INSTRUCTOR'S SOLUTIONS MANUAL

PART ONE

ARDIS • BORZELLINO • BUCHANAN • KOUBA • MOGILL • NELSON

THOMAS' CALCULUS
EARLY TRANSCENDENTALS

ELEVENTH EDITION

BASED ON THE ORIGINAL WORK BY

George B. Thomas, Jr.

Massachusetts Institute of Technology

AS REVISED BY

Maurice D. Weir

Naval Postgraduate School

Joel Hass

University of California, Davis

Frank R. Giordano

Naval Postgraduate School

Boston San Francisco New York
London Toronto Sydney Tokyo Singapore Madrid
Mexico City Munich Paris Cape Town Hong Kong Montreal

Reproduced by Pearson Addison-Wesley from electronic files supplied by the author.

Publishing as Pearson Addison-Wesley, 75 Arlington Street, Boston, MA 02116.

ISBN 0-321-22634-8

3 4 5 6 VHG 08 07 06 05

PREFACE TO THE INSTRUCTOR

This Instructor's Solutions Manual contains the solutions to every exercise in the 11th Edition of THOMAS' CALCULUS: EARLY TRANSCENDENTALS by Maurice Weir, Joel Hass and Frank Giordano, including the Computer Algebra System (CAS) exercises. The corresponding Student's Solutions Manual omits the solutions to the even-numbered exercises as well as the solutions to the CAS exercises (because the CAS command templates would give them all away).

In addition to including the solutions to all of the new exercises in this edition of Thomas, we have carefully revised or rewritten every solution which appeared in previous solutions manuals to ensure that each solution

- conforms exactly to the methods, procedures and steps presented in the text
- is mathematically correct
- includes all of the steps necessary so a typical calculus student can follow the logical argument and algebra
- includes a graph or figure whenever called for by the exercise, or if needed to help with the explanation
- is formatted in an appropriate style to aid in its understanding

Every CAS exercise is solved in both the MAPLE and *MATHEMATICA* computer algebra systems. A template showing an example of the CAS commands needed to execute the solution is provided for each exercise type. Similar exercises within the text grouping require a change only in the input function or other numerical input parameters associated with the problem (such as the interval endpoints or the number of iterations).

Acknowledgments

Solutions Writers

William Ardis, Collin County Community College-Preston Ridge Campus
Joseph Borzellino, California Polytechnic State University
Linda Buchanan, Howard College
Duane Kouba, University of California-Davis
Tim Mogill
Patricia Nelson, University of Wisconsin-La Crosse

Accuracy Checkers

Karl Kattchee, University of Wisconsin-La Crosse
Marie Vanisko, California State University, Stanislaus
Tom Weigleitner, VISTA Information Technologies

Thanks to Rachel Reeve, Christine O'Brien, Sheila Spinney, Elka Block, and Joe Vetere for all their guidance and help at every step.

TABLE OF CONTENTS

1 Functions 1

1.1 Functions and Their Graphs 1
1.2 Identifying Functions; Mathematical Models 6
1.3 Combining Functions; Shifting and Scaling Graphs 11
1.4 Graphing with Calculators and Computers 25
1.5 Exponential Functions 32
1.6 Inverse Functions and Logarithms 35
Practice Exercises 42
Additional and Advanced Exercises 50

2 Limits and Continuity 57

2.1 Rates of Change and Limits 57
2.2 Calculating Limits Using the Limit Laws 64
2.3 The Precise Definition of a Limit 68
2.4 One-Sided Limits and Limits at Infinity 76
2.5 Infinite Limits and Vertical Asymptotes 82
2.6 Continuity 90
2.7 Tangents and Derivatives 95
Practice Exercises 101
Additional and Advanced Exercises 106

3 Differentiation 113

3.1 The Derivative as a Function 113
3.2 Differentiation Rules for Polynomials, Exponentials, Products, and Quotients 122
3.3 The Derivative as a Rate of Change 127
3.4 Derivatives of Trigonometric Functions 133
3.5 The Chain Rule and Parametric Equations 140
3.6 Implicit Differentiation 152
3.7 Derivatives of Inverse Functions and Logarithms 161
3.8 Inverse Trigonometric Functions 169
3.9 Related Rates 176
3.10 Linearizations and Differentials 180
Practice Exercises 189
Additional and Advanced Exercises 203

4 Applications of Derivatives 209

4.1 Extreme Values of Functions 209
4.2 The Mean Value Theorem 225
4.3 Monotonic Functions and the First Derivative Test 230
4.4 Concavity and Curve Sketching 243
4.5 Applied Optimization Problems 261
4.6 Indeterminate Forms and L'Hôpital's Rule 276
4.7 Newton's Method 283
4.8 Antiderivatives 287
Practice Exercises 295
Additional and Advanced Exercises 313

5 Integration 319

5.1 Estimating with Finite Sums 319
5.2 Sigma Notation and Limits of Finite Sums 324
5.3 The Definite Integral 329
5.4 The Fundamental Theorem of Calculus 341
5.5 Indefinite Integrals and the Substitution Rule 350
5.6 Substitution and Area Between Curves 357
Practice Exercises 374
Additional and Advanced Exercises 388

6 Applications of Definite Integrals 395

6.1 Volumes by Slicing and Rotation About an Axis 395
6.2 Volumes by Cylindrical Shells 406
6.3 Lengths of Plane Curves 414
6.4 Moments and Centers of Mass 420
6.5 Areas of Surfaces of Revolution and the Theorems of Pappus 430
6.6 Work 438
6.7 Fluid Pressures and Forces 444
Practice Exercises 448
Additional and Advanced Exercises 457

7 Integrals and Transcendental Functions 463

7.1 The Logarithm Defined as an Inegral 463
7.2 Exponential Growth and Decay 469
7.3 Relative Rates of Growth 472
7.4 Hyperbolic Functions 476
Practice Exercises 483
Additional and Advanced Exercises 487

8 Techniques of Integration 489

8.1 Basic Integration Formulas 489
8.2 Integration by Parts 497
8.3 Integration of Rational Functions by Partial Fractions 504
8.4 Trigonometric Integrals 511
8.5 Trigonometric Substitutions 515
8.5 Integral Tables and Computer Algebra Systems 520
8.6 Numerical Integration 534
8.7 Improper Integrals 545
Practice Exercises 555
Additional and Advanced Exercises 571

9 Further Applications of Integration 581

9.1 Slope Fields and Separable Differential Equations 581
9.2 First-Order Linear Differential Equations 585
9.3 Euler's Method 589
9.4 Graphical Solutions of Autonomous Differential Equations 595
9.5 Applications of First-Order Differential Equations 602
Practice Exercises 608
Additional and Advanced Exercises 613

10 Conic Sections and Polar Coordinates 615

10.1 Conic Sections and Quadratic Equations 615
10.2 Classifying Conic Sections by Eccentricity 626
10.3 Quadratic Equations and Rotations 633
10.4 Conics and Parametric Equations; The Cycloid 638
10.5 Polar Coordinates 643
10.6 Graphing in Polar Coordinates 647
10.7 Areas and Lengths in Polar Coordinates 657
10.8 Conic Sections in Polar Coordinates 663
Practice Exercises 671
Additional and Advanced Exercises 682

11 Infinite Sequences and Series 693

11.1 Sequences 693
11.2 Infinite Series 704
11.3 The Integral Test 709
11.4 Comparison Tests 713
11.5 The Ratio and Root Tests 718
11.6 Alternating Series, Absolute and Conditional Convergence 721
11.7 Power Series 727
11.8 Taylor and Maclaurin Series 735
11.9 Convergence of Taylor Series; Error Estimates 739
11.10 Applications of Power Series 746
11.11 Fourier Series 753
Practice Exercises 757
Additional and Advanced Exercises 769

CHAPTER 1 FUNCTIONS

1.1 FUNCTIONS AND THEIR GRAPHS

1. domain $= (-\infty, \infty)$; range $= [1, \infty)$

2. domain $= [0, \infty)$; range $= (-\infty, 1]$

3. domain $= (0, \infty)$; y in range $\Rightarrow\ y = \frac{1}{\sqrt{t}},\ t > 0 \ \Rightarrow\ y^2 = \frac{1}{t}$ and $y > 0 \ \Rightarrow$ y can be any positive real number $\Rightarrow$ range $= (0, \infty)$.

4. domain $= [0, \infty)$; y in range $\Rightarrow\ y = \frac{1}{1+\sqrt{t}},\ t > 0$. If $t = 0$, then $y = 1$ and as t increases, y becomes a smaller and smaller positive real number $\Rightarrow$ range $= (0, 1]$.

5. $4 - z^2 = (2 - z)(2 + z) \ge 0 \ \Leftrightarrow\ z \in [-2, 2] =$ domain. Largest value is $g(0) = \sqrt{4} = 2$ and smallest value is $g(-2) = g(2) = \sqrt{0} = 0 \ \Rightarrow$ range $= [0, 2]$.

6. domain $= (-2, 2)$ from Exercise 5; smallest value is $g(0) = \frac{1}{2}$ and as $0 < z$ increases to 2, g(z) gets larger and larger (also true as $z < 0$ decreases to -2) $\Rightarrow$ range $= \left[\frac{1}{2}, \infty\right)$.

7. (a) Not the graph of a function of x since it fails the vertical line test.
   (b) Is the graph of a function of x since any vertical line intersects the graph at most once.

8. (a) Not the graph of a function of x since it fails the vertical line test.
   (b) Not the graph of a function of x since it fails the vertical line test.

9. $y = \sqrt{\left(\frac{1}{x}\right) - 1} \Rightarrow \frac{1}{x} - 1 \ge 0 \Rightarrow x \le 1$ and $x > 0$. So,
   (a) No $(x > 0)$; (b) No; division by 0 undefined;
   (c) No; if $x \ge 1,\ \frac{1}{x} < 1 \Rightarrow \frac{1}{x} - 1 < 0$; (d) $(0, 1]$

10. $y = \sqrt{2 - \sqrt{x}} \Rightarrow 2 - \sqrt{x} \ge 0 \Rightarrow \sqrt{x} \ge 0$ and $\sqrt{x} \le 2$. $\sqrt{x} \ge 0 \Rightarrow x \ge 0$ and $\sqrt{x} \le 2 \ \Rightarrow x \le 4$. So, $0 \le x \le 4$.
    (a) No; (b) No; (c) $[0, 4]$

11. base $= x$; $(\text{height})^2 + \left(\frac{x}{2}\right)^2 = x^2 \ \Rightarrow$ height $= \frac{\sqrt{3}}{2}\,x$; area is $a(x) = \frac{1}{2}(\text{base})(\text{height}) = \frac{1}{2}(x)\left(\frac{\sqrt{3}}{2}x\right) = \frac{\sqrt{3}}{4}\,x^2$; perimeter is $p(x) = x + x + x = 3x$.

12. $s =$ side length $\Rightarrow\ s^2 + s^2 = d^2 \ \Rightarrow\ s = \frac{d}{\sqrt{2}}$; and area is $a = s^2 \ \Rightarrow\ a = \frac{1}{2}\,d^2$

13. Let $D =$ diagonal of a face of the cube and $\ell =$ the length of an edge. Then $\ell^2 + D^2 = d^2$ and $D^2 = 2\ell^2 \ \Rightarrow\ 3\ell^2 = d^2 \ \Rightarrow\ \ell = \frac{d}{\sqrt{3}}$. The surface area is $6\ell^2 = \frac{6d^2}{3} = 2d^2$ and the volume is $\ell^3 = \left(\frac{d^2}{3}\right)^{3/2} = \frac{d^3}{3\sqrt{3}}$.

14. The coordinates of P are $\left(x, \sqrt{x}\right)$ so the slope of the line joining P to the origin is $m = \frac{\sqrt{x}}{x} = \frac{1}{\sqrt{x}}$ $(x > 0)$. Thus, $\left(x, \sqrt{x}\right) = \left(\frac{1}{m^2}, \frac{1}{m}\right)$.

15. The domain is $(-\infty, \infty)$.

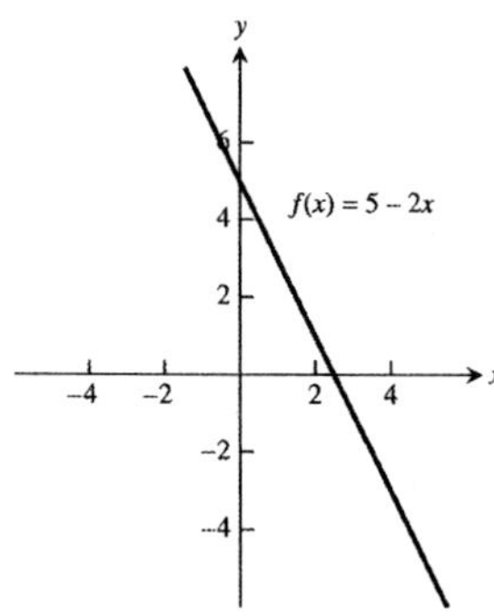

16. The domain is $(-\infty, \infty)$.

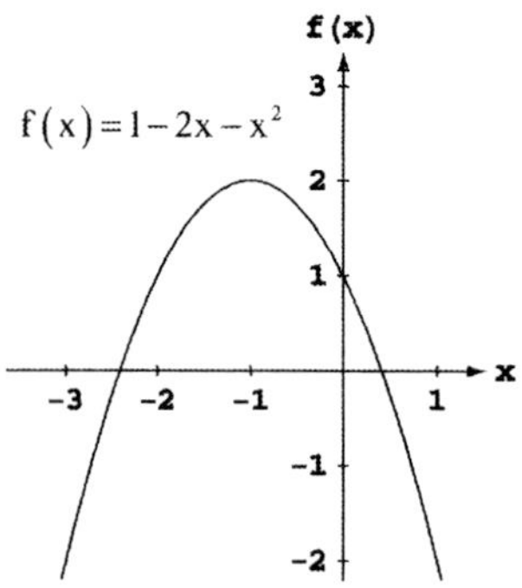

17. The domain is $(-\infty, \infty)$.

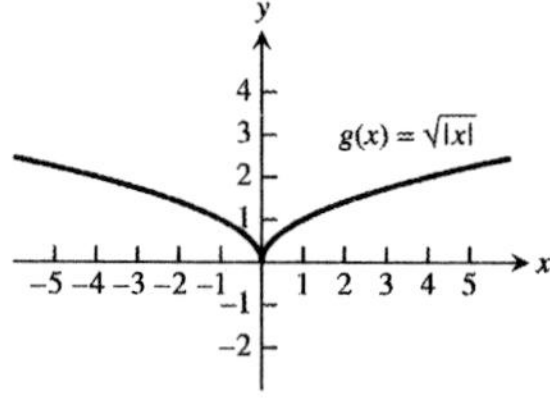

18. The domain is $(-\infty, 0]$.

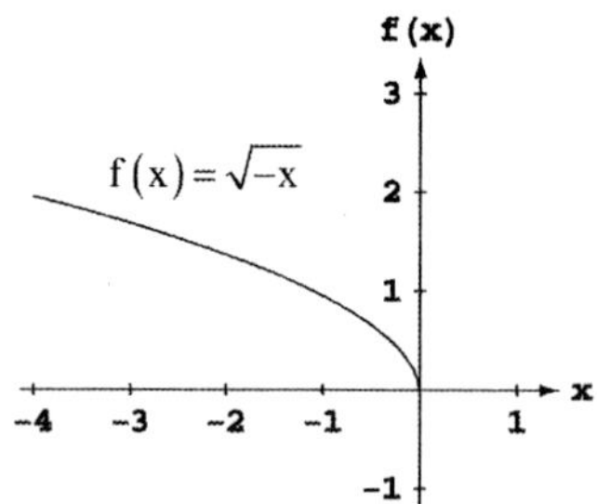

19. The domain is $(-\infty, 0) \cup (0, \infty)$.

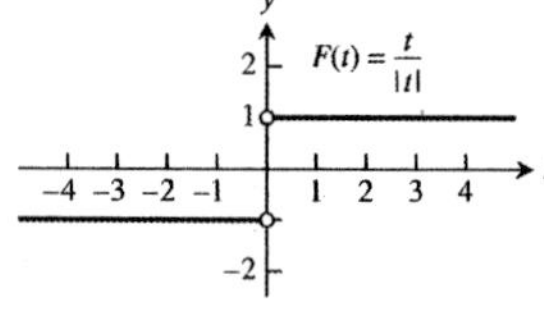

20. The domain is $(-\infty, 0) \cup (0, \infty)$.

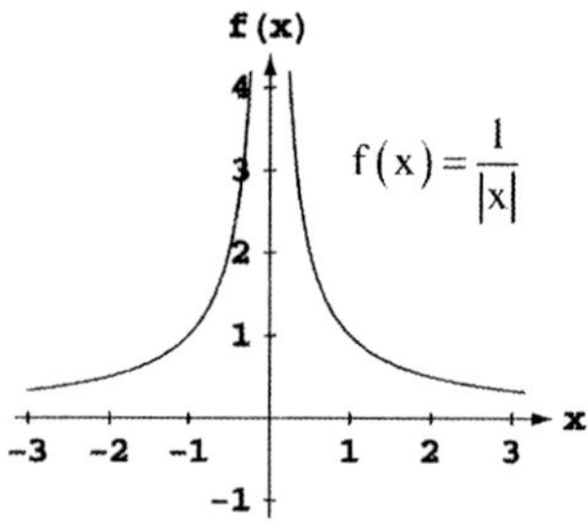

21. Neither graph passes the vertical line test

(a)

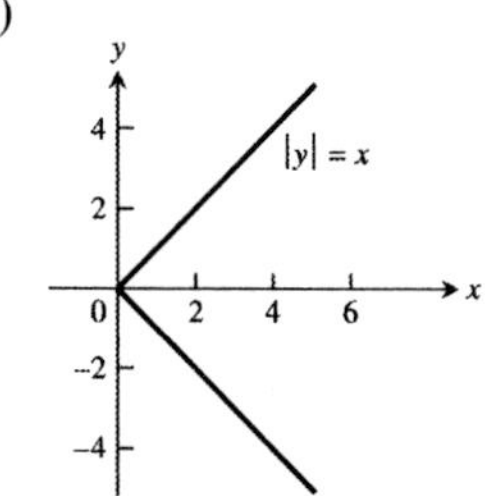

(b)

22. Neither graph passes the vertical line test

(a)

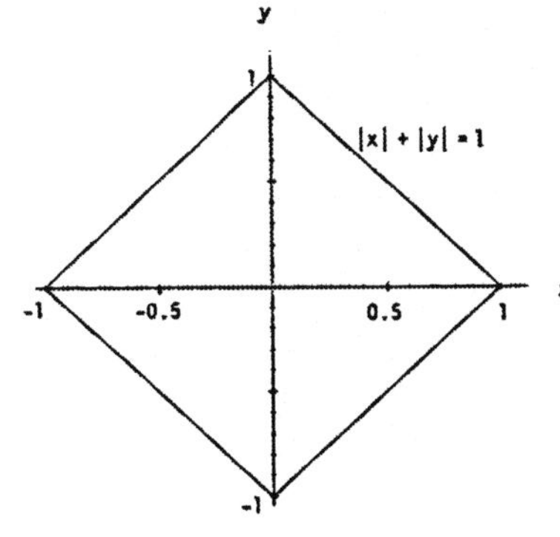

(b)

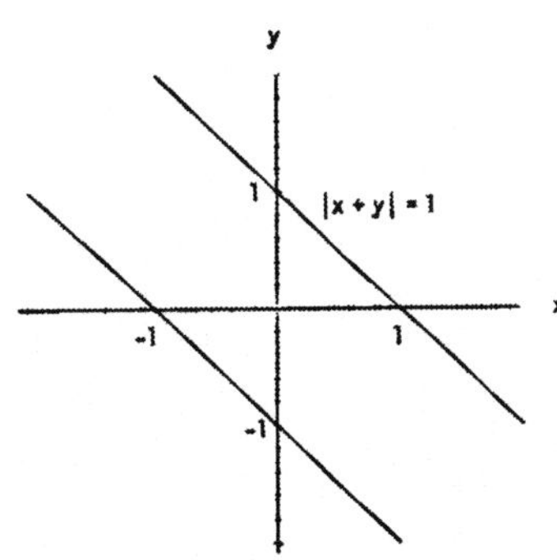

$$|x + y| = 1 \Leftrightarrow \left\{ \begin{matrix} x + y = 1 \\ \text{or} \\ x + y = -1 \end{matrix} \right\} \Leftrightarrow \left\{ \begin{matrix} y = 1 - x \\ \text{or} \\ y = -1 - x \end{matrix} \right\}$$

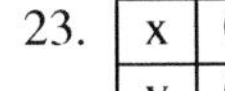

23.

x	0	1	2
y	0	1	0

$$f(x) = \begin{cases} x, & 0 \le x \le 1 \\ 2 - x, & 1 < x \le 2 \end{cases}$$

24.

x	0	1	2
y	1	0	0

$$y = \begin{cases} 1 - x, & 0 \le x \le 1 \\ 2 - x, & 1 < x \le 2 \end{cases}$$

25. $y = \begin{cases} 3 - x, & x \le 1 \\ 2x, & 1 < x \end{cases}$

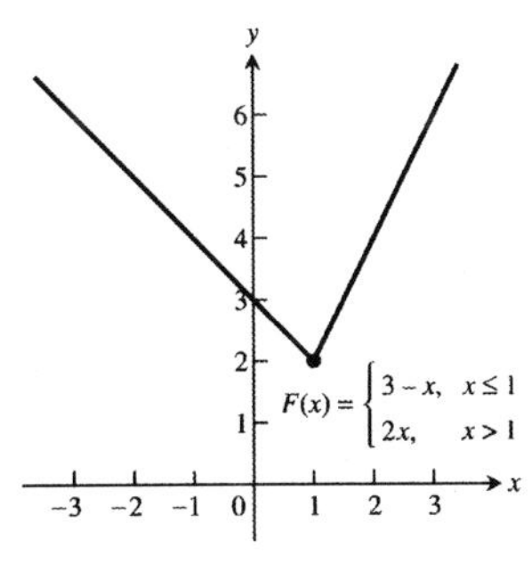

26. $y = \begin{cases} \frac{1}{x}, & x < 0 \\ x, & 0 \le x \end{cases}$

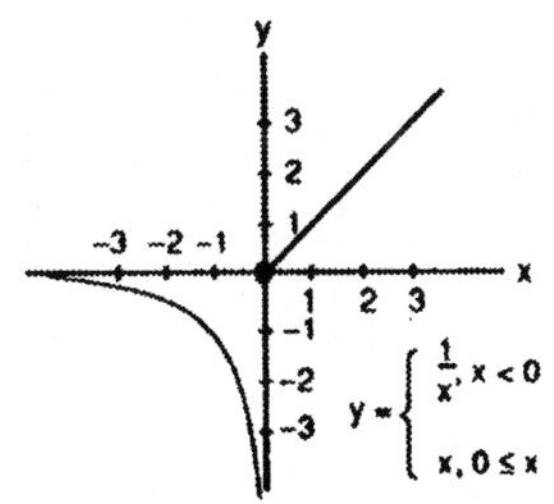

27. (a) Line through $(0, 0)$ and $(1, 1)$: $y = x$

Line through $(1, 1)$ and $(2, 0)$: $y = -x + 2$

$$f(x) = \begin{cases} x, & 0 \le x \le 1 \\ -x + 2, & 1 < x \le 2 \end{cases}$$

(b) $f(x) = \begin{cases} 2, & 0 \le x < 1 \\ 0, & 1 \le x < 2 \\ 2, & 2 \le x < 3 \\ 0, & 3 \le x \le 4 \end{cases}$

28. (a) Line through $(0, 2)$ and $(2, 0)$: $y = -x + 2$

Line through $(2, 1)$ and $(5, 0)$: $m = \frac{0-1}{5-2} = \frac{-1}{3} = -\frac{1}{3}$, so $y = -\frac{1}{3}(x - 2) + 1 = -\frac{1}{3}x + \frac{5}{3}$

$$f(x) = \begin{cases} -x + 2, & 0 < x \le 2 \\ -\frac{1}{3}x + \frac{5}{3}, & 2 < x \le 5 \end{cases}$$

(b) Line through $(-1, 0)$ and $(0, -3)$: $m = \frac{-3-0}{0-(-1)} = -3$, so $y = -3x - 3$

Line through $(0, 3)$ and $(2, -1)$: $m = \frac{-1-3}{2-0} = \frac{-4}{2} = -2$, so $y = -2x + 3$

$$f(x) = \begin{cases} -3x - 3, & -1 < x \le 0 \\ -2x + 3, & 0 < x \le 2 \end{cases}$$

29. (a) Line through $(-1, 1)$ and $(0, 0)$: $y = -x$

Line through $(0, 1)$ and $(1, 1)$: $y = 1$

Line through $(1, 1)$ and $(3, 0)$: $m = \frac{0-1}{3-1} = \frac{-1}{2} = -\frac{1}{2}$, so $y = -\frac{1}{2}(x - 1) + 1 = -\frac{1}{2}x + \frac{3}{2}$

$$f(x) = \begin{cases} -x & -1 \le x < 0 \\ 1 & 0 < x \le 1 \\ -\frac{1}{2}x + \frac{3}{2} & 1 < x < 3 \end{cases}$$

(b) Line through $(-2, -1)$ and $(0, 0)$: $y = \frac{1}{2}x$

Line through $(0, 2)$ and $(1, 0)$: $y = -2x + 2$

Line through $(1, -1)$ and $(3, -1)$: $y = -1$

$$f(x) = \begin{cases} \frac{1}{2}x & -2 \le x \le 0 \\ -2x + 2 & 0 < x \le 1 \\ -1 & 1 < x \le 3 \end{cases}$$

30. (a) Line through $\left(\frac{T}{2}, 0\right)$ and $(T, 1)$: $m = \frac{1-0}{T-(T/2)} = \frac{2}{T}$, so $y = \frac{2}{T}\left(x - \frac{T}{2}\right) + 0 = \frac{2}{T}x - 1$

$$f(x) = \begin{cases} 0, & 0 \le x \le \frac{T}{2} \\ \frac{2}{T}x - 1, & \frac{T}{2} < x \le T \end{cases}$$

(b) $$f(x) = \begin{cases} A, & 0 \le x < \frac{T}{2} \\ -A, & \frac{T}{2} \le x < T \\ A, & T \le x < \frac{3T}{2} \\ -A, & \frac{3T}{2} \le x \le 2T \end{cases}$$

31. (a) From the graph, $\frac{x}{2} > 1 + \frac{4}{x} \Rightarrow x \in (-2, 0) \cup (4, \infty)$

(b) $\frac{x}{2} > 1 + \frac{4}{x} \Rightarrow \frac{x}{2} - 1 - \frac{4}{x} > 0$

$x > 0$: $\frac{x}{2} - 1 - \frac{4}{x} > 0 \Rightarrow \frac{x^2-2x-8}{2x} > 0 \Rightarrow \frac{(x-4)(x+2)}{2x} > 0$

$\Rightarrow x > 4$ since x is positive;

$x < 0$: $\frac{x}{2} - 1 - \frac{4}{x} > 0 \Rightarrow \frac{x^2-2x-8}{2x} < 0 \Rightarrow \frac{(x-4)(x+2)}{2x} < 0$

$\Rightarrow x < -2$ since x is negative;

sign of $(x - 4)(x + 2)$

+   −2   −   4   +

Solution interval: $(-2, 0) \cup (4, \infty)$

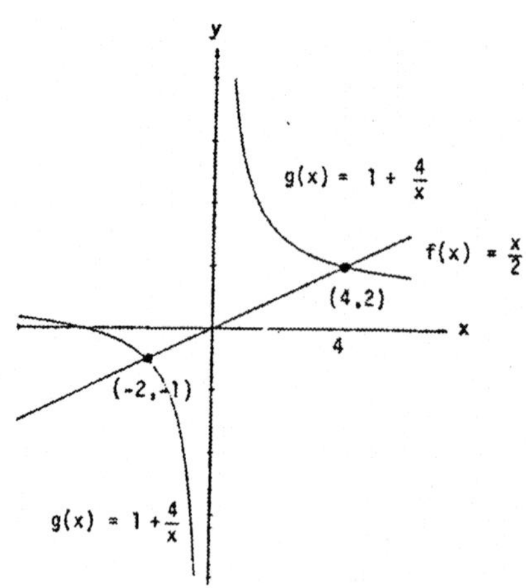

32. (a) From the graph, $\frac{3}{x-1} < \frac{2}{x+1} \Rightarrow x \in (-\infty, -5) \cup (-1, 1)$

(b) Case $x < -1$: $\frac{3}{x-1} < \frac{2}{x+1} \Rightarrow \frac{3(x+1)}{x-1} > 2$
$\Rightarrow 3x + 3 < 2x - 2 \Rightarrow x < -5.$
Thus, $x \in (-\infty, -5)$ solves the inequality.

Case $-1 < x < 1$: $\frac{3}{x-1} < \frac{2}{x+1} \Rightarrow \frac{3(x+1)}{x-1} < 2$
$\Rightarrow 3x + 3 > 2x - 2 \Rightarrow x > -5$ which is true if $x > -1$. Thus, $x \in (-1, 1)$ solves the inequality.

Case $1 < x$: $\frac{3}{x-1} < \frac{2}{x+1} \Rightarrow 3x + 3 < 2x - 2 \Rightarrow x < -5$ which is never true if $1 < x$, so no solution here.

In conclusion, $x \in (-\infty, -5) \cup (-1, 1)$.

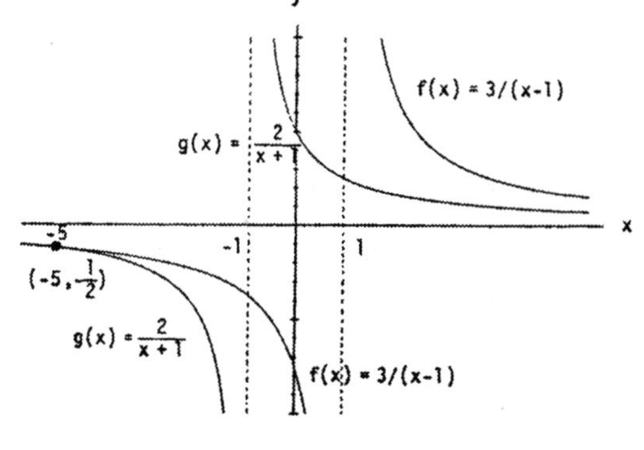

33. (a) $\lfloor x \rfloor = 0$ for $x \in [0, 1)$ (b) $\lceil x \rceil = 0$ for $x \in (-1, 0]$

34. $\lfloor x \rfloor = \lceil x \rceil$ only when x is an integer.

35. For any real number x, $n \le x \le n + 1$, where n is an integer. Now: $n \le x \le n + 1 \Rightarrow -(n + 1) \le -x \le -n$. By definition: $\lceil -x \rceil = -n$ and $\lfloor x \rfloor = n \Rightarrow -\lfloor x \rfloor = -n$. So $\lceil -x \rceil = -\lfloor x \rfloor$ for all $x \in \Re$.

36. To find f(x) you delete the decimal or fractional portion of x, leaving only the integer part.

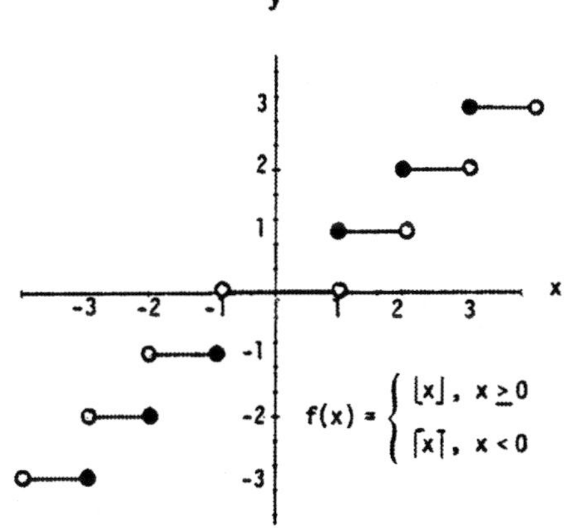

37. $v = f(x) = x(14 - 2x)(22 - 2x) = 4x^3 - 72x^2 + 308x;\ 0 < x < 7.$

38. (a) Let h = height of the triangle. Since the triangle is isosceles, $\overline{AB}^2 + \overline{AB}^2 = 2^2 \Rightarrow \overline{AB} = \sqrt{2}$. So, $h^2 + 1^2 = \left(\sqrt{2}\right)^2 \Rightarrow h = 1 \Rightarrow$ B is at (0, 1) $\Rightarrow$ slope of AB $= -1 \Rightarrow$ The equation of AB is $y = f(x) = -x + 1;\ x \in [0, 1]$.

(b) $A(x) = 2x\,y = 2x(-x + 1) = -2x^2 + 2x;\ x \in [0, 1]$.

39. (a) Because the circumference of the original circle was 8π and a piece of length x was removed.

(b) $r = \frac{8\pi - x}{2\pi} = 4 - \frac{x}{2\pi}$

(c) $h = \sqrt{16 - r^2} = \sqrt{16 - \left(4 - \frac{x}{2\pi}\right)^2} = \sqrt{16 - \left(16 - \frac{4x}{\pi} + \frac{x^2}{4\pi^2}\right)} = \sqrt{\frac{4x}{\pi} - \frac{x^2}{4\pi^2}} = \sqrt{\frac{16\pi x}{4\pi^2} - \frac{x^2}{4\pi^2}} = \frac{\sqrt{16\pi x - x^2}}{2\pi}$

(d) $V = \frac{1}{3}\pi r^2 h = \frac{1}{3}\pi\left(\frac{8\pi - x}{2\pi}\right)^2 \cdot \frac{\sqrt{16\pi x - x^2}}{2\pi} = \frac{(8\pi - x)^2\sqrt{16\pi x - x^2}}{24\pi^2}$

40. (a) Note that 2 mi = 10,560 ft, so there are $\sqrt{800^2 + x^2}$ feet of river cable at \$180 per foot and $(10,560 - x)$ feet of land cable at \$100 per foot. The cost is $C(x) = 180\sqrt{800^2 + x^2} + 100(10,560 - x)$.

(b) $C(0) = \$1,200,000$
$C(500) \approx \$1,175,812$
$C(1000) \approx \$1,186,512$
$C(1500) \approx \$1,212,000$
$C(2000) \approx \$1,243,732$
$C(2500) \approx \$1,278,479$
$C(3000) \approx \$1,314,870$
Values beyond this are all larger. It would appear that the least expensive location is less than 2000 feet from the point P.

41. A curve symmetric about the x-axis will not pass the vertical line test because the points (x, y) and $(x, -y)$ lie on the same vertical line. The graph of the function $y = f(x) = 0$ is the x-axis, a horizontal line for which there is a single y-value, 0, for any x.

42. Pick 11, for example: $11 + 5 = 16 \to 2 \cdot 16 = 32 \to 32 - 6 = 26 \to \frac{26}{2} = 13 \to 13 - 2 = 11$, the original number. $f(x) = \frac{2(x+5)-6}{2} - 2 = x$, the number you started with.

1.2 IDENTIFYING FUNCTIONS; MATHEMATICAL MODELS

1. (a) linear, polynomial of degree 1, algebraic. (b) power, algebraic.
(c) rational, algebraic. (d) exponential.

2. (a) polynomial of degree 4, algebraic. (b) exponential.
(c) algebraic. (d) power, algebraic.

3. (a) rational, algebraic. (b) algebraic.
(c) trigonometric. (d) logarithmic.

4. (a) logarithmic. (b) algebraic.
(c) exponential. (d) trigonometric.

5. (a) Graph h because it is an even function and rises less rapidly than does Graph g.
(b) Graph f because it is an odd function.
(c) Graph g because it is an even function and rises more rapidly than does Graph h.

6. (a) Graph f because it is linear.
(b) Graph g because it contains $(0, 1)$.
(c) Graph h because it is a nonlinear odd function.

7. Symmetric about the origin
Dec: $-\infty < x < \infty$
Inc: nowhere

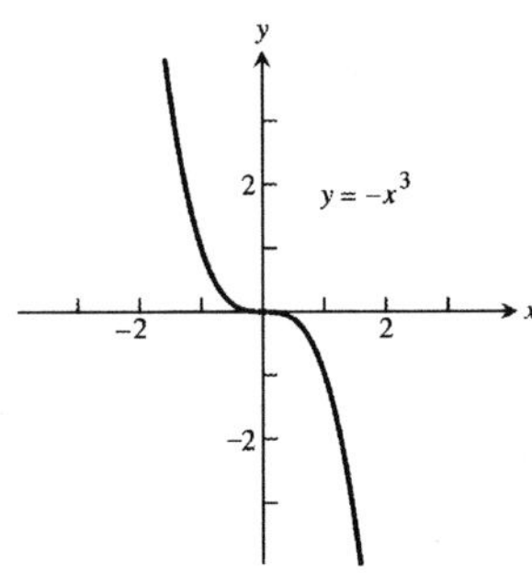

8. Symmetric about the y-axis
Dec: $-\infty < x < 0$
Inc: $0 < x < \infty$

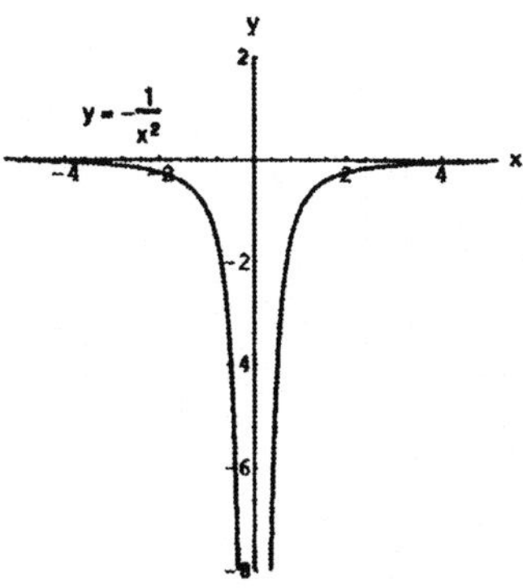

9. Symmetric about the origin
Dec: nowhere
Inc: $-\infty < x < 0$
$0 < x < \infty$

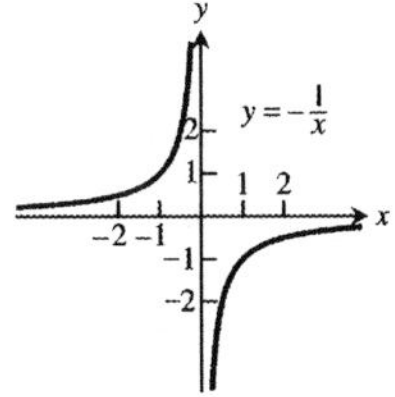

10. Symmetric about the y-axis
Dec: $0 < x < \infty$
Inc: $-\infty < x < 0$

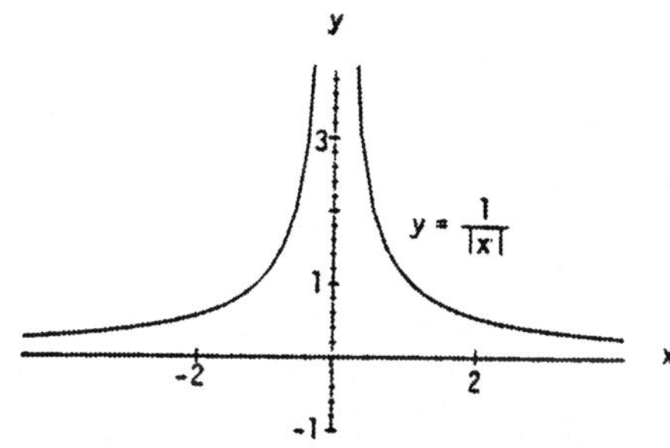

11. Symmetric about the y-axis
Dec: $-\infty < x \le 0$
Inc: $0 < x < \infty$

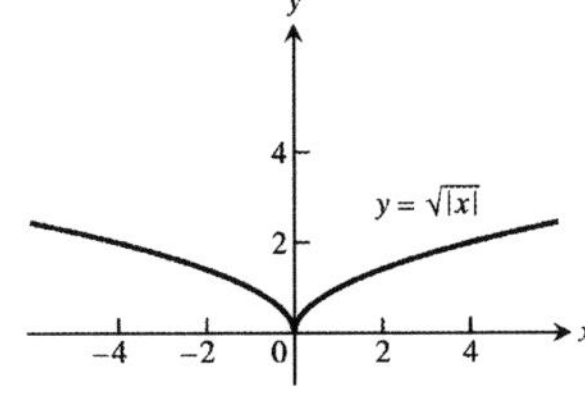

12. No symmetry
Dec: $-\infty < x \le 0$
Inc: nowhere

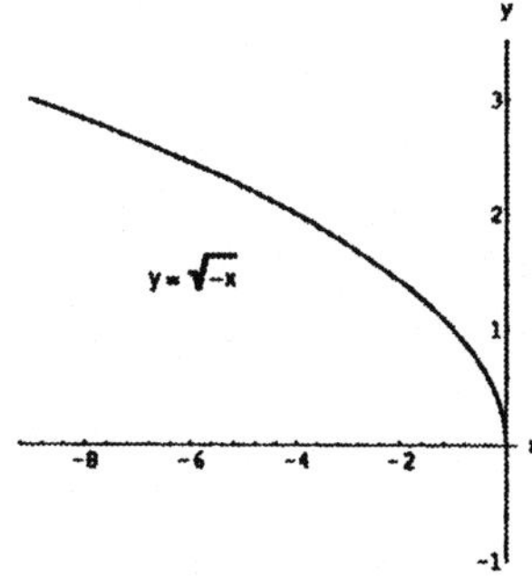

13. Symmetric about the origin

Dec: nowhere

Inc: $-\infty < x < \infty$

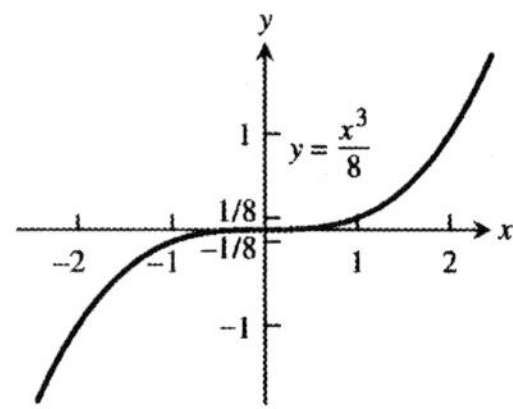

14. No symmetry

Dec: $0 \le x < \infty$

Inc: nowhere

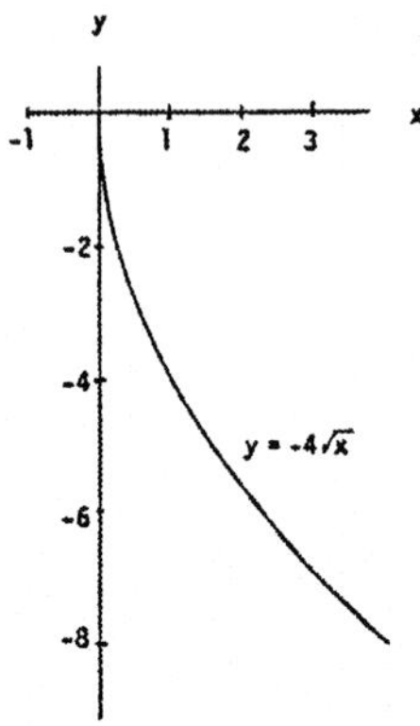

15. No symmetry

Dec: $0 \le x < \infty$

Inc: nowhere

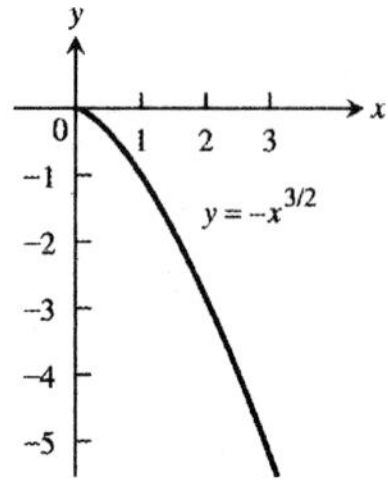

16. No symmetry

Dec: $-\infty < x \le 0$

Inc: nowhere

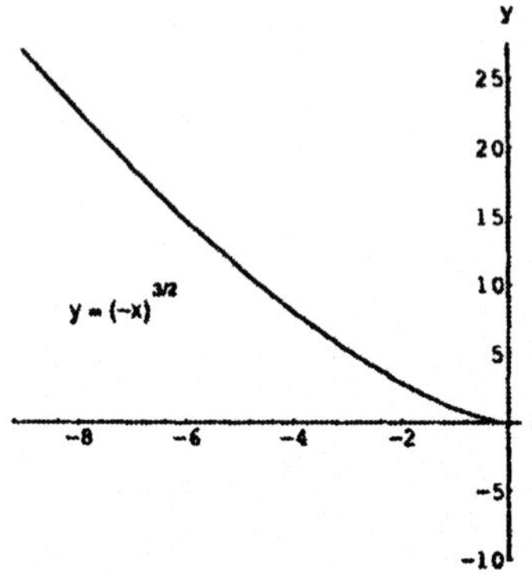

17. Symmetric about the y-axis

Dec: $-\infty < x \le 0$

Inc: $0 < x < \infty$

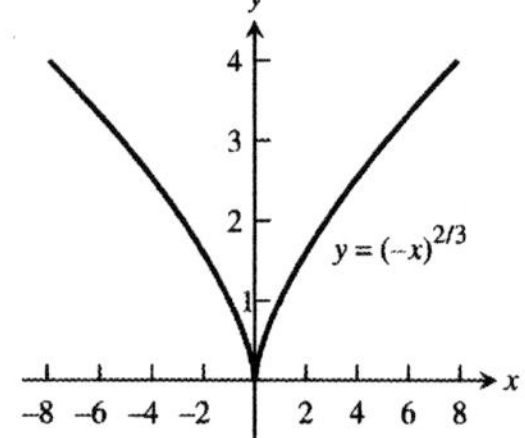

18. Symmetric about the y-axis

Dec: $0 \le x < \infty$

Inc: $-\infty < x < 0$

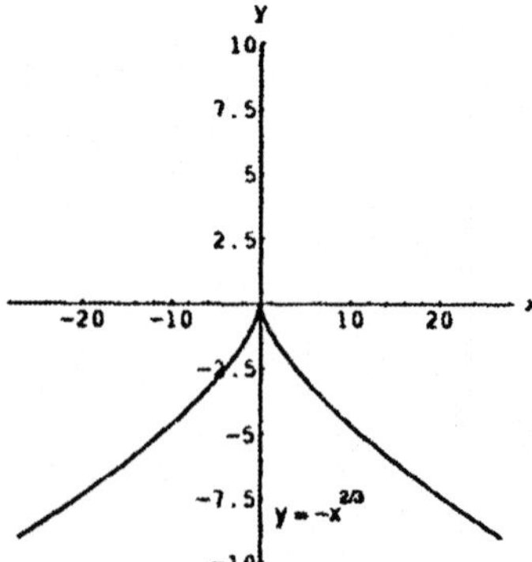

19. Since a horizontal line not through the origin is symmetric with respect to the y-axis, but not with respect to the origin, the function is even.

20. $f(x) = x^{-5} = \frac{1}{x^5}$ and $f(-x) = (-x)^{-5} = \frac{1}{(-x)^5} = -\left(\frac{1}{x^5}\right) = -f(x)$. Thus the function is odd.

21. Since $f(x) = x^2 + 1 = (-x)^2 + 1 = -f(x)$. The function is even.

22. Since $[f(x) = x^2 + x] \neq [f(-x) = (-x)^2 - x]$ and $[f(x) = x^2 + x] \neq [-f(x) = -(x)^2 - x]$ the function is neither even nor odd.

23. Since $g(x) = x^3 + x$, $g(-x) = -x^3 - x = -(x^3 + x) = -g(x)$. So the function is odd.

24. $g(x) = x^4 + 3x^2 - 1 = (-x)^4 + 3(-x)^2 - 1 = g(-x)$, thus the function is even.

25. $g(x) = \frac{1}{x^2 - 1} = \frac{1}{(-x)^2 - 1} = g(-x)$. Thus the function is even.

26. $g(x) = \frac{x}{x^2 - 1}$; $g(-x) = -\frac{x}{x^2 - 1} = g(-x)$. So the function is odd.

27. $h(t) = \frac{1}{t - 1}$; $h(-t) = \frac{1}{-t - 1}$; $-h(t) = \frac{1}{1 - t}$. Since $h(t) \neq -h(t)$ and $h(t) \neq h(-t)$, the function is neither even nor odd.

28. Since $|t^3| = |(-t)^3|$, $h(t) = h(-t)$ and the function is even.

29. $h(t) = 2t + 1$, $h(-t) = -2t + 1$. So $h(t) \neq h(-t)$. $-h(t) = -2t - 1$, so $h(t) \neq -h(t)$. The function is neither even nor odd.

30. $h(t) = 2|t| + 1$ and $h(-t) = 2|-t| + 1 = 2|t| + 1$. So $h(t) = h(-t)$ and the function is even.

31. (a)

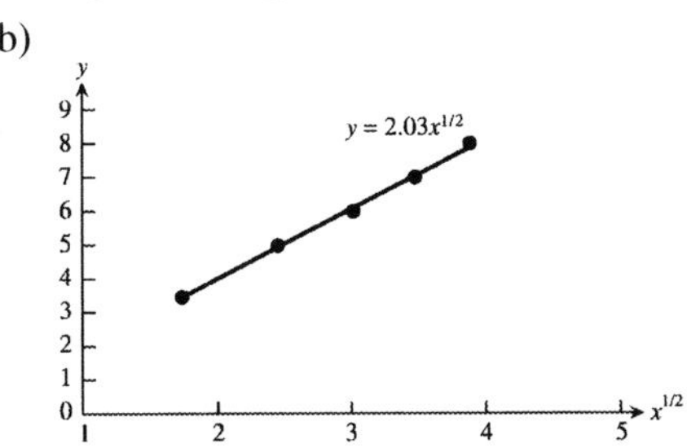

The graph supports the assumption that y is proportional to x. The constant of proportionality is estimated from the slope of the regression line, which is 0.166.

(b)

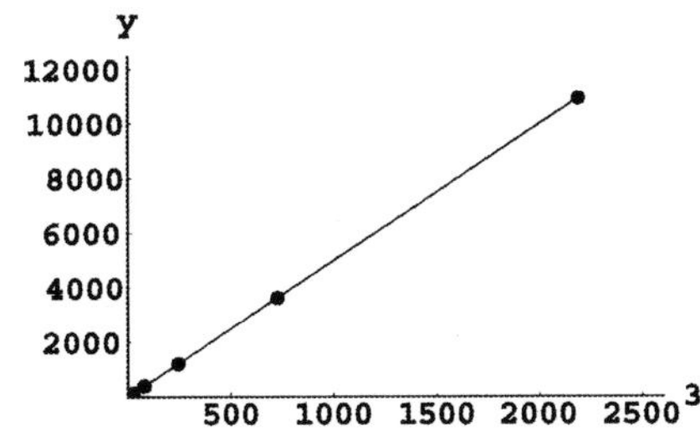

The graph supports the assumption that y is proportional to $x^{1/2}$. The constant of proportionality is estimated from the slope of the regression line, which is 2.03.

32. (a) Because of the wide range of values of the data, two graphs are needed to observe all of the points in relation to the regression line.

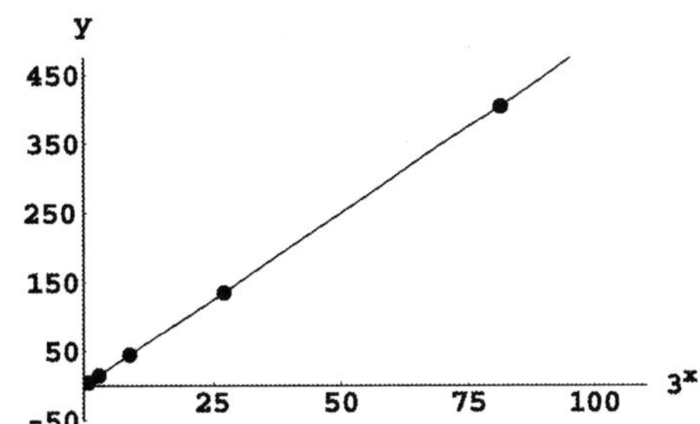

The graphs support the assumption that y is proportional to 3^x. The constant of proportionality is estimated from the slope of the regression line, which is 5.00.

(b) The graph supports the assumption that y is proportional to ln x. The constant of proportionality is extimated from the slope of the regression line, which is 2.99.

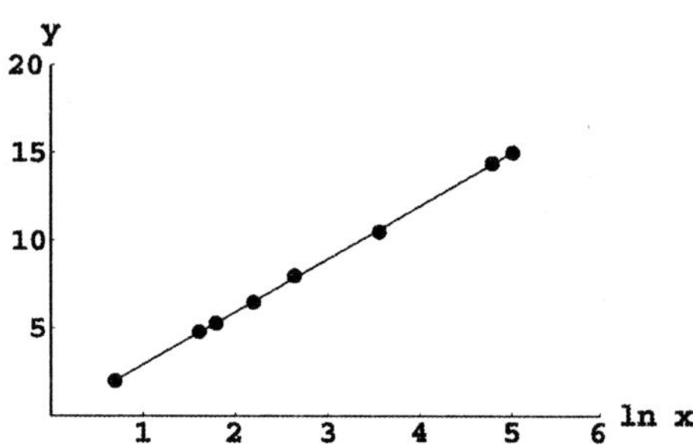

33. (a) The scatterplot of y = reaction distance versus x = speed is

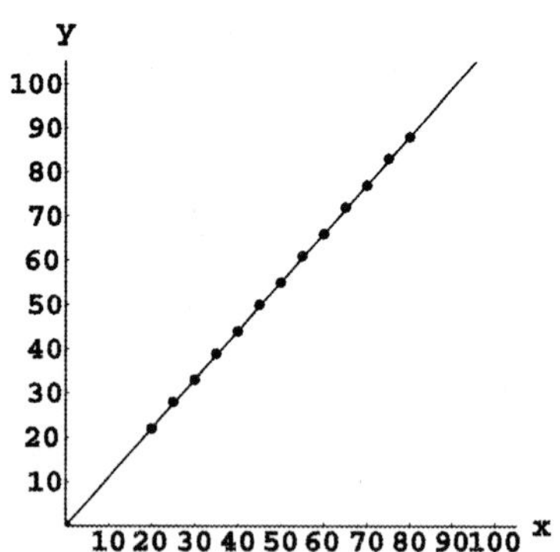

Answers for the constant of proportionality may vary. The constant of proportionality is the slope of the line, which is approximately 1.1.

(b) Calculate $x' =$ speed squared. The scatterplot of x'versus y = braking distance is:

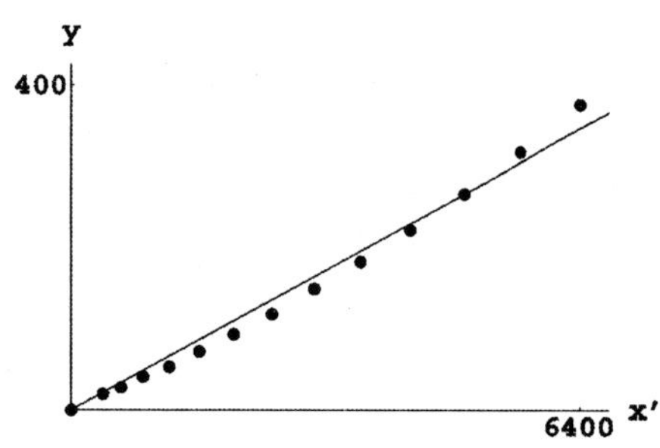

Answers for the constant of proportionality may vary. The constant of proportionality is the slope of the line, which is approximately 0.059.

34. Kepler's 3rd Law is $T(\text{days}) = 0.41R^{3/2}$, R in millions of miles. "Quaoar" is 4×10^9 miles from Earth, or about $4 \times 10^9 + 93 \times 10^6 \approx 4 \times 10^9$ miles from the sun. Let R = 4000 (millions of miles) and $T = (0.41)(4000)^{3/2}$ days $\approx 103,723$ days.

35. (a)

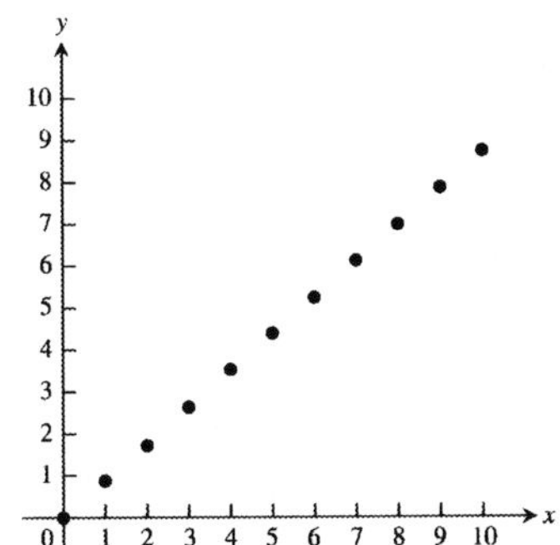

The hypothesis is reasonable.

(b) The constant of proportionality is the slope of the line $\approx \frac{8.741-0}{10-0}$ in./unit mass $= 0.874$ in./unit mass.

(c) $y(\text{in.}) = (0.87 \text{ in./unit mass})(13 \text{ unit mass}) = 11.31$ in.

36. (a)

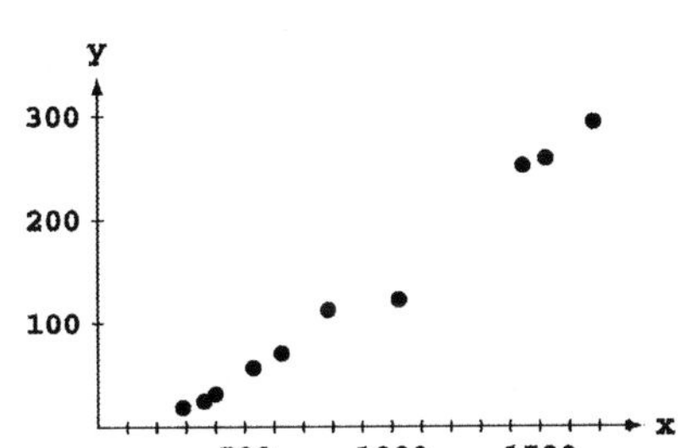

(b)

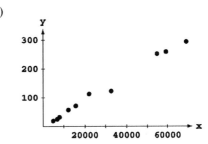

Graph (b) suggests that $y = k\,x^3$ is the better model. This graph is more linear than is graph (a).

1.3 COMBINING FUNCTIONS; SHIFTING AND SCALING GRAPHS

1. $D_f: -\infty < x < \infty$, $D_g: x \ge 1 \Rightarrow D_{f+g} = D_{fg}: x \ge 1$. $R_f: -\infty < y < \infty$, $R_g: y \ge 0$, $R_{f+g}: y \ge 1$, $R_{fg}: y \ge 0$

2. $D_f: x+1 \ge 0 \Rightarrow x \ge -1$, $D_g: x-1 \ge 0 \Rightarrow x \ge 1$. Therefore $D_{f+g} = D_{fg}: x \ge 1$.
$R_f = R_g: y \ge 0$, $R_{f+g}: y \ge \sqrt{2}$, $R_{fg}: y \ge 0$

3. $D_f: -\infty < x < \infty$, $D_g: -\infty < x < \infty$, $D_{f/g}: -\infty < x < \infty$, $D_{g/f}: -\infty < x < \infty$, $R_f: y = 2$, $R_g: y \ge 1$,
$R_{f/g}: 0 < y \le 2$, $R_{g/f}: \frac{1}{2} \le y < \infty$

4. $D_f: -\infty < x < \infty$, $D_g: x \ge 0$, $D_{f/g}: x \ge 0$, $D_{g/f}: x \ge 0$; $R_f: y = 1$, $R_g: y \ge 1$, $R_{f/g}: 0 < y \le 1$, $R_{g/f}: 1 \le y < \infty$

5. (a) 2 (b) 22 (c) $x^2 + 2$
(d) $(x+5)^2 - 3 = x^2 + 10x + 22$ (e) 5 (f) -2
(g) $x + 10$ (h) $(x^2-3)^2 - 3 = x^4 - 6x^2 + 6$

6. (a) $-\frac{1}{3}$ (b) 2 (c) $\frac{1}{x+1} - 1 = \frac{-x}{x+1}$
(d) $\frac{1}{x}$ (e) 0 (f) $\frac{3}{4}$
(g) $x - 2$ (h) $\frac{1}{\frac{1}{x+1}+1} = \frac{1}{\frac{x+2}{x+1}} = \frac{x+1}{x+2}$

7. (a) $\frac{4}{x^2} - 5$ (b) $\frac{4}{x^2} - 5$ (c) $\left(\frac{4}{x} - 5\right)^2$
(d) $\frac{1}{(4x-5)^2}$ (e) $\frac{1}{4x^2-5}$ (f) $\frac{1}{(4x-5)^2}$

8. (a) $\sqrt{x} - 8$ (b) $4\left(\sqrt{\frac{x}{4}}\right) - 8 = 4\frac{\sqrt{x}}{2} - 8 = 2\sqrt{x} - 8$ (c) $\frac{4\sqrt{x}-8}{4} = \sqrt{x} - 2$
(d) $\frac{\sqrt{4x-8}}{4} = \frac{2\sqrt{x-2}}{4} = \frac{\sqrt{x-2}}{2}$ (e) $\sqrt{\frac{4x-8}{4}} = \sqrt{x-2}$ (f) $\sqrt{x-8}$

9. (a) $(f \circ g)(x)$ (b) $(j \circ g)(x)$ (c) $(g \circ g)(x)$
(d) $(j \circ j)(x)$ (e) $(g \circ h \circ f)(x)$ (f) $(h \circ j \circ f)(x)$

10. (a) $(f \circ j)(x)$ (b) $(g \circ h)(x)$ (c) $(h \circ h)(x)$
(d) $(f \circ f)(x)$ (e) $(j \circ g \circ f)(x)$ (f) $(g \circ f \circ h)(x)$

11.

	g(x)	f(x)	(f ∘ g)(x)
(a)	$x - 7$	$\sqrt{x}$	$\sqrt{x-7}$
(b)	$x + 2$	$3x$	$3(x+2) = 3x + 6$
(c)	x^2	$\sqrt{x-5}$	$\sqrt{x^2-5}$
(d)	$\frac{x}{x-1}$	$\frac{x}{x-1}$	$\frac{\frac{x}{x-1}}{\frac{x}{x-1}-1} = \frac{x}{x-(x-1)} = x$
(e)	$\frac{1}{x-1}$	$1 + \frac{1}{x}$	x
(f)	$\frac{1}{x}$	$\frac{1}{x}$	x

12. (a) $(f \circ g)(x) = |g(x)| = \frac{1}{|x-1|}$.
(b) $(f \circ g)(x) = \frac{g(x)-1}{g(x)} = \frac{x}{x+1} \Rightarrow 1 - \frac{1}{g(x)} = \frac{x}{x+1} \Rightarrow 1 - \frac{x}{x+1} = \frac{1}{g(x)} \Rightarrow \frac{1}{x+1} = \frac{1}{g(x)}$, so $g(x) = x + 1$.
(c) Since $(f \circ g)(x) = \sqrt{g(x)} = |x|$, $g(x) = x^2$.
(d) Since $(f \circ g)(x) = f(\sqrt{x}) = |x|$, $f(x) = x^2$. (Note that the domain of the composite is $[0, \infty)$.)
The completed table is shown. Note that the absolute value sign in part (d) is optional.

g(x)	f(x)	(f∘g)(x)
$\frac{1}{x-1}$	$\|x\|$	$\frac{1}{\|x-1\|}$
$x + 1$	$\frac{x-1}{x}$	$\frac{x}{x+1}$
x^2	$\sqrt{x}$	$\|x\|$
$\sqrt{x}$	x^2	$\|x\|$

13. (a) $f(g(x)) = \sqrt{\frac{1}{x} + 1} = \sqrt{\frac{1+x}{x}}$
$g(f(x)) = \frac{1}{\sqrt{x+1}}$
(b) Domain $(f \circ g)$: $(0, \infty)$, domain $(g \circ f)$: $(-1, \infty)$
(c) Range $(f \circ g)$: $(1, \infty)$, range $(g \circ f)$: $(0, \infty)$

14. (a) $f(g(x)) = 1 - 2\sqrt{x} + x$
$g(f(x)) = 1 - |x|$
(b) Domain $(f \circ g)$: $(0, \infty)$, domain $(g \circ f)$: $(0, \infty)$
(c) Range $(f \circ g)$: $(0, \infty)$, range $(g \circ f)$: $(-\infty, 1)$

15. (a) $y = -(x+7)^2$ (b) $y = -(x-4)^2$

16. (a) $y = x^2 + 3$ (b) $y = x^2 - 5$

17. (a) Position 4 (b) Position 1 (c) Position 2 (d) Position 3

18. (a) $y = -(x-1)^2 + 4$ (b) $y = -(x+2)^2 + 3$ (c) $y = -(x+4)^2 - 1$ (d) $y = -(x-2)^2$

19.

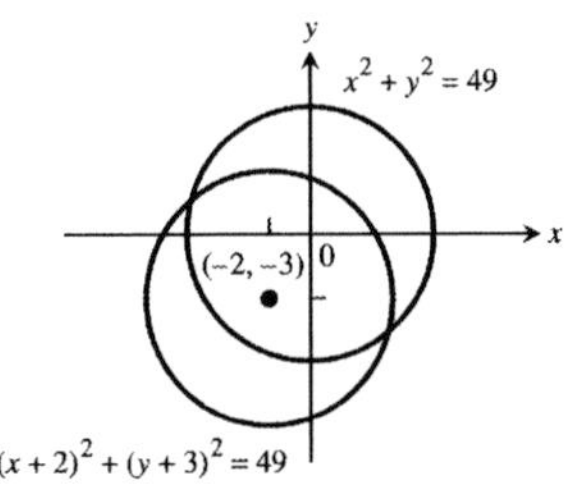

20.

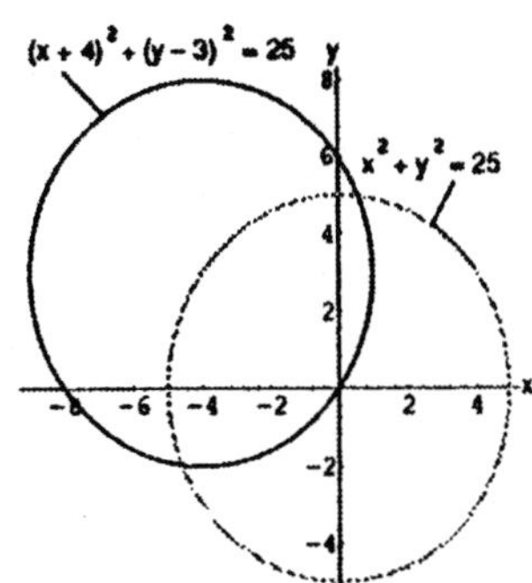

21.

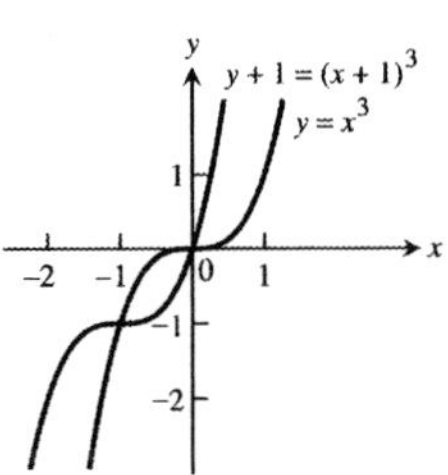

22.

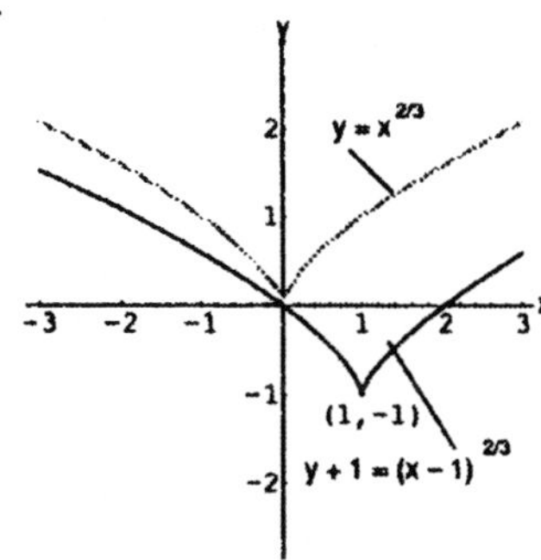

23.

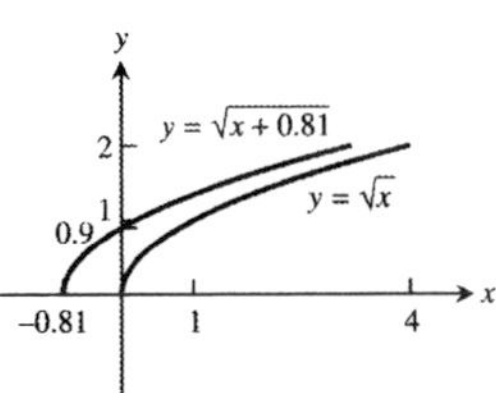

24.

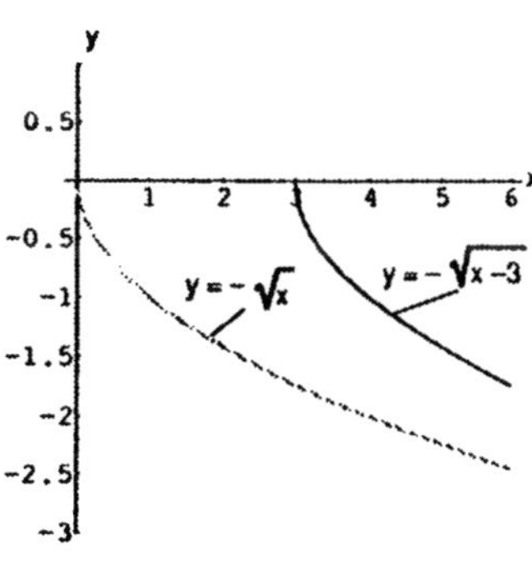

25.

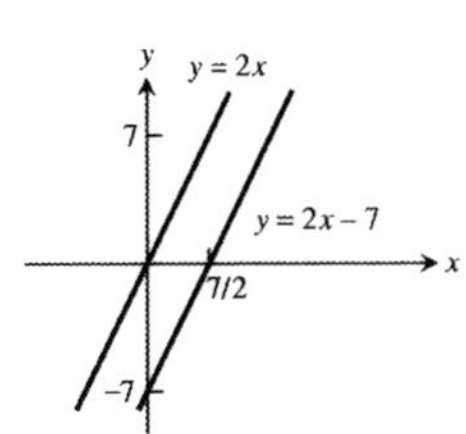

26.

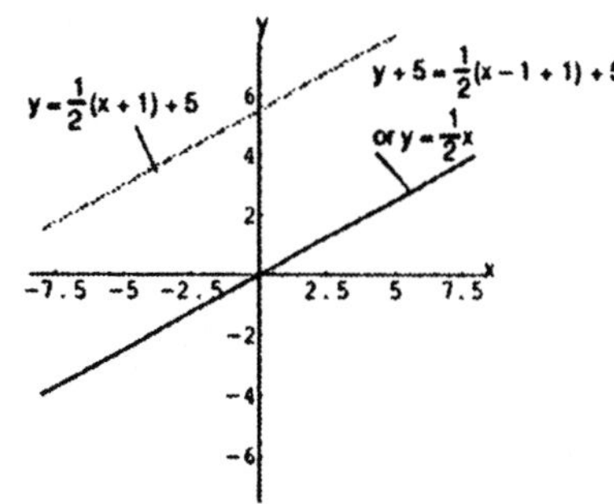

27.

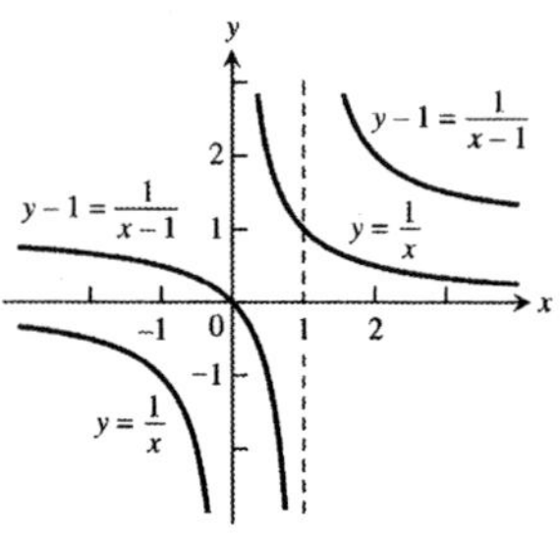

28.

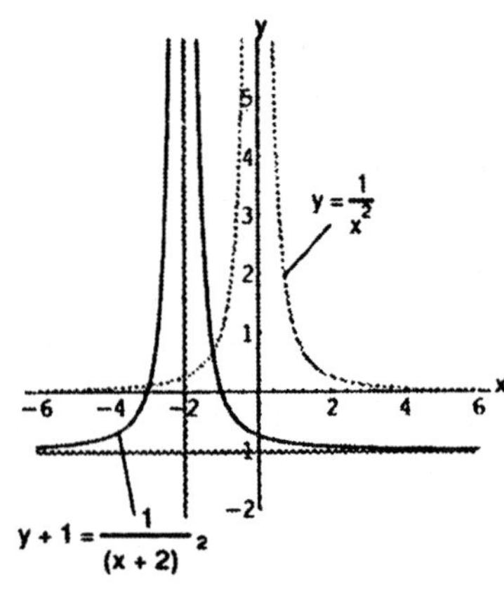

29.

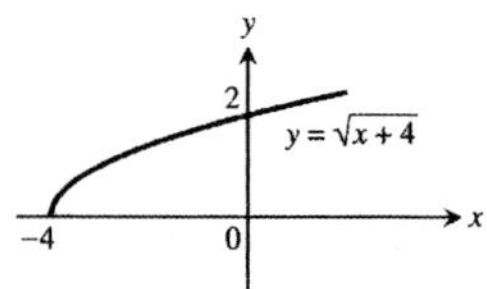

30.

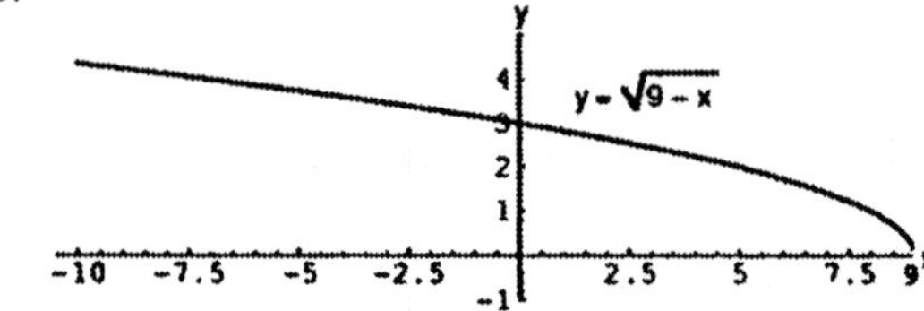

31.

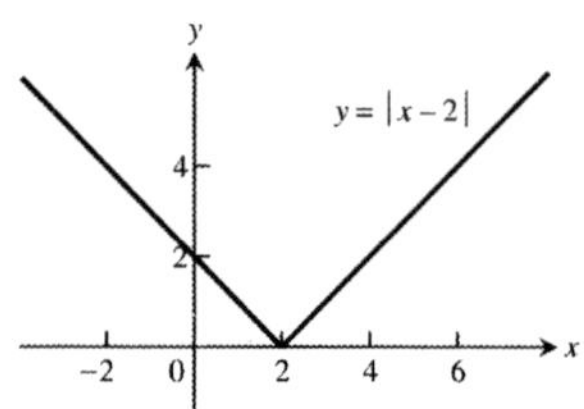

32.

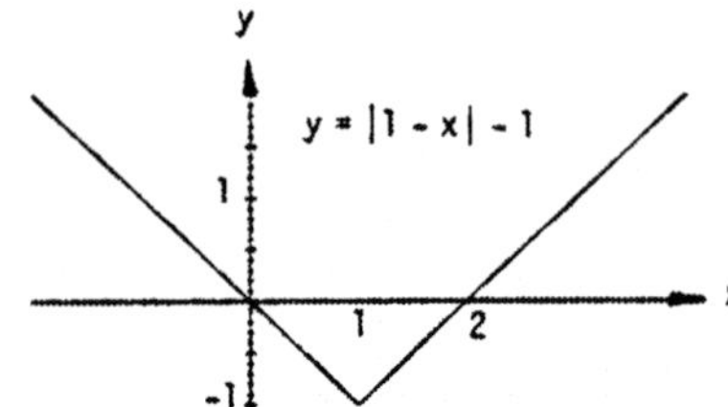

33.

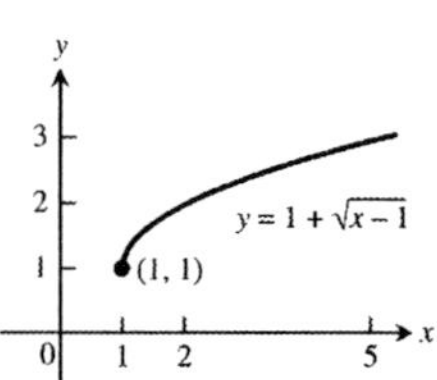

34.

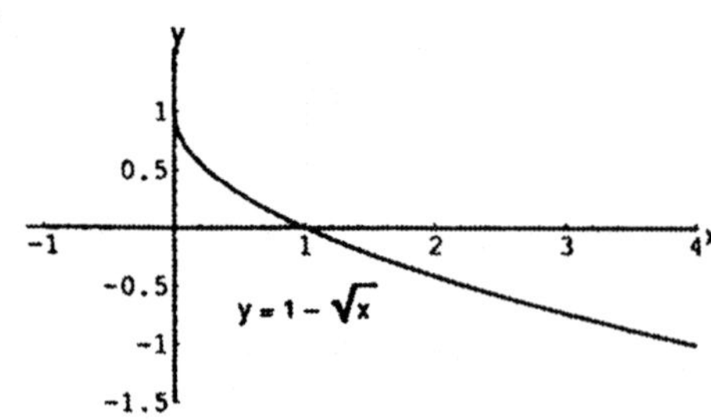

35.

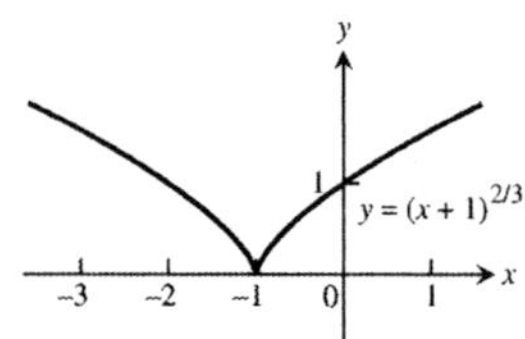

36.

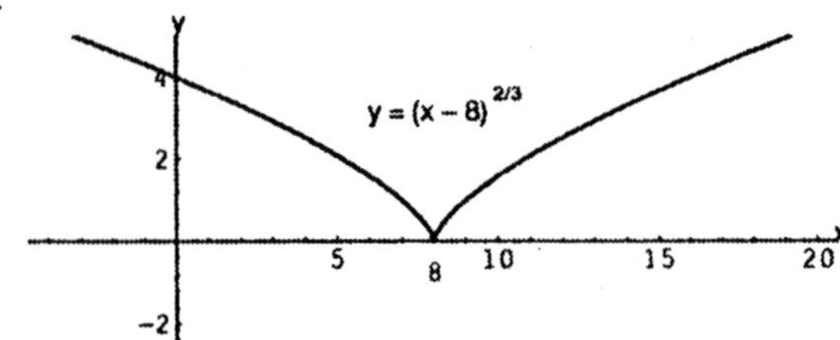

37.

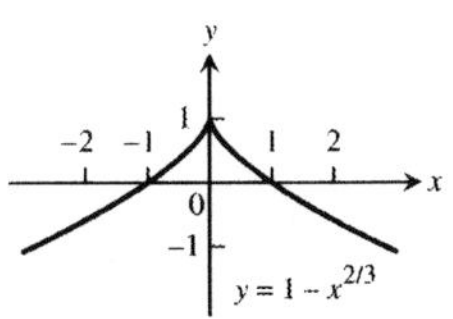

38.

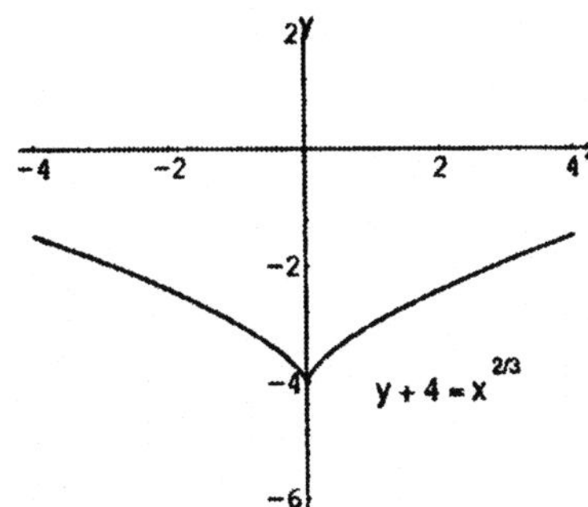

39.

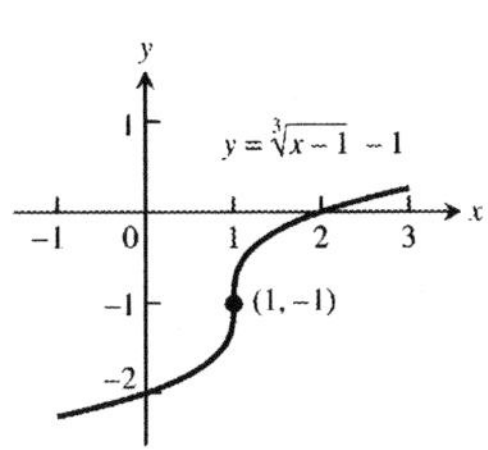

40.

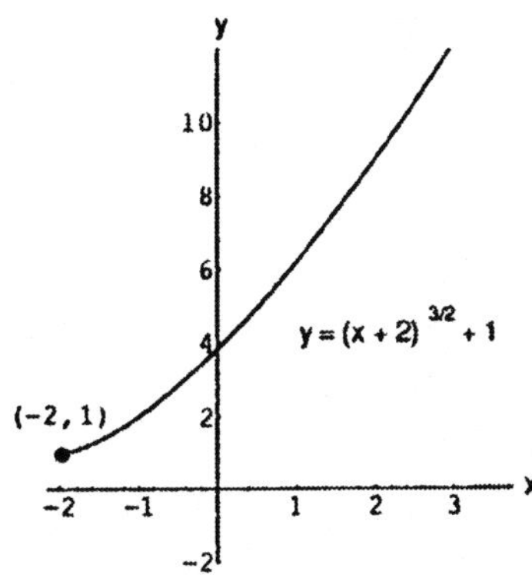

41.

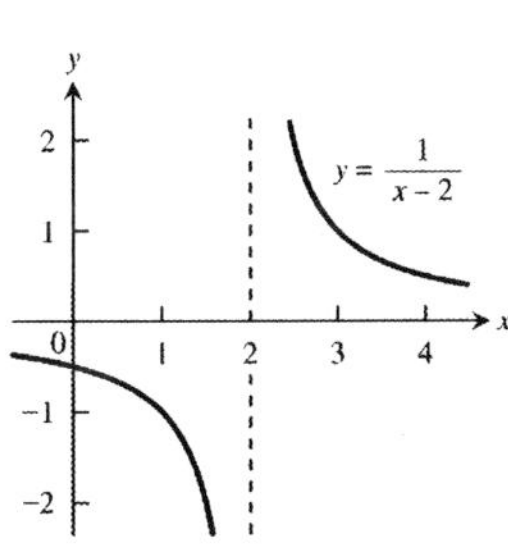

42.

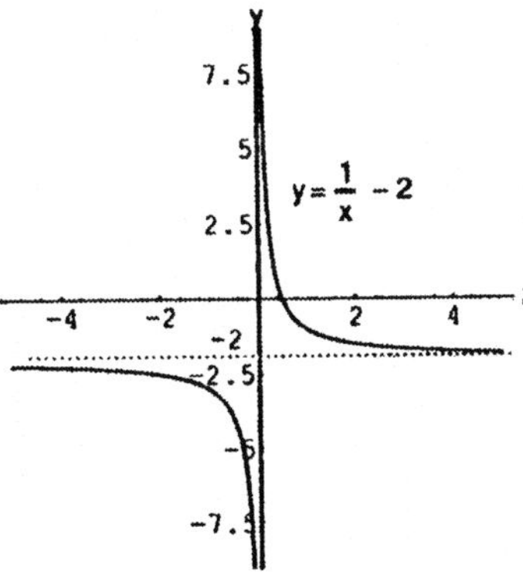

43.

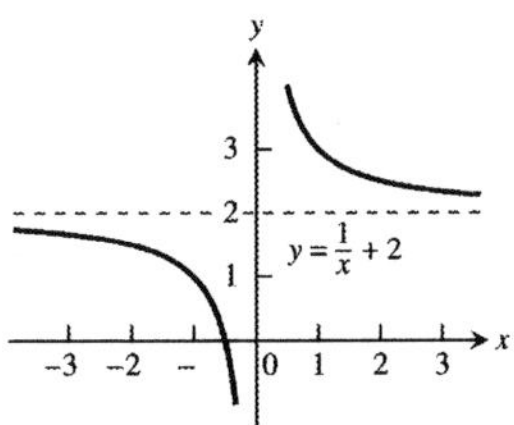

44.

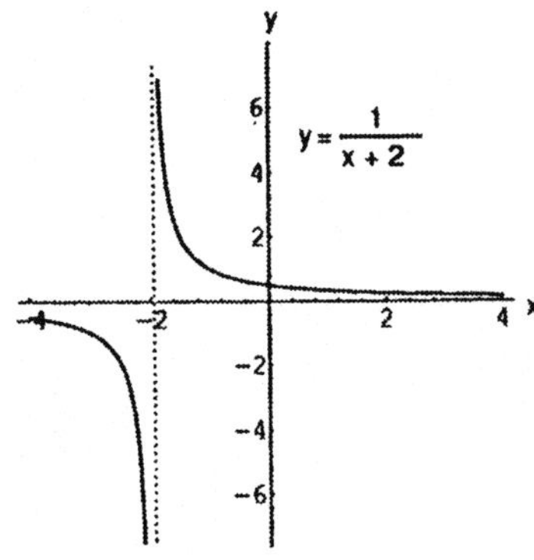

45.

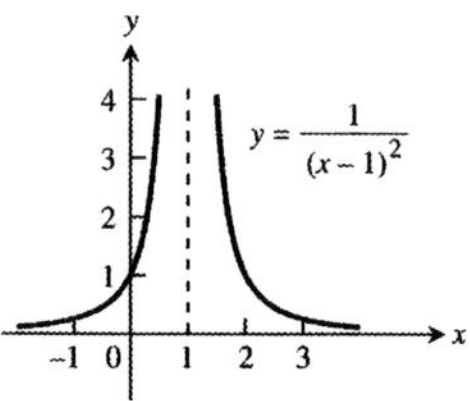

46.

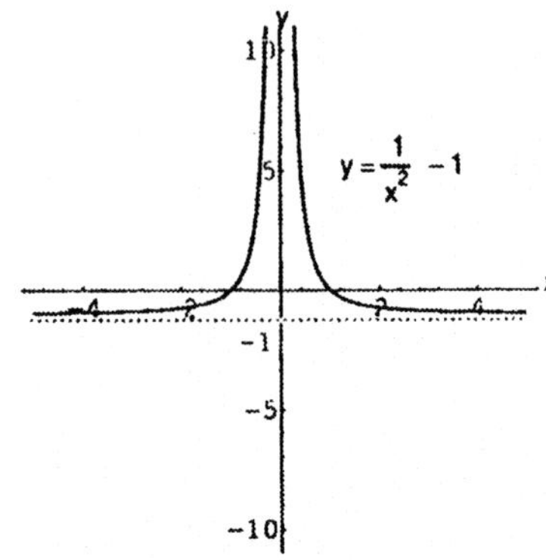

47.

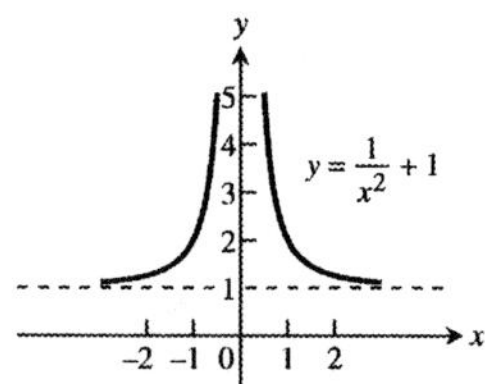

48.

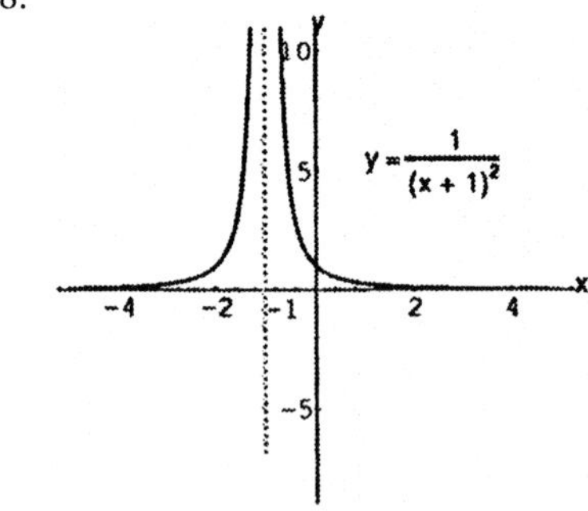

49. (a) domain: $[0, 2]$; range: $[2, 3]$

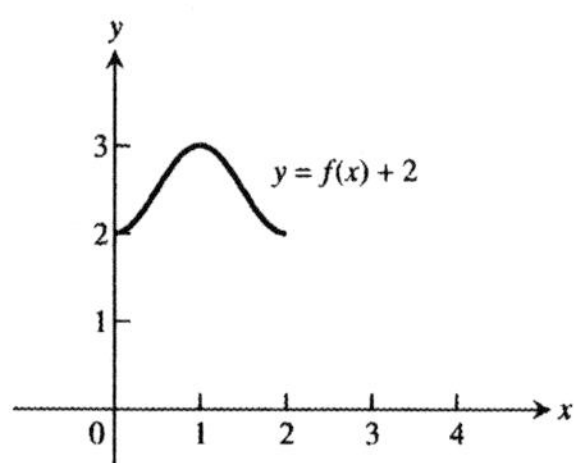

(b) domain: $[0, 2]$; range: $[-1, 0]$

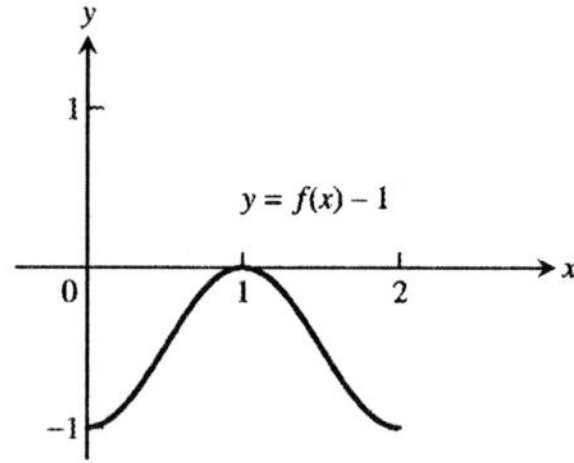

(c) domain: $[0, 2]$; range: $[0, 2]$

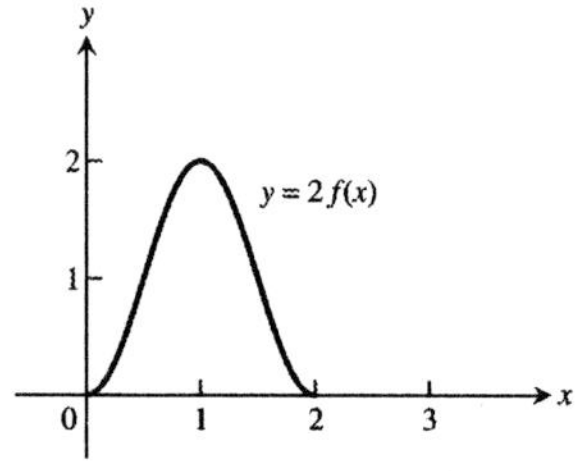

(d) domain: $[0, 2]$; range: $[-1, 0]$

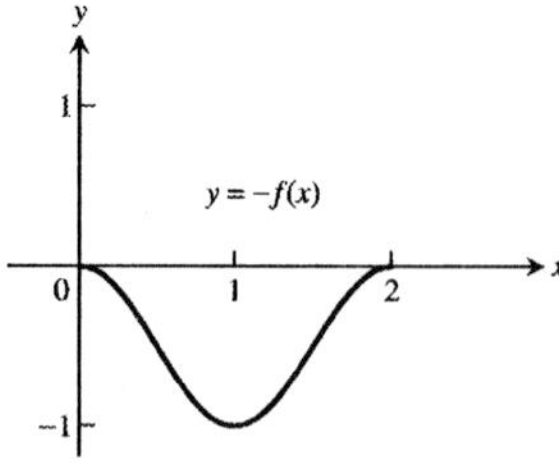

(e) domain: $[-2, 0]$; range: $[0, 1]$

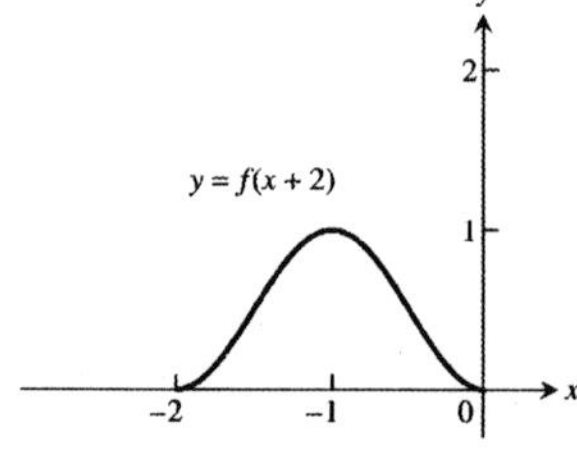

(f) domain: $[1, 3]$; range: $[0, 1]$

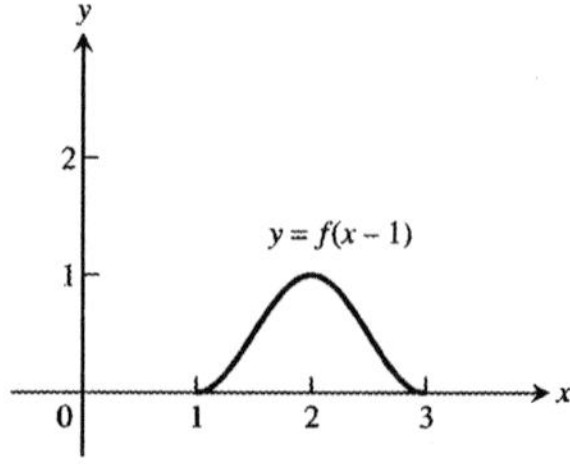

(g) domain: $[-2, 0]$; range: $[0, 1]$

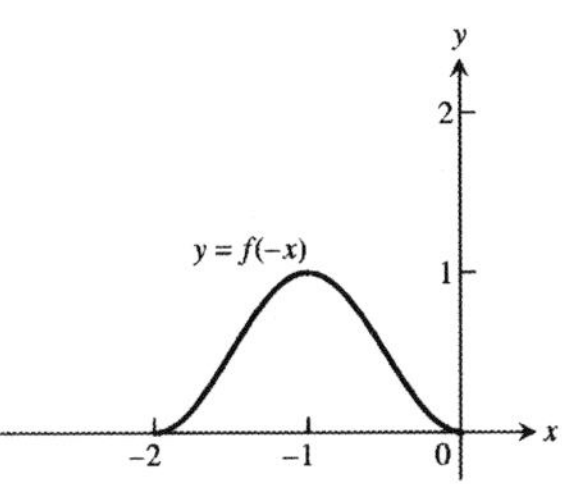

(h) domain: $[-1, 1]$; range: $[0, 1]$

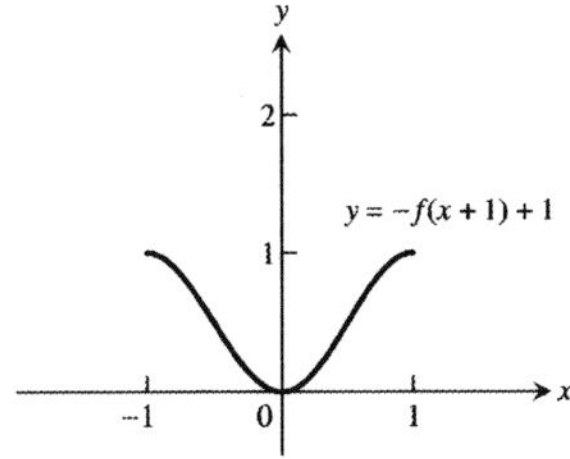

50. (a) domain: $[0, 4]$; range: $[-3, 0]$

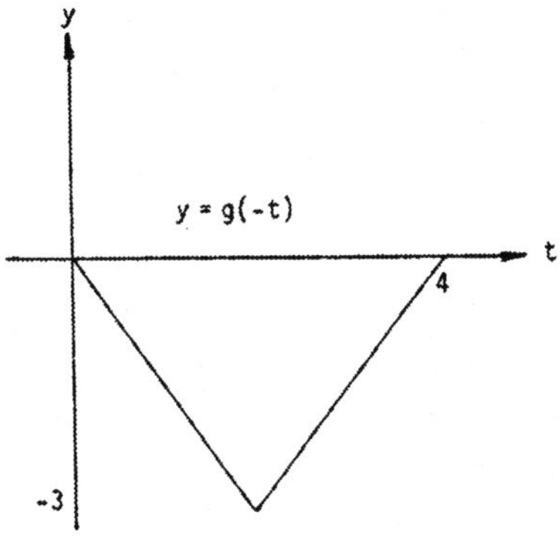

(b) domain: $[-4, 0]$; range: $[0, 3]$

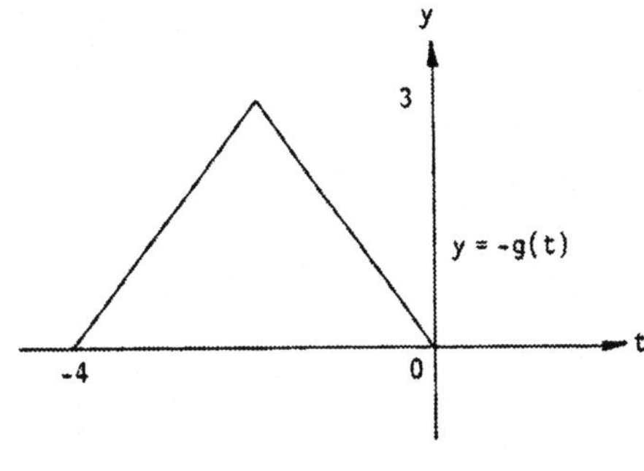

(c) domain: $[-4, 0]$; range: $[0, 3]$

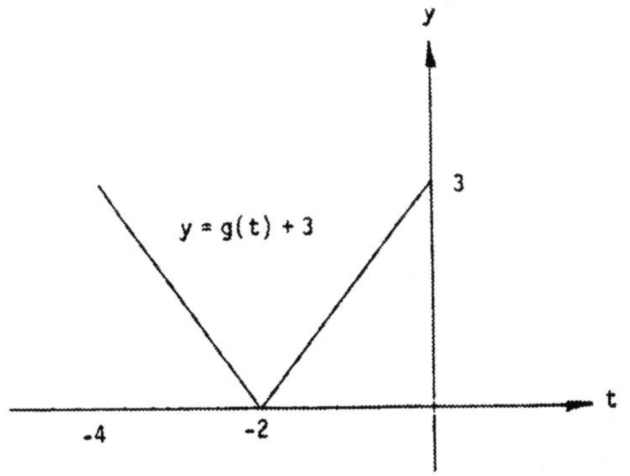

(d) domain: $[-4, 0]$; range: $[1, 4]$

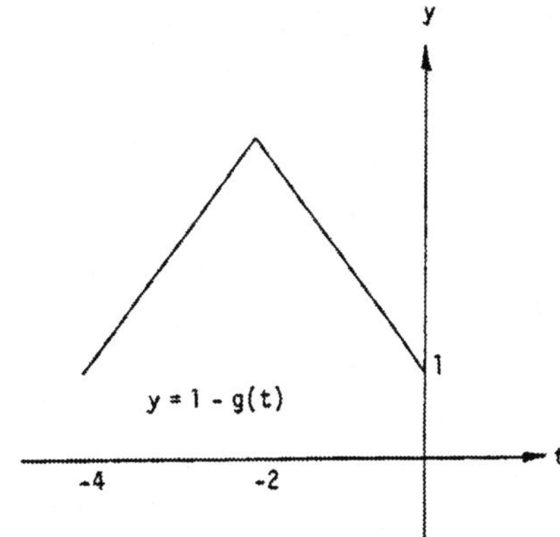

(e) domain: $[2, 4]$; range: $[-3, 0]$

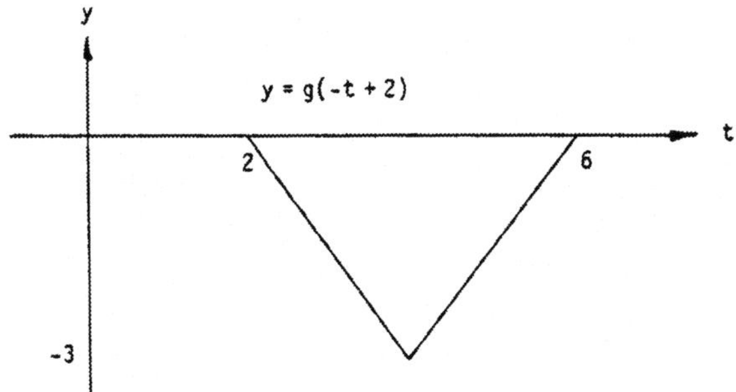

(f) domain: $[-2, 2]$; range: $[-3, 0]$

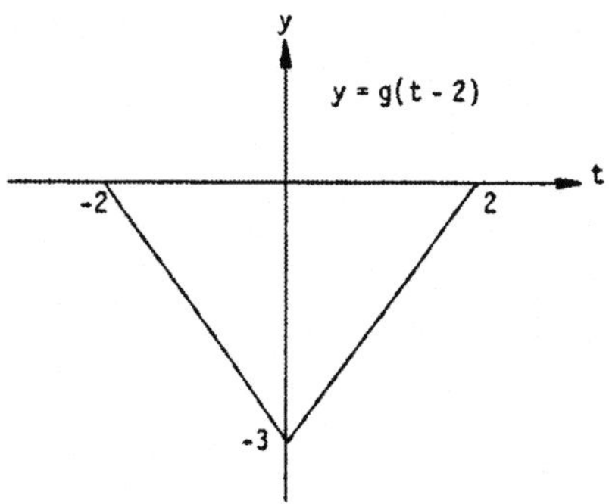

(g) domain: $[1, 5]$; range: $[-3, 0]$

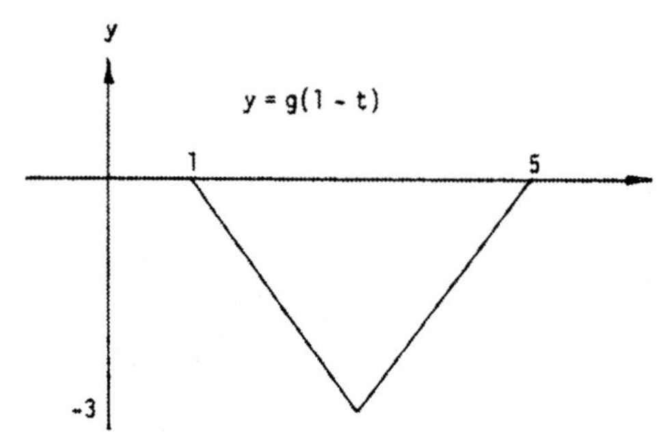

(h) domain: $[0, 4]$; range: $[0, 3]$

51. $y = 3x^2 - 3$

52. $y = (2x)^2 - 1 = 4x^2 - 1$

53. $y = \frac{1}{2}\left(1 + \frac{1}{x^2}\right) = \frac{1}{2} + \frac{1}{2x^2}$

54. $y = 1 + \frac{1}{(x/3)^2} = 1 + \frac{9}{x^2}$

55. $y = \sqrt{4x + 1}$

56. $y = 3\sqrt{x + 1}$

57. $y = \sqrt{4 - \left(\frac{x}{2}\right)^2} = \frac{1}{2}\sqrt{16 - x^2}$

58. $y = \frac{1}{3}\sqrt{4 - x^2}$

59. $y = 1 - (3x)^3 = 1 - 27x^3$

60. $y = 1 - \left(\frac{x}{2}\right)^3 = 1 - \frac{x^3}{8}$

61. Let $y = -\sqrt{2x + 1} = f(x)$ and let $g(x) = x^{1/2}$, $h(x) = \left(x + \frac{1}{2}\right)^{1/2}$, $i(x) = \sqrt{2}\left(x + \frac{1}{2}\right)^{1/2}$, and $j(x) = -\left[\sqrt{2}\left(x + \frac{1}{2}\right)^{1/2}\right] = f(x)$. The graph of $h(x)$ is the graph of $g(x)$ shifted left $\frac{1}{2}$ unit; the graph of $i(x)$ is the graph of $h(x)$ stretched vertically by a factor of $\sqrt{2}$; and the graph of $j(x) = f(x)$ is the graph of $i(x)$ reflected across the x-axis.

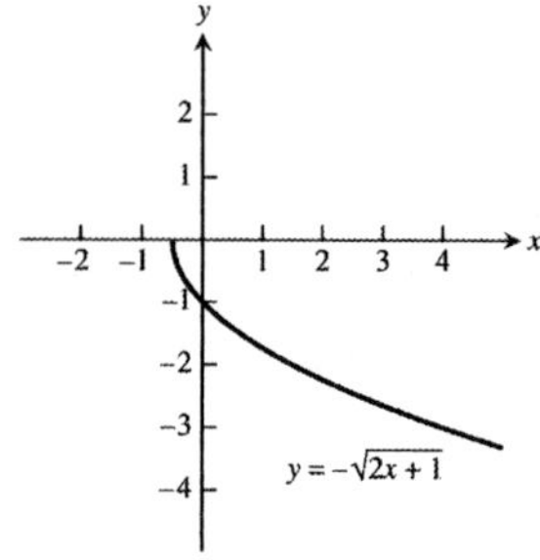

62. Let $y = \sqrt{1 - \frac{x}{2}} = f(x)$. Let $g(x) = (-x)^{1/2}$, $h(x) = (-x + 2)^{1/2}$, and $i(x) = \frac{1}{\sqrt{2}}(-x + 2)^{1/2} = \sqrt{1 - \frac{x}{2}} = f(x)$. The graph of $g(x)$ is the graph of $y = \sqrt{x}$ reflected across the x-axis. The graph of $h(x)$ is the graph of $g(x)$ shifted right two units. And the graph of $i(x)$ is the graph of $h(x)$ compressed vertically by a factor of $\sqrt{2}$.

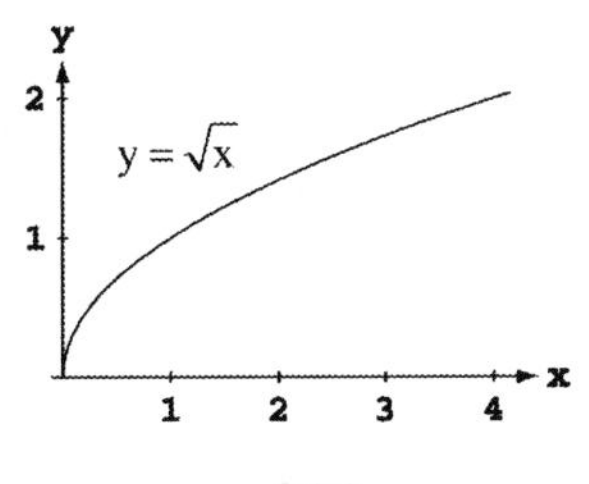

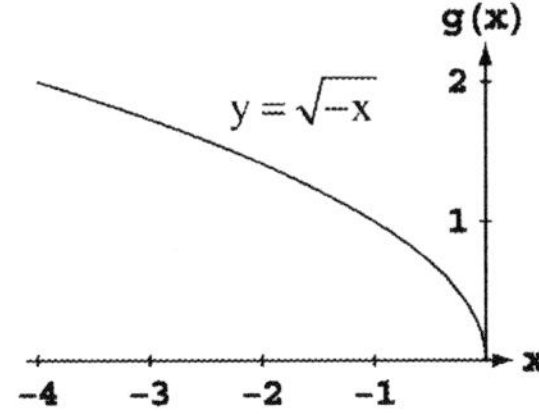

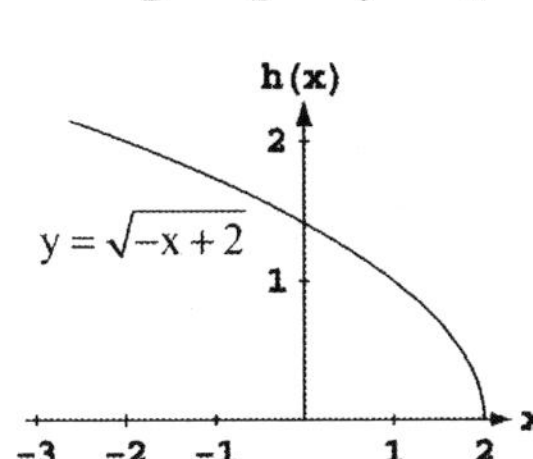

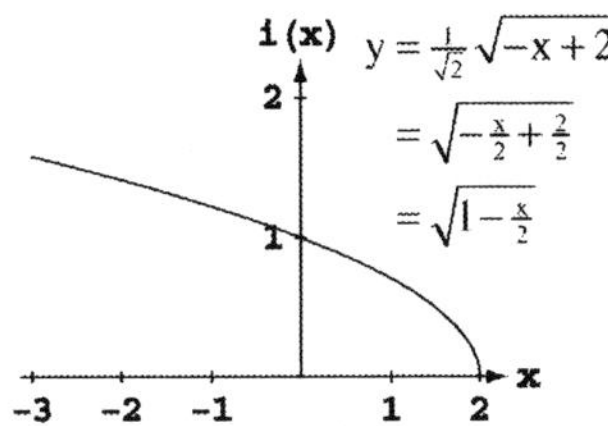

63. $y = f(x) = x^3$. Shift $f(x)$ one unit right followed by a shift two units up to get $g(x) = (x - 1)^3 + 2$.

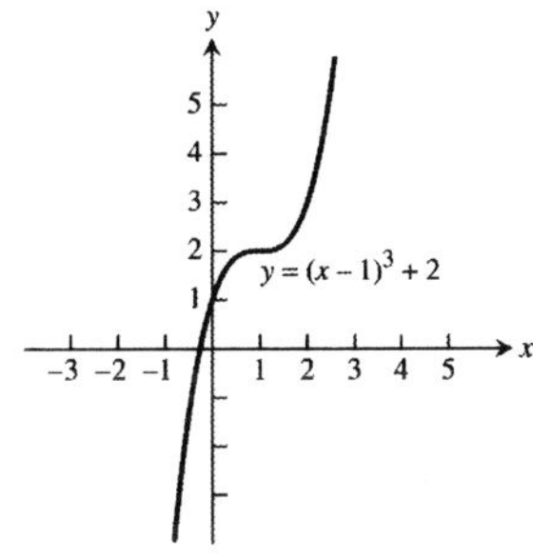

64. $y = (1 - x)^3 + 2 = -[(x - 1)^3 + (-2)] = f(x)$. Let $g(x) = x^3$, $h(x) = (x - 1)^3$, $i(x) = (x - 1)^3 + (-2)$, and $j(x) = -[(x - 1)^3 + (-2)]$. The graph of $h(x)$ is the graph of $g(x)$ shifted right one unit; the graph of $i(x)$ is the graph of $h(x)$ shifted down two units; and the graph of $f(x)$ is the graph of $i(x)$ reflected across the x-axis.

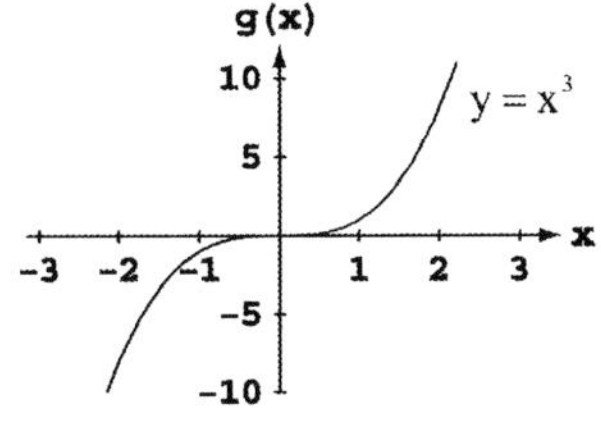

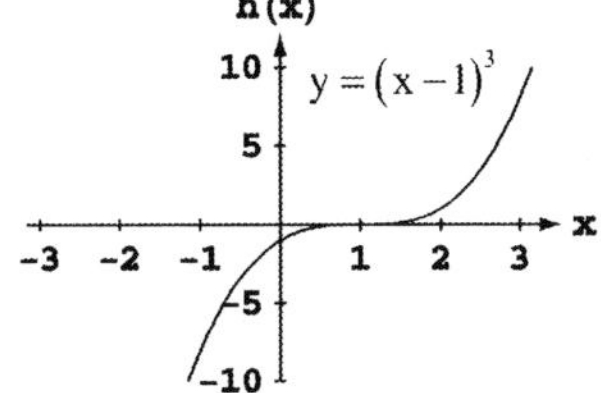

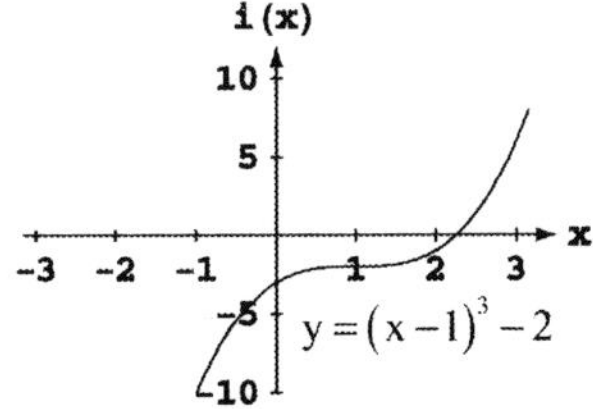

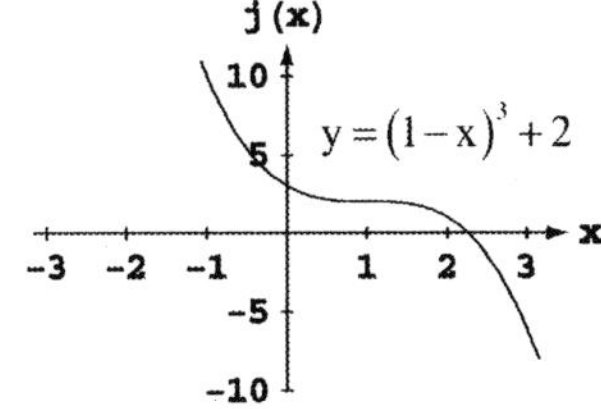

65. Compress the graph of $f(x) = \frac{1}{x}$ horizontally by a factor of 2 to get $g(x) = \frac{1}{2x}$. Then shift $g(x)$ vertically down 1 unit to get $h(x) = \frac{1}{2x} - 1$.

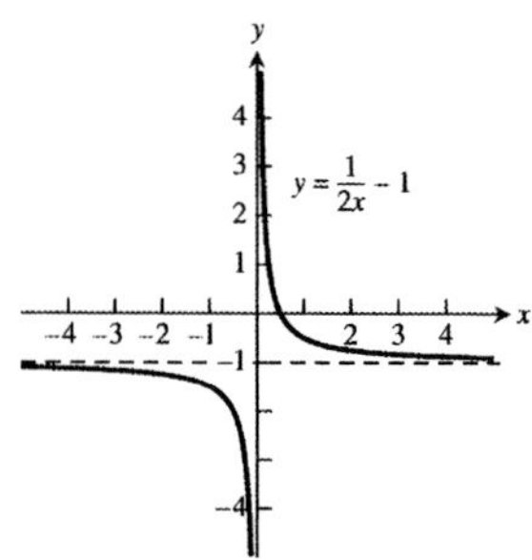

66. Let $f(x) = \frac{1}{x^2}$ and $g(x) = \frac{2}{x^2} + 1 = \frac{1}{\left(\frac{x^2}{2}\right)} + 1 = \frac{1}{\left(x/\sqrt{2}\right)^2} + 1 = \frac{1}{\left[\left(1/\sqrt{2}\right)x\right]^2} + 1$. Since $\sqrt{2} \approx 1.4$, we see that the graph of $f(x)$ stretched horizontally by a factor of 1.4 and shifted up 1 unit is the graph of $g(x)$.

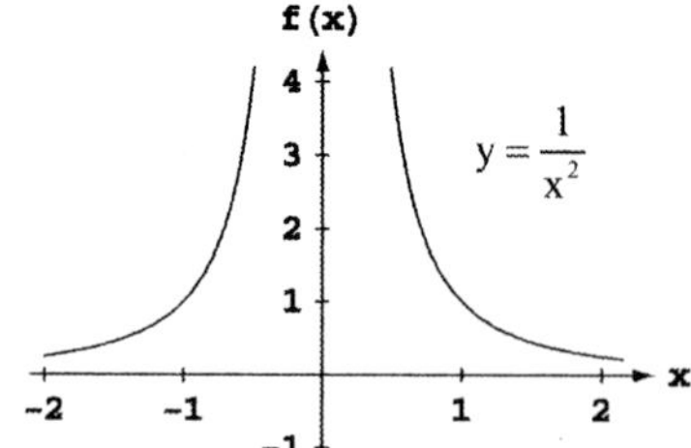

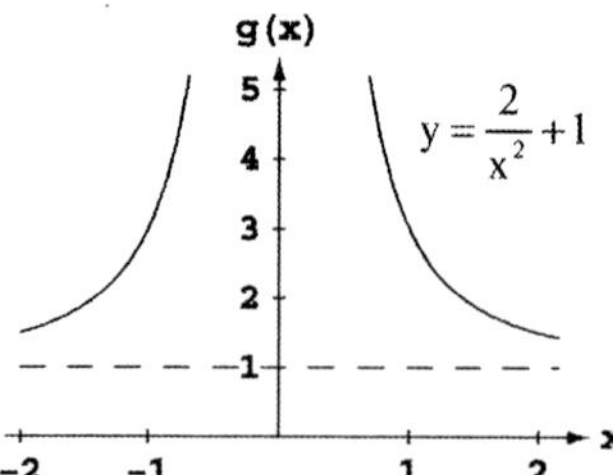

67. Reflect the graph of $y = f(x) = \sqrt[3]{x}$ across the x-axis to get $g(x) = -\sqrt[3]{x}$.

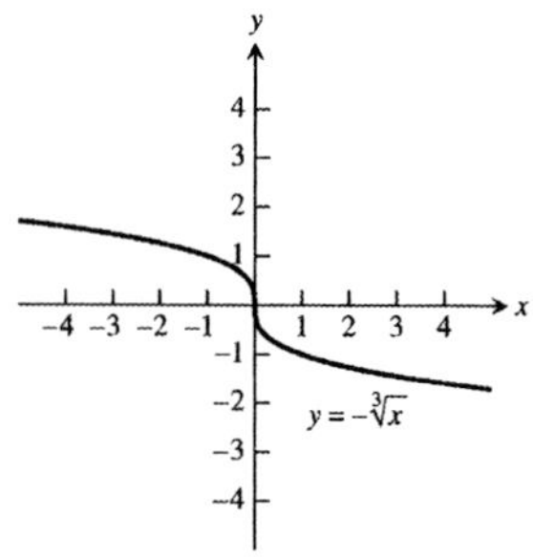

68. $y = f(x) = (-2x)^{2/3} = [(-1)(2)x]^{2/3} = (-1)^{2/3}(2x)^{2/3} = (2x)^{2/3}$. So the graph of $f(x)$ is the graph of $g(x) = x^{2/3}$ compressed horizontally by a factor of 2.

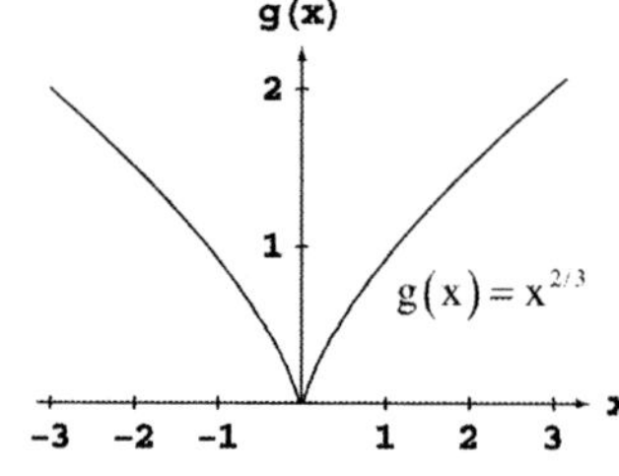

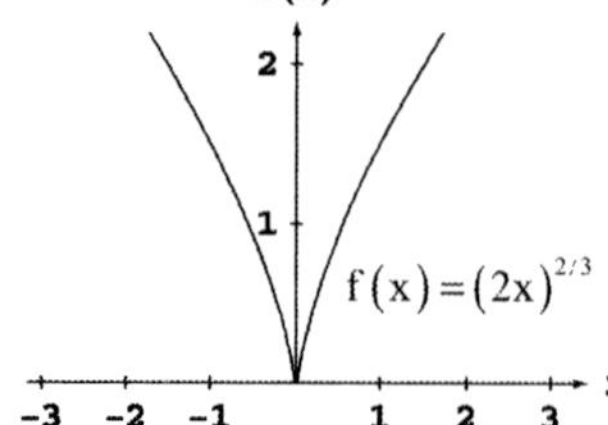

69.

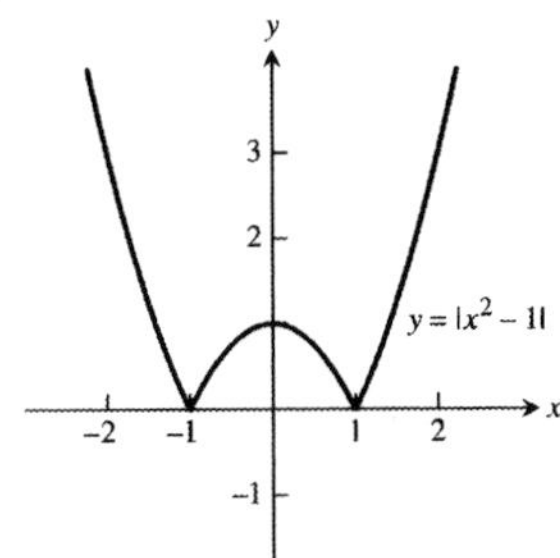

70.

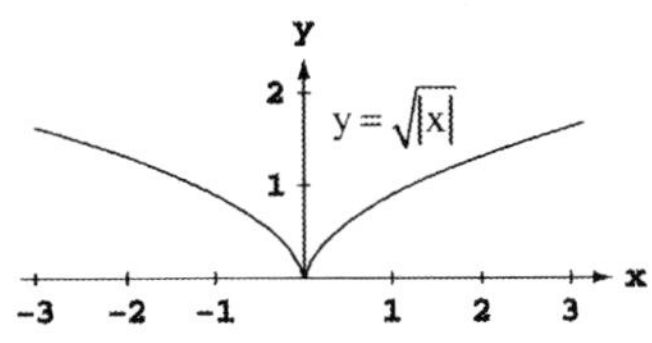

71. $9x^2 + 25y^2 = 225 \Rightarrow \frac{x^2}{5^2} + \frac{y^2}{3^2} = 1$

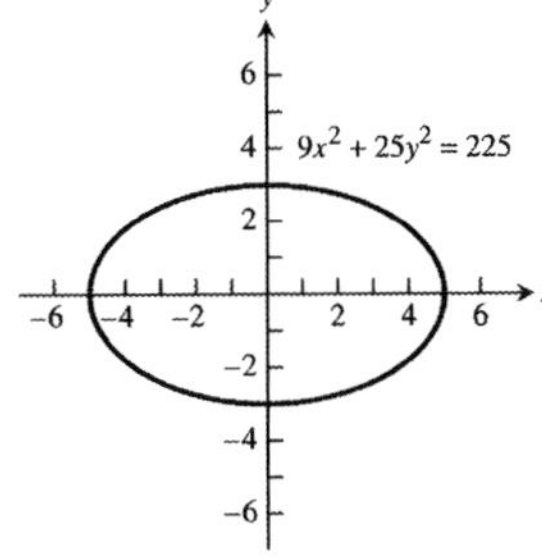

72. $16x^2 + 7y^2 = 112 \Rightarrow \frac{x^2}{\left(\sqrt{7}\right)^2} + \frac{y^2}{4^2} = 1$

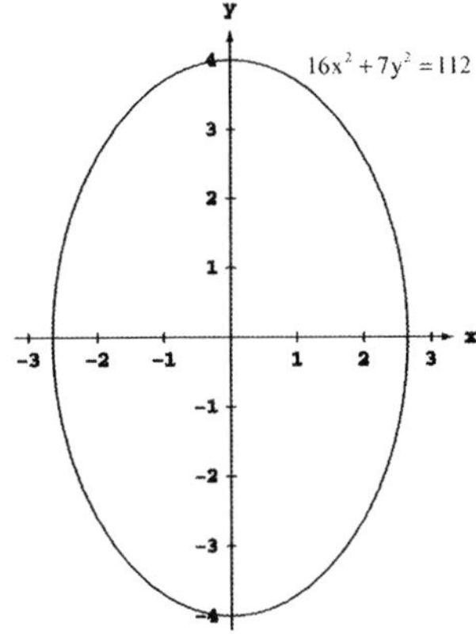

73. $3x^2 + (y-2)^2 = 3 \Rightarrow \frac{x^2}{1^2} + \frac{(y-2)^2}{\left(\sqrt{3}\right)^2} = 1$

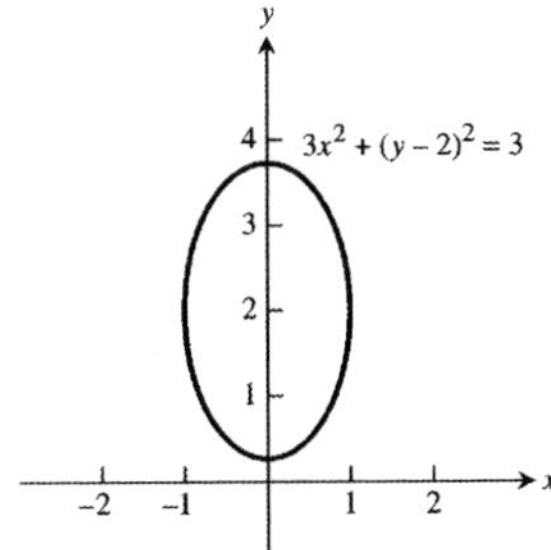

74. $(x+1)^2 + 2y^2 = 4 \Rightarrow \frac{[x-(-1)]^2}{2^2} + \frac{y^2}{\left(\sqrt{2}\right)^2} = 1$

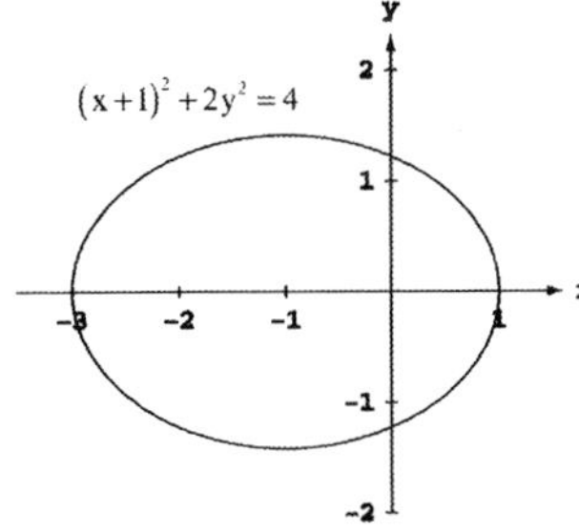

75. $3(x-1)^2 + 2(y+2)^2 = 6$

$\Rightarrow \frac{(x-1)^2}{\left(\sqrt{2}\right)^2} + \frac{[y-(-2)]^2}{\left(\sqrt{3}\right)^2} = 1$

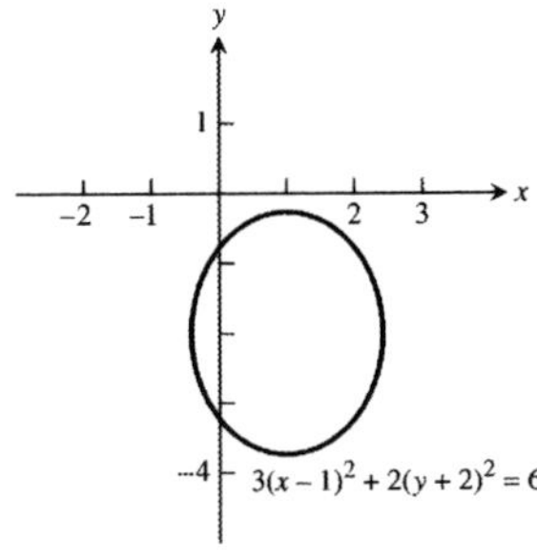

76. $6\left(x+\frac{3}{2}\right)^2 + 9\left(y-\frac{1}{2}\right)^2 = 54$

$\Rightarrow \frac{\left[x-\left(-\frac{3}{2}\right)\right]^2}{3^2} + \frac{\left(y-\frac{1}{2}\right)^2}{\left(\sqrt{6}\right)^2} = 1$

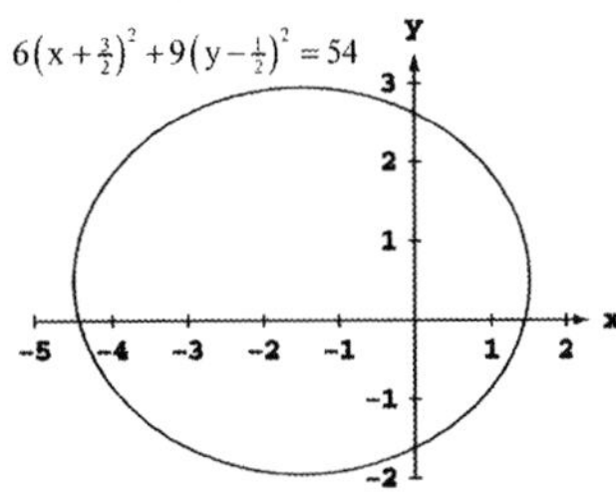

77. $\frac{x^2}{16} + \frac{y^2}{9} = 1$ has its center at $(0, 0)$. Shiftinig 4 units left and 3 units up gives the center at $(h, k) = (-4, 3)$. So the equation is $\frac{[x-(-4)]^2}{4^2} + \frac{(y-3)^2}{3^2} = 1 \Rightarrow \frac{(x+4)^2}{4^2} + \frac{(y-3)^2}{3^2} = 1$. Center, C, is $(-4, 3)$, and major axis, $\overline{AB}$, is the segment from $(-8, 3)$ to $(0, 3)$.

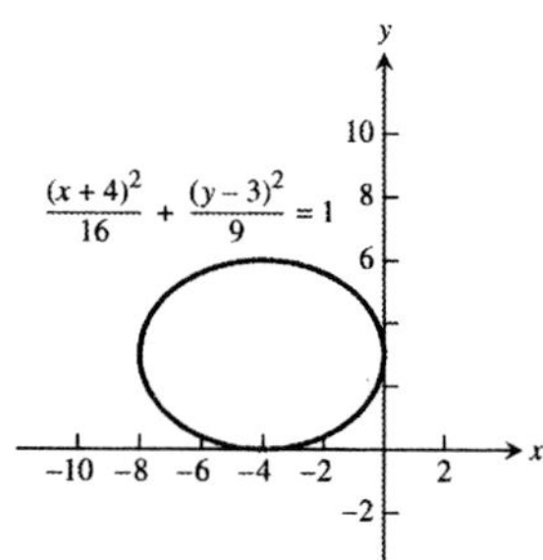

78. The ellipse $\frac{x^2}{4} + \frac{y^2}{25} = 1$ has center $(h, k) = (0, 0)$. Shifting the ellipse 3 units right and 2 units down produces an ellipse with center at $(h, k) = (3, -2)$ and an equation $\frac{(x-3)^2}{4} + \frac{[y-(-2)]^2}{25} = 1$. Center, C, is $(3, -2)$, and $\overline{AB}$, the segment from $(3, 3)$ to $(3, -7)$ is the major axis.

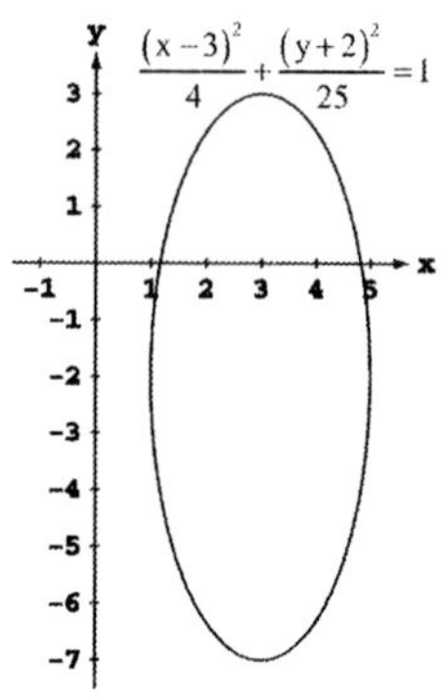

79. (a) $(fg)(-x) = f(-x)g(-x) = f(x)(-g(x)) = -(fg)(x)$, odd
    (b) $\left(\frac{f}{g}\right)(-x) = \frac{f(-x)}{g(-x)} = \frac{f(x)}{-g(x)} = -\left(\frac{f}{g}\right)(x)$, odd
    (c) $\left(\frac{g}{f}\right)(-x) = \frac{g(-x)}{f(-x)} = \frac{-g(x)}{f(x)} = -\left(\frac{g}{f}\right)(x)$, odd
    (d) $f^2(-x) = f(-x)f(-x) = f(x)f(x) = f^2(x)$, even
    (e) $g^2(-x) = (g(-x))^2 = (-g(x))^2 = g^2(x)$, even
    (f) $(f \circ g)(-x) = f(g(-x)) = f(-g(x)) = f(g(x)) = (f \circ g)(x)$, even
    (g) $(g \circ f)(-x) = g(f(-x)) = g(f(x)) = (g \circ f)(x)$, even
    (h) $(f \circ f)(-x) = f(f(-x)) = f(f(x)) = (f \circ f)(x)$, even
    (i) $(g \circ g)(-x) = g(g(-x)) = g(-g(x)) = -g(g(x)) = -(g \circ g)(x)$, odd

80. Yes, $f(x) = 0$ is both even and odd since $f(-x) = 0 = f(x)$ and $f(-x) = 0 = -f(x)$.

81. (a) 

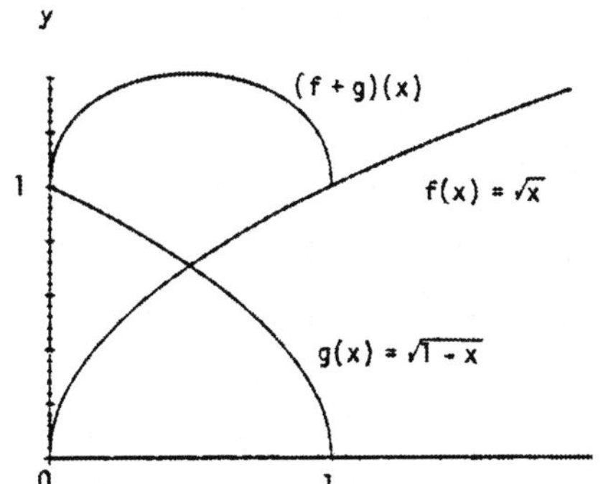

(b) 

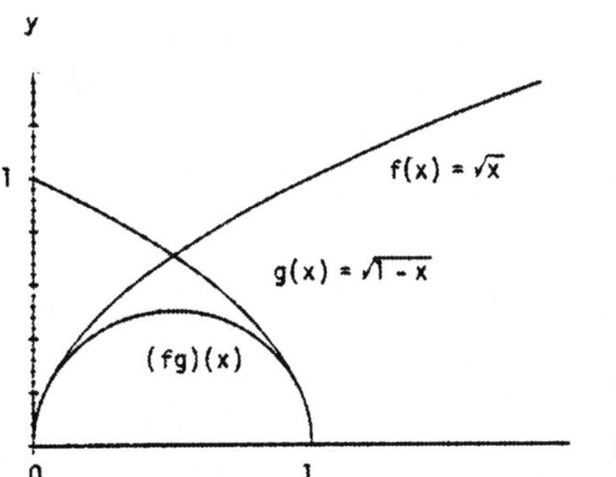

(c)

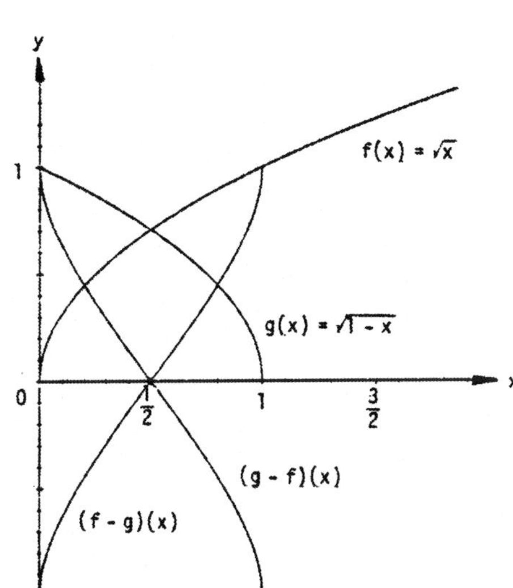

(d) 

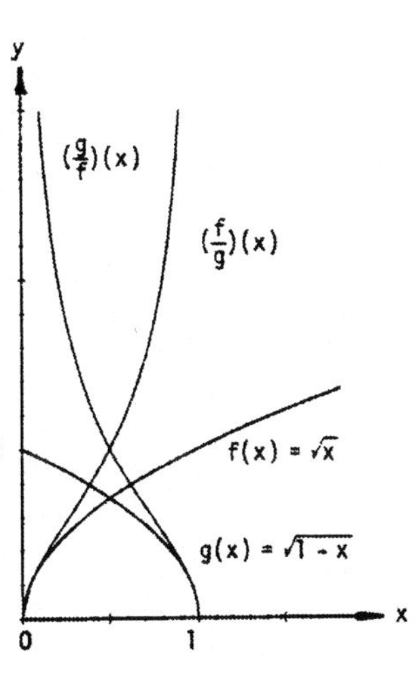

82.

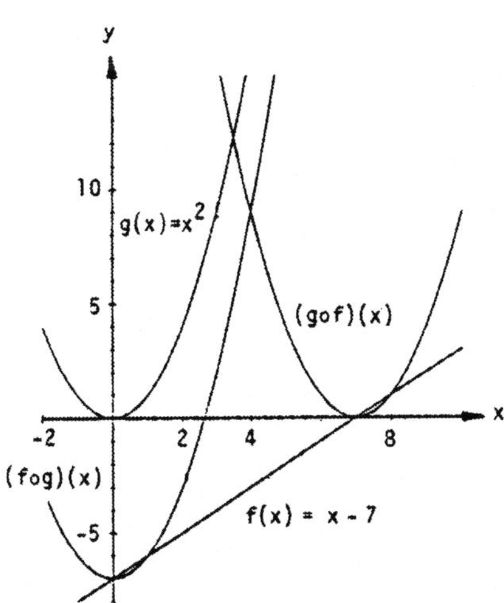

83. $A = 2$, $B = 2\pi$, $C = -\pi$, $D = -1$

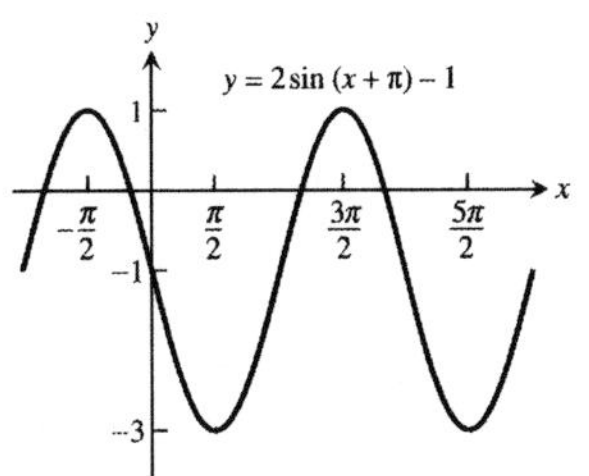

84. $A = \frac{1}{2}$, $B = 2$, $C = 1$, $D = \frac{1}{2}$

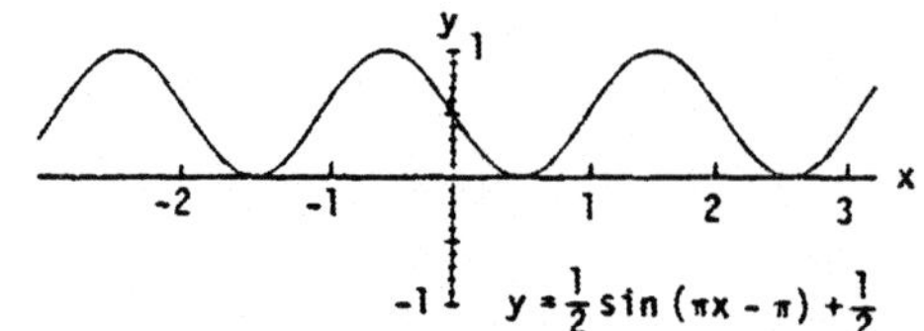

85. $A = -\frac{2}{\pi}$, $B = 4$, $C = 0$, $D = \frac{1}{\pi}$

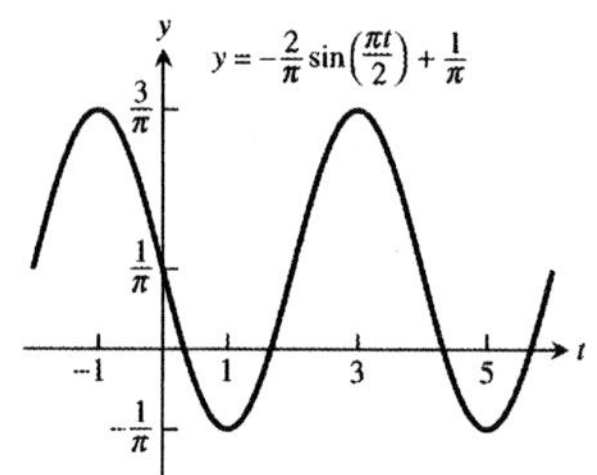

86. $A = \frac{L}{2\pi}$, $B = L$, $C = 0$, $D = 0$

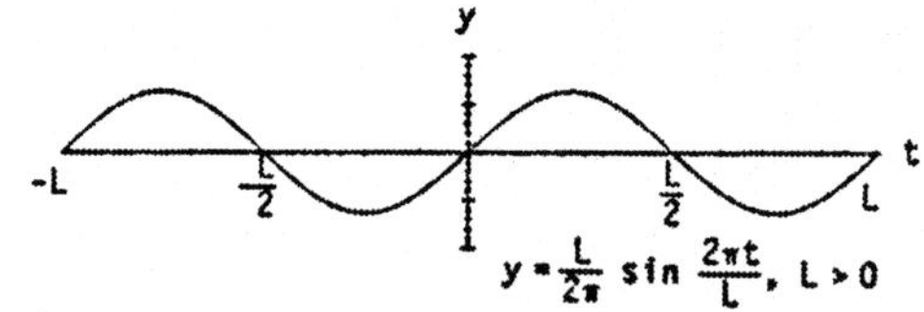

87-90. Example CAS commands:

Maple

```
f := x -> A*sin((2*Pi/B)*(x-C))+D1;
A:=3; C:=0; D1:=0;
f_list := [seq( f(x), B=[1,3,2*Pi,5*Pi] )];
plot( f_list, x=-4*Pi..4*Pi, scaling=constrained,
      color=[red,blue,green,cyan], linestyle=[1,3,4,7],
      legend=["B=1","B=3","B=2*Pi","B=3*Pi"],
      title="#67 (Section 1.6)" );
```

Mathematica

```
Clear[a, b, c, d, f, x]
f[x_]:=a Sin[2π/b (x – c)] + d
Plot[f[x]/.{a → 3, b → 1, c → 0, d → 0}, {x, –4π, 4π }]
```

87. (a) The graph stretches horizontally.

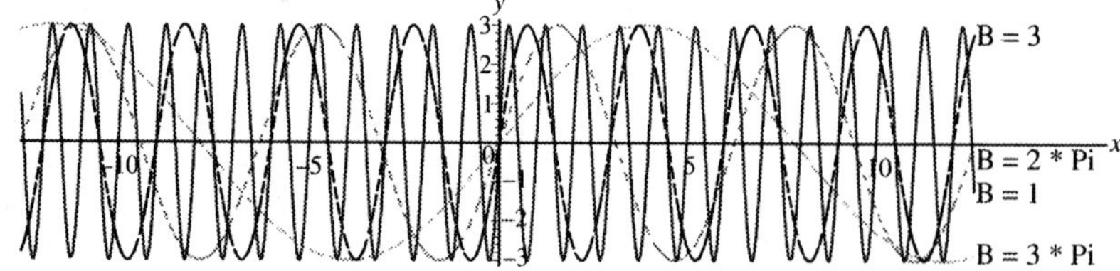

(b) The period remains the same: period $= |B|$. The graph has a horizontal shift of $\frac{1}{2}$ period.

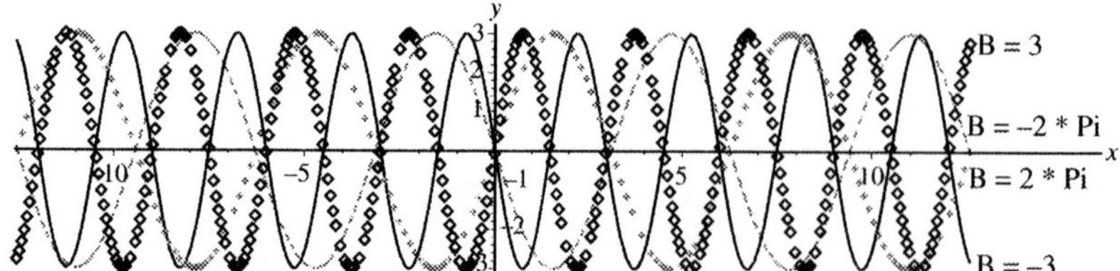

88. (a) The graph is shifted right C units.

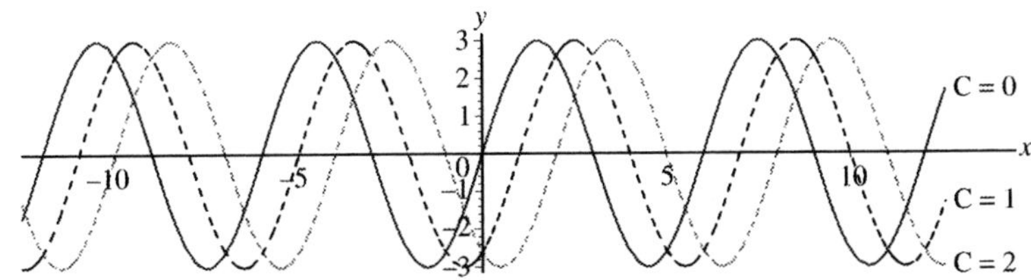

(b) The graph is shifted left C units.

(c) A shift of $\pm$ one period will produce no apparent shift. $|C| = 6$

89. The graph shifts upwards $|D|$ units for $D > 0$ and down $|D|$ units for $D < 0$.

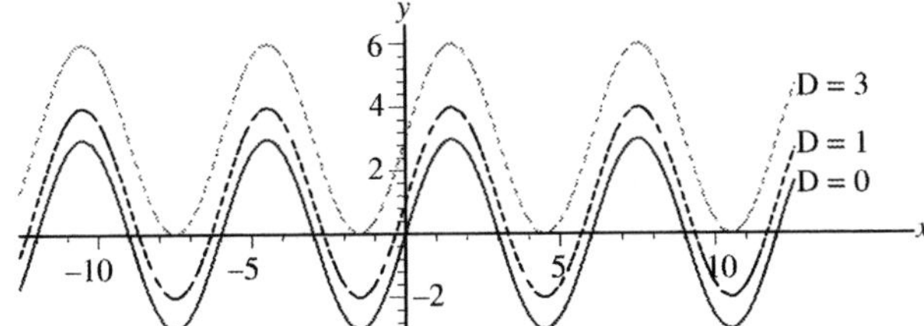

90. (a) The graph stretches $|A|$ units.

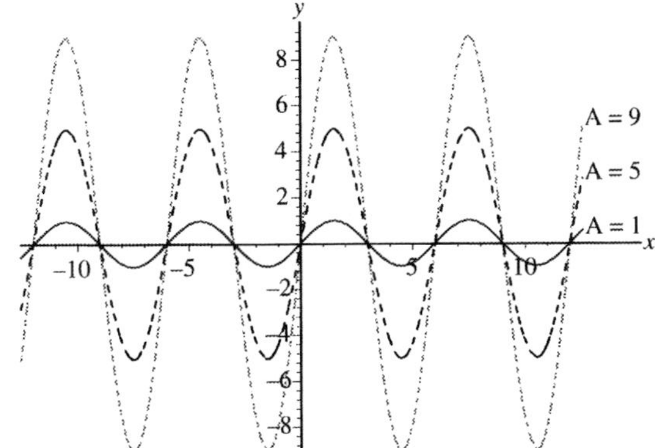

(b) For $A < 0$, the graph is inverted.

1.4 GRAPHING WITH CALCULATORS AND COMPUTERS

1-4. The most appropriate viewing window displays the maxima, minima, intercepts, and end behavior of the graphs and has little unused space.

1. d.

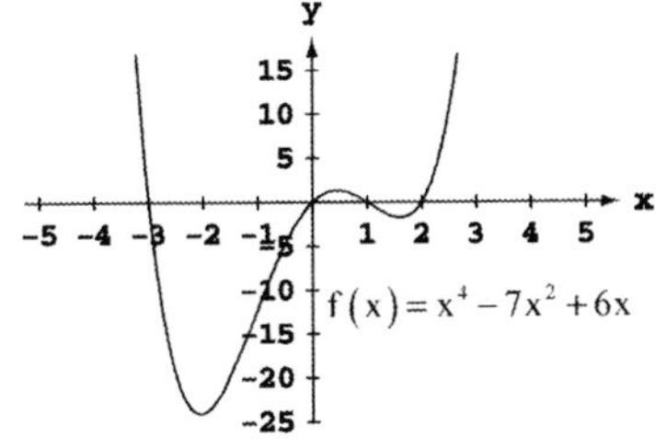

2. c.

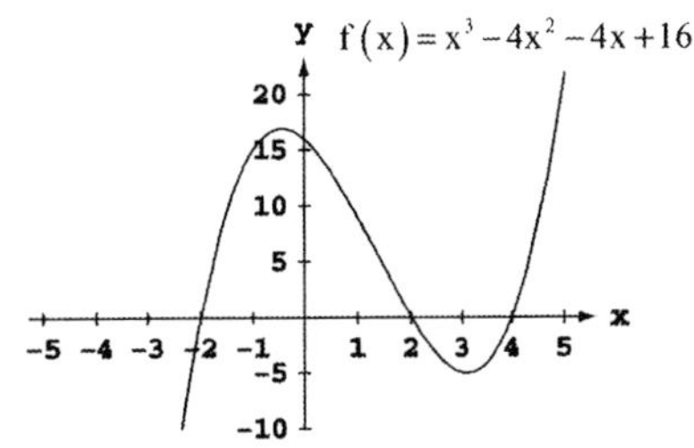

3. d.

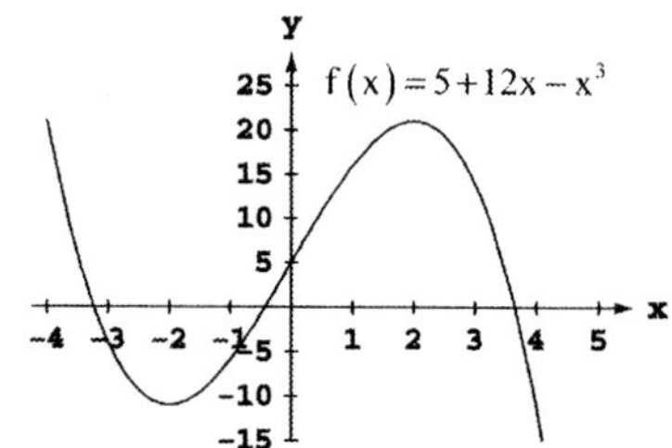

4. b.

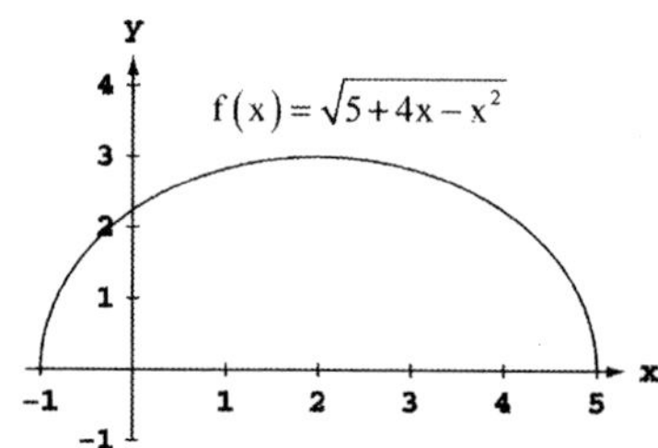

5-30. For any display there are many appropriate display widows. The graphs given as answers in Exercises 5–30 are not unique in appearance.

5. $[-2, 5]$ by $[-15, 40]$

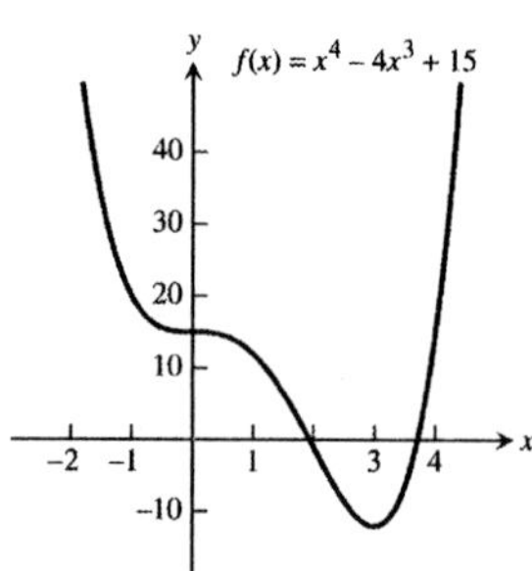

6. $[-4, 4]$ by $[-4, 4]$

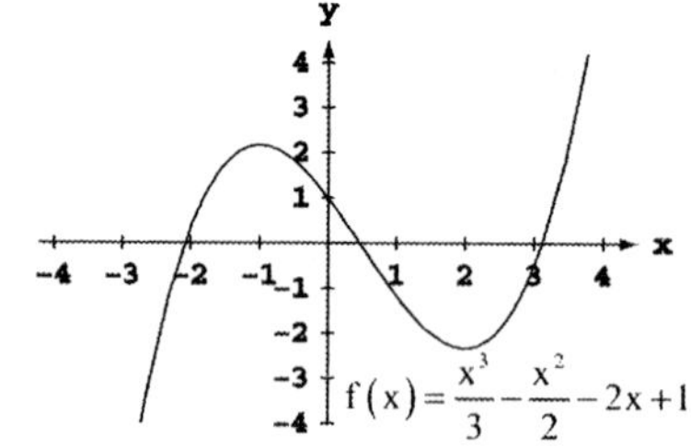

7. $[-2, 6]$ by $[-250, 50]$

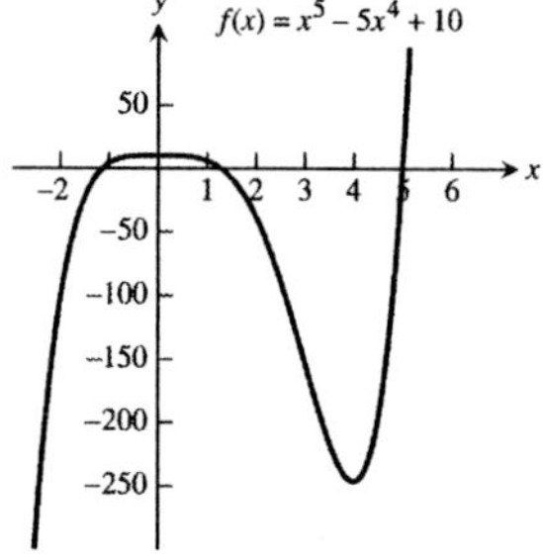

8. $[-1, 5]$ by $[-5, 30]$

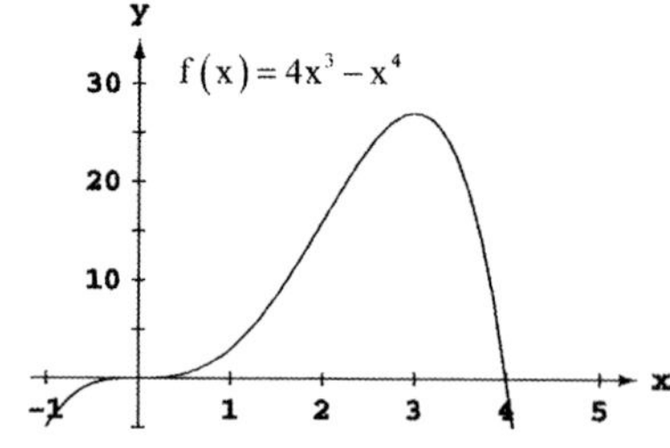

9. $[-4, 4]$ by $[-5, 5]$

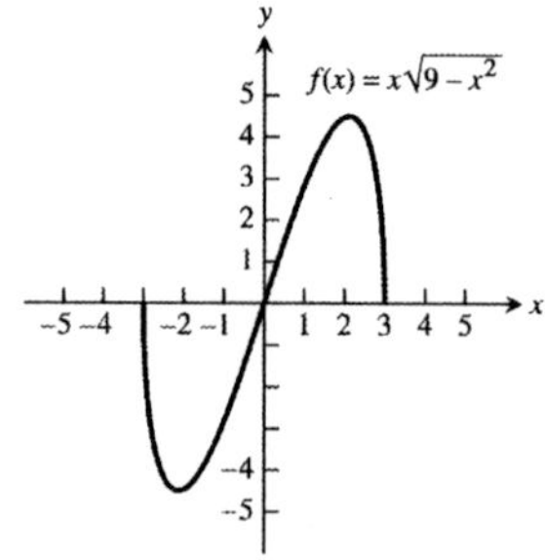

10. $[-2, 2]$ by $[-2, 8]$

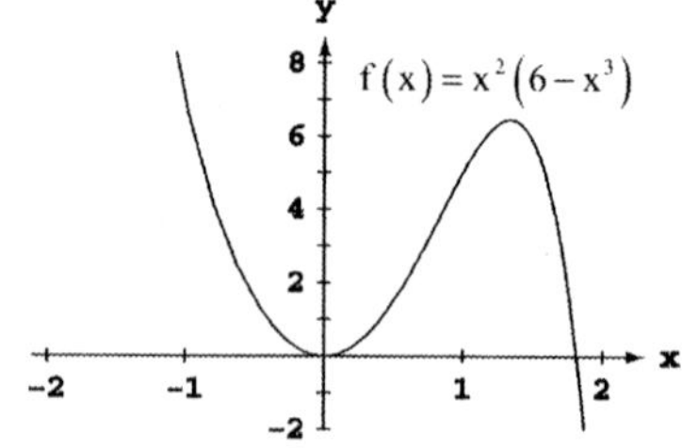

11. $[-2, 6]$ by $[-5, 4]$

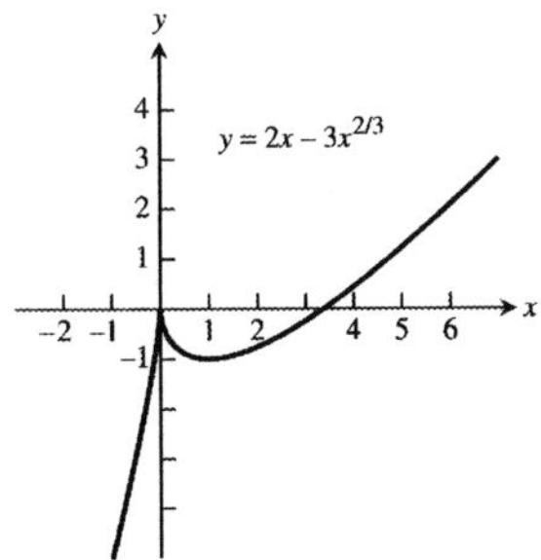

12. $[-4, 4]$ by $[-8, 8]$

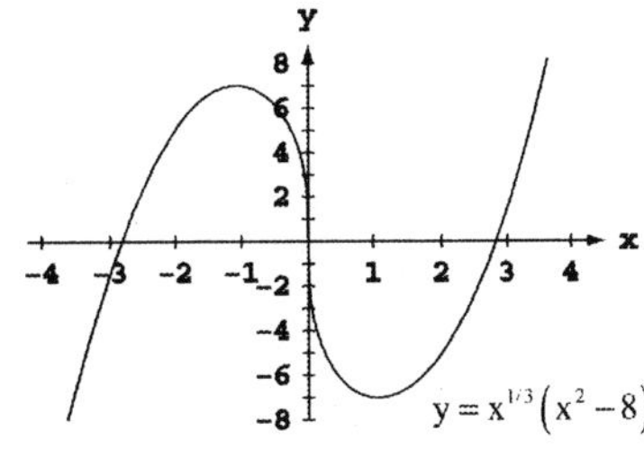

13. $[-1, 6]$ by $[-1, 4]$

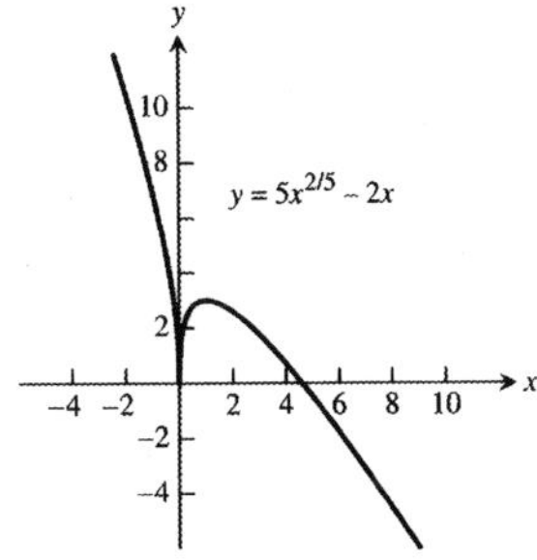

14. $[-1, 6]$ by $[-1, 5]$

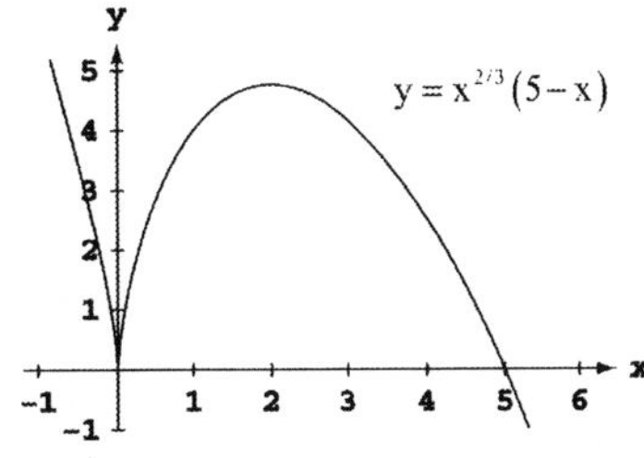

15. $[-3, 3]$ by $[0, 10]$

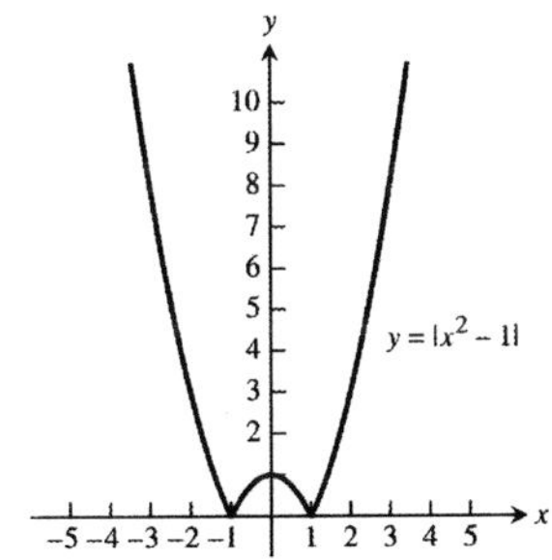

16. $[-1, 2]$ by $[0, 1]$

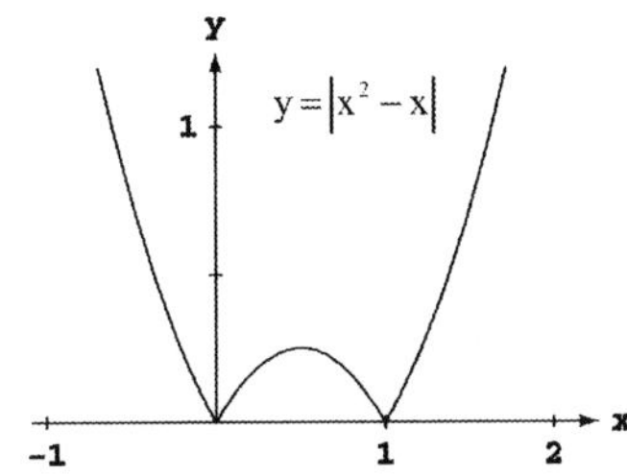

17. $[-5, 1]$ by $[-5, 5]$

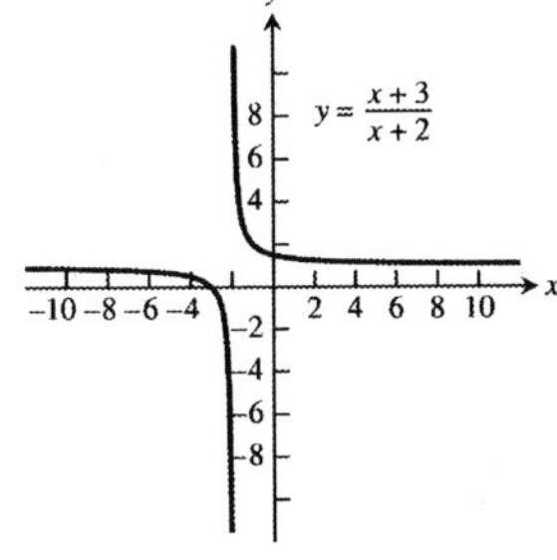

18. $[-5, 1]$ by $[-2, 4]$

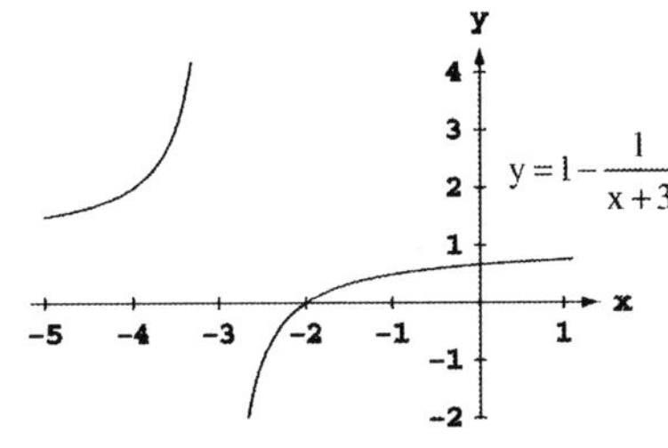

19. $[-4, 4]$ by $[0, 3]$

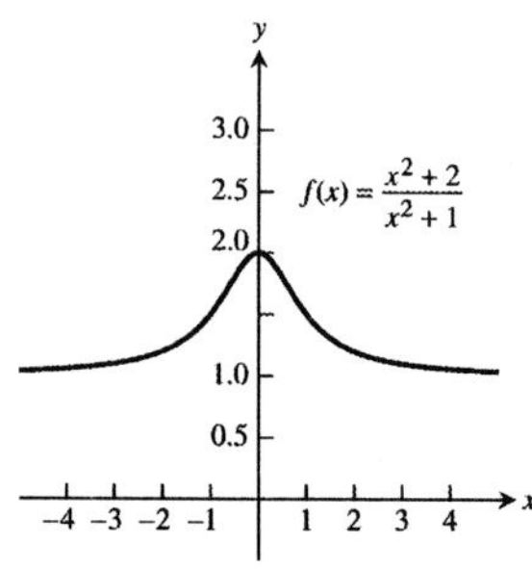

20. $[-5, 5]$ by $[-2, 2]$

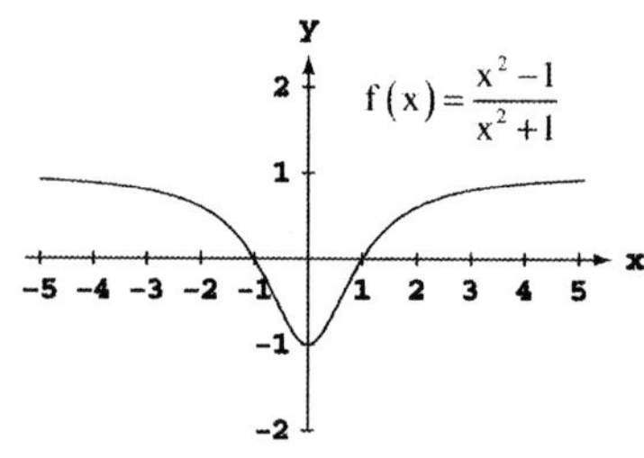

21. $[-10, 10]$ by $[-6, 6]$

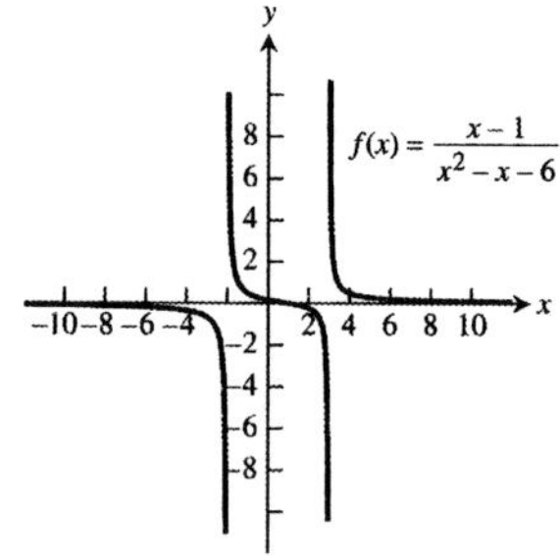

22. $[-5, 5]$ by $[-2, 2]$

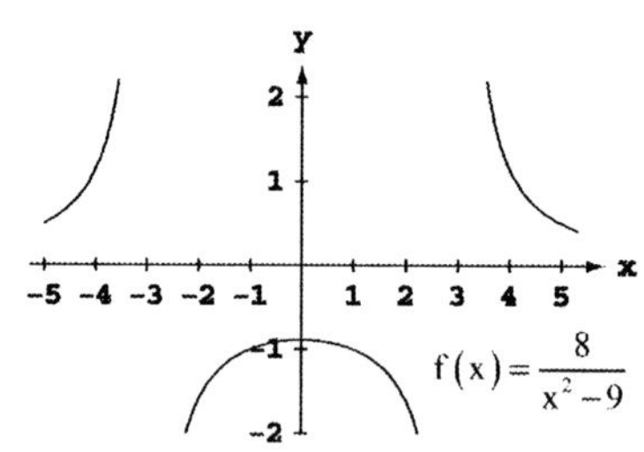

23. $[-6, 10]$ by $[-6, 6]$

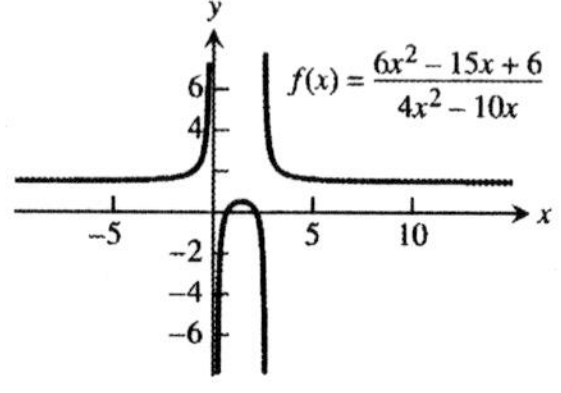

24. $[-3, 5]$ by $[-2, 10]$

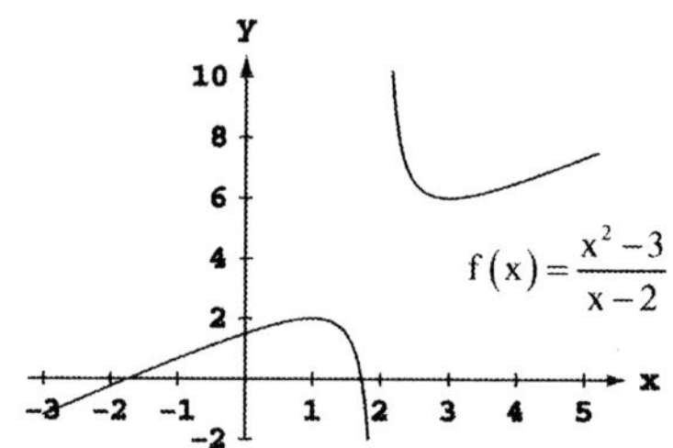

25. $[-0.03, 0.03]$ by $[-1.25, 1.25]$

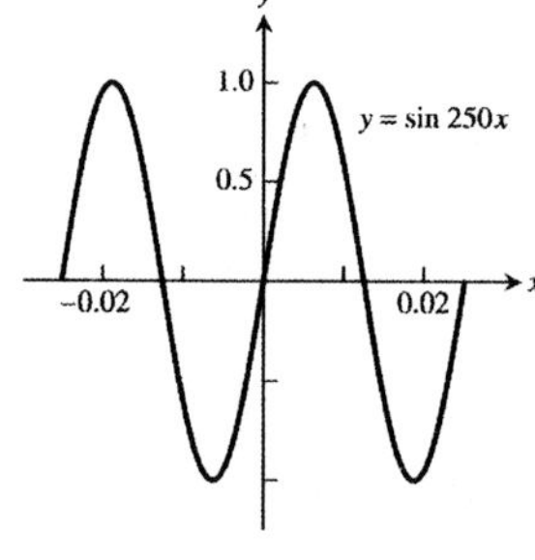

26. $[-0.1, 0.1]$ by $[-3, 3]$

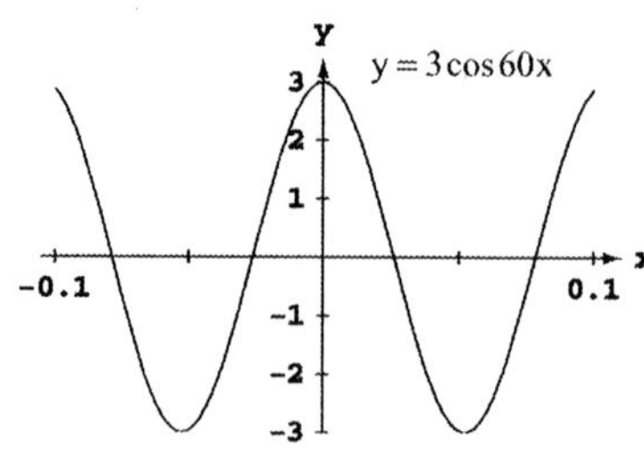

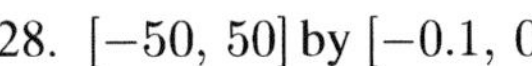

27. $[-300, 300]$ by $[-1.25, 1.25]$

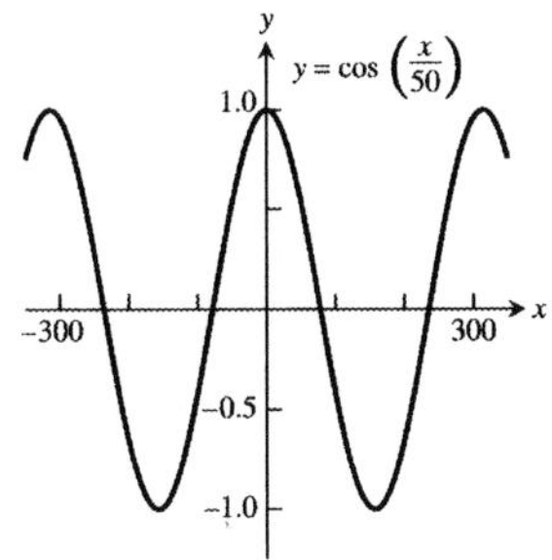

28. $[-50, 50]$ by $[-0.1, 0.1]$

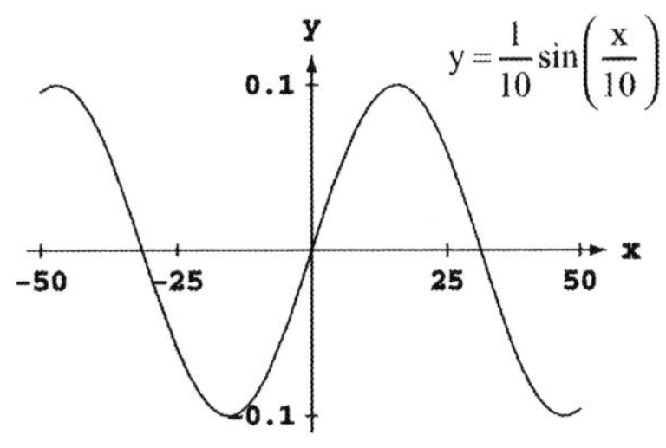

29. $[-0.25, 0.25]$ by $[-0.3, 0.3]$

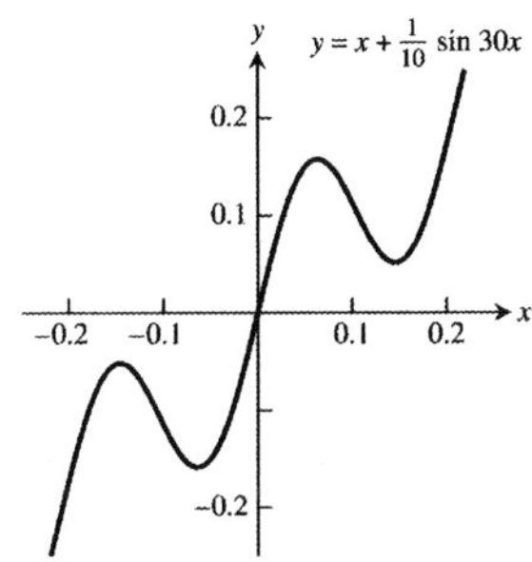

30. $[-0.15, 0.15]$ by $[-0.02, 0.05]$

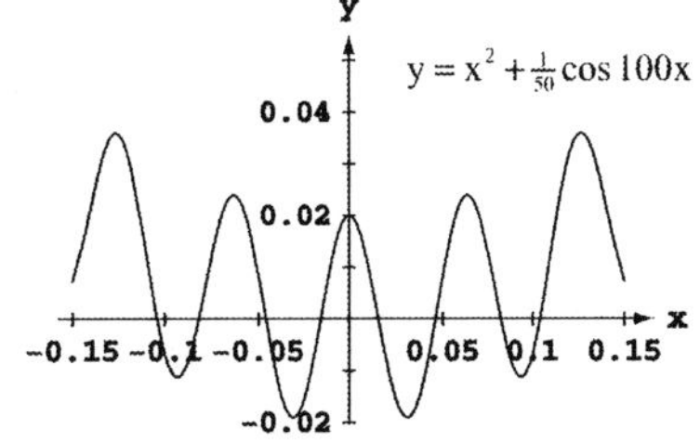

31. $x^2 + 2x = 4 + 4y - y^2 \Rightarrow y = 2 \pm \sqrt{-x^2 - 2x + 8}$.
The lower half is produced by graphing
$y = 2 - \sqrt{-x^2 - 2x + 8}$.

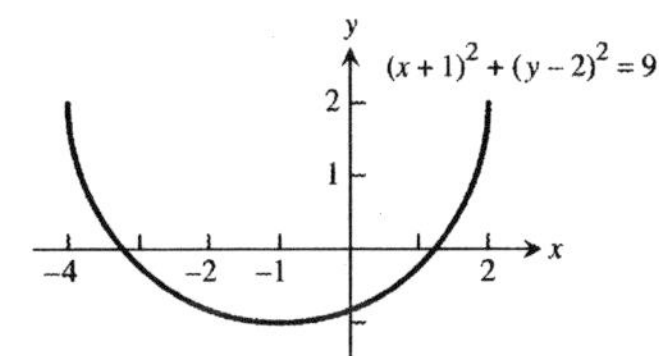

32. $y^2 - 16x^2 = 1 \Rightarrow y = \pm\sqrt{1 + 16x^2}$. The upper branch is produced by graphing $y = \sqrt{1 + 16x^2}$.

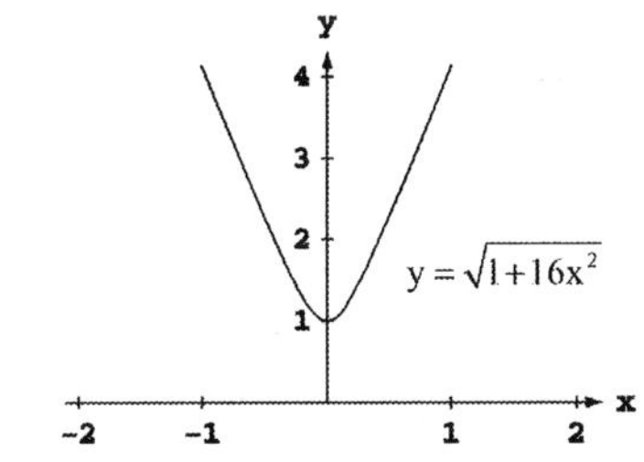

33.

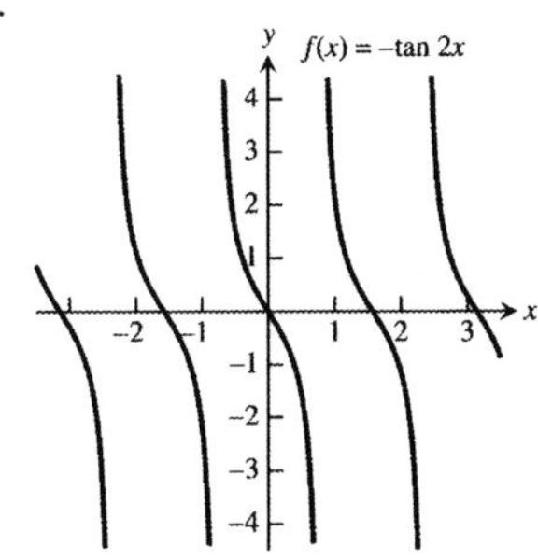

34.

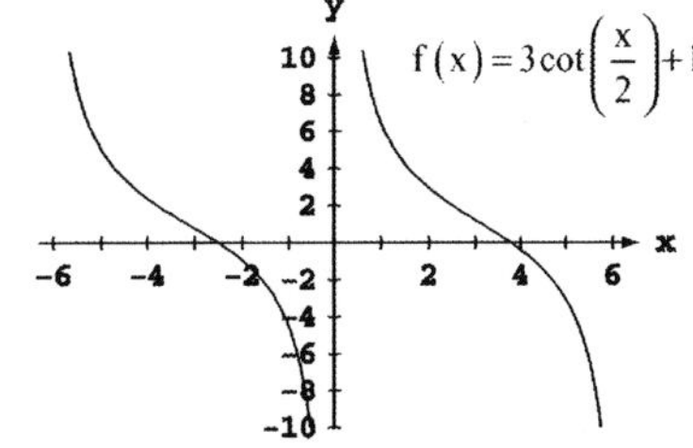

35.

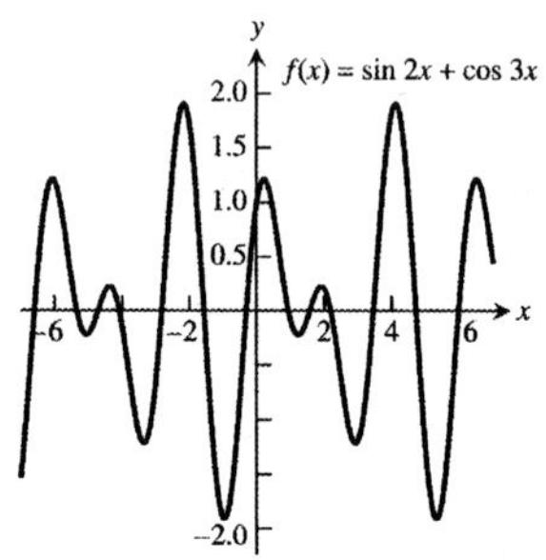

36.

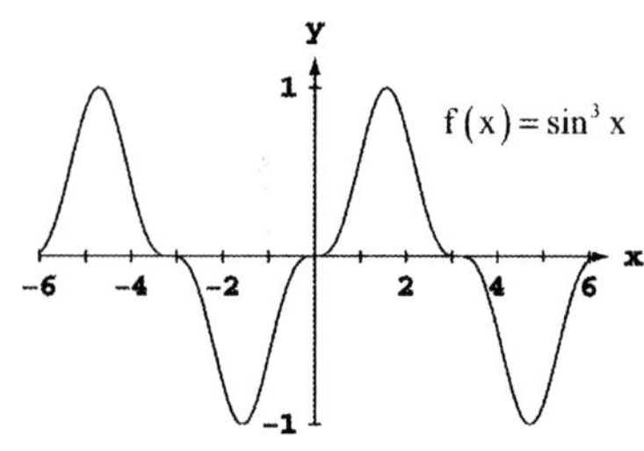

37.

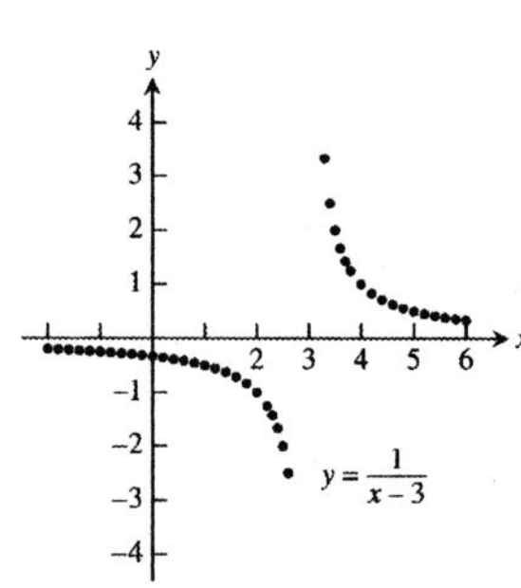

38.

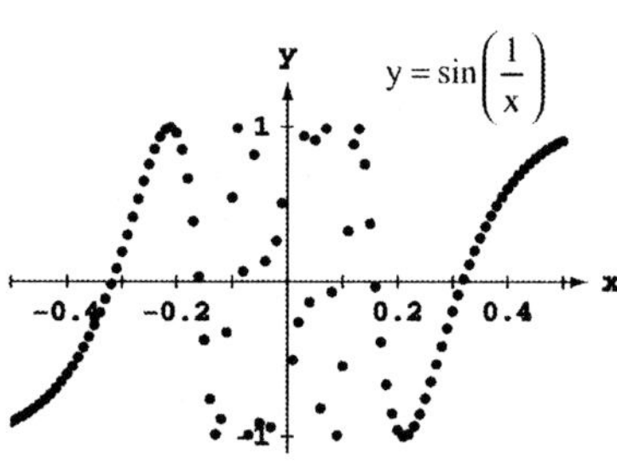

39.

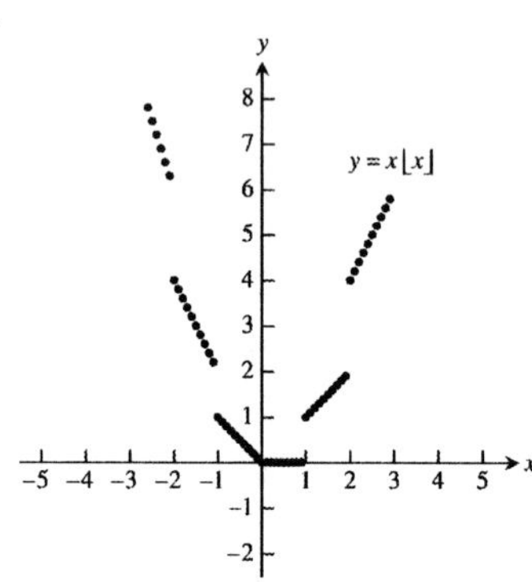

40.

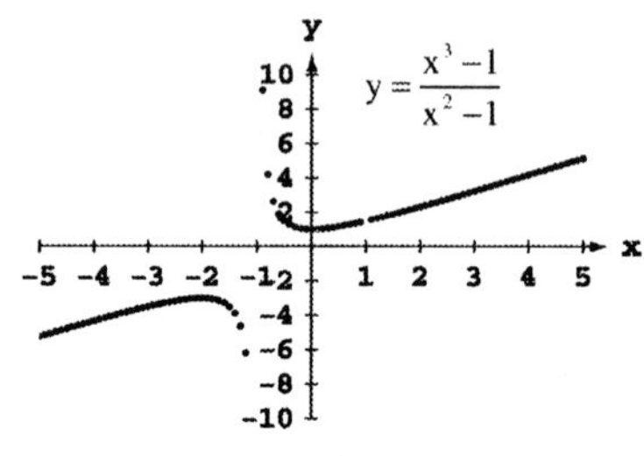

41. (a) $y = 1059.14x - 2074972$

(b) $m = 1059.14$ dollars/year, which is the yearly increase in compensation.

(c)

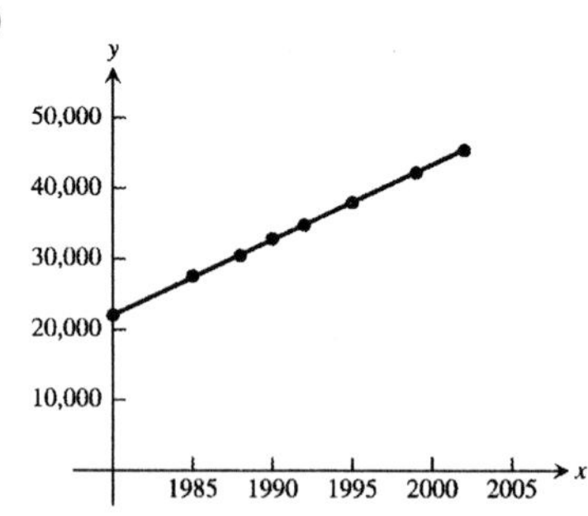

(d) Answers may vary slightly. $y = (1059.14)(2010) - 2074972 = \$53,899$

42. (a) Let C = cost and x = year.

$C = (7960.71)x - 1.6 \times 10^7$

(b) Slope represents increase in cost per year

(c) $C = (2637.14)x - 5.2 \times 10^6$

(d) The median price is rising faster in the northeast (the slope is larger).

43. (a) Let x represent the speed in miles per hour and d the stopping distance in feet. The quadratic regression function is $d = 0.0866x^2 - 1.97x + 50.1$.

(b)

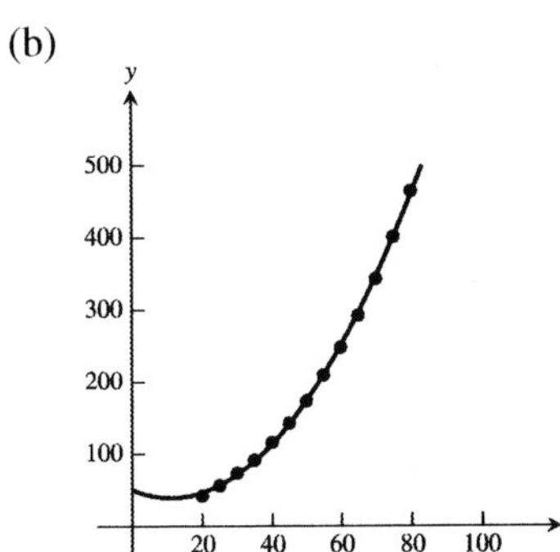

(c) From the graph in part (b), the stopping distance is about 370 feet when the vehicle is 72 mph and it is about 525 feet when the speed is 85 mph.

Algebraically: $d_{\text{quadratic}}(72) = 0.0866(72)^2 - 1.97(72) + 50.1 = 367.6$ ft.

$d_{\text{quadratic}}(85) = 0.0866(85)^2 - 1.97(85) + 50.1 = 522.8$ ft.

(d) The linear regression function is $d = 6.89x - 140.4 \Rightarrow d_{\text{linear}}(72) = 6.89(72) - 140.4 = 355.7$ ft and $d_{\text{linear}}(85) = 6.89(85) - 140.4 = 445.2$ ft. The linear regression line is shown on the graph in part (b). The quadratic regression curve clearly gives the better fit.

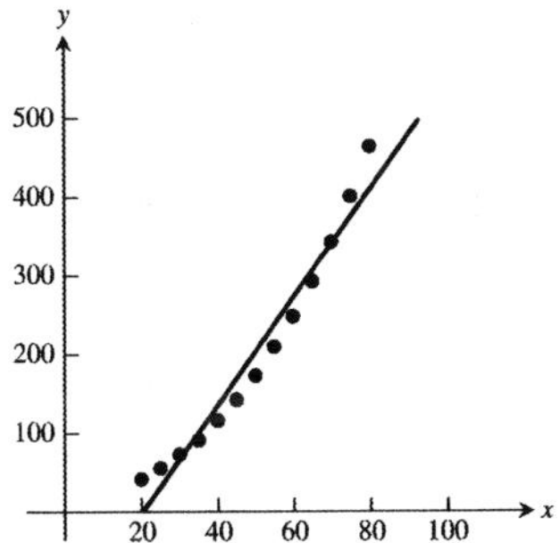

44. (a) The power regression function is $y = 4.44647x^{0.511414}$.

(b)

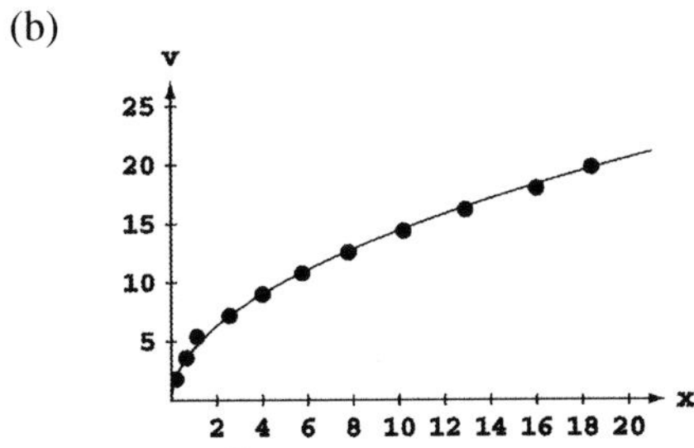

(c) 15.2 km/h

(d) The linear regression function is $y = 0.913675x + 4.189976$ and it is shown on the graph in part (b). The linear regession function gives a speed of 14.2 km/h when $y = 11$ m. The power regression curve in part (a) better fits the data.

1.5 EXPONENTIAL FUNCTIONS

1.

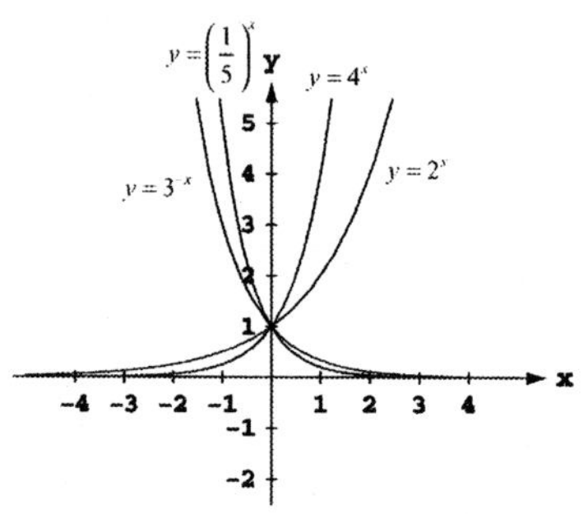

2.

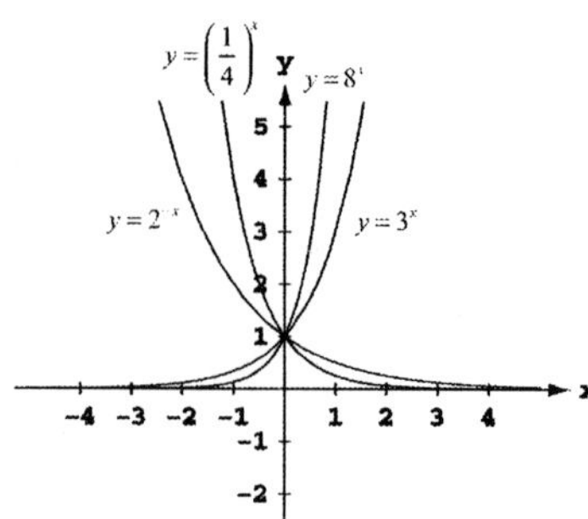

3.

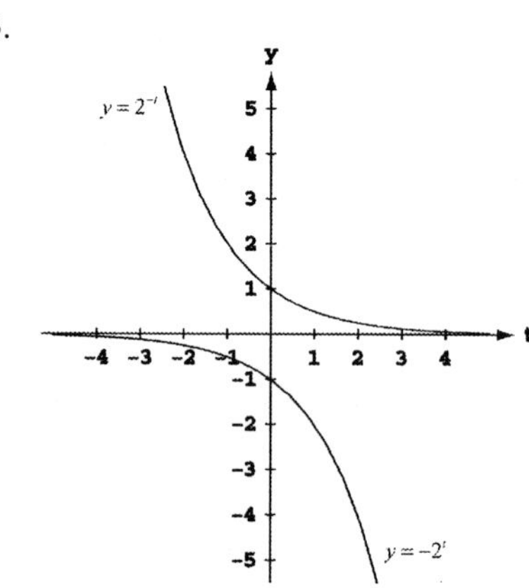

4.

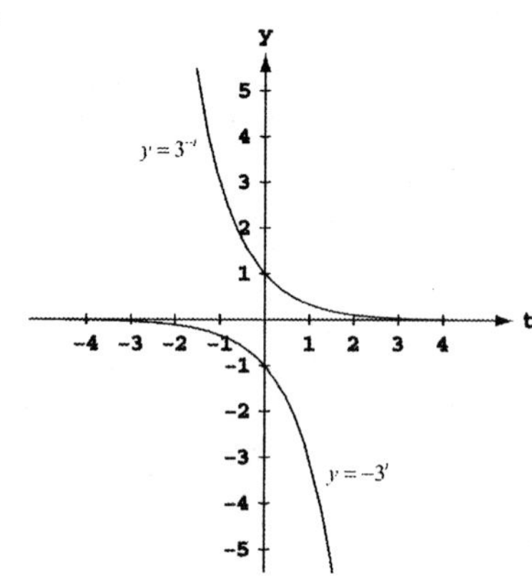

5.

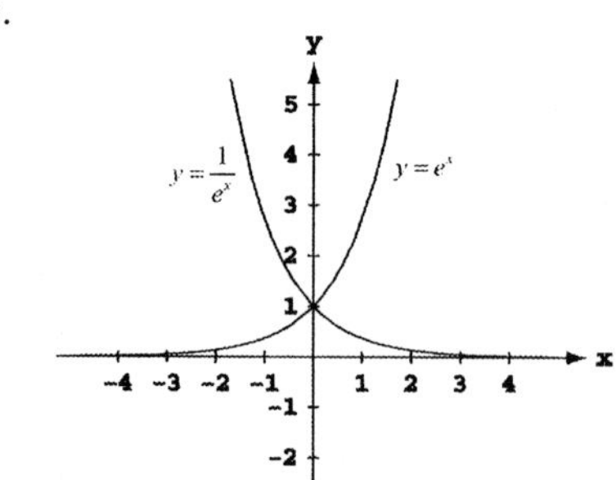

6.

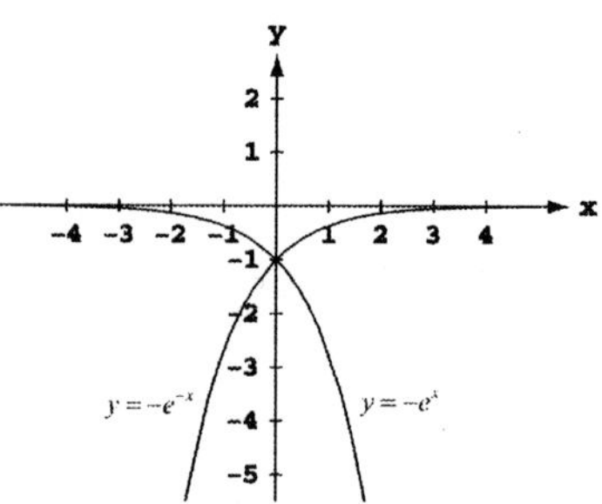

7.

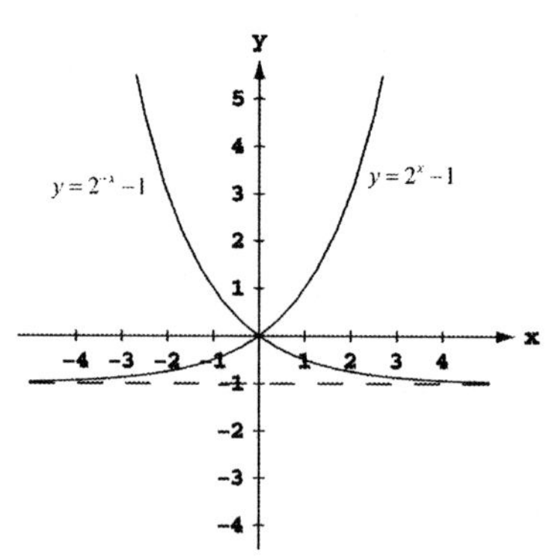

8.

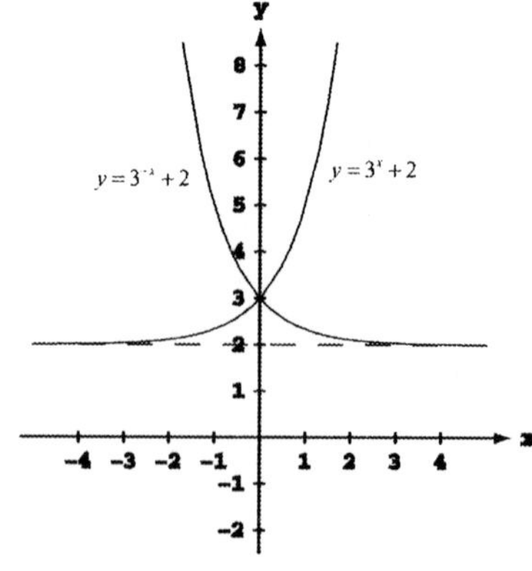

9.

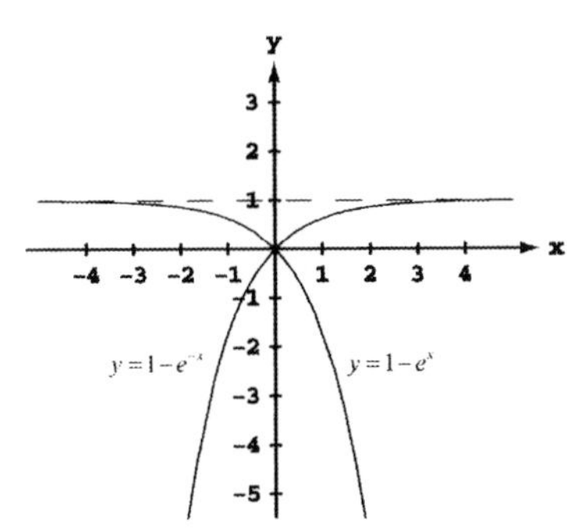

10.

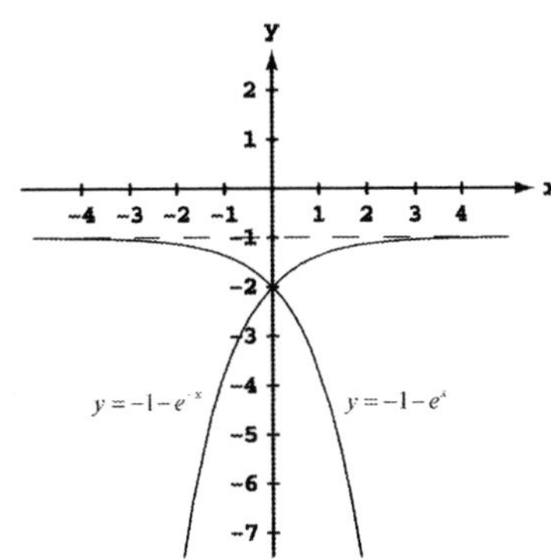

11. $16^2 \cdot 16^{-1.75} = 16^{2+(-1.75)} = 16^{0.25} = 16^{1/4} = 2$

12. $9^{1/3} \cdot 9^{1/6} = 9^{\frac{1}{3}+\frac{1}{6}} = 9^{1/2} = 3$

13. $\frac{4^{4.2}}{4^{3.7}} = 4^{4.2-3.7} = 4^{0.5} = 4^{1/2} = 2$

14. $\frac{3^{5/3}}{3^{2/3}} = 3^{\frac{5}{3}-\frac{2}{3}} = 3^1 = 3$

15. $\left(25^{1/8}\right)^4 = 25^{4/8} = 25^{1/2} = 5$

16. $\left(13^{\sqrt{2}}\right)^{\sqrt{2}/2} = 13^{2/2} = 13$

17. $2^{\sqrt{3}} \cdot 7^{\sqrt{3}} = (2 \cdot 7)^{\sqrt{3}} = 14^{\sqrt{3}}$

18. $\left(\sqrt{3}\right)^{1/2}\left(\sqrt{12}\right)^{1/2} = \left(\sqrt{3} \cdot \sqrt{12}\right)^{1/2} = \left(\sqrt{36}\right)^{1/2} = 6^{1/2}$

19. $\left(\frac{2}{\sqrt{2}}\right)^4 = \frac{2^4}{\left(2^{1/2}\right)^4} = \frac{16}{2^2} = 4$

20. $\left(\frac{\sqrt{6}}{3}\right)^2 = \frac{\left(6^{1/2}\right)^2}{3^2} = \frac{6}{9} = \frac{2}{3}$

21. Domain: $(-\infty, \infty)$; y in range $\Rightarrow y = \frac{1}{2+e^x}$. As x increases, e^x becomes infinitely large and y becomes a smaller and smaller positive real number. As x decreases, e^x becomes a smaller and smaller positive real number, $y < \frac{1}{2}$, and y gets arbitrarily close to $\frac{1}{2} \Rightarrow$ Range: $\left(0, \frac{1}{2}\right)$.

22. Domain: $(-\infty, \infty)$; y in range $\Rightarrow y = \cos(e^{-t})$. Since the values of e^{-t} are $(0, \infty)$ and $-1 \le \cos x \le 1 \Rightarrow$ Range: $[-1, 1]$.

23. Domain: $(-\infty, \infty)$; y in range $\Rightarrow y = \sqrt{1 + 3^{-t}}$. Since the values of 3^{-t} are $(0, \infty) \Rightarrow$ Range: $(1, \infty)$.

24. If $e^{2x} = 1$, then $x = 0 \Rightarrow$ Domain: $(-\infty, 0) \cup (0, \infty)$; y in range $\Rightarrow y = \frac{3}{1-e^{2x}}$. If $x > 0$, then $1 < e^{2x} < \infty$ $\Rightarrow -\infty < y < 0$. If $x < 0$, then $0 < e^{2x} < 1 \Rightarrow 3 < y < \infty \Rightarrow$ Range: $(-\infty, 0) \cup (3, \infty)$.

25.

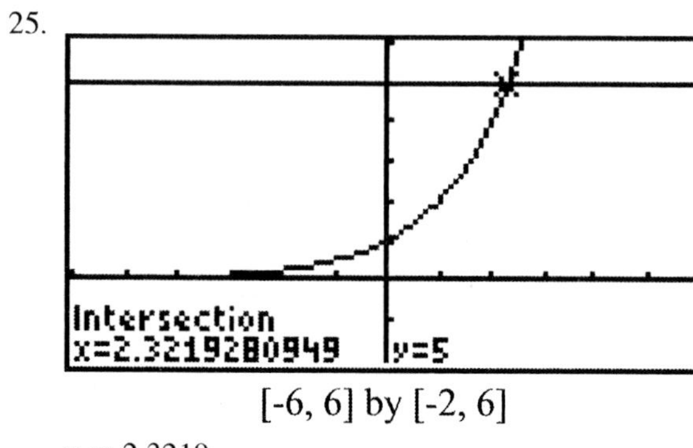

[-6, 6] by [-2, 6]

$x \approx 2.3219$

26.

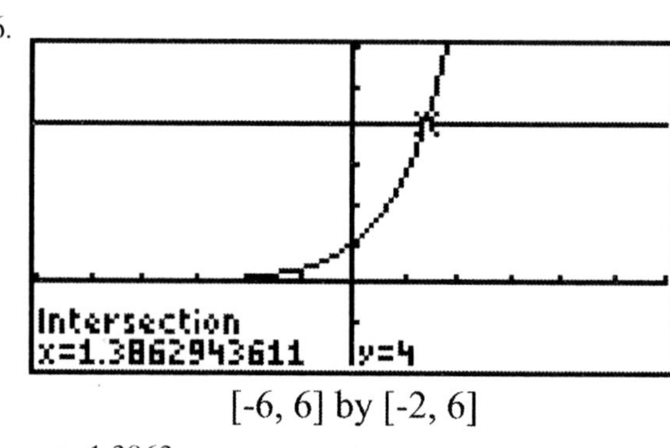

[-6, 6] by [-2, 6]

$x \approx 1.3863$

27.

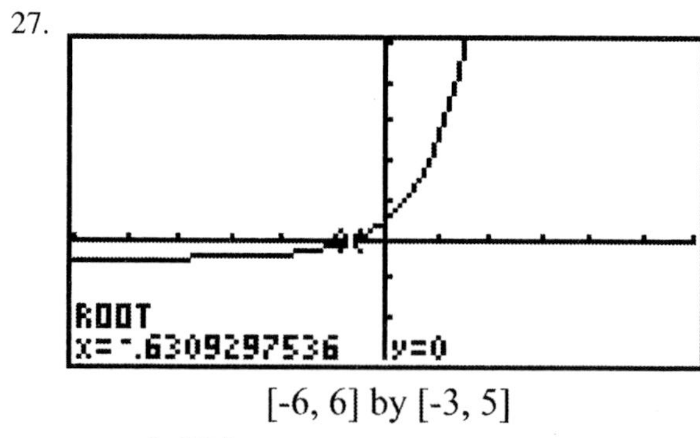

[-6, 6] by [-3, 5]

$x \approx -0.6309$

28.

ROOT
X=-1.584962501 Y=-1E-13

[-6, 6] by [-3, 5]

$x \approx -1.5850$

29. Let t be the number of years. Solving $500{,}000(1.0375)^t = 1{,}000{,}000$ graphically, we find that $t \approx 18.828$. The population will reach 1 million in about 19 years.

30. (a) The population is given by $P(t) = 6250(1.0275)^t$, where t is the number of years after 1890.
Population in 1915: $P(25) \approx 12{,}315$
Population in 1940: $P(50) \approx 24{,}265$
(b) Solving $P(t) = 50{,}000$ graphically, we find that $t \approx 76.651$. The population reached 50,000 about 77 years after 1890, in 1967.

31. (a) $A(t) = 6.6\left(\frac{1}{2}\right)^{t/14}$
(b) Solving $A(t) = 1$ graphically, we find that $t \approx 38$. There will be 1 gram remaining after about 38.1145 days.

32. Let t be the number of years. Solving $2300(1.06)^t = 4150$ graphically, we find that $t \approx 10.129$. It will take about 10.129 years. (If the interest is not credited to the account until the end of each year, it will take 11 years.)

33. Let A be the amount of the initial investment, and let t be the number of years. We wish to solve $A(1.0625)^t = 2A$, which is equivalent to $1.0625^t = 2$. Solving graphically, we find that $t \approx 11.433$. It will take about 11.433 years. (If the interest is credited at the end of each year, it will take 12 years.)

34. Let A be the amount of the initial investment, and let t be the number of years. We wish to solve $A\left(1 + \frac{0.0625}{12}\right)^{12t} = 2A$, which is equivalent to $\left(1 + \frac{0.0625}{12}\right)^{12t} = 2$. Solving graphically, we find that $t \approx 11.119$. It will take about 11.119 years. (If the interest is credited at the end of each month, it will take 11 years 2 months.)

35. Let A be the amount of the initial investment, and let t be the number of years. We wish to solve $Ae^{0.0625t} = 2A$, which is equivalent to $e^{0.0625t} = 2$. Solving graphically, we find that $t \approx 11.090$. It will take about 11.090 years.

36. Let A be the amount of the initial investment, and let t be the number of years. We wish to solve $A(1.0575)^t = 3A$, which is equivalent to $(1.0575)^t = 3$. Solving graphically, we find that $t \approx 19.650$. It will take abou 19.650 years. (If the interest is credited at the end of each year, it will take 20 years.)

37. Let A be the amount of the initial investment, and let t be the number of years. We wish to solve $A\left(1 + \frac{0.0575}{365}\right)^{365t} = 3A$, which is equivalent to $\left(1 + \frac{0.0575}{365}\right)^{365t} = 3$. Solving graphically, we find that $t \approx 19.108$. It will take about 19.108 years.

38. Let A be the amount of the initial investment, and let t be the number of years. We wish to solve $Ae^{0.0575t} = 3A$, which is equivalent to $e^{0.0575t} = 3$. Solving graphically, we find that $t \approx 19.106$. It will take about 19.106 years.

39. After t hours, the population is $P(t) = 2^{t/0.5}$, or equivalently, $P(t) = 2^{2t}$. After 24 hours, the population is $P(24) = 2^{48} \approx 2.815 \times 10^{14}$ bacteria.

40. (a) Each year, the number of cases is $100\% - 20\% = 80\%$ of the previous year's number of cases. After t years, the number of cases will be $C(t) = 10{,}000(0.8)^t$. Solving $C(t) = 1000$ graphically, we find that $t \approx 10.319$. It will take 10.319 years.
    (b) Solving $C(t) = 1$ graphically, we find that $t \approx 41.275$. It will take about 41.275 years.

41. (a) Let $x = 0$ represent 1900, $x = 1$ prepresent 1901, and so on. The regression equation is $P(x) = 6.033(1.030)^x$.

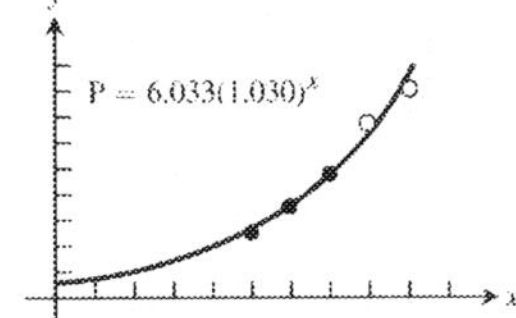

    (b) The regression equation gives an estimate of $P(0) \approx 6.03$ million, which is not very close to the actual population.
    (c) Since the equation is of the form $P(x) = P(0) \cdot 1.030^x$, the annual rate of growth is about 3%.

42. (a) The regression equation is $P(x) = 4.831(1.019)^x$.

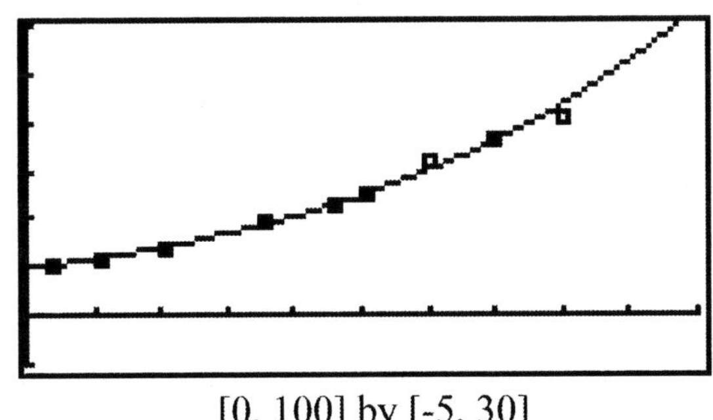

[0, 100] by [-5, 30]

    (b) $P90 \approx 26.3$ million.
    (c) Since the equation is of the form $P(x) = P(0) \cdot 1.019^x$, the annual rate of growth is about 1.9%.

1.6 INVERSE FUNCTIONS AND LOGARITHMS

1. Yes one-to-one, the graph passes the horizontal test.

2. Not one-to-one, the graph fails the horizontal test.

3. Not one-to-one since (for example) the horizontal line $y = 2$ intersects the graph twice.

4. Not one-to-one, the graph fails the horizontal test.

5. Yes one-to-one, the graph passes the horizontal test

6. Yes one-to-one, the graph passes the horizontal test

7. Domain: $0 < x \le 1$, Range: $0 \le y$

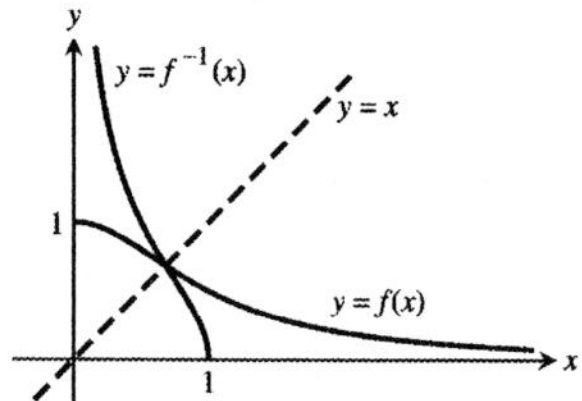

8. Domain: $x < 1$, Range: $y > 0$

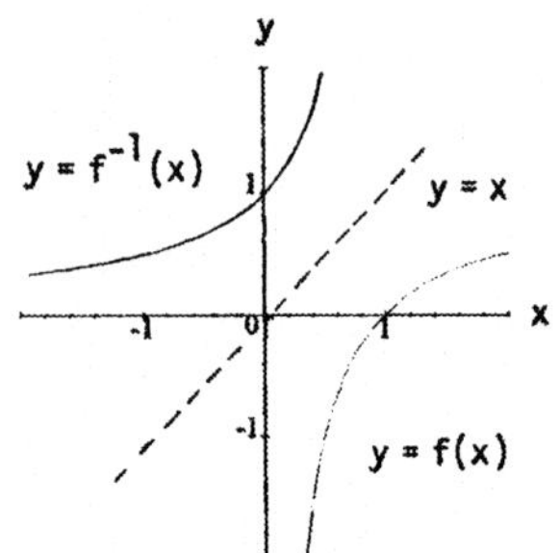

9. Domain: $-1 \le x \le 1$, Range: $-\frac{\pi}{2} \le y \le \frac{\pi}{2}$

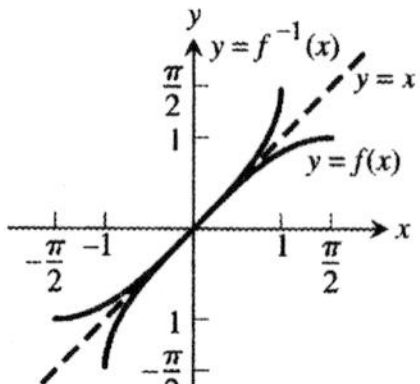

10. Domain: $-\infty < x < \infty$, Range: $-\frac{\pi}{2} < y \le \frac{\pi}{2}$

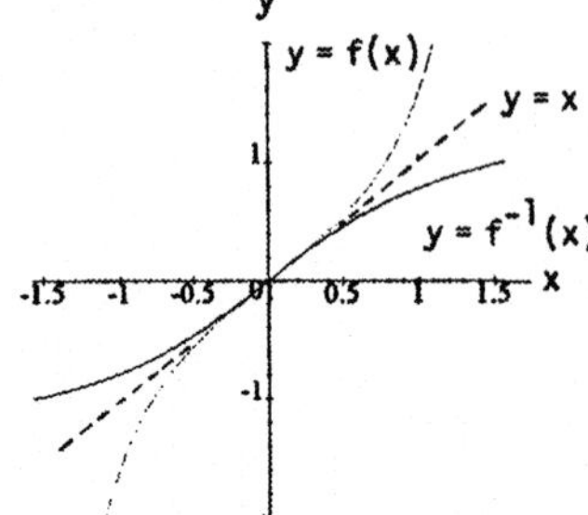

11. (a) The graph is symmetric about $y = x$.

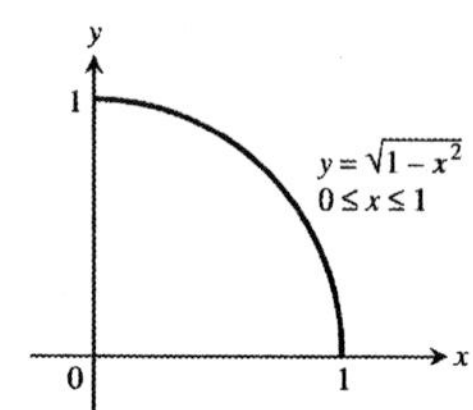

(b) $y = \sqrt{1 - x^2} \Rightarrow y^2 = 1 - x^2 \Rightarrow x^2 = 1 - y^2 \Rightarrow x = \sqrt{1 - y^2} \Rightarrow y = \sqrt{1 - x^2} = f^{-1}(x)$

12. (a) The graph is symmetric about $y = x$.

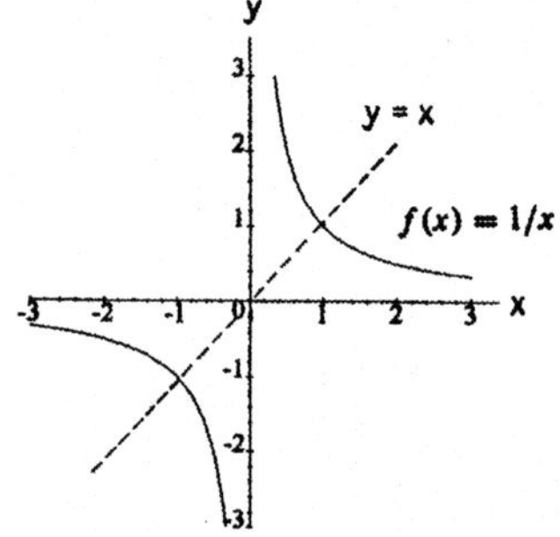

(b) $y = \frac{1}{x} \Rightarrow x = \frac{1}{y} \Rightarrow y = \frac{1}{x} = f^{-1}(x)$

13. Step 1: $y = x^2 + 1 \Rightarrow x^2 = y - 1 \Rightarrow x = \sqrt{y-1}$
Step 2: $y = \sqrt{x-1} = f^{-1}(x)$

14. Step 1: $y = x^2 \Rightarrow x = -\sqrt{y}$, since $x \le 0$.
Step 2: $y = -\sqrt{x} = f^{-1}(x)$

15. Step 1: $y = x^3 - 1 \Rightarrow x^3 = y + 1 \Rightarrow x = (y+1)^{1/3}$
Step 2: $y = \sqrt[3]{x+1} = f^{-1}(x)$

16. Step 1: $y = x^2 - 2x + 1 \Rightarrow y = (x-1)^2 \Rightarrow \sqrt{y} = x - 1$, since $x \ge 1 \Rightarrow x = 1 + \sqrt{y}$
Step 2: $y = 1 + \sqrt{x} = f^{-1}(x)$

17. Step 1: $y = (x+1)^2 \Rightarrow \sqrt{y} = x + 1$, since $x \ge -1 \Rightarrow x = \sqrt{y} - 1$
Step 2: $y = \sqrt{x} - 1 = f^{-1}(x)$

18. Step 1: $y = x^{2/3} \Rightarrow x = y^{3/2}$
Step 2: $y = x^{3/2} = f^{-1}(x)$

19. Step 1: $y = x^5 \Rightarrow x = y^{1/5}$
Step 2: $y = \sqrt[5]{x} = f^{-1}(x)$;
Domain and Range of f^{-1}: all reals;
$f(f^{-1}(x)) = \left(x^{1/5}\right)^5 = x$ and $f^{-1}(f(x)) = \left(x^5\right)^{1/5} = x$

20. Step 1: $y = x^4 \Rightarrow x = y^{1/4}$
Step 2: $y = \sqrt[4]{x} = f^{-1}(x)$;
Domain of f^{-1}: $x \ge 0$, Range of f^{-1}: $y \ge 0$;
$f(f^{-1}(x)) = \left(x^{1/4}\right)^4 = x$ and $f^{-1}(f(x)) = \left(x^4\right)^{1/4} = x$

21. Step 1: $y = x^3 + 1 \Rightarrow x^3 = y - 1 \Rightarrow x = (y-1)^{1/3}$
Step 2: $y = \sqrt[3]{x-1} = f^{-1}(x)$;
Domain and Range of f^{-1}: all reals;
$f(f^{-1}(x)) = \left((x-1)^{1/3}\right)^3 + 1 = (x-1) + 1 = x$ and $f^{-1}(f(x)) = \left((x^3+1) - 1\right)^{1/3} = \left(x^3\right)^{1/3} = x$

22. Step 1: $y = \frac{1}{2}x - \frac{7}{2} \Rightarrow \frac{1}{2}x = y + \frac{7}{2} \Rightarrow x = 2y + 7$
Step 2: $y = 2x + 7 = f^{-1}(x)$;
Domain and Range of f^{-1}: all reals;
$f(f^{-1}(x)) = \frac{1}{2}(2x+7) - \frac{7}{2} = \left(x + \frac{7}{2}\right) - \frac{7}{2} = x$ and $f^{-1}(f(x)) = 2\left(\frac{1}{2}x - \frac{7}{2}\right) + 7 = (x-7) + 7 = x$

23. Step 1: $y = \frac{1}{x^2} \Rightarrow x^2 = \frac{1}{y} \Rightarrow x = \frac{1}{\sqrt{y}}$
Step 2: $y = \frac{1}{\sqrt{x}} = f^{-1}(x)$
Domain of f^{-1}: $x > 0$, Range of f^{-1}: $y > 0$;
$f(f^{-1}(x)) = \frac{1}{\left(\frac{1}{\sqrt{x}}\right)^2} = \frac{1}{\left(\frac{1}{x}\right)} = x$ and $f^{-1}(f(x)) = \frac{1}{\sqrt{\frac{1}{x^2}}} = \frac{1}{\left(\frac{1}{x}\right)} = x$ since $x > 0$

24. Step 1: $y = \frac{1}{x^3} \Rightarrow x^3 = \frac{1}{y} \Rightarrow x = \frac{1}{y^{1/3}}$

Step 2: $y = \frac{1}{x^{1/3}} = \sqrt[3]{\frac{1}{x}} = f^{-1}(x)$;

Domain of f^{-1}: $x \neq 0$, Range of f^{-1}: $y \neq 0$;

$f(f^{-1}(x)) = \frac{1}{(x^{-1/3})^3} = \frac{1}{x^{-1}} = x$ and $f^{-1}(f(x)) = \left(\frac{1}{x^3}\right)^{-1/3} = \left(\frac{1}{x}\right)^{-1} = x$

25. (a) $\ln 0.75 = \ln \frac{3}{4} = \ln 3 - \ln 4 = \ln 3 - \ln 2^2 = \ln 3 - 2 \ln 2$

(b) $\ln \frac{4}{9} = \ln 4 - \ln 9 = \ln 2^2 - \ln 3^2 = 2 \ln 2 - 2 \ln 3$

(c) $\ln \frac{1}{2} = \ln 1 - \ln 2 = -\ln 2$

(d) $\ln \sqrt[3]{9} = \frac{1}{3} \ln 9 = \frac{1}{3} \ln 3^2 = \frac{2}{3} \ln 3$

(e) $\ln 3\sqrt{2} = \ln 3 + \ln 2^{1/2} = \ln 3 + \frac{1}{2} \ln 2$

(f) $\ln \sqrt{13.5} = \frac{1}{2} \ln 13.5 = \frac{1}{2} \ln \frac{27}{2} = \frac{1}{2}(\ln 3^3 - \ln 2) = \frac{1}{2}(3 \ln 3 - \ln 2)$

26. (a) $\ln \frac{1}{125} = \ln 1 - 3 \ln 5 = -3 \ln 5$

(b) $\ln 9.8 = \ln \frac{49}{5} = \ln 7^2 - \ln 5 = 2 \ln 7 - \ln 5$

(c) $\ln 7\sqrt{7} = \ln 7^{3/2} = \frac{3}{2} \ln 7$

(d) $\ln 1225 = \ln 35^2 = 2 \ln 35 = 2 \ln 5 + 2 \ln 7$

(e) $\ln 0.056 = \ln \frac{7}{125} = \ln 7 - \ln 5^3 = \ln 7 - 3 \ln 5$

(f) $\frac{\ln 35 + \ln \frac{1}{7}}{\ln 25} = \frac{\ln 5 + \ln 7 - \ln 7}{2 \ln 5} = \frac{1}{2}$

27. (a) $\ln \sin\theta - \ln\left(\frac{\sin\theta}{5}\right) = \ln\left(\frac{\sin\theta}{\left(\frac{\sin\theta}{5}\right)}\right) = \ln 5$

(b) $\ln(3x^2 - 9x) + \ln\left(\frac{1}{3x}\right) = \ln\left(\frac{3x^2 - 9x}{3x}\right) = \ln(x - 3)$

(c) $\frac{1}{2}\ln(4t^4) - \ln 2 = \ln\sqrt{4t^4} - \ln 2 = \ln 2t^2 - \ln 2 = \ln\left(\frac{2t^2}{2}\right) = \ln(t^2)$

28. (a) $\ln \sec\theta + \ln \cos\theta = \ln[(\sec\theta)(\cos\theta)] = \ln 1 = 0$

(b) $\ln(8x + 4) - \ln 2^2 = \ln(8x + 4) - \ln 4 = \ln\left(\frac{8x+4}{4}\right) = \ln(2x + 1)$

(c) $3 \ln \sqrt[3]{t^2 - 1} - \ln(t + 1) = 3 \ln(t^2 - 1)^{1/3} - \ln(t + 1) = 3\left(\frac{1}{3}\right)\ln(t^2 - 1) - \ln(t + 1) = \ln\left(\frac{(t+1)(t-1)}{(t+1)}\right)$

$= \ln(t - 1)$

29. (a) $e^{\ln 7.2} = 7.2$

(b) $e^{-\ln x^2} = \frac{1}{e^{\ln x^2}} = \frac{1}{x^2}$

(c) $e^{\ln x - \ln y} = e^{\ln(x/y)} = \frac{x}{y}$

30. (a) $e^{\ln(x^2+y^2)} = x^2 + y^2$

(b) $e^{-\ln 0.3} = \frac{1}{e^{\ln 0.3}} = \frac{1}{0.3}$

(c) $e^{\ln \pi x - \ln 2} = e^{\ln(\pi x/2)} = \frac{\pi x}{2}$

31. (a) $2 \ln \sqrt{e} = 2 \ln e^{1/2} = (2)\left(\frac{1}{2}\right) \ln e = 1$

(b) $\ln(\ln e^e) = \ln(e \ln e) = \ln e = 1$

(c) $\ln e^{(-x^2-y^2)} = (-x^2 - y^2) \ln e = -x^2 - y^2$

32. (a) $\ln\left(e^{\sec\theta}\right) = (\sec\theta)(\ln e) = \sec\theta$

(b) $\ln e^{(e^x)} = (e^x)(\ln e) = e^x$

(c) $\ln\left(e^{2\ln x}\right) = \ln\left(e^{\ln x^2}\right) = \ln x^2 = 2 \ln x$

33. $\ln y = 2t + 4 \Rightarrow e^{\ln y} = e^{2t+4} \Rightarrow y = e^{2t+4}$

34. $\ln y = -t + 5 \Rightarrow e^{\ln y} = e^{-t+5} \Rightarrow y = e^{-t+5}$

35. $\ln(y - 40) = 5t \Rightarrow e^{\ln(y-40)} = e^{5t} \Rightarrow y - 40 = e^{5t} \Rightarrow y = e^{5t} + 40$

36. $\ln(1 - 2y) = t \Rightarrow e^{\ln(1-2y)} = e^t \Rightarrow 1 - 2y = e^t \Rightarrow -2y = e^t - 1 \Rightarrow y = -\left(\frac{e^t - 1}{2}\right)$

37. $\ln(y - 1) - \ln 2 = x + \ln x \Rightarrow \ln(y - 1) - \ln 2 - \ln x = x \Rightarrow \ln\left(\frac{y-1}{2x}\right) = x \Rightarrow e^{\ln\left(\frac{y-1}{2x}\right)} = e^x \Rightarrow \frac{y-1}{2x} = e^x$

$\Rightarrow y - 1 = 2xe^x \Rightarrow y = 2xe^x + 1$

38. $\ln(y^2-1) - \ln(y+1) = \ln(\sin x) \Rightarrow \ln\left(\frac{y^2-1}{y+1}\right) = \ln(\sin x) \Rightarrow \ln(y-1) = \ln(\sin x) \Rightarrow e^{\ln(y-1)} = e^{\ln(\sin x)}$
$\Rightarrow y - 1 = \sin x \Rightarrow y = \sin x + 1$

39. (a) $e^{2k} = 4 \Rightarrow \ln e^{2k} = \ln 4 \Rightarrow 2k \ln e = \ln 2^2 \Rightarrow 2k = 2\ln 2 \Rightarrow k = \ln 2$
(b) $100e^{10k} = 200 \Rightarrow e^{10k} = 2 \Rightarrow \ln e^{10k} = \ln 2 \Rightarrow 10k \ln e = \ln 2 \Rightarrow 10k = \ln 2 \Rightarrow k = \frac{\ln 2}{10}$
(c) $e^{k/1000} = a \Rightarrow \ln e^{k/1000} = \ln a \Rightarrow \frac{k}{1000} \ln e = \ln a \Rightarrow \frac{k}{1000} = \ln a \Rightarrow k = 1000 \ln a$

40. (a) $e^{5k} = \frac{1}{4} \Rightarrow \ln e^{5k} = \ln 4^{-1} \Rightarrow 5k \ln e = -\ln 4 \Rightarrow 5k = -\ln 4 \Rightarrow k = -\frac{\ln 4}{5}$
(b) $80e^{k} = 1 \Rightarrow e^{k} = 80^{-1} \Rightarrow \ln e^{k} = \ln 80^{-1} \Rightarrow k \ln e = -\ln 80 \Rightarrow k = -\ln 80$
(c) $e^{(\ln 0.8)k} = 0.8 \Rightarrow \left(e^{\ln 0.8}\right)^k = 0.8 \Rightarrow (0.8)^k = 0.8 \Rightarrow k = 1$

41. (a) $e^{-0.3t} = 27 \Rightarrow \ln e^{-0.3t} = \ln 3^3 \Rightarrow (-0.3t)\ln e = 3\ln 3 \Rightarrow -0.3t = 3\ln 3 \Rightarrow t = -10\ln 3$
(b) $e^{kt} = \frac{1}{2} \Rightarrow \ln e^{kt} = \ln 2^{-1} = kt \ln e = -\ln 2 \Rightarrow t = -\frac{\ln 2}{k}$
(c) $e^{(\ln 0.2)t} = 0.4 \Rightarrow \left(e^{\ln 0.2}\right)^t = 0.4 \Rightarrow 0.2^t = 0.4 \Rightarrow \ln 0.2^t = \ln 0.4 \Rightarrow t \ln 0.2 = \ln 0.4 \Rightarrow t = \frac{\ln 0.4}{\ln 0.2}$

42. (a) $e^{-0.01t} = 1000 \Rightarrow \ln e^{-0.01t} = \ln 1000 \Rightarrow (-0.01t)\ln e = \ln 1000 \Rightarrow -0.01t = \ln 1000 \Rightarrow t = -100\ln 1000$
(b) $e^{kt} = \frac{1}{10} \Rightarrow \ln e^{kt} = \ln 10^{-1} = kt \ln e = -\ln 10 \Rightarrow kt = -\ln 10 \Rightarrow t = -\frac{\ln 10}{k}$
(c) $e^{(\ln 2)t} = \frac{1}{2} \Rightarrow \left(e^{\ln 2}\right)^t = 2^{-1} \Rightarrow 2^t = 2^{-1} \Rightarrow t = -1$

43. $e^{\sqrt{t}} = x^2 \Rightarrow \ln e^{\sqrt{t}} = \ln x^2 \Rightarrow \sqrt{t} = 2\ln x \Rightarrow t = 4(\ln x)^2$

44. $e^{x^2} e^{2x+1} = e^t \Rightarrow e^{x^2+2x+1} = e^t \Rightarrow \ln e^{x^2+2x+1} = \ln e^t \Rightarrow t = x^2 + 2x + 1$

45. (a) $5^{\log_5 7} = 7$ (b) $8^{\log_8 \sqrt{2}} = \sqrt{2}$ (c) $1.3^{\log_{1.3} 75} = 75$
(d) $\log_4 16 = \log_4 4^2 = 2\log_4 4 = 2 \cdot 1 = 2$ (e) $\log_3 \sqrt{3} = \log_3 3^{1/2} = \frac{1}{2}\log_3 3 = \frac{1}{2} \cdot 1 = \frac{1}{2} = 0.5$
(f) $\log_4\left(\frac{1}{4}\right) = \log_4 4^{-1} = -1\log_4 4 = -1 \cdot 1 = -1$

46. (a) $2^{\log_2 3} = 3$ (b) $10^{\log_{10}(1/2)} = \frac{1}{2}$ (c) $\pi^{\log_\pi 7} = 7$
(d) $\log_{11} 121 = \log_{11} 11^2 = 2\log_{11} 11 = 2 \cdot 1 = 2$
(e) $\log_{121} 11 = \log_{121} 121^{1/2} = \left(\frac{1}{2}\right)\log_{121} 121 = \left(\frac{1}{2}\right) \cdot 1 = \frac{1}{2}$
(f) $\log_3\left(\frac{1}{9}\right) = \log_3 3^{-2} = -2\log_3 3 = -2 \cdot 1 = -2$

47. (a) Let $z = \log_4 x \Rightarrow 4^z = x \Rightarrow 2^{2z} = x \Rightarrow (2^z)^2 = x \Rightarrow 2^z = \sqrt{x}$
(b) Let $z = \log_3 x \Rightarrow 3^z = x \Rightarrow (3^z)^2 = x^2 \Rightarrow 3^{2z} = x^2 \Rightarrow 9^z = x^2$
(c) $\log_2\left(e^{(\ln 2)\sin x}\right) = \log_2 2^{\sin x} = \sin x$

48. (a) Let $z = \log_5(3x^2) \Rightarrow 5^z = 3x^2 \Rightarrow 25^z = 9x^4$
(b) $\log_e(e^x) = x$
(c) $\log_4\left(2^{e^x \sin x}\right) = \log_4 4^{(e^x \sin x)/2} = \frac{e^x \sin x}{2}$

49. (a) $\frac{\log_2 x}{\log_3 x} = \frac{\ln x}{\ln 2} \div \frac{\ln x}{\ln 3} = \frac{\ln x}{\ln 2} \cdot \frac{\ln 3}{\ln x} = \frac{\ln 3}{\ln 2}$ (b) $\frac{\log_2 x}{\log_8 x} = \frac{\ln x}{\ln 2} \div \frac{\ln x}{\ln 8} = \frac{\ln x}{\ln 2} \cdot \frac{\ln 8}{\ln x} = \frac{3\ln 2}{\ln 2} = 3$
(c) $\frac{\log_x a}{\log_{x^2} a} = \frac{\ln a}{\ln x} \div \frac{\ln a}{\ln x^2} = \frac{\ln a}{\ln x} \cdot \frac{\ln x^2}{\ln a} = \frac{2\ln x}{\ln x} = 2$

50. (a) $\frac{\log_9 x}{\log_3 x} = \frac{\ln x}{\ln 9} \div \frac{\ln x}{\ln 3} = \frac{\ln x}{2\ln 3} \cdot \frac{\ln 3}{\ln x} = \frac{1}{2}$
(b) $\frac{\log_{\sqrt{10}} x}{\log_{\sqrt{2}} x} = \frac{\ln x}{\ln \sqrt{10}} \div \frac{\ln x}{\ln \sqrt{2}} = \frac{\ln x}{\left(\frac{1}{2}\right)\ln 10} \cdot \frac{\left(\frac{1}{2}\right)\ln 2}{\ln x} = \frac{\ln 2}{\ln 10}$

(c) $\frac{\log_a b}{\log_b a} = \frac{\ln b}{\ln a} \div \frac{\ln a}{\ln b} = \frac{\ln b}{\ln a} \cdot \frac{\ln b}{\ln a} = \left(\frac{\ln b}{\ln a}\right)^2$

51. $3^{\log_3(7)} + 2^{\log_2(5)} = 5^{\log_5(x)} \Rightarrow 7 + 5 = x \Rightarrow x = 12$

52. $8^{\log_8(3)} - e^{\ln 5} = x^2 - 7^{\log_7(3x)} \Rightarrow 3 - 5 = x^2 - 3x \Rightarrow 0 = x^2 - 3x + 2 = (x-1)(x-2) \Rightarrow x = 1$ or $x = 2$

53. $3^{\log_3(x^2)} = 5e^{\ln x} - 3 \cdot 10^{\log_{10}(2)} \Rightarrow x^2 = 5x - 6 \Rightarrow x^2 - 5x + 6 = 0 \Rightarrow (x-2)(x-3) = 0 \Rightarrow x = 2$ or $x = 3$

54. $\ln e + 4^{-2\log_4(x)} = \frac{1}{x}\log_{10} 100 \Rightarrow 1 + 4^{\log_4(x^{-2})} = \frac{1}{x}\log_{10} 10^2 \Rightarrow 1 + x^{-2} = \left(\frac{1}{x}\right)(2) \Rightarrow 1 + \frac{1}{x^2} - \frac{2}{x} = 0$
$\Rightarrow x^2 - 2x + 1 = 0 \Rightarrow (x-1)^2 = 0 \Rightarrow x = 1$

55. (a) $\log_3 8 = \frac{\ln 8}{\ln 3} \approx 1.89279$
(b) $\log_7 0.5 = \frac{\ln 0.5}{\ln 7} \approx -0.35621$
(c) $\log_{20} 17 = \frac{\ln 17}{\ln 20} \approx 0.94575$
(d) $\log_{0.5} 7 = \frac{\ln 7}{\ln 0.5} \approx -2.80735$
(e) $\ln x = (\log_{10} x)(\ln 10) = 2.3 \ln 10 \approx 5.29595$
(f) $\ln x = (\log_2 x)(\ln 2) = 1.4 \ln 2 \approx 0.97041$
(g) $\ln x = (\log_2 x)(\ln 2) = -1.5 \ln 2 \approx -1.03972$
(h) $\ln x = (\log_{10} x)(\ln 10) = -0.7 \ln 10 \approx -1.61181$

56. (a) $\frac{\ln 10}{\ln 2} \cdot \log_{10} x = \frac{\ln 10}{\ln 2} \cdot \frac{\ln x}{\ln 10} = \frac{\ln x}{\ln 2} = \log_2 x$
(b) $\frac{\ln a}{\ln b} \cdot \log_a x = \frac{\ln a}{\ln b} \cdot \frac{\ln x}{\ln a} = \frac{\ln x}{\ln b} = \log_b x$

57. $e^{\ln x} = x$ for $x > 0$ and $\ln(e^x) = x$ for all x

58. Graph $y = \ln x - 1$ and determine the zero of y to be $x = 2.7182818$.

59. (a) $-\frac{\pi}{6}$ (b) $\frac{\pi}{4}$ (c) $-\frac{\pi}{3}$

60. (a) $\frac{\pi}{3}$ (b) $\frac{3\pi}{4}$ (c) $\frac{\pi}{6}$

61. (a) $\arccos(-1) = \pi$ since $\cos(\pi) = 1$ and $0 \le \pi \le \pi$.
(b) $\arccos(0) = \frac{\pi}{2}$ since $\cos\left(\frac{\pi}{2}\right) = 0$ and $0 \le \frac{\pi}{2} \le \pi$.

62. (a) $\arcsin(-1) = -\frac{\pi}{2}$ since $\sin\left(-\frac{\pi}{2}\right) = -1$ and $-\frac{\pi}{2} \le -\frac{\pi}{2} \le \frac{\pi}{2}$.
(a) $\arcsin\left(-\frac{1}{\sqrt{2}}\right) = -\frac{\pi}{4}$ since $\sin\left(-\frac{\pi}{4}\right) = -\frac{1}{\sqrt{2}}$ and $-\frac{\pi}{2} \le -\frac{\pi}{4} \le \frac{\pi}{2}$.

63. The function g(x) is also one-to-one. The reasoning: f(x) is one-to-one means that if $x_1 \ne x_2$ then $f(x_1) \ne f(x_2)$, so $-f(x_1) \ne -f(x_2)$ and therefore $g(x_1) \ne g(x_2)$. Therefore g(x) is one-to-one as well.

64. The function h(x) is also one-to-one. The reasoning: f(x) is one-to-one means that if $x_1 \ne x_2$ then $f(x_1) \ne f(x_2)$, so $\frac{1}{f(x_1)} \ne \frac{1}{f(x_2)}$, and therefore $h(x_1) \ne h(x_2)$.

65. The composite is one-to-one also. The reasoning: If $x_1 \ne x_2$ then $g(x_1) \ne g(x_2)$ because g is one-to-one. Since $g(x_1) \ne g(x_2)$, we also have $f(g(x_1)) \ne f(g(x_2))$ because f is one-to-one. Thus, $f \circ g$ is one-to-one because $x_1 \ne x_2 \Rightarrow f(g(x_1)) \ne f(g(x_2))$.

66. Yes, g must be one-to-one. If g were not one-to-one, there would exist numbers $x_1 \ne x_2$ in the domain of g with $g(x_1) = g(x_2)$. For these numbers we would also have $f(g(x_1)) = f(g(x_2))$, contradicting the assumption that $f \circ g$ is one-to-one.

67. (a) $y = \frac{100}{1+2^{-x}} \to 1 + 2^{-x} = \frac{100}{y} \to 2^{-x} = \frac{100}{y} - 1 \to \log_2(2^{-x}) = \log_2\left(\frac{100}{y} - 1\right) \to -x = \log_2\left(\frac{100}{y} - 1\right)$

$x = -\log_2\left(\frac{100}{y} - 1\right) = -\log_2\left(\frac{100-y}{y}\right) = \log_2\left(\frac{y}{100-y}\right)$.

Interchange x and y: $y = \log_2\left(\frac{x}{100-x}\right) \to f^{-1}(x) = \log_2\left(\frac{x}{100-x}\right)$

Verify.

$(f \circ f^{-1})(x) = f\left(\log_2\left(\frac{x}{100-x}\right)\right) = \frac{100}{1+2^{-\log_2\left(\frac{x}{100-x}\right)}} = \frac{100}{1+2^{\log_2\left(\frac{100-x}{x}\right)}} = \frac{100}{1+\frac{100-x}{x}} = \frac{100x}{x+100-x} = \frac{100x}{100} = x$

$(f^{-1} \circ f)(x) = f^{-1}\left(\frac{100}{1+2^{-x}}\right) = \log_2\left(\frac{\frac{100}{1+2^{-x}}}{100 - \frac{100}{1+2^{-x}}}\right) = \log_2\left(\frac{100}{100(1+2^{-x}) - 100}\right) = \log_2\left(\frac{1}{2^{-x}}\right) = \log_2(2^x) = x$

(b) $y = \frac{50}{1+1.1^{-x}} \to 1 + 1.1^{-x} = \frac{50}{y} \to 1.1^{-x} = \frac{50}{y} - 1 \to \log_{1.1}(1.1^{-x}) = \log_{1.1}\left(\frac{50}{y} - 1\right) \to -x = \log_{1.1}\left(\frac{50}{y} - 1\right)$

$x = -\log_{1.1}\left(\frac{50}{y} - 1\right) = -\log_{1.1}\left(\frac{50-y}{y}\right) = \log_{1.1}\left(\frac{y}{50-y}\right)$.

Interchange x and y: $y = \log_{1.1}\left(\frac{x}{50-x}\right) \to f^{-1}(x) = \log_{1.1}\left(\frac{x}{50-x}\right)$

Verify.

$(f \circ f^{-1})(x) = f\left(\log_{1.1}\left(\frac{x}{50-x}\right)\right) = \frac{50}{1+1.1^{-\log_{1.1}\left(\frac{x}{50-x}\right)}} = \frac{50}{1+1.1^{\log_{1.1}\left(\frac{50-x}{x}\right)}} = \frac{50}{1+\frac{50-x}{x}} = \frac{50x}{x+50-x} = \frac{50x}{50} = x$

$(f^{-1} \circ f)(x) = f^{-1}\left(\frac{50}{1+1.1^{-x}}\right) = \log_{1.1}\left(\frac{\frac{50}{1+1.1^{-x}}}{50 - \frac{50}{1+1.1^{-x}}}\right) = \log_{1.1}\left(\frac{50}{50(1+1.1^{-x}) - 50}\right) = \log_{1.1}\left(\frac{1}{1.1^{-x}}\right) = \log_{1.1}(1.1^x) = x$

68. (a) Suppose that $f(x_1) = f(x_2)$. Then $mx_1 + b = mx_2 + b$ so $mx_1 = mx_2$. Since $m \neq 0$, this gives $x_1 = x_2$.

(b) $y = mx + b \to y - b = mx \to \frac{y-b}{m} = x$.

Interchange x and y: $\frac{x-b}{m} = y \to f^{-1}(x) = \frac{x-b}{m}$

The slopes are reciprocals.

(c) If the original functions both have slope m, each of the inverse functions will have slope $\frac{1}{m}$. The graphs of the inverses will be paraallel lines with nonzero slopes.

(d) If the original functions have slopes m and $-\frac{1}{m}$, respectively, then the inverse functions will have slopes $\frac{1}{m}$ and $-m$, respectively. Since each of $\frac{1}{m}$ and $-m$ is the negative reciprocal of the other, the graphs of the inverses will be perpendicular lines with nonzero slopes.

69. (a) Amount $= 8\left(\frac{1}{2}\right)^{t/12}$

(b) $8\left(\frac{1}{2}\right)^{t/12} = 1 \to \left(\frac{1}{2}\right)^{t/12} = \frac{1}{8} \to \left(\frac{1}{2}\right)^{t/12} = \left(\frac{1}{2}\right)^3 \to \frac{t}{12} = 3 \to t = 36$

There will be 1 gram remaining after 36 hours.

70. $500(1.0475)^t = 1000 \to 1.0475^t = 2 \to \ln(1.0475^t) = \ln(2) \to t\ln(1.0475) = \ln(2) \to t = \frac{\ln(2)}{\ln(1.0475)} \approx 14.936$

It will take about 14.936 years. (If the interest is paid at the end of each year, it will take 15 years.)

71. $375{,}000(1.0225)^t = 1{,}000{,}000 \to 1.0225^t = \frac{8}{3} \to \ln(1.0225^t) = \ln\left(\frac{8}{3}\right) \to t\ln(1.0225) = \ln\left(\frac{8}{3}\right) \to t = \frac{\ln\left(\frac{8}{3}\right)}{\ln(1.0225)} \approx 44.081$

It will take about 44.081 years.

72. $y = y_0e^{-0.18t}$ represents the decay equation; solving $(0.9)y_0 = y_0e^{-0.18t} \Rightarrow t = \frac{\ln(0.9)}{-0.18} \approx 0.585$ days

73. From zooming in on the graph at the right, we estimate the third root to be $x \approx -0.76666$

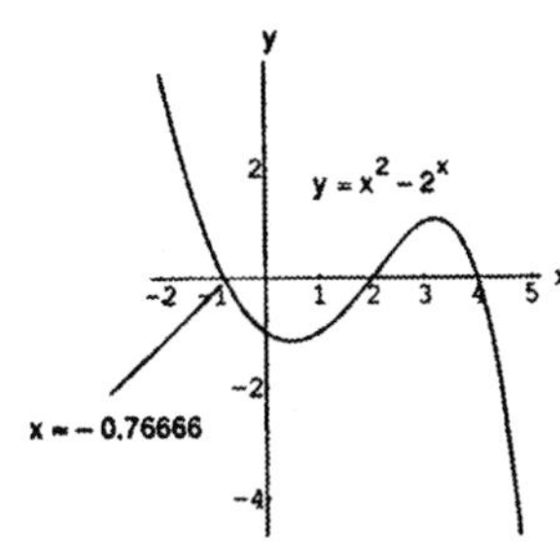

74. The functions $f(x) = x^{\ln 2}$ and $g(x) = 2^{\ln x}$ appear to have identical graphs for $x > 0$. This is no accident, because $x^{\ln 2} = e^{\ln 2 \cdot \ln x} = (e^{\ln 2})^{\ln x} = 2^{\ln x}$.

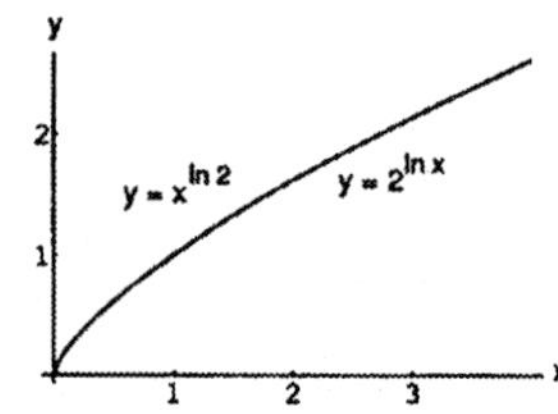

75. (a) Begin with $y = \ln x$ and reduce the y-value by $3 \Rightarrow y = \ln x - 3$.
(b) Begin with $y = \ln x$ and replace x with $x - 1 \Rightarrow y = \ln(x - 1)$.
(c) Begin with $y = \ln x$, replace x with $x + 1$, and increase the y-value by $3 \Rightarrow y = \ln(x + 1) + 3$.
(d) Begin with $y = \ln x$, reduce the y-value by 4, and replace x with $x - 2 \Rightarrow y = \ln(x - 2) - 4$.
(e) Begin with $y = \ln x$ and replace x with $-x \Rightarrow y = \ln(-x)$.
(f) Begin with $y = \ln x$ and switch x and $y \Rightarrow x = \ln y$ or $y = e^x$.

76. (a) Begin with $y = \ln x$ and multiply the y-value by $2 \Rightarrow y = 2 \ln x$.
(b) Begin with $y = \ln x$ and replace x with $\frac{x}{3} \Rightarrow y = \ln\left(\frac{x}{3}\right)$.
(c) Begin with $y = \ln x$ and multiply the y-value by $\frac{1}{4} \Rightarrow y = \frac{1}{4}\ln x$.
(d) Begin with $y = \ln x$ and replace x with $2x \Rightarrow y = \ln 2x$.

77. There are a few cases to consider. First, if $x = 0$ then $\cos(\sin^{-1}(x)) = \cos(\sin^{-1}(0)) = \cos(0) = 1 = \sqrt{1 - 0}$ $= \sqrt{1 - x^2}$. Note that if $x \neq 0$ then either $x > 0$ or $x < 0$. However, since $\sin(x)$ is an odd function, so is $\sin^{-1}(x)$ odd. Now $\cos(x)$ is even so that $\cos(\sin^{-1}(x)) = \cos(-\sin^{-1}(-x)) = \cos(\sin^{-1}(-x))$. Thus, it suffices to assume only that $x > 0$; the case $x < 0$ follows. So consider the given right triangle with $\theta = \sin^{-1}(x)$. We have $\sin\theta = x = \frac{x}{1}$ $= \frac{\text{opposite}}{\text{hypotenuse}}$. Thus $\cos(\sin^{-1}(x)) = \cos\theta = \frac{\text{adjacent}}{\text{hypotenuse}} = \sqrt{1 - x^2}$.

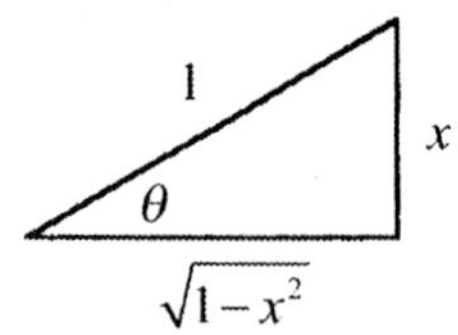

78. $\sin^{-1}(1) + \cos^{-1}(1) = \frac{\pi}{2} + 0 = \frac{\pi}{2}$; $\sin^{-1}(0) + \cos^{-1}(0) = 0 + \frac{\pi}{2} = \frac{\pi}{2}$; and $\sin^{-1}(-1) + \cos^{-1}(-1) = -\frac{\pi}{2} + \pi = \frac{\pi}{2}$. If $x \in (-1, 0)$ and $x = -a$, then $\sin^{-1}(x) + \cos^{-1}(x) = \sin^{-1}(-a) + \cos^{-1}(-a) = -\sin^{-1} a + (\pi - \cos^{-1} a)$ $= \pi - (\sin^{-1} a + \cos^{-1} a) = \pi - \frac{\pi}{2} = \frac{\pi}{2}$ from Equations (3) and (4) in the text.

CHAPTER 1 PRACTICE EXERCISES

1. The area is $A = \pi r^2$ and the circumference is $C = 2\pi r$. Thus, $r = \frac{C}{2\pi} \Rightarrow A = \pi\left(\frac{C}{2\pi}\right)^2 = \frac{C^2}{4\pi}$.

2. The surface area is $S = 4\pi r^2 \Rightarrow r = \left(\frac{S}{4\pi}\right)^{1/2}$. The volume is $V = \frac{4}{3}\pi r^3 \Rightarrow r = \sqrt[3]{\frac{3V}{4\pi}}$. Substitution into the formula for surface area gives $S = 4\pi r^2 = 4\pi\left(\frac{3V}{4\pi}\right)^{2/3}$.

3. The coordinates of a point on the parabola are (x, x^2). The angle of inclination θ joining this point to the origin satisfies the equation $\tan\theta = \frac{x^2}{x} = x$. Thus the point has coordinates $(x, x^2) = (\tan\theta, \tan^2\theta)$.

4. $\tan\theta = \frac{\text{rise}}{\text{run}} = \frac{h}{500} \Rightarrow h = 500 \tan\theta$ ft.

5.

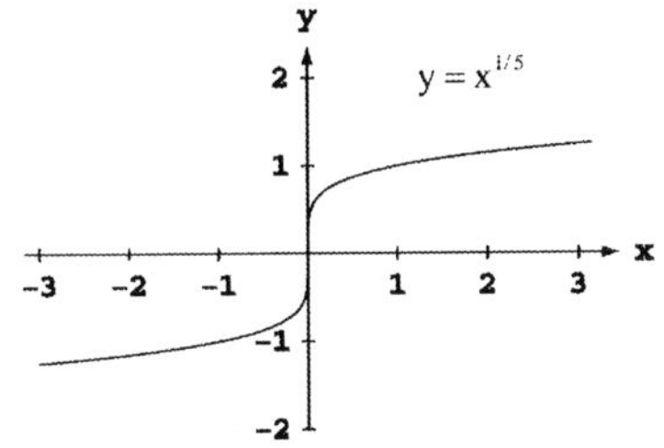

Symmetric about the origin.

6.

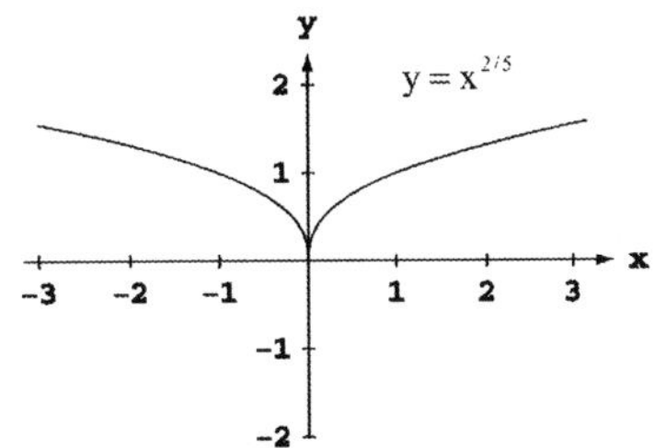

Symmetric about the y-axis.

7.

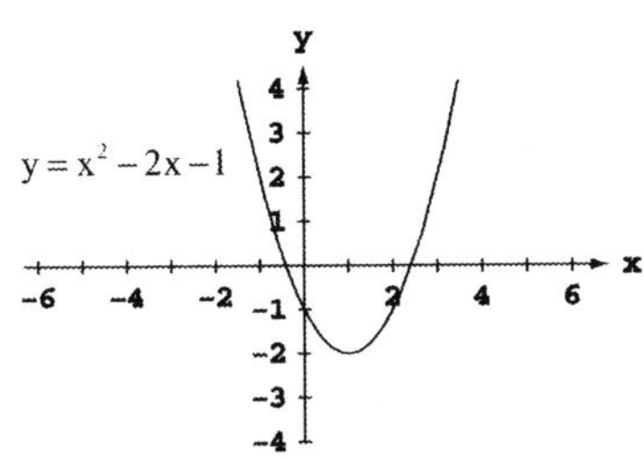

Neither

8.

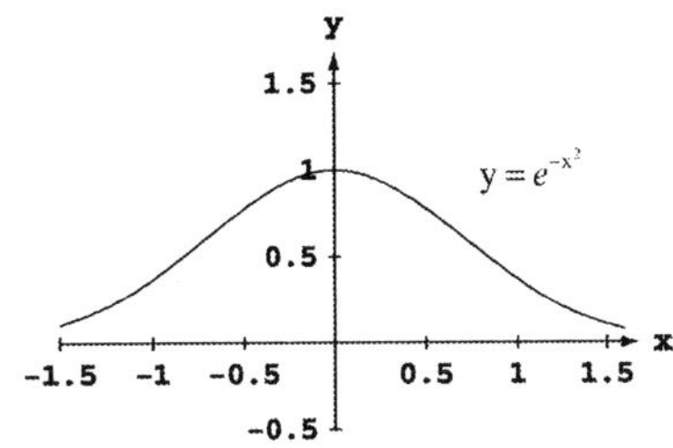

Symmetric about the y-axis.

9. $y(-x) = (-x)^2 + 1 = x^2 + 1 = y(x)$. Even.

10. $y(-x) = (-x)^5 - (-x)^3 - (-x) = -x^5 + x^3 + x = -y(x)$. Odd.

11. $y(-x) = 1 - \cos(-x) = 1 - \cos x = y(x)$. Even.

12. $y(-x) = \sec(-x)\tan(-x) = \frac{\sin(-x)}{\cos^2(-x)} = \frac{-\sin x}{\cos^2 x} = -\sec x \tan x = -y(x)$. Odd.

13. $y(-x) = \frac{(-x)^4+1}{(-x)^3-2(-x)} = \frac{x^4+1}{-x^3+2x} = -\frac{x^4+1}{x^3-2x} = -y(x)$. Odd.

14. $y(-x) = 1 - \sin(-x) = 1 + \sin x$. Neither even nor odd.

15. $y(-x) = -x + \cos(-x) = -x + \cos x$. Neither even nor odd.

16. $y(-x) = \sqrt{(-x)^4 - 1} = \sqrt{x^4 - 1} = y(x)$. Even.

17. (a) The function is defined for all values of x, so the domain is $(-\infty, \infty)$.
    (b) Since $|x|$ attains all nonnegative values, the range is $[-2, \infty)$.

18. (a) Since the square root requires $1 - x \geq 0$, the domain is $(-\infty, 1]$.
    (b) Since $\sqrt{1 - x}$ attains all nonnegative values, the range is $[-2, \infty)$.

19. (a) Since the square root requires $16 - x^2 \geq 0$, the domain is $[-4, 4]$.
    (b) For values of x in the domain, $0 \leq 16 - x^2 \leq 16$, so $0 \leq \sqrt{16 - x^2} \leq 4$. The range is $[0, 4]$.

20. (a) The function is defined for all values of x, so the domain is $(-\infty, \infty)$.
(b) Since 3^{2-x} attains all positive values, the range is $(1, \infty)$.

21. (a) The function is defined for all values of x, so the domain is $(-\infty, \infty)$.
(b) Since $2e^{-x}$ attains all positive values, the range is $(-3, \infty)$.

22. (a) The function is equivalent to $y = \tan 2x$, so we require $2x \neq \frac{k\pi}{2}$ for odd integers k. The domain is given by $x \neq \frac{k\pi}{4}$ for odd integers k.
(b) Since the tangent function attains all values, the range is $(-\infty, \infty)$.

23. (a) The function is defined for all values of x, so the domain is $(-\infty, \infty)$.
(b) The sine function attains values from -1 to 1, so $-2 \leq 2\sin(3x + \pi) \leq 2$ and hence $-3 \leq 2\sin(3x + \pi) - 1 \leq 1$. The range is $[-3, 1]$.

24. (a) The function is defined for all values of x, so the domain is $(-\infty, \infty)$.
(b) The function is equivalent to $y = \sqrt[5]{x^2}$, which attains all nonnegative values. The range is $[0, \infty)$.

25. (a) The logarithm requires $x - 3 > 0$, so the domain is $(3, \infty)$.
(b) The logarithm attains all real values, so the range is $(-\infty, \infty)$.

26. (a) The function is defined for all values of x, so the domain is $(-\infty, \infty)$.
(b) The cube root attains all real values, so the range is $(-\infty, \infty)$.

27. (a) The function is defined for $-4 \leq x \leq 4$, so the domain is $[-4, 4]$.
(b) The function is equivalent to $y = \sqrt{|x|}$, $-4 \leq x \leq 4$, which attains values from 0 to 2 for x in the domain. The range is $[0, 2]$.

28. (a) The function is defined for $-2 \leq x \leq 2$, so the domain is $[-2, 2]$.
(b) The range is $[-1, 1]$.

29. First piece: Line through (0, 1) and (1, 0). $m = \frac{0-1}{1-0} = \frac{-1}{1} = -1 \Rightarrow y = -x + 1 = 1 - x$
Second piece: Line through (1, 1) and (2, 0). $m = \frac{0-1}{2-1} = \frac{-1}{1} = -1 \Rightarrow y = -(x - 1) + 1 = -x + 2 = 2 - x$

$$f(x) = \begin{cases} 1 - x, & 0 \leq x < 1 \\ 2 - x, & 1 \leq x \leq 2 \end{cases}$$

30. First piece: Line through (0, 0) and (2, 5). $m = \frac{5-0}{2-0} = \frac{5}{2} \Rightarrow y = \frac{5}{2}x$
Second piece: Line through (2, 5) and (4, 0). $m = \frac{0-5}{4-2} = \frac{-5}{2} = -\frac{5}{2} \Rightarrow y = -\frac{5}{2}(x - 2) + 5 = -\frac{5}{2}x + 10 = 10 - \frac{5x}{2}$

$$f(x) = \begin{cases} \frac{5}{2}x, & 0 \leq x < 2 \\ 10 - \frac{5x}{2}, & 2 \leq x \leq 4 \end{cases}$$ (Note: $x = 2$ can be included on either piece.)

31. (a) $(f \circ g)(-1) = f(g(-1)) = f\left(\frac{1}{\sqrt{-1+2}}\right) = f(1) = \frac{1}{1} = 1$
(b) $(g \circ f)(2) = g(f(2)) = g\left(\frac{1}{2}\right) = \frac{1}{\sqrt{\frac{1}{2}+2}} = \frac{1}{\sqrt{2.5}}$ or $\sqrt{\frac{2}{5}}$
(c) $(f \circ f)(x) = f(f(x)) = f\left(\frac{1}{x}\right) = \frac{1}{1/x} = x,\ x \neq 0$
(d) $(g \circ g)(x) = g(g(x)) = g\left(\frac{1}{\sqrt{x+2}}\right) = \frac{1}{\sqrt{\frac{1}{\sqrt{x+2}}+2}} = \frac{\sqrt[4]{x+2}}{\sqrt{1+2\sqrt{x+2}}}$

32. (a) $(f\circ g)(-1) = f(g(-1)) = f(\sqrt[3]{-1+1}) = f(0) = 2 - 0 = 2$
    (b) $(g\circ f)(2) = f(g(2)) = g(2-2) = g(0) = \sqrt[3]{0+1} = 1$
    (c) $(f\circ f)(x) = f(f(x)) = f(2-x) = 2-(2-x) = x$
    (d) $(g\circ g)(x) = g(g(x)) = g(\sqrt[3]{x+1}) = \sqrt[3]{\sqrt[3]{x+1}+1}$

33. (a) $(f\circ g)(x) = f(g(x)) = f(\sqrt{x+2}) = 2 - (\sqrt{x+2})^2 = -x,\ x \geq -2.$
    $(g\circ f)(x) = f(g(x)) = g(2-x^2) = \sqrt{(2-x^2)+2} = \sqrt{4-x^2}$
    (b) Domain of $f\circ g$: $[-2, \infty)$.
    Domain of $g\circ f$: $[-2, 2]$.
    (c) Range of $f\circ g$: $(-\infty, 2]$.
    Range of $g\circ f$: $[0, 2]$.

34. (a) $(f\circ g)(x) = f(g(x)) = f\left(\sqrt{1-x}\right) = \sqrt{\sqrt{1-x}} = \sqrt[4]{1-x}.$
    $(g\circ f)(x) = f(g(x)) = g(\sqrt{x}) = \sqrt{1-\sqrt{x}}$
    (b) Domain of $f\circ g$: $(-\infty, 1]$.
    Domain of $g\circ f$: $[0, 1]$.
    (c) Range of $f\circ g$: $[0, \infty)$.
    Range of $g\circ f$: $[0, 1]$.

35.

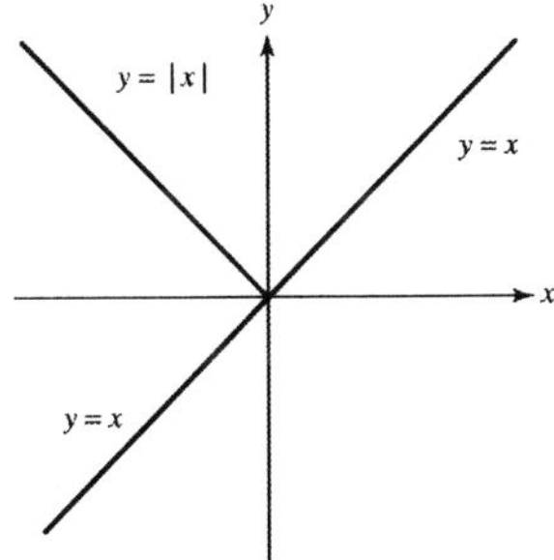

The graph of $f_2(x) = f_1(|x|)$ is the same as the graph of $f_1(x)$ to the right of the y-axis. The graph of $f_2(x)$ to the left of the y-axis is the reflection of $y = f_1(x),\ x \geq 0$ across the y-axis.

36.

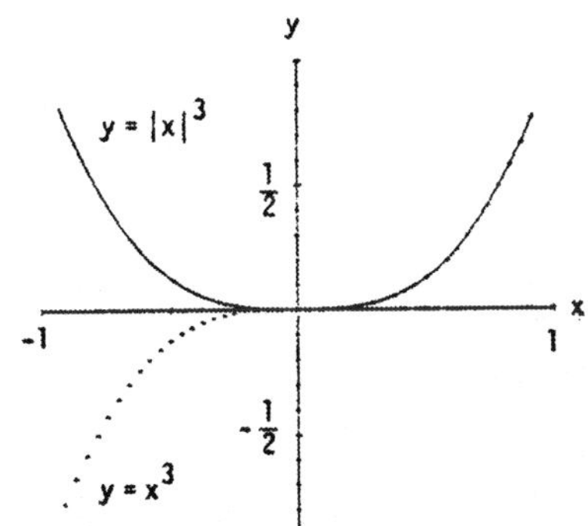

The graph of $f_2(x) = f_1(|x|)$ is the same as the graph of $f_1(x)$ to the right of the y-axis. The graph of $f_2(x)$ to the left of the y-axis is the reflection of $y = f_1(x),\ x \geq 0$ across the y-axis.

37.

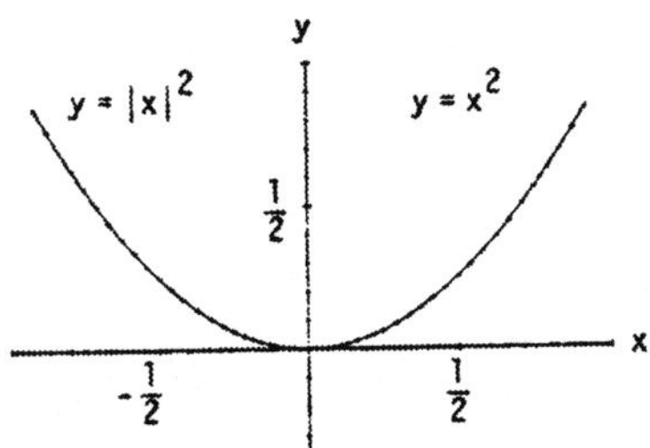

It does not change the graph.

38.

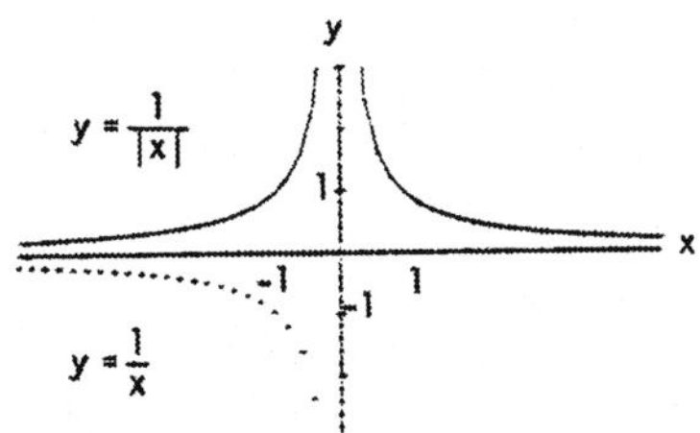

The graph of $f_2(x) = f_1(|x|)$ is the same as the graph of $f_1(x)$ to the right of the y-axis. The graph of $f_2(x)$ to the left of the y-axis is the reflection of $y = f_1(x),\ x \geq 0$ across the y-axis.

39.

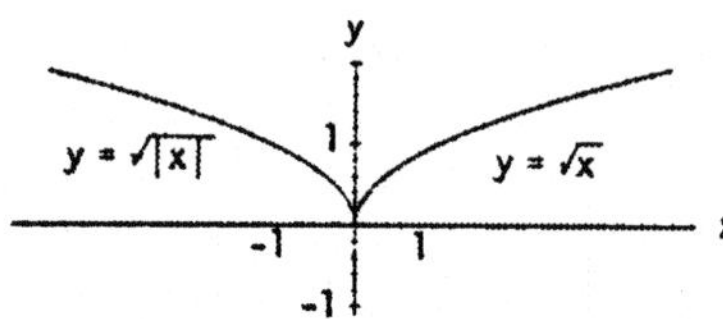

The graph of $f_2(x) = f_1(|x|)$ is the same as the graph of $f_1(x)$ to the right of the y-axis. The graph of $f_2(x)$ to the left of the y-axis is the reflection of $y = f_1(x)$, $x \geq 0$ across the y-axis.

40.

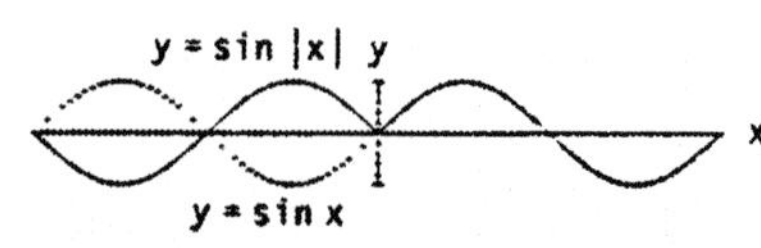

The graph of $f_2(x) = f_1(|x|)$ is the same as the graph of $f_1(x)$ to the right of the y-axis. The graph of $f_2(x)$ to the left of the y-axis is the reflection of $y = f_1(x)$, $x \geq 0$ across the y-axis.

41.

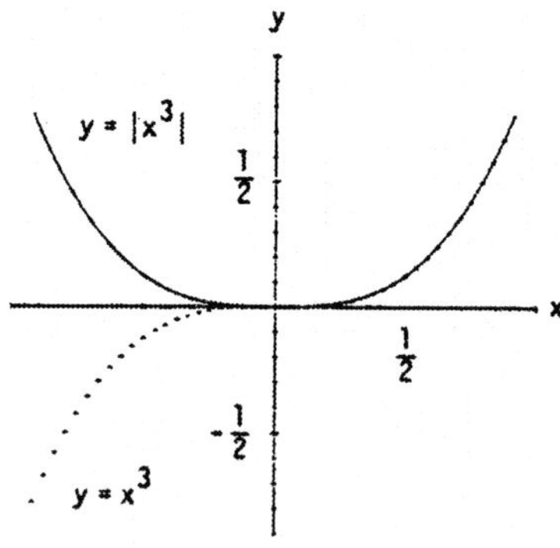

Whenever $g_1(x)$ is positive, the graph of $y = g_2(x) = |g_1(x)|$ is the same as the graph of $y = g_1(x)$. When $g_1(x)$ is negative, the graph of $y = g_2(x)$ is the reflection of the graph of $y = g_1(x)$ across the x-axis.

42.

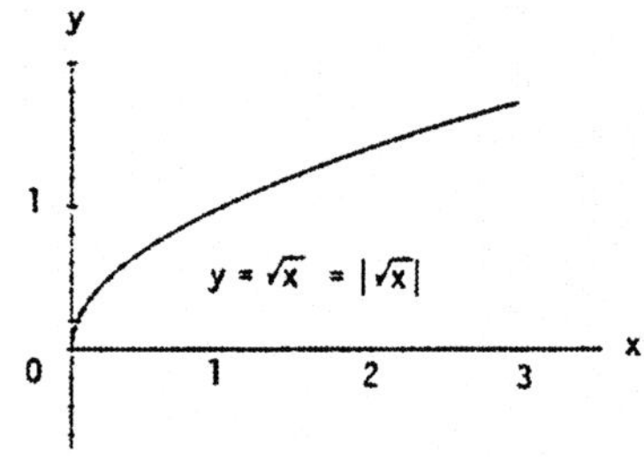

It does not change the graph.

43.

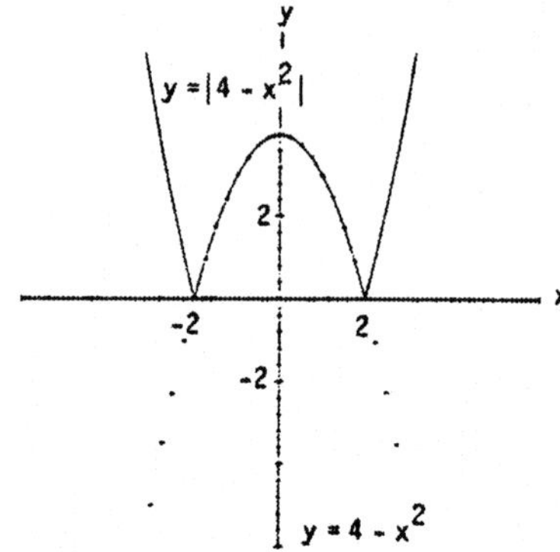

Whenever $g_1(x)$ is positive, the graph of $y = g_2(x) = |g_1(x)|$ is the same as the graph of $y = g_1(x)$. When $g_1(x)$ is negative, the graph of $y = g_2(x)$ is the reflection of the graph of $y = g_1(x)$ across the x-axis.

44.

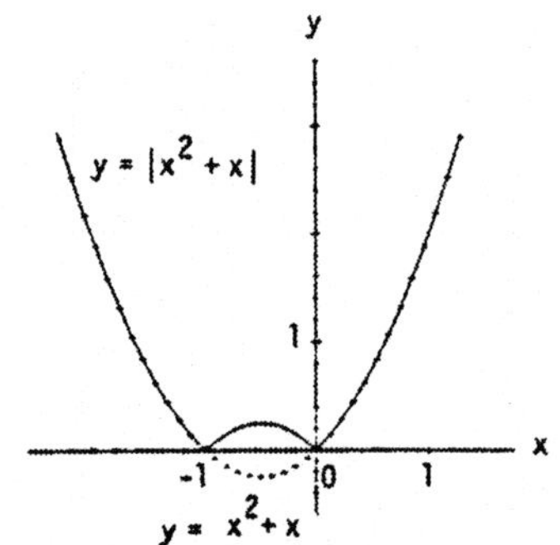

Whenever $g_1(x)$ is positive, the graph of $y = g_2(x) = |g_1(x)|$ is the same as the graph of $y = g_1(x)$. When $g_1(x)$ is negative, the graph of $y = g_2(x)$ is the reflection of the graph of $y = g_1(x)$ across the x-axis.

45. (a) Domain: $(-\infty, \infty)$

(b) Range: $(0, \infty)$

46. (a) Domain: $(-\infty, 0) \cup (0, \infty)$
(b) If $4 - x^2 = (2 - x)(2 + x) = 0$, then $x = 2$ or $x = -2 \Rightarrow$ Domain: $(-\infty, 2) \cup (-2, 2) \cup (2, \infty)$.

47. (a) If $-1 \le \frac{x}{3} \le 1$, then $-3 \le x \le 3 \Rightarrow$ Domain: $[-3, 3]$.
(b) If $-1 \le \sqrt{x} - 1 \le 1$, then $0 \le \sqrt{x} \le 2 \Rightarrow 0 \le x \le 4 \Rightarrow$ Domain: $[0, 4]$.

48. (a) If $\cos^{-1} x = 0$, then $x = 1 \Rightarrow$ Domain: $[-1, 1)$.
(b) Since $\frac{\pi}{2} \le \sin^{-1} x \le \frac{\pi}{2} \Rightarrow$ Domain: $[-1, 1]$.

49. $(f \circ g)(x) = f(g(x)) = f(4 - x^2) = \ln(4 - x^2) \Rightarrow$ Domain of $f \circ g$: $(-2, 2)$; $(g \circ f)(x) = g(f(x)) = g(\ln x) = 4 - (\ln x)^2$ $\Rightarrow$ Domain of $g \circ f$: $(0, \infty)$; $(f \circ f)(x) = f(f(x)) = f(\ln x) = \ln(\ln x)$ and since $\ln 1 = 0 \Rightarrow$ Domain of $f \circ f$: $(1, \infty)$; $(g \circ g)(x) = g(g(x)) = g(4 - x^2) = 4 - (4 - x^2)^2 = -x^4 + 8x^2 - 12 \Rightarrow$ Domain of $g \circ g$: $(-\infty, \infty)$.

50. (a) Since $f(-x) = e^{-(-x)^2} = e^{-x^2} = f(x) \Rightarrow f$ is even.
(b) Since $f(-x) = 1 + \sin^{-1}(-(-x)) = 1 + \sin^{-1} x$, $-f(-x) = -1 - \sin^{-1} x$, and $f(x) = 1 + \sin^{-1}(-x) = 1 - \sin^{-1} x$ $\Rightarrow f$ is neither even nor odd.
(c) Since $f(-x) = |e^{-x}| = e^{-x}$, $-f(-x) = -e^{-x}$ and $f(x) = |e^x| = e^x \Rightarrow f$ is neither even nor odd.
(d) Since $f(-x) = e^{\ln|-x|+1} = e^{\ln|x|+1} = f(x) \Rightarrow f$ is even.

51. $y = \ln kx \Rightarrow y = \ln x + \ln k$; thus the graph of $y = \ln kx$ is the graph of $y = \ln x$ shifted vertically by $\ln k$, $k > 0$.

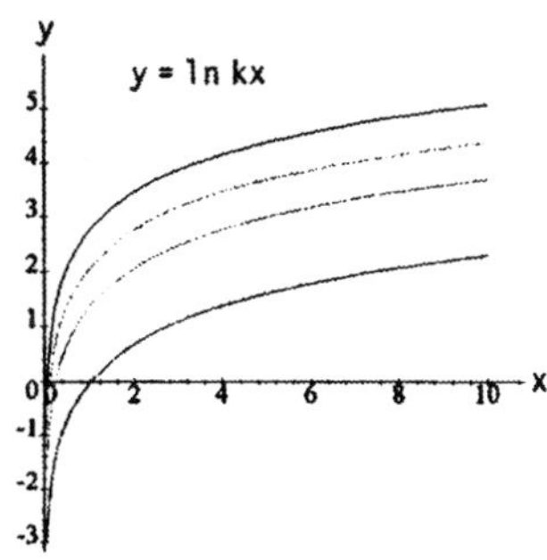

52. For $c > 0$, the domain is $(-\infty, \infty)$
For $c \le 0$, the domain is $(-\infty, -\sqrt{c}) \cup (\sqrt{c}, \infty)$.

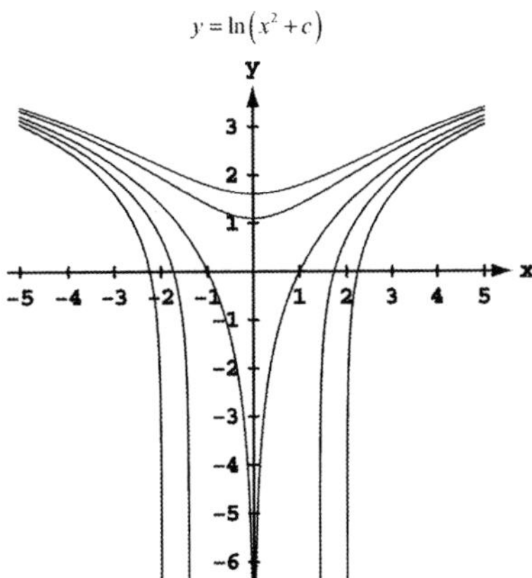

53. To turn the arches upside down we would use the formula $y = -\ln|\sin x| = \ln \frac{1}{|\sin x|}$.

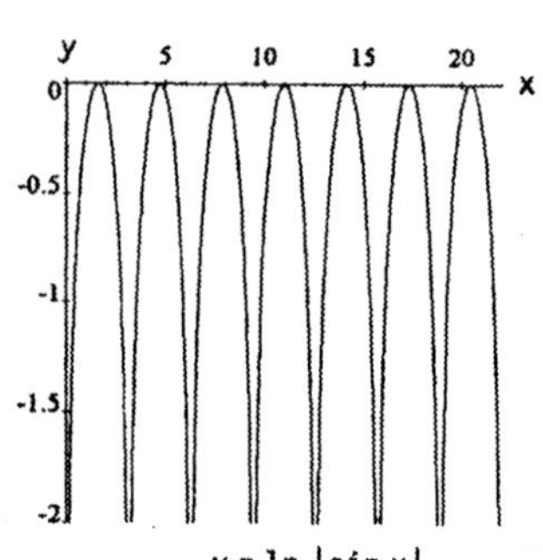

54. For large values of x, $y = a^x$ has the largest values; $y = \log_a x$ the smallest.

55. (a) Domain: $-\infty < x < \infty$; Range: $-\frac{\pi}{2} \le y \le \frac{\pi}{2}$

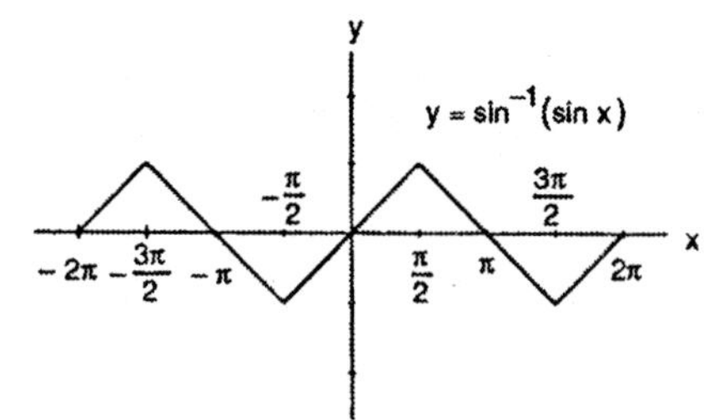

(b) Domain: $-1 \le x \le 1$; Range: $-1 \le y \le 1$
The graph of $y = \sin^{-1}(\sin x)$ is periodic; the graph of $y = \sin(\sin^{-1} x) = x$ for $-1 \le x \le 1$.

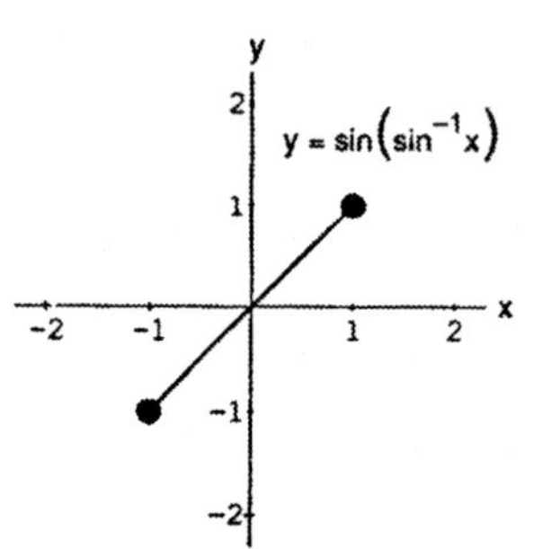

56. (a) Domain: $-\infty < x < \infty$; Range: $0 \le y \le \pi$

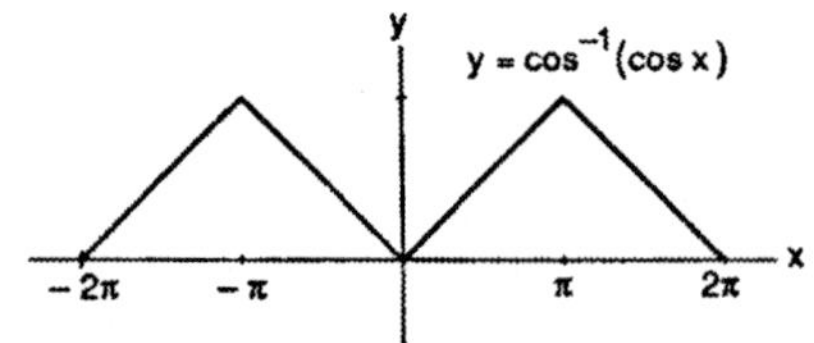

(b) Domain: $-1 \le x \le 1$; Range: $-1 \le y \le 1$
The graph of $y = \cos^{-1}(\cos x)$ is periodic; the graph of $y = \cos(\cos^{-1} x) = x$ for $-1 \le x \le 1$.

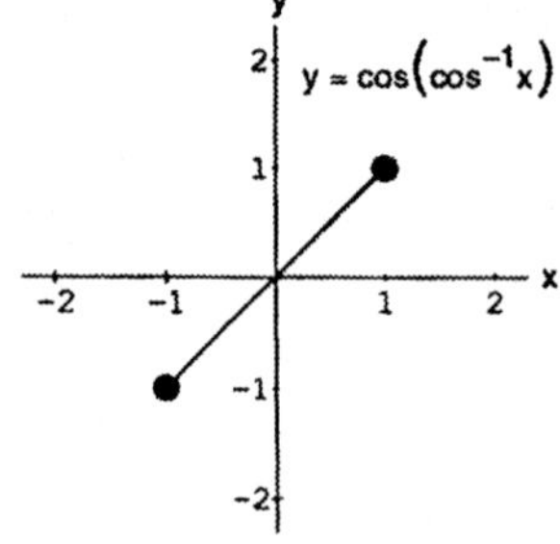

57. (a) $f(x) = x^3 - \frac{x}{2} = x\left(x^2 - \frac{1}{2}\right) \Rightarrow f(x) = 0 \Rightarrow x = 0, \pm \frac{1}{\sqrt{2}}$
Thus the graph of $y = f(x)$ resembles the graph shown at the right.

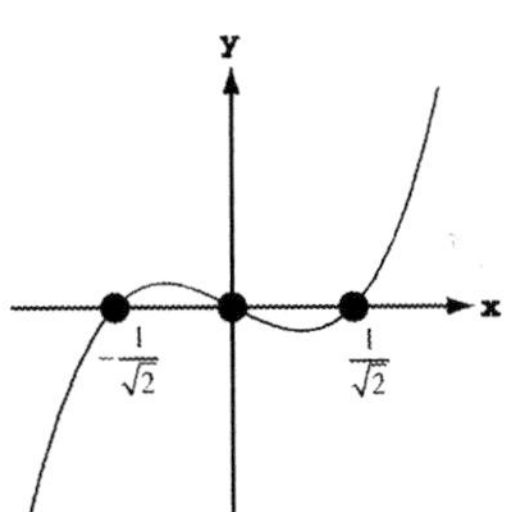

Clearly, f is not one-to-one.

(b) $f(x) = x^3 + \frac{x}{2} = x\left(x^2 + \frac{1}{2}\right) \Rightarrow f(x) = 0 \Rightarrow x = 0.$
Thus the graph of $y = f(x)$ resembles the graph shown at the right.

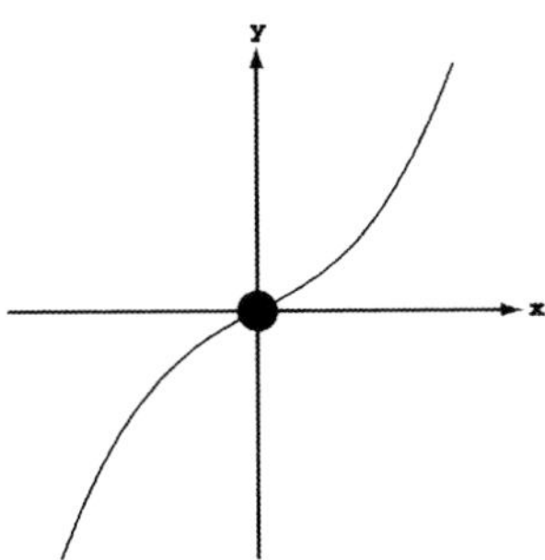

We see that f is one-to-one.

58. Answers depend on the view screen used. For $[15, 17] \times [5 \cdot 10^6, 10^7]$ it appears that $e^x > 10^7$ for $x \geq 16.128$, for example.

59. (a) $f(g(x)) = \left(\sqrt[3]{x}\right)^3 = x,\ g(f(x)) = \sqrt[3]{x^3} = x$

(b)

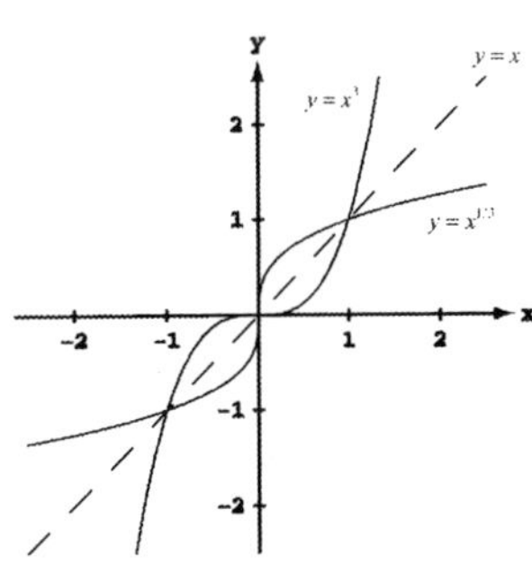

60. (a) $h(k(x)) = \frac{1}{4}\left(\left(4x\right)^{1/3}\right)^3 = x,$
$k(h(x)) = \left(4 \cdot \frac{x^3}{4}\right)^{1/3} = x$

(b)

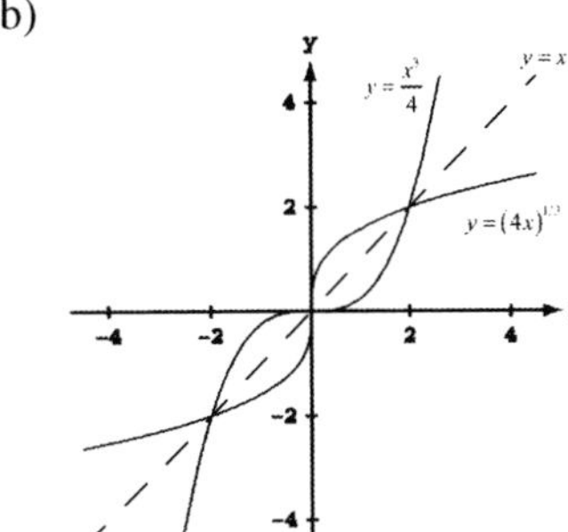

61. (a) $y = 20.627x + 338.622$

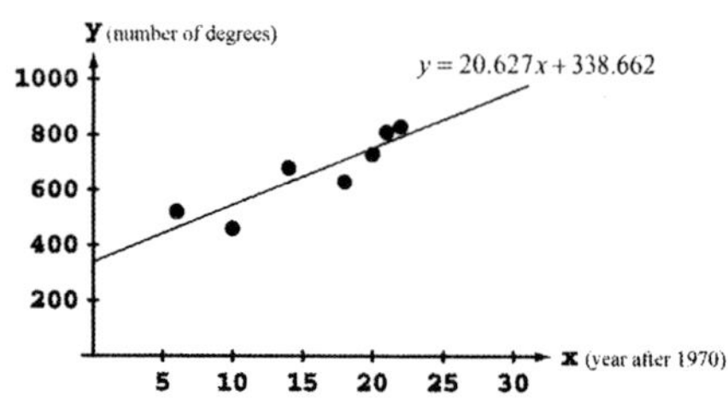

(b) When $x = 30$, $y = 957.445$. According to the regression equation, abut 957 degrees will be earned.

(c) The slope is 20.627. It represents the approximate annual increase in the number of doctorates earned by Hispanic Americans per year.

62. (a) The TI-83 Plus calculator gives $Q = 1.00(2.0138^x) = 1.00e^{0.7x}$

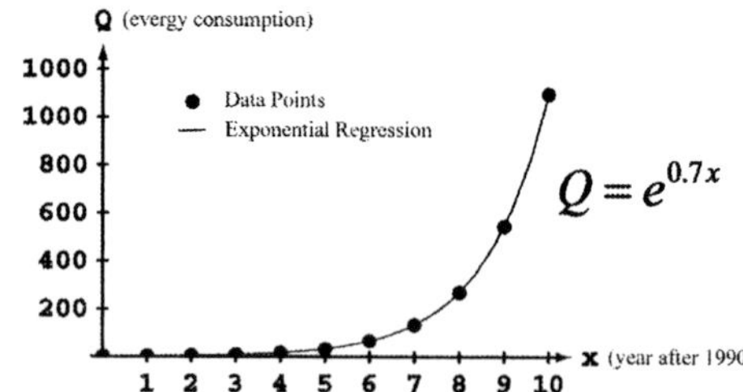

(b) For 1996, $x = 9.6 \Rightarrow Q(9.6) = e^{0.7(9.6)} = 828.82$ units of energy consumed that year as estimated by the exponential regression. The exponential regression shows that energy consumption has doubled (i.e., increased by 100%) each decade during the 20th cntury. The annual rate of increase during this time is $e^{0.7(0.1)} - e^{0.7(0)} = 0.0725 = 7.25\%$,

CHAPTER 1 ADDITIONAL AND ADVANCED EXERCISES

1. (a) The given graph is reflected about the y-axis.

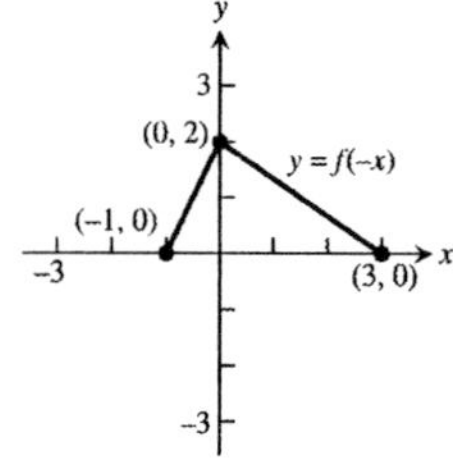

(b) The given graph is reflected about the x-axis.

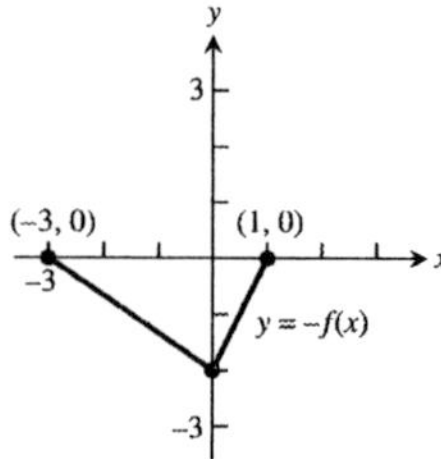

(c) The given graph is shifted left 1 unit, stretched vertically by a factor of 2, reflected about the x-axis, and then shifted upward 1 unit.

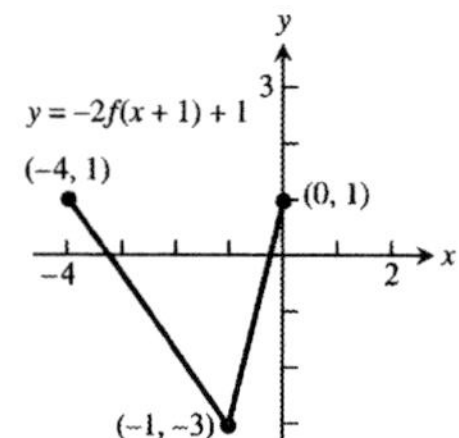

(d) The given graph is shifted right 2 units, stretched vertically by a factor of 3, and then shifted downward 2 units.

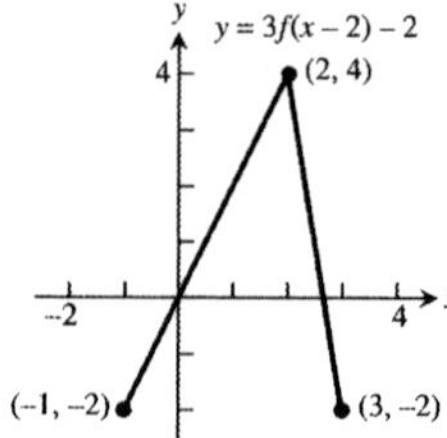

2. (a)

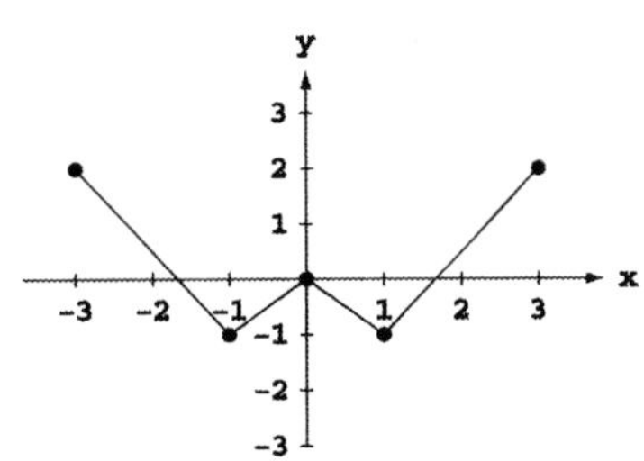

(b)

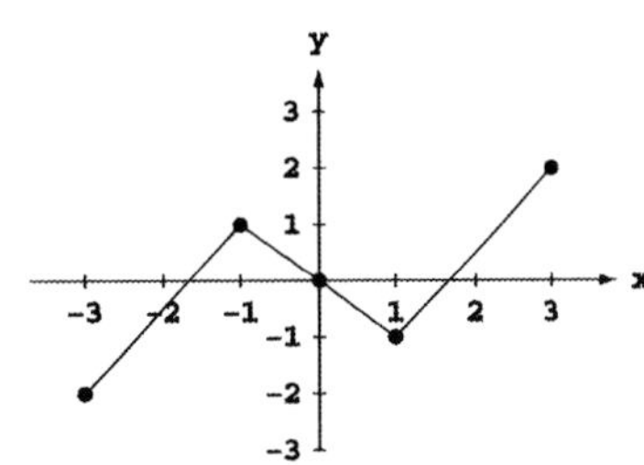

3. There are (infinitely) many such function pairs. For example, $f(x) = 3x$ and $g(x) = 4x$ satisfy $f(g(x)) = f(4x) = 3(4x) = 12x = 4(3x) = g(3x) = g(f(x))$.

4. Yes, there are many such function pairs. For example, if $g(x) = (2x + 3)^3$ and $f(x) = x^{1/3}$, then $(f \circ g)(x) = f(g(x)) = f\left((2x + 3)^3\right) = \left((2x + 3)^3\right)^{1/3} = 2x + 3$.

5. If f is odd and defined at x, then $f(-x) = -f(x)$. Thus $g(-x) = f(-x) - 2 = -f(x) - 2$ whereas $-g(x) = -(f(x) - 2) = -f(x) + 2$. Then g cannot be odd because $g(-x) = -g(x) \Rightarrow -f(x) - 2 = -f(x) + 2$ $\Rightarrow 4 = 0$, which is a contradiction. Also, g(x) is not even unless $f(x) = 0$ for all x. On the other hand, if f is even, then $g(x) = f(x) - 2$ is also even: $g(-x) = f(-x) - 2 = f(x) - 2 = g(x)$.

6. If g is odd and g(0) is defined, then $g(0) = g(-0) = -g(0)$. Therefore, $2g(0) = 0 \Rightarrow g(0) = 0$.

7. For (x, y) in the 1st quadrant, $|x| + |y| = 1 + x$ $\Leftrightarrow x + y = 1 + x \Leftrightarrow y = 1$. For (x, y) in the 2nd quadrant, $|x| + |y| = x + 1 \Leftrightarrow -x + y = x + 1$ $\Leftrightarrow y = 2x + 1$. In the 3rd quadrant, $|x| + |y| = x + 1$ $\Leftrightarrow -x - y = x + 1 \Leftrightarrow y = -2x - 1$. In the 4th quadrant, $|x| + |y| = x + 1 \Leftrightarrow x + (-y) = x + 1$ $\Leftrightarrow y = -1$. The graph is given at the right.

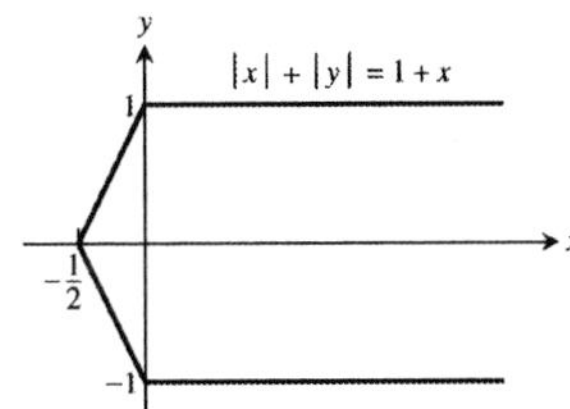

8. We use reasoning similar to Exercise 7.
   (1) 1st quadrant: $y + |y| = x + |x|$ $\Leftrightarrow 2y = 2x \Leftrightarrow y = x$.
   (2) 2nd quadrant: $y + |y| = x + |x|$ $\Leftrightarrow 2y = x + (-x) = 0 \Leftrightarrow y = 0$.
   (3) 3rd quadrant: $y + |y| = x + |x|$ $\Leftrightarrow y + (-y) = x + (-x) \Leftrightarrow 0 = 0$ $\Rightarrow$ all points in the 3rd quadrant satisfy the equation.
   (4) 4th quadrant: $y + |y| = x + |x|$ $\Leftrightarrow y + (-y) = 2x \Leftrightarrow 0 = x$. Combining these results we have the graph given at the right:

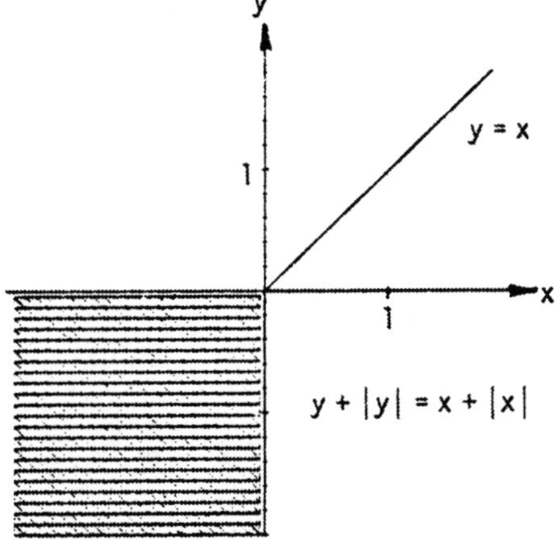

9. If f is even and odd, then $f(-x) = -f(x)$ and $f(-x) = f(x) \Rightarrow f(x) = -f(x)$ for all x in the domain of f. Thus $2f(x) = 0 \Rightarrow f(x) = 0$.

10. (a) As suggested, let $E(x) = \frac{f(x)+f(-x)}{2} \Rightarrow E(-x) = \frac{f(-x)+f(-(-x))}{2} = \frac{f(x)+f(-x)}{2} = E(x) \Rightarrow$ E is an even function. Define $O(x) = f(x) - E(x) = f(x) - \frac{f(x)+f(-x)}{2} = \frac{f(x)-f(-x)}{2}$. Then

$O(-x) = \frac{f(-x) - f(-(-x))}{2} = \frac{f(-x) - f(x)}{2} = -\left(\frac{f(x) - f(-x)}{2}\right) = -O(x) \Rightarrow$ O is an odd function

$\Rightarrow f(x) = E(x) + O(x)$ is the sum of an even and an odd function.

(b) Part (a) shows that $f(x) = E(x) + O(x)$ is the sum of an even and an odd function. If also $f(x) = E_1(x) + O_1(x)$, where E_1 is even and O_1 is odd, then $f(x) - f(x) = 0 = (E_1(x) + O_1(x)) - (E(x) + O(x))$. Thus, $E(x) - E_1(x) = O_1(x) - O(x)$ for all x in the domain of f (which is the same as the domain of $E - E_1$ and $O - O_1$). Now $(E - E_1)(-x) = E(-x) - E_1(-x) = E(x) - E_1(x)$ (since E and E_1 are even) $= (E - E_1)(x) \Rightarrow E - E_1$ is even. Likewise, $(O_1 - O)(-x) = O_1(-x) - O(-x) = -O_1(x) - (-O(x))$ (since O and O_1 are odd) $= -(O_1(x) - O(x)) = -(O_1 - O)(x) \Rightarrow O_1 - O$ is odd. Therefore, $E - E_1$ and $O_1 - O$ are both even and odd so they must be zero at each x in the domain of f by Exercise 9. That is, $E_1 = E$ and $O_1 = O$, so the decomposition of f found in part (a) is unique.

11. (a) If f is even, then $f(x) = f(-x)$ and $h(-x) = g(f(-x)) = g(f(x)) = h(x)$.
If f is odd, then $f(-x) = -f(x)$ and $h(-x) = g(f(-x)) = g(f(x)) = h(x)$ because g is even.
If f is neither, then h may not be even. For example, if $f(x) = x^2 + x$ and $g(x) = x^2$, then $h(x) = x^4 + 2x^3 + x^2$ and $h(-x) = x^4 - 2x^3 + x^2 \neq h(x)$. Therefore, h need not be even.

(b) No, h is not always odd. Let $g(t) = t$ and $f(x) = x^2$. Then, $h(x) = g(f(x)) = f(x) = x^2$ s even although g is odd.
If f is odd, then $f(-x) = -f(x)$ and $h(-x) = g(f(-x)) = g(-f(x)) = -g(f(x)) = -h(x)$ because g is odd.
In this case, h is odd. However, if f is even, as in the above counter example, we see that h need not be odd.

12. If f is one-to-one and an odd function then we know $f(a) = b$ and $f(-a) = -b$, by definition of odd function. Hence, using the inverse function, $f^{-1}(b) = a$ and $f^{-1}(-b) = -a$. This fufills the definition of an odd function.
On the other hand, if f is an even function, then $f(a) = b$ and $f(-a) = b$ by definition. Then in the inverse function, we have $f^{-1}(b) = \pm a$. Thus, f^{-1} can only exist if f is the function which has domain $\{0\}$ and range $\{0\}$. There is no other function which is even and has an inverse.

13. If the graph of f(x) passes the horizontal line test, so will the graph of $g(x) = -f(x)$ since it's the same graph reflected about the x-axis.
Alternate answer: If $g(x_1) = g(x_2)$ then $-f(x_1) = -f(x_2)$, $f(x_1) = f(x_2)$, and $x_1 = x_2$ since f is one-to-one.

14. Suppose that $g(x_1) = g(x_2)$. Then $\frac{1}{f(x_1)} = \frac{1}{f(x_2)}$, $f(x_1) = f(x_2)$, and $x_1 = x_2$ since f is one-to-one.

15. (a) The expression $a(b^{c-x}) + d$ is defined for all values of x, so the domain is $(-\infty, \infty)$. Since b^{c-x} attains all positive values, the range is (d, ∞) if $a > 0$ and the range is $(-d, \infty)$ if $a < 0$.

(b) The expression $a \log_b(x - c) + d$ is defined when $x - c > 0$, so the domain is (c, ∞).
Since $a \log_b(x - c) + d$ attains every real value for some value of x, the range is $(-\infty, \infty)$.

16. (a) Suppose $f(x_1) = f(x_2)$. Then:

$\frac{ax_1 + b}{cx_1 + d} = \frac{ax_2 + b}{cx_2 + d}$

$(ax_1 + b)(cx_2 + d) = (ax_2 + b)(cx_1 + d)$

$acx_1x_2 + adx_1 + bcx_2 + bd = acx_1x_2 + adx_2 + bcx_1 + bd$

$adx_1 + bcx_2 = adx_2 + bcx_1$

$(ad - bc)x_1 = (ad - bc)x_2$

Since $ad - bc \neq 0$, this means that $x_1 = x_2$.

(b) $y = \frac{ax+b}{cx+d}$

$cxy + dy = ax + b$

$(cy - a)x = -dy + b$

$x = \frac{-dy+b}{cy-a}$

Interchange x and y.

$y = \frac{-dx+b}{cx-a}$

$f^{-1}(x) = \frac{-dx+b}{cx-a}$

(c) As $x \to \pm\infty$, $f(x) = \frac{ax+b}{cx+d} \to \frac{a}{c}$, so the horizontal asymptote is $y = \frac{a}{c}$ $(c \neq 0)$. Since f(x) is undefined at $x = -\frac{d}{c}$, the vertical asymptote is $x = -\frac{d}{c}$ provided $c \neq 0$.

(d) As $x \to \pm\infty$, $f^{-1}(x) = \frac{-dx+b}{cx-a} \to -\frac{d}{c}$, so the horizontal asymptote is $y = -\frac{d}{c}$ $(c \neq 0)$. Since f(x) is undefined at $x = \frac{a}{c}$, the vertical asymptote is $x = \frac{a}{c}$. The horizontal asymptote of f becomes the vertical asyptote of f^{-1} and vice versa due to the reflection of the graph about the line $y = x$.

17. (a) $y = 100{,}000 - 10{,}000x$, $0 \le x \le 10$

(b)
$$y = 55{,}000$$
$$100{,}000 - 10{,}000x = 55{,}000$$
$$-10{,}000x = 55{,}000$$
$$x = 4.5$$

The value is \$55,000 after 4.5 years.

18. (a) $f(0) = 90$ units.

(b) $f(2) = 90 - 52\ln 3 \approx 32.8722$ units.

(c)

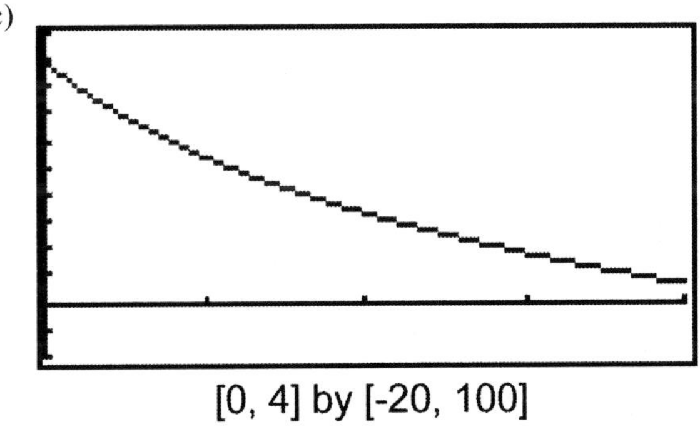

[0, 4] by [-20, 100]

19. $1500(1.08)^t = 5000 \to 1.08^t = \frac{5000}{1500} = \frac{10}{3} \to \ln(1.08)^t = \ln\frac{10}{3} \to t\ln 1.08 = \ln\frac{10}{3} \to t = \frac{\ln(10/3)}{\ln 1.08} \approx 15.6439$.
It will take about 15.6439 years. (If the bank only pays interest at the end of the year, it will take 16 years.)

20. $A(t) = A_0e^{rt}$; $A(t) = 2A_0 \Rightarrow 2A_0 = A_0e^{rt} \Rightarrow e^{rt} = 2 \Rightarrow rt = \ln 2 \Rightarrow t = \frac{\ln 2}{r} \Rightarrow t \approx \frac{0.7}{r} = \frac{70}{100r} = \frac{70}{(r\%)}$

21. $\ln x^{(x^x)} = x^x \ln x$ and $\ln(x^x)^x = x\ln(x^x) = x^2\ln x$; then, $x^x \ln x = x^2 \ln x \Rightarrow x^x = x^2 \Rightarrow x\ln x = 2\ln x \Rightarrow x = 2$.
Therefore, $x^{(x^x)} = (x^x)^x$ when $x = 2$.

22. (a) No, there are two intersections: one at $x = 2$ and the other at $x = 4$

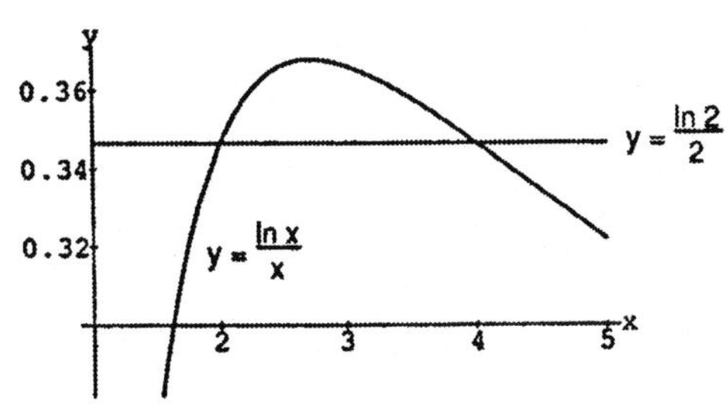

(b) Yes, because there is only one intersection

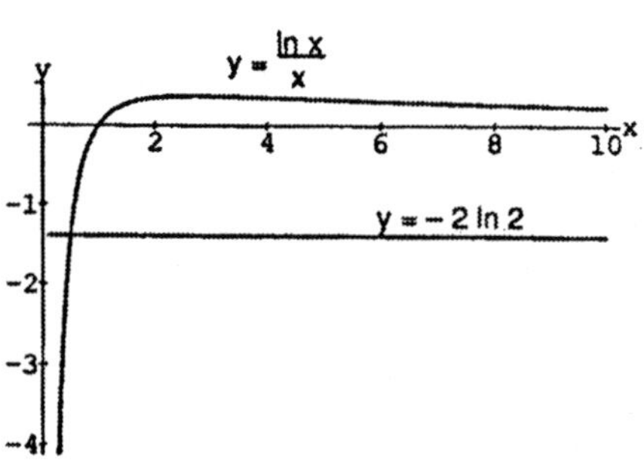

23. $\frac{\log_4 x}{\log_2 x} = \frac{\left(\frac{\ln x}{\ln 4}\right)}{\left(\frac{\ln x}{\ln 2}\right)} = \frac{\ln x}{\ln 4} \cdot \frac{\ln 2}{\ln x} = \frac{\ln 2}{\ln 4} = \frac{\ln 2}{2 \ln 2} = \frac{1}{2}$

24. (a) $f(x) = \frac{\ln 2}{\ln x}$, $g(x) = \frac{\ln x}{\ln 2}$

(b) f is negative when g is negative, positive when g is positive, and undefined when $g = 0$; the values of f decrease as those of g increase

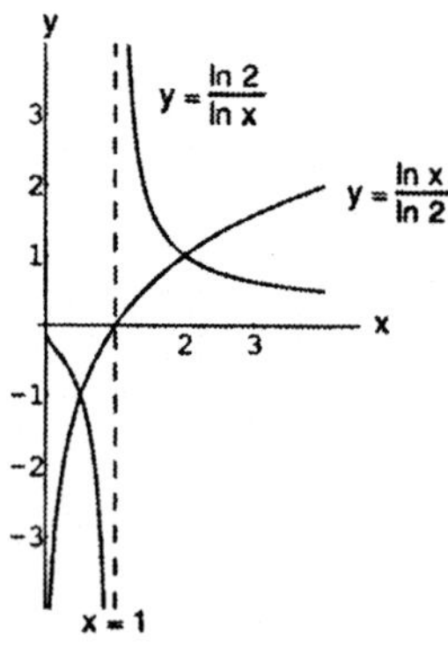

(b) $\frac{\ln 2}{\ln x} = \frac{\ln x}{\ln 2} \Rightarrow (\ln 2)^2 = (\ln x)^2$
$\Rightarrow (\ln 2 - \ln x)(\ln 2 + \ln x) = 0 \Rightarrow \ln x = \ln 2$ or $\ln x = -\ln 2 \Rightarrow e^{\ln x} = e^{\ln 2}$ or $e^{\ln x} = e^{\ln(1/2)}$
$\Rightarrow x = 2$ or $x = \frac{1}{2}$. Therefore, the two curves cross at the two points $\left(\frac{1}{2}, \frac{\ln(1/2)}{\ln 2}\right) = \left(\frac{1}{2}, -1\right)$ and $\left(2, \frac{\ln 2}{\ln 2}\right) = (2, 1)$.

25. $y = ax^2 + bx + c = a\left(x^2 + \frac{b}{a}x + \frac{b^2}{4a^2}\right) - \frac{b^2}{4a} + c = a\left(x + \frac{b}{2a}\right)^2 - \frac{b^2}{4a} + c$

(a) If $a > 0$ the graph is a parabola that opens upward. Increasing a causes a vertical stretching and a shift of the vertex toward the y-axis and upward. If $a < 0$ the graph is a parabola that opens downward. Decreasing a causes a vertical stretching and a shift of the vertex toward the y-axis and downward.

(b) If $a > 0$ the graph is a parabola that opens upward. If also $b > 0$, then increasing b causes a shift of the graph downward to the left; if $b < 0$, then decreasing b causes a shift of the graph downward and to the right.

If $a < 0$ the graph is a parabola that opens downward. If $b > 0$, increasing b shifts the graph upward to the right. If $b < 0$, decreasing b shifts the graph upward to the left.

(c) Changing c (for fixed a and b) by Δc shifts the graph upward Δc units if $\Delta c > 0$, and downward $-\Delta c$ units if $\Delta c < 0$.

26. (a) If $a > 0$, the graph rises to the right of the vertical line $x = -b$ and falls to the left. If $a < 0$, the graph falls to the right of the line $x = -b$ and rises to the left. If $a = 0$, the graph reduces to the horizontal line $y = c$. As $|a|$ increases, the slope at any given point $x = x_0$ increases in magnitude and the graph becomes steeper. As $|a|$ decreases, the slope at x_0 decreases in magnitude and the graph rises or falls more gradually.

(b) Increasing b shifts the graph to the left; decreasing b shifts it to the right.

(c) Increasing c shifts the graph upward; decreasing c shifts it downward.

27. If $m > 0$, the x-intercept of $y = mx + 2$ must be negative. If $m < 0$, then the x-intercept exceeds $\frac{1}{2}$ $\Rightarrow 0 = mx + 2$ and $x > \frac{1}{2} \Rightarrow x = -\frac{2}{m} > \frac{1}{2} \Rightarrow 0 > m > -4$.

28. $y = 80.21 \cdot (1.01)^x$.
In 1935 the population was about $y(35) = 113.6$ million. In 2012 the population will be, according to this model, $y(112) = 244.6$ million, in 2025 it should be $y(125) = 278.2$ million.

29. Each of the triangles pictured has the same base $b = v\Delta t = v(1 \text{ sec})$. Moreover, the height of each triangle is the same value h. Thus $\frac{1}{2}$ (base)(height) $= \frac{1}{2}bh = A_1 = A_2 = A_3 = \ldots$. In conclusion, the object sweeps out equal areas in each one second interval.

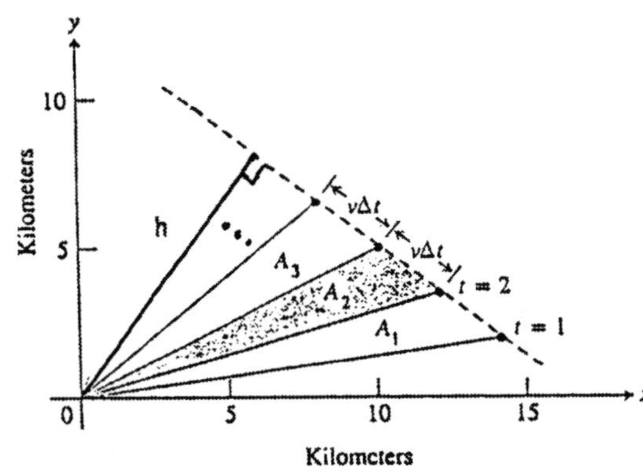

30. (a) The coordinates of P are $\left(\frac{a+0}{2}, \frac{b+0}{2}\right) = \left(\frac{a}{2}, \frac{b}{2}\right)$. Thus the slope of OP $= \frac{\Delta y}{\Delta x} = \frac{b/2}{a/2} = \frac{b}{a}$.
(b) The slope of AB $= \frac{b-0}{0-a} = -\frac{b}{a}$. The line segments AB and OP are perpendicular when the product of their slopes is $-1 = \left(\frac{b}{a}\right)\left(-\frac{b}{a}\right) = -\frac{b^2}{a^2}$. Thus, $b^2 = a^2 \Rightarrow a = b$ (since both are positive). Therefore, AB is perpendicular to OP when $a = b$.

NOTES:

CHAPTER 2 LIMITS AND CONTINUITY

2.1 RATES OF CHANGE AND LIMITS

1. (a) Does not exist. As x approaches 1 from the right, g(x) approaches 0. As x approaches 1 from the left, g(x) approaches 1. There is no single number L that all the values g(x) get arbitrarily close to as $x \to 1$.
   (b) 1
   (c) 0

2. (a) 0
   (b) -1
   (c) Does not exist. As t approaches 0 from the left, f(t) approaches -1. As t approaches 0 from the right, f(t) approaches 1. There is no single number L that f(t) gets arbitrarily close to as $t \to 0$.

3. (a) True (b) True (c) False
   (d) False (e) False (f) True

4. (a) False (b) False (c) True
   (d) True (e) True

5. $\lim\limits_{x \to 0} \frac{x}{|x|}$ does not exist because $\frac{x}{|x|} = \frac{x}{x} = 1$ if $x > 0$ and $\frac{x}{|x|} = \frac{x}{-x} = -1$ if $x < 0$. As x approaches 0 from the left, $\frac{x}{|x|}$ approaches -1. As x approaches 0 from the right, $\frac{x}{|x|}$ approaches 1. There is no single number L that all the function values get arbitrarily close to as $x \to 0$.

6. As x approaches 1 from the left, the values of $\frac{1}{x-1}$ become increasingly large and negative. As x approaches 1 from the right, the values become increasingly large and positive. There is no one number L that all the function values get arbitrarily close to as $x \to 1$, so $\lim\limits_{x \to 1} \frac{1}{x-1}$ does not exist.

7. Nothing can be said about f(x) because the existence of a limit as $x \to x_0$ does not depend on how the function is defined at x_0. In order for a limit to exist, f(x) must be arbitrarily close to a single real number L when x is close enough to x_0. That is, the existence of a limit depends on the values of f(x) for x <u>near</u> x_0, not on the definition of f(x) at x_0 itself.

8. Nothing can be said. In order for $\lim\limits_{x \to 0} f(x)$ to exist, f(x) must close to a single value for x near 0 regardless of the value f(0) itself.

9. No, the definition does not require that f be defined at $x = 1$ in order for a limiting value to exist there. If f(1) is defined, it can be any real number, so we can conclude nothing about f(1) from $\lim\limits_{x \to 1} f(x) = 5$.

10. No, because the existence of a limit depends on the values of f(x) when x is near 1, not on f(1) itself. If $\lim\limits_{x \to 1} f(x)$ exists, its value may be some number other than $f(1) = 5$. We can conclude nothing about $\lim\limits_{x \to 1} f(x)$, whether it exists or what its value is if it does exist, from knowing the value of f(1) alone.

11. (a) $f(x) = (x^2 - 9)/(x + 3)$

x	−3.1	−3.01	−3.001	−3.0001	−3.00001	−3.000001
f(x)	−6.1	−6.01	−6.001	−6.0001	−6.00001	−6.000001

x	−2.9	−2.99	−2.999	−2.9999	−2.99999	−2.999999
f(x)	−5.9	−5.99	−5.999	−5.9999	−5.99999	−5.999999

The estimate is $\lim_{x \to -3} f(x) = -6.$

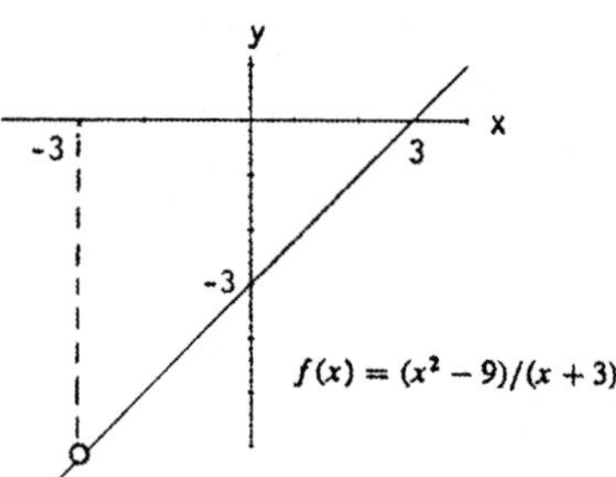

(c) $f(x) = \frac{x^2-9}{x+3} = \frac{(x+3)(x-3)}{x+3} = x - 3$ if $x \neq -3$, and $\lim_{x \to -3} (x - 3) = -3 - 3 = -6.$

12. (a) $g(x) = (x^2 - 2)/\left(x - \sqrt{2}\right)$

x	1.4	1.41	1.414	1.4142	1.41421	1.414213
g(x)	2.81421	2.82421	2.82821	2.828413	2.828423	2.828426

(b)

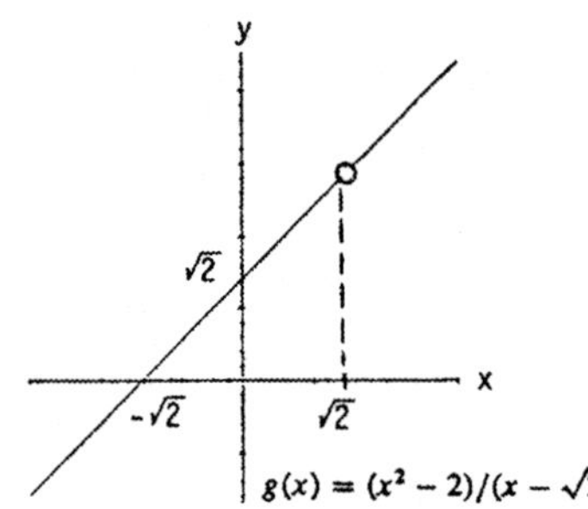

(c) $g(x) = \frac{x^2-2}{x-\sqrt{2}} = \frac{\left(x+\sqrt{2}\right)\left(x-\sqrt{2}\right)}{\left(x-\sqrt{2}\right)} = x + \sqrt{2}$ if $x \neq \sqrt{2}$, and $\lim_{x \to \sqrt{2}} \left(x + \sqrt{2}\right) = \sqrt{2} + \sqrt{2} = 2\sqrt{2}.$

13. (a) $G(x) = (x + 6)/\left(x^2 + 4x - 12\right)$

x	−5.9	−5.99	−5.999	−5.9999	−5.99999	−5.999999
G(x)	−.126582	−.1251564	−.1250156	−.1250015	−.1250001	−.1250000

x	−6.1	−6.01	−6.001	−6.0001	−6.00001	−6.000001
G(x)	−.123456	−.124843	−.124984	−.124998	−.124999	−.124999

(b)

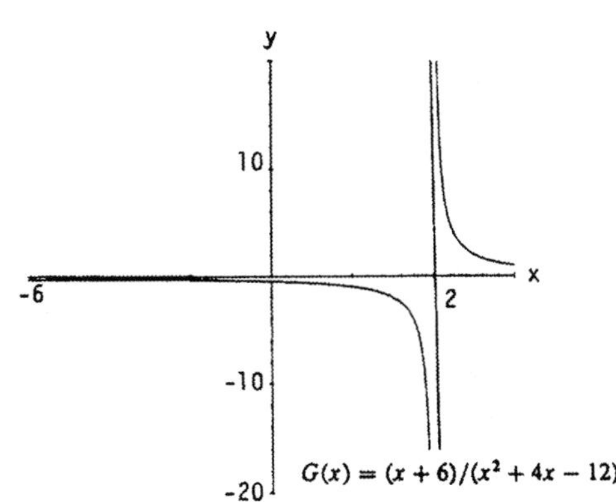

(c) $G(x) = \frac{x+6}{(x^2+4x-12)} = \frac{x+6}{(x+6)(x-2)} = \frac{1}{x-2}$ if $x \neq -6$, and $\lim_{x \to -6} \frac{1}{x-2} = \frac{1}{-6-2} = -\frac{1}{8} = -0.125$.

14. (a) $h(x) = (x^2 - 2x - 3)/(x^2 - 4x + 3)$

x	2.9	2.99	2.999	2.9999	2.99999	2.999999
h(x)	2.052631	2.005025	2.000500	2.000050	2.000005	2.0000005

x	3.1	3.01	3.001	3.0001	3.00001	3.000001
h(x)	1.952380	1.995024	1.999500	1.999950	1.999995	1.999999

(b)

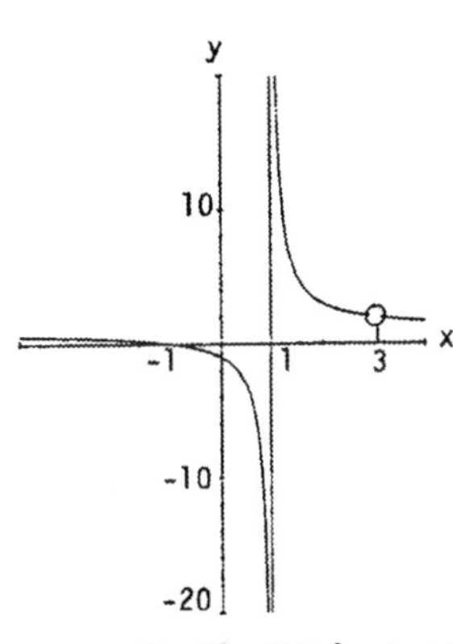

$h(x) = (x^2 - 2x - 3)/(x^2 - 4x + 3)$

(c) $h(x) = \frac{x^2-2x-3}{x^2-4x+3} = \frac{(x-3)(x+1)}{(x-3)(x-1)} = \frac{x+1}{x-1}$ if $x \neq 3$, and $\lim_{x \to 3} \frac{x+1}{x-1} = \frac{3+1}{3-1} = \frac{4}{2} = 2$.

15. (a) $f(x) = (x^2 - 1)/(|x| - 1)$

x	−1.1	−1.01	−1.001	−1.0001	−1.00001	−1.000001
f(x)	2.1	2.01	2.001	2.0001	2.00001	2.000001

x	−.9	−.99	−.999	−.9999	−.99999	−.999999
f(x)	1.9	1.99	1.999	1.9999	1.99999	1.999999

(b)

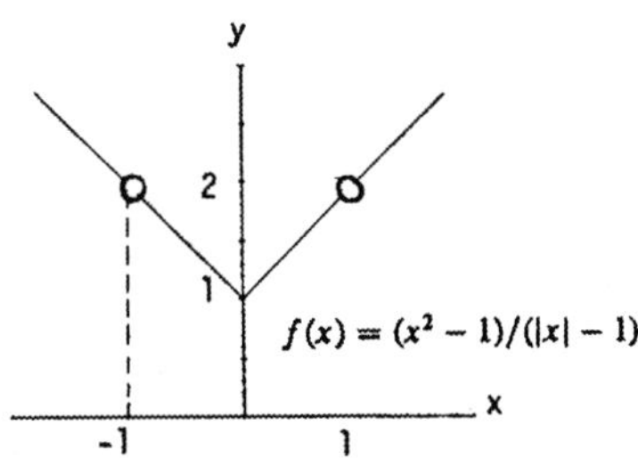

(c) $f(x) = \frac{x^2-1}{|x|-1} = \begin{cases} \frac{(x+1)(x-1)}{x-1} = x+1, & x \ge 0 \text{ and } x \ne 1 \\ \frac{(x+1)(x-1)}{-(x+1)} = 1-x, & x < 0 \text{ and } x \ne -1 \end{cases}$, and $\lim\limits_{x \to -1} (1-x) = 1-(-1) = 2.$

16. (a) $F(x) = (x^2+3x+2)/(2-|x|)$

x	−2.1	−2.01	−2.001	−2.0001	−2.00001	−2.000001
F(x)	−1.1	−1.01	−1.001	−1.0001	−1.00001	−1.000001

x	−1.9	−1.99	−1.999	−1.9999	−1.99999	−1.999999
F(x)	−.9	−.99	−.999	−.9999	−.99999	−.999999

(b)

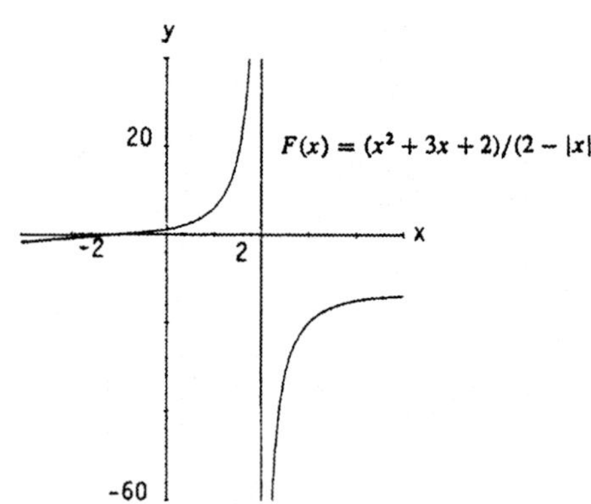

(c) $F(x) = \frac{x^2+3x+2}{2-|x|} = \begin{cases} \frac{(x+2)(x+1)}{2-x}, & x \ge 0 \\ \frac{(x+2)(x+1)}{2+x} = x+1, & x < 0 \text{ and } x \ne -2 \end{cases}$, and $\lim\limits_{x \to -2} (x+1) = -2+1 = -1.$

17. (a) $g(\theta) = (\sin\theta)/\theta$

θ	.1	.01	.001	.0001	.00001	.000001
$g(\theta)$	.998334	.999983	.999999	.999999	.999999	.999999

θ	−.1	−.01	−.001	−.0001	−.00001	−.000001
$g(\theta)$	.998334	.999983	.999999	.999999	.999999	.999999

$\lim\limits_{\theta \to 0} g(\theta) = 1$

(b)

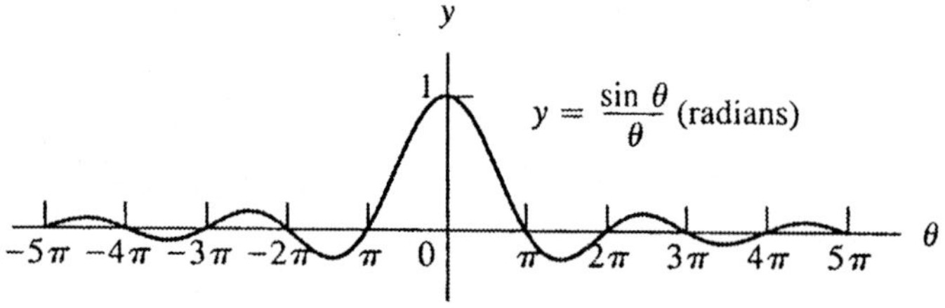

18. (a) $G(t) = (1-\cos t)/t^2$

t	.1	.01	.001	.0001	.00001	.000001
G(t)	.499583	.499995	.499999	.5	.5	.5

t	−.1	−.01	−.001	−.0001	−.00001	−.000001
G(t)	.499583	.499995	.499999	.5	.5	.5

$\lim\limits_{t \to 0} G(t) = 0.5$

(b)

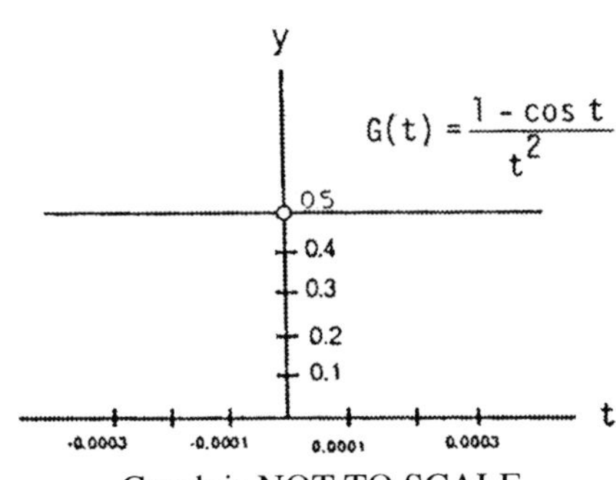

Graph is NOT TO SCALE

19. (a) $f(x) = x^{1/(1-x)}$

x	.9	.99	.999	.9999	.99999	.999999
f(x)	.348678	.366032	.367695	.367861	.367877	.367879

x	1.1	1.01	1.001	1.0001	1.00001	1.000001
f(x)	.385543	.369711	.368063	.367897	.367881	.367878

$\lim_{x \to 1} f(x) \approx 0.36788$

(b)

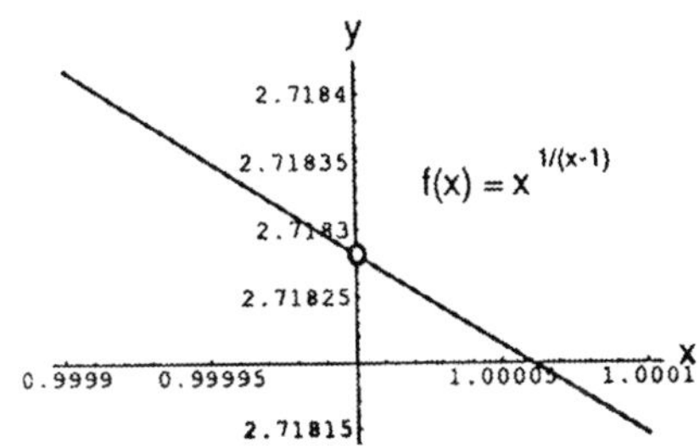

Graph is NOT TO SCALE. Also the intersection of the axes is not the origin: the axes intersect at the point $(1, 2.71820)$.

20. (a) $f(x) = (3^x - 1)/x$

x	.1	.01	.001	.0001	.00001	.000001
f(x)	1.161231	1.104669	1.099215	1.098672	1.098618	1.098612

x	−.1	−.01	−.001	−.0001	−.00001	−.000001
f(x)	1.040415	1.092599	1.098009	1.098551	1.098606	1.098611

$\lim_{x \to 0} f(x) \approx 1.0986$

(b)

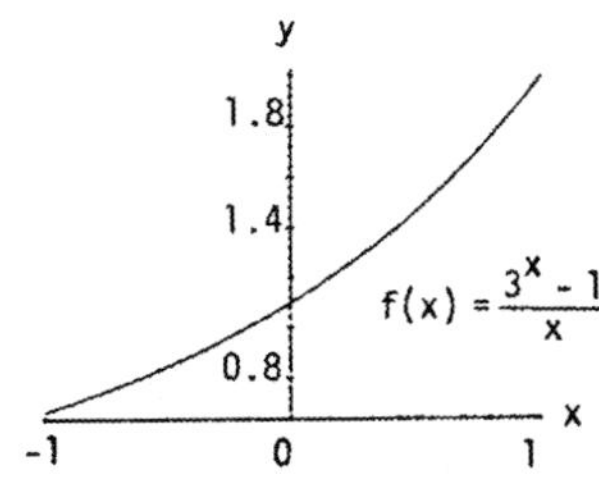

21. $\lim_{x \to 2} 2x = 2(2) = 4$

22. $\lim_{x \to 0} 2x = 2(0) = 0$

23. $\lim_{x \to \frac{1}{3}} (3x - 1) = 3\left(\frac{1}{3}\right) - 1 = 0$

24. $\lim_{x \to 1} \frac{-1}{3x-1} = \frac{-1}{3(1)-1} = -\frac{1}{2}$

25. $\lim_{x \to -1} 3x(2x-1) = 3(-1)(2(-1)-1) = 9$

26. $\lim_{x \to -1} \frac{3x^2}{2x-1} = \frac{3(-1)^2}{2(-1)-1} = \frac{3}{-3} = -1$

27. $\lim_{x \to \frac{\pi}{2}} x \sin x = \frac{\pi}{2} \sin \frac{\pi}{2} = \frac{\pi}{2}$

28. $\lim_{x \to \pi} \frac{\cos x}{1-\pi} = \frac{\cos \pi}{1-\pi} = \frac{-1}{1-\pi} = \frac{1}{\pi-1}$

29. (a) $\frac{\Delta f}{\Delta x} = \frac{f(3)-f(2)}{3-2} = \frac{28-9}{1} = 19$

(b) $\frac{\Delta f}{\Delta x} = \frac{f(1)-f(-1)}{1-(-1)} = \frac{2-0}{2} = 1$

30. (a) $\frac{\Delta g}{\Delta x} = \frac{g(1)-g(-1)}{1-(-1)} = \frac{1-1}{2} = 0$

(b) $\frac{\Delta g}{\Delta x} = \frac{g(0)-g(-2)}{0-(-2)} = \frac{0-4}{2} = -2$

31. (a) $\frac{\Delta h}{\Delta t} = \frac{h\left(\frac{3\pi}{4}\right)-h\left(\frac{\pi}{4}\right)}{\frac{3\pi}{4}-\frac{\pi}{4}} = \frac{-1-1}{\frac{\pi}{2}} = -\frac{4}{\pi}$

(b) $\frac{\Delta h}{\Delta t} = \frac{h\left(\frac{\pi}{2}\right)-h\left(\frac{\pi}{6}\right)}{\frac{\pi}{2}-\frac{\pi}{6}} = \frac{0-\sqrt{3}}{\frac{\pi}{3}} = \frac{-3\sqrt{3}}{\pi}$

32. (a) $\frac{\Delta g}{\Delta t} = \frac{g(\pi)-g(0)}{\pi-0} = \frac{(2-1)-(2+1)}{\pi-0} = -\frac{2}{\pi}$

(b) $\frac{\Delta g}{\Delta t} = \frac{g(\pi)-g(-\pi)}{\pi-(-\pi)} = \frac{(2-1)-(2-1)}{2\pi} = 0$

33. $\frac{\Delta R}{\Delta \theta} = \frac{R(2)-R(0)}{2-0} = \frac{\sqrt{8+1}-\sqrt{1}}{2} = \frac{3-1}{2} = 1$

34. $\frac{\Delta P}{\Delta \theta} = \frac{P(2)-P(1)}{2-1} = \frac{(8-16+10)-(1-4+5)}{1} = 2-2 = 0$

35. (a)

Q	Slope of PQ $= \frac{\Delta p}{\Delta t}$
$Q_1(10, 225)$	$\frac{650-225}{20-10} = 42.5$ m/sec
$Q_2(14, 375)$	$\frac{650-375}{20-14} = 45.83$ m/sec
$Q_3(16.5, 475)$	$\frac{650-475}{20-16.5} = 50.00$ m/sec
$Q_4(18, 550)$	$\frac{650-550}{20-18} = 50.00$ m/sec

(b) At $t = 20$, the Cobra was traveling approximately 50 m/sec or 180 km/h.

36. (a)

Q	Slope of PQ $= \frac{\Delta p}{\Delta t}$
$Q_1(5, 20)$	$\frac{80-20}{10-5} = 12$ m/sec
$Q_2(7, 39)$	$\frac{80-39}{10-7} = 13.7$ m/sec
$Q_3(8.5, 58)$	$\frac{80-58}{10-8.5} = 14.7$ m/sec
$Q_4(9.5, 72)$	$\frac{80-72}{10-9.5} = 16$ m/sec

(b) Approximately 16 m/sec

37. (a)

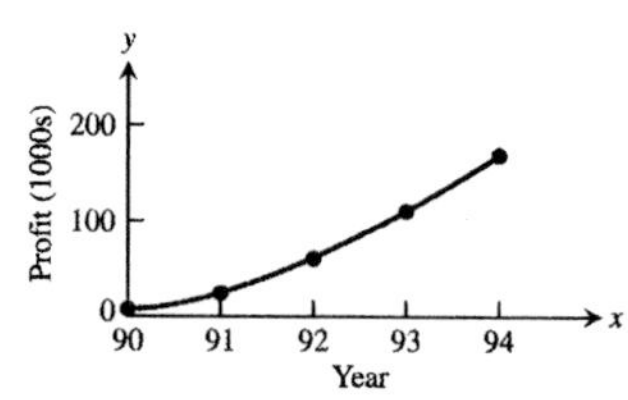

(b) $\frac{\Delta p}{\Delta t} = \frac{174-62}{1994-1992} = \frac{112}{2} = 56$ thousand dollars per year

(c) The average rate of change from 1991 to 1992 is $\frac{\Delta p}{\Delta t} = \frac{62-27}{1992-1991} = 35$ thousand dollars per year.

The average rate of change from 1992 to 1993 is $\frac{\Delta p}{\Delta t} = \frac{111-62}{1993-1992} = 49$ thousand dollars per year.

So, the rate at which profits were changing in 1992 is approximatley $\frac{1}{2}(35+49) = 42$ thousand dollars per year.

38. (a) $F(x) = (x+2)/(x-2)$

x	1.2	1.1	1.01	1.001	1.0001	1
F(x)	-4.0	$-3.\overline{4}$	$-3.\overline{04}$	$-3.\overline{004}$	$-3.\overline{0004}$	-3

$\frac{\Delta F}{\Delta x} = \frac{-4.0-(-3)}{1.2-1} = -5.0$; $\frac{\Delta F}{\Delta x} = \frac{-3.\overline{4}-(-3)}{1.1-1} = -4.\overline{4}$;

$\frac{\Delta F}{\Delta x} = \frac{-3.\overline{04}-(-3)}{1.01-1} = -4.\overline{04}$; $\frac{\Delta F}{\Delta x} = \frac{-3.\overline{004}-(-3)}{1.001-1} = -4.\overline{004}$;

$\frac{\Delta F}{\Delta x} = \frac{-3.\overline{0004}-(-3)}{1.0001-1} = -4.\overline{0004}$;

(b) The rate of change of F(x) at $x = 1$ is -4.

39. (a) $\frac{\Delta g}{\Delta x} = \frac{g(2)-g(1)}{2-1} = \frac{\sqrt{2}-1}{2-1} \approx 0.414213$ $\frac{\Delta g}{\Delta x} = \frac{g(1.5)-g(1)}{1.5-1} = \frac{\sqrt{1.5}-1}{0.5} \approx 0.449489$

$\frac{\Delta g}{\Delta x} = \frac{g(1+h)-g(1)}{(1+h)-1} = \frac{\sqrt{1+h}-1}{h}$

(b) $g(x) = \sqrt{x}$

$1+h$	1.1	1.01	1.001	1.0001	1.00001	1.000001
$\sqrt{1+h}$	1.04880	1.004987	1.0004998	1.0000499	1.000005	1.0000005
$\left(\sqrt{1+h}-1\right)/h$	0.4880	0.4987	0.4998	0.499	0.5	0.5

(c) The rate of change of g(x) at $x = 1$ is 0.5.

(d) The calculator gives $\lim_{h \to 0} \frac{\sqrt{1+h}-1}{h} = \frac{1}{2}$.

40. (a) i) $\frac{f(3)-f(2)}{3-2} = \frac{\frac{1}{3}-\frac{1}{2}}{1} = \frac{\frac{-1}{6}}{1} = -\frac{1}{6}$

ii) $\frac{f(T)-f(2)}{T-2} = \frac{\frac{1}{T}-\frac{1}{2}}{T-2} = \frac{\frac{2}{2T}-\frac{T}{2T}}{T-2} = \frac{2-T}{2T(T-2)} = \frac{2-T}{-2T(2-T)} = -\frac{1}{2T},\ T \neq 2$

(b)

T	2.1	2.01	2.001	2.0001	2.00001	2.000001
f(T)	0.476190	0.497512	0.499750	0.4999750	0.499997	0.499999
$(f(T)-f(2))/(T-2)$	-0.2381	-0.2488	-0.2500	-0.2500	-0.2500	-0.2500

(c) The table indicates the rate of change is -0.25 at $t = 2$.

(d) $\lim_{T \to 2} \left(\frac{1}{-2T}\right) = -\frac{1}{4}$

41-46. Example CAS commands:

Maple:

```
f := x -> (x^4 – 16)/(x – 2);
x0 := 2;
plot( f(x), x = x0-1..x0+1, color = black,
      title = "Section 2.1, #41(a)" );
limit( f(x), x = x0 );
```

In Exercise 43, note that the standard cube root, x^(1/3), is not defined for x<0 in many CASs. This can be overcome in Maple by entering the function as f := x -> (surd(x+1, 3) – 1)/x.

Mathematica: (assigned function and values for x0 and h may vary)

```
Clear[f, x]
f[x_]:=(x^3 – x^2 – 5x – 3)/(x + 1)^2
x0= –1; h= 0.1;
Plot[f[x],{x, x0 – h, x0 + h}]
Limit[f[x], x → x0]
```

2.2 CALCULATING LIMITS USING THE LIMIT LAWS

1. $\lim\limits_{x \to -7} (2x + 5) = 2(-7) + 5 = -14 + 5 = -9$

2. $\lim\limits_{x \to 12} (10 - 3x) = 10 - 3(12) = 10 - 36 = -26$

3. $\lim\limits_{x \to 2} (-x^2 + 5x - 2) = -(2)^2 + 5(2) - 2 = -4 + 10 - 2 = 4$

4. $\lim\limits_{x \to -2} (x^3 - 2x^2 + 4x + 8) = (-2)^3 - 2(-2)^2 + 4(-2) + 8 = -8 - 8 - 8 + 8 = -16$

5. $\lim\limits_{t \to 6} 8(t - 5)(t - 7) = 8(6 - 5)(6 - 7) = -8$

6. $\lim\limits_{s \to \frac{2}{3}} 3s(2s - 1) = 3\left(\frac{2}{3}\right)\left[2\left(\frac{2}{3}\right) - 1\right] = 2\left(\frac{4}{3} - 1\right) = \frac{2}{3}$

7. $\lim\limits_{x \to 2} \frac{x+3}{x+6} = \frac{2+3}{2+6} = \frac{5}{8}$

8. $\lim\limits_{x \to 5} \frac{4}{x-7} = \frac{4}{5-7} = \frac{4}{-2} = -2$

9. $\lim\limits_{y \to -5} \frac{y^2}{5-y} = \frac{(-5)^2}{5-(-5)} = \frac{25}{10} = \frac{5}{2}$

10. $\lim\limits_{y \to 2} \frac{y+2}{y^2+5y+6} = \frac{2+2}{(2)^2+5(2)+6} = \frac{4}{4+10+6} = \frac{4}{20} = \frac{1}{5}$

11. $\lim\limits_{x \to -1} 3(2x - 1)^2 = 3(2(-1) - 1)^2 = 3(-3)^2 = 27$

12. $\lim\limits_{x \to -4} (x + 3)^{1984} = (-4 + 3)^{1984} = (-1)^{1984} = 1$

13. $\lim\limits_{y \to -3} (5 - y)^{4/3} = [5 - (-3)]^{4/3} = (8)^{4/3} = \left((8)^{1/3}\right)^4 = 2^4 = 16$

14. $\lim\limits_{z \to 0} (2z - 8)^{1/3} = (2(0) - 8)^{1/3} = (-8)^{1/3} = -2$

15. $\lim\limits_{h \to 0} \frac{3}{\sqrt{3h+1}+1} = \frac{3}{\sqrt{3(0)+1}+1} = \frac{3}{\sqrt{1}+1} = \frac{3}{2}$

16. $\lim\limits_{h \to 0} \frac{5}{\sqrt{5h+4}+2} = \frac{5}{\sqrt{5(0)+4}+2} = \frac{5}{\sqrt{4}+2} = \frac{5}{4}$

17. $\lim\limits_{h \to 0} \frac{\sqrt{3h+1}-1}{h} = \lim\limits_{h \to 0} \frac{\sqrt{3h+1}-1}{h} \cdot \frac{\sqrt{3h+1}+1}{\sqrt{3h+1}+1} = \lim\limits_{h \to 0} \frac{(3h+1)-1}{h\left(\sqrt{3h+1}+1\right)} = \lim\limits_{h \to 0} \frac{3h}{h\left(\sqrt{3h+1}+1\right)} = \lim\limits_{h \to 0} \frac{3}{\sqrt{3h+1}+1}$
$= \frac{3}{\sqrt{1}+1} = \frac{3}{2}$

18. $\lim\limits_{h \to 0} \frac{\sqrt{5h+4}-2}{h} = \lim\limits_{h \to 0} \frac{\sqrt{5h+4}-2}{h} \cdot \frac{\sqrt{5h+4}+2}{\sqrt{5h+4}+2} = \lim\limits_{h \to 0} \frac{(5h+4)-4}{h\left(\sqrt{5h+4}+2\right)} = \lim\limits_{h \to 0} \frac{5h}{h\left(\sqrt{5h+4}+2\right)} = \lim\limits_{h \to 0} \frac{5}{\sqrt{5h+4}+2}$
$= \frac{5}{\sqrt{4}+2} = \frac{5}{4}$

19. $\lim\limits_{x \to 5} \frac{x-5}{x^2-25} = \lim\limits_{x \to 5} \frac{x-5}{(x+5)(x-5)} = \lim\limits_{x \to 5} \frac{1}{x+5} = \frac{1}{5+5} = \frac{1}{10}$

20. $\lim\limits_{x \to -3} \frac{x+3}{x^2+4x+3} = \lim\limits_{x \to -3} \frac{x+3}{(x+3)(x+1)} = \lim\limits_{x \to -3} \frac{1}{x+1} = \frac{1}{-3+1} = -\frac{1}{2}$

21. $\lim\limits_{x \to -5} \frac{x^2+3x-10}{x+5} = \lim\limits_{x \to -5} \frac{(x+5)(x-2)}{x+5} = \lim\limits_{x \to -5} (x - 2) = -5 - 2 = -7$

22. $\lim\limits_{x \to 2} \frac{x^2 - 7x + 10}{x - 2} = \lim\limits_{x \to 2} \frac{(x-5)(x-2)}{x-2} = \lim\limits_{x \to 2} (x - 5) = 2 - 5 = -3$

23. $\lim\limits_{t \to 1} \frac{t^2 + t - 2}{t^2 - 1} = \lim\limits_{t \to 1} \frac{(t+2)(t-1)}{(t-1)(t+1)} = \lim\limits_{t \to 1} \frac{t+2}{t+1} = \frac{1+2}{1+1} = \frac{3}{2}$

24. $\lim\limits_{t \to -1} \frac{t^2 + 3t + 2}{t^2 - t - 2} = \lim\limits_{t \to -1} \frac{(t+2)(t+1)}{(t-2)(t+1)} = \lim\limits_{t \to -1} \frac{t+2}{t-2} = \frac{-1+2}{-1-2} = -\frac{1}{3}$

25. $\lim\limits_{x \to -2} \frac{-2x - 4}{x^3 + 2x^2} = \lim\limits_{x \to -2} \frac{-2(x+2)}{x^2(x+2)} = \lim\limits_{x \to -2} \frac{-2}{x^2} = \frac{-2}{4} = -\frac{1}{2}$

26. $\lim\limits_{y \to 0} \frac{5y^3 + 8y^2}{3y^4 - 16y^2} = \lim\limits_{y \to 0} \frac{y^2(5y+8)}{y^2(3y^2 - 16)} = \lim\limits_{y \to 0} \frac{5y+8}{3y^2 - 16} = \frac{8}{-16} = -\frac{1}{2}$

27. $\lim\limits_{u \to 1} \frac{u^4 - 1}{u^3 - 1} = \lim\limits_{u \to 1} \frac{(u^2+1)(u+1)(u-1)}{(u^2+u+1)(u-1)} = \lim\limits_{u \to 1} \frac{(u^2+1)(u+1)}{u^2+u+1} = \frac{(1+1)(1+1)}{1+1+1} = \frac{4}{3}$

28. $\lim\limits_{v \to 2} \frac{v^3 - 8}{v^4 - 16} = \lim\limits_{v \to 2} \frac{(v-2)(v^2+2v+4)}{(v-2)(v+2)(v^2+4)} = \lim\limits_{v \to 2} \frac{v^2+2v+4}{(v+2)(v^2+4)} = \frac{4+4+4}{(4)(8)} = \frac{12}{32} = \frac{3}{8}$

29. $\lim\limits_{x \to 9} \frac{\sqrt{x} - 3}{x - 9} = \lim\limits_{x \to 9} \frac{\sqrt{x} - 3}{(\sqrt{x}-3)(\sqrt{x}+3)} = \lim\limits_{x \to 9} \frac{1}{\sqrt{x}+3} = \frac{1}{\sqrt{9}+3} = \frac{1}{6}$

30. $\lim\limits_{x \to 4} \frac{4x - x^2}{2 - \sqrt{x}} = \lim\limits_{x \to 4} \frac{x(4-x)}{2-\sqrt{x}} = \lim\limits_{x \to 4} \frac{x(2+\sqrt{x})(2-\sqrt{x})}{2-\sqrt{x}} = \lim\limits_{x \to 4} x(2+\sqrt{x}) = 4(2+2) = 16$

31. $\lim\limits_{x \to 1} \frac{x - 1}{\sqrt{x+3} - 2} = \lim\limits_{x \to 1} \frac{(x-1)(\sqrt{x+3}+2)}{(\sqrt{x+3}-2)(\sqrt{x+3}+2)} = \lim\limits_{x \to 1} \frac{(x-1)(\sqrt{x+3}+2)}{(x+3)-4} = \lim\limits_{x \to 1} \left(\sqrt{x+3}+2\right)$
$= \sqrt{4} + 2 = 4$

32. $\lim\limits_{x \to -1} \frac{\sqrt{x^2+8} - 3}{x+1} = \lim\limits_{x \to -1} \frac{\left(\sqrt{x^2+8}-3\right)\left(\sqrt{x^2+8}+3\right)}{(x+1)\left(\sqrt{x^2+8}+3\right)} = \lim\limits_{x \to -1} \frac{(x^2+8)-9}{(x+1)\left(\sqrt{x^2+8}+3\right)}$
$= \lim\limits_{x \to -1} \frac{(x+1)(x-1)}{(x+1)\left(\sqrt{x^2+8}+3\right)} = \lim\limits_{x \to -1} \frac{x-1}{\sqrt{x^2+8}+3} = \frac{-2}{3+3} = -\frac{1}{3}$

33. $\lim\limits_{x \to 2} \frac{\sqrt{x^2+12} - 4}{x-2} = \lim\limits_{x \to 2} \frac{\left(\sqrt{x^2+12}-4\right)\left(\sqrt{x^2+12}+4\right)}{(x-2)\left(\sqrt{x^2+12}+4\right)} = \lim\limits_{x \to 2} \frac{(x^2+12)-16}{(x-2)\left(\sqrt{x^2+12}+4\right)}$
$= \lim\limits_{x \to 2} \frac{(x-2)(x+2)}{(x-2)\left(\sqrt{x^2+12}+4\right)} = \lim\limits_{x \to 2} \frac{x+2}{\sqrt{x^2+12}+4} = \frac{4}{\sqrt{16}+4} = \frac{1}{2}$

34. $\lim\limits_{x \to -2} \frac{x+2}{\sqrt{x^2+5} - 3} = \lim\limits_{x \to -2} \frac{(x+2)\left(\sqrt{x^2+5}+3\right)}{\left(\sqrt{x^2+5}-3\right)\left(\sqrt{x^2+5}+3\right)} = \lim\limits_{x \to -2} \frac{(x+2)\left(\sqrt{x^2+5}+3\right)}{(x^2+5)-9}$
$= \lim\limits_{x \to -2} \frac{(x+2)\left(\sqrt{x^2+5}+3\right)}{(x+2)(x-2)} = \lim\limits_{x \to -2} \frac{\sqrt{x^2+5}+3}{x-2} = \frac{\sqrt{9}+3}{-4} = -\frac{3}{2}$

35. $\lim\limits_{x \to -3} \frac{2 - \sqrt{x^2-5}}{x+3} = \lim\limits_{x \to -3} \frac{\left(2-\sqrt{x^2-5}\right)\left(2+\sqrt{x^2-5}\right)}{(x+3)\left(2+\sqrt{x^2-5}\right)} = \lim\limits_{x \to -3} \frac{4-(x^2-5)}{(x+3)\left(2+\sqrt{x^2-5}\right)}$
$= \lim\limits_{x \to -3} \frac{9-x^2}{(x+3)\left(2+\sqrt{x^2-5}\right)} = \lim\limits_{x \to -3} \frac{(3-x)(3+x)}{(x+3)\left(2+\sqrt{x^2-5}\right)} = \lim\limits_{x \to -3} \frac{3-x}{2+\sqrt{x^2-5}} = \frac{6}{2+\sqrt{4}} = \frac{3}{2}$

36. $\lim_{x \to 4} \frac{4-x}{5-\sqrt{x^2+9}} = \lim_{x \to 4} \frac{(4-x)\left(5+\sqrt{x^2+9}\right)}{\left(5-\sqrt{x^2+9}\right)\left(5+\sqrt{x^2+9}\right)} = \lim_{x \to 4} \frac{(4-x)\left(5+\sqrt{x^2+9}\right)}{25-(x^2+9)}$

$= \lim_{x \to 4} \frac{(4-x)\left(5+\sqrt{x^2+9}\right)}{16-x^2} = \lim_{x \to 4} \frac{(4-x)\left(5+\sqrt{x^2+9}\right)}{(4-x)(4+x)} = \lim_{x \to 4} \frac{5+\sqrt{x^2+9}}{4+x} = \frac{5+\sqrt{25}}{8} = \frac{5}{4}$

37. (a) quotient rule
(b) difference and power rules
(c) sum and constant multiple rules

38. (a) quotient rule
(b) power and product rules
(c) difference and constant multiple rules

39. (a) $\lim_{x \to c} f(x)\,g(x) = \left[\lim_{x \to c} f(x)\right]\left[\lim_{x \to c} g(x)\right] = (5)(-2) = -10$
(b) $\lim_{x \to c} 2f(x)\,g(x) = 2\left[\lim_{x \to c} f(x)\right]\left[\lim_{x \to c} g(x)\right] = 2(5)(-2) = -20$
(c) $\lim_{x \to c} [f(x) + 3g(x)] = \lim_{x \to c} f(x) + 3\lim_{x \to c} g(x) = 5 + 3(-2) = -1$
(d) $\lim_{x \to c} \frac{f(x)}{f(x)-g(x)} = \frac{\lim_{x \to c} f(x)}{\lim_{x \to c} f(x) - \lim_{x \to c} g(x)} = \frac{5}{5-(-2)} = \frac{5}{7}$

40. (a) $\lim_{x \to 4} [g(x)+3] = \lim_{x \to 4} g(x) + \lim_{x \to 4} 3 = -3 + 3 = 0$
(b) $\lim_{x \to 4} xf(x) = \lim_{x \to 4} x \cdot \lim_{x \to 4} f(x) = (4)(0) = 0$
(c) $\lim_{x \to 4} [g(x)]^2 = \left[\lim_{x \to 4} g(x)\right]^2 = [-3]^2 = 9$
(d) $\lim_{x \to 4} \frac{g(x)}{f(x)-1} = \frac{\lim_{x \to 4} g(x)}{\lim_{x \to 4} f(x) - \lim_{x \to 4} 1} = \frac{-3}{0-1} = 3$

41. (a) $\lim_{x \to b} [f(x)+g(x)] = \lim_{x \to b} f(x) + \lim_{x \to b} g(x) = 7 + (-3) = 4$
(b) $\lim_{x \to b} f(x) \cdot g(x) = \left[\lim_{x \to b} f(x)\right]\left[\lim_{x \to b} g(x)\right] = (7)(-3) = -21$
(c) $\lim_{x \to b} 4g(x) = \left[\lim_{x \to b} 4\right]\left[\lim_{x \to b} g(x)\right] = (4)(-3) = -12$
(d) $\lim_{x \to b} f(x)/g(x) = \lim_{x \to b} f(x)/\lim_{x \to b} g(x) = \frac{7}{-3} = -\frac{7}{3}$

42. (a) $\lim_{x \to -2} [p(x)+r(x)+s(x)] = \lim_{x \to -2} p(x) + \lim_{x \to -2} r(x) + \lim_{x \to -2} s(x) = 4 + 0 + (-3) = 1$
(b) $\lim_{x \to -2} p(x) \cdot r(x) \cdot s(x) = \left[\lim_{x \to -2} p(x)\right]\left[\lim_{x \to -2} r(x)\right]\left[\lim_{x \to -2} s(x)\right] = (4)(0)(-3) = 0$
(c) $\lim_{x \to -2} [-4p(x)+5r(x)]/s(x) = \left[-4\lim_{x \to -2} p(x) + 5\lim_{x \to -2} r(x)\right] \Big/ \lim_{x \to -2} s(x) = [-4(4)+5(0)]/-3 = \frac{16}{3}$

43. $\lim_{h \to 0} \frac{(1+h)^2-1^2}{h} = \lim_{h \to 0} \frac{1+2h+h^2-1}{h} = \lim_{h \to 0} \frac{h(2+h)}{h} = \lim_{h \to 0} (2+h) = 2$

44. $\lim_{h \to 0} \frac{(-2+h)^2-(-2)^2}{h} = \lim_{h \to 0} \frac{4-4h+h^2-4}{h} = \lim_{h \to 0} \frac{h(h-4)}{h} = \lim_{h \to 0} (h-4) = -4$

45. $\lim_{h \to 0} \frac{[3(2+h)-4]-[3(2)-4]}{h} = \lim_{h \to 0} \frac{3h}{h} = 3$

46. $\lim_{h \to 0} \frac{\left(\frac{1}{-2+h}\right)-\left(\frac{1}{-2}\right)}{h} = \lim_{h \to 0} \frac{\frac{-2}{-2+h}-1}{-2h} = \lim_{h \to 0} \frac{-2-(-2+h)}{-2h(-2+h)} = \lim_{h \to 0} \frac{-h}{h(4-2h)} = -\frac{1}{4}$

47. $\lim\limits_{h \to 0} \frac{\sqrt{7+h}-\sqrt{7}}{h} = \lim\limits_{h \to 0} \frac{\left(\sqrt{7+h}-\sqrt{7}\right)\left(\sqrt{7+h}+\sqrt{7}\right)}{h\left(\sqrt{7+h}+\sqrt{7}\right)} = \lim\limits_{h \to 0} \frac{(7+h)-7}{h\left(\sqrt{7+h}+\sqrt{7}\right)}$

$= \lim\limits_{h \to 0} \frac{h}{h\left(\sqrt{7+h}+\sqrt{7}\right)} = \lim\limits_{h \to 0} \frac{1}{\sqrt{7+h}+\sqrt{7}} = \frac{1}{2\sqrt{7}}$

48. $\lim\limits_{h \to 0} \frac{\sqrt{3(0+h)+1}-\sqrt{3(0)+1}}{h} = \lim\limits_{h \to 0} \frac{\left(\sqrt{3h+1}-1\right)\left(\sqrt{3h+1}+1\right)}{h\left(\sqrt{3h+1}+1\right)} = \lim\limits_{h \to 0} \frac{(3h+1)-1}{h\left(\sqrt{3h+1}+1\right)}$

$= \lim\limits_{h \to 0} \frac{3h}{h\left(\sqrt{3h+1}+1\right)} = \lim\limits_{h \to 0} \frac{3}{\sqrt{3h+1}+1} = \frac{3}{2}$

49. $\lim\limits_{x \to 0} \sqrt{5-2x^2} = \sqrt{5-2(0)^2} = \sqrt{5}$ and $\lim\limits_{x \to 0} \sqrt{5-x^2} = \sqrt{5-(0)^2} = \sqrt{5}$; by the sandwich theorem,

$\lim\limits_{x \to 0} f(x) = \sqrt{5}$

50. $\lim\limits_{x \to 0} (2-x^2) = 2-0 = 2$ and $\lim\limits_{x \to 0} 2\cos x = 2(1) = 2$; by the sandwich theorem, $\lim\limits_{x \to 0} g(x) = 2$

51. (a) $\lim\limits_{x \to 0} \left(1-\frac{x^2}{6}\right) = 1-\frac{0}{6} = 1$ and $\lim\limits_{x \to 0} 1 = 1$; by the sandwich theorem, $\lim\limits_{x \to 0} \frac{x \sin x}{2-2\cos x} = 1$

(b) For $x \neq 0$, $y = (x \sin x)/(2-2\cos x)$ lies between the other two graphs in the figure, and the graphs converge as $x \to 0$.

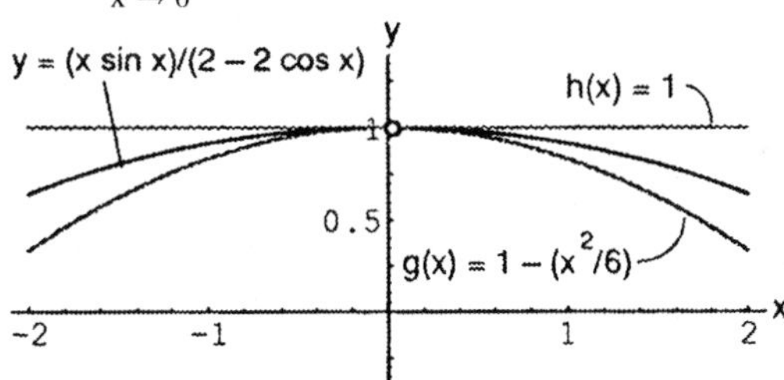

52. (a) $\lim\limits_{x \to 0} \left(\frac{1}{2}-\frac{x^2}{24}\right) = \lim\limits_{x \to 0} \frac{1}{2} - \lim\limits_{x \to 0} \frac{x^2}{24} = \frac{1}{2} - 0 = \frac{1}{2}$ and $\lim\limits_{x \to 0} \frac{1}{2} = \frac{1}{2}$; by the sandwich theorem,

$\lim\limits_{x \to 0} \frac{1-\cos x}{x^2} = \frac{1}{2}$.

(b) For all $x \neq 0$, the graph of $f(x) = (1-\cos x)/x^2$ lies between the line $y = \frac{1}{2}$ and the parabola $y = \frac{1}{2} - x^2/24$, and the graphs converge as $x \to 0$.

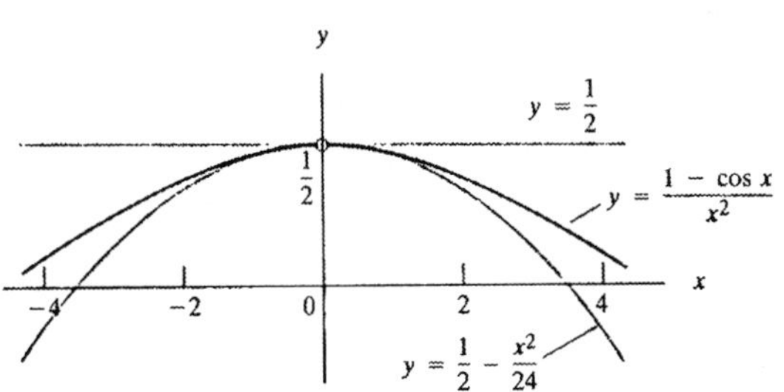

53. $\lim\limits_{x \to c} f(x)$ exists at those points c where $\lim\limits_{x \to c} x^4 = \lim\limits_{x \to c} x^2$. Thus, $c^4 = c^2 \Rightarrow c^2(1-c^2) = 0$

$\Rightarrow c = 0, 1$, or -1. Moreover, $\lim\limits_{x \to 0} f(x) = \lim\limits_{x \to 0} x^2 = 0$ and $\lim\limits_{x \to -1} f(x) = \lim\limits_{x \to 1} f(x) = 1$.

54. Nothing can be concluded about the values of f, g, and h at $x = 2$. Yes, f(2) could be 0. Since the conditions of the sandwich theorem are satisfied, $\lim\limits_{x \to 2} f(x) = -5 \neq 0$.

55. $1 = \lim\limits_{x \to 4} \frac{f(x)-5}{x-2} = \frac{\lim\limits_{x \to 4} f(x) - \lim\limits_{x \to 4} 5}{\lim\limits_{x \to 4} x - \lim\limits_{x \to 4} 2} = \frac{\lim\limits_{x \to 4} f(x) - 5}{4-2} \Rightarrow \lim\limits_{x \to 4} f(x) - 5 = 2(1) \Rightarrow \lim\limits_{x \to 4} f(x) = 2+5 = 7.$

56. (a) $1 = \lim\limits_{x \to -2} \frac{f(x)}{x^2} = \frac{\lim\limits_{x \to -2} f(x)}{\lim\limits_{x \to -2} x^2} = \frac{\lim\limits_{x \to -2} f(x)}{4} \Rightarrow \lim\limits_{x \to -2} f(x) = 4.$

(b) $1 = \lim\limits_{x \to -2} \frac{f(x)}{x^2} = \left[\lim\limits_{x \to -2} \frac{f(x)}{x}\right]\left[\lim\limits_{x \to -2} \frac{1}{x}\right] = \left[\lim\limits_{x \to -2} \frac{f(x)}{x}\right]\left(\frac{1}{-2}\right) \Rightarrow \lim\limits_{x \to -2} \frac{f(x)}{x} = -2.$

57. (a) $0 = 3 \cdot 0 = \left[\lim\limits_{x \to 2} \frac{f(x)-5}{x-2}\right]\left[\lim\limits_{x \to 2} (x-2)\right] = \lim\limits_{x \to 2} \left[\left(\frac{f(x)-5}{x-2}\right)(x-2)\right] = \lim\limits_{x \to 2} [f(x)-5] = \lim\limits_{x \to 2} f(x) - 5$

$\Rightarrow \lim\limits_{x \to 2} f(x) = 5.$

(b) $0 = 4 \cdot 0 = \left[\lim\limits_{x \to 2} \frac{f(x)-5}{x-2}\right]\left[\lim\limits_{x \to 2} (x-2)\right] \Rightarrow \lim\limits_{x \to 2} f(x) = 5$ as in part (a).

58. (a) $0 = 1 \cdot 0 = \left[\lim\limits_{x \to 0} \frac{f(x)}{x^2}\right]\left[\lim\limits_{x \to 0} x\right]^2 = \left[\lim\limits_{x \to 0} \frac{f(x)}{x^2}\right]\left[\lim\limits_{x \to 0} x^2\right] = \lim\limits_{x \to 0} \left[\frac{f(x)}{x^2} \cdot x^2\right] = \lim\limits_{x \to 0} f(x).$ That is, $\lim\limits_{x \to 0} f(x) = 0.$

(b) $0 = 1 \cdot 0 = \left[\lim\limits_{x \to 0} \frac{f(x)}{x^2}\right]\left[\lim\limits_{x \to 0} x\right] = \lim\limits_{x \to 0} \left[\frac{f(x)}{x^2} \cdot x\right] = \lim\limits_{x \to 0} \frac{f(x)}{x}.$ That is, $\lim\limits_{x \to 0} \frac{f(x)}{x} = 0.$

59. (a) $\lim\limits_{x \to 0} x \sin \frac{1}{x} = 0$

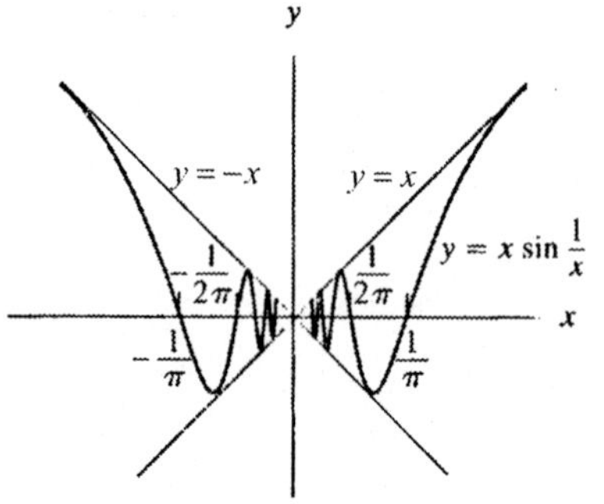

(b) $-1 \le \sin \frac{1}{x} \le 1$ for $x \ne 0$:

$x > 0 \Rightarrow -x \le x \sin \frac{1}{x} \le x \Rightarrow \lim\limits_{x \to 0} x \sin \frac{1}{x} = 0$ by the sandwich theorem;

$x < 0 \Rightarrow -x \ge x \sin \frac{1}{x} \ge x \Rightarrow \lim\limits_{x \to 0} x \sin \frac{1}{x} = 0$ by the sandwich theorem.

60. (a) $\lim\limits_{x \to 0} x^2 \cos\left(\frac{1}{x^3}\right) = 0$

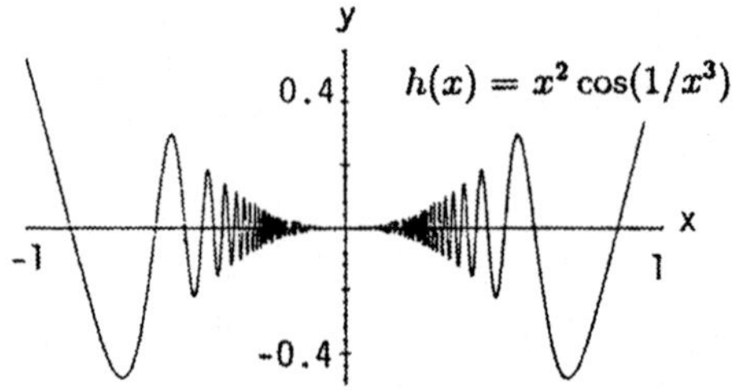

(b) $-1 \le \cos\left(\frac{1}{x^3}\right) \le 1$ for $x \ne 0 \Rightarrow -x^2 \le x^2 \cos\left(\frac{1}{x^3}\right) \le x^2 \Rightarrow \lim\limits_{x \to 0} x^2 \cos\left(\frac{1}{x^3}\right) = 0$ by the sandwich theorem since $\lim\limits_{x \to 0} x^2 = 0.$

2.3 THE PRECISE DEFINITION OF A LIMIT

1. 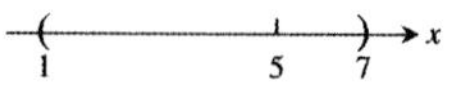

Step 1: $|x-5| < \delta \Rightarrow -\delta < x-5 < \delta \Rightarrow -\delta + 5 < x < \delta + 5$

Step 2: $\delta + 5 = 7 \Rightarrow \delta = 2$, or $-\delta + 5 = 1 \Rightarrow \delta = 4$.

The value of δ which assures $|x-5| < \delta \Rightarrow 1 < x < 7$ is the smaller value, $\delta = 2$.

2. 1 2 7 x

Step 1: $|x - 2| < \delta \Rightarrow -\delta < x - 2 < \delta \Rightarrow -\delta + 2 < x < \delta + 2$

Step 2: $-\delta + 2 = 1 \Rightarrow \delta = 1$, or $\delta + 2 = 7 \Rightarrow \delta = 5$.

The value of δ which assures $|x - 2| < \delta \Rightarrow 1 < x < 7$ is the smaller value, $\delta = 1$.

3. −7/2 −3 −1/2 x

Step 1: $|x - (-3)| < \delta \Rightarrow -\delta < x + 3 < \delta \Rightarrow -\delta - 3 < x < \delta - 3$

Step 2: $-\delta - 3 = -\frac{7}{2} \Rightarrow \delta = \frac{1}{2}$, or $\delta - 3 = -\frac{1}{2} \Rightarrow \delta = \frac{5}{2}$.

The value of δ which assures $|x - (-3)| < \delta \Rightarrow -\frac{7}{2} < x < -\frac{1}{2}$ is the smaller value, $\delta = \frac{1}{2}$.

4.

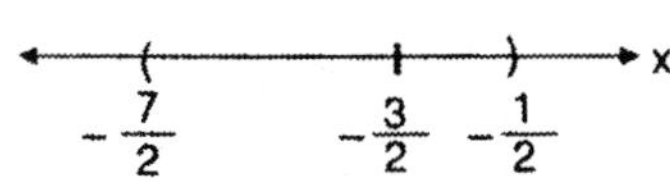

Step 1: $\left|x - \left(-\frac{3}{2}\right)\right| < \delta \Rightarrow -\delta < x + \frac{3}{2} < \delta \Rightarrow -\delta - \frac{3}{2} < x < \delta - \frac{3}{2}$

Step 2: $-\delta - \frac{3}{2} = -\frac{7}{2} \Rightarrow \delta = 2$, or $\delta - \frac{3}{2} = -\frac{1}{2} \Rightarrow \delta = 1$.

The value of δ which assures $\left|x - \left(-\frac{3}{2}\right)\right| < \delta \Rightarrow -\frac{7}{2} < x < -\frac{1}{2}$ is the smaller value, $\delta = 1$.

5. 4/9 1/2 4/7 x

Step 1: $\left|x - \frac{1}{2}\right| < \delta \Rightarrow -\delta < x - \frac{1}{2} < \delta \Rightarrow -\delta + \frac{1}{2} < x < \delta + \frac{1}{2}$

Step 2: $-\delta + \frac{1}{2} = \frac{4}{9} \Rightarrow \delta = \frac{1}{18}$, or $\delta + \frac{1}{2} = \frac{4}{7} \Rightarrow \delta = \frac{1}{14}$.

The value of δ which assures $\left|x - \frac{1}{2}\right| < \delta \Rightarrow \frac{4}{9} < x < \frac{4}{7}$ is the smaller value, $\delta = \frac{1}{18}$.

6. 2.7591 3 3.2391 x

Step 1: $|x - 3| < \delta \Rightarrow -\delta < x - 3 < \delta \Rightarrow -\delta + 3 < x < \delta + 3$

Step 2: $-\delta + 3 = 2.7591 \Rightarrow \delta = 0.2409$, or $\delta + 3 = 3.2391 \Rightarrow \delta = 0.2391$.

The value of δ which assures $|x - 3| < \delta \Rightarrow 2.7591 < x < 3.2391$ is the smaller value, $\delta = 0.2391$.

7. Step 1: $|x - 5| < \delta \Rightarrow -\delta < x - 5 < \delta \Rightarrow -\delta + 5 < x < \delta + 5$

Step 2: From the graph, $-\delta + 5 = 4.9 \Rightarrow \delta = 0.1$, or $\delta + 5 = 5.1 \Rightarrow \delta = 0.1$; thus $\delta = 0.1$ in either case.

8. Step 1: $|x - (-3)| < \delta \Rightarrow -\delta < x + 3 < \delta \Rightarrow -\delta - 3 < x < \delta - 3$

Step 2: From the graph, $-\delta - 3 = -3.1 \Rightarrow \delta = 0.1$, or $\delta - 3 = -2.9 \Rightarrow \delta = 0.1$; thus $\delta = 0.1$.

9. Step 1: $|x - 1| < \delta \Rightarrow -\delta < x - 1 < \delta \Rightarrow -\delta + 1 < x < \delta + 1$

Step 2: From the graph, $-\delta + 1 = \frac{9}{16} \Rightarrow \delta = \frac{7}{16}$, or $\delta + 1 = \frac{25}{16} \Rightarrow \delta = \frac{9}{16}$; thus $\delta = \frac{7}{16}$.

10. Step 1: $|x - 3| < \delta \Rightarrow -\delta < x - 3 < \delta \Rightarrow -\delta + 3 < x < \delta + 3$

Step 2: From the graph, $-\delta + 3 = 2.61 \Rightarrow \delta = 0.39$, or $\delta + 3 = 3.41 \Rightarrow \delta = 0.41$; thus $\delta = 0.39$.

11. Step 1: $|x - 2| < \delta \Rightarrow -\delta < x - 2 < \delta \Rightarrow -\delta + 2 < x < \delta + 2$

Step 2: From the graph, $-\delta + 2 = \sqrt{3} \Rightarrow \delta = 2 - \sqrt{3} \approx 0.2679$, or $\delta + 2 = \sqrt{5} \Rightarrow \delta = \sqrt{5} - 2 \approx 0.2361$; thus $\delta = \sqrt{5} - 2$.

12. Step 1: $|x-(-1)| < \delta \Rightarrow -\delta < x+1 < \delta \Rightarrow -\delta - 1 < x < \delta - 1$

 Step 2: From the graph, $-\delta - 1 = -\frac{\sqrt{5}}{2} \Rightarrow \delta = \frac{\sqrt{5}-2}{2} \approx 0.1180$, or $\delta - 1 = -\frac{\sqrt{3}}{2} \Rightarrow \delta = \frac{2-\sqrt{3}}{2} \approx 0.1340$; thus $\delta = \frac{\sqrt{5}-2}{2}$.

13. Step 1: $|x-(-1)| < \delta \Rightarrow -\delta < x+1 < \delta \Rightarrow -\delta - 1 < x < \delta - 1$

 Step 2: From the graph, $-\delta - 1 = -\frac{16}{9} \Rightarrow \delta = \frac{7}{9} \approx 0.77$, or $\delta - 1 = -\frac{16}{25} \Rightarrow \frac{9}{25} = 0.36$; thus $\delta = \frac{9}{25} = 0.36$.

14. Step 1: $\left|x - \frac{1}{2}\right| < \delta \Rightarrow -\delta < x - \frac{1}{2} < \delta \Rightarrow -\delta + \frac{1}{2} < x < \delta + \frac{1}{2}$

 Step 2: From the graph, $-\delta + \frac{1}{2} = \frac{1}{2.01} \Rightarrow \delta = \frac{1}{2} - \frac{1}{2.01} \approx 0.00248$, or $\delta + \frac{1}{2} = \frac{1}{1.99} \Rightarrow \delta = \frac{1}{1.99} - \frac{1}{2} \approx 0.00251$; thus $\delta = 0.00248$.

15. Step 1: $|(x+1) - 5| < 0.01 \Rightarrow |x-4| < 0.01 \Rightarrow -0.01 < x - 4 < 0.01 \Rightarrow 3.99 < x < 4.01$

 Step 2: $|x-4| < \delta \Rightarrow -\delta < x - 4 < \delta \Rightarrow -\delta + 4 < x < \delta + 4 \Rightarrow \delta = 0.01$.

16. Step 1: $|(2x-2) - (-6)| < 0.02 \Rightarrow |2x+4| < 0.02 \Rightarrow -0.02 < 2x + 4 < 0.02 \Rightarrow -4.02 < 2x < -3.98$ $\Rightarrow -2.01 < x < -1.99$

 Step 2: $|x-(-2)| < \delta \Rightarrow -\delta < x + 2 < \delta \Rightarrow -\delta - 2 < x < \delta - 2 \Rightarrow \delta = 0.01$.

17. Step 1: $\left|\sqrt{x+1} - 1\right| < 0.1 \Rightarrow -0.1 < \sqrt{x+1} - 1 < 0.1 \Rightarrow 0.9 < \sqrt{x+1} < 1.1 \Rightarrow 0.81 < x + 1 < 1.21$ $\Rightarrow -0.19 < x < 0.21$

 Step 2: $|x-0| < \delta \Rightarrow -\delta < x < \delta$. Then, $-\delta = -0.19 \Rightarrow \delta = 0.19$ or $\delta = 0.21$; thus, $\delta = 0.19$.

18. Step 1: $\left|\sqrt{x} - \frac{1}{2}\right| < 0.1 \Rightarrow -0.1 < \sqrt{x} - \frac{1}{2} < 0.1 \Rightarrow 0.4 < \sqrt{x} < 0.6 \Rightarrow 0.16 < x < 0.36$

 Step 2: $\left|x - \frac{1}{4}\right| < \delta \Rightarrow -\delta < x - \frac{1}{4} < \delta \Rightarrow -\delta + \frac{1}{4} < x < \delta + \frac{1}{4}$.

 Then, $-\delta + \frac{1}{4} = 0.16 \Rightarrow \delta = 0.09$ or $\delta + \frac{1}{4} = 0.36 \Rightarrow \delta = 0.11$; thus $\delta = 0.09$.

19. Step 1: $\left|\sqrt{19-x} - 3\right| < 1 \Rightarrow -1 < \sqrt{19-x} - 3 < 1 \Rightarrow 2 < \sqrt{19-x} < 4 \Rightarrow 4 < 19 - x < 16$ $\Rightarrow -4 > x - 19 > -16 \Rightarrow 15 > x > 3$ or $3 < x < 15$

 Step 2: $|x-10| < \delta \Rightarrow -\delta < x - 10 < \delta \Rightarrow -\delta + 10 < x < \delta + 10$.

 Then $-\delta + 10 = 3 \Rightarrow \delta = 7$, or $\delta + 10 = 15 \Rightarrow \delta = 5$; thus $\delta = 5$.

20. Step 1: $\left|\sqrt{x-7} - 4\right| < 1 \Rightarrow -1 < \sqrt{x-7} - 4 < 1 \Rightarrow 3 < \sqrt{x-7} < 5 \Rightarrow 9 < x - 7 < 25 \Rightarrow 16 < x < 32$

 Step 2: $|x-23| < \delta \Rightarrow -\delta < x - 23 < \delta \Rightarrow -\delta + 23 < x < \delta + 23$.

 Then $-\delta + 23 = 16 \Rightarrow \delta = 7$, or $\delta + 23 = 32 \Rightarrow \delta = 9$; thus $\delta = 7$.

21. Step 1: $\left|\frac{1}{x} - \frac{1}{4}\right| < 0.05 \Rightarrow -0.05 < \frac{1}{x} - \frac{1}{4} < 0.05 \Rightarrow 0.2 < \frac{1}{x} < 0.3 \Rightarrow \frac{10}{2} > x > \frac{10}{3}$ or $\frac{10}{3} < x < 5$.

 Step 2: $|x-4| < \delta \Rightarrow -\delta < x - 4 < \delta \Rightarrow -\delta + 4 < x < \delta + 4$.

 Then $-\delta + 4 = \frac{10}{3}$ or $\delta = \frac{2}{3}$, or $\delta + 4 = 5$ or $\delta = 1$; thus $\delta = \frac{2}{3}$.

22. Step 1: $|x^2 - 3| < 0.1 \Rightarrow -0.1 < x^2 - 3 < 0.1 \Rightarrow 2.9 < x^2 < 3.1 \Rightarrow \sqrt{2.9} < x < \sqrt{3.1}$

 Step 2: $\left|x - \sqrt{3}\right| < \delta \Rightarrow -\delta < x - \sqrt{3} < \delta \Rightarrow -\delta + \sqrt{3} < x < \delta + \sqrt{3}$.

 Then $-\delta + \sqrt{3} = \sqrt{2.9} \Rightarrow \delta = \sqrt{3} - \sqrt{2.9} \approx 0.0291$, or $\delta + \sqrt{3} = \sqrt{3.1} \Rightarrow \delta = \sqrt{3.1} - \sqrt{3} \approx 0.0286$; thus $\delta = 0.0286$.

23. Step 1: $|x^2 - 4| < 0.5 \Rightarrow -0.5 < x^2 - 4 < 0.5 \Rightarrow 3.5 < x^2 < 4.5 \Rightarrow \sqrt{3.5} < |x| < \sqrt{4.5} \Rightarrow -\sqrt{4.5} < x < -\sqrt{3.5}$, for x near -2.
Step 2: $|x - (-2)| < \delta \Rightarrow -\delta < x + 2 < \delta \Rightarrow -\delta - 2 < x < \delta - 2$.
Then $-\delta - 2 = -\sqrt{4.5} \Rightarrow \delta = \sqrt{4.5} - 2 \approx 0.1213$, or $\delta - 2 = -\sqrt{3.5} \Rightarrow \delta = 2 - \sqrt{3.5} \approx 0.1292$; thus $\delta = \sqrt{4.5} - 2 \approx 0.12$.

24. Step 1: $\left|\frac{1}{x} - (-1)\right| < 0.1 \Rightarrow -0.1 < \frac{1}{x} + 1 < 0.1 \Rightarrow -\frac{11}{10} < \frac{1}{x} < -\frac{9}{10} \Rightarrow -\frac{10}{11} > x > -\frac{10}{9}$ or $-\frac{10}{9} < x < -\frac{10}{11}$.
Step 2: $|x - (-1)| < \delta \Rightarrow -\delta < x + 1 < \delta \Rightarrow -\delta - 1 < x < \delta - 1$.
Then $-\delta - 1 = -\frac{10}{9} \Rightarrow \delta = \frac{1}{9}$, or $\delta - 1 = -\frac{10}{11} \Rightarrow \delta = \frac{1}{11}$; thus $\delta = \frac{1}{11}$.

25. Step 1: $|(x^2 - 5) - 11| < 1 \Rightarrow |x^2 - 16| < 1 \Rightarrow -1 < x^2 - 16 < 1 \Rightarrow 15 < x^2 < 17 \Rightarrow \sqrt{15} < x < \sqrt{17}$.
Step 2: $|x - 4| < \delta \Rightarrow -\delta < x - 4 < \delta \Rightarrow -\delta + 4 < x < \delta + 4$.
Then $-\delta + 4 = \sqrt{15} \Rightarrow \delta = 4 - \sqrt{15} \approx 0.1270$, or $\delta + 4 = \sqrt{17} \Rightarrow \delta = \sqrt{17} - 4 \approx 0.1231$; thus $\delta = \sqrt{17} - 4 \approx 0.12$.

26. Step 1: $\left|\frac{120}{x} - 5\right| < 1 \Rightarrow -1 < \frac{120}{x} - 5 < 1 \Rightarrow 4 < \frac{120}{x} < 6 \Rightarrow \frac{1}{4} > \frac{x}{120} > \frac{1}{6} \Rightarrow 30 > x > 20$ or $20 < x < 30$.
Step 2: $|x - 24| < \delta \Rightarrow -\delta < x - 24 < \delta \Rightarrow -\delta + 24 < x < \delta + 24$.
Then $-\delta + 24 = 20 \Rightarrow \delta = 4$, or $\delta + 24 = 30 \Rightarrow \delta = 6$; thus $\Rightarrow \delta = 4$.

27. Step 1: $|mx - 2m| < 0.03 \Rightarrow -0.03 < mx - 2m < 0.03 \Rightarrow -0.03 + 2m < mx < 0.03 + 2m \Rightarrow 2 - \frac{0.03}{m} < x < 2 + \frac{0.03}{m}$.
Step 2: $|x - 2| < \delta \Rightarrow -\delta < x - 2 < \delta \Rightarrow -\delta + 2 < x < \delta + 2$.
Then $-\delta + 2 = 2 - \frac{0.03}{m} \Rightarrow \delta = \frac{0.03}{m}$, or $\delta + 2 = 2 + \frac{0.03}{m} \Rightarrow \delta = \frac{0.03}{m}$. In either case, $\delta = \frac{0.03}{m}$.

28. Step 1: $|mx - 3m| < c \Rightarrow -c < mx - 3m < c \Rightarrow -c + 3m < mx < c + 3m \Rightarrow 3 - \frac{c}{m} < x < 3 + \frac{c}{m}$
Step 2: $|x - 3| < \delta \Rightarrow -\delta < x - 3 < \delta \Rightarrow -\delta + 3 < x < \delta + 3$.
Then $-\delta + 3 = 3 - \frac{c}{m} \Rightarrow \delta = \frac{c}{m}$, or $\delta + 3 = 3 + \frac{c}{m} \Rightarrow \delta = \frac{c}{m}$. In either case, $\delta = \frac{c}{m}$.

29. Step 1: $\left|(mx + b) - \left(\frac{m}{2} + b\right)\right| < c \Rightarrow -c < mx - \frac{m}{2} < c \Rightarrow -c + \frac{m}{2} < mx < c + \frac{m}{2} \Rightarrow \frac{1}{2} - \frac{c}{m} < x < \frac{1}{2} + \frac{c}{m}$.
Step 2: $\left|x - \frac{1}{2}\right| < \delta \Rightarrow -\delta < x - \frac{1}{2} < \delta \Rightarrow -\delta + \frac{1}{2} < x < \delta + \frac{1}{2}$.
Then $-\delta + \frac{1}{2} = \frac{1}{2} - \frac{c}{m} \Rightarrow \delta = \frac{c}{m}$, or $\delta + \frac{1}{2} = \frac{1}{2} + \frac{c}{m} \Rightarrow \delta = \frac{c}{m}$. In either case, $\delta = \frac{c}{m}$.

30. Step 1: $|(mx + b) - (m + b)| < 0.05 \Rightarrow -0.05 < mx - m < 0.05 \Rightarrow -0.05 + m < mx < 0.05 + m \Rightarrow 1 - \frac{0.05}{m} < x < 1 + \frac{0.05}{m}$.
Step 2: $|x - 1| < \delta \Rightarrow -\delta < x - 1 < \delta \Rightarrow -\delta + 1 < x < \delta + 1$.
Then $-\delta + 1 = 1 - \frac{0.05}{m} \Rightarrow \delta = \frac{0.05}{m}$, or $\delta + 1 = 1 + \frac{0.05}{m} \Rightarrow \delta = \frac{0.05}{m}$. In either case, $\delta = \frac{0.05}{m}$.

31. $\lim_{x \to 3} (3 - 2x) = 3 - 2(3) = -3$
Step 1: $|(3 - 2x) - (-3)| < 0.02 \Rightarrow -0.02 < 6 - 2x < 0.02 \Rightarrow -6.02 < -2x < -5.98 \Rightarrow 3.01 > x > 2.99$ or $2.99 < x < 3.01$.
Step 2: $0 < |x - 3| < \delta \Rightarrow -\delta < x - 3 < \delta \Rightarrow -\delta + 3 < x < \delta + 3$.
Then $-\delta + 3 = 2.99 \Rightarrow \delta = 0.01$, or $\delta + 3 = 3.01 \Rightarrow \delta = 0.01$; thus $\delta = 0.01$.

32. $\lim_{x \to -1} (-3x - 2) = (-3)(-1) - 2 = 1$
Step 1: $|(-3x - 2) - 1| < 0.03 \Rightarrow -0.03 < -3x - 3 < 0.03 \Rightarrow 0.01 > x + 1 > -0.01 \Rightarrow -1.01 < x < -0.99$.

Step 2: $|x-(-1)| < \delta \Rightarrow -\delta < x+1 < \delta \Rightarrow -\delta - 1 < x < \delta - 1.$
Then $-\delta - 1 = -1.01 \Rightarrow \delta = 0.01$, or $\delta - 1 = -0.99 \Rightarrow \delta = 0.01$; thus $\delta = 0.01$.

33. $\lim\limits_{x \to 2} \frac{x^2-4}{x-2} = \lim\limits_{x \to 2} \frac{(x+2)(x-2)}{(x-2)} = \lim\limits_{x \to 2} (x+2) = 2+2 = 4,\ x \neq 2$

Step 1: $\left|\left(\frac{x^2-4}{x-2}\right) - 4\right| < 0.05 \Rightarrow -0.05 < \frac{(x+2)(x-2)}{(x-2)} - 4 < 0.05 \Rightarrow 3.95 < x+2 < 4.05,\ x \neq 2$
$\Rightarrow 1.95 < x < 2.05,\ x \neq 2.$

Step 2: $|x-2| < \delta \Rightarrow -\delta < x-2 < \delta \Rightarrow -\delta + 2 < x < \delta + 2.$
Then $-\delta + 2 = 1.95 \Rightarrow \delta = 0.05$, or $\delta + 2 = 2.05 \Rightarrow \delta = 0.05$; thus $\delta = 0.05$.

34. $\lim\limits_{x \to -5} \frac{x^2+6x+5}{x+5} = \lim\limits_{x \to -5} \frac{(x+5)(x+1)}{(x+5)} = \lim\limits_{x \to -5} (x+1) = -4,\ x \neq -5.$

Step 1: $\left|\left(\frac{x^2+6x+5}{x+5}\right) - (-4)\right| < 0.05 \Rightarrow -0.05 < \frac{(x+5)(x+1)}{(x+5)} + 4 < 0.05 \Rightarrow -4.05 < x+1 < -3.95,\ x \neq -5$
$\Rightarrow -5.05 < x < -4.95,\ x \neq -5.$

Step 2: $|x-(-5)| < \delta \Rightarrow -\delta < x+5 < \delta \Rightarrow -\delta - 5 < x < \delta - 5.$
Then $-\delta - 5 = -5.05 \Rightarrow \delta = 0.05$, or $\delta - 5 = -4.95 \Rightarrow \delta = 0.05$; thus $\delta = 0.05$.

35. $\lim\limits_{x \to -3} \sqrt{1-5x} = \sqrt{1-5(-3)} = \sqrt{16} = 4$

Step 1: $\left|\sqrt{1-5x} - 4\right| < 0.5 \Rightarrow -0.5 < \sqrt{1-5x} - 4 < 0.5 \Rightarrow 3.5 < \sqrt{1-5x} < 4.5 \Rightarrow 12.25 < 1-5x < 20.25$
$\Rightarrow 11.25 < -5x < 19.25 \Rightarrow -3.85 < x < -2.25.$

Step 2: $|x-(-3)| < \delta \Rightarrow -\delta < x+3 < \delta \Rightarrow -\delta - 3 < x < \delta - 3.$
Then $-\delta - 3 = -3.85 \Rightarrow \delta = 0.85$, or $\delta - 3 = -2.25 \Rightarrow 0.75$; thus $\delta = 0.75$.

36. $\lim\limits_{x \to 2} \frac{4}{x} = \frac{4}{2} = 2$

Step 1: $\left|\frac{4}{x} - 2\right| < 0.4 \Rightarrow -0.4 < \frac{4}{x} - 2 < 0.4 \Rightarrow 1.6 < \frac{4}{x} < 2.4 \Rightarrow \frac{10}{16} > \frac{x}{4} > \frac{10}{24} \Rightarrow \frac{10}{4} > x > \frac{10}{6}$ or $\frac{5}{3} < x < \frac{5}{2}$.

Step 2: $|x-2| < \delta \Rightarrow -\delta < x-2 < \delta \Rightarrow -\delta + 2 < x < \delta + 2.$
Then $-\delta + 2 = \frac{5}{3} \Rightarrow \delta = \frac{1}{3}$, or $\delta + 2 = \frac{5}{2} \Rightarrow \delta = \frac{1}{2}$; thus $\delta = \frac{1}{3}$.

37. Step 1: $|(9-x)-5| < \epsilon \Rightarrow -\epsilon < 4-x < \epsilon \Rightarrow -\epsilon - 4 < -x < \epsilon - 4 \Rightarrow \epsilon + 4 > x > 4 - \epsilon \Rightarrow 4 - \epsilon < x < 4 + \epsilon.$

Step 2: $|x-4| < \delta \Rightarrow -\delta < x-4 < \delta \Rightarrow -\delta + 4 < x < \delta + 4.$
Then $-\delta + 4 = -\epsilon + 4 \Rightarrow \delta = \epsilon$, or $\delta + 4 = \epsilon + 4 \Rightarrow \delta = \epsilon$. Thus choose $\delta = \epsilon$.

38. Step 1: $|(3x-7)-2| < \epsilon \Rightarrow -\epsilon < 3x-9 < \epsilon \Rightarrow 9 - \epsilon < 3x < 9 + \epsilon \Rightarrow 3 - \frac{\epsilon}{3} < x < 3 + \frac{\epsilon}{3}.$

Step 2: $|x-3| < \delta \Rightarrow -\delta < x-3 < \delta \Rightarrow -\delta + 3 < x < \delta + 3.$
Then $-\delta + 3 = 3 - \frac{\epsilon}{3} \Rightarrow \delta = \frac{\epsilon}{3}$, or $\delta + 3 = 3 + \frac{\epsilon}{3} \Rightarrow \delta = \frac{\epsilon}{3}$. Thus choose $\delta = \frac{\epsilon}{3}$.

39. Step 1: $\left|\sqrt{x-5} - 2\right| < \epsilon \Rightarrow -\epsilon < \sqrt{x-5} - 2 < \epsilon \Rightarrow 2 - \epsilon < \sqrt{x-5} < 2 + \epsilon \Rightarrow (2-\epsilon)^2 < x-5 < (2+\epsilon)^2$
$\Rightarrow (2-\epsilon)^2 + 5 < x < (2+\epsilon)^2 + 5.$

Step 2: $|x-9| < \delta \Rightarrow -\delta < x-9 < \delta \Rightarrow -\delta + 9 < x < \delta + 9.$
Then $-\delta + 9 = \epsilon^2 - 4\epsilon + 9 \Rightarrow \delta = 4\epsilon - \epsilon^2$, or $\delta + 9 = \epsilon^2 + 4\epsilon + 9 \Rightarrow \delta = 4\epsilon + \epsilon^2$. Thus choose the smaller distance, $\delta = 4\epsilon - \epsilon^2$.

40. Step 1: $\left|\sqrt{4-x} - 2\right| < \epsilon \Rightarrow -\epsilon < \sqrt{4-x} - 2 < \epsilon \Rightarrow 2 - \epsilon < \sqrt{4-x} < 2 + \epsilon \Rightarrow (2-\epsilon)^2 < 4-x < (2+\epsilon)^2$
$\Rightarrow -(2+\epsilon)^2 < x-4 < -(2-\epsilon)^2 \Rightarrow -(2+\epsilon)^2 + 4 < x < -(2-\epsilon)^2 + 4.$

Step 2: $|x-0| < \delta \Rightarrow -\delta < x < \delta.$

Then $-\delta = -(2+\epsilon)^2 + 4 = -\epsilon^2 - 4\epsilon \Rightarrow \delta = 4\epsilon + \epsilon^2$, or $\delta = -(2-\epsilon)^2 + 4 = 4\epsilon - \epsilon^2$. Thus choose the smaller distance, $\delta = 4\epsilon - \epsilon^2$.

41. Step 1: For $x \neq 1$, $|x^2 - 1| < \epsilon \Rightarrow -\epsilon < x^2 - 1 < \epsilon \Rightarrow 1 - \epsilon < x^2 < 1 + \epsilon \Rightarrow \sqrt{1-\epsilon} < |x| < \sqrt{1+\epsilon}$ $\Rightarrow \sqrt{1-\epsilon} < x < \sqrt{1+\epsilon}$ near $x = 1$.

Step 2: $|x - 1| < \delta \Rightarrow -\delta < x - 1 < \delta \Rightarrow -\delta + 1 < x < \delta + 1$.
Then $-\delta + 1 = \sqrt{1-\epsilon} \Rightarrow \delta = 1 - \sqrt{1-\epsilon}$, or $\delta + 1 = \sqrt{1+\epsilon} \Rightarrow \delta = \sqrt{1+\epsilon} - 1$. Choose $\delta = \min\left\{1 - \sqrt{1-\epsilon}, \sqrt{1+\epsilon} - 1\right\}$, that is, the smaller of the two distances.

42. Step 1: For $x \neq -2$, $|x^2 - 4| < \epsilon \Rightarrow -\epsilon < x^2 - 4 < \epsilon \Rightarrow 4 - \epsilon < x^2 < 4 + \epsilon \Rightarrow \sqrt{4-\epsilon} < |x| < \sqrt{4+\epsilon}$ $\Rightarrow -\sqrt{4+\epsilon} < x < -\sqrt{4-\epsilon}$ near $x = -2$.

Step 2: $|x - (-2)| < \delta \Rightarrow -\delta < x + 2 < \delta \Rightarrow -\delta - 2 < x < \delta - 2$.
Then $-\delta - 2 = -\sqrt{4+\epsilon} \Rightarrow \delta = \sqrt{4+\epsilon} - 2$, or $\delta - 2 = -\sqrt{4-\epsilon} \Rightarrow \delta = 2 - \sqrt{4-\epsilon}$. Choose $\delta = \min\left\{\sqrt{4+\epsilon} - 2, 2 - \sqrt{4-\epsilon}\right\}$.

43. Step 1: $\left|\frac{1}{x} - 1\right| < \epsilon \Rightarrow -\epsilon < \frac{1}{x} - 1 < \epsilon \Rightarrow 1 - \epsilon < \frac{1}{x} < 1 + \epsilon \Rightarrow \frac{1}{1+\epsilon} < x < \frac{1}{1-\epsilon}$.

Step 2: $|x - 1| < \delta \Rightarrow -\delta < x - 1 < \delta \Rightarrow 1 - \delta < x < 1 + \delta$.
Then $1 - \delta = \frac{1}{1+\epsilon} \Rightarrow \delta = 1 - \frac{1}{1+\epsilon} = \frac{\epsilon}{1+\epsilon}$, or $1 + \delta = \frac{1}{1-\epsilon} \Rightarrow \delta = \frac{1}{1-\epsilon} - 1 = \frac{\epsilon}{1-\epsilon}$.
Choose $\delta = \frac{\epsilon}{1+\epsilon}$, the smaller of the two distances.

44. Step 1: $\left|\frac{1}{x^2} - \frac{1}{3}\right| < \epsilon \Rightarrow -\epsilon < \frac{1}{x^2} - \frac{1}{3} < \epsilon \Rightarrow \frac{1}{3} - \epsilon < \frac{1}{x^2} < \frac{1}{3} + \epsilon \Rightarrow \frac{1-3\epsilon}{3} < \frac{1}{x^2} < \frac{1+3\epsilon}{3} \Rightarrow \frac{3}{1-3\epsilon} > x^2 > \frac{3}{1+3\epsilon}$ $\Rightarrow \sqrt{\frac{3}{1+3\epsilon}} < |x| < \sqrt{\frac{3}{1-3\epsilon}}$, or $\sqrt{\frac{3}{1+3\epsilon}} < x < \sqrt{\frac{3}{1-3\epsilon}}$ for x near $\sqrt{3}$.

Step 2: $\left|x - \sqrt{3}\right| < \delta \Rightarrow -\delta < x - \sqrt{3} < \delta \Rightarrow \sqrt{3} - \delta < x < \sqrt{3} + \delta$.
Then $\sqrt{3} - \delta = \sqrt{\frac{3}{1+3\epsilon}} \Rightarrow \delta = \sqrt{3} - \sqrt{\frac{3}{1+3\epsilon}}$, or $\sqrt{3} + \delta = \sqrt{\frac{3}{1-3\epsilon}} \Rightarrow \delta = \sqrt{\frac{3}{1-3\epsilon}} - \sqrt{3}$.
Choose $\delta = \min\left\{\sqrt{3} - \sqrt{\frac{3}{1+3\epsilon}}, \sqrt{\frac{3}{1-3\epsilon}} - \sqrt{3}\right\}$.

45. Step 1: $\left|\left(\frac{x^2-9}{x+3}\right) - (-6)\right| < \epsilon \Rightarrow -\epsilon < (x-3) + 6 < \epsilon, x \neq -3 \Rightarrow -\epsilon < x + 3 < \epsilon \Rightarrow -\epsilon - 3 < x < \epsilon - 3$.

Step 2: $|x - (-3)| < \delta \Rightarrow -\delta < x + 3 < \delta \Rightarrow -\delta - 3 < x < \delta - 3$.
Then $-\delta - 3 = -\epsilon - 3 \Rightarrow \delta = \epsilon$, or $\delta - 3 = \epsilon - 3 \Rightarrow \delta = \epsilon$. Choose $\delta = \epsilon$.

46. Step 1: $\left|\left(\frac{x^2-1}{x-1}\right) - 2\right| < \epsilon \Rightarrow -\epsilon < (x+1) - 2 < \epsilon, x \neq 1 \Rightarrow 1 - \epsilon < x < 1 + \epsilon$.

Step 2: $|x - 1| < \delta \Rightarrow -\delta < x - 1 < \delta \Rightarrow 1 - \delta < x < 1 + \delta$.
Then $1 - \delta = 1 - \epsilon \Rightarrow \delta = \epsilon$, or $1 + \delta = 1 + \epsilon \Rightarrow \delta = \epsilon$. Choose $\delta = \epsilon$.

47. Step 1: $x < 1$: $|(4 - 2x) - 2| < \epsilon \Rightarrow 0 < 2 - 2x < \epsilon$ since $x < 1$. Thus, $1 - \frac{\epsilon}{2} < x < 0$;
$x \geq 1$: $|(6x - 4) - 2| < \epsilon \Rightarrow 0 \leq 6x - 6 < \epsilon$ since $x \geq 1$. Thus, $1 \leq x < 1 + \frac{\epsilon}{6}$.

Step 2: $|x - 1| < \delta \Rightarrow -\delta < x - 1 < \delta \Rightarrow 1 - \delta < x < 1 + \delta$.
Then $1 - \delta = 1 - \frac{\epsilon}{2} \Rightarrow \delta = \frac{\epsilon}{2}$, or $1 + \delta = 1 + \frac{\epsilon}{6} \Rightarrow \delta = \frac{\epsilon}{6}$. Choose $\delta = \frac{\epsilon}{6}$.

48. Step 1: $x < 0$: $|2x - 0| < \epsilon \Rightarrow -\epsilon < 2x < 0 \Rightarrow -\frac{\epsilon}{2} < x < 0$;
$x \geq 0$: $\left|\frac{x}{2} - 0\right| < \epsilon \Rightarrow 0 \leq x < 2\epsilon$.

Step 2: $|x - 0| < \delta \Rightarrow -\delta < x < \delta.$
Then $-\delta = -\frac{\epsilon}{2} \Rightarrow \delta = \frac{\epsilon}{2}$, or $\delta = 2\epsilon \Rightarrow \delta = 2\epsilon$. Choose $\delta = \frac{\epsilon}{2}$.

49. By the figure, $-x \le x \sin \frac{1}{x} \le x$ for all $x > 0$ and $-x \ge x \sin \frac{1}{x} \ge x$ for $x < 0$. Since $\lim_{x \to 0} (-x) = \lim_{x \to 0} x = 0$, then by the sandwich theorem, in either case, $\lim_{x \to 0} x \sin \frac{1}{x} = 0$.

50. By the figure, $-x^2 \le x^2 \sin \frac{1}{x} \le x^2$ for all x except possibly at $x = 0$. Since $\lim_{x \to 0} (-x^2) = \lim_{x \to 0} x^2 = 0$, then by the sandwich theorem, $\lim_{x \to 0} x^2 \sin \frac{1}{x} = 0$.

51. As x approaches the value 0, the values of g(x) approach k. Thus for every number $\epsilon > 0$, there exists a $\delta > 0$ such that $0 < |x - 0| < \delta \Rightarrow |g(x) - k| < \epsilon$.

52. Write $x = h + c$. Then $0 < |x - c| < \delta \Leftrightarrow -\delta < x - c < \delta, x \ne c \Leftrightarrow -\delta < (h + c) - c < \delta, h + c \ne c$
$\Leftrightarrow -\delta < h < \delta, h \ne 0 \Leftrightarrow 0 < |h - 0| < \delta$.
Thus, $\lim_{x \to c} f(x) = L \Leftrightarrow$ for any $\epsilon > 0$, there exists $\delta > 0$ such that $|f(x) - L| < \epsilon$ whenever $0 < |x - c| < \delta$
$\Leftrightarrow |f(h + c) - L| < \epsilon$ whenever $0 < |h - 0| < \delta \Leftrightarrow \lim_{h \to 0} f(h + c) = L$.

53. Let $f(x) = x^2$. The function values do get closer to -1 as x approaches 0, but $\lim_{x \to 0} f(x) = 0$, not -1. The function $f(x) = x^2$ never gets arbitrarily close to -1 for x near 0.

54. Let $f(x) = \sin x$, $L = \frac{1}{2}$, and $x_0 = 0$. There exists a value of x (namely, $x = \frac{\pi}{6}$) for which $\left|\sin x - \frac{1}{2}\right| < \epsilon$ for any given $\epsilon > 0$. However, $\lim_{x \to 0} \sin x = 0$, not $\frac{1}{2}$. The wrong statement does not require x to be arbitrarily close to x_0. As another example, let $g(x) = \sin \frac{1}{x}$, $L = \frac{1}{2}$, and $x_0 = 0$. We can choose infinitely many values of x near 0 such that $\sin \frac{1}{x} = \frac{1}{2}$ as you can see from the accompanying figure. However, $\lim_{x \to 0} \sin \frac{1}{x}$ fails to exist. The wrong statement does not require all values of x arbitrarily close to $x_0 = 0$ to lie within $\epsilon > 0$ of $L = \frac{1}{2}$. Again you can see from the figure that there are also infinitely many values of x near 0 such that $\sin \frac{1}{x} = 0$. If we choose $\epsilon < \frac{1}{4}$ we cannot satisfy the inequality $\left|\sin \frac{1}{x} - \frac{1}{2}\right| < \epsilon$ for all values of x sufficiently near $x_0 = 0$.

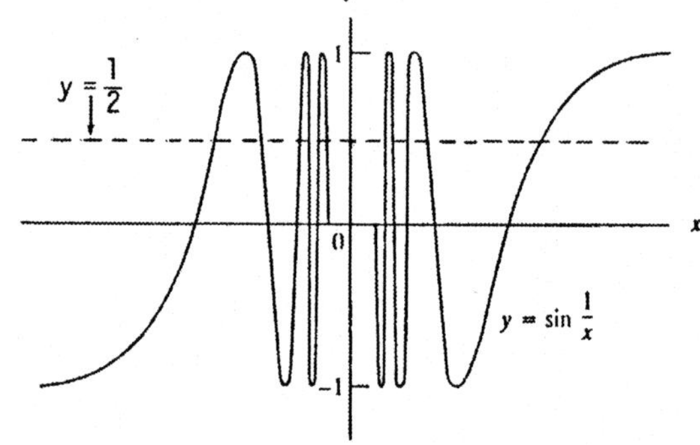

55. $|A - 9| \le 0.01 \Rightarrow -0.01 \le \pi \left(\frac{x}{2}\right)^2 - 9 \le 0.01 \Rightarrow 8.99 \le \frac{\pi x^2}{4} \le 9.01 \Rightarrow \frac{4}{\pi}(8.99) \le x^2 \le \frac{4}{\pi}(9.01)$
$\Rightarrow 2\sqrt{\frac{8.99}{\pi}} \le x \le 2\sqrt{\frac{9.01}{\pi}}$ or $3.384 \le x \le 3.387$. To be safe, the left endpoint was rounded up and the right endpoint was rounded down.

56. $V = RI \Rightarrow \frac{V}{R} = I \Rightarrow \left|\frac{V}{R} - 5\right| \le 0.1 \Rightarrow -0.1 \le \frac{120}{R} - 5 \le 0.1 \Rightarrow 4.9 \le \frac{120}{R} \le 5.1 \Rightarrow \frac{10}{49} \ge \frac{R}{120} \ge \frac{10}{51} \Rightarrow$
$\frac{(120)(10)}{51} \le R \le \frac{(120)(10)}{49} \Rightarrow 23.53 \le R \le 24.48$.

To be safe, the left endpoint was rounded up and the right endpoint was rounded down.

57. (a) $-\delta < x - 1 < 0 \Rightarrow 1 - \delta < x < 1 \Rightarrow f(x) = x$. Then $|f(x) - 2| = |x - 2| = 2 - x > 2 - 1 = 1$. That is, $|f(x) - 2| \ge 1 \ge \frac{1}{2}$ no matter how small δ is taken when $1 - \delta < x < 1 \Rightarrow \lim_{x \to 1} f(x) \ne 2$.

(b) $0 < x - 1 < \delta \Rightarrow 1 < x < 1 + \delta \Rightarrow f(x) = x + 1$. Then $|f(x) - 1| = |(x + 1) - 1| = |x| = x > 1$. That is, $|f(x) - 1| \ge 1$ no matter how small δ is taken when $1 < x < 1 + \delta \Rightarrow \lim_{x \to 1} f(x) \ne 1$.

(c) $-\delta < x - 1 < 0 \Rightarrow 1 - \delta < x < 1 \Rightarrow f(x) = x$. Then $|f(x) - 1.5| = |x - 1.5| = 1.5 - x > 1.5 - 1 = 0.5$. Also, $0 < x - 1 < \delta \Rightarrow 1 < x < 1 + \delta \Rightarrow f(x) = x + 1$. Then $|f(x) - 1.5| = |(x + 1) - 1.5| = |x - 0.5| = x - 0.5 > 1 - 0.5 = 0.5$. Thus, no matter how small δ is taken, there exists a value of x such that $-\delta < x - 1 < \delta$ but $|f(x) - 1.5| \ge \frac{1}{2} \Rightarrow \lim_{x \to 1} f(x) \ne 1.5$.

58. (a) For $2 < x < 2 + \delta \Rightarrow h(x) = 2 \Rightarrow |h(x) - 4| = 2$. Thus for $\epsilon < 2$, $|h(x) - 4| \ge \epsilon$ whenever $2 < x < 2 + \delta$ no matter how small we choose $\delta > 0 \Rightarrow \lim_{x \to 2} h(x) \ne 4$.

(b) For $2 < x < 2 + \delta \Rightarrow h(x) = 2 \Rightarrow |h(x) - 3| = 1$. Thus for $\epsilon < 1$, $|h(x) - 3| \ge \epsilon$ whenever $2 < x < 2 + \delta$ no matter how small we choose $\delta > 0 \Rightarrow \lim_{x \to 2} h(x) \ne 3$.

(c) For $2 - \delta < x < 2 \Rightarrow h(x) = x^2$ so $|h(x) - 2| = |x^2 - 2|$. No matter how small $\delta > 0$ is chosen, x^2 is close to 4 when x is near 2 and to the left on the real line $\Rightarrow |x^2 - 2|$ will be close to 2. Thus if $\epsilon < 1$, $|h(x) - 2| \ge \epsilon$ whenever $2 - \delta < x < 2$ no mater how small we choose $\delta > 0 \Rightarrow \lim_{x \to 2} h(x) \ne 2$.

59. (a) For $3 - \delta < x < 3 \Rightarrow f(x) > 4.8 \Rightarrow |f(x) - 4| \ge 0.8$. Thus for $\epsilon < 0.8$, $|f(x) - 4| \ge \epsilon$ whenever $3 - \delta < x < 3$ no matter how small we choose $\delta > 0 \Rightarrow \lim_{x \to 3} f(x) \ne 4$.

(b) For $3 < x < 3 + \delta \Rightarrow f(x) < 3 \Rightarrow |f(x) - 4.8| \ge 1.8$. Thus for $\epsilon < 1.8$, $|f(x) - 4.8| \ge \epsilon$ whenever $3 < x < 3 + \delta$ no matter how small we choose $\delta > 0 \Rightarrow \lim_{x \to 3} f(x) \ne 4.8$.

(c) For $3 - \delta < x < 3 \Rightarrow f(x) > 4.8 \Rightarrow |f(x) - 3| \ge 1.8$. Again, for $\epsilon < 1.8$, $|f(x) - 3| \ge \epsilon$ whenever $3 - \delta < x < 3$ no matter how small we choose $\delta > 0 \Rightarrow \lim_{x \to 3} f(x) \ne 3$.

60. (a) No matter how small we choose $\delta > 0$, for x near -1 satisfying $-1 - \delta < x < -1 + \delta$, the values of g(x) are near $1 \Rightarrow |g(x) - 2|$ is near 1. Then, for $\epsilon = \frac{1}{2}$ we have $|g(x) - 2| \ge \frac{1}{2}$ for some x satisfying $-1 - \delta < x < -1 + \delta$, or $0 < |x + 1| < \delta \Rightarrow \lim_{x \to -1} g(x) \ne 2$.

(b) Yes, $\lim_{x \to -1} g(x) = 1$ because from the graph we can find a $\delta > 0$ such that $|g(x) - 1| < \epsilon$ if $0 < |x - (-1)| < \delta$.

61-66. Example CAS commands (values of del may vary for a specified eps):

<u>Maple</u>:

```
f := x -> (x^4-81)/(x-3);x0 := 3;
plot( f(x), x=x0-1..x0+1, color=black,                # (a)
      title="Section 2.3, #61(a)" );
L := limit( f(x), x=x0 );                             # (b)
epsilon := 0.2;                                       # (c)
plot( [f(x),L-epsilon,L+epsilon], x=x0-0.01..x0+0.01,
      color=black, linestyle=[1,3,3], title="Section 2.3, #61(c)" );
q := fsolve( abs( f(x)-L ) = epsilon, x=x0-1..x0+1 );  # (d)
delta := abs(x0-q);
plot( [f(x),L-epsilon,L+epsilon], x=x0-delta..x0+delta, color=black, title="Section 2.3, #61(d)" );
for eps in [0.1, 0.005, 0.001 ] do                    # (e)
 q := fsolve( abs( f(x)-L ) = eps, x=x0-1..x0+1 );
```

```
    delta := abs(x0-q);
    head := sprintf("Section 2.3, #61(e)\n epsilon = %5f, delta = %5f\n", eps, delta );
    print(plot( [f(x),L-eps,L+eps], x=x0-delta..x0+delta,
              color=black, linestyle=[1,3,3], title=head ));
  end do:
```

Mathematica (assigned function and values for x0, eps and del may vary):

```
Clear[f, x]
y1: = L − eps; y2: = L + eps; x0 = 1;
f[x_]: = (3x^2 − (7x + 1)Sqrt[x] + 5)/(x − 1)
Plot[f[x], {x, x0 − 0.2, x0 + 0.2}]
L: = Limit[f[x], x → x0]
eps = 0.1; del = 0.2;
Plot[{f[x], y1, y2},{x, x0 − del, x0 + del}, PlotRange → {L − 2eps, L + 2eps}]
```

2.4 ONE-SIDED LIMITS AND LIMITS AT INFINITY

1. (a) True (b) True (c) False (d) True
   (e) True (f) True (g) False (h) False
   (i) False (j) False (k) True (l) False

2. (a) True (b) False (c) False (d) True
   (e) True (f) True (g) True (h) True
   (i) True (j) False (k) True

3. (a) $\lim\limits_{x \to 2^+} f(x) = \frac{2}{2} + 1 = 2,\ \lim\limits_{x \to 2^-} f(x) = 3 - 2 = 1$
   (b) No, $\lim\limits_{x \to 2} f(x)$ does not exist because $\lim\limits_{x \to 2^+} f(x) \neq \lim\limits_{x \to 2^-} f(x)$
   (c) $\lim\limits_{x \to 4^-} f(x) = \frac{4}{2} + 1 = 3,\ \lim\limits_{x \to 4^+} f(x) = \frac{4}{2} + 1 = 3$
   (d) Yes, $\lim\limits_{x \to 4} f(x) = 3$ because $3 = \lim\limits_{x \to 4^-} f(x) = \lim\limits_{x \to 4^+} f(x)$

4. (a) $\lim\limits_{x \to 2^+} f(x) = \frac{2}{2} = 1,\ \lim\limits_{x \to 2^-} f(x) = 3 - 2 = 1,\ f(2) = 2$
   (b) Yes, $\lim\limits_{x \to 2} f(x) = 1$ because $1 = \lim\limits_{x \to 2^+} f(x) = \lim\limits_{x \to 2^-} f(x)$
   (c) $\lim\limits_{x \to -1^-} f(x) = 3 - (-1) = 4,\ \lim\limits_{x \to -1^+} f(x) = 3 - (-1) = 4$
   (d) Yes, $\lim\limits_{x \to -1} f(x) = 4$ because $4 = \lim\limits_{x \to -1^-} f(x) = \lim\limits_{x \to -1^+} f(x)$

5. (a) No, $\lim\limits_{x \to 0^+} f(x)$ does not exist since $\sin\left(\frac{1}{x}\right)$ does not approach any single value as x approaches 0
   (b) $\lim\limits_{x \to 0^-} f(x) = \lim\limits_{x \to 0^-} 0 = 0$
   (c) $\lim\limits_{x \to 0} f(x)$ does not exist because $\lim\limits_{x \to 0^+} f(x)$ does not exist

6. (a) Yes, $\lim\limits_{x \to 0^+} g(x) = 0$ by the sandwich theorem since $-\sqrt{x} \le g(x) \le \sqrt{x}$ when $x > 0$
   (b) No, $\lim\limits_{x \to 0^-} g(x)$ does not exist since $\sqrt{x}$ is not defined for $x < 0$
   (c) No, $\lim\limits_{x \to 0} g(x)$ does not exist since $\lim\limits_{x \to 0^-} g(x)$ does not exist

7. (a)

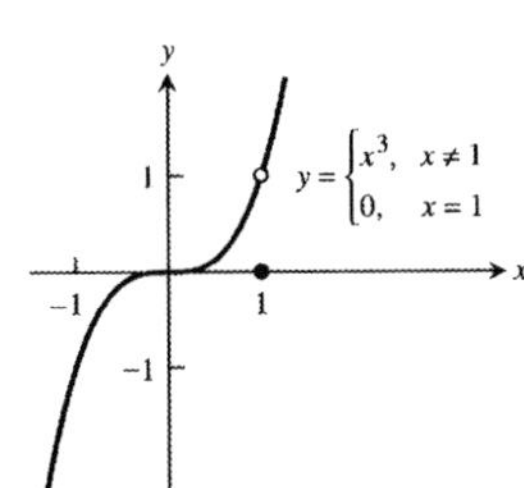

(b) $\lim_{x \to 1^-} f(x) = 1 = \lim_{x \to 1^+} f(x)$

(c) Yes, $\lim_{x \to 1} f(x) = 1$ since the right-hand and left-hand limits exist and equal 1

8. (a)

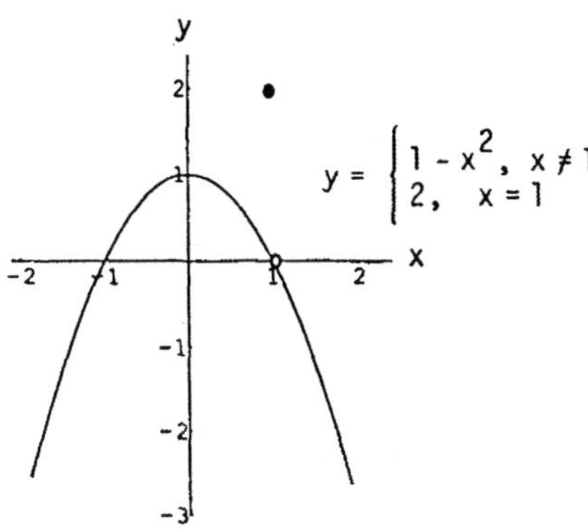

(b) $\lim_{x \to 1^+} f(x) = 0 = \lim_{x \to 1^-} f(x)$

(c) Yes, $\lim_{x \to 1} f(x) = 0$ since the right-hand and left-hand limits exist and equal 0

9. (a) domain: $0 \le x \le 2$
range: $0 < y \le 1$ and $y = 2$

(b) $\lim_{x \to c} f(x)$ exists for c belonging to $(0, 1) \cup (1, 2)$

(c) $x = 2$

(d) $x = 0$

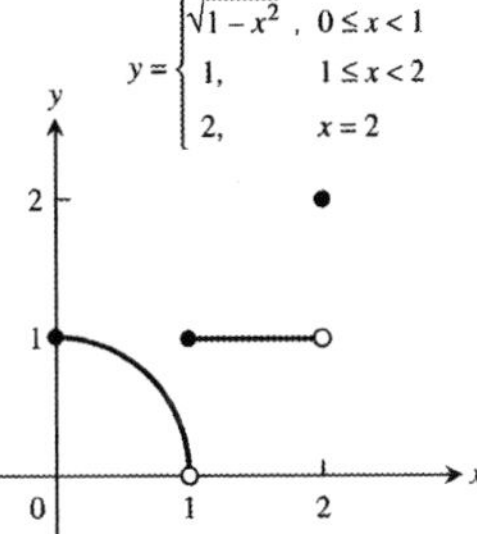

10. (a) domain: $-\infty < x < \infty$
range: $-1 \le y \le 1$

(b) $\lim_{x \to c} f(x)$ exists for c belonging to $(-\infty, -1) \cup (-1, 1) \cup (1, \infty)$

(c) none

(d) none

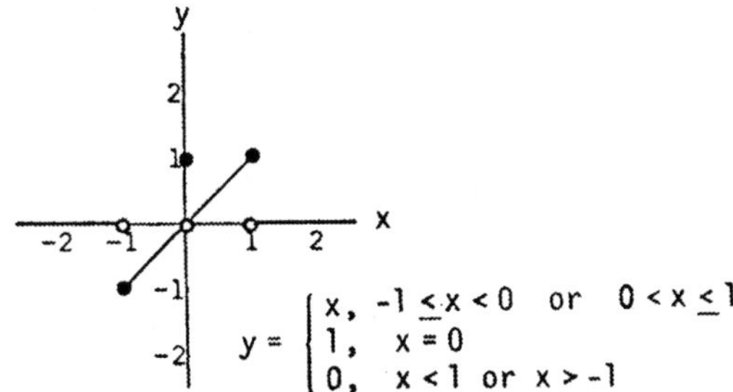

11. $\lim_{x \to -0.5^-} \sqrt{\frac{x+2}{x-1}} = \sqrt{\frac{-0.5+2}{-0.5+1}} = \sqrt{\frac{3/2}{1/2}} = \sqrt{3}$

12. $\lim_{x \to 1^+} \sqrt{\frac{x-1}{x+2}} = \sqrt{\frac{1-1}{1+2}} = \sqrt{0} = 0$

13. $\lim_{x \to -2^+} \left(\frac{x}{x+1}\right)\left(\frac{2x+5}{x^2+x}\right) = \left(\frac{-2}{-2+1}\right)\left(\frac{2(-2)+5}{(-2)^2+(-2)}\right) = (2)\left(\frac{1}{2}\right) = 1$

14. $\lim_{x \to 1^-} \left(\frac{1}{x+1}\right)\left(\frac{x+6}{x}\right)\left(\frac{3-x}{7}\right) = \left(\frac{1}{1+1}\right)\left(\frac{1+6}{1}\right)\left(\frac{3-1}{7}\right) = \left(\frac{1}{2}\right)\left(\frac{7}{1}\right)\left(\frac{2}{7}\right) = 1$

15. $\lim_{h \to 0^+} \frac{\sqrt{h^2+4h+5}-\sqrt{5}}{h} = \lim_{h \to 0^+} \left(\frac{\sqrt{h^2+4h+5}-\sqrt{5}}{h}\right)\left(\frac{\sqrt{h^2+4h+5}+\sqrt{5}}{\sqrt{h^2+4h+5}+\sqrt{5}}\right)$

$= \lim_{h \to 0^+} \frac{(h^2+4h+5)-5}{h\left(\sqrt{h^2+4h+5}+\sqrt{5}\right)} = \lim_{h \to 0^+} \frac{h(h+4)}{h\left(\sqrt{h^2+4h+5}+\sqrt{5}\right)} = \frac{0+4}{\sqrt{5}+\sqrt{5}} = \frac{2}{\sqrt{5}}$

16. $\lim_{h \to 0^-} \frac{\sqrt{6}-\sqrt{5h^2+11h+6}}{h} = \lim_{h \to 0^-} \left(\frac{\sqrt{6}-\sqrt{5h^2+11h+6}}{h}\right)\left(\frac{\sqrt{6}+\sqrt{5h^2+11h+6}}{\sqrt{6}+\sqrt{5h^2+11h+6}}\right)$
$= \lim_{h \to 0^-} \frac{6-(5h^2+11h+6)}{h\left(\sqrt{6}+\sqrt{5h^2+11h+6}\right)} = \lim_{h \to 0^-} \frac{-h(5h+11)}{h\left(\sqrt{6}+\sqrt{5h^2+11h+6}\right)} = \frac{-(0+11)}{\sqrt{6}+\sqrt{6}} = -\frac{11}{2\sqrt{6}}$

17. (a) $\lim_{x \to -2^+} (x+3)\frac{|x+2|}{x+2} = \lim_{x \to -2^+} (x+3)\frac{(x+2)}{(x+2)}$ $\quad (|x+2| = x+2 \text{ for } x > -2)$
$= \lim_{x \to -2^+} (x+3) = (-2)+3 = 1$

(b) $\lim_{x \to -2^-} (x+3)\frac{|x+2|}{x+2} = \lim_{x \to -2^-} (x+3)\left[\frac{-(x+2)}{(x+2)}\right]$ $\quad (|x+2| = -(x+2) \text{ for } x < -2)$
$= \lim_{x \to -2^-} (x+3)(-1) = -(-2+3) = -1$

18. (a) $\lim_{x \to 1^+} \frac{\sqrt{2x}(x-1)}{|x-1|} = \lim_{x \to 1^+} \frac{\sqrt{2x}(x-1)}{(x-1)}$ $\quad (|x-1| = x-1 \text{ for } x > 1)$
$= \lim_{x \to 1^+} \sqrt{2x} = \sqrt{2}$

(b) $\lim_{x \to 1^-} \frac{\sqrt{2x}(x-1)}{|x-1|} = \lim_{x \to 1^-} \frac{\sqrt{2x}(x-1)}{-(x-1)}$ $\quad (|x-1| = -(x-1) \text{ for } x < 1)$
$= \lim_{x \to 1^-} -\sqrt{2x} = -\sqrt{2}$

19. (a) $\lim_{\theta \to 3^+} \frac{\lfloor\theta\rfloor}{\theta} = \frac{3}{3} = 1$ (b) $\lim_{\theta \to 3^-} \frac{\lfloor\theta\rfloor}{\theta} = \frac{2}{3}$

20. (a) $\lim_{t \to 4^+} (t - \lfloor t \rfloor) = 4 - 4 = 0$ (b) $\lim_{t \to 4^-} (t - \lfloor t \rfloor) = 4 - 3 = 1$

21. $\lim_{\theta \to 0} \frac{\sin\sqrt{2}\theta}{\sqrt{2}\theta} = \lim_{x \to 0} \frac{\sin x}{x} = 1$ $\quad$ (where $x = \sqrt{2}\theta$)

22. $\lim_{t \to 0} \frac{\sin kt}{t} = \lim_{t \to 0} \frac{k \sin kt}{kt} = \lim_{\theta \to 0} \frac{k \sin\theta}{\theta} = k \lim_{\theta \to 0} \frac{\sin\theta}{\theta} = k \cdot 1 = k$ $\quad$ (where $\theta = kt$)

23. $\lim_{y \to 0} \frac{\sin 3y}{4y} = \frac{1}{4} \lim_{y \to 0} \frac{3 \sin 3y}{3y} = \frac{3}{4} \lim_{y \to 0} \frac{\sin 3y}{3y} = \frac{3}{4} \lim_{\theta \to 0} \frac{\sin\theta}{\theta} = \frac{3}{4}$ $\quad$ (where $\theta = 3y$)

24. $\lim_{h \to 0^-} \frac{h}{\sin 3h} = \lim_{h \to 0^-} \left(\frac{1}{3} \cdot \frac{3h}{\sin 3h}\right) = \frac{1}{3} \lim_{h \to 0^-} \frac{1}{\left(\frac{\sin 3h}{3h}\right)} = \frac{1}{3}\left(\frac{1}{\lim_{\theta \to 0^-} \frac{\sin\theta}{\theta}}\right) = \frac{1}{3} \cdot 1 = \frac{1}{3}$ $\quad$ (where $\theta = 3h$)

25. $\lim_{x \to 0} \frac{\tan 2x}{x} = \lim_{x \to 0} \frac{\left(\frac{\sin 2x}{\cos 2x}\right)}{x} = \lim_{x \to 0} \frac{\sin 2x}{x \cos 2x} = \left(\lim_{x \to 0} \frac{1}{\cos 2x}\right)\left(\lim_{x \to 0} \frac{2 \sin 2x}{2x}\right) = 1 \cdot 2 = 2$

26. $\lim_{t \to 0} \frac{2t}{\tan t} = 2 \lim_{t \to 0} \frac{t}{\left(\frac{\sin t}{\cos t}\right)} = 2 \lim_{t \to 0} \frac{t \cos t}{\sin t} = 2\left(\lim_{t \to 0} \cos t\right)\left(\frac{1}{\lim_{t \to 0} \frac{\sin t}{t}}\right) = 2 \cdot 1 \cdot 1 = 2$

27. $\lim_{x \to 0} \frac{x \csc 2x}{\cos 5x} = \lim_{x \to 0} \left(\frac{x}{\sin 2x} \cdot \frac{1}{\cos 5x}\right) = \left(\frac{1}{2} \lim_{x \to 0} \frac{2x}{\sin 2x}\right)\left(\lim_{x \to 0} \frac{1}{\cos 5x}\right) = \left(\frac{1}{2} \cdot 1\right)(1) = \frac{1}{2}$

28. $\lim_{x \to 0} 6x^2(\cot x)(\csc 2x) = \lim_{x \to 0} \frac{6x^2 \cos x}{\sin x \sin 2x} = \lim_{x \to 0} \left(3 \cos x \cdot \frac{x}{\sin x} \cdot \frac{2x}{\sin 2x}\right) = 3 \cdot 1 \cdot 1 = 3$

29. $\lim_{x \to 0} \frac{x + x\cos x}{\sin x \cos x} = \lim_{x \to 0} \left(\frac{x}{\sin x \cos x} + \frac{x \cos x}{\sin x \cos x}\right) = \lim_{x \to 0} \left(\frac{x}{\sin x} \cdot \frac{1}{\cos x}\right) + \lim_{x \to 0} \frac{x}{\sin x}$
$= \lim_{x \to 0} \left(\frac{1}{\frac{\sin x}{x}}\right) \cdot \lim_{x \to 0} \left(\frac{1}{\cos x}\right) + \lim_{x \to 0} \left(\frac{1}{\frac{\sin x}{x}}\right) = (1)(1) + 1 = 2$

30. $\lim_{x \to 0} \frac{x^2 - x + \sin x}{2x} = \lim_{x \to 0} \left(\frac{x}{2} - \frac{1}{2} + \frac{1}{2}\left(\frac{\sin x}{x}\right)\right) = 0 - \frac{1}{2} + \frac{1}{2}(1) = 0$

31. $\lim\limits_{t \to 0} \frac{\sin(1-\cos t)}{1-\cos t} = \lim\limits_{\theta \to 0} \frac{\sin\theta}{\theta} = 1$ since $\theta = 1 - \cos t \to 0$ as $t \to 0$

32. $\lim\limits_{h \to 0} \frac{\sin(\sin h)}{\sin h} = \lim\limits_{\theta \to 0} \frac{\sin\theta}{\theta} = 1$ since $\theta = \sin h \to 0$ as $h \to 0$

33. $\lim\limits_{\theta \to 0} \frac{\sin\theta}{\sin 2\theta} = \lim\limits_{\theta \to 0} \left(\frac{\sin\theta}{\sin 2\theta} \cdot \frac{2\theta}{2\theta}\right) = \frac{1}{2} \lim\limits_{\theta \to 0} \left(\frac{\sin\theta}{\theta} \cdot \frac{2\theta}{\sin 2\theta}\right) = \frac{1}{2} \cdot 1 \cdot 1 = \frac{1}{2}$

34. $\lim\limits_{x \to 0} \frac{\sin 5x}{\sin 4x} = \lim\limits_{x \to 0} \left(\frac{\sin 5x}{\sin 4x} \cdot \frac{4x}{5x} \cdot \frac{5}{4}\right) = \frac{5}{4} \lim\limits_{x \to 0} \left(\frac{\sin 5x}{5x} \cdot \frac{4x}{\sin 4x}\right) = \frac{5}{4} \cdot 1 \cdot 1 = \frac{5}{4}$

35. $\lim\limits_{x \to 0} \frac{\tan 3x}{\sin 8x} = \lim\limits_{x \to 0} \left(\frac{\sin 3x}{\cos 3x} \cdot \frac{1}{\sin 8x}\right) = \lim\limits_{x \to 0} \left(\frac{\sin 3x}{\cos 3x} \cdot \frac{1}{\sin 8x} \cdot \frac{8x}{3x} \cdot \frac{3}{8}\right)$
$= \frac{3}{8} \lim\limits_{x \to 0} \left(\frac{1}{\cos 3x}\right)\left(\frac{\sin 3x}{3x}\right)\left(\frac{8x}{\sin 8x}\right) = \frac{3}{8} \cdot 1 \cdot 1 \cdot 1 = \frac{3}{8}$

36. $\lim\limits_{y \to 0} \frac{\sin 3y \cot 5y}{y \cot 4y} = \lim\limits_{y \to 0} \frac{\sin 3y \sin 4y \cos 5y}{y \cos 4y \sin 5y} = \lim\limits_{y \to 0} \left(\frac{\sin 3y}{y}\right)\left(\frac{\sin 4y}{\cos 4y}\right)\left(\frac{\cos 5y}{\sin 5y}\right)\left(\frac{3 \cdot 4 \cdot 5y}{3 \cdot 4 \cdot 5y}\right)$
$= \lim\limits_{y \to 0} \left(\frac{\sin 3y}{3y}\right)\left(\frac{\sin 4y}{4y}\right)\left(\frac{5y}{\sin 5y}\right)\left(\frac{\cos 5y}{\cos 4y}\right)\left(\frac{3 \cdot 4}{5}\right) = 1 \cdot 1 \cdot 1 \cdot 1 \cdot \frac{12}{5} = \frac{12}{5}$

Note: In these exercises we use the result $\lim\limits_{x \to \pm\infty} \frac{1}{x^{m/n}} = 0$ whenever $\frac{m}{n} > 0$. This result follows immediately from Example 6 and the power rule in Theorem 8: $\lim\limits_{x \to \pm\infty} \left(\frac{1}{x^{m/n}}\right) = \lim\limits_{x \to \pm\infty} \left(\frac{1}{x}\right)^{m/n} = \left(\lim\limits_{x \to \pm\infty} \frac{1}{x}\right)^{m/n} = 0^{m/n} = 0.$

37. (a) -3 (b) -3

38. (a) π (b) π

39. (a) $\frac{1}{2}$ (b) $\frac{1}{2}$

40. (a) $\frac{1}{8}$ (b) $\frac{1}{8}$

41. (a) $-\frac{5}{3}$ (b) $-\frac{5}{3}$

42. (a) $\frac{3}{4}$ (b) $\frac{3}{4}$

43. $-\frac{1}{x} \le \frac{\sin 2x}{x} \le \frac{1}{x} \Rightarrow \lim\limits_{x \to \infty} \frac{\sin 2x}{x} = 0$ by the Sandwich Theorem

44. $-\frac{1}{3\theta} \le \frac{\cos\theta}{3\theta} \le \frac{1}{3\theta} \Rightarrow \lim\limits_{\theta \to -\infty} \frac{\cos\theta}{3\theta} = 0$ by the Sandwich Theorem

45. $\lim\limits_{t \to \infty} \frac{2-t+\sin t}{t+\cos t} = \lim\limits_{t \to \infty} \frac{\frac{2}{t} - 1 + \left(\frac{\sin t}{t}\right)}{1 + \left(\frac{\cos t}{t}\right)} = \frac{0-1+0}{1+0} = -1$

46. $\lim\limits_{r \to \infty} \frac{r+\sin r}{2r+7-5\sin r} = \lim\limits_{r \to \infty} \frac{1+\left(\frac{\sin r}{r}\right)}{2+\frac{7}{r}-5\left(\frac{\sin r}{r}\right)} = \lim\limits_{r \to \infty} \frac{1+0}{2+0-0} = \frac{1}{2}$

47. $-\frac{1}{e^x} \le \frac{\sin x}{e^x} \le \frac{1}{e^x} \Rightarrow \lim\limits_{x \to \infty} e^{-x}\sin x = 0$ by the Sandwich Theorem.

48. $\lim\limits_{x \to -\infty} (e^x)\left(\cos^{-1}\left(\frac{1}{x}\right)\right) = 0 \cdot \cos^{-1} 0 = 0 \cdot \frac{\pi}{2} = 0$

49. $\lim\limits_{x \to -\infty} \frac{e^x - e^{-x}}{e^x + e^{-x}} = \lim\limits_{x \to -\infty} \frac{e^x - \frac{1}{e^x}}{e^x + \frac{1}{e^x}} = \lim\limits_{x \to -\infty} \frac{\frac{e^{2x}-1}{e^x}}{\frac{e^{2x}+1}{e^x}} = \lim\limits_{x \to -\infty} \frac{e^{2x}-1}{e^{2x}+1} = \frac{0-1}{0+1} = -1$

50. $\lim_{x \to \infty} \frac{3x^2 + e^{-x}}{\sin(1/x) - 2x^2} = \lim_{x \to \infty} \frac{3 + \frac{1}{x^2 e^x}}{\frac{\sin(1/x)}{x^2} - 2} = \frac{3+0}{0-2} = -\frac{3}{2}$

51. (a) $\lim_{x \to \infty} \frac{2x+3}{5x+7} = \lim_{x \to \infty} \frac{2 + \frac{3}{x}}{5 + \frac{7}{x}} = \frac{2}{5}$ (b) $\frac{2}{5}$ (same process as part (a))

52. (a) $\lim_{x \to \infty} \frac{2x^3+7}{x^3 - x^2 + x + 7} = \lim_{x \to \infty} \frac{2 + \left(\frac{7}{x^3}\right)}{1 - \frac{1}{x} + \frac{1}{x^2} + \frac{7}{x^3}} = 2$
(b) 2 (same process as part (a))

53. (a) $\lim_{x \to \infty} \frac{x+1}{x^2+3} = \lim_{x \to \infty} \frac{\frac{1}{x} + \frac{1}{x^2}}{1 + \frac{3}{x^2}} = 0$ (b) 0 (same process as part (a))

54. (a) $\lim_{x \to \infty} \frac{3x+7}{x^2-2} = \lim_{x \to \infty} \frac{\frac{3}{x} + \frac{7}{x^2}}{1 - \frac{2}{x^2}} = 0$ (b) 0 (same process as part (a))

55. (a) $\lim_{x \to \infty} \frac{7x^3}{x^3 - 3x^2 + 6x} = \lim_{x \to \infty} \frac{7}{1 - \frac{3}{x} + \frac{6}{x^2}} = 7$ (b) 7 (same process as part (a))

56. (a) $\lim_{x \to \infty} \frac{1}{x^3 - 4x + 1} = \lim_{x \to \infty} \frac{\frac{1}{x^3}}{1 - \frac{4}{x^2} + \frac{1}{x^3}} = 0$ (b) 0 (same process as part (a))

57. (a) $\lim_{x \to \infty} \frac{10x^5 + x^4 + 31}{x^6} = \lim_{x \to \infty} \frac{\frac{10}{x} + \frac{1}{x^2} + \frac{31}{x^6}}{1} = 0$
(b) 0 (same process as part (a))

58. (a) $\lim_{x \to \infty} \frac{9x^4 + x}{2x^4 + 5x^2 - x + 6} = \lim_{x \to \infty} \frac{9 + \frac{1}{x^3}}{2 + \frac{5}{x^2} - \frac{1}{x^3} + \frac{6}{x^4}} = \frac{9}{2}$
(b) $\frac{9}{2}$ (same process as part (a))

59. (a) $\lim_{x \to \infty} \frac{-2x^3 - 2x + 3}{3x^3 + 3x^2 - 5x} = \lim_{x \to \infty} \frac{-2 - \frac{2}{x^2} + \frac{3}{x^3}}{3 + \frac{3}{x} - \frac{5}{x^2}} = -\frac{2}{3}$
(b) $-\frac{2}{3}$ (same process as part (a))

60. (a) $\lim_{x \to \infty} \frac{-x^4}{x^4 - 7x^3 + 7x^2 + 9} = \lim_{x \to \infty} \frac{-1}{1 - \frac{7}{x} + \frac{7}{x^2} + \frac{9}{x^4}} = -1$
(b) -1 (same process as part (a))

61. $\lim_{x \to \infty} \frac{2\sqrt{x} + x^{-1}}{3x - 7} = \lim_{x \to \infty} \frac{\left(\frac{2}{x^{1/2}}\right) + \left(\frac{1}{x^2}\right)}{3 - \frac{7}{x}} = 0$

62. $\lim_{x \to \infty} \frac{2 + \sqrt{x}}{2 - \sqrt{x}} = \lim_{x \to \infty} \frac{\left(\frac{2}{x^{1/2}}\right) + 1}{\left(\frac{2}{x^{1/2}}\right) - 1} = -1$

63. $\lim_{x \to -\infty} \frac{\sqrt[3]{x} - \sqrt[5]{x}}{\sqrt[3]{x} + \sqrt[5]{x}} = \lim_{x \to -\infty} \frac{1 - x^{(1/5)-(1/3)}}{1 + x^{(1/5)-(1/3)}} = \lim_{x \to -\infty} \frac{1 - \left(\frac{1}{x^{2/15}}\right)}{1 + \left(\frac{1}{x^{2/15}}\right)} = 1$

64. $\lim_{x \to \infty} \frac{x^{-1} + x^{-4}}{x^{-2} - x^{-3}} = \lim_{x \to \infty} \frac{x + \frac{1}{x^2}}{1 - \frac{1}{x}} = \infty$

65. $\lim_{x \to \infty} \frac{2x^{5/3} - x^{1/3} + 7}{x^{8/5} + 3x + \sqrt{x}} = \lim_{x \to \infty} \frac{2x^{1/15} - \frac{1}{x^{19/15}} + \frac{7}{x^{8/5}}}{1 + \frac{3}{x^{3/5}} + \frac{1}{x^{11/10}}} = \infty$

66. $\lim_{x \to -\infty} \frac{\sqrt[3]{x} - 5x + 3}{2x + x^{2/3} - 4} = \lim_{x \to -\infty} \frac{\frac{1}{x^{2/3}} - 5 + \frac{3}{x}}{2 + \frac{1}{x^{1/3}} - \frac{4}{x}} = -\frac{5}{2}$

67. Yes. If $\lim_{x \to a^+} f(x) = L = \lim_{x \to a^-} f(x)$, then $\lim_{x \to a} f(x) = L$. If $\lim_{x \to a^+} f(x) \neq \lim_{x \to a^-} f(x)$, then $\lim_{x \to a} f(x)$ does not exist.

68. Since $\lim_{x \to c} f(x) = L$ if and only if $\lim_{x \to c^+} f(x) = L$ and $\lim_{x \to c^-} f(x) = L$, then $\lim_{x \to c} f(x)$ can be found by calculating $\lim_{x \to c^+} f(x)$.

69. If f is an odd function of x, then $f(-x) = -f(x)$. Given $\lim_{x \to 0^+} f(x) = 3$, then $\lim_{x \to 0^-} f(x) = -3$.

70. If f is an even function of x, then $f(-x) = f(x)$. Given $\lim_{x \to 2^-} f(x) = 7$ then $\lim_{x \to -2^+} f(x) = 7$. However, nothing can be said about $\lim_{x \to -2^-} f(x)$ because we don't know $\lim_{x \to 2^+} f(x)$.

71. Yes. If $\lim_{x \to \infty} \frac{f(x)}{g(x)} = 2$ then the ratio of the polynomials' leading coefficients is 2, so $\lim_{x \to -\infty} \frac{f(x)}{g(x)} = 2$ as well.

72. Yes, it can have a horizontal or oblique asymptote.

73. At most 1 horizontal asymptote: If $\lim_{x \to \infty} \frac{f(x)}{g(x)} = L$, then the ratio of the polynomials' leading coefficients is L, so $\lim_{x \to -\infty} \frac{f(x)}{g(x)} = L$ as well.

74. $\lim_{x \to \infty} \sqrt{x^2+x} - \sqrt{x^2-x} = \lim_{x \to \infty} \left[\sqrt{x^2+x} - \sqrt{x^2-x}\right] \cdot \left[\frac{\sqrt{x^2+x}+\sqrt{x^2-x}}{\sqrt{x^2+x}+\sqrt{x^2-x}}\right] = \lim_{x \to \infty} \frac{(x^2+x)-(x^2-x)}{\sqrt{x^2+x}+\sqrt{x^2-x}}$
$= \lim_{x \to \infty} \frac{2x}{\sqrt{x^2+x}+\sqrt{x^2-x}} = \lim_{x \to \infty} \frac{2}{\sqrt{1+\frac{1}{x}}+\sqrt{1-\frac{1}{x}}} = \frac{2}{1+1} = 1$

75. For any $\epsilon > 0$, take $N = 1$. Then for all $x > N$ we have that $|f(x) - k| = |k - k| = 0 < \epsilon$.

76. For any $\epsilon > 0$, take $N = 1$. Then for all $y < -N$ we have that $|f(x) - k| = |k - k| = 0 < \epsilon$.

77. $I = (5, 5+\delta) \Rightarrow 5 < x < 5+\delta$. Also, $\sqrt{x-5} < \epsilon \Rightarrow x - 5 < \epsilon^2 \Rightarrow x < 5 + \epsilon^2$. Choose $\delta = \epsilon^2$
$\Rightarrow \lim_{x \to 5^+} \sqrt{x-5} = 0$.

78. $I = (4-\delta, 4) \Rightarrow 4 - \delta < x < 4$. Also, $\sqrt{4-x} < \epsilon \Rightarrow 4 - x < \epsilon^2 \Rightarrow x > 4 - \epsilon^2$. Choose $\delta = \epsilon^2$
$\Rightarrow \lim_{x \to 4^-} \sqrt{4-x} = 0$.

79. As $x \to 0^-$ the number x is always negative. Thus, $\left|\frac{x}{|x|} - (-1)\right| < \epsilon \Rightarrow \left|\frac{x}{-x} + 1\right| < \epsilon \Rightarrow 0 < \epsilon$ which is always true independent of the value of x. Hence we can choose any $\delta > 0$ with $-\delta < x < 0 \Rightarrow \lim_{x \to 0^-} \frac{x}{|x|} = -1$.

80. Since $x \to 2^+$ we have $x > 2$ and $|x - 2| = x - 2$. Then, $\left|\frac{x-2}{|x-2|} - 1\right| = \left|\frac{x-2}{x-2} - 1\right| < \epsilon \Rightarrow 0 < \epsilon$ which is always true so long as $x > 2$. Hence we can choose any $\delta > 0$, and thus $2 < x < 2 + \delta$
$\Rightarrow \left|\frac{x-2}{|x-2|} - 1\right| < \epsilon$. Thus, $\lim_{x \to -2^+} \frac{x-2}{|x-2|} = 1$.

81. (a) $\lim_{x \to 400^+} \lfloor x \rfloor = 400$. Just observe that if $400 < x < 401$, then $\lfloor x \rfloor = 400$. Thus if we choose $\delta = 1$, we have for any number $\epsilon > 0$ that $400 < x < 400 + \delta \Rightarrow |\lfloor x \rfloor - 400| = |400 - 400| = 0 < \epsilon$.
(b) $\lim_{x \to 400^-} \lfloor x \rfloor = 399$. Just observe that if $399 < x < 400$ then $\lfloor x \rfloor = 399$. Thus if we choose $\delta = 1$, we have for any number $\epsilon > 0$ that $400 - \delta < x < 400 \Rightarrow |\lfloor x \rfloor - 399| = |399 - 399| = 0 < \epsilon$.

(c) Since $\lim_{x \to 400^+} \lfloor x \rfloor \neq \lim_{x \to 400^-} \lfloor x \rfloor$ we conclude that $\lim_{x \to 400} \lfloor x \rfloor$ does not exist.

82. (a) $\lim_{x \to 0^+} f(x) = \lim_{x \to 0^+} \sqrt{x} = \sqrt{0} = 0$; $\left|\sqrt{x} - 0\right| < \epsilon \Rightarrow -\epsilon < \sqrt{x} < \epsilon \Rightarrow 0 < x < \epsilon^2$ for x positive. Choose $\delta = \epsilon^2$ $\Rightarrow \lim_{x \to 0^+} f(x) = 0$.

(b) $\lim_{x \to 0^-} f(x) = \lim_{x \to 0^-} x^2 \sin\left(\frac{1}{x}\right) = 0$ by the sandwich theorem since $-x^2 \le x^2 \sin\left(\frac{1}{x}\right) \le x^2$ for all $x \neq 0$. Since $|x^2 - 0| = |-x^2 - 0| = x^2 < \epsilon$ whenever $|x| < \sqrt{\epsilon}$, we choose $\delta = \sqrt{\epsilon}$ and obtain $\left|x^2 \sin\left(\frac{1}{x}\right) - 0\right| < \epsilon$ if $-\delta < x < 0$.

(c) The function f has limit 0 at $x_0 = 0$ since both the right-hand and left-hand limits exist and equal 0.

83. $\lim_{x \to \pm\infty} x \sin \frac{1}{x} = \lim_{\theta \to 0} \frac{1}{\theta} \sin \theta = 1, \left(\theta = \frac{1}{x}\right)$

84. $\lim_{x \to -\infty} \frac{\cos \frac{1}{x}}{1 + \frac{1}{x}} = \lim_{\theta \to 0^-} \frac{\cos \theta}{1 + \theta} = \frac{1}{1} = 1, \left(\theta = \frac{1}{x}\right)$

85. $\lim_{x \to \pm\infty} \frac{3x + 4}{2x - 5} = \lim_{x \to \pm\infty} \frac{3 + \frac{4}{x}}{2 - \frac{5}{x}} = \lim_{t \to 0} \frac{3 + 4t}{2 - 5t} = \frac{3}{2}, \left(t = \frac{1}{x}\right)$

86. $\lim_{x \to \infty} \left(\frac{1}{x}\right)^{1/x} = \lim_{z \to 0^+} z^z = 1, \left(z = \frac{1}{x}\right)$

87. $\lim_{x \to \pm\infty} \left(3 + \frac{2}{x}\right)\left(\cos \frac{1}{x}\right) = \lim_{\theta \to 0} (3 + 2\theta)(\cos \theta) = (3)(1) = 3, \left(\theta = \frac{1}{x}\right)$

88. $\lim_{x \to \infty} \left(\frac{3}{x^2} - \cos \frac{1}{x}\right)\left(1 + \sin \frac{1}{x}\right) = \lim_{\theta \to 0^+} (3\theta^2 - \cos \theta)(1 + \sin \theta) = (0 - 1)(1 + 0) = -1, \left(\theta = \frac{1}{x}\right)$

2.5 INFINITE LIMITS AND VERTICAL ASYMPTOTES

1. $\lim_{x \to 0^+} \frac{1}{3x} = \infty \quad \left(\frac{\text{positive}}{\text{positive}}\right)$

2. $\lim_{x \to 0^-} \frac{5}{2x} = -\infty \quad \left(\frac{\text{positive}}{\text{negative}}\right)$

3. $\lim_{x \to 2^-} \frac{3}{x-2} = -\infty \quad \left(\frac{\text{positive}}{\text{negative}}\right)$

4. $\lim_{x \to 3^+} \frac{1}{x-3} = \infty \quad \left(\frac{\text{positive}}{\text{positive}}\right)$

5. $\lim_{x \to -8^+} \frac{2x}{x+8} = -\infty \quad \left(\frac{\text{negative}}{\text{positive}}\right)$

6. $\lim_{x \to -5^-} \frac{3x}{2x+10} = \infty \quad \left(\frac{\text{negative}}{\text{negative}}\right)$

7. $\lim_{x \to 7} \frac{4}{(x-7)^2} = \infty \quad \left(\frac{\text{positive}}{\text{positive}}\right)$

8. $\lim_{x \to 0} \frac{-1}{x^2(x+1)} = -\infty \quad \left(\frac{\text{negative}}{\text{positive} \cdot \text{positive}}\right)$

9. (a) $\lim_{x \to 0^+} \frac{2}{3x^{1/3}} = \infty$

(b) $\lim_{x \to 0^-} \frac{2}{3x^{1/3}} = -\infty$

10. (a) $\lim_{x \to 0^+} \frac{2}{x^{1/5}} = \infty$

(b) $\lim_{x \to 0^-} \frac{2}{x^{1/5}} = -\infty$

11. $\lim_{x \to 0} \frac{4}{x^{2/5}} = \lim_{x \to 0} \frac{4}{(x^{1/5})^2} = \infty$

12. $\lim_{x \to 0} \frac{1}{x^{2/3}} = \lim_{x \to 0} \frac{1}{(x^{1/3})^2} = \infty$

13. $\lim_{x \to \left(\frac{\pi}{2}\right)^-} \tan x = \infty$

14. $\lim_{x \to \left(\frac{-\pi}{2}\right)^+} \sec x = \infty$

15. $\lim_{\theta \to 0^-} (1 + \csc \theta) = -\infty$

16. $\lim_{\theta \to 0^+} (2 - \cot \theta) = -\infty$ and $\lim_{\theta \to 0^-} (2 - \cot \theta) = \infty$, so the limit does not exist

17. (a) $\lim_{x \to 2^+} \frac{1}{x^2-4} = \lim_{x \to 2^+} \frac{1}{(x+2)(x-2)} = \infty$ $\left(\frac{1}{\text{positive}\cdot\text{positive}}\right)$

(b) $\lim_{x \to 2^-} \frac{1}{x^2-4} = \lim_{x \to 2^-} \frac{1}{(x+2)(x-2)} = -\infty$ $\left(\frac{1}{\text{positive}\cdot\text{negative}}\right)$

(c) $\lim_{x \to -2^+} \frac{1}{x^2-4} = \lim_{x \to -2^+} \frac{1}{(x+2)(x-2)} = -\infty$ $\left(\frac{1}{\text{positive}\cdot\text{negative}}\right)$

(d) $\lim_{x \to -2^-} \frac{1}{x^2-4} = \lim_{x \to -2^-} \frac{1}{(x+2)(x-2)} = \infty$ $\left(\frac{1}{\text{negative}\cdot\text{negative}}\right)$

18. (a) $\lim_{x \to 1^+} \frac{x}{x^2-1} = \lim_{x \to 1^+} \frac{x}{(x+1)(x-1)} = \infty$ $\left(\frac{\text{positive}}{\text{positive}\cdot\text{positive}}\right)$

(b) $\lim_{x \to 1^-} \frac{x}{x^2-1} = \lim_{x \to 1^-} \frac{x}{(x+1)(x-1)} = -\infty$ $\left(\frac{\text{positive}}{\text{positive}\cdot\text{negative}}\right)$

(c) $\lim_{x \to -1^+} \frac{x}{x^2-1} = \lim_{x \to -1^+} \frac{x}{(x+1)(x-1)} = \infty$ $\left(\frac{\text{negative}}{\text{positive}\cdot\text{negative}}\right)$

(d) $\lim_{x \to -1^-} \frac{x}{x^2-1} = \lim_{x \to -1^-} \frac{x}{(x+1)(x-1)} = -\infty$ $\left(\frac{\text{negative}}{\text{negative}\cdot\text{negative}}\right)$

19. (a) $\lim_{x \to 0^+} \frac{x^2}{2} - \frac{1}{x} = 0 + \lim_{x \to 0^+} \frac{1}{-x} = -\infty$ $\left(\frac{1}{\text{negative}}\right)$

(b) $\lim_{x \to 0^-} \frac{x^2}{2} - \frac{1}{x} = 0 + \lim_{x \to 0^-} \frac{1}{-x} = \infty$ $\left(\frac{1}{\text{positive}}\right)$

(c) $\lim_{x \to \sqrt[3]{2}} \frac{x^2}{2} - \frac{1}{x} = \frac{2^{2/3}}{2} - \frac{1}{2^{1/3}} = 2^{-1/3} - 2^{-1/3} = 0$

(d) $\lim_{x \to -1} \frac{x^2}{2} - \frac{1}{x} = \frac{1}{2} - \left(\frac{1}{-1}\right) = \frac{3}{2}$

20. (a) $\lim_{x \to -2^+} \frac{x^2-1}{2x+4} = \infty$ $\left(\frac{\text{positive}}{\text{positive}}\right)$ (b) $\lim_{x \to -2^-} \frac{x^2-1}{2x+4} = -\infty$ $\left(\frac{\text{positive}}{\text{negative}}\right)$

(c) $\lim_{x \to 1^+} \frac{x^2-1}{2x+4} = \lim_{x \to 1^+} \frac{(x+1)(x-1)}{2x+4} = \frac{2\cdot 0}{2+4} = 0$

(d) $\lim_{x \to 0^-} \frac{x^2-1}{2x+4} = \frac{-1}{4}$

21. (a) $\lim_{x \to 0^+} \frac{x^2-3x+2}{x^3-2x^2} = \lim_{x \to 0^+} \frac{(x-2)(x-1)}{x^2(x-2)} = -\infty$ $\left(\frac{\text{negative}\cdot\text{negative}}{\text{positive}\cdot\text{negative}}\right)$

(b) $\lim_{x \to 2^+} \frac{x^2-3x+2}{x^3-2x^2} = \lim_{x \to 2^+} \frac{(x-2)(x-1)}{x^2(x-2)} = \lim_{x \to 2^+} \frac{x-1}{x^2} = \frac{1}{4}, x \neq 2$

(c) $\lim_{x \to 2^-} \frac{x^2-3x+2}{x^3-2x^2} = \lim_{x \to 2^-} \frac{(x-2)(x-1)}{x^2(x-2)} = \lim_{x \to 2^-} \frac{x-1}{x^2} = \frac{1}{4}, x \neq 2$

(d) $\lim_{x \to 2} \frac{x^2-3x+2}{x^3-2x^2} = \lim_{x \to 2} \frac{(x-2)(x-1)}{x^2(x-2)} = \lim_{x \to 2} \frac{x-1}{x^2} = \frac{1}{4}, x \neq 2$

(e) $\lim_{x \to 0} \frac{x^2-3x+2}{x^3-2x^2} = \lim_{x \to 0} \frac{(x-2)(x-1)}{x^2(x-2)} = -\infty$ $\left(\frac{\text{negative}\cdot\text{negative}}{\text{positive}\cdot\text{negative}}\right)$

22. (a) $\lim_{x \to 2^+} \frac{x^2-3x+2}{x^3-4x} = \lim_{x \to 2^+} \frac{(x-2)(x-1)}{x(x-2)(x+2)} = \lim_{x \to 2^+} \frac{(x-1)}{x(x+2)} = \frac{1}{2(4)} = \frac{1}{8}$

(b) $\lim_{x \to -2^+} \frac{x^2-3x+2}{x^3-4x} = \lim_{x \to -2^+} \frac{(x-2)(x-1)}{x(x-2)(x+2)} = \lim_{x \to -2^+} \frac{(x-1)}{x(x+2)} = \infty$ $\left(\frac{\text{negative}}{\text{negative}\cdot\text{positive}}\right)$

(c) $\lim_{x \to 0^-} \frac{x^2-3x+2}{x^3-4x} = \lim_{x \to 0^-} \frac{(x-2)(x-1)}{x(x-2)(x+2)} = \lim_{x \to 0^-} \frac{(x-1)}{x(x+2)} = \infty$ $\left(\frac{\text{negative}}{\text{negative}\cdot\text{positive}}\right)$

(d) $\lim_{x \to 1^+} \frac{x^2-3x+2}{x^3-4x} = \lim_{x \to 1^+} \frac{(x-2)(x-1)}{x(x-2)(x+2)} = \lim_{x \to 1^+} \frac{(x-1)}{x(x+2)} = \frac{0}{(1)(3)} = 0$

(e) $\lim_{x \to 0^+} \frac{x-1}{x(x+2)} = -\infty$ $\left(\frac{\text{negative}}{\text{positive}\cdot\text{positive}}\right)$

and $\lim_{x \to 0^-} \frac{x-1}{x(x+2)} = \infty$ $\left(\frac{\text{negative}}{\text{negative}\cdot\text{positive}}\right)$

so the function has no limit as $x \to 0$.

23. (a) $\lim_{t \to 0^+} \left[2 - \frac{3}{t^{1/3}}\right] = -\infty$ (b) $\lim_{t \to 0^-} \left[2 - \frac{3}{t^{1/3}}\right] = \infty$

24. (a) $\lim_{t \to 0^+} \left[\frac{1}{t^{3/5}} + 7\right] = \infty$ (b) $\lim_{t \to 0^-} \left[\frac{1}{t^{3/5}} + 7\right] = -\infty$

25. (a) $\lim_{x \to 0^+} \left[\frac{1}{x^{2/3}} + \frac{2}{(x-1)^{2/3}}\right] = \infty$ (b) $\lim_{x \to 0^-} \left[\frac{1}{x^{2/3}} + \frac{2}{(x-1)^{2/3}}\right] = \infty$

(c) $\lim_{x \to 1^+} \left[\frac{1}{x^{2/3}} + \frac{2}{(x-1)^{2/3}}\right] = \infty$ (d) $\lim_{x \to 1^-} \left[\frac{1}{x^{2/3}} + \frac{2}{(x-1)^{2/3}}\right] = \infty$

26. (a) $\lim_{x \to 0^+} \left[\frac{1}{x^{1/3}} - \frac{1}{(x-1)^{4/3}}\right] = \infty$ (b) $\lim_{x \to 0^-} \left[\frac{1}{x^{1/3}} - \frac{1}{(x-1)^{4/3}}\right] = -\infty$

(c) $\lim_{x \to 1^+} \left[\frac{1}{x^{1/3}} - \frac{1}{(x-1)^{4/3}}\right] = -\infty$ (d) $\lim_{x \to 1^-} \left[\frac{1}{x^{1/3}} - \frac{1}{(x-1)^{4/3}}\right] = -\infty$

27. $y = \frac{1}{x-1}$

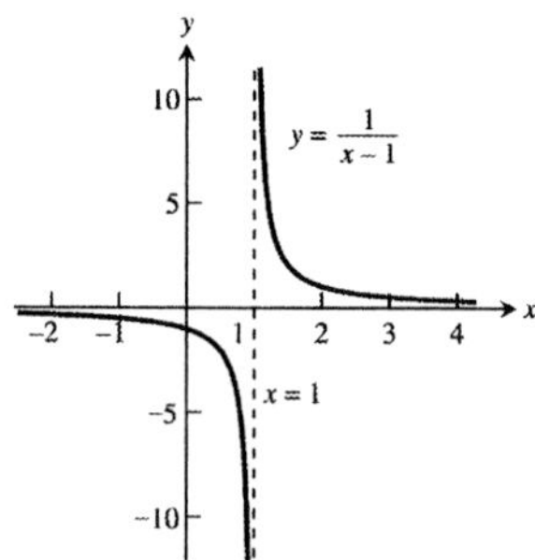

28. $y = \frac{1}{x+1}$

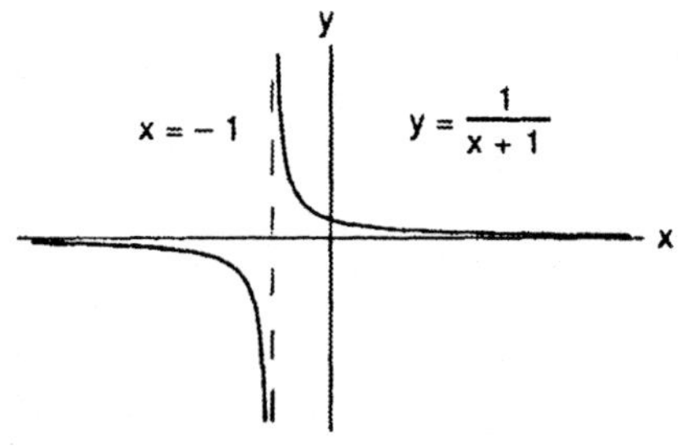

29. $y = \frac{1}{2x+4}$

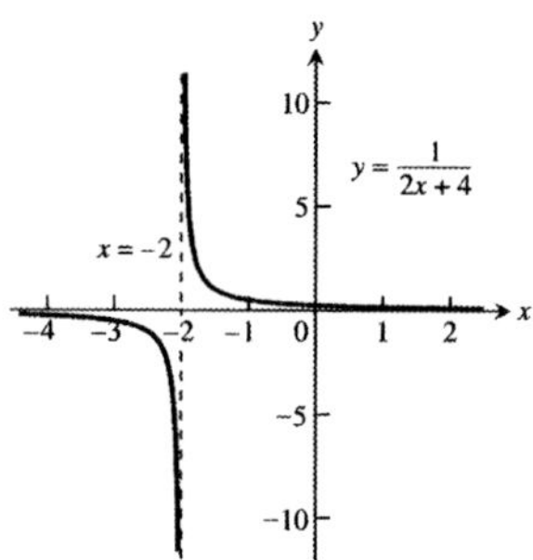

30. $y = \frac{-3}{x-3}$

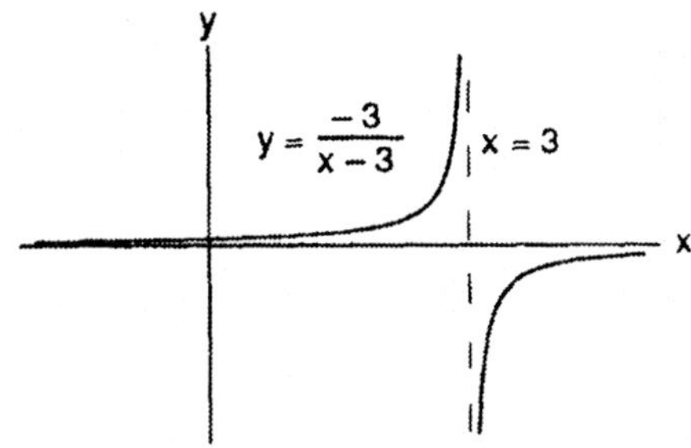

31. $y = \frac{x+3}{x+2} = 1 + \frac{1}{x+2}$

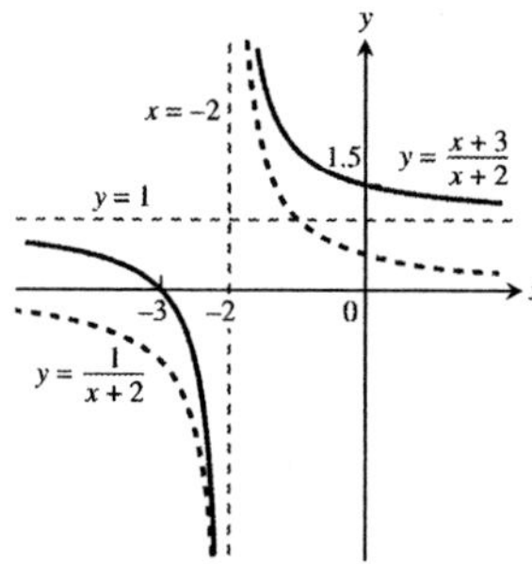

32. $y = \frac{2x}{x+1} = 2 - \frac{2}{x+1}$

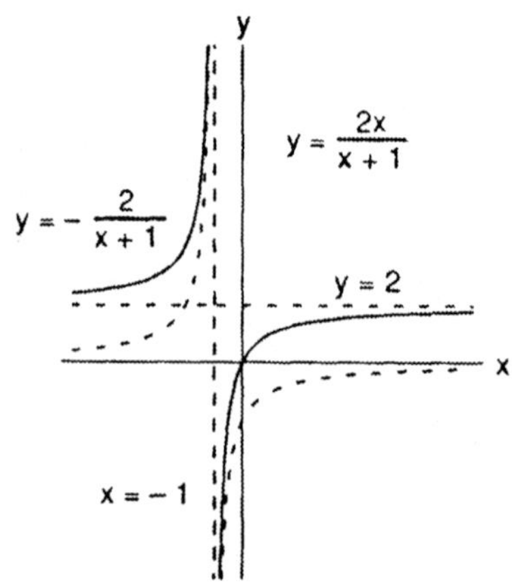

33. $y = \frac{x^2}{x-1} = x + 1 + \frac{1}{x-1}$

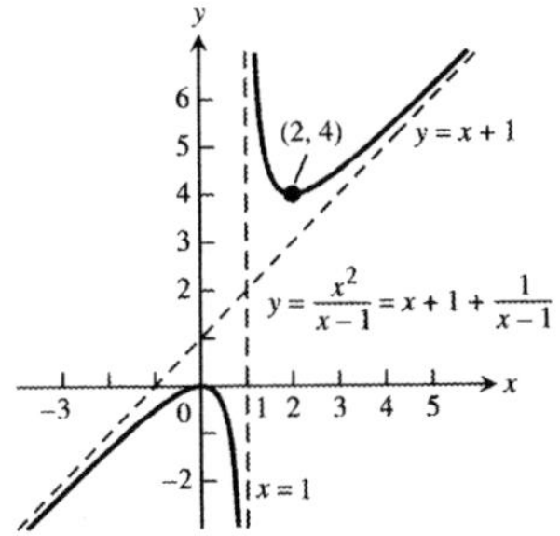

34. $y = \frac{x^2+1}{x-1} = x + 1 + \frac{2}{x-1}$

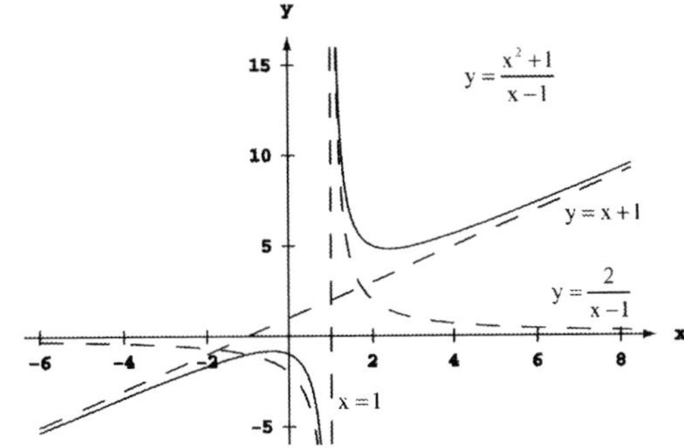

35. $y = \frac{x^2-4}{x-1} = x + 1 - \frac{3}{x-1}$

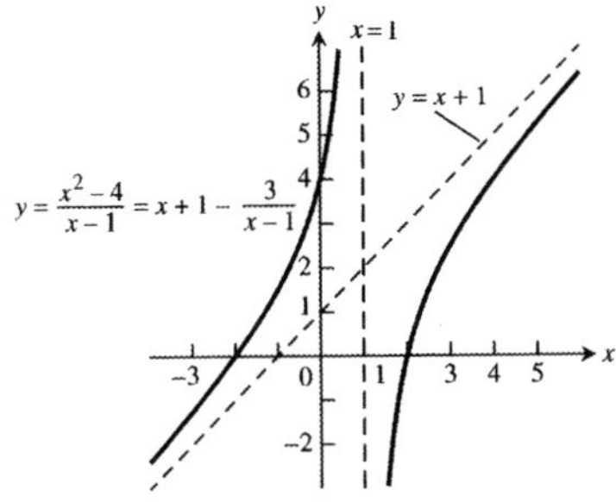

36. $y = \frac{x^2-1}{2x+4} = \frac{1}{2}x - 1 + \frac{3}{2x+4}$

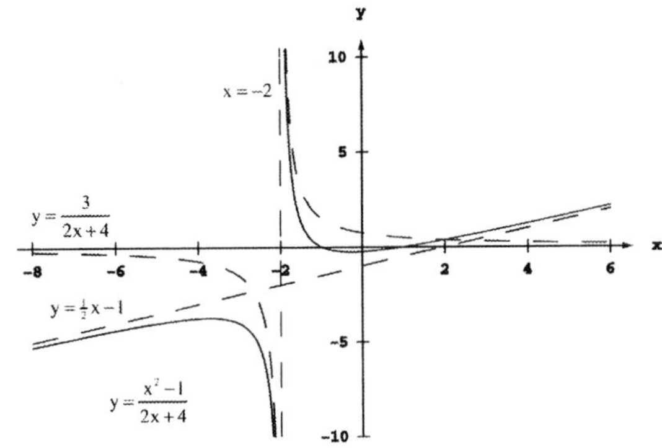

37. $y = \frac{x^2-1}{x} = x - \frac{1}{x}$

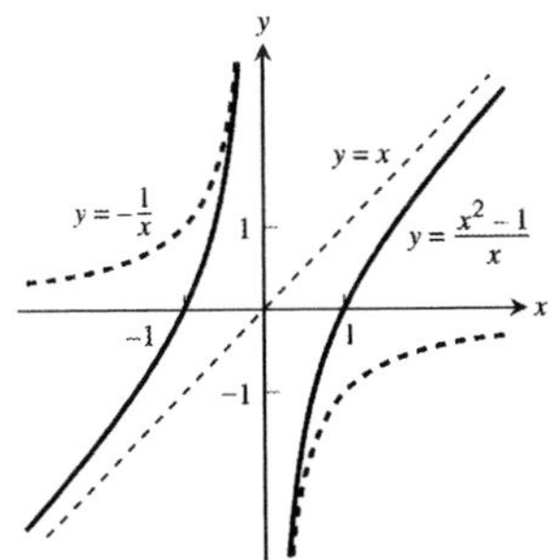

38. $y = \frac{x^3+1}{x^2} = x + \frac{1}{x^2}$

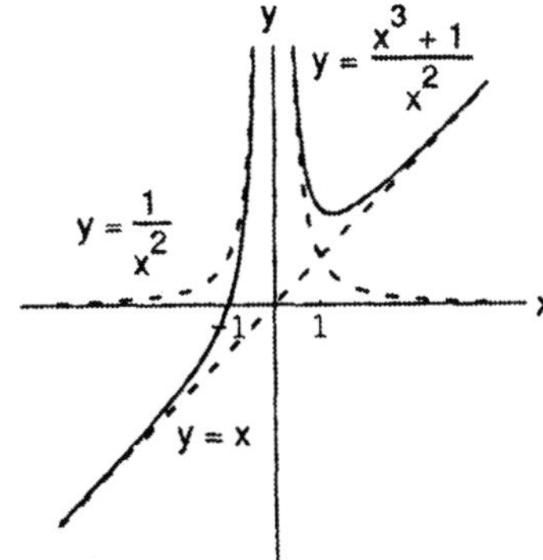

39. Here is one possibility.

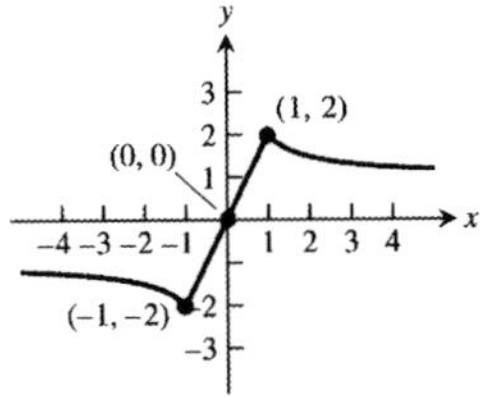

40. Here is one possibility.

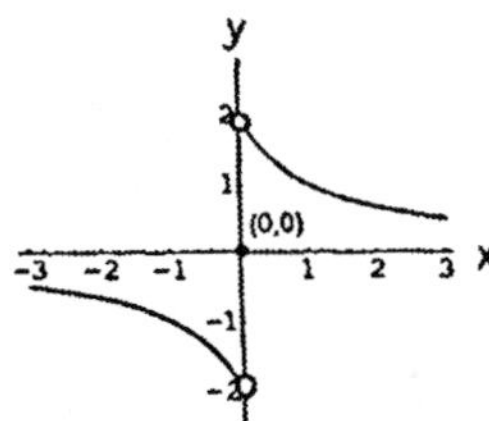

41. Here is one possibility.

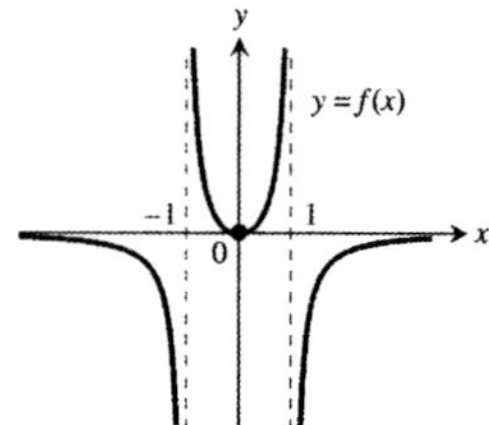

42. Here is one possibility.

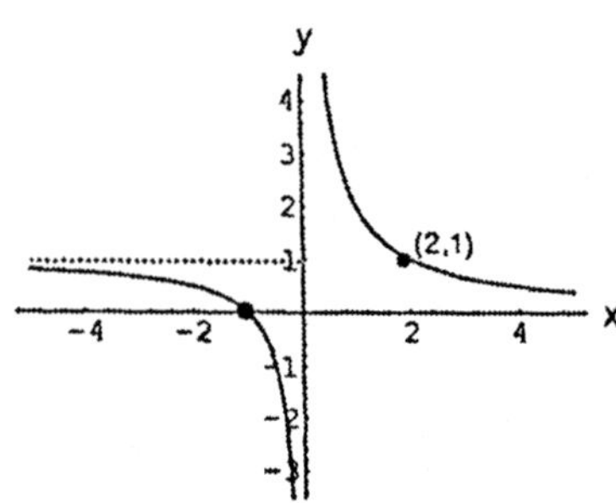

43. Here is one possibility.

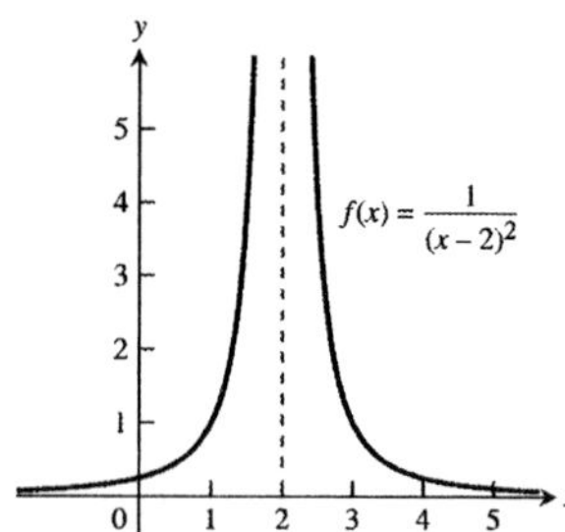

44. Here is one possibility.

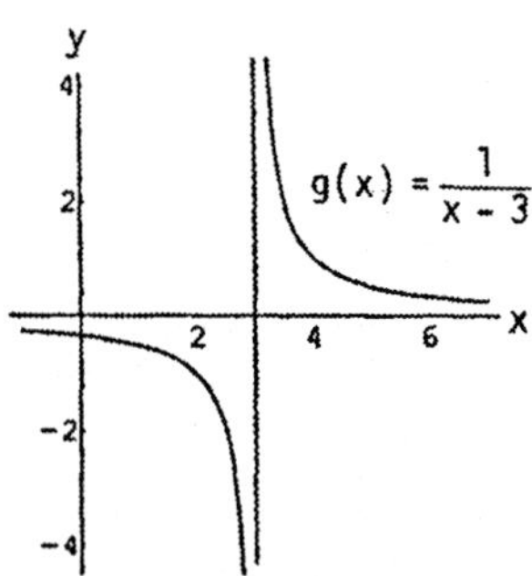

45. Here is one possibility.

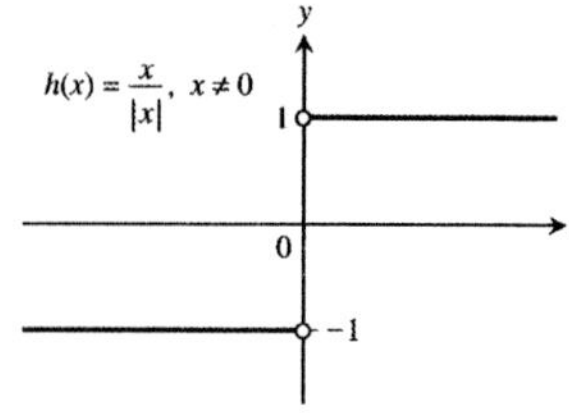

46. Here is one possibility.

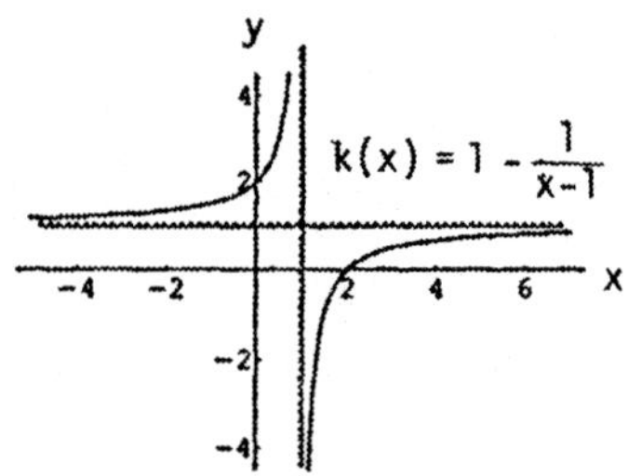

47. For every real number $-B < 0$, we must find a $\delta > 0$ such that for all x, $0 < |x - 0| < \delta \Rightarrow \frac{-1}{x^2} < -B$. Now, $-\frac{1}{x^2} < -B < 0 \Leftrightarrow \frac{1}{x^2} > B > 0 \Leftrightarrow x^2 < \frac{1}{B} \Leftrightarrow |x| < \frac{1}{\sqrt{B}}$. Choose $\delta = \frac{1}{\sqrt{B}}$, then $0 < |x| < \delta \Rightarrow |x| < \frac{1}{\sqrt{B}}$ $\Rightarrow \frac{-1}{x^2} < -B$ so that $\lim_{x \to 0} -\frac{1}{x^2} = -\infty$.

48. For every real number $B > 0$, we must find a $\delta > 0$ such that for all x, $0 < |x - 0| < \delta \Rightarrow \frac{1}{|x|} > B$. Now, $\frac{1}{|x|} > B > 0 \Leftrightarrow |x| < \frac{1}{B}$. Choose $\delta = \frac{1}{B}$. Then $0 < |x - 0| < \delta \Rightarrow |x| < \frac{1}{B} \Rightarrow \frac{1}{|x|} > B$ so that $\lim_{x \to 0} \frac{1}{|x|} = \infty$.

49. For every real number $-B < 0$, we must find a $\delta > 0$ such that for all x, $0 < |x - 3| < \delta \Rightarrow \frac{-2}{(x-3)^2} < -B$. Now, $\frac{-2}{(x-3)^2} < -B < 0 \Leftrightarrow \frac{2}{(x-3)^2} > B > 0 \Leftrightarrow \frac{(x-3)^2}{2} < \frac{1}{B} \Leftrightarrow (x-3)^2 < \frac{2}{B} \Leftrightarrow 0 < |x - 3| < \sqrt{\frac{2}{B}}$. Choose $\delta = \sqrt{\frac{2}{B}}$, then $0 < |x - 3| < \delta \Rightarrow \frac{-2}{(x-3)^2} < -B < 0$ so that $\lim_{x \to 3} \frac{-2}{(x-3)^2} = -\infty$.

50. For every real number $B > 0$, we must find a $\delta > 0$ such that for all x, $0 < |x - (-5)| < \delta \Rightarrow \frac{1}{(x+5)^2} > B$. Now, $\frac{1}{(x+5)^2} > B > 0 \Leftrightarrow (x+5)^2 < \frac{1}{B} \Leftrightarrow |x+5| < \frac{1}{\sqrt{B}}$. Choose $\delta = \frac{1}{\sqrt{B}}$. Then $0 < |x - (-5)| < \delta$ $\Rightarrow |x+5| < \frac{1}{\sqrt{B}} \Rightarrow \frac{1}{(x+5)^2} > B$ so that $\lim_{x \to -5} \frac{1}{(x+5)^2} = \infty$.

51. (a) We say that f(x) approaches infinity as x approaches x_0 from the left, and write $\lim_{x \to x_0^-} f(x) = \infty$, if for every positive number B, there exists a corresponding number $\delta > 0$ such that for all x, $x_0 - \delta < x < x_0 \Rightarrow f(x) > B$.

(b) We say that f(x) approaches minus infinity as x approaches x_0 from the right, and write $\lim_{x \to x_0^+} f(x) = -\infty$, if for every positive number B (or negative number $-B$) there exists a corresponding number $\delta > 0$ such that for all x, $x_0 < x < x_0 + \delta \Rightarrow f(x) < -B$.

(c) We say that f(x) approaches minus infinity as x approaches x_0 from the left, and write $\lim_{x \to x_0^-} f(x) = -\infty$, if for every positive number B (or negative number $-B$) there exists a corresponding number $\delta > 0$ such that for all x, $x_0 - \delta < x < x_0 \Rightarrow f(x) < -B$.

52. For $B > 0$, $\frac{1}{x} > B > 0 \Leftrightarrow x < \frac{1}{B}$. Choose $\delta = \frac{1}{B}$. Then $0 < x < \delta \Rightarrow 0 < x < \frac{1}{B} \Rightarrow \frac{1}{x} > B$ so that $\lim_{x \to 0^+} \frac{1}{x} = \infty$.

53. For $B > 0$, $\frac{1}{x} < -B < 0 \Leftrightarrow -\frac{1}{x} > B > 0 \Leftrightarrow -x < \frac{1}{B} \Leftrightarrow -\frac{1}{B} < x$. Choose $\delta = \frac{1}{B}$. Then $-\delta < x < 0$ $\Rightarrow -\frac{1}{B} < x \Rightarrow \frac{1}{x} < -B$ so that $\lim_{x \to 0^-} \frac{1}{x} = -\infty$.

54. For $B > 0$, $\frac{1}{x-2} < -B \Leftrightarrow -\frac{1}{x-2} > B \Leftrightarrow -(x-2) < \frac{1}{B} \Leftrightarrow x - 2 > -\frac{1}{B} \Leftrightarrow x > 2 - \frac{1}{B}$. Choose $\delta = \frac{1}{B}$. Then $2 - \delta < x < 2 \Rightarrow -\delta < x - 2 < 0 \Rightarrow -\frac{1}{B} < x - 2 < 0 \Rightarrow \frac{1}{x-2} < -B < 0$ so that $\lim_{x \to 2^-} \frac{1}{x-2} = -\infty$.

55. For $B > 0$, $\frac{1}{x-2} > B \Leftrightarrow 0 < x - 2 < \frac{1}{B}$. Choose $\delta = \frac{1}{B}$. Then $2 < x < 2 + \delta \Rightarrow 0 < x - 2 < \delta \Rightarrow 0 < x - 2 < \frac{1}{B}$ $\Rightarrow \frac{1}{x-2} > B > 0$ so that $\lim_{x \to 2^+} \frac{1}{x-2} = \infty$.

56. For $B > 0$ and $0 < x < 1$, $\frac{1}{1-x^2} > B \Leftrightarrow 1 - x^2 < \frac{1}{B} \Leftrightarrow (1-x)(1+x) < \frac{1}{B}$. Now $\frac{1+x}{2} < 1$ since $x < 1$. Choose $\delta < \frac{1}{2B}$. Then $1 - \delta < x < 1 \Rightarrow -\delta < x - 1 < 0 \Rightarrow 1 - x < \delta < \frac{1}{2B} \Rightarrow (1-x)(1+x) < \frac{1}{B}\left(\frac{1+x}{2}\right) < \frac{1}{B}$ $\Rightarrow \frac{1}{1-x^2} > B$ for $0 < x < 1$ and x near 1 $\Rightarrow \lim_{x \to 1^-} \frac{1}{1-x^2} = \infty$.

57. $y = \sec x + \frac{1}{x}$

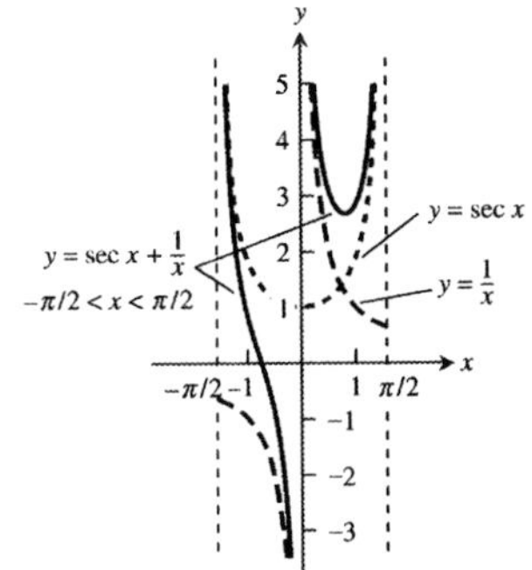

58. $y = \sec x - \frac{1}{x^2}$

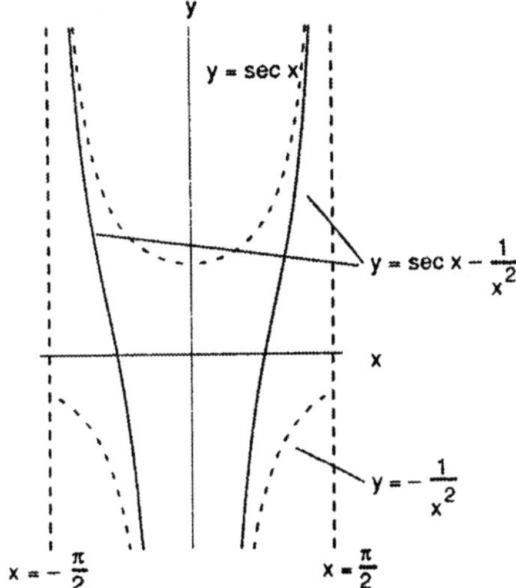

59. $y = \tan x + \frac{1}{x^2}$

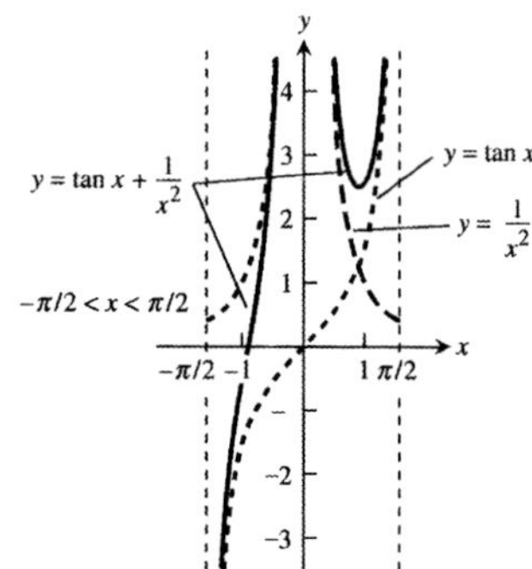

60. $y = \frac{1}{x} - \tan x$

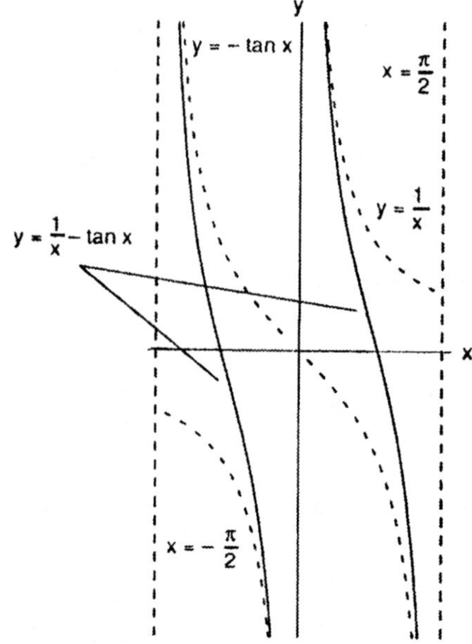

61. $y = \frac{x}{\sqrt{4 - x^2}}$

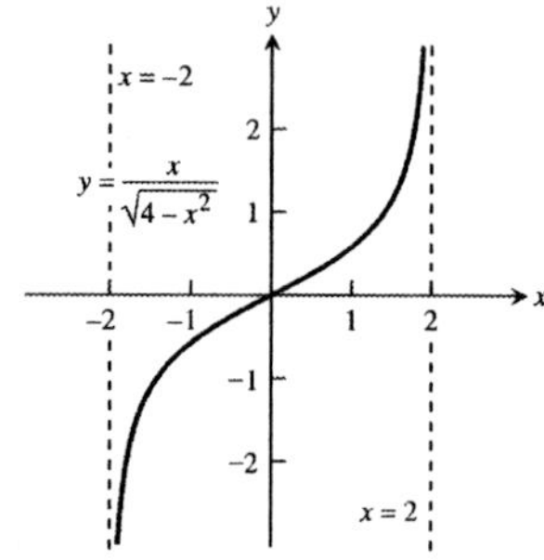

62. $y = \frac{-1}{\sqrt{4 - x^2}}$

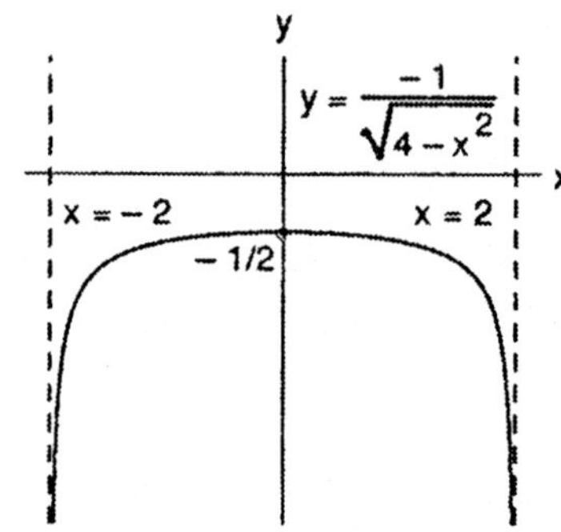

63. $y = x^{2/3} + \frac{1}{x^{1/3}}$

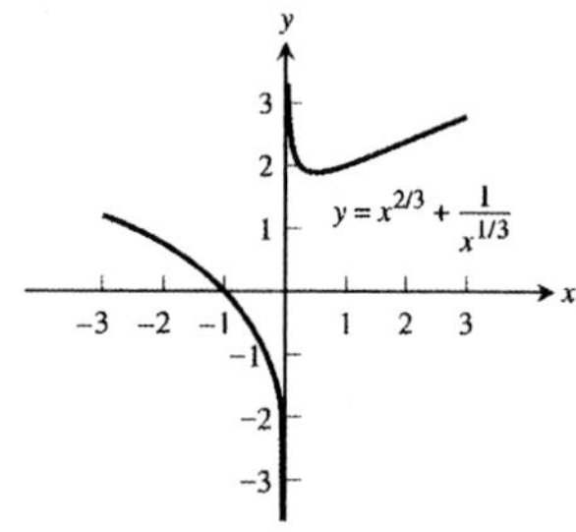

64. $y = \sin\left(\frac{\pi}{x^2+1}\right)$

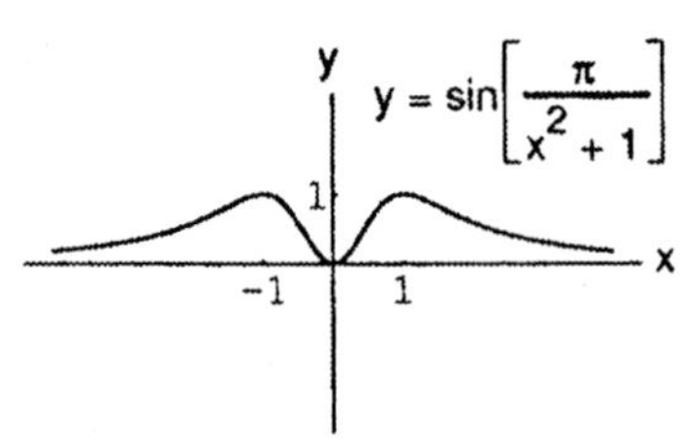

65. The graph of $y = f\left(\frac{1}{x}\right) = \frac{1}{x}e^{1/x}$ is shown.

$\lim_{x \to \infty} f(x) = \lim_{x \to 0^+} f\left(\frac{1}{x}\right) = \infty$

$\lim_{x \to -\infty} f(x) = \lim_{x \to 0^-} f\left(\frac{1}{x}\right) = 0$

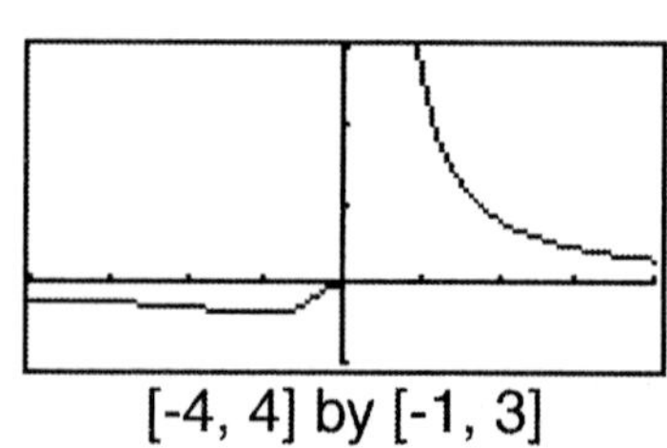

[-4, 4] by [-1, 3]

66. The graph of $y = f\left(\frac{1}{x}\right) = \frac{1}{x^2}e^{-1/x}$ is shown.

$\lim_{x \to \infty} f(x) = \lim_{x \to 0^+} f\left(\frac{1}{x}\right) = 0$

$\lim_{x \to -\infty} f(x) = \lim_{x \to 0^-} f\left(\frac{1}{x}\right) = \infty$

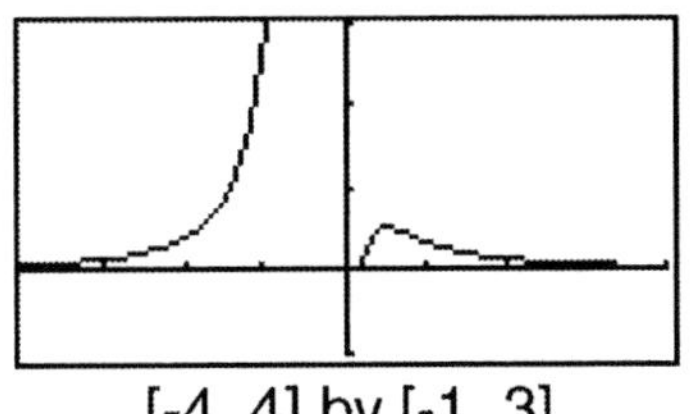

[-4, 4] by [-1, 3]

67. The graph of $y = f\left(\frac{1}{x}\right) = x \ln\left|\frac{1}{x}\right|$ is shown.

$\lim_{x \to \infty} f(x) = \lim_{x \to 0^+} f\left(\frac{1}{x}\right) = 0$

$\lim_{x \to -\infty} f(x) = \lim_{x \to 0^-} f\left(\frac{1}{x}\right) = 0$

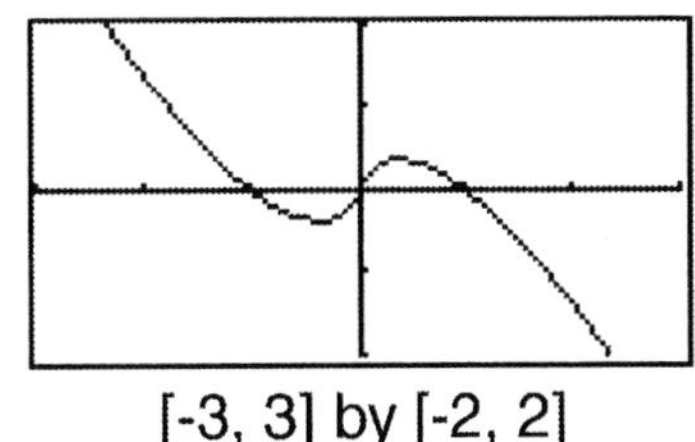

[-3, 3] by [-2, 2]

68. Let $t = \frac{1}{x} \Rightarrow \lim_{x \to \infty} \frac{e^{1/x}}{\ln|x|} = \lim_{t \to 0^+} \frac{e^t}{\ln(1/t)} = \lim_{t \to 0^+} \frac{e^t}{\ln 1 - \ln t} = \lim_{t \to 0^+} -\frac{e^t}{\ln t} = 0$ (since $\lim_{t \to 0^+} e^t = 1$ and $\lim_{t \to 0^+} \ln t = -\infty$);

similarly, $\lim_{x \to -\infty} \frac{e^{1/x}}{\ln|x|} = 0$

69. (a) $y \to \infty$ (see accompanying graph)
(b) $y \to \infty$ (see accompanying graph)
(c) cusps at $x = \pm 1$ (see accompanying graph)

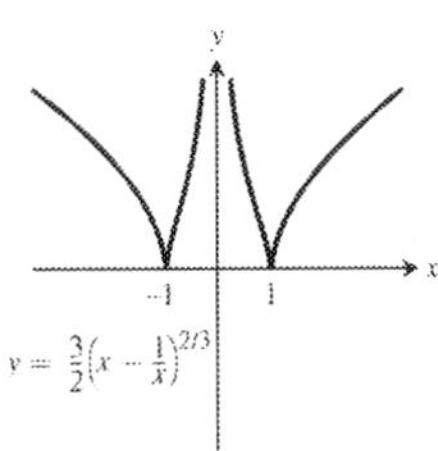

70. (a) $y \to 0$ and a cusp at $x = 0$ (see the accompanying graph)
(b) $y \to \frac{3}{2}$ (see accompanying graph)
(c) a vertical asymptote at $x = 1$ and contains the point $\left(-1, \frac{3}{2\sqrt[3]{4}}\right)$ (see accompanying graph)

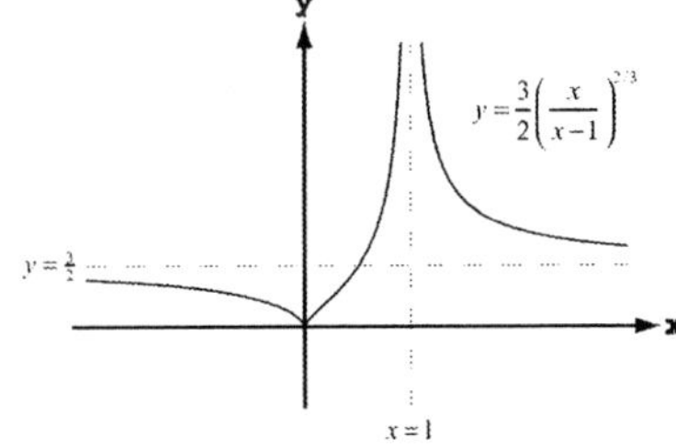

71. (a) The function $y = e^x$ is a right end behavior model because $\lim_{x \to \infty} \frac{e^x - 2x}{e^x} = \lim_{x \to \infty} \left(1 - \frac{2x}{e^x}\right) = 1 - 0 = 1$.
(b) The function $y = -2x$ is a left end behavior model because $\lim_{x \to -\infty} \frac{e^x - 2x}{-2x} = \lim_{x \to -\infty} \left(-\frac{e^x}{2x} + 1\right) = 0 + 1 = 1$.

72. (a) The function $y = x^2$ is a right end behavior model because $\lim_{x \to \infty} \frac{x^2 + e^{-x}}{x^2} = \lim_{x \to \infty} \left(1 + \frac{e^{-x}}{x^2}\right) = 1 - 0 = 1$.
(b) The function $y = e^{-x}$ is a left end behavior model because $\lim_{x \to -\infty} \frac{x^2 + e^{-x}}{e^{-x}} = \lim_{x \to -\infty} \left(\frac{x^2}{e^{-x}} + 1\right)$
$= \lim_{x \to -\infty} (x^2 e^x + 1) = 0 + 1 = 1$.

73. (a, b) The function $y = x$ is both a right end behavior model and a a left end behavior model because $\lim_{x \to \pm\infty} \frac{x + \ln x}{x} = \lim_{x \to \pm\infty} \left(1 + \frac{\ln x}{x}\right) = 1 - 0 = 1$.

74. (a, b) The function $y = x^2$ is both a right end behavior model and a a left end behavior model because $\lim_{x \to \pm\infty} \frac{x^2 + \sin x}{x^2} = \lim_{x \to \pm\infty} \left(1 + \frac{\sin x}{x^2}\right) = 1$.

75. $y = \frac{2x^3 - 3x^2 + 1}{x+3} = 2x^2 - 9x + 27 - \frac{80}{x+3} \Rightarrow$
   (a) $g(x) = 2x^2 - 9x + 27$ is a right end behavior model for y.
   (b) $h(x) = 2x^2 - 9x + 27$ is a left end behavior model for y.

76. $y = \frac{2x^4 - x^3 + x^2 - 1}{2-x} = -2x^3 - 3x^2 - 7x - 14 + \frac{27}{2-x} \Rightarrow$
   (a) $g(x) = -2x^3 - 3x^2 - 7x - 14$ is a right end behavior model for y.
   (b) $h(x) = -2x^3 - 3x^2 - 7x - 14$ is a left end behavior model for y.

2.6 CONTINUITY

1. No, discontinuous at $x = 2$, not defined at $x = 2$

2. No, discontinuous at $x = 3$, $1 = \lim_{x \to 3^-} g(x) \neq g(3) = 1.5$

3. Continuous on $[-1, 3]$

4. No, discontinuous at $x = 1$, $1.5 = \lim_{x \to 1^-} k(x) \neq \lim_{x \to 1^+} k(x) = 0$

5. (a) Yes (b) Yes, $\lim_{x \to -1^+} f(x) = 0$
   (c) Yes (d) Yes

6. (a) Yes, $f(1) = 1$ (b) Yes, $\lim_{x \to 1} f(x) = 2$
   (c) No (d) No

7. (a) No (b) No

8. $[-1, 0) \cup (0, 1) \cup (1, 2) \cup (2, 3)$

9. $f(2) = 0$, since $\lim_{x \to 2^-} f(x) = -2(2) + 4 = 0 = \lim_{x \to 2^+} f(x)$

10. $f(1)$ should be changed to $2 = \lim_{x \to 1} f(x)$

11. Nonremovable discontinuity at $x = 1$ because $\lim_{x \to 1} f(x)$ fails to exist ($\lim_{x \to 1^-} f(x) = 1$ and $\lim_{x \to 1^+} f(x) = 0$). Removable discontinuity at $x = 0$ by assigning the number $\lim_{x \to 0} f(x) = 0$ to be the value of $f(0)$ rather than $f(0) = 1$.

12. Nonremovable discontinuity at $x = 1$ because $\lim_{x \to 1} f(x)$ fails to exist ($\lim_{x \to 1^-} f(x) = 2$ and $\lim_{x \to 1^+} f(x) = 1$). Removable discontinuity at $x = 2$ by assigning the number $\lim_{x \to 2} f(x) = 1$ to be the value of $f(2)$ rather than $f(2) = 2$.

13. Discontinuous only when $x - 2 = 0 \Rightarrow x = 2$

14. Discontinuous only when $(x + 2)^2 = 0 \Rightarrow x = -2$

15. Discontinuous only when $x^2 - 4x + 3 = 0 \Rightarrow (x - 3)(x - 1) = 0 \Rightarrow x = 3$ or $x = 1$

16. Discontinuous only when $x^2 - 3x - 10 = 0 \Rightarrow (x - 5)(x + 2) = 0 \Rightarrow x = 5$ or $x = -2$

17. Continuous everywhere. ($|x - 1| + \sin x$ defined for all x; limits exist and are equal to function values.)

18. Continuous everywhere. ($|x| + 1 \neq 0$ for all x; limits exist and are equal to function values.)

19. Discontinuous only at $x = 0$

20. Discontinuous at odd integer multiples of $\frac{\pi}{2}$, i.e., $x = (2n - 1)\frac{\pi}{2}$, n an integer, but continuous at all other x.

21. Discontinuous when 2x is an integer multiple of π, i.e., $2x = n\pi$, n an integer $\Rightarrow x = \frac{n\pi}{2}$, n an integer, but continuous at all other x.

22. Discontinuous when $\frac{\pi x}{2}$ is an odd integer multiple of $\frac{\pi}{2}$, i.e., $\frac{\pi x}{2} = (2n - 1)\frac{\pi}{2}$, n an integer $\Rightarrow x = 2n - 1$, n an integer (i.e., x is an odd integer). Continuous everywhere else.

23. Discontinuous at odd integer multiples of $\frac{\pi}{2}$, i.e., $x = (2n - 1)\frac{\pi}{2}$, n an integer, but continuous at all other x.

24. Continuous everywhere since $x^4 + 1 \geq 1$ and $-1 \leq \sin x \leq 1 \Rightarrow 0 \leq \sin^2 x \leq 1 \Rightarrow 1 + \sin^2 x \geq 1$; limits exist and are equal to the function values.

25. Discontinuous when $2x + 3 < 0$ or $x < -\frac{3}{2} \Rightarrow$ continuous on the interval $\left[-\frac{3}{2}, \infty\right)$.

26. Discontinuous when $3x - 1 < 0$ or $x < \frac{1}{3} \Rightarrow$ continuous on the interval $\left[\frac{1}{3}, \infty\right)$.

27. Continuous everywhere: $(2x - 1)^{1/3}$ is defined for all x; limits exist and are equal to function values.

28. Continuous everywhere: $(2 - x)^{1/5}$ is defined for all x; limits exist and are equal to function values.

29. $\lim\limits_{x \to \pi} \sin(x - \sin x) = \sin(\pi - \sin \pi) = \sin(\pi - 0) = \sin \pi = 0$, and function continuous at $x = \pi$.

30. $\lim\limits_{t \to 0} \sin\left(\frac{\pi}{2}\cos(\tan t)\right) = \sin\left(\frac{\pi}{2}\cos(\tan(0))\right) = \sin\left(\frac{\pi}{2}\cos(0)\right) = \sin\left(\frac{\pi}{2}\right) = 1$, and function continuous at $t = 0$.

31. $\lim\limits_{y \to 1} \sec(y \sec^2 y - \tan^2 y - 1) = \lim\limits_{y \to 1} \sec(y \sec^2 y - \sec^2 y) = \lim\limits_{y \to 1} \sec((y - 1)\sec^2 y) = \sec((1 - 1)\sec^2 1)$ $= \sec 0 = 1$, , and function continuous at $y = 1$.

32. $\lim\limits_{x \to 0} \tan\left[\frac{\pi}{4}\cos\left(\sin x^{1/3}\right)\right] = \tan\left[\frac{\pi}{4}\cos(\sin(0))\right] = \tan\left(\frac{\pi}{4}\cos(0)\right) = \tan\left(\frac{\pi}{4}\right) = 1$, and function continuous at $x = 0$.

33. $\lim\limits_{x \to 0^+} \sin\left(\frac{\pi}{2}e^{\sqrt{x}}\right) = \sin\left(\frac{\pi}{2}e^0\right) = \sin\left(\frac{\pi}{2}\right) = 1$, and function is continuous at $x = 0$.

34. $\lim\limits_{x \to 1} \cos^{-1}\left(\ln \sqrt{x}\right) = \cos^{-1}\left(\ln \sqrt{1}\right) = \cos^{-1}(0) = \frac{\pi}{2}$, and function is continuous at $x = 1$.

35. $g(x) = \frac{x^2 - 9}{x - 3} = \frac{(x + 3)(x - 3)}{(x - 3)} = x + 3, x \neq 3 \Rightarrow g(3) = \lim\limits_{x \to 3}(x + 3) = 6$

36. $h(t) = \frac{t^2 + 3t - 10}{t - 2} = \frac{(t + 5)(t - 2)}{t - 2} = t + 5, t \neq 2 \Rightarrow h(2) = \lim\limits_{t \to 2}(t + 5) = 7$

37. $f(s) = \frac{s^3-1}{s^2-1} = \frac{(s^2+s+1)(s-1)}{(s+1)(s-1)} = \frac{s^2+s+1}{s+1}, s \neq 1 \Rightarrow f(1) = \lim_{s \to 1} \left(\frac{s^2+s+1}{s+1}\right) = \frac{3}{2}$

38. $g(x) = \frac{x^2-16}{x^2-3x-4} = \frac{(x+4)(x-4)}{(x-4)(x+1)} = \frac{x+4}{x+1}, x \neq 4 \Rightarrow g(4) = \lim_{x \to 4} \left(\frac{x+4}{x+1}\right) = \frac{8}{5}$

39. As defined, $\lim_{x \to 3^-} f(x) = (3)^2 - 1 = 8$ and $\lim_{x \to 3^+} (2a)(3) = 6a$. For f(x) to be continuous we must have $6a = 8 \Rightarrow a = \frac{4}{3}$.

40. As defined, $\lim_{x \to -2^-} g(x) = -2$ and $\lim_{x \to -2^+} g(x) = b(-2)^2 = 4b$. For g(x) to be continuous we must have $4b = -2 \Rightarrow b = -\frac{1}{2}$.

41. The function can be extended: $f(0) \approx 2.3$.

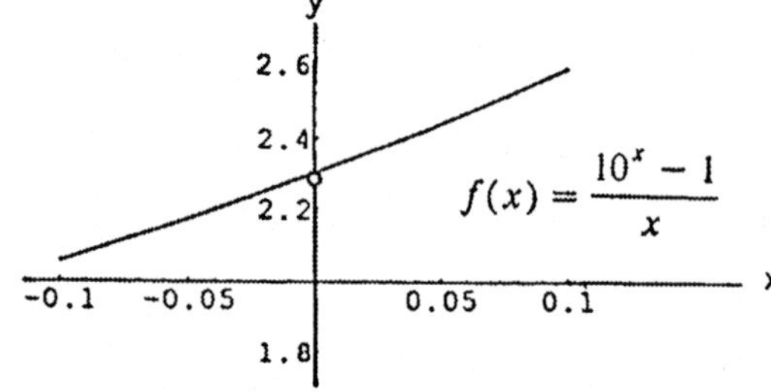

42. The function cannot be extended to be continuous at $x = 0$. If $f(0) \approx 2.3$, it will be continuous from the right. Or if $f(0) \approx -2.3$, it will be continuous from the left.

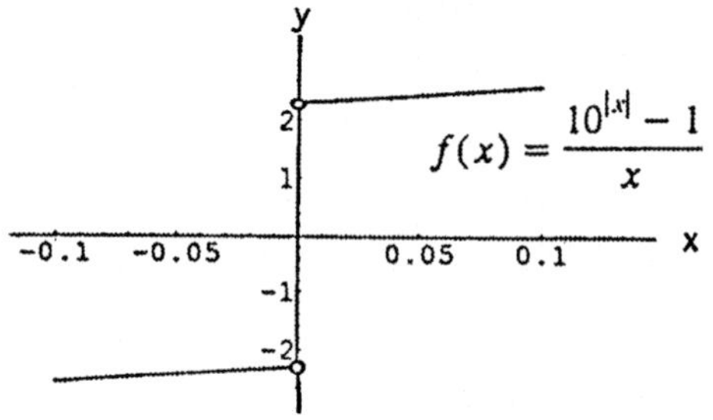

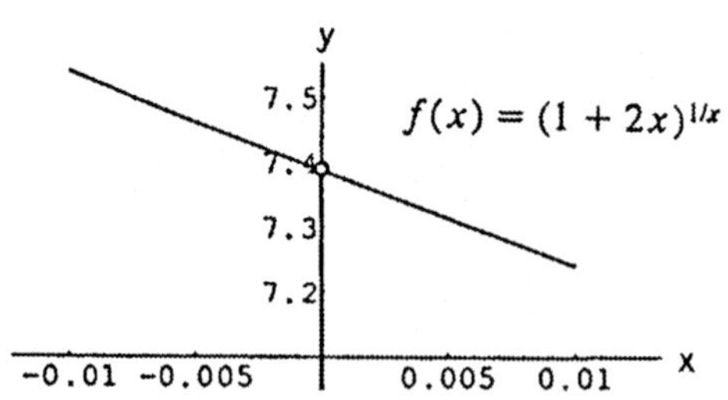

43. The function cannot be extended to be continuous at $x = 0$. If $f(0) = 1$, it will be continuous from the right. Or if $f(0) = -1$, it will be continuous from the left.

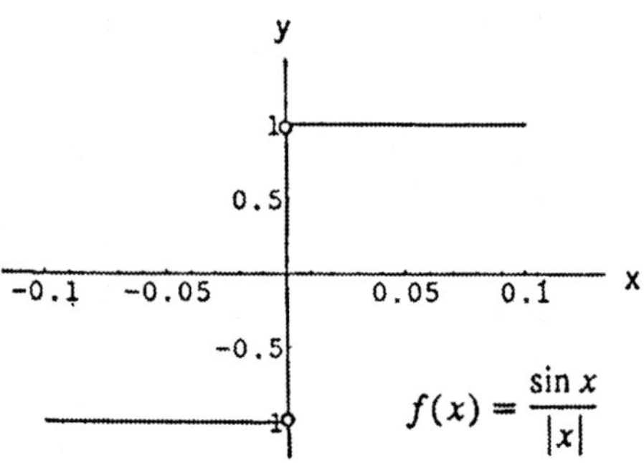

44. The function can be extended: $f(0) \approx 7.39$.

45. f(x) is continuous on [0, 1] and $f(0) < 0$, $f(1) > 0$ $\Rightarrow$ by the Intermediate Value Theorem f(x) takes on every value between f(0) and f(1) $\Rightarrow$ the equation $f(x) = 0$ has at least one solution between $x = 0$ and $x = 1$.

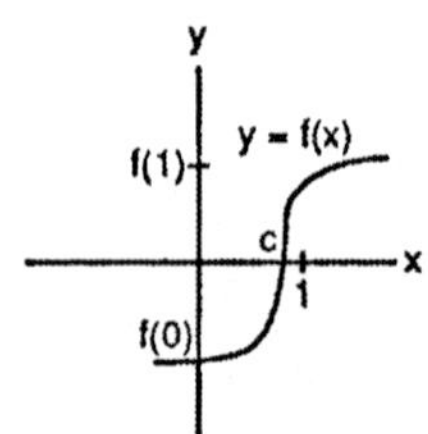

46. $\cos x = x \Rightarrow (\cos x) - x = 0$. If $x = -\frac{\pi}{2}$, $\cos\left(-\frac{\pi}{2}\right) - \left(-\frac{\pi}{2}\right) > 0$. If $x = \frac{\pi}{2}$, $\cos\left(\frac{\pi}{2}\right) - \frac{\pi}{2} < 0$. Thus $\cos x - x = 0$ for some x between $-\frac{\pi}{2}$ and $\frac{\pi}{2}$ according to the Intermediate Value Theorem, since the function $\cos x - x$ is continuous.

47. Let $f(x) = x^3 - 15x + 1$, which is continuous on $[-4, 4]$. Then $f(-4) = -3$, $f(-1) = 15$, $f(1) = -13$, and $f(4) = 5$. By the Intermediate Value Theorem, $f(x) = 0$ for some x in each of the intervals $-4 < x < -1$, $-1 < x < 1$, and $1 < x < 4$. That is, $x^3 - 15x + 1 = 0$ has three solutions in $[-4, 4]$. Since a polynomial of degree 3 can have at most 3 solutions, these are the only solutions.

48. Without loss of generality, assume that $a < b$. Then $F(x) = (x - a)^2 (x - b)^2 + x$ is continuous for all values of x, so it is continuous on the interval $[a, b]$. Moreover $F(a) = a$ and $F(b) = b$. By the Intermediate Value Theorem, since $a < \frac{a+b}{2} < b$, there is a number c between a and b such that $F(x) = \frac{a+b}{2}$.

49. Answers may vary. Note that f is continuous for every value of x.
    (a) $f(0) = 10$, $f(1) = 1^3 - 8(1) + 10 = 3$. Since $3 < \pi < 10$, by the Intermediate Value Theorem, there exists a c so that $0 < c < 1$ and $f(c) = \pi$.
    (b) $f(0) = 10$, $f(-4) = (-4)^3 - 8(-4) + 10 = -22$. Since $-22 < -\sqrt{3} < 10$, by the Intermediate Value Theorem, there exists a c so that $-4 < c < 0$ and $f(c) = -\sqrt{3}$.
    (c) $f(0) = 10$, $f(1000) = (1000)^3 - 8(1000) + 10 = 999{,}992{,}010$. Since $10 < 5{,}000{,}000 < 999{,}992{,}010$, by the Intermediate Value Theorem, there exists a c so that $0 < c < 1000$ and $f(c) = 5{,}000{,}000$.

50. All five statements ask for the same information because of the intermediate value property of continuous functions.
    (a) A root of $f(x) = x^3 - 3x - 1$ is a point c where $f(c) = 0$.
    (b) The points where $y = x^3$ crosses $y = 3x + 1$ have the same y-coordinate, or $y = x^3 = 3x + 1$ $\Rightarrow f(x) = x^3 - 3x - 1 = 0$.
    (c) $x^3 - 3x = 1 \Rightarrow x^3 - 3x - 1 = 0$. The solutions to the equation are the roots of $f(x) = x^3 - 3x - 1$.
    (d) The points where $y = x^3 - 3x$ crosses $y = 1$ have common y-coordinates, or $y = x^3 - 3x = 1$ $\Rightarrow f(x) = x^3 - 3x - 1 = 0$.
    (e) The solutions of $x^3 - 3x - 1 = 0$ are those points where $f(x) = x^3 - 3x - 1$ has value 0.

51. Answers may vary. For example, $f(x) = \frac{\sin(x-2)}{x-2}$ is discontinuous at $x = 2$ because it is not defined there. However, the discontinuity can be removed because f has a limit (namely 1) as $x \to 2$.

52. Answers may vary. For example, $g(x) = \frac{1}{x+1}$ has a discontinuity at $x = -1$ because $\lim_{x \to -1} g(x)$ does not exist. $\left(\lim_{x \to -1^-} g(x) = -\infty \text{ and } \lim_{x \to -1^+} g(x) = +\infty.\right)$

53. (a) Suppose x_0 is rational $\Rightarrow f(x_0) = 1$. Choose $\epsilon = \frac{1}{2}$. For any $\delta > 0$ there is an irrational number x (actually infinitely many) in the interval $(x_0 - \delta, x_0 + \delta) \Rightarrow f(x) = 0$. Then $0 < |x - x_0| < \delta$ but $|f(x) - f(x_0)| = 1 > \frac{1}{2} = \epsilon$, so $\lim_{x \to x_0} f(x)$ fails to exist $\Rightarrow$ f is discontinuous at x_0 rational.
    On the other hand, x_0 irrational $\Rightarrow f(x_0) = 0$ and there is a rational number x in $(x_0 - \delta, x_0 + \delta) \Rightarrow f(x) = 1$. Again $\lim_{x \to x_0} f(x)$ fails to exist $\Rightarrow$ f is discontinuous at x_0 irrational. That is, f is discontinuous at every point.
    (b) f is neither right-continuous nor left-continuous at any point x_0 because in every interval $(x_0 - \delta, x_0)$ or $(x_0, x_0 + \delta)$ there exist both rational and irrational real numbers. Thus neither limits $\lim_{x \to x_0^-} f(x)$ and $\lim_{x \to x_0^+} f(x)$ exist by the same arguments used in part (a).

54. Yes. Both $f(x) = x$ and $g(x) = x - \frac{1}{2}$ are continuous on $[0, 1]$. However $\frac{f(x)}{g(x)}$ is undefined at $x = \frac{1}{2}$ since $g\left(\frac{1}{2}\right) = 0 \Rightarrow \frac{f(x)}{g(x)}$ is discontinuous at $x = \frac{1}{2}$.

55. No. For instance, if $f(x) = 0$, $g(x) = \lceil x \rceil$, then $h(x) = 0(\lceil x \rceil) = 0$ is continuous at $x = 0$ and $g(x)$ is not.

56. Let $f(x) = \frac{1}{x-1}$ and $g(x) = x + 1$. Both functions are continuous at $x = 0$. The composition $f \circ g = f(g(x))$ $= \frac{1}{(x+1)-1} = \frac{1}{x}$ is discontinuous at $x = 0$, since it is not defined there. Theorem 10 requires that $f(x)$ be continuous at $g(0)$, which is not the case here since $g(0) = 1$ and f is undefined at 1.

57. Yes, because of the Intermediate Value Theorem. If f(a) and f(b) did have different signs then f would have to equal zero at some point between a and b since f is continuous on [a, b].

58. Let f(x) be the new position of point x and let $d(x) = f(x) - x$. The displacement function d is negative if x is the left-hand point of the rubber band and positive if x is the right-hand point of the rubber band. By the Intermediate Value Theorem, $d(x) = 0$ for some point in between. That is, $f(x) = x$ for some point x, which is then in its original position.

59. If $f(0) = 0$ or $f(1) = 1$, we are done (i.e., $c = 0$ or $c = 1$ in those cases). Then let $f(0) = a > 0$ and $f(1) = b < 1$ because $0 \le f(x) \le 1$. Define $g(x) = f(x) - x \Rightarrow$ g is continuous on [0, 1]. Moreover, $g(0) = f(0) - 0 = a > 0$ and $g(1) = f(1) - 1 = b - 1 < 0 \Rightarrow$ by the Intermediate Value Theorem there is a number c in (0, 1) such that $g(c) = 0 \Rightarrow f(c) - c = 0$ or $f(c) = c$.

60. Let $\epsilon = \frac{|f(c)|}{2} > 0$. Since f is continuous at $x = c$ there is a $\delta > 0$ such that $|x - c| < \delta \Rightarrow |f(x) - f(c)| < \epsilon$ $\Rightarrow f(c) - \epsilon < f(x) < f(c) + \epsilon$.
If $f(c) > 0$, then $\epsilon = \frac{1}{2} f(c) \Rightarrow \frac{1}{2} f(c) < f(x) < \frac{3}{2} f(c) \Rightarrow f(x) > 0$ on the interval $(c - \delta, c + \delta)$.
If $f(c) < 0$, then $\epsilon = -\frac{1}{2} f(c) \Rightarrow \frac{3}{2} f(c) < f(x) < \frac{1}{2} f(c) \Rightarrow f(x) < 0$ on the interval $(c - \delta, c + \delta)$.

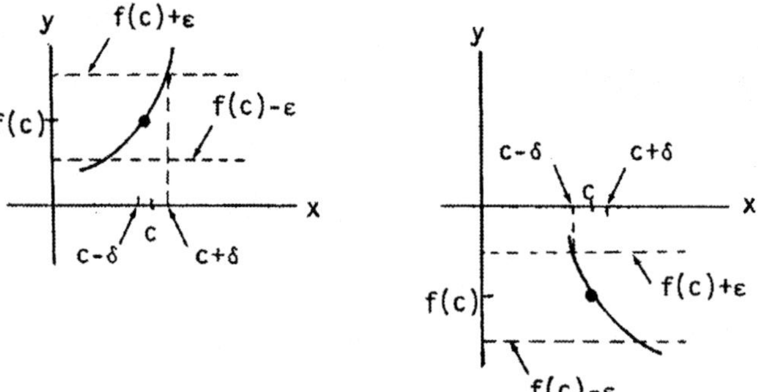

61. By Exercises 52 in Section 2.3, we have $\lim_{x \to c} f(x) = L \Leftrightarrow \lim_{h \to 0} f(c + h) = L$.
Thus, f(x) is continuous at $x = c \Leftrightarrow \lim_{x \to c} f(x) = f(c) \Leftrightarrow \lim_{h \to 0} f(c + h) = f(c)$.

62. By Exercise 61, it suffices to show that $\lim_{h \to 0} \sin(c + h) = \sin c$ and $\lim_{h \to 0} \cos(c + h) = \cos c$.
Now $\lim_{h \to 0} \sin(c + h) = \lim_{h \to 0} [(\sin c)(\cos h) + (\cos c)(\sin h)] = (\sin c)\left(\lim_{h \to 0} \cos h\right) + (\cos c)\left(\lim_{h \to 0} \sin h\right)$
By Example 6 Section 2.2, $\lim_{h \to 0} \cos h = 1$ and $\lim_{h \to 0} \sin h = 0$. So $\lim_{h \to 0} \sin(c + h) = \sin c$ and thus $f(x) = \sin x$ is continuous at $x = c$. Similarly,
$\lim_{h \to 0} \cos(c + h) = \lim_{h \to 0} [(\cos c)(\cos h) - (\sin c)(\sin h)] = (\cos c)\left(\lim_{h \to 0} \cos h\right) - (\sin c)\left(\lim_{h \to 0} \sin h\right) = \cos c$.
Thus, $g(x) = \cos x$ is continuous at $x = c$.

63. $x \approx 1.8794, -1.5321, -0.3473$

64. $x \approx 1.4516, -0.8547, 0.4030$

65. $x \approx 1.7549$

66. $x \approx 1.5596$

67. $x \approx 3.5156$

68. $x \approx -3.9058, 3.8392, 0.0667$

69. $x \approx 0.7391$

70. $x \approx -1.8955, 0, 1.8955$

2.7 TANGENTS AND DERIVATIVES

1. P_1: $m_1 = 1$, P_2: $m_2 = 5$

2. P_1: $m_1 = -2$, P_2: $m_2 = 0$

3. P_1: $m_1 = \frac{5}{2}$, P_2: $m_2 = -\frac{1}{2}$

4. P_1: $m_1 = 3$, P_2: $m_2 = -3$

5. $m = \lim\limits_{h \to 0} \frac{[4-(-1+h)^2]-(4-(-1)^2)}{h}$
$= \lim\limits_{h \to 0} \frac{-(1-2h+h^2)+1}{h} = \lim\limits_{h \to 0} \frac{h(2-h)}{h} = 2;$
at $(-1, 3)$: $y = 3 + 2(x-(-1)) \Rightarrow y = 2x + 5$, tangent line

6. $m = \lim\limits_{h \to 0} \frac{[(1+h-1)^2+1]-[(1-1)^2+1]}{h} = \lim\limits_{h \to 0} \frac{h^2}{h}$
$= \lim\limits_{h \to 0} h = 0$; at $(1, 1)$: $y = 1 + 0(x-1) \Rightarrow y = 1$, tangent line

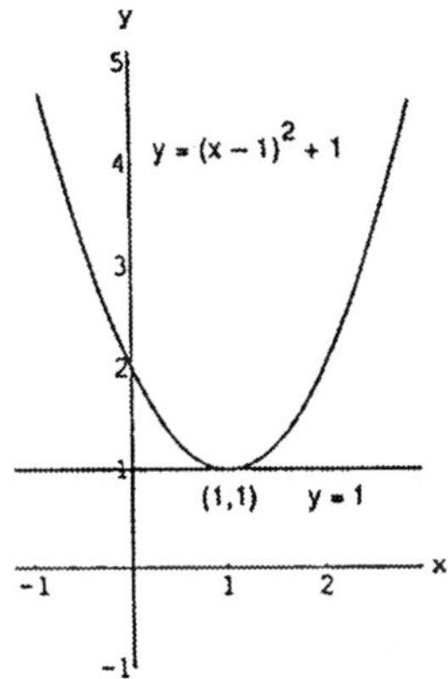

7. $m = \lim\limits_{h \to 0} \frac{2\sqrt{1+h}-2\sqrt{1}}{h} = \lim\limits_{h \to 0} \frac{2\sqrt{1+h}-2}{h} \cdot \frac{2\sqrt{1+h}+2}{2\sqrt{1+h}+2}$
$= \lim\limits_{h \to 0} \frac{4(1+h)-4}{2h\left(\sqrt{1+h}+1\right)} = \lim\limits_{h \to 0} \frac{2}{\sqrt{1+h}+1} = 1;$
at $(1, 2)$: $y = 2 + 1(x-1) \Rightarrow y = x + 1$, tangent line

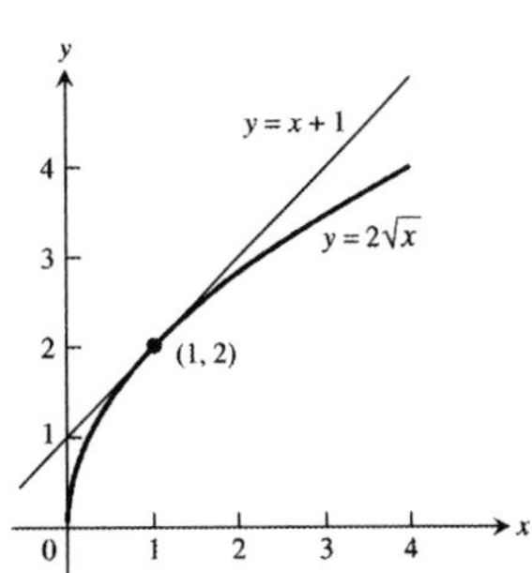

8. $m = \lim_{h \to 0} \frac{\frac{1}{(-1+h)^2} - \frac{1}{(-1)^2}}{h} = \lim_{h \to 0} \frac{1-(-1+h)^2}{h(-1+h)^2}$
$= \lim_{h \to 0} \frac{-(-2h+h^2)}{h(-1+h)^2} = \lim_{h \to 0} \frac{2-h}{(-1+h)^2} = 2;$
at $(-1, 1)$: $y = 1 + 2(x - (-1)) \Rightarrow y = 2x + 3$, tangent line

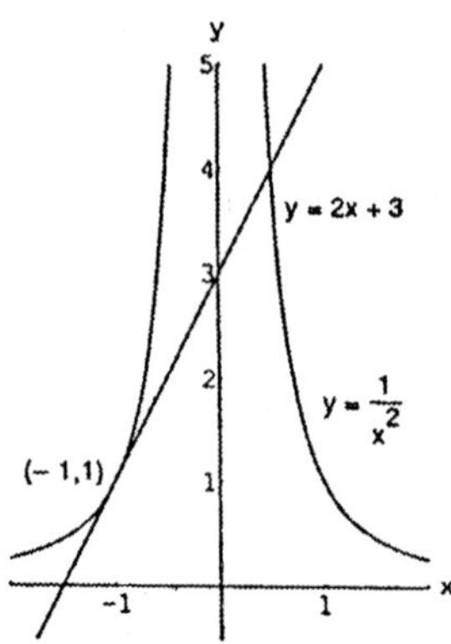

9. $m = \lim_{h \to 0} \frac{(-2+h)^3 - (-2)^3}{h} = \lim_{h \to 0} \frac{-8+12h-6h^2+h^3+8}{h}$
$= \lim_{h \to 0} (12 - 6h + h^2) = 12;$
at $(-2, -8)$: $y = -8 + 12(x - (-2)) \Rightarrow y = 12x + 16$, tangent line

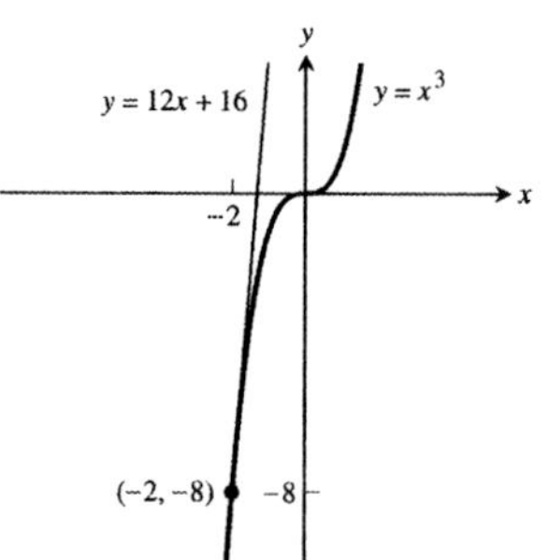

10. $m = \lim_{h \to 0} \frac{\frac{1}{(-2+h)^3} - \frac{1}{(-2)^3}}{h} = \lim_{h \to 0} \frac{-8-(-2+h)^3}{-8h(-2+h)^3}$
$= \lim_{h \to 0} \frac{-(12h-6h^2+h^3)}{-8h(-2+h)^3} = \lim_{h \to 0} \frac{12-6h+h^2}{8(-2+h)^3}$
$= \frac{12}{8(-8)} = -\frac{3}{16};$
at $\left(-2, -\frac{1}{8}\right)$: $y = -\frac{1}{8} - \frac{3}{16}(x - (-2))$
$\Rightarrow y = -\frac{3}{16}x - \frac{1}{2}$, tangent line

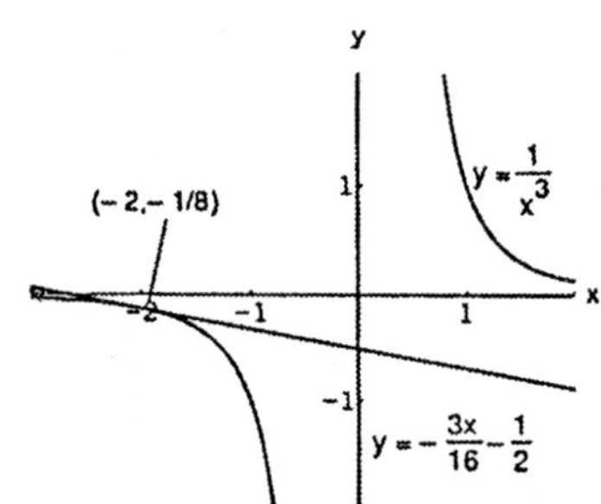

11. $m = \lim_{h \to 0} \frac{[(2+h)^2+1]-5}{h} = \lim_{h \to 0} \frac{(5+4h+h^2)-5}{h} = \lim_{h \to 0} \frac{h(4+h)}{h} = 4;$
at $(2, 5)$: $y - 5 = 4(x - 2)$, tangent line

12. $m = \lim_{h \to 0} \frac{[(1+h)-2(1+h)^2]-(-1)}{h} = \lim_{h \to 0} \frac{(1+h-2-4h-2h^2)+1}{h} = \lim_{h \to 0} \frac{h(-3-2h)}{h} = -3;$
at $(1, -1)$: $y + 1 = -3(x - 1)$, tangent line

13. $m = \lim_{h \to 0} \frac{\frac{3+h}{(3+h)-2} - 3}{h} = \lim_{h \to 0} \frac{(3+h)-3(h+1)}{h(h+1)} = \lim_{h \to 0} \frac{-2h}{h(h+1)} = -2;$
at $(3, 3)$: $y - 3 = -2(x - 3)$, tangent line

14. $m = \lim_{h \to 0} \frac{\frac{8}{(2+h)^2} - 2}{h} = \lim_{h \to 0} \frac{8-2(2+h)^2}{h(2+h)^2} = \lim_{h \to 0} \frac{8-2(4+4h+h^2)}{h(2+h)^2} = \lim_{h \to 0} \frac{-2h(4+h)}{h(2+h)^2} = \frac{-8}{4} = -2;$
at $(2, 2)$: $y - 2 = -2(x - 2)$

15. $m = \lim_{h \to 0} \frac{(2+h)^3-8}{h} = \lim_{h \to 0} \frac{(8+12h+6h^2+h^3)-8}{h} = \lim_{h \to 0} \frac{h(12+6h+h^2)}{h} = 12;$
at $(2, 8)$: $y - 8 = 12(t - 2)$, tangent line

16. $m = \lim_{h \to 0} \frac{[(1+h)^3 + 3(1+h)] - 4}{h} = \lim_{h \to 0} \frac{(1 + 3h + 3h^2 + h^3 + 3 + 3h) - 4}{h} = \lim_{h \to 0} \frac{h(6 + 3h + h^2)}{h} = 6;$
at $(1, 4)$: $y - 4 = 6(t - 1)$, tangent line

17. $m = \lim_{h \to 0} \frac{\sqrt{4+h} - 2}{h} = \lim_{h \to 0} \frac{\sqrt{4+h} - 2}{h} \cdot \frac{\sqrt{4+h}+2}{\sqrt{4+h}+2} = \lim_{h \to 0} \frac{(4+h) - 4}{h\left(\sqrt{4+h}+2\right)} = \lim_{h \to 0} \frac{h}{h\left(\sqrt{4+h}+2\right)} = \frac{1}{\sqrt{4}+2}$
$= \frac{1}{4}$; at $(4, 2)$: $y - 2 = \frac{1}{4}(x - 4)$, tangent line

18. $m = \lim_{h \to 0} \frac{\sqrt{(8+h)+1} - 3}{h} = \lim_{h \to 0} \frac{\sqrt{9+h} - 3}{h} \cdot \frac{\sqrt{9+h}+3}{\sqrt{9+h}+3} = \lim_{h \to 0} \frac{(9+h) - 9}{h\left(\sqrt{9+h}+3\right)} = \lim_{h \to 0} \frac{h}{h\left(\sqrt{9+h}+3\right)}$
$= \frac{1}{\sqrt{9}+3} = \frac{1}{6}$; at $(8, 3)$: $y - 3 = \frac{1}{6}(x - 8)$, tangent line

19. At $x = -1, y = 5 \Rightarrow m = \lim_{h \to 0} \frac{5(-1+h)^2 - 5}{h} = \lim_{h \to 0} \frac{5(1 - 2h + h^2) - 5}{h} = \lim_{h \to 0} \frac{5h(-2+h)}{h} = -10$, slope

20. At $x = 2, y = -3 \Rightarrow m = \lim_{h \to 0} \frac{[1 - (2+h)^2] - (-3)}{h} = \lim_{h \to 0} \frac{(1 - 4 - 4h - h^2) + 3}{h} = \lim_{h \to 0} \frac{-h(4+h)}{h} = -4$, slope

21. At $x = 3, y = \frac{1}{2} \Rightarrow m = \lim_{h \to 0} \frac{\frac{1}{(3+h)-1} - \frac{1}{2}}{h} = \lim_{h \to 0} \frac{2 - (2+h)}{2h(2+h)} = \lim_{h \to 0} \frac{-h}{2h(2+h)} = -\frac{1}{4}$, slope

22. At $x = 0, y = -1 \Rightarrow m = \lim_{h \to 0} \frac{\frac{h-1}{h+1} - (-1)}{h} = \lim_{h \to 0} \frac{(h-1) + (h+1)}{h(h+1)} = \lim_{h \to 0} \frac{2h}{h(h+1)} = 2$, slope

23. At a horizontal tangent the slope $m = 0 \Rightarrow 0 = m = \lim_{h \to 0} \frac{[(x+h)^2 + 4(x+h) - 1] - (x^2 + 4x - 1)}{h}$
$= \lim_{h \to 0} \frac{(x^2 + 2xh + h^2 + 4x + 4h - 1) - (x^2 + 4x - 1)}{h} = \lim_{h \to 0} \frac{(2xh + h^2 + 4h)}{h} = \lim_{h \to 0} (2x + h + 4) = 2x + 4;$
$2x + 4 = 0 \Rightarrow x = -2$. Then $f(-2) = 4 - 8 - 1 = -5 \Rightarrow (-2, -5)$ is the point on the graph where there is a horizontal tangent.

24. $0 = m = \lim_{h \to 0} \frac{[(x+h)^3 - 3(x+h)] - (x^3 - 3x)}{h} = \lim_{h \to 0} \frac{(x^3 + 3x^2h + 3xh^2 + h^3 - 3x - 3h) - (x^3 - 3x)}{h}$
$= \lim_{h \to 0} \frac{3x^2h + 3xh^2 + h^3 - 3h}{h} = \lim_{h \to 0} (3x^2 + 3xh + h^2 - 3) = 3x^2 - 3$; $3x^2 - 3 = 0 \Rightarrow x = -1$ or $x = 1$. Then $f(-1) = 2$ and $f(1) = -2 \Rightarrow (-1, 2)$ and $(1, -2)$ are the points on the graph where a horizontal tangent exists.

25. $-1 = m = \lim_{h \to 0} \frac{\frac{1}{(x+h)-1} - \frac{1}{x-1}}{h} = \lim_{h \to 0} \frac{(x-1) - (x+h-1)}{h(x-1)(x+h-1)} = \lim_{h \to 0} \frac{-h}{h(x-1)(x+h-1)} = -\frac{1}{(x-1)^2}$
$\Rightarrow (x-1)^2 = 1 \Rightarrow x^2 - 2x = 0 \Rightarrow x(x-2) = 0 \Rightarrow x = 0$ or $x = 2$. If $x = 0$, then $y = -1$ and $m = -1$
$\Rightarrow y = -1 - (x - 0) = -(x + 1)$. If $x = 2$, then $y = 1$ and $m = -1 \Rightarrow y = 1 - (x - 2) = -(x - 3)$.

26. $\frac{1}{4} = m = \lim_{h \to 0} \frac{\sqrt{x+h} - \sqrt{x}}{h} = \lim_{h \to 0} \frac{\sqrt{x+h} - \sqrt{x}}{h} \cdot \frac{\sqrt{x+h} + \sqrt{x}}{\sqrt{x+h} + \sqrt{x}} = \lim_{h \to 0} \frac{(x+h) - x}{h\left(\sqrt{x+h} + \sqrt{x}\right)}$
$= \lim_{h \to 0} \frac{h}{h\left(\sqrt{x+h} + \sqrt{x}\right)} = \frac{1}{2\sqrt{x}}$. Thus, $\frac{1}{4} = \frac{1}{2\sqrt{x}} \Rightarrow \sqrt{x} = 2 \Rightarrow x = 4 \Rightarrow y = 2$. The tangent line is
$y = 2 + \frac{1}{4}(x - 4) = \frac{x}{4} + 1$.

27. $\lim_{h \to 0} \frac{f(2+h) - f(2)}{h} = \lim_{h \to 0} \frac{(100 - 4.9(2+h)^2) - (100 - 4.9(2)^2)}{h} = \lim_{h \to 0} \frac{-4.9(4 + 4h + h^2) + 4.9(4)}{h}$
$= \lim_{h \to 0} (-19.6 - 4.9h) = -19.6$. The minus sign indicates the object is falling downward at a speed of 19.6 m/sec.

28. $\lim_{h \to 0} \frac{f(10+h) - f(10)}{h} = \lim_{h \to 0} \frac{3(10+h)^2 - 3(10)^2}{h} = \lim_{h \to 0} \frac{3(20h + h^2)}{h} = 60$ ft/sec.

29. $\lim_{h \to 0} \frac{f(3+h)-f(3)}{h} = \lim_{h \to 0} \frac{\pi(3+h)^2 - \pi(3)^2}{h} = \lim_{h \to 0} \frac{\pi[9+6h+h^2-9]}{h} = \lim_{h \to 0} \pi(6+h) = 6\pi$

30. $\lim_{h \to 0} \frac{f(2+h)-f(2)}{h} = \lim_{h \to 0} \frac{\frac{4\pi}{3}(2+h)^3 - \frac{4\pi}{3}(2)^3}{h} = \lim_{h \to 0} \frac{\frac{4\pi}{3}[12h+6h^2+h^3]}{h} = \lim_{h \to 0} \frac{4\pi}{3}[12+6h+h^2] = 16\pi$

31. Slope at origin $= \lim_{h \to 0} \frac{f(0+h)-f(0)}{h} = \lim_{h \to 0} \frac{h^2 \sin\left(\frac{1}{h}\right)}{h} = \lim_{h \to 0} h \sin\left(\frac{1}{h}\right) = 0 \Rightarrow$ yes, f(x) does have a tangent at the origin with slope 0.

32. $\lim_{h \to 0} \frac{g(0+h)-g(0)}{h} = \lim_{h \to 0} \frac{h \sin\left(\frac{1}{h}\right)}{h} = \lim_{h \to 0} \sin \frac{1}{h}$. Since $\lim_{h \to 0} \sin \frac{1}{h}$ does not exist, f(x) has no tangent at the origin.

33. $\lim_{h \to 0^-} \frac{f(0+h)-f(0)}{h} = \lim_{h \to 0^-} \frac{-1-0}{h} = \infty$, and $\lim_{h \to 0^+} \frac{f(0+h)-f(0)}{h} = \lim_{h \to 0^+} \frac{1-0}{h} = \infty$. Therefore, $\lim_{h \to 0} \frac{f(0+h)-f(0)}{h} = \infty \Rightarrow$ yes, the graph of f has a vertical tangent at the origin.

34. $\lim_{h \to 0^-} \frac{U(0+h)-U(0)}{h} = \lim_{h \to 0^-} \frac{0-1}{h} = \infty$, and $\lim_{h \to 0^+} \frac{U(0+h)-U(0)}{h} = \lim_{h \to 0^+} \frac{1-1}{h} = 0 \Rightarrow$ no, the graph of f does not have a vertical tangent at (0, 1) because the limit does not exist.

35. (a) The graph appears to have a cusp at x = 0.

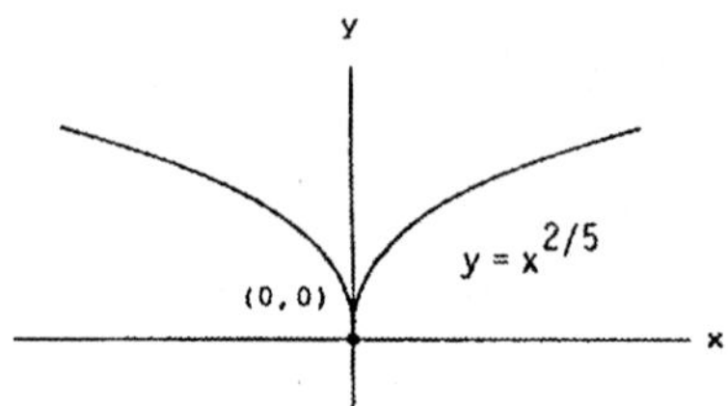

(b) $\lim_{h \to 0^-} \frac{f(0+h)-f(0)}{h} = \lim_{h \to 0^-} \frac{h^{2/5}-0}{h} = \lim_{h \to 0^-} \frac{1}{h^{3/5}} = -\infty$ and $\lim_{h \to 0^+} \frac{1}{h^{3/5}} = \infty \Rightarrow$ limit does not exist $\Rightarrow$ the graph of $y = x^{2/5}$ does not have a vertical tangent at x = 0.

36. (a) The graph appears to have a cusp at x = 0.

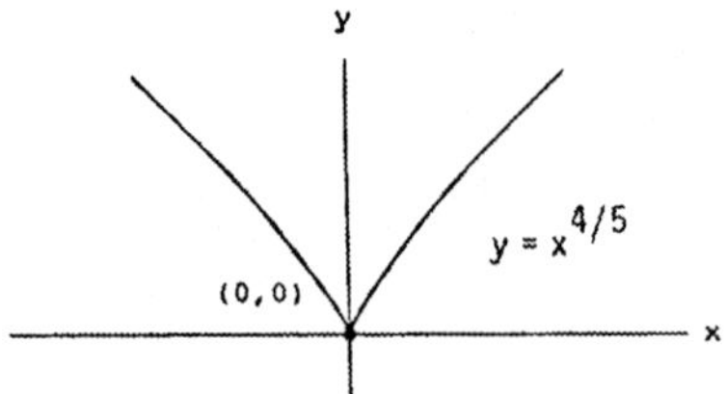

(b) $\lim_{h \to 0^-} \frac{f(0+h)-f(0)}{h} = \lim_{h \to 0^-} \frac{h^{4/5}-0}{h} = \lim_{h \to 0^-} \frac{1}{h^{1/5}} = -\infty$ and $\lim_{h \to 0^+} \frac{1}{h^{1/5}} = \infty \Rightarrow$ limit does not exist $\Rightarrow y = x^{4/5}$ does not have a vertical tangent at x = 0.

37. (a) The graph appears to have a vertical tangent at x = 0.

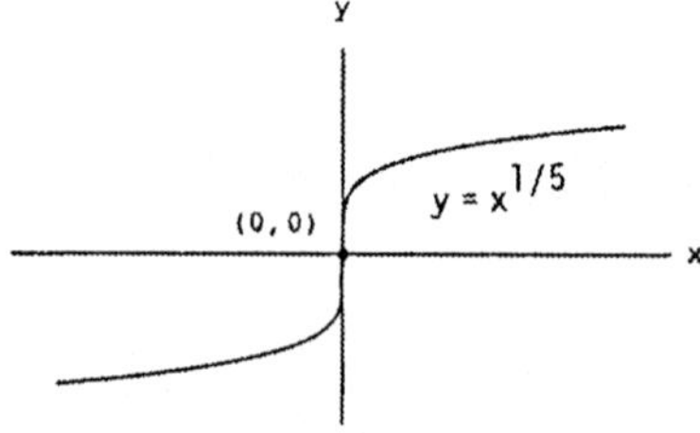

(b) $\lim_{h \to 0} \frac{f(0+h)-f(0)}{h} = \lim_{h \to 0} \frac{h^{1/5}-0}{h} = \lim_{h \to 0} \frac{1}{h^{4/5}} = \infty \Rightarrow y = x^{1/5}$ has a vertical tangent at x = 0.

38. (a) The graph appears to have a vertical tangent at $x = 0$.

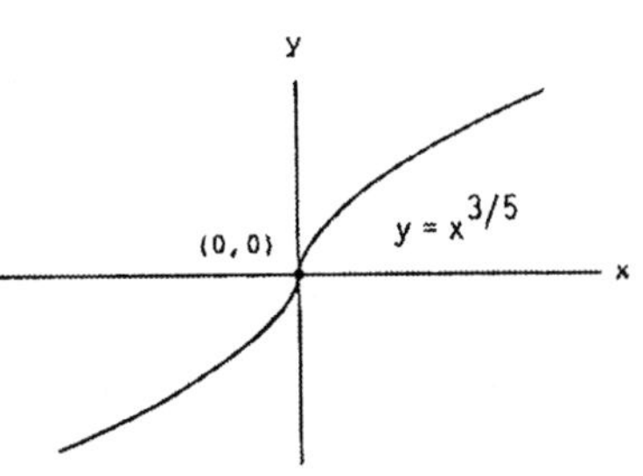

(b) $\lim\limits_{h \to 0} \frac{f(0+h) - f(0)}{h} = \lim\limits_{h \to 0} \frac{h^{3/5} - 0}{h} = \lim\limits_{h \to 0} \frac{1}{h^{2/5}} = \infty \Rightarrow$ the graph of $y = x^{3/5}$ has a vertical tangent at $x = 0$.

39. (a) The graph appears to have a cusp at $x = 0$.

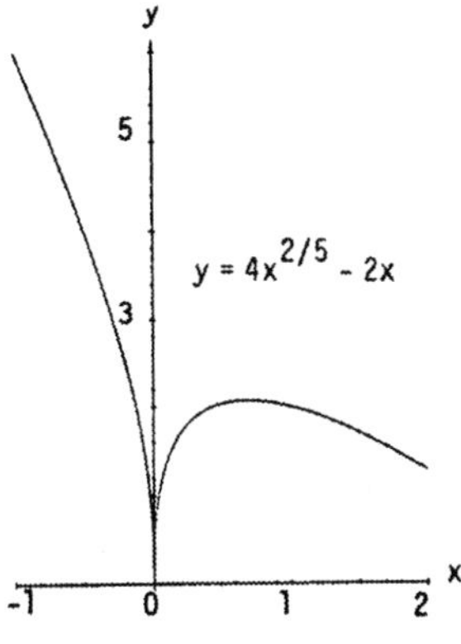

(b) $\lim\limits_{h \to 0^-} \frac{f(0+h) - f(0)}{h} = \lim\limits_{h \to 0^-} \frac{4h^{2/5} - 2h}{h} = \lim\limits_{h \to 0^-} \frac{4}{h^{3/5}} - 2 = -\infty$ and $\lim\limits_{h \to 0^+} \frac{4}{h^{3/5}} - 2 = \infty$
$\Rightarrow$ limit does not exist $\Rightarrow$ the graph of $y = 4x^{2/5} - 2x$ does not have a vertical tangent at $x = 0$.

40. (a) The graph appears to have a cusp at $x = 0$.

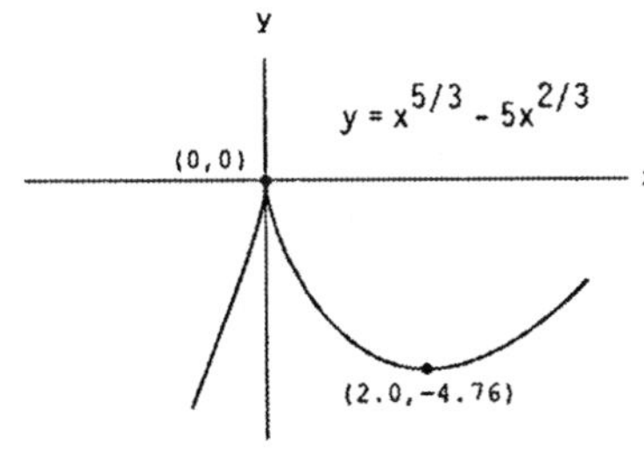

(b) $\lim\limits_{h \to 0} \frac{f(0+h) - f(0)}{h} = \lim\limits_{h \to 0} \frac{h^{5/3} - 5h^{2/3}}{h} = \lim\limits_{h \to 0} h^{2/3} - \frac{5}{h^{1/3}} = 0 - \lim\limits_{h \to 0} \frac{5}{h^{1/3}}$ does not exist $\Rightarrow$ the graph of $y = x^{5/3} - 5x^{2/3}$ does not have a vertical tangent at $x = 0$.

41. (a) The graph appears to have a vertical tangent at $x = 1$ and a cusp at $x = 0$.

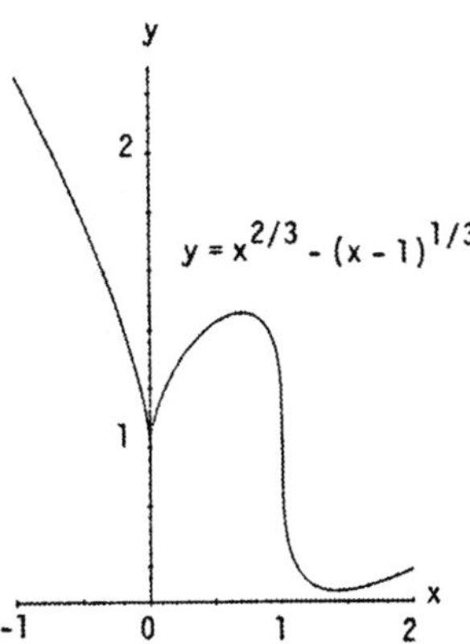

(b) $x = 1$: $\lim\limits_{h \to 0} \frac{(1+h)^{2/3} - (1+h-1)^{1/3} - 1}{h} = \lim\limits_{h \to 0} \frac{(1+h)^{2/3} - h^{1/3} - 1}{h} = -\infty$
$\Rightarrow y = x^{2/3} - (x-1)^{1/3}$ has a vertical tangent at $x = 1$;

$x = 0$: $\lim\limits_{h \to 0} \frac{f(0+h)-f(0)}{h} = \lim\limits_{h \to 0} \frac{h^{2/3}-(h-1)^{1/3}-(-1)^{1/3}}{h} = \lim\limits_{h \to 0} \left[\frac{1}{h^{1/3}} - \frac{(h-1)^{1/3}}{h} + \frac{1}{h}\right]$ does not exist $\Rightarrow$ $y = x^{2/3} - (x-1)^{1/3}$ does not have a vertical tangent at $x = 0$.

42. (a) The graph appears to have vertical tangents at $x = 0$ and $x = 1$.

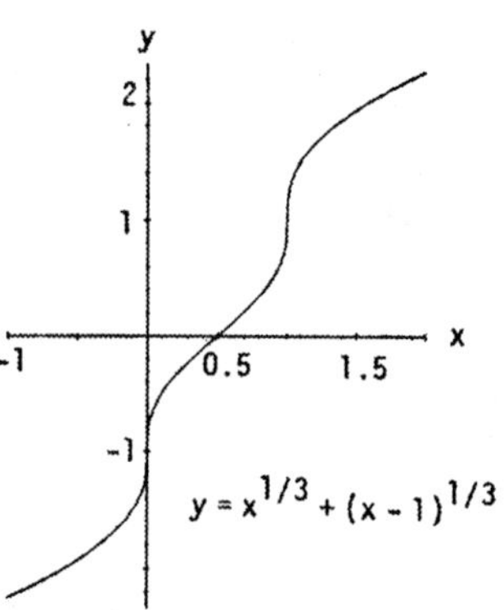

(b) $x = 0$: $\lim\limits_{h \to 0} \frac{f(0+h)-f(0)}{h} = \lim\limits_{h \to 0} \frac{h^{1/3}+(h-1)^{1/3}-(-1)^{1/3}}{h} = \infty \Rightarrow y = x^{1/3} + (x-1)^{1/3}$ has a vertical tangent at $x = 0$;

$x = 1$: $\lim\limits_{h \to 0} \frac{f(1+h)-f(1)}{h} = \lim\limits_{h \to 0} \frac{(1+h)^{1/3}+(1+h-1)^{1/3}-1}{h} = \infty \Rightarrow y = x^{1/3} + (x-1)^{1/3}$ has a vertical tangent at $x = 1$.

43. (a) The graph appears to have a vertical tangent at $x = 0$.

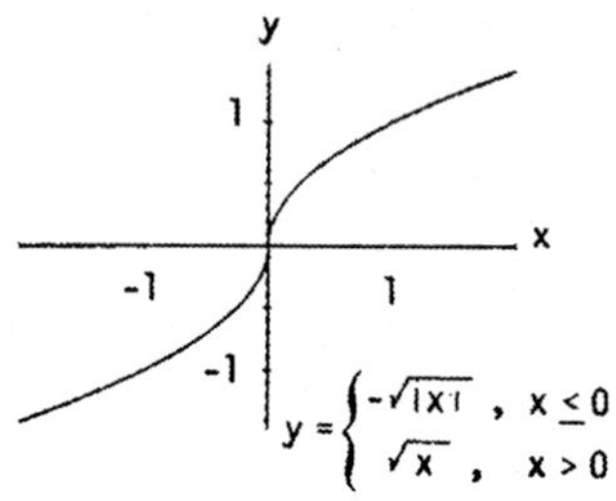

(b) $\lim\limits_{h \to 0^+} \frac{f(0+h)-f(0)}{h} = \lim\limits_{x \to 0^+} \frac{\sqrt{h}-0}{h} = \lim\limits_{h \to 0} \frac{1}{\sqrt{h}} = \infty$;

$\lim\limits_{h \to 0^-} \frac{f(0+h)-f(0)}{h} = \lim\limits_{h \to 0^-} \frac{-\sqrt{|h|}-0}{h} = \lim\limits_{h \to 0^-} \frac{-\sqrt{|h|}}{-|h|} = \lim\limits_{h \to 0^-} \frac{1}{\sqrt{|h|}} = \infty$

$\Rightarrow$ y has a vertical tangent at $x = 0$.

44. (a) The graph appears to have a cusp at $x = 4$.

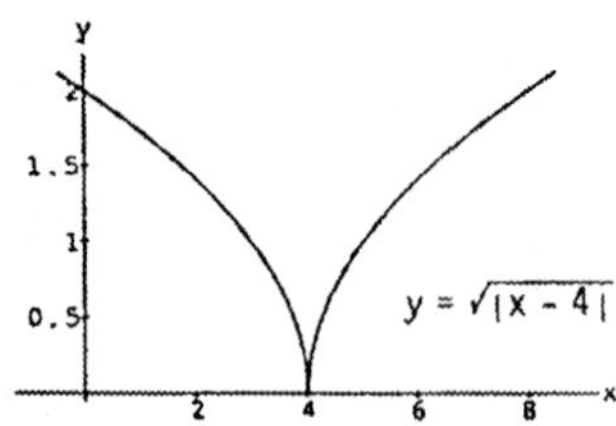

(b) $\lim\limits_{h \to 0^+} \frac{f(4+h)-f(4)}{h} = \lim\limits_{h \to 0^+} \frac{\sqrt{|4-(4+h)|}-0}{h} = \lim\limits_{h \to 0^+} \frac{\sqrt{|h|}}{h} = \lim\limits_{h \to 0^+} \frac{1}{\sqrt{h}} = \infty$;

$\lim\limits_{h \to 0^-} \frac{f(4+h)-f(4)}{h} = \lim\limits_{h \to 0^-} \frac{\sqrt{|4-(4+h)|}}{h} = \lim\limits_{h \to 0^-} \frac{\sqrt{|h|}}{-|h|} = \lim\limits_{h \to 0^-} \frac{-1}{\sqrt{|h|}} = -\infty$

$\Rightarrow$ $y = \sqrt{4-x}$ does not have a vertical tangent at $x = 4$.

45-48. Example CAS commands:

Maple:

```
f := x -> x^3 + 2*x;x0 := 0;
plot( f(x), x=x0-1/2..x0+3, color=black,              # part (a)
      title="Section 2.7, #45(a)" );
q := unapply( (f(x0+h)-f(x0))/h, h );                 # part (b)
```

```
L := limit( q(h), h=0 );                                    # part (c)
sec_lines := seq( f(x0)+q(h)*(x-x0), h=1..3 );              # part (d)
tan_line := f(x0) + L*(x-x0);
plot( [f(x),tan_line,sec_lines], x=x0-1/2..x0+3, color=black,
      linestyle=[1,2,5,6,7], title="Section 2.7, #45(d)",
      legend=["y=f(x)","Tangent line at x=0","Secant line (h=1)",
              "Secant line (h=2)","Secant line (h=3)"] );
```

Mathematica: (function and value for x0 may change)

```
Clear[f, m, x, h]
x0 = p;
f[x_]: = Cos[x] + 4Sin[2x]
Plot[f[x], {x, x0 − 1, x0 + 3}]
dq[h_]: = (f[x0+h] − f[x0])/h
m = Limit[dq[h], h → 0]
ytan: = f[x0] + m(x − x0)
y1: = f[x0] + dq[1](x − x0)
y2: = f[x0] + dq[2](x − x0)
y3: = f[x0] + dq[3](x − x0)
Plot[{f[x], ytan, y1, y2, y3}, {x, x0 − 1, x0 + 3}]
```

CHAPTER 2 PRACTICE EXERCISES

1. At $x = -1$: $\lim_{x \to -1^-} f(x) = \lim_{x \to -1^+} f(x) = 1$

   $\Rightarrow \lim_{x \to -1} f(x) = 1 = f(-1)$

   $\Rightarrow$ f is continuous at $x = -1$.

   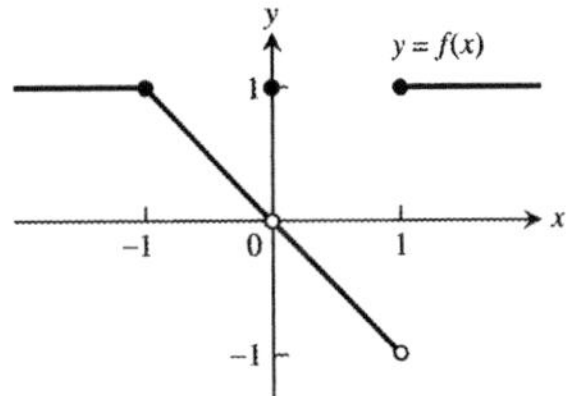

   At $x = 0$: $\lim_{x \to 0^-} f(x) = \lim_{x \to 0^+} f(x) = 0 \Rightarrow \lim_{x \to 0} f(x) = 0$.

   But $f(0) = 1 \neq \lim_{x \to 0} f(x)$

   $\Rightarrow$ f is discontinuous at $x = 0$.

   If we define $f(0) = 0$, then the discontinuity at $x = 0$ is removable.

   At $x = 1$: $\lim_{x \to 1^-} f(x) = -1$ and $\lim_{x \to 1^+} f(x) = 1$

   $\Rightarrow \lim_{x \to 1} f(x)$ does not exist

   $\Rightarrow$ f is discontinuous at $x = 1$.

2. At $x = -1$: $\lim_{x \to -1^-} f(x) = 0$ and $\lim_{x \to -1^+} f(x) = -1$

   $\Rightarrow \lim_{x \to -1} f(x)$ does not exist

   $\Rightarrow$ f is discontinuous at $x = -1$.

   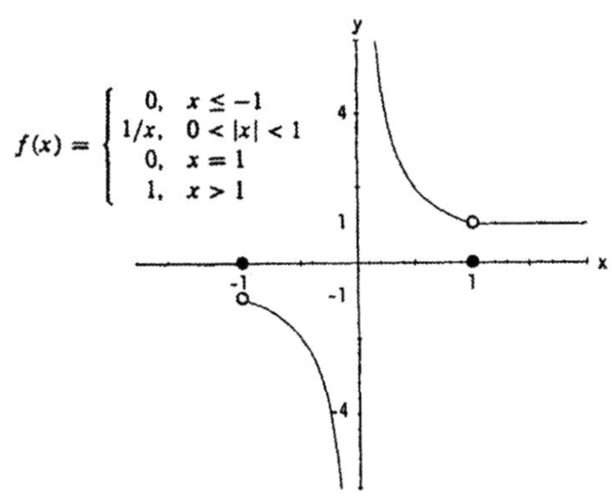

   At $x = 0$: $\lim_{x \to 0^-} f(x) = -\infty$ and $\lim_{x \to 0^+} f(x) = \infty$

   $\Rightarrow \lim_{x \to 0} f(x)$ does not exist

   $\Rightarrow$ f is discontinuous at $x = 0$.

   At $x = 1$: $\lim_{x \to 1^-} f(x) = \lim_{x \to 1^+} f(x) = 1 \Rightarrow \lim_{x \to 1} f(x) = 1$.

   But $f(1) = 0 \neq \lim_{x \to 1} f(x)$

   $\Rightarrow$ f is discontinuous at $x = 1$.

   If we define $f(1) = 1$, then the discontinuity at $x = 1$ is removable.

3. (a) $\lim\limits_{t \to t_0} (3f(t)) = 3 \lim\limits_{t \to t_0} f(t) = 3(-7) = -21$

(b) $\lim\limits_{t \to t_0} (f(t))^2 = \left(\lim\limits_{t \to t_0} f(t)\right)^2 = (-7)^2 = 49$

(c) $\lim\limits_{t \to t_0} (f(t) \cdot g(t)) = \lim\limits_{t \to t_0} f(t) \cdot \lim\limits_{t \to t_0} g(t) = (-7)(0) = 0$

(d) $\lim\limits_{t \to t_0} \frac{f(t)}{g(t)-7} = \frac{\lim\limits_{t \to t_0} f(t)}{\lim\limits_{t \to t_0} (g(t)-7)} = \frac{\lim\limits_{t \to t_0} f(t)}{\lim\limits_{t \to t_0} g(t) - \lim\limits_{t \to t_0} 7} = \frac{-7}{0-7} = 1$

(e) $\lim\limits_{t \to t_0} \cos(g(t)) = \cos\left(\lim\limits_{t \to t_0} g(t)\right) = \cos 0 = 1$

(f) $\lim\limits_{t \to t_0} |f(t)| = \left|\lim\limits_{t \to t_0} f(t)\right| = |-7| = 7$

(g) $\lim\limits_{t \to t_0} (f(t) + g(t)) = \lim\limits_{t \to t_0} f(t) + \lim\limits_{t \to t_0} g(t) = -7 + 0 = -7$

(h) $\lim\limits_{t \to t_0} \left(\frac{1}{f(t)}\right) = \frac{1}{\lim\limits_{t \to t_0} f(t)} = \frac{1}{-7} = -\frac{1}{7}$

4. (a) $\lim\limits_{x \to 0} -g(x) = -\lim\limits_{x \to 0} g(x) = -\sqrt{2}$

(b) $\lim\limits_{x \to 0} (g(x) \cdot f(x)) = \lim\limits_{x \to 0} g(x) \cdot \lim\limits_{x \to 0} f(x) = \left(\sqrt{2}\right)\left(\frac{1}{2}\right) = \frac{\sqrt{2}}{2}$

(c) $\lim\limits_{x \to 0} (f(x) + g(x)) = \lim\limits_{x \to 0} f(x) + \lim\limits_{x \to 0} g(x) = \frac{1}{2} + \sqrt{2}$

(d) $\lim\limits_{x \to 0} \frac{1}{f(x)} = \frac{1}{\lim\limits_{x \to 0} f(x)} = \frac{1}{\frac{1}{2}} = 2$

(e) $\lim\limits_{x \to 0} (x + f(x)) = \lim\limits_{x \to 0} x + \lim\limits_{x \to 0} f(x) = 0 + \frac{1}{2} = \frac{1}{2}$

(f) $\lim\limits_{x \to 0} \frac{f(x) \cdot \cos x}{x-1} = \frac{\lim\limits_{x \to 0} f(x) \cdot \lim\limits_{x \to 0} \cos x}{\lim\limits_{x \to 0} x - \lim\limits_{x \to 0} 1} = \frac{\left(\frac{1}{2}\right)(1)}{0-1} = -\frac{1}{2}$

5. Since $\lim\limits_{x \to 0} x = 0$ we must have that $\lim\limits_{x \to 0} (4 - g(x)) = 0$. Otherwise, if $\lim\limits_{x \to 0} (4 - g(x))$ is a finite positive number, we would have $\lim\limits_{x \to 0^-} \left[\frac{4-g(x)}{x}\right] = -\infty$ and $\lim\limits_{x \to 0^+} \left[\frac{4-g(x)}{x}\right] = \infty$ so the limit could not equal 1 as $x \to 0$. Similar reasoning holds if $\lim\limits_{x \to 0} (4 - g(x))$ is a finite negative number. We conclude that $\lim\limits_{x \to 0} g(x) = 4$.

6. $2 = \lim\limits_{x \to -4} \left[x \lim\limits_{x \to 0} g(x)\right] = \lim\limits_{x \to -4} x \cdot \lim\limits_{x \to -4} \left[\lim\limits_{x \to 0} g(x)\right] = -4 \lim\limits_{x \to -4} \left[\lim\limits_{x \to 0} g(x)\right] = -4 \lim\limits_{x \to 0} g(x)$ (since $\lim\limits_{x \to 0} g(x)$ is a constant) $\Rightarrow \lim\limits_{x \to 0} g(x) = \frac{2}{-4} = -\frac{1}{2}$.

7. (a) $\lim\limits_{x \to c} f(x) = \lim\limits_{x \to c} x^{1/3} = c^{1/3} = f(c)$ for every real number $c \Rightarrow f$ is continuous on $(-\infty, \infty)$.

(b) $\lim\limits_{x \to c} g(x) = \lim\limits_{x \to c} x^{3/4} = c^{3/4} = g(c)$ for every nonnegative real number $c \Rightarrow g$ is continuous on $[0, \infty)$.

(c) $\lim\limits_{x \to c} h(x) = \lim\limits_{x \to c} x^{-2/3} = \frac{1}{c^{2/3}} = h(c)$ for every nonzero real number $c \Rightarrow h$ is continuous on $(-\infty, 0)$ and $(-\infty, \infty)$.

(d) $\lim\limits_{x \to c} k(x) = \lim\limits_{x \to c} x^{-1/6} = \frac{1}{c^{1/6}} = k(c)$ for every positive real number $c \Rightarrow k$ is continuous on $(0, \infty)$

8. (a) $\bigcup\limits_{n \in I} \left(\left(n - \frac{1}{2}\right)\pi, \left(n + \frac{1}{2}\right)\pi\right)$, where I = the set of all integers.

(b) $\bigcup\limits_{n \in I} (n\pi, (n+1)\pi)$, where I = the set of all integers.

(c) $(-\infty, \pi) \cup (\pi, \infty)$

(d) $(-\infty, 0) \cup (0, \infty)$

9. (a) $\lim\limits_{x \to 0} \frac{x^2 - 4x + 4}{x^3 + 5x^2 - 14x} = \lim\limits_{x \to 0} \frac{(x-2)(x-2)}{x(x+7)(x-2)} = \lim\limits_{x \to 0} \frac{x-2}{x(x+7)}$, $x \neq 2$; the limit does not exist because $\lim\limits_{x \to 0^-} \frac{x-2}{x(x+7)} = \infty$ and $\lim\limits_{x \to 0^+} \frac{x-2}{x(x+7)} = -\infty$

(b) $\lim\limits_{x \to 2} \frac{x^2 - 4x + 4}{x^3 + 5x^2 - 14x} = \lim\limits_{x \to 2} \frac{(x-2)(x-2)}{x(x+7)(x-2)} = \lim\limits_{x \to 2} \frac{x-2}{x(x+7)}$, $x \neq 2$, and $\lim\limits_{x \to 2} \frac{x-2}{x(x+7)} = \frac{0}{2(9)} = 0$

10. (a) $\lim\limits_{x\to 0} \frac{x^2+x}{x^5+2x^4+x^3} = \lim\limits_{x\to 0} \frac{x(x+1)}{x^3(x^2+2x+1)} = \lim\limits_{x\to 0} \frac{x+1}{x^2(x+1)(x+1)} = \lim\limits_{x\to 0} \frac{1}{x^2(x+1)}$, $x \neq 0$ and $x \neq -1$.
Now $\lim\limits_{x\to 0^-} \frac{1}{x^2(x+1)} = \infty$ and $\lim\limits_{x\to 0^+} \frac{1}{x^2(x+1)} = \infty \Rightarrow \lim\limits_{x\to 0} \frac{x^2+x}{x^5+2x^4+x^3} = \infty$.

(b) $\lim\limits_{x\to -1} \frac{x^2+x}{x^5+2x^4+x^3} = \lim\limits_{x\to -1} \frac{x(x+1)}{x^3(x^2+2x+1)} = \lim\limits_{x\to -1} \frac{1}{x^2(x+1)}$, $x \neq 0$ and $x \neq -1$. The limit does not exist because $\lim\limits_{x\to -1^-} \frac{1}{x^2(x+1)} = -\infty$ and $\lim\limits_{x\to -1^+} \frac{1}{x^2(x+1)} = \infty$.

11. $\lim\limits_{x\to 1} \frac{1-\sqrt{x}}{1-x} = \lim\limits_{x\to 1} \frac{1-\sqrt{x}}{(1-\sqrt{x})(1+\sqrt{x})} = \lim\limits_{x\to 1} \frac{1}{1+\sqrt{x}} = \frac{1}{2}$

12. $\lim\limits_{x\to a} \frac{x^2-a^2}{x^4-a^4} = \lim\limits_{x\to a} \frac{(x^2-a^2)}{(x^2+a^2)(x^2-a^2)} = \lim\limits_{x\to a} \frac{1}{x^2+a^2} = \frac{1}{2a^2}$

13. $\lim\limits_{h\to 0} \frac{(x+h)^2-x^2}{h} = \lim\limits_{h\to 0} \frac{(x^2+2hx+h^2)-x^2}{h} = \lim\limits_{h\to 0} (2x+h) = 2x$

14. $\lim\limits_{x\to 0} \frac{(x+h)^2-x^2}{h} = \lim\limits_{x\to 0} \frac{(x^2+2hx+h^2)-x^2}{h} = \lim\limits_{x\to 0} (2x+h) = h$

15. $\lim\limits_{x\to 0} \frac{\frac{1}{2+x}-\frac{1}{2}}{x} = \lim\limits_{x\to 0} \frac{2-(2+x)}{2x(2+x)} = \lim\limits_{x\to 0} \frac{-1}{4+2x} = -\frac{1}{4}$

16. $\lim\limits_{x\to 0} \frac{(2+x)^3-8}{x} = \lim\limits_{x\to 0} \frac{(x^3+6x^2+12x+8)-8}{x} = \lim\limits_{x\to 0} (x^2+6x+12) = 12$

17. $\lim\limits_{x\to 0} \frac{\tan 2x}{\tan \pi x} = \lim\limits_{x\to 0} \frac{\sin 2x}{\cos 2x} \cdot \frac{\cos \pi x}{\sin \pi x} = \lim\limits_{x\to 0} \left(\frac{\sin 2x}{2x}\right)\left(\frac{\cos \pi x}{\cos 2x}\right)\left(\frac{\pi x}{\sin \pi x}\right)\left(\frac{2x}{\pi x}\right) = 1\cdot 1\cdot 1\cdot \frac{2}{\pi} = \frac{2}{\pi}$

18. $\lim\limits_{x\to \pi^-} \csc x = \lim\limits_{x\to \pi^-} \frac{1}{\sin x} = \infty$ and $\lim\limits_{x\to \pi^+} \csc x = \lim\limits_{x\to \pi^+} \frac{1}{\sin x} = -\infty \Rightarrow \lim\limits_{x\to \pi} \csc x =$ does not exist

19. $\lim\limits_{x\to \pi} \sin\left(\frac{x}{2} + \sin x\right) = \sin\left(\frac{\pi}{2} + \sin \pi\right) = \sin\left(\frac{\pi}{2}\right) = 1$

20. $\lim\limits_{x\to 1} e^{(x^2+x-2)} = e^0 = 1$

21. Let $x = t-3 \Rightarrow \lim\limits_{t\to 3^+} \ln(t-3) = \lim\limits_{x\to 0^+} \ln x = -\infty$

22. $\lim\limits_{t\to 1} t^2 \ln\left(2-\sqrt{t}\right) = \ln 1 = 0$

23. $-1 \le \cos\left(\frac{\pi}{\theta}\right) \le 1 \Rightarrow e^{-1} \le e^{\cos(\pi/\theta)} \le e^1 \Rightarrow \sqrt{\theta}\, e^{-1} \le \sqrt{\theta}\, e^{\cos(\pi/\theta)} \le \sqrt{\theta}\, e \Rightarrow \lim\limits_{\theta\to 0^+} \sqrt{\theta}\, e^{\cos(\pi/\theta)} = 0$ by the Sandwich Theorem

24. $\lim\limits_{z\to 0^+} \frac{2e^{1/z}}{e^{1/z}+1} = \lim\limits_{z\to 0^+} \frac{2}{1+e^{-1/z}} = \frac{2}{1+0} = 2$

25. $\lim\limits_{x\to 0^+} [4\,g(x)]^{1/3} = 2 \Rightarrow \left[\lim\limits_{x\to 0^+} 4\,g(x)\right]^{1/3} = 2 \Rightarrow \lim\limits_{x\to 0^+} 4\,g(x) = 8$, since $2^3 = 8$. Then $\lim\limits_{x\to 0^+} g(x) = 2$.

26. $\lim\limits_{x\to \sqrt{5}} \frac{1}{x+g(x)} = 2 \Rightarrow \lim\limits_{x\to \sqrt{5}} (x+g(x)) = \frac{1}{2} \Rightarrow \sqrt{5} + \lim\limits_{x\to \sqrt{5}} g(x) = \frac{1}{2} \Rightarrow \lim\limits_{x\to \sqrt{5}} g(x) = \frac{1}{2} - \sqrt{5}$

27. $\lim\limits_{x\to 1} \frac{3x^2+1}{g(x)} = \infty \Rightarrow \lim\limits_{x\to 1} g(x) = 0$ since $\lim\limits_{x\to 1} (3x^2+1) = 4$

28. $\lim\limits_{x \to -2} \frac{5-x^2}{\sqrt{g(x)}} = 0 \Rightarrow \lim\limits_{x \to -2} g(x) = \infty$ since $\lim\limits_{x \to -2} (5 - x^2) = 1$

29. $\lim\limits_{x \to \infty} \frac{2x+3}{5x+7} = \lim\limits_{x \to \infty} \frac{2+\frac{3}{x}}{5+\frac{7}{x}} = \frac{2+0}{5+0} = \frac{2}{5}$

30. $\lim\limits_{x \to -\infty} \frac{2x^2+3}{5x^2+7} = \lim\limits_{x \to -\infty} \frac{2+\frac{3}{x^2}}{5+\frac{7}{x^2}} = \frac{2+0}{5+0} = \frac{2}{5}$

31. $\lim\limits_{x \to -\infty} \frac{x^2-4x+8}{3x^3} = \lim\limits_{x \to -\infty} \left(\frac{1}{3x} - \frac{4}{3x^2} + \frac{8}{3x^3}\right) = 0 - 0 + 0 = 0$

32. $\lim\limits_{x \to \infty} \frac{1}{x^2-7x+1} = \lim\limits_{x \to \infty} \frac{\frac{1}{x^2}}{1-\frac{7}{x}+\frac{1}{x^2}} = \frac{0}{1-0+0} = 0$

33. $\lim\limits_{x \to -\infty} \frac{x^2-7x}{x+1} = \lim\limits_{x \to -\infty} \frac{x-7}{1+\frac{1}{x}} = -\infty$

34. $\lim\limits_{x \to \infty} \frac{x^4+x^3}{12x^3+128} = \lim\limits_{x \to -\infty} \frac{x+1}{12+\frac{128}{x^3}} = \infty$

35. $\lim\limits_{x \to \infty} \frac{|\sin x|}{\lfloor x \rfloor} \le \lim\limits_{x \to \infty} \frac{1}{\lfloor x \rfloor} = 0$ since int $x \to \infty$ as $x \to \infty \Rightarrow \lim\limits_{x \to \infty} \frac{|\sin x|}{\lfloor x \rfloor} = 0.$

36. $\lim\limits_{\theta \to \infty} \frac{|\cos \theta - 1|}{\theta} \le \lim\limits_{\theta \to \infty} \frac{|-2|}{\theta} = 0 \Rightarrow \lim\limits_{\theta \to \infty} \frac{|\cos \theta - 1|}{\theta} = 0.$

37. $\lim\limits_{x \to \infty} \frac{x + \sin x + 2\sqrt{x}}{x + \sin x} = \lim\limits_{x \to \infty} \frac{1 + \frac{\sin x}{x} + \frac{2}{\sqrt{x}}}{1 + \frac{\sin x}{x}} = \frac{1+0+0}{1+0} = 1$

38. $\lim\limits_{x \to \infty} \frac{x^{2/3} + x^{-1}}{x^{2/3} + \cos^2 x} = \lim\limits_{x \to \infty} \left(\frac{1 + x^{-5/3}}{1 + \frac{\cos^2 x}{x^{2/3}}}\right) = \frac{1+0}{1+0} = 1$

39. $\lim\limits_{x \to \infty} e^{1/x} \cos\left(\frac{1}{x}\right) = e^0 \cdot \cos(0) = 1 \cdot 1 = 1$

40. $\lim\limits_{t \to \infty} \ln\left(1 + \frac{1}{t}\right) = \ln 1 = 0$

41. $\lim\limits_{x \to -\infty} \tan^{-1} x = -\frac{\pi}{2}$

42. $\lim\limits_{t \to -\infty} e^{3t} \sin^{-1}\left(\frac{1}{t}\right) = 0 \cdot \sin^{-1}(0) = 0 \cdot 0 = 0$

43. At $x = -1$: $\lim\limits_{x \to -1^-} f(x) = \lim\limits_{x \to -1^-} \frac{x(x^2-1)}{|x^2-1|}$
$= \lim\limits_{x \to -1^-} \frac{x(x^2-1)}{x^2-1} = \lim\limits_{x \to -1^-} x = -1$, and
$\lim\limits_{x \to -1^+} f(x) = \lim\limits_{x \to -1^+} \frac{x(x^2-1)}{|x^2-1|} = \lim\limits_{x \to -1^+} \frac{x(x^2-1)}{-(x^2-1)}$
$= \lim\limits_{x \to -1} (-x) = -(-1) = 1$. Since
$\lim\limits_{x \to -1^-} f(x) \ne \lim\limits_{x \to -1^+} f(x)$
$\Rightarrow \lim\limits_{x \to -1} f(x)$ does not exist, the function f cannot be extended to a continuous function at $x = -1$.

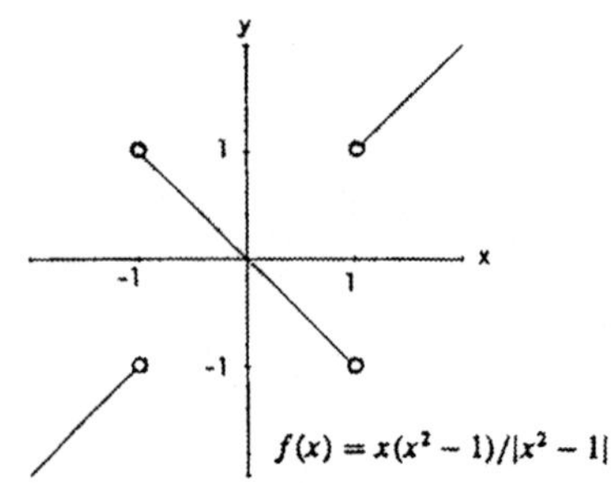

At $x = 1$: $\lim\limits_{x \to 1^-} f(x) = \lim\limits_{x \to 1^-} \frac{x(x^2-1)}{|x^2-1|} = \lim\limits_{x \to 1^-} \frac{x(x^2-1)}{-(x^2-1)} = \lim\limits_{x \to 1^-} (-x) = -1$, and
$\lim\limits_{x \to 1^+} f(x) = \lim\limits_{x \to 1^+} \frac{x(x^2-1)}{|x^2-1|} = \lim\limits_{x \to 1^+} \frac{x(x^2-1)}{x^2-1} = \lim\limits_{x \to 1^+} x = 1$. Again $\lim\limits_{x \to 1} f(x)$ does not exist so f cannot be extended to a continuous function at $x = 1$ either.

44. The discontinuity at $x = 0$ of $f(x) = \sin\left(\frac{1}{x}\right)$ is nonremovable because $\lim\limits_{x \to 0} \sin \frac{1}{x}$ does not exist.

45. Yes, f does have a continuous extension to $a = 1$:
define $f(1) = \lim\limits_{x \to 1} \frac{x-1}{x - \sqrt[4]{x}} = \frac{4}{3}$.

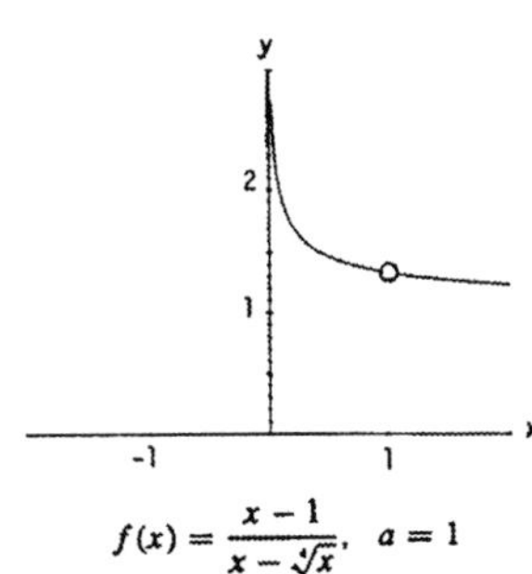

$f(x) = \frac{x-1}{x - \sqrt[4]{x}}, \quad a = 1$

46. Yes, g does have a continuous extension to $a = \frac{\pi}{2}$:
$g\left(\frac{\pi}{2}\right) = \lim\limits_{\theta \to \frac{\pi}{2}} \frac{5\cos\theta}{4\theta - 2\pi} = -\frac{5}{4}$.

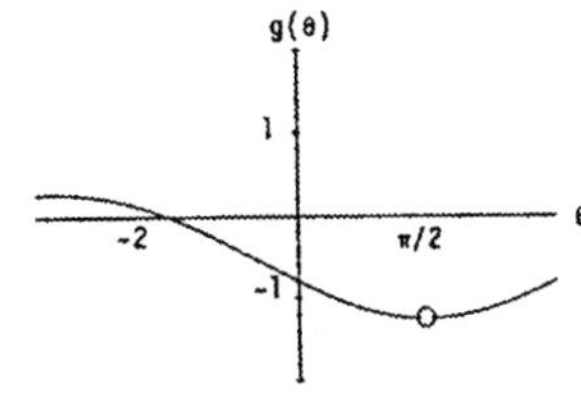

$g(\theta) = \frac{5\cos\theta}{4\theta - 2\pi}, \quad a = \pi/2$

47. From the graph we see that $\lim\limits_{t \to 0^-} h(t) \neq \lim\limits_{t \to 0^+} h(t)$
so h cannot be extended to a continuous function
at $a = 0$.

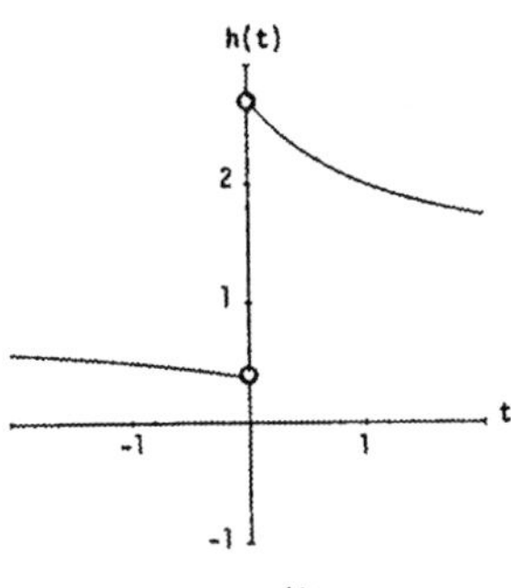

$h(t) = (1 + |t|)^{1/t}, \quad a = 0$

48. From the graph we see that $\lim\limits_{x \to 0^-} k(x) \neq \lim\limits_{x \to 0^+} k(x)$
so k cannot be extended to a continuous function
at $a = 0$.

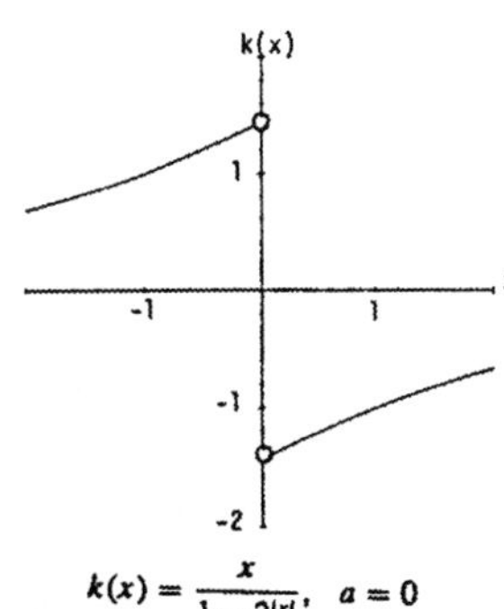

$k(x) = \frac{x}{1 - 2^{|x|}}, \quad a = 0$

49. (a) $f(-1) = -1$ and $f(2) = 5 \Rightarrow$ f has a root between -1 and 2 by the Intermediate Value Theorem.
(b), (c) root is 1.32471795724

50. (a) $f(-2) = -2$ and $f(0) = 2 \Rightarrow$ f has a root between -2 and 0 by the Intermediate Value Theorem.
(b), (c) root is -1.76929235424

CHAPTER 2 ADDITIONAL AND ADVANCED EXERCISES

1. (a)

x	0.1	0.01	0.001	0.0001	0.00001
x^x	0.7943	0.9550	0.9931	0.9991	0.9999

Apparently, $\lim_{x \to 0^+} x^x = 1$

(b)

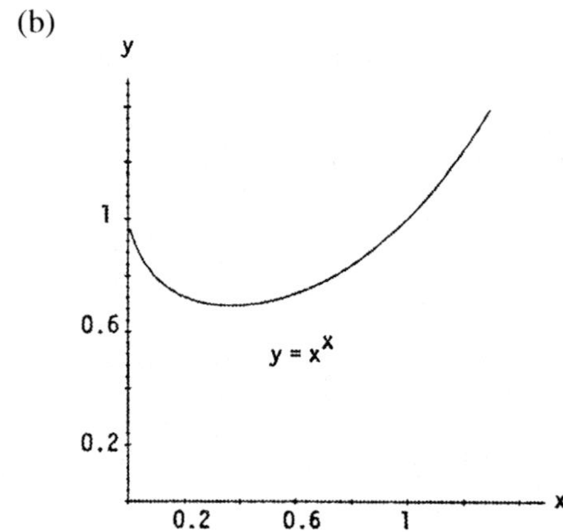

2. (a)

x	10	100	1000
$\left(\frac{1}{x}\right)^{1/(\ln x)}$	0.3679	0.3679	0.3679

Apparently, $\lim_{x \to \infty} \left(\frac{1}{x}\right)^{1/(\ln x)} = 0.3678 = \frac{1}{e}$

(b)

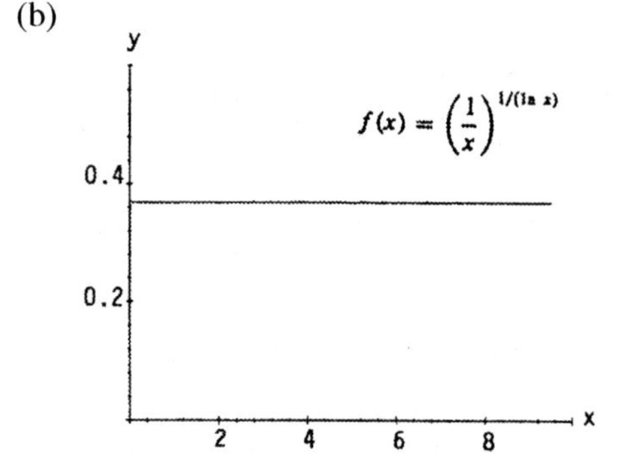

3. $\lim_{v \to c^-} L = \lim_{v \to c^-} L_0 \sqrt{1 - \frac{v^2}{c^2}} = L_0 \sqrt{1 - \frac{\lim_{v \to c^-} v^2}{c^2}} = L_0 \sqrt{1 - \frac{c^2}{c^2}} = 0$

The left-hand limit was needed because the function L is undefined if $v > c$ (the rocket cannot move faster than the speed of light).

4. $\left|\frac{\sqrt{x}}{2} - 1\right| < 0.2 \Rightarrow -0.2 < \frac{\sqrt{x}}{2} - 1 < 0.2 \Rightarrow 0.8 < \frac{\sqrt{x}}{2} < 1.2 \Rightarrow 1.6 < \sqrt{x} < 2.4 \Rightarrow 2.56 < x < 5.76.$

$\left|\frac{\sqrt{x}}{2} - 1\right| < 0.1 \Rightarrow -0.1 < \frac{\sqrt{x}}{2} - 1 < 0.1 \Rightarrow 0.9 < \frac{\sqrt{x}}{2} < 1.1 \Rightarrow 1.8 < \sqrt{x} < 2.2 \Rightarrow 3.24 < x < 4.84.$

5. $|10 + (t - 70) \times 10^{-4} - 10| < 0.0005 \Rightarrow |(t - 70) \times 10^{-4}| < 0.0005 \Rightarrow -0.0005 < (t - 70) \times 10^{-4} < 0.0005$
$\Rightarrow -5 < t - 70 < 5 \Rightarrow 65° < t < 75° \Rightarrow$ Within 5° F.

6. We want to know in what interval to hold values of h to make V satisfy the inequality $|V - 1000| = |36\pi h - 1000| \le 10$. To find out, we solve the inequality:
$|36\pi h - 1000| \le 10 \Rightarrow -10 \le 36\pi h - 1000 \le 10 \Rightarrow 990 \le 36\pi h \le 1010 \Rightarrow \frac{990}{36\pi} \le h \le \frac{1010}{36\pi}$
$\Rightarrow 8.8 \le h \le 8.9$. where 8.8 was rounded up, to be safe, and 8.9 was rounded down, to be safe.

The interval in which we should hold h is about $8.9 - 8.8 = 0.1$ cm wide (1 mm). With stripes 1 mm wide, we can expect to measure a liter of water with an accuracy of 1%, which is more than enough accuracy for cooking.

7. Show $\lim_{x \to 1} f(x) = \lim_{x \to 1} (x^2 - 7) = -6 = f(1)$.
Step 1: $|(x^2 - 7) + 6| < \epsilon \Rightarrow -\epsilon < x^2 - 1 < \epsilon \Rightarrow 1 - \epsilon < x^2 < 1 + \epsilon \Rightarrow \sqrt{1-\epsilon} < x < \sqrt{1+\epsilon}$.
Step 2: $|x - 1| < \delta \Rightarrow -\delta < x - 1 < \delta \Rightarrow -\delta + 1 < x < \delta + 1$.
Then $-\delta + 1 = \sqrt{1-\epsilon}$ or $\delta + 1 = \sqrt{1+\epsilon}$. Choose $\delta = \min\left\{1 - \sqrt{1-\epsilon}, \sqrt{1+\epsilon} - 1\right\}$, then
$0 < |x - 1| < \delta \Rightarrow |(x^2 - 7) - 6| < \epsilon$ and $\lim_{x \to 1} f(x) = -6$. By the continuity test, f(x) is continuous at $x = 1$.

8. Show $\lim_{x \to \frac{1}{4}} g(x) = \lim_{x \to \frac{1}{4}} \frac{1}{2x} = 2 = g\left(\frac{1}{4}\right)$.
Step 1: $\left|\frac{1}{2x} - 2\right| < \epsilon \Rightarrow -\epsilon < \frac{1}{2x} - 2 < \epsilon \Rightarrow 2 - \epsilon < \frac{1}{2x} < 2 + \epsilon \Rightarrow \frac{1}{4-2\epsilon} > x > \frac{1}{4+2\epsilon}$.
Step 2: $\left|x - \frac{1}{4}\right| < \delta \Rightarrow -\delta < x - \frac{1}{4} < \delta \Rightarrow -\delta + \frac{1}{4} < x < \delta + \frac{1}{4}$.
Then $-\delta + \frac{1}{4} = \frac{1}{4+2\epsilon} \Rightarrow \delta = \frac{1}{4} - \frac{1}{4+2\epsilon} = \frac{\epsilon}{4(2+\epsilon)}$, or $\delta + \frac{1}{4} = \frac{1}{4-2\epsilon} \Rightarrow \delta = \frac{1}{4-2\epsilon} - \frac{1}{4} = \frac{\epsilon}{4(2-\epsilon)}$.
Choose $\delta = \frac{\epsilon}{4(2+\epsilon)}$, the smaller of the two values. Then $0 < \left|x - \frac{1}{4}\right| < \delta \Rightarrow \left|\frac{1}{2x} - 2\right| < \epsilon$ and $\lim_{x \to \frac{1}{4}} \frac{1}{2x} = 2$.

By the continuity test, g(x) is continuous at $x = \frac{1}{4}$.

9. Show $\lim_{x \to 2} h(x) = \lim_{x \to 2} \sqrt{2x - 3} = 1 = h(2)$.
Step 1: $\left|\sqrt{2x-3} - 1\right| < \epsilon \Rightarrow -\epsilon < \sqrt{2x-3} - 1 < \epsilon \Rightarrow 1 - \epsilon < \sqrt{2x-3} < 1 + \epsilon \Rightarrow \frac{(1-\epsilon)^2+3}{2} < x < \frac{(1+\epsilon)^2+3}{2}$.
Step 2: $|x - 2| < \delta \Rightarrow -\delta < x - 2 < \delta$ or $-\delta + 2 < x < \delta + 2$.
Then $-\delta + 2 = \frac{(1-\epsilon)^2+3}{2} \Rightarrow \delta = 2 - \frac{(1-\epsilon)^2+3}{2} = \frac{1-(1-\epsilon)^2}{2} = \epsilon - \frac{\epsilon^2}{2}$, or $\delta + 2 = \frac{(1+\epsilon)^2+3}{2}$
$\Rightarrow \delta = \frac{(1+\epsilon)^2+3}{2} - 2 = \frac{(1+\epsilon)^2-1}{2} = \epsilon + \frac{\epsilon^2}{2}$. Choose $\delta = \epsilon - \frac{\epsilon^2}{2}$, the smaller of the two values. Then,
$0 < |x - 2| < \delta \Rightarrow \left|\sqrt{2x-3} - 1\right| < \epsilon$, so $\lim_{x \to 2} \sqrt{2x-3} = 1$. By the continuity test, h(x) is continuous at $x = 2$.

10. Show $\lim_{x \to 5} F(x) = \lim_{x \to 5} \sqrt{9 - x} = 2 = F(5)$.
Step 1: $\left|\sqrt{9-x} - 2\right| < \epsilon \Rightarrow -\epsilon < \sqrt{9-x} - 2 < \epsilon \Rightarrow 9 - (2-\epsilon)^2 > x > 9 - (2+\epsilon)^2$.
Step 2: $0 < |x - 5| < \delta \Rightarrow -\delta < x - 5 < \delta \Rightarrow -\delta + 5 < x < \delta + 5$.
Then $-\delta + 5 = 9 - (2+\epsilon)^2 \Rightarrow \delta = (2+\epsilon)^2 - 4 = \epsilon^2 + 2\epsilon$, or $\delta + 5 = 9 - (2-\epsilon)^2 \Rightarrow \delta = 4 - (2-\epsilon)^2 = \epsilon^2 - 2\epsilon$.
Choose $\delta = \epsilon^2 - 2\epsilon$, the smaller of the two values. Then, $0 < |x - 5| < \delta \Rightarrow \left|\sqrt{9-x} - 2\right| < \epsilon$, so
$\lim_{x \to 5} \sqrt{9-x} = 2$. By the continuity test, F(x) is continuous at $x = 5$.

11. Suppose L_1 and L_2 are two different limits. Without loss of generality assume $L_2 > L_1$. Let $\epsilon = \frac{1}{3}(L_2 - L_1)$.
Since $\lim_{x \to x_0} f(x) = L_1$ there is a $\delta_1 > 0$ such that $0 < |x - x_0| < \delta_1 \Rightarrow |f(x) - L_1| < \epsilon \Rightarrow -\epsilon < f(x) - L_1 < \epsilon$
$\Rightarrow -\frac{1}{3}(L_2 - L_1) + L_1 < f(x) < \frac{1}{3}(L_2 - L_1) + L_1 \Rightarrow 4L_1 - L_2 < 3f(x) < 2L_1 + L_2$. Likewise, $\lim_{x \to x_0} f(x) = L_2$
so there is a δ_2 such that $0 < |x - x_0| < \delta_2 \Rightarrow |f(x) - L_2| < \epsilon \Rightarrow -\epsilon < f(x) - L_2 < \epsilon$
$\Rightarrow -\frac{1}{3}(L_2 - L_1) + L_2 < f(x) < \frac{1}{3}(L_2 - L_1) + L_2 \Rightarrow 2L_2 + L_1 < 3f(x) < 4L_2 - L_1$
$\Rightarrow L_1 - 4L_2 < -3f(x) < -2L_2 - L_1$. If $\delta = \min\{\delta_1, \delta_2\}$ both inequalities must hold for $0 < |x - x_0| < \delta$:
$$\left.\begin{array}{l} 4L_1 - L_2 < 3f(x) < 2L_1 + L_2 \\ L_1 - 4L_2 < -3f(x) < -2L_2 - L_1 \end{array}\right\} \Rightarrow 5(L_1 - L_2) < 0 < L_1 - L_2.$$
That is, $L_1 - L_2 < 0$ and $L_1 - L_2 > 0$, a contradiction.

12. Suppose $\lim_{x \to c} f(x) = L$. If $k = 0$, then $\lim_{x \to c} kf(x) = \lim_{x \to c} 0 = 0 = 0 \cdot \lim_{x \to c} f(x)$ and we are done.
If $k \neq 0$, then given any $\epsilon > 0$, there is a $\delta > 0$ so that $0 < |x - c| < \delta \Rightarrow |f(x) - L| < \frac{\epsilon}{|k|} \Rightarrow |k||f(x) - L| < \epsilon$
$\Rightarrow |k(f(x) - L)| < \epsilon \Rightarrow |(kf(x)) - (kL)| < \epsilon$. Thus, $\lim_{x \to c} kf(x) = kL = k\left(\lim_{x \to c} f(x)\right)$.

13. (a) Since $x \to 0^+, 0 < x^3 < x < 1 \Rightarrow (x^3 - x) \to 0^- \Rightarrow \lim_{x \to 0^+} f(x^3 - x) = \lim_{y \to 0^-} f(y) = B$ where $y = x^3 - x$.
(b) Since $x \to 0^-, -1 < x < x^3 < 0 \Rightarrow (x^3 - x) \to 0^+ \Rightarrow \lim_{x \to 0^-} f(x^3 - x) = \lim_{y \to 0^+} f(y) = A$ where $y = x^3 - x$.
(c) Since $x \to 0^+, 0 < x^4 < x^2 < 1 \Rightarrow (x^2 - x^4) \to 0^+ \Rightarrow \lim_{x \to 0^+} f(x^2 - x^4) = \lim_{y \to 0^+} f(y) = A$ where $y = x^2 - x^4$.
(d) Since $x \to 0^-, -1 < x < 0 \Rightarrow 0 < x^4 < x^2 < 1 \Rightarrow (x^2 - x^4) \to 0^+ \Rightarrow \lim_{x \to 0^+} f(x^2 - x^4) = A$ as in part (c).

14. (a) True, because if $\lim_{x \to a} (f(x) + g(x))$ exists then $\lim_{x \to a} (f(x) + g(x)) - \lim_{x \to a} f(x) = \lim_{x \to a} [(f(x) + g(x)) - f(x)]$
$= \lim_{x \to a} g(x)$ exists, contrary to assumption.
(b) False; for example take $f(x) = \frac{1}{x}$ and $g(x) = -\frac{1}{x}$. Then neither $\lim_{x \to 0} f(x)$ nor $\lim_{x \to 0} g(x)$ exists, but
$\lim_{x \to 0} (f(x) + g(x)) = \lim_{x \to 0} \left(\frac{1}{x} - \frac{1}{x}\right) = \lim_{x \to 0} 0 = 0$ exists.
(c) True, because $g(x) = |x|$ is continuous $\Rightarrow g(f(x)) = |f(x)|$ is continuous (it is the composite of continuous functions).
(d) False; for example let $f(x) = \begin{cases} -1, & x \le 0 \\ 1, & x > 0 \end{cases} \Rightarrow f(x)$ is discontinuous at $x = 0$. However $|f(x)| = 1$ is continuous at $x = 0$.

15. Show $\lim_{x \to -1} f(x) = \lim_{x \to -1} \frac{x^2 - 1}{x + 1} = \lim_{x \to -1} \frac{(x + 1)(x - 1)}{(x + 1)} = -2, x \neq -1$.
Define the continuous extension of f(x) as $F(x) = \begin{cases} \frac{x^2 - 1}{x + 1}, & x \neq -1 \\ -2, & x = -1 \end{cases}$. We now prove the limit of f(x) as $x \to -1$ exists and has the correct value.
Step 1: $\left|\frac{x^2 - 1}{x + 1} - (-2)\right| < \epsilon \Rightarrow -\epsilon < \frac{(x + 1)(x - 1)}{(x + 1)} + 2 < \epsilon \Rightarrow -\epsilon < (x - 1) + 2 < \epsilon, x \neq -1 \Rightarrow -\epsilon - 1 < x < \epsilon - 1$.
Step 2: $|x - (-1)| < \delta \Rightarrow -\delta < x + 1 < \delta \Rightarrow -\delta - 1 < x < \delta - 1$.
Then $-\delta - 1 = -\epsilon - 1 \Rightarrow \delta = \epsilon$, or $\delta - 1 = \epsilon - 1 \Rightarrow \delta = \epsilon$. Choose $\delta = \epsilon$. Then $0 < |x - (-1)| < \delta$
$\Rightarrow \left|\frac{x^2 - 1}{x + 1} - (-2)\right| < \epsilon \Rightarrow \lim_{x \to -1} F(x) = -2$. Since the conditions of the continuity test are met by F(x), then f(x) has a continuous extension to F(x) at $x = -1$.

16. Show $\lim_{x \to 3} g(x) = \lim_{x \to 3} \frac{x^2 - 2x - 3}{2x - 6} = \lim_{x \to 3} \frac{(x - 3)(x + 1)}{2(x - 3)} = 2, x \neq 3$.
Define the continuous extension of g(x) as $G(x) = \begin{cases} \frac{x^2 - 2x - 3}{2x - 6}, & x \neq 3 \\ 2, & x = 3 \end{cases}$. We now prove the limit of g(x) as $x \to 3$ exists and has the correct value.
Step 1: $\left|\frac{x^2 - 2x - 3}{2x - 6} - 2\right| < \epsilon \Rightarrow -\epsilon < \frac{(x - 3)(x + 1)}{2(x - 3)} - 2 < \epsilon \Rightarrow -\epsilon < \frac{x + 1}{2} - 2 < \epsilon, x \neq 3 \Rightarrow 3 - 2\epsilon < x < 3 + 2\epsilon$.
Step 2: $|x - 3| < \delta \Rightarrow -\delta < x - 3 < \delta \Rightarrow 3 - \delta < x < \delta + 3$.
Then, $3 - \delta = 3 - 2\epsilon \Rightarrow \delta = 2\epsilon$, or $\delta + 3 = 3 + 2\epsilon \Rightarrow \delta = 2\epsilon$. Choose $\delta = 2\epsilon$. Then $0 < |x - 3| < \delta$
$\Rightarrow \left|\frac{x^2 - 2x - 3}{2x - 6} - 2\right| < \epsilon \Rightarrow \lim_{x \to 3} \frac{(x - 3)(x + 1)}{2(x - 3)} = 2$. Since the conditions of the continuity test hold for G(x), g(x) can be continuously extended to G(x) at $x = 3$.

17. (a) Let $\epsilon > 0$ be given. If x is rational, then $f(x) = x \Rightarrow |f(x) - 0| = |x - 0| < \epsilon \Leftrightarrow |x - 0| < \epsilon$; i.e., choose $\delta = \epsilon$. Then $|x - 0| < \delta \Rightarrow |f(x) - 0| < \epsilon$ for x rational. If x is irrational, then $f(x) = 0 \Rightarrow |f(x) - 0| < \epsilon$ $\Leftrightarrow 0 < \epsilon$ which is true no matter how close irrational x is to 0, so again we can choose $\delta = \epsilon$. In either case,

given $\epsilon > 0$ there is a $\delta = \epsilon > 0$ such that $0 < |x - 0| < \delta \Rightarrow |f(x) - 0| < \epsilon$. Therefore, f is continuous at $x = 0$.

(b) Choose $x = c > 0$. Then within any interval $(c - \delta, c + \delta)$ there are both rational and irrational numbers. If c is rational, pick $\epsilon = \frac{c}{2}$. No matter how small we choose $\delta > 0$ there is an irrational number x in $(c - \delta, c + \delta) \Rightarrow |f(x) - f(c)| = |0 - c| = c > \frac{c}{2} = \epsilon$. That is, f is not continuous at any rational $c > 0$. On the other hand, suppose c is irrational $\Rightarrow f(c) = 0$. Again pick $\epsilon = \frac{c}{2}$. No matter how small we choose $\delta > 0$ there is a rational number x in $(c - \delta, c + \delta)$ with $|x - c| < \frac{c}{2} = \epsilon \Leftrightarrow \frac{c}{2} < x < \frac{3c}{2}$. Then $|f(x) - f(c)| = |x - 0| = |x| > \frac{c}{2} = \epsilon \Rightarrow$ f is not continuous at any irrational $c > 0$.

If $x = c < 0$, repeat the argument picking $\epsilon = \frac{|c|}{2} = \frac{-c}{2}$. Therefore f fails to be continuous at any nonzero value $x = c$.

18. (a) Let $c = \frac{m}{n}$ be a rational number in [0, 1] reduced to lowest terms $\Rightarrow f(c) = \frac{1}{n}$. Pick $\epsilon = \frac{1}{2n}$. No matter how small $\delta > 0$ is taken, there is an irrational number x in the interval $(c - \delta, c + \delta) \Rightarrow |f(x) - f(c)| = \left|0 - \frac{1}{n}\right| = \frac{1}{n} > \frac{1}{2n} = \epsilon$. Therefore f is discontinuous at $x = c$, a rational number.

(b) Now suppose c is an irrational number $\Rightarrow f(c) = 0$. Let $\epsilon > 0$ be given. Notice that $\frac{1}{2}$ is the only rational number reduced to lowest terms with denominator 2 and belonging to [0, 1]; $\frac{1}{3}$ and $\frac{2}{3}$ the only rationals with denominator 3 belonging to [0, 1]; $\frac{1}{4}$ and $\frac{3}{4}$ with denominator 4 in [0, 1]; $\frac{1}{5}, \frac{2}{5}, \frac{3}{5}$ and $\frac{4}{5}$ with denominator 5 in [0, 1]; etc. In general, choose N so that $\frac{1}{N} < \epsilon \Rightarrow$ there exist only finitely many rationals in [0, 1] having denominator $\le N$, say $r_1, r_2, \ldots, r_p$. Let $\delta = \min\{|c - r_i| : i = 1, \ldots, p\}$. Then the interval $(c - \delta, c + \delta)$ contains no rational numbers with denominator $\le N$. Thus, $0 < |x - c| < \delta \Rightarrow |f(x) - f(c)| = |f(x) - 0| = |f(x)| \le \frac{1}{N} < \epsilon \Rightarrow$ f is continuous at $x = c$ irrational.

(c) The graph looks like the markings on a typical ruler when the points $(x, f(x))$ on the graph of $f(x)$ are connected to the x-axis with vertical lines.

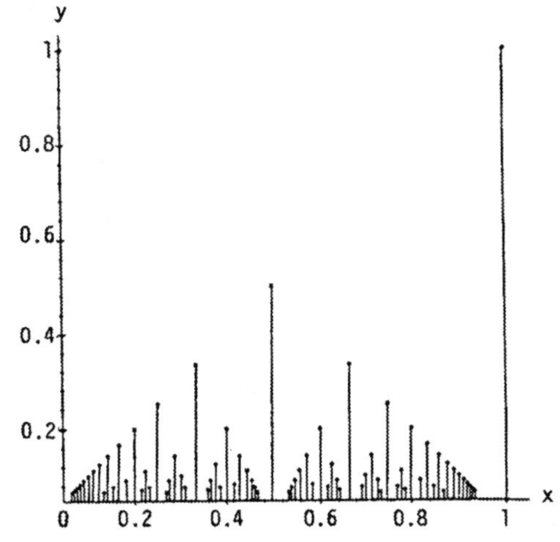

19. Yes. Let R be the radius of the equator (earth) and suppose at a fixed instant of time we label noon as the zero point, 0, on the equator $\Rightarrow 0 + \pi R$ represents the midnight point (at the same exact time). Suppose x_1 is a point on the equator "just after" noon $\Rightarrow x_1 + \pi R$ is simultaneously "just after" midnight. It seems reasonable that the temperature T at a point just after noon is hotter than it would be at the diametrically opposite point just after midnight: That is, $T(x_1) - T(x_1 + \pi R) > 0$. At exactly the same moment in time pick x_2 to be a point just before midnight $\Rightarrow x_2 + \pi R$ is just before noon. Then $T(x_2) - T(x_2 + \pi R) < 0$. Assuming the temperature function T is continuous along the equator (which is reasonable), the Intermediate Value Theorem says there is a point c between 0 (noon) and πR (simultaneously midnight) such that $T(c) - T(c + \pi R) = 0$; i.e., there is always a pair of antipodal points on the earth's equator where the temperatures are the same.

20. $\lim_{x\to c} f(x)g(x) = \lim_{x\to c} \frac{1}{4}\left[(f(x)+g(x))^2 - (f(x)-g(x))^2\right] = \frac{1}{4}\left[\left(\lim_{x\to c}(f(x)+g(x))\right)^2 - \left(\lim_{x\to c}(f(x)-g(x))\right)^2\right]$
$= \frac{1}{4}(3^2 - (-1)^2) = 2.$

21. (a) At x = 0: $\lim_{a\to 0} r_+(a) = \lim_{a\to 0} \frac{-1+\sqrt{1+a}}{a} = \lim_{a\to 0} \left(\frac{-1+\sqrt{1+a}}{a}\right)\left(\frac{-1-\sqrt{1+a}}{-1-\sqrt{1+a}}\right)$
$= \lim_{a\to 0} \frac{1-(1+a)}{a(-1-\sqrt{1+a})} = \frac{-1}{-1-\sqrt{1+0}} = \frac{1}{2}$

At x = −1: $\lim_{a\to -1^+} r_+(a) = \lim_{a\to -1^+} \frac{1-(1+a)}{a(-1-\sqrt{1+a})} = \lim_{a\to -1} \frac{-a}{a(-1-\sqrt{1+a})} = \frac{-1}{-1-\sqrt{0}} = 1$

(b) At x = 0: $\lim_{a\to 0^-} r_-(a) = \lim_{a\to 0^-} \frac{-1-\sqrt{1+a}}{a} = \lim_{a\to 0^-} \left(\frac{-1-\sqrt{1+a}}{a}\right)\left(\frac{-1+\sqrt{1+a}}{-1+\sqrt{1+a}}\right)$
$= \lim_{a\to 0^-} \frac{1-(1+a)}{a(-1+\sqrt{1+a})} = \lim_{a\to 0^-} \frac{-a}{a(-1+\sqrt{1+a})} = \lim_{a\to 0^-} \frac{-1}{-1+\sqrt{1+a}} = \infty$ (because the denominator is always negative); $\lim_{a\to 0^+} r_-(a) = \lim_{a\to 0^+} \frac{-1}{-1+\sqrt{1+a}} = -\infty$ (because the denominator is always positive). Therefore, $\lim_{a\to 0} r_-(a)$ does not exist.

At x = −1: $\lim_{a\to -1^+} r_-(a) = \lim_{a\to -1^+} \frac{-1-\sqrt{1+a}}{a} = \lim_{a\to -1^+} \frac{-1}{-1+\sqrt{1+a}} = 1$

(c)

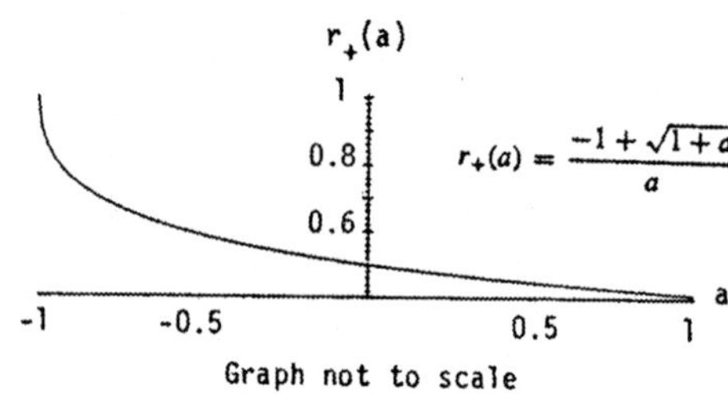

r_(a)
$r_-(a) = \frac{-1-\sqrt{1+a}}{a}$
2
a
-1
2
4
6
8
-2
-4

(d)

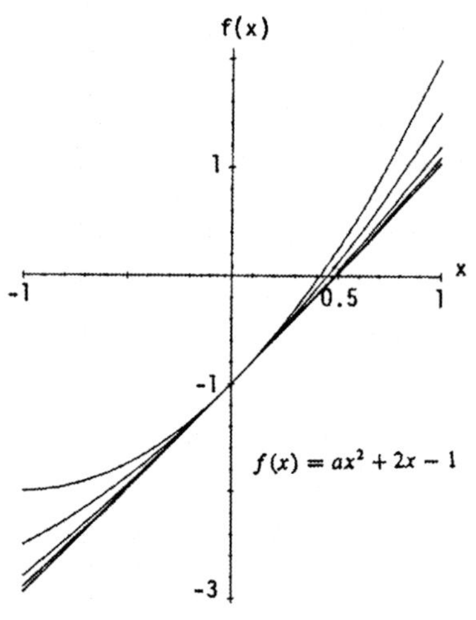

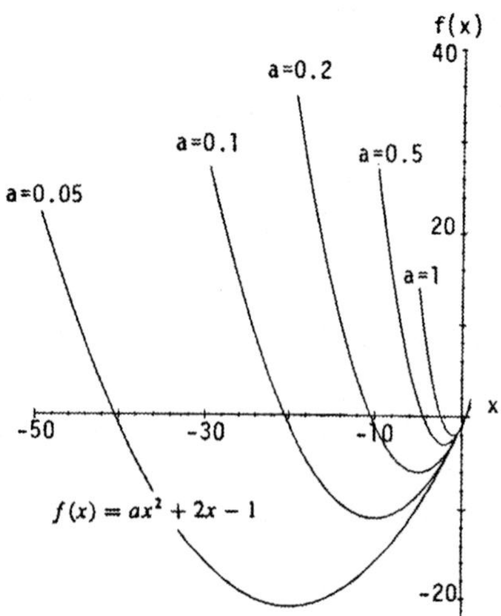

22. $f(x) = x + 2\cos x \Rightarrow f(0) = 0 + 2\cos 0 = 2 > 0$ and $f(-\pi) = -\pi + 2\cos(-\pi) = -\pi - 2 < 0$. Since f(x) is continuous on $[-\pi, 0]$, by the Intermediate Value Theorem, f(x) must take on every value between $[-\pi - 2, 2]$. Thus there is some number c in $[-\pi, 0]$ such that $f(c) = 0$; i.e., c is a solution to $x + 2\cos x = 0$.

23. (a) The function f is bounded on D if $f(x) \geq M$ and $f(x) \leq N$ for all x in D. This means $M \leq f(x) \leq N$ for all x in D. Choose B to be $\max\{|M|, |N|\}$. Then $|f(x)| \leq B$. On the other hand, if $|f(x)| \leq B$, then $-B \leq f(x) \leq B \Rightarrow f(x) \geq -B$ and $f(x) \leq B \Rightarrow f(x)$ is bounded on D with N = B an upper bound and M = −B a lower bound.

(b) Assume $f(x) \le N$ for all x and that $L > N$. Let $\epsilon = \frac{L-N}{2}$. Since $\lim_{x \to x_0} f(x) = L$ there is a $\delta > 0$ such that $0 < |x - x_0| < \delta \Rightarrow |f(x) - L| < \epsilon \Leftrightarrow L - \epsilon < f(x) < L + \epsilon \Leftrightarrow L - \frac{L-N}{2} < f(x) < L + \frac{L-N}{2}$ $\Leftrightarrow \frac{L+N}{2} < f(x) < \frac{3L-N}{2}$. But $L > N \Rightarrow \frac{L+N}{2} > N \Rightarrow N < f(x)$ contrary to the boundedness assumption $f(x) \le N$. This contradiction proves $L \le N$.

(c) Assume $M \le f(x)$ for all x and that $L < M$. Let $\epsilon = \frac{M-L}{2}$. As in part (b), $0 < |x - x_0| < \delta$ $\Rightarrow L - \frac{M-L}{2} < f(x) < L + \frac{M-L}{2} \Leftrightarrow \frac{3L-M}{2} < f(x) < \frac{M+L}{2} < M$, a contradiction.

24. (a) If $a \ge b$, then $a - b \ge 0 \Rightarrow |a - b| = a - b \Rightarrow \max(a, b) = \frac{a+b}{2} + \frac{|a-b|}{2} = \frac{a+b}{2} + \frac{a-b}{2} = \frac{2a}{2} = a$.
If $a \le b$, then $a - b \le 0 \Rightarrow |a - b| = -(a - b) = b - a \Rightarrow \max(a, b) = \frac{a+b}{2} + \frac{|a-b|}{2} = \frac{a+b}{2} + \frac{b-a}{2}$ $= \frac{2b}{2} = b$.

(b) Let $\min(a, b) = \frac{a+b}{2} - \frac{|a-b|}{2}$.

25. $\lim_{x \to 0} \frac{\sin(1-\cos x)}{x} = \lim_{x \to 0} \frac{\sin(1-\cos x)}{1-\cos x} \cdot \frac{1-\cos x}{x} \cdot \frac{1+\cos x}{1+\cos x} = \lim_{x \to 0} \frac{\sin(1-\cos x)}{1-\cos x} \cdot \lim_{x \to 0} \frac{1-\cos^2 x}{x(1+\cos x)} = 1 \cdot \lim_{x \to 0} \frac{\sin^2 x}{x(1+\cos x)}$
$= \lim_{x \to 0} \frac{\sin x}{x} \cdot \frac{\sin x}{1+\cos x} = 1 \cdot \left(\frac{0}{2}\right) = 0$.

26. $\lim_{x \to 0^+} \frac{\sin x}{\sin \sqrt{x}} = \lim_{x \to 0^+} \frac{\sin x}{x} \cdot \frac{\sqrt{x}}{\sin \sqrt{x}} \cdot \frac{x}{\sqrt{x}} = 1 \cdot \lim_{x \to 0^+} \frac{1}{\left(\frac{\sin \sqrt{x}}{\sqrt{x}}\right)} \cdot \lim_{x \to 0^+} \sqrt{x} = 1 \cdot 0 \cdot 0 = 0$.

27. $\lim_{x \to 0} \frac{\sin(\sin x)}{x} = \lim_{x \to 0} \frac{\sin(\sin x)}{\sin x} \cdot \frac{\sin x}{x} = \lim_{x \to 0} \frac{\sin(\sin x)}{\sin x} \cdot \lim_{x \to 0} \frac{\sin x}{x} = 1 \cdot 1 = 1$.

28. $\lim_{x \to 0} \frac{\sin(x^2+x)}{x} = \lim_{x \to 0} \frac{\sin(x^2+x)}{x^2+x} \cdot (x+1) = \lim_{x \to 0} \frac{\sin(x^2+x)}{x^2+x} \cdot \lim_{x \to 0} (x+1) = 1 \cdot 1 = 1$

29. $\lim_{x \to 2} \frac{\sin(x^2-4)}{x-2} = \lim_{x \to 2} \frac{\sin(x^2-4)}{x^2-4} \cdot (x+2) = \lim_{x \to 2} \frac{\sin(x^2-4)}{x^2-4} \cdot \lim_{x \to 2} (x+2) = 1 \cdot 4 = 4$

30. $\lim_{x \to 9} \frac{\sin(\sqrt{x}-3)}{x-9} = \lim_{x \to 9} \frac{\sin(\sqrt{x}-3)}{\sqrt{x}-3} \cdot \frac{1}{\sqrt{x}+3} = \lim_{x \to 9} \frac{\sin(\sqrt{x}-3)}{\sqrt{x}-3} \cdot \lim_{x \to 9} \frac{1}{\sqrt{x}+3} = 1 \cdot \frac{1}{6} = \frac{1}{6}$

NOTES:

CHAPTER 3 DIFFERENTIATION

3.1 THE DERIVATIVE AS A FUNCTION

1. Step 1: $f(x) = 4 - x^2$ and $f(x+h) = 4 - (x+h)^2$

   Step 2: $\frac{f(x+h)-f(x)}{h} = \frac{[4-(x+h)^2]-(4-x^2)}{h} = \frac{(4-x^2-2xh-h^2)-4+x^2}{h} = \frac{-2xh-h^2}{h} = \frac{h(-2x-h)}{h}$
   $= -2x - h$

   Step 3: $f'(x) = \lim_{h \to 0} (-2x-h) = -2x$; $f'(-3) = 6$, $f'(0) = 0$, $f'(1) = -2$

2. $F(x) = (x-1)^2 + 1$ and $F(x+h) = (x+h-1)^2 + 1 \Rightarrow F'(x) = \lim_{h \to 0} \frac{[(x+h-1)^2+1]-[(x-1)^2+1]}{h}$
   $= \lim_{h \to 0} \frac{(x^2+2xh+h^2-2x-2h+1+1)-(x^2-2x+1+1)}{h} = \lim_{h \to 0} \frac{2xh+h^2-2h}{h} = \lim_{h \to 0} (2x+h-2)$
   $= 2(x-1)$; $F'(-1) = -4$, $F'(0) = -2$, $F'(2) = 2$

3. Step 1: $g(t) = \frac{1}{t^2}$ and $g(t+h) = \frac{1}{(t+h)^2}$

   Step 2: $\frac{g(t+h)-g(t)}{h} = \frac{\frac{1}{(t+h)^2}-\frac{1}{t^2}}{h} = \frac{\left(\frac{t^2-(t+h)^2}{(t+h)^2 \cdot t^2}\right)}{h} = \frac{t^2-(t^2+2th+h^2)}{(t+h)^2 \cdot t^2 \cdot h} = \frac{-2th-h^2}{(t+h)^2 t^2 h}$
   $= \frac{h(-2t-h)}{(t+h)^2 t^2 h} = \frac{-2t-h}{(t+h)^2 t^2}$

   Step 3: $g'(t) = \lim_{h \to 0} \frac{-2t-h}{(t+h)^2 t^2} = \frac{-2t}{t^2 \cdot t^2} = \frac{-2}{t^3}$; $g'(-1) = 2$, $g'(2) = -\frac{1}{4}$, $g'\left(\sqrt{3}\right) = -\frac{2}{3\sqrt{3}}$

4. $k(z) = \frac{1-z}{2z}$ and $k(z+h) = \frac{1-(z+h)}{2(z+h)} \Rightarrow k'(z) = \lim_{h \to 0} \frac{\left(\frac{1-(z+h)}{2(z+h)} - \frac{1-z}{2z}\right)}{h}$
   $= \lim_{h \to 0} \frac{(1-z-h)z-(1-z)(z+h)}{2(z+h)zh} = \lim_{h \to 0} \frac{z-z^2-zh-z-h+z^2+zh}{2(z+h)zh} = \lim_{h \to 0} \frac{-h}{2(z+h)zh} = \lim_{h \to 0} \frac{-1}{2(z+h)z}$
   $= \frac{-1}{2z^2}$; $k'(-1) = -\frac{1}{2}$, $k'(1) = -\frac{1}{2}$, $k'\left(\sqrt{2}\right) = -\frac{1}{4}$

5. Step 1: $p(\theta) = \sqrt{3\theta}$ and $p(\theta+h) = \sqrt{3(\theta+h)}$

   Step 2: $\frac{p(\theta+h)-p(\theta)}{h} = \frac{\sqrt{3(\theta+h)}-\sqrt{3\theta}}{h} = \frac{\left(\sqrt{3\theta+3h}-\sqrt{3\theta}\right)}{h} \cdot \frac{\left(\sqrt{3\theta+3h}+\sqrt{3\theta}\right)}{\left(\sqrt{3\theta+3h}+\sqrt{3\theta}\right)} = \frac{(3\theta+3h)-3\theta}{h\left(\sqrt{3\theta+3h}+\sqrt{3\theta}\right)}$
   $= \frac{3h}{h\left(\sqrt{3\theta+3h}+\sqrt{3\theta}\right)} = \frac{3}{\sqrt{3\theta+3h}+\sqrt{3\theta}}$

   Step 3: $p'(\theta) = \lim_{h \to 0} \frac{3}{\sqrt{3\theta+3h}+\sqrt{3\theta}} = \frac{3}{\sqrt{3\theta}+\sqrt{3\theta}} = \frac{3}{2\sqrt{3\theta}}$; $p'(1) = \frac{3}{2\sqrt{3}}$, $p'(3) = \frac{1}{2}$, $p'\left(\frac{2}{3}\right) = \frac{3}{2\sqrt{2}}$

6. $r(s) = \sqrt{2s+1}$ and $r(s+h) = \sqrt{2(s+h)+1} \Rightarrow r'(s) = \lim_{h \to 0} \frac{\sqrt{2s+2h+1}-\sqrt{2s+1}}{h}$
   $= \lim_{h \to 0} \frac{\left(\sqrt{2s+h+1}-\sqrt{2s+1}\right)}{h} \cdot \frac{\left(\sqrt{2s+2h+1}+\sqrt{2s+1}\right)}{\left(\sqrt{2s+2h+1}+\sqrt{2s+1}\right)} = \lim_{h \to 0} \frac{(2s+2h+1)-(2s+1)}{h\left(\sqrt{2s+2h+1}+\sqrt{2s+1}\right)}$
   $= \lim_{h \to 0} \frac{2h}{h\left(\sqrt{2s+2h+1}+\sqrt{2s+1}\right)} = \lim_{h \to 0} \frac{2}{\sqrt{2s+2h+1}+\sqrt{2s+1}} = \frac{2}{\sqrt{2s+1}+\sqrt{2s+1}} = \frac{2}{2\sqrt{2s+1}}$
   $= \frac{1}{\sqrt{2s+1}}$; $r'(0) = 1$, $r'(1) = \frac{1}{\sqrt{3}}$, $r'\left(\frac{1}{2}\right) = \frac{1}{\sqrt{2}}$

7. $y = f(x) = 2x^3$ and $f(x+h) = 2(x+h)^3 \Rightarrow \frac{dy}{dx} = \lim_{h \to 0} \frac{2(x+h)^3-2x^3}{h} = \lim_{h \to 0} \frac{2(x^3+3x^2h+3xh^2+h^3)-2x^3}{h}$
   $= \lim_{h \to 0} \frac{6x^2h+6xh^2+2h^3}{h} = \lim_{h \to 0} \frac{h(6x^2+6xh+2h^2)}{h} = \lim_{h \to 0} (6x^2+6xh+2h^2) = 6x^2$

8. $r = \frac{s^3}{2} + 1 \Rightarrow \frac{dr}{ds} = \lim_{h \to 0} \frac{\left[\frac{(s+h)^3}{2} + 1\right] - \left[\frac{s^3}{2} + 1\right]}{h} = \frac{1}{2} \lim_{h \to 0} \frac{[(s+h)^3 + 2] - [s^3 + 2]}{h}$

$= \frac{1}{2} \lim_{h \to 0} \frac{s^3 + 3s^2h + 3sh^2 + h^3 + 2 - s^3 - 2}{h} = \frac{1}{2} \lim_{h \to 0} \frac{h[3s^2 + 3sh + h^2]}{h} = \frac{1}{2} \lim_{h \to 0} (3s^2 + 3sh + h^2) = \frac{3}{2} s^2$

9. $s = r(t) = \frac{t}{2t+1}$ and $r(t+h) = \frac{t+h}{2(t+h)+1} \Rightarrow \frac{ds}{dt} = \lim_{h \to 0} \frac{\left(\frac{t+h}{2(t+h)+1}\right) - \left(\frac{t}{2t+1}\right)}{h}$

$= \lim_{h \to 0} \frac{\left(\frac{(t+h)(2t+1) - t(2t+2h+1)}{(2t+2h+1)(2t+1)}\right)}{h} = \lim_{h \to 0} \frac{(t+h)(2t+1) - t(2t+2h+1)}{(2t+2h+1)(2t+1)h}$

$= \lim_{h \to 0} \frac{2t^2 + t + 2ht + h - 2t^2 - 2ht - t}{(2t+2h+1)(2t+1)h} = \lim_{h \to 0} \frac{h}{(2t+2h+1)(2t+1)h} = \lim_{h \to 0} \frac{1}{(2t+2h+1)(2t+1)}$

$= \frac{1}{(2t+1)(2t+1)} = \frac{1}{(2t+1)^2}$

10. $\frac{dv}{dt} = \lim_{h \to 0} \frac{\left[(t+h) - \frac{1}{t+h}\right] - \left(t - \frac{1}{t}\right)}{h} = \lim_{h \to 0} \frac{h - \frac{1}{t+h} + \frac{1}{t}}{h} = \lim_{h \to 0} \frac{\left(\frac{h(t+h)t - t + (t+h)}{(t+h)t}\right)}{h}$

$= \lim_{h \to 0} \frac{ht^2 + h^2t + h}{h(t+h)t} = \lim_{h \to 0} \frac{t^2 + ht + 1}{(t+h)t} = \frac{t^2+1}{t^2} = 1 + \frac{1}{t^2}$

11. $p = f(q) = \frac{1}{\sqrt{q+1}}$ and $f(q+h) = \frac{1}{\sqrt{(q+h)+1}} \Rightarrow \frac{dp}{dq} = \lim_{h \to 0} \frac{\left(\frac{1}{\sqrt{(q+h)+1}}\right) - \left(\frac{1}{\sqrt{q+1}}\right)}{h}$

$= \lim_{h \to 0} \frac{\left(\frac{\sqrt{q+1} - \sqrt{q+h+1}}{\sqrt{q+h+1}\sqrt{q+1}}\right)}{h} = \lim_{h \to 0} \frac{\sqrt{q+1} - \sqrt{q+h+1}}{h\sqrt{q+h+1}\sqrt{q+1}}$

$= \lim_{h \to 0} \frac{\left(\sqrt{q+1} - \sqrt{q+h+1}\right)}{h\sqrt{q+h+1}\sqrt{q+1}} \cdot \frac{\left(\sqrt{q+1} + \sqrt{q+h+1}\right)}{\left(\sqrt{q+1} + \sqrt{q+h+1}\right)} = \lim_{h \to 0} \frac{(q+1) - (q+h+1)}{h\sqrt{q+h+1}\sqrt{q+1}\left(\sqrt{q+1} + \sqrt{q+h+1}\right)}$

$= \lim_{h \to 0} \frac{-h}{h\sqrt{q+h+1}\sqrt{q+1}\left(\sqrt{q+1} + \sqrt{q+h+1}\right)} = \lim_{h \to 0} \frac{-1}{\sqrt{q+h+1}\sqrt{q+1}\left(\sqrt{q+1} + \sqrt{q+h+1}\right)}$

$= \frac{-1}{\sqrt{q+1}\sqrt{q+1}\left(\sqrt{q+1} + \sqrt{q+1}\right)} = \frac{-1}{2(q+1)\sqrt{q+1}}$

12. $\frac{dz}{dw} = \lim_{h \to 0} \frac{\left(\frac{1}{\sqrt{3(w+h)-2}} - \frac{1}{\sqrt{3w-2}}\right)}{h} = \lim_{h \to 0} \frac{\sqrt{3w-2} - \sqrt{3w+3h-2}}{h\sqrt{3w+3h-2}\sqrt{3w-2}}$

$= \lim_{h \to 0} \frac{\left(\sqrt{3w-2} - \sqrt{3w+3h-2}\right)}{h\sqrt{3w+3h-2}\sqrt{3w-2}} \cdot \frac{\left(\sqrt{3w-2} + \sqrt{3w+3h-2}\right)}{\left(\sqrt{3w-2} + \sqrt{3w+3h-2}\right)}$

$= \lim_{h \to 0} \frac{(3w-2) - (3w+3h-2)}{h\sqrt{3w+3h-2}\sqrt{3w-2}\left(\sqrt{3w-2} + \sqrt{3w+3h-2}\right)}$

$= \lim_{h \to 0} \frac{-3}{\sqrt{3w+3h-2}\sqrt{3w-2}\left(\sqrt{3w-2} + \sqrt{3w+3h-2}\right)} = \frac{-3}{\sqrt{3w-2}\sqrt{3w-2}\left(\sqrt{3w-2} + \sqrt{3w-2}\right)}$

$= \frac{-3}{2(3w-2)\sqrt{3w-2}}$

13. $f(x) = x + \frac{9}{x}$ and $f(x+h) = (x+h) + \frac{9}{(x+h)} \Rightarrow \frac{f(x+h) - f(x)}{h} = \frac{\left[(x+h) + \frac{9}{(x+h)}\right] - \left[x + \frac{9}{x}\right]}{h}$

$= \frac{x(x+h)^2 + 9x - x^2(x+h) - 9(x+h)}{x(x+h)h} = \frac{x^3 + 2x^2h + xh^2 + 9x - x^3 - x^2h - 9x - 9h}{x(x+h)h} = \frac{x^2h + xh^2 - 9h}{x(x+h)h}$

$= \frac{h(x^2 + xh - 9)}{x(x+h)h} = \frac{x^2 + xh - 9}{x(x+h)}$; $f'(x) = \lim_{h \to 0} \frac{x^2 + xh - 9}{x(x+h)} = \frac{x^2 - 9}{x^2} = 1 - \frac{9}{x^2}$; $m = f'(-3) = 0$

14. $k(x) = \frac{1}{2+x}$ and $k(x+h) = \frac{1}{2+(x+h)} \Rightarrow k'(x) = \lim_{h \to 0} \frac{k(x+h) - k(x)}{h} = \lim_{h \to 0} \frac{\left(\frac{1}{2+x+h} - \frac{1}{2+x}\right)}{h}$

$= \lim_{h \to 0} \frac{(2+x) - (2+x+h)}{h(2+x)(2+x+h)} = \lim_{h \to 0} \frac{-h}{h(2+x)(2+x+h)} = \lim_{h \to 0} \frac{-1}{(2+x)(2+x+h)} = \frac{-1}{(2+x)^2}$;

$k'(2) = -\frac{1}{16}$

15. $\frac{ds}{dt} = \lim_{h \to 0} \frac{[(t+h)^3 - (t+h)^2] - (t^3 - t^2)}{h} = \lim_{h \to 0} \frac{(t^3 + 3t^2h + 3th^2 + h^3) - (t^2 + 2th + h^2) - t^3 + t^2}{h}$

$= \lim_{h \to 0} \frac{3t^2h + 3th^2 + h^3 - 2th - h^2}{h} = \lim_{h \to 0} \frac{h(3t^2 + 3th + h^2 - 2t - h)}{h} = \lim_{h \to 0} (3t^2 + 3th + h^2 - 2t - h)$

$= 3t^2 - 2t$; $m = \left.\frac{ds}{dt}\right|_{t=-1} = 5$

16. $\frac{dy}{dx} = \lim\limits_{h \to 0} \frac{(x+h+1)^3 - (x+1)^3}{h} = \lim\limits_{h \to 0} \frac{(x+1)^3 + 3(x+1)^2h + 3(x+1)h^2 + h^3 - (x+1)^3}{h}$
$= \lim\limits_{h \to 0} [3(x+1)^2 + 3(x+1)h + h^2] = 3(x+1)^2$; $m = \left.\frac{dy}{dx}\right|_{x=-2} = 3$

17. $f(x) = \frac{8}{\sqrt{x-2}}$ and $f(x+h) = \frac{8}{\sqrt{(x+h)-2}} \Rightarrow \frac{f(x+h)-f(x)}{h} = \frac{\frac{8}{\sqrt{(x+h)-2}} - \frac{8}{\sqrt{x-2}}}{h}$
$= \frac{8\left(\sqrt{x-2} - \sqrt{x+h-2}\right)}{h\sqrt{x+h-2}\sqrt{x-2}} \cdot \frac{\left(\sqrt{x-2} + \sqrt{x+h-2}\right)}{\left(\sqrt{x-2} + \sqrt{x+h-2}\right)} = \frac{8[(x-2)-(x+h-2)]}{h\sqrt{x+h-2}\sqrt{x-2}\left(\sqrt{x-2}+\sqrt{x+h-2}\right)}$
$= \frac{-8h}{h\sqrt{x+h-2}\sqrt{x-2}\left(\sqrt{x-2}+\sqrt{x+h-2}\right)} \Rightarrow f'(x) = \lim\limits_{h \to 0} \frac{-8}{\sqrt{x+h-2}\sqrt{x-2}\left(\sqrt{x-2}+\sqrt{x+h-2}\right)}$
$= \frac{-8}{\sqrt{x-2}\sqrt{x-2}\left(\sqrt{x-2}+\sqrt{x-2}\right)} = \frac{-4}{(x-2)\sqrt{x-2}}$; $m = f'(6) = \frac{-4}{4\sqrt{4}} = -\frac{1}{2} \Rightarrow$ the equation of the tangent line at $(6, 4)$ is $y - 4 = -\frac{1}{2}(x-6) \Rightarrow y = -\frac{1}{2}x + 3 + 4 \Rightarrow y = -\frac{1}{2}x + 7$.

18. $g'(z) = \lim\limits_{h \to 0} \frac{\left(1+\sqrt{4-(z+h)}\right) - \left(1+\sqrt{4-z}\right)}{h} = \lim\limits_{h \to 0} \frac{\left(\sqrt{4-z-h} - \sqrt{4-z}\right)}{h} \cdot \frac{\left(\sqrt{4-z-h}+\sqrt{4-z}\right)}{\left(\sqrt{4-z-h}+\sqrt{4-z}\right)}$
$= \lim\limits_{h \to 0} \frac{(4-z-h)-(4-z)}{h\left(\sqrt{4-z-h}+\sqrt{4-z}\right)} = \lim\limits_{h \to 0} \frac{-h}{h\left(\sqrt{4-z-h}+\sqrt{4-z}\right)} = \lim\limits_{h \to 0} \frac{-1}{\left(\sqrt{4-z-h}+\sqrt{4-z}\right)} = \frac{-1}{2\sqrt{4-z}}$;
$m = g'(3) = \frac{-1}{2\sqrt{4-3}} = -\frac{1}{2} \Rightarrow$ the equation of the tangent line at $(3, 2)$ is $w - 2 = -\frac{1}{2}(z-3)$
$\Rightarrow w = -\frac{1}{2}z + \frac{3}{2} + 2 \Rightarrow w = -\frac{1}{2}z + \frac{7}{2}$.

19. $s = f(t) = 1 - 3t^2$ and $f(t+h) = 1 - 3(t+h)^2 = 1 - 3t^2 - 6th - 3h^2 \Rightarrow \frac{ds}{dt} = \lim\limits_{h \to 0} \frac{f(t+h)-f(t)}{h}$
$= \lim\limits_{h \to 0} \frac{(1-3t^2-6th-3h^2)-(1-3t^2)}{h} = \lim\limits_{h \to 0} (-6t - 3h) = -6t \Rightarrow \left.\frac{ds}{dt}\right|_{t=-1} = 6$

20. $y = f(x) = 1 - \frac{1}{x}$ and $f(x+h) = 1 - \frac{1}{x+h} \Rightarrow \frac{dy}{dx} = \lim\limits_{h \to 0} \frac{f(x+h)-f(x)}{h} = \lim\limits_{h \to 0} \frac{\left(1-\frac{1}{x+h}\right)-\left(1-\frac{1}{x}\right)}{h}$
$= \lim\limits_{h \to 0} \frac{\frac{1}{x} - \frac{1}{x+h}}{h} = \lim\limits_{h \to 0} \frac{h}{x(x+h)h} = \lim\limits_{h \to 0} \frac{1}{x(x+h)} = \frac{1}{x^2} \Rightarrow \left.\frac{dy}{dx}\right|_{x=\sqrt{3}} = \frac{1}{3}$

21. $r = f(\theta) = \frac{2}{\sqrt{4-\theta}}$ and $f(\theta+h) = \frac{2}{\sqrt{4-(\theta+h)}} \Rightarrow \frac{dr}{d\theta} = \lim\limits_{h \to 0} \frac{f(\theta+h)-f(\theta)}{h} = \lim\limits_{h \to 0} \frac{\frac{2}{\sqrt{4-\theta-h}} - \frac{2}{\sqrt{4-\theta}}}{h}$
$= \lim\limits_{h \to 0} \frac{2\sqrt{4-\theta} - 2\sqrt{4-\theta-h}}{h\sqrt{4-\theta}\sqrt{4-\theta-h}} = \lim\limits_{h \to 0} \frac{2\sqrt{4-\theta} - 2\sqrt{4-\theta-h}}{h\sqrt{4-\theta}\sqrt{4-\theta-h}} \cdot \frac{\left(2\sqrt{4-\theta}+2\sqrt{4-\theta-h}\right)}{\left(2\sqrt{4-\theta}+2\sqrt{4-\theta-h}\right)}$
$= \lim\limits_{h \to 0} \frac{4(4-\theta)-4(4-\theta-h)}{2h\sqrt{4-\theta}\sqrt{4-\theta-h}\left(\sqrt{4-\theta}+\sqrt{4-\theta-h}\right)} = \lim\limits_{h \to 0} \frac{2}{\sqrt{4-\theta}\sqrt{4-\theta-h}\left(\sqrt{4-\theta}+\sqrt{4-\theta-h}\right)}$
$= \frac{2}{(4-\theta)\left(2\sqrt{4-\theta}\right)} = \frac{1}{(4-\theta)\sqrt{4-\theta}} \Rightarrow \left.\frac{dr}{d\theta}\right|_{\theta=0} = \frac{1}{8}$

22. $w = f(z) = z + \sqrt{z}$ and $f(z+h) = (z+h) + \sqrt{z+h} \Rightarrow \frac{dw}{dz} = \lim\limits_{h \to 0} \frac{f(z+h)-f(z)}{h}$
$= \lim\limits_{h \to 0} \frac{\left(z+h+\sqrt{z+h}\right)-\left(z+\sqrt{z}\right)}{h} = \lim\limits_{h \to 0} \frac{h+\sqrt{z+h}-\sqrt{z}}{h} = \lim\limits_{h \to 0} \left[1 + \frac{\sqrt{z+h}-\sqrt{z}}{h} \cdot \frac{\left(\sqrt{z+h}+\sqrt{z}\right)}{\left(\sqrt{z+h}+\sqrt{z}\right)}\right]$
$= 1 + \lim\limits_{h \to 0} \frac{(z+h)-z}{h\left(\sqrt{z+h}+\sqrt{z}\right)} = 1 + \lim\limits_{h \to 0} \frac{1}{\sqrt{z+h}+\sqrt{z}} = 1 + \frac{1}{2\sqrt{z}} \Rightarrow \left.\frac{dw}{dz}\right|_{z=4} = \frac{5}{4}$

23. $f'(x) = \lim\limits_{z \to x} \frac{f(z)-f(x)}{z-x} = \lim\limits_{z \to x} \frac{\frac{1}{z+2} - \frac{1}{x+2}}{z-x} = \lim\limits_{z \to x} \frac{(x+2)-(z+2)}{(z-x)(z+2)(x+2)} = \lim\limits_{z \to x} \frac{x-z}{(z-x)(z+2)(x+2)} = \lim\limits_{z \to x} \frac{-1}{(z+2)(x+2)}$
$= \frac{-1}{(x+2)^2}$

24. $f'(x) = \lim\limits_{z \to x} \frac{f(z)-f(x)}{z-x} = \lim\limits_{z \to x} \frac{\frac{1}{(z-1)^2} - \frac{1}{(x-1)^2}}{z-x} = \lim\limits_{z \to x} \frac{(x-1)^2-(z-1)^2}{(z-x)(z-1)^2(x-1)^2} = \lim\limits_{z \to x} \frac{[(x-1)-(z-1)][(x-1)+(z-1)]}{(z-x)(z-1)^2(x-1)^2}$
$= \lim\limits_{z \to x} \frac{(x-z)(x+z-2)}{(z-x)(z-1)^2(x-1)^2} = \lim\limits_{z \to x} \frac{-1(x+z-2)}{(z-1)^2(x-1)^2} = \frac{-1(2x-2)}{(x-1)^4} = \frac{-2(x-1)}{(x-1)^4} = \frac{-2}{(x-1)^3}$

25. $g'(x) = \lim\limits_{z \to x} \frac{g(z)-g(x)}{z-x} = \lim\limits_{z \to x} \frac{\frac{z}{z-1} - \frac{x}{x-1}}{z-x} = \lim\limits_{z \to x} \frac{z(x-1)-x(z-1)}{(z-x)(z-1)(x-1)} = \lim\limits_{z \to x} \frac{-z+x}{(z-x)(z-1)(x-1)} = \lim\limits_{z \to x} \frac{-1}{(z-1)(x-1)}$
$= \frac{-1}{(x-1)^2}$

26. $g'(x) = \lim\limits_{z \to x} \frac{g(z)-g(x)}{z-x} = \lim\limits_{z \to x} \frac{(1+\sqrt{z})-(1+\sqrt{x})}{z-x} = \lim\limits_{z \to x} \frac{\sqrt{z}-\sqrt{x}}{z-x} \cdot \frac{\sqrt{z}+\sqrt{x}}{\sqrt{z}+\sqrt{x}} = \lim\limits_{z \to x} \frac{z-x}{(z-x)(\sqrt{z}+\sqrt{x})}$
$= \lim\limits_{z \to x} \frac{1}{\sqrt{z}+\sqrt{x}} = \frac{1}{2\sqrt{x}}$

27. Note that as x increases, the slope of the tangent line to the curve is first negative, then zero (when $x = 0$), then positive $\Rightarrow$ the slope is always increasing which matches (b).

28. Note that the slope of the tangent line is never negative. For x negative, $f_2'(x)$ is positive but decreasing as x increases. When $x = 0$, the slope of the tangent line to x is 0. For $x > 0$, $f_2'(x)$ is positive and increasing. This graph matches (a).

29. $f_3(x)$ is an oscillating function like the cosine. Everywhere that the graph of f_3 has a horizontal tangent we expect f_3' to be zero, and (d) matches this condition.

30. The graph matches with (c).

31. (a) f' is not defined at $x = 0, 1, 4$. At these points, the left-hand and right-hand derivatives do not agree. For example, $\lim\limits_{x \to 0^-} \frac{f(x)-f(0)}{x-0}$ = slope of line joining $(-4, 0)$ and $(0, 2) = \frac{1}{2}$ but $\lim\limits_{x \to 0^+} \frac{f(x)-f(0)}{x-0}$ = slope of line joining $(0, 2)$ and $(1, -2) = -4$. Since these values are not equal, $f'(0) = \lim\limits_{x \to 0} \frac{f(x)-f(0)}{x-0}$ does not exist.

(b)

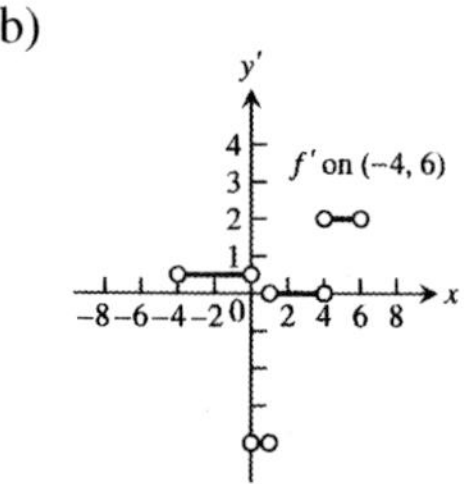

32. (a)

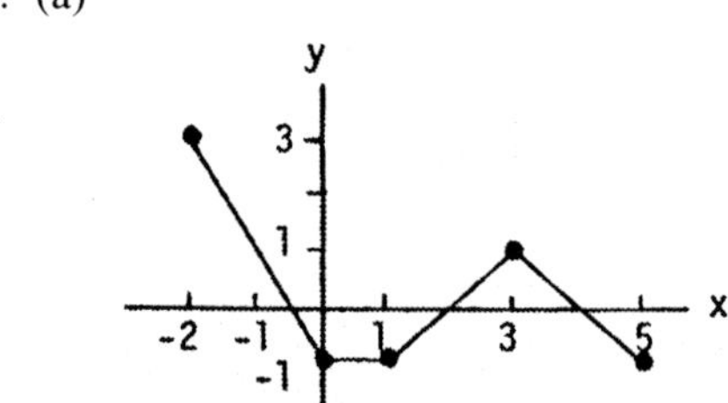

(b) Shift the graph in (a) down 3 units

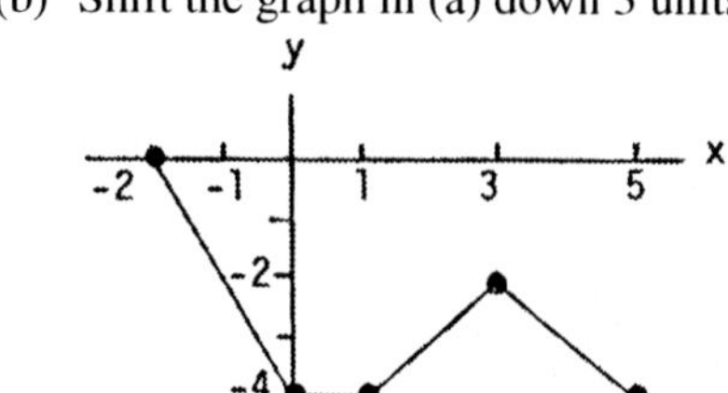

33.

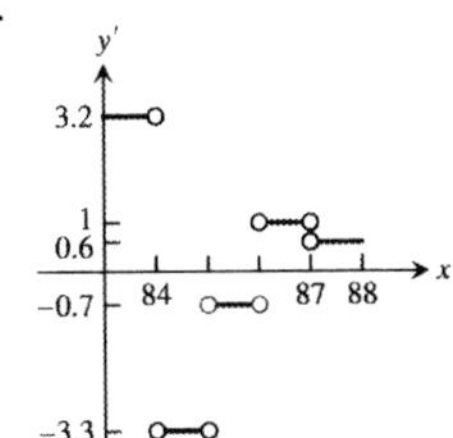

34. (a)

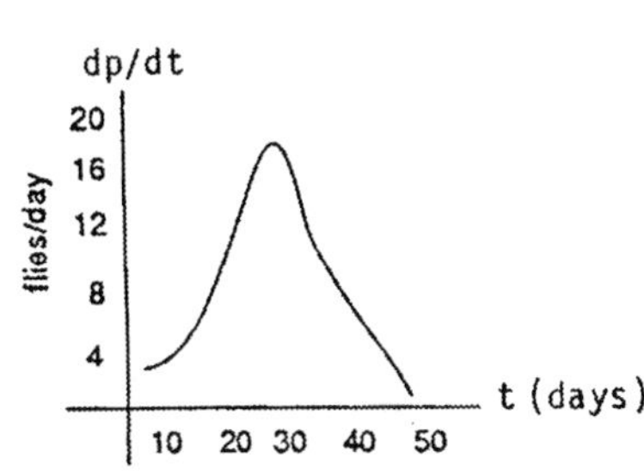

(b) The fastest is between the 20th and 30th days; slowest is between the 40th and 50th days.

35. Left-hand derivative: For $h < 0$, $f(0+h) = f(h) = h^2$ (using $y = x^2$ curve) $\Rightarrow \lim_{h \to 0^-} \frac{f(0+h)-f(0)}{h}$ $= \lim_{h \to 0^-} \frac{h^2 - 0}{h} = \lim_{h \to 0^-} h = 0$;
Right-hand derivative: For $h > 0$, $f(0+h) = f(h) = h$ (using $y = x$ curve) $\Rightarrow \lim_{h \to 0^+} \frac{f(0+h)-f(0)}{h}$ $= \lim_{h \to 0^+} \frac{h-0}{h} = \lim_{h \to 0^+} 1 = 1$;
Then $\lim_{h \to 0^-} \frac{f(0+h)-f(0)}{h} \neq \lim_{h \to 0^+} \frac{f(0+h)-f(0)}{h} \Rightarrow$ the derivative $f'(0)$ does not exist.

36. Left-hand derivative: When $h < 0$, $1 + h < 1 \Rightarrow f(1+h) = 2 \Rightarrow \lim_{h \to 0^-} \frac{f(1+h)-f(1)}{h} = \lim_{h \to 0^-} \frac{2-2}{h}$ $= \lim_{h \to 0^-} 0 = 0$;
Right-hand derivative: When $h > 0$, $1 + h > 1 \Rightarrow f(1+h) = 2(1+h) = 2 + 2h \Rightarrow \lim_{h \to 0^+} \frac{f(1+h)-f(1)}{h}$ $= \lim_{h \to 0^+} \frac{(2+2h)-2}{h} = \lim_{h \to 0^+} \frac{2h}{h} = \lim_{h \to 0^+} 2 = 2$;
Then $\lim_{h \to 0^-} \frac{f(1+h)-f(1)}{h} \neq \lim_{h \to 0^+} \frac{f(1+h)-f(1)}{h} \Rightarrow$ the derivative $f'(1)$ does not exist.

37. Left-hand derivative: When $h < 0$, $1 + h < 1 \Rightarrow f(1+h) = \sqrt{1+h} \Rightarrow \lim_{h \to 0^-} \frac{f(1+h)-f(1)}{h}$ $= \lim_{h \to 0^-} \frac{\sqrt{1+h}-1}{h} = \lim_{h \to 0^-} \frac{\left(\sqrt{1+h}-1\right)}{h} \cdot \frac{\left(\sqrt{1+h}+1\right)}{\left(\sqrt{1+h}+1\right)} = \lim_{h \to 0^-} \frac{(1+h)-1}{h\left(\sqrt{1+h}+1\right)} = \lim_{h \to 0^-} \frac{1}{\sqrt{1+h}+1} = \frac{1}{2}$;
Right-hand derivative: When $h > 0$, $1 + h > 1 \Rightarrow f(1+h) = 2(1+h) - 1 = 2h + 1 \Rightarrow \lim_{h \to 0^+} \frac{f(1+h)-f(1)}{h}$ $= \lim_{h \to 0^+} \frac{(2h+1)-1}{h} = \lim_{h \to 0^+} 2 = 2$;
Then $\lim_{h \to 0^-} \frac{f(1+h)-f(1)}{h} \neq \lim_{h \to 0^+} \frac{f(1+h)-f(1)}{h} \Rightarrow$ the derivative $f'(1)$ does not exist.

38. Left-hand derivative: $\lim_{h \to 0^-} \frac{f(1+h)-f(1)}{h} = \lim_{h \to 0^-} \frac{(1+h)-1}{h} = \lim_{h \to 0^-} 1 = 1$;
Right-hand derivative: $\lim_{h \to 0^+} \frac{f(1+h)-f(1)}{h} = \lim_{h \to 0^+} \frac{\left(\frac{1}{1+h}-1\right)}{h} = \lim_{h \to 0^+} \frac{\left(\frac{1-(1+h)}{1+h}\right)}{h}$ $= \lim_{h \to 0^+} \frac{-h}{h(1+h)} = \lim_{h \to 0^+} \frac{-1}{1+h} = -1$;
Then $\lim_{h \to 0^-} \frac{f(1+h)-f(1)}{h} \neq \lim_{h \to 0^+} \frac{f(1+h)-f(1)}{h} \Rightarrow$ the derivative $f'(1)$ does not exist.

39. (a) The function is differentiable on its domain $-3 \le x \le 2$ (it is smooth)
(b) none
(c) none

40. (a) The function is differentiable on its domain $-2 \le x \le 3$ (it is smooth)
(b) none
(c) none

41. (a) The function is differentiable on $-3 \le x < 0$ and $0 < x \le 3$
(b) none
(c) The function is neither continuous nor differentiable at $x = 0$ since $\lim_{x \to 0^-} f(x) \neq \lim_{x \to 0^+} f(x)$

42. (a) f is differentiable on $-2 \le x < -1, -1 < x < 0, 0 < x < 2$, and $2 < x \le 3$
(b) f is continuous but not differentiable at $x = -1$: $\lim_{x \to -1} f(x) = 0$ exists but there is a corner at $x = -1$ since $\lim_{h \to 0^-} \frac{f(-1+h)-f(-1)}{h} = -3$ and $\lim_{h \to 0^+} \frac{f(-1+h)-f(-1)}{h} = 3 \Rightarrow f'(-1)$ does not exist
(c) f is neither continuous nor differentiable at $x = 0$ and $x = 2$:
at $x = 0$, $\lim_{x \to 0^-} f(x) = 3$ but $\lim_{x \to 0^+} f(x) = 0 \Rightarrow \lim_{x \to 0} f(x)$ does not exist;
at $x = 2$, $\lim_{x \to 2} f(x)$ exists but $\lim_{x \to 2} f(x) \neq f(2)$

43. (a) f is differentiable on $-1 \le x < 0$ and $0 < x \le 2$
(b) f is continuous but not differentiable at $x = 0$: $\lim_{x \to 0} f(x) = 0$ exists but there is a cusp at $x = 0$, so $f'(0) = \lim_{h \to 0} \frac{f(0+h)-f(0)}{h}$ does not exist
(c) none

44. (a) f is differentiable on $-3 \le x < -2, -2 < x < 2$, and $2 < x \le 3$
(b) f is continuous but not differentiable at $x = -2$ and $x = 2$: there are corners at those points
(c) none

45. (a) $f'(x) = \lim_{h \to 0} \frac{f(x+h)-f(x)}{h} = \lim_{h \to 0} \frac{-(x+h)^2-(-x^2)}{h} = \lim_{h \to 0} \frac{-x^2-2xh-h^2+x^2}{h} = \lim_{h \to 0} (-2x - h) = -2x$
(b)

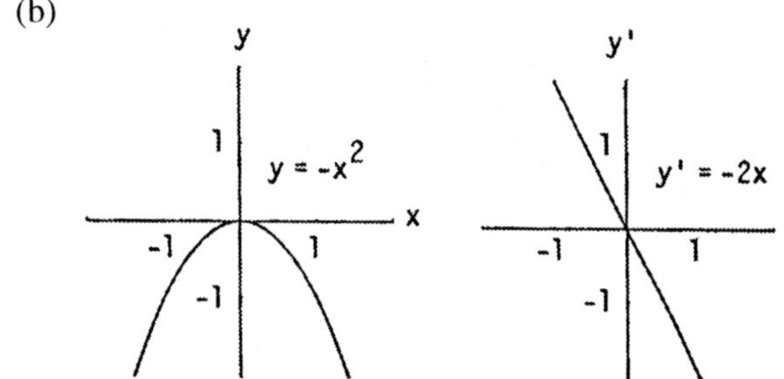

(c) $y' = -2x$ is positive for $x < 0$, y' is zero when $x = 0$, y' is negative when $x > 0$
(d) $y = -x^2$ is increasing for $-\infty < x < 0$ and decreasing for $0 < x < \infty$; the function is increasing on intervals where $y' > 0$ and decreasing on intervals where $y' < 0$

46. (a) $f'(x) = \lim_{h \to 0} \frac{f(x+h)-f(x)}{h} = \lim_{h \to 0} \frac{\left(\frac{-1}{x+h} - \frac{-1}{x}\right)}{h} = \lim_{h \to 0} \frac{-x+(x+h)}{x(x+h)h} = \lim_{h \to 0} \frac{1}{x(x+h)} = \frac{1}{x^2}$

(b)

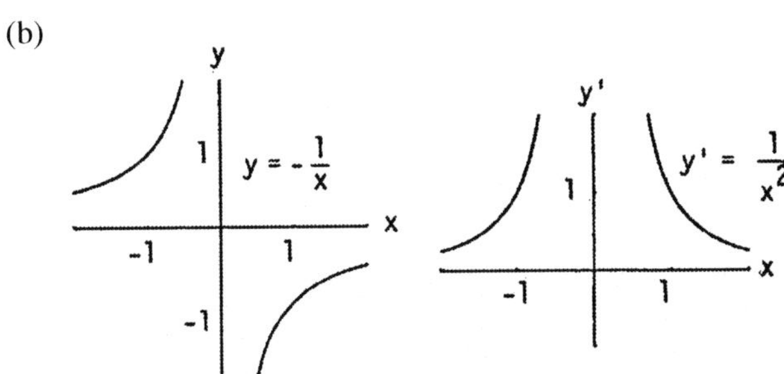

(c) y' is positive for all $x \neq 0$, y' is never 0, y' is never negative

(d) $y = -\frac{1}{x}$ is increasing for $-\infty < x < 0$ and $0 < x < \infty$

47. (a) Using the alternate formula for calculating derivatives: $f'(x) = \lim\limits_{z \to x} \frac{f(z) - f(x)}{z - x} = \lim\limits_{z \to x} \frac{\left(\frac{z^3}{3} - \frac{x^3}{3}\right)}{z - x}$
$= \lim\limits_{z \to x} \frac{z^3 - x^3}{3(z - x)} = \lim\limits_{z \to x} \frac{(z - x)(z^2 + zx + x^2)}{3(z - x)} = \lim\limits_{z \to x} \frac{z^2 + zx + x^2}{3} = x^2 \Rightarrow f'(x) = x^2$

(b)

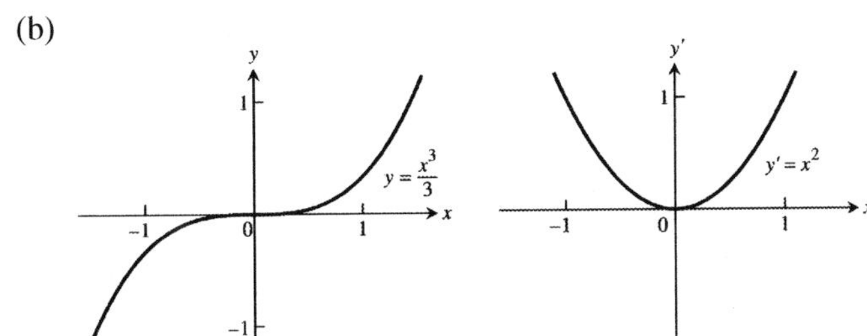

(c) y' is positive for all $x \neq 0$, and $y' = 0$ when $x = 0$; y' is never negative

(d) $y = \frac{x^3}{3}$ is increasing for all $x \neq 0$ (the graph is horizontal at $x = 0$) because y is increasing where $y' > 0$; y is never decreasing

48. (a) Using the alternate form for calculating derivatives: $f'(x) = \lim\limits_{z \to x} \frac{f(z) - f(x)}{z - x} = \lim\limits_{z \to x} \frac{\left(\frac{z^4}{4} - \frac{x^4}{4}\right)}{z - x}$
$= \lim\limits_{z \to x} \frac{z^4 - x^4}{4(z - x)} = \lim\limits_{z \to x} \frac{(z - x)(z^3 + xz^2 + x^2z + x^3)}{4(z - x)} = \lim\limits_{z \to x} \frac{z^3 + xz^2 + x^2z + x^3}{4} = x^3 \Rightarrow f'(x) = x^3$

(b)

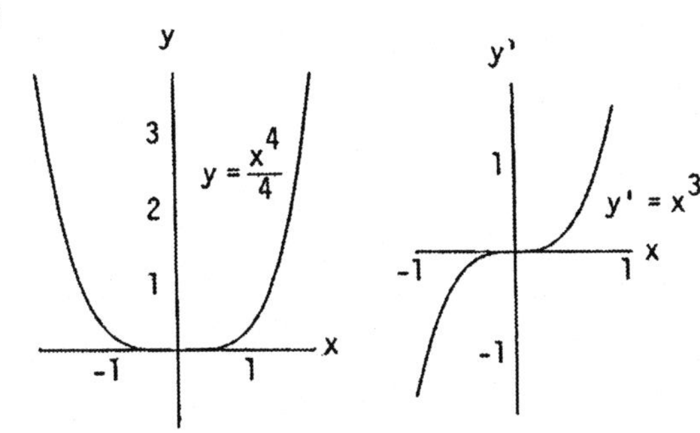

(c) y' is positive for $x > 0$, y' is zero for $x = 0$, y' is negative for $x < 0$

(d) $y = \frac{x^4}{4}$ is increasing on $0 < x < \infty$ and decreasing on $-\infty < x < 0$

49. $y' = \lim\limits_{x \to c} \frac{f(x) - f(c)}{x - c} = \lim\limits_{x \to c} \frac{x^3 - c^3}{x - c} = \lim\limits_{x \to c} \frac{(x - c)(x^2 + xc + c^2)}{x - c} = \lim\limits_{x \to c} (x^2 + xc + c^2) = 3c^2$.
The slope of the curve $y = x^3$ at $x = c$ is $y' = 3c^2$. Notice that $3c^2 \geq 0$ for all $c \Rightarrow y = x^3$ never has a negative slope.

50. Horizontal tangents occur where $y' = 0$. Thus, $y' = \lim\limits_{h \to 0} \frac{2\sqrt{x+h} - 2\sqrt{x}}{h}$
$= \lim\limits_{h \to 0} \frac{2\left(\sqrt{x+h} - \sqrt{x}\right)}{h} \cdot \frac{\left(\sqrt{x+h} + \sqrt{x}\right)}{\left(\sqrt{x+h} + \sqrt{x}\right)} = \lim\limits_{h \to 0} \frac{2((x+h) - x))}{h\left(\sqrt{x+h} + \sqrt{x}\right)} = \lim\limits_{h \to 0} \frac{2}{\sqrt{x+h} + \sqrt{x}} = \frac{1}{\sqrt{x}}$.

Then $y' = 0$ when $\frac{1}{\sqrt{x}} = 0$ which is never true $\Rightarrow$ the curve has no horizontal tangents.

51. $y' = \lim_{h \to 0} \frac{(2(x+h)^2 - 13(x+h) + 5) - (2x^2 - 13x + 5)}{h} = \lim_{h \to 0} \frac{2x^2 + 4xh + 2h^2 - 13x - 13h + 5 - 2x^2 + 13x - 5}{h}$
$= \lim_{h \to 0} \frac{4xh + 2h^2 - 13h}{h} = \lim_{h \to 0} (4x + 2h - 13) = 4x - 13$, slope at x. The slope is -1 when $4x - 13 = -1$
$\Rightarrow 4x = 12 \Rightarrow x = 3 \Rightarrow y = 2 \cdot 3^2 - 13 \cdot 3 + 5 = -16$. Thus the tangent line is $y + 16 = (-1)(x - 3)$
$\Rightarrow y = -x - 13$ and the point of tangency is $(3, -16)$.

52. For the curve $y = \sqrt{x}$, we have $y' = \lim_{h \to 0} \frac{\left(\sqrt{x+h} - \sqrt{x}\right)}{h} \cdot \frac{\left(\sqrt{x+h} + \sqrt{x}\right)}{\left(\sqrt{x+h} + \sqrt{x}\right)} = \lim_{h \to 0} \frac{(x+h) - x}{\left(\sqrt{x+h} + \sqrt{x}\right) h}$
$= \lim_{h \to 0} \frac{1}{\sqrt{x+h} + \sqrt{x}} = \frac{1}{2\sqrt{x}}$. Suppose $\left(a, \sqrt{a}\right)$ is the point of tangency of such a line and $(-1, 0)$ is the point on the line where it crosses the x-axis. Then the slope of the line is $\frac{\sqrt{a} - 0}{a - (-1)} = \frac{\sqrt{a}}{a+1}$ which must also equal $\frac{1}{2\sqrt{a}}$; using the derivative formula at $x = a \Rightarrow \frac{\sqrt{a}}{a+1} = \frac{1}{2\sqrt{a}} \Rightarrow 2a = a + 1 \Rightarrow a = 1$. Thus such a line does exist: its point of tangency is $(1, 1)$, its slope is $\frac{1}{2\sqrt{a}} = \frac{1}{2}$; and an equation of the line is $y - 1 = \frac{1}{2}(x - 1)$
$\Rightarrow y = \frac{1}{2}x + \frac{1}{2}$.

53. No. Derivatives of functions have the intermediate value property. The function $f(x) = \lfloor x \rfloor$ satisfies $f(0) = 0$ and $f(1) = 1$ but does not take on the value $\frac{1}{2}$ anywhere in $[0, 1] \Rightarrow$ f does not have the intermediate value property. Thus f cannot be the derivative of any function on $[0, 1] \Rightarrow$ f cannot be the derivative of any function on $(-\infty, \infty)$.

54. The graphs are the same. So we know that for $f(x) = |x|$, we have $f'(x) = \frac{|x|}{x}$.

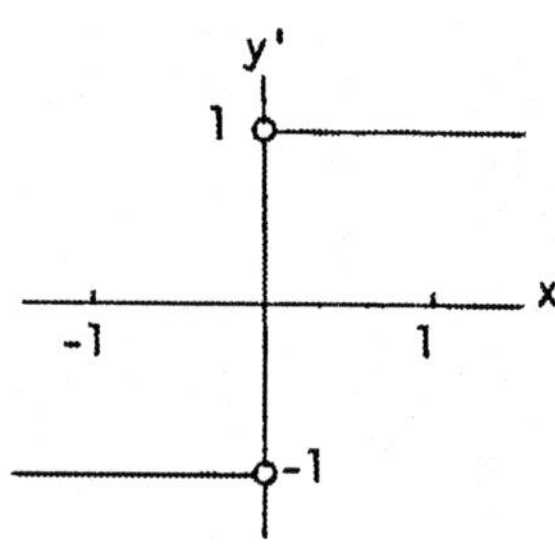

55. Yes; the derivative of $-f$ is $-f'$ so that $f'(x_0)$ exists $\Rightarrow -f'(x_0)$ exists as well.

56. Yes; the derivative of 3g is 3g′ so that $g'(7)$ exists $\Rightarrow 3g'(7)$ exists as well.

57. Yes, $\lim_{t \to 0} \frac{g(t)}{h(t)}$ can exist but it need not equal zero. For example, let $g(t) = mt$ and $h(t) = t$. Then $g(0) = h(0)$ $= 0$, but $\lim_{t \to 0} \frac{g(t)}{h(t)} = \lim_{t \to 0} \frac{mt}{t} = \lim_{t \to 0} m = m$, which need not be zero.

58. (a) Suppose $|f(x)| \le x^2$ for $-1 \le x \le 1$. Then $|f(0)| \le 0^2 \Rightarrow f(0) = 0$. Then $f'(0) = \lim_{h \to 0} \frac{f(0+h) - f(0)}{h}$
$= \lim_{h \to 0} \frac{f(h) - 0}{h} = \lim_{h \to 0} \frac{f(h)}{h}$. For $|h| \le 1, -h^2 \le f(h) \le h^2 \Rightarrow -h \le \frac{f(h)}{h} \le h \Rightarrow f'(0) = \lim_{h \to 0} \frac{f(h)}{h} = 0$ by the Sandwich Theorem for limits.

(b) Note that for $x \ne 0$, $|f(x)| = \left|x^2 \sin \frac{1}{x}\right| = |x^2| \, |\sin x| \le |x^2| \cdot 1 = x^2$ (since $-1 \le \sin x \le 1$). By part (a), f is differentiable at $x = 0$ and $f'(0) = 0$.

59. The graphs are shown below for $h = 1, 0.5, 0.1$. The function $y = \frac{1}{2\sqrt{x}}$ is the derivative of the function $y = \sqrt{x}$ so that $\frac{1}{2\sqrt{x}} = \lim_{h \to 0} \frac{\sqrt{x+h} - \sqrt{x}}{h}$. The graphs reveal that $y = \frac{\sqrt{x+h} - \sqrt{x}}{h}$ gets closer to $y = \frac{1}{2\sqrt{x}}$

as h gets smaller and smaller.

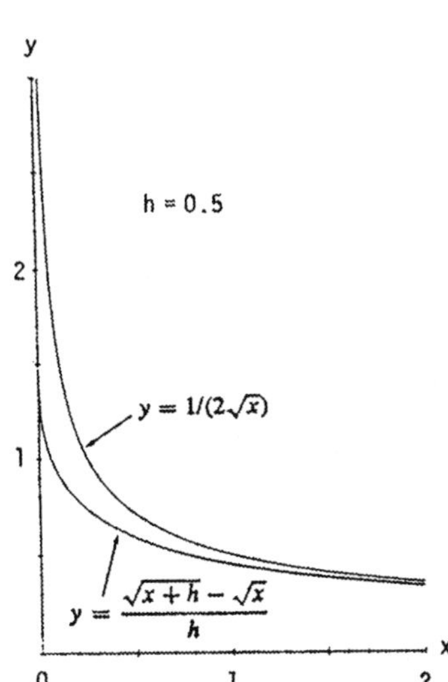

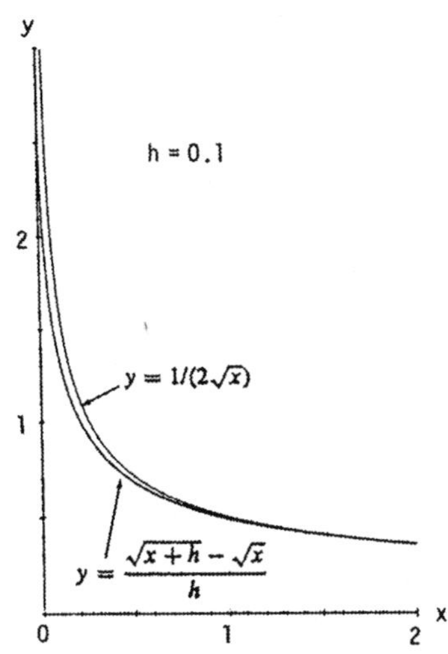

60. The graphs are shown below for h = 2, 1, 0.5. The function $y = 3x^2$ is the derivative of the function $y = x^3$ so that $3x^2 = \lim_{h \to 0} \frac{(x+h)^3 - x^3}{h}$. The graphs reveal that $y = \frac{(x+h)^3 - x^3}{h}$ gets closer to $y = 3x^2$ as h gets smaller and smaller.

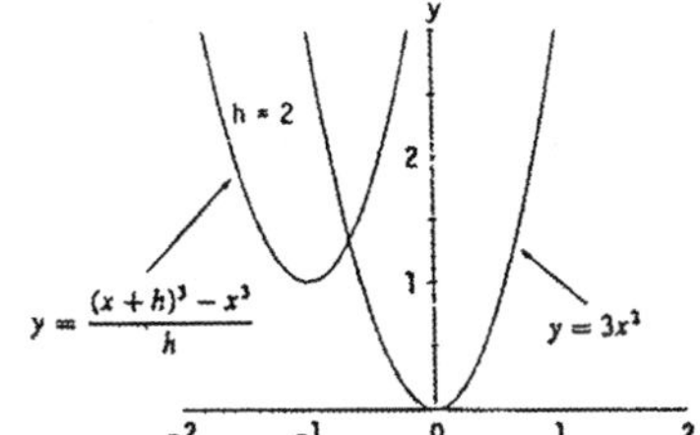

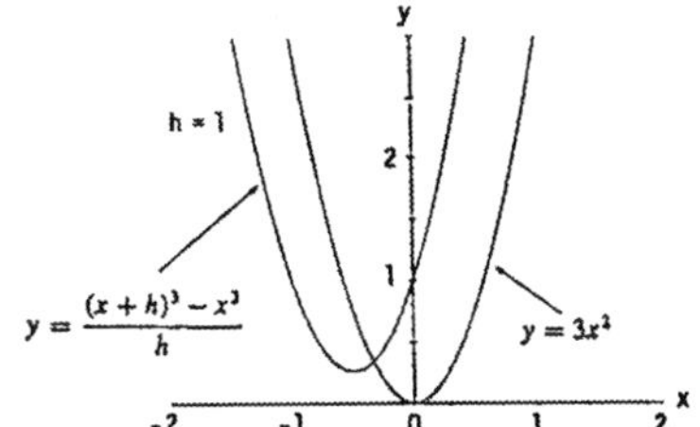

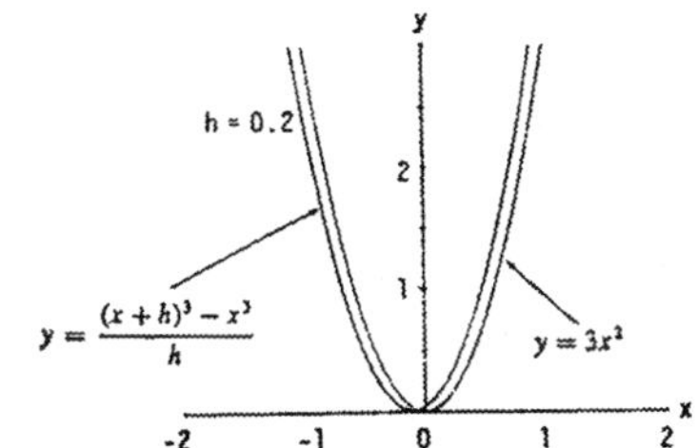

61. Weierstrass's nowhere differentiable continuous function.

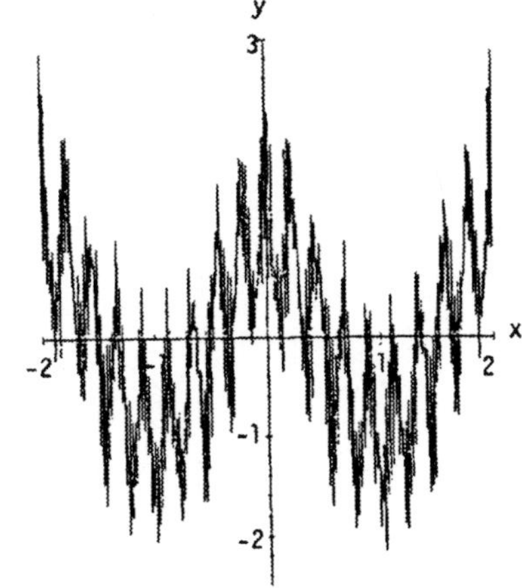

$$g(x) = \cos(\pi x) + \left(\frac{2}{3}\right)^1 \cos(9\pi x) + \left(\frac{2}{3}\right)^2 \cos(9^2 \pi x) + \left(\frac{2}{3}\right)^3 \cos(9^3 \pi x) + \cdots + \left(\frac{2}{3}\right)^7 \cos(9^7 \pi x)$$

62-67. Example CAS commands:

<u>Maple</u>:

```
f := x -> x^3 + x^2 - x;
```

```
x0 := 1;
plot( f(x), x=x0-5..x0+2, color=black,
      title="Section 3_1, #62(a)" );
q := unapply( (f(x+h)-f(x))/h, (x,h) );                         # (b)
L := limit( q(x,h), h=0 );                                      # (c)
m := eval( L, x=x0 );
tan_line := f(x0) + m*(x-x0);
plot( [f(x),tan_line], x=x0-2..x0+3, color=black,
     linestyle=[1,7], title="Section 3.1 #62(d)",
     legend=["y=f(x)","Tangent line at x=1"] );
Xvals := sort( [ x0+2^(-k) $ k=0..5, x0-2^(-k) $ k=0..5 ] ):    # (e)
Yvals := map( f, Xvals ):
evalf[4](< convert(Xvals,Matrix) , convert(Yvals,Matrix) >);
plot( L, x=x0-5..x0+3, color=black, title="Section 3.1 #62(f)" );
```

Mathematica: (functions and x0 may vary) (see section 2.5 re. RealOnly):

```
<<Miscellaneous`RealOnly`
Clear[f, m, x, y, h]
x0= π /4;
f[x_]:=x^2 Cos[x]
Plot[f[x], {x, x0 − 3, x0 + 3}]
q[x_, h_]:=(f[x + h] − f[x])/h
m[x_]:=Limit[q[x, h], h → 0]
ytan:=f[x0] + m[x0] (x − x0)
Plot[{f[x], ytan},{x, x0 − 3, x0 + 3}]
m[x0 − 1]//N
m[x0 + 1]//N
Plot[{f[x], m[x]},{x, x0 − 3, x0 + 3}]
```

3.2 DIFFERENTIATION RULES FOR POLYNOMIALS, EXPONENTIALS, PRODUCTS, AND QUOTIENTS

1. $y = -x^2 + 3 \Rightarrow \frac{dy}{dx} = \frac{d}{dx}(-x^2) + \frac{d}{dx}(3) = -2x + 0 = -2x \Rightarrow \frac{d^2y}{dx^2} = -2$

2. $y = x^2 + x + 8 \Rightarrow \frac{dy}{dx} = 2x + 1 + 0 = 2x + 1 \Rightarrow \frac{d^2y}{dx^2} = 2$

3. $s = 5t^3 - 3t^5 \Rightarrow \frac{ds}{dt} = \frac{d}{dt}(5t^3) - \frac{d}{dt}(3t^5) = 15t^2 - 15t^4 \Rightarrow \frac{d^2s}{dt^2} = \frac{d}{dt}(15t^2) - \frac{d}{dt}(15t^4) = 30t - 60t^3$

4. $w = 3z^7 - 7z^3 + 21z^2 \Rightarrow \frac{dw}{dz} = 21z^6 - 21z^2 + 42z \Rightarrow \frac{d^2w}{dz^2} = 126z^5 - 42z + 42$

5. $y = \frac{4x^3}{3} - x + 2e^x \Rightarrow \frac{dy}{dx} = 4x^2 - 1 + 2e^x \Rightarrow \frac{d^2y}{dx^2} = 8x + 2e^x$

6. $y = \frac{x^3}{3} + \frac{x^2}{2} + \frac{x}{4} \Rightarrow \frac{dy}{dx} = x^2 + x + \frac{1}{4} \Rightarrow \frac{d^2y}{dx^2} = 2x + 1 + 0 = 2x + 1$

7. $w = 3z^{-2} - z^{-1} \Rightarrow \frac{dw}{dz} = -6z^{-3} + z^{-2} = \frac{-6}{z^3} + \frac{1}{z^2} \Rightarrow \frac{d^2w}{dz^2} = 18z^{-4} - 2z^{-3} = \frac{18}{z^4} - \frac{2}{z^3}$

8. $s = -2t^{-1} + 4t^{-2} \Rightarrow \frac{ds}{dt} = 2t^{-2} - 8t^{-3} = \frac{2}{t^2} - \frac{8}{t^3} \Rightarrow \frac{d^2s}{dt^2} = -4t^{-3} + 24t^{-4} = \frac{-4}{t^3} + \frac{24}{t^4}$

9. $y = 6x^2 - 10x - 5x^{-2} \Rightarrow \frac{dy}{dx} = 12x - 10 + 10x^{-3} = 12x - 10 + \frac{10}{x^3} \Rightarrow \frac{d^2y}{dx^2} = 12 - 0 - 30x^{-4} = 12 - \frac{30}{x^4}$

10. $y = 4 - 2x - x^{-3} \Rightarrow \frac{dy}{dx} = -2 + 3x^{-4} = -2 + \frac{3}{x^4} \Rightarrow \frac{d^2y}{dx^2} = 0 - 12x^{-5} = \frac{-12}{x^5}$

11. $r = \frac{1}{3}s^{-2} - \frac{5}{2}s^{-1} \Rightarrow \frac{dr}{ds} = -\frac{2}{3}s^{-3} + \frac{5}{2}s^{-2} = \frac{-2}{3s^3} + \frac{5}{2s^2} \Rightarrow \frac{d^2r}{ds^2} = 2s^{-4} - 5s^{-3} = \frac{2}{s^4} - \frac{5}{s^3}$

12. $r = 12\theta^{-1} - 4\theta^{-3} + \theta^{-4} \Rightarrow \frac{dr}{d\theta} = -12\theta^{-2} + 12\theta^{-4} - 4\theta^{-5} = \frac{-12}{\theta^2} + \frac{12}{\theta^4} - \frac{4}{\theta^5} \Rightarrow \frac{d^2r}{d\theta^2} = 24\theta^{-3} - 48\theta^{-5} + 20\theta^{-6}$
$= \frac{24}{\theta^3} - \frac{48}{\theta^5} + \frac{20}{\theta^6}$

13. (a) $y = (3 - x^2)(x^3 - x + 1) \Rightarrow y' = (3 - x^2) \cdot \frac{d}{dx}(x^3 - x + 1) + (x^3 - x + 1) \cdot \frac{d}{dx}(3 - x^2)$
$= (3 - x^2)(3x^2 - 1) + (x^3 - x + 1)(-2x) = -5x^4 + 12x^2 - 2x - 3$
(b) $y = -x^5 + 4x^3 - x^2 - 3x + 3 \Rightarrow y' = -5x^4 + 12x^2 - 2x - 3$

14. (a) $y = (x - 1)(x^2 + x + 1) \Rightarrow y' = (x - 1)(2x + 1) + (x^2 + x + 1)(1) = 3x^2$
(b) $y = (x - 1)(x^2 + x + 1) = x^3 - 1 \Rightarrow y' = 3x^2$

15. (a) $y = (x^2 + 1)\left(x + 5 + \frac{1}{x}\right) \Rightarrow y' = (x^2 + 1) \cdot \frac{d}{dx}\left(x + 5 + \frac{1}{x}\right) + \left(x + 5 + \frac{1}{x}\right) \cdot \frac{d}{dx}(x^2 + 1)$
$= (x^2 + 1)(1 - x^{-2}) + (x + 5 + x^{-1})(2x) = (x^2 - 1 + 1 - x^{-2}) + (2x^2 + 10x + 2) = 3x^2 + 10x + 2 - \frac{1}{x^2}$
(b) $y = x^3 + 5x^2 + 2x + 5 + \frac{1}{x} \Rightarrow y' = 3x^2 + 10x + 2 - \frac{1}{x^2}$

16. $y = \left(x + \frac{1}{x}\right)\left(x - \frac{1}{x} + 1\right)$
(a) $y' = (x + x^{-1}) \cdot (1 + x^{-2}) + (x - x^{-1} + 1)(1 - x^{-2}) = 2x + 1 - \frac{1}{x^2} + \frac{2}{x^3}$
(b) $y = x^2 + x + \frac{1}{x} - \frac{1}{x^2} \Rightarrow y' = 2x + 1 - \frac{1}{x^2} + \frac{2}{x^3}$

17. $y = \frac{2x+5}{3x-2}$; use the quotient rule: $u = 2x + 5$ and $v = 3x - 2 \Rightarrow u' = 2$ and $v' = 3 \Rightarrow y' = \frac{vu' - uv'}{v^2}$
$= \frac{(3x-2)(2) - (2x+5)(3)}{(3x-2)^2} = \frac{6x - 4 - 6x - 15}{(3x-2)^2} = \frac{-19}{(3x-2)^2}$

18. $z = \frac{2x+1}{x^2-1} \Rightarrow \frac{dz}{dx} = \frac{(x^2-1)(2) - (2x+1)(2x)}{(x^2-1)^2} = \frac{2x^2 - 2 - 4x^2 - 2x}{(x^2-1)^2} = \frac{-2x^2 - 2x - 2}{(x^2-1)^2} = \frac{-2(x^2+x+1)}{(x^2-1)^2}$

19. $g(x) = \frac{x^2-4}{x+0.5}$; use the quotient rule: $u = x^2 - 4$ and $v = x + 0.5 \Rightarrow u' = 2x$ and $v' = 1 \Rightarrow g'(x) = \frac{vu' - uv'}{v^2}$
$= \frac{(x+0.5)(2x) - (x^2-4)(1)}{(x+0.5)^2} = \frac{2x^2 + x - x^2 + 4}{(x+0.5)^2} = \frac{x^2 + x + 4}{(x+0.5)^2}$

20. $f(t) = \frac{t^2-1}{t^2+t-2} = \frac{(t-1)(t+1)}{(t+2)(t-1)} = \frac{t+1}{t+2}, t \neq 1 \Rightarrow f'(t) = \frac{(t+2)(1) - (t+1)(1)}{(t+2)^2} = \frac{t+2-t-1}{(t+2)^2} = \frac{1}{(t+2)^2}$

21. $v = (1 - t)(1 + t^2)^{-1} = \frac{1-t}{1+t^2} \Rightarrow \frac{dv}{dt} = \frac{(1+t^2)(-1) - (1-t)(2t)}{(1+t^2)^2} = \frac{-1 - t^2 - 2t + 2t^2}{(1+t^2)^2} = \frac{t^2 - 2t - 1}{(1+t^2)^2}$

22. $w = \frac{x+5}{2x-7} \Rightarrow w' = \frac{(2x-7)(1) - (x+5)(2)}{(2x-7)^2} = \frac{2x - 7 - 2x - 10}{(2x-7)^2} = \frac{-17}{(2x-7)^2}$

23. $y = 2e^{-x} \Rightarrow y' = 2e^{-x} \cdot (-1) = -2e^{-x}$

24. $y = \frac{x^2 + 3e^x}{2e^x - x} \Rightarrow y' = \frac{(2e^x - x)(2x + 3e^x) - (x^2 + 3e^x)(2e^x - 1)}{(2e^x - x)^2} = \frac{(4xe^x + 6e^{2x} - 2x^2 - 3xe^x) - (2x^2e^x - x^2 + 6e^{2x} - 3e^x)}{(2e^x - x)^2}$
$y' = \frac{xe^x - x^2 - 2x^2e^x + 3e^x}{(2e^x - x)^2}$

25. $v = \frac{1 + x - 4\sqrt{x}}{x} \Rightarrow v' = \frac{x\left(1 - \frac{2}{\sqrt{x}}\right) - (1 + x - 4\sqrt{x})}{x^2} = \frac{2\sqrt{x} - 1}{x^2}$

26. $r = 2\left(\frac{1}{\sqrt{\theta}} + \sqrt{\theta}\right) \Rightarrow r' = 2\left(\frac{\sqrt{\theta}(0) - 1\left(\frac{1}{2\sqrt{\theta}}\right)}{\theta} + \frac{1}{2\sqrt{\theta}}\right) = -\frac{1}{\theta^{3/2}} + \frac{1}{\theta^{1/2}}$

27. $y = x^3e^x \Rightarrow y' = x^3 \cdot e^x + 3x^2 \cdot e^x = (x^3 + 3x^2)e^x$

28. $w = re^{-r} \Rightarrow w' = r \cdot e^{-r}(-1) + (1) \cdot e^{-r} = (1 - r)e^{-r}$

29. $y = \frac{1}{2}x^4 - \frac{3}{2}x^2 - x \Rightarrow y' = 2x^3 - 3x - 1 \Rightarrow y'' = 6x^2 - 3 \Rightarrow y''' = 12x \Rightarrow y^{(4)} = 12 \Rightarrow y^{(n)} = 0$ for all $n \ge 5$

30. $y = \frac{1}{120}x^5 \Rightarrow y' = \frac{1}{24}x^4 \Rightarrow y'' = \frac{1}{6}x^3 \Rightarrow y''' = \frac{1}{2}x^2 \Rightarrow y^{(4)} = x \Rightarrow y^{(5)} = 1 \Rightarrow y^{(n)} = 0$ for all $n \ge 6$

31. $y = \frac{x^3+7}{x} = x^2 + 7x^{-1} \Rightarrow \frac{dy}{dx} = 2x - 7x^{-2} = 2x - \frac{7}{x^2} \Rightarrow \frac{d^2y}{dx^2} = 2 + 14x^{-3} = 2 + \frac{14}{x^3}$

32. $s = \frac{t^2+5t-1}{t^2} = 1 + \frac{5}{t} - \frac{1}{t^2} = 1 + 5t^{-1} - t^{-2} \Rightarrow \frac{ds}{dt} = 0 - 5t^{-2} + 2t^{-3} = -5t^{-2} + 2t^{-3} = \frac{-5}{t^2} + \frac{2}{t^3}$
$\Rightarrow \frac{d^2s}{dt^2} = 10t^{-3} - 6t^{-4} = \frac{10}{t^3} - \frac{6}{t^4}$

33. $r = \frac{(\theta-1)(\theta^2+\theta+1)}{\theta^3} = \frac{\theta^3-1}{\theta^3} = 1 - \frac{1}{\theta^3} = 1 - \theta^{-3} \Rightarrow \frac{dr}{d\theta} = 0 + 3\theta^{-4} = 3\theta^{-4} = \frac{3}{\theta^4} \Rightarrow \frac{d^2r}{d\theta^2} = -12\theta^{-5} = \frac{-12}{\theta^5}$

34. $u = \frac{(x^2+x)(x^2-x+1)}{x^4} = \frac{x(x+1)(x^2-x+1)}{x^4} = \frac{x(x^3+1)}{x^4} = \frac{x^4+x}{x^4} = 1 + \frac{x}{x^4} = 1 + x^{-3}$
$\Rightarrow \frac{du}{dx} = 0 - 3x^{-4} = -3x^{-4} = \frac{-3}{x^4} \Rightarrow \frac{d^2u}{dx^2} = 12x^{-5} = \frac{12}{x^5}$

35. $w = 3z^2e^z \Rightarrow w' = 3z^2 \cdot e^z + 6z \cdot e^z = (3z^2 + 6z)e^z$ and $w'' = (3z^2 + 6z)e^z + (6z + 6)e^z = (3z^2 + 12z + 6)e^z$

36. $w = e^z(z-1)(z^2+1) = e^z(z^3 - z^2 + z - 1) \Rightarrow w' = e^z \cdot (z^3 - z^2 + z - 1) + e^z \cdot (3z^2 - 2z + 1) = e^z(z^3 + 2z^2 - z)$
and $w'' = e^z \cdot (z^3 + 2z^2 - z) + e^z \cdot (3z^2 + 4z - 1) = e^z(z^3 + 5z^2 + 3z - 1)$

37. $p = \left(\frac{q^2+3}{12q}\right)\left(\frac{q^4-1}{q^3}\right) = \frac{q^6 - q^2 + 3q^4 - 3}{12q^4} = \frac{1}{12}q^2 - \frac{1}{12}q^{-2} + \frac{1}{4} - \frac{1}{4}q^{-4} \Rightarrow \frac{dp}{dq} = \frac{1}{6}q + \frac{1}{6}q^{-3} + q^{-5} = \frac{1}{6}q + \frac{1}{6q^3} + \frac{1}{q^5}$
$\Rightarrow \frac{d^2p}{dq^2} = \frac{1}{6} - \frac{1}{2}q^{-4} - 5q^{-6} = \frac{1}{6} - \frac{1}{2q^4} - \frac{5}{q^6}$

38. $p = \frac{q^2+3}{(q-1)^3 + (q+1)^3} = \frac{q^2+3}{(q^3 - 3q^2 + 3q - 1) + (q^3 + 3q^2 + 3q + 1)} = \frac{q^2+3}{2q^3+6q} = \frac{q^2+3}{2q(q^2+3)} = \frac{1}{2q} = \frac{1}{2}q^{-1}$
$\Rightarrow \frac{dp}{dq} = -\frac{1}{2}q^{-2} = -\frac{1}{2q^2} \Rightarrow \frac{d^2p}{dq^2} = q^{-3} = \frac{1}{q^3}$

39. $u(0) = 5,\ u'(0) = -3,\ v(0) = -1,\ v'(0) = 2$

(a) $\frac{d}{dx}(uv) = uv' + vu' \Rightarrow \frac{d}{dx}(uv)\big|_{x=0} = u(0)v'(0) + v(0)u'(0) = 5 \cdot 2 + (-1)(-3) = 13$

(b) $\frac{d}{dx}\left(\frac{u}{v}\right) = \frac{vu' - uv'}{v^2} \Rightarrow \frac{d}{dx}\left(\frac{u}{v}\right)\big|_{x=0} = \frac{v(0)u'(0) - u(0)v'(0)}{(v(0))^2} = \frac{(-1)(-3) - (5)(2)}{(-1)^2} = -7$

(c) $\frac{d}{dx}\left(\frac{v}{u}\right) = \frac{uv' - vu'}{u^2} \Rightarrow \frac{d}{dx}\left(\frac{v}{u}\right)\big|_{x=0} = \frac{u(0)v'(0) - v(0)u'(0)}{(u(0))^2} = \frac{(5)(2) - (-1)(-3)}{(5)^2} = \frac{7}{25}$

(d) $\frac{d}{dx}(7v - 2u) = 7v' - 2u' \Rightarrow \frac{d}{dx}(7v - 2u)\big|_{x=0} = 7v'(0) - 2u'(0) = 7 \cdot 2 - 2(-3) = 20$

40. $u(1) = 2,\ u'(1) = 0,\ v(1) = 5,\ v'(1) = -1$

(a) $\frac{d}{dx}(uv)\big|_{x=1} = u(1)v'(1) + v(1)u'(1) = 2 \cdot (-1) + 5 \cdot 0 = -2$

(b) $\frac{d}{dx}\left(\frac{u}{v}\right)\big|_{x=1} = \frac{v(1)u'(1) - u(1)v'(1)}{(v(1))^2} = \frac{5 \cdot 0 - 2 \cdot (-1)}{(5)^2} = \frac{2}{25}$

(c) $\frac{d}{dx}\left(\frac{v}{u}\right)\big|_{x=1} = \frac{u(1)v'(1) - v(1)u'(1)}{(u(1))^2} = \frac{2 \cdot (-1) - 5 \cdot 0}{(2)^2} = -\frac{1}{2}$

(d) $\frac{d}{dx}(7v - 2u)\big|_{x=1} = 7v'(1) - 2u'(1) = 7 \cdot (-1) - 2 \cdot 0 = -7$

41. $y = x^3 - 4x + 1$. Note that $(2, 1)$ is on the curve: $1 = 2^3 - 4(2) + 1$

(a) Slope of the tangent at (x, y) is $y' = 3x^2 - 4 \Rightarrow$ slope of the tangent at $(2, 1)$ is $y'(2) = 3(2)^2 - 4 = 8$. Thus the slope of the line perpendicular to the tangent at $(2, 1)$ is $-\frac{1}{8} \Rightarrow$ the equation of the line perpendicular to to the tangent line at $(2, 1)$ is $y - 1 = -\frac{1}{8}(x - 2)$ or $y = -\frac{x}{8} + \frac{5}{4}$.

(b) The slope of the curve at x is $m = 3x^2 - 4$ and the smallest value for m is -4 when $x = 0$ and $y = 1$.

(c) We want the slope of the curve to be $8 \Rightarrow y' = 8 \Rightarrow 3x^2 - 4 = 8 \Rightarrow 3x^2 = 12 \Rightarrow x^2 = 4 \Rightarrow x = \pm 2$. When $x = 2$, $y = 1$ and the tangent line has equation $y - 1 = 8(x - 2)$ or $y = 8x - 15$; when $x = -2$, $y = (-2)^3 - 4(-2) + 1 = 1$, and the tangent line has equation $y - 1 = 8(x + 2)$ or $y = 8x + 17$.

42. (a) $y = x^3 - 3x - 2 \Rightarrow y' = 3x^2 - 3$. For the tangent to be horizontal, we need $m = y' = 0 \Rightarrow 0 = 3x^2 - 3$ $\Rightarrow 3x^2 = 3 \Rightarrow x = \pm 1$. When $x = -1$, $y = 0 \Rightarrow$ the tangent line has equation $y = 0$. The line perpendicular to this line at $(-1, 0)$ is $x = -1$. When $x = 1$, $y = -4 \Rightarrow$ the tangent line has equation $y = -4$. The line perpendicular to this line at $(1, -4)$ is $x = 1$.

(b) The smallest value of y' is -3, and this occurs when $x = 0$ and $y = -2$. The tangent to the curve at $(0, -2)$ has slope $-3 \Rightarrow$ the line perpendicular to the tangent at $(0, -2)$ has slope $\frac{1}{3} \Rightarrow y + 2 = \frac{1}{3}(x - 0)$ or $y = \frac{1}{3}x - 2$ is an equation of the perpendicular line.

43. $y = \frac{4x}{x^2+1} \Rightarrow \frac{dy}{dx} = \frac{(x^2+1)(4) - (4x)(2x)}{(x^2+1)^2} = \frac{4x^2+4-8x^2}{(x^2+1)^2} = \frac{4(-x^2+1)}{(x^2+1)^2}$. When $x = 0$, $y = 0$ and $y' = \frac{4(0+1)}{1}$ $= 4$, so the tangent to the curve at $(0, 0)$ is the line $y = 4x$. When $x = 1$, $y = 2 \Rightarrow y' = 0$, so the tangent to the curve at $(1, 2)$ is the line $y = 2$.

44. $y = \frac{8}{x^2+4} \Rightarrow y' = \frac{(x^2+4)(0) - 8(2x)}{(x^2+4)^2} = \frac{-16x}{(x^2+4)^2}$. When $x = 2$, $y = 1$ and $y' = \frac{-16(2)}{(2^2+4)^2} = -\frac{1}{2}$, so the tangent line to the curve at $(2, 1)$ has the equation $y - 1 = -\frac{1}{2}(x - 2)$, or $y = -\frac{x}{2} + 2$.

45. $y = ax^2 + bx + c$ passes through $(0, 0) \Rightarrow 0 = a(0) + b(0) + c \Rightarrow c = 0$; $y = ax^2 + bx$ passes through $(1, 2)$ $\Rightarrow 2 = a + b$; $y' = 2ax + b$ and since the curve is tangent to $y = x$ at the origin, its slope is 1 at $x = 0$ $\Rightarrow y' = 1$ when $x = 0 \Rightarrow 1 = 2a(0) + b \Rightarrow b = 1$. Then $a + b = 2 \Rightarrow a = 1$. In summary $a = b = 1$ and $c = 0$ so the curve is $y = x^2 + x$.

46. $y = cx - x^2$ passes through $(1, 0) \Rightarrow 0 = c(1) - 1 \Rightarrow c = 1 \Rightarrow$ the curve is $y = x - x^2$. For this curve, $y' = 1 - 2x$ and $x = 1 \Rightarrow y' = -1$. Since $y = x - x^2$ and $y = x^2 + ax + b$ have common tangents at $x = 0$, $y = x^2 + ax + b$ must also have slope -1 at $x = 1$. Thus $y' = 2x + a \Rightarrow -1 = 2 \cdot 1 + a \Rightarrow a = -3$ $\Rightarrow y = x^2 - 3x + b$. Since this last curve passes through $(1, 0)$, we have $0 = 1 - 3 + b \Rightarrow b = 2$. In summary, $a = -3$, $b = 2$ and $c = 1$ so the curves are $y = x^2 - 3x + 2$ and $y = x - x^2$.

47. (a) $y = x^3 - x \Rightarrow y' = 3x^2 - 1$. When $x = -1$, $y = 0$ and $y' = 2 \Rightarrow$ the tangent line to the curve at $(-1, 0)$ is $y = 2(x + 1)$ or $y = 2x + 2$.

(b)

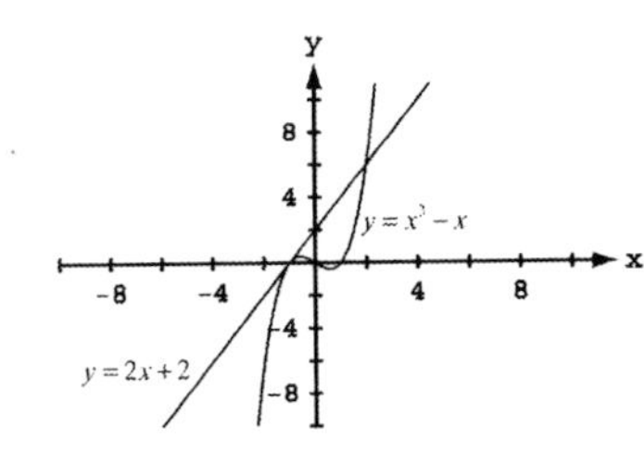

(c) $\left.\begin{array}{l} y = x^3 - x \\ y = 2x + 2 \end{array}\right\} \Rightarrow x^3 - x = 2x + 2 \Rightarrow x^3 - 3x - 2 = (x - 2)(x + 1)^2 = 0 \Rightarrow x = 2$ or $x = -1$. Since $y = 2(2) + 2 = 6$; the other intersection point is $(2, 6)$

48. (a) $y = x^3 - 6x^2 + 5x \Rightarrow y' = 3x^2 - 12x + 5$. When $x = 0$, $y = 0$ and $y' = 5 \Rightarrow$ the tangent line to the curve at $(0,0)$ is $y = 5x$.

(b)

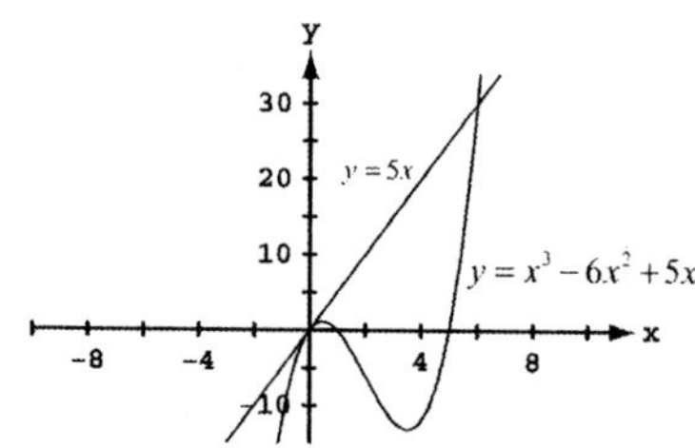

(c) $\left.\begin{array}{l} y = x^3 - 6x^2 + 5x \\ y = 5x \end{array}\right\} \Rightarrow x^3 - 6x^2 + 5x = 5x \Rightarrow x^3 - 6x^2 = 0 \Rightarrow x^2(x-6) = 0 \Rightarrow x = 0$ or $x = 6$.

Since $y = 5(6) = 30$, the other intersection point is $(6, 30)$.

49. $P(x) = a_n x^n + a_{n-1}x^{n-1} + \cdots + a_2x^2 + a_1x + a_0 \Rightarrow P'(x) = na_n x^{n-1} + (n-1)a_{n-1}x^{n-2} + \cdots + 2a_2x + a_1$

50. $R = M^2\left(\frac{C}{2} - \frac{M}{3}\right) = \frac{C}{2}M^2 - \frac{1}{3}M^3$, where C is a constant $\Rightarrow \frac{dR}{dM} = CM - M^2$

51. Let c be a constant $\Rightarrow \frac{dc}{dx} = 0 \Rightarrow \frac{d}{dx}(u \cdot c) = u \cdot \frac{dc}{dx} + c \cdot \frac{du}{dx} = u \cdot 0 + c\,\frac{du}{dx} = c\,\frac{du}{dx}$. Thus when one of the functions is a constant, the Product Rule is just the Constant Multiple Rule $\Rightarrow$ the Constant Multiple Rule is a special case of the Product Rule.

52. (a) We use the Quotient rule to derive the Reciprocal Rule (with $u = 1$): $\frac{d}{dx}\left(\frac{1}{v}\right) = \frac{v \cdot 0 - 1 \cdot \frac{dv}{dx}}{v^2} = \frac{-1 \cdot \frac{dv}{dx}}{v^2}$
$= -\frac{1}{v^2} \cdot \frac{dv}{dx}$.

(b) Now, using the Reciprocal Rule and the Product Rule, we'll derive the Quotient Rule: $\frac{d}{dx}\left(\frac{u}{v}\right) = \frac{d}{dx}\left(u \cdot \frac{1}{v}\right)$
$= u \cdot \frac{d}{dx}\left(\frac{1}{v}\right) + \frac{1}{v} \cdot \frac{du}{dx}$ (Product Rule) $= u \cdot \left(\frac{-1}{v^2}\right)\frac{dv}{dx} + \frac{1}{v}\frac{du}{dx}$ (Reciprocal Rule) $\Rightarrow \frac{d}{dx}\left(\frac{u}{v}\right) = \frac{-u\frac{dv}{dx} + v\frac{du}{dx}}{v^2}$
$= \frac{v\frac{du}{dx} - u\frac{dv}{dx}}{v^2}$, the Quotient Rule.

53. (a) $\frac{d}{dx}(uvw) = \frac{d}{dx}((uv) \cdot w) = (uv)\frac{dw}{dx} + w \cdot \frac{d}{dx}(uv) = uv\frac{dw}{dx} + w\left(u\frac{dv}{dx} + v\frac{du}{dx}\right) = uv\frac{dw}{dx} + wu\frac{dv}{dx} + wv\frac{du}{dx}$
$= uvw' + uv'w + u'vw$

(b) $\frac{d}{dx}(u_1u_2u_3u_4) = \frac{d}{dx}((u_1u_2u_3)\,u_4) = (u_1u_2u_3)\frac{du_4}{dx} + u_4\frac{d}{dx}(u_1u_2u_3) \Rightarrow \frac{d}{dx}(u_1u_2u_3u_4)$
$= u_1u_2u_3\frac{du_4}{dx} + u_4\left(u_1u_2\frac{du_3}{dx} + u_3u_1\frac{du_2}{dx} + u_3u_2\frac{du_1}{dx}\right)$ (using (a) above)
$\Rightarrow \frac{d}{dx}(u_1u_2u_3u_4) = u_1u_2u_3\frac{du_4}{dx} + u_1u_2u_4\frac{du_3}{dx} + u_1u_3u_4\frac{du_2}{dx} + u_2u_3u_4\frac{du_1}{dx}$
$= u_1u_2u_3u_4' + u_1u_2u_3'u_4 + u_1u_2'u_3u_4 + u_1'u_2u_3u_4$

(c) Generalizing (a) and (b) above, $\frac{d}{dx}(u_1 \cdots u_n) = u_1u_2 \cdots u_{n-1}u_n' + u_1u_2 \cdots u_{n-2}u_{n-1}'u_n + \ldots + u_1'u_2 \cdots u_n$

54. In this problem we don't know the Power Rule works with fractional powers so we can't use it. Remember $\frac{d}{dx}\left(\sqrt{x}\right) = \frac{1}{2\sqrt{x}}$ (from Example 2 in Section 2.1)

(a) $\frac{d}{dx}\left(x^{3/2}\right) = \frac{d}{dx}\left(x \cdot x^{1/2}\right) = x \cdot \frac{d}{dx}\left(\sqrt{x}\right) + \sqrt{x}\,\frac{d}{dx}(x) = x \cdot \frac{1}{2\sqrt{x}} + \sqrt{x} \cdot 1 = \frac{\sqrt{x}}{2} + \sqrt{x} = \frac{3\sqrt{x}}{2} = \frac{3}{2}x^{1/2}$

(b) $\frac{d}{dx}\left(x^{5/2}\right) = \frac{d}{dx}\left(x^2 \cdot x^{1/2}\right) = x^2\frac{d}{dx}\left(\sqrt{x}\right) + \sqrt{x}\,\frac{d}{dx}\left(x^2\right) = x^2 \cdot \left(\frac{1}{2\sqrt{x}}\right) + \sqrt{x} \cdot 2x = \frac{1}{2}x^{3/2} + 2x^{3/2} = \frac{5}{2}x^{3/2}$

(c) $\frac{d}{dx}\left(x^{7/2}\right) = \frac{d}{dx}\left(x^3 \cdot x^{1/2}\right) = x^3\frac{d}{dx}\left(\sqrt{x}\right) + \sqrt{x}\,\frac{d}{dx}\left(x^3\right) = x^3 \cdot \left(\frac{1}{2\sqrt{x}}\right) + \sqrt{x} \cdot 3x^2 = \frac{1}{2}x^{5/2} + 3x^{5/2} = \frac{7}{2}x^{5/2}$

(d) We have $\frac{d}{dx}\left(x^{3/2}\right) = \frac{3}{2}x^{1/2}$, $\frac{d}{dx}\left(x^{5/2}\right) = \frac{5}{2}x^{3/2}$, $\frac{d}{dx}\left(x^{7/2}\right) = \frac{7}{2}x^{5/2}$ so it appears that $\frac{d}{dx}\left(x^{n/2}\right) = \frac{n}{2}x^{(n/2)-1}$ whenever n is an odd positive integer ≥ 3.

55. $p = \frac{nRT}{V-nb} - \frac{an^2}{V^2}$. We are holding T constant, and a, b, n, R are also constant so their derivatives are zero
$\Rightarrow \frac{dP}{dV} = \frac{(V-nb)\cdot 0 - (nRT)(1)}{(V-nb)^2} - \frac{V^2(0) - (an^2)(2V)}{(V^2)^2} = \frac{-nRT}{(V-nb)^2} + \frac{2an^2}{V^3}$

56. $A(q) = \frac{km}{q} + cm + \frac{hq}{2} = (km)q^{-1} + cm + \left(\frac{h}{2}\right)q \Rightarrow \frac{dA}{dq} = -(km)q^{-2} + \left(\frac{h}{2}\right) = -\frac{km}{q^2} + \frac{h}{2} \Rightarrow \frac{d^2A}{dt^2} = 2(km)q^{-3} = \frac{2km}{q^3}$

3.3 THE DERIVATIVE AS A RATE OF CHANGE

1. $s = t^2 - 3t + 2, 0 \le t \le 2$
   (a) displacement $= \Delta s = s(2) - s(0) = 0m - 2m = -2$ m, $v_{av} = \frac{\Delta s}{\Delta t} = \frac{-2}{2} = -1$ m/sec
   (b) $v = \frac{ds}{dt} = 2t - 3 \Rightarrow |v(0)| = |-3| = 3$ m/sec and $|v(2)| = 1$ m/sec;
   $a = \frac{d^2s}{dt^2} = 2 \Rightarrow a(0) = 2$ m/sec^2 and $a(2) = 2$ m/sec^2
   (c) $v = 0 \Rightarrow 2t - 3 = 0 \Rightarrow t = \frac{3}{2}$. v is negative in the interval $0 < t < \frac{3}{2}$ and v is positive when $\frac{3}{2} < t < 2 \Rightarrow$ the body changes direction at $t = \frac{3}{2}$.

2. $s = 6t - t^2, 0 \le t \le 6$
   (a) displacement $= \Delta s = s(6) - s(0) = 0$ m, $v_{av} = \frac{\Delta s}{\Delta t} = \frac{0}{6} = 0$ m/sec
   (b) $v = \frac{ds}{dt} = 6 - 2t \Rightarrow |v(0)| = |6| = 6$ m/sec and $|v(6)| = |-6| = 6$ m/sec;
   $a = \frac{d^2s}{dt^2} = -2 \Rightarrow a(0) = -2$ m/sec^2 and $a(6) = -2$ m/sec^2
   (c) $v = 0 \Rightarrow 6 - 2t = 0 \Rightarrow t = 3$. v is positive in the interval $0 < t < 3$ and v is negative when $3 < t < 6 \Rightarrow$ the body changes direction at $t = 3$.

3. $s = -t^3 + 3t^2 - 3t, 0 \le t \le 3$
   (a) displacement $= \Delta s = s(3) - s(0) = -9$ m, $v_{av} = \frac{\Delta s}{\Delta t} = \frac{-9}{3} = -3$ m/sec
   (b) $v = \frac{ds}{dt} = -3t^2 + 6t - 3 \Rightarrow |v(0)| = |-3| = 3$ m/sec and $|v(3)| = |-12| = 12$ m/sec; $a = \frac{d^2s}{dt^2} = -6t + 6$
   $\Rightarrow a(0) = 6$ m/sec^2 and $a(3) = -12$ m/sec^2
   (c) $v = 0 \Rightarrow -3t^2 + 6t - 3 = 0 \Rightarrow t^2 - 2t + 1 = 0 \Rightarrow (t-1)^2 = 0 \Rightarrow t = 1$. For all other values of t in the interval the velocity v is negative (the graph of $v = -3t^2 + 6t - 3$ is a parabola with vertex at $t = 1$ which opens downward $\Rightarrow$ the body never changes direction).

4. $s = \frac{t^4}{4} - t^3 + t^2, 0 \le t \le 3$
   (a) $\Delta s = s(3) - s(0) = \frac{9}{4}$ m, $v_{av} = \frac{\Delta s}{\Delta t} = \frac{\frac{9}{4}}{3} = \frac{3}{4}$ m/sec
   (b) $v = t^3 - 3t^2 + 2t \Rightarrow |v(0)| = 0$ m/sec and $|v(3)| = 6$ m/sec; $a = 3t^2 - 6t + 2 \Rightarrow a(0) = 2$ m/sec^2 and $a(3) = 11$ m/sec^2
   (c) $v = 0 \Rightarrow t^3 - 3t^2 + 2t = 0 \Rightarrow t(t-2)(t-1) = 0 \Rightarrow t = 0, 1, 2 \Rightarrow v = t(t-2)(t-1)$ is positive in the interval for $0 < t < 1$ and v is negative for $1 < t < 2$ and v is positive for $2 < t < 3 \Rightarrow$ the body changes direction at $t = 1$ and at $t = 2$.

5. $s = \frac{25}{t^2} - \frac{5}{t}, 1 \le t \le 5$
   (a) $\Delta s = s(5) - s(1) = -20$ m, $v_{av} = \frac{-20}{4} = -5$ m/sec
   (b) $v = \frac{-50}{t^3} + \frac{5}{t^2} \Rightarrow |v(1)| = 45$ m/sec and $|v(5)| = \frac{1}{5}$ m/sec; $a = \frac{150}{t^4} - \frac{10}{t^3} \Rightarrow a(1) = 140$ m/sec^2 and $a(5) = \frac{4}{25}$ m/sec^2
   (c) $v = 0 \Rightarrow \frac{-50 + 5t}{t^3} = 0 \Rightarrow -50 + 5t = 0 \Rightarrow t = 10 \Rightarrow$ the body does not change direction in the interval

6. $s = \frac{25}{t+5}, -4 \le t \le 0$
   (a) $\Delta s = s(0) - s(-4) = -20$ m, $v_{av} = -\frac{20}{4} = -5$ m/sec

(b) $v = \frac{-25}{(t+5)^2} \Rightarrow |v(-4)| = 25$ m/sec and $|v(0)| = 1$ m/sec; $a = \frac{50}{(t+5)^3} \Rightarrow a(-4) = 50$ m/sec^2 and $a(0) = \frac{2}{5}$ m/sec^2

(c) $v = 0 \Rightarrow \frac{-25}{(t+5)^2} = 0 \Rightarrow$ v is never 0 $\Rightarrow$ the body never changes direction

7. $s = t^3 - 6t^2 + 9t$ and let the positive direction be to the right on the s-axis.
   (a) $v = 3t^2 - 12t + 9$ so that $v = 0 \Rightarrow t^2 - 4t + 3 = (t-3)(t-1) = 0 \Rightarrow t = 1$ or 3; $a = 6t - 12 \Rightarrow a(1) = -6$ m/sec^2 and $a(3) = 6$ m/sec^2. Thus the body is motionless but being accelerated left when $t = 1$, and motionless but being accelerated right when $t = 3$.
   (b) $a = 0 \Rightarrow 6t - 12 = 0 \Rightarrow t = 2$ with speed $|v(2)| = |12 - 24 + 9| = 3$ m/sec
   (c) The body moves to the right or forward on $0 \le t < 1$, and to the left or backward on $1 < t < 2$. The positions are $s(0) = 0$, $s(1) = 4$ and $s(2) = 2 \Rightarrow$ total distance $= |s(1) - s(0)| + |s(2) - s(1)| = |4| + |-2| = 6$ m.

8. $v = t^2 - 4t + 3 \Rightarrow a = 2t - 4$
   (a) $v = 0 \Rightarrow t^2 - 4t + 3 = 0 \Rightarrow t = 1$ or $3 \Rightarrow a(1) = -2$ m/sec^2 and $a(3) = 2$ m/sec^2
   (b) $v > 0 \Rightarrow (t-3)(t-1) > 0 \Rightarrow 0 \le t < 1$ or $t > 3$ and the body is moving forward; $v < 0 \Rightarrow (t-3)(t-1) < 0 \Rightarrow 1 < t < 3$ and the body is moving backward
   (c) velocity increasing $\Rightarrow a > 0 \Rightarrow 2t - 4 > 0 \Rightarrow t > 2$; velocity decreasing $\Rightarrow a < 0 \Rightarrow 2t - 4 < 0 \Rightarrow 0 \le t < 2$

9. $s_m = 1.86t^2 \Rightarrow v_m = 3.72t$ and solving $3.72t = 27.8 \Rightarrow t \approx 7.5$ sec on Mars; $s_j = 11.44t^2 \Rightarrow v_j = 22.88t$ and solving $22.88t = 27.8 \Rightarrow t \approx 1.2$ sec on Jupiter.

10. (a) $v(t) = s'(t) = 24 - 1.6t$ m/sec, and $a(t) = v'(t) = s''(t) = -1.6$ m/sec^2
   (b) Solve $v(t) = 0 \Rightarrow 24 - 1.6t = 0 \Rightarrow t = 15$ sec
   (c) $s(15) = 24(15) - .8(15)^2 = 180$ m
   (d) Solve $s(t) = 90 \Rightarrow 24t - .8t^2 = 90 \Rightarrow t = \frac{30 \pm 15\sqrt{2}}{2} \approx 4.39$ sec going up and 25.6 sec going down
   (e) Twice the time it took to reach its highest point or 30 sec

11. $s = 15t - \frac{1}{2} g_s t^2 \Rightarrow v = 15 - g_s t$ so that $v = 0 \Rightarrow 15 - g_s t = 0 \Rightarrow g_s = \frac{15}{t}$. Therefore $g_s = \frac{15}{20} = \frac{3}{4} = 0.75$ m/sec^2

12. Solving $s_m = 832t - 2.6t^2 = 0 \Rightarrow t(832 - 2.6t) = 0 \Rightarrow t = 0$ or $320 \Rightarrow 320$ sec on the moon; solving $s_e = 832t - 16t^2 = 0 \Rightarrow t(832 - 16t) = 0 \Rightarrow t = 0$ or $52 \Rightarrow 52$ sec on the earth. Also, $v_m = 832 - 5.2t = 0 \Rightarrow t = 160$ and $s_m(160) = 66{,}560$ ft, the height it reaches above the moon's surface; $v_e = 832 - 32t = 0 \Rightarrow t = 26$ and $s_e(26) = 10{,}816$ ft, the height it reaches above the earth's surface.

13. (a) $s = 179 - 16t^2 \Rightarrow v = -32t \Rightarrow$ speed $= |v| = 32t$ ft/sec and $a = -32$ ft/sec^2
   (b) $s = 0 \Rightarrow 179 - 16t^2 = 0 \Rightarrow t = \sqrt{\frac{179}{16}} \approx 3.3$ sec
   (c) When $t = \sqrt{\frac{179}{16}}$, $v = -32\sqrt{\frac{179}{16}} = -8\sqrt{179} \approx -107.0$ ft/sec

14. (a) $\lim_{\theta \to \frac{\pi}{2}} v = \lim_{\theta \to \frac{\pi}{2}} 9.8(\sin\theta)t = 9.8t$ so we expect $v = 9.8t$ m/sec in free fall
   (b) $a = \frac{dv}{dt} = 9.8$ m/sec^2

15. (a) at 2 and 7 seconds

(b) between 3 and 6 seconds: $3 \le t \le 6$

(c)

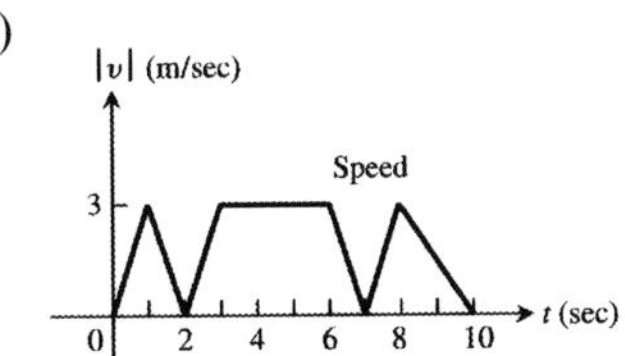

(d)

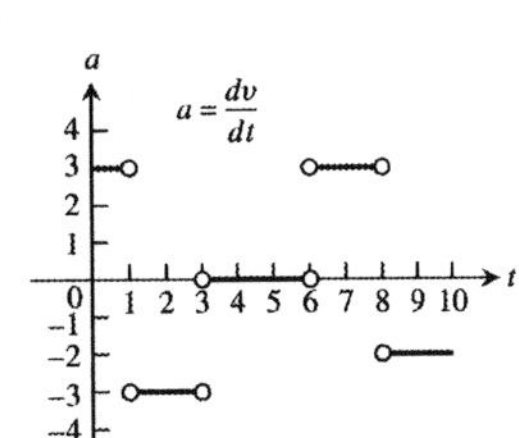

16. (a) P is moving to the left when $2 < t < 3$ or $5 < t < 6$; P is moving to the right when $0 < t < 1$; P is standing still when $1 < t < 2$ or $3 < t < 5$

(b)

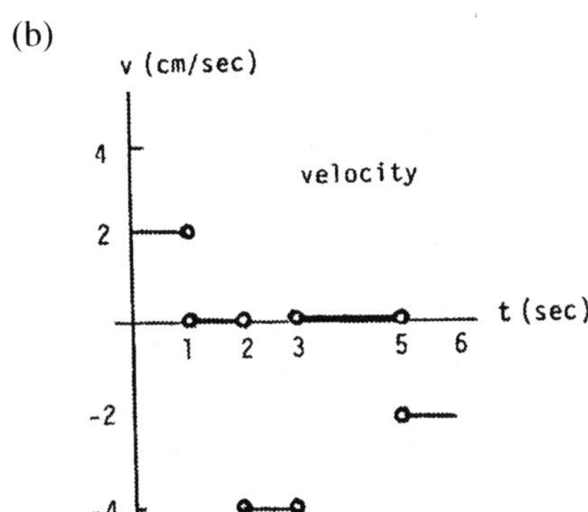

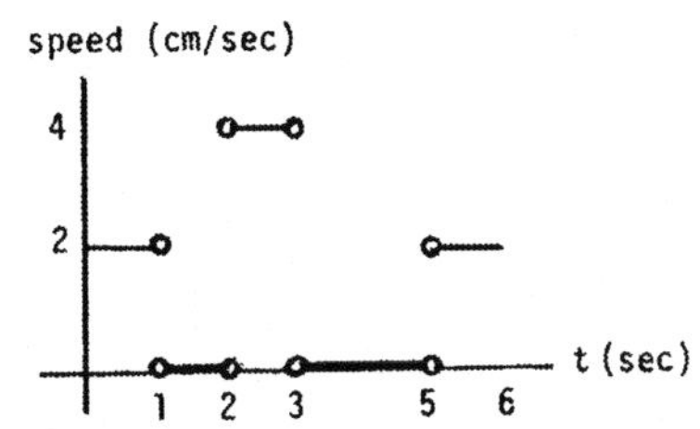

17. (a) 190 ft/sec

(b) 2 sec

(c) at 8 sec, 0 ft/sec

(d) 10.8 sec, 90 ft/sec

(e) From $t = 8$ until $t = 10.8$ sec, a total of 2.8 sec

(f) Greatest acceleration happens 2 sec after launch

(g) From $t = 2$ to $t = 10.8$ sec; during this period, $a = \frac{v(10.8) - v(2)}{10.8 - 2} \approx -32$ ft/sec^2

18. (a) Forward: $0 \le t < 1$ and $5 < t < 7$; Backward: $1 < t < 5$; Speeds up: $1 < t < 2$ and $5 < t < 6$; Slows down: $0 \le t < 1$, $3 < t < 5$, and $6 < t < 7$

(b) Positive: $3 < t < 6$; negative: $0 \le t < 2$ and $6 < t < 7$; zero: $2 < t < 3$ and $7 < t < 9$

(c) $t = 0$ and $2 \le t \le 3$

(d) $7 \le t \le 9$

19. $s = 490t^2 \Rightarrow v = 980t \Rightarrow a = 980$

(a) Solving $160 = 490t^2 \Rightarrow t = \frac{4}{7}$ sec. The average velocity was $\frac{s(4/7) - s(0)}{4/7} = 280$ cm/sec.

(b) At the 160 cm mark the balls are falling at $v(4/7) = 560$ cm/sec. The acceleration at the 160 cm mark was 980 cm/sec^2.

(c) The light was flashing at a rate of $\frac{17}{4/7} = 29.75$ flashes per second.

20. (a)

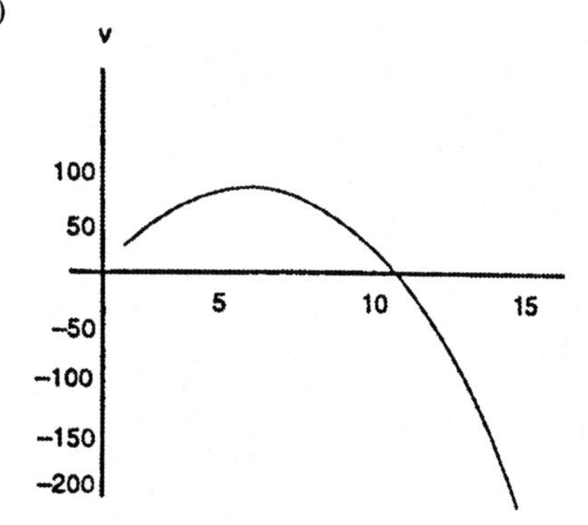

(b)

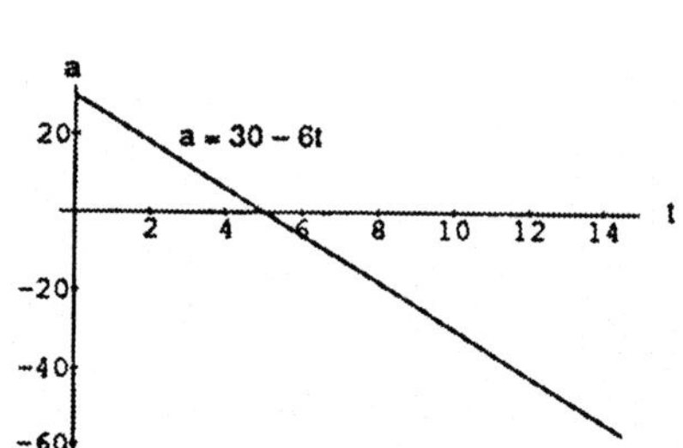

21. C = position, A = velocity, and B = acceleration. Neither A nor C can be the derivative of B because B's derivative is constant. Graph C cannot be the derivative of A either, because A has some negative slopes while C has only positive values. So, C (being the derivative of neither A nor B) must be the graph of position. Curve C has both positive and negative slopes, so its derivative, the velocity, must be A and not B. That leaves B for acceleration.

22. C = position, B = velocity, and A = acceleration. Curve C cannot be the derivative of either A or B because C has only negative values while both A and B have some positive slopes. So, C represents position. Curve C has no positive slopes, so its derivative, the velocity, must be B. That leaves A for acceleration. Indeed, A is negative where B has negative slopes and positive where B has positive slopes.

23. (a) $c(100) = 11{,}000 \Rightarrow c_{av} = \frac{11{,}000}{100} = \$110$

(b) $c(x) = 2000 + 100x - .1x^2 \Rightarrow c'(x) = 100 - .2x$. Marginal cost $= c'(x) \Rightarrow$ the marginal cost of producing 100 machines is $c'(100) = \$80$

(c) The cost of producing the 101st machine is $c(101) - c(100) = 100 - \frac{201}{10} = \$79.90$

24. (a) $r(x) = 20000\left(1 - \frac{1}{x}\right) \Rightarrow r'(x) = \frac{20000}{x^2}$, which is marginal revenue.

(b) $r'(100) = \frac{20000}{100^2} = \$2$.

(c) $\lim_{x \to \infty} r'(x) = \lim_{x \to \infty} \frac{20000}{x^2} = 0$. The increase in revenue as the number of items increases without bound will approach zero.

25. $b(t) = 10^6 + 10^4 t - 10^3 t^2 \Rightarrow b'(t) = 10^4 - (2)(10^3 t) = 10^3(10 - 2t)$

(a) $b'(0) = 10^4$ bacteria/hr

(b) $b'(5) = 0$ bacteria/hr

(c) $b'(10) = -10^4$ bacteria/hr

26. $Q(t) = 200(30 - t)^2 = 200(900 - 60t + t^2) \Rightarrow Q'(t) = 200(-60 + 2t) \Rightarrow Q'(10) = -8{,}000$ gallons/min is the rate the water is running at the end of 10 min. Then $\frac{Q(10) - Q(0)}{10} = -10{,}000$ gallons/min is the average rate the water flows during the first 10 min. The negative signs indicate water is <u>leaving</u> the tank.

27. (a) $y = 6\left(1 - \frac{t}{12}\right)^2 = 6\left(1 - \frac{t}{6} + \frac{t^2}{144}\right) \Rightarrow \frac{dy}{dt} = \frac{t}{12} - 1$

(b) The largest value of $\frac{dy}{dt}$ is 0 m/h when $t = 12$ and the fluid level is falling the slowest at that time. The smallest value of $\frac{dy}{dt}$ is -1 m/h, when $t = 0$, and the fluid level is falling the fastest at that time.

(c) In this situation, $\frac{dy}{dt} \le 0 \Rightarrow$ the graph of y is always decreasing. As $\frac{dy}{dt}$ increases in value, the slope of the graph of y increases from -1 to 0 over the interval $0 \le t \le 12$.

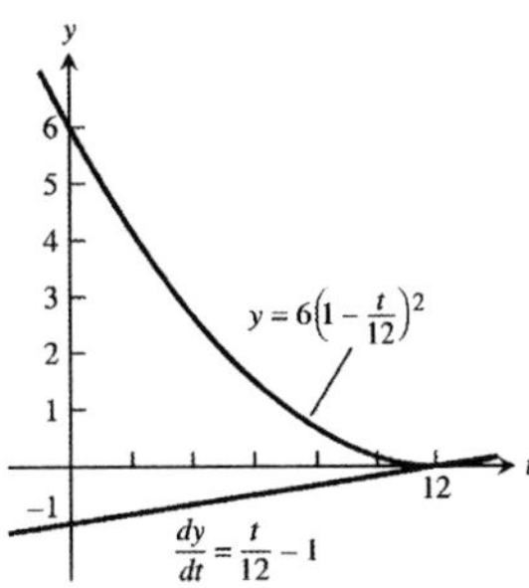

28. (a) $V = \frac{4}{3}\pi r^3 \Rightarrow \frac{dV}{dr} = 4\pi r^2 \Rightarrow \frac{dV}{dr}\Big|_{r=2} = 4\pi(2)^2 = 16\pi \text{ ft}^3/\text{ft}$

(b) When $r = 2$, $\frac{dV}{dr} = 16\pi$ so that when r changes by 1 unit, we expect V to change by approximately 16π. Therefore when r changes by 0.2 units V changes by approximately $(16\pi)(0.2) = 3.2\pi \approx 10.05 \text{ ft}^3$. Note that $V(2.2) - V(2) \approx 11.09 \text{ ft}^3$.

29. $200 \text{ km/hr} = 55\frac{5}{9} \text{m/sec} = \frac{500}{9}$ m/sec, and $D = \frac{10}{9}t^2 \Rightarrow V = \frac{20}{9}t$. Thus $V = \frac{500}{9} \Rightarrow \frac{20}{9}t = \frac{500}{9} \Rightarrow t = 25$ sec. When $t = 25$, $D = \frac{10}{9}(25)^2 = \frac{6250}{9}$ m

30. $s = v_0 t - 16t^2 \Rightarrow v = v_0 - 32t$; $v = 0 \Rightarrow t = \frac{v_0}{32}$; $1900 = v_0 t - 16t^2$ so that $t = \frac{v_0}{32} \Rightarrow 1900 = \frac{v_0^2}{32} - \frac{v_0^2}{64}$ $\Rightarrow v_0 = \sqrt{(64)(1900)} = 80\sqrt{19}$ ft/sec and, finally, $\frac{80\sqrt{19}\text{ ft}}{\text{sec}} \cdot \frac{60 \text{ sec}}{1 \text{ min}} \cdot \frac{60 \text{ min}}{1 \text{ hr}} \cdot \frac{1 \text{ mi}}{5280 \text{ ft}} \approx 238$ mph.

31.

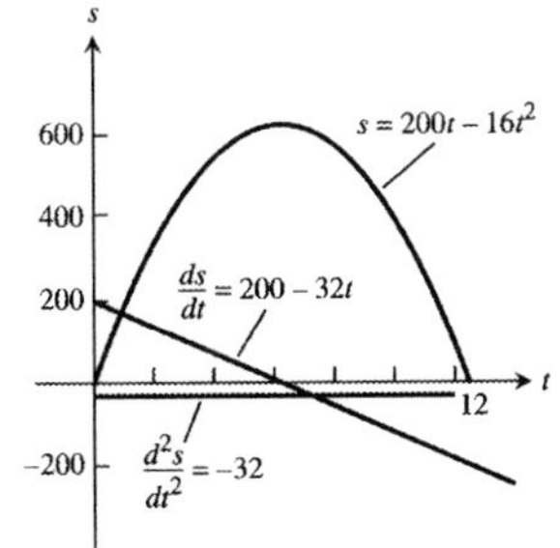

(a) $v = 0$ when $t = 6.25$ sec

(b) $v > 0$ when $0 \le t < 6.25 \Rightarrow$ body moves up; $v < 0$ when $6.25 < t \le 12.5 \Rightarrow$ body moves down

(c) body changes direction at $t = 6.25$ sec

(d) body speeds up on $(6.25, 12.5]$ and slows down on $[0, 6.25)$

(e) The body is moving fastest at the endpoints $t = 0$ and $t = 12.5$ when it is traveling 200 ft/sec. It's moving slowest at $t = 6.25$ when the speed is 0.

(f) When $t = 6.25$ the body is $s = 625$ m from the origin and farthest away.

32.

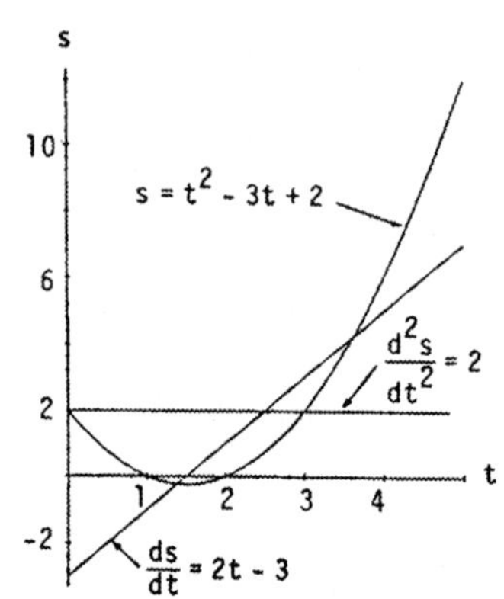

(a) $v = 0$ when $t = \frac{3}{2}$ sec

(b) $v < 0$ when $0 \le t < 1.5 \Rightarrow$ body moves down; $v > 0$ when $1.5 < t \le 5 \Rightarrow$ body moves up

(c) body changes direction at $t = \frac{3}{2}$ sec

(d) body speeds up on $\left(\frac{3}{2}, 5\right]$ and slows down on $\left[0, \frac{3}{2}\right)$

(e) body is moving fastest at $t = 5$ when the speed $= |v(5)| = 7$ units/sec; it is moving slowest at $t = \frac{3}{2}$ when the speed is 0

(f) When $t = 5$ the body is $s = 12$ units from the origin and farthest away.

33.

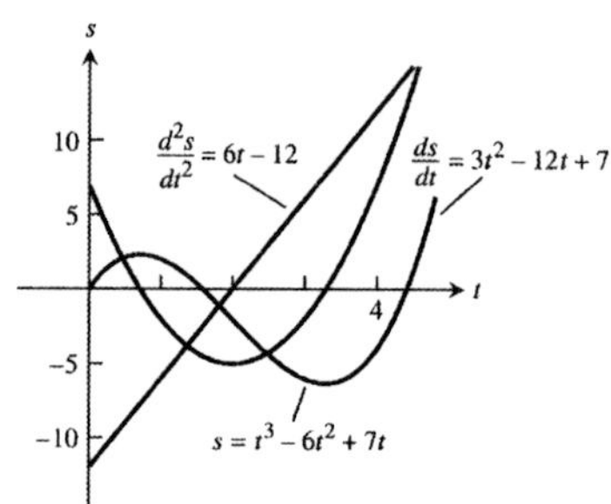

(a) $v = 0$ when $t = \frac{6 \pm \sqrt{15}}{3}$ sec

(b) $v < 0$ when $\frac{6 - \sqrt{15}}{3} < t < \frac{6 + \sqrt{15}}{3} \Rightarrow$ body moves left; $v > 0$ when $0 \le t < \frac{6 - \sqrt{15}}{3}$ or $\frac{6 + \sqrt{15}}{3} < t \le 4$ $\Rightarrow$ body moves right

(c) body changes direction at $t = \frac{6 \pm \sqrt{15}}{3}$ sec

(d) body speeds up on $\left(\frac{6 - \sqrt{15}}{3}, 2\right) \cup \left(\frac{6 + \sqrt{15}}{3}, 4\right]$ and slows down on $\left[0, \frac{6 - \sqrt{15}}{3}\right) \cup \left(2, \frac{6 + \sqrt{15}}{3}\right)$.

(e) The body is moving fastest at $t = 0$ and $t = 4$ when it is moving 7 units/sec and slowest at $t = \frac{6 \pm \sqrt{15}}{3}$ sec

(f) When $t = \frac{6 + \sqrt{15}}{3}$ the body is at position $s \approx -6.303$ units and farthest from the origin.

34.

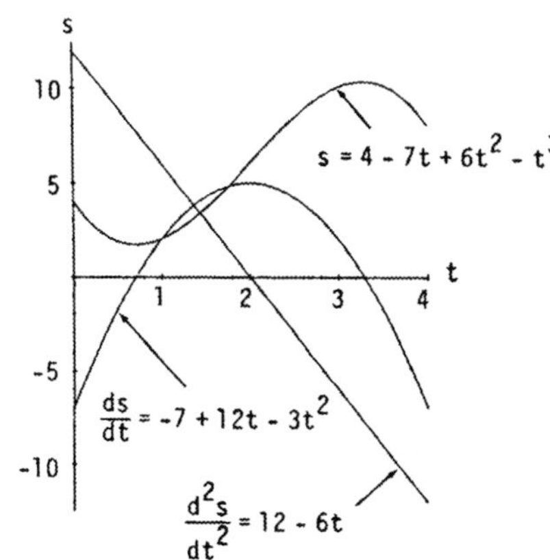

(a) $v = 0$ when $t = \frac{6 \pm \sqrt{15}}{3}$

(b) $v < 0$ when $0 \le t < \frac{6-\sqrt{15}}{3}$ or $\frac{6+\sqrt{15}}{3} < t \le 4 \Rightarrow$ body is moving left; $v > 0$ when $\frac{6-\sqrt{15}}{3} < t < \frac{6+\sqrt{15}}{3} \Rightarrow$ body is moving right

(c) body changes direction at $t = \frac{6 \pm \sqrt{15}}{3}$ sec

(d) body speeds up on $\left(\frac{6-\sqrt{15}}{3}, 2\right) \cup \left(\frac{6+\sqrt{15}}{3}, 4\right]$ and slows down on $\left[0, \frac{6-\sqrt{15}}{3}\right) \cup \left(2, \frac{6+\sqrt{15}}{3}\right)$

(e) The body is moving fastest at 7 units/sec when $t = 0$ and $t = 4$; it is moving slowest and stationary at $t = \frac{6 \pm \sqrt{15}}{3}$

(f) When $t = \frac{6+\sqrt{15}}{3}$ the position is $s \approx 10.303$ units and the body is farthest from the origin.

35. (a) It takes 135 seconds.

(b) Average speed $= \frac{\Delta F}{\Delta t} = \frac{5-0}{73-0} = \frac{5}{73} \approx 0.068$ furlongs/sec.

(c) Using a symmetric difference quotient, the horse's speed is approximately $\frac{\Delta F}{\Delta t} = \frac{4-2}{59-33} = \frac{2}{26} \approx 0.077$ furlongs/sec.

(d) The horse is running the fastest during the last furlong (between the 9th and 10th furlong markers). This furlong takes only 11 seconds to run, which is the least amount of time for a furlong.

(e) The horse accelerates the fastest during the first furlong (between markers 0 and 1).

3.4 DERIVATIVES OF TRIGONOMETRIC FUNCTIONS

1. $y = -10x + 3\cos x \Rightarrow \frac{dy}{dx} = -10 + 3\frac{d}{dx}(\cos x) = -10 - 3\sin x$

2. $y = \frac{3}{x} + 5\sin x \Rightarrow \frac{dy}{dx} = \frac{-3}{x^2} + 5\frac{d}{dx}(\sin x) = \frac{-3}{x^2} + 5\cos x$

3. $y = \csc x - 4\sqrt{x} + 7 \Rightarrow \frac{dy}{dx} = -\csc x \cot x - \frac{4}{2\sqrt{x}} + 0 = -\csc x \cot x - \frac{2}{\sqrt{x}}$

4. $y = x^2 \cot x - \frac{1}{x^2} \Rightarrow \frac{dy}{dx} = x^2 \frac{d}{dx}(\cot x) + \cot x \cdot \frac{d}{dx}(x^2) + \frac{2}{x^3} = -x^2\csc^2 x + (\cot x)(2x) + \frac{2}{x^3}$
$= -x^2 \csc^2 x + 2x\cot x + \frac{2}{x^3}$

5. $y = (\sec x + \tan x)(\sec x - \tan x) \Rightarrow \frac{dy}{dx} = (\sec x + \tan x)\frac{d}{dx}(\sec x - \tan x) + (\sec x - \tan x)\frac{d}{dx}(\sec x + \tan x)$
$= (\sec x + \tan x)(\sec x \tan x - \sec^2 x) + (\sec x - \tan x)(\sec x \tan x + \sec^2 x)$
$= (\sec^2 x \tan x + \sec x \tan^2 x - \sec^3 x - \sec^2 x \tan x) + (\sec^2 x \tan x - \sec x \tan^2 x + \sec^3 x - \tan x \sec^2 x) = 0.$
$\left(\text{Note also that } y = \sec^2 x - \tan^2 x = (\tan^2 x + 1) - \tan^2 x = 1 \Rightarrow \frac{dy}{dx} = 0.\right)$

6. $y = (\sin x + \cos x)\sec x \Rightarrow \frac{dy}{dx} = (\sin x + \cos x)\frac{d}{dx}(\sec x) + \sec x \frac{d}{dx}(\sin x + \cos x)$
$= (\sin x + \cos x)(\sec x \tan x) + (\sec x)(\cos x - \sin x) = \frac{(\sin x + \cos x)\sin x}{\cos^2 x} + \frac{\cos x - \sin x}{\cos x}$
$= \frac{\sin^2 x + \cos x \sin x + \cos^2 x - \cos x \sin x}{\cos^2 x} = \frac{1}{\cos^2 x} = \sec^2 x$
$\left(\text{Note also that } y = \sin x \sec x + \cos x \sec x = \tan x + 1 \Rightarrow \frac{dy}{dx} = \sec^2 x.\right)$

7. $y = \frac{\cot x}{1 + \cot x} \Rightarrow \frac{dy}{dx} = \frac{(1 + \cot x)\frac{d}{dx}(\cot x) - (\cot x)\frac{d}{dx}(1 + \cot x)}{(1 + \cot x)^2} = \frac{(1 + \cot x)(-\csc^2 x) - (\cot x)(-\csc^2 x)}{(1 + \cot x)^2}$
$= \frac{-\csc^2 x - \csc^2 x \cot x + \csc^2 x \cot x}{(1 + \cot x)^2} = \frac{-\csc^2 x}{(1 + \cot x)^2}$

8. $y = \frac{\cos x}{1 + \sin x} \Rightarrow \frac{dy}{dx} = \frac{(1 + \sin x)\frac{d}{dx}(\cos x) - (\cos x)\frac{d}{dx}(1 + \sin x)}{(1 + \sin x)^2} = \frac{(1 + \sin x)(-\sin x) - (\cos x)(\cos x)}{(1 + \sin x)^2}$
$= \frac{-\sin x - \sin^2 x - \cos^2 x}{(1 + \sin x)^2} = \frac{-\sin x - 1}{(1 + \sin x)^2} = \frac{-(1 + \sin x)}{(1 + \sin x)^2} = \frac{-1}{1 + \sin x}$

9. $y = \frac{4}{\cos x} + \frac{1}{\tan x} = 4 \sec x + \cot x \Rightarrow \frac{dy}{dx} = 4 \sec x \tan x - \csc^2 x$

10. $y = \frac{\cos x}{x} + \frac{x}{\cos x} \Rightarrow \frac{dy}{dx} = \frac{x(-\sin x) - (\cos x)(1)}{x^2} + \frac{(\cos x)(1) - x(-\sin x)}{\cos^2 x} = \frac{-x \sin x - \cos x}{x^2} + \frac{\cos x + x \sin x}{\cos^2 x}$

11. $y = x^2 \sin x + 2x \cos x - 2 \sin x \Rightarrow \frac{dy}{dx} = (x^2 \cos x + (\sin x)(2x)) + ((2x)(-\sin x) + (\cos x)(2)) - 2 \cos x$
$= x^2 \cos x + 2x \sin x - 2x \sin x + 2 \cos x - 2 \cos x = x^2 \cos x$

12. $y = x^2 \cos x - 2x \sin x - 2 \cos x \Rightarrow \frac{dy}{dx} = (x^2(-\sin x) + (\cos x)(2x)) - (2x \cos x + (\sin x)(2)) - 2(-\sin x)$
$= -x^2 \sin x + 2x \cos x - 2x \cos x - 2 \sin x + 2 \sin x = -x^2 \sin x$

13. $s = \tan t - e^{-t} \Rightarrow \frac{ds}{dt} = \sec^2 t + e^{-t}$

14. $s = t^2 - \sec t + 5e^t \Rightarrow \frac{ds}{dt} = 2t - \sec t \tan t + 5e^t$

15. $s = \frac{1 + \csc t}{1 - \csc t} \Rightarrow \frac{ds}{dt} = \frac{(1 - \csc t)(-\csc t \cot t) - (1 + \csc t)(\csc t \cot t)}{(1 - \csc t)^2}$
$= \frac{-\csc t \cot t + \csc^2 t \cot t - \csc t \cot t - \csc^2 t \cot t}{(1 - \csc t)^2} = \frac{-2 \csc t \cot t}{(1 - \csc t)^2}$

16. $s = \frac{\sin t}{1 - \cos t} \Rightarrow \frac{ds}{dt} = \frac{(1 - \cos t)(\cos t) - (\sin t)(\sin t)}{(1 - \cos t)^2} = \frac{\cos t - \cos^2 t - \sin^2 t}{(1 - \cos t)^2} = \frac{\cos t - 1}{(1 - \cos t)^2} = -\frac{1}{1 - \cos t}$
$= \frac{1}{\cos t - 1}$

17. $r = 4 - \theta^2 \sin \theta \Rightarrow \frac{dr}{d\theta} = -\left(\theta^2 \frac{d}{d\theta}(\sin \theta) + (\sin \theta)(2\theta)\right) = -(\theta^2 \cos \theta + 2\theta \sin \theta) = -\theta(\theta \cos \theta + 2 \sin \theta)$

18. $r = \theta \sin \theta + \cos \theta \Rightarrow \frac{dr}{d\theta} = (\theta \cos \theta + (\sin \theta)(1)) - \sin \theta = \theta \cos \theta$

19. $r = \sec \theta \csc \theta \Rightarrow \frac{dr}{d\theta} = (\sec \theta)(-\csc \theta \cot \theta) + (\csc \theta)(\sec \theta \tan \theta)$
$= \left(\frac{-1}{\cos \theta}\right)\left(\frac{1}{\sin \theta}\right)\left(\frac{\cos \theta}{\sin \theta}\right) + \left(\frac{1}{\sin \theta}\right)\left(\frac{1}{\cos \theta}\right)\left(\frac{\sin \theta}{\cos \theta}\right) = \frac{-1}{\sin^2 \theta} + \frac{1}{\cos^2 \theta} = \sec^2 \theta - \csc^2 \theta$

20. $r = (1 + \sec \theta)\sin \theta \Rightarrow \frac{dr}{d\theta} = (1 + \sec \theta)\cos \theta + (\sin \theta)(\sec \theta \tan \theta) = (\cos \theta + 1) + \tan^2 \theta = \cos \theta + \sec^2 \theta$

21. $p = 5 + \frac{1}{\cot q} = 5 + \tan q \Rightarrow \frac{dp}{dq} = \sec^2 q$

22. $p = (1 + \csc q)\cos q \Rightarrow \frac{dp}{dq} = (1 + \csc q)(-\sin q) + (\cos q)(-\csc q \cot q) = (-\sin q - 1) - \cot^2 q = -\sin q - \csc^2 q$

23. $p = \frac{\sin q + \cos q}{\cos q} \Rightarrow \frac{dp}{dq} = \frac{(\cos q)(\cos q - \sin q) - (\sin q + \cos q)(-\sin q)}{\cos^2 q}$
$= \frac{\cos^2 q - \cos q \sin q + \sin^2 q + \cos q \sin q}{\cos^2 q} = \frac{1}{\cos^2 q} = \sec^2 q$

24. $p = \frac{\tan q}{1 + \tan q} \Rightarrow \frac{dp}{dq} = \frac{(1 + \tan q)(\sec^2 q) - (\tan q)(\sec^2 q)}{(1 + \tan q)^2} = \frac{\sec^2 q + \tan q \sec^2 q - \tan q \sec^2 q}{(1 + \tan q)^2} = \frac{\sec^2 q}{(1 + \tan q)^2}$

25. (a) $y = \csc x \Rightarrow y' = -\csc x \cot x \Rightarrow y'' = -((\csc x)(-\csc^2 x) + (\cot x)(-\csc x \cot x)) = \csc^3 x + \csc x \cot^2 x$
$= (\csc x)(\csc^2 x + \cot^2 x) = (\csc x)(\csc^2 x + \csc^2 x - 1) = 2\csc^3 x - \csc x$
(b) $y = \sec x \Rightarrow y' = \sec x \tan x \Rightarrow y'' = (\sec x)(\sec^2 x) + (\tan x)(\sec x \tan x) = \sec^3 x + \sec x \tan^2 x$
$= (\sec x)(\sec^2 x + \tan^2 x) = (\sec x)(\sec^2 x + \sec^2 x - 1) = 2\sec^3 x - \sec x$

26. (a) $y = -2\sin x \Rightarrow y' = -2\cos x \Rightarrow y'' = -2(-\sin x) = 2\sin x \Rightarrow y''' = 2\cos x \Rightarrow y^{(4)} = -2\sin x$
(b) $y = 9\cos x \Rightarrow y' = -9\sin x \Rightarrow y'' = -9\cos x \Rightarrow y''' = -9(-\sin x) = 9\sin x \Rightarrow y^{(4)} = 9\cos x$

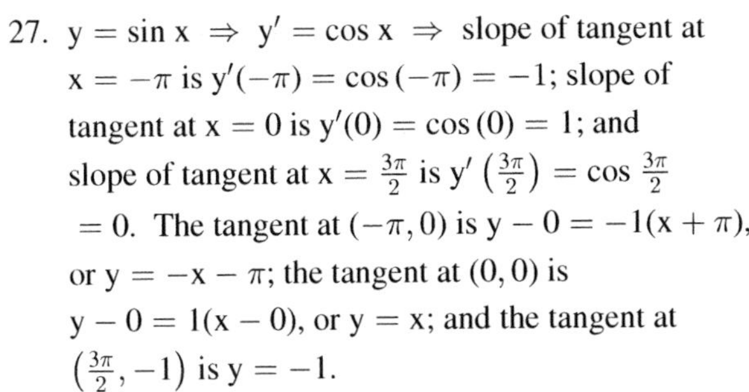

27. $y = \sin x \Rightarrow y' = \cos x \Rightarrow$ slope of tangent at $x = -\pi$ is $y'(-\pi) = \cos(-\pi) = -1$; slope of tangent at $x = 0$ is $y'(0) = \cos(0) = 1$; and slope of tangent at $x = \frac{3\pi}{2}$ is $y'\left(\frac{3\pi}{2}\right) = \cos\frac{3\pi}{2} = 0$. The tangent at $(-\pi, 0)$ is $y - 0 = -1(x + \pi)$, or $y = -x - \pi$; the tangent at $(0, 0)$ is $y - 0 = 1(x - 0)$, or $y = x$; and the tangent at $\left(\frac{3\pi}{2}, -1\right)$ is $y = -1$.

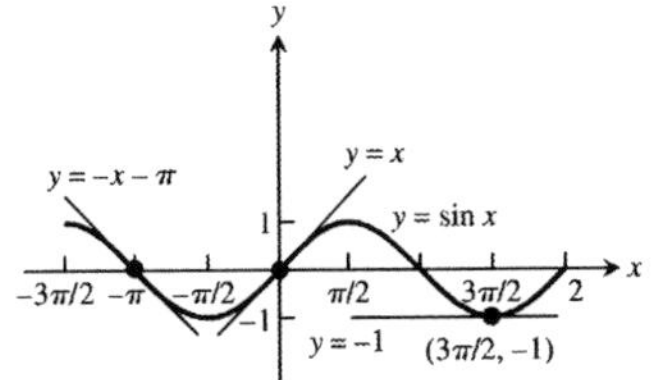

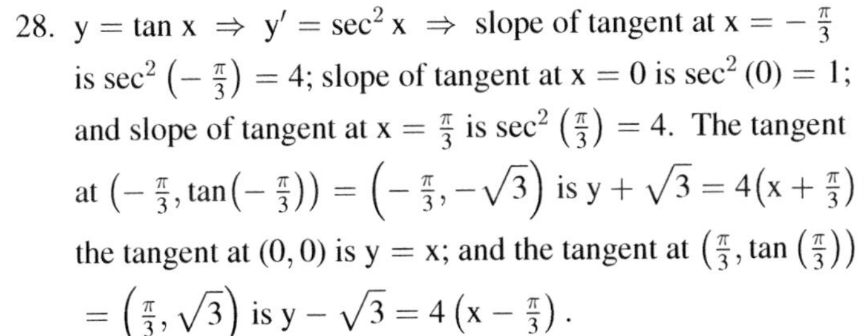

28. $y = \tan x \Rightarrow y' = \sec^2 x \Rightarrow$ slope of tangent at $x = -\frac{\pi}{3}$ is $\sec^2\left(-\frac{\pi}{3}\right) = 4$; slope of tangent at $x = 0$ is $\sec^2(0) = 1$; and slope of tangent at $x = \frac{\pi}{3}$ is $\sec^2\left(\frac{\pi}{3}\right) = 4$. The tangent at $\left(-\frac{\pi}{3}, \tan\left(-\frac{\pi}{3}\right)\right) = \left(-\frac{\pi}{3}, -\sqrt{3}\right)$ is $y + \sqrt{3} = 4\left(x + \frac{\pi}{3}\right)$; the tangent at $(0, 0)$ is $y = x$; and the tangent at $\left(\frac{\pi}{3}, \tan\left(\frac{\pi}{3}\right)\right) = \left(\frac{\pi}{3}, \sqrt{3}\right)$ is $y - \sqrt{3} = 4\left(x - \frac{\pi}{3}\right)$.

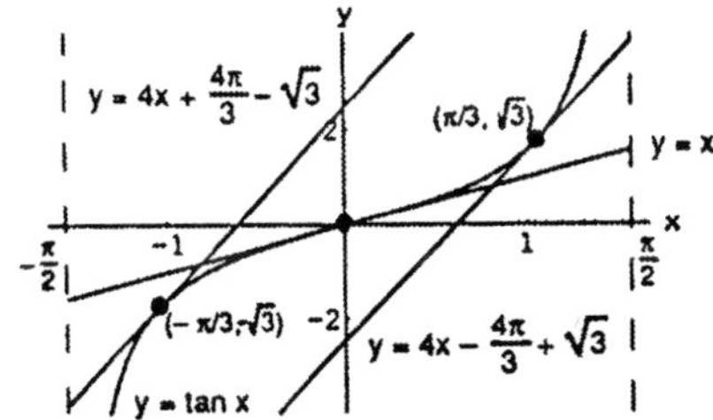

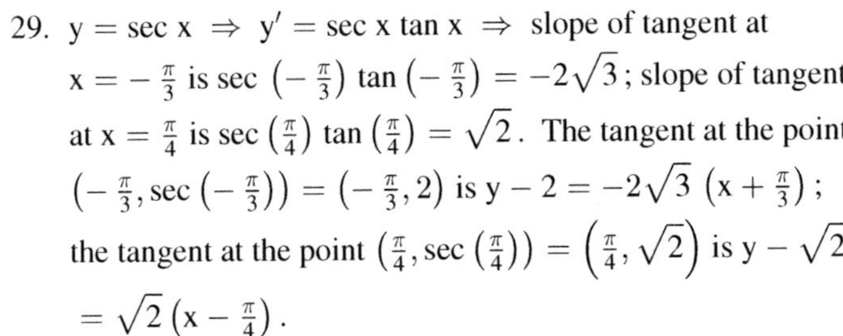

29. $y = \sec x \Rightarrow y' = \sec x \tan x \Rightarrow$ slope of tangent at $x = -\frac{\pi}{3}$ is $\sec\left(-\frac{\pi}{3}\right)\tan\left(-\frac{\pi}{3}\right) = -2\sqrt{3}$; slope of tangent at $x = \frac{\pi}{4}$ is $\sec\left(\frac{\pi}{4}\right)\tan\left(\frac{\pi}{4}\right) = \sqrt{2}$. The tangent at the point $\left(-\frac{\pi}{3}, \sec\left(-\frac{\pi}{3}\right)\right) = \left(-\frac{\pi}{3}, 2\right)$ is $y - 2 = -2\sqrt{3}\left(x + \frac{\pi}{3}\right)$; the tangent at the point $\left(\frac{\pi}{4}, \sec\left(\frac{\pi}{4}\right)\right) = \left(\frac{\pi}{4}, \sqrt{2}\right)$ is $y - \sqrt{2} = \sqrt{2}\left(x - \frac{\pi}{4}\right)$.

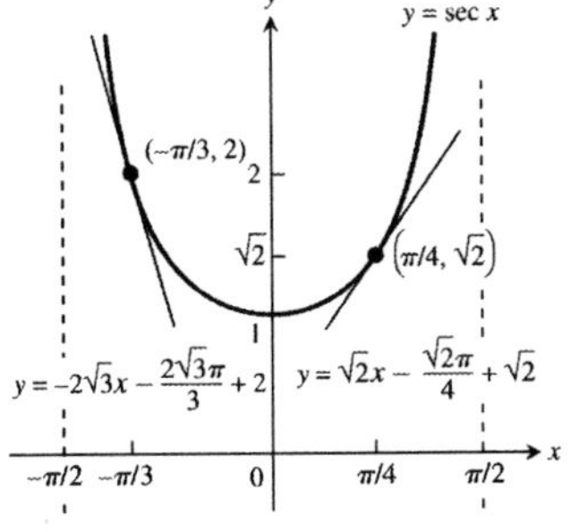

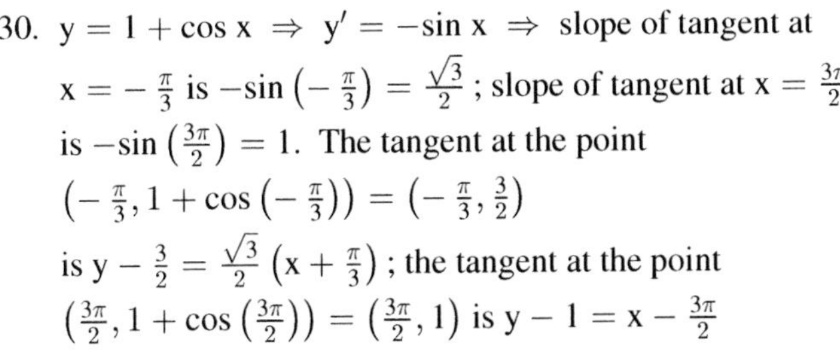

30. $y = 1 + \cos x \Rightarrow y' = -\sin x \Rightarrow$ slope of tangent at $x = -\frac{\pi}{3}$ is $-\sin\left(-\frac{\pi}{3}\right) = \frac{\sqrt{3}}{2}$; slope of tangent at $x = \frac{3\pi}{2}$ is $-\sin\left(\frac{3\pi}{2}\right) = 1$. The tangent at the point $\left(-\frac{\pi}{3}, 1 + \cos\left(-\frac{\pi}{3}\right)\right) = \left(-\frac{\pi}{3}, \frac{3}{2}\right)$ is $y - \frac{3}{2} = \frac{\sqrt{3}}{2}\left(x + \frac{\pi}{3}\right)$; the tangent at the point $\left(\frac{3\pi}{2}, 1 + \cos\left(\frac{3\pi}{2}\right)\right) = \left(\frac{3\pi}{2}, 1\right)$ is $y - 1 = x - \frac{3\pi}{2}$

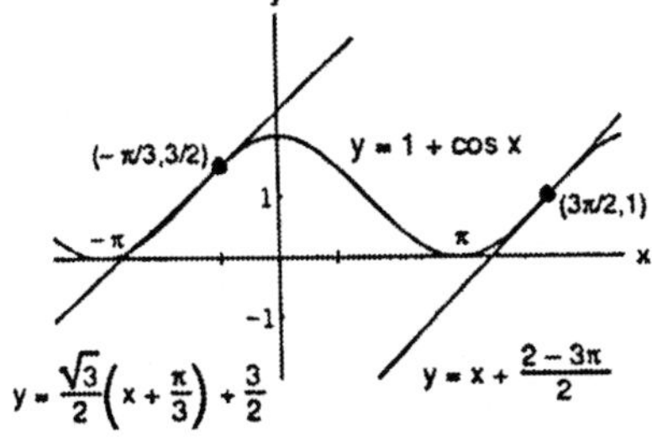

31. Yes, $y = x + \sin x \Rightarrow y' = 1 + \cos x$; horizontal tangent occurs where $1 + \cos x = 0 \Rightarrow \cos x = -1$ $\Rightarrow x = \pi$

32. No, $y = 2x + \sin x \Rightarrow y' = 2 + \cos x$; horizontal tangent occurs where $2 + \cos x = 0 \Rightarrow \cos x = -2$. But there are no x-values for which $\cos x = -2$.

33. No, $y = x - \cot x \Rightarrow y' = 1 + \csc^2 x$; horizontal tangent occurs where $1 + \csc^2 x = 0 \Rightarrow \csc^2 x = -1$. But there are no x-values for which $\csc^2 x = -1$.

34. Yes, $y = x + 2\cos x \Rightarrow y' = 1 - 2\sin x$; horizontal tangent occurs where $1 - 2\sin x = 0 \Rightarrow 1 = 2\sin x$ $\Rightarrow \frac{1}{2} = \sin x \Rightarrow x = \frac{\pi}{6}$ or $x = \frac{5\pi}{6}$

35. We want all points on the curve where the tangent line has slope 2. Thus, $y = \tan x \Rightarrow y' = \sec^2 x$ so that $y' = 2 \Rightarrow \sec^2 x = 2 \Rightarrow \sec x = \pm\sqrt{2}$ $\Rightarrow x = \pm\frac{\pi}{4}$. Then the tangent line at $\left(\frac{\pi}{4}, 1\right)$ has equation $y - 1 = 2\left(x - \frac{\pi}{4}\right)$; the tangent line at $\left(-\frac{\pi}{4}, -1\right)$ has equation $y + 1 = 2\left(x + \frac{\pi}{4}\right)$.

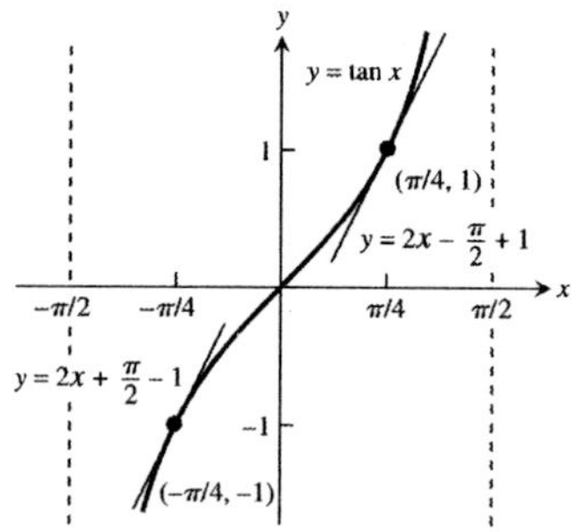

36. We want all points on the curve $y = \cot x$ where the tangent line has slope -1. Thus $y = \cot x$ $\Rightarrow y' = -\csc^2 x$ so that $y' = -1 \Rightarrow -\csc^2 x = -1$ $\Rightarrow \csc^2 x = 1 \Rightarrow \csc x = \pm 1 \Rightarrow x = \frac{\pi}{2}$. The tangent line at $\left(\frac{\pi}{2}, 0\right)$ is $y = -x + \frac{\pi}{2}$.

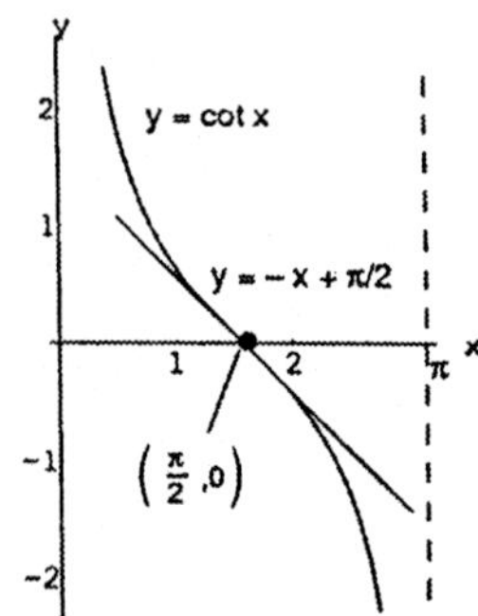

37. $y = 4 + \cot x - 2\csc x \Rightarrow y' = -\csc^2 x + 2\csc x \cot x = -\left(\frac{1}{\sin x}\right)\left(\frac{1 - 2\cos x}{\sin x}\right)$
   (a) When $x = \frac{\pi}{2}$, then $y' = -1$; the tangent line is $y = -x + \frac{\pi}{2} + 2$.
   (b) To find the location of the horizontal tangent set $y' = 0 \Rightarrow 1 - 2\cos x = 0 \Rightarrow x = \frac{\pi}{3}$ radians. When $x = \frac{\pi}{3}$, then $y = 4 - \sqrt{3}$ is the horizontal tangent.

38. $y = 1 + \sqrt{2}\csc x + \cot x \Rightarrow y' = -\sqrt{2}\csc x \cot x - \csc^2 x = -\left(\frac{1}{\sin x}\right)\left(\frac{\sqrt{2}\cos x + 1}{\sin x}\right)$
   (a) If $x = \frac{\pi}{4}$, then $y' = -4$; the tangent line is $y = -4x + \pi + 4$.
   (b) To find the location of the horizontal tangent set $y' = 0 \Rightarrow \sqrt{2}\cos x + 1 = 0 \Rightarrow x = \frac{3\pi}{4}$ radians. When $x = \frac{3\pi}{4}$, then $y = 2$ is the horizontal tangent.

39. $\lim_{x \to 2} \sin\left(\frac{1}{x} - \frac{1}{2}\right) = \sin\left(\frac{1}{2} - \frac{1}{2}\right) = \sin 0 = 0$

40. $\lim_{x \to -\frac{\pi}{6}} \sqrt{1 + \cos(\pi \csc x)} = \sqrt{1 + \cos\left(\pi \csc\left(-\frac{\pi}{6}\right)\right)} = \sqrt{1 + \cos(\pi \cdot (-2))} = \sqrt{2}$

41. $\lim_{x \to 0} \sec\left[e^x + \pi \tan\left(\frac{\pi}{4\sec x}\right) - 1\right] = \sec\left[1 + \pi \tan\left(\frac{\pi}{4\sec 0}\right) - 1\right] = = \sec\left[\pi \tan\left(\frac{\pi}{4}\right)\right] = \sec \pi = -1$

42. $\lim\limits_{x \to 0} \sin\left(\frac{\pi + \tan x}{\tan x - 2 \sec x}\right) = \sin\left(\frac{\pi + \tan 0}{\tan 0 - 2 \sec 0}\right) = \sin\left(-\frac{\pi}{2}\right) = -1$

43. $\lim\limits_{t \to 0} \tan\left(1 - \frac{\sin t}{t}\right) = \tan\left(1 - \lim\limits_{t \to 0} \frac{\sin t}{t}\right) = \tan(1 - 1) = 0$

44. $\lim\limits_{\theta \to 0} \cos\left(\frac{\pi\theta}{\sin\theta}\right) = \cos\left(\pi \lim\limits_{\theta \to 0} \frac{\theta}{\sin\theta}\right) = \cos\left(\pi \cdot \frac{1}{\lim\limits_{\theta \to 0} \frac{\sin\theta}{\theta}}\right) = \cos\left(\pi \cdot \frac{1}{1}\right) = -1$

45. $s = 2 - 2\sin t \Rightarrow v = \frac{ds}{dt} = -2\cos t \Rightarrow a = \frac{dv}{dt} = 2\sin t \Rightarrow j = \frac{da}{dt} = 2\cos t$. Therefore, velocity $= v\left(\frac{\pi}{4}\right)$ $= -\sqrt{2}$ m/sec; speed $= \left|v\left(\frac{\pi}{4}\right)\right| = \sqrt{2}$ m/sec; acceleration $= a\left(\frac{\pi}{4}\right) = \sqrt{2}$ m/sec^2; jerk $= j\left(\frac{\pi}{4}\right) = \sqrt{2}$ m/sec^3.

46. $s = \sin t + \cos t \Rightarrow v = \frac{ds}{dt} = \cos t - \sin t \Rightarrow a = \frac{dv}{dt} = -\sin t - \cos t \Rightarrow j = \frac{da}{dt} = -\cos t + \sin t$. Therefore velocity $= v\left(\frac{\pi}{4}\right) = 0$ m/sec; speed $= \left|v\left(\frac{\pi}{4}\right)\right| = 0$ m/sec; acceleration $= a\left(\frac{\pi}{4}\right) = -\sqrt{2}$ m/sec^2; jerk $= j\left(\frac{\pi}{4}\right) = 0$ m/sec^3.

47. $\lim\limits_{x \to 0} f(x) = \lim\limits_{x \to 0} \frac{\sin^2 3x}{x^2} = \lim\limits_{x \to 0} 9\left(\frac{\sin 3x}{3x}\right)\left(\frac{\sin 3x}{3x}\right) = 9$ so that f is continuous at $x = 0 \Rightarrow \lim\limits_{x \to 0} f(x) = f(0)$ $\Rightarrow 9 = c$.

48. $\lim\limits_{x \to 0^-} g(x) = \lim\limits_{x \to 0^-} (x + b) = b$ and $\lim\limits_{x \to 0^+} g(x) = \lim\limits_{x \to 0^+} \cos x = 1$ so that g is continuous at $x = 0 \Rightarrow \lim\limits_{x \to 0^-} g(x)$ $= \lim\limits_{x \to 0^+} g(x) \Rightarrow b = 1$. Now g is not differentiable at $x = 0$: At $x = 0$, the left-hand derivative is $\frac{d}{dx}(x + b)\Big|_{x=0} = 1$, but the right-hand derivative is $\frac{d}{dx}(\cos x)\Big|_{x=0} = -\sin 0 = 0$. The left- and right-hand derivatives can never agree at $x = 0$, so g is not differentiable at $x = 0$ for any value of b (including $b = 1$).

49. $\frac{d^{999}}{dx^{999}}(\cos x) = \sin x$ because $\frac{d^4}{dx^4}(\cos x) = \cos x \Rightarrow$ the derivative of cos x any number of times that is a multiple of 4 is cos x. Thus, dividing 999 by 4 gives $999 = 249 \cdot 4 + 3 \Rightarrow \frac{d^{999}}{dx^{999}}(\cos x)$ $= \frac{d^3}{dx^3}\left[\frac{d^{249\cdot 4}}{dx^{249\cdot 4}}(\cos x)\right] = \frac{d^3}{dx^3}(\cos x) = \sin x$.

50. (a) $y = \sec x = \frac{1}{\cos x} \Rightarrow \frac{dy}{dx} = \frac{(\cos x)(0) - (1)(-\sin x)}{(\cos x)^2} = \frac{\sin x}{\cos^2 x} = \left(\frac{1}{\cos x}\right)\left(\frac{\sin x}{\cos x}\right) = \sec x \tan x$ $\Rightarrow \frac{d}{dx}(\sec x) = \sec x \tan x$

(b) $y = \csc x = \frac{1}{\sin x} \Rightarrow \frac{dy}{dx} = \frac{(\sin x)(0) - (1)(\cos x)}{(\sin x)^2} = \frac{-\cos x}{\sin^2 x} = \left(\frac{-1}{\sin x}\right)\left(\frac{\cos x}{\sin x}\right) = -\csc x \cot x$ $\Rightarrow \frac{d}{dx}(\csc x) = -\csc x \cot x$

(c) $y = \cot x = \frac{\cos x}{\sin x} \Rightarrow \frac{dy}{dx} = \frac{(\sin x)(-\sin x) - (\cos x)(\cos x)}{(\sin x)^2} = \frac{-\sin^2 x - \cos^2 x}{\sin^2 x} = \frac{-1}{\sin^2 x} = -\csc^2 x$ $\Rightarrow \frac{d}{dx}(\cot x) = -\csc^2 x$

51.

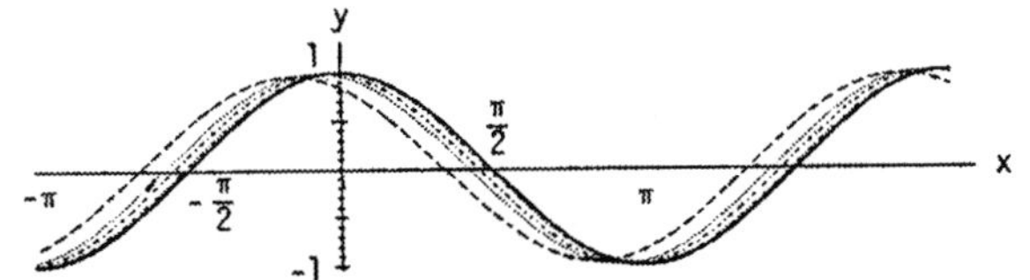

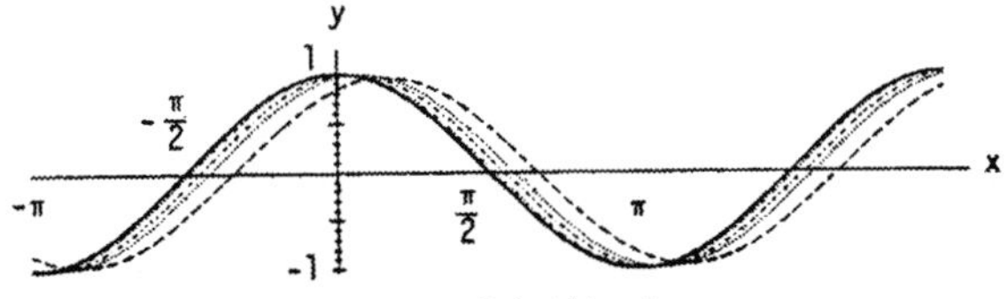

As h takes on the values of 1, 0.5, 0.3 and 0.1 the corresponding dashed curves of $y = \frac{\sin(x+h) - \sin x}{h}$ get closer and closer to the black curve $y = \cos x$ because $\frac{d}{dx}(\sin x) = \lim\limits_{h \to 0} \frac{\sin(x+h) - \sin x}{h} = \cos x$. The same is true as h takes on the values of $-1, -0.5, -0.3$ and -0.1.

52.

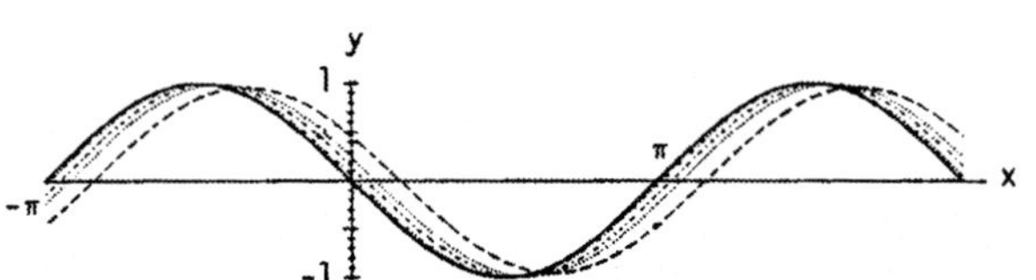

As h takes on the values of 1, 0.5, 0.3, and 0.1 the corresponding dashed curves of $y = \frac{\cos(x+h) - \cos x}{h}$ get closer and closer to the black curve $y = -\sin x$ because $\frac{d}{dx}(\cos x) = \lim_{h \to 0} \frac{\cos(x+h) - \cos x}{h} = -\sin x$. The same is true as h takes on the values of $-1, -0.5, -0.3$, and -0.1.

53. (a)

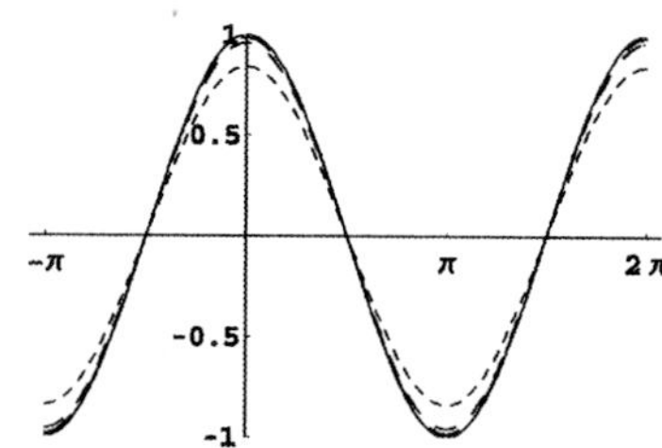

The dashed curves of $y = \frac{\sin(x+h) - \sin(x-h)}{2h}$ are closer to the black curve $y = \cos x$ than the corresponding dashed curves in Exercise 51 illustrating that the centered difference quotient is a better approximation of the derivative of this function.

(b)

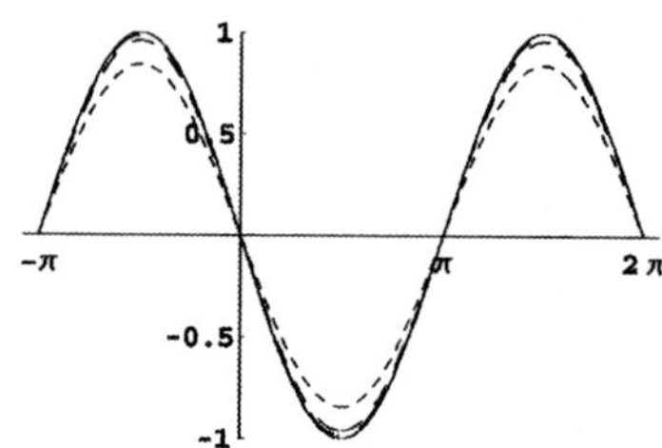

The dashed curves of $y = \frac{\cos(x+h) - \cos(x-h)}{2h}$ are closer to the black curve $y = -\sin x$ than the corresponding dashed curves in Exercise 52 illustrating that the centered difference quotient is a better approximation of the derivative of this function.

54. $\lim_{h \to 0} \frac{|0+h| - |0-h|}{2h} = \lim_{x \to 0} \frac{|h| - |h|}{2h} = \lim_{h \to 0} 0 = 0 \Rightarrow$ the limits of the centered difference quotient exists even though the derivative of $f(x) = |x|$ does not exist at $x = 0$.

55. $y = \tan x \Rightarrow y' = \sec^2 x$, so the smallest value $y' = \sec^2 x$ takes on is $y' = 1$ when $x = 0$; y' has no maximum value since $\sec^2 x$ has no largest value on $\left(-\frac{\pi}{2}, \frac{\pi}{2}\right)$; y' is never negative since $\sec^2 x \geq 1$.

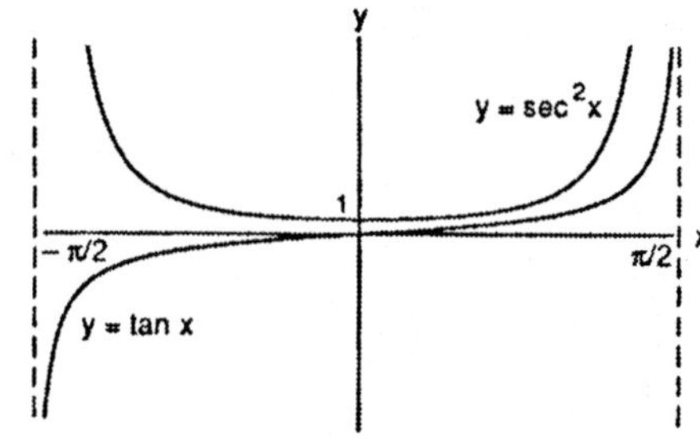

56. $y = \cot x \Rightarrow y' = -\csc^2 x$ so y' has no smallest value since $-\csc^2 x$ has no minimum value on $(0, \pi)$; the largest value of y' is -1, when $x = \frac{\pi}{2}$; the slope is never positive since the largest value $y' = -\csc^2 x$ takes on is -1.

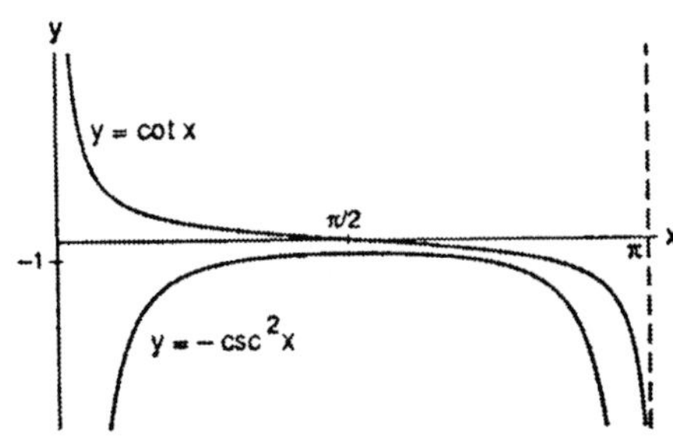

57. $y = \frac{\sin x}{x}$ appears to cross the y-axis at $y = 1$, since $\lim\limits_{x \to 0} \frac{\sin x}{x} = 1$; $y = \frac{\sin 2x}{x}$ appears to cross the y-axis at $y = 2$, since $\lim\limits_{x \to 0} \frac{\sin 2x}{x} = 2$; $y = \frac{\sin 4x}{x}$ appears to cross the y-axis at $y = 4$, since $\lim\limits_{x \to 0} \frac{\sin 4x}{x} = 4$. However, none of these graphs actually cross the y-axis since $x = 0$ is not in the domain of the functions. Also, $\lim\limits_{x \to 0} \frac{\sin 5x}{x} = 5$, $\lim\limits_{x \to 0} \frac{\sin(-3x)}{x} = -3$, and $\lim\limits_{x \to 0} \frac{\sin kx}{x} = k \Rightarrow$ the graphs of $y = \frac{\sin 5x}{x}$, $y = \frac{\sin(-3x)}{x}$, and $y = \frac{\sin kx}{x}$ approach 5, -3, and k, respectively, as $x \to 0$. However, the graphs do not actually cross the y-axis.

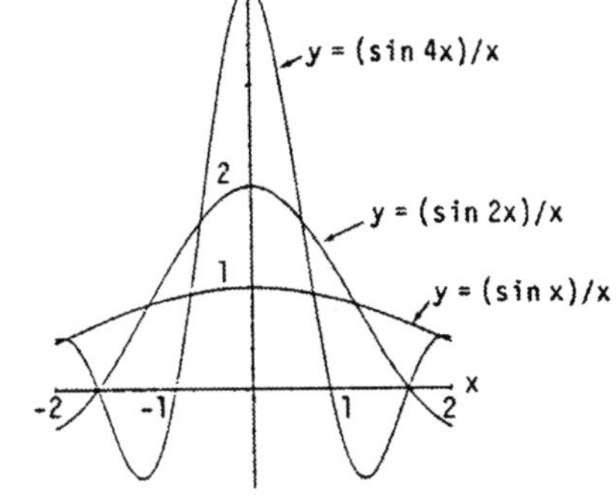

58. (a)

h	$\frac{\sin h}{h}$	$\left(\frac{\sin h}{h}\right)\left(\frac{180}{\pi}\right)$
1	.017452406	.99994923
0.01	.017453292	1
0.001	.017453292	1
0.0001	.017453292	1

$\lim\limits_{h \to 0} \frac{\sin h^\circ}{h} = \lim\limits_{x \to 0} \frac{\sin\left(h \cdot \frac{\pi}{180}\right)}{h} = \lim\limits_{h \to 0} \frac{\frac{\pi}{180} \sin\left(h \cdot \frac{\pi}{180}\right)}{\frac{\pi}{180} \cdot h} = \lim\limits_{\theta \to 0} \frac{\frac{\pi}{180} \sin\theta}{\theta} = \frac{\pi}{180}$ $\quad \left(\theta = h \cdot \frac{\pi}{180}\right)$

(converting to radians)

(b)

h	$\frac{\cos h - 1}{h}$
1	-0.0001523
0.01	-0.0000015
0.001	-0.0000001
0.0001	0

$\lim\limits_{h \to 0} \frac{\cos h - 1}{h} = 0$, whether h is measured in degrees or radians.

(c) In degrees, $\frac{d}{dx}(\sin x) = \lim\limits_{h \to 0} \frac{\sin(x+h) - \sin x}{h} = \lim\limits_{h \to 0} \frac{(\sin x \cos h + \cos x \sin h) - \sin x}{h}$

$= \lim\limits_{h \to 0} \left(\sin x \cdot \frac{\cos h - 1}{h}\right) + \lim\limits_{h \to 0} \left(\cos x \cdot \frac{\sin h}{h}\right) = (\sin x) \cdot \lim\limits_{h \to 0} \left(\frac{\cos h - 1}{h}\right) + (\cos x) \cdot \lim\limits_{h \to 0} \left(\frac{\sin h}{h}\right)$

$= (\sin x)(0) + (\cos x)\left(\frac{\pi}{180}\right) = \frac{\pi}{180} \cos x$

(d) In degrees, $\frac{d}{dx}(\cos x) = \lim\limits_{h \to 0} \frac{\cos(x+h) - \cos x}{h} = \lim\limits_{h \to 0} \frac{(\cos x \cos h - \sin x \sin h) - \cos x}{h}$

$= \lim\limits_{h \to 0} \frac{(\cos x)(\cos h - 1) - \sin x \sin h}{h} = \lim\limits_{h \to 0} \left(\cos x \cdot \frac{\cos h - 1}{h}\right) - \lim\limits_{h \to 0} \left(\sin x \cdot \frac{\sin h}{h}\right)$

$= (\cos x) \lim\limits_{h \to 0} \left(\frac{\cos h - 1}{h}\right) - (\sin x) \lim\limits_{h \to 0} \left(\frac{\sin h}{h}\right) = (\cos x)(0) - (\sin x)\left(\frac{\pi}{180}\right) = -\frac{\pi}{180} \sin x$

(e) $\frac{d^2}{dx^2}(\sin x) = \frac{d}{dx}\left(\frac{\pi}{180} \cos x\right) = -\left(\frac{\pi}{180}\right)^2 \sin x$; $\frac{d^3}{dx^3}(\sin x) = \frac{d}{dx}\left(-\left(\frac{\pi}{180}\right)^2 \sin x\right) = -\left(\frac{\pi}{180}\right)^3 \cos x$;

$\frac{d^2}{dx^2}(\cos x) = \frac{d}{dx}\left(-\frac{\pi}{180} \sin x\right) = -\left(\frac{\pi}{180}\right)^2 \cos x$; $\frac{d^3}{dx^3}(\cos x) = \frac{d}{dx}\left(-\left(\frac{\pi}{180}\right)^2 \cos x\right) = \left(\frac{\pi}{180}\right)^3 \sin x$

3.5 THE CHAIN RULE AND PARAMETRIC EQUATIONS

1. $f(u) = 6u - 9 \Rightarrow f'(u) = 6 \Rightarrow f'(g(x)) = 6$; $g(x) = \frac{1}{2}x^4 \Rightarrow g'(x) = 2x^3$; therefore $\frac{dy}{dx} = f'(g(x))g'(x)$ $= 6 \cdot 2x^3 = 12x^3$

2. $f(u) = 2u^3 \Rightarrow f'(u) = 6u^2 \Rightarrow f'(g(x)) = 6(8x-1)^2$; $g(x) = 8x - 1 \Rightarrow g'(x) = 8$; therefore $\frac{dy}{dx} = f'(g(x))g'(x)$ $= 6(8x-1)^2 \cdot 8 = 48(8x-1)^2$

3. $f(u) = \sin u \Rightarrow f'(u) = \cos u \Rightarrow f'(g(x)) = \cos(3x+1)$; $g(x) = 3x + 1 \Rightarrow g'(x) = 3$; therefore $\frac{dy}{dx} = f'(g(x))g'(x)$ $= (\cos(3x+1))(3) = 3\cos(3x+1)$

4. $f(u) = \cos u \Rightarrow f'(u) = -\sin u \Rightarrow f'(g(x)) = -\sin\left(\frac{-x}{3}\right)$; $g(x) = \frac{-x}{3} \Rightarrow g'(x) = -\frac{1}{3}$; therefore $\frac{dy}{dx} = f'(g(x))g'(x)$ $= -\sin\left(\frac{-x}{3}\right) \cdot \left(\frac{-1}{3}\right) = \frac{1}{3}\sin\left(\frac{-x}{3}\right)$

5. $f(u) = \cos u \Rightarrow f'(u) = -\sin u \Rightarrow f'(g(x)) = -\sin(\sin x)$; $g(x) = \sin x \Rightarrow g'(x) = \cos x$; therefore $\frac{dy}{dx} = f'(g(x))g'(x) = -(\sin(\sin x))\cos x$

6. $f(u) = \sin u \Rightarrow f'(u) = \cos u \Rightarrow f'(g(x)) = \cos(x - \cos x)$; $g(x) = x - \cos x \Rightarrow g'(x) = 1 + \sin x$; therefore $\frac{dy}{dx} = f'(g(x))g'(x) = (\cos(x - \cos x))(1 + \sin x)$

7. $f(u) = \tan u \Rightarrow f'(u) = \sec^2 u \Rightarrow f'(g(x)) = \sec^2(10x-5)$; $g(x) = 10x - 5 \Rightarrow g'(x) = 10$; therefore $\frac{dy}{dx} = f'(g(x))g'(x) = (\sec^2(10x-5))(10) = 10\sec^2(10x-5)$

8. $f(u) = -\sec u \Rightarrow f'(u) = -\sec u \tan u \Rightarrow f'(g(x)) = -\sec(x^2+7x)\tan(x^2+7x)$; $g(x) = x^2 + 7x$ $\Rightarrow g'(x) = 2x + 7$; therefore $\frac{dy}{dx} = f'(g(x))g'(x) = -(2x+7)\sec(x^2+7x)\tan(x^2+7x)$

9. With $u = (2x+1)$, $y = u^5$: $\frac{dy}{dx} = \frac{dy}{du}\frac{du}{dx} = 5u^4 \cdot 2 = 10(2x+1)^4$

10. With $u = (4-3x)$, $y = u^9$: $\frac{dy}{dx} = \frac{dy}{du}\frac{du}{dx} = 9u^8 \cdot (-3) = -27(4-3x)^8$

11. With $u = \left(1 - \frac{x}{7}\right)$, $y = u^{-7}$: $\frac{dy}{dx} = \frac{dy}{du}\frac{du}{dx} = -7u^{-8} \cdot \left(-\frac{1}{7}\right) = \left(1 - \frac{x}{7}\right)^{-8}$

12. With $u = \left(\frac{x}{2} - 1\right)$, $y = u^{-10}$: $\frac{dy}{dx} = \frac{dy}{du}\frac{du}{dx} = -10u^{-11} \cdot \left(\frac{1}{2}\right) = -5\left(\frac{x}{2} - 1\right)^{-11}$

13. With $u = \left(\frac{x^2}{8} + x - \frac{1}{x}\right)$, $y = u^4$: $\frac{dy}{dx} = \frac{dy}{du}\frac{du}{dx} = 4u^3 \cdot \left(\frac{x}{4} + 1 + \frac{1}{x^2}\right) = 4\left(\frac{x^2}{8} + x - \frac{1}{x}\right)^3\left(\frac{x}{4} + 1 + \frac{1}{x^2}\right)$

14. With $u = \left(\frac{x}{5} + \frac{1}{5x}\right)$, $y = u^5$: $\frac{dy}{dx} = \frac{dy}{du}\frac{du}{dx} = 5u^4 \cdot \left(\frac{1}{5} - \frac{1}{5x^2}\right) = \left(\frac{x}{5} + \frac{1}{5x}\right)^4\left(1 - \frac{1}{x^2}\right)$

15. With $u = \tan x$, $y = \sec u$: $\frac{dy}{dx} = \frac{dy}{du}\frac{du}{dx} = (\sec u \tan u)(\sec^2 x) = (\sec(\tan x)\tan(\tan x))\sec^2 x$

16. With $u = \pi - \frac{1}{x}$, $y = \cot u$: $\frac{dy}{dx} = \frac{dy}{du}\frac{du}{dx} = (-\csc^2 u)\left(\frac{1}{x^2}\right) = -\frac{1}{x^2}\csc^2\left(\pi - \frac{1}{x}\right)$

17. With $u = \sin x$, $y = u^3$: $\frac{dy}{dx} = \frac{dy}{du}\frac{du}{dx} = 3u^2\cos x = 3(\sin^2 x)(\cos x)$

18. With $u = \cos x$, $y = 5u^{-4}$: $\frac{dy}{dx} = \frac{dy}{du}\frac{du}{dx} = (-20u^{-5})(-\sin x) = 20(\cos^{-5} x)(\sin x)$

19. $y = e^{-5x} \Rightarrow y' = e^{-5x} \frac{d}{dx}(-5x) \Rightarrow y' = -5e^{-5x}$

20. $y = e^{2x/3} \Rightarrow y' = e^{2x/3} \frac{d}{dx}\left(\frac{2x}{3}\right) \Rightarrow y' = \frac{2}{3} e^{2x/3}$

21. $y = e^{5-7x} \Rightarrow y' = e^{5-7x} \frac{d}{dx}(5-7x) \Rightarrow y' = -7e^{5-7x}$

22. $y = e^{\left(4\sqrt{x}+x^2\right)} \Rightarrow y' = e^{\left(4\sqrt{x}+x^2\right)} \frac{d}{dx}\left(4\sqrt{x}+x^2\right) \Rightarrow y' = \left(\frac{2}{\sqrt{x}} + 2x\right) e^{\left(4\sqrt{x}+x^2\right)}$

23. $p = \sqrt{3-t} = (3-t)^{1/2} \Rightarrow \frac{dp}{dt} = \frac{1}{2}(3-t)^{-1/2} \cdot \frac{d}{dt}(3-t) = -\frac{1}{2}(3-t)^{-1/2} = \frac{-1}{2\sqrt{3-t}}$

24. $q = \sqrt{2r-r^2} = \left(2r-r^2\right)^{1/2} \Rightarrow \frac{dq}{dr} = \frac{1}{2}\left(2r-r^2\right)^{-1/2} \cdot \frac{d}{dr}\left(2r-r^2\right) = \frac{1}{2}\left(2r-r^2\right)^{-1/2}(2-2r) = \frac{1-r}{\sqrt{2r-r^2}}$

25. $s = \frac{4}{3\pi} \sin 3t + \frac{4}{5\pi} \cos 5t \Rightarrow \frac{ds}{dt} = \frac{4}{3\pi} \cos 3t \cdot \frac{d}{dt}(3t) + \frac{4}{5\pi}(-\sin 5t) \cdot \frac{d}{dt}(5t) = \frac{4}{\pi} \cos 3t - \frac{4}{\pi} \sin 5t$
$= \frac{4}{\pi}(\cos 3t - \sin 5t)$

26. $s = \sin\left(\frac{3\pi t}{2}\right) + \cos\left(\frac{3\pi t}{2}\right) \Rightarrow \frac{ds}{dt} = \cos\left(\frac{3\pi t}{2}\right) \cdot \frac{d}{dt}\left(\frac{3\pi t}{2}\right) - \sin\left(\frac{3\pi t}{2}\right) \cdot \frac{d}{dt}\left(\frac{3\pi t}{2}\right) = \frac{3\pi}{2} \cos\left(\frac{3\pi t}{2}\right) - \frac{3\pi}{2} \sin\left(\frac{3\pi t}{2}\right)$
$= \frac{3\pi}{2}\left(\cos \frac{3\pi t}{2} - \sin \frac{3\pi t}{2}\right)$

27. $r = (\csc\theta + \cot\theta)^{-1} \Rightarrow \frac{dr}{d\theta} = -(\csc\theta + \cot\theta)^{-2} \frac{d}{d\theta}(\csc\theta + \cot\theta) = \frac{\csc\theta\cot\theta + \csc^2\theta}{(\csc\theta + \cot\theta)^2} = \frac{\csc\theta(\cot\theta + \csc\theta)}{(\csc\theta + \cot\theta)^2}$
$= \frac{\csc\theta}{\csc\theta + \cot\theta}$

28. $r = -(\sec\theta + \tan\theta)^{-1} \Rightarrow \frac{dr}{d\theta} = (\sec\theta + \tan\theta)^{-2} \frac{d}{d\theta}(\sec\theta + \tan\theta) = \frac{\sec\theta\tan\theta + \sec^2\theta}{(\sec\theta + \tan\theta)^2} = \frac{\sec\theta(\tan\theta + \sec\theta)}{(\sec\theta + \tan\theta)^2}$
$= \frac{\sec\theta}{\sec\theta + \tan\theta}$

29. $y = x^2 \sin^4 x + x \cos^{-2} x \Rightarrow \frac{dy}{dx} = x^2 \frac{d}{dx}(\sin^4 x) + \sin^4 x \cdot \frac{d}{dx}(x^2) + x \frac{d}{dx}(\cos^{-2} x) + \cos^{-2} x \cdot \frac{d}{dx}(x)$
$= x^2\left(4 \sin^3 x \frac{d}{dx}(\sin x)\right) + 2x \sin^4 x + x\left(-2 \cos^{-3} x \cdot \frac{d}{dx}(\cos x)\right) + \cos^{-2} x$
$= x^2(4 \sin^3 x \cos x) + 2x \sin^4 x + x((-2 \cos^{-3} x)(-\sin x)) + \cos^{-2} x$
$= 4x^2 \sin^3 x \cos x + 2x \sin^4 x + 2x \sin x \cos^{-3} x + \cos^{-2} x$

30. $y = \frac{1}{x} \sin^{-5} x - \frac{x}{3} \cos^3 x \Rightarrow y' = \frac{1}{x} \frac{d}{dx}(\sin^{-5} x) + \sin^{-5} x \cdot \frac{d}{dx}\left(\frac{1}{x}\right) - \frac{x}{3} \frac{d}{dx}(\cos^3 x) - \cos^3 x \cdot \frac{d}{dx}\left(\frac{x}{3}\right)$
$= \frac{1}{x}(-5 \sin^{-6} x \cos x) + (\sin^{-5} x)\left(-\frac{1}{x^2}\right) - \frac{x}{3}((3 \cos^2 x)(-\sin x)) - (\cos^3 x)\left(\frac{1}{3}\right)$
$= -\frac{5}{x} \sin^{-6} x \cos x - \frac{1}{x^2} \sin^{-5} x + x \cos^2 x \sin x - \frac{1}{3} \cos^3 x$

31. $y = \frac{1}{21}(3x-2)^7 + \left(4 - \frac{1}{2x^2}\right)^{-1} \Rightarrow \frac{dy}{dx} = \frac{7}{21}(3x-2)^6 \cdot \frac{d}{dx}(3x-2) + (-1)\left(4 - \frac{1}{2x^2}\right)^{-2} \cdot \frac{d}{dx}\left(4 - \frac{1}{2x^2}\right)$
$= \frac{7}{21}(3x-2)^6 \cdot 3 + (-1)\left(4 - \frac{1}{2x^2}\right)^{-2}\left(\frac{1}{x^3}\right) = (3x-2)^6 - \frac{1}{x^3\left(4 - \frac{1}{2x^2}\right)^2}$

32. $y = (5-2x)^{-3} + \frac{1}{8}\left(\frac{2}{x} + 1\right)^4 \Rightarrow \frac{dy}{dx} = -3(5-2x)^{-4}(-2) + \frac{4}{8}\left(\frac{2}{x} + 1\right)^3\left(-\frac{2}{x^2}\right) = 6(5-2x)^{-4} - \left(\frac{1}{x^2}\right)\left(\frac{2}{x} + 1\right)^3$
$= \frac{6}{(5-2x)^4} - \frac{\left(\frac{2}{x}+1\right)^3}{x^2}$

33. $y = (4x+3)^4(x+1)^{-3} \Rightarrow \frac{dy}{dx} = (4x+3)^4(-3)(x+1)^{-4} \cdot \frac{d}{dx}(x+1) + (x+1)^{-3}(4)(4x+3)^3 \cdot \frac{d}{dx}(4x+3)$
$= (4x+3)^4(-3)(x+1)^{-4}(1) + (x+1)^{-3}(4)(4x+3)^3(4) = -3(4x+3)^4(x+1)^{-4} + 16(4x+3)^3(x+1)^{-3}$
$= \frac{(4x+3)^3}{(x+1)^4}[-3(4x+3) + 16(x+1)] = \frac{(4x+3)^3(4x+7)}{(x+1)^4}$

34. $y=(2x-5)^{-1}\left(x^2-5x\right)^6 \Rightarrow \frac{dy}{dx}=(2x-5)^{-1}(6)\left(x^2-5x\right)^5(2x-5)+\left(x^2-5x\right)^6(-1)(2x-5)^{-2}(2)$
$=6\left(x^2-5x\right)^5-\frac{2\left(x^2-5x\right)^6}{(2x-5)^2}$

35. $y=xe^{-x}+e^{3x} \Rightarrow y'=x\cdot e^{-x}(-1)+(1)\cdot e^{-x}+3e^{3x}=(1-x)e^{-x}+3e^{3x}$

36. $y=(1+2x)e^{-2x} \Rightarrow y'=(1+2x)\cdot e^{-2x}(-2)+(2)\cdot e^{-2x}=-4xe^{-2x}$

37. $y=\left(x^2-2x+2\right)e^{5x/2} \Rightarrow y'=\left(x^2-2x+2\right)\cdot e^{5x/2}\left(\frac{5}{2}\right)+(2x-2)\cdot e^{5x/2}=\left(\frac{5}{2}x^2-3x+3\right)e^{5x/2}$

38. $y=\left(9x^2-6x+2\right)e^{x^3} \Rightarrow y'=\left(9x^2-6x+2\right)\cdot e^{x^3}\left(3x^2\right)+(18x-6)\cdot e^{x^3}=\left(27x^4-18x^3+6x^2+18x-6\right)e^{x^3}$

39. $h(x)=x\tan\left(2\sqrt{x}\right)+7 \Rightarrow h'(x)=x\frac{d}{dx}\left(\tan\left(2x^{1/2}\right)\right)+\tan\left(2x^{1/2}\right)\cdot\frac{d}{dx}(x)+0$
$=x\sec^2\left(2x^{1/2}\right)\cdot\frac{d}{dx}\left(2x^{1/2}\right)+\tan\left(2x^{1/2}\right)=x\sec^2\left(2\sqrt{x}\right)\cdot\frac{1}{\sqrt{x}}+\tan\left(2\sqrt{x}\right)=\sqrt{x}\sec^2\left(2\sqrt{x}\right)+\tan\left(2\sqrt{x}\right)$

40. $k(x)=x^2\sec\left(\frac{1}{x}\right) \Rightarrow k'(x)=x^2\frac{d}{dx}\left(\sec\frac{1}{x}\right)+\sec\left(\frac{1}{x}\right)\cdot\frac{d}{dx}\left(x^2\right)=x^2\sec\left(\frac{1}{x}\right)\tan\left(\frac{1}{x}\right)\cdot\frac{d}{dx}\left(\frac{1}{x}\right)+2x\sec\left(\frac{1}{x}\right)$
$=x^2\sec\left(\frac{1}{x}\right)\tan\left(\frac{1}{x}\right)\cdot\left(-\frac{1}{x^2}\right)+2x\sec\left(\frac{1}{x}\right)=2x\sec\left(\frac{1}{x}\right)-\sec\left(\frac{1}{x}\right)\tan\left(\frac{1}{x}\right)$

41. $f(\theta)=\left(\frac{\sin\theta}{1+\cos\theta}\right)^2 \Rightarrow f'(\theta)=2\left(\frac{\sin\theta}{1+\cos\theta}\right)\cdot\frac{d}{d\theta}\left(\frac{\sin\theta}{1+\cos\theta}\right)=\frac{2\sin\theta}{1+\cos\theta}\cdot\frac{(1+\cos\theta)(\cos\theta)-(\sin\theta)(-\sin\theta)}{(1+\cos\theta)^2}$
$=\frac{(2\sin\theta)\left(\cos\theta+\cos^2\theta+\sin^2\theta\right)}{(1+\cos\theta)^3}=\frac{(2\sin\theta)(\cos\theta+1)}{(1+\cos\theta)^3}=\frac{2\sin\theta}{(1+\cos\theta)^2}$

42. $g(t)=\left(\frac{1+\cos t}{\sin t}\right)^{-1} \Rightarrow g'(t)=-\left(\frac{1+\cos t}{\sin t}\right)^{-2}\cdot\frac{d}{dt}\left(\frac{1+\cos t}{\sin t}\right)=-\frac{\sin^2 t}{(1+\cos t)^2}\cdot\frac{(\sin t)(-\sin t)-(1+\cos t)(\cos t)}{(\sin t)^2}$
$=\frac{-\left(-\sin^2 t-\cos t-\cos^2 t\right)}{(1+\cos t)^2}=\frac{1}{1+\cos t}$

43. $r=\sin\left(\theta^2\right)\cos(2\theta) \Rightarrow \frac{dr}{d\theta}=\sin\left(\theta^2\right)(-\sin 2\theta)\frac{d}{d\theta}(2\theta)+\cos(2\theta)\left(\cos\left(\theta^2\right)\right)\cdot\frac{d}{d\theta}\left(\theta^2\right)$
$=\sin\left(\theta^2\right)(-\sin 2\theta)(2)+(\cos 2\theta)\left(\cos\left(\theta^2\right)\right)(2\theta)=-2\sin\left(\theta^2\right)\sin(2\theta)+2\theta\cos(2\theta)\cos\left(\theta^2\right)$

44. $r=\left(\sec\sqrt{\theta}\right)\tan\left(\frac{1}{\theta}\right) \Rightarrow \frac{dr}{d\theta}=\left(\sec\sqrt{\theta}\right)\left(\sec^2\frac{1}{\theta}\right)\left(-\frac{1}{\theta^2}\right)+\tan\left(\frac{1}{\theta}\right)\left(\sec\sqrt{\theta}\tan\sqrt{\theta}\right)\left(\frac{1}{2\sqrt{\theta}}\right)$
$=-\frac{1}{\theta^2}\sec\sqrt{\theta}\sec^2\left(\frac{1}{\theta}\right)+\frac{1}{2\sqrt{\theta}}\tan\left(\frac{1}{\theta}\right)\sec\sqrt{\theta}\tan\sqrt{\theta}=\left(\sec\sqrt{\theta}\right)\left[\frac{\tan\sqrt{\theta}\tan\left(\frac{1}{\theta}\right)}{2\sqrt{\theta}}-\frac{\sec^2\left(\frac{1}{\theta}\right)}{\theta^2}\right]$

45. $q=\sin\left(\frac{t}{\sqrt{t+1}}\right) \Rightarrow \frac{dq}{dt}=\cos\left(\frac{t}{\sqrt{t+1}}\right)\cdot\frac{d}{dt}\left(\frac{t}{\sqrt{t+1}}\right)=\cos\left(\frac{t}{\sqrt{t+1}}\right)\cdot\frac{\sqrt{t+1}(1)-t\cdot\frac{d}{dt}\left(\sqrt{t+1}\right)}{\left(\sqrt{t+1}\right)^2}$
$=\cos\left(\frac{t}{\sqrt{t+1}}\right)\cdot\frac{\sqrt{t+1}-\frac{t}{2\sqrt{t+1}}}{t+1}=\cos\left(\frac{t}{\sqrt{t+1}}\right)\left(\frac{2(t+1)-t}{2(t+1)^{3/2}}\right)=\left(\frac{t+2}{2(t+1)^{3/2}}\right)\cos\left(\frac{t}{\sqrt{t+1}}\right)$

46. $q=\cot\left(\frac{\sin t}{t}\right) \Rightarrow \frac{dq}{dt}=-\csc^2\left(\frac{\sin t}{t}\right)\cdot\frac{d}{dt}\left(\frac{\sin t}{t}\right)=\left(-\csc^2\left(\frac{\sin t}{t}\right)\right)\left(\frac{t\cos t-\sin t}{t^2}\right)$

47. $y=\cos\left(e^{-\theta^2}\right) \Rightarrow \frac{dy}{d\theta}=-\sin\left(e^{-\theta^2}\right)\frac{d}{d\theta}\left(e^{-\theta^2}\right)=\left(-\sin\left(e^{-\theta^2}\right)\right)\left(e^{-\theta^2}\right)\frac{d}{d\theta}\left(-\theta^2\right)=2\theta e^{-\theta^2}\sin\left(e^{-\theta^2}\right)$

48. $y=\theta^3e^{-2\theta}\cos 5\theta \Rightarrow \frac{dy}{d\theta}=\left(3\theta^2\right)\left(e^{-2\theta}\cos 5\theta\right)+\left(\theta^3\cos 5\theta\right)e^{-2\theta}\frac{d}{d\theta}(-2\theta)-5(\sin 5\theta)\left(\theta^3e^{-2\theta}\right)$
$=\theta^2e^{-2\theta}(3\cos 5\theta-2\theta\cos 5\theta-5\theta\sin 5\theta)$

49. $y=\sin^2(\pi t-2) \Rightarrow \frac{dy}{dt}=2\sin(\pi t-2)\cdot\frac{d}{dt}\sin(\pi t-2)=2\sin(\pi t-2)\cdot\cos(\pi t-2)\cdot\frac{d}{dt}(\pi t-2)$
$=2\pi\sin(\pi t-2)\cos(\pi t-2)$

50. $y = \sec^2 \pi t \Rightarrow \frac{dy}{dt} = (2 \sec \pi t) \cdot \frac{d}{dt}(\sec \pi t) = (2 \sec \pi t)(\sec \pi t \tan \pi t) \cdot \frac{d}{dt}(\pi t) = 2\pi \sec^2 \pi t \tan \pi t$

51. $y = (1 + \cos 2t)^{-4} \Rightarrow \frac{dy}{dt} = -4(1 + \cos 2t)^{-5} \cdot \frac{d}{dt}(1 + \cos 2t) = -4(1 + \cos 2t)^{-5}(-\sin 2t) \cdot \frac{d}{dt}(2t) = \frac{8 \sin 2t}{(1 + \cos 2t)^5}$

52. $y = \left(1 + \cot\left(\frac{t}{2}\right)\right)^{-2} \Rightarrow \frac{dy}{dt} = -2\left(1 + \cot\left(\frac{t}{2}\right)\right)^{-3} \cdot \frac{d}{dt}\left(1 + \cot\left(\frac{t}{2}\right)\right) = -2\left(1 + \cot\left(\frac{t}{2}\right)\right)^{-3} \cdot \left(-\csc^2\left(\frac{t}{2}\right)\right) \cdot \frac{d}{dt}\left(\frac{t}{2}\right)$
$= \frac{\csc^2\left(\frac{t}{2}\right)}{\left(1 + \cot\left(\frac{t}{2}\right)\right)^3}$

53. $y = e^{\cos^2(\pi t - 1)} \Rightarrow \frac{dy}{dt} = e^{\cos^2(\pi t - 1)} \cdot 2\cos(\pi t - 1) \cdot (-\sin(\pi t - 1)) \cdot \pi = -2\pi \sin(\pi t - 1)\cos(\pi t - 1)e^{\cos^2(\pi t - 1)}$

54. $y = \left(e^{\sin(t/2)}\right)^3 \Rightarrow \frac{dy}{dt} = 3\left(e^{\sin(t/2)}\right)^2 \cdot e^{\sin(t/2)} \cdot \cos\left(\frac{t}{2}\right) \cdot \frac{1}{2} = \frac{3}{2}\cos\left(\frac{t}{2}\right)e^{\sin(t/2)}e^{2\sin(t/2)} = \frac{3}{2}\cos\left(\frac{t}{2}\right)e^{3\sin(t/2)}$

55. $y = \sin(\cos(2t - 5)) \Rightarrow \frac{dy}{dt} = \cos(\cos(2t - 5)) \cdot \frac{d}{dt}\cos(2t - 5) = \cos(\cos(2t - 5)) \cdot (-\sin(2t - 5)) \cdot \frac{d}{dt}(2t - 5)$
$= -2\cos(\cos(2t - 5))(\sin(2t - 5))$

56. $y = \cos\left(5 \sin\left(\frac{t}{3}\right)\right) \Rightarrow \frac{dy}{dt} = -\sin\left(5 \sin\left(\frac{t}{3}\right)\right) \cdot \frac{d}{dt}\left(5 \sin\left(\frac{t}{3}\right)\right) = -\sin\left(5 \sin\left(\frac{t}{3}\right)\right)\left(5 \cos\left(\frac{t}{3}\right)\right) \cdot \frac{d}{dt}\left(\frac{t}{3}\right)$
$= -\frac{5}{3}\sin\left(5 \sin\left(\frac{t}{3}\right)\right)\left(\cos\left(\frac{t}{3}\right)\right)$

57. $y = \left[1 + \tan^4\left(\frac{t}{12}\right)\right]^3 \Rightarrow \frac{dy}{dt} = 3\left[1 + \tan^4\left(\frac{t}{12}\right)\right]^2 \cdot \frac{d}{dt}\left[1 + \tan^4\left(\frac{t}{12}\right)\right] = 3\left[1 + \tan^4\left(\frac{t}{12}\right)\right]^2\left[4 \tan^3\left(\frac{t}{12}\right) \cdot \frac{d}{dt}\tan\left(\frac{t}{12}\right)\right]$
$= 12\left[1 + \tan^4\left(\frac{t}{12}\right)\right]^2\left[\tan^3\left(\frac{t}{12}\right)\sec^2\left(\frac{t}{12}\right) \cdot \frac{1}{12}\right] = \left[1 + \tan^4\left(\frac{t}{12}\right)\right]^2\left[\tan^3\left(\frac{t}{12}\right)\sec^2\left(\frac{t}{12}\right)\right]$

58. $y = \frac{1}{6}\left[1 + \cos^2(7t)\right]^3 \Rightarrow \frac{dy}{dt} = \frac{3}{6}\left[1 + \cos^2(7t)\right]^2 \cdot 2\cos(7t)(-\sin(7t))(7) = -7\left[1 + \cos^2(7t)\right]^2(\cos(7t)\sin(7t))$

59. $y = (1 + \cos(t^2))^{1/2} \Rightarrow \frac{dy}{dt} = \frac{1}{2}(1 + \cos(t^2))^{-1/2} \cdot \frac{d}{dt}(1 + \cos(t^2)) = \frac{1}{2}(1 + \cos(t^2))^{-1/2}\left(-\sin(t^2) \cdot \frac{d}{dt}(t^2)\right)$
$= -\frac{1}{2}(1 + \cos(t^2))^{-1/2}(\sin(t^2)) \cdot 2t = -\frac{t \sin(t^2)}{\sqrt{1 + \cos(t^2)}}$

60. $y = 4 \sin\left(\sqrt{1 + \sqrt{t}}\right) \Rightarrow \frac{dy}{dt} = 4 \cos\left(\sqrt{1 + \sqrt{t}}\right) \cdot \frac{d}{dt}\left(\sqrt{1 + \sqrt{t}}\right) = 4 \cos\left(\sqrt{1 + \sqrt{t}}\right) \cdot \frac{1}{2\sqrt{1 + \sqrt{t}}} \cdot \frac{d}{dt}(1 + \sqrt{t})$
$= \frac{2 \cos\left(\sqrt{1 + \sqrt{t}}\right)}{\sqrt{1 + \sqrt{t}} \cdot 2\sqrt{t}} = \frac{\cos\left(\sqrt{1 + \sqrt{t}}\right)}{\sqrt{t + \sqrt{t}}}$

61. $y = \left(1 + \frac{1}{x}\right)^3 \Rightarrow y' = 3\left(1 + \frac{1}{x}\right)^2\left(-\frac{1}{x^2}\right) = -\frac{3}{x^2}\left(1 + \frac{1}{x}\right)^2 \Rightarrow y'' = \left(-\frac{3}{x^2}\right) \cdot \frac{d}{dx}\left(1 + \frac{1}{x}\right)^2 - \left(1 + \frac{1}{x}\right)^2 \cdot \frac{d}{dx}\left(\frac{3}{x^2}\right)$
$= \left(-\frac{3}{x^2}\right)\left(2\left(1 + \frac{1}{x}\right)\left(-\frac{1}{x^2}\right)\right) + \left(\frac{6}{x^3}\right)\left(1 + \frac{1}{x}\right)^2 = \frac{6}{x^4}\left(1 + \frac{1}{x}\right) + \frac{6}{x^3}\left(1 + \frac{1}{x}\right)^2 = \frac{6}{x^3}\left(1 + \frac{1}{x}\right)\left(\frac{1}{x} + 1 + \frac{1}{x}\right)$
$= \frac{6}{x^3}\left(1 + \frac{1}{x}\right)\left(1 + \frac{2}{x}\right)$

62. $y = \left(1 - \sqrt{x}\right)^{-1} \Rightarrow y' = -\left(1 - \sqrt{x}\right)^{-2}\left(-\frac{1}{2}x^{-1/2}\right) = \frac{1}{2}\left(1 - \sqrt{x}\right)^{-2}x^{-1/2}$
$\Rightarrow y'' = \frac{1}{2}\left[\left(1 - \sqrt{x}\right)^{-2}\left(-\frac{1}{2}x^{-3/2}\right) + x^{-1/2}(-2)\left(1 - \sqrt{x}\right)^{-3}\left(-\frac{1}{2}x^{-1/2}\right)\right]$
$= \frac{1}{2}\left[\frac{-1}{2}x^{-3/2}\left(1 - \sqrt{x}\right)^{-2} + x^{-1}\left(1 - \sqrt{x}\right)^{-3}\right] = \frac{1}{2}x^{-1}\left(1 - \sqrt{x}\right)^{-3}\left[-\frac{1}{2}x^{-1/2}\left(1 - \sqrt{x}\right) + 1\right]$
$= \frac{1}{2x}\left(1 - \sqrt{x}\right)^{-3}\left(-\frac{1}{2\sqrt{x}} + \frac{1}{2} + 1\right) = \frac{1}{2x}\left(1 - \sqrt{x}\right)^{-3}\left(\frac{3}{2} - \frac{1}{2\sqrt{x}}\right)$

63. $y = \frac{1}{9}\cot(3x - 1) \Rightarrow y' = -\frac{1}{9}\csc^2(3x - 1)(3) = -\frac{1}{3}\csc^2(3x - 1) \Rightarrow y'' = \left(-\frac{2}{3}\right)\left(\csc(3x - 1) \cdot \frac{d}{dx}\csc(3x - 1)\right)$
$= -\frac{2}{3}\csc(3x - 1)\left(-\csc(3x - 1)\cot(3x - 1) \cdot \frac{d}{dx}(3x - 1)\right) = 2\csc^2(3x - 1)\cot(3x - 1)$

64. $y = 9\tan\left(\frac{x}{3}\right) \Rightarrow y' = 9\left(\sec^2\left(\frac{x}{3}\right)\right)\left(\frac{1}{3}\right) = 3\sec^2\left(\frac{x}{3}\right) \Rightarrow y'' = 3\cdot 2\sec\left(\frac{x}{3}\right)\left(\sec\left(\frac{x}{3}\right)\tan\left(\frac{x}{3}\right)\right)\left(\frac{1}{3}\right) = 2\sec^2\left(\frac{x}{3}\right)\tan\left(\frac{x}{3}\right)$

65. $y = e^{x^2} + 5x \Rightarrow y' = 2xe^{x^2} + 5 \Rightarrow y'' = 2x\cdot e^{x^2}(2x) + 2e^{x^2} = (4x^2+2)e^{x^2}$

66. $y = \sin(x^2e^x) \Rightarrow y' = \cos(x^2e^x)\cdot(x^2e^x + 2xe^x) = (x^2+2x)e^x\cos(x^2e^x)$
$\Rightarrow$ (Use triple product rule: $D(fgh) = f'gh + fg'h + fgh'$)
$y'' = (2x+2)e^x\cos(x^2e^x) + (x^2+2x)e^x\cos(x^2e^x) + (x^2+2x)e^x(-\sin(x^2e^x)\cdot(x^2e^x+2xe^x))$
$= (x^2+4x+2)e^x\cos(x^2e^x) - xe^{2x}(x^3+4x^2+4x)\sin(x^2e^x)$

67. $g(x) = \sqrt{x} \Rightarrow g'(x) = \frac{1}{2\sqrt{x}} \Rightarrow g(1) = 1$ and $g'(1) = \frac{1}{2}$; $f(u) = u^5 + 1 \Rightarrow f'(u) = 5u^4 \Rightarrow f'(g(1)) = f'(1) = 5$; therefore, $(f\circ g)'(1) = f'(g(1))\cdot g'(1) = 5\cdot\frac{1}{2} = \frac{5}{2}$

68. $g(x) = (1-x)^{-1} \Rightarrow g'(x) = -(1-x)^{-2}(-1) = \frac{1}{(1-x)^2} \Rightarrow g(-1) = \frac{1}{2}$ and $g'(-1) = \frac{1}{4}$; $f(u) = 1 - \frac{1}{u}$
$\Rightarrow f'(u) = \frac{1}{u^2} \Rightarrow f'(g(-1)) = f'\left(\frac{1}{2}\right) = 4$; therefore, $(f\circ g)'(-1) = f'(g(-1))g'(-1) = 4\cdot\frac{1}{4} = 1$

69. $g(x) = 5\sqrt{x} \Rightarrow g'(x) = \frac{5}{2\sqrt{x}} \Rightarrow g(1) = 5$ and $g'(1) = \frac{5}{2}$; $f(u) = \cot\left(\frac{\pi u}{10}\right) \Rightarrow f'(u) = -\csc^2\left(\frac{\pi u}{10}\right)\left(\frac{\pi}{10}\right)$
$= \frac{-\pi}{10}\csc^2\left(\frac{\pi u}{10}\right) \Rightarrow f'(g(1)) = f'(5) = -\frac{\pi}{10}\csc^2\left(\frac{\pi}{2}\right) = -\frac{\pi}{10}$; therefore, $(f\circ g)'(1) = f'(g(1))g'(1) = -\frac{\pi}{10}\cdot\frac{5}{2}$
$= -\frac{\pi}{4}$

70. $g(x) = \pi x \Rightarrow g'(x) = \pi \Rightarrow g\left(\frac{1}{4}\right) = \frac{\pi}{4}$ and $g'\left(\frac{1}{4}\right) = \pi$; $f(u) = u + \sec^2 u \Rightarrow f'(u) = 1 + 2\sec u\cdot\sec u\tan u$
$= 1 + 2\sec^2 u\tan u \Rightarrow f'\left(g\left(\frac{1}{4}\right)\right) = f'\left(\frac{\pi}{4}\right) = 1 + 2\sec^2\frac{\pi}{4}\tan\frac{\pi}{4} = 5$; therefore, $(f\circ g)'\left(\frac{1}{4}\right) = f'\left(g\left(\frac{1}{4}\right)\right)g'\left(\frac{1}{4}\right) = 5\pi$

71. $g(x) = 10x^2 + x + 1 \Rightarrow g'(x) = 20x + 1 \Rightarrow g(0) = 1$ and $g'(0) = 1$; $f(u) = \frac{2u}{u^2+1} \Rightarrow f'(u) = \frac{(u^2+1)(2)-(2u)(2u)}{(u^2+1)^2}$
$= \frac{-2u^2+2}{(u^2+1)^2} \Rightarrow f'(g(0)) = f'(1) = 0$; therefore, $(f\circ g)'(0) = f'(g(0))g'(0) = 0\cdot 1 = 0$

72. $g(x) = \frac{1}{x^2} - 1 \Rightarrow g'(x) = -\frac{2}{x^3} \Rightarrow g(-1) = 0$ and $g'(-1) = 2$; $f(u) = \left(\frac{u-1}{u+1}\right)^2 \Rightarrow f'(u) = 2\left(\frac{u-1}{u+1}\right)\frac{d}{du}\left(\frac{u-1}{u+1}\right)$
$= 2\left(\frac{u-1}{u+1}\right)\cdot\frac{(u+1)(1)-(u-1)(1)}{(u+1)^2} = \frac{2(u-1)(2)}{(u+1)^3} = \frac{4(u-1)}{(u+1)^3} \Rightarrow f'(g(-1)) = f'(0) = -4$; therefore,
$(f\circ g)'(-1) = f'(g(-1))g'(-1) = (-4)(2) = -8$

73. (a) $y = 2f(x) \Rightarrow \frac{dy}{dx} = 2f'(x) \Rightarrow \left.\frac{dy}{dx}\right|_{x=2} = 2f'(2) = 2\left(\frac{1}{3}\right) = \frac{2}{3}$

(b) $y = f(x) + g(x) \Rightarrow \frac{dy}{dx} = f'(x) + g'(x) \Rightarrow \left.\frac{dy}{dx}\right|_{x=3} = f'(3) + g'(3) = 2\pi + 5$

(c) $y = f(x)\cdot g(x) \Rightarrow \frac{dy}{dx} = f(x)g'(x) + g(x)f'(x) \Rightarrow \left.\frac{dy}{dx}\right|_{x=3} = f(3)g'(3) + g(3)f'(3) = 3\cdot 5 + (-4)(2\pi) = 15 - 8\pi$

(d) $y = \frac{f(x)}{g(x)} \Rightarrow \frac{dy}{dx} = \frac{g(x)f'(x) - f(x)g'(x)}{[g(x)]^2} \Rightarrow \left.\frac{dy}{dx}\right|_{x=2} = \frac{g(2)f'(2) - f(2)g'(2)}{[g(2)]^2} = \frac{(2)\left(\frac{1}{3}\right) - (8)(-3)}{2^2} = \frac{37}{6}$

(e) $y = f(g(x)) \Rightarrow \frac{dy}{dx} = f'(g(x))g'(x) \Rightarrow \left.\frac{dy}{dx}\right|_{x=2} = f'(g(2))g'(2) = f'(2)(-3) = \frac{1}{3}(-3) = -1$

(f) $y = (f(x))^{1/2} \Rightarrow \frac{dy}{dx} = \frac{1}{2}(f(x))^{-1/2}\cdot f'(x) = \frac{f'(x)}{2\sqrt{f(x)}} \Rightarrow \left.\frac{dy}{dx}\right|_{x=2} = \frac{f'(2)}{2\sqrt{f(2)}} = \frac{\left(\frac{1}{3}\right)}{2\sqrt{8}} = \frac{1}{6\sqrt{8}} = \frac{1}{12\sqrt{2}} = \frac{\sqrt{2}}{24}$

(g) $y = (g(x))^{-2} \Rightarrow \frac{dy}{dx} = -2(g(x))^{-3}\cdot g'(x) \Rightarrow \left.\frac{dy}{dx}\right|_{x=3} = -2(g(3))^{-3}g'(3) = -2(-4)^{-3}\cdot 5 = \frac{5}{32}$

(h) $y = \left((f(x))^2 + (g(x))^2\right)^{1/2} \Rightarrow \frac{dy}{dx} = \frac{1}{2}\left((f(x))^2 + (g(x))^2\right)^{-1/2}(2f(x)\cdot f'(x) + 2g(x)\cdot g'(x))$
$\Rightarrow \left.\frac{dy}{dx}\right|_{x=2} = \frac{1}{2}\left((f(2))^2 + (g(2))^2\right)^{-1/2}(2f(2)f'(2) + 2g(2)g'(2)) = \frac{1}{2}\left(8^2 + 2^2\right)^{-1/2}\left(2\cdot 8\cdot\frac{1}{3} + 2\cdot 2\cdot(-3)\right)$
$= -\frac{5}{3\sqrt{17}}$

74. (a) $y = 5f(x) - g(x) \Rightarrow \frac{dy}{dx} = 5f'(x) - g'(x) \Rightarrow \left.\frac{dy}{dx}\right|_{x=1} = 5f'(1) - g'(1) = 5\left(-\frac{1}{3}\right) - \left(\frac{-8}{3}\right) = 1$

(b) $y = f(x)(g(x))^3 \Rightarrow \frac{dy}{dx} = f(x)\left(3(g(x))^2 g'(x)\right) + (g(x))^3 f'(x) \Rightarrow \left.\frac{dy}{dx}\right|_{x=0} = 3f(0)(g(0))^2 g'(0) + (g(0))^3 f'(0)$
$= 3(1)(1)^2\left(\frac{1}{3}\right) + (1)^3(5) = 6$

(c) $y = \frac{f(x)}{g(x)+1} \Rightarrow \frac{dy}{dx} = \frac{(g(x)+1)f'(x) - f(x)g'(x)}{(g(x)+1)^2} \Rightarrow \left.\frac{dy}{dx}\right|_{x=1} = \frac{(g(1)+1)f'(1) - f(1)g'(1)}{(g(1)+1)^2}$
$= \frac{(-4+1)\left(-\frac{1}{3}\right) - (3)\left(-\frac{8}{3}\right)}{(-4+1)^2} = 1$

(d) $y = f(g(x)) \Rightarrow \frac{dy}{dx} = f'(g(x))g'(x) \Rightarrow \left.\frac{dy}{dx}\right|_{x=0} = f'(g(0))g'(0) = f'(1)\left(\frac{1}{3}\right) = \left(-\frac{1}{3}\right)\left(\frac{1}{3}\right) = -\frac{1}{9}$

(e) $y = g(f(x)) \Rightarrow \frac{dy}{dx} = g'(f(x))f'(x) \Rightarrow \left.\frac{dy}{dx}\right|_{x=0} = g'(f(0))f'(0) = g'(1)(5) = \left(-\frac{8}{3}\right)(5) = -\frac{40}{3}$

(f) $y = \left(x^{11} + f(x)\right)^{-2} \Rightarrow \frac{dy}{dx} = -2\left(x^{11} + f(x)\right)^{-3}\left(11x^{10} + f'(x)\right) \Rightarrow \left.\frac{dy}{dx}\right|_{x=1} = -2(1 + f(1))^{-3}\left(11 + f'(1)\right)$
$= -2(1+3)^{-3}\left(11 - \frac{1}{3}\right) = \left(-\frac{2}{4^3}\right)\left(\frac{32}{3}\right) = -\frac{1}{3}$

(g) $y = f(x + g(x)) \Rightarrow \frac{dy}{dx} = f'(x + g(x))\left(1 + g'(x)\right) \Rightarrow \left.\frac{dy}{dx}\right|_{x=0} = f'(0 + g(0))\left(1 + g'(0)\right) = f'(1)\left(1 + \frac{1}{3}\right)$
$= \left(-\frac{1}{3}\right)\left(\frac{4}{3}\right) = -\frac{4}{9}$

75. $\frac{ds}{dt} = \frac{ds}{d\theta} \cdot \frac{d\theta}{dt}$: $s = \cos\theta \Rightarrow \frac{ds}{d\theta} = -\sin\theta \Rightarrow \left.\frac{ds}{d\theta}\right|_{\theta=\frac{3\pi}{2}} = -\sin\left(\frac{3\pi}{2}\right) = 1$ so that $\frac{ds}{dt} = \frac{ds}{d\theta} \cdot \frac{d\theta}{dt} = 1 \cdot 5 = 5$

76. $\frac{dy}{dt} = \frac{dy}{dx} \cdot \frac{dx}{dt}$: $y = x^2 + 7x - 5 \Rightarrow \frac{dy}{dx} = 2x + 7 \Rightarrow \left.\frac{dy}{dx}\right|_{x=1} = 9$ so that $\frac{dy}{dt} = \frac{dy}{dx} \cdot \frac{dx}{dt} = 9 \cdot \frac{1}{3} = 3$

77. With $y = x$, we should get $\frac{dy}{dx} = 1$ for both (a) and (b):

(a) $y = \frac{u}{5} + 7 \Rightarrow \frac{dy}{du} = \frac{1}{5}$; $u = 5x - 35 \Rightarrow \frac{du}{dx} = 5$; therefore, $\frac{dy}{dx} = \frac{dy}{du} \cdot \frac{du}{dx} = \frac{1}{5} \cdot 5 = 1$, as expected

(b) $y = 1 + \frac{1}{u} \Rightarrow \frac{dy}{du} = -\frac{1}{u^2}$; $u = (x-1)^{-1} \Rightarrow \frac{du}{dx} = -(x-1)^{-2}(1) = \frac{-1}{(x-1)^2}$; therefore $\frac{dy}{dx} = \frac{dy}{du} \cdot \frac{du}{dx}$
$= \frac{-1}{u^2} \cdot \frac{-1}{(x-1)^2} = \frac{-1}{\left((x-1)^{-1}\right)^2} \cdot \frac{-1}{(x-1)^2} = (x-1)^2 \cdot \frac{1}{(x-1)^2} = 1$, again as expected

78. With $y = x^{3/2}$, we should get $\frac{dy}{dx} = \frac{3}{2}x^{1/2}$ for both (a) and (b):

(a) $y = u^3 \Rightarrow \frac{dy}{du} = 3u^2$; $u = \sqrt{x} \Rightarrow \frac{du}{dx} = \frac{1}{2\sqrt{x}}$; therefore, $\frac{dy}{dx} = \frac{dy}{du} \cdot \frac{du}{dx} = 3u^2 \cdot \frac{1}{2\sqrt{x}} = 3\left(\sqrt{x}\right)^2 \cdot \frac{1}{2\sqrt{x}} = \frac{3}{2}\sqrt{x}$, as expected.

(b) $y = \sqrt{u} \Rightarrow \frac{dy}{du} = \frac{1}{2\sqrt{u}}$; $u = x^3 \Rightarrow \frac{du}{dx} = 3x^2$; therefore, $\frac{dy}{dx} = \frac{dy}{du} \cdot \frac{du}{dx} = \frac{1}{2\sqrt{u}} \cdot 3x^2 = \frac{1}{2\sqrt{x^3}} \cdot 3x^2 = \frac{3}{2}x^{1/2}$, again as expected.

79. $y = 2\tan\left(\frac{\pi x}{4}\right) \Rightarrow \frac{dy}{dx} = \left(2\sec^2\frac{\pi x}{4}\right)\left(\frac{\pi}{4}\right) = \frac{\pi}{2}\sec^2\frac{\pi x}{4}$

(a) $\left.\frac{dy}{dx}\right|_{x=1} = \frac{\pi}{2}\sec^2\left(\frac{\pi}{4}\right) = \pi \Rightarrow$ slope of tangent is 2; thus, $y(1) = 2\tan\left(\frac{\pi}{4}\right) = 2$ and $y'(1) = \pi \Rightarrow$ tangent line is given by $y - 2 = \pi(x - 1) \Rightarrow y = \pi x + 2 - \pi$

(b) $y' = \frac{\pi}{2}\sec^2\left(\frac{\pi x}{4}\right)$ and the smallest value the secant function can have in $-2 < x < 2$ is 1 $\Rightarrow$ the minimum value of y' is $\frac{\pi}{2}$ and that occurs when $\frac{\pi}{2} = \frac{\pi}{2}\sec^2\left(\frac{\pi x}{4}\right) \Rightarrow 1 = \sec^2\left(\frac{\pi x}{4}\right) \Rightarrow \pm 1 = \sec\left(\frac{\pi x}{4}\right) \Rightarrow x = 0$.

80. (a) $y = \sin 2x \Rightarrow y' = 2\cos 2x \Rightarrow y'(0) = 2\cos(0) = 2 \Rightarrow$ tangent to $y = \sin 2x$ at the origin is $y = 2x$; $y = -\sin\left(\frac{x}{2}\right) \Rightarrow y' = -\frac{1}{2}\cos\left(\frac{x}{2}\right) \Rightarrow y'(0) = -\frac{1}{2}\cos 0 = -\frac{1}{2} \Rightarrow$ tangent to $y = -\sin\left(\frac{x}{2}\right)$ at the origin is $y = -\frac{1}{2}x$. The tangents are perpendicular to each other at the origin since the product of their slopes is -1.

(b) $y = \sin(mx) \Rightarrow y' = m\cos(mx) \Rightarrow y'(0) = m\cos 0 = m$; $y = -\sin\left(\frac{x}{m}\right) \Rightarrow y' = -\frac{1}{m}\cos\left(\frac{x}{m}\right)$
$\Rightarrow y'(0) = -\frac{1}{m}\cos(0) = -\frac{1}{m}$. Since $m \cdot \left(-\frac{1}{m}\right) = -1$, the tangent lines are perpendicular at the origin.

(c) $y = \sin(mx) \Rightarrow y' = m\cos(mx)$. The largest value $\cos(mx)$ can attain is 1 at $x = 0 \Rightarrow$ the largest value y' can attain is $|m|$ because $|y'| = |m\cos(mx)| = |m|\,|\cos mx| \le |m| \cdot 1 = |m|$. Also, $y = -\sin\left(\frac{x}{m}\right)$ $\Rightarrow y' = -\frac{1}{m}\cos\left(\frac{x}{m}\right) \Rightarrow |y'| = \left|\frac{-1}{m}\cos\left(\frac{x}{m}\right)\right| \le \left|\frac{1}{m}\right|\left|\cos\left(\frac{x}{m}\right)\right| \le \frac{1}{|m|} \Rightarrow$ the largest value y' can attain is $\left|\frac{1}{m}\right|$.

(d) $y = \sin(mx) \Rightarrow y' = m\cos(mx) \Rightarrow y'(0) = m \Rightarrow$ slope of curve at the origin is m. Also, $\sin(mx)$ completes m periods on $[0, 2\pi]$. Therefore the slope of the curve $y = \sin(mx)$ at the origin is the same as the number of periods it completes on $[0, 2\pi]$. In particular, for large m, we can think of "compressing" the graph of $y = \sin x$ horizontally which gives more periods completed on $[0, 2\pi]$, but also increases the slope of the graph at the origin.

81. $x = \cos 2t,\ y = \sin 2t,\ 0 \le t \le \pi$
$\Rightarrow \cos^2 2t + \sin^2 2t = 1 \Rightarrow x^2 + y^2 = 1$

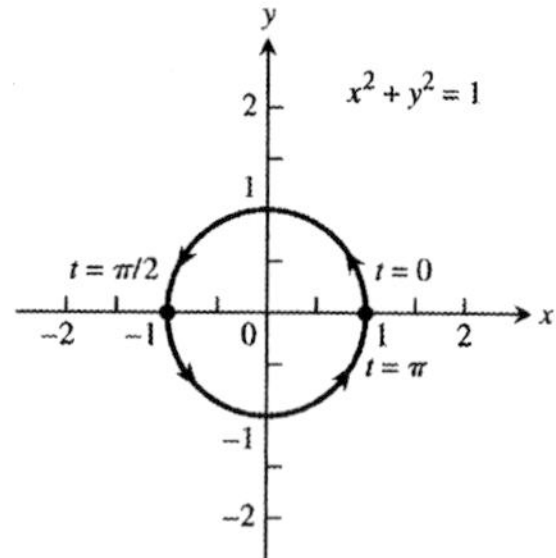

82. $x = \cos(\pi - t),\ y = \sin(\pi - t),\ 0 \le t \le \pi$
$\Rightarrow \cos^2(\pi - t) + \sin^2(\pi - t) = 1$
$\Rightarrow x^2 + y^2 = 1,\ y \ge 0$

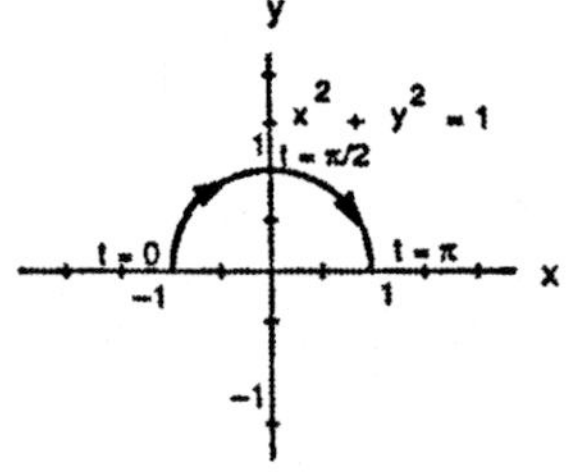

83. $x = 4\cos t,\ y = 2\sin t,\ 0 \le t \le 2\pi$
$\Rightarrow \frac{16\cos^2 t}{16} + \frac{4\sin^2 t}{4} = 1 \Rightarrow \frac{x^2}{16} + \frac{y^2}{4} = 1$

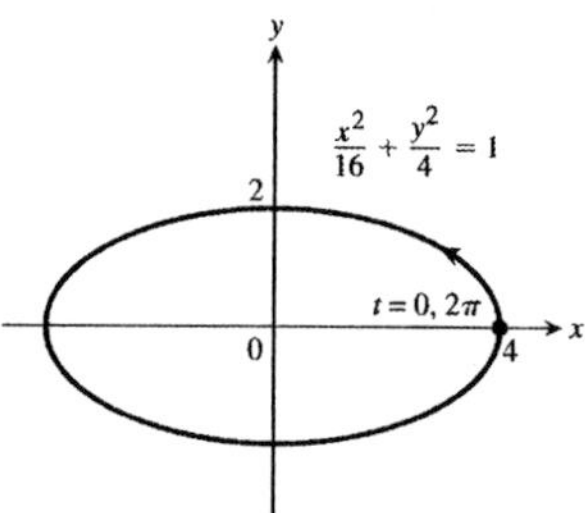

84. $x = 4\sin t,\ y = 5\cos t,\ 0 \le t \le 2\pi$
$\Rightarrow \frac{16\sin^2 t}{16} + \frac{25\cos^2 t}{25} = 1 \Rightarrow \frac{x^2}{16} + \frac{y^2}{25} = 1$

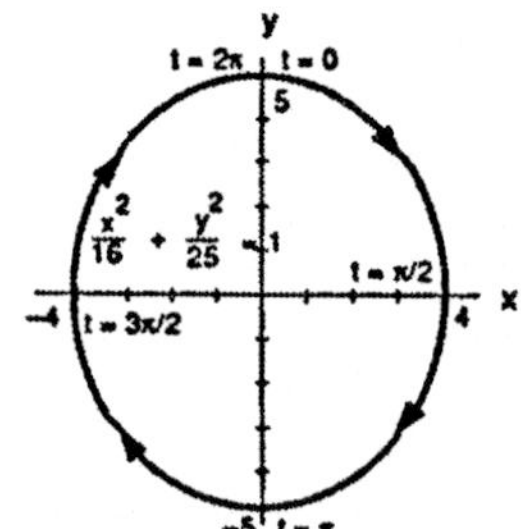

85. $x = 3t,\ y = 9t^2,\ -\infty < t < \infty \Rightarrow y = x^2$

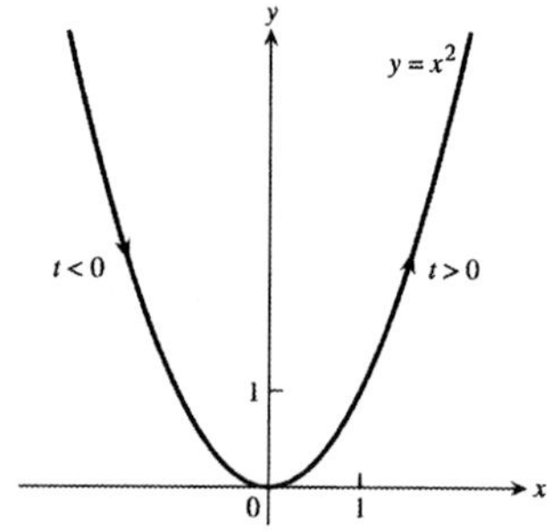

86. $x = -\sqrt{t},\ y = t,\ t \ge 0 \Rightarrow x = -\sqrt{y}$
or $y = x^2,\ x \le 0$

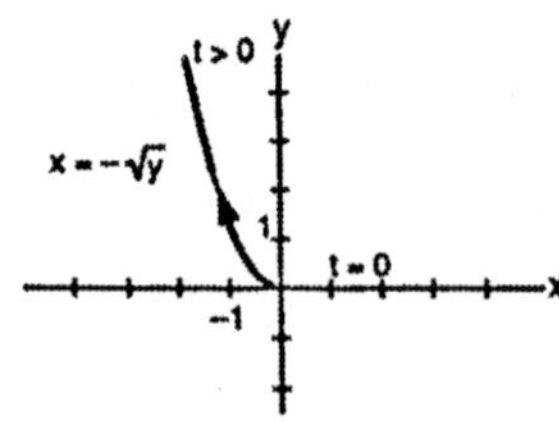

87. $x = 2t - 5,\ y = 4t - 7,\ -\infty < t < \infty$
$\Rightarrow x + 5 = 2t \Rightarrow 2(x + 5) = 4t$
$\Rightarrow y = 2(x + 5) - 7 \Rightarrow y = 2x + 3$

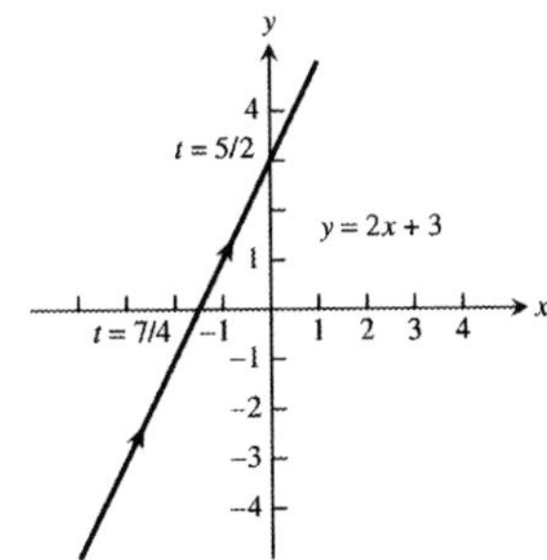

88. $x = 3 - 3t,\ y = 2t,\ 0 \le t \le 1 \Rightarrow \frac{y}{2} = t$
$\Rightarrow x = 3 - 3\left(\frac{y}{2}\right) \Rightarrow 2x = 6 - 3y$
$\Rightarrow y = 2 - \frac{2}{3}x,\ 0 \le x \le 3$

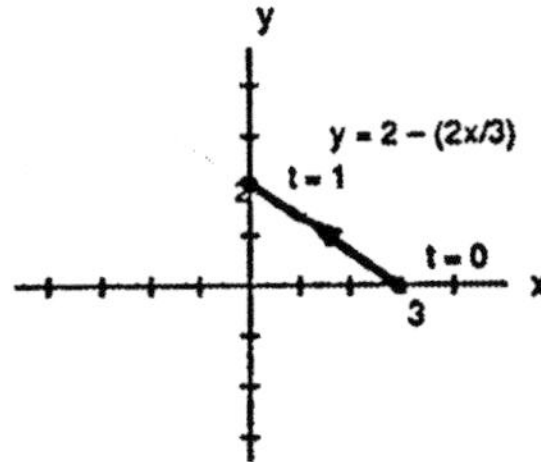

89. $x = t,\ y = \sqrt{1 - t^2},\ -1 \le t \le 0$
$\Rightarrow y = \sqrt{1 - x^2}$

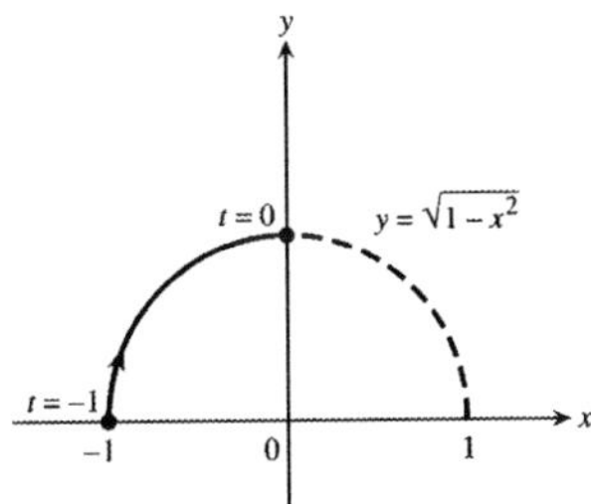

90. $x = \sqrt{t + 1},\ y = \sqrt{t},\ t \ge 0$
$\Rightarrow y^2 = t \Rightarrow x = \sqrt{y^2 + 1},\ y \ge 0$

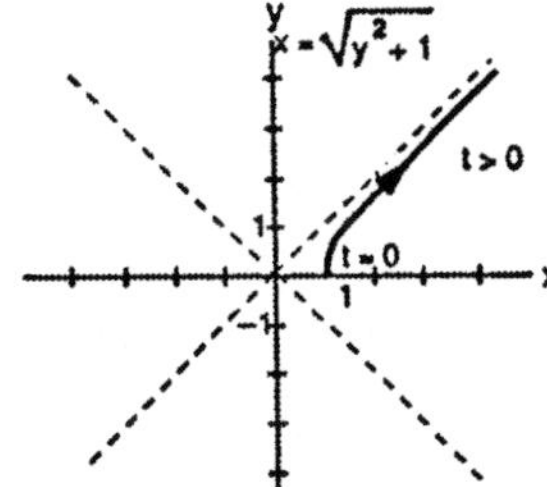

91. $x = \sec^2 t - 1,\ y = \tan t,\ -\frac{\pi}{2} < t < \frac{\pi}{2}$
$\Rightarrow \sec^2 t - 1 = \tan^2 t \Rightarrow x = y^2$

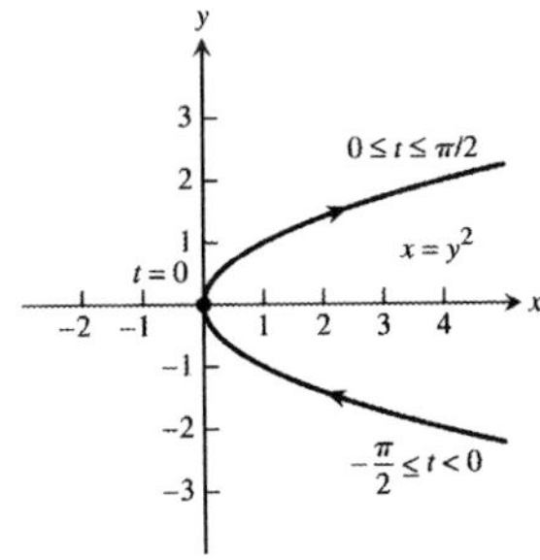

92. $x = -\sec t,\ y = \tan t,\ -\frac{\pi}{2} < t < \frac{\pi}{2}$
$\Rightarrow \sec^2 t - \tan^2 t = 1 \Rightarrow x^2 - y^2 = 1$

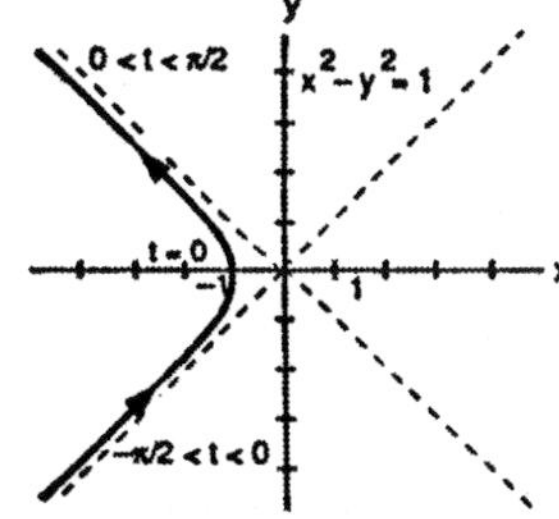

93. (a) $x = a\cos t,\ y = -a\sin t,\ 0 \le t \le 2\pi$
(b) $x = a\cos t,\ y = a\sin t,\ 0 \le t \le 2\pi$
(c) $x = a\cos t,\ y = -a\sin t,\ 0 \le t \le 4\pi$
(d) $x = a\cos t,\ y = a\sin t,\ 0 \le t \le 4\pi$

94. (a) $x = a\sin t,\ y = b\cos t,\ \frac{\pi}{2} \le t \le \frac{5\pi}{2}$
(b) $x = a\cos t,\ y = b\sin t,\ 0 \le t \le 2\pi$
(c) $x = a\sin t,\ y = b\cos t,\ \frac{\pi}{2} \le t \le \frac{9\pi}{2}$
(d) $x = a\cos t,\ y = b\sin t,\ 0 \le t \le 4\pi$

95. Using $(-1, -3)$ we create the parametric equations $x = -1 + at$ and $y = -3 + bt$, representing a line which goes through $(-1, -3)$ at $t = 0$. We determine a and b so that the line goes through $(4, 1)$ when $t = 1$.
Since $4 = -1 + a \Rightarrow a = 5$. Since $1 = -3 + b \Rightarrow b = 4$. Therefore, one possible parameterization is $x = -1 + 5t$, $y = -3 - 4t,\ 0 \le t \le 1$.

96. Using $(-1, 3)$ we create the parametric equations $x = -1 + at$ and $y = 3 + bt$, representing a line which goes through $(-1, 3)$ at $t = 0$. We determine a and b so that the line goes through $(3, -2)$ when $t = 1$. Since $3 = -1 + a \Rightarrow a = 4$. Since $-2 = 3 + b \Rightarrow b = -5$. Therefore, one possible parameterization is $x = -1 + 4t$, $y = -3 - 5t$, $0 \le t \le 1$.

97. The lower half of the parabola is given by $x = y^2 + 1$ for $y \le 0$. Substituting t for y, we obtain one possible parameterization $x = t^2 + 1$, $y = t$, $t \le 0$.

98. The vertex of the parabola is at $(-1, -1)$, so the left half of the parabola is given by $y = x^2 + 2x$ for $x \le -1$. Substituting t for x, we obtain one possible parametrization: $x = t$, $y = t^2 + 2t$, $t \le -1$.

99. For simplicity, we assume that x and y are linear functions of t and that the point(x, y) starts at $(2, 3)$ for $t = 0$ and passes through $(-1, -1)$ at $t = 1$. Then $x = f(t)$, where $f(0) = 2$ and $f(1) = -1$.
Since slope $= \frac{\Delta x}{\Delta t} = \frac{-1-2}{1-0} = -3$, $x = f(t) = -3t + 2 = 2 - 3t$. Also, $y = g(t)$, where $g(0) = 3$ and $g(1) = -1$.
Since slope $= \frac{\Delta y}{\Delta t} = \frac{-1-3}{1-0} = -4$. $y = g(t) = -4t + 3 = 3 - 4t$.
One possible parameterization is: $x = 2 - 3t$, $y = 3 - 4t$, $t \ge 0$.

100. For simplicity, we assume that x and y are linear functions of t and that the point(x, y) starts at $(-1, 2)$ for $t = 0$ and passes through $(0, 0)$ at $t = 1$. Then $x = f(t)$, where $f(0) = -1$ and $f(1) = 0$.
Since slope $= \frac{\Delta x}{\Delta t} = \frac{0-(-1)}{1-0} = 1$, $x = f(t) = 1t + (-1) = -1 + t$. Also, $y = g(t)$, where $g(0) = 2$ and $g(1) = 0$.
Since slope $= \frac{\Delta y}{\Delta t} = \frac{0-2}{1-0} = -2$. $y = g(t) = -2t + 2 = 2 - 2t$.
One possible parameterization is: $x = -1 + t$, $y = 2 - 2t$, $t \ge 0$.

101. $t = \frac{\pi}{4} \Rightarrow x = 2\cos\frac{\pi}{4} = \sqrt{2}$, $y = 2\sin\frac{\pi}{4} = \sqrt{2}$; $\frac{dx}{dt} = -2\sin t$, $\frac{dy}{dt} = 2\cos t \Rightarrow \frac{dy}{dx} = \frac{dy/dt}{dx/dt} = \frac{2\cos t}{-2\sin t} = -\cot t$
$\Rightarrow \left.\frac{dy}{dx}\right|_{t=\frac{\pi}{4}} = -\cot\frac{\pi}{4} = -1$; tangent line is $y - \sqrt{2} = -1\left(x - \sqrt{2}\right)$ or $y = -x + 2\sqrt{2}$; $\frac{dy'}{dt} = \csc^2 t$
$\Rightarrow \frac{d^2y}{dx^2} = \frac{dy'/dt}{dx/dt} = \frac{\csc^2 t}{-2\sin t} = -\frac{1}{2\sin^3 t} \Rightarrow \left.\frac{d^2y}{dx^2}\right|_{t=\frac{\pi}{4}} = -\sqrt{2}$

102. $t = \frac{2\pi}{3} \Rightarrow x = \cos\frac{2\pi}{3} = -\frac{1}{2}$, $y = \sqrt{3}\cos\frac{2\pi}{3} = -\frac{\sqrt{3}}{2}$; $\frac{dx}{dt} = -\sin t$, $\frac{dy}{dt} = -\sqrt{3}\sin t \Rightarrow \frac{dy}{dx} = \frac{-\sqrt{3}\sin t}{-\sin t} = \sqrt{3}$
$\Rightarrow \left.\frac{dy}{dx}\right|_{t=\frac{2\pi}{3}} = \sqrt{3}$; tangent line is $y - \left(-\frac{\sqrt{3}}{2}\right) = \sqrt{3}\left[x - \left(-\frac{1}{2}\right)\right]$ or $y = \sqrt{3}x$; $\frac{dy'}{dt} = 0 \Rightarrow \frac{d^2y}{dx^2} = \frac{0}{-\sin t} = 0$
$\Rightarrow \left.\frac{d^2y}{dx^2}\right|_{t=\frac{2\pi}{3}} = 0$

103. $t = \frac{1}{4} \Rightarrow x = \frac{1}{4}$, $y = \frac{1}{2}$; $\frac{dx}{dt} = 1$, $\frac{dy}{dt} = \frac{1}{2\sqrt{t}} \Rightarrow \frac{dy}{dx} = \frac{dy/dt}{dx/dt} = \frac{1}{2\sqrt{t}} \Rightarrow \left.\frac{dy}{dx}\right|_{t=\frac{1}{4}} = \frac{1}{2\sqrt{\frac{1}{4}}} = 1$; tangent line is
$y - \frac{1}{2} = 1 \cdot \left(x - \frac{1}{4}\right)$ or $y = x + \frac{1}{4}$; $\frac{dy'}{dt} = -\frac{1}{4}t^{-3/2} \Rightarrow \frac{d^2y}{dx^2} = \frac{dy'/dt}{dx/dt} = -\frac{1}{4}t^{-3/2} \Rightarrow \left.\frac{d^2y}{dx^2}\right|_{t=\frac{1}{4}} = -2$

104. $t = 3 \Rightarrow x = -\sqrt{3+1} = -2$, $y = \sqrt{3(3)} = 3$; $\frac{dx}{dt} = -\frac{1}{2}(t+1)^{-1/2}$, $\frac{dy}{dt} = \frac{3}{2}(3t)^{-1/2} \Rightarrow \frac{dy}{dx} = \frac{\left(\frac{3}{2}\right)(3t)^{-1/2}}{\left(-\frac{1}{2}\right)(t+1)^{-1/2}}$
$= -\frac{3\sqrt{t+1}}{\sqrt{3t}} = \left.\frac{dy}{dx}\right|_{t=3} = \frac{-3\sqrt{3+1}}{\sqrt{3(3)}} = -2$; tangent line is $y - 3 = -2[x - (-2)]$ or $y = -2x - 1$;
$\frac{dy'}{dt} = \frac{\sqrt{3t}\left[-\frac{3}{2}(t+1)^{-1/2}\right] + 3\sqrt{t+1}\left[\frac{3}{2}(3t)^{-1/2}\right]}{3t} = \frac{3}{2t\sqrt{3t}\sqrt{t+1}} \Rightarrow \frac{d^2y}{dx^2} = \frac{\left(\frac{3}{2t\sqrt{3t}\sqrt{t+1}}\right)}{\left(\frac{-1}{2\sqrt{t+1}}\right)} = -\frac{3}{t\sqrt{3t}}$
$\Rightarrow \left.\frac{d^2y}{dx^2}\right|_{t=3} = -\frac{1}{3}$

105. $t = -1 \Rightarrow x = 5, y = 1; \frac{dx}{dt} = 4t, \frac{dy}{dt} = 4t^3 \Rightarrow \frac{dy}{dx} = \frac{dy/dt}{dx/dt} = \frac{4t^3}{4t} = t^2 \Rightarrow \left.\frac{dy}{dx}\right|_{t=-1} = (-1)^2 = 1$; tangent line is $y - 1 = 1 \cdot (x - 5)$ or $y = x - 4$; $\frac{dy'}{dt} = 2t \Rightarrow \frac{d^2y}{dx^2} = \frac{dy'/dt}{dx/dt} = \frac{2t}{4t} = \frac{1}{2} \Rightarrow \left.\frac{d^2y}{dx^2}\right|_{t=-1} = \frac{1}{2}$

106. $t = \frac{\pi}{3} \Rightarrow x = \frac{\pi}{3} - \sin\frac{\pi}{3} = \frac{\pi}{3} - \frac{\sqrt{3}}{2}, y = 1 - \cos\frac{\pi}{3} = 1 - \frac{1}{2} = \frac{1}{2}; \frac{dx}{dt} = 1 - \cos t, \frac{dy}{dt} = \sin t \Rightarrow \frac{dy}{dx} = \frac{dy/dt}{dx/dt}$ $= \frac{\sin t}{1-\cos t} \Rightarrow \left.\frac{dy}{dx}\right|_{t=\frac{\pi}{3}} = \frac{\sin\left(\frac{\pi}{3}\right)}{1-\cos\left(\frac{\pi}{3}\right)} = \frac{\left(\frac{\sqrt{3}}{2}\right)}{\left(\frac{1}{2}\right)} = \sqrt{3}$; tangent line is $y - \frac{1}{2} = \sqrt{3}\left(x - \frac{\pi}{3} + \frac{\sqrt{3}}{2}\right)$ $\Rightarrow y = \sqrt{3}x - \frac{\pi\sqrt{3}}{3} + 2; \frac{dy'}{dt} = \frac{(1-\cos t)(\cos t) - (\sin t)(\sin t)}{(1-\cos t)^2} = \frac{-1}{1-\cos t} \Rightarrow \frac{d^2y}{dx^2} = \frac{dy'/dt}{dx/dt} = \frac{\left(\frac{-1}{1-\cos t}\right)}{1-\cos t}$ $= \frac{-1}{(1-\cos t)^2} \Rightarrow \left.\frac{d^2y}{dx^2}\right|_{t=\frac{\pi}{3}} = -4$

107. $t = \frac{\pi}{2} \Rightarrow x = \cos\frac{\pi}{2} = 0, y = 1 + \sin\frac{\pi}{2} = 2; \frac{dx}{dt} = -\sin t, \frac{dy}{dt} = \cos t \Rightarrow \frac{dy}{dx} = \frac{\cos t}{-\sin t} = -\cot t$ $\Rightarrow \left.\frac{dy}{dx}\right|_{t=\frac{\pi}{2}} = -\cot\frac{\pi}{2} = 0$; tangent line is $y = 2$; $\frac{dy'}{dt} = \csc^2 t \Rightarrow \frac{d^2y}{dx^2} = \frac{\csc^2 t}{-\sin t} = -\csc^3 t \Rightarrow \left.\frac{d^2y}{dx^2}\right|_{t=\frac{\pi}{2}} = -1$

108. $t = -\frac{\pi}{4} \Rightarrow x = \sec^2\left(-\frac{\pi}{4}\right) - 1 = 1, y = \tan\left(-\frac{\pi}{4}\right) = -1; \frac{dx}{dt} = 2\sec^2 t \tan t, \frac{dy}{dt} = \sec^2 t$ $\Rightarrow \frac{dy}{dx} = \frac{\sec^2 t}{2\sec^2 t \tan t} = \frac{1}{2\tan t} = \frac{1}{2}\cot t \Rightarrow \left.\frac{dy}{dx}\right|_{t=-\frac{\pi}{4}} = \frac{1}{2}\cot\left(-\frac{\pi}{4}\right) = -\frac{1}{2}$; tangent line is $y - (-1) = -\frac{1}{2}(x - 1)$ or $y = -\frac{1}{2}x - \frac{1}{2}$; $\frac{dy'}{dt} = -\frac{1}{2}\csc^2 t \Rightarrow \frac{d^2y}{dx^2} = \frac{-\frac{1}{2}\csc^2 t}{2\sec^2 t \tan t} = -\frac{1}{4}\cot^3 t$ $\Rightarrow \left.\frac{d^2y}{dx^2}\right|_{t=-\frac{\pi}{4}} = \frac{1}{4}$

109. $s = A\cos(2\pi bt) \Rightarrow v = \frac{ds}{dt} = -A\sin(2\pi bt)(2\pi b) = -2\pi bA\sin(2\pi bt)$. If we replace b with 2b to double the frequency, the velocity formula gives $v = -4\pi bA\sin(4\pi bt) \Rightarrow$ doubling the frequency causes the velocity to double. Also $v = -2\pi bA\sin(2\pi bt) \Rightarrow a = \frac{dv}{dt} = -4\pi^2b^2A\cos(2\pi bt)$. If we replace b with 2b in the acceleration formula, we get $a = -16\pi^2b^2A\cos(4\pi bt) \Rightarrow$ doubling the frequency causes the acceleration to quadruple. Finally, $a = -4\pi^2b^2A\cos(2\pi bt) \Rightarrow j = \frac{da}{dt} = 8\pi^3b^3A\sin(2\pi bt)$. If we replace b with 2b in the jerk formula, we get $j = 64\pi^3b^3A\sin(4\pi bt) \Rightarrow$ doubling the frequency multiplies the jerk by a factor of 8.

110. (a) $y = 37\sin\left[\frac{2\pi}{365}(x - 101)\right] + 25 \Rightarrow y' = 37\cos\left[\frac{2\pi}{365}(x - 101)\right]\left(\frac{2\pi}{365}\right) = \frac{74\pi}{365}\cos\left[\frac{2\pi}{365}(x - 101)\right]$.
The temperature is increasing the fastest when y' is as large as possible. The largest value of $\cos\left[\frac{2\pi}{365}(x - 101)\right]$ is 1 and occurs when $\frac{2\pi}{365}(x - 101) = 0 \Rightarrow x = 101 \Rightarrow$ on day 101 of the year ($\sim$ April 11), the temperature is increasing the fastest.
(b) $y'(101) = \frac{74\pi}{365}\cos\left[\frac{2\pi}{365}(101 - 101)\right] = \frac{74\pi}{365}\cos(0) = \frac{74\pi}{365} \approx 0.64$ °F/day

111. $s = (1 + 4t)^{1/2} \Rightarrow v = \frac{ds}{dt} = \frac{1}{2}(1 + 4t)^{-1/2}(4) = 2(1 + 4t)^{-1/2} \Rightarrow v(6) = 2(1 + 4 \cdot 6)^{-1/2} = \frac{2}{5}$ m/sec; $v = 2(1 + 4t)^{-1/2} \Rightarrow a = \frac{dv}{dt} = -\frac{1}{2} \cdot 2(1 + 4t)^{-3/2}(4) = -4(1 + 4t)^{-3/2} \Rightarrow a(6) = -4(1 + 4 \cdot 6)^{-3/2} = -\frac{4}{125}$ m/sec^2

112. We need to show $a = \frac{dv}{dt}$ is constant: $a = \frac{dv}{dt} = \frac{dv}{ds} \cdot \frac{ds}{dt}$ and $\frac{dv}{ds} = \frac{d}{ds}\left(k\sqrt{s}\right) = \frac{k}{2\sqrt{s}} \Rightarrow a = \frac{dv}{ds} \cdot \frac{ds}{dt} = \frac{dv}{ds} \cdot v$ $= \frac{k}{2\sqrt{s}} \cdot k\sqrt{s} = \frac{k^2}{2}$ which is a constant.

113. v proportional to $\frac{1}{\sqrt{s}} \Rightarrow v = \frac{k}{\sqrt{s}}$ for some constant k $\Rightarrow \frac{dv}{ds} = -\frac{k}{2s^{3/2}}$. Thus, $a = \frac{dv}{dt} = \frac{dv}{ds} \cdot \frac{ds}{dt} = \frac{dv}{ds} \cdot v$ $= -\frac{k}{2s^{3/2}} \cdot \frac{k}{\sqrt{s}} = -\frac{k^2}{2}\left(\frac{1}{s^2}\right) \Rightarrow$ acceleration is a constant times $\frac{1}{s^2}$ so a is inversely proportional to s^2.

114. Let $\frac{dx}{dt} = f(x)$. Then, $a = \frac{dv}{dt} = \frac{dv}{dx} \cdot \frac{dx}{dt} = \frac{dv}{dx} \cdot f(x) = \frac{d}{dx}\left(\frac{dx}{dt}\right) \cdot f(x) = \frac{d}{dx}(f(x)) \cdot f(x) = f'(x)f(x)$, as required.

115. $T = 2\pi\sqrt{\frac{L}{g}} \Rightarrow \frac{dT}{dL} = 2\pi \cdot \frac{1}{2\sqrt{\frac{L}{g}}} \cdot \frac{1}{g} = \frac{\pi}{g\sqrt{\frac{L}{g}}} = \frac{\pi}{\sqrt{gL}}$. Therefore, $\frac{dT}{du} = \frac{dT}{dL} \cdot \frac{dL}{du} = \frac{\pi}{\sqrt{gL}} \cdot kL = \frac{\pi k\sqrt{L}}{\sqrt{g}} = \frac{1}{2} \cdot 2\pi k\sqrt{\frac{L}{g}}$
$= \frac{kT}{2}$, as required.

116. No. The chain rule says that when g is differentiable at 0 and f is differentiable at g(0), then $f \circ g$ is differentiable at 0. But the chain rule says nothing about what happens when g is not differentiable at 0 so there is no contradiction.

117. The graph of $y = (f \circ g)(x)$ has a horizontal tangent at $x = 1$ provided that $(f \circ g)'(1) = 0 \Rightarrow f'(g(1))g'(1) = 0$ $\Rightarrow$ either $f'(g(1)) = 0$ or $g'(1) = 0$ (or both) $\Rightarrow$ either the graph of f has a horizontal tangent at $u = g(1)$, or the graph of g has a horizontal tangent at $x = 1$ (or both).

118. $(f \circ g)'(-5) < 0 \Rightarrow f'(g(-5)) \cdot g'(-5) < 0 \Rightarrow f'(g(-5))$ and $g'(-5)$ are both nonzero and have opposite signs. That is, either $[f'(g(-5)) > 0$ and $g'(-5) < 0]$ or $[f'(g(-5)) < 0$ and $g'(-5) > 0]$.

119. As $h \to 0$, the graph of $y = \frac{\sin 2(x+h) - \sin 2x}{h}$ approaches the graph of $y = 2\cos 2x$ because $\lim_{h \to 0} \frac{\sin 2(x+h) - \sin 2x}{h} = \frac{d}{dx}(\sin 2x) = 2\cos 2x$.

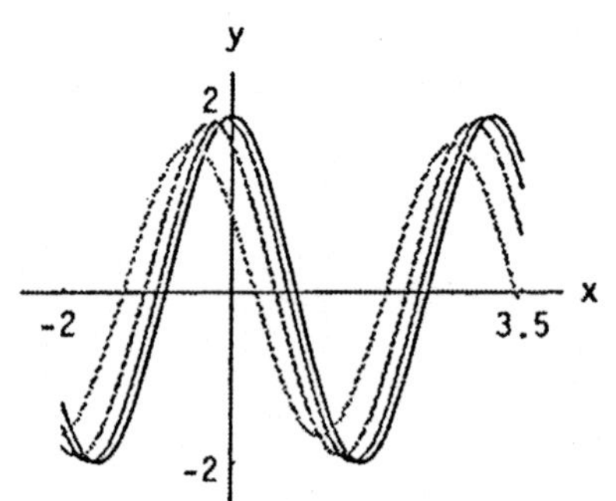

120. As $h \to 0$, the graph of $y = \frac{\cos[(x+h)^2] - \cos(x^2)}{h}$ approaches the graph of $y = -2x\sin(x^2)$ because $\lim_{h \to 0} \frac{\cos[(x+h)^2] - \cos(x^2)}{h} = \frac{d}{dx}[\cos(x^2)] = -2x\sin(x^2)$.

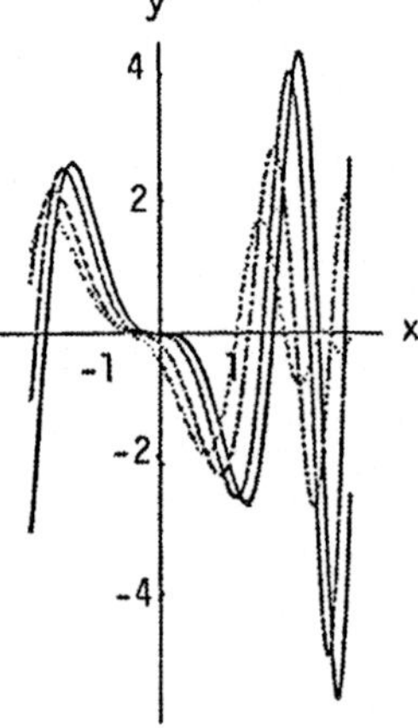

121. $\frac{dx}{dt} = \cos t$ and $\frac{dy}{dt} = 2\cos 2t \Rightarrow \frac{dy}{dx} = \frac{dy/dt}{dx/dt} = \frac{2\cos 2t}{\cos t} = \frac{2(2\cos^2 t - 1)}{\cos t}$; then $\frac{dy}{dx} = 0 \Rightarrow \frac{2(2\cos^2 t - 1)}{\cos t} = 0$
$\Rightarrow 2\cos^2 t - 1 = 0 \Rightarrow \cos t = \pm\frac{1}{\sqrt{2}} \Rightarrow t = \frac{\pi}{4}, \frac{3\pi}{4}, \frac{5\pi}{4}, \frac{7\pi}{4}$. In the 1st quadrant: $t = \frac{\pi}{4} \Rightarrow x = \sin\frac{\pi}{4} = \frac{\sqrt{2}}{2}$ and $y = \sin 2\left(\frac{\pi}{4}\right) = 1 \Rightarrow \left(\frac{\sqrt{2}}{2}, 1\right)$ is the point where the tangent line is horizontal. At the origin: $x = 0$ and $y = 0$ $\Rightarrow \sin t = 0 \Rightarrow t = 0$ or $t = \pi$ and $\sin 2t = 0 \Rightarrow t = 0, \frac{\pi}{2}, \pi, \frac{3\pi}{2}$; thus $t = 0$ and $t = \pi$ give the tangent lines at the origin. Tangents at origin: $\left.\frac{dy}{dx}\right|_{t=0} = 2 \Rightarrow y = 2x$ and $\left.\frac{dy}{dx}\right|_{t=\pi} = -2 \Rightarrow y = -2x$

122. $\frac{dx}{dt} = 2\cos 2t$ and $\frac{dy}{dt} = 3\cos 3t \Rightarrow \frac{dy}{dx} = \frac{dy/dt}{dx/dt} = \frac{3\cos 3t}{2\cos 2t} = \frac{3(\cos 2t\cos t - \sin 2t\sin t)}{2(2\cos^2 t - 1)}$
$= \frac{3[(2\cos^2 t - 1)(\cos t) - 2\sin t\cos t\sin t]}{2(2\cos^2 t - 1)} = \frac{(3\cos t)(2\cos^2 t - 1 - 2\sin^2 t)}{2(2\cos^2 t - 1)} = \frac{(3\cos t)(4\cos^2 t - 3)}{2(2\cos^2 t - 1)}$; then

$\frac{dy}{dx} = 0 \Rightarrow \frac{(3\cos t)(4\cos^2 t - 3)}{2(2\cos^2 t - 1)} = 0 \Rightarrow 3\cos t = 0$ or $4\cos^2 t - 3 = 0$: $3\cos t = 0 \Rightarrow t = \frac{\pi}{2}, \frac{3\pi}{2}$ and $4\cos^2 t - 3 = 0 \Rightarrow \cos t = \pm\frac{\sqrt{3}}{2} \Rightarrow t = \frac{\pi}{6}, \frac{5\pi}{6}, \frac{7\pi}{6}, \frac{11\pi}{6}$. In the 1st quadrant: $t = \frac{\pi}{6} \Rightarrow x = \sin 2\left(\frac{\pi}{6}\right) = \frac{\sqrt{3}}{2}$ and $y = \sin 3\left(\frac{\pi}{6}\right) = 1 \Rightarrow \left(\frac{\sqrt{3}}{2}, 1\right)$ is the point where the graph has a horizontal tangent. At the origin: $x = 0$ and $y = 0 \Rightarrow \sin 2t = 0$ and $\sin 3t = 0 \Rightarrow t = 0, \frac{\pi}{2}, \pi, \frac{3\pi}{2}$ and $t = 0, \frac{\pi}{3}, \frac{2\pi}{3}, \pi, \frac{4\pi}{3}, \frac{5\pi}{3} \Rightarrow t = 0$ and $t = \pi$ give the tangent lines at the origin. Tangents at the origin: $\left.\frac{dy}{dx}\right|_{t=0} = \frac{3\cos 0}{2\cos 0} = \frac{3}{2} \Rightarrow y = \frac{3}{2}x$, and $\left.\frac{dy}{dx}\right|_{t=\pi} = \frac{3\cos(3\pi)}{2\cos(2\pi)} = -\frac{3}{2} \Rightarrow y = -\frac{3}{2}x$

123. From the power rule, with $y = x^{1/4}$, we get $\frac{dy}{dx} = \frac{1}{4}x^{-3/4}$. From the chain rule, $y = \sqrt{\sqrt{x}}$ $\Rightarrow \frac{dy}{dx} = \frac{1}{2\sqrt{\sqrt{x}}} \cdot \frac{d}{dx}\left(\sqrt{x}\right) = \frac{1}{2\sqrt{\sqrt{x}}} \cdot \frac{1}{2\sqrt{x}} = \frac{1}{4}x^{-3/4}$, in agreement.

124. From the power rule, with $y = x^{3/4}$, we get $\frac{dy}{dx} = \frac{3}{4}x^{-1/4}$. From the chain rule, $y = \sqrt{x\sqrt{x}}$ $\Rightarrow \frac{dy}{dx} = \frac{1}{2\sqrt{x\sqrt{x}}} \cdot \frac{d}{dx}\left(x\sqrt{x}\right) \Rightarrow \frac{dy}{dx} = \frac{1}{2\sqrt{x\sqrt{x}}} \cdot \left(x \cdot \frac{1}{2\sqrt{x}} + \sqrt{x}\right) = \frac{1}{2\sqrt{x\sqrt{x}}} \cdot \left(\frac{3}{2}\sqrt{x}\right) = \frac{3\sqrt{x}}{4\sqrt{x\sqrt{x}}}$ $= \frac{3\sqrt{x}}{4\sqrt{x}\sqrt{\sqrt{x}}} = \frac{3}{4}x^{-1/4}$, in agreement.

125. (a)

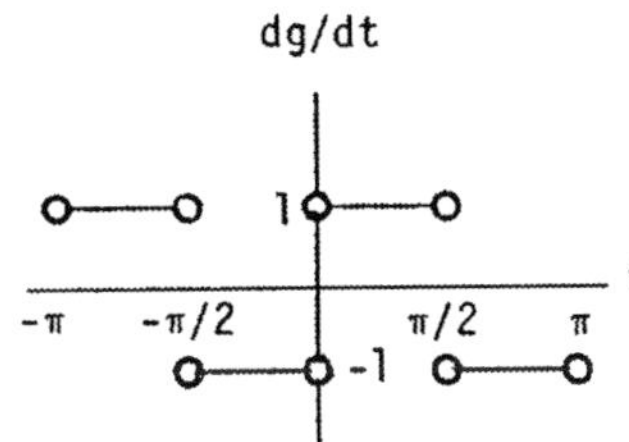

(b) $\frac{df}{dt} = 1.27324\sin 2t + 0.42444\sin 6t + 0.2546\sin 10t + 0.18186\sin 14t$

(c) The curve of $y = \frac{df}{dt}$ approximates $y = \frac{dg}{dt}$ the best when t is not $-\pi, -\frac{\pi}{2}, 0, \frac{\pi}{2}$, nor π.

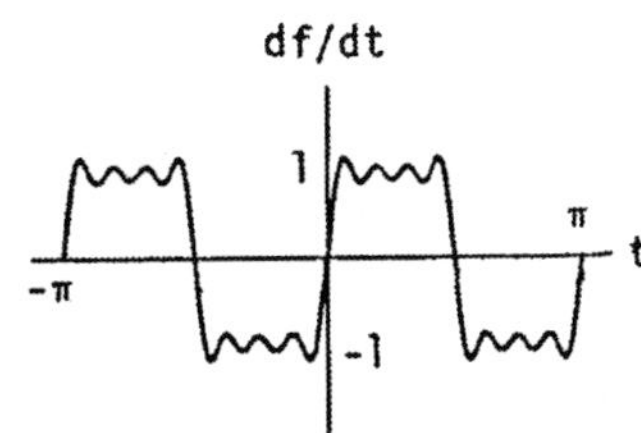

126. (a)

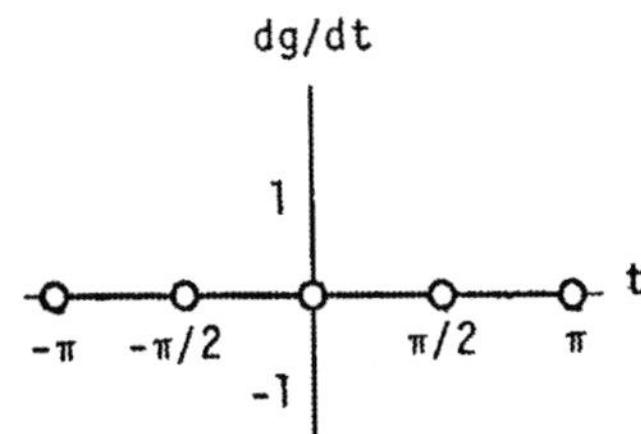

(b) $\frac{dh}{dt} = 2.5464\cos(2t) + 2.5464\cos(6t) + 2.5465\cos(10t) + 2.54646\cos(14t) + 2.54646\cos(18t)$

(c)

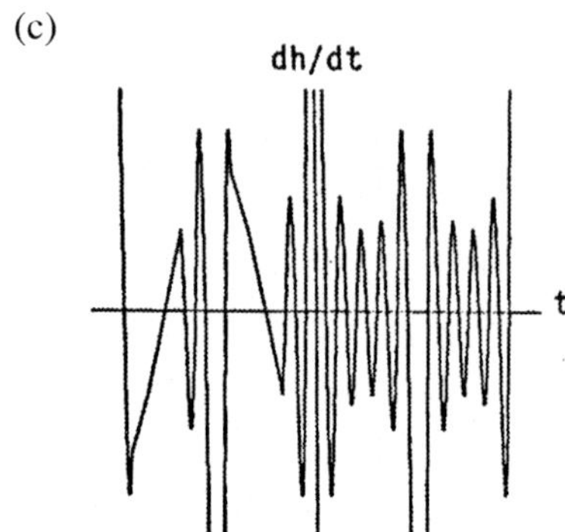

125-130. Example CAS commands:

Maple:

```
f := t -> 0.78540 - 0.63662*cos(2*t) - 0.07074*cos(6*t)
    - 0.02546*cos(10*t) - 0.01299*cos(14*t);
g := t -> piecewise( t<-Pi/2, t+Pi, t<0, -t, t<Pi/2, t, Pi-t );
plot( [f(t),g(t)], t=-Pi..Pi );
Df := D(f);
Dg := D(g);
plot( [Df(t),Dg(t)], t=-Pi..Pi );
```

Mathematica: (functions, domains, and value for t0 may change):

To see the relationship between f[t] and f'[t] in 111 and h[t] in 112

```
Clear[t, f]
f[t_] = 0.78540 – 0.63662 Cos[2t] – 0.07074 Cos[6t] – 0.02546 Cos[10t] – 0.01299 Cos[14t]
f'[t]
Plot[{f[t], f'[t]},{t, –π, π}]
```

For the parametric equations in 113 - 116, do the following. Do NOT use the colon when defining tanline.

```
Clear[x, y, t]
t0 = p/4;
x[t_]:=1–Cos[t]
y[t_]:=1 + Sin[t]
p1=ParametricPlot[{x[t], y[t]},{t, –π, π}]
yp[t_]:=y'[t]/x'[t]
ypp[t_]:=yp'[t]/x'[t]
yp[t0]//N
ypp[t0]//N
tanline[x_]=y[t0] + yp[t0] (x – x[t0])
p2=Plot[tanline[x], {x, 0, 1}]
Show[p1, p2]
```

3.6 IMPLICIT DIFFERENTIATION

1. $y = x^{9/4} \Rightarrow \frac{dy}{dx} = \frac{9}{4}x^{5/4}$

2. $y = x^{-3/5} \Rightarrow \frac{dy}{dx} = -\frac{3}{5}x^{-8/5}$

3. $y = \sqrt[3]{2x} = (2x)^{1/3} \Rightarrow \frac{dy}{dx} = \frac{1}{3}(2x)^{-2/3} \cdot 2 = \frac{2^{1/3}}{3x^{2/3}}$

4. $y = \sqrt[4]{5x} = (5x)^{1/4} \Rightarrow \frac{dy}{dx} = \frac{1}{4}(5x)^{-3/4} \cdot 5 = \frac{5^{1/4}}{4x^{3/4}}$

5. $y = 7\sqrt{x+6} = 7(x+6)^{1/2} \Rightarrow \frac{dy}{dx} = \frac{7}{2}(x+6)^{-1/2} = \frac{7}{2\sqrt{x+6}}$

6. $y = -2\sqrt{x-1} = -2(x-1)^{1/2} \Rightarrow \frac{dy}{dx} = -1(x-1)^{-1/2} = -\frac{1}{\sqrt{x-1}}$

7. $y = (2x+5)^{-1/2} \Rightarrow \frac{dy}{dx} = -\frac{1}{2}(2x+5)^{-3/2} \cdot 2 = -(2x+5)^{-3/2}$

8. $y = (1-6x)^{2/3} \Rightarrow \frac{dy}{dx} = \frac{2}{3}(1-6x)^{-1/3}(-6) = -4(1-6x)^{-1/3}$

9. $y = x\left(x^2+1\right)^{1/2} \Rightarrow y' = x \cdot \frac{1}{2}\left(x^2+1\right)^{-1/2}(2x) + \left(x^2+1\right)^{1/2} \cdot 1 = \left(x^2+1\right)^{-1/2}\left(x^2+x^2+1\right) = \frac{2x^2+1}{\sqrt{x^2+1}}$

10. $y = x\left(x^2+1\right)^{-1/2} \Rightarrow y' = x \cdot \left(-\frac{1}{2}\right)\left(x^2+1\right)^{-3/2}(2x) + \left(x^2+1\right)^{-1/2} \cdot 1 = \left(x^2+1\right)^{-3/2}\left(-x^2+x^2+1\right) = \frac{1}{\left(x^2+1\right)^{3/2}}$

11. $s = \sqrt[7]{t^2} = t^{2/7} \Rightarrow \frac{ds}{dt} = \frac{2}{7}t^{-5/7}$

12. $r = \sqrt[4]{\theta^{-3}} = \theta^{-3/4} \Rightarrow \frac{dr}{d\theta} = -\frac{3}{4}\theta^{-7/4}$

13. $y = \sin\left((2t+5)^{-2/3}\right) \Rightarrow \frac{dy}{dt} = \cos\left((2t+5)^{-2/3}\right) \cdot \left(-\frac{2}{3}\right)(2t+5)^{-5/3} \cdot 2 = -\frac{4}{3}(2t+5)^{-5/3}\cos\left((2t+5)^{-2/3}\right)$

14. $z = \cos\left((1-6t)^{2/3}\right) \Rightarrow \frac{dz}{dt} = -\sin\left((1-6t)^{2/3}\right) \cdot \frac{2}{3}(1-6t)^{-1/3}(-6) = 4(1-6t)^{-1/3}\sin\left((1-6t)^{2/3}\right)$

15. $f(x) = \sqrt{1-\sqrt{x}} = \left(1-x^{1/2}\right)^{1/2} \Rightarrow f'(x) = \frac{1}{2}\left(1-x^{1/2}\right)^{-1/2}\left(-\frac{1}{2}x^{-1/2}\right) = \frac{-1}{4\left(\sqrt{1-\sqrt{x}}\right)\sqrt{x}} = \frac{-1}{4\sqrt{x\left(1-\sqrt{x}\right)}}$

16. $g(x) = 2\left(2x^{-1/2}+1\right)^{-1/3} \Rightarrow g'(x) = -\frac{2}{3}\left(2x^{-1/2}+1\right)^{-4/3} \cdot (-1)x^{-3/2} = \frac{2}{3}\left(2x^{-1/2}+1\right)^{-4/3}x^{-3/2}$

17. $h(\theta) = \sqrt[3]{1+\cos(2\theta)} = (1+\cos 2\theta)^{1/3} \Rightarrow h'(\theta) = \frac{1}{3}(1+\cos 2\theta)^{-2/3} \cdot (-\sin 2\theta) \cdot 2 = -\frac{2}{3}(\sin 2\theta)(1+\cos 2\theta)^{-2/3}$

18. $k(\theta) = (\sin(\theta+5))^{5/4} \Rightarrow k'(\theta) = \frac{5}{4}(\sin(\theta+5))^{1/4} \cdot \cos(\theta+5) = \frac{5}{4}\cos(\theta+5)(\sin(\theta+5))^{1/4}$

19. $x^2y + xy^2 = 6$:

Step 1: $\left(x^2\frac{dy}{dx} + y \cdot 2x\right) + \left(x \cdot 2y\frac{dy}{dx} + y^2 \cdot 1\right) = 0$

Step 2: $x^2\frac{dy}{dx} + 2xy\frac{dy}{dx} = -2xy - y^2$

Step 3: $\frac{dy}{dx}\left(x^2+2xy\right) = -2xy - y^2$

Step 4: $\frac{dy}{dx} = \frac{-2xy-y^2}{x^2+2xy}$

20. $x^3 + y^3 = 18xy \Rightarrow 3x^2 + 3y^2\frac{dy}{dx} = 18y + 18x\frac{dy}{dx} \Rightarrow \left(3y^2-18x\right)\frac{dy}{dx} = 18y - 3x^2 \Rightarrow \frac{dy}{dx} = \frac{6y-x^2}{y^2-6x}$

21. $2xy + y^2 = x + y$:

Step 1: $\left(2x\frac{dy}{dx} + 2y\right) + 2y\frac{dy}{dx} = 1 + \frac{dy}{dx}$

Step 2: $2x\frac{dy}{dx} + 2y\frac{dy}{dx} - \frac{dy}{dx} = 1 - 2y$

Step 3: $\frac{dy}{dx}(2x+2y-1) = 1 - 2y$

Step 4: $\frac{dy}{dx} = \frac{1-2y}{2x+2y-1}$

22. $x^3 - xy + y^3 = 1 \Rightarrow 3x^2 - y - x\frac{dy}{dx} + 3y^2\frac{dy}{dx} = 0 \Rightarrow \left(3y^2-x\right)\frac{dy}{dx} = y - 3x^2 \Rightarrow \frac{dy}{dx} = \frac{y-3x^2}{3y^2-x}$

23. $x^2(x-y)^2 = x^2 - y^2$:

Step 1: $x^2\left[2(x-y)\left(1-\frac{dy}{dx}\right)\right] + (x-y)^2(2x) = 2x - 2y\frac{dy}{dx}$

Step 2: $-2x^2(x-y)\frac{dy}{dx} + 2y\frac{dy}{dx} = 2x - 2x^2(x-y) - 2x(x-y)^2$

Step 3: $\frac{dy}{dx}\left[-2x^2(x-y) + 2y\right] = 2x\left[1 - x(x-y) - (x-y)^2\right]$

Step 4: $\frac{dy}{dx} = \frac{2x\left[1-x(x-y)-(x-y)^2\right]}{-2x^2(x-y)+2y} = \frac{x\left[1-x(x-y)-(x-y)^2\right]}{y-x^2(x-y)} = \frac{x\left(1-x^2+xy-x^2+2xy-y^2\right)}{x^2y-x^3+y}$

$= \frac{x-2x^3+3x^2y-xy^2}{x^2y-x^3+y}$

24. $(3xy+7)^2 = 6y \Rightarrow 2(3xy+7)\cdot\left(3x\frac{dy}{dx}+3y\right) = 6\frac{dy}{dx} \Rightarrow 2(3xy+7)(3x)\frac{dy}{dx} - 6\frac{dy}{dx} = -6y(3xy+7)$
$\Rightarrow \frac{dy}{dx}\left[6x(3xy+7)-6\right] = -6y(3xy+7) \Rightarrow \frac{dy}{dx} = -\frac{y(3xy+7)}{x(3xy+7)-1} = \frac{3xy^2+7y}{1-3x^2y-7x}$

25. $y^2 = \frac{x-1}{x+1} \Rightarrow 2y\frac{dy}{dx} = \frac{(x+1)-(x-1)}{(x+1)^2} = \frac{2}{(x+1)^2} \Rightarrow \frac{dy}{dx} = \frac{1}{y(x+1)^2}$

26. $x^2 = \frac{x-y}{x+y} \Rightarrow x^3 + x^2y = x - y \Rightarrow 3x^2 + 2xy + x^2y' = 1 - y' \Rightarrow (x^2+1)\,y' = 1 - 3x^2 - 2xy \Rightarrow y' = \frac{1-3x^2-2xy}{x^2+1}$

27. $x = \tan y \Rightarrow 1 = (\sec^2 y)\frac{dy}{dx} \Rightarrow \frac{dy}{dx} = \frac{1}{\sec^2 y} = \cos^2 y$

28. $xy = \cot(xy) \Rightarrow x\frac{dy}{dx} + y = -\csc^2(xy)\left(x\frac{dy}{dx}+y\right) \Rightarrow x\frac{dy}{dx} + x\csc^2(xy)\frac{dy}{dx} = -y\csc^2(xy) - y$
$\Rightarrow \frac{dy}{dx}\left[x + x\csc^2(xy)\right] = -y\left[\csc^2(xy)+1\right] \Rightarrow \frac{dy}{dx} = \frac{-y\left[\csc^2(xy)+1\right]}{x\left[1+\csc^2(xy)\right]} = -\frac{y}{x}$

29. $e^{2x} = \sin(x+3y) \Rightarrow 2e^{2x} = (1+3y')\cos(x+3y) \Rightarrow 1+3y' = \frac{2e^{2x}}{\cos(x+3y)} \Rightarrow 3y' = \frac{2e^{2x}}{\cos(x+3y)} - 1$
$\Rightarrow y' = \frac{2e^{2x}-\cos(x+3y)}{3\cos(x+3y)}$

30. $x + \sin y = xy \Rightarrow 1 + (\cos y)\frac{dy}{dx} = y + x\frac{dy}{dx} \Rightarrow (\cos y - x)\frac{dy}{dx} = y - 1 \Rightarrow \frac{dy}{dx} = \frac{y-1}{\cos y - x}$

31. $y\sin\left(\frac{1}{y}\right) = 1 - xy \Rightarrow y\left[\cos\left(\frac{1}{y}\right)\cdot(-1)\frac{1}{y^2}\cdot\frac{dy}{dx}\right] + \sin\left(\frac{1}{y}\right)\cdot\frac{dy}{dx} = -x\frac{dy}{dx} - y \Rightarrow$
$\frac{dy}{dx}\left[-\frac{1}{y}\cos\left(\frac{1}{y}\right) + \sin\left(\frac{1}{y}\right) + x\right] = -y \Rightarrow \frac{dy}{dx} = \frac{-y}{-\frac{1}{y}\cos\left(\frac{1}{y}\right)+\sin\left(\frac{1}{y}\right)+x} = \frac{-y^2}{y\sin\left(\frac{1}{y}\right)-\cos\left(\frac{1}{y}\right)+xy}$

32. $e^{x^2y} = 2x + 2y \Rightarrow e^{x^2y}(x^2y' + 2xy) = 2 + 2y' \Rightarrow x^2e^{x^2y}y' + 2xye^{x^2y} = 2 + 2y' \Rightarrow x^2e^{x^2y}y' - 2y' = 2 - 2xye^{x^2y}$
$\Rightarrow y' = \frac{2-2xye^{x^2y}}{x^2e^{x^2y}-2}$

33. $\theta^{1/2} + r^{1/2} = 1 \Rightarrow \frac{1}{2}\theta^{-1/2} + \frac{1}{2}r^{-1/2}\cdot\frac{dr}{d\theta} = 0 \Rightarrow \frac{dr}{d\theta}\left[\frac{1}{2\sqrt{r}}\right] = \frac{-1}{2\sqrt{\theta}} \Rightarrow \frac{dr}{d\theta} = -\frac{2\sqrt{r}}{2\sqrt{\theta}} = -\frac{\sqrt{r}}{\sqrt{\theta}}$

34. $r - 2\sqrt{\theta} = \frac{3}{2}\theta^{2/3} + \frac{4}{3}\theta^{3/4} \Rightarrow \frac{dr}{d\theta} - \theta^{-1/2} = \theta^{-1/3} + \theta^{-1/4} \Rightarrow \frac{dr}{d\theta} = \theta^{-1/2} + \theta^{-1/3} + \theta^{-1/4}$

35. $\sin(r\theta) = \frac{1}{2} \Rightarrow [\cos(r\theta)]\left(r + \theta\frac{dr}{d\theta}\right) = 0 \Rightarrow \frac{dr}{d\theta}[\theta\cos(r\theta)] = -r\cos(r\theta) \Rightarrow \frac{dr}{d\theta} = \frac{-r\cos(r\theta)}{\theta\cos(r\theta)} = -\frac{r}{\theta}$,
$\cos(r\theta) \neq 0$

36. $\cos r + \cot\theta = e^{r\theta} \Rightarrow -\sin r\cdot\frac{dr}{d\theta} - \csc^2\theta = e^{r\theta}\left(r + \theta\frac{dr}{d\theta}\right) \Rightarrow -\sin r\cdot\frac{dr}{d\theta} - \csc^2\theta = re^{r\theta} + \theta e^{r\theta}\frac{dr}{d\theta}$
$\Rightarrow -\sin r\frac{dr}{d\theta} - \theta e^{r\theta}\frac{dr}{d\theta} = re^{r\theta} + \csc^2\theta \Rightarrow \frac{dr}{d\theta} = -\frac{re^{r\theta}+\csc^2\theta}{\theta e^{r\theta}+\sin r}$

37. $x^2 + y^2 = 1 \Rightarrow 2x + 2yy' = 0 \Rightarrow 2yy' = -2x \Rightarrow \frac{dy}{dx} = y' = -\frac{x}{y}$; now to find $\frac{d^2y}{dx^2}$, $\frac{d}{dx}(y') = \frac{d}{dx}\left(-\frac{x}{y}\right)$
$\Rightarrow y'' = \frac{y(-1)+xy'}{y^2} = \frac{-y+x\left(-\frac{x}{y}\right)}{y^2}$ since $y' = -\frac{x}{y} \Rightarrow \frac{d^2y}{dx^2} = y'' = \frac{-y^2-x^2}{y^3} = \frac{-y^2-(1-y^2)}{y^3} = \frac{-1}{y^3}$

38. $x^{2/3} + y^{2/3} = 1 \Rightarrow \frac{2}{3}x^{-1/3} + \frac{2}{3}y^{-1/3}\frac{dy}{dx} = 0 \Rightarrow \frac{dy}{dx}\left[\frac{2}{3}y^{-1/3}\right] = -\frac{2}{3}x^{-1/3} \Rightarrow y' = \frac{dy}{dx} = -\frac{x^{-1/3}}{y^{-1/3}} = -\left(\frac{y}{x}\right)^{1/3}$;

Differentiating again, $y'' = \frac{x^{1/3}\cdot\left(-\frac{1}{3}y^{-2/3}\right)y' + y^{1/3}\left(\frac{1}{3}x^{-2/3}\right)}{x^{2/3}} = \frac{x^{1/3}\cdot\left(-\frac{1}{3}y^{-2/3}\right)\left(-\frac{y^{1/3}}{x^{1/3}}\right) + y^{1/3}\left(\frac{1}{3}x^{-2/3}\right)}{x^{2/3}}$

$\Rightarrow \frac{d^2y}{dx^2} = \frac{1}{3}x^{-2/3}y^{-1/3} + \frac{1}{3}y^{1/3}x^{-4/3} = \frac{y^{1/3}}{3x^{4/3}} + \frac{1}{3y^{1/3}x^{2/3}}$

39. $y^2 = e^{x^2} + 2x \Rightarrow 2yy' = 2x e^{x^2} + 2 \Rightarrow \frac{dy}{dx} = \frac{x e^{x^2}+1}{y} \Rightarrow \frac{d^2y}{dx^2} = \frac{y\left(2x^2 e^{x^2} + e^{x^2}\right) - \left(x e^{x^2}+1\right)y'}{y^2}$

$= \frac{y\left(2x^2 e^{x^2} + e^{x^2}\right) - \left(x e^{x^2}+1\right)\cdot\frac{x e^{x^2}+1}{y}}{y^2} = \frac{y^2\left(2x^2 e^{x^2} + e^{x^2}\right) - \left(x^2 e^{2x^2} + 2x e^{x^2} + 1\right)}{y^3}$

$= \frac{(2x^2y^2 + y^2 - 2x)e^{x^2} - x^2 e^{2x^2} - 1}{y^3}$

40. $y^2 - 2x = 1 - 2y \Rightarrow 2y\cdot y' - 2 = -2y' \Rightarrow y'(2y+2) = 2 \Rightarrow y' = \frac{1}{y+1} = (y+1)^{-1}$; then $y'' = -(y+1)^{-2}\cdot y'$

$= -(y+1)^{-2}(y+1)^{-1} \Rightarrow \frac{d^2y}{dx^2} = y'' = \frac{-1}{(y+1)^3}$

41. $2\sqrt{y} = x - y \Rightarrow y^{-1/2}y' = 1 - y' \Rightarrow y'\left(y^{-1/2}+1\right) = 1 \Rightarrow \frac{dy}{dx} = y' = \frac{1}{y^{-1/2}+1} = \frac{\sqrt{y}}{\sqrt{y}+1}$; we can differentiate the equation $y'\left(y^{-1/2}+1\right) = 1$ again to find y'': $y'\left(-\frac{1}{2}y^{-3/2}y'\right) + \left(y^{-1/2}+1\right)y'' = 0$

$\Rightarrow \left(y^{-1/2}+1\right)y'' = \frac{1}{2}[y']^2 y^{-3/2} \Rightarrow \frac{d^2y}{dx^2} = y'' = \frac{\frac{1}{2}\left(\frac{1}{y^{-1/2}+1}\right)^2 y^{-3/2}}{\left(y^{-1/2}+1\right)} = \frac{1}{2y^{3/2}\left(y^{-1/2}+1\right)^3} = \frac{1}{2\left(1+\sqrt{y}\right)^3}$

42. $xy + y^2 = 1 \Rightarrow xy' + y + 2yy' = 0 \Rightarrow xy' + 2yy' = -y \Rightarrow y'(x+2y) = -y \Rightarrow y' = \frac{-y}{(x+2y)}$; $\frac{d^2y}{dx^2} = y''$

$= \frac{-(x+2y)y' + y(1+2y')}{(x+2y)^2} = \frac{-(x+2y)\left[\frac{-y}{(x+2y)}\right] + y\left[1 + 2\left(\frac{-y}{(x+2y)}\right)\right]}{(x+2y)^2} = \frac{\frac{1}{(x+2y)}\left[y(x+2y) + y(x+2y) - 2y^2\right]}{(x+2y)^2}$

$= \frac{2y(x+2y) - 2y^2}{(x+2y)^3} = \frac{2y^2 + 2xy}{(x+2y)^3} = \frac{2y(x+y)}{(x+2y)^3}$

43. $x^3 + y^3 = 16 \Rightarrow 3x^2 + 3y^2y' = 0 \Rightarrow 3y^2y' = -3x^2 \Rightarrow y' = -\frac{x^2}{y^2}$; we differentiate $y^2y' = -x^2$ to find y'':

$y^2y'' + y'[2y\cdot y'] = -2x \Rightarrow y^2y'' = -2x - 2y[y']^2 \Rightarrow y'' = \frac{-2x - 2y\left(-\frac{x^2}{y^2}\right)^2}{y^2} = \frac{-2x - \frac{2x^4}{y^3}}{y^2}$

$= \frac{-2xy^3 - 2x^4}{y^5} \Rightarrow \left.\frac{d^2y}{dx^2}\right|_{(2,2)} = \frac{-32-32}{32} = -2$

44. $xy + y^2 = 1 \Rightarrow xy' + y + 2yy' = 0 \Rightarrow y'(x+2y) = -y \Rightarrow y' = \frac{-y}{(x+2y)} \Rightarrow y'' = \frac{(x+2y)(-y') - (-y)(1+2y')}{(x+2y)^2}$;

since $y'|_{(0,-1)} = -\frac{1}{2}$ we obtain $y''|_{(0,-1)} = \frac{(-2)\left(\frac{1}{2}\right) - (1)(0)}{4} = -\frac{1}{4}$

45. $y^2 + x^2 = y^4 - 2x$ at $(-2,1)$ and $(-2,-1) \Rightarrow 2y\frac{dy}{dx} + 2x = 4y^3\frac{dy}{dx} - 2 \Rightarrow 2y\frac{dy}{dx} - 4y^3\frac{dy}{dx} = -2 - 2x$

$\Rightarrow \frac{dy}{dx}(2y - 4y^3) = -2 - 2x \Rightarrow \frac{dy}{dx} = \frac{x+1}{2y^3-y} \Rightarrow \left.\frac{dy}{dx}\right|_{(-2,1)} = -1$ and $\left.\frac{dy}{dx}\right|_{(-2,-1)} = 1$

46. $\left(x^2+y^2\right)^2 = (x-y)^2$ at $(1,0)$ and $(1,-1) \Rightarrow 2\left(x^2+y^2\right)\left(2x + 2y\frac{dy}{dx}\right) = 2(x-y)\left(1 - \frac{dy}{dx}\right)$

$\Rightarrow \frac{dy}{dx}\left[2y\left(x^2+y^2\right) + (x-y)\right] = -2x\left(x^2+y^2\right) + (x-y) \Rightarrow \frac{dy}{dx} = \frac{-2x\left(x^2+y^2\right) + (x-y)}{2y\left(x^2+y^2\right) + (x-y)} \Rightarrow \left.\frac{dy}{dx}\right|_{(1,0)} = -1$

and $\left.\frac{dy}{dx}\right|_{(1,-1)} = 1$

47. $x^2 + xy - y^2 = 1 \Rightarrow 2x + y + xy' - 2yy' = 0 \Rightarrow (x-2y)y' = -2x - y \Rightarrow y' = \frac{2x+y}{2y-x}$;

(a) the slope of the tangent line $m = y'|_{(2,3)} = \frac{7}{4} \Rightarrow$ the tangent line is $y - 3 = \frac{7}{4}(x-2) \Rightarrow y = \frac{7}{4}x - \frac{1}{2}$

(b) the normal line is $y - 3 = -\frac{4}{7}(x-2) \Rightarrow y = -\frac{4}{7}x + \frac{29}{7}$

48. $x^2 + y^2 = 25 \Rightarrow 2x + 2yy' = 0 \Rightarrow y' = -\frac{x}{y}$;

(a) the slope of the tangent line $m = y'|_{(3,-4)} = -\frac{x}{y}\Big|_{(3,-4)} = \frac{3}{4} \Rightarrow$ the tangent line is $y + 4 = \frac{3}{4}(x - 3)$ $\Rightarrow y = \frac{3}{4}x - \frac{25}{4}$

(b) the normal line is $y + 4 = -\frac{4}{3}(x - 3) \Rightarrow y = -\frac{4}{3}x$

49. $x^2y^2 = 9 \Rightarrow 2xy^2 + 2x^2yy' = 0 \Rightarrow x^2yy' = -xy^2 \Rightarrow y' = -\frac{y}{x}$;

(a) the slope of the tangent line $m = y'|_{(-1,3)} = -\frac{y}{x}\Big|_{(-1,3)} = 3 \Rightarrow$ the tangent line is $y - 3 = 3(x + 1)$ $\Rightarrow y = 3x + 6$

(b) the normal line is $y - 3 = -\frac{1}{3}(x + 1) \Rightarrow y = -\frac{1}{3}x + \frac{8}{3}$

50. $y^2 - 2x - 4y - 1 = 0 \Rightarrow 2yy' - 2 - 4y' = 0 \Rightarrow 2(y - 2)y' = 2 \Rightarrow y' = \frac{1}{y-2}$;

(a) the slope of the tangent line $m = y'|_{(-2,1)} = -1 \Rightarrow$ the tangent line is $y - 1 = -1(x + 2) \Rightarrow y = -x - 1$

(b) the normal line is $y - 1 = 1(x + 2) \Rightarrow y = x + 3$

51. $6x^2 + 3xy + 2y^2 + 17y - 6 = 0 \Rightarrow 12x + 3y + 3xy' + 4yy' + 17y' = 0 \Rightarrow y'(3x + 4y + 17) = -12x - 3y$ $\Rightarrow y' = \frac{-12x - 3y}{3x + 4y + 17}$;

(a) the slope of the tangent line $m = y'|_{(-1,0)} = \frac{-12x - 3y}{3x + 4y + 17}\Big|_{(-1,0)} = \frac{6}{7} \Rightarrow$ the tangent line is $y - 0 = \frac{6}{7}(x + 1)$ $\Rightarrow y = \frac{6}{7}x + \frac{6}{7}$

(b) the normal line is $y - 0 = -\frac{7}{6}(x + 1) \Rightarrow y = -\frac{7}{6}x - \frac{7}{6}$

52. $x^2 - \sqrt{3}xy + 2y^2 = 5 \Rightarrow 2x - \sqrt{3}xy' - \sqrt{3}y + 4yy' = 0 \Rightarrow y'\left(4y - \sqrt{3}x\right) = \sqrt{3}y - 2x \Rightarrow y' = \frac{\sqrt{3}y - 2x}{4y - \sqrt{3}x}$;

(a) the slope of the tangent line $m = y'|_{(\sqrt{3},2)} = \frac{\sqrt{3}y - 2x}{4y - \sqrt{3}x}\Big|_{(\sqrt{3},2)} = 0 \Rightarrow$ the tangent line is $y = 2$

(b) the normal line is $x = \sqrt{3}$

53. $2xy + \pi \sin y = 2\pi \Rightarrow 2xy' + 2y + \pi(\cos y)y' = 0 \Rightarrow y'(2x + \pi \cos y) = -2y \Rightarrow y' = \frac{-2y}{2x + \pi \cos y}$;

(a) the slope of the tangent line $m = y'|_{(1,\frac{\pi}{2})} = \frac{-2y}{2x + \pi \cos y}\Big|_{(1,\frac{\pi}{2})} = -\frac{\pi}{2} \Rightarrow$ the tangent line is $y - \frac{\pi}{2} = -\frac{\pi}{2}(x - 1) \Rightarrow y = -\frac{\pi}{2}x + \pi$

(b) the normal line is $y - \frac{\pi}{2} = \frac{2}{\pi}(x - 1) \Rightarrow y = \frac{2}{\pi}x - \frac{2}{\pi} + \frac{\pi}{2}$

54. $x \sin 2y = y \cos 2x \Rightarrow x(\cos 2y)2y' + \sin 2y = -2y \sin 2x + y' \cos 2x \Rightarrow y'(2x \cos 2y - \cos 2x)$ $= -\sin 2y - 2y \sin 2x \Rightarrow y' = \frac{\sin 2y + 2y \sin 2x}{\cos 2x - 2x \cos 2y}$;

(a) the slope of the tangent line $m = y'|_{(\frac{\pi}{4},\frac{\pi}{2})} = \frac{\sin 2y + 2y \sin 2x}{\cos 2x - 2x \cos 2y}\Big|_{(\frac{\pi}{4},\frac{\pi}{2})} = \frac{\pi}{\frac{\pi}{2}} = 2 \Rightarrow$ the tangent line is $y - \frac{\pi}{2} = 2\left(x - \frac{\pi}{4}\right) \Rightarrow y = 2x$

(b) the normal line is $y - \frac{\pi}{2} = -\frac{1}{2}\left(x - \frac{\pi}{4}\right) \Rightarrow y = -\frac{1}{2}x + \frac{5\pi}{8}$

55. $y = 2\sin(\pi x - y) \Rightarrow y' = 2[\cos(\pi x - y)] \cdot (\pi - y') \Rightarrow y'[1 + 2\cos(\pi x - y)] = 2\pi \cos(\pi x - y)$ $\Rightarrow y' = \frac{2\pi \cos(\pi x - y)}{1 + 2\cos(\pi x - y)}$;

(a) the slope of the tangent line $m = y'|_{(1,0)} = \frac{2\pi \cos(\pi x - y)}{1 + 2\cos(\pi x - y)}\Big|_{(1,0)} = 2\pi \Rightarrow$ the tangent line is $y - 0 = 2\pi(x - 1) \Rightarrow y = 2\pi x - 2\pi$

(b) the normal line is $y - 0 = -\frac{1}{2\pi}(x - 1) \Rightarrow y = -\frac{x}{2\pi} + \frac{1}{2\pi}$

56. $x^2 \cos^2 y - \sin y = 0 \Rightarrow x^2(2 \cos y)(-\sin y)y' + 2x \cos^2 y - y' \cos y = 0 \Rightarrow y'[-2x^2 \cos y \sin y - \cos y]$ $= -2x \cos^2 y \Rightarrow y' = \frac{2x \cos^2 y}{2x^2 \cos y \sin y + \cos y}$;

(a) the slope of the tangent line $m = y'|_{(0,\pi)} = \left.\frac{2x \cos^2 y}{2x^2 \cos y \sin y + \cos y}\right|_{(0,\pi)} = 0 \Rightarrow$ the tangent line is $y = \pi$

(b) the normal line is $x = 0$

57. Solving $x^2 + xy + y^2 = 7$ and $y = 0 \Rightarrow x^2 = 7 \Rightarrow x = \pm\sqrt{7} \Rightarrow \left(-\sqrt{7}, 0\right)$ and $\left(\sqrt{7}, 0\right)$ are the points where the curve crosses the x-axis. Now $x^2 + xy + y^2 = 7 \Rightarrow 2x + y + xy' + 2yy' = 0 \Rightarrow (x + 2y)y' = -2x - y$ $\Rightarrow y' = -\frac{2x+y}{x+2y} \Rightarrow m = -\frac{2x+y}{x+2y} \Rightarrow$ the slope at $\left(-\sqrt{7}, 0\right)$ is $m = -\frac{-2\sqrt{7}}{-\sqrt{7}} = -2$ and the slope at $\left(\sqrt{7}, 0\right)$ is $m = -\frac{2\sqrt{7}}{\sqrt{7}} = -2$. Since the slope is -2 in each case, the corresponding tangents must be parallel.

58. $x^2 + xy + y^2 = 7 \Rightarrow 2x + y + x\frac{dy}{dx} + 2y\frac{dy}{dx} = 0 \Rightarrow (x + 2y)\frac{dy}{dx} = -2x - y \Rightarrow \frac{dy}{dx} = \frac{-2x-y}{x+2y}$ and $\frac{dx}{dy} = \frac{x+2y}{-2x-y}$;

(a) Solving $\frac{dy}{dx} = 0 \Rightarrow -2x - y = 0 \Rightarrow y = -2x$ and substitution into the original equation gives $x^2 + x(-2x) + (-2x)^2 = 7 \Rightarrow 3x^2 = 7 \Rightarrow x = \pm\sqrt{\frac{7}{3}}$ and $y = \mp 2\sqrt{\frac{7}{3}}$ when the tangents are parallel to the x-axis.

(b) Solving $\frac{dx}{dy} = 0 \Rightarrow x + 2y = 0 \Rightarrow y = -\frac{x}{2}$ and substitution gives $x^2 + x\left(-\frac{x}{2}\right) + \left(-\frac{x}{2}\right)^2 = 7 \Rightarrow \frac{3x^2}{4} = 7$ $\Rightarrow x = \pm 2\sqrt{\frac{7}{3}}$ and $y = \mp\sqrt{\frac{7}{3}}$ when the tangents are parallel to the y-axis.

59. $y^4 = y^2 - x^2 \Rightarrow 4y^3y' = 2yy' - 2x \Rightarrow 2(2y^3 - y)y' = -2x \Rightarrow y' = \frac{x}{y-2y^3}$; the slope of the tangent line at $\left(\frac{\sqrt{3}}{4}, \frac{\sqrt{3}}{2}\right)$ is $\left.\frac{x}{y-2y^3}\right|_{\left(\frac{\sqrt{3}}{4}, \frac{\sqrt{3}}{2}\right)} = \frac{\frac{\sqrt{3}}{4}}{\frac{\sqrt{3}}{2} - \frac{6\sqrt{3}}{8}} = \frac{\frac{1}{4}}{\frac{1}{2} - \frac{3}{4}} = \frac{1}{2-3} = -1$; the slope of the tangent line at $\left(\frac{\sqrt{3}}{4}, \frac{1}{2}\right)$ is $\left.\frac{x}{y-2y^3}\right|_{\left(\frac{\sqrt{3}}{4}, \frac{1}{2}\right)} = \frac{\frac{\sqrt{3}}{4}}{\frac{1}{2} - \frac{2}{8}} = \frac{2\sqrt{3}}{4-2} = \sqrt{3}$

60. $y^2(2 - x) = x^3 \Rightarrow 2yy'(2 - x) + y^2(-1) = 3x^2 \Rightarrow y' = \frac{y^2+3x^2}{2y(2-x)}$; the slope of the tangent line is $m = \left.\frac{y^2+3x^2}{2y(2-x)}\right|_{(1,1)} = \frac{4}{2} = 2 \Rightarrow$ the tangent line is $y - 1 = 2(x - 1) \Rightarrow y = 2x - 1$; the normal line is $y - 1 = -\frac{1}{2}(x - 1) \Rightarrow y = -\frac{1}{2}x + \frac{3}{2}$

61. $y^4 - 4y^2 = x^4 - 9x^2 \Rightarrow 4y^3y' - 8yy' = 4x^3 - 18x \Rightarrow y'(4y^3 - 8y) = 4x^3 - 18x \Rightarrow y' = \frac{4x^3-18x}{4y^3-8y} = \frac{2x^3-9x}{2y^3-4y}$ $= \frac{x(2x^2-9)}{y(2y^2-4)} = m$; $(-3, 2)$: $m = \frac{(-3)(18-9)}{2(8-4)} = -\frac{27}{8}$; $(-3, -2)$: $m = \frac{27}{8}$; $(3, 2)$: $m = \frac{27}{8}$; $(3, -2)$: $m = -\frac{27}{8}$

62. $x^3 + y^3 - 9xy = 0 \Rightarrow 3x^2 + 3y^2y' - 9xy' - 9y = 0 \Rightarrow y'(3y^2 - 9x) = 9y - 3x^2 \Rightarrow y' = \frac{9y-3x^2}{3y^2-9x} = \frac{3y-x^2}{y^2-3x}$

(a) $y'|_{(4,2)} = \frac{5}{4}$ and $y'|_{(2,4)} = \frac{4}{5}$;

(b) $y' = 0 \Rightarrow \frac{3y-x^2}{y^2-3x} = 0 \Rightarrow 3y - x^2 = 0 \Rightarrow y = \frac{x^2}{3} \Rightarrow x^3 + \left(\frac{x^2}{3}\right)^3 - 9x\left(\frac{x^2}{3}\right) = 0 \Rightarrow x^6 - 54x^3 = 0$ $\Rightarrow x^3(x^3 - 54) = 0 \Rightarrow x = 0$ or $x = \sqrt[3]{54} = 3\sqrt[3]{2} \Rightarrow$ there is a horizontal tangent at $x = 3\sqrt[3]{2}$. To find the corresponding y-value, we will use part (c).

(c) $\frac{dx}{dy} = 0 \Rightarrow \frac{y^2-3x}{3y-x^2} = 0 \Rightarrow y^2 - 3x = 0 \Rightarrow y = \pm\sqrt{3x}$; $y = \sqrt{3x} \Rightarrow x^3 + \left(\sqrt{3x}\right)^3 - 9x\sqrt{3x} = 0$ $\Rightarrow x^3 - 6\sqrt{3}x^{3/2} = 0 \Rightarrow x^{3/2}\left(x^{3/2} - 6\sqrt{3}\right) = 0 \Rightarrow x^{3/2} = 0$ or $x^{3/2} = 6\sqrt{3} \Rightarrow x = 0$ or $x = \sqrt[3]{108} = 3\sqrt[3]{4}$.

Since the equation $x^3 + y^3 - 9xy = 0$ is symmetric in x and y, the graph is symmetric about the line $y = x$. That is, if (a, b) is a point on the folium, then so is (b, a). Moreover, if $y'|_{(a,b)} = m$, then $y'|_{(b,a)} = \frac{1}{m}$. Thus, if the folium has a horizontal tangent at (a, b), it has a vertical tangent at (b, a) so one might expect that with a horizontal tangent at $x = \sqrt[3]{54}$ and a vertical tangent at $x = 3\sqrt[3]{4}$, the points of tangency are $\left(\sqrt[3]{54}, 3\sqrt[3]{4}\right)$ and $\left(3\sqrt[3]{4}, \sqrt[3]{54}\right)$, respectively. One can check that these points do satisfy the equation $x^3 + y^3 - 9xy = 0$.

63. $x^2 - 2tx + 2t^2 = 4 \Rightarrow 2x\frac{dx}{dt} - 2x - 2t\frac{dx}{dt} + 4t = 0 \Rightarrow (2x - 2t)\frac{dx}{dt} = 2x - 4t \Rightarrow \frac{dx}{dt} = \frac{2x-4t}{2x-2t} = \frac{x-2t}{x-t}$;
$2y^3 - 3t^2 = 4 \Rightarrow 6y^2\frac{dy}{dt} - 6t = 0 \Rightarrow \frac{dy}{dt} = \frac{6t}{6y^2} = \frac{t}{y^2}$; thus $\frac{dy}{dx} = \frac{dy/dt}{dx/dt} = \frac{\left(\frac{t}{y^2}\right)}{\left(\frac{x-2t}{x-t}\right)} = \frac{t(x-t)}{y^2(x-2t)}$; $t = 2$
$\Rightarrow x^2 - 2(2)x + 2(2)^2 = 4 \Rightarrow x^2 - 4x + 4 = 0 \Rightarrow (x-2)^2 = 0 \Rightarrow x = 2$; $t = 2 \Rightarrow 2y^3 - 3(2)^2 = 4$
$\Rightarrow 2y^3 = 16 \Rightarrow y^3 = 8 \Rightarrow y = 2$; therefore $\left.\frac{dy}{dx}\right|_{t=2} = \frac{2(2-2)}{(2)^2(2-2(2))} = 0$

64. $x = \sqrt{5 - \sqrt{t}} \Rightarrow \frac{dx}{dt} = \frac{1}{2}\left(5 - \sqrt{t}\right)^{-1/2}\left(-\frac{1}{2}t^{-1/2}\right) = -\frac{1}{4\sqrt{t}\sqrt{5-\sqrt{t}}}$; $y(t-1) = \sqrt{t} \Rightarrow y + (t-1)\frac{dy}{dt} = \frac{1}{2}t^{-1/2}$
$\Rightarrow (t-1)\frac{dy}{dt} = \frac{1}{2\sqrt{t}} - y \Rightarrow \frac{dy}{dt} = \frac{\frac{1}{2\sqrt{t}} - y}{(t-1)} = \frac{1-2y\sqrt{t}}{2t\sqrt{t}-2\sqrt{t}}$; thus $\frac{dy}{dx} = \frac{\frac{dy}{dt}}{\frac{dx}{dt}} = \frac{\frac{1-2y\sqrt{t}}{2t\sqrt{t}-2\sqrt{t}}}{\frac{-1}{4\sqrt{t}\sqrt{5-\sqrt{t}}}} = \frac{1-2y\sqrt{t}}{2\sqrt{t}(t-1)} \cdot \frac{4\sqrt{t}\sqrt{5-\sqrt{t}}}{-1}$
$= \frac{2\left(1-2y\sqrt{t}\right)\sqrt{5-\sqrt{t}}}{1-t}$; $t = 4 \Rightarrow x = \sqrt{5-\sqrt{4}} = \sqrt{3}$; $t = 4 \Rightarrow y(3) = \sqrt{4} = 2$
therefore, $\left.\frac{dy}{dx}\right|_{t=4} = \frac{2\left(1-2(2)\sqrt{4}\right)\sqrt{5-\sqrt{4}}}{1-4} = \frac{14}{3}$

65. $x + 2x^{3/2} = t^2 + t \Rightarrow \frac{dx}{dt} + 3x^{1/2}\frac{dx}{dt} = 2t + 1 \Rightarrow \left(1 + 3x^{1/2}\right)\frac{dx}{dt} = 2t + 1 \Rightarrow \frac{dx}{dt} = \frac{2t+1}{1+3x^{1/2}}$; $y\sqrt{t+1} + 2t\sqrt{y} = 4$
$\Rightarrow \frac{dy}{dt}\sqrt{t+1} + y\left(\frac{1}{2}\right)(t+1)^{-1/2} + 2\sqrt{y} + 2t\left(\frac{1}{2}y^{-1/2}\right)\frac{dy}{dt} = 0 \Rightarrow \frac{dy}{dt}\sqrt{t+1} + \frac{y}{2\sqrt{t+1}} + 2\sqrt{y} + \left(\frac{t}{\sqrt{y}}\right)\frac{dy}{dt} = 0$
$\Rightarrow \left(\sqrt{t+1} + \frac{t}{\sqrt{y}}\right)\frac{dy}{dt} = \frac{-y}{2\sqrt{t+1}} - 2\sqrt{y} \Rightarrow \frac{dy}{dt} = \frac{\left(\frac{-y}{2\sqrt{t+1}} - 2\sqrt{y}\right)}{\left(\sqrt{t+1} + \frac{t}{\sqrt{y}}\right)} = \frac{-y\sqrt{y} - 4y\sqrt{t+1}}{2\sqrt{y}(t+1) + 2t\sqrt{t+1}}$; thus
$\frac{dy}{dx} = \frac{dy/dt}{dx/dt} = \frac{\left(\frac{-y\sqrt{y}-4y\sqrt{t+1}}{2\sqrt{y}(t+1)+2t\sqrt{t+1}}\right)}{\left(\frac{2t+1}{1+3x^{1/2}}\right)}$; $t = 0 \Rightarrow x + 2x^{3/2} = 0 \Rightarrow x\left(1 + 2x^{1/2}\right) = 0 \Rightarrow x = 0$; $t = 0$
$\Rightarrow y\sqrt{0+1} + 2(0)\sqrt{y} = 4 \Rightarrow y = 4$; therefore $\left.\frac{dy}{dx}\right|_{t=0} = \frac{\left(\frac{-4\sqrt{4}-4(4)\sqrt{0+1}}{2\sqrt{4}(0+1)+2(0)\sqrt{0+1}}\right)}{\left(\frac{2(0)+1}{1+3(0)^{1/2}}\right)} = -6$

66. $x \sin t + 2x = t \Rightarrow \frac{dx}{dt}\sin t + x\cos t + 2\frac{dx}{dt} = 1 \Rightarrow (\sin t + 2)\frac{dx}{dt} = 1 - x\cos t \Rightarrow \frac{dx}{dt} = \frac{1 - x\cos t}{\sin t + 2}$;
$t \sin t - 2t = y \Rightarrow \sin t + t\cos t - 2 = \frac{dy}{dt}$; thus $\frac{dy}{dx} = \frac{\sin t + t\cos t - 2}{\left(\frac{1 - x\cos t}{\sin t + 2}\right)}$; $t = \pi \Rightarrow x\sin\pi + 2x = \pi$
$\Rightarrow x = \frac{\pi}{2}$; therefore $\left.\frac{dy}{dx}\right|_{t=\pi} = \frac{\sin\pi + \pi\cos\pi - 2}{\left[\frac{1-\left(\frac{\pi}{2}\right)\cos\pi}{\sin\pi + 2}\right]} = \frac{-4\pi - 8}{2 + \pi} = -4$

67. (a) if $f(x) = \frac{3}{2}x^{2/3} - 3$, then $f'(x) = x^{-1/3}$ and $f''(x) = -\frac{1}{3}x^{-4/3}$ so the claim $f''(x) = x^{-1/3}$ is false
(b) if $f(x) = \frac{9}{10}x^{5/3} - 7$, then $f'(x) = \frac{3}{2}x^{2/3}$ and $f''(x) = x^{-1/3}$ is true
(c) $f''(x) = x^{-1/3} \Rightarrow f'''(x) = -\frac{1}{3}x^{-4/3}$ is true
(d) if $f'(x) = \frac{3}{2}x^{2/3} + 6$, then $f''(x) = x^{-1/3}$ is true

68. $2x^2 + 3y^2 = 5 \Rightarrow 4x + 6yy' = 0 \Rightarrow y' = -\frac{2x}{3y} \Rightarrow y'|_{(1,1)} = -\left.\frac{2x}{3y}\right|_{(1,1)} = -\frac{2}{3}$ and $y'|_{(1,-1)} = -\left.\frac{2x}{3y}\right|_{(1,-1)} = \frac{2}{3}$; also,
$y^2 = x^3 \Rightarrow 2yy' = 3x^2 \Rightarrow y' = \frac{3x^2}{2y} \Rightarrow y'|_{(1,1)} = \left.\frac{3x^2}{2y}\right|_{(1,1)} = \frac{3}{2}$ and $y'|_{(1,-1)} = \left.\frac{3x^2}{2y}\right|_{(1,-1)} = -\frac{3}{2}$. Therefore the tangents to the curves are perpendicular at $(1, 1)$ and $(1, -1)$ (i.e., the curves are orthogonal at these two points of intersection).

69. $x^2 + 2xy - 3y^2 = 0 \Rightarrow 2x + 2xy' + 2y - 6yy' = 0 \Rightarrow y'(2x - 6y) = -2x - 2y \Rightarrow y' = \frac{x+y}{3y-x} \Rightarrow$ the slope of the tangent line $m = y'|_{(1,1)} = \left.\frac{x+y}{3y-x}\right|_{(1,1)} = 1 \Rightarrow$ the equation of the normal line at $(1, 1)$ is $y - 1 = -1(x - 1)$ $\Rightarrow y = -x + 2$. To find where the normal line intersects the curve we substitute into its equation: $x^2 + 2x(2 - x) - 3(2 - x)^2 = 0 \Rightarrow x^2 + 4x - 2x^2 - 3(4 - 4x + x^2) = 0 \Rightarrow -4x^2 + 16x - 12 = 0$ $\Rightarrow x^2 - 4x + 3 = 0 \Rightarrow (x - 3)(x - 1) = 0 \Rightarrow x = 3$ and $y = -x + 2 = -1$. Therefore, the normal to the curve at $(1, 1)$ intersects the curve at the point $(3, -1)$. Note that it also intersects the curve at $(1, 1)$.

70. $xy + 2x - y = 0 \Rightarrow x\frac{dy}{dx} + y + 2 - \frac{dy}{dx} = 0 \Rightarrow \frac{dy}{dx} = \frac{y+2}{1-x}$; the slope of the line $2x + y = 0$ is -2. In order to be parallel, the normal lines must also have slope of -2. Since a normal is perpendicular to a tangent, the slope of the tangent is $\frac{1}{2}$. Therefore, $\frac{y+2}{1-x} = \frac{1}{2} \Rightarrow 2y + 4 = 1 - x \Rightarrow x = -3 - 2y$. Substituting in the original equation, $y(-3 - 2y) + 2(-3 - 2y) - y = 0 \Rightarrow y^2 + 4y + 3 = 0 \Rightarrow y = -3$ or $y = -1$. If $y = -3$, then $x = 3$ and $y + 3 = -2(x - 3) \Rightarrow y = -2x + 3$. If $y = -1$, then $x = -1$ and $y + 1 = -2(x + 1) \Rightarrow y = -2x - 3$.

71. $y^2 = x \Rightarrow \frac{dy}{dx} = \frac{1}{2y}$. If a normal is drawn from $(a, 0)$ to (x_1, y_1) on the curve its slope satisfies $\frac{y_1 - 0}{x_1 - a} = -2y_1$ $\Rightarrow y_1 = -2y_1(x_1 - a)$ or $a = x_1 + \frac{1}{2}$. Since $x_1 \ge 0$ on the curve, we must have that $a \ge \frac{1}{2}$. By symmetry, the two points on the parabola are $\left(x_1, \sqrt{x_1}\right)$ and $\left(x_1, -\sqrt{x_1}\right)$. For the normal to be perpendicular, $\left(\frac{\sqrt{x_1}}{x_1 - a}\right)\left(\frac{\sqrt{x_1}}{a - x_1}\right) = -1 \Rightarrow \frac{x_1}{(a - x_1)^2} = 1 \Rightarrow x_1 = (a - x_1)^2 \Rightarrow x_1 = \left(x_1 + \frac{1}{2} - x_1\right)^2 \Rightarrow x_1 = \frac{1}{4}$ and $y_1 = \pm\frac{1}{2}$. Therefore, $\left(\frac{1}{4}, \pm\frac{1}{2}\right)$ and $a = \frac{3}{4}$.

72. Ex. 6b.) $y = x^{1/2}$ has no derivative at $x = 0$ because the slope of the graph becomes vertical at $x = 0$.
Ex. 7a.) $y = \left(1 - x^2\right)^{1/4}$ has a derivative only on $(-1, 1)$ because the function is defined only on $[-1, 1]$ and the slope of the tangent becomes vertical at both $x = -1$ and $x = 1$.

73. $xy^3 + x^2y = 6 \Rightarrow x\left(3y^2\frac{dy}{dx}\right) + y^3 + x^2\frac{dy}{dx} + 2xy = 0 \Rightarrow \frac{dy}{dx}\left(3xy^2 + x^2\right) = -y^3 - 2xy \Rightarrow \frac{dy}{dx} = \frac{-y^3 - 2xy}{3xy^2 + x^2}$ $= -\frac{y^3 + 2xy}{3xy^2 + x^2}$; also, $xy^3 + x^2y = 6 \Rightarrow x\left(3y^2\right) + y^3\frac{dx}{dy} + x^2 + y\left(2x\frac{dx}{dy}\right) = 0 \Rightarrow \frac{dx}{dy}\left(y^3 + 2xy\right) = -3xy^2 - x^2$ $\Rightarrow \frac{dx}{dy} = -\frac{3xy^2 + x^2}{y^3 + 2xy}$; thus $\frac{dx}{dy}$ appears to equal $\frac{1}{\frac{dy}{dx}}$. The two different treatments view the graphs as functions symmetric across the line $y = x$, so their slopes are reciprocals of one another at the corresponding points (a, b) and (b, a).

74. $x^3 + y^2 = \sin^2 y \Rightarrow 3x^2 + 2y\frac{dy}{dx} = (2\sin y)(\cos y)\frac{dy}{dx} \Rightarrow \frac{dy}{dx}(2y - 2\sin y\cos y) = -3x^2 \Rightarrow \frac{dy}{dx} = \frac{-3x^2}{2y - 2\sin y\cos y}$ $= \frac{3x^2}{2\sin y\cos y - 2y}$; also, $x^3 + y^2 = \sin^2 y \Rightarrow 3x^2\frac{dx}{dy} + 2y = 2\sin y\cos y \Rightarrow \frac{dx}{dy} = \frac{2\sin y\cos y - 2y}{3x^2}$; thus $\frac{dx}{dy}$ appears to equal $\frac{1}{\frac{dy}{dx}}$. The two different treatments view the graphs as functions symmetric across the line $y = x$ so their slopes are reciprocals of one another at the corresponding points (a, b) and (b, a).

75. $x^4 + 4y^2 = 1$:

(a) $y^2 = \frac{1-x^4}{4} \Rightarrow y = \pm\frac{1}{2}\sqrt{1-x^4}$

$\Rightarrow \frac{dy}{dx} = \pm\frac{1}{4}\left(1-x^4\right)^{-1/2}\left(-4x^3\right) = \frac{\pm x^3}{\left(1-x^4\right)^{1/2}}$;

differentiating implicitly, we find, $4x^3 + 8y\frac{dy}{dx} = 0$

$\Rightarrow \frac{dy}{dx} = \frac{-4x^3}{8y} = \frac{-4x^3}{8\left(\pm\frac{1}{2}\sqrt{1-x^4}\right)} = \frac{\pm x^3}{\left(1-x^4\right)^{1/2}}$.

(b)

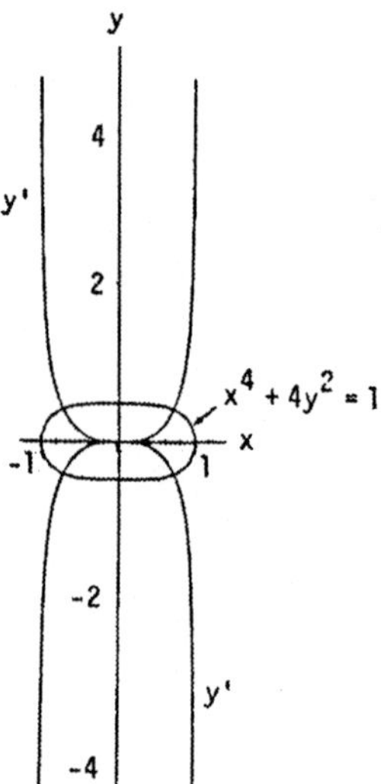

76. $(x-2)^2 + y^2 = 4$:

(a) $y = \pm\sqrt{4-(x-2)^2}$

$\Rightarrow \frac{dy}{dx} = \pm\frac{1}{2}\left(4-(x-2)^2\right)^{-1/2}(-2(x-2))$

$= \frac{\pm(x-2)}{\left[4-(x-2)^2\right]^{1/2}}$; differentiating implicitly,

$2(x-2) + 2y\frac{dy}{dx} = 0 \Rightarrow \frac{dy}{dx} = \frac{-2(x-2)}{2y}$

$= \frac{-(x-2)}{y} = \frac{-(x-2)}{\pm\left[4-(x-2)^2\right]^{1/2}} = \frac{\pm(x-2)}{\left[4-(x-2)^2\right]^{1/2}}$.

(b)

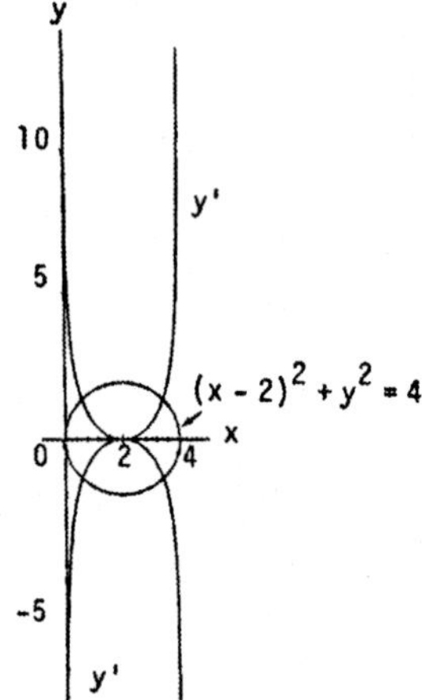

77-84. Example CAS commands:

Maple:

```
q1 := x^3-x*y+y^3 = 7;
pt := [x=2,y=1];
p1 := implicitplot( q1, x=-3..3, y=-3..3 ):
p1;
eval( q1, pt );
q2 := implicitdiff( q1, y, x );
m := eval( q2, pt );
tan_line := y = 1 + m*(x-2);
p2 := implicitplot( tan_line, x=-5..5, y=-5..5, color=green ):
p3 := pointplot( eval([x,y],pt), color=blue ):
display( [p1,p2,p3], ="Section 3.6 #77(c)" );
```

Mathematica: (functions and x0 may vary):

Note use of double equal sign (logic statement) in definition of eqn and tanline.

```
<<Graphics`ImplicitPlot`
```

```
Clear[x, y]
{x0, y0}={1, π/4};
eqn=x + Tan[y/x]==2;
ImplicitPlot[eqn,{ x, x0 − 3, x0 + 3},{y, y0 − 3, y0 + 3}]
eqn/.{x → x0, y → y0}
eqn/.{ y → y[x]}
D[%, x]
Solve[%, y'[x]]
slope=y'[x]/.First[%]
m=slope/.{x → x0, y[x] → y0}
tanline=y==y0 + m (x − x0)
ImplicitPlot[{eqn, tanline}, {x, x0 − 3, x0 + 3},{y, y0 − 3, y0 + 3}]
```

3.7 DERIVATIVES OF INVERSE FUNCTIONS AND LOGARITHMS

1. (a) $y = 2x + 3 \Rightarrow 2x = y - 3$
$\Rightarrow x = \frac{y}{2} - \frac{3}{2} \Rightarrow f^{-1}(x) = \frac{x}{2} - \frac{3}{2}$
(c) $\left.\frac{df}{dx}\right|_{x=-1} = 2, \left.\frac{df^{-1}}{dx}\right|_{x=1} = \frac{1}{2}$

(b)

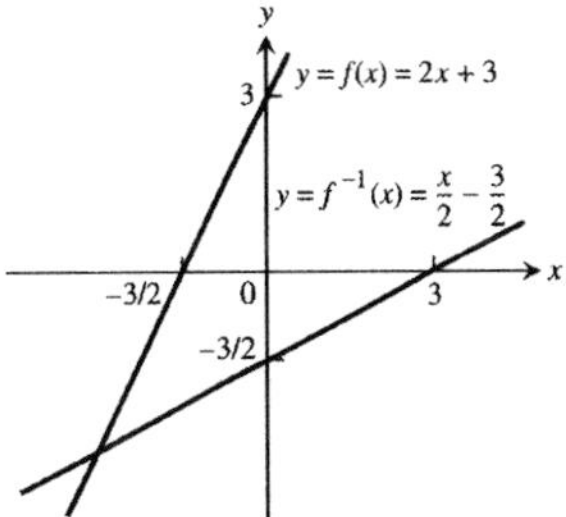

2. (a) $y = \frac{1}{5}x + 7 \Rightarrow \frac{1}{5}x = y - 7$
$\Rightarrow x = 5y - 35 \Rightarrow f^{-1}(x) = 5x - 35$
(c) $\left.\frac{df}{dx}\right|_{x=-1} = \frac{1}{5}, \left.\frac{df^{-1}}{dx}\right|_{x=34/5} = 5$

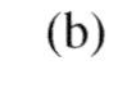
(b)

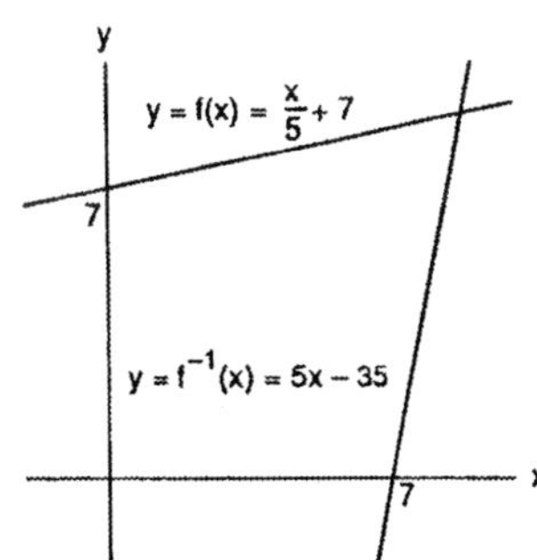

3. (a) $y = 5 - 4x \Rightarrow 4x = 5 - y$
$\Rightarrow x = \frac{5}{4} - \frac{y}{4} \Rightarrow f^{-1}(x) = \frac{5}{4} - \frac{x}{4}$
(c) $\left.\frac{df}{dx}\right|_{x=1/2} = -4, \left.\frac{df^{-1}}{dx}\right|_{x=3} = -\frac{1}{4}$

(b)

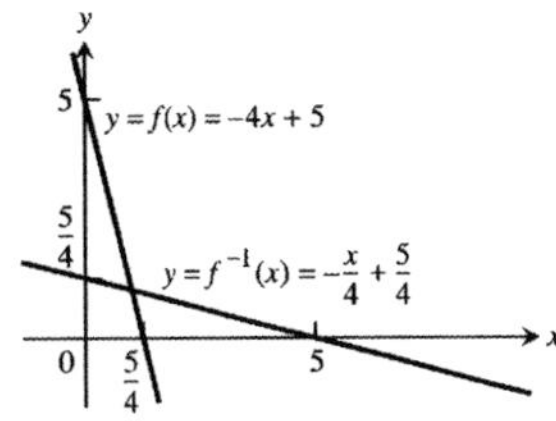

4. (a) $y = 2x^2 \Rightarrow x^2 = \frac{1}{2}y$
$\Rightarrow x = \frac{1}{\sqrt{2}}\sqrt{y} \Rightarrow f^{-1}(x) = \sqrt{\frac{x}{2}}$

(b)

(c) $\left.\frac{df}{dx}\right|_{x=5} = 4x\Big|_{x=5} = 20,$
$\left.\frac{df^{-1}}{dx}\right|_{x=50} = \left.\frac{1}{2\sqrt{2}}x^{-1/2}\right|_{x=50} = \frac{1}{20}$

5. (a) $f(g(x)) = \left(\sqrt[3]{x}\right)^3 = x,\ g(f(x)) = \sqrt[3]{x^3} = x$

(b)

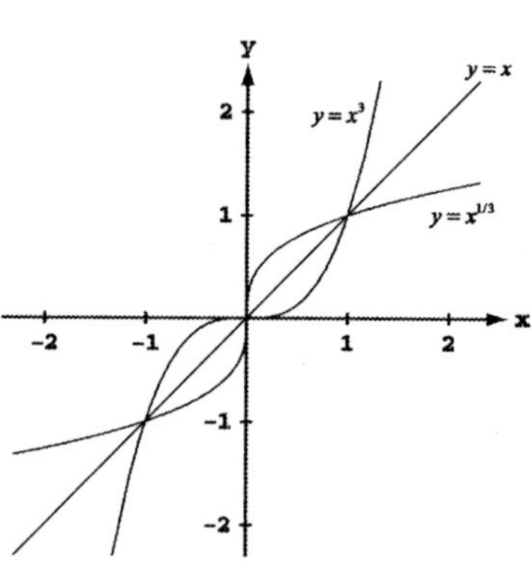

(c) $f'(x) = 3x^2 \Rightarrow f'(1) = 3,\ f'(-1) = 3;$
$g'(x) = \frac{1}{3}x^{-2/3} \Rightarrow g'(1) = \frac{1}{3},\ g'(-1) = \frac{1}{3}$

(d) The line $y = 0$ is tangent to $f(x) = x^3$ at $(0, 0)$; the line $x = 0$ is tangent to $g(x) = \sqrt[3]{x}$ at $(0, 0)$

6. (a) $h(k(x)) = \frac{1}{4}\left((4x)^{1/3}\right)^3 = x,$
$k(h(x)) = \left(4 \cdot \frac{x^3}{4}\right)^{1/3} = x$

(b)

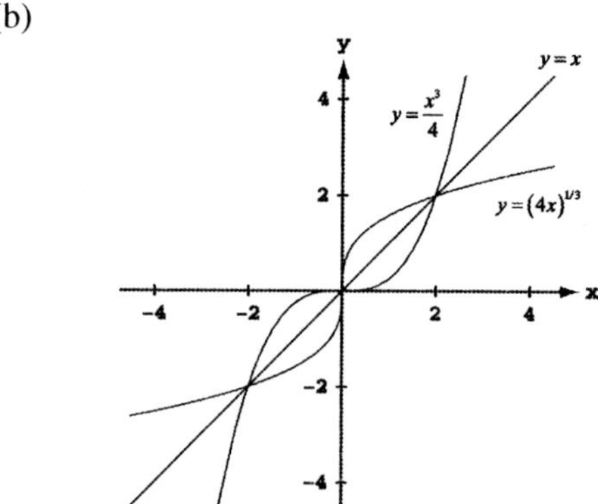

(c) $h'(x) = \frac{3x^2}{4} \Rightarrow h'(2) = 3,\ h'(-2) = 3;$
$k'(x) = \frac{4}{3}(4x)^{-2/3} \Rightarrow k'(2) = \frac{1}{3},\ k'(-2) = \frac{1}{3}$

(d) The line $y = 0$ is tangent to $h(x) = \frac{x^3}{4}$ at $(0, 0)$; the line $x = 0$ is tangent to $k(x) = (4x)^{1/3}$ at $(0, 0)$

7. $\frac{df}{dx} = 3x^2 - 6x \Rightarrow \left.\frac{df^{-1}}{dx}\right|_{x=f(3)} = \left.\frac{1}{\frac{df}{dx}}\right|_{x=3} = \frac{1}{9}$

8. $\frac{df}{dx} = 2x - 4 \Rightarrow \left.\frac{df^{-1}}{dx}\right|_{x=f(5)} = \left.\frac{1}{\frac{df}{dx}}\right|_{x=5} = \frac{1}{6}$

9. $\left.\frac{df^{-1}}{dx}\right|_{x=4} = \left.\frac{df^{-1}}{dx}\right|_{x=f(2)} = \left.\frac{1}{\frac{df}{dx}}\right|_{x=2} = \frac{1}{\left(\frac{1}{3}\right)} = 3$

10. $\left.\frac{dg^{-1}}{dx}\right|_{x=0} = \left.\frac{dg^{-1}}{dx}\right|_{x=f(0)} = \left.\frac{1}{\frac{dg}{dx}}\right|_{x=0} = \frac{1}{2}$

11. $y = \ln 3x \Rightarrow y' = \left(\frac{1}{3x}\right)(3) = \frac{1}{x}$

12. $y = \ln kx \Rightarrow y' = \left(\frac{1}{kx}\right)(k) = x$

13. $y = \ln(t^2) \Rightarrow \frac{dy}{dt} = \left(\frac{1}{t^2}\right)(2t) = \frac{2}{t}$

14. $y = \ln\left(t^{3/2}\right) \Rightarrow \frac{dy}{dt} = \left(\frac{1}{t^{3/2}}\right)\left(\frac{3}{2}t^{1/2}\right) = \frac{3}{2t}$

15. $y = \ln\frac{3}{x} = \ln 3x^{-1} \Rightarrow \frac{dy}{dx} = \left(\frac{1}{3x^{-1}}\right)(-3x^{-2}) = -\frac{1}{x}$

16. $y = \ln\frac{10}{x} = \ln 10x^{-1} \Rightarrow \frac{dy}{dx} = \left(\frac{1}{10x^{-1}}\right)(-10x^{-2}) = -\frac{1}{x}$

17. $y = \ln(\theta + 1) \Rightarrow \frac{dy}{d\theta} = \left(\frac{1}{\theta+1}\right)(1) = \frac{1}{\theta+1}$

18. $y = \ln(2\theta + 2) \Rightarrow \frac{dy}{d\theta} = \left(\frac{1}{2\theta+2}\right)(2) = \frac{1}{\theta+1}$

19. $y = \ln x^3 \Rightarrow \frac{dy}{dx} = \left(\frac{1}{x^3}\right)(3x^2) = \frac{3}{x}$

20. $y = (\ln x)^3 \Rightarrow \frac{dy}{dx} = 3(\ln x)^2 \cdot \frac{d}{dx}(\ln x) = \frac{3(\ln x)^2}{x}$

21. $y = t(\ln t)^2 \Rightarrow \frac{dy}{dt} = (\ln t)^2 + 2t(\ln t) \cdot \frac{d}{dt}(\ln t) = (\ln t)^2 + \frac{2t \ln t}{t} = (\ln t)^2 + 2 \ln t$

22. $y = t\sqrt{\ln t} = t(\ln t)^{1/2} \Rightarrow \frac{dy}{dt} = (\ln t)^{1/2} + \frac{1}{2} t(\ln t)^{-1/2} \cdot \frac{d}{dt}(\ln t) = (\ln t)^{1/2} + \frac{t(\ln t)^{-1/2}}{2t}$
$= (\ln t)^{1/2} + \frac{1}{2(\ln t)^{1/2}}$

23. $y = \frac{x^4}{4} \ln x - \frac{x^4}{16} \Rightarrow \frac{dy}{dx} = x^3 \ln x + \frac{x^4}{4} \cdot \frac{1}{x} - \frac{4x^3}{16} = x^3 \ln x$

24. $y = \frac{x^3}{3} \ln x - \frac{x^3}{9} \Rightarrow \frac{dy}{dx} = x^2 \ln x + \frac{x^3}{3} \cdot \frac{1}{x} - \frac{3x^2}{9} = x^2 \ln x$

25. $y = \frac{\ln t}{t} \Rightarrow \frac{dy}{dt} = \frac{t\left(\frac{1}{t}\right) - (\ln t)(1)}{t^2} = \frac{1 - \ln t}{t^2}$

26. $y = \frac{1 + \ln t}{t} \Rightarrow \frac{dy}{dt} = \frac{t\left(\frac{1}{t}\right) - (1 + \ln t)(1)}{t^2} = \frac{1 - 1 - \ln t}{t^2} = -\frac{\ln t}{t^2}$

27. $y = \frac{\ln x}{1 + \ln x} \Rightarrow y' = \frac{(1 + \ln x)\left(\frac{1}{x}\right) - (\ln x)\left(\frac{1}{x}\right)}{(1 + \ln x)^2} = \frac{\frac{1}{x} + \frac{\ln x}{x} - \frac{\ln x}{x}}{(1 + \ln x)^2} = \frac{1}{x(1 + \ln x)^2}$

28. $y = \frac{x \ln x}{1 + \ln x} \Rightarrow y' = \frac{(1+\ln x)\left(\ln x + x \cdot \frac{1}{x}\right) - (x \ln x)\left(\frac{1}{x}\right)}{(1+\ln x)^2} = \frac{(1 + \ln x)^2 - \ln x}{(1 + \ln x)^2} = 1 - \frac{\ln x}{(1 + \ln x)^2}$

29. $y = \ln(\ln x) \Rightarrow y' = \left(\frac{1}{\ln x}\right)\left(\frac{1}{x}\right) = \frac{1}{x \ln x}$

30. $y = \ln(\ln(\ln x)) \Rightarrow y' = \frac{1}{\ln(\ln x)} \cdot \frac{d}{dx}(\ln(\ln x)) = \frac{1}{\ln(\ln x)} \cdot \frac{1}{\ln x} \cdot \frac{d}{dx}(\ln x) = \frac{1}{x(\ln x)\ln(\ln x)}$

31. $y = \theta[\sin(\ln\theta) + \cos(\ln\theta)] \Rightarrow \frac{dy}{d\theta} = [\sin(\ln\theta) + \cos(\ln\theta)] + \theta\left[\cos(\ln\theta) \cdot \frac{1}{\theta} - \sin(\ln\theta) \cdot \frac{1}{\theta}\right]$
$= \sin(\ln\theta) + \cos(\ln\theta) + \cos(\ln\theta) - \sin(\ln\theta) = 2\cos(\ln\theta)$

32. $y = \ln(\sec\theta + \tan\theta) \Rightarrow \frac{dy}{d\theta} = \frac{\sec\theta\tan\theta + \sec^2\theta}{\sec\theta + \tan\theta} = \frac{\sec\theta(\tan\theta + \sec\theta)}{\tan\theta + \sec\theta} = \sec\theta$

33. $y = \ln\frac{1}{x\sqrt{x+1}} = -\ln x - \frac{1}{2}\ln(x+1) \Rightarrow y' = -\frac{1}{x} - \frac{1}{2}\left(\frac{1}{x+1}\right) = -\frac{2(x+1)+x}{2x(x+1)} = -\frac{3x+2}{2x(x+1)}$

34. $y = \frac{1}{2}\ln\frac{1+x}{1-x} = \frac{1}{2}[\ln(1+x) - \ln(1-x)] \Rightarrow y' = \frac{1}{2}\left[\frac{1}{1+x} - \left(\frac{1}{1-x}\right)(-1)\right] = \frac{1}{2}\left[\frac{1-x+1+x}{(1+x)(1-x)}\right] = \frac{1}{1-x^2}$

35. $y = \frac{1+\ln t}{1-\ln t} \Rightarrow \frac{dy}{dt} = \frac{(1-\ln t)\left(\frac{1}{t}\right) - (1+\ln t)\left(\frac{-1}{t}\right)}{(1-\ln t)^2} = \frac{\frac{1}{t} - \frac{\ln t}{t} + \frac{1}{t} + \frac{\ln t}{t}}{(1-\ln t)^2} = \frac{2}{t(1-\ln t)^2}$

36. $y = \sqrt{\ln\sqrt{t}} = \left(\ln t^{1/2}\right)^{1/2} \Rightarrow \frac{dy}{dt} = \frac{1}{2}\left(\ln t^{1/2}\right)^{-1/2} \cdot \frac{d}{dt}\left(\ln t^{1/2}\right) = \frac{1}{2}\left(\ln t^{1/2}\right)^{-1/2} \cdot \frac{1}{t^{1/2}} \cdot \frac{d}{dt}\left(t^{1/2}\right)$
$= \frac{1}{2}\left(\ln t^{1/2}\right)^{-1/2} \cdot \frac{1}{t^{1/2}} \cdot \frac{1}{2}t^{-1/2} = \frac{1}{4t\sqrt{\ln\sqrt{t}}}$

37. $y = \ln(\sec(\ln\theta)) \Rightarrow \frac{dy}{d\theta} = \frac{1}{\sec(\ln\theta)} \cdot \frac{d}{d\theta}(\sec(\ln\theta)) = \frac{\sec(\ln\theta)\tan(\ln\theta)}{\sec(\ln\theta)} \cdot \frac{d}{d\theta}(\ln\theta) = \frac{\tan(\ln\theta)}{\theta}$

38. $y = \ln\frac{\sqrt{\sin\theta\cos\theta}}{1+2\ln\theta} = \frac{1}{2}(\ln\sin\theta + \ln\cos\theta) - \ln(1+2\ln\theta) \Rightarrow \frac{dy}{d\theta} = \frac{1}{2}\left(\frac{\cos\theta}{\sin\theta} - \frac{\sin\theta}{\cos\theta}\right) - \frac{\frac{2}{\theta}}{1+2\ln\theta}$
$= \frac{1}{2}\left[\cot\theta - \tan\theta - \frac{4}{\theta(1+2\ln\theta)}\right]$

39. $y = \ln\left(\frac{(x^2+1)^5}{\sqrt{1-x}}\right) = 5\ln(x^2+1) - \frac{1}{2}\ln(1-x) \Rightarrow y' = \frac{5\cdot 2x}{x^2+1} - \frac{1}{2}\left(\frac{1}{1-x}\right)(-1) = \frac{10x}{x^2+1} + \frac{1}{2(1-x)}$

40. $y = \ln\sqrt{\frac{(x+1)^5}{(x+2)^{20}}} = \frac{1}{2}[5\ln(x+1) - 20\ln(x+2)] \Rightarrow y' = \frac{1}{2}\left(\frac{5}{x+1} - \frac{20}{x+2}\right) = \frac{5}{2}\left[\frac{(x+2)-4(x+1)}{(x+1)(x+2)}\right]$
$= -\frac{5}{2}\left[\frac{3x+2}{(x+1)(x+2)}\right]$

41. $y = \sqrt{x(x+1)} = (x(x+1))^{1/2} \Rightarrow \ln y = \frac{1}{2}\ln(x(x+1)) \Rightarrow 2\ln y = \ln(x) + \ln(x+1) \Rightarrow \frac{2y'}{y} = \frac{1}{x} + \frac{1}{x+1}$
$\Rightarrow y' = \left(\frac{1}{2}\right)\sqrt{x(x+1)}\left(\frac{1}{x} + \frac{1}{x+1}\right) = \frac{\sqrt{x(x+1)}\,(2x+1)}{2x(x+1)} = \frac{2x+1}{2\sqrt{x(x+1)}}$

42. $y = \sqrt{(x^2+1)(x-1)^2} \Rightarrow \ln y = \frac{1}{2}[\ln(x^2+1) + 2\ln(x-1)] \Rightarrow \frac{y'}{y} = \frac{1}{2}\left(\frac{2x}{x^2+1} + \frac{2}{x-1}\right)$
$\Rightarrow y' = \sqrt{(x^2+1)(x-1)^2}\left(\frac{x}{x^2+1} + \frac{1}{x-1}\right) = \sqrt{(x^2+1)(x-1)^2}\left[\frac{x^2-x+x^2+1}{(x^2+1)(x-1)}\right] = \frac{(2x^2-x+1)\,|x-1|}{\sqrt{x^2+1}\,(x-1)}$

43. $y = \sqrt{\frac{t}{t+1}} = \left(\frac{t}{t+1}\right)^{1/2} \Rightarrow \ln y = \frac{1}{2}[\ln t - \ln(t+1)] \Rightarrow \frac{1}{y}\frac{dy}{dt} = \frac{1}{2}\left(\frac{1}{t} - \frac{1}{t+1}\right)$
$\Rightarrow \frac{dy}{dt} = \frac{1}{2}\sqrt{\frac{t}{t+1}}\left(\frac{1}{t} - \frac{1}{t+1}\right) = \frac{1}{2}\sqrt{\frac{t}{t+1}}\left[\frac{1}{t(t+1)}\right] = \frac{1}{2\sqrt{t}\,(t+1)^{3/2}}$

44. $y = \sqrt{\frac{1}{t(t+1)}} = [t(t+1)]^{-1/2} \Rightarrow \ln y = \frac{1}{2}[\ln t + \ln(t+1)] \Rightarrow \frac{1}{y}\frac{dy}{dt} = -\frac{1}{2}\left(\frac{1}{t} + \frac{1}{t+1}\right)$
$\Rightarrow \frac{dy}{dt} = -\frac{1}{2}\sqrt{\frac{1}{t(t+1)}}\left[\frac{2t+1}{t(t+1)}\right] = -\frac{2t+1}{2\,(t^2+t)^{3/2}}$

45. $y = \sqrt{\theta+3}\,(\sin\theta) = (\theta+3)^{1/2}\sin\theta \Rightarrow \ln y = \frac{1}{2}\ln(\theta+3) + \ln(\sin\theta) \Rightarrow \frac{1}{y}\frac{dy}{d\theta} = \frac{1}{2(\theta+3)} + \frac{\cos\theta}{\sin\theta}$
$\Rightarrow \frac{dy}{d\theta} = \sqrt{\theta+3}\,(\sin\theta)\left[\frac{1}{2(\theta+3)} + \cot\theta\right]$

46. $y = (\tan\theta)\sqrt{2\theta+1} = (\tan\theta)(2\theta+1)^{1/2} \Rightarrow \ln y = \ln(\tan\theta) + \frac{1}{2}\ln(2\theta+1) \Rightarrow \frac{1}{y}\frac{dy}{d\theta} = \frac{\sec^2\theta}{\tan\theta} + \left(\frac{1}{2}\right)\left(\frac{2}{2\theta+1}\right)$
$\Rightarrow \frac{dy}{d\theta} = (\tan\theta)\sqrt{2\theta+1}\left(\frac{\sec^2\theta}{\tan\theta} + \frac{1}{2\theta+1}\right) = (\sec^2\theta)\sqrt{2\theta+1} + \frac{\tan\theta}{\sqrt{2\theta+1}}$

47. $y = t(t+1)(t+2) \Rightarrow \ln y = \ln t + \ln(t+1) + \ln(t+2) \Rightarrow \frac{1}{y}\frac{dy}{dt} = \frac{1}{t} + \frac{1}{t+1} + \frac{1}{t+2}$
$\Rightarrow \frac{dy}{dt} = t(t+1)(t+2)\left(\frac{1}{t} + \frac{1}{t+1} + \frac{1}{t+2}\right) = t(t+1)(t+2)\left[\frac{(t+1)(t+2)+t(t+2)+t(t+1)}{t(t+1)(t+2)}\right] = 3t^2 + 6t + 2$

48. $y = \frac{1}{t(t+1)(t+2)} \Rightarrow \ln y = \ln 1 - \ln t - \ln(t+1) - \ln(t+2) \Rightarrow \frac{1}{y}\frac{dy}{dt} = -\frac{1}{t} - \frac{1}{t+1} - \frac{1}{t+2}$
$\Rightarrow \frac{dy}{dt} = \frac{1}{t(t+1)(t+2)}\left[-\frac{1}{t} - \frac{1}{t+1} - \frac{1}{t+2}\right] = \frac{-1}{t(t+1)(t+2)}\left[\frac{(t+1)(t+2)+t(t+2)+t(t+1)}{t(t+1)(t+2)}\right]$
$= -\frac{3t^2+6t+2}{(t^3+3t^2+2t)^2}$

49. $y = \frac{\theta+5}{\theta\cos\theta} \Rightarrow \ln y = \ln(\theta+5) - \ln\theta - \ln(\cos\theta) \Rightarrow \frac{1}{y}\frac{dy}{d\theta} = \frac{1}{\theta+5} - \frac{1}{\theta} + \frac{\sin\theta}{\cos\theta}$
$\Rightarrow \frac{dy}{d\theta} = \left(\frac{\theta+5}{\theta\cos\theta}\right)\left(\frac{1}{\theta+5} - \frac{1}{\theta} + \tan\theta\right)$

50. $y = \frac{\theta\sin\theta}{\sqrt{\sec\theta}} \Rightarrow \ln y = \ln\theta + \ln(\sin\theta) - \frac{1}{2}\ln(\sec\theta) \Rightarrow \frac{1}{y}\frac{dy}{d\theta} = \left[\frac{1}{\theta} + \frac{\cos\theta}{\sin\theta} - \frac{(\sec\theta)(\tan\theta)}{2\sec\theta}\right]$
$\Rightarrow \frac{dy}{d\theta} = \frac{\theta\sin\theta}{\sqrt{\sec\theta}}\left(\frac{1}{\theta} + \cot\theta - \frac{1}{2}\tan\theta\right)$

51. $y = \frac{x\sqrt{x^2+1}}{(x+1)^{2/3}} \Rightarrow \ln y = \ln x + \frac{1}{2}\ln(x^2+1) - \frac{2}{3}\ln(x+1) \Rightarrow \frac{y'}{y} = \frac{1}{x} + \frac{x}{x^2+1} - \frac{2}{3(x+1)}$
$\Rightarrow y' = \frac{x\sqrt{x^2+1}}{(x+1)^{2/3}}\left[\frac{1}{x} + \frac{x}{x^2+1} - \frac{2}{3(x+1)}\right]$

52. $y = \sqrt{\frac{(x+1)^{10}}{(2x+1)^5}} \Rightarrow \ln y = \frac{1}{2}\left[10 \ln(x+1) - 5 \ln(2x+1)\right] \Rightarrow \frac{y'}{y} = \frac{5}{x+1} - \frac{5}{2x+1}$
$\Rightarrow y' = \sqrt{\frac{(x+1)^{10}}{(2x+1)^5}}\left(\frac{5}{x+1} - \frac{5}{2x+1}\right)$

53. $y = \sqrt[3]{\frac{x(x-2)}{x^2+1}} \Rightarrow \ln y = \frac{1}{3}\left[\ln x + \ln(x-2) - \ln(x^2+1)\right] \Rightarrow \frac{y'}{y} = \frac{1}{3}\left(\frac{1}{x} + \frac{1}{x-2} - \frac{2x}{x^2+1}\right)$
$\Rightarrow y' = \frac{1}{3}\sqrt[3]{\frac{x(x-2)}{x^2+1}}\left(\frac{1}{x} + \frac{1}{x-2} - \frac{2x}{x^2+1}\right)$

54. $y = \sqrt[3]{\frac{x(x+1)(x-2)}{(x^2+1)(2x+3)}} \Rightarrow \ln y = \frac{1}{3}\left[\ln x + \ln(x+1) + \ln(x-2) - \ln(x^2+1) - \ln(2x+3)\right]$
$\Rightarrow y' = \frac{1}{3}\sqrt[3]{\frac{x(x+1)(x-2)}{(x^2+1)(2x+3)}}\left(\frac{1}{x} + \frac{1}{x+1} + \frac{1}{x-2} - \frac{2x}{x^2+1} - \frac{2}{2x+3}\right)$

55. $y = \ln(\cos^2\theta) \Rightarrow \frac{dy}{d\theta} = \frac{1}{\cos^2\theta} \cdot 2\cos\theta \cdot (-\sin\theta) = -2\tan\theta$

56. $y = \ln\left(3\theta e^{-\theta}\right) = \ln 3 + \ln\theta + \ln e^{-\theta} = \ln 3 + \ln\theta - \theta \Rightarrow \frac{dy}{d\theta} = \frac{1}{\theta} - 1$

57. $y = \ln(3te^{-t}) = \ln 3 + \ln t + \ln e^{-t} = \ln 3 + \ln t - t \Rightarrow \frac{dy}{dt} = \frac{1}{t} - 1 = \frac{1-t}{t}$

58. $y = \ln(2e^{-t}\sin t) = \ln 2 + \ln e^{-t} + \ln \sin t = \ln 2 - t + \ln\sin t \Rightarrow \frac{dy}{dt} = -1 + \left(\frac{1}{\sin t}\right)\frac{d}{dt}(\sin t) = -1 + \frac{\cos t}{\sin t}$
$= \frac{\cos t - \sin t}{\sin t}$

59. $y = \ln\frac{e^\theta}{1+e^\theta} = \ln e^\theta - \ln\left(1+e^\theta\right) = \theta - \ln\left(1+e^\theta\right) \Rightarrow \frac{dy}{d\theta} = 1 - \left(\frac{1}{1+e^\theta}\right)\frac{d}{d\theta}\left(1+e^\theta\right) = 1 - \frac{e^\theta}{1+e^\theta} = \frac{1}{1+e^\theta}$

60. $y = \ln\frac{\sqrt{\theta}}{1+\sqrt{\theta}} = \ln\sqrt{\theta} - \ln\left(1+\sqrt{\theta}\right) \Rightarrow \frac{dy}{d\theta} = \left(\frac{1}{\sqrt{\theta}}\right)\frac{d}{d\theta}\left(\sqrt{\theta}\right) - \left(\frac{1}{1+\sqrt{\theta}}\right)\frac{d}{d\theta}\left(1+\sqrt{\theta}\right)$
$= \left(\frac{1}{\sqrt{\theta}}\right)\left(\frac{1}{2\sqrt{\theta}}\right) - \left(\frac{1}{1+\sqrt{\theta}}\right)\left(\frac{1}{2\sqrt{\theta}}\right) = \frac{\left(1+\sqrt{\theta}\right) - \sqrt{\theta}}{2\theta\left(1+\sqrt{\theta}\right)} = \frac{1}{2\theta\left(1+\sqrt{\theta}\right)} = \frac{1}{2\theta\left(1+\theta^{1/2}\right)}$

61. $y = e^{(\cos t + \ln t)} = e^{\cos t} e^{\ln t} = te^{\cos t} \Rightarrow \frac{dy}{dt} = e^{\cos t} + te^{\cos t}\frac{d}{dt}(\cos t) = (1 - t\sin t)e^{\cos t}$

62. $y = e^{\sin t}(\ln t^2 + 1) \Rightarrow \frac{dy}{dt} = e^{\sin t}(\cos t)(\ln t^2 + 1) + \frac{2}{t}e^{\sin t} = e^{\sin t}\left[(\ln t^2 + 1)(\cos t) + \frac{2}{t}\right]$

63. $\ln y = e^y \sin x \Rightarrow \left(\frac{1}{y}\right)y' = (y'e^y)(\sin x) + e^y\cos x \Rightarrow y'\left(\frac{1}{y} - e^y\sin x\right) = e^y\cos x$
$\Rightarrow y'\left(\frac{1 - ye^y\sin x}{y}\right) = e^y\cos x \Rightarrow y' = \frac{ye^y\cos x}{1 - ye^y\sin x}$

64. $\ln xy = e^{x+y} \Rightarrow \ln x + \ln y = e^{x+y} \Rightarrow \frac{1}{x} + \left(\frac{1}{y}\right)y' = (1+y')e^{x+y} \Rightarrow y'\left(\frac{1}{y} - e^{x+y}\right) = e^{x+y} - \frac{1}{x}$
$\Rightarrow y'\left(\frac{1 - ye^{x+y}}{y}\right) = \frac{xe^{x+y} - 1}{x} \Rightarrow y' = \frac{y(xe^{x+y} - 1)}{x(1 - ye^{x+y})}$

65. $x^y = y^x \Rightarrow \ln x^y = \ln y^x \Rightarrow y\ln x = x\ln y \Rightarrow y \cdot \frac{1}{x} + y' \cdot \ln x = x \cdot \frac{1}{y} \cdot y' + (1) \cdot \ln y \Rightarrow \ln x \cdot y' - \frac{x}{y} \cdot y' = \ln y - \frac{y}{x}$
$\Rightarrow y' = \frac{\ln y - \frac{y}{x}}{\ln x - \frac{x}{y}} = \frac{xy\ln y - y^2}{xy\ln x - x^2} = \frac{y}{x}\left(\frac{x\ln y - y}{y\ln x - x}\right)$

66. $\tan y = e^x + \ln x \Rightarrow (\sec^2 y)y' = e^x + \frac{1}{x} \Rightarrow y' = \frac{(xe^x + 1)\cos^2 y}{x}$

67. $y = 2^x \Rightarrow y' = 2^x \ln 2$

68. $y = 3^{-x} \Rightarrow y' = 3^{-x}(\ln 3)(-1) = -3^{-x}\ln 3$

69. $y = 5^{\sqrt{s}} \Rightarrow \frac{dy}{ds} = 5^{\sqrt{s}}(\ln 5)\left(\frac{1}{2}s^{-1/2}\right) = \left(\frac{\ln 5}{2\sqrt{s}}\right)5^{\sqrt{s}}$

70. $y = 2^{s^2} \Rightarrow \frac{dy}{ds} = 2^{s^2}(\ln 2)2s = (\ln 2^2)\left(s2^{s^2}\right) = (\ln 4)s2^{s^2}$

71. $y = x^{\pi} \Rightarrow y' = \pi x^{(\pi-1)}$

72. $y = t^{1-e} \Rightarrow \frac{dy}{dt} = (1-e)t^{-e}$

73. $y = \log_2 5\theta = \frac{\ln 5\theta}{\ln 2} \Rightarrow \frac{dy}{d\theta} = \left(\frac{1}{\ln 2}\right)\left(\frac{1}{5\theta}\right)(5) = \frac{1}{\theta \ln 2}$

74. $y = \log_3(1+\theta\ln 3) = \frac{\ln(1+\theta\ln 3)}{\ln 3} \Rightarrow \frac{dy}{d\theta} = \left(\frac{1}{\ln 3}\right)\left(\frac{1}{1+\theta\ln 3}\right)(\ln 3) = \frac{1}{1+\theta\ln 3}$

75. $y = \frac{\ln x}{\ln 4} + \frac{\ln x^2}{\ln 4} = \frac{\ln x}{\ln 4} + 2\frac{\ln x}{\ln 4} = 3\frac{\ln x}{\ln 4} \Rightarrow y' = \frac{3}{x\ln 4}$

76. $y = \frac{x\ln e}{\ln 25} - \frac{\ln x}{2\ln 5} = \frac{x}{2\ln 5} - \frac{\ln x}{2\ln 5} = \left(\frac{1}{2\ln 5}\right)(x - \ln x) \Rightarrow y' = \left(\frac{1}{2\ln 5}\right)\left(1 - \frac{1}{x}\right) = \frac{x-1}{2x\ln 5}$

77. $y = \log_2 r \cdot \log_4 r = \left(\frac{\ln r}{\ln 2}\right)\left(\frac{\ln r}{\ln 4}\right) = \frac{\ln^2 r}{(\ln 2)(\ln 4)} \Rightarrow \frac{dy}{dr} = \left[\frac{1}{(\ln 2)(\ln 4)}\right](2\ln r)\left(\frac{1}{r}\right) = \frac{2\ln r}{r(\ln 2)(\ln 4)}$

78. $y = \log_3 r \cdot \log_9 r = \left(\frac{\ln r}{\ln 3}\right)\left(\frac{\ln r}{\ln 9}\right) = \frac{\ln^2 r}{(\ln 3)(\ln 9)} \Rightarrow \frac{dy}{dr} = \left[\frac{1}{(\ln 3)(\ln 9)}\right](2\ln r)\left(\frac{1}{r}\right) = \frac{2\ln r}{r(\ln 3)(\ln 9)}$

79. $y = \log_3\left(\left(\frac{x+1}{x-1}\right)^{\ln 3}\right) = \frac{\ln\left(\frac{x+1}{x-1}\right)^{\ln 3}}{\ln 3} = \frac{(\ln 3)\ln\left(\frac{x+1}{x-1}\right)}{\ln 3} = \ln\left(\frac{x+1}{x-1}\right) = \ln(x+1) - \ln(x-1)$
$\Rightarrow \frac{dy}{dx} = \frac{1}{x+1} - \frac{1}{x-1} = \frac{-2}{(x+1)(x-1)}$

80. $y = \log_5\sqrt{\left(\frac{7x}{3x+2}\right)^{\ln 5}} = \log_5\left(\frac{7x}{3x+2}\right)^{(\ln 5)/2} = \frac{\ln\left(\frac{7x}{3x+2}\right)^{(\ln 5)/2}}{\ln 5} = \left(\frac{\ln 5}{2}\right)\left[\frac{\ln\left(\frac{7x}{3x+2}\right)}{\ln 5}\right] = \frac{1}{2}\ln\left(\frac{7x}{3x+2}\right)$
$= \frac{1}{2}\ln 7x - \frac{1}{2}\ln(3x+2) \Rightarrow \frac{dy}{dx} = \frac{7}{2\cdot 7x} - \frac{3}{2\cdot(3x+2)} = \frac{(3x+2)-3x}{2x(3x+2)} = \frac{1}{x(3x+2)}$

81. $y = \theta\sin(\log_7\theta) = \theta\sin\left(\frac{\ln\theta}{\ln 7}\right) \Rightarrow \frac{dy}{d\theta} = \sin\left(\frac{\ln\theta}{\ln 7}\right) + \theta\left[\cos\left(\frac{\ln\theta}{\ln 7}\right)\right]\left(\frac{1}{\theta\ln 7}\right) = \sin(\log_7\theta) + \frac{1}{\ln 7}\cos(\log_7\theta)$

82. $y = \log_7\left(\frac{\sin\theta\cos\theta}{e^{\theta}2^{\theta}}\right) = \frac{\ln(\sin\theta) + \ln(\cos\theta) - \ln e^{\theta} - \ln 2^{\theta}}{\ln 7} = \frac{\ln(\sin\theta) + \ln(\cos\theta) - \theta - \theta\ln 2}{\ln 7}$
$\Rightarrow \frac{dy}{d\theta} = \frac{\cos\theta}{(\sin\theta)(\ln 7)} - \frac{\sin\theta}{(\cos\theta)(\ln 7)} - \frac{1}{\ln 7} - \frac{\ln 2}{\ln 7} = \left(\frac{1}{\ln 7}\right)(\cot\theta - \tan\theta - 1 - \ln 2)$

83. $y = \log_5 e^x = \frac{\ln e^x}{\ln 5} = \frac{x}{\ln 5} \Rightarrow y' = \frac{1}{\ln 5}$

84. $y = \log_2\left(\frac{x^2e^2}{2\sqrt{x+1}}\right) = \frac{\ln x^2 + \ln e^2 - \ln 2 - \ln\sqrt{x+1}}{\ln 2} = \frac{2\ln x + 2 - \ln 2 - \frac{1}{2}\ln(x+1)}{\ln 2}$
$\Rightarrow y' = \frac{2}{x\ln 2} - \frac{1}{2(\ln 2)(x+1)} = \frac{4(x+1) - x}{2x(x+1)(\ln 2)} = \frac{3x+4}{2x(x+1)\ln 2}$

85. $y = 3^{\log_2 t} = 3^{(\ln t)/(\ln 2)} \Rightarrow \frac{dy}{dt} = \left[3^{(\ln t)/(\ln 2)}(\ln 3)\right]\left(\frac{1}{t\ln 2}\right) = \frac{1}{t}(\log_2 3)3^{\log_2 t}$

86. $y = 3\log_8(\log_2 t) = \frac{3\ln(\log_2 t)}{\ln 8} = \frac{3\ln\left(\frac{\ln t}{\ln 2}\right)}{\ln 8} \Rightarrow \frac{dy}{dt} = \left(\frac{3}{\ln 8}\right)\left[\frac{1}{(\ln t)/(\ln 2)}\right]\left(\frac{1}{t\ln 2}\right) = \frac{3}{t(\ln t)(\ln 8)}$
$= \frac{1}{t(\ln t)(\ln 2)}$

87. $y = \log_2(8t^{\ln 2}) = \frac{\ln 8 + \ln(t^{\ln 2})}{\ln 2} = \frac{3\ln 2 + (\ln 2)(\ln t)}{\ln 2} = 3 + \ln t \Rightarrow \frac{dy}{dt} = \frac{1}{t}$

88. $y = \frac{t\ln\left(\left(e^{\ln 3}\right)^{\sin t}\right)}{\ln 3} = \frac{t\ln\left(3^{\sin t}\right)}{\ln 3} = \frac{t(\sin t)(\ln 3)}{\ln 3} = t\sin t \Rightarrow \frac{dy}{dt} = \sin t + t\cos t$

89. $y = (x+1)^x \Rightarrow \ln y = \ln(x+1)^x = x\ln(x+1) \Rightarrow \frac{y'}{y} = \ln(x+1) + x\cdot\frac{1}{(x+1)} \Rightarrow y' = (x+1)^x\left[\frac{x}{x+1} + \ln(x+1)\right]$

90. $y = x^{(x+1)} \Rightarrow \ln y = \ln x^{(x+1)} = (x+1)\ln x \Rightarrow \frac{y'}{y} = \ln x + (x+1)\left(\frac{1}{x}\right) = \ln x + 1 + \frac{1}{x}$
$\Rightarrow y' = x^{(x+1)}\left(1 + \frac{1}{x} + \ln x\right)$

91. $y = \left(\sqrt{t}\right)^t = \left(t^{1/2}\right)^t = t^{t/2} \Rightarrow \ln y = \ln t^{t/2} = \left(\frac{t}{2}\right)\ln t \Rightarrow \frac{1}{y}\frac{dy}{dt} = \left(\frac{1}{2}\right)(\ln t) + \left(\frac{t}{2}\right)\left(\frac{1}{t}\right) = \frac{\ln t}{2} + \frac{1}{2}$
$\Rightarrow \frac{dy}{dt} = \left(\sqrt{t}\right)^t\left(\frac{\ln t}{2} + \frac{1}{2}\right)$

92. $y = t^{\sqrt{t}} = t^{\left(t^{1/2}\right)} \Rightarrow \ln y = \ln t^{\left(t^{1/2}\right)} = \left(t^{1/2}\right)(\ln t) \Rightarrow \frac{1}{y}\frac{dy}{dt} = \left(\frac{1}{2}t^{-1/2}\right)(\ln t) + t^{1/2}\left(\frac{1}{t}\right) = \frac{\ln t + 2}{2\sqrt{t}}$
$\Rightarrow \frac{dy}{dt} = \left(\frac{\ln t + 2}{2\sqrt{t}}\right)t^{\sqrt{t}}$

93. $y = (\sin x)^x \Rightarrow \ln y = \ln(\sin x)^x = x\ln(\sin x) \Rightarrow \frac{y'}{y} = \ln(\sin x) + x\left(\frac{\cos x}{\sin x}\right) \Rightarrow y' = (\sin x)^x\left[\ln(\sin x) + x\cot x\right]$

94. $y = x^{\sin x} \Rightarrow \ln y = \ln x^{\sin x} = (\sin x)(\ln x) \Rightarrow \frac{y'}{y} = (\cos x)(\ln x) + (\sin x)\left(\frac{1}{x}\right) = \frac{\sin x + x(\ln x)(\cos x)}{x}$
$\Rightarrow y' = x^{\sin x}\left[\frac{\sin x + x(\ln x)(\cos x)}{x}\right]$

95. $y = x^{\ln x}, x > 0 \Rightarrow \ln y = (\ln x)^2 \Rightarrow \frac{y'}{y} = 2(\ln x)\left(\frac{1}{x}\right) \Rightarrow y' = \left(x^{\ln x}\right)\left(\frac{\ln x^2}{x}\right)$

96. $y = (\ln x)^{\ln x} \Rightarrow \ln y = (\ln x)\ln(\ln x) \Rightarrow \frac{y'}{y} = \left(\frac{1}{x}\right)\ln(\ln x) + (\ln x)\left(\frac{1}{\ln x}\right)\frac{d}{dx}(\ln x) = \frac{\ln(\ln x)}{x} + \frac{1}{x}$
$\Rightarrow y' = \left(\frac{\ln(\ln x) + 1}{x}\right)(\ln x)^{\ln x}$

97. $(g \circ f)(x) = x \Rightarrow g(f(x)) = x \Rightarrow g'(f(x))f'(x) = 1$

98. $\lim_{n\to\infty}\left(1 + \frac{x}{n}\right)^n = \lim_{n\to\infty}\left[\left(1 + \frac{1}{(n/x)}\right)^{(n/x)}\right]^x = e^x$ for any $x > 0$.

99. $y = A\sin(\ln x) + B\cos(\ln x) \Rightarrow y' = A\cos(\ln x)\cdot\frac{1}{x} - B\sin(\ln x)\cdot\frac{1}{x} = (A\cos(\ln x) - B\sin(\ln x))\cdot\frac{1}{x}$
$\Rightarrow y'' = (A\cos(\ln x) - B\sin(\ln x))\cdot\frac{-1}{x^2} + \left(-A\sin(\ln x)\cdot\frac{1}{x} - B\cos(\ln x)\cdot\frac{1}{x}\right)\cdot\frac{1}{x}$
$= (-A(\cos(\ln x) + \sin(\ln x)) + B(\sin(\ln x) - \cos(\ln x)))\cdot\frac{1}{x^2}$
$\Rightarrow x^2y'' + xy' + y = (-A(\cos(\ln x) + \sin(\ln x)) + B(\sin(\ln x) - \cos(\ln x))) + (A\cos(\ln x) - B\sin(\ln x))$
$+ (A\sin(\ln x) + B\cos(\ln x)) = 0$

100. Suppose $n = 1$. Then $\frac{d}{dx}\ln x = \frac{1}{x} = (-1)^0\cdot\frac{0!}{x^1}$ and so the base case is established. Now if the statement holds for $n = k$ we have that, for $n = k+1$, the following holds:
$\frac{d^n}{dx^n}(\ln x) = \frac{d^{k+1}}{dx^{k+1}}(\ln x) = \frac{d}{dx}\left(\frac{d^k}{dx^k}(\ln x)\right) = \frac{d}{dx}\left((-1)^{k-1}\frac{(k-1)!}{x^k}\right) = (-1)^{k-1}(k-1)!\cdot(-k)x^{-k-1} = \frac{(-1)^k k!}{x^{k+1}} = \frac{(-1)^{n-1}(n-1)!}{x^n}$
Thus by mathematical induction the result is established for all $n \geq 1$.

101-108.Example CAS commands:

Maple:

```
with( plots );#101
f := x -> sqrt(3*x-2);
```

```
domain := 2/3 .. 4;
x0 := 3;
Df := D(f);                             # (a)
plot( [f(x),Df(x)], x=domain, color=[red,blue], linestyle=[1,3], legend=["y=f(x)","y=f '(x)"],
    title="#53(a) (Section 7.1)" );
q1 := solve( y=f(x), x );               # (b)
g := unapply( q1, y );
m1 := Df(x0);                           # (c)
t1 := f(x0)+m1*(x-x0);
y=t1;
m2 := 1/Df(x0);                         # (d)
t2 := g(f(x0)) + m2*(x-f(x0));
y=t2;
domaing := map(f,domain);       # (e)
p1 := plot( [f(x),x], x=domain, color=[pink,green], linestyle=[1,9], thickness=[3,0] ):
p2 := plot( g(x), x=domaing, color=cyan, linestyle=3, thickness=4 ):
p3 := plot( t1, x=x0-1..x0+1, color=red, linestyle=4, thickness=0 ):
p4 := plot( t2, x=f(x0)-1..f(x0)+1, color=blue, linestyle=7, thickness=1 ):
p5 := plot( [ [x0,f(x0)], [f(x0),x0] ], color=green ):
display( [p1,p2,p3,p4,p5], scaling=constrained, title="#53(e) (Section 7.1)" );
```

<u>Mathematica:</u> (assigned function and values for a, b, and x0 may vary)

If a function requires the odd root of a negative number, begin by loading the RealOnly package that allows Mathematica to do this. See section 2.5 for details.

```
<<Miscellaneous `RealOnly`
Clear[x, y]
{a,b} = {−2, 1}; x0 = 1/2 ;
f[x_] = (3x + 2) / (2x − 11)
Plot[{f[x], f'[x]}, {x, a, b}]
solx = Solve[y == f[x], x]
g[y_] = x /. solx[[1]]
y0 = f[x0]
ftan[x_] = y0 + f'[x0] (x-x0)
gtan[y_] = x0 + 1/ f'[x0] (y − y0)
Plot[{f[x], ftan[x], g[x], gtan[x], Identity[x]},{x, a, b},
Epilog → Line[{{x0, y0},{y0, x0}}], PlotRange → {{a,b},{a,b}}, AspectRatio → Automatic]
```

109-110.Example CAS commands:

<u>Maple</u>:

```
with( plots );
eq := cos(y) = x^(1/5);
domain := 0 .. 1;
x0 := 1/2;
f := unapply( solve( eq, y ), x );  # (a)
Df := D(f);
plot( [f(x),Df(x)], x=domain, color=[red,blue], linestyle=[1,3], legend=["y=f(x)","y=f '(x)"],
    title="#62(a) (Section 7.1)" );
q1 := solve( eq, x );                   # (b)
g := unapply( q1, y );
m1 := Df(x0);                           # (c)
```

```
t1 := f(x0)+m1*(x-x0);
y=t1;
m2 := 1/Df(x0);                    # (d)
t2 := g(f(x0)) + m2*(x-f(x0));
y=t2;
domaing := map(f,domain);          # (e)
p1 := plot( [f(x),x], x=domain, color=[pink,green], linestyle=[1,9], thickness=[3,0] ):
p2 := plot( g(x), x=domaing, color=cyan, linestyle=3, thickness=4 ):
p3 := plot( t1, x=x0-1..x0+1, color=red, linestyle=4, thickness=0 ):
p4 := plot( t2, x=f(x0)-1..f(x0)+1, color=blue, linestyle=7, thickness=1 ):
p5 := plot( [ [x0,f(x0)], [f(x0),x0] ], color=green ):
display( [p1,p2,p3,p4,p5], scaling=constrained, title="#62(e) (Section 7.1)" );
```

Mathematica: (assigned function and values for a, b, and x0 may vary)

For problems 61 and 62, the code is just slightly altered. At times, different "parts" of solutions need to be used, as in the definitions of f[x] and g[y]

```
Clear[x, y]
{a,b} = {0, 1}; x0 = 1/2 ;
eqn = Cos[y] == x^(1/5)
soly = Solve[eqn, y]
f[x_] = y /. soly[[2]]
Plot[{f[x], f'[x]}, {x, a, b}]
solx = Solve[eqn, x]
g[y_] = x /. solx[[1]]
y0 = f[x0]
ftan[x_] = y0 + f'[x0] (x − x0)
gtan[y_] = x0 + 1/ f'[x0] (y − y0)
Plot[{f[x], ftan[x], g[x], gtan[x], Identity[x]},{x, a, b},
Epilog → Line[{{x0, y0},{y0, x0}}], PlotRange → {{a, b}, {a, b}}, AspectRatio → Automatic]
```

3.8 INVERSE TRIGONOMETRIC FUNCTIONS

1. (a) $\frac{\pi}{4}$ (b) $-\frac{\pi}{3}$ (c) $\frac{\pi}{6}$

2. (a) $-\frac{\pi}{4}$ (b) $\frac{\pi}{3}$ (c) $-\frac{\pi}{6}$

3. (a) $-\frac{\pi}{6}$ (b) $\frac{\pi}{4}$ (c) $-\frac{\pi}{3}$

4. (a) $\frac{\pi}{6}$ (b) $-\frac{\pi}{4}$ (c) $\frac{\pi}{3}$

5. (a) $\frac{\pi}{3}$ (b) $\frac{3\pi}{4}$ (c) $\frac{\pi}{6}$

6. (a) $\frac{2\pi}{3}$ (b) $\frac{\pi}{4}$ (c) $\frac{5\pi}{6}$

7. (a) $\frac{3\pi}{4}$ (b) $\frac{\pi}{6}$ (c) $\frac{2\pi}{3}$

8. (a) $\frac{\pi}{4}$ (b) $\frac{5\pi}{6}$ (c) $\frac{\pi}{3}$

9. (a) $\frac{\pi}{4}$ (b) $-\frac{\pi}{3}$ (c) $\frac{\pi}{6}$

10. (a) $-\frac{\pi}{4}$ (b) $\frac{\pi}{3}$ (c) $-\frac{\pi}{6}$

11. (a) $\frac{3\pi}{4}$ (b) $\frac{\pi}{6}$ (c) $\frac{2\pi}{3}$

12. (a) $\frac{\pi}{4}$ (b) $\frac{5\pi}{6}$ (c) $\frac{\pi}{3}$

13. $\alpha = \sin^{-1}\left(\frac{5}{13}\right) \Rightarrow \cos\alpha = \frac{12}{13}$, $\tan\alpha = \frac{5}{12}$, $\sec\alpha = \frac{13}{12}$, $\csc\alpha = \frac{13}{5}$, and $\cot\alpha = \frac{12}{5}$

14. $\alpha = \tan^{-1}\left(\frac{4}{3}\right) \Rightarrow \sin\alpha = \frac{4}{5}$, $\cos\alpha = \frac{3}{5}$, $\sec\alpha = \frac{5}{3}$, $\csc\alpha = \frac{5}{4}$, and $\cot\alpha = \frac{3}{4}$

15. $\alpha = \sec^{-1}\left(-\sqrt{5}\right) \Rightarrow \sin\alpha = \frac{2}{\sqrt{5}}$, $\cos\alpha = -\frac{1}{\sqrt{5}}$, $\tan\alpha = -2$, $\csc\alpha = \frac{\sqrt{5}}{2}$, and $\cot\alpha = -\frac{1}{2}$

16. $\alpha = \sec^{-1}\left(-\frac{\sqrt{13}}{2}\right) \Rightarrow \sin\alpha = \frac{3}{\sqrt{13}}, \cos\alpha = -\frac{2}{\sqrt{13}}, \tan\alpha = -\frac{3}{2}, \csc\alpha = \frac{\sqrt{13}}{3}$, and $\cot\alpha = -\frac{2}{3}$

17. $\sin\left(\cos^{-1}\frac{\sqrt{2}}{2}\right) = \sin\left(\frac{\pi}{4}\right) = \frac{1}{\sqrt{2}}$

18. $\sec\left(\cos^{-1}\frac{1}{2}\right) = \sec\left(\frac{\pi}{3}\right) = 2$

19. $\tan\left(\sin^{-1}\left(-\frac{1}{2}\right)\right) = \tan\left(-\frac{\pi}{6}\right) = -\frac{1}{\sqrt{3}}$

20. $\cot\left(\sin^{-1}\left(-\frac{\sqrt{3}}{2}\right)\right) = \cot\left(-\frac{\pi}{3}\right) = -\frac{1}{\sqrt{3}}$

21. $\csc\left(\sec^{-1}2\right) + \cos\left(\tan^{-1}\left(-\sqrt{3}\right)\right) = \csc\left(\cos^{-1}\left(\frac{1}{2}\right)\right) + \cos\left(-\frac{\pi}{3}\right) = \csc\left(\frac{\pi}{3}\right) + \cos\left(-\frac{\pi}{3}\right) = \frac{2}{\sqrt{3}} + \frac{1}{2} = \frac{4+\sqrt{3}}{2\sqrt{3}}$

22. $\tan\left(\sec^{-1}1\right) + \sin\left(\csc^{-1}(-2)\right) = \tan\left(\cos^{-1}\frac{1}{1}\right) + \sin\left(\sin^{-1}\left(-\frac{1}{2}\right)\right) = \tan(0) + \sin\left(-\frac{\pi}{6}\right) = 0 + \left(-\frac{1}{2}\right) = -\frac{1}{2}$

23. $\sin\left(\sin^{-1}\left(-\frac{1}{2}\right) + \cos^{-1}\left(-\frac{1}{2}\right)\right) = \sin\left(-\frac{\pi}{6} + \frac{2\pi}{3}\right) = \sin\left(\frac{\pi}{2}\right) = 1$

24. $\cot\left(\sin^{-1}\left(-\frac{1}{2}\right) - \sec^{-1}2\right) = \cot\left(-\frac{\pi}{6} - \cos^{-1}\left(\frac{1}{2}\right)\right) = \cot\left(-\frac{\pi}{6} - \frac{\pi}{3}\right) = \cot\left(-\frac{\pi}{2}\right) = 0$

25. $\sec\left(\tan^{-1}1 + \csc^{-1}1\right) = \sec\left(\frac{\pi}{4} + \sin^{-1}\frac{1}{1}\right) = \sec\left(\frac{\pi}{4} + \frac{\pi}{2}\right) = \sec\left(\frac{3\pi}{4}\right) = -\sqrt{2}$

26. $\sec\left(\cot^{-1}\sqrt{3} + \csc^{-1}(-1)\right) = \sec\left(\frac{\pi}{6} + \sin^{-1}\left(\frac{1}{-1}\right)\right) = \sec\left(\frac{\pi}{2} - \frac{\pi}{3} - \frac{\pi}{2}\right) = \sec\left(-\frac{\pi}{3}\right) = 2$

27. $\sec^{-1}\left(\sec\left(-\frac{\pi}{6}\right)\right) = \sec^{-1}\left(\frac{2}{\sqrt{3}}\right) = \cos^{-1}\left(\frac{\sqrt{3}}{2}\right) = \frac{\pi}{6}$

28. $\cot^{-1}\left(\cot\left(-\frac{\pi}{4}\right)\right) = \cot^{-1}(-1) = \frac{3\pi}{4}$

29. $\alpha = \tan^{-1}\frac{x}{2}$ indicates the diagram

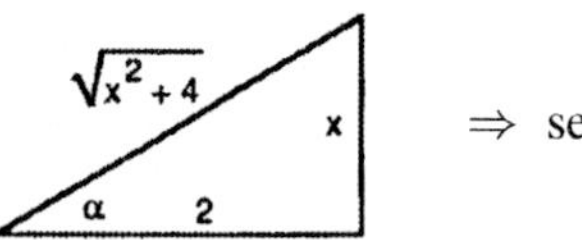

$\Rightarrow \sec\left(\tan^{-1}\frac{x}{2}\right) = \sec\alpha = \frac{\sqrt{x^2+4}}{2}$

30. $\alpha = \tan^{-1}2x$ indicates the diagram

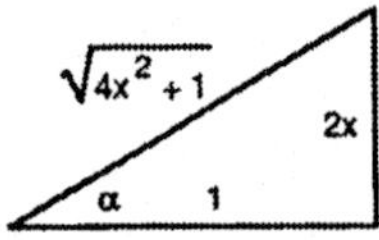

$\Rightarrow \sec\left(\tan^{-1}2x\right) = \sec\alpha = \sqrt{4x^2+1}$

31. $\alpha = \sec^{-1}3y$ indicates the diagram

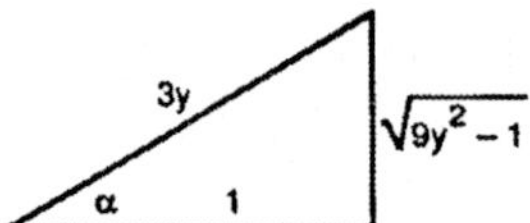

$\Rightarrow \tan\left(\sec^{-1}3y\right) = \tan\alpha = \sqrt{9y^2-1}$

32. $\alpha = \sec^{-1}\frac{y}{5}$ indicates the diagram

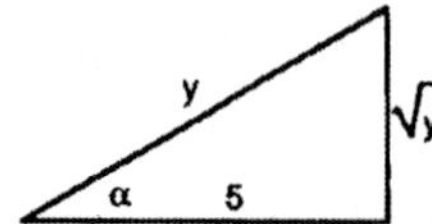

$\Rightarrow \tan\left(\sec^{-1}\frac{y}{5}\right) = \tan\alpha = \frac{\sqrt{y^2-25}}{5}$

33. $\alpha = \sin^{-1}x$ indicates the diagram

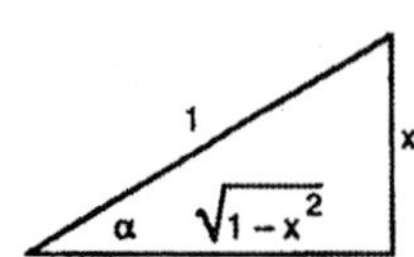

$\Rightarrow \cos\left(\sin^{-1}x\right) = \cos\alpha = \sqrt{1-x^2}$

34. $\alpha = \cos^{-1} x$ indicates the diagram

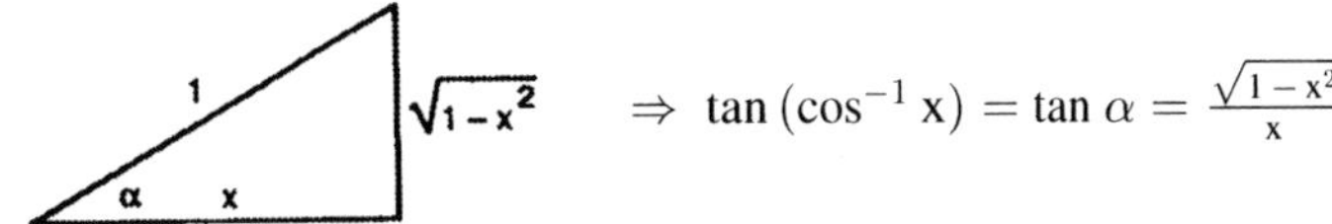

$\Rightarrow \tan(\cos^{-1} x) = \tan \alpha = \frac{\sqrt{1-x^2}}{x}$

35. $\alpha = \tan^{-1} \sqrt{x^2 - 2x}$ indicates the diagram $\Rightarrow \sin\left(\tan^{-1} \sqrt{x^2 - 2x}\right)$

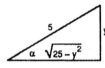

$= \sin \alpha = \frac{\sqrt{x^2-2x}}{x-1}$

36. $\alpha = \tan^{-1} \frac{x}{\sqrt{x^2+1}}$ indicates the diagram $\Rightarrow \sin\left(\tan^{-1} \frac{x}{\sqrt{x^2+1}}\right) = \sin \alpha = \frac{x}{\sqrt{2x^2+1}}$

37. $\alpha = \sin^{-1} \frac{2y}{3}$ indicates the diagram $\Rightarrow \cos\left(\sin^{-1} \frac{2y}{3}\right) = \cos \alpha = \frac{\sqrt{9-4y^2}}{3}$

38. $\alpha = \sin^{-1} \frac{y}{5}$ indicates the diagram $\Rightarrow \cos\left(\sin^{-1} \frac{y}{5}\right) = \cos \alpha = \frac{\sqrt{25-y^2}}{5}$

39. $\alpha = \sec^{-1} \frac{x}{4}$ indicates the diagram

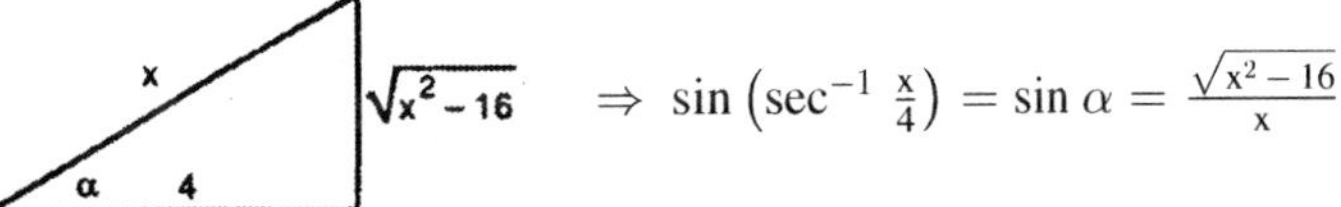

$\Rightarrow \sin\left(\sec^{-1} \frac{x}{4}\right) = \sin \alpha = \frac{\sqrt{x^2-16}}{x}$

40. $\alpha = \sec^{-1} \frac{\sqrt{x^2+4}}{x}$ indicates the diagram

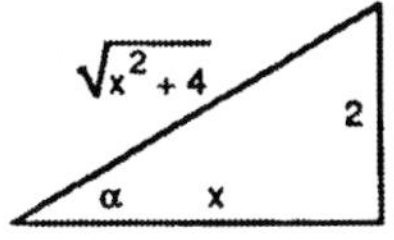

$\Rightarrow \sin\left(\sec^{-1} \frac{\sqrt{x^2+4}}{x}\right) = \sin \alpha = \frac{2}{\sqrt{x^2+4}}$

41. $\lim\limits_{x \to 1^-} \sin^{-1} x = \frac{\pi}{2}$

42. $\lim\limits_{x \to -1^+} \cos^{-1} x = \pi$

43. $\lim\limits_{x \to \infty} \tan^{-1} x = \frac{\pi}{2}$

44. $\lim\limits_{x \to -\infty} \tan^{-1} x = -\frac{\pi}{2}$

45. $\lim\limits_{x \to \infty} \sec^{-1} x = \frac{\pi}{2}$

46. $\lim\limits_{x \to -\infty} \sec^{-1} x = \lim\limits_{x \to -\infty} \cos^{-1}\left(\frac{1}{x}\right) = \frac{\pi}{2}$

47. $\lim\limits_{x \to \infty} \csc^{-1} x = \lim\limits_{x \to \infty} \sin^{-1}\left(\frac{1}{x}\right) = 0$

48. $\lim\limits_{x \to -\infty} \csc^{-1} x = \lim\limits_{x \to -\infty} \sin^{-1}\left(\frac{1}{x}\right) = 0$

49. $y = \cos^{-1}(x^2) \Rightarrow \frac{dy}{dx} = -\frac{2x}{\sqrt{1-(x^2)^2}} = \frac{-2x}{\sqrt{1-x^4}}$

50. $y = \cos^{-1}\left(\frac{1}{x}\right) = \sec^{-1} x \Rightarrow \frac{dy}{dx} = \frac{1}{|x|\sqrt{x^2-1}}$

51. $y = \sin^{-1} \sqrt{2}t \Rightarrow \frac{dy}{dt} = \frac{\sqrt{2}}{\sqrt{1-\left(\sqrt{2}t\right)^2}} = \frac{\sqrt{2}}{\sqrt{1-2t^2}}$

52. $y = \sin^{-1}(1-t) \Rightarrow \frac{dy}{dt} = \frac{-1}{\sqrt{1-(1-t)^2}} = \frac{-1}{\sqrt{2t-t^2}}$

53. $y = \sec^{-1}(2s+1) \Rightarrow \frac{dy}{ds} = \frac{2}{|2s+1|\sqrt{(2s+1)^2-1}} = \frac{2}{|2s+1|\sqrt{4s^2+4s}} = \frac{1}{|2s+1|\sqrt{s^2+s}}$

54. $y = \sec^{-1} 5s \Rightarrow \frac{dy}{ds} = \frac{5}{|5s|\sqrt{(5s)^2 - 1}} = \frac{1}{|s|\sqrt{25s^2 - 1}}$

55. $y = \csc^{-1}(x^2 + 1) \Rightarrow \frac{dy}{dx} = -\frac{2x}{|x^2 + 1|\sqrt{(x^2 + 1)^2 - 1}} = \frac{-2x}{(x^2 + 1)\sqrt{x^4 + 2x^2}}$

56. $y = \csc^{-1}\left(\frac{x}{2}\right) \Rightarrow \frac{dy}{dx} = -\frac{\left(\frac{1}{2}\right)}{\left|\frac{x}{2}\right|\sqrt{\left(\frac{x}{2}\right)^2 - 1}} = \frac{-1}{|x|\sqrt{\frac{x^2 - 4}{4}}} = \frac{-2}{|x|\sqrt{x^2 - 4}}$

57. $y = \sec^{-1}\left(\frac{1}{t}\right) = \cos^{-1} t \Rightarrow \frac{dy}{dt} = \frac{-1}{\sqrt{1 - t^2}}$

58. $y = \sin^{-1}\left(\frac{3}{t^2}\right) = \csc^{-1}\left(\frac{t^2}{3}\right) \Rightarrow \frac{dy}{dt} = -\frac{\left(\frac{2t}{3}\right)}{\left|\frac{t^2}{3}\right|\sqrt{\left(\frac{t^2}{3}\right)^2 - 1}} = \frac{-2t}{t^2\sqrt{\frac{t^4 - 9}{9}}} = \frac{-6}{t\sqrt{t^4 - 9}}$

59. $y = \cot^{-1}\sqrt{t} = \cot^{-1} t^{1/2} \Rightarrow \frac{dy}{dt} = -\frac{\left(\frac{1}{2}\right)t^{-1/2}}{1 + \left(t^{1/2}\right)^2} = \frac{-1}{2\sqrt{t}(1 + t)}$

60. $y = \cot^{-1}\sqrt{t - 1} = \cot^{-1}(t - 1)^{1/2} \Rightarrow \frac{dy}{dt} = -\frac{\left(\frac{1}{2}\right)(t - 1)^{-1/2}}{1 + \left[(t - 1)^{1/2}\right]^2} = \frac{-1}{2\sqrt{t - 1}(1 + t - 1)} = \frac{-1}{2t\sqrt{t - 1}}$

61. $y = \ln(\tan^{-1} x) \Rightarrow \frac{dy}{dx} = \frac{\left(\frac{1}{1 + x^2}\right)}{\tan^{-1} x} = \frac{1}{(\tan^{-1} x)(1 + x^2)}$

62. $y = \tan^{-1}(\ln x) \Rightarrow \frac{dy}{dx} = \frac{\left(\frac{1}{x}\right)}{1 + (\ln x)^2} = \frac{1}{x\left[1 + (\ln x)^2\right]}$

63. $y = \csc^{-1}(e^t) \Rightarrow \frac{dy}{dt} = -\frac{e^t}{|e^t|\sqrt{(e^t)^2 - 1}} = \frac{-1}{\sqrt{e^{2t} - 1}}$

64. $y = \cos^{-1}(e^{-t}) \Rightarrow \frac{dy}{dt} = -\frac{-e^{-t}}{\sqrt{1 - (e^{-t})^2}} = \frac{e^{-t}}{\sqrt{1 - e^{-2t}}}$

65. $y = s\sqrt{1 - s^2} + \cos^{-1} s = s\left(1 - s^2\right)^{1/2} + \cos^{-1} s \Rightarrow \frac{dy}{ds} = \left(1 - s^2\right)^{1/2} + s\left(\frac{1}{2}\right)\left(1 - s^2\right)^{-1/2}(-2s) - \frac{1}{\sqrt{1 - s^2}}$
$= \sqrt{1 - s^2} - \frac{s^2}{\sqrt{1 - s^2}} - \frac{1}{\sqrt{1 - s^2}} = \sqrt{1 - s^2} - \frac{s^2 + 1}{\sqrt{1 - s^2}} = \frac{1 - s^2 - s^2 - 1}{\sqrt{1 - s^2}} = \frac{-2s^2}{\sqrt{1 - s^2}}$

66. $y = \sqrt{s^2 - 1} - \sec^{-1} s = \left(s^2 - 1\right)^{1/2} - \sec^{-1} s \Rightarrow \frac{dy}{dx} = \left(\frac{1}{2}\right)\left(s^2 - 1\right)^{-1/2}(2s) - \frac{1}{|s|\sqrt{s^2 - 1}} = \frac{s}{\sqrt{s^2 - 1}} - \frac{1}{|s|\sqrt{s^2 - 1}}$
$= \frac{s|s| - 1}{|s|\sqrt{s^2 - 1}}$

67. $y = \tan^{-1}\sqrt{x^2 - 1} + \csc^{-1} x = \tan^{-1}\left(x^2 - 1\right)^{1/2} + \csc^{-1} x \Rightarrow \frac{dy}{dx} = \frac{\left(\frac{1}{2}\right)\left(x^2 - 1\right)^{-1/2}(2x)}{1 + \left[\left(x^2 - 1\right)^{1/2}\right]^2} - \frac{1}{|x|\sqrt{x^2 - 1}}$
$= \frac{1}{x\sqrt{x^2 - 1}} - \frac{1}{|x|\sqrt{x^2 - 1}} = 0$, for $x > 1$

68. $y = \cot^{-1}\left(\frac{1}{x}\right) - \tan^{-1} x = \frac{\pi}{2} - \tan^{-1}(x^{-1}) - \tan^{-1} x \Rightarrow \frac{dy}{dx} = 0 - \frac{-x^{-2}}{1 + (x^{-1})^2} - \frac{1}{1 + x^2} = \frac{1}{x^2 + 1} - \frac{1}{1 + x^2} = 0$

69. $y = x\sin^{-1} x + \sqrt{1 - x^2} = x\sin^{-1} x + \left(1 - x^2\right)^{1/2} \Rightarrow \frac{dy}{dx} = \sin^{-1} x + x\left(\frac{1}{\sqrt{1 - x^2}}\right) + \left(\frac{1}{2}\right)\left(1 - x^2\right)^{-1/2}(-2x)$
$= \sin^{-1} x + \frac{x}{\sqrt{1 - x^2}} - \frac{x}{\sqrt{1 - x^2}} = \sin^{-1} x$

70. $y = \ln(x^2+4) - x\tan^{-1}\left(\frac{x}{2}\right) \Rightarrow \frac{dy}{dx} = \frac{2x}{x^2+4} - \tan^{-1}\left(\frac{x}{2}\right) - x\left[\frac{\left(\frac{1}{2}\right)}{1+\left(\frac{x}{2}\right)^2}\right] = \frac{2x}{x^2+4} - \tan^{-1}\left(\frac{x}{2}\right) - \frac{2x}{4+x^2}$

$= -\tan^{-1}\left(\frac{x}{2}\right)$

71. (a) $y = \frac{\pi}{2}$
    (b) $y = -\frac{\pi}{2}$
    (c) None, since $\frac{d}{dx}\tan^{-1}x = \frac{1}{1+x^2} \neq 0$.

72. (a) $y = 0$
    (b) $y = \pi$
    (c) None, since $\frac{d}{dx}\cot^{-1}x = -\frac{1}{1+x^2} \neq 0$.

73. (a) $y = \frac{\pi}{2}$
    (b) $y = \frac{\pi}{2}$
    (c) None, since $\frac{d}{dx}\sec^{-1}x = \frac{1}{|x|\sqrt{x^2-1}} \neq 0$.

74. (a) $y = 0$
    (b) $y = \pi$
    (c) None, since $\frac{d}{dx}\csc^{-1}x = -\frac{1}{|x|\sqrt{x^2-1}} \neq 0$.

75. The angle α is the large angle between the wall and the right end of the blackboard minus the small angle between the left end of the blackboard and the wall $\Rightarrow \alpha = \cot^{-1}\left(\frac{x}{15}\right) - \cot^{-1}\left(\frac{x}{3}\right)$.

76. $65° + (90° - \beta) + (90° - \alpha) = 180° \Rightarrow \alpha = 65° - \beta = 65° - \tan^{-1}\left(\frac{21}{50}\right) \approx 65° - 22.78° \approx 42.22°$

77. Take each square as a unit square. From the diagram we have the following: the smallest angle α has a tangent of 1 $\Rightarrow \alpha = \tan^{-1} 1$; the middle angle β has a tangent of 2 $\Rightarrow \beta = \tan^{-1} 2$; and the largest angle γ has a tangent of 3 $\Rightarrow \gamma = \tan^{-1} 3$. The sum of these three angles is $\pi \Rightarrow \alpha + \beta + \gamma = \pi$
$\Rightarrow \tan^{-1} 1 + \tan^{-1} 2 + \tan^{-1} 3 = \pi$.

78. (a) From the symmetry of the diagram, we see that $\pi - \sec^{-1} x$ is the vertical distance from the graph of $y = \sec^{-1} x$ to the line $y = \pi$ and this distance is the same as the height of $y = \sec^{-1} x$ above the x-axis at $-x$; i.e., $\pi - \sec^{-1} x = \sec^{-1}(-x)$.
    (b) $\cos^{-1}(-x) = \pi - \cos^{-1} x$, where $-1 \le x \le 1 \Rightarrow \cos^{-1}\left(-\frac{1}{x}\right) = \pi - \cos^{-1}\left(\frac{1}{x}\right)$, where $x \ge 1$ or $x \le -1$
    $\Rightarrow \sec^{-1}(-x) = \pi - \sec^{-1} x$

79. If $x = 1$: $\sin^{-1}(1) + \cos^{-1}(1) = \frac{\pi}{2} + 0 = \frac{\pi}{2}$.
If $x = 0$: $\sin^{-1}(0) + \cos^{-1}(0) = 0 + \frac{\pi}{2} = \frac{\pi}{2}$.
If $x = -1$: $\sin^{-1}(-1) + \cos^{-1}(-1) = -\frac{\pi}{2} + \pi = \frac{\pi}{2}$.
The identity $\sin^{-1}(x) + \cos^{-1}(x) = \frac{\pi}{2}$ has been established for x in (0, 1), by Figure 1.6.7. So now if x is in (−1, 0), note that −x is in (0, 1), and we have that

$\sin^{-1}(x) + \cos^{-1}(x) = -\sin^{-1}(-x) + \cos^{-1}(x)$	since $\sin^{-1}$ is odd
$= -\sin^{-1}(-x) + \pi - \cos^{-1}(-x)$	by Eq. 3, Section 1.6
$= -(\sin^{-1}(-x) + \cos^{-1}(-x)) + \pi$	
$= -\frac{\pi}{2} + \pi$	
$= \frac{\pi}{2}$	

This establishes the identity for all x in [−1, 1].

80. 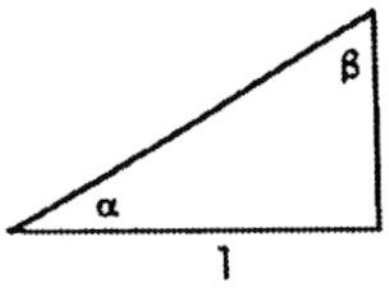

x $\Rightarrow \tan\alpha = x$ and $\tan\beta = \frac{1}{x} \Rightarrow \frac{\pi}{2} = \alpha + \beta = \tan^{-1} x + \tan^{-1}\frac{1}{x}$.

81. (a) Defined; there is an angle whose tangent is 2.
    (b) Not defined; there is no angle whose cosine is 2.

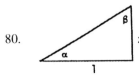

82. (a) Not defined; there is no angle whose cosecant is $\frac{1}{2}$.
(b) Defined; there is an angle whose cosecant is 2.

83. (a) Not defined; there is no angle whose secant is 0.
(b) Not defined; there is no angle whose sine is $\sqrt{2}$.

84. (a) Defined; there is an angle whose cotangent is $-\frac{1}{2}$.
(b) Not defined; there is no angle whose cosine is -5.

85. $\csc^{-1} u = \frac{\pi}{2} - \sec^{-1} u \Rightarrow \frac{d}{dx}(\csc^{-1} u) = \frac{d}{dx}\left(\frac{\pi}{2} - \sec^{-1} u\right) = 0 - \frac{\frac{du}{dx}}{|u|\sqrt{u^2-1}} = -\frac{\frac{du}{dx}}{|u|\sqrt{u^2-1}}$, $|u| > 1$

86. $y = \tan^{-1} x \Rightarrow \tan y = x \Rightarrow \frac{d}{dx}(\tan y) = \frac{d}{dx}(x)$
$\Rightarrow (\sec^2 y)\frac{dy}{dx} = 1 \Rightarrow \frac{dy}{dx} = \frac{1}{\sec^2 y} = \frac{1}{\left(\sqrt{1+x^2}\right)^2}$
$= \frac{1}{1+x^2}$, as indicated by the triangle

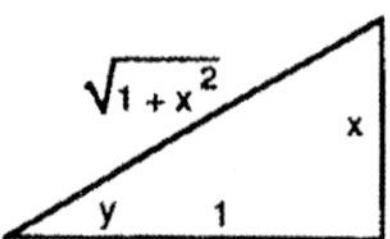

87. $f(x) = \sec x \Rightarrow f'(x) = \sec x \tan x \Rightarrow \left.\frac{df^{-1}}{dx}\right|_{x=b} = \frac{1}{\left.\frac{df}{dx}\right|_{x=f^{-1}(b)}} = \frac{1}{\sec(\sec^{-1} b)\tan(\sec^{-1} b)} = \frac{1}{b\left(\pm\sqrt{b^2-1}\right)}$.
Since the slope of $\sec^{-1} x$ is always positive, we the right sign by writing $\frac{d}{dx}\sec^{-1} x = \frac{1}{|x|\sqrt{x^2-1}}$.

88. $\cot^{-1} u = \frac{\pi}{2} - \tan^{-1} u \Rightarrow \frac{d}{dx}(\cot^{-1} u) = \frac{d}{dx}\left(\frac{\pi}{2} - \tan^{-1} u\right) = 0 - \frac{\frac{du}{dx}}{1+u^2} = -\frac{\frac{du}{dx}}{1+u^2}$

89. The functions f and g have the same derivative (for $x \geq 0$), namely $\frac{1}{\sqrt{x}(x+1)}$. The functions therefore differ by a constant. To identify the constant we can set x equal to 0 in the equation $f(x) = g(x) + C$, obtaining $\sin^{-1}(-1) = 2\tan^{-1}(0) + C \Rightarrow -\frac{\pi}{2} = 0 + C \Rightarrow C = -\frac{\pi}{2}$. For $x \geq 0$, we have $\sin^{-1}\left(\frac{x-1}{x+1}\right) = 2\tan^{-1}\sqrt{x} - \frac{\pi}{2}$.

90. The functions f and g have the same derivative for $x > 0$, namely $\frac{-1}{1+x^2}$. The functions therefore differ by a constant for $x > 0$. To identify the constant we can set x equal to 1 in the equation $f(x) = g(x) + C$, obtaining $\sin^{-1}\left(\frac{1}{\sqrt{2}}\right) = \tan^{-1} 1 + C \Rightarrow \frac{\pi}{4} = \frac{\pi}{4} + C \Rightarrow C = 0$. For $x > 0$, we have $\sin^{-1}\frac{1}{\sqrt{x^2+1}} = \tan^{-1}\frac{1}{x}$.

91. (a) $\sec^{-1} 1.5 = \cos^{-1}\frac{1}{1.5} \approx 0.84107$
(b) $\csc^{-1}(-1.5) = \sin^{-1}\left(-\frac{1}{1.5}\right) \approx -0.72973$
(c) $\cot^{-1} 2 = \frac{\pi}{2} - \tan^{-1} 2 \approx 0.46365$

92. (a) $\sec^{-1}(-3) = \cos^{-1}\left(-\frac{1}{3}\right) \approx 1.91063$
(b) $\csc^{-1} 1.7 = \sin^{-1}\left(\frac{1}{1.7}\right) \approx 0.62887$
(c) $\cot^{-1}(-2) = \frac{\pi}{2} - \tan^{-1}(-2) \approx 2.67795$

93. (a) Domain: all real numbers except those having the form $\frac{\pi}{2} + k\pi$ where k is an integer.
Range: $-\frac{\pi}{2} < y < \frac{\pi}{2}$

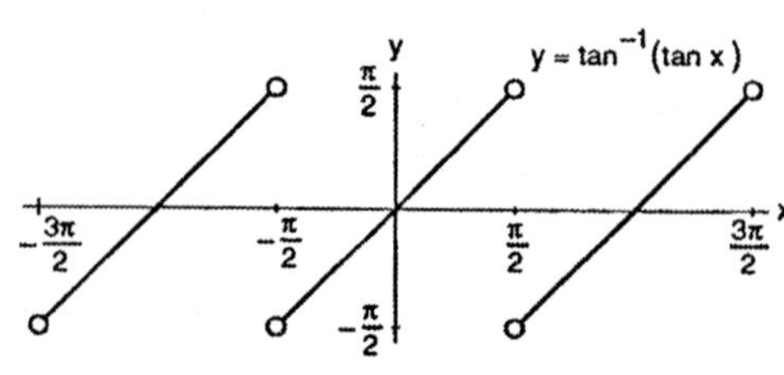

(b) Domain: $-\infty < x < \infty$; Range: $-\infty < y < \infty$
The graph of $y = \tan^{-1}(\tan x)$ is periodic, the graph of $y = \tan(\tan^{-1} x) = x$ for $-\infty \le x < \infty$.

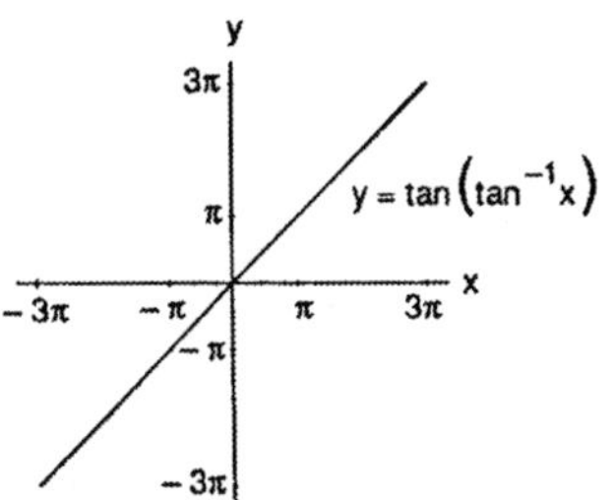

94. (a) Domain: $-\infty < x < \infty$; Range: $-\frac{\pi}{2} \le y \le \frac{\pi}{2}$

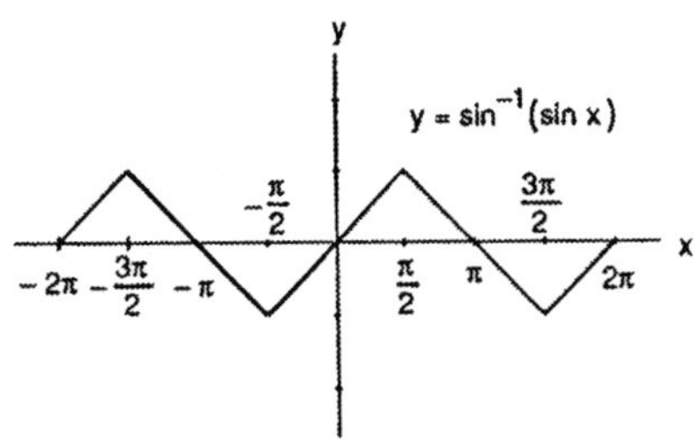

(b) Domain: $-1 \le x \le 1$; Range: $-1 \le y \le 1$
The graph of $y = \sin^{-1}(\sin x)$ is periodic; the graph of $y = \sin(\sin^{-1} x) = x$ for $-1 \le x \le 1$.

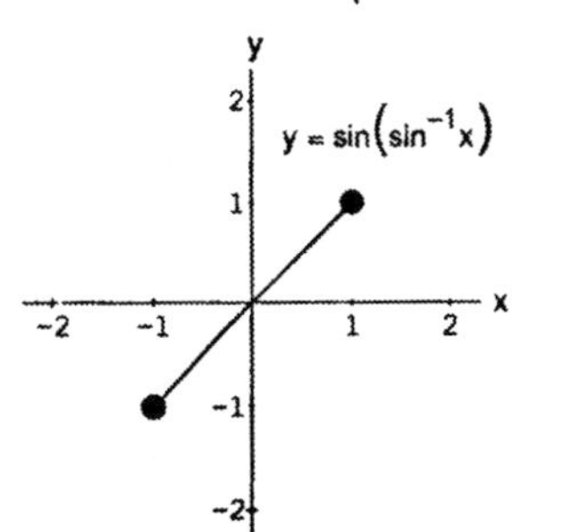

95. (a) Domain: $-\infty < x < \infty$; Range: $0 \le y \le \pi$

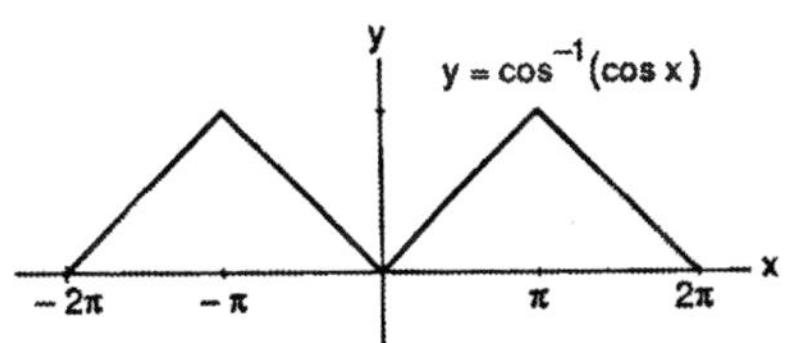

(b) Domain: $-1 \le x \le 1$; Range: $-1 \le y \le 1$
The graph of $y = \cos^{-1}(\cos x)$ is periodic; the graph of $y = \cos(\cos^{-1} x) = x$ for $-1 \le x \le 1$.

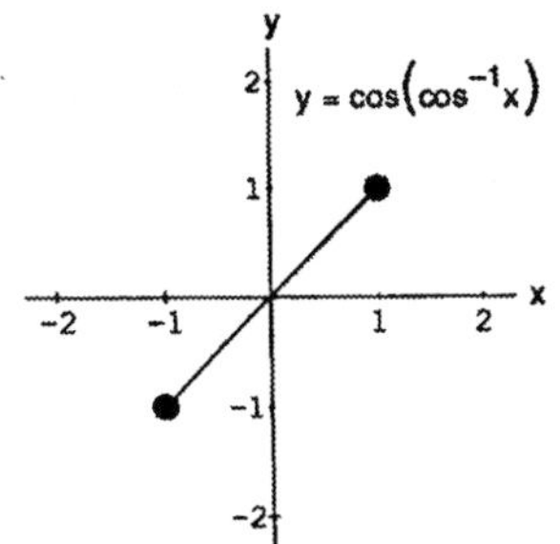

96. Since the domain of $\sec^{-1} x$ is $(-\infty, -1] \cup [1, \infty)$, we have $\sec(\sec^{-1} x) = x$ for $|x| \ge 1$. The graph of $y = \sec(\sec^{-1} x)$ is the line $y = x$ with the open line segment from $(-1, -1)$ to $(1, 1)$ removed.

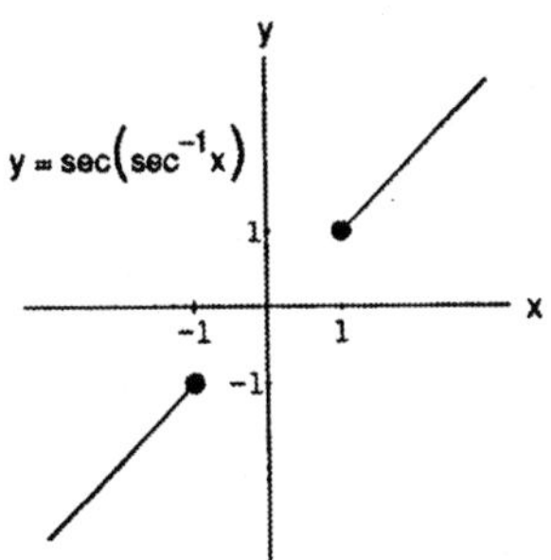

97. The graphs are identical for $y = 2\sin(2\tan^{-1}x)$
$= 4[\sin(\tan^{-1}x)][\cos(\tan^{-1}x)] = 4\left(\frac{x}{\sqrt{x^2+1}}\right)\left(\frac{1}{\sqrt{x^2+1}}\right)$
$= \frac{4x}{x^2+1}$ from the triangle

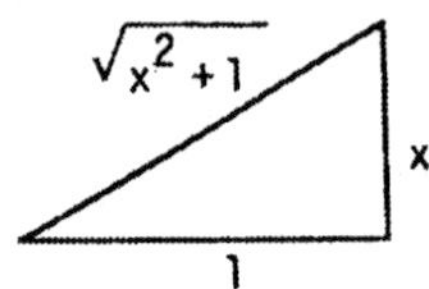

98. The graphs are identical for $y = \cos(2\sec^{-1}x)$
$= \cos^2(\sec^{-1}x) - \sin^2(\sec^{-1}x) = \frac{1}{x^2} - \frac{x^2-1}{x^2}$
$= \frac{2-x^2}{x^2}$ from the triangle

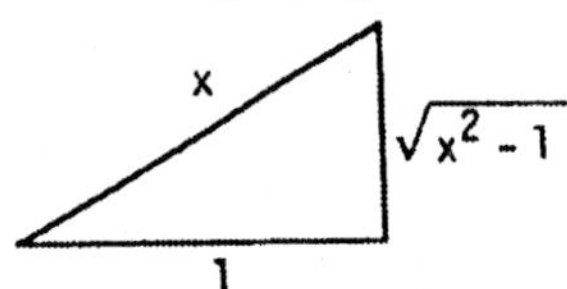

99. The values of f increase over the interval $[-1, 1]$ because $f' > 0$, and the graph of f steepens as the values of f' increase towards the ends of the interval. The graph of f is concave down to the left of the origin where $f'' < 0$, and concave up to the right of the origin where $f'' > 0$. There is an inflection point at $x = 0$ where $f'' = 0$ and f' has a local minimum value.

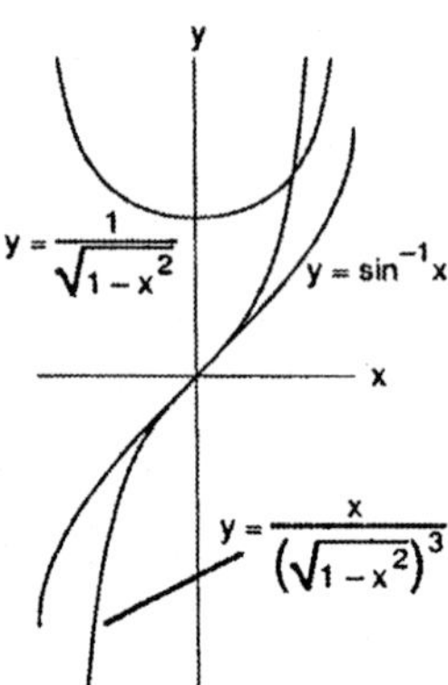

100. The values of f increase throughout the interval $(-\infty, \infty)$ because $f' > 0$, and they increase most rapidly near the origin where the values of f' are relatively large. The graph of f is concave up to the left of the origin where $f'' > 0$, and concave down to the right of the origin where $f'' < 0$. There is an inflection point at $x = 0$ where $f'' = 0$ and f' has a local maximum value.

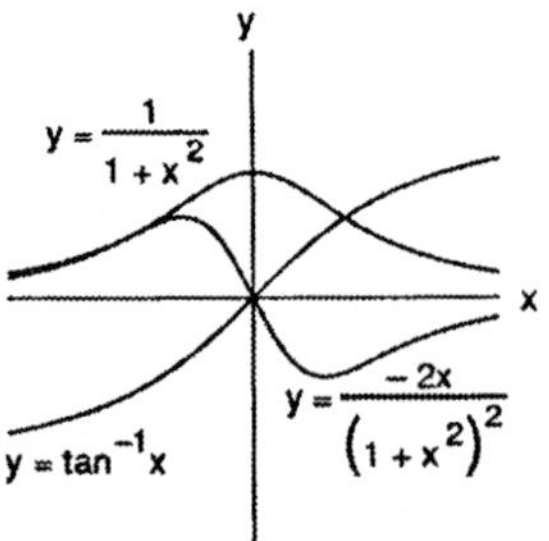

3.9 RELATED RATES

1. $A = \pi r^2 \Rightarrow \frac{dA}{dt} = 2\pi r\,\frac{dr}{dt}$

2. $S = 4\pi r^2 \Rightarrow \frac{dS}{dt} = 8\pi r\,\frac{dr}{dt}$

3. (a) $V = \pi r^2 h \Rightarrow \frac{dV}{dt} = \pi r^2\,\frac{dh}{dt}$
(b) $V = \pi r^2 h \Rightarrow \frac{dV}{dt} = 2\pi rh\,\frac{dr}{dt}$
(c) $V = \pi r^2 h \Rightarrow \frac{dV}{dt} = \pi r^2\,\frac{dh}{dt} + 2\pi rh\,\frac{dr}{dt}$

4. (a) $V = \frac{1}{3}\pi r^2 h \Rightarrow \frac{dV}{dt} = \frac{1}{3}\pi r^2\,\frac{dh}{dt}$
(b) $V = \frac{1}{3}\pi r^2 h \Rightarrow \frac{dV}{dt} = \frac{2}{3}\pi rh\,\frac{dr}{dt}$
(c) $\frac{dV}{dt} = \frac{1}{3}\pi r^2\,\frac{dh}{dt} + \frac{2}{3}\pi rh\,\frac{dr}{dt}$

5. (a) $\frac{dV}{dt} = 1$ volt/sec (b) $\frac{dI}{dt} = -\frac{1}{3}$ amp/sec
(c) $\frac{dV}{dt} = R\left(\frac{dI}{dt}\right) + I\left(\frac{dR}{dt}\right) \Rightarrow \frac{dR}{dt} = \frac{1}{I}\left(\frac{dV}{dt} - R\frac{dI}{dt}\right) \Rightarrow \frac{dR}{dt} = \frac{1}{I}\left(\frac{dV}{dt} - \frac{V}{I}\frac{dI}{dt}\right)$
(d) $\frac{dR}{dt} = \frac{1}{2}\left[1 - \frac{12}{2}\left(-\frac{1}{3}\right)\right] = \left(\frac{1}{2}\right)(3) = \frac{3}{2}$ ohms/sec, R is increasing

6. (a) $P = RI^2 \Rightarrow \frac{dP}{dt} = I^2\frac{dR}{dt} + 2RI\frac{dI}{dt}$
(b) $P = RI^2 \Rightarrow 0 = \frac{dP}{dt} = I^2\frac{dR}{dt} + 2RI\frac{dI}{dt} \Rightarrow \frac{dR}{dt} = -\frac{2RI}{I^2}\frac{dI}{dt} = -\frac{2\left(\frac{P}{I^2}\right)}{I}\frac{dI}{dt} = -\frac{2P}{I^3}\frac{dI}{dt}$

7. (a) $s = \sqrt{x^2+y^2} = (x^2+y^2)^{1/2} \Rightarrow \frac{ds}{dt} = \frac{x}{\sqrt{x^2+y^2}}\frac{dx}{dt}$
(b) $s = \sqrt{x^2+y^2} = (x^2+y^2)^{1/2} \Rightarrow \frac{ds}{dt} = \frac{x}{\sqrt{x^2+y^2}}\frac{dx}{dt} + \frac{y}{\sqrt{x^2+y^2}}\frac{dy}{dt}$
(c) $s = \sqrt{x^2+y^2} \Rightarrow s^2 = x^2+y^2 \Rightarrow 2s\frac{ds}{dt} = 2x\frac{dx}{dt} + 2y\frac{dy}{dt} \Rightarrow 2s\cdot 0 = 2x\frac{dx}{dt} + 2y\frac{dy}{dt} \Rightarrow \frac{dx}{dt} = -\frac{y}{x}\frac{dy}{dt}$

8. (a) $s = \sqrt{x^2+y^2+z^2} \Rightarrow s^2 = x^2+y^2+z^2 \Rightarrow 2s\frac{ds}{dt} = 2x\frac{dx}{dt} + 2y\frac{dy}{dt} + 2z\frac{dz}{dt}$
$\Rightarrow \frac{ds}{dt} = \frac{x}{\sqrt{x^2+y^2+z^2}}\frac{dx}{dt} + \frac{y}{\sqrt{x^2+y^2+z^2}}\frac{dy}{dt} + \frac{z}{\sqrt{x^2+y^2+z^2}}\frac{dz}{dt}$
(b) From part (a) with $\frac{dx}{dt} = 0 \Rightarrow \frac{ds}{dt} = \frac{y}{\sqrt{x^2+y^2+z^2}}\frac{dy}{dt} + \frac{z}{\sqrt{x^2+y^2+z^2}}\frac{dz}{dt}$
(c) From part (a) with $\frac{ds}{dt} = 0 \Rightarrow 0 = 2x\frac{dx}{dt} + 2y\frac{dy}{dt} + 2z\frac{dz}{dt} \Rightarrow \frac{dx}{dt} + \frac{y}{x}\frac{dy}{dt} + \frac{z}{x}\frac{dz}{dt} = 0$

9. (a) $A = \frac{1}{2}ab\sin\theta \Rightarrow \frac{dA}{dt} = \frac{1}{2}ab\cos\theta\frac{d\theta}{dt}$ (b) $A = \frac{1}{2}ab\sin\theta \Rightarrow \frac{dA}{dt} = \frac{1}{2}ab\cos\theta\frac{d\theta}{dt} + \frac{1}{2}b\sin\theta\frac{da}{dt}$
(c) $A = \frac{1}{2}ab\sin\theta \Rightarrow \frac{dA}{dt} = \frac{1}{2}ab\cos\theta\frac{d\theta}{dt} + \frac{1}{2}b\sin\theta\frac{da}{dt} + \frac{1}{2}a\sin\theta\frac{db}{dt}$

10. Given $A = \pi r^2$, $\frac{dr}{dt} = 0.01$ cm/sec, and r = 50 cm. Since $\frac{dA}{dt} = 2\pi r\frac{dr}{dt}$, then $\left.\frac{dA}{dt}\right|_{r=50} = 2\pi(50)\left(\frac{1}{100}\right)$
$= \pi$ cm^2/min.

11. Given $\frac{d\ell}{dt} = -2$ cm/sec, $\frac{dw}{dt} = 2$ cm/sec, $\ell = 12$ cm and w = 5 cm.
(a) $A = \ell w \Rightarrow \frac{dA}{dt} = \ell\frac{dw}{dt} + w\frac{d\ell}{dt} \Rightarrow \frac{dA}{dt} = 12(2) + 5(-2) = 14$ cm^2/sec, increasing
(b) $P = 2\ell + 2w \Rightarrow \frac{dP}{dt} = 2\frac{d\ell}{dt} + 2\frac{dw}{dt} = 2(-2) + 2(2) = 0$ cm/sec, constant
(c) $D = \sqrt{w^2+\ell^2} = (w^2+\ell^2)^{1/2} \Rightarrow \frac{dD}{dt} = \frac{1}{2}(w^2+\ell^2)^{-1/2}\left(2w\frac{dw}{dt} + 2\ell\frac{d\ell}{dt}\right) \Rightarrow \frac{dD}{dt} = \frac{w\frac{dw}{dt} + \ell\frac{d\ell}{dt}}{\sqrt{w^2+\ell^2}}$
$= \frac{(5)(2) + (12)(-2)}{\sqrt{25+144}} = -\frac{14}{13}$ cm/sec, decreasing

12. (a) $V = xyz \Rightarrow \frac{dV}{dt} = yz\frac{dx}{dt} + xz\frac{dy}{dt} + xy\frac{dz}{dt} \Rightarrow \left.\frac{dV}{dt}\right|_{(4,3,2)} = (3)(2)(1) + (4)(2)(-2) + (4)(3)(1) = 2$ m^3/sec
(b) $S = 2xy + 2xz + 2yz \Rightarrow \frac{dS}{dt} = (2y+2z)\frac{dx}{dt} + (2x+2z)\frac{dy}{dt} + (2x+2y)\frac{dz}{dt}$
$\Rightarrow \left.\frac{dS}{dt}\right|_{(4,3,2)} = (10)(1) + (12)(-2) + (14)(1) = 0$ m^2/sec
(c) $\ell = \sqrt{x^2+y^2+z^2} = (x^2+y^2+z^2)^{1/2} \Rightarrow \frac{d\ell}{dt} = \frac{x}{\sqrt{x^2+y^2+z^2}}\frac{dx}{dt} + \frac{y}{\sqrt{x^2+y^2+z^2}}\frac{dy}{dt} + \frac{z}{\sqrt{x^2+y^2+z^2}}\frac{dz}{dt}$
$\Rightarrow \left.\frac{d\ell}{dt}\right|_{(4,3,2)} = \left(\frac{4}{\sqrt{29}}\right)(1) + \left(\frac{3}{\sqrt{29}}\right)(-2) + \left(\frac{2}{\sqrt{29}}\right)(1) = 0$ m/sec

13. Given: $\frac{dx}{dt} = 5$ ft/sec, the ladder is 13 ft long, and x = 12, y = 5 at the instant of time
(a) Since $x^2 + y^2 = 169 \Rightarrow \frac{dy}{dt} = -\frac{x}{y}\frac{dx}{dt} = -\left(\frac{12}{5}\right)(5) = -12$ ft/sec, the ladder is sliding down the wall
(b) The area of the triangle formed by the ladder and walls is $A = \frac{1}{2}xy \Rightarrow \frac{dA}{dt} = \left(\frac{1}{2}\right)\left(x\frac{dy}{dt} + y\frac{dx}{dt}\right)$. The area is changing at $\frac{1}{2}[12(-12) + 5(5)] = -\frac{119}{2} = -59.5$ ft^2/sec.
(c) $\cos\theta = \frac{x}{13} \Rightarrow -\sin\theta\frac{d\theta}{dt} = \frac{1}{13}\cdot\frac{dx}{dt} \Rightarrow \frac{d\theta}{dt} = -\frac{1}{13\sin\theta}\cdot\frac{dx}{dt} = -\left(\frac{1}{5}\right)(5) = -1$ rad/sec

14. $s^2 = y^2 + x^2 \Rightarrow 2s\,\frac{ds}{dt} = 2x\,\frac{dx}{dt} + 2y\,\frac{dy}{dt} \Rightarrow \frac{ds}{dt} = \frac{1}{s}\left(x\,\frac{dx}{dt} + y\,\frac{dy}{dt}\right) \Rightarrow \frac{ds}{dt} = \frac{1}{\sqrt{169}}\,[5(-442) + 12(-481)]$ $= -614$ knots

15. Let s represent the distance between the girl and the kite and x represents the horizontal distance between the girl and kite $\Rightarrow s^2 = (300)^2 + x^2 \Rightarrow \frac{ds}{dt} = \frac{x}{s}\,\frac{dx}{dt} = \frac{400(25)}{500} = 20$ ft/sec.

16. When the diameter is 3.8 in., the radius is 1.9 in. and $\frac{dr}{dt} = \frac{1}{3000}$ in/min. Also $V = 6\pi r^2 \Rightarrow \frac{dV}{dt} = 12\pi r\,\frac{dr}{dt}$ $\Rightarrow \frac{dV}{dt} = 12\pi(1.9)\left(\frac{1}{3000}\right) = 0.0076\pi$. The volume is changing at about 0.0239 in^3/min.

17. $V = \frac{1}{3}\,\pi r^2 h,\ h = \frac{3}{8}\,(2r) = \frac{3r}{4} \Rightarrow r = \frac{4h}{3} \Rightarrow V = \frac{1}{3}\,\pi\left(\frac{4h}{3}\right)^2 h = \frac{16\pi h^3}{27} \Rightarrow \frac{dV}{dt} = \frac{16\pi h^2}{9}\,\frac{dh}{dt}$
   (a) $\left.\frac{dh}{dt}\right|_{h=4} = \left(\frac{9}{16\pi 4^2}\right)(10) = \frac{90}{256\pi} \approx 0.1119$ m/sec $= 11.19$ cm/sec
   (b) $r = \frac{4h}{3} \Rightarrow \frac{dr}{dt} = \frac{4}{3}\,\frac{dh}{dt} = \frac{4}{3}\left(\frac{90}{256\pi}\right) = \frac{15}{32\pi} \approx 0.1492$ m/sec $= 14.92$ cm/sec

18. (a) $V = \frac{1}{3}\,\pi r^2 h$ and $r = \frac{15h}{2} \Rightarrow V = \frac{1}{3}\,\pi\left(\frac{15h}{2}\right)^2 h = \frac{75\pi h^3}{4} \Rightarrow \frac{dV}{dt} = \frac{225\pi h^2}{4}\,\frac{dh}{dt} \Rightarrow \left.\frac{dh}{dt}\right|_{h=5} = \frac{4(-50)}{225\pi(5)^2} = \frac{-8}{225\pi}$ ≈ -0.0113 m/min $= -1.13$ cm/min
   (b) $r = \frac{15h}{2} \Rightarrow \frac{dr}{dt} = \frac{15}{2}\,\frac{dh}{dt} \Rightarrow \left.\frac{dr}{dt}\right|_{h=5} = \left(\frac{15}{2}\right)\left(\frac{-8}{225\pi}\right) = \frac{-4}{15\pi} \approx -0.0849$ m/sec $= -8.49$ cm/sec

19. (a) $V = \frac{\pi}{3}\,y^2(3R - y) \Rightarrow \frac{dV}{dt} = \frac{\pi}{3}\,[2y(3R - y) + y^2(-1)]\,\frac{dy}{dt} \Rightarrow \frac{dy}{dt} = \left[\frac{\pi}{3}\,(6Ry - 3y^2)\right]^{-1}\frac{dV}{dt} \Rightarrow$ at $R = 13$ and $y = 8$ we have $\frac{dy}{dt} = \frac{1}{144\pi}(-6) = \frac{-1}{24\pi}$ m/min
   (b) The hemisphere is on the circle $r^2 + (13 - y)^2 = 169 \Rightarrow r = \sqrt{26y - y^2}$ m
   (c) $r = (26y - y^2)^{1/2} \Rightarrow \frac{dr}{dt} = \frac{1}{2}\,(26y - y^2)^{-1/2}(26 - 2y)\,\frac{dy}{dt} \Rightarrow \frac{dr}{dt} = \frac{13 - y}{\sqrt{26y - y^2}}\,\frac{dy}{dt} \Rightarrow \left.\frac{dr}{dt}\right|_{y=8} = \frac{13 - 8}{\sqrt{26\cdot 8 - 64}}\left(\frac{-1}{24\pi}\right)$ $= \frac{-5}{288\pi}$ m/min

20. If $V = \frac{4}{3}\,\pi r^3$, $S = 4\pi r^2$, and $\frac{dV}{dt} = kS = 4k\pi r^2$, then $\frac{dV}{dt} = 4\pi r^2\,\frac{dr}{dt} \Rightarrow 4k\pi r^2 = 4\pi r^2\,\frac{dr}{dt} \Rightarrow \frac{dr}{dt} = k$, a constant. Therefore, the radius is increasing at a constant rate.

21. If $V = \frac{4}{3}\,\pi r^3$, $r = 5$, and $\frac{dV}{dt} = 100\pi$ ft^3/min, then $\frac{dV}{dt} = 4\pi r^2\,\frac{dr}{dt} \Rightarrow \frac{dr}{dt} = 1$ ft/min. Then $S = 4\pi r^2 \Rightarrow \frac{dS}{dt}$ $= 8\pi r\,\frac{dr}{dt} = 8\pi(5)(1) = 40\pi$ ft^2/min, the rate at which the surface area is increasing.

22. Let s represent the length of the rope and x the horizontal distance of the boat from the dock.
   (a) We have $s^2 = x^2 + 36 \Rightarrow \frac{dx}{dt} = \frac{s}{x}\,\frac{ds}{dt} = \frac{s}{\sqrt{s^2 - 36}}\,\frac{ds}{dt}$. Therefore, the boat is approaching the dock at $\left.\frac{dx}{dt}\right|_{s=10} = \frac{10}{\sqrt{10^2 - 36}}(-2) = -2.5$ ft/sec.
   (b) $\cos\theta = \frac{6}{r} \Rightarrow -\sin\theta\,\frac{d\theta}{dt} = -\frac{6}{r^2}\,\frac{dr}{dt} \Rightarrow \frac{d\theta}{dt} = \frac{6}{r^2\sin\theta}\,\frac{dr}{dt}$. Thus, $r = 10$, $x = 8$, and $\sin\theta = \frac{8}{10}$ $\Rightarrow \frac{d\theta}{dt} = \frac{6}{10^2\left(\frac{8}{10}\right)}\cdot(-2) = -\frac{3}{20}$ rad/sec

23. Let s represent the distance between the bicycle and balloon, h the height of the balloon and x the horizontal distance between the balloon and the bicycle. The relationship between the variables is $s^2 = h^2 + x^2$ $\Rightarrow \frac{ds}{dt} = \frac{1}{s}\left(h\,\frac{dh}{dt} + x\,\frac{dx}{dt}\right) \Rightarrow \frac{ds}{dt} = \frac{1}{85}\,[68(1) + 51(17)] = 11$ ft/sec.

24. (a) Let h be the height of the coffee in the pot. Since the radius of the pot is 3, the volume of the coffee is $V = 9\pi h \Rightarrow \frac{dV}{dt} = 9\pi\,\frac{dh}{dt} \Rightarrow$ the rate the coffee is rising is $\frac{dh}{dt} = \frac{1}{9\pi}\,\frac{dV}{dt} = \frac{10}{9\pi}$ in/min.
   (b) Let h be the height of the coffee in the pot. From the figure, the radius of the filter $r = \frac{h}{2} \Rightarrow V = \frac{1}{3}\,\pi r^2 h$ $= \frac{\pi h^3}{12}$, the volume of the filter. The rate the coffee is falling is $\frac{dh}{dt} = \frac{4}{\pi h^2}\,\frac{dV}{dt} = \frac{4}{25\pi}(-10) = -\frac{8}{5\pi}$ in/min.

25. $y = QD^{-1} \Rightarrow \frac{dy}{dt} = D^{-1}\frac{dQ}{dt} - QD^{-2}\frac{dD}{dt} = \frac{1}{41}(0) - \frac{233}{(41)^2}(-2) = \frac{466}{1681}$ L/min $\Rightarrow$ increasing about 0.2772 L/min

26. (a) $\frac{dc}{dt} = (3x^2 - 12x + 15)\frac{dx}{dt} = (3(2)^2 - 12(2) + 15)(0.1) = 0.3$, $\frac{dr}{dt} = 9\frac{dx}{dt} = 9(0.1) = 0.9$, $\frac{dp}{dt} = 0.9 - 0.3 = 0.6$

(b) $\frac{dc}{dt} = (3x^2 - 12x - 45x^{-2})\frac{dx}{dt} = (3(1.5)^2 - 12(1.5) - 45(1.5)^{-2})(0.05) = -1.5625$, $\frac{dr}{dt} = 70\frac{dx}{dt} = 70(0.05) = 3.5$, $\frac{dp}{dt} = 3.5 - (-1.5625) = 5.0625$

27. Let P(x, y) represent a point on the curve $y = x^2$ and θ the angle of inclination of a line containing P and the origin. Consequently, $\tan\theta = \frac{y}{x} \Rightarrow \tan\theta = \frac{x^2}{x} = x \Rightarrow \sec^2\theta\frac{d\theta}{dt} = \frac{dx}{dt} \Rightarrow \frac{d\theta}{dt} = \cos^2\theta\frac{dx}{dt}$. Since $\frac{dx}{dt} = 10$ m/sec and $\cos^2\theta\big|_{x=3} = \frac{x^2}{y^2+x^2} = \frac{3^2}{9^2+3^2} = \frac{1}{10}$, we have $\frac{d\theta}{dt}\big|_{x=3} = 1$ rad/sec.

28. $y = (-x)^{1/2}$ and $\tan\theta = \frac{y}{x} \Rightarrow \tan\theta = \frac{(-x)^{1/2}}{x} \Rightarrow \sec^2\theta\frac{d\theta}{dt} = \frac{\left(\frac{1}{2}\right)(-x)^{-1/2}(-1)x - (-x)^{1/2}(1)}{x^2}\frac{dx}{dt}$
$\Rightarrow \frac{d\theta}{dt} = \left(\frac{\frac{-x}{2\sqrt{-x}} - \sqrt{-x}}{x^2}\right)(\cos^2\theta)\left(\frac{dx}{dt}\right)$. Now, $\tan\theta = \frac{2}{-4} = -\frac{1}{2} \Rightarrow \cos\theta = -\frac{2}{\sqrt{5}} \Rightarrow \cos^2\theta = \frac{4}{5}$. Then $\frac{d\theta}{dt} = \left(\frac{\frac{4}{4} - 2}{16}\right)\left(\frac{4}{5}\right)(-8) = \frac{2}{5}$ rad/sec.

29. The distance from the origin is $s = \sqrt{x^2 + y^2}$ and we wish to find $\frac{ds}{dt}\big|_{(5,12)}$
$= \frac{1}{2}(x^2 + y^2)^{-1/2}\left(2x\frac{dx}{dt} + 2y\frac{dy}{dt}\right)\Big|_{(5,12)} = \frac{(5)(-1) + (12)(-5)}{\sqrt{25 + 144}} = -5$ m/sec

30. When s represents the length of the shadow and x the distance of the man from the streetlight, then $s = \frac{3}{5}x$.

(a) If I represents the distance of the tip of the shadow from the streetlight, then $I = s + x \Rightarrow \frac{dI}{dt} = \frac{ds}{dt} + \frac{dx}{dt}$ (which is velocity not speed) $\Rightarrow \left|\frac{dI}{dt}\right| = \left|\frac{3}{5}\frac{dx}{dt} + \frac{dx}{dt}\right| = \left|\frac{8}{5}\right|\left|\frac{dx}{dt}\right| = \frac{8}{5}|-5| = 8$ ft/sec, the speed the tip of the shadow is moving along the ground.

(b) $\frac{ds}{dt} = \frac{3}{5}\frac{dx}{dt} = \frac{3}{5}(-5) = -3$ ft/sec, so the length of the shadow is decreasing at a rate of 3 ft/sec.

31. Let $s = 16t^2$ represent the distance the ball has fallen, h the distance between the ball and the ground, and I the distance between the shadow and the point directly beneath the ball. Accordingly, $s + h = 50$ and since the triangle LOQ and triangle PRQ are similar we have $I = \frac{30h}{50 - h} \Rightarrow h = 50 - 16t^2$ and $I = \frac{30(50 - 16t^2)}{50 - (50 - 16t^2)}$
$= \frac{1500}{16t^2} - 30 \Rightarrow \frac{dI}{dt} = -\frac{1500}{8t^3} \Rightarrow \frac{dI}{dt}\big|_{t=\frac{1}{2}} = -1500$ ft/sec.

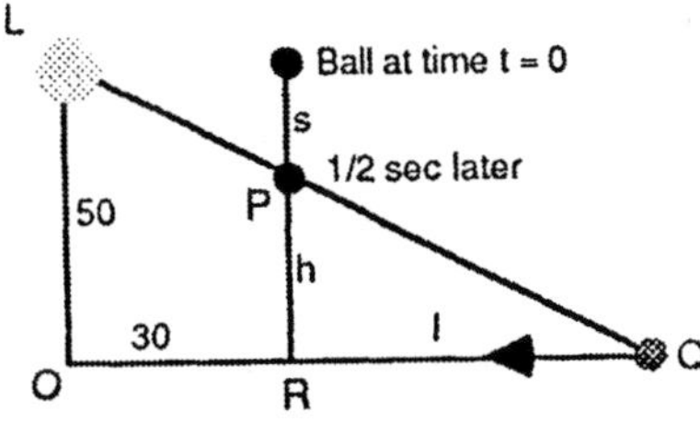

32. Let s = distance of car from foot of perpendicular in the textbook diagram $\Rightarrow \tan\theta = \frac{s}{132} \Rightarrow \sec^2\theta\frac{d\theta}{dt} = \frac{1}{132}\frac{ds}{dt}$
$\Rightarrow \frac{d\theta}{dt} = \frac{\cos^2\theta}{132}\frac{ds}{dt}$; $\frac{ds}{dt} = -264$ and $\theta = 0 \Rightarrow \frac{d\theta}{dt} = -2$ rad/sec. A half second later the car has traveled 132 ft right of the perpendicular $\Rightarrow |\theta| = \frac{\pi}{4}$, $\cos^2\theta = \frac{1}{2}$, and $\frac{ds}{dt} = 264$ (since s increases) $\Rightarrow \frac{d\theta}{dt} = \frac{\left(\frac{1}{2}\right)}{132}(264) = 1$ rad/sec.

33. The volume of the ice is $V = \frac{4}{3}\pi r^3 - \frac{4}{3}\pi 4^3 \Rightarrow \frac{dV}{dt} = 4\pi r^2\frac{dr}{dt} \Rightarrow \frac{dr}{dt}\big|_{r=6} = \frac{-5}{72\pi}$ in./min when $\frac{dV}{dt} = -10$ in^3/min, the thickness of the ice is decreasing at $\frac{5}{72\pi}$ in/min. The surface area is $S = 4\pi r^2 \Rightarrow \frac{dS}{dt} = 8\pi r\frac{dr}{dt} \Rightarrow \frac{dS}{dt}\big|_{r=6} = 48\pi\left(\frac{-5}{72\pi}\right)$ $= -\frac{10}{3}$ in^2/min, the outer surface area of the ice is decreasing at $\frac{10}{3}$ in^2/min.

34. Let s represent the horizontal distance between the car and plane while r is the line-of-sight distance between the car and plane $\Rightarrow 9 + s^2 = r^2 \Rightarrow \frac{ds}{dt} = \frac{r}{\sqrt{r^2 - 9}}\frac{dr}{dt} \Rightarrow \frac{ds}{dt}\big|_{r=5} = \frac{5}{\sqrt{16}}(-160) = -200$ mph $\Rightarrow$ speed of plane + speed of car = 200 mph $\Rightarrow$ the speed of the car is 80 mph.

35. When x represents the length of the shadow, then $\tan\theta = \frac{80}{x} \Rightarrow \sec^2\theta \frac{d\theta}{dt} = -\frac{80}{x^2}\frac{dx}{dt} \Rightarrow \frac{dx}{dt} = \frac{-x^2\sec^2\theta}{80}\frac{d\theta}{dt}$. We are given that $\frac{d\theta}{dt} = 0.27° = \frac{3\pi}{2000}$ rad/min. At x = 60, $\cos\theta = \frac{3}{5} \Rightarrow$ $\left|\frac{dx}{dt}\right| = \left|\frac{-x^2\sec^2\theta}{80}\frac{d\theta}{dt}\right|\Big|_{\left(\frac{d\theta}{dt} = \frac{3\pi}{2000} \text{ and } \sec\theta = \frac{5}{3}\right)} = \frac{3\pi}{16}$ ft/min ≈ 0.589 ft/min ≈ 7.1 in./min.

36. Let A represent the side opposite θ and B represent the side adjacent θ. $\tan\theta = \frac{A}{B} \Rightarrow \sec^2\theta \frac{d\theta}{dt} = \frac{1}{B}\frac{dA}{dt} - \frac{A}{B^2}\frac{dB}{dt}$ $\Rightarrow$ at A = 10 m and B = 20 m we have $\cos\theta = \frac{20}{10\sqrt{5}} = \frac{2}{\sqrt{5}}$ and $\frac{d\theta}{dt} = \left[\left(\frac{1}{20}\right)(-2) - \left(\frac{10}{400}(1)\right)\right]\left(\frac{4}{5}\right)$ $= \left(\frac{-1}{10} - \frac{1}{40}\right)\left(\frac{4}{5}\right) = -\frac{1}{10}$ rad/sec $= -\frac{18°}{\pi}$/sec $\approx -6°$/sec

37. Let x represent distance of the player from second base and s the distance to third base. Then $\frac{dx}{dt} = -16$ ft/sec
(a) $s^2 = x^2 + 8100 \Rightarrow 2s\frac{ds}{dt} = 2x\frac{dx}{dt} \Rightarrow \frac{ds}{dt} = \frac{x}{s}\frac{dx}{dt}$. When the player is 30 ft from first base, x = 60 $\Rightarrow s = 30\sqrt{13}$ and $\frac{ds}{dt} = \frac{60}{30\sqrt{13}}(-16) = \frac{-32}{\sqrt{13}} \approx -8.875$ ft/sec
(b) $\cos\theta_1 = \frac{90}{s} \Rightarrow -\sin\theta_1\frac{d\theta_1}{dt} = -\frac{90}{s^2}\cdot\frac{ds}{dt} \Rightarrow \frac{d\theta_1}{dt} = \frac{90}{s^2\sin\theta_1}\cdot\frac{ds}{dt} = \frac{90}{sx}\cdot\frac{ds}{dt}$. Therefore, x = 60 and $s = 30\sqrt{13}$ $\Rightarrow \frac{d\theta_1}{dt} = \frac{90}{\left(30\sqrt{13}\right)(60)}\cdot\left(\frac{-32}{\sqrt{13}}\right) = \frac{-8}{65}$ rad/sec; $\sin\theta_2 = \frac{90}{s} \Rightarrow \cos\theta_2\frac{d\theta_2}{dt} = -\frac{90}{s^2}\cdot\frac{ds}{dt} \Rightarrow \frac{d\theta_2}{dt} = \frac{-90}{s^2\cos\theta_2}\cdot\frac{ds}{dt}$ $= \frac{-90}{sx}\cdot\frac{ds}{dt}$. Therefore, x = 60 and $s = 30\sqrt{13} \Rightarrow \frac{d\theta_2}{dt} = \frac{8}{65}$ rad/sec.
(c) $\frac{d\theta_1}{dt} = \frac{90}{s^2\sin\theta_1}\cdot\frac{ds}{dt} = \frac{90}{\left(s^2\cdot\frac{x}{s}\right)}\cdot\left(\frac{x}{s}\right)\cdot\left(\frac{dx}{dt}\right) = \left(\frac{90}{s^2}\right)\left(\frac{dx}{dt}\right) = \left(\frac{90}{x^2+8100}\right)\frac{dx}{dt} \Rightarrow \lim_{x\to 0}\frac{d\theta_1}{dt}$ $= \lim_{x\to 0}\left(\frac{90}{x^2+8100}\right)(-15) = -\frac{1}{6}$ rad/sec; $\frac{d\theta_2}{dt} = \frac{-90}{s^2\cos\theta_2}\cdot\frac{ds}{dt} = \left(\frac{-90}{s^2\cdot\frac{x}{s}}\right)\left(\frac{x}{s}\right)\left(\frac{dx}{dt}\right) = \left(\frac{-90}{s^2}\right)\left(\frac{dx}{dt}\right)$ $= \left(\frac{-90}{x^2+8100}\right)\frac{dx}{dt} \Rightarrow \lim_{x\to 0}\frac{d\theta_2}{dt} = \frac{1}{6}$ rad/sec

38. Let a represent the distance between point O and ship A, b the distance between point O and ship B, and D the distance between the ships. By the Law of Cosines, $D^2 = a^2 + b^2 - 2ab\cos 120°$ $\Rightarrow \frac{dD}{dt} = \frac{1}{2D}\left[2a\frac{da}{dt} + 2b\frac{db}{dt} + a\frac{db}{dt} + b\frac{da}{dt}\right]$. When a = 5, $\frac{da}{dt} = 14$, b = 3, and $\frac{db}{dt} = 21$, then $\frac{dD}{dt} = \frac{413}{2D}$ where D = 7. The ships are moving $\frac{dD}{dt} = 29.5$ knots apart.

3.10 LINEARIZATION AND DIFFERENTIALS

1. $f(x) = x^3 - 2x + 3 \Rightarrow f'(x) = 3x^2 - 2 \Rightarrow L(x) = f'(2)(x-2) + f(2) = 10(x-2) + 7 \Rightarrow L(x) = 10x - 13$ at x = 2

2. $f(x) = \sqrt{x^2+9} = (x^2+9)^{1/2} \Rightarrow f'(x) = \left(\frac{1}{2}\right)(x^2+9)^{-1/2}(2x) = \frac{x}{\sqrt{x^2+9}} \Rightarrow L(x) = f'(-4)(x+4) + f(-4)$ $= -\frac{4}{5}(x+4) + 5 \Rightarrow L(x) = -\frac{4}{5}x + \frac{9}{5}$ at x = −4

3. $f(x) = x + \frac{1}{x} \Rightarrow f'(x) = 1 - x^{-2} \Rightarrow L(x) = f(1) + f'(1)(x-1) = 2 + 0(x-1) = 2$

4. $f(x) = x^{1/3} \Rightarrow f'(x) = \frac{1}{3x^{2/3}} \Rightarrow L(x) = f'(-8)(x-(-8)) + f(-8) = \frac{1}{12}(x+8) - 2 \Rightarrow L(x) = \frac{1}{12}x - \frac{4}{3}$

5. $f(x) = \tan x \Rightarrow f'(x) = \sec^2 x \Rightarrow L(x) = f(\pi) + f'(\pi)(x-\pi) = 0 + 1(x-\pi) = x - \pi$

6. (a) $f(x) = \sin x \Rightarrow f'(x) = \cos x \Rightarrow L(x) = f(0) + f'(0)(x-0) = x \Rightarrow L(x) = x$
(b) $f(x) = \cos x \Rightarrow f'(x) = -\sin x \Rightarrow L(x) = f(0) + f'(0)(x-0) = 1 \Rightarrow L(x) = 1$
(c) $f(x) = \tan x \Rightarrow f'(x) = \sec^2 x \Rightarrow L(x) = f(0) + f'(0)(x-0) = x \Rightarrow L(x) = x$
(d) $f(x) = e^x \Rightarrow f'(x) = e^x \Rightarrow L(x) = f(0) + f'(0)(x-0) = x + 1 \Rightarrow L(x) = x + 1$
(e) $f(x) = \ln(1+x) \Rightarrow f'(x) = \frac{1}{1+x} \Rightarrow L(x) = f(0) + f'(0)(x-0) = x \Rightarrow L(x) = x$

7. $f(x) = x^2 + 2x \Rightarrow f'(x) = 2x + 2 \Rightarrow L(x) = f'(0)(x-0) + f(0) = 2(x-0) + 0 \Rightarrow L(x) = 2x$ at x = 0

8. $f(x) = x^{-1} \Rightarrow f'(x) = -x^{-2} \Rightarrow L(x) = f'(1)(x-1) + f(1) = (-1)(x-1) + 1 \Rightarrow L(x) = -x + 2$ at $x = 1$

9. $f(x) = 2x^2 + 4x - 3 \Rightarrow f'(x) = 4x + 4 \Rightarrow L(x) = f'(-1)(x+1) + f(-1) = 0(x+1) + (-5) \Rightarrow L(x) = -5$ at $x = -1$

10. $f(x) = 1 + x \Rightarrow f'(x) = 1 \Rightarrow L(x) = f'(8)(x-8) + f(8) = 1(x-8) + 9 \Rightarrow L(x) = x + 1$ at $x = 8$

11. $f(x) = \sqrt[3]{x} = x^{1/3} \Rightarrow f'(x) = \left(\frac{1}{3}\right) x^{-2/3} \Rightarrow L(x) = f'(8)(x-8) + f(8) = \frac{1}{12}(x-8) + 2 \Rightarrow L(x) = \frac{1}{12}x + \frac{4}{3}$ at $x = 8$

12. $f(x) = \frac{x}{x+1} \Rightarrow f'(x) = \frac{(1)(x+1)-(1)(x)}{(x+1)^2} = \frac{1}{(x+1)^2} \Rightarrow L(x) = f'(1)(x-1) + f(1) = \frac{1}{4}(x-1) + \frac{1}{2}$
$\Rightarrow L(x) = \frac{1}{4}x + \frac{1}{4}$ at $x = 1$

13. $f(x) = e^{-x} \Rightarrow f'(x) = -e^{-x} \Rightarrow L(x) = f(0) + f'(0)(x-0) = -x + 1$

14. $f(x) = \sin^{-1}x \Rightarrow f'(x) = \frac{1}{\sqrt{1-x^2}} \Rightarrow L(x) = f(0) + f'(0)(x-0) = x$

15. $f'(x) = k(1+x)^{k-1}$. We have $f(0) = 1$ and $f'(0) = k$. $L(x) = f(0) + f'(0)(x-0) = 1 + k(x-0) = 1 + kx$

16. (a) $f(x) = (1-x)^6 = \left[1 + (-x)\right]^6 \approx 1 + 6(-x) = 1 - 6x$
(b) $f(x) = \frac{2}{1-x} = 2\left[1 + (-x)\right]^{-1} \approx 2\left[1 + (-1)(-x)\right] = 2 + 2x$
(c) $f(x) = (1+x)^{-1/2} \approx 1 + \left(-\frac{1}{2}\right)x = 1 - \frac{x}{2}$
(d) $f(x) = \sqrt{1+x^2} = \sqrt{2}\left(1 + \frac{x^2}{2}\right)^{1/2} \approx \sqrt{2}\left(1 + \frac{1}{2}\frac{x^2}{2}\right) = \sqrt{2}\left(1 + \frac{x^2}{4}\right)$
(e) $f(x) = (4+3x)^{1/3} = 4^{1/3}\left(1 + \frac{3x}{4}\right)^{1/3} \approx 4^{1/3}\left(1 + \frac{1}{3}\frac{3x}{4}\right) = 4^{1/3}\left(1 + \frac{x}{4}\right)$
(f) $f(x) = \left(1 - \frac{1}{2+x}\right)^{2/3} = \left[1 + \left(-\frac{1}{2+x}\right)\right]^{2/3} \approx 1 + \frac{2}{3}\left(-\frac{1}{2+x}\right) = 1 - \frac{2}{6+3x}$

17. (a) $(1.0002)^{50} = (1 + 0.0002)^{50} \approx 1 + 50(0.0002) = 1 + .01 = 1.01$
(b) $\sqrt[3]{1.009} = (1 + 0.009)^{1/3} \approx 1 + \left(\frac{1}{3}\right)(0.009) = 1 + 0.003 = 1.003$

18. $f(x) = \sqrt{x+1} + \sin x = (x+1)^{1/2} + \sin x \Rightarrow f'(x) = \left(\frac{1}{2}\right)(x+1)^{-1/2} + \cos x \Rightarrow L_f(x) = f'(0)(x-0) + f(0)$
$= \frac{3}{2}(x-0) + 1 \Rightarrow L_f(x) = \frac{3}{2}x + 1$, the linearization of $f(x)$; $g(x) = \sqrt{x+1} = (x+1)^{1/2} \Rightarrow g'(x)$
$= \left(\frac{1}{2}\right)(x+1)^{-1/2} \Rightarrow L_g(x) = g'(0)(x-0) + g(0) = \frac{1}{2}(x-0) + 1 \Rightarrow L_g(x) = \frac{1}{2}x + 1$, the linearization of $g(x)$;
$h(x) = \sin x \Rightarrow h'(x) = \cos x \Rightarrow L_h(x) = h'(0)(x-0) + h(0) = (1)(x-0) + 0 \Rightarrow L_h(x) = x$, the linearization of
$h(x)$. $L_f(x) = L_g(x) + L_h(x)$ implies that the linearization of a sum is equal to the sum of the linearizations.

19. $y = x^3 - 3\sqrt{x} = x^3 - 3x^{1/2} \Rightarrow dy = \left(3x^2 - \frac{3}{2}x^{-1/2}\right)dx \Rightarrow dy = \left(3x^2 - \frac{3}{2\sqrt{x}}\right)dx$

20. $y = x\sqrt{1-x^2} = x\left(1-x^2\right)^{1/2} \Rightarrow dy = \left[(1)\left(1-x^2\right)^{1/2} + (x)\left(\frac{1}{2}\right)\left(1-x^2\right)^{-1/2}(-2x)\right]dx$
$= \left(1-x^2\right)^{-1/2}\left[\left(1-x^2\right) - x^2\right]dx = \frac{(1-2x^2)}{\sqrt{1-x^2}}\,dx$

21. $y = \frac{2x}{1+x^2} \Rightarrow dy = \left(\frac{(2)(1+x^2) - (2x)(2x)}{(1+x^2)^2}\right)dx = \frac{2-2x^2}{(1+x^2)^2}\,dx$

22. $y = \frac{2\sqrt{x}}{3(1+\sqrt{x})} = \frac{2x^{1/2}}{3(1+x^{1/2})} \Rightarrow dy = \left(\frac{x^{-1/2}(3(1+x^{1/2})) - 2x^{1/2}\left(\frac{3}{2}x^{-1/2}\right)}{9(1+x^{1/2})^2}\right) dx = \frac{3x^{-1/2}+3-3}{9(1+x^{1/2})^2}\, dx$
$\Rightarrow dy = \frac{1}{3\sqrt{x}(1+\sqrt{x})^2}\, dx$

23. $2y^{3/2} + xy - x = 0 \Rightarrow 3y^{1/2}\, dy + y\, dx + x\, dy - dx = 0 \Rightarrow \left(3y^{1/2} + x\right) dy = (1-y)\, dx \Rightarrow dy = \frac{1-y}{3\sqrt{y}+x}\, dx$

24. $xy^2 - 4x^{3/2} - y = 0 \Rightarrow y^2\, dx + 2xy\, dy - 6x^{1/2}\, dx - dy = 0 \Rightarrow (2xy-1)\, dy = \left(6x^{1/2} - y^2\right) dx$
$\Rightarrow dy = \frac{6\sqrt{x} - y^2}{2xy-1}\, dx$

25. $y = \sin\left(5\sqrt{x}\right) = \sin\left(5x^{1/2}\right) \Rightarrow dy = \left(\cos\left(5x^{1/2}\right)\right)\left(\frac{5}{2}x^{-1/2}\right) dx \Rightarrow dy = \frac{5\cos\left(5\sqrt{x}\right)}{2\sqrt{x}}\, dx$

26. $y = \cos(x^2) \Rightarrow dy = [-\sin(x^2)](2x)\, dx = -2x\sin(x^2)\, dx$

27. $y = 4\tan\left(\frac{x^3}{3}\right) \Rightarrow dy = 4\left(\sec^2\left(\frac{x^3}{3}\right)\right)(x^2)\, dx \Rightarrow dy = 4x^2\sec^2\left(\frac{x^3}{3}\right) dx$

28. $y = \sec(x^2-1) \Rightarrow dy = [\sec(x^2-1)\tan(x^2-1)](2x)\, dx = 2x[\sec(x^2-1)\tan(x^2-1)]\, dx$

29. $y = 3\csc\left(1-2\sqrt{x}\right) = 3\csc\left(1-2x^{1/2}\right) \Rightarrow dy = 3\left(-\csc\left(1-2x^{1/2}\right)\right)\cot\left(1-2x^{1/2}\right)\left(-x^{-1/2}\right) dx$
$\Rightarrow dy = \frac{3}{\sqrt{x}}\csc\left(1-2\sqrt{x}\right)\cot\left(1-2\sqrt{x}\right) dx$

30. $y = 2\cot\left(\frac{1}{\sqrt{x}}\right) = 2\cot\left(x^{-1/2}\right) \Rightarrow dy = -2\csc^2\left(x^{-1/2}\right)\left(-\frac{1}{2}\right)\left(x^{-3/2}\right) dx \Rightarrow dy = \frac{1}{\sqrt{x^3}}\csc^2\left(\frac{1}{\sqrt{x}}\right) dx$

31. $y = e^{\sqrt{x}} \Rightarrow dy = \frac{e^{\sqrt{x}}}{2\sqrt{x}}\, dx$

32. $y = x\, e^{-x} \Rightarrow dy = \left(-x\, e^{-x} + e^{-x}\right) dx = (1-x)\, e^{-x}\, dx$

33. $y = \ln(1+x^2) \Rightarrow dy = \frac{2x}{1+x^2}\, dx$

34. $y = \ln\left(\frac{x+1}{\sqrt{x-1}}\right) = \ln(x+1) - \frac{1}{2}\ln(x-1) \Rightarrow dy = \left(\frac{1}{x+1} - \frac{1}{2}\cdot\frac{1}{x-1}\right) dx = \frac{x-3}{2(x^2-1)}\, dx$

35. $y = \tan^{-1}\left(e^{x^2}\right) \Rightarrow dy = \frac{1}{1+\left(e^{x^2}\right)^2}\cdot e^{x^2}\cdot 2x\, dx = \frac{2x\, e^{x^2}}{1+e^{2x^2}}\, dx$

36. $y = \cot^{-1}\left(\frac{1}{x^2}\right) + \cos^{-1}(2x)$ Note: $\frac{d}{d\theta}\cot^{-1}\theta = \frac{-1\, d\theta}{1+\theta^2}$, so that $\frac{d}{dx}\left(\cot^{-1}\left(\frac{1}{x^2}\right)\right) = \frac{-1\cdot\left(-\frac{2}{x^3}\right)}{1+\frac{1}{x^4}} = \frac{-1\cdot\left(-\frac{2}{x^3}\right)}{\frac{x^4+1}{x^4}} = \frac{2x}{x^4+1}$
Note: $\frac{d}{d\theta}\cos^{-1}\theta = \frac{-1}{\sqrt{1-\theta^2}}d\theta$, so that $\frac{d}{dx}(\cos^{-1}(2x)) = \frac{-1\cdot 2}{\sqrt{1-4x^2}} = \frac{-2}{\sqrt{1-4x^2}}$. Thus $dy = \left(\frac{2x}{x^4+1} - \frac{2}{\sqrt{1-4x^2}}\right) dx$

37. $y = \sec^{-1}(e^{-x}) \Rightarrow dy = \frac{1}{e^{-x}\sqrt{(e^{-x})^2-1}}\cdot(-e^{-x})\, dx = \frac{-1}{\sqrt{\left(\frac{1}{e^x}\right)^2-1}}\, dx = \frac{-e^x}{\sqrt{1-e^{2x}}}\, dx$

38. $y = e^{\tan^{-1}\sqrt{x^2+1}} \Rightarrow dy = e^{\tan^{-1}\sqrt{x^2+1}}\cdot\frac{1}{1+\left(\sqrt{x^2+1}\right)^2}\cdot\frac{1}{2}(x^2+1)^{-1/2}\cdot 2x\, dx = \frac{x\, e^{\tan^{-1}\sqrt{x^2+1}}}{(x^2+2)\sqrt{x^2+1}}\, dx$

39. $f(x) = x^2 + 2x$, $x_0 = 1$, $dx = 0.1 \Rightarrow f'(x) = 2x + 2$
(a) $\Delta f = f(x_0 + dx) - f(x_0) = f(1.1) - f(1) = 3.41 - 3 = 0.41$
(b) $df = f'(x_0)\,dx = [2(1) + 2](0.1) = 0.4$
(c) $|\Delta f - df| = |0.41 - 0.4| = 0.01$

40. $f(x) = 2x^2 + 4x - 3$, $x_0 = -1$, $dx = 0.1 \Rightarrow f'(x) = 4x + 4$
(a) $\Delta f = f(x_0 + dx) - f(x_0) = f(-.9) - f(-1) = .02$
(b) $df = f'(x_0)\,dx = [4(-1) + 4](.1) = 0$
(c) $|\Delta f - df| = |.02 - 0| = .02$

41. $f(x) = x^3 - x$, $x_0 = 1$, $dx = 0.1 \Rightarrow f'(x) = 3x^2 - 1$
(a) $\Delta f = f(x_0 + dx) - f(x_0) = f(1.1) - f(1) = .231$
(b) $df = f'(x_0)\,dx = [3(1)^2 - 1](.1) = .2$
(c) $|\Delta f - df| = |.231 - .2| = .031$

42. $f(x) = x^4$, $x_0 = 1$, $dx = 0.1 \Rightarrow f'(x) = 4x^3$
(a) $\Delta f = f(x_0 + dx) - f(x_0) = f(1.1) - f(1) = .4641$
(b) $df = f'(x_0)\,dx = 4(1)^3(.1) = .4$
(c) $|\Delta f - df| = |.4641 - .4| = .0641$

43. $f(x) = x^{-1}$, $x_0 = 0.5$, $dx = 0.1 \Rightarrow f'(x) = -x^{-2}$
(a) $\Delta f = f(x_0 + dx) - f(x_0) = f(.6) - f(.5) = -\frac{1}{3}$
(b) $df = f'(x_0)\,dx = (-4)\left(\frac{1}{10}\right) = -\frac{2}{5}$
(c) $|\Delta f - df| = \left|-\frac{1}{3} + \frac{2}{5}\right| = \frac{1}{15}$

44. $f(x) = x^3 - 2x + 3$, $x_0 = 2$, $dx = 0.1 \Rightarrow f'(x) = 3x^2 - 2$
(a) $\Delta f = f(x_0 + dx) - f(x_0) = f(2.1) - f(2) = 1.061$
(b) $df = f'(x_0)\,dx = (10)(0.10) = 1$
(c) $|\Delta f - df| = |1.061 - 1| = .061$

45. $V = \frac{4}{3}\pi r^3 \Rightarrow dV = 4\pi r_0^2\,dr$

46. $V = x^3 \Rightarrow dV = 3x_0^2\,dx$

47. $S = 6x^2 \Rightarrow dS = 12x_0\,dx$

48. $S = \pi r\sqrt{r^2 + h^2} = \pi r\left(r^2 + h^2\right)^{1/2}$, h constant $\Rightarrow \frac{dS}{dr} = \pi\left(r^2 + h^2\right)^{1/2} + \pi r \cdot r\left(r^2 + h^2\right)^{-1/2}$
$\Rightarrow \frac{dS}{dr} = \frac{\pi\left(r^2 + h^2\right) + \pi r^2}{\sqrt{r^2 + h^2}} \Rightarrow dS = \frac{\pi\left(2r_0^2 + h^2\right)}{\sqrt{r_0^2 + h^2}}\,dr$, h constant

49. $V = \pi r^2 h$, height constant $\Rightarrow dV = 2\pi r_0 h\,dr$

50. $S = 2\pi rh \Rightarrow dS = 2\pi r\,dh$

51. Given $r = 2$ m, $dr = .02$ m
(a) $A = \pi r^2 \Rightarrow dA = 2\pi r\,dr = 2\pi(2)(.02) = .08\pi\ m^2$
(b) $\left(\frac{.08\pi}{4\pi}\right)(100\%) = 2\%$

52. $C = 2\pi r$ and $dC = 2$ in. $\Rightarrow dC = 2\pi\,dr \Rightarrow dr = \frac{1}{\pi} \Rightarrow$ the diameter grew about $\frac{2}{\pi}$ in.; $A = \pi r^2 \Rightarrow dA = 2\pi r\,dr$
$= 2\pi(5)\left(\frac{1}{\pi}\right) = 10\ in.^2$

53. The volume of a cylinder is $V = \pi r^2 h$. When h is held fixed, we have $\frac{dV}{dr} = 2\pi rh$, and so $dV = 2\pi rh\, dr$. For $h = 30$ in., $r = 6$ in., and $dr = 0.5$ in., the volume of the material in the shell is approximately $dV = 2\pi rh\, dr = 2\pi(6)(30)(0.5)$ $= 180\pi \approx 565.5\, in^3$.

54. Let θ = angle of elevation and h = height of building. Then $h = 30\tan\theta$, so $dh = 30\sec^2\theta\, d\theta$. We want $|dh| < 0.04h$, which gives: $|30\sec^2\theta\, d\theta| < 0.04|30\tan\theta| \Rightarrow \frac{1}{\cos^2\theta}|d\theta| < \frac{0.04\sin\theta}{\cos\theta} \Rightarrow |d\theta| < 0.04\sin\theta\cos\theta \Rightarrow |d\theta| < 0.04\sin\frac{5\pi}{12}\cos\frac{5\pi}{12}$ $= 0.01$ radian. The angle should be measured with an error of less than 0.01 radian (or approximatley 0.57 degrees), which is a percentage error of approximately 0.76%.

55. $V = \pi h^3 \Rightarrow dV = 3\pi h^2\, dh$; recall that $\Delta V \approx dV$. Then $|\Delta V| \le (1\%)(V) = \frac{(1)(\pi h^3)}{100} \Rightarrow |dV| \le \frac{(1)(\pi h^3)}{100}$ $\Rightarrow |3\pi h^2\, dh| \le \frac{(1)(\pi h^3)}{100} \Rightarrow |dh| \le \frac{1}{300} h = \left(\frac{1}{3}\%\right) h$. Therefore the greatest tolerated error in the measurement of h is $\frac{1}{3}$ %.

56. (a) Let D_i represent the inside diameter. Then $V = \pi r^2 h = \pi\left(\frac{D_i}{2}\right)^2 h = \frac{\pi D_i^2 h}{4}$ and $h = 10 \Rightarrow V = \frac{5\pi D_i^2}{2} \Rightarrow$ $dV = 5\pi D_i\, dD_i$. Recall that $\Delta V \approx dV$. We want $|\Delta V| \le (1\%)(V) \Rightarrow |dV| \le \left(\frac{1}{100}\right)\left(\frac{5\pi D_i^2}{2}\right) = \frac{\pi D_i^2}{40}$ $\Rightarrow 5\pi D_i\, dD_i \le \frac{\pi D_i^2}{40} \Rightarrow \frac{dD_i}{D_i} \le 200$. The inside diameter must be measured to within 0.5%.

(b) Let D_e represent the exterior diameter, h the height and S the area of the painted surface. $S = \pi D_e h \Rightarrow dS = \pi h dD_e$ $\Rightarrow \frac{dS}{S} = \frac{dD_e}{D_e}$. Thus for small changes in exterior diameter, the approximate percentage change in the exterior diameter is equal to the approximate percentage change in the area painted, and to estimate the amount of paint required to within 5%, the tanks's exterior diameter must be measured to within 5%.

57. $V = \pi r^2 h$, h is constant $\Rightarrow dV = 2\pi rh\, dr$; recall that $\Delta V \approx dV$. We want $|\Delta V| \le \frac{1}{1000} V \Rightarrow |dV| \le \frac{\pi r^2 h}{1000}$ $\Rightarrow |2\pi rh\, dr| \le \frac{\pi r^2 h}{1000} \Rightarrow |dr| \le \frac{r}{2000} = (.05\%)r \Rightarrow$ a .05% variation in the radius can be tolerated.

58. $\frac{\Delta P}{P} \times 100\% \approx \frac{dP}{P} \times 100\% = \frac{\left(200-\frac{x}{2}\right)e^{-x/400}}{200xe^{-x/400}}dx$. As sales change from $x = 145$ to $x = 150$, $\Delta x = dx = 5$ and $\frac{\Delta P}{P} \times 100\% \approx \frac{\left(200-\frac{145}{2}\right)e^{-145/400}}{200(145)e^{-145/400}}(5) \times 100\% = 2.2\%$.

59. $W = a + \frac{b}{g} = a + bg^{-1} \Rightarrow dW = -bg^{-2}\, dg = -\frac{b\,dg}{g^2} \Rightarrow \frac{dW_{moon}}{dW_{earth}} = \frac{\left(-\frac{b\,dg}{(5.2)^2}\right)}{\left(-\frac{b\,dg}{(32)^2}\right)} = \left(\frac{32}{5.2}\right)^2 = 37.87$, so a change of gravity on the moon has about 38 times the effect that a change of the same magnitude has on Earth.

60. (a) $T = 2\pi\left(\frac{L}{g}\right)^{1/2} \Rightarrow dT = 2\pi\sqrt{L}\left(-\frac{1}{2}g^{-3/2}\right)dg = -\pi\sqrt{L}\,g^{-3/2}\,dg$

(b) If g increases, then $dg > 0 \Rightarrow dT < 0$. The period T decreases and the clock ticks more frequently. Both the pendulum speed and clock speed increase.

(c) $0.001 = -\pi\sqrt{100}\left(980^{-3/2}\right) dg \Rightarrow dg \approx -0.977$ cm/sec^2 $\Rightarrow$ the new $g \approx 979$ cm/sec^2

61. The error in measurement $dx = (1\%)(10) = 0.1$ cm; $V = x^3 \Rightarrow dV = 3x^2\, dx = 3(10)^2(0.1) = 30\ cm^3 \Rightarrow$ the percentage error in the volume calculation is $\left(\frac{30}{1000}\right)(100\%) = 3\%$

62. $A = s^2 \Rightarrow dA = 2s\, ds$; recall that $\Delta A \approx dA$. Then $|\Delta A| \le (2\%)A = \frac{2s^2}{100} = \frac{s^2}{50} \Rightarrow |dA| \le \frac{s^2}{50} \Rightarrow |2s\, ds| \le \frac{s^2}{50}$ $\Rightarrow |ds| \le \frac{s^2}{(2s)(50)} = \frac{s}{100} = (1\%)s \Rightarrow$ the error must be no more than 1% of the true value.

63. Given D = 100 cm, dD = 1 cm, $V = \frac{4}{3}\pi\left(\frac{D}{2}\right)^3 = \frac{\pi D^3}{6} \Rightarrow dV = \frac{\pi}{2} D^2\,dD = \frac{\pi}{2}(100)^2(1) = \frac{10^4\pi}{2}$. Then $\frac{dV}{V}(100\%)$ $= \left[\frac{\frac{10^4\pi}{2}}{\frac{10^6\pi}{6}}\right](10^2\%) = \left[\frac{\frac{10^6\pi}{2}}{\frac{10^6\pi}{6}}\right]\% = 3\%$

64. $V = \frac{4}{3}\pi r^3 = \frac{4}{3}\pi\left(\frac{D}{2}\right)^3 = \frac{\pi D^3}{6} \Rightarrow dV = \frac{\pi D^2}{2}\,dD$; recall that $\Delta V \approx dV$. Then $|\Delta V| \le (3\%)V = \left(\frac{3}{100}\right)\left(\frac{\pi D^3}{6}\right)$ $= \frac{\pi D^3}{200} \Rightarrow |dV| \le \frac{\pi D^3}{200} \Rightarrow \left|\frac{\pi D^2}{2}\,dD\right| \le \frac{\pi D^3}{200} \Rightarrow |dD| \le \frac{D}{100} = (1\%)\,D \Rightarrow$ the allowable percentage error in measuring the diameter is 1%.

65. A 5% error in measuring t $\Rightarrow dt = (5\%)t = \frac{t}{20}$. Then $s = 16t^2 \Rightarrow ds = 32t\,dt = 32t\left(\frac{t}{20}\right) = \frac{32t^2}{20} = \frac{16t^2}{10} = \left(\frac{1}{10}\right)s$ $= (10\%)s \Rightarrow$ a 10% error in the calculation of s.

66. From Example 8 we have $\frac{dV}{V} = 4\,\frac{dr}{r}$. An increase of 12.5% in r will give a 50% increase in V.

67. $\lim\limits_{x\to 0} \frac{\sqrt{1+x}}{1+\frac{x}{2}} = \frac{\sqrt{1+0}}{1+\frac{0}{2}} = 1$

68. $\lim\limits_{x\to 0} \frac{\tan x}{x} = \lim\limits_{x\to 0} \left(\frac{\sin x}{x}\right)\left(\frac{1}{\cos x}\right) = (1)(1) = 1$

69. $E(x) = f(x) - g(x) \Rightarrow E(x) = f(x) - m(x-a) - c$. Then $E(a) = 0 \Rightarrow f(a) - m(a-a) - c = 0 \Rightarrow c = f(a)$. Next we calculate m: $\lim\limits_{x\to a} \frac{E(x)}{x-a} = 0 \Rightarrow \lim\limits_{x\to a} \frac{f(x)-m(x-a)-c}{x-a} = 0 \Rightarrow \lim\limits_{x\to a}\left[\frac{f(x)-f(a)}{x-a} - m\right] = 0$ (since $c = f(a)$) $\Rightarrow f'(a) - m = 0 \Rightarrow m = f'(a)$. Therefore, $g(x) = m(x-a) + c = f'(a)(x-a) + f(a)$ is the linear approximation, as claimed.

70. (a) i. $Q(a) = f(a)$ implies that $b_0 = f(a)$.

ii. Since $Q'(x) = b_1 + 2b_2(x-a)$, $Q'(a) = f'(a)$ implies that $b_1 = f'(a)$.

iii. Since $Q''(x) = 2b_2$, $Q''(a) = f''(a)$ implies that $b_2 = \frac{f''(a)}{2}$.

In summary, $b_0 = f(a)$, $b_1 = f'(a)$, and $b_2 = \frac{f''(a)}{2}$.

(b) $f(x) = (1-x)^{-1}$

$f'(x) = -1(1-x)^{-2}(-1) = (1-x)^{-2}$

$f''(x) = -2(1-x)^{-3}(-1) = 2(1-x)^{-3}$

Since $f(0) = 1$, $f'(0) = 1$, and $f''(0) = 2$, the coefficients are $b_0 = 1$, $b_1 = 1$, $b_2 = \frac{2}{2} = 1$. The quadratic approximation is $Q(x) = 1 + x + x^2$.

(c)

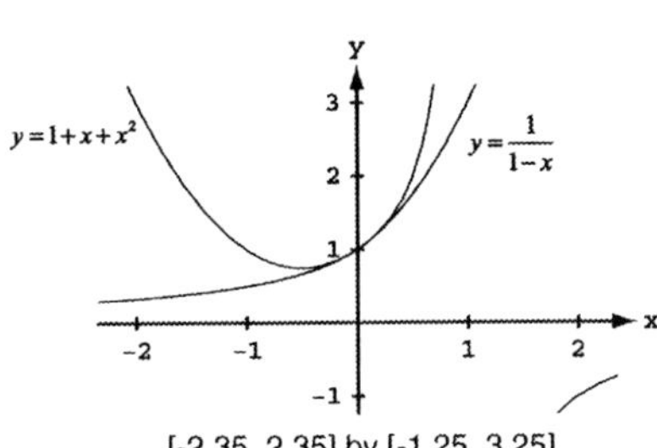

[-2.35, 2.35] by [-1.25, 3.25]

As one zooms in, the two graphs quickly become indistinguishable. They appear to be identical.

(d) $g(x) = x^{-1}$

$g'(x) = -1x^{-2}$

$g''(x) = 2x^{-3}$

Since $g(1) = 1$, $g'(1) = -1$, and $g''(1) = 2$, the coefficients are $b_0 = 1$, $b_1 = -1$, $b_2 = \frac{2}{2} = 1$. The quadratic approximation is $Q(x) = 1 - (x-1) + (x-1)^2$.

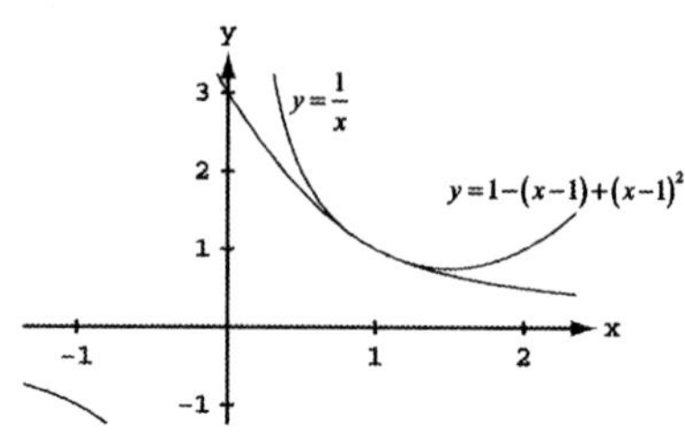

As one zooms in, the two graphs quickly become indistinguishable. They appear to be identical.

(e) $h(x) = (1+x)^{1/2}$

$h'(x) = \frac{1}{2}(1+x)^{-1/2}$

$h''(x) = -\frac{1}{4}(1+x)^{-3/2}$

Since $h(0) = 1$, $h'(0) = \frac{1}{2}$, and $h''(0) = -\frac{1}{4}$, the coefficients are $b_0 = 1$, $b_1 = \frac{1}{2}$, $b_2 = \frac{-\frac{1}{4}}{2} = -\frac{1}{8}$. The quadratic approximation is $Q(x) = 1 + \frac{x}{2} - \frac{x^2}{8}$.

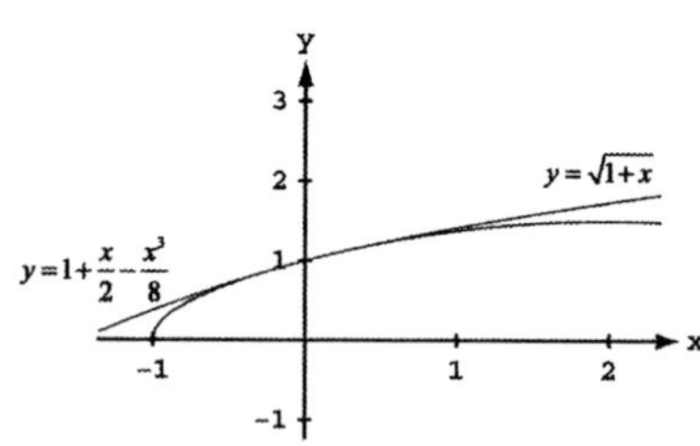

As one zooms in, the two graphs quickly become indistinguishable. They appear to be identical.

(f) The linearization of any differentiable function $u(x)$ at $x = a$ is $L(x) = u(a) + u'(a)(x-a) = b_0 + b_1(x-a)$, where b_0 and b_1 are the coefficients of the constant and linear terms of the quadratic approximation. Thus, the linearization for $f(x)$ at $x = 0$ is $1 + x$; the linearization for $g(x)$ at $x = 1$ is $1 - (x-1)$ or $2 - x$; and the linearization for $h(x)$ at $x = 0$ is $1 + \frac{x}{2}$.

71. (a) $f(x) = 2^x \Rightarrow f'(x) = 2^x \ln 2$; $L(x) = (2^0 \ln 2)x + 2^0 = x \ln 2 + 1 \approx 0.69x + 1$

(b)

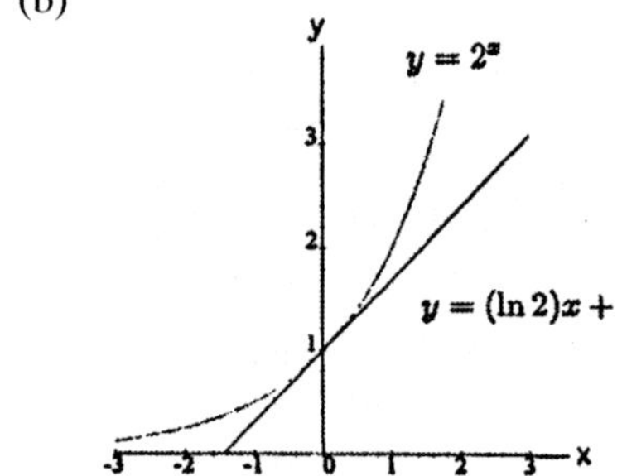

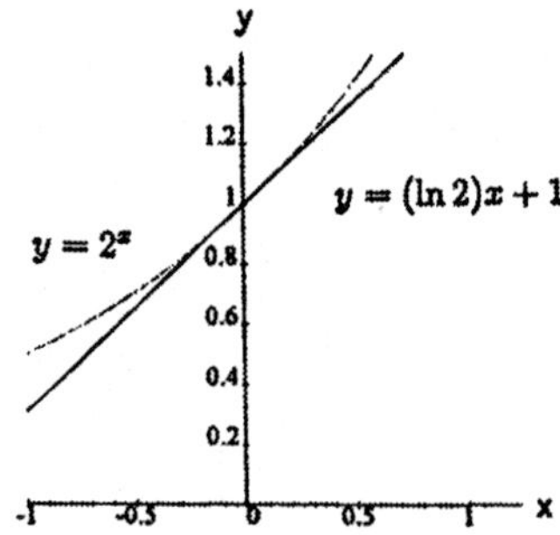

72. (a) $f(x) = \log_3 x \Rightarrow f'(x) = \frac{1}{x \ln 3}$, and $f(3) = \frac{\ln 3}{\ln 3} \Rightarrow L(x) = \frac{1}{3 \ln 3}(x-3) + \frac{\ln 3}{\ln 3} = \frac{x}{3 \ln 3} - \frac{1}{\ln 3} + 1$
$\approx 0.30x + 0.09$

(b)

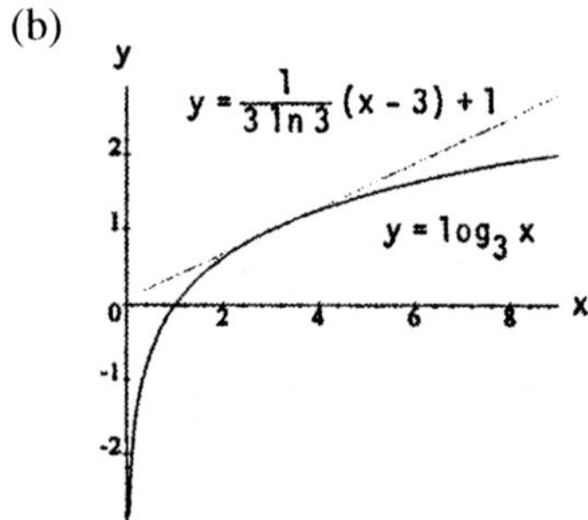

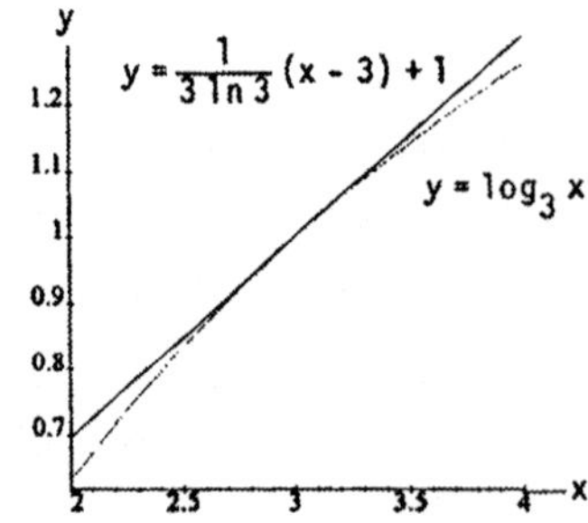

73. (a) $x = 1$

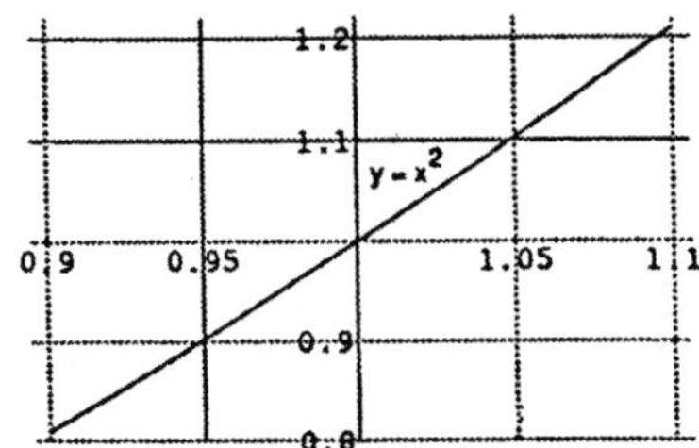

(b) $x = 1$; $m = 2.5$, $e^1 \approx 2.7$ $\qquad$ $x = 0$; $m = 1$, $e^0 = 1$ $\qquad$ $x = -1$; $m = 0.3$, $e^{-1} \approx 0.4$

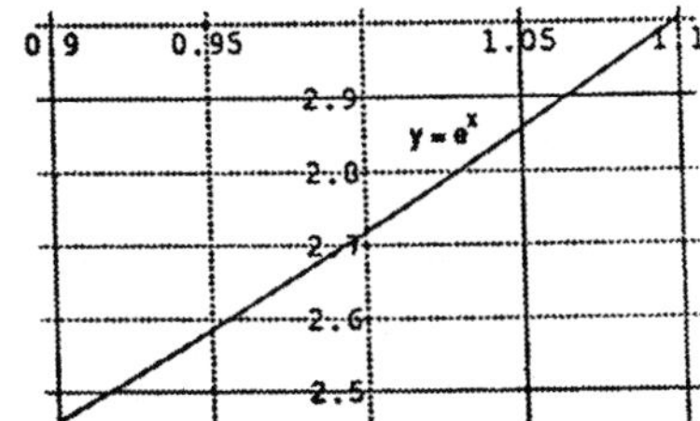

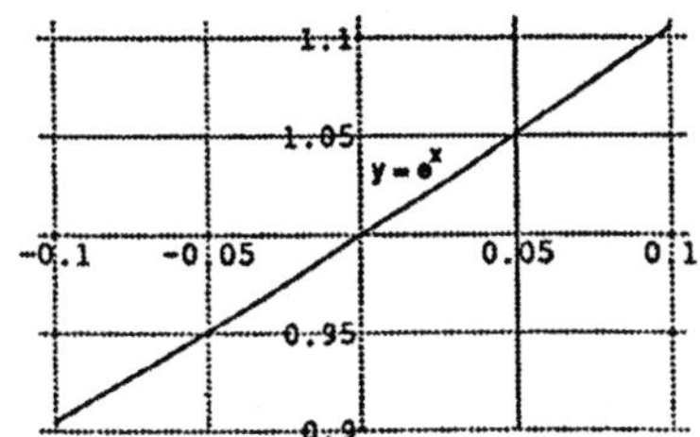

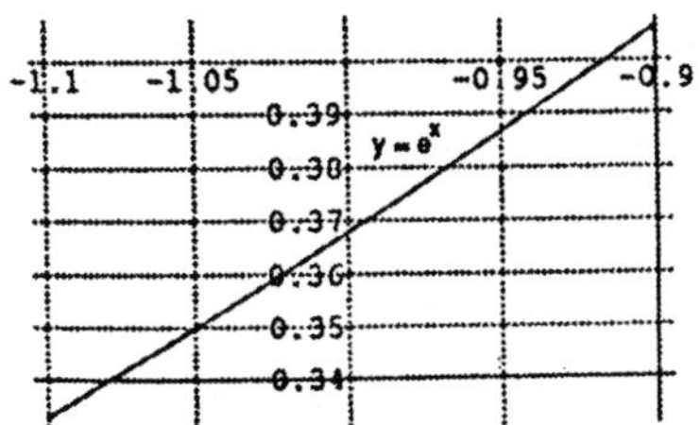

74. If f has a horizontal tangent at $x = a$, then $f'(a) = 0$ and the linearization of f at $x = a$ is $L(x) = f(a) + f'(a)(x - a) = f(a) + 0 \cdot (x - a) = f(a)$. The linearization is a constant.

75. Find $|v|$ when $m = 1.01m_0$. $m = \frac{m_0}{\sqrt{1-\frac{v^2}{c^2}}} \Rightarrow m\sqrt{1 - \frac{v^2}{c^2}} = m_0 \Rightarrow \sqrt{1 - \frac{v^2}{c^2}} = \frac{m_0}{m} \Rightarrow 1 - \frac{v^2}{c^2} = \frac{m_0^2}{m^2} \Rightarrow v^2 = c^2\left(1 - \frac{m_0^2}{m^2}\right)$

$\Rightarrow |v| = c\sqrt{1 - \frac{m_0^2}{m^2}} \Rightarrow dv = c \cdot \frac{1}{2}\left(1 - \frac{m_0^2}{m^2}\right)^{-1/2}\left(\frac{2m_0^2}{m^3}\right)dm$, $dm = 0.01m_0 \Rightarrow dv = \frac{c\,m_0^2}{m^3\sqrt{1-\frac{m_0^2}{m^2}}}\left(\frac{m_0}{100}\right)$. $m = \frac{101}{100}m_0$,

$dv = \frac{c \cdot m_0^2}{\frac{101^3}{100^3}m_0^3\sqrt{1 - \frac{m_0^2}{\frac{101^2}{100^2}m_0^2}}}\left(\frac{m_0}{100}\right) = \frac{1000}{101^3\sqrt{1-\frac{100^2}{101^2}}} \approx 0.69c$. Body at rest $\Rightarrow v_0 = 0$ and $v = v_0 + dv$

$\Rightarrow v = 0.69c$.

76. (a) The successive square roots of 2 appear to converge to the number 1. For tenth roots the convergence is more rapid.

(b) Successive square roots of 0.5 also converge to 1. In fact, successive square roots of any positive number converge to 1.

A graph indicates what is going on:

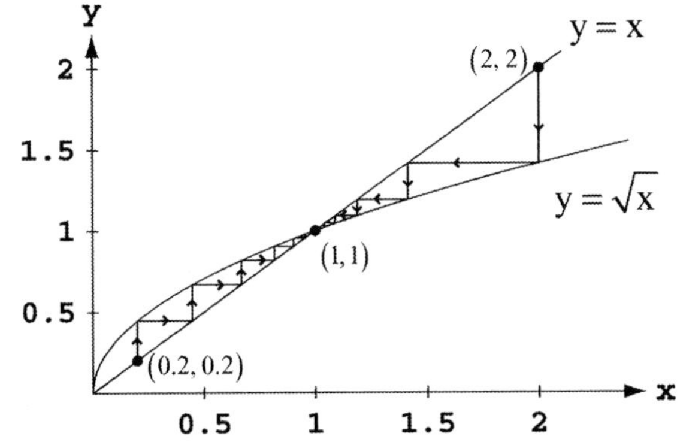

Starting on the line $y = x$, the successive square roots are found by moving to the graph of $y = \sqrt{x}$ and then across to the line $y = x$ again. From any positive starting value x, the iterates converge to 1.

77. (a) Window: $-0.00006 \le x \le 0.00006,\ 0.9999 \le y \le 1.0001$

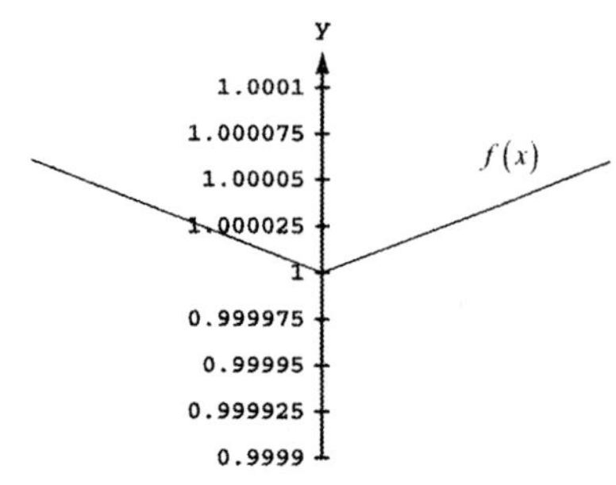

After zooming in seven times, starting with the window $-1 \le x \le 1$ and $0 \le y \le 2$ on a TI-92 Plus calculator, the graph of $f(x)$ shows no signs of straightening out.

(b) Window: $-0.01 \le x \le 0.01,\ 0.98 \le y \le 1.02$

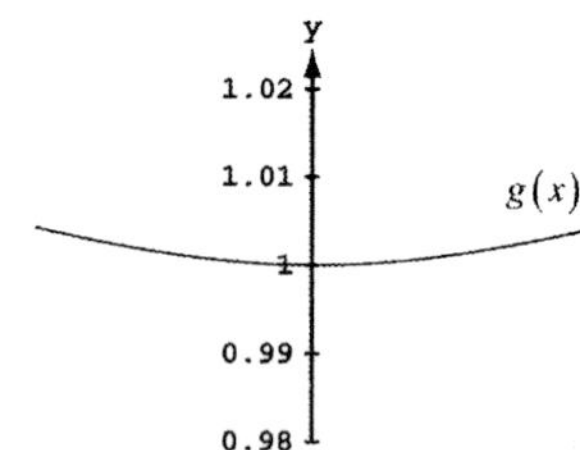

After zooming in only twice, starting with the window $-1 \le x \le 1$ and $0 \le y \le 2$ on a TI-92 Plus calculator, the graph of $g(x)$ already appears to be smoothing toward a horizontal straight line.

(c) After seven zooms, starting with the window $-1 \le x \le 1$ and $0 \le y \le 2$ on a TI-92 Plus calculator, the graph of $g(x)$ looks exactly like a horizontal straight line.

(d) Window: $-1 \le x \le 1,\ 0 \le y \le 2$

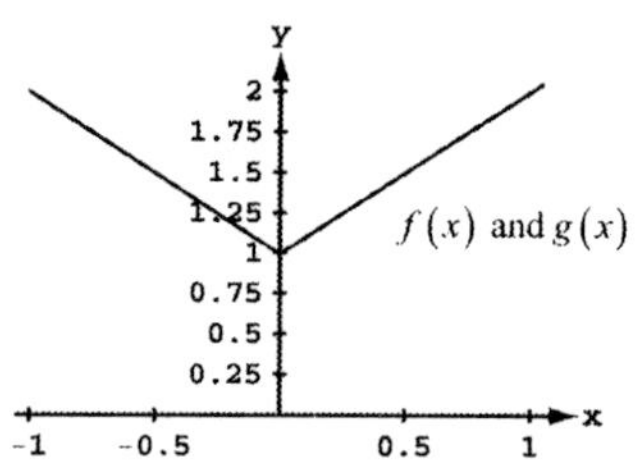

78. Volume $= (x + \Delta x)^3 = x^3 + 3x^2(\Delta x) + 3x(\Delta x)^2 + (\Delta x)^3$

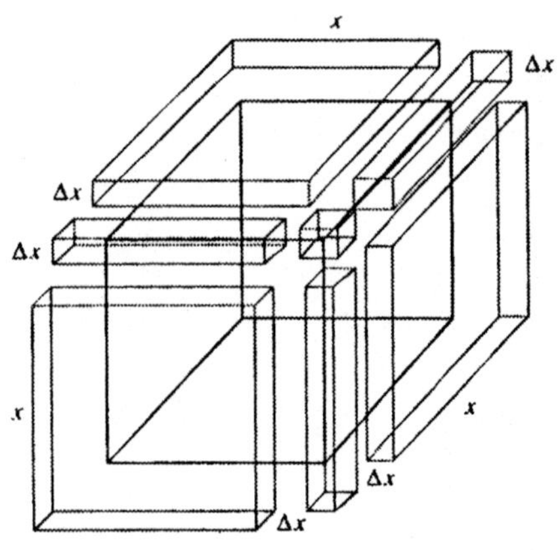

79-84. Example CAS commands:

Maple:

```
with(plots):
a:= 1: f:=x -> x ∧ 3 + x ∧ 2 − 2*x;
plot(f(x), x=−1..2);
diff(f(x),x);
fp := unapply (″,x);
```

```
L:=x -> f(a) + fp(a)*(x − a);
plot({f(x), L(x)}, x=−1..2);
err:=x -> abs(f(x) − L(x));
plot(err(x), x=−1..2, title = #absolute error function#);
err(−1);
```

Mathematica: (function, x1, x2, and a may vary):

```
Clear[f, x]
{x1, x2} = {−1, 2}; a = 1;
f[x_]:=x^3 + x^2 − 2x
Plot[f[x], {x, x1, x2}]
lin[x_]=f[a] + f'[a](x − a)
Plot[{f[x], lin[x]}, {x, x1, x2}]
err[x_]=Abs[f[x] − lin[x]]
Plot[err[x], {x, x1,x 2}]
err//N
```

After reviewing the error function, plot the error function and epsilon for differing values of epsilon (eps) and delta (del)

```
eps = 0.5; del = 0.4
Plot[{err[x], eps},{x, a − del, a + del}]
```

CHAPTER 3 PRACTICE EXERCISES

1. $y = x^5 - 0.125x^2 + 0.25x \Rightarrow \frac{dy}{dx} = 5x^4 - 0.25x + 0.25$

2. $y = 3 - 0.7x^3 + 0.3x^7 \Rightarrow \frac{dy}{dx} = -2.1x^2 + 2.1x^6$

3. $y = x^3 - 3(x^2 + \pi^2) \Rightarrow \frac{dy}{dx} = 3x^2 - 3(2x + 0) = 3x^2 - 6x = 3x(x - 2)$

4. $y = x^7 + \sqrt{7}x - \frac{1}{\pi+1} \Rightarrow \frac{dy}{dx} = 7x^6 + \sqrt{7}$

5. $y = (x+1)^2(x^2+2x) \Rightarrow \frac{dy}{dx} = (x+1)^2(2x+2) + (x^2+2x)(2(x+1)) = 2(x+1)\left[(x+1)^2 + x(x+2)\right]$
$= 2(x+1)(2x^2+4x+1)$

6. $y = (2x-5)(4-x)^{-1} \Rightarrow \frac{dy}{dx} = (2x-5)(-1)(4-x)^{-2}(-1) + (4-x)^{-1}(2) = (4-x)^{-2}\left[(2x-5) + 2(4-x)\right]$
$= 3(4-x)^{-2}$

7. $y = (\theta^2 + \sec\theta + 1)^3 \Rightarrow \frac{dy}{d\theta} = 3(\theta^2 + \sec\theta + 1)^2(2\theta + \sec\theta\tan\theta)$

8. $y = \left(-1 - \frac{\csc\theta}{2} - \frac{\theta^2}{4}\right)^2 \Rightarrow \frac{dy}{d\theta} = 2\left(-1 - \frac{\csc\theta}{2} - \frac{\theta^2}{4}\right)\left(\frac{\csc\theta\cot\theta}{2} - \frac{\theta}{2}\right) = \left(-1 - \frac{\csc\theta}{2} - \frac{\theta^2}{4}\right)(\csc\theta\cot\theta - \theta)$

9. $s = \frac{\sqrt{t}}{1+\sqrt{t}} \Rightarrow \frac{ds}{dt} = \frac{(1+\sqrt{t})\cdot\frac{1}{2\sqrt{t}} - \sqrt{t}\left(\frac{1}{2\sqrt{t}}\right)}{(1+\sqrt{t})^2} = \frac{(1+\sqrt{t}) - \sqrt{t}}{2\sqrt{t}(1+\sqrt{t})^2} = \frac{1}{2\sqrt{t}(1+\sqrt{t})^2}$

10. $s = \frac{1}{\sqrt{t}-1} \Rightarrow \frac{ds}{dt} = \frac{(\sqrt{t}-1)(0) - 1\left(\frac{1}{2\sqrt{t}}\right)}{(\sqrt{t}-1)^2} = \frac{-1}{2\sqrt{t}(\sqrt{t}-1)^2}$

11. $y = 2\tan^2 x - \sec^2 x \Rightarrow \frac{dy}{dx} = (4\tan x)(\sec^2 x) - (2\sec x)(\sec x\tan x) = 2\sec^2 x\tan x$

12. $y = \frac{1}{\sin^2 x} - \frac{2}{\sin x} = \csc^2 x - 2\csc x \Rightarrow \frac{dy}{dx} = (2\csc x)(-\csc x \cot x) - 2(-\csc x \cot x) = (2\csc x \cot x)(1 - \csc x)$

13. $s = \cos^4(1-2t) \Rightarrow \frac{ds}{dt} = 4\cos^3(1-2t)(-\sin(1-2t))(-2) = 8\cos^3(1-2t)\sin(1-2t)$

14. $s = \cot^3\left(\frac{2}{t}\right) \Rightarrow \frac{ds}{dt} = 3\cot^2\left(\frac{2}{t}\right)\left(-\csc^2\left(\frac{2}{t}\right)\right)\left(\frac{-2}{t^2}\right) = \frac{6}{t^2}\cot^2\left(\frac{2}{t}\right)\csc^2\left(\frac{2}{t}\right)$

15. $s = (\sec t + \tan t)^5 \Rightarrow \frac{ds}{dt} = 5(\sec t + \tan t)^4(\sec t \tan t + \sec^2 t) = 5(\sec t)(\sec t + \tan t)^5$

16. $s = \csc^5(1 - t + 3t^2) \Rightarrow \frac{ds}{dt} = 5\csc^4(1-t+3t^2)(-\csc(1-t+3t^2)\cot(1-t+3t^2))(-1+6t)$
$= -5(6t-1)\csc^5(1-t+3t^2)\cot(1-t+3t^2)$

17. $r = \sqrt{2\theta \sin\theta} = (2\theta\sin\theta)^{1/2} \Rightarrow \frac{dr}{d\theta} = \frac{1}{2}(2\theta\sin\theta)^{-1/2}(2\theta\cos\theta + 2\sin\theta) = \frac{\theta\cos\theta + \sin\theta}{\sqrt{2\theta\sin\theta}}$

18. $r = 2\theta\sqrt{\cos\theta} = 2\theta(\cos\theta)^{1/2} \Rightarrow \frac{dr}{d\theta} = 2\theta\left(\frac{1}{2}\right)(\cos\theta)^{-1/2}(-\sin\theta) + 2(\cos\theta)^{1/2} = \frac{-\theta\sin\theta}{\sqrt{\cos\theta}} + 2\sqrt{\cos\theta}$
$= \frac{2\cos\theta - \theta\sin\theta}{\sqrt{\cos\theta}}$

19. $r = \sin\sqrt{2\theta} = \sin(2\theta)^{1/2} \Rightarrow \frac{dr}{d\theta} = \cos(2\theta)^{1/2}\left(\frac{1}{2}(2\theta)^{-1/2}(2)\right) = \frac{\cos\sqrt{2\theta}}{\sqrt{2\theta}}$

20. $r = \sin\left(\theta + \sqrt{\theta+1}\right) \Rightarrow \frac{dr}{d\theta} = \cos\left(\theta + \sqrt{\theta+1}\right)\left(1 + \frac{1}{2\sqrt{\theta+1}}\right) = \frac{2\sqrt{\theta+1}+1}{2\sqrt{\theta+1}}\cos\left(\theta + \sqrt{\theta+1}\right)$

21. $y = \frac{1}{2}x^2\csc\frac{2}{x} \Rightarrow \frac{dy}{dx} = \frac{1}{2}x^2\left(-\csc\frac{2}{x}\cot\frac{2}{x}\right)\left(\frac{-2}{x^2}\right) + \left(\csc\frac{2}{x}\right)\left(\frac{1}{2}\cdot 2x\right) = \csc\frac{2}{x}\cot\frac{2}{x} + x\csc\frac{2}{x}$

22. $y = 2\sqrt{x}\sin\sqrt{x} \Rightarrow \frac{dy}{dx} = 2\sqrt{x}\left(\cos\sqrt{x}\right)\left(\frac{1}{2\sqrt{x}}\right) + \left(\sin\sqrt{x}\right)\left(\frac{2}{2\sqrt{x}}\right) = \cos\sqrt{x} + \frac{\sin\sqrt{x}}{\sqrt{x}}$

23. $y = x^{-1/2}\sec(2x)^2 \Rightarrow \frac{dy}{dx} = x^{-1/2}\sec(2x)^2\tan(2x)^2(2(2x)\cdot 2) + \sec(2x)^2\left(-\frac{1}{2}x^{-3/2}\right)$
$= 8x^{1/2}\sec(2x)^2\tan(2x)^2 - \frac{1}{2}x^{-3/2}\sec(2x)^2 = \frac{1}{2}x^{1/2}\sec(2x)^2\left[16\tan(2x)^2 - x^{-2}\right]$ or $\frac{1}{2x^{3/2}}\sec(2x)^2\left[16x^2\tan(2x)^2 - 1\right]$

24. $y = \sqrt{x}\csc(x+1)^3 = x^{1/2}\csc(x+1)^3$
$\Rightarrow \frac{dy}{dx} = x^{1/2}\left(-\csc(x+1)^3\cot(x+1)^3\right)\left(3(x+1)^2\right) + \csc(x+1)^3\left(\frac{1}{2}x^{-1/2}\right)$
$= -3\sqrt{x}(x+1)^2\csc(x+1)^3\cot(x+1)^3 + \frac{\csc(x+1)^3}{2\sqrt{x}} = \frac{1}{2}\sqrt{x}\csc(x+1)^3\left[\frac{1}{x} - 6(x+1)^2\cot(x+1)^3\right]$
or $\frac{1}{2\sqrt{x}}\csc(x+1)^3\left[1 - 6x(x+1)^2\cot(x+1)^3\right]$

25. $y = 5\cot x^2 \Rightarrow \frac{dy}{dx} = 5(-\csc^2 x^2)(2x) = -10x\csc^2(x^2)$

26. $y = x^2\cot 5x \Rightarrow \frac{dy}{dx} = x^2(-\csc^2 5x)(5) + (\cot 5x)(2x) = -5x^2\csc^2 5x + 2x\cot 5x$

27. $y = x^2\sin^2(2x^2) \Rightarrow \frac{dy}{dx} = x^2(2\sin(2x^2))(\cos(2x^2))(4x) + \sin^2(2x^2)(2x) = 8x^3\sin(2x^2)\cos(2x^2) + 2x\sin^2(2x^2)$

28. $y = x^{-2}\sin^2(x^3) \Rightarrow \frac{dy}{dx} = x^{-2}(2\sin(x^3))(\cos(x^3))(3x^2) + \sin^2(x^3)(-2x^{-3}) = 6\sin(x^3)\cos(x^3) - 2x^{-3}\sin^2(x^3)$

29. $s = \left(\frac{4t}{t+1}\right)^{-2} \Rightarrow \frac{ds}{dt} = -2\left(\frac{4t}{t+1}\right)^{-3}\left(\frac{(t+1)(4)-(4t)(1)}{(t+1)^2}\right) = -2\left(\frac{4t}{t+1}\right)^{-3}\frac{4}{(t+1)^2} = -\frac{(t+1)}{8t^3}$

30. $s = \frac{-1}{15(15t-1)^3} = -\frac{1}{15}(15t-1)^{-3} \Rightarrow \frac{ds}{dt} = -\frac{1}{15}(-3)(15t-1)^{-4}(15) = \frac{3}{(15t-1)^4}$

31. $y = \left(\frac{\sqrt{x}}{x+1}\right)^2 \Rightarrow \frac{dy}{dx} = 2\left(\frac{\sqrt{x}}{x+1}\right) \cdot \frac{(x+1)\left(\frac{1}{2\sqrt{x}}\right) - (\sqrt{x})(1)}{(x+1)^2} = \frac{(x+1)-2x}{(x+1)^3} = \frac{1-x}{(x+1)^3}$

32. $y = \left(\frac{2\sqrt{x}}{2\sqrt{x}+1}\right)^2 \Rightarrow \frac{dy}{dx} = 2\left(\frac{2\sqrt{x}}{2\sqrt{x}+1}\right)\left(\frac{(2\sqrt{x}+1)\left(\frac{1}{\sqrt{x}}\right) - (2\sqrt{x})\left(\frac{1}{\sqrt{x}}\right)}{(2\sqrt{x}+1)^2}\right) = \frac{4\sqrt{x}\left(\frac{1}{\sqrt{x}}\right)}{(2\sqrt{x}+1)^3} = \frac{4}{(2\sqrt{x}+1)^3}$

33. $y = \sqrt{\frac{x^2+x}{x^2}} = \left(1+\frac{1}{x}\right)^{1/2} \Rightarrow \frac{dy}{dx} = \frac{1}{2}\left(1+\frac{1}{x}\right)^{-1/2}\left(-\frac{1}{x^2}\right) = -\frac{1}{2x^2\sqrt{1+\frac{1}{x}}}$

34. $y = 4x\sqrt{x+\sqrt{x}} = 4x\left(x+x^{1/2}\right)^{1/2} \Rightarrow \frac{dy}{dx} = 4x\left(\frac{1}{2}\right)\left(x+x^{1/2}\right)^{-1/2}\left(1+\frac{1}{2}x^{-1/2}\right) + \left(x+x^{1/2}\right)^{1/2}(4)$
$= \left(x+\sqrt{x}\right)^{-1/2}\left[2x\left(1+\frac{1}{2\sqrt{x}}\right) + 4\left(x+\sqrt{x}\right)\right] = \left(x+\sqrt{x}\right)^{-1/2}\left(2x+\sqrt{x}+4x+4\sqrt{x}\right) = \frac{6x+5\sqrt{x}}{\sqrt{x+\sqrt{x}}}$

35. $r = \left(\frac{\sin\theta}{\cos\theta-1}\right)^2 \Rightarrow \frac{dr}{d\theta} = 2\left(\frac{\sin\theta}{\cos\theta-1}\right)\left[\frac{(\cos\theta-1)(\cos\theta)-(\sin\theta)(-\sin\theta)}{(\cos\theta-1)^2}\right]$
$= 2\left(\frac{\sin\theta}{\cos\theta-1}\right)\left(\frac{\cos^2\theta-\cos\theta+\sin^2\theta}{(\cos\theta-1)^2}\right) = \frac{(2\sin\theta)(1-\cos\theta)}{(\cos\theta-1)^3} = \frac{-2\sin\theta}{(\cos\theta-1)^2}$

36. $r = \left(\frac{\sin\theta+1}{1-\cos\theta}\right)^2 \Rightarrow \frac{dr}{d\theta} = 2\left(\frac{\sin\theta+1}{1-\cos\theta}\right)\left[\frac{(1-\cos\theta)(\cos\theta)-(\sin\theta+1)(\sin\theta)}{(1-\cos\theta)^2}\right]$
$= \frac{2(\sin\theta+1)}{(1-\cos\theta)^3}\left(\cos\theta-\cos^2\theta-\sin^2\theta-\sin\theta\right) = \frac{2(\sin\theta+1)(\cos\theta-\sin\theta-1)}{(1-\cos\theta)^3}$

37. $y = (2x+1)\sqrt{2x+1} = (2x+1)^{3/2} \Rightarrow \frac{dy}{dx} = \frac{3}{2}(2x+1)^{1/2}(2) = 3\sqrt{2x+1}$

38. $y = 20(3x-4)^{1/4}(3x-4)^{-1/5} = 20(3x-4)^{1/20} \Rightarrow \frac{dy}{dx} = 20\left(\frac{1}{20}\right)(3x-4)^{-19/20}(3) = \frac{3}{(3x-4)^{19/20}}$

39. $y = 3\left(5x^2+\sin 2x\right)^{-3/2} \Rightarrow \frac{dy}{dx} = 3\left(-\frac{3}{2}\right)\left(5x^2+\sin 2x\right)^{-5/2}[10x+(\cos 2x)(2)] = \frac{-9(5x+\cos 2x)}{(5x^2+\sin 2x)^{5/2}}$

40. $y = \left(3+\cos^3 3x\right)^{-1/3} \Rightarrow \frac{dy}{dx} = -\frac{1}{3}\left(3+\cos^3 3x\right)^{-4/3}\left(3\cos^2 3x\right)(-\sin 3x)(3) = \frac{3\cos^2 3x\sin 3x}{(3+\cos^3 3x)^{4/3}}$

41. $y = 10e^{-x/5} \Rightarrow \frac{dy}{dx} = (10)\left(-\frac{1}{5}\right)e^{-x/5} = -2e^{-x/5}$

42. $y = \sqrt{2}e^{\sqrt{2}x} \Rightarrow \frac{dy}{dx} = \left(\sqrt{2}\right)\left(\sqrt{2}\right)e^{\sqrt{2}x} = 2e^{\sqrt{2}x}$

43. $y = \frac{1}{4}xe^{4x} - \frac{1}{16}e^{4x} \Rightarrow \frac{dy}{dx} = \frac{1}{4}\left[x\left(4e^{4x}\right)+e^{4x}(1)\right] - \frac{1}{16}\left(4e^{4x}\right) = xe^{4x} + \frac{1}{4}e^{4x} - \frac{1}{4}e^{4x} = xe^{4x}$

44. $y = x^2e^{-2/x} = x^2e^{-2x^{-1}} \Rightarrow \frac{dy}{dx} = x^2\left[\left(2x^{-2}\right)e^{-2x^{-1}}\right] + e^{-2x^{-1}}(2x) = (2+2x)e^{-2x^{-1}} = 2e^{-2/x}(1+x)$

45. $y = \ln\left(\sin^2\theta\right) \Rightarrow \frac{dy}{d\theta} = \frac{2(\sin\theta)(\cos\theta)}{\sin^2\theta} = \frac{2\cos\theta}{\sin\theta} = 2\cot\theta$

46. $y = \ln\left(\sec^2\theta\right) \Rightarrow \frac{dy}{d\theta} = \frac{2(\sec\theta)(\sec\theta\tan\theta)}{\sec^2\theta} = 2\tan\theta$

47. $y = \log_2\left(\frac{x^2}{2}\right) = \frac{\ln\left(\frac{x^2}{2}\right)}{\ln 2} \Rightarrow \frac{dy}{dx} = \frac{1}{\ln 2}\left(\frac{x}{\left(\frac{x^2}{2}\right)}\right) = \frac{2}{(\ln 2)x}$

48. $y = \log_5(3x-7) = \frac{\ln(3x-7)}{\ln 5} \Rightarrow \frac{dy}{dx} = \left(\frac{1}{\ln 5}\right)\left(\frac{3}{3x-7}\right) = \frac{3}{(\ln 5)(3x-7)}$

49. $y = 8^{-t} \Rightarrow \frac{dy}{dt} = 8^{-t}(\ln 8)(-1) = -8^{-t}(\ln 8)$

50. $y = 9^{2t} \Rightarrow \frac{dy}{dt} = 9^{2t}(\ln 9)(2) = 9^{2t}(2\ln 9)$

51. $y = 5x^{3.6} \Rightarrow \frac{dy}{dx} = 5(3.6)x^{2.6} = 18x^{2.6}$

52. $y = \sqrt{2}\,x^{-\sqrt{2}} \Rightarrow \frac{dy}{dx} = \left(\sqrt{2}\right)\left(-\sqrt{2}\right)x^{\left(-\sqrt{2}-1\right)} = -2x^{\left(-\sqrt{2}-1\right)}$

53. $y = (x+2)^{x+2} \Rightarrow \ln y = \ln(x+2)^{x+2} = (x+2)\ln(x+2) \Rightarrow \frac{y'}{y} = (x+2)\left(\frac{1}{x+2}\right) + (1)\ln(x+2)$
$\Rightarrow \frac{dy}{dx} = (x+2)^{x+2}\left[\ln(x+2)+1\right]$

54. $y = 2(\ln x)^{x/2} \Rightarrow \ln y = \ln\left[2(\ln x)^{x/2}\right] = \ln(2) + \left(\frac{x}{2}\right)\ln(\ln x) \Rightarrow \frac{y'}{y} = 0 + \left(\frac{x}{2}\right)\left[\frac{\left(\frac{1}{x}\right)}{\ln x}\right] + (\ln(\ln x))\left(\frac{1}{2}\right)$
$\Rightarrow y' = \left[\frac{1}{2\ln x} + \left(\frac{1}{2}\right)\ln(\ln x)\right]2(\ln x)^{x/2} = (\ln x)^{x/2}\left[\ln(\ln x) + \frac{1}{\ln x}\right]$

55. $y = \sin^{-1}\sqrt{1-u^2} = \sin^{-1}\left(1-u^2\right)^{1/2} \Rightarrow \frac{dy}{du} = \frac{\frac{1}{2}(1-u^2)^{-1/2}(-2u)}{\sqrt{1-\left[(1-u^2)^{1/2}\right]^2}} = \frac{-u}{\sqrt{1-u^2}\sqrt{1-(1-u^2)}} = \frac{-u}{|u|\sqrt{1-u^2}}$
$= \frac{-u}{u\sqrt{1-u^2}} = \frac{-1}{\sqrt{1-u^2}},\ 0 < u < 1$

56. $y = \sin^{-1}\left(\frac{1}{\sqrt{v}}\right) = \sin^{-1}v^{-1/2} \Rightarrow \frac{dy}{dv} = \frac{-\frac{1}{2}v^{-3/2}}{\sqrt{1-(v^{-1/2})^2}} = \frac{-1}{2v^{3/2}\sqrt{1-v^{-1}}} = \frac{-1}{2v^{3/2}\sqrt{\frac{v-1}{v}}} = \frac{-\sqrt{v}}{2v^{3/2}\sqrt{v-1}}$
$= \frac{-1}{2v\sqrt{v-1}}$

57. $y = \ln\left(\cos^{-1}x\right) \Rightarrow y' = \frac{\left(\frac{-1}{\sqrt{1-x^2}}\right)}{\cos^{-1}x} = \frac{-1}{\sqrt{1-x^2}\cos^{-1}x}$

58. $y = z\cos^{-1}z - \sqrt{1-z^2} = z\cos^{-1}z - \left(1-z^2\right)^{1/2} \Rightarrow \frac{dy}{dz} = \cos^{-1}z - \frac{z}{\sqrt{1-z^2}} - \left(\frac{1}{2}\right)\left(1-z^2\right)^{-1/2}(-2z)$
$= \cos^{-1}z - \frac{z}{\sqrt{1-z^2}} + \frac{z}{\sqrt{1-z^2}} = \cos^{-1}z$

59. $y = t\tan^{-1}t - \left(\frac{1}{2}\right)\ln t \Rightarrow \frac{dy}{dt} = \tan^{-1}t + t\left(\frac{1}{1+t^2}\right) - \left(\frac{1}{2}\right)\left(\frac{1}{t}\right) = \tan^{-1}t + \frac{t}{1+t^2} - \frac{1}{2t}$

60. $y = \left(1+t^2\right)\cot^{-1}2t \Rightarrow \frac{dy}{dt} = 2t\cot^{-1}2t + \left(1+t^2\right)\left(\frac{-2}{1+4t^2}\right)$

61. $y = z\sec^{-1}z - \sqrt{z^2-1} = z\sec^{-1}z - \left(z^2-1\right)^{1/2} \Rightarrow \frac{dy}{dz} = z\left(\frac{1}{|z|\sqrt{z^2-1}}\right) + \left(\sec^{-1}z\right)(1) - \frac{1}{2}\left(z^2-1\right)^{-1/2}(2z)$
$= \frac{z}{|z|\sqrt{z^2-1}} - \frac{z}{\sqrt{z^2-1}} + \sec^{-1}z = \frac{1-z}{\sqrt{z^2-1}} + \sec^{-1}z,\ z > 1$

62. $y = 2\sqrt{x-1}\sec^{-1}\sqrt{x} = 2(x-1)^{1/2}\sec^{-1}\left(x^{1/2}\right)$
$\Rightarrow \frac{dy}{dx} = 2\left[\left(\frac{1}{2}\right)(x-1)^{-1/2}\sec^{-1}\left(x^{1/2}\right) + (x-1)^{1/2}\left(\frac{\left(\frac{1}{2}\right)x^{-1/2}}{\sqrt{x}\sqrt{x-1}}\right)\right] = 2\left(\frac{\sec^{-1}\sqrt{x}}{2\sqrt{x-1}} + \frac{1}{2x}\right) = \frac{\sec^{-1}\sqrt{x}}{\sqrt{x-1}} + \frac{1}{x}$

63. $y = \csc^{-1}(\sec\theta) \Rightarrow \frac{dy}{d\theta} = \frac{-\sec\theta\tan\theta}{|\sec\theta|\sqrt{\sec^2\theta - 1}} = -\frac{\tan\theta}{|\tan\theta|} = -1,\ 0 < \theta < \frac{\pi}{2}$

64. $y = \left(1+x^2\right)e^{\tan^{-1}x} \Rightarrow y' = 2xe^{\tan^{-1}x} + \left(1+x^2\right)\left(\frac{e^{\tan^{-1}x}}{1+x^2}\right) = 2xe^{\tan^{-1}x} + e^{\tan^{-1}x}$

65. $xy + 2x + 3y = 1 \Rightarrow (xy' + y) + 2 + 3y' = 0 \Rightarrow xy' + 3y' = -2 - y \Rightarrow y'(x+3) = -2 - y \Rightarrow y' = -\frac{y+2}{x+3}$

66. $x^2 + xy + y^2 - 5x = 2 \Rightarrow 2x + \left(x\frac{dy}{dx} + y\right) + 2y\frac{dy}{dx} - 5 = 0 \Rightarrow x\frac{dy}{dx} + 2y\frac{dy}{dx} = 5 - 2x - y \Rightarrow \frac{dy}{dx}(x+2y)$
$= 5 - 2x - y \Rightarrow \frac{dy}{dx} = \frac{5-2x-y}{x+2y}$

67. $x^3 + 4xy - 3y^{4/3} = 2x \Rightarrow 3x^2 + \left(4x\frac{dy}{dx} + 4y\right) - 4y^{1/3}\frac{dy}{dx} = 2 \Rightarrow 4x\frac{dy}{dx} - 4y^{1/3}\frac{dy}{dx} = 2 - 3x^2 - 4y$
$\Rightarrow \frac{dy}{dx}\left(4x - 4y^{1/3}\right) = 2 - 3x^2 - 4y \Rightarrow \frac{dy}{dx} = \frac{2-3x^2-4y}{4x-4y^{1/3}}$

68. $5x^{4/5} + 10y^{6/5} = 15 \Rightarrow 4x^{-1/5} + 12y^{1/5}\frac{dy}{dx} = 0 \Rightarrow 12y^{1/5}\frac{dy}{dx} = -4x^{-1/5} \Rightarrow \frac{dy}{dx} = -\frac{1}{3}x^{-1/5}y^{-1/5} = -\frac{1}{3(xy)^{1/5}}$

69. $(xy)^{1/2} = 1 \Rightarrow \frac{1}{2}(xy)^{-1/2}\left(x\frac{dy}{dx} + y\right) = 0 \Rightarrow x^{1/2}y^{-1/2}\frac{dy}{dx} = -x^{-1/2}y^{1/2} \Rightarrow \frac{dy}{dx} = -x^{-1}y \Rightarrow \frac{dy}{dx} = -\frac{y}{x}$

70. $x^2y^2 = 1 \Rightarrow x^2\left(2y\frac{dy}{dx}\right) + y^2(2x) = 0 \Rightarrow 2x^2y\frac{dy}{dx} = -2xy^2 \Rightarrow \frac{dy}{dx} = -\frac{y}{x}$

71. $y^2 = \frac{x}{x+1} \Rightarrow 2y\frac{dy}{dx} = \frac{(x+1)(1)-(x)(1)}{(x+1)^2} \Rightarrow \frac{dy}{dx} = \frac{1}{2y(x+1)^2}$

72. $y^2 = \left(\frac{1+x}{1-x}\right)^{1/2} \Rightarrow y^4 = \frac{1+x}{1-x} \Rightarrow 4y^3\frac{dy}{dx} = \frac{(1-x)(1)-(1+x)(-1)}{(1-x)^2} \Rightarrow \frac{dy}{dx} = \frac{1}{2y^3(1-x)^2}$

73. $e^{x+2y} = 1 \Rightarrow e^{x+2y}\left(1 + 2\frac{dy}{dx}\right) = 0 \Rightarrow \frac{dy}{dx} = -\frac{1}{2}$

74. $y^2 = 2e^{-1/x} \Rightarrow 2y\frac{dy}{dx} = 2e^{-1/x}\frac{d}{dx}(-x^{-1}) = \frac{2e^{-1/x}}{x^2} \Rightarrow \frac{dy}{dx} = \frac{e^{-1/x}}{yx^2}$

75. $\ln\left(\frac{x}{y}\right) = 1 \Rightarrow \frac{1}{x/y}\frac{d}{dx}\left(\frac{x}{y}\right) = 0 \Rightarrow \frac{y(1) - x\frac{dy}{dx}}{y^2} = 0 \Rightarrow \frac{dy}{dx} = \frac{y}{x}$

76. $x\sin^{-1}y = 1 + x^2 \Rightarrow y = \sin(x^{-1} + x) \Rightarrow \frac{dy}{dx} = \cos(x^{-1}+x)\frac{d}{dx}(x^{-1}+x) = (1-x^{-2})\cos(x^{-1}+x)$
$= \left(\frac{x^2-1}{x}\right)\cos\left(\frac{x^2+1}{x}\right)$

77. $y\,e^{\tan^{-1}x} = 2 \Rightarrow y = 2e^{-\tan^{-1}x} \Rightarrow \frac{dy}{dx} = 2e^{-\tan^{-1}x}\frac{d}{dx}(-\tan^{-1}x) = -2e^{-\tan^{-1}x}\left(\frac{1}{1+x^2}\right) = -\frac{2e^{-\tan^{-1}x}}{1+x^2}$

78. $x^y = \sqrt{2} \Rightarrow \ln(x^y) = \ln\left(2^{1/2}\right) \Rightarrow y\ln x = \frac{\ln 2}{2} \Rightarrow \frac{d}{dx}(y\ln x) = 0 \Rightarrow y\left(\frac{1}{x}\right) + \ln x\frac{dy}{dx} = 0 \Rightarrow \frac{dy}{dx} = -\frac{y}{x\ln x}$

79. $p^3 + 4pq - 3q^2 = 2 \Rightarrow 3p^2\frac{dp}{dq} + 4\left(p + q\frac{dp}{dq}\right) - 6q = 0 \Rightarrow 3p^2\frac{dp}{dq} + 4q\frac{dp}{dq} = 6q - 4p \Rightarrow \frac{dp}{dq}(3p^2 + 4q) = 6q - 4p$
$\Rightarrow \frac{dp}{dq} = \frac{6q-4p}{3p^2+4q}$

80. $q = (5p^2 + 2p)^{-3/2} \Rightarrow 1 = -\frac{3}{2}(5p^2+2p)^{-5/2}\left(10p\frac{dp}{dq} + 2\frac{dp}{dq}\right) \Rightarrow -\frac{2}{3}(5p^2+2p)^{5/2} = \frac{dp}{dq}(10p+2)$
$\Rightarrow \frac{dp}{dq} = -\frac{(5p^2+2p)^{5/2}}{3(5p+1)}$

81. $r\cos 2s + \sin^2 s = \pi \Rightarrow r(-\sin 2s)(2) + (\cos 2s)\left(\frac{dr}{ds}\right) + 2\sin s\cos s = 0 \Rightarrow \frac{dr}{ds}(\cos 2s) = 2r\sin 2s - 2\sin s\cos s$
$\Rightarrow \frac{dr}{ds} = \frac{2r\sin 2s - \sin 2s}{\cos 2s} = \frac{(2r-1)(\sin 2s)}{\cos 2s} = (2r-1)(\tan 2s)$

82. $2rs - r - s + s^2 = -3 \Rightarrow 2\left(r + s\,\frac{dr}{ds}\right) - \frac{dr}{ds} - 1 + 2s = 0 \Rightarrow \frac{dr}{ds}(2s - 1) = 1 - 2s - 2r \Rightarrow \frac{dr}{ds} = \frac{1-2s-2r}{2s-1}$

83. (a) $x^3 + y^3 = 1 \Rightarrow 3x^2 + 3y^2\,\frac{dy}{dx} = 0 \Rightarrow \frac{dy}{dx} = -\frac{x^2}{y^2} \Rightarrow \frac{d^2y}{dx^2} = \frac{y^2(-2x) - (-x^2)\left(2y\,\frac{dy}{dx}\right)}{y^4}$

$\Rightarrow \frac{d^2y}{dx^2} = \frac{-2xy^2 + (2yx^2)\left(-\frac{x^2}{y^2}\right)}{y^4} = \frac{-2xy^2 - \frac{2x^4}{y}}{y^4} = \frac{-2xy^3 - 2x^4}{y^5}$

(b) $y^2 = 1 - \frac{2}{x} \Rightarrow 2y\,\frac{dy}{dx} = \frac{2}{x^2} \Rightarrow \frac{dy}{dx} = \frac{1}{yx^2} \Rightarrow \frac{dy}{dx} = (yx^2)^{-1} \Rightarrow \frac{d^2y}{dx^2} = -(yx^2)^{-2}\left[y(2x) + x^2\,\frac{dy}{dx}\right]$

$\Rightarrow \frac{d^2y}{dx^2} = \frac{-2xy - x^2\left(\frac{1}{yx^2}\right)}{y^2x^4} = \frac{-2xy^2 - 1}{y^3x^4}$

84. (a) $x^2 - y^2 = 1 \Rightarrow 2x - 2y\,\frac{dy}{dx} = 0 \Rightarrow -2y\,\frac{dy}{dx} = -2x \Rightarrow \frac{dy}{dx} = \frac{x}{y}$

(b) $\frac{dy}{dx} = \frac{x}{y} \Rightarrow \frac{d^2y}{dx^2} = \frac{y(1) - x\,\frac{dy}{dx}}{y^2} = \frac{y - x\left(\frac{x}{y}\right)}{y^2} = \frac{y^2 - x^2}{y^3} = \frac{-1}{y^3}$ (since $y^2 - x^2 = -1$)

85. (a) Let $h(x) = 6f(x) - g(x) \Rightarrow h'(x) = 6f'(x) - g'(x) \Rightarrow h'(1) = 6f'(1) - g'(1) = 6\left(\frac{1}{2}\right) - (-4) = 7$

(b) Let $h(x) = f(x)g^2(x) \Rightarrow h'(x) = f(x)\,(2g(x))\,g'(x) + g^2(x)f'(x) \Rightarrow h'(0) = 2f(0)g(0)g'(0) + g^2(0)f'(0)$
$= 2(1)(1)\left(\frac{1}{2}\right) + (1)^2(-3) = -2$

(c) Let $h(x) = \frac{f(x)}{g(x)+1} \Rightarrow h'(x) = \frac{(g(x)+1)f'(x) - f(x)g'(x)}{(g(x)+1)^2} \Rightarrow h'(1) = \frac{(g(1)+1)f'(1) - f(1)g'(1)}{(g(1)+1)^2}$
$= \frac{(5+1)\left(\frac{1}{2}\right) - 3(-4)}{(5+1)^2} = \frac{5}{12}$

(d) Let $h(x) = f(g(x)) \Rightarrow h'(x) = f'(g(x))g'(x) \Rightarrow h'(0) = f'(g(0))g'(0) = f'(1)\left(\frac{1}{2}\right) = \left(\frac{1}{2}\right)\left(\frac{1}{2}\right) = \frac{1}{4}$

(e) Let $h(x) = g(f(x)) \Rightarrow h'(x) = g'(f(x))f'(x) \Rightarrow h'(0) = g'(f(0))f'(0) = g'(1)f'(0) = (-4)(-3) = 12$

(f) Let $h(x) = (x + f(x))^{3/2} \Rightarrow h'(x) = \frac{3}{2}(x + f(x))^{1/2}(1 + f'(x)) \Rightarrow h'(1) = \frac{3}{2}(1 + f(1))^{1/2}(1 + f'(1))$
$= \frac{3}{2}(1+3)^{1/2}\left(1 + \frac{1}{2}\right) = \frac{9}{2}$

(g) Let $h(x) = f(x + g(x)) \Rightarrow h'(x) = f'(x + g(x))(1 + g'(x)) \Rightarrow h'(0) = f'(g(0))(1 + g'(0))$
$= f'(1)\left(1 + \frac{1}{2}\right) = \left(\frac{1}{2}\right)\left(\frac{3}{2}\right) = \frac{3}{4}$

86. (a) Let $h(x) = \sqrt{x}\,f(x) \Rightarrow h'(x) = \sqrt{x}\,f'(x) + f(x)\cdot\frac{1}{2\sqrt{x}} \Rightarrow h'(1) = \sqrt{1}\,f'(1) + f(1)\cdot\frac{1}{2\sqrt{1}} = \frac{1}{5} + (-3)\left(\frac{1}{2}\right) = -\frac{13}{10}$

(b) Let $h(x) = (f(x))^{1/2} \Rightarrow h'(x) = \frac{1}{2}(f(x))^{-1/2}(f'(x)) \Rightarrow h'(0) = \frac{1}{2}(f(0))^{-1/2}f'(0) = \frac{1}{2}(9)^{-1/2}(-2) = -\frac{1}{3}$

(c) Let $h(x) = f\left(\sqrt{x}\right) \Rightarrow h'(x) = f'\left(\sqrt{x}\right)\cdot\frac{1}{2\sqrt{x}} \Rightarrow h'(1) = f'\left(\sqrt{1}\right)\cdot\frac{1}{2\sqrt{1}} = \frac{1}{5}\cdot\frac{1}{2} = \frac{1}{10}$

(d) Let $h(x) = f(1 - 5\tan x) \Rightarrow h'(x) = f'(1 - 5\tan x)(-5\sec^2 x) \Rightarrow h'(0) = f'(1 - 5\tan 0)(-5\sec^2 0)$
$= f'(1)(-5) = \frac{1}{5}(-5) = -1$

(e) Let $h(x) = \frac{f(x)}{2 + \cos x} \Rightarrow h'(x) = \frac{(2 + \cos x)f'(x) - f(x)(-\sin x)}{(2 + \cos x)^2} \Rightarrow h'(0) = \frac{(2+1)f'(0) - f(0)(0)}{(2+1)^2} = \frac{3(-2)}{9} = -\frac{2}{3}$

(f) Let $h(x) = 10\sin\left(\frac{\pi x}{2}\right)f^2(x) \Rightarrow h'(x) = 10\sin\left(\frac{\pi x}{2}\right)(2f(x)f'(x)) + f^2(x)\left(10\cos\left(\frac{\pi x}{2}\right)\right)\left(\frac{\pi}{2}\right)$
$\Rightarrow h'(1) = 10\sin\left(\frac{\pi}{2}\right)(2f(1)f'(1)) + f^2(1)\left(10\cos\left(\frac{\pi}{2}\right)\right)\left(\frac{\pi}{2}\right) = 20(-3)\left(\frac{1}{5}\right) + 0 = -12$

87. $x = t^2 + \pi \Rightarrow \frac{dx}{dt} = 2t;\ y = 3\sin 2x \Rightarrow \frac{dy}{dx} = 3(\cos 2x)(2) = 6\cos 2x = 6\cos(2t^2 + 2\pi) = 6\cos(2t^2)$; thus,
$\frac{dy}{dt} = \frac{dy}{dx}\cdot\frac{dx}{dt} = 6\cos(2t^2)\cdot 2t \Rightarrow \left.\frac{dy}{dt}\right|_{t=0} = 6\cos(0)\cdot 0 = 0$

88. $t = (u^2 + 2u)^{1/3} \Rightarrow \frac{dt}{du} = \frac{1}{3}(u^2 + 2u)^{-2/3}(2u + 2) = \frac{2}{3}(u^2 + 2u)^{-2/3}(u + 1);\ s = t^2 + 5t \Rightarrow \frac{ds}{dt} = 2t + 5$
$= 2(u^2 + 2u)^{1/3} + 5$; thus $\frac{ds}{du} = \frac{ds}{dt}\cdot\frac{dt}{du} = \left[2(u^2 + 2u)^{1/3} + 5\right]\left(\frac{2}{3}\right)(u^2 + 2u)^{-2/3}(u + 1)$
$\Rightarrow \left.\frac{ds}{du}\right|_{u=2} = \left[2(2^2 + 2(2))^{1/3} + 5\right]\left(\frac{2}{3}\right)(2^2 + 2(2))^{-2/3}(2 + 1) = 2\left(2\cdot 8^{1/3} + 5\right)\left(8^{-2/3}\right) = 2(2\cdot 2 + 5)\left(\frac{1}{4}\right) = \frac{9}{2}$

89. $\frac{dw}{ds} = \frac{dw}{dr} \cdot \frac{dr}{ds} = \left[\cos\left(e^{\sqrt{r}}\right)\left(e^{\sqrt{r}}\frac{1}{2\sqrt{r}}\right)\right]\left[3\cos\left(s + \frac{\pi}{6}\right)\right]$ at $x = 0$, $r = 3\sin\frac{\pi}{6} = \frac{3}{2}$

$\Rightarrow \frac{dw}{ds} = \cos\left(e^{\sqrt{3/2}}\right)\left(e^{\sqrt{3/2}}\frac{1}{2\sqrt{3/2}}\right)\left(3\cos\left(\frac{\pi}{6}\right)\right) = \frac{3\sqrt{3}e^{\sqrt{3/2}}}{4\sqrt{3/2}}\cos\left(e^{\sqrt{3/2}}\right) = \frac{3\sqrt{2}e^{\sqrt{3/2}}}{4}\cos\left(e^{\sqrt{3/2}}\right)$

90. $\frac{dr}{dt} = \frac{dr}{d\theta}\frac{d\theta}{dt}$; $\frac{dr}{d\theta} = \frac{1}{3}(\theta^2 + 7)^{-2/3}(2\theta)$; $\theta^2 e^t + \theta = 1 \Rightarrow \frac{d}{dt}(\theta^2 e^t + \theta) = \frac{d}{dt}(1) \Rightarrow \theta^2 e^t + 2\theta\frac{d\theta}{dt}e^t + \frac{d\theta}{dt} = 0$

$\Rightarrow (1 + 2\theta e^t)\frac{d\theta}{dt} = -\theta^2 e^t \Rightarrow \frac{d\theta}{dt} = -\frac{\theta^2 e^t}{1+2\theta e^t} \Rightarrow \frac{dr}{dt} = \left[\frac{2\theta}{3(\theta^2+7)^{2/3}}\right]\left[-\frac{\theta^2 e^t}{1+2\theta e^t}\right] = -\frac{2\theta^3 e^t}{3(1+2\theta e^t)(\theta^2+7)^{2/3}}$

At $t = 0$, $\theta^2 + \theta - 1 = 0 \Rightarrow \theta = \frac{-1 \pm \sqrt{5}}{2} \Rightarrow \frac{dr}{dt} = -\frac{2\left(\frac{-1\pm\sqrt{5}}{2}\right)^3}{3\left(1+\left(-1\pm\sqrt{5}\right)\right)\left(\left(\frac{-1\pm\sqrt{5}}{2}\right)^2+7\right)^{2/3}}$

91. $y^3 + y = 2\cos x \Rightarrow 3y^2\frac{dy}{dx} + \frac{dy}{dx} = -2\sin x \Rightarrow \frac{dy}{dx}(3y^2 + 1) = -2\sin x \Rightarrow \frac{dy}{dx} = \frac{-2\sin x}{3y^2+1} \Rightarrow \left.\frac{dy}{dx}\right|_{(0,1)}$

$= \frac{-2\sin(0)}{3+1} = 0$; $\frac{d^2y}{dx^2} = \frac{(3y^2+1)(-2\cos x) - (-2\sin x)\left(6y\frac{dy}{dx}\right)}{(3y^2+1)^2}$

$\Rightarrow \left.\frac{d^2y}{dx^2}\right|_{(0,1)} = \frac{(3+1)(-2\cos 0) - (-2\sin 0)(6\cdot 0)}{(3+1)^2} = -\frac{1}{2}$

92. $x^{1/3} + y^{1/3} = 4 \Rightarrow \frac{1}{3}x^{-2/3} + \frac{1}{3}y^{-2/3}\frac{dy}{dx} = 0 \Rightarrow \frac{dy}{dx} = -\frac{y^{2/3}}{x^{2/3}} \Rightarrow \left.\frac{dy}{dx}\right|_{(8,8)} = -1$; $\frac{dy}{dx} = \frac{-y^{2/3}}{x^{2/3}}$

$\Rightarrow \frac{d^2y}{dx^2} = \frac{\left(x^{2/3}\right)\left(-\frac{2}{3}y^{-1/3}\frac{dy}{dx}\right) - \left(-y^{2/3}\right)\left(\frac{2}{3}x^{-1/3}\right)}{\left(x^{2/3}\right)^2} \Rightarrow \left.\frac{d^2y}{dx^2}\right|_{(8,8)} = \frac{\left(8^{2/3}\right)\left[-\frac{2}{3}\cdot 8^{-1/3}\cdot(-1)\right] + \left(8^{2/3}\right)\left(\frac{2}{3}\cdot 8^{-1/3}\right)}{8^{4/3}}$

$= \frac{\frac{1}{3}+\frac{1}{3}}{8^{2/3}} = \frac{\frac{2}{3}}{4} = \frac{1}{6}$

93. $f(t) = \frac{1}{2t+1}$ and $f(t+h) = \frac{1}{2(t+h)+1} \Rightarrow \frac{f(t+h) - f(t)}{h} = \frac{\frac{1}{2(t+h)+1} - \frac{1}{2t+1}}{h} = \frac{2t+1-(2t+2h+1)}{(2t+2h+1)(2t+1)h}$

$= \frac{-2h}{(2t+2h+1)(2t+1)h} = \frac{-2}{(2t+2h+1)(2t+1)} \Rightarrow f'(t) = \lim\limits_{h\to 0}\frac{f(t+h) - f(t)}{h} = \lim\limits_{h\to 0}\frac{-2}{(2t+2h+1)(2t+1)}$

$= \frac{-2}{(2t+1)^2}$

94. $g(x) = 2x^2 + 1$ and $g(x+h) = 2(x+h)^2 + 1 = 2x^2 + 4xh + 2h^2 + 1 \Rightarrow \frac{g(x+h) - g(x)}{h}$

$= \frac{(2x^2 + 4xh + 2h^2 + 1) - (2x^2 + 1)}{h} = \frac{4xh + 2h^2}{h} = 4x + 2h \Rightarrow g'(x) = \lim\limits_{h\to 0}\frac{g(x+h) - g(x)}{h} = \lim\limits_{h\to 0}(4x + 2h)$

$= 4x$

95. (a)

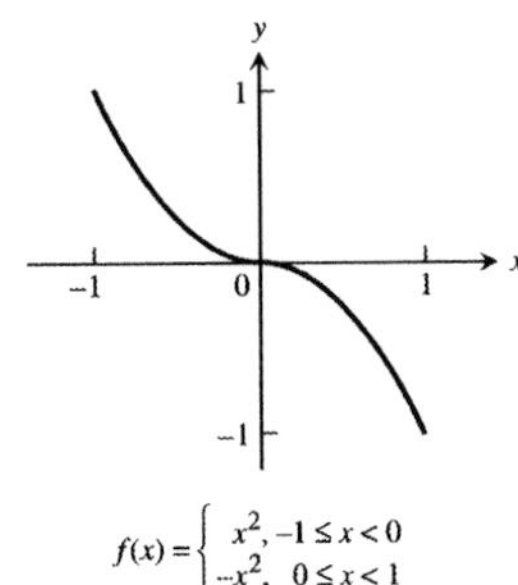

$f(x) = \begin{cases} x^2, & -1 \le x < 0 \\ -x^2, & 0 \le x < 1 \end{cases}$

(b) $\lim\limits_{x\to 0^-} f(x) = \lim\limits_{x\to 0^-} x^2 = 0$ and $\lim\limits_{x\to 0^+} f(x) = \lim\limits_{x\to 0^+} -x^2 = 0 \Rightarrow \lim\limits_{x\to 0} f(x) = 0$. Since $\lim\limits_{x\to 0} f(x) = 0 = f(0)$ it follows that f is continuous at $x = 0$.

(c) $\lim\limits_{x\to 0^-} f'(x) = \lim\limits_{x\to 0^-} (2x) = 0$ and $\lim\limits_{x\to 0^+} f'(x) = \lim\limits_{x\to 0^+} (-2x) = 0 \Rightarrow \lim\limits_{x\to 0} f'(x) = 0$. Since this limit exists, it follows that f is differentiable at $x = 0$.

96. (a)

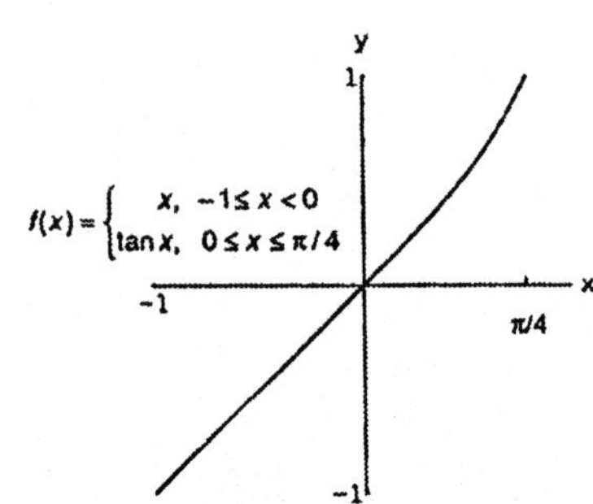

(b) $\lim_{x \to 0^-} f(x) = \lim_{x \to 0^-} x = 0$ and $\lim_{x \to 0^+} f(x) = \lim_{x \to 0^+} \tan x = 0 \Rightarrow \lim_{x \to 0} f(x) = 0$. Since $\lim_{x \to 0} f(x) = 0 = f(0)$, it follows that f is continuous at $x = 0$.

(c) $\lim_{x \to 0^-} f'(x) = \lim_{x \to 0^-} 1 = 1$ and $\lim_{x \to 0^+} f'(x) = \lim_{x \to 0^+} \sec^2 x = 1 \Rightarrow \lim_{x \to 0} f'(x) = 1$. Since this limit exists it follows that f is differentiable at $x = 0$.

97. (a)

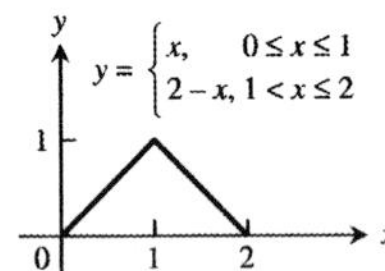

(b) $\lim_{x \to 1^-} f(x) = \lim_{x \to 1^-} x = 1$ and $\lim_{x \to 1^+} f(x) = \lim_{x \to 1^+} (2 - x) = 1 \Rightarrow \lim_{x \to 1} f(x) = 1$. Since $\lim_{x \to 1} f(x) = 1 = f(1)$, it follows that f is continuous at $x = 1$.

(c) $\lim_{x \to 1^-} f'(x) = \lim_{x \to 1^-} 1 = 1$ and $\lim_{x \to 1^+} f'(x) = \lim_{x \to 1^+} -1 = -1 \Rightarrow \lim_{x \to 1^-} f'(x) \neq \lim_{x \to 1^+} f'(x)$, so $\lim_{x \to 1} f'(x)$ does not exist $\Rightarrow$ f is not differentiable at $x = 1$.

98. (a) $\lim_{x \to 0^-} f(x) = \lim_{x \to 0^-} \sin 2x = 0$ and $\lim_{x \to 0^+} f(x) = \lim_{x \to 0^+} mx = 0 \Rightarrow \lim_{x \to 0} f(x) = 0$, independent of m; since $f(0) = 0 = \lim_{x \to 0} f(x)$ it follows that f is continuous at $x = 0$ for all values of m.

(b) $\lim_{x \to 0^-} f'(x) = \lim_{x \to 0^-} (\sin 2x)' = \lim_{x \to 0^-} 2\cos 2x = 2$ and $\lim_{x \to 0^+} f'(x) = \lim_{x \to 0^+} (mx)' = \lim_{x \to 0^+} m = m \Rightarrow$ f is differentiable at $x = 0$ provided that $\lim_{x \to 0^-} f'(x) = \lim_{x \to 0^+} f'(x) \Rightarrow m = 2$.

99. $y = \frac{x}{2} + \frac{1}{2x-4} = \frac{1}{2}x + (2x-4)^{-1} \Rightarrow \frac{dy}{dx} = \frac{1}{2} - 2(2x-4)^{-2}$; the slope of the tangent is $-\frac{3}{2} \Rightarrow -\frac{3}{2} = \frac{1}{2} - 2(2x-4)^{-2} \Rightarrow -2 = -2(2x-4)^{-2} \Rightarrow 1 = \frac{1}{(2x-4)^2} \Rightarrow (2x-4)^2 = 1 \Rightarrow 4x^2 - 16x + 16 = 1$ $\Rightarrow 4x^2 - 16x + 15 = 0 \Rightarrow (2x-5)(2x-3) = 0 \Rightarrow x = \frac{5}{2}$ or $x = \frac{3}{2} \Rightarrow \left(\frac{5}{2}, \frac{9}{4}\right)$ and $\left(\frac{3}{2}, -\frac{1}{4}\right)$ are points on the curve where the slope is $-\frac{3}{2}$.

100. $y = x - e^{-x}$; $\frac{dy}{dx} = 1 + e^{-x} = 2 \Rightarrow e^{-x} = 1 \Rightarrow x = 0 \Rightarrow y = 0 - e^0 = -1$. Therefore, the curve has a tangent with a slope of 2 at the point $(0, -1)$.

101. $y = 2x^3 - 3x^2 - 12x + 20 \Rightarrow \frac{dy}{dx} = 6x^2 - 6x - 12$; the tangent is parallel to the x-axis when $\frac{dy}{dx} = 0$ $\Rightarrow 6x^2 - 6x - 12 = 0 \Rightarrow x^2 - x - 2 = 0 \Rightarrow (x-2)(x+1) = 0 \Rightarrow x = 2$ or $x = -1 \Rightarrow (2, 0)$ and $(-1, 27)$ are points on the curve where the tangent is parallel to the x-axis.

102. $y = x^3 \Rightarrow \frac{dy}{dx} = 3x^2 \Rightarrow \left.\frac{dy}{dx}\right|_{(-2,-8)} = 12$; an equation of the tangent line at $(-2, -8)$ is $y + 8 = 12(x + 2)$ $\Rightarrow y = 12x + 16$; x-intercept: $0 = 12x + 16 \Rightarrow x = -\frac{4}{3} \Rightarrow \left(-\frac{4}{3}, 0\right)$; y-intercept: $y = 12(0) + 16 = 16 \Rightarrow (0, 16)$

103. $y = 2x^3 - 3x^2 - 12x + 20 \Rightarrow \frac{dy}{dx} = 6x^2 - 6x - 12$

(a) The tangent is perpendicular to the line $y = 1 - \frac{x}{24}$ when $\frac{dy}{dx} = -\left(\frac{1}{-\left(\frac{1}{24}\right)}\right) = 24$; $6x^2 - 6x - 12 = 24$ $\Rightarrow x^2 - x - 2 = 4 \Rightarrow x^2 - x - 6 = 0 \Rightarrow (x-3)(x+2) = 0 \Rightarrow x = -2$ or $x = 3 \Rightarrow (-2, 16)$ and $(3, 11)$ are points where the tangent is perpendicular to $y = 1 - \frac{x}{24}$.

(b) The tangent is parallel to the line $y = \sqrt{2} - 12x$ when $\frac{dy}{dx} = -12 \Rightarrow 6x^2 - 6x - 12 = -12 \Rightarrow x^2 - x = 0$ $\Rightarrow x(x-1) = 0 \Rightarrow x = 0$ or $x = 1 \Rightarrow (0, 20)$ and $(1, 7)$ are points where the tangent is parallel to $y = \sqrt{2} - 12x$.

104. $y = \frac{\pi \sin x}{x} \Rightarrow \frac{dy}{dx} = \frac{x(\pi \cos x) - (\pi \sin x)(1)}{x^2} \Rightarrow m_1 = \left.\frac{dy}{dx}\right|_{x=\pi} = \frac{-\pi^2}{\pi^2} = -1$ and $m_2 = \left.\frac{dy}{dx}\right|_{x=-\pi} \frac{\pi^2}{\pi^2} = 1$.
Since $m_1 = -\frac{1}{m_2}$ the tangents intersect at right angles.

105. $y = \tan x, -\frac{\pi}{2} < x < \frac{\pi}{2} \Rightarrow \frac{dy}{dx} = \sec^2 x$; now the slope of $y = -\frac{x}{2}$ is $-\frac{1}{2} \Rightarrow$ the normal line is parallel to $y = -\frac{x}{2}$ when $\frac{dy}{dx} = 2$. Thus, $\sec^2 x = 2 \Rightarrow \frac{1}{\cos^2 x} = 2$ $\Rightarrow \cos^2 x = \frac{1}{2} \Rightarrow \cos x = \frac{\pm 1}{\sqrt{2}} \Rightarrow x = -\frac{\pi}{4}$ and $x = \frac{\pi}{4}$ for $-\frac{\pi}{2} < x < \frac{\pi}{2} \Rightarrow \left(-\frac{\pi}{4}, -1\right)$ and $\left(\frac{\pi}{4}, 1\right)$ are points where the normal is parallel to $y = -\frac{x}{2}$.

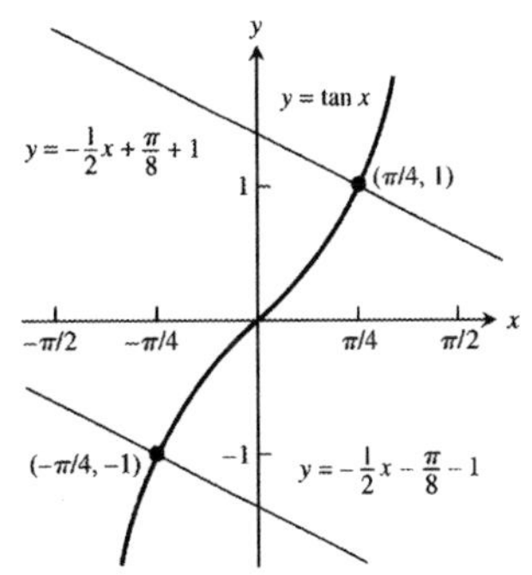

106. $y = 1 + \cos x \Rightarrow \frac{dy}{dx} = -\sin x \Rightarrow \left.\frac{dy}{dx}\right|_{\left(\frac{\pi}{2}, 1\right)} = -1$ $\Rightarrow$ the tangent at $\left(\frac{\pi}{2}, 1\right)$ is the line $y - 1 = -\left(x - \frac{\pi}{2}\right)$ $\Rightarrow y = -x + \frac{\pi}{2} + 1$; the normal at $\left(\frac{\pi}{2}, 1\right)$ is $y - 1 = (1)\left(x - \frac{\pi}{2}\right) \Rightarrow y = x - \frac{\pi}{2} + 1$

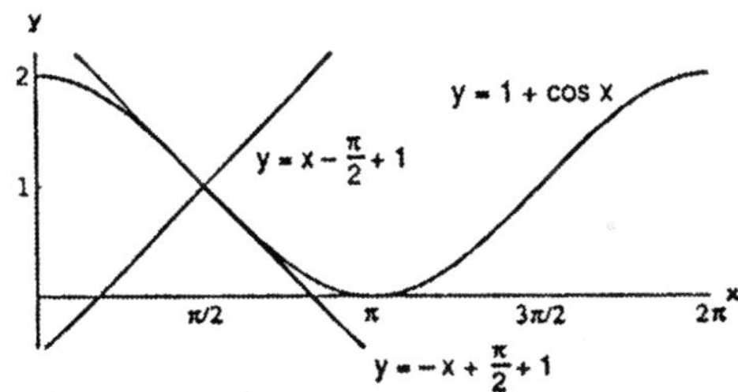

107. $y = x^2 + C \Rightarrow \frac{dy}{dx} = 2x$ and $y = x \Rightarrow \frac{dy}{dx} = 1$; the parabola is tangent to $y = x$ when $2x = 1 \Rightarrow x = \frac{1}{2} \Rightarrow y = \frac{1}{2}$; thus, $\frac{1}{2} = \left(\frac{1}{2}\right)^2 + C \Rightarrow C = \frac{1}{4}$

108. $y = x^3 \Rightarrow \frac{dy}{dx} = 3x^2 \Rightarrow \left.\frac{dy}{dx}\right|_{x=a} = 3a^2 \Rightarrow$ the tangent line at (a, a^3) is $y - a^3 = 3a^2(x - a)$. The tangent line intersects $y = x^3$ when $x^3 - a^3 = 3a^2(x - a) \Rightarrow (x - a)(x^2 + xa + a^2) = 3a^2(x - a) \Rightarrow (x - a)(x^2 + xa - 2a^2) = 0$ $\Rightarrow (x - a)^2(x + 2a) = 0 \Rightarrow x = a$ or $x = -2a$. Now $\left.\frac{dy}{dx}\right|_{x=-2a} = 3(-2a)^2 = 12a^2 = 4(3a^2)$, so the slope at $x = -2a$ is 4 times as large as the slope at (a, a^3) where $x = a$.

109. The line through $(0, 3)$ and $(5, -2)$ has slope $m = \frac{3-(-2)}{0-5} = -1 \Rightarrow$ the line through $(0, 3)$ and $(5, -2)$ is $y = -x + 3$; $y = \frac{c}{x+1} \Rightarrow \frac{dy}{dx} = \frac{-c}{(x+1)^2}$, so the curve is tangent to $y = -x + 3 \Rightarrow \frac{dy}{dx} = -1 = \frac{-c}{(x+1)^2}$ $\Rightarrow (x+1)^2 = c, x \neq -1$. Moreover, $y = \frac{c}{x+1}$ intersects $y = -x + 3 \Rightarrow \frac{c}{x+1} = -x + 3, x \neq -1$ $\Rightarrow c = (x+1)(-x+3), x \neq -1$. Thus $c = c \Rightarrow (x+1)^2 = (x+1)(-x+3) \Rightarrow (x+1)[x+1-(-x+3)]$ $= 0, x \neq -1 \Rightarrow (x+1)(2x-2) = 0 \Rightarrow x = 1$ (since $x \neq -1$) $\Rightarrow c = 4$.

110. Let $\left(b, \pm\sqrt{a^2-b^2}\right)$ be a point on the circle $x^2+y^2=a^2$. Then $x^2+y^2=a^2 \Rightarrow 2x+2y\frac{dy}{dx}=0 \Rightarrow \frac{dy}{dx}=-\frac{x}{y}$ $\Rightarrow \left.\frac{dy}{dx}\right|_{x=b}=\frac{-b}{\pm\sqrt{a^2-b^2}} \Rightarrow$ normal line through $\left(b, \pm\sqrt{a^2-b^2}\right)$ has slope $\frac{\pm\sqrt{a^2-b^2}}{b} \Rightarrow$ normal line is $y-\left(\pm\sqrt{a^2-b^2}\right)=\frac{\pm\sqrt{a^2-b^2}}{b}(x-b) \Rightarrow y\mp\sqrt{a^2-b^2}=\frac{\pm\sqrt{a^2-b^2}}{b}x\mp\sqrt{a^2-b^2} \Rightarrow y=\pm\frac{\sqrt{a^2-b^2}}{b}x$ which passes through the origin.

111. $x^2+2y^2=9 \Rightarrow 2x+4y\frac{dy}{dx}=0 \Rightarrow \frac{dy}{dx}=-\frac{x}{2y} \Rightarrow \left.\frac{dy}{dx}\right|_{(1,2)}=-\frac{1}{4} \Rightarrow$ the tangent line is $y=2-\frac{1}{4}(x-1)$ $=-\frac{1}{4}x+\frac{9}{4}$ and the normal line is $y=2+4(x-1)=4x-2$.

112. $e^x+y^2=2 \Rightarrow \frac{d}{dx}(e^x+y^2)=\frac{d}{dx}(2) \Rightarrow e^x+2y\frac{dy}{dx}=0 \Rightarrow \frac{dy}{dx}=-\frac{e^x}{2y} \Rightarrow m_{\tan}=\left.\frac{dy}{dx}\right|_{(0,\,1)}=-\frac{e^0}{2(1)}=-\frac{1}{2}$; $m_\perp=-\frac{1}{m_{\tan}}=2$; tangent line: $y-1=-\frac{1}{2}(x-0) \Rightarrow y=1-\frac{x}{2}$; normal line: $y-1=2(x-0) \Rightarrow y=2x+1$

113. $xy+2x-5y=2 \Rightarrow \left(x\frac{dy}{dx}+y\right)+2-5\frac{dy}{dx}=0 \Rightarrow \frac{dy}{dx}(x-5)=-y-2 \Rightarrow \frac{dy}{dx}=\frac{-y-2}{x-5} \Rightarrow \left.\frac{dy}{dx}\right|_{(3,2)}=2$ $\Rightarrow$ the tangent line is $y=2+2(x-3)=2x-4$ and the normal line is $y=2+\frac{-1}{2}(x-3)=-\frac{1}{2}x+\frac{7}{2}$.

114. $(y-x)^2=2x+4 \Rightarrow 2(y-x)\left(\frac{dy}{dx}-1\right)=2 \Rightarrow (y-x)\frac{dy}{dx}=1+(y-x) \Rightarrow \frac{dy}{dx}=\frac{1+y-x}{y-x} \Rightarrow \left.\frac{dy}{dx}\right|_{(6,2)}=\frac{3}{4}$ $\Rightarrow$ the tangent line is $y=2+\frac{3}{4}(x-6)=\frac{3}{4}x-\frac{5}{2}$ and the normal line is $y=2-\frac{4}{3}(x-6)=-\frac{4}{3}x+10$.

115. $x+\sqrt{xy}=6 \Rightarrow 1+\frac{1}{2\sqrt{xy}}\left(x\frac{dy}{dx}+y\right)=0 \Rightarrow x\frac{dy}{dx}+y=-2\sqrt{xy} \Rightarrow \frac{dy}{dx}=\frac{-2\sqrt{xy}-y}{x} \Rightarrow \left.\frac{dy}{dx}\right|_{(4,1)}=\frac{-5}{4}$ $\Rightarrow$ the tangent line is $y=1-\frac{5}{4}(x-4)=-\frac{5}{4}x+6$ and the normal line is $y=1+\frac{4}{5}(x-4)=\frac{4}{5}x-\frac{11}{5}$.

116. $x^{3/2}+2y^{3/2}=17 \Rightarrow \frac{3}{2}x^{1/2}+3y^{1/2}\frac{dy}{dx}=0 \Rightarrow \frac{dy}{dx}=\frac{-x^{1/2}}{2y^{1/2}} \Rightarrow \left.\frac{dy}{dx}\right|_{(1,4)}=-\frac{1}{4} \Rightarrow$ the tangent line is $y=4-\frac{1}{4}(x-1)=-\frac{1}{4}x+\frac{17}{4}$ and the normal line is $y=4+4(x-1)=4x$.

117. $x^3y^3+y^2=x+y \Rightarrow \left[x^3\left(3y^2\frac{dy}{dx}\right)+y^3(3x^2)\right]+2y\frac{dy}{dx}=1+\frac{dy}{dx} \Rightarrow 3x^3y^2\frac{dy}{dx}+2y\frac{dy}{dx}-\frac{dy}{dx}=1-3x^2y^3$ $\Rightarrow \frac{dy}{dx}(3x^3y^2+2y-1)=1-3x^2y^3 \Rightarrow \frac{dy}{dx}=\frac{1-3x^2y^3}{3x^3y^2+2y-1} \Rightarrow \left.\frac{dy}{dx}\right|_{(1,1)}=-\frac{2}{4}$, but $\left.\frac{dy}{dx}\right|_{(1,-1)}$ is undefined. Therefore, the curve has slope $-\frac{1}{2}$ at $(1,1)$ but the slope is undefined at $(1,-1)$.

118. $y=\sin(x-\sin x) \Rightarrow \frac{dy}{dx}=[\cos(x-\sin x)](1-\cos x)$; $y=0 \Rightarrow \sin(x-\sin x)=0 \Rightarrow x-\sin x=k\pi$, $k=-2,-1,0,1,2$ (for our interval) $\Rightarrow \cos(x-\sin x)=\cos(k\pi)=\pm 1$. Therefore, $\frac{dy}{dx}=0$ and $y=0$ when $1-\cos x=0$ and $x=k\pi$. For $-2\pi\le x\le 2\pi$, these equations hold when $k=-2$, 0, and 2 (since $\cos(-\pi)=\cos\pi=-1$). Thus the curve has horizontal tangents at the x-axis for the x-values -2π, 0, and 2π (which are even integer multiples of π) $\Rightarrow$ the curve has an infinite number of horizontal tangents.

119. $x=\frac{1}{2}\tan t,\ y=\frac{1}{2}\sec t \Rightarrow \frac{dy}{dx}=\frac{dy/dt}{dx/dt}=\frac{\frac{1}{2}\sec t\tan t}{\frac{1}{2}\sec^2 t}=\frac{\tan t}{\sec t}=\sin t \Rightarrow \left.\frac{dy}{dx}\right|_{t=\pi/3}=\sin\frac{\pi}{3}=\frac{\sqrt{3}}{2}$; $t=\frac{\pi}{3}$ $\Rightarrow x=\frac{1}{2}\tan\frac{\pi}{3}=\frac{\sqrt{3}}{2}$ and $y=\frac{1}{2}\sec\frac{\pi}{3}=1 \Rightarrow y=\frac{\sqrt{3}}{2}x+\frac{1}{4}$; $\frac{d^2y}{dx^2}=\frac{dy'/dt}{dx/dt}=\frac{\cos t}{\frac{1}{2}\sec^2 t}=2\cos^3 t \Rightarrow \left.\frac{d^2y}{dx^2}\right|_{t=\pi/3}$ $=2\cos^3\left(\frac{\pi}{3}\right)=\frac{1}{4}$

120. $x = 1 + \frac{1}{t^2}$, $y = 1 - \frac{3}{t} \Rightarrow \frac{dy}{dx} = \frac{dy/dt}{dx/dt} = \frac{\left(\frac{3}{t^2}\right)}{\left(-\frac{2}{t^3}\right)} = -\frac{3}{2}t \Rightarrow \left.\frac{dy}{dx}\right|_{t=2} = -\frac{3}{2}(2) = -3$; $t = 2 \Rightarrow x = 1 + \frac{1}{2^2} = \frac{5}{4}$ and $y = 1 - \frac{3}{2} = -\frac{1}{2} \Rightarrow y = -3x + \frac{13}{4}$; $\frac{d^2y}{dx^2} = \frac{dy'/dt}{dx/dt} = \frac{\left(-\frac{3}{2}\right)}{\left(-\frac{2}{t^3}\right)} = \frac{3}{4}t^3 \Rightarrow \left.\frac{d^2y}{dx^2}\right|_{t=2} = \frac{3}{4}(2)^3 = 6$

121. B = graph of f, A = graph of f′. Curve B cannot be the derivative of A because A has only negative slopes while some of B's values are positive.

122. A = graph of f, B = graph of f′. Curve A cannot be the derivative of B because B has only negative slopes while A has positive values for $x > 0$.

123.

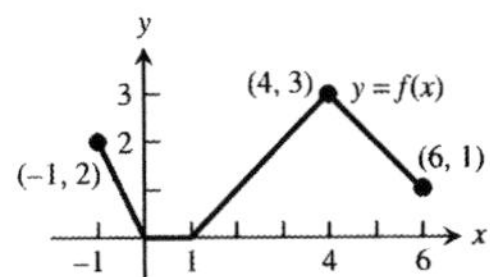

124.

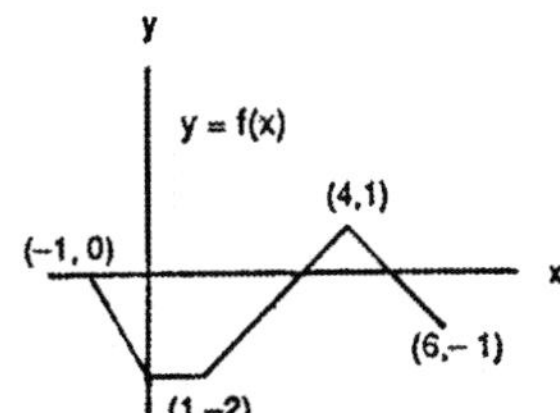

125. (a) 0, 0 (b) largest 1700, smallest about 1400

126. rabbits/day and foxes/day

127. $\lim\limits_{x \to 0} \frac{\sin x}{2x^2 - x} = \lim\limits_{x \to 0} \left[\left(\frac{\sin x}{x}\right) \cdot \frac{1}{(2x-1)}\right] = (1)\left(\frac{1}{-1}\right) = -1$

128. $\lim\limits_{x \to 0} \frac{3x - \tan 7x}{2x} = \lim\limits_{x \to 0} \left(\frac{3x}{2x} - \frac{\sin 7x}{2x \cos 7x}\right) = \frac{3}{2} - \lim\limits_{x \to 0} \left(\frac{1}{\cos 7x} \cdot \frac{\sin 7x}{7x} \cdot \frac{1}{\left(\frac{2}{7}\right)}\right) = \frac{3}{2} - \left(1 \cdot 1 \cdot \frac{7}{2}\right) = -2$

129. $\lim\limits_{r \to 0} \frac{\sin r}{\tan 2r} = \lim\limits_{r \to 0} \left(\frac{\sin r}{r} \cdot \frac{2r}{\tan 2r} \cdot \frac{1}{2}\right) = \left(\frac{1}{2}\right)(1) \lim\limits_{r \to 0} \frac{\cos 2r}{\left(\frac{\sin 2r}{2r}\right)} = \left(\frac{1}{2}\right)(1)\left(\frac{1}{1}\right) = \frac{1}{2}$

130. $\lim\limits_{\theta \to 0} \frac{\sin(\sin\theta)}{\theta} = \lim\limits_{\theta \to 0} \left(\frac{\sin(\sin\theta)}{\sin\theta}\right)\left(\frac{\sin\theta}{\theta}\right) = \lim\limits_{\theta \to 0} \frac{\sin(\sin\theta)}{\sin\theta}$. Let $x = \sin\theta$. Then $x \to 0$ as $\theta \to 0$ $\Rightarrow \lim\limits_{\theta \to 0} \frac{\sin(\sin\theta)}{\sin\theta} = \lim\limits_{x \to 0} \frac{\sin x}{x} = 1$

131. $\lim\limits_{\theta \to \left(\frac{\pi}{2}\right)^-} \frac{4\tan^2\theta + \tan\theta + 1}{\tan^2\theta + 5} = \lim\limits_{\theta \to \left(\frac{\pi}{2}\right)^-} \frac{\left(4 + \frac{1}{\tan\theta} + \frac{1}{\tan^2\theta}\right)}{\left(1 + \frac{5}{\tan^2\theta}\right)} = \frac{(4+0+0)}{(1+0)} = 4$

132. $\lim\limits_{\theta \to 0^+} \frac{1 - 2\cot^2\theta}{5\cot^2\theta - 7\cot\theta - 8} = \lim\limits_{\theta \to 0^+} \frac{\left(\frac{1}{\cot^2\theta} - 2\right)}{\left(5 - \frac{7}{\cot\theta} - \frac{8}{\cot^2\theta}\right)} = \frac{(0-2)}{(5-0-0)} = -\frac{2}{5}$

133. $\lim\limits_{x \to 0} \frac{x\sin x}{2 - 2\cos x} = \lim\limits_{x \to 0} \frac{x\sin x}{2(1-\cos x)} = \lim\limits_{x \to 0} \frac{x\sin x}{2\left(2\sin^2\left(\frac{x}{2}\right)\right)} = \lim\limits_{x \to 0} \left[\frac{\frac{x}{2}\cdot\frac{x}{2}}{\sin^2\left(\frac{x}{2}\right)} \cdot \frac{\sin x}{x}\right]$ $= \lim\limits_{x \to 0} \left[\frac{\left(\frac{x}{2}\right)}{\sin\left(\frac{x}{2}\right)} \cdot \frac{\left(\frac{x}{2}\right)}{\sin\left(\frac{x}{2}\right)} \cdot \frac{\sin x}{x}\right] = (1)(1)(1) = 1$

134. $\lim\limits_{\theta \to 0} \frac{1-\cos\theta}{\theta^2} = \lim\limits_{\theta \to 0} \frac{2\sin^2\left(\frac{\theta}{2}\right)}{\theta^2} = \lim\limits_{\theta \to 0} \left[\frac{\sin\left(\frac{\theta}{2}\right)}{\left(\frac{\theta}{2}\right)} \cdot \frac{\sin\left(\frac{\theta}{2}\right)}{\left(\frac{\theta}{2}\right)} \cdot \frac{1}{2}\right] = (1)(1)\left(\frac{1}{2}\right) = \frac{1}{2}$

135. $\lim_{x\to 0} \frac{\tan x}{x} = \lim_{x\to 0} \left(\frac{1}{\cos x} \cdot \frac{\sin x}{x}\right) = 1$; let $\theta = \tan x \Rightarrow \theta \to 0$ as $x \to 0 \Rightarrow \lim_{x\to 0} g(x) = \lim_{x\to 0} \frac{\tan(\tan x)}{\tan x}$ $= \lim_{\theta\to 0} \frac{\tan\theta}{\theta} = 1$. Therefore, to make g continuous at the origin, define $g(0) = 1$.

136. $\lim_{x\to 0} f(x) = \lim_{x\to 0} \frac{\tan(\tan x)}{\sin(\sin x)} = \lim_{x\to 0} \left[\frac{\tan(\tan x)}{\tan x} \cdot \frac{\sin x}{\sin(\sin x)} \cdot \frac{1}{\cos x}\right] = 1 \cdot \lim_{x\to 0} \frac{\sin x}{\sin(\sin x)}$ (using the result of #135); let $\theta = \sin x \Rightarrow \theta \to 0$ as $x \to 0 \Rightarrow \lim_{x\to 0} \frac{\sin x}{\sin(\sin x)} = \lim_{\theta\to 0} \frac{\theta}{\sin\theta} = 1$. Therefore, to make f continuous at the origin, define $f(0) = 1$.

137. $y = \frac{2(x^2+1)}{\sqrt{\cos 2x}} \Rightarrow \ln y = \ln\left(\frac{2(x^2+1)}{\sqrt{\cos 2x}}\right) = \ln(2) + \ln(x^2+1) - \frac{1}{2}\ln(\cos 2x) \Rightarrow \frac{y'}{y} = 0 + \frac{2x}{x^2+1} - \left(\frac{1}{2}\right)\frac{(-2\sin 2x)}{\cos 2x}$ $\Rightarrow y' = \left(\frac{2x}{x^2+1} + \tan 2x\right) y = \frac{2(x^2+1)}{\sqrt{\cos 2x}}\left(\frac{2x}{x^2+1} + \tan 2x\right)$

138. $y = \sqrt[10]{\frac{3x+4}{2x-4}} \Rightarrow \ln y = \ln\sqrt[10]{\frac{3x+4}{2x-4}} = \frac{1}{10}[\ln(3x+4) - \ln(2x-4)] \Rightarrow \frac{y'}{y} = \frac{1}{10}\left(\frac{3}{3x+4} - \frac{2}{2x-4}\right)$ $\Rightarrow y' = \frac{1}{10}\left(\frac{3}{3x+4} - \frac{1}{x-2}\right) y = \sqrt[10]{\frac{3x+4}{2x-4}}\left(\frac{1}{10}\right)\left(\frac{3}{3x+4} - \frac{1}{x-2}\right)$

139. $y = \left[\frac{(t+1)(t-1)}{(t-2)(t+3)}\right]^5 \Rightarrow \ln y = 5[\ln(t+1) + \ln(t-1) - \ln(t-2) - \ln(t+3)] \Rightarrow \left(\frac{1}{y}\right)\left(\frac{dy}{dt}\right)$ $= 5\left(\frac{1}{t+1} + \frac{1}{t-1} - \frac{1}{t-2} - \frac{1}{t+3}\right) \Rightarrow \frac{dy}{dt} = 5\left[\frac{(t+1)(t-1)}{(t-2)(t+3)}\right]^5\left(\frac{1}{t+1} + \frac{1}{t-1} - \frac{1}{t-2} - \frac{1}{t+3}\right)$

140. $y = \frac{2u2^u}{\sqrt{u^2+1}} \Rightarrow \ln y = \ln 2 + \ln u + u\ln 2 - \frac{1}{2}\ln(u^2+1) \Rightarrow \left(\frac{1}{y}\right)\left(\frac{dy}{du}\right) = \frac{1}{u} + \ln 2 - \frac{1}{2}\left(\frac{2u}{u^2+1}\right)$ $\Rightarrow \frac{dy}{du} = \frac{2u2^u}{\sqrt{u^2+1}}\left(\frac{1}{u} + \ln 2 - \frac{u}{u^2+1}\right)$

141. $y = (\sin\theta)^{\sqrt{\theta}} \Rightarrow \ln y = \sqrt{\theta}\ln(\sin\theta) \Rightarrow \left(\frac{1}{y}\right)\left(\frac{dy}{d\theta}\right) = \sqrt{\theta}\left(\frac{\cos\theta}{\sin\theta}\right) + \frac{1}{2}\theta^{-1/2}\ln(\sin\theta)$ $\Rightarrow \frac{dy}{d\theta} = (\sin\theta)^{\sqrt{\theta}}\left(\sqrt{\theta}\cot\theta + \frac{\ln(\sin\theta)}{2\sqrt{\theta}}\right)$

142. $y = (\ln x)^{1/\ln x} \Rightarrow \ln y = \left(\frac{1}{\ln x}\right)\ln(\ln x) \Rightarrow \frac{y'}{y} = \left(\frac{1}{\ln x}\right)\left(\frac{1}{\ln x}\right)\left(\frac{1}{x}\right) + \ln(\ln x)\left[\frac{-1}{(\ln x)^2}\right]\left(\frac{1}{x}\right)$ $\Rightarrow y' = (\ln x)^{1/\ln x}\left[\frac{1-\ln(\ln x)}{x(\ln x)^2}\right]$

143. (a) $S = 2\pi r^2 + 2\pi rh$ and h constant $\Rightarrow \frac{dS}{dt} = 4\pi r\frac{dr}{dt} + 2\pi h\frac{dr}{dt} = (4\pi r + 2\pi h)\frac{dr}{dt}$

(b) $S = 2\pi r^2 + 2\pi rh$ and r constant $\Rightarrow \frac{dS}{dt} = 2\pi r\frac{dh}{dt}$

(c) $S = 2\pi r^2 + 2\pi rh \Rightarrow \frac{dS}{dt} = 4\pi r\frac{dr}{dt} + 2\pi\left(r\frac{dh}{dt} + h\frac{dr}{dt}\right) = (4\pi r + 2\pi h)\frac{dr}{dt} + 2\pi r\frac{dh}{dt}$

(d) S constant $\Rightarrow \frac{dS}{dt} = 0 \Rightarrow 0 = (4\pi r + 2\pi h)\frac{dr}{dt} + 2\pi r\frac{dh}{dt} \Rightarrow (2r+h)\frac{dr}{dt} = -r\frac{dh}{dt} \Rightarrow \frac{dr}{dt} = \frac{-r}{2r+h}\frac{dh}{dt}$

144. $S = \pi r\sqrt{r^2+h^2} \Rightarrow \frac{dS}{dt} = \pi r \cdot \frac{\left(r\frac{dr}{dt} + h\frac{dh}{dt}\right)}{\sqrt{r^2+h^2}} + \pi\sqrt{r^2+h^2}\frac{dr}{dt}$;

(a) h constant $\Rightarrow \frac{dh}{dt} = 0 \Rightarrow \frac{dS}{dt} = \frac{\pi r^2\frac{dr}{dt}}{\sqrt{r^2+h^2}} + \pi\sqrt{r^2+h^2}\frac{dr}{dt} = \left[\pi\sqrt{r^2+h^2} + \frac{\pi r^2}{\sqrt{r^2+h^2}}\right]\frac{dr}{dt}$

(b) r constant $\Rightarrow \frac{dr}{dt} = 0 \Rightarrow \frac{dS}{dt} = \frac{\pi rh}{\sqrt{r^2+h^2}}\frac{dh}{dt}$

(c) In general, $\frac{dS}{dt} = \left[\pi\sqrt{r^2+h^2} + \frac{\pi r^2}{\sqrt{r^2+h^2}}\right]\frac{dr}{dt} + \frac{\pi rh}{\sqrt{r^2+h^2}}\frac{dh}{dt}$

145. $A = \pi r^2 \Rightarrow \frac{dA}{dt} = 2\pi r\frac{dr}{dt}$; so $r = 10$ and $\frac{dr}{dt} = -\frac{2}{\pi}$ m/sec $\Rightarrow \frac{dA}{dt} = (2\pi)(10)\left(-\frac{2}{\pi}\right) = -40\ \text{m}^2/\text{sec}$

146. $V = s^3 \Rightarrow \frac{dV}{dt} = 3s^2 \cdot \frac{ds}{dt} \Rightarrow \frac{ds}{dt} = \frac{1}{3s^2}\frac{dV}{dt}$; so $s = 20$ and $\frac{dV}{dt} = 1200\ \text{cm}^3/\text{min} \Rightarrow \frac{ds}{dt} = \frac{1}{3(20)^2}(1200) = 1$ cm/min

147. $\frac{dR_1}{dt} = -1$ ohm/sec, $\frac{dR_2}{dt} = 0.5$ ohm/sec; and $\frac{1}{R} = \frac{1}{R_1} + \frac{1}{R_2} \Rightarrow \frac{-1}{R^2}\frac{dR}{dt} = \frac{-1}{R_1^2}\frac{dR_1}{dt} - \frac{1}{R_2^2}\frac{dR_2}{dt}$. Also, $R_1 = 75$ ohms and $R_2 = 50$ ohms $\Rightarrow \frac{1}{R} = \frac{1}{75} + \frac{1}{50} \Rightarrow R = 30$ ohms. Therefore, from the derivative equation, $\frac{-1}{(30)^2}\frac{dR}{dt} = \frac{-1}{(75)^2}(-1) - \frac{1}{(50)^2}(0.5) = \left(\frac{1}{5625} - \frac{1}{5000}\right) \Rightarrow \frac{dR}{dt} = (-900)\left(\frac{5000-5625}{5625 \cdot 5000}\right) = \frac{9(625)}{50(5625)} = \frac{1}{50}$ $= 0.02$ ohm/sec.

148. $\frac{dR}{dt} = 3$ ohms/sec and $\frac{dX}{dt} = -2$ ohms/sec; $Z = \sqrt{R^2 + X^2} \Rightarrow \frac{dZ}{dt} = \frac{R\frac{dR}{dt} + X\frac{dX}{dt}}{\sqrt{R^2+X^2}}$ so that $R = 10$ ohms and $X = 20$ ohms $\Rightarrow \frac{dZ}{dt} = \frac{(10)(3)+(20)(-2)}{\sqrt{10^2+20^2}} = \frac{-1}{\sqrt{5}} \approx -0.45$ ohm/sec.

149. Given $\frac{dx}{dt} = 10$ m/sec and $\frac{dy}{dt} = 5$ m/sec, let D be the distance from the origin $\Rightarrow D^2 = x^2 + y^2 \Rightarrow 2D\frac{dD}{dt} = 2x\frac{dx}{dt} + 2y\frac{dy}{dt} \Rightarrow D\frac{dD}{dt} = x\frac{dx}{dt} + y\frac{dy}{dt}$. When $(x, y) = (3, -4)$, $D = \sqrt{3^2 + (-4)^2} = 5$ and $5\frac{dD}{dt} = (5)(10) + (12)(5) \Rightarrow \frac{dD}{dt} = \frac{110}{5} = 22$. Therefore, the particle is moving away from the origin at 22 m/sec (because the distance D is increasing).

150. Let D be the distance from the origin. We are given that $\frac{dD}{dt} = 11$ units/sec. Then $D^2 = x^2 + y^2 = x^2 + \left(x^{3/2}\right)^2 = x^2 + x^3 \Rightarrow 2D\frac{dD}{dt} = 2x\frac{dx}{dt} + 3x^2\frac{dx}{dt} = x(2 + 3x)\frac{dx}{dt}$; $x = 3 \Rightarrow D = \sqrt{3^2 + 3^3} = 6$ and substitution in the derivative equation gives $(2)(6)(11) = (3)(2 + 9)\frac{dx}{dt} \Rightarrow \frac{dx}{dt} = 4$ units/sec.

151. (a) From the diagram we have $\frac{10}{h} = \frac{4}{r} \Rightarrow r = \frac{2}{5}h$.

(b) $V = \frac{1}{3}\pi r^2 h = \frac{1}{3}\pi\left(\frac{2}{5}h\right)^2 h = \frac{4\pi h^3}{75} \Rightarrow \frac{dV}{dt} = \frac{4\pi h^2}{25}\frac{dh}{dt}$, so $\frac{dV}{dt} = -5$ and $h = 6 \Rightarrow \frac{dh}{dt} = -\frac{125}{144\pi}$ ft/min.

152. From the sketch in the text, $s = r\theta \Rightarrow \frac{ds}{dt} = r\frac{d\theta}{dt} + \theta\frac{dr}{dt}$. Also $r = 1.2$ is constant $\Rightarrow \frac{dr}{dt} = 0 \Rightarrow \frac{ds}{dt} = r\frac{d\theta}{dt} = (1.2)\frac{d\theta}{dt}$. Therefore, $\frac{ds}{dt} = 6$ ft/sec and $r = 1.2$ ft $\Rightarrow \frac{d\theta}{dt} = 5$ rad/sec

153. (a) From the sketch in the text, $\frac{d\theta}{dt} = -0.6$ rad/sec and $x = \tan\theta$. Also $x = \tan\theta \Rightarrow \frac{dx}{dt} = \sec^2\theta\frac{d\theta}{dt}$; at point A, $x = 0 \Rightarrow \theta = 0 \Rightarrow \frac{dx}{dt} = (\sec^2 0)(-0.6) = -0.6$. Therefore the speed of the light is $0.6 = \frac{3}{5}$ km/sec when it reaches point A.

(b) $\frac{(3/5)\text{ rad}}{\text{sec}} \cdot \frac{1\text{ rev}}{2\pi\text{ rad}} \cdot \frac{60\text{ sec}}{\text{min}} = \frac{18}{\pi}$ revs/min

154. From the figure, $\frac{a}{r} = \frac{b}{BC} \Rightarrow \frac{a}{r} = \frac{b}{\sqrt{b^2 - r^2}}$. We are given that r is constant. Differentiation gives,

$$\frac{1}{r} \cdot \frac{da}{dt} = \frac{\left(\sqrt{b^2 - r^2}\right)\left(\frac{db}{dt}\right) - (b)\left(\frac{b}{\sqrt{b^2 - r^2}}\right)\left(\frac{db}{dt}\right)}{b^2 - r^2}.$$

Then, $b = 2r$ and $\frac{db}{dt} = -0.3r$

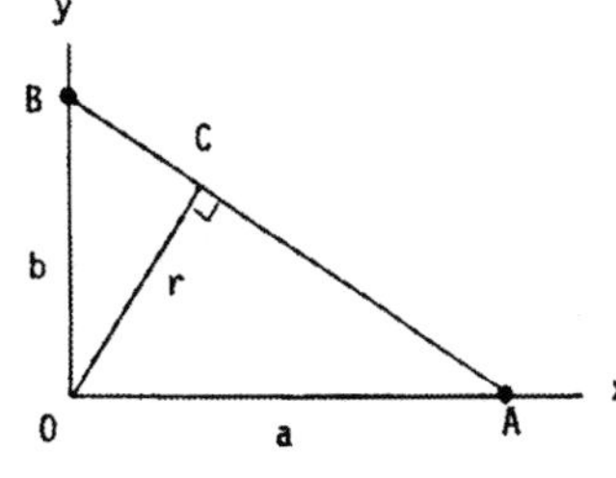

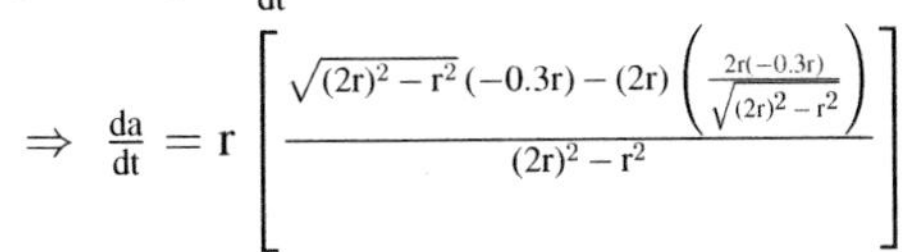

$$\Rightarrow \frac{da}{dt} = r\left[\frac{\sqrt{(2r)^2 - r^2}\,(-0.3r) - (2r)\left(\frac{2r(-0.3r)}{\sqrt{(2r)^2 - r^2}}\right)}{(2r)^2 - r^2}\right]$$

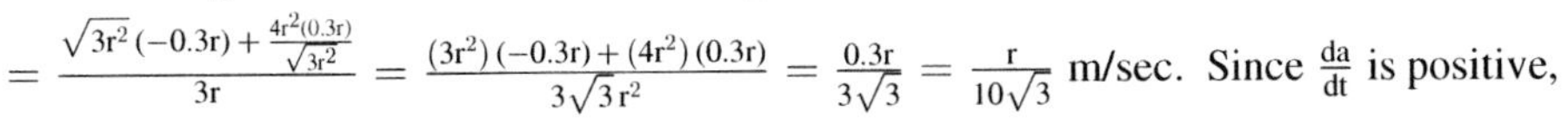

$$= \frac{\sqrt{3r^2}\,(-0.3r) + \frac{4r^2(0.3r)}{\sqrt{3r^2}}}{3r} = \frac{(3r^2)(-0.3r) + (4r^2)(0.3r)}{3\sqrt{3}\,r^2} = \frac{0.3r}{3\sqrt{3}} = \frac{r}{10\sqrt{3}}$$ m/sec. Since $\frac{da}{dt}$ is positive,

the distance OA is increasing when OB = 2r, and B is moving toward O at the rate of 0.3r m/sec.

155. (a) If $f(x) = \tan x$ and $x = -\frac{\pi}{4}$, then $f'(x) = \sec^2 x$, $f\left(-\frac{\pi}{4}\right) = -1$ and $f'\left(-\frac{\pi}{4}\right) = 2$. The linearization of $f(x)$ is $L(x) = 2\left(x + \frac{\pi}{4}\right) + (-1) = 2x + \frac{\pi - 2}{2}$.

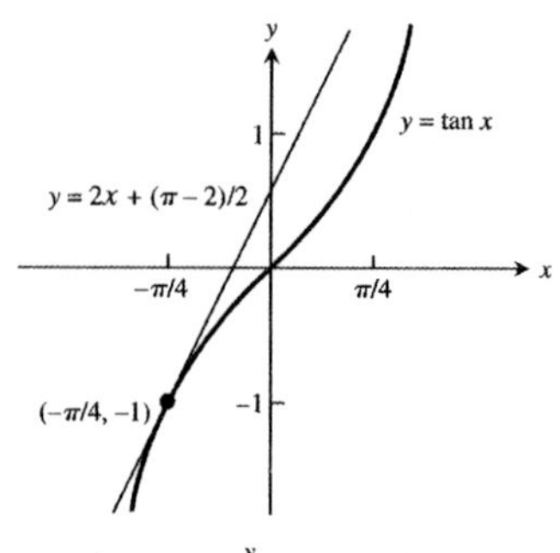

(b) If $f(x) = \sec x$ and $x = -\frac{\pi}{4}$, then $f'(x) = \sec x \tan x$, $f\left(-\frac{\pi}{4}\right) = \sqrt{2}$ and $f'\left(-\frac{\pi}{4}\right) = -\sqrt{2}$. The linearization of $f(x)$ is $L(x) = -\sqrt{2}\left(x + \frac{\pi}{4}\right) + \sqrt{2}$ $= -\sqrt{2}x + \frac{\sqrt{2}(4 - \pi)}{4}$.

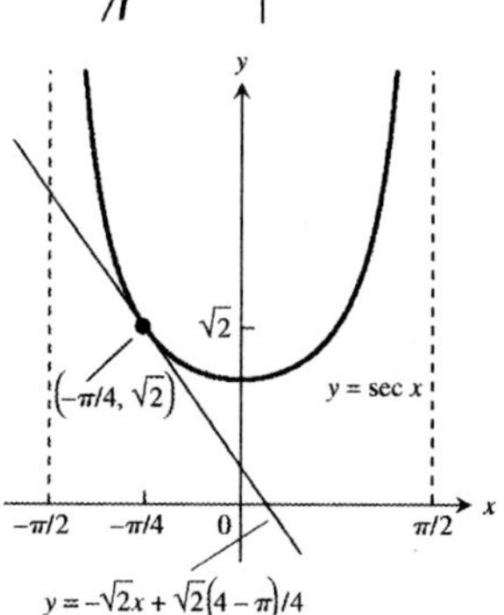

156. $f(x) = \frac{1}{1 + \tan x} \Rightarrow f'(x) = \frac{-\sec^2 x}{(1 + \tan x)^2}$. The linearization at $x = 0$ is $L(x) = f'(0)(x - 0) + f(0) = 1 - x$.

157. $f(x) = \sqrt{x + 1} + \sin x - 0.5 = (x + 1)^{1/2} + \sin x - 0.5 \Rightarrow f'(x) = \left(\frac{1}{2}\right)(x + 1)^{-1/2} + \cos x$ $\Rightarrow L(x) = f'(0)(x - 0) + f(0) = 1.5(x - 0) + 0.5 \Rightarrow L(x) = 1.5x + 0.5$, the linearization of $f(x)$.

158. $f(x) = \frac{2}{1 - x} + \sqrt{1 + x} - 3.1 = 2(1 - x)^{-1} + (1 + x)^{1/2} - 3.1 \Rightarrow f'(x) = -2(1 - x)^{-2}(-1) + \frac{1}{2}(1 + x)^{-1/2}$ $= \frac{2}{(1 - x)^2} + \frac{1}{2\sqrt{1 + x}} \Rightarrow L(x) = f'(0)(x - 0) + f(0) = 2.5x - 0.1$, the linearization of $f(x)$.

159. $S = \pi r \sqrt{r^2 + h^2}$, r constant $\Rightarrow dS = \pi r \cdot \frac{1}{2}\left(r^2 + h^2\right)^{-1/2} 2h\, dh = \frac{\pi r h}{\sqrt{r^2 + h^2}} dh$. Height changes from h_0 to $h_0 + dh$ $\Rightarrow dS = \frac{\pi r h_0 (dh)}{\sqrt{r^2 + h_0^2}}$

160. (a) $S = 6r^2 \Rightarrow dS = 12r\, dr$. We want $|dS| \le (2\%) S \Rightarrow |12r\, dr| \le \frac{12r^2}{100} \Rightarrow |dr| \le \frac{r}{100}$. The measurement of the edge r must have an error less than 1%.

(b) When $V = r^3$, then $dV = 3r^2\, dr$. The accuracy of the volume is $\left(\frac{dV}{V}\right)(100\%) = \left(\frac{3r^2\, dr}{r^3}\right)(100\%)$ $= \left(\frac{3}{r}\right)(dr)(100\%) = \left(\frac{3}{r}\right)\left(\frac{r}{100}\right)(100\%) = 3\%$

161. $C = 2\pi r \Rightarrow r = \frac{C}{2\pi}$, $S = 4\pi r^2 = \frac{C^2}{\pi}$, and $V = \frac{4}{3}\pi r^3 = \frac{C^3}{6\pi^2}$. It also follows that $dr = \frac{1}{2\pi} dC$, $dS = \frac{2C}{\pi} dC$ and $dV = \frac{C^2}{2\pi^2} dC$. Recall that $C = 10$ cm and $dC = 0.4$ cm.

(a) $dr = \frac{0.4}{2\pi} = \frac{0.2}{\pi}$ cm $\Rightarrow \left(\frac{dr}{r}\right)(100\%) = \left(\frac{0.2}{\pi}\right)\left(\frac{2\pi}{10}\right)(100\%) = (.04)(100\%) = 4\%$

(b) $dS = \frac{20}{\pi}(0.4) = \frac{8}{\pi}$ cm $\Rightarrow \left(\frac{dS}{S}\right)(100\%) = \left(\frac{8}{\pi}\right)\left(\frac{\pi}{100}\right)(100\%) = 8\%$

(c) $dV = \frac{10^2}{2\pi^2}(0.4) = \frac{20}{\pi^2}$ cm $\Rightarrow \left(\frac{dV}{V}\right)(100\%) = \left(\frac{20}{\pi^2}\right)\left(\frac{6\pi^2}{1000}\right)(100\%) = 12\%$

162. Similar triangles yield $\frac{35}{h} = \frac{15}{6} \Rightarrow h = 14$ ft. The same triangles imply that $\frac{20 + a}{h} = \frac{a}{6} \Rightarrow h = 120a^{-1} + 6$ $\Rightarrow dh = -120a^{-2}\, da = -\frac{120}{a^2}\, da = \left(-\frac{120}{a^2}\right)\left(\pm\frac{1}{12}\right) = \left(-\frac{120}{15^2}\right)\left(\pm\frac{1}{12}\right) = \pm\frac{2}{45} \approx \pm .0444$ ft $= \pm 0.53$ inches.

CHAPTER 3 ADDITIONAL AND ADVANCED EXERCISES

1. (a) $\sin 2\theta = 2 \sin \theta \cos \theta \Rightarrow \frac{d}{d\theta}(\sin 2\theta) = \frac{d}{d\theta}(2 \sin \theta \cos \theta) \Rightarrow 2 \cos 2\theta = 2[(\sin \theta)(-\sin \theta) + (\cos \theta)(\cos \theta)]$
$\Rightarrow \cos 2\theta = \cos^2 \theta - \sin^2 \theta$

(b) $\cos 2\theta = \cos^2 \theta - \sin^2 \theta \Rightarrow \frac{d}{d\theta}(\cos 2\theta) = \frac{d}{d\theta}(\cos^2 \theta - \sin^2 \theta) \Rightarrow -2 \sin 2\theta = (2 \cos \theta)(-\sin \theta) - (2 \sin \theta)(\cos \theta)$
$\Rightarrow \sin 2\theta = \cos \theta \sin \theta + \sin \theta \cos \theta \Rightarrow \sin 2\theta = 2 \sin \theta \cos \theta$

2. The derivative of $\sin(x + a) = \sin x \cos a + \cos x \sin a$ with respect to x is $\cos(x + a) = \cos x \cos a - \sin x \sin a$, which is also an identity. This principle does not apply to the equation $x^2 - 2x - 8 = 0$, since $x^2 - 2x - 8 = 0$ is not an identity: it holds for 2 values of x (-2 and 4), but not for all x.

3. (a) $f(x) = \cos x \Rightarrow f'(x) = -\sin x \Rightarrow f''(x) = -\cos x$, and $g(x) = a + bx + cx^2 \Rightarrow g'(x) = b + 2cx \Rightarrow g''(x) = 2c$; also, $f(0) = g(0) \Rightarrow \cos(0) = a \Rightarrow a = 1$; $f'(0) = g'(0) \Rightarrow -\sin(0) = b \Rightarrow b = 0$; $f''(0) = g''(0)$ $\Rightarrow -\cos(0) = 2c \Rightarrow c = -\frac{1}{2}$. Therefore, $g(x) = 1 - \frac{1}{2}x^2$.

(b) $f(x) = \sin(x + a) \Rightarrow f'(x) = \cos(x + a)$, and $g(x) = b \sin x + c \cos x \Rightarrow g'(x) = b \cos x - c \sin x$; also, $f(0) = g(0) \Rightarrow \sin(a) = b \sin(0) + c \cos(0) \Rightarrow c = \sin a$; $f'(0) = g'(0) \Rightarrow \cos(a) = b \cos(0) - c \sin(0)$ $\Rightarrow b = \cos a$. Therefore, $g(x) = \sin x \cos a + \cos x \sin a$.

(c) When $f(x) = \cos x$, $f'''(x) = \sin x$ and $f^{(4)}(x) = \cos x$; when $g(x) = 1 - \frac{1}{2}x^2$, $g'''(x) = 0$ and $g^{(4)}(x) = 0$. Thus $f'''(0) = 0 = g'''(0)$ so the third derivatives agree at $x = 0$. However, the fourth derivatives do not agree since $f^{(4)}(0) = 1$ but $g^{(4)}(0) = 0$. In case (b), when $f(x) = \sin(x + a)$ and $g(x)$ $= \sin x \cos a + \cos x \sin a$, notice that $f(x) = g(x)$ for all x, not just $x = 0$. Since this is an identity, we have $f^{(n)}(x) = g^{(n)}(x)$ for any x and any positive integer n.

4. (a) $y = \sin x \Rightarrow y' = \cos x \Rightarrow y'' = -\sin x \Rightarrow y'' + y = -\sin x + \sin x = 0$; $y = \cos x \Rightarrow y' = -\sin x$ $\Rightarrow y'' = -\cos x \Rightarrow y'' + y = -\cos x + \cos x = 0$; $y = a \cos x + b \sin x \Rightarrow y' = -a \sin x + b \cos x$ $\Rightarrow y'' = -a \cos x - b \sin x \Rightarrow y'' + y = (-a \cos x - b \sin x) + (a \cos x + b \sin x) = 0$

(b) $y = \sin(2x) \Rightarrow y' = 2 \cos(2x) \Rightarrow y'' = -4 \sin(2x) \Rightarrow y'' + 4y = -4 \sin(2x) + 4 \sin(2x) = 0$. Similarly, $y = \cos(2x)$ and $y = a \cos(2x) + b \sin(2x)$ satisfy the differential equation $y'' + 4y = 0$. In general, $y = \cos(mx)$, $y = \sin(mx)$ and $y = a \cos(mx) + b \sin(mx)$ satisfy the differential equation $y'' + m^2 y = 0$.

5. If the circle $(x - h)^2 + (y - k)^2 = a^2$ and $y = x^2 + 1$ are tangent at $(1, 2)$, then the slope of this tangent is $m = 2x|_{(1,2)} = 2$ and the tangent line is $y = 2x$. The line containing (h, k) and $(1, 2)$ is perpendicular to $y = 2x \Rightarrow \frac{k-2}{h-1} = -\frac{1}{2} \Rightarrow h = 5 - 2k \Rightarrow$ the location of the center is $(5 - 2k, k)$. Also, $(x - h)^2 + (y - k)^2 = a^2$ $\Rightarrow x - h + (y - k)y' = 0 \Rightarrow 1 + (y')^2 + (y - k)y'' = 0 \Rightarrow y'' = \frac{1 + (y')^2}{k - y}$. At the point $(1, 2)$ we know $y' = 2$ from the tangent line and that $y'' = 2$ from the parabola. Since the second derivatives are equal at $(1, 2)$ we obtain $2 = \frac{1 + (2)^2}{k - 2} \Rightarrow k = \frac{9}{2}$. Then $h = 5 - 2k = -4 \Rightarrow$ the circle is $(x + 4)^2 + \left(y - \frac{9}{2}\right)^2 = a^2$. Since $(1, 2)$ lies on the circle we have that $a = \frac{5\sqrt{5}}{2}$.

6. The total revenue is the number of people times the price of the fare: $r(x) = xp = x\left(3 - \frac{x}{40}\right)^2$, where $0 \le x \le 60$. The marginal revenue is $\frac{dr}{dx} = \left(3 - \frac{x}{40}\right)^2 + 2x\left(3 - \frac{x}{40}\right)\left(-\frac{1}{40}\right) \Rightarrow \frac{dr}{dx} = \left(3 - \frac{x}{40}\right)\left[\left(3 - \frac{x}{40}\right) - \frac{2x}{40}\right]$ $= 3\left(3 - \frac{x}{40}\right)\left(1 - \frac{x}{40}\right)$. Then $\frac{dr}{dx} = 0 \Rightarrow x = 40$ (since $x = 120$ does not belong to the domain). When 40 people are on the bus the marginal revenue is zero and the fare is $p(40) = \left(3 - \frac{x}{40}\right)^2\Big|_{x=40} = \$4.00$.

7. (a) $y = uv \Rightarrow \frac{dy}{dt} = \frac{du}{dt}v + u\frac{dv}{dt} = (0.04u)v + u(0.05v) = 0.09uv = 0.09y \Rightarrow$ the rate of growth of the total production is 9% per year.

(b) If $\frac{du}{dt} = -0.02u$ and $\frac{dv}{dt} = 0.03v$, then $\frac{dy}{dt} = (-0.02u)v + (0.03v)u = 0.01uv = 0.01y$, increasing at 1% per year.

8. When $x^2 + y^2 = 225$, then $y' = -\frac{x}{y}$. The tangent line to the balloon at $(12, -9)$ is $y + 9 = \frac{4}{3}(x - 12)$ $\Rightarrow y = \frac{4}{3}x - 25$. The top of the gondola is $15 + 8 = 23$ ft below the center of the balloon. The intersection of $y = -23$ and $y = \frac{4}{3}x - 25$ is at the far right edge of the gondola $\Rightarrow -23 = \frac{4}{3}x - 25$ $\Rightarrow x = \frac{3}{2}$. Thus the gondola is $2x = 3$ ft wide.

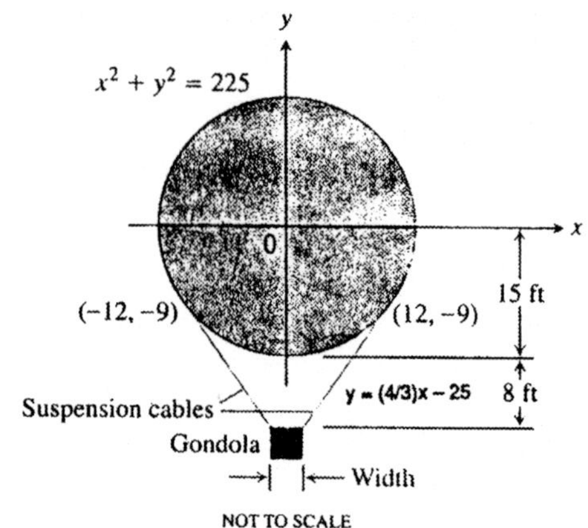

9. Answers will vary. Here is one possibility.

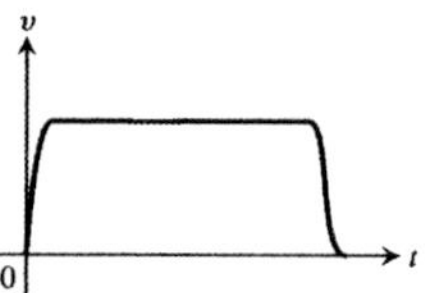

10. $s(t) = 10\cos\left(t + \frac{\pi}{4}\right) \Rightarrow v(t) = \frac{ds}{dt} = -10\sin\left(t + \frac{\pi}{4}\right) \Rightarrow a(t) = \frac{dv}{dt} = \frac{d^2s}{dt^2} = -10\cos\left(t + \frac{\pi}{4}\right)$

(a) $s(0) = 10\cos\left(\frac{\pi}{4}\right) = \frac{10}{\sqrt{2}}$

(b) Left: -10, Right: 10

(c) Solving $10\cos\left(t + \frac{\pi}{4}\right) = -10 \Rightarrow \cos\left(t + \frac{\pi}{4}\right) = -1 \Rightarrow t = \frac{3\pi}{4}$ when the particle is farthest to the left. Solving $10\cos\left(t + \frac{\pi}{4}\right) = 10 \Rightarrow \cos\left(t + \frac{\pi}{4}\right) = 1 \Rightarrow t = -\frac{\pi}{4}$, but $t \ge 0 \Rightarrow t = 2\pi + \frac{-\pi}{4} = \frac{7\pi}{4}$ when the particle is farthest to the right. Thus, $v\left(\frac{3\pi}{4}\right) = 0$, $v\left(\frac{7\pi}{4}\right) = 0$, $a\left(\frac{3\pi}{4}\right) = 10$, and $a\left(\frac{7\pi}{4}\right) = -10$.

(d) Solving $10\cos\left(t + \frac{\pi}{4}\right) = 0 \Rightarrow t = \frac{\pi}{4} \Rightarrow v\left(\frac{\pi}{4}\right) = -10$, $\left|v\left(\frac{\pi}{4}\right)\right| = 10$ and $a\left(\frac{\pi}{4}\right) = 0$.

11. (a) $s(t) = 64t - 16t^2 \Rightarrow v(t) = \frac{ds}{dt} = 64 - 32t = 32(2 - t)$. The maximum height is reached when $v(t) = 0$ $\Rightarrow t = 2$ sec. The velocity when it leaves the hand is $v(0) = 64$ ft/sec.

(b) $s(t) = 64t - 2.6t^2 \Rightarrow v(t) = \frac{ds}{dt} = 64 - 5.2t$. The maximum height is reached when $v(t) = 0 \Rightarrow t \approx 12.31$ sec. The maximum height is about $s(12.31) = 393.85$ ft.

12. $s_1 = 3t^3 - 12t^2 + 18t + 5$ and $s_2 = -t^3 + 9t^2 - 12t \Rightarrow v_1 = 9t^2 - 24t + 18$ and $v_2 = -3t^2 + 18t - 12$; $v_1 = v_2$ $\Rightarrow 9t^2 - 24t + 18 = -3t^2 + 18t - 12 \Rightarrow 2t^2 - 7t + 5 = 0 \Rightarrow (t - 1)(2t - 5) = 0 \Rightarrow t = 1$ sec and $t = 2.5$ sec.

13. $m\left(v^2 - v_0^2\right) = k\left(x_0^2 - x^2\right) \Rightarrow m\left(2v\frac{dv}{dt}\right) = k\left(-2x\frac{dx}{dt}\right) \Rightarrow m\frac{dv}{dt} = k\left(-\frac{2x}{2v}\right)\frac{dx}{dt} \Rightarrow m\frac{dv}{dt} = -kx\left(\frac{1}{v}\right)\frac{dx}{dt}$. Then substituting $\frac{dx}{dt} = v \Rightarrow m\frac{dv}{dt} = -kx$, as claimed.

14. (a) $x = At^2 + Bt + C$ on $[t_1, t_2] \Rightarrow v = \frac{dx}{dt} = 2At + B \Rightarrow v\left(\frac{t_1 + t_2}{2}\right) = 2A\left(\frac{t_1 + t_2}{2}\right) + B = A(t_1 + t_2) + B$ is the instantaneous velocity at the midpoint. The average velocity over the time interval is $v_{av} = \frac{\Delta x}{\Delta t}$ $= \frac{\left(At_2^2 + Bt_2 + C\right) - \left(At_1^2 + Bt_1 + C\right)}{t_2 - t_1} = \frac{(t_2 - t_1)\left[A(t_2 + t_1) + B\right]}{t_2 - t_1} = A(t_2 + t_1) + B.$

(b) On the graph of the parabola $x = At^2 + Bt + C$, the slope of the curve at the midpoint of the interval $[t_1, t_2]$ is the same as the average slope of the curve over the interval.

15. (a) To be continuous at $x = \pi$ requires that $\lim\limits_{x \to \pi^-} \sin x = \lim\limits_{x \to \pi^+} (mx + b) \Rightarrow 0 = m\pi + b \Rightarrow m = -\frac{b}{\pi}$;

(b) If $y' = \begin{cases} \cos x, & x < \pi \\ m, & x \ge \pi \end{cases}$ is differentiable at $x = \pi$, then $\lim\limits_{x \to \pi^-} \cos x = m \Rightarrow m = -1$ and $b = \pi$.

16. $f(x)$ is continuous at 0 because $\lim\limits_{x \to 0} \frac{1-\cos x}{x} = 0 = f(0)$. $f'(0) = \lim\limits_{x \to 0} \frac{f(x) - f(0)}{x - 0} = \lim\limits_{x \to 0} \frac{\frac{1-\cos x}{x} - 0}{x}$
$= \lim\limits_{x \to 0} \left(\frac{1-\cos x}{x^2}\right)\left(\frac{1+\cos x}{1+\cos x}\right) = \lim\limits_{x \to 0} \left(\frac{\sin x}{x}\right)^2 \left(\frac{1}{1+\cos x}\right) = \frac{1}{2}$. Therefore $f'(0)$ exists with value $\frac{1}{2}$.

17. (a) For all a, b and for all $x \ne 2$, f is differentiable at x. Next, f differentiable at $x = 2 \Rightarrow$ f continuous at $x = 2$ $\Rightarrow \lim\limits_{x \to 2^-} f(x) = f(2) \Rightarrow 2a = 4a - 2b + 3 \Rightarrow 2a - 2b + 3 = 0$. Also, f differentiable at $x \ne 2$ $\Rightarrow f'(x) = \begin{cases} a, & x < 2 \\ 2ax - b, & x > 2 \end{cases}$. In order that $f'(2)$ exist we must have $a = 2a(2) - b \Rightarrow a = 4a - b \Rightarrow 3a = b$. Then $2a - 2b + 3 = 0$ and $3a = b \Rightarrow a = \frac{3}{4}$ and $b = \frac{9}{4}$.

(b) For $x < 2$, the graph of f is a straight line having a slope of $\frac{3}{4}$ and passing through the origin; for $x \ge 2$, the graph of f is a parabola. At $x = 2$, the value of the y-coordinate on the parabola is $\frac{3}{2}$ which matches the y-coordinate of the point on the straight line at $x = 2$. In addition, the slope of the parabola at the match up point is $\frac{3}{4}$ which is equal to the slope of the straight line. Therefore, since the graph is differentiable at the match up point, the graph is smooth there.

18. (a) For any a, b and for any $x \ne -1$, g is differentiable at x. Next, g differentiable at $x = -1 \Rightarrow$ g continuous at $x = -1 \Rightarrow \lim\limits_{x \to -1^+} g(x) = g(-1) \Rightarrow -a - 1 + 2b = -a + b \Rightarrow b = 1$. Also, g differentiable at $x \ne -1$ $\Rightarrow g'(x) = \begin{cases} a, & x < -1 \\ 3ax^2 + 1, & x > -1 \end{cases}$. In order that $g'(-1)$ exist we must have $a = 3a(-1)^2 + 1 \Rightarrow a = 3a + 1$ $\Rightarrow a = -\frac{1}{2}$.

(b) For $x \le -1$, the graph of f is a straight line having a slope of $-\frac{1}{2}$ and a y-intercept of 1. For $x > -1$, the graph of f is a parabola. At $x = -1$, the value of the y-coordinate on the parabola is $\frac{3}{2}$ which matches the y-coordinate of the point on the straight line at $x = -1$. In addition, the slope of the parabola at the match up point is $-\frac{1}{2}$ which is equal to the slope of the straight line. Therefore, since the graph is differentiable at the match up point, the graph is smooth there.

19. f odd $\Rightarrow f(-x) = -f(x) \Rightarrow \frac{d}{dx}(f(-x)) = \frac{d}{dx}(-f(x)) \Rightarrow f'(-x)(-1) = -f'(x) \Rightarrow f'(-x) = f'(x) \Rightarrow f'$ is even.

20. f even $\Rightarrow f(-x) = f(x) \Rightarrow \frac{d}{dx}(f(-x)) = \frac{d}{dx}(f(x)) \Rightarrow f'(-x)(-1) = f'(x) \Rightarrow f'(-x) = -f'(x) \Rightarrow f'$ is odd.

21. Let $h(x) = (fg)(x) = f(x)\,g(x) \Rightarrow h'(x) = \lim\limits_{x \to x_0} \frac{h(x) - h(x_0)}{x - x_0} = \lim\limits_{x \to x_0} \frac{f(x)\,g(x) - f(x_0)\,g(x_0)}{x - x_0}$
$= \lim\limits_{x \to x_0} \frac{f(x)\,g(x) - f(x)\,g(x_0) + f(x)\,g(x_0) - f(x_0)\,g(x_0)}{x - x_0} = \lim\limits_{x \to x_0} \left[f(x)\left[\frac{g(x) - g(x_0)}{x - x_0}\right]\right] + \lim\limits_{x \to x_0} \left[g(x_0)\left[\frac{f(x) - f(x_0)}{x - x_0}\right]\right]$
$= f(x_0) \lim\limits_{x \to x_0} \left[\frac{g(x) - g(x_0)}{x - x_0}\right] + g(x_0)\,f'(x_0) = 0 \cdot \lim\limits_{x \to x_0} \left[\frac{g(x) - g(x_0)}{x - x_0}\right] + g(x_0)\,f'(x_0) = g(x_0)\,f'(x_0)$, if g is continuous at x_0. Therefore $(fg)(x)$ is differentiable at x_0 if $f(x_0) = 0$, and $(fg)'(x_0) = g(x_0)\,f'(x_0)$.

22. From Exercise 21 we have that fg is differentiable at 0 if f is differentiable at 0, $f(0) = 0$ and g is continuous at 0.

(a) If $f(x) = \sin x$ and $g(x) = |x|$, then $|x| \sin x$ is differentiable because $f'(0) = \cos(0) = 1$, $f(0) = \sin(0) = 0$ and $g(x) = |x|$ is continuous at $x = 0$.

(b) If $f(x) = \sin x$ and $g(x) = x^{2/3}$, then $x^{2/3} \sin x$ is differentiable because $f'(0) = \cos(0) = 1$, $f(0) = \sin(0) = 0$ and $g(x) = x^{2/3}$ is continuous at $x = 0$.

(c) If $f(x) = 1 - \cos x$ and $g(x) = \sqrt[3]{x}$, then $\sqrt[3]{x}(1 - \cos x)$ is differentiable because $f'(0) = \sin(0) = 0$, $f(0) = 1 - \cos(0) = 0$ and $g(x) = x^{1/3}$ is continuous at $x = 0$.

(d) If $f(x) = x$ and $g(x) = x \sin\left(\frac{1}{x}\right)$, then $x^2 \sin\left(\frac{1}{x}\right)$ is differentiable because $f'(0) = 1$, $f(0) = 0$ and $\lim_{x \to 0} x \sin\left(\frac{1}{x}\right) = \lim_{x \to 0} \frac{\sin\left(\frac{1}{x}\right)}{\frac{1}{x}} = \lim_{t \to \infty} \frac{\sin t}{t} = 0$ (so g is continuous at $x = 0$).

23. If $f(x) = x$ and $g(x) = x \sin\left(\frac{1}{x}\right)$, then $x^2 \sin\left(\frac{1}{x}\right)$ is differentiable at $x = 0$ because $f'(0) = 1$, $f(0) = 0$ and $\lim_{x \to 0} x \sin\left(\frac{1}{x}\right) = \lim_{x \to 0} \frac{\sin\left(\frac{1}{x}\right)}{\frac{1}{x}} = \lim_{t \to \infty} \frac{\sin t}{t} = 0$ (so g is continuous at $x = 0$). In fact, from Exercise 21, $h'(0) = g(0)f'(0) = 0$. However, for $x \neq 0$, $h'(x) = \left[x^2 \cos\left(\frac{1}{x}\right)\right]\left(-\frac{1}{x^2}\right) + 2x \sin\left(\frac{1}{x}\right)$. But $\lim_{x \to 0} h'(x) = \lim_{x \to 0} \left[-\cos\left(\frac{1}{x}\right) + 2x \sin\left(\frac{1}{x}\right)\right]$ does not exist because $\cos\left(\frac{1}{x}\right)$ has no limit as $x \to 0$. Therefore, the derivative is not continuous at $x = 0$ because it has no limit there.

24. From the given conditions we have $f(x + h) = f(x) f(h)$, $f(h) - 1 = hg(h)$ and $\lim_{h \to 0} g(h) = 1$. Therefore,
$f'(x) = \lim_{h \to 0} \frac{f(x+h) - f(x)}{h} = \lim_{h \to 0} \frac{f(x) f(h) - f(x)}{h} = \lim_{h \to 0} f(x) \left[\frac{f(h) - 1}{h}\right] = f(x) \left[\lim_{h \to 0} g(h)\right] = f(x) \cdot 1 = f(x)$
$\Rightarrow f'(x) = f(x)$ and $f'(x)$exists at every value of x.

25. Step 1: The formula holds for $n = 2$ (a single product) since $y = u_1u_2 \Rightarrow \frac{dy}{dx} = \frac{du_1}{dx} u_2 + u_1 \frac{du_2}{dx}$.
Step 2: Assume the formula holds for $n = k$:
$$y = u_1u_2 \cdots u_k \Rightarrow \frac{dy}{dx} = \frac{du_1}{dx} u_2u_3 \cdots u_k + u_1 \frac{du_2}{dx} u_3 \cdots u_k + \ldots + u_1u_2 \cdots u_{k-1} \frac{du_k}{dx}.$$
If $y = u_1u_2 \cdots u_ku_{k+1} = (u_1u_2 \cdots u_k)\, u_{k+1}$, then $\frac{dy}{dx} = \frac{d(u_1u_2 \cdots u_k)}{dx} u_{k+1} + u_1u_2 \cdots u_k \frac{du_{k+1}}{dx}$
$= \left(\frac{du_1}{dx} u_2u_3 \cdots u_k + u_1 \frac{du_2}{dx} u_3 \cdots u_k + \cdots + u_1u_2 \cdots u_{k-1} \frac{du_k}{dx}\right) u_{k+1} + u_1u_2 \cdots u_k \frac{du_{k+1}}{dx}$
$= \frac{du_1}{dx} u_2u_3 \cdots u_{k+1} + u_1 \frac{du_2}{dx} u_3 \cdots u_{k+1} + \cdots + u_1u_2 \cdots u_{k-1} \frac{du_k}{dx} u_{k+1} + u_1u_2 \cdots u_k \frac{du_{k+1}}{dx}$.
Thus the original formula holds for $n = (k+1)$ whenever it holds for $n = k$.

26. Recall $\binom{m}{k} = \frac{m!}{k!\,(m-k)!}$. Then $\binom{m}{1} = \frac{m!}{1!\,(m-1)!} = m$ and $\binom{m}{k} + \binom{m}{k+1} = \frac{m!}{k!\,(m-k)!} + \frac{m!}{(k+1)!\,(m-k-1)!}$
$= \frac{m!\,(k+1) + m!\,(m-k)}{(k+1)!\,(m-k)!} = \frac{m!\,(m+1)}{(k+1)!\,(m-k)!} = \frac{(m+1)!}{(k+1)!\,((m+1)-(k+1))!} = \binom{m+1}{k+1}$. Now, we prove Leibniz's rule by mathematical induction.
Step 1: If $n = 1$, then $\frac{d(uv)}{dx} = u \frac{dv}{dx} + v \frac{du}{dx}$. Assume that the statement is true for $n = k$, that is:
$$\frac{d^k(uv)}{dx^k} = \frac{d^ku}{dx^k} v + k \frac{d^{k-1}u}{dx^{k-1}} \frac{dv}{dx} + \binom{k}{2} \frac{d^{k-2}u}{dx^{k-2}} \frac{d^2v}{dx^2} + \ldots + \binom{k}{k-1} \frac{du}{dv} \frac{d^{k-1}v}{dx^{k-1}} + u \frac{d^kv}{dx^k}.$$
Step 2: If $n = k + 1$, then $\frac{d^{k+1}(uv)}{dx^{k+1}} = \frac{d}{dx}\left(\frac{d^k(uv)}{dx^k}\right) = \left[\frac{d^{k+1}u}{dx^{k+1}} v + \frac{d^ku}{dx^k} \frac{dv}{dx}\right] + \left[k \frac{d^ku}{dx^k} \frac{dv}{dx} + k \frac{d^{k-1}u}{dx^{k-1}} \frac{d^2v}{dx^2}\right]$
$+ \left[\binom{k}{2} \frac{d^{k-1}u}{dx^{k-1}} \frac{d^2v}{dx^2} + \binom{k}{2} \frac{d^{k-2}u}{dx^{k-2}} \frac{d^3v}{dx^3}\right] + \ldots + \left[\binom{k}{k-1} \frac{d^2u}{dx^2} \frac{d^{k-1}v}{dx^{k-1}} + \binom{k}{k-1} \frac{du}{dx} \frac{d^ku}{dx^k} v\right]$
$+ \left[\frac{du}{dx} \frac{d^kv}{dx^k} + u \frac{d^{k+1}u}{dx^{k+1}}\right] = \frac{d^{k+1}u}{dx^{k+1}} v + (k + 1) \frac{d^ku}{dx^k} \frac{dv}{dx} + \left[\binom{k}{1} + \binom{k}{2}\right] \frac{d^{k-1}u}{dx^{k-1}} \frac{d^2v}{dx^2} + \ldots$
$+ \left[\binom{k}{k-1} + \binom{k}{k}\right] \frac{du}{dx} \frac{d^kv}{dx^k} + u \frac{d^{k+1}v}{dx^{k+1}} = \frac{d^{k+1}u}{dx^{k+1}} v + (k + 1) \frac{d^ku}{dx^k} \frac{dv}{dx} + \binom{k+1}{2} \frac{d^{k-1}u}{dx^{k-1}} \frac{d^2v}{dx^2} + \ldots$
$+ \binom{k+1}{k} \frac{du}{dx} \frac{d^kv}{dx^k} + u \frac{d^{k+1}v}{dx^{k+1}}$.
Therefore the formula (c) holds for $n = (k + 1)$ whenever it holds for $n = k$.

27. (a) $T^2 = \frac{4\pi^2L}{g} \Rightarrow L = \frac{T^2g}{4\pi^2} \Rightarrow L = \frac{(1\ \text{sec}^2)(32.2\ \text{ft/sec}^2)}{4\pi^2} \Rightarrow L \approx 0.8156$ ft

(b) $T^2 = \frac{4\pi^2L}{g} \Rightarrow T = \frac{2\pi}{\sqrt{g}} \sqrt{L}$; $dT = \frac{2\pi}{\sqrt{g}} \cdot \frac{1}{2\sqrt{L}} dL = \frac{\pi}{\sqrt{Lg}} dL$; $dT = \frac{\pi}{\sqrt{(0.8156\ \text{ft})(32.2\ \text{ft/sec}^2)}} (0.01\ \text{ft}) \approx 0.00613$ sec.

(c) Since there are 86,400 sec in a day, we have (0.00613 sec)(86,400 sec/day) $\approx$ 529.6 sec/day, or 8.83 min/day; the clock will lose about 8.83 min/day.

28. $v = s^3 \Rightarrow \frac{dv}{dt} = 3s^2 \frac{ds}{dt} = -k(6s^2) \Rightarrow \frac{ds}{dt} = -2k$. If $s_0 =$ the initial length of the cube's side, then $s_1 = s_0 - 2k$ $\Rightarrow 2k = s_0 - s_1$. Let $t =$ the time it will take the ice cube to melt. Now, $t = \frac{s_0}{2k} = \frac{s_0}{s_0 - s_1} = \frac{(v_0)^{1/3}}{(v_0)^{1/3} - \left(\frac{3}{4}v_0\right)^{1/3}}$ $= \frac{1}{1 - \left(\frac{3}{4}\right)^{1/3}} \approx 11$ hr.

NOTES:

CHAPTER 4 APPLICATIONS OF DERIVATIVES

4.1 EXTREME VALUES OF FUNCTIONS

1. An absolute minimum at $x = c_2$, an absolute maximum at $x = b$. Theorem 1 guarantees the existence of such extreme values because h is continuous on [a, b].

2. An absolute minimum at $x = b$, an absolute maximum at $x = c$. Theorem 1 guarantees the existence of such extreme values because f is continuous on [a, b].

3. No absolute minimum. An absolute maximum at $x = c$. Since the function's domain is an open interval, the function does not satisfy the hypotheses of Theorem 1 and need not have absolute extreme values.

4. No absolute extrema. The function is neither continuous nor defined on a closed interval, so it need not fulfill the conclusions of Theorem 1.

5. An absolute minimum at $x = a$ and an absolute maximum at $x = c$. Note that $y = g(x)$ is not continuous but still has extrema. When the hypothesis of Theorem 1 is satisfied then extrema are guaranteed, but when the hypothesis is not satisfied, absolute extrema may or may not occur.

6. Absolute minimum at $x = c$ and an absolute maximum at $x = a$. Note that $y = g(x)$ is not continuous but still has absolute extrema. When the hypothesis of Theorem 1 is satisfied then extrema are guaranteed, but when the hypothesis is not satisfied, absolute extrema may or may not occur.

7. Local minimum at $(-1, 0)$, local maximum at $(1, 0)$

8. Minima at $(-2, 0)$ and $(2, 0)$, maximum at $(0, 2)$

9. Maximum at $(0, 5)$. Note that there is no minimum since the endpoint $(2, 0)$ is excluded from the graph.

10. Local maximum at $(-3, 0)$, local minimum at $(2, 0)$, maximum at $(1, 2)$, minimum at $(0, -1)$

11. Graph (c), since this the only graph that has positive slope at c.

12. Graph (b), since this is the only graph that represents a differentiable function at a and b and has negative slope at c.

13. Graph (d), since this is the only graph representing a funtion that is differentiable at b but not at a.

14. Graph (a), since this is the only graph that represents a function that is not differentiable at a or b.

15. $f(x) = \frac{2}{3}x - 5 \Rightarrow f'(x) = \frac{2}{3} \Rightarrow$ no critical points; $f(-2) = -\frac{19}{3}$, $f(3) = -3 \Rightarrow$ the absolute maximum is -3 at $x = 3$ and the absolute minimum is $-\frac{19}{3}$ at $x = -2$

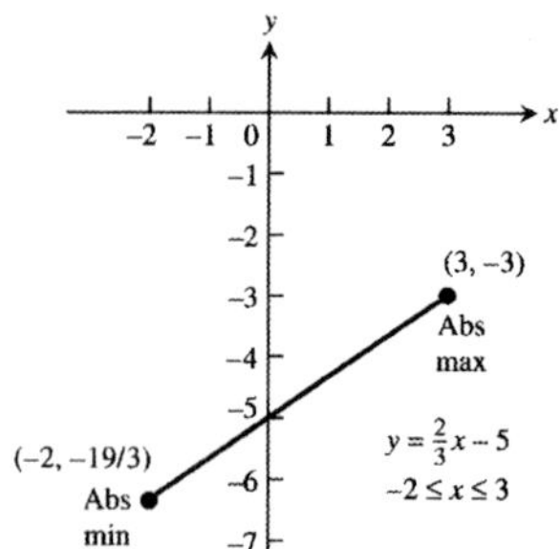

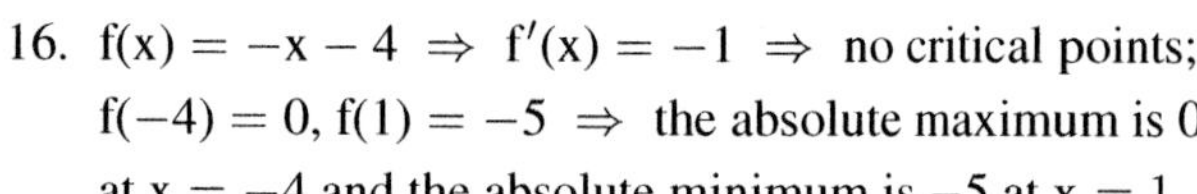

16. $f(x) = -x - 4 \Rightarrow f'(x) = -1 \Rightarrow$ no critical points; $f(-4) = 0$, $f(1) = -5 \Rightarrow$ the absolute maximum is 0 at $x = -4$ and the absolute minimum is -5 at $x = 1$

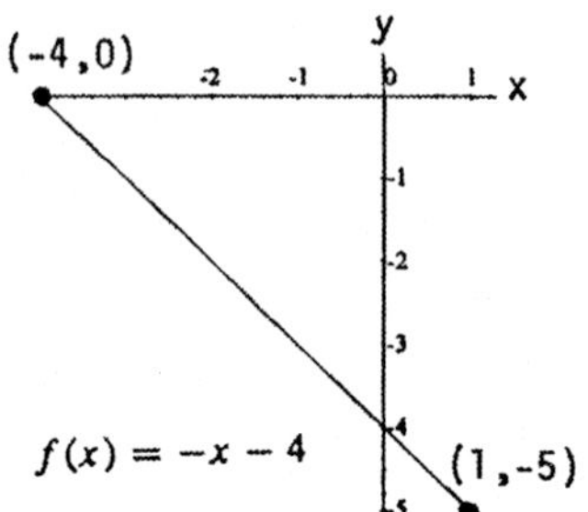

17. $f(x) = x^2 - 1 \Rightarrow f'(x) = 2x \Rightarrow$ a critical point at $x = 0$; $f(-1) = 0$, $f(0) = -1$, $f(2) = 3 \Rightarrow$ the absolute maximum is 3 at $x = 2$ and the absolute minimum is -1 at $x = 0$

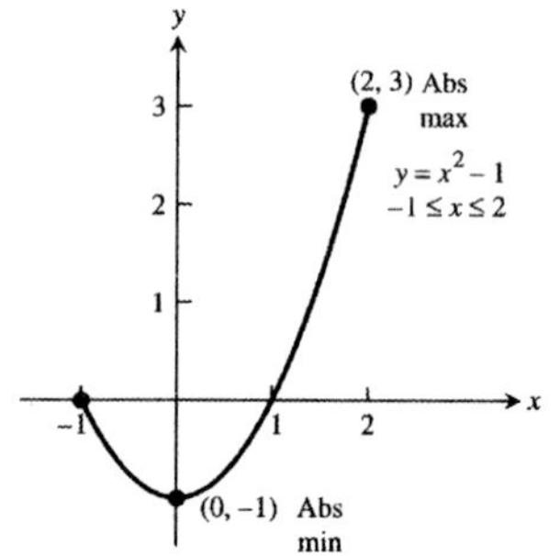

18. $f(x) = 4 - x^2 \Rightarrow f'(x) = -2x \Rightarrow$ a critical point at $x = 0$; $f(-3) = -5$, $f(0) = 4$, $f(1) = 3 \Rightarrow$ the absolute maximum is 4 at $x = 0$ and the absolute minimum is -5 at $x = -3$

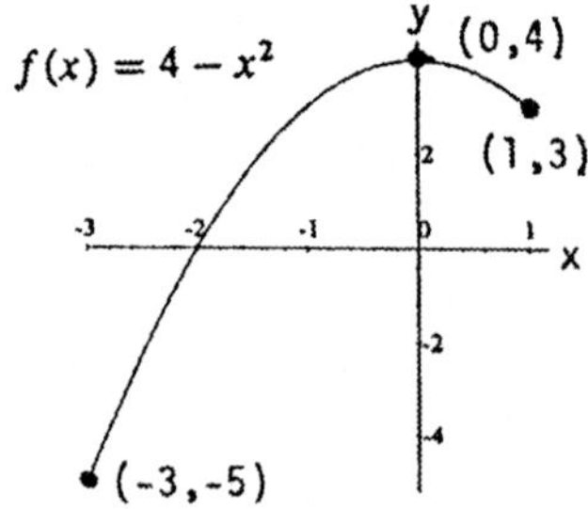

19. $F(x) = -\frac{1}{x^2} = -x^{-2} \Rightarrow F'(x) = 2x^{-3} = \frac{2}{x^3}$, however $x = 0$ is not a critical point since 0 is not in the domain; $F(0.5) = -4$, $F(2) = -0.25 \Rightarrow$ the absolute maximum is -0.25 at $x = 2$ and the absolute minimum is -4 at $x = 0.5$

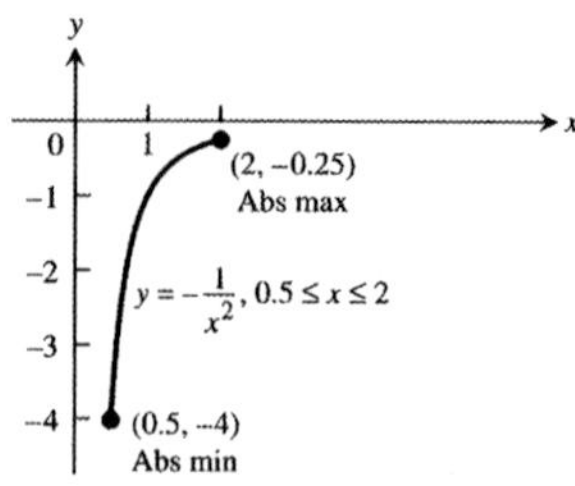

20. $F(x) = -\frac{1}{x} = -x^{-1} \Rightarrow F'(x) = x^{-2} = \frac{1}{x^2}$, however $x = 0$ is not a critical point since 0 is not in the domain; $F(-2) = \frac{1}{2}$, $F(-1) = 1 \Rightarrow$ the absolute maximum is 1 at $x = -1$ and the absolute minimum is $\frac{1}{2}$ at $x = -2$

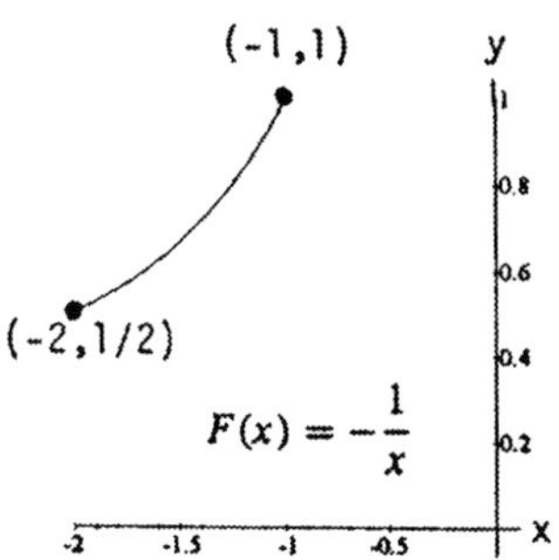

21. $h(x) = \sqrt[3]{x} = x^{1/3} \Rightarrow h'(x) = \frac{1}{3}x^{-2/3} \Rightarrow$ a critical point at $x = 0$; $h(-1) = -1$, $h(0) = 0$, $h(8) = 2 \Rightarrow$ the absolute maximum is 2 at $x = 8$ and the absolute minimum is -1 at $x = -1$

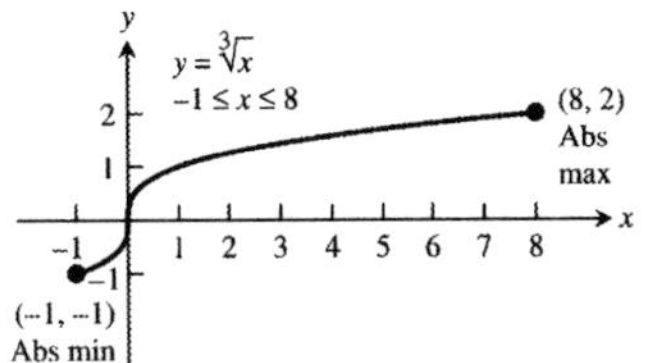

22. $h(x) = -3x^{2/3} \Rightarrow h'(x) = -2x^{-1/3} \Rightarrow$ a critical point at $x = 0$; $h(-1) = -3$, $h(0) = 0$, $h(1) = -3 \Rightarrow$ the absolute maximum is 0 at $x = 0$ and the absolute minimum is -3 at $x = 1$ and at $x = -1$

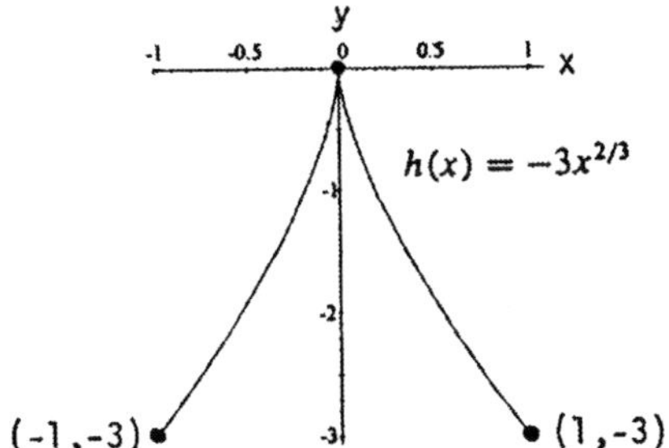

23. $g(x) = \sqrt{4 - x^2} = (4 - x^2)^{1/2}$
$\Rightarrow g'(x) = \frac{1}{2}(4 - x^2)^{-1/2}(-2x) = \frac{-x}{\sqrt{4 - x^2}}$
$\Rightarrow$ critical points at $x = -2$ and $x = 0$, but not at $x = 2$ because 2 is not in the domain; $g(-2) = 0$, $g(0) = 2$, $g(1) = \sqrt{3} \Rightarrow$ the absolute maximum is 2 at $x = 0$ and the absolute minimum is 0 at $x = -2$

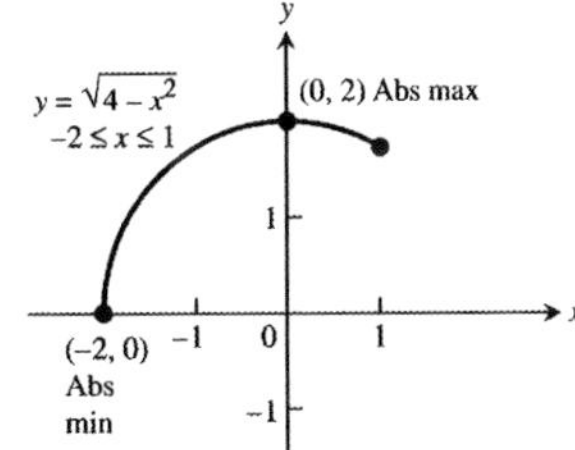

24. $g(x) = -\sqrt{5 - x^2} = -(5 - x^2)^{1/2}\ (5 - x^2)^{-1/2}(-2x)$
$\Rightarrow g'(x) = -\left(\frac{1}{2}\right) = \frac{x}{\sqrt{5 - x^2}} \Rightarrow$ critical points at $x = -\sqrt{5}$ and $x = 0$, but not at $x = \sqrt{5}$ because $\sqrt{5}$ is not in the domain; $f\left(-\sqrt{5}\right) = 0$, $f(0) = -\sqrt{5}$
$\Rightarrow$ the absolute maximum is 0 at $x = -\sqrt{5}$ and the absolute minimum is $-\sqrt{5}$ at $x = 0$

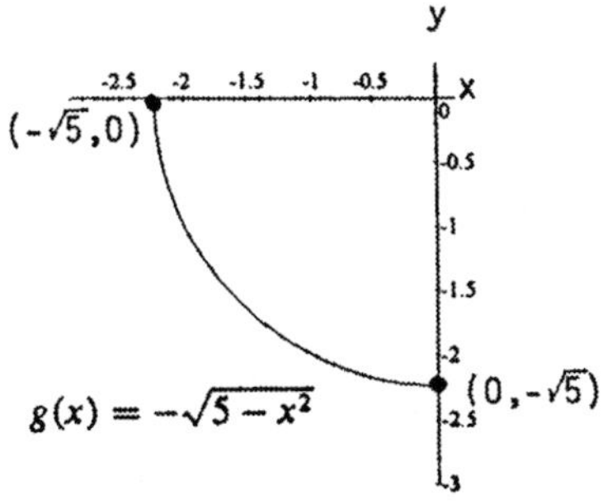

25. $f(\theta) = \sin\theta \Rightarrow f'(\theta) = \cos\theta \Rightarrow \theta = \frac{\pi}{2}$ is a critical point, but $\theta = \frac{-\pi}{2}$ is not a critical point because $\frac{-\pi}{2}$ is not interior to the domain; $f\left(\frac{-\pi}{2}\right) = -1$, $f\left(\frac{\pi}{2}\right) = 1$, $f\left(\frac{5\pi}{6}\right) = \frac{1}{2}$
$\Rightarrow$ the absolute maximum is 1 at $\theta = \frac{\pi}{2}$ and the absolute minimum is -1 at $\theta = \frac{-\pi}{2}$

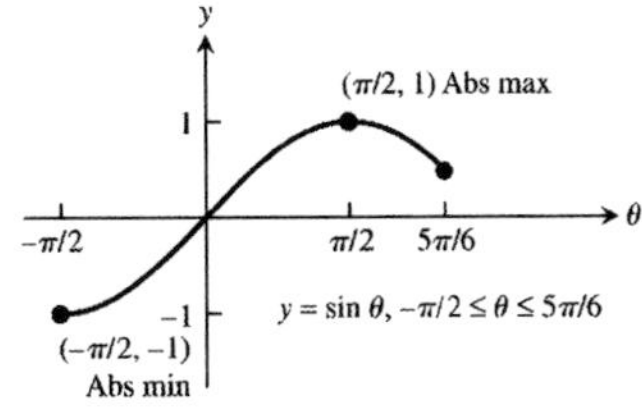

26. $f(\theta) = \tan\theta \Rightarrow f'(\theta) = \sec^2\theta \Rightarrow$ f has no critical points in $\left(\frac{-\pi}{3}, \frac{\pi}{4}\right)$. The extreme values therefore occur at the endpoints: $f\left(\frac{-\pi}{3}\right) = -\sqrt{3}$ and $f\left(\frac{\pi}{4}\right) = 1 \Rightarrow$ the absolute maximum is 1 at $\theta = \frac{\pi}{4}$ and the absolute minimum is $-\sqrt{3}$ at $\theta = \frac{-\pi}{3}$

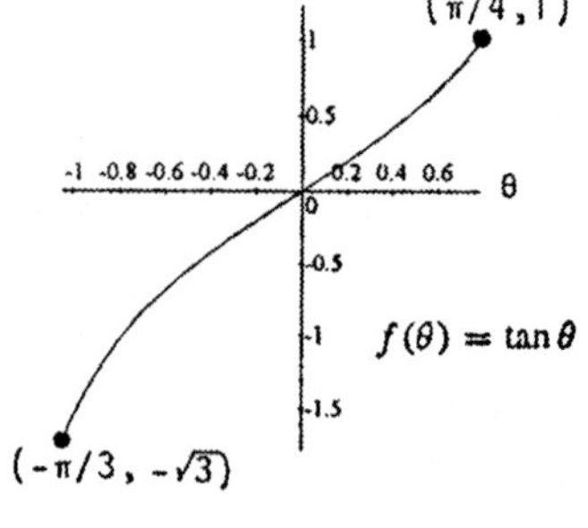

27. $g(x) = \csc x \Rightarrow g'(x) = -(\csc x)(\cot x) \Rightarrow$ a critical point at $x = \frac{\pi}{2}$; $g\left(\frac{\pi}{3}\right) = \frac{2}{\sqrt{3}}$, $g\left(\frac{\pi}{2}\right) = 1$, $g\left(\frac{2\pi}{3}\right) = \frac{2}{\sqrt{3}} \Rightarrow$ the absolute maximum is $\frac{2}{\sqrt{3}}$ at $x = \frac{\pi}{3}$ and $x = \frac{2\pi}{3}$, and the absolute minimum is 1 at $x = \frac{\pi}{2}$

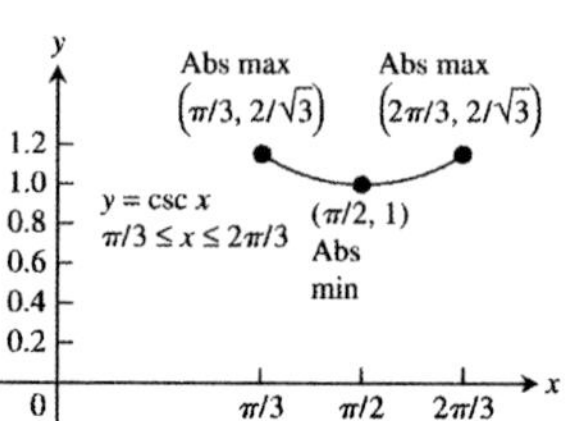

28. $g(x) = \sec x \Rightarrow g'(x) = (\sec x)(\tan x) \Rightarrow$ a critical point at $x = 0$; $g\left(-\frac{\pi}{3}\right) = 2$, $g(0) = 1$, $g\left(\frac{\pi}{6}\right) = \frac{2}{\sqrt{3}} \Rightarrow$ the absolute maximum is 2 at $x = -\frac{\pi}{3}$ and the absolute minimum is 1 at $x = 0$

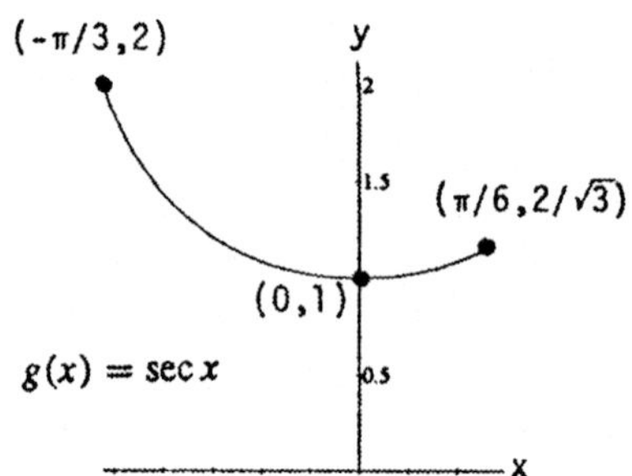

29. $f(t) = 2 - |t| = 2 - \sqrt{t^2} = 2 - (t^2)^{1/2}$
$\Rightarrow f'(t) = -\frac{1}{2}(t^2)^{-1/2}(2t) = -\frac{t}{\sqrt{t^2}} = -\frac{t}{|t|}$
$\Rightarrow$ a critical point at $t = 0$; $f(-1) = 1$,
$f(0) = 2$, $f(3) = -1 \Rightarrow$ the absolute maximum is 2 at $t = 0$ and the absolute minimum is -1 at $t = 3$

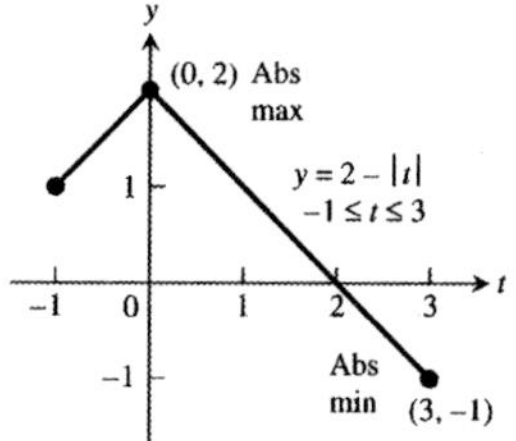

30. $f(t) = |t - 5| = \sqrt{(t-5)^2} = ((t-5)^2)^{1/2} \Rightarrow f'(t)$
$= \frac{1}{2}((t-5)^2)^{-1/2}(2(t-5)) = \frac{t-5}{\sqrt{(t-5)^2}} = \frac{t-5}{|t-5|}$
$\Rightarrow$ a critical point at $t = 5$; $f(4) = 1$, $f(5) = 0$, $f(7) = 2$
$\Rightarrow$ the absolute maximum is 2 at $t = 7$ and the absolute minimum is 0 at $t = 5$

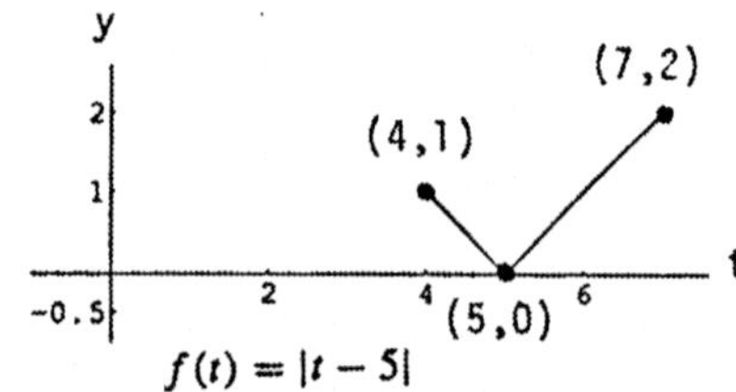

31. $g(x) = xe^{-x} \Rightarrow g'(x) = e^{-x} - xe^{-x}$
$\Rightarrow$ a critical point at $x = 1$; $g(-1) = -e$, and $g(1) = \frac{1}{e}$,
$\Rightarrow$ the absolute maximum is $\frac{1}{e}$ at $x = 1$ and the absolute minimum is $-e$ at $x = -1$

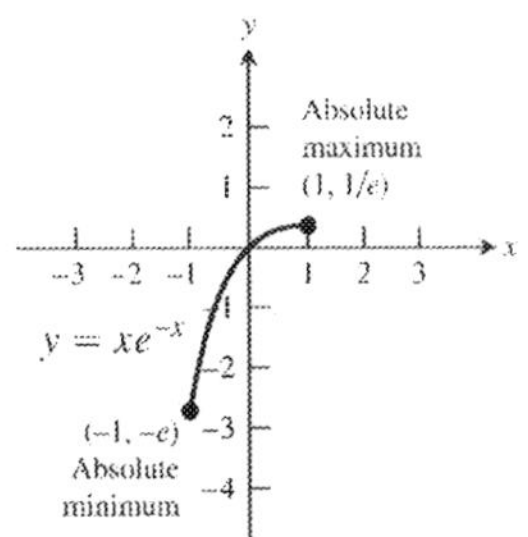

32. The first derivative $h'(x) = \frac{1}{x+1}$ has no zeros, so we need only consider the endpoints. $h(0) = \ln 1$; $h(3) = \ln 4$
Maximum value is $\ln 4$ at $x = 3$;
Minimum value is $\ln 1$ at $x = 0$.

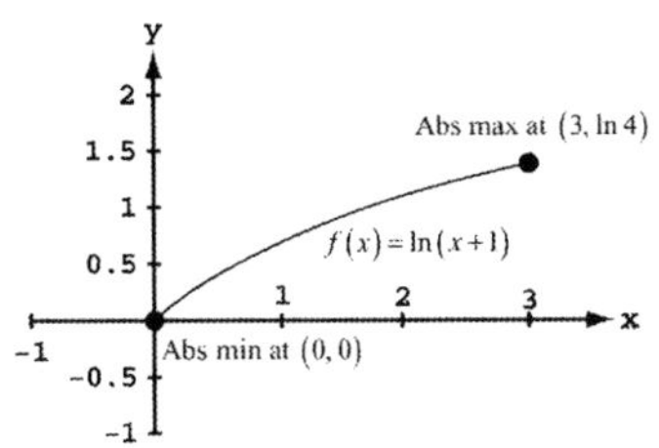

33. The first derivative $f'(x) = -\frac{1}{x^2} + \frac{1}{x}$ has a zero at $x = 1$.
Critical point value: $f(1) = 1 + \ln 1 = 1$
Endpoint values: $f(0.5) = 2 + \ln 0.5 \approx 1.307$;
$f(4) = \frac{1}{4} + \ln 4 \approx 1.636$;
Absolute maximum value is $\frac{1}{4} + \ln 4$ at $x = 4$;
Absolute Minimum value is 1 at $x = 1$;
Local maximum at $\left(\frac{1}{2}, 2 - \ln 2\right)$

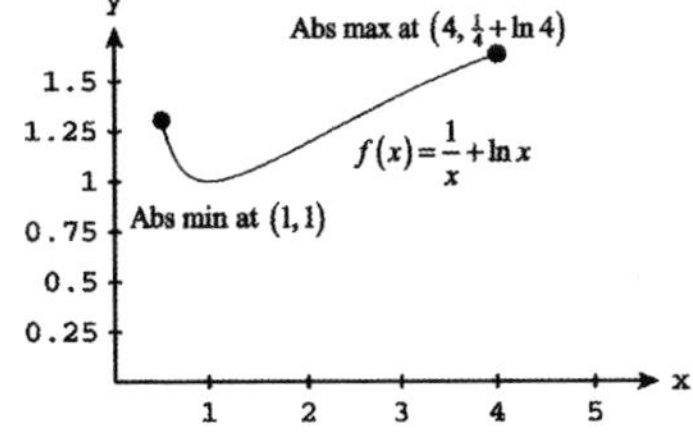

34. $g(x) = e^{-x^2} \Rightarrow g'(x) = -2xe^{-x^2} \Rightarrow$ a critical point at $x = 0$; $g(-2) = e^{-4}$, $g(0) = 1$, and $g(1) = e^{-1}$
$\Rightarrow$ the absolute maximum is 1 at $x = 0$ and the absolute minimum is e^{-4} at $x = -2$

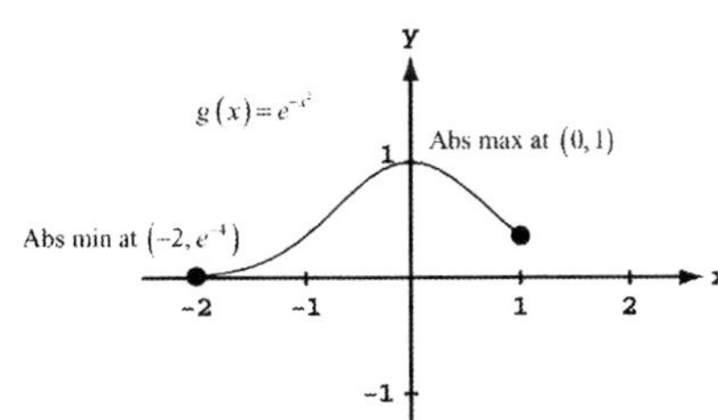

35. $f(x) = x^{4/3} \Rightarrow f'(x) = \frac{4}{3}x^{1/3} \Rightarrow$ a critical point at $x = 0$; $f(-1) = 1$, $f(0) = 0$, $f(8) = 16 \Rightarrow$ the absolute maximum is 16 at $x = 8$ and the absolute minimum is 0 at $x = 0$

36. $f(x) = x^{5/3} \Rightarrow f'(x) = \frac{5}{3}x^{2/3} \Rightarrow$ a critical point at $x = 0$; $f(-1) = -1$, $f(0) = 0$, $f(8) = 32 \Rightarrow$ the absolute maximum is 32 at $x = 8$ and the absolute minimum is -1 at $x = -1$

37. $g(\theta) = \theta^{3/5} \Rightarrow g'(\theta) = \frac{3}{5}\theta^{-2/5} \Rightarrow$ a critical point at $\theta = 0$; $g(-32) = -8$, $g(0) = 0$, $g(1) = 1 \Rightarrow$ the absolute maximum is 1 at $\theta = 1$ and the absolute minimum is -8 at $\theta = -32$

38. $h(\theta) = 3\theta^{2/3} \Rightarrow h'(\theta) = 2\theta^{-1/3} \Rightarrow$ a critical point at $\theta = 0$; $h(-27) = 27$, $h(0) = 0$, $h(8) = 12 \Rightarrow$ the absolute maximum is 27 at $\theta = -27$ and the absolute minimum is 0 at $\theta = 0$

39. Minimum value is 1 at x = 2.

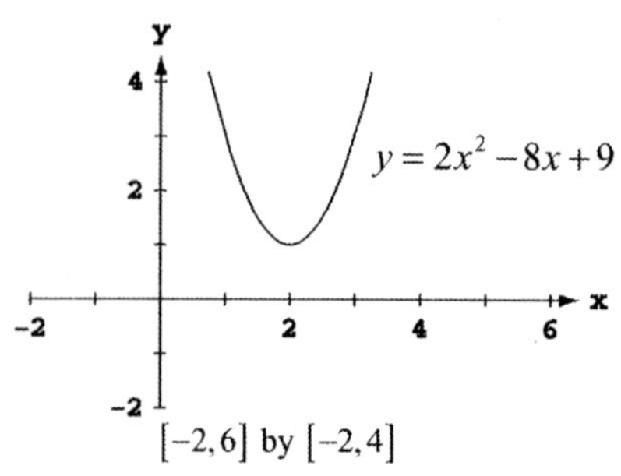

40. To find the exact values, note that $y' = 3x^2 - 2$, which is zero when $x = \pm\sqrt{\frac{2}{3}}$. Local maximum at $\left(-\sqrt{\frac{2}{3}}, 4 + \frac{4\sqrt{6}}{9}\right) \approx (-0.816, 5.089)$; local minimum at $\left(\sqrt{\frac{2}{3}}, 4 - \frac{4\sqrt{6}}{9}\right) \approx (0.816, 2.911)$

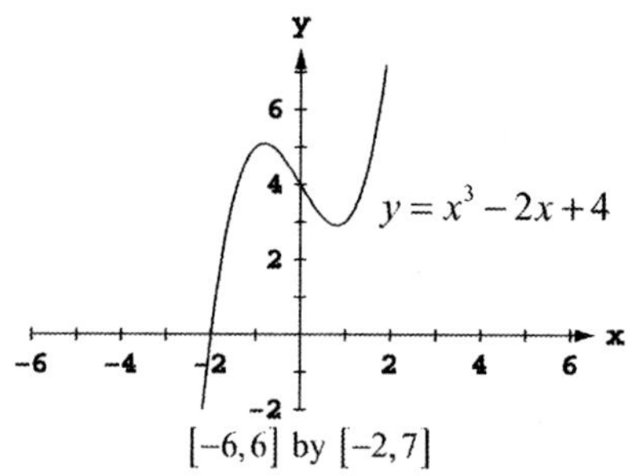

41. To find the exact values, note that that $y' = 3x^2 + 2x - 8$ $= (3x - 4)(x + 2)$, which is zero when $x = -2$ or $x = \frac{4}{3}$. Local maximum at $(-1, 17)$; local minimum at $\left(\frac{4}{3}, -\frac{41}{27}\right)$

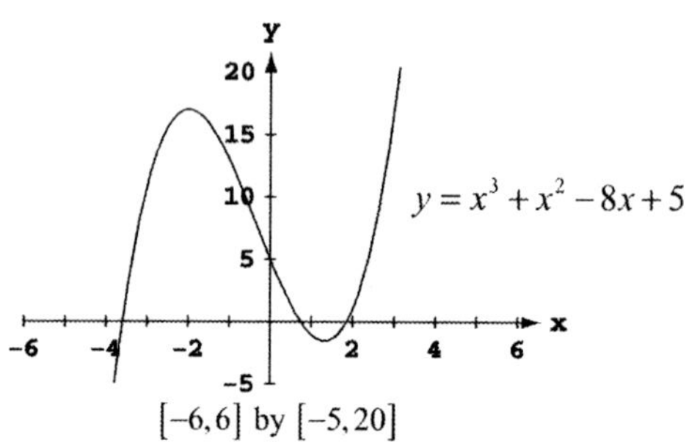

42. Note that $y' = 3x^2 - 6x + 3 = 3(x - 1)^2$, which is zero at $x = 1$. The graph shows that the function assumes lower values to the left and higher values to the right of this point, so the function has no local or global extreme values.

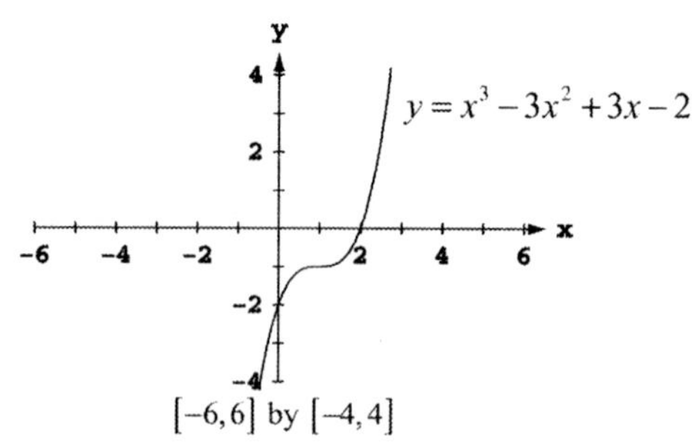

43. Minimum value is 0 when $x = -1$ or $x = 1$.

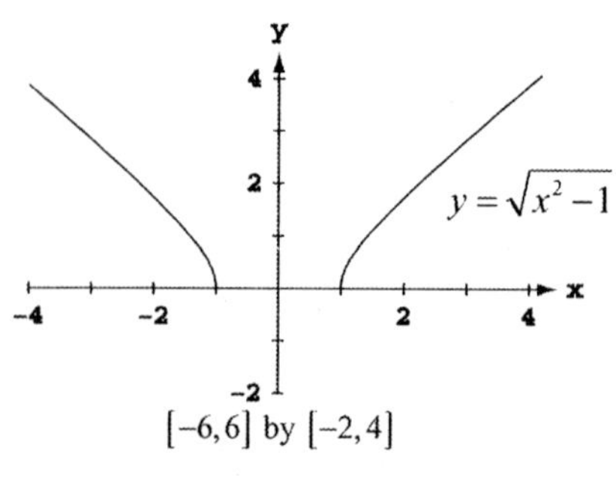

44. The minimum value is 1 at $x = 0$.

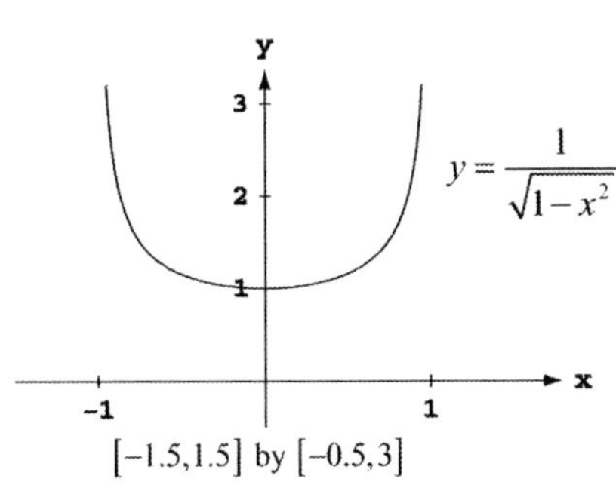

$[-1.5, 1.5]$ by $[-0.5, 3]$

45. The actual graph of the function has asymptotes at $x = \pm 1$, so there are no extrema near these values. (This is an example of grapher failure.) There is a local minimum at $(0, 1)$.

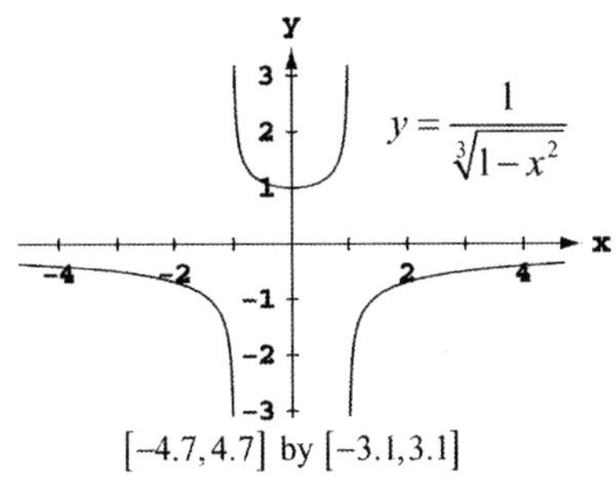

$[-4.7, 4.7]$ by $[-3.1, 3.1]$

46. Maximum value is 2 at $x = 1$;
minimum value is 0 at $x = -1$ and $x = 3$.

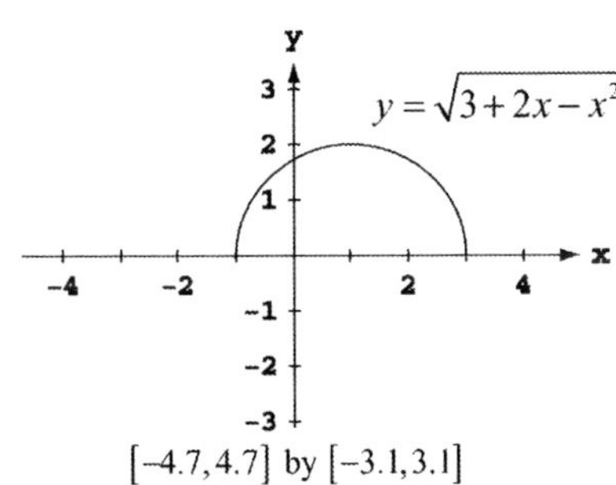

$[-4.7, 4.7]$ by $[-3.1, 3.1]$

47. Maximum value is $\frac{1}{2}$ at $x = 1$;
minimum value is $-\frac{1}{2}$ as $x = -1$.

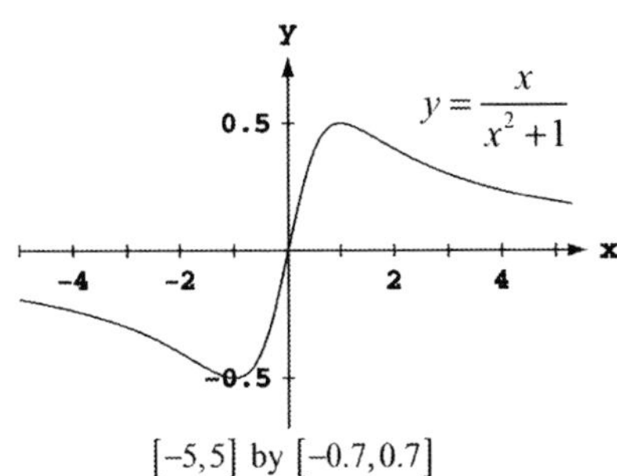

$[-5, 5]$ by $[-0.7, 0.7]$

48. Maximum value is $\frac{1}{2}$ at $x = 0$;
minimum value is $-\frac{1}{2}$ as $x = -2$.

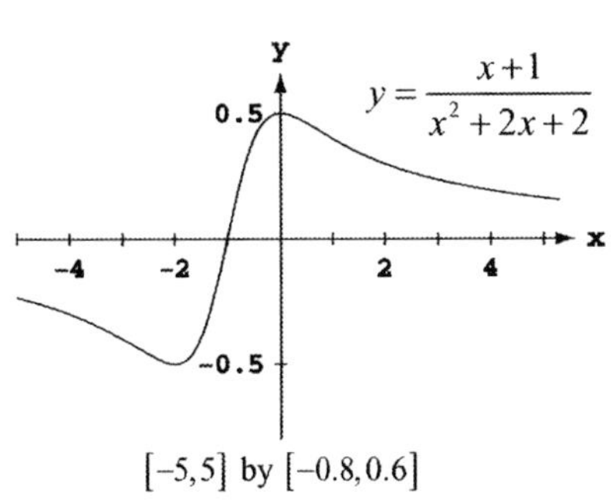

$[-5, 5]$ by $[-0.8, 0.6]$

49. $y = e^x + e^{-x} \Rightarrow y' = e^x - e^{-x} = e^x - \frac{1}{e^x} = \frac{e^{2x}-1}{e^x}$, which is 0 at $x = 0$; an absolute minimum value is 2 at $x = 0$.

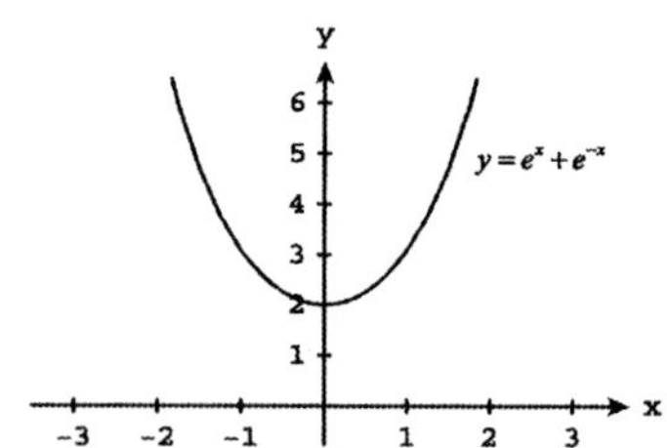

50. $y = e^x - e^{-x} \Rightarrow y' = e^x - (-e^{-x}) = e^x + \frac{1}{e^x} = \frac{e^{2x}+1}{e^x}$, which is never zero; there are no extreme values.

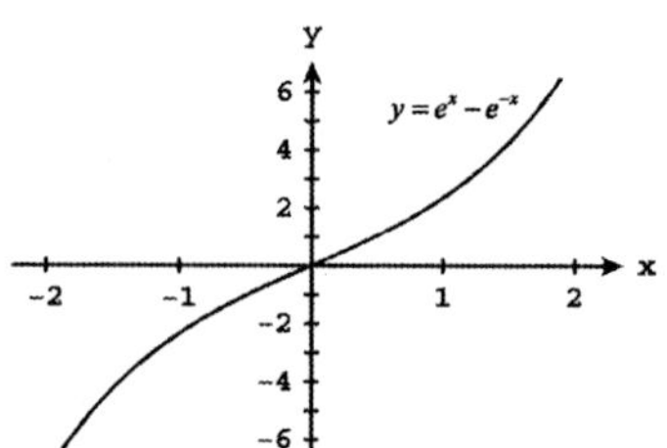

51. $y = x \ln x \Rightarrow y' = x \cdot \frac{1}{x} + (1) \cdot \ln x = 1 + \ln x$, which is zero at $x = e^{-1}$; an absolute minimum value is $-\frac{1}{e}$ at $x = \frac{1}{e}$

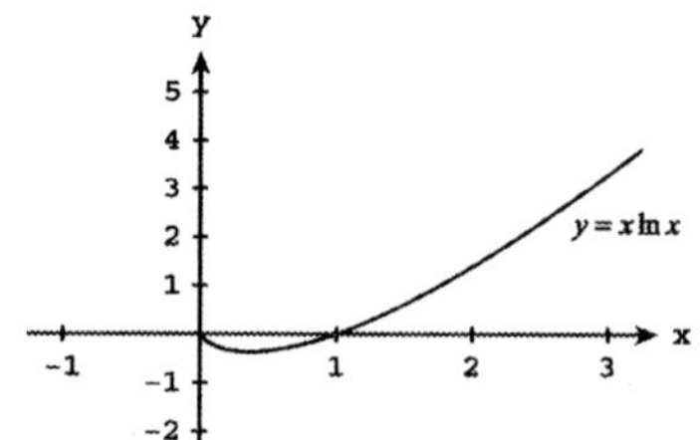

52. $y = x^2 \ln x \Rightarrow y' = x^2 \cdot \frac{1}{x} + 2x \cdot \ln x = x(1 + 2 \ln x)$, which is zero at $x = e^{-1/2}$; an absolute minimum value is $-\frac{1}{2e}$ at $x = \frac{1}{\sqrt{e}}$

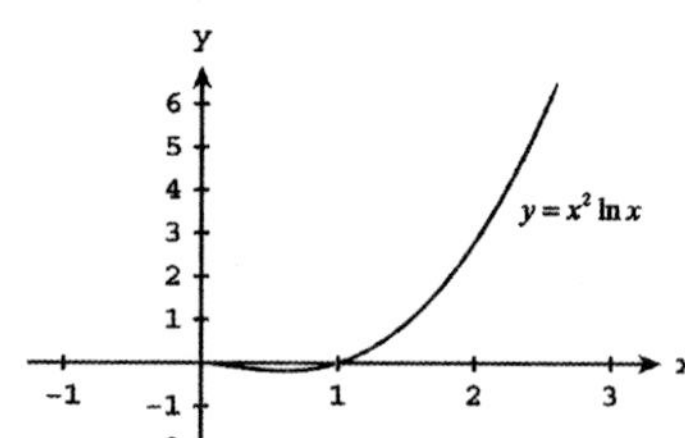

53. $y = \cos^{-1}(x^2) \Rightarrow y' = \frac{-1}{\sqrt{1-(x^2)^2}} \cdot (2x) = \frac{-2x}{\sqrt{1-x^4}}$, which is zero at $x = 0$; an absolute maximum value is $\frac{\pi}{2}$ at $x = 0$; an absolute minimum value is 0 at $x = 1$ and $x = -1$.

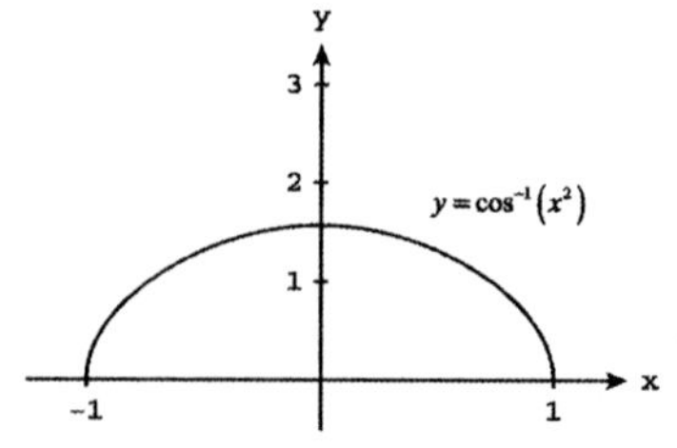

54. $y = \sin^{-1}(e^x) \Rightarrow y' = \frac{1}{\sqrt{1-(e^x)^2}} \cdot (e^x) = \frac{e^x}{\sqrt{1-e^{2x}}}$, which is never zero; an absolute maximum value is $\frac{\pi}{2}$ at $x = 0$.

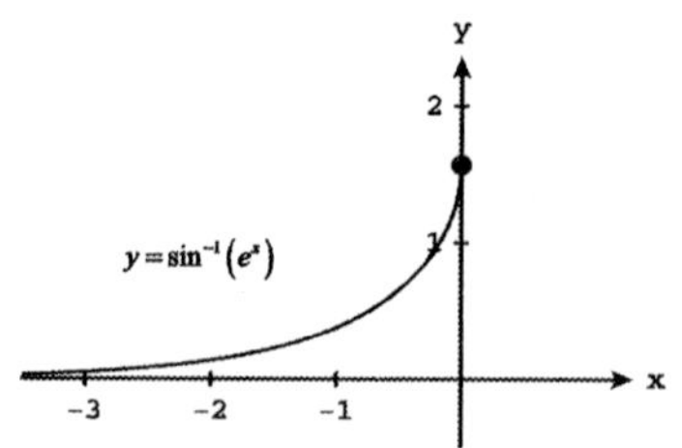

55. $y' = x^{2/3}(1) + \frac{2}{3}x^{-1/3}(x+2) = \frac{5x+4}{3\sqrt[3]{x}}$

crit. pt.	derivative	extremum	value
$x = -\frac{4}{5}$	0	local max	$\frac{12}{25}10^{1/3} = 1.034$
$x = 0$	undefined	local min	0

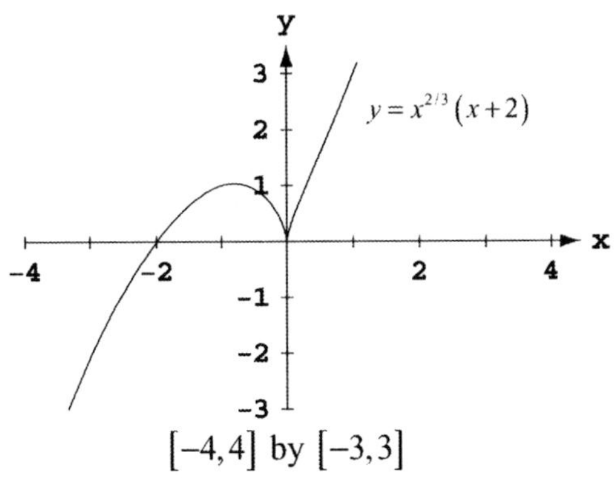

$[-4, 4]$ by $[-3, 3]$

56. $y' = x^{2/3}(2x) + \frac{2}{3}x^{-1/3}(x^2-4) = \frac{8x^2-8}{3\sqrt[3]{x}}$

crit. pt.	derivative	extremum	value
$x = -1$	0	minimum	-3
$x = 0$	undefined	local max	0
$x = 1$	0	minimum	3

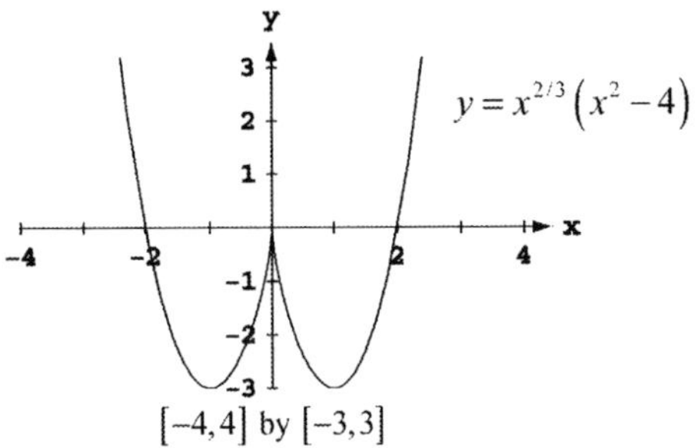

$[-4, 4]$ by $[-3, 3]$

57. $y' = x\frac{1}{2\sqrt{4-x^2}}(-2x) + (1)\sqrt{4-x^2}$

$= \frac{-x^2+(4-x^2)}{\sqrt{4-x^2}} = \frac{4-2x^2}{\sqrt{4-x^2}}$

crit. pt.	derivative	extremum	value
$x = -2$	undefined	local max	0
$x = -\sqrt{2}$	0	minimum	-2
$x = \sqrt{2}$	0	maximum	2
$x = 2$	undefined	local min	0

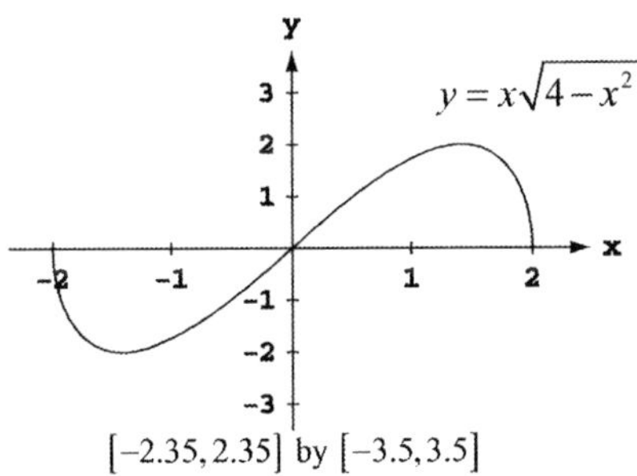

$[-2.35, 2.35]$ by $[-3.5, 3.5]$

58. $y' = x^2\frac{1}{2\sqrt{3-x}}(-1) + 2x\sqrt{3-x}$

$= \frac{-x^2+(4x)(3-x)}{2\sqrt{3-x}} = \frac{-5x^2+12x}{2\sqrt{3-x}}$

crit. pt.	derivative	extremum	value
$x = 0$	0	minimum	0
$x = \frac{12}{5}$	0	local max	$\frac{144}{125}15^{1/2} \approx 4.462$
$x = 3$	undefined	minimum	0

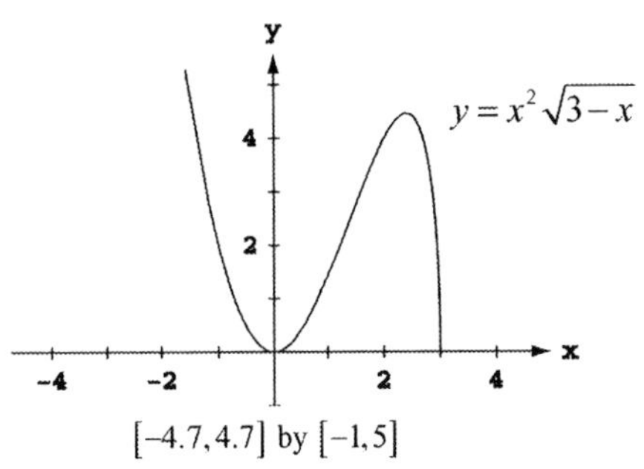

$[-4.7, 4.7]$ by $[-1, 5]$

59. $y' = \begin{cases} -2, & x < 1 \\ 1, & x > 1 \end{cases}$

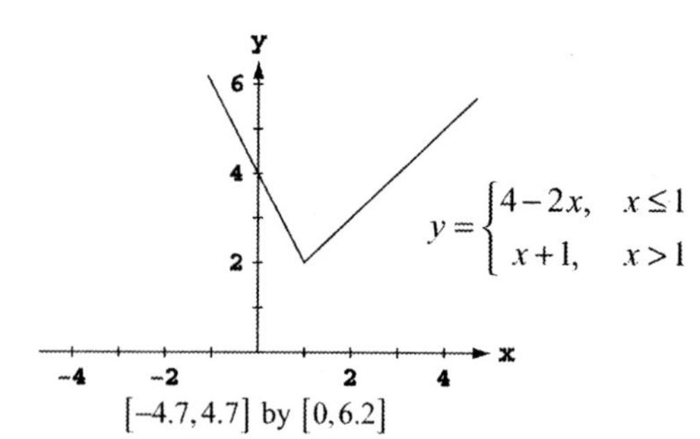

crit. pt.	derivative	extremum	value
x = 1	undefined	minimum	2

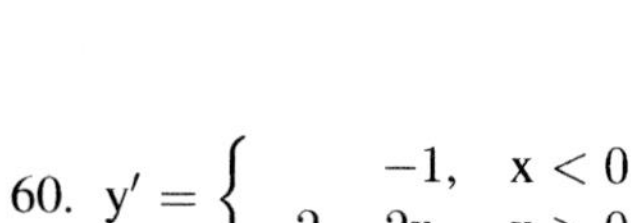

60. $y' = \begin{cases} -1, & x < 0 \\ 2 - 2x, & x > 0 \end{cases}$

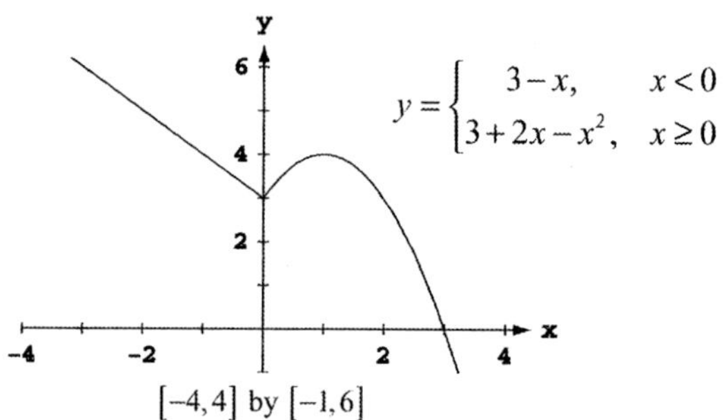

crit. pt.	derivative	extremum	value
x = 0	undefined	local min	3
x = 1	0	local max	4

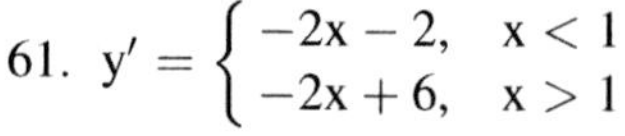

61. $y' = \begin{cases} -2x - 2, & x < 1 \\ -2x + 6, & x > 1 \end{cases}$

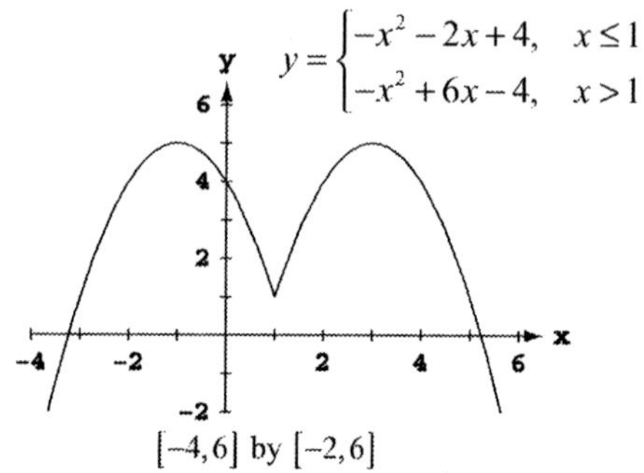

crit. pt.	derivative	extremum	value
x = −1	0	maximum	5
x = 1	undefined	local min	1
x = 3	0	maximum	5

62. We begin by determining whether $f'(x)$ is defined at $x = 1$, where $f(x) = \begin{cases} -\frac{1}{4}x^2 - \frac{1}{2}x + \frac{15}{4}, & x \le 1 \\ x^3 - 6x^2 + 8x, & x > 1 \end{cases}$

Clearly, $f'(x) = -\frac{1}{2}x - \frac{1}{2}$ if $x < 1$, and $\lim_{h\to 0^-} f'(1+h) = -1$. Also, $f'(x) = 3x^2 - 12x + 8$ if $x > 1$, and $\lim_{h\to 0^+} f'(1+h) = -1$. Since f is continuous at $x = 1$, we have that $f'(1) = -1$. Thus,

$$f'(x) = \begin{cases} -\frac{1}{2}x - \frac{1}{2}, & x \le 1 \\ 3x^2 - 12x + 8, & x > 1 \end{cases}$$

Note that $-\frac{1}{2}x - \frac{1}{2} = 0$ when $x = -1$, and $3x^2 - 12x + 8 = 0$ when $x = \frac{12 \pm \sqrt{12^2 - 4(3)(8)}}{2(3)} = \frac{12 \pm \sqrt{48}}{6} = 2 \pm \frac{2\sqrt{3}}{3}$.

But $2 - \frac{2\sqrt{3}}{3} \approx 0.845 < 1$, so the critical points occur at $x = -1$ and $x = 2 + \frac{2\sqrt{3}}{3} \approx 3.155$.

crit. pt.	derivative	extremum	value
$x = -1$	0	local max	4
$x \approx 3.155$	0	local min	≈ -3.079

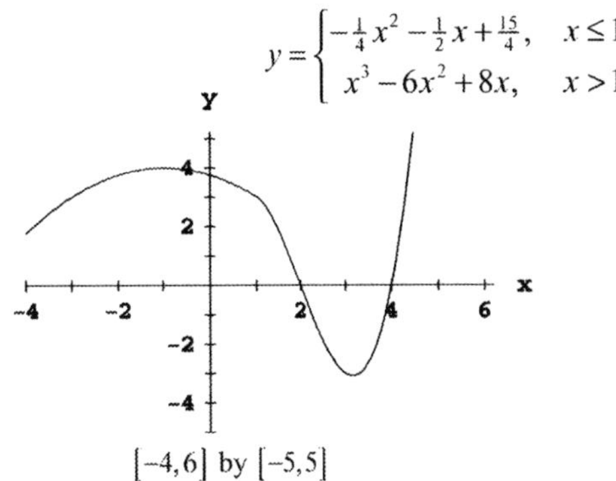

63. (a) No, since $f'(x) = \frac{2}{3}(x-2)^{-1/3}$, which is undefined at $x = 2$.
(b) The derivative is defined and nonzero for all $x \neq 2$. Also, $f(2) = 0$ and $f(x) > 0$ for all $x \neq 2$.
(c) No, $f(x)$ need not have a global maximum because its domain is all real numbers. Any restriction of f to a closed interval of the form [a, b] would have both a maximum value and minimum value on the interval.
(d) The answers are the same as (a) and (b) with 2 replaced by a.

64. Note that $f(x) = \begin{cases} -x^3 + 9x, & x \le -3 \text{ or } 0 \le x < 3 \\ x^3 - 9x, & -3 < x < 0 \text{ or } x \ge 3 \end{cases}$. Therefore, $f'(x) = \begin{cases} -3x^3 + 9, & x < -3 \text{ or } 0 < x < 3 \\ 3x^3 - 9, & -3 < x < 0 \text{ or } x > 3 \end{cases}$.
(a) No, since the left- and right-hand derivatives at $x = 0$, are -9 and 9, respectively.
(b) No, since the left- and right-hand derivatives at $x = 3$, are -18 and 18, respectively.
(c) No, since the left- and right-hand derivatives at $x = -3$, are 18 and -18, respectively.
(d) The critical points occur when $f'(x) = 0$ (at $x = \pm\sqrt{3}$) and when $f'(x)$ is undefined (at $x = 0$ and $x = \pm 3$). The minimum value is 0 at $x = -3$, at $x = 0$, and at $x = 3$; local maxima occur at $\left(-\sqrt{3}, 6\sqrt{3}\right)$ and $\left(\sqrt{3}, 6\sqrt{3}\right)$.

65.

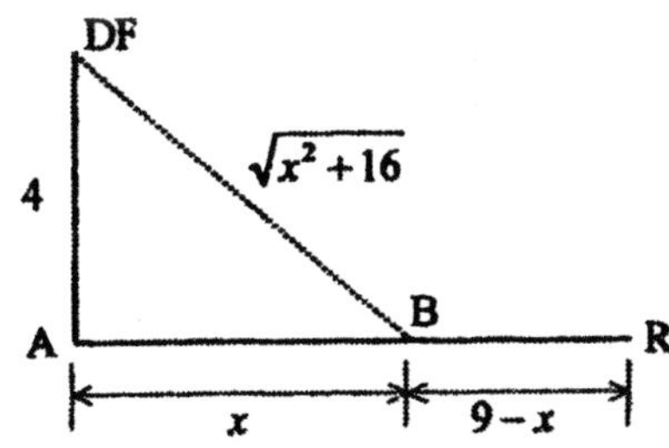

(a) The construction cost is $C(x) = 0.3\sqrt{16 + x^2} + 0.2(9 - x)$ million dollars, where $0 \le x \le 9$ miles. The following is a graph of $C(x)$.

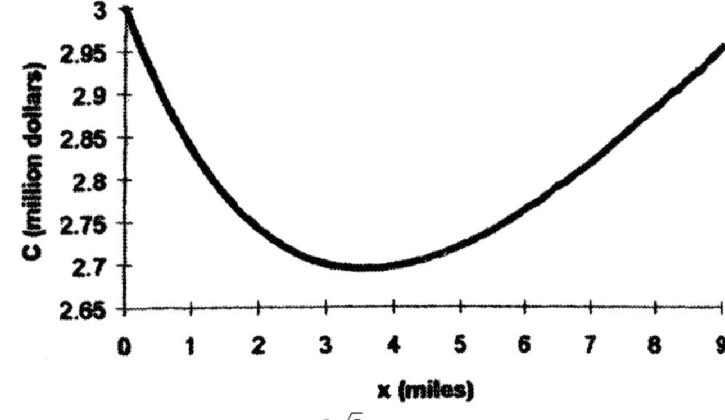

Solving $C'(x) = \frac{0.3x}{\sqrt{16+x^2}} - 0.2 = 0$ gives $x = \pm\frac{8\sqrt{5}}{5} \approx \pm 3.58$ miles, but only $x = 3.58$ miles is a critical point is the specified domain. Evaluating the costs at the critical and endpoints gives $C(0) = \$3$ million, $C\left(\frac{8\sqrt{5}}{5}\right) \approx \$2.694$ million, and $C(9) \approx \$2.955$ million. Therefore, to minimize the cost of construction, the pipeline should be placed from the docking facility to point B, 3.58 miles along the shore from point A, and then along the shore from B to the refinery.

(b) If the per mile cost of underwater construction is p, then $C(x) = p\sqrt{16 + x^2} + 0.2(9 - x)$ and $C'(x) = \frac{0.3x}{\sqrt{16+x^2}} - 0.2 = 0$ gives $x_c = \frac{0.8}{\sqrt{p^2 - 0.04}}$, which minimizes the construction cost provided $x_c \le 9$. The value of p that gives $x_c = 9$ miles is 0.218864. Consequently, if the underwater construction costs \$218,864 per mile or less, then running the pipeline along a straight line directly from the docking facility to the refinery will minimize the cost of construction.

In theory, p would have to be infinite to justify running the pipe directly from the docking facility to point A (i.e., for x_c to be zero). For all values of $p > 0.218864$ there is always an $x_c \in (0, 9)$ that will give a minimum value for C. This is proved by looking at $C''(x_c) = \frac{16p}{(16+x_c^2)^{3/2}}$ which is always positive for $p > 0$.

66. There are two options to consider. The first is to build a new road straight from Village A to Village B. The second is to build a new highway segment from Village A to the Old Road, reconstruct a segment of Old Road, and build a new highway segment from Old Road to Village B, as shown in the figure. The cost of the first option is $C_1 = 0.5(150)$ million dollars = 75 million dollars.

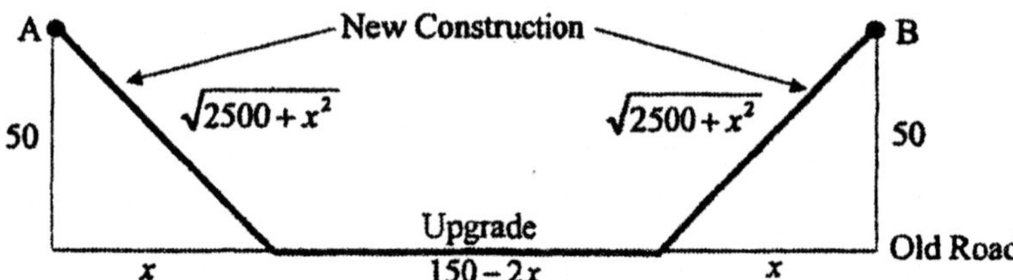

The construction cost for the second option is $C_2(x) = 0.5\left(2\sqrt{2500 + x^2}\right) + 0.3(150 - 2x)$ million dollars for $0 \le x \le 75$ miles. The following is a graph of $C_2(x)$.

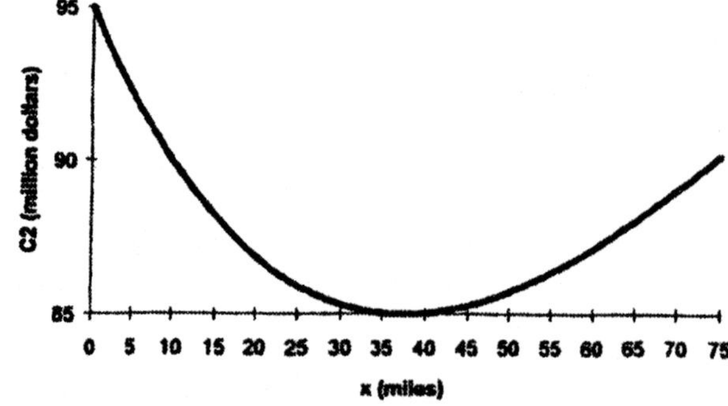

Solving $C_2'(x) = \frac{x}{\sqrt{2500+x^2}} - 0.6 = 0$ give $x = \pm 37.5$ miles, but only $x = 37.5$ miles is in the specified domain. In summary, $C_1 = \$75$ million, $C_2(0) = \$95$ million, $C_2(37.5) = \$85$ million, and $C_2(75) = \$90.139$ million. Consequently, a new road straight from village A to village B is the least expensive option.

67.

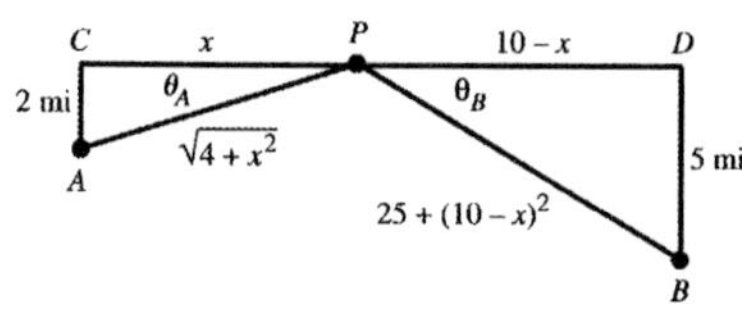

The length of pipeline is $L(x) = \sqrt{4 + x^2} + \sqrt{25 + (10 - x)^2}$ for $0 \le x \le 10$. The following is a graph of $L(x)$.

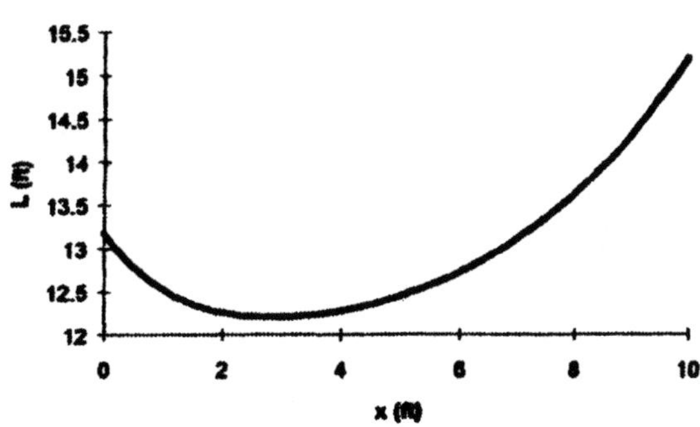

Setting the derivative of $L(x)$ equal to zero gives $L'(x) = \frac{x}{\sqrt{4+x^2}} - \frac{(10-x)}{\sqrt{25+(10-x)^2}} = 0$. Note that $\frac{x}{\sqrt{4+x^2}} = \cos\theta_A$ and $\frac{10-x}{\sqrt{25+(10-x)^2}} = \cos\theta_B$, therefore, $L'(x) = 0$ when $\cos\theta_A = \cos\theta_B$, or $\theta_A = \theta_B$ and $\triangle ACP$ is similar to $\triangle BDP$. Use simple proportions to determine x as follows: $\frac{x}{2} = \frac{10-x}{5} \Rightarrow x = \frac{20}{7} \approx 2.857$ miles along the coast from town A to town B. If the two towns were on opposite sides of the river, the obvious solution would be to place the pump station on a straight line (the shortest distance) between two towns, again forcing $\theta_A = \theta_B$. The shortest length of pipe is the same regardless of whether the towns are on thee same or opposite sides of the river.

68.

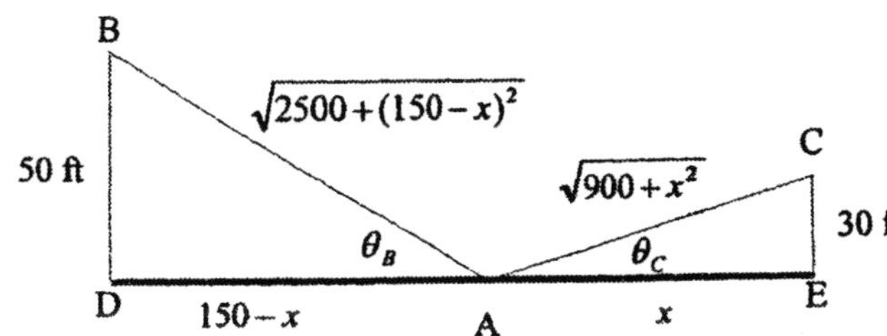

(a) The length of guy wire is $L(x) = \sqrt{900+x^2} + \sqrt{2500+(150-x)^2}$ for $0 \le x \le 150$. The following is a graph of $L(x)$.

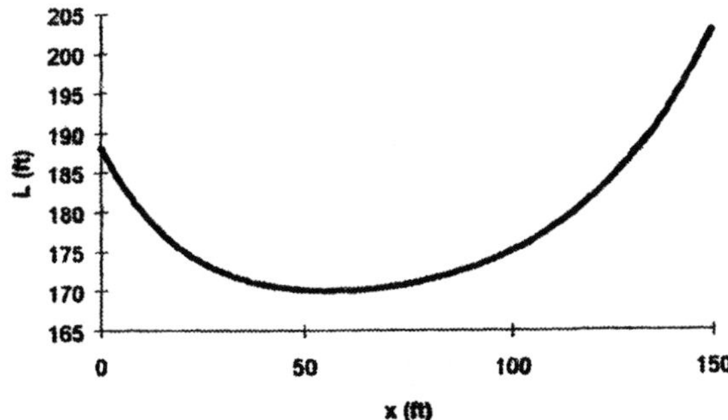

Setting $L'(x)$ equal to zero gives $L'(x) = \frac{x}{\sqrt{900+x^2}} - \frac{(150-x)}{\sqrt{2500+(150-x)^2}} = 0$. Note that $\frac{x}{\sqrt{900+x^2}} = \cos\theta_A$ and $\frac{(150-x)}{\sqrt{2500+(150-x)^2}} = \cos\theta_B$. Therefore, $L'(x) = 0$ when $\cos\theta_A = \cos\theta_B$, or $\theta_A = \theta_B$ and $\triangle ACE$ is similar to $\triangle ABD$.

Use simple proportions to determine x: $\frac{x}{30} = \frac{150-x}{50} \Rightarrow x = \frac{225}{4} = 56.25$ feet.

(b) If the heights of the towers are h_B and h_C, and the horizontal distance between them is s, then $L(x) = \sqrt{h_C^2 + x^2} + \sqrt{h_B^2 + (s-x)^2}$ and $L'(x) = \frac{x}{\sqrt{h_C^2+x^2}} - \frac{(s-x)}{\sqrt{h_B+(s-x)^2}}$. However, $\frac{x}{\sqrt{h_C^2+x^2}} = \cos\theta_C$ and $\frac{(s-x)}{\sqrt{h_B+(s-x)^2}} = \cos\theta_B$. Therefore, $L'(x) = 0$ when $\cos\theta_C = \cos\theta_B$, or $\theta_C = \theta_B$ and $\triangle ACE$ is similar to $\triangle ABD$.

Simple proportions can again be used to determine the optimum x: $\frac{x}{h_c} = \frac{s-x}{h_B} \Rightarrow x = \left(\frac{h_c}{h_B+h_c}\right)s$.

69. (a) $V(x) = 160x - 52x^2 + 4x^3$

$V'(x) = 160 - 104x + 12x^2 = 4(x-2)(3x-20)$

The only critical point in the interval $(0, 5)$ is at $x = 2$. The maximum value of $V(x)$ is 144 at $x = 2$.

(b) The largest possible volume of the box is 144 cubic units, and it occurs when $x = 2$ units.

70. (a) $P'(x) = 2 - 200x^{-2}$

The only critical point in the interval $(0, \infty)$ is at $x = 10$. The minimum value of $P(x)$ is 40 at $x = 10$.

(b) The smallest possible perimeter of the rectangel is 40 units and it occurs at $x = 10$ units which makes the rectangle a 10 by 10 square.

71. Let x represent the length of the base and $\sqrt{25 - x^2}$ the height of the triangle. The area of the triangle is represented by $A(x) = \frac{x}{2}\sqrt{25 - x^2}$ where $0 \le x \le 5$. Consequently, solving $A'(x) = 0 \Rightarrow \frac{25 - 2x^2}{2\sqrt{25 - x^2}} = 0 \Rightarrow x = \frac{5}{\sqrt{2}}$. Since $A(0) = A(5) = 0$, $A(x)$ is maximized at $x = \frac{5}{\sqrt{2}}$. The largest possible area is $A\left(\frac{5}{\sqrt{2}}\right) = \frac{25}{4}$ cm^2.

72. (a) From the diagram the perimeter $P = 2x + 2\pi r = 400$
$\Rightarrow x = 200 - \pi r$. The area A is $2rx$
$\Rightarrow A(r) = 400r - 2\pi r^2$ where $0 \le r \le \frac{200}{\pi}$.

(b) $A'(r) = 400 - 4\pi r$ so the only critical point is $r = \frac{100}{\pi}$. Since $A(r) = 0$ if $r = 0$ and $x = 200 - \pi r = 0$, the values $r = \frac{100}{\pi} \approx 31.83$ m and $x = 100$ m maximize the area over the interval $0 \le r \le \frac{200}{\pi}$.

73. $s = -\frac{1}{2}gt^2 + v_0 t + s_0 \Rightarrow \frac{ds}{dt} = -gt + v_0 = 0 \Rightarrow t = \frac{v_0}{g}$. Now $s(t) = s_0 \Leftrightarrow t\left(-\frac{gt}{2} + v_0\right) = 0 \Leftrightarrow t = 0$ or $t = \frac{2v_0}{g}$. Thus $s\left(\frac{v_0}{g}\right) = -\frac{1}{2}g\left(\frac{v_0}{g}\right)^2 + v_0\left(\frac{v_0}{g}\right) + s_0 = \frac{v_0^2}{2g} + s_0 > s_0$ is the maximum height over the interval $0 \le t \le \frac{2v_0}{g}$.

74. $\frac{dI}{dt} = -2\sin t + 2\cos t$, solving $\frac{dI}{dt} = 0 \Rightarrow \tan t = 1 \Rightarrow t = \frac{\pi}{4} + n\pi$ where n is a nonnegative integer (in this exercise t is never negative) $\Rightarrow$ the peak current is $2\sqrt{2}$ amps.

75. Yes, since $f(x) = |x| = \sqrt{x^2} = (x^2)^{1/2} \Rightarrow f'(x) = \frac{1}{2}(x^2)^{-1/2}(2x) = \frac{x}{(x^2)^{1/2}} = \frac{x}{|x|}$ is not defined at $x = 0$. Thus it is not required that f' be zero at a local extreme point since f' may be undefined there.

76. If $f(c)$ is a local maximum value of f, then $f(x) \le f(c)$ for all x in some open interval (a, b) containing c. Since f is even, $f(-x) = f(x) \le f(c) = f(-c)$ for all $-x$ in the open interval $(-b, -a)$ containing $-c$. That is, f assumes a local maximum at the point $-c$. This is also clear from the graph of f because the graph of an even function is symmetric about the y-axis.

77. If $g(c)$ is a local minimum value of g, then $g(x) \ge g(c)$ for all x in some open interval (a, b) containing c. Since g is odd, $g(-x) = -g(x) \le -g(c) = g(-c)$ for all $-x$ in the open interval $(-b, -a)$ containing $-c$. That is, g assumes a local maximum at the point $-c$. This is also clear from the graph of g because the graph of an odd function is symmetric about the origin.

78. If there are no boundary points or critical points the function will have no extreme values in its domain. Such functions do indeed exist, for example $f(x) = x$ for $-\infty < x < \infty$. (Any other linear function $f(x) = mx + b$ with $m \ne 0$ will do as well.)

79. (a) $f'(x) = 3ax^2 + 2bx + c$ is a quadratic, so it can have 0, 1, or 2 zeros, which would be the critical points of f. The function $f(x) = x^3 - 3x$ has two critical points at $x = -1$ and $x = 1$. The function $f(x) = x^3 - 1$ has one critical point at $x = 0$. The function $f(x) = x^3 + x$ has no critical points.

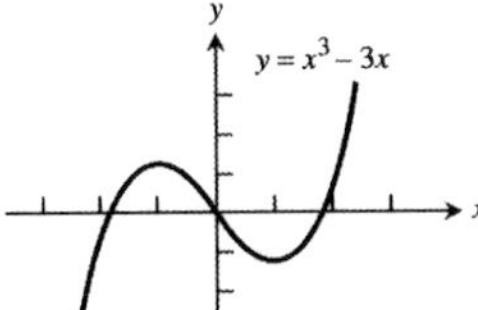

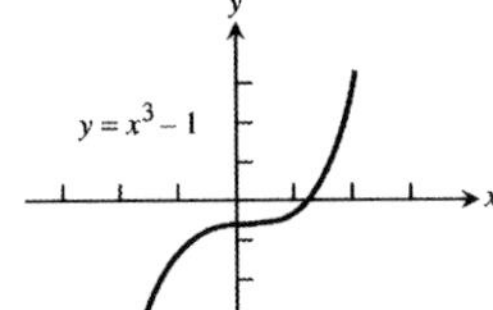

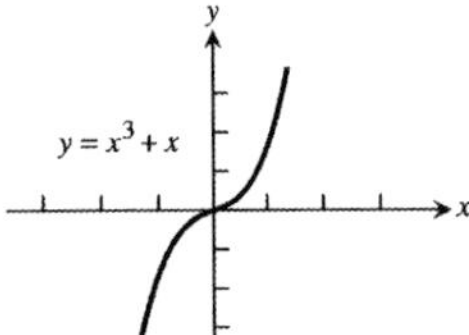

(b) The function can have either two local extreme values or no extreme values. (If there is only one critical point, the cubic function has no extreme values.)

80. (a)

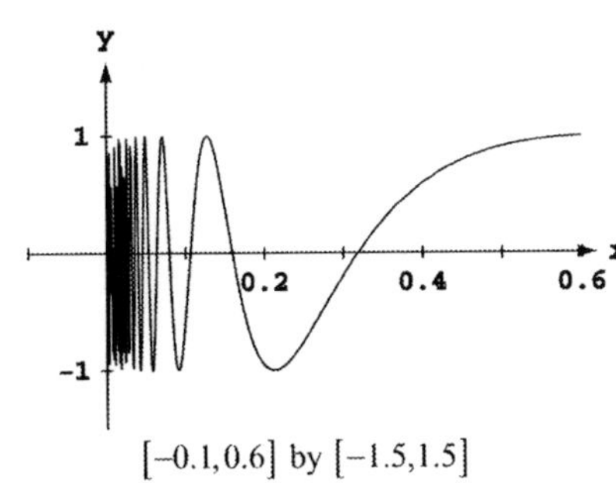

$[-0.1, 0.6]$ by $[-1.5, 1.5]$

$f(0) = 0$ is not a local extreme value because in any open interval containing $x = 0$, there are infinitely many points where $f(x) = 1$ and where $f(x) = -1$.

(b) One possible answer, on the interval $[0, 1]$:

$$f(x) = \begin{cases} (1 - x)\cos \frac{1}{1-x}, & 0 \le x < 1 \\ 0, & x = 1 \end{cases}$$

This function has no local extreme value at $x = 1$. Note that it is continuous on $[0, 1]$.

81. Maximum value is 11 at $x = 5$;
minimum value is 5 on the interval $[-3, 2]$;
local maximum at $(-5, 9)$

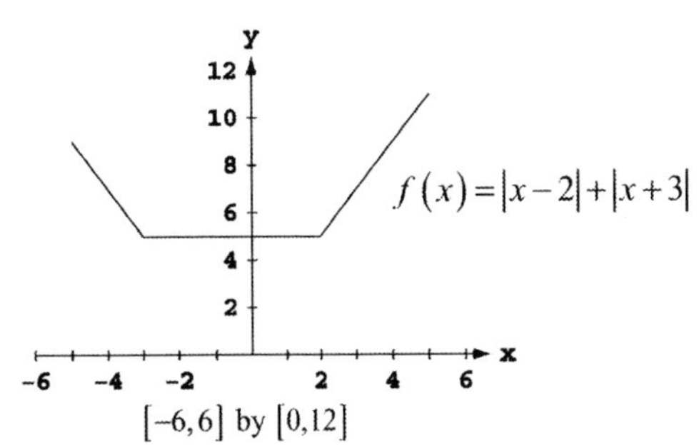

$[-6, 6]$ by $[0, 12]$

82. Maximum value is 4 on the interval $[5, 7]$;
minimum value is -4 on the interval $[-2, 1]$.

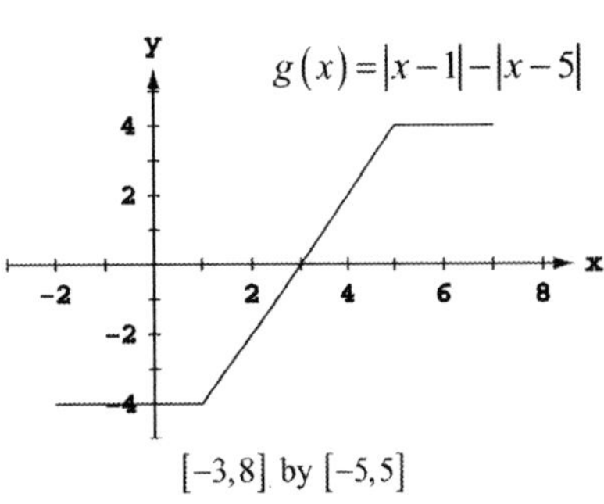

$[-3, 8]$ by $[-5, 5]$

83. Maximum value is 5 on the interval $[3, \infty)$;
minimum value is -5 on the interval $(-\infty, -2]$.

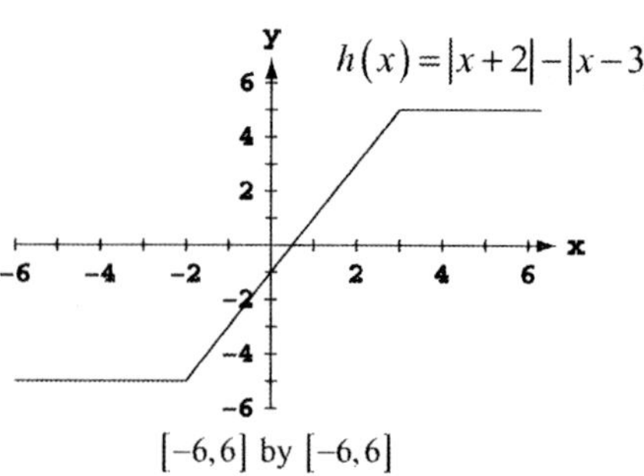

$[-6,6]$ by $[-6,6]$

84. Minimum value is 4 on the interval $[-1, 3]$

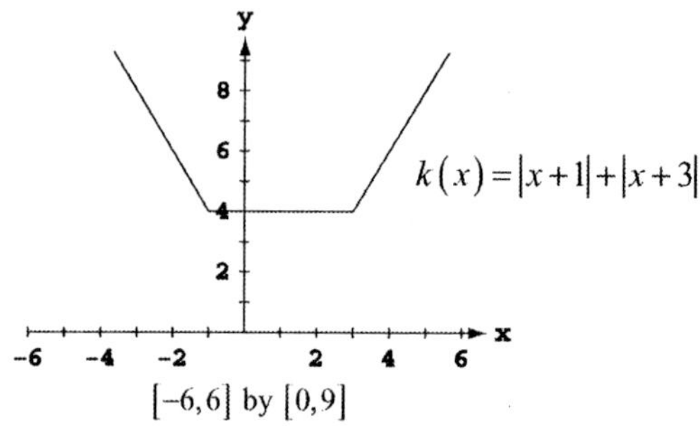

$[-6,6]$ by $[0,9]$

85-92. Example CAS commands:

Maple:

```
with(student):
f := x -> x^4 - 8*x^2 + 4*x + 2;
domain := x=-20/25..64/25;
plot( f(x), domain, color=black, title="Section 4.1 #85(a)" );
Df := D(f);
plot( Df(x), domain, color=black, title="Section 4.1 # 85(b)" )
StatPt := fsolve( Df(x)=0, domain )
SingPt := NULL;
EndPt := op(rhs(domain));
Pts :=evalf([EndPt,StatPt,SingPt]);
Values := [seq( f(x), x=Pts )];
```

Maximum value is 2.7608 and occurs at x=2.56 (right endpoint).

Minimum value [34] is -6.2680 and occurs at x=1.86081 (singular point).

Mathematica: (functions may vary) (see section 2.5 re. RealsOnly):

```
<<Miscellaneous `RealOnly`
Clear[f,x]
a = −1; b = 10/3;
f[x_] =2 + 2x − 3 x^(2/3)
f'[x]
Plot[{f[x], f'[x]}, {x, a, b}]
NSolve[f'[x]==0, x]
{f[a], f[0], f[x]/.%, f[b]//N
```

In more complicated expressions, NSolve may not yield results. In this case, an approximate solution (say 1.1 here) is observed from the graph and the following command is used:

```
FindRoot[f'[x]==0,{x, 1.1}]
```

4.2 THE MEAN VALUE THEOREM

1. When $f(x) = x^2 + 2x - 1$ for $0 \le x \le 1$, then $\frac{f(1) - f(0)}{1 - 0} = f'(c) \Rightarrow 3 = 2c + 2 \Rightarrow c = \frac{1}{2}$.

2. When $f(x) = x^{2/3}$ for $0 \le x \le 1$, then $\frac{f(1) - f(0)}{1 - 0} = f'(c) \Rightarrow 1 = \left(\frac{2}{3}\right) c^{-1/3} \Rightarrow c = \frac{8}{27}$.

3. When $f(x) = \sin^{-1}(x)$ for $-1 \le x \le 1$, then $f'(c) = \frac{f(1) - f(-1)}{1 - (-1)} \Rightarrow \frac{1}{\sqrt{1 - c^2}} = \frac{\frac{\pi}{2} - \left(-\frac{\pi}{2}\right)}{2} \Rightarrow \sqrt{1 - c^2} = \frac{2}{\pi} \Rightarrow 1 - c^2 = \frac{4}{\pi^2}$ $\Rightarrow c^2 = 1 - \frac{4}{\pi^2} \Rightarrow c = \pm\sqrt{1 - \frac{4}{\pi^2}} \approx \pm 0.771$

4. When $f(x) = \ln(x - 1)$ for $2 \le x \le 4$, then $f'(c) = \frac{f(4) - f(2)}{4 - 2} \Rightarrow \frac{1}{c - 1} = \frac{\ln 3 - \ln 1}{2} \Rightarrow c - 1 = \frac{2}{\ln 3} \Rightarrow c = 1 + \frac{2}{\ln 3} \approx 2.820$

5. Does not; f(x) is not differentiable at $x = 0$ in $(-1, 8)$.

6. Does; f(x) is continuous for every point of [0, 1] and differentiable for every point in (0, 1).

7. Does; f(x) is continuous for every point of [0, 1] and differentiable for every point in (0, 1).

8. Does not; f(x) is not continuous at $x = 0$ because $\lim_{x \to 0^-} f(x) = 1 \ne 0 = f(0)$.

9. Since f(x) is not continuous on $0 \le x \le 1$, Rolle's Theorem does not apply: $\lim_{x \to 1^-} f(x) = \lim_{x \to 1^-} x = 1$ $\ne 0 = f(1)$.

10. Since f(x) must be continuous at $x = 0$ and $x = 1$ we have $\lim_{x \to 0^+} f(x) = a = f(0) \Rightarrow a = 3$ and $\lim_{x \to 1^-} f(x) = \lim_{x \to 1^+} f(x) \Rightarrow -1 + 3 + a = m + b \Rightarrow 5 = m + b$. Since f(x) must also be differentiable at $x = 1$ we have $\lim_{x \to 1^-} f'(x) = \lim_{x \to 1^+} f'(x) \Rightarrow -2x + 3\big|_{x=1} = m\big|_{x=1} \Rightarrow 1 = m$. Therefore, $a = 3$, $m = 1$ and $b = 4$.

11. (a) i ii iii iv

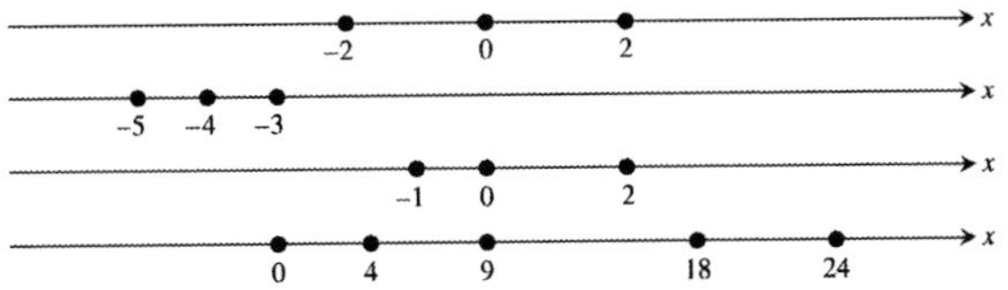

(b) Let r_1 and r_2 be zeros of the polynomial $P(x) = x^n + a_{n-1}x^{n-1} + \ldots + a_1x + a_0$, then $P(r_1) = P(r_2) = 0$. Since polynomials are everywhere continuous and differentiable, by Rolle's Theorem $P'(r) = 0$ for some r between r_1 and r_2, where $P'(x) = nx^{n-1} + (n-1)\,a_{n-1}x^{n-2} + \ldots + a_1$.

12. With f both differentiable and continuous on [a, b] and $f(r_1) = f(r_2) = f(r_3) = 0$ where r_1, r_2 and r_3 are in [a, b], then by Rolle's Theorem there exists a c_1 between r_1 and r_2 such that $f'(c_1) = 0$ and a c_2 between r_2 and r_3 such that $f'(c_2) = 0$. Since f' is both differentiable and continuous on [a, b], Rolle's Theorem again applies and we have a c_3 between c_1 and c_2 such that $f''(c_3) = 0$. To generalize, if f has n+1 zeros in [a, b] and $f^{(n)}$ is continuous on [a, b], then $f^{(n)}$ has at least one zero between a and b.

13. Since f'' exists throughout [a, b] the derivative function f' is continuous there. If f' has more than one zero in [a, b], say $f'(r_1) = f'(r_2) = 0$ for $r_1 \ne r_2$, then by Rolle's Theorem there is a c between r_1 and r_2 such that $f''(c) = 0$, contrary to $f'' > 0$ throughout [a, b]. Therefore f' has at most one zero in [a, b]. The same argument holds if $f'' < 0$ throughout [a, b].

14. If f(x) is a cubic polynomial with four or more zeros, then by Rolle's Theorem $f'(x)$ has three or more zeros, $f''(x)$ has 2 or more zeros and $f'''(x)$ has at least one zero. This is a contradiction since $f'''(x)$ is a non-zero constant when f(x) is a cubic polynomial.

15. With $f(-2) = 11 > 0$ and $f(-1) = -1 < 0$ we conclude from the Intermediate Value Theorem that $f(x) = x^4 + 3x + 1$ has at least one zero between -2 and -1. Then $-2 < x < -1 \Rightarrow -8 < x^3 < -1$ $\Rightarrow -32 < 4x^3 < -4 \Rightarrow -29 < 4x^3 + 3 < -1 \Rightarrow f'(x) < 0$ for $-2 < x < -1 \Rightarrow$ f(x) is decreasing on $[-2, -1]$ $\Rightarrow f(x) = 0$ has exactly one solution in the interval $(-2, -1)$.

16. $f(x) = x^3 + \frac{4}{x^2} + 7 \Rightarrow f'(x) = 3x^2 - \frac{8}{x^3} > 0$ on $(-\infty, 0) \Rightarrow$ f(x) is increasing on $(-\infty, 0)$. Also, $f(x) < 0$ if $x < -2$ and $f(x) > 0$ if $-2 < x < 0 \Rightarrow$ f(x) has exactly one zero in $(-\infty, 0)$.

17. $g(t) = \sqrt{t} + \sqrt{t+1} - 4 \Rightarrow g'(t) = \frac{1}{2\sqrt{t}} + \frac{1}{2\sqrt{t+1}} > 0 \Rightarrow$ g(t) is increasing for t in $(0, \infty)$; $g(3) = \sqrt{3} - 2 < 0$ and $g(15) = \sqrt{15} > 0 \Rightarrow$ g(t) has exactly one zero in $(0, \infty)$.

18. $g(t) = \frac{1}{1-t} + \sqrt{1+t} - 3.1 \Rightarrow g'(t) = \frac{1}{(1-t)^2} + \frac{1}{2\sqrt{1+t}} > 0 \Rightarrow$ g(t) is increasing for t in $(-1, 1)$; $g(-0.99) = -2.5$ and $g(0.99) = 98.3 \Rightarrow$ g(t) has exactly one zero in $(-1, 1)$.

19. $r(\theta) = \theta + \sin^2\left(\frac{\theta}{3}\right) - 8 \Rightarrow r'(\theta) = 1 + \frac{2}{3}\sin\left(\frac{\theta}{3}\right)\cos\left(\frac{\theta}{3}\right) = 1 + \frac{1}{3}\sin\left(\frac{2\theta}{3}\right) > 0$ on $(-\infty, \infty) \Rightarrow r(\theta)$ is increasing on $(-\infty, \infty)$; $r(0) = -8$ and $r(8) = \sin^2\left(\frac{8}{3}\right) > 0 \Rightarrow r(\theta)$ has exactly one zero in $(-\infty, \infty)$.

20. $r(\theta) = 2\theta - \cos^2\theta + \sqrt{2} \Rightarrow r'(\theta) = 2 + 2\sin\theta\cos\theta = 2 + \sin 2\theta > 0$ on $(-\infty, \infty) \Rightarrow r(\theta)$ is increasing on $(-\infty, \infty)$; $r(-2\pi) = -4\pi - \cos(-2\pi) + \sqrt{2} = -4\pi - 1 + \sqrt{2} < 0$ and $r(2\pi) = 4\pi - 1 + \sqrt{2} > 0 \Rightarrow r(\theta)$ has exactly one zero in $(-\infty, \infty)$.

21. $r(\theta) = \sec\theta - \frac{1}{\theta^3} + 5 \Rightarrow r'(\theta) = (\sec\theta)(\tan\theta) + \frac{3}{\theta^4} > 0$ on $\left(0, \frac{\pi}{2}\right) \Rightarrow r(\theta)$ is increasing on $\left(0, \frac{\pi}{2}\right)$; $r(0.1) \approx -994$ and $r(1.57) \approx 1260.5 \Rightarrow r(\theta)$ has exactly one zero in $\left(0, \frac{\pi}{2}\right)$.

22. $r(\theta) = \tan\theta - \cot\theta - \theta \Rightarrow r'(\theta) = \sec^2\theta + \csc^2\theta - 1 = \sec^2\theta + \cot^2\theta > 0$ on $\left(0, \frac{\pi}{2}\right) \Rightarrow r(\theta)$ is increasing on $\left(0, \frac{\pi}{2}\right)$; $r\left(\frac{\pi}{4}\right) = -\frac{\pi}{4} < 0$ and $r(1.57) \approx 1254.2 \Rightarrow r(\theta)$ has exactly one zero in $\left(0, \frac{\pi}{2}\right)$.

23. By Corollary 1, $f'(x) = 0$ for all x $\Rightarrow f(x) = C$, where C is a constant. Since $f(-1) = 3$ we have $C = 3$ $\Rightarrow f(x) = 3$ for all x.

24. $g(x) = 2x + 5 \Rightarrow g'(x) = 2 = f'(x)$ for all x. By Corollary 2, $f(x) = g(x) + C$ for some constant C. Then $f(0) = g(0) + C \Rightarrow 5 = 5 + C \Rightarrow C = 0 \Rightarrow f(x) = g(x) = 2x + 5$ for all x.

25. $g(x) = x^2 \Rightarrow g'(x) = 2x = f'(x)$ for all x. By Corollary 2, $f(x) = g(x) + C$.
(a) $f(0) = 0 \Rightarrow 0 = g(0) + C = 0 + C \Rightarrow C = 0 \Rightarrow f(x) = x^2 \Rightarrow f(2) = 4$
(b) $f(1) = 0 \Rightarrow 0 = g(1) + C = 1 + C \Rightarrow C = -1 \Rightarrow f(x) = x^2 - 1 \Rightarrow f(2) = 3$
(c) $f(-2) = 3 \Rightarrow 3 = g(-2) + C \Rightarrow 3 = 4 + C \Rightarrow C = -1 \Rightarrow f(x) = x^2 - 1 \Rightarrow f(2) = 3$

26. $g(x) = mx \Rightarrow g'(x) = m$, a constant. If $f'(x) = m$, then by Corollary 2, $f(x) = g(x) + b = mx + b$ where b is a constant. Therefore all functions whose derivatives are constant can be graphed as straight lines $y = mx + b$.

27. (a) $y = \frac{x^2}{2} + C$ (b) $y = \frac{x^3}{3} + C$ (c) $y = \frac{x^4}{4} + C$

28. (a) $y = x^2 + C$ (b) $y = x^2 - x + C$ (c) $y = x^3 + x^2 - x + C$

29. (a) $y = \ln\theta + C$ if $\theta > 0$ and $y = \ln(-\theta) + C$ if $\theta < 0$, where C is a constant. (These functions can be combined as $y = \ln|\theta| + C$.)
(a) $y = \theta - \ln\theta + C$ if $\theta > 0$ and $y = \theta - \ln(-\theta) + C$ if $\theta < 0$, where C is a constant. (These functions can be combined as $y = \theta - \ln|\theta| + C$.)
(a) $y = 5\theta + \ln\theta + C$ if $\theta > 0$ and $y = 5\theta + \ln(-\theta) + C$ if $\theta < 0$, where C is a constant. (These functions can be combined as $y = 5\theta + \ln|\theta| + C$.)

30. (a) $y' = \frac{1}{2}x^{-1/2} \Rightarrow y = x^{1/2} + C \Rightarrow y = \sqrt{x} + C$ (b) $y = 2\sqrt{x} + C$
(c) $y = 2x^2 - 2\sqrt{x} + C$

31. (a) $y = -\frac{1}{2}\cos 2t + C$ (b) $y = 2\sin\frac{t}{2} + C$
(c) $y = -\frac{1}{2}\cos 2t + 2\sin\frac{t}{2} + C$

32. (a) $y = \tan\theta + C$ (b) $y' = \theta^{1/2} \Rightarrow y = \frac{2}{3}\theta^{3/2} + C$ (c) $y = \frac{2}{3}\theta^{3/2} - \tan\theta + C$

33. $f(x) = x^2 - x + C;\ 0 = f(0) = 0^2 - 0 + C \Rightarrow C = 0 \Rightarrow f(x) = x^2 - x$

34. $g(x) = \begin{Bmatrix} x^2 + \ln x + C & \text{if } x > 0 \\ x^2 + \ln(-x) + C & \text{if } x < 0 \end{Bmatrix} = x^2 + \ln|x| + C;\ g(1) = -1 \Rightarrow 1^2 + \ln 1 + C = -1 \Rightarrow C = -2$
$\Rightarrow g(x) = x^2 + \ln|x| - 2$

35. $f(x) = \frac{e^{2x}}{2} + C;\ f(0) = \frac{3}{2} \Rightarrow \frac{e^{2(0)}}{2} + C = \frac{3}{2} \Rightarrow C = 1 \Rightarrow f(x) = 1 + \frac{e^{2x}}{2}$

36. $r(t) = \sec t - t + C;\ 0 = r(0) = \sec(0) - 0 + C \Rightarrow C = -1 \Rightarrow r(t) = \sec t - t - 1$

37. $v = \frac{ds}{dt} = 9.8t + 5 \Rightarrow s = 4.9t^2 + 5t + C$; at $s = 10$ and $t = 0$ we have $C = 10 \Rightarrow s = 4.9t^2 + 5t + 10$

38. $v = \frac{ds}{dt} = 32t - 2 \Rightarrow s = 16t^2 - 2t + C$; at $s = 4$ and $t = \frac{1}{2}$ we have $C = 1 \Rightarrow s = 6t^2 - 2t + 1$

39. $v = \frac{ds}{dt} = \sin(\pi t) \Rightarrow s = -\frac{1}{\pi}\cos(\pi t) + C$; at $s = 0$ and $t = 0$ we have $C = \frac{1}{\pi} \Rightarrow s = \frac{1 - \cos(\pi t)}{\pi}$

40. $v = \frac{ds}{dt} = \frac{1}{t+2} \Rightarrow s = \ln(t+2) + C$; at $s = \frac{1}{2}$ and $t = -1$ we have $C = \frac{1}{2} \Rightarrow s = \frac{1}{2} + \ln(t+2)$

41. $a = \frac{dv}{dt} = e^t \Rightarrow v = e^t + C$; at $v = 20$ and $t = 0$ we have $C = 19 \Rightarrow v = e^t + 19$
$v = \frac{ds}{dt} = e^t + 19 \Rightarrow s = e^t + 19t + C$; at $s = 5$ and $t = 0$ we have $C = 4 \Rightarrow s = e^t + 19t + 4$

42. $a = 9.8 \Rightarrow v = 9.8t + C_1$; at $v = -3$ and $t = 0$ we have $C_1 = -3 \Rightarrow v = 9.8t - 3 \Rightarrow s = 4.9t^2 - 3t + C_2$; at $s = 0$ and $t = 0$ we have $C_2 = 0 \Rightarrow s = 4.9t^2 - 3t$

43. $a = -4\sin(2t) \Rightarrow v = 2\cos(2t) + C_1$; at $v = 2$ and $t = 0$ we have $C_1 = 0 \Rightarrow v = 2\cos(2t) \Rightarrow s = \sin(2t) + C_2$; at $s = -3$ and $t = 0$ we have $C_2 = -3 \Rightarrow s = \sin(2t) - 3$

44. $a = \frac{9}{\pi^2}\cos\left(\frac{3t}{\pi}\right) \Rightarrow v = \frac{3}{\pi}\sin\left(\frac{3t}{\pi}\right) + C_1$; at $v = 0$ and $t = 0$ we have $C_1 = 0 \Rightarrow v = \frac{3}{\pi}\sin\left(\frac{3t}{\pi}\right) \Rightarrow s = -\cos\left(\frac{3t}{\pi}\right) + C_2$; at $s = -1$ and $t = 0$ we have $C_2 = 0 \Rightarrow s = -\cos\left(\frac{3t}{\pi}\right)$

45. If T(t) is the temperature of the thermometer at time t, then $T(0) = -19°$ C and $T(14) = 100°$ C. From the Mean Value Theorem there exists a $0 < t_0 < 14$ such that $\frac{T(14) - T(0)}{14 - 0} = 8.5°$ C/sec $= T'(t_0)$, the rate at which the temperature was changing at $t = t_0$ as measured by the rising mercury on the thermometer.

46. Because the trucker's average speed was 79.5 mph, by the Mean Value Theorem, the trucker must have been going that speed at least once during the trip.

47. Because its average speed was approximately 7.667 knots, and by the Mean Value Theorem, it must have been going that speed at least once during the trip.

48. The runner's average speed for the marathon was approximately 11.909 mph. Therefore, by the Mean Value Theorem, the runner must have been going that speed at least once during the marathon. Since the initial speed and final speed are both 0 mph and the runner's speed is continuous, by the Intermediate Value Theorem, the runner's speed must have been 11 mph at least twice.

49. Let d(t) represent the distance the automobile traveled in time t. The average speed over $0 \le t \le 2$ is $\frac{d(2) - d(0)}{2 - 0}$. The Mean Value Theorem says that for some $0 < t_0 < 2$, $d'(t_0) = \frac{d(2) - d(0)}{2 - 0}$. The value $d'(t_0)$ is the speed of the automobile at time t_0 (which is read on the speedometer).

50. $a(t) = v'(t) = 1.6 \Rightarrow v(t) = 1.6t + C$; at $(0, 0)$ we have $C = 0 \Rightarrow v(t) = 1.6t$. When $t = 30$, then $v(30) = 48$ m/sec.

51. The conclusion of the Mean Value Theorem yields $\frac{\frac{1}{b} - \frac{1}{a}}{b - a} = -\frac{1}{c^2} \Rightarrow c^2\left(\frac{a - b}{ab}\right) = a - b \Rightarrow c = \sqrt{ab}$.

52. The conclusion of the Mean Value Theorem yields $\frac{b^2 - a^2}{b - a} = 2c \Rightarrow c = \frac{a+b}{2}$.

53. $f'(x) = [\cos x \sin(x + 2) + \sin x \cos(x + 2)] - 2\sin(x + 1)\cos(x + 1) = \sin(x + x + 2) - \sin 2(x + 1)$
$= \sin(2x + 2) - \sin(2x + 2) = 0$. Therefore, the function has the constant value $f(0) = -\sin^2 1 \approx -0.7081$ which explains why the graph is a horizontal line.

54. (a) $f(x) = (x + 2)(x + 1)x(x - 1)(x - 2) = x^5 - 5x^3 + 4x$ is one possibility.
(b) Graphing $f(x) = x^5 - 5x^3 + 4x$ and $f'(x) = 5x^4 - 15x^2 + 4$ on $[-3, 3]$ by $[-7, 7]$ we see that each x-intercept of $f'(x)$ lies between a pair of x-intercepts of f(x), as expected by Rolle's Theorem.

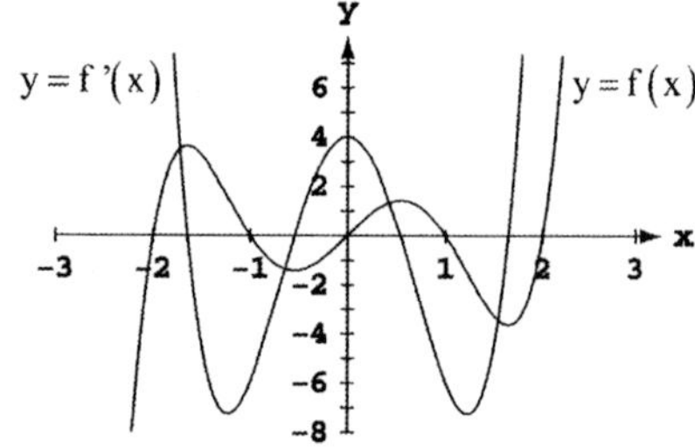

(c) Yes, since sin is continuous and differentiable on $(-\infty, \infty)$.

55. f(x) must be zero at least once between a and b by the Intermediate Value Theorem. Now suppose that f(x) is zero twice between a and b. Then by the Mean Value Theorem, $f'(x)$ would have to be zero at least once between the two zeros of

f(x), but this can't be true since we are given that $f'(x) \neq 0$ on this interval. Therefore, f(x) is zero once and only once between a and b.

56. Consider the function $k(x) = f(x) - g(x)$. k(x) is continuous and differentiable on [a, b], and since $k(a) = f(a) - g(a)$ and $k(b) = f(b) - g(b)$, by the Mean Value Theorem, there must be a point c in (a, b) where $k'(c) = 0$. But since $k'(c) = f'(c) - g'(c)$, this means that $f'(c) = g'(c)$, and c is a point where the graphs of f and g have tangent lines with the same slope, so these lines are either parallel or are the same line.

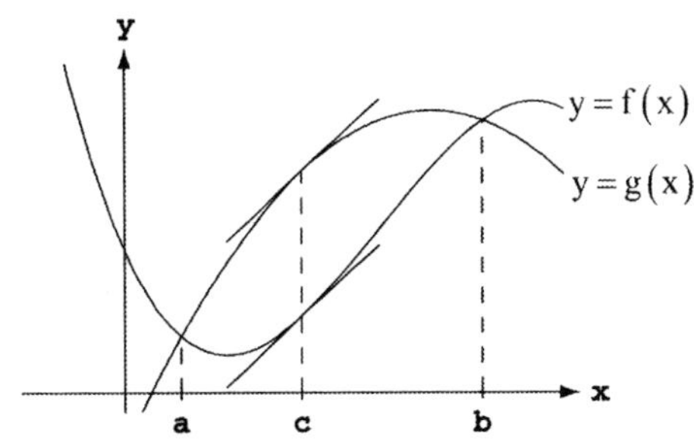

57. Yes. By Corollary 2 we have $f(x) = g(x) + c$ since $f'(x) = g'(x)$. If the graphs start at the same point $x = a$, then $f(a) = g(a) \Rightarrow c = 0 \Rightarrow f(x) = g(x)$.

58. Let $f(x) = \sin x$ for $a \le x \le b$. From the Mean Value Theorem there exists a c between a and b such that $\frac{\sin b - \sin a}{b - a} = \cos c \Rightarrow -1 \le \frac{\sin b - \sin a}{b - a} \le 1 \Rightarrow \left|\frac{\sin b - \sin a}{b - a}\right| \le 1 \Rightarrow |\sin b - \sin a| \le |b - a|$.

59. By the Mean Value Theorem we have $\frac{f(b) - f(a)}{b - a} = f'(c)$ for some point c between a and b. Since $b - a > 0$ and $f(b) < f(a)$, we have $f(b) - f(a) < 0 \Rightarrow f'(c) < 0$.

60. The condition is that f' should be continuous over [a, b]. The Mean Value Theorem then guarantees the existence of a point c in (a, b) such that $\frac{f(b) - f(a)}{b - a} = f'(c)$. If f' is continuous, then it has a minimum and maximum value on [a, b], and $\min f' \le f'(c) \le \max f'$, as required.

61. $f'(x) = (1 + x^4 \cos x)^{-1} \Rightarrow f''(x) = -(1 + x^4 \cos x)^{-2}(4x^3 \cos x - x^4 \sin x)$
$= -x^3(1 + x^4 \cos x)^{-2}(4 \cos x - x \sin x) < 0$ for $0 \le x \le 0.1 \Rightarrow f'(x)$ is decreasing when $0 \le x \le 0.1$
$\Rightarrow \min f' \approx 0.9999$ and $\max f' = 1$. Now we have $0.9999 \le \frac{f(0.1) - 1}{0.1} \le 1 \Rightarrow 0.09999 \le f(0.1) - 1 \le 0.1$
$\Rightarrow 1.09999 \le f(0.1) \le 1.1$.

62. $f'(x) = (1 - x^4)^{-1} \Rightarrow f''(x) = -(1 - x^4)^{-2}(-4x^3) = \frac{4x^3}{(1 - x^4)^3} > 0$ for $0 < x \le 0.1 \Rightarrow f'(x)$ is increasing when $0 \le x \le 0.1 \Rightarrow \min f' = 1$ and $\max f' = 1.0001$. Now we have $1 \le \frac{f(0.1) - 2}{0.1} \le 1.0001$
$\Rightarrow 0.1 \le f(0.1) - 2 \le 0.10001 \Rightarrow 2.1 \le f(0.1) \le 2.10001$.

63. (a) Suppose $x < 1$, then by the Mean Value Theorem $\frac{f(x) - f(1)}{x - 1} < 0 \Rightarrow f(x) > f(1)$. Suppose $x > 1$, then by the Mean Value Theorem $\frac{f(x) - f(1)}{x - 1} > 0 \Rightarrow f(x) > f(1)$. Therefore $f(x) \ge 1$ for all x since $f(1) = 1$.
(b) Yes. From part (a), $\lim_{x \to 1^-} \frac{f(x) - f(1)}{x - 1} \le 0$ and $\lim_{x \to 1^+} \frac{f(x) - f(1)}{x - 1} \ge 0$. Since $f'(1)$ exists, these two one-sided limits are equal and have the value $f'(1) \Rightarrow f'(1) \le 0$ and $f'(1) \ge 0 \Rightarrow f'(1) = 0$.

64. From the Mean Value Theorem we have $\frac{f(b) - f(a)}{b - a} = f'(c)$ where c is between a and b. But $f'(c) = 2pc + q = 0$ has only one solution $c = -\frac{q}{2p}$. (Note: $p \neq 0$ since f is a quadratic function.)

65. Proof that $\ln\left(\frac{b}{x}\right) = \ln b - \ln x$: Note that $\frac{d}{dx}\ln\left(\frac{b}{x}\right) = \frac{1}{b/x} \cdot \frac{-b}{x^2} = \frac{-1}{x}$ and $\frac{d}{dx}(\ln b - \ln x) = \frac{-1}{x}$; by Corollary 2 of the Mean Value Theorem there is a constant C so that $\ln\left(\frac{b}{x}\right) = \ln b - \ln x + C$; if $x = b$, then $\ln 1 = \ln b - \ln b + C \Rightarrow C = 0$
$\Rightarrow \ln\left(\frac{b}{x}\right) = \ln b - \ln x$.

66. (a) $\frac{d}{dx}(\tan^{-1}x + \cot^{-1}x) = \frac{1}{1+x^2} + \frac{-1}{1+x^2} = 0 \Rightarrow$ by Corollary 2 of the Mean Value Theorem that $\tan^{-1}x + \cot^{-1}x = C$ for some constant C; if x = 1, then $\tan^{-1}1 + \cot^{-1}1 = \frac{\pi}{4} + \frac{\pi}{4} = \frac{\pi}{2} = C \Rightarrow \tan^{-1}x + \cot^{-1}x = \frac{\pi}{2}$.

(b) $\frac{d}{dx}(\sec^{-1}x + \csc^{-1}x) = \frac{1}{|x|\sqrt{x^2-1}} + \frac{-1}{|x|\sqrt{x^2-1}} = 0 \Rightarrow$ by Corollary 2 of the Mean Value Theorem that $\sec^{-1}x + \csc^{-1}x = C$ for some constant C; if $x = \sqrt{2}$, then $\sec^{-1}\sqrt{2} + \csc^{-1}\sqrt{2} = \frac{\pi}{4} + \frac{\pi}{4} = \frac{\pi}{2} = C$ $\Rightarrow \sec^{-1}x + \csc^{-1}x = \frac{\pi}{2}$.

67. $e^x e^{-x} = e^{(x-x)} = e^0 = 1 \Rightarrow e^{-x} = \frac{1}{e^x}$ for all x; $\frac{e^{x_1}}{e^{x_2}} = e^{x_1}\left(\frac{1}{e^{x_2}}\right) = e^{x_1}e^{-x_2} = e^{x_1 - x_2}$

68. $y = (e^{x_1})^{x_2} \Rightarrow \ln y = x_2 \ln e^{x_1} = x_2x_1 = x_1x_2 \Rightarrow e^{\ln y} = e^{x_1x_2} \Rightarrow y = e^{x_1x_2} \Rightarrow (e^{x_1})^{x_2} = e^{x_1x_2}$

4.3 MONOTONIC FUNCTIONS AND THE FIRST DERIVATIVE TEST

1. (a) $f'(x) = x(x-1) \Rightarrow$ critical points at 0 and 1
   (b) $f' = +++ \underset{0}{|} --- \underset{1}{|} +++ \Rightarrow$ increasing on $(-\infty, 0)$ and $(1, \infty)$, decreasing on $(0, 1)$
   (c) Local maximum at x = 0 and a local minimum at x = 1

2. (a) $f'(x) = (x-1)(x+2) \Rightarrow$ critical points at -2 and 1
   (b) $f' = +++ \underset{-2}{|} --- \underset{1}{|} +++ \Rightarrow$ increasing on $(-\infty, -2)$ and $(1, \infty)$, decreasing on $(-2, 1)$
   (c) Local maximum at $x = -2$ and a local minimum at x = 1

3. (a) $f'(x) = (x-1)^2(x+2) \Rightarrow$ critical points at -2 and 1
   (b) $f' = --- \underset{-2}{|} +++ \underset{1}{|} +++ \Rightarrow$ increasing on $(-2, 1)$ and $(1, \infty)$, decreasing on $(-\infty, -2)$
   (c) No local maximum and a local minimum at $x = -2$

4. (a) $f'(x) = (x-1)^2(x+2)^2 \Rightarrow$ critical points at -2 and 1
   (b) $f' = +++ \underset{-2}{|} +++ \underset{1}{|} +++ \Rightarrow$ increasing on $(-\infty, -2) \cup (-2, 1) \cup (1, \infty)$, never decreasing
   (c) No local extrema

5. (a) $f'(x) = (x-1)e^{-x} \Rightarrow$ critical point at x = 1
   (b) $f' = ---- \underset{1}{|} ++++ \Rightarrow$ decreasing on $(-\infty, 1]$, increasing on $[1, \infty)$
   (c) Local (and absolute) minimum at x = 1

6. (a) $f'(x) = (x-7)(x+1)(x+5) \Rightarrow$ critical points at -5, -1 and 7
   (b) $f' = --- \underset{-5}{|} +++ \underset{-1}{|} --- \underset{7}{|} +++ \Rightarrow$ increasing on $(-5, -1)$ and $(7, \infty)$, decreasing on $(-\infty, -5)$ and $(-1, 7)$
   (c) Local maximum at $x = -1$, local minima at $x = -5$ and $x = 7$

7. (a) $f'(x) = x^{-1/3}(x+2) \Rightarrow$ critical points at -2 and 0
   (b) $f' = +++ \underset{-2}{|} --- \underset{0}{)(} +++ \Rightarrow$ increasing on $(-\infty, -2)$ and $(0, \infty)$, decreasing on $(-2, 0)$
   (c) Local maximum at $x = -2$, local minimum at x = 0

8. (a) $f'(x) = x^{-1/2}(x-3) \Rightarrow$ critical points at 0 and 3
   (b) $f' = \underset{0}{(} --- \underset{3}{|} +++ \Rightarrow$ increasing on $(3, \infty)$, decreasing on $(0, 3)$
   (c) No local maximum and a local minimum at x = 3

9. (a) $g(t) = -t^2 - 3t + 3 \Rightarrow g'(t) = -2t - 3 \Rightarrow$ a critical point at $t = -\frac{3}{2}$; $g' = +++\underset{-3/2}{|}---$, increasing on $\left(-\infty, -\frac{3}{2}\right)$, decreasing on $\left(-\frac{3}{2}, \infty\right)$

(b) local maximum value of $g\left(-\frac{3}{2}\right) = \frac{21}{4}$ at $t = -\frac{3}{2}$

(c) absolute maximum is $\frac{21}{4}$ at $t = -\frac{3}{2}$

(d)

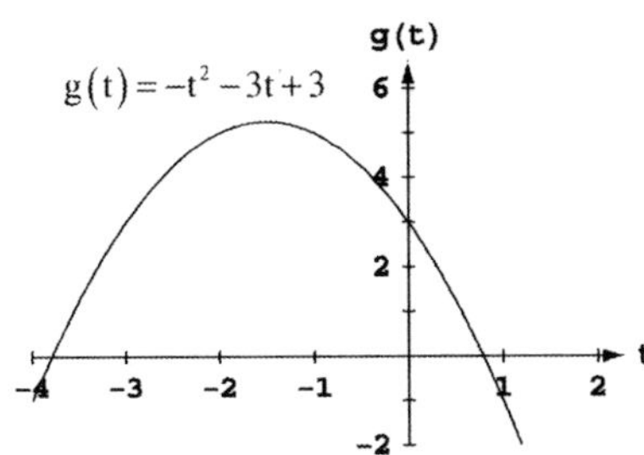

10. (a) $g(t) = -3t^2 + 9t + 5 \Rightarrow g'(t) = -6t + 9 \Rightarrow$ a critical point at $t = \frac{3}{2}$; $g' = +++\underset{3/2}{|}---$, increasing on $\left(-\infty, \frac{3}{2}\right)$, decreasing on $\left(\frac{3}{2}, \infty\right)$

(b) local maximum value of $g\left(\frac{3}{2}\right) = \frac{47}{4}$ at $t = \frac{3}{2}$

(c) absolute maximum is $\frac{47}{4}$ at $t = \frac{3}{2}$

(d)

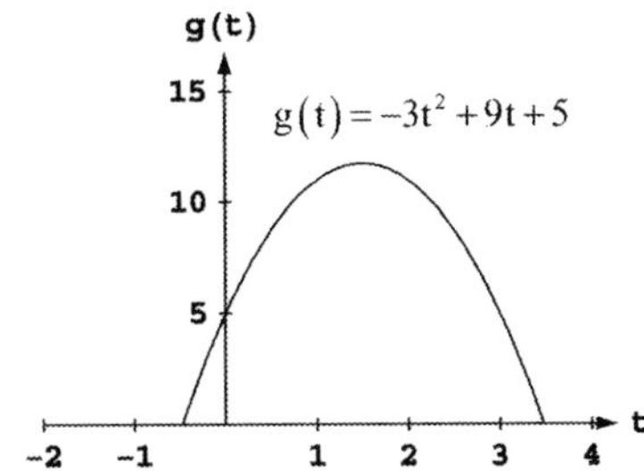

11. (a) $h(x) = -x^3 + 2x^2 \Rightarrow h'(x) = -3x^2 + 4x = x(4 - 3x) \Rightarrow$ critical points at $x = 0, \frac{4}{3}$
$\Rightarrow h' = ---\underset{0}{|}+++\underset{4/3}{|}---$, increasing on $\left(0, \frac{4}{3}\right)$, decreasing on $(-\infty, 0)$ and $\left(\frac{4}{3}, \infty\right)$

(b) local maximum value of $h\left(\frac{4}{3}\right) = \frac{32}{27}$ at $x = \frac{4}{3}$; local minimum value of $h(0) = 0$ at $x = 0$

(c) no absolute extrema

(d)

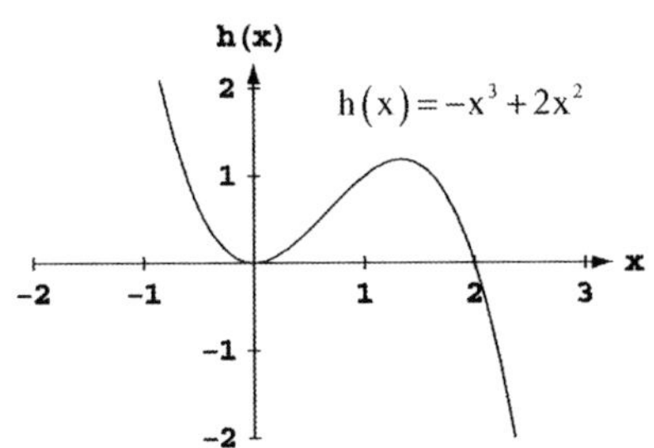

12. (a) $h(x) = 2x^3 - 18x \Rightarrow h'(x) = 6x^2 - 18 = 6\left(x + \sqrt{3}\right)\left(x - \sqrt{3}\right) \Rightarrow$ critical points at $x = \pm\sqrt{3}$
$\Rightarrow h' = +++\underset{-\sqrt{3}}{|}---\underset{\sqrt{3}}{|}+++$, increasing on $\left(-\infty, -\sqrt{3}\right)$ and $\left(\sqrt{3}, \infty\right)$, decreasing on $\left(-\sqrt{3}, \sqrt{3}\right)$

(b) a local maximum is $h\left(-\sqrt{3}\right) = 12\sqrt{3}$ at $x = -\sqrt{3}$; local minimum is $h\left(\sqrt{3}\right) = -12\sqrt{3}$ at $x = \sqrt{3}$

(c) no absolute extrema

(d)

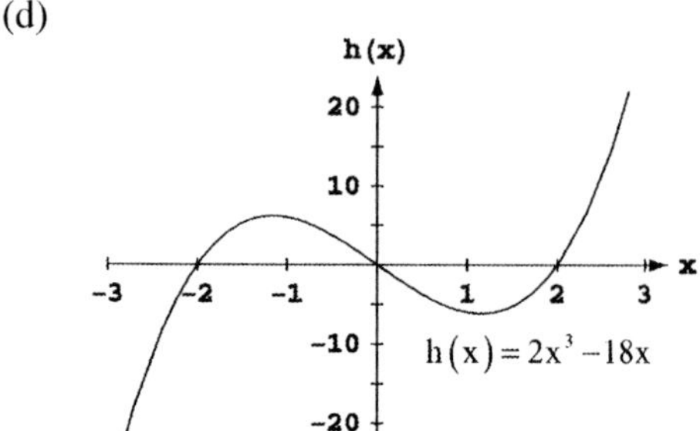

13. (a) $f(\theta) = 3\theta^2 - 4\theta^3 \Rightarrow f'(\theta) = 6\theta - 12\theta^2 = 6\theta(1 - 2\theta) \Rightarrow$ critical points at $\theta = 0, \frac{1}{2} \Rightarrow f' = ---\underset{0}{|}+++\underset{1/2}{|}---$, increasing on $\left(0, \frac{1}{2}\right)$, decreasing on $(-\infty, 0)$ and $\left(\frac{1}{2}, \infty\right)$

(b) a local maximum is $f\left(\frac{1}{2}\right) = \frac{1}{4}$ at $\theta = \frac{1}{2}$, a local minimum is $f(0) = 0$ at $\theta = 0$

(c) no absolute extrema

(d)

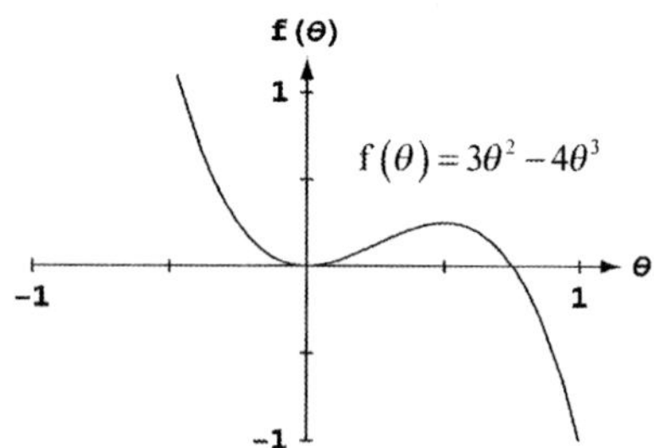

14. (a) $f(\theta) = 6\theta - \theta^3 \Rightarrow f'(\theta) = 6 - 3\theta^2 = 3\left(\sqrt{2} - \theta\right)\left(\sqrt{2} + \theta\right) \Rightarrow$ critical points at $\theta = \pm\sqrt{2} \Rightarrow$ $f' = ---\underset{-\sqrt{2}}{|}+++\underset{\sqrt{2}}{|}---$, increasing on $\left(-\sqrt{2}, \sqrt{2}\right)$, decreasing on $\left(-\infty, -\sqrt{2}\right)$ and $\left(\sqrt{2}, \infty\right)$

(b) a local maximum is $f\left(\sqrt{2}\right) = 4\sqrt{2}$ at $\theta = \sqrt{2}$, a local minimum is $f\left(-\sqrt{2}\right) = -4\sqrt{2}$ at $\theta = -\sqrt{2}$

(c) no absolute extrema

(d)

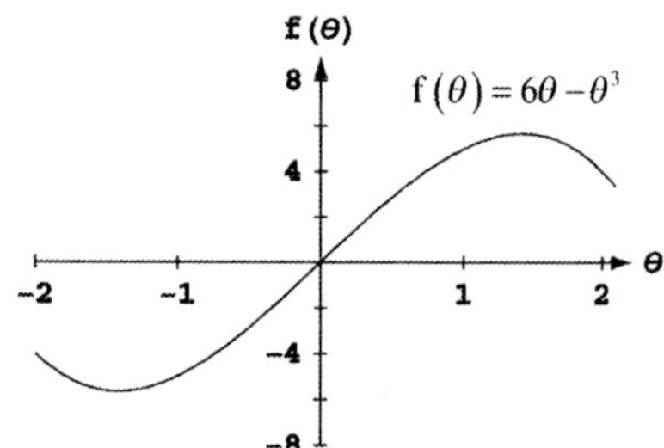

15. (a) $f(r) = 3r^3 + 16r \Rightarrow f'(r) = 9r^2 + 16 \Rightarrow$ no critical points $\Rightarrow f' = +++++$, increasing on $(-\infty, \infty)$, never decreasing

(b) no local extrema

(c) no absolute extrema

(d)

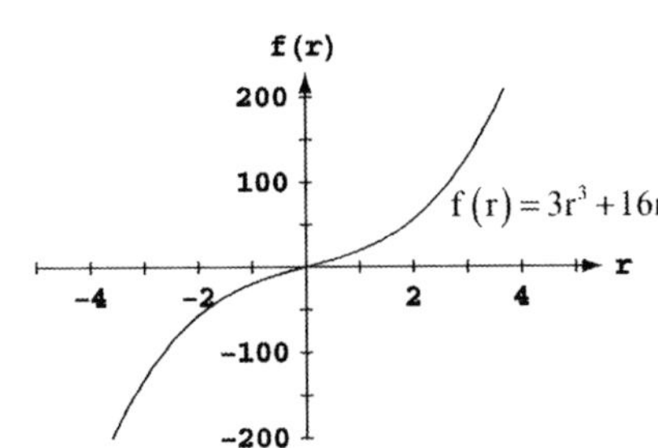

16. (a) $h(r) = (r+7)^3 \Rightarrow h'(r) = 3(r+7)^2 \Rightarrow$ a critical point at $r = -7 \Rightarrow h' = \underset{-7}{+++|+++}$, increasing on $(-\infty, -7) \cup (-7, \infty)$, never decreasing

(b) no local extrema

(c) no absolute extrema

(d)

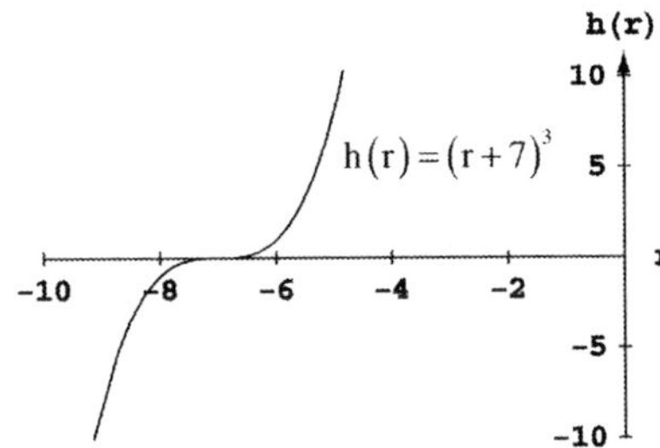

17. (a) $f(x) = x^4 - 8x^2 + 16 \Rightarrow f'(x) = 4x^3 - 16x = 4x(x+2)(x-2) \Rightarrow$ critical points at $x = 0$ and $x = \pm 2$ $\Rightarrow f' = \underset{-2}{---|}\underset{0}{+++|}\underset{2}{---|}+++$, increasing on $(-2, 0)$ and $(2, \infty)$, decreasing on $(-\infty, -2)$ and $(0, 2)$

(b) a local maximum is $f(0) = 16$ at $x = 0$, local minima are $f(\pm 2) = 0$ at $x = \pm 2$

(c) no absolute maximum; absolute minimum is 0 at $x = \pm 2$

(d)

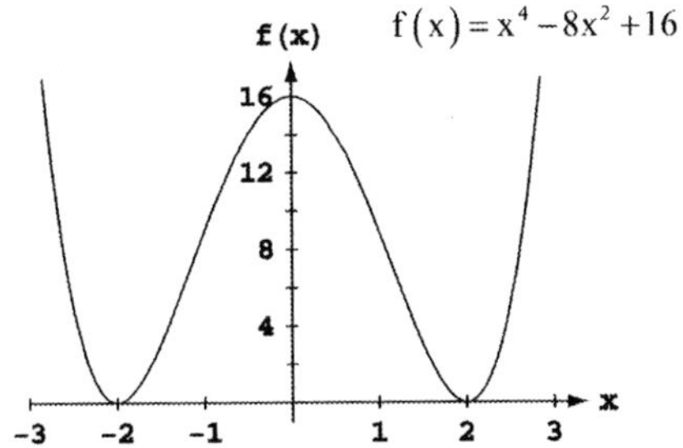

18. (a) $g(x) = x^4 - 4x^3 + 4x^2 \Rightarrow g'(x) = 4x^3 - 12x^2 + 8x = 4x(x-2)(x-1) \Rightarrow$ critical points at $x = 0, 1, 2$ $\Rightarrow g' = \underset{0}{---|}\underset{1}{+++|}\underset{2}{---|}+++$, increasing on $(0, 1)$ and $(2, \infty)$, decreasing on $(-\infty, 0)$ and $(1, 2)$

(b) a local maximum is $g(1) = 1$ at $x = 1$, local minima are $g(0) = 0$ at $x = 0$ and $g(2) = 0$ at $x = 2$

(c) no absolute maximum; absolute minimum is 0 at $x = 0, 2$

(d)

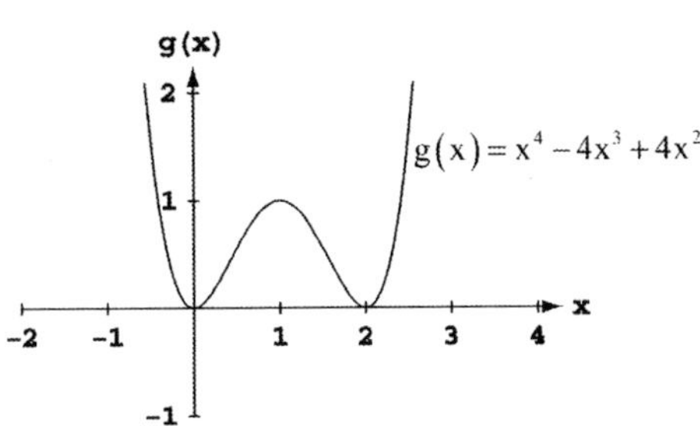

19. (a) $H(t) = \frac{3}{2}t^4 - t^6 \Rightarrow H'(t) = 6t^3 - 6t^5 = 6t^3(1+t)(1-t) \Rightarrow$ critical points at $t = 0, \pm 1$
$\Rightarrow H' = \underset{-1}{+++\,|}\,\underset{0}{---\,|}\,\underset{1}{+++\,|}\,---$, increasing on $(-\infty, -1)$ and $(0, 1)$, decreasing on $(-1, 0)$ and $(1, \infty)$

(b) the local maxima are $H(-1) = \frac{1}{2}$ at $t = -1$ and $H(1) = \frac{1}{2}$ at $t = 1$, the local minimum is $H(0) = 0$ at $t = 0$

(c) absolute maximum is $\frac{1}{2}$ at $t = \pm 1$; no absolute minimum

(d)

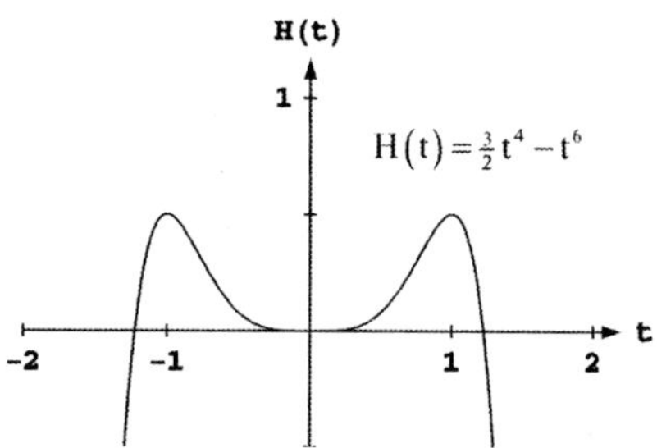

20. (a) $K(t) = 15t^3 - t^5 \Rightarrow K'(t) = 45t^2 - 5t^4 = 5t^2(3+t)(3-t) \Rightarrow$ critical points at $t = 0, \pm 3$
$\Rightarrow K' = \underset{-3}{---\,|}\,\underset{0}{+++\,|}\,\underset{3}{+++\,|}\,---$, increasing on $(-3, 0) \cup (0, 3)$, decreasing on $(-\infty, -3)$ and $(3, \infty)$

(b) a local maximum is $K(3) = 162$ at $t = 3$, a local minimum is $K(-3) = -162$ at $t = -3$

(c) no absolute extrema

(d)

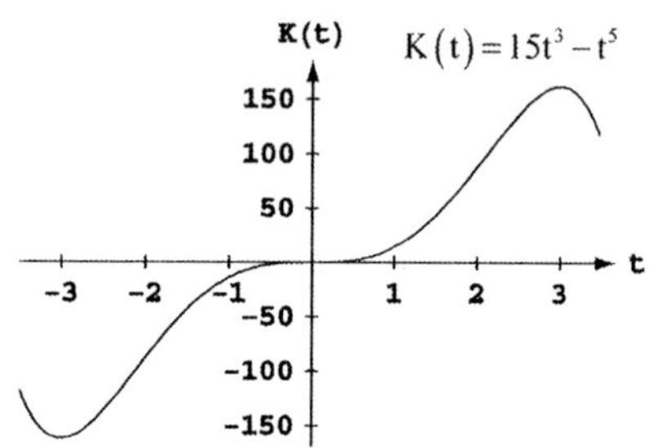

21. (a) $g(x) = x\sqrt{8 - x^2} = x(8 - x^2)^{1/2} \Rightarrow g'(x) = (8 - x^2)^{1/2} + x\left(\frac{1}{2}\right)(8 - x^2)^{-1/2}(-2x) = \frac{2(2-x)(2+x)}{\sqrt{\left(2\sqrt{2}-x\right)\left(2\sqrt{2}+x\right)}}$

$\Rightarrow$ critical points at $x = \pm 2, \pm 2\sqrt{2} \Rightarrow g' = \underset{-2\sqrt{2}}{(}\ \underset{-2}{---\,|}\,\underset{2}{+++\,|}\,\underset{2\sqrt{2}}{---\,)}$, increasing on $(-2, 2)$, decreasing on $\left(-2\sqrt{2}, -2\right)$ and $\left(2, 2\sqrt{2}\right)$

(b) local maxima are $g(2) = 4$ at $x = 2$ and $g\left(-2\sqrt{2}\right) = 0$ at $x = -2\sqrt{2}$, local minima are $g(-2) = -4$ at $x = -2$ and $g\left(2\sqrt{2}\right) = 0$ at $x = 2\sqrt{2}$

(c) absolute maximum is 4 at $x = 2$; absolute minimum is -4 at $x = -2$

(d)

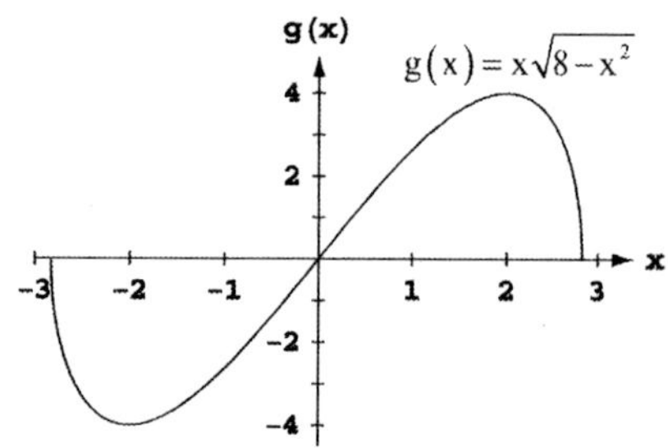

22. (a) $g(x) = x^2\sqrt{5 - x} = x^2(5 - x)^{1/2} \Rightarrow g'(x) = 2x(5 - x)^{1/2} + x^2\left(\frac{1}{2}\right)(5 - x)^{-1/2}(-1) = \frac{5x(4-x)}{2\sqrt{5-x}} \Rightarrow$ critical points at $x = 0, 4$ and $5 \Rightarrow g' = \underset{0}{---\,|}\,\underset{4}{+++\,|}\,\underset{5}{---\,)}$, increasing on $(0, 4)$, decreasing on $(-\infty, 0)$ and $(4, 5)$

(b) a local maximum is $g(4) = 16$ at $x = 4$, a local minimum is 0 at $x = 0$ and $x = 5$

(c) no absolute maximum; absolute minimum is 0 at $x = 0, 5$

(d)

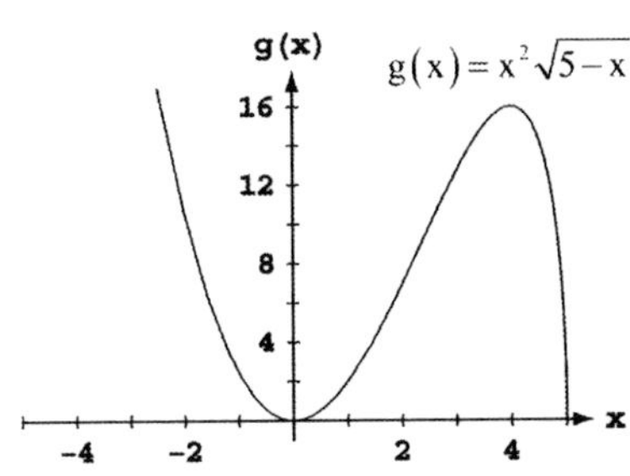

23. (a) $f(x) = \frac{x^2-3}{x-2} \Rightarrow f'(x) = \frac{2x(x-2)-(x^2-3)(1)}{(x-2)^2} = \frac{(x-3)(x-1)}{(x-2)^2} \Rightarrow$ critical points at $x = 1, 3$

$\Rightarrow f' = +++ \underset{1}{|} --- \underset{2}{)(} --- \underset{3}{|} +++$, increasing on $(-\infty, 1)$ and $(3, \infty)$, decreasing on $(1, 2)$ and $(2, 3)$, discontinuous at $x = 2$

(b) a local maximum is $f(1) = 2$ at $x = 1$, a local minimum is $f(3) = 6$ at $x = 3$

(c) no absolute extrema

(d)

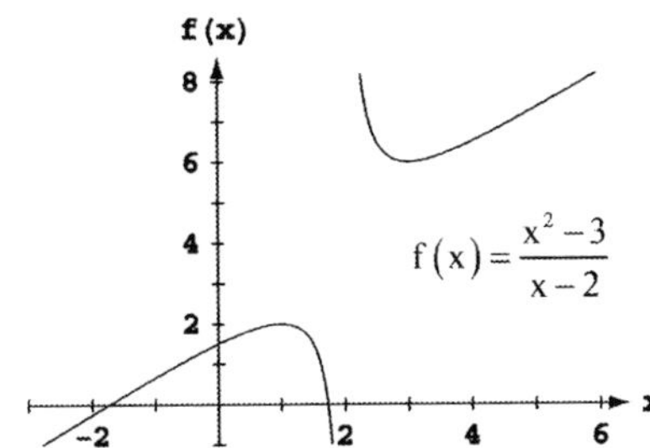

24. (a) $f(x) = \frac{x^3}{3x^2+1} \Rightarrow f'(x) = \frac{3x^2(3x^2+1)-x^3(6x)}{(3x^2+1)^2} = \frac{3x^2(x^2+1)}{(3x^2+1)^2} \Rightarrow$ a critical point at $x = 0$

$\Rightarrow f' = +++ \underset{0}{|} +++$, increasing on $(-\infty, 0) \cup (0, \infty)$, and never decreasing

(b) no local extrema

(c) no absolute extrema

(d)

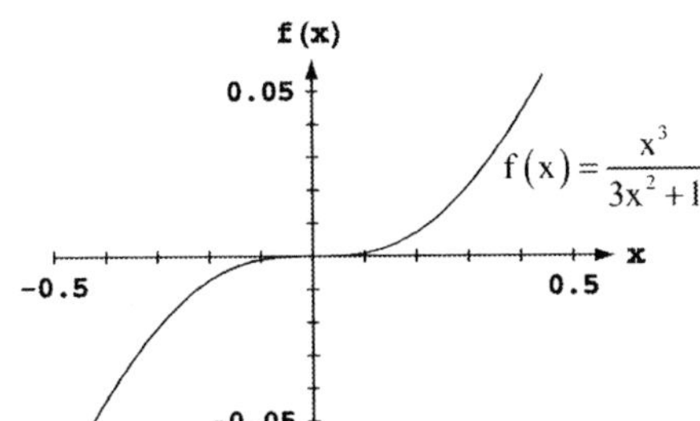

25. (a) $f(x) = x^{1/3}(x+8) = x^{4/3} + 8x^{1/3} \Rightarrow f'(x) = \frac{4}{3}x^{1/3} + \frac{8}{3}x^{-2/3} = \frac{4(x+2)}{3x^{2/3}} \Rightarrow$ critical points at $x = 0, -2$

$\Rightarrow f' = --- \underset{-2}{|} +++ \underset{0}{)(} +++$, increasing on $(-2, 0) \cup (0, \infty)$, decreasing on $(-\infty, -2)$

(b) no local maximum, a local minimum is $f(-2) = -6\sqrt[3]{2} \approx -7.56$ at $x = -2$

(c) no absolute maximum; absolute minimum is $-6\sqrt[3]{2}$ at $x = -2$

(d)

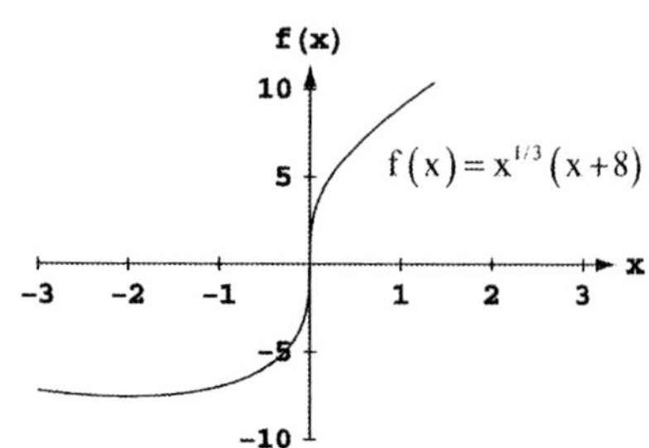

26. (a) $g(x) = x^{2/3}(x+5) = x^{5/3} + 5x^{2/3} \Rightarrow g'(x) = \frac{5}{3}x^{2/3} + \frac{10}{3}x^{-1/3} = \frac{5(x+2)}{3\sqrt[3]{x}} \Rightarrow$ critical points at $x = -2$ and $x = 0 \Rightarrow g' = +++\underset{-2}{|} ---\underset{0}{)(} +++$, increasing on $(-\infty, -2)$ and $(0, \infty)$, decreasing on $(-2, 0)$

(b) local maximum is $g(-2) = 3\sqrt[3]{4} \approx 4.762$ at $x = -2$, a local minimum is $g(0) = 0$ at $x = 0$

(c) no absolute extrema

(d)

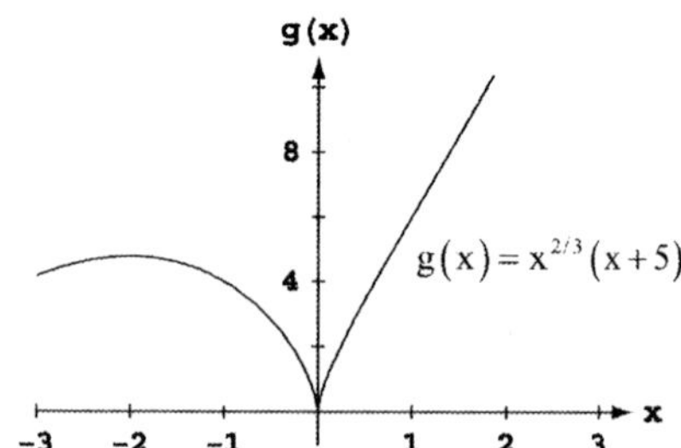

27. (a) $h(x) = x^{1/3}(x^2-4) = x^{7/3} - 4x^{1/3} \Rightarrow h'(x) = \frac{7}{3}x^{4/3} - \frac{4}{3}x^{-2/3} = \frac{(\sqrt{7}x+2)(\sqrt{7}x-2)}{3\sqrt[3]{x^2}} \Rightarrow$ critical points at $x = 0, \frac{\pm 2}{\sqrt{7}} \Rightarrow h' = +++\underset{-2/\sqrt{7}}{|} ---\underset{0}{)(}---\underset{2/\sqrt{7}}{|} +++$, increasing on $\left(-\infty, \frac{-2}{\sqrt{7}}\right)$ and $\left(\frac{2}{\sqrt{7}}, \infty\right)$, decreasing on $\left(\frac{-2}{\sqrt{7}}, 0\right)$ and $\left(0, \frac{2}{\sqrt{7}}\right)$

(b) local maximum is $h\left(\frac{-2}{\sqrt{7}}\right) = \frac{24\sqrt[3]{2}}{7^{7/6}} \approx 3.12$ at $x = \frac{-2}{\sqrt{7}}$, the local minimum is $h\left(\frac{2}{\sqrt{7}}\right) = -\frac{24\sqrt[3]{2}}{7^{7/6}} \approx -3.12$

(c) no absolute extrema

(d)

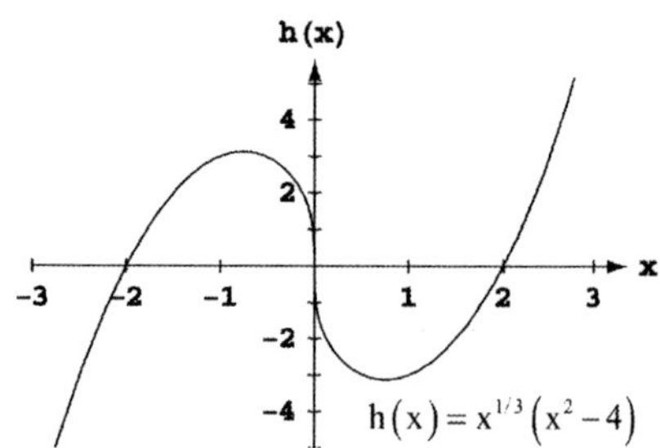

28. (a) $k(x) = x^{2/3}(x^2-4) = x^{8/3} - 4x^{2/3} \Rightarrow k'(x) = \frac{8}{3}x^{5/3} - \frac{8}{3}x^{-1/3} = \frac{8(x+1)(x-1)}{3\sqrt[3]{x}} \Rightarrow$ critical points at $x = 0, \pm 1 \Rightarrow k' = ---\underset{-1}{|} +++\underset{0}{)(}---\underset{1}{|} +++$, increasing on $(-1, 0)$ and $(1, \infty)$, decreasing on $(-\infty, -1)$ and $(0, 1)$

(b) local maximum is $k(0) = 0$ at $x = 0$, local minima are $k(\pm 1) = -3$ at $x = \pm 1$

(c) no absolute maximum; absolute minimum is -3 at $x = \pm 1$

(d)

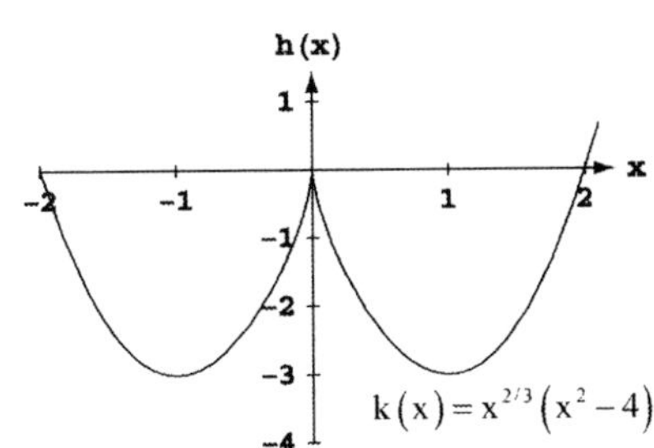

29. (a) $f(x) = e^{2x} + e^{-x} \Rightarrow f'(x) = 2e^{2x} - e^{-x} = 0 \Rightarrow e^{3x} = \frac{1}{2} \Rightarrow$ a critical point at $x = \frac{1}{3}\ln\left(\frac{1}{2}\right)$

$\Rightarrow f' = \underset{\frac{1}{3}\ln\left(\frac{1}{2}\right)}{- - - - |+ + + +}$, increasing on $\left(\frac{1}{3}\ln\left(\frac{1}{2}\right), \infty\right)$, decreasing on $\left(-\infty, \frac{1}{3}\ln\left(\frac{1}{2}\right)\right)$

(b) a local minimum is $\frac{3}{2^{2/3}}$ at $x = \frac{1}{3}\ln\left(\frac{1}{2}\right)$; no local maximum

(c) an absolute minimum $\frac{3}{2^{2/3}}$ at $x = \frac{1}{3}\ln\left(\frac{1}{2}\right)$; no absolute maximum

(d)

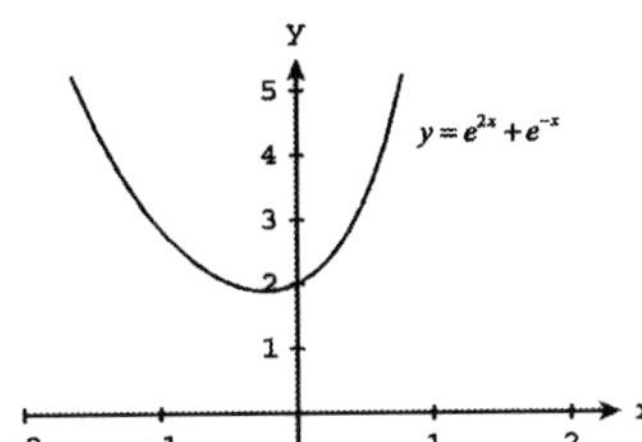

30. (a) $f(x) = e^{\sqrt{x}} \Rightarrow f'(x) = \frac{e^{\sqrt{x}}}{2\sqrt{x}} \Rightarrow$ no critical points $\Rightarrow f' = \underset{0}{|+ + +}$, increasing on $(0, \infty)$

(b) A local minimum is 1 at x = 0, no local maximum

(c) An absolute minimum is 1 at x = 0, no absolute maximum

(d)

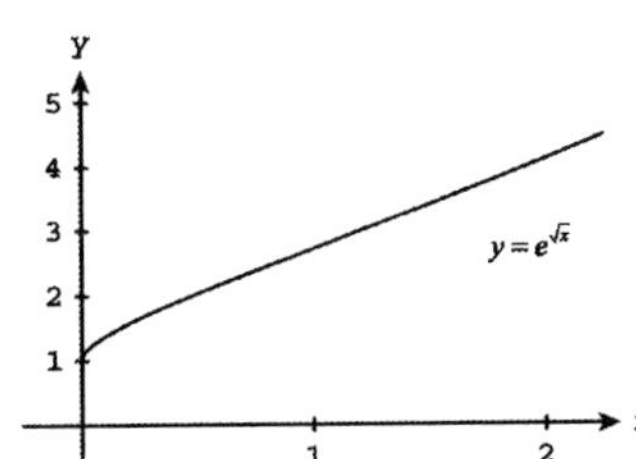

31. (a) $f(x) = x \ln x \Rightarrow f'(x) = 1 + \ln x \Rightarrow$ a critical point at $x = e^{-1} \Rightarrow f' = \underset{0}{[} - - - \underset{e^{-1}}{|} + + +$, increasing on (e^{-1}, ∞),

decreasing on $(0, e^{-1})$

(b) A local minimum is $-e^{-1}$ at $x = e^{-1}$, no local maximum

(c) An absolute minimum is $-e^{-1}$ at $x = e^{-1}$, no absolute maximum

(d)

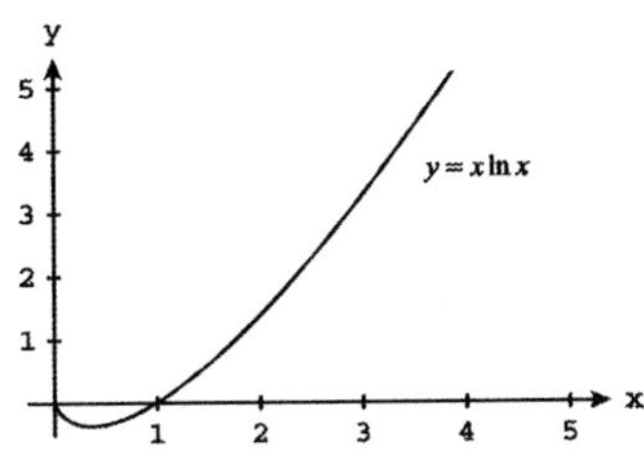

32. (a) $f(x) = x^2 \ln x \Rightarrow f'(x) = x + 2x \ln x = x(1 + 2\ln x) \Rightarrow$ a critical point at $x = e^{-1/2} \Rightarrow f' = \underset{0}{[} --- \underset{e^{-1/2}}{|} +++$,

increasing on $(e^{-1/2}, \infty)$, decreasing on $(0, e^{-1/2})$

(b) A local minimum is $-\frac{e^{-1}}{2}$ at $x = e^{-1/2}$, no local maximum

(c) An absolute minimum is $-\frac{e^{-1}}{2}$ at $x = e^{-1/2}$, no absolute maximum

(d)

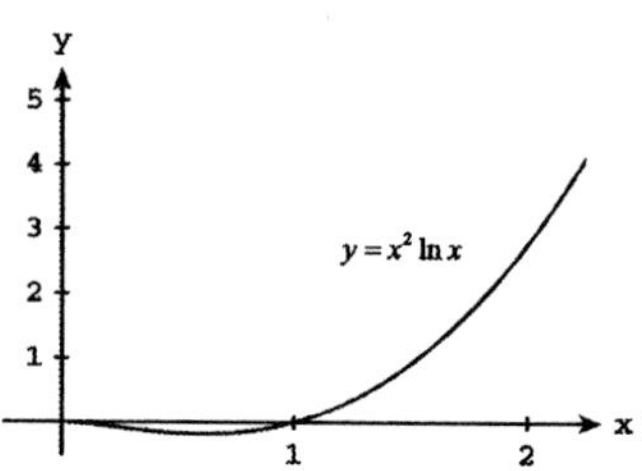

33. (a) $f(x) = 2x - x^2 \Rightarrow f'(x) = 2 - 2x \Rightarrow$ a critical point at $x = 1 \Rightarrow f' = +++ \underset{1}{|} --- \underset{2}{]}$ and $f(1) = 1$ and $f(2) = 0$

a local maximum is 1 at $x = 1$, a local minimum is 0 at $x = 2$.

(b) There is an absolute maximum of 1 at $x = 1$; no absolute minimum.

(c)

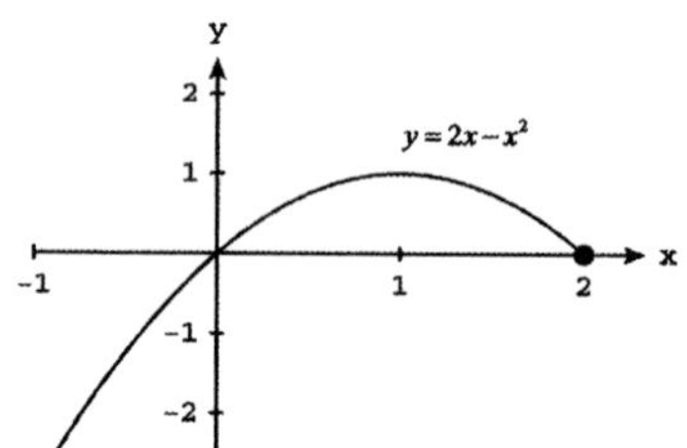

34. (a) $f(x) = (x + 1)^2 \Rightarrow f'(x) = 2(x + 1) \Rightarrow$ a critical point at $x = -1 \Rightarrow f' = --- \underset{-1}{|} +++ \underset{0}{]}$ and $f(-1) = 0$, $f(0) = 1$

$\Rightarrow$ a local maximum is 1 at $x = 0$, a local minimum is 0 at $x = -1$

(b) no absolute maximum; absolute minimum is 0 at $x = -1$

(c)

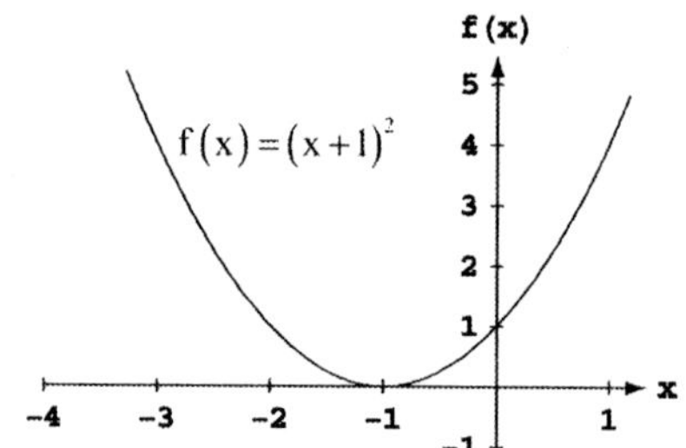

35. (a) $g(x) = x^2 - 4x + 4 \Rightarrow g'(x) = 2x - 4 = 2(x - 2) \Rightarrow$ a critical point at $x = 2 \Rightarrow g' = \underset{1}{[} --- \underset{2}{|} +++$ and

$g(1) = 1$, $g(2) = 0 \Rightarrow$ a local maximum is 1 at $x = 1$, a local minimum is $g(2) = 0$ at $x = 2$

(b) no absolute maximum; absolute minimum is 0 at $x = 2$

(c)

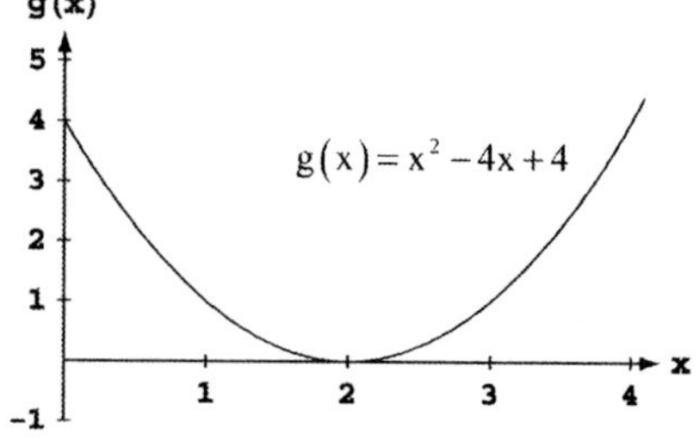

36. (a) $g(x) = -x^2 - 6x - 9 \Rightarrow g'(x) = -2x - 6 = -2(x+3) \Rightarrow$ a critical point at $x = -3 \Rightarrow g' = \underset{-4}{[}\ +++\ \underset{-3}{|}\ ---$ and

$g(-4) = -1,\ g(-3) = 0 \Rightarrow$ a local maximum is 0 at $x = -3$, a local minimum is -1 at $x = -4$

(b) absolute maximum is 0 at $x = -3$; no absolute minimum

(c)

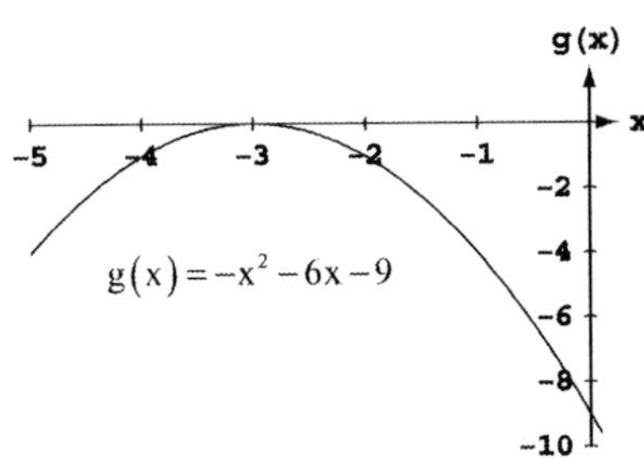

37. (a) $f(t) = 12t - t^3 \Rightarrow f'(t) = 12 - 3t^2 = 3(2+t)(2-t) \Rightarrow$ critical points at $t = \pm 2 \Rightarrow f' = \underset{-3}{[}\ ---\ \underset{-2}{|}\ +++\ \underset{2}{|}\ ---$

and $f(-3) = -9,\ f(-2) = -16,\ f(2) = 16 \Rightarrow$ local maxima are -9 at $t = -3$ and 16 at $t = -2$, a local minimum is -16 at $t = -2$

(b) absolute maximum is 16 at $t = 2$; no absolute minimum

(c)

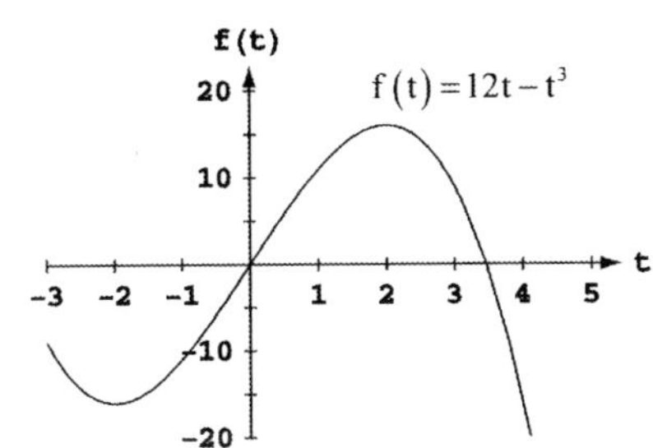

38. (a) $f(t) = t^3 - 3t^2 \Rightarrow f'(t) = 3t^2 - 6t = 3t(t-2) \Rightarrow$ critical points at $t = 0$ and $t = 2$

$\Rightarrow f' = +++\ \underset{0}{|}\ ---\ \underset{2}{|}\ +++\ \underset{3}{]}$ and $f(0) = 0,\ f(2) = -4,\ f(3) = 0 \Rightarrow$ a local maximum is 0 at $t = 0$ and $t = 3$, a local minimum is -4 at $t = 2$

(b) absolute maximum is 0 at $t = 0, 3$; no absolute minimum

(c)

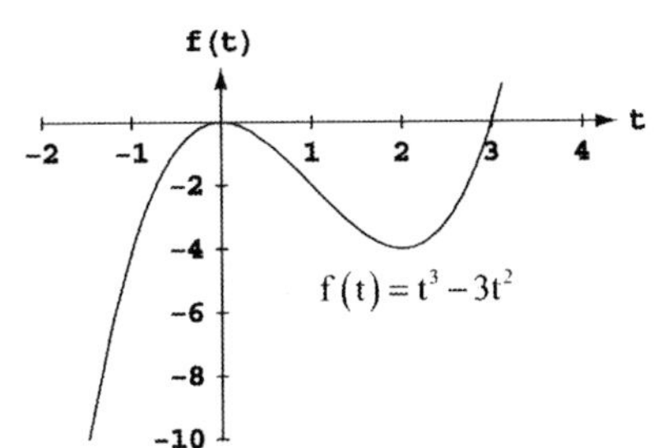

39. (a) $h(x) = \frac{x^3}{3} - 2x^2 + 4x \Rightarrow h'(x) = x^2 - 4x + 4 = (x-2)^2 \Rightarrow$ a critical point at $x = 2 \Rightarrow h' = \underset{0}{[}\ +++\ \underset{2}{|}\ +++$ and

$h(0) = 0 \Rightarrow$ no local maximum, a local minimum is 0 at $x = 0$

(b) no absolute maximum; absolute minimum is 0 at $x = 0$

(c)

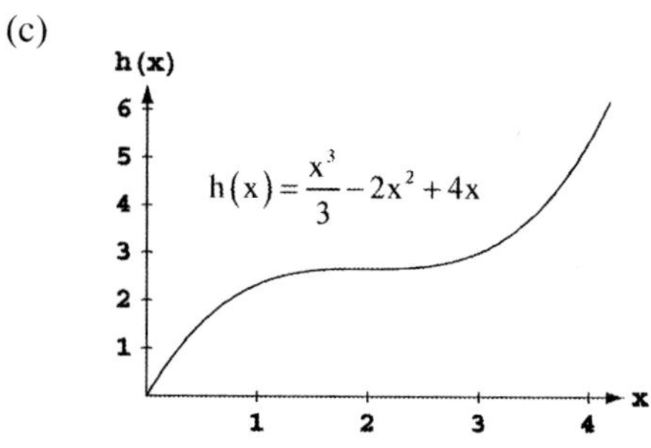

40. (a) $k(x) = x^3 + 3x^2 + 3x + 1 \Rightarrow k'(x) = 3x^2 + 6x + 3 = 3(x+1)^2 \Rightarrow$ a critical point at $x = -1$
$\Rightarrow k' = \underset{}{+++} \underset{-1}{|} \underset{}{+++} \underset{0}{]}$ and $k(-1) = 0$, $k(0) = 1 \Rightarrow$ a local maximum is 1 at $x = 0$, no local minimum

(b) absolute maximum is 1 at $x = 0$; no absolute minimum

(c)

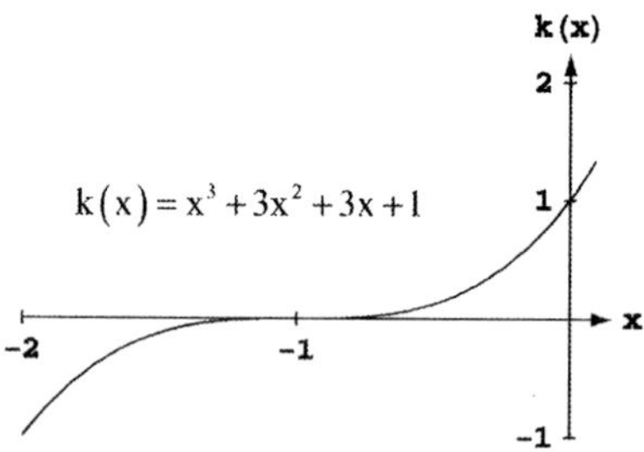

41. (a) $f(x) = \frac{x}{2} - 2\sin\left(\frac{x}{2}\right) \Rightarrow f'(x) = \frac{1}{2} - \cos\left(\frac{x}{2}\right)$, $f'(x) = 0 \Rightarrow \cos\left(\frac{x}{2}\right) = \frac{1}{2} \Rightarrow$ a critical point at $x = \frac{2\pi}{3}$
$\Rightarrow f' = \underset{0}{[} --- \underset{2\pi/3}{|} +++ \underset{2\pi}{]}$ and $f(0) = 0$, $f\left(\frac{2\pi}{3}\right) = \frac{\pi}{3} - \sqrt{3}$, $f(2\pi) = \pi \Rightarrow$ local maxima are 0 at $x = 0$ and π at $x = 2\pi$, a local minimum is $\frac{\pi}{3} - \sqrt{3}$ at $x = \frac{2\pi}{3}$

(b) The graph of f rises when $f' > 0$, falls when $f' < 0$, and has a local minimum value at the point where f' changes from negative to positive.

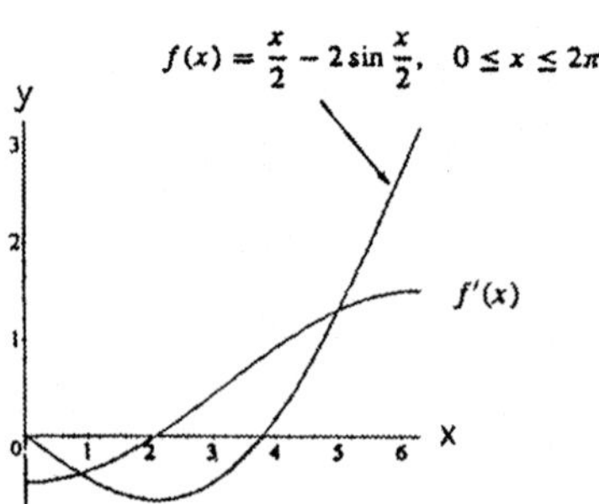

42. (a) $f(x) = -2\cos x - \cos^2 x \Rightarrow f'(x) = 2\sin x + 2\cos x \sin x = 2(\sin x)(1 + \cos x) \Rightarrow$ critical points at $x = -\pi, 0, \pi \Rightarrow f' = \underset{-\pi}{[} --- \underset{0}{|} +++ \underset{\pi}{]}$ and $f(-\pi) = 1$, $f(0) = -3$, $f(\pi) = 1 \Rightarrow$ a local maximum is 1 at $x = \pm\pi$, a local minimum is -3 at $x = 0$

(b) The graph of f rises when $f' > 0$, falls when $f' < 0$, and has local extreme values where $f' = 0$. The function f has a local minimum value at $x = 0$, where the values of f' change from negative to positive.

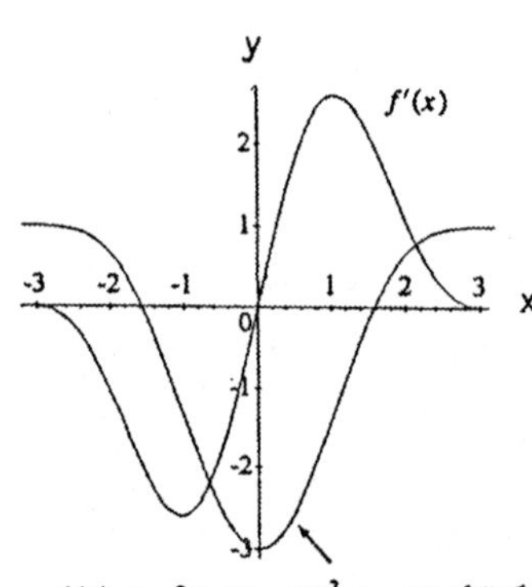

43. (a) $f(x) = \csc^2 x - 2\cot x \Rightarrow f'(x) = 2(\csc x)(-\csc x)(\cot x) - 2(-\csc^2 x) = -2(\csc^2 x)(\cot x - 1) \Rightarrow$ a critical point at $x = \frac{\pi}{4} \Rightarrow f' = \underset{0}{(} --- \underset{\pi/4}{|} +++ \underset{\pi}{)}$ and $f\left(\frac{\pi}{4}\right) = 0 \Rightarrow$ no local maximum, a local minimum is 0 at $x = \frac{\pi}{4}$

(b) The graph of f rises when $f' > 0$, falls when $f' < 0$, and has a local minimum value at the point where $f' = 0$ and the values of f' change from negative to positive. The graph of f steepens as $f'(x) \to \pm\infty$.

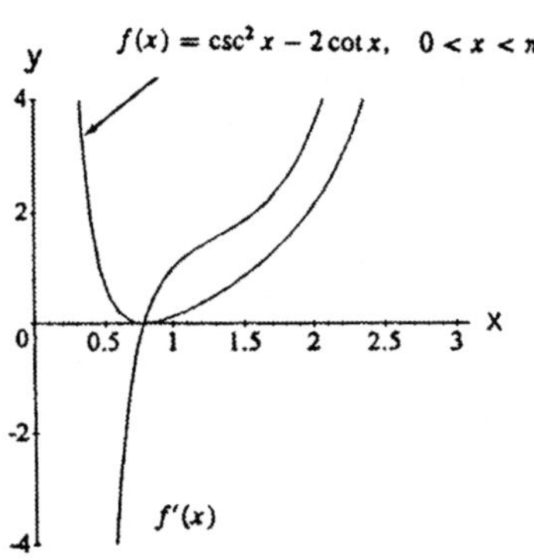

44. (a) $f(x) = \sec^2 x - 2\tan x \Rightarrow f'(x) = 2(\sec x)(\sec x)(\tan x) - 2\sec^2 x = (2\sec^2 x)(\tan x - 1) \Rightarrow$ a critical point at $x = \frac{\pi}{4} \Rightarrow f' = \underset{-\pi/2}{(} --- \underset{\pi/4}{|} +++ \underset{\pi/2}{)}$ and $f\left(\frac{\pi}{4}\right) = 0 \Rightarrow$ no local maximum, a local minimum is 0 at $x = \frac{\pi}{4}$

(b) The graph of f rises when $f' > 0$, falls when $f' < 0$, and has a local minimum value where $f' = 0$ and the values of f' change from negative to positive.

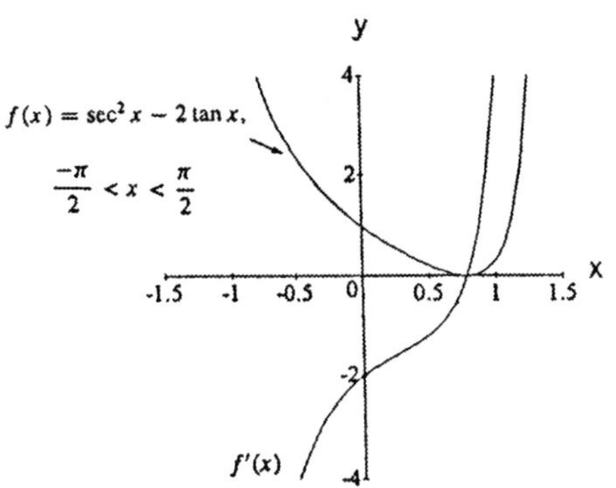

45. $h(\theta) = 3\cos\left(\frac{\theta}{2}\right) \Rightarrow h'(\theta) = -\frac{3}{2}\sin\left(\frac{\theta}{2}\right) \Rightarrow h' = \underset{0}{[} --- \underset{2\pi}{]}$, $(0, 3)$ and $(2\pi, -3) \Rightarrow$ a local maximum is 3 at $\theta = 0$, a local minimum is -3 at $\theta = 2\pi$

46. $h(\theta) = 5\sin\left(\frac{\theta}{2}\right) \Rightarrow h'(\theta) = \frac{5}{2}\cos\left(\frac{\theta}{2}\right) \Rightarrow h' = \underset{0}{[} +++ \underset{\pi}{]}$, $(0, 0)$ and $(\pi, 5) \Rightarrow$ a local maximum is 5 at $\theta = \pi$, a local minimum is 0 at $\theta = 0$

47. (a)

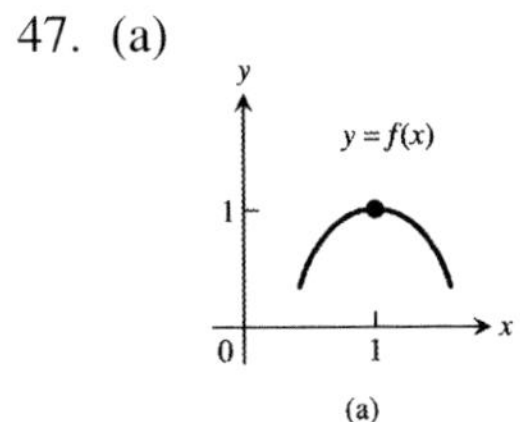

(b)

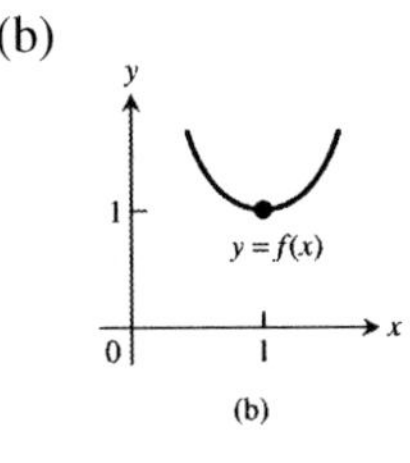

(c)

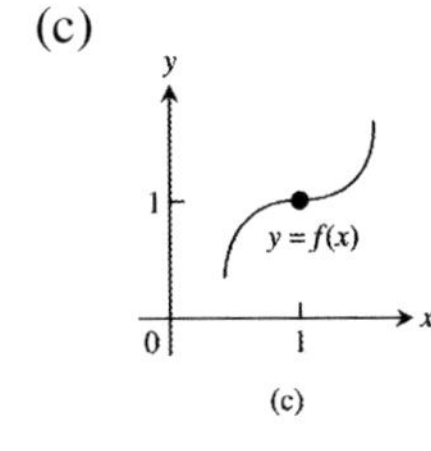

(d)

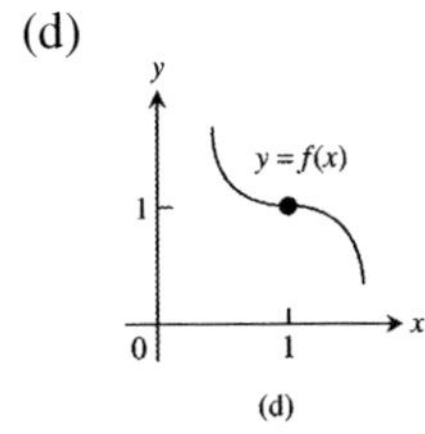

48. (a)

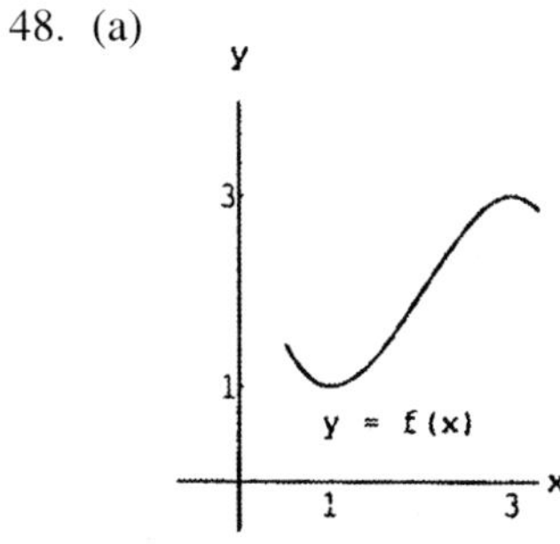

(b)

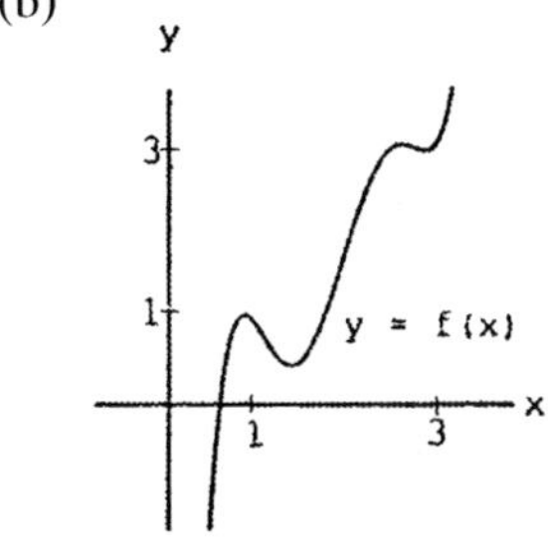

(c)

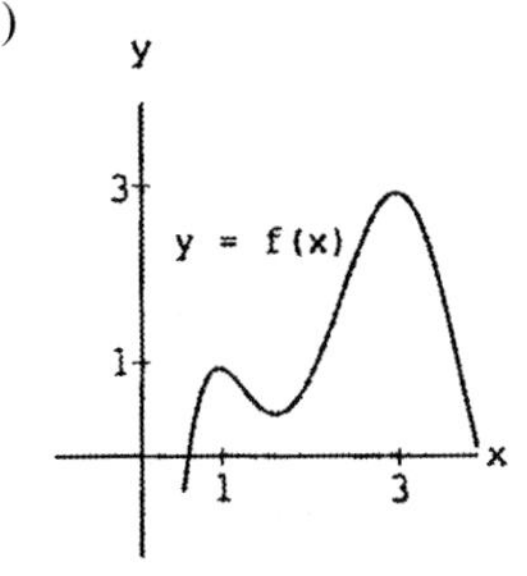

(d)

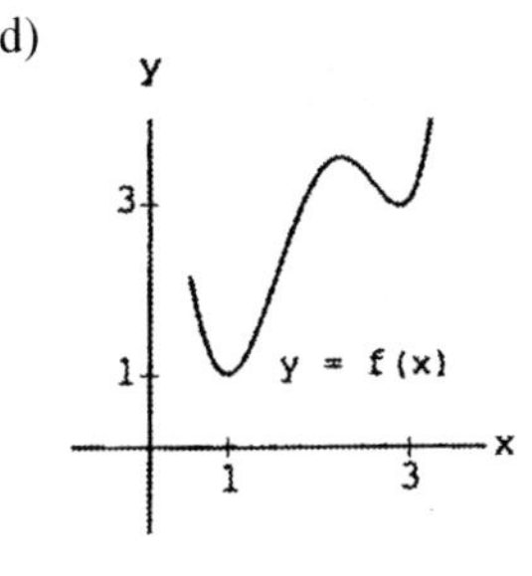

49. (a)

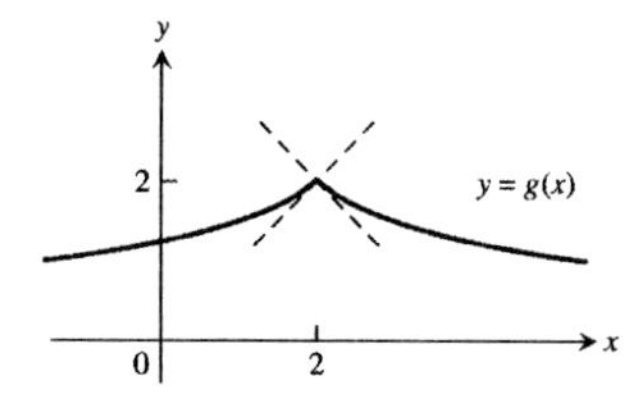

(b)

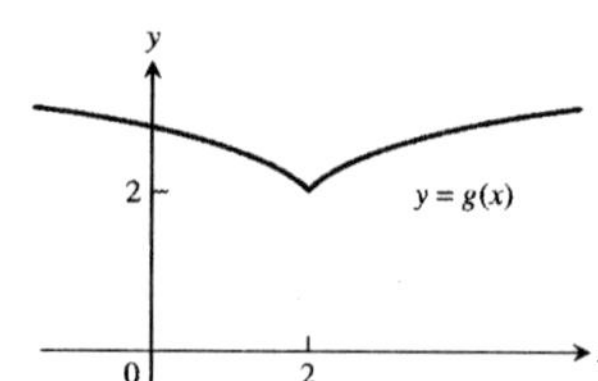

50. (a)

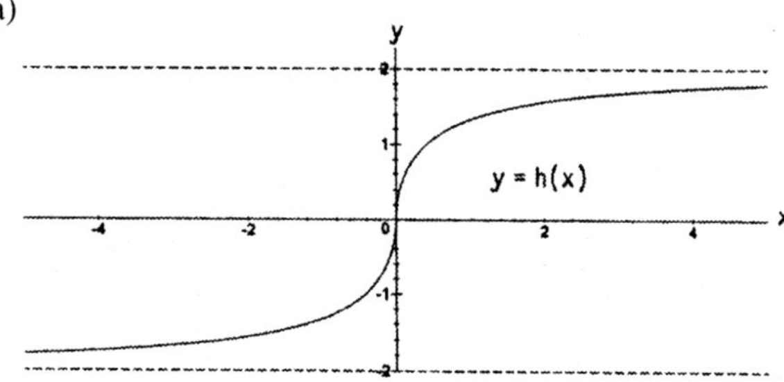

(b)

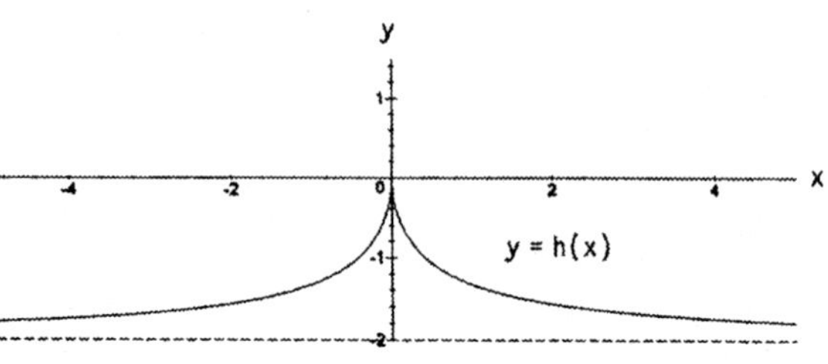

51. (a) $f(x) = \ln(\cos x) \Rightarrow f'(x) = -\frac{\sin x}{\cos x} = -\tan x = 0 \Rightarrow x = 0$; $f'(x) > 0$ for $-\frac{\pi}{4} \le x < 0$ and $f'(x) < 0$ for $0 < x \le \frac{\pi}{3} \Rightarrow$ there is a relative maximum at $x = 0$ with $f(0) = \ln(\cos 0) = \ln 1 = 0$; $f\left(-\frac{\pi}{4}\right) = \ln\left(\cos\left(-\frac{\pi}{4}\right)\right) = \ln\left(\frac{1}{\sqrt{2}}\right) = -\frac{1}{2}\ln 2$ and $f\left(\frac{\pi}{3}\right) = \ln\left(\cos\left(\frac{\pi}{3}\right)\right) = \ln\frac{1}{2} = -\ln 2$. Therefore, the absolute minimum occurs at $x = \frac{\pi}{3}$ with $f\left(\frac{\pi}{3}\right) = -\ln 2$ and the absolute maximum occurs at $x = 0$ with $f(0) = 0$.

(b) $f(x) = \cos(\ln x) \Rightarrow f'(x) = \frac{-\sin(\ln x)}{x} = 0 \Rightarrow x = 1$; $f'(x) > 0$ for $\frac{1}{2} \le x < 1$ and $f'(x) < 0$ for $1 < x \le 2$ $\Rightarrow$ there is a relative maximum at $x = 1$ with $f(1) = \cos(\ln 1) = \cos 0 = 1$; $f\left(\frac{1}{2}\right) = \cos\left(\ln\left(\frac{1}{2}\right)\right) = \cos(-\ln 2) = \cos(\ln 2)$ and $f(2) = \cos(\ln 2)$. Therefore, the absolute minimum occurs at $x = \frac{1}{2}$ and $x = 2$ with $f\left(\frac{1}{2}\right) = f(2) = \cos(\ln 2)$, and the absolute maximum occurs at $x = 1$ with $f(1) = 1$.

52. (a) $f(x) = x - \ln x \Rightarrow f'(x) = 1 - \frac{1}{x}$; if $x > 1$, then $f'(x) > 0$ which means that $f(x)$ is increasing

(b) $f(1) = 1 - \ln 1 = 1 \Rightarrow f(x) = x - \ln x > 0$, if $x > 1$ by part (a) $\Rightarrow x > \ln x$ if $x > 1$

53. $f(x) = e^x - 2x \Rightarrow f'(x) = e^x - 2$; $f'(x) = 0 \Rightarrow e^x = 2 \Rightarrow x = \ln 2$; $f(0) = 1$, the absolute maximum; $f(\ln 2) = 2 - 2\ln 2 \approx 0.613706$, the absolute minimum; $f(1) = e - 2 \approx 0.71828$, a relative or local maximum since $f''(x) = e^x$ is always positive.

54. The function $f(x) = 2e^{\sin(x/2)}$ has a maximum whenever $\sin\frac{x}{2} = 1$ and a minimum whenever $\sin\frac{x}{2} = -1$. Therefore the maximums occur at $x = \pi + 2k(2\pi)$ and the minimums occur at $x = 3\pi + 2k(2\pi)$, where k is any integer. The maximum is $2e \approx 5.43656$ and the minimum is $\frac{2}{e} \approx 0.73576$.

55. $f(x) = x^2 \ln\frac{1}{x} \Rightarrow f'(x) = 2x\ln\frac{1}{x} + x^2\left(\frac{1}{\frac{1}{x}}\right)(-x^{-2}) = 2x\ln\frac{1}{x} - x = -x(2\ln x + 1)$; $f'(x) = 0 \Rightarrow x = 0$ or $\ln x = -\frac{1}{2}$. Since $x = 0$ is not in the domain of f, $x = e^{-1/2} = \frac{1}{\sqrt{e}}$. Also, $f'(x) > 0$ for $0 < x < \frac{1}{\sqrt{e}}$ and $f'(x) < 0$ for $x > \frac{1}{\sqrt{e}}$. Therefore, $f\left(\frac{1}{\sqrt{e}}\right) = \frac{1}{e}\ln\sqrt{e} = \frac{1}{e}\ln e^{1/2} = \frac{1}{2e}\ln e = \frac{1}{2e}$ is the absolute maximum value of f assumed at $x = \frac{1}{\sqrt{e}}$.

56. $f(x) = ax^2 + bx + c = a\left(x^2 + \frac{b}{a}x\right) + c = a\left(x^2 + \frac{b}{a}x + \frac{b^2}{4a^2}\right) - \frac{b^2}{4a} + c = a\left(x + \frac{b}{2a}\right)^2 - \frac{b^2 - 4ac}{4a}$, a parabola whose vertex is at $x = -\frac{b}{2a}$. Thus when $a > 0$, f is increasing on $\left(\frac{-b}{2a}, \infty\right)$ and decreasing on $\left(-\infty, \frac{-b}{2a}\right)$; when $a < 0$, f is increasing on $\left(-\infty, \frac{-b}{2a}\right)$ and decreasing on $\left(\frac{-b}{2a}, \infty\right)$. Also note that $f'(x) = 2ax + b = 2a\left(x + \frac{b}{2a}\right) \Rightarrow$ for $a > 0$, $f' = \underset{-b/2a}{---|}\ +++$; for $a < 0$, $f' = \underset{-b/2a}{+++|}\ ---$.

57. $f(x) = x^3 - 3x + 2 \Rightarrow f'(x) = 3x^2 - 3 = 3(x-1)(x+1) \Rightarrow f' = +++\underset{-1}{|}---\underset{1}{|}+++ \Rightarrow$ rising for $x = c = 2$ since $f'(x) > 0$ for $x = c = 2$.

58. (a) Let $f(x) = e^x - x - 1 \Rightarrow f'(x) = e^x - 1 \Rightarrow$ a critical point at $x = 0 \Rightarrow f' = ---\underset{0}{|}+++$, so f is increasing on $(0, \infty)$; since $f(0) = 0$ it follows that $f(x) = e^x - x - 1 \geq 0$ for $x \geq 0 \Rightarrow e^x \geq x + 1$ for $x \geq 0$.
(b) Let $f(x) = e^x - \frac{1}{2}x^2 - x - 1 \Rightarrow f'(x) = e^x - x - 1 \geq 0$ for $x \geq 0$ by part (a), so f is increasing on $(0, \infty)$; since $f(0) = 0$ it follows that $f(x) = e^x - \frac{1}{2}x^2 - x - 1 \geq 0$ for $x \geq 0 \Rightarrow e^x \geq \frac{1}{2}x^2 + x + 1$ for $x \geq 0$.

59. Let $x_1 \neq x_2$ be two numbers in the domain of an increasing function f. Then, either $x_1 < x_2$ or $x_1 > x_2$ which implies $f(x_1) < f(x_2)$ or $f(x_1) > f(x_2)$, since f(x) is increasing. In either case, $f(x_1) \neq f(x_2)$ and f is one-to-one. Similar arguments hold if f is decreasing.

60. f(x) is increasing since $x_2 > x_1 \Rightarrow \frac{1}{3}x_2 + \frac{5}{6} > \frac{1}{3}x_1 + \frac{5}{6}$; $\frac{df}{dx} = \frac{1}{3} \Rightarrow \frac{df^{-1}}{dx} = \frac{1}{\left(\frac{1}{3}\right)} = 3$

61. f(x) is increasing since $x_2 > x_1 \Rightarrow 27x_2^3 > 27x_1^3$; $y = 27x^3 \Rightarrow x = \frac{1}{3}y^{1/3} \Rightarrow f^{-1}(x) = \frac{1}{3}x^{1/3}$;
$\frac{df}{dx} = 81x^2 \Rightarrow \frac{df^{-1}}{dx} = \frac{1}{81x^2}\Big|_{\frac{1}{3}x^{1/3}} = \frac{1}{9x^{2/3}} = \frac{1}{9}x^{-2/3}$

62. f(x) is decreasing since $x_2 > x_1 \Rightarrow 1 - 8x_2^3 < 1 - 8x_1^3$; $y = 1 - 8x^3 \Rightarrow x = \frac{1}{2}(1-y)^{1/3} \Rightarrow f^{-1}(x) = \frac{1}{2}(1-x)^{1/3}$;
$\frac{df}{dx} = -24x^2 \Rightarrow \frac{df^{-1}}{dx} = \frac{1}{-24x^2}\Big|_{\frac{1}{2}(1-x)^{1/3}} = \frac{-1}{6(1-x)^{2/3}} = -\frac{1}{6}(1-x)^{-2/3}$

63. f(x) is decreasing since $x_2 > x_1 \Rightarrow (1-x_2)^3 < (1-x_1)^3$; $y = (1-x)^3 \Rightarrow x = 1 - y^{1/3} \Rightarrow f^{-1}(x) = 1 - x^{1/3}$;
$\frac{df}{dx} = -3(1-x)^2 \Rightarrow \frac{df^{-1}}{dx} = \frac{1}{-3(1-x)^2}\Big|_{1-x^{1/3}} = \frac{-1}{3x^{2/3}} = -\frac{1}{3}x^{-2/3}$

64. f(x) is increasing since $x_2 > x_1 \Rightarrow x_2^{5/3} > x_1^{5/3}$; $y = x^{5/3} \Rightarrow x = y^{3/5} \Rightarrow f^{-1}(x) = x^{3/5}$;
$\frac{df}{dx} = \frac{5}{3}x^{2/3} \Rightarrow \frac{df^{-1}}{dx} = \frac{1}{\frac{5}{3}x^{2/3}}\Big|_{x^{3/5}} = \frac{3}{5x^{2/5}} = \frac{3}{5}x^{-2/5}$

4.4 CONCAVITY AND CURVE SKETCHING

1. $y = \frac{x^3}{3} - \frac{x^2}{2} - 2x + \frac{1}{3} \Rightarrow y' = x^2 - x - 2 = (x-2)(x+1) \Rightarrow y'' = 2x - 1 = 2\left(x - \frac{1}{2}\right)$. The graph is rising on $(-\infty, -1)$ and $(2, \infty)$, falling on $(-1, 2)$, concave up on $\left(\frac{1}{2}, \infty\right)$ and concave down on $\left(-\infty, \frac{1}{2}\right)$. Consequently, a local maximum is $\frac{3}{2}$ at $x = -1$, a local minimum is -3 at $x = 2$, and $\left(\frac{1}{2}, -\frac{3}{4}\right)$ is a point of inflection.

2. $y = \frac{x^4}{4} - 2x^2 + 4 \Rightarrow y' = x^3 - 4x = x(x^2 - 4) = x(x+2)(x-2) \Rightarrow y'' = 3x^2 - 4 = \left(\sqrt{3}x + 2\right)\left(\sqrt{3}x - 2\right)$. The graph is rising on $(-2, 0)$ and $(2, \infty)$, falling on $(-\infty, -2)$ and $(0, 2)$, concave up on $\left(-\infty, -\frac{2}{\sqrt{3}}\right)$ and $\left(\frac{2}{\sqrt{3}}, \infty\right)$ and concave down on $\left(-\frac{2}{\sqrt{3}}, \frac{2}{\sqrt{3}}\right)$. Consequently, a local maximum is 4 at $x = 0$, local minima are 0 at $x = \pm 2$, and $\left(-\frac{2}{\sqrt{3}}, \frac{16}{9}\right)$ and $\left(\frac{2}{\sqrt{3}}, \frac{16}{9}\right)$ are points of inflection.

3. $y = \frac{3}{4}(x^2 - 1)^{2/3} \Rightarrow y' = \left(\frac{3}{4}\right)\left(\frac{2}{3}\right)(x^2 - 1)^{-1/3}(2x) = x(x^2 - 1)^{-1/3}$, $y' = ---\underset{-1}{)}\,(+++\underset{0}{|}---\underset{1}{)}(+++$
$\Rightarrow$ the graph is rising on $(-1, 0)$ and $(1, \infty)$, falling on $(-\infty, -1)$ and $(0, 1)$ $\Rightarrow$ a local maximum is $\frac{3}{4}$ at $x = 0$, local minima are 0 at $x = \pm 1$; $y'' = (x^2 - 1)^{-1/3} + (x)\left(-\frac{1}{3}\right)(x^2 - 1)^{-4/3}(2x) = \frac{x^2 - 3}{3\sqrt[3]{(x^2-1)^4}}$,

$y'' = +++\underset{-\sqrt{3}}{|}\ ---\underset{-1}{)}\ (---\underset{1}{)(}---\underset{\sqrt{3}}{|}\ +++ \Rightarrow$ the graph is concave up on $\left(-\infty, -\sqrt{3}\right)$ and $\left(\sqrt{3}, \infty\right)$, concave down on $\left(-\sqrt{3}, \sqrt{3}\right) \Rightarrow$ points of inflection at $\left(\pm\sqrt{3}, \frac{3\sqrt[3]{4}}{4}\right)$

4. $y = \frac{9}{14}x^{1/3}(x^2 - 7) \Rightarrow y' = \frac{3}{14}x^{-2/3}(x^2 - 7) + \frac{9}{14}x^{1/3}(2x) = \frac{3}{2}x^{-2/3}(x^2 - 1)$, $y' = +++\underset{-1}{|}\ ---\underset{0}{)(}---\underset{1}{|}+++$ $\Rightarrow$ the graph is rising on $(-\infty, -1)$ and $(1, \infty)$, falling on $(-1, 1) \Rightarrow$ a local maximum is $\frac{27}{7}$ at $x = -1$, a local minimum is $-\frac{27}{7}$ at $x = 1$; $y'' = -x^{-5/3}(x^2 - 1) + 3x^{1/3} = 2x^{1/3} + x^{-5/3} = x^{-5/3}(2x^2 + 1)$, $y'' = ---\underset{0}{)(}+++ \Rightarrow$ the graph is concave up on $(0, \infty)$, concave down on $(-\infty, 0) \Rightarrow$ a point of inflection at $(0, 0)$

5. $y = x + \sin 2x \Rightarrow y' = 1 + 2\cos 2x$, $y' = \underset{-2\pi/3}{[}---\underset{-\pi/3}{|}\ +++\underset{\pi/3}{|}\ ---\underset{2\pi/3}{]}$ $\Rightarrow$ the graph is rising on $\left(-\frac{\pi}{3}, \frac{\pi}{3}\right)$, falling on $\left(-\frac{2\pi}{3}, -\frac{\pi}{3}\right)$ and $\left(\frac{\pi}{3}, \frac{2\pi}{3}\right) \Rightarrow$ local maxima are $-\frac{2\pi}{3} + \frac{\sqrt{3}}{2}$ at $x = -\frac{2\pi}{3}$ and $\frac{\pi}{3} + \frac{\sqrt{3}}{2}$ at $x = \frac{\pi}{3}$, local minima are $-\frac{\pi}{3} - \frac{\sqrt{3}}{2}$ at $x = -\frac{\pi}{3}$ and $\frac{2\pi}{3} - \frac{\sqrt{3}}{2}$ at $x = \frac{2\pi}{3}$; $y'' = -4\sin 2x$, $y'' = \underset{-2\pi/3}{[}\ ---\underset{-\pi/2}{|}\ +++\underset{0}{|}---\underset{\pi/2}{|}\ +++\underset{2\pi/3}{]}$ $\Rightarrow$ the graph is concave up on $\left(-\frac{\pi}{2}, 0\right)$ and $\left(\frac{\pi}{2}, \frac{2\pi}{3}\right)$, concave down on $\left(-\frac{2\pi}{3}, -\frac{\pi}{2}\right)$ and $\left(0, \frac{\pi}{2}\right) \Rightarrow$ points of inflection at $\left(-\frac{\pi}{2}, -\frac{\pi}{2}\right)$, $(0, 0)$, and $\left(\frac{\pi}{2}, \frac{\pi}{2}\right)$

6. $y = \tan x - 4x \Rightarrow y' = \sec^2 x - 4$, $y' = \underset{-\pi/2}{(}\ +++\underset{-\pi/3}{|}\ ---\underset{\pi/3}{|}\ +++\underset{\pi/2}{)}$ $\Rightarrow$ the graph is rising on $\left(-\frac{\pi}{2}, -\frac{\pi}{3}\right)$ and $\left(\frac{\pi}{3}, \frac{\pi}{2}\right)$, falling on $\left(-\frac{\pi}{3}, \frac{\pi}{3}\right) \Rightarrow$ a local maximum is $-\sqrt{3} + \frac{4\pi}{3}$ at $x = -\frac{\pi}{3}$, a local minimum is $\sqrt{3} - \frac{4\pi}{3}$ at $x = \frac{\pi}{3}$; $y'' = 2(\sec x)(\sec x)(\tan x) = 2(\sec^2 x)(\tan x)$, $y'' = \underset{-\pi/2}{(}---\underset{0}{|}+++\underset{\pi/2}{)}$ $\Rightarrow$ the graph is concave up on $\left(0, \frac{\pi}{2}\right)$, concave down on $\left(-\frac{\pi}{2}, 0\right) \Rightarrow$ a point of inflection at $(0, 0)$

7. If $x \geq 0$, $\sin|x| = \sin x$ and if $x < 0$, $\sin|x| = \sin(-x)$ $= -\sin x$. From the sketch the graph is rising on $\left(-\frac{3\pi}{2}, -\frac{\pi}{2}\right)$, $\left(0, \frac{\pi}{2}\right)$ and $\left(\frac{3\pi}{2}, 2\pi\right)$, falling on $\left(-2\pi, -\frac{3\pi}{2}\right)$, $\left(-\frac{\pi}{2}, 0\right)$ and $\left(\frac{\pi}{2}, \frac{3\pi}{2}\right)$; local minima are -1 at $x = \pm\frac{3\pi}{2}$ and 0 at $x = 0$; local maxima are 1 at $x = \pm\frac{\pi}{2}$ and 0 at $x = \pm 2\pi$; concave up on $(-2\pi, -\pi)$ and $(\pi, 2\pi)$, and concavedown on $(-\pi, 0)$ and $(0, \pi) \Rightarrow$ points of inflection are $(-\pi, 0)$ and $(\pi, 0)$

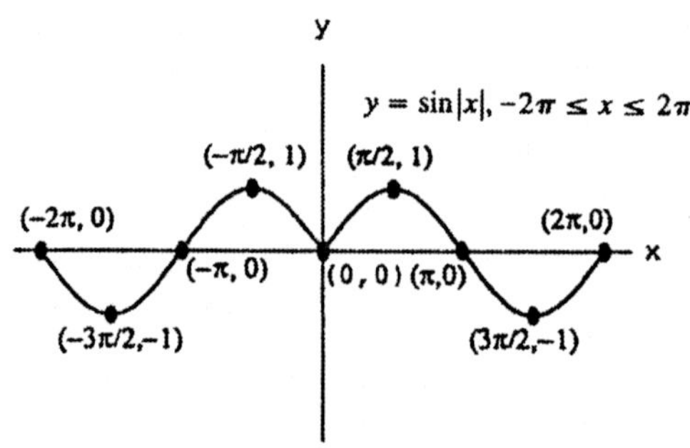

8. $y = 2\cos x - \sqrt{2}x \Rightarrow y' = -2\sin x - \sqrt{2}$, $y' = \underset{-\pi}{[}\ ---\underset{-3\pi/4}{|}\ +++\underset{-\pi/4}{|}\ ---\underset{5\pi/4}{|}\ +++\underset{3\pi/2}{]}$ $\Rightarrow$ rising on $\left(-\frac{3\pi}{4}, -\frac{\pi}{4}\right)$and $\left(\frac{5\pi}{4}, \frac{3\pi}{2}\right)$, falling on $\left(-\pi, -\frac{3\pi}{4}\right)$ and $\left(-\frac{\pi}{4}, \frac{5\pi}{4}\right) \Rightarrow$ local maxima are $-2 + \pi\sqrt{2}$ at $x = -\pi$, $\sqrt{2} + \frac{\pi\sqrt{2}}{4}$ at $x = -\frac{\pi}{4}$ and $-\frac{3\pi\sqrt{2}}{2}$ at $x = \frac{3\pi}{2}$, and local minima are $-\sqrt{2} + \frac{3\pi\sqrt{2}}{4}$ at $x = -\frac{3\pi}{4}$ and $-\sqrt{2} - \frac{5\pi\sqrt{2}}{4}$ at $x = \frac{5\pi}{4}$; $y'' = -2\cos x$, $y'' = \underset{-\pi}{[}\ +++\underset{-\pi/2}{|}\ ---\underset{\pi/2}{|}\ +++\underset{3\pi/2}{]}$ $\Rightarrow$ concave up on $\left(-\pi, -\frac{\pi}{2}\right)$ and $\left(\frac{\pi}{2}, \frac{3\pi}{2}\right)$, concave down on $\left(-\frac{\pi}{2}, \frac{\pi}{2}\right) \Rightarrow$ points of inflection at $\left(-\frac{\pi}{2}, \frac{\sqrt{2}\pi}{2}\right)$ and $\left(\frac{\pi}{2}, -\frac{\sqrt{2}\pi}{2}\right)$

9. When $y = x^2 - 4x + 3$, then $y' = 2x - 4 = 2(x - 2)$ and $y'' = 2$. The curve rises on $(2, \infty)$ and falls on $(-\infty, 2)$. At $x = 2$ there is a minimum. Since $y'' > 0$, the curve is concave up for all x.

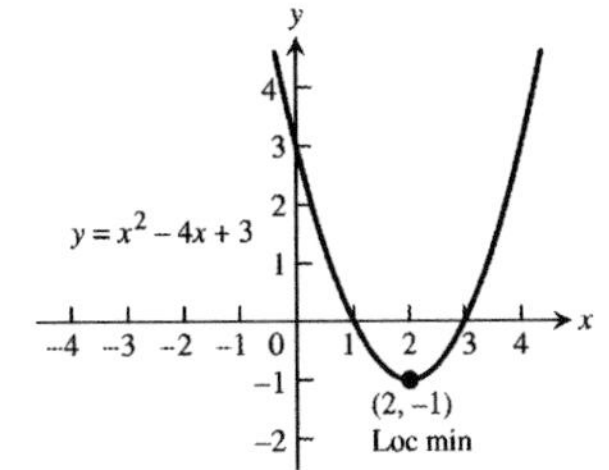

10. When $y = 6 - 2x - x^2$, then $y' = -2 - 2x = -2(1 + x)$ and $y'' = -2$. The curve rises on $(-\infty, -1)$ and falls on $(-1, \infty)$. At $x = -1$ there is a maximum. Since $y'' < 0$, the curve is concave down for all x.

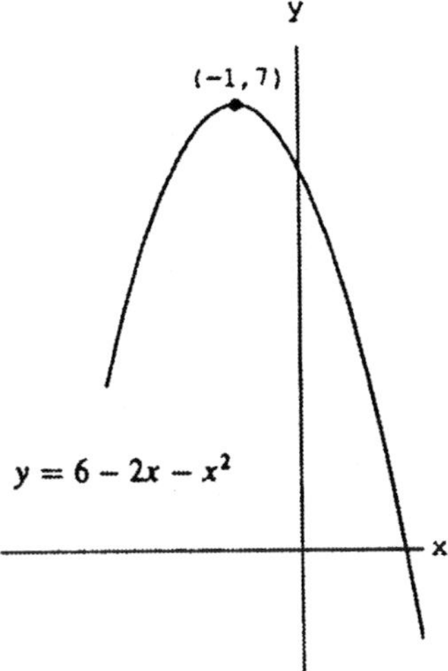

11. When $y = x^3 - 3x + 3$, then $y' = 3x^2 - 3 = 3(x - 1)(x + 1)$ and $y'' = 6x$. The curve rises on $(-\infty, -1) \cup (1, \infty)$ and falls on $(-1, 1)$. At $x = -1$ there is a local maximum and at $x = 1$ a local minimum. The curve is concave down on $(-\infty, 0)$ and concave up on $(0, \infty)$. There is a point of inflection at $x = 0$.

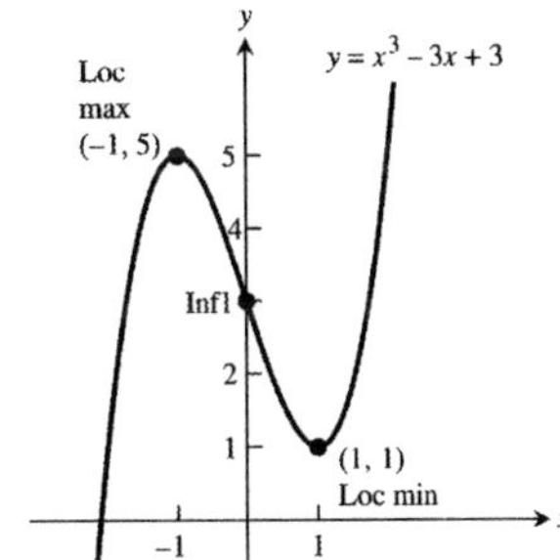

12. When $y = x(6 - 2x)^2$, then $y' = -4x(6 - 2x) + (6 - 2x)^2$ $= 12(3 - x)(1 - x)$ and $y'' = -12(3 - x) - 12(1 - x)$ $= 24(x - 2)$. The curve rises on $(-\infty, 1) \cup (3, \infty)$ and falls on $(1, 3)$. The curve is concave down on $(-\infty, 2)$ and concave up on $(2, \infty)$. At $x = 2$ there is a point of inflection.

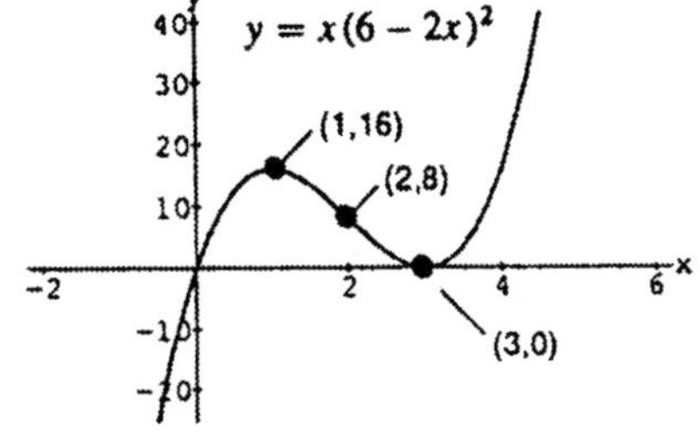

13. When $y = -2x^3 + 6x^2 - 3$, then $y' = -6x^2 + 12x$ $= -6x(x - 2)$ and $y'' = -12x + 12 = -12(x - 1)$. The curve rises on $(0, 2)$ and falls on $(-\infty, 0)$ and $(2, \infty)$. At $x = 0$ there is a local minimum and at $x = 2$ a local maximum. The curve is concave up on $(-\infty, 1)$ and concave down on $(1, \infty)$. At $x = 1$ there is a point of inflection.

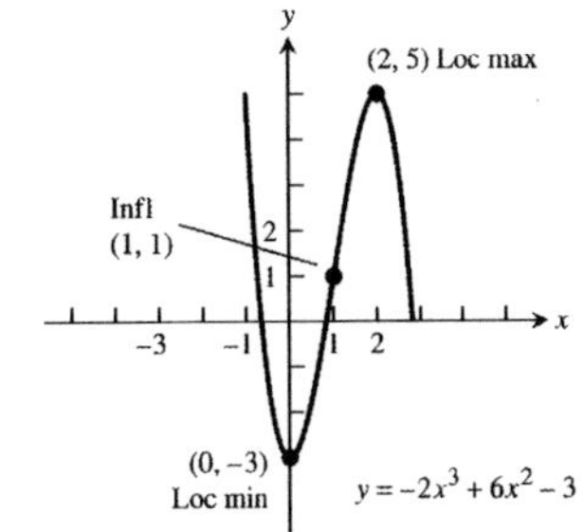

14. When $y = (x-2)^3 + 1$, then $y' = 3(x-2)^2$ and $y'' = 6(x-2)$. The curve never falls and there are no local extrema. The curve is concave down on $(-\infty, 2)$ and concave up on $(2, \infty)$. At $x = 2$ there is a point of inflection.

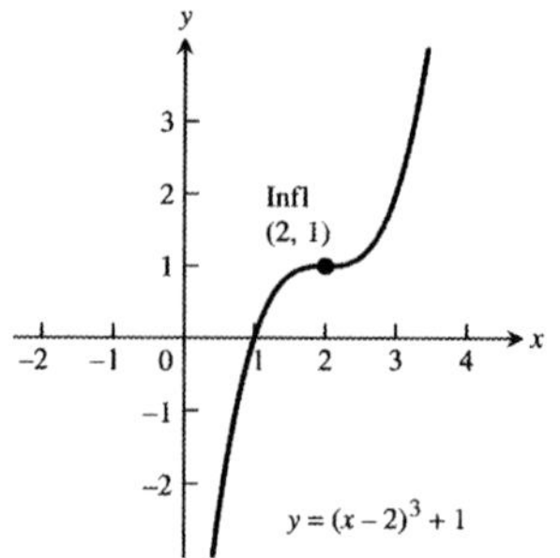

15. When $y = x^4 - 2x^2$, then $y' = 4x^3 - 4x = 4x(x+1)(x-1)$ and $y'' = 12x^2 - 4 = 12\left(x + \frac{1}{\sqrt{3}}\right)\left(x - \frac{1}{\sqrt{3}}\right)$. The curve rises on $(-1, 0)$ and $(1, \infty)$ and falls on $(-\infty, -1)$ and $(0, 1)$. At $x = \pm 1$ there are local minima and at $x = 0$ a local maximum. The curve is concave up on $\left(-\infty, -\frac{1}{\sqrt{3}}\right)$ and $\left(\frac{1}{\sqrt{3}}, \infty\right)$ and concave down on $\left(-\frac{1}{\sqrt{3}}, \frac{1}{\sqrt{3}}\right)$. At $x = \frac{\pm 1}{\sqrt{3}}$ there are points of inflection.

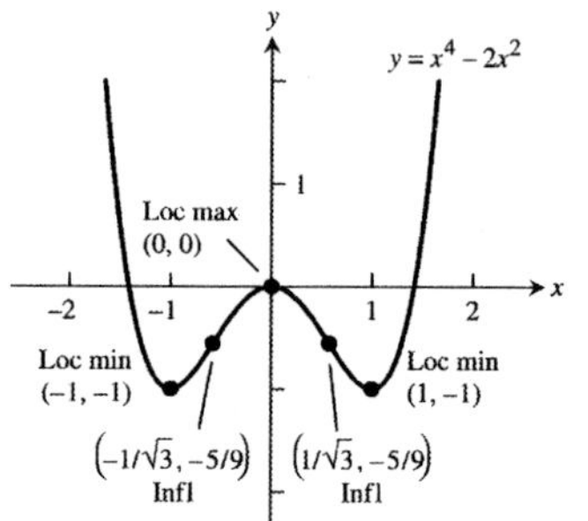

16. When $y = -x^4 + 6x^2 - 4$, then $y' = -4x^3 + 12x$ $= -4x\left(x + \sqrt{3}\right)\left(x - \sqrt{3}\right)$ and $y'' = -12x^2 + 12$ $= -12(x+1)(x-1)$. The curve rises on $\left(-\infty, -\sqrt{3}\right)$ and $\left(0, \sqrt{3}\right)$, and falls on $\left(-\sqrt{3}, 0\right)$ and $\left(\sqrt{3}, \infty\right)$. At $x = \pm\sqrt{3}$there are local maxima and at $x = 0$ a local minimum. The curve is concave up on $(-1, 1)$ and concave down on $(-\infty, -1)$ and $(1, \infty)$. At $x = \pm 1$ there are points of inflection.

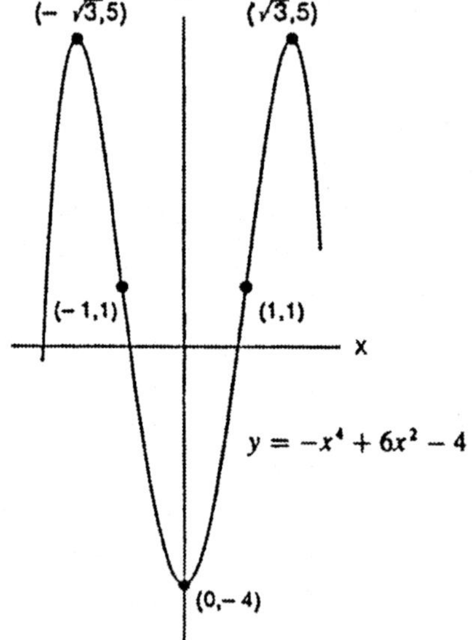

17. When $y = 4x^3 - x^4$, then $y' = 12x^2 - 4x^3 = 4x^2(3-x)$ and $y'' = 24x - 12x^2 = 12x(2-x)$. The curve rises on $(-\infty, 3)$ and falls on $(3, \infty)$. At $x = 3$ there is a local maximum, but there is no local minimum. The graph is concave up on $(0, 2)$ and concave down on $(-\infty, 0)$ and $(2, \infty)$. There are inflection points at $x = 0$ and $x = 2$.

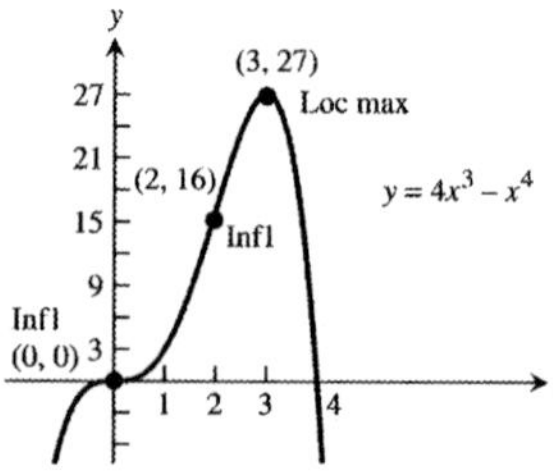

18. When $y = x^4 + 2x^3$, then $y' = 4x^3 + 6x^2 = 2x^2(2x + 3)$ and $y'' = 12x^2 + 12x = 12x(x + 1)$. The curve rises on $\left(-\frac{3}{2}, \infty\right)$ and falls on $\left(-\infty, -\frac{3}{2}\right)$. There is a local minimum at $x = -\frac{3}{2}$, but no local maximum. The curve is concave up on $(-\infty, -1)$ and $(0, \infty)$, and concave down on $(-1, 0)$. At $x = -1$ and $x = 0$ there are points of inflection.

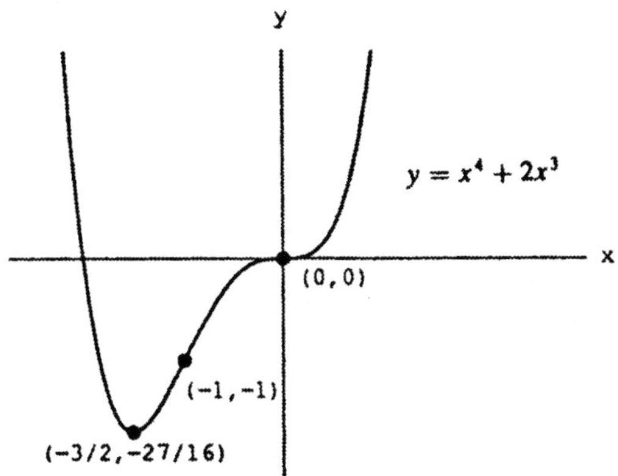

19. When $y = x^5 - 5x^4$, then $y' = 5x^4 - 20x^3 = 5x^3(x - 4)$ and $y'' = 20x^3 - 60x^2 = 20x^2(x - 3)$. The curve rises on $(-\infty, 0)$ and $(4, \infty)$, and falls on $(0, 4)$. There is a local maximum at $x = 0$, and a local minimum at $x = 4$. The curve is concave down on $(-\infty, 3)$ and concave up on $(3, \infty)$. At $x = 3$ there is a point of inflection.

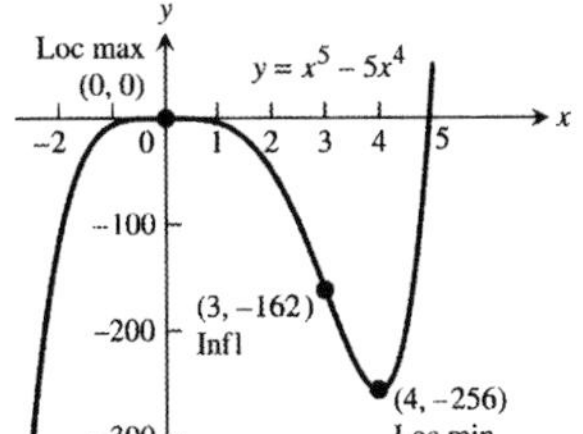

20. When $y = x\left(\frac{x}{2} - 5\right)^4$, then $y' = \left(\frac{x}{2} - 5\right)^4 + x(4)\left(\frac{x}{2} - 5\right)^3\left(\frac{1}{2}\right)$ $= \left(\frac{x}{2} - 5\right)^3\left(\frac{5x}{2} - 5\right)$, and $y'' = 3\left(\frac{x}{2} - 5\right)^2\left(\frac{1}{2}\right)\left(\frac{5x}{2} - 5\right)$ $+ \left(\frac{x}{2} - 5\right)^3\left(\frac{5}{2}\right) = 5\left(\frac{x}{2} - 5\right)^2(x - 4)$. The curve is rising on $(-\infty, 2)$ and $(10, \infty)$, and falling on $(2, 10)$. There is a local maximum at $x = 2$ and a local minimum at $x = 10$. The curve is concave down on $(-\infty, 4)$ and concave up on $(4, \infty)$. At $x = 4$ there is a point of inflection.

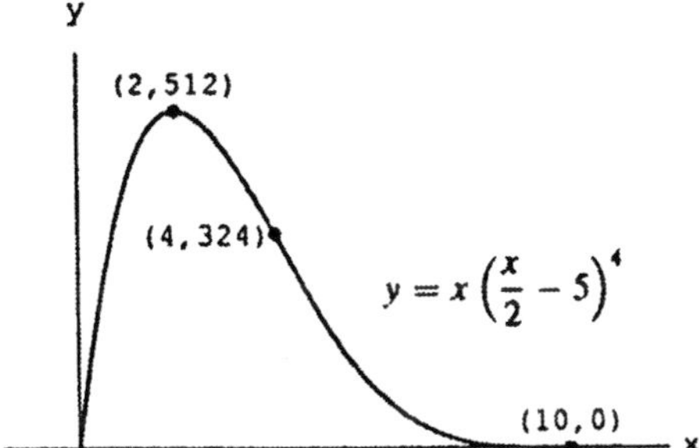

21. When $y = x + \sin x$, then $y' = 1 + \cos x$ and $y'' = -\sin x$. The curve rises on $(0, 2\pi)$. At $x = 0$ there is a local and absolute minimum and at $x = 2\pi$ there is a local and absolute maximum. The curve is concave down on $(0, \pi)$ and concave up on $(\pi, 2\pi)$. At $x = \pi$ there is a point of inflection.

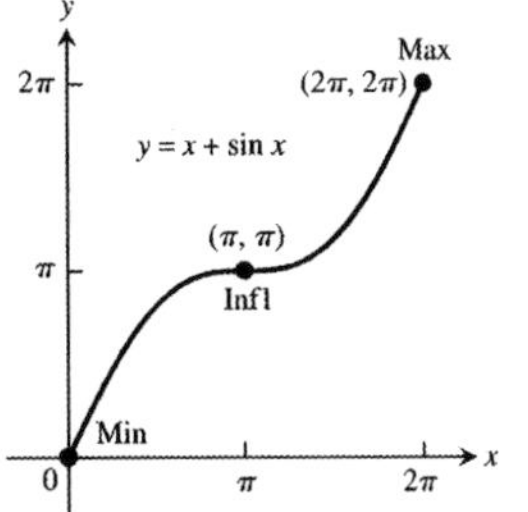

22. When $y = x - \sin x$, then $y' = 1 - \cos x$ and $y'' = \sin x$. The curve rises on $(0, 2\pi)$. At $x = 0$ there is a local and absolute minimum and at $x = 2\pi$ there is a local and absolute maximum. The curve is concave up on $(0, \pi)$ and concave down on $(\pi, 2\pi)$. At $x = \pi$ there is a point of inflection.

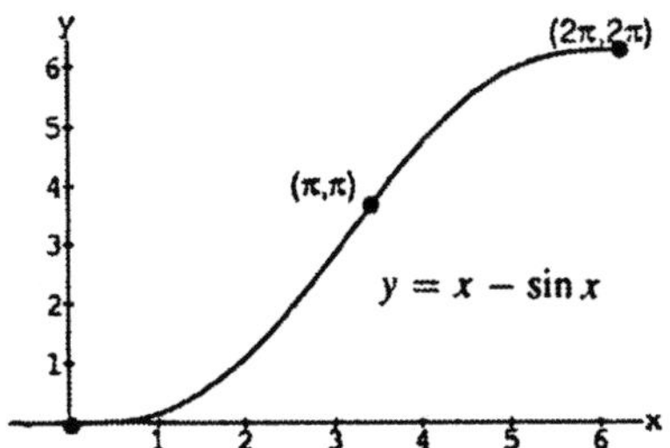

23. When $y = x^{1/5}$, then $y' = \frac{1}{5}x^{-4/5}$ and $y'' = -\frac{4}{25}x^{-9/5}$. The curve rises on $(-\infty, \infty)$ and there are no extrema. The curve is concave up on $(-\infty, 0)$ and concave down on $(0, \infty)$. At $x = 0$ there is a point of inflection.

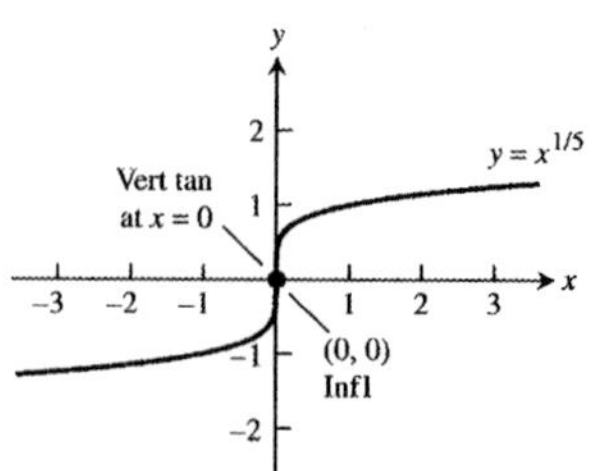

24. When $y = x^{2/5}$, then $y' = \frac{2}{5}x^{-3/5}$ and $y'' = -\frac{6}{25}x^{-8/5}$. The curve is rising on $(0, \infty)$ and falling on $(-\infty, 0)$. At $x = 0$ there is a local and absolute minimum. There is no local or absolute maximum. The curve is concave down on $(-\infty, 0)$ and $(0, \infty)$. There are no points of inflection, but a cusp exists at $x = 0$.

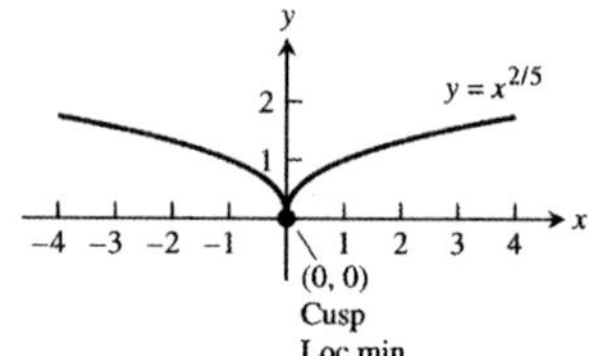

25. When $y = 2x - 3x^{2/3}$, then $y' = 2 - 2x^{-1/3}$ and $y'' = \frac{2}{3}x^{-4/3}$. The curve is rising on $(-\infty, 0)$ and $(1, \infty)$, and falling on $(0, 1)$. There is a local maximum at $x = 0$ and a local minimum at $x = 1$. The curve is concave up on $(-\infty, 0)$ and $(0, \infty)$. There are no points of inflection, but a cusp exists at $x = 0$.

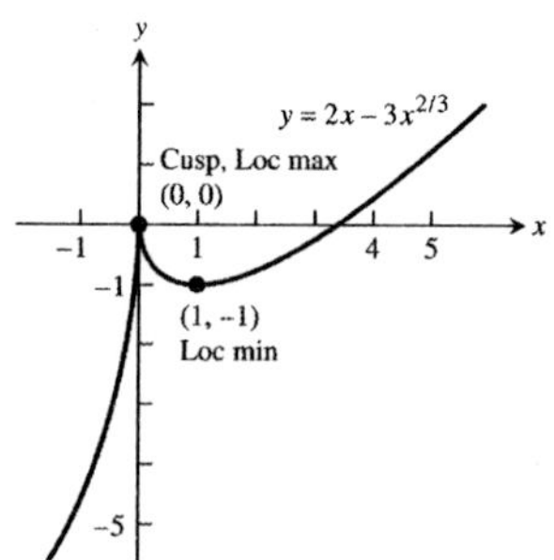

26. When $y = x^{2/3}\left(\frac{5}{2} - x\right) = \frac{5}{2}x^{2/3} - x^{5/3}$, then $y' = \frac{5}{3}x^{-1/3} - \frac{5}{3}x^{2/3} = \frac{5}{3}x^{-1/3}(1 - x)$ and $y'' = -\frac{5}{9}x^{-4/3} - \frac{10}{9}x^{-1/3} = -\frac{5}{9}x^{-4/3}(1 + 2x)$. The curve is rising on $(0, 1)$ and falling on $(-\infty, 0)$ and $(1, \infty)$. There is a local minimum at $x = 0$ and a local maximum at $x = 1$. The curve is concave up on $\left(-\infty, -\frac{1}{2}\right)$ and concave down on $\left(-\frac{1}{2}, 0\right)$ and $(0, \infty)$. There is a point of inflection at $x = -\frac{1}{2}$ and a cusp at $x = 0$.

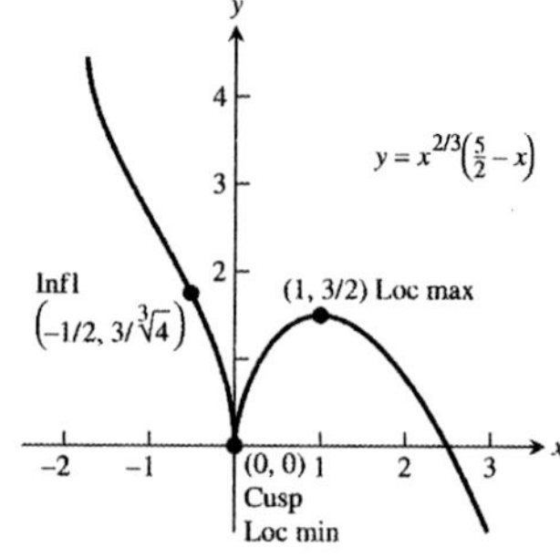

27. When $y = x\sqrt{8 - x^2} = x(8 - x^2)^{1/2}$, then

$$y' = (8 - x^2)^{1/2} + (x)\left(\tfrac{1}{2}\right)(8 - x^2)^{-1/2}(-2x)$$

$$= (8 - x^2)^{-1/2}(8 - 2x^2) = \frac{2(2-x)(2+x)}{\sqrt{\left(2\sqrt{2}+x\right)\left(2\sqrt{2}-x\right)}} \text{ and}$$

$$y'' = \left(-\tfrac{1}{2}\right)(8 - x^2)^{-\frac{3}{2}}(-2x)(8 - 2x^2) + (8 - x^2)^{-\frac{1}{2}}(-4x)$$

$$= \frac{2x(x^2 - 12)}{\sqrt{(8 - x^2)^3}}.$$

The curve is rising on $(-2, 2)$, and falling on $\left(-2\sqrt{2}, -2\right)$ and $\left(2, 2\sqrt{2}\right)$. There are local minima $x = -2$ and $x = 2\sqrt{2}$, and local maxima at $x = -2\sqrt{2}$ and $x = 2$. The curve is concave up on $\left(-2\sqrt{2}, 0\right)$ and concave down on $\left(0, 2\sqrt{2}\right)$. There is a point of inflection at $x = 0$.

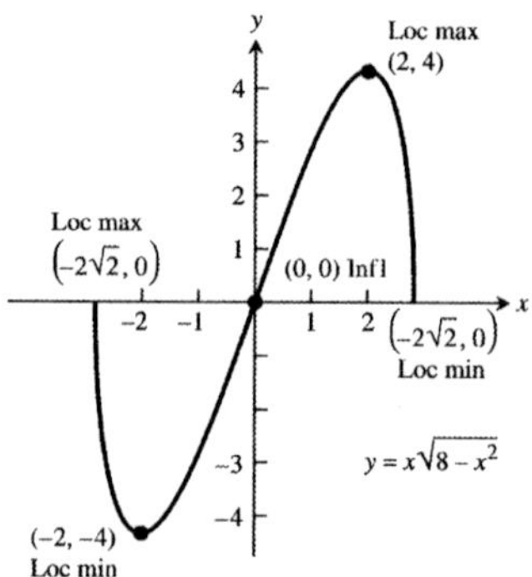

28. When $y = (2 - x^2)^{3/2}$, then $y' = \left(\frac{3}{2}\right)(2 - x^2)^{1/2}(-2x)$
$= -3x\sqrt{2 - x^2} = -3x\sqrt{\left(\sqrt{2} - x\right)\left(\sqrt{2} + x\right)}$ and
$y'' = (-3)(2 - x^2)^{1/2} + (-3x)\left(\frac{1}{2}\right)(2 - x^2)^{-1/2}(-2x)$
$= \frac{-6(1 - x)(1 + x)}{\sqrt{\left(\sqrt{2} - x\right)\left(\sqrt{2} + x\right)}}$. The curve is rising on $\left(-\sqrt{2}, 0\right)$ and falling on $\left(0, \sqrt{2}\right)$. There is a local maximum at $x = 0$, and local minima at $x = \pm\sqrt{2}$. The curve is concave down on $(-1, 1)$ and concave up on $\left(-\sqrt{2}, -1\right)$ and $\left(1, \sqrt{2}\right)$. There are points of inflection at $x = \pm 1$.

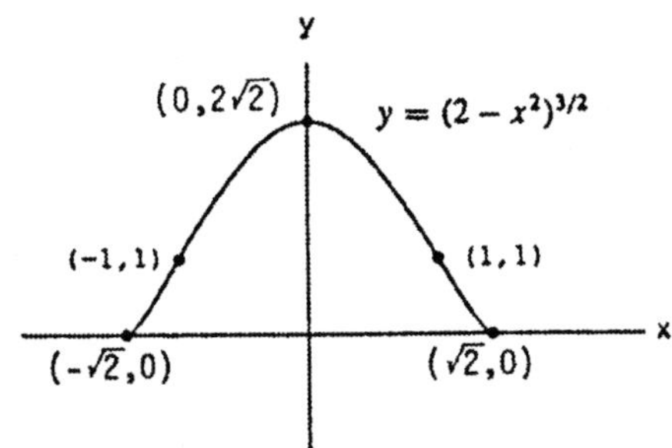

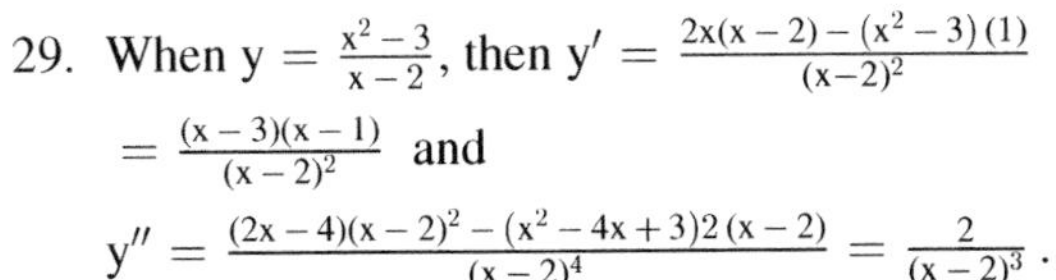

29. When $y = \frac{x^2 - 3}{x - 2}$, then $y' = \frac{2x(x - 2) - (x^2 - 3)(1)}{(x - 2)^2}$
$= \frac{(x - 3)(x - 1)}{(x - 2)^2}$ and
$y'' = \frac{(2x - 4)(x - 2)^2 - (x^2 - 4x + 3)2(x - 2)}{(x - 2)^4} = \frac{2}{(x - 2)^3}$.
The curve is rising on $(-\infty, 1)$ and $(3, \infty)$, and falling on $(1, 2)$ and $(2, 3)$. There is a local maximum at $x = 1$ and a local minimum at $x = 3$. The curve is concave down on $(-\infty, 2)$ and concave up on $(2, \infty)$. There are no points of inflection because $x = 2$ is not in the domain.

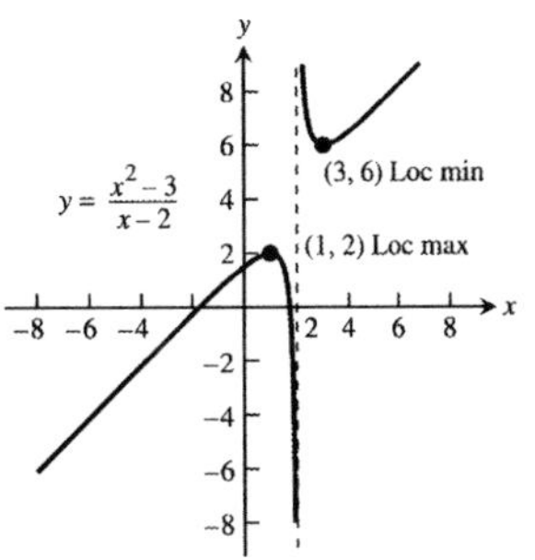

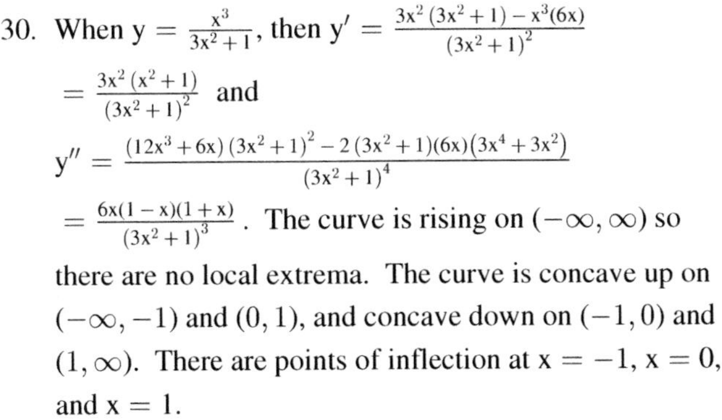

30. When $y = \frac{x^3}{3x^2 + 1}$, then $y' = \frac{3x^2(3x^2 + 1) - x^3(6x)}{(3x^2 + 1)^2}$
$= \frac{3x^2(x^2 + 1)}{(3x^2 + 1)^2}$ and
$y'' = \frac{(12x^3 + 6x)(3x^2 + 1)^2 - 2(3x^2 + 1)(6x)(3x^4 + 3x^2)}{(3x^2 + 1)^4}$
$= \frac{6x(1 - x)(1 + x)}{(3x^2 + 1)^3}$. The curve is rising on $(-\infty, \infty)$ so there are no local extrema. The curve is concave up on $(-\infty, -1)$ and $(0, 1)$, and concave down on $(-1, 0)$ and $(1, \infty)$. There are points of inflection at $x = -1$, $x = 0$, and $x = 1$.

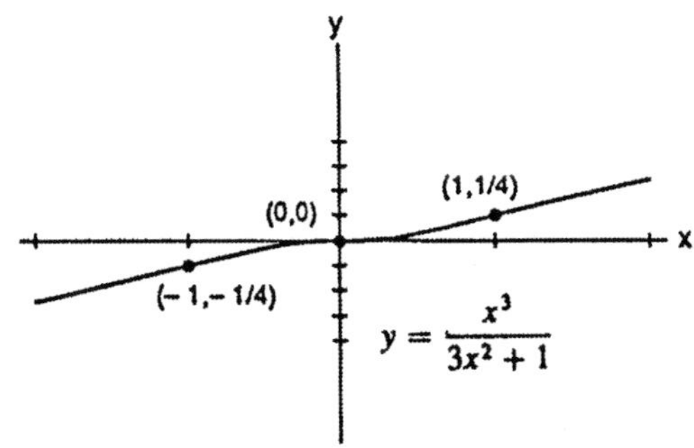

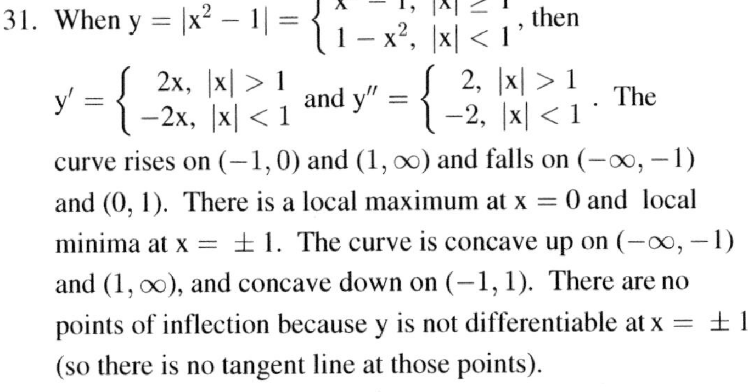

31. When $y = |x^2 - 1| = \begin{cases} x^2 - 1, & |x| \geq 1 \\ 1 - x^2, & |x| < 1 \end{cases}$, then
$y' = \begin{cases} 2x, & |x| > 1 \\ -2x, & |x| < 1 \end{cases}$ and $y'' = \begin{cases} 2, & |x| > 1 \\ -2, & |x| < 1 \end{cases}$. The curve rises on $(-1, 0)$ and $(1, \infty)$ and falls on $(-\infty, -1)$ and $(0, 1)$. There is a local maximum at $x = 0$ and local minima at $x = \pm 1$. The curve is concave up on $(-\infty, -1)$ and $(1, \infty)$, and concave down on $(-1, 1)$. There are no points of inflection because y is not differentiable at $x = \pm 1$ (so there is no tangent line at those points).

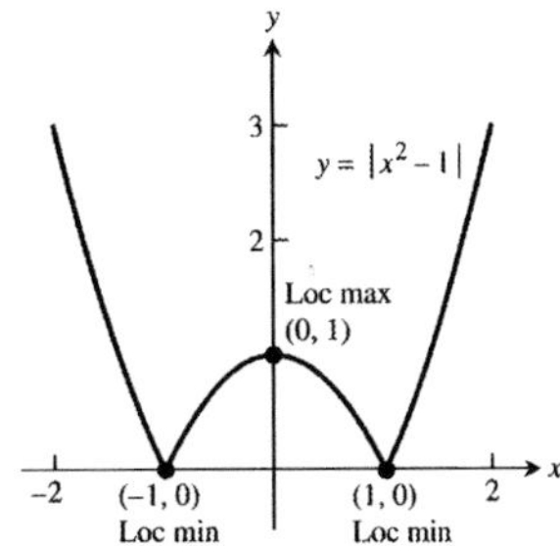

32. When $y = \sqrt{|x|} = \begin{cases} \sqrt{-x}, & x \le 0 \\ \sqrt{x}, & x > 0 \end{cases}$, then

$$y' = \begin{cases} -\frac{1}{2\sqrt{-x}} & x < 0 \\ \frac{1}{2\sqrt{x}} & x > 0 \end{cases} \text{ and } y'' = \begin{cases} -\frac{1}{4}(-x)^{-3/2} & x < 0 \\ -\frac{1}{4}x^{-3/2} & x > 0 \end{cases}.$$

The curve is rising on $(0, \infty)$ and falling on $(-\infty, 0)$. There is a absolute minimum of 0 at $x = 0$. The curve is concave down on $(-\infty, 0)$ and $(0, \infty)$.

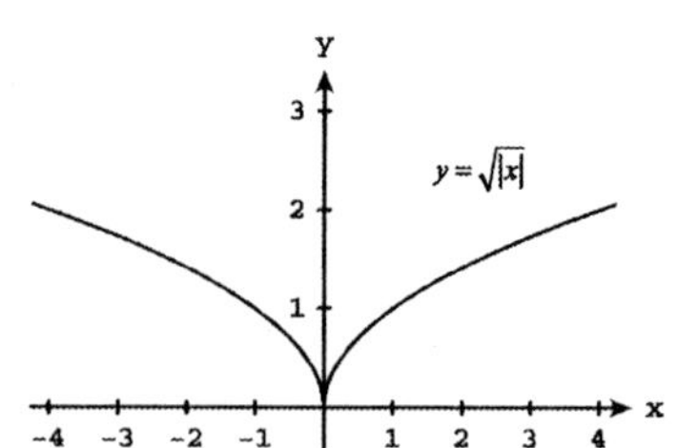

33. When $y = x e^{1/x}$, then $y' = -\frac{x e^{1/x}}{x^2} + e^{1/x} = e^{1/x}\left(1 - \frac{1}{x}\right)$

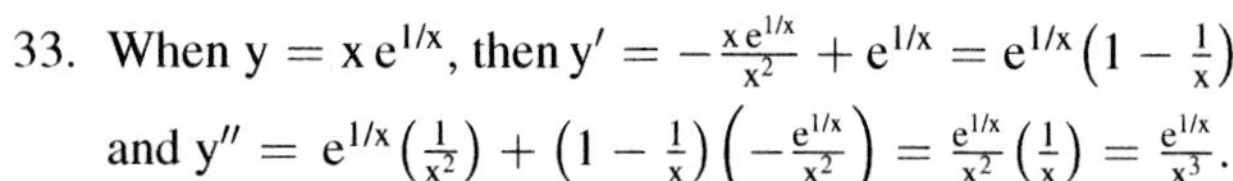

and $y'' = e^{1/x}\left(\frac{1}{x^2}\right) + \left(1 - \frac{1}{x}\right)\left(-\frac{e^{1/x}}{x^2}\right) = \frac{e^{1/x}}{x^2}\left(\frac{1}{x}\right) = \frac{e^{1/x}}{x^3}$.

The curve is rising on $(1, \infty)$ and $(-\infty, 0)$ and falling on $(0, 1)$. The curve is concave down on $(-\infty, 0)$ and concave up on $(0, \infty)$. There is a local minimum of e at $x = 1$, but there are no inflection points.

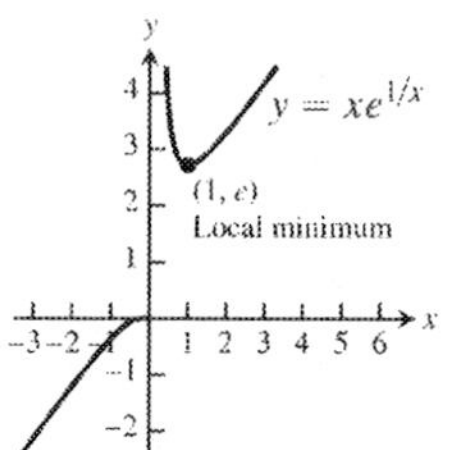

34. $y = \frac{e^x}{x}, \Rightarrow y' = \frac{x e^x - e^x}{x^2} = \frac{(x-1)e^x}{x^2}$

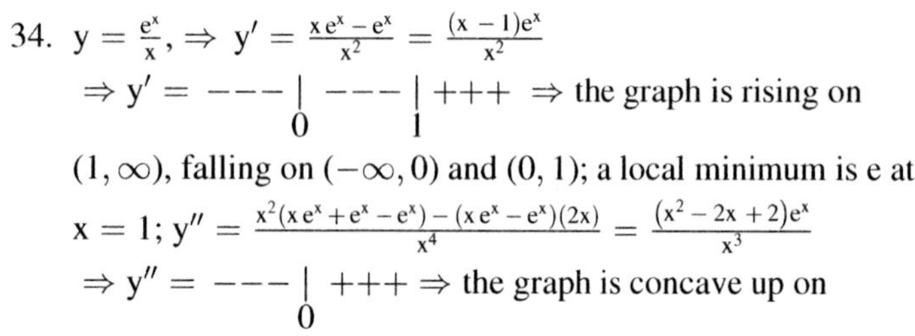

$\Rightarrow y' = \underset{0}{---|}\underset{1}{---|}+++ \Rightarrow$ the graph is rising on $(1, \infty)$, falling on $(-\infty, 0)$ and $(0, 1)$; a local minimum is e at $x = 1$; $y'' = \frac{x^2(x e^x + e^x - e^x) - (x e^x - e^x)(2x)}{x^4} = \frac{(x^2 - 2x + 2)e^x}{x^3}$

$\Rightarrow y'' = \underset{0}{---|}\, +++ \Rightarrow$ the graph is concave up on $(0, \infty)$, concave down on $(-\infty, 0)$, but has no inflection points.

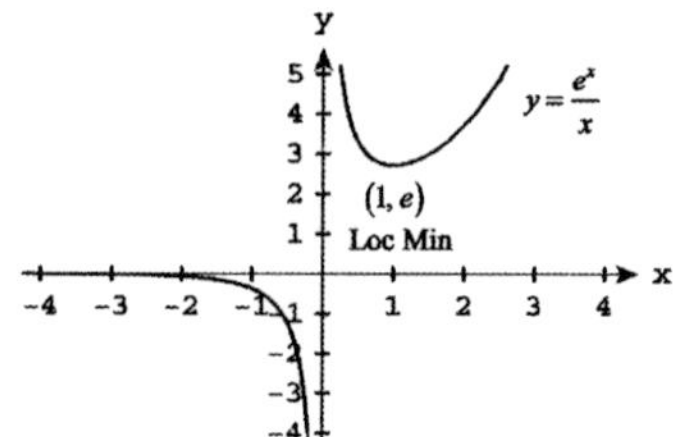

35. $y = \ln(3 - x^2), \Rightarrow y' = \frac{-2x}{3 - x^2} = \frac{2x}{x^2 - 3}$

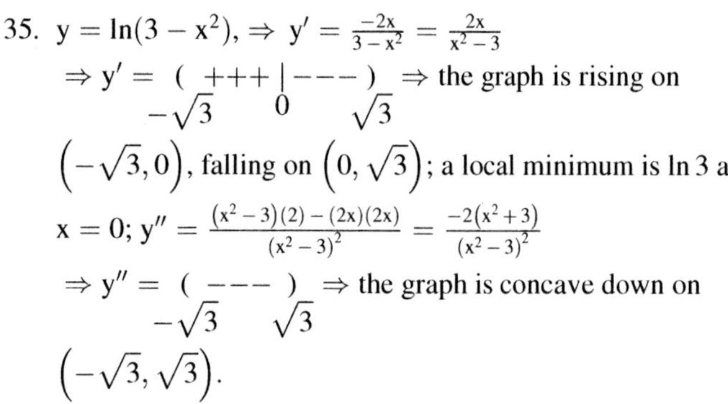

$\Rightarrow y' = \underset{-\sqrt{3}}{(}\ +++\underset{0}{|}---\underset{\sqrt{3}}{)} \Rightarrow$ the graph is rising on $\left(-\sqrt{3}, 0\right)$, falling on $\left(0, \sqrt{3}\right)$; a local minimum is $\ln 3$ at $x = 0$; $y'' = \frac{(x^2-3)(2) - (2x)(2x)}{(x^2-3)^2} = \frac{-2(x^2+3)}{(x^2-3)^2}$

$\Rightarrow y'' = \underset{-\sqrt{3}}{(}\ ---\ \underset{\sqrt{3}}{)} \Rightarrow$ the graph is concave down on $\left(-\sqrt{3}, \sqrt{3}\right)$.

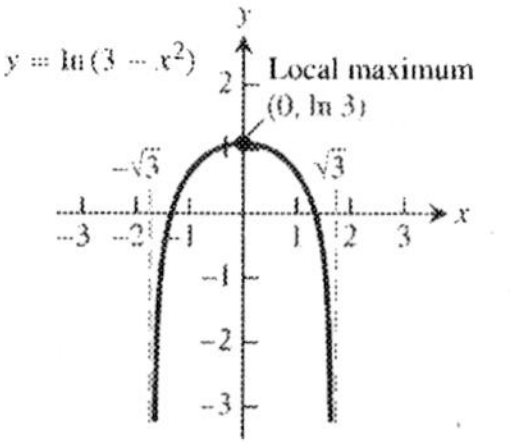

36. $y = x(\ln x)^2, \Rightarrow y' = x \cdot 2\ln x \cdot \left(\frac{1}{x}\right) + (1) \cdot (\ln x)^2$

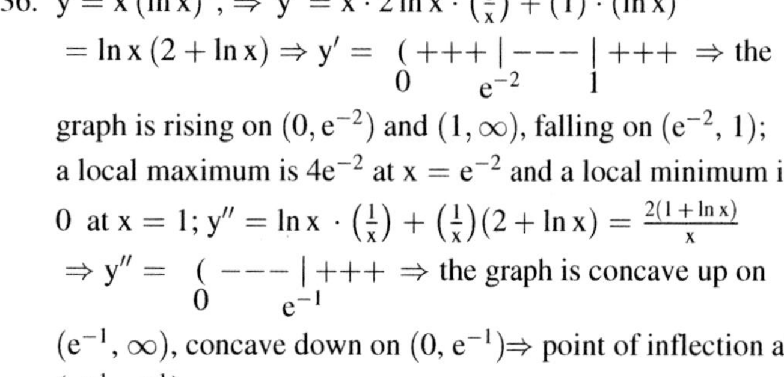

$= \ln x\,(2 + \ln x) \Rightarrow y' = \underset{0}{(}+++\underset{e^{-2}}{|}---\underset{1}{|}+++ \Rightarrow$ the graph is rising on $(0, e^{-2})$ and $(1, \infty)$, falling on $(e^{-2}, 1)$; a local maximum is $4e^{-2}$ at $x = e^{-2}$ and a local minimum is 0 at $x = 1$; $y'' = \ln x \cdot \left(\frac{1}{x}\right) + \left(\frac{1}{x}\right)(2 + \ln x) = \frac{2(1 + \ln x)}{x}$

$\Rightarrow y'' = \underset{0}{(}---\underset{e^{-1}}{|}+++ \Rightarrow$ the graph is concave up on (e^{-1}, ∞), concave down on $(0, e^{-1}) \Rightarrow$ point of inflection at (e^{-1}, e^{-1}).

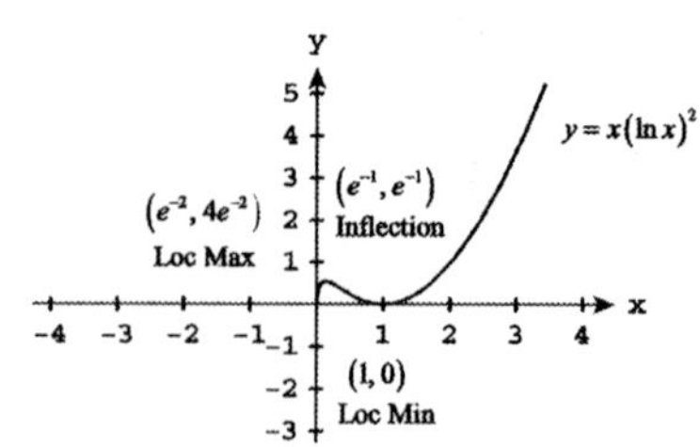

37. $y = e^x - 2e^{-x} - 3x \Rightarrow y' = e^x + 2e^{-x} - 3 = \frac{(e^x)^2 - 3e^x + 2}{e^x}$
$= \frac{(e^x - 2)(e^x - 1)}{e^x} \Rightarrow y' = \underset{0}{+++\ |}\underset{\ln 2}{---\ |}\ +++ \Rightarrow$
the graph is incrreasing on $(-\infty, 0)$ and $(\ln 2, \infty)$,
decreasing on $(0, \ln 2)$; a local maximum is -1 at $x = 0$ and
a local minimum is $1 - 3\ln 2$ at $x = \ln 2$; $y'' = e^x - 2e^{-x}$
$= \frac{(e^x)^2 - 2}{e^x} \Rightarrow y'' = \underset{\frac{1}{2}\ln 2}{---\ |}\ +++ \Rightarrow$ the graph is
concave up on $\left(\frac{1}{2}\ln 2, \infty\right)$, concave down on $\left(-\infty, \frac{1}{2}\ln 2\right)$
$\Rightarrow$ point of inflection at $\left(\frac{1}{2}\ln 2, -\frac{3}{2}\ln 2\right)$.

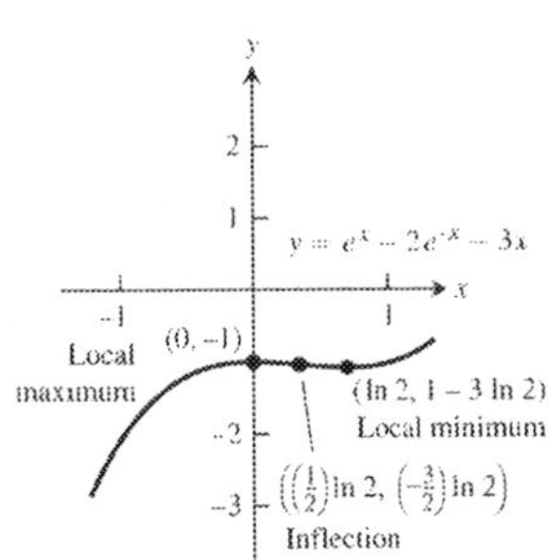

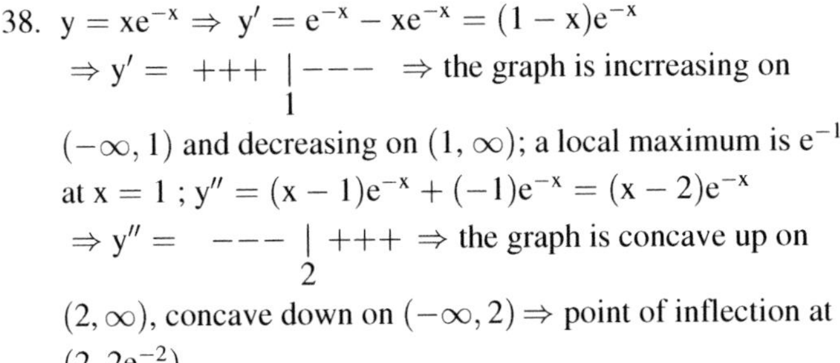
38. $y = xe^{-x} \Rightarrow y' = e^{-x} - xe^{-x} = (1 - x)e^{-x}$
$\Rightarrow y' = \underset{1}{+++\ |}\ --- \Rightarrow$ the graph is incrreasing on
$(-\infty, 1)$ and decreasing on $(1, \infty)$; a local maximum is e^{-1}
at $x = 1$; $y'' = (x - 1)e^{-x} + (-1)e^{-x} = (x - 2)e^{-x}$
$\Rightarrow y'' = \underset{2}{---\ |}\ +++ \Rightarrow$ the graph is concave up on
$(2, \infty)$, concave down on $(-\infty, 2) \Rightarrow$ point of inflection at
$(2, 2e^{-2})$.

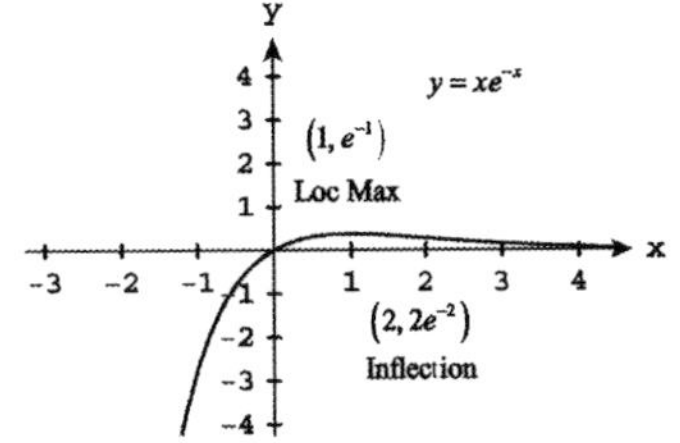

39. $y = \ln(\cos x) \Rightarrow y' = \frac{-\sin x}{\cos x} = -\tan x$
$\Rightarrow y' = \ \ldots\ \underset{-\frac{7\pi}{2}}{)}\ \text{none}\ \underset{-\frac{5\pi}{2}}{(}\ +++\ \underset{-2\pi}{|}\ ---\ \underset{-\frac{3\pi}{2}}{)}\ \text{none}\ \underset{-\frac{\pi}{2}}{(}\ +++\ \underset{0}{|}\ ---\ \underset{\frac{\pi}{2}}{)}\ \text{none}\ \underset{\frac{3\pi}{2}}{(}\ +++\ \underset{2\pi}{|}\ ---\ \underset{\frac{5\pi}{2}}{)}\ \text{none}\ \underset{\frac{7\pi}{2}}{(}\ \ldots$
$\Rightarrow$ the graph is incrreasing on $\ldots, \left(-\frac{5\pi}{2}, -2\pi\right), \left(-\frac{\pi}{2}, 0\right)$,
$\left(\frac{3\pi}{2}, 2\pi\right), \ldots$, decreasing on $\left(-2\pi, -\frac{3\pi}{2}\right), \left(0, \frac{\pi}{2}\right), \left(2\pi, \frac{5\pi}{2}\right)$;
local maxima are 0 at $x = 0, \ \pm 2\pi, \ \pm 4\pi, \ldots$; $y'' = -\sec^2 x$
$\frac{-1}{\cos^2 x} \Rightarrow$ the graph is concave down on $\left(-\frac{5\pi}{2}, -\frac{3\pi}{2}\right), \left(-\frac{\pi}{2}, \frac{\pi}{2}\right)$,
$\left(\frac{3\pi}{2}, \frac{5\pi}{2}\right), \ldots$

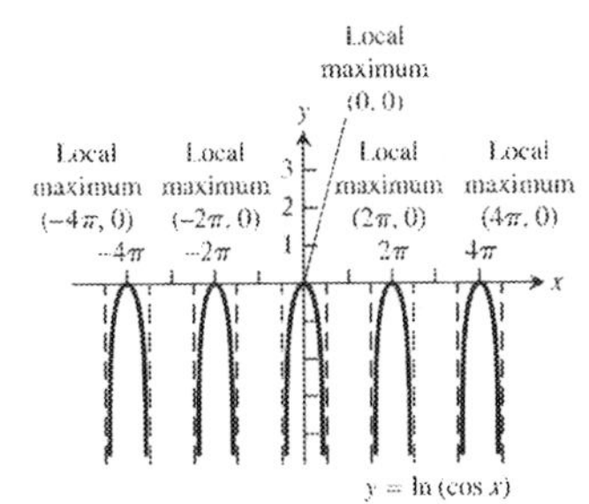

40. $y = \frac{\ln x}{\sqrt{x}} \Rightarrow y' = \frac{\sqrt{x}\cdot\left(\frac{1}{x}\right) - \ln x\cdot\frac{1}{2\sqrt{x}}}{\left(\sqrt{x}\right)^2} = \frac{2 - \ln x}{2x^{3/2}}$
$\Rightarrow y' = \underset{0}{(}\ +++\ \underset{e^2}{|}\ --- \Rightarrow$ the graph is incrreasing on
$(0, e^2)$, decreasing on (e^2, ∞); a local maximum is $\frac{2}{e}$ at
$x = e^2$; $y'' = \frac{2x^{3/2}\left(-\frac{1}{x}\right) - (2 - \ln x)\cdot 2\cdot\frac{3}{2}x^{1/2}}{\left(2x^{3/2}\right)^2} = \frac{3\ln x - 8}{4x^{5/2}}$
$\Rightarrow y'' = \underset{0}{(}\ ---\ \underset{e^{8/3}}{|}\ +++ \Rightarrow$ the graph is concave up on
$\left(e^{8/3}, \infty\right)$, concave down on $\left(0, e^{8/3}\right) \Rightarrow$ point of inflection
is $\left(e^{8/3}, \frac{8}{3e^{4/3}}\right)$.

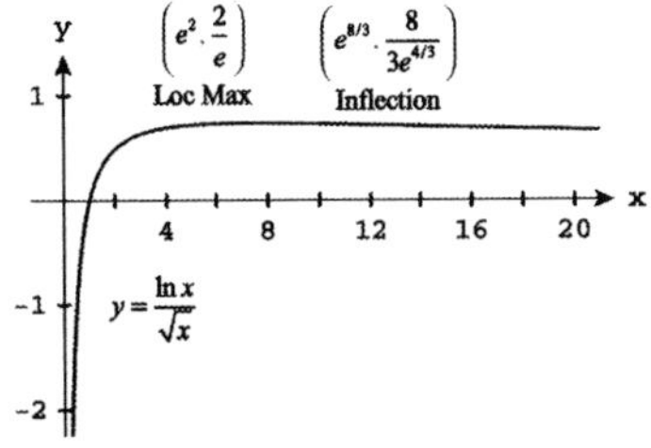

41. $y = \frac{1}{1+e^{-x}} = \frac{e^x}{e^x+1} \Rightarrow y' = \frac{(e^x+1)e^x - e^x \cdot e^x}{(e^x+1)^2} = \frac{e^x}{(e^x+1)^2}$
$\Rightarrow y' = +++ \Rightarrow$ the graph is incrreasing on $(-\infty, \infty)$;
$y'' = \frac{(e^x+1)^2 \cdot e^x - e^x \cdot 2(e^x+1)e^x}{(e^x+1)^4} = \frac{e^x(1-e^x)}{(e^x+1)^3}$
$\Rightarrow y'' = +++ \underset{0}{|} --- \Rightarrow$ the graph is concave up on $(-\infty, 0)$, concave down on $(0, \infty) \Rightarrow$ point of inflection is $\left(0, \frac{1}{2}\right)$.

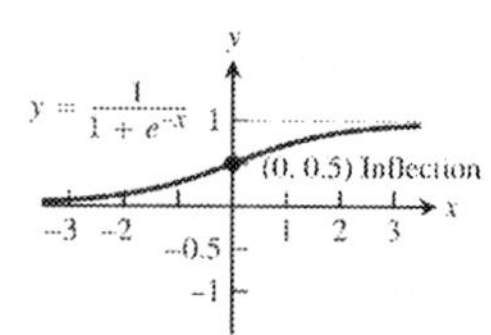

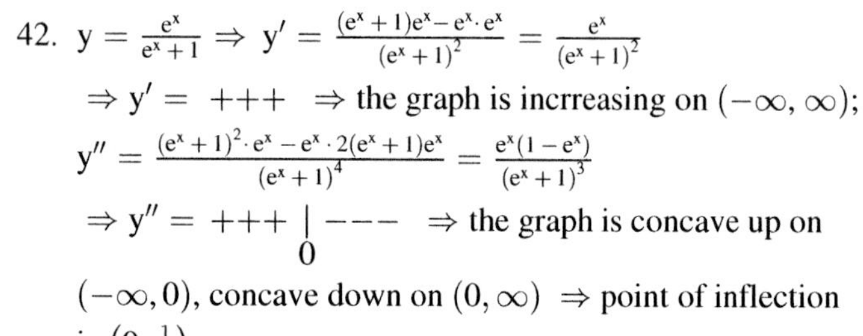

42. $y = \frac{e^x}{e^x+1} \Rightarrow y' = \frac{(e^x+1)e^x - e^x \cdot e^x}{(e^x+1)^2} = \frac{e^x}{(e^x+1)^2}$
$\Rightarrow y' = +++ \Rightarrow$ the graph is incrreasing on $(-\infty, \infty)$;
$y'' = \frac{(e^x+1)^2 \cdot e^x - e^x \cdot 2(e^x+1)e^x}{(e^x+1)^4} = \frac{e^x(1-e^x)}{(e^x+1)^3}$
$\Rightarrow y'' = +++ \underset{0}{|} --- \Rightarrow$ the graph is concave up on $(-\infty, 0)$, concave down on $(0, \infty) \Rightarrow$ point of inflection is $\left(0, \frac{1}{2}\right)$.

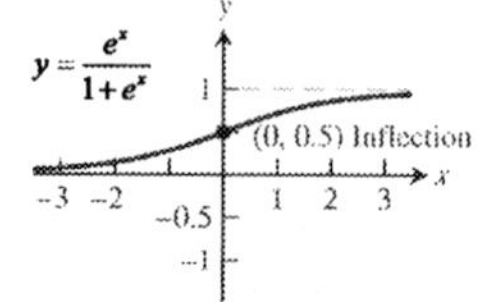

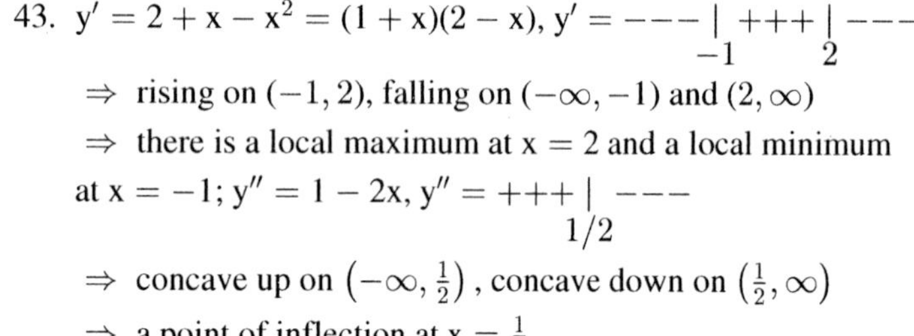

43. $y' = 2 + x - x^2 = (1+x)(2-x)$, $y' = --- \underset{-1}{|} +++ \underset{2}{|} ---$
$\Rightarrow$ rising on $(-1, 2)$, falling on $(-\infty, -1)$ and $(2, \infty)$
$\Rightarrow$ there is a local maximum at $x = 2$ and a local minimum at $x = -1$; $y'' = 1 - 2x$, $y'' = +++ \underset{1/2}{|} ---$
$\Rightarrow$ concave up on $\left(-\infty, \frac{1}{2}\right)$, concave down on $\left(\frac{1}{2}, \infty\right)$
$\Rightarrow$ a point of inflection at $x = \frac{1}{2}$

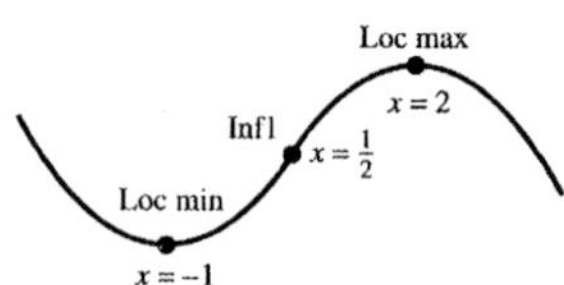

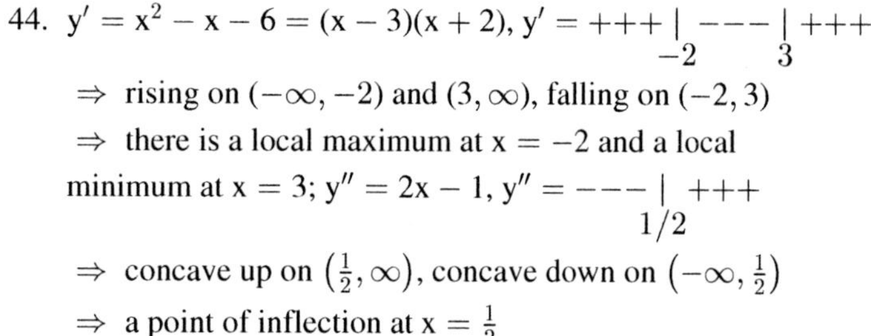

44. $y' = x^2 - x - 6 = (x-3)(x+2)$, $y' = +++ \underset{-2}{|} --- \underset{3}{|} +++$
$\Rightarrow$ rising on $(-\infty, -2)$ and $(3, \infty)$, falling on $(-2, 3)$
$\Rightarrow$ there is a local maximum at $x = -2$ and a local minimum at $x = 3$; $y'' = 2x - 1$, $y'' = --- \underset{1/2}{|} +++$
$\Rightarrow$ concave up on $\left(\frac{1}{2}, \infty\right)$, concave down on $\left(-\infty, \frac{1}{2}\right)$
$\Rightarrow$ a point of inflection at $x = \frac{1}{2}$

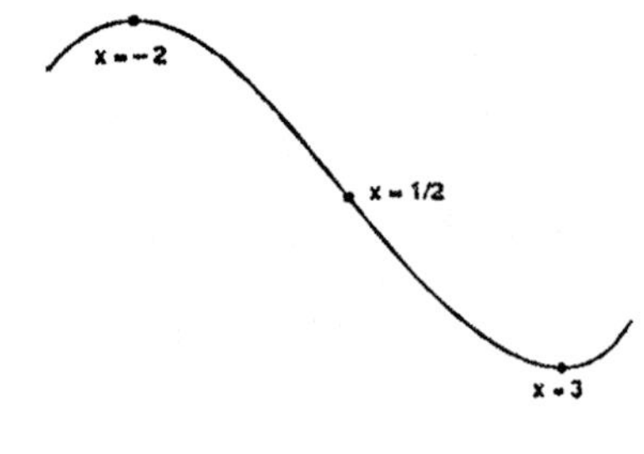

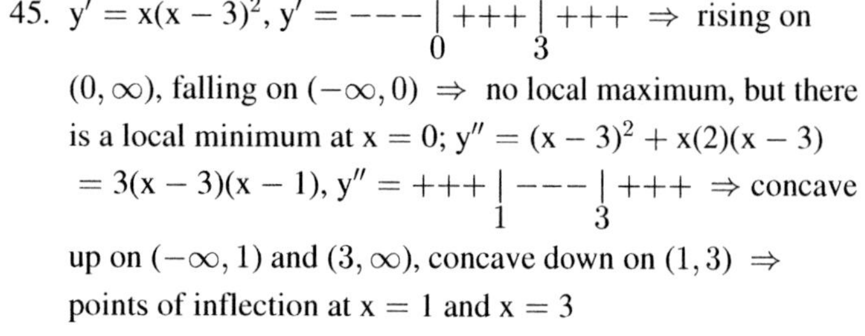

45. $y' = x(x-3)^2$, $y' = --- \underset{0}{|} +++ \underset{3}{|} +++ \Rightarrow$ rising on $(0, \infty)$, falling on $(-\infty, 0) \Rightarrow$ no local maximum, but there is a local minimum at $x = 0$; $y'' = (x-3)^2 + x(2)(x-3)$
$= 3(x-3)(x-1)$, $y'' = +++ \underset{1}{|} --- \underset{3}{|} +++ \Rightarrow$ concave up on $(-\infty, 1)$ and $(3, \infty)$, concave down on $(1, 3) \Rightarrow$ points of inflection at $x = 1$ and $x = 3$

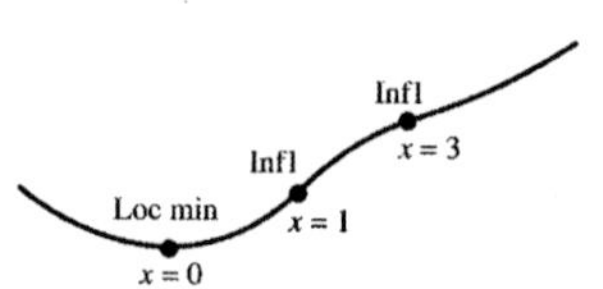

46. $y' = x^2(2 - x)$, $y' = \underset{0}{+++|}\;\underset{2}{+++|}\;---$ $\Rightarrow$ rising on $(-\infty, 2)$, falling on $(2, \infty)$ $\Rightarrow$ there is a local maximum at $x = 2$, but no local minimum; $y'' = 2x(2 - x) + x^2(-1)$ $= x(4 - 3x)$, $y'' = \underset{0}{---|}\;\underset{4/3}{+++|}\;---$ $\Rightarrow$ concave up on $\left(0, \frac{4}{3}\right)$, concave down on $(-\infty, 0)$ and $\left(\frac{4}{3}, \infty\right)$ $\Rightarrow$ points of inflection at $x = 0$ and $x = \frac{4}{3}$

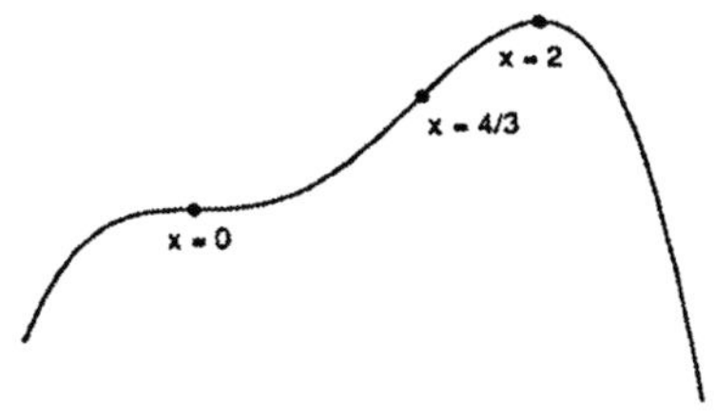

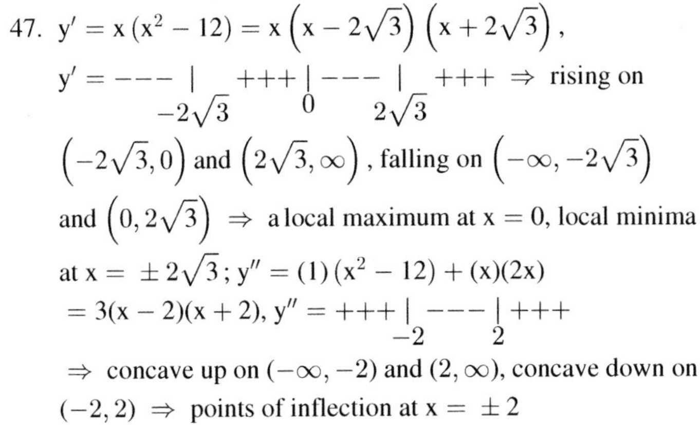

47. $y' = x\,(x^2 - 12) = x\left(x - 2\sqrt{3}\right)\left(x + 2\sqrt{3}\right)$,
$y' = \underset{-2\sqrt{3}}{---|}\;\underset{0}{+++|}\;\underset{2\sqrt{3}}{---|}\;+++$ $\Rightarrow$ rising on $\left(-2\sqrt{3}, 0\right)$ and $\left(2\sqrt{3}, \infty\right)$, falling on $\left(-\infty, -2\sqrt{3}\right)$ and $\left(0, 2\sqrt{3}\right)$ $\Rightarrow$ a local maximum at $x = 0$, local minima at $x = \pm 2\sqrt{3}$; $y'' = (1)\,(x^2 - 12) + (x)(2x)$ $= 3(x - 2)(x + 2)$, $y'' = \underset{-2}{+++|}\;\underset{2}{---|}\;+++$ $\Rightarrow$ concave up on $(-\infty, -2)$ and $(2, \infty)$, concave down on $(-2, 2)$ $\Rightarrow$ points of inflection at $x = \pm 2$

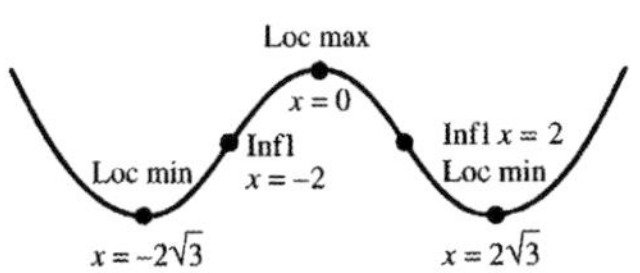

48. $y' = (x^2 - 2x)\,(x - 5)^2 = x(x - 2)(x - 5)^2$,
$y' = \underset{0}{+++|}\;\underset{2}{---|}\;\underset{5}{+++|}\;+++$ $\Rightarrow$ rising on $(-\infty, 0)$ and $(2, \infty)$, falling on $(0, 2)$ $\Rightarrow$ a local maximum at $x = 0$, a local minimum at $x = 2$;
$y'' = (2x - 2)(x - 5)^2 + (x^2 - 2x)\,(2)(x - 5)$
$= 2(x - 5)\,(2x^2 - 8x + 5)$,
$y'' = \underset{\frac{4-\sqrt{6}}{2}}{---|}\;\underset{\frac{4+\sqrt{6}}{2}}{+++|}\;\underset{5}{---|}\;+++$ $\Rightarrow$ concave up on $\left(\frac{4-\sqrt{6}}{2}, \frac{4+\sqrt{6}}{2}\right)$ and $(5, \infty)$, concave down on $\left(-\infty, \frac{4-\sqrt{6}}{2}\right)$ and $\left(\frac{4+\sqrt{6}}{2}, 5\right)$ $\Rightarrow$ points of inflection at $x = \frac{4\pm\sqrt{6}}{2}$ and $x = 5$

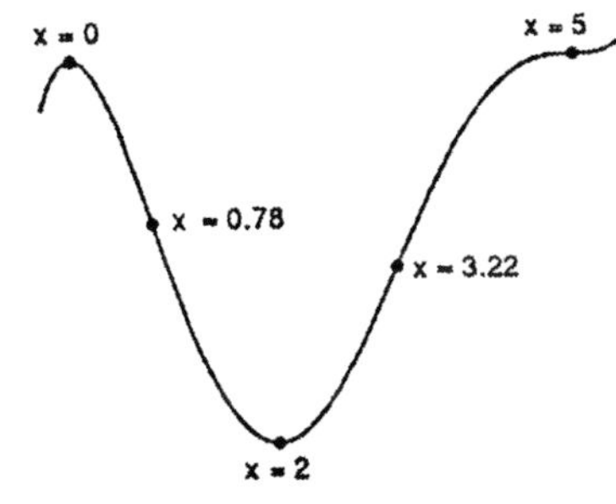

49. $y' = \sec^2 x$, $y' = \underset{-\pi/2}{(}\;+++\;\underset{\pi/2}{)}$ $\Rightarrow$ rising on $\left(-\frac{\pi}{2}, \frac{\pi}{2}\right)$, never falling $\Rightarrow$ no local extrema;
$y'' = 2(\sec x)(\sec x)(\tan x) = 2\,(\sec^2 x)\,(\tan x)$,
$y'' = \underset{-\pi/2}{(}\;\underset{0}{---|}\;+++\;\underset{\pi/2}{)}$ $\Rightarrow$ concave up on $\left(0, \frac{\pi}{2}\right)$, concave down on $\left(-\frac{\pi}{2}, 0\right)$, 0 is a opoint of inflection.

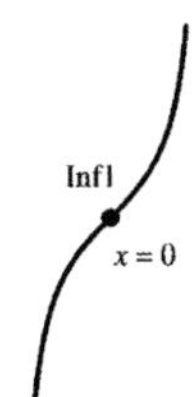

50. $y' = \tan x$, $y' = \underset{-\pi/2}{(}\;\underset{0}{---|}\;+++\;\underset{\pi/2}{)}$ $\Rightarrow$ rising on $\left(0, \frac{\pi}{2}\right)$, falling on $\left(-\frac{\pi}{2}, 0\right)$ $\Rightarrow$ no local maximum, a local minimum at $x = 0$; $y'' = \sec^2 x$, $y'' = \underset{-\pi/2}{(}\;+++\;\underset{\pi/2}{)}$ $\Rightarrow$ concave up on $\left(-\frac{\pi}{2}, \frac{\pi}{2}\right)$ $\Rightarrow$ no points of inflection

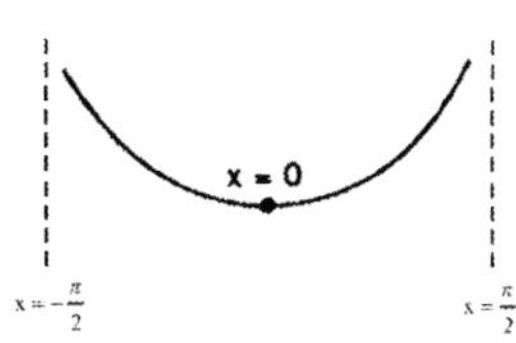

51. $y' = \cot \frac{\theta}{2}$, $y' = \underset{0}{(}+++\underset{\pi}{|}---\underset{2\pi}{)}$ $\Rightarrow$ rising on $(0, \pi)$, falling on $(\pi, 2\pi)$ $\Rightarrow$ a local maximum at $\theta = \pi$, no local minimum; $y'' = -\frac{1}{2}\csc^2 \frac{\theta}{2}$, $y'' = \underset{0}{(}---\underset{2\pi}{)}$ $\Rightarrow$ never concave up, concave down on $(0, 2\pi)$ $\Rightarrow$ no points of inflection

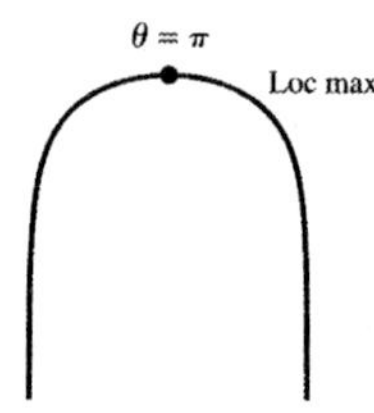

52. $y' = \csc^2 \frac{\theta}{2}$, $y' = \underset{0}{(}+++\underset{2\pi}{)}$ $\Rightarrow$ rising on $(0, 2\pi)$, never falling $\Rightarrow$ no local extrema;

$y'' = 2\left(\csc \frac{\theta}{2}\right)\left(-\csc \frac{\theta}{2}\right)\left(\cot \frac{\theta}{2}\right)\left(\frac{1}{2}\right)$

$= -\left(\csc^2 \frac{\theta}{2}\right)\left(\cot \frac{\theta}{2}\right)$, $y'' = \underset{0}{(}---\underset{\pi}{|}+++\underset{2\pi}{)}$

$\Rightarrow$ concave up on $(\pi, 2\pi)$, concave down on $(0, \pi)$

$\Rightarrow$ a point of inflection at $\theta = \pi$

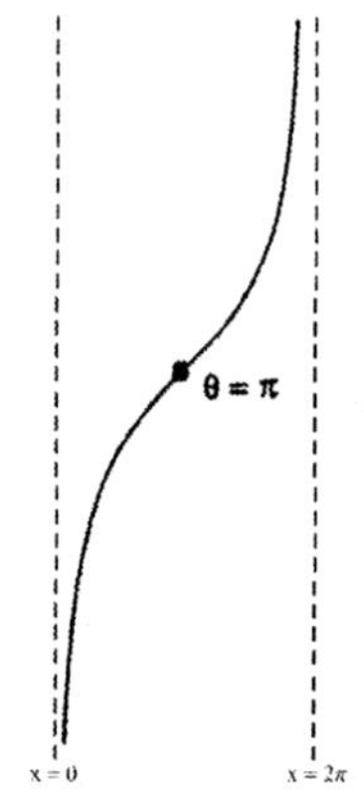

53. $y' = \tan^2 \theta - 1 = (\tan \theta - 1)(\tan \theta + 1)$,

$y' = \underset{-\pi/2}{(}\ +++\underset{-\pi/4}{|}\ ---\underset{\pi/4}{|}\ +++\underset{\pi/2}{)}$ $\Rightarrow$ rising on

$\left(-\frac{\pi}{2}, -\frac{\pi}{4}\right)$ and $\left(\frac{\pi}{4}, \frac{\pi}{2}\right)$, falling on $\left(-\frac{\pi}{4}, \frac{\pi}{4}\right)$

$\Rightarrow$ a local maximum at $\theta = -\frac{\pi}{4}$, a local minimum at $\theta = \frac{\pi}{4}$;

$y'' = 2 \tan \theta \sec^2 \theta$, $y'' = \underset{-\pi/2}{(}\ ---\underset{0}{|}+++\underset{\pi/2}{)}$

$\Rightarrow$ concave up on $\left(0, \frac{\pi}{2}\right)$, concave down on $\left(-\frac{\pi}{2}, 0\right)$

$\Rightarrow$ a point of inflection at $\theta = 0$

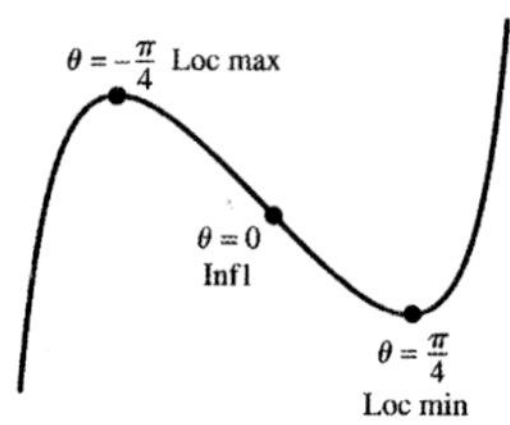

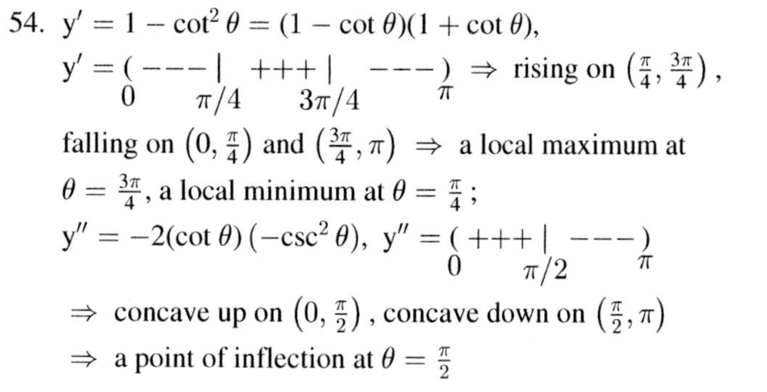

54. $y' = 1 - \cot^2 \theta = (1 - \cot \theta)(1 + \cot \theta)$,

$y' = \underset{0}{(}---\underset{\pi/4}{|}\ +++\underset{3\pi/4}{|}\ ---\underset{\pi}{)}$ $\Rightarrow$ rising on $\left(\frac{\pi}{4}, \frac{3\pi}{4}\right)$,

falling on $\left(0, \frac{\pi}{4}\right)$ and $\left(\frac{3\pi}{4}, \pi\right)$ $\Rightarrow$ a local maximum at $\theta = \frac{3\pi}{4}$, a local minimum at $\theta = \frac{\pi}{4}$;

$y'' = -2(\cot \theta)(-\csc^2 \theta)$, $y'' = \underset{0}{(}+++\underset{\pi/2}{|}\ ---\underset{\pi}{)}$

$\Rightarrow$ concave up on $\left(0, \frac{\pi}{2}\right)$, concave down on $\left(\frac{\pi}{2}, \pi\right)$

$\Rightarrow$ a point of inflection at $\theta = \frac{\pi}{2}$

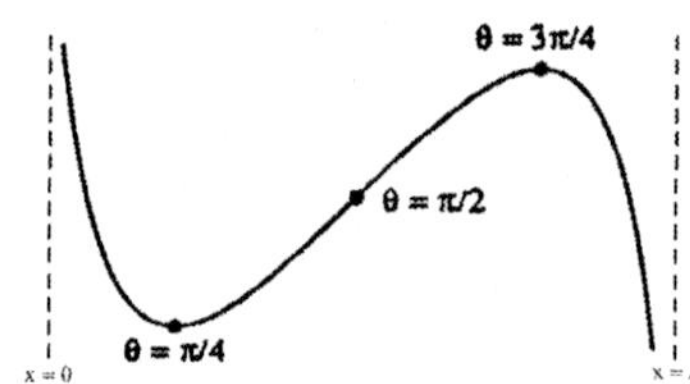

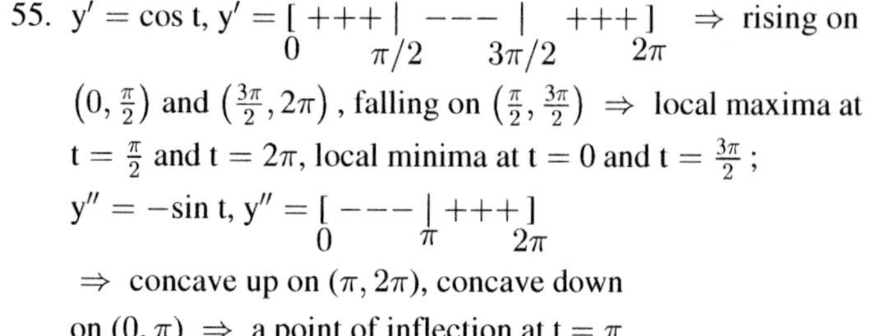

55. $y' = \cos t$, $y' = \underset{0}{[}\ +++\underset{\pi/2}{|}\ ---\underset{3\pi/2}{|}\ +++\underset{2\pi}{]}$ $\Rightarrow$ rising on

$\left(0, \frac{\pi}{2}\right)$ and $\left(\frac{3\pi}{2}, 2\pi\right)$, falling on $\left(\frac{\pi}{2}, \frac{3\pi}{2}\right)$ $\Rightarrow$ local maxima at $t = \frac{\pi}{2}$ and $t = 2\pi$, local minima at $t = 0$ and $t = \frac{3\pi}{2}$;

$y'' = -\sin t$, $y'' = \underset{0}{[}\ ---\underset{\pi}{|}+++\underset{2\pi}{]}$

$\Rightarrow$ concave up on $(\pi, 2\pi)$, concave down on $(0, \pi)$ $\Rightarrow$ a point of inflection at $t = \pi$

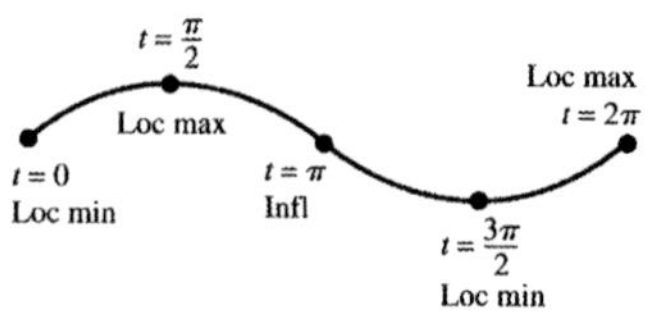

56. $y' = \sin t$, $y' = \underset{0}{[}\,+++\,\underset{\pi}{|}\,---\,\underset{2\pi}{]}$ $\Rightarrow$ rising on $(0, \pi)$, falling on $(\pi, 2\pi)$ $\Rightarrow$ a local maximum at $t = \pi$, local minima at $t = 0$ and $t = 2\pi$; $y'' = \cos t$, $y'' = \underset{0}{[}\,+++\,\underset{\pi/2}{|}\,---\,\underset{3\pi/2}{|}\,+++\,\underset{2\pi}{]}$ $\Rightarrow$ concave up on $\left(0, \frac{\pi}{2}\right)$ and $\left(\frac{3\pi}{2}, 2\pi\right)$, concave down on $\left(\frac{\pi}{2}, \frac{3\pi}{2}\right)$ $\Rightarrow$ points of inflection at $t = \frac{\pi}{2}$ and $t = \frac{3\pi}{2}$

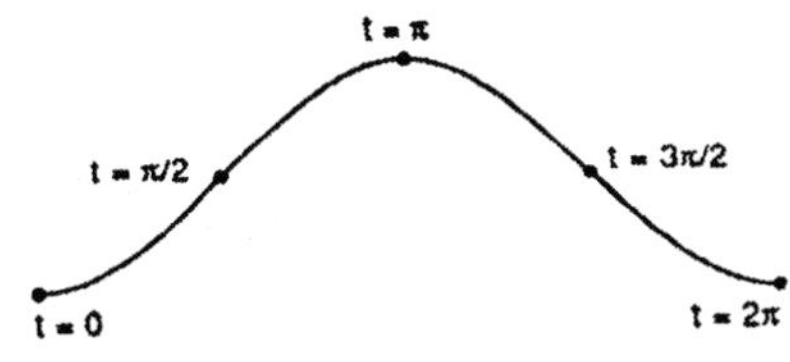

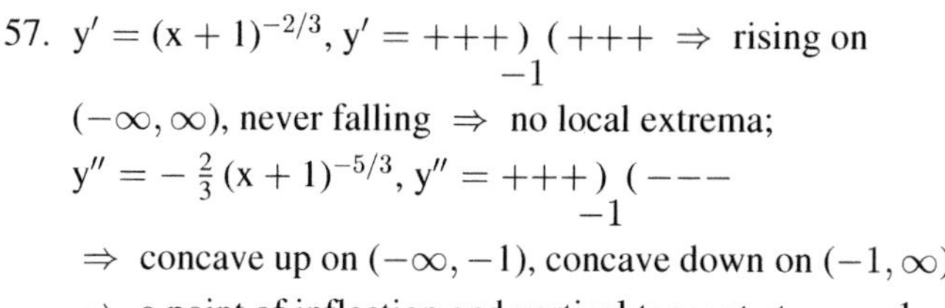

57. $y' = (x+1)^{-2/3}$, $y' = +++\,\underset{-1}{)(}\,+++$ $\Rightarrow$ rising on $(-\infty, \infty)$, never falling $\Rightarrow$ no local extrema; $y'' = -\frac{2}{3}(x+1)^{-5/3}$, $y'' = +++\,\underset{-1}{)(}\,---$ $\Rightarrow$ concave up on $(-\infty, -1)$, concave down on $(-1, \infty)$ $\Rightarrow$ a point of inflection and vertical tangent at $x = -1$

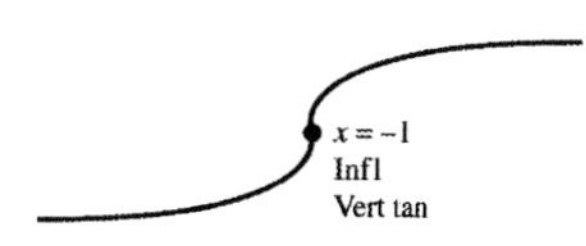

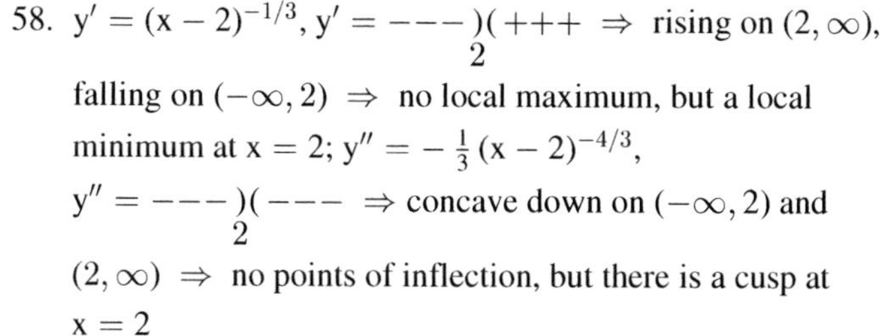

58. $y' = (x-2)^{-1/3}$, $y' = ---\,\underset{2}{)(}\,+++$ $\Rightarrow$ rising on $(2, \infty)$, falling on $(-\infty, 2)$ $\Rightarrow$ no local maximum, but a local minimum at $x = 2$; $y'' = -\frac{1}{3}(x-2)^{-4/3}$, $y'' = ---\,\underset{2}{)(}\,---$ $\Rightarrow$ concave down on $(-\infty, 2)$ and $(2, \infty)$ $\Rightarrow$ no points of inflection, but there is a cusp at $x = 2$

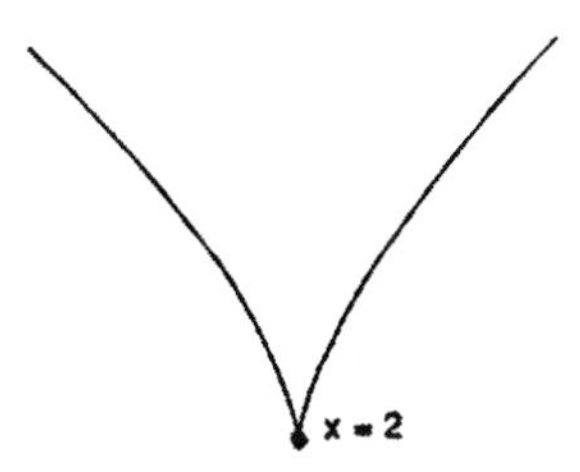

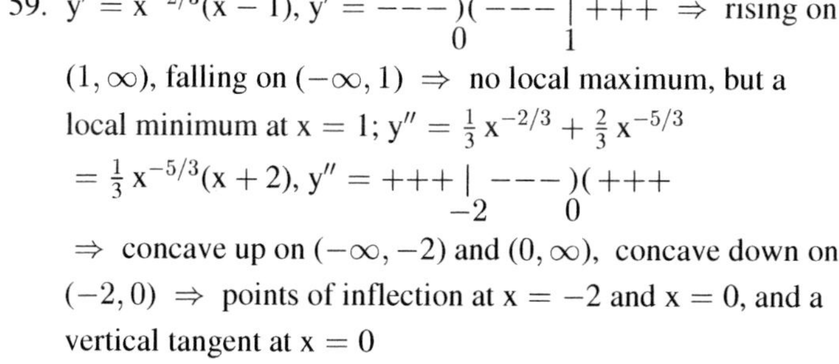

59. $y' = x^{-2/3}(x-1)$, $y' = ---\,\underset{0}{)(}\,---\,\underset{1}{|}\,+++$ $\Rightarrow$ rising on $(1, \infty)$, falling on $(-\infty, 1)$ $\Rightarrow$ no local maximum, but a local minimum at $x = 1$; $y'' = \frac{1}{3}x^{-2/3} + \frac{2}{3}x^{-5/3}$ $= \frac{1}{3}x^{-5/3}(x+2)$, $y'' = +++\,\underset{-2}{|}\,---\,\underset{0}{)(}\,+++$ $\Rightarrow$ concave up on $(-\infty, -2)$ and $(0, \infty)$, concave down on $(-2, 0)$ $\Rightarrow$ points of inflection at $x = -2$ and $x = 0$, and a vertical tangent at $x = 0$

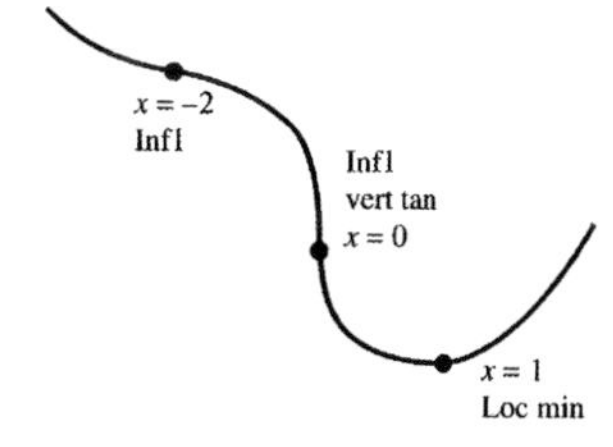

60. $y' = x^{-4/5}(x+1)$, $y' = ---\,\underset{-1}{|}\,+++\,\underset{0}{)(}\,+++$ $\Rightarrow$ rising on $(-1, 0)$ and $(0, \infty)$, falling on $(-\infty, -1)$ $\Rightarrow$ no local maximum, but a local minimum at $x = -1$; $y'' = \frac{1}{5}x^{-4/5} - \frac{4}{5}x^{-9/5} = \frac{1}{5}x^{-9/5}(x-4)$, $y'' = +++\,\underset{0}{)(}\,---\,\underset{4}{|}\,+++$ $\Rightarrow$ concave up on $(-\infty, 0)$ and $(4, \infty)$, concave down on $(0, 4)$ $\Rightarrow$ points of inflection at $x = 0$ and $x = 4$, and a vertical tangent at $x = 0$

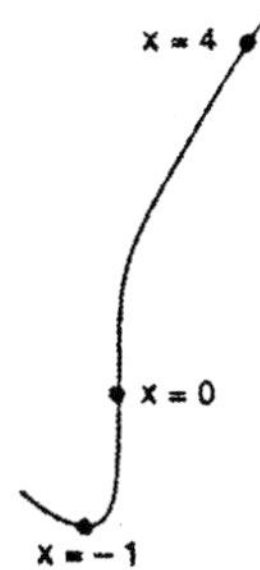

61. $y' = \begin{cases} -2x, & x \le 0 \\ 2x, & x > 0 \end{cases}$, $y' = +++\,\underset{0}{|}\,+++$ $\Rightarrow$ rising on $(-\infty, \infty)$ $\Rightarrow$ no local extrema; $y'' = \begin{cases} -2, & x < 0 \\ 2, & x > 0 \end{cases}$, $y'' = ---\,\underset{0}{)(}\,+++$ $\Rightarrow$ concave up on $(0, \infty)$, concave down on $(-\infty, 0)$ $\Rightarrow$ a point of inflection at $x = 0$

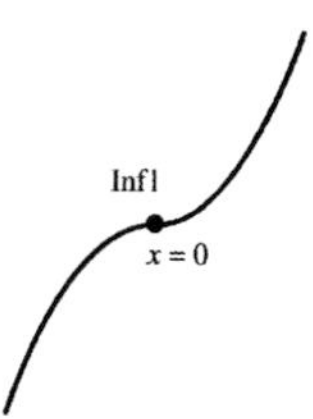

62. $y' = \begin{cases} -x^2, & x \le 0 \\ x^2, & x > 0 \end{cases}$, $y' = \underset{0}{---|+++} \Rightarrow$ rising on $(0, \infty)$, falling on $(-\infty, 0) \Rightarrow$ no local maximum, but a local minimum at $x = 0$; $y'' = \begin{cases} -2x, & x \le 0 \\ 2x, & x > 0 \end{cases}$, $y'' = \underset{0}{+++|+++} \Rightarrow$ concave up on $(-\infty, \infty)$ $\Rightarrow$ no point of inflection

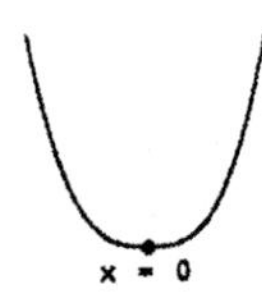

63. The graph of $y = f''(x) \Rightarrow$ the graph of $y = f(x)$ is concave up on $(0, \infty)$, concave down on $(-\infty, 0) \Rightarrow$ a point of inflection at $x = 0$; the graph of $y = f'(x)$ $\Rightarrow y' = +++|---|+++ \Rightarrow$ the graph $y = f(x)$ has both a local maximum and a local minimum

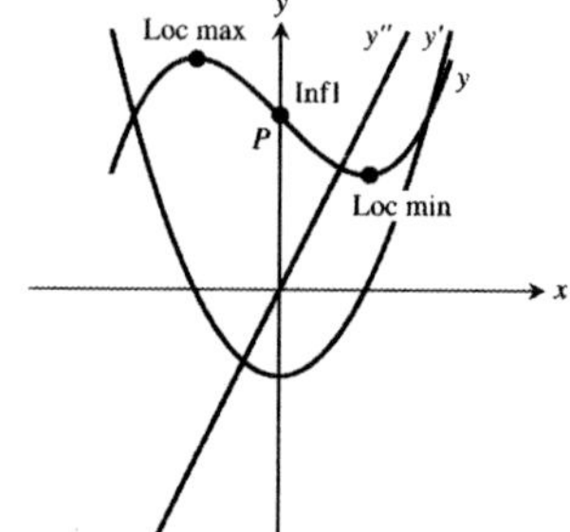

64. The graph of $y = f''(x) \Rightarrow y'' = +++|--- \Rightarrow$ the graph of $y = f(x)$ has a point of inflection, the graph of $y = f'(x) \Rightarrow y' = ---|+++|--- \Rightarrow$ the graph of $y = f(x)$ has both a local maximum and a local minimum

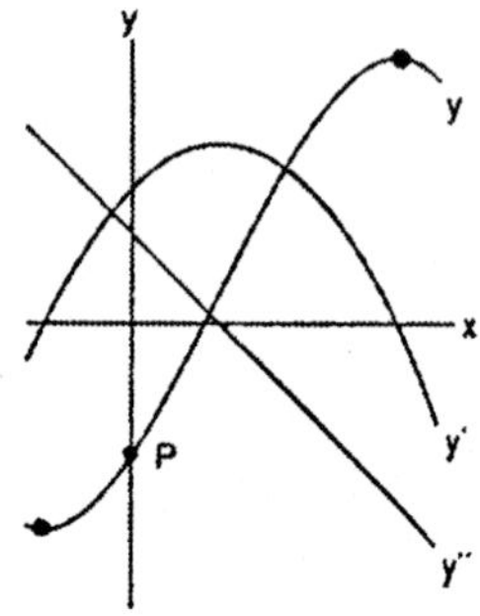

65. The graph of $y = f''(x) \Rightarrow y'' = ---|+++|---$ $\Rightarrow$ the graph of $y = f(x)$ has two points of inflection, the graph of $y = f'(x) \Rightarrow y' = ---|+++ \Rightarrow$ the graph of $y = f(x)$ has a local minimum

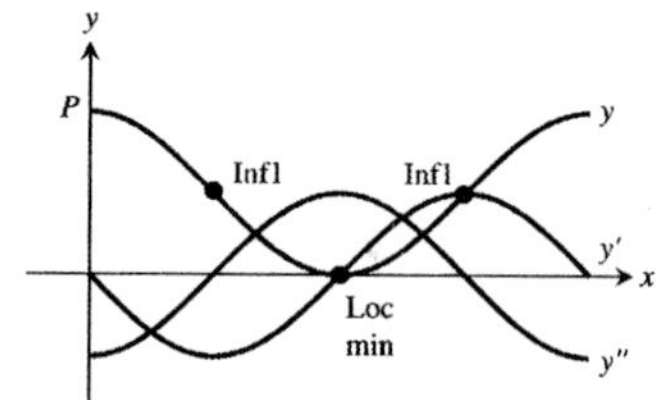

66. The graph of $y = f''(x) \Rightarrow y'' = +++|--- \Rightarrow$ the graph of $y = f(x)$ has a point of inflection; the graph of $y = f'(x) \Rightarrow y' = ---|+++|--- \Rightarrow$ the graph of $y = f(x)$ has both a local maximum and a local minimum

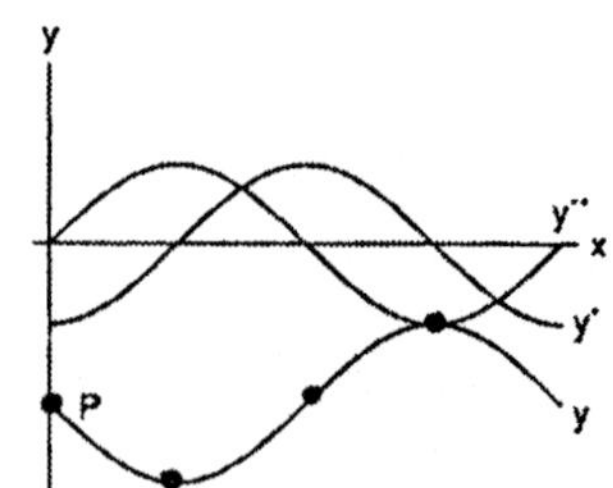

67.

Point	y'	y''
P	−	+
Q	+	0
R	+	−
S	0	−
T	−	−

68.

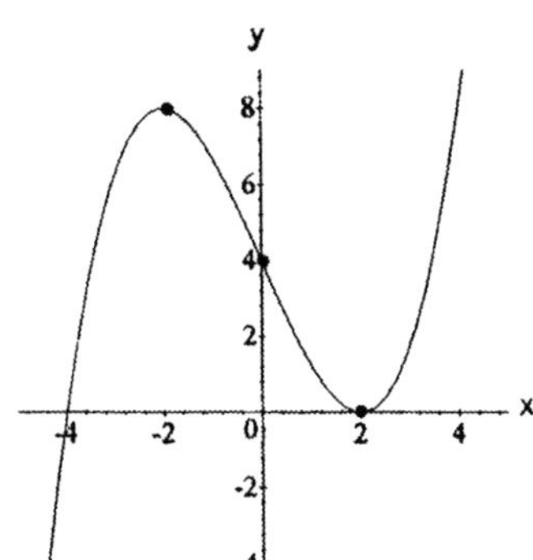

69.

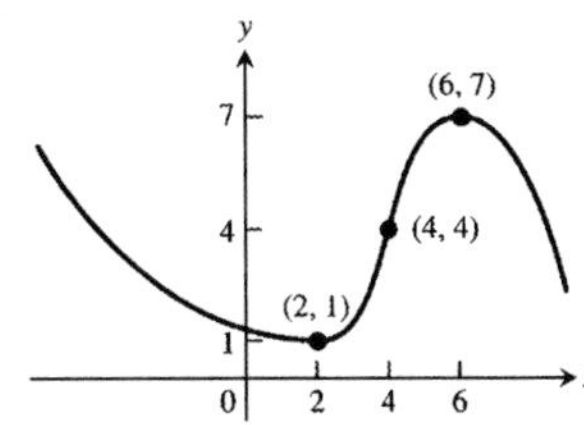

70.

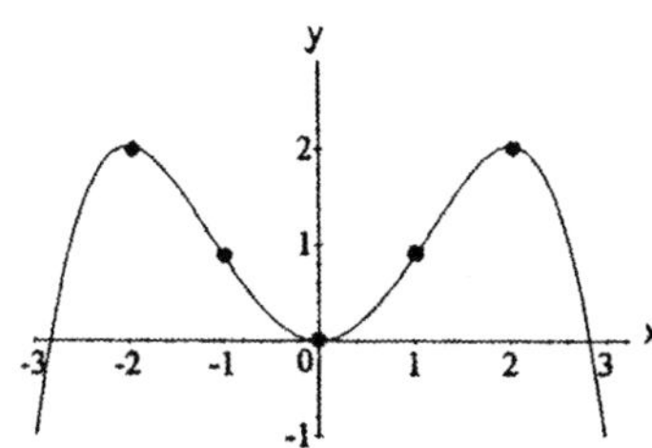

71. Graphs printed in color can shift during a press run, so your values may differ somewhat from those given here.

(a) The body is moving away from the origin when |displacement| is increasing as t increases, $0 < t < 2$ and $6 < t < 9.5$; the body is moving toward the origin when |displacement| is decreasing as t increases, $2 < t < 6$ and $9.5 < t < 15$

(b) The velocity will be zero when the slope of the tangent line for $y = s(t)$ is horizontal. The velocity is zero when t is approximately 2, 6, or 9.5 sec.

(c) The acceleration will be zero at those values of t where the curve $y = s(t)$ has points of inflection. The acceleration is zero when t is approximately 4, 7.5, or 12.5 sec.

(d) The acceleration is positive when the concavity is up, $4 < t < 7.5$ and $12.5 < t < 15$; the acceleration is negative when the concavity is down, $0 < t < 4$ and $7.5 < t < 12.5$

72. (a) The body is moving away from the origin when |displacement| is increasing as t increases, $1.5 < t < 4$, $10 < t < 12$ and $13.5 < t < 16$; the body is moving toward the origin when |displacement| is decreasing as t increases, $0 < t < 1.5$, $4 < t < 10$ and $12 < t < 13.5$

(b) The velocity will be zero when the slope of the tangent line for $y = s(t)$ is horizontal. The velocity is zero when t is approximately 0, 4, 12 or 16 sec.

(c) The acceleration will be zero at those values of t where the curve $y = s(t)$ has points of inflection. The acceleration is zero when t is approximately 1.5, 6, 8, 10.5, or 13.5 sec.

(d) The acceleration is positive when the concavity is up, $0 < t < 1.5$, $6 < t < 8$ and $10 < t < 13.5$, the acceleration is negative when the concavity is down, $1.5 < t < 6$, $8 < t < 10$ and $13.5 < t < 16$.

73. The marginal cost is $\frac{dc}{dx}$ which changes from decreasing to increasing when its derivative $\frac{d^2c}{dx^2}$ is zero. This is a point of inflection of the cost curve and occurs when the production level x is approximately 60 thousand units.

74. The marginal revenue is $\frac{dy}{dx}$ and it is increasing when its derivative $\frac{d^2y}{dx^2}$ is positive $\Rightarrow$ the curve is concave up $\Rightarrow 0 < t < 2$ and $5 < t < 9$; marginal revenue is decreasing when $\frac{d^2y}{dx^2} < 0 \Rightarrow$ the curve is concave down $\Rightarrow 2 < t < 5$ and $9 < t < 12$.

75. When $y' = (x-1)^2(x-2)$, then $y'' = 2(x-1)(x-2) + (x-1)^2$. The curve falls on $(-\infty, 2)$ and rises on $(2, \infty)$. At $x = 2$ there is a local minimum. There is no local maximum. The curve is concave upward on $(-\infty, 1)$ and

$\left(\frac{5}{3}, \infty\right)$, and concave downward on $\left(1, \frac{5}{3}\right)$. At $x = 1$ or $x = \frac{5}{3}$ there are inflection points.

76. When $y' = (x-1)^2(x-2)(x-4)$, then $y'' = 2(x-1)(x-2)(x-4) + (x-1)^2(x-4) + (x-1)^2(x-2)$ $= (x-1)\left[2\left(x^2-6x+8\right) + \left(x^2-5x+4\right) + \left(x^2-3x+2\right)\right] = 2(x-1)\left(2x^2-10x+11\right)$. The curve rises on $(-\infty, 2)$ and $(4, \infty)$ and falls on $(2, 4)$. At $x = 2$ there is a local maximum and at $x = 4$ a local minimum. The curve is concave downward on $(-\infty, 1)$ and $\left(\frac{5-\sqrt{3}}{2}, \frac{5+\sqrt{3}}{2}\right)$ and concave upward on $\left(1, \frac{5-\sqrt{3}}{2}\right)$ and $\left(\frac{5+\sqrt{3}}{2}, \infty\right)$. At $x = 1$, $\frac{5-\sqrt{3}}{2}$ and $\frac{5+\sqrt{3}}{2}$ there are inflection points.

77. The graph must be concave down for $x > 0$ because $f''(x) = -\frac{1}{x^2} < 0$.

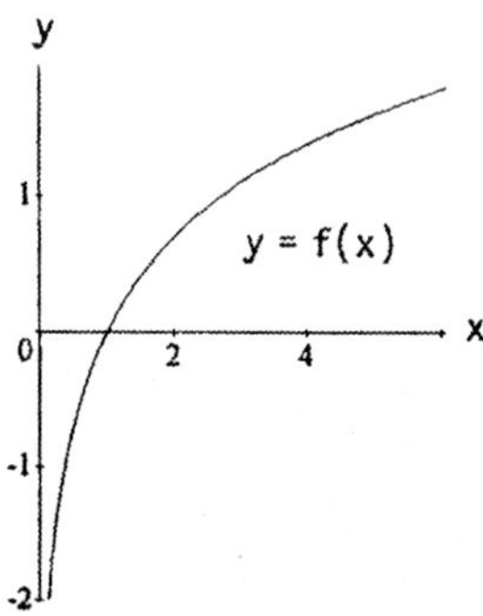

78. The second derivative, being continuous and never zero, cannot change sign. Therefore the graph will always be concave up or concave down so it will have no inflection points and no cusps or corners.

79. The curve will have a point of inflection at $x = 1$ if 1 is a solution of $y'' = 0$; $y = x^3 + bx^2 + cx + d$ $\Rightarrow y' = 3x^2 + 2bx + c \Rightarrow y'' = 6x + 2b$ and $6(1) + 2b = 0 \Rightarrow b = -3$.

80. (a) True. If f(x) is a polynomial of even degree then f' is of odd degree. Every polynomial of odd degree has at least one real root $\Rightarrow f'(x) = 0$ for some $x = r \Rightarrow$ f has a horizontal tangent at $x = r$.
    (b) False. For example, $f(x) = x - 1$ is a polynomial of odd degree but $f'(x) = 1$ is never 0. As another example, $y = \frac{1}{3}x^3 + x^2 + x$ is a polynomial of odd degree, but $y' = x^2 + 2x + 1 = (x+1)^2 > 0$ for all x.

81. (a) $f(x) = ax^2 + bx + c = a\left(x^2 + \frac{b}{a}x\right) + c = a\left(x^2 + \frac{b}{a}x + \frac{b^2}{4a^2}\right) - \frac{b^2}{4a} + c = a\left(x + \frac{b}{2a}\right)^2 - \frac{b^2-4ac}{4a}$ a parabola whose vertex is at $x = -\frac{b}{2a} \Rightarrow$ the coordinates of the vertex are $\left(-\frac{b}{2a}, -\frac{b^2-4ac}{4a}\right)$
    (b) The second derivative, $f''(x) = 2a$, describes concavity $\Rightarrow$ when $a > 0$ the parabola is concave up and when $a < 0$ the parabola is concave down.

82. No, $f''(x)$ could be decreasing to zero at $x = c$ and then increase again so it would be concave up on every interval even though $f''(x) = 0$. For example $f(x) = x^4$ is always concave up even though $f''(0) = 0$.

83. A quadratic curve never has an inflection point. If $y = ax^2 + bx + c$ where $a \neq 0$, then $y' = 2ax + b$ and $y'' = 2a$. Since 2a is a constant, it is not possible for y'' to change signs.

84. A cubic curve always has exactly one inflection point. If $y = ax^3 + bx^2 + cx + d$ where $a \neq 0$, then $y' = 3ax^2 + 2bx + c$ and $y'' = 6ax + 2b$. Since $\frac{-b}{3a}$ is a solution of $y'' = 0$, we have that y'' changes its sign at $x = -\frac{b}{3a}$ and y' exists everywhere (so there is a tangent at $x = -\frac{b}{3a}$). Thus the curve has an inflection point at $x = -\frac{b}{3a}$. There are no other inflection points because y'' changes sign only at this zero.

85. If $y = x^5 - 5x^4 - 240$, then $y' = 5x^3(x - 4)$ and $y'' = 20x^2(x - 3)$. The zeros of y' are extrema, and there is a point of inflection at $x = 3$.

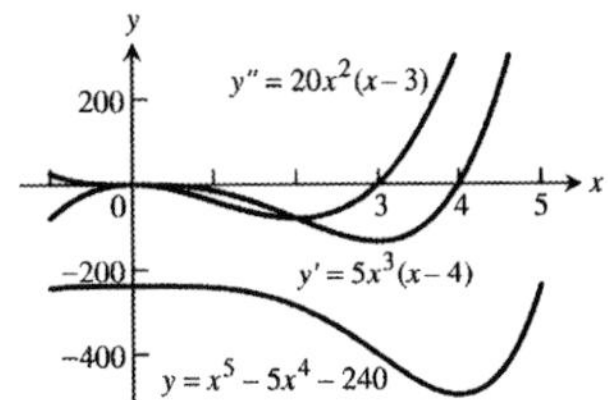

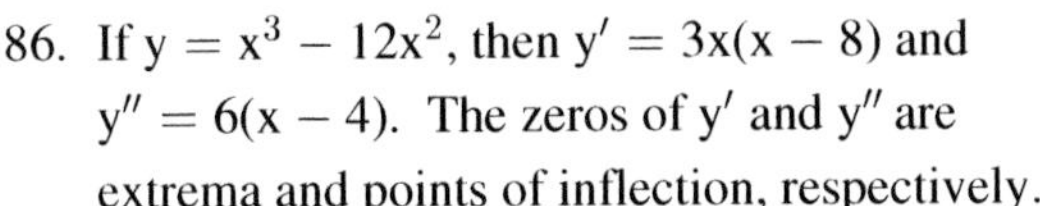

86. If $y = x^3 - 12x^2$, then $y' = 3x(x - 8)$ and $y'' = 6(x - 4)$. The zeros of y' and y'' are extrema and points of inflection, respectively.

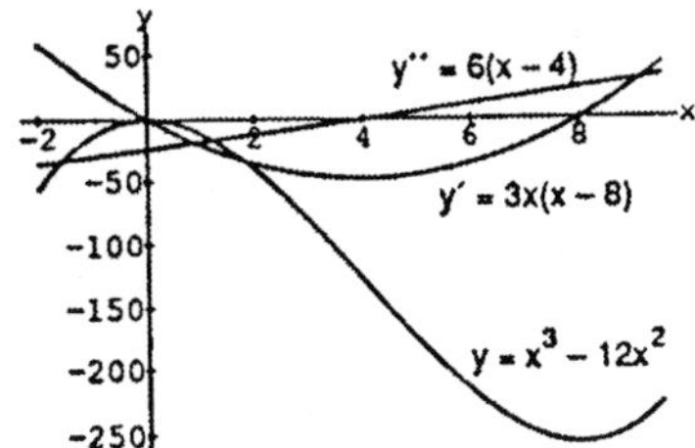

87. If $y = \frac{4}{5}x^5 + 16x^2 - 25$, then $y' = 4x\left(x^3 + 8\right)$ and $y'' = 16\left(x^3 + 2\right)$. The zeros of y' and y'' are extrema and points of inflection, respectively.

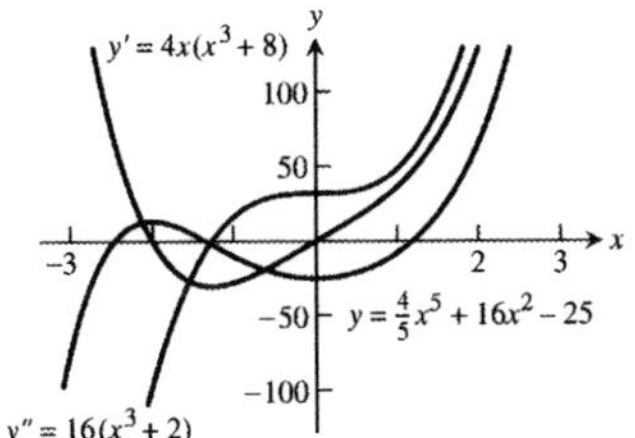

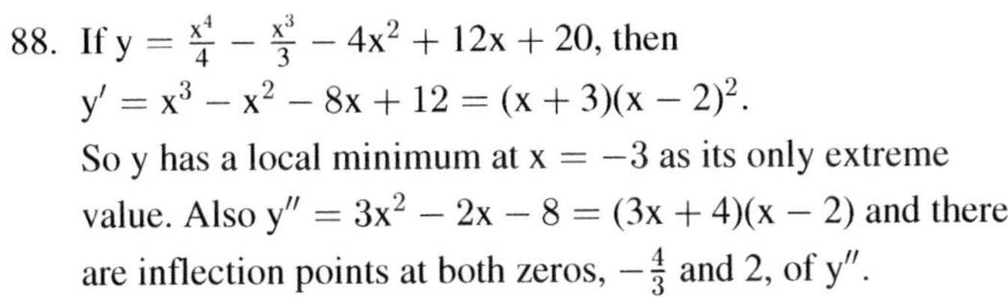

88. If $y = \frac{x^4}{4} - \frac{x^3}{3} - 4x^2 + 12x + 20$, then $y' = x^3 - x^2 - 8x + 12 = (x + 3)(x - 2)^2$. So y has a local minimum at $x = -3$ as its only extreme value. Also $y'' = 3x^2 - 2x - 8 = (3x + 4)(x - 2)$ and there are inflection points at both zeros, $-\frac{4}{3}$ and 2, of y''.

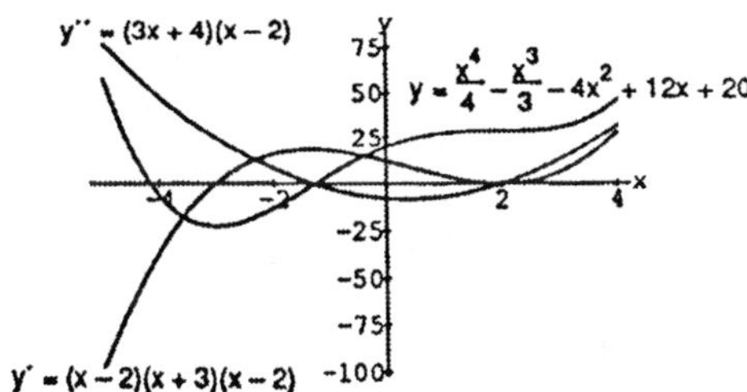

89. The graph of f falls where $f' < 0$, rises where $f' > 0$, and has horizontal tangents where $f' = 0$. It has local minima at points where f' changes from negative to positive and local maxima where f' changes from positive to negative. The graph of f is concave down where $f'' < 0$ and concave up where $f'' > 0$. It has an inflection point each time f''changes sign, provided a tangent line exists there.

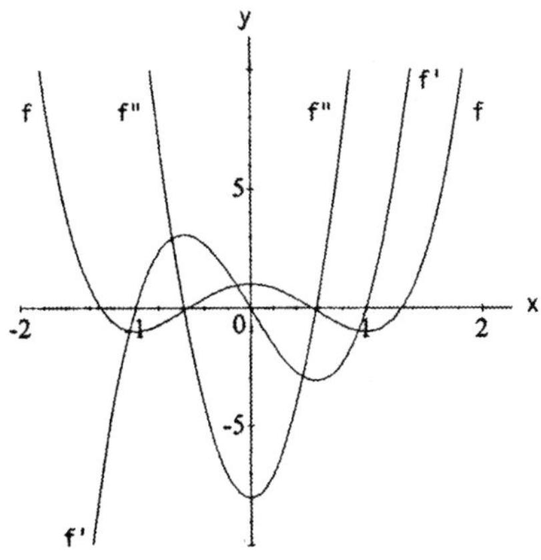

90. The graph f is concave down where $f'' < 0$, and concave up where $f'' > 0$. It has an inflection point each time f'' changes sign, provided a tangent line exists there.

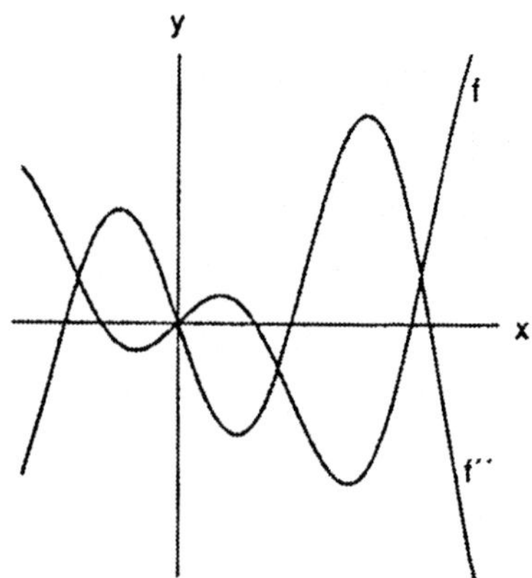

91. (a) It appears to control the number and magnitude of the local extrema. If $k < 0$, there is a local maximum to the left of the origin and a local minimum to the right. The larger the magnitude of k $(k < 0)$, the greater the magnitude of the extrema. If $k > 0$, the graph has only positive slopes and lies entirely in the first and third quadrants with no local extrema. The graph becomes increasingly steep and straight as $k \to \infty$.

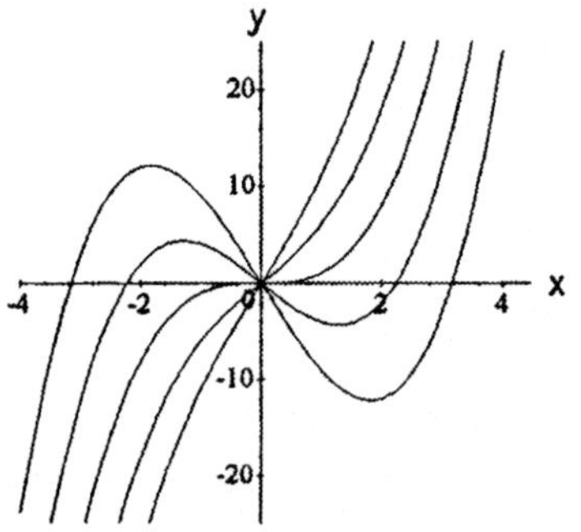

(b) $f'(x) = 3x^2 + k \Rightarrow$ the discriminant $0^2 - 4(3)(k) = -12k$ is positive for $k < 0$, zero for $k = 0$, and negative for $k > 0$; f' has two zeros $x = \pm\sqrt{-\frac{k}{3}}$ when $k < 0$, one zero $x = 0$ when $k = 0$ and no real zeros when $k > 0$; the sign of k controls the number of local extrema.

(c) As $k \to \infty$, $f'(x) \to \infty$ and the graph becomes increasingly steep and straight. As $k \to -\infty$, the crest of the graph (local maximum) in the second quadrant becomes increasingly high and the trough (local minimum) in the fourth quadrant becomes increasingly deep.

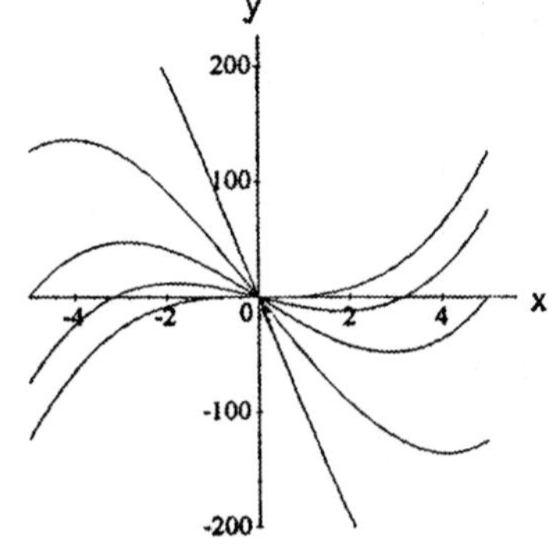

92. (a) It appears to control the concavity and the number of local extrema.

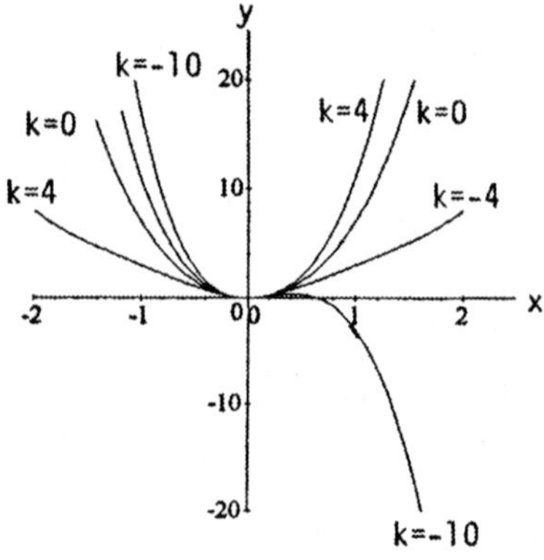

(b) $f(x) = x^4 + kx^3 + 6x^2 \Rightarrow f'(x) = 4x^3 + 3kx^2 + 12x$ $\Rightarrow f''(x) = 12x^2 + 6kx + 12 \Rightarrow$ the discriminant is $36k^2 - 4(12)(12) = 36(k+4)(k-4)$, so the sign line of the discriminant is $\underset{-4}{+++\,|}\;---\;\underset{4}{|\,+++} \Rightarrow$ the discriminant is positive when $|k| > 4$, zero when $k = \pm 4$, and negative when $|k| < 4$; $f''(x) = 0$ has two zeros when $|k| > 4$, one zero when $k = \pm 4$, and no real zeros for $|k| < 4$; the value of k controls the number of possible points of inflection.

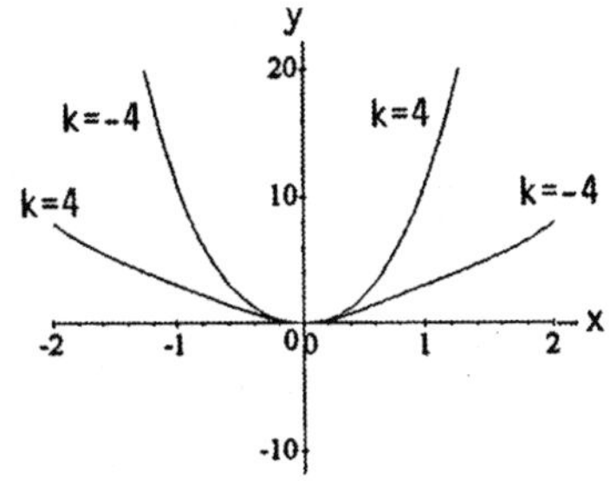

93. (a) If $y = x^{2/3}(x^2 - 2)$, then $y' = \frac{4}{3}x^{-1/3}(2x^2 - 1)$ and $y'' = \frac{4}{9}x^{-4/3}(10x^2 + 1)$. The curve rises on $\left(-\frac{1}{\sqrt{2}}, 0\right)$ and $\left(\frac{1}{\sqrt{2}}, \infty\right)$ and falls on $\left(-\infty, -\frac{1}{\sqrt{2}}\right)$ and $\left(0, \frac{1}{\sqrt{2}}\right)$. The curve is concave up on $(-\infty, 0)$ and $(0, \infty)$.

(b) A cusp since $\lim\limits_{x \to 0^-} y' = \infty$ and $\lim\limits_{x \to 0^+} y' = -\infty$.

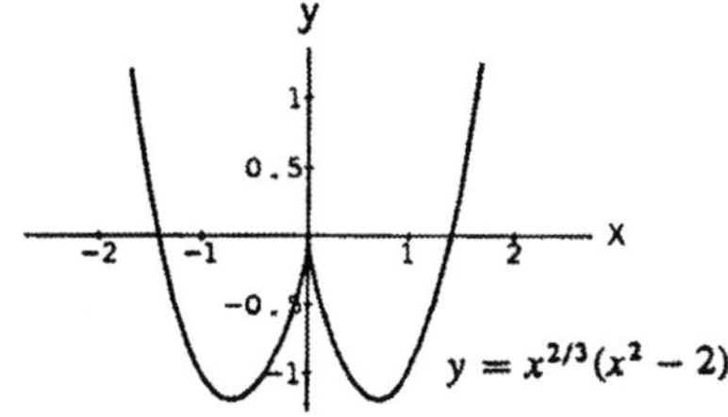

94. (a) If $y = 9x^{2/3}(x - 1)$, then $y' = \frac{15\left(x - \frac{2}{5}\right)}{x^{1/3}}$ and $y'' = \frac{10\left(x + \frac{1}{5}\right)}{x^{4/3}}$. The curve rises on $(-\infty, 0)$ and $\left(\frac{2}{5}, \infty\right)$ and falls on $\left(0, \frac{2}{5}\right)$. The curve is concave down on $\left(-\infty, -\frac{1}{5}\right)$ and concave up on $\left(-\frac{1}{5}, 0\right)$ and $(0, \infty)$.

(b) A cusp since $\lim\limits_{x \to 0^-} y' = \infty$ and $\lim\limits_{x \to 0^+} y' = -\infty$.

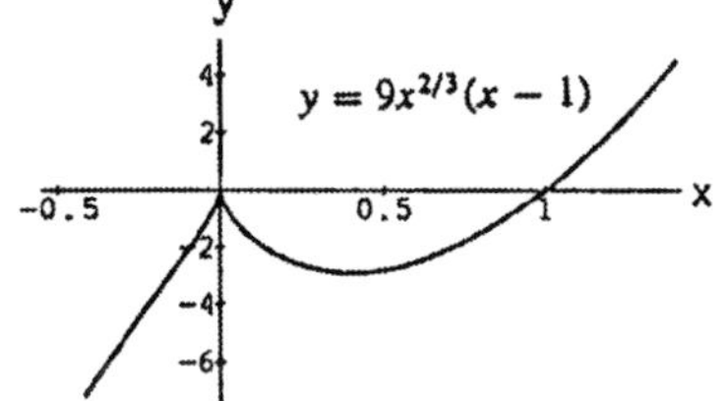

95. Yes: $y = x^2 + 3\sin 2x \Rightarrow y' = 2x + 6\cos 2x$. The graph of y' is zero near -3 and this indicates a horizontal tangent near $x = -3$.

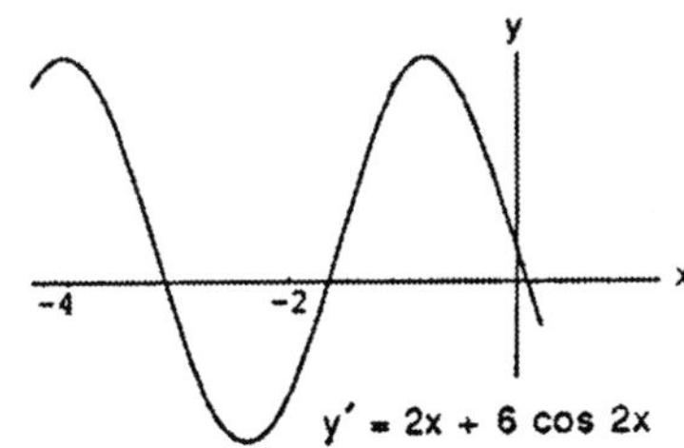

4.5 APPLIED OPTIMIZATION PROBLEMS

1. Let ℓ and w represent the length and width of the rectangle, respectively. With an area of 16 in.^{2}, we have that $(\ell)(w) = 16 \Rightarrow w = 16\ell^{-1} \Rightarrow$ the perimeter is $P = 2\ell + 2w = 2\ell + 32\ell^{-1}$ and $P'(\ell) = 2 - \frac{32}{\ell^2} = \frac{2(\ell^2 - 16)}{\ell^2}$. Solving $P'(\ell) = 0 \Rightarrow \frac{2(\ell+4)(\ell-4)}{\ell^2} = 0 \Rightarrow \ell = -4, 4$. Since $\ell > 0$ for the length of a rectangle, ℓ must be 4 and $w = 4 \Rightarrow$ the perimeter is 16 in., a minimum since $P''(\ell) = \frac{16}{\ell^3} > 0$.

2. Let x represent the length of the rectangle in meters $(0 < x < 4)$ Then the width is $4 - x$ and the area is $A(x) = x(4 - x) = 4x - x^2$. Since $A'(x) = 4 - 2x$, the critical point occurs at $x = 2$. Since, $A'(x) > 0$ for $0 < x < 2$ and $A'(x) < 0$ for $2 < x < 4$, this critical point corresponds to the maximum area. The rectangle with the largest area measures 2 m by $4 - 2 = 2$ m, so it is a square.

Graphical Support:

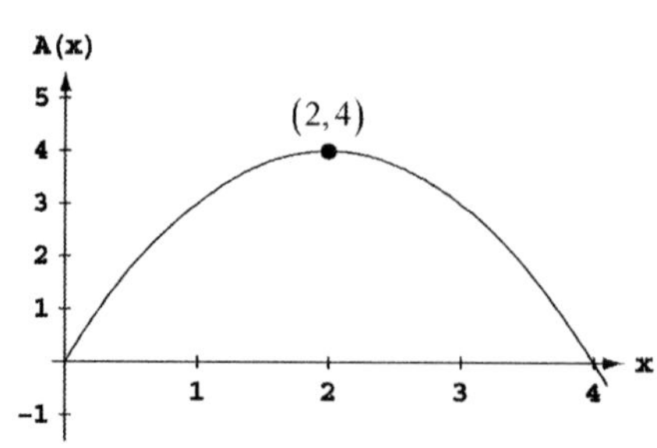

3. (a) The line containing point P also contains the points $(0, 1)$ and $(1, 0)$ $\Rightarrow$ the line containing P is $y = 1 - x$ $\Rightarrow$ a general point on that line is $(x, 1 - x)$.
   (b) The area $A(x) = 2x(1 - x)$, where $0 \le x \le 1$.
   (c) When $A(x) = 2x - 2x^2$, then $A'(x) = 0 \Rightarrow 2 - 4x = 0 \Rightarrow x = \frac{1}{2}$. Since $A(0) = 0$ and $A(1) = 0$, we conclude that $A\left(\frac{1}{2}\right) = \frac{1}{2}$ sq units is the largest area. The dimensions are 1 unit by $\frac{1}{2}$ unit.

4. The area of the rectangle is $A = 2xy = 2x\left(12 - x^2\right)$, where $0 \le x \le \sqrt{12}$. Solving $A'(x) = 0 \Rightarrow 24 - 6x^2 = 0$ $\Rightarrow x = -2$ or 2. Now -2 is not in the domain, and since $A(0) = 0$ and $A\left(\sqrt{12}\right) = 0$, we conclude that $A(2) = 32$ square units is the maximum area. The dimensions are 4 units by 8 units.

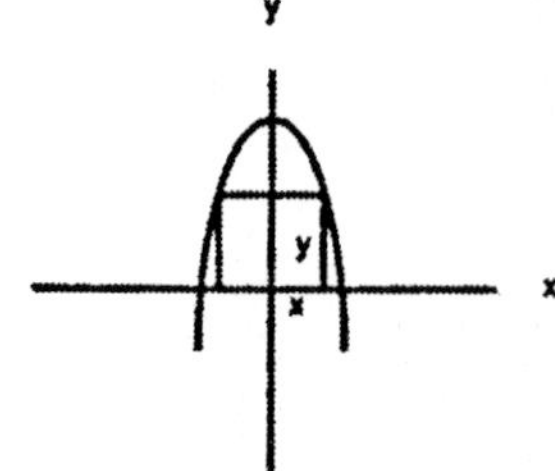

5. The volume of the box is $V(x) = x(15 - 2x)(8 - 2x)$ $= 120x - 46x^2 + 4x^3$, where $0 \le x \le 4$. Solving $V'(x) = 0$ $\Rightarrow 120 - 92x + 12x^2 = 4(6 - x)(5 - 3x) = 0 \Rightarrow x = \frac{5}{3}$ or 6, but 6 is not in the domain. Since $V(0) = V(4) = 0$, $V\left(\frac{5}{3}\right) = \frac{2450}{27} \approx 91 \text{ in}^3$ must be the maximum volume of the box with dimensions $\frac{14}{3} \times \frac{35}{3} \times \frac{5}{3}$ inches.

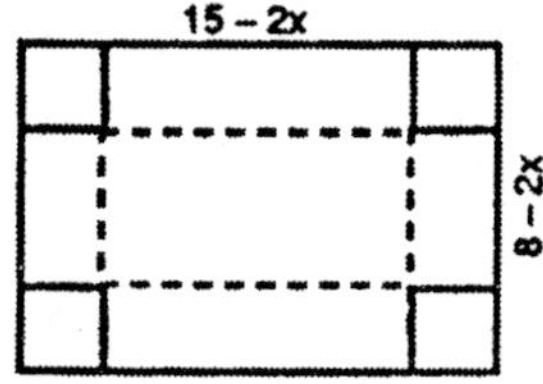

6. The area of the triangle is $A = \frac{1}{2}ba = \frac{b}{2}\sqrt{400 - b^2}$, where $0 \le b \le 20$. Then $\frac{dA}{db} = \frac{1}{2}\sqrt{400 - b^2} - \frac{b^2}{2\sqrt{400 - b^2}}$ $= \frac{200 - b^2}{\sqrt{400 - b^2}} = 0 \Rightarrow$ the interior critical point is $b = 10\sqrt{2}$. When $b = 0$ or 20, the area is zero $\Rightarrow A\left(10\sqrt{2}\right)$ is the maximum area. When $a^2 + b^2 = 400$ and $b = 10\sqrt{2}$, the value of a is also $10\sqrt{2}$ $\Rightarrow$ the maximum area occurs when $a = b$.

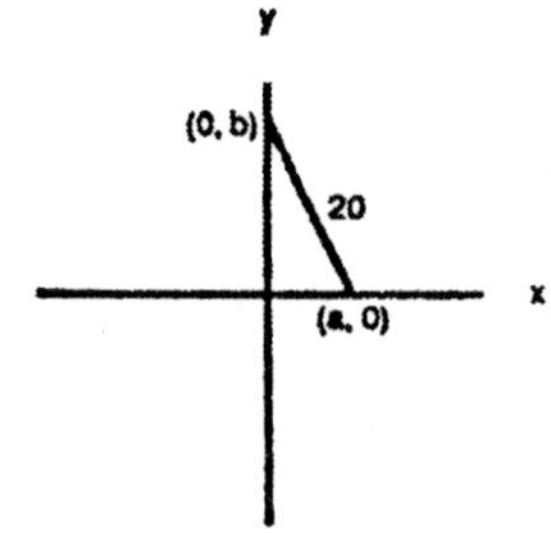

7. The area is $A(x) = x(800 - 2x)$, where $0 \le x \le 400$. Solving $A'(x) = 800 - 4x = 0 \Rightarrow x = 200$. With $A(0) = A(400) = 0$, the maximum area is $A(200) = 80{,}000 \text{ m}^2$. The dimensions are 200 m by 400 m.

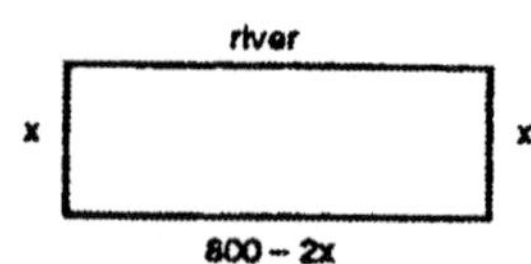

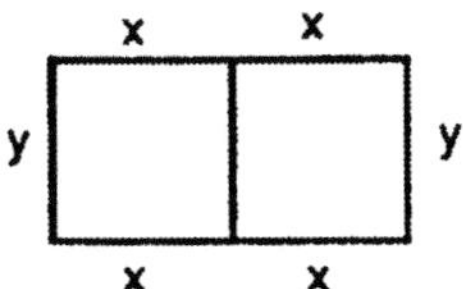

8. The area is $2xy = 216 \Rightarrow y = \frac{108}{x}$. The amount of fence needed is $P = 4x + 3y = 4x + 324x^{-1}$, where $0 < x$; $\frac{dP}{dx} = 4 - \frac{324}{x^2} = 0 \Rightarrow x^2 - 81 = 0 \Rightarrow$ the critical points are 0 and ± 9, but 0 and -9 are not in the domain. Then $P''(9) > 0 \Rightarrow$ at $x = 9$ there is a minimum $\Rightarrow$ the dimensions of the outer rectangle are 18 m by 12 m $\Rightarrow$ 72 meters of fence will be needed.

9. (a) We minimize the weight = tS where S is the surface area, and t is the thickness of the steel walls of the tank. The surace area is $S = x^2 + 4xy$ where x is the length of a side of the square base of the tank, and y is its depth. The volume of the tank must be $500\text{ft}^3 \Rightarrow y = \frac{500}{x^2}$. Therefore, the weight of the tank is $w(x) = t\left(x^2 + \frac{2000}{x}\right)$. Treating the thickness as a constant gives $w'(x) = t\left(2x - \frac{2000}{x^2}\right)$ for x.0. The critical value is at $x = 10$. Since $w''(10) = t\left(2 + \frac{4000}{10^3}\right) > 0$, there is a minimum at $x = 10$. Therefore, the optimum dimensions of the tank are 10 ft on the base edges and 5 ft deep.

   (b) Minimizing the surface area of the tank minimizes its weight for a given wall thickness. The thickness of the steel walls would likely be determined by other considerations such as structural requirements.

10. (a) The volume of the tank being 1125 ft^3, we have that $yx^2 = 1125 \Rightarrow y = \frac{1125}{x^2}$. The cost of building the tank is $c(x) = 5x^2 + 30x\left(\frac{1125}{x^2}\right)$, where $0 < x$. Then $c'(x) = 10x - \frac{33750}{x^2} = 0 \Rightarrow$ the critical points are 0 and 15, but 0 is not in the domain. Thus, $c''(15) > 0 \Rightarrow$ at $x = 15$ we have a minimum. The values of $x = 15$ ft and $y = 5$ ft will minimize the cost.

   (b) The cost function $c = 5(x^2 + 4xy) + 10xy$, can be separated into two items: (1) the cost of the materials and labor to fabricate the tank, and (2) the cost for the excavation. Since the area of the sides and bottom of the tanks is $(x^2 + 4xy)$, it can be deduced that the unit cost to fabricate the tanks is \$5/ft^2. Normally, excavation costs are per unit volume of excavated material. Consequently, the total excavation cost can be taken as $10xy = \left(\frac{10}{x}\right)(x^2y)$. This suggests that the unit cost of excavation is $\frac{\$10/\text{ft}^2}{x}$ where x is the length of a side of the square base of the tank in feet. For the least expensive tank, the unit cost for the excavation is $\frac{\$10/\text{ft}^2}{15\text{ ft}} = \frac{\$0.67}{\text{ft}^3} = \frac{\$18}{\text{yd}^3}$. The total cost of the least expensive tank is \$3375, which is the sum of \$2625 for fabrication and \$750 for the excavation.

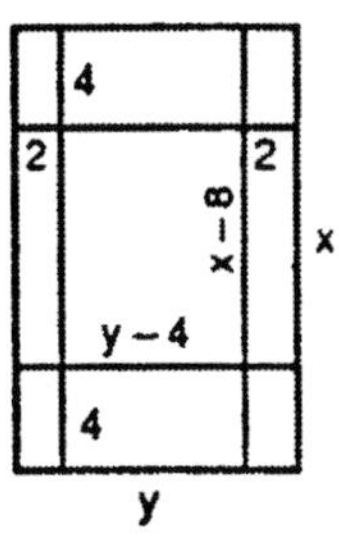

11. The area of the printing is $(y - 4)(x - 8) = 50$. Consequently, $y = \left(\frac{50}{x-8}\right) + 4$. The area of the paper is $A(x) = x\left(\frac{50}{x-8} + 4\right)$, where $8 < x$. Then $A'(x) = \left(\frac{50}{x-8} + 4\right) - x\left(\frac{50}{(x-8)^2}\right) = \frac{4(x-8)^2 - 400}{(x-8)^2} = 0$ $\Rightarrow$ the critical points are -2 and 18, but -2 is not in the domain. Thus $A''(18) > 0 \Rightarrow$ at $x = 18$ we have a minimum. Therefore the dimensions 18 by 9 inches minimize the amount of paper.

12. The volume of the cone is $V = \frac{1}{3}\pi r^2 h$, where $r = x = \sqrt{9 - y^2}$ and $h = y + 3$ (from the figure in the text). Thus, $V(y) = \frac{\pi}{3}(9 - y^2)(y + 3) = \frac{\pi}{3}(27 + 9y - 3y^2 - y^3) \Rightarrow V'(y) = \frac{\pi}{3}(9 - 6y - 3y^2) = \pi(1 - y)(3 + y)$. The critical points are -3 and 1, but -3 is not in the domain. Thus $V''(1) = \frac{\pi}{3}(-6 - 6(1)) < 0 \Rightarrow$ at $y = 1$ we have a maximum volume of $V(1) = \frac{\pi}{3}(8)(4) = \frac{32\pi}{3}$ cubic units.

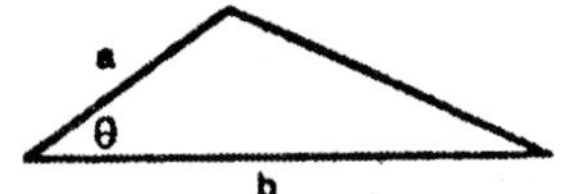

13. The area of the triangle is $A(\theta) = \frac{ab\sin\theta}{2}$, where $0 < \theta < \pi$. Solving $A'(\theta) = 0 \Rightarrow \frac{ab\cos\theta}{2} = 0 \Rightarrow \theta = \frac{\pi}{2}$. Since $A''(\theta) = -\frac{ab\sin\theta}{2} \Rightarrow A''\left(\frac{\pi}{2}\right) < 0$, there is a maximum at $\theta = \frac{\pi}{2}$.

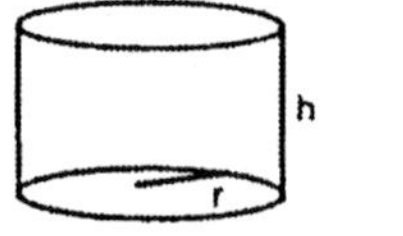

14. A volume $V = \pi r^2 h = 1000 \Rightarrow h = \frac{1000}{\pi r^2}$. The amount of material is the surface area given by the sides and bottom of the can $\Rightarrow S = 2\pi rh + \pi r^2 = \frac{2000}{r} + \pi r^2$, $0 < r$. Then $\frac{dS}{dr} = -\frac{2000}{r^2} + 2\pi r = 0 \Rightarrow \frac{\pi r^3 - 1000}{r^2} = 0$. The critical points are 0 and $\frac{10}{\sqrt[3]{\pi}}$, but 0 is not in the domain. Since $\frac{d^2S}{dr^2} = \frac{4000}{r^3} + 2\pi > 0$, we have a minimum surface area when $r = \frac{10}{\sqrt[3]{\pi}}$ cm and $h = \frac{1000}{\pi r^2} = \frac{10}{\sqrt[3]{\pi}}$ cm. Comparing this result to the result found in Example 2, if we include both ends of the can, then we have a minimum surface area when the can is shorter-specifically, when the height of the can is the same as its diameter.

15. With a volume of 1000 cm and $V = \pi r^2 h$, then $h = \frac{1000}{\pi r^2}$. The amount of aluminum used per can is $A = 8r^2 + 2\pi rh = 8r^2 + \frac{2000}{r}$. Then $A'(r) = 16r - \frac{2000}{r^2} = 0 \Rightarrow \frac{8r^3 - 1000}{r^2} = 0 \Rightarrow$ the critical points are 0 and 5, but $r = 0$ results in no can. Since $A''(r) = 16 + \frac{1000}{r^3} > 0$ we have a minimum at $r = 5 \Rightarrow h = \frac{40}{\pi}$ and $h{:}r = 8{:}\pi$.

16. (a) The base measures $10 - 2x$ in. by $\frac{15-2x}{2}$ in., so the volume formula is $V(x) = \frac{x(10-2x)(15-2x)}{2} = 2x^3 - 25x^2 + 75x$.

(b) We require $x > 0$, $2x < 10$, and $2x < 15$. Combining these requirements, the domain is the interval (0, 5).

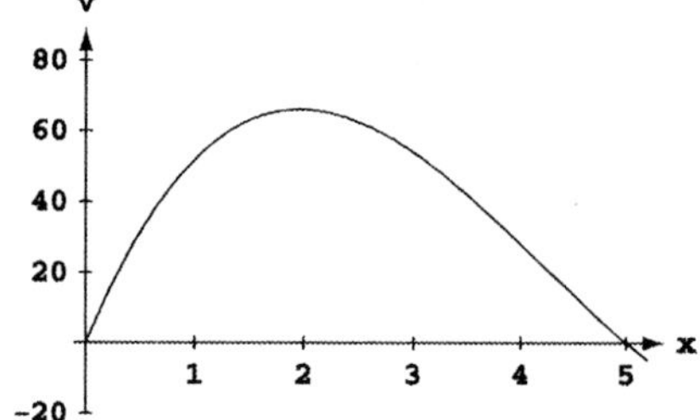

(c) The maximum volume is approximately 66.02 in.^{3} when $x \approx 1.96$ in.

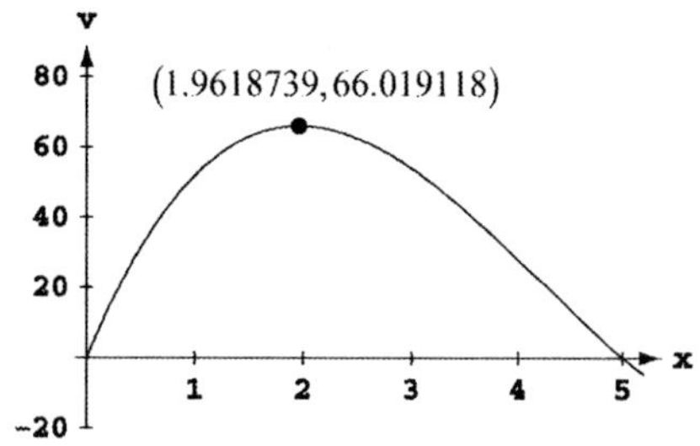

(d) $V'(x) = 6x^2 - 50x + 75$. The critical point occurs when $V'(x) = 0$, at $x = \frac{50 \pm \sqrt{(-50)^2 - 4(6)(75)}}{2(6)} = \frac{50 \pm \sqrt{700}}{12} = \frac{25 \pm 5\sqrt{7}}{6}$, that is, $x \approx 1.96$ or $x \approx 6.37$. We discard the larger value because it is not in the domain. Since $V''(x) = 12x - 50$, which is negative when $x \approx 1.96$, the critical point corresponds to the maximum volume. The maximum volume occurs when $x = \frac{25 - 5\sqrt{7}}{6} \approx 1.96$, which comfimrs the result in (c).

17. (a) The" sides" of the suitcase will measure $24 - 2x$ in. by $18 - 2x$ in. and will be $2x$ in. apart, so the volume formula is $V(x) = 2x(24 - 2x)(18 - 2x) = 8x^3 - 168x^2 + 862x$.

(b) We require $x > 0$, $2x < 18$, and $2x < 24$. Combining these requirements, the domain is the interval $(0, 9)$.

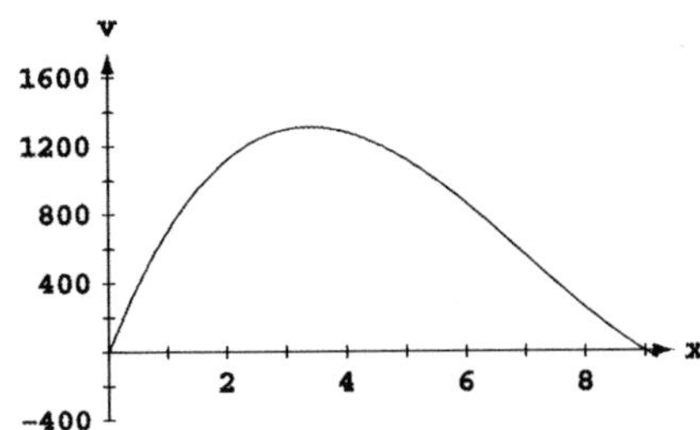

(c) The maximum volume is approximately 1309.95 in.^{3} when $x \approx 3.39$ in.

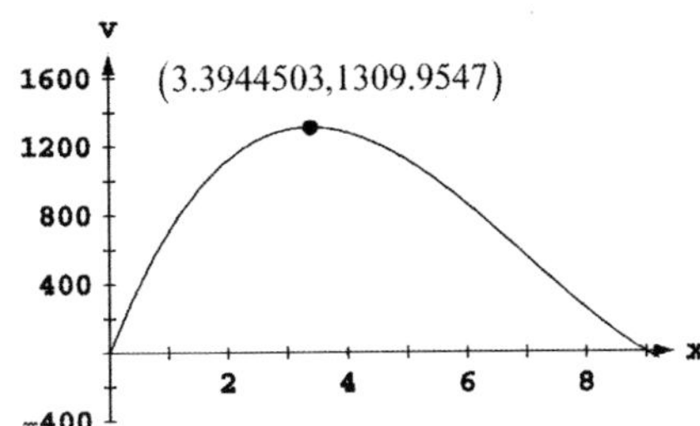

(d) $V'(x) = 24x^2 - 336x + 864 = 24(x^2 - 14x + 36)$. The critical point is at $x = \frac{14 \pm \sqrt{(-14)^2 - 4(1)(36)}}{2(1)} = \frac{14 \pm \sqrt{52}}{2}$ $= 7 \pm \sqrt{13}$, that is, $x \approx 3.39$ or $x \approx 10.61$. We discard the larger value because it is not in the domain. Since $V''(x) = 24(2x - 14)$ which is negative when $x \approx 3.39$, the critical point corresponds to the maximum volume. The maximum value occurs at $x = 7 - \sqrt{13} \approx 3.39$, which confirms the results in (c).

(e) $8x^3 - 168x^2 + 862x = 1120 \Rightarrow 8(x^3 - 21x^2 + 108x - 140) = 0 \Rightarrow 8(x - 2)(x - 5)(x - 14) = 0$. Since 14 is not in the fomain, the possible values of x are $x = 2$ in. or $x = 5$ in.

(f) The dimensions of the resulting box are 2x in., $(24 - 2x)$ in., and $(18 - 2x)$. Each of these measurements must be positive, so that gives the domain of $(0, 9)$.

18. If the upper right vertex of the rectangle is located at $(x, 4\cos 0.5x)$ for $0 < x < \pi$, then the rectangle has width 2x and height $4\cos 0.5x$, so the area is $A(x) = 8x\cos 0.5x$. Solving $A'(x) = 0$ graphically for $0 < x < \pi$, we find that $x \approx 2.214$. Evaluating 2x and $4\cos 0.5x$ for $x \approx 2.214$, the dimensions of the rectangle are approximately 4.43 (width) by 1.79 (height), and the maximum area is approximately 7.923.

19. Let the radius of the cylinder be r cm, $0 < r < 10$. Then the height is $2\sqrt{100 - r^2}$ and the volume is $V(r) = 2\pi r^2\sqrt{100 - r^2}$ cm^3. Then, $V'(r) = 2\pi r^2\left(\frac{1}{\sqrt{100 - r^2}}\right)(-2r) + \left(2\pi\sqrt{100 - r^2}\right)(2r)$ $= \frac{-2\pi r^3 + 4\pi r(100 - r^2)}{\sqrt{100 - r^2}} = \frac{2\pi r(200 - 3r^2)}{\sqrt{100 - r^2}}$. The critical point for $0 < r < 10$ occurs at $r = \sqrt{\frac{200}{3}} = 10\sqrt{\frac{2}{3}}$. Since $V'(r) > 0$ for $0 < r < 10\sqrt{\frac{2}{3}}$ and $V'(r) < 0$ for $10\sqrt{\frac{2}{3}} < r < 10$, the critical point corresponds to the maximum volume. The dimensions are $r = 10\sqrt{\frac{2}{3}} \approx 8.16$ cm and $h = \frac{20}{\sqrt{3}} \approx 11.55$ cm, and the volume is $\frac{4000\pi}{3\sqrt{3}} \approx 2418.40$ cm^3.

20. (a) From the diagram we have $4x + \ell = 108$ and $V = x^2\ell$. The volume of the box is $V(x) = x^2(108 - 4x)$, where $0 \le x < 27$. Then
$V'(x) = 216x - 12x^2 = 12x(18 - x) = 0$
$\Rightarrow$ the critical points are 0 and 18, but $x = 0$ results in no box. Since $V''(x) = 216 - 24x < 0$ at $x = 18$ we have a maximum. The dimensions of the box are $18 \times 18 \times 36$ in.

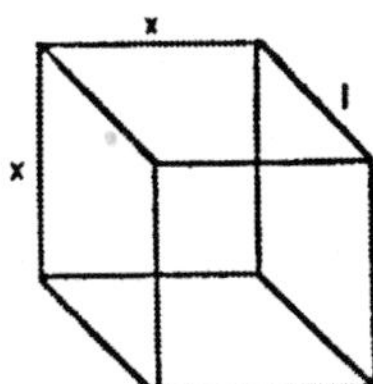

(b) In terms of length, $V(\ell) = x^2\ell = \left(\frac{108-\ell}{4}\right)^2 \ell$. The graph indicates that the maximum volume occurs near $\ell = 36$, which is consistent with the result of part (a).

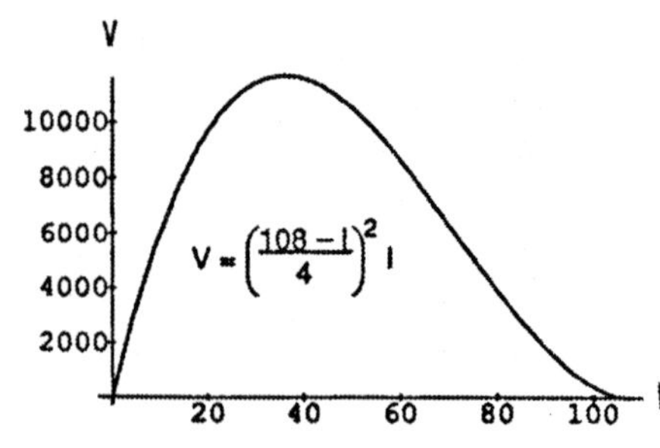

21. (a) From the diagram we have $3h + 2w = 108$ and $V = h^2 w \Rightarrow V(h) = h^2\left(54 - \frac{3}{2}h\right) = 54h^2 - \frac{3}{2}h^3$. Then $V'(h) = 108h - \frac{9}{2}h^2 = \frac{9}{2}h(24 - h) = 0$ $\Rightarrow h = 0$ or $h = 24$, but $h = 0$ results in no box. Since $V''(h) = 108 - 9h < 0$ at $h = 24$, we have a maximum volume at $h = 24$ and $w = 54 - \frac{3}{2}h = 18$.

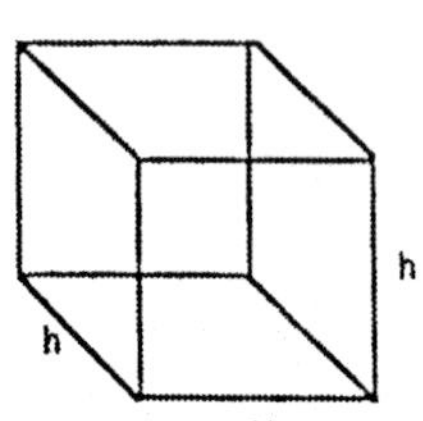

(b)

(24, 10368)
Abs max

$\left(v = 54h^2 - \frac{3}{2}h^3\right)$

22. From the diagram the perimeter is $P = 2r + 2h + \pi r$, where r is the radius of the semicircle and h is the height of the rectangle. The amount of light transmitted proportional to
$A = 2rh + \frac{1}{4}\pi r^2 = r(P - 2r - \pi r) + \frac{1}{4}\pi r^2$
$= rP - 2r^2 - \frac{3}{4}\pi r^2$. Then $\frac{dA}{dr} = P - 4r - \frac{3}{2}\pi r = 0$
$\Rightarrow r = \frac{2P}{8+3\pi} \Rightarrow 2h = P - \frac{4P}{8+3\pi} - \frac{2\pi P}{8+3\pi} = \frac{(4+\pi)P}{8+3\pi}$.
Therefore, $\frac{2r}{h} = \frac{8}{4+\pi}$ gives the proportions that admit the <u>most</u> light since $\frac{d^2A}{dr^2} = -4 - \frac{3}{2}\pi < 0$.

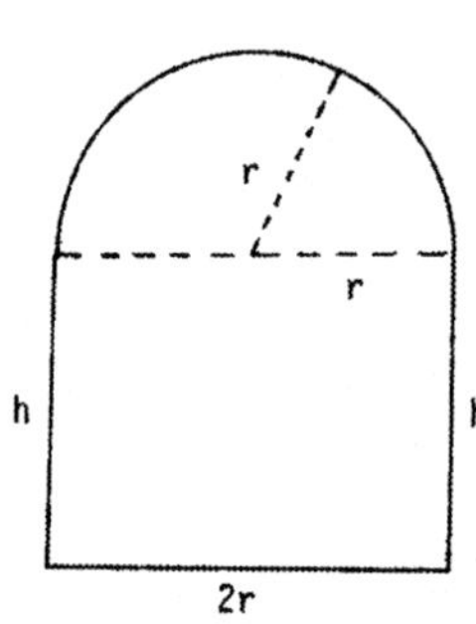

23. The fixed volume is $V = \pi r^2 h + \frac{2}{3}\pi r^3 \Rightarrow h = \frac{V}{\pi r^2} - \frac{2r}{3}$, where h is the height of the cylinder and r is the radius of the hemisphere. To minimize the cost we must minimize surface area of the cylinder added to twice the surface area of the hemisphere. Thus, we minimize $C = 2\pi rh + 4\pi r^2 = 2\pi r\left(\frac{V}{\pi r^2} - \frac{2r}{3}\right) + 4\pi r^2 = \frac{2V}{r} + \frac{8}{3}\pi r^2$. Then $\frac{dC}{dr} = -\frac{2V}{r^2} + \frac{16}{3}\pi r = 0 \Rightarrow V = \frac{8}{3}\pi r^3 \Rightarrow r = \left(\frac{3V}{8\pi}\right)^{1/3}$. From the volume equation, $h = \frac{V}{\pi r^2} - \frac{2r}{3}$ $= \frac{4V^{1/3}}{\pi^{1/3}\cdot 3^{2/3}} - \frac{2\cdot 3^{1/3}\cdot V^{1/3}}{3\cdot 2\cdot \pi^{1/3}} = \frac{3^{1/3}\cdot 2\cdot 4\cdot V^{1/3} - 2\cdot 3^{1/3}\cdot V^{1/3}}{3\cdot 2\cdot \pi^{1/3}} = \left(\frac{3V}{\pi}\right)^{1/3}$. Since $\frac{d^2C}{dr^2} = \frac{4V}{r^3} + \frac{16}{3}\pi > 0$, these dimensions do minimize the cost.

24. The volume of the trough is maximized when the area of the cross section is maximized. From the diagram the area of the cross section is $A(\theta) = \cos\theta + \sin\theta\cos\theta$, $0 < \theta < \frac{\pi}{2}$. Then $A'(\theta) = -\sin\theta + \cos^2\theta - \sin^2\theta$ $= -\left(2\sin^2\theta + \sin\theta - 1\right) = -(2\sin\theta - 1)(\sin\theta + 1)$ so $A'(\theta) = 0 \Rightarrow \sin\theta = \frac{1}{2}$ or $\sin\theta = -1 \Rightarrow \theta = \frac{\pi}{6}$ because $\sin\theta \neq -1$ when $0 < \theta < \frac{\pi}{2}$. Also, $A'(\theta) > 0$ for $0 < \theta < \frac{\pi}{6}$ and $A'(\theta) < 0$ for $\frac{\pi}{6} < \theta < \frac{\pi}{2}$. Therefore, at $\theta = \frac{\pi}{6}$ there is a maximum.

25. (a) From the diagram we have: $\overline{AP} = x$, $\overline{RA} = \sqrt{L - x^2}$, $\overline{PB} = 8.5 - x$, $\overline{CH} = \overline{DR} = 11 - \overline{RA} = 11 - \sqrt{L - x^2}$, $\overline{QB} = \sqrt{x^2 - (8.5 - x)^2}$, $\overline{HQ} = 11 - \overline{CH} - \overline{QB}$ $= 11 - \left[11 - \sqrt{L - x^2} + \sqrt{x^2 - (8.5 - x)^2}\right]$ $= \sqrt{L - x^2} - \sqrt{x^2 - (8.5 - x)^2}$, $\overline{RQ}^2 = \overline{RH}^2 + \overline{HQ}^2$ $= (8.5)^2 + \left(\sqrt{L - x^2} - \sqrt{x^2 - (8.5 - x)^2}\right)^2$. It

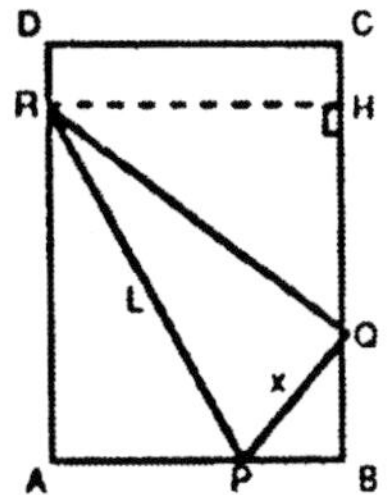

follows that $\overline{RP}^2 = \overline{PQ}^2 + \overline{RQ}^2 \Rightarrow L^2 = x^2 + \left(\sqrt{L^2 - x^2} - \sqrt{x^2 - (x - 8.5)^2}\right)^2 + (8.5)^2$

$\Rightarrow L^2 = x^2 + L^2 - x^2 - 2\sqrt{L^2 - x^2}\,\sqrt{17x - (8.5)^2} + 17x - (8.5)^2 + (8.5)^2$

$\Rightarrow 17^2x^2 = 4\left(L^2 - x^2\right)\left(17x - (8.5)^2\right) \Rightarrow L^2 = x^2 + \frac{17^2x^2}{4\left[17x - (8.5)^2\right]} = \frac{17x^3}{17x - (8.5)^2}$

$= \frac{17x^3}{17x - \left(\frac{17}{2}\right)^2} = \frac{4x^3}{4x - 17} = \frac{2x^3}{2x - 8.5}$.

(b) If $f(x) = \frac{4x^3}{4x-17}$ is minimized, then L^2 is minimized. Now $f'(x) = \frac{4x^2(8x - 51)}{(4x - 17)^2} \Rightarrow f'(x) < 0$ when $x < \frac{51}{8}$ and $f'(x) > 0$ when $x > \frac{51}{8}$. Thus L^2 is minimized when $x = \frac{51}{8}$.

(c) When $x = \frac{51}{8}$, then $L \approx 11.0$ in.

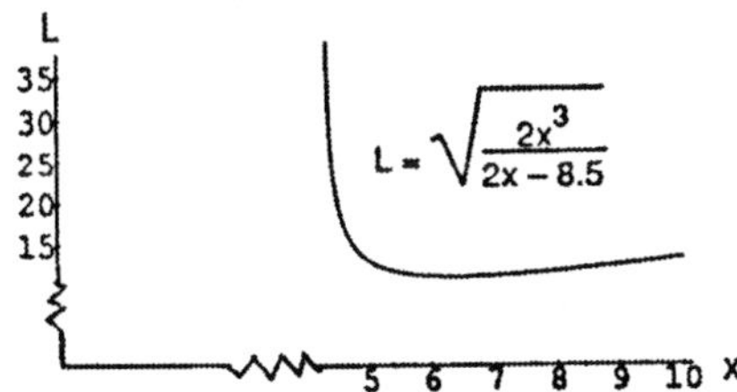

26. (a) From the figure in the text we have $P = 2x + 2y \Rightarrow y = \frac{P}{2} - x$. If $P = 36$, then $y = 18 - x$. When the cylinder is formed, $x = 2\pi r \Rightarrow r = \frac{x}{2\pi}$ and $h = y \Rightarrow h = 18 - x$. The volume of the cylinder is $V = \pi r^2 h$ $\Rightarrow V(x) = \frac{18x^2 - x^3}{4\pi}$. Solving $V'(x) = \frac{3x(12 - x)}{4\pi} = 0 \Rightarrow x = 0$ or 12; but when $x = 0$, there is no cylinder. Then $V''(x) = \frac{3}{\pi}\left(3 - \frac{x}{2}\right) \Rightarrow V''(12) < 0 \Rightarrow$ there is a maximum at $x = 12$. The values of $x = 12$ cm and $y = 6$ cm give the largest volume.

(b) In this case $V(x) = \pi x^2(18 - x)$. Solving $V'(x) = 3\pi x(12 - x) = 0 \Rightarrow x = 0$ or 12; but $x = 0$ would result in no cylinder. Then $V''(x) = 6\pi(6 - x) \Rightarrow V''(12) < 0 \Rightarrow$ there is a maximum at $x = 12$. The values of $x = 12$ cm and $y = 6$ cm give the largest volume.

27. Note that $h^2 + r^2 = 3$ and so $r = \sqrt{3 - h^2}$. Then the volume is given by $V = \frac{\pi}{3}r^2h = \frac{\pi}{3}(3 - h^2)h = \pi h - \frac{\pi}{3}h^3$ for $0 < h < \sqrt{3}$, and so $\frac{dV}{dh} = \pi - \pi r^2 = \pi(1 - r^2)$. The critical point (for $h > 0$) occurs at $h = 1$. Since $\frac{dV}{dh} > 0$ for $0 < h < 1$, and $\frac{dV}{dh} < 0$ for $1 < h < \sqrt{3}$, the critical point corresponds to the maximum volume. The cone of greatest volume has radius $\sqrt{2}$ m, height 1m, and volume $\frac{2\pi}{3}$ m^3.

28. (a) $f(x) = x^2 + \frac{a}{x} \Rightarrow f'(x) = x^{-2}\left(2x^3 - a\right)$, so that $f'(x) = 0$ when $x = 2$ implies $a = 16$

(b) $f(x) = x^2 + \frac{a}{x} \Rightarrow f''(x) = 2x^{-3}\left(x^3 + a\right)$, so that $f''(x) = 0$ when $x = 1$ implies $a = -1$

29. If $f(x) = x^2 + \frac{a}{x}$, then $f'(x) = 2x - ax^{-2}$ and $f''(x) = 2 + 2ax^{-3}$. The critical points are 0 and $\sqrt[3]{\frac{a}{2}}$, but $x \neq 0$. Now $f''\left(\sqrt[3]{\frac{a}{2}}\right) = 6 > 0 \Rightarrow$ at $x = \sqrt[3]{\frac{a}{2}}$ there is a local minimum. However, no local maximum exists for any a.

30. If $f(x) = x^3 + ax^2 + bx$, then $f'(x) = 3x^2 + 2ax + b$ and $f''(x) = 6x + 2a$.

(a) A local maximum at $x = -1$ and local minimum at $x = 3 \Rightarrow f'(-1) = 0$ and $f'(3) = 0 \Rightarrow 3 - 2a + b = 0$ and $27 + 6a + b = 0 \Rightarrow a = -3$ and $b = -9$.

(b) A local minimum at $x = 4$ and a point of inflection at $x = 1 \Rightarrow f'(4) = 0$ and $f''(1) = 0 \Rightarrow 48 + 8a + b = 0$ and $6 + 2a = 0 \Rightarrow a = -3$ and $b = -24$.

31. (a) $s(t) = -16t^2 + 96t + 112 \Rightarrow v(t) = s'(t) = -32t + 96$. At $t = 0$, the velocity is $v(0) = 96$ ft/sec.
(b) The maximum height ocurs when $v(t) = 0$, when $t = 3$. The maximum height is $s(3) = 256$ ft and it occurs at $t = 3$ sec.
(c) Note that $s(t) = -16t^2 + 96t + 112 = -16(t+1)(t-7)$, so $s = 0$ at $t = -1$ or $t = 7$. Choosing the positive value of t, the velocity when $s = 0$ is $v(7) = -128$ ft/sec.

32.
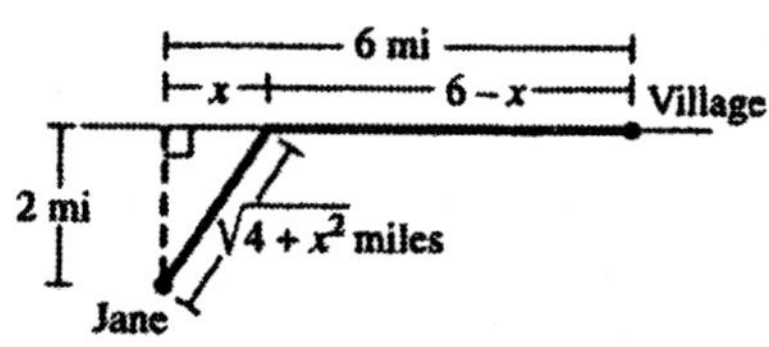

Let x be the distance from the point on the shoreline nearest Jane's boat to the point where she lands her boat. Then she needs to row $\sqrt{4 + x^2}$ mi at 2 mph and walk $6 - x$ mi at 5 mph. The total amount of time to reach the village is $f(x) = \frac{\sqrt{4+x^2}}{2} + \frac{6-x}{5}$ hours $(0 \le x \le 6)$. Then $f'(x) = \frac{1}{2}\frac{1}{2\sqrt{4+x^2}}(2x) - \frac{1}{5} = \frac{x}{2\sqrt{4+x^2}} - \frac{1}{5}$. Solving $f'(x) = 0$, we have: $\frac{x}{2\sqrt{4+x^2}} = \frac{1}{5} \Rightarrow 5x = 2\sqrt{4+x^2} \Rightarrow 25x^2 = 4(4+x^2) \Rightarrow 21x^2 = 16 \Rightarrow x = \pm\frac{4}{\sqrt{21}}$. We discard the negative value of x because it is not in the domain. Checking the endpoints and critical point, we have $f(0) = 2.2$, $f\left(\frac{4}{\sqrt{21}}\right) \approx 2.12$, and $f(6) \approx 3.16$. Jane should land her boat $\frac{4}{\sqrt{21}} \approx 0.87$ miles donw the shoreline from the point nearest her boat.

33. $\frac{8}{x} = \frac{h}{x+27} \Rightarrow h = 8 + \frac{216}{x}$ and $L(x) = \sqrt{h^2 + (x+27)^2}$
$= \sqrt{\left(8 + \frac{216}{x}\right)^2 + (x+27)^2}$ when $x \ge 0$. Note that $L(x)$ is minimized when $f(x) = \left(8 + \frac{216}{x}\right)^2 + (x+27)^2$ is minimized. If $f'(x) = 0$, then
$2\left(8 + \frac{216}{x}\right)\left(-\frac{216}{x^2}\right) + 2(x+27) = 0$
$\Rightarrow (x+27)\left(1 - \frac{1728}{x^3}\right) = 0 \Rightarrow x = -27$ (not acceptable since distance is never negative or $x = 12$. Then $L(12) = \sqrt{2197} \approx 46.87$ ft.

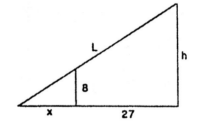

34. (a) From the diagram we have $d^2 = 4r^2 - w^2$. The strength of the beam is $S = kwd^2 = kw\left(4r^2 - w^2\right)$. When $r = 6$, then $S = 144kw - kw^3$. Also, $S'(w) = 144k - 3kw^2 = 3k\left(48 - w^2\right)$ so $S'(w) = 0 \Rightarrow w = \pm 4\sqrt{3}$; $S''\left(4\sqrt{3}\right) < 0$ and $-4\sqrt{3}$ is not acceptable. Therefore $S\left(4\sqrt{3}\right)$ is the maximum strength. The dimensions of the strongest beam are $4\sqrt{3}$ by $4\sqrt{6}$ inches.

(b)
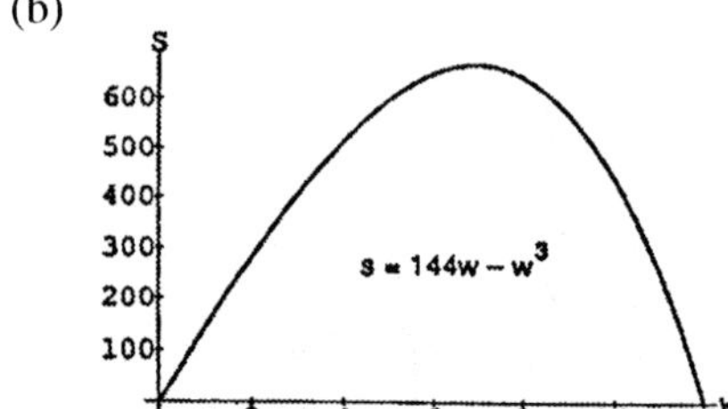

(c)
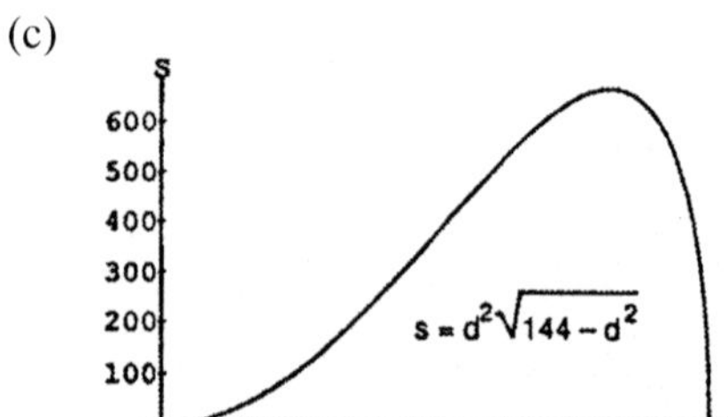

Both graphs indicate the same maximum value and are consistent with each other. Changing k does not change the dimensions that give the strongest beam (i.e., do not change the values of w and d that produce the strongest beam).

35. (a) From the situation we have $w^2 = 144 - d^2$. The stiffness of the beam is $S = kwd^3 = kd^3 (144 - d^2)^{1/2}$, where $0 \le d \le 12$. Also, $S'(d) = \frac{4kd^2(108 - d^2)}{\sqrt{144 - d^2}} \Rightarrow$ critical points at 0, 12, and $6\sqrt{3}$. Both $d = 0$ and $d = 12$ cause $S = 0$. The maximum occurs at $d = 6\sqrt{3}$. The dimensions are 6 by $6\sqrt{3}$ inches.

(b)

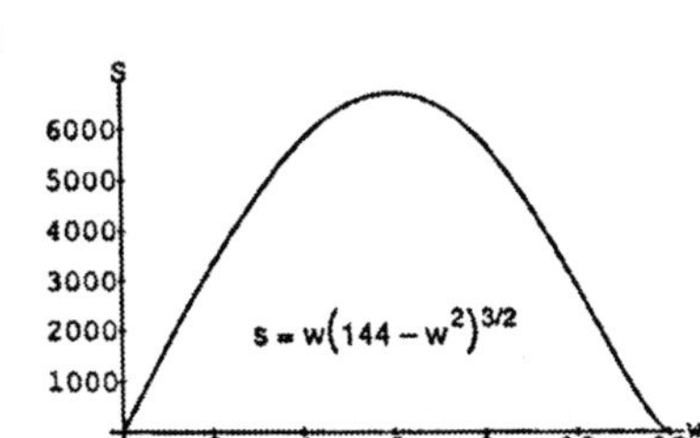

(c)

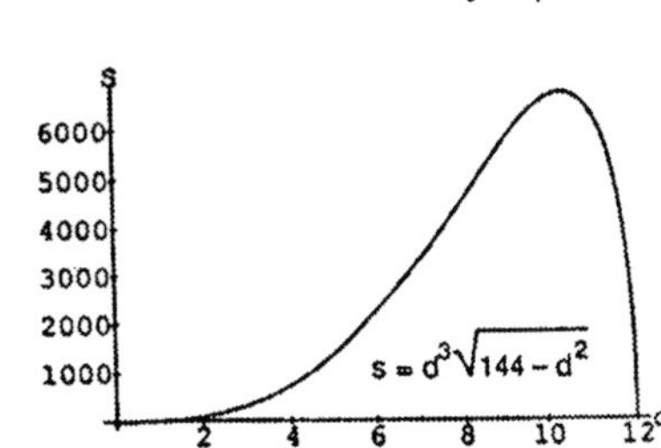

Both graphs indicate the same maximum value and are consistent with each other. The changing of k has no effect.

36. (a) $s_1 = s_2 \Rightarrow \sin t = \sin\left(t + \frac{\pi}{3}\right) \Rightarrow \sin t = \sin t \cos\frac{\pi}{3} + \sin\frac{\pi}{3}\cos t \Rightarrow \sin t = \frac{1}{2}\sin t + \frac{\sqrt{3}}{2}\cos t \Rightarrow \tan t = \sqrt{3}$ $\Rightarrow t = \frac{\pi}{3}$ or $\frac{4\pi}{3}$

(b) The distance between the particles is $s(t) = |s_1 - s_2| = \left|\sin t - \sin\left(t + \frac{\pi}{3}\right)\right| = \frac{1}{2}\left|\sin t - \sqrt{3}\cos t\right|$

$\Rightarrow s'(t) = \frac{\left(\sin t - \sqrt{3}\cos t\right)\left(\cos t + \sqrt{3}\sin t\right)}{2\left|\sin t - \sqrt{3}\cos t\right|}$ since $\frac{d}{dx}|x| = \frac{x}{|x|} \Rightarrow$ critical times and endpoints are $0, \frac{\pi}{3}, \frac{5\pi}{6}, \frac{4\pi}{3}, \frac{11\pi}{6}, 2\pi$; then $s(0) = \frac{\sqrt{3}}{2}$, $s\left(\frac{\pi}{3}\right) = 0$, $s\left(\frac{5\pi}{6}\right) = 1$, $s\left(\frac{4\pi}{3}\right) = 0$, $s\left(\frac{11\pi}{6}\right) = 1$, $s(2\pi) = \frac{\sqrt{3}}{2} \Rightarrow$ the greatest distance between the particles is 1.

(c) Since $s'(t) = \frac{\left(\sin t - \sqrt{3}\cos t\right)\left(\cos t + \sqrt{3}\sin t\right)}{2\left|\sin t - \sqrt{3}\cos t\right|}$ we can conclude that at $t = \frac{\pi}{3}$ and $\frac{4\pi}{3}$, $s'(t)$ has cusps and the distance between the particles is changing the fastest near these points.

37. (a) $s = 10\cos(\pi t) \Rightarrow v = -10\pi\sin(\pi t) \Rightarrow \text{speed} = |10\pi\sin(\pi t)| = 10\pi|\sin(\pi t)| \Rightarrow$ the maximum speed is $10\pi \approx 31.42$ cm/sec since the maximum value of $|\sin(\pi t)|$ is 1; the cart is moving the fastest at t = 0.5 sec, 1.5 sec, 2.5 sec and 3.5 sec when $|\sin(\pi t)|$ is 1. At these times the distance is $s = 10\cos\left(\frac{\pi}{2}\right) = 0$ cm and $a = -10\pi^2\cos(\pi t) \Rightarrow |a| = 10\pi^2|\cos(\pi t)| \Rightarrow |a| = 0\ \text{cm/sec}^2$

(b) $|a| = 10\pi^2|\cos(\pi t)|$ is greatest at t = 0.0 sec, 1.0 sec, 2.0 sec, 3.0 sec and 4.0 sec, and at these times the magnitude of the cart's position is $|s| = 10$ cm from the rest position and the speed is 0 cm/sec.

38. (a) $2\sin t = \sin 2t \Rightarrow 2\sin t - 2\sin t\cos t = 0 \Rightarrow (2\sin t)(1 - \cos t) = 0 \Rightarrow t = k\pi$ where k is a positive integer

(b) The vertical distance between the masses is $s(t) = |s_1 - s_2| = \left((s_1 - s_2)^2\right)^{1/2} = \left((\sin 2t - 2\sin t)^2\right)^{1/2}$

$\Rightarrow s'(t) = \left(\frac{1}{2}\right)\left((\sin 2t - 2\sin t)^2\right)^{-1/2}(2)(\sin 2t - 2\sin t)(2\cos 2t - 2\cos t)$

$= \frac{2(\cos 2t - \cos t)(\sin 2t - 2\sin t)}{|\sin 2t - 2\sin t|} = \frac{4(2\cos t + 1)(\cos t - 1)(\sin t)(\cos t - 1)}{|\sin 2t - 2\sin t|} \Rightarrow$ critical times at $0, \frac{2\pi}{3}, \pi, \frac{4\pi}{3}, 2\pi$; then $s(0) = 0$, $s\left(\frac{2\pi}{3}\right) = \left|\sin\left(\frac{4\pi}{3}\right) - 2\sin\left(\frac{2\pi}{3}\right)\right| = \frac{3\sqrt{3}}{2}$, $s(\pi) = 0$, $s\left(\frac{4\pi}{3}\right)$ $= \left|\sin\left(\frac{8\pi}{3}\right) - 2\sin\left(\frac{4\pi}{3}\right)\right| = \frac{3\sqrt{3}}{2}$, $s(2\pi) = 0 \Rightarrow$ the greatest distance is $\frac{3\sqrt{3}}{2}$ at $t = \frac{2\pi}{3}$ and $\frac{4\pi}{3}$

39.

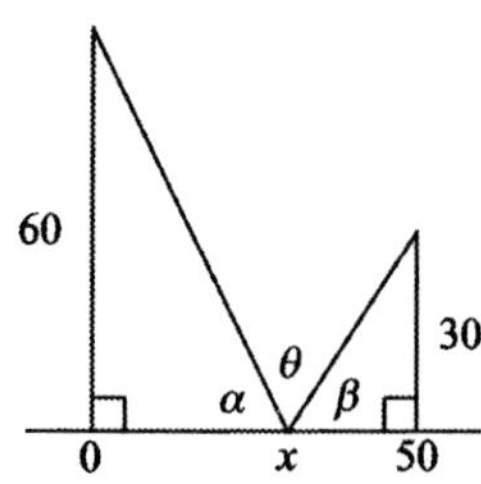

From the diagram above we have $\alpha + \beta + \theta = \pi \Rightarrow \theta = \pi - \alpha - \beta$. From the left triangle we have $\cot\alpha = \frac{x}{60} \Rightarrow \cot^{-1}\left(\frac{x}{60}\right) = \alpha$, and from the triangle on the right side we have $\cot\beta = \frac{50-x}{60} \Rightarrow \cot^{-1}\left(\frac{50-x}{30}\right) = \beta$

$\Rightarrow \theta = \pi - \cot^{-1}\left(\frac{x}{60}\right) - \cot^{-1}\left(\frac{50-x}{30}\right) \Rightarrow \frac{d\theta}{dx} = 0 - \left(-\frac{1}{1+\left(\frac{x}{60}\right)^2}\cdot\frac{1}{60}\right) - \left(-\frac{1}{1+\left(\frac{50-x}{30}\right)^2}\cdot\frac{-1}{30}\right)$

$\frac{d\theta}{dx} = \frac{1}{60\left(1+\left(\frac{x}{60}\right)^2\right)} - \frac{1}{30\left(1+\left(\frac{50-x}{30}\right)^2\right)} = \frac{1}{60+\frac{x^2}{60}} - \frac{1}{30+\frac{(50-x)^2}{30}} = \frac{60}{60^2+x^2} - \frac{30}{30^2+(50-x)^2}$

$\frac{60}{60^2+x^2} - \frac{30}{30^2+(50-x)^2} = 0 \Rightarrow 60\left(30^2 + (50-x)^2\right) - 30(60^2+x^2) \Rightarrow 30x^2 - 6000x + 96000 = 0$

$\Rightarrow x = 20\left(5 \pm \sqrt{17}\right)$. Since $0 \le x \le 50 \Rightarrow x = 20\left(5-\sqrt{17}\right) \approx 17.54$. $\theta'(10) = \frac{60}{60^2+10^2} - \frac{30}{30^2+(50-10)^2}$

$= \frac{60}{3700} - \frac{30}{2500} = \frac{39}{9250} > 0$ and $\theta'(20) = \frac{60}{60^2+20^2} - \frac{30}{30^2+(50-20)^2} = \frac{60}{4000} - \frac{30}{1800} = -\frac{1}{600}$.

$\Rightarrow$ the maximum angle is $\theta = \pi - \cot^{-1}\left(\frac{20\left(5-\sqrt{17}\right)}{60}\right) - \cot^{-1}\left(\frac{50-20\left(5-\sqrt{17}\right)}{30}\right) \approx 1.10917$ rad or

$63.55°$ when the solar station is ≈ 17.54 m west of the left (60 m) building,

40. The distance $\overline{OT} + \overline{TB}$ is minimized when $\overline{OB}$ is a straight line. Hence $\angle\alpha = \angle\beta \Rightarrow \theta_1 = \theta_2$.

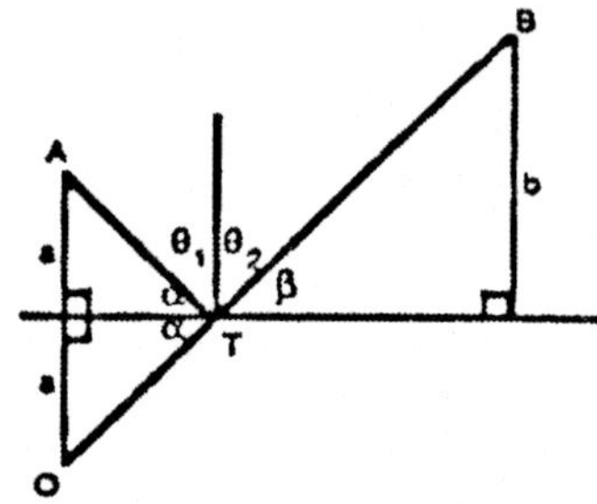

41. If $v = kax - kx^2$, then $v' = ka - 2kx$ and $v'' = -2k$, so $v' = 0 \Rightarrow x = \frac{a}{2}$. At $x = \frac{a}{2}$ there is a maximum since $v''\left(\frac{a}{2}\right) = -2k < 0$. The maximum value of v is $\frac{ka^2}{4}$.

42. (a) According to the graph, $y'(0) = 0$.
(b) According to the graph, $y'(-L) = 0$.
(c) $y(0) = 0$, so $d = 0$. Now $y'(x) = 3ax^2 + 2bx + c$, so $y'(0) = 0$ implies that $c = 0$. There fore, $y(x) = ax^3 + bx^2$ and $y'(x) = 3ax^2 + 2bx$. Then $y(-L) = -aL^3 + bL^2 = H$ and $y'(-L) = 3aL^2 - 2bL = 0$, so we have two linear equations in two unknowns a and b. The second equation gives $b = \frac{3aL}{2}$. Substituting into the first equation, we have $-aL^3 + \frac{3aL^3}{2} = H$, or $\frac{aL^3}{2} = H$, so $a = 2\frac{H}{L^3}$. Therefore, $b = 3\frac{H}{L^2}$ and the equation for y is $y(x) = 2\frac{H}{L^3}x^3 + 3\frac{H}{L^2}x^2$, or $y(x) = H\left[2\left(\frac{x}{L}\right)^3 + 3\left(\frac{x}{L}\right)^2\right]$.

43. The profit is $p = nx - nc = n(x-c) = \left[a(x-c)^{-1} + b(100-x)\right](x-c) = a + b(100-x)(x-c)$ $= a + (bc + 100b)x - 100bc - bx^2$. Then $p'(x) = bc + 100b - 2bx$ and $p''(x) = -2b$. Solving $p'(x) = 0 \Rightarrow$ $x = \frac{c}{2} + 50$. At $x = \frac{c}{2} + 50$ there is a maximum profit since $p''(x) = -2b < 0$ for all x.

44. Let x represent the number of people over 50. The profit is $p(x) = (50+x)(200-2x) - 32(50+x) - 6000$ $= -2x^2 + 68x + 2400$. Then $p'(x) = -4x + 68$ and $p'' = -4$. Solving $p'(x) = 0 \Rightarrow x = 17$. At $x = 17$ there is a maximum since $p''(17) < 0$. It would take 67 people to maximize the profit.

45. (a) $A(q) = kmq^{-1} + cm + \frac{h}{2}q$, where $q > 0 \Rightarrow A'(q) = -kmq^{-2} + \frac{h}{2} = \frac{hq^2 - 2km}{2q^2}$ and $A''(q) = 2kmq^{-3}$. The critical points are $-\sqrt{\frac{2km}{h}}$, 0, and $\sqrt{\frac{2km}{h}}$, but only $\sqrt{\frac{2km}{h}}$ is in the domain. Then $A''\left(\sqrt{\frac{2km}{h}}\right) > 0 \Rightarrow$ at $q = \sqrt{\frac{2km}{h}}$ there is a minimum average weekly cost.

(b) $A(q) = \frac{(k+bq)m}{q} + cm + \frac{h}{2}q = kmq^{-1} + bm + cm + \frac{h}{2}q$, where $q > 0 \Rightarrow A'(q) = 0$ at $q = \sqrt{\frac{2km}{h}}$ as in (a). Also $A''(q) = 2kmq^{-3} > 0$ so the most economical quantity to order is still $q = \sqrt{\frac{2km}{h}}$ which minimizes the average weekly cost.

46. We start with $c(x) =$ the cost of producing x items, $x > 0$, and $\frac{c(x)}{x} =$ the average cost of producing x items, assumed to be differentiable. If the average cost can be minimized, it will be at a production level at which $\frac{d}{dx}\left(\frac{c(x)}{x}\right) = 0$ $\Rightarrow \frac{x\,c'(x) - c(x)}{x^2} = 0$ (by the quotient rule) $\Rightarrow x\,c'(x) - c(x) = 0$ (multiply both sides by x^2) $\Rightarrow c'(x) = \frac{c(x)}{x}$ where $c'(x)$ is the marginal cost. This concludes the proof. (Note: The theorem does not assure a production level that will give a minimum cost, but rather, it indicates where to look to see if there is one. Find the production levels where the average cost equals the marginal cost, then check to see if any of them give a mimimum.)

47. The profit $p(x) = r(x) - c(x) = 6x - (x^3 - 6x^2 + 15x) = -x^3 + 6x^2 - 9x$, where $x \geq 0$. Then $p'(x) = -3x^2 + 12x - 9 = -3(x-3)(x-1)$ and $p''(x) = -6x + 12$. The critical points are 1 and 3. Thus $p''(1) = 6 > 0 \Rightarrow$ at $x = 1$ there is a local minimum, and $p''(3) = -6 < 0 \Rightarrow$ at $x = 3$ there is a local maximum. But $p(3) = 0 \Rightarrow$ the best you can do is break even.

48. The average cost of producing x items is $\bar{c}(x) = \frac{c(x)}{x} = x^2 - 20x + 20{,}000 \Rightarrow \bar{c}'(x) = 2x - 20 = 0 \Rightarrow x = 10$, the only critical value. The average cost is $\bar{c}(10) = \$19{,}900$ per item is a minimum cost because $\bar{c}''(10) = 2 > 0$.

49. (a) The artisan should order px units of material in order to have enough until the next delivery.

(b) The average number of units in storage until the next delivery is $\frac{px}{2}$ and so the cost of storing then is $s\left(\frac{px}{2}\right)$ per day, and the total cost for x days is $\left(\frac{px}{2}\right)sx$. When added to the delivery cost, the total cost for delivery and storage for each cycle is: cost per cycle $= d + \frac{px}{2}sx$.

(c) The average cost per day for storage and delivery of materials is: average cost per day $= \frac{\left(d + \frac{ps}{2}x^2\right)}{x} + \frac{d}{x} + \frac{ps}{2}x$. To minimize the average cost per day, set the derivative equal to zero. $\frac{d}{dx}\left(d(x)^{-1} + \frac{ps}{2}x\right) = -d(x)^{-2} + \frac{ps}{2} = 0$ $\Rightarrow x = \pm\sqrt{\frac{2d}{ps}}$. Only the positive root makes sense in this context so that $x^* = \sqrt{\frac{2d}{ps}}$. To verify that x^* gives a minimum, check the second derivative $\left[\frac{d}{dx}\left(-d(x)^{-2} + \frac{ps}{2}\right)\right]\Bigg|_{\sqrt{\frac{2d}{ps}}} = \frac{2d}{x^3}\Bigg|_{\sqrt{\frac{2d}{ps}}} = \frac{2d}{\left(\sqrt{\frac{2d}{ps}}\right)^3} > 0 \Rightarrow$ a minimum.

The amount to deliver is $px^* = \sqrt{\frac{2pd}{s}}$.

(d) The line and the hyperbola intersect when $\frac{d}{x} = \frac{ps}{2}x$. Solving for x gives $x_{\text{intersection}} = \pm\sqrt{\frac{2d}{ps}}$. For $x > 0$, $x_{\text{intersection}} = \sqrt{\frac{2d}{ps}} = x^*$. From this result, the average cost per day is minimized when the average daily cost of delivery is equal to the average daily cost of storage.

50. Average Cost: $\frac{c(x)}{x} = \frac{2000}{x} + 96 + 4x^{1/2} \Rightarrow \frac{d}{dx}\left(\frac{c(x)}{x}\right) = -\frac{2000}{x^2} + 2x^{-1/2} = 0 \Rightarrow x = 100$. Check for a minimum: $\frac{d^2}{dx^2}\left(\frac{c(x)}{x}\right)\Bigg|_{x=100} = \frac{4000}{100^3} - 100^{-3/2} = 0.003 > 0 \Rightarrow$ a minimum at $x = 100$. At a production level of 100,000 units, the average cost will be minimized at \$156 per unit.

51. We have $\frac{dR}{dM} = CM - M^2$. Solving $\frac{d^2R}{dM^2} = C - 2M = 0 \Rightarrow M = \frac{C}{2}$. Also, $\frac{d^3R}{dM^3} = -2 < 0 \Rightarrow$ at $M = \frac{C}{2}$ there is a maximum.

52. (a) If $v = cr_0r^2 - cr^3$, then $v' = 2cr_0r - 3cr^2 = cr(2r_0 - 3r)$ and $v'' = 2cr_0 - 6cr = 2c(r_0 - 3r)$. The solution of $v' = 0$ is $r = 0$ or $\frac{2r_0}{3}$, but 0 is not in the domain. Also, $v' > 0$ for $r < \frac{2r_0}{3}$ and $v' < 0$ for $r > \frac{2r_0}{3} \Rightarrow$ at $r = \frac{2r_0}{3}$ there is a maximum.

(b) The graph confirms the findings in (a).

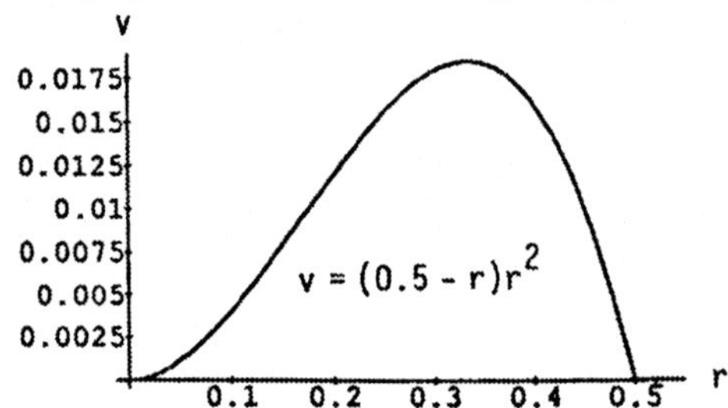

53. If $x > 0$, then $(x-1)^2 \ge 0 \Rightarrow x^2 + 1 \ge 2x \Rightarrow \frac{x^2+1}{x} \ge 2$. In particular if a, b, c and d are positive integers, then $\left(\frac{a^2+1}{a}\right)\left(\frac{b^2+1}{b}\right)\left(\frac{c^2+1}{c}\right)\left(\frac{d^2+1}{d}\right) \ge 16$.

54. (a) $f(x) = \frac{x}{\sqrt{a^2+x^2}} \Rightarrow f'(x) = \frac{(a^2+x^2)^{1/2} - x^2(a^2+x^2)^{-1/2}}{(a^2+x^2)} = \frac{a^2+x^2-x^2}{(a^2+x^2)^{3/2}} = \frac{a^2}{(a^2+x^2)^{3/2}} > 0$

$\Rightarrow$ f(x) is an increasing function of x

(b) $g(x) = \frac{d-x}{\sqrt{b^2+(d-x)^2}} \Rightarrow g'(x) = \frac{-(b^2+(d-x)^2)^{1/2} + (d-x)^2(b^2+(d-x)^2)^{-1/2}}{b^2+(d-x)^2}$

$= \frac{-(b^2+(d-x)^2)+(d-x)^2}{(b^2+(d-x)^2)^{3/2}} = \frac{-b^2}{(b^2+(d-x)^2)^{3/2}} < 0 \Rightarrow$ g(x) is a decreasing function of x

(c) Since $c_1, c_2 > 0$, the derivative $\frac{dt}{dx}$ is an increasing function of x (from part (a)) minus a decreasing function of x (from part (b)): $\frac{dt}{dx} = \frac{1}{c_1}f(x) - \frac{1}{c_2}g(x) \Rightarrow \frac{d^2t}{dx^2} = \frac{1}{c_1}f'(x) - \frac{1}{c_2}g'(x) > 0$ since $f'(x) > 0$ and $g'(x) < 0 \Rightarrow \frac{dt}{dx}$ is an increasing function of x.

55. At x = c, the tangents to the curves are parallel. Justification: The vertical distance between the curves is $D(x) = f(x) - g(x)$, so $D'(x) = f'(x) - g'(x)$. The maximum value of D will occur at a point c where $D' = 0$. At such a point, $f'(c) - g'(c) = 0$, or $f'(c) = g'(c)$.

56. (a) $f(x) = 3 + 4\cos x + \cos 2x$ is a periodic function with period 2π

(b) No, $f(x) = 3 + 4\cos x + \cos 2x = 3 + 4\cos x + (2\cos^2 x - 1) = 2(1 + 2\cos x + \cos^2 x) = 2(1+\cos x)^2 \ge 0$ $\Rightarrow$ f(x) is never negative

57. (a) If $y = \cot x - \sqrt{2}\csc x$ where $0 < x < \pi$, then $y' = (\csc x)\left(\sqrt{2}\cot x - \csc x\right)$. Solving $y' = 0$ $\Rightarrow \cos x = \frac{1}{\sqrt{2}} \Rightarrow x = \frac{\pi}{4}$. For $0 < x < \frac{\pi}{4}$ we have $y' > 0$, and $y' < 0$ when $\frac{\pi}{4} < x < \pi$. Therefore, at $x = \frac{\pi}{4}$ there is a maximum value of $y = -1$.

(b)

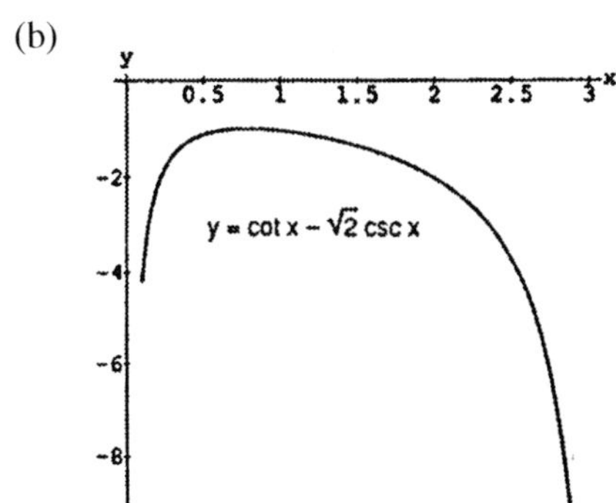

The graph confirms the findings in (a).

58. (a) If $y = \tan x + 3 \cot x$ where $0 < x < \frac{\pi}{2}$, then $y' = \sec^2 x - 3 \csc^2 x$. Solving $y' = 0 \Rightarrow \tan x = \pm\sqrt{3}$ $\Rightarrow x = \pm\frac{\pi}{3}$, but $-\frac{\pi}{3}$ is not in the domain. Also, $y'' = 2 \sec^2 x \tan x + 3 \csc^2 x \cot x > 0$ for all $0 < x < \frac{\pi}{2}$. Therefore at $x = \frac{\pi}{3}$ there is a minimum value of $y = 2\sqrt{3}$.

(b)

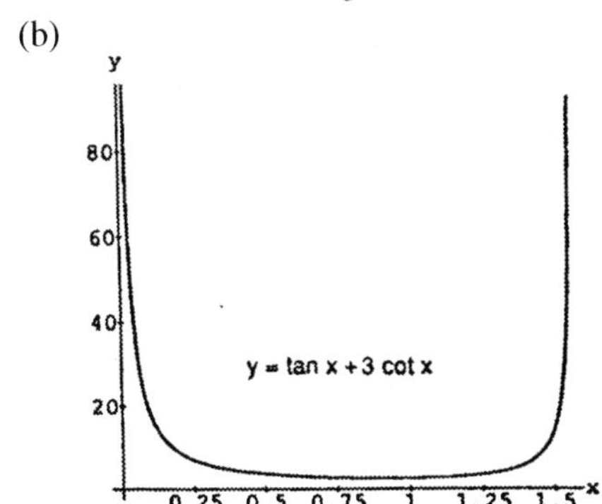

The graph confirms the findings in (a).

59. (a) The square of the distance is $D(x) = \left(x - \frac{3}{2}\right)^2 + \left(\sqrt{x} + 0\right)^2 = x^2 - 2x + \frac{9}{4}$, so $D'(x) = 2x - 2$ and the critical point occurs at $x = 1$. Since $D'(x) < 0$ for $x < 1$ and $D'(x) > 0$ for $x > 1$, the critical point corresponds to the minimum distance. The minimum distance is $\sqrt{D(1)} = \frac{\sqrt{5}}{2}$.

(b)

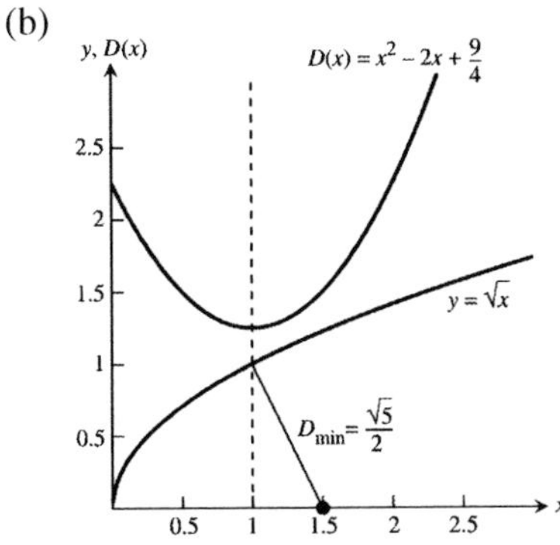

The minimum distance is from the point $\left(\frac{3}{2}, 0\right)$ to the point $(1, 1)$ on the graph of $y = \sqrt{x}$, and this occurs at the value $x = 1$ where $D(x)$, the distance squared, has its minimum value.

60. (a) Calculus Method:

The square of the distance from the point $\left(1, \sqrt{3}\right)$ to $\left(x, \sqrt{16 - x^2}\right)$ is given by

$D(x) = (x - 1)^2 + \left(\sqrt{16 - x^2} - \sqrt{3}\right)^2 = x^2 - 2x + 1 + 16 - x^2 - 2\sqrt{48 - 3x^2} + 3 = -2x + 20 - 2\sqrt{48 - 3x^2}$.

Then $D'(x) = -2 - \frac{2}{\sqrt{48 - 3x^2}}(-6x) = -2 + \frac{6x}{\sqrt{48 - 3x^2}}$. Solving $D'(x) = 0$ we have: $6x = 2\sqrt{48 - 3x^2}$

$\Rightarrow 36x^2 = 4(48 - 3x^2) \Rightarrow 9x^2 = 48 - 3x^2 \Rightarrow 12x^2 = 48 \Rightarrow x = \pm 2$. We discard $x = -2$ as an extraneous solution, leaving $x = 2$. Since $D'(x) < 0$ for $-4 < x < 2$ and $D'(x) > 0$ for $2 < x < 4$, the critical point corresponds to the minimum distance. The minimum distance is $\sqrt{D(2)} = 2$.

Geometry Method:

The semicircle is centered at the origin and has radius 4. The distance from the origin to $\left(1, \sqrt{3}\right)$ is $\sqrt{1^2 + \left(\sqrt{3}\right)^2} = 2$. The shortest distance from the point to the semicircle is the distance along the radius containing the point $\left(1, \sqrt{3}\right)$. That distance is $4 - 2 = 2$.

(b)

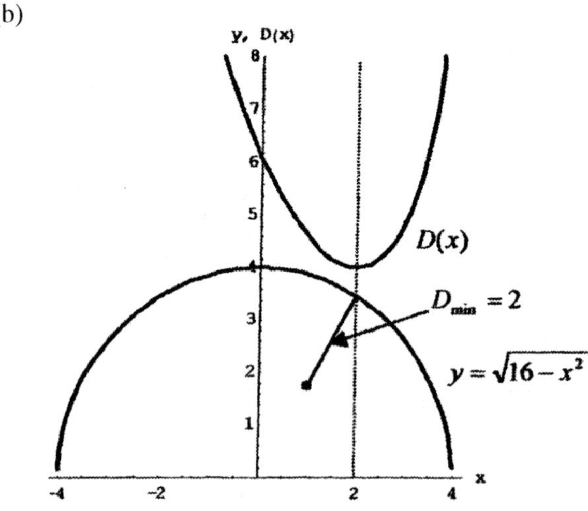

The minimum distance is from the point $\left(1, \sqrt{3}\right)$ to the point $\left(2, 2\sqrt{3}\right)$ on the graph of $y = \sqrt{16 - x^2}$, and this occurs at the value $x = 2$ where $D(x)$, the distance squared, has its minimum value.

61. (a) The base radius of the cone is $r = \frac{2\pi a - x}{2\pi}$ and so the height is $h = \sqrt{a^2 - r^2} = \sqrt{a^2 - \left(\frac{2\pi a - x}{2\pi}\right)^2}$. Therefore, $V(x) = \frac{\pi}{3} r^2 h = \frac{\pi}{3}\left(\frac{2\pi a - x}{2\pi}\right)^2 \sqrt{a^2 - \left(\frac{2\pi a - x}{2\pi}\right)^2}$.

(b) To simplify the calculations, we shall consider the volume as a function of r: volume $= f(r) = \frac{\pi}{3} r^2 \sqrt{a^2 - r^2}$, where $0 < r < a$. $f'(r) = \frac{\pi}{3}\frac{d}{dr}\left(r^2\sqrt{a^2 - r^2}\right) = \frac{\pi}{3}\left[r^2 \cdot \frac{1}{2\sqrt{a^2 - r^2}}(-2r) + \left(\sqrt{a^2 - r^2}\right)(2r)\right] = \frac{\pi}{3}\left[\frac{-r^3 + 2r(a^2 - r^2)}{\sqrt{a^2 - r^2}}\right]$ $= \frac{\pi}{3}\left[\frac{2a^2 r - 3r^3}{\sqrt{a^2 - r^2}}\right] = \frac{\pi r(2a^2 - 3r^2)}{3\sqrt{a^2 - r^2}}$. The critical point occurs when $r^2 = \frac{2a^2}{3}$, which gives $r = a\sqrt{\frac{2}{3}} = \frac{a\sqrt{6}}{3}$. Then $h = \sqrt{a^2 - r^2} = \sqrt{a^2 - \frac{2a^2}{3}} = \sqrt{\frac{a^2}{3}} = \frac{a\sqrt{3}}{3}$. Using $r = \frac{a\sqrt{6}}{3}$ and $h = \frac{a\sqrt{3}}{3}$, we may now find the values of r and h for the given values of a.

When $a = 4$: $r = \frac{4\sqrt{6}}{3}$, $h = \frac{4\sqrt{3}}{3}$;

When $a = 5$: $r = \frac{5\sqrt{6}}{3}$, $h = \frac{5\sqrt{3}}{3}$;

When $a = 6$: $r = 2\sqrt{6}$, $h = 2\sqrt{3}$;

When $a = 8$: $r = \frac{8\sqrt{6}}{3}$, $h = \frac{8\sqrt{3}}{3}$;

(c) Since $r = \frac{a\sqrt{6}}{3}$ and $h = \frac{a\sqrt{3}}{3}$, the relationship is $\frac{r}{h} = \sqrt{2}$.

62. (a) Let x_0 represent the fixed value of x at the point P, so that P has the coordinates (x_0, a), and let $m = f'(x_0)$ be the slope of the line RT. Then the equation of the line RT is $y = m(x - x_0) + a$. The y-intercept of this line is $m(0 - x_0) + a = a - mx_0$, and the x-intercept is the solution of $m(x - x_0) + a = 0$, or $x = \frac{mx_0 - a}{m}$. Let O designate the origin. Then

(Area of triangle RST)

$= 2$(Area of triangle ORT)

$= 2 \cdot \frac{1}{2}$(x-intercept of line RT)(y-intercept of line RT)

$= 2 \cdot \frac{1}{2}\left(\frac{mx_0 - a}{m}\right)(a - mx_0)$

$= -m\left(\frac{mx_0 - a}{m}\right)\left(\frac{mx_0 - a}{m}\right)$

$= -m\left(\frac{mx_0 - a}{m}\right)^2$

$= -m\left(x_0 - \frac{a}{m}\right)^2$

Substituting x for x_0, $f'(x)$ for m, and $f(x)$ for a, we have $A(x) = -f'(x)\left[x - \frac{f(x)}{f'(x)}\right]^2$.

(b) The domain is the open interval (0, 10). To graph, let $y_1 = f(x) = 5 + 5\sqrt{1 - \frac{x^2}{100}}$, $y_2 = f'(x) = \text{NDER}(y_1)$, and $y_3 = A(x) = -y_2\left(x - \frac{y_1}{y_2}\right)^2$. The graph of the area function $y_3 = A(x)$ is shown below.

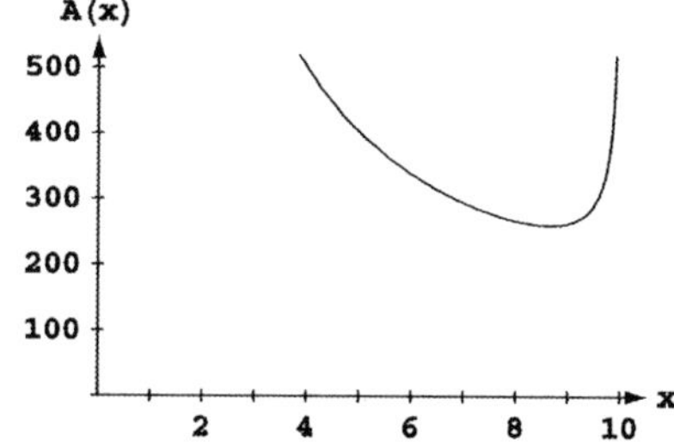

The vertical asymptotes at $x = 0$ and $x = 10$ correspond to horizontal or vertical tangent lines, which do not form triangles.

(c) Using our expression for the y-intercept of the tangent line, the height of the triangle is
$a - mx = f(x) - f'(x) \cdot x = 5 + \frac{1}{2}\sqrt{100 - x^2} - \frac{-x}{2\sqrt{100 - x^2}}x = 5 + \frac{1}{2}\sqrt{100 - x^2} + \frac{x^2}{2\sqrt{100 - x^2}}$
We may use graphing methods or the analytic method in part (d) to find that the minimum value of $A(x)$ occurs at $x \approx 8.66$. Substituting this value into the expression above, the height of the triangle is 15. This is 3 times the y-coordinate of the center of the ellipse.

(d) Part (a) remains unchanged. Assuming $C \geq B$, the domain is $(0, C)$. To graph, note that
$f(x) = B + B\sqrt{1 - \frac{x^2}{C^2}} = B + \frac{B}{C}\sqrt{C^2 - x^2}$ and $f'(x) = \frac{B}{C}\frac{1}{2\sqrt{C^2 - x^2}}(-2x) = \frac{-Bx}{C\sqrt{C^2 - x^2}}$. Therefore we have

$$A(x) = -f'(x)\left[x - \frac{f(x)}{f'(x)}\right]^2 = \frac{Bx}{C\sqrt{C^2 - x^2}}\left(x - \frac{B + \frac{B}{C}\sqrt{C^2 - x^2}}{\frac{-Bx}{C\sqrt{C^2 - x^2}}}\right)^2 = \frac{Bx}{C\sqrt{C^2 - x^2}}\left(x - \frac{\left(BC + B\sqrt{C^2 - x^2}\right)\left(\sqrt{C^2 - x^2}\right)}{-Bx}\right)^2$$

$$= \frac{1}{BCx\sqrt{C^2 - x^2}}\left[Bx^2 + \left(BC + B\sqrt{C^2 - x^2}\right)\left(\sqrt{C^2 - x^2}\right)\right]^2 = \frac{1}{BCx\sqrt{C^2 - x^2}}\left[Bx^2 + BC\sqrt{C^2 - x^2} + B(C^2 - x^2)\right]^2$$

$$= \frac{1}{BCx\sqrt{C^2 - x^2}}\left[BC\left(C + \sqrt{C^2 - x^2}\right)\right]^2 = \frac{BC\left(C + \sqrt{C^2 - x^2}\right)^2}{x\sqrt{C^2 - x^2}}$$

$$A'(x) = BC \cdot \frac{\left(x\sqrt{C^2 - x^2}\right)(2)\left(C + \sqrt{C^2 - x^2}\right)\left(\frac{-x}{\sqrt{C^2 - x^2}}\right) - \left(C + \sqrt{C^2 - x^2}\right)^2\left(x\frac{-x}{\sqrt{C^2 - x^2}} + \sqrt{C^2 - x^2}(1)\right)}{x^2(C^2 - x^2)}$$

$$= \frac{BC\left(C + \sqrt{C^2 - x^2}\right)}{x^2(C^2 - x^2)}\left[-2x^2 - \left(C + \sqrt{C^2 - x^2}\right)\left(\frac{-x^2}{\sqrt{C^2 - x^2}} + \sqrt{C^2 - x^2}\right)\right]$$

$$= \frac{BC\left(C + \sqrt{C^2 - x^2}\right)}{x^2(C^2 - x^2)}\left[-2x^2 + \frac{Cx^2}{\sqrt{C^2 - x^2}} - C\sqrt{C^2 - x^2} + x^2 - (C^2 - x^2)\right]$$

$$= \frac{BC\left(C + \sqrt{C^2 - x^2}\right)}{x^2(C^2 - x^2)}\left(\frac{Cx^2}{\sqrt{C^2 - x^2}} - C\sqrt{C^2 - x^2} - C^2\right) = \frac{BC\left(C + \sqrt{C^2 - x^2}\right)}{x^2(C^2 - x^2)^{3/2}}\left[Cx^2 - C(C^2 - x^2) - C^2\sqrt{C^2 - x^2}\right]$$

$$= \frac{BC^2\left(C + \sqrt{C^2 - x^2}\right)}{x^2(C^2 - x^2)^{3/2}}\left(2x^2 - C^2 - C\sqrt{C^2 - x^2}\right)$$

To find the critical points for $0 < x < C$, we solve: $2x^2 - C^2 = C\sqrt{C^2 - x^2} \Rightarrow 4x^4 - 4C^2x^2 + C^4 = C^4 - C^2x^2$ $\Rightarrow 4x^4 - 3C^2x^2 = 0 \Rightarrow x^2(4x^2 - 3C^2) = 0$. The minimum value of $A(x)$ for $0 < x < C$ occurs at the critical point $x = \frac{C\sqrt{3}}{2}$, or $x^2 = \frac{3C^2}{4}$. The corresponding triangle height is

$$\begin{aligned} a - mx &= f(x) - f'(x) \cdot x \\ &= B + \tfrac{B}{C}\sqrt{C^2 - x^2} + \frac{Bx^2}{C\sqrt{C^2 - \frac{3C^2}{4}}} \\ &= B + \tfrac{B}{C}\sqrt{C^2 - x^2} + \frac{B\left(\frac{3C^2}{4}\right)}{C\sqrt{C^2 - \frac{3C^2}{4}}} \\ &= B + \tfrac{B}{C}\left(\tfrac{C}{2}\right) + \frac{\frac{3BC^2}{4}}{\frac{C^2}{2}} \\ &= B + \tfrac{B}{2} + \tfrac{3B}{2} \\ &= 3B \end{aligned}$$

This shows that the traingle has minimum arrea when its height is 3B.

4.6 INDETERMINATE FORMS AND L'HÔPITAL'S RULE

1. l'Hôpital: $\lim\limits_{x \to 2} \frac{x-2}{x^2-4} = \frac{1}{2x}\Big|_{x=2} = \frac{1}{4}$ or $\lim\limits_{x \to 2} \frac{x-2}{x^2-4} = \lim\limits_{x \to 2} \frac{x-2}{(x-2)(x+2)} = \lim\limits_{x \to 2} \frac{1}{x+2} = \frac{1}{4}$

2. l'Hôpital: $\lim\limits_{x \to 0} \frac{\sin 5x}{x} = \frac{5\cos 5x}{1}\Big|_{x=0} = 5$ or $\lim\limits_{x \to 0} \frac{\sin 5x}{x} = 5 \lim\limits_{5x \to 0} \frac{\sin 5x}{5x} = 5 \cdot 1 = 5$

3. l'Hôpital: $\lim\limits_{x \to \infty} \frac{5x^2-3x}{7x^2+1} = \lim\limits_{x \to \infty} \frac{10x-3}{14x} = \lim\limits_{x \to \infty} \frac{10}{14} = \frac{5}{7}$ or $\lim\limits_{x \to \infty} \frac{5x^2-3x}{7x^2+1} = \lim\limits_{x \to \infty} \frac{5-\frac{3}{x}}{7+\frac{1}{x}} = \frac{5}{7}$

4. l'Hôpital: $\lim\limits_{x \to 1} \frac{x^3-1}{4x^3-x-3} = \lim\limits_{x \to 1} \frac{3x^2}{12x^2-1} = \frac{3}{11}$ or $\lim\limits_{x \to 1} \frac{x^3-1}{4x^3-x-3} = \lim\limits_{x \to 1} \frac{(x-1)(x^2+x+1)}{(x-1)(4x^2+4x+3)}$ $= \lim\limits_{x \to 1} \frac{(x^2+x+1)}{(4x^2+4x+3)} = \frac{3}{11}$

5. l'Hôpital: $\lim\limits_{x \to 0} \frac{1-\cos x}{x^2} = \lim\limits_{x \to 0} \frac{\sin x}{2x} = \lim\limits_{x \to 0} \frac{\cos x}{2} = \frac{1}{2}$ or $\lim\limits_{x \to 0} \frac{1-\cos x}{x^2} = \lim\limits_{x \to 0} \left[\frac{(1-\cos x)}{x^2}\left(\frac{1+\cos x}{1+\cos x}\right)\right]$ $= \lim\limits_{x \to 0} \frac{\sin^2 x}{x^2(1+\cos x)} = \lim\limits_{x \to 0} \left[\left(\frac{\sin x}{x}\right)\left(\frac{\sin x}{x}\right)\left(\frac{1}{1+\cos x}\right)\right] = \frac{1}{2}$

6. l'Hôpital: $\lim\limits_{x \to \infty} \frac{2x^2+3x}{x^3+x+1} = \lim\limits_{x \to \infty} \frac{4x+3}{3x^2+1} = \lim\limits_{x \to \infty} \frac{4}{6x} = 0$ or $\lim\limits_{x \to \infty} \frac{2x^2+3x}{x^3+x+1} = \lim\limits_{x \to \infty} \frac{\frac{2}{x}+\frac{3}{x^2}}{1+\frac{1}{x^2}+\frac{1}{x^3}} = \frac{0}{1} = 0$

7. $\lim\limits_{x \to 2} \frac{x-2}{x^2-4} = \lim\limits_{x \to 2} \frac{1}{2x} = \frac{1}{4}$

8. $\lim\limits_{x \to -5} \frac{x^2-25}{x+5} = \lim\limits_{x \to -5} \frac{2x}{1} = -10$

9. $\lim\limits_{t \to -3} \frac{t^3-4t+15}{t^2-t-12} = \lim\limits_{t \to -3} \frac{3t^2-4}{2t-1} = \frac{3(-3)^2-4}{2(-3)-1} = -\frac{23}{7}$

10. $\lim\limits_{t \to 1} \frac{t^3-1}{4t^3-t-3} = \lim\limits_{t \to 1} \frac{3t^2}{12t^2-1} = \frac{3}{11}$

11. $\lim\limits_{x \to \infty} \frac{5x^2-3x}{7x^2+1} = \lim\limits_{x \to \infty} \frac{10x-3}{14x} = \lim\limits_{x \to \infty} \frac{10}{14} = \frac{5}{7}$

12. $\lim\limits_{x \to \infty} \frac{x - 8x^2}{12x^2 + 5x} = \lim\limits_{x \to \infty} \frac{1 - 16x}{24x + 5} = \lim\limits_{x \to \infty} \frac{-16}{24} = -\frac{2}{3}$

13. $\lim\limits_{t \to 0} \frac{\sin t^2}{t} = \lim\limits_{t \to 0} \frac{(\cos t^2)(2t)}{1} = 0$

14. $\lim\limits_{t \to 0} \frac{\sin 5t}{t} = \lim\limits_{t \to 0} \frac{5 \cos 5t}{1} = 5$

15. $\lim\limits_{x \to 0} \frac{8x^2}{\cos x - 1} = \lim\limits_{x \to 0} \frac{16x}{-\sin x} = \lim\limits_{x \to 0} \frac{16}{-\cos x} = \frac{16}{-1} = -16$

16. $\lim\limits_{x \to 0} \frac{\sin x - x}{x^3} = \lim\limits_{x \to 0} \frac{\cos x - 1}{3x^2} = \lim\limits_{x \to 0} \frac{-\sin x}{6x} = \lim\limits_{x \to 0} \frac{-\cos x}{6} = -\frac{1}{6}$

17. $\lim\limits_{\theta \to \pi/2} \frac{2\theta - \pi}{\cos(2\pi - \theta)} = \lim\limits_{\theta \to \pi/2} \frac{2}{\sin(2\pi - \theta)} = \frac{2}{\sin\left(\frac{3\pi}{2}\right)} = -2$

18. $\lim\limits_{\theta \to -\pi/3} \frac{3\theta + \pi}{\sin\left(\theta + \frac{\pi}{3}\right)} = \lim\limits_{\theta \to -\pi/3} \frac{3}{\cos\left(\theta + \frac{\pi}{3}\right)} = 3$

19. $\lim\limits_{\theta \to \pi/2} \frac{1 - \sin\theta}{1 + \cos 2\theta} = \lim\limits_{\theta \to \pi/2} \frac{-\cos\theta}{-2\sin 2\theta} = \lim\limits_{\theta \to \pi/2} \frac{\sin\theta}{-4\cos 2\theta} = \frac{1}{(-4)(-1)} = \frac{1}{4}$

20. $\lim\limits_{x \to 1} \frac{x - 1}{\ln x - \sin(\pi x)} = \lim\limits_{x \to 1} \frac{1}{\frac{1}{x} - \pi\cos(\pi x)} = \frac{1}{1 + \pi}$

21. $\lim\limits_{x \to 0} \frac{x^2}{\ln(\sec x)} = \lim\limits_{x \to 0} \frac{2x}{\left(\frac{\sec x \tan x}{\sec x}\right)} = \lim\limits_{x \to 0} \frac{2x}{\tan x} = \lim\limits_{x \to 0} \frac{2}{\sec^2 x} = \frac{2}{1^2} = 2$

22. $\lim\limits_{x \to \pi/2} \frac{\ln(\csc x)}{\left(x - \left(\frac{\pi}{2}\right)\right)^2} = \lim\limits_{x \to \pi/2} \frac{-\left(\frac{\csc x \cot x}{\csc x}\right)}{2\left(x - \left(\frac{\pi}{2}\right)\right)} = \lim\limits_{x \to \pi/2} \frac{-\cot x}{2\left(x - \left(\frac{\pi}{2}\right)\right)} = \lim\limits_{x \to \pi/2} \frac{\csc^2 x}{2} = \frac{1^2}{2} = \frac{1}{2}$

23. $\lim\limits_{t \to 0} \frac{t(1 - \cos t)}{t - \sin t} = \lim\limits_{t \to 0} \frac{(1 - \cos t) + t(\sin t)}{1 - \cos t} = \lim\limits_{t \to 0} \frac{\sin t + (\sin t + t\cos t)}{\sin t}$
$= \lim\limits_{t \to 0} \frac{\cos t + \cos t + \cos t - t\sin t}{\cos t} = \frac{1 + 1 + 1 - 0}{1} = 3$

24. $\lim\limits_{t \to 0} \frac{t\sin t}{1 - \cos t} = \lim\limits_{t \to 0} \frac{\sin t + t\cos t}{\sin t} = \lim\limits_{t \to 0} \frac{\cos t + (\cos t - t\sin t)}{\cos t} = \frac{1 + (1 - 0)}{1} = 2$

25. $\lim\limits_{x \to (\pi/2)^-} \left(x - \frac{\pi}{2}\right)\sec x = \lim\limits_{x \to (\pi/2)^-} \frac{\left(x - \frac{\pi}{2}\right)}{\cos x} = \lim\limits_{x \to (\pi/2)^-} \left(\frac{1}{-\sin x}\right) = \frac{1}{-1} = -1$

26. $\lim\limits_{x \to (\pi/2)^-} \left(\frac{\pi}{2} - x\right)\tan x = \lim\limits_{x \to (\pi/2)^-} \frac{\left(\frac{\pi}{2} - x\right)}{\cot x} = \lim\limits_{x \to (\pi/2)^-} \left(\frac{-1}{-\csc^2 x}\right) = \lim\limits_{x \to (\pi/2)^-} \sin^2 x = 1$

27. $\lim\limits_{\theta \to 0} \frac{3^{\sin\theta} - 1}{\theta} = \lim\limits_{\theta \to 0} \frac{3^{\sin\theta}(\ln 3)(\cos\theta)}{1} = \frac{(3^0)(\ln 3)(1)}{1} = \ln 3$

28. $\lim\limits_{\theta \to 0} \frac{\left(\frac{1}{2}\right)^\theta - 1}{\theta} = \lim\limits_{\theta \to 0} \frac{\left(\ln\left(\frac{1}{2}\right)\right)\left(\frac{1}{2}\right)^\theta}{1} = \ln\left(\frac{1}{2}\right) = \ln 1 - \ln 2 = -\ln 2$

29. $\lim\limits_{x \to 0} \frac{x\,2^x}{2^x - 1} = \lim\limits_{x \to 0} \frac{(1)(2^x) + (x)(\ln 2)(2^x)}{(\ln 2)(2^x)} = \frac{1 \cdot 2^0 + 0}{(\ln 2) \cdot 2^0} = \frac{1}{\ln 2}$

30. $\lim\limits_{x \to 0} \frac{3^x - 1}{2^x - 1} = \lim\limits_{x \to 0} \frac{3^x \ln 3}{2^x \ln 2} = \frac{3^0 \cdot \ln 3}{2^0 \cdot \ln 2} = \frac{\ln 3}{\ln 2}$

31. $\lim\limits_{x \to \infty} \frac{\ln(x+1)}{\log_2 x} = \lim\limits_{x \to \infty} \frac{\ln(x+1)}{\left(\frac{\ln x}{\ln 2}\right)} = (\ln 2)\lim\limits_{x \to \infty} \frac{\left(\frac{1}{x+1}\right)}{\left(\frac{1}{x}\right)} = (\ln 2)\lim\limits_{x \to \infty} \frac{x}{x+1} = (\ln 2)\lim\limits_{x \to \infty} \frac{1}{1} = \ln 2$

32. $\lim\limits_{x\to\infty} \frac{\log_2 x}{\log_3 (x+3)} = \lim\limits_{x\to\infty} \frac{\left(\frac{\ln x}{\ln 2}\right)}{\left(\frac{\ln (x+3)}{\ln 3}\right)} = \left(\frac{\ln 3}{\ln 2}\right) \lim\limits_{x\to\infty} \frac{\ln x}{\ln (x+3)} = \left(\frac{\ln 3}{\ln 2}\right) \lim\limits_{x\to\infty} \frac{\left(\frac{1}{x}\right)}{\left(\frac{1}{x+3}\right)} = \left(\frac{\ln 3}{\ln 2}\right) \lim\limits_{x\to\infty} \frac{x+3}{x}$
$= \left(\frac{\ln 3}{\ln 2}\right) \lim\limits_{x\to\infty} \frac{1}{1} = \frac{\ln 3}{\ln 2}$

33. $\lim\limits_{x\to 0^+} \frac{\ln (x^2+2x)}{\ln x} = \lim\limits_{x\to 0^+} \frac{\left(\frac{2x+2}{x^2+2x}\right)}{\left(\frac{1}{x}\right)} = \lim\limits_{x\to 0^+} \frac{2x^2+2x}{x^2+2x} = \lim\limits_{x\to 0^+} \frac{4x+2}{2x+2} = \lim\limits_{x\to 0^+} \frac{2}{2} = 1$

34. $\lim\limits_{x\to 0^+} \frac{\ln (e^x-1)}{\ln x} = \lim\limits_{x\to 0^+} \frac{\left(\frac{e^x}{e^x-1}\right)}{\left(\frac{1}{x}\right)} = \lim\limits_{x\to 0^+} \frac{xe^x}{e^x-1} = \lim\limits_{x\to 0^+} \frac{e^x+xe^x}{e^x} = \frac{1+0}{1} = 1$

35. $\lim\limits_{y\to 0} \frac{\sqrt{5y+25}-5}{y} = \lim\limits_{y\to 0} \frac{(5y+25)^{1/2}-5}{y} = \lim\limits_{y\to 0} \frac{\left(\frac{1}{2}\right)(5y+25)^{-1/2}(5)}{1} = \lim\limits_{y\to 0} \frac{5}{2\sqrt{5y+25}} = \frac{1}{2}$

36. $\lim\limits_{y\to 0} \frac{\sqrt{ay+a^2}-a}{y} = \lim\limits_{y\to 0} \frac{(ay+a^2)^{1/2}-a}{y} = \lim\limits_{y\to 0} \frac{\left(\frac{1}{2}\right)(ay+a^2)^{-1/2}(a)}{1} = \lim\limits_{y\to 0} \frac{a}{2\sqrt{ay+a^2}} = \frac{1}{2},\ a>0$

37. $\lim\limits_{x\to\infty} [\ln 2x - \ln (x+1)] = \lim\limits_{x\to\infty} \ln \left(\frac{2x}{x+1}\right) = \ln \left(\lim\limits_{x\to\infty} \frac{2x}{x+1}\right) = \ln \left(\lim\limits_{x\to\infty} \frac{2}{1}\right) = \ln 2$

38. $\lim\limits_{x\to 0^+} (\ln x - \ln \sin x) = \lim\limits_{x\to 0^+} \ln \left(\frac{x}{\sin x}\right) = \ln \left(\lim\limits_{x\to 0^+} \frac{x}{\sin x}\right) = \ln \left(\lim\limits_{x\to 0^+} \frac{1}{\cos x}\right) = \ln 1 = 0$

39. $\lim\limits_{h\to 0} \frac{\sin(a+h)-\sin a}{h} = \lim\limits_{h\to 0} \frac{\cos(a+h)-\cos a}{1} = 0$

40. $\lim\limits_{x\to 0^+} \left(\frac{3x+1}{x} - \frac{1}{\sin x}\right) = \lim\limits_{x\to 0^+} \left(\frac{(3x+1)(\sin x)-x}{x \sin x}\right) = \lim\limits_{x\to 0^+} \frac{3 \sin x + (3x+1)(\cos x) - 1}{\sin x + x \cos x}$
$= \lim\limits_{x\to 0^+} \left(\frac{3 \cos x + 3 \cos x + (3x+1)(-\sin x)}{\cos x + \cos x - x \sin x}\right) = \frac{3+3+(1)(0)}{1+1-0} = \frac{6}{2} = 3$

41. $\lim\limits_{x\to 1^+} \left(\frac{1}{x-1} - \frac{1}{\ln x}\right) = \lim\limits_{x\to 1^+} \left(\frac{\ln x - (x-1)}{(x-1)(\ln x)}\right) = \lim\limits_{x\to 1^+} \left(\frac{\frac{1}{x}-1}{(\ln x)+(x-1)\left(\frac{1}{x}\right)}\right) = \lim\limits_{x\to 1^+} \left(\frac{1-x}{(x \ln x)+x-1}\right)$
$= \lim\limits_{x\to 1^+} \left(\frac{-1}{(\ln x+1)+1}\right) = \frac{-1}{(0+1)+1} = -\frac{1}{2}$

42. $\lim\limits_{x\to 0^+} (\csc x - \cot x + \cos x) = \lim\limits_{x\to 0^+} \left(\frac{1}{\sin x} - \frac{\cos x}{\sin x} + \cos x\right) = \lim\limits_{x\to 0^+} \left(\frac{(1-\cos x)+(\sin x)(\cos x)}{\sin x}\right)$
$= \lim\limits_{x\to 0^+} \left(\frac{\sin x + \cos^2 x - \sin^2 x}{\cos x}\right) = \frac{0+1-0}{1} = 1$

43. $\lim\limits_{\theta\to 0} \frac{\cos \theta - 1}{e^\theta - \theta - 1} = \lim\limits_{\theta\to 0} \frac{-\sin \theta}{e^\theta - 1} = \lim\limits_{\theta\to 0} \frac{-\cos \theta}{e^\theta} = -1$

44. $\lim\limits_{h\to 0} \frac{e^h-(1+h)}{h^2} = \lim\limits_{h\to 0} \frac{e^h-1}{2h} = \lim\limits_{h\to 0} \frac{e^h}{2} = \frac{1}{2}$

45. $\lim\limits_{t\to\infty} \frac{e^t+t^2}{e^t-1} = \lim\limits_{t\to\infty} \frac{e^t+2t}{e^t} = \lim\limits_{t\to\infty} \frac{e^t+2}{e^t} = \lim\limits_{t\to\infty} \frac{e^t}{e^t} = 1$

46. $\lim\limits_{x\to\infty} x^2e^{-x} = \lim\limits_{x\to\infty} \frac{x^2}{e^x} = \lim\limits_{x\to\infty} \frac{2x}{e^x} = \lim\limits_{x\to\infty} \frac{2}{e^x} = 0$

47. The limit leads to the indeterminate form 1^∞. Let $f(x) = x^{1/(1-x)} \Rightarrow \ln f(x) = \ln \left(x^{1/(1-x)}\right) = \frac{\ln x}{1-x}$. Now
$\lim\limits_{x\to 1^+} \ln f(x) = \lim\limits_{x\to 1^+} \frac{\ln x}{1-x} = \lim\limits_{x\to 1^+} \frac{\left(\frac{1}{x}\right)}{-1} = -1$. Therefore $\lim\limits_{x\to 1^+} x^{1/(1-x)} = \lim\limits_{x\to 1^+} f(x) = \lim\limits_{x\to 1^+} e^{\ln f(x)} = e^{-1} = \frac{1}{e}$

48. The limit leads to the indeterminate form 1^∞. Let $f(x) = x^{1/(x-1)} \Rightarrow \ln f(x) = \ln\left(x^{1/(x-1)}\right) = \frac{\ln x}{x-1}$. Now $\lim\limits_{x \to 1^+} \ln f(x) = \lim\limits_{x \to 1^+} \frac{\ln x}{x-1} = \lim\limits_{x \to 1^+} \frac{\left(\frac{1}{x}\right)}{1} = 1$. Therefore $\lim\limits_{x \to 1^+} x^{1/(x-1)} = \lim\limits_{x \to 1^+} f(x) = \lim\limits_{x \to 1^+} e^{\ln f(x)} = e^1 = e$

49. The limit leads to the indeterminate form ∞^0. Let $f(x) = (\ln x)^{1/x} \Rightarrow \ln f(x) = \ln(\ln x)^{1/x} = \frac{\ln(\ln x)}{x}$. Now $\lim\limits_{x \to \infty} \ln f(x) = \lim\limits_{x \to \infty} \frac{\ln(\ln x)}{x} = \lim\limits_{x \to \infty} \frac{\left(\frac{1}{x \ln x}\right)}{1} = 0$. Therefore $\lim\limits_{x \to \infty} (\ln x)^{1/x} = \lim\limits_{x \to \infty} f(x)$ $= \lim\limits_{x \to \infty} e^{\ln f(x)} = e^0 = 1$

50. The limit leads to the indeterminate form 1^∞. Let $f(x) = (\ln x)^{1/(x-e)} \Rightarrow \ln f(x) = \frac{\ln(\ln x)}{x-e} = \lim\limits_{x \to e^+} \ln f(x)$ $= \lim\limits_{x \to e^+} \frac{\ln(\ln x)}{x-e} = \lim\limits_{x \to e^+} \frac{\left(\frac{1}{x \ln x}\right)}{1} = \frac{1}{e}$. Therefore $(\ln x)^{1/(x-e)} = \lim\limits_{x \to e^+} f(x) = \lim\limits_{x \to e^+} e^{\ln f(x)} = e^{1/e}$

51. The limit leads to the indeterminate form 0^0. Let $f(x) = x^{-1/\ln x} \Rightarrow \ln f(x) = -\frac{\ln x}{\ln x} = -1$. Therefore $\lim\limits_{x \to 0^+} x^{-1/\ln x} = \lim\limits_{x \to 0^+} f(x) = \lim\limits_{x \to 0^+} e^{\ln f(x)} = e^{-1} = \frac{1}{e}$

52. The limit leads to the indeterminate form ∞^0. Let $f(x) = x^{1/\ln x} \Rightarrow \ln f(x) = \frac{\ln x}{\ln x} = 1$. Therefore $\lim\limits_{x \to \infty} x^{1/\ln x}$ $= \lim\limits_{x \to \infty} f(x) = \lim\limits_{x \to \infty} e^{\ln f(x)} = e^1 = e$

53. The limit leads to the indeterminate form ∞^0. Let $f(x) = (1+2x)^{1/(2\ln x)} \Rightarrow \ln f(x) = \frac{\ln(1+2x)}{2\ln x}$ $\Rightarrow \lim\limits_{x \to \infty} \ln f(x) = \lim\limits_{x \to \infty} \frac{\ln(1+2x)}{2\ln x} = \lim\limits_{x \to \infty} \frac{x}{1+2x} = \lim\limits_{x \to \infty} \frac{1}{2} = \frac{1}{2}$. Therefore $\lim\limits_{x \to \infty} (1+2x)^{1/(2\ln x)}$ $= \lim\limits_{x \to \infty} f(x) = \lim\limits_{x \to \infty} e^{\ln f(x)} = e^{1/2}$

54. The limit leads to the indeterminate form 1^∞. Let $f(x) = (e^x + x)^{1/x} \Rightarrow \ln f(x) = \frac{\ln(e^x + x)}{x}$ $\Rightarrow \lim\limits_{x \to 0} \ln f(x) = \lim\limits_{x \to 0} \frac{\ln(e^x + x)}{x} = \lim\limits_{x \to 0} \frac{e^x + 1}{e^x + x} = 2$. Therefore $\lim\limits_{x \to 0} (e^x + x)^{1/x} = \lim\limits_{x \to 0} f(x)$ $= \lim\limits_{x \to 0} e^{\ln f(x)} = e^2$

55. The limit leads to the indeterminate form 0^0. Let $f(x) = x^x \Rightarrow \ln f(x) = x \ln x \Rightarrow \ln f(x) = \frac{\ln x}{\left(\frac{1}{x}\right)}$ $= \lim\limits_{x \to 0^+} \ln f(x) = \lim\limits_{x \to 0^+} \frac{\ln x}{\left(\frac{1}{x}\right)} = \lim\limits_{x \to 0^+} \frac{\left(\frac{1}{x}\right)}{\left(-\frac{1}{x^2}\right)} = \lim\limits_{x \to 0^+} (-x) = 0$. Therefore $\lim\limits_{x \to 0^+} x^x = \lim\limits_{x \to 0^+} f(x)$ $= \lim\limits_{x \to 0^+} e^{\ln f(x)} = e^0 = 1$

56. The limit leads to the indeterminate form ∞^0. Let $f(x) = \left(1 + \frac{1}{x}\right)^x \Rightarrow \ln f(x) = \frac{\ln(1+x^{-1})}{x^{-1}} \Rightarrow \lim\limits_{x \to 0^+} \ln f(x)$ $= \lim\limits_{x \to 0^+} \frac{\left(\frac{-x^{-2}}{1+x^{-1}}\right)}{-x^{-2}} = \lim\limits_{x \to 0^+} \frac{1}{1+x^{-1}} = \lim\limits_{x \to 0^+} \frac{x}{x+1} = 0$. Therefore $\lim\limits_{x \to 0^+} \left(1 + \frac{1}{x}\right)^x = \lim\limits_{x \to 0^+} f(x)$ $= \lim\limits_{x \to 0^+} e^{\ln f(x)} = e^0 = 1$

57. $\lim\limits_{x \to \infty} \frac{\sqrt{9x+1}}{\sqrt{x+1}} = \sqrt{\lim\limits_{x \to \infty} \frac{9x+1}{x+1}} = \sqrt{\lim\limits_{x \to \infty} \frac{9}{1}} = \sqrt{9} = 3$

58. $\lim\limits_{x \to 0^+} \frac{\sqrt{x}}{\sqrt{\sin x}} = \sqrt{\frac{1}{\lim\limits_{x \to 0^+} \frac{\sin x}{x}}} = \sqrt{\frac{1}{1}} = 1$

59. $\lim\limits_{x \to \pi/2^-} \frac{\sec x}{\tan x} = \lim\limits_{x \to \pi/2^-} \left(\frac{1}{\cos x}\right)\left(\frac{\cos x}{\sin x}\right) = \lim\limits_{x \to \pi/2^-} \frac{1}{\sin x} = 1$

60. $\lim\limits_{x \to 0^+} \frac{\cot x}{\csc x} = \lim\limits_{x \to 0^+} \frac{\left(\frac{\cos x}{\sin x}\right)}{\left(\frac{1}{\sin x}\right)} = \lim\limits_{x \to 0^+} \cos x = 1$

61. Part (b) is correct because part (a) is neither in the $\frac{0}{0}$ nor $\frac{\infty}{\infty}$ form and so l'Hôpital's rule may not be used.

62. Part (b) is correct; the step $\lim\limits_{x \to 0} \frac{2x-2}{2x-\cos x} = \lim\limits_{x \to 0} \frac{2}{2+\sin x}$ in part (a) is false because $\lim\limits_{x \to 0} \frac{2x-2}{2x-\cos x}$ is not an indeterminate quotient form.

63. Part (d) is correct, the other parts are indeterminate forms and cannot be calculated by the incorrect arithmetic

64. (a) For $x \neq 0$, $f'(x) = \frac{d}{dx}(x+2) = 1$ and $g'(x) = \frac{d}{dx}(x+1) = 1$. Therefore, $\lim\limits_{x \to 0} \frac{f'(x)}{g'(x)} = \frac{1}{1} = 1$, while $\lim\limits_{x \to 0} \frac{f(x)}{g(x)} = \frac{x+2}{x+1} = \frac{0+2}{0+1} = 2$.

(b) This does not contradict l'Hôpital's rule because neither f nor g is differentiable at $x = 0$ (as evidenced by the fact that neither is continuous at $x = 0$), so l'Hôpital's rule does not apply.

65. If f(x) is to be continuous at $x = 0$, then $\lim\limits_{x \to 0} f(x) = f(0) \Rightarrow c = f(0) = \lim\limits_{x \to 0} \frac{9x - 3\sin 3x}{5x^3} = \lim\limits_{x \to 0} \frac{9 - 9\cos 3x}{15x^2}$ $= \lim\limits_{x \to 0} \frac{27 \sin 3x}{30x} = \lim\limits_{x \to 0} \frac{81 \cos 3x}{30} = \frac{27}{10}$.

66. (a)

(b) The limit leads to the indeterminate form $\infty - \infty$:

$$\lim_{x \to \infty} \left(x - \sqrt{x^2+x}\right) = \lim_{x \to \infty} \left(x - \sqrt{x^2+x}\right)\left(\frac{x+\sqrt{x^2+x}}{x+\sqrt{x^2+x}}\right) = \lim_{x \to \infty} \left(\frac{x^2-(x^2+x)}{x+\sqrt{x^2+x}}\right) = \lim_{x \to \infty} \frac{-x}{x+\sqrt{x^2+x}}$$

$$= \lim_{x \to \infty} \frac{-1}{1+\sqrt{1+\frac{1}{x}}} = \frac{-1}{1+\sqrt{1+0}} = -\frac{1}{2}$$

67. The graph indicates a limit near -1. The limit leads to the indeterminate form $\frac{0}{0}$: $\lim\limits_{x \to 1} \frac{2x^2 - (3x+1)\sqrt{x}+2}{x-1}$

$$= \lim_{x \to 1} \frac{2x^2 - 3x^{3/2} - x^{1/2} + 2}{x-1} = \lim_{x \to 1} \frac{4x - \frac{9}{2}x^{1/2} - \frac{1}{2}x^{-1/2}}{1}$$

$$= \frac{4 - \frac{9}{2} - \frac{1}{2}}{1} = \frac{4-5}{1} = -1$$

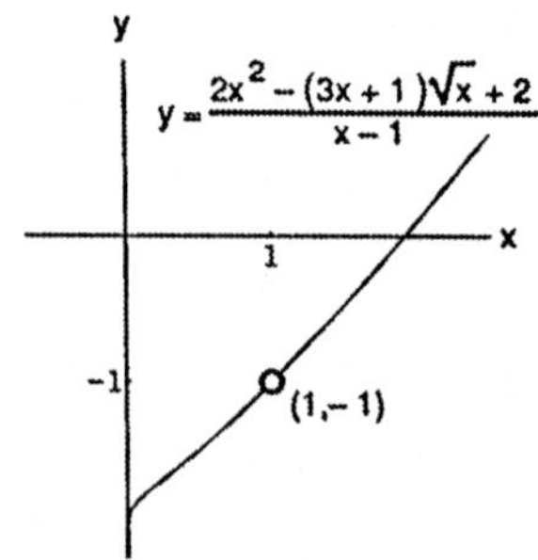

68. (a) The limit leads to the indeterminate form 1^∞. Let $f(x) = \left(1+\frac{1}{x}\right)^x \Rightarrow \ln f(x) = x \ln\left(1+\frac{1}{x}\right) \Rightarrow \lim\limits_{x \to \infty} \ln f(x)$

$$= \lim_{x \to \infty} \frac{\ln\left(1+\frac{1}{x}\right)}{\left(\frac{1}{x}\right)} = \lim_{x \to \infty} \frac{\ln\left(1+x^{-1}\right)}{x^{-1}} = \lim_{x \to \infty} \frac{\left(\frac{-x^{-2}}{1+x^{-1}}\right)}{-x^{-2}} = \lim_{x \to \infty} \frac{1}{1+\left(\frac{1}{x}\right)} = \frac{1}{1+0} = 1$$

$$\Rightarrow \lim_{x \to \infty} \left(1+\frac{1}{x}\right)^x = \lim_{x \to \infty} f(x) = \lim_{x \to \infty} e^{\ln f(x)} = e^1 = e$$

(b)

x	$\left(1+\frac{1}{x}\right)^x$
10	2.5937424601
100	2.70481382942
1000	2.71692393224
10,000	2.71814592683
100,000	2.71826823717

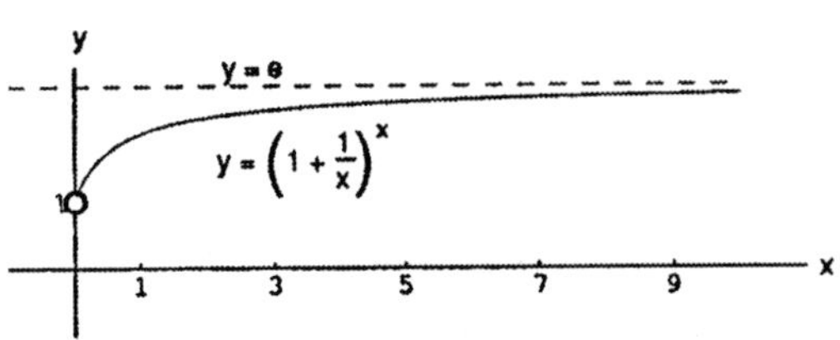

Both functions have limits as x approaches infinity. The function f has a maximum but no minimum while g has no extrema. The limit of f(x) leads to the indeterminate form 1^∞.

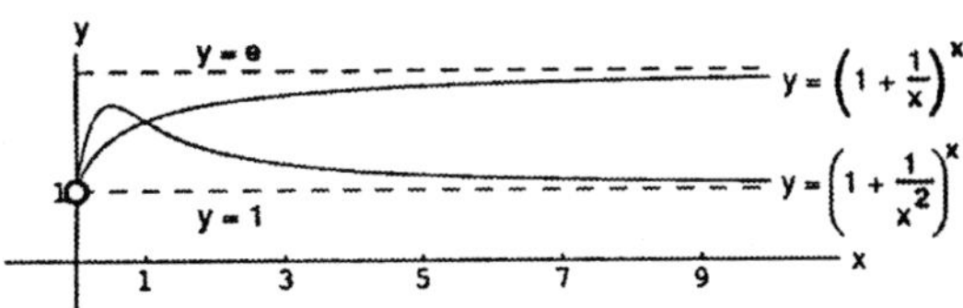

(c) Let $f(x) = \left(1+\frac{1}{x^2}\right)^x \Rightarrow \ln f(x) = x \ln\left(1+x^{-2}\right)$

$\Rightarrow \lim\limits_{x\to\infty} \ln f(x) = \lim\limits_{x\to\infty} \frac{\ln\left(1+x^{-2}\right)}{x^{-1}} = \lim\limits_{x\to\infty} \frac{\left(\frac{-2x^{-3}}{1+x^{-2}}\right)}{-x^{-2}} = \lim\limits_{x\to\infty} \frac{2x^2}{(x^3+x)} = \lim\limits_{x\to\infty} \frac{4x}{(3x^2+1)} = \lim\limits_{x\to\infty} \frac{4}{6x} = 0.$

Therefore $\lim\limits_{x\to\infty} \left(1+\frac{1}{x^2}\right)^x = \lim\limits_{x\to\infty} f(x) = \lim\limits_{x\to\infty} e^{\ln f(x)} = e^0 = 1$

69. Let $f(k) = \left(1+\frac{r}{k}\right)^k \Rightarrow \ln f(k) = \frac{\ln\left(1+rk^{-1}\right)}{k^{-1}} \Rightarrow \lim\limits_{k\to\infty} \frac{\ln\left(1+rk^{-1}\right)}{k^{-1}} = \lim\limits_{k\to\infty} \frac{\left(\frac{-rk^{-2}}{1+rk^{-1}}\right)}{-k^{-2}} = \lim\limits_{k\to\infty} \frac{r}{1+rk^{-1}}$

$= \lim\limits_{k\to\infty} \frac{rk}{k+r} = \lim\limits_{k\to\infty} \frac{r}{1} = r$. Therefore $\lim\limits_{k\to\infty} \left(1+\frac{r}{k}\right)^k = \lim\limits_{k\to\infty} f(k) = \lim\limits_{k\to\infty} e^{\ln f(k)} = e^r$.

70. (a) $y = x^{1/x} \Rightarrow \ln y = \frac{\ln x}{x} \Rightarrow \frac{y'}{y} = \frac{\left(\frac{1}{x}\right)(x) - \ln x}{x^2} \Rightarrow y' = \left(\frac{1-\ln x}{x^2}\right)\left(x^{1/x}\right)$. The sign pattern is

$y' = \underset{0}{|} + + + + + \underset{e}{|} - - - -$ which indicates a maximum value of $y = e^{1/e}$ when $x = e$

(b) $y = x^{1/x^2} \Rightarrow \ln y = \frac{\ln x}{x^2} \Rightarrow \frac{y'}{y} = \frac{\left(\frac{1}{x}\right)(x^2) - 2x\ln x}{x^4} \Rightarrow y' = \left(\frac{1-2\ln x}{x^3}\right)\left(x^{1/x^2}\right)$. The sign pattern is

$y' = \underset{0}{|} + + + \underset{\sqrt{e}}{|} - - - -$ which indicates a maximum of $y = e^{1/2e}$ when $x = \sqrt{e}$

(c) $y = x^{1/x^n} \Rightarrow \ln y = \frac{\ln x}{x^n} = \frac{\left(\frac{1}{x}\right)(x^n) - (\ln x)\left(nx^{n-1}\right)}{x^{2n}} \Rightarrow y' = \frac{x^{n-1}(1 - n\ln x)}{x^{2n}} \cdot x^{1/x^n}$. The sign pattern is

$y' = \underset{0}{|} + + + \underset{\sqrt[n]{e}}{|} - - - -$ which indicates a maximum of $y = e^{1/ne}$ when $x = \sqrt[n]{e}$

(d) $\lim\limits_{x\to\infty} x^{1/x^n} = \lim\limits_{x\to\infty} \left(e^{\ln x}\right)^{1/x^n} = \lim\limits_{x\to\infty} e^{(\ln x)/x^n} = \exp\left(\lim\limits_{x\to\infty} \frac{\ln x}{x^n}\right) = \exp\left(\lim\limits_{x\to\infty} \left(\frac{1}{nx^n}\right)\right) = e^0 = 1$

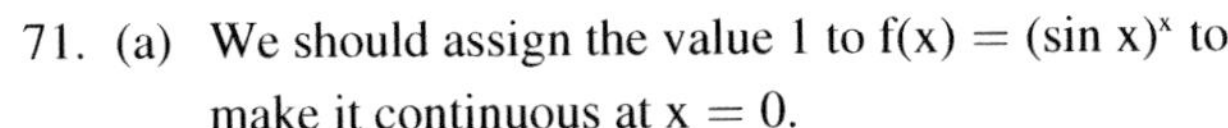

71. (a) We should assign the value 1 to $f(x) = (\sin x)^x$ to make it continuous at $x = 0$.

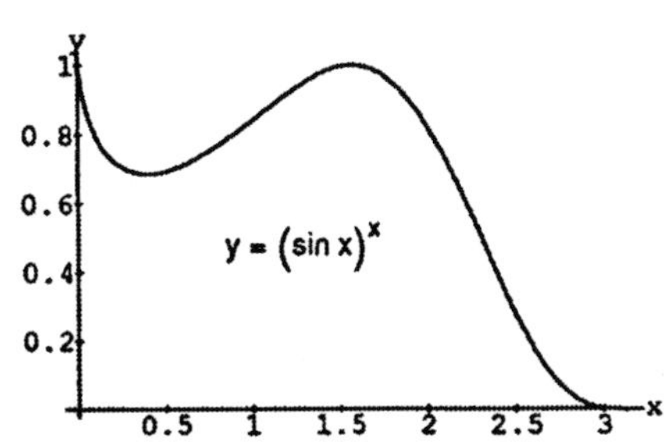

(b) $\ln f(x) = x \ln(\sin x) = \frac{\ln(\sin x)}{\left(\frac{1}{x}\right)} \Rightarrow \lim\limits_{x\to 0^+} \ln f(x) = \lim\limits_{x\to 0^+} \frac{\ln(\sin x)}{\left(\frac{1}{x}\right)} = \lim\limits_{x\to 0^+} \frac{\left(\frac{1}{\sin x}\right)(\cos x)}{\left(-\frac{1}{x^2}\right)}$

$= \lim\limits_{x\to 0} \frac{-x^2}{\tan x} = \lim\limits_{x\to 0} \frac{-2x}{\sec^2 x} = 0 \Rightarrow \lim\limits_{x\to 0} f(x) = e^0 = 1$

(c) The maximum value of f(x) is close to 1 near the point $x \approx 1.55$ (see the graph in part (a)).

(d) The root in question is near 1.57.

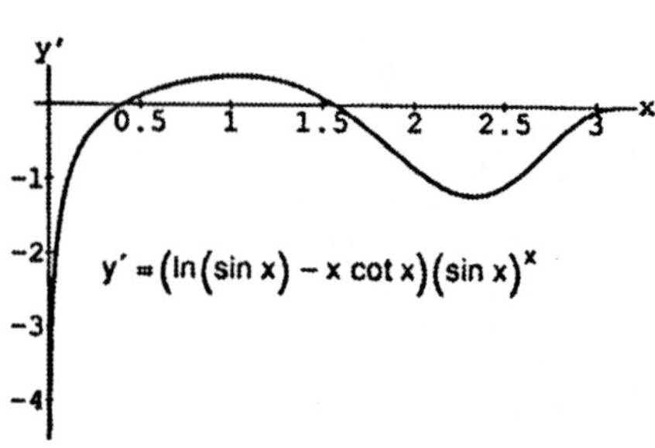

72. (a) When sin x $<$ 0 there are gaps in the sketch. The width of each gap is π.

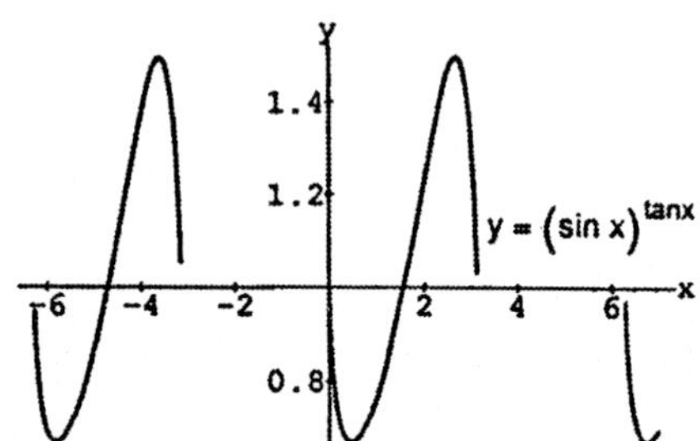

(b) Let $f(x) = (\sin x)^{\tan x} \Rightarrow \ln f(x) = (\tan x) \ln (\sin x)$

$\Rightarrow \lim\limits_{x \to \pi/2^-} \ln f(x) = \lim\limits_{x \to \pi/2^-} \frac{\ln (\sin x)}{\cot x}$

$= \lim\limits_{x \to \pi/2^-} \frac{\left(\frac{1}{\sin x}\right)(\cos x)}{-\csc^2 x} = \lim\limits_{x \to \pi/2^-} \frac{\cos x}{(-\sin x)} = 0$

$\Rightarrow \lim\limits_{x \to \pi/2^-} f(x) = e^0 = 1$. Similarly,

$\lim\limits_{x \to \pi/2^+} f(x) = e^0 = 1$. Therefore, $\lim\limits_{x \to \pi/2} f(x) = 1$.

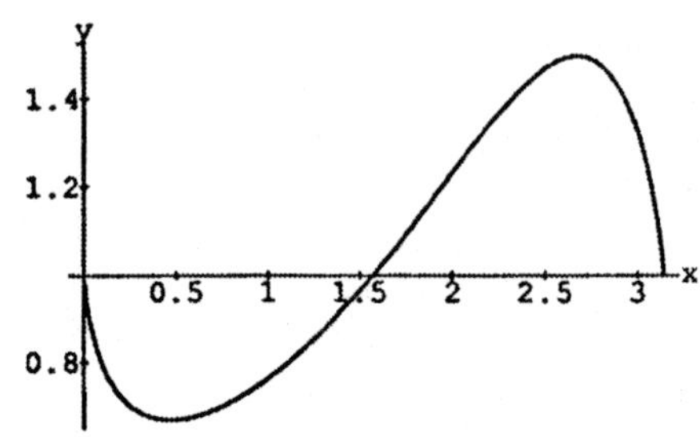

(c) From the graph in part (b) we have a minimum of about 0.665 at $x \approx 0.47$ and the maximum is about 1.491 at $x \approx 2.66$.

73. Graphing $f(x) = \frac{1 - \cos x^6}{x^{12}}$ on th window $[-1, 1]$ by $[-0.5, 1]$ it appears that $\lim\limits_{x \to 0} f(x) = 0$. However, we see that if we let $u = x^6$, then $\lim\limits_{x \to 0} f(x) = \lim\limits_{u \to 0} \frac{1 - \cos u}{u^2} = \lim\limits_{u \to 0} \frac{\sin u}{2u} = \lim\limits_{u \to 0} \frac{\cos u}{2} = \frac{1}{2}$.

74. (a) We seek c in $(-2, 0)$ so that $\frac{f'(c)}{g'(c)} = \frac{f(0) - f(-2)}{g(0) - g(-2)} = \frac{0+2}{0-4} = -\frac{1}{2}$. Since $f'(c) = 1$ and $g'(c) = 2c$ we have that $\frac{1}{2c} = -\frac{1}{2}$ $\Rightarrow c = -1$.

(b) We seek c in (a, b) so that $\frac{f'(c)}{g'(c)} = \frac{f(b) - f(a)}{g(b) - g(a)} = \frac{b-a}{b^2 - a^2} = \frac{1}{b+a}$. Since $f'(c) = 1$ and $g'(c) = 2c$ we have that $\frac{1}{2c} = \frac{1}{b+a}$ $\Rightarrow c = \frac{b+a}{2}$.

(c) We seek c in $(0, 3)$ so that $\frac{f'(c)}{g'(c)} = \frac{f(3) - f(0)}{g(3) - g(0)} = \frac{-3-0}{9-0} = -\frac{1}{3}$. Since $f'(c) = c^2 - 4$ and $g'(c) = 2c$ we have that $\frac{c^2 - 4}{2c} = -\frac{1}{3} \Rightarrow c = \frac{-1 \pm \sqrt{37}}{3} \Rightarrow c = \frac{-1 + \sqrt{37}}{3}$.

75. (a) By similar triangles, $\frac{PA}{AB} = \frac{CE}{EB}$ where E is the point on $\overleftrightarrow{AB}$ such that $\overleftrightarrow{CE} \perp \overleftrightarrow{AB}$:

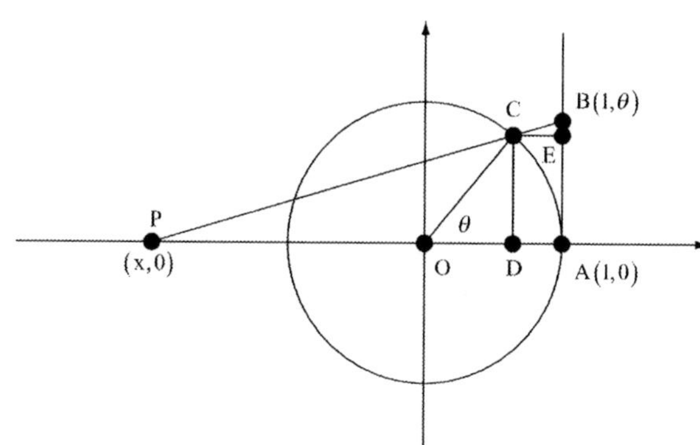

Thus $\frac{1-x}{\theta} = \frac{1-\cos\theta}{\theta - \sin\theta}$, since the coordinates of C are $(\cos\theta, \sin\theta)$. Hence, $1 - x = \frac{\theta(1-\cos\theta)}{\theta - \sin\theta}$.

(b) $\lim\limits_{\theta \to 0} (1-x) = \lim\limits_{\theta \to 0} \frac{\theta(1-\cos\theta)}{\theta - \sin\theta} = \lim\limits_{\theta \to 0} \frac{\theta \sin\theta + 1 - \cos\theta}{1-\cos\theta} = \lim\limits_{\theta \to 0} \frac{\theta\cos\theta + \sin\theta + \sin\theta}{\sin\theta} = \lim\limits_{\theta \to 0} \frac{\theta\cos\theta + 2\sin\theta}{\sin\theta}$
$= \lim\limits_{\theta \to 0} \frac{\theta(-\sin\theta) + \cos\theta + 2\cos\theta}{\cos\theta} = \lim\limits_{\theta \to 0} \frac{-\theta\sin\theta + 3\cos\theta}{\cos\theta} = \frac{0+3}{1} = 3$

(c) We have that $\lim\limits_{\theta \to \infty} [(1-x) - (1-\cos\theta)] = \lim\limits_{\theta \to \infty} \left[\frac{\theta(1-\cos\theta)}{\theta - \sin\theta} - (1-\cos\theta)\right] = \lim\limits_{\theta \to \infty} (1-\cos\theta)\left[\frac{\theta}{\theta - \sin\theta} - 1\right]$
As $\theta \to \infty$, $(1-\cos\theta)$ oscillates between 0 and 2, and so it is bounded. Since $\lim\limits_{\theta \to \infty} \left(\frac{\theta}{\theta - \sin\theta} - 1\right) = 1 - 1 = 0$,
$\lim\limits_{\theta \to \infty} (1-\cos\theta)\left[\frac{\theta}{\theta - \sin\theta} - 1\right] = 0$. Geometrically, this means that as $\theta \to \infty$, the distance between points P and D approaches 0.

76. Throughout this problem note that $r^2 = y^2 + 1$, $r > y$ and that both $r \to \infty$ and $y \to \infty$ as $\theta \to \frac{\pi}{2}$.

(a) $\lim\limits_{\theta \to \pi/2} r - y = \lim\limits_{\theta \to \pi/2} \frac{1}{r+y} = 0$

(b) $\lim\limits_{\theta \to \pi/2} r^2 - y^2 = \lim\limits_{\theta \to \pi/2} 1 = 1$

(c) We have that $r^3 - y^3 = (r-y)(r^2 + ry + y^2) = \frac{r^2 + ry + y^2}{r+y} > \frac{y^2 + y \cdot y + y^2}{r} = \frac{3y^2}{r} = 3y \cdot \frac{y}{r}$.
Since $\lim\limits_{\theta \to \pi/2} 3y \cdot \frac{y}{r} = \lim\limits_{\theta \to \pi/2} 3\sin\theta \cdot y = \infty$ we have that $\lim\limits_{\theta \to \pi/2} r^3 - y^3 = \infty$.

4.7 NEWTON'S METHOD

1. $y = x^2 + x - 1 \Rightarrow y' = 2x + 1 \Rightarrow x_{n+1} = x_n - \frac{x_n^2 + x_n - 1}{2x_n + 1}$; $x_0 = 1 \Rightarrow x_1 = 1 - \frac{1+1-1}{2+1} = \frac{2}{3}$
$\Rightarrow x_2 = \frac{2}{3} - \frac{\frac{4}{9} + \frac{2}{3} - 1}{\frac{4}{3} + 1} \Rightarrow x_2 = \frac{2}{3} - \frac{4+6-9}{12+9} = \frac{2}{3} - \frac{1}{21} = \frac{13}{21} \approx .61905$; $x_0 = -1 \Rightarrow x_1 = 1 - \frac{1-1-1}{-2+1} = -2$
$\Rightarrow x_2 = -2 - \frac{4-2-1}{-4+1} = -\frac{5}{3} \approx -1.66667$

2. $y = x^3 + 3x + 1 \Rightarrow y' = 3x^2 + 3 \Rightarrow x_{n+1} = x_n - \frac{x_n^3 + 3x_n + 1}{3x_n^2 + 3}$; $x_0 = 0 \Rightarrow x_1 = 0 - \frac{1}{3} = -\frac{1}{3}$
$\Rightarrow x_2 = -\frac{1}{3} - \frac{-\frac{1}{27} - 1 + 1}{\frac{1}{3} + 3} = -\frac{1}{3} + \frac{1}{90} = -\frac{29}{90} \approx -0.32222$

3. $y = x^4 + x - 3 \Rightarrow y' = 4x^3 + 1 \Rightarrow x_{n+1} = x_n - \frac{x_n^4 + x_n - 3}{4x_n^3 + 1}$; $x_0 = 1 \Rightarrow x_1 = 1 - \frac{1+1-3}{4+1} = \frac{6}{5}$
$\Rightarrow x_2 = \frac{6}{5} - \frac{\frac{1296}{625} + \frac{6}{5} - 3}{\frac{864}{125} + 1} = \frac{6}{5} - \frac{1296 + 750 - 1875}{4320 + 625} = \frac{6}{5} - \frac{171}{4945} = \frac{5763}{4945} \approx 1.16542$; $x_0 = -1 \Rightarrow x_1 = -1 - \frac{1-1-3}{-4+1}$
$= -2 \Rightarrow x_2 = -2 - \frac{16-2-3}{-32+1} = -2 + \frac{11}{31} = -\frac{51}{31} \approx -1.64516$

4. $y = 2x - x^2 + 1 \Rightarrow y' = 2 - 2x \Rightarrow x_{n+1} = x_n - \frac{2x_n - x_n^2 + 1}{2 - 2x_n}$; $x_0 = 0 \Rightarrow x_1 = 0 - \frac{0-0+1}{2-0} = -\frac{1}{2}$
$\Rightarrow x_2 = -\frac{1}{2} - \frac{-1-\frac{1}{4}+1}{2+1} = -\frac{1}{2} + \frac{1}{12} = -\frac{5}{12} \approx -.41667$; $x_0 = 2 \Rightarrow x_1 = 2 - \frac{4-4+1}{2-4} = \frac{5}{2} \Rightarrow x_2 = \frac{5}{2} - \frac{5-\frac{25}{4}+1}{2-5}$
$= \frac{5}{2} - \frac{20-25+4}{-12} = \frac{5}{2} - \frac{1}{12} = \frac{29}{12} \approx 2.41667$

5. One obvious root is $x = 0$. Graphing e^{-x} and $2x + 1$ shows that $x = 0$ is the only root. Taking a naive approach we can use Newton's Method to estimate the root as follows: Let $f(x) = e^{-x} - 2x - 1$, $x_0 = 1$, and $x_{n+1} = x_n - \frac{f(x_n)}{f'(x_n)}$ $= x_n + \frac{e^{-x_n} - 2x_n - 1}{e^{-x_n} + 2}$. Performing iterations on a calculator, spreadsheet, or CAS gives $x_1 = -0.111594$, $x_2 = -0.00215192$, $x_3 = -0.000000773248$. You may get different results depending upon what you select for f(x) and x_0, and what calculator or computer you may use.

6. Graphing $\tan^{-1}(x)$ and $1 - 2x$ shows that there is only one root and it is between $x = 0.3$ and $x = 0.4$. Let $f(x) = \tan^{-1}(x) + 2x - 1$, $x_1 = 0.3$, and $x_{n+1} = x_n - \frac{f(x_n)}{f'(x_n)} = x_n - \frac{\tan^{-1}(x_n) + 2x_n - 1}{\frac{1}{1+x_n^2} + 2}$. Performing iterations on a calculator, spreadsheet, or CAS gives $x_2 = 0.337205$, $x_3 = 0.337329$, $x_4 = 0.337329$. You may get different results depending upon what you select for f(x) and x_1, and what calculator or computer you may use.

7. $f(x_0) = 0$ and $f'(x_0) \neq 0 \Rightarrow x_{n+1} = x_n - \frac{f(x_n)}{f'(x_n)}$ gives $x_1 = x_0 \Rightarrow x_2 = x_0 \Rightarrow x_n = x_0$ for all $n \geq 0$. That is, all of the approximations in Newton's method will be the root of $f(x) = 0$.

8. It does matter. If you start too far away from $x = \frac{\pi}{2}$, the calculated values may approach some other root. Starting with $x_0 = -0.5$, for instance, leads to $x = -\frac{\pi}{2}$ as the root, not $x = \frac{\pi}{2}$.

9. If $x_0 = h > 0 \Rightarrow x_1 = x_0 - \frac{f(x_0)}{f'(x_0)} = h - \frac{f(h)}{f'(h)}$
$= h - \frac{\sqrt{h}}{\left(\frac{1}{2\sqrt{h}}\right)} = h - \left(\sqrt{h}\right)\left(2\sqrt{h}\right) = -h;$
if $x_0 = -h < 0 \Rightarrow x_1 = x_0 - \frac{f(x_0)}{f'(x_0)} = -h - \frac{f(-h)}{f'(-h)}$
$= -h - \frac{\sqrt{h}}{\left(\frac{-1}{2\sqrt{h}}\right)} = -h + \left(\sqrt{h}\right)\left(2\sqrt{h}\right) = h.$

$y = \begin{cases} \sqrt{x}, & x \geq 0 \\ \sqrt{-x}, & x < 0 \end{cases}$

10. $f(x) = x^{1/3} \Rightarrow f'(x) = \left(\frac{1}{3}\right)x^{-2/3} \Rightarrow x_{n+1} = x_n - \frac{x_n^{1/3}}{\left(\frac{1}{3}\right)x_n^{-2/3}}$
$= -2x_n$; $x_0 = 1 \Rightarrow x_1 = -2$, $x_2 = 4$, $x_3 = -8$, and $x_4 = 16$ and so forth. Since $|x_n| = 2|x_{n-1}|$ we may conclude that $n \to \infty \Rightarrow |x_n| \to \infty$.

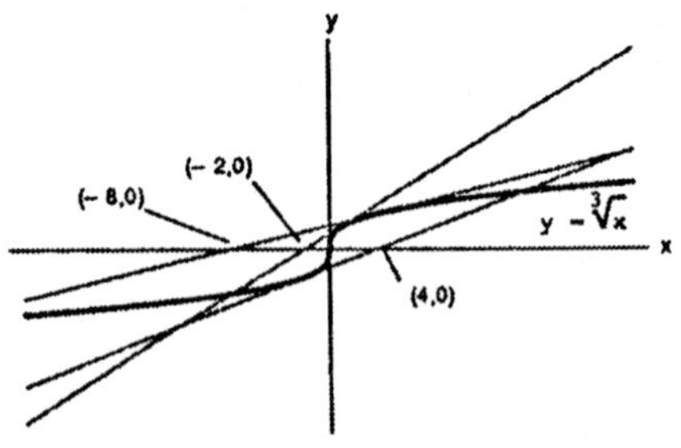

11. i) is equivalent to solving $x^3 - 3x - 1 = 0$.
ii) is equivalent to solving $x^3 - 3x - 1 = 0$.
iii) is equivalent to solving $x^3 - 3x - 1 = 0$.
iv) is equivalent to solving $x^3 - 3x - 1 = 0$.
All four equations are equivalent.

12. $f(x) = x - 1 - 0.5 \sin x \Rightarrow f'(x) = 1 - 0.5 \cos x \Rightarrow x_{n+1} = x_n - \frac{x_n - 1 - 0.5 \sin x_n}{1 - 0.5 \cos x_n}$; if $x_0 = 1.5$, then $x_1 = 1.49870$

13. For $x_0 = -0.3$, the procedure converges to the root $-0.32218535....$

(a)

```
Plot1 Plot2 Plot3
\y1=x^3+3x+1
\y2=nDer(y1,x,x)
```

(b)

```
-.3→x
                 -.3
```

(c)

```
x-y1/y2→x
       -.322324152194
       -.322185360292
       -.322185354626
       -.322185354626
```

(d) Values for x will vary. One possible choice is $x_0 = 0.1$.

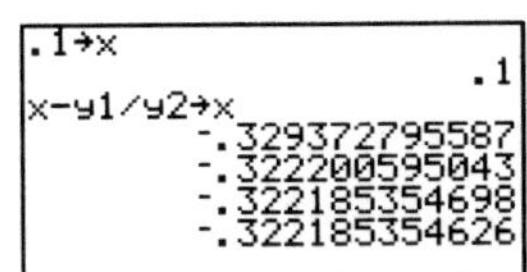
```
.1→x
                   .1
x-y1/y2→x
       -.329372795587
       -.322200595043
       -.322185354698
       -.322185354626
```

(e) Values for x will vary.

14. (a) $f(x) = x^3 - 3x - 1 \Rightarrow f'(x) = 3x^2 - 3 \Rightarrow x_{n+1} = x_n - \frac{x_n^3 - 3x_n - 1}{3x_n^2 - 3} \Rightarrow$ the two negative zeros are -1.53209 and -0.34730

(b) The estimated solutions of $x^3 - 3x - 1 = 0$ are $-1.53209, -0.34730, 1.87939$.

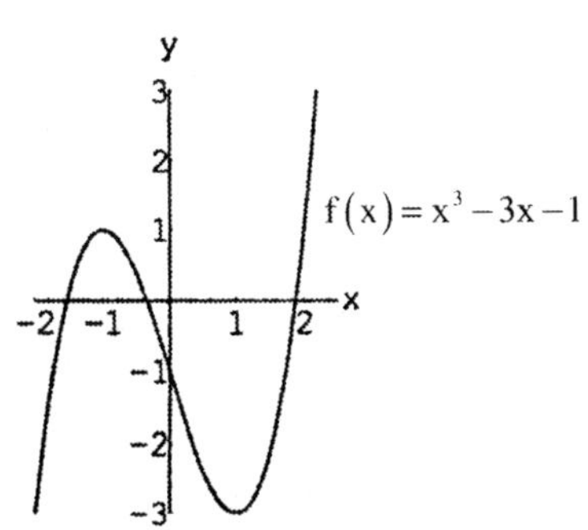

(c) The estimated x-values where $g(x) = 0.25x^4 - 1.5x^2 - x + 5$ has horizontal tangents are the roots of $g'(x) = x^3 - 3x - 1$, and these are $-1.53209, -0.34730, 1.87939$.

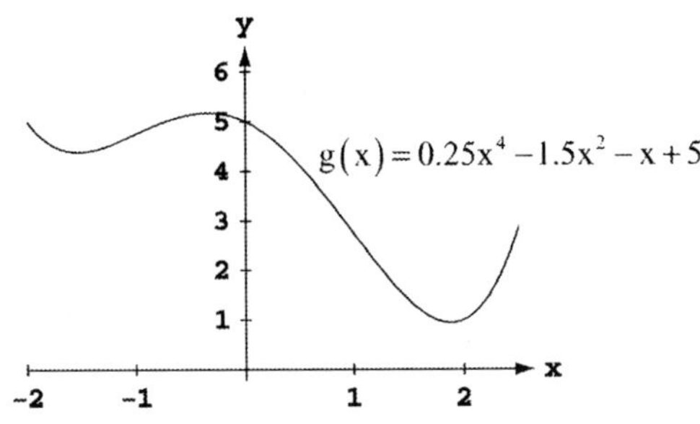

15. $f(x) = \tan x - 2x \Rightarrow f'(x) = \sec^2 x - 2 \Rightarrow x_{n+1} = x_n - \frac{\tan(x_n) - 2x_n}{\sec^2(x_n)}$; $x_0 = 1 \Rightarrow x_1 = 12920445$ $\Rightarrow x_2 = 1.155327774 \Rightarrow x_{16} = x_{17} = 1.165561185$

16. $f(x) = x^4 - 2x^3 - x^2 - 2x + 2 \Rightarrow f'(x) = 4x^3 - 6x^2 - 2x - 2 \Rightarrow x_{n+1} = x_n - \frac{x_n^4 - 2x_n^3 - x_n^2 - 2x_n + 2}{4x_n^3 - 6x_n^2 - 2x_n - 2}$; if $x_0 = 0.5$, then $x_4 = 0.630115396$; if $x_0 = 2.5$, then $x_4 = 2.57327196$

17. (a) The graph of $f(x) = \sin 3x - 0.99 + x^2$ in the window $-2 \le x \le 2$, $-2 \le y \le 3$ suggests three roots. However, when you zoom in on the x-axis near $x = 1.2$, you can see that the graph lies above the axis there. There are only two roots, one near $x = -1$, the other near $x = 0.4$.

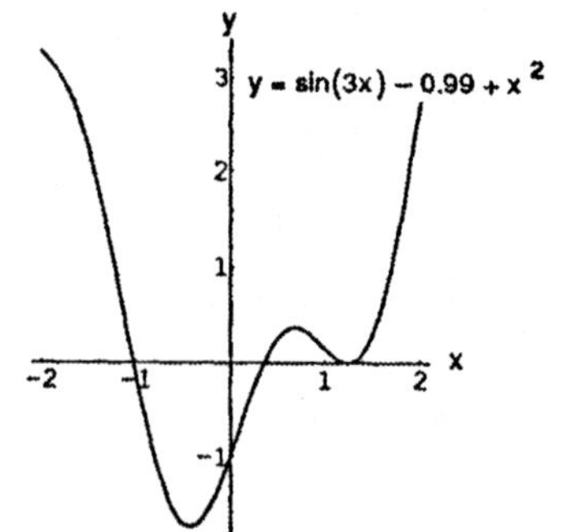

(b) $f(x) = \sin 3x - 0.99 + x^2 \Rightarrow f'(x) = 3\cos 3x + 2x$

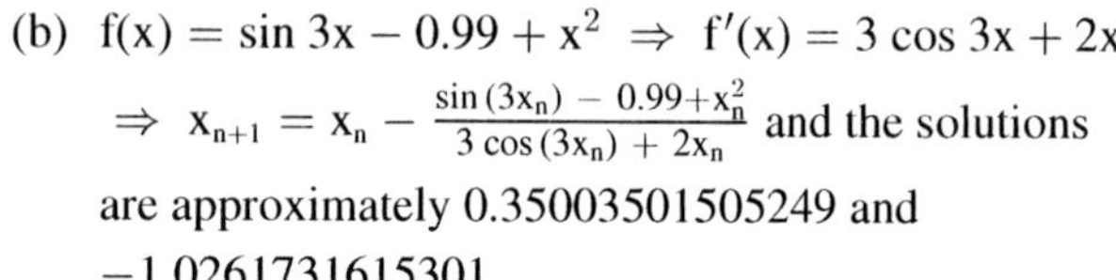

$\Rightarrow x_{n+1} = x_n - \frac{\sin(3x_n) - 0.99 + x_n^2}{3\cos(3x_n) + 2x_n}$ and the solutions are approximately 0.35003501505249 and -1.0261731615301

18. (a) Yes, three times as indicted by the graphs

(b) $f(x) = \cos 3x - x \Rightarrow f'(x) = -3\sin 3x - 1 \Rightarrow x_{n+1} = x_n - \frac{\cos(3x_n) - x_n}{-3\sin(3x_n) - 1}$; at approximately -0.979367, -0.887726, and 0.39004 we have $\cos 3x = x$

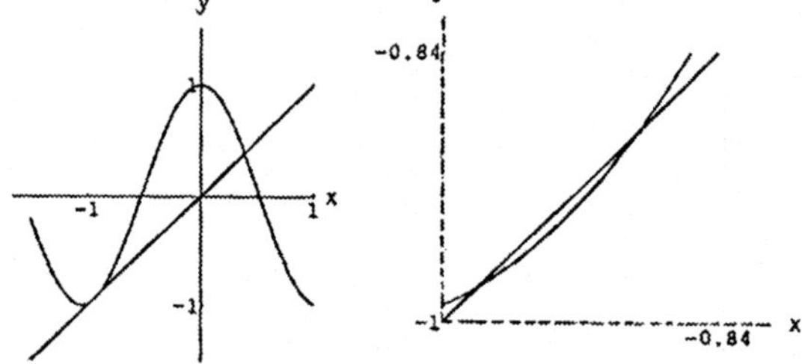

19. $f(x) = 2x^4 - 4x^2 + 1 \Rightarrow f'(x) = 8x^3 - 8x \Rightarrow x_{n+1} = x_n - \frac{2x_n^1 - 4x_n^2 + 1}{8x_n^3 - 8x_n}$; if $x_0 = -2$, then $x_6 = -1.30656296$; if $x_0 = -0.5$, then $x_3 = -0.5411961$; the roots are approximately ± 0.5411961 and ± 1.30656296 because $f(x)$ is an even function.

20. $f(x) = \tan x \Rightarrow f'(x) = \sec^2 x \Rightarrow x_{n+1} = x_n - \frac{\tan(x_n)}{\sec^2(x_n)}$; $x_0 = 3 \Rightarrow x_1 = 3.13971 \Rightarrow x_2 = 3.14159$ and we approximate π to be 3.14159.

21. Graphing e^{-x^2} and $x^2 - x + 1$ shows that there are two places where the curves intersect, one at $x = 0$ and the other between $x = 0.5$ and $x = 0.6$. Let $f(x) = e^{-x^2} - x^2 + x - 1$, $x_0 = 0.5$, and $x_{n+1} = x_n - \frac{f(x_n)}{f'(x_n)} = x_n - \frac{e^{-x_n^2} - x_n^2 + x_n - 1}{1 - 2x_n - 2x_n e^{-x_n^2}}$. Performing iterations on a calculator, spreadsheet, or CAS gives $x_1 = 0.536981$, $x_2 = 0.534856$, $x_3 = 0.53485$, $x_4 = 0.53485$. (You may get different results depending upon what you select for $f(x)$ and x_0, and what calculator or computer you may use.) Therefore, the two curves intersect at $x = 0$ and $x = 0.53485$.

22. Graphing $\ln(1 - x^2)$ and $x - 1$ shows that there are two places where the curves intersect, one between $x = -1$ and $x = -0.9$, and the other between $x = 0.5$ and $x = 0.6$. Let $f(x) = \ln(1 - x^2) - x + 1$, and $x_{n+1} = x_n - \frac{f(x_n)}{f'(x_n)}$ $= x_n - \frac{\ln(1 - x_n^2) - x_n + 1}{\frac{-2x_n}{1 - x_n^2} - 1}$. Performing iterations on a calculator, spreadsheet, or CAS with $x_0 = 0.5$ gives $x_1 = 0.590992$, $x_2 = 0.583658$, $x_3 = 0.583597$, $x_4 = 0.583597$ and with $x_0 = -0.9$ gives $x_1 = -0.928237$, $x_2 = -0.924247$, $x_3 = -0.924119$, $x_4 = -0.924119$. (You may get different results depending upon what you select for $f(x)$ and x_0, and what calculator or computer you may use.) Therefore, the two curves intersect at $x = -0.924119$ and $x = 0.583597$.

23. If $f(x) = x^3 + 2x - 4$, then $f(1) = -1 < 0$ and $f(2) = 8 > 0 \Rightarrow$ by the Intermediate Value Theorem the equation $x^3 + 2x - 4 = 0$ has a solution between 1 and 2. Consequently, $f'(x) = 3x^2 + 2$ and $x_{n+1} = x_n - \frac{x_n^3 + 2x_n - 4}{3x_n^2 + 2}$. Then $x_0 = 1 \Rightarrow x_1 = 1.2 \Rightarrow x_2 = 1.17975 \Rightarrow x_3 = 1.179509 \Rightarrow x_4 = 1.1795090 \Rightarrow$ the root is approximately 1.17951.

24. We wish to solve $8x^4 - 14x^3 - 9x^2 + 11x - 1 = 0$. Let $f(x) = 8x^4 - 14x^3 - 9x^2 + 11x - 1$, then $f'(x) = 32x^3 - 42x^2 - 18x + 11 \Rightarrow x_{n+1} = x_n - \frac{8x_n^4 - 14x_n^3 - 9x_n^2 + 11x_n - 1}{32x_n^3 - 42x_n^2 - 18x_n + 11}$.

x_0	approximation of corresponding root
−1.0	−0.976823589
0.1	0.100363332
0.6	0.642746671
2.0	1.983713587

25. $f(x) = 4x^4 - 4x^2 \Rightarrow f'(x) = 16x^3 - 8x \Rightarrow x_{i+1} = x_i - \frac{f(x_i)}{f'(x_i)} = x_i - \frac{x_i^3 - x_i}{4x_i^2 - 2}$. Iterations are performed using the procedure in problem 13 in this section.

(a) For $x_0 = 2$ or $x_0 = -0.8$, $x_i \to -1$ as i gets large.

(b) For $x_0 = -0.5$ or $x_0 = 0.25$, $x_i \to 0$ as i gets large.

(c) For $x_0 = 0.8$ or $x_0 = 2$, $x_i \to 1$ as i gets large.

(d) (If your calculator has a CAS, put it in exact mode, otherwise approximate the radicals with a decimal value.) For $x_0 = -\frac{\sqrt{21}}{7}$ or $x_0 = -\frac{\sqrt{21}}{7}$, Newton's method does not converge. The values of x_i alternate between $x_0 = -\frac{\sqrt{21}}{7}$ or $x_0 = -\frac{\sqrt{21}}{7}$ as i increases.

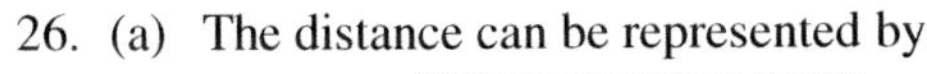

26. (a) The distance can be represented by $D(x) = \sqrt{(x-2)^2 + \left(x^2 + \frac{1}{2}\right)^2}$, where $x \ge 0$. The distance D(x) is minimized when $f(x) = (x-2)^2 + \left(x^2 + \frac{1}{2}\right)^2$ is minimized. If $f(x) = (x-2)^2 + \left(x^2 + \frac{1}{2}\right)^2$, then $f'(x) = 4\left(x^3 + x - 1\right)$ and $f''(x) = 4\left(3x^2 + 1\right) > 0$. Now $f'(x) = 0 \Rightarrow x^3 + x - 1 = 0 \Rightarrow x\left(x^2 + 1\right) = 1 \Rightarrow x = \frac{1}{x^2+1}$.

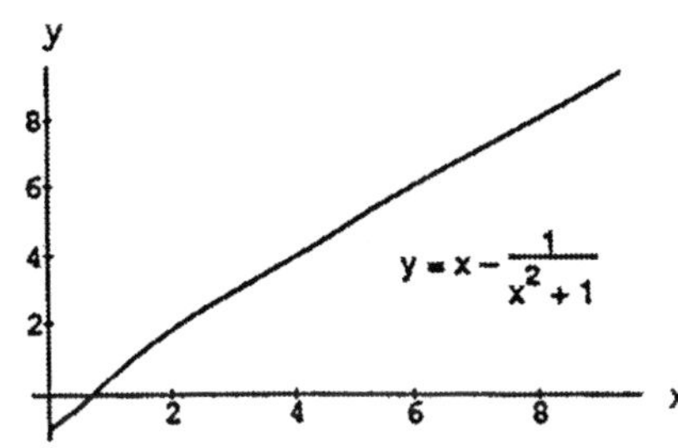

(b) Let $g(x) = \frac{1}{x^2+1} - x = \left(x^2+1\right)^{-1} - x \Rightarrow g'(x) = -\left(x^2+1\right)^{-2}(2x) - 1 = \frac{-2x}{\left(x^2+1\right)^2} - 1$

$\Rightarrow x_{n+1} = x_n - \frac{\left(\frac{1}{x_n^2+1} - x_n\right)}{\left(\frac{-2x_n}{\left(x_n^2+1\right)^2} - 1\right)}$; $x_0 = 1 \Rightarrow x_4 = 0.68233$ to five decimal places.

27. $f(x) = (x-1)^{40} \Rightarrow f'(x) = 40(x-1)^{39} \Rightarrow x_{n+1} = x_n - \frac{(x_n - 1)^{40}}{40(x_n - 1)^{39}} = \frac{39x_n + 1}{40}$. With $x_0 = 2$, our computer gave $x_{87} = x_{88} = x_{89} = \cdots = x_{200} = 1.11051$, coming within 0.11051 of the root $x = 1$.

28. $f(x) = x^3 + 3.6x^2 - 36.4 \Rightarrow f'(x) = 3x^2 + 7.2x \Rightarrow x_{n+1} = x_n - \frac{x_n^3 + 3.6x_n^2 - 36.4}{3x_n^2 + 7.2x_n}$; $x_0 = 2 \Rightarrow x_1 = 2.53\overline{03}$ $\Rightarrow x_2 = 2.45418225 \Rightarrow x_3 = 2.45238021 \Rightarrow x_4 = 2.45237921$ which is 2.45 to two decimal places. Recall that $x = 10^4\,[H_3O^+] \Rightarrow [H_3O^+] = (x)\left(10^{-4}\right) = (2.45)\left(10^{-4}\right) = 0.000245$

4.8 ANTIDERIVATIVES

1. (a) x^2 (b) $\frac{x^3}{3}$ (c) $\frac{x^3}{3} - x^2 + x$

2. (a) $3x^2$ (b) $\frac{x^8}{8}$ (c) $\frac{x^8}{8} - 3x^2 + 8x$

3. (a) x^{-3} (b) $-\frac{x^{-3}}{3}$ (c) $-\frac{x^{-3}}{3} + x^2 + 3x$

4. (a) $-x^{-2}$ (b) $-\frac{x^{-2}}{4}+\frac{x^3}{3}$ (c) $\frac{x^{-2}}{2}+\frac{x^2}{2}-x$

5. (a) $\frac{-1}{x}$ (b) $\frac{-5}{x}$ (c) $2x+\frac{5}{x}$

6. (a) $\frac{1}{x^2}$ (b) $\frac{-1}{4x^2}$ (c) $\frac{x^4}{4}+\frac{1}{2x^2}$

7. (a) $\sqrt{x^3}$ (b) $\sqrt{x}$ (c) $\frac{2}{3}\sqrt{x^3}+2\sqrt{x}$

8. (a) $x^{4/3}$ (b) $\frac{1}{2}x^{2/3}$ (c) $\frac{3}{4}x^{4/3}+\frac{3}{2}x^{2/3}$

9. (a) $x^{2/3}$ (b) $x^{1/3}$ (c) $x^{-1/3}$

10. (a) $x^{1/2}$ (b) $x^{-1/2}$ (c) $x^{-3/2}$

11. (a) $\ln|x|$ (b) $7\ln|x|$ (c) $x-5\ln|x|$

12. (a) $\frac{1}{3}\ln|x|$ (b) $\frac{2}{5}\ln|x|$ (c) $x+\frac{4}{3}\ln|x|+\frac{1}{x}$

13. (a) $\cos(\pi x)$ (b) $-3\cos x$ (c) $\frac{-\cos(\pi x)}{\pi}+\cos(3x)$

14. (a) $\sin(\pi x)$ (b) $\sin\left(\frac{\pi x}{2}\right)$ (c) $\left(\frac{2}{\pi}\right)\sin\left(\frac{\pi x}{2}\right)+\pi\sin x$

15. (a) $\tan x$ (b) $2\tan\left(\frac{x}{3}\right)$ (c) $-\frac{2}{3}\tan\left(\frac{3x}{2}\right)$

16. (a) $-\cot x$ (b) $\cot\left(\frac{3x}{2}\right)$ (c) $x+4\cot(2x)$

17. (a) $-\csc x$ (b) $\frac{1}{5}\csc(5x)$ (c) $2\csc\left(\frac{\pi x}{2}\right)$

18. (a) $\sec x$ (b) $\frac{4}{3}\sec(3x)$ (c) $\frac{2}{\pi}\sec\left(\frac{\pi x}{2}\right)$

19. (a) $\frac{1}{3}e^{3x}$ (b) $-e^{-x}$ (c) $2e^{x/2}$

20. (a) $-\frac{1}{2}e^{-2x}$ (b) $\frac{3}{4}e^{4x/3}$ (c) $-5e^{-x/5}$

21. (a) $\frac{1}{\ln 3}\cdot 3^x$ (b) $\frac{-1}{\ln 2}\cdot 2^{-x}$ (c) $\frac{1}{\ln(5/3)}\cdot\left(\frac{5}{3}\right)^x$

22. (a) $\frac{1}{\sqrt{3}+1}x^{\sqrt{3}+1}$ (b) $\frac{1}{\pi+1}x^{\pi+1}$ (c) $\frac{1}{\sqrt{2}}x^{\sqrt{2}}$

23. (a) $2\sin^{-1}x$ (b) $\frac{1}{2}\tan^{-1}x$ (c) $\frac{1}{2}\tan^{-1}(2x)$

24. (a) $\frac{1}{2}x^2-\frac{1}{\ln(1/2)}\cdot\left(\frac{1}{2}\right)^x$ (b) $\frac{1}{3}x^3+\frac{1}{\ln 2}\cdot 2^x$ (c) $\frac{1}{\ln\pi}\cdot\pi^x-\ln|x|$

25. $\int(x+1)\,dx=\frac{x^2}{2}+x+C$

26. $\int(5-6x)\,dx=5x-3x^2+C$

27. $\int\left(3t^2+\frac{t}{2}\right)dt=t^3+\frac{t^2}{4}+C$

28. $\int\left(\frac{t^2}{2}+4t^3\right)dt=\frac{t^3}{6}+t^4+C$

29. $\int (2x^3 - 5x + 7)\,dx = \frac{1}{2}x^4 - \frac{5}{2}x^2 + 7x + C$

30. $\int (1 - x^2 - 3x^5)\,dx = x - \frac{1}{3}x^3 - \frac{1}{2}x^6 + C$

31. $\int \left(\frac{1}{x^2} - x^2 - \frac{1}{3}\right) dx = \int \left(x^{-2} - x^2 - \frac{1}{3}\right) dx = \frac{x^{-1}}{-1} - \frac{x^3}{3} - \frac{1}{3}x + C = -\frac{1}{x} - \frac{x^3}{3} - \frac{x}{3} + C$

32. $\int \left(\frac{1}{5} - \frac{2}{x^3} + 2x\right) dx = \int \left(\frac{1}{5} - 2x^{-3} + 2x\right) dx = \frac{1}{5}x - \left(\frac{2x^{-2}}{-2}\right) + \frac{2x^2}{2} + C = \frac{x}{5} + \frac{1}{x^2} + x^2 + C$

33. $\int x^{-1/3}\,dx = \frac{x^{2/3}}{\frac{2}{3}} + C = \frac{3}{2}x^{2/3} + C$

34. $\int x^{-5/4}\,dx = \frac{x^{-1/4}}{-\frac{1}{4}} + C = \frac{-4}{\sqrt[4]{x}} + C$

35. $\int \left(\sqrt{x} + \sqrt[3]{x}\right) dx = \int \left(x^{1/2} + x^{1/3}\right) dx = \frac{x^{3/2}}{\frac{3}{2}} + \frac{x^{4/3}}{\frac{4}{3}} + C = \frac{2}{3}x^{3/2} + \frac{3}{4}x^{4/3} + C$

36. $\int \left(\frac{\sqrt{x}}{2} + \frac{2}{\sqrt{x}}\right) dx = \int \left(\frac{1}{2}x^{1/2} + 2x^{-1/2}\right) dx = \frac{1}{2}\left(\frac{x^{3/2}}{\frac{3}{2}}\right) + 2\left(\frac{x^{1/2}}{\frac{1}{2}}\right) + C = \frac{1}{3}x^{3/2} + 4x^{1/2} + C$

37. $\int \left(8y - \frac{2}{y^{1/4}}\right) dy = \int \left(8y - 2y^{-1/4}\right) dy = \frac{8y^2}{2} - 2\left(\frac{y^{3/4}}{\frac{3}{4}}\right) + C = 4y^2 - \frac{8}{3}y^{3/4} + C$

38. $\int \left(\frac{1}{7} - \frac{1}{y^{5/4}}\right) dy = \int \left(\frac{1}{7} - y^{-5/4}\right) dy = \frac{1}{7}y - \left(\frac{y^{-1/4}}{-\frac{1}{4}}\right) + C = \frac{y}{7} + \frac{4}{y^{1/4}} + C$

39. $\int 2x\left(1 - x^{-3}\right) dx = \int \left(2x - 2x^{-2}\right) dx = \frac{2x^2}{2} - 2\left(\frac{x^{-1}}{-1}\right) + C = x^2 + \frac{2}{x} + C$

40. $\int x^{-3}(x+1)\,dx = \int \left(x^{-2} + x^{-3}\right) dx = \frac{x^{-1}}{-1} + \left(\frac{x^{-2}}{-2}\right) + C = -\frac{1}{x} - \frac{1}{2x^2} + C$

41. $\int \frac{t\sqrt{t}+\sqrt{t}}{t^2}\,dt = \int \left(\frac{t^{3/2}}{t^2} + \frac{t^{1/2}}{t^2}\right) dt = \int \left(t^{-1/2} + t^{-3/2}\right) dt = \frac{t^{1/2}}{\frac{1}{2}} + \left(\frac{t^{-1/2}}{-\frac{1}{2}}\right) + C = 2\sqrt{t} - \frac{2}{\sqrt{t}} + C$

42. $\int \frac{4+\sqrt{t}}{t^3}\,dt = \int \left(\frac{4}{t^3} + \frac{t^{1/2}}{t^3}\right) dt = \int \left(4t^{-3} + t^{-5/2}\right) dt = 4\left(\frac{t^{-2}}{-2}\right) + \left(\frac{t^{-3/2}}{-\frac{3}{2}}\right) + C = -\frac{2}{t^2} - \frac{2}{3t^{3/2}} + C$

43. $\int -2\cos t\,dt = -2\sin t + C$

44. $\int -5\sin t\,dt = 5\cos t + C$

45. $\int 7\sin\frac{\theta}{3}\,d\theta = -21\cos\frac{\theta}{3} + C$

46. $\int 3\cos 5\theta\,d\theta = \frac{3}{5}\sin 5\theta + C$

47. $\int -3\csc^2 x\,dx = 3\cot x + C$

48. $\int -\frac{\sec^2 x}{3}\,dx = -\frac{\tan x}{3} + C$

49. $\int \frac{\csc\theta\cot\theta}{2}\,d\theta = -\frac{1}{2}\csc\theta + C$

50. $\int \frac{2}{5}\sec\theta\tan\theta\,d\theta = \frac{2}{5}\sec\theta + C$

51. $\int \left(e^{3x} + 5e^{-x}\right) dx = \frac{e^{3x}}{3} - 5e^{-x} + C$

52. $\int \left(2e^x - 3e^{-2x}\right) dx = 2e^x + \frac{3}{2}e^{-2x} + C$

53. $\int \left(e^{-x} + 4^x\right) dx = -e^{-x} + \frac{4^x}{\ln 4} + C$

54. $\int (1.3)^x\,dx = \frac{(1.3)^x}{\ln(1.3)} + C$

55. $\int \left(4\sec x\tan x - 2\sec^2 x\right) dx = 4\sec x - 2\tan x + C$

56. $\int \frac{1}{2}\left(\csc^2 x - \csc x\cot x\right) dx = -\frac{1}{2}\cot x + \frac{1}{2}\csc x + C$

57. $\int \left(\sin 2x - \csc^2 x\right) dx = -\frac{1}{2}\cos 2x + \cot x + C$

58. $\int (2\cos 2x - 3\sin 3x)\,dx = \sin 2x + \cos 3x + C$

59. $\int \frac{1+\cos 4t}{2}\,dt = \int \left(\frac{1}{2} + \frac{1}{2}\cos 4t\right) dt = \frac{1}{2}t + \frac{1}{2}\left(\frac{\sin 4t}{4}\right) + C = \frac{t}{2} + \frac{\sin 4t}{8} + C$

60. $\int \frac{1-\cos 6t}{2}\,dt = \int \left(\frac{1}{2} - \frac{1}{2}\cos 6t\right) dt = \frac{1}{2}t - \frac{1}{2}\left(\frac{\sin 6t}{6}\right) + C = \frac{t}{2} - \frac{\sin 6t}{12} + C$

61. $\int \left(\frac{1}{x} - \frac{5}{x^2+1}\right) dx = \ln|x| - 5\tan^{-1}x + C$

62. $\int \left(\frac{2}{\sqrt{1-y^2}} - \frac{1}{y^{1/4}}\right) dy = 2\sin^{-1}y - \frac{4}{3}y^{3/4} + C$

63. $\int 3x^{\sqrt{3}}\,dx = \frac{3x^{(\sqrt{3}+1)}}{\sqrt{3}+1} + C$

64. $\int x^{(\sqrt{2}-1)}\,dx = \frac{x^{\sqrt{2}}}{\sqrt{2}} + C$

65. $\int (1+\tan^2\theta)\,d\theta = \int \sec^2\theta\,d\theta = \tan\theta + C$

66. $\int (2+\tan^2\theta)\,d\theta = \int (1+1+\tan^2\theta)\,d\theta = \int (1+\sec^2\theta)\,d\theta = \theta + \tan\theta + C$

67. $\int \cot^2 x\,dx = \int (\csc^2 x - 1)\,dx = -\cot x - x + C$

68. $\int (1-\cot^2 x)\,dx = \int (1-(\csc^2 x - 1))\,dx = \int (2-\csc^2 x)\,dx = 2x + \cot x + C$

69. $\int \cos\theta\,(\tan\theta + \sec\theta)\,d\theta = \int (\sin\theta + 1)\,d\theta = -\cos\theta + \theta + C$

70. $\int \frac{\csc\theta}{\csc\theta - \sin\theta}\,d\theta = \int \left(\frac{\csc\theta}{\csc\theta - \sin\theta}\right)\left(\frac{\sin\theta}{\sin\theta}\right) d\theta = \int \frac{1}{1-\sin^2\theta}\,d\theta = \int \frac{1}{\cos^2\theta}\,d\theta = \int \sec^2\theta\,d\theta = \tan\theta + C$

71. $\frac{d}{dx}\left(\frac{(7x-2)^4}{28} + C\right) = \frac{4(7x-2)^3(7)}{28} = (7x-2)^3$

72. $\frac{d}{dx}\left(-\frac{(3x+5)^{-1}}{3} + C\right) = -\left(-\frac{(3x+5)^{-2}(3)}{3}\right) = (3x+5)^{-2}$

73. $\frac{d}{dx}\left(\frac{1}{5}\tan(5x-1) + C\right) = \frac{1}{5}\left(\sec^2(5x-1)\right)(5) = \sec^2(5x-1)$

74. $\frac{d}{dx}\left(-3\cot\left(\frac{x-1}{3}\right) + C\right) = -3\left(-\csc^2\left(\frac{x-1}{3}\right)\right)\left(\frac{1}{3}\right) = \csc^2\left(\frac{x-1}{3}\right)$

75. $\frac{d}{dx}\left(\frac{-1}{x+1} + C\right) = (-1)(-1)(x+1)^{-2} = \frac{1}{(x+1)^2}$

76. $\frac{d}{dx}\left(\frac{x}{x+1} + C\right) = \frac{(x+1)(1) - x(1)}{(x+1)^2} = \frac{1}{(x+1)^2}$

77. $\frac{d}{dx}\left(\ln(x+1) + C\right) = \frac{1}{x+1}$

78. $\frac{d}{dx}\left(x e^x - e^x\right) = x \cdot e^x + (1) \cdot e^x - e^x = x e^x$

79. $\frac{d}{dx}\left(\frac{1}{a}\tan^{-1}\left(\frac{x}{a}\right)\right) = \frac{1}{a} \cdot \frac{1}{1+\left(\frac{x}{a}\right)^2} \cdot \frac{d}{dx}\left(\frac{x}{a}\right) = \frac{1}{a^2\left(1+\frac{x^2}{a^2}\right)} = \frac{1}{a^2+x^2}$

80. $\frac{d}{dx}\left(\sin^{-1}\left(\frac{x}{a}\right)\right) = \frac{1}{\sqrt{1-\left(\frac{x}{a}\right)^2}} \cdot \frac{d}{dx}\left(\frac{x}{a}\right) = \frac{1}{a\sqrt{1-\left(\frac{x}{a}\right)^2}} = \frac{1}{\sqrt{a^2-x^2}}$

81. If $y = \ln x - \frac{1}{2}\ln(1+x^2) - \frac{\tan^{-1}x}{x} + C$, then $dy = \left[\frac{1}{x} - \frac{x}{1+x^2} - \frac{\left(\frac{x}{1+x^2}\right) - \tan^{-1}x}{x^2}\right] dx$

$= \left(\frac{1}{x} - \frac{x}{1+x^2} - \frac{1}{x(1+x^2)} + \frac{\tan^{-1}x}{x^2}\right) dx = \frac{x(1+x^2) - x^3 - x + (\tan^{-1}x)(1+x^2)}{x^2(1+x^2)}\,dx = \frac{\tan^{-1}x}{x^2}\,dx,$

which verifies the formula

82. If $y = x(\sin^{-1} x)^2 - 2x + 2\sqrt{1-x^2}\sin^{-1} x + C$, then
$dy = \left[(\sin^{-1} x)^2 + \frac{2x(\sin^{-1} x)}{\sqrt{1-x^2}} - 2 + \frac{-2x}{\sqrt{1-x^2}}\sin^{-1} x + 2\sqrt{1-x^2}\left(\frac{1}{\sqrt{1-x^2}}\right)\right] dx = (\sin^{-1} x)^2\, dx$, which verifies the formula

83. (a) Wrong: $\frac{d}{dx}\left(\frac{x^2}{2}\sin x + C\right) = \frac{2x}{2}\sin x + \frac{x^2}{2}\cos x = x\sin x + \frac{x^2}{2}\cos x \neq x\sin x$
(b) Wrong: $\frac{d}{dx}(-x\cos x + C) = -\cos x + x\sin x \neq x\sin x$
(c) Right: $\frac{d}{dx}(-x\cos x + \sin x + C) = -\cos x + x\sin x + \cos x = x\sin x$

84. (a) Wrong: $\frac{d}{d\theta}\left(\frac{\sec^3\theta}{3} + C\right) = \frac{3\sec^2\theta}{3}(\sec\theta\tan\theta) = \sec^3\theta\tan\theta \neq \tan\theta\sec^2\theta$
(b) Right: $\frac{d}{d\theta}\left(\frac{1}{2}\tan^2\theta + C\right) = \frac{1}{2}(2\tan\theta)\sec^2\theta = \tan\theta\sec^2\theta$
(c) Right: $\frac{d}{d\theta}\left(\frac{1}{2}\sec^2\theta + C\right) = \frac{1}{2}(2\sec\theta)\sec\theta\tan\theta = \tan\theta\sec^2\theta$

85. (a) Wrong: $\frac{d}{dx}\left(\frac{(2x+1)^3}{3} + C\right) = \frac{3(2x+1)^2(2)}{3} = 2(2x+1)^2 \neq (2x+1)^2$
(b) Wrong: $\frac{d}{dx}((2x+1)^3 + C) = 3(2x+1)^2(2) = 6(2x+1)^2 \neq 3(2x+1)^2$
(c) Right: $\frac{d}{dx}((2x+1)^3 + C) = 6(2x+1)^2$

86. (a) Wrong: $\frac{d}{dx}(x^2+x+C)^{1/2} = \frac{1}{2}(x^2+x+C)^{-1/2}(2x+1) = \frac{2x+1}{2\sqrt{x^2+x+C}} \neq \sqrt{2x+1}$
(b) Wrong: $\frac{d}{dx}\left((x^2+x)^{1/2} + C\right) = \frac{1}{2}(x^2+x)^{-1/2}(2x+1) = \frac{2x+1}{2\sqrt{x^2+x}} \neq \sqrt{2x+1}$
(c) Right: $\frac{d}{dx}\left(\frac{1}{3}\left(\sqrt{2x+1}\right)^3 + C\right) = \frac{d}{dx}\left(\frac{1}{3}(2x+1)^{3/2} + C\right) = \frac{3}{6}(2x+1)^{1/2}(2) = \sqrt{2x+1}$

87. Graph (b), because $\frac{dy}{dx} = 2x \Rightarrow y = x^2 + C$. Then $y(1) = 4 \Rightarrow C = 3$.

88. Graph (b), because $\frac{dy}{dx} = -x \Rightarrow y = -\frac{1}{2}x^2 + C$. Then $y(-1) = 1 \Rightarrow C = \frac{3}{2}$.

89. $\frac{dy}{dx} = 2x - 7 \Rightarrow y = x^2 - 7x + C$; at $x = 2$ and $y = 0$ we have $0 = 2^2 - 7(2) + C \Rightarrow C = 10 \Rightarrow y = x^2 - 7x + 10$

90. $\frac{dy}{dx} = 10 - x \Rightarrow y = 10x - \frac{x^2}{2} + C$; at $x = 0$ and $y = -1$ we have $-1 = 10(0) - \frac{0^2}{2} + C \Rightarrow C = -1$
$\Rightarrow y = 10x - \frac{x^2}{2} - 1$

91. $\frac{dy}{dx} = \frac{1}{x^2} + x = x^{-2} + x \Rightarrow y = -x^{-1} + \frac{x^2}{2} + C$; at $x = 2$ and $y = 1$ we have $1 = -2^{-1} + \frac{2^2}{2} + C \Rightarrow C = -\frac{1}{2}$
$\Rightarrow y = -x^{-1} + \frac{x^2}{2} - \frac{1}{2}$ or $y = -\frac{1}{x} + \frac{x^2}{2} - \frac{1}{2}$

92. $\frac{dy}{dx} = 9x^2 - 4x + 5 \Rightarrow y = 3x^3 - 2x^2 + 5x + C$; at $x = -1$ and $y = 0$ we have $0 = 3(-1)^3 - 2(-1)^2 + 5(-1) + C$
$\Rightarrow C = 10 \Rightarrow y = 3x^3 - 2x^2 + 5x + 10$

93. $\frac{dy}{dx} = 3x^{-2/3} \Rightarrow y = \frac{3x^{1/3}}{\frac{1}{3}} + C = 9$; at $x = 9x^{1/3} + C$; at $x = -1$ and $y = -5$ we have $-5 = 9(-1)^{1/3} + C \Rightarrow C = 4$
$\Rightarrow y = 9x^{1/3} + 4$

94. $\frac{dy}{dx} = \frac{1}{2\sqrt{x}} = \frac{1}{2}x^{-1/2} \Rightarrow y = x^{1/2} + C$; at $x = 4$ and $y = 0$ we have $0 = 4^{1/2} + C \Rightarrow C = -2 \Rightarrow y = x^{1/2} - 2$

95. $\frac{ds}{dt} = 1 + \cos t \Rightarrow s = t + \sin t + C$; at $t = 0$ and $s = 4$ we have $4 = 0 + \sin 0 + C \Rightarrow C = 4 \Rightarrow s = t + \sin t + 4$

96. $\frac{ds}{dt} = \cos t + \sin t \Rightarrow s = \sin t - \cos t + C$; at $t = \pi$ and $s = 1$ we have $1 = \sin \pi - \cos \pi + C \Rightarrow C = 0$ $\Rightarrow s = \sin t - \cos t$

97. $\frac{dr}{d\theta} = -\pi \sin \pi\theta \Rightarrow r = \cos(\pi\theta) + C$; at $r = 0$ and $\theta = 0$ we have $0 = \cos(\pi 0) + C \Rightarrow C = -1 \Rightarrow r = \cos(\pi\theta) - 1$

98. $\frac{dr}{d\theta} = \cos \pi\theta \Rightarrow r = \frac{1}{\pi}\sin(\pi\theta) + C$; at $r = 1$ and $\theta = 0$ we have $1 = \frac{1}{\pi}\sin(\pi 0) + C \Rightarrow C = 1 \Rightarrow r = \frac{1}{\pi}\sin(\pi\theta) + 1$

99. $\frac{dv}{dt} = \frac{1}{2}\sec t \tan t \Rightarrow v = \frac{1}{2}\sec t + C$; at $v = 1$ and $t = 0$ we have $1 = \frac{1}{2}\sec(0) + C \Rightarrow C = \frac{1}{2} \Rightarrow v = \frac{1}{2}\sec t + \frac{1}{2}$

100. $\frac{dv}{dt} = 8t + \csc^2 t \Rightarrow v = 4t^2 - \cot t + C$; at $v = -7$ and $t = \frac{\pi}{2}$ we have $-7 = 4\left(\frac{\pi}{2}\right)^2 - \cot\left(\frac{\pi}{2}\right) + C \Rightarrow C = -7 - \pi^2$ $\Rightarrow v = 4t^2 - \cot t - 7 - \pi^2$

101. $\frac{dv}{dt} = \frac{3}{t\sqrt{t^2-1}}, t > 1 \Rightarrow v = 3\sec^{-1} t + C$; at $t = 2$ and $v = 0$ we have $0 = 3\sec^{-1} 2 + C \Rightarrow C = -\pi \Rightarrow v = 3\sec^{-1} t - \pi$

102. $\frac{dv}{dt} = \frac{8}{1+t^2} + \sec^2 t \Rightarrow v = 8\tan^{-1} t + \tan t + C$; at $t = 0$ and $v = 1$ we have $1 = 8\tan^{-1}(0) + \tan(0) + C \Rightarrow C = 1$ $\Rightarrow v = 8\tan^{-1} t + \tan t + 1$

103. $\frac{d^2y}{dx^2} = 2 - 6x \Rightarrow \frac{dy}{dx} = 2x - 3x^2 + C_1$; at $\frac{dy}{dx} = 4$ and $x = 0$ we have $4 = 2(0) - 3(0)^2 + C_1 \Rightarrow C_1 = 4$ $\Rightarrow \frac{dy}{dx} = 2x - 3x^2 + 4 \Rightarrow y = x^2 - x^3 + 4x + C_2$; at $y = 1$ and $x = 0$ we have $1 = 0^2 - 0^3 + 4(0) + C_2 \Rightarrow C_2 = 1$ $\Rightarrow y = x^2 - x^3 + 4x + 1$

104. $\frac{d^2y}{dx^2} = 0 \Rightarrow \frac{dy}{dx} = C_1$; at $\frac{dy}{dx} = 2$ and $x = 0$ we have $C_1 = 2 \Rightarrow \frac{dy}{dx} = 2 \Rightarrow y = 2x + C_2$; at $y = 0$ and $x = 0$ we have $0 = 2(0) + C_2 \Rightarrow C_2 = 0 \Rightarrow y = 2x$

105. $\frac{d^2r}{dt^2} = \frac{2}{t^3} = 2t^{-3} \Rightarrow \frac{dr}{dt} = -t^{-2} + C_1$; at $\frac{dr}{dt} = 1$ and $t = 1$ we have $1 = -(1)^{-2} + C_1 \Rightarrow C_1 = 2 \Rightarrow \frac{dr}{dt} = -t^{-2} + 2$ $\Rightarrow r = t^{-1} + 2t + C_2$; at $r = 1$ and $t = 1$ we have $1 = 1^{-1} + 2(1) + C_2 \Rightarrow C_2 = -2 \Rightarrow r = t^{-1} + 2t - 2$ or $r = \frac{1}{t} + 2t - 2$

106. $\frac{d^2s}{dt^2} = \frac{3t}{8} \Rightarrow \frac{ds}{dt} = \frac{3t^2}{16} + C_1$; at $\frac{ds}{dt} = 3$ and $t = 4$ we have $3 = \frac{3(4)^2}{16} + C_1 \Rightarrow C_1 = 0 \Rightarrow \frac{ds}{dt} = \frac{3t^2}{16} \Rightarrow s = \frac{t^3}{16} + C_2$; at $s = 4$ and $t = 4$ we have $4 = \frac{4^3}{16} + C_2 \Rightarrow C_2 = 0 \Rightarrow s = \frac{t^3}{16}$

107. $\frac{d^3y}{dx^3} = 6 \Rightarrow \frac{d^2y}{dx^2} = 6x + C_1$; at $\frac{d^2y}{dx^2} = -8$ and $x = 0$ we have $-8 = 6(0) + C_1 \Rightarrow C_1 = -8 \Rightarrow \frac{d^2y}{dx^2} = 6x - 8$ $\Rightarrow \frac{dy}{dx} = 3x^2 - 8x + C_2$; at $\frac{dy}{dx} = 0$ and $x = 0$ we have $0 = 3(0)^2 - 8(0) + C_2 \Rightarrow C_2 = 0 \Rightarrow \frac{dy}{dx} = 3x^2 - 8x$ $\Rightarrow y = x^3 - 4x^2 + C_3$; at $y = 5$ and $x = 0$ we have $5 = 0^3 - 4(0)^2 + C_3 \Rightarrow C_3 = 5 \Rightarrow y = x^3 - 4x^2 + 5$

108. $\frac{d^3\theta}{dt^3} = 0 \Rightarrow \frac{d^2\theta}{dt^2} = C_1$; at $\frac{d^2\theta}{dt^2} = -2$ and $t = 0$ we have $\frac{d^2\theta}{dt^2} = -2 \Rightarrow \frac{d\theta}{dt} = -2t + C_2$; at $\frac{d\theta}{dt} = -\frac{1}{2}$ and $t = 0$ we have $-\frac{1}{2} = -2(0) + C_2 \Rightarrow C_2 = -\frac{1}{2} \Rightarrow \frac{d\theta}{dt} = -2t - \frac{1}{2} \Rightarrow \theta = -t^2 - \frac{1}{2}t + C_3$; at $\theta = \sqrt{2}$ and $t = 0$ we have $\sqrt{2} = -0^2 - \frac{1}{2}(0) + C_3 \Rightarrow C_3 = \sqrt{2} \Rightarrow \theta = -t^2 - \frac{1}{2}t + \sqrt{2}$

109. $y^{(4)} = -\sin t + \cos t \Rightarrow y''' = \cos t + \sin t + C_1$; at $y''' = 7$ and $t = 0$ we have $7 = \cos(0) + \sin(0) + C_1$ $\Rightarrow C_1 = 6 \Rightarrow y''' = \cos t + \sin t + 6 \Rightarrow y'' = \sin t - \cos t + 6t + C_2$; at $y'' = -1$ and $t = 0$ we have $-1 = \sin(0) - \cos(0) + 6(0) + C_2 \Rightarrow C_2 = 0 \Rightarrow y'' = \sin t - \cos t + 6t \Rightarrow y' = -\cos t - \sin t + 3t^2 + C_3$; at $y' = -1$ and $t = 0$ we have $-1 = -\cos(0) - \sin(0) + 3(0)^2 + C_3 \Rightarrow C_3 = 0 \Rightarrow y' = -\cos t - \sin t + 3t^2$ $\Rightarrow y = -\sin t + \cos t + t^3 + C_4$; at $y = 0$ and $t = 0$ we have $0 = -\sin(0) + \cos(0) + 0^3 + C_4 \Rightarrow C_4 = -1$ $\Rightarrow y = -\sin t + \cos t + t^3 - 1$

110. $y^{(4)} = -\cos x + 8\sin(2x) \Rightarrow y''' = -\sin x - 4\cos(2x) + C_1$; at $y''' = 0$ and $x = 0$ we have $0 = -\sin(0) - 4\cos(2(0)) + C_1 \Rightarrow C_1 = 4 \Rightarrow y''' = -\sin x - 4\cos(2x) + 4 \Rightarrow y'' = \cos x - 2\sin(2x) + 4x + C_2$; at $y'' = 1$ and $x = 0$ we have $1 = \cos(0) - 2\sin(2(0)) + 4(0) + C_2 \Rightarrow C_2 = 0 \Rightarrow y'' = \cos x - 2\sin(2x) + 4x$ $\Rightarrow y' = \sin x + \cos(2x) + 2x^2 + C_3$; at $y' = 1$ and $x = 0$ we have $1 = \sin(0) + \cos(2(0)) + 2(0)^2 + C_3 \Rightarrow C_3 = 0$ $\Rightarrow y' = \sin x + \cos(2x) + 2x^2 \Rightarrow y = -\cos x + \frac{1}{2}\sin(2x) + \frac{2}{3}x^3 + C_4$; at $y = 3$ and $x = 0$ we have $3 = -\cos(0) + \frac{1}{2}\sin(2(0)) + \frac{2}{3}(0)^3 + C_4 \Rightarrow C_4 = 4 \Rightarrow y = -\cos x + \frac{1}{2}\sin(2x) + \frac{2}{3}x^3 + 4$

111. $m = y' = 3\sqrt{x} = 3x^{1/2} \Rightarrow y = 2x^{3/2} + C$; at $(9, 4)$ we have $4 = 2(9)^{3/2} + C \Rightarrow C = -50 \Rightarrow y = 2x^{3/2} - 50$

112. (a) $\frac{d^2y}{dx^2} = 6x \Rightarrow \frac{dy}{dx} = 3x^2 + C_1$; at $y' = 0$ and $x = 0$ we have $0 = 3(0)^2 + C_1 \Rightarrow C_1 = 0 \Rightarrow \frac{dy}{dx} = 3x^2$ $\Rightarrow y = x^3 + C_2$; at $y = 1$ and $x = 0$ we have $C_2 = 1 \Rightarrow y = x^3 + 1$

(b) One, because any other possible function would differ from $x^3 + 1$ by a constant that must be zero because of the initial conditions

113. $\frac{dy}{dx} = 1 - \frac{4}{3}x^{1/3} \Rightarrow y = \int\left(1 - \frac{4}{3}x^{1/3}\right)dx = x - x^{4/3} + C$; at $(1, 0.5)$ on the curve we have $0.5 = 1 - 1^{4/3} + C$ $\Rightarrow C = 0.5 \Rightarrow y = x - x^{4/3} + \frac{1}{2}$

114. $\frac{dy}{dx} = x - 1 \Rightarrow y = \int(x-1)\,dx = \frac{x^2}{2} - x + C$; at $(-1, 1)$ on the curve we have $1 = \frac{(-1)^2}{2} - (-1) + C$ $\Rightarrow C = -\frac{1}{2} \Rightarrow y = \frac{x^2}{2} - x - \frac{1}{2}$

115. $\frac{dy}{dx} = \sin x - \cos x \Rightarrow y = \int(\sin x - \cos x)\,dx = -\cos x - \sin x + C$; at $(-\pi, -1)$ on the curve we have $-1 = -\cos(-\pi) - \sin(-\pi) + C \Rightarrow C = -2 \Rightarrow y = -\cos x - \sin x - 2$

116. $\frac{dy}{dx} = \frac{1}{2\sqrt{x}} + \pi\sin\pi x = \frac{1}{2}x^{-1/2} + \pi\sin\pi x \Rightarrow y = \int\left(\frac{1}{2}x^{-1/2} + \sin\pi x\right)dx = x^{1/2} - \cos\pi x + C$; at $(1, 2)$ on the curve we have $2 = 1^{1/2} - \cos\pi(1) + C \Rightarrow C = 0 \Rightarrow y = \sqrt{x} - \cos\pi x$

117. (a) $\frac{ds}{dt} = 9.8t - 3 \Rightarrow s = 4.9t^2 - 3t + C$; (i) at $s = 5$ and $t = 0$ we have $C = 5 \Rightarrow s = 4.9t^2 - 3t + 5$; displacement $= s(3) - s(1) = ((4.9)(9) - 9 + 5) - (4.9 - 3 + 5) = 33.2$ units; (ii) at $s = -2$ and $t = 0$ we have $C = -2 \Rightarrow s = 4.9t^2 - 3t - 2$; displacement $= s(3) - s(1) = ((4.9)(9) - 9 - 2) - (4.9 - 3 - 2) = 33.2$ units; (iii) at $s = s_0$ and $t = 0$ we have $C = s_0 \Rightarrow s = 4.9t^2 - 3t + s_0$; displacement $= s(3) - s(1)$ $= ((4.9)(9) - 9 + s_0) - (4.9 - 3 + s_0) = 33.2$ units

(b) True. Given an antiderivative f(t) of the velocity function, we know that the body's position function is $s = f(t) + C$ for some constant C. Therefore, the displacement from $t = a$ to $t = b$ is $(f(b) + C) - (f(a) + C)$ $= f(b) - f(a)$. Thus we can find the displacement from any antiderivative f as the numerical difference $f(b) - f(a)$ without knowing the exact values of C and s.

118. $a(t) = v'(t) = 20 \Rightarrow v(t) = 20t + C$; at $(0, 0)$ we have $C = 0 \Rightarrow v(t) = 20t$. When $t = 60$, then $v(60) = 20(60)$ $= 1200$ m/sec.

119. Step 1: $\frac{d^2s}{dt^2} = -k \Rightarrow \frac{ds}{dt} = -kt + C_1$; at $\frac{ds}{dt} = 88$ and $t = 0$ we have $C_1 = 88 \Rightarrow \frac{ds}{dt} = -kt + 88 \Rightarrow$ $s = -k\left(\frac{t^2}{2}\right) + 88t + C_2$; at $s = 0$ and $t = 0$ we have $C_2 = 0 \Rightarrow s = -\frac{kt^2}{2} + 88t$

Step 2: $\frac{ds}{dt} = 0 \Rightarrow 0 = -kt + 88 \Rightarrow t = \frac{88}{k}$

Step 3: $242 = \frac{-k\left(\frac{88}{k}\right)^2}{2} + 88\left(\frac{88}{k}\right) \Rightarrow 242 = -\frac{(88)^2}{2k} + \frac{(88)^2}{k} \Rightarrow 242 = \frac{(88)^2}{2k} \Rightarrow k = 16$

120. $\frac{d^2s}{dt^2} = -k \Rightarrow \frac{ds}{dt} = \int -k\,dt = -kt + C$; at $\frac{ds}{dt} = 44$ when $t = 0$ we have $44 = -k(0) + C \Rightarrow C = 44$
$\Rightarrow \frac{ds}{dt} = -kt + 44 \Rightarrow s = -\frac{kt^2}{2} + 44t + C_1$; at $s = 0$ when $t = 0$ we have $0 = -\frac{k(0)^2}{2} + 44(0) + C_1 \Rightarrow C_1 = 0$
$\Rightarrow s = -\frac{kt^2}{2} + 44t$. Then $\frac{ds}{dt} = 0 \Rightarrow -kt + 44 = 0 \Rightarrow t = \frac{44}{k}$ and $s\left(\frac{44}{k}\right) = -\frac{k\left(\frac{44}{k}\right)^2}{2} + 44\left(\frac{44}{k}\right) = 45$
$\Rightarrow -\frac{968}{k} + \frac{1936}{k} = 45 \Rightarrow \frac{968}{k} = 45 \Rightarrow k = \frac{968}{45} \approx 21.5\ \frac{ft}{sec^2}$.

121. (a) $v = \int a\,dt = \int \left(15t^{1/2} - 3t^{-1/2}\right) dt = 10t^{3/2} - 6t^{1/2} + C$; $\frac{ds}{dt}(1) = 4 \Rightarrow 4 = 10(1)^{3/2} - 6(1)^{1/2} + C \Rightarrow C = 0$
$\Rightarrow v = 10t^{3/2} - 6t^{1/2}$

(b) $s = \int v\,dt = \int \left(10t^{3/2} - 6t^{1/2}\right) dt = 4t^{5/2} - 4t^{3/2} + C$; $s(1) = 0 \Rightarrow 0 = 4(1)^{5/2} - 4(1)^{3/2} + C \Rightarrow C = 0$
$\Rightarrow s = 4t^{5/2} - 4t^{3/2}$

122. $\frac{d^2s}{dt^2} = -5.2 \Rightarrow \frac{ds}{dt} = -5.2t + C_1$; at $\frac{ds}{dt} = 0$ and $t = 0$ we have $C_1 = 0 \Rightarrow \frac{ds}{dt} = -5.2t \Rightarrow s = -2.6t^2 + C_2$; at $s = 4$ and $t = 0$ we have $C_2 = 4 \Rightarrow s = -2.6t^2 + 4$. Then $s = 0 \Rightarrow 0 = -2.6t^2 + 4 \Rightarrow t = \sqrt{\frac{4}{2.6}} \approx 1.24$ sec, since $t > 0$

123. $\frac{d^2s}{dt^2} = a \Rightarrow \frac{ds}{dt} = \int a\,dt = at + C$; $\frac{ds}{dt} = v_0$ when $t = 0 \Rightarrow C = v_0 \Rightarrow \frac{ds}{dt} = at + v_0 \Rightarrow s = \frac{at^2}{2} + v_0t + C_1$; $s = s_0$ when $t = 0 \Rightarrow s_0 = \frac{a(0)^2}{2} + v_0(0) + C_1 \Rightarrow C_1 = s_0 \Rightarrow s = \frac{at^2}{2} + v_0t + s_0$

124. The appropriate initial value problem is: Differential Equation: $\frac{d^2s}{dt^2} = -g$ with Initial Conditions: $\frac{ds}{dt} = v_0$ and $s = s_0$ when $t = 0$. Thus, $\frac{ds}{dt} = \int -g\,dt = -gt + C_1$; $\frac{ds}{dt}(0) = v_0 \Rightarrow v_0 = (-g)(0) + C_1 \Rightarrow C_1 = v_0$
$\Rightarrow \frac{ds}{dt} = -gt + v_0$. Thus $s = \int(-gt + v_0)\,dt = -\frac{1}{2}gt^2 + v_0t + C_2$; $s(0) = s_0 = -\frac{1}{2}(g)(0)^2 + v_0(0) + C_2 \Rightarrow C_2 = s_0$
Thus $s = -\frac{1}{2}gt^2 + v_0t + s_0$.

125. (a) $\int f(x)\,dx = 1 - \sqrt{x} + C_1 = -\sqrt{x} + C$ (b) $\int g(x)\,dx = x + 2 + C_1 = x + C$
(c) $\int -f(x)\,dx = -\left(1 - \sqrt{x}\right) + C_1 = \sqrt{x} + C$ (d) $\int -g(x)\,dx = -(x + 2) + C_1 = -x + C$
(e) $\int [f(x) + g(x)]\,dx = \left(1 - \sqrt{x}\right) + (x + 2) + C_1 = x - \sqrt{x} + C$
(f) $\int [f(x) - g(x)]\,dx = \left(1 - \sqrt{x}\right) - (x + 2) + C_1 = -x - \sqrt{x} + C$

126. Yes. If F(x) and G(x) both solve the initial value problem on an interval I then they both have the same first derivative. Therefore, by Corollary 2 of the Mean Value Theorem there is a constant C such that $F(x) = G(x) + C$ for all x. In particular, $F(x_0) = G(x_0) + C$, so $C = F(x_0) - G(x_0) = 0$. Hence $F(x) = G(x)$ for all x.

127 – 130 Example CAS commands:

Maple:

```
with(student):
f := x -> cos(x)^2 + sin(x);
ic := [x=Pi,y=1];
F := unapply( int( f(x), x ) + C, x );
eq := eval( y=F(x), ic );
solnC := solve( eq, {C} );
Y := unapply( eval( F(x), solnC ), x );
DEplot( diff(y(x),x) = f(x), y(x), x=0..2*Pi, [[y(Pi)=1]],
    color=black, linecolor=black, stepsize=0.05, title="Section 4.8 #127" );
```

Mathematica: (functions and values may vary)

The following commands use the definite integral and the Fundamental Theorem of calculus to construct the solution of the initial value problems for exercises 127 - 130.

```
Clear[x, y, yprime]
yprime[x_] = Cos[x]^2 + Sin[x];
initxvalue = π; inityvalue = 1;
y[x_] = Integrate[yprime[t], {t, initxvalue, x}] + inityvalue
```

If the solution satisfies the differential equation and initial condition, the following yield True

```
yprime[x]==D[y[x], x] //Simplify
y[initxvalue]==inityvalue
```

Since exercise 106 is a second order differential equation, two integrations will be required.

```
Clear[x, y, yprime]
y2prime[x_] = 3 Exp[x/2] + 1;
initxval = 0; inityval = 4; inityprimeval = -1;
yprime[x_] = Integrate[y2prime[t],{t, initxval, x}] + inityprimeval
y[x_] = Integrate[yprime[t], {t, initxval, x}] + inityval
```

Verify that y[x] solves the differential equation and initial condition and plot the solution (red) and its derivative (blue).

```
y2prime[x]==D[y[x], {x, 2}]//Simplify
y[initxval]==inityval
yprime[initxval]==inityprimeval
Plot[{y[x], yprime[x]}, {x, initxval - 3, initxval + 3}, PlotStyle → {RGBColor[1,0,0], RGBColor[0,0,1]}]
```

CHAPTER 4 PRACTICE EXERCISES

1. No, since $f(x) = x^3 + 2x + \tan x \Rightarrow f'(x) = 3x^2 + 2 + \sec^2 x > 0 \Rightarrow f(x)$ is always increasing on its domain

2. No, since $g(x) = \csc x + 2\cot x \Rightarrow g'(x) = -\csc x \cot x - 2\csc^2 x = -\frac{\cos x}{\sin^2 x} - \frac{2}{\sin^2 x} = -\frac{1}{\sin^2 x}(\cos x + 2) < 0$ $\Rightarrow g(x)$ is always decreasing on its domain

3. No absolute minimum because $\lim_{x \to \infty} (7 + x)(11 - 3x)^{1/3} = -\infty$. Next $f'(x) = (11 - 3x)^{1/3} - (7 + x)(11 - 3x)^{-2/3} = \frac{(11-3x)-(7+x)}{(11-3x)^{2/3}} = \frac{4(1-x)}{(11-3x)^{2/3}} \Rightarrow x = 1$ and $x = \frac{11}{3}$ are critical points. Since $f' > 0$ if $x < 1$ and $f' < 0$ if $x > 1$, $f(1) = 16$ is the absolute maximum.

4. $f(x) = \frac{ax+b}{x^2-1} \Rightarrow f'(x) = \frac{a(x^2-1)-2x(ax+b)}{(x^2-1)^2} = \frac{-(ax^2+2bx+a)}{(x^2-1)^2}$; $f'(3) = 0 \Rightarrow -\frac{1}{64}(9a + 6b + a) = 0 \Rightarrow 5a + 3b = 0$. We require also that $f(3) = 1$. Thus $1 = \frac{3a+b}{8} \Rightarrow 3a + b = 8$. Solving both equations yields $a = 6$ and $b = -10$. Now, $f'(x) = \frac{-2(3x-1)(x-3)}{(x^2-1)^2}$ so that $f' = ---\underset{-1}{|}---\underset{1/3}{|}+++\underset{1}{|}+++\underset{3}{|}---$. Thus f' changes sign at $x = 3$ from positive to negative so there is a local maximum at $x = 3$ which has a value $f(3) = 1$.

5. $g(x) = e^x - x \Rightarrow g'(x) = e^x - 1 \Rightarrow g' = ---\underset{0}{|}+++ \Rightarrow$ the graph is decreasing on $(-\infty, 0)$, increasing on $(0, \infty)$; an absolute minimum value is 1 at $x = 0$; $x = 0$ is the only critical point of g; there is no absolute maximum value

6. $f(x) = \frac{2e^x}{1+x^2} \Rightarrow f'(x) = \frac{(1+x^2)\cdot 2e^x - 2e^x \cdot 2x}{(1+x^2)^2} = \frac{2e^x(1-x)^2}{(1+x^2)^2} \Rightarrow f' = +++\underset{1}{|}+++ \Rightarrow$ the graph is increasing on $(-\infty, \infty)$; $x = 1$ is the only critical point of f; there are no absolute maximum values or absolute minimum values.

7. $f(x) = x - 2\ln x$ on $1 \le x \le 3 \Rightarrow f'(x) = 1 - \frac{2}{x} \Rightarrow f' = \underset{1}{|}---\underset{2}{|}+++\underset{3}{|} \Rightarrow$ the graph is decreasing on $(1, 2)$, increasing on $(2, 3)$; an absolute minimum value is $2 - 2\ln 2$ at $x = 2$; an absolute maximum value is 1 at $x = 1$.

8. $f(x) = \frac{4}{x} + \ln x^2$ on $1 \le x \le 4 \Rightarrow f'(x) = -\frac{4}{x^2} + \frac{2}{x} = \frac{2x-4}{x^2} \Rightarrow f' = \underset{1}{|}\,---\underset{2}{|}\,+++\underset{4}{|}$ $\Rightarrow$ the graph is decreasing on $(1, 2)$, increasing on $(2, 4)$; an absolute minimum value is $2 + \ln 4$ at $x = 2$; an absolute maximum value is 4 at $x = 1$.

9. Yes, because at each point of $[0, 1)$ except $x = 0$, the function's value is a local minimum value as well as a local maximum value. At $x = 0$ the function's value, 0, is not a local minimum value because each open interval around $x = 0$ on the x-axis contains points to the left of 0 where f equals -1.

10. (a) The first derivative of the function $f(x) = x^3$ is zero at $x = 0$ even though f has no local extreme value at $x = 0$.
    (b) Theorem 2 says only that if f is differentiable and f has a local extreme at $x = c$ then $f'(c) = 0$. It does not assert the (false) reverse implication $f'(c) = 0 \Rightarrow$ f has a local extreme at $x = c$.

11. No, because the interval $0 < x < 1$ fails to be closed. The Extreme Value Theorem says that if the function is continuous throughout a finite closed interval $a \le x \le b$ then the existence of absolute extrema is guaranteed on that interval.

12. The absolute maximum is $|-1| = 1$ and the absolute minimum is $|0| = 0$. This is not inconsistent with the Extreme Value Theorem for continuous functions, which says a continuous function on a closed interval attains its extreme values on that interval. The theorem says nothing about the behavior of a continuous function on an interval which is half open and half closed, such as $[-1, 1)$, so there is nothing to contradict.

13. (a) There appear to be local minima at $x = -1.75$ and 1.8. Points of inflection are indicated at approximately $x = 0$ and $x = \pm 1$.

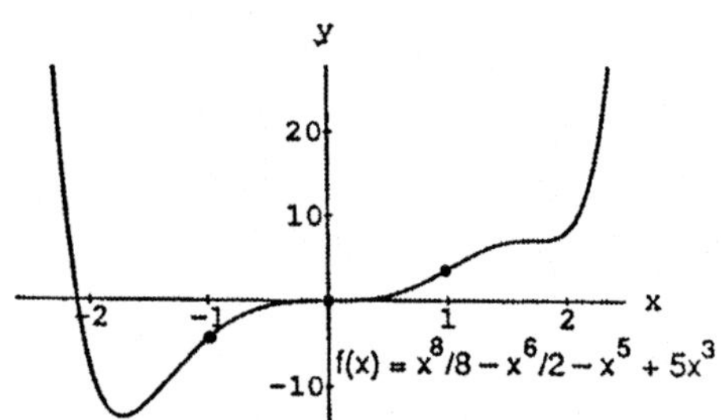

    (b) $f'(x) = x^7 - 3x^5 - 5x^4 + 15x^2 = x^2(x^2 - 3)(x^3 - 5)$. The pattern $y' = ---\underset{-\sqrt{3}}{|}\,+++\underset{0}{|}\,+++\underset{\sqrt[3]{5}}{|}\,---\underset{\sqrt{3}}{|}\,+++$ indicates a local maximum at $x = \sqrt[3]{5}$ and local minima at $x = \pm\sqrt{3}$.

    (c)

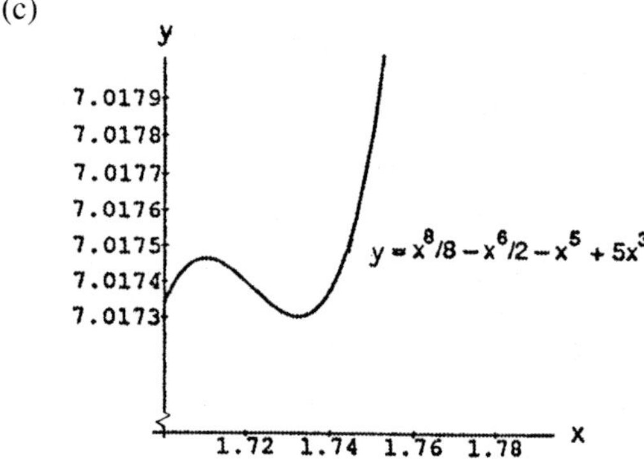

14. (a) The graph does not indicate any local extremum. Points of inflection are indicated at approximately $x = -\frac{3}{4}$ and $x = 1$.

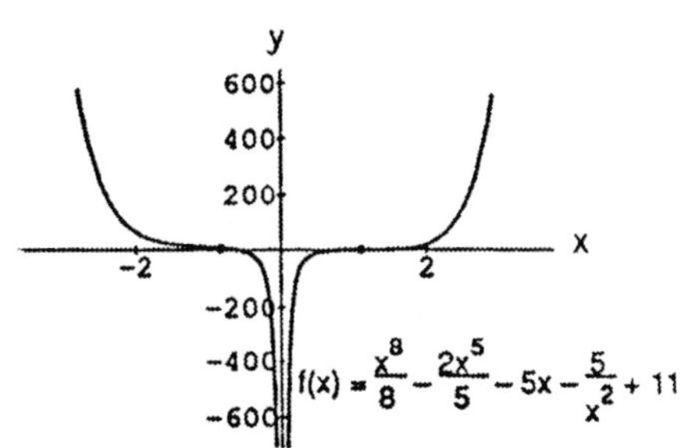

(b) $f'(x) = x^7 - 2x^4 - 5 + \frac{10}{x^3} = x^{-3}(x^3 - 2)(x^7 - 5)$. The pattern $f' = ---\underset{0}{)}(+++\underset{\sqrt[7]{5}}{|}---\underset{\sqrt[3]{2}}{|}+++$ indicates a local maximum at $x = \sqrt[7]{5}$ and a local minimum at $x = \sqrt[3]{2}$.

(c)

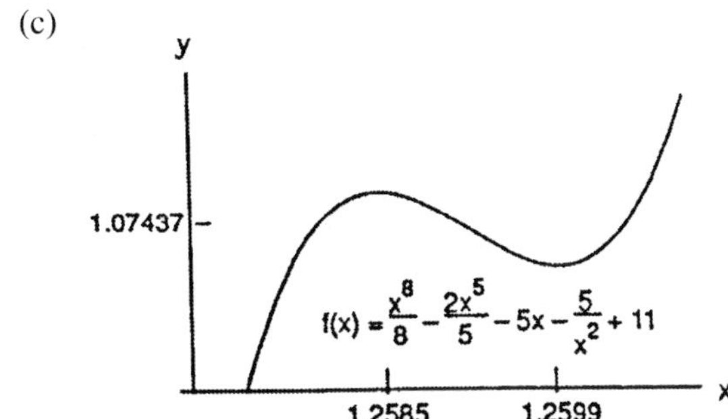

15. (a) $g(t) = \sin^2 t - 3t \Rightarrow g'(t) = 2 \sin t \cos t - 3 = \sin(2t) - 3 \Rightarrow g' < 0 \Rightarrow g(t)$ is always falling and hence must decrease on every interval in its domain.

(b) One, since $\sin^2 t - 3t - 5 = 0$ and $\sin^2 t - 3t = 5$ have the same solutions: $f(t) = \sin^2 t - 3t - 5$ has the same derivative as $g(t)$ in part (a) and is always decreasing with $f(-3) > 0$ and $f(0) < 0$. The Intermediate Value Theorem guarantees the continuous function f has a root in $[-3, 0]$.

16. (a) $y = \tan\theta \Rightarrow \frac{dy}{d\theta} = \sec^2\theta > 0 \Rightarrow y = \tan\theta$ is always rising on its domain $\Rightarrow y = \tan\theta$ increases on every interval in its domain

(b) The interval $\left[\frac{\pi}{4}, \pi\right]$ is not in the tangent's domain because $\tan\theta$ is undefined at $\theta = \frac{\pi}{2}$. Thus the tangent need not increase on this interval.

17. (a) $f(x) = x^4 + 2x^2 - 2 \Rightarrow f'(x) = 4x^3 + 4x$. Since $f(0) = -2 < 0$, $f(1) = 1 > 0$ and $f'(x) \geq 0$ for $0 \leq x \leq 1$, we may conclude from the Intermediate Value Theorem that $f(x)$ has exactly one solution when $0 \leq x \leq 1$.

(b) $x^2 = \frac{-2 \pm \sqrt{4+8}}{2} > 0 \Rightarrow x^2 = \sqrt{3} - 1$ and $x \geq 0 \Rightarrow x \approx \sqrt{.7320508076} \approx .8555996772$

18. (a) $y = \frac{x}{x+1} \Rightarrow y' = \frac{1}{(x+1)^2} > 0$, for all x in the domain of $\frac{x}{x+1} \Rightarrow y = \frac{x}{x+1}$ is increasing in every interval in its domain

(b) $y = x^3 + 2x \Rightarrow y' = 3x^2 + 2 > 0$ for all $x \Rightarrow$ the graph of $y = x^3 + 2x$ is always increasing and can never have a local maximum or minimum

19. Let $V(t)$ represent the volume of the water in the reservoir at time t, in minutes, let $V(0) = a_0$ be the initial amount and $V(1440) = a_0 + (1400)(43{,}560)(7.48)$ gallons be the amount of water contained in the reservoir after the rain, where 24 hr = 1440 min. Assume that $V(t)$ is continuous on $[0, 1440]$ and differentiable on $(0, 1440)$. The Mean Value Theorem says that for some t_0 in $(0, 1440)$ we have $V'(t_0) = \frac{V(1440) - V(0)}{1440 - 0}$ $= \frac{a_0 + (1400)(43{,}560)(7.48) - a_0}{1440} = \frac{456{,}160{,}320 \text{ gal}}{1440 \text{ min}} = 316{,}778$ gal/min. Therefore at t_0 the reservoir's volume was increasing at a rate in excess of 225,000 gal/min.

20. Yes, all differentiable functions g(x) having 3 as a derivative differ by only a constant. Consequently, the difference $3x - g(x)$ is a constant K because $g'(x) = 3 = \frac{d}{dx}(3x)$. Thus $g(x) = 3x + K$, the same form as F(x).

21. No, $\frac{x}{x+1} = 1 + \frac{-1}{x+1} \Rightarrow \frac{x}{x+1}$ differs from $\frac{-1}{x+1}$ by the constant 1. Both functions have the same derivative $\frac{d}{dx}\left(\frac{x}{x+1}\right) = \frac{(x+1) - x(1)}{(x+1)^2} = \frac{1}{(x+1)^2} = \frac{d}{dx}\left(\frac{-1}{x+1}\right)$.

22. $f'(x) = g'(x) = \frac{2x}{(x^2+1)^2} \Rightarrow f(x) - g(x) = C$ for some constant $C \Rightarrow$ the graphs differ by a vertical shift.

23. The global minimum value of $\frac{1}{2}$ occurs at $x = 2$.

24. (a) The function is increasing on the intervals $[-3, -2]$ and $[1, 2]$.
(b) The function is decreasing on the intervals $[-2, 0)$ and $(0, 1]$.
(c) The local maximum values occur only at $x = -2$, and at $x = 2$; local minimum values occur at $x = -3$ and at $x = 1$ provided f is continuous at $x = 0$.

25. (a) $t = 0, 6, 12$ (b) $t = 3, 9$ (c) $6 < t < 12$ (d) $0 < t < 6,\ 12 < t < 14$

26. (a) $t = 4$ (b) at no time (c) $0 < t < 4$ (d) $4 < t < 8$

27.

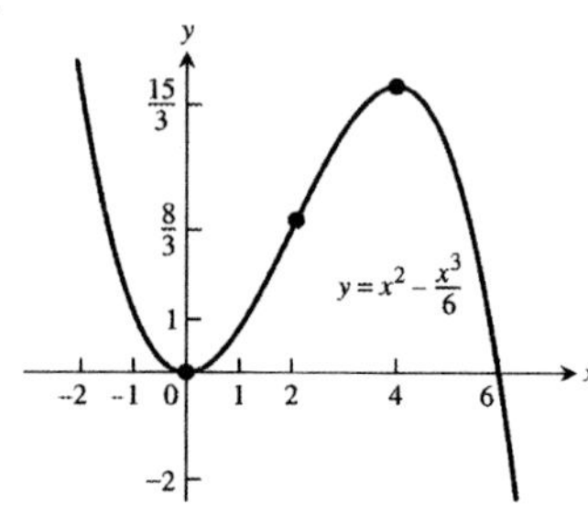

28.

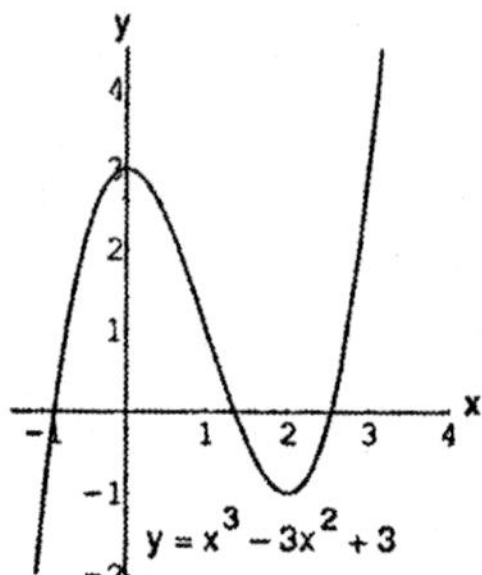

29.

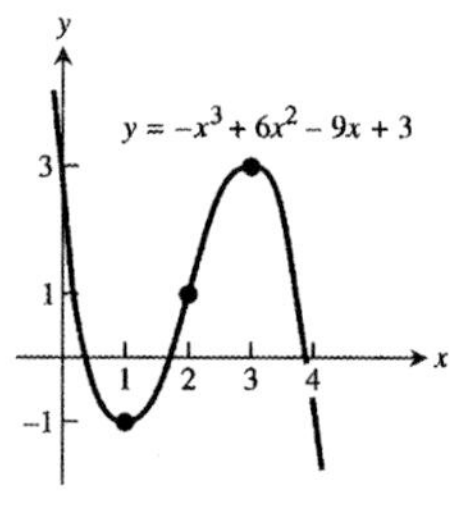

30.

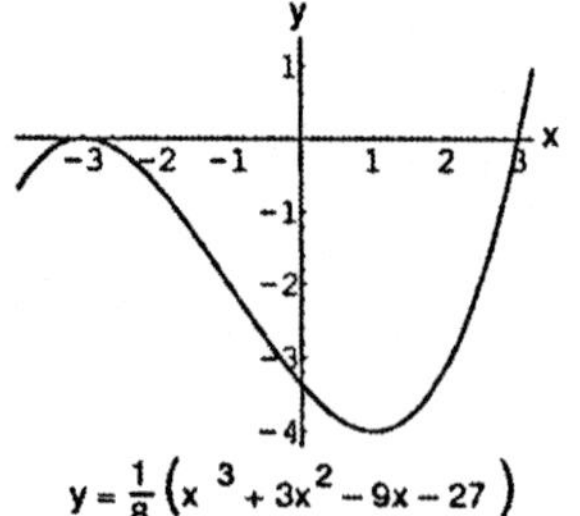

31.

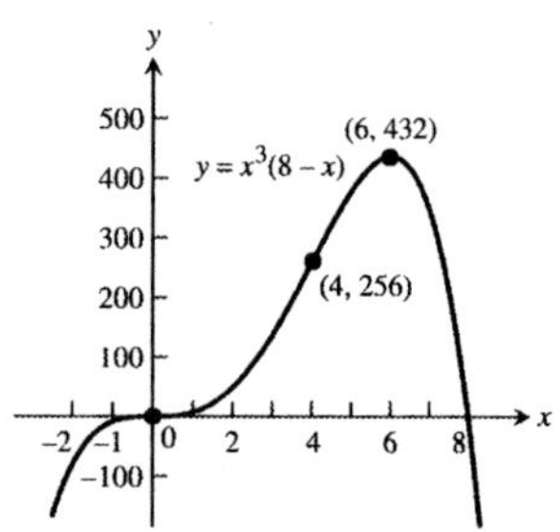

32.

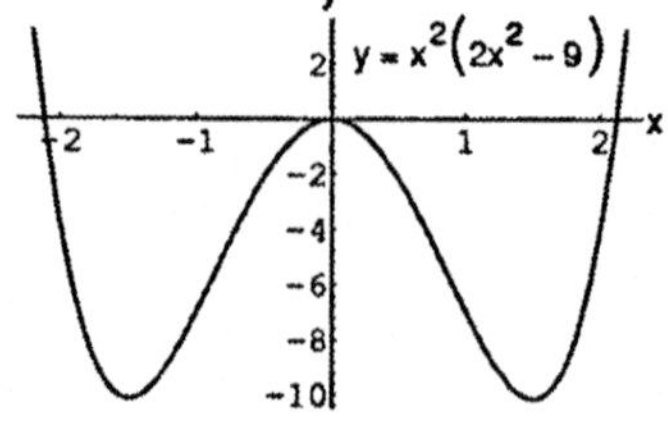

33.

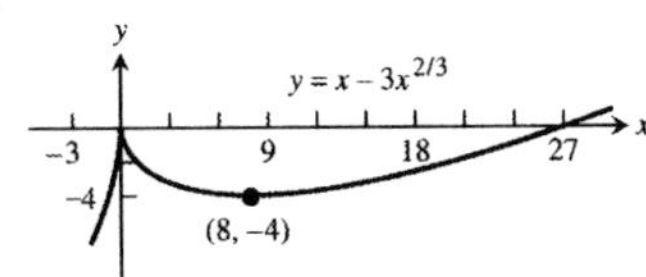

34.

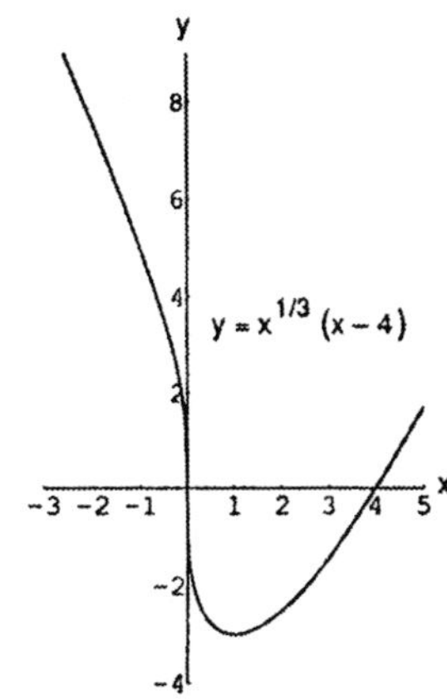

35.

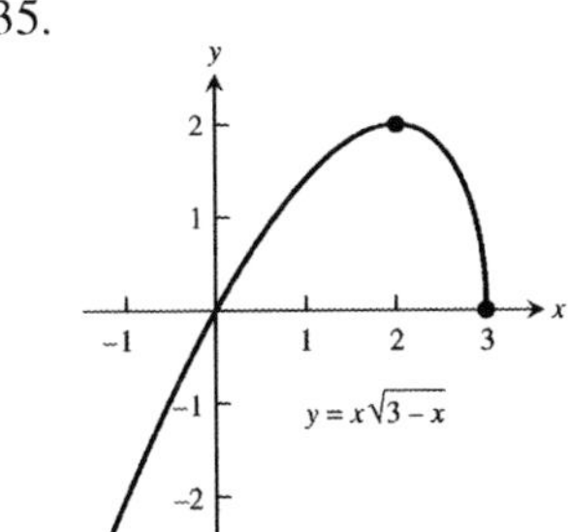

36.

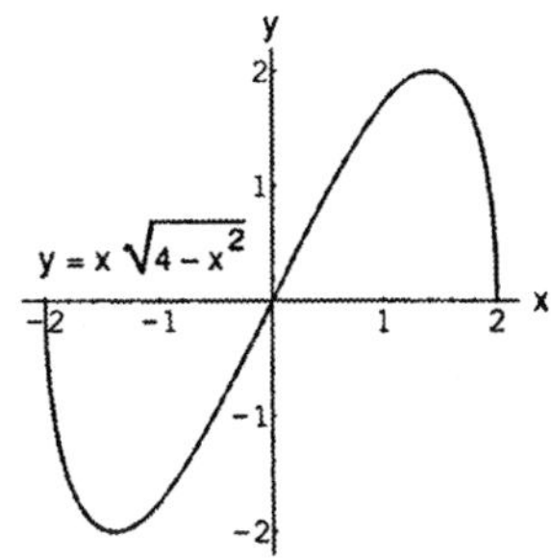

37.

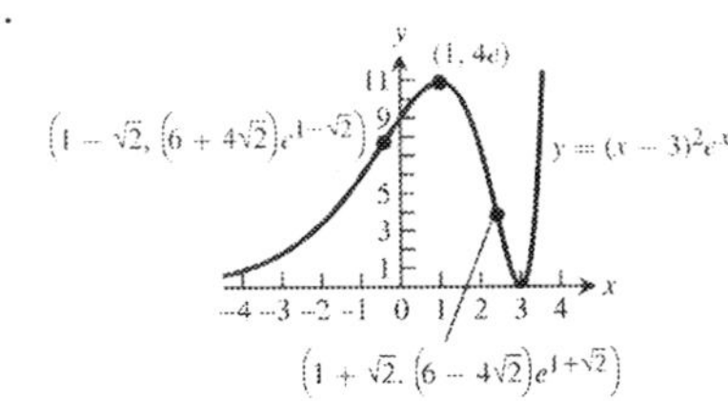

38.

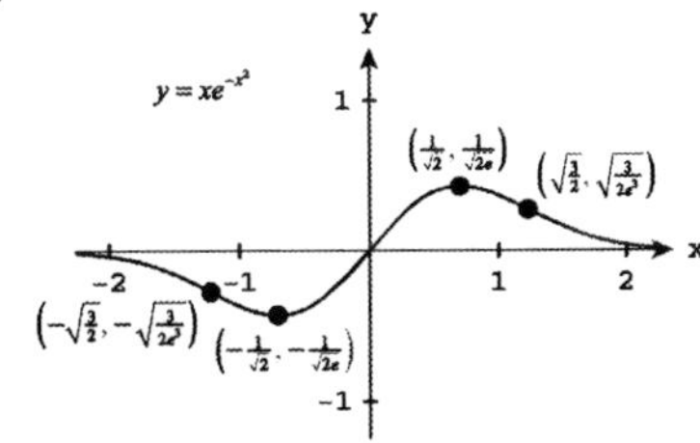

39.

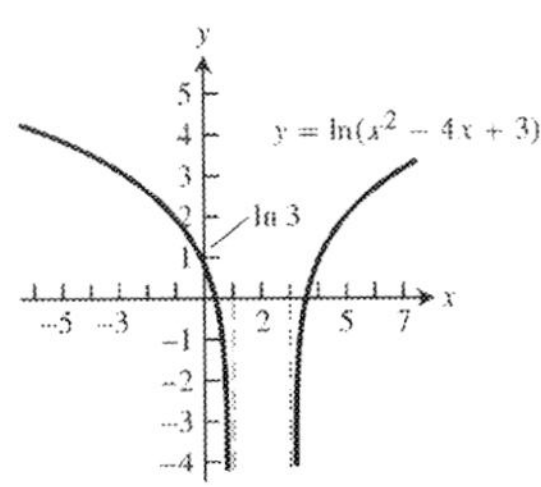

40.

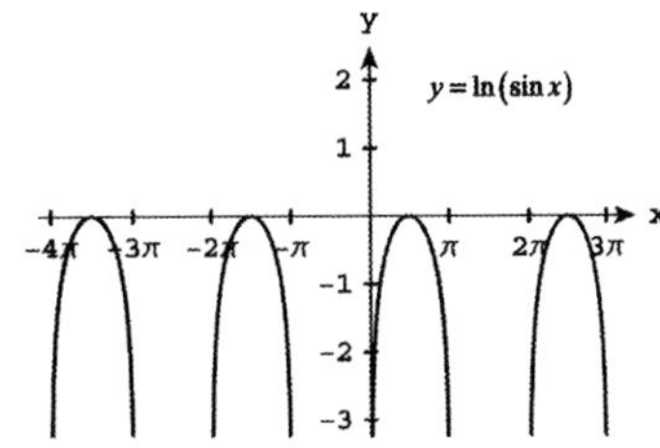

41.

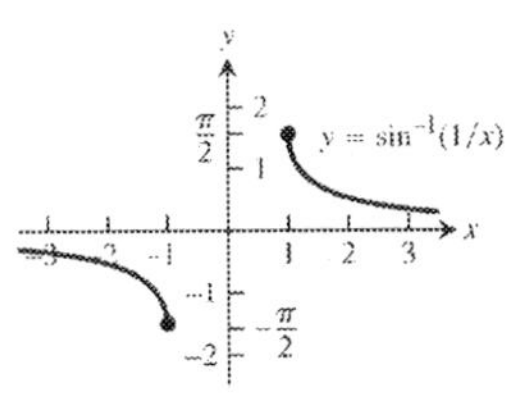

42.

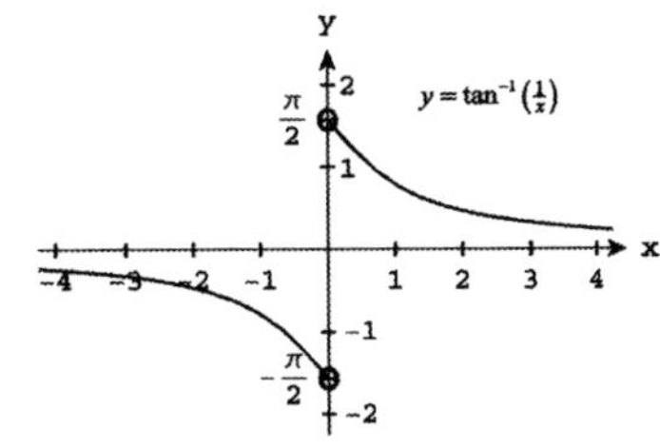

43. (a) $y' = 16 - x^2 \Rightarrow y' = \underset{-4}{---|}\underset{4}{+++|}---$ $\Rightarrow$ the curve is rising on $(-4, 4)$, falling on $(-\infty, -4)$ and $(4, \infty)$ $\Rightarrow$ a local maximum at $x = 4$ and a local minimum at $x = -4$; $y'' = -2x \Rightarrow y'' = \underset{0}{+++|}---$ $\Rightarrow$ the curve is concave up on $(-\infty, 0)$, concave down on $(0, \infty)$ $\Rightarrow$ a point of inflection at $x = 0$

(b)

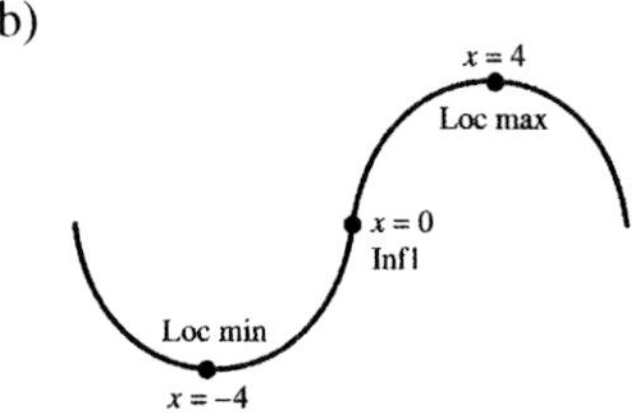

44. (a) $y' = x^2 - x - 6 = (x - 3)(x + 2) \Rightarrow y' = \underset{-2}{+++|}\underset{3}{---|}+++$ $\Rightarrow$ the curve is rising on $(-\infty, -2)$ and $(3, \infty)$, falling on $(-2, 3)$ $\Rightarrow$ local maximum at $x = -2$ and a local minimum at $x = 3$; $y'' = 2x - 1$ $\Rightarrow y'' = \underset{1/2}{---|}+++$ $\Rightarrow$ concave up on $\left(\frac{1}{2}, \infty\right)$, concave down on $\left(-\infty, \frac{1}{2}\right)$ $\Rightarrow$ a point of inflection at $x = \frac{1}{2}$

(b)

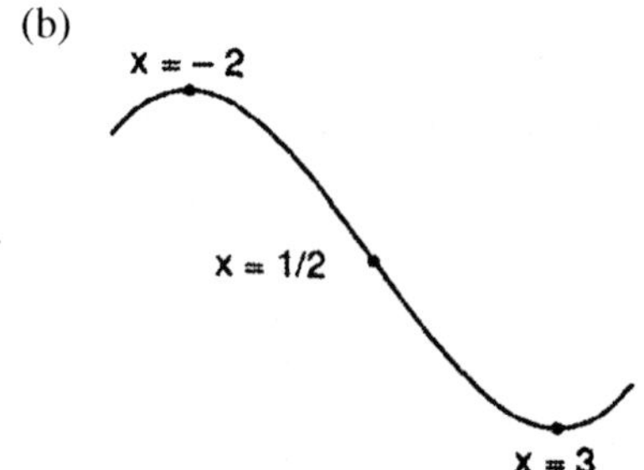

45. (a) $y' = 6x(x + 1)(x - 2) = 6x^3 - 6x^2 - 12x \Rightarrow y' = \underset{-1}{---|}\underset{0}{+++|}\underset{2}{---|}+++$ $\Rightarrow$ the graph is rising on $(-1, 0)$ and $(2, \infty)$, falling on $(-\infty, -1)$ and $(0, 2)$ $\Rightarrow$ a local maximum at $x = 0$, local minima at $x = -1$ and $x = 2$; $y'' = 18x^2 - 12x - 12 = 6(3x^2 - 2x - 2) = 6\left(x - \frac{1-\sqrt{7}}{3}\right)\left(x - \frac{1+\sqrt{7}}{3}\right) \Rightarrow$ $y'' = \underset{\frac{1-\sqrt{7}}{3}}{+++|}\underset{\frac{1+\sqrt{7}}{3}}{---|}+++$ $\Rightarrow$ the curve is concave up on $\left(-\infty, \frac{1-\sqrt{7}}{3}\right)$ and $\left(\frac{1+\sqrt{7}}{3}, \infty\right)$, concave down on $\left(\frac{1-\sqrt{7}}{3}, \frac{1+\sqrt{7}}{3}\right)$ $\Rightarrow$ points of inflection at $x = \frac{1 \pm \sqrt{7}}{3}$

(b)

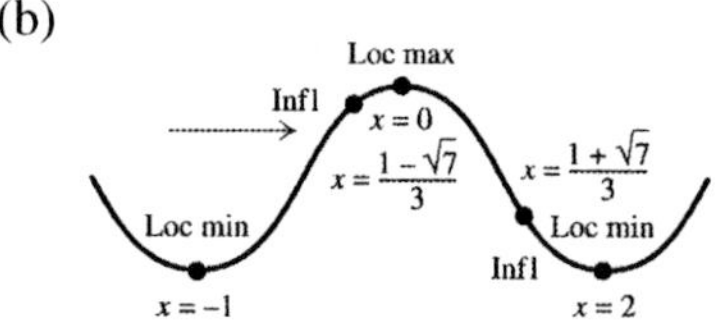

46. (a) $y' = x^2(6 - 4x) = 6x^2 - 4x^3 \Rightarrow y' = \underset{0}{+++|}\underset{3/2}{+++|}---$ $\Rightarrow$ the curve is rising on $\left(-\infty, \frac{3}{2}\right)$, falling on $\left(\frac{3}{2}, \infty\right)$ $\Rightarrow$ a local maximum at $x = \frac{3}{2}$; $y'' = 12x - 12x^2 = 12x(1 - x) \Rightarrow y'' = \underset{0}{---|}\underset{1}{+++|}---$ $\Rightarrow$ concave up on $(0, 1)$, concave down on $(-\infty, 0)$ and $(1, \infty)$ $\Rightarrow$ points of inflection at $x = 0$ and $x = 1$

(b)

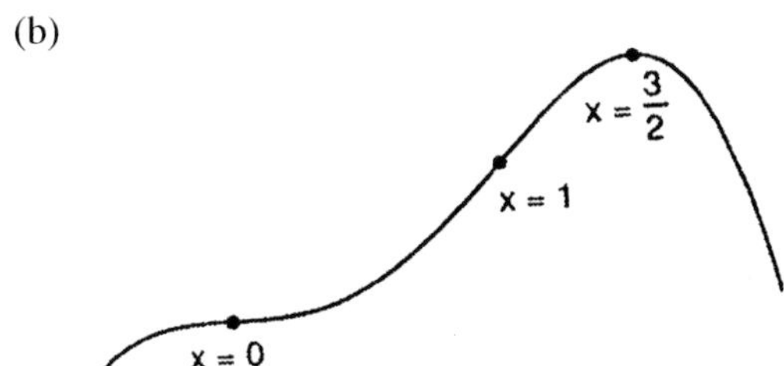

47. (a) $y' = x^4 - 2x^2 = x^2(x^2 - 2) \Rightarrow y' = \underset{}{+++} \underset{-\sqrt{2}}{|} --- \underset{0}{|} --- \underset{\sqrt{2}}{|} +++ \Rightarrow$ the curve is rising on $\left(-\infty, -\sqrt{2}\right)$ and $\left(\sqrt{2}, \infty\right)$, falling on $\left(-\sqrt{2}, \sqrt{2}\right) \Rightarrow$ a local maximum at $x = -\sqrt{2}$ and a local minimum at $x = \sqrt{2}$; $y'' = 4x^3 - 4x = 4x(x-1)(x+1) \Rightarrow y'' = --- \underset{-1}{|} +++ \underset{0}{|} --- \underset{1}{|} +++ \Rightarrow$ concave up on $(-1, 0)$ and $(1, \infty)$, concave down on $(-\infty, -1)$ and $(0, 1) \Rightarrow$ points of inflection at $x = 0$ and $x = \pm 1$

(b)

Loc max
$x = -\sqrt{2}$
Infl
$x = -1$
Infl
$x = 0$
Infl
$x = 1$
$x = \sqrt{2}$
Loc min

48. (a) $y' = 4x^2 - x^4 = x^2(4 - x^2) \Rightarrow y' = --- \underset{-2}{|} +++ \underset{0}{|} +++ \underset{2}{|} --- \Rightarrow$ the curve is rising on $(-2, 0)$ and $(0, 2)$, falling on $(-\infty, -2)$ and $(2, \infty) \Rightarrow$ a local maximum at $x = 2$, a local minimum at $x = -2$; $y'' = 8x - 4x^3 = 4x(2 - x^2) \Rightarrow y'' = +++ \underset{-\sqrt{2}}{|} --- \underset{0}{|} +++ \underset{\sqrt{2}}{|} --- \Rightarrow$ concave up on $\left(-\infty, -\sqrt{2}\right)$ and $\left(0, \sqrt{2}\right)$, concave down on $\left(-\sqrt{2}, 0\right)$ and $\left(\sqrt{2}, \infty\right) \Rightarrow$ points of inflection at $x = 0$ and $x = \pm\sqrt{2}$

(b)

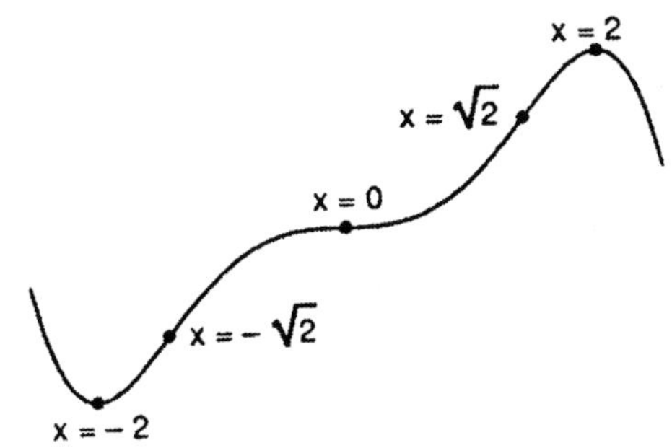

49. The values of the first derivative indicate that the curve is rising on $(0, \infty)$ and falling on $(-\infty, 0)$. The slope of the curve approaches $-\infty$ as $x \to 0^-$, and approaches ∞ as $x \to 0^+$ and $x \to 1$. The curve should therefore have a cusp and local minimum at $x = 0$, and a vertical tangent at $x = 1$.

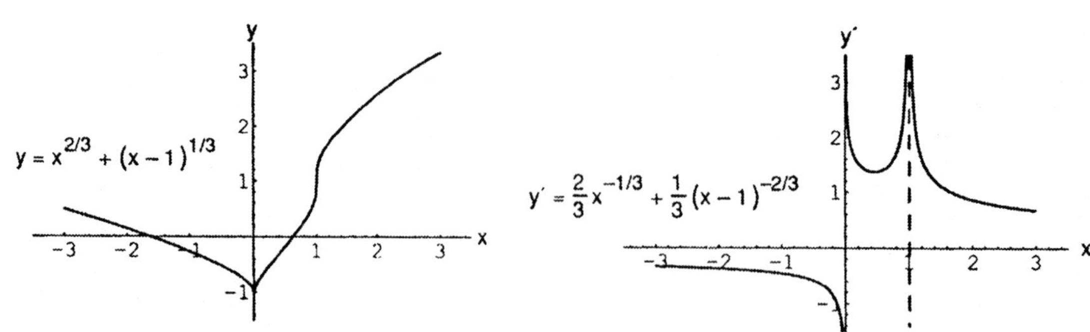

50. The values of the first derivative indicate that the curve is rising on $\left(0, \frac{1}{2}\right)$ and $(1, \infty)$, and falling on $(-\infty, 0)$ and $\left(\frac{1}{2}, 1\right)$. The derivative changes from positive to negative at $x = \frac{1}{2}$, indicating a local maximum there. The slope of the curve approaches $-\infty$ as $x \to 0^-$ and $x \to 1^-$, and approaches ∞ as $x \to 0^+$ and as $x \to 1^+$, indicating cusps and local minima at both $x = 0$ and $x = 1$.

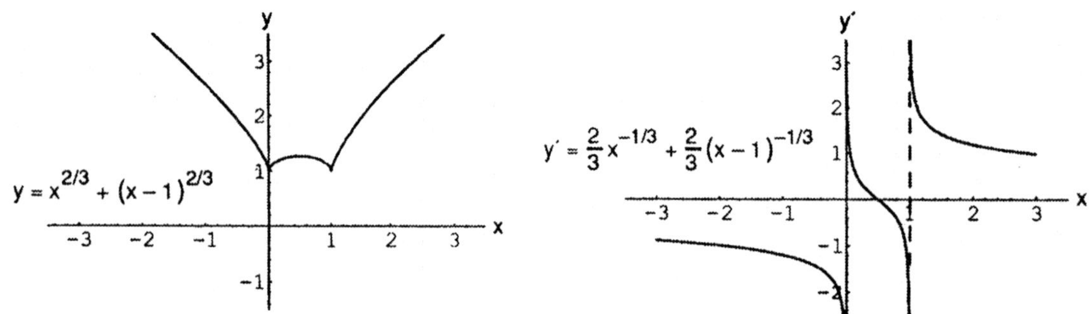

51. The values of the first derivative indicate that the curve is always rising. The slope of the curve approaches ∞ as $x \to 0$ and as $x \to 1$, indicating vertical tangents at both $x = 0$ and $x = 1$.

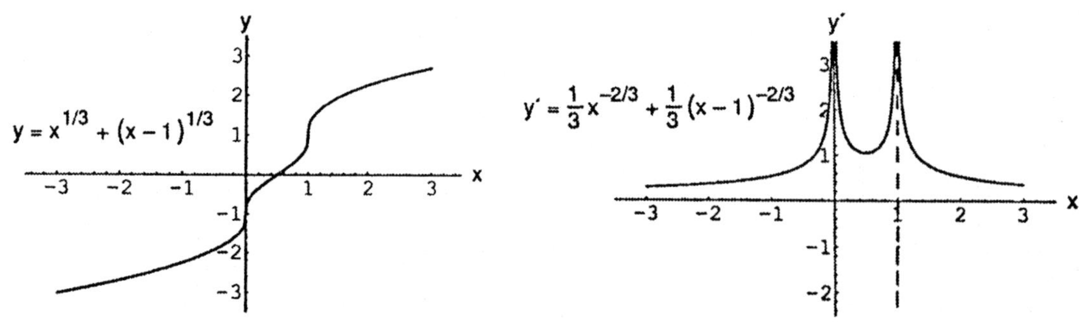

52. The graph of the first derivative indicates that the curve is rising on $\left(0, \frac{17-\sqrt{33}}{16}\right)$ and $\left(\frac{17+\sqrt{33}}{16}, \infty\right)$, falling on $(-\infty, 0)$ and $\left(\frac{17-\sqrt{33}}{16}, \frac{17+\sqrt{33}}{16}\right) \Rightarrow$ a local maximum at $x = \frac{17-\sqrt{33}}{16}$, a local minimum at $x = \frac{17+\sqrt{33}}{16}$. The derivative approaches $-\infty$ as $x \to 0^-$ and $x \to 1$, and approaches ∞ as $x \to 0^+$, indicating a cusp and local minimum at $x = 0$ and a vertical tangent at $x = 1$.

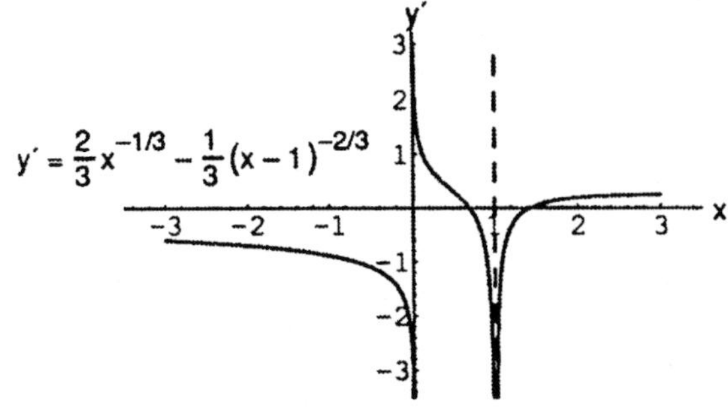

53. $y = \frac{x+1}{x-3} = 1 + \frac{4}{x-3}$

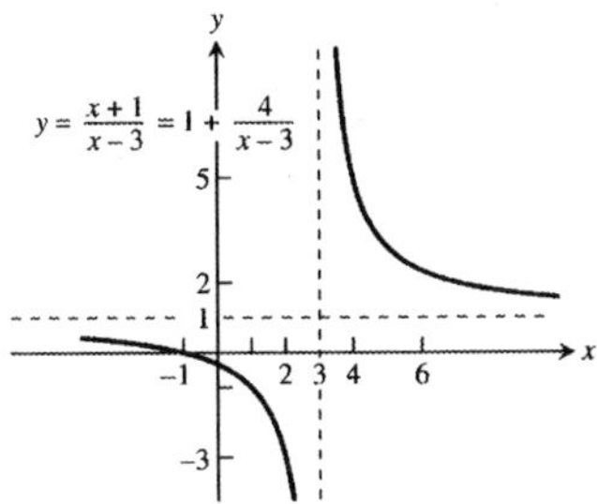

54. $y = \frac{2x}{x+5} = 2 - \frac{10}{x+5}$

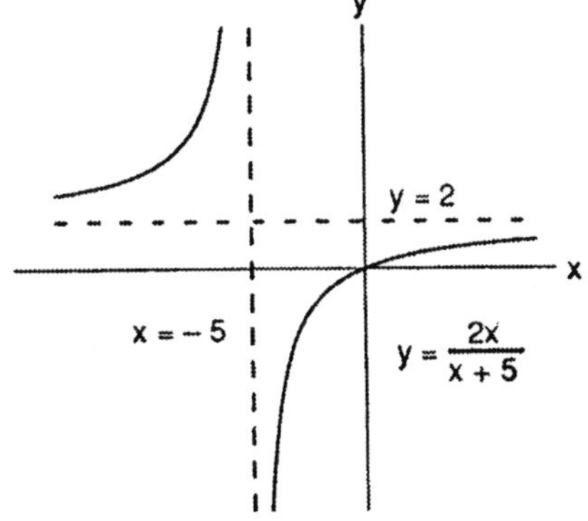

55. $y = \frac{x^2+1}{x} = x + \frac{1}{x}$

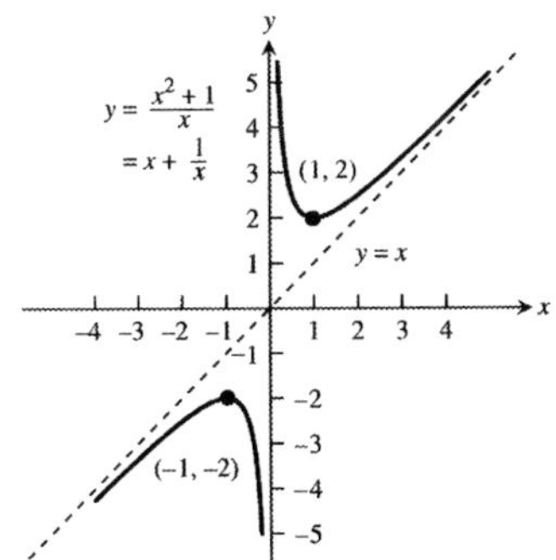

56. $y = \frac{x^2-x+1}{x} = x - 1 + \frac{1}{x}$

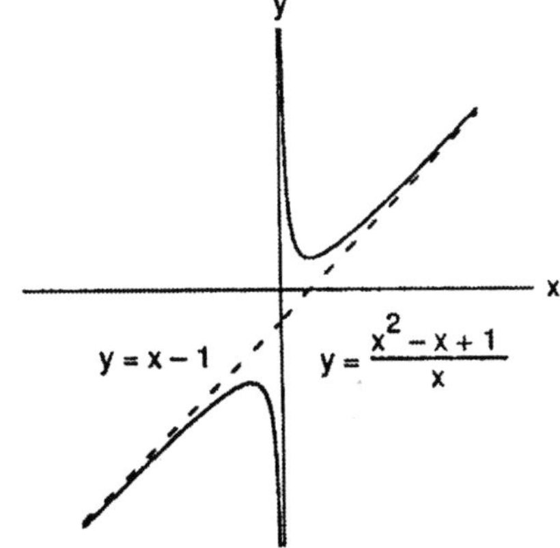

57. $y = \frac{x^3+2}{2x} = \frac{x^2}{2} + \frac{1}{x}$

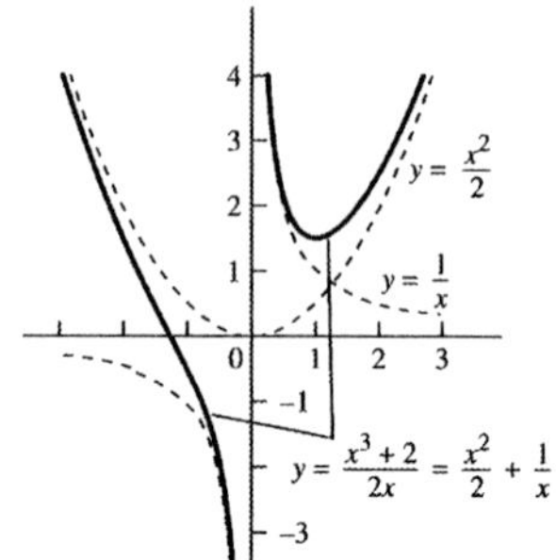

58. $y = \frac{x^4-1}{x^2} = x^2 - \frac{1}{x^2}$

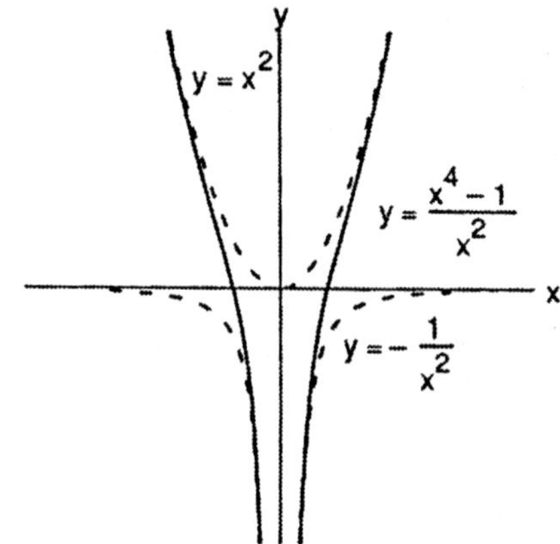

59. $y = \frac{x^2-4}{x^2-3} = 1 - \frac{1}{x^2-3}$

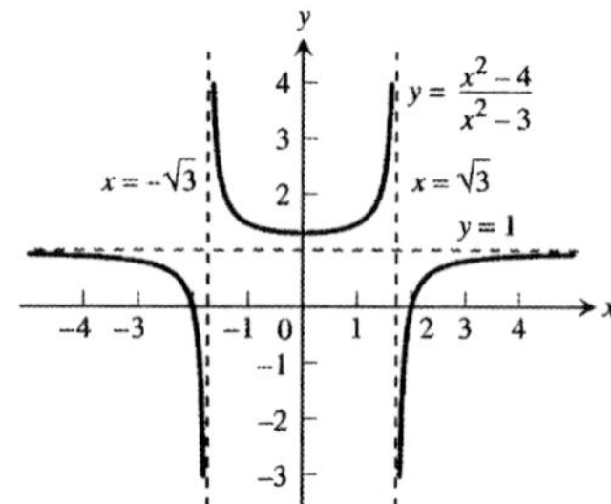

60. $y = \frac{x^2}{x^2-4} = 1 + \frac{4}{x^2-4}$

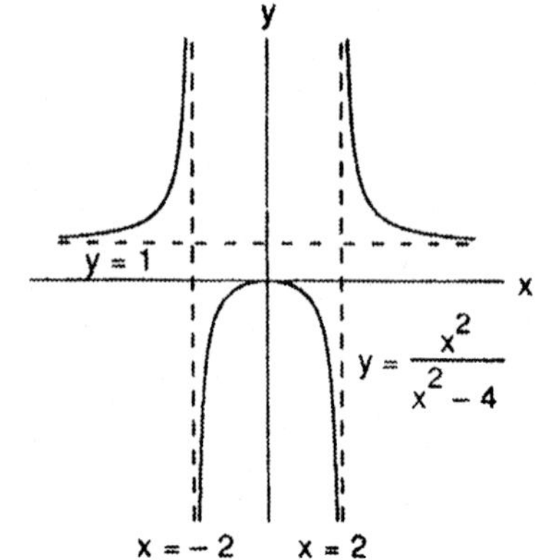

61. $\lim\limits_{x \to 1} \frac{x^2+3x-4}{x-1} = \lim\limits_{x \to 1} \frac{2x+3}{1} = 5$

62. $\lim\limits_{x \to 1} \frac{x^a-1}{x^b-1} = \lim\limits_{x \to 1} \frac{ax^{a-1}}{bx^{b-1}} = \frac{a}{b}$

63. $\lim\limits_{x \to \pi} \frac{\tan x}{x} = \frac{\tan \pi}{\pi} = 0$

64. $\lim\limits_{x \to 0} \frac{\tan x}{x + \sin x} = \lim\limits_{x \to 0} \frac{\sec^2 x}{1 + \cos x} = \frac{1}{1+1} = \frac{1}{2}$

65. $\lim\limits_{x \to 0} \frac{\sin^2 x}{\tan(x^2)} = \lim\limits_{x \to 0} \frac{2\sin x \cdot \cos x}{2x \sec^2(x^2)} = \lim\limits_{x \to 0} \frac{\sin(2x)}{2x \sec^2(x^2)} = \lim\limits_{x \to 0} \frac{2\cos(2x)}{2x\,(2\sec^2(x^2)\tan(x^2) \cdot 2x) + 2\sec^2(x^2)} = \frac{2}{0+2 \cdot 1} = 1$

66. $\lim\limits_{x \to 0} \frac{\sin(mx)}{\sin(nx)} = \lim\limits_{x \to 0} \frac{m\cos(mx)}{n\cos(nx)} = \frac{m}{n}$

67. $\lim\limits_{x \to \pi/2^-} \sec(7x)\cos(3x) = \lim\limits_{x \to \pi/2^-} \frac{\cos(3x)}{\cos(7x)} = \lim\limits_{x \to \pi/2^-} \frac{-3\sin(3x)}{-7\sin(7x)} = \frac{3}{7}$

68. $\lim\limits_{x \to 0^+} \sqrt{x} \sec x = \lim\limits_{x \to 0^+} \frac{\sqrt{x}}{\cos x} = \frac{0}{1} = 0$

69. $\lim\limits_{x \to 0} (\csc x - \cot x) = \lim\limits_{x \to 0} \frac{1 - \cos x}{\sin x} = \lim\limits_{x \to 0} \frac{\sin x}{\cos x} = \frac{0}{1} = 0$

70. $\lim_{x \to 0} \left(\frac{1}{x^4} - \frac{1}{x^2}\right) = \lim_{x \to 0} \left(\frac{1-x^2}{x^4}\right) = \lim_{x \to 0} (1-x^2) \cdot \frac{1}{x^4} = \lim_{x \to 0} (1-x^2) = \lim_{x \to 0} \frac{1}{x^4} = 1 \cdot \infty = \infty$

71. $\lim_{x \to \infty} \left(\sqrt{x^2+x+1} - \sqrt{x^2-x}\right) = \lim_{x \to \infty} \left(\sqrt{x^2+x+1} - \sqrt{x^2-x}\right) \cdot \frac{\sqrt{x^2+x+1}+\sqrt{x^2-x}}{\sqrt{x^2+x+1}+\sqrt{x^2-x}}$

$= \lim_{x \to \infty} \frac{2x+1}{\sqrt{x^2+x+1}+\sqrt{x^2-x}}$

Notice that $x = \sqrt{x^2}$ for $x > 0$ so this is equivalent to

$= \lim_{x \to \infty} \frac{\frac{2x+1}{x}}{\sqrt{\frac{x^2+x+1}{x^2}}+\sqrt{\frac{x^2-x}{x^2}}} = \lim_{x \to \infty} \frac{2+\frac{1}{x}}{\sqrt{1+\frac{1}{x}+\frac{1}{x^2}}+\sqrt{1-\frac{1}{x}}} = \frac{2}{\sqrt{1}+\sqrt{1}} = 1$

72. $\lim_{x \to \infty} \left(\frac{x^3}{x^2-1} - \frac{x^3}{x^2+1}\right) = \lim_{x \to \infty} \frac{x^3(x^2+1) - x^3(x^2-1)}{(x^2-1)(x^2+1)} = \lim_{x \to \infty} \frac{2x^3}{x^4-1} = \lim_{x \to \infty} \frac{6x^2}{4x^3} = \lim_{x \to \infty} \frac{12x}{12x^2}$

$= \lim_{x \to \infty} \frac{12}{24x} = \lim_{x \to \infty} \frac{1}{2x} = 0$

73. The limit leads to the indeterminate form $\frac{0}{0}$: $\lim_{x \to 0} \frac{10^x - 1}{x} = \lim_{x \to 0} \frac{(\ln 10)10^x}{1} = \ln 10$

74. The limit leads to the indeterminate form $\frac{0}{0}$: $\lim_{\theta \to 0} \frac{3^\theta - 1}{\theta} = \lim_{\theta \to 0} \frac{(\ln 3)3^\theta}{1} = \ln 3$

75. The limit leads to the indeterminate form $\frac{0}{0}$: $\lim_{x \to 0} \frac{2^{\sin x} - 1}{e^x - 1} = \lim_{x \to 0} \frac{2^{\sin x}(\ln 2)(\cos x)}{e^x} = \ln 2$

76. The limit leads to the indeterminate form $\frac{0}{0}$: $\lim_{x \to 0} \frac{2^{-\sin x} - 1}{e^x - 1} = \lim_{x \to 0} \frac{2^{-\sin x}(\ln 2)(-\cos x)}{e^x} = -\ln 2$

77. The limit leads to the indeterminate form $\frac{0}{0}$: $\lim_{x \to 0} \frac{5 - 5\cos x}{e^x - x - 1} = \lim_{x \to 0} \frac{5\sin x}{e^x - 1} = \lim_{x \to 0} \frac{5\cos x}{e^x} = 5$

78. The limit leads to the indeterminate form $\frac{0}{0}$: $\lim_{x \to 0} \frac{4 - 4e^x}{xe^x} = \lim_{x \to 0} \frac{-4e^x}{e^x + xe^x} = -4$

79. The limit leads to the indeterminate form $\frac{0}{0}$: $\lim_{t \to 0^+} \frac{t - \ln(1+2t)}{t^2} = \lim_{t \to 0^+} \frac{\left(1 - \frac{2}{1+2t}\right)}{2t} = -\infty$

80. The limit leads to the indeterminate form $\frac{0}{0}$: $\lim_{x \to 4} \frac{\sin^2(\pi x)}{e^{x-4}+3-x} = \lim_{x \to 4} \frac{2\pi(\sin \pi x)(\cos \pi x)}{e^{x-4}-1}$

$= \lim_{x \to 4} \frac{\pi \sin(2\pi x)}{e^{x-4}-1} = \lim_{x \to 4} \frac{2\pi^2 \cos(2\pi x)}{e^{x-4}} = 2\pi^2$

81. The limit leads to the indeterminate form $\frac{0}{0}$: $\lim_{t \to 0^+} \left(\frac{e^t}{t} - \frac{1}{t}\right) = \lim_{t \to 0^+} \left(\frac{e^t - 1}{t}\right) = \lim_{t \to 0^+} \frac{e^t}{1} = 1$

82. The limit leads to the indeterminate form $\frac{\infty}{\infty}$: $\lim_{y \to 0^+} e^{-1/y} \ln y = \lim_{y \to 0^+} \frac{\ln y}{e^{y^{-1}}} = \lim_{y \to 0^+} \frac{y^{-1}}{-e^{y^{-1}}(y^{-2})}$

$= -\lim_{y \to 0^+} \frac{y}{e^{y^{-1}}} = 0$

83. $\lim_{x \to \infty} \left(1 + \frac{b}{x}\right)^{kx} = \lim_{x \to \infty} \left[\left(1 + \frac{1}{x/b}\right)^{x/b}\right]^{bk} = e^{bk}$

84. $\lim_{x \to \infty} \left(1 + \frac{2}{x} + \frac{7}{x^2}\right) = 1 + 0 + 0 = 1$

85. (a) Maximize $f(x) = \sqrt{x} - \sqrt{36-x} = x^{1/2} - (36-x)^{1/2}$ where $0 \le x \le 36$

$\Rightarrow f'(x) = \frac{1}{2}x^{-1/2} - \frac{1}{2}(36-x)^{-1/2}(-1) = \frac{\sqrt{36-x}+\sqrt{x}}{2\sqrt{x}\sqrt{36-x}} \Rightarrow$ derivative fails to exist at 0 and 36; $f(0) = -6$, and $f(36) = 6 \Rightarrow$ the numbers are 0 and 36

(b) Maximize $g(x) = \sqrt{x} + \sqrt{36-x} = x^{1/2} + (36-x)^{1/2}$ where $0 \le x \le 36$

$\Rightarrow g'(x) = \frac{1}{2}x^{-1/2} + \frac{1}{2}(36-x)^{-1/2}(-1) = \frac{\sqrt{36-x}-\sqrt{x}}{2\sqrt{x}\sqrt{36-x}} \Rightarrow$ critical points at 0, 18 and 36; $g(0) = 6$, $g(18) = 2\sqrt{18} = 6\sqrt{2}$ and $g(36) = 6 \Rightarrow$ the numbers are 18 and 18

86. (a) Maximize $f(x) = \sqrt{x}(20-x) = 20x^{1/2} - x^{3/2}$ where $0 \le x \le 20 \Rightarrow f'(x) = 10x^{-1/2} - \frac{3}{2}x^{1/2}$ $= \frac{20-3x}{2\sqrt{x}} = 0 \Rightarrow x = 0$ and $x = \frac{20}{3}$ are critical points; $f(0) = f(20) = 0$ and $f\left(\frac{20}{3}\right) = \sqrt{\frac{20}{3}}\left(20 - \frac{20}{3}\right)$ $= \frac{40\sqrt{20}}{3\sqrt{3}} \Rightarrow$ the numbers are $\frac{20}{3}$ and $\frac{40}{3}$.

(b) Maximize $g(x) = x + \sqrt{20-x} = x + (20-x)^{1/2}$ where $0 \le x \le 20 \Rightarrow g'(x) = \frac{2\sqrt{20-x}-1}{2\sqrt{20-x}} = 0$ $\Rightarrow \sqrt{20-x} = \frac{1}{2} \Rightarrow x = \frac{79}{4}$. The critical points are $x = \frac{79}{4}$ and $x = 20$. Since $g\left(\frac{79}{4}\right) = \frac{81}{4}$ and $g(20) = 20$, the numbers must be $\frac{79}{4}$ and $\frac{1}{4}$.

87. $A(x) = \frac{1}{2}(2x)(27-x^2)$ for $0 \le x \le \sqrt{27}$ $\Rightarrow A'(x) = 3(3+x)(3-x)$ and $A''(x) = -6x$. The critical points are -3 and 3, but -3 is not in the domain. Since $A''(3) = -18 < 0$ and $A\left(\sqrt{27}\right) = 0$, the maximum occurs at $x = 3 \Rightarrow$ the largest area is $A(3) = 54$ sq units.

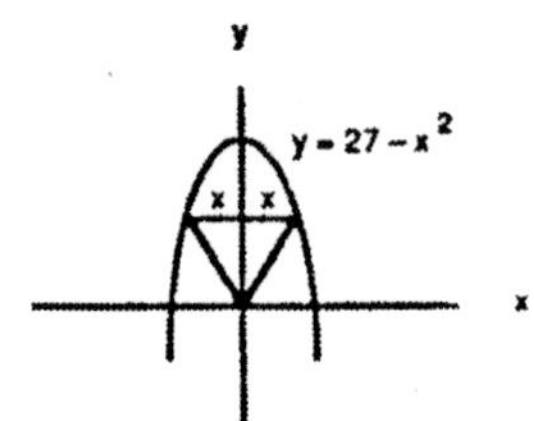

88. The volume is $V = x^2h = 32 \Rightarrow h = \frac{32}{x^2}$. The surface area is $S(x) = x^2 + 4x\left(\frac{32}{x^2}\right) = x^2 + \frac{128}{x}$, where $x > 0 \Rightarrow S'(x) = \frac{2(x-4)(x^2+4x+16)}{x^2}$ $\Rightarrow$ the critical points are 0 and 4, but 0 is not in the domain. Now $S''(4) = 2 + \frac{256}{4^3} > 0 \Rightarrow$ at $x = 4$ there is a minimum. The dimensions 4 ft by 4 ft by 2 ft minimize the surface area.

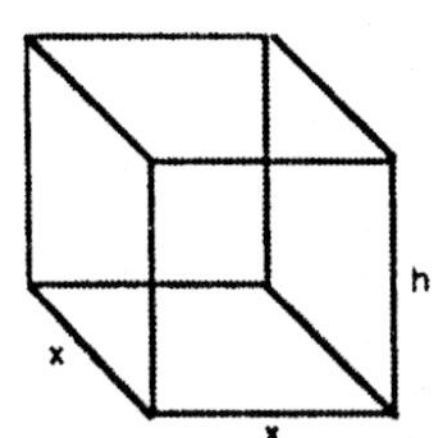

89. From the diagram we have $\left(\frac{h}{2}\right)^2 + r^2 = \left(\sqrt{3}\right)^2$ $\Rightarrow r^2 = \frac{12-h^2}{4}$. The volume of the cylinder is $V = \pi r^2 h = \pi\left(\frac{12-h^2}{4}\right)h = \frac{\pi}{4}(12h - h^3)$, where $0 \le h \le 2\sqrt{3}$. Then $V'(h) = \frac{3\pi}{4}(2+h)(2-h)$ $\Rightarrow$ the critical points are -2 and 2, but -2 is not in the domain. At $h = 2$ there is a maximum since $V''(2) = -3\pi < 0$. The dimensions of the largest cylinder are radius $= \sqrt{2}$ and height $= 2$.

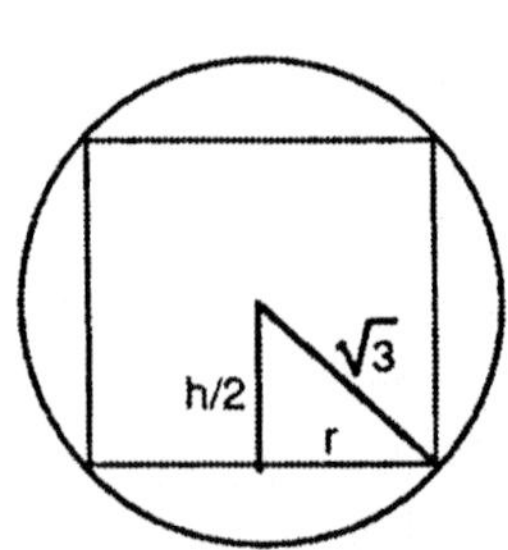

90. From the diagram we have x = radius and y = height = 12 − 2x and $V(x) = \frac{1}{3}\pi x^2(12 - 2x)$, where $0 \le x \le 6 \Rightarrow V'(x) = 2\pi x(4 - x)$ and $V''(4) = -8\pi$. The critical points are 0 and 4; $V(0) = V(6) = 0 \Rightarrow x = 4$ gives the maximum. Thus the values of r = 4 and h = 4 yield the largest volume for the smaller cone.

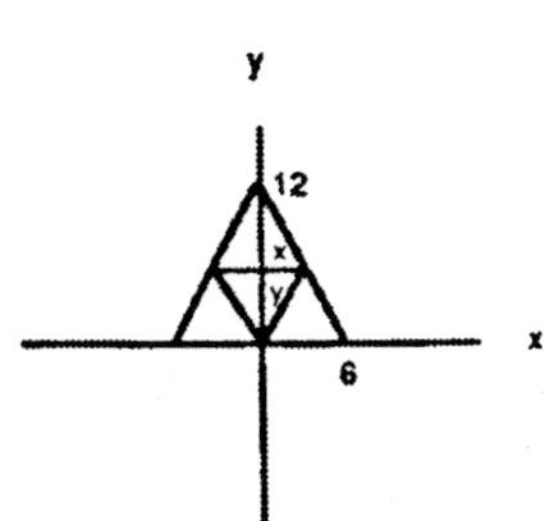

91. The profit $P = 2px + py = 2px + p\left(\frac{40 - 10x}{5 - x}\right)$, where p is the profit on grade B tires and $0 \le x \le 4$. Thus $P'(x) = \frac{2p}{(5 - x)^2}(x^2 - 10x + 20) \Rightarrow$ the critical points are $\left(5 - \sqrt{5}\right)$, 5, and $\left(5 + \sqrt{5}\right)$, but only $\left(5 - \sqrt{5}\right)$ is in the domain. Now $P'(x) > 0$ for $0 < x < \left(5 - \sqrt{5}\right)$ and $P'(x) < 0$ for $\left(5 - \sqrt{5}\right) < x < 4 \Rightarrow$ at $x = \left(5 - \sqrt{5}\right)$ there is a local maximum. Also $P(0) = 8p$, $P\left(5 - \sqrt{5}\right) = 4p\left(5 - \sqrt{5}\right) \approx 11p$, and $P(4) = 8p \Rightarrow$ at $x = \left(5 - \sqrt{5}\right)$ there is an absolute maximum. The maximum occurs when $x = \left(5 - \sqrt{5}\right)$ and $y = 2\left(5 - \sqrt{5}\right)$, the units are hundreds of tires, i.e., $x \approx 276$ tires and $y \approx 553$ tires.

92. (a) The distance between the particles is $|f(t)|$ where $f(t) = -\cos t + \cos\left(t + \frac{\pi}{4}\right)$. Then, $f'(t) = \sin t - \sin\left(t + \frac{\pi}{4}\right)$. Solving $f'(t) = 0$ graphically, we obtain $t \approx 1.178$, $t \approx 4.320$, and so on.

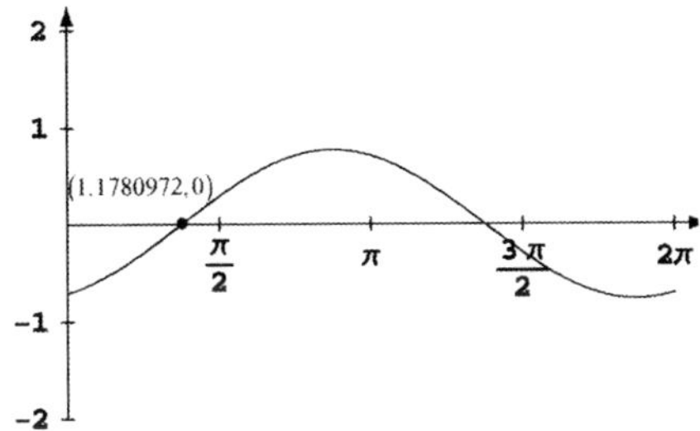

Alternatively, $f'(t) = 0$ may be solved analytically as follows. $f'(t) = \sin\left[\left(t + \frac{\pi}{8}\right) - \frac{\pi}{8}\right] - \sin\left[\left(t + \frac{\pi}{8}\right) + \frac{\pi}{8}\right]$
$= \left[\sin\left(t + \frac{\pi}{8}\right)\cos\frac{\pi}{8} - \cos\left(t + \frac{\pi}{8}\right)\sin\frac{\pi}{8}\right] - \left[\sin\left(t + \frac{\pi}{8}\right)\cos\frac{\pi}{8} + \cos\left(t + \frac{\pi}{8}\right)\sin\frac{\pi}{8}\right] = -2\sin\frac{\pi}{8}\cos\left(t + \frac{\pi}{8}\right)$
so the critical points occur when $\cos\left(t + \frac{\pi}{8}\right) = 0$, or $t = \frac{3\pi}{8} + k\pi$. At each of these values, $f(t) = \pm\cos\frac{3\pi}{8}$ $\approx \pm 0.765$ units, so the maximum distance between the particles is 0.765 units.

(b) Solving $\cos t = \cos\left(t + \frac{\pi}{4}\right)$ graphically, we obtain $t \approx 2.749$, $t \approx 5.890$, and so on.

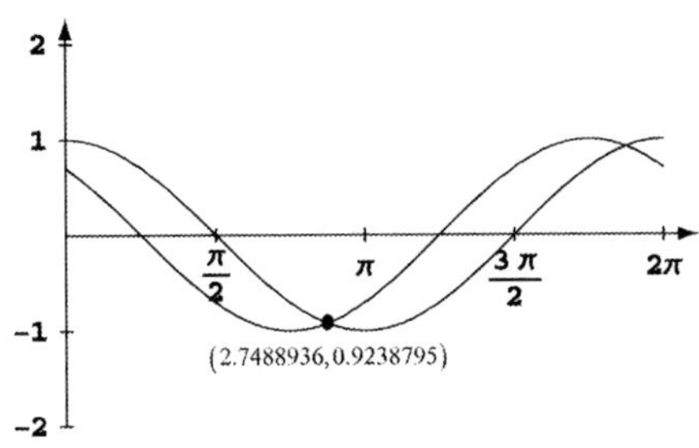

Alternatively, this problem can be solved analytically as follows.

$$\cos t = \cos\left(t + \tfrac{\pi}{4}\right)$$

$$\cos\left[\left(t + \tfrac{\pi}{8}\right) - \tfrac{\pi}{8}\right] = \cos\left[\left(t + \tfrac{\pi}{8}\right) + \tfrac{\pi}{8}\right]$$

$$\cos\left(t + \tfrac{\pi}{8}\right)\cos\tfrac{\pi}{8} + \sin\left(t + \tfrac{\pi}{8}\right)\sin\tfrac{\pi}{8} = \cos\left(t + \tfrac{\pi}{8}\right)\cos\tfrac{\pi}{8} - \sin\left(t + \tfrac{\pi}{8}\right)\sin\tfrac{\pi}{8}$$

$$2\sin\left(t + \tfrac{\pi}{8}\right)\sin\tfrac{\pi}{8} = 0$$

$$\sin\left(t + \tfrac{\pi}{8}\right) = 0$$

$$t = \frac{7\pi}{8} + k\pi$$

The particles collide when $t = \frac{7\pi}{8} \approx 2.749$. (plus multiples of π if they keep going.)

93. The dimensions will be x in. by $10 - 2x$ in. by $16 - 2x$ in., so $V(x) = x(10 - 2x)(16 - 2x) = 4x^3 - 52x^2 + 160x$ for $0 < x < 5$. Then $V'(x) = 12x^2 - 104x + 160 = 4(x - 2)(3x - 20)$, so the critical point in the correct domain is $x = 2$. This critical point corresponds to the maximum possible volume because $V'(x) > 0$ for $0 < x < 2$ and $V'(x) < 0$ for $2 < x < 5$. The box of largest volume has a height of 2 in. and a base measuring 6 in. by 12 in., and its volume is 144 in.^{3} Graphical support:

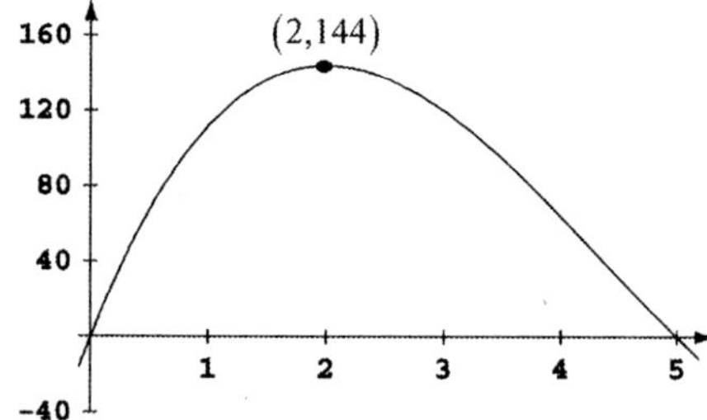

94. The length of the ladder is $d_1 + d_2 = 8 \sec\theta + 6 \csc\theta$. We wish to maximize $I(\theta) = 8 \sec\theta + 6 \csc\theta \Rightarrow I'(\theta) = 8 \sec\theta \tan\theta - 6 \csc\theta \cot\theta$. Then $I'(\theta) = 0 \Rightarrow 8 \sin^3\theta - 6 \cos^3\theta = 0 \Rightarrow \tan\theta = \frac{\sqrt[3]{6}}{2} \Rightarrow$ $d_1 = 4\sqrt{4 + \sqrt[3]{36}}$ and $d_2 = \sqrt[3]{36}\sqrt{4 + \sqrt[3]{36}}$ $\Rightarrow$ the length of the ladder is about $\left(4 + \sqrt[3]{36}\right)\sqrt{4 + \sqrt[3]{36}} = \left(4 + \sqrt[3]{36}\right)^{3/2} \approx 19.7$ ft.

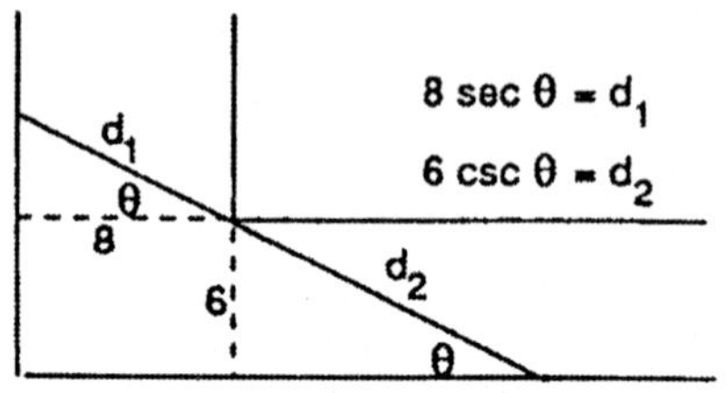

95. $g(x) = 3x - x^3 + 4 \Rightarrow g(2) = 2 > 0$ and $g(3) = -14 < 0 \Rightarrow g(x) = 0$ in the interval $[2, 3]$ by the Intermediate Value Theorem. Then $g'(x) = 3 - 3x^2 \Rightarrow x_{n+1} = x_n - \frac{3x_n - x_n^3 + 4}{3 - 3x_n^2}$; $x_0 = 2 \Rightarrow x_1 = 2.\overline{22} \Rightarrow x_2 = 2.196215$, and so forth to $x_5 = 2.195823345$.

96. $g(x) = x^4 - x^3 - 75 \Rightarrow g(3) = -21 < 0$ and $g(4) = 117 > 0 \Rightarrow g(x) = 0$ in the interval $[3, 4]$ by the Intermediate Value Theorem. Then $g'(x) = 4x^3 - 3x^2 \Rightarrow x_{n+1} = x_n - \frac{x_n^4 - x_n^3 - 75}{4x_n^3 - 3x_n^2}$; $x_0 = 3 \Rightarrow x_1 = 3.259259$ $\Rightarrow x_2 = 3.229050$, and so forth to $x_5 = 3.22857729$.

97. $\int (x^3 + 5x - 7)\, dx = \frac{x^4}{4} + \frac{5x^2}{2} - 7x + C$

98. $\int \left(8t^3 - \frac{t^2}{2} + t\right) dt = \frac{8t^4}{4} - \frac{t^3}{6} + \frac{t^2}{2} + C = 2t^4 - \frac{t^3}{6} + \frac{t^2}{2} + C$

99. $\int \left(3\sqrt{t} + \frac{4}{t^2}\right) dt = \int \left(3t^{1/2} + 4t^{-2}\right) dt = \frac{3t^{3/2}}{\left(\frac{3}{2}\right)} + \frac{4t^{-1}}{-1} + C = 2t^{3/2} - \frac{4}{t} + C$

100. $\int \left(\frac{1}{2\sqrt{t}} - \frac{3}{t^4}\right) dt = \int \left(\frac{1}{2}t^{-1/2} - 3t^{-4}\right) dt = \frac{1}{2}\left(\frac{t^{1/2}}{\frac{1}{2}}\right) - \frac{3t^{-3}}{(-3)} + C = \sqrt{t} + \frac{1}{t^3} + C$

101. Let $u = r + 5 \Rightarrow du = dr$

$\int \frac{dr}{(r+5)^2} = \int \frac{du}{u^2} = \int u^{-2}\,du = \frac{u^{-1}}{-1} + C = -u^{-1} + C = -\frac{1}{(r+5)} + C$

102. Let $u = r - \sqrt{2} \Rightarrow du = dr$

$\int \frac{6\,dr}{\left(r-\sqrt{2}\right)^3} = 6\int \frac{dr}{\left(r-\sqrt{2}\right)^3} = 6\int \frac{du}{u^3} = 6\int u^{-3}\,du = 6\left(\frac{u^{-2}}{-2}\right) + C = -3u^{-2} + C = -\frac{3}{\left(r-\sqrt{2}\right)^2} + C$

103. Let $u = \theta^2 + 1 \Rightarrow du = 2\theta\,d\theta \Rightarrow \frac{1}{2}\,du = \theta\,d\theta$

$\int 3\theta\sqrt{\theta^2+1}\,d\theta = \int \sqrt{u}\left(\frac{3}{2}\,du\right) = \frac{3}{2}\int u^{1/2}\,du = \frac{3}{2}\left(\frac{u^{3/2}}{\frac{3}{2}}\right) + C = u^{3/2} + C = \left(\theta^2+1\right)^{3/2} + C$

104. Let $u = 7 + \theta^2 \Rightarrow du = 2\theta\,d\theta \Rightarrow \frac{1}{2}\,du = \theta\,d\theta$

$\int \frac{\theta}{\sqrt{7+\theta^2}}\,d\theta = \int \frac{1}{\sqrt{u}}\left(\frac{1}{2}\,du\right) = \frac{1}{2}\int u^{-1/2}\,du = \frac{1}{2}\left(\frac{u^{1/2}}{\frac{1}{2}}\right) + C = u^{1/2} + C = \sqrt{7+\theta^2} + C$

105. Let $u = 1 + x^4 \Rightarrow du = 4x^3\,dx \Rightarrow \frac{1}{4}\,du = x^3\,dx$

$\int x^3\left(1+x^4\right)^{-1/4}\,dx = \int u^{-1/4}\left(\frac{1}{4}\,du\right) = \frac{1}{4}\int u^{-1/4}\,du = \frac{1}{4}\left(\frac{u^{3/4}}{\frac{3}{4}}\right) + C = \frac{1}{3}u^{3/4} + C = \frac{1}{3}\left(1+x^4\right)^{3/4} + C$

106. Let $u = 2 - x \Rightarrow du = -\,dx \Rightarrow -\,du = dx$

$\int (2-x)^{3/5}\,dx = \int u^{3/5}(-\,du) = -\int u^{3/5}\,du = -\frac{u^{8/5}}{\left(\frac{8}{5}\right)} + C = -\frac{5}{8}u^{8/5} + C = -\frac{5}{8}(2-x)^{8/5} + C$

107. Let $u = \frac{s}{10} \Rightarrow du = \frac{1}{10}\,ds \Rightarrow 10\,du = ds$

$\int \sec^2\frac{s}{10}\,ds = \int \left(\sec^2 u\right)(10\,du) = 10\int \sec^2 u\,du = 10\tan u + C = 10\tan\frac{s}{10} + C$

108. Let $u = \pi s \Rightarrow du = \pi\,ds \Rightarrow \frac{1}{\pi}\,du = ds$

$\int \csc^2 \pi s\,ds = \int \left(\csc^2 u\right)\left(\frac{1}{\pi}\,du\right) = \frac{1}{\pi}\int \csc^2 u\,du = -\frac{1}{\pi}\cot u + C = -\frac{1}{\pi}\cot \pi s + C$

109. Let $u = \sqrt{2}\,\theta \Rightarrow du = \sqrt{2}\,d\theta \Rightarrow \frac{1}{\sqrt{2}}\,du = d\theta$

$\int \csc\sqrt{2}\theta\cot\sqrt{2}\theta\,d\theta = \int (\csc u\cot u)\left(\frac{1}{\sqrt{2}}\,du\right) = \frac{1}{\sqrt{2}}(-\csc u) + C = -\frac{1}{\sqrt{2}}\csc\sqrt{2}\theta + C$

110. Let $u = \frac{\theta}{3} \Rightarrow du = \frac{1}{3}\,d\theta \Rightarrow 3\,du = d\theta$

$\int \sec\frac{\theta}{3}\tan\frac{\theta}{3}\,d\theta = \int (\sec u\tan u)(3\,du) = 3\sec u + C = 3\sec\frac{\theta}{3} + C$

111. Let $u = \frac{x}{4} \Rightarrow du = \frac{1}{4}\,dx \Rightarrow 4\,du = dx$

$\int \sin^2\frac{x}{4}\,dx = \int \left(\sin^2 u\right)(4\,du) = \int 4\left(\frac{1-\cos 2u}{2}\right)du = 2\int (1-\cos 2u)\,du = 2\left(u - \frac{\sin 2u}{2}\right) + C$

$= 2u - \sin 2u + C = 2\left(\frac{x}{4}\right) - \sin 2\left(\frac{x}{4}\right) + C = \frac{x}{2} - \sin\frac{x}{2} + C$

112. Let $u = \frac{x}{2} \Rightarrow du = \frac{1}{2}\,dx \Rightarrow 2\,du = dx$

$\int \cos^2\frac{x}{2}\,dx = \int \left(\cos^2 u\right)(2\,du) = \int 2\left(\frac{1+\cos 2u}{2}\right)du = \int (1+\cos 2u)\,du = u + \frac{\sin 2u}{2} + C$

$= \frac{x}{2} + \frac{1}{2}\sin x + C$

113. $\int \left(\frac{3}{x} - x\right)dx = 3\ln|x| - \frac{x^2}{2} + C$

114. $\int \left(\frac{5}{x^2} + \frac{2}{x^2+1}\right)dx = \int \left(5x^{-2} + \frac{2}{x^2+1}\right)dx = -5x^{-1} + 2\tan^{-1}x + C$

115. $\int \left(\frac{1}{2}e^t - e^{-t}\right)dt = \frac{1}{2}e^t - \frac{e^{-t}}{-1} + C = \frac{1}{2}e^t + e^{-t} + C$

116. $\int (5^s + s^5)ds = \frac{5^s}{\ln 5} + \frac{s^6}{6} + C$

117. $\int (\theta^{1-\pi})d\theta = \frac{\theta^{2-\pi}}{2-\pi} + C$

118. $\int (2^{\pi+r})dr = \frac{2^{\pi+r}}{\ln 2} + C$

119. $\int \frac{3}{2x\sqrt{x^2-1}}dx = \frac{3}{2}\int \frac{1}{x\sqrt{x^2-1}}dx = \frac{3}{2}\sec^{-1}x + C$

120. $\int \frac{d\theta}{\sqrt{16-\theta^2}} = \int \frac{d\theta}{\sqrt{16\left(1-\frac{\theta^2}{16}\right)}} = \int \frac{\frac{1}{4}d\theta}{\sqrt{1-\frac{\theta^2}{16}}} = \sin^{-1}\left(\frac{\theta}{4}\right) + C$

121. $y = \int \frac{x^2+1}{x^2}\,dx = \int (1 + x^{-2})\,dx = x - x^{-1} + C = x - \frac{1}{x} + C$; $y = -1$ when $x = 1 \Rightarrow 1 - \frac{1}{1} + C = -1$ $\Rightarrow C = -1 \Rightarrow y = x - \frac{1}{x} - 1$

122. $y = \int \left(x + \frac{1}{x}\right)^2 dx = \int \left(x^2 + 2 + \frac{1}{x^2}\right) dx = \int (x^2 + 2 + x^{-2})\,dx = \frac{x^3}{3} + 2x - x^{-1} + C = \frac{x^3}{3} + 2x - \frac{1}{x} + C$; $y = 1$ when $x = 1 \Rightarrow \frac{1}{3} + 2 - \frac{1}{1} + C = 1 \Rightarrow C = -\frac{1}{3} \Rightarrow y = \frac{x^3}{3} + 2x - \frac{1}{x} - \frac{1}{3}$

123. $\frac{dr}{dt} = \int \left(15\sqrt{t} + \frac{3}{\sqrt{t}}\right) dt = \int \left(15t^{1/2} + 3t^{-1/2}\right) dt = 10t^{3/2} + 6t^{1/2} + C$; $\frac{dr}{dt} = 8$ when $t = 1$ $\Rightarrow 10(1)^{3/2} + 6(1)^{1/2} + C = 8 \Rightarrow C = -8$. Thus $\frac{dr}{dt} = 10t^{3/2} + 6t^{1/2} - 8 \Rightarrow r = \int \left(10t^{3/2} + 6t^{1/2} - 8\right) dt$ $= 4t^{5/2} + 4t^{3/2} - 8t + C$; $r = 0$ when $t = 1 \Rightarrow 4(1)^{5/2} + 4(1)^{3/2} - 8(1) + C_1 = 0 \Rightarrow C_1 = 0$. Therefore, $r = 4t^{5/2} + 4t^{3/2} - 8t$

124. $\frac{d^2r}{dt^2} = \int -\cos t\,dt = -\sin t + C$; $r'' = 0$ when $t = 0 \Rightarrow -\sin 0 + C = 0 \Rightarrow C = 0$. Thus, $\frac{d^2r}{dt^2} = -\sin t$ $\Rightarrow \frac{dr}{dt} = \int -\sin t\,dt = \cos t + C_1$; $r' = 0$ when $t = 0 \Rightarrow 1 + C_1 = 0 \Rightarrow C_1 = -1$. Then $\frac{dr}{dt} = \cos t - 1$ $\Rightarrow r = \int (\cos t - 1)\,dt = \sin t - t + C_2$; $r = -1$ when $t = 0 \Rightarrow 0 - 0 + C_2 = -1 \Rightarrow C_2 = -1$. Therefore, $r = \sin t - t - 1$

125. Yes, $\sin^{-1} x$ and $-\cos^{-1} x$ differ by the constant $\frac{\pi}{2}$

126. Yes, the derivatives of $y = -\cos^{-1} x + C$ and $y = \cos^{-1}(-x) + C$ are both $\frac{1}{\sqrt{1-x^2}}$

127. $A = xy = xe^{-x^2} \Rightarrow \frac{dA}{dx} = e^{-x^2} + (x)(-2x)e^{-x^2} = e^{-x^2}\left(1 - 2x^2\right)$. Solving $\frac{dA}{dx} = 0 \Rightarrow 1 - 2x^2 = 0$ $\Rightarrow x = \frac{1}{\sqrt{2}}$; $\frac{dA}{dx} < 0$ for $x > \frac{1}{\sqrt{2}}$ and $\frac{dA}{dx} > 0$ for $0 < x < \frac{1}{\sqrt{2}} \Rightarrow$ absolute maximum of $\frac{1}{\sqrt{2}}e^{-1/2} = \frac{1}{\sqrt{2e}}$ at $x = \frac{1}{\sqrt{2}}$ units long by $y = e^{-1/2} = \frac{1}{\sqrt{e}}$ units high.

128. $A = xy = x\left(\frac{\ln x}{x^2}\right) = \frac{\ln x}{x} \Rightarrow \frac{dA}{dx} = \frac{1}{x^2} - \frac{\ln x}{x^2} = \frac{1-\ln x}{x^2}$. Solving $\frac{dA}{dx} = 0 \Rightarrow 1 - \ln x = 0 \Rightarrow x = e$; $\frac{dA}{dx} < 0$ for $x > e$ and $\frac{dA}{dx} > 0$ for $x < e \Rightarrow$ absolute maximum of $\frac{\ln e}{e} = \frac{1}{e}$ at $x = e$ units long and $y = \frac{1}{e^2}$ units high.

129. $y = x \ln 2x - x \Rightarrow y' = x\left(\frac{2}{2x}\right) + \ln(2x) - 1 = \ln 2x$; solving $y' = 0 \Rightarrow x = \frac{1}{2}$; $y' > 0$ for $x > \frac{1}{2}$ and $y' < 0$ for $x < \frac{1}{2} \Rightarrow$ relative minimum of $-\frac{1}{2}$ at $x = \frac{1}{2}$; $f\left(\frac{1}{2e}\right) = -\frac{1}{e}$ and $f\left(\frac{e}{2}\right) = 0 \Rightarrow$ absolute minimum is $-\frac{1}{2}$ at $x = \frac{1}{2}$ and the absolute maximum is 0 at $x = \frac{e}{2}$

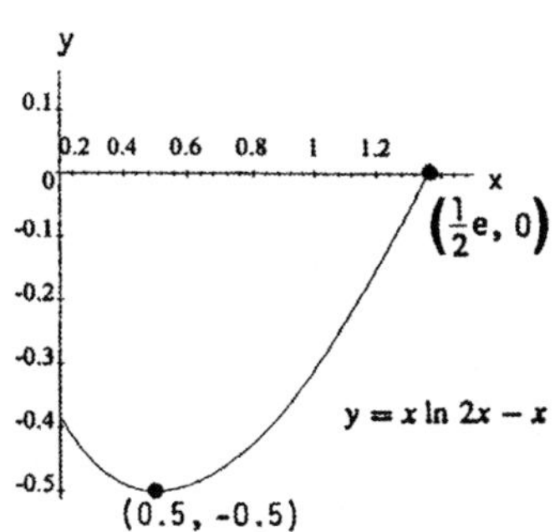

130. $y = 10x(2 - \ln x) \Rightarrow y' = 10(2 - \ln x) - 10x\left(\frac{1}{x}\right)$ $= 20 - 10 \ln x - 10 = 10(1 - \ln x)$; solving $y' = 0$ $\Rightarrow x = e$; $y' < 0$ for $x > e$ and $y' > 0$ for $x < e$ $\Rightarrow$ relative maximum at $x = e$ of $10e$; $y \geq 0$ on $(0, e^2]$ and $y(e^2) = 10e^2(2 - 2\ln e) = 0 \Rightarrow$ absolute minimum is 0 at $x = e^2$ and the absolute maximum is $10e$ at $x = e$

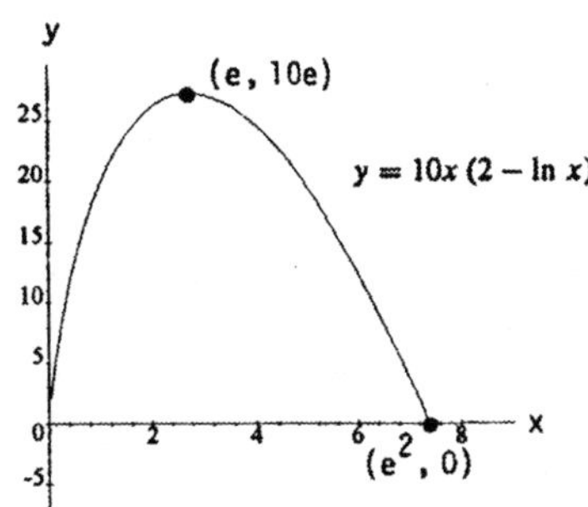

131. $f(x) = e^{x/\sqrt{x^4+1}}$ for all x in $(-\infty, \infty)$; $f'(x) = \left[\frac{\left(\sqrt{x^4+1}\right)\cdot 1 - x\left(\frac{2x^3}{\sqrt{x^4+1}}\right)}{\left(\sqrt{x^4+1}\right)^2}\right] e^{x/\sqrt{x^4+1}} = \frac{1-x^4}{\left(\sqrt{x^4+1}\right)^3} e^{x/\sqrt{x^4+1}}$

$= \frac{(1-x^2)(1+x^2)}{(x^4+1)^{3/2}} e^{x/\sqrt{x^4+1}} = 0 \Rightarrow 1 - x^2 = 0 \Rightarrow x = \pm 1$ are critical points. Consider the behavior of f as $x \to \pm\infty$; $\lim\limits_{x \to \infty} e^{x/\sqrt{x^4+1}} = \lim\limits_{x \to -\infty} e^{x/\sqrt{x^4+1}} = 1$ as suggested by the following table (14 digit precision, 12 digits displayed):

x	$x/\sqrt{x^4+1}$	$e^{x/\sqrt{x^4+1}}$
$-\infty$	0	1
⋮	⋮	⋮
−100000	−0.0000 10000 0000 00000	0.9999 9000 0050
−10000	−0.0001 0000 0000 000	0.9999 0000 5000
−1000	−0.0010 0000 0000 00	0.9990 0049 9833
−100	−0.0099 9999 9950 00	0.9900 4983 3799
−10	−0.0999 9500 0375 0	0.9048 4194 1895
0	0	1
10	0.0999 9500 0375 0	1.1051 6539 265
100	0.0099 9999 9950 00	1.0100 5016 703
1000	0.0010 0000 0000 00	1.0010 0050 017
10000	0.0001 0000 0000 000	1.0001 0000 500
100000	0.0000 10000 0000 00000	1.0000 1000 005
⋮	⋮	⋮
∞	0	1

Therefore, $y = 1$ is a horizontal asymptote in both directions. Check the critical points for absolute extreme values: $f(-1) = e^{-\sqrt{2}/2} \approx 0.4931$, $f(1) = e^{\sqrt{2}/2} \approx 2.0281 \Rightarrow$ the absolute minimum value of the function is $e^{-\sqrt{2}/2}$ at $x = -1$, and the absolute maximum value is $e^{\sqrt{2}/2}$ at $x = 1$.

132. $f(x) = e^{\sqrt{3-2x-x^2}}$; The domain of g is all x such that $3 - 2x - x^2 \geq 0$. The parabola $y = 3 - 2x - x^2$ is concave down with x-intercepts at $x = -3$ and $x = 1$, therefore $3 - 2x - x^2 \geq 0$ if $-3 \leq x \leq 1$, and this interval is the domain of g; $g'(x) = -\frac{1+x}{\sqrt{3-2x-x^2}} e^{\sqrt{3-2x-x^2}} = 0 \Rightarrow 1 + x = 0 \Rightarrow x = -1$ is a critical point; $g(-3) = g(1) = e^0 = 1$,

$g(1) = e^2 \approx 7.3891 \Rightarrow$ the absolute minimum value of the function is 1 at $x = -3$ and $x = 1$, and the aboslute maximum value is e^2 at $x = -1$.

133. (a) $y = \frac{\ln x}{\sqrt{x}} \Rightarrow y' = \frac{1}{x\sqrt{x}} - \frac{\ln x}{2x^{3/2}} = \frac{2 - \ln x}{2x\sqrt{x}}$
$\Rightarrow y'' = -\frac{3}{4}x^{-5/2}(2 - \ln x) - \frac{1}{2}x^{-5/2} = x^{-5/2}\left(\frac{3}{4}\ln x - 2\right)$; solving $y' = 0 \Rightarrow \ln x = 2 \Rightarrow x = e^2$; $y' < 0$ for $x > e^2$ and and $y' > 0$ for $x < e^2 \Rightarrow$ a maximum of $\frac{2}{e}$; $y'' = 0$ $\Rightarrow \ln x = \frac{8}{3} \Rightarrow x = e^{8/3}$; the curve is concave down on $\left(0, e^{8/3}\right)$ and concave up on $\left(e^{8/3}, \infty\right)$; so there is an inflection point at $\left(e^{8/3}, \frac{8}{3e^{4/3}}\right)$.

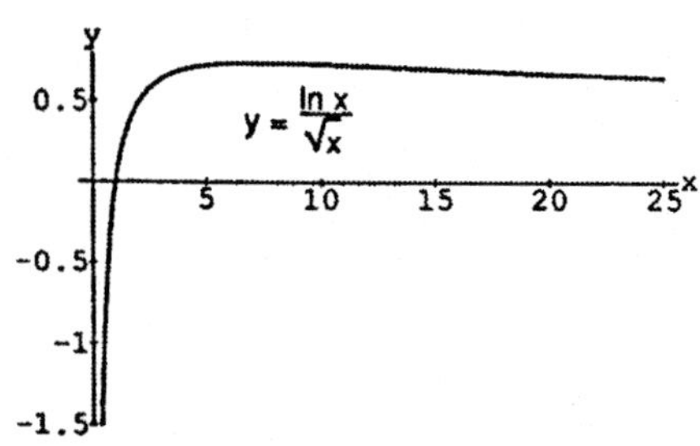

(b) $y = e^{-x^2} \Rightarrow y' = -2xe^{-x^2} \Rightarrow y'' = -2e^{-x^2} + 4x^2e^{-x^2}$ $= (4x^2 - 2)e^{-x^2}$; solving $y' = 0 \Rightarrow x = 0$; $y' < 0$ for $x > 0$ and $y' > 0$ for $x < 0 \Rightarrow$ a maximum at $x = 0$ of $e^0 = 1$; there are points of inflection at $x = \pm\frac{1}{\sqrt{2}}$; the curve is concave down for $-\frac{1}{\sqrt{2}} < x < \frac{1}{\sqrt{2}}$ and concave up otherwise.

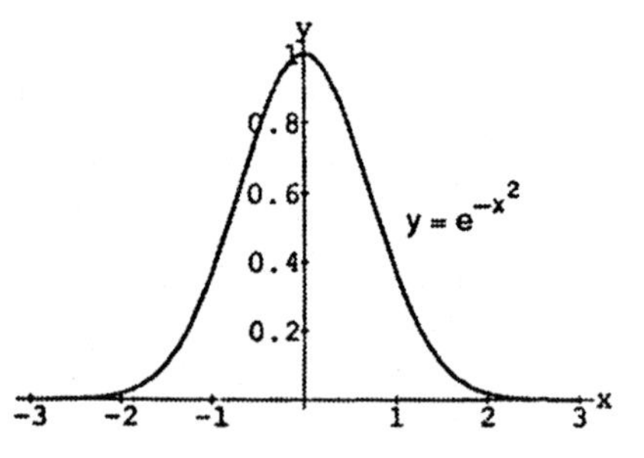

(c) $y = (1 + x)e^{-x} \Rightarrow y' = e^{-x} - (1 + x)e^{-x} = -xe^{-x}$
$\Rightarrow y'' = -e^{-x} + xe^{-x} = (x - 1)e^{-x}$; solving $y' = 0$ $\Rightarrow -xe^{-x} = 0 \Rightarrow x = 0$; $y' < 0$ for $x > 0$ and $y' > 0$ for $x < 0 \Rightarrow$ a maximum at $x = 0$ of $(1 + 0)e^0 = 1$; there is a point of inflection at $x = 1$ and the curve is concave up for $x > 1$ and concave down for $x < 1$.

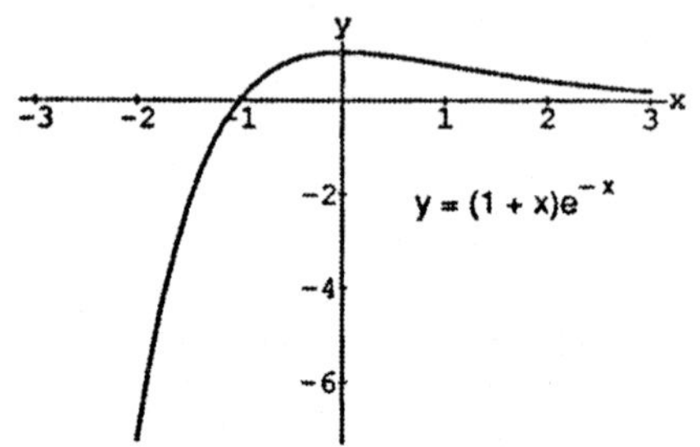

134. $y = x \ln x \Rightarrow y' = \ln x + x\left(\frac{1}{x}\right) = \ln x + 1$; solving $y' = 0$ $\Rightarrow \ln x + 1 = 0 \Rightarrow \ln x = -1 \Rightarrow x = e^{-1}$; $y' > 0$ for $x > e^{-1}$ and $y' < 0$ for $x < e^{-1} \Rightarrow$ a minimum of $e^{-1}\ln e^{-1}$ $= -\frac{1}{e}$ at $x = e^{-1}$. This minimum is an absolute minimum since $y'' = \frac{1}{x}$ is positive for all $x > 0$.

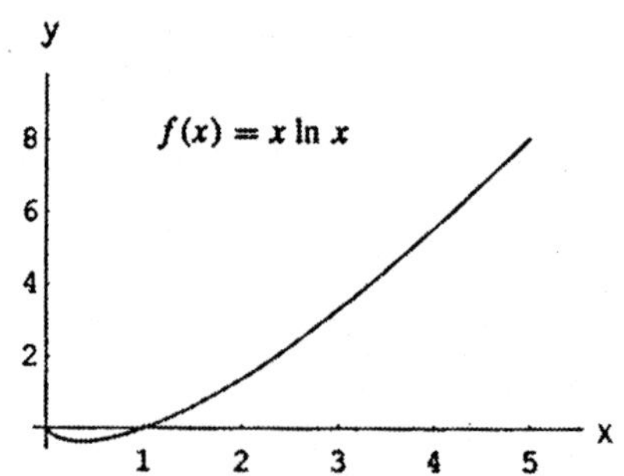

135 In the interval $\pi < x < 2\pi$ the function $\sin x < 0$ $\Rightarrow (\sin x)^{\sin x}$ is not defined for all values in that interval or its translation by 2π.

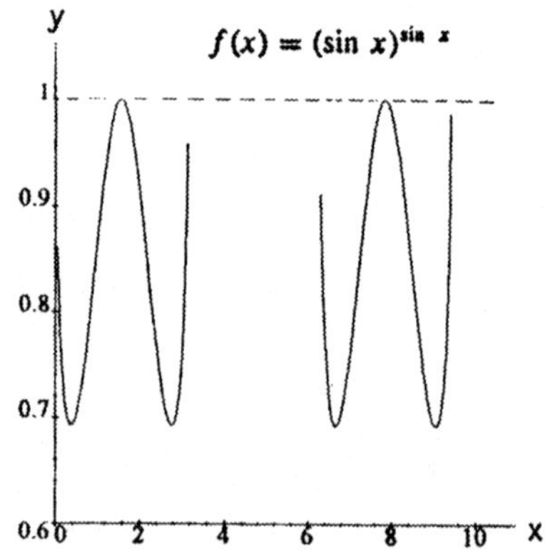

136. $v = x^2 \ln\left(\frac{1}{x}\right) = x^2(\ln 1 - \ln x) = -x^2 \ln x \Rightarrow \frac{dv}{dx} = -2x \ln x - x^2\left(\frac{1}{x}\right) = -x(2\ln x + 1)$; solving $\frac{dv}{dx} = 0$ $\Rightarrow 2\ln x + 1 = 0 \Rightarrow \ln x = -\frac{1}{2} \Rightarrow x = e^{-1/2}$; $\frac{dv}{dx} < 0$ for $x > e^{-1/2}$ and $\frac{dv}{dx} > 0$ for $x < e^{-1/2} \Rightarrow$ a relative maximum at $x = e^{-1/2}$; $\frac{r}{h} = x$ and $r = 1 \Rightarrow h = e^{1/2} = \sqrt{e} \approx 1.65$ cm

CHAPTER 4 ADDITIONAL AND ADVANCED EXERCISES

1. If M and m are the maximum and minimum values, respectively, then $m \le f(x) \le M$ for all $x \in I$. If $m = M$ then f is constant on I.

2. No, the function $f(x) = \begin{cases} 3x+6, & -2 \le x < 0 \\ 9 - x^2, & 0 \le x \le 2 \end{cases}$ has an absolute minimum value of 0 at $x = -2$ and an absolute maximum value of 9 at $x = 0$, but it is discontinuous at $x = 0$.

3. On an open interval the extreme values of a continuous function (if any) must occur at an interior critical point. On a half-open interval the extreme values of a continuous function may be at a critical point or at the closed endpoint. Extreme values occur only where $f' = 0$, f' does not exist, or at the endpoints of the interval. Thus the extreme points will not be at the ends of an open interval.

4. The pattern $f' = +++\underset{1}{|}----\underset{2}{|}----\underset{3}{|}++++\underset{4}{|}+++$ indicates a local maximum at $x = 1$ and a local minimum at $x = 3$.

5. (a) If $y' = 6(x+1)(x-2)^2$, then $y' < 0$ for $x < -1$ and $y' > 0$ for $x > -1$. The sign pattern is $f' = ---\underset{-1}{|}+++\underset{2}{|}+++ \Rightarrow$ f has a local minimum at $x = -1$. Also $y'' = 6(x-2)^2 + 12(x+1)(x-2)$ $= 6(x-2)(3x) \Rightarrow y'' > 0$ for $x < 0$ or $x > 2$, while $y'' < 0$ for $0 < x < 2$. Therefore f has points of inflection at $x = 0$ and $x = 2$. There is no local maximum.
   (b) If $y' = 6x(x+1)(x-2)$, then $y' < 0$ for $x < -1$ and $0 < x < 2$; $y' > 0$ for $-1 < x < 0$ and $x > 2$. The sign sign pattern is $y' = ---\underset{-1}{|}+++\underset{0}{|}---\underset{2}{|}+++$. Therefore f has a local maximum at $x = 0$ and local minima at $x = -1$ and $x = 2$. Also, $y'' = 18\left[x - \left(\frac{1-\sqrt{7}}{3}\right)\right]\left[x - \left(\frac{1+\sqrt{7}}{3}\right)\right]$, so $y'' < 0$ for $\frac{1-\sqrt{7}}{3} < x < \frac{1+\sqrt{7}}{3}$ and $y'' > 0$ for all other x $\Rightarrow$ f has points of inflection at $x = \frac{1 \pm \sqrt{7}}{3}$.

6. The Mean Value Theorem indicates that $\frac{f(6) - f(0)}{6 - 0} = f'(c) \le 2$ for some c in $(0, 6)$. Then $f(6) - f(0) \le 12$ indicates the most that f can increase is 12.

7. If f is continuous on $[a, c)$ and $f'(x) \le 0$ on $[a, c)$, then by the Mean Value Theorem for all $x \in [a, c)$ we have $\frac{f(c) - f(x)}{c - x} \le 0 \Rightarrow f(c) - f(x) \le 0 \Rightarrow f(x) \ge f(c)$. Also if f is continuous on $(c, b]$ and $f'(x) \ge 0$ on $(c, b]$, then for all $x \in (c, b]$ we have $\frac{f(x) - f(c)}{x - c} \ge 0 \Rightarrow f(x) - f(c) \ge 0 \Rightarrow f(x) \ge f(c)$. Therefore $f(x) \ge f(c)$ for all $x \in [a, b]$.

8. (a) For all x, $-(x+1)^2 \le 0 \le (x-1)^2 \Rightarrow -(1+x^2) \le 2x \le (1+x^2) \Rightarrow -\frac{1}{2} \le \frac{x}{1+x^2} \le \frac{1}{2}$.
   (b) There exists $c \in (a, b)$ such that $\frac{c}{1+c^2} = \frac{f(b) - f(a)}{b - a} \Rightarrow \left|\frac{f(b) - f(a)}{b - a}\right| = \left|\frac{c}{1+c^2}\right| \le \frac{1}{2}$, from part (a) $\Rightarrow |f(b) - f(a)| \le \frac{1}{2}|b - a|$.

9. No. Corollary 1 requires that $f'(x) = 0$ for <u>all</u> x in some interval I, not $f'(x) = 0$ at a single point in I.

10. (a) $h(x) = f(x)g(x) \Rightarrow h'(x) = f'(x)g(x) + f(x)g'(x)$ which changes signs at $x = a$ since $f'(x), g'(x) > 0$ when $x < a$, $f'(x), g'(x) < 0$ when $x > a$ and $f(x), g(x) > 0$ for all x. Therefore h(x) does have a local maximum at $x = a$.

(b) No, let $f(x) = g(x) = x^3$ which have points of inflection at $x = 0$, but $h(x) = x^6$ has no point of inflection (it has a local minimum at $x = 0$).

11. From (ii), $f(-1) = \frac{-1+a}{b-c+2} = 0 \Rightarrow a = 1$; from (iii), either $1 = \lim_{x \to \infty} f(x)$ or $1 = \lim_{x \to -\infty} f(x)$. In either case, $\lim_{x \to \pm\infty} f(x) = \lim_{x \to \pm\infty} \frac{x+1}{bx^2+cx+2} = \lim_{x \to \pm\infty} \frac{1+\frac{1}{x}}{bx+c+\frac{2}{x}} = 1 \Rightarrow b = 0$ and $c = 1$. For if $b = 1$, then $\lim_{x \to \pm\infty} \frac{1+\frac{1}{x}}{x+c+\frac{2}{x}} = 0$ and if $c = 0$, then $\lim_{x \to \pm\infty} \frac{1+\frac{1}{x}}{bx+\frac{2}{x}} = \lim_{x \to \pm\infty} \frac{1+\frac{1}{x}}{\frac{2}{x}} = \pm\infty$. Thus $a = 1$, $b = 0$, and $c = 1$.

12. $\frac{dy}{dx} = 3x^2 + 2kx + 3 = 0 \Rightarrow x = \frac{-2k \pm \sqrt{4k^2 - 36}}{6} \Rightarrow x$ has only one value when $4k^2 - 36 = 0 \Rightarrow k^2 = 9$ or $k = \pm 3$.

13. The area of the ΔABC is $A(x) = \frac{1}{2}(2)\sqrt{1-x^2} = (1-x^2)^{1/2}$, where $0 \le x \le 1$. Thus $A'(x) = \frac{-x}{\sqrt{1-x^2}} \Rightarrow 0$ and ± 1 are critical points. Also $A(\pm 1) = 0$ so $A(0) = 1$ is the maximum. When $x = 0$ the ΔABC is isosceles since $AC = BC = \sqrt{2}$.

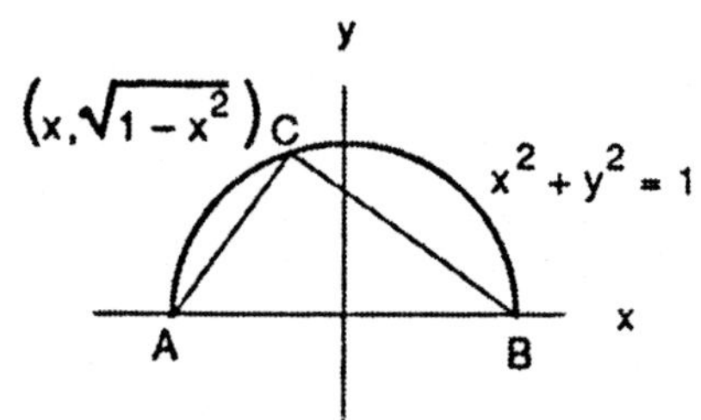

14. $\lim_{h \to 0} \frac{f'(c+h) - f'(c)}{h} = f''(c) \Rightarrow$ for $\epsilon = \frac{1}{2}|f''(c)| > 0$ there exists a $\delta > 0$ such that $0 < |h| < \delta$
$\Rightarrow \left|\frac{f'(c+h) - f'(c)}{h} - f''(c)\right| < \frac{1}{2}|f''(c)|$. Then $f'(c) = 0 \Rightarrow -\frac{1}{2}|f''(c)| < \frac{f'(c+h)}{h} - f''(c) < \frac{1}{2}|f''(c)|$
$\Rightarrow f''(c) - \frac{1}{2}|f''(c)| < \frac{f'(c+h)}{h} < f''(c) + \frac{1}{2}|f''(c)|$. If $f''(c) < 0$, then $|f''(c)| = -f''(c)$
$\Rightarrow \frac{3}{2}f''(c) < \frac{f'(c+h)}{h} < \frac{1}{2}f''(c) < 0$; likewise if $f''(c) > 0$, then $0 < \frac{1}{2}f''(c) < \frac{f'(c+h)}{h} < \frac{3}{2}f''(c)$.

(a) If $f''(c) < 0$, then $-\delta < h < 0 \Rightarrow f'(c+h) > 0$ and $0 < h < \delta \Rightarrow f'(c+h) < 0$. Therefore, $f(c)$ is a local maximum.

(b) If $f''(c) > 0$, then $-\delta < h < 0 \Rightarrow f'(c+h) < 0$ and $0 < h < \delta \Rightarrow f'(c+h) > 0$. Therefore, $f(c)$ is a local minimum.

15. The time it would take the water to hit the ground from height y is $\sqrt{\frac{2y}{g}}$, where g is the acceleration of gravity. The product of time and exit velocity (rate) yields the distance the water travels: $D(y) = \sqrt{\frac{2y}{g}}\sqrt{64(h-y)} = 8\sqrt{\frac{2}{g}}(hy-y^2)^{1/2}, 0 \le y \le h \Rightarrow D'(y) = -4\sqrt{\frac{2}{g}}(hy-y^2)^{-1/2}(h-2y) \Rightarrow 0, \frac{h}{2}$ and h are critical points. Now $D(0) = 0$, $D\left(\frac{h}{2}\right) = \frac{8h}{\sqrt{g}}$ and $D(h) = 0 \Rightarrow$ the best place to drill the hole is at $y = \frac{h}{2}$.

16. From the figure in the text, $\tan(\beta + \theta) = \frac{b+a}{h}$; $\tan(\beta+\theta) = \frac{\tan\beta + \tan\theta}{1 - \tan\beta\tan\theta}$; and $\tan\theta = \frac{a}{h}$. These equations give $\frac{b+a}{h} = \frac{\tan\beta + \frac{a}{h}}{1 - \frac{a}{h}\tan\beta} = \frac{h\tan\beta + a}{h - a\tan\beta}$. Solving for $\tan\beta$ gives $\tan\beta = \frac{bh}{h^2 + a(b+a)}$ or $(h^2 - a(b+a))\tan\beta = bh$. Differentiating both sides with respect to h gives $2h\tan\beta + (h^2 + a(b+a))\sec^2\beta\,\frac{d\beta}{dh} = b$. Then $\frac{d\beta}{dh} = 0 \Rightarrow 2h\tan\beta = b \Rightarrow 2h\left(\frac{bh}{h^2+a(b+a)}\right) = b$
$\Rightarrow 2bh^2 = bh^2 + ab(b+a) \Rightarrow h^2 = a(b+a) \Rightarrow h = \sqrt{a(a+b)}$.

17. The surface area of the cylinder is $S = 2\pi r^2 + 2\pi rh$. From the diagram we have $\frac{r}{R} = \frac{H-h}{H} \Rightarrow h = \frac{RH - rH}{R}$ and $S(r) = 2\pi r(r+h) = 2\pi r\left(r + H - r\frac{H}{R}\right)$
$= 2\pi\left(1 - \frac{H}{R}\right)r^2 + 2\pi Hr$, where $0 \le r \le R$.

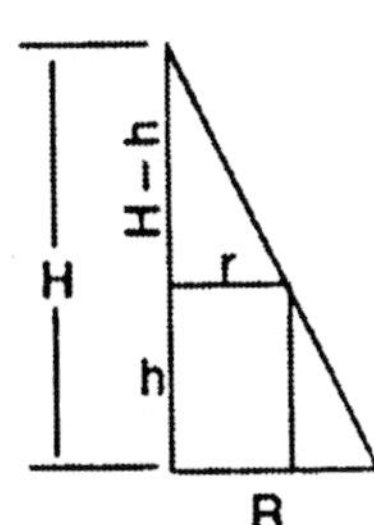

Case 1: $H < R \Rightarrow S(r)$ is a quadratic equation containing the origin and concave upward $\Rightarrow S(r)$ is maximum at $r = R$.

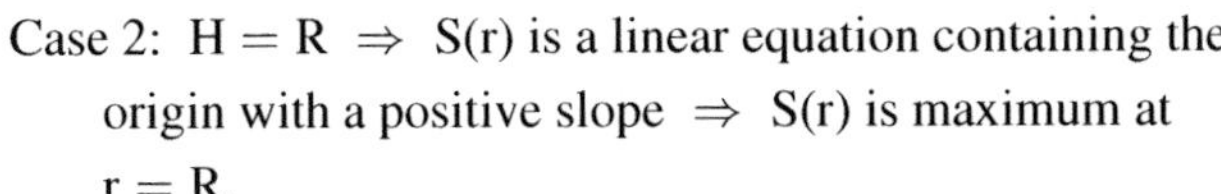

Case 2: $H = R \Rightarrow S(r)$ is a linear equation containing the origin with a positive slope $\Rightarrow S(r)$ is maximum at $r = R$.

Case 3: $H > R \Rightarrow S(r)$ is a quadratic equation containing the origin and concave downward. Then $\frac{dS}{dr} = 4\pi\left(1 - \frac{H}{R}\right)r + 2\pi H$ and $\frac{dS}{dr} = 0 \Rightarrow 4\pi\left(1 - \frac{H}{R}\right)r + 2\pi H = 0 \Rightarrow r = \frac{RH}{2(H-R)}$. For simplification we let $r^* = \frac{RH}{2(H-R)}$.

(a) If $R < H < 2R$, then $0 > H - 2R \Rightarrow H > 2(H-R) \Rightarrow \frac{RH}{2(H-R)} > R$ which is impossible.

(b) If $H = 2R$, then $r^* = \frac{2R^2}{2R} = R \Rightarrow S(r)$ is maximum at $r = R$.

(c) If $H > 2R$, then $2R + H < 2H \Rightarrow H < 2(H-R) \Rightarrow \frac{H}{2(H-R)} < 1 \Rightarrow \frac{RH}{2(H-R)} < R \Rightarrow r^* < R$. Therefore, $S(r)$ is a maximum at $r = r^* = \frac{RH}{2(H-R)}$.

Conclusion: If $H \in (0, R]$ or $H = 2R$, then the maximum surface area is at $r = R$. If $H \in (R, 2R)$, then $r > R$ which is not possible. If $H \in (2R, \infty)$, then the maximum is at $r = r^* = \frac{RH}{2(H-R)}$.

18. $f(x) = mx - 1 + \frac{1}{x} \Rightarrow f'(x) = m - \frac{1}{x^2}$ and $f''(x) = \frac{2}{x^3} > 0$ when $x > 0$. Then $f'(x) = 0 \Rightarrow x = \frac{1}{\sqrt{m}}$ yields a minimum. If $f\left(\frac{1}{\sqrt{m}}\right) \ge 0$, then $\sqrt{m} - 1 + \sqrt{m} = 2\sqrt{m} - 1 \ge 0 \Rightarrow m \ge \frac{1}{4}$. Thus the smallest acceptable value for m is $\frac{1}{4}$.

19. (a) $\lim\limits_{x \to 0} \frac{2\sin(5x)}{3x} = \lim\limits_{x \to 0} \frac{2\sin(5x)}{\frac{3}{5}(5x)} = \lim\limits_{x \to 0} \frac{10}{3}\frac{\sin(5x)}{(5x)} = \frac{10}{3} \cdot 1 = \frac{10}{3}$

(b) $\lim\limits_{x \to 0} \sin(5x)\cot(3x) = \lim\limits_{x \to 0} \frac{\sin(5x)\cos(3x)}{\sin(3x)} = \lim\limits_{x \to 0} \frac{-3\sin(5x)\sin(3x) + 5\cos(5x)\cos(3x)}{3\cos(3x)} = \frac{5}{3}$

(c) $\lim\limits_{x \to 0} x\csc^2\sqrt{2x} = \lim\limits_{x \to 0} \frac{x}{\sin^2\sqrt{2x}} = \lim\limits_{x \to 0} \frac{1}{\frac{2\sin\sqrt{2x}\cos\sqrt{2x}}{\sqrt{2x}}} = \lim\limits_{x \to 0} \frac{\sqrt{2x}}{\sin\left(2\sqrt{2x}\right)} = \lim\limits_{x \to 0} \frac{\frac{1}{\sqrt{2x}}}{\cos\left(2\sqrt{2x}\right)\frac{2}{\sqrt{2x}}}$
$= \lim\limits_{x \to 0} \frac{1}{\cos\left(2\sqrt{2x}\right) \cdot 2} = \frac{1}{2}$

(d) $\lim\limits_{x \to \pi/2} (\sec x - \tan x) = \lim\limits_{x \to \pi/2} \frac{1 - \sin x}{\cos x} = \lim\limits_{x \to \pi/2} \frac{-\cos x}{-\sin x} = 0.$

(e) $\lim\limits_{x \to 0} \frac{x - \sin x}{x - \tan x} = \lim\limits_{x \to 0} \frac{1 - \cos x}{1 - \sec^2 x} = \lim\limits_{x \to 0} \frac{1 - \cos x}{-\tan^2 x} = \lim\limits_{x \to 0} \frac{\cos x - 1}{\tan^2 x} = \lim\limits_{x \to 0} \frac{-\sin x}{2\tan x \sec^2 x} = \lim\limits_{x \to 0} \frac{-\sin x}{\frac{2\sin x}{\cos^3 x}} =$
$= \lim\limits_{x \to 0} \frac{\cos^3 x}{-2} = -\frac{1}{2}$

(f) $\lim\limits_{x \to 0} \frac{\sin(x^2)}{x\sin x} = \lim\limits_{x \to 0} \frac{2x\cos(x^2)}{x\cos x + \sin x} = \lim\limits_{x \to 0} \frac{-(2x^2)\sin(x^2) + 2\cos(x^2)}{-x\sin x + 2\cos x} = \frac{2}{2} = 1$

(g) $\lim\limits_{x \to 0} \frac{\sec x - 1}{x^2} = \lim\limits_{x \to 0} \frac{\sec x \tan x}{2x} = \lim\limits_{x \to 0} \frac{\sec^3 x + \tan^2 x \sec x}{2} = \frac{1+0}{2} = \frac{1}{2}$

(h) $\lim\limits_{x \to 2} \frac{x^3 - 8}{x^2 - 4} = \lim\limits_{x \to 2} \frac{(x-2)(x^2+2x+4)}{(x-2)(x+2)} = \lim\limits_{x \to 2} \frac{x^2+2x+4}{x+2} = \frac{4+4+4}{4} = 3$

20. (a) $\lim\limits_{x \to \infty} \frac{\sqrt{x+5}}{\sqrt{x}+5} = \lim\limits_{x \to \infty} \frac{\frac{\sqrt{x+5}}{\sqrt{x}}}{\frac{\sqrt{x}+5}{\sqrt{x}}} = \lim\limits_{x \to \infty} \frac{\sqrt{1+\frac{5}{x}}}{1+\frac{5}{\sqrt{x}}} = \frac{1}{1} = 1$

(b) $\lim\limits_{x \to \infty} \frac{2x}{x + 7\sqrt{x}} = \lim\limits_{x \to \infty} \frac{\frac{2x}{x}}{\frac{x+7\sqrt{x}}{x}} = \lim\limits_{x \to \infty} \frac{2}{1 + 7\sqrt{\frac{1}{x}}} = \frac{2}{1+0} = 2$

21. (a) The profit function is $P(x) = (c - ex)x - (a + bx) = -ex^2 + (c - b)x - a$. $P'(x) = -2ex + c - b = 0$ $\Rightarrow x = \frac{c-b}{2e}$. $P''(x) = -2e < 0$ if $e > 0$ so that the profit function is maximized at $x = \frac{c-b}{2e}$.

(b) The price therefore that corresponds to a production level yeilding a maximum profit is $p\Big|_{x=\frac{c-b}{2e}} = c - e\left(\frac{c-b}{2e}\right) = \frac{c+b}{2}$ dollars.

(c) The weekly profit at this production level is $P(x) = -e\left(\frac{c-b}{2e}\right)^2 + (c - b)\left(\frac{c-b}{2e}\right) - a = \frac{(c-b)^2}{4e} - a$.

(d) The tax increases cost to the new profit function is $F(x) = (c - ex)x - (a + bx + tx) = -ex^2 + (c - b - t)x - a$. Now $F'(x) = -2ex + c - b - t = 0$ when $x = \frac{t+b-c}{-2e} = \frac{c-b-t}{2e}$. Since $F''(x) = -2e < 0$ if $e > 0$, F is maximized when $x = \frac{c-b-t}{2e}$ units per week. Thus the price per unit is $p = c - e\left(\frac{c-b-t}{2e}\right) = \frac{c+b+t}{2}$ dollars. Thus, such a tax increases the cost per unit by $\frac{c+b+t}{2} - \frac{c+b}{2} = \frac{t}{2}$ dollars if units are priced to maximize profit.

22. (a)

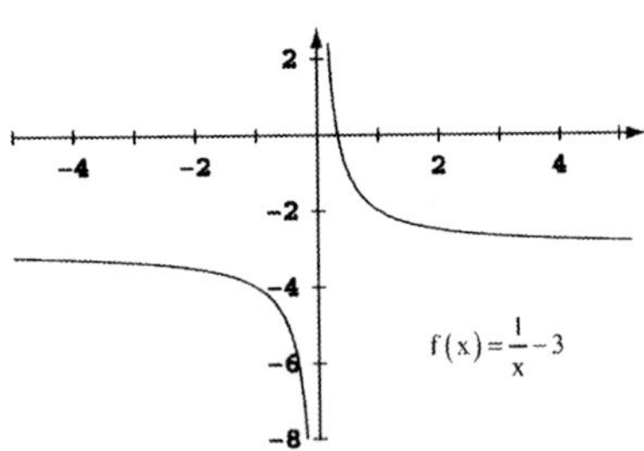

The x-intercept occurs when $\frac{1}{x} - 3 = 0 \Rightarrow \frac{1}{x} = 3 \Rightarrow x = \frac{1}{3}$.

(b) By Newton's method, $x_{n+1} = x_n - \frac{f(x_n)}{f'(x_n)}$. Here $f'(x_n) = -x_n^{-2} = \frac{-1}{x_n^2}$. So $x_{n+1} = x_n - \frac{\frac{1}{x_n} - 3}{\frac{-1}{x_n^2}} = x_n + \left(\frac{1}{x_n} - 3\right)x_n^2$ $= x_n + x_n - 3x_n^2 = 2x_n - 3x_n^2 = x_n(2 - 3x_n)$.

23. $x_1 = x_0 - \frac{f(x_0)}{f'(x_0)} = x_0 - \frac{x_0^q - a}{qx_0^{q-1}} = \frac{qx_0^q - x_0^q - a}{qx_0^{q-1}} = \frac{x_0^q(q-1) - a}{qx_0^{q-1}} = x_0\left(\frac{q-1}{q}\right) + \frac{a}{x_0^{q-1}}\left(\frac{1}{q}\right)$ so that x_1 is a weighted average of x_0 and $\frac{a}{x_0^{q-1}}$ with weights $m_0 = \frac{q-1}{q}$ and $m_1 = \frac{1}{q}$.

In the case where $x_0 = \frac{a}{x_0^{q-1}}$ we have $x_0^q = a$ and $x_1 = \frac{a}{x_0^{q-1}}\left(\frac{q-1}{q}\right) + \frac{a}{x_0^{q-1}}\left(\frac{1}{q}\right) = \frac{a}{x_0^{q-1}}\left(\frac{q-1}{q} + \frac{1}{q}\right) = \frac{a}{x_0^{q-1}}$.

24. We have that $(x - h)^2 + (y - h)^2 = r^2$ and so $2(x - h) + 2(y - h)\frac{dy}{dx} = 0$ and $2 + 2\frac{dy}{dx} + 2(y - h)\frac{d^2y}{dx^2} = 0$ hold. Thus $2x + 2y\frac{dy}{dx} = 2h + 2h\frac{dy}{dx}$, by the former. Solving for h, we obtain $h = \frac{x + y\frac{dy}{dx}}{1 + \frac{dy}{dx}}$. Substituting this into the second equation yields $2 + 2\frac{dy}{dx} + 2y\frac{d^2y}{dx^2} - 2\left(\frac{x + y\frac{dy}{dx}}{1 + \frac{dy}{dx}}\right) = 0$. Dividing by 2 results in $1 + \frac{dy}{dx} + y\frac{d^2y}{dx^2} - \left(\frac{x + y\frac{dy}{dx}}{1 + \frac{dy}{dx}}\right) = 0$.

25. $\frac{ds}{dt} = ks \Rightarrow \frac{ds}{s} = k\,dt \Rightarrow \ln s = kt + C \Rightarrow s = s_0e^{kt}$ $\Rightarrow$ the 14th century model of free fall was exponential; note that the motion starts too slowly at first and then becomes too fast after about 7 seconds

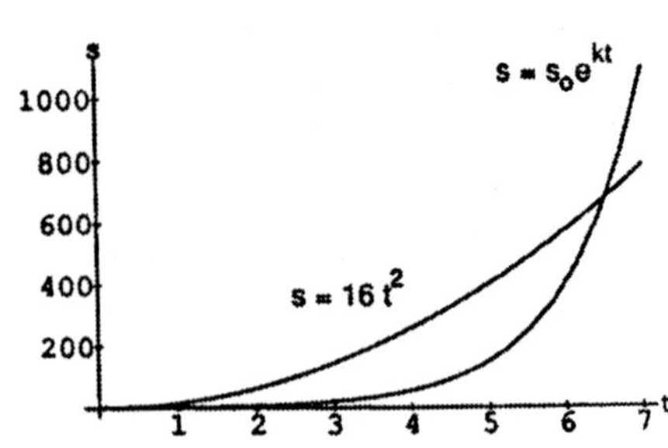

26. Two views of the graph of $y = 1000\left[1 - (.99)^x + \frac{1}{x}\right]$ are shown below.

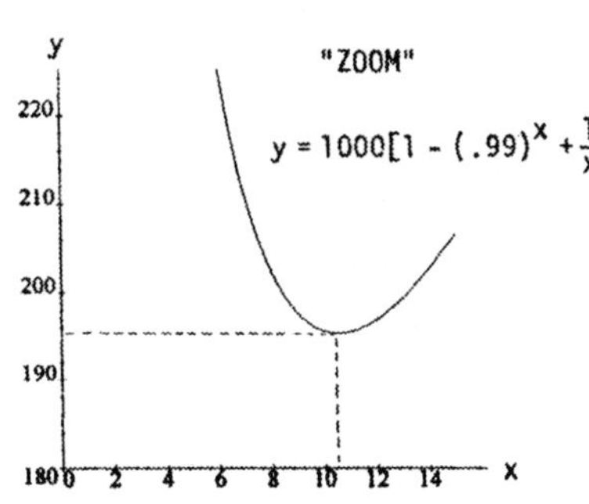

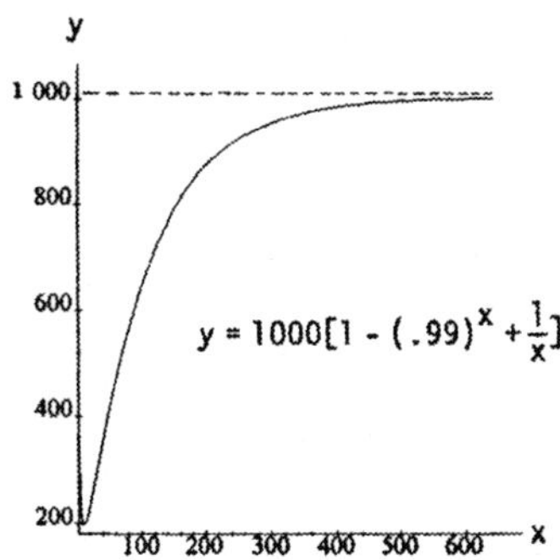

(a) At about $x = 11$ there is a minimum

(b) There is no maximum; however, the curve is asymptotic to $y = 1000$. The curve is near 1000 when $x \geq 643$.

27. (a) $a(t) = s''(t) = -k \quad (k > 0) \Rightarrow s'(t) = -kt + C_1$, where $s'(0) = 88 \Rightarrow C_1 = 88 \Rightarrow s'(t) = -kt + 88$. So $s(t) = \frac{-kt^2}{2} + 88t + C_2$ where $s(0) = 0 \Rightarrow C_2 = 0$ so $s(t) = \frac{-kt^2}{2} + 88t$. Now $s(t) = 100$ when $\frac{-kt^2}{2} + 88t = 100$. Solving for t we obtain $t = \frac{88 \pm \sqrt{88^2 - 200k}}{k}$. At such t we want $s'(t) = 0$, thus $-k\left(\frac{88 + \sqrt{88^2 - 200k}}{k}\right) + 88 = 0$ or $-k\left(\frac{88 - \sqrt{88^2 - 200k}}{k}\right) + 88 = 0$. In either case we obtain $88^2 - 200k = 0$ so that $k = \frac{88^2}{200} \approx 38.72$ ft/sec^2.

(b) The initial condition that $s'(0) = 44$ ft/sec implies that $s'(t) = -kt + 44$ and $s(t) = \frac{-kt^2}{2} + 44t$ where k is as above. The car is stopped at a time t such that $s'(t) = -kt + 44 = 0 \Rightarrow t = \frac{44}{k}$. At this time the car has traveled a distance $s\left(\frac{44}{k}\right) = \frac{-k}{2}\left(\frac{44}{k}\right)^2 + 44\left(\frac{44}{k}\right) = \frac{44^2}{2k} = \frac{968}{k} = 968\left(\frac{200}{88^2}\right) = 25$ feet. Thus halving the initial velocity quarters stopping distance.

28. $h(x) = f^2(x) + g^2(x) \Rightarrow h'(x) = 2f(x)f'(x) + 2g(x)g'(x) = 2\left[f(x)f'(x) + g(x)g'(x)\right] = 2\left[f(x)g(x) + g(x)(-f(x))\right] = 2 \cdot 0 = 0$. Thus $h(x) = c$, a constant. Since $h(0) = 5$, $h(x) = 5$ for all x in the domain of h. Thus $h(10) = 5$.

29. Yes. The curve $y = x$ satisfies all three conditions since $\frac{dy}{dx} = 1$ everywhere, when $x = 0$, $y = 0$, and $\frac{d^2y}{dx^2} = 0$ everywhere.

30. $y' = 3x^2 + 2$ for all $x \Rightarrow y = x^3 + 2x + C$ where $-1 = 1^3 + 2 \cdot 1 + C \Rightarrow C = -4 \Rightarrow y = x^3 + 2x - 4$.

31. $s''(t) = a = -t^2 \Rightarrow v = s'(t) = \frac{-t^3}{3} + C$. We seek $v_0 = s'(0) = C$. We know that $s(t^*) = b$ for some t^* and s is at a maximum for this t^*. Since $s(t) = \frac{-t^4}{12} + Ct + k$ and $s(0) = 0$ we have that $s(t) = \frac{-t^4}{12} + Ct$ and also $s'(t^*) = 0$ so that $t^* = (3C)^{1/3}$. So $\frac{\left[-(3C)^{1/3}\right]^4}{12} + C(3C)^{1/3} = b \Rightarrow (3C)^{1/3}\left(C - \frac{3C}{12}\right) = b \Rightarrow (3C)^{1/3}\left(\frac{3C}{4}\right) = b \Rightarrow 3^{1/3}C^{4/3} = \frac{4b}{3}$ $\Rightarrow C = \frac{(4b)^{3/4}}{3}$. Thus $v_0 = s'(0) = \frac{(4b)^{3/4}}{3} = \frac{2\sqrt{2}}{3}b^{3/4}$.

32. (a) $s''(t) = t^{1/2} - t^{-1/2} \Rightarrow v(t) = s'(t) = \frac{2}{3}t^{3/2} - 2t^{1/2} + k$ where $v(0) = k = \frac{4}{3} \Rightarrow v(t) = \frac{2}{3}t^{3/2} - 2t^{1/2} + \frac{4}{3}$.

(b) $s(t) = \frac{4}{15}t^{5/2} - \frac{4}{3}t^{3/2} + \frac{4}{3}t + k_2$ where $s(0) = k_2 = -\frac{4}{15}$. Thus $s(t) = \frac{4}{15}t^{5/2} - \frac{4}{3}t^{3/2} + \frac{4}{3}t - \frac{4}{15}$.

33. (a) $L = k\left(\frac{a - b\cot\theta}{R^4} + \frac{b\csc\theta}{r^4}\right) \Rightarrow \frac{dL}{d\theta} = k\left(\frac{b\csc^2\theta}{R^4} - \frac{b\csc\theta\cot\theta}{r^4}\right)$; solving $\frac{dL}{d\theta} = 0$ $\Rightarrow r^4 b\csc^2\theta - bR^4\csc\theta\cot\theta = 0 \Rightarrow (b\csc\theta)(r^4\csc\theta - R^4\cot\theta) = 0$; but $b\csc\theta \neq 0$ since $\theta \neq \frac{\pi}{2} \Rightarrow r^4\csc\theta - R^4\cot\theta = 0 \Rightarrow \cos\theta = \frac{r^4}{R^4} \Rightarrow \theta = \cos^{-1}\left(\frac{r^4}{R^4}\right)$, the critical value of θ

(b) $\theta = \cos^{-1}\left(\frac{5}{6}\right)^4 \approx \cos^{-1}(0.48225) \approx 61°$

34. (a) If $\frac{1}{x} - 3 = 0$, then $\frac{1-3x}{x} = 0 \Rightarrow x = \frac{1}{3}$.

(b) $f(x) = \frac{1}{x} - 3$ and $f'(x) = -\frac{1}{x^2}$

$\Rightarrow x_{n+1} = x_n - \frac{f(x_n)}{f'(x_n)} = x_n - \frac{\frac{1}{x_n} - 3}{-\frac{1}{x_n^2}}$

$= 2x_n - 3x_n^2 = x_n(2 - 3x_n)$

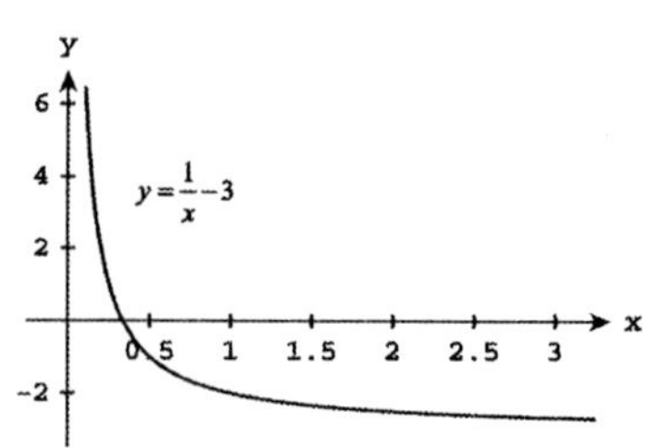

35. The graph of $f(x) = ax^2 + bx + c$ with $a > 0$ is a parabola opening upwards. Thus $f(x) \geq 0$ for all x if $f(x) = 0$ for at most one real value of x. The solutions to $f(x) = 0$ are, by the quadratic equation $\frac{-2b \pm \sqrt{(2b)^2 - 4ac}}{2a}$. Thus we require $(2b)^2 - 4ac \leq 0 \Rightarrow b^2 - ac \leq 0$.

36. (a) Clearly $f(x) = (a_1x + b_1)^2 + \ldots + (a_nx + b_n)^2 \geq 0$ for all x. Expanding we see

$f(x) = (a_1^2x^2 + 2a_1b_1x + b_1^2) + \ldots + (a_n^2x^2 + 2a_nb_nx + b_n^2)$

$= (a_1^2 + a_2^2 + \ldots + a_n^2)x^2 + 2(a_1b_1 + a_2b_2 + \ldots + a_nb_n)x + (b_1^2 + b_2^2 + \ldots + b_n^2) \geq 0$.

Thus $(a_1b_1 + a_2b_2 + \ldots + a_nb_n)^2 - (a_1^2 + a_2^2 + \ldots + a_n^2)(b_1^2 + b_2^2 + \ldots + b_n^2) \leq 0$ by Exercise 35.

Thus $(a_1b_1 + a_2b_2 + \ldots + a_nb_n)^2 \leq (a_1^2 + a_2^2 + \ldots + a_n^2)(b_1^2 + b_2^2 + \ldots + b_n^2)$.

(b) Referring to Exercise 35: It is clear that $f(x) = 0$ for some real $x \Leftrightarrow b^2 - 4ac = 0$, by quadratic formula. Now notice that this implies that

$f(x) = (a_1x + b_1)^2 + \ldots + (a_nx + b_n)^2$

$= (a_1^2 + a_2^2 + \ldots + a_n^2)x^2 + 2(a_1b_1 + a_2b_2 + \ldots + a_nb_n)x + (b_1^2 + b_2^2 + \ldots + b_n^2) = 0$

$\Leftrightarrow (a_1b_1 + a_2b_2 + \ldots + a_nb_n)^2 - (a_1^2 + a_2^2 + \ldots + a_n^2)(b_1^2 + b_2^2 + \ldots + b_n^2) = 0$

$\Leftrightarrow (a_1b_1 + a_2b_2 + \ldots + a_nb_n)^2 = (a_1^2 + a_2^2 + \ldots + a_n^2)(b_1^2 + b_2^2 + \ldots + b_n^2)$

But now $f(x) = 0 \Leftrightarrow a_ix + b_i = 0$ for all $i = 1, 2, \ldots, n \Leftrightarrow a_ix = -b_i = 0$ for all $i = 1, 2, \ldots, n$.

CHAPTER 5 INTEGRATION

5.1 ESTIMATING WITH FINITE SUMS

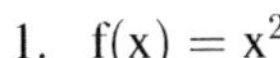

1. $f(x) = x^2$

Since f is increasing on [0, 1], we use left endpoints to obtain lower sums and right endpoints to obtain upper sums.

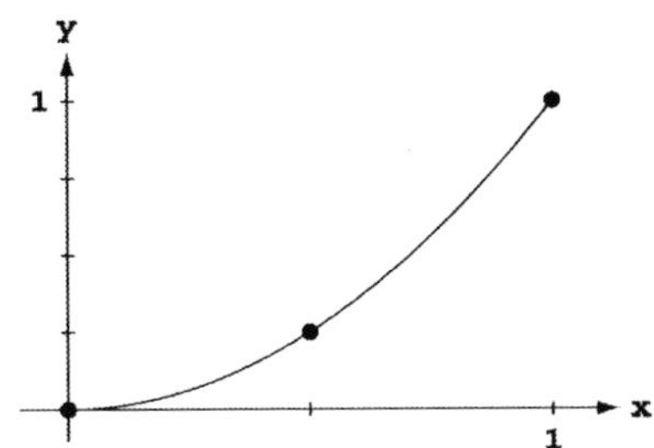

(a) $\triangle x = \frac{1-0}{2} = \frac{1}{2}$ and $x_i = i\triangle x = \frac{i}{2} \Rightarrow$ a lower sum is $\sum_{i=0}^{1} \left(\frac{i}{2}\right)^2 \cdot \frac{1}{2} = \frac{1}{2}\left(0^2 + \left(\frac{1}{2}\right)^2\right) = \frac{1}{8}$

(b) $\triangle x = \frac{1-0}{4} = \frac{1}{4}$ and $x_i = i\triangle x = \frac{i}{4} \Rightarrow$ a lower sum is $\sum_{i=0}^{3} \left(\frac{i}{4}\right)^2 \cdot \frac{1}{4} = \frac{1}{4}\left(0^2 + \left(\frac{1}{4}\right)^2 + \left(\frac{1}{2}\right)^2 + \left(\frac{3}{4}\right)^2\right) = \frac{1}{4} \cdot \frac{7}{8} = \frac{7}{32}$

(c) $\triangle x = \frac{1-0}{2} = \frac{1}{2}$ and $x_i = i\triangle x = \frac{i}{2} \Rightarrow$ an upper sum is $\sum_{i=1}^{2} \left(\frac{i}{2}\right)^2 \cdot \frac{1}{2} = \frac{1}{2}\left(\left(\frac{1}{2}\right)^2 + 1^2\right) = \frac{5}{8}$

(d) $\triangle x = \frac{1-0}{4} = \frac{1}{4}$ and $x_i = i\triangle x = \frac{i}{4} \Rightarrow$ an upper sum is $\sum_{i=1}^{4} \left(\frac{i}{4}\right)^2 \cdot \frac{1}{4} = \frac{1}{4}\left(\left(\frac{1}{4}\right)^2 + \left(\frac{1}{2}\right)^2 + \left(\frac{3}{4}\right)^2 + 1^2\right) = \frac{1}{4} \cdot \left(\frac{30}{16}\right) = \frac{15}{32}$

2. $f(x) = x^3$

Since f is increasing on [0, 1], we use left endpoints to obtain lower sums and right endpoints to obtain upper sums.

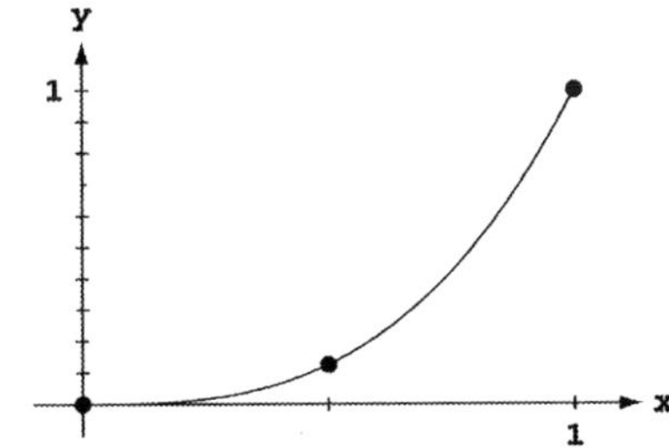

(a) $\triangle x = \frac{1-0}{2} = \frac{1}{2}$ and $x_i = i\triangle x = \frac{i}{2} \Rightarrow$ a lower sum is $\sum_{i=0}^{1} \left(\frac{i}{2}\right)^3 \cdot \frac{1}{2} = \frac{1}{2}\left(0^3 + \left(\frac{1}{2}\right)^3\right) = \frac{1}{16}$

(b) $\triangle x = \frac{1-0}{4} = \frac{1}{4}$ and $x_i = i\triangle x = \frac{i}{4} \Rightarrow$ a lower sum is $\sum_{i=0}^{3} \left(\frac{i}{4}\right)^3 \cdot \frac{1}{4} = \frac{1}{4}\left(0^3 + \left(\frac{1}{4}\right)^3 + \left(\frac{1}{2}\right)^3 + \left(\frac{3}{4}\right)^3\right) = \frac{36}{256} = \frac{9}{64}$

(c) $\triangle x = \frac{1-0}{2} = \frac{1}{2}$ and $x_i = i\triangle x = \frac{i}{2} \Rightarrow$ an upper sum is $\sum_{i=1}^{2} \left(\frac{i}{2}\right)^3 \cdot \frac{1}{2} = \frac{1}{2}\left(\left(\frac{1}{2}\right)^3 + 1^3\right) = \frac{1}{2} \cdot \frac{9}{8} = \frac{9}{16}$

(d) $\triangle x = \frac{1-0}{4} = \frac{1}{4}$ and $x_i = i\triangle x = \frac{i}{4} \Rightarrow$ an upper sum is $\sum_{i=1}^{4} \left(\frac{i}{4}\right)^3 \cdot \frac{1}{4} = \frac{1}{4}\left(\left(\frac{1}{4}\right)^3 + \left(\frac{1}{2}\right)^3 + \left(\frac{3}{4}\right)^3 + 1^3\right) = = \frac{100}{256} = \frac{25}{64}$

3. $f(x) = \frac{1}{x}$

Since f is decreasing on [0, 1], we use left endpoints to obtain upper sums and right endpoints to obtain lower sums.

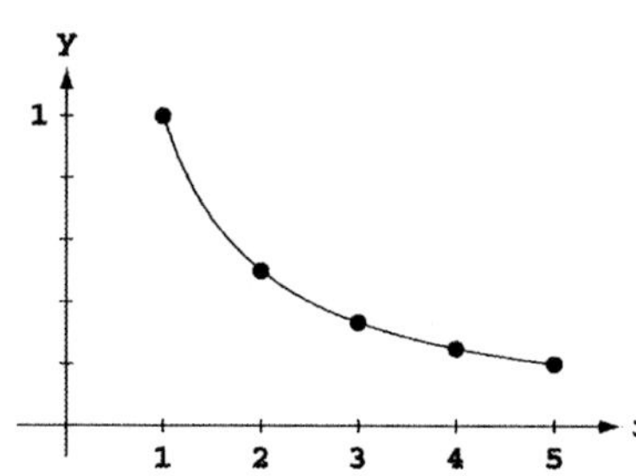

(a) $\triangle x = \frac{5-1}{2} = 2$ and $x_i = 1 + i\triangle x = 1 + 2i \Rightarrow$ a lower sum is $\sum_{i=1}^{2} \frac{1}{x_i} \cdot 2 = 2\left(\frac{1}{3} + \frac{1}{5}\right) = \frac{16}{15}$

(b) $\triangle x = \frac{5-1}{4} = 1$ and $x_i = 1 + i\triangle x = 1 + i \Rightarrow$ a lower sum is $\sum_{i=1}^{4} \frac{1}{x_i} \cdot 1 = 1\left(\frac{1}{2} + \frac{1}{3} + \frac{1}{4} + \frac{1}{5}\right) = \frac{77}{60}$

(c) $\triangle x = \frac{5-1}{2} = 2$ and $x_i = 1 + i\triangle x = 1 + 2i \Rightarrow$ an upper sum is $\sum_{i=0}^{1} \frac{1}{x_i} \cdot 2 = 2\left(1 + \frac{1}{3}\right) = \frac{8}{3}$

(d) $\triangle x = \frac{5-1}{4} = 1$ and $x_i = 1 + i\triangle x = 1 + i \Rightarrow$ an upper sum is $\sum_{i=0}^{3} \frac{1}{x_i} \cdot 1 = 1\left(1 + \frac{1}{2} + \frac{1}{3} + \frac{1}{4}\right) = \frac{25}{12}$

4. $f(x) = 4 - x^2$

Since f is increasing on [−2, 0] and decreasing on [0, 2], we use left endpoints on [−2, 0] and right endpoints on [0, 2] to obtain lower sums and use right endpoints on [−2, 0] and left endpoints on [0, 2] to obtain upper sums.

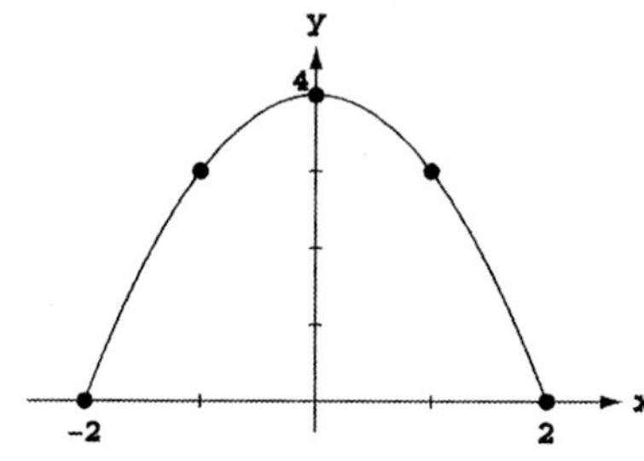

(a) $\triangle x = \frac{2-(-2)}{2} = 2$ and $x_i = -2 + i\triangle x = -2 + 2i \Rightarrow$ a lower sum is $2 \cdot \left(4 - (-2)^2\right) + 2 \cdot (4 - 2^2) = 0$

(b) $\triangle x = \frac{2-(-2)}{4} = 1$ and $x_i = -2 + i\triangle x = -2 + i \Rightarrow$ a lower sum is $\sum_{i=0}^{1}\left(4 - (x_i)^2\right) \cdot 1 + \sum_{i=3}^{4}\left(4 - (x_i)^2\right) \cdot 1$

$= 1\left(\left(4 - (-2)^2\right) + \left(4 - (-1)^2\right) + (4 - 1^2) + (4 - 2^2)\right) = 6$

(c) $\triangle x = \frac{2-(-2)}{2} = 2$ and $x_i = -2 + i\triangle x = -2 + 2i \Rightarrow$ a upper sum is $2 \cdot \left(4 - (0)^2\right) + 2 \cdot (4 - 0^2) = 16$

(d) $\triangle x = \frac{2-(-2)}{4} = 1$ and $x_i = -2 + i\triangle x = -2 + i \Rightarrow$ a upper sum is $\sum_{i=1}^{2}\left(4 - (x_i)^2\right) \cdot 1 + \sum_{i=2}^{3}\left(4 - (x_i)^2\right) \cdot 1$

$= 1\left(\left(4 - (-1)^2\right) + (4 - 0^2) + (4 - 0^2) + (4 - 1^2)\right) = 14$

5. $f(x) = x^2$

Using 2 rectangles $\Rightarrow \triangle x = \frac{1-0}{2} = \frac{1}{2} \Rightarrow \frac{1}{2}\left(f\left(\frac{1}{4}\right) + f\left(\frac{3}{4}\right)\right)$
$= \frac{1}{2}\left(\left(\frac{1}{4}\right)^2 + \left(\frac{3}{4}\right)^2\right) = \frac{10}{32} = \frac{5}{16}$

Using 4 rectangles $\Rightarrow \triangle x = \frac{1-0}{4} = \frac{1}{4}$
$\Rightarrow \frac{1}{4}\left(f\left(\frac{1}{8}\right) + f\left(\frac{3}{8}\right) + f\left(\frac{5}{8}\right) + f\left(\frac{7}{8}\right)\right)$
$= \frac{1}{4}\left(\left(\frac{1}{8}\right)^2 + \left(\frac{3}{8}\right)^2 + \left(\frac{5}{8}\right)^2 + \left(\frac{7}{8}\right)^2\right) = \frac{21}{64}$

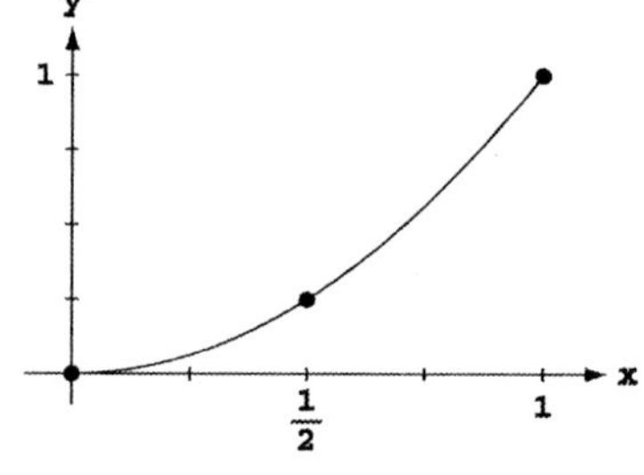

6. $f(x) = x^3$

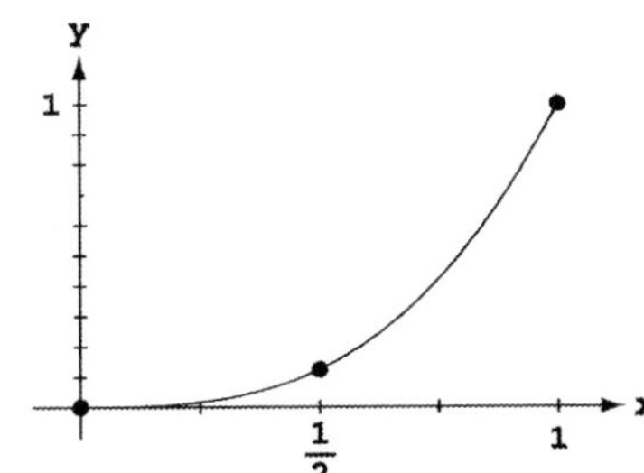

Using 2 rectangles $\Rightarrow \triangle x = \frac{1-0}{2} = \frac{1}{2} \Rightarrow \frac{1}{2}\left(f\left(\frac{1}{4}\right) + f\left(\frac{3}{4}\right)\right)$
$= \frac{1}{2}\left(\left(\frac{1}{4}\right)^3 + \left(\frac{3}{4}\right)^3\right) = \frac{28}{2 \cdot 64} = \frac{7}{32}$

Using 4 rectangles $\Rightarrow \triangle x = \frac{1-0}{4} = \frac{1}{4}$
$\Rightarrow \frac{1}{4}\left(f\left(\frac{1}{8}\right) + f\left(\frac{3}{8}\right) + f\left(\frac{5}{8}\right) + f\left(\frac{7}{8}\right)\right)$
$= \frac{1}{4}\left(\frac{1^3+3^3+5^3+7^3}{8^3}\right) = \frac{496}{4 \cdot 8^3} = \frac{124}{8^3} = \frac{31}{128}$

7. $f(x) = \frac{1}{x}$

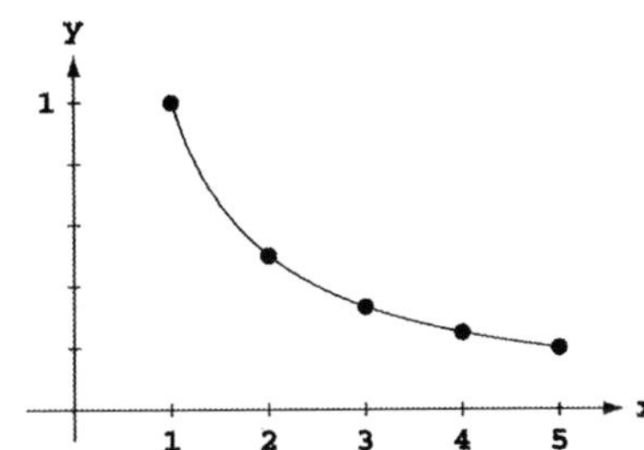

Using 2 rectangles $\Rightarrow \triangle x = \frac{5-1}{2} = 2 \Rightarrow 2(f(2) + f(4))$
$= 2\left(\frac{1}{2} + \frac{1}{4}\right) = \frac{3}{2}$

Using 4 rectangles $\Rightarrow \triangle x = \frac{5-1}{4} = 1$
$\Rightarrow 1\left(f\left(\frac{3}{2}\right) + f\left(\frac{5}{2}\right) + f\left(\frac{7}{2}\right) + f\left(\frac{9}{2}\right)\right)$
$= 1\left(\frac{2}{3} + \frac{2}{5} + \frac{2}{7} + \frac{2}{9}\right) = \frac{1488}{3 \cdot 5 \cdot 7 \cdot 9} = \frac{496}{5 \cdot 7 \cdot 9} = \frac{496}{315}$

8. $f(x) = 4 - x^2$

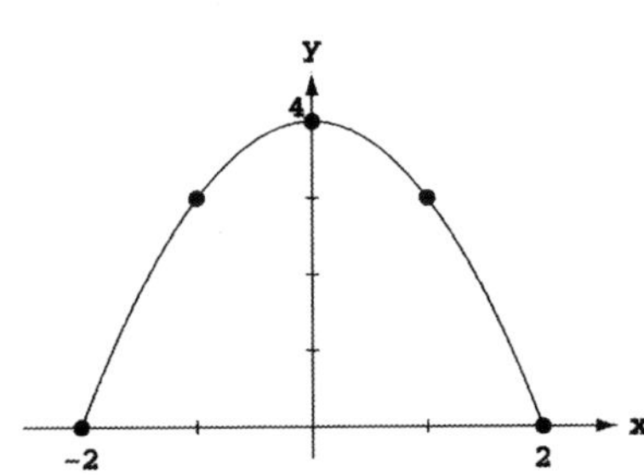

Using 2 rectangles $\Rightarrow \triangle x = \frac{2-(-2)}{2} = 2 \Rightarrow 2(f(-1) + f(1))$
$= 2(3 + 3) = 12$

Using 4 rectangles $\Rightarrow \triangle x = \frac{2-(-2)}{4} = 1$
$\Rightarrow 1\left(f\left(-\frac{3}{2}\right) + f\left(-\frac{1}{2}\right) + f\left(\frac{1}{2}\right) + f\left(\frac{3}{2}\right)\right)$
$= 1\left(\left(4 - \left(-\frac{3}{2}\right)^2\right) + \left(4 - \left(-\frac{1}{2}\right)^2\right) + \left(4 - \left(\frac{1}{2}\right)^2\right) + \left(4 - \left(\frac{3}{2}\right)^2\right)\right)$
$= 16 - \left(\frac{9}{4} \cdot 2 + \frac{1}{4} \cdot 2\right) = 16 - \frac{10}{2} = 11$

9. (a) $D \approx (0)(1) + (12)(1) + (22)(1) + (10)(1) + (5)(1) + (13)(1) + (11)(1) + (6)(1) + (2)(1) + (6)(1) = 87$ inches
   (b) $D \approx (12)(1) + (22)(1) + (10)(1) + (5)(1) + (13)(1) + (11)(1) + (6)(1) + (2)(1) + (6)(1) + (0)(1) = 87$ inches

10. (a) $D \approx (1)(300) + (1.2)(300) + (1.7)(300) + (2.0)(300) + (1.8)(300) + (1.6)(300) + (1.4)(300) + (1.2)(300)$ $+ (1.0)(300) + (1.8)(300) + (1.5)(300) + (1.2)(300) = 5220$ meters (NOTE: 5 minutes = 300 seconds)
    (b) $D \approx (1.2)(300) + (1.7)(300) + (2.0)(300) + (1.8)(300) + (1.6)(300) + (1.4)(300) + (1.2)(300) + (1.0)(300)$ $+ (1.8)(300) + (1.5)(300) + (1.2)(300) + (0)(300) = 4920$ meters (NOTE: 5 minutes = 300 seconds)

11. (a) $D \approx (0)(10) + (44)(10) + (15)(10) + (35)(10) + (30)(10) + (44)(10) + (35)(10) + (15)(10) + (22)(10)$ $+ (35)(10) + (44)(10) + (30)(10) = 3490$ feet ≈ 0.66 miles
    (b) $D \approx (44)(10) + (15)(10) + (35)(10) + (30)(10) + (44)(10) + (35)(10) + (15)(10) + (22)(10) + (35)(10)$ $+ (44)(10) + (30)(10) + (35)(10) = 3840$ feet ≈ 0.73 miles

12. (a) The distance traveled will be the area under the curve. We will use the approximate velocities at the midpoints of each time interval to approximate this area using rectangles. Thus,
    $D \approx (20)(0.001) + (50)(0.001) + (72)(0.001) + (90)(0.001) + (102)(0.001) + (112)(0.001) + (120)(0.001)$ $+ (128)(0.001) + (134)(0.001) + (139)(0.001) \approx 0.967$ miles
    (b) Roughly, after 0.0063 hours, the car would have gone 0.484 miles, where 0.0060 hours = 22.7 sec. At 22.7 sec, the velocity was approximately 120 mi/hr.

13. (a) Because the acceleration is decreasing, an upper estimate is obtained using left end-points in summing acceleration $\cdot \Delta t$. Thus, $\Delta t = 1$ and speed $\approx [32.00 + 19.41 + 11.77 + 7.14 + 4.33](1) = 74.65$ ft/sec

(b) Using right end-points we obtain a lower estimate: speed $\approx [19.41 + 11.77 + 7.14 + 4.33 + 2.63](1) = 45.28$ ft/sec

(c) Upper estimates for the speed at each second are:

t	0	1	2	3	4	5
v	0	32.00	51.41	63.18	70.32	74.65

Thus, the distance fallen when t = 3 seconds is $s \approx [32.00 + 51.41 + 63.18](1) = 146.59$ ft.

14. (a) The speed is a decreasing function of time $\Rightarrow$ right end-points give an lower estimate for the height (distance) attained. Also

t	0	1	2	3	4	5
v	400	368	336	304	272	240

gives the time-velocity table by subtracting the constant $g = 32$ from the speed at each time increment $\Delta t = 1$ sec. Thus, the speed ≈ 240 ft/sec after 5 seconds.

(b) A lower estimate for height attained is $h \approx [368 + 336 + 304 + 272 + 240](1) = 1520$ ft.

15. Partition $[0,2]$ into the four subintervals $[0,0.5]$, $[0.5,1]$, $[1,1.5]$, and $[1.5,2]$. The midpoints of these subintervals are $m_1 = 0.25$, $m_2 = 0.75$, $m_3 = 1.25$, and $m_4 = 1.75$. The heights of the four approximating rectangles are $f(m_1) = (0.25)^3 = \frac{1}{64}$, $f(m_2) = (0.75)^3 = \frac{27}{64}$, $f(m_3) = (1.25)^3 = \frac{125}{64}$, and $f(m_4) = (1.75)^3 = \frac{343}{64}$. Notice that the average value is approximated by $\frac{1}{2}\left[\left(\frac{1}{4}\right)^3\left(\frac{1}{2}\right) + \left(\frac{3}{4}\right)^3\left(\frac{1}{2}\right) + \left(\frac{5}{4}\right)^3\left(\frac{1}{2}\right) + \left(\frac{7}{4}\right)^3\left(\frac{1}{2}\right)\right] = \frac{31}{16}$ $= \frac{1}{\text{length of }[0,2]} \cdot \left[\begin{matrix}\text{approximate area under}\\ \text{curve } f(x) = x^3\end{matrix}\right]$. We use this observation in solving the next several exercises.

16. Partition $[1,9]$ into the four subintervals $[1,3]$, $[3,5]$, $[5,7]$, and $[7,9]$. The midpoints of these subintervals are $m_1 = 2$, $m_2 = 4$, $m_3 = 6$, and $m_4 = 8$. The heights of the four approximating rectangles are $f(m_1) = \frac{1}{2}$, $f(m_2) = \frac{1}{4}$, $f(m_3) = \frac{1}{6}$, and $f(m_4) = \frac{1}{8}$. The width of each rectangle is $\Delta x = 2$. Thus, Area $\approx 2\left(\frac{1}{2}\right) + 2\left(\frac{1}{4}\right) + 2\left(\frac{1}{6}\right) + 2\left(\frac{1}{8}\right) = \frac{25}{12} \Rightarrow$ average value $\approx \frac{\text{area}}{\text{length of }[1,9]} = \frac{\left(\frac{25}{12}\right)}{8} = \frac{25}{96}$.

17. Partition $[0,2]$ into the four subintervals $[0,0.5]$, $[0.5,1]$, $[1,1.5]$, and $[1.5,2]$. The midpoints of the subintervals are $m_1 = 0.25$, $m_2 = 0.75$, $m_3 = 1.25$, and $m_4 = 1.75$. The heights of the four approximating rectangles are $f(m_1) = \frac{1}{2} + \sin^2\frac{\pi}{4} = \frac{1}{2} + \frac{1}{2} = 1$, $f(m_2) = \frac{1}{2} + \sin^2\frac{3\pi}{4} = \frac{1}{2} + \frac{1}{2} = 1$, $f(m_3) = \frac{1}{2} + \sin^2\frac{5\pi}{4} = \frac{1}{2} + \left(-\frac{1}{\sqrt{2}}\right)^2$ $= \frac{1}{2} + \frac{1}{2} = 1$, and $f(m_4) = \frac{1}{2} + \sin^2\frac{7\pi}{4} = \frac{1}{2} + \left(-\frac{1}{\sqrt{2}}\right)^2 = 1$. The width of each rectangle is $\Delta x = \frac{1}{2}$. Thus, Area $\approx (1 + 1 + 1 + 1)\left(\frac{1}{2}\right) = 2 \Rightarrow$ average value $\approx \frac{\text{area}}{\text{length of }[0,2]} = \frac{2}{2} = 1$.

18. Partition $[0,4]$ into the four subintervals $[0,1]$, $[1,2,]$, $[2,3]$, and $[3,4]$. The midpoints of the subintervals are $m_1 = \frac{1}{2}$, $m_2 = \frac{3}{2}$, $m_3 = \frac{5}{2}$, and $m_4 = \frac{7}{2}$. The heights of the four approximating rectangles are $f(m_1) = 1 - \left(\cos\left(\frac{\pi\left(\frac{1}{2}\right)}{4}\right)\right)^4 = 1 - \left(\cos\left(\frac{\pi}{8}\right)\right)^4 = 0.27145$ (to 5 decimal places), $f(m_2) = 1 - \left(\cos\left(\frac{\pi\left(\frac{3}{2}\right)}{4}\right)\right)^4 = 1 - \left(\cos\left(\frac{3\pi}{8}\right)\right)^4 = 0.97855$, $f(m_3) = 1 - \left(\cos\left(\frac{\pi\left(\frac{5}{2}\right)}{4}\right)\right)^4 = 1 - \left(\cos\left(\frac{5\pi}{8}\right)\right)^4$ $= 0.97855$, and $f(m_4) = 1 - \left(\cos\left(\frac{\pi\left(\frac{7}{2}\right)}{4}\right)\right)^4 = 1 - \left(\cos\left(\frac{7\pi}{8}\right)\right)^4 = 0.27145$. The width of each rectangle is $\Delta x = 1$. Thus, Area $\approx (0.27145)(1) + (0.97855)(1) + (0.97855)(1) + (0.27145)(1) = 2.5 \Rightarrow$ average value $\approx \frac{\text{area}}{\text{length of }[0,4]} = \frac{2.5}{4} = \frac{5}{8}$.

19. Since the leakage is increasing, an upper estimate uses right endpoints and a lower estimate uses left endpoints:

(a) upper estimate $= (70)(1) + (97)(1) + (136)(1) + (190)(1) + (265)(1) = 758$ gal,
lower estimate $= (50)(1) + (70)(1) + (97)(1) + (136)(1) + (190)(1) = 543$ gal.

(b) upper estimate $= (70 + 97 + 136 + 190 + 265 + 369 + 516 + 720) = 2363$ gal,
lower estimate $= (50 + 70 + 97 + 136 + 190 + 265 + 369 + 516) = 1693$ gal.

(c) worst case: $2363 + 720t = 25{,}000 \Rightarrow t \approx 31.4$ hrs;
best case: $1693 + 720t = 25{,}000 \Rightarrow t \approx 32.4$ hrs

20. Since the pollutant release increases over time, an upper estimate uses right endpoints and a lower estimate uses left endpoints:

(a) upper estimate $= (0.2)(30) + (0.25)(30) + (0.27)(30) + (0.34)(30) + (0.45)(30) + (0.52)(30) = 60.9$ tons
lower estimate $= (0.05)(30) + (0.2)(30) + (0.25)(30) + (0.27)(30) + (0.34)(30) + (0.45)(30) = 46.8$ tons

(b) Using the lower (best case) estimate: $46.8 + (0.52)(30) + (0.63)(30) + (0.70)(30) + (0.81)(30) = 126.6$ tons, so near the end of September 125 tons of pollutants will have been released.

21. (a) The diagonal of the square has length 2, so the side length is $\sqrt{2}$. Area $= \left(\sqrt{2}\right)^2 = 2$

(b) Think of the octagon as a collection of 16 right triangles with a hypotenuse of length 1 and an acute angle measuring $\frac{2\pi}{16} = \frac{\pi}{8}$.
Area $= 16\left(\frac{1}{2}\right)\left(\sin\frac{\pi}{8}\right)\left(\cos\frac{\pi}{8}\right) = 4\sin\frac{\pi}{4} = 2\sqrt{2} \approx 2.828$

(c) Think of the 16-gon as a collection of 32 right triangles with a hypotenuse of length 1 and an acute angle measuring $\frac{2\pi}{32} = \frac{\pi}{16}$.
Area $= 32\left(\frac{1}{2}\right)\left(\sin\frac{\pi}{16}\right)\left(\cos\frac{\pi}{16}\right) = 8\sin\frac{\pi}{8} = 2\sqrt{2} \approx 3.061$

(d) Each area is less than the area of the circle, π. As n increases, the area approaches π.

22. (a) Each of the isosceles triangles is made up of two right triangles having hypotenuse 1 and an acute angle measuring $\frac{2\pi}{2n} = \frac{\pi}{n}$. The area of each isosceles triangle is $A_T = 2\left(\frac{1}{2}\right)\left(\sin\frac{\pi}{n}\right)\left(\cos\frac{\pi}{n}\right) = \frac{1}{2}\sin\frac{2\pi}{n}$.

(b) The area of the polygon is $A_P = nA_T = \frac{n}{2}\sin\frac{2\pi}{n}$, so $\lim\limits_{n\to\infty} \frac{n}{2}\sin\frac{2\pi}{n} = \lim\limits_{n\to\infty} \pi \cdot \frac{\sin\frac{2\pi}{n}}{\left(\frac{2\pi}{n}\right)} = \pi$

(c) Multiply each area by r^2.
$A_T = \frac{1}{2}r^2\sin\frac{2\pi}{n}$
$A_P = \frac{n}{2}r^2\sin\frac{2\pi}{n}$
$\lim\limits_{n\to\infty} A_P = \pi r^2$

23-26. Example CAS commands:

Maple:

```
with( Student[Calculus1] );
f := x -> sin(x);
a := 0;
b := Pi;
plot( f(x), x=a..b, title="#23(a) (Section 5.1)" );
N := [ 100, 200, 1000 ];                          # (b)
for n in N do
  Xlist := [ a+1.*(b-a)/n*i $ i=0..n ];
  Ylist := map( f, Xlist );
end do:
for n in N do                                      # (c)
  Avg[n] := evalf(add(y,y=Ylist)/nops(Ylist));
```

```
end do;
avg := FunctionAverage( f(x), x=a..b, output=value );
evalf( avg );
FunctionAverage(f(x),x=a..b,output=plot);     # (d)
fsolve( f(x)=avg, x=0.5 );
fsolve( f(x)=avg, x=2.5 );
fsolve( f(x)=Avg[1000], x=0.5 );
fsolve( f(x)=Avg[1000], x=2.5 );
```

Mathematica: (assigned function and values for a and b may vary):

Symbols for π, $\rightarrow$, powers, roots, fractions, etc. are available in Palettes (under File).

Never insert a space between the name of a function and its argument.

```
Clear[x]
f[x_]:=x Sin[1/x]
{a,b}={π/4, π}
Plot[f[x],{x, a, b}]
```

The following code computes the value of the function for each interval midpoint and then finds the average. Each sequence of commands for a different value of n (number of subdivisions) should be placed in a separate cell.

```
n =100; dx = (b − a) /n;
values = Table[N[f[x]], {x, a + dx/2, b, dx}]
average=Sum[values[[i]],{i, 1, Length[values]}] / n
n =200; dx = (b − a) /n;
values = Table[N[f[x]],{x, a + dx/2, b, dx}]
average=Sum[values[[i]],{i, 1, Length[values]}] / n
n =1000; dx = (b − a) /n;
values = Table[N[f[x]],{x, a + dx/2, b, dx}]
average=Sum[values[[i]],{i, 1, Length[values]}] / n
FindRoot[f[x] == average,{x, a}]
```

5.2 SIGMA NOTATION AND LIMITS OF FINITE SUMS

1. $\sum\limits_{k=1}^{2} \frac{6k}{k+1} = \frac{6(1)}{1+1} + \frac{6(2)}{2+1} = \frac{6}{2} + \frac{12}{3} = 7$

2. $\sum\limits_{k=1}^{3} \frac{k-1}{k} = \frac{1-1}{1} + \frac{2-1}{2} + \frac{3-1}{3} = 0 + \frac{1}{2} + \frac{2}{3} = \frac{7}{6}$

3. $\sum\limits_{k=1}^{4} \cos k\pi = \cos(1\pi) + \cos(2\pi) + \cos(3\pi) + \cos(4\pi) = -1 + 1 - 1 + 1 = 0$

4. $\sum\limits_{k=1}^{5} \sin k\pi = \sin(1\pi) + \sin(2\pi) + \sin(3\pi) + \sin(4\pi) + \sin(5\pi) = 0 + 0 + 0 + 0 + 0 = 0$

5. $\sum\limits_{k=1}^{3} (-1)^{k+1} \sin\frac{\pi}{k} = (-1)^{1+1}\sin\frac{\pi}{1} + (-1)^{2+1}\sin\frac{\pi}{2} + (-1)^{3+1}\sin\frac{\pi}{3} = 0 - 1 + \frac{\sqrt{3}}{2} = \frac{\sqrt{3}-2}{2}$

6. $\sum\limits_{k=1}^{4} (-1)^{k} \cos k\pi = (-1)^{1}\cos(1\pi) + (-1)^{2}\cos(2\pi) + (-1)^{3}\cos(3\pi) + (-1)^{4}\cos(4\pi)$

$= -(-1) + 1 - (-1) + 1 = 4$

7. (a) $\sum_{k=1}^{6} 2^{k-1} = 2^{1-1} + 2^{2-1} + 2^{3-1} + 2^{4-1} + 2^{5-1} + 2^{6-1} = 1 + 2 + 4 + 8 + 16 + 32$

(b) $\sum_{k=0}^{5} 2^k = 2^0 + 2^1 + 2^2 + 2^3 + 2^4 + 2^5 = 1 + 2 + 4 + 8 + 16 + 32$

(c) $\sum_{k=-1}^{4} 2^{k+1} = 2^{-1+1} + 2^{0+1} + 2^{1+1} + 2^{2+1} + 2^{3+1} + 2^{4+1} = 1 + 2 + 4 + 8 + 16 + 32$

All of them represent $1 + 2 + 4 + 8 + 16 + 32$

8. (a) $\sum_{k=1}^{6} (-2)^{k-1} = (-2)^{1-1} + (-2)^{2-1} + (-2)^{3-1} + (-2)^{4-1} + (-2)^{5-1} + (-2)^{6-1} = 1 - 2 + 4 - 8 + 16 - 32$

(b) $\sum_{k=0}^{5} (-1)^k 2^k = (-1)^0 2^0 + (-1)^1 2^1 + (-1)^2 2^2 + (-1)^3 2^3 + (-1)^4 2^4 + (-1)^5 2^5 = 1 - 2 + 4 - 8 + 16 - 32$

(c) $\sum_{k=-2}^{3} (-1)^{k+1} 2^{k+2} = (-1)^{-2+1} 2^{-2+2} + (-1)^{-1+1} 2^{-1+2} + (-1)^{0+1} 2^{0+2} + (-1)^{1+1} 2^{1+2} + (-1)^{2+1} 2^{2+2}$
$+ (-1)^{3+1} 2^{3+2} = -1 + 2 - 4 + 8 - 16 + 32;$

(a) and (b) represent $1 - 2 + 4 - 8 + 16 - 32$; (c) is not equivalent to the other two

9. (a) $\sum_{k=2}^{4} \frac{(-1)^{k-1}}{k-1} = \frac{(-1)^{2-1}}{2-1} + \frac{(-1)^{3-1}}{3-1} + \frac{(-1)^{4-1}}{4-1} = -1 + \frac{1}{2} - \frac{1}{3}$

(b) $\sum_{k=0}^{2} \frac{(-1)^k}{k+1} = \frac{(-1)^0}{0+1} + \frac{(-1)^1}{1+1} + \frac{(-1)^2}{2+1} = 1 - \frac{1}{2} + \frac{1}{3}$

(c) $\sum_{k=-1}^{1} \frac{(-1)^k}{k+2} = \frac{(-1)^{-1}}{-1+2} + \frac{(-1)^0}{0+2} + \frac{(-1)^1}{1+2} = -1 + \frac{1}{2} - \frac{1}{3}$

(a) and (c) are equivalent; (b) is not equivalent to the other two.

10. (a) $\sum_{k=1}^{4} (k-1)^2 = (1-1)^2 + (2-1)^2 + (3-1)^2 + (4-1)^2 = 0 + 1 + 4 + 9$

(b) $\sum_{k=-1}^{3} (k+1)^2 = (-1+1)^2 + (0+1)^2 + (1+1)^2 + (2+1)^2 + (3+1)^2 = 0 + 1 + 4 + 9 + 16$

(c) $\sum_{k=-3}^{-1} k^2 = (-3)^2 + (-2)^2 + (-1)^2 = 9 + 4 + 1$

(a) and (c) are equivalent to each other; (b) is not equivalent to the other two.

11. $\sum_{k=1}^{6} k$

12. $\sum_{k=1}^{4} k^2$

13. $\sum_{k=1}^{4} \frac{1}{2^k}$

14. $\sum_{k=1}^{5} 2k$

15. $\sum_{k=1}^{5} (-1)^{k+1} \frac{1}{k}$

16. $\sum_{k=1}^{5} (-1)^k \frac{k}{5}$

17. (a) $\sum_{k=1}^{n} 3a_k = 3 \sum_{k=1}^{n} a_k = 3(-5) = -15$

(b) $\sum_{k=1}^{n} \frac{b_k}{6} = \frac{1}{6} \sum_{k=1}^{n} b_k = \frac{1}{6}(6) = 1$

(c) $\sum_{k=1}^{n} (a_k + b_k) = \sum_{k=1}^{n} a_k + \sum_{k=1}^{n} b_k = -5 + 6 = 1$

(d) $\sum_{k=1}^{n} (a_k - b_k) = \sum_{k=1}^{n} a_k - \sum_{k=1}^{n} b_k = -5 - 6 = -11$

(e) $\sum_{k=1}^{n} (b_k - 2a_k) = \sum_{k=1}^{n} b_k - 2 \sum_{k=1}^{n} a_k = 6 - 2(-5) = 16$

18. (a) $\sum_{k=1}^{n} 8a_k = 8 \sum_{k=1}^{n} a_k = 8(0) = 0$ (b) $\sum_{k=1}^{n} 250b_k = 250 \sum_{k=1}^{n} b_k = 250(1) = 250$

(c) $\sum_{k=1}^{n} (a_k + 1) = \sum_{k=1}^{n} a_k + \sum_{k=1}^{n} 1 = 0 + n = n$ (d) $\sum_{k=1}^{n} (b_k - 1) = \sum_{k=1}^{n} b_k - \sum_{k=1}^{n} 1 = 1 - n$

19. (a) $\sum_{k=1}^{10} k = \frac{10(10+1)}{2} = 55$ (b) $\sum_{k=1}^{10} k^2 = \frac{10(10+1)(2(10)+1)}{6} = 385$

(c) $\sum_{k=1}^{10} k^3 = \left[\frac{10(10+1)}{2}\right]^2 = 55^2 = 3025$

20. (a) $\sum_{k=1}^{13} k = \frac{13(13+1)}{2} = 91$ (b) $\sum_{k=1}^{13} k^2 = \frac{13(13+1)(2(13)+1)}{6} = 819$

(c) $\sum_{k=1}^{13} k^3 = \left[\frac{13(13+1)}{2}\right]^2 = 91^2 = 8281$

21. $\sum_{k=1}^{7} -2k = -2 \sum_{k=1}^{7} k = -2\left(\frac{7(7+1)}{2}\right) = -56$

22. $\sum_{k=1}^{5} \frac{\pi k}{15} = \frac{\pi}{15} \sum_{k=1}^{5} k = \frac{\pi}{15}\left(\frac{5(5+1)}{2}\right) = \pi$

23. $\sum_{k=1}^{6} (3 - k^2) = \sum_{k=1}^{6} 3 - \sum_{k=1}^{6} k^2 = 3(6) - \frac{6(6+1)(2(6)+1)}{6} = -73$

24. $\sum_{k=1}^{6} (k^2 - 5) = \sum_{k=1}^{6} k^2 - \sum_{k=1}^{6} 5 = \frac{6(6+1)(2(6)+1)}{6} - 5(6) = 61$

25. $\sum_{k=1}^{5} k(3k+5) = \sum_{k=1}^{5} (3k^2 + 5k) = 3\sum_{k=1}^{5} k^2 + 5\sum_{k=1}^{5} k = 3\left(\frac{5(5+1)(2(5)+1)}{6}\right) + 5\left(\frac{5(5+1)}{2}\right) = 240$

26. $\sum_{k=1}^{7} k(2k+1) = \sum_{k=1}^{7} (2k^2 + k) = 2\sum_{k=1}^{7} k^2 + \sum_{k=1}^{7} k = 2\left(\frac{7(7+1)(2(7)+1)}{6}\right) + \frac{7(7+1)}{2} = 308$

27. $\sum_{k=1}^{5} \frac{k^3}{225} + \left(\sum_{k=1}^{5} k\right)^3 = \frac{1}{225}\sum_{k=1}^{5} k^3 + \left(\sum_{k=1}^{5} k\right)^3 = \frac{1}{225}\left(\frac{5(5+1)}{2}\right)^2 + \left(\frac{5(5+1)}{2}\right)^3 = 3376$

28. $\left(\sum_{k=1}^{7} k\right)^2 - \sum_{k=1}^{7} \frac{k^3}{4} = \left(\sum_{k=1}^{7} k\right)^2 - \frac{1}{4}\sum_{k=1}^{7} k^3 = \left(\frac{7(7+1)}{2}\right)^2 - \frac{1}{4}\left(\frac{7(7+1)}{2}\right)^2 = 588$

29. (a)

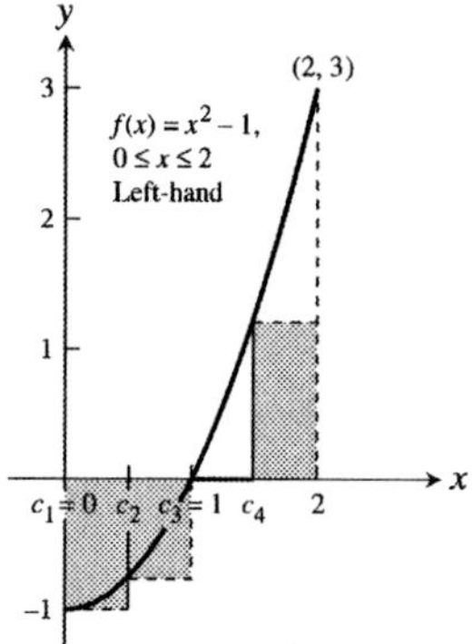

(b)

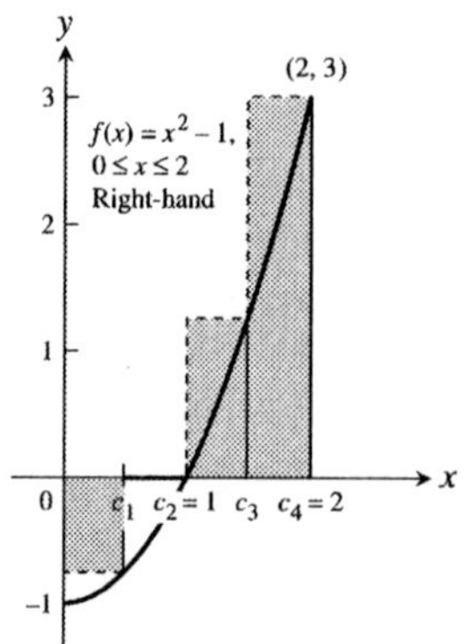

(c)

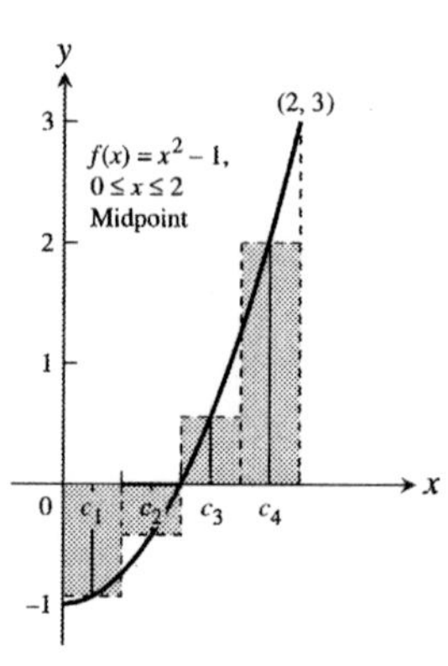

30. (a) 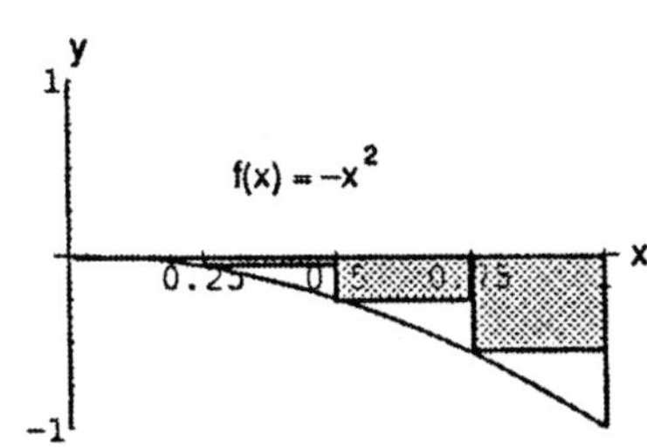

(b) 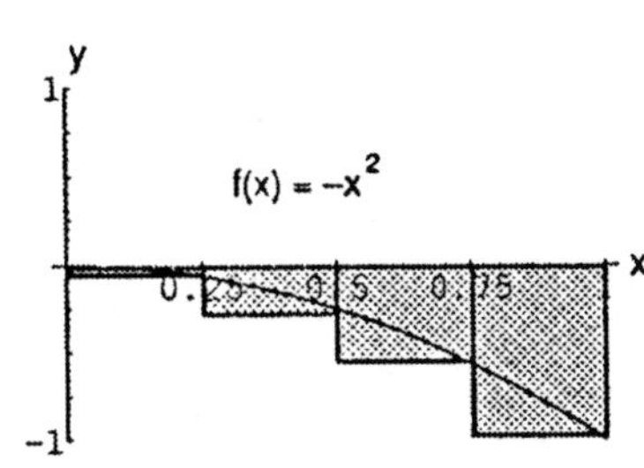

(c) 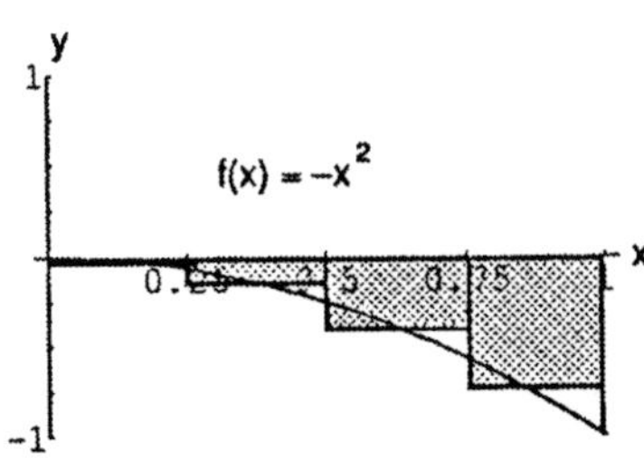

31. (a) 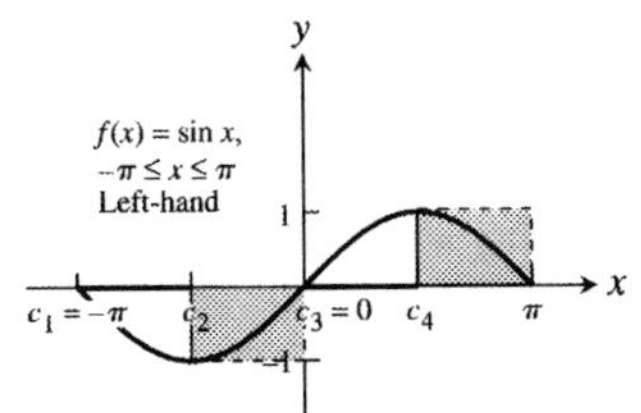

(b) 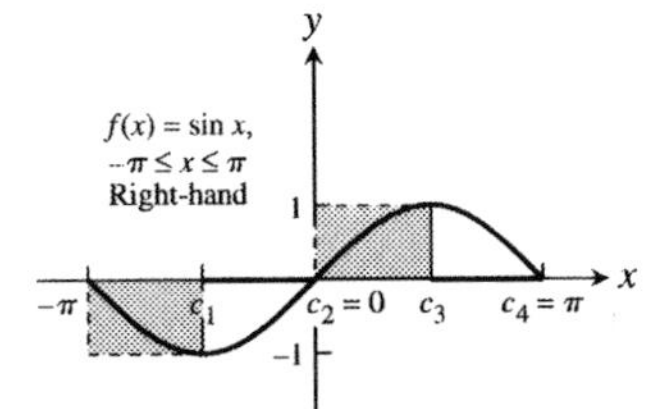

(c) 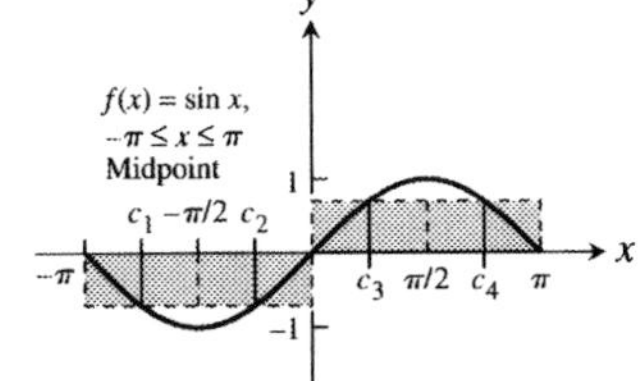

32. (a) 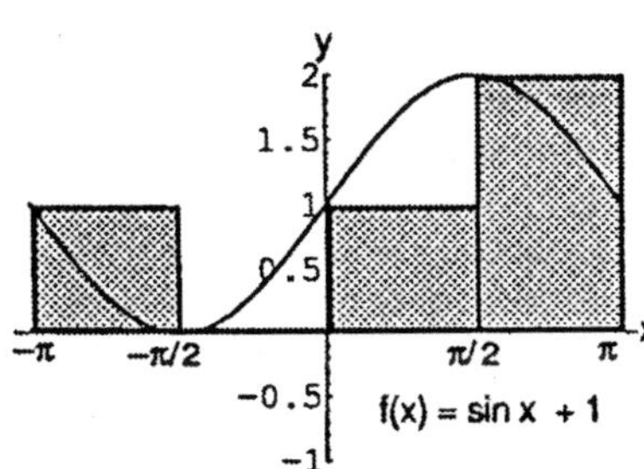

(b) 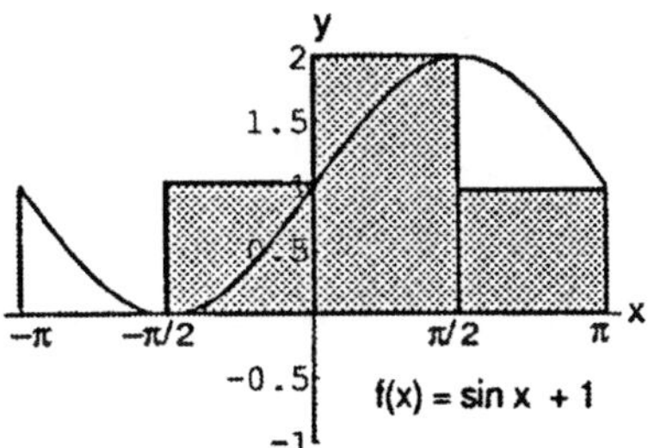

(c) 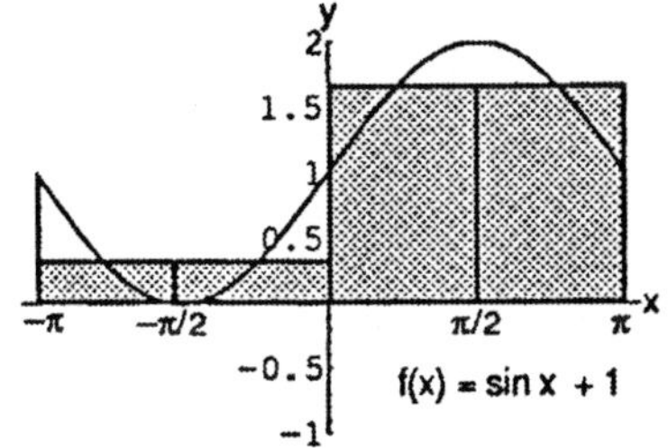

33. $|x_1 - x_0| = |1.2 - 0| = 1.2$, $|x_2 - x_1| = |1.5 - 1.2| = 0.3$, $|x_3 - x_2| = |2.3 - 1.5| = 0.8$, $|x_4 - x_3| = |2.6 - 2.3| = 0.3$, and $|x_5 - x_4| = |3 - 2.6| = 0.4$; the largest is $\|P\| = 1.2$.

34. $|x_1 - x_0| = |-1.6 - (-2)| = 0.4$, $|x_2 - x_1| = |-0.5 - (-1.6)| = 1.1$, $|x_3 - x_2| = |0 - (-0.5)| = 0.5$, $|x_4 - x_3| = |0.8 - 0| = 0.8$, and $|x_5 - x_4| = |1 - 0.8| = 0.2$; the largest is $\|P\| = 1.1$.

35. $f(x) = 1 - x^2$

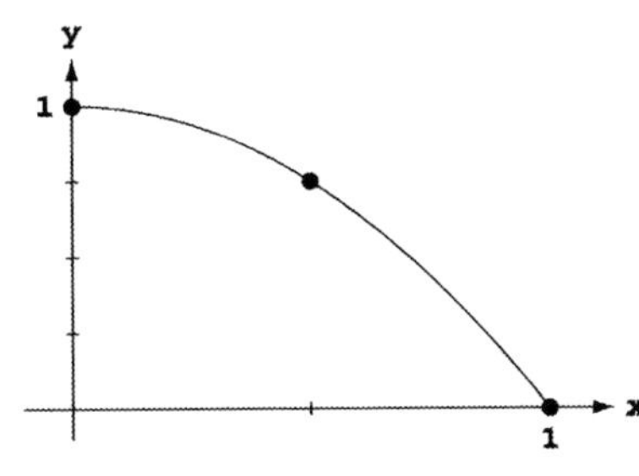

Since f is decreasing on $[0, 1]$ we use left endpoints to obtain upper sums. $\triangle x = \frac{1-0}{n} = \frac{1}{n}$ and $x_i = i\triangle x = \frac{i}{n}$. So an upper sum is $\sum_{i=0}^{n-1}(1 - x_i^2)\frac{1}{n} = \frac{1}{n}\sum_{i=0}^{n-1}\left(1 - \left(\frac{i}{n}\right)^2\right) = \frac{1}{n^3}\sum_{i=0}^{n-1}(n^2 - i^2)$

$= \frac{n^3}{n^3} - \frac{1}{n^3}\sum_{i=0}^{n} i^2 = 1 - \frac{(n-1)n(2(n-1)+1)}{6n^3} = 1 - \frac{2n^3 - 3n^2 + n}{6n^3}$

$= 1 - \frac{2 - \frac{3}{n} + \frac{1}{n^2}}{6}$. Thus,

$\lim_{n\to\infty}\sum_{i=0}^{n-1}(1 - x_i^2)\frac{1}{n} = \lim_{n\to\infty}\left(1 - \frac{2 - \frac{3}{n} + \frac{1}{n^2}}{6}\right) = 1 - \frac{1}{3} = \frac{2}{3}$

36. $f(x) = 2x$

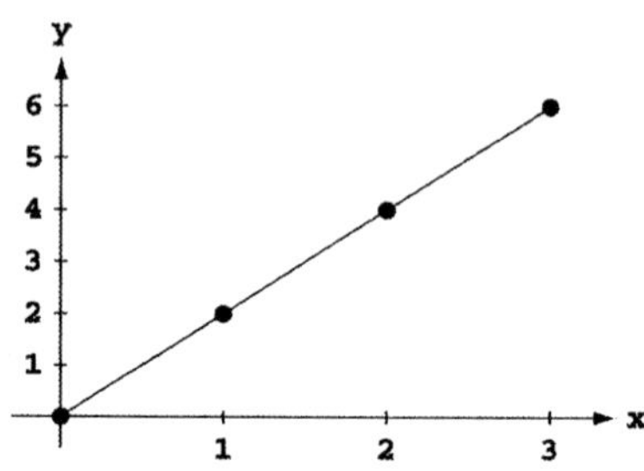

Since f is increasing on $[0, 3]$ we use right endpoints to obtain upper sums. $\triangle x = \frac{3-0}{n} = \frac{3}{n}$ and $x_i = i\triangle x = \frac{3i}{n}$. So an upper sum is $\sum_{i=1}^{n} 2x_i\left(\frac{3}{n}\right) = \sum_{i=1}^{n} \frac{6i}{n} \cdot \frac{3}{n} = \frac{18}{n^2} \sum_{i=1}^{n} i = \frac{18}{n^2} \cdot \frac{n(n+1)}{2} = \frac{9n^2+9n}{n^2}$

Thus, $\lim_{n\to\infty} \sum_{i=1}^{n} \frac{6i}{n} \cdot \frac{3}{n} = \lim_{n\to\infty} \frac{9n^2+9n}{n^2} = \lim_{n\to\infty} \left(9 + \frac{9}{n}\right) = 9.$

37. $f(x) = x^2 + 1$

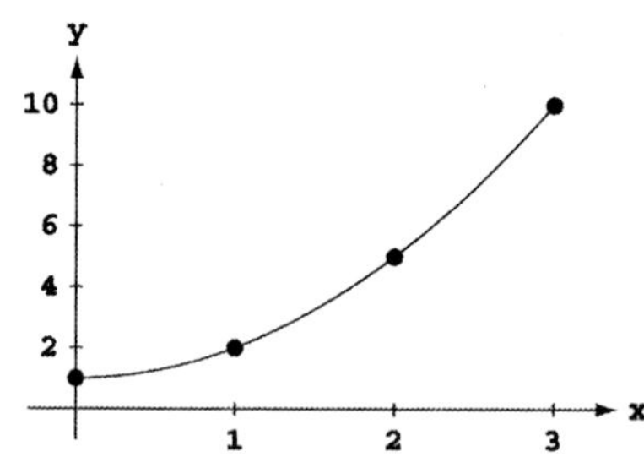

Since f is increasing on $[0, 3]$ we use right endpoints to obtain upper sums. $\triangle x = \frac{3-0}{n} = \frac{3}{n}$ and $x_i = i\triangle x = \frac{3i}{n}$. So an upper sum is $\sum_{i=1}^{n} (x_i^2 + 1)\frac{3}{n} = \sum_{i=1}^{n} \left(\left(\frac{3i}{n}\right)^2 + 1\right)\frac{3}{n} = \frac{3}{n} \sum_{i=1}^{n} \left(\frac{9i^2}{n^2} + 1\right)$

$= \frac{27}{n}\sum_{i=1}^{n} i^2 + \frac{3}{n} \cdot n = \frac{27}{n^3}\left(\frac{n(n+1)(2n+1)}{6}\right) + 3$

$= \frac{9(2n^3+3n^2+n)}{2n^3} + 3 = \frac{18 + \frac{27}{n} + \frac{9}{n^2}}{2} + 3.$ Thus,

$\lim_{n\to\infty} \sum_{i=1}^{n} (x_i^2 + 1)\frac{3}{n} = \lim_{n\to\infty} \left(\frac{18 + \frac{27}{n} + \frac{9}{n^2}}{2} + 3\right) = 9 + 3 = 12.$

38. $f(x) = 3x^2$

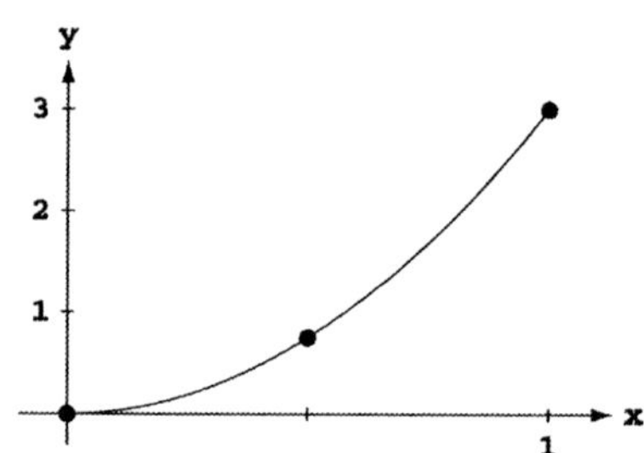

Since f is increasing on $[0, 1]$ we use right endpoints to obtain upper sums. $\triangle x = \frac{1-0}{n} = \frac{1}{n}$ and $x_i = i\triangle x = \frac{i}{n}$. So an upper sum is $\sum_{i=1}^{n} 3x_i^2\left(\frac{1}{n}\right) = \sum_{i=1}^{n} 3\left(\frac{i}{n}\right)^2\left(\frac{1}{n}\right) = \frac{3}{n^3} \sum_{i=1}^{n} i^2 = \frac{3}{n^3} \cdot \left(\frac{n(n+1)(2n+1)}{6}\right)$

$= \frac{2n^3+3n^2+n}{2n^3} = \frac{2 + \frac{3}{n} + \frac{1}{n^2}}{2}$. Thus, $\lim_{n\to\infty} \sum_{i=1}^{n} 3x_i^2\left(\frac{1}{n}\right)$

$= \lim_{n\to\infty} \left(\frac{2 + \frac{3}{n} + \frac{1}{n^2}}{2}\right) = \frac{2}{2} = 1.$

39. $f(x) = x + x^2 = x(1 + x)$

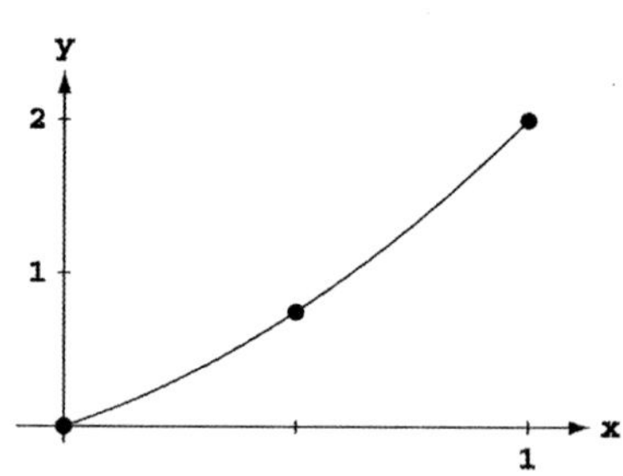

Since f is increasing on $[0, 1]$ we use right endpoints to obtain upper sums. $\triangle x = \frac{1-0}{n} = \frac{1}{n}$ and $x_i = i\triangle x = \frac{i}{n}$. So an upper sum is $\sum_{i=1}^{n} (x_i + x_i^2)\frac{1}{n} = \sum_{i=1}^{n} \left(\frac{i}{n} + \left(\frac{i}{n}\right)^2\right)\frac{1}{n} = \frac{1}{n^2}\sum_{i=1}^{n} i + \frac{1}{n^3}\sum_{i=1}^{n} i^2$

$= \frac{1}{n^2}\left(\frac{n(n+1)}{2}\right) + \frac{1}{n^3}\left(\frac{n(n+1)(2n+1)}{6}\right) = \frac{n^2+n}{2n^2} + \frac{2n^3+3n^2+n}{6n^3}$

$= \frac{1+\frac{1}{n}}{2} + \frac{2 + \frac{3}{n} + \frac{1}{n^2}}{6}$. Thus, $\lim_{n\to\infty} \sum_{i=1}^{n} (x_i + x_i^2)\frac{1}{n}$

$= \lim_{n\to\infty} \left[\left(\frac{1+\frac{1}{n}}{2}\right) + \left(\frac{2 + \frac{3}{n} + \frac{1}{n^2}}{6}\right)\right] = \frac{1}{2} + \frac{2}{6} = \frac{5}{6}.$

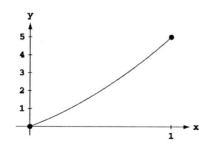

40. $f(x) = 3x + 2x^2$

Since f is increasing on $[0, 1]$ we use right endpoints to obtain upper sums. $\triangle x = \frac{1-0}{n} = \frac{1}{n}$ and $x_i = i\triangle x = \frac{i}{n}$. So an upper sum is $\sum_{i=1}^{n} (3x_i + 2x_i^2)\frac{1}{n} = \sum_{i=1}^{n} \left(\frac{3i}{n} + 2\left(\frac{i}{n}\right)^2\right)\frac{1}{n} = \frac{3}{n^2}\sum_{i=1}^{n} i + \frac{2}{n^3}\sum_{i=1}^{n} i^2$

$= \frac{3}{n^2}\left(\frac{n(n+1)}{2}\right) + \frac{2}{n^3}\left(\frac{n(n+1)(2n+1)}{6}\right) = \frac{3n^2+3n}{2n^2} + \frac{2n^2+3n+1}{3n^2}$

$= \frac{3+\frac{3}{n}}{2} + \frac{2 + \frac{3}{n} + \frac{1}{n^2}}{3}$. Thus, $\lim_{n\to\infty} \sum_{i=1}^{n} (3x_i + 2x_i^2)\frac{1}{n}$

$= \lim_{n\to\infty} \left[\left(\frac{3+\frac{3}{n}}{2}\right) + \left(\frac{2 + \frac{3}{n} + \frac{1}{n^2}}{3}\right)\right] = \frac{3}{2} + \frac{2}{3} = \frac{13}{6}.$

5.3 THE DEFINITE INTEGRAL

1. $\int_0^2 x^2\,dx$
2. $\int_{-1}^0 2x^3\,dx$
3. $\int_{-7}^5 (x^2 - 3x)\,dx$
4. $\int_1^4 \frac{1}{x}\,dx$
5. $\int_2^3 \frac{1}{1-x}\,dx$
6. $\int_0^1 \sqrt{4 - x^2}\,dx$
7. $\int_{-\pi/4}^0 (\sec x)\,dx$
8. $\int_0^{\pi/4} (\tan x)\,dx$

9. (a) $\int_2^2 g(x)\,dx = 0$

   (b) $\int_5^1 g(x)\,dx = -\int_1^5 g(x)\,dx = -8$

   (c) $\int_1^2 3f(x)\,dx = 3\int_1^2 f(x)\,dx = 3(-4) = -12$

   (d) $\int_2^5 f(x)\,dx = \int_1^5 f(x)\,dx - \int_1^2 f(x)\,dx = 6 - (-4) = 10$

   (e) $\int_1^5 [f(x) - g(x)]\,dx = \int_1^5 f(x)\,dx - \int_1^5 g(x)\,dx = 6 - 8 = -2$

   (f) $\int_1^5 [4f(x) - g(x)]\,dx = 4\int_1^5 f(x)\,dx - \int_1^5 g(x)\,dx = 4(6) - 8 = 16$

10. (a) $\int_1^9 -2f(x)\,dx = -2\int_1^9 f(x)\,dx = -2(-1) = 2$

    (b) $\int_7^9 [f(x) + h(x)]\,dx = \int_7^9 f(x)\,dx + \int_7^9 h(x)\,dx = 5 + 4 = 9$

    (c) $\int_7^9 [2f(x) - 3h(x)]\,dx = 2\int_7^9 f(x)\,dx - 3\int_7^9 h(x)\,dx = 2(5) - 3(4) = -2$

    (d) $\int_9^1 f(x)\,dx = -\int_1^9 f(x)\,dx = -(-1) = 1$

    (e) $\int_1^7 f(x)\,dx = \int_1^9 f(x)\,dx - \int_7^9 f(x)\,dx = -1 - 5 = -6$

    (f) $\int_9^7 [h(x) - f(x)]\,dx = \int_7^9 [f(x) - h(x)]\,dx = \int_7^9 f(x)\,dx - \int_7^9 h(x)\,dx = 5 - 4 = 1$

11. (a) $\int_1^2 f(u)\,du = \int_1^2 f(x)\,dx = 5$

    (b) $\int_1^2 \sqrt{3}\,f(z)\,dz = \sqrt{3}\int_1^2 f(z)\,dz = 5\sqrt{3}$

    (c) $\int_2^1 f(t)\,dt = -\int_1^2 f(t)\,dt = -5$

    (d) $\int_1^2 [-f(x)]\,dx = -\int_1^2 f(x)\,dx = -5$

12. (a) $\int_0^{-3} g(t)\,dt = -\int_{-3}^0 g(t)\,dt = -\sqrt{2}$

    (b) $\int_{-3}^0 g(u)\,du = \int_{-3}^0 g(t)\,dt = \sqrt{2}$

    (c) $\int_{-3}^0 [-g(x)]\,dx = -\int_{-3}^0 g(x)\,dx = -\sqrt{2}$

    (d) $\int_{-3}^0 \frac{g(r)}{\sqrt{2}}\,dr = \frac{1}{\sqrt{2}}\int_{-3}^0 g(t)\,dt = \left(\frac{1}{\sqrt{2}}\right)\left(\sqrt{2}\right) = 1$

13. (a) $\int_3^4 f(z)\,dz = \int_0^4 f(z)\,dz - \int_0^3 f(z)\,dz = 7 - 3 = 4$

    (b) $\int_4^3 f(t)\,dt = -\int_3^4 f(t)\,dt = -4$

14. (a) $\int_1^3 h(r)\,dr = \int_{-1}^3 h(r)\,dr - \int_{-1}^1 h(r)\,dr = 6 - 0 = 6$

    (b) $-\int_3^1 h(u)\,du = -\left(-\int_1^3 h(u)\,du\right) = \int_1^3 h(u)\,du = 6$

15. The area of the trapezoid is $A = \frac{1}{2}(B + b)h$
$= \frac{1}{2}(5 + 2)(6) = 21 \Rightarrow \int_{-2}^{4} \left(\frac{x}{2} + 3\right) dx$
$= 21$ square units

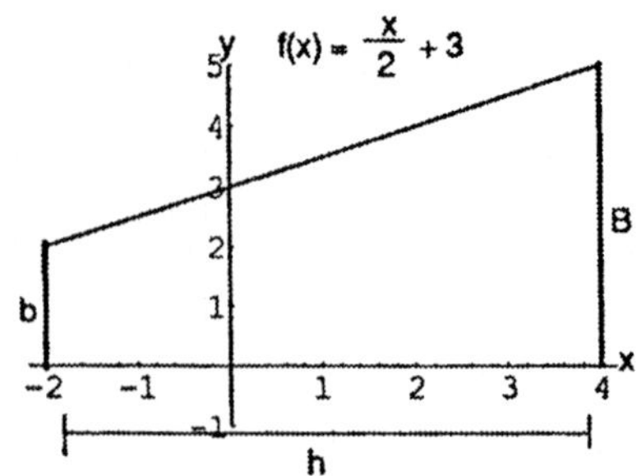

16. The area of the trapezoid is $A = \frac{1}{2}(B + b)h$
$= \frac{1}{2}(3 + 1)(1) = 2 \Rightarrow \int_{1/2}^{3/2} (-2x + 4)\, dx$
$= 2$ square units

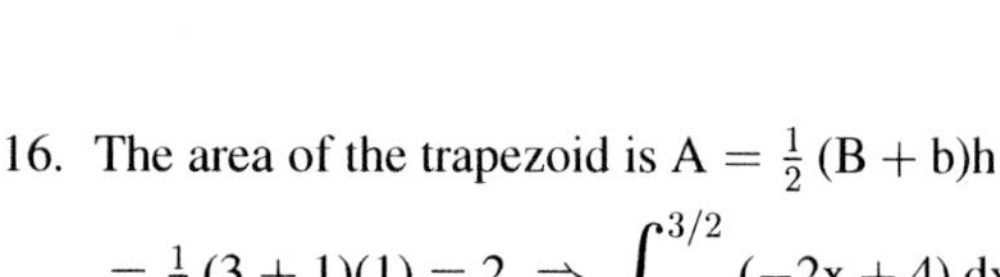

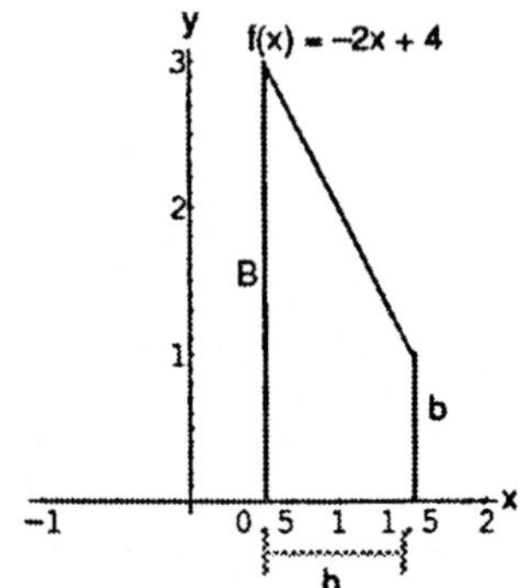

17. The area of the semicircle is $A = \frac{1}{2}\pi r^2 = \frac{1}{2}\pi(3)^2$
$= \frac{9}{2}\pi \Rightarrow \int_{-3}^{3} \sqrt{9 - x^2}\, dx = \frac{9}{2}\pi$ square units

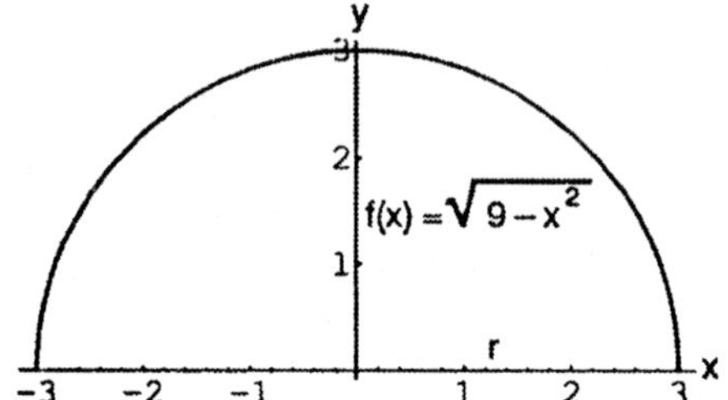

18. The graph of the quarter circle is $A = \frac{1}{4}\pi r^2 = \frac{1}{4}\pi(4)^2$
$= 4\pi \Rightarrow \int_{-4}^{0} \sqrt{16 - x^2}\, dx = 4\pi$ square units

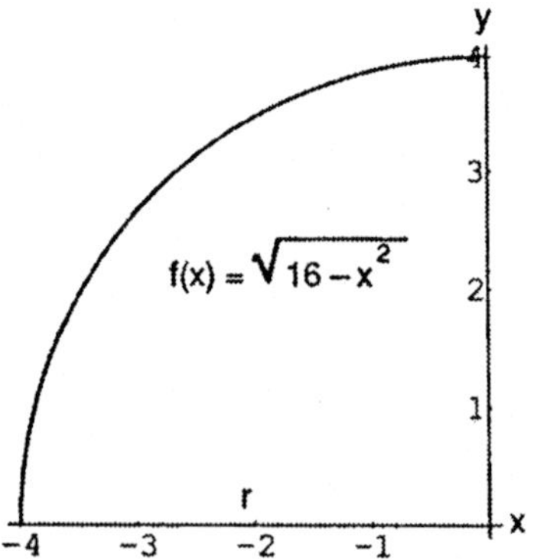

19. The area of the triangle on the left is $A = \frac{1}{2}bh = \frac{1}{2}(2)(2)$
$= 2$. The area of the triangle on the right is $A = \frac{1}{2}bh$
$= \frac{1}{2}(1)(1) = \frac{1}{2}$. Then, the total area is 2.5
$\Rightarrow \int_{-2}^{1} |x|\, dx = 2.5$ square units

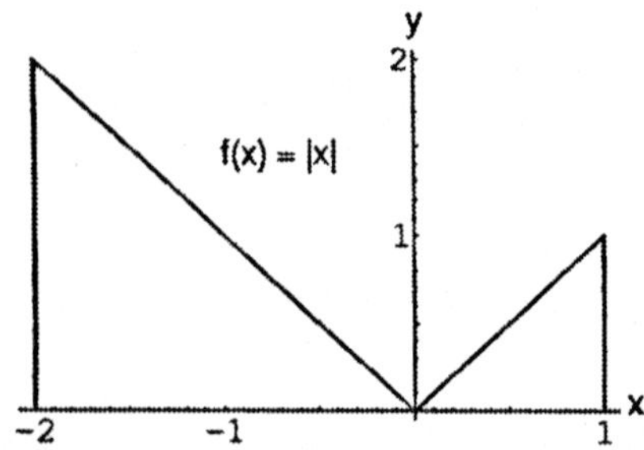

20. The area of the triangle is $A = \frac{1}{2}\,bh = \frac{1}{2}\,(2)(1) = 1$
$\Rightarrow \int_{-1}^{1} (1 - |x|)\,dx = 1$ square unit

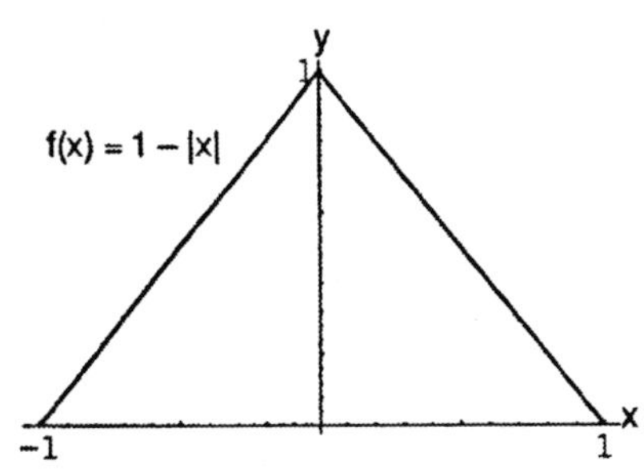

21. The area of the triangular peak is $A = \frac{1}{2}\,bh = \frac{1}{2}\,(2)(1) = 1$.
The area of the rectangular base is $S = \ell w = (2)(1) = 2$.
Then the total area is $3 \Rightarrow \int_{-1}^{1} (2 - |x|)\,dx = 3$ square units

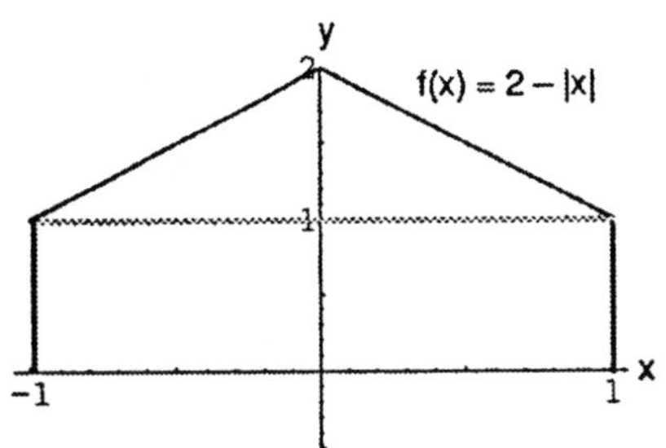

22. $y = 1 + \sqrt{1 - x^2} \Rightarrow y - 1 = \sqrt{1 - x^2}$
$\Rightarrow (y - 1)^2 = 1 - x^2 \Rightarrow x^2 + (y - 1)^2 = 1$, a circle with center $(0, 1)$ and radius of $1 \Rightarrow y = 1 + \sqrt{1 - x^2}$ is the upper semicircle. The area of this semicircle is $A = \frac{1}{2}\,\pi r^2 = \frac{1}{2}\,\pi(1)^2 = \frac{\pi}{2}$. The area of the rectangular base is $A = \ell w = (2)(1) = 2$. Then the total area is $2 + \frac{\pi}{2}$
$\Rightarrow \int_{-1}^{1} \left(1 + \sqrt{1 - x^2}\right) dx = 2 + \frac{\pi}{2}$ square units

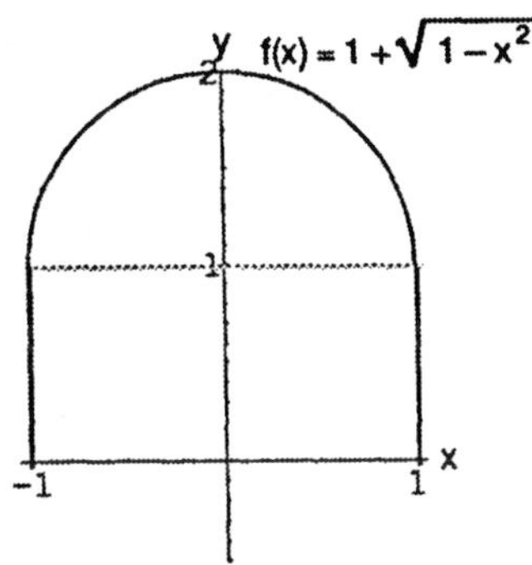

23. $\int_0^b \frac{x}{2}\,dx = \frac{1}{2}\,(b)\left(\frac{b}{2}\right) = \frac{b^2}{4}$

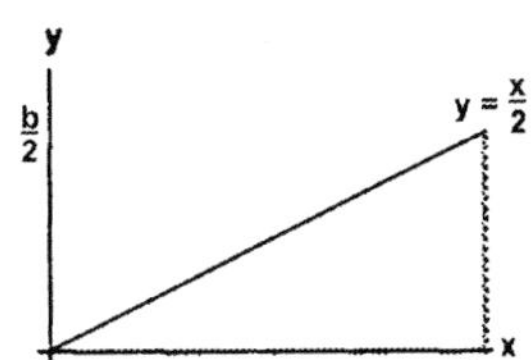

24. $\int_0^b 4x\,dx = \frac{1}{2}\,b(4b) = 2b^2$

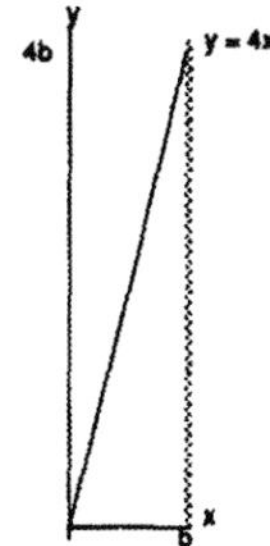

25. $\int_a^b 2s\,ds = \frac{1}{2}b(2b) - \frac{1}{2}a(2a) = b^2 - a^2$

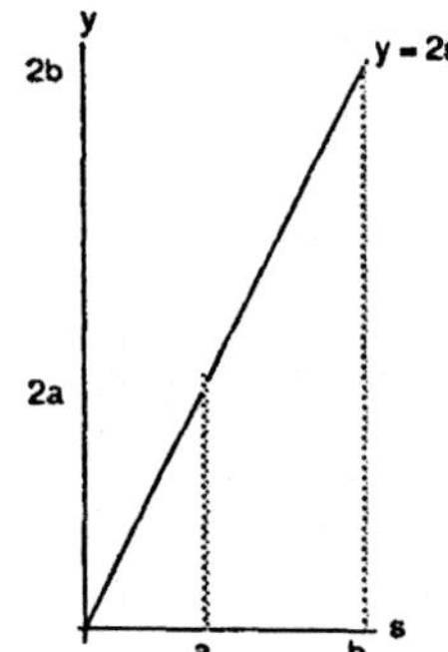

26. $\int_a^b 3t\,dt = \frac{1}{2}b(3b) - \frac{1}{2}a(3a) = \frac{3}{2}(b^2 - a^2)$

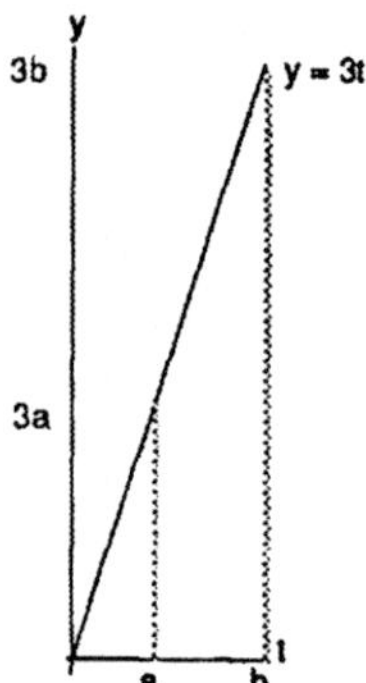

27. $\int_1^{\sqrt{2}} x\,dx = \frac{(\sqrt{2})^2}{2} - \frac{(1)^2}{2} = \frac{1}{2}$

28. $\int_{0.5}^{2.5} x\,dx = \frac{(2.5)^2}{2} - \frac{(0.5)^2}{2} = 3$

29. $\int_\pi^{2\pi} \theta\,d\theta = \frac{(2\pi)^2}{2} - \frac{\pi^2}{2} = \frac{3\pi^2}{2}$

30. $\int_{\sqrt{2}}^{5\sqrt{2}} r\,dr = \frac{(5\sqrt{2})^2}{2} - \frac{(\sqrt{2})^2}{2} = 24$

31. $\int_0^{\sqrt[3]{7}} x^2\,dx = \frac{(\sqrt[3]{7})^3}{3} = \frac{7}{3}$

32. $\int_0^{0.3} s^2\,ds = \frac{(0.3)^3}{3} = 0.009$

33. $\int_0^{1/2} t^2\,dt = \frac{(\frac{1}{2})^3}{3} = \frac{1}{24}$

34. $\int_0^{\pi/2} \theta^2\,d\theta = \frac{(\frac{\pi}{2})^3}{3} = \frac{\pi^3}{24}$

35. $\int_a^{2a} x\,dx = \frac{(2a)^2}{2} - \frac{a^2}{2} = \frac{3a^2}{2}$

36. $\int_a^{\sqrt{3}a} x\,dx = \frac{(\sqrt{3}a)^2}{2} - \frac{a^2}{2} = a^2$

37. $\int_0^{\sqrt[3]{b}} x^2\,dx = \frac{(\sqrt[3]{b})^3}{3} = \frac{b}{3}$

38. $\int_0^{3b} x^2\,dx = \frac{(3b)^3}{3} = 9b^3$

39. $\int_3^1 7\,dx = 7(1-3) = -14$

40. $\int_0^{-2} \sqrt{2}\,dx = \sqrt{2}(-2-0) = -2\sqrt{2}$

41. $\int_0^2 5x\,dx = 5\int_0^2 x\,dx = 5\left[\frac{2^2}{2} - \frac{0^2}{2}\right] = 10$

42. $\int_3^5 \frac{x}{8}\,dx = \frac{1}{8}\int_3^5 x\,dx = \frac{1}{8}\left[\frac{5^2}{2} - \frac{3^2}{2}\right] = \frac{16}{16} = 1$

43. $\int_0^2 (2t-3)\,dt = 2\int_1^1 t\,dt - \int_0^2 3\,dt = 2\left[\frac{2^2}{2} - \frac{0^2}{2}\right] - 3(2-0) = 4 - 6 = -2$

44. $\int_0^{\sqrt{2}} \left(t - \sqrt{2}\right) dt = \int_0^{\sqrt{2}} t\,dt - \int_0^{\sqrt{2}} \sqrt{2}\,dt = \left[\frac{(\sqrt{2})^2}{2} - \frac{0^2}{2}\right] - \sqrt{2}\left[\sqrt{2} - 0\right] = 1 - 2 = -1$

45. $\int_2^1 \left(1 + \frac{z}{2}\right) dz = \int_2^1 1\,dz + \int_2^1 \frac{z}{2}\,dz = \int_2^1 1\,dz - \frac{1}{2}\int_1^2 z\,dz = 1[1-2] - \frac{1}{2}\left[\frac{2^2}{2} - \frac{1^2}{2}\right] = -1 - \frac{1}{2}\left(\frac{3}{2}\right) = -\frac{7}{4}$

46. $\int_3^0 (2z-3)\,dz = \int_3^0 2z\,dz - \int_3^0 3\,dz = -2\int_0^3 z\,dz - \int_3^0 3\,dz = -2\left[\frac{3^2}{2} - \frac{0^2}{2}\right] - 3[0-3] = -9 + 9 = 0$

47. $\int_1^2 3u^2\,du = 3\int_1^2 u^2\,du = 3\left[\int_0^2 u^2\,du - \int_0^1 u^2\,du\right] = 3\left(\left[\frac{2^3}{3} - \frac{0^3}{3}\right] - \left[\frac{1^3}{3} - \frac{0^3}{3}\right]\right) = 3\left[\frac{2^3}{3} - \frac{1^3}{3}\right] = 3\left(\frac{7}{3}\right) = 7$

48. $\int_{1/2}^{1} 24u^2\,du = 24\int_{1/2}^{1} u^2\,du = 24\left[\int_0^1 u^2\,du - \int_0^{1/2} u^2\,du\right] = 24\left[\frac{1^3}{3} - \frac{\left(\frac{1}{2}\right)^3}{3}\right] = 24\left[\frac{\left(\frac{7}{8}\right)}{3}\right] = 7$

49. $\int_0^2 (3x^2 + x - 5)\,dx = 3\int_0^2 x^2\,dx + \int_0^2 x\,dx - \int_0^2 5\,dx = 3\left[\frac{2^3}{3} - \frac{0^3}{3}\right] + \left[\frac{2^2}{2} - \frac{0^2}{2}\right] - 5[2-0] = (8+2) - 10 = 0$

50. $\int_1^0 (3x^2 + x - 5)\,dx = -\int_0^1 (3x^2 + x - 5)\,dx = -\left[3\int_0^1 x^2\,dx + \int_0^1 x\,dx - \int_0^1 5\,dx\right]$
$= -\left[3\left(\frac{1^3}{3} - \frac{0^3}{3}\right) + \left(\frac{1^2}{2} - \frac{0^2}{2}\right) - 5(1-0)\right] = -\left(\frac{3}{2} - 5\right) = \frac{7}{2}$

51. Let $\Delta x = \frac{b-0}{n} = \frac{b}{n}$ and let $x_0 = 0$, $x_1 = \Delta x$, $x_2 = 2\Delta x, \ldots, x_{n-1} = (n-1)\Delta x$, $x_n = n\Delta x = b$. Let the c_k's be the right end-points of the subintervals $\Rightarrow c_1 = x_1$, $c_2 = x_2$, and so on. The rectangles defined have areas:

$f(c_1)\,\Delta x = f(\Delta x)\,\Delta x = 3(\Delta x)^2\,\Delta x = 3(\Delta x)^3$
$f(c_2)\,\Delta x = f(2\Delta x)\,\Delta x = 3(2\Delta x)^2\,\Delta x = 3(2)^2(\Delta x)^3$
$f(c_3)\,\Delta x = f(3\Delta x)\,\Delta x = 3(3\Delta x)^2\,\Delta x = 3(3)^2(\Delta x)^3$
$\vdots$
$f(c_n)\,\Delta x = f(n\Delta x)\,\Delta x = 3(n\Delta x)^2\,\Delta x = 3(n)^2(\Delta x)^3$

Then $S_n = \sum_{k=1}^{n} f(c_k)\,\Delta x = \sum_{k=1}^{n} 3k^2(\Delta x)^3$
$= 3(\Delta x)^3 \sum_{k=1}^{n} k^2 = 3\left(\frac{b^3}{n^3}\right)\left(\frac{n(n+1)(2n+1)}{6}\right)$
$= \frac{b^3}{2}\left(2 + \frac{3}{n} + \frac{1}{n^2}\right) \Rightarrow \int_0^b 3x^2\,dx = \lim_{n\to\infty} \frac{b^3}{2}\left(2 + \frac{3}{n} + \frac{1}{n^2}\right) = b^3.$

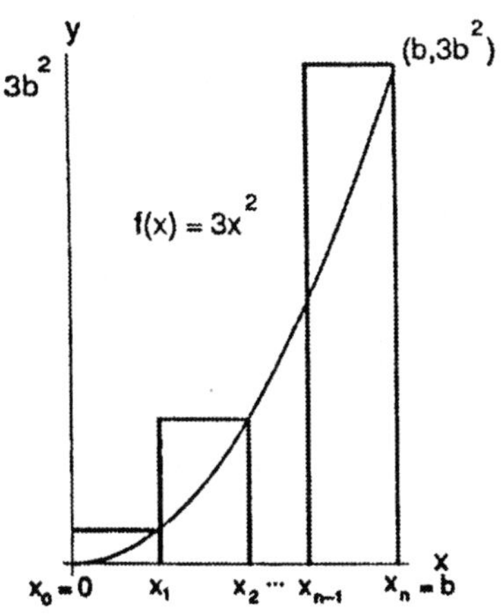

52. Let $\Delta x = \frac{b-0}{n} = \frac{b}{n}$ and let $x_0 = 0$, $x_1 = \Delta x$, $x_2 = 2\Delta x, \ldots, x_{n-1} = (n-1)\Delta x$, $x_n = n\Delta x = b$. Let the c_k's be the right end-points of the subintervals $\Rightarrow c_1 = x_1$, $c_2 = x_2$, and so on. The rectangles defined have areas:

$f(c_1)\,\Delta x = f(\Delta x)\,\Delta x = \pi(\Delta x)^2\,\Delta x = \pi(\Delta x)^3$
$f(c_2)\,\Delta x = f(2\Delta x)\,\Delta x = \pi(2\Delta x)^2\,\Delta x = \pi(2)^2(\Delta x)^3$
$f(c_3)\,\Delta x = f(3\Delta x)\,\Delta x = \pi(3\Delta x)^2\,\Delta x = \pi(3)^2(\Delta x)^3$
$\vdots$
$f(c_n)\,\Delta x = f(n\Delta x)\,\Delta x = \pi(n\Delta x)^2\,\Delta x = \pi(n)^2(\Delta x)^3$

Then $S_n = \sum_{k=1}^{n} f(c_k)\,\Delta x = \sum_{k=1}^{n} \pi k^2(\Delta x)^3$
$= \pi(\Delta x)^3 \sum_{k=1}^{n} k^2 = \pi\left(\frac{b^3}{n^3}\right)\left(\frac{n(n+1)(2n+1)}{6}\right)$
$= \frac{\pi b^3}{6}\left(2 + \frac{3}{n} + \frac{1}{n^2}\right) \Rightarrow \int_0^b \pi x^2\,dx = \lim_{n\to\infty} \frac{\pi b^3}{6}\left(2 + \frac{3}{n} + \frac{1}{n^2}\right) = \frac{\pi b^3}{3}.$

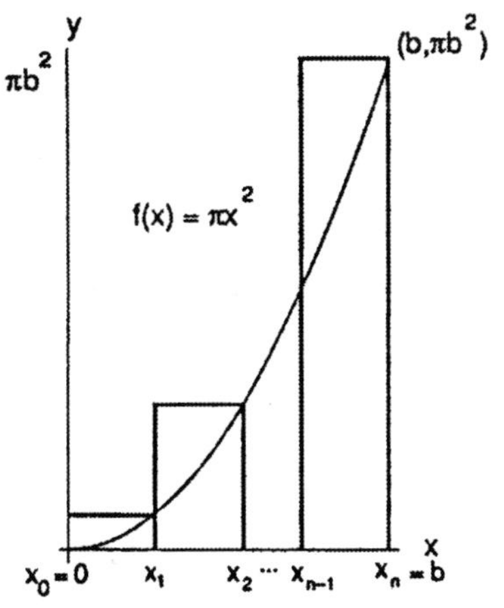

53. Let $\Delta x = \frac{b-0}{n} = \frac{b}{n}$ and let $x_0 = 0$, $x_1 = \Delta x$, $x_2 = 2\Delta x, \ldots, x_{n-1} = (n-1)\Delta x$, $x_n = n\Delta x = b$. Let the c_k's be the right end-points of the subintervals $\Rightarrow c_1 = x_1$, $c_2 = x_2$, and so on. The rectangles defined have areas:

$$f(c_1)\,\Delta x = f(\Delta x)\,\Delta x = 2(\Delta x)(\Delta x) = 2(\Delta x)^2$$
$$f(c_2)\,\Delta x = f(2\Delta x)\,\Delta x = 2(2\Delta x)(\Delta x) = 2(2)(\Delta x)^2$$
$$f(c_3)\,\Delta x = f(3\Delta x)\,\Delta x = 2(3\Delta x)(\Delta x) = 2(3)(\Delta x)^2$$
$$\vdots$$
$$f(c_n)\,\Delta x = f(n\Delta x)\,\Delta x = 2(n\Delta x)(\Delta x) = 2(n)(\Delta x)^2$$

Then $S_n = \sum_{k=1}^{n} f(c_k)\,\Delta x = \sum_{k=1}^{n} 2k(\Delta x)^2$

$= 2(\Delta x)^2 \sum_{k=1}^{n} k = 2\left(\frac{b^2}{n^2}\right)\left(\frac{n(n+1)}{2}\right)$

$= b^2\left(1 + \frac{1}{n}\right) \Rightarrow \int_0^b 2x\,dx = \lim_{n\to\infty} b^2\left(1 + \frac{1}{n}\right) = b^2.$

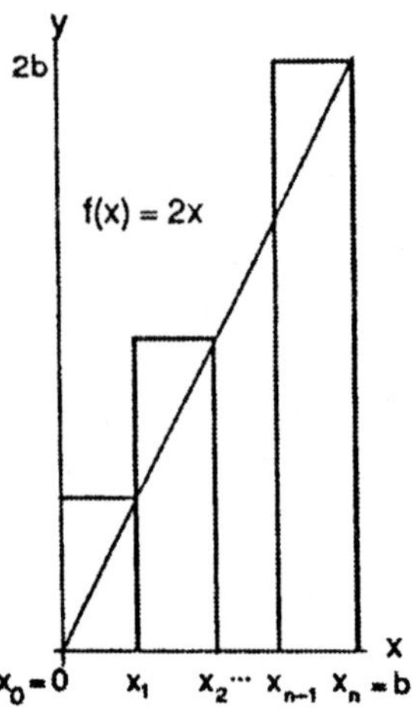

54. Let $\Delta x = \frac{b-0}{n} = \frac{b}{n}$ and let $x_0 = 0$, $x_1 = \Delta x$, $x_2 = 2\Delta x, \ldots, x_{n-1} = (n-1)\Delta x$, $x_n = n\Delta x = b$. Let the c_k's be the right end-points of the subintervals $\Rightarrow c_1 = x_1$, $c_2 = x_2$, and so on. The rectangles defined have areas:

$$f(c_1)\,\Delta x = f(\Delta x)\,\Delta x = \left(\frac{\Delta x}{2} + 1\right)(\Delta x) = \frac{1}{2}(\Delta x)^2 + \Delta x$$
$$f(c_2)\,\Delta x = f(2\Delta x)\,\Delta x = \left(\frac{2\Delta x}{2} + 1\right)(\Delta x) = \frac{1}{2}(2)(\Delta x)^2 + \Delta x$$
$$f(c_3)\,\Delta x = f(3\Delta x)\,\Delta x = \left(\frac{3\Delta x}{2} + 1\right)(\Delta x) = \frac{1}{2}(3)(\Delta x)^2 + \Delta x$$
$$\vdots$$
$$f(c_n)\,\Delta x = f(n\Delta x)\,\Delta x = \left(\frac{n\Delta x}{2} + 1\right)(\Delta x) = \frac{1}{2}(n)(\Delta x)^2 + \Delta x$$

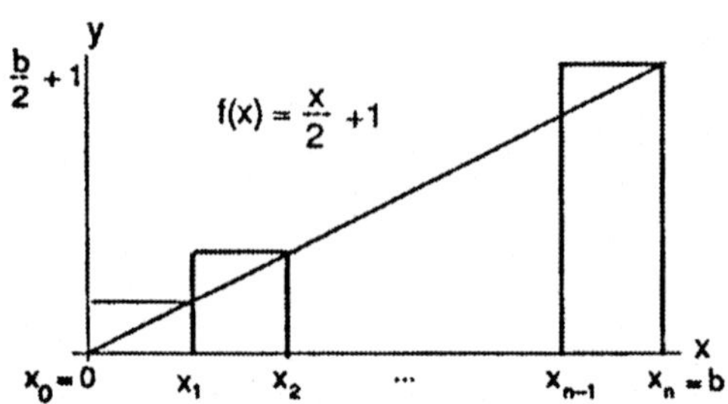

Then $S_n = \sum_{k=1}^{n} f(c_k)\,\Delta x = \sum_{k=1}^{n}\left(\frac{1}{2}k(\Delta x)^2 + \Delta x\right) = \frac{1}{2}(\Delta x)^2 \sum_{k=1}^{n} k + \Delta x \sum_{k=1}^{n} 1 = \frac{1}{2}\left(\frac{b^2}{n^2}\right)\left(\frac{n(n+1)}{2}\right) + \left(\frac{b}{n}\right)(n)$

$= \frac{1}{4}b^2\left(1 + \frac{1}{n}\right) + b \Rightarrow \int_0^b \left(\frac{x}{2} + 1\right)dx = \lim_{n\to\infty}\left(\frac{1}{4}b^2\left(1 + \frac{1}{n}\right) + b\right) = \frac{1}{4}b^2 + b.$

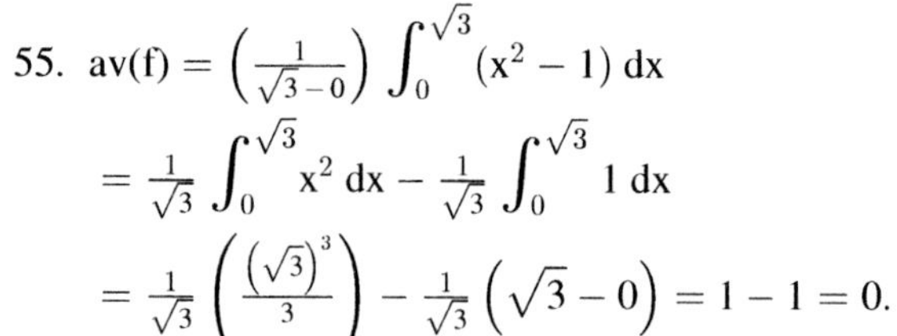

55. $\text{av}(f) = \left(\frac{1}{\sqrt{3}-0}\right)\int_0^{\sqrt{3}} (x^2 - 1)\,dx$

$= \frac{1}{\sqrt{3}}\int_0^{\sqrt{3}} x^2\,dx - \frac{1}{\sqrt{3}}\int_0^{\sqrt{3}} 1\,dx$

$= \frac{1}{\sqrt{3}}\left(\frac{\left(\sqrt{3}\right)^3}{3}\right) - \frac{1}{\sqrt{3}}\left(\sqrt{3} - 0\right) = 1 - 1 = 0.$

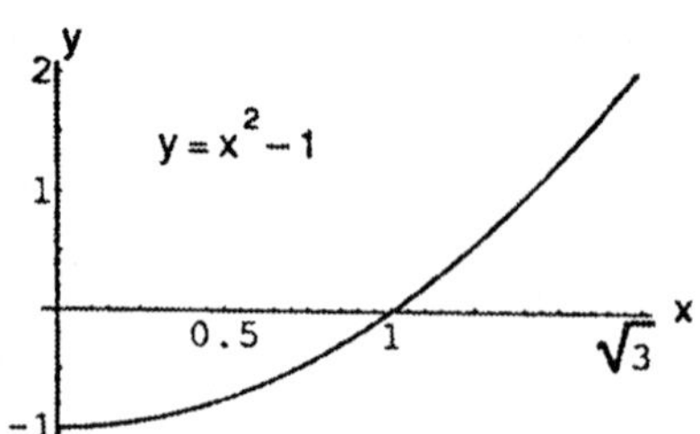

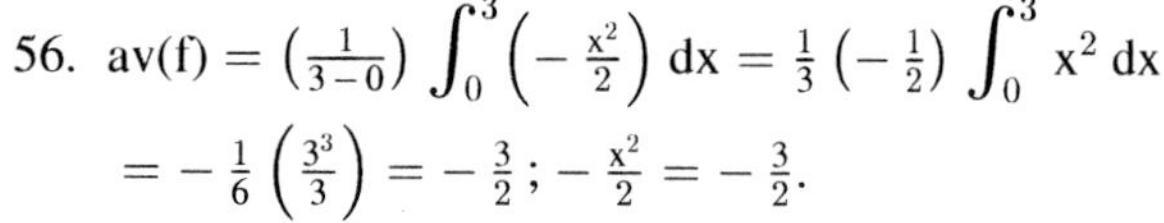

56. $\text{av}(f) = \left(\frac{1}{3-0}\right)\int_0^3 \left(-\frac{x^2}{2}\right)dx = \frac{1}{3}\left(-\frac{1}{2}\right)\int_0^3 x^2\,dx$

$= -\frac{1}{6}\left(\frac{3^3}{3}\right) = -\frac{3}{2};\ -\frac{x^2}{2} = -\frac{3}{2}.$

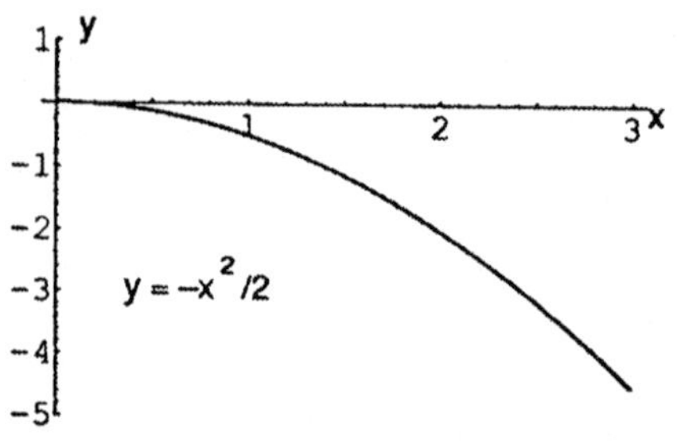

57. $\text{av}(f) = \left(\frac{1}{1-0}\right) \int_0^1 (-3x^2 - 1)\, dx =$
$$= -3 \int_0^1 x^2\, dx - \int_0^1 1\, dx = -3\left(\frac{1^3}{3}\right) - (1-0)$$
$$= -2.$$

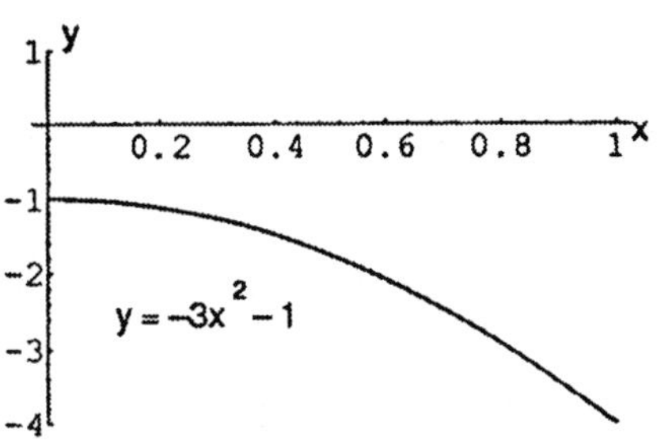

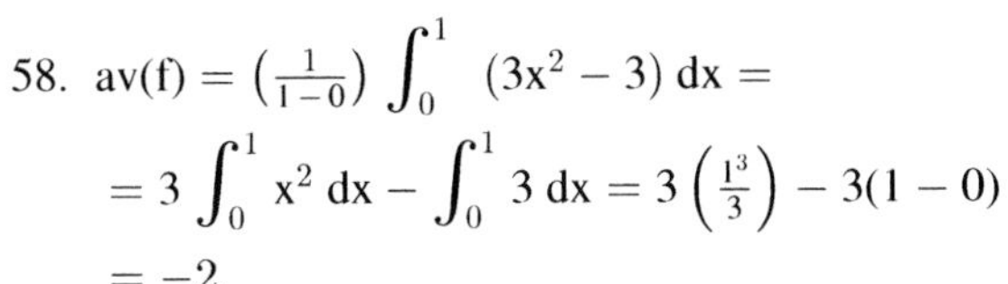

58. $\text{av}(f) = \left(\frac{1}{1-0}\right) \int_0^1 (3x^2 - 3)\, dx =$
$$= 3 \int_0^1 x^2\, dx - \int_0^1 3\, dx = 3\left(\frac{1^3}{3}\right) - 3(1-0)$$
$$= -2.$$

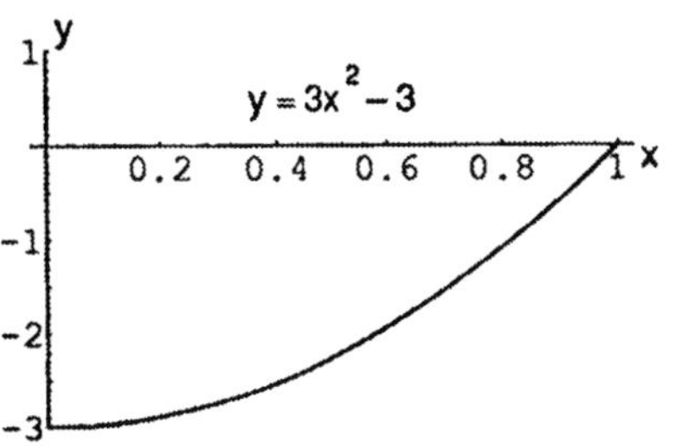

59. $\text{av}(f) = \left(\frac{1}{3-0}\right) \int_0^3 (t-1)^2\, dt$
$$= \frac{1}{3} \int_0^3 t^2\, dt - \frac{2}{3} \int_0^3 t\, dt + \frac{1}{3} \int_0^3 1\, dt$$
$$= \frac{1}{3}\left(\frac{3^3}{3}\right) - \frac{2}{3}\left(\frac{3^2}{2} - \frac{0^2}{2}\right) + \frac{1}{3}(3-0) = 1.$$

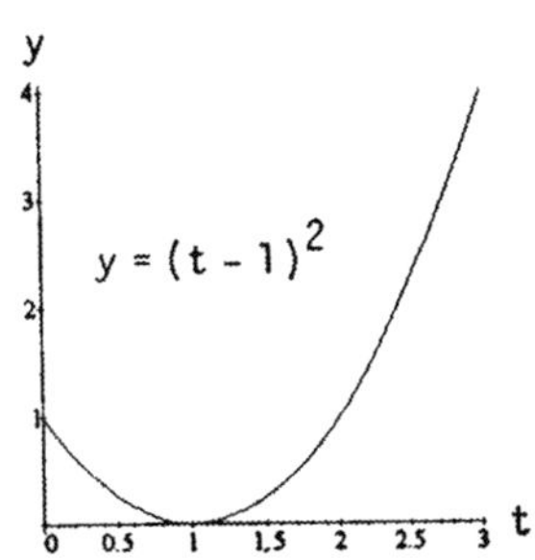

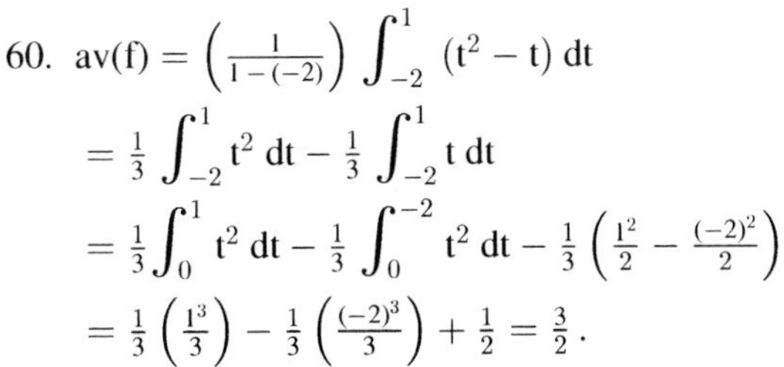

60. $\text{av}(f) = \left(\frac{1}{1-(-2)}\right) \int_{-2}^1 (t^2 - t)\, dt$
$$= \frac{1}{3} \int_{-2}^1 t^2\, dt - \frac{1}{3} \int_{-2}^1 t\, dt$$
$$= \frac{1}{3} \int_0^1 t^2\, dt - \frac{1}{3} \int_0^{-2} t^2\, dt - \frac{1}{3}\left(\frac{1^2}{2} - \frac{(-2)^2}{2}\right)$$
$$= \frac{1}{3}\left(\frac{1^3}{3}\right) - \frac{1}{3}\left(\frac{(-2)^3}{3}\right) + \frac{1}{2} = \frac{3}{2}.$$

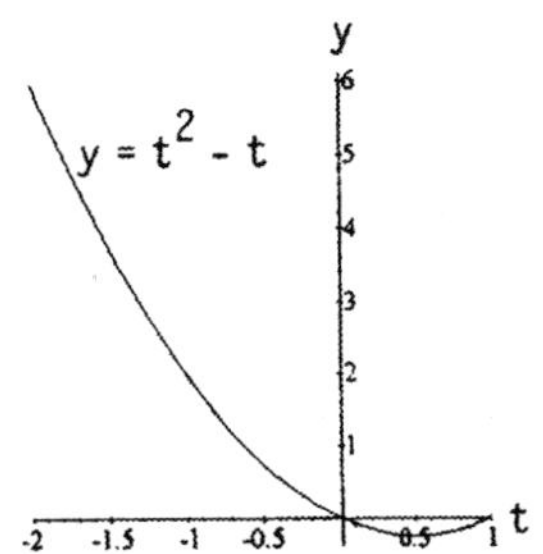

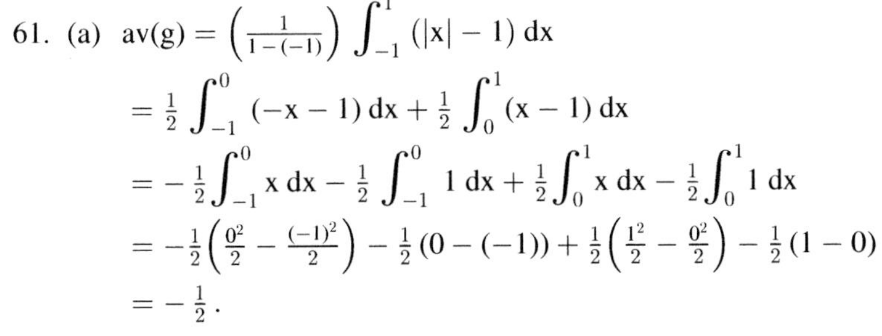

61. (a) $\text{av}(g) = \left(\frac{1}{1-(-1)}\right) \int_{-1}^1 (|x| - 1)\, dx$
$$= \frac{1}{2} \int_{-1}^0 (-x-1)\, dx + \frac{1}{2} \int_0^1 (x-1)\, dx$$
$$= -\frac{1}{2} \int_{-1}^0 x\, dx - \frac{1}{2} \int_{-1}^0 1\, dx + \frac{1}{2} \int_0^1 x\, dx - \frac{1}{2} \int_0^1 1\, dx$$
$$= -\frac{1}{2}\left(\frac{0^2}{2} - \frac{(-1)^2}{2}\right) - \frac{1}{2}(0 - (-1)) + \frac{1}{2}\left(\frac{1^2}{2} - \frac{0^2}{2}\right) - \frac{1}{2}(1-0)$$
$$= -\frac{1}{2}.$$

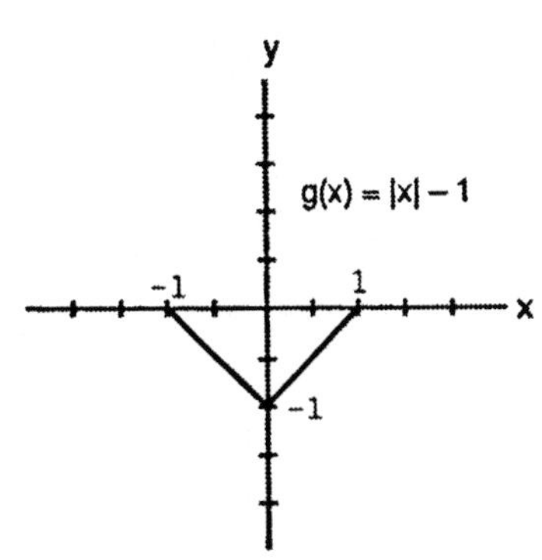

(b) $\text{av}(g) = \left(\frac{1}{3-1}\right)\int_1^3 (|x| - 1)\,dx = \frac{1}{2}\int_1^3 (x-1)\,dx$
$= \frac{1}{2}\int_1^3 x\,dx - \frac{1}{2}\int_1^3 1\,dx = \frac{1}{2}\left(\frac{3^2}{2} - \frac{1^2}{2}\right) - \frac{1}{2}(3-1)$
$= 1.$

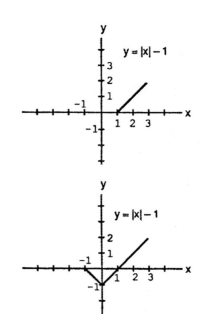

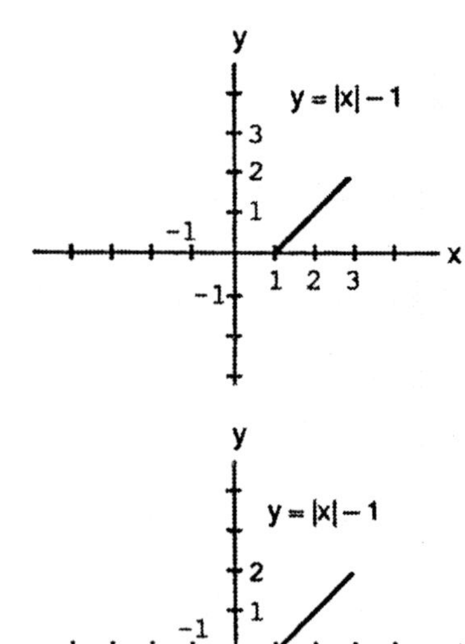

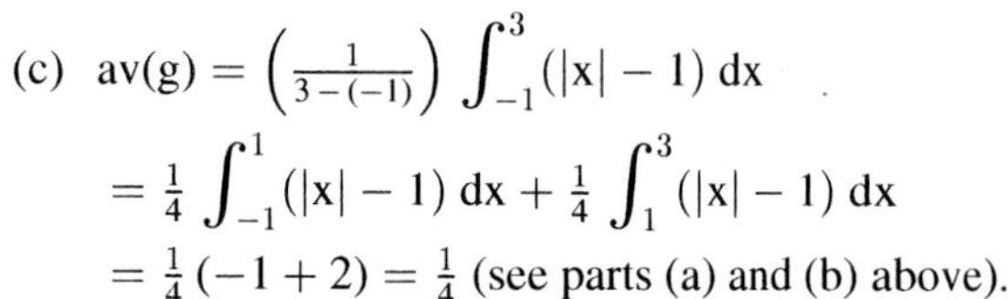

(c) $\text{av}(g) = \left(\frac{1}{3-(-1)}\right)\int_{-1}^3 (|x| - 1)\,dx$
$= \frac{1}{4}\int_{-1}^1 (|x| - 1)\,dx + \frac{1}{4}\int_1^3 (|x| - 1)\,dx$
$= \frac{1}{4}(-1 + 2) = \frac{1}{4}$ (see parts (a) and (b) above).

62. (a) $\text{av}(h) = \left(\frac{1}{0-(-1)}\right)\int_{-1}^0 -|x|\,dx = \int_{-1}^0 -(-x)\,dx$
$= \int_{-1}^0 x\,dx = \frac{0^2}{2} - \frac{(-1)^2}{2} = -\frac{1}{2}.$

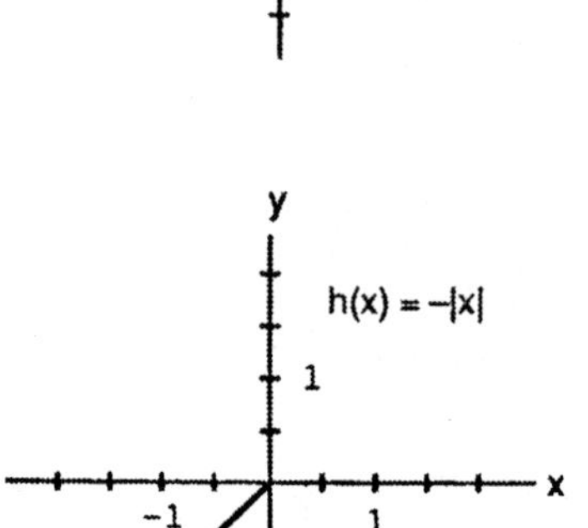

(b) $\text{av}(h) = \left(\frac{1}{1-0}\right)\int_0^1 -|x|\,dx = -\int_0^1 x\,dx$
$= -\left(\frac{1^2}{2} - \frac{0^2}{2}\right) = -\frac{1}{2}.$

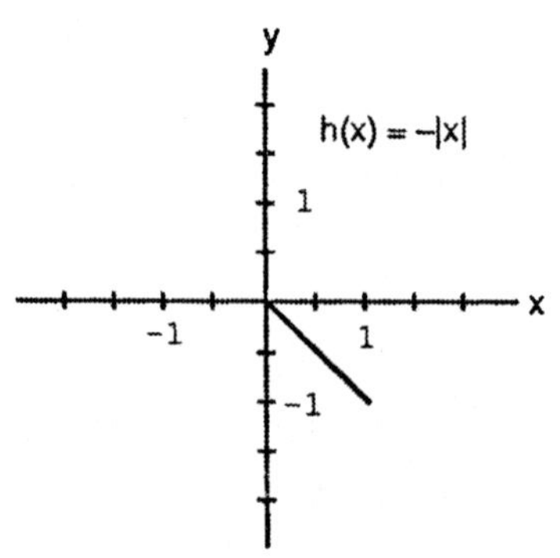

(c) $\text{av}(h) = \left(\frac{1}{1-(-1)}\right)\int_{-1}^1 -|x|\,dx$
$= \frac{1}{2}\left(\int_{-1}^0 -|x|\,dx + \int_0^1 -|x|\,dx\right)$
$= \frac{1}{2}\left(-\frac{1}{2} + \left(-\frac{1}{2}\right)\right) = -\frac{1}{2}$ (see parts (a) and (b) above).

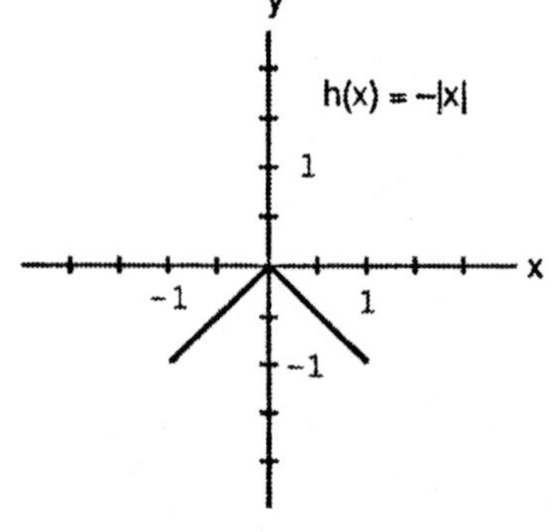

63. To find where $x - x^2 \geq 0$, let $x - x^2 = 0 \Rightarrow x(1-x) = 0 \Rightarrow x = 0$ or $x = 1$. If $0 < x < 1$, then $0 < x - x^2 \Rightarrow a = 0$ and $b = 1$ maximize the integral.

64. To find where $x^4 - 2x^2 \leq 0$, let $x^4 - 2x^2 = 0 \Rightarrow x^2(x^2 - 2) = 0 \Rightarrow x = 0$ or $x = \pm\sqrt{2}$. By the sign graph, $+++++ \underset{-\sqrt{2}}{0} -- \underset{0}{0} -- \underset{\sqrt{2}}{0} +++++++$, we can see that $x^4 - 2x^2 \leq 0$ on $\left[-\sqrt{2}, \sqrt{2}\right] \Rightarrow a = -\sqrt{2}$ and $b = \sqrt{2}$ minimize the integral.

65. $f(x) = \frac{1}{1+x^2}$ is decreasing on $[0,1] \Rightarrow$ maximum value of f occurs at $0 \Rightarrow \max f = f(0) = 1$; minimum value of f occurs at $1 \Rightarrow \min f = f(1) = \frac{1}{1+1^2} = \frac{1}{2}$. Therefore, $(1-0)\min f \leq \int_0^1 \frac{1}{1+x^2}\,dx \leq (1-0)\max f$ $\Rightarrow \frac{1}{2} \leq \int_0^1 \frac{1}{1+x^2}\,dx \leq 1$. That is, an upper bound $= 1$ and a lower bound $= \frac{1}{2}$.

66. See Exercise 65 above. On $[0, 0.5]$, $\max f = \frac{1}{1+0^2} = 1$, $\min f = \frac{1}{1+(0.5)^2} = 0.8$. Therefore $(0.5-0)\min f \leq \int_0^{0.5} f(x)\,dx \leq (0.5-0)\max f \Rightarrow \frac{2}{5} \leq \int_0^{0.5} \frac{1}{1+x^2}\,dx \leq \frac{1}{2}$. On $[0.5, 1]$, $\max f = \frac{1}{1+(0.5)^2} = 0.8$ and $\min f = \frac{1}{1+1^2} = 0.5$. Therefore $(1-0.5)\min f \leq \int_{0.5}^1 \frac{1}{1+x^2}\,dx \leq (1-0.5)\max f \Rightarrow \frac{1}{4} \leq \int_{0.5}^1 \frac{1}{1+x^2}\,dx \leq \frac{2}{5}$. Then $\frac{1}{4} + \frac{2}{5} \leq \int_0^{0.5} \frac{1}{1+x^2}\,dx + \int_{0.5}^1 \frac{1}{1+x^2}\,dx \leq \frac{1}{2} + \frac{2}{5} \Rightarrow \frac{13}{20} \leq \int_0^1 \frac{1}{1+x^2}\,dx \leq \frac{9}{10}$.

67. $-1 \leq \sin(x^2) \leq 1$ for all $x \Rightarrow (1-0)(-1) \leq \int_0^1 \sin(x^2)\,dx \leq (1-0)(1)$ or $\int_0^1 \sin x^2\,dx \leq 1 \Rightarrow \int_0^1 \sin x^2\,dx$ cannot equal 2.

68. $f(x) = \sqrt{x+8}$ is increasing on $[0,1] \Rightarrow \max f = f(1) = \sqrt{1+8} = 3$ and $\min f = f(0) = \sqrt{0+8} = 2\sqrt{2}$. Therefore, $(1-0)\min f \leq \int_0^1 \sqrt{x+8}\,dx \leq (1-0)\max f \Rightarrow 2\sqrt{2} \leq \int_0^1 \sqrt{x+8}\,dx \leq 3$.

69. If $f(x) \geq 0$ on $[a,b]$, then $\min f \geq 0$ and $\max f \geq 0$ on $[a,b]$. Now, $(b-a)\min f \leq \int_a^b f(x)\,dx \leq (b-a)\max f$. Then $b \geq a \Rightarrow b - a \geq 0 \Rightarrow (b-a)\min f \geq 0 \Rightarrow \int_a^b f(x)\,dx \geq 0$.

70. If $f(x) \leq 0$ on $[a,b]$, then $\min f \leq 0$ and $\max f \leq 0$. Now, $(b-a)\min f \leq \int_a^b f(x)\,dx \leq (b-a)\max f$. Then $b \geq a \Rightarrow b - a \geq 0 \Rightarrow (b-a)\max f \leq 0 \Rightarrow \int_a^b f(x)\,dx \leq 0$.

71. $\sin x \leq x$ for $x \geq 0 \Rightarrow \sin x - x \leq 0$ for $x \geq 0 \Rightarrow \int_0^1 (\sin x - x)\,dx \leq 0$ (see Exercise 70) $\Rightarrow \int_0^1 \sin x\,dx - \int_0^1 x\,dx$ $\leq 0 \Rightarrow \int_0^1 \sin x\,dx \leq \int_0^1 x\,dx \Rightarrow \int_0^1 \sin x\,dx \leq \left(\frac{1^2}{2} - \frac{0^2}{2}\right) \Rightarrow \int_0^1 \sin x\,dx \leq \frac{1}{2}$. Thus an upper bound is $\frac{1}{2}$.

72. $\sec x \geq 1 + \frac{x^2}{2}$ on $\left(-\frac{\pi}{2}, \frac{\pi}{2}\right) \Rightarrow \sec x - \left(1 + \frac{x^2}{2}\right) \geq 0$ on $\left(-\frac{\pi}{2}, \frac{\pi}{2}\right) \Rightarrow \int_0^1 \left[\sec x - \left(1 + \frac{x^2}{2}\right)\right] dx \geq 0$ (see Exercise 69) since $[0,1]$ is contained in $\left(-\frac{\pi}{2}, \frac{\pi}{2}\right) \Rightarrow \int_0^1 \sec x\,dx - \int_0^1 \left(1 + \frac{x^2}{2}\right) dx \geq 0 \Rightarrow \int_0^1 \sec x\,dx$ $\geq \int_0^1 \left(1 + \frac{x^2}{2}\right) dx \Rightarrow \int_0^1 \sec x\,dx \geq \int_0^1 1\,dx + \frac{1}{2}\int_0^1 x^2\,dx \Rightarrow \int_0^1 \sec x\,dx \geq (1-0) + \frac{1}{2}\left(\frac{1^3}{3}\right) \Rightarrow \int_0^1 \sec x\,dx \geq \frac{7}{6}$. Thus a lower bound is $\frac{7}{6}$.

73. Yes, for the following reasons: $\text{av}(f) = \frac{1}{b-a}\int_a^b f(x)\,dx$ is a constant K. Thus $\int_a^b \text{av}(f)\,dx = \int_a^b K\,dx$ $= K(b-a) \Rightarrow \int_a^b \text{av}(f)\,dx = (b-a)K = (b-a)\cdot\frac{1}{b-a}\int_a^b f(x)\,dx = \int_a^b f(x)\,dx.$

74. All three rules hold. The reasons: On any interval [a, b] on which f and g are integrable, we have:
(a) $\text{av}(f+g) = \frac{1}{b-a}\int_a^b [f(x)+g(x)]\,dx = \frac{1}{b-a}\left[\int_a^b f(x)\,dx + \int_a^b g(x)\,dx\right] = \frac{1}{b-a}\int_a^b f(x)\,dx + \frac{1}{b-a}\int_a^b g(x)\,dx$ $= \text{av}(f) + \text{av}(g)$
(b) $\text{av}(kf) = \frac{1}{b-a}\int_a^b kf(x)\,dx = \frac{1}{b-a}\left[k\int_a^b f(x)\,dx\right] = k\left[\frac{1}{b-a}\int_a^b f(x)\,dx\right] = k\,\text{av}(f)$
(c) $\text{av}(f) = \frac{1}{b-a}\int_a^b f(x)\,dx \le \frac{1}{b-a}\int_a^b g(x)\,dx$ since $f(x) \le g(x)$ on [a, b], and $\frac{1}{b-a}\int_a^b g(x)\,dx = \text{av}(g)$. Therefore, $\text{av}(f) \le \text{av}(g)$.

75. Consider the partition P that subdivides the interval [a, b] into n subintervals of width $\triangle x = \frac{b-a}{n}$ and let c_k be the right endpoint of each subinterval. So the partition is $P = \{a, a + \frac{b-a}{n}, a + \frac{2(b-a)}{n}, \ldots, a + \frac{n(b-a)}{n}\}$ and $c_k = a + \frac{k(b-a)}{n}$. We get the Riemann sum $\sum_{k=1}^{n} f(c_k)\triangle x = \sum_{k=1}^{n} c\cdot\frac{b-a}{n} = \frac{c(b-a)}{n}\sum_{k=1}^{n} 1 = \frac{c(b-a)}{n}\cdot n = c(b-a)$. As $n\to\infty$ and $\|P\|\to 0$ this expression remains $c(b-a)$. Thus, $\int_a^b c\,dx = c(b-a)$.

76. Consider the partition P that subdivides the interval [a, b] into n subintervals of width $\triangle x = \frac{b-a}{n}$ and let c_k be the right endpoint of each subinterval. So the partition is $P = \{a, a + \frac{b-a}{n}, a + \frac{2(b-a)}{n}, \ldots, a + \frac{n(b-a)}{n}\}$ and $c_k = a + \frac{k(b-a)}{n}$. We get the Riemann sum $\sum_{k=1}^{n} f(c_k)\triangle x = \sum_{k=1}^{n} c_k^2\left(\frac{b-a}{n}\right) = \frac{b-a}{n}\sum_{k=1}^{n}\left(a + \frac{k(b-a)}{n}\right)^2 = \frac{b-a}{n}\sum_{k=1}^{n}\left(a^2 + \frac{2ak(b-a)}{n} + \frac{k^2(b-a)^2}{n^2}\right)$
$= \frac{b-a}{n}\left(\sum_{k=1}^{n} a^2 + \frac{2a(b-a)}{n}\sum_{k=1}^{n} k + \frac{(b-a)^2}{n^2}\sum_{k=1}^{n} k^2\right) = \frac{b-a}{n}\cdot na^2 + \frac{2a(b-a)^2}{n^2}\cdot\frac{n(n+1)}{2} + \frac{(b-a)^3}{n^3}\cdot\frac{n(n+1)(2n+1)}{6}$
$= (b-a)a^2 + a(b-a)^2\cdot\frac{n+1}{n} + \frac{(b-a)^3}{6}\cdot\frac{(n+1)(2n+1)}{n^2} = (b-a)a^2 + a(b-a)^2\cdot\frac{1+\frac{1}{n}}{1} + \frac{(b-a)^3}{6}\cdot\frac{2+\frac{3}{n}+\frac{1}{n^2}}{1}$
As $n\to\infty$ and $\|P\|\to 0$ this expression has value $(b-a)a^2 + a(b-a)^2\cdot 1 + \frac{(b-a)^3}{6}\cdot 2$
$= ba^2 - a^3 + ab^2 - 2a^2b + a^3 + \frac{1}{3}(b^3 - 3b^2a + 3ba^2 - a^3) = \frac{b^3}{3} - \frac{a^3}{3}$. Thus, $\int_a^b x^2dx = \frac{b^3}{3} - \frac{a^3}{3}$.

77. (a) $U = \max_1 \Delta x + \max_2 \Delta x + \ldots + \max_n \Delta x$ where $\max_1 = f(x_1)$, $\max_2 = f(x_2), \ldots, \max_n = f(x_n)$ since f is increasing on [a, b]; $L = \min_1 \Delta x + \min_2 \Delta x + \ldots + \min_n \Delta x$ where $\min_1 = f(x_0)$, $\min_2 = f(x_1), \ldots,$ $\min_n = f(x_{n-1})$ since f is increasing on [a, b]. Therefore
$U - L = (\max_1 - \min_1)\Delta x + (\max_2 - \min_2)\Delta x + \ldots + (\max_n - \min_n)\Delta x$
$= (f(x_1) - f(x_0))\Delta x + (f(x_2) - f(x_1))\Delta x + \ldots + (f(x_n) - f(x_{n-1}))\Delta x = (f(x_n) - f(x_0))\Delta x = (f(b) - f(a))\Delta x.$
(b) $U = \max_1 \Delta x_1 + \max_2 \Delta x_2 + \ldots + \max_n \Delta x_n$ where $\max_1 = f(x_1)$, $\max_2 = f(x_2), \ldots, \max_n = f(x_n)$ since f is increasing on[a, b]; $L = \min_1 \Delta x_1 + \min_2 \Delta x_2 + \ldots + \min_n \Delta x_n$ where
$\min_1 = f(x_0)$, $\min_2 = f(x_1), \ldots, \min_n = f(x_{n-1})$ since f is increasing on [a, b]. Therefore
$U - L = (\max_1 - \min_1)\Delta x_1 + (\max_2 - \min_2)\Delta x_2 + \ldots + (\max_n - \min_n)\Delta x_n$
$= (f(x_1) - f(x_0))\Delta x_1 + (f(x_2) - f(x_1))\Delta x_2 + \ldots + (f(x_n) - f(x_{n-1}))\Delta x_n$
$\le (f(x_1) - f(x_0))\Delta x_{max} + (f(x_2) - f(x_1))\Delta x_{max} + \ldots + (f(x_n) - f(x_{n-1}))\Delta x_{max}$. Then
$U - L \le (f(x_n) - f(x_0))\Delta x_{max} = (f(b) - f(a))\Delta x_{max} = |f(b) - f(a)|\,\Delta x_{max}$ since $f(b) \ge f(a)$. Thus
$\lim_{\|P\|\to 0}(U - L) = \lim_{\|P\|\to 0}(f(b) - f(a))\Delta x_{max} = 0$, since $\Delta x_{max} = \|P\|$.

78. (a) $U = \max_1 \Delta x + \max_2 \Delta x + \ldots + \max_n \Delta x$ where $\max_1 = f(x_0)$, $\max_2 = f(x_1)$, $\ldots$, $\max_n = f(x_{n-1})$ since f is decreasing on [a, b]; $L = \min_1 \Delta x + \min_2 \Delta x + \ldots + \min_n \Delta x$ where $\min_1 = f(x_1)$, $\min_2 = f(x_2)$, $\ldots$, $\min_n = f(x_n)$ since f is decreasing on [a, b]. Therefore $U - L = (\max_1 - \min_1)\Delta x + (\max_2 - \min_2)\Delta x + \ldots + (\max_n - \min_n)\Delta x$ $= (f(x_0) - f(x_1))\Delta x + (f(x_1) - f(x_2))\Delta x + \ldots + (f(x_{n-1}) - f(x_n))\Delta x = (f(x_0) - f(x_n))\Delta x$ $= (f(a) - f(b))\Delta x$.

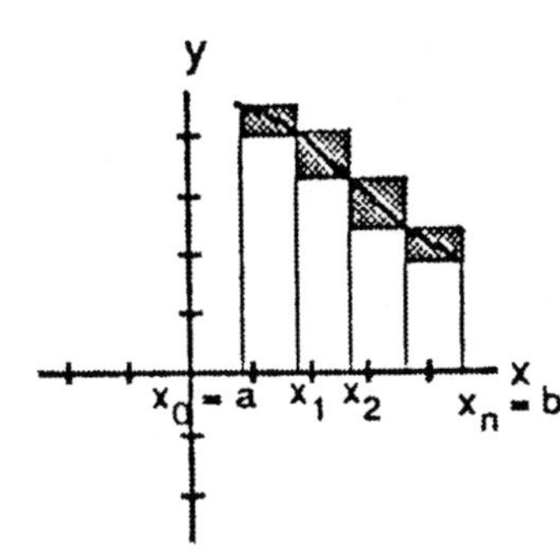

(b) $U = \max_1 \Delta x_1 + \max_2 \Delta x_2 + \ldots + \max_n \Delta x_n$ where $\max_1 = f(x_0)$, $\max_2 = f(x_1)$, $\ldots$, $\max_n = f(x_{n-1})$ since f is decreasing on[a, b]; $L = \min_1 \Delta x_1 + \min_2 \Delta x_2 + \ldots + \min_n \Delta x_n$ where $\min_1 = f(x_1)$, $\min_2 = f(x_2)$, $\ldots$, $\min_n = f(x_n)$ since f is decreasing on [a, b]. Therefore $U - L = (\max_1 - \min_1)\Delta x_1 + (\max_2 - \min_2)\Delta x_2 + \ldots + (\max_n - \min_n)\Delta x_n$ $= (f(x_0) - f(x_1))\Delta x_1 + (f(x_1) - f(x_2))\Delta x_2 + \ldots + (f(x_{n-1}) - f(x_n))\Delta x_n$ $\le (f(x_0) - f(x_n))\Delta x_{max} = (f(a) - f(b)\Delta x_{max} = |f(b) - f(a)|\Delta x_{max}$ since $f(b) \le f(a)$. Thus $\lim\limits_{\|P\| \to 0} (U - L) = \lim\limits_{\|P\| \to 0} |f(b) - f(a)|\Delta x_{max} = 0$, since $\Delta x_{max} = \|P\|$.

79. (a) Partition $\left[0, \frac{\pi}{2}\right]$ into n subintervals, each of length $\Delta x = \frac{\pi}{2n}$ with points $x_0 = 0$, $x_1 = \Delta x$, $x_2 = 2\Delta x, \ldots, x_n = n\Delta x = \frac{\pi}{2}$. Since sin x is increasing on $\left[0, \frac{\pi}{2}\right]$, the upper sum U is the sum of the areas of the circumscribed rectangles of areas $f(x_1)\Delta x = (\sin \Delta x)\Delta x$, $f(x_2)\Delta x = (\sin 2\Delta x)\Delta x, \ldots, f(x_n)\Delta x$ $= (\sin n\Delta x)\Delta x$. Then $U = (\sin \Delta x + \sin 2\Delta x + \ldots + \sin n\Delta x)\Delta x = \left[\frac{\cos \frac{\Delta x}{2} - \cos\left(\left(n + \frac{1}{2}\right)\Delta x\right)}{2 \sin \frac{\Delta x}{2}}\right]\Delta x$

$$= \left[\frac{\cos \frac{\pi}{4n} - \cos\left(\left(n + \frac{1}{2}\right)\frac{\pi}{2n}\right)}{2 \sin \frac{\pi}{4n}}\right]\left(\frac{\pi}{2n}\right) = \frac{\pi\left(\cos \frac{\pi}{4n} - \cos\left(\frac{\pi}{2} + \frac{\pi}{4n}\right)\right)}{4n \sin \frac{\pi}{4n}} = \frac{\cos \frac{\pi}{4n} - \cos\left(\frac{\pi}{2} + \frac{\pi}{4n}\right)}{\left(\frac{\sin \frac{\pi}{4n}}{\frac{\pi}{4n}}\right)}$$

(b) The area is $\int_0^{\pi/2} \sin x \, dx = \lim\limits_{n \to \infty} \frac{\cos \frac{\pi}{4n} - \cos\left(\frac{\pi}{2} + \frac{\pi}{4n}\right)}{\left(\frac{\sin \frac{\pi}{4n}}{\frac{\pi}{4n}}\right)} = \frac{1 - \cos \frac{\pi}{2}}{1} = 1.$

80. (a) The area of the shaded region is $\sum\limits_{i=1}^{n} \Delta x_i \cdot m_i$ which is equal to L.

(b) The area of the shaded region is $\sum\limits_{i=1}^{n} \Delta x_i \cdot M_i$ which is equal to U.

(c) The area of the shaded region is the difference in the areas of the shaded regions shown in the second part of the figure and the first part of the figure. Thus this area is U – L.

81. By Exercise 80, $U - L = \sum\limits_{i=1}^{n} \Delta x_i \cdot M_i - \sum\limits_{i=1}^{n} \Delta x_i \cdot m_i$ where $M_i = \max\{f(x) \text{ on the ith subinterval}\}$ and $m_i = \min\{f(x) \text{ on the ith subinterval}\}$. Thus $U - L = \sum\limits_{i=1}^{n} (M_i - m_i)\Delta x_i < \sum\limits_{i=1}^{n} \epsilon \cdot \Delta x_i$ provided $\Delta x_i < \delta$ for each $i = 1, \ldots, n$. Since $\sum\limits_{i=1}^{n} \epsilon \cdot \Delta x_i = \epsilon \sum\limits_{i=1}^{n} \Delta x_i = \epsilon(b - a)$ the result, $U - L < \epsilon(b - a)$ follows.

82. The car drove the first 150 miles in 5 hours and the second 150 miles in 3 hours, which means it drove 300 miles in 8 hours, for an average of $\frac{300}{8}$ mi/hr $= 37.5$ mi/hr. In terms of average values of functions, the function whose average value we seek is $v(t) = \begin{cases} 30, & 0 \le t \le 5 \\ 50, & 5 < 1 \le 8 \end{cases}$, and the average value is $\frac{(30)(5) + (50)(3)}{8} = 37.5$.

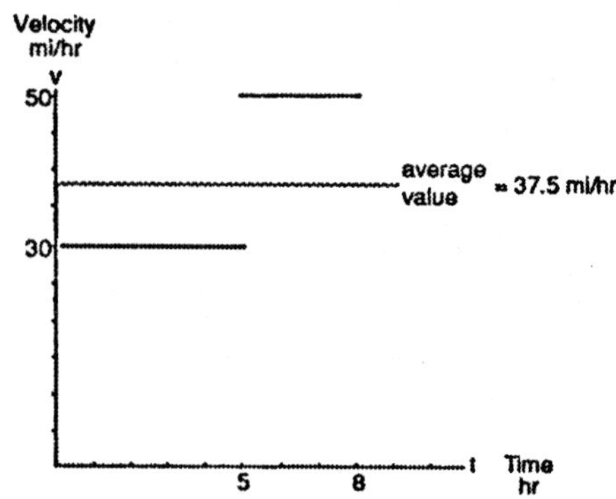

83-88. Example CAS commands:

Maple:

```
with( plots );
with( Student[Calculus1] );
f := x -> 1-x;
a := 0;
b := 1;
N :=[ 4, 10, 20, 50 ];
P := [seq( RiemannSum( f(x), x=a..b, partition=n, method=random, output=plot ), n=N )]:
display( P, insequence=true );
```

89-96. Example CAS commands:

Maple:

```
with( Student[Calculus1] );
f := x -> sin(x);
a := 0;
b := Pi;
plot( f(x), x=a..b, title="#89(a) (Section 5.1)" );
N := [ 100, 200, 1000 ];                          # (b)
for n in N do
  Xlist := [ a+1.*(b-a)/n*i $ i=0..n ];
  Ylist := map( f, Xlist );
end do:
for n in N do                                     # (c)
  Avg[n] := evalf(add(y,y=Ylist)/nops(Ylist));
end do;
avg := FunctionAverage( f(x), x=a..b, output=value );
evalf( avg );
FunctionAverage(f(x),x=a..b,output=plot);      # (d)
fsolve( f(x)=avg, x=0.5 );
fsolve( f(x)=avg, x=2.5 );
fsolve( f(x)=Avg[1000], x=0.5 );
fsolve( f(x)=Avg[1000], x=2.5 );
```

83-96. Example CAS commands:

Mathematica: (assigned function and values for a, b, and n may vary)

Sums of rectangles evaluated at left-hand endpoints can be represented and evaluated by this set of commands

```
Clear[x, f, a, b, n]
```

```
{a, b}={0, π}; n =10; dx = (b − a)/n;
f = Sin[x]^2;
xvals =Table[N[x], {x, a, b − dx, dx}];
yvals = f /.x → xvals;
boxes = MapThread[Line[{{#1,0},{#1, #3},{#2, #3},{#2, 0}]&,{xvals, xvals + dx, yvals}];
Plot[f, {x, a, b}, Epilog → boxes];
Sum[yvals[[i]] dx, {i, 1, Length[yvals]}]//N
```

Sums of rectangles evaluated at right-hand endpoints can be represented and evaluated by this set of commands.

```
Clear[x, f, a, b, n]
{a, b}={0, π}; n =10; dx = (b − a)/n;
f = Sin[x]^2;
xvals =Table[N[x], {x, a + dx, b, dx}];
yvals = f /.x → xvals;
boxes = MapThread[Line[{{#1,0},{#1, #3},{#2, #3},{#2, 0}]&,{xvals − dx,xvals, yvals}];
Plot[f, {x, a, b}, Epilog → boxes];
Sum[yvals[[i]] dx, {i, 1,Length[yvals]}]//N
```

Sums of rectangles evaluated at midpoints can be represented and evaluated by this set of commands.

```
Clear[x, f, a, b, n]
{a, b}={0, π}; n =10; dx = (b − a)/n;
f = Sin[x]^2;
xvals =Table[N[x], {x, a + dx/2, b − dx/2, dx}];
yvals = f /.x → xvals;
boxes = MapThread[Line[{{#1,0},{#1, #3},{#2, #3},{#2, 0}]&,{xvals − dx/2, xvals + dx/2, yvals}];
Plot[f, {x, a, b},Epilog → boxes];
Sum[yvals[[i]] dx, {i, 1, Length[yvals]}]//N
```

5.4 THE FUNDAMENTAL THEOREM OF CALCULUS

1. $\int_{-2}^{0}(2x+5)\,dx = \left[x^2+5x\right]_{-2}^{0} = (0^2+5(0)) - ((-2)^2+5(-2)) = 6$

2. $\int_{-3}^{4}\left(5-\frac{x}{2}\right)dx = \left[5x-\frac{x^2}{4}\right]_{-3}^{4} = \left(5(4)-\frac{4^2}{4}\right) - \left(5(-3)-\frac{(-3)^2}{4}\right) = \frac{133}{4}$

3. $\int_{0}^{4}\left(3x-\frac{x^3}{4}\right)dx = \left[\frac{3x^2}{2}-\frac{x^4}{16}\right]_{0}^{4} = \left(\frac{3(4)^2}{2}-\frac{4^4}{16}\right) - \left(\frac{3(0)^2}{2}-\frac{(0)^4}{16}\right) = 8$

4. $\int_{-2}^{2}(x^3-2x+3)\,dx = \left[\frac{x^4}{4}-x^2+3x\right]_{-2}^{2} = \left(\frac{2^4}{4}-2^2+3(2)\right) - \left(\frac{(-2)^4}{4}-(-2)^2+3(-2)\right) = 12$

5. $\int_{0}^{1}\left(x^2+\sqrt{x}\right)dx = \left[\frac{x^3}{3}+\frac{2}{3}x^{3/2}\right]_{0}^{1} = \left(\frac{1}{3}+\frac{2}{3}\right) - 0 = 1$

6. $\int_{0}^{5}x^{3/2}\,dx = \left[\frac{2}{5}x^{5/2}\right]_{0}^{5} = \frac{2}{5}(5)^{5/2} - 0 = 2(5)^{3/2} = 10\sqrt{5}$

7. $\int_{1}^{32}x^{-6/5}\,dx = \left[-5x^{-1/5}\right]_{1}^{32} = \left(-\frac{5}{2}\right) - (-5) = \frac{5}{2}$

8. $\int_{-2}^{-1}\frac{2}{x^2}\,dx = \int_{-2}^{-1}2x^{-2}\,dx = \left[-2x^{-1}\right]_{-2}^{-1} = \left(\frac{-2}{-1}\right) - \left(\frac{-2}{-2}\right) = 1$

9. $\int_0^{\pi} \sin x \, dx = [-\cos x]_0^{\pi} = (-\cos \pi) - (-\cos 0) = -(-1) - (-1) = 2$

10. $\int_0^{\pi} (1 + \cos x) \, dx = [x + \sin x]_0^{\pi} = (\pi + \sin \pi) - (0 + \sin 0) = \pi$

11. $\int_0^{\pi/3} 2 \sec^2 x \, dx = [2 \tan x]_0^{\pi/3} = \left(2 \tan\left(\frac{\pi}{3}\right)\right) - (2 \tan 0) = 2\sqrt{3} - 0 = 2\sqrt{3}$

12. $\int_{\pi/6}^{5\pi/6} \csc^2 x \, dx = [-\cot x]_{\pi/6}^{5\pi/6} = \left(-\cot\left(\frac{5\pi}{6}\right)\right) - \left(-\cot\left(\frac{\pi}{6}\right)\right) = -\left(-\sqrt{3}\right) - \left(-\sqrt{3}\right) = 2\sqrt{3}$

13. $\int_{\pi/4}^{3\pi/4} \csc \theta \cot \theta \, d\theta = [-\csc \theta]_{\pi/4}^{3\pi/4} = \left(-\csc\left(\frac{3\pi}{4}\right)\right) - \left(-\csc\left(\frac{\pi}{4}\right)\right) = -\sqrt{2} - \left(-\sqrt{2}\right) = 0$

14. $\int_0^{\pi/3} 4 \sec u \tan u \, du = [4 \sec u]_0^{\pi/3} = 4 \sec\left(\frac{\pi}{3}\right) - 4 \sec 0 = 4(2) - 4(1) = 4$

15. $\int_{\pi/2}^{0} \frac{1+\cos 2t}{2} \, dt = \int_{\pi/2}^{0} \left(\frac{1}{2} + \frac{1}{2} \cos 2t\right) dt = \left[\frac{1}{2} t + \frac{1}{4} \sin 2t\right]_{\pi/2}^{0} = \left(\frac{1}{2}(0) + \frac{1}{4} \sin 2(0)\right) - \left(\frac{1}{2}\left(\frac{\pi}{2}\right) + \frac{1}{4} \sin 2\left(\frac{\pi}{2}\right)\right)$
$= -\frac{\pi}{4}$

16. $\int_{-\pi/3}^{\pi/3} \frac{1-\cos 2t}{2} \, dt = \int_{-\pi/3}^{\pi/3} \left(\frac{1}{2} - \frac{1}{2} \cos 2t\right) dt = \left[\frac{1}{2} t - \frac{1}{4} \sin 2t\right]_{-\pi/3}^{\pi/3}$
$= \left(\frac{1}{2}\left(\frac{\pi}{3}\right) - \frac{1}{4} \sin 2\left(\frac{\pi}{3}\right)\right) - \left(\frac{1}{2}\left(-\frac{\pi}{3}\right) - \frac{1}{4} \sin 2\left(-\frac{\pi}{3}\right)\right) = \frac{\pi}{6} - \frac{1}{4} \sin \frac{2\pi}{3} + \frac{\pi}{6} + \frac{1}{4} \sin\left(\frac{-2\pi}{3}\right) = \frac{\pi}{3} - \frac{\sqrt{3}}{4}$

17. $\int_{-\pi/2}^{\pi/2} (8y^2 + \sin y) \, dy = \left[\frac{8y^3}{3} - \cos y\right]_{-\pi/2}^{\pi/2} = \left(\frac{8\left(\frac{\pi}{2}\right)^3}{3} - \cos \frac{\pi}{2}\right) - \left(\frac{8\left(-\frac{\pi}{2}\right)^3}{3} - \cos\left(-\frac{\pi}{2}\right)\right) = \frac{2\pi^3}{3}$

18. $\int_{-\pi/3}^{-\pi/4} \left(4 \sec^2 t + \frac{\pi}{t^2}\right) dt = \int_{-\pi/3}^{-\pi/4} (4 \sec^2 t + \pi t^{-2}) \, dt = \left[4 \tan t - \frac{\pi}{t}\right]_{-\pi/3}^{-\pi/4}$
$= \left(4 \tan\left(-\frac{\pi}{4}\right) - \frac{\pi}{\left(-\frac{\pi}{4}\right)}\right) - \left(4 \tan\left(\frac{\pi}{3}\right) - \frac{\pi}{\left(-\frac{\pi}{3}\right)}\right) = (4(-1) + 4) - \left(4\left(-\sqrt{3}\right) + 3\right) = 4\sqrt{3} - 3$

19. $\int_1^{-1} (r + 1)^2 \, dr = \int_1^{-1} (r^2 + 2r + 1) \, dr = \left[\frac{r^3}{3} + r^2 + r\right]_1^{-1} = \left(\frac{(-1)^3}{3} + (-1)^2 + (-1)\right) - \left(\frac{1^3}{3} + 1^2 + 1\right) = -\frac{8}{3}$

20. $\int_{-\sqrt{3}}^{\sqrt{3}} (t + 1)(t^2 + 4) \, dt = \int_{-\sqrt{3}}^{\sqrt{3}} (t^3 + t^2 + 4t + 4) \, dt = \left[\frac{t^4}{4} + \frac{t^3}{3} + 2t^2 + 4t\right]_{-\sqrt{3}}^{\sqrt{3}}$
$= \left(\frac{\left(\sqrt{3}\right)^4}{4} + \frac{\left(\sqrt{3}\right)^3}{3} + 2\left(\sqrt{3}\right)^2 + 4\sqrt{3}\right) - \left(\frac{\left(-\sqrt{3}\right)^4}{4} + \frac{\left(-\sqrt{3}\right)^3}{3} + 2\left(-\sqrt{3}\right)^2 + 4\left(-\sqrt{3}\right)\right) = 10\sqrt{3}$

21. $\int_{\sqrt{2}}^{1} \left(\frac{u^7}{2} - \frac{1}{u^5}\right) du = \int_{\sqrt{2}}^{1} \left(\frac{u^7}{2} - u^{-5}\right) du = \left[\frac{u^8}{16} + \frac{1}{4u^4}\right]_{\sqrt{2}}^{1} = \left(\frac{1^8}{16} + \frac{1}{4(1)^4}\right) - \left(\frac{\left(\sqrt{2}\right)^8}{16} + \frac{1}{4\left(\sqrt{2}\right)^4}\right) = -\frac{3}{4}$

22. $\int_{1/2}^{1} \left(\frac{1}{v^3} - \frac{1}{v^4}\right) dv = \int_{1/2}^{1} (v^{-3} - v^{-4}) \, dv = \left[\frac{-1}{2v^2} + \frac{1}{3v^3}\right]_{1/2}^{1} = \left(\frac{-1}{2(1)^2} + \frac{1}{3(1)^3}\right) - \left(\frac{-1}{2\left(\frac{1}{2}\right)^2} + \frac{1}{3\left(\frac{1}{2}\right)^3}\right) = -\frac{5}{6}$

23. $\int_1^{\sqrt{2}} \frac{s^2 + \sqrt{s}}{s^2} \, ds = \int_1^{\sqrt{2}} \left(1 + s^{-3/2}\right) ds = \left[s - \frac{2}{\sqrt{s}}\right]_1^{\sqrt{2}} = \left(\sqrt{2} - \frac{2}{\sqrt{\sqrt{2}}}\right) - \left(1 - \frac{2}{\sqrt{1}}\right) = \sqrt{2} - 2^{3/4} + 1$
$= \sqrt{2} - \sqrt[4]{8} + 1$

24. $\int_9^4 \frac{1-\sqrt{u}}{\sqrt{u}}\,du = \int_9^4 \left(u^{-1/2} - 1\right) du = \left[2\sqrt{u} - u\right]_9^4 = \left(2\sqrt{4} - 4\right) - \left(2\sqrt{9} - 9\right) = 3$

25. $\int_{-4}^4 |x|\,dx = \int_{-4}^0 |x|\,dx + \int_0^4 |x|\,dx = -\int_{-4}^0 x\,dx + \int_0^4 x\,dx = \left[-\frac{x^2}{2}\right]_{-4}^0 + \left[\frac{x^2}{2}\right]_0^4 = \left(-\frac{0^2}{2} + \frac{(-4)^2}{2}\right) + \left(\frac{4^2}{2} - \frac{0^2}{2}\right)$
$= 16$

26. $\int_0^\pi \frac{1}{2}\left(\cos x + |\cos x|\right) dx = \int_0^{\pi/2} \frac{1}{2}(\cos x + \cos x)\,dx + \int_{\pi/2}^\pi \frac{1}{2}(\cos x - \cos x)\,dx = \int_0^{\pi/2} \cos x\,dx = [\sin x]_0^{\pi/2}$
$= \sin\frac{\pi}{2} - \sin 0 = 1$

27. $\int_0^{\ln 2} e^{3x}dx = \frac{1}{3}e^{3x}\Big|_0^{\ln 2} = \frac{1}{3}e^{3\ln 2} - \frac{1}{3}e^0 = \frac{1}{3}e^{\ln 8} - \frac{1}{3} = \frac{8}{3} - \frac{1}{3} = \frac{7}{3}$

28. $\int_1^2 \left(\frac{1}{x} - e^{-x}\right)dx = \left(\ln x + e^{-x}\right)\Big|_1^2 = \left(\ln 2 + e^{-2}\right) - \left(\ln 1 + e^{-1}\right) = \ln 2 + \frac{1}{e^2} - \frac{1}{e}$

29. $\int_0^1 \frac{4}{1+x^2}dx = 4\tan^{-1}x\Big|_0^1 = 4\tan^{-1}1 - 4\tan^{-1}0 = 4\left(\frac{\pi}{4}\right) - 4(0) = \pi$

30. $\int_2^5 \frac{x}{\sqrt{1+x^2}}dx = \int_2^5 x\left(1 + x^2\right)^{1/2}dx = \sqrt{1 + x^2}\Big|_2^5 = \sqrt{26} - \sqrt{5}$

31. $\int_2^4 x^{\pi-1}dx = \frac{x^\pi}{\pi}\Big|_2^4 = \frac{1}{\pi}\left(4^\pi - 2^\pi\right)$

32. $\int_{-1}^0 \pi^{x-1}dx = \frac{1}{\ln \pi}\cdot \pi^{x-1}\Big|_{-1}^0 = \frac{1}{\ln \pi}\left(\pi^{-1} - \pi^{-2}\right)$

33. $\int_0^1 x\,e^{x^2}dx = \frac{1}{2}e^{x^2}\Big|_0^1 = \frac{1}{2}e^1 - \frac{1}{2}e^0 = \frac{1}{2}(e - 1)$

34. $\int_1^2 \frac{\ln x}{x}dx = \frac{1}{2}(\ln x)^2\Big|_1^2 = \frac{1}{2}(\ln 2)^2 - \frac{1}{2}(\ln 1)^2 = \frac{1}{2}(\ln 2)^2$

35. (a) $\int_0^{\sqrt{x}} \cos t\,dt = [\sin t]_0^{\sqrt{x}} = \sin\sqrt{x} - \sin 0 = \sin\sqrt{x} \Rightarrow \frac{d}{dx}\left(\int_0^{\sqrt{x}} \cos t\,dt\right) = \frac{d}{dx}\left(\sin\sqrt{x}\right) = \cos\sqrt{x}\left(\frac{1}{2}x^{-1/2}\right)$
$= \frac{\cos\sqrt{x}}{2\sqrt{x}}$

(b) $\frac{d}{dx}\left(\int_0^{\sqrt{x}} \cos t\,dt\right) = \left(\cos\sqrt{x}\right)\left(\frac{d}{dx}\left(\sqrt{x}\right)\right) = \left(\cos\sqrt{x}\right)\left(\frac{1}{2}x^{-1/2}\right) = \frac{\cos\sqrt{x}}{2\sqrt{x}}$

36. (a) $\int_1^{\sin x} 3t^2\,dt = \left[t^3\right]_1^{\sin x} = \sin^3 x - 1 \Rightarrow \frac{d}{dx}\left(\int_1^{\sin x} 3t^2\,dt\right) = \frac{d}{dx}\left(\sin^3 x - 1\right) = 3\sin^2 x\cos x$

(b) $\frac{d}{dx}\left(\int_1^{\sin x} 3t^2\,dt\right) = \left(3\sin^2 x\right)\left(\frac{d}{dx}(\sin x)\right) = 3\sin^2 x\cos x$

37. (a) $\int_0^{t^4} \sqrt{u}\,du = \int_0^{t^4} u^{1/2}\,du = \left[\frac{2}{3}u^{3/2}\right]_0^{t^4} = \frac{2}{3}\left(t^4\right)^{3/2} - 0 = \frac{2}{3}t^6 \Rightarrow \frac{d}{dt}\left(\int_0^{t^4} \sqrt{u}\,du\right) = \frac{d}{dt}\left(\frac{2}{3}t^6\right) = 4t^5$

(b) $\frac{d}{dt}\left(\int_0^{t^4} \sqrt{u}\,du\right) = \sqrt{t^4}\left(\frac{d}{dt}\left(t^4\right)\right) = t^2\left(4t^3\right) = 4t^5$

38. (a) $\int_0^{\tan\theta} \sec^2 y\, dy = [\tan y]_0^{\tan\theta} = \tan(\tan\theta) - 0 = \tan(\tan\theta) \Rightarrow \frac{d}{d\theta}\left(\int_0^{\tan\theta} \sec^2 y\, dy\right) = \frac{d}{d\theta}(\tan(\tan\theta))$
$= (\sec^2(\tan\theta))\sec^2\theta$

(b) $\frac{d}{d\theta}\left(\int_0^{\tan\theta} \sec^2 y\, dy\right) = (\sec^2(\tan\theta))\left(\frac{d}{d\theta}(\tan\theta)\right) = (\sec^2(\tan\theta))\sec^2\theta$

39. (a) $\int_0^{x^3} e^{-t} dt = -e^{-t}\Big|_0^{x^3} = -e^{-x^3} + 1 \Rightarrow \frac{d}{dx}\left(\int_0^{x^3} e^{-t} dt\right) = \frac{d}{dx}\left(-e^{-x^3} + 1\right) = 3x^2e^{-x^3}$

(b) $\frac{d}{dx}\left(\int_0^{x^3} e^{-t} dt\right) = e^{-x^3} \cdot \frac{d}{dx}(x^3) = 3x^2e^{-x^3}$

40. (a) $\int_0^{\sqrt{t}} \left(x^4 + \frac{3}{\sqrt{1-x^2}}\right) dx = \frac{x^5}{5} + 3\sin^{-1}x\Big|_0^{\sqrt{t}} = \frac{1}{5}t^{5/2} + 3\sin^{-1}\sqrt{t} \Rightarrow \frac{d}{dt}\int_0^{\sqrt{t}} \left(x^4 + \frac{3}{\sqrt{1-x^2}}\right) dx$
$= \frac{d}{dt}\left(\frac{1}{5}t^{5/2} + 3\sin^{-1}\sqrt{t}\right) = \frac{1}{5} \cdot \frac{5}{2}t^{3/2} + \frac{3}{\sqrt{1-t}} \cdot \frac{1}{2\sqrt{t}} = \frac{1}{2}t^{3/2} + \frac{3}{2\sqrt{t-t^2}}$

(b) $\frac{d}{dt}\int_0^{\sqrt{t}} \left(x^4 + \frac{3}{\sqrt{1-x^2}}\right) dx = \left(t^2 + \frac{3}{\sqrt{1-t}}\right) \cdot \frac{d}{dt}\left(\sqrt{t}\right) = \left(t^2 + \frac{3}{\sqrt{1-t}}\right) \cdot \frac{1}{2\sqrt{t}} = \frac{1}{2}t^{3/2} + \frac{3}{2\sqrt{t-t^2}}$

41. $y = \int_0^x \sqrt{1+t^2}\, dt \Rightarrow \frac{dy}{dx} = \sqrt{1+x^2}$

42. $y = \int_1^x \frac{1}{t}\, dt \Rightarrow \frac{dy}{dx} = \frac{1}{x},\ x > 0$

43. $y = \int_{\sqrt{x}}^0 \sin t^2\, dt = -\int_0^{\sqrt{x}} \sin t^2\, dt \Rightarrow \frac{dy}{dx} = -\left(\sin\left(\sqrt{x}\right)^2\right)\left(\frac{d}{dx}\left(\sqrt{x}\right)\right) = -(\sin x)\left(\frac{1}{2}x^{-1/2}\right) = -\frac{\sin x}{2\sqrt{x}}$

44. $y = \int_0^{x^2} \cos\sqrt{t}\, dt \Rightarrow \frac{dy}{dx} = \left(\cos\sqrt{x^2}\right)\left(\frac{d}{dx}(x^2)\right) = 2x\cos|x|$

45. $y = \int_0^{\sin x} \frac{dt}{\sqrt{1-t^2}},\ |x| < \frac{\pi}{2} \Rightarrow \frac{dy}{dx} = \frac{1}{\sqrt{1-\sin^2 x}}\left(\frac{d}{dx}(\sin x)\right) = \frac{1}{\sqrt{\cos^2 x}}(\cos x) = \frac{\cos x}{|\cos x|} = \frac{\cos x}{\cos x} = 1$ since $|x| < \frac{\pi}{2}$

46. $y = \int_{\tan x}^0 \frac{dt}{1+t^2} = -\int_0^{\tan x} \frac{dt}{1+t^2} \Rightarrow \frac{dy}{dx} = \left(-\frac{1}{1+\tan^2 x}\right)\left(\frac{d}{dx}(\tan x)\right) = \left(-\frac{1}{\sec^2 x}\right)(\sec^2 x) = -1$

47. $y = \int_0^{e^{x^2}} \frac{1}{\sqrt{t}} dt \Rightarrow \frac{dy}{dx} = \frac{1}{\sqrt{e^{x^2}}} \cdot \frac{d}{dx}\left(e^{x^2}\right) = \frac{1}{e^{\frac{1}{2}x^2}} \cdot 2x\, e^{x^2} = 2x\, e^{\frac{1}{2}x^2}$

48. $y = \int_{2^x}^1 \sqrt[3]{t}\, dt = -\int_1^{2^x} t^{1/3}\, dt \Rightarrow \frac{dy}{dx} = -(2^x)^{1/3} \cdot \frac{d}{dx}(2^x) = -2^{x/3} \cdot 2^x \ln 2 = -2^{4x/3}\ln 2$

49. $y = \int_0^{\sin^{-1}t} \cos t\, dt \Rightarrow \frac{dy}{dx} = \cos(\sin^{-1}x) \cdot \frac{d}{dx}(\sin^{-1}x) = \sqrt{1-x^2} \cdot \frac{1}{\sqrt{1-x^2}} = 1$

50. $y = \int_{-1}^{x^{1/\pi}} \sin^{-1} t\, dt \Rightarrow \frac{dy}{dx} = \sin^{-1}\left(x^{\frac{1}{\pi}}\right) \cdot \frac{d}{dx}\left(x^{\frac{1}{\pi}}\right) = \sin^{-1}\left(x^{\frac{1}{\pi}}\right) \cdot \frac{1}{\pi}x^{\frac{1}{\pi}-1}$

51. $-x^2 - 2x = 0 \Rightarrow -x(x+2) = 0 \Rightarrow x = 0$ or $x = -2$; Area

$$= -\int_{-3}^{-2}(-x^2 - 2x)dx + \int_{-2}^{0}(-x^2 - 2x)dx - \int_{0}^{2}(-x^2 - 2x)dx$$

$$= -\left[-\tfrac{x^3}{3} - x^2\right]_{-3}^{-2} + \left[-\tfrac{x^3}{3} - x^2\right]_{-2}^{0} - \left[-\tfrac{x^3}{3} - x^2\right]_{0}^{2}$$

$$= -\left(\left(-\tfrac{(-2)^3}{3} - (-2)^2\right) - \left(-\tfrac{(-3)^3}{3} - (-3)^2\right)\right)$$

$$+ \left(\left(-\tfrac{0^3}{3} - 0^2\right) - \left(-\tfrac{(-2)^3}{3} - (-2)^2\right)\right)$$

$$- \left(\left(-\tfrac{2^3}{3} - 2^2\right) - \left(-\tfrac{0^3}{3} - 0^2\right)\right) = \tfrac{28}{3}$$

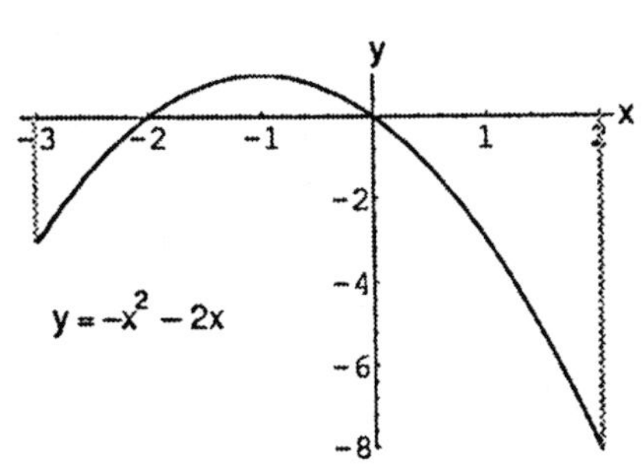

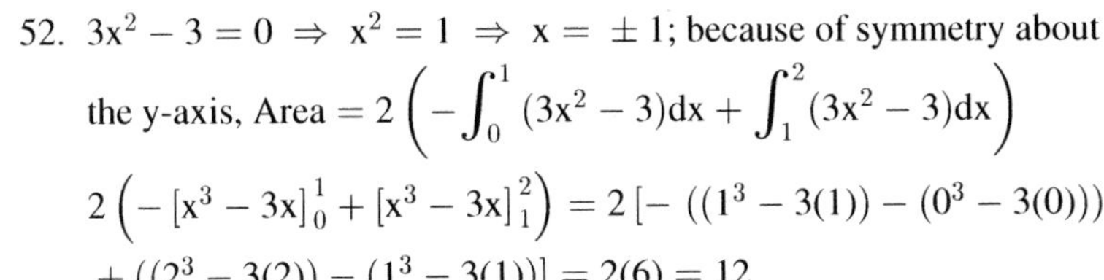

52. $3x^2 - 3 = 0 \Rightarrow x^2 = 1 \Rightarrow x = \pm 1$; because of symmetry about the y-axis, Area $= 2\left(-\int_{0}^{1}(3x^2 - 3)dx + \int_{1}^{2}(3x^2 - 3)dx\right)$

$$2\left(-\left[x^3 - 3x\right]_0^1 + \left[x^3 - 3x\right]_1^2\right) = 2\left[-\left((1^3 - 3(1)) - (0^3 - 3(0))\right)\right.$$

$$\left. + \left((2^3 - 3(2)) - (1^3 - 3(1))\right)\right] = 2(6) = 12$$

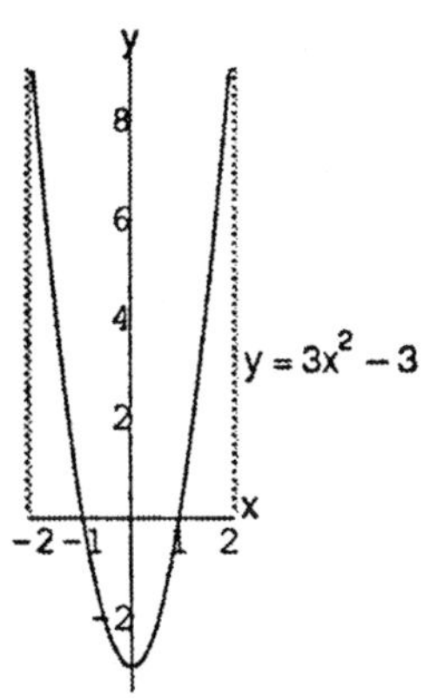

53. $x^3 - 3x^2 + 2x = 0 \Rightarrow x(x^2 - 3x + 2) = 0$

$\Rightarrow x(x-2)(x-1) = 0 \Rightarrow x = 0, 1$, or 2;

$$\text{Area} = \int_{0}^{1}(x^3 - 3x^2 + 2x)dx - \int_{1}^{2}(x^3 - 3x^2 + 2x)dx$$

$$= \left[\tfrac{x^4}{4} - x^3 + x^2\right]_0^1 - \left[\tfrac{x^4}{4} - x^3 + x^2\right]_1^2$$

$$= \left(\tfrac{1^4}{4} - 1^3 + 1^2\right) - \left(\tfrac{0^4}{4} - 0^3 + 0^2\right)$$

$$- \left[\left(\tfrac{2^4}{4} - 2^3 + 2^2\right) - \left(\tfrac{1^4}{4} - 1^3 + 1^2\right)\right] = \tfrac{1}{2}$$

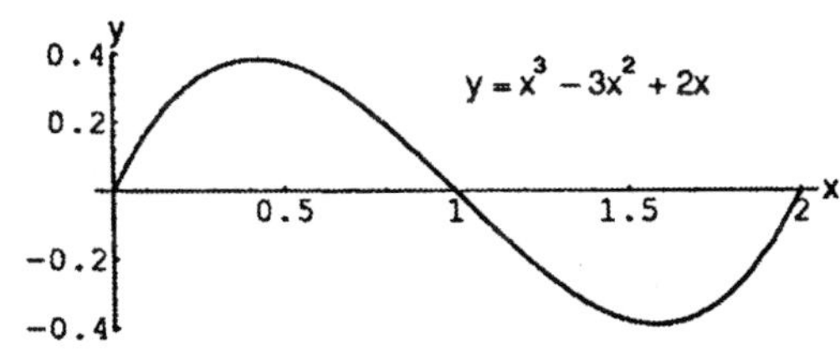

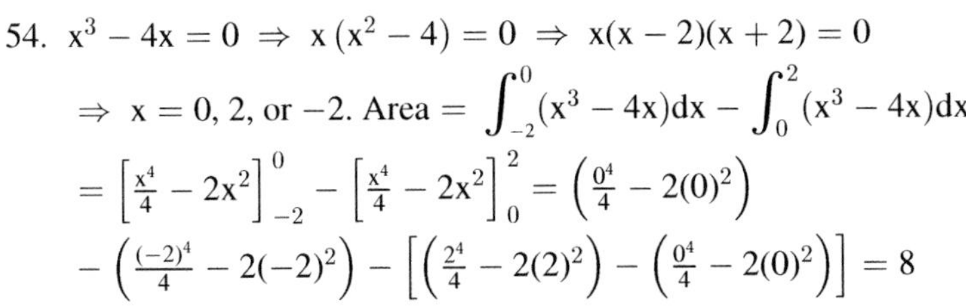

54. $x^3 - 4x = 0 \Rightarrow x(x^2 - 4) = 0 \Rightarrow x(x-2)(x+2) = 0$

$\Rightarrow x = 0, 2$, or -2. Area $= \int_{-2}^{0}(x^3 - 4x)dx - \int_{0}^{2}(x^3 - 4x)dx$

$$= \left[\tfrac{x^4}{4} - 2x^2\right]_{-2}^{0} - \left[\tfrac{x^4}{4} - 2x^2\right]_{0}^{2} = \left(\tfrac{0^4}{4} - 2(0)^2\right)$$

$$- \left(\tfrac{(-2)^4}{4} - 2(-2)^2\right) - \left[\left(\tfrac{2^4}{4} - 2(2)^2\right) - \left(\tfrac{0^4}{4} - 2(0)^2\right)\right] = 8$$

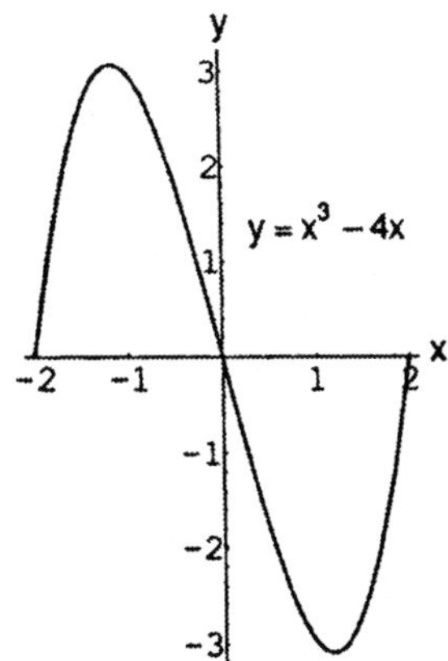

55. $x^{1/3} = 0 \Rightarrow x = 0$; Area $= -\int_{-1}^{0} x^{1/3}\,dx + \int_{0}^{8} x^{1/3}\,dx$
$= \left[-\frac{3}{4}x^{4/3}\right]_{-1}^{0} + \left[\frac{3}{4}x^{4/3}\right]_{0}^{8}$
$= \left(-\frac{3}{4}(0)^{4/3}\right) - \left(-\frac{3}{4}(-1)^{4/3}\right) + \left(\frac{3}{4}(8)^{4/3}\right) - \left(\frac{3}{4}(0)^{4/3}\right)$
$= \frac{51}{4}$

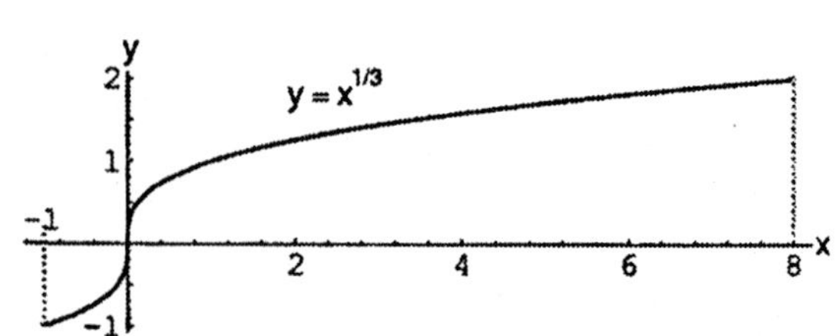

56. $x^{1/3} - x = 0 \Rightarrow x^{1/3}\left(1 - x^{2/3}\right) = 0 \Rightarrow x^{1/3} = 0$ or
$1 - x^{2/3} = 0 \Rightarrow x = 0$ or $1 = x^{2/3} \Rightarrow x = 0$ or
$1 = x^2 \Rightarrow x = 0$ or ± 1;
Area $= -\int_{-1}^{0}\left(x^{1/3} - x\right)dx + \int_{0}^{1}\left(x^{1/3} - x\right)dx - \int_{1}^{8}\left(x^{1/3} - x\right)dx$
$= -\left[\frac{3}{4}x^{4/3} - \frac{x^2}{2}\right]_{-1}^{0} + \left[\frac{3}{4}x^{4/3} - \frac{x^2}{2}\right]_{0}^{1} - \left[\frac{3}{4}x^{4/3} - \frac{x^2}{2}\right]_{1}^{8}$
$= -\left[\left(\frac{3}{4}(0)^{4/3} - \frac{0^2}{2}\right) - \left(\frac{3}{4}(-1)^{4/3} - \frac{(-1)^2}{2}\right)\right]$
$+ \left[\left(\frac{3}{4}(1)^{4/3} - \frac{1^2}{2}\right) - \left(\frac{3}{4}(0)^{4/3} - \frac{0^2}{2}\right)\right]$
$- \left[\left(\frac{3}{4}(8)^{4/3} - \frac{8^2}{2}\right) - \left(\frac{3}{4}(1)^{4/3} - \frac{1^2}{2}\right)\right]$
$= \frac{1}{4} + \frac{1}{4} - \left(-20 - \frac{3}{4} + \frac{1}{2}\right) = \frac{83}{4}$

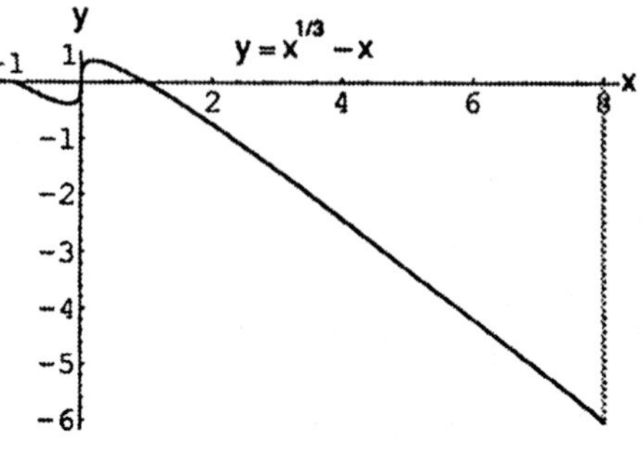

57. The area of the rectangle bounded by the lines $y = 2$, $y = 0$, $x = \pi$, and $x = 0$ is 2π. The area under the curve $y = 1 + \cos x$ on $[0, \pi]$ is $\int_{0}^{\pi}(1 + \cos x)\,dx = [x + \sin x]_0^{\pi} = (\pi + \sin \pi) - (0 + \sin 0) = \pi$. Therefore the area of the shaded region is $2\pi - \pi = \pi$.

58. The area of the rectangle bounded by the lines $x = \frac{\pi}{6}$, $x = \frac{5\pi}{6}$, $y = \sin\frac{\pi}{6} = \frac{1}{2} = \sin\frac{5\pi}{6}$, and $y = 0$ is $\frac{1}{2}\left(\frac{5\pi}{6} - \frac{\pi}{6}\right) = \frac{\pi}{3}$. The area under the curve $y = \sin x$ on $\left[\frac{\pi}{6}, \frac{5\pi}{6}\right]$ is $\int_{\pi/6}^{5\pi/6} \sin x\,dx = [-\cos x]_{\pi/6}^{5\pi/6}$
$= \left(-\cos\frac{5\pi}{6}\right) - \left(-\cos\frac{\pi}{6}\right) = -\left(-\frac{\sqrt{3}}{2}\right) + \frac{\sqrt{3}}{2} = \sqrt{3}$. Therefore the area of the shaded region is $\sqrt{3} - \frac{\pi}{3}$.

59. On $\left[-\frac{\pi}{4}, 0\right]$: The area of the rectangle bounded by the lines $y = \sqrt{2}$, $y = 0$, $\theta = 0$, and $\theta = -\frac{\pi}{4}$ is $\sqrt{2}\left(\frac{\pi}{4}\right)$ $= \frac{\pi\sqrt{2}}{4}$. The area between the curve $y = \sec\theta\tan\theta$ and $y = 0$ is $-\int_{-\pi/4}^{0} \sec\theta\tan\theta\,d\theta = [-\sec\theta]_{-\pi/4}^{0}$ $= (-\sec 0) - \left(-\sec\left(-\frac{\pi}{4}\right)\right) = \sqrt{2} - 1$. Therefore the area of the shaded region on $\left[-\frac{\pi}{4}, 0\right]$ is $\frac{\pi\sqrt{2}}{4} + \left(\sqrt{2} - 1\right)$.
On $\left[0, \frac{\pi}{4}\right]$: The area of the rectangle bounded by $\theta = \frac{\pi}{4}$, $\theta = 0$, $y = \sqrt{2}$, and $y = 0$ is $\sqrt{2}\left(\frac{\pi}{4}\right) = \frac{\pi\sqrt{2}}{4}$. The area under the curve $y = \sec\theta\tan\theta$ is $\int_{0}^{\pi/4} \sec\theta\tan\theta\,d\theta = [\sec\theta]_0^{\pi/4} = \sec\frac{\pi}{4} - \sec 0 = \sqrt{2} - 1$. Therefore the area of the shaded region on $\left[0, \frac{\pi}{4}\right]$ is $\frac{\pi\sqrt{2}}{4} - \left(\sqrt{2} - 1\right)$. Thus, the area of the total shaded region is $\left(\frac{\pi\sqrt{2}}{4} + \sqrt{2} - 1\right) + \left(\frac{\pi\sqrt{2}}{4} - \sqrt{2} + 1\right) = \frac{\pi\sqrt{2}}{2}$.

60. The area of the rectangle bounded by the lines $y = 2$, $y = 0$, $t = -\frac{\pi}{4}$, and $t = 1$ is $2\left(1 - \left(-\frac{\pi}{4}\right)\right) = 2 + \frac{\pi}{2}$. The area under the curve $y = \sec^2 t$ on $\left[-\frac{\pi}{4}, 0\right]$ is $\int_{-\pi/4}^{0} \sec^2 t\,dt = [\tan t]_{-\pi/4}^{0} = \tan 0 - \tan\left(-\frac{\pi}{4}\right) = 1$. The area under the curve $y = 1 - t^2$ on $[0, 1]$ is $\int_{0}^{1}\left(1 - t^2\right)dt = \left[t - \frac{t^3}{3}\right]_0^1 = \left(1 - \frac{1^3}{3}\right) - \left(0 - \frac{0^3}{3}\right) = \frac{2}{3}$. Thus, the total area under the curves on $\left[-\frac{\pi}{4}, 1\right]$ is $1 + \frac{2}{3} = \frac{5}{3}$. Therefore the area of the shaded region is $\left(2 + \frac{\pi}{2}\right) - \frac{5}{3} = \frac{1}{3} + \frac{\pi}{2}$.

61. $y = \int_{\pi}^{x} \frac{1}{t}\,dt - 3 \Rightarrow \frac{dy}{dx} = \frac{1}{x}$ and $y(\pi) = \int_{\pi}^{\pi} \frac{1}{t}\,dt - 3 = 0 - 3 = -3 \Rightarrow$ (d) is a solution to this problem.

62. $y = \int_{-1}^{x} \sec t\,dt + 4 \Rightarrow \frac{dy}{dx} = \sec x$ and $y(-1) = \int_{-1}^{-1} \sec t\,dt + 4 = 0 + 4 = 4 \Rightarrow$ (c) is a solution to this problem.

63. $y = \int_{0}^{x} \sec t\,dt + 4 \Rightarrow \frac{dy}{dx} = \sec x$ and $y(0) = \int_{0}^{0} \sec t\,dt + 4 = 0 + 4 = 4 \Rightarrow$ (b) is a solution to this problem.

64. $y = \int_{1}^{x} \frac{1}{t}\,dt - 3 \Rightarrow \frac{dy}{dx} = \frac{1}{x}$ and $y(1) = \int_{1}^{1} \frac{1}{t}\,dt - 3 = 0 - 3 = -3 \Rightarrow$ (a) is a solution to this problem.

65. $y = \int_{2}^{x} \sec t\,dt + 3$

66. $y = \int_{1}^{x} \sqrt{1+t^2}\,dt - 2$

67. $s = \int_{t_0}^{t} f(x)\,dx + s_0$

68. $v = \int_{t_0}^{t} g(x)\,dx + v_0$

69. $\text{Area} = \int_{-b/2}^{b/2} \left(h - \left(\frac{4h}{b^2}\right)x^2\right) dx = \left[hx - \frac{4hx^3}{3b^2}\right]_{-b/2}^{b/2}$
$= \left(h\left(\frac{b}{2}\right) - \frac{4h\left(\frac{b}{2}\right)^3}{3b^2}\right) - \left(h\left(-\frac{b}{2}\right) - \frac{4h\left(-\frac{b}{2}\right)^3}{3b^2}\right)$
$= \left(\frac{bh}{2} - \frac{bh}{6}\right) - \left(-\frac{bh}{2} + \frac{bh}{6}\right) = bh - \frac{bh}{3} = \frac{2}{3}bh$

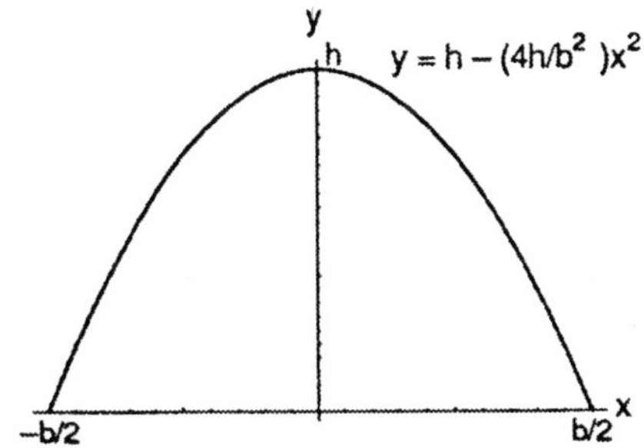

70. $r = \int_{0}^{3} \left(2 - \frac{2}{(x+1)^2}\right) dx = 2\int_{0}^{3} \left(1 - \frac{1}{(x+1)^2}\right) dx = 2\left[x - \left(\frac{-1}{x+1}\right)\right]_0^3 = 2\left[\left(3 + \frac{1}{(3+1)}\right) - \left(0 + \frac{1}{(0+1)}\right)\right]$
$= 2\left[3\frac{1}{4} - 1\right] = 2\left(2\frac{1}{4}\right) = 4.5$ or \$4500

71. $\frac{dc}{dx} = \frac{1}{2\sqrt{x}} = \frac{1}{2}x^{-1/2} \Rightarrow c = \int_{0}^{x} \frac{1}{2}t^{-1/2}dt = \left[t^{1/2}\right]_0^x = \sqrt{x}$
$c(100) - c(1) = \sqrt{100} - \sqrt{1} = \$9.00$

72. By Exercise 71, $c(400) - c(100) = \sqrt{400} - \sqrt{100} = 20 - 10 = \$10.00$

73. (a) $v = \frac{ds}{dt} = \frac{d}{dt}\int_{0}^{t} f(x)\,dx = f(t) \Rightarrow v(5) = f(5) = 2$ m/sec

(b) $a = \frac{df}{dt}$ is negative since the slope of the tangent line at $t = 5$ is negative

(c) $s = \int_{0}^{3} f(x)\,dx = \frac{1}{2}(3)(3) = \frac{9}{2}$ m since the integral is the area of the triangle formed by $y = f(x)$, the x-axis, and $x = 3$

(d) $t = 6$ since from $t = 6$ to $t = 9$, the region lies below the x-axis

(e) At $t = 4$ and $t = 7$, since there are horizontal tangents there

(f) Toward the origin between $t = 6$ and $t = 9$ since the velocity is negative on this interval. Away from the origin between $t = 0$ and $t = 6$ since the velocity is positive there.

(g) Right or positive side, because the integral of f from 0 to 9 is positive, there being more area above the x-axis than below it.

74. (a) $v = \frac{dg}{dt} = \frac{d}{dt}\int_0^t g(x)\,dx = g(t) \Rightarrow v(3) = g(3) = 0$ m/sec.

(b) $a = \frac{df}{dt}$ is positive, since the slope of the tangent line at $t = 3$ is positive

(c) At $t = 3$, the particle's position is $\int_0^3 g(x)\,dx = \frac{1}{2}(3)(-6) = -9$

(d) The particle passes through the origin at $t = 6$ because $s(6) = \int_0^6 g(x)\,dx = 0$

(e) At $t = 7$, since there is a horizontal tangent there

(f) The particle starts at the origin and moves away to the left for $0 < t < 3$. It moves back toward the origin for $3 < t < 6$, passes through the origin at $t = 6$, and moves away to the right for $t > 6$.

(g) Right side, since its position at $t = 9$ is positive, there being more area above the x-axis than below it at $t = 9$.

75. $k > 0 \Rightarrow$ one arch of $y = \sin kx$ will occur over the interval $\left[0, \frac{\pi}{k}\right] \Rightarrow$ the area $= \int_0^{\pi/k} \sin kx\,dx = \left[-\frac{1}{k}\cos kx\right]_0^{\pi/k}$ $= -\frac{1}{k}\cos\left(k\left(\frac{\pi}{k}\right)\right) - \left(-\frac{1}{k}\cos(0)\right) = \frac{2}{k}$

76. $\lim\limits_{x\to 0}\frac{1}{x^3}\int_0^x \frac{t^2}{t^4+1}\,dt = \lim\limits_{x\to 0}\frac{\int_0^x \frac{t^2}{t^4+1}dt}{x^3} = \lim\limits_{x\to 0}\frac{\frac{x^2}{x^4+1}}{3x^2} = \lim\limits_{x\to 0}\frac{1}{3(x^4+1)} = \infty.$

77. $\int_1^x f(t)\,dt = x^2 - 2x + 1 \Rightarrow f(x) = \frac{d}{dx}\int_1^x f(t)\,dt = \frac{d}{dx}(x^2 - 2x + 1) = 2x - 2$

78. $\int_0^x f(t)\,dt = x\cos \pi x \Rightarrow f(x) = \frac{d}{dx}\int_0^x f(t)\,dt = \cos \pi x - \pi x \sin \pi x \Rightarrow f(4) = \cos \pi(4) - \pi(4)\sin \pi(4) = 1$

79. $f(x) = 2 - \int_2^{x+1} \frac{9}{1+t}\,dt \Rightarrow f'(x) = -\frac{9}{1+(x+1)} = \frac{-9}{x+2} \Rightarrow f'(1) = -3;\ f(1) = 2 - \int_2^{1+1}\frac{9}{1+t}\,dt = 2 - 0 = 2;$ $L(x) = -3(x-1) + f(1) = -3(x-1) + 2 = -3x + 5$

80. $g(x) = 3 + \int_1^{x^2} \sec(t-1)\,dt \Rightarrow g'(x) = (\sec(x^2-1))(2x) = 2x\sec(x^2-1) \Rightarrow g'(-1) = 2(-1)\sec((-1)^2-1)$ $= -2;\ g(-1) = 3 + \int_1^{(-1)^2}\sec(t-1)\,dt = 3 + \int_1^1 \sec(t-1)\,dt = 3 + 0 = 3;\ L(x) = -2(x-(-1)) + g(-1)$ $= -2(x+1) + 3 = -2x + 1$

81. (a) True: since f is continuous, g is differentiable by Part 1 of the Fundamental Theorem of Calculus.

(b) True: g is continuous because it is differentiable.

(c) True, since $g'(1) = f(1) = 0$.

(d) False, since $g''(1) = f'(1) > 0$.

(e) True, since $g'(1) = 0$ and $g''(1) = f'(1) > 0$.

(f) False: $g''(x) = f'(x) > 0$, so g'' never changes sign.

(g) True, since $g'(1) = f(1) = 0$ and $g'(x) = f(x)$ is an increasing function of x (because $f'(x) > 0$).

82. (a) True: by Part 1 of the Fundamental Theorem of Calculus, $h'(x) = f(x)$. Since f is differentiable for all x, h has a second derivative for all x.

(b) True: they are continuous because they are differentiable.

(c) True, since $h'(1) = f(1) = 0$.

(d) True, since $h'(1) = 0$ and $h''(1) = f'(1) < 0$.

(e) False, since $h''(1) = f'(1) < 0$.

(f) False, since $h''(x) = f'(x) < 0$ never changes sign.

(g) True, since $h'(1) = f(1) = 0$ and $h'(x) = f(x)$ is a decreasing function of x (because $f'(x) < 0$).

83.

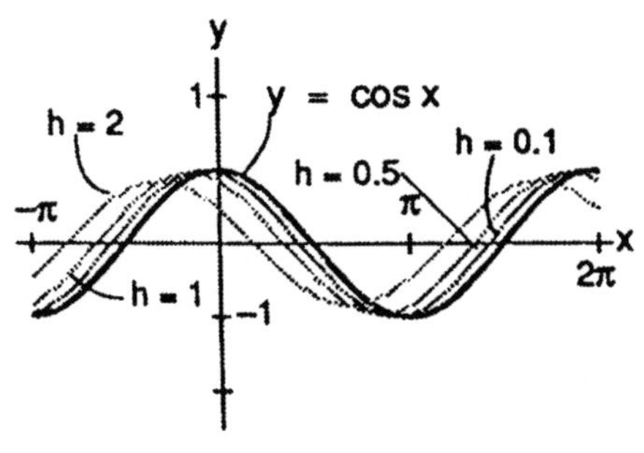

84. The limit is $3x^2$

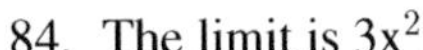

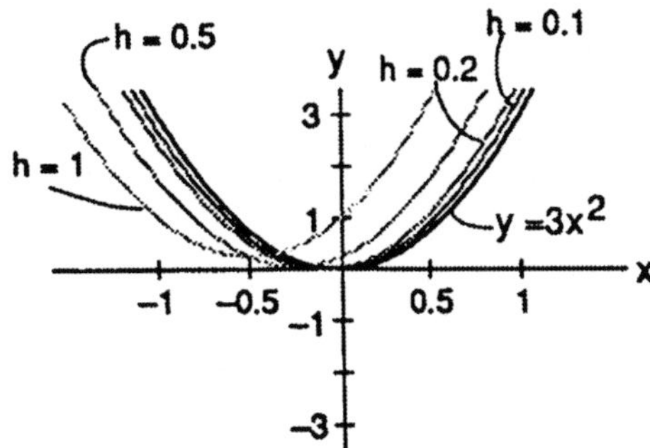

85-88. Example CAS commands:

Maple:

```
with( plots );
f := x -> x^3-4*x^2+3*x;
a := 0;
b := 4;
F := unapply( int(f(t),t=a..x), x );                     # (a)
p1 := plot( [f(x),F(x)], x=a..b, legend=["y = f(x)","y = F(x)"], title="#85(a) (Section 5.4)" ):
p1;
dF := D(F);                                              # (b)
q1 := solve( dF(x)=0, x );
pts1 := [ seq( [x,f(x)], x=remove(has,evalf([q1]),I) ) ];
p2 := plot( pts1, style=point, color=blue, symbolsize=18, symbol=diamond, legend="(x,f(x)) where F '(x)=0" ):
display( [p1,p2], title="85(b) (Section 5.4)" );
incr := solve( dF(x)>0, x );                             # (c)
decr := solve( dF(x)<0, x );
df := D(f);                                              # (d)
p3 := plot( [df(x),F(x)], x=a..b, legend=["y = f '(x)","y = F(x)"], title="#85(d) (Section 5.4)" ):
p3;
q2 := solve( df(x)=0, x );
pts2 := [ seq( [x,F(x)], x=remove(has,evalf([q2]),I) ) ];
p4 := plot( pts2, style=point, color=blue, symbolsize=18, symbol=diamond, legend="(x,f(x)) where f '(x)=0" ):
display( [p3,p4], title="85(d) (Section 5.4)" );
```

89-92. Example CAS commands:

Maple:

```
a := 1;
u := x -> x^2;
f := x -> sqrt(1-x^2);
F := unapply( int( f(t), t=a..u(x) ), x );
dF := D(F);                 # (b)
cp := solve( dF(x)=0, x );
solve( dF(x)>0, x );
solve( dF(x)<0, x );
d2F := D(dF);               # (c)
solve( d2F(x)=0, x );
plot( F(x), x=-1..1, title="#89(d) (Section 5.4)" );
```

93. Example CAS commands:

Maple:

```
f := `f`;
q1 := Diff( Int( f(t), t=a..u(x) ), x );
d1 := value( q1 );
```

94. Example CAS commands:

Maple:

```
f := `f`;
q2 := Diff( Int( f(t), t=a..u(x) ), x,x );
value( q2 );
```

85-94. Example CAS commands:

Mathematica: (assigned function and values for a, and b may vary)

For transcendental functions the FindRoot is needed instead of the Solve command.

The Map command executes FindRoot over a set of initial guesses

Initial guesses will vary as the functions vary.

```
Clear[x, f, F]
{a, b}= {0, 2π}; f[x_] = Sin[2x] Cos[x/3]
F[x_] = Integrate[f[t], {t, a, x}]
Plot[{f[x], F[x]},{x, a, b}]
x/.Map[FindRoot[F'[x]==0, {x, #}] &,{2, 3, 5, 6}]
x/.Map[FindRoot[f'[x]==0, {x, #}] &,{1, 2, 4, 5, 6}]
```

Slightly alter above commands for 75 - 80.

```
Clear[x, f, F, u]
a=0;  f[x_] = x^2 − 2x − 3
u[x_] = 1 − x^2
F[x_] = Integrate[f[t], {t, a, u(x)}]
x/.Map[FindRoot[F'[x]==0,{x,#}] &,{1,2,3,4}]
x/.Map[FindRoot[F''[x]==0,{x,#}] &,{1,2,3,4}]
```

After determining an appropriate value for b, the following can be entered

```
b = 4;
Plot[{F[x], {x, a, b}]
```

5.5 INDEFINTE INTEGRALS AND THE SUBSTITUTION RULE

1. Let $u = 3x \Rightarrow du = 3\,dx \Rightarrow \frac{1}{3}\,du = dx$

$\int \sin 3x\,dx = \int \frac{1}{3}\sin u\,du = -\frac{1}{3}\cos u + C = -\frac{1}{3}\cos 3x + C$

2. Let $u = 2x^2 \Rightarrow du = 4x\,dx \Rightarrow \frac{1}{4}\,du = x\,dx$

$\int x\sin(2x^2)\,dx = \int \frac{1}{4}\sin u\,du = -\frac{1}{4}\cos u + C = -\frac{1}{4}\cos 2x^2 + C$

3. Let $u = 2t \Rightarrow du = 2\,dt \Rightarrow \frac{1}{2}\,du = dt$

$\int \sec 2t\tan 2t\,dt = \int \frac{1}{2}\sec u\tan u\,du = \frac{1}{2}\sec u + C = \frac{1}{2}\sec 2t + C$

4. Let $u = 1 - \cos\frac{t}{2} \Rightarrow du = \frac{1}{2}\sin\frac{t}{2}\,dt \Rightarrow 2\,du = \sin\frac{t}{2}\,dt$

$\int \left(1 - \cos\frac{t}{2}\right)^2\left(\sin\frac{t}{2}\right)dt = \int 2u^2\,du = \frac{2}{3}u^3 + C = \frac{2}{3}\left(1 - \cos\frac{t}{2}\right)^3 + C$

5. Let $u = 7x - 2 \Rightarrow du = 7\,dx \Rightarrow \frac{1}{7}\,du = dx$

$\int 28(7x-2)^{-5}\,dx = \int \frac{1}{7}(28)u^{-5}\,du = \int 4u^{-5}\,du = -u^{-4} + C = -(7x-2)^{-4} + C$

6. Let $u = x^4 - 1 \Rightarrow du = 4x^3\,dx \Rightarrow \frac{1}{4}\,du = x^3\,dx$

$\int x^3\left(x^4-1\right)^2 dx = \int \frac{1}{4}u^2\,du = \frac{u^3}{12} + C = \frac{1}{12}\left(x^4-1\right)^3 + C$

7. Let $u = 1 - r^3 \Rightarrow du = -3r^2\,dr \Rightarrow -3\,du = 9r^2\,dr$

$\int \frac{9r^2\,dr}{\sqrt{1-r^3}} = \int -3u^{-1/2}\,du = -3(2)u^{1/2} + C = -6\left(1-r^3\right)^{1/2} + C$

8. Let $u = y^4 + 4y^2 + 1 \Rightarrow du = (4y^3 + 8y)\,dy \Rightarrow 3\,du = 12\left(y^3 + 2y\right)dy$

$\int 12\left(y^4+4y^2+1\right)^2\left(y^3+2y\right)dy = \int 3u^2\,du = u^3 + C = \left(y^4+4y^2+1\right)^3 + C$

9. Let $u = x^{3/2} - 1 \Rightarrow du = \frac{3}{2}x^{1/2}\,dx \Rightarrow \frac{2}{3}\,du = \sqrt{x}\,dx$

$\int \sqrt{x}\sin^2\left(x^{3/2}-1\right)dx = \int \frac{2}{3}\sin^2 u\,du = \frac{2}{3}\left(\frac{u}{2} - \frac{1}{4}\sin 2u\right) + C = \frac{1}{3}\left(x^{3/2}-1\right) - \frac{1}{6}\sin\left(2x^{3/2}-2\right) + C$

10. Let $u = -\frac{1}{x} \Rightarrow du = \frac{1}{x^2}\,dx$

$\int \frac{1}{x^2}\cos^2\left(\frac{1}{x}\right)dx = \int \cos^2(-u)\,du = \int \cos^2(u)\,du = \left(\frac{u}{2} + \frac{1}{4}\sin 2u\right) + C = -\frac{1}{2x} + \frac{1}{4}\sin\left(-\frac{2}{x}\right) + C$
$= -\frac{1}{2x} - \frac{1}{4}\sin\left(\frac{2}{x}\right) + C$

11. (a) Let $u = \cot 2\theta \Rightarrow du = -2\csc^2 2\theta\,d\theta \Rightarrow -\frac{1}{2}\,du = \csc^2 2\theta\,d\theta$

$\int \csc^2 2\theta\cot 2\theta\,d\theta = -\int \frac{1}{2}u\,du = -\frac{1}{2}\left(\frac{u^2}{2}\right) + C = -\frac{u^2}{4} + C = -\frac{1}{4}\cot^2 2\theta + C$

(b) Let $u = \csc 2\theta \Rightarrow du = -2\csc 2\theta\cot 2\theta\,d\theta \Rightarrow -\frac{1}{2}\,du = \csc 2\theta\cot 2\theta\,d\theta$

$\int \csc^2 2\theta\cot 2\theta\,d\theta = \int -\frac{1}{2}u\,du = -\frac{1}{2}\left(\frac{u^2}{2}\right) + C = -\frac{u^2}{4} + C = -\frac{1}{4}\csc^2 2\theta + C$

12. (a) Let $u = 5x + 8 \Rightarrow du = 5\,dx \Rightarrow \frac{1}{5}\,du = dx$

$\int \frac{dx}{\sqrt{5x+8}} = \int \frac{1}{5}\left(\frac{1}{\sqrt{u}}\right)du = \frac{1}{5}\int u^{-1/2}\,du = \frac{1}{5}\left(2u^{1/2}\right) + C = \frac{2}{5}u^{1/2} + C = \frac{2}{5}\sqrt{5x+8} + C$

(b) Let $u = \sqrt{5x+8} \Rightarrow du = \frac{1}{2}(5x+8)^{-1/2}(5)\,dx \Rightarrow \frac{2}{5}\,du = \frac{dx}{\sqrt{5x+8}}$

$\int \frac{dx}{\sqrt{5x+8}} = \int \frac{2}{5}\,du = \frac{2}{5}u + C = \frac{2}{5}\sqrt{5x+8} + C$

13. Let $u = 3 - 2s \Rightarrow du = -2\,ds \Rightarrow -\frac{1}{2}\,du = ds$

$\int \sqrt{3-2s}\,ds = \int \sqrt{u}\left(-\frac{1}{2}\,du\right) = -\frac{1}{2}\int u^{1/2}\,du = \left(-\frac{1}{2}\right)\left(\frac{2}{3}u^{3/2}\right) + C = -\frac{1}{3}(3-2s)^{3/2} + C$

14. Let $u = 2x + 1 \Rightarrow du = 2\,dx \Rightarrow \frac{1}{2}\,du = dx$

$\int (2x+1)^3\,dx = \int u^3\left(\frac{1}{2}\,du\right) = \frac{1}{2}\int u^3\,du = \left(\frac{1}{2}\right)\left(\frac{u^4}{4}\right) + C = \frac{1}{8}(2x+1)^4 + C$

15. Let $u = 5s + 4 \Rightarrow du = 5\,ds \Rightarrow \frac{1}{5}\,du = ds$

$\int \frac{1}{\sqrt{5s+4}}\,ds = \int \frac{1}{\sqrt{u}}\left(\frac{1}{5}\,du\right) = \frac{1}{5}\int u^{-1/2}\,du = \left(\frac{1}{5}\right)\left(2u^{1/2}\right) + C = \frac{2}{5}\sqrt{5s+4} + C$

16. Let $u = 2 - x \Rightarrow du = -dx \Rightarrow -du = dx$

$\int \frac{3}{(2-x)^2}\,dx = \int \frac{3(-du)}{u^2} = -3\int u^{-2}\,du = -3\left(\frac{u^{-1}}{-1}\right) + C = \frac{3}{2-x} + C$

17. Let $u = 1 - \theta^2 \Rightarrow du = -2\theta\, d\theta \Rightarrow -\frac{1}{2}\, du = \theta\, d\theta$

$\int \theta \sqrt[4]{1-\theta^2}\, d\theta = \int \sqrt[4]{u}\left(-\frac{1}{2}\, du\right) = -\frac{1}{2}\int u^{1/4}\, du = \left(-\frac{1}{2}\right)\left(\frac{4}{5}u^{5/4}\right) + C = -\frac{2}{5}\left(1-\theta^2\right)^{5/4} + C$

18. Let $u = 2y^2 + 1 \Rightarrow du = 4y\, dy$

$\int \frac{4y\, dy}{\sqrt{2y^2+1}} = \int \frac{1}{\sqrt{u}}\, du = \int u^{-1/2}\, du = 2u^{1/2} + C = 2\sqrt{2y^2+1} + C$

19. Let $u = 1 + \sqrt{x} \Rightarrow du = \frac{1}{2\sqrt{x}}\, dx \Rightarrow 2\, du = \frac{1}{\sqrt{x}}\, dx$

$\int \frac{1}{\sqrt{x}\left(1+\sqrt{x}\right)^2}\, dx = \int \frac{2\, du}{u^2} = -\frac{2}{u} + C = \frac{-2}{1+\sqrt{x}} + C$

20. Let $u = 1 + \sqrt{x} \Rightarrow du = \frac{1}{2\sqrt{x}}\, dx \Rightarrow 2\, du = \frac{1}{\sqrt{x}}\, dx$

$\int \frac{\left(1+\sqrt{x}\right)^3}{\sqrt{x}}\, dx = \int u^3\, (2\, du) = 2\left(\frac{1}{4}u^4\right) + C = \frac{1}{2}\left(1+\sqrt{x}\right)^4 + C$

21. Let $u = 3z + 4 \Rightarrow du = 3\, dz \Rightarrow \frac{1}{3}\, du = dz$

$\int \cos(3z+4)\, dz = \int (\cos u)\left(\frac{1}{3}\, du\right) = \frac{1}{3}\int \cos u\, du = \frac{1}{3}\sin u + C = \frac{1}{3}\sin(3z+4) + C$

22. Let $u = \tan x \Rightarrow du = \sec^2 x\, dx$

$\int \tan^2 x \sec^2 x\, dx = \int u^2\, du = \frac{1}{3}u^3 + C = \frac{1}{3}\tan^3 x + C$

23. Let $u = \cos x \Rightarrow du = -\sin x\, dx \Rightarrow -du = \sin x\, dx$

$\int \tan x\, dx = \int \frac{\sin x}{\cos x}\, dx = -\int \frac{1}{u}\, du = -\ln|u| + C = -\ln|\cos x| + C = \ln|\cos x|^{-1} + C = \ln|\sec x| + C$

24. Let $u = \tan\left(\frac{x}{2}\right) \Rightarrow du = \frac{1}{2}\sec^2\left(\frac{x}{2}\right) dx \Rightarrow 2\, du = \sec^2\left(\frac{x}{2}\right) dx$

$\int \tan^7\left(\frac{x}{2}\right)\sec^2\left(\frac{x}{2}\right) dx = \int u^7\, (2\, du) = 2\left(\frac{1}{8}u^8\right) + C = \frac{1}{4}\tan^8\left(\frac{x}{2}\right) + C$

25. Let $u = \frac{r^3}{18} - 1 \Rightarrow du = \frac{r^2}{6}\, dr \Rightarrow 6\, du = r^2\, dr$

$\int r^2\left(\frac{r^3}{18} - 1\right)^5 dr = \int u^5\, (6\, du) = 6\int u^5\, du = 6\left(\frac{u^6}{6}\right) + C = \left(\frac{r^3}{18} - 1\right)^6 + C$

26. Let $u = 7 - \frac{r^5}{10} \Rightarrow du = -\frac{1}{2}r^4\, dr \Rightarrow -2\, du = r^4\, dr$

$\int r^4\left(7 - \frac{r^5}{10}\right)^3 dr = \int u^3\, (-2\, du) = -2\int u^3\, du = -2\left(\frac{u^4}{4}\right) + C = -\frac{1}{2}\left(7 - \frac{r^5}{10}\right)^4 + C$

27. Let $u = x^{3/2} + 1 \Rightarrow du = \frac{3}{2}x^{1/2}\, dx \Rightarrow \frac{2}{3}\, du = x^{1/2}\, dx$

$\int x^{1/2}\sin\left(x^{3/2}+1\right) dx = \int (\sin u)\left(\frac{2}{3}\, du\right) = \frac{2}{3}\int \sin u\, du = \frac{2}{3}(-\cos u) + C = -\frac{2}{3}\cos\left(x^{3/2}+1\right) + C$

28. Let $u = x^{4/3} - 8 \Rightarrow du = \frac{4}{3}x^{1/3}\, dx \Rightarrow \frac{3}{4}\, du = x^{1/3}\, dx$

$\int x^{1/3}\sin\left(x^{4/3}-8\right) dx = \int (\sin u)\left(\frac{3}{4}\, du\right) = \frac{3}{4}\int \sin u\, du = \frac{3}{4}(-\cos u) + C = -\frac{3}{4}\cos\left(x^{4/3}-8\right) + C$

29. Let $u = \cos(2t+1) \Rightarrow du = -2\sin(2t+1)\, dt \Rightarrow -\frac{1}{2}\, du = \sin(2t+1)\, dt$

$\int \frac{\sin(2t+1)}{\cos^2(2t+1)}\, dt = \int -\frac{1}{2}\frac{du}{u^2} = \frac{1}{2u} + C = \frac{1}{2\cos(2t+1)} + C$

30. Let $u = 2 + \sin t \Rightarrow du = \cos t\, dt$

$\int \frac{6 \cos t}{(2+\sin t)^3}\, dt = \int \frac{6}{u^3}\, du = 6 \int u^{-3}\, du = 6\left(\frac{u^{-2}}{-2}\right) + C = -3(2+\sin t)^{-2} + C$

31. Let $u = \sin \frac{1}{\theta} \Rightarrow du = \left(\cos \frac{1}{\theta}\right)\left(-\frac{1}{\theta^2}\right) d\theta \Rightarrow -du = \frac{1}{\theta^2} \cos \frac{1}{\theta}\, d\theta$

$\int \frac{1}{\theta^2} \sin \frac{1}{\theta} \cos \frac{1}{\theta}\, d\theta = \int -u\, du = -\frac{1}{2} u^2 + C = -\frac{1}{2} \sin^2 \frac{1}{\theta} + C$

32. Let $u = \sec z \Rightarrow du = \sec z \tan z\, dz$

$\int \frac{\sec z \tan z}{\sqrt{\sec z}}\, dz = \int \frac{1}{\sqrt{u}}\, du = \int u^{-1/2}\, du = 2u^{1/2} + C = 2\sqrt{\sec z} + C$

33. Let $u = \frac{1}{t} - 1 = t^{-1} - 1 \Rightarrow du = -t^{-2}\, dt \Rightarrow -du = \frac{1}{t^2}\, dt$

$\int \frac{1}{t^2} \cos\left(\frac{1}{t} - 1\right) dt = \int (\cos u)(-du) = -\int \cos u\, du = -\sin u + C = -\sin\left(\frac{1}{t} - 1\right) + C$

34. Let $u = \csc \sqrt{\theta} \Rightarrow du = \left(-\csc \sqrt{\theta} \cot \sqrt{\theta}\right)\left(\frac{1}{2\sqrt{\theta}}\right) d\theta \Rightarrow -2\, du = \frac{1}{\sqrt{\theta}} \cot \sqrt{\theta} \csc \sqrt{\theta}\, d\theta$

$\int \frac{\cos \sqrt{\theta}}{\sqrt{\theta} \sin^2 \sqrt{\theta}}\, d\theta = \int \frac{1}{\sqrt{\theta}} \cot \sqrt{\theta} \csc \sqrt{\theta}\, d\theta = \int -2\, du = -2u + C = -2 \csc \sqrt{\theta} + C = -\frac{2}{\sin \sqrt{\theta}} + C$

35. Let $u = s^3 + 2s^2 - 5s + 5 \Rightarrow du = (3s^2 + 4s - 5)\, ds$

$\int (s^3 + 2s^2 - 5s + 5)(3s^2 + 4s - 5)\, ds = \int u\, du = \frac{u^2}{2} + C = \frac{(s^3 + 2s^2 - 5s + 5)^2}{2} + C$

36. Let $u = 1 + t^4 \Rightarrow du = 4t^3\, dt \Rightarrow \frac{1}{4}\, du = t^3\, dt$

$\int t^3 (1 + t^4)^3\, dt = \int u^3 \left(\frac{1}{4}\, du\right) = \frac{1}{4}\left(\frac{1}{4} u^4\right) + C = \frac{1}{16}(1 + t^4)^4 + C$

37. Let $u = 1 - \frac{1}{x} \Rightarrow du = \frac{1}{x^2}\, dx$

$\int \sqrt{\frac{x-1}{x^5}}\, dx = \int \frac{1}{x^2} \sqrt{\frac{x-1}{x}}\, dx = \int \frac{1}{x^2} \sqrt{1 - \frac{1}{x}}\, dx = \int \sqrt{u}\, du = \int u^{1/2}\, du = \frac{2}{3} u^{3/2} + C = \frac{2}{3}\left(1 - \frac{1}{x}\right)^{3/2} + C$

38. Let $u = x^2 + 1$. Then $du = 2x dx$ and $\frac{1}{2} du = x dx$ and $x^2 = u - 1$. Thus $\int x^3 \sqrt{x^2 + 1}\, dx = \int (u - 1) \frac{1}{2} \sqrt{u}\, du$
$= \frac{1}{2} \int \left(u^{3/2} - u^{1/2}\right) du = \frac{1}{2}\left[\frac{2}{5} u^{5/2} - \frac{2}{3} u^{3/2}\right] + C = \frac{1}{5} u^{5/2} - \frac{1}{3} u^{3/2} + C = \frac{1}{5}(x^2 + 1)^{5/2} - \frac{1}{3}(x^2 + 1)^{3/2} + C$

39. Let $u = \sin x \Rightarrow du = \cos x\, dx$

$\int (\cos x) e^{\sin x}\, dx = \int e^u\, du = e^u + C = e^{\sin x} + C$

40. Let $u = \sin^2\theta \Rightarrow du = 2 \sin\theta \cos\theta\, d\theta = \sin 2\theta\, d\theta$

$\int (\sin 2\theta) e^{\sin^2\theta}\, d\theta = \int e^u\, du = e^u + C = e^{\sin^2\theta} + C$

41. Let $u = e^{\sqrt{x}} + 1 \Rightarrow du = \frac{1}{2\sqrt{x}} e^{\sqrt{x}} dx \Rightarrow 2\, du = \frac{1}{\sqrt{x}} e^{\sqrt{x}} dx = \frac{1}{\sqrt{x}\, e^{-\sqrt{x}}}\, dx$

$\int \frac{1}{\sqrt{x}\, e^{-\sqrt{x}}} \sec^2\left(e^{\sqrt{x}} + 1\right) dx = 2 \int \sec^2 u\, du = 2 \tan u + C = 2 \tan\left(e^{\sqrt{x}} + 1\right) + C$

42. Let $u = 1 + e^{\frac{1}{x}} \Rightarrow du = e^{\frac{1}{x}} \cdot \frac{-1}{x^2} dx \Rightarrow -du = e^{\frac{1}{x}} \cdot \frac{1}{x^2} dx$

$\int \frac{1}{x^2} e^{\frac{1}{x}} \sec\left(1 + e^{\frac{1}{x}}\right) \tan\left(1 + e^{\frac{1}{x}}\right) dx = -\int \sec u \tan u\, du = -\sec u + C = -\sec\left(1 + e^{\frac{1}{x}}\right) + C$

43. Let $u = \ln x \Rightarrow du = \frac{1}{x}dx$

$\int \frac{1}{x \ln x} dx = \int \frac{1}{u} du = \ln|u| + C = \ln|\ln x| + C$

44. Let $u = \ln t \Rightarrow du = \frac{1}{t}dt$

$\int \frac{\ln \sqrt{t}}{t} dt = \int \frac{\ln t^{1/2}}{t} dt = \frac{1}{2}\int \frac{\ln t}{t} dt = \frac{1}{2}\int u\, du = \frac{1}{2} \cdot \frac{u^2}{2} + C = \frac{1}{4}(\ln t)^2 + C$

45. Let $u = e^{-z} + 1 \Rightarrow du = -e^{-z}\, dz \Rightarrow -du = e^{-z}\, dz$

$\int \frac{dz}{1+e^z} = \int \frac{1}{1+e^z} \cdot \frac{e^{-z}}{e^{-z}} dz = \int \frac{e^{-z}}{e^{-z}+1} dz = -\int \frac{1}{u} du = -\ln|u| + C = -\ln(e^{-z}+1) + C = -\ln\left(\frac{1+e^z}{e^z}\right) + C$
$= -(\ln(1+e^z) - \ln e^z) + C = z - \ln(1+e^z) + C$

46. Let $u = x^2 \Rightarrow du = 2x\, dx \Rightarrow \frac{1}{2}du = x\, dx$

$\int \frac{dx}{x\sqrt{x^4-1}} = \int \frac{x\, dx}{x^2\sqrt{(x^2)^2-1}} = \frac{1}{2}\int \frac{du}{u\sqrt{u^2-1}} = \frac{1}{2}\sec^{-1} u + C = \frac{1}{2}\sec^{-1}(x^2) + C$

47. Let $u = \frac{2}{3}r \Rightarrow du = \frac{2}{3}dr \Rightarrow \frac{3}{2}du = dr$

$\int \frac{5}{9+4r^2} dr = \frac{5}{9}\int \frac{1}{1+\left(\frac{2}{3}r\right)^2} dr = \frac{5}{9}\int \frac{\frac{3}{2}}{1+u^2} du = \frac{5}{6}\tan^{-1} u + C = \frac{5}{6}\tan^{-1}\left(\frac{2}{3}r\right) + C$

48. Let $u = e^\theta \Rightarrow du = e^\theta\, d\theta$

$\int \frac{1}{\sqrt{e^{2\theta}-1}} d\theta = \int \frac{e^\theta\, d\theta}{e^\theta\sqrt{(e^\theta)^2-1}} = \int \frac{du}{u\sqrt{u^2-1}} = \sec^{-1} u + C = \sec^{-1}\left(e^\theta\right) + C$

49. $\int \frac{e^{\sin^{-1} x}}{\sqrt{1-x^2}} dx = \int e^u\, du$, where $u = \sin^{-1} x$ and $du = \frac{dx}{\sqrt{1-x^2}}$
$= e^u + C = e^{\sin^{-1} x} + C$

50. $\int \frac{e^{\cos^{-1} x}}{\sqrt{1-x^2}} dx = -\int e^u\, du$, where $u = \cos^{-1} x$ and $du = \frac{-dx}{\sqrt{1-x^2}}$
$= -e^u + C = -e^{\cos^{-1} x} + C$

51. $\int \frac{(\sin^{-1} x)^2}{\sqrt{1-x^2}} dx = \int u^2\, du$, where $u = \sin^{-1} x$ and $du = \frac{dx}{\sqrt{1-x^2}}$
$= \frac{u^3}{3} + C = \frac{(\sin^{-1} x)^3}{3} + C$

52. $\int \frac{\sqrt{\tan^{-1} x}}{1+x^2} dx = \int u^{1/2}\, du$, where $u = \tan^{-1} x$ and $du = \frac{dx}{1+x^2}$
$= \frac{2}{3}u^{3/2} + C = \frac{2}{3}(\tan^{-1} x)^{3/2} + C = \frac{2}{3}\sqrt{(\tan^{-1} x)^3} + C$

53. $\int \frac{1}{(\tan^{-1} y)(1+y^2)} dy = \int \frac{\left(\frac{1}{1+y^2}\right)}{\tan^{-1} y} dy = \int \frac{1}{u} du$, where $u = \tan^{-1} y$ and $du = \frac{dy}{1+y^2}$
$= \ln|u| + C = \ln|\tan^{-1} y| + C$

54. $\int \frac{1}{(\sin^{-1} y)\sqrt{1+y^2}} dy = \int \frac{\left(\frac{1}{\sqrt{1-y^2}}\right)}{\sin^{-1} y} dy = \int \frac{1}{u} du$, where $u = \sin^{-1} y$ and $du = \frac{dy}{\sqrt{1-y^2}}$
$= \ln|u| + C = \ln|\sin^{-1} y| + C$

55. (a) Let $u = \tan x \Rightarrow du = \sec^2 x\, dx$; $v = u^3 \Rightarrow dv = 3u^2\, du \Rightarrow 6\, dv = 18u^2\, du$; $w = 2 + v \Rightarrow dw = dv$

$\int \frac{18 \tan^2 x \sec^2 x}{(2+\tan^3 x)^2}\, dx = \int \frac{18u^2}{(2+u^3)^2}\, du = \int \frac{6\, dv}{(2+v)^2} = \int \frac{6\, dw}{w^2} = 6 \int w^{-2}\, dw = -6w^{-1} + C = -\frac{6}{2+v} + C$
$= -\frac{6}{2+u^3} + C = -\frac{6}{2+\tan^3 x} + C$

(b) Let $u = \tan^3 x \Rightarrow du = 3 \tan^2 x \sec^2 x\, dx \Rightarrow 6\, du = 18 \tan^2 x \sec^2 x\, dx$; $v = 2 + u \Rightarrow dv = du$

$\int \frac{18 \tan^2 x \sec^2 x}{(2+\tan^3 x)^2}\, dx = \int \frac{6\, du}{(2+u)^2} = \int \frac{6\, dv}{v^2} = -\frac{6}{v} + C = -\frac{6}{2+u} + C = -\frac{6}{2+\tan^3 x} + C$

(c) Let $u = 2 + \tan^3 x \Rightarrow du = 3 \tan^2 x \sec^2 x\, dx \Rightarrow 6\, du = 18 \tan^2 x \sec^2 x\, dx$

$\int \frac{18 \tan^2 x \sec^2 x}{(2+\tan^3 x)^2}\, dx = \int \frac{6\, du}{u^2} = -\frac{6}{u} + C = -\frac{6}{2+\tan^3 x} + C$

56. (a) Let $u = x - 1 \Rightarrow du = dx$; $v = \sin u \Rightarrow dv = \cos u\, du$; $w = 1 + v^2 \Rightarrow dw = 2v\, dv \Rightarrow \frac{1}{2}\, dw = v\, dv$

$\int \sqrt{1 + \sin^2 (x-1)} \sin (x-1) \cos (x-1)\, dx = \int \sqrt{1 + \sin^2 u} \sin u \cos u\, du = \int v\sqrt{1 + v^2}\, dv$
$= \int \frac{1}{2} \sqrt{w}\, dw = \frac{1}{3} w^{3/2} + C = \frac{1}{3} (1 + v^2)^{3/2} + C = \frac{1}{3} (1 + \sin^2 u)^{3/2} + C = \frac{1}{3} (1 + \sin^2 (x-1))^{3/2} + C$

(b) Let $u = \sin (x-1) \Rightarrow du = \cos (x-1)\, dx$; $v = 1 + u^2 \Rightarrow dv = 2u\, du \Rightarrow \frac{1}{2}\, dv = u\, du$

$\int \sqrt{1 + \sin^2 (x-1)} \sin (x-1) \cos (x-1)\, dx = \int u \sqrt{1 + u^2}\, du = \int \frac{1}{2} \sqrt{v}\, dv = \int \frac{1}{2} v^{1/2}\, dv$
$= \left(\frac{1}{2}\left(\frac{2}{3}\right) v^{3/2}\right) + C = \frac{1}{3} v^{3/2} + C = \frac{1}{3} (1 + u^2)^{3/2} + C = \frac{1}{3} (1 + \sin^2 (x-1))^{3/2} + C$

(c) Let $u = 1 + \sin^2 (x-1) \Rightarrow du = 2 \sin (x-1) \cos (x-1)\, dx \Rightarrow \frac{1}{2}\, du = \sin (x-1) \cos (x-1)\, dx$

$\int \sqrt{1 + \sin^2 (x-1)} \sin (x-1) \cos (x-1)\, dx = \int \frac{1}{2} \sqrt{u}\, du = \int \frac{1}{2} u^{1/2}\, du = \frac{1}{2} \left(\frac{2}{3} u^{3/2}\right) + C$
$= \frac{1}{3} (1 + \sin^2 (x-1))^{3/2} + C$

57. Let $u = 3(2r-1)^2 + 6 \Rightarrow du = 6(2r-1)(2)\, dr \Rightarrow \frac{1}{12}\, du = (2r-1)\, dr$; $v = \sqrt{u} \Rightarrow dv = \frac{1}{2\sqrt{u}}\, du \Rightarrow \frac{1}{6}\, dv = \frac{1}{12\sqrt{u}}\, du$

$\int \frac{(2r-1) \cos \sqrt{3(2r-1)^2+6}}{\sqrt{3(2r-1)^2+6}}\, dr = \int \left(\frac{\cos \sqrt{u}}{\sqrt{u}}\right)\left(\frac{1}{12}\, du\right) = \int (\cos v)\left(\frac{1}{6}\, dv\right) = \frac{1}{6} \sin v + C = \frac{1}{6} \sin \sqrt{u} + C$
$= \frac{1}{6} \sin \sqrt{3(2r-1)^2 + 6} + C$

58. Let $u = \cos \sqrt{\theta} \Rightarrow du = \left(-\sin \sqrt{\theta}\right)\left(\frac{1}{2\sqrt{\theta}}\right) d\theta \Rightarrow -2\, du = \frac{\sin \sqrt{\theta}}{\sqrt{\theta}}\, d\theta$

$\int \frac{\sin \sqrt{\theta}}{\sqrt{\theta \cos^3 \sqrt{\theta}}}\, d\theta = \int \frac{\sin \sqrt{\theta}}{\sqrt{\theta}\sqrt{\cos^3 \sqrt{\theta}}}\, d\theta = \int \frac{-2\, du}{u^{3/2}} = -2 \int u^{-3/2}\, du = -2\left(-2u^{-1/2}\right) + C = \frac{4}{\sqrt{u}} + C$
$= \frac{4}{\sqrt{\cos \sqrt{\theta}}} + C$

59. Let $u = 3t^2 - 1 \Rightarrow du = 6t\, dt \Rightarrow 2\, du = 12t\, dt$

$s = \int 12t\, (3t^2 - 1)^3\, dt = \int u^3\, (2\, du) = 2\left(\frac{1}{4} u^4\right) + C = \frac{1}{2} u^4 + C = \frac{1}{2} (3t^2 - 1)^4 + C$;

$s = 3$ when $t = 1 \Rightarrow 3 = \frac{1}{2} (3-1)^4 + C \Rightarrow 3 = 8 + C \Rightarrow C = -5 \Rightarrow s = \frac{1}{2} (3t^2 - 1)^4 - 5$

60. Let $u = x^2 + 8 \Rightarrow du = 2x\, dx \Rightarrow 2\, du = 4x\, dx$

$y = \int 4x\, (x^2 + 8)^{-1/3}\, dx = \int u^{-1/3}\, (2\, du) = 2\left(\frac{3}{2} u^{2/3}\right) + C = 3u^{2/3} + C = 3\,(x^2 + 8)^{2/3} + C$;

$y = 0$ when $x = 0 \Rightarrow 0 = 3(8)^{2/3} + C \Rightarrow C = -12 \Rightarrow y = 3\,(x^2 + 8)^{2/3} - 12$

61. Let $u = t + \frac{\pi}{12} \Rightarrow du = dt$

$s = \int 8 \sin^2\left(t + \frac{\pi}{12}\right) dt = \int 8 \sin^2 u\, du = 8\left(\frac{u}{2} - \frac{1}{4} \sin 2u\right) + C = 4\left(t + \frac{\pi}{12}\right) - 2 \sin\left(2t + \frac{\pi}{6}\right) + C$;

$s = 8$ when $t = 0 \Rightarrow 8 = 4\left(\frac{\pi}{12}\right) - 2 \sin\left(\frac{\pi}{6}\right) + C \Rightarrow C = 8 - \frac{\pi}{3} + 1 = 9 - \frac{\pi}{3}$

$\Rightarrow s = 4\left(t + \frac{\pi}{12}\right) - 2 \sin\left(2t + \frac{\pi}{6}\right) + 9 - \frac{\pi}{3} = 4t - 2 \sin\left(2t + \frac{\pi}{6}\right) + 9$

62. Let $u = \frac{\pi}{4} - \theta \Rightarrow -du = d\theta$

$r = \int 3\cos^2\left(\frac{\pi}{4} - \theta\right) d\theta = -\int 3\cos^2 u\, du = -3\left(\frac{u}{2} + \frac{1}{4}\sin 2u\right) + C = -\frac{3}{2}\left(\frac{\pi}{4} - \theta\right) - \frac{3}{4}\sin\left(\frac{\pi}{2} - 2\theta\right) + C;$
$r = \frac{\pi}{8}$ when $\theta = 0 \Rightarrow \frac{\pi}{8} = -\frac{3\pi}{8} - \frac{3}{4}\sin\frac{\pi}{2} + C \Rightarrow C = \frac{\pi}{2} + \frac{3}{4} \Rightarrow r = -\frac{3}{2}\left(\frac{\pi}{4} - \theta\right) - \frac{3}{4}\sin\left(\frac{\pi}{2} - 2\theta\right) + \frac{\pi}{2} + \frac{3}{4}$
$\Rightarrow r = \frac{3}{2}\theta - \frac{3}{4}\sin\left(\frac{\pi}{2} - 2\theta\right) + \frac{\pi}{8} + \frac{3}{4} \Rightarrow r = \frac{3}{2}\theta - \frac{3}{4}\cos 2\theta + \frac{\pi}{8} + \frac{3}{4}$

63. Let $u = 2t - \frac{\pi}{2} \Rightarrow du = 2\, dt \Rightarrow -2\, du = -4\, dt$

$\frac{ds}{dt} = \int -4\sin\left(2t - \frac{\pi}{2}\right) dt = \int (\sin u)(-2\, du) = 2\cos u + C_1 = 2\cos\left(2t - \frac{\pi}{2}\right) + C_1;$
at $t = 0$ and $\frac{ds}{dt} = 100$ we have $100 = 2\cos\left(-\frac{\pi}{2}\right) + C_1 \Rightarrow C_1 = 100 \Rightarrow \frac{ds}{dt} = 2\cos\left(2t - \frac{\pi}{2}\right) + 100$
$\Rightarrow s = \int \left(2\cos\left(2t - \frac{\pi}{2}\right) + 100\right) dt = \int (\cos u + 50)\, du = \sin u + 50u + C_2 = \sin\left(2t - \frac{\pi}{2}\right) + 50\left(2t - \frac{\pi}{2}\right) + C_2;$
at $t = 0$ and $s = 0$ we have $0 = \sin\left(-\frac{\pi}{2}\right) + 50\left(-\frac{\pi}{2}\right) + C_2 \Rightarrow C_2 = 1 + 25\pi$
$\Rightarrow s = \sin\left(2t - \frac{\pi}{2}\right) + 100t - 25\pi + (1 + 25\pi) \Rightarrow s = \sin\left(2t - \frac{\pi}{2}\right) + 100t + 1$

64. Let $u = \tan 2x \Rightarrow du = 2\sec^2 2x\, dx \Rightarrow 2\, du = 4\sec^2 2x\, dx;\ v = 2x \Rightarrow dv = 2\, dx \Rightarrow \frac{1}{2}\, dv = dx$

$\frac{dy}{dx} = \int 4\sec^2 2x \tan 2x\, dx = \int u(2\, du) = u^2 + C_1 = \tan^2 2x + C_1;$
at $x = 0$ and $\frac{dy}{dx} = 4$ we have $4 = 0 + C_1 \Rightarrow C_1 = 4 \Rightarrow \frac{dy}{dx} = \tan^2 2x + 4 = (\sec^2 2x - 1) + 4 = \sec^2 2x + 3$
$\Rightarrow y = \int (\sec^2 2x + 3)\, dx = \int (\sec^2 v + 3)\left(\frac{1}{2}\, dv\right) = \frac{1}{2}\tan v + \frac{3}{2}v + C_2 = \frac{1}{2}\tan 2x + 3x + C_2;$
at $x = 0$ and $y = -1$ we have $-1 = \frac{1}{2}(0) + 0 + C_2 \Rightarrow C_2 = -1 \Rightarrow y = \frac{1}{2}\tan 2x + 3x - 1$

65. Let $u = 2t \Rightarrow du = 2\, dt \Rightarrow 3\, du = 6\, dt$

$s = \int 6\sin 2t\, dt = \int (\sin u)(3\, du) = -3\cos u + C = -3\cos 2t + C;$
at $t = 0$ and $s = 0$ we have $0 = -3\cos 0 + C \Rightarrow C = 3 \Rightarrow s = 3 - 3\cos 2t \Rightarrow s\left(\frac{\pi}{2}\right) = 3 - 3\cos(\pi) = 6$ m

66. Let $u = \pi t \Rightarrow du = \pi\, dt \Rightarrow \pi\, du = \pi^2\, dt$

$v = \int \pi^2 \cos \pi t\, dt = \int (\cos u)(\pi\, du) = \pi \sin u + C_1 = \pi \sin(\pi t) + C_1;$
at $t = 0$ and $v = 8$ we have $8 = \pi(0) + C_1 \Rightarrow C_1 = 8 \Rightarrow v = \frac{ds}{dt} = \pi\sin(\pi t) + 8 \Rightarrow s = \int (\pi \sin(\pi t) + 8)\, dt$
$= \int \sin u\, du + 8t + C_2 = -\cos(\pi t) + 8t + C_2;$ at $t = 0$ and $s = 0$ we have $0 = -1 + C_2 \Rightarrow C_2 = 1$
$\Rightarrow s = 8t - \cos(\pi t) + 1 \Rightarrow s(1) = 8 - \cos\pi + 1 = 10$ m

67. All three integrations are correct. In each case, the derivative of the function on the right is the integrand on the left, and each formula has an arbitrary constant for generating the remaining antiderivatives. Moreover, $\sin^2 x + C_1 = 1 - \cos^2 x + C_1 \Rightarrow C_2 = 1 + C_1$; also $-\cos^2 x + C_2 = -\frac{\cos 2x}{2} - \frac{1}{2} + C_2 \Rightarrow C_3 = C_2 - \frac{1}{2} = C_1 + \frac{1}{2}$.

68. Both integrations are correct. In each case, the derivative of the function on the right is the integrand on the left, and each formula has an arbitrary constant for generating the remaining antiderivatives. Moreover,
$\frac{\tan^2 x}{2} + C = \frac{\sec^2 x - 1}{2} + C = \frac{\sec^2 x}{2} + \underbrace{\left(C - \frac{1}{2}\right)}_{\text{a constant}}$

69. (a) $\left(\frac{1}{\frac{1}{60} - 0}\right)\int_0^{1/60} V_{max}\sin 120\pi t\, dt = 60\left[-V_{max}\left(\frac{1}{120\pi}\right)\cos(120\pi t)\right]_0^{1/60} = -\frac{V_{max}}{2\pi}[\cos 2\pi - \cos 0]$
$= -\frac{V_{max}}{2\pi}[1 - 1] = 0$

(b) $V_{max} = \sqrt{2}\, V_{rms} = \sqrt{2}(240) \approx 339$ volts

(c) $\int_0^{1/60} (V_{max})^2 \sin^2 120\pi t\, dt = (V_{max})^2 \int_0^{1/60} \left(\frac{1-\cos 240\pi t}{2}\right) dt = \frac{(V_{max})^2}{2} \int_0^{1/60} (1-\cos 240\pi t)\, dt$
$= \frac{(V_{max})^2}{2} \left[t - \left(\frac{1}{240\pi}\right) \sin 240\pi t\right]_0^{1/60} = \frac{(V_{max})^2}{2} \left[\left(\frac{1}{60} - \left(\frac{1}{240\pi}\right) \sin(4\pi)\right) - \left(0 - \left(\frac{1}{240\pi}\right) \sin(0)\right)\right] = \frac{(V_{max})^2}{120}$

5.6 SUBSTITUTION AND AREA BETWEEN CURVES

1. (a) Let $u = y + 1 \Rightarrow du = dy;\ y = 0 \Rightarrow u = 1,\ y = 3 \Rightarrow u = 4$
$\int_0^3 \sqrt{y+1}\, dy = \int_1^4 u^{1/2}\, du = \left[\frac{2}{3} u^{3/2}\right]_1^4 = \left(\frac{2}{3}\right)(4)^{3/2} - \left(\frac{2}{3}\right)(1)^{3/2} = \left(\frac{2}{3}\right)(8) - \left(\frac{2}{3}\right)(1) = \frac{14}{3}$
(b) Use the same substitution for u as in part (a); $y = -1 \Rightarrow u = 0,\ y = 0 \Rightarrow u = 1$
$\int_{-1}^0 \sqrt{y+1}\, dy = \int_0^1 u^{1/2}\, du = \left[\frac{2}{3} u^{3/2}\right]_0^1 = \left(\frac{2}{3}\right)(1)^{3/2} - 0 = \frac{2}{3}$

2. (a) Let $u = 1 - r^2 \Rightarrow du = -2r\, dr \Rightarrow -\frac{1}{2}\, du = r\, dr;\ r = 0 \Rightarrow u = 1,\ r = 1 \Rightarrow u = 0$
$\int_0^1 r\sqrt{1-r^2}\, dr = \int_1^0 -\frac{1}{2}\sqrt{u}\, du = \left[-\frac{1}{3} u^{3/2}\right]_1^0 = 0 - \left(-\frac{1}{3}\right)(1)^{3/2} = \frac{1}{3}$
(b) Use the same substitution for u as in part (a); $r = -1 \Rightarrow u = 0,\ r = 1 \Rightarrow u = 0$
$\int_{-1}^1 r\sqrt{1-r^2}\, dr = \int_0^0 -\frac{1}{2}\sqrt{u}\, du = 0$

3. (a) Let $u = \tan x \Rightarrow du = \sec^2 x\, dx;\ x = 0 \Rightarrow u = 0,\ x = \frac{\pi}{4} \Rightarrow u = 1$
$\int_0^{\pi/4} \tan x \sec^2 x\, dx = \int_0^1 u\, du = \left[\frac{u^2}{2}\right]_0^1 = \frac{1^2}{2} - 0 = \frac{1}{2}$
(b) Use the same substitution as in part (a); $x = -\frac{\pi}{4} \Rightarrow u = -1,\ x = 0 \Rightarrow u = 0$
$\int_{-\pi/4}^0 \tan x \sec^2 x\, dx = \int_{-1}^0 u\, du = \left[\frac{u^2}{2}\right]_{-1}^0 = 0 - \frac{1}{2} = -\frac{1}{2}$

4. (a) Let $u = \cos x \Rightarrow du = -\sin x\, dx \Rightarrow -du = \sin x\, dx;\ x = 0 \Rightarrow u = 1,\ x = \pi \Rightarrow u = -1$
$\int_0^\pi 3\cos^2 x \sin x\, dx = \int_1^{-1} -3u^2\, du = \left[-u^3\right]_1^{-1} = -(-1)^3 - \left(-(1)^3\right) = 2$
(b) Use the same substitution as in part (a); $x = 2\pi \Rightarrow u = 1,\ x = 3\pi \Rightarrow u = -1$
$\int_{2\pi}^{3\pi} 3\cos^2 x \sin x\, dx = \int_1^{-1} -3u^2\, du = 2$

5. (a) $u = 1 + t^4 \Rightarrow du = 4t^3\, dt \Rightarrow \frac{1}{4}\, du = t^3\, dt;\ t = 0 \Rightarrow u = 1,\ t = 1 \Rightarrow u = 2$
$\int_0^1 t^3 \left(1+t^4\right)^3 dt = \int_1^2 \frac{1}{4} u^3\, du = \left[\frac{u^4}{16}\right]_1^2 = \frac{2^4}{16} - \frac{1^4}{16} = \frac{15}{16}$
(b) Use the same substitution as in part (a); $t = -1 \Rightarrow u = 2,\ t = 1 \Rightarrow u = 2$
$\int_{-1}^1 t^3 \left(1+t^4\right)^3 dt = \int_2^2 \frac{1}{4} u^3\, du = 0$

6. (a) Let $u = t^2 + 1 \Rightarrow du = 2t\, dt \Rightarrow \frac{1}{2}\, du = t\, dt;\ t = 0 \Rightarrow u = 1,\ t = \sqrt{7} \Rightarrow u = 8$
$\int_0^{\sqrt{7}} t\left(t^2+1\right)^{1/3} dt = \int_1^8 \frac{1}{2} u^{1/3}\, du = \left[\left(\frac{1}{2}\right)\left(\frac{3}{4}\right) u^{4/3}\right]_1^8 = \left(\frac{3}{8}\right)(8)^{4/3} - \left(\frac{3}{8}\right)(1)^{4/3} = \frac{45}{8}$
(b) Use the same substitution as in part (a); $t = -\sqrt{7} \Rightarrow u = 8,\ t = 0 \Rightarrow u = 1$
$\int_{-\sqrt{7}}^0 t\left(t^2+1\right)^{1/3} dt = \int_8^1 \frac{1}{2} u^{1/3}\, du = -\int_1^8 \frac{1}{2} u^{1/3}\, du = -\frac{45}{8}$

7. (a) Let $u = 4 + r^2 \Rightarrow du = 2r\, dr \Rightarrow \frac{1}{2}\, du = r\, dr;\ r = -1 \Rightarrow u = 5,\ r = 1 \Rightarrow u = 5$
$\int_{-1}^1 \frac{5r}{\left(4+r^2\right)^2}\, dr = 5\int_5^5 \frac{1}{2} u^{-2}\, du = 0$

(b) Use the same substitution as in part (a); $r = 0 \Rightarrow u = 4$, $r = 1 \Rightarrow u = 5$

$$\int_0^1 \frac{5r}{(4+r^2)^2}\,dr = 5\int_4^5 \tfrac{1}{2}u^{-2}\,du = 5\left[-\tfrac{1}{2}u^{-1}\right]_4^5 = 5\left(-\tfrac{1}{2}(5)^{-1}\right) - 5\left(-\tfrac{1}{2}(4)^{-1}\right) = \tfrac{1}{8}$$

8. (a) Let $u = 1 + v^{3/2} \Rightarrow du = \frac{3}{2}v^{1/2}\,dv \Rightarrow \frac{20}{3}\,du = 10\sqrt{v}\,dv$; $v = 0 \Rightarrow u = 1$, $v = 1 \Rightarrow u = 2$

$$\int_0^1 \frac{10\sqrt{v}}{(1+v^{3/2})^2}\,dv = \int_1^2 \tfrac{1}{u^2}\left(\tfrac{20}{3}\,du\right) = \tfrac{20}{3}\int_1^2 u^{-2}\,du = -\tfrac{20}{3}\left[\tfrac{1}{u}\right]_1^2 = -\tfrac{20}{3}\left[\tfrac{1}{2} - \tfrac{1}{1}\right] = \tfrac{10}{3}$$

(b) Use the same substitution as in part (a); $v = 1 \Rightarrow u = 2$, $v = 4 \Rightarrow u = 1 + 4^{3/2} = 9$

$$\int_1^4 \frac{10\sqrt{v}}{(1+v^{3/2})^2}\,dv = \int_2^9 \tfrac{1}{u^2}\left(\tfrac{20}{3}\,du\right) = -\tfrac{20}{3}\left[\tfrac{1}{u}\right]_2^9 = -\tfrac{20}{3}\left(\tfrac{1}{9} - \tfrac{1}{2}\right) = -\tfrac{20}{3}\left(-\tfrac{7}{18}\right) = \tfrac{70}{27}$$

9. (a) Let $u = x^2 + 1 \Rightarrow du = 2x\,dx \Rightarrow 2\,du = 4x\,dx$; $x = 0 \Rightarrow u = 1$, $x = \sqrt{3} \Rightarrow u = 4$

$$\int_0^{\sqrt{3}} \frac{4x}{\sqrt{x^2+1}}\,dx = \int_1^4 \tfrac{2}{\sqrt{u}}\,du = \int_1^4 2u^{-1/2}\,du = \left[4u^{1/2}\right]_1^4 = 4(4)^{1/2} - 4(1)^{1/2} = 4$$

(b) Use the same substitution as in part (a); $x = -\sqrt{3} \Rightarrow u = 4$, $x = \sqrt{3} \Rightarrow u = 4$

$$\int_{-\sqrt{3}}^{\sqrt{3}} \frac{4x}{\sqrt{x^2+1}}\,dx = \int_4^4 \tfrac{2}{\sqrt{u}}\,du = 0$$

10. (a) Let $u = x^4 + 9 \Rightarrow du = 4x^3\,dx \Rightarrow \frac{1}{4}\,du = x^3\,dx$; $x = 0 \Rightarrow u = 9$, $x = 1 \Rightarrow u = 10$

$$\int_0^1 \frac{x^3}{\sqrt{x^4+9}}\,dx = \int_9^{10} \tfrac{1}{4}u^{-1/2}\,du = \left[\tfrac{1}{4}(2)u^{1/2}\right]_9^{10} = \tfrac{1}{2}(10)^{1/2} - \tfrac{1}{2}(9)^{1/2} = \tfrac{\sqrt{10}-3}{2}$$

(b) Use the same substitution as in part (a); $x = -1 \Rightarrow u = 10$, $x = 0 \Rightarrow u = 9$

$$\int_{-1}^0 \frac{x^3}{\sqrt{x^4+9}}\,dx = \int_{10}^9 \tfrac{1}{4}u^{-1/2}\,du = -\int_9^{10} \tfrac{1}{4}u^{-1/2}\,du = \tfrac{3-\sqrt{10}}{2}$$

11. (a) Let $u = 1 - \cos 3t \Rightarrow du = 3\sin 3t\,dt \Rightarrow \frac{1}{3}\,du = \sin 3t\,dt$; $t = 0 \Rightarrow u = 0$, $t = \frac{\pi}{6} \Rightarrow u = 1 - \cos\frac{\pi}{2} = 1$

$$\int_0^{\pi/6} (1 - \cos 3t)\sin 3t\,dt = \int_0^1 \tfrac{1}{3}u\,du = \left[\tfrac{1}{3}\left(\tfrac{u^2}{2}\right)\right]_0^1 = \tfrac{1}{6}(1)^2 - \tfrac{1}{6}(0)^2 = \tfrac{1}{6}$$

(b) Use the same substitution as in part (a); $t = \frac{\pi}{6} \Rightarrow u = 1$, $t = \frac{\pi}{3} \Rightarrow u = 1 - \cos\pi = 2$

$$\int_{\pi/6}^{\pi/3} (1 - \cos 3t)\sin 3t\,dt = \int_1^2 \tfrac{1}{3}u\,du = \left[\tfrac{1}{3}\left(\tfrac{u^2}{2}\right)\right]_1^2 = \tfrac{1}{6}(2)^2 - \tfrac{1}{6}(1)^2 = \tfrac{1}{2}$$

12. (a) Let $u = 2 + \tan\frac{t}{2} \Rightarrow du = \frac{1}{2}\sec^2\frac{t}{2}\,dt \Rightarrow 2\,du = \sec^2\frac{t}{2}\,dt$; $t = \frac{-\pi}{2} \Rightarrow u = 2 + \tan\left(\frac{-\pi}{4}\right) = 1$, $t = 0 \Rightarrow u = 2$

$$\int_{-\pi/2}^0 \left(2 + \tan\tfrac{t}{2}\right)\sec^2\tfrac{t}{2}\,dt = \int_1^2 u\,(2\,du) = \left[u^2\right]_1^2 = 2^2 - 1^2 = 3$$

(b) Use the same substitution as in part (a); $t = \frac{-\pi}{2} \Rightarrow u = 1$, $t = \frac{\pi}{2} \Rightarrow u = 3$

$$\int_{-\pi/2}^{\pi/2} \left(2 + \tan\tfrac{t}{2}\right)\sec^2\tfrac{t}{2}\,dt = 2\int_1^3 u\,du = \left[u^2\right]_1^3 = 3^2 - 1^2 = 8$$

13. (a) Let $u = 4 + 3\sin z \Rightarrow du = 3\cos z\,dz \Rightarrow \frac{1}{3}\,du = \cos z\,dz$; $z = 0 \Rightarrow u = 4$, $z = 2\pi \Rightarrow u = 4$

$$\int_0^{2\pi} \frac{\cos z}{\sqrt{4+3\sin z}}\,dz = \int_4^4 \tfrac{1}{\sqrt{u}}\left(\tfrac{1}{3}\,du\right) = 0$$

(b) Use the same substitution as in part (a); $z = -\pi \Rightarrow u = 4 + 3\sin(-\pi) = 4$, $z = \pi \Rightarrow u = 4$

$$\int_{-\pi}^{\pi} \frac{\cos z}{\sqrt{4+3\sin z}}\,dz = \int_4^4 \tfrac{1}{\sqrt{u}}\left(\tfrac{1}{3}\,du\right) = 0$$

14. (a) Let $u = 3 + 2\cos w \Rightarrow du = -2\sin w\,dw \Rightarrow -\frac{1}{2}\,du = \sin w\,dw$; $w = -\frac{\pi}{2} \Rightarrow u = 3$, $w = 0 \Rightarrow u = 5$

$$\int_{-\pi/2}^0 \frac{\sin w}{(3+2\cos w)^2}\,dw = \int_3^5 u^{-2}\left(-\tfrac{1}{2}\,du\right) = \tfrac{1}{2}\left[u^{-1}\right]_3^5 = \tfrac{1}{2}\left(\tfrac{1}{5} - \tfrac{1}{3}\right) = -\tfrac{1}{15}$$

(b) Use the same substitution as in part (a); $w = 0 \Rightarrow u = 5$, $w = \frac{\pi}{2} \Rightarrow u = 3$

$$\int_0^{\pi/2} \frac{\sin w}{(3 + 2\cos w)^2}\, dw = \int_5^3 u^{-2}\left(-\frac{1}{2}\, du\right) = \frac{1}{2}\int_3^5 u^{-2}\, du = \frac{1}{15}$$

15. Let $u = t^5 + 2t \Rightarrow du = (5t^4 + 2)\, dt$; $t = 0 \Rightarrow u = 0$, $t = 1 \Rightarrow u = 3$

$$\int_0^1 \sqrt{t^5 + 2t}\,(5t^4 + 2)\, dt = \int_0^3 u^{1/2}\, du = \left[\frac{2}{3}u^{3/2}\right]_0^3 = \frac{2}{3}(3)^{3/2} - \frac{2}{3}(0)^{3/2} = 2\sqrt{3}$$

16. Let $u = 1 + \sqrt{y} \Rightarrow du = \frac{dy}{2\sqrt{y}}$; $y = 1 \Rightarrow u = 2$, $y = 4 \Rightarrow u = 3$

$$\int_1^4 \frac{dy}{2\sqrt{y}\,(1 + \sqrt{y})^2} = \int_2^3 \frac{1}{u^2}\, du = \int_2^3 u^{-2}\, du = \left[-u^{-1}\right]_2^3 = \left(-\frac{1}{3}\right) - \left(-\frac{1}{2}\right) = \frac{1}{6}$$

17. Let $u = \cos 2\theta \Rightarrow du = -2\sin 2\theta\, d\theta \Rightarrow -\frac{1}{2}\, du = \sin 2\theta\, d\theta$; $\theta = 0 \Rightarrow u = 1$, $\theta = \frac{\pi}{6} \Rightarrow u = \cos 2\left(\frac{\pi}{6}\right) = \frac{1}{2}$

$$\int_0^{\pi/6} \cos^{-3} 2\theta \sin 2\theta\, d\theta = \int_1^{1/2} u^{-3}\left(-\frac{1}{2}\, du\right) = -\frac{1}{2}\int_1^{1/2} u^{-3}\, du = \left[-\frac{1}{2}\left(\frac{u^{-2}}{-2}\right)\right]_1^{1/2} = \frac{1}{4\left(\frac{1}{2}\right)^2} - \frac{1}{4(1)^2} = \frac{3}{4}$$

18. Let $u = \tan\left(\frac{\theta}{6}\right) \Rightarrow du = \frac{1}{6}\sec^2\left(\frac{\theta}{6}\right) d\theta \Rightarrow 6\, du = \sec^2\left(\frac{\theta}{6}\right) d\theta$; $\theta = \pi \Rightarrow u = \tan\left(\frac{\pi}{6}\right) = \frac{1}{\sqrt{3}}$, $\theta = \frac{3\pi}{2} \Rightarrow$ $u = \tan\frac{\pi}{4} = 1$

$$\int_\pi^{3\pi/2} \cot^5\left(\frac{\theta}{6}\right)\sec^2\left(\frac{\theta}{6}\right) d\theta = \int_{1/\sqrt{3}}^1 u^{-5}(6\, du) = \left[6\left(\frac{u^{-4}}{-4}\right)\right]_{1/\sqrt{3}}^1 = \left[-\frac{3}{2u^4}\right]_{1/\sqrt{3}}^1 = -\frac{3}{2(1)^4} - \left(-\frac{3}{2\left(\frac{1}{\sqrt{3}}\right)^4}\right) = 12$$

19. Let $u = 5 - 4\cos t \Rightarrow du = 4\sin t\, dt \Rightarrow \frac{1}{4}\, du = \sin t\, dt$; $t = 0 \Rightarrow u = 5 - 4\cos 0 = 1$, $t = \pi \Rightarrow$ $u = 5 - 4\cos\pi = 9$

$$\int_0^\pi 5(5 - 4\cos t)^{1/4}\sin t\, dt = \int_1^9 5u^{1/4}\left(\frac{1}{4}\, du\right) = \frac{5}{4}\int_1^9 u^{1/4}\, du = \left[\frac{5}{4}\left(\frac{4}{5}u^{5/4}\right)\right]_1^9 = 9^{5/4} - 1 = 3^{5/2} - 1$$

20. Let $u = 1 - \sin 2t \Rightarrow du = -2\cos 2t\, dt \Rightarrow -\frac{1}{2}\, du = \cos 2t\, dt$; $t = 0 \Rightarrow u = 1$, $t = \frac{\pi}{4} \Rightarrow u = 0$

$$\int_0^{\pi/4} (1 - \sin 2t)^{3/2}\cos 2t\, dt = \int_1^0 -\frac{1}{2}u^{3/2}\, du = \left[-\frac{1}{2}\left(\frac{2}{5}u^{5/2}\right)\right]_1^0 = \left(-\frac{1}{5}(0)^{5/2}\right) - \left(-\frac{1}{5}(1)^{5/2}\right) = \frac{1}{5}$$

21. Let $u = 4y - y^2 + 4y^3 + 1 \Rightarrow du = (4 - 2y + 12y^2)\, dy$; $y = 0 \Rightarrow u = 1$, $y = 1 \Rightarrow u = 4(1) - (1)^2 + 4(1)^3 + 1 = 8$

$$\int_0^1 (4y - y^2 + 4y^3 + 1)^{-2/3}(12y^2 - 2y + 4)\, dy = \int_1^8 u^{-2/3}\, du = \left[3u^{1/3}\right]_1^8 = 3(8)^{1/3} - 3(1)^{1/3} = 3$$

22. Let $u = y^3 + 6y^2 - 12y + 9 \Rightarrow du = (3y^2 + 12y - 12)\, dy \Rightarrow \frac{1}{3}\, du = (y^2 + 4y - 4)\, dy$; $y = 0 \Rightarrow u = 9$, $y = 1$ $\Rightarrow u = 4$

$$\int_0^1 (y^3 + 6y^2 - 12y + 9)^{-1/2}(y^2 + 4y - 4)\, dy = \int_9^4 \frac{1}{3}u^{-1/2}\, du = \left[\frac{1}{3}\left(2u^{1/2}\right)\right]_9^4 = \frac{2}{3}(4)^{1/2} - \frac{2}{3}(9)^{1/2} = \frac{2}{3}(2 - 3)$$
$$= -\frac{2}{3}$$

23. Let $u = \theta^{3/2} \Rightarrow du = \frac{3}{2}\theta^{1/2}\, d\theta \Rightarrow \frac{2}{3}\, du = \sqrt{\theta}\, d\theta$; $\theta = 0 \Rightarrow u = 0$, $\theta = \sqrt[3]{\pi^2} \Rightarrow u = \pi$

$$\int_0^{\sqrt[3]{\pi^2}} \sqrt{\theta}\cos^2\left(\theta^{3/2}\right) d\theta = \int_0^\pi \cos^2 u\left(\frac{2}{3}\, du\right) = \left[\frac{2}{3}\left(\frac{u}{2} + \frac{1}{4}\sin 2u\right)\right]_0^\pi = \frac{2}{3}\left(\frac{\pi}{2} + \frac{1}{4}\sin 2\pi\right) - \frac{2}{3}(0) = \frac{\pi}{3}$$

24. Let $u = 1 + \frac{1}{t} \Rightarrow du = -t^{-2}\, dt$; $t = -1 \Rightarrow u = 0$, $t = -\frac{1}{2} \Rightarrow u = -1$

$$\int_{-1}^{-1/2} t^{-2}\sin^2\left(1 + \frac{1}{t}\right) dt = \int_0^{-1} -\sin^2 u\, du = \left[-\left(\frac{u}{2} - \frac{1}{4}\sin 2u\right)\right]_0^{-1} = -\left[\left(-\frac{1}{2} - \frac{1}{4}\sin(-2)\right) - \left(\frac{0}{2} - \frac{1}{4}\sin 0\right)\right]$$
$$= \frac{1}{2} - \frac{1}{4}\sin 2$$

25. Let $u = \tan\theta \Rightarrow du = \sec^2\theta\, d\theta$; $\theta = 0 \Rightarrow u = 0$, $\theta = \frac{\pi}{4} \Rightarrow u = 1$;
$\int_0^{\pi/4} \left(1 + e^{\tan\theta}\right) \sec^2\theta\, d\theta = \int_0^{\pi/4} \sec^2\theta\, d\theta + \int_0^1 e^u\, du = [\tan\theta]_0^{\pi/4} + [e^u]_0^1 = \left[\tan\left(\frac{\pi}{4}\right) - \tan(0)\right] + (e^1 - e^0)$
$= (1 - 0) + (e - 1) = e$

26. Let $u = \cot\theta \Rightarrow du = -\csc^2\theta\, d\theta$; $\theta = \frac{\pi}{4} \Rightarrow u = 1$, $\theta = \frac{\pi}{2} \Rightarrow u = 0$;
$\int_{\pi/4}^{\pi/2} \left(1 + e^{\cot\theta}\right) \csc^2\theta\, d\theta = \int_{\pi/4}^{\pi/2} \csc^2\theta\, d\theta - \int_1^0 e^u\, du = [-\cot\theta]_{\pi/4}^{\pi/2} - [e^u]_1^0 = \left[-\cot\left(\frac{\pi}{2}\right) + \cot\left(\frac{\pi}{4}\right)\right] - (e^0 - e^1)$
$= (0 + 1) - (1 - e) = e$

27. $\int_0^\pi \frac{\sin t}{2 - \cos t}\, dt = [\ln|2 - \cos t|]_0^\pi = \ln 3 - \ln 1 = \ln 3$; or let $u = 2 - \cos t \Rightarrow du = \sin t\, dt$ with $t = 0$
$\Rightarrow u = 1$ and $t = \pi \Rightarrow u = 3 \Rightarrow \int_0^\pi \frac{\sin t}{2 - \cos t}\, dt = \int_1^3 \frac{1}{u}\, du = [\ln|u|]_1^3 = \ln 3 - \ln 1 = \ln 3$

28. $\int_0^{\pi/3} \frac{4\sin\theta}{1 - 4\cos\theta}\, d\theta = [\ln|1 - 4\cos\theta|]_0^{\pi/3} = \ln|1 - 2| = -\ln 3 = \ln\frac{1}{3}$; or let $u = 1 - 4\cos\theta \Rightarrow du = 4\sin\theta\, d\theta$
with $\theta = 0 \Rightarrow u = -3$ and $\theta = \frac{\pi}{3} \Rightarrow u = -1 \Rightarrow \int_0^{\pi/3} \frac{4\sin\theta}{1 - 4\cos\theta}\, d\theta = \int_{-3}^{-1} \frac{1}{u}\, du = [\ln|u|]_{-3}^{-1} = -\ln 3 = \ln\frac{1}{3}$

29. Let $u = \ln x \Rightarrow du = \frac{1}{x}\, dx$; $x = 1 \Rightarrow u = 0$ and $x = 2 \Rightarrow u = \ln 2$;
$\int_1^2 \frac{2\ln x}{x}\, dx = \int_0^{\ln 2} 2u\, du = \left[u^2\right]_0^{\ln 2} = (\ln 2)^2$

30. Let $u = \ln x \Rightarrow du = \frac{1}{x}\, dx$; $x = 2 \Rightarrow u = \ln 2$ and $x = 4 \Rightarrow u = \ln 4$;
$\int_2^4 \frac{dx}{x\ln x} = \int_{\ln 2}^{\ln 4} \frac{1}{u}\, du = [\ln u]_{\ln 2}^{\ln 4} = \ln(\ln 4) - \ln(\ln 2) = \ln\left(\frac{\ln 4}{\ln 2}\right) = \ln\left(\frac{\ln 2^2}{\ln 2}\right) = \ln\left(\frac{2\ln 2}{\ln 2}\right) = \ln 2$

31. Let $u = \ln x \Rightarrow du = \frac{1}{x}\, dx$; $x = 2 \Rightarrow u = \ln 2$ and $x = 4 \Rightarrow u = \ln 4$;
$\int_2^4 \frac{dx}{x(\ln x)^2} = \int_{\ln 2}^{\ln 4} u^{-2}\, du = \left[-\frac{1}{u}\right]_{\ln 2}^{\ln 4} = -\frac{1}{\ln 4} + \frac{1}{\ln 2} = -\frac{1}{\ln 2^2} + \frac{1}{\ln 2} = -\frac{1}{2\ln 2} + \frac{1}{\ln 2} = \frac{1}{2\ln 2} = \frac{1}{\ln 4}$

32. Let $u = \ln x \Rightarrow du = \frac{1}{x}\, dx$; $x = 2 \Rightarrow u = \ln 2$ and $x = 16 \Rightarrow u = \ln 16$;
$\int_2^{16} \frac{dx}{2x\sqrt{\ln x}} = \frac{1}{2}\int_{\ln 2}^{\ln 16} u^{-1/2}\, du = \left[u^{1/2}\right]_{\ln 2}^{\ln 16} = \sqrt{\ln 16} - \sqrt{\ln 2} = \sqrt{4\ln 2} - \sqrt{\ln 2} = 2\sqrt{\ln 2} - \sqrt{\ln 2} = \sqrt{\ln 2}$

33. Let $u = \cos\frac{x}{2} \Rightarrow du = -\frac{1}{2}\sin\frac{x}{2}\, dx \Rightarrow -2\, du = \sin\frac{x}{2}\, dx$; $x = 0 \Rightarrow u = 1$ and $x = \frac{\pi}{2} \Rightarrow u = \frac{1}{\sqrt{2}}$;
$\int_0^{\pi/2} \tan\frac{x}{2}\, dx = \int_0^{\pi/2} \frac{\sin\frac{x}{2}}{\cos\frac{x}{2}}\, dx = -2\int_1^{1/\sqrt{2}} \frac{du}{u} = [-2\ln|u|]_1^{1/\sqrt{2}} = -2\ln\frac{1}{\sqrt{2}} = 2\ln\sqrt{2} = \ln 2$

34. Let $u = \sin t \Rightarrow du = \cos t\, dt$; $t = \frac{\pi}{4} \Rightarrow u = \frac{1}{\sqrt{2}}$ and $t = \frac{\pi}{2} \Rightarrow u = 1$;
$\int_{\pi/4}^{\pi/2} \cot t\, dt = \int_{\pi/4}^{\pi/2} \frac{\cos t}{\sin t}\, dt = \int_{1/\sqrt{2}}^1 \frac{du}{u} = [\ln|u|]_{1/\sqrt{2}}^1 = -\ln\frac{1}{\sqrt{2}} = \ln\sqrt{2}$

35. Let $u = \sin\frac{\theta}{3} \Rightarrow du = \frac{1}{3}\cos\frac{\theta}{3}\, d\theta \Rightarrow 6\, du = 2\cos\frac{\theta}{3}\, d\theta$; $\theta = \frac{\pi}{2} \Rightarrow u = \frac{1}{2}$ and $\theta = \pi \Rightarrow u = \frac{\sqrt{3}}{2}$;
$\int_{\pi/2}^{\pi} 2\cot\frac{\theta}{3}\, d\theta = \int_{\pi/2}^{\pi} \frac{2\cos\frac{\theta}{3}}{\sin\frac{\theta}{3}}\, d\theta = 6\int_{1/2}^{\sqrt{3}/2} \frac{du}{u} = 6[\ln|u|]_{1/2}^{\sqrt{3}/2} = 6\left(\ln\frac{\sqrt{3}}{2} - \ln\frac{1}{2}\right) = 6\ln\sqrt{3} = \ln 27$

36. Let $u = \cos 3x \Rightarrow du = -3\sin 3x\, dx \Rightarrow -2\, du = 6\sin 3x\, dx$; $x = 0 \Rightarrow u = 1$ and $x = \frac{\pi}{12} \Rightarrow u = \frac{1}{\sqrt{2}}$;
$\int_0^{\pi/12} 6\tan 3x\, dx = \int_0^{\pi/12} \frac{6\sin 3x}{\cos 3x}\, dx = -2\int_1^{1/\sqrt{2}} \frac{du}{u} = -2[\ln|u|]_1^{1/\sqrt{2}} = -2\ln\frac{1}{\sqrt{2}} - \ln 1 = 2\ln\sqrt{2} = \ln 2$

37. $\int_{-\pi/2}^{\pi/2} \frac{2\cos\theta\, d\theta}{1+(\sin\theta)^2} = 2\int_{-1}^{1} \frac{du}{1+u^2}$, where $u = \sin\theta$ and $du = \cos\theta\, d\theta$; $\theta = -\frac{\pi}{2} \Rightarrow u = -1, \theta = \frac{\pi}{2} \Rightarrow u = 1$
$= \left[2\tan^{-1} u\right]_{-1}^{1} = 2\left(\tan^{-1} 1 - \tan^{-1}(-1)\right) = 2\left[\frac{\pi}{4} - \left(-\frac{\pi}{4}\right)\right] = \pi$

38. $\int_{\pi/6}^{\pi/4} \frac{\csc^2 x\, dx}{1+(\cot x)^2} = -\int_{\sqrt{3}}^{1} \frac{du}{1+u^2}$, where $u = \cot x$ and $du = -\csc^2 x\, dx$; $x = \frac{\pi}{6} \Rightarrow u = \sqrt{3}, x = \frac{\pi}{4} \Rightarrow u = 1$
$= \left[-\tan^{-1} u\right]_{\sqrt{3}}^{1} = -\tan^{-1} 1 + \tan^{-1}\sqrt{3} = -\frac{\pi}{4} + \frac{\pi}{3} = \frac{\pi}{12}$

39. $\int_{0}^{\ln\sqrt{3}} \frac{e^x\, dx}{1+e^{2x}} = \int_{1}^{\sqrt{3}} \frac{du}{1+u^2}$, where $u = e^x$ and $du = e^x\, dx$; $x = 0 \Rightarrow u = 1, x = \ln\sqrt{3} \Rightarrow u = \sqrt{3}$
$= \left[\tan^{-1} u\right]_{1}^{\sqrt{3}} = \tan^{-1}\sqrt{3} - \tan^{-1} 1 = \frac{\pi}{3} - \frac{\pi}{4} = \frac{\pi}{12}$

40. $\int_{1}^{e^{\pi/4}} \frac{4\, dt}{t(1+\ln^2 t)} = 4\int_{0}^{\pi/4} \frac{du}{1+u^2}$, where $u = \ln t$ and $du = \frac{1}{t}\, dt$; $t = 1 \Rightarrow u = 0, t = e^{\pi/4} \Rightarrow u = \frac{\pi}{4}$
$= \left[4\tan^{-1} u\right]_{0}^{\pi/4} = 4\left(\tan^{-1}\frac{\pi}{4} - \tan^{-1} 0\right) = 4\tan^{-1}\frac{\pi}{4}$

41. $\int_{0}^{1} \frac{4\, ds}{\sqrt{4-s^2}} = \left[4\sin^{-1}\frac{s}{2}\right]_{0}^{1} = 4\left(\sin^{-1}\frac{1}{2} - \sin^{-1} 0\right) = 4\left(\frac{\pi}{6} - 0\right) = \frac{2\pi}{3}$

42. $\int_{0}^{3\sqrt{2}/4} \frac{ds}{\sqrt{9-4s^2}} = \frac{1}{2}\int_{0}^{3\sqrt{2}/4} \frac{du}{\sqrt{9-u^2}}$, where $u = 2s$ and $du = 2\, ds$; $s = 0 \Rightarrow u = 0, s = \frac{3\sqrt{2}}{4} \Rightarrow u = \frac{3\sqrt{2}}{2}$
$= \left[\frac{1}{2}\sin^{-1}\frac{u}{3}\right]_{0}^{3\sqrt{2}/2} = \frac{1}{2}\left(\sin^{-1}\frac{\sqrt{2}}{2} - \sin^{-1} 0\right) = \frac{1}{2}\left(\frac{\pi}{4} - 0\right) = \frac{\pi}{8}$

43. $\int_{\sqrt{2}}^{2} \frac{\sec^2(\sec^{-1} x)}{x\sqrt{x^2-1}}\, dx = \int_{\pi/4}^{\pi/3} \sec^2 u\, du$, where $u = \sec^{-1} x$ and $du = \frac{dx}{x\sqrt{x^2-1}}$; $x = \sqrt{2} \Rightarrow u = \frac{\pi}{4}, x = 2 \Rightarrow u = \frac{\pi}{3}$
$= \left[\tan u\right]_{\pi/4}^{\pi/3} = \tan\frac{\pi}{3} - \tan\frac{\pi}{4} = \sqrt{3} - 1$

44. $\int_{2/\sqrt{3}}^{2} \frac{\cos(\sec^{-1} x)}{x\sqrt{x^2-1}}\, dx = \int_{\pi/6}^{\pi/3} \cos u\, du$, where $u = \sec^{-1} x$ and $du = \frac{dx}{x\sqrt{x^2-1}}$; $x = \frac{2}{\sqrt{3}} \Rightarrow u = \frac{\pi}{6}, x = 2 \Rightarrow u = \frac{\pi}{3}$
$= \left[\sin u\right]_{\pi/6}^{\pi/3} = \sin\frac{\pi}{3} - \sin\frac{\pi}{6} = \frac{\sqrt{3}-1}{2}$

45. $\int_{-1}^{-\sqrt{2}/2} \frac{dy}{y\sqrt{4y^2-1}} = \int_{-2}^{-\sqrt{2}} \frac{du}{u\sqrt{u^2-1}}$, where $u = 2y$ and $du = 2\, dy$; $y = -1 \Rightarrow u = -2, y = -\frac{\sqrt{2}}{2} \Rightarrow u = -\sqrt{2}$
$= \left[\sec^{-1}|u|\right]_{-2}^{-\sqrt{2}} = \sec^{-1}\left|-\sqrt{2}\right| - \sec^{-1}|-2| = \frac{\pi}{4} - \frac{\pi}{3} = -\frac{\pi}{12}$

46. $\int_{-2/3}^{-\sqrt{2}/3} \frac{dy}{y\sqrt{9y^2-1}} = \int_{-2}^{-\sqrt{2}} \frac{du}{u\sqrt{u^2-1}}$, where $u = 3y$ and $du = 3\, dy$; $y = -\frac{2}{3} \Rightarrow u = -2, y = -\frac{\sqrt{2}}{3} \Rightarrow u = -\sqrt{2}$
$= \left[\sec^{-1}|u|\right]_{-2}^{-\sqrt{2}} = \sec^{-1}\left|-\sqrt{2}\right| - \sec^{-1}|-2| = \frac{\pi}{4} - \frac{\pi}{3} = -\frac{\pi}{12}$

47. Let $u = 4 - x^2 \Rightarrow du = -2x\, dx \Rightarrow -\frac{1}{2}\, du = x\, dx$; $x = -2 \Rightarrow u = 0, x = 0 \Rightarrow u = 4, x = 2 \Rightarrow u = 0$
$A = -\int_{-2}^{0} x\sqrt{4-x^2}\, dx + \int_{0}^{2} x\sqrt{4-x^2}\, dx = -\int_{0}^{4} -\frac{1}{2}u^{1/2}\, du + \int_{4}^{0} -\frac{1}{2}u^{1/2}\, du = 2\int_{0}^{4} \frac{1}{2}u^{1/2}\, du = \int_{0}^{4} u^{1/2}\, du$
$= \left[\frac{2}{3}u^{3/2}\right]_{0}^{4} = \frac{2}{3}(4)^{3/2} - \frac{2}{3}(0)^{3/2} = \frac{16}{3}$

48. Let $u = 1 - \cos x \Rightarrow du = \sin x\, dx$; $x = 0 \Rightarrow u = 0, x = \pi \Rightarrow u = 2$
$\int_{0}^{\pi} (1 - \cos x)\sin x\, dx = \int_{0}^{2} u\, du = \left[\frac{u^2}{2}\right]_{0}^{2} = \frac{2^2}{2} - \frac{0^2}{2} = 2$

49. Let $u = 1 + \cos x \Rightarrow du = -\sin x\,dx \Rightarrow -du = \sin x\,dx$; $x = -\pi \Rightarrow u = 1 + \cos(-\pi) = 0$, $x = 0$ $\Rightarrow u = 1 + \cos 0 = 2$

$A = -\int_{-\pi}^{0} 3(\sin x)\sqrt{1+\cos x}\,dx = -\int_{0}^{2} 3u^{1/2}(-du) = 3\int_{0}^{2} u^{1/2}\,du = \left[2u^{3/2}\right]_0^2 = 2(2)^{3/2} - 2(0)^{3/2} = 2^{5/2}$

50. Let $u = \pi + \pi \sin x \Rightarrow du = \pi \cos x\,dx \Rightarrow \frac{1}{\pi}\,du = \cos x\,dx$; $x = -\frac{\pi}{2} \Rightarrow u = \pi + \pi\sin\left(-\frac{\pi}{2}\right) = 0$, $x = 0 \Rightarrow u = \pi$

Because of symmetry about $x = -\frac{\pi}{2}$, $A = 2\int_{-\pi/2}^{0} \frac{\pi}{2}(\cos x)(\sin(\pi + \pi\sin x))\,dx = 2\int_{0}^{\pi} \frac{\pi}{2}(\sin u)\left(\frac{1}{\pi}\,du\right)$

$= \int_0^{\pi} \sin u\,du = [-\cos u]_0^{\pi} = (-\cos \pi) - (-\cos 0) = 2$

51. For the sketch given, $a = 0$, $b = \pi$; $f(x) - g(x) = 1 - \cos^2 x = \sin^2 x = \frac{1-\cos 2x}{2}$;

$A = \int_0^{\pi} \frac{(1-\cos 2x)}{2}\,dx = \frac{1}{2}\int_0^{\pi}(1-\cos 2x)\,dx = \frac{1}{2}\left[x - \frac{\sin 2x}{2}\right]_0^{\pi} = \frac{1}{2}[(\pi - 0) - (0 - 0)] = \frac{\pi}{2}$

52. For the sketch given, $a = -\frac{\pi}{3}$, $b = \frac{\pi}{3}$; $f(t) - g(t) = \frac{1}{2}\sec^2 t - (-4\sin^2 t) = \frac{1}{2}\sec^2 t + 4\sin^2 t$;

$A = \int_{-\pi/3}^{\pi/3}\left(\frac{1}{2}\sec^2 t + 4\sin^2 t\right)dt = \frac{1}{2}\int_{-\pi/3}^{\pi/3}\sec^2 t\,dt + 4\int_{-\pi/3}^{\pi/3}\sin^2 t\,dt = \frac{1}{2}\int_{-\pi/3}^{\pi/3}\sec^2 t\,dt + 4\int_{-\pi/3}^{\pi/3}\frac{(1-\cos 2t)}{2}\,dt$

$= \frac{1}{2}\int_{-\pi/3}^{\pi/3}\sec^2 t\,dt + 2\int_{-\pi/3}^{\pi/3}(1-\cos 2t)\,dt = \frac{1}{2}[\tan t]_{-\pi/3}^{\pi/3} + 2\left[t - \frac{\sin 2t}{2}\right]_{-\pi/3}^{\pi/3} = \sqrt{3} + 4\cdot\frac{\pi}{3} - \sqrt{3} = \frac{4\pi}{3}$

53. For the sketch given, $a = -2$, $b = 2$; $f(x) - g(x) = 2x^2 - (x^4 - 2x^2) = 4x^2 - x^4$;

$A = \int_{-2}^{2}(4x^2 - x^4)\,dx = \left[\frac{4x^3}{3} - \frac{x^5}{5}\right]_{-2}^{2} = \left(\frac{32}{3} - \frac{32}{5}\right) - \left[-\frac{32}{3} - \left(-\frac{32}{5}\right)\right] = \frac{64}{3} - \frac{64}{5} = \frac{320-192}{15} = \frac{128}{15}$

54. For the sketch given, $c = 0$, $d = 1$; $f(y) - g(y) = y^2 - y^3$;

$A = \int_0^1 (y^2 - y^3)\,dy = \int_0^1 y^2\,dy - \int_0^1 y^3\,dy = \left[\frac{y^3}{3}\right]_0^1 - \left[\frac{y^4}{4}\right]_0^1 = \frac{(1-0)}{3} - \frac{(1-0)}{4} = \frac{1}{3} - \frac{1}{4} = \frac{1}{12}$

55. For the sketch given, $c = 0$, $d = 1$; $f(y) - g(y) = (12y^2 - 12y^3) - (2y^2 - 2y) = 10y^2 - 12y^3 + 2y$;

$A = \int_0^1 (10y^2 - 12y^3 + 2y)\,dy = \int_0^1 10y^2\,dy - \int_0^1 12y^3\,dy + \int_0^1 2y\,dy = \left[\frac{10}{3}y^3\right]_0^1 - \left[\frac{12}{4}y^4\right]_0^1 + \left[\frac{2}{2}y^2\right]_0^1$

$= \left(\frac{10}{3} - 0\right) - (3 - 0) + (1 - 0) = \frac{4}{3}$

56. For the sketch given, $a = -1$, $b = 1$; $f(x) - g(x) = x^2 - (-2x^4) = x^2 + 2x^4$;

$A = \int_{-1}^{1}(x^2 + 2x^4)\,dx = \left[\frac{x^3}{3} + \frac{2x^5}{5}\right]_{-1}^{1} = \left(\frac{1}{3} + \frac{2}{5}\right) - \left[-\frac{1}{3} + \left(-\frac{2}{5}\right)\right] = \frac{2}{3} + \frac{4}{5} = \frac{10+12}{15} = \frac{22}{15}$

57. We want the area between the line $y = 1$, $0 \le x \le 2$, and the curve $y = \frac{x^2}{4}$, *minus* the area of a triangle (formed by $y = x$ and $y = 1$) with base 1 and height 1. Thus, $A = \int_0^2\left(1 - \frac{x^2}{4}\right)dx - \frac{1}{2}(1)(1) = \left[x - \frac{x^3}{12}\right]_0^2 - \frac{1}{2}$

$= \left(2 - \frac{8}{12}\right) - \frac{1}{2} = 2 - \frac{2}{3} - \frac{1}{2} = \frac{5}{6}$

58. We want the area between the x-axis and the curve $y = x^2$, $0 \le x \le 1$ *plus* the area of a triangle (formed by $x = 1$, $x + y = 2$, and the x-axis) with base 1 and height 1. Thus, $A = \int_0^1 x^2\,dx + \frac{1}{2}(1)(1) = \left[\frac{x^3}{3}\right]_0^1 + \frac{1}{2} = \frac{1}{3} + \frac{1}{2} = \frac{5}{6}$

59. AREA = A1 + A2

A1: For the sketch given, $a = -3$ and we find b by solving the equations $y = x^2 - 4$ and $y = -x^2 - 2x$ simultaneously for x: $x^2 - 4 = -x^2 - 2x \Rightarrow 2x^2 + 2x - 4 = 0 \Rightarrow 2(x+2)(x-1) \Rightarrow x = -2$ or $x = 1$ so $b = -2$: $f(x) - g(x) = (x^2 - 4) - (-x^2 - 2x) = 2x^2 + 2x - 4 \Rightarrow A1 = \int_{-3}^{-2}(2x^2 + 2x - 4)\,dx$

$= \left[\frac{2x^3}{3} + \frac{2x^2}{2} - 4x\right]_{-3}^{-2} = \left(-\frac{16}{3} + 4 + 8\right) - (-18 + 9 + 12) = 9 - \frac{16}{3} = \frac{11}{3};$

A2: For the sketch given, a = −2 and b = 1: $f(x) - g(x) = (-x^2 - 2x) - (x^2 - 4) = -2x^2 - 2x + 4$

$\Rightarrow A2 = -\int_{-2}^{1} (2x^2 + 2x - 4)\,dx = -\left[\frac{2x^3}{3} + x^2 - 4x\right]_{-2}^{1} = -\left(\frac{2}{3} + 1 - 4\right) + \left(-\frac{16}{3} + 4 + 8\right)$

$= -\frac{2}{3} - 1 + 4 - \frac{16}{3} + 4 + 8 = 9;$

Therefore, AREA = A1 + A2 = $\frac{11}{3} + 9 = \frac{38}{3}$

60. AREA = A1 + A2

A1: For the sketch given, a = −2 and b = 0: $f(x) - g(x) = (2x^3 - x^2 - 5x) - (-x^2 + 3x) = 2x^3 - 8x$

$\Rightarrow A1 = \int_{-2}^{0} (2x^3 - 8x)\,dx = \left[\frac{2x^4}{4} - \frac{8x^2}{2}\right]_{-2}^{0} = 0 - (8 - 16) = 8;$

A2: For the sketch given, a = 0 and b = 2: $f(x) - g(x) = (-x^2 + 3x) - (2x^3 - x^2 - 5x) = 8x - 2x^3$

$\Rightarrow A2 = \int_{0}^{2} (8x - 2x^3)\,dx = \left[\frac{8x^2}{2} - \frac{2x^4}{4}\right]_{0}^{2} = (16 - 8) = 8;$

Therefore, AREA = A1 + A2 = 16

61. AREA = A1 + A2 + A3

A1: For the sketch given, a = −2 and b = −1: $f(x) - g(x) = (-x + 2) - (4 - x^2) = x^2 - x - 2$

$\Rightarrow A1 = \int_{-2}^{-1} (x^2 - x - 2)\,dx = \left[\frac{x^3}{3} - \frac{x^2}{2} - 2x\right]_{-2}^{-1} = \left(-\frac{1}{3} - \frac{1}{2} + 2\right) - \left(-\frac{8}{3} - \frac{4}{2} + 4\right) = \frac{7}{3} - \frac{1}{2} = \frac{14-3}{6} = \frac{11}{6};$

A2: For the sketch given, a = −1 and b = 2: $f(x) - g(x) = (4 - x^2) - (-x + 2) = -(x^2 - x - 2)$

$\Rightarrow A2 = -\int_{-1}^{2} (x^2 - x - 2)\,dx = -\left[\frac{x^3}{3} - \frac{x^2}{2} - 2x\right]_{-1}^{2} = -\left(\frac{8}{3} - \frac{4}{2} - 4\right) + \left(-\frac{1}{3} - \frac{1}{2} + 2\right) = -3 + 8 - \frac{1}{2} = \frac{9}{2};$

A3: For the sketch given, a = 2 and b = 3: $f(x) - g(x) = (-x + 2) - (4 - x^2) = x^2 - x - 2$

$\Rightarrow A3 = \int_{2}^{3} (x^2 - x - 2)\,dx = \left[\frac{x^3}{3} - \frac{x^2}{2} - 2x\right]_{2}^{3} = \left(\frac{27}{3} - \frac{9}{2} - 6\right) - \left(\frac{8}{3} - \frac{4}{2} - 4\right) = 9 - \frac{9}{2} - \frac{8}{3};$

Therefore, AREA = A1 + A2 + A3 = $\frac{11}{6} + \frac{9}{2} + \left(9 - \frac{9}{2} - \frac{8}{3}\right) = 9 - \frac{5}{6} = \frac{49}{6}$

62. AREA = A1 + A2 + A3

A1: For the sketch given, a = −2 and b = 0: $f(x) - g(x) = \left(\frac{x^3}{3} - x\right) - \frac{x}{3} = \frac{x^3}{3} - \frac{4}{3}x = \frac{1}{3}(x^3 - 4x)$

$\Rightarrow A1 = \frac{1}{3}\int_{-2}^{0} (x^3 - 4x)\,dx = \frac{1}{3}\left[\frac{x^4}{4} - 2x^2\right]_{-2}^{0} = 0 - \frac{1}{3}(4 - 8) = \frac{4}{3};$

A2: For the sketch given, a = 0 and we find b by solving the equations $y = \frac{x^3}{3} - x$ and $y = \frac{x}{3}$ simultaneously for x: $\frac{x^3}{3} - x = \frac{x}{3} \Rightarrow \frac{x^3}{3} - \frac{4}{3}x = 0 \Rightarrow \frac{x}{3}(x - 2)(x + 2) = 0 \Rightarrow x = -2,\ x = 0,$ or $x = 2$ so b = 2:

$f(x) - g(x) = \frac{x}{3} - \left(\frac{x^3}{3} - x\right) = -\frac{1}{3}(x^3 - 4x) \Rightarrow A2 = -\frac{1}{3}\int_{0}^{2} (x^3 - 4x)\,dx = \frac{1}{3}\int_{0}^{2} (4x - x^3) = \frac{1}{3}\left[2x^2 - \frac{x^4}{4}\right]_{0}^{2}$

$= \frac{1}{3}(8 - 4) = \frac{4}{3};$

A3: For the sketch given, a = 2 and b = 3: $f(x) - g(x) = \left(\frac{x^3}{3} - x\right) - \frac{x}{3} = \frac{1}{3}(x^3 - 4x)$

$\Rightarrow A3 = \frac{1}{3}\int_{2}^{3} (x^3 - 4x)\,dx = \frac{1}{3}\left[\frac{x^4}{4} - 2x^2\right]_{2}^{3} = \frac{1}{3}\left[\left(\frac{81}{4} - 2 \cdot 9\right) - \left(\frac{16}{4} - 8\right)\right] = \frac{1}{3}\left(\frac{81}{4} - 14\right) = \frac{25}{12};$

Therefore, AREA = A1 + A2 + A3 = $\frac{4}{3} + \frac{4}{3} + \frac{25}{12} = \frac{32+25}{12} = \frac{19}{4}$

63. $a = -2,\ b = 2;$
$f(x) - g(x) = 2 - (x^2 - 2) = 4 - x^2$
$\Rightarrow A = \int_{-2}^{2} (4 - x^2)dx = \left[4x - \frac{x^3}{3}\right]_{-2}^{2} = \left(8 - \frac{8}{3}\right) - \left(-8 + \frac{8}{3}\right)$
$= 2 \cdot \left(\frac{24}{3} - \frac{8}{3}\right) = \frac{32}{3}$

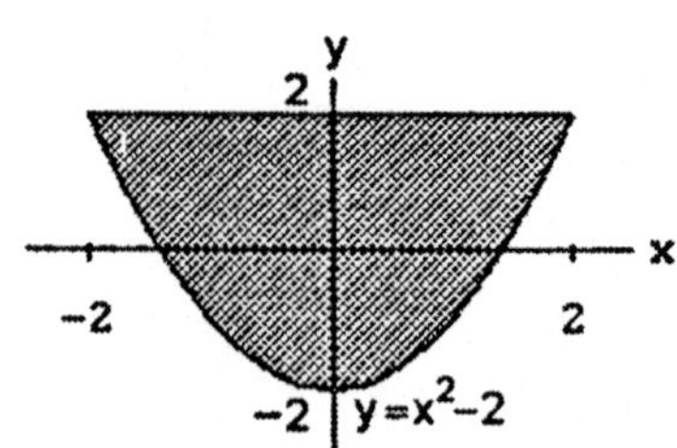

64. $a = -1,\ b = 3;$
$f(x) - g(x) = (2x - x^2) - (-3) = 2x - x^2 + 3$
$\Rightarrow A = \int_{-1}^{3} (2x - x^2 + 3)\, dx = \left[x^2 - \frac{x^3}{3} + 3x\right]_{-1}^{3}$
$= \left(9 - \frac{27}{3} + 9\right) - \left(1 + \frac{1}{3} - 3\right) = 11 - \frac{1}{3} = \frac{32}{3}$

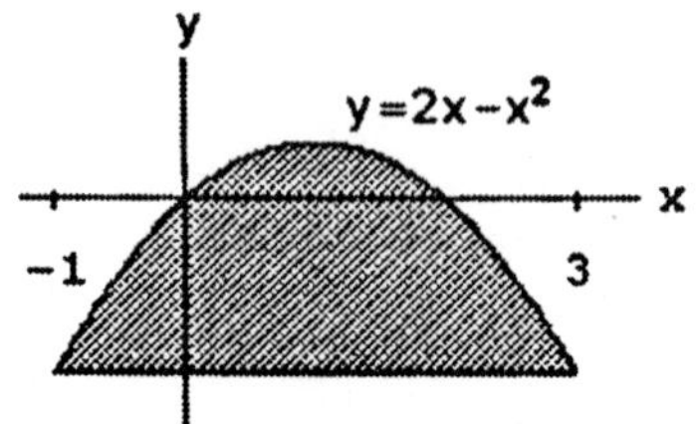

65. $a = 0,\ b = 2;$
$f(x) - g(x) = 8x - x^4 \Rightarrow A = \int_{0}^{2} (8x - x^4)\, dx$
$= \left[\frac{8x^2}{2} - \frac{x^5}{5}\right]_{0}^{2} = 16 - \frac{32}{5} = \frac{80 - 32}{5} = \frac{48}{5}$

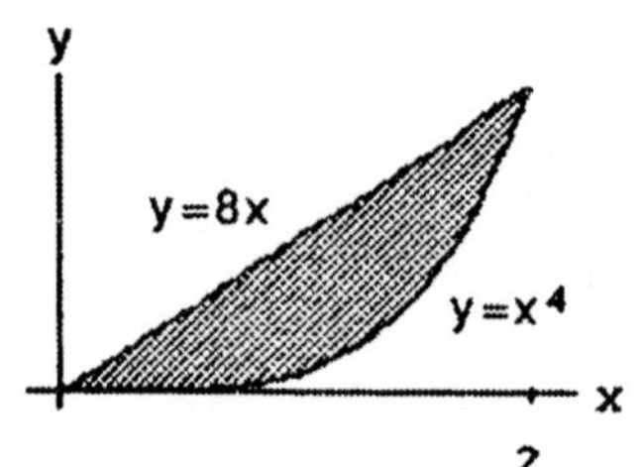

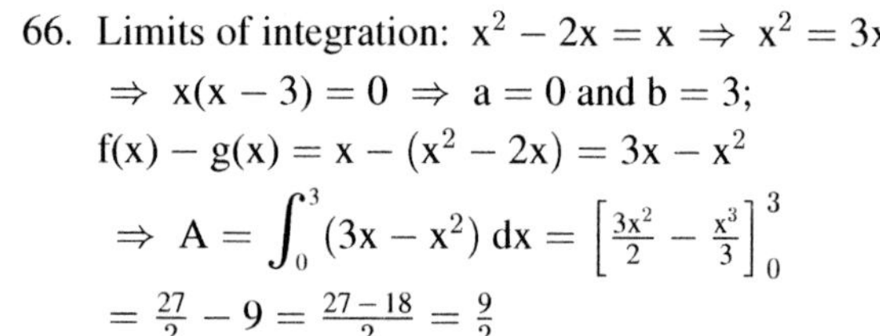

66. Limits of integration: $x^2 - 2x = x \Rightarrow x^2 = 3x$
$\Rightarrow x(x - 3) = 0 \Rightarrow a = 0$ and $b = 3;$
$f(x) - g(x) = x - (x^2 - 2x) = 3x - x^2$
$\Rightarrow A = \int_{0}^{3} (3x - x^2)\, dx = \left[\frac{3x^2}{2} - \frac{x^3}{3}\right]_{0}^{3}$
$= \frac{27}{2} - 9 = \frac{27 - 18}{2} = \frac{9}{2}$

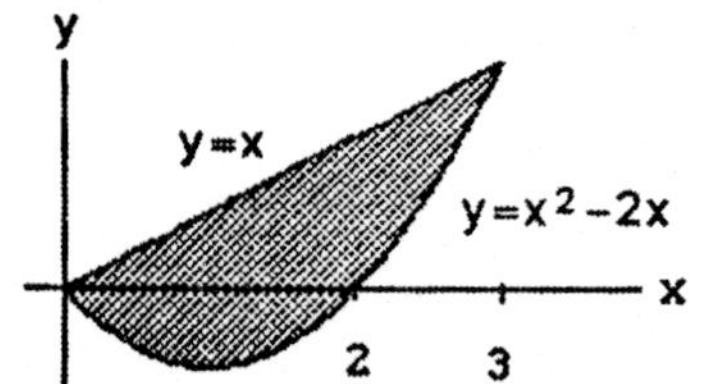

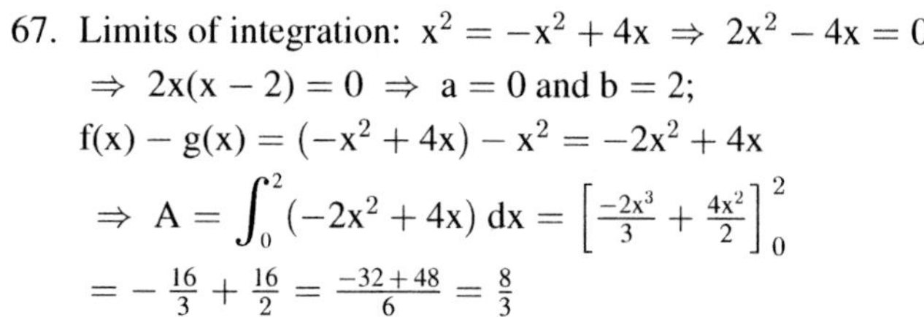

67. Limits of integration: $x^2 = -x^2 + 4x \Rightarrow 2x^2 - 4x = 0$
$\Rightarrow 2x(x - 2) = 0 \Rightarrow a = 0$ and $b = 2;$
$f(x) - g(x) = (-x^2 + 4x) - x^2 = -2x^2 + 4x$
$\Rightarrow A = \int_{0}^{2} (-2x^2 + 4x)\, dx = \left[\frac{-2x^3}{3} + \frac{4x^2}{2}\right]_{0}^{2}$
$= -\frac{16}{3} + \frac{16}{2} = \frac{-32 + 48}{6} = \frac{8}{3}$

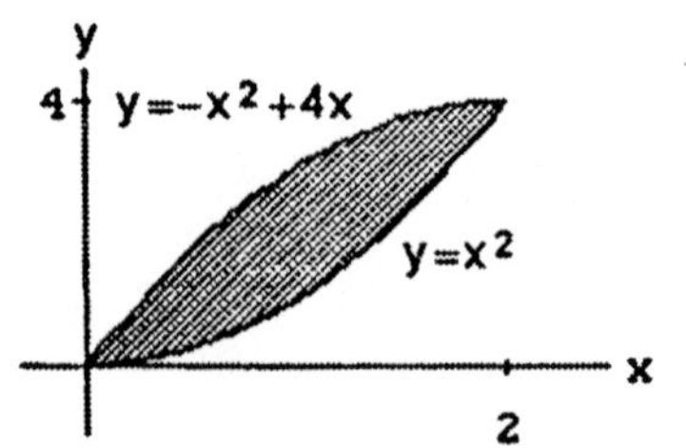

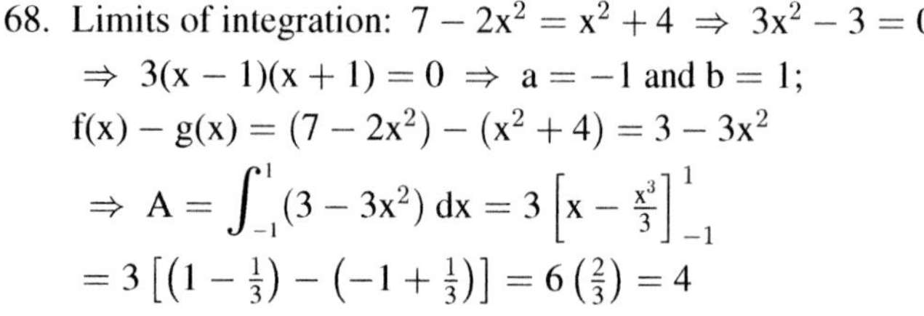

68. Limits of integration: $7 - 2x^2 = x^2 + 4 \Rightarrow 3x^2 - 3 = 0$
$\Rightarrow 3(x - 1)(x + 1) = 0 \Rightarrow a = -1$ and $b = 1;$
$f(x) - g(x) = (7 - 2x^2) - (x^2 + 4) = 3 - 3x^2$
$\Rightarrow A = \int_{-1}^{1} (3 - 3x^2)\, dx = 3\left[x - \frac{x^3}{3}\right]_{-1}^{1}$
$= 3\left[\left(1 - \frac{1}{3}\right) - \left(-1 + \frac{1}{3}\right)\right] = 6\left(\frac{2}{3}\right) = 4$

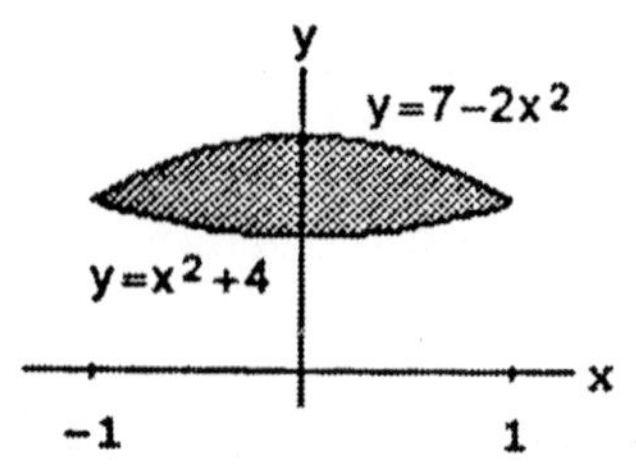

69. Limits of integration: $x^4 - 4x^2 + 4 = x^2$
$\Rightarrow x^4 - 5x^2 + 4 = 0 \Rightarrow (x^2 - 4)(x^2 - 1) = 0$
$\Rightarrow (x+2)(x-2)(x+1)(x-1) = 0 \Rightarrow x = -2, -1, 1, 2;$
$f(x) - g(x) = (x^4 - 4x^2 + 4) - x^2 = x^4 - 5x^2 + 4$ and
$g(x) - f(x) = x^2 - (x^4 - 4x^2 + 4) = -x^4 + 5x^2 - 4$
$\Rightarrow A = \int_{-2}^{-1} (-x^4 + 5x^2 - 4)dx + \int_{-1}^{1} (x^4 - 5x^2 + 4)dx$
$+ \int_{1}^{2} (-x^4 + 5x^2 - 4)dx$
$= \left[-\frac{x^5}{5} + \frac{5x^3}{3} - 4x\right]_{-2}^{-1} + \left[\frac{x^5}{5} - \frac{5x^3}{3} + 4x\right]_{-1}^{1} + \left[\frac{-x^5}{5} + \frac{5x^3}{3} - 4x\right]_{1}^{2}$
$= \left(\frac{1}{5} - \frac{5}{3} + 4\right) - \left(\frac{32}{5} - \frac{40}{3} + 8\right) + \left(\frac{1}{5} - \frac{5}{3} + 4\right) - \left(-\frac{1}{5} + \frac{5}{3} - 4\right) + \left(-\frac{32}{5} + \frac{40}{3} - 8\right) - \left(-\frac{1}{5} + \frac{5}{3} - 4\right)$
$= -\frac{60}{5} + \frac{60}{3} = \frac{300 - 180}{15} = 8$

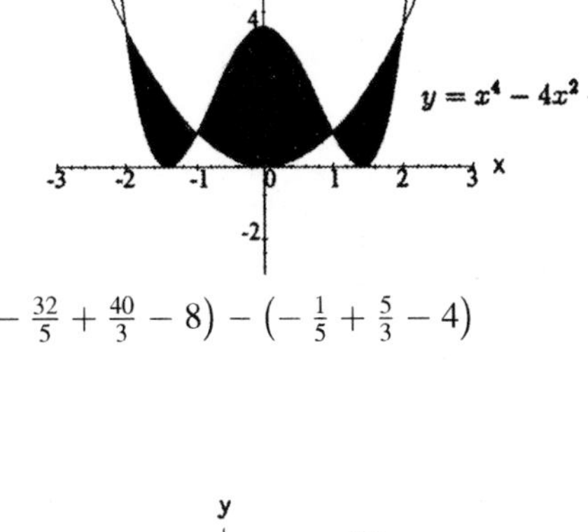

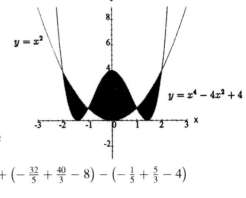

70. Limits of integration: $x\sqrt{a^2 - x^2} = 0 \Rightarrow x = 0$ or
$\sqrt{a^2 - x^2} = 0 \Rightarrow x = 0$ or $a^2 - x^2 = 0 \Rightarrow x = -a, 0, a;$

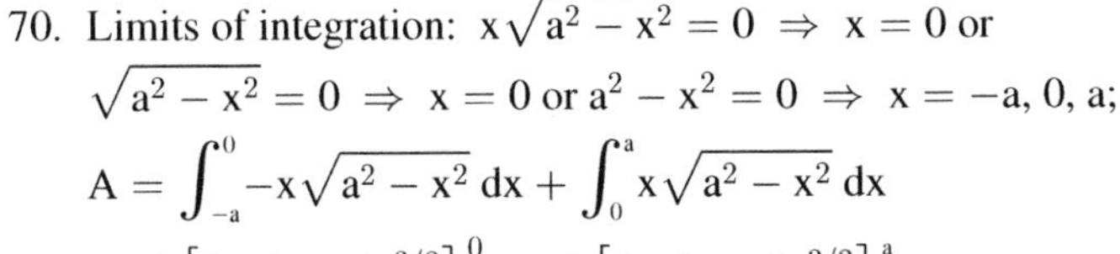

$A = \int_{-a}^{0} -x\sqrt{a^2 - x^2}\, dx + \int_{0}^{a} x\sqrt{a^2 - x^2}\, dx$
$= \frac{1}{2}\left[\frac{2}{3}(a^2 - x^2)^{3/2}\right]_{-a}^{0} - \frac{1}{2}\left[\frac{2}{3}(a^2 - x^2)^{3/2}\right]_{0}^{a}$
$= \frac{1}{3}(a^2)^{3/2} - \left[-\frac{1}{3}(a^2)^{3/2}\right] = \frac{2a^3}{3}$

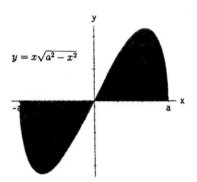

71. Limits of integration: $y = \sqrt{|x|} = \begin{cases} \sqrt{-x}, & x \le 0 \\ \sqrt{x}, & x \ge 0 \end{cases}$ and
$5y = x + 6$ or $y = \frac{x}{5} + \frac{6}{5}$; for $x \le 0$: $\sqrt{-x} = \frac{x}{5} + \frac{6}{5}$
$\Rightarrow 5\sqrt{-x} = x + 6 \Rightarrow 25(-x) = x^2 + 12x + 36$
$\Rightarrow x^2 + 37x + 36 = 0 \Rightarrow (x+1)(x+36) = 0$
$\Rightarrow x = -1, -36$ (but $x = -36$ is not a solution);
for $x \ge 0$: $5\sqrt{x} = x + 6 \Rightarrow 25x = x^2 + 12x + 36$
$\Rightarrow x^2 - 13x + 36 = 0 \Rightarrow (x-4)(x-9) = 0$
$\Rightarrow x = 4, 9$; there are three intersection points and

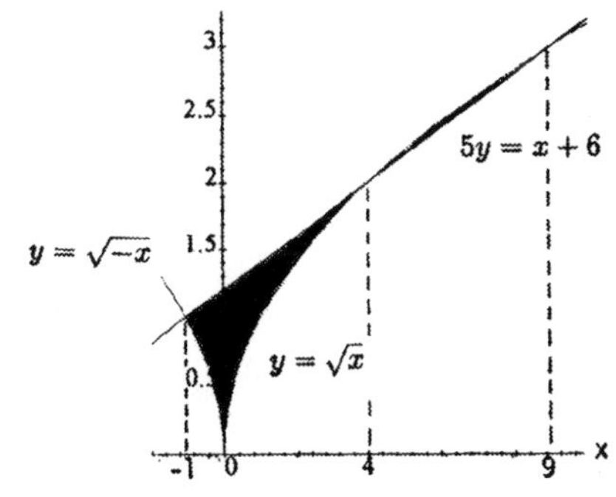

$A = \int_{-1}^{0} \left(\frac{x+6}{5} - \sqrt{-x}\right)dx + \int_{0}^{4} \left(\frac{x+6}{5} - \sqrt{x}\right)dx + \int_{4}^{9} \left(\sqrt{x} - \frac{x+6}{5}\right)dx$
$= \left[\frac{(x+6)^2}{10} + \frac{2}{3}(-x)^{3/2}\right]_{-1}^{0} + \left[\frac{(x+6)^2}{10} - \frac{2}{3}x^{3/2}\right]_{0}^{4} + \left[\frac{2}{3}x^{3/2} - \frac{(x+6)^2}{10}\right]_{4}^{9}$
$= \left(\frac{36}{10} - \frac{25}{10} - \frac{2}{3}\right) + \left(\frac{100}{10} - \frac{2}{3} \cdot 4^{3/2} - \frac{36}{10} + 0\right) + \left(\frac{2}{3} \cdot 9^{3/2} - \frac{225}{10} - \frac{2}{3} \cdot 4^{3/2} + \frac{100}{10}\right) = -\frac{50}{10} + \frac{20}{3} = \frac{5}{3}$

72. Limits of integration:
$y = |x^2 - 4| = \begin{cases} x^2 - 4, & x \le -2 \text{ or } x \ge 2 \\ 4 - x^2, & -2 \le x \le 2 \end{cases}$
for $x \le -2$ and $x \ge 2$: $x^2 - 4 = \frac{x^2}{2} + 4$
$\Rightarrow 2x^2 - 8 = x^2 + 8 \Rightarrow x^2 = 16 \Rightarrow x = \pm 4;$
for $-2 \le x \le 2$: $4 - x^2 = \frac{x^2}{2} + 4 \Rightarrow 8 - 2x^2 = x^2 + 8$
$\Rightarrow x^2 = 0 \Rightarrow x = 0$; by symmetry of the graph,

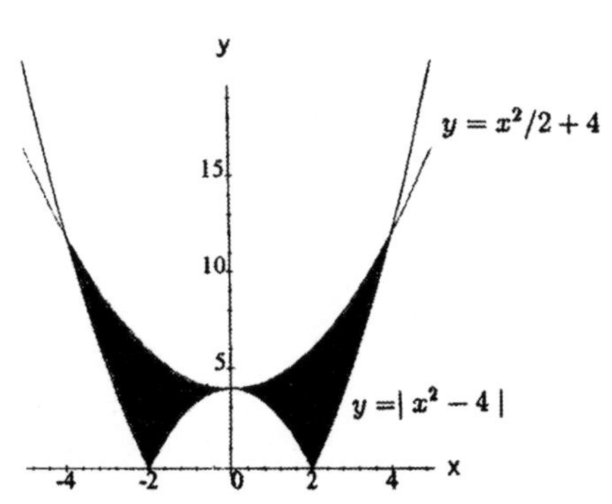

$A = 2\int_0^2 \left[\left(\frac{x^2}{2}+4\right) - (4 - x^2)\right]dx + 2\int_2^4 \left[\left(\frac{x^2}{2}+4\right) - (x^2 - 4)\right]dx = 2\left[\frac{x^3}{2}\right]_0^2 + 2\left[8x - \frac{x^3}{6}\right]_2^4$
$= 2\left(\frac{8}{2} - 0\right) + 2\left(32 - \frac{64}{6} - 16 + \frac{8}{6}\right) = 40 - \frac{56}{3} = \frac{64}{3}$

73. Limits of integration: $c = 0$ and $d = 3$;
$f(y) - g(y) = 2y^2 - 0 = 2y^2$
$\Rightarrow A = \int_0^3 2y^2\,dy = \left[\frac{2y^3}{3}\right]_0^3 = 2 \cdot 9 = 18$

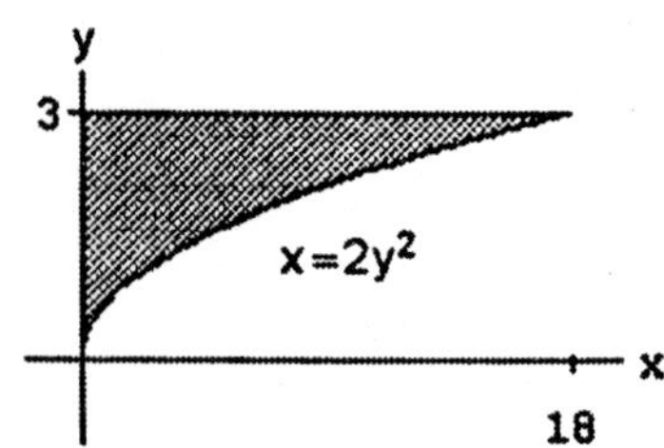

74. Limits of integration: $y^2 = y + 2 \Rightarrow (y+1)(y-2) = 0$
$\Rightarrow c = -1$ and $d = 2$; $f(y) - g(y) = (y + 2) - y^2$
$\Rightarrow A = \int_{-1}^2 (y + 2 - y^2)\,dy = \left[\frac{y^2}{2} + 2y - \frac{y^3}{3}\right]_{-1}^2$
$= \left(\frac{4}{2} + 4 - \frac{8}{3}\right) - \left(\frac{1}{2} - 2 + \frac{1}{3}\right) = 6 - \frac{8}{3} - \frac{1}{2} + 2 - \frac{1}{3} = \frac{9}{2}$

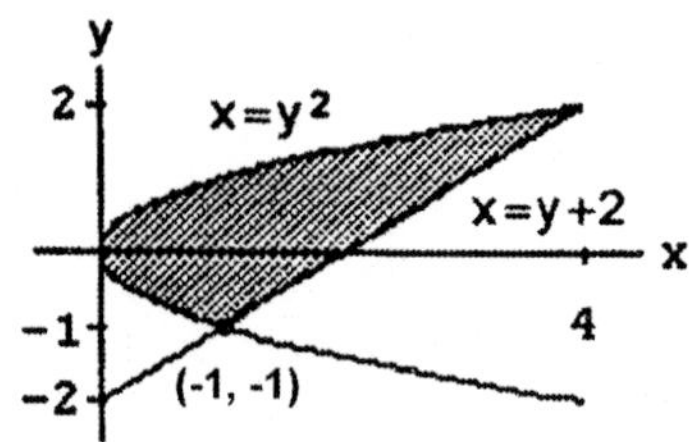

75. Limits of integration: $4x = y^2 - 4$ and $4x = 16 + y$
$\Rightarrow y^2 - 4 = 16 + y \Rightarrow y^2 - y - 20 = 0 \Rightarrow$
$(y-5)(y+4) = 0 \Rightarrow c = -4$ and $d = 5$;
$f(y) - g(y) = \left(\frac{16+y}{4}\right) - \left(\frac{y^2-4}{4}\right) = \frac{-y^2+y+20}{4}$
$\Rightarrow A = \frac{1}{4}\int_{-4}^5 (-y^2 + y + 20)\,dy$
$= \frac{1}{4}\left[-\frac{y^3}{3} + \frac{y^2}{2} + 20y\right]_{-4}^5$
$= \frac{1}{4}\left(-\frac{125}{3} + \frac{25}{2} + 100\right) - \frac{1}{4}\left(\frac{64}{3} + \frac{16}{2} - 80\right)$
$= \frac{1}{4}\left(-\frac{189}{3} + \frac{9}{2} + 180\right) = \frac{243}{8}$

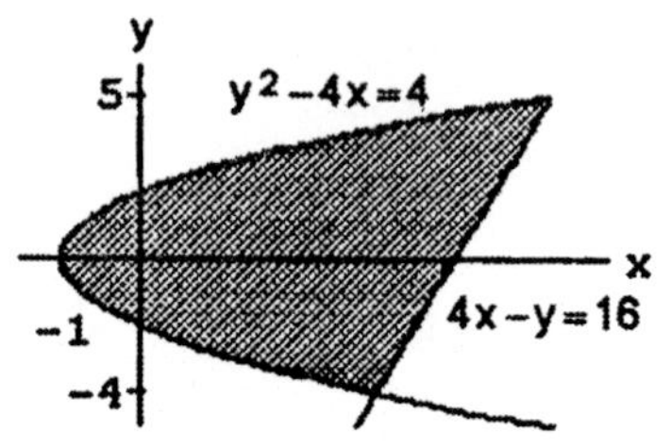

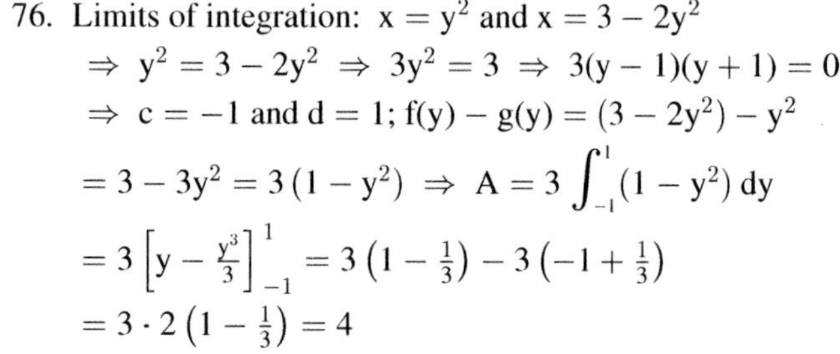

76. Limits of integration: $x = y^2$ and $x = 3 - 2y^2$
$\Rightarrow y^2 = 3 - 2y^2 \Rightarrow 3y^2 = 3 \Rightarrow 3(y-1)(y+1) = 0$
$\Rightarrow c = -1$ and $d = 1$; $f(y) - g(y) = (3 - 2y^2) - y^2$
$= 3 - 3y^2 = 3(1 - y^2) \Rightarrow A = 3\int_{-1}^1 (1 - y^2)\,dy$
$= 3\left[y - \frac{y^3}{3}\right]_{-1}^1 = 3\left(1 - \frac{1}{3}\right) - 3\left(-1 + \frac{1}{3}\right)$
$= 3 \cdot 2\left(1 - \frac{1}{3}\right) = 4$

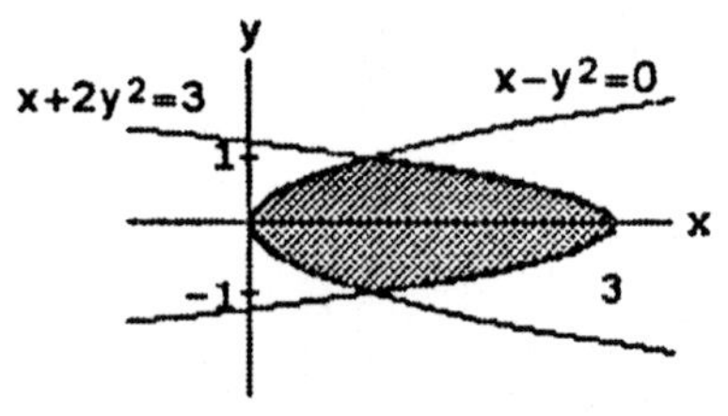

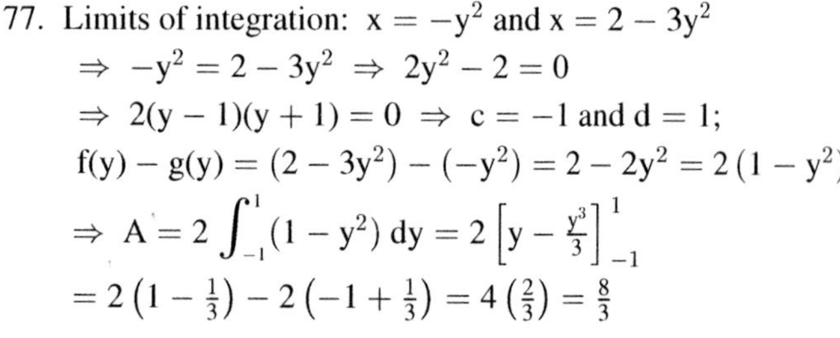

77. Limits of integration: $x = -y^2$ and $x = 2 - 3y^2$
$\Rightarrow -y^2 = 2 - 3y^2 \Rightarrow 2y^2 - 2 = 0$
$\Rightarrow 2(y-1)(y+1) = 0 \Rightarrow c = -1$ and $d = 1$;
$f(y) - g(y) = (2 - 3y^2) - (-y^2) = 2 - 2y^2 = 2(1 - y^2)$
$\Rightarrow A = 2\int_{-1}^1 (1 - y^2)\,dy = 2\left[y - \frac{y^3}{3}\right]_{-1}^1$
$= 2\left(1 - \frac{1}{3}\right) - 2\left(-1 + \frac{1}{3}\right) = 4\left(\frac{2}{3}\right) = \frac{8}{3}$

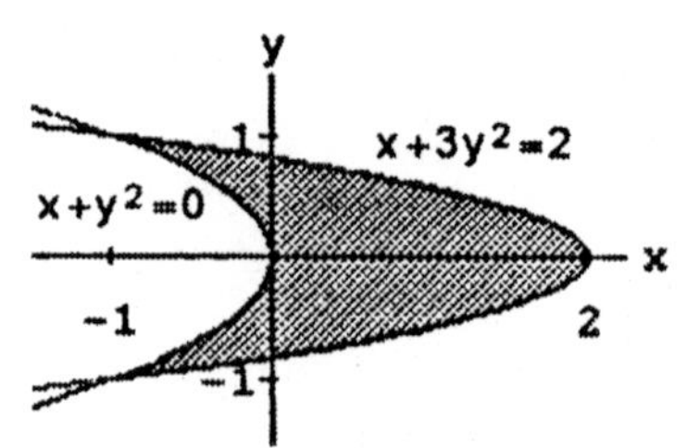

78. Limits of integration: $x = y^{2/3}$ and $x = 2 - y^4$
$\Rightarrow y^{2/3} = 2 - y^4 \Rightarrow c = -1$ and $d = 1$;
$f(y) - g(y) = (2 - y^4) - y^{2/3}$
$\Rightarrow A = \int_{-1}^{1} (2 - y^4 - y^{2/3})\,dy$
$= \left[2y - \frac{y^5}{5} - \frac{3}{5}y^{5/3}\right]_{-1}^{1}$
$= \left(2 - \frac{1}{5} - \frac{3}{5}\right) - \left(-2 + \frac{1}{5} + \frac{3}{5}\right)$
$= 2\left(2 - \frac{1}{5} - \frac{3}{5}\right) = \frac{12}{5}$

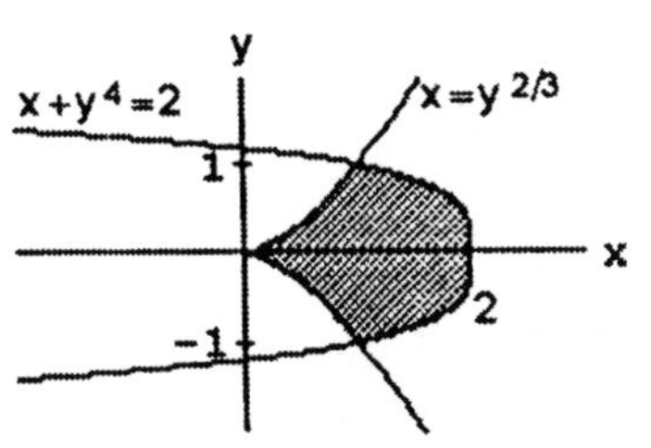

79. Limits of integration: $x = y^2 - 1$ and $x = |y|\sqrt{1-y^2}$
$\Rightarrow y^2 - 1 = |y|\sqrt{1-y^2} \Rightarrow y^4 - 2y^2 + 1 = y^2(1 - y^2)$
$\Rightarrow y^4 - 2y^2 + 1 = y^2 - y^4 \Rightarrow 2y^4 - 3y^2 + 1 = 0$
$\Rightarrow (2y^2 - 1)(y^2 - 1) = 0 \Rightarrow 2y^2 - 1 = 0$ or $y^2 - 1 = 0$
$\Rightarrow y^2 = \frac{1}{2}$ or $y^2 = 1 \Rightarrow y = \pm\frac{\sqrt{2}}{2}$ or $y = \pm 1$.
Substitution shows that $\frac{\pm\sqrt{2}}{2}$ are not solutions $\Rightarrow y = \pm 1$;
for $-1 \le y \le 0$, $f(x) - g(x) = -y\sqrt{1-y^2} - (y^2 - 1)$
$= 1 - y^2 - y(1 - y^2)^{1/2}$, and by symmetry of the graph,
$A = 2\int_{-1}^{0}\left[1 - y^2 - y(1 - y^2)^{1/2}\right] dy$
$= 2\int_{-1}^{0}(1 - y^2)\,dy - 2\int_{-1}^{0} y(1 - y^2)^{1/2}\,dy$
$= 2\left[y - \frac{y^3}{3}\right]_{-1}^{0} + 2\left(\frac{1}{2}\right)\left[\frac{2(1-y^2)^{3/2}}{3}\right]_{-1}^{0} = 2\left[(0 - 0) - \left(-1 + \frac{1}{3}\right)\right] + \left(\frac{2}{3} - 0\right) = 2$

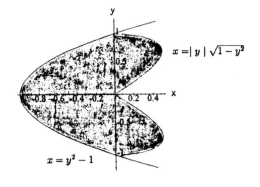

80. AREA = A1 + A2
Limits of integration: $x = 2y$ and $x = y^3 - y^2 \Rightarrow$
$y^3 - y^2 = 2y \Rightarrow y(y^2 - y - 2) = y(y + 1)(y - 2) = 0$
$\Rightarrow y = -1, 0, 2$:
for $-1 \le y \le 0$, $f(y) - g(y) = y^3 - y^2 - 2y$
$\Rightarrow A1 = \int_{-1}^{0}(y^3 - y^2 - 2y)\,dy = \left[\frac{y^4}{4} - \frac{y^3}{3} - y^2\right]_{-1}^{0}$
$= 0 - \left(\frac{1}{4} + \frac{1}{3} - 1\right) = \frac{5}{12}$;
for $0 \le y \le 2$, $f(y) - g(y) = 2y - y^3 + y^2$
$\Rightarrow A2 = \int_{0}^{2}(2y - y^3 + y^2)\,dy = \left[y^2 - \frac{y^4}{4} + \frac{y^3}{3}\right]_{0}^{2}$
$\Rightarrow \left(4 - \frac{16}{4} + \frac{8}{3}\right) - 0 = \frac{8}{3}$;
Therefore, $A1 + A2 = \frac{5}{12} + \frac{8}{3} = \frac{37}{12}$

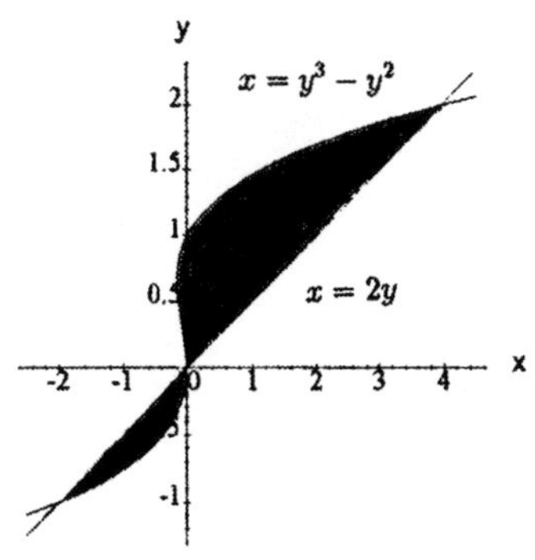

81. Limits of integration: $y = -4x^2 + 4$ and $y = x^4 - 1$
$\Rightarrow x^4 - 1 = -4x^2 + 4 \Rightarrow x^4 + 4x^2 - 5 = 0$
$\Rightarrow (x^2 + 5)(x - 1)(x + 1) = 0 \Rightarrow a = -1$ and $b = 1$;
$f(x) - g(x) = -4x^2 + 4 - x^4 + 1 = -4x^2 - x^4 + 5$
$\Rightarrow A = \int_{-1}^{1}(-4x^2 - x^4 + 5)\,dx = \left[-\frac{4x^3}{3} - \frac{x^5}{5} + 5x\right]_{-1}^{1}$
$= \left(-\frac{4}{3} - \frac{1}{5} + 5\right) - \left(\frac{4}{3} + \frac{1}{5} - 5\right) = 2\left(-\frac{4}{3} - \frac{1}{5} + 5\right) = \frac{104}{15}$

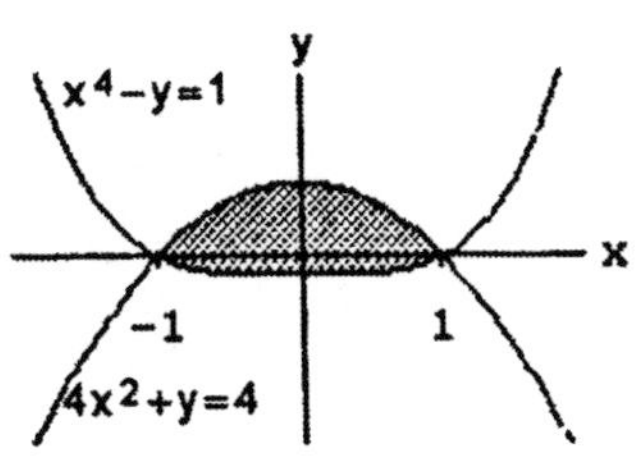

82. Limits of integration: $y = x^3$ and $y = 3x^2 - 4$
$\Rightarrow x^3 - 3x^2 + 4 = 0 \Rightarrow (x^2 - x - 2)(x - 2) = 0$
$\Rightarrow (x + 1)(x - 2)^2 = 0 \Rightarrow a = -1$ and $b = 2$;
$f(x) - g(x) = x^3 - (3x^2 - 4) = x^3 - 3x^2 + 4$
$\Rightarrow A = \int_{-1}^{2} (x^3 - 3x^2 + 4)\,dx = \left[\frac{x^4}{4} - \frac{3x^3}{3} + 4x\right]_{-1}^{2}$
$= \left(\frac{16}{4} - \frac{24}{3} + 8\right) - \left(\frac{1}{4} + 1 - 4\right) = \frac{27}{4}$

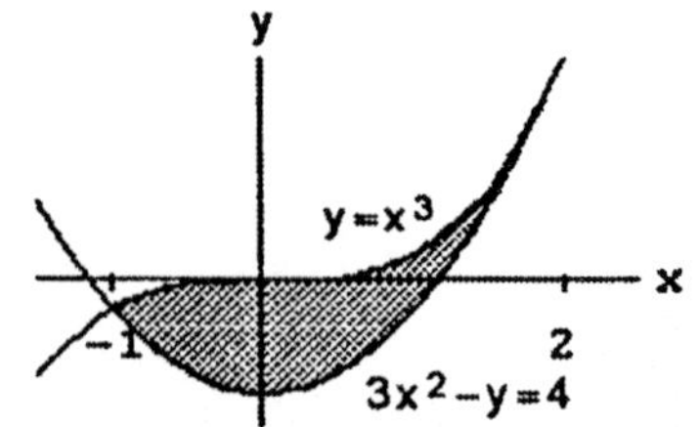

83. Limits of integration: $x = 4 - 4y^2$ and $x = 1 - y^4$
$\Rightarrow 4 - 4y^2 = 1 - y^4 \Rightarrow y^4 - 4y^2 + 3 = 0$
$\Rightarrow \left(y - \sqrt{3}\right)\left(y + \sqrt{3}\right)(y - 1)(y + 1) = 0 \Rightarrow c = -1$
and $d = 1$ since $x \geq 0$; $f(y) - g(y) = (4 - 4y^2) - (1 - y^4)$
$= 3 - 4y^2 + y^4 \Rightarrow A = \int_{-1}^{1} (3 - 4y^2 + y^4)\,dy$
$= \left[3y - \frac{4y^3}{3} + \frac{y^5}{5}\right]_{-1}^{1} = 2\left(3 - \frac{4}{3} + \frac{1}{5}\right) = \frac{56}{15}$

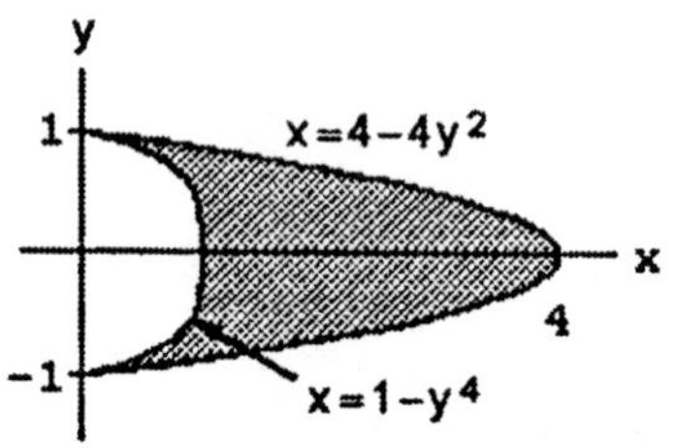

84. Limits of integration: $x = 3 - y^2$ and $x = -\frac{y^2}{4}$
$\Rightarrow 3 - y^2 = -\frac{y^2}{4} \Rightarrow \frac{3y^2}{4} - 3 = 0 \Rightarrow \frac{3}{4}(y - 2)(y + 2) = 0$
$\Rightarrow c = -2$ and $d = 2$; $f(y) - g(y) = (3 - y^2) - \left(\frac{-y^2}{4}\right)$
$= 3\left(1 - \frac{y^2}{4}\right) \Rightarrow A = 3\int_{-2}^{2} \left(1 - \frac{y^2}{4}\right) dy = 3\left[y - \frac{y^3}{12}\right]_{-2}^{2}$
$= 3\left[\left(2 - \frac{8}{12}\right) - \left(-2 + \frac{8}{12}\right)\right] = 3\left(4 - \frac{16}{12}\right) = 12 - 4 = 8$

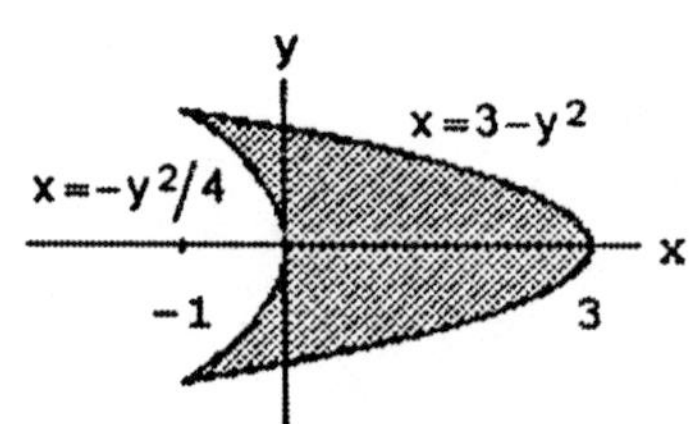

85. $a = 0$, $b = \pi$; $f(x) - g(x) = 2\sin x - \sin 2x$
$\Rightarrow A = \int_{0}^{\pi} (2\sin x - \sin 2x)\,dx = \left[-2\cos x + \frac{\cos 2x}{2}\right]_{0}^{\pi}$
$= \left[-2(-1) + \frac{1}{2}\right] - \left(-2 \cdot 1 + \frac{1}{2}\right) = 4$

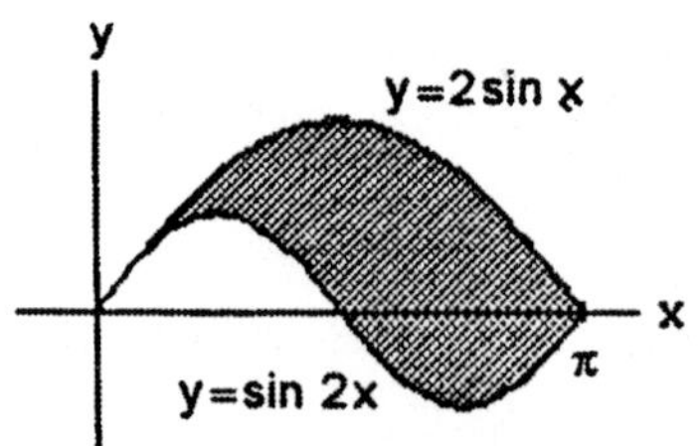

86. $a = -\frac{\pi}{3}$, $b = \frac{\pi}{3}$; $f(x) - g(x) = 8\cos x - \sec^2 x$
$\Rightarrow A = \int_{-\pi/3}^{\pi/3} (8\cos x - \sec^2 x)\,dx = [8\sin x - \tan x]_{-\pi/3}^{\pi/3}$
$= \left(8 \cdot \frac{\sqrt{3}}{2} - \sqrt{3}\right) - \left(-8 \cdot \frac{\sqrt{3}}{2} + \sqrt{3}\right) = 6\sqrt{3}$

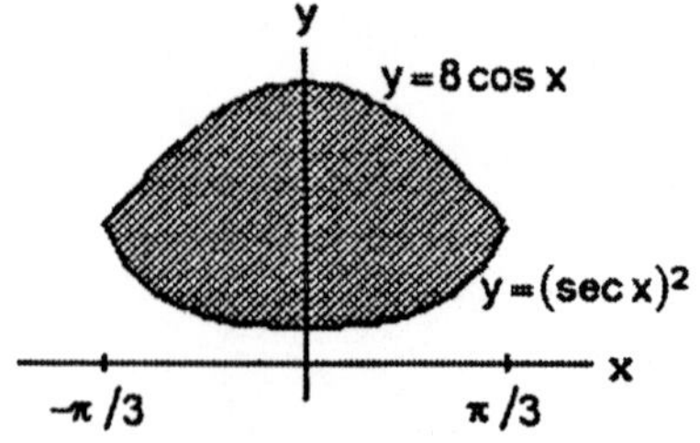

87. $a = -1$, $b = 1$; $f(x) - g(x) = (1 - x^2) - \cos\left(\frac{\pi x}{2}\right)$
$\Rightarrow A = \int_{-1}^{1} \left[1 - x^2 - \cos\left(\frac{\pi x}{2}\right)\right] dx = \left[x - \frac{x^3}{3} - \frac{2}{\pi}\sin\left(\frac{\pi x}{2}\right)\right]_{-1}^{1}$
$= \left(1 - \frac{1}{3} - \frac{2}{\pi}\right) - \left(-1 + \frac{1}{3} + \frac{2}{\pi}\right) = 2\left(\frac{2}{3} - \frac{2}{\pi}\right) = \frac{4}{3} - \frac{4}{\pi}$

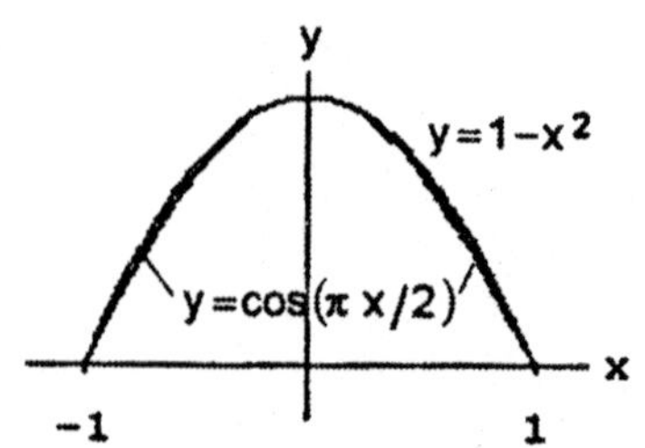

88. $A = A1 + A2$
$a_1 = -1, b_1 = 0$ and $a_2 = 0, b_2 = 1;$
$f_1(x) - g_1(x) = x - \sin\left(\frac{\pi x}{2}\right)$ and $f_2(x) - g_2(x) = \sin\left(\frac{\pi x}{2}\right) - x$
$\Rightarrow$ by symmetry about the origin,
$A_1 + A_2 = 2A_1 \Rightarrow A = 2\int_0^1 \left[\sin\left(\frac{\pi x}{2}\right) - x\right] dx$
$= 2\left[-\frac{2}{\pi}\cos\left(\frac{\pi x}{2}\right) - \frac{x^2}{2}\right]_0^1 = 2\left[\left(-\frac{2}{\pi}\cdot 0 - \frac{1}{2}\right) - \left(-\frac{2}{\pi}\cdot 1 - 0\right)\right]$
$= 2\left(\frac{2}{\pi} - \frac{1}{2}\right) = 2\left(\frac{4-\pi}{2\pi}\right) = \frac{4-\pi}{\pi}$

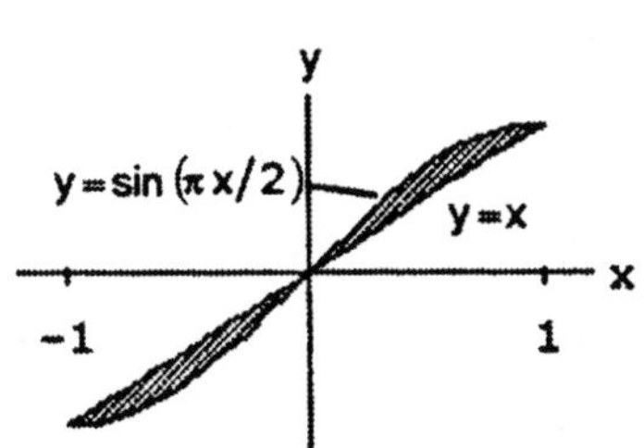

89. $a = -\frac{\pi}{4}, b = \frac{\pi}{4}; f(x) - g(x) = \sec^2 x - \tan^2 x$
$\Rightarrow A = \int_{-\pi/4}^{\pi/4} (\sec^2 x - \tan^2 x)\, dx$
$= \int_{-\pi/4}^{\pi/4} [\sec^2 x - (\sec^2 x - 1)]\, dx$
$= \int_{-\pi/4}^{\pi/4} 1 \cdot dx = [x]_{-\pi/4}^{\pi/4} = \frac{\pi}{4} - \left(-\frac{\pi}{4}\right) = \frac{\pi}{2}$

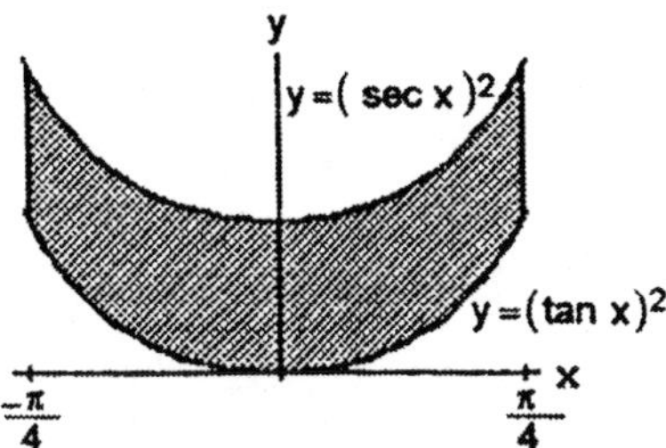

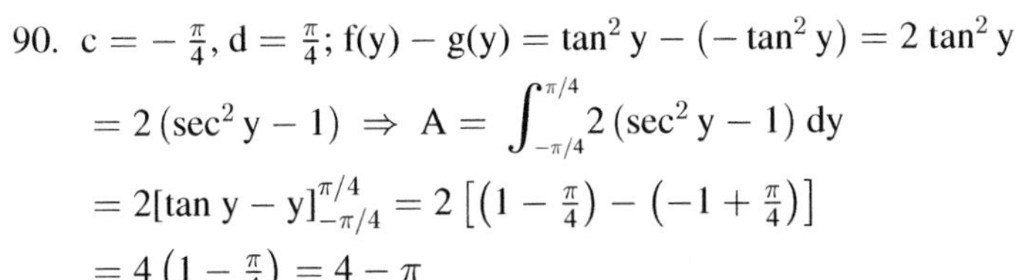

90. $c = -\frac{\pi}{4}, d = \frac{\pi}{4}; f(y) - g(y) = \tan^2 y - (-\tan^2 y) = 2\tan^2 y$
$= 2(\sec^2 y - 1) \Rightarrow A = \int_{-\pi/4}^{\pi/4} 2(\sec^2 y - 1)\, dy$
$= 2[\tan y - y]_{-\pi/4}^{\pi/4} = 2\left[\left(1 - \frac{\pi}{4}\right) - \left(-1 + \frac{\pi}{4}\right)\right]$
$= 4\left(1 - \frac{\pi}{4}\right) = 4 - \pi$

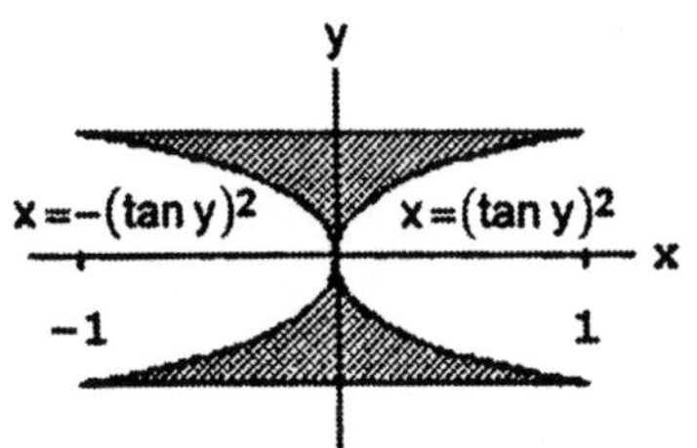

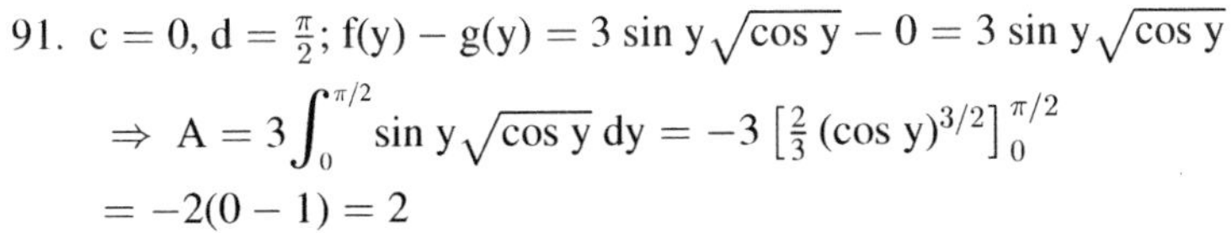

91. $c = 0, d = \frac{\pi}{2}; f(y) - g(y) = 3\sin y\sqrt{\cos y} - 0 = 3\sin y\sqrt{\cos y}$
$\Rightarrow A = 3\int_0^{\pi/2} \sin y\sqrt{\cos y}\, dy = -3\left[\frac{2}{3}(\cos y)^{3/2}\right]_0^{\pi/2}$
$= -2(0 - 1) = 2$

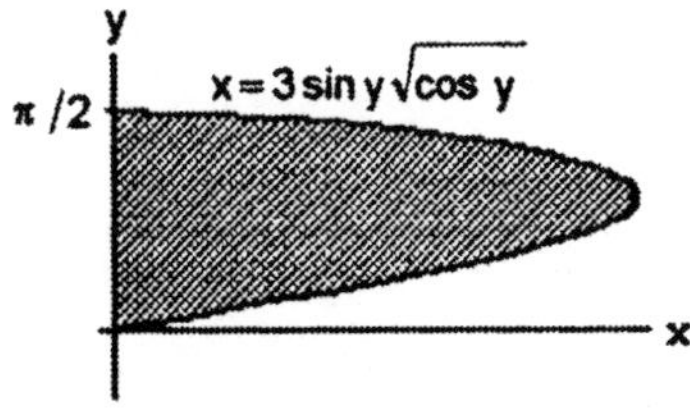

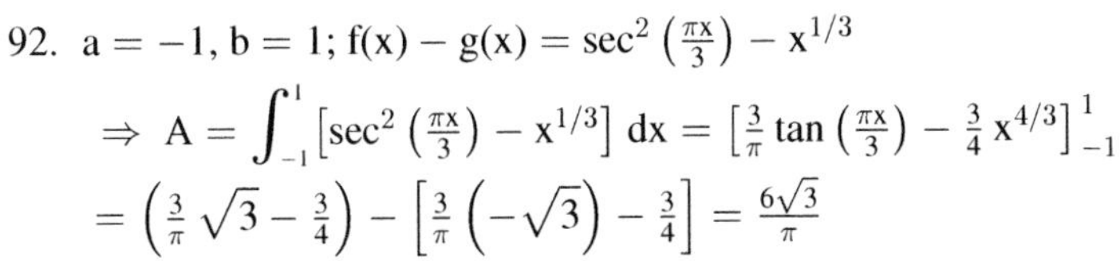

92. $a = -1, b = 1; f(x) - g(x) = \sec^2\left(\frac{\pi x}{3}\right) - x^{1/3}$
$\Rightarrow A = \int_{-1}^{1} \left[\sec^2\left(\frac{\pi x}{3}\right) - x^{1/3}\right] dx = \left[\frac{3}{\pi}\tan\left(\frac{\pi x}{3}\right) - \frac{3}{4}x^{4/3}\right]_{-1}^{1}$
$= \left(\frac{3}{\pi}\sqrt{3} - \frac{3}{4}\right) - \left[\frac{3}{\pi}\left(-\sqrt{3}\right) - \frac{3}{4}\right] = \frac{6\sqrt{3}}{\pi}$

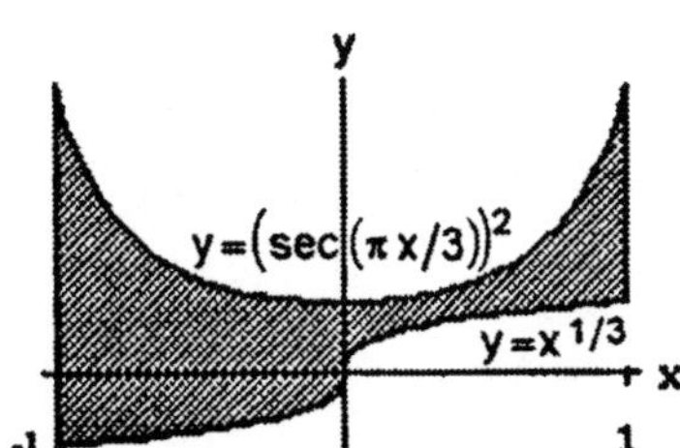

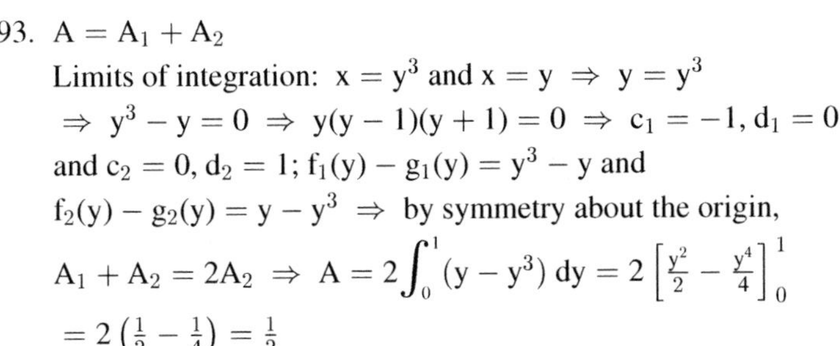

93. $A = A_1 + A_2$
Limits of integration: $x = y^3$ and $x = y \Rightarrow y = y^3$
$\Rightarrow y^3 - y = 0 \Rightarrow y(y-1)(y+1) = 0 \Rightarrow c_1 = -1, d_1 = 0$
and $c_2 = 0, d_2 = 1; f_1(y) - g_1(y) = y^3 - y$ and
$f_2(y) - g_2(y) = y - y^3 \Rightarrow$ by symmetry about the origin,
$A_1 + A_2 = 2A_2 \Rightarrow A = 2\int_0^1 (y - y^3)\, dy = 2\left[\frac{y^2}{2} - \frac{y^4}{4}\right]_0^1$
$= 2\left(\frac{1}{2} - \frac{1}{4}\right) = \frac{1}{2}$

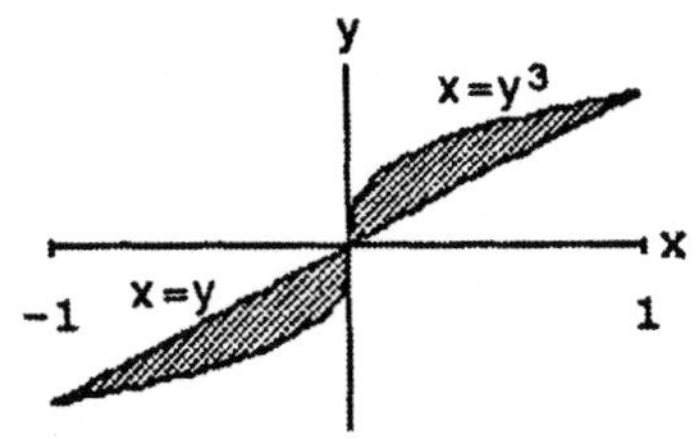

94. $A = A_1 + A_2$

Limits of integration: $y = x^3$ and $y = x^5 \Rightarrow x^3 = x^5$
$\Rightarrow x^5 - x^3 = 0 \Rightarrow x^3(x-1)(x+1) = 0 \Rightarrow a_1 = -1, b_1 = 0$
and $a_2 = 0, b_2 = 1$; $f_1(x) - g_1(x) = x^3 - x^5$ and
$f_2(x) - g_2(x) = x^5 - x^3 \Rightarrow$ by symmetry about the origin,
$A_1 + A_2 = 2A_2 \Rightarrow A = 2\int_0^1 (x^3 - x^5)\,dx = 2\left[\frac{x^4}{4} - \frac{x^6}{6}\right]_0^1$
$= 2\left(\frac{1}{4} - \frac{1}{6}\right) = \frac{1}{6}$

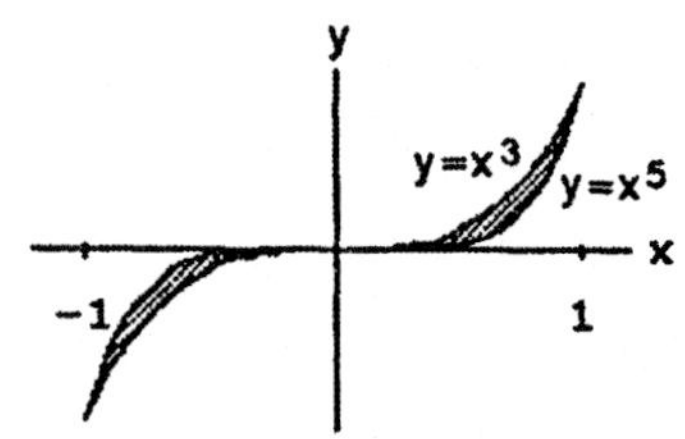

95. $A = A_1 + A_2$

Limits of integration: $y = x$ and $y = \frac{1}{x^2} \Rightarrow x = \frac{1}{x^2}, x \neq 0$
$\Rightarrow x^3 = 1 \Rightarrow x = 1$, $f_1(x) - g_1(x) = x - 0 = x$
$\Rightarrow A_1 = \int_0^1 x\,dx = \left[\frac{x^2}{2}\right]_0^1 = \frac{1}{2}$; $f_2(x) - g_2(x) = \frac{1}{x^2} - 0$
$= x^{-2} \Rightarrow A_2 = \int_1^2 x^{-2}\,dx = \left[\frac{-1}{x}\right]_1^2 = -\frac{1}{2} + 1 = \frac{1}{2}$;
$A = A_1 + A_2 = \frac{1}{2} + \frac{1}{2} = 1$

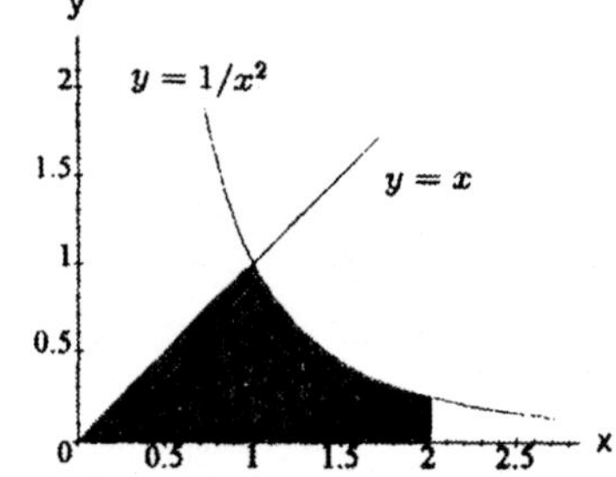

96. Limits of integration: $\sin x = \cos x \Rightarrow x = \frac{\pi}{4} \Rightarrow a = 0$
and $b = \frac{\pi}{4}$; $f(x) - g(x) = \cos x - \sin x$
$\Rightarrow A = \int_0^{\pi/4} (\cos x - \sin x)\,dx = [\sin x + \cos x]_0^{\pi/4}$
$= \left(\frac{\sqrt{2}}{2} + \frac{\sqrt{2}}{2}\right) - (0 + 1) = \sqrt{2} - 1$

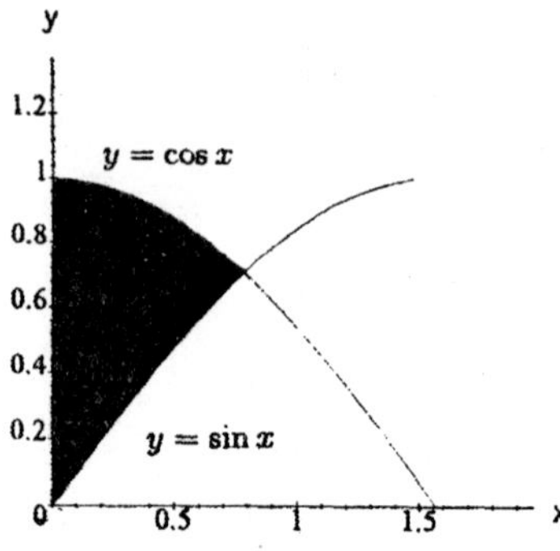

97. $\int_1^5 (\ln 2x - \ln x)\,dx = \int_1^5 (-\ln x + \ln 2 + \ln x)\,dx = (\ln 2)\int_1^5 dx = (\ln 2)(5-1) = \ln 2^4 = \ln 16$

98. $A = \int_{-\pi/4}^0 -\tan x\,dx + \int_0^{\pi/3} \tan x\,dx = \int_{-\pi/4}^0 \frac{-\sin x}{\cos x}\,dx - \int_0^{\pi/3} \frac{-\sin x}{\cos x}\,dx = [\ln|\cos x|]_{-\pi/4}^0 - [\ln|\cos x|]_0^{\pi/3}$
$= \left(\ln 1 - \ln \frac{1}{\sqrt{2}}\right) - \left(\ln \frac{1}{2} - \ln 1\right) = \ln \sqrt{2} + \ln 2 = \frac{3}{2}\ln 2$

99. $\int_0^{\ln 3} (e^{2x} - e^x)\,dx = \left[\frac{e^{2x}}{2} - e^x\right]_0^{\ln 3} = \left(\frac{e^{2\ln 3}}{2} - e^{\ln 3}\right) - \left(\frac{e^0}{2} - e^0\right) = \left(\frac{9}{2} - 3\right) - \left(\frac{1}{2} - 1\right) = \frac{8}{2} - 2 = 2$

100. $\int_0^{2\ln 2} \left(e^{x/2} - e^{-x/2}\right) dx = \left[2e^{x/2} + 2e^{-x/2}\right]_0^{2\ln 2} = \left(2e^{\ln 2} + 2e^{-\ln 2}\right) - \left(2e^0 + 2e^0\right) = (4+1) - (2+2) = 5 - 4 = 1$

101. $A = \int_{-2}^2 \frac{2x}{1+x^2}\,dx = 2\int_0^2 \frac{2x}{1+x^2}\,dx$; $[u = 1 + x^2 \Rightarrow du = 2x\,dx;\ x = 0 \Rightarrow u = 1,\ x = 2 \Rightarrow u = 5]$
$\rightarrow A = 2\int_1^5 \frac{1}{u}\,du = 2\,[\ln|u|]_1^5 = 2(\ln 5 - \ln 1) = 2\ln 5$

102. $A = \int_{-1}^1 2^{(1-x)}\,dx = 2\int_{-1}^1 \left(\frac{1}{2}\right)^x dx = 2\left[\frac{\left(\frac{1}{2}\right)^x}{\ln\left(\frac{1}{2}\right)}\right]_{-1}^1 = -\frac{2}{\ln 2}\left(\frac{1}{2} - 2\right) = \left(-\frac{2}{\ln 2}\right)\left(-\frac{3}{2}\right) = \frac{3}{\ln 2}$

103. (a) The coordinates of the points of intersection of the line and parabola are $c = x^2 \Rightarrow x = \pm\sqrt{c}$ and $y = c$

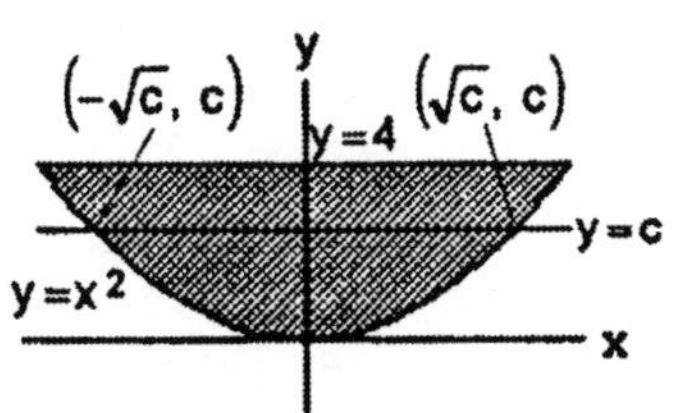

(b) $f(y) - g(y) = \sqrt{y} - (-\sqrt{y}) = 2\sqrt{y} \Rightarrow$ the area of the lower section is, $A_L = \int_0^c [f(y) - g(y)]\,dy$ $= 2\int_0^c \sqrt{y}\,dy = 2\left[\frac{2}{3}y^{3/2}\right]_0^c = \frac{4}{3}c^{3/2}$. The area of the entire shaded region can be found by setting $c = 4$: $A = \left(\frac{4}{3}\right)4^{3/2} = \frac{4\cdot 8}{3} = \frac{32}{3}$. Since we want c to divide the region into subsections of equal area we have $A = 2A_L \Rightarrow \frac{32}{3} = 2\left(\frac{4}{3}c^{3/2}\right) \Rightarrow c = 4^{2/3}$

(c) $f(x) - g(x) = c - x^2 \Rightarrow A_L = \int_{-\sqrt{c}}^{\sqrt{c}} [f(x) - g(x)]\,dx = \int_{-\sqrt{c}}^{\sqrt{c}} (c - x^2)\,dx = \left[cx - \frac{x^3}{3}\right]_{-\sqrt{c}}^{\sqrt{c}} = 2\left[c^{3/2} - \frac{c^{3/2}}{3}\right]$ $= \frac{4}{3}c^{3/2}$. Again, the area of the whole shaded region can be found by setting $c = 4 \Rightarrow A = \frac{32}{3}$. From the condition $A = 2A_L$, we get $\frac{4}{3}c^{3/2} = \frac{32}{3} \Rightarrow c = 4^{2/3}$ as in part (b).

104. (a) Limits of integration: $y = 3 - x^2$ and $y = -1$
$\Rightarrow 3 - x^2 = -1 \Rightarrow x^2 = 4 \Rightarrow a = -2$ and $b = 2$;
$f(x) - g(x) = (3 - x^2) - (-1) = 4 - x^2$
$\Rightarrow A = \int_{-1}^{2} (4 - x^2)\,dx = \left[4x - \frac{x^3}{3}\right]_{-2}^{2}$
$= \left(8 - \frac{8}{3}\right) - \left(-8 + \frac{8}{3}\right) = 16 - \frac{16}{3} = \frac{32}{3}$

(b) Limits of integration: let $x = 0$ in $y = 3 - x^2$
$\Rightarrow y = 3$; $f(y) - g(y) = \sqrt{3 - y} - \left(-\sqrt{3 - y}\right)$
$= 2(3 - y)^{1/2}$
$\Rightarrow A = 2\int_{-1}^{3} (3 - y)^{1/2}\,dy = -2\int_{-1}^{3} (3 - y)^{1/2}(-1)\,dy = (-2)\left[\frac{2(3 - y)^{3/2}}{3}\right]_{-1}^{3} = \left(-\frac{4}{3}\right)\left[0 - (3 + 1)^{3/2}\right]$
$= \left(\frac{4}{3}\right)(8) = \frac{32}{3}$

105. Limits of integration: $y = 1 + \sqrt{x}$ and $y = \frac{2}{\sqrt{x}}$
$\Rightarrow 1 + \sqrt{x} = \frac{2}{\sqrt{x}}, x \neq 0 \Rightarrow \sqrt{x} + x = 2 \Rightarrow x = (2 - x)^2$
$\Rightarrow x = 4 - 4x + x^2 \Rightarrow x^2 - 5x + 4 = 0$
$\Rightarrow (x - 4)(x - 1) = 0 \Rightarrow x = 1, 4$ (but $x = 4$ does not satisfy the equation); $y = \frac{2}{\sqrt{x}}$ and $y = \frac{x}{4} \Rightarrow \frac{2}{\sqrt{x}} = \frac{x}{4}$
$\Rightarrow 8 = x\sqrt{x} \Rightarrow 64 = x^3 \Rightarrow x = 4.$

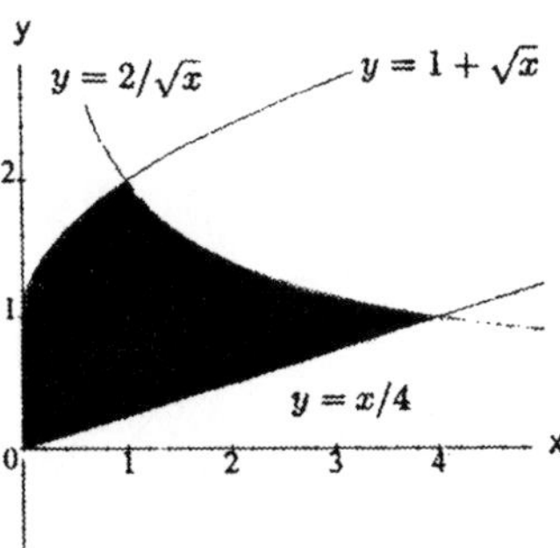

Therefore, AREA $= A_1 + A_2$: $f_1(x) - g_1(x)$
$= \left(1 + x^{1/2}\right) - \frac{x}{4} \Rightarrow A_1 = \int_0^1 \left(1 + x^{1/2} - \frac{x}{4}\right)dx$
$= \left[x + \frac{2}{3}x^{3/2} - \frac{x^2}{8}\right]_0^1$
$= \left(1 + \frac{2}{3} - \frac{1}{8}\right) - 0 = \frac{37}{24}$; $f_2(x) - g_2(x) = 2x^{-1/2} - \frac{x}{4} \Rightarrow A_2 = \int_1^4 \left(2x^{-1/2} - \frac{x}{4}\right)dx = \left[4x^{1/2} - \frac{x^2}{8}\right]_1^4$
$= \left(4\cdot 2 - \frac{16}{8}\right) - \left(4 - \frac{1}{8}\right) = 4 - \frac{15}{8} = \frac{17}{8}$; Therefore, AREA $= A_1 + A_2 = \frac{37}{24} + \frac{17}{8} = \frac{37+51}{24} = \frac{88}{24} = \frac{11}{3}$

106. Limits of integration: $(y-1)^2 = 3-y \Rightarrow y^2-2y+1 = 3-y \Rightarrow y^2-y-2=0 \Rightarrow (y-2)(y+1)=0 \Rightarrow y=2$ since $y>0$; also, $2\sqrt{y} = 3-y \Rightarrow 4y = 9-6y+y^2 \Rightarrow y^2-10y+9=0 \Rightarrow (y-9)(y-1)=0 \Rightarrow y=1$ since $y=9$ does not satisfy the equation;

AREA $= A_1 + A_2$

$f_1(y) - g_1(y) = 2\sqrt{y} - 0 = 2y^{1/2}$

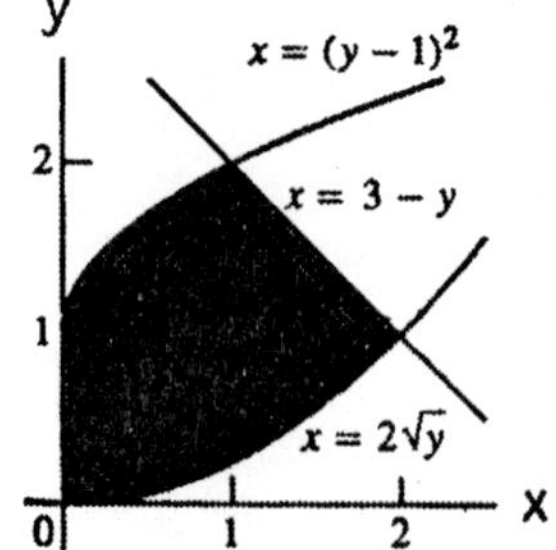

$\Rightarrow A_1 = 2\int_0^1 y^{1/2}\,dy = 2\left[\frac{2y^{3/2}}{3}\right]_0^1 = \frac{4}{3}$; $f_2(y) - g_2(y) = (3-y) - (y-1)^2$

$\Rightarrow A_2 = \int_1^2 [3-y-(y-1)^2]\,dy = \left[3y - \frac{1}{2}y^2 - \frac{1}{3}(y-1)^3\right]_1^2 = \left(6-2-\frac{1}{3}\right) - \left(3-\frac{1}{2}+0\right) = 1 - \frac{1}{3} + \frac{1}{2} = \frac{7}{6}$;

Therefore, $A_1 + A_2 = \frac{4}{3} + \frac{7}{6} = \frac{15}{6} = \frac{5}{2}$

107. Area between parabola and $y = a^2$: $A = 2\int_0^a (a^2 - x^2)\,dx = 2\left[a^2x - \frac{1}{3}x^3\right]_0^a = 2\left(a^3 - \frac{a^3}{3}\right) - 0 = \frac{4a^3}{3}$;

Area of triangle AOC: $\frac{1}{2}(2a)(a^2) = a^3$; limit of ratio $= \lim_{a\to 0^+} \frac{a^3}{\left(\frac{4a^3}{3}\right)} = \frac{3}{4}$ which is independent of a.

108. $A = \int_a^b 2f(x)\,dx - \int_a^b f(x)\,dx = 2\int_a^b f(x)\,dx - \int_a^b f(x)\,dx = \int_a^b f(x)\,dx = 4$

109. Neither one; they are both zero. Neither integral takes into account the changes in the formulas for the region's upper and lower bounding curves at $x = 0$. The area of the shaded region is actually $A = \int_{-1}^0 [-x-(x)]\,dx + \int_0^1 [x-(-x)]\,dx = \int_{-1}^0 -2x\,dx + \int_0^1 2x\,dx = 2.$

110. It is sometimes true. It is true if $f(x) \ge g(x)$ for all x between a and b. Otherwise it is false. If the graph of f lies below the graph of g for a portion of the interval of integration, the integral over that portion will be negative and the integral over [a, b] will be less than the area between the curves.

111. Let $u = 2x \Rightarrow du = 2\,dx \Rightarrow \frac{1}{2}\,du = dx$; $x = 1 \Rightarrow u = 2$, $x = 3 \Rightarrow u = 6$

$\int_1^3 \frac{\sin 2x}{x}\,dx = \int_2^6 \frac{\sin u}{\left(\frac{u}{2}\right)}\left(\frac{1}{2}\,du\right) = \int_2^6 \frac{\sin u}{u}\,du = [F(u)]_2^6 = F(6) - F(2)$

112. Let $u = 1-x \Rightarrow du = -dx \Rightarrow -du = dx$; $x = 0 \Rightarrow u = 1$, $x = 1 \Rightarrow u = 0$

$\int_0^1 f(1-x)\,dx = \int_1^0 f(u)(-du) = -\int_1^0 f(u)\,du = \int_0^1 f(u)\,du = \int_0^1 f(x)\,dx$

113. (a) Let $u = -x \Rightarrow du = -dx$; $x = -1 \Rightarrow u = 1$, $x = 0 \Rightarrow u = 0$

f odd $\Rightarrow f(-x) = -f(x)$. Then $\int_{-1}^0 f(x)\,dx = \int_1^0 f(-u)(-du) = \int_1^0 -f(u)(-du) = \int_1^0 f(u)\,du = -\int_0^1 f(u)\,du = -3$

(b) Let $u = -x \Rightarrow du = -dx$; $x = -1 \Rightarrow u = 1$, $x = 0 \Rightarrow u = 0$

f even $\Rightarrow f(-x) = f(x)$. Then $\int_{-1}^0 f(x)\,dx = \int_1^0 f(-u)(-du) = -\int_1^0 f(u)\,du = \int_0^1 f(u)\,du = 3$

114. (a) Consider $\int_{-a}^0 f(x)\,dx$ when f is odd. Let $u = -x \Rightarrow du = -dx \Rightarrow -du = dx$ and $x = -a \Rightarrow u = a$ and $x = 0 \Rightarrow u = 0$. Thus $\int_{-a}^0 f(x)\,dx = \int_a^0 -f(-u)\,du = \int_a^0 f(u)\,du = -\int_0^a f(u)\,du = -\int_0^a f(x)\,dx.$

Thus $\int_{-a}^a f(x)\,dx = \int_{-a}^0 f(x)\,dx + \int_0^a f(x)\,dx = -\int_0^a f(x)\,dx + \int_0^a f(x)\,dx = 0.$

(b) $\int_{-\pi/2}^{\pi/2} \sin x\, dx = [-\cos x]_{-\pi/2}^{\pi/2} = -\cos\left(\frac{\pi}{2}\right) + \cos\left(-\frac{\pi}{2}\right) = 0 + 0 = 0.$

115. Let $u = a - x \Rightarrow du = -dx$; $x = 0 \Rightarrow u = a$, $x = a \Rightarrow u = 0$

$$I = \int_0^a \frac{f(x)\, dx}{f(x)+f(a-x)} = \int_a^0 \frac{f(a-u)}{f(a-u)+f(u)}\,(-du) = \int_0^a \frac{f(a-u)\, du}{f(u)+f(a-u)} = \int_0^a \frac{f(a-x)\, dx}{f(x)+f(a-x)}$$

$$\Rightarrow I + I = \int_0^a \frac{f(x)\, dx}{f(x)+f(a-x)} + \int_0^a \frac{f(a-x)\, dx}{f(x)+f(a-x)} = \int_0^a \frac{f(x)+f(a-x)}{f(x)+f(a-x)}\, dx = \int_0^a dx = [x]_0^a = a - 0 = a.$$

Therefore, $2I = a \Rightarrow I = \frac{a}{2}$.

116. Let $u = \frac{xy}{t} \Rightarrow du = -\frac{xy}{t^2}\, dt \Rightarrow -\frac{t}{xy}\, du = \frac{1}{t}\, dt \Rightarrow -\frac{1}{u}\, du = \frac{1}{t}\, dt$; $t = x \Rightarrow u = y$, $t = xy \Rightarrow u = 1$. Therefore,

$$\int_x^{xy} \frac{1}{t}\, dt = \int_y^1 -\frac{1}{u}\, du = -\int_y^1 \frac{1}{u}\, du = \int_1^y \frac{1}{u}\, du = \int_1^y \frac{1}{t}\, dt$$

117. Let $u = x + c \Rightarrow du = dx$; $x = a - c \Rightarrow u = a$, $x = b - c \Rightarrow u = b$

$$\int_{a-c}^{b-c} f(x+c)\, dx = \int_a^b f(u)\, du = \int_a^b f(x)\, dx$$

118. (a)

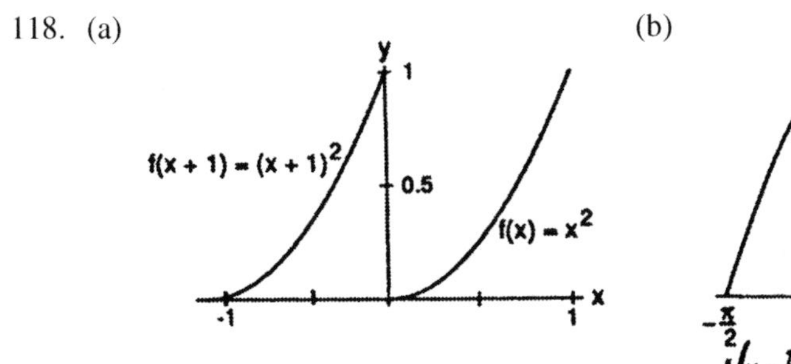

(b)

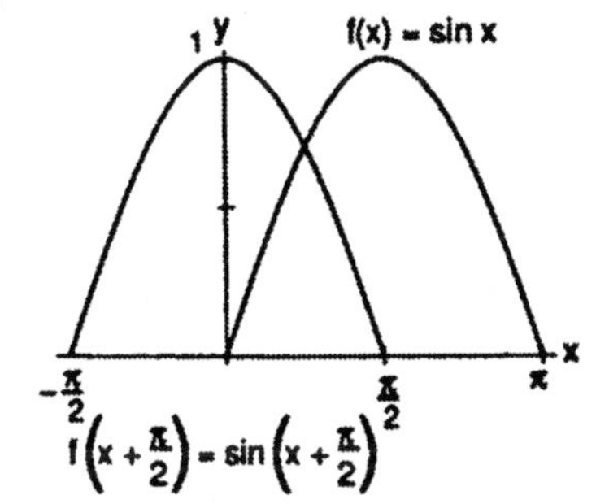

(c)

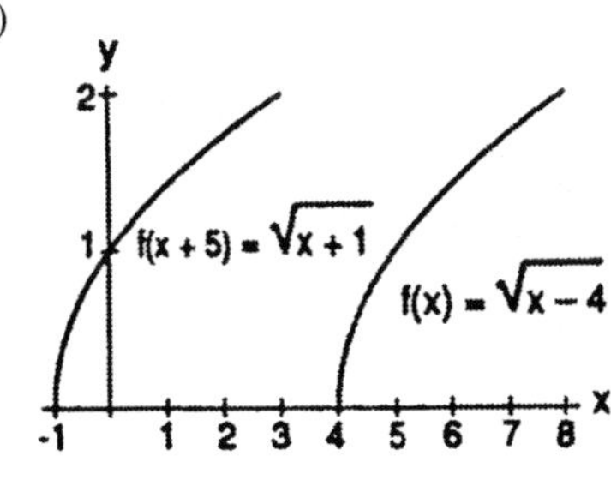

119-122.Example CAS commands:

Maple:

```
f := x -> x^3/3-x^2/2-2*x+1/3;
g := x -> x-1;
plot( [f(x),g(x)], x=-5..5, legend=["y = f(x)","y = g(x)"], title="#119(a) (Section 5.6)" );
q1 := [ -5, -2, 1, 4 ];                    # (b)
q2 := [seq( fsolve( f(x)=g(x), x=q1[i]..q1[i+1] ), i=1..nops(q1)-1 )];
for i from 1 to nops(q2)-1 do        # (c)
  area[i] := int( abs(f(x)-g(x)),x=q2[i]..q2[i+1] );
end do;
add( area[i], i=1..nops(q2)-1 );     # (d)
```

Mathematica: (assigned functions may vary)

```
Clear[x, f, g]
f[x_] = x^2 Cos[x]
g[x_] = x^3 - x
Plot[{f[x], g[x]}, {x, -2, 2}]
```

After examining the plots, the initial guesses for FindRoot can be determined.

```
pts = x/.Map[FindRoot[f[x]==g[x],{x, #}]&, {-1, 0, 1}]
i1=NIntegrate[f[x] - g[x], {x, pts[[1]], pts[[2]]}]
i2=NIntegrate[f[x] - g[x], {x, pts[[2]], pts[[3]]}]
i1 + i2
```

CHAPTER 5 PRACTICE EXERCISES

1. (a) Each time subinterval is of length $\Delta t = 0.4$ sec. The distance traveled over each subinterval, using the midpoint rule, is $\Delta h = \frac{1}{2}(v_i + v_{i+1})\Delta t$, where v_i is the velocity at the left endpoint and v_{i+1} the velocity at the right endpoint of the subinterval. We then add Δh to the height attained so far at the left endpoint v_i to arrive at the height associated with velocity v_{i+1} at the right endpoint. Using this methodology we build the following table based on the figure in the text:

t (sec)	0	0.4	0.8	1.2	1.6	2.0	2.4	2.8	3.2	3.6	4.0	4.4	4.8	5.2	5.6	6.0
v (fps)	0	10	25	55	100	190	180	165	150	140	130	115	105	90	76	65
h (ft)	0	2	9	25	56	114	188	257	320	378	432	481	525	564	592	620.2

t (sec)	6.4	6.8	7.2	7.6	8.0
v (fps)	50	37	25	12	0
h (ft)	643.2	660.6	672	679.4	681.8

NOTE: Your table values may vary slightly from ours depending on the v-values you read from the graph. Remember that some shifting of the graph occurs in the printing process.
The total height attained is about 680 ft.

(b) The graph is based on the table in part (a).

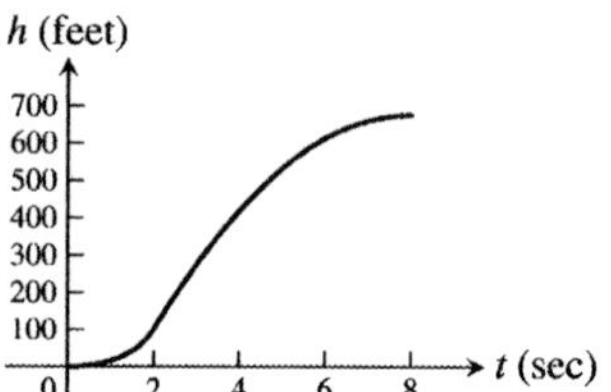

2. (a) Each time subinterval is of length $\Delta t = 1$ sec. The distance traveled over each subinterval, using the midpoint rule, is $\Delta s = \frac{1}{2}(v_i + v_{i+1})\Delta t$, where v_i is the velocity at the left, and v_{i+1} the velocity at the right, endpoint of the subinterval. We then add Δs to the distance attained so far at the left endpoint v_i to arrive at the distance associated with velocity v_{i+1} at the right endpoint. Using this methodology we build the table given below based on the figure in the text, obtaining approximately 26 m for the total distance traveled:

t (sec)	0	1	2	3	4	5	6	7	8	9	10
v (m/sec)	0	0.5	1.2	2	3.4	4.5	4.8	4.5	3.5	2	0
s (m)	0	0.25	1.1	2.7	5.4	9.35	14	18.65	22.65	25.4	26.4

(b) The graph shows the distance traveled by the moving body as a function of time for $0 \le t \le 10$.

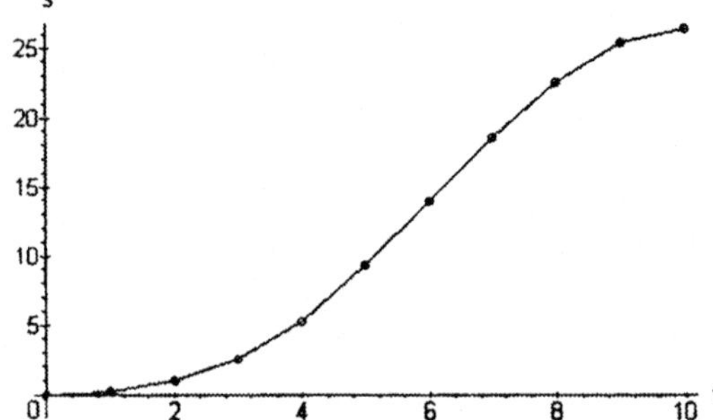

3. (a) $\sum_{k=1}^{10} \frac{a_k}{4} = \frac{1}{4}\sum_{k=1}^{10} a_k = \frac{1}{4}(-2) = -\frac{1}{2}$

(b) $\sum_{k=1}^{10} (b_k - 3a_k) = \sum_{k=1}^{10} b_k - 3\sum_{k=1}^{10} a_k = 25 - 3(-2) = 31$

(c) $\sum_{k=1}^{10} (a_k + b_k - 1) = \sum_{k=1}^{10} a_k + \sum_{k=1}^{10} b_k - \sum_{k=1}^{10} 1 = -2 + 25 - (1)(10) = 13$

(d) $\sum_{k=1}^{10} \left(\frac{5}{2} - b_k\right) = \sum_{k=1}^{10} \frac{5}{2} - \sum_{k=1}^{10} b_k = \frac{5}{2}(10) - 25 = 0$

4. (a) $\sum_{k=1}^{20} 3a_k = 3 \sum_{k=1}^{20} a_k = 3(0) = 0$ (b) $\sum_{k=1}^{20} (a_k + b_k) = \sum_{k=1}^{20} a_k + \sum_{k=1}^{20} b_k = 0 + 7 = 7$

(c) $\sum_{k=1}^{20} \left(\frac{1}{2} - \frac{2b_k}{7}\right) = \sum_{k=1}^{20} \frac{1}{2} - \frac{2}{7} \sum_{k=1}^{20} b_k = \frac{1}{2}(20) - \frac{2}{7}(7) = 8$

(d) $\sum_{k=1}^{20} (a_k - 2) = \sum_{k=1}^{20} a_k - \sum_{k=1}^{20} 2 = 0 - 2(20) = -40$

5. Let $u = 2x - 1 \Rightarrow du = 2\,dx \Rightarrow \frac{1}{2}\,du = dx$; $x = 1 \Rightarrow u = 1$, $x = 5 \Rightarrow u = 9$

$\int_1^5 (2x-1)^{-1/2}\,dx = \int_1^9 u^{-1/2}\left(\frac{1}{2}\,du\right) = \left[u^{1/2}\right]_1^9 = 3 - 1 = 2$

6. Let $u = x^2 - 1 \Rightarrow du = 2x\,dx \Rightarrow \frac{1}{2}\,du = x\,dx$; $x = 1 \Rightarrow u = 0$, $x = 3 \Rightarrow u = 8$

$\int_1^3 x\left(x^2-1\right)^{1/3} dx = \int_0^8 u^{1/3}\left(\frac{1}{2}\,du\right) = \left[\frac{3}{8}u^{4/3}\right]_0^8 = \frac{3}{8}(16 - 0) = 6$

7. Let $u = \frac{x}{2} \Rightarrow 2\,du = dx$; $x = -\pi \Rightarrow u = -\frac{\pi}{2}$, $x = 0 \Rightarrow u = 0$

$\int_{-\pi}^0 \cos\left(\frac{x}{2}\right) dx = \int_{-\pi/2}^0 (\cos u)(2\,du) = [2 \sin u]_{-\pi/2}^0 = 2\sin 0 - 2\sin\left(-\frac{\pi}{2}\right) = 2(0 - (-1)) = 2$

8. Let $u = \sin x \Rightarrow du = \cos x\,dx$; $x = 0 \Rightarrow u = 0$, $x = \frac{\pi}{2} \Rightarrow u = 1$

$\int_0^{\pi/2} (\sin x)(\cos x)\,dx = \int_0^1 u\,du = \left[\frac{u^2}{2}\right]_0^1 = \frac{1}{2}$

9. (a) $\int_{-2}^2 f(x)\,dx = \frac{1}{3}\int_{-2}^2 3\,f(x)\,dx = \frac{1}{3}(12) = 4$ (b) $\int_2^5 f(x)\,dx = \int_{-2}^5 f(x)\,dx - \int_{-2}^2 f(x)\,dx = 6 - 4 = 2$

(c) $\int_5^{-2} g(x)\,dx = -\int_{-2}^5 g(x)\,dx = -2$ (d) $\int_{-2}^5 (-\pi\,g(x))\,dx = -\pi\int_{-2}^5 g(x)\,dx = -\pi(2) = -2\pi$

(e) $\int_{-2}^5 \left(\frac{f(x)+g(x)}{5}\right) dx = \frac{1}{5}\int_{-2}^5 f(x)\,dx + \frac{1}{5}\int_{-2}^5 g(x)\,dx = \frac{1}{5}(6) + \frac{1}{5}(2) = \frac{8}{5}$

10. (a) $\int_0^2 g(x)\,dx = \frac{1}{7}\int_0^2 7\,g(x)\,dx = \frac{1}{7}(7) = 1$ (b) $\int_1^2 g(x)\,dx = \int_0^2 g(x)\,dx - \int_0^1 g(x)\,dx = 1 - 2 = -1$

(c) $\int_2^0 f(x)\,dx = -\int_0^2 f(x)\,dx = -\pi$ (d) $\int_0^2 \sqrt{2}\,f(x)\,dx = \sqrt{2}\int_0^2 f(x)\,dx = \sqrt{2}(\pi) = \pi\sqrt{2}$

(e) $\int_0^2 [g(x) - 3\,f(x)]\,dx = \int_0^2 g(x)\,dx - 3\int_0^2 f(x)\,dx = 1 - 3\pi$

11. $x^2 - 4x + 3 = 0 \Rightarrow (x-3)(x-1) = 0 \Rightarrow x = 3$ or $x = 1$;

Area $= \int_0^1 (x^2 - 4x + 3)\,dx - \int_1^3 (x^2 - 4x + 3)\,dx$

$= \left[\frac{x^3}{3} - 2x^2 + 3x\right]_0^1 - \left[\frac{x^3}{3} - 2x^2 + 3x\right]_1^3$

$= \left[\left(\frac{1^3}{3} - 2(1)^2 + 3(1)\right) - 0\right]$

$- \left[\left(\frac{3^3}{3} - 2(3)^2 + 3(3)\right) - \left(\frac{1^3}{3} - 2(1)^2 + 3(1)\right)\right]$

$= \left(\frac{1}{3} + 1\right) - \left[0 - \left(\frac{1}{3} + 1\right)\right] = \frac{8}{3}$

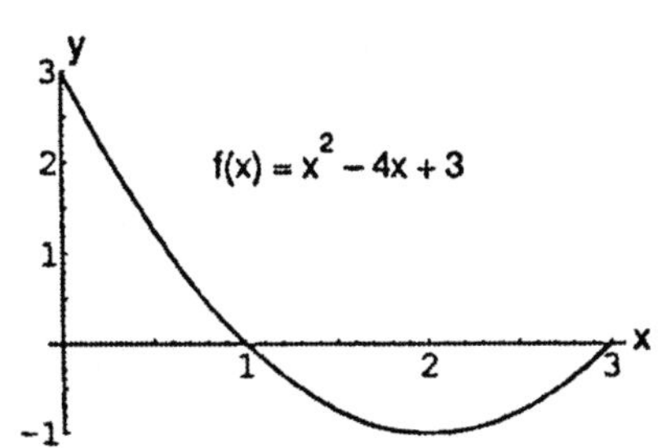

12. $1 - \frac{x^2}{4} = 0 \Rightarrow 4 - x^2 - 0 \Rightarrow x = \pm 2;$

$$\text{Area} = \int_{-2}^{2}\left(1 - \frac{x^2}{4}\right)dx - \int_{2}^{3}\left(1 - \frac{x^2}{4}\right)dx$$
$$= \left[x - \frac{x^3}{12}\right]_{-2}^{2} - \left[x - \frac{x^3}{12}\right]_{2}^{3}$$
$$= \left[\left(2 - \frac{2^3}{12}\right) - \left(-2 - \frac{(-2)^3}{12}\right)\right] - \left[\left(3 - \frac{3^3}{12}\right) - \left(2 - \frac{2^3}{12}\right)\right]$$
$$= \left[\frac{4}{3} - \left(-\frac{4}{3}\right)\right] - \left(\frac{3}{4} - \frac{4}{3}\right) = \frac{13}{4}$$

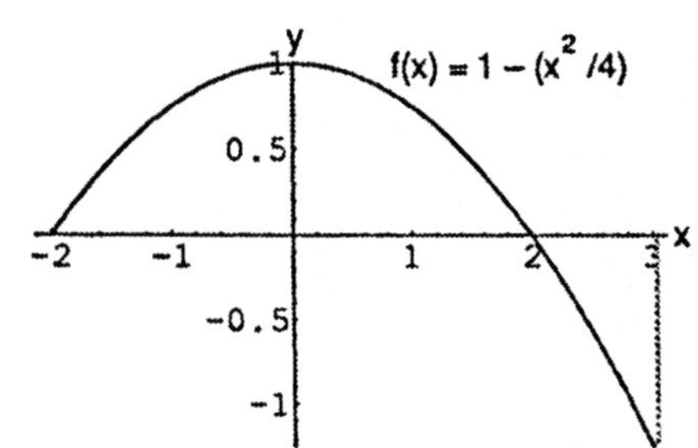

13. $5 - 5x^{2/3} = 0 \Rightarrow 1 - x^{2/3} = 0 \Rightarrow x = \pm 1;$

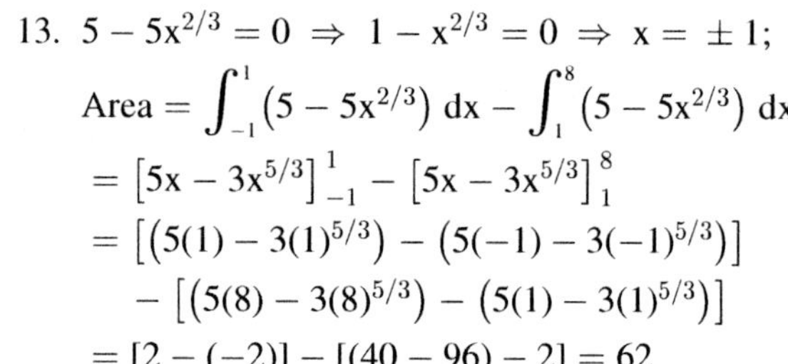

$$\text{Area} = \int_{-1}^{1}\left(5 - 5x^{2/3}\right)dx - \int_{1}^{8}\left(5 - 5x^{2/3}\right)dx$$
$$= \left[5x - 3x^{5/3}\right]_{-1}^{1} - \left[5x - 3x^{5/3}\right]_{1}^{8}$$
$$= \left[\left(5(1) - 3(1)^{5/3}\right) - \left(5(-1) - 3(-1)^{5/3}\right)\right]$$
$$- \left[\left(5(8) - 3(8)^{5/3}\right) - \left(5(1) - 3(1)^{5/3}\right)\right]$$
$$= [2 - (-2)] - [(40 - 96) - 2] = 62$$

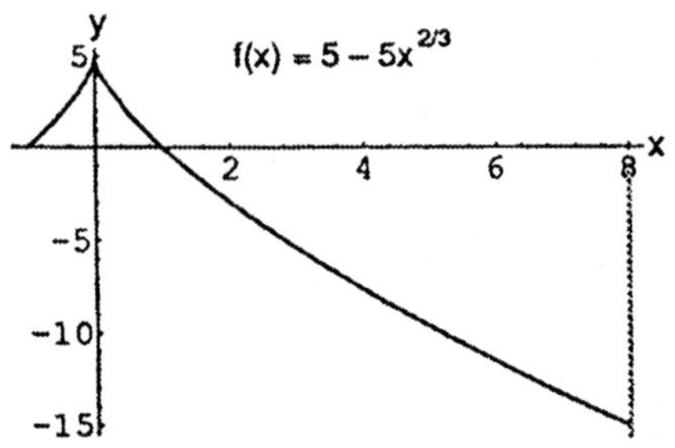

14. $1 - \sqrt{x} = 0 \Rightarrow x = 1;$

$$\text{Area} = \int_{0}^{1}\left(1 - \sqrt{x}\right)dx - \int_{1}^{4}\left(1 - \sqrt{x}\right)dx$$
$$= \left[x - \frac{2}{3}x^{3/2}\right]_{0}^{1} - \left[x - \frac{2}{3}x^{3/2}\right]_{1}^{4}$$
$$= \left[\left(1 - \frac{2}{3}(1)^{3/2}\right) - 0\right] - \left[\left(4 - \frac{2}{3}(4)^{3/2}\right) - \left(1 - \frac{2}{3}(1)^{3/2}\right)\right]$$
$$= \frac{1}{3} - \left[\left(4 - \frac{16}{3}\right) - \frac{1}{3}\right] = 2$$

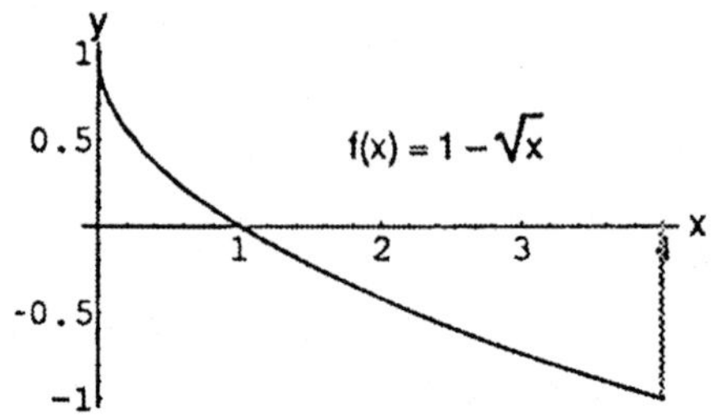

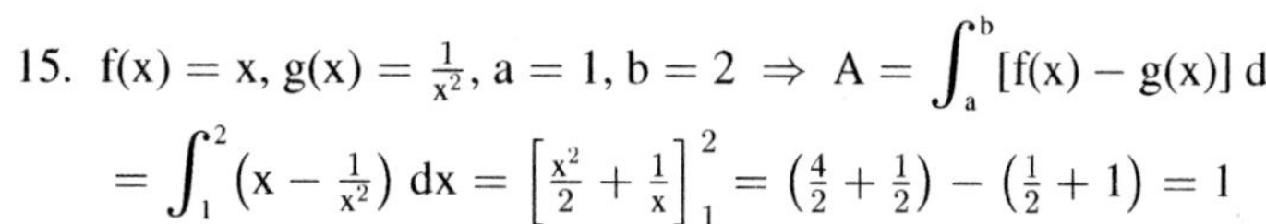

15. $f(x) = x,\ g(x) = \frac{1}{x^2},\ a = 1,\ b = 2 \Rightarrow A = \int_{a}^{b}[f(x) - g(x)]\,dx$
$$= \int_{1}^{2}\left(x - \frac{1}{x^2}\right)dx = \left[\frac{x^2}{2} + \frac{1}{x}\right]_{1}^{2} = \left(\frac{4}{2} + \frac{1}{2}\right) - \left(\frac{1}{2} + 1\right) = 1$$

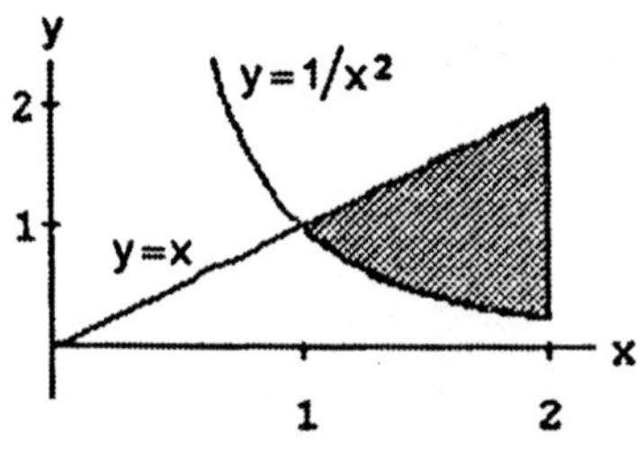

16. $f(x) = x,\ g(x) = \frac{1}{\sqrt{x}},\ a = 1,\ b = 2 \Rightarrow A = \int_{a}^{b}[f(x) - g(x)]\,dx$
$$= \int_{1}^{2}\left(x - \frac{1}{\sqrt{x}}\right)dx = \left[\frac{x^2}{2} - 2\sqrt{x}\right]_{1}^{2}$$
$$= \left(\frac{4}{2} - 2\sqrt{2}\right) - \left(\frac{1}{2} - 2\right) = \frac{7 - 4\sqrt{2}}{2}$$

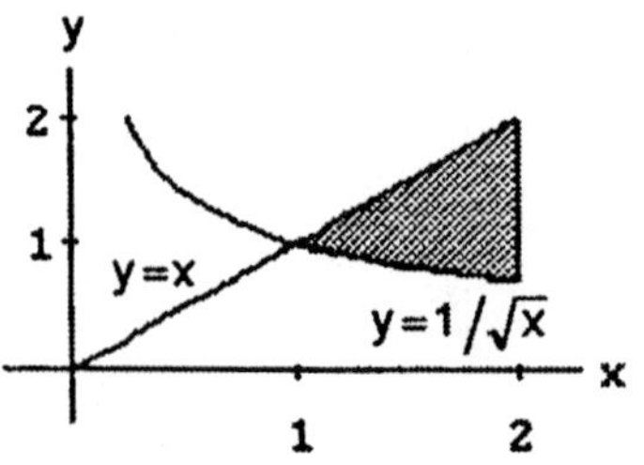

17. $f(x) = \left(1 - \sqrt{x}\right)^2,\ g(x) = 0,\ a = 0,\ b = 1 \Rightarrow A = \int_{a}^{b}[f(x) - g(x)]\,dx = \int_{0}^{1}\left(1 - \sqrt{x}\right)^2 dx = \int_{0}^{1}\left(1 - 2\sqrt{x} + x\right)dx$
$$= \int_{0}^{1}\left(1 - 2x^{1/2} + x\right)dx = \left[x - \frac{4}{3}x^{3/2} + \frac{x^2}{2}\right]_{0}^{1} = 1 - \frac{4}{3} + \frac{1}{2} = \frac{1}{6}(6 - 8 + 3) = \frac{1}{6}$$

18. $f(x) = (1 - x^3)^2$, $g(x) = 0$, $a = 0$, $b = 1 \Rightarrow A = \int_a^b [f(x) - g(x)]\,dx = \int_0^1 (1 - x^3)^2\,dx = \int_0^1 (1 - 2x^3 + x^6)\,dx$
$= \left[x - \frac{x^4}{2} + \frac{x^7}{7}\right]_0^1 = 1 - \frac{1}{2} + \frac{1}{7} = \frac{9}{14}$

19. $f(y) = 2y^2$, $g(y) = 0$, $c = 0$, $d = 3$
$\Rightarrow A = \int_c^d [f(y) - g(y)]\,dy = \int_0^3 (2y^2 - 0)\,dy$
$= 2\int_0^3 y^2\,dy = \frac{2}{3}\left[y^3\right]_0^3 = 18$

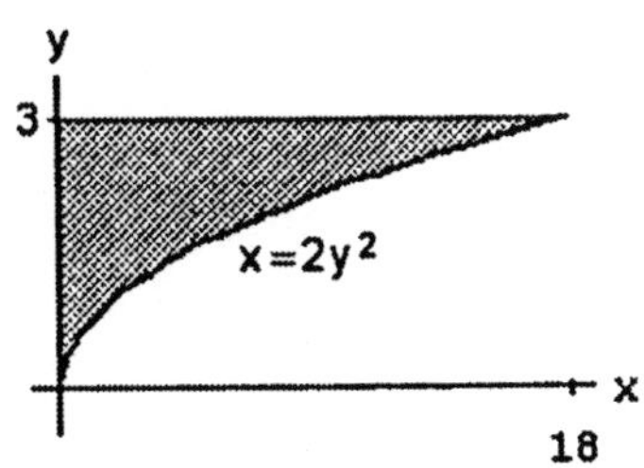

20. $f(y) = 4 - y^2$, $g(y) = 0$, $c = -2$, $d = 2$
$\Rightarrow A = \int_c^d [f(y) - g(y)]\,dy = \int_{-2}^2 (4 - y^2)\,dy$
$= \left[4y - \frac{y^3}{3}\right]_{-2}^2 = 2\left(8 - \frac{8}{3}\right) = \frac{32}{3}$

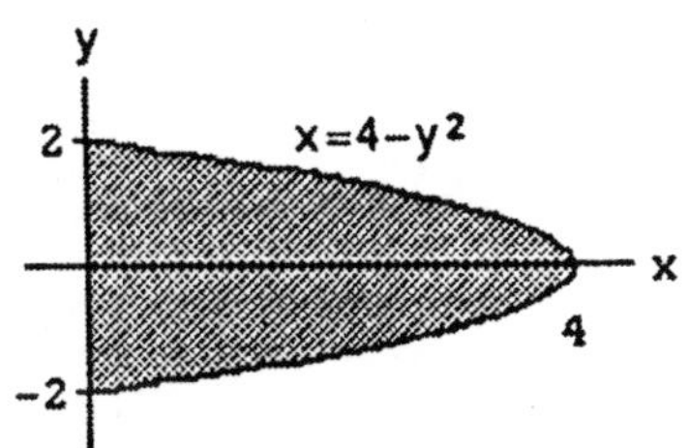

21. Let us find the intersection points: $\frac{y^2}{4} = \frac{y+2}{4}$
$\Rightarrow y^2 - y - 2 = 0 \Rightarrow (y-2)(y+1) = 0 \Rightarrow y = -1$
or $y = 2 \Rightarrow c = -1$, $d = 2$; $f(y) = \frac{y+2}{4}$, $g(y) = \frac{y^2}{4}$
$\Rightarrow A = \int_c^d [f(y) - g(y)]\,dy = \int_{-1}^2 \left(\frac{y+2}{4} - \frac{y^2}{4}\right)dy$
$= \frac{1}{4}\int_{-1}^2 (y + 2 - y^2)\,dy = \frac{1}{4}\left[\frac{y^2}{2} + 2y - \frac{y^3}{3}\right]_{-1}^2$
$= \frac{1}{4}\left[\left(\frac{4}{2} + 4 - \frac{8}{3}\right) - \left(\frac{1}{2} - 2 + \frac{1}{3}\right)\right] = \frac{9}{8}$

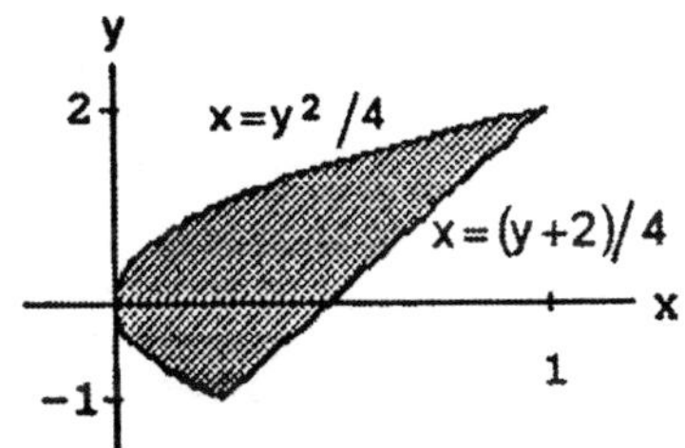

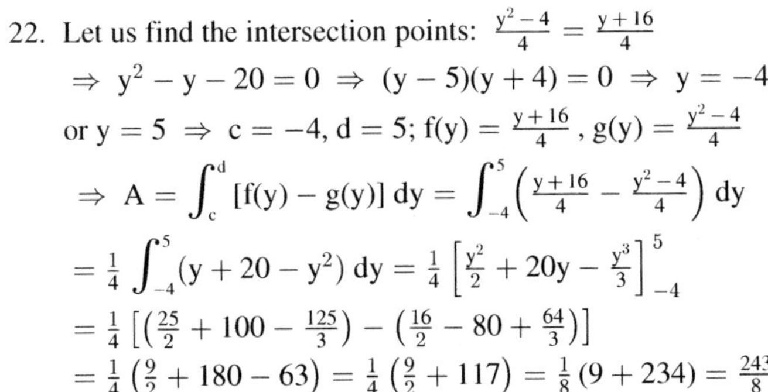

22. Let us find the intersection points: $\frac{y^2-4}{4} = \frac{y+16}{4}$
$\Rightarrow y^2 - y - 20 = 0 \Rightarrow (y-5)(y+4) = 0 \Rightarrow y = -4$
or $y = 5 \Rightarrow c = -4$, $d = 5$; $f(y) = \frac{y+16}{4}$, $g(y) = \frac{y^2-4}{4}$
$\Rightarrow A = \int_c^d [f(y) - g(y)]\,dy = \int_{-4}^5 \left(\frac{y+16}{4} - \frac{y^2-4}{4}\right)dy$
$= \frac{1}{4}\int_{-4}^5 (y + 20 - y^2)\,dy = \frac{1}{4}\left[\frac{y^2}{2} + 20y - \frac{y^3}{3}\right]_{-4}^5$
$= \frac{1}{4}\left[\left(\frac{25}{2} + 100 - \frac{125}{3}\right) - \left(\frac{16}{2} - 80 + \frac{64}{3}\right)\right]$
$= \frac{1}{4}\left(\frac{9}{2} + 180 - 63\right) = \frac{1}{4}\left(\frac{9}{2} + 117\right) = \frac{1}{8}(9 + 234) = \frac{243}{8}$

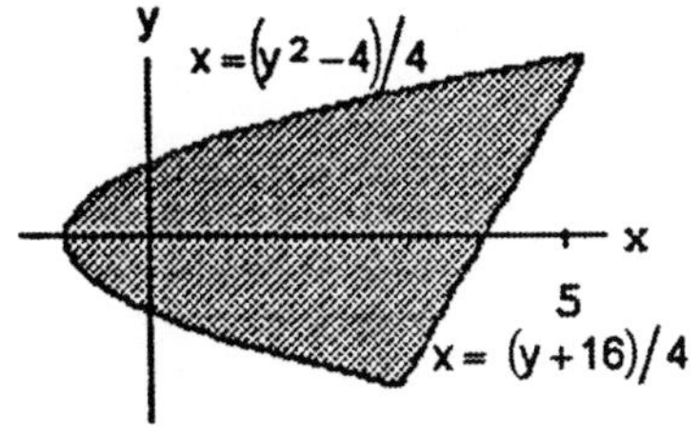

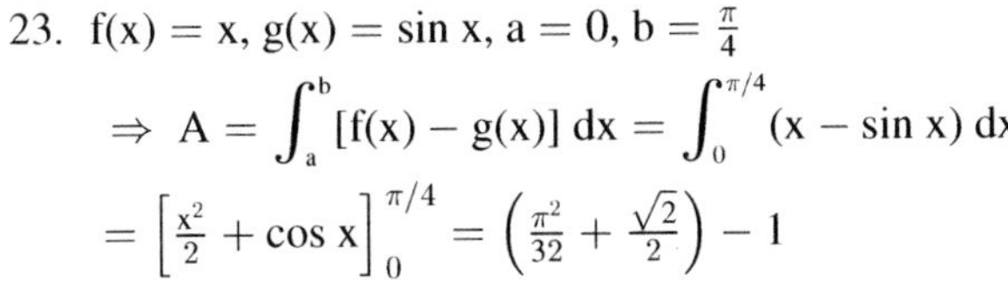

23. $f(x) = x$, $g(x) = \sin x$, $a = 0$, $b = \frac{\pi}{4}$
$\Rightarrow A = \int_a^b [f(x) - g(x)]\,dx = \int_0^{\pi/4} (x - \sin x)\,dx$
$= \left[\frac{x^2}{2} + \cos x\right]_0^{\pi/4} = \left(\frac{\pi^2}{32} + \frac{\sqrt{2}}{2}\right) - 1$

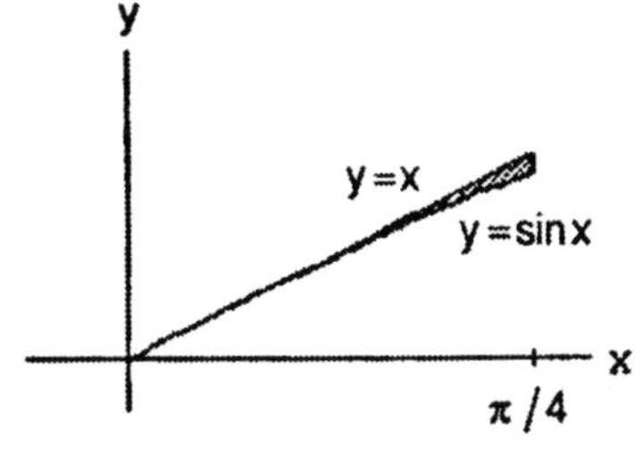

24. $f(x) = 1,\ g(x) = |\sin x|,\ a = -\frac{\pi}{2},\ b = \frac{\pi}{2}$

$\Rightarrow A = \int_a^b [f(x) - g(x)]\,dx = \int_{-\pi/2}^{\pi/2} (1 - |\sin x|)\,dx$

$= \int_{-\pi/2}^{0} (1 + \sin x)\,dx + \int_0^{\pi/2} (1 - \sin x)\,dx$

$= 2\int_0^{\pi/2} (1 - \sin x)\,dx = 2[x + \cos x]_0^{\pi/2}$

$= 2\left(\frac{\pi}{2} - 1\right) = \pi - 2$

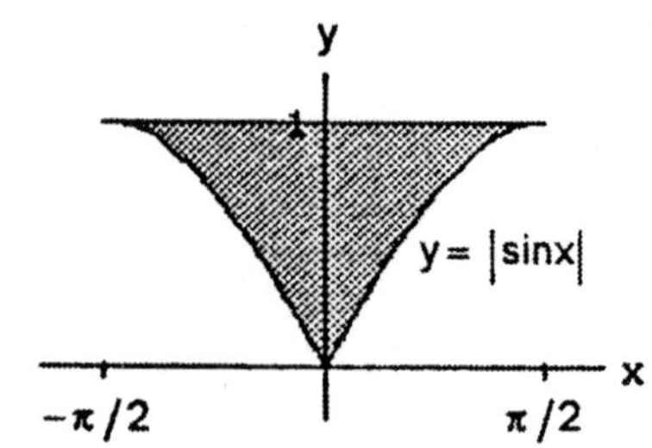

25. $a = 0,\ b = \pi,\ f(x) - g(x) = 2\sin x - \sin 2x$

$\Rightarrow A = \int_0^{\pi} (2\sin x - \sin 2x)\,dx = \left[-2\cos x + \frac{\cos 2x}{2}\right]_0^{\pi}$

$= \left[-2\cdot(-1) + \frac{1}{2}\right] - \left(-2\cdot 1 + \frac{1}{2}\right) = 4$

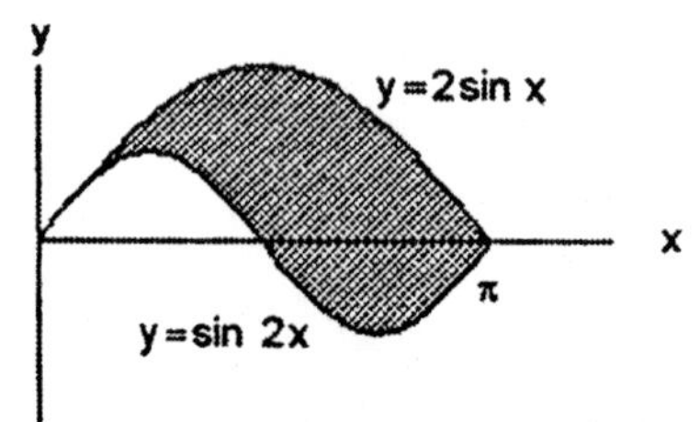

26. $a = -\frac{\pi}{3},\ b = \frac{\pi}{3},\ f(x) - g(x) = 8\cos x - \sec^2 x$

$\Rightarrow A = \int_{-\pi/3}^{\pi/3} (8\cos x - \sec^2 x)\,dx = [8\sin x - \tan x]_{-\pi/3}^{\pi/3}$

$= \left(8\cdot\frac{\sqrt{3}}{2} - \sqrt{3}\right) - \left(-8\cdot\frac{\sqrt{3}}{2} + \sqrt{3}\right) = 6\sqrt{3}$

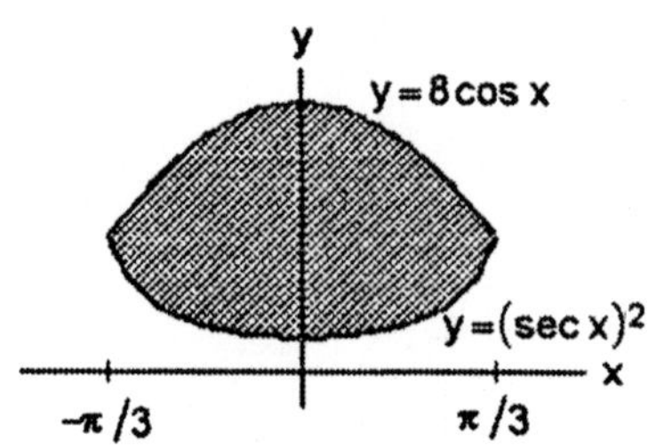

27. $f(y) = \sqrt{y},\ g(y) = 2 - y,\ c = 1,\ d = 2$

$\Rightarrow A = \int_c^d [f(y) - g(y)]\,dy = \int_1^2 \left[\sqrt{y} - (2 - y)\right]dy$

$= \int_1^2 \left(\sqrt{y} - 2 + y\right)dy = \left[\frac{2}{3}y^{3/2} - 2y + \frac{y^2}{2}\right]_1^2$

$= \left(\frac{4}{3}\sqrt{2} - 4 + 2\right) - \left(\frac{2}{3} - 2 + \frac{1}{2}\right) = \frac{4}{3}\sqrt{2} - \frac{7}{6} = \frac{8\sqrt{2}-7}{6}$

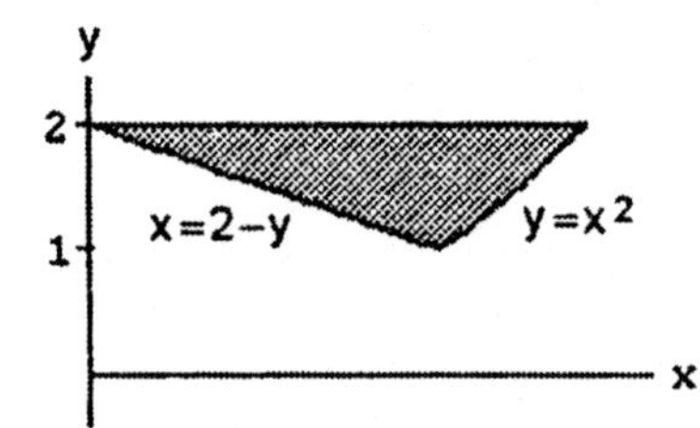

28. $f(y) = 6 - y,\ g(y) = y^2,\ c = 1,\ d = 2$

$\Rightarrow A = \int_c^d [f(y) - g(y)]\,dy = \int_1^2 (6 - y - y^2)\,dy$

$= \left[6y - \frac{y^2}{2} - \frac{y^3}{3}\right]_1^2 = \left(12 - 2 - \frac{8}{3}\right) - \left(6 - \frac{1}{2} - \frac{1}{3}\right)$

$= 4 - \frac{7}{3} + \frac{1}{2} = \frac{24 - 14 + 3}{6} = \frac{13}{6}$

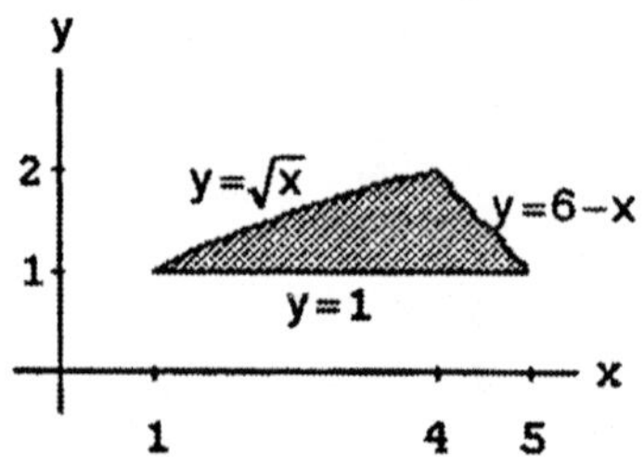

29. $f(x) = x^3 - 3x^2 = x^2(x - 3) \Rightarrow f'(x) = 3x^2 - 6x = 3x(x - 2) \Rightarrow f' = +++\underset{0}{|}----\underset{2}{|}+++$

$\Rightarrow f(0) = 0$ is a maximum and $f(2) = -4$ is a minimum. $A = -\int_0^3 (x^3 - 3x^2)\,dx = -\left[\frac{x^4}{4} - x^3\right]_0^3$

$= -\left(\frac{81}{4} - 27\right) = \frac{27}{4}$

30. $A = \int_0^a \left(a^{1/2} - x^{1/2}\right)^2 dx = \int_0^a \left(a - 2\sqrt{a}\,x^{1/2} + x\right)dx = \left[ax - \frac{4}{3}\sqrt{a}\,x^{3/2} + \frac{x^2}{2}\right]_0^a = a^2 - \frac{4}{3}\sqrt{a}\cdot a\sqrt{a} + \frac{a^2}{2}$

$= a^2\left(1 - \frac{4}{3} + \frac{1}{2}\right) = \frac{a^2}{6}(6 - 8 + 3) = \frac{a^2}{6}$

31. The area above the x-axis is $A_1 = \int_0^1 \left(y^{2/3} - y\right) dy$
$= \left[\frac{3y^{5/3}}{5} - \frac{y^2}{2}\right]_0^1 = \frac{1}{10}$; the area below the x-axis is
$A_2 = \int_{-1}^{0} \left(y^{2/3} - y\right) dy = \left[\frac{3y^{5/3}}{5} - \frac{y^2}{2}\right]_{-1}^{0} = \frac{11}{10}$
$\Rightarrow$ the total area is $A_1 + A_2 = \frac{6}{5}$

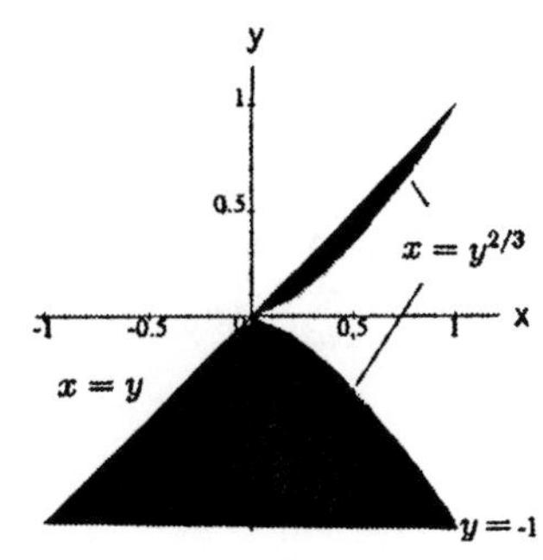

32. $A = \int_0^{\pi/4} (\cos x - \sin x)\, dx + \int_{\pi/4}^{5\pi/4} (\sin x - \cos x)\, dx$
$+ \int_{5\pi/4}^{3\pi/2} (\cos x - \sin x)\, dx = [\sin x + \cos x]_0^{\pi/4}$
$+ [-\cos x - \sin x]_{\pi/4}^{5\pi/4} + [\sin x + \cos x]_{5\pi/4}^{3\pi/2}$
$= \left[\left(\frac{\sqrt{2}}{2} + \frac{\sqrt{2}}{2}\right) - (0 + 1)\right] + \left[\left(\frac{\sqrt{2}}{2} + \frac{\sqrt{2}}{2}\right) - \left(-\frac{\sqrt{2}}{2} - \frac{\sqrt{2}}{2}\right)\right]$
$+ \left[(-1 + 0) - \left(-\frac{\sqrt{2}}{2} - \frac{\sqrt{2}}{2}\right)\right] = \frac{8\sqrt{2}}{2} - 2 = 4\sqrt{2} - 2$

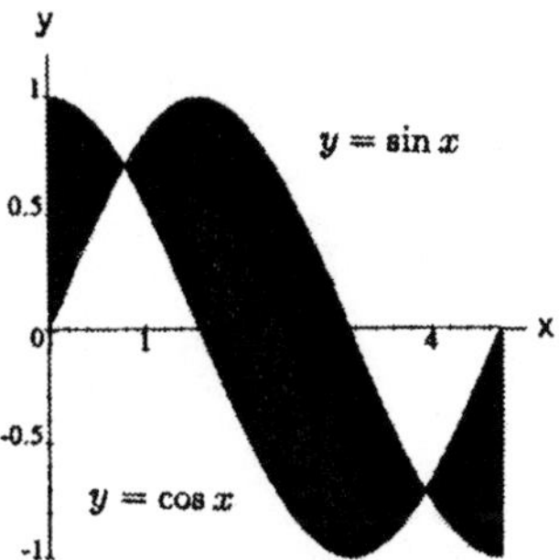

33. $A = \int_1^e \frac{2 \ln x}{x}\, dx = \int_0^1 2u\, du = \left[u^2\right]_0^1 = 1$, where
$u = \ln x$ and $du = \frac{1}{x}\, dx$; $x = 1 \Rightarrow u = 0$, $x = e \Rightarrow u = 1$

34. (a) $A_1 = \int_{10}^{20} \frac{1}{x}\, dx = [\ln |x|]_{10}^{20} = \ln 20 - \ln 10 = \ln \frac{20}{10} = \ln 2$, and $A_2 = \int_1^2 \frac{1}{x}\, dx = [\ln |x|]_1^2 = \ln 2 - \ln 1 = \ln 2$
(b) $A_1 = \int_{ka}^{kb} \frac{1}{x}\, dx = [\ln |x|]_{ka}^{kb} = \ln kb - \ln ka = \ln \frac{kb}{ka} = \ln \frac{b}{a} = \ln b - \ln a$, and $A_2 = \int_a^b \frac{1}{x}\, dx = [\ln |x|]_a^b$
$= \ln b - \ln a$

35. $y = x^2 + \int_1^x \frac{1}{t}\, dt \Rightarrow \frac{dy}{dx} = 2x + \frac{1}{x} \Rightarrow \frac{d^2y}{dx^2} = 2 - \frac{1}{x^2}$; $y(1) = 1 + \int_1^1 \frac{1}{t}\, dt = 1$ and $y'(1) = 2 + 1 = 3$

36. $y = \int_0^x \left(1 + 2\sqrt{\sec t}\right) dt \Rightarrow \frac{dy}{dx} = 1 + 2\sqrt{\sec x} \Rightarrow \frac{d^2y}{dx^2} = 2\left(\frac{1}{2}\right)(\sec x)^{-1/2}(\sec x \tan x) = \sqrt{\sec x}\,(\tan x)$;
$x = 0 \Rightarrow y = \int_0^0 \left(1 + 2\sqrt{\sec t}\right) dt = 0$ and $x = 0 \Rightarrow \frac{dy}{dx} = 1 + 2\sqrt{\sec 0} = 3$

37. $y = \int_5^x \frac{\sin t}{t}\, dt - 3 \Rightarrow \frac{dy}{dx} = \frac{\sin x}{x}$; $x = 5 \Rightarrow y = \int_5^5 \frac{\sin t}{t}\, dt - 3 = -3$

38. $y = \int_{-1}^{x} \sqrt{2 - \sin^2 t}\, dt + 2$ so that $\frac{dy}{dx} = \sqrt{2 - \sin^2 x}$; $x = -1 \Rightarrow y = \int_{-1}^{-1} \sqrt{2 - \sin^2 t}\, dt + 2 = 2$

39. $\frac{dy}{dx} = \frac{1}{\sqrt{1 - x^2}} \Rightarrow dy = \frac{dx}{\sqrt{1 - x^2}} \Rightarrow y = \sin^{-1} x + C$; $x = 0$ and $y = 0 \Rightarrow 0 = \sin^{-1} 0 + C \Rightarrow C = 0 \Rightarrow y = \sin^{-1} x$

40. $\frac{dy}{dx} = \frac{1}{x^2 + 1} - 1 \Rightarrow dy = \left(\frac{1}{1 + x^2} - 1\right) dx \Rightarrow y = \tan^{-1}(x) - x + C$; $x = 0$ and $y = 1 \Rightarrow 1 = \tan^{-1} 0 - 0 + C$
$\Rightarrow C = 1 \Rightarrow y = \tan^{-1}(x) - x + 1$

41. $\frac{dy}{dx} = \frac{1}{x\sqrt{x^2 - 1}} \Rightarrow dy = \frac{dx}{x\sqrt{x^2 - 1}} \Rightarrow y = \sec^{-1} |x| + C$; $x = 2$ and $y = \pi \Rightarrow \pi = \sec^{-1} 2 + C \Rightarrow C = \pi - \sec^{-1} 2$
$= \pi - \frac{\pi}{3} = \frac{2\pi}{3} \Rightarrow y = \sec^{-1}(x) + \frac{2\pi}{3}$, $x > 1$

42. $\frac{dy}{dx} = \frac{1}{1+x^2} - \frac{2}{\sqrt{1-x^2}} \Rightarrow dy = \left(\frac{1}{1+x^2} - \frac{2}{\sqrt{1-x^2}}\right) dx \Rightarrow y = \tan^{-1} x - 2\sin^{-1} x + C$; $x = 0$ and $y = 2$
$\Rightarrow 2 = \tan^{-1} 0 - 2\sin^{-1} 0 + C \Rightarrow C = 2 \Rightarrow y = \tan^{-1} x - 2\sin^{-1} x + 2$

43. Let $u = \cos x \Rightarrow du = -\sin x\, dx \Rightarrow -du = \sin x\, dx$
$\int 2(\cos x)^{-1/2} \sin x\, dx = \int 2u^{-1/2}(-du) = -2\int u^{-1/2}\, du = -2\left(\frac{u^{1/2}}{\frac{1}{2}}\right) + C = -4u^{1/2} + C$
$= -4(\cos x)^{1/2} + C$

44. Let $u = \tan x \Rightarrow du = \sec^2 x\, dx$
$\int (\tan x)^{-3/2} \sec^2 x\, dx = \int u^{-3/2}\, du = \frac{u^{-1/2}}{\left(-\frac{1}{2}\right)} + C = -2u^{-1/2} + C = \frac{-2}{(\tan x)^{1/2}} + C$

45. Let $u = 2\theta + 1 \Rightarrow du = 2\, d\theta \Rightarrow \frac{1}{2}\, du = d\theta$
$\int [2\theta + 1 + 2\cos(2\theta + 1)]\, d\theta = \int (u + 2\cos u)\left(\frac{1}{2}\, du\right) = \frac{u^2}{4} + \sin u + C_1 = \frac{(2\theta+1)^2}{4} + \sin(2\theta + 1) + C_1$
$= \theta^2 + \theta + \sin(2\theta + 1) + C$, where $C = C_1 + \frac{1}{4}$ is still an arbitrary constant

46. Let $u = 2\theta - \pi \Rightarrow du = 2\, d\theta \Rightarrow \frac{1}{2}\, du = d\theta$
$\int \left(\frac{1}{\sqrt{2\theta - \pi}} + 2\sec^2(2\theta - \pi)\right) d\theta = \int \left(\frac{1}{\sqrt{u}} + 2\sec^2 u\right)\left(\frac{1}{2}\, du\right) = \frac{1}{2}\int \left(u^{-1/2} + 2\sec^2 u\right) du$
$= \frac{1}{2}\left(\frac{u^{1/2}}{\frac{1}{2}}\right) + \frac{1}{2}(2\tan u) + C = u^{1/2} + \tan u + C = (2\theta - \pi)^{1/2} + \tan(2\theta - \pi) + C$

47. $\int \left(t - \frac{2}{t}\right)\left(t + \frac{2}{t}\right) dt = \int \left(t^2 - \frac{4}{t^2}\right) dt = \int \left(t^2 - 4t^{-2}\right) dt = \frac{t^3}{3} - 4\left(\frac{t^{-1}}{-1}\right) + C = \frac{t^3}{3} + \frac{4}{t} + C$

48. $\int \frac{(t+1)^2 - 1}{t^4}\, dt = \int \frac{t^2 + 2t}{t^4}\, dt = \int \left(\frac{1}{t^2} + \frac{2}{t^3}\right) dt = \int \left(t^{-2} + 2t^{-3}\right) dt = \frac{t^{-1}}{(-1)} + 2\left(\frac{t^{-2}}{-2}\right) + C = -\frac{1}{t} - \frac{1}{t^2} + C$

49. Let $u = 2t^{3/2} \Rightarrow du = 3\sqrt{t}\, dt \Rightarrow \frac{1}{3} du = \sqrt{t}\, dt$
$\int \sqrt{t} \sin\left(2t^{3/2}\right) dt = \frac{1}{3}\int \sin u\, du = -\frac{1}{3}\cos u + C = -\frac{1}{3}\cos\left(2t^{3/2}\right) + C$

50. Let $u = 1 + \sec\theta \Rightarrow du = \sec\theta \tan\theta\, d\theta \Rightarrow \int \sec\theta \tan\theta \sqrt{1 + \sec\theta}\, d\theta = \int u^{1/2} du = \frac{2}{3}u^{3/2} + C$
$= \frac{2}{3}(1 + \sec\theta)^{3/2} + C$

51. $\int e^x \sec^2(e^x - 7)\, dx = \int \sec^2 u\, du$, where $u = e^x - 7$ and $du = e^x\, dx$
$= \tan u + C = \tan(e^x - 7) + C$

52. $\int e^y \csc(e^y + 1)\cot(e^y + 1)\, dy = \int \csc u \cot u\, du$, where $u = e^y + 1$ and $du = e^y\, dy$
$= -\csc u + C = -\csc(e^y + 1) + C$

53. $\int (\sec^2 x)\, e^{\tan x}\, dx = \int e^u\, du$, where $u = \tan x$ and $du = \sec^2 x\, dx$
$= e^u + C = e^{\tan x} + C$

54. $\int (\csc^2 x)\, e^{\cot x}\, dx = -\int e^u\, du$, where $u = \cot x$ and $du = -\csc^2 x\, dx$
$= -e^u + C = -e^{\cot x} + C$

55. $\int_{-1}^{1} \frac{1}{3x-4}\, dx = \frac{1}{3}\int_{-7}^{-1} \frac{1}{u}\, du$, where $u = 3x - 4$, $du = 3\, dx$; $x = -1 \Rightarrow u = -7$, $x = 1 \Rightarrow u = -1$
$= \frac{1}{3}\left[\ln|u|\right]_{-7}^{-1} = \frac{1}{3}\left[\ln|-1| - \ln|-7|\right] = \frac{1}{3}[0 - \ln 7] = -\frac{\ln 7}{3}$

56. $\int_{1}^{e} \frac{\sqrt{\ln x}}{x}\, dx = \int_{0}^{1} u^{1/2}\, du$, where $u = \ln x$, $du = \frac{1}{x}\, dx$; $x = 1 \Rightarrow u = 0$, $x = e \Rightarrow u = 1$
$= \left[\frac{2}{3} u^{3/2}\right]_{0}^{1} = \left[\frac{2}{3} 1^{3/2} - \frac{2}{3} 0^{3/2}\right] = \frac{2}{3}$

57. $\int_{0}^{4} \frac{2t}{t^2-25}\, dt = \int_{-25}^{-9} \frac{1}{u}\, du$, where $u = t^2 - 25$, $du = 2t\, dt$; $t = 0 \Rightarrow u = -25$, $t = 4 \Rightarrow u = -9$
$= \left[\ln|u|\right]_{-25}^{-9} = \ln|-9| - \ln|-25| = \ln 9 - \ln 25 = \ln \frac{9}{25}$

58. $\int \frac{\tan(\ln v)}{v}\, dv = \int \tan u\, du = \int \frac{\sin u}{\cos u}\, du$, where $u = \ln v$ and $du = \frac{1}{v}\, dv$
$= -\ln|\cos u| + C = -\ln|\cos(\ln v)| + C$

59. $\int \frac{(\ln x)^{-3}}{x}\, dx = \int u^{-3}\, du$, where $u = \ln x$ and $du = \frac{1}{x}\, dx$
$= \frac{u^{-2}}{-2} + C = -\frac{1}{2}(\ln x)^{-2} + C$

60. $\int \frac{1}{r} \csc^2(1 + \ln r)\, dr = \int \csc^2 u\, du$, where $u = 1 + \ln r$ and $du = \frac{1}{r}\, dr$
$= -\cot u + C = -\cot(1 + \ln r) + C$

61. $\int x3^{x^2}\, dx = \frac{1}{2}\int 3^u\, du$, where $u = x^2$ and $du = 2x\, dx$
$= \frac{1}{2\ln 3}(3^u) + C = \frac{1}{2\ln 3}\left(3^{x^2}\right) + C$

62. $\int 2^{\tan x} \sec^2 x\, dx = \int 2^u\, du$, where $u = \tan x$ and $du = \sec^2 x\, dx$
$= \frac{1}{\ln 2}(2^u) + C = \frac{2^{\tan x}}{\ln 2} + C$

63. $\int \frac{3\, dr}{\sqrt{1-4(r-1)^2}} = \frac{3}{2}\int \frac{du}{\sqrt{1-u^2}}$, where $u = 2(r - 1)$ and $du = 2\, dr$
$= \frac{3}{2}\sin^{-1} u + C = \frac{3}{2}\sin^{-1} 2(r - 1) + C$

64. $\int \frac{6\, dr}{\sqrt{4-(r+1)^2}} = 6\int \frac{du}{\sqrt{4-u^2}}$, where $u = r + 1$ and $du = dr$
$= 6\sin^{-1}\frac{u}{2} + C = 6\sin^{-1}\left(\frac{r+1}{2}\right) + C$

65. $\int \frac{dx}{2+(x-1)^2} = \int \frac{du}{2+u^2}$, where $u = x - 1$ and $du = dx$
$= \frac{1}{\sqrt{2}}\tan^{-1}\frac{u}{\sqrt{2}} + C = \frac{1}{\sqrt{2}}\tan^{-1}\left(\frac{x-1}{\sqrt{2}}\right) + C$

66. $\int \frac{dx}{1+(3x+1)^2} = \frac{1}{3}\int \frac{du}{1+u^2}$, where $u = 3x + 1$ and $du = 3\, dx$
$= \frac{1}{3}\tan^{-1} u + C = \frac{1}{3}\tan^{-1}(3x + 1) + C$

67. $\int \frac{dx}{(2x-1)\sqrt{(2x-1)^2-4}} = \frac{1}{2}\int \frac{du}{u\sqrt{u^2-4}}$, where $u = 2x - 1$ and $du = 2\, dx$
$= \frac{1}{2}\cdot\frac{1}{2}\sec^{-1}\left|\frac{u}{2}\right| + C = \frac{1}{4}\sec^{-1}\left|\frac{2x-1}{2}\right| + C$

68. $\int \frac{dx}{(x+3)\sqrt{(x+3)^2-25}} = \int \frac{du}{u\sqrt{u^2-25}}$, where $u = x+3$ and $du = dx$
$= \frac{1}{5}\sec^{-1}\left|\frac{u}{5}\right| + C = \frac{1}{5}\sec^{-1}\left|\frac{x+3}{5}\right| + C$

69. $\int \frac{e^{\sin^{-1}\sqrt{x}}}{2\sqrt{x-x^2}}\,dx = \int e^u\,du$, where $u = \sin^{-1}\sqrt{x}$ and $du = \frac{dx}{2\sqrt{x-x^2}}$
$= e^u + C = e^{\sin^{-1}\sqrt{x}} + C$

70. $\int \frac{\sqrt{\sin^{-1}x}}{\sqrt{1-x^2}}\,dx = \int u^{1/2}\,du$, where $u = \sin^{-1}x$ and $du = \frac{dx}{\sqrt{1-x^2}}$
$= \frac{2}{3}u^{3/2} + C = \frac{2}{3}\left(\sin^{-1}x\right)^{3/2} + C$

71. $\int \frac{1}{\sqrt{\tan^{-1}y}\,(1+y^2)}\,dy = \int \frac{\left(\frac{1}{1+y^2}\right)}{\sqrt{\tan^{-1}y}}\,dy = \int u^{-1/2}\,du$, where $u = \tan^{-1}y$ and $du = \frac{dy}{1+y^2}$
$= 2\,u^{1/2} + C = 2\sqrt{\tan^{-1}y} + C$

72. $\int \frac{(\tan^{-1}x)^2}{1+x^2}\,dx = \int u^2\,du$, where $u = \tan^{-1}x$ and $du = \frac{dx}{1+x^2}$
$= \frac{1}{3}u^3 + C = \frac{1}{3}\left(\tan^{-1}x\right)^3 + C$

73. $\int_{-1}^{1}(3x^2-4x+7)\,dx = \left[x^3-2x^2+7x\right]_{-1}^{1} = \left[1^3-2(1)^2+7(1)\right] - \left[(-1)^3-2(-1)^2+7(-1)\right] = 6-(-10) = 16$

74. $\int_0^1 (8s^3-12s^2+5)\,ds = \left[2s^4-4s^3+5s\right]_0^1 = \left[2(1)^4-4(1)^3+5(1)\right] - 0 = 3$

75. $\int_1^2 \frac{4}{v^2}\,dv = \int_1^2 4v^{-2}\,dv = \left[-4v^{-1}\right]_1^2 = \left(\frac{-4}{2}\right) - \left(\frac{-4}{1}\right) = 2$

76. $\int_1^{27} x^{-4/3}\,dx = \left[-3x^{-1/3}\right]_1^{27} = -3(27)^{-1/3} - \left(-3(1)^{-1/3}\right) = -3\left(\frac{1}{3}\right) + 3(1) = 2$

77. $\int_1^4 \frac{dt}{t\sqrt{t}} = \int_1^4 \frac{dt}{t^{3/2}} = \int_1^4 t^{-3/2}\,dt = \left[-2t^{-1/2}\right]_1^4 = \frac{-2}{\sqrt{4}} - \frac{(-2)}{\sqrt{1}} = 1$

78. Let $x = 1+\sqrt{u} \Rightarrow dx = \frac{1}{2}u^{-1/2}\,du \Rightarrow 2\,dx = \frac{du}{\sqrt{u}}$; $u = 1 \Rightarrow x = 2$, $u = 4 \Rightarrow x = 3$
$\int_1^4 \frac{\left(1+\sqrt{u}\right)^{1/2}}{\sqrt{u}}\,du = \int_2^3 x^{1/2}(2\,dx) = \left[2\left(\frac{2}{3}\right)x^{3/2}\right]_2^3 = \frac{4}{3}\left(3^{3/2}\right) - \frac{4}{3}\left(2^{3/2}\right) = 4\sqrt{3} - \frac{8}{3}\sqrt{2} = \frac{4}{3}\left(3\sqrt{3}-2\sqrt{2}\right)$

79. Let $u = 2x+1 \Rightarrow du = 2\,dx \Rightarrow 18\,du = 36\,dx$; $x = 0 \Rightarrow u = 1$, $x = 1 \Rightarrow u = 3$
$\int_0^1 \frac{36\,dx}{(2x+1)^3} = \int_1^3 18u^{-3}\,du = \left[\frac{18u^{-2}}{-2}\right]_1^3 = \left[\frac{-9}{u^2}\right]_1^3 = \left(\frac{-9}{3^2}\right) - \left(\frac{-9}{1^2}\right) = 8$

80. Let $u = 7-5r \Rightarrow du = -5\,dr \Rightarrow -\frac{1}{5}\,du = dr$; $r = 0 \Rightarrow u = 7$, $r = 1 \Rightarrow u = 2$
$\int_0^1 \frac{dr}{\sqrt[3]{(7-5r)^2}} = \int_0^1 (7-5r)^{-2/3}\,dr = \int_7^2 u^{-2/3}\left(-\frac{1}{5}\,du\right) = -\frac{1}{5}\left[3u^{1/3}\right]_7^2 = \frac{3}{5}\left(\sqrt[3]{7}-\sqrt[3]{2}\right)$

81. Let $u = 1-x^{2/3} \Rightarrow du = -\frac{2}{3}x^{-1/3}\,dx \Rightarrow -\frac{3}{2}\,du = x^{-1/3}\,dx$; $x = \frac{1}{8} \Rightarrow u = 1-\left(\frac{1}{8}\right)^{2/3} = \frac{3}{4}$,
$x = 1 \Rightarrow u = 1-1^{2/3} = 0$
$\int_{1/8}^{1} x^{-1/3}\left(1-x^{2/3}\right)^{3/2}\,dx = \int_{3/4}^{0} u^{3/2}\left(-\frac{3}{2}\,du\right) = \left[\left(-\frac{3}{2}\right)\left(\frac{u^{5/2}}{\frac{5}{2}}\right)\right]_{3/4}^{0} = \left[-\frac{3}{5}u^{5/2}\right]_{3/4}^{0} = -\frac{3}{5}(0)^{5/2} - \left(-\frac{3}{5}\right)\left(\frac{3}{4}\right)^{5/2}$
$= \frac{27\sqrt{3}}{160}$

82. Let $u = 1 + 9x^4 \Rightarrow du = 36x^3\,dx \Rightarrow \frac{1}{36}\,du = x^3\,dx$; $x = 0 \Rightarrow u = 1$, $x = \frac{1}{2} \Rightarrow u = 1 + 9\left(\frac{1}{2}\right)^4 = \frac{25}{16}$

$$\int_0^{1/2} x^3\left(1+9x^4\right)^{-3/2} dx = \int_1^{25/16} u^{-3/2}\left(\tfrac{1}{36}\,du\right) = \left[\tfrac{1}{36}\left(\tfrac{u^{-1/2}}{-\frac{1}{2}}\right)\right]_1^{25/16} = \left[-\tfrac{1}{18}u^{-1/2}\right]_1^{25/16}$$
$$= -\tfrac{1}{18}\left(\tfrac{25}{16}\right)^{-1/2} - \left(-\tfrac{1}{18}(1)^{-1/2}\right) = \tfrac{1}{90}$$

83. Let $u = 5r \Rightarrow du = 5\,dr \Rightarrow \frac{1}{5}\,du = dr$; $r = 0 \Rightarrow u = 0$, $r = \pi \Rightarrow u = 5\pi$

$$\int_0^{\pi} \sin^2 5r\,dr = \int_0^{5\pi} \left(\sin^2 u\right)\left(\tfrac{1}{5}\,du\right) = \tfrac{1}{5}\left[\tfrac{u}{2} - \tfrac{\sin 2u}{4}\right]_0^{5\pi} = \left(\tfrac{\pi}{2} - \tfrac{\sin 10\pi}{20}\right) - \left(0 - \tfrac{\sin 0}{20}\right) = \tfrac{\pi}{2}$$

84. Let $u = 4t - \frac{\pi}{4} \Rightarrow du = 4\,dt \Rightarrow \frac{1}{4}\,du = dt$; $t = 0 \Rightarrow u = -\frac{\pi}{4}$, $t = \frac{\pi}{4} \Rightarrow u = \frac{3\pi}{4}$

$$\int_0^{\pi/4} \cos^2\left(4t - \tfrac{\pi}{4}\right) dt = \int_{-\pi/4}^{3\pi/4} \left(\cos^2 u\right)\left(\tfrac{1}{4}\,du\right) = \tfrac{1}{4}\left[\tfrac{u}{2} + \tfrac{\sin 2u}{4}\right]_{-\pi/4}^{3\pi/4} = \tfrac{1}{4}\left(\tfrac{3\pi}{8} + \tfrac{\sin\left(\frac{3\pi}{2}\right)}{4}\right) - \tfrac{1}{4}\left(-\tfrac{\pi}{8} + \tfrac{\sin\left(-\frac{\pi}{2}\right)}{4}\right)$$
$$= \tfrac{\pi}{8} - \tfrac{1}{16} + \tfrac{1}{16} = \tfrac{\pi}{8}$$

85. $$\int_0^{\pi/3} \sec^2\theta\,d\theta = [\tan\theta]_0^{\pi/3} = \tan\tfrac{\pi}{3} - \tan 0 = \sqrt{3}$$

86. $$\int_{\pi/4}^{3\pi/4} \csc^2 x\,dx = [-\cot x]_{\pi/4}^{3\pi/4} = \left(-\cot\tfrac{3\pi}{4}\right) - \left(-\cot\tfrac{\pi}{4}\right) = 2$$

87. Let $u = \frac{x}{6} \Rightarrow du = \frac{1}{6}\,dx \Rightarrow 6\,du = dx$; $x = \pi \Rightarrow u = \frac{\pi}{6}$, $x = 3\pi \Rightarrow u = \frac{\pi}{2}$

$$\int_{\pi}^{3\pi} \cot^2\tfrac{x}{6}\,dx = \int_{\pi/6}^{\pi/2} 6\cot^2 u\,du = 6\int_{\pi/6}^{\pi/2} \left(\csc^2 u - 1\right) du = [6(-\cot u - u)]_{\pi/6}^{\pi/2} = 6\left(-\cot\tfrac{\pi}{2} - \tfrac{\pi}{2}\right) - 6\left(-\cot\tfrac{\pi}{6} - \tfrac{\pi}{6}\right)$$
$$= 6\sqrt{3} - 2\pi$$

88. Let $u = \frac{\theta}{3} \Rightarrow du = \frac{1}{3}\,d\theta \Rightarrow 3\,du = d\theta$; $\theta = 0 \Rightarrow u = 0$, $\theta = \pi \Rightarrow u = \frac{\pi}{3}$

$$\int_0^{\pi} \tan^2\tfrac{\theta}{3}\,d\theta = \int_0^{\pi}\left(\sec^2\tfrac{\theta}{3} - 1\right) d\theta = \int_0^{\pi/3} 3\left(\sec^2 u - 1\right) du = [3\tan u - 3u]_0^{\pi/3}$$
$$= \left[3\tan\tfrac{\pi}{3} - 3\left(\tfrac{\pi}{3}\right)\right] - (3\tan 0 - 0) = 3\sqrt{3} - \pi$$

89. $$\int_{-\pi/3}^{0} \sec x\tan x\,dx = [\sec x]_{-\pi/3}^{0} = \sec 0 - \sec\left(-\tfrac{\pi}{3}\right) = 1 - 2 = -1$$

90. $$\int_{\pi/4}^{3\pi/4} \csc z\cot z\,dz = [-\csc z]_{\pi/4}^{3\pi/4} = \left(-\csc\tfrac{3\pi}{4}\right) - \left(-\csc\tfrac{\pi}{4}\right) = -\sqrt{2} + \sqrt{2} = 0$$

91. Let $u = \sin x \Rightarrow du = \cos x\,dx$; $x = 0 \Rightarrow u = 0$, $x = \frac{\pi}{2} \Rightarrow u = 1$

$$\int_0^{\pi/2} 5(\sin x)^{3/2}\cos x\,dx = \int_0^1 5u^{3/2}\,du = \left[5\left(\tfrac{2}{5}\right)u^{5/2}\right]_0^1 = \left[2u^{5/2}\right]_0^1 = 2(1)^{5/2} - 2(0)^{5/2} = 2$$

92. Let $u = \sin 3x \Rightarrow du = 3\cos 3x\,dx \Rightarrow \frac{1}{3}\,du = \cos 3x\,dx$; $x = -\frac{\pi}{2} \Rightarrow u = \sin\left(-\frac{3\pi}{2}\right) = 1$, $x = \frac{\pi}{2} \Rightarrow u = \sin\left(\frac{3\pi}{2}\right)$
$= -1$

$$\int_{-\pi/2}^{\pi/2} 15\sin^4 3x\cos 3x\,dx = \int_1^{-1} 15u^4\left(\tfrac{1}{3}\,du\right) = \int_1^{-1} 5u^4\,du = \left[u^5\right]_1^{-1} = (-1)^5 - (1)^5 = -2$$

93. Let $u = 1 + 3\sin^2 x \Rightarrow du = 6\sin x\cos x\,dx \Rightarrow \frac{1}{2}\,du = 3\sin x\cos x\,dx$; $x = 0 \Rightarrow u = 1$, $x = \frac{\pi}{2}$
$\Rightarrow u = 1 + 3\sin^2\frac{\pi}{2} = 4$

$$\int_0^{\pi/2} \frac{3\sin x\cos x}{\sqrt{1+3\sin^2 x}}\,dx = \int_1^4 \tfrac{1}{\sqrt{u}}\left(\tfrac{1}{2}\,du\right) = \int_1^4 \tfrac{1}{2}u^{-1/2}\,du = \left[\tfrac{1}{2}\left(\tfrac{u^{1/2}}{\frac{1}{2}}\right)\right]_1^4 = \left[u^{1/2}\right]_1^4 = 4^{1/2} - 1^{1/2} = 1$$

94. Let $u = 1 + 7\tan x \Rightarrow du = 7\sec^2 x\, dx \Rightarrow \frac{1}{7}\, du = \sec^2 x\, dx$; $x = 0 \Rightarrow u = 1 + 7\tan 0 = 1$,
$x = \frac{\pi}{4} \Rightarrow u = 1 + 7\tan\frac{\pi}{4} = 8$

$$\int_0^{\pi/4} \frac{\sec^2 x}{(1+7\tan x)^{2/3}}\, dx = \int_1^8 \frac{1}{u^{2/3}}\left(\frac{1}{7}\, du\right) = \int_1^8 \frac{1}{7} u^{-2/3}\, du = \left[\frac{1}{7}\left(\frac{u^{1/3}}{\frac{1}{3}}\right)\right]_1^8 = \left[\frac{3}{7} u^{1/3}\right]_1^8 = \frac{3}{7}(8)^{1/3} - \frac{3}{7}(1)^{1/3} = \frac{3}{7}$$

95. $\int_1^4 \left(\frac{x}{8} + \frac{1}{2x}\right) dx = \frac{1}{2}\int_1^4 \left(\frac{1}{4}x + \frac{1}{x}\right) dx = \frac{1}{2}\left[\frac{1}{8}x^2 + \ln|x|\right]_1^4 = \frac{1}{2}\left[\left(\frac{16}{8} + \ln 4\right) - \left(\frac{1}{8} + \ln 1\right)\right] = \frac{15}{16} + \frac{1}{2}\ln 4$
$= \frac{15}{16} + \ln\sqrt{4} = \frac{15}{16} + \ln 2$

96. $\int_1^8 \left(\frac{2}{3x} - \frac{8}{x^2}\right) dx = \frac{2}{3}\int_1^8 \left(\frac{1}{x} - 12x^{-2}\right) dx = \frac{2}{3}\left[\ln|x| + 12x^{-1}\right]_1^8 = \frac{2}{3}\left[\left(\ln 8 + \frac{12}{8}\right) - (\ln 1 + 12)\right]$
$= \frac{2}{3}\left(\ln 8 + \frac{3}{2} - 12\right) = \frac{2}{3}\left(\ln 8 - \frac{21}{2}\right) = \frac{2}{3}(\ln 8) - 7 = \ln\left(8^{2/3}\right) - 7 = \ln 4 - 7$

97. $\int_{-2}^{-1} e^{-(x+1)}\, dx = -\int_1^0 e^u\, du$, where $u = -(x+1)$, $du = -\, dx$; $x = -2 \Rightarrow u = 1$, $x = -1 \Rightarrow u = 0$
$= -\left[e^u\right]_1^0 = -\left(e^0 - e^1\right) = e - 1$

98. $\int_{-\ln 2}^0 e^{2w}\, dw = \frac{1}{2}\int_{\ln(1/4)}^0 e^u\, du$, where $u = 2w$, $du = 2\, dw$; $w = -\ln 2 \Rightarrow u = \ln\frac{1}{4}$, $w = 0 \Rightarrow u = 0$
$= \frac{1}{2}\left[e^u\right]_{\ln(1/4)}^0 = \frac{1}{2}\left[e^0 - e^{\ln(1/4)}\right] = \frac{1}{2}\left(1 - \frac{1}{4}\right) = \frac{3}{8}$

99. $\int_1^{\ln 5} e^r\left(3e^r + 1\right)^{-3/2} dr = \frac{1}{3}\int_4^{16} u^{-3/2}\, du$, where $u = 3e^r + 1$, $du = 3e^r dr$; $r = 0 \Rightarrow u = 4$, $r = \ln 5 \Rightarrow u = 16$
$= -\frac{2}{3}\left[u^{-1/2}\right]_4^{16} = -\frac{2}{3}\left(16^{-1/2} - 4^{-1/2}\right) = \left(-\frac{2}{3}\right)\left(\frac{1}{4} - \frac{1}{2}\right) = \left(-\frac{2}{3}\right)\left(-\frac{1}{4}\right) = \frac{1}{6}$

100. $\int_0^{\ln 9} e^\theta\left(e^\theta - 1\right)^{1/2} d\theta = \int_0^8 u^{1/2}\, du$, where $u = e^\theta - 1$, $du = e^\theta\, d\theta$; $\theta = 0 \Rightarrow u = 0$, $\theta = \ln 9 \Rightarrow u = 8$
$= \frac{2}{3}\left[u^{3/2}\right]_0^8 = \frac{2}{3}\left(8^{3/2} - 0^{3/2}\right) = \frac{2}{3}\left(2^{9/2} - 0\right) = \frac{2^{11/2}}{3} = \frac{32\sqrt{2}}{3}$

101. $\int_1^e \frac{1}{x}(1 + 7\ln x)^{-1/3}\, dx = \frac{1}{7}\int_1^8 u^{-1/3}\, du$, where $u = 1 + 7\ln x$, $du = \frac{7}{x}\, dx$, $x = 1 \Rightarrow u = 1$, $x = e \Rightarrow u = 8$
$= \frac{3}{14}\left[u^{2/3}\right]_1^8 = \frac{3}{14}\left(8^{2/3} - 1^{2/3}\right) = \left(\frac{3}{14}\right)(4 - 1) = \frac{9}{14}$

102. $\int_1^3 \frac{[\ln(v+1)]^2}{v+1}\, dv = \int_1^3 [\ln(v+1)]^2 \frac{1}{v+1}\, dv = \int_{\ln 2}^{\ln 4} u^2\, du$, where $u = \ln(v+1)$, $du = \frac{1}{v+1}\, dv$;
$v = 1 \Rightarrow u = \ln 2$, $v = 3 \Rightarrow u = \ln 4$;
$= \frac{1}{3}\left[u^3\right]_{\ln 2}^{\ln 4} = \frac{1}{3}\left[(\ln 4)^3 - (\ln 2)^3\right] = \frac{1}{3}\left[(2\ln 2)^3 - (\ln 2)^3\right] = \frac{(\ln 2)^3}{3}(8 - 1) = \frac{7}{3}(\ln 2)^3$

103. $\int_1^8 \frac{\log_4 \theta}{\theta}\, d\theta = \frac{1}{\ln 4}\int_1^8 (\ln\theta)\left(\frac{1}{\theta}\right) d\theta = \frac{1}{\ln 4}\int_0^{\ln 8} u\, du$, where $u = \ln\theta$, $du = \frac{1}{\theta}\, d\theta$, $\theta = 1 \Rightarrow u = 0$, $\theta = 8 \Rightarrow u = \ln 8$
$= \frac{1}{2\ln 4}\left[u^2\right]_0^{\ln 8} = \frac{1}{\ln 16}\left[(\ln 8)^2 - 0^2\right] = \frac{(3\ln 2)^2}{4\ln 2} = \frac{9\ln 2}{4}$

104. $\int_1^e \frac{8(\ln 3)(\log_3\theta)}{\theta}\, d\theta = \int_1^e \frac{8(\ln 3)(\ln\theta)}{\theta(\ln 3)}\, d\theta = 8\int_1^e (\ln\theta)\left(\frac{1}{\theta}\right) d\theta = 8\int_0^1 u\, du$, where $u = \ln\theta$, $du = \frac{1}{\theta}\, d\theta$;
$\theta = 1 \Rightarrow u = 0$, $\theta = e \Rightarrow u = 1$
$= 4\left[u^2\right]_0^1 = 4\left(1^2 - 0^2\right) = 4$

105. $\int_{-3/4}^{3/4} \frac{6}{\sqrt{9 - 4x^2}}\, dx = 3\int_{-3/4}^{3/4} \frac{2}{\sqrt{3^2 - (2x)^2}}\, dx = 3\int_{-3/2}^{3/2} \frac{1}{\sqrt{3^2 - u^2}}\, du$, where $u = 2x$, $du = 2\, dx$;
$x = -\frac{3}{4} \Rightarrow u = -\frac{3}{2}$, $x = \frac{3}{4} \Rightarrow u = \frac{3}{2}$
$= 3\left[\sin^{-1}\left(\frac{u}{3}\right)\right]_{-3/2}^{3/2} = 3\left[\sin^{-1}\left(\frac{1}{2}\right) - \sin^{-1}\left(-\frac{1}{2}\right)\right] = 3\left[\frac{\pi}{6} - \left(-\frac{\pi}{6}\right)\right] = 3\left(\frac{\pi}{3}\right) = \pi$

106. $\int_{-1/5}^{1/5} \frac{6}{\sqrt{4-25x^2}}\,dx = \frac{6}{5}\int_{-1/5}^{1/5} \frac{5}{\sqrt{2^2-(5x)^2}}\,dx = \frac{6}{5}\int_{-1}^{1} \frac{1}{\sqrt{2^2-u^2}}\,du$, where $u = 5x$, $du = 5\,dx$;

$x = -\frac{1}{5} \Rightarrow u = -1, x = \frac{1}{5} \Rightarrow u = 1$

$= \frac{6}{5}\left[\sin^{-1}\left(\frac{u}{2}\right)\right]_{-1}^{1} = \frac{6}{5}\left[\sin^{-1}\left(\frac{1}{2}\right) - \sin^{-1}\left(-\frac{1}{2}\right)\right] = \frac{6}{5}\left[\frac{\pi}{6} - \left(-\frac{\pi}{6}\right)\right] = \frac{6}{5}\left(\frac{\pi}{3}\right) = \frac{2\pi}{5}$

107. $\int_{-2}^{2} \frac{3}{4+3t^2}\,dt = \sqrt{3}\int_{-2}^{2} \frac{\sqrt{3}}{2^2+\left(\sqrt{3}t\right)^2}\,dt = \sqrt{3}\int_{-2\sqrt{3}}^{2\sqrt{3}} \frac{1}{2^2+u^2}\,du$, where $u = \sqrt{3}t$, $du = \sqrt{3}\,dt$;

$t = -2 \Rightarrow u = -2\sqrt{3}, t = 2 \Rightarrow u = 2\sqrt{3}$

$= \sqrt{3}\left[\frac{1}{2}\tan^{-1}\left(\frac{u}{2}\right)\right]_{-2\sqrt{3}}^{2\sqrt{3}} = \frac{\sqrt{3}}{2}\left[\tan^{-1}\left(\sqrt{3}\right) - \tan^{-1}\left(-\sqrt{3}\right)\right] = \frac{\sqrt{3}}{2}\left[\frac{\pi}{3} - \left(-\frac{\pi}{3}\right)\right] = \frac{\pi}{\sqrt{3}}$

108. $\int_{\sqrt{3}}^{3} \frac{1}{3+t^2}\,dt = \int_{\sqrt{3}}^{3} \frac{1}{\left(\sqrt{3}\right)^2+t^2}\,dt = \left[\frac{1}{\sqrt{3}}\tan^{-1}\left(\frac{t}{\sqrt{3}}\right)\right]_{\sqrt{3}}^{3} = \frac{1}{\sqrt{3}}\left(\tan^{-1}\sqrt{3} - \tan^{-1}1\right) = \frac{1}{\sqrt{3}}\left(\frac{\pi}{3} - \frac{\pi}{4}\right) = \frac{\sqrt{3}\pi}{36}$

109. $\int \frac{1}{y\sqrt{4y^2-1}}\,dy = \int \frac{2}{(2y)\sqrt{(2y)^2-1}}\,dy = \int \frac{1}{u\sqrt{u^2-1}}\,du$, where $u = 2y$ and $du = 2\,dy$
$= \sec^{-1}|u| + C = \sec^{-1}|2y| + C$

110. $\int \frac{24}{y\sqrt{y^2-16}}\,dy = 24\int \frac{1}{y\sqrt{y^2-4^2}}\,dy = 24\left(\frac{1}{4}\sec^{-1}\left|\frac{y}{4}\right|\right) + C = 6\sec^{-1}\left|\frac{y}{4}\right| + C$

111. $\int_{\sqrt{2}/3}^{2/3} \frac{1}{|y|\sqrt{9y^2-1}}\,dy = \int_{\sqrt{2}/3}^{2/3} \frac{3}{|3y|\sqrt{(3y)^2-1}}\,dy = \int_{\sqrt{2}}^{2} \frac{1}{|u|\sqrt{u^2-1}}\,du$, where $u = 3y$, $du = 3\,dy$;

$y = \frac{\sqrt{2}}{3} \Rightarrow u = \sqrt{2}, y = \frac{2}{3} \Rightarrow u = 2$

$= \left[\sec^{-1}u\right]_{\sqrt{2}}^{2} = \left[\sec^{-1}2 - \sec^{-1}\sqrt{2}\right] = \frac{\pi}{3} - \frac{\pi}{4} = \frac{\pi}{12}$

112. $\int_{-2/\sqrt{5}}^{-\sqrt{6}/\sqrt{5}} \frac{1}{|y|\sqrt{5y^2-3}}\,dy = \int_{-2/\sqrt{5}}^{-\sqrt{6}/\sqrt{5}} \frac{\sqrt{5}}{-\sqrt{5}y\sqrt{\left(\sqrt{5}y\right)^2-\left(\sqrt{3}\right)^2}}\,dy = \int_{-2}^{-\sqrt{6}} \frac{1}{-u\sqrt{u^2-\left(\sqrt{3}\right)^2}}\,du$,

where $u = \sqrt{5}y$, $du = \sqrt{5}\,dy$; $y = -\frac{2}{\sqrt{5}} \Rightarrow u = -2, y = -\frac{\sqrt{6}}{\sqrt{5}} \Rightarrow u = -\sqrt{6}$

$= \left[-\frac{1}{\sqrt{3}}\sec^{-1}\left|\frac{u}{\sqrt{3}}\right|\right]_{-2}^{-\sqrt{6}} = \frac{-1}{\sqrt{3}}\left[\sec^{-1}\sqrt{2} - \sec^{-1}\frac{2}{\sqrt{3}}\right] = \frac{-1}{\sqrt{3}}\left(\frac{\pi}{4} - \frac{\pi}{6}\right) = \frac{-1}{\sqrt{3}}\left[\frac{3\pi}{12} - \frac{2\pi}{12}\right] = \frac{-\pi}{12\sqrt{3}} = \frac{-\sqrt{3}\pi}{36}$

113. (a) $av(f) = \frac{1}{1-(-1)}\int_{-1}^{1}(mx+b)\,dx = \frac{1}{2}\left[\frac{mx^2}{2} + bx\right]_{-1}^{1} = \frac{1}{2}\left[\left(\frac{m(1)^2}{2} + b(1)\right) - \left(\frac{m(-1)^2}{2} + b(-1)\right)\right] = \frac{1}{2}(2b) = b$

(b) $av(f) = \frac{1}{k-(-k)}\int_{-k}^{k}(mx+b)\,dx = \frac{1}{2k}\left[\frac{mx^2}{2} + bx\right]_{-k}^{k} = \frac{1}{2k}\left[\left(\frac{m(k)^2}{2} + b(k)\right) - \left(\frac{m(-k)^2}{2} + b(-k)\right)\right]$
$= \frac{1}{2k}(2bk) = b$

114. (a) $y_{av} = \frac{1}{3-0}\int_0^3 \sqrt{3x}\,dx = \frac{1}{3}\int_0^3 \sqrt{3}x^{1/2}\,dx = \frac{\sqrt{3}}{3}\left[\frac{2}{3}x^{3/2}\right]_0^3 = \frac{\sqrt{3}}{3}\left[\frac{2}{3}(3)^{3/2} - \frac{2}{3}(0)^{3/2}\right] = \frac{\sqrt{3}}{3}\left(2\sqrt{3}\right) = 2$

(b) $y_{av} = \frac{1}{a-0}\int_0^a \sqrt{ax}\,dx = \frac{1}{a}\int_0^a \sqrt{a}x^{1/2}\,dx = \frac{\sqrt{a}}{a}\left[\frac{2}{3}x^{3/2}\right]_0^a = \frac{\sqrt{a}}{a}\left(\frac{2}{3}(a)^{3/2} - \frac{2}{3}(0)^{3/2}\right) = \frac{\sqrt{a}}{a}\left(\frac{2}{3}a\sqrt{a}\right) = \frac{2}{3}a$

115. $f'_{av} = \frac{1}{b-a}\int_a^b \sqrt{axf'(x)}\,dx = \frac{1}{b-a}[f(x)]_a^b = \frac{1}{b-a}[f(b) - f(a)] = \frac{f(b)-f(a)}{b-a}$ so the average value of f' over $[a, b]$ is the slope of the secant line joining the points $(a, f(a))$ and $(b, f(b))$, which is the average rate of change of f over $[a, b]$.

116. Yes, because the average value of f on $[a, b]$ is $\frac{1}{b-a}\int_a^b f(x)\,dx$. If the length of the interval is 2, then $b - a = 2$ and the average value of the function is $\frac{1}{2}\int_a^b f(x)\,dx$.

117. (a) $\frac{d}{dx}(x \ln x - x + C) = x \cdot \frac{1}{x} + \ln x - 1 + 0 = \ln x$

(b) average value $= \frac{1}{e-1}\int_1^e \ln x\, dx = \frac{1}{e-1}[x \ln x - x]_1^e = \frac{1}{e-1}[(e \ln e - e) - (1 \ln 1 - 1)]$
$= \frac{1}{e-1}(e - e + 1) = \frac{1}{e-1}$

118. average value $= \frac{1}{2-1}\int_1^2 \frac{1}{x}\, dx = [\ln|x|]_1^2 = \ln 2 - \ln 1 = \ln 2$

119. We want to evaluate

$$\frac{1}{365-0}\int_0^{365} f(x)\, dx = \frac{1}{365}\int_0^{365}\left(37\sin\left[\frac{2\pi}{365}(x-101)\right] + 25\right)dx = \frac{37}{365}\int_0^{365}\sin\left[\frac{2\pi}{365}(x-101)\right]dx + \frac{25}{365}\int_0^{365} dx$$

Notice that the period of $y = \sin\left[\frac{2\pi}{365}(x-101)\right]$ is $\frac{2\pi}{\frac{2\pi}{365}} = 365$ and that we are integrating this function over an iterval of length 365. Thus the value of $\frac{37}{365}\int_0^{365}\sin\left[\frac{2\pi}{365}(x-101)\right]dx + \frac{25}{365}\int_0^{365} dx$ is $\frac{37}{365}\cdot 0 + \frac{25}{365}\cdot 365 = 25$.

120. $\frac{1}{675-20}\int_{20}^{675}(8.27 + 10^{-5}(26T - 1.87T^2))dT = \frac{1}{655}\left[8.27T + \frac{26T^2}{2\cdot 10^5} - \frac{1.87T^3}{3\cdot 10^5}\right]_{20}^{675}$

$= \frac{1}{655}\left(\left[8.27(675) + \frac{26(675)^2}{2\cdot 10^5} - \frac{1.87(675)^3}{3\cdot 10^5}\right] - \left[8.27(20) + \frac{26(20)^2}{2\cdot 10^5} - \frac{1.87(20)^3}{3\cdot 10^5}\right]\right) \approx \frac{1}{655}(3724.44 - 165.40)$

$= 5.43 =$ the average value of C_v on [20, 675]. To find the temperature T at which $C_v = 5.43$, solve $5.43 = 8.27 + 10^{-5}(26T - 1.87T^2)$ for T. We obtain $1.87T^2 - 26T - 284000 = 0$

$\Rightarrow T = \frac{26 \pm \sqrt{(26)^2 - 4(1.87)(-284000)}}{2(1.87)} = \frac{26 \pm \sqrt{2124996}}{3.74}$. So $T = -382.82$ or $T = 396.72$. Only $T = 396.72$ lies in the interval [20, 675], so $T = 396.72°C$.

121. $\frac{dy}{dx} = \sqrt{2 + \cos^3 x}$

122. $\frac{dy}{dx} = \sqrt{2 + \cos^3(7x^2)} \cdot \frac{d}{dx}(7x^2) = 14x\sqrt{2 + \cos^3(7x^2)}$

123. $\frac{dy}{dx} = \frac{d}{dx}\left(-\int_1^x \frac{6}{3+t^4}dt\right) = -\frac{6}{3+x^4}$

124. $\frac{dy}{dx} = \frac{d}{dx}\left(\int_{\sec x}^2 \frac{1}{t^2+1}dt\right) = -\frac{d}{dx}\left(\int_2^{\sec x}\frac{1}{t^2+1}dt\right) = -\frac{1}{\sec^2 x + 1}\frac{d}{dx}(\sec x) = -\frac{\sec x \tan x}{1 + \sec^2 x}$

125. $y = \int_{\ln x^2}^0 e^{\cos t}\, dt = -\int_0^{\ln x^2} e^{\cos t}\, dt \Rightarrow \frac{dy}{dx} = -e^{\cos(\ln x^2)} \cdot \frac{d}{dx}(\ln x^2) = -\frac{2}{x}e^{\cos(\ln x^2)}$

126. $y = \int_1^{e^{\sqrt{x}}} \ln(t^2 + 1)dt \Rightarrow \frac{dy}{dx} = \ln\left(e^{2\sqrt{x}} + 1\right) \cdot \frac{d}{dx}\left(e^{\sqrt{x}}\right) = \frac{e^{\sqrt{x}}}{2\sqrt{x}}\ln\left(e^{2\sqrt{x}} + 1\right)$

127. $y = \int_0^{\sin^{-1}x} \frac{dt}{\sqrt{1-2t^2}} \Rightarrow \frac{dy}{dx} = \frac{1}{\sqrt{1 - 2(\sin^{-1}x)^2}} \cdot \frac{d}{dx}(\sin^{-1}x) = \frac{1}{\sqrt{1-2(\sin^{-1}x)^2}} \cdot \frac{1}{\sqrt{1-x^2}}$

128. $y = \int_{\tan^{-1}x}^{\pi/4} e^{\sqrt{t}}\, dt = -\int_{\pi/4}^{\tan^{-1}x} e^{\sqrt{t}}\, dt \Rightarrow \frac{dy}{dx} = -e^{\sqrt{\tan^{-1}x}} \cdot \frac{d}{dx}(\tan^{-1}x) = \frac{-e^{\sqrt{\tan^{-1}x}}}{1+x^2}$

129. Yes. The function f, being differentiable on [a, b], is then continuous on [a, b]. The Fundamental Theorem of Calculus says that every continuous function on [a, b] is the derivative of a function on [a, b].

130. The second part of the Fundamental Theorem of Calculus states that if F(x) is an antiderivative of f(x) on [a, b], then $\int_a^b f(x)\,dx = F(b) - F(a)$. In particular, if F(x) is an antiderivaitve of $\sqrt{1+x^4}$ on [0, 1], then $\int_0^1 \sqrt{1+x^4}\,dx = F(1) - F(0)$.

131. $y = \int_x^1 \sqrt{1+t^2}\,dt = -\int_1^x \sqrt{1+t^2}\,dt \Rightarrow \frac{dy}{dx} = \frac{d}{dx}\left[-\int_1^x \sqrt{1+t^2}\,dt\right] = -\frac{d}{dx}\left[\int_1^x \sqrt{1+t^2}\,dt\right] = -\sqrt{1+x^2}$

132. $y = \int_{\cos x}^0 \frac{1}{1-t^2}\,dt = -\int_0^{\cos x} \frac{1}{1-t^2}\,dt \Rightarrow \frac{dy}{dx} = \frac{d}{dx}\left[-\int_0^{\cos x} \frac{1}{1-t^2}\,dt\right] = -\frac{d}{dx}\left[\int_0^{\cos x} \frac{1}{1-t^2}\,dt\right]$
$= -\left(\frac{1}{1-\cos^2 x}\right)\left(\frac{d}{dx}(\cos x)\right) = -\left(\frac{1}{\sin^2 x}\right)(-\sin x) = \frac{1}{\sin x} = \csc x$

133. We estimate the area A using midpoints of the vertical intervals, and we will estimate the width of the parking lot on each interval by averaging the widths at top and bottom. This gives the estimate
$A \approx 15 \cdot \left(\frac{0+36}{2} + \frac{36+54}{2} + \frac{54+51}{2} + \frac{51+49.5}{2} + \frac{49.5+54}{2} + \frac{54+64.4}{2} + \frac{64.4+67.5}{2} + \frac{67.5+42}{2}\right) \approx 5961 \text{ ft}^2$
The cost is Area · ($\$2.10/\text{ft}^2$) $\approx$ (5961 ft^2) ($\$2.10/\text{ft}^2$) = \$12,518.10 $\Rightarrow$ the job cannot be done for \$11,000.

134. (a) Before the chute opens for A, $a = -32 \text{ ft/sec}^2$. Since the helicopter is hovering, $v_0 = 0$ ft/sec $\Rightarrow v = \int -32\,dt = -32t + v_0 = -32t$. Then $s_0 = 6400$ ft $\Rightarrow s = \int -32t\,dt = -16t^2 + s_0 = -16t^2 + 6400$. At t = 4 sec, $s = -16(4)^2 + 6400 = 6144$ ft when A's chute opens;

(b) For B, $s_0 = 7000$ ft, $v_0 = 0$, $a = -32 \text{ ft/sec}^2 \Rightarrow v = \int -32\,dt = -32t + v_0 = -32t \Rightarrow s = \int -32t\,dt$ $= -16t^2 + s_0 = -16t^2 + 7000$. At t = 13 sec, $s = -16(13)^2 + 7000 = 4296$ ft when B's chute opens;

(c) After the chutes open, $v = -16$ ft/sec $\Rightarrow s = \int -16\,dt = -16t + s_0$. For A, $s_0 = 6144$ ft and for B, $s_0 = 4296$ ft. Therefore, for A, $s = -16t + 6144$ and for B, $s = -16t + 4296$. When they hit the ground, $s = 0 \Rightarrow$ for A, $0 = -16t + 6144 \Rightarrow t = \frac{6144}{16} = 384$ seconds, and for B, $0 = -16t + 4296 \Rightarrow t = \frac{4296}{16}$ $= 268.5$ seconds to hit the ground after the chutes open. Since B's chute opens 58 seconds after A's opens $\Rightarrow$ B hits the ground first.

135. $av(I) = \frac{1}{30}\int_0^{30} (1200 - 40t)\,dt = \frac{1}{30}\left[1200t - 20t^2\right]_0^{30} = \frac{1}{30}\left[\left(\left(1200(30) - 20(30)^2\right) - \left(1200(0) - 20(0)^2\right)\right)\right]$
$= \frac{1}{30}(18{,}000) = 600$; Average Daily Holding Cost = (600)(\$0.03) = \$18

136. $av(I) = \frac{1}{14}\int_0^{14} (600 + 600t)\,dt = \frac{1}{14}\left[600t + 300t^2\right]_0^{14} = \frac{1}{14}\left[600(14) + 300(14)^2 - 0\right] = 4800$; Average Daily Holding Cost = (4800)(\$0.04) = \$192

137. $av(I) = \frac{1}{30}\int_0^{30}\left(450 - \frac{t^2}{2}\right)dt = \frac{1}{30}\left[450t - \frac{t^3}{6}\right]_0^{30} = \frac{1}{30}\left[450(30) - \frac{30^3}{6} - 0\right] = 300$; Average Daily Holding Cost
$= (300)(\$0.02) = \$6$

138. $av(I) = \frac{1}{60}\int_0^{60}\left(600 - 20\sqrt{15t}\right)dt = \frac{1}{60}\int_0^{60}\left(600 - 20\sqrt{15}\,t^{1/2}\right)dt = \frac{1}{60}\left[600t - 20\sqrt{15}\left(\frac{2}{3}\right)t^{3/2}\right]_0^{60}$
$= \frac{1}{60}\left[600(60) - \frac{40\sqrt{15}}{3}(60)^{3/2} - 0\right] = \frac{1}{60}\left(36{,}000 - \left(\frac{320}{3}\right)15^2\right) = 200$; Average Daily Holding Cost
$= (200)(\$0.005) = \$1.00$

CHAPTER 5 ADDITIONAL AND ADVANCED EXERCISES

1. (a) Yes, because $\int_0^1 f(x)\,dx = \frac{1}{7}\int_0^1 7f(x)\,dx = \frac{1}{7}(7) = 1$

   (b) No. For example, $\int_0^1 8x\,dx = \left[4x^2\right]_0^1 = 4$, but $\int_0^1 \sqrt{8x}\,dx = \left[2\sqrt{2}\left(\frac{x^{3/2}}{\frac{3}{2}}\right)\right]_0^1 = \frac{4\sqrt{2}}{3}\left(1^{3/2} - 0^{3/2}\right)$
   $= \frac{4\sqrt{2}}{3} \neq \sqrt{4}$

2. (a) True: $\int_5^2 f(x)\,dx = -\int_2^5 f(x)\,dx = -3$

   (b) True: $\int_{-2}^5 [f(x) + g(x)]\,dx = \int_{-2}^5 f(x)\,dx + \int_{-2}^5 g(x)\,dx = \int_{-2}^2 f(x)\,dx + \int_2^5 f(x)\,dx + \int_{-2}^5 g(x)\,dx$
   $= 4 + 3 + 2 = 9$

   (c) False: $\int_{-2}^5 f(x)\,dx = 4 + 3 = 7 > 2 = \int_{-2}^5 g(x)\,dx \Rightarrow \int_{-2}^5 [f(x) - g(x)]\,dx > 0 \Rightarrow \int_{-2}^5 [g(x) - f(x)]\,dx < 0.$
   On the other hand, $f(x) \le g(x) \Rightarrow [g(x) - f(x)] \ge 0 \Rightarrow \int_{-2}^5 [g(x) - f(x)]\,dx \ge 0$ which is a contradiction.

3. $y = \frac{1}{a}\int_0^x f(t)\sin a(x-t)\,dt = \frac{1}{a}\int_0^x f(t)\sin ax\cos at\,dt - \frac{1}{a}\int_0^x f(t)\cos ax\sin at\,dt$
   $= \frac{\sin ax}{a}\int_0^x f(t)\cos at\,dt - \frac{\cos ax}{a}\int_0^x f(t)\sin at\,dt \Rightarrow \frac{dy}{dx} = \cos ax\left(\int_0^x f(t)\cos at\,dt\right)$
   $+ \frac{\sin ax}{a}\left(\frac{d}{dx}\int_0^x f(t)\cos at\,dt\right) + \sin ax\int_0^x f(t)\sin at\,dt - \frac{\cos ax}{a}\left(\frac{d}{dx}\int_0^x f(t)\sin at\,dt\right)$
   $= \cos ax\int_0^x f(t)\cos at\,dt + \frac{\sin ax}{a}(f(x)\cos ax) + \sin ax\int_0^x f(t)\sin at\,dt - \frac{\cos ax}{a}(f(x)\sin ax)$
   $\Rightarrow \frac{dy}{dx} = \cos ax\int_0^x f(t)\cos at\,dt + \sin ax\int_0^x f(t)\sin at\,dt$. Next,
   $\frac{d^2y}{dx^2} = -a\sin ax\int_0^x f(t)\cos at\,dt + (\cos ax)\left(\frac{d}{dx}\int_0^x f(t)\cos at\,dt\right) + a\cos ax\int_0^x f(t)\sin at\,dt$
   $+ (\sin ax)\left(\frac{d}{dx}\int_0^x f(t)\sin at\,dt\right) = -a\sin ax\int_0^x f(t)\cos at\,dt + (\cos ax)f(x)\cos ax$
   $+ a\cos ax\int_0^x f(t)\sin at\,dt + (\sin ax)f(x)\sin ax = -a\sin ax\int_0^x f(t)\cos at\,dt + a\cos ax\int_0^x f(t)\sin at\,dt + f(x).$
   Therefore, $y'' + a^2y = a\cos ax\int_0^x f(t)\sin at\,dt - a\sin ax\int_0^x f(t)\cos at\,dt + f(x)$
   $+ a^2\left(\frac{\sin ax}{a}\int_0^x f(t)\cos at\,dt - \frac{\cos ax}{a}\int_0^x f(t)\sin at\,dt\right) = f(x)$. Note also that $y'(0) = y(0) = 0$.

4. $x = \int_0^y \frac{1}{\sqrt{1+4t^2}}\,dt \Rightarrow \frac{d}{dx}(x) = \frac{d}{dx}\int_0^y \frac{1}{\sqrt{1+4t^2}}\,dt = \frac{d}{dy}\left[\int_0^y \frac{1}{\sqrt{1+4t^2}}\,dt\right]\left(\frac{dy}{dx}\right)$ from the chain rule
   $\Rightarrow 1 = \frac{1}{\sqrt{1+4y^2}}\left(\frac{dy}{dx}\right) \Rightarrow \frac{dy}{dx} = \sqrt{1+4y^2}$. Then $\frac{d^2y}{dx^2} = \frac{d}{dx}\left(\sqrt{1+4y^2}\right) = \frac{d}{dy}\left(\sqrt{1+4y^2}\right)\left(\frac{dy}{dx}\right)$
   $= \frac{1}{2}\left(1+4y^2\right)^{-1/2}(8y)\left(\frac{dy}{dx}\right) = \frac{4y\left(\frac{dy}{dx}\right)}{\sqrt{1+4y^2}} = \frac{4y\left(\sqrt{1+4y^2}\right)}{\sqrt{1+4y^2}} = 4y$. Thus $\frac{d^2y}{dx^2} = 4y$, and the constant of proportionality is 4.

5. (a) $\int_0^{x^2} f(t)\,dt = x\cos\pi x \Rightarrow \frac{d}{dx}\int_0^{x^2} f(t)\,dt = \cos\pi x - \pi x\sin\pi x \Rightarrow f(x^2)(2x) = \cos\pi x - \pi x\sin\pi x$
   $\Rightarrow f(x^2) = \frac{\cos\pi x - \pi x\sin\pi x}{2x}$. Thus, $x = 2 \Rightarrow f(4) = \frac{\cos 2\pi - 2\pi\sin 2\pi}{4} = \frac{1}{4}$

   (b) $\int_0^{f(x)} t^2\,dt = \left[\frac{t^3}{3}\right]_0^{f(x)} = \frac{1}{3}(f(x))^3 \Rightarrow \frac{1}{3}(f(x))^3 = x\cos\pi x \Rightarrow (f(x))^3 = 3x\cos\pi x \Rightarrow f(x) = \sqrt[3]{3x\cos\pi x}$
   $\Rightarrow f(4) = \sqrt[3]{3(4)\cos 4\pi} = \sqrt[3]{12}$

6. $\int_0^a f(x)\,dx = \frac{a^2}{2} + \frac{a}{2}\sin a + \frac{\pi}{2}\cos a$. Let $F(a) = \int_0^a f(t)\,dt \Rightarrow f(a) = F'(a)$. Now $F(a) = \frac{a^2}{2} + \frac{a}{2}\sin a + \frac{\pi}{2}\cos a$
$\Rightarrow f(a) = F'(a) = a + \frac{1}{2}\sin a + \frac{a}{2}\cos a - \frac{\pi}{2}\sin a \Rightarrow f\left(\frac{\pi}{2}\right) = \frac{\pi}{2} + \frac{1}{2}\sin\frac{\pi}{2} + \frac{\left(\frac{\pi}{2}\right)}{2}\cos\frac{\pi}{2} - \frac{\pi}{2}\sin\frac{\pi}{2} = \frac{\pi}{2} + \frac{1}{2} - \frac{\pi}{2} = \frac{1}{2}$

7. $\int_1^b f(x)\,dx = \sqrt{b^2+1} - \sqrt{2} \Rightarrow f(b) = \frac{d}{db}\int_1^b f(x)\,dx = \frac{1}{2}(b^2+1)^{-1/2}(2b) = \frac{b}{\sqrt{b^2+1}} \Rightarrow f(x) = \frac{x}{\sqrt{x^2+1}}$

8. The derivative of the left side of the equation is: $\frac{d}{dx}\left[\int_0^x\left[\int_0^u f(t)\,dt\right]du\right] = \int_0^x f(t)\,dt$; the derivative of the right side of the equation is: $\frac{d}{dx}\left[\int_0^x f(u)(x-u)\,du\right] = \frac{d}{dx}\int_0^x f(u)\,x\,du - \frac{d}{dx}\int_0^x u\,f(u)\,du$
$= \frac{d}{dx}\left[x\int_0^x f(u)\,du\right] - \frac{d}{dx}\int_0^x u\,f(u)\,du = \int_0^x f(u)\,du + x\left[\frac{d}{dx}\int_0^x f(u)\,du\right] - xf(x) = \int_0^x f(u)\,du + xf(x) - xf(x)$
$= \int_0^x f(u)\,du$. Since each side has the same derivative, they differ by a constant, and since both sides equal 0 when $x = 0$, the constant must be 0. Therefore, $\int_0^x\left[\int_0^u f(t)\,dt\right]du = \int_0^x f(u)(x-u)\,du$.

9. $\frac{dy}{dx} = 3x^2 + 2 \Rightarrow y = \int (3x^2+2)\,dx = x^3 + 2x + C$. Then $(1,-1)$ on the curve $\Rightarrow 1^3 + 2(1) + C = -1 \Rightarrow C = -4$
$\Rightarrow y = x^3 + 2x - 4$

10. The acceleration due to gravity downward is -32 ft/sec^2 $\Rightarrow v = \int -32\,dt = -32t + v_0$, where v_0 is the initial velocity $\Rightarrow v = -32t + 32 \Rightarrow s = \int (-32t+32)\,dt = -16t^2 + 32t + C$. If the release point, at $t = 0$, is $s = 0$, then $C = 0 \Rightarrow s = -16t^2 + 32t$. Then $s = 17 \Rightarrow 17 = -16t^2 + 32t \Rightarrow 16t^2 - 32t + 17 = 0$. The discriminant of this quadratic equation is -64 which says there is no real time when $s = 17$ ft. You had better duck.

11. $\int_{-8}^3 f(x)\,dx = \int_{-8}^0 x^{2/3}\,dx + \int_0^3 -4\,dx$
$= \left[\frac{3}{5}x^{5/3}\right]_{-8}^0 + [-4x]_0^3$
$= \left(0 - \frac{3}{5}(-8)^{5/3}\right) + (-4(3) - 0) = \frac{96}{5} - 12$
$= \frac{36}{5}$

12. $\int_{-4}^3 f(x)\,dx = \int_{-4}^0 \sqrt{-x}\,dx + \int_0^3 (x^2-4)\,dx$
$= \left[-\frac{2}{3}(-x)^{3/2}\right]_{-4}^0 + \left[\frac{x^3}{3} - 4x\right]_0^3$
$= \left[0 - \left(-\frac{2}{3}(4)^{3/2}\right)\right] + \left[\left(\frac{3^3}{3} - 4(3)\right) - 0\right]$
$= \frac{16}{3} - 3 = \frac{7}{3}$

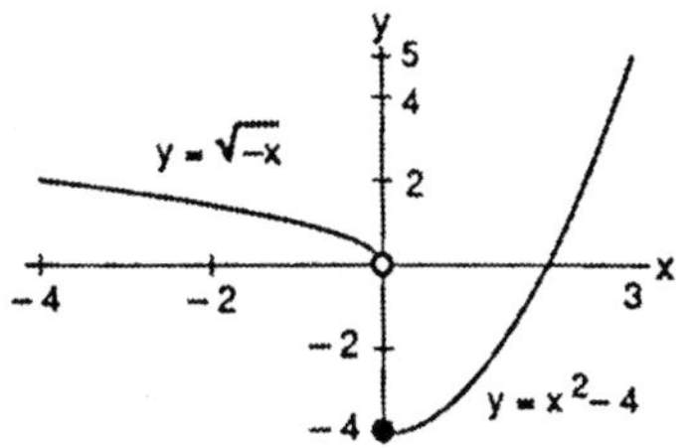

13. $\int_0^2 g(t)\,dt = \int_0^1 t\,dt + \int_1^2 \sin\pi t\,dt$
$= \left[\frac{t^2}{2}\right]_0^1 + \left[-\frac{1}{\pi}\cos\pi t\right]_1^2$
$= \left(\frac{1}{2} - 0\right) + \left[-\frac{1}{\pi}\cos 2\pi - \left(-\frac{1}{\pi}\cos\pi\right)\right]$
$= \frac{1}{2} - \frac{2}{\pi}$

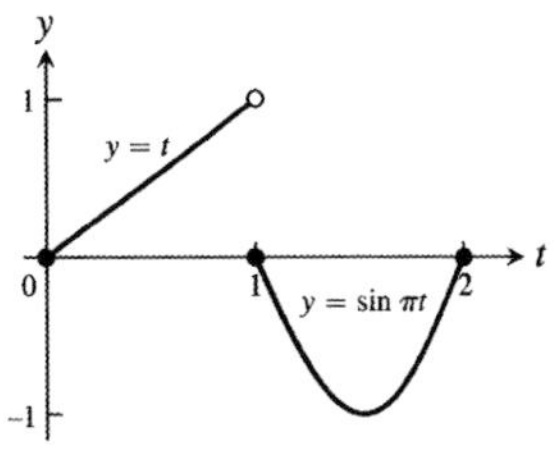

14. $\int_0^2 h(z)\,dz = \int_0^1 \sqrt{1-z}\,dz + \int_1^2 (7z-6)^{-1/3}\,dz$
$= \left[-\frac{2}{3}(1-z)^{3/2}\right]_0^1 + \left[\frac{3}{14}(7z-6)^{2/3}\right]_1^2$
$= \left[-\frac{2}{3}(1-1)^{3/2} - \left(-\frac{2}{3}(1-0)^{3/2}\right)\right]$
$+ \left[\frac{3}{14}(7(2)-6)^{2/3} - \frac{3}{14}(7(1)-6)^{2/3}\right]$
$= \frac{2}{3} + \left(\frac{6}{7} - \frac{3}{14}\right) = \frac{55}{42}$

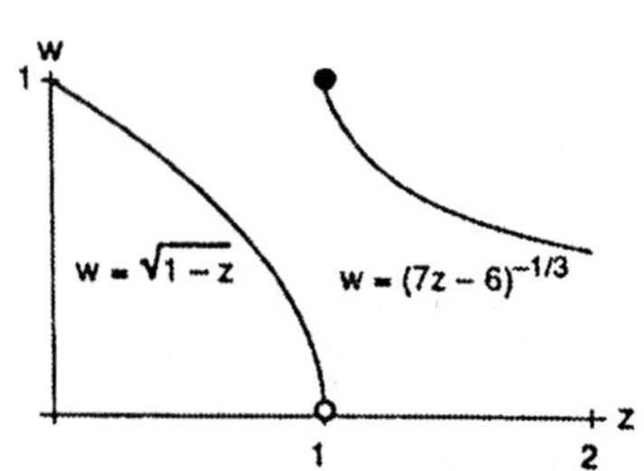

15. $\int_{-2}^2 f(x)\,dx = \int_{-2}^{-1} dx + \int_{-1}^1 (1-x^2)\,dx + \int_1^2 2\,dx$
$= [x]_{-2}^{-1} + \left[x - \frac{x^3}{3}\right]_{-1}^1 + [2x]_1^2$
$= (-1-(-2)) + \left[\left(1-\frac{1^3}{3}\right) - \left(-1 - \frac{(-1)^3}{3}\right)\right] + \left[2(2) - 2(1)\right]$
$= 1 + \frac{2}{3} - \left(-\frac{2}{3}\right) + 4 - 2 = \frac{13}{3}$

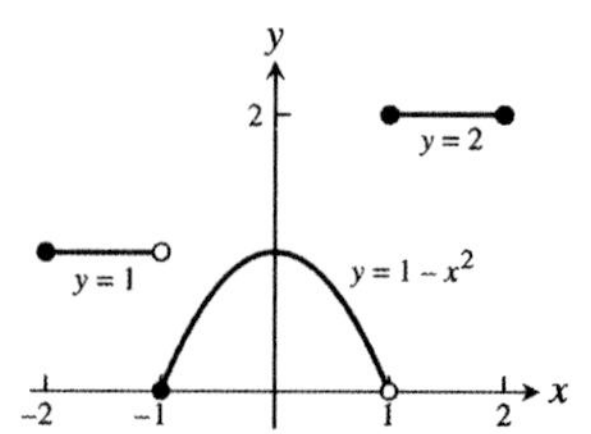

16. $\int_{-1}^2 h(r)\,dr = \int_{-1}^0 r\,dr + \int_0^1 (1-r^2)\,dr + \int_1^2 dr$
$= \left[\frac{r^2}{2}\right]_{-1}^0 + \left[r - \frac{r^3}{3}\right]_0^1 + [r]_1^2$
$= \left(0 - \frac{(-1)^2}{2}\right) + \left(\left(1 - \frac{1^3}{3}\right) - 0\right) + (2-1)$
$= -\frac{1}{2} + \frac{2}{3} + 1 = \frac{7}{6}$

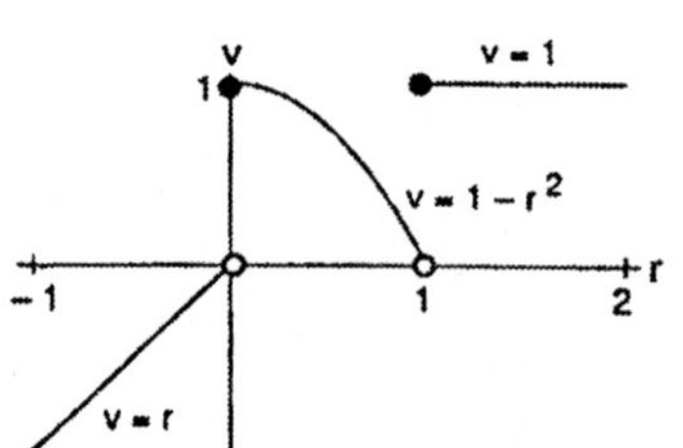

17. Ave. value $= \frac{1}{b-a}\int_a^b f(x)\,dx = \frac{1}{2-0}\int_0^2 f(x)\,dx = \frac{1}{2}\left[\int_0^1 x\,dx + \int_1^2 (x-1)\,dx\right] = \frac{1}{2}\left[\frac{x^2}{2}\right]_0^1 + \frac{1}{2}\left[\frac{x^2}{2} - x\right]_1^2$
$= \frac{1}{2}\left[\left(\frac{1^2}{2} - 0\right) + \left(\frac{2^2}{2} - 2\right) - \left(\frac{1^2}{2} - 1\right)\right] = \frac{1}{2}$

18. Ave. value $= \frac{1}{b-a}\int_a^b f(x)\,dx = \frac{1}{3-0}\int_0^3 f(x)\,dx = \frac{1}{3}\left[\int_0^1 dx + \int_1^2 0\,dx + \int_2^3 dx\right] = \frac{1}{3}[1 - 0 + 0 + 3 - 2] = \frac{2}{3}$

19. $\lim\limits_{b\to 1^-} \int_0^b \frac{1}{\sqrt{1-x^2}}\,dx = \lim\limits_{b\to 1^-} \left[\sin^{-1} x\right]_0^b = \lim\limits_{b\to 1^-} \left(\sin^{-1} b - \sin^{-1} 0\right) = \lim\limits_{b\to 1^-} \left(\sin^{-1} b - 0\right) = \lim\limits_{b\to 1^-} \sin^{-1} b = \frac{\pi}{2}$

20. $\lim\limits_{x\to\infty} \frac{1}{x}\int_0^x \tan^{-1} t\,dt = \lim\limits_{x\to\infty} \frac{\int_0^x \tan^{-1} t\,dt}{x}$ $\left(\frac{\infty}{\infty}\text{ form}\right)$
$= \lim\limits_{x\to\infty} \frac{\tan^{-1} x}{1} = \frac{\pi}{2}$

21. $\lim\limits_{x\to\infty} \left(\frac{1}{n+1} + \frac{1}{n+2} + \ldots + \frac{1}{2n}\right) = \lim\limits_{x\to\infty} \left(\left(\frac{1}{n}\right)\left[\frac{1}{1+\left(\frac{1}{n}\right)}\right] + \left(\frac{1}{n}\right)\left[\frac{1}{1+2\left(\frac{1}{n}\right)}\right] + \ldots + \left(\frac{1}{n}\right)\left[\frac{1}{1+n\left(\frac{1}{n}\right)}\right]\right)$
which can be interpreted as a Riemann sum with partitioning $\Delta x = \frac{1}{n} \Rightarrow \lim\limits_{x\to\infty} \left(\frac{1}{n+1} + \frac{1}{n+2} + \ldots + \frac{1}{2n}\right)$
$= \int_0^1 \frac{1}{1+x}\,dx = [\ln(1+x)]_0^1 = \ln 2$

22. $\lim\limits_{x\to\infty} \frac{1}{n}\left[e^{1/n} + e^{2/n} + \ldots + e\right] = \lim\limits_{x\to\infty} \left[\left(\frac{1}{n}\right)e^{(1/n)} + \left(\frac{1}{n}\right)e^{2(1/n)} + \ldots + \left(\frac{1}{n}\right)e^{n(1/n)}\right]$ which can be interpreted as a Riemann sum with partitioning $\Delta x = \frac{1}{n} \Rightarrow \lim\limits_{x\to\infty} \frac{1}{n}\left[e^{1/n} + e^{2/n} + \ldots + e\right] = \int_0^1 e^x\,dx = \left[e^x\right]_0^1 = e - 1$

23. Let $f(x) = x^5$ on $[0, 1]$. Partition $[0, 1]$ into n subintervals with $\Delta x = \frac{1-0}{n} = \frac{1}{n}$. Then $\frac{1}{n}, \frac{2}{n}, \ldots, \frac{n}{n}$ are the right-hand endpoints of the subintervals. Since f is increasing on $[0, 1]$, $U = \sum_{j=1}^{\infty} \left(\frac{j}{n}\right)^5 \left(\frac{1}{n}\right)$ is the upper sum for $f(x) = x^5$ on $[0, 1] \Rightarrow \lim_{n \to \infty} \sum_{j=1}^{\infty} \left(\frac{j}{n}\right)^5 \left(\frac{1}{n}\right) = \lim_{n \to \infty} \frac{1}{n}\left[\left(\frac{1}{n}\right)^5 + \left(\frac{2}{n}\right)^5 + \ldots + \left(\frac{n}{n}\right)^5\right] = \lim_{n \to \infty} \left[\frac{1^5 + 2^5 + \ldots + n^5}{n^6}\right]$ $= \int_0^1 x^5\, dx = \left[\frac{x^6}{6}\right]_0^1 = \frac{1}{6}$

24. Let $f(x) = x^3$ on $[0, 1]$. Partition $[0, 1]$ into n subintervals with $\Delta x = \frac{1-0}{n} = \frac{1}{n}$. Then $\frac{1}{n}, \frac{2}{n}, \ldots, \frac{n}{n}$ are the right-hand endpoints of the subintervals. Since f is increasing on $[0, 1]$, $U = \sum_{j=1}^{\infty} \left(\frac{j}{n}\right)^3 \left(\frac{1}{n}\right)$ is the upper sum for $f(x) = x^3$ on $[0, 1] \Rightarrow \lim_{n \to \infty} \sum_{j=1}^{\infty} \left(\frac{j}{n}\right)^3 \left(\frac{1}{n}\right) = \lim_{n \to \infty} \frac{1}{n}\left[\left(\frac{1}{n}\right)^3 + \left(\frac{2}{n}\right)^3 + \ldots + \left(\frac{n}{n}\right)^3\right] = \lim_{n \to \infty} \left[\frac{1^3 + 2^3 + \ldots + n^3}{n^4}\right]$ $= \int_0^1 x^3\, dx = \left[\frac{x^4}{4}\right]_0^1 = \frac{1}{4}$

25. Let $y = f(x)$ on $[0, 1]$. Partition $[0, 1]$ into n subintervals with $\Delta x = \frac{1-0}{n} = \frac{1}{n}$. Then $\frac{1}{n}, \frac{2}{n}, \ldots, \frac{n}{n}$ are the right-hand endpoints of the subintervals. Since f is continuous on $[0, 1]$, $\sum_{j=1}^{\infty} f\left(\frac{j}{n}\right)\left(\frac{1}{n}\right)$ is a Riemann sum of $y = f(x)$ on $[0, 1] \Rightarrow \lim_{n \to \infty} \sum_{j=1}^{\infty} f\left(\frac{j}{n}\right)\left(\frac{1}{n}\right) = \lim_{n \to \infty} \frac{1}{n}\left[f\left(\frac{1}{n}\right) + f\left(\frac{2}{n}\right) + \ldots + f\left(\frac{n}{n}\right)\right] = \int_0^1 f(x)\, dx$

26. (a) $\lim_{n \to \infty} \frac{1}{n^2}[2 + 4 + 6 + \ldots + 2n] = \lim_{n \to \infty} \frac{1}{n}\left[\frac{2}{n} + \frac{4}{n} + \frac{6}{n} + \ldots + \frac{2n}{n}\right] = \int_0^1 2x\, dx = \left[x^2\right]_0^1 = 1$, where $f(x) = 2x$ on $[0, 1]$ (see Exercise 25)

(b) $\lim_{n \to \infty} \frac{1}{n^{16}}\left[1^{15} + 2^{15} + \ldots + n^{15}\right] = \lim_{n \to \infty} \frac{1}{n}\left[\left(\frac{1}{n}\right)^{15} + \left(\frac{2}{n}\right)^{15} + \ldots + \left(\frac{n}{n}\right)^{15}\right] = \int_0^1 x^{15}\, dx = \left[\frac{x^{16}}{16}\right]_0^1 = \frac{1}{16}$, where $f(x) = x^{15}$ on $[0, 1]$ (see Exercise 25)

(c) $\lim_{n \to \infty} \frac{1}{n}\left[\sin\frac{\pi}{n} + \sin\frac{2\pi}{n} + \ldots + \sin\frac{n\pi}{n}\right] = \int_0^1 \sin n\pi\, dx = \left[-\frac{1}{\pi}\cos \pi x\right]_0^1 = -\frac{1}{\pi}\cos\pi - \left(-\frac{1}{\pi}\cos 0\right)$ $= \frac{2}{\pi}$, where $f(x) = \sin \pi x$ on $[0, 1]$ (see Exercise 25)

(d) $\lim_{n \to \infty} \frac{1}{n^{17}}\left[1^{15} + 2^{15} + \ldots + n^{15}\right] = \left(\lim_{n \to \infty} \frac{1}{n}\right)\left(\lim_{n \to \infty} \frac{1}{n^{16}}\left[1^{15} + 2^{15} + \ldots + n^{15}\right]\right) = \left(\lim_{n \to \infty} \frac{1}{n}\right)\int_0^1 x^{15}\, dx$ $= 0\left(\frac{1}{16}\right) = 0$ (see part (b) above)

(e) $\lim_{n \to \infty} \frac{1}{n^{15}}\left[1^{15} + 2^{15} + \ldots + n^{15}\right] = \lim_{n \to \infty} \frac{n}{n^{16}}\left[1^{15} + 2^{15} + \ldots + n^{15}\right]$ $= \left(\lim_{n \to \infty} n\right)\left(\lim_{n \to \infty} \frac{1}{n^{16}}\left[1^{15} + 2^{15} + \ldots + n^{15}\right]\right) = \left(\lim_{n \to \infty} n\right)\int_0^1 x^{15}\, dx = \infty$ (see part (b) above)

27. (a) Let the polygon be inscribed in a circle of radius r. If we draw a radius from the center of the circle (and the polygon) to each vertex of the polygon, we have n isosceles triangles formed (the equal sides are equal to r, the radius of the circle) and a vertex angle of θ_n where $\theta_n = \frac{2\pi}{n}$. The area of each triangle is $A_n = \frac{1}{2}r^2 \sin\theta_n \Rightarrow$ the area of the polygon is $A = nA_n = \frac{nr^2}{2}\sin\theta_n = \frac{nr^2}{2}\sin\frac{2\pi}{n}$.

(b) $\lim_{n \to \infty} A = \lim_{n \to \infty} \frac{nr^2}{2}\sin\frac{2\pi}{n} = \lim_{n \to \infty} \frac{n\pi r^2}{2\pi}\sin\frac{2\pi}{n} = \lim_{n \to \infty} (\pi r^2)\frac{\sin\left(\frac{2\pi}{n}\right)}{\left(\frac{2\pi}{n}\right)} = (\pi r^2)\lim_{2\pi/n \to 0}\frac{\sin\left(\frac{2\pi}{n}\right)}{\left(\frac{2\pi}{n}\right)} = \pi r^2$

28. Partition $[0, 1]$ into n subintervals, each of length $\Delta x = \frac{1}{n}$ with the points $x_0 = 0$, $x_1 = \frac{1}{n}$, $x_2 = \frac{2}{n}, \ldots, x_n = \frac{n}{n} = 1$. The inscribed rectangles so determined have areas $f(x_0)\,\Delta x = (0)^2\,\Delta x$, $f(x_1)\,\Delta x = \left(\frac{1}{n}\right)^2\Delta x$, $f(x_2)\,\Delta x = \left(\frac{2}{n}\right)^2\Delta x, \ldots, f(x_{n-1}) = \left(\frac{n-1}{n}\right)^2\Delta x$. The sum of these areas is $S_n = \left(0^2 + \left(\frac{1}{n}\right)^2 + \left(\frac{2}{n}\right)^2 + \ldots + \left(\frac{n-1}{n}\right)^2\right)\Delta x = \left(\frac{1^2}{n^2} + \frac{2^2}{n^2} + \ldots + \frac{(n-1)^2}{n^2}\right)\frac{1}{n} = \frac{1^2}{n^3} + \frac{2^2}{n^3} + \ldots + \frac{(n-1)^2}{n^3}$. Then $\lim_{n \to \infty} S_n = \lim_{n \to \infty}\left(\frac{1^2}{n^3} + \frac{2^2}{n^3} + \ldots + \frac{(n-1)^2}{n^3}\right) = \int_0^1 x^2\, dx = \frac{1^3}{3} = \frac{1}{3}$.

29. $A_1 = \int_1^e \frac{2 \log_2 x}{x}\,dx = \frac{2}{\ln 2}\int_1^e \frac{\ln x}{x}\,dx = \left[\frac{(\ln x)^2}{\ln 2}\right]_1^e = \frac{1}{\ln 2}$; $A_2 = \int_1^e \frac{2 \log_4 x}{4}\,dx = \frac{2}{\ln 4}\int_1^e \frac{\ln x}{x}\,dx$ $= \left[\frac{(\ln x)^2}{2\ln 2}\right]_1^e = \frac{1}{2\ln 2} \Rightarrow A_1 : A_2 = 2:1$

30. $\ln x^{(x^x)} = x^x \ln x$ and $\ln (x^x)^x = x \ln x^x = x^2 \ln x$; then, $x^x \ln x = x^2 \ln x \Rightarrow (x^x - x^2)\ln x = 0 \Rightarrow x^x = x^2$ or $\ln x = 0$. $\ln x = 0 \Rightarrow x = 1$; $x^x = x^2 \Rightarrow x \ln x = 2 \ln x \Rightarrow x = 2$. Therefore, $x^{(x^x)} = (x^x)^x$ when $x = 2$ or $x = 1$.

31. $f(x) = e^{g(x)} \Rightarrow f'(x) = e^{g(x)}\, g'(x)$, where $g'(x) = \frac{x}{1+x^4} \Rightarrow f'(2) = e^0\left(\frac{2}{1+16}\right) = \frac{2}{17}$

32. (a) $\frac{df}{dx} = \frac{2\ln e^x}{e^x} \cdot e^x = 2x$

(b) $f(0) = \int_1^1 \frac{2\ln t}{t}\,dt = 0$

(c) $\frac{df}{dx} = 2x \Rightarrow f(x) = x^2 + C$; $f(0) = 0 \Rightarrow C = 0 \Rightarrow f(x) = x^2 \Rightarrow$ the graph of f(x) is a parabola

33. (a) $g(1) = \int_1^1 f(t)\,dt = 0$

(b) $g(3) = \int_1^3 f(t)\,dt = -\frac{1}{2}(2)(1) = -1$

(c) $g(-1) = \int_1^{-1} f(t)\,dt = -\int_{-1}^1 f(t)\,dt = -\frac{1}{4}(\pi\, 2^2) = -\pi$

(d) $g'(x) = f(x) = 0 \Rightarrow x = -3, 1, 3$ and the sign chart for $g'(x) = f(x)$ is $\underset{-3}{|}+++\underset{1}{|}---\underset{3}{|}+++$. So g has a relative maximum at $x = 1$.

(e) $g'(-1) = f(-1) = 2$ is the slope and $g(-1) = \int_1^{-1} f(t)\,dt = -\pi$, by (c). Thus the equation is $y + \pi = 2(x+1)$ $y = 2x + 2 - \pi$.

(f) $g''(x) = f'(x) = 0$ at $x = -1$ and $g''(x) = f'(x)$ is negative on $(-3, -1)$ and positive on $(-1, 1)$ so there is an inflection point for g at $x = -1$. We notice that $g''(x) = f'(x) < 0$ for x on $(-1, 2)$ and $g''(x) = f'(x) > 0$ for x on $(2, 4)$, even though $g''(2)$ does not exist, g has a tangent line at $x = 2$, so there is an inflection point at $x = 2$.

(g) g is continuous on $[-3, 4]$ and so it attains its absolute maximum and minimum values on this interval. We saw in (d) that $g'(x) = 0 \Rightarrow x = -3, 1, 3$. We have that

$$g(-3) = \int_1^{-3} f(t)\,dt = -\int_{-3}^1 f(t)\,dt = -\frac{\pi 2^2}{2} = -2\pi$$

$$g(1) = \int_1^1 f(t)\,dt = 0$$

$$g(3) = \int_1^3 f(t)\,dt = -1$$

$$g(4) = \int_1^4 f(t)\,dt = -1 + \frac{1}{2}\cdot 1 \cdot 1 = -\frac{1}{2}$$

Thus, the absolute minimum is -2π and the absolute maximum is 0. Thus, the range is $[-2\pi, 0]$.

34. $y = \sin x + \int_x^\pi \cos 2t\,dt + 1 = \sin x - \int_\pi^x \cos 2t\,dt + 1 \Rightarrow y' = \cos x - \cos(2x)$; when $x = \pi$ we have $y' = \cos\pi - \cos(2\pi) = -1 - 1 = -2$. And $y'' = -\sin x + 2\sin(2x)$; when $x = \pi$, $y = \sin\pi + \int_x^\pi \cos 2t\,dt + 1$ $= 0 + 0 + 1 = 1$.

35. The area of the shaded region is $\int_0^1 \sin^{-1} x\,dx = \int_0^1 \sin^{-1} y\,dy$, which is the same as the area of the region to the left of the curve $y = \sin x$ (and part of the rectangle formed by the coordinate axes and dashed lines $y = 1$, $x = \frac{\pi}{2}$). The area of the rectangle is $\frac{\pi}{2} = \int_0^1 \sin^{-1} y\,dy + \int_0^{\pi/2} \sin x\,dx$, so we have $\frac{\pi}{2} = \int_0^1 \sin^{-1} x\,dx + \int_0^{\pi/2} \sin x\,dx \Rightarrow \int_0^{\pi/2} \sin x\,dx = \frac{\pi}{2} - \int_0^1 \sin^{-1} x\,dx$.

36. (a) slope of L_3 < slope of L_2 < slope of L_1 $\Rightarrow \frac{1}{b} < \frac{\ln b - \ln a}{b - a} < \frac{1}{a}$

(b) area of small (shaded) rectangle < area under curve < area of large rectangle

$\Rightarrow \frac{1}{b}(b-a) < \int_a^b \frac{1}{x}\,dx < \frac{1}{a}(b-a) \Rightarrow \frac{1}{b} < \frac{\ln b - \ln a}{b-a} < \frac{1}{a}$

37. $f(x) = \int_{1/x}^{x} \frac{1}{t}\,dt \Rightarrow f'(x) = \frac{1}{x}\left(\frac{dx}{dx}\right) - \left(\frac{1}{\frac{1}{x}}\right)\left(\frac{d}{dx}\left(\frac{1}{x}\right)\right) = \frac{1}{x} - x\left(-\frac{1}{x^2}\right) = \frac{1}{x} + \frac{1}{x} = \frac{2}{x}$

38. $f(x) = \int_{\cos x}^{\sin x} \frac{1}{1-t^2}\,dt \Rightarrow f'(x) = \left(\frac{1}{1-\sin^2 x}\right)\left(\frac{d}{dx}(\sin x)\right) - \left(\frac{1}{1-\cos^2 x}\right)\left(\frac{d}{dx}(\cos x)\right) = \frac{\cos x}{\cos^2 x} + \frac{\sin x}{\sin^2 x}$
$= \frac{1}{\cos x} + \frac{1}{\sin x}$

39. $g(y) = \int_{\sqrt{y}}^{2\sqrt{y}} \sin t^2\,dt \Rightarrow g'(y) = \left(\sin\left(2\sqrt{y}\right)^2\right)\left(\frac{d}{dy}\left(2\sqrt{y}\right)\right) - \left(\sin\left(\sqrt{y}\right)^2\right)\left(\frac{d}{dy}\left(\sqrt{y}\right)\right) = \frac{\sin 4y}{\sqrt{y}} - \frac{\sin y}{2\sqrt{y}}$

40. $g(y) = \int_{\sqrt{y}}^{y^2} \frac{e^t}{t}\,dt \Rightarrow g'(y) = \frac{e^{y^2}}{y^2} \cdot \frac{d}{dy}(y^2) - \frac{e^{\sqrt{y}}}{\sqrt{y}} \cdot \frac{d}{dy}\left(\sqrt{y}\right) = \frac{e^{y^2}}{y^2}(2y) - \frac{e^{\sqrt{y}}}{\sqrt{y}} \cdot \frac{1}{2\sqrt{y}} = \frac{4e^{y^2}}{2y} - \frac{e^{\sqrt{y}}}{2y} = \frac{4e^{y^2} - e^{\sqrt{y}}}{2y}$

41. $y = \int_{x^2/2}^{x^2} \ln\sqrt{t}\,dt \Rightarrow \frac{dy}{dx} = \left(\ln\sqrt{x^2}\right) \cdot \frac{d}{dx}(x^2) - \left(\ln\sqrt{\frac{x^2}{2}}\right) \cdot \frac{d}{dx}\left(\frac{x^2}{2}\right) = 2x \ln|x| - x \ln\frac{|x|}{\sqrt{2}}$

42. $y = \int_{\sqrt{x}}^{\sqrt[3]{x}} \ln t\,dt \Rightarrow \frac{dy}{dx} = \left(\ln\sqrt[3]{x}\right) \cdot \frac{d}{dx}\left(\sqrt[3]{x}\right) - \left(\ln\sqrt{x}\right) \cdot \frac{d}{dx}\left(\sqrt{x}\right) = \left(\ln\sqrt[3]{x}\right)\left(\frac{1}{3}x^{-2/3}\right) - \left(\ln\sqrt{x}\right)\left(\frac{1}{2}x^{-1/2}\right)$
$= \frac{\ln\sqrt[3]{x}}{3\sqrt[3]{x^2}} - \frac{\ln\sqrt{x}}{2\sqrt{x}}$

43. $\int_0^{\ln x} \sin e^t\,dt \Rightarrow y' = \left(\sin e^{\ln x}\right) \cdot \frac{d}{dx}(\ln x) = \frac{\sin x}{x}$

44. $y = \int_{e^{4\sqrt{x}}}^{e^{2x}} \ln t\,dt \Rightarrow y' = \left(\ln e^{2x}\right) \cdot \frac{d}{dx}\left(e^{2x}\right) - \left(\ln e^{4\sqrt{x}}\right) \cdot \frac{d}{dx}\left(e^{4\sqrt{x}}\right) = (2x)\left(2e^{2x}\right) - \left(4\sqrt{x}\right)\left(e^{4\sqrt{x}}\right) \cdot \frac{d}{dx}\left(4\sqrt{x}\right)$
$= 4xe^{2x} - 4\sqrt{x}\,e^{4\sqrt{x}}\left(\frac{2}{\sqrt{x}}\right) = 4xe^{2x} - 8e^{4\sqrt{x}}$

45. $f(x) = \int_x^{x+3} t(5-t)\,dt \Rightarrow f'(x) = (x+3)(5-(x+3))\left(\frac{d}{dx}(x+3)\right) - x(5-x)\left(\frac{dx}{dx}\right) = (x+3)(2-x) - x(5-x)$
$= 6 - x - x^2 - 5x + x^2 = 6 - 6x$. Thus $f'(x) = 0 \Rightarrow 6 - 6x = 0 \Rightarrow x = 1$. Also, $f''(x) = -6 < 0 \Rightarrow x = 1$ gives a maximum.

NOTES:

CHAPTER 6 APPLICATIONS OF DEFINITE INTEGRALS

6.1 VOLUMES BY SLICING AND ROTATION ABOUT AN AXIS

1. (a) $A = \pi(\text{radius})^2$ and radius $= \sqrt{1-x^2} \Rightarrow A(x) = \pi(1-x^2)$
   (b) $A = \text{width} \cdot \text{height}$, width $=$ height $= 2\sqrt{1-x^2} \Rightarrow A(x) = 4(1-x^2)$
   (c) $A = (\text{side})^2$ and diagonal $= \sqrt{2}(\text{side}) \Rightarrow A = \frac{(\text{diagonal})^2}{2}$; diagonal $= 2\sqrt{1-x^2} \Rightarrow A(x) = 2(1-x^2)$
   (d) $A = \frac{\sqrt{3}}{4}(\text{side})^2$ and side $= 2\sqrt{1-x^2} \Rightarrow A(x) = \sqrt{3}(1-x^2)$

2. (a) $A = \pi(\text{radius})^2$ and radius $= \sqrt{x} \Rightarrow A(x) = \pi x$
   (b) $A = \text{width} \cdot \text{height}$, width $=$ height $= 2\sqrt{x} \Rightarrow A(x) = 4x$
   (c) $A = (\text{side})^2$ and diagonal $= \sqrt{2}(\text{side}) \Rightarrow A = \frac{(\text{diagonal})^2}{2}$; diagonal $= 2\sqrt{x} \Rightarrow A(x) = 2x$
   (d) $A = \frac{\sqrt{3}}{4}(\text{side})^2$ and side $= 2\sqrt{x} \Rightarrow A(x) = \sqrt{3}x$

3. $A(x) = \frac{(\text{diagonal})^2}{2} = \frac{\left(\sqrt{x}-(-\sqrt{x})\right)^2}{2} = 2x$ (see Exercise 1c); $a = 0$, $b = 4$;
   $V = \int_a^b A(x)\,dx = \int_0^4 2x\,dx = \left[x^2\right]_0^4 = 16$

4. $A(x) = \frac{\pi(\text{diameter})^2}{4} = \frac{\pi[(2-x^2)-x^2]^2}{4} = \frac{\pi[2(1-x^2)]^2}{4} = \pi(1-2x^2+x^4)$; $a = -1$, $b = 1$;
   $V = \int_a^b A(x)\,dx = \int_{-1}^1 \pi(1-2x^2+x^4)\,dx = \pi\left[x - \frac{2}{3}x^3 + \frac{x^5}{5}\right]_{-1}^1 = 2\pi\left(1-\frac{2}{3}+\frac{1}{5}\right) = \frac{16\pi}{15}$

5. $A(x) = (\text{edge})^2 = \left[\sqrt{1-x^2} - \left(-\sqrt{1-x^2}\right)\right]^2 = \left(2\sqrt{1-x^2}\right)^2 = 4(1-x^2)$; $a = -1$, $b = 1$;
   $V = \int_a^b A(x)\,dx = \int_{-1}^1 4(1-x^2)\,dx = 4\left[x - \frac{x^3}{3}\right]_{-1}^1 = 8\left(1-\frac{1}{3}\right) = \frac{16}{3}$

6. $A(x) = \frac{(\text{diagonal})^2}{2} = \frac{\left[\sqrt{1-x^2}-\left(-\sqrt{1-x^2}\right)\right]^2}{2} = \frac{\left(2\sqrt{1-x^2}\right)^2}{2} = 2(1-x^2)$ (see Exercise 1c); $a = -1$, $b = 1$;
   $V = \int_a^b A(x)\,dx = 2\int_{-1}^1 (1-x^2)\,dx = 2\left[x - \frac{x^3}{3}\right]_{-1}^1 = 4\left(1-\frac{1}{3}\right) = \frac{8}{3}$

7. (a) STEP 1) $A(x) = \frac{1}{2}(\text{side}) \cdot (\text{side}) \cdot \left(\sin\frac{\pi}{3}\right) = \frac{1}{2} \cdot \left(2\sqrt{\sin x}\right) \cdot \left(2\sqrt{\sin x}\right)\left(\sin\frac{\pi}{3}\right) = \sqrt{3}\sin x$
   STEP 2) $a = 0$, $b = \pi$
   STEP 3) $V = \int_a^b A(x)\,dx = \sqrt{3}\int_0^\pi \sin x\,dx = \left[-\sqrt{3}\cos x\right]_0^\pi = \sqrt{3}(1+1) = 2\sqrt{3}$
   (b) STEP 1) $A(x) = (\text{side})^2 = \left(2\sqrt{\sin x}\right)\left(2\sqrt{\sin x}\right) = 4\sin x$
   STEP 2) $a = 0$, $b = \pi$
   STEP 3) $V = \int_a^b A(x)\,dx = \int_0^\pi 4\sin x\,dx = \left[-4\cos x\right]_0^\pi = 8$

8. (a) STEP 1) $A(x) = \frac{\pi(\text{diameter})^2}{4} = \frac{\pi}{4}(\sec x - \tan x)^2 = \frac{\pi}{4}(\sec^2 x + \tan^2 x - 2\sec x\tan x)$
   $= \frac{\pi}{4}\left[\sec^2 x + (\sec^2 x - 1) - 2\,\frac{\sin x}{\cos^2 x}\right]$
   STEP 2) $a = -\frac{\pi}{3}$, $b = \frac{\pi}{3}$
   STEP 3) $V = \int_a^b A(x)\,dx = \int_{-\pi/3}^{\pi/3} \frac{\pi}{4}\left(2\sec^2 x - 1 - \frac{2\sin x}{\cos^2 x}\right) dx = \frac{\pi}{4}\left[2\tan x - x + 2\left(-\frac{1}{\cos x}\right)\right]_{-\pi/3}^{\pi/3}$

$= \frac{\pi}{4}\left[2\sqrt{3} - \frac{\pi}{3} + 2\left(-\frac{1}{\left(\frac{1}{2}\right)}\right) - \left(-2\sqrt{3} + \frac{\pi}{3} + 2\left(-\frac{1}{\left(\frac{1}{2}\right)}\right)\right)\right] = \frac{\pi}{4}\left(4\sqrt{3} - \frac{2\pi}{3}\right)$

(b) STEP 1) $A(x) = (\text{edge})^2 = (\sec x - \tan x)^2 = \left(2\sec^2 x - 1 - 2\,\frac{\sin x}{\cos^2 x}\right)$

STEP 2) $a = -\frac{\pi}{3},\ b = \frac{\pi}{3}$

STEP 3) $V = \int_a^b A(x)\,dx = \int_{-\pi/3}^{\pi/3} \left(2\sec^2 x - 1 - \frac{2\sin x}{\cos^2 x}\right) dx = 2\left(2\sqrt{3} - \frac{\pi}{3}\right) = 4\sqrt{3} - \frac{2\pi}{3}$

9. $A(y) = \frac{\pi}{4}(\text{diameter})^2 = \frac{\pi}{4}\left(\sqrt{5}y^2 - 0\right)^2 = \frac{5\pi}{4}y^4;$

$c = 0,\ d = 2;\ V = \int_c^d A(y)\,dy = \int_0^2 \frac{5\pi}{4} y^4\,dy$

$= \left[\left(\frac{5\pi}{4}\right)\left(\frac{y^5}{5}\right)\right]_0^2 = \frac{\pi}{4}\left(2^5 - 0\right) = 8\pi$

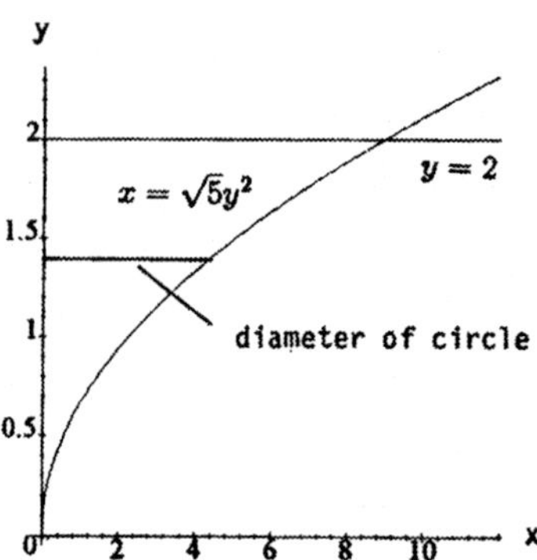

10. $A(y) = \frac{1}{2}(\text{leg})(\text{leg}) = \frac{1}{2}\left[\sqrt{1-y^2} - \left(-\sqrt{1-y^2}\right)\right]^2 = \frac{1}{2}\left(2\sqrt{1-y^2}\right)^2 = 2\left(1-y^2\right);\ c = -1,\ d = 1;$

$V = \int_c^d A(y)\,dy = \int_{-1}^1 2\left(1-y^2\right) dy = 2\left[y - \frac{y^3}{3}\right]_{-1}^1 = 4\left(1 - \frac{1}{3}\right) = \frac{8}{3}$

11. (a) $A(x) = \frac{\pi}{4}(\text{diameter})^2 = \frac{\pi}{4}\left[\frac{1}{\sqrt{1+x^2}} - \left(-\frac{1}{\sqrt{1+x^2}}\right)\right]^2 = \frac{\pi}{1+x^2} \Rightarrow V = \int_a^b A(x)\,dx = \int_{-1}^1 \frac{\pi\,dx}{1+x^2}$

$= \pi\left[\tan^{-1} x\right]_{-1}^1 = (\pi)(2)\left(\frac{\pi}{4}\right) = \frac{\pi^2}{2}$

(b) $A(x) = (\text{edge})^2 = \left[\frac{1}{\sqrt{1+x^2}} - \left(-\frac{1}{\sqrt{1+x^2}}\right)\right]^2 = \frac{4}{1+x^2} \Rightarrow V = \int_a^b A(x)\,dx = \int_{-1}^1 \frac{4\,dx}{1+x^2}$

$= 4\left[\tan^{-1} x\right]_{-1}^1 = 4\left[\tan^{-1}(1) - \tan^{-1}(-1)\right] = 4\left[\frac{\pi}{4} - \left(-\frac{\pi}{4}\right)\right] = 2\pi$

12. (a) $A(x) = \frac{\pi}{4}(\text{diameter})^2 = \frac{\pi}{4}\left(\frac{2}{\sqrt[4]{1-x^2}} - 0\right)^2 = \frac{\pi}{4}\left(\frac{4}{\sqrt{1-x^2}}\right) = \frac{\pi}{\sqrt{1-x^2}} \Rightarrow V = \int_a^b A(x)\,dx$

$= \int_{-\sqrt{2}/2}^{\sqrt{2}/2} \frac{\pi}{\sqrt{1-x^2}}\,dx = \pi\left[\sin^{-1} x\right]_{-\sqrt{2}/2}^{\sqrt{2}/2} = \pi\left[\sin^{-1}\left(\frac{\sqrt{2}}{2}\right) - \sin^{-1}\left(-\frac{\sqrt{2}}{2}\right)\right] = \pi\left[\frac{\pi}{4} - \left(-\frac{\pi}{4}\right)\right] = \frac{\pi^2}{2}$

(b) $A(x) = \frac{(\text{diagonal})^2}{2} = \frac{1}{2}\left(\frac{2}{\sqrt[4]{1-x^2}} - 0\right)^2 = \frac{2}{\sqrt{1-x^2}} \Rightarrow V = \int_a^b A(x)\,dx = \int_{-\sqrt{2}/2}^{\sqrt{2}/2} \frac{2}{\sqrt{1-x^2}}\,dx$

$= 2\left[\sin^{-1} x\right]_{-\sqrt{2}/2}^{\sqrt{2}/2} = 2\left(\frac{\pi}{4}\cdot 2\right) = \pi$

13. (a) It follows from Cavalieri's Principle that the volume of a column is the same as the volume of a right prism with a square base of side length s and altitude h. Thus, STEP 1) $A(x) = (\text{side length})^2 = s^2$; STEP 2) $a = 0,\ b = h$; STEP 3) $V = \int_a^b A(x)\,dx = \int_0^h s^2\,dx = s^2h$

(b) From Cavalieri's Principle we conclude that the volume of the column is the same as the volume of the prism described above, regardless of the number of turns $\Rightarrow V = s^2h$

14. 1) The solid and the cone have the same altitude of 12.
    2) The cross sections of the solid are disks of diameter $x - \left(\frac{x}{2}\right) = \frac{x}{2}$. If we place the vertex of the cone at the origin of the coordinate system and make its axis of symmetry coincide with the x-axis then the cone's cross sections will be circular disks of diameter $\frac{x}{4} - \left(-\frac{x}{4}\right) = \frac{x}{2}$ (see accompanying figure).
    3) The solid and the cone have equal altitudes and identical parallel cross sections. From Cavalieri's Principle we conclude that the solid and the cone have the same volume.

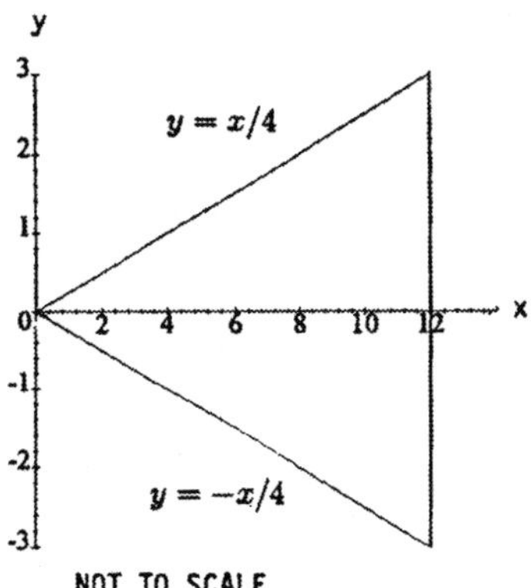

15. $R(x) = y = 1 - \frac{x}{2} \Rightarrow V = \int_0^2 \pi[R(x)]^2\,dx = \pi\int_0^2 \left(1 - \frac{x}{2}\right)^2 dx = \pi\int_0^2 \left(1 - x + \frac{x^2}{4}\right) dx = \pi\left[x - \frac{x^2}{2} + \frac{x^3}{12}\right]_0^2$
$= \pi\left(2 - \frac{4}{2} + \frac{8}{12}\right) = \frac{2\pi}{3}$

16. $R(y) = x = \frac{3y}{2} \Rightarrow V = \int_0^2 \pi[R(y)]^2\,dy = \pi\int_0^2 \left(\frac{3y}{2}\right)^2 dy = \pi\int_0^2 \frac{9}{4}y^2\,dy = \pi\left[\frac{3}{4}y^3\right]_0^2 = \pi \cdot \frac{3}{4} \cdot 8 = 6\pi$

17. $R(x) = \tan\left(\frac{\pi}{4}y\right)$; $u = \frac{\pi}{4}y \Rightarrow du = \frac{\pi}{4}\,dy \Rightarrow 4\,du = \pi\,dy$; $y = 0 \Rightarrow u = 0$, $y = 1 \Rightarrow u = \frac{\pi}{4}$;
$V = \int_0^1 \pi[R(y)]^2\,dy = \pi\int_0^1 \left[\tan\left(\frac{\pi}{4}y\right)\right]^2 dy = 4\int_0^{\pi/4} \tan^2 u\,du = 4\int_0^{\pi/4} (-1 + \sec^2 u)\,du = 4[-u + \tan u]_0^{\pi/4}$
$= 4\left(-\frac{\pi}{4} + 1 - 0\right) = 4 - \pi$

18. $R(x) = \sin x \cos x$; $R(x) = 0 \Rightarrow a = 0$ and $b = \frac{\pi}{2}$ are the limits of integration; $V = \int_0^{\pi/2} \pi[R(x)]^2\,dx$
$= \pi\int_0^{\pi/2} (\sin x \cos x)^2\,dx = \pi\int_0^{\pi/2} \frac{(\sin 2x)^2}{4}\,dx$; $\left[u = 2x \Rightarrow du = 2\,dx \Rightarrow \frac{du}{8} = \frac{dx}{4}; x = 0 \Rightarrow u = 0,\right.$
$\left.x = \frac{\pi}{2} \Rightarrow u = \pi\right] \rightarrow V = \pi\int_0^{\pi} \frac{1}{8}\sin^2 u\,du = \frac{\pi}{8}\left[\frac{u}{2} - \frac{1}{4}\sin 2u\right]_0^{\pi} = \frac{\pi}{8}\left[\left(\frac{\pi}{2} - 0\right) - 0\right] = \frac{\pi^2}{16}$

19. $R(x) = x^2 \Rightarrow V = \int_0^2 \pi[R(x)]^2\,dx = \pi\int_0^2 (x^2)^2\,dx$
$= \pi\int_0^2 x^4\,dx = \pi\left[\frac{x^5}{5}\right]_0^2 = \frac{32\pi}{5}$

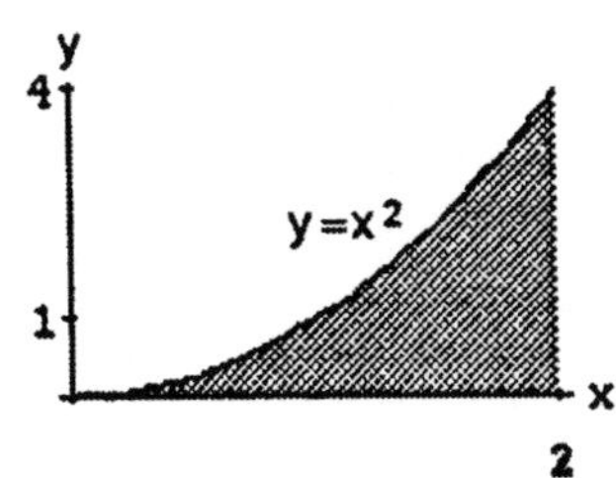

20. $R(x) = x^3 \Rightarrow V = \int_0^2 \pi[R(x)]^2\,dx = \pi\int_0^2 (x^3)^2\,dx$
$= \pi\int_0^2 x^6\,dx = \pi\left[\frac{x^7}{7}\right]_0^2 = \frac{128\pi}{7}$

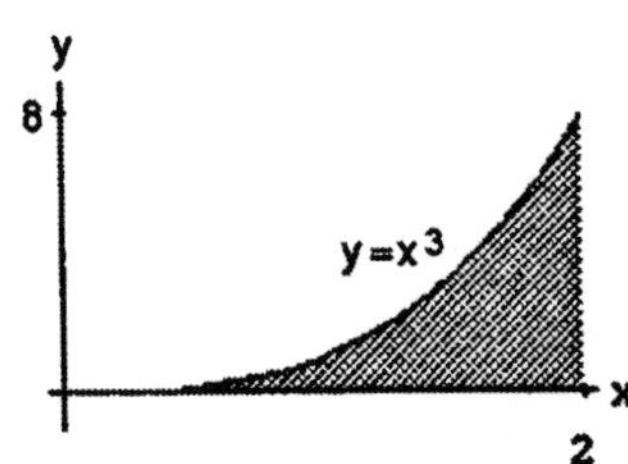

21. $R(x) = \sqrt{9 - x^2} \Rightarrow V = \int_{-3}^{3} \pi[R(x)]^2\,dx = \pi \int_{-3}^{3} (9 - x^2)\,dx$
$= \pi \left[9x - \frac{x^3}{3}\right]_{-3}^{3} = 2\pi \left[9(3) - \frac{27}{3}\right] = 2 \cdot \pi \cdot 18 = 36\pi$

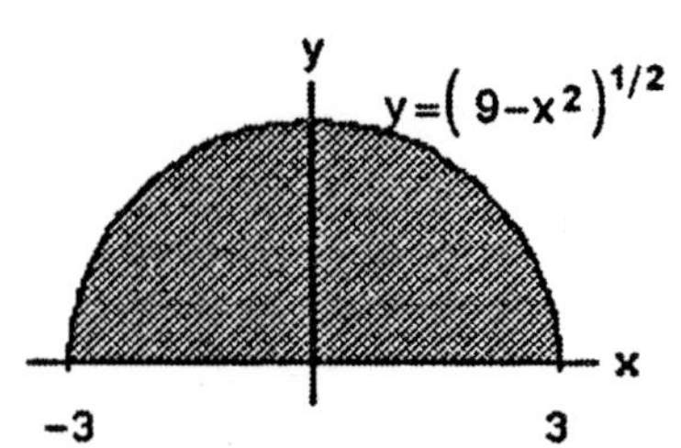

22. $R(x) = x - x^2 \Rightarrow V = \int_0^1 \pi[R(x)]^2\,dx = \pi \int_0^1 (x - x^2)^2\,dx$
$= \pi \int_0^1 (x^2 - 2x^3 + x^4)\,dx = \pi \left[\frac{x^3}{3} - \frac{2x^4}{4} + \frac{x^5}{5}\right]_0^1$
$= \pi\left(\frac{1}{3} - \frac{1}{2} + \frac{1}{5}\right) = \frac{\pi}{30}(10 - 15 + 6) = \frac{\pi}{30}$

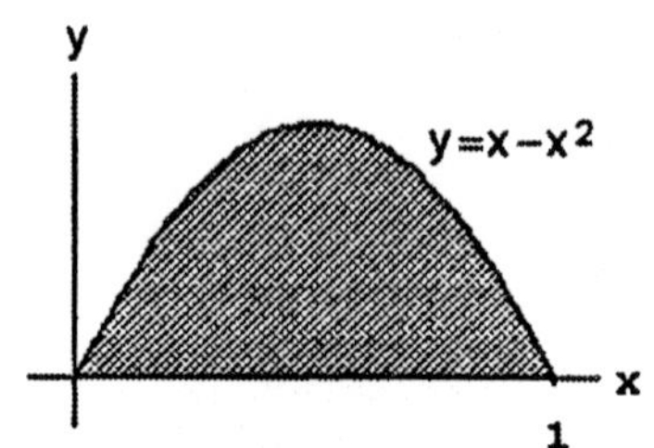

23. $R(x) = \sqrt{\cos x} \Rightarrow V = \int_0^{\pi/2} \pi[R(x)]^2\,dx = \pi \int_0^{\pi/2} \cos x\,dx$
$= \pi\,[\sin x]_0^{\pi/2} = \pi(1 - 0) = \pi$

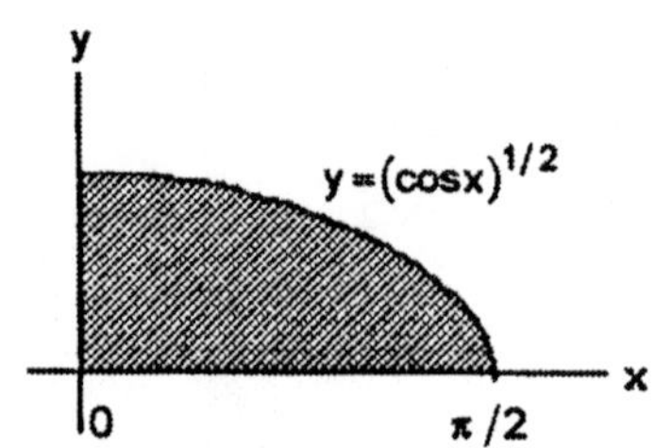

24. $R(x) = \sec x \Rightarrow V = \int_{-\pi/4}^{\pi/4} \pi[R(x)]^2\,dx = \pi \int_{-\pi/4}^{\pi/4} \sec^2 x\,dx$
$= \pi\,[\tan x]_{-\pi/4}^{\pi/4} = \pi[1 - (-1)] = 2\pi$

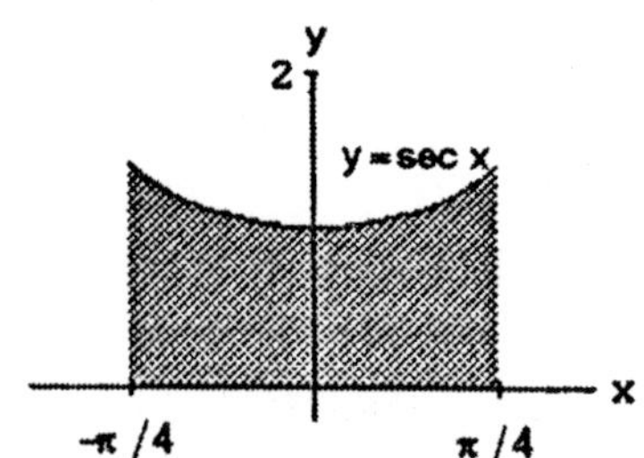

25. $R(x) = e^{-x} \Rightarrow V = \int_0^1 \pi[R(x)]^2\,dx = \pi \int_0^1 (e^{-x})^2\,dx$
$= \pi \int_0^1 e^{-2x}\,dx = -\frac{\pi}{2} e^{-2x}\Big|_0^1 = -\frac{\pi}{2}(e^{-2} - 1)$
$= \frac{\pi}{2}\left(1 - \frac{1}{e^2}\right) = \frac{\pi(e^2 - 1)}{2e^2}$

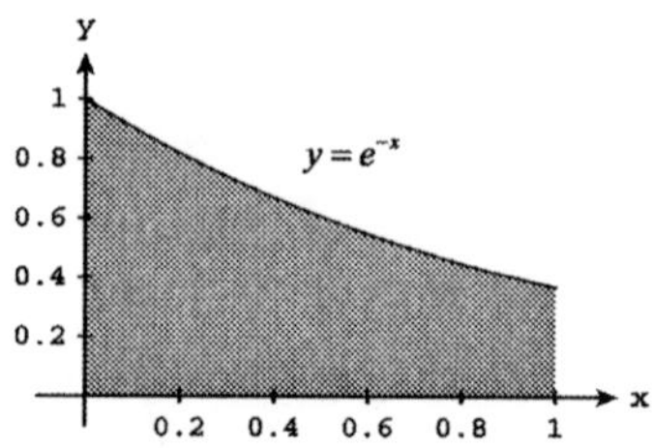

26. $V = \pi \int_{\pi/6}^{\pi/2} \cot x\,dx = \pi \int_{\pi/6}^{\pi/2} \frac{\cos x}{\sin x}\,dx = \pi\,[\ln(\sin x)]_{\pi/6}^{\pi/2} = \pi\left(\ln 1 - \ln \frac{1}{2}\right) = \pi \ln 2$

27. $V = \frac{\pi}{4} \int_{1/4}^{4} \frac{1}{x}\,dx = \frac{\pi}{4}\,[\ln x]_{1/4}^{4} = \frac{\pi}{4}\left(\ln 4 - \ln \frac{1}{4}\right) = \frac{\pi}{2} \ln 4$

28. $V = \pi \int_0^{2\pi} y^2\,dx = \pi \int_0^{2\pi} y^2 \left(\frac{dx}{d\theta}\right) d\theta = \pi \int_0^{2\pi} (1 - \cos\theta)^2(1 - \cos\theta)\,d\theta = \pi \int_0^{2\pi} (1 - 3\cos\theta + 3\cos^2\theta - \cos^3\theta)\,d\theta$;
evaluating each integral: $I_1 = \pi \int_0^{2\pi} d\theta = 2\pi^2$; $I_2 = \pi \int_0^{2\pi} (-3\cos\theta)\,d\theta = -3\pi\,[\sin\theta]_0^{2\pi} = 0$; $I_3 = \pi \int_0^{2\pi} 3\cos^2\theta\,d\theta$

$= 3\pi\left[\frac{1}{2}\theta + \frac{1}{4}\sin 2\theta\right]_0^{2\pi} = 3\pi^2$; $I_4 = \pi\int_0^{2\pi} \cos^3\theta\, d\theta = \pi\int_0^{2\pi} (1 - \sin^2\theta)(\cos\theta)\, d\theta = 0$; therefore
$V = I_1 + I_2 + I_3 + I_4 = 5\pi^2$

29. $R(x) = \sqrt{2} - \sec x \tan x \Rightarrow V = \int_0^{\pi/4} \pi[R(x)]^2\, dx$
$= \pi \int_0^{\pi/4} \left(\sqrt{2} - \sec x \tan x\right)^2 dx$
$= \pi \int_0^{\pi/4} \left(2 - 2\sqrt{2}\sec x \tan x + \sec^2 x \tan^2 x\right) dx$
$= \pi \left(\int_0^{\pi/4} 2\, dx - 2\sqrt{2}\int_0^{\pi/4} \sec x \tan x\, dx + \int_0^{\pi/4} (\tan x)^2 \sec^2 x\, dx\right)$
$= \pi \left([2x]_0^{\pi/4} - 2\sqrt{2}\,[\sec x]_0^{\pi/4} + \left[\frac{\tan^3 x}{3}\right]_0^{\pi/4}\right)$
$= \pi \left[\left(\frac{\pi}{2} - 0\right) - 2\sqrt{2}\left(\sqrt{2} - 1\right) + \frac{1}{3}(1^3 - 0)\right] = \pi\left(\frac{\pi}{2} + 2\sqrt{2} - \frac{11}{3}\right)$

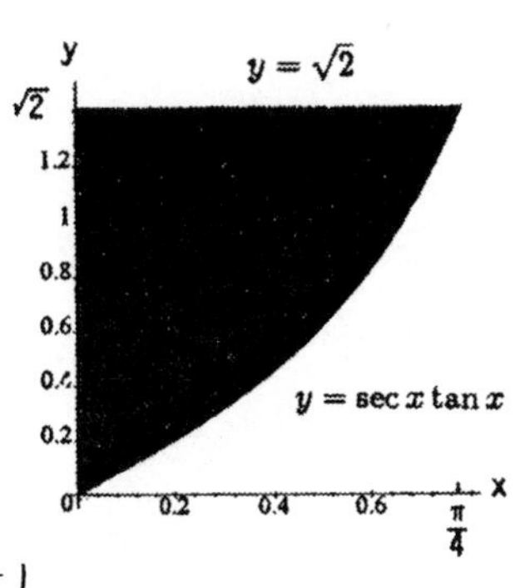

30. $R(x) = 2 - 2\sin x = 2(1 - \sin x) \Rightarrow V = \int_0^{\pi/2} \pi[R(x)]^2\, dx$
$= \pi \int_0^{\pi/2} 4(1 - \sin x)^2\, dx = 4\pi \int_0^{\pi/2} (1 + \sin^2 x - 2\sin x)\, dx$
$= 4\pi \int_0^{\pi/2} \left[1 + \frac{1}{2}(1 - \cos 2x) - 2\sin x\right] dx$
$= 4\pi \int_0^{\pi/2} \left(\frac{3}{2} - \frac{\cos 2x}{2} - 2\sin x\right)$
$= 4\pi \left[\frac{3}{2}x - \frac{\sin 2x}{4} + 2\cos x\right]_0^{\pi/2}$
$= 4\pi\left[\left(\frac{3\pi}{4} - 0 + 0\right) - (0 - 0 + 2)\right] = \pi(3\pi - 8)$

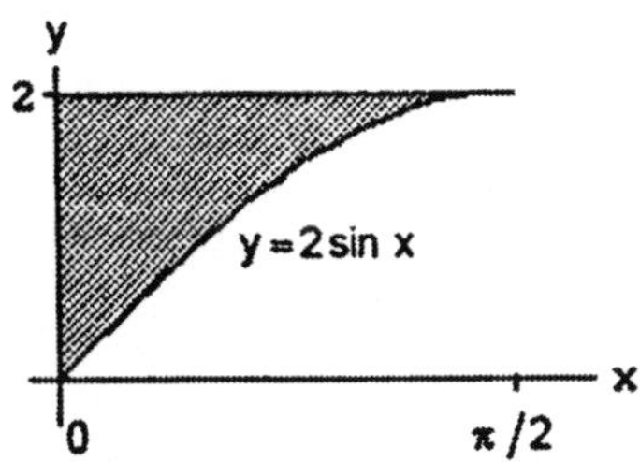

31. $R(y) = \sqrt{5} \cdot y^2 \Rightarrow V = \int_{-1}^{1} \pi[R(y)]^2\, dy = \pi \int_{-1}^{1} 5y^4\, dy$
$= \pi\left[y^5\right]_{-1}^{1} = \pi[1 - (-1)] = 2\pi$

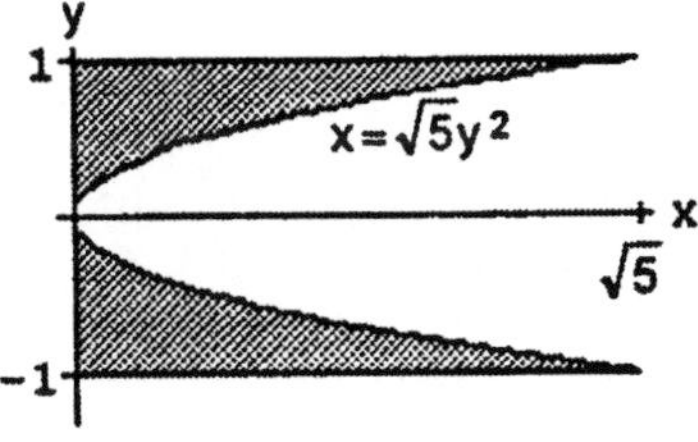

32. $R(y) = y^{3/2} \Rightarrow V = \int_0^2 \pi[R(y)]^2\, dy = \pi\int_0^2 y^3\, dy$
$= \pi\left[\frac{y^4}{4}\right]_0^2 = 4\pi$

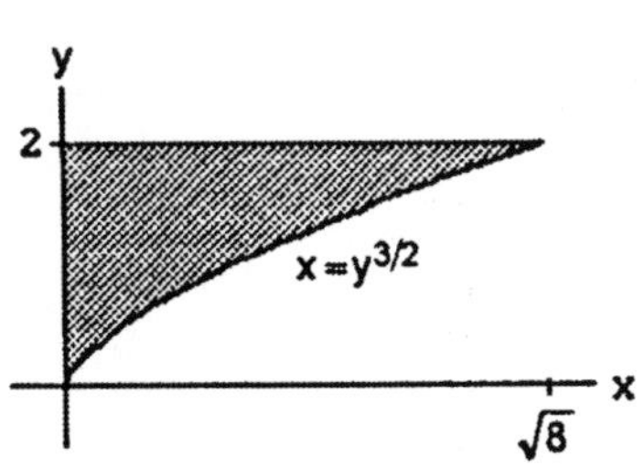

33. $R(y) = \sqrt{2\sin 2y} \Rightarrow V = \int_0^{\pi/2} \pi[R(y)]^2\, dy$
$= \pi\int_0^{\pi/2} 2\sin 2y\, dy = \pi\left[-\cos 2y\right]_0^{\pi/2}$
$= \pi[1 - (-1)] = 2\pi$

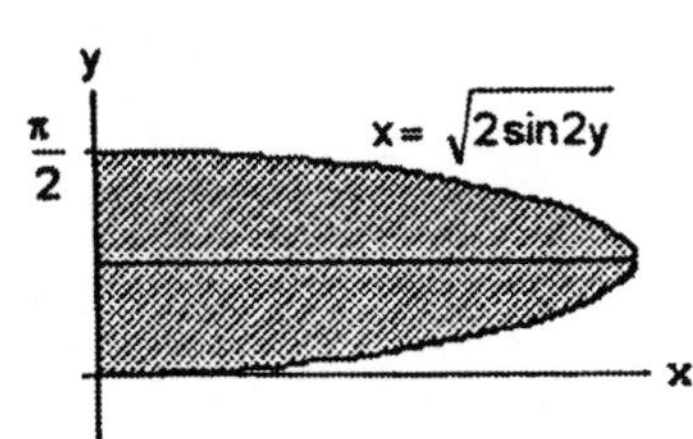

34. $R(y) = \sqrt{\cos \frac{\pi y}{4}} \Rightarrow V = \int_{-2}^{0} \pi [R(y)]^2 \, dy$
$= \pi \int_{-2}^{0} \cos\left(\frac{\pi y}{4}\right) dy = 4\left[\sin \frac{\pi y}{4}\right]_{-2}^{0} = 4[0 - (-1)] = 4$

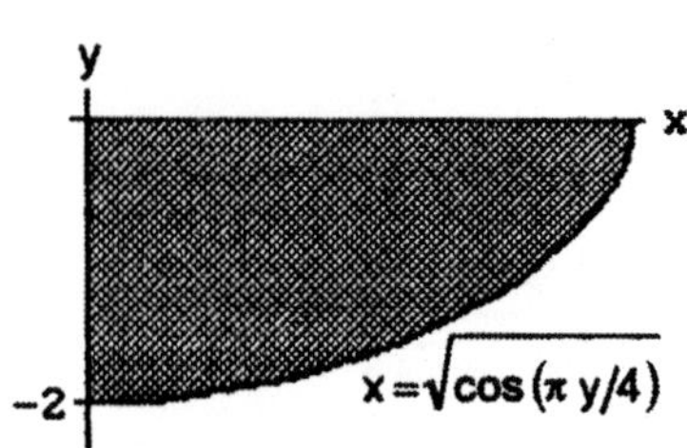

35. $R(y) = \frac{2}{y+1} \Rightarrow V = \int_{0}^{3} \pi [R(y)]^2 \, dy = 4\pi \int_{0}^{3} \frac{1}{(y+1)^2} \, dy$
$= 4\pi \left[\frac{-1}{y+1}\right]_{0}^{3} = 4\pi\left[-\frac{1}{4} - (-1)\right] = 3\pi$

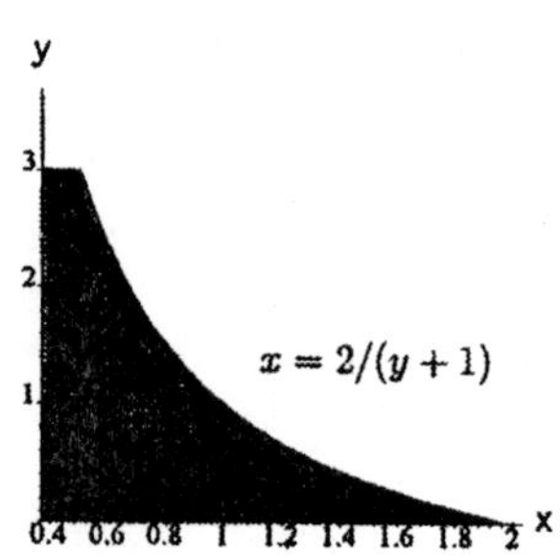

36. $R(y) = \frac{\sqrt{2y}}{y^2+1} \Rightarrow V = \int_{0}^{1} \pi [R(y)]^2 \, dy = \pi \int_{0}^{1} 2y\left(y^2+1\right)^{-2} dy;$
$[u = y^2 + 1 \Rightarrow du = 2y \, dy; \; y = 0 \Rightarrow u = 1, \; y = 1 \Rightarrow u = 2]$
$\rightarrow V = \pi \int_{1}^{2} u^{-2} \, du = \pi\left[-\frac{1}{u}\right]_{1}^{2} = \pi\left[-\frac{1}{2} - (-1)\right] = \frac{\pi}{2}$

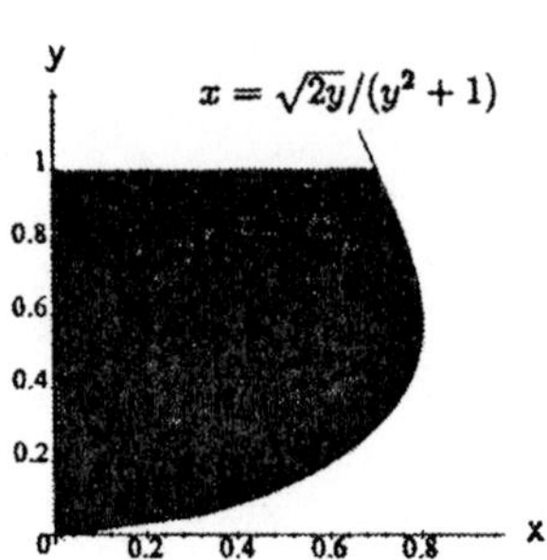

37. For the sketch given, $a = -\frac{\pi}{2}, \; b = \frac{\pi}{2}; \; R(x) = 1, \; r(x) = \sqrt{\cos x}; \; V = \int_{a}^{b} \pi\left([R(x)]^2 - [r(x)]^2\right) dx$
$= \int_{-\pi/2}^{\pi/2} \pi(1 - \cos x) \, dx = 2\pi \int_{0}^{\pi/2} (1 - \cos x) \, dx = 2\pi[x - \sin x]_{0}^{\pi/2} = 2\pi\left(\frac{\pi}{2} - 1\right) = \pi^2 - 2\pi$

38. For the sketch given, $c = 0, \; d = \frac{\pi}{4}; \; R(y) = 1, \; r(y) = \tan y; \; V = \int_{c}^{d} \pi\left([R(y)]^2 - [r(y)]^2\right) dy$
$= \pi \int_{0}^{\pi/4} \left(1 - \tan^2 y\right) dy = \pi \int_{0}^{\pi/4} \left(2 - \sec^2 y\right) dy = \pi[2y - \tan y]_{0}^{\pi/4} = \pi\left(\frac{\pi}{2} - 1\right) = \frac{\pi^2}{2} - \pi$

39. $r(x) = x$ and $R(x) = 1 \Rightarrow V = \int_{0}^{1} \pi\left([R(x)]^2 - [r(x)]^2\right) dx$
$= \int_{0}^{1} \pi\left(1 - x^2\right) dx = \pi\left[x - \frac{x^3}{3}\right]_{0}^{1} = \pi\left[\left(1 - \frac{1}{3}\right) - 0\right] = \frac{2\pi}{3}$

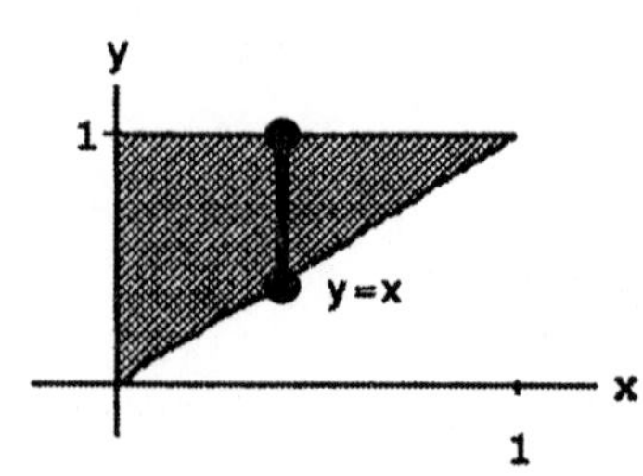

40. $r(x) = 2\sqrt{x}$ and $R(x) = 2 \Rightarrow V = \int_0^1 \pi\left([R(x)]^2 - [r(x)]^2\right) dx$
$= \pi \int_0^1 (4 - 4x)\, dx = 4\pi\left[x - \frac{x^2}{2}\right]_0^1 = 4\pi\left(1 - \frac{1}{2}\right) = 2\pi$

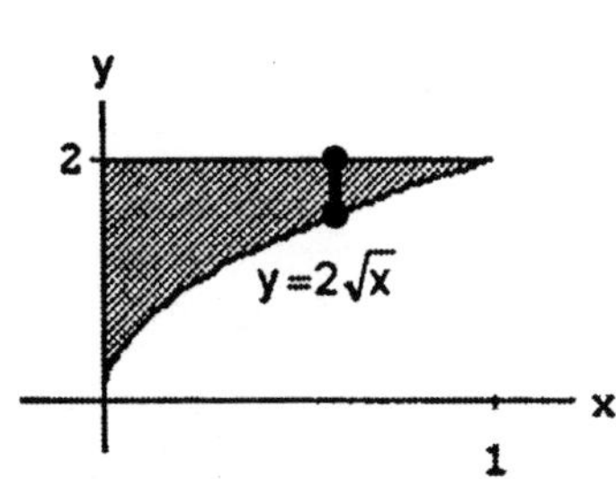

41. $r(x) = x^2 + 1$ and $R(x) = x + 3$
$\Rightarrow V = \int_{-1}^2 \pi\left([R(x)]^2 - [r(x)]^2\right) dx$
$= \pi \int_{-1}^2 \left[(x+3)^2 - (x^2+1)^2\right] dx$
$= \pi \int_{-1}^2 \left[(x^2 + 6x + 9) - (x^4 + 2x^2 + 1)\right] dx$
$= \pi \int_{-1}^2 (-x^4 - x^2 + 6x + 8)\, dx$
$= \pi \left[-\frac{x^5}{5} - \frac{x^3}{3} + \frac{6x^2}{2} + 8x\right]_{-1}^2$
$= \pi\left[\left(-\frac{32}{5} - \frac{8}{3} + \frac{24}{2} + 16\right) - \left(\frac{1}{5} + \frac{1}{3} + \frac{6}{2} - 8\right)\right] = \pi\left(-\frac{33}{5} - 3 + 28 - 3 + 8\right) = \pi\left(\frac{5 \cdot 30 - 33}{5}\right) = \frac{117\pi}{5}$

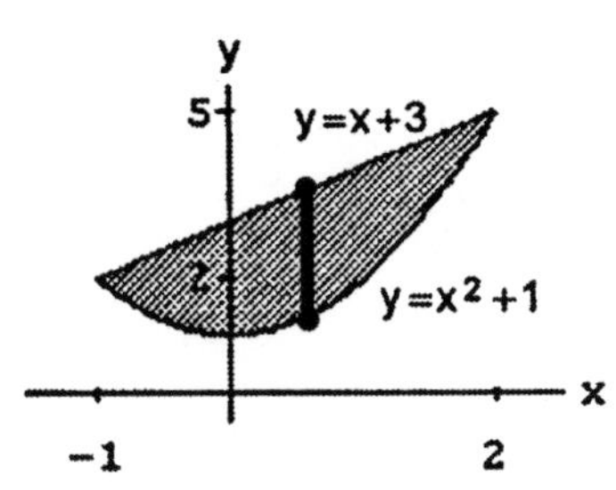

42. $r(x) = 2 - x$ and $R(x) = 4 - x^2$
$\Rightarrow V = \int_{-1}^2 \pi\left([R(x)]^2 - [r(x)]^2\right) dx$
$= \pi \int_{-1}^2 \left[(4 - x^2)^2 - (2 - x)^2\right] dx$
$= \pi \int_{-1}^2 \left[(16 - 8x^2 + x^4) - (4 - 4x + x^2)\right] dx$
$= \pi \int_{-1}^2 (12 + 4x - 9x^2 + x^4)\, dx$
$= \pi \left[12x + 2x^2 - 3x^3 + \frac{x^5}{5}\right]_{-1}^2$
$= \pi\left[\left(24 + 8 - 24 + \frac{32}{5}\right) - \left(-12 + 2 + 3 - \frac{1}{5}\right)\right] = \pi\left(15 + \frac{33}{5}\right) = \frac{108\pi}{5}$

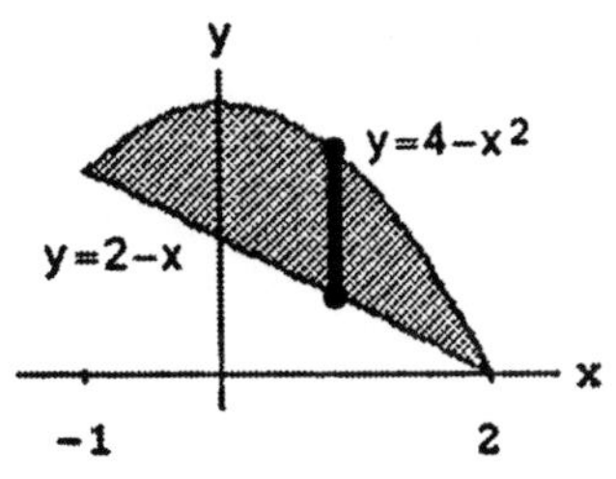

43. $r(x) = \sec x$ and $R(x) = \sqrt{2}$
$\Rightarrow V = \int_{-\pi/4}^{\pi/4} \pi\left([R(x)]^2 - [r(x)]^2\right) dx$
$= \pi \int_{-\pi/4}^{\pi/4} (2 - \sec^2 x)\, dx = \pi[2x - \tan x]_{-\pi/4}^{\pi/4}$
$= \pi\left[\left(\frac{\pi}{2} - 1\right) - \left(-\frac{\pi}{2} + 1\right)\right] = \pi(\pi - 2)$

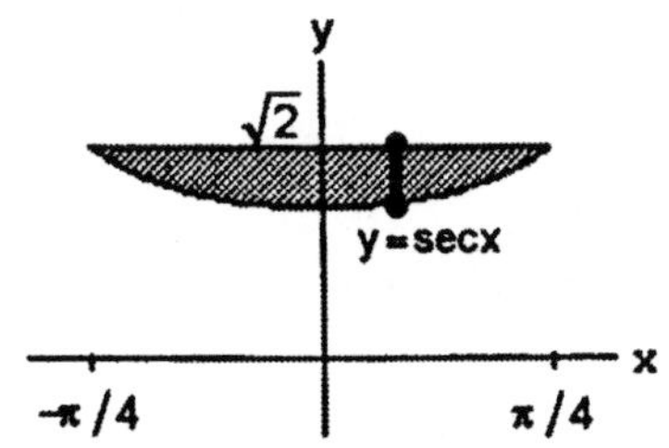

44. $R(x) = \sec x$ and $r(x) = \tan x$
$\Rightarrow V = \int_0^1 \pi\left([R(x)]^2 - [r(x)]^2\right) dx$
$= \pi \int_0^1 (\sec^2 x - \tan^2 x)\, dx = \pi \int_0^1 1\, dx = \pi[x]_0^1 = \pi$

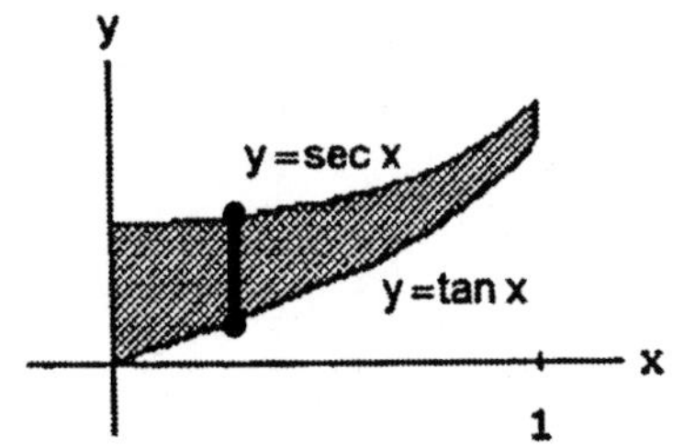

45. $r(y) = 1$ and $R(y) = 1 + y$

$$\Rightarrow V = \int_0^1 \pi \left([R(y)]^2 - [r(y)]^2\right) dy$$
$$= \pi \int_0^1 \left[(1+y)^2 - 1\right] dy = \pi \int_0^1 (1 + 2y + y^2 - 1)\, dy$$
$$= \pi \int_0^1 (2y + y^2)\, dy = \pi \left[y^2 + \tfrac{y^3}{3}\right]_0^1 = \pi\left(1 + \tfrac{1}{3}\right) = \tfrac{4\pi}{3}$$

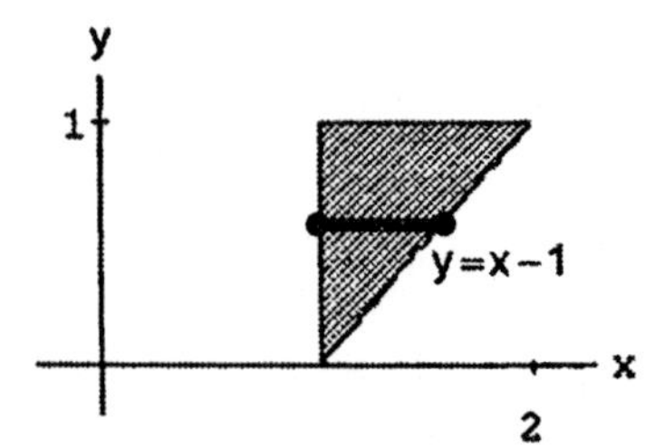

46. $R(y) = 1$ and $r(y) = 1 - y \Rightarrow V = \int_0^1 \pi \left([R(y)]^2 - [r(y)]^2\right) dy$

$$= \pi \int_0^1 \left[1 - (1-y)^2\right] dy = \pi \int_0^1 \left[1 - (1 - 2y + y^2)\right] dy$$
$$= \pi \int_0^1 (2y - y^2)\, dy = \pi \left[y^2 - \tfrac{y^3}{3}\right]_0^1 = \pi\left(1 - \tfrac{1}{3}\right) = \tfrac{2\pi}{3}$$

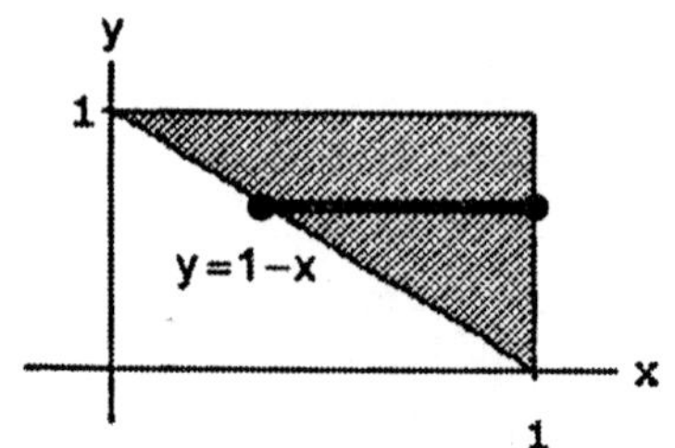

47. $R(y) = 2$ and $r(y) = \sqrt{y}$

$$\Rightarrow V = \int_0^4 \pi \left([R(y)]^2 - [r(y)]^2\right) dy$$
$$= \pi \int_0^4 (4 - y)\, dy = \pi \left[4y - \tfrac{y^2}{2}\right]_0^4 = \pi(16 - 8) = 8\pi$$

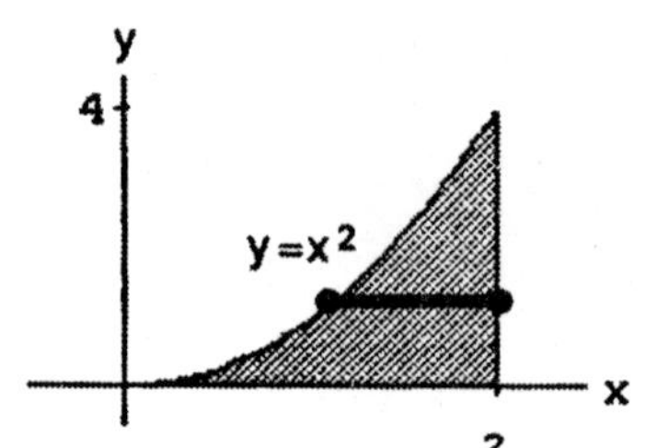

48. $R(y) = \sqrt{3}$ and $r(y) = \sqrt{3 - y^2}$

$$\Rightarrow V = \int_0^{\sqrt{3}} \pi \left([R(y)]^2 - [r(y)]^2\right) dy$$
$$= \pi \int_0^{\sqrt{3}} \left[3 - (3 - y^2)\right] dy = \pi \int_0^{\sqrt{3}} y^2\, dy$$
$$= \pi \left[\tfrac{y^3}{3}\right]_0^{\sqrt{3}} = \pi\sqrt{3}$$

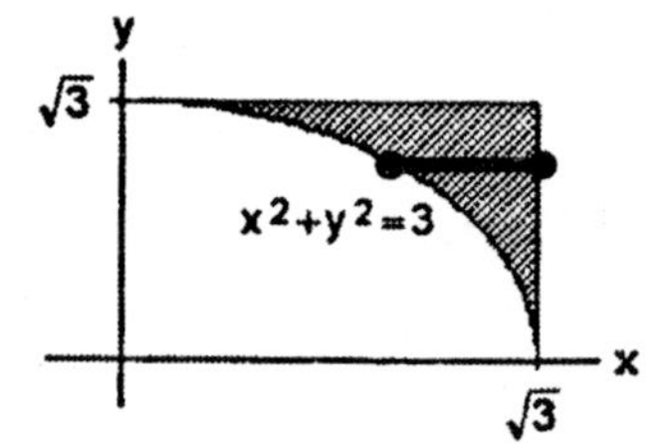

49. $R(y) = 2$ and $r(y) = 1 + \sqrt{y}$

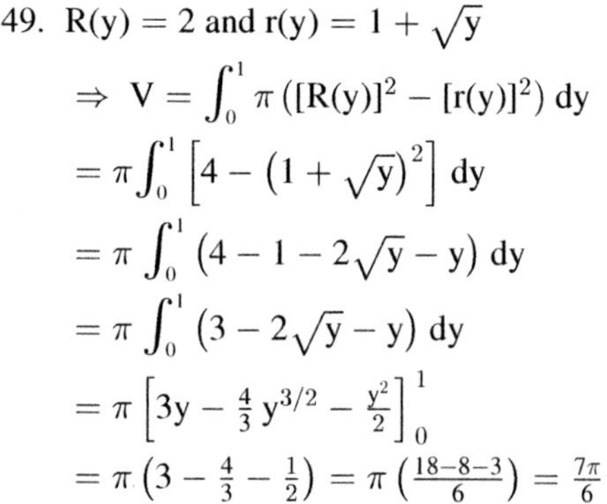

$$\Rightarrow V = \int_0^1 \pi \left([R(y)]^2 - [r(y)]^2\right) dy$$
$$= \pi \int_0^1 \left[4 - \left(1 + \sqrt{y}\right)^2\right] dy$$
$$= \pi \int_0^1 \left(4 - 1 - 2\sqrt{y} - y\right) dy$$
$$= \pi \int_0^1 \left(3 - 2\sqrt{y} - y\right) dy$$
$$= \pi \left[3y - \tfrac{4}{3} y^{3/2} - \tfrac{y^2}{2}\right]_0^1$$
$$= \pi\left(3 - \tfrac{4}{3} - \tfrac{1}{2}\right) = \pi\left(\tfrac{18-8-3}{6}\right) = \tfrac{7\pi}{6}$$

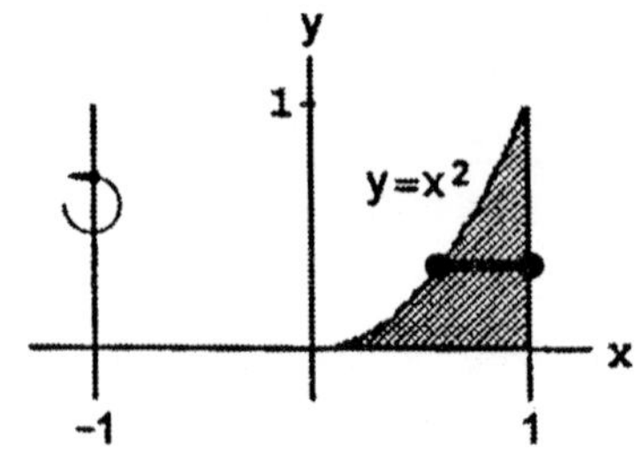

50. $R(y) = 2 - y^{1/3}$ and $r(y) = 1$

$$\Rightarrow V = \int_0^1 \pi\left([R(y)]^2 - [r(y)]^2\right) dy$$
$$= \pi \int_0^1 \left[\left(2 - y^{1/3}\right)^2 - 1\right] dy$$
$$= \pi \int_0^1 \left(4 - 4y^{1/3} + y^{2/3} - 1\right) dy$$
$$= \pi \int_0^1 \left(3 - 4y^{1/3} + y^{2/3}\right) dy$$
$$= \pi \left[3y - 3y^{4/3} + \frac{3y^{5/3}}{5}\right]_0^1 = \pi\left(3 - 3 + \frac{3}{5}\right) = \frac{3\pi}{5}$$

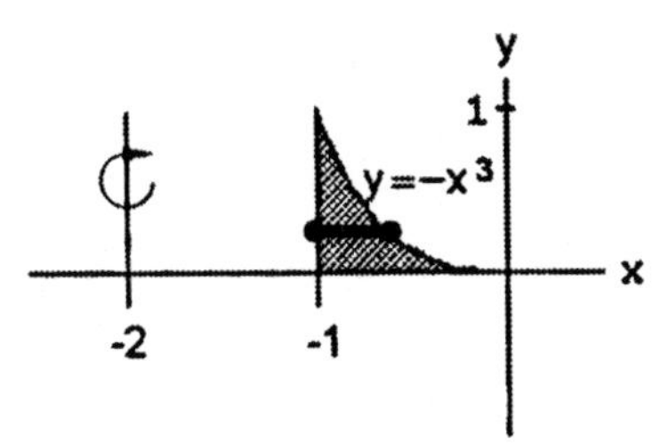

51. (a) $r(x) = \sqrt{x}$ and $R(x) = 2$

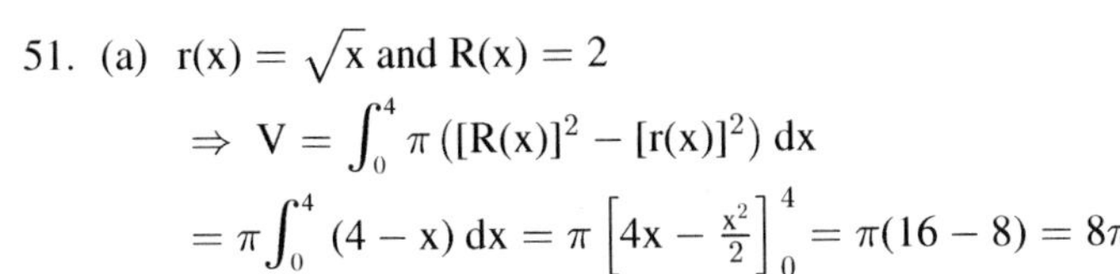

$$\Rightarrow V = \int_0^4 \pi\left([R(x)]^2 - [r(x)]^2\right) dx$$
$$= \pi \int_0^4 (4 - x)\, dx = \pi\left[4x - \frac{x^2}{2}\right]_0^4 = \pi(16 - 8) = 8\pi$$

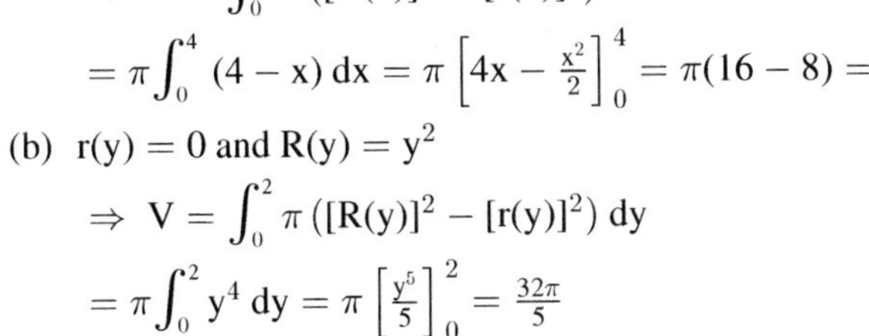

(b) $r(y) = 0$ and $R(y) = y^2$

$$\Rightarrow V = \int_0^2 \pi\left([R(y)]^2 - [r(y)]^2\right) dy$$
$$= \pi \int_0^2 y^4\, dy = \pi\left[\frac{y^5}{5}\right]_0^2 = \frac{32\pi}{5}$$

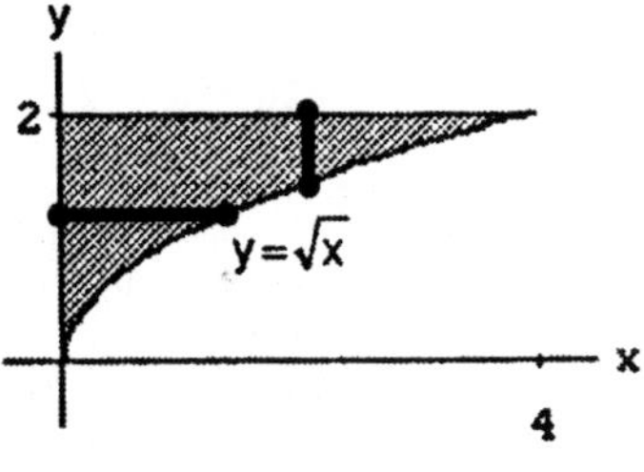

(c) $r(x) = 0$ and $R(x) = 2 - \sqrt{x} \Rightarrow V = \int_0^4 \pi\left([R(x)]^2 - [r(x)]^2\right) dx = \pi \int_0^4 \left(2 - \sqrt{x}\right)^2 dx$

$$= \pi \int_0^4 \left(4 - 4\sqrt{x} + x\right) dx = \pi\left[4x - \frac{8x^{3/2}}{3} + \frac{x^2}{2}\right]_0^4 = \pi\left(16 - \frac{64}{3} + \frac{16}{2}\right) = \frac{8\pi}{3}$$

(d) $r(y) = 4 - y^2$ and $R(y) = 4 \Rightarrow V = \int_0^2 \pi\left([R(y)]^2 - [r(y)]^2\right) dy = \pi \int_0^2 \left[16 - \left(4 - y^2\right)^2\right] dy$

$$= \pi \int_0^2 \left(16 - 16 + 8y^2 - y^4\right) dy = \pi \int_0^2 \left(8y^2 - y^4\right) dy = \pi\left[\frac{8}{3}y^3 - \frac{y^5}{5}\right]_0^2 = \pi\left(\frac{64}{3} - \frac{32}{5}\right) = \frac{224\pi}{15}$$

52. (a) $r(y) = 0$ and $R(y) = 1 - \frac{y}{2}$

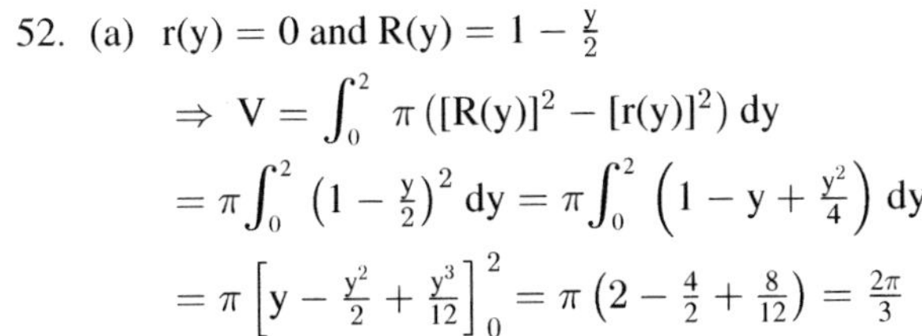

$$\Rightarrow V = \int_0^2 \pi\left([R(y)]^2 - [r(y)]^2\right) dy$$
$$= \pi \int_0^2 \left(1 - \frac{y}{2}\right)^2 dy = \pi \int_0^2 \left(1 - y + \frac{y^2}{4}\right) dy$$
$$= \pi\left[y - \frac{y^2}{2} + \frac{y^3}{12}\right]_0^2 = \pi\left(2 - \frac{4}{2} + \frac{8}{12}\right) = \frac{2\pi}{3}$$

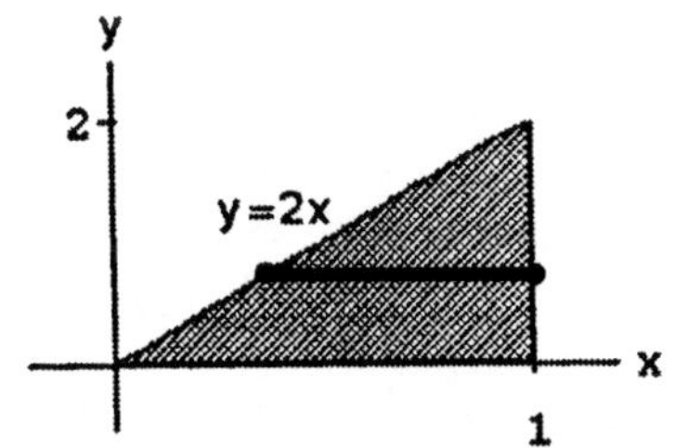

(b) $r(y) = 1$ and $R(y) = 2 - \frac{y}{2}$

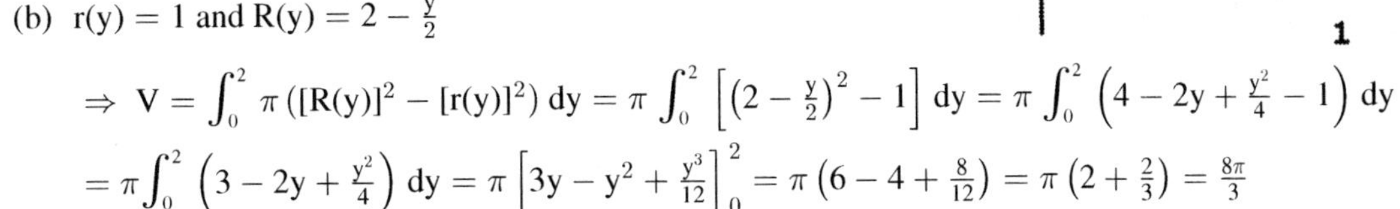

$$\Rightarrow V = \int_0^2 \pi\left([R(y)]^2 - [r(y)]^2\right) dy = \pi \int_0^2 \left[\left(2 - \frac{y}{2}\right)^2 - 1\right] dy = \pi \int_0^2 \left(4 - 2y + \frac{y^2}{4} - 1\right) dy$$
$$= \pi \int_0^2 \left(3 - 2y + \frac{y^2}{4}\right) dy = \pi\left[3y - y^2 + \frac{y^3}{12}\right]_0^2 = \pi\left(6 - 4 + \frac{8}{12}\right) = \pi\left(2 + \frac{2}{3}\right) = \frac{8\pi}{3}$$

53. (a) $r(x) = 0$ and $R(x) = 1 - x^2$

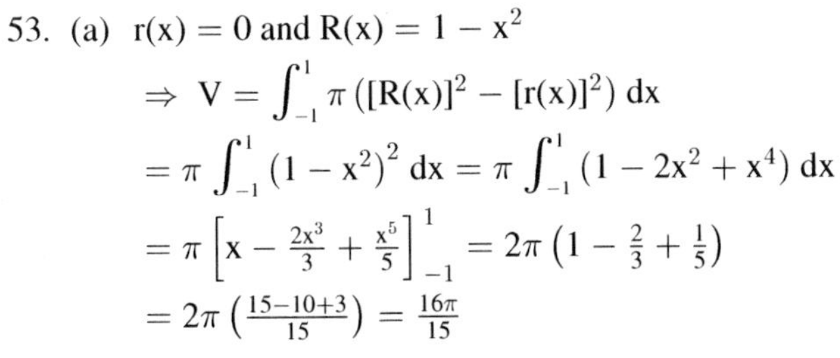

$$\Rightarrow V = \int_{-1}^1 \pi\left([R(x)]^2 - [r(x)]^2\right) dx$$
$$= \pi \int_{-1}^1 \left(1 - x^2\right)^2 dx = \pi \int_{-1}^1 \left(1 - 2x^2 + x^4\right) dx$$
$$= \pi\left[x - \frac{2x^3}{3} + \frac{x^5}{5}\right]_{-1}^1 = 2\pi\left(1 - \frac{2}{3} + \frac{1}{5}\right)$$
$$= 2\pi\left(\frac{15-10+3}{15}\right) = \frac{16\pi}{15}$$

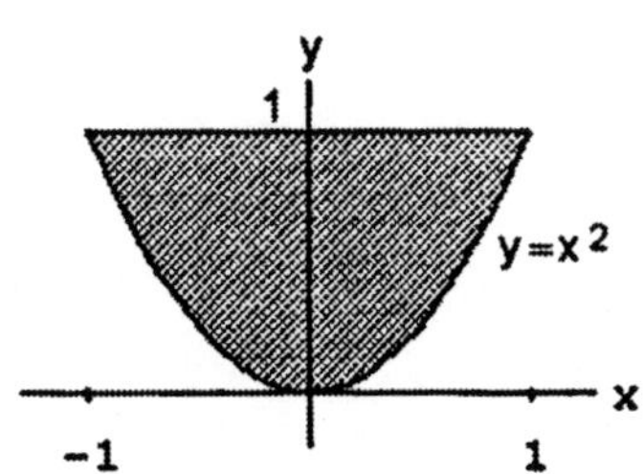

(b) $r(x) = 1$ and $R(x) = 2 - x^2 \Rightarrow V = \int_{-1}^1 \pi\left([R(x)]^2 - [r(x)]^2\right) dx = \pi \int_{-1}^1 \left[\left(2 - x^2\right)^2 - 1\right] dx$

$$= \pi \int_{-1}^1 \left(4 - 4x^2 + x^4 - 1\right) dx = \pi \int_{-1}^1 \left(3 - 4x^2 + x^4\right) dx = \pi\left[3x - \frac{4}{3}x^3 + \frac{x^5}{5}\right]_{-1}^1 = 2\pi\left(3 - \frac{4}{3} + \frac{1}{5}\right)$$
$$= \frac{2\pi}{15}(45 - 20 + 3) = \frac{56\pi}{15}$$

(c) $r(x) = 1 + x^2$ and $R(x) = 2 \Rightarrow V = \int_{-1}^{1} \pi \left([R(x)]^2 - [r(x)]^2\right) dx = \pi \int_{-1}^{1} \left[4 - (1+x^2)^2\right] dx$
$= \pi \int_{-1}^{1} (4 - 1 - 2x^2 - x^4)\, dx = \pi \int_{-1}^{1} (3 - 2x^2 - x^4)\, dx = \pi \left[3x - \frac{2}{3}x^3 - \frac{x^5}{5}\right]_{-1}^{1} = 2\pi \left(3 - \frac{2}{3} - \frac{1}{5}\right)$
$= \frac{2\pi}{15}(45 - 10 - 3) = \frac{64\pi}{15}$

54. (a) $r(x) = 0$ and $R(x) = -\frac{h}{b}x + h$
$\Rightarrow V = \int_0^b \pi \left([R(x)]^2 - [r(x)]^2\right) dx$
$= \pi \int_0^b \left(-\frac{h}{b}x + h\right)^2 dx$
$= \pi \int_0^b \left(\frac{h^2}{b^2}x^2 - \frac{2h^2}{b}x + h^2\right) dx$
$= \pi h^2 \left[\frac{x^3}{3b^2} - \frac{x^2}{b} + x\right]_0^b = \pi h^2 \left(\frac{b}{3} - b + b\right) = \frac{\pi h^2 b}{3}$

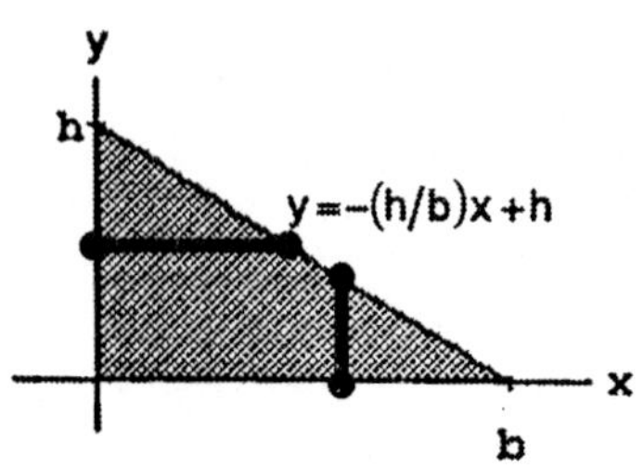

(b) $r(y) = 0$ and $R(y) = b\left(1 - \frac{y}{h}\right) \Rightarrow V = \int_0^h \pi \left([R(y)]^2 - [r(y)]^2\right) dy = \pi b^2 \int_0^h \left(1 - \frac{y}{h}\right)^2 dy$
$= \pi b^2 \int_0^h \left(1 - \frac{2y}{h} + \frac{y^2}{h^2}\right) dy = \pi b^2 \left[y - \frac{y^2}{h} + \frac{y^3}{3h^2}\right]_0^h = \pi b^2 \left(h - h + \frac{h}{3}\right) = \frac{\pi b^2 h}{3}$

55. $R(y) = b + \sqrt{a^2 - y^2}$ and $r(y) = b - \sqrt{a^2 - y^2}$
$\Rightarrow V = \int_{-a}^{a} \pi \left([R(y)]^2 - [r(y)]^2\right) dy$
$= \pi \int_{-a}^{a} \left[\left(b + \sqrt{a^2 - y^2}\right)^2 - \left(b - \sqrt{a^2 - y^2}\right)^2\right] dy$
$= \pi \int_{-a}^{a} 4b\sqrt{a^2 - y^2}\, dy = 4b\pi \int_{-a}^{a} \sqrt{a^2 - y^2}\, dy$
$= 4b\pi \cdot$ area of semicircle of radius $a = 4b\pi \cdot \frac{\pi a^2}{2} = 2a^2 b\pi^2$

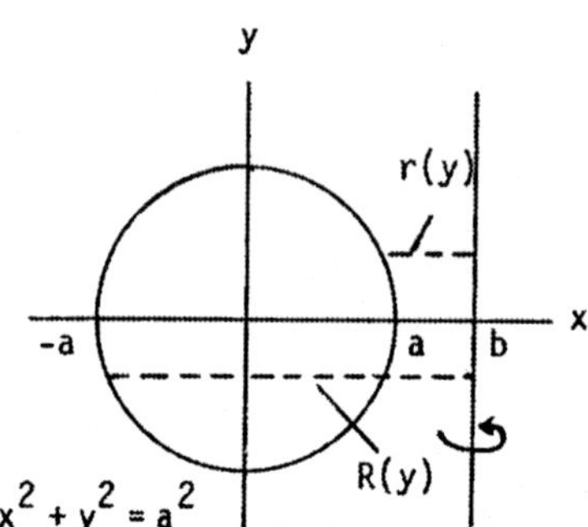

56. (a) A cross section has radius $r = \sqrt{2y}$ and area $\pi r^2 = 2\pi y$. The volume is $\int_0^5 2\pi y\, dy = \pi \left[y^2\right]_0^5 = 25\pi$.
(b) $V(h) = \int A(h)\, dh$, so $\frac{dV}{dh} = A(h)$. Therefore $\frac{dV}{dt} = \frac{dV}{dh} \cdot \frac{dh}{dt} = A(h) \cdot \frac{dh}{dt}$, so $\frac{dh}{dt} = \frac{1}{A(h)} \cdot \frac{dV}{dt}$.
For $h = 4$, the area is $2\pi(4) = 8\pi$, so $\frac{dh}{dt} = \frac{1}{8\pi} \cdot 3 \frac{\text{units}^3}{\text{sec}} = \frac{3}{8\pi} \cdot \frac{\text{units}^3}{\text{sec}}$.

57. (a) $R(y) = \sqrt{a^2 - y^2} \Rightarrow V = \pi \int_{-a}^{h-a} (a^2 - y^2)\, dy = \pi \left[a^2 y - \frac{y^3}{3}\right]_{-a}^{h-a} = \pi \left[a^2 h - a^3 - \frac{(h-a)^3}{3} - \left(-a^3 + \frac{a^3}{3}\right)\right]$
$= \pi \left[a^2 h - \frac{1}{3}(h^3 - 3h^2 a + 3ha^2 - a^3) - \frac{a^3}{3}\right] = \pi \left(a^2 h - \frac{h^3}{3} + h^2 a - ha^2\right) = \frac{\pi h^2 (3a - h)}{3}$
(b) Given $\frac{dV}{dt} = 0.2\ \text{m}^3/\text{sec}$ and $a = 5$ m, find $\left.\frac{dh}{dt}\right|_{h=4}$. From part (a), $V(h) = \frac{\pi h^2 (15 - h)}{3} = 5\pi h^2 - \frac{\pi h^3}{3}$
$\Rightarrow \frac{dV}{dh} = 10\pi h - \pi h^2 \Rightarrow \frac{dV}{dt} = \frac{dV}{dh} \cdot \frac{dh}{dt} = \pi h(10 - h) \frac{dh}{dt} \Rightarrow \left.\frac{dh}{dt}\right|_{h=4} = \frac{0.2}{4\pi(10-4)} = \frac{1}{(20\pi)(6)} = \frac{1}{120\pi}$ m/sec.

58. $R(y) = -\frac{r}{h}(y - h)$ so $V = \int_0^h \pi [R(y)]^2 dy$
$= \int_0^h \pi \left[-\frac{r}{h}(y - h)\right]^2 dy$
$= \frac{\pi r^2}{h^2} \int_0^h (y^2 - 2hy + h^2)\, dy = \frac{\pi r^3}{3}$

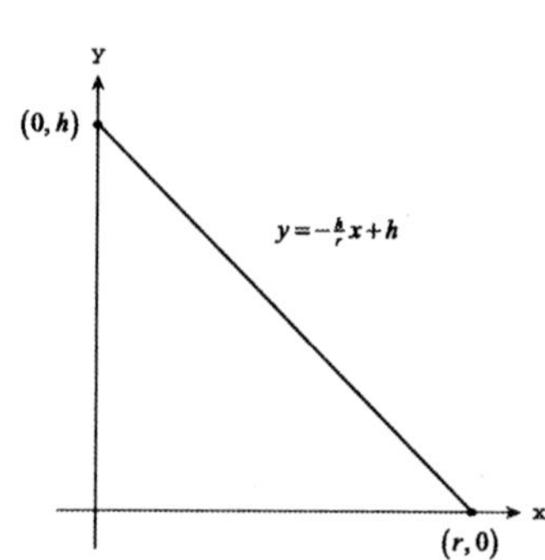

59. $V = \pi \int_{-\sqrt{3}/3}^{\sqrt{3}} \left(\frac{1}{\sqrt{1+x^2}}\right)^2 dx = \pi \int_{-\sqrt{3}/3}^{\sqrt{3}} \frac{1}{1+x^2}\, dx = \pi \left[\tan^{-1} x\right]_{-\sqrt{3}/3}^{\sqrt{3}} = \pi \left[\tan^{-1} \sqrt{3} - \tan^{-1}\left(-\frac{\sqrt{3}}{3}\right)\right]$
$= \pi \left[\frac{\pi}{3} - \left(-\frac{\pi}{6}\right)\right] = \frac{\pi^2}{2}$

60. The cross section of a solid right circular cylinder with a cone removed is a disk with radius R from which a disk of radius h has been removed. Thus its area is $A_1 = \pi R^2 - \pi h^2 = \pi\left(R^2 - h^2\right)$. The cross section of the hemisphere is a disk of radius $\sqrt{R^2 - h^2}$. Therefore its area is $A_2 = \pi\left(\sqrt{R^2 - h^2}\right)^2 = \pi\left(R^2 - h^2\right)$. We can see that $A_1 = A_2$. The altitudes of both solids are R. Applying Cavalieri's Principle we find
Volume of Hemisphere $=$ (Volume of Cylinder) $-$ (Volume of Cone) $= \left(\pi R^2\right) R - \frac{1}{3}\pi\left(R^2\right) R = \frac{2}{3}\pi R^3$.

61. $R(y) = \sqrt{256 - y^2} \Rightarrow V = \int_{-16}^{-7} \pi[R(y)]^2\, dy = \pi \int_{-16}^{-7} \left(256 - y^2\right) dy = \pi \left[256y - \frac{y^3}{3}\right]_{-16}^{-7}$
$= \pi \left[(256)(-7) + \frac{7^3}{3} - \left((256)(-16) + \frac{16^3}{3}\right)\right] = \pi \left(\frac{7^3}{3} + 256(16 - 7) - \frac{16^3}{3}\right) = 1053\pi \text{ cm}^3 \approx 3308 \text{ cm}^3$

62. $R(x) = \frac{x}{12}\sqrt{36 - x^2} \Rightarrow V = \int_0^6 \pi[R(x)]^2\, dx = \pi \int_0^6 \frac{x^2}{144}\left(36 - x^2\right) dx = \frac{\pi}{144} \int_0^6 \left(36x^2 - x^4\right) dx$
$= \frac{\pi}{144}\left[12x^3 - \frac{x^5}{5}\right]_0^6 = \frac{\pi}{144}\left(12 \cdot 6^3 - \frac{6^5}{5}\right) = \frac{\pi \cdot 6^3}{144}\left(12 - \frac{36}{5}\right) = \left(\frac{196\pi}{144}\right)\left(\frac{60-36}{5}\right) = \frac{36\pi}{5} \text{ cm}^3$. The plumb bob will weigh about $W = (8.5)\left(\frac{36\pi}{5}\right) \approx 192$ gm, to the nearest gram.

63. (a) $R(x) = |c - \sin x|$, so $V = \pi \int_0^\pi [R(x)]^2\, dx = \pi \int_0^\pi (c - \sin x)^2\, dx = \pi \int_0^\pi \left(c^2 - 2c \sin x + \sin^2 x\right) dx$
$= \pi \int_0^\pi \left(c^2 - 2c \sin x + \frac{1 - \cos 2x}{2}\right) dx = \pi \int_0^\pi \left(c^2 + \frac{1}{2} - 2c \sin x - \frac{\cos 2x}{2}\right) dx$
$= \pi \left[\left(c^2 + \frac{1}{2}\right) x + 2c \cos x - \frac{\sin 2x}{4}\right]_0^\pi = \pi \left[\left(c^2 \pi + \frac{\pi}{2} - 2c - 0\right) - (0 + 2c - 0)\right] = \pi\left(c^2\pi + \frac{\pi}{2} - 4c\right)$. Let $V(c) = \pi\left(c^2\pi + \frac{\pi}{2} - 4c\right)$. We find the extreme values of V(c): $\frac{dV}{dc} = \pi(2c\pi - 4) = 0 \Rightarrow c = \frac{2}{\pi}$ is a critical point, and $V\left(\frac{2}{\pi}\right) = \pi\left(\frac{4}{\pi} + \frac{\pi}{2} - \frac{8}{\pi}\right) = \pi\left(\frac{\pi}{2} - \frac{4}{\pi}\right) = \frac{\pi^2}{2} - 4$; Evaluate V at the endpoints: $V(0) = \frac{\pi^2}{2}$ and $V(1) = \pi\left(\frac{3}{2}\pi - 4\right) = \frac{\pi^2}{2} - (4 - \pi)\pi$. Now we see that the function's absolute minimum value is $\frac{\pi^2}{2} - 4$, taken on at the critical point $c = \frac{2}{\pi}$. (See also the accompanying graph.)

(b) From the discussion in part (a) we conclude that the function's absolute maximum value is $\frac{\pi^2}{2}$, taken on at the endpoint $c = 0$.

(c) The graph of the solid's volume as a function of c for $0 \le c \le 1$ is given at the right. As c moves away from [0, 1] the volume of the solid increases without bound. If we approximate the solid as a set of solid disks, we can see that the radius of a typical disk increases without bounds as c moves away from [0, 1].

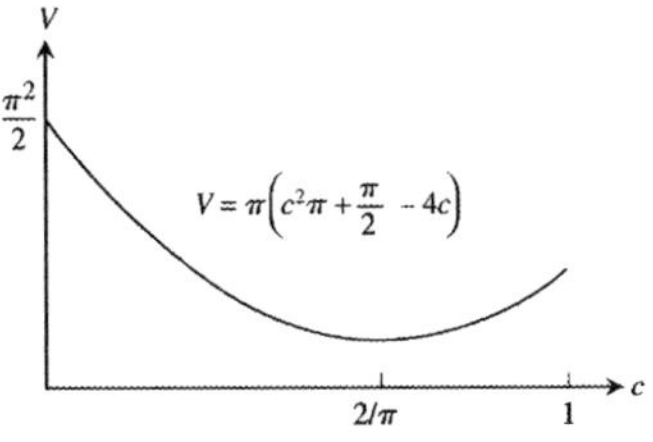

64. (a) $R(x) = 1 - \frac{x^2}{16} \Rightarrow V = \int_{-4}^{4} \pi[R(x)]^2\, dx$
$= \pi \int_{-4}^{4} \left(1 - \frac{x^2}{16}\right)^2 dx = \pi \int_{-4}^{4} \left(1 - \frac{x^2}{8} + \frac{x^4}{16^2}\right) dx$
$= \pi \left[x - \frac{x^3}{24} + \frac{x^5}{5 \cdot 16^2}\right]_{-4}^{4} = 2\pi\left(4 - \frac{4^3}{24} + \frac{4^5}{5 \cdot 16^2}\right)$
$= 2\pi\left(4 - \frac{8}{3} + \frac{4}{5}\right) = \frac{2\pi}{15}(60 - 40 + 12) = \frac{64\pi}{15} \text{ ft}^3$

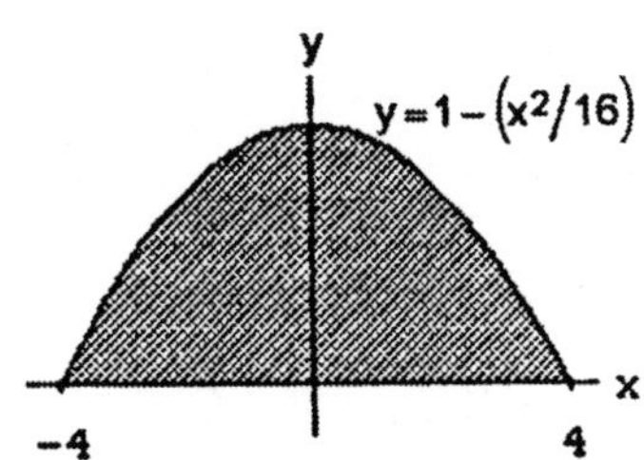

(b) The helicopter will be able to fly $\left(\frac{64\pi}{15}\right)(7.481)(2) \approx 201$ additional miles.

6.2 VOLUME BY CYLINDRICAL SHELLS

1. For the sketch given, a = 0, b = 2;
$V = \int_a^b 2\pi \left(\begin{smallmatrix}\text{shell}\\ \text{radius}\end{smallmatrix}\right)\left(\begin{smallmatrix}\text{shell}\\ \text{height}\end{smallmatrix}\right) dx = \int_0^2 2\pi x \left(1+\frac{x^2}{4}\right) dx = 2\pi \int_0^2 \left(x+\frac{x^3}{4}\right) dx = 2\pi \left[\frac{x^2}{2}+\frac{x^4}{16}\right]_0^2 = 2\pi\left(\frac{4}{2}+\frac{16}{16}\right)$
$= 2\pi \cdot 3 = 6\pi$

2. For the sketch given, a = 0, b = 2;
$V = \int_a^b 2\pi \left(\begin{smallmatrix}\text{shell}\\ \text{radius}\end{smallmatrix}\right)\left(\begin{smallmatrix}\text{shell}\\ \text{height}\end{smallmatrix}\right) dx = \int_0^2 2\pi x \left(2-\frac{x^2}{4}\right) dx = 2\pi \int_0^2 \left(2x-\frac{x^3}{4}\right) dx = 2\pi \left[x^2-\frac{x^4}{16}\right]_0^2 = 2\pi(4-1) = 6\pi$

3. For the sketch given, c = 0, d = $\sqrt{2}$;
$V = \int_c^d 2\pi \left(\begin{smallmatrix}\text{shell}\\ \text{radius}\end{smallmatrix}\right)\left(\begin{smallmatrix}\text{shell}\\ \text{height}\end{smallmatrix}\right) dy = \int_0^{\sqrt{2}} 2\pi y \cdot (y^2)\, dy = 2\pi \int_0^{\sqrt{2}} y^3\, dy = 2\pi \left[\frac{y^4}{4}\right]_0^{\sqrt{2}} = 2\pi$

4. For the sketch given, c = 0, d = $\sqrt{3}$;
$V = \int_c^d 2\pi \left(\begin{smallmatrix}\text{shell}\\ \text{radius}\end{smallmatrix}\right)\left(\begin{smallmatrix}\text{shell}\\ \text{height}\end{smallmatrix}\right) dy = \int_0^{\sqrt{3}} 2\pi y \cdot [3-(3-y^2)]\, dy = 2\pi \int_0^{\sqrt{3}} y^3\, dy = 2\pi \left[\frac{y^4}{4}\right]_0^{\sqrt{3}} = \frac{9\pi}{2}$

5. For the sketch given, a = 0, b = $\sqrt{3}$;
$V = \int_a^b 2\pi \left(\begin{smallmatrix}\text{shell}\\ \text{radius}\end{smallmatrix}\right)\left(\begin{smallmatrix}\text{shell}\\ \text{height}\end{smallmatrix}\right) dx = \int_0^{\sqrt{3}} 2\pi x \cdot \left(\sqrt{x^2+1}\right) dx;$
$\left[u = x^2+1 \Rightarrow du = 2x\, dx;\ x = 0 \Rightarrow u = 1,\ x = \sqrt{3} \Rightarrow u = 4\right]$
$\rightarrow V = \pi \int_1^4 u^{1/2}\, du = \pi \left[\frac{2}{3} u^{3/2}\right]_1^4 = \frac{2\pi}{3}\left(4^{3/2}-1\right) = \left(\frac{2\pi}{3}\right)(8-1) = \frac{14\pi}{3}$

6. For the sketch given, a = 0, b = 3;
$V = \int_a^b 2\pi \left(\begin{smallmatrix}\text{shell}\\ \text{radius}\end{smallmatrix}\right)\left(\begin{smallmatrix}\text{shell}\\ \text{height}\end{smallmatrix}\right) dx = \int_0^3 2\pi x \left(\frac{9x}{\sqrt{x^3+9}}\right) dx;$
$[u = x^3+9 \Rightarrow du = 3x^2\, dx \Rightarrow 3\, du = 9x^2\, dx;\ x = 0 \Rightarrow u = 9,\ x = 3 \Rightarrow u = 36]$
$\rightarrow V = 2\pi \int_9^{36} 3u^{-1/2}\, du = 6\pi \left[2u^{1/2}\right]_9^{36} = 12\pi\left(\sqrt{36}-\sqrt{9}\right) = 36\pi$

7. a = 0, b = 2;
$V = \int_a^b 2\pi \left(\begin{smallmatrix}\text{shell}\\ \text{radius}\end{smallmatrix}\right)\left(\begin{smallmatrix}\text{shell}\\ \text{height}\end{smallmatrix}\right) dx = \int_0^2 2\pi x \left[x - \left(-\frac{x}{2}\right)\right] dx$
$= \int_0^2 2\pi x^2 \cdot \frac{3}{2}\, dx = \pi \int_0^2 3x^2\, dx = \pi \left[x^3\right]_0^2 = 8\pi$

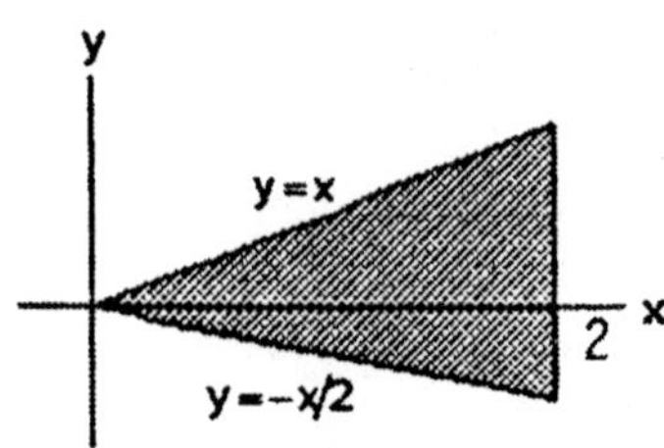

8. a = 0, b = 1;
$V = \int_a^b 2\pi \left(\begin{smallmatrix}\text{shell}\\ \text{radius}\end{smallmatrix}\right)\left(\begin{smallmatrix}\text{shell}\\ \text{height}\end{smallmatrix}\right) dx = \int_0^1 2\pi x \left(2x - \frac{x}{2}\right) dx$
$= \pi \int_0^1 2\left(\frac{3x^2}{2}\right) dx = \pi \int_0^1 3x^2\, dx = \pi \left[x^3\right]_0^1 = \pi$

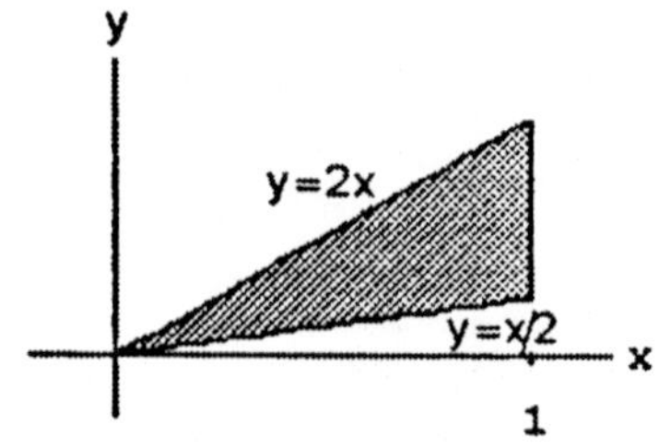

9. $a = 0, b = 1$;

$V = \int_a^b 2\pi \left(\begin{smallmatrix}\text{shell}\\ \text{radius}\end{smallmatrix}\right)\left(\begin{smallmatrix}\text{shell}\\ \text{height}\end{smallmatrix}\right) dx = \int_0^1 2\pi x\,[(2-x) - x^2]\,dx$

$= 2\pi \int_0^1 (2x - x^2 - x^3)\,dx = 2\pi \left[x^2 - \frac{x^3}{3} - \frac{x^4}{4}\right]_0^1$

$= 2\pi\left(1 - \frac{1}{3} - \frac{1}{4}\right) = 2\pi\left(\frac{12-4-3}{12}\right) = \frac{10\pi}{12} = \frac{5\pi}{6}$

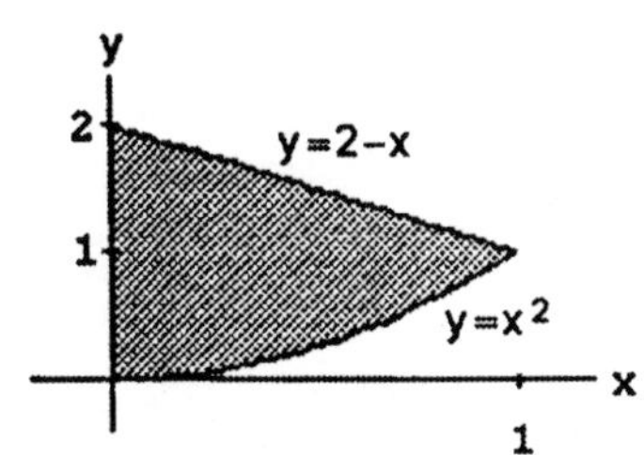

10. $a = 0, b = 1$;

$V = \int_a^b 2\pi \left(\begin{smallmatrix}\text{shell}\\ \text{radius}\end{smallmatrix}\right)\left(\begin{smallmatrix}\text{shell}\\ \text{height}\end{smallmatrix}\right) dx = \int_0^1 2\pi x\,[(2-x^2) - x^2]\,dx$

$= 2\pi \int_0^1 x\,(2 - 2x^2)\,dx = 4\pi \int_0^1 (x - x^3)\,dx$

$= 4\pi \left[\frac{x^2}{2} - \frac{x^4}{4}\right]_0^1 = 4\pi\left(\frac{1}{2} - \frac{1}{4}\right) = \pi$

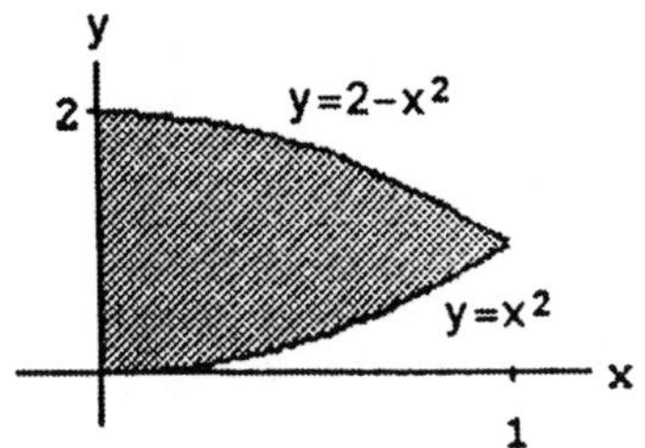

11. $a = 0, b = 1$;

$V = \int_a^b 2\pi \left(\begin{smallmatrix}\text{shell}\\ \text{radius}\end{smallmatrix}\right)\left(\begin{smallmatrix}\text{shell}\\ \text{height}\end{smallmatrix}\right) dx = \int_0^1 2\pi x\left[\sqrt{x} - (2x-1)\right] dx$

$= 2\pi \int_0^1 \left(x^{3/2} - 2x^2 + x\right) dx = 2\pi\left[\frac{2}{5}x^{5/2} - \frac{2}{3}x^3 + \frac{1}{2}x^2\right]_0^1$

$= 2\pi\left(\frac{2}{5} - \frac{2}{3} + \frac{1}{2}\right) = 2\pi\left(\frac{12-20+15}{30}\right) = \frac{7\pi}{15}$

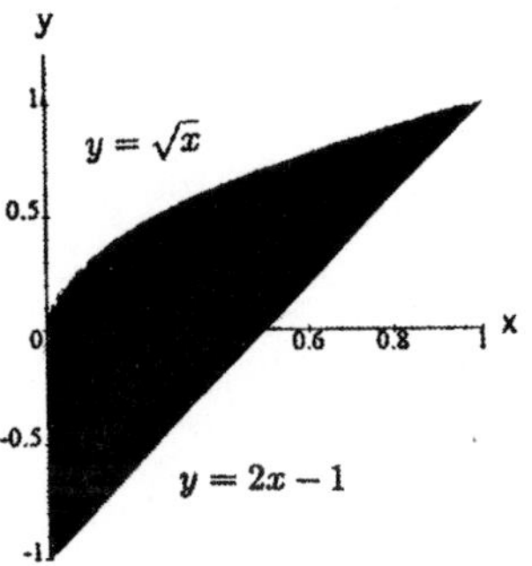

12. $a = 1, b = 4$;

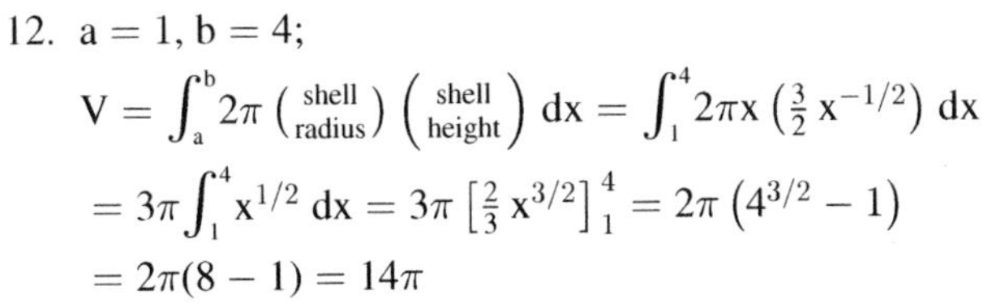

$V = \int_a^b 2\pi \left(\begin{smallmatrix}\text{shell}\\ \text{radius}\end{smallmatrix}\right)\left(\begin{smallmatrix}\text{shell}\\ \text{height}\end{smallmatrix}\right) dx = \int_1^4 2\pi x\left(\frac{3}{2}x^{-1/2}\right) dx$

$= 3\pi \int_1^4 x^{1/2}\,dx = 3\pi\left[\frac{2}{3}x^{3/2}\right]_1^4 = 2\pi\left(4^{3/2} - 1\right)$

$= 2\pi(8-1) = 14\pi$

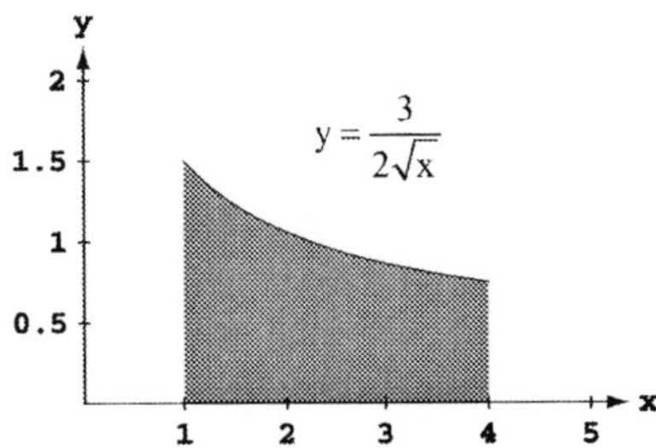

13. (a) $xf(x) = \begin{cases} x \cdot \frac{\sin x}{x}, & 0 < x \le \pi \\ x, & x = 0 \end{cases} \Rightarrow xf(x) = \begin{cases} \sin x, & 0 < x \le \pi \\ 0, & x = 0 \end{cases}$; since $\sin 0 = 0$ we have

$xf(x) = \begin{cases} \sin x, & 0 < x \le \pi \\ \sin x, & x = 0 \end{cases} \Rightarrow xf(x) = \sin x,\ 0 \le x \le \pi$

(b) $V = \int_a^b 2\pi \left(\begin{smallmatrix}\text{shell}\\ \text{radius}\end{smallmatrix}\right)\left(\begin{smallmatrix}\text{shell}\\ \text{height}\end{smallmatrix}\right) dx = \int_0^\pi 2\pi x \cdot f(x)\,dx$ and $x \cdot f(x) = \sin x,\ 0 \le x \le \pi$ by part (a)

$\Rightarrow V = 2\pi \int_0^\pi \sin x\,dx = 2\pi[-\cos x]_0^\pi = 2\pi(-\cos\pi + \cos 0) = 4\pi$

14. (a) $xg(x) = \begin{cases} x \cdot \frac{\tan^2 x}{x}, & 0 < x \le \frac{\pi}{4} \\ x \cdot 0, & x = 0 \end{cases} \Rightarrow xg(x) = \begin{cases} \tan^2 x, & 0 < x \le \pi/4 \\ 0, & x = 0 \end{cases}$; since $\tan 0 = 0$ we have

$xg(x) = \begin{cases} \tan^2 x, & 0 < x \le \pi/4 \\ \tan^2 x, & x = 0 \end{cases} \Rightarrow xg(x) = \tan^2 x,\ 0 \le x \le \pi/4$

(b) $V = \int_a^b 2\pi \left(\begin{smallmatrix}\text{shell}\\ \text{radius}\end{smallmatrix}\right) \left(\begin{smallmatrix}\text{shell}\\ \text{height}\end{smallmatrix}\right) dx = \int_0^{\pi/4} 2\pi x \cdot g(x)\, dx$ and $x \cdot g(x) = \tan^2 x,\ 0 \le x \le \pi/4$ by part (a)

$\Rightarrow V = 2\pi \int_0^{\pi/4} \tan^2 x\, dx = 2\pi \int_0^{\pi/4} (\sec^2 x - 1)\, dx = 2\pi[\tan x - x]_0^{\pi/4} = 2\pi\left(1 - \frac{\pi}{4}\right) = \frac{4\pi - \pi^2}{2}$

15. $c = 0, d = 2$;

$V = \int_c^d 2\pi \left(\begin{smallmatrix}\text{shell}\\ \text{radius}\end{smallmatrix}\right) \left(\begin{smallmatrix}\text{shell}\\ \text{height}\end{smallmatrix}\right) dy = \int_0^2 2\pi y \left[\sqrt{y} - (-y)\right] dy$

$= 2\pi \int_0^2 \left(y^{3/2} + y^2\right) dy = 2\pi \left[\frac{2y^{5/2}}{5} + \frac{y^3}{3}\right]_0^2$

$= 2\pi \left[\frac{2}{5}\left(\sqrt{2}\right)^5 + \frac{2^3}{3}\right] = 2\pi \left(\frac{8\sqrt{2}}{5} + \frac{8}{3}\right) = 16\pi \left(\frac{\sqrt{2}}{5} + \frac{1}{3}\right)$

$= \frac{16\pi}{15}\left(3\sqrt{2} + 5\right)$

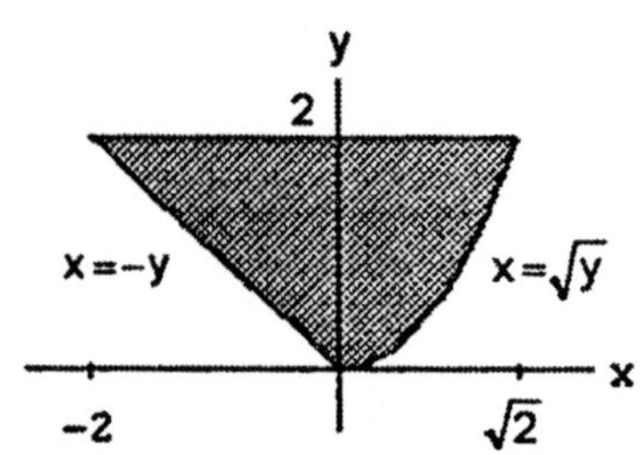

16. $c = 0, d = 2$;

$V = \int_c^d 2\pi \left(\begin{smallmatrix}\text{shell}\\ \text{radius}\end{smallmatrix}\right) \left(\begin{smallmatrix}\text{shell}\\ \text{height}\end{smallmatrix}\right) dy = \int_0^2 2\pi y \left[y^2 - (-y)\right] dy$

$= 2\pi \int_0^2 \left(y^3 + y^2\right) dy = 2\pi \left[\frac{y^4}{4} + \frac{y^3}{3}\right]_0^2 = 16\pi\left(\frac{2}{4} + \frac{1}{3}\right)$

$= 16\pi\left(\frac{5}{6}\right) = \frac{40\pi}{3}$

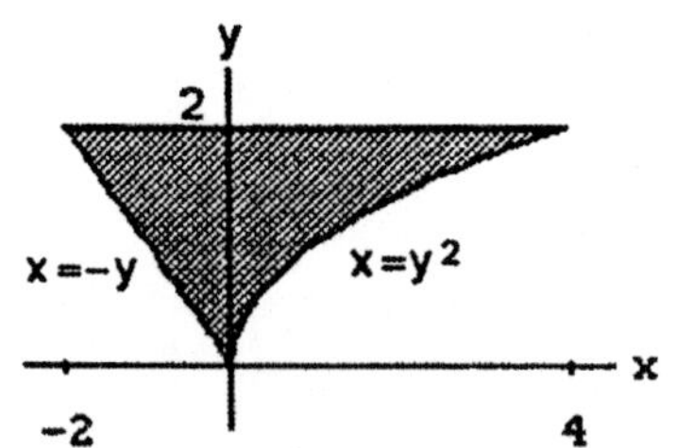

17. $c = 0, d = 2$;

$V = \int_c^d 2\pi \left(\begin{smallmatrix}\text{shell}\\ \text{radius}\end{smallmatrix}\right) \left(\begin{smallmatrix}\text{shell}\\ \text{height}\end{smallmatrix}\right) dy = \int_0^2 2\pi y \left(2y - y^2\right) dy$

$= 2\pi \int_0^2 \left(2y^2 - y^3\right) dy = 2\pi \left[\frac{2y^3}{3} - \frac{y^4}{4}\right]_0^2 = 2\pi\left(\frac{16}{3} - \frac{16}{4}\right)$

$= 32\pi\left(\frac{1}{3} - \frac{1}{4}\right) = \frac{32\pi}{12} = \frac{8\pi}{3}$

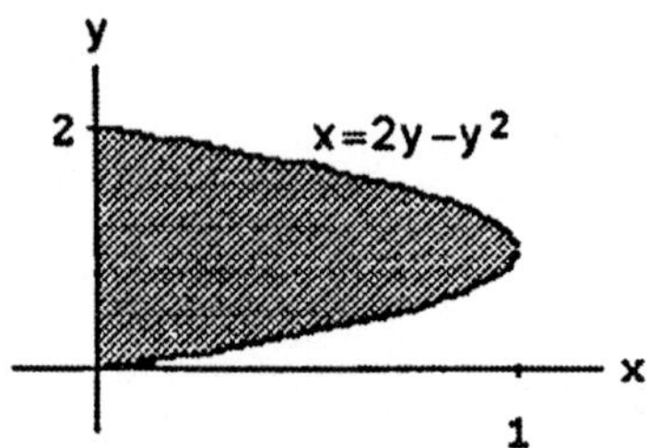

18. $c = 0, d = 1$;

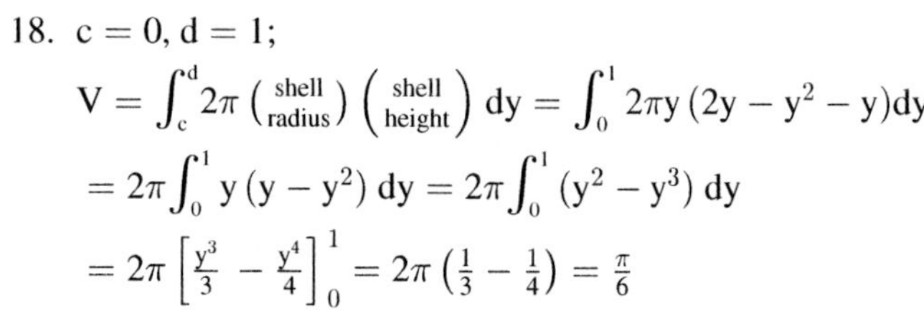

$V = \int_c^d 2\pi \left(\begin{smallmatrix}\text{shell}\\ \text{radius}\end{smallmatrix}\right) \left(\begin{smallmatrix}\text{shell}\\ \text{height}\end{smallmatrix}\right) dy = \int_0^1 2\pi y \left(2y - y^2 - y\right) dy$

$= 2\pi \int_0^1 y\left(y - y^2\right) dy = 2\pi \int_0^1 \left(y^2 - y^3\right) dy$

$= 2\pi \left[\frac{y^3}{3} - \frac{y^4}{4}\right]_0^1 = 2\pi\left(\frac{1}{3} - \frac{1}{4}\right) = \frac{\pi}{6}$

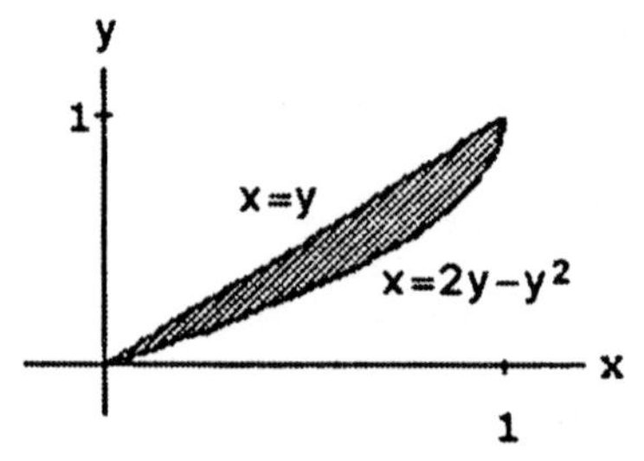

19. $c = 0, d = 1$;

$V = \int_c^d 2\pi \left(\begin{smallmatrix}\text{shell}\\ \text{radius}\end{smallmatrix}\right) \left(\begin{smallmatrix}\text{shell}\\ \text{height}\end{smallmatrix}\right) dy = 2\pi \int_0^1 y\left[y - (-y)\right] dy$

$= 2\pi \int_0^1 2y^2\, dy = \frac{4\pi}{3}\left[y^3\right]_0^1 = \frac{4\pi}{3}$

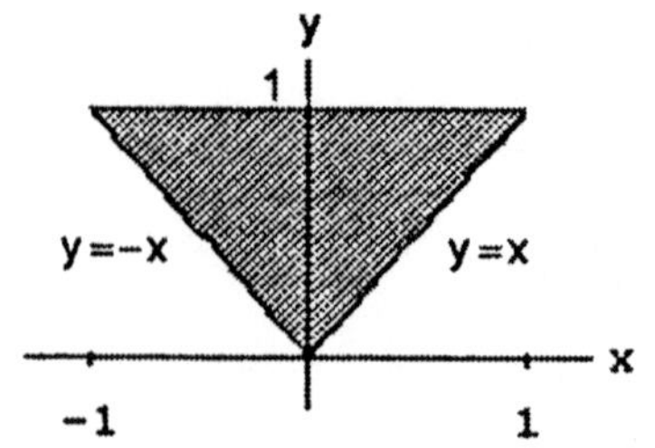

20. $c = 0, d = 2;$

$V = \int_c^d 2\pi \left(\frac{\text{shell}}{\text{radius}}\right)\left(\frac{\text{shell}}{\text{height}}\right) dy = \int_0^2 2\pi\, y\left(y - \frac{y}{2}\right)dy$

$= 2\pi \int_0^2 \frac{y^2}{2}dy = \frac{\pi}{3}\left[y^3\right]_0^1 = \frac{8\pi}{3}$

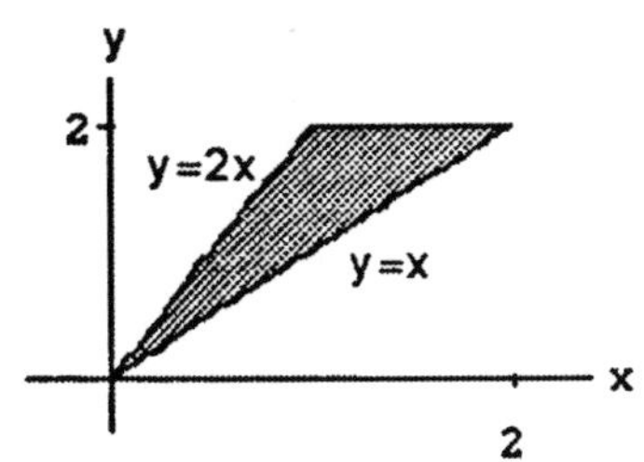

21. $c = 0, d = 2;$

$V = \int_c^d 2\pi \left(\frac{\text{shell}}{\text{radius}}\right)\left(\frac{\text{shell}}{\text{height}}\right) dy = \int_0^2 2\pi y\left[(2 + y) - y^2\right] dy$

$= 2\pi \int_0^2 \left(2y + y^2 - y^3\right) dy = 2\pi \left[y^2 + \frac{y^3}{3} - \frac{y^4}{4}\right]_0^2$

$= 2\pi\left(4 + \frac{8}{3} - \frac{16}{4}\right) = \frac{\pi}{6}(48 + 32 - 48) = \frac{16\pi}{3}$

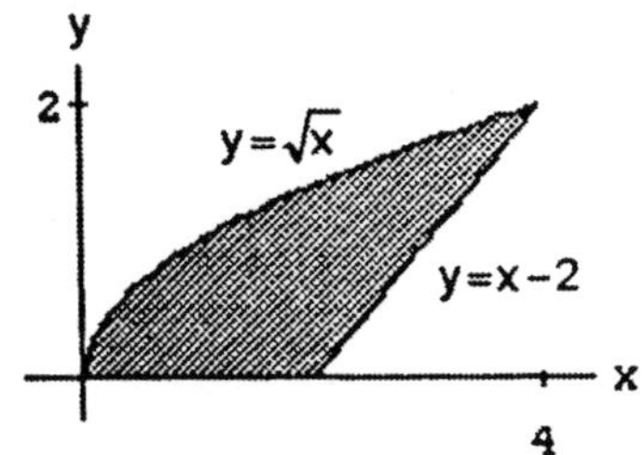

22. $c = 0, d = 1;$

$V = \int_c^d 2\pi \left(\frac{\text{shell}}{\text{radius}}\right)\left(\frac{\text{shell}}{\text{height}}\right) dy = \int_0^1 2\pi y\left[(2 - y) - y^2\right] dy$

$= 2\pi\int_0^1 \left(2y - y^2 - y^3\right) dy = 2\pi \left[y^2 - \frac{y^3}{3} - \frac{y^4}{4}\right]_0^1$

$= 2\pi\left(1 - \frac{1}{3} - \frac{1}{4}\right) = \frac{\pi}{6}(12 - 4 - 3) = \frac{5\pi}{6}$

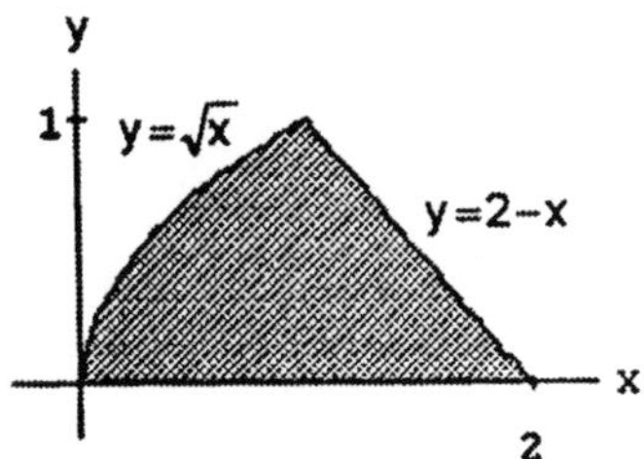

23. (a) $V = \int_c^d 2\pi \left(\frac{\text{shell}}{\text{radius}}\right)\left(\frac{\text{shell}}{\text{height}}\right) dy = \int_0^1 2\pi y \cdot 12\left(y^2 - y^3\right) dy = 24\pi \int_0^1 \left(y^3 - y^4\right) dy = 24\pi \left[\frac{y^4}{4} - \frac{y^5}{5}\right]_0^1$

$= 24\pi\left(\frac{1}{4} - \frac{1}{5}\right) = \frac{24\pi}{20} = \frac{6\pi}{5}$

(b) $V = \int_c^d 2\pi \left(\frac{\text{shell}}{\text{radius}}\right)\left(\frac{\text{shell}}{\text{height}}\right) dy = \int_0^1 2\pi(1 - y)\left[12\left(y^2 - y^3\right)\right] dy = 24\pi\int_0^1 (1 - y)\left(y^2 - y^3\right) dy$

$= 24\pi\int_0^1 \left(y^2 - 2y^3 + y^4\right) dy = 24\pi\left[\frac{y^3}{3} - \frac{y^4}{2} + \frac{y^5}{5}\right]_0^1 = 24\pi\left(\frac{1}{3} - \frac{1}{2} + \frac{1}{5}\right) = 24\pi\left(\frac{1}{30}\right) = \frac{4\pi}{5}$

(c) $V = \int_c^d 2\pi \left(\frac{\text{shell}}{\text{radius}}\right)\left(\frac{\text{shell}}{\text{height}}\right) dy = \int_0^1 2\pi\left(\frac{8}{5} - y\right)\left[12\left(y^2 - y^3\right)\right] dy = 24\pi \int_0^1 \left(\frac{8}{5} - y\right)\left(y^2 - y^3\right) dy$

$= 24\pi\int_0^1 \left(\frac{8}{5}y^2 - \frac{13}{5}y^3 + y^4\right) dy = 24\pi\left[\frac{8}{15}y^3 - \frac{13}{20}y^4 + \frac{y^5}{5}\right]_0^1 = 24\pi\left(\frac{8}{15} - \frac{13}{20} + \frac{1}{5}\right) = \frac{24\pi}{60}(32 - 39 + 12)$

$= \frac{24\pi}{12} = 2\pi$

(d) $V = \int_c^d 2\pi \left(\frac{\text{shell}}{\text{radius}}\right)\left(\frac{\text{shell}}{\text{height}}\right) dy = \int_0^1 2\pi\left(y + \frac{2}{5}\right)\left[12\left(y^2 - y^3\right)\right] dy = 24\pi\int_0^1 \left(y + \frac{2}{5}\right)\left(y^2 - y^3\right) dy$

$= 24\pi\int_0^1 \left(y^3 - y^4 + \frac{2}{5}y^2 - \frac{2}{5}y^3\right) dy = 24\pi\int_0^1 \left(\frac{2}{5}y^2 + \frac{3}{5}y^3 - y^4\right) dy = 24\pi\left[\frac{2}{15}y^3 + \frac{3}{20}y^4 - \frac{y^5}{5}\right]_0^1$

$= 24\pi\left(\frac{2}{15} + \frac{3}{20} - \frac{1}{5}\right) = \frac{24\pi}{60}(8 + 9 - 12) = \frac{24\pi}{12} = 2\pi$

24. (a) $V = \int_c^d 2\pi \left(\frac{\text{shell}}{\text{radius}}\right)\left(\frac{\text{shell}}{\text{height}}\right) dy = \int_0^2 2\pi y\left[\frac{y^2}{2} - \left(\frac{y^4}{4} - \frac{y^2}{2}\right)\right] dy = \int_0^2 2\pi y\left(y^2 - \frac{y^4}{4}\right) dy = 2\pi\int_0^2 \left(y^3 - \frac{y^5}{4}\right) dy$

$= 2\pi\left[\frac{y^4}{4} - \frac{y^6}{24}\right]_0^2 = 2\pi\left(\frac{2^4}{4} - \frac{2^6}{24}\right) = 32\pi\left(\frac{1}{4} - \frac{4}{24}\right) = 32\pi\left(\frac{1}{4} - \frac{1}{6}\right) = 32\pi\left(\frac{2}{24}\right) = \frac{8\pi}{3}$

(b) $V = \int_c^d 2\pi \left(\frac{\text{shell}}{\text{radius}}\right)\left(\frac{\text{shell}}{\text{height}}\right) dy = \int_0^2 2\pi(2 - y)\left[\frac{y^2}{2} - \left(\frac{y^4}{4} - \frac{y^2}{2}\right)\right] dy = \int_0^2 2\pi(2 - y)\left(y^2 - \frac{y^4}{4}\right) dy$

$= 2\pi\int_0^2 \left(2y^2 - \frac{y^4}{2} - y^3 + \frac{y^5}{4}\right) dy = 2\pi\left[\frac{2y^3}{3} - \frac{y^5}{10} - \frac{y^4}{4} + \frac{y^6}{24}\right]_0^2 = 2\pi\left(\frac{16}{3} - \frac{32}{10} - \frac{16}{4} + \frac{64}{24}\right) = \frac{8\pi}{5}$

(c) $V = \int_c^d 2\pi \left(\begin{smallmatrix}\text{shell}\\\text{radius}\end{smallmatrix}\right)\left(\begin{smallmatrix}\text{shell}\\\text{height}\end{smallmatrix}\right) dy = \int_0^2 2\pi(5-y)\left[\frac{y^2}{2} - \left(\frac{y^4}{4} - \frac{y^2}{2}\right)\right] dy = \int_0^2 2\pi(5-y)\left(y^2 - \frac{y^4}{4}\right) dy$

$= 2\pi \int_0^2 \left(5y^2 - \frac{5}{4}y^4 - y^3 + \frac{y^5}{4}\right) dy = 2\pi\left[\frac{5y^3}{3} - \frac{5y^5}{20} - \frac{y^4}{4} + \frac{y^6}{24}\right]_0^2 = 2\pi\left(\frac{40}{3} - \frac{160}{20} - \frac{16}{4} + \frac{64}{24}\right) = 8\pi$

(d) $V = \int_c^d 2\pi \left(\begin{smallmatrix}\text{shell}\\\text{radius}\end{smallmatrix}\right)\left(\begin{smallmatrix}\text{shell}\\\text{height}\end{smallmatrix}\right) dy = \int_0^2 2\pi\left(y + \frac{5}{8}\right)\left[\frac{y^2}{2} - \left(\frac{y^4}{4} - \frac{y^2}{2}\right)\right] dy = \int_0^2 2\pi\left(y + \frac{5}{8}\right)\left(y^2 - \frac{y^4}{4}\right) dy$

$= 2\pi \int_0^2 \left(y^3 - \frac{y^5}{4} + \frac{5}{8}y^2 - \frac{5}{32}y^4\right) dy = 2\pi\left[\frac{y^4}{4} - \frac{y^6}{24} + \frac{5y^3}{24} - \frac{5y^5}{160}\right]_0^2 = 2\pi\left(\frac{16}{4} - \frac{64}{24} + \frac{40}{24} - \frac{160}{160}\right) = 4\pi$

25. (a) About x-axis: $V = \int_c^d 2\pi \left(\begin{smallmatrix}\text{shell}\\\text{radius}\end{smallmatrix}\right)\left(\begin{smallmatrix}\text{shell}\\\text{height}\end{smallmatrix}\right) dy$

$= \int_0^1 2\pi y(\sqrt{y} - y)dy = 2\pi \int_0^1 (y^{3/2} - y^2)dy$

$= 2\pi\left[\frac{2}{5}y^{5/2} - \frac{1}{3}y^3\right]_0^1 = 2\pi\left(\frac{2}{5} - \frac{1}{3}\right) = \frac{2\pi}{15}$

About y-axis: $V = \int_a^b 2\pi \left(\begin{smallmatrix}\text{shell}\\\text{radius}\end{smallmatrix}\right)\left(\begin{smallmatrix}\text{shell}\\\text{height}\end{smallmatrix}\right) dx$

$= \int_0^1 2\pi x(x - x^2)dx = 2\pi \int_0^1 (x^2 - x^3)dx$

$= 2\pi\left[\frac{x^3}{3} - \frac{x^4}{4}\right]_0^1 = 2\pi\left(\frac{1}{3} - \frac{1}{4}\right) = \frac{\pi}{6}$

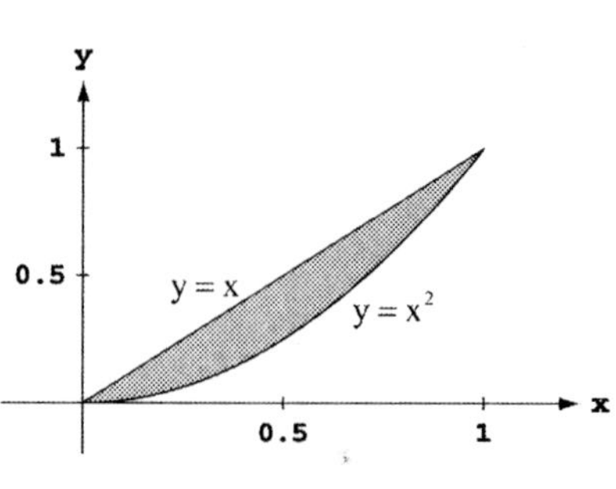

(b) About x-axis: $R(x) = x$ and $r(x) = x^2 \Rightarrow V = \int_a^b \pi\left[R(x)^2 - r(x)^2\right]dx = \int_0^1 \pi[x^2 - x^4]dx$

$= \pi\left[\frac{x^3}{3} - \frac{x^5}{5}\right]_0^1 = \pi\left(\frac{1}{3} - \frac{1}{5}\right) = \frac{2\pi}{15}$

About y-axis: $R(y) = \sqrt{y}$ and $r(y) = y \Rightarrow V = \int_c^d \pi\left[R(y)^2 - r(y)^2\right]dy = \int_0^1 \pi[y - y^2]dy$

$= \pi\left[\frac{y^2}{2} - \frac{y^3}{3}\right]_0^1 = \pi\left(\frac{1}{2} - \frac{1}{3}\right) = \frac{\pi}{6}$

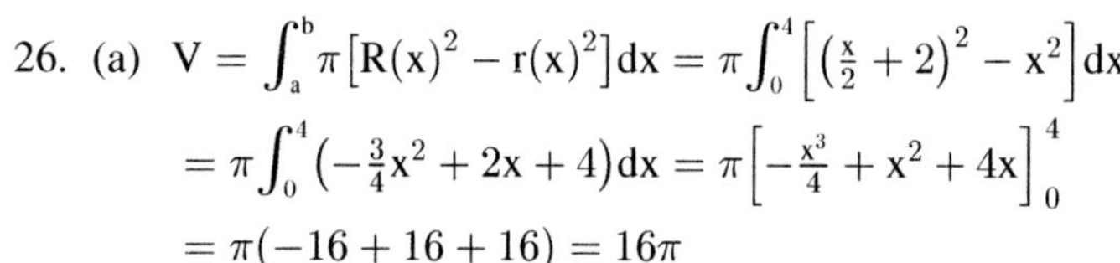

26. (a) $V = \int_a^b \pi\left[R(x)^2 - r(x)^2\right]dx = \pi \int_0^4 \left[\left(\frac{x}{2} + 2\right)^2 - x^2\right]dx$

$= \pi \int_0^4 \left(-\frac{3}{4}x^2 + 2x + 4\right)dx = \pi\left[-\frac{x^3}{4} + x^2 + 4x\right]_0^4$

$= \pi(-16 + 16 + 16) = 16\pi$

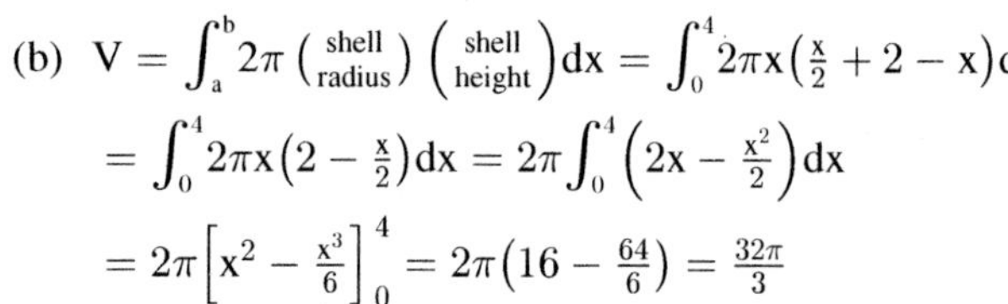

(b) $V = \int_a^b 2\pi \left(\begin{smallmatrix}\text{shell}\\\text{radius}\end{smallmatrix}\right)\left(\begin{smallmatrix}\text{shell}\\\text{height}\end{smallmatrix}\right) dx = \int_0^4 2\pi x\left(\frac{x}{2} + 2 - x\right)dx$

$= \int_0^4 2\pi x\left(2 - \frac{x}{2}\right)dx = 2\pi \int_0^4 \left(2x - \frac{x^2}{2}\right)dx$

$= 2\pi\left[x^2 - \frac{x^3}{6}\right]_0^4 = 2\pi\left(16 - \frac{64}{6}\right) = \frac{32\pi}{3}$

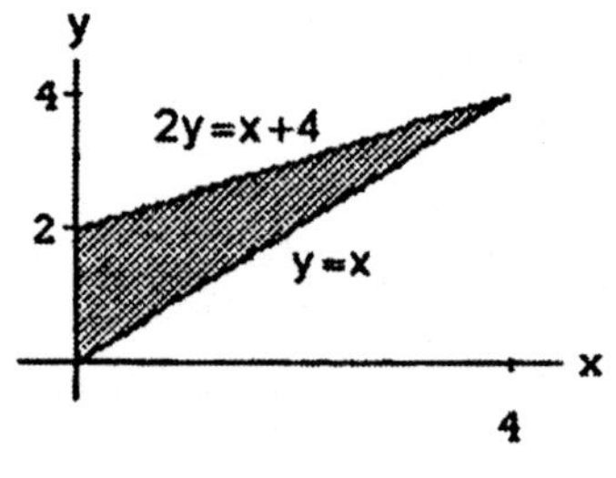

(c) $V = \int_a^b 2\pi \left(\begin{smallmatrix}\text{shell}\\\text{radius}\end{smallmatrix}\right)\left(\begin{smallmatrix}\text{shell}\\\text{height}\end{smallmatrix}\right) dx = \int_0^4 2\pi(4 - x)\left(\frac{x}{2} + 2 - x\right)dx = \int_0^4 2\pi(4 - x)\left(2 - \frac{x}{2}\right)dx = 2\pi \int_0^4 \left(8 - 4x + \frac{x^2}{2}\right)dx$

$= 2\pi\left[8x - 2x^2 + \frac{x^3}{6}\right]_0^4 = 2\pi\left(32 - 32 + \frac{64}{6}\right) = \frac{64\pi}{3}$

(d) $V = \int_a^b \pi\left[R(x)^2 - r(x)^2\right]dx = \pi \int_0^4 \left[(8 - x)^2 - \left(6 - \frac{x}{2}\right)^2\right]dx = \pi \int_0^4 \left[(64 - 16x + x^2) - \left(36 - 6x + \frac{x^2}{4}\right)\right]dx$

$\pi \int_0^4 \left(\frac{3}{4}x^2 - 10x + 28\right)dx = \pi\left[\frac{x^3}{4} - 5x^2 + 28x\right]_0^4 = \pi\left[16 - (5)(16) + (7)(16)\right] = \pi(3)(16) = 48\pi$

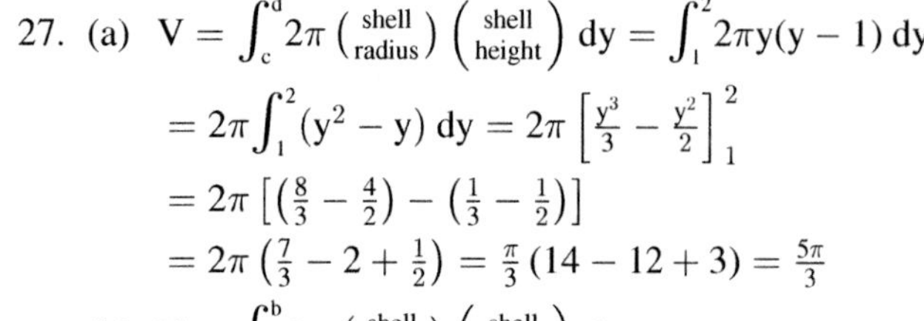

27. (a) $V = \int_c^d 2\pi \left(\begin{smallmatrix}\text{shell}\\\text{radius}\end{smallmatrix}\right)\left(\begin{smallmatrix}\text{shell}\\\text{height}\end{smallmatrix}\right) dy = \int_1^2 2\pi y(y - 1)\, dy$

$= 2\pi \int_1^2 (y^2 - y)\, dy = 2\pi\left[\frac{y^3}{3} - \frac{y^2}{2}\right]_1^2$

$= 2\pi\left[\left(\frac{8}{3} - \frac{4}{2}\right) - \left(\frac{1}{3} - \frac{1}{2}\right)\right]$

$= 2\pi\left(\frac{7}{3} - 2 + \frac{1}{2}\right) = \frac{\pi}{3}(14 - 12 + 3) = \frac{5\pi}{3}$

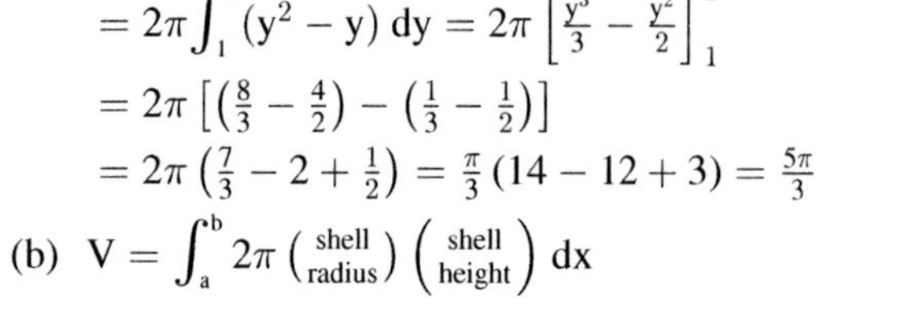

(b) $V = \int_a^b 2\pi \left(\begin{smallmatrix}\text{shell}\\\text{radius}\end{smallmatrix}\right)\left(\begin{smallmatrix}\text{shell}\\\text{height}\end{smallmatrix}\right) dx$

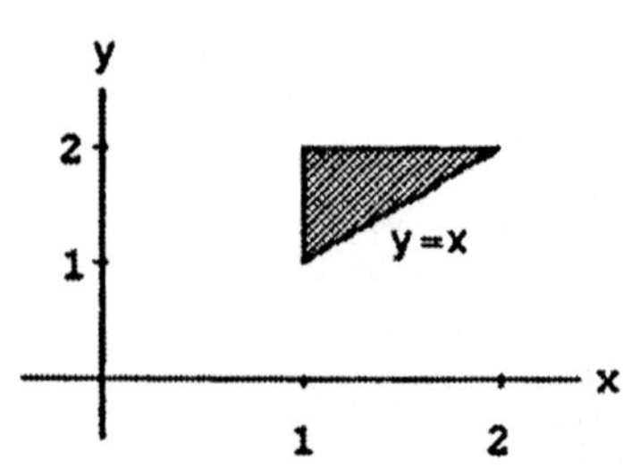

$= \int_1^2 2\pi x(2-x)\,dx = 2\pi \int_1^2 (2x - x^2)\,dx = 2\pi \left[x^2 - \frac{x^3}{3}\right]_1^2 = 2\pi\left[\left(4 - \frac{8}{3}\right) - \left(1 - \frac{1}{3}\right)\right]$

$= 2\pi\left[\left(\frac{12-8}{3}\right) - \left(\frac{3-1}{3}\right)\right] = 2\pi\left(\frac{4}{3} - \frac{2}{3}\right) = \frac{4\pi}{3}$

(c) $V = \int_a^b 2\pi \left(\begin{smallmatrix}\text{shell}\\ \text{radius}\end{smallmatrix}\right)\left(\begin{smallmatrix}\text{shell}\\ \text{height}\end{smallmatrix}\right) dx = \int_1^2 2\pi\left(\frac{10}{3} - x\right)(2-x)\,dx = 2\pi \int_1^2 \left(\frac{20}{3} - \frac{16}{3}x + x^2\right) dx$

$= 2\pi\left[\frac{20}{3}x - \frac{8}{3}x^2 + \frac{1}{3}x^3\right]_1^2 = 2\pi\left[\left(\frac{40}{3} - \frac{32}{3} + \frac{8}{3}\right) - \left(\frac{20}{3} - \frac{8}{3} + \frac{1}{3}\right)\right] = 2\pi\left(\frac{3}{3}\right) = 2\pi$

(d) $V = \int_c^d 2\pi \left(\begin{smallmatrix}\text{shell}\\ \text{radius}\end{smallmatrix}\right)\left(\begin{smallmatrix}\text{shell}\\ \text{height}\end{smallmatrix}\right) dy = \int_1^2 2\pi(y-1)(y-1)\,dy = 2\pi \int_1^2 (y-1)^2 = 2\pi\left[\frac{(y-1)^3}{3}\right]_1^2 = \frac{2\pi}{3}$

28. (a) $V = \int_c^d 2\pi \left(\begin{smallmatrix}\text{shell}\\ \text{radius}\end{smallmatrix}\right)\left(\begin{smallmatrix}\text{shell}\\ \text{height}\end{smallmatrix}\right) dy = \int_0^2 2\pi y(y^2 - 0)\,dy$

$= 2\pi \int_0^2 y^3\,dy = 2\pi\left[\frac{y^4}{4}\right]_0^2 = 2\pi\left(\frac{2^4}{4}\right) = 8\pi$

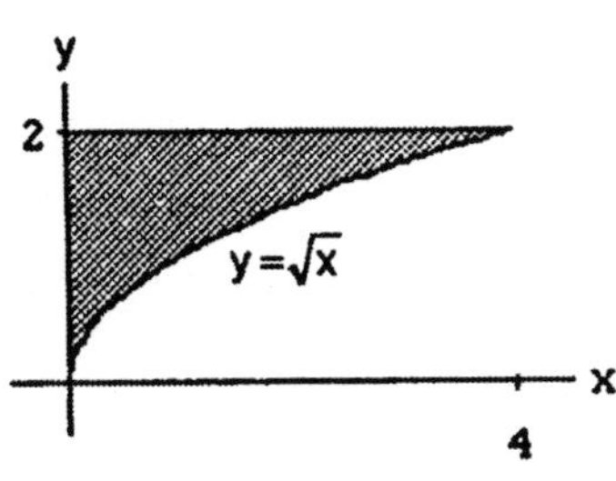

(b) $V = \int_a^b 2\pi \left(\begin{smallmatrix}\text{shell}\\ \text{radius}\end{smallmatrix}\right)\left(\begin{smallmatrix}\text{shell}\\ \text{height}\end{smallmatrix}\right) dx$

$= \int_0^4 2\pi x\left(2 - \sqrt{x}\right) dx = 2\pi \int_0^4 \left(2x - x^{3/2}\right) dx$

$= 2\pi\left[x^2 - \frac{2}{5}x^{5/2}\right]_0^4 = 2\pi\left(16 - \frac{2\cdot 2^5}{5}\right)$

$= 2\pi\left(16 - \frac{64}{5}\right) = \frac{2\pi}{5}(80 - 64) = \frac{32\pi}{5}$

(c) $V = \int_a^b 2\pi \left(\begin{smallmatrix}\text{shell}\\ \text{radius}\end{smallmatrix}\right)\left(\begin{smallmatrix}\text{shell}\\ \text{height}\end{smallmatrix}\right) dx = \int_0^4 2\pi(4-x)\left(2 - \sqrt{x}\right) dx = 2\pi \int_0^4 \left(8 - 4x^{1/2} - 2x + x^{3/2}\right) dx$

$= 2\pi\left[8x - \frac{8}{3}x^{3/2} - x^2 + \frac{2}{5}x^{5/2}\right]_0^4 = 2\pi\left(32 - \frac{64}{3} - 16 + \frac{64}{5}\right) = \frac{2\pi}{15}(240 - 320 + 192) = \frac{2\pi}{15}(112) = \frac{224\pi}{15}$

(d) $V = \int_c^d 2\pi \left(\begin{smallmatrix}\text{shell}\\ \text{radius}\end{smallmatrix}\right)\left(\begin{smallmatrix}\text{shell}\\ \text{height}\end{smallmatrix}\right) dy = \int_0^2 2\pi(2-y)(y^2)\,dy = 2\pi \int_0^2 (2y^2 - y^3)\,dy = 2\pi\left[\frac{2}{3}y^3 - \frac{y^4}{4}\right]_0^2$

$= 2\pi\left(\frac{16}{3} - \frac{16}{4}\right) = \frac{32\pi}{12}(4-3) = \frac{8\pi}{3}$

29. (a) $V = \int_c^d 2\pi \left(\begin{smallmatrix}\text{shell}\\ \text{radius}\end{smallmatrix}\right)\left(\begin{smallmatrix}\text{shell}\\ \text{height}\end{smallmatrix}\right) dy = \int_0^1 2\pi y(y - y^3)\,dy$

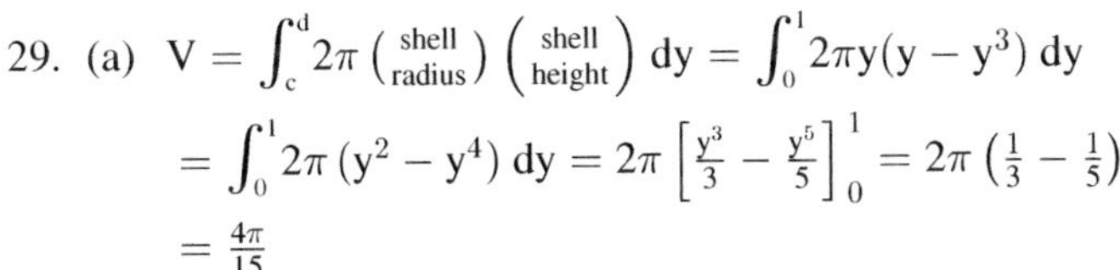

$= \int_0^1 2\pi(y^2 - y^4)\,dy = 2\pi\left[\frac{y^3}{3} - \frac{y^5}{5}\right]_0^1 = 2\pi\left(\frac{1}{3} - \frac{1}{5}\right)$

$= \frac{4\pi}{15}$

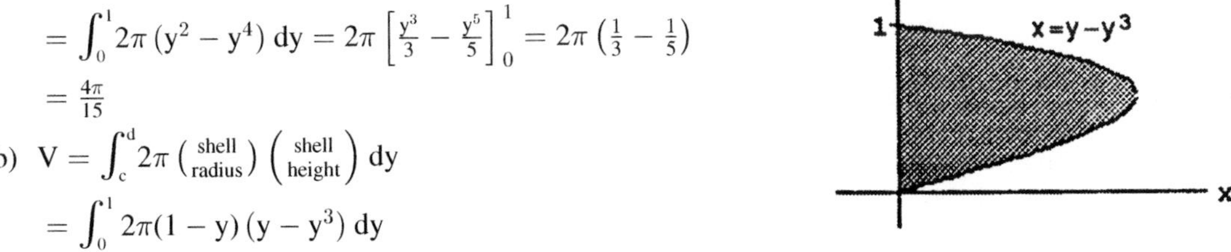

(b) $V = \int_c^d 2\pi \left(\begin{smallmatrix}\text{shell}\\ \text{radius}\end{smallmatrix}\right)\left(\begin{smallmatrix}\text{shell}\\ \text{height}\end{smallmatrix}\right) dy$

$= \int_0^1 2\pi(1-y)(y - y^3)\,dy$

$= 2\pi \int_0^1 (y - y^2 - y^3 + y^4)\,dy = 2\pi\left[\frac{y^2}{2} - \frac{y^3}{3} - \frac{y^4}{4} + \frac{y^5}{5}\right]_0^1 = 2\pi\left(\frac{1}{2} - \frac{1}{3} - \frac{1}{4} + \frac{1}{5}\right) = \frac{2\pi}{60}(30 - 20 - 15 + 12) = \frac{7\pi}{30}$

30. (a) $V = \int_c^d 2\pi \left(\begin{smallmatrix}\text{shell}\\ \text{radius}\end{smallmatrix}\right)\left(\begin{smallmatrix}\text{shell}\\ \text{height}\end{smallmatrix}\right) dy$

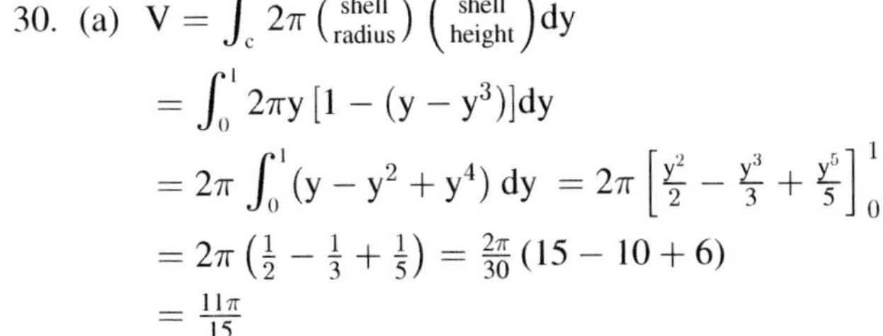

$= \int_0^1 2\pi y\left[1 - (y - y^3)\right] dy$

$= 2\pi \int_0^1 (y - y^2 + y^4)\,dy = 2\pi\left[\frac{y^2}{2} - \frac{y^3}{3} + \frac{y^5}{5}\right]_0^1$

$= 2\pi\left(\frac{1}{2} - \frac{1}{3} + \frac{1}{5}\right) = \frac{2\pi}{30}(15 - 10 + 6)$

$= \frac{11\pi}{15}$

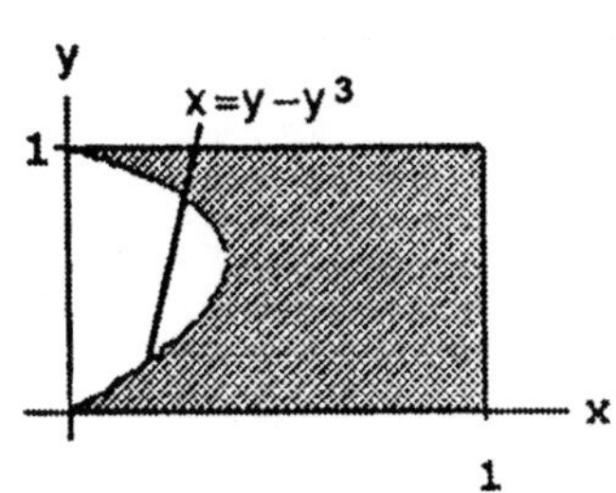

(b) Use the washer method:

$V = \int_c^d \pi\left[R^2(y) - r^2(y)\right] dy = \int_0^1 \pi\left[1^2 - (y - y^3)^2\right] dy = \pi \int_0^1 (1 - y^2 - y^6 + 2y^4)\,dy = \pi\left[y - \frac{y^3}{3} - \frac{y^7}{7} + \frac{2y^5}{5}\right]_0^1$

$= \pi\left(1 - \frac{1}{3} - \frac{1}{7} + \frac{2}{5}\right) = \frac{\pi}{105}(105 - 35 - 15 + 42) = \frac{97\pi}{105}$

(c) Use the washer method:

$V = \int_c^d \pi\left[R^2(y) - r^2(y)\right] dy = \int_0^1 \pi\left[\left[1 - (y - y^3)\right]^2 - 0\right] dy = \pi \int_0^1 \left[1 - 2(y - y^3) + (y - y^3)^2\right] dy$

$= \pi \int_0^1 (1 + y^2 + y^6 - 2y + 2y^3 - 2y^4)\, dy = \pi \left[y + \frac{y^3}{3} + \frac{y^7}{7} - y^2 + \frac{y^4}{2} - \frac{2y^5}{5} \right]_0^1 = \pi \left(1 + \frac{1}{3} + \frac{1}{7} - 1 + \frac{1}{2} - \frac{2}{5}\right)$

$= \frac{\pi}{210}(70 + 30 + 105 - 2 \cdot 42) = \frac{121\pi}{210}$

(d) $V = \int_c^d 2\pi \left(\begin{smallmatrix}\text{shell}\\ \text{radius}\end{smallmatrix}\right) \left(\begin{smallmatrix}\text{shell}\\ \text{height}\end{smallmatrix}\right) dy = \int_0^1 2\pi(1 - y)\left[1 - (y - y^3)\right] dy = 2\pi \int_0^1 (1 - y)(1 - y + y^3)\, dy$

$= 2\pi \int_0^1 (1 - y + y^3 - y + y^2 - y^4)\, dy = 2\pi \int_0^1 (1 - 2y + y^2 + y^3 - y^4)\, dy = 2\pi \left[y - y^2 + \frac{y^3}{3} + \frac{y^4}{4} - \frac{y^5}{5} \right]_0^1$

$= 2\pi \left(1 - 1 + \frac{1}{3} + \frac{1}{4} - \frac{1}{5}\right) = \frac{2\pi}{60}(20 + 15 - 12) = \frac{23\pi}{30}$

31. (a) $V = \int_c^d 2\pi \left(\begin{smallmatrix}\text{shell}\\ \text{radius}\end{smallmatrix}\right) \left(\begin{smallmatrix}\text{shell}\\ \text{height}\end{smallmatrix}\right) dy = \int_0^2 2\pi y \left(\sqrt{8y} - y^2\right) dy$

$= 2\pi \int_0^2 \left(2\sqrt{2}\, y^{3/2} - y^3\right) dy = 2\pi \left[\frac{4\sqrt{2}}{5} y^{5/2} - \frac{y^4}{4} \right]_0^2$

$= 2\pi \left(\frac{4\sqrt{2} \cdot \left(\sqrt{2}\right)^5}{5} - \frac{2^4}{4} \right) = 2\pi \left(\frac{4 \cdot 2^3}{5} - \frac{4 \cdot 4}{4} \right)$

$= 2\pi \cdot 4 \left(\frac{8}{5} - 1\right) = \frac{8\pi}{5}(8 - 5) = \frac{24\pi}{5}$

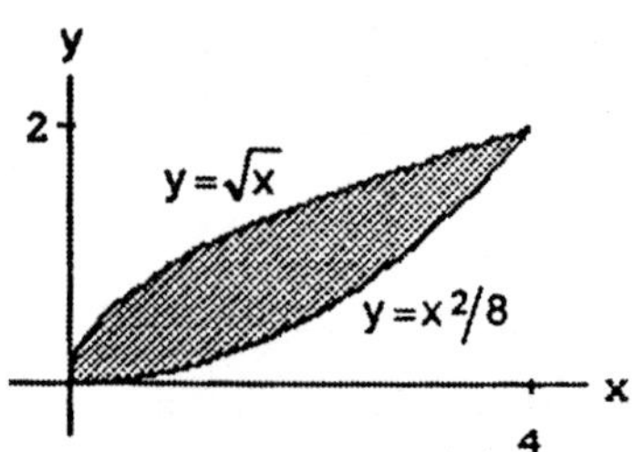

(b) $V = \int_a^b 2\pi \left(\begin{smallmatrix}\text{shell}\\ \text{radius}\end{smallmatrix}\right) \left(\begin{smallmatrix}\text{shell}\\ \text{height}\end{smallmatrix}\right) dx = \int_0^4 2\pi x \left(\sqrt{x} - \frac{x^2}{8}\right) dx = 2\pi \int_0^4 \left(x^{3/2} - \frac{x^3}{8}\right) dx = 2\pi \left[\frac{2}{5} x^{5/2} - \frac{x^4}{32} \right]_0^4$

$= 2\pi \left(\frac{2 \cdot 2^5}{5} - \frac{4^4}{32} \right) = 2\pi \left(\frac{2^6}{5} - \frac{2^8}{32} \right) = \frac{\pi \cdot 2^7}{160}(32 - 20) = \frac{\pi \cdot 2^9 \cdot 3}{160} = \frac{\pi \cdot 2^4 \cdot 3}{5} = \frac{48\pi}{5}$

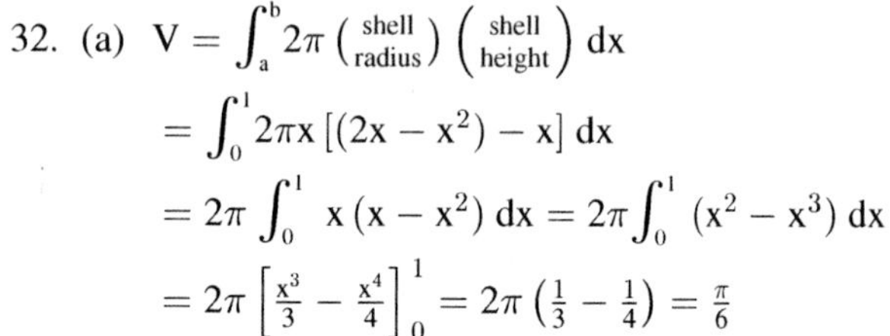

32. (a) $V = \int_a^b 2\pi \left(\begin{smallmatrix}\text{shell}\\ \text{radius}\end{smallmatrix}\right) \left(\begin{smallmatrix}\text{shell}\\ \text{height}\end{smallmatrix}\right) dx$

$= \int_0^1 2\pi x \left[(2x - x^2) - x\right] dx$

$= 2\pi \int_0^1 x(x - x^2)\, dx = 2\pi \int_0^1 (x^2 - x^3)\, dx$

$= 2\pi \left[\frac{x^3}{3} - \frac{x^4}{4} \right]_0^1 = 2\pi \left(\frac{1}{3} - \frac{1}{4}\right) = \frac{\pi}{6}$

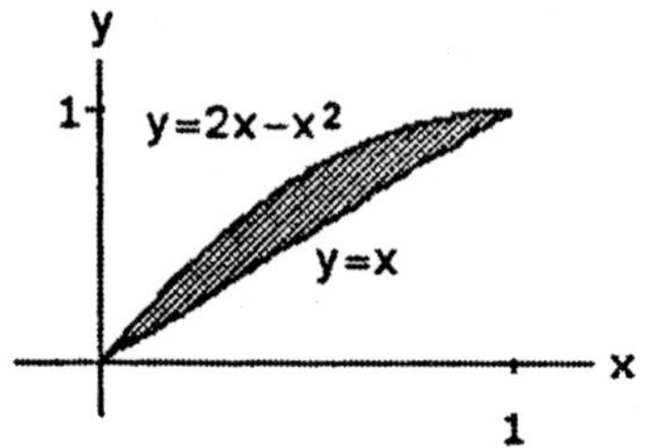

(b) $V = \int_a^b 2\pi \left(\begin{smallmatrix}\text{shell}\\ \text{radius}\end{smallmatrix}\right) \left(\begin{smallmatrix}\text{shell}\\ \text{height}\end{smallmatrix}\right) dx = \int_0^1 2\pi(1 - x)\left[(2x - x^2) - x\right] dx = 2\pi \int_0^1 (1 - x)(x - x^2)\, dx$

$= 2\pi \int_0^1 (x - 2x^2 + x^3)\, dx = 2\pi \left[\frac{x^2}{2} - \frac{2}{3}x^3 + \frac{x^4}{4} \right]_0^1 = 2\pi \left(\frac{1}{2} - \frac{2}{3} + \frac{1}{4}\right) = \frac{2\pi}{12}(6 - 8 + 3) = \frac{\pi}{6}$

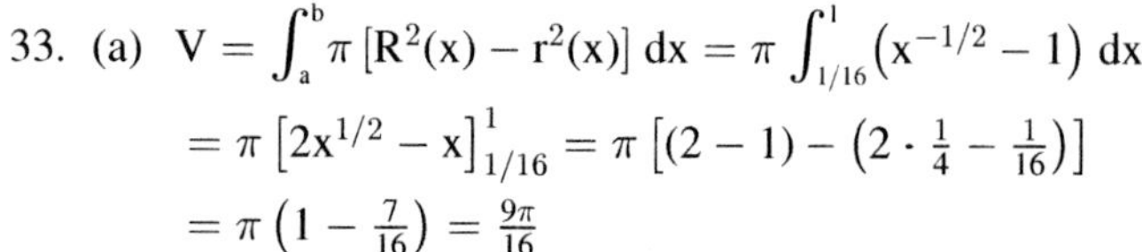

33. (a) $V = \int_a^b \pi \left[R^2(x) - r^2(x)\right] dx = \pi \int_{1/16}^1 \left(x^{-1/2} - 1\right) dx$

$= \pi \left[2x^{1/2} - x\right]_{1/16}^1 = \pi \left[(2 - 1) - \left(2 \cdot \frac{1}{4} - \frac{1}{16}\right)\right]$

$= \pi \left(1 - \frac{7}{16}\right) = \frac{9\pi}{16}$

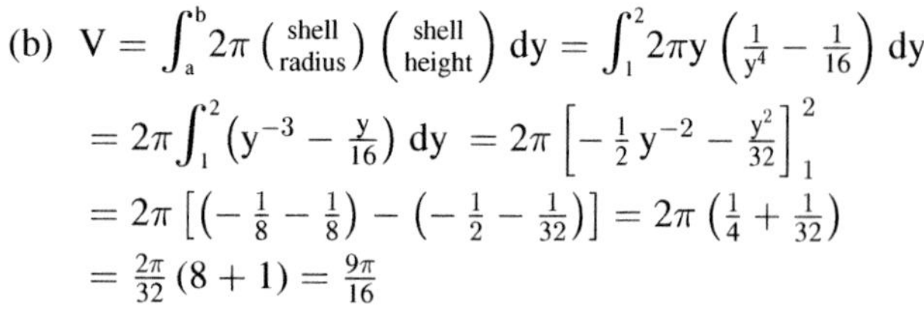

(b) $V = \int_a^b 2\pi \left(\begin{smallmatrix}\text{shell}\\ \text{radius}\end{smallmatrix}\right) \left(\begin{smallmatrix}\text{shell}\\ \text{height}\end{smallmatrix}\right) dy = \int_1^2 2\pi y \left(\frac{1}{y^4} - \frac{1}{16}\right) dy$

$= 2\pi \int_1^2 \left(y^{-3} - \frac{y}{16}\right) dy = 2\pi \left[-\frac{1}{2} y^{-2} - \frac{y^2}{32} \right]_1^2$

$= 2\pi \left[\left(-\frac{1}{8} - \frac{1}{8}\right) - \left(-\frac{1}{2} - \frac{1}{32}\right)\right] = 2\pi \left(\frac{1}{4} + \frac{1}{32}\right)$

$= \frac{2\pi}{32}(8 + 1) = \frac{9\pi}{16}$

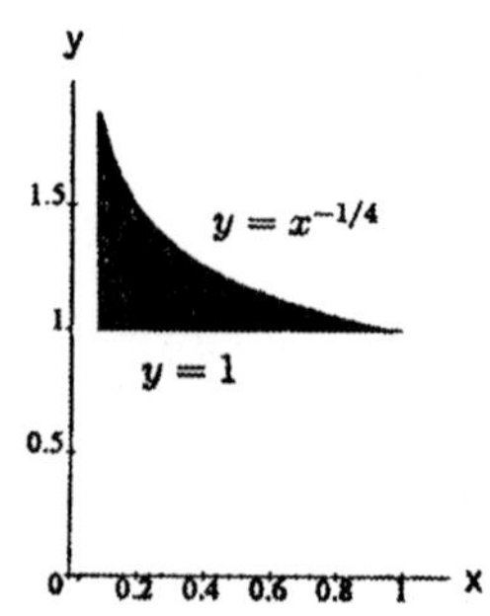

34. (a) $V = \int_c^d \pi \left[R^2(y) - r^2(y)\right] dy = \int_1^2 \pi \left(\frac{1}{y^4} - \frac{1}{16}\right) dy$

$= \pi \left[-\frac{1}{3} y^{-3} - \frac{y}{16} \right]_1^2 = \pi \left[\left(-\frac{1}{24} - \frac{1}{8}\right) - \left(-\frac{1}{3} - \frac{1}{16}\right)\right]$

$= \frac{\pi}{48}(-2 - 6 + 16 + 3) = \frac{11\pi}{48}$

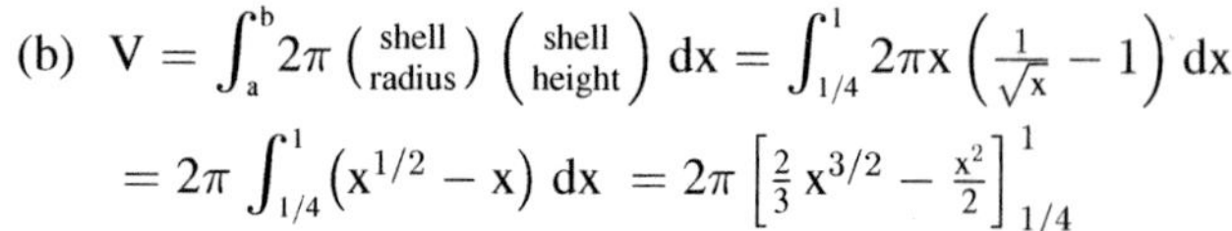

(b) $V = \int_a^b 2\pi \left(\begin{smallmatrix}\text{shell}\\ \text{radius}\end{smallmatrix}\right) \left(\begin{smallmatrix}\text{shell}\\ \text{height}\end{smallmatrix}\right) dx = \int_{1/4}^1 2\pi x \left(\frac{1}{\sqrt{x}} - 1\right) dx$

$= 2\pi \int_{1/4}^1 \left(x^{1/2} - x\right) dx = 2\pi \left[\frac{2}{3} x^{3/2} - \frac{x^2}{2} \right]_{1/4}^1$

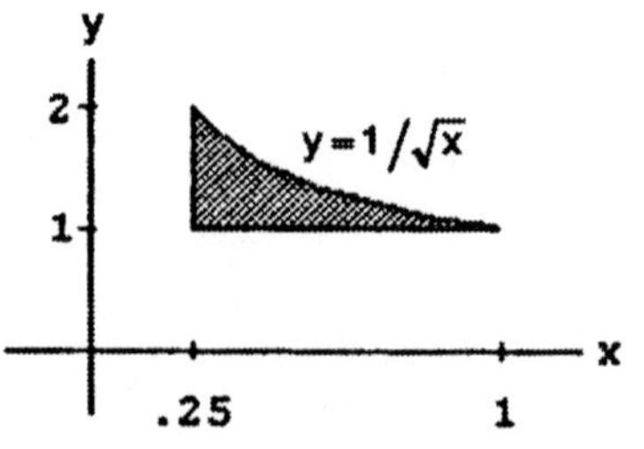

$= 2\pi\left[\left(\frac{2}{3}-\frac{1}{2}\right)-\left(\frac{2}{3}\cdot\frac{1}{8}-\frac{1}{32}\right)\right] = \pi\left(\frac{4}{3}-1-\frac{1}{6}+\frac{1}{16}\right) = \frac{\pi}{48}(4\cdot 16-48-8+3) = \frac{11\pi}{48}$

35. (a) *Disk*: $V = V_1 - V_2$

$V_1 = \int_{a_1}^{b_1} \pi[R_1(x)]^2\, dx$ and $V_2 = \int_{a_2}^{b_2} \pi[R_2(x)]^2$ with $R_1(x) = \sqrt{\frac{x+2}{3}}$ and $R_2(x) = \sqrt{x}$,

$a_1 = -2,\ b_1 = 1;\ a_2 = 0,\ b_2 = 1 \Rightarrow$ two integrals are required

(b) *Washer*: $V = V_1 - V_2$

$V_1 = \int_{a_1}^{b_1} \pi\left([R_1(x)]^2 - [r_1(x)]^2\right) dx$ with $R_1(x) = \sqrt{\frac{x+2}{3}}$ and $r_1(x) = 0$; $a_1 = -2$ and $b_1 = 0$;

$V_2 = \int_{a_2}^{b_2} \pi\left([R_2(x)]^2 - [r_2(x)]^2\right) dx$ with $R_2(x) = \sqrt{\frac{x+2}{3}}$ and $r_2(x) = \sqrt{x}$; $a_2 = 0$ and $b_2 = 1$

$\Rightarrow$ two integrals are required

(c) *Shell*: $V = \int_c^d 2\pi\left(\frac{\text{shell}}{\text{radius}}\right)\left(\frac{\text{shell}}{\text{height}}\right) dy = \int_c^d 2\pi y\left(\frac{\text{shell}}{\text{height}}\right) dy$ where shell height $= y^2 - (3y^2 - 2) = 2 - 2y^2$;

$c = 0$ and $d = 1$. Only *one* integral is required. It is, therefore preferable to use the *shell* method. However, whichever method you use, you will get $V = \pi$.

36. (a) *Disk*: $V = V_1 - V_2 - V_3$

$V_i = \int_{c_i}^{d_i} \pi[R_i(y)]^2\, dy$, $i = 1, 2, 3$ with $R_1(y) = 1$ and $c_1 = -1,\ d_1 = 1$; $R_2(y) = \sqrt{y}$ and $c_2 = 0$ and $d_2 = 1$;

$R_3(y) = (-y)^{1/4}$ and $c_3 = -1,\ d_3 = 0 \Rightarrow$ three integrals are required

(b) *Washer*: $V = V_1 + V_2$

$V_i = \int_{c_i}^{d_i} \pi\left([R_i(y)]^2 - [r_i(y)]^2\right) dy$, $i = 1, 2$ with $R_1(y) = 1$, $r_1(y) = \sqrt{y}$, $c_1 = 0$ and $d_1 = 1$;

$R_2(y) = 1$, $r_2(y) = (-y)^{1/4}$, $c_2 = -1$ and $d_2 = 0 \Rightarrow$ two integrals are required

(c) *Shell*: $V = \int_a^b 2\pi\left(\frac{\text{shell}}{\text{radius}}\right)\left(\frac{\text{shell}}{\text{height}}\right) dx = \int_a^b 2\pi x\left(\frac{\text{shell}}{\text{height}}\right) dx$, where shell height $= x^2 - (-x^4) = x^2 + x^4$,

$a = 0$ and $b = 1 \Rightarrow$ only one integral is required. It is, therefore preferable to use the *shell* method. However, whichever method you use, you will get $V = \frac{5\pi}{6}$.

37. $W(a) = \int_{f(a)}^{f(a)} \pi\left[\left(f^{-1}(y)\right)^2 - a^2\right] dy = 0 = \int_a^a 2\pi x[f(a) - f(x)]\, dx = S(a)$; $W'(t) = \pi\left[\left(f^{-1}(f(t))\right)^2 - a^2\right] f'(t)$

$= \pi\left(t^2 - a^2\right) f'(t)$; also $S(t) = 2\pi f(t)\int_a^t x\, dx - 2\pi\int_a^t xf(x)\, dx = \left[\pi f(t)t^2 - \pi f(t)a^2\right] - 2\pi\int_a^t xf(x)\, dx$

$\Rightarrow S'(t) = \pi t^2 f'(t) + 2\pi t f(t) - \pi a^2 f'(t) - 2\pi t f(t) = \pi\left(t^2 - a^2\right) f'(t) \Rightarrow W'(t) = S'(t)$. Therefore, $W(t) = S(t)$ for all $t \in [a, b]$.

38. $V = \pi\int_0^{\pi/3} \left[2^2 - (\sec y)^2\right] dy = \pi\left[4y - \tan y\right]_0^{\pi/3} = \pi\left(\frac{4\pi}{3} - \sqrt{3}\right)$

39. $V = \int_a^b 2\pi\left(\frac{\text{shell}}{\text{radius}}\right)\left(\frac{\text{shell}}{\text{height}}\right) dx = \int_0^1 2\pi\, x\, e^{-x^2} dx = -\pi\, e^{-x^2}\Big|_0^1 = -\pi\left(e^{-1} - e^0\right) = \pi\left(1 - \frac{1}{e}\right)$

40. Use washer cross sections. A washer has inner radius $r = 1$, outer radius $R = e^{x/2}$, and area $\pi\left(R^2 - r^2\right) = \pi\left(e^x - 1\right)$.

The volume is $V = \int_0^{\ln 3} \pi\left(e^x - 1\right) dx = \pi\left[e^x - x\right]_0^{\ln 3}$

$= \pi(3 - \ln 3 - 1) = \pi(2 - \ln 3)$

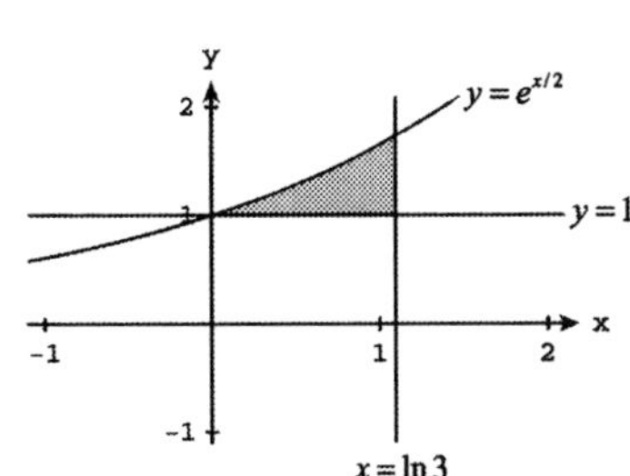

6.3 LENGTHS OF PLANE CURVES

1. $\frac{dx}{dt} = -1$ and $\frac{dy}{dt} = 3 \Rightarrow \sqrt{\left(\frac{dx}{dt}\right)^2 + \left(\frac{dy}{dt}\right)^2} = \sqrt{(-1)^2 + (3)^2} = \sqrt{10}$
$\Rightarrow \text{Length} = \int_{-2/3}^{1} \sqrt{10}\, dt = \sqrt{10}\, [t]_{-2/3}^{1} = \sqrt{10} - \left(-\frac{2}{3}\sqrt{10}\right) = \frac{5\sqrt{10}}{3}$

2. $\frac{dx}{dt} = -\sin t$ and $\frac{dy}{dt} = 1 + \cos t \Rightarrow \sqrt{\left(\frac{dx}{dt}\right)^2 + \left(\frac{dy}{dt}\right)^2} = \sqrt{(-\sin t)^2 + (1 + \cos t)^2} = \sqrt{2 + 2\cos t}$
$\Rightarrow \text{Length} = \int_0^{\pi} \sqrt{2 + 2\cos t}\, dt = \sqrt{2} \int_0^{\pi} \sqrt{\left(\frac{1-\cos t}{1-\cos t}\right)(1 + \cos t)}\, dt = \sqrt{2} \int_0^{\pi} \sqrt{\frac{\sin^2 t}{1 - \cos t}}\, dt$
$= \sqrt{2} \int_0^{\pi} \frac{\sin t}{\sqrt{1 - \cos t}}\, dt$ (since $\sin t \geq 0$ on $[0, \pi]$); $[u = 1 - \cos t \Rightarrow du = \sin t\, dt;\ t = 0 \Rightarrow u = 0,$
$t = \pi \Rightarrow u = 2] \rightarrow \sqrt{2} \int_0^{2} u^{-1/2}\, du = \sqrt{2}\, \left[2u^{1/2}\right]_0^2 = 4$

3. $\frac{dx}{dt} = 3t^2$ and $\frac{dy}{dt} = 3t \Rightarrow \sqrt{\left(\frac{dx}{dt}\right)^2 + \left(\frac{dy}{dt}\right)^2} = \sqrt{(3t^2)^2 + (3t)^2} = \sqrt{9t^4 + 9t^2} = 3t\sqrt{t^2 + 1}$ $\left(\text{since } t \geq 0 \text{ on } \left[0, \sqrt{3}\right]\right)$
$\Rightarrow \text{Length} = \int_0^{\sqrt{3}} 3t\sqrt{t^2 + 1}\, dt;$ $\left[u = t^2 + 1 \Rightarrow \frac{3}{2}\, du = 3t\, dt;\ t = 0 \Rightarrow u = 1, t = \sqrt{3} \Rightarrow u = 4\right]$
$\rightarrow \int_1^4 \frac{3}{2} u^{1/2}\, du = \left[u^{3/2}\right]_1^4 = (8 - 1) = 7$

4. $\frac{dx}{dt} = t$ and $\frac{dy}{dt} = (2t + 1)^{1/2} \Rightarrow \sqrt{\left(\frac{dx}{dt}\right)^2 + \left(\frac{dy}{dt}\right)^2} = \sqrt{t^2 + (2t + 1)} = \sqrt{(t + 1)^2} = |t + 1| = t + 1$ since $0 \leq t \leq 4$
$\Rightarrow \text{Length} = \int_0^4 (t + 1)\, dt = \left[\frac{t^2}{2} + t\right]_0^4 = (8 + 4) = 12$

5. $\frac{dx}{dt} = (2t + 3)^{1/2}$ and $\frac{dy}{dt} = 1 + t \Rightarrow \sqrt{\left(\frac{dx}{dt}\right)^2 + \left(\frac{dy}{dt}\right)^2} = \sqrt{(2t + 3) + (1 + t)^2} = \sqrt{t^2 + 4t + 4} = |t + 2| = t + 2$
since $0 \leq t \leq 3 \Rightarrow \text{Length} = \int_0^3 (t + 2)\, dt = \left[\frac{t^2}{2} + 2t\right]_0^3 = \frac{21}{2}$

6. $\frac{dx}{dt} = 8t\cos t$ and $\frac{dy}{dt} = 8t\sin t \Rightarrow \sqrt{\left(\frac{dx}{dt}\right)^2 + \left(\frac{dy}{dt}\right)^2} = \sqrt{(8t\cos t)^2 + (8t\sin t)^2} = \sqrt{64t^2\cos^2 t + 64t^2\sin^2 t}$
$= |8t| = 8t$ since $0 \leq t \leq \frac{\pi}{2} \Rightarrow \text{Length} = \int_0^{\pi/2} 8t\, dt = \left[4t^2\right]_0^{\pi/2} = \pi^2$

7. $\frac{dx}{dt} = e^t - 1$ and $\frac{dy}{dt} = 2e^{t/2} \Rightarrow \sqrt{\left(\frac{dx}{dt}\right)^2 + \left(\frac{dy}{dt}\right)^2} = \sqrt{(e^t - 1)^2 + (2e^{t/2})^2} = \sqrt{e^{2t} + 2e^t + 1} = \sqrt{(e^t + 1)^2} = |e^t + 1|$
$= e^t + 1$ (since $e^t + 1 > 0$ for all t) $\Rightarrow L = \int_0^3 (e^t + 1)\, dt = \left[e^t + t\right]_0^3 = e^3 + 2$

8. $\frac{dx}{dt} = e^t(\cos t - \sin t)$ and $\frac{dy}{dt} = e^t(\sin t + \cos t) \Rightarrow \sqrt{\left(\frac{dx}{dt}\right)^2 + \left(\frac{dy}{dt}\right)^2} = \sqrt{\left[e^t(\cos t - \sin t)\right]^2 + \left[e^t(\sin t + \cos t)\right]^2}$
$= \sqrt{2e^{2t}} = \sqrt{2}e^t \Rightarrow L = \int_0^{\pi} \sqrt{2}e^t dt = \left[\sqrt{2}e^t\right]_0^{\pi} = \sqrt{2}(e^{\pi} - 1)$

9. $\frac{dy}{dx} = \frac{1}{3} \cdot \frac{3}{2}\left(x^2+2\right)^{1/2} \cdot 2x = \sqrt{(x^2+2)} \cdot x$
$\Rightarrow L = \int_0^3 \sqrt{1+(x^2+2)\,x^2}\,dx = \int_0^3 \sqrt{1+2x^2+x^4}\,dx$
$= \int_0^3 \sqrt{(1+x^2)^2}\,dx = \int_0^3 (1+x^2)\,dx = \left[x+\frac{x^3}{3}\right]_0^3$
$= 3 + \frac{27}{3} = 12$

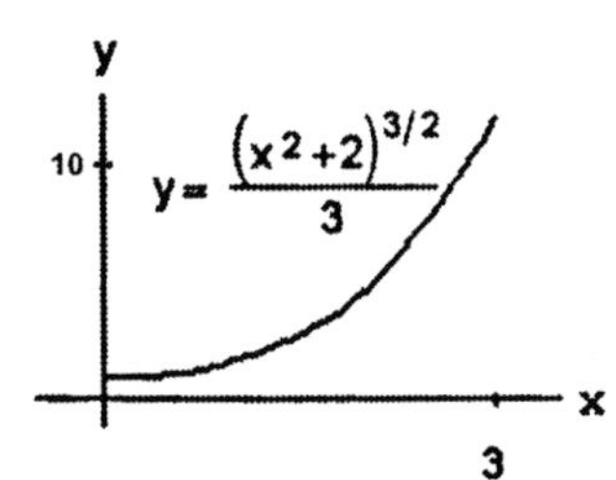

10. $\frac{dy}{dx} = \frac{3}{2}\sqrt{x} \Rightarrow L = \int_0^4 \sqrt{1+\frac{9}{4}x}\,dx$; $\left[u = 1+\frac{9}{4}x\right.$
$\Rightarrow du = \frac{9}{4}\,dx \Rightarrow \frac{4}{9}\,du = dx;\ x = 0 \Rightarrow u = 1;\ x = 4$
$\left.\Rightarrow u = 10\right] \rightarrow L = \int_1^{10} u^{1/2}\left(\frac{4}{9}\,du\right) = \frac{4}{9}\left[\frac{2}{3}u^{3/2}\right]_1^{10}$
$= \frac{8}{27}\left(10\sqrt{10}-1\right)$

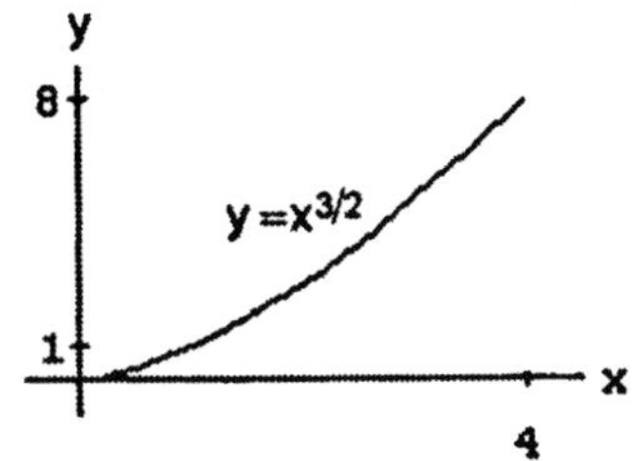

11. $\frac{dx}{dy} = y^2 - \frac{1}{4y^2} \Rightarrow \left(\frac{dx}{dy}\right)^2 = y^4 - \frac{1}{2} + \frac{1}{16y^4}$
$\Rightarrow L = \int_1^3 \sqrt{1+y^4-\frac{1}{2}+\frac{1}{16y^4}}\,dy$
$= \int_1^3 \sqrt{y^4+\frac{1}{2}+\frac{1}{16y^4}}\,dy$
$= \int_1^3 \sqrt{\left(y^2+\frac{1}{4y^2}\right)^2}\,dy = \int_1^3 \left(y^2+\frac{1}{4y^2}\right)dy$
$= \left[\frac{y^3}{3} - \frac{y^{-1}}{4}\right]_1^3 = \left(\frac{27}{3}-\frac{1}{12}\right) - \left(\frac{1}{3}-\frac{1}{4}\right) = 9 - \frac{1}{12} - \frac{1}{3} + \frac{1}{4} = 9 + \frac{(-1-4+3)}{12} = 9 + \frac{(-2)}{12} = \frac{53}{6}$

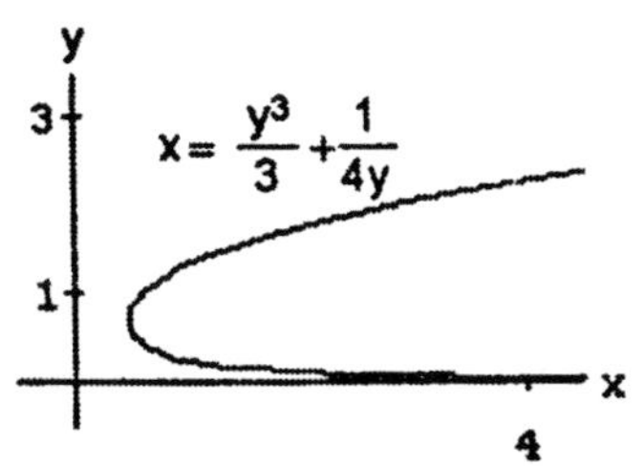

12. $\frac{dx}{dy} = \frac{1}{2}y^{1/2} - \frac{1}{2}y^{-1/2} \Rightarrow \left(\frac{dx}{dy}\right)^2 = \frac{1}{4}\left(y-2+\frac{1}{y}\right)$
$\Rightarrow L = \int_1^9 \sqrt{1+\frac{1}{4}\left(y-2+\frac{1}{y}\right)}\,dy$
$= \int_1^9 \sqrt{\frac{1}{4}\left(y+2+\frac{1}{y}\right)}\,dy = \int_1^9 \frac{1}{2}\sqrt{\left(\sqrt{y}+\frac{1}{\sqrt{y}}\right)^2}\,dy$
$= \frac{1}{2}\int_1^9 \left(y^{1/2}+y^{-1/2}\right)dy = \frac{1}{2}\left[\frac{2}{3}y^{3/2}+2y^{1/2}\right]_1^9$
$= \left[\frac{y^{3/2}}{3}+y^{1/2}\right]_1^9 = \left(\frac{3^3}{3}+3\right) - \left(\frac{1}{3}+1\right) = 11 - \frac{1}{3} = \frac{32}{3}$

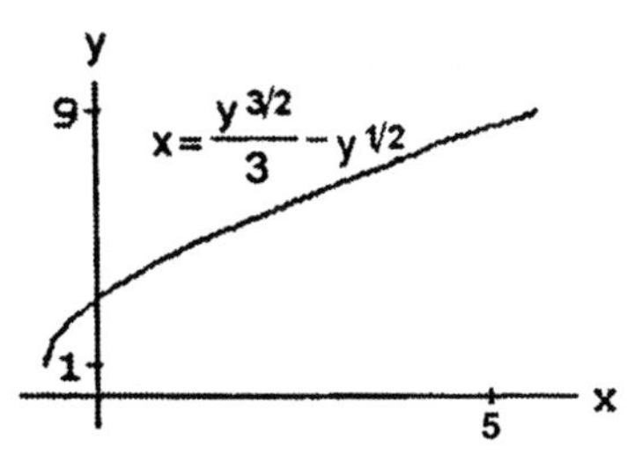

13. $\frac{dx}{dy} = y^3 - \frac{1}{4y^3} \Rightarrow \left(\frac{dx}{dy}\right)^2 = y^6 - \frac{1}{2} + \frac{1}{16y^6}$
$\Rightarrow L = \int_1^2 \sqrt{1+y^6-\frac{1}{2}+\frac{1}{16y^6}}\,dy$
$= \int_1^2 \sqrt{y^6+\frac{1}{2}+\frac{1}{16y^6}}\,dy = \int_1^2 \sqrt{\left(y^3+\frac{y^{-3}}{4}\right)^2}\,dy$
$= \int_1^2 \left(y^3+\frac{y^{-3}}{4}\right)dy = \left[\frac{y^4}{4} - \frac{y^{-2}}{8}\right]_1^2$
$= \left(\frac{16}{4} - \frac{1}{(16)(2)}\right) - \left(\frac{1}{4}-\frac{1}{8}\right) = 4 - \frac{1}{32} - \frac{1}{4} + \frac{1}{8} = \frac{128-1-8+4}{32} = \frac{123}{32}$

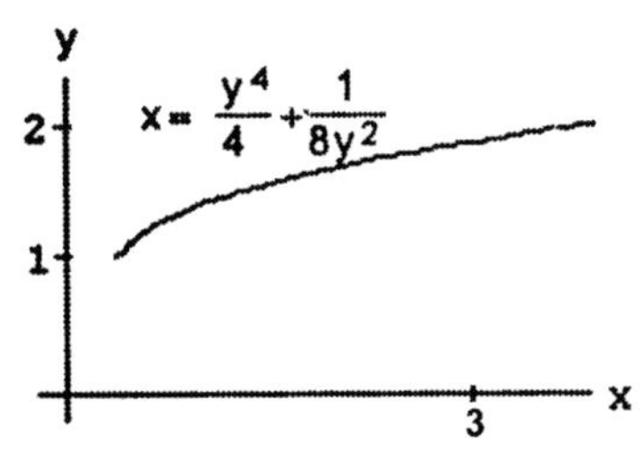

14. $\frac{dx}{dy} = \frac{y^2}{2} - \frac{1}{2y^2} \Rightarrow \left(\frac{dx}{dy}\right)^2 = \frac{1}{4}\left(y^4 - 2 + y^{-4}\right)$

$\Rightarrow L = \int_2^3 \sqrt{1 + \frac{1}{4}\left(y^4 - 2 + y^{-4}\right)}\, dy$

$= \int_2^3 \sqrt{\frac{1}{4}\left(y^4 + 2 + y^{-4}\right)}\, dy$

$= \frac{1}{2}\int_2^3 \sqrt{\left(y^2 + y^{-2}\right)^2}\, dy = \frac{1}{2}\int_2^3 \left(y^2 + y^{-2}\right) dy$

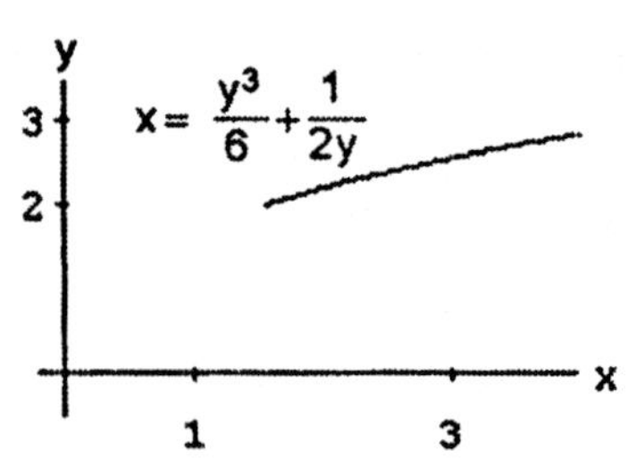

$= \frac{1}{2}\left[\frac{y^3}{3} - y^{-1}\right]_2^3 = \frac{1}{2}\left[\left(\frac{27}{3} - \frac{1}{3}\right) - \left(\frac{8}{3} - \frac{1}{2}\right)\right] = \frac{1}{2}\left(\frac{26}{3} - \frac{8}{3} + \frac{1}{2}\right) = \frac{1}{2}\left(6 + \frac{1}{2}\right) = \frac{13}{4}$

15. $\frac{dy}{dx} = x^{1/3} - \frac{1}{4}x^{-1/3} \Rightarrow \left(\frac{dy}{dx}\right)^2 = x^{2/3} - \frac{1}{2} + \frac{x^{-2/3}}{16}$

$\Rightarrow L = \int_1^8 \sqrt{1 + x^{2/3} - \frac{1}{2} + \frac{x^{-2/3}}{16}}\, dx$

$= \int_1^8 \sqrt{x^{2/3} + \frac{1}{2} + \frac{x^{-2/3}}{16}}\, dx$

$= \int_1^8 \sqrt{\left(x^{1/3} + \frac{1}{4}x^{-1/3}\right)^2}\, dx = \int_1^8 \left(x^{1/3} + \frac{1}{4}x^{-1/3}\right) dx$

$= \left[\frac{3}{4}x^{4/3} + \frac{3}{8}x^{2/3}\right]_1^8 = \frac{3}{8}\left[2x^{4/3} + x^{2/3}\right]_1^8$

$= \frac{3}{8}\left[\left(2 \cdot 2^4 + 2^2\right) - (2 + 1)\right] = \frac{3}{8}(32 + 4 - 3) = \frac{99}{8}$

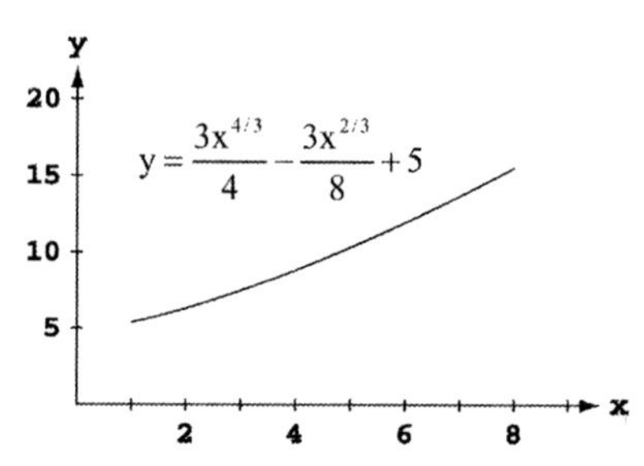

16. $\frac{dy}{dx} = x^2 + 2x + 1 - \frac{4}{(4x+4)^2} = x^2 + 2x + 1 - \frac{1}{4}\frac{1}{(1+x)^2}$

$= (1+x)^2 - \frac{1}{4}\frac{1}{(1+x)^2} \Rightarrow \left(\frac{dy}{dx}\right)^2 = (1+x)^4 - \frac{1}{2} + \frac{1}{16(1+x)^4}$

$\Rightarrow L = \int_0^2 \sqrt{1 + (1+x)^4 - \frac{1}{2} + \frac{(1+x)^{-4}}{16}}\, dx$

$= \int_0^2 \sqrt{(1+x)^4 + \frac{1}{2} + \frac{(1+x)^{-4}}{16}}\, dx$

$= \int_0^2 \sqrt{\left[(1+x)^2 + \frac{(1+x)^{-2}}{4}\right]^2}\, dx$

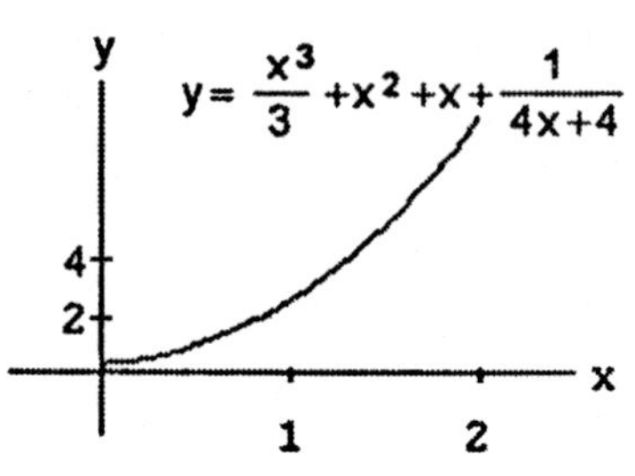

$= \int_0^2 \left[(1+x)^2 + \frac{(1+x)^{-2}}{4}\right] dx$; $[u = 1 + x \Rightarrow du = dx;\ x = 0 \Rightarrow u = 1,\ x = 2 \Rightarrow u = 3]$

$\rightarrow L = \int_1^3 \left(u^2 + \frac{1}{4}u^{-2}\right) du = \left[\frac{u^3}{3} - \frac{1}{4}u^{-1}\right]_1^3 = \left(9 - \frac{1}{12}\right) - \left(\frac{1}{3} - \frac{1}{4}\right) = \frac{108 - 1 - 4 + 3}{12} = \frac{106}{12} = \frac{53}{6}$

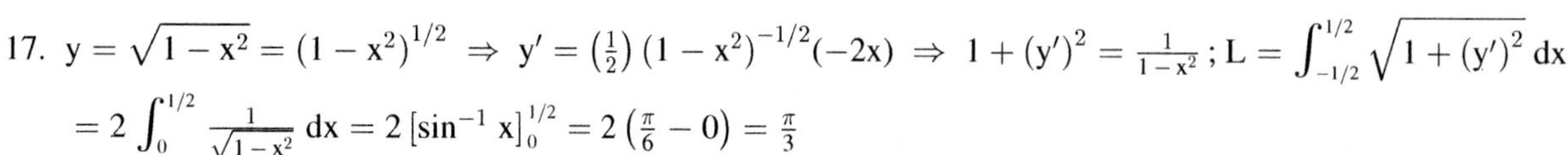

17. $y = \sqrt{1 - x^2} = \left(1 - x^2\right)^{1/2} \Rightarrow y' = \left(\frac{1}{2}\right)\left(1 - x^2\right)^{-1/2}(-2x) \Rightarrow 1 + (y')^2 = \frac{1}{1 - x^2}$; $L = \int_{-1/2}^{1/2} \sqrt{1 + (y')^2}\, dx$

$= 2\int_0^{1/2} \frac{1}{\sqrt{1 - x^2}}\, dx = 2\left[\sin^{-1} x\right]_0^{1/2} = 2\left(\frac{\pi}{6} - 0\right) = \frac{\pi}{3}$

18. $\frac{dx}{dy} = \sqrt{\sec^4 y - 1} \Rightarrow \left(\frac{dx}{dy}\right)^2 = \sec^4 y - 1$

$\Rightarrow L = \int_{-\pi/4}^{\pi/4} \sqrt{1 + \left(\sec^4 y - 1\right)}\, dy = \int_{-\pi/4}^{\pi/4} \sec^2 y\, dy$

$= \left[\tan y\right]_{-\pi/4}^{\pi/4} = 1 - (-1) = 2$

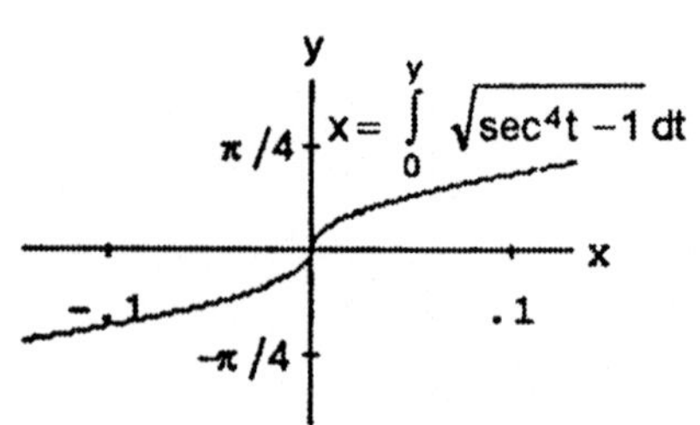

19. (a) $\frac{dy}{dx} = 2x \Rightarrow \left(\frac{dy}{dx}\right)^2 = 4x^2$
$\Rightarrow L = \int_{-1}^{2} \sqrt{1 + \left(\frac{dy}{dx}\right)^2}\, dx$
$= \int_{-1}^{2} \sqrt{1 + 4x^2}\, dx$

(b)

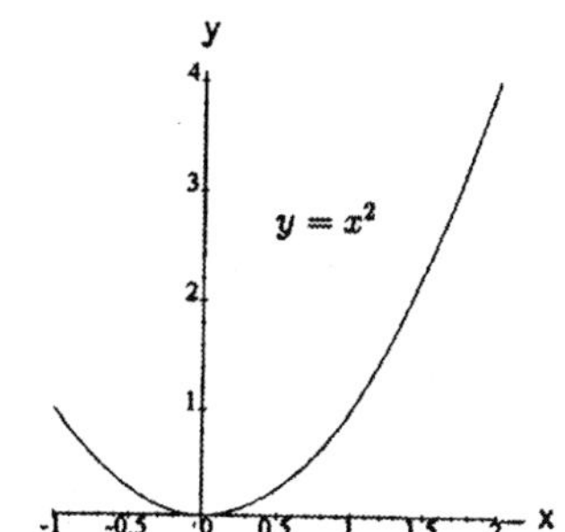

(c) $L \approx 6.13$

20. (a) $\frac{dy}{dx} = \sec^2 x \Rightarrow \left(\frac{dy}{dx}\right)^2 = \sec^4 x$
$\Rightarrow L = \int_{-\pi/3}^{0} \sqrt{1 + \sec^4 x}\, dx$

(b)

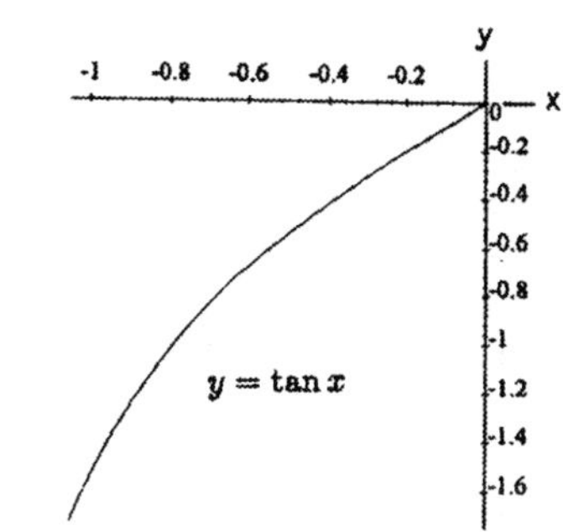

(c) $L \approx 2.06$

21. (a) $\frac{dx}{dy} = \cos y \Rightarrow \left(\frac{dx}{dy}\right)^2 = \cos^2 y$
$\Rightarrow L = \int_{0}^{\pi} \sqrt{1 + \cos^2 y}\, dy$

(b)

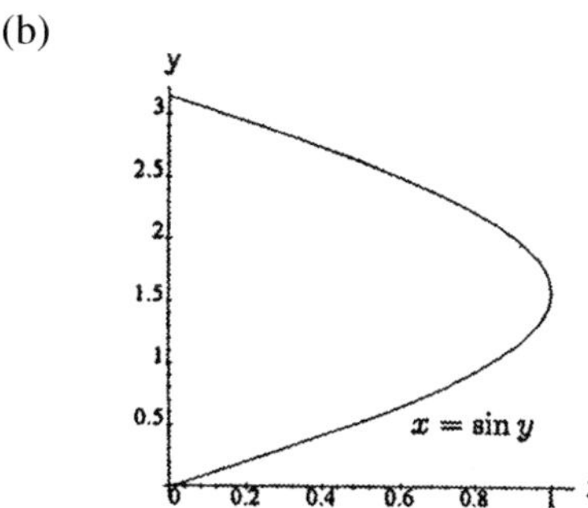

(c) $L \approx 3.82$

22. (a) $\frac{dx}{dy} = -\frac{y}{\sqrt{1-y^2}} \Rightarrow \left(\frac{dx}{dy}\right)^2 = \frac{y^2}{1-y^2}$
$\Rightarrow L = \int_{-1/2}^{1/2} \sqrt{1 + \frac{y^2}{(1-y^2)}}\, dy = \int_{-1/2}^{1/2} \sqrt{\frac{1}{1-y^2}}\, dy$
$= \int_{-1/2}^{1/2} (1 - y^2)^{-1/2}\, dy$

(b)

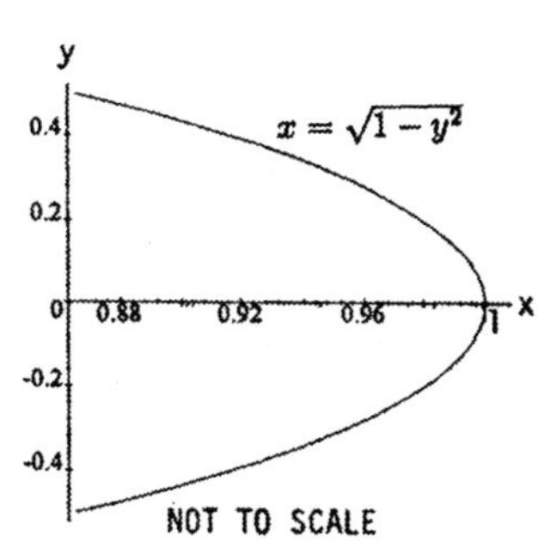

(c) $L \approx 1.05$

23. (a) $2y + 2 = 2\,\frac{dx}{dy} \Rightarrow \left(\frac{dx}{dy}\right)^2 = (y+1)^2$
$\Rightarrow L = \int_{-1}^{3} \sqrt{1 + (y+1)^2}\, dy$

(b)

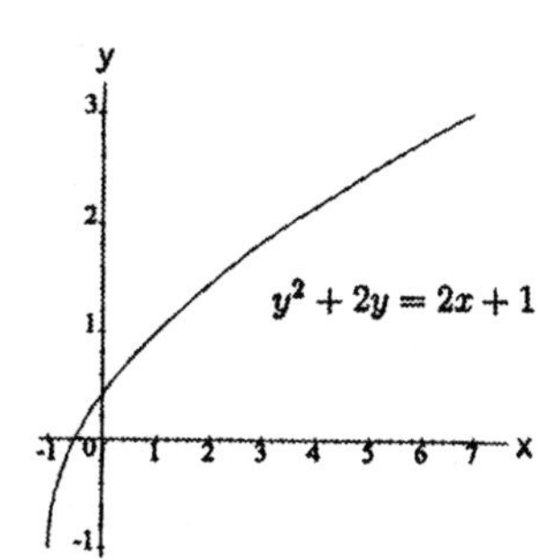

(c) $L \approx 9.29$

24. (a) $\frac{dy}{dx} = \cos x - \cos x + x \sin x \Rightarrow \left(\frac{dy}{dx}\right)^2 = x^2 \sin^2 x$
$\Rightarrow L = \int_0^{\pi} \sqrt{1 + x^2 \sin^2 x}\, dx$
(b)
(c) $L \approx 4.70$

25. (a) $\frac{dy}{dx} = \tan x \Rightarrow \left(\frac{dy}{dx}\right)^2 = \tan^2 x$
$\Rightarrow L = \int_0^{\pi/6} \sqrt{1 + \tan^2 x}\, dx = \int_0^{\pi/6} \sqrt{\frac{\sin^2 x + \cos^2 x}{\cos^2 x}}\, dx$
$= \int_0^{\pi/6} \frac{dx}{\cos x} = \int_0^{\pi/6} \sec x\, dx$
(c) $L \approx 0.55$

(b)

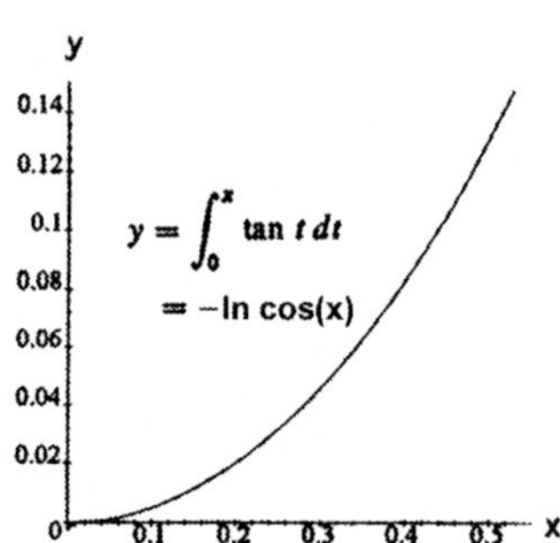

26. (a) $\frac{dx}{dy} = \sqrt{\sec^2 y - 1} \Rightarrow \left(\frac{dx}{dy}\right)^2 = \sec^2 y - 1$
$\Rightarrow L = \int_{-\pi/3}^{\pi/4} \sqrt{1 + (\sec^2 y - 1)}\, dy$
$= \int_{-\pi/3}^{\pi/4} |\sec y|\, dy = \int_{-\pi/3}^{\pi/4} \sec y\, dy$
(c) $L \approx 2.20$

(b)

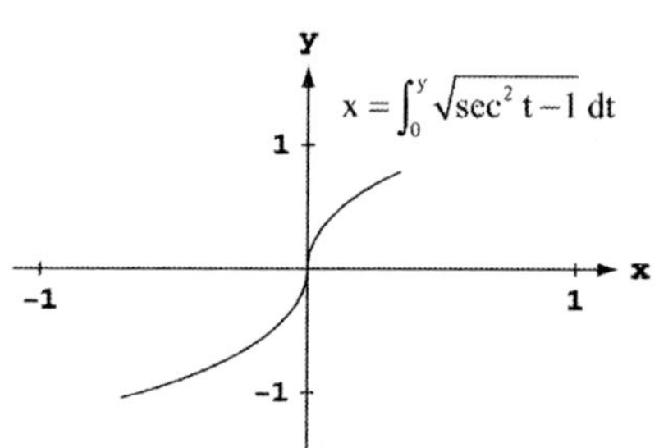

27. $\sqrt{2}\, x = \int_0^x \sqrt{1 + \left(\frac{dy}{dt}\right)^2}\, dt,\ x \ge 0 \Rightarrow \sqrt{2} = \sqrt{1 + \left(\frac{dy}{dx}\right)^2} \Rightarrow \frac{dy}{dx} = \pm 1 \Rightarrow y = f(x) = \pm x + C$ where C is any real number.

28. (a) From the accompanying figure and definition of the differential (change along the tangent line) we see that $dy = f'(x_{k-1})\,\Delta x_k \Rightarrow$ length of kth tangent fin is $\sqrt{(\Delta x_k)^2 + (dy)^2} = \sqrt{(\Delta x_k)^2 + [f'(x_{k-1})\,\Delta x_k]^2}$.

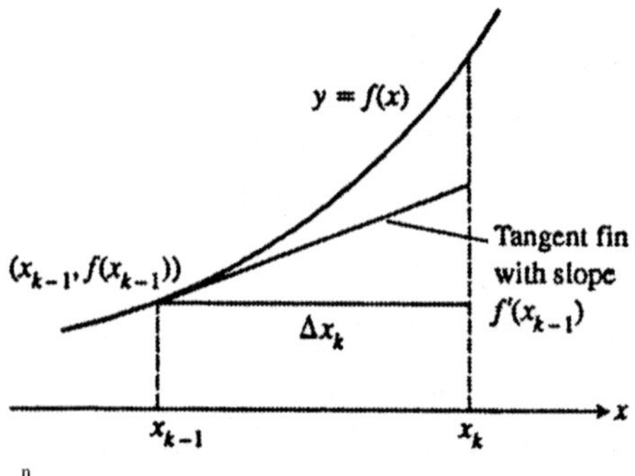

(b) Length of curve $= \lim_{n\to\infty} \sum_{k=1}^{n}$ (length of kth tangent fin) $= \lim_{n\to\infty} \sum_{k=1}^{n} \sqrt{(\Delta x_k)^2 + [f'(x_{k-1})\,\Delta x_k]^2}$
$= \lim_{n\to\infty} \sum_{k=1}^{n} \sqrt{1 + [f'(x_{k-1})]^2}\,\Delta x_k = \int_a^b \sqrt{1 + [f'(x)]^2}\, dx$

29. (a) $\left(\frac{dy}{dx}\right)^2$ corresponds to $\frac{1}{4x}$ here, so take $\frac{dy}{dx}$ as $\frac{1}{2\sqrt{x}}$. Then $y = \sqrt{x} + C$ and since (1, 1) lies on the curve, C = 0. So $y = \sqrt{x}$ from (1, 1) to (4, 2).

(b) Only one. We know the derivative of the function and the value of the function at one value of x.

30. (a) $\left(\frac{dx}{dy}\right)^2$ corresponds to $\frac{1}{y^4}$ here, so take $\frac{dy}{dx}$ as $\frac{1}{y^2}$. Then $x = -\frac{1}{y} + C$ and, since (0, 1) lies on the curve, $C = 1$

So $y = \frac{1}{1-x}$.

(b) Only one. We know the derivative of the function and the value of the function at one value of x.

31. $L = \int_0^1 \sqrt{1 + \frac{e^x}{4}}\, dx \Rightarrow \frac{dy}{dx} = \frac{e^{x/2}}{2} \Rightarrow y = e^{x/2} + C;\ y(0) = 0 \Rightarrow 0 = e^0 + C \Rightarrow C = -1 \Rightarrow y = e^{x/2} - 1$

32. $L = \int_1^2 \sqrt{1 + \frac{1}{x^2}}\, dx \Rightarrow \frac{dy}{dx} = \frac{1}{x} \Rightarrow y = \ln|x| + C = \ln x + C$ since $x > 0 \Rightarrow 0 = \ln 1 + C \Rightarrow C = 0 \Rightarrow y = \ln x$

33. $\frac{dx}{dt} = \left(\frac{1}{\sec t + \tan t}\right)(\sec t \tan t + \sec^2 t) - \cos t = \sec t - \cos t$ and $\frac{dy}{dt} = -\sin t \Rightarrow \sqrt{\left(\frac{dx}{dt}\right)^2 + \left(\frac{dy}{dt}\right)^2}$

$= \sqrt{(\sec t - \cos t)^2 + (-\sin t)^2} = \sqrt{\sec^2 t - 1} = \sqrt{\tan^2 t} = |\tan t| = \tan t$ since $0 \le t \le \frac{\pi}{3}$

$\Rightarrow \text{Length} = \int_0^{\pi/3} \tan t\, dt = \int_0^{\pi/3} \frac{\sin t}{\cos t}\, dt = [-\ln|\cos t|]_0^{\pi/3} = -\ln\frac{1}{2} + \ln 1 = \ln 2$

34. $x = a(\theta - \sin\theta) \Rightarrow \frac{dx}{d\theta} = a(1 - \cos\theta) \Rightarrow \left(\frac{dx}{d\theta}\right)^2 = a^2(1 - 2\cos\theta + \cos^2\theta)$ and $y = a(1 - \cos\theta)$

$\Rightarrow \frac{dy}{d\theta} = a\sin\theta \Rightarrow \left(\frac{dy}{d\theta}\right)^2 = a^2\sin^2\theta \Rightarrow \text{Length} = \int_0^{2\pi} \sqrt{\left(\frac{dx}{d\theta}\right)^2 + \left(\frac{dy}{d\theta}\right)^2}\, d\theta = \int_0^{2\pi} \sqrt{2a^2(1 - \cos\theta)}\, d\theta$

$= a\sqrt{2}\int_0^{2\pi} \sqrt{2}\sqrt{\frac{1-\cos\theta}{2}}\, d\theta = 2a\int_0^{2\pi} \left|\sin\frac{\theta}{2}\right| d\theta = 2a\int_0^{2\pi} \sin\frac{\theta}{2}\, d\theta = -4a\left[\cos\frac{\theta}{2}\right]_0^{2\pi} = 8a$

35. $\frac{dx}{dt} = e^t - e^{-t}$ and $\frac{dy}{dt} = -2 \Rightarrow \sqrt{\left(\frac{dx}{dt}\right)^2 + \left(\frac{dy}{dt}\right)^2} = \sqrt{(e^t - e^{-t})^2 + (-2)^2} = \sqrt{e^{2t} + 2 + e^{-2t}} = \sqrt{\frac{e^{4t} + 2e^{2t} + 1}{e^{2t}}}$

$= \sqrt{\frac{(e^{2t}+1)^2}{(e^t)^2}} = \frac{e^{2t}+1}{e^t} = e^t + e^{-t} \Rightarrow \text{Length} = \int_0^3 (e^t + e^{-t})dt = (e^t - e^{-t})\Big|_0^3 = (e^3 - e^{-3}) - (e^0 - e^0) = e^3 - \frac{1}{e^3}$

36. (a) $\frac{dx}{dt} = -2\sin 2t$ and $\frac{dy}{dt} = 2\cos 2t \Rightarrow \sqrt{\left(\frac{dx}{dt}\right)^2 + \left(\frac{dy}{dt}\right)^2} = \sqrt{(-2\sin 2t)^2 + (2\cos 2t)^2} = 2$

$\Rightarrow \text{Length} = \int_0^{\pi/2} 2\, dt = [2t]_0^{\pi/2} = \pi$

(b) $\frac{dx}{dt} = \pi\cos\pi t$ and $\frac{dy}{dt} = -\pi\sin\pi t = \sqrt{\left(\frac{dx}{dt}\right)^2 + \left(\frac{dy}{dt}\right)^2} = \sqrt{(\pi\cos\pi t)^2 + (-\pi\sin\pi t)^2} = \pi$

$\Rightarrow \text{Length} = \int_{-1/2}^{1/2} \pi\, dt = [\pi t]_{-1/2}^{1/2} = \pi$

37-42. Example CAS commands:

Maple:

```
with( plots );
with( Student[Calculus1] );
with( student );
f := x -> sqrt(1-x^2);a := -1;
b := 1;
N := [2, 4, 8 ];
for n in N do
  xx := [seq( a+i*(b-a)/n, i=0..n )];
  pts := [seq([x,f(x)],x=xx)];
  L := simplify(add( distance(pts[i+1],pts[i]), i=1..n ));                # (b)
  T := sprintf("#31(a) (Section 6.3)\nn=%3d  L=%8.5f\n", n, L );
  P[n] := plot( [f(x),pts], x=a..b, title=T ):                             # (a)
```

```
end do:
display( [seq(P[n],n=N)], insequence=true, scaling=constrained );
L := ArcLength( f(x), x=a..b, output=integral ):
L = evalf( L );                                                    # (c)
```

43-46. Example CAS commands:

Maple:

```
with( plots );
with( student );
x := t -> t^3/3;
y := t -> t^2/2;
a := 0;
b := 1;
N := [2, 4, 8 ];
for n in N do
  tt := [seq( a+i*(b-a)/n, i=0..n )];
  pts := [seq([x(t),y(t)],t=tt)];
  L := simplify(add( student[distance](pts[i+1],pts[i]), i=1..n ));   # (b)
  T := sprintf("#37(a) (Section 6.3)\nn=%3d  L=%8.5f\n", n, L );
  P[n] := plot( [[x(t),y(t),t=a..b],pts], title=T ):                   # (a)
end do:
display( [seq(P[n],n=N)], insequence=true );
ds := t ->sqrt( simplify(D(x)(t)^2 + D(y)(t)^2) ):                     # (c)
L := Int( ds(t), t=a..b ):
L = evalf(L);
```

37-46. Example CAS commands:

Mathematica: (assigned function and values for a, b, and n may vary)

```
Clear[x, f]
{a, b} = {−1, 1}; f[x_] = Sqrt[1 − x^2]
p1 = Plot[f[x], {x, a, b}]
n = 8;
pts = Table[{xn, f[xn]}, {xn, a, b, (b − a)/n}]// N
Show[{p1,Graphics[{Line[pts]}]}]
Sum[ Sqrt[ (pts[[i + 1, 1]] − pts[[i, 1]])^2 + (pts[[i + 1, 2]] − pts[[i, 2]])^2], {i, 1, n}]
NIntegrate[ Sqrt[ 1 + f'[x]^2],{x, a, b}]
```

6.4 MOMENTS AND CENTERS OF MASS

1. Because the children are balanced, the moment of the system about the origin must be equal to zero: $5 \cdot 80 = x \cdot 100 \Rightarrow x = 4$ ft, the distance of the 100-lb child from the fulcrum.

2. Suppose the log has length 2a. Align the log along the x-axis so the 100-lb end is placed at $x = -a$ and the 200-lb end at $x = a$. Then the center of mass $\overline{x}$ satisfies $\overline{x} = \frac{100(-a) + 200(a)}{300} \Rightarrow \overline{x} = \frac{a}{3}$. That is, $\overline{x}$ is located at a distance $a - \frac{a}{3} = \frac{2a}{3} = \frac{1}{3}(2a)$ which is $\frac{1}{3}$ of the length of the log from the 200-lb (heavier) end (see figure) or $\frac{2}{3}$ of the way from the lighter end toward the heavier end.

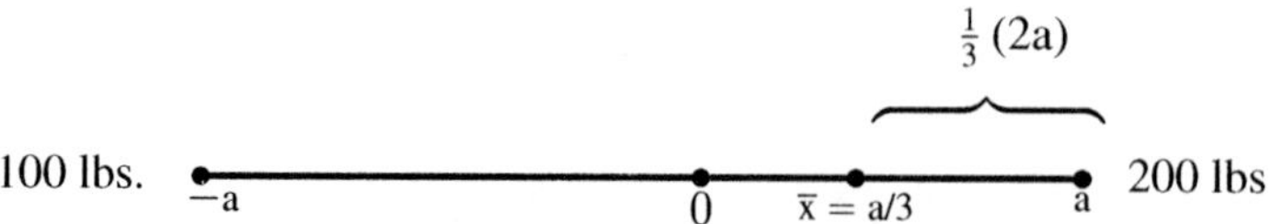

3. The center of mass of each rod is in its center (see Example 1). The rod system is equivalent to two point masses located at the centers of the rods at coordinates $\left(\frac{L}{2}, 0\right)$ and $\left(0, \frac{L}{2}\right)$. Therefore $\overline{x} = \frac{m_y}{m}$ $= \frac{x_1 m_1 + x_2 m_2}{m_1 + m_2} = \frac{\frac{L}{2} \cdot m + 0}{m + m} = \frac{L}{4}$ and $\overline{y} = \frac{m_x}{m} = \frac{y_1 m_2 + y_2 m_2}{m_1 + m_2} = \frac{0 + \frac{L}{2} \cdot m}{m + m} = \frac{L}{4} \Rightarrow \left(\frac{L}{4}, \frac{L}{4}\right)$ is the center of mass location.

4. Let the rods have lengths $x = L$ and $y = 2L$. The center of mass of each rod is in its center (see Example 1). The rod system is equivalent to two point masses located at the centers of the rods at coordinates $\left(\frac{L}{2}, 0\right)$ and $(0, L)$. Therefore $\overline{x} = \frac{\frac{L}{2} \cdot m + 0 \cdot 2m}{m + 2m} = \frac{L}{6}$ and $\overline{y} = \frac{0 \cdot m + L \cdot 2m}{m + 2m} = \frac{2L}{3} \Rightarrow \left(\frac{L}{6}, \frac{2L}{3}\right)$ is the center of mass location.

5. $M_0 = \int_0^2 x \cdot 4\, dx = \left[4 \frac{x^2}{2}\right]_0^2 = 4 \cdot \frac{4}{2} = 8$; $M = \int_0^2 4\, dx = [4x]_0^2 = 4 \cdot 2 = 8 \Rightarrow \overline{x} = \frac{M_0}{M} = 1$

6. $M_0 = \int_1^3 x \cdot 4\, dx = \left[4 \frac{x^2}{2}\right]_1^3 = \frac{4}{2}(9 - 1) = 16$; $M = \int_1^3 4\, dx = [4x]_1^3 = 12 - 4 = 8 \Rightarrow \overline{x} = \frac{M_0}{M} = \frac{16}{8} = 2$

7. $M_0 = \int_0^3 x\left(1 + \frac{x}{3}\right) dx = \int_0^3 \left(x + \frac{x^2}{3}\right) dx = \left[\frac{x^2}{2} + \frac{x^3}{9}\right]_0^3 = \left(\frac{9}{2} + \frac{27}{9}\right) = \frac{15}{2}$; $M = \int_0^3 \left(1 + \frac{x}{3}\right) dx = \left[x + \frac{x^2}{6}\right]_0^3$ $= 3 + \frac{9}{6} = \frac{9}{2} \Rightarrow \overline{x} = \frac{M_0}{M} = \frac{\left(\frac{15}{2}\right)}{\left(\frac{9}{2}\right)} = \frac{15}{9} = \frac{5}{3}$

8. $M_0 = \int_0^4 x\left(2 - \frac{x}{4}\right) dx = \int_0^4 \left(2x - \frac{x^2}{4}\right) dx = \left[x^2 - \frac{x^3}{12}\right]_0^4 = \left(16 - \frac{64}{12}\right) = 16 - \frac{16}{3} = 16 \cdot \frac{2}{3} = \frac{32}{3}$;
$M = \int_0^4 \left(2 - \frac{x}{4}\right) dx = \left[2x - \frac{x^2}{8}\right]_0^4 = 8 - \frac{16}{8} = 6 \Rightarrow \overline{x} = \frac{M_0}{M} = \frac{32}{3 \cdot 6} = \frac{16}{9}$

9. $M_0 = \int_1^4 x\left(1 + \frac{1}{\sqrt{x}}\right) dx = \int_1^4 \left(x + x^{1/2}\right) dx = \left[\frac{x^2}{2} + \frac{2x^{3/2}}{3}\right]_1^4 = \left(8 + \frac{16}{3}\right) - \left(\frac{1}{2} + \frac{2}{3}\right) = \frac{15}{2} + \frac{14}{3} = \frac{45+28}{6} = \frac{73}{6}$;
$M = \int_1^4 \left(1 + x^{-1/2}\right) dx = \left[x + 2x^{1/2}\right]_1^4 = (4 + 4) - (1 + 2) = 5 \Rightarrow \overline{x} = \frac{M_0}{M} = \frac{\left(\frac{73}{6}\right)}{5} = \frac{73}{30}$

10. $M_0 = \int_{1/4}^1 x \cdot 3\left(x^{-3/2} + x^{-5/2}\right) dx = 3\int_{1/4}^1 \left(x^{-1/2} + x^{-3/2}\right) dx = 3\left[2x^{1/2} - \frac{2}{x^{1/2}}\right]_{1/4}^1 = 3\left[(2 - 2) - \left(2 \cdot \frac{1}{2} - \frac{2}{\left(\frac{1}{2}\right)}\right)\right]$
$= 3(4 - 1) = 9$; $M = 3\int_{1/4}^1 \left(x^{-3/2} + x^{-5/2}\right) dx = 3\left[\frac{-2}{x^{1/2}} - \frac{2}{3x^{3/2}}\right]_{1/4}^1 = 3\left[\left(-2 - \frac{2}{3}\right) - \left(-4 - \frac{16}{3}\right)\right] = 3\left(2 + \frac{14}{3}\right)$
$= 6 + 14 = 20 \Rightarrow \overline{x} = \frac{M_0}{M} = \frac{9}{20}$

11. $M_0 = \int_0^1 x(2 - x)\, dx + \int_1^2 x \cdot x\, dx = \int_0^1 (2x - x^2)\, dx + \int_1^2 x^2\, dx = \left[\frac{2x^2}{2} - \frac{x^3}{3}\right]_0^1 + \left[\frac{x^3}{3}\right]_1^2 = \left(1 - \frac{1}{3}\right) + \left(\frac{8}{3} - \frac{1}{3}\right)$
$= \frac{9}{3} = 3$; $M = \int_0^1 (2 - x)\, dx + \int_1^2 x\, dx = \left[2x - \frac{x^2}{2}\right]_0^1 + \left[\frac{x^2}{2}\right]_1^2 = \left(2 - \frac{1}{2}\right) + \left(\frac{4}{2} - \frac{1}{2}\right) = 3 \Rightarrow \overline{x} = \frac{M_0}{M} = 1$

12. $M_0 = \int_0^1 x(x + 1)\, dx + \int_1^2 2x\, dx = \int_0^1 (x^2 + x)\, dx + \int_1^2 2x\, dx = \left[\frac{x^3}{3} + \frac{x^2}{2}\right]_0^1 + [x^2]_1^2 = \left(\frac{1}{3} + \frac{1}{2}\right) + (4 - 1)$
$= 3 + \frac{5}{6} = \frac{23}{6}$; $M = \int_0^1 (x + 1)\, dx + \int_1^2 2\, dx = \left[\frac{x^2}{2} + x\right]_0^1 + [2x]_1^2 = \left(\frac{1}{2} + 1\right) + (4 - 2) = 2 + \frac{3}{2} = \frac{7}{2}$
$\Rightarrow \overline{x} = \frac{M_0}{M} = \left(\frac{23}{6}\right)\left(\frac{2}{7}\right) = \frac{23}{21}$

13. Since the plate is symmetric about the y-axis and its density is constant, the distribution of mass is symmetric about the y-axis and the center of mass lies on the y-axis. This means that $\overline{x} = 0$. It remains to find $\overline{y} = \frac{M_x}{M}$. We model the distribution of mass with *vertical* strips. The typical strip has center of mass: $(\widetilde{x}, \widetilde{y}) = \left(x, \frac{x^2+4}{2}\right)$, length: $4 - x^2$, width: dx, area:

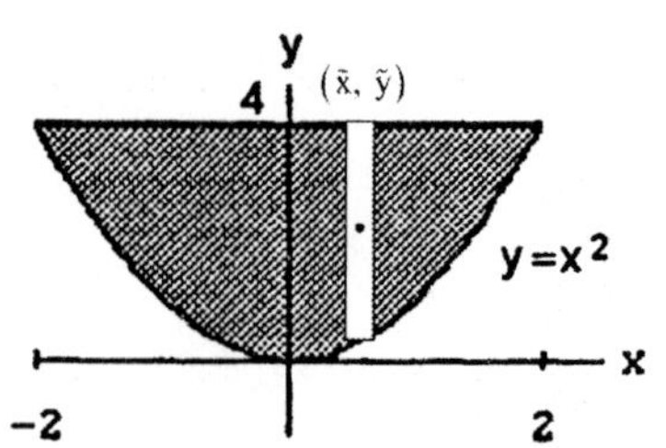

$dA = (4 - x^2)\,dx$, mass: $dm = \delta\,dA = \delta\,(4 - x^2)\,dx$. The moment of the strip about the x-axis is $\widetilde{y}\ dm = \left(\frac{x^2+4}{2}\right)\delta\,(4 - x^2)\,dx = \frac{\delta}{2}\,(16 - x^4)\,dx$. The moment of the plate about the x-axis is $M_x = \int \widetilde{y}\ dm$ $= \int_{-2}^{2} \frac{\delta}{2}\,(16 - x^4)\,dx = \frac{\delta}{2}\left[16x - \frac{x^5}{5}\right]_{-2}^{2} = \frac{\delta}{2}\left[\left(16 \cdot 2 - \frac{2^5}{5}\right) - \left(-16 \cdot 2 + \frac{2^5}{5}\right)\right] = \frac{\delta \cdot 2}{2}\left(32 - \frac{32}{5}\right) = \frac{128\delta}{5}$. The mass of the plate is $M = \int \delta\,(4 - x^2)\,dx = \delta\left[4x - \frac{x^3}{3}\right]_{-2}^{2} = 2\delta\left(8 - \frac{8}{3}\right) = \frac{32\delta}{3}$. Therefore $\overline{y} = \frac{M_x}{M} = \frac{\left(\frac{128\delta}{5}\right)}{\left(\frac{32\delta}{3}\right)} = \frac{12}{5}$. The plate's center of mass is the point $(\overline{x}, \overline{y}) = \left(0, \frac{12}{5}\right)$.

14. Applying the symmetry argument analogous to the one in Exercise 13, we find $\overline{x} = 0$. To find $\overline{y} = \frac{M_x}{M}$, we use the *vertical* strips technique. The typical strip has center of mass: $(\widetilde{x}, \widetilde{y}) = \left(x, \frac{25 - x^2}{2}\right)$, length: $25 - x^2$, width: dx, area: $dA = (25 - x^2)dx$, mass: $dm = \delta\,dA = \delta\,(25 - x^2)\,dx$. The moment of the strip about the x-axis is

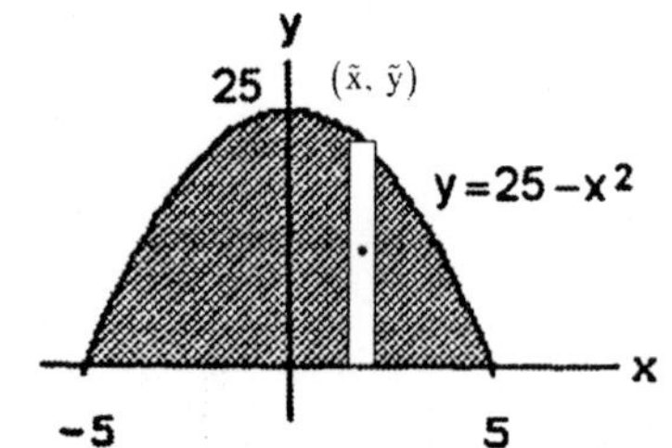

$\widetilde{y}\ dm = \left(\frac{25 - x^2}{2}\right)\delta\,(25 - x^2)\,dx = \frac{\delta}{2}\,(25 - x^2)^2\,dx$. The moment of the plate about the x-axis is $M_x = \int \widetilde{y}\ dm$ $= \int_{-5}^{5} \frac{\delta}{2}\,(25 - x^2)^2\,dx = \frac{\delta}{2}\int_{-5}^{5}(625 - 50x^2 + x^4)\,dx = \frac{\delta}{2}\left[625x - \frac{50}{3}x^3 + \frac{x^5}{5}\right]_{-5}^{5} = 2 \cdot \frac{\delta}{2}\left(625 \cdot 5 - \frac{50}{3} \cdot 5^3 + \frac{5^5}{5}\right)$ $= \delta \cdot 625\left(5 - \frac{10}{3} + 1\right) = \delta \cdot 625 \cdot \left(\frac{8}{3}\right)$. The mass of the plate is $M = \int dm = \int_{-5}^{5} \delta\,(25 - x^2)\,dx = \delta\left[25x - \frac{x^3}{3}\right]_{-5}^{5}$ $= 2\delta\left(5^3 - \frac{5^3}{3}\right) = \frac{4}{3}\delta \cdot 5^3$. Therefore $\overline{y} = \frac{M_x}{M} = \frac{\delta \cdot 5^4 \cdot \left(\frac{8}{3}\right)}{\delta \cdot 5^3 \cdot \left(\frac{4}{3}\right)} = 10$. The plate's center of mass is the point $(\overline{x}, \overline{y}) = (0, 10)$.

15. Intersection points: $x - x^2 = -x \Rightarrow 2x - x^2 = 0$ $\Rightarrow x(2 - x) = 0 \Rightarrow x = 0$ or $x = 2$. The typical *vertical* strip has center of mass: $(\widetilde{x}, \widetilde{y}) = \left(x, \frac{(x - x^2) + (-x)}{2}\right)$ $= \left(x, -\frac{x^2}{2}\right)$, length: $(x - x^2) - (-x) = 2x - x^2$, width: dx, area: $dA = (2x - x^2)\,dx$, mass: $dm = \delta\,dA = \delta\,(2x - x^2)\,dx$. The moment of the strip about the x-axis is

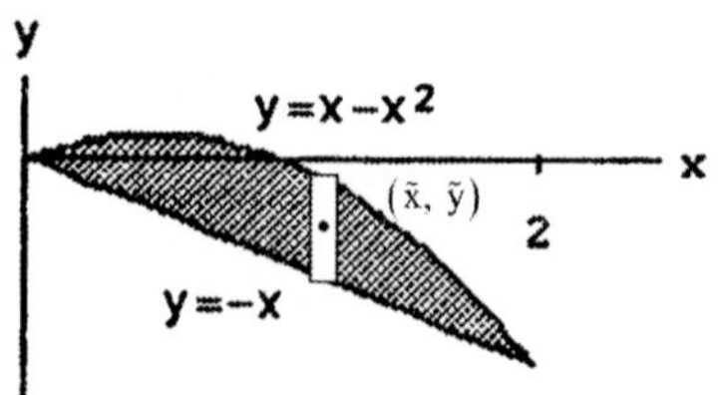

$\widetilde{y}\ dm = \left(-\frac{x^2}{2}\right)\delta\,(2x - x^2)\,dx$; about the y-axis it is $\widetilde{x}\ dm = x \cdot \delta\,(2x - x^2)\,dx$. Thus, $M_x = \int \widetilde{y}\ dm$ $= -\int_0^2 \left(\frac{\delta}{2}x^2\right)(2x - x^2)\,dx = -\frac{\delta}{2}\int_0^2 (2x^3 - x^4)\,dx = -\frac{\delta}{2}\left[\frac{x^4}{2} - \frac{x^5}{5}\right]_0^2 = -\frac{\delta}{2}\left(2^3 - \frac{2^5}{5}\right) = -\frac{\delta}{2} \cdot 2^3\left(1 - \frac{4}{5}\right)$ $= -\frac{4\delta}{5}$; $M_y = \int \widetilde{x}\ dm = \int_0^2 x \cdot \delta\,(2x - x^2)\,dx = \delta\int_0^2 (2x^2 - x^3) = \delta\left[\frac{2}{3}x^3 - \frac{x^4}{4}\right]_0^2 = \delta\left(2 \cdot \frac{2^3}{3} - \frac{2^4}{4}\right) = \frac{\delta \cdot 2^4}{12} = \frac{4\delta}{3}$; $M = \int dm = \int_0^2 \delta\,(2x - x^2)\,dx = \delta\int_0^2 (2x - x^2)\,dx = \delta\left[x^2 - \frac{x^3}{3}\right]_0^2 = \delta\left(4 - \frac{8}{3}\right) = \frac{4\delta}{3}$. Therefore, $\overline{x} = \frac{M_y}{M}$ $= \left(\frac{4\delta}{3}\right)\left(\frac{3}{4\delta}\right) = 1$ and $\overline{y} = \frac{M_x}{M} = \left(-\frac{4\delta}{5}\right)\left(\frac{3}{4\delta}\right) = -\frac{3}{5} \Rightarrow (\overline{x}, \overline{y}) = \left(1, -\frac{3}{5}\right)$ is the center of mass.

16. Intersection points: $x^2 - 3 = -2x^2 \Rightarrow 3x^2 - 3 = 0$ $\Rightarrow 3(x-1)(x+1) = 0 \Rightarrow x = -1$ or $x = 1$. Applying the symmetry argument analogous to the one in Exercise 13, we find $\bar{x} = 0$. The typical *vertical* strip has center of mass:

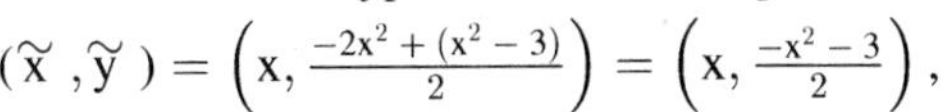

$(\tilde{x}, \tilde{y}) = \left(x, \frac{-2x^2 + (x^2 - 3)}{2}\right) = \left(x, \frac{-x^2 - 3}{2}\right)$,

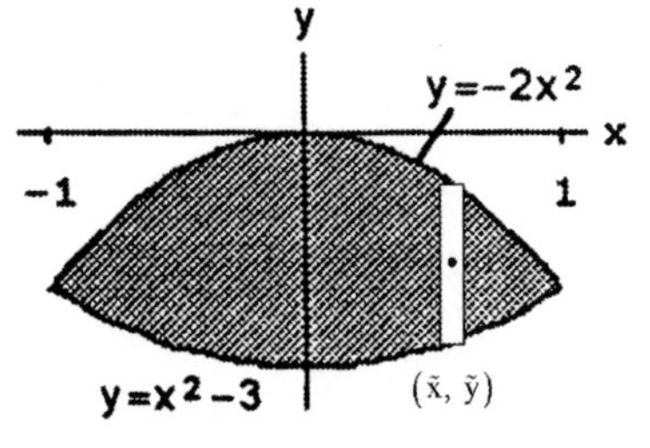

length: $-2x^2 - (x^2 - 3) = 3(1 - x^2)$, width: dx,
area: $dA = 3(1 - x^2)\,dx$, mass: $dm = \delta\,dA = 3\delta(1 - x^2)\,dx$.
The moment of the strip about the x-axis is
$\tilde{y}\,dm = \frac{3}{2}\delta(-x^2 - 3)(1 - x^2)\,dx = \frac{3}{2}\delta(x^4 + 3x^2 - x^2 - 3)\,dx = \frac{3}{2}\delta(x^4 + 2x^2 - 3)\,dx$; $M_x = \int \tilde{y}\,dm$
$= \frac{3}{2}\delta \int_{-1}^{1}(x^4 + 2x^2 - 3)\,dx = \frac{3}{2}\delta\left[\frac{x^5}{5} + \frac{2x^3}{3} - 3x\right]_{-1}^{1} = \frac{3}{2}\cdot\delta\cdot 2\left(\frac{1}{5} + \frac{2}{3} - 3\right) = 3\delta\left(\frac{3+10-45}{15}\right) = -\frac{32\delta}{5}$;
$M = \int dm = 3\delta\int_{-1}^{1}(1 - x^2)\,dx = 3\delta\left[x - \frac{x^3}{3}\right]_{-1}^{1} = 3\delta\cdot 2\left(1 - \frac{1}{3}\right) = 4\delta$. Therefore, $\bar{y} = \frac{M_x}{M} = -\frac{\delta\cdot 32}{5\cdot\delta\cdot 4} = -\frac{8}{5}$
$\Rightarrow (\bar{x}, \bar{y}) = \left(0, -\frac{8}{5}\right)$ is the center of mass.

17. The typical *horizontal* strip has center of mass:
$(\tilde{x}, \tilde{y}) = \left(\frac{y - y^3}{2}, y\right)$, length: $y - y^3$, width: dy,
area: $dA = (y - y^3)\,dy$, mass: $dm = \delta\,dA = \delta(y - y^3)\,dy$.
The moment of the strip about the y-axis is
$\tilde{x}\,dm = \delta\left(\frac{y - y^3}{2}\right)(y - y^3)\,dy = \frac{\delta}{2}(y - y^3)^2\,dy$
$= \frac{\delta}{2}(y^2 - 2y^4 + y^6)\,dy$; the moment about the x-axis is

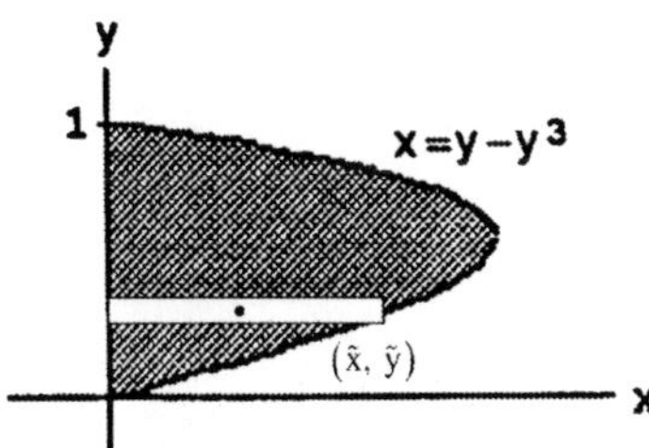

$\tilde{y}\,dm = \delta y(y - y^3)\,dy = \delta(y^2 - y^4)\,dy$. Thus, $M_x = \int \tilde{y}\,dm = \delta\int_0^1 (y^2 - y^4)\,dy = \delta\left[\frac{y^3}{3} - \frac{y^5}{5}\right]_0^1 = \delta\left(\frac{1}{3} - \frac{1}{5}\right) = \frac{2\delta}{15}$;
$M_y = \int \tilde{x}\,dm = \frac{\delta}{2}\int_0^1 (y^2 - 2y^4 + y^6)\,dy = \frac{\delta}{2}\left[\frac{y^3}{3} - \frac{2y^5}{5} + \frac{y^7}{7}\right]_0^1 = \frac{\delta}{2}\left(\frac{1}{3} - \frac{2}{5} + \frac{1}{7}\right) = \frac{\delta}{2}\left(\frac{35 - 42 + 15}{3\cdot 5\cdot 7}\right) = \frac{4\delta}{105}$; $M = \int dm$
$= \delta\int_0^1 (y - y^3)\,dy = \delta\left[\frac{y^2}{2} - \frac{y^4}{4}\right]_0^1 = \delta\left(\frac{1}{2} - \frac{1}{4}\right) = \frac{\delta}{4}$. Therefore, $\bar{x} = \frac{M_y}{M} = \left(\frac{4\delta}{105}\right)\left(\frac{4}{\delta}\right) = \frac{16}{105}$ and $\bar{y} = \frac{M_x}{M} = \left(\frac{2\delta}{15}\right)\left(\frac{4}{\delta}\right)$
$= \frac{8}{15} \Rightarrow (\bar{x}, \bar{y}) = \left(\frac{16}{105}, \frac{8}{15}\right)$ is the center of mass.

18. Intersection points: $y = y^2 - y \Rightarrow y^2 - 2y = 0$
$\Rightarrow y(y - 2) = 0 \Rightarrow y = 0$ or $y = 2$. The typical *horizontal* strip has center of mass:
$(\tilde{x}, \tilde{y}) = \left(\frac{(y^2 - y) + y}{2}, y\right) = \left(\frac{y^2}{2}, y\right)$,

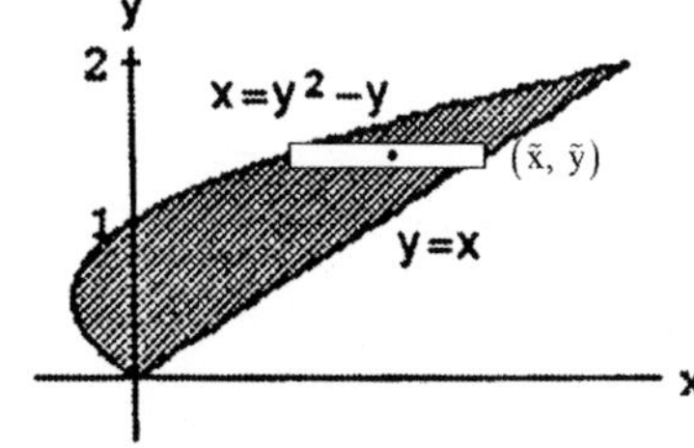

length: $y - (y^2 - y) = 2y - y^2$, width: dy,
area: $dA = (2y - y^2)\,dy$, mass: $dm = \delta\,dA = \delta(2y - y^2)\,dy$.
The moment about the y-axis is $\tilde{x}\,dm = \frac{\delta}{2}\cdot y^2(2y - y^2)\,dy$
$= \frac{\delta}{2}(2y^3 - y^4)\,dy$; the moment about the x-axis is $\tilde{y}\,dm = \delta y(2y - y^2)\,dy = \delta(2y^2 - y^3)\,dy$. Thus,
$M_x = \int \tilde{y}\,dm = \int_0^2 \delta(2y^2 - y^3)\,dy = \delta\left[\frac{2y^3}{3} - \frac{y^4}{4}\right]_0^2 = \delta\left(\frac{16}{3} - \frac{16}{4}\right) = \frac{16\delta}{12}(4 - 3) = \frac{4\delta}{3}$; $M_y = \int \tilde{x}\,dm$
$= \int_0^2 \frac{\delta}{2}(2y^3 - y^4)\,dy = \frac{\delta}{2}\left[\frac{y^4}{2} - \frac{y^5}{5}\right]_0^2 = \frac{\delta}{2}\left(8 - \frac{32}{5}\right) = \frac{\delta}{2}\left(\frac{40 - 32}{5}\right) = \frac{4\delta}{5}$; $M = \int dm = \int_0^2 \delta(2y - y^2)\,dy$
$= \delta\left[y^2 - \frac{y^3}{3}\right]_0^2 = \delta\left(4 - \frac{8}{3}\right) = \frac{4\delta}{3}$. Therefore, $\bar{x} = \frac{M_y}{M} = \left(\frac{4\delta}{5}\right)\left(\frac{3}{4\delta}\right) = \frac{3}{5}$ and $\bar{y} = \frac{M_x}{M} = \left(\frac{4\delta}{3}\right)\left(\frac{3}{4\delta}\right) = 1$
$\Rightarrow (\bar{x}, \bar{y}) = \left(\frac{3}{5}, 1\right)$ is the center of mass.

19. Applying the symmetry argument analogous to the one used in Exercise 13, we find $\bar{x} = 0$. The typical *vertical* strip has center of mass: $(\tilde{x}, \tilde{y}) = \left(x, \frac{\cos x}{2}\right)$, length: $\cos x$, width: dx, area: $dA = \cos x \, dx$, mass: $dm = \delta \, dA = \delta \cos x \, dx$. The moment of the strip about the x-axis is $\tilde{y} \, dm = \delta \cdot \frac{\cos x}{2} \cdot \cos x \, dx = \frac{\delta}{2} \cos^2 x \, dx = \frac{\delta}{2}\left(\frac{1+\cos 2x}{2}\right) dx = \frac{\delta}{4}(1+\cos 2x)\, dx$; thus,

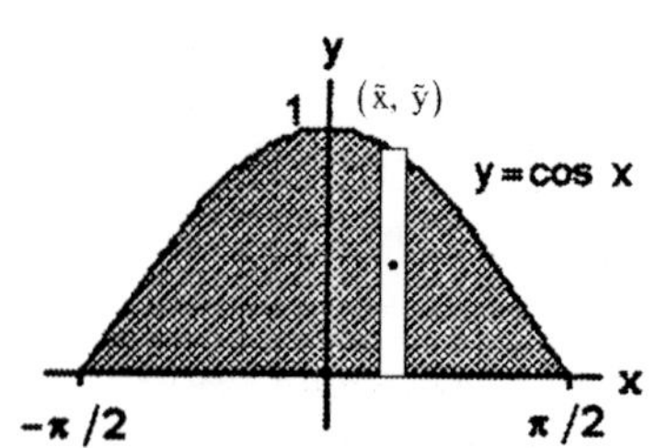

$M_x = \int \tilde{y} \, dm = \int_{-\pi/2}^{\pi/2} \frac{\delta}{4}(1+\cos 2x)\, dx = \frac{\delta}{4}\left[x + \frac{\sin 2x}{2}\right]_{-\pi/2}^{\pi/2} = \frac{\delta}{4}\left[\left(\frac{\pi}{2}+0\right) - \left(-\frac{\pi}{2}\right)\right] = \frac{\delta\pi}{4}$; $M = \int dm = \delta \int_{-\pi/2}^{\pi/2} \cos x \, dx$ $= \delta[\sin x]_{-\pi/2}^{\pi/2} = 2\delta$. Therefore, $\bar{y} = \frac{M_x}{M} = \frac{\delta\pi}{4 \cdot 2\delta} = \frac{\pi}{8} \Rightarrow (\bar{x}, \bar{y}) = \left(0, \frac{\pi}{8}\right)$ is the center of mass.

20. Applying the symmetry argument analogous to the one used in Exercise 13, we find $\bar{x} = 0$. The typical vertical strip has center of mass: $(\tilde{x}, \tilde{y}) = \left(x, \frac{\sec^2 x}{2}\right)$, length: $\sec^2 x$, width: dx, area: $dA = \sec^2 x \, dx$, mass: $dm = \delta \, dA = \delta \sec^2 x \, dx$. The moment about the x-axis is $\tilde{y} \, dm = \left(\frac{\sec^2 x}{2}\right)(\delta \sec^2 x)\, dx$ $= \frac{\delta}{2}\sec^4 x \, dx$. $M_x = \int_{-\pi/4}^{\pi/4} \tilde{y} \, dm = \frac{\delta}{2}\int_{-\pi/4}^{\pi/4} \sec^4 x \, dx$

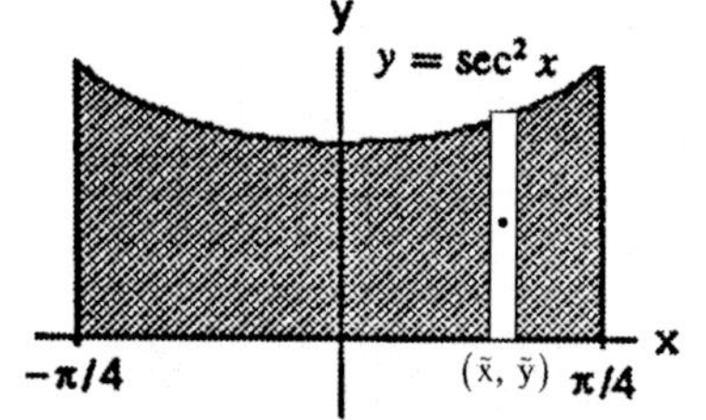

$= \frac{\delta}{2}\int_{-\pi/4}^{\pi/4} (\tan^2 x + 1)(\sec^2 x)\, dx = \frac{\delta}{2}\int_{-\pi/4}^{\pi/4} (\tan x)^2 (\sec^2 x)\, dx + \frac{\delta}{2}\int_{-\pi/4}^{\pi/4} \sec^2 x \, dx = \frac{\delta}{2}\left[\frac{(\tan x)^3}{3}\right]_{-\pi/4}^{\pi/4} + \frac{\delta}{2}[\tan x]_{-\pi/4}^{\pi/4}$

$= \frac{\delta}{2}\left[\frac{1}{3} - \left(-\frac{1}{3}\right)\right] + \frac{\delta}{2}[1-(-1)] = \frac{\delta}{3} + \delta = \frac{4\delta}{3}$; $M = \int dm = \delta \int_{-\pi/4}^{\pi/4} \sec^2 x \, dx = \delta[\tan x]_{-\pi/4}^{\pi/4} = \delta[1-(-1)] = 2\delta$.

Therefore, $\bar{y} = \frac{M_x}{M} = \left(\frac{4\delta}{3}\right)\left(\frac{1}{2\delta}\right) = \frac{2}{3} \Rightarrow (\bar{x}, \bar{y}) = \left(0, \frac{2}{3}\right)$ is the center of mass.

21. (a) $M_y = \int_1^2 x\left(\frac{1}{x}\right) dx = 1$, $M_x = \int_1^2 \left(\frac{1}{2x}\right)\left(\frac{1}{x}\right) dx$ $= \frac{1}{2}\int_1^2 \frac{1}{x^2}\, dx = \left[-\frac{1}{2x}\right]_1^2 = \frac{1}{4}$, $M = \int_1^2 \frac{1}{x}\, dx$ $= [\ln |x|]_1^2 = \ln 2 \Rightarrow \bar{x} = \frac{M_y}{M} = \frac{1}{\ln 2} \approx 1.44$ and $\bar{y} = \frac{M_x}{M} = \frac{\left(\frac{1}{4}\right)}{\ln 2} \approx 0.36$

(b)

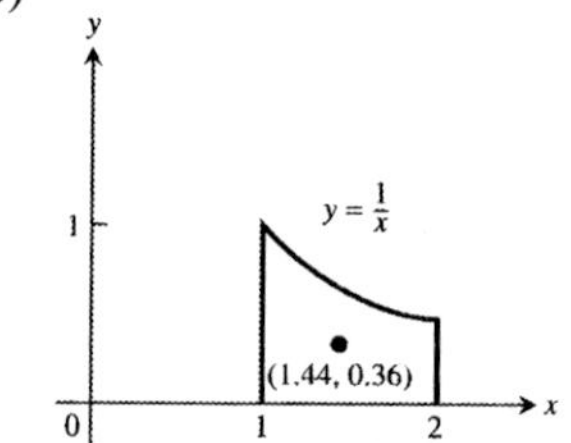

22. (a) Since the plate is symmetric about the line $x = y$ and its density is constant, the distribution of mass is symmetric about this line. This means that $\bar{x} = \bar{y}$. The typical *vertical* strip has center of mass: $(\tilde{x}, \tilde{y}) = \left(x, \frac{\sqrt{9-x^2}}{2}\right)$, length: $\sqrt{9-x^2}$, width: dx, area: $dA = \sqrt{9-x^2}\, dx$, mass: $dm = \delta \, dA = \delta\sqrt{9-x^2}\, dx$. The moment about the x-axis is

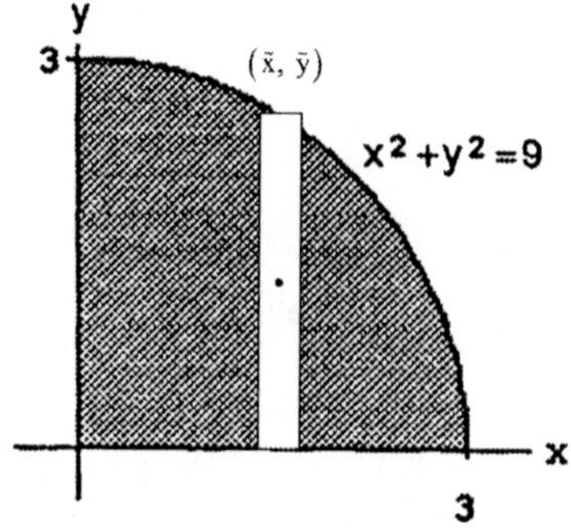

$\tilde{y} \, dm = \delta\left(\frac{\sqrt{9-x^2}}{2}\right)\sqrt{9-x^2}\, dx = \frac{\delta}{2}(9-x^2)\, dx$. Thus, $M_x = \int \tilde{y} \, dm = \int_0^3 \frac{\delta}{2}(9-x^2)\, dx = \frac{\delta}{2}\left[9x - \frac{x^3}{3}\right]_0^3$ $= \frac{\delta}{2}(27-9) = 9\delta$; $M = \int dm = \int \delta \, dA = \delta \int dA = \delta$(Area of a quarter of a circle of radius 3) $= \delta\left(\frac{9\pi}{4}\right) = \frac{9\pi\delta}{4}$.
Therefore, $\bar{y} = \frac{M_x}{M} = (9\delta)\left(\frac{4}{9\pi\delta}\right) = \frac{4}{\pi} \Rightarrow (\bar{x}, \bar{y}) = \left(\frac{4}{\pi}, \frac{4}{\pi}\right)$ is the center of mass.

(b) Applying the symmetry argument analogous to the one used in Exercise 13, we find that $\overline{x} = 0$. The typical vertical strip has the same parameters as in part (a). Thus, $M_x = \int \tilde{y}\, dm = \int_{-3}^{3} \frac{\delta}{2}(9 - x^2)\, dx$ $= 2\int_0^3 \frac{\delta}{2}(9 - x^2)\, dx = 2(9\delta) = 18\delta$; $M = \int dm = \int \delta\, dA = \delta \int dA$ $= \delta(\text{Area of a semi-circle of radius } 3) = \delta\left(\frac{9\pi}{2}\right) = \frac{9\pi\delta}{2}$. Therefore, $\overline{y} = \frac{M_x}{M} = (18\delta)\left(\frac{2}{9\pi\delta}\right) = \frac{4}{\pi}$, the same $\overline{y}$ as in part (a) $\Rightarrow$ $(\overline{x}, \overline{y}) = \left(0, \frac{4}{\pi}\right)$ is the center of mass.

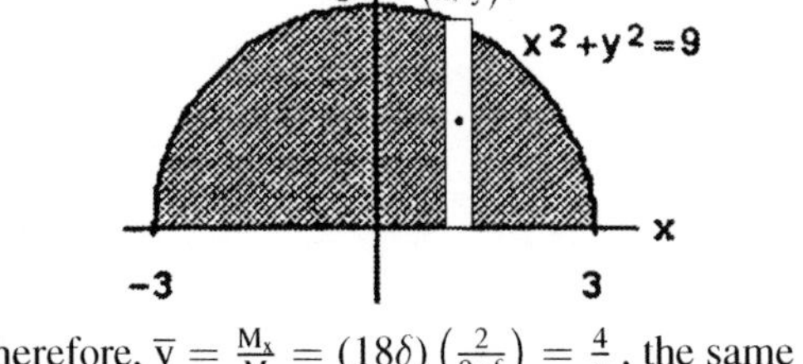

23. $M = \int_0^1 \frac{2}{1+x^2}\, dx = 2\left[\tan^{-1} x\right]_0^1 = \frac{\pi}{2}$ and $M_y = \int_0^1 \frac{2x}{1+x^2}\, dx = \left[\ln(1+x^2)\right]_0^1 = \ln 2 \Rightarrow \overline{x} = \frac{M_y}{M} = \frac{\ln 2}{\left(\frac{\pi}{2}\right)} = \frac{\ln 4}{\pi}$; $\overline{y} = 0$ by symmetry

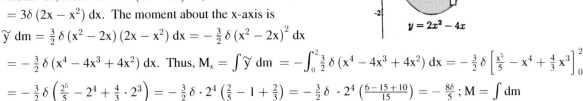

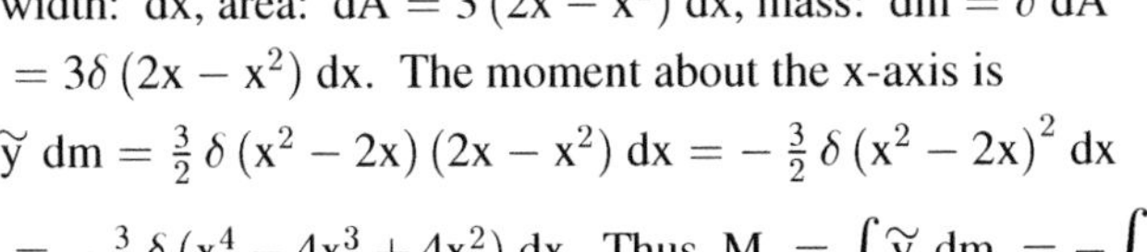

24. Since the plate is symmetric about the line $x = 1$ and its density is constant, the distribution of mass is symmetric about this line and the center of mass lies on it. This means that $\overline{x} = 1$. The typical *vertical* strip has center of mass: $(\tilde{x}, \tilde{y}) = \left(x, \frac{(2x - x^2) + (2x^2 - 4x)}{2}\right) = \left(x, \frac{x^2 - 2x}{2}\right)$, length: $(2x - x^2) - (2x^2 - 4x) = -3x^2 + 6x = 3(2x - x^2)$, width: dx, area: $dA = 3(2x - x^2)\, dx$, mass: $dm = \delta\, dA$ $= 3\delta(2x - x^2)\, dx$. The moment about the x-axis is $\tilde{y}\, dm = \frac{3}{2}\delta(x^2 - 2x)(2x - x^2)\, dx = -\frac{3}{2}\delta(x^2 - 2x)^2\, dx$ $= -\frac{3}{2}\delta(x^4 - 4x^3 + 4x^2)\, dx$. Thus, $M_x = \int \tilde{y}\, dm = -\int_0^2 \frac{3}{2}\delta(x^4 - 4x^3 + 4x^2)\, dx = -\frac{3}{2}\delta\left[\frac{x^5}{5} - x^4 + \frac{4}{3}x^3\right]_0^2$ $= -\frac{3}{2}\delta\left(\frac{2^5}{5} - 2^4 + \frac{4}{3}\cdot 2^3\right) = -\frac{3}{2}\delta \cdot 2^4\left(\frac{2}{5} - 1 + \frac{2}{3}\right) = -\frac{3}{2}\delta \cdot 2^4\left(\frac{6 - 15 + 10}{15}\right) = -\frac{8\delta}{5}$; $M = \int dm$ $= \int_0^2 3\delta(2x - x^2)\, dx = 3\delta\left[x^2 - \frac{x^3}{3}\right]_0^2 = 3\delta\left(4 - \frac{8}{3}\right) = 4\delta$. Therefore, $\overline{y} = \frac{M_x}{M} = \left(-\frac{8\delta}{5}\right)\left(\frac{1}{4\delta}\right) = -\frac{2}{5}$ $\Rightarrow$ $(\overline{x}, \overline{y}) = \left(1, -\frac{2}{5}\right)$ is the center of mass.

25. $M_y = \int_1^{16} x\left(\frac{1}{\sqrt{x}}\right) dx = \int_1^{16} x^{1/2}\, dx = \frac{2}{3}\left[x^{3/2}\right]_1^{16} = 42$; $M_x = \int_1^{16}\left(\frac{1}{2\sqrt{x}}\right)\left(\frac{1}{\sqrt{x}}\right) dx = \frac{1}{2}\int_1^{16} \frac{1}{x}\, dx$ $= \frac{1}{2}\left[\ln|x|\right]_1^{16} = \ln 4$, $M = \int_1^{16} \frac{1}{\sqrt{x}}\, dx = \left[2x^{1/2}\right]_1^{16} = 6 \Rightarrow \overline{x} = \frac{M_y}{M} = 7$ and $\overline{y} = \frac{M_x}{M} = \frac{\ln 4}{6}$

26. Applying the symmetry argument analogous to the one used in Exercise 13, we find that $\overline{y} = 0$. The typical *vertical* strip has center of mass: $(\tilde{x}, \tilde{y}) = \left(x, \frac{\frac{1}{x^3} - \frac{1}{x^3}}{2}\right) = (x, 0)$, length: $\frac{1}{x^3} - \left(-\frac{1}{x^3}\right) = \frac{2}{x^3}$, width: dx, area: $dA = \frac{2}{x^3}\, dx$, mass: $dm = \delta\, dA = \frac{2\delta}{x^3}\, dx$. The moment about the y-axis is $\tilde{x}\, dm = x \cdot \frac{2\delta}{x^3}\, dx = \frac{2\delta}{x^2}\, dx$. Thus, $M_y = \int \tilde{x}\, dm = \int_1^a \frac{2\delta}{x^2}\, dx$ $= 2\delta\left[-\frac{1}{x}\right]_1^a = 2\delta\left(-\frac{1}{a} + 1\right) = \frac{2\delta(a-1)}{a}$; $M = \int dm = \int_1^a \frac{2\delta}{x^3}\, dx = \delta\left[-\frac{1}{x^2}\right]_1^a = \delta\left(-\frac{1}{a^2} + 1\right) = \frac{\delta(a^2 - 1)}{a^2}$. Therefore, $\overline{x} = \frac{M_y}{M} = \left[\frac{2\delta(a-1)}{a}\right]\left[\frac{a^2}{\delta(a^2 - 1)}\right] = \frac{2a}{a+1} \Rightarrow (\overline{x}, \overline{y}) = \left(\frac{2a}{a+1}, 0\right)$. Also, $\lim_{a \to \infty} \overline{x} = 2$.

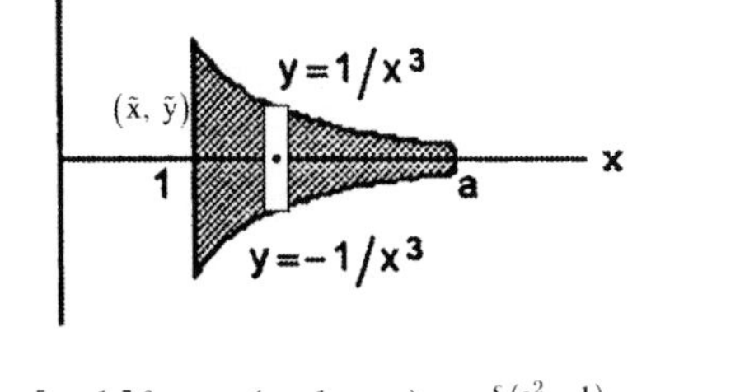

27. $M_x = \int \tilde{y}\, dm = \int_1^2 \frac{\left(\frac{2}{x^2}\right)}{2} \cdot \delta \cdot \left(\frac{2}{x^2}\right) dx$
$= \int_1^2 \left(\frac{1}{x^2}\right)(x^2)\left(\frac{2}{x^2}\right) dx = \int_1^2 \frac{2}{x^2}\, dx = 2\int_1^2 x^{-2}\, dx$
$= 2\left[-x^{-1}\right]_1^2 = 2\left[\left(-\frac{1}{2}\right) - (-1)\right] = 2\left(\frac{1}{2}\right) = 1;$
$M_y = \int \tilde{x}\, dm = \int_1^2 x \cdot \delta \cdot \left(\frac{2}{x^2}\right) dx$
$= \int_1^2 x\,(x^2)\left(\frac{2}{x^2}\right) dx = 2\int_1^2 x\, dx = 2\left[\frac{x^2}{2}\right]_1^2$
$= 2\left(2 - \frac{1}{2}\right) = 4 - 1 = 3;$ $M = \int dm = \int_1^2 \delta\left(\frac{2}{x^2}\right) dx = \int_1^2 x^2\left(\frac{2}{x^2}\right) dx = 2\int_1^2 dx = 2[x]_1^2 = 2(2-1) = 2.$ So $\overline{x} = \frac{M_y}{M} = \frac{3}{2}$ and $\overline{y} = \frac{M_x}{M} = \frac{1}{2} \Rightarrow (\overline{x}, \overline{y}) = \left(\frac{3}{2}, \frac{1}{2}\right)$ is the center of mass.

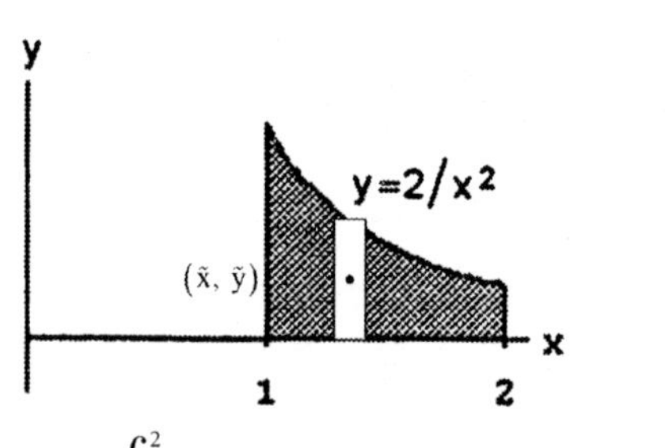

28. We use the *vertical* strip approach:
$M_x = \int \tilde{y}\, dm = \int_0^1 \frac{(x+x^2)}{2}(x - x^2) \cdot \delta\, dx$
$= \frac{1}{2}\int_0^1 (x^2 - x^4) \cdot 12x\, dx$
$= 6\int_0^1 (x^3 - x^5)\, dx = 6\left[\frac{x^4}{4} - \frac{x^6}{6}\right]_0^1$
$= 6\left(\frac{1}{4} - \frac{1}{6}\right) = \frac{6}{4} - 1 = \frac{1}{2};$

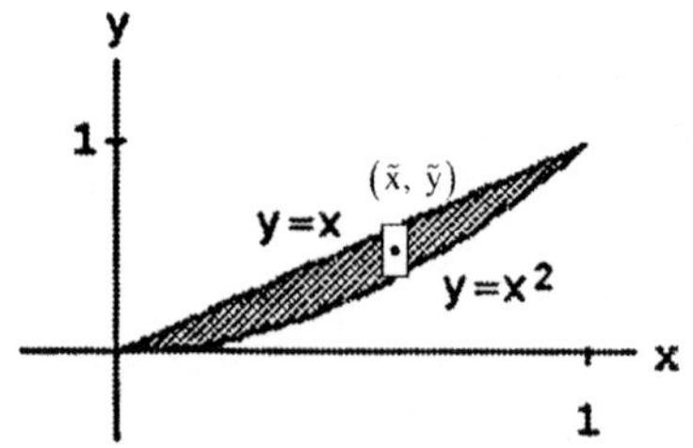

$M_y = \int \tilde{x}\, dm = \int_0^1 x\,(x - x^2) \cdot \delta\, dx = \int_0^1 (x^2 - x^3) \cdot 12x\, dx = 12\int_0^1 (x^3 - x^4)\, dx = 12\left[\frac{x^4}{4} - \frac{x^5}{5}\right]_0^1 = 12\left(\frac{1}{4} - \frac{1}{5}\right)$
$= \frac{12}{20} = \frac{3}{5};$ $M = \int dm = \int_0^1 (x - x^2) \cdot \delta\, dx = 12\int_0^1 (x^2 - x^3)\, dx = 12\left[\frac{x^3}{3} - \frac{x^4}{4}\right]_0^1 = 12\left(\frac{1}{3} - \frac{1}{4}\right) = \frac{12}{12} = 1.$ So
$\overline{x} = \frac{M_y}{M} = \frac{3}{5}$ and $\overline{y} = \frac{M_x}{M} = \frac{1}{2} \Rightarrow \left(\frac{3}{5}, \frac{1}{2}\right)$ is the center of mass.

29. $M_y = \int_1^{16} x\left(\frac{1}{\sqrt{x}}\right)\left(\frac{4}{\sqrt{x}}\right) dx = 4\int_1^{16} dx = 60,$ $M_x = \int_1^{16} \left(\frac{1}{2\sqrt{x}}\right)\left(\frac{1}{\sqrt{x}}\right)\left(\frac{4}{\sqrt{x}}\right) dx = 2\int_1^{16} x^{-3/2}\, dx$
$= -4\left[x^{-1/2}\right]_1^{16} = 3,$ $M = \int_1^{16} \left(\frac{1}{\sqrt{x}}\right)\left(\frac{4}{\sqrt{x}}\right) dx = 4\int_1^{16} \frac{1}{x}\, dx = [4 \ln |x|]_1^{16} = 4 \ln 16 \Rightarrow \overline{x} = \frac{M_y}{M} = \frac{15}{\ln 16}$ and
$\overline{y} = \frac{M_x}{M} = \frac{3}{4 \ln 16}$

30. (a) We use the disk method: $V = \int_a^b \pi R^2(x)\, dx = \int_1^4 \pi\left(\frac{4}{x^2}\right) dx = 4\pi\int_1^4 x^{-2}\, dx = 4\pi\left[-\frac{1}{x}\right]_1^4$
$= 4\pi\left[\frac{-1}{4} - (-1)\right] = \pi[-1 + 4] = 3\pi$

(b) We model the distribution of mass with vertical strips: $M_x = \int \tilde{y}\, dm = \int_1^4 \frac{\left(\frac{2}{x}\right)}{2} \cdot \left(\frac{2}{x}\right) \cdot \delta\, dx = \int_1^4 \frac{2}{x^2} \cdot \sqrt{x}\, dx$
$= 2\int_1^4 x^{-3/2}\, dx = 2\left[\frac{-2}{\sqrt{x}}\right]_1^4 = 2[-1 - (-2)] = 2;$ $M_y = \int \tilde{x}\, dm = \int_1^4 x \cdot \frac{2}{x} \cdot \delta\, dx = 2\int_1^4 x^{1/2}\, dx$
$= 2\left[\frac{2x^{3/2}}{3}\right]_1^4 = 2\left[\frac{16}{3} - \frac{2}{3}\right] = \frac{28}{3};$ $M = \int dm = \int_1^4 \frac{2}{x} \cdot \delta\, dx = 2\int_1^4 \frac{\sqrt{x}}{x}\, dx = 2\int_1^4 x^{-1/2}\, dx = 2\left[2x^{1/2}\right]_1^4$
$= 2(4 - 2) = 4.$ So $\overline{x} = \frac{M_y}{M} = \frac{\left(\frac{28}{3}\right)}{4} = \frac{7}{3}$ and $\overline{y} = \frac{M_x}{M} = \frac{2}{4} = \frac{1}{2} \Rightarrow (\overline{x}, \overline{y}) = \left(\frac{7}{3}, \frac{1}{2}\right)$ is the center of mass.

(c)

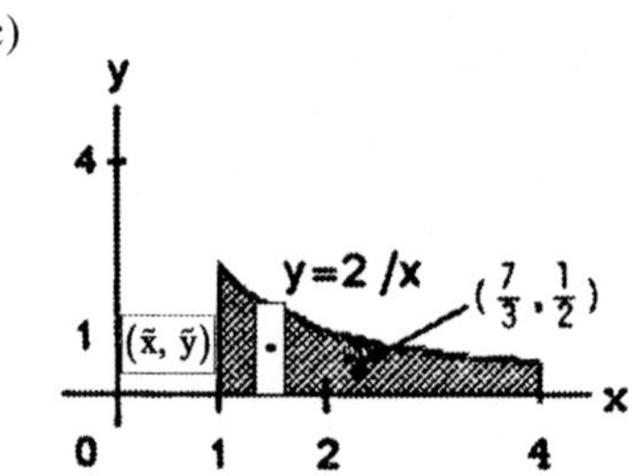

31. (a) We use the shell method: $V = \int_a^b 2\pi \left(\substack{\text{shell}\\ \text{radius}}\right)\left(\substack{\text{shell}\\ \text{height}}\right) dx = \int_1^4 2\pi x \left[\frac{4}{\sqrt{x}} - \left(-\frac{4}{\sqrt{x}}\right)\right] dx$

$= 16\pi \int_1^4 \frac{x}{\sqrt{x}}\, dx = 16\pi \int_1^4 x^{1/2}\, dx = 16\pi \left[\frac{2}{3} x^{3/2}\right]_1^4 = 16\pi \left(\frac{2}{3}\cdot 8 - \frac{2}{3}\right) = \frac{32\pi}{3}(8-1) = \frac{224\pi}{3}$

(b) Since the plate is symmetric about the x-axis and its density $\delta(x) = \frac{1}{x}$ is a function of x alone, the distribution of its mass is symmetric about the x-axis. This means that $\bar{y} = 0$. We use the vertical strip approach to find $\bar{x}$: $M_y = \int \tilde{x}\, dm = \int_1^4 x \cdot \left[\frac{4}{\sqrt{x}} - \left(-\frac{4}{\sqrt{x}}\right)\right] \cdot \delta\, dx = \int_1^4 x \cdot \frac{8}{\sqrt{x}} \cdot \frac{1}{x}\, dx = 8\int_1^4 x^{-1/2}\, dx$

$= 8\left[2x^{1/2}\right]_1^4 = 8(2\cdot 2 - 2) = 16$; $M = \int dm = \int_1^4 \left[\frac{4}{\sqrt{x}} - \left(\frac{-4}{\sqrt{x}}\right)\right] \cdot \delta\, dx = 8\int_1^4 \left(\frac{1}{\sqrt{x}}\right)\left(\frac{1}{x}\right) dx = 8\int_1^4 x^{-3/2}\, dx$

$= 8\left[-2x^{-1/2}\right]_1^4 = 8[-1-(-2)] = 8$. So $\bar{x} = \frac{M_y}{M} = \frac{16}{8} = 2 \Rightarrow (\bar{x}, \bar{y}) = (2, 0)$ is the center of mass.

(c)

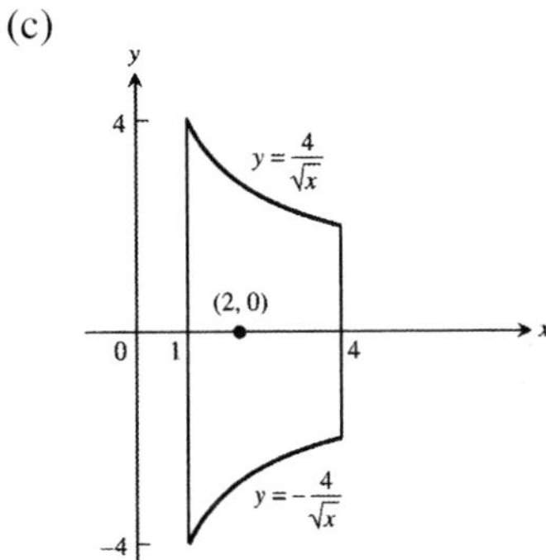

32. (a) $V = \pi \int_{1/4}^4 \left(\frac{1}{2\sqrt{x}}\right)^2 dx = \frac{\pi}{4}\int_{1/4}^4 \frac{1}{x}\, dx = \frac{\pi}{4}\left[\ln|x|\right]_{1/4}^4 = \frac{\pi}{4}\left(\ln 4 - \ln\frac{1}{4}\right) = \frac{\pi}{4}\ln 16 = \frac{\pi}{4}\ln(2^4) = \pi \ln 2$

(b) $M_y = \int_{1/4}^4 x\left(\frac{1}{2\sqrt{x}}\right) dx = \frac{1}{2}\int_{1/4}^4 x^{1/2}\, dx = \left[\frac{1}{3}x^{3/2}\right]_{1/4}^4 = \left(\frac{8}{3} - \frac{1}{24}\right) = \frac{64-1}{24} = \frac{63}{24}$;

$M_x = \int_{1/4}^4 \frac{1}{2}\left(\frac{1}{2\sqrt{x}}\right)\left(\frac{1}{2\sqrt{x}}\right) dx = \frac{1}{8}\int_{1/4}^4 \frac{1}{x}\, dx = \left[\frac{1}{8}\ln|x|\right]_{1/4}^4 = \frac{1}{8}\ln 16 = \frac{1}{2}\ln 2$;

$M = \int_{1/4}^4 \frac{1}{2\sqrt{x}}\, dx = \int_{1/4}^4 \frac{1}{2}x^{-1/2}\, dx = \left[x^{1/2}\right]_{1/4}^4 = 2 - \frac{1}{2} = \frac{3}{2}$; therefore, $\bar{x} = \frac{M_y}{M} = \left(\frac{63}{24}\right)\left(\frac{2}{3}\right) = \frac{21}{12} = \frac{7}{4}$ and $\bar{y} = \frac{M_x}{M} = \left(\frac{1}{2}\ln 2\right)\left(\frac{2}{3}\right) = \frac{\ln 2}{3}$

33. The mass of a horizontal strip is $dm = \delta\, dA = \delta L\, dy$, where L is the width of the triangle at a distance of y above its base on the x-axis as shown in the figure in the text. Also, by similar triangles we have $\frac{L}{b} = \frac{h-y}{h}$ $\Rightarrow L = \frac{b}{h}(h-y)$. Thus, $M_x = \int \tilde{y}\, dm = \int_0^h \delta y \left(\frac{b}{h}\right)(h-y)\, dy = \frac{\delta b}{h}\int_0^h (hy - y^2)\, dy = \frac{\delta b}{h}\left[\frac{hy^2}{2} - \frac{y^3}{3}\right]_0^h$

$= \frac{\delta b}{h}\left(\frac{h^3}{2} - \frac{h^3}{3}\right) = \delta b h^2 \left(\frac{1}{2} - \frac{1}{3}\right) = \frac{\delta b h^2}{6}$; $M = \int dm = \int_0^h \delta\left(\frac{b}{h}\right)(h-y)\, dy = \frac{\delta b}{h}\int_0^h (h-y)\, dy = \frac{\delta b}{h}\left[hy - \frac{y^2}{2}\right]_0^h$

$= \frac{\delta b}{h}\left(h^2 - \frac{h^2}{2}\right) = \frac{\delta b h}{2}$. So $\bar{y} = \frac{M_x}{M} = \left(\frac{\delta b h^2}{6}\right)\left(\frac{2}{\delta b h}\right) = \frac{h}{3} \Rightarrow$ the center of mass lies above the base of the triangle one-third of the way toward the opposite vertex. Similarly the other two sides of the triangle can be placed on the x-axis and the same results will occur. Therefore the centroid does lie at the intersection of the medians, as claimed.

34. From the symmetry about the y-axis it follows that $\bar{x} = 0$. It also follows that the line through the points (0, 0) and (0, 3) is a median $\Rightarrow \bar{y} = \frac{1}{3}(3-0) = 1 \Rightarrow (\bar{x}, \bar{y}) = (0, 1)$.

(0,3)
(-1,0) (+1,0)

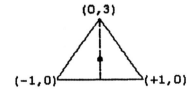

35. From the symmetry about the line $x = y$ it follows that $\bar{x} = \bar{y}$. It also follows that the line through the points (0, 0) and $\left(\frac{1}{2}, \frac{1}{2}\right)$ is a median $\Rightarrow \bar{y} = \bar{x} = \frac{2}{3}\cdot\left(\frac{1}{2} - 0\right) = \frac{1}{3}$ $\Rightarrow (\bar{x}, \bar{y}) = \left(\frac{1}{3}, \frac{1}{3}\right)$.

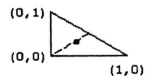

(0,1)
(0,0)
(1,0)

36. From the symmetry about the line $x = y$ it follows that $\overline{x} = \overline{y}$. It also follows that the line through the point $(0, 0)$ and $\left(\frac{a}{2}, \frac{a}{2}\right)$ is a median $\Rightarrow \overline{y} = \overline{x} = \frac{2}{3}\left(\frac{a}{2} - 0\right) = \frac{1}{3}a$ $\Rightarrow (\overline{x}, \overline{y}) = \left(\frac{a}{3}, \frac{a}{3}\right)$.

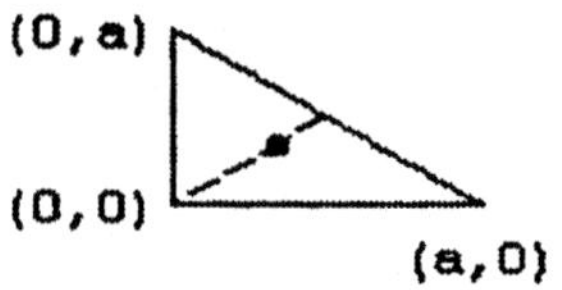

37. The point of intersection of the median from the vertex $(0, b)$ to the opposite side has coordinates $\left(0, \frac{a}{2}\right)$
$\Rightarrow \overline{y} = (b - 0) \cdot \frac{1}{3} = \frac{b}{3}$ and $\overline{x} = \left(\frac{a}{2} - 0\right) \cdot \frac{2}{3} = \frac{a}{3}$
$\Rightarrow (\overline{x}, \overline{y}) = \left(\frac{a}{3}, \frac{b}{3}\right)$.

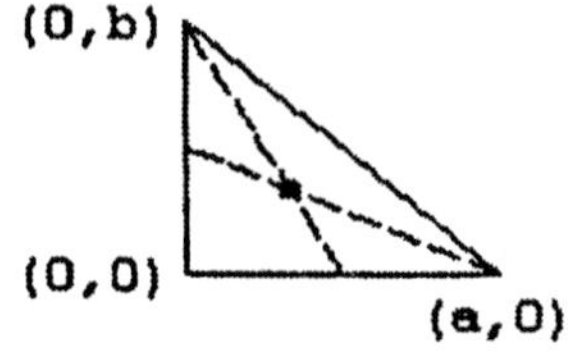

38. From the symmetry about the line $x = \frac{a}{2}$ it follows that $\overline{x} = \frac{a}{2}$. It also follows that the line through the points $\left(\frac{a}{2}, 0\right)$ and $\left(\frac{a}{2}, b\right)$ is a median $\Rightarrow \overline{y} = \frac{1}{3}(b - 0) = \frac{b}{3}$
$\Rightarrow (\overline{x}, \overline{y}) = \left(\frac{a}{2}, \frac{b}{3}\right)$.

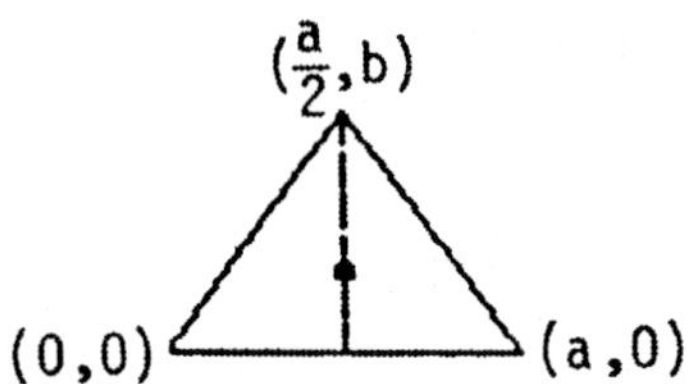

39. $y = x^{1/2} \Rightarrow dy = \frac{1}{2}x^{-1/2}\,dx$
$\Rightarrow ds = \sqrt{(dx)^2 + (dy)^2} = \sqrt{1 + \frac{1}{4x}}\,dx;$
$M_x = \delta\int_0^2 \sqrt{x}\sqrt{1 + \frac{1}{4x}}\,dx$
$= \delta\int_0^2 \sqrt{x + \frac{1}{4}}\,dx = \frac{2\delta}{3}\left[\left(x + \frac{1}{4}\right)^{3/2}\right]_0^2$
$= \frac{2\delta}{3}\left[\left(2 + \frac{1}{4}\right)^{3/2} - \left(\frac{1}{4}\right)^{3/2}\right]$
$= \frac{2\delta}{3}\left[\left(\frac{9}{4}\right)^{3/2} - \left(\frac{1}{4}\right)^{3/2}\right] = \frac{2\delta}{3}\left(\frac{27}{8} - \frac{1}{8}\right) = \frac{13\delta}{6}$

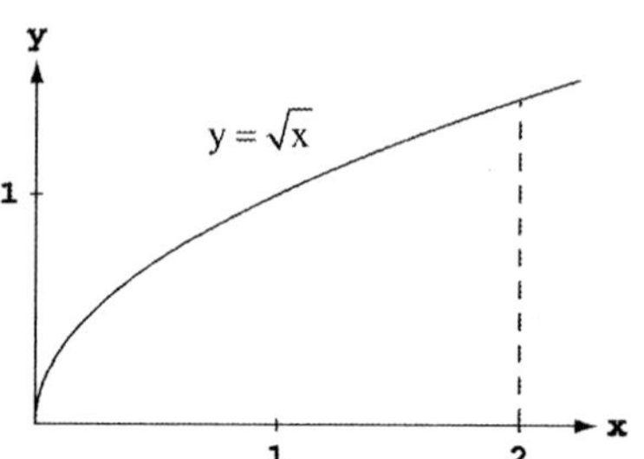

40. $y = x^3 \Rightarrow dy = 3x^2\,dx$
$\Rightarrow dx = \sqrt{(dx)^2 + (3x^2\,dx)^2} = \sqrt{1 + 9x^4}\,dx;$
$M_x = \delta\int_0^1 x^3\sqrt{1 + 9x^4}\,dx;$
$[u = 1 + 9x^4 \Rightarrow du = 36x^3\,dx \Rightarrow \frac{1}{36}\,du = x^3\,dx;$
$x = 0 \Rightarrow u = 1, x = 1 \Rightarrow u = 10]$
$\rightarrow M_x = \delta\int_1^{10} \frac{1}{36}u^{1/2}\,du = \frac{\delta}{36}\left[\frac{2}{3}u^{3/2}\right]_1^{10} = \frac{\delta}{54}\left(10^{3/2} - 1\right)$

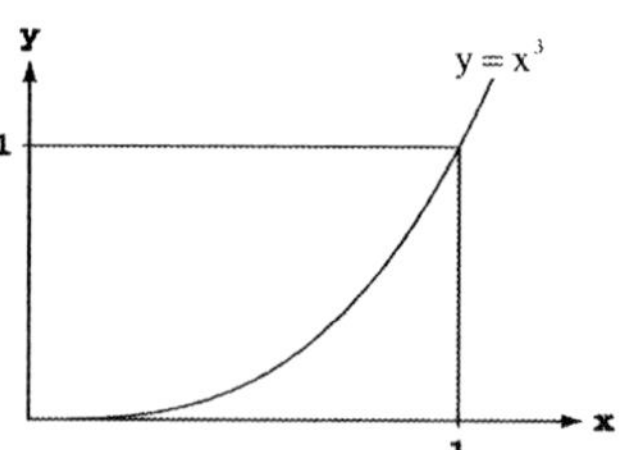

41. From Example 6 we have $M_x = \int_0^\pi a(a\sin\theta)(k\sin\theta)\,d\theta = a^2k\int_0^\pi \sin^2\theta\,d\theta = \frac{a^2k}{2}\int_0^\pi (1 - \cos 2\theta)\,d\theta$
$= \frac{a^2k}{2}\left[\theta - \frac{\sin 2\theta}{2}\right]_0^\pi = \frac{a^2k\pi}{2}$; $M_y = \int_0^\pi a(a\cos\theta)(k\sin\theta)\,d\theta = a^2k\int_0^\pi \sin\theta\cos\theta\,d\theta = \frac{a^2k}{2}\left[\sin^2\theta\right]_0^\pi = 0;$
$M = \int_0^\pi ak\sin\theta\,d\theta = ak[-\cos\theta]_0^\pi = 2ak$. Therefore, $\overline{x} = \frac{M_y}{M} = 0$ and $\overline{y} = \frac{M_x}{M} = \left(\frac{a^2k\pi}{2}\right)\left(\frac{1}{2ak}\right) = \frac{a\pi}{4} \Rightarrow \left(0, \frac{a\pi}{4}\right)$ is the center of mass.

42. $M_x = \int \widetilde{y}\,dm = \int_0^\pi (a\sin\theta) \cdot \delta \cdot a\,d\theta$
$= \int_0^\pi (a^2\sin\theta)(1 + k|\cos\theta|)\,d\theta$
$= a^2\int_0^{\pi/2} (\sin\theta)(1 + k\cos\theta)\,d\theta$
$+ a^2\int_{\pi/2}^{\pi} (\sin\theta)(1 - k\cos\theta)\,d\theta$

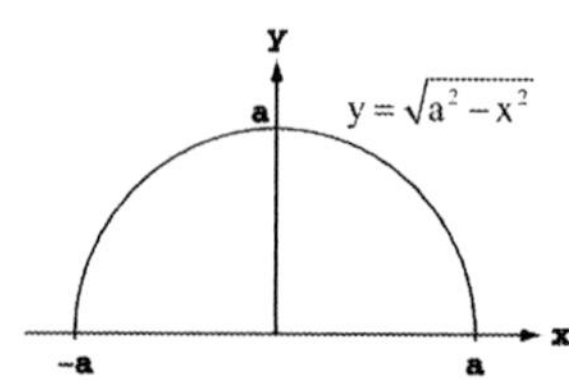

$= a^2\int_0^{\pi/2} \sin\theta\, d\theta + a^2k\int_0^{\pi/2} \sin\theta \cos\theta\, d\theta + a^2\int_{\pi/2}^{\pi} \sin\theta\, d\theta - a^2k \int_{\pi/2}^{\pi} \sin\theta\cos\theta\, d\theta$

$= a^2[-\cos\theta]_0^{\pi/2} + a^2k\left[\frac{\sin^2\theta}{2}\right]_0^{\pi/2} + a^2[-\cos\theta]_{\pi/2}^{\pi} - a^2k\left[\frac{\sin^2\theta}{2}\right]_{\pi/2}^{\pi}$

$= a^2[0-(-1)] + a^2k\left(\frac{1}{2}-0\right) + a^2[-(-1)-0] - a^2k\left(0-\frac{1}{2}\right) = a^2 + \frac{a^2k}{2} + a^2 + \frac{a^2k}{2}$

$= 2a^2 + a^2k = a^2(2+k)$;

$M_y = \int \widetilde{x}\, dm = \int_0^{\pi} (a\cos\theta)\cdot\delta\cdot a\, d\theta = \int_0^{\pi} (a^2\cos\theta)(1+k|\cos\theta|)\, d\theta$

$= a^2\int_0^{\pi/2} (\cos\theta)(1+k\cos\theta)\, d\theta + a^2\int_{\pi/2}^{\pi} (\cos\theta)(1-k\cos\theta)\, d\theta$

$= a^2\int_0^{\pi/2} \cos\theta\, d\theta + a^2k\int_0^{\pi/2} \left(\frac{1+\cos 2\theta}{2}\right) d\theta + a^2\int_{\pi/2}^{\pi} \cos\theta\, d\theta - a^2k\int_{\pi/2}^{\pi} \left(\frac{1+\cos 2\theta}{2}\right) d\theta$

$= a^2[\sin\theta]_0^{\pi/2} + \frac{a^2k}{2}\left[\theta + \frac{\sin 2\theta}{2}\right]_0^{\pi/2} + a^2[\sin\theta]_{\pi/2}^{\pi} - \frac{a^2k}{2}\left[\theta + \frac{\sin 2\theta}{2}\right]_{\pi/2}^{\pi}$

$= a^2(1-0) + \frac{a^2k}{2}\left[\left(\frac{\pi}{2}-0\right)-(0+0)\right] + a^2(0-1) - \frac{a^2k}{2}\left[(\pi+0)-\left(\frac{\pi}{2}+0\right)\right] = a^2 + \frac{a^2k\pi}{4} - a^2 - \frac{a^2k\pi}{4} = 0$;

$M = \int_0^{\pi} \delta\cdot a\, d\theta = a\int_0^{\pi} (1+k|\cos\theta|)\, d\theta = a\int_0^{\pi/2} (1+k\cos\theta)\, d\theta + a\int_{\pi/2}^{\pi} (1-k\cos\theta)\, d\theta$

$= a[\theta + k\sin\theta]_0^{\pi/2} + a[\theta - k\sin\theta]_{\pi/2}^{\pi} = a\left[\left(\frac{\pi}{2}+k\right)-0\right] + a\left[(\pi+0)-\left(\frac{\pi}{2}-k\right)\right]$

$= \frac{a\pi}{2} + ak + a\left(\frac{\pi}{2}+k\right) = a\pi + 2ak = a(\pi+2k)$. So $\overline{x} = \frac{M_y}{M} = 0$ and $\overline{y} = \frac{M_x}{M} = \frac{a^2(2+k)}{a(\pi+2k)} = \frac{a(2+k)}{\pi+2k}$

$\Rightarrow \left(0, \frac{2a+ka}{\pi+2k}\right)$ is the center of mass.

43. Consider the curve as an infinite number of line segments joined together. From the derivation of arc length we have that the length of a particular segment is $ds = \sqrt{(dx)^2 + (dy)^2}$. This implies that $M_x = \int \delta y\, ds$, $M_y = \int \delta x\, ds$ and $M = \int \delta\, ds$. If δ is constant, then $\overline{x} = \frac{M_y}{M} = \frac{\int x\, ds}{\int ds} = \frac{\int x\, ds}{\text{length}}$ and $\overline{y} = \frac{M_x}{M} = \frac{\int y\, ds}{\int ds} = \frac{\int y\, ds}{\text{length}}$.

44. Applying the symmetry argument analogous to the one used in Exercise 13, we find that $\overline{x} = 0$. The typical vertical strip has center of mass: $(\widetilde{x}, \widetilde{y}) = \left(x, \frac{a+\frac{x^2}{4p}}{2}\right)$, length: $a - \frac{x^2}{4p}$, width: dx, area: $dA = \left(a - \frac{x^2}{4p}\right) dx$, mass: $dm = \delta\, dA = \delta\left(a - \frac{x^2}{4p}\right) dx$. Thus, $M_x = \int \widetilde{y}\, dm = \int_{-2\sqrt{pa}}^{2\sqrt{pa}} \frac{1}{2}\left(a + \frac{x^2}{4p}\right)\left(a - \frac{x^2}{4p}\right)\delta\, dx$

$= \frac{\delta}{2}\int_{-2\sqrt{pa}}^{2\sqrt{pa}} \left(a^2 - \frac{x^4}{16p^2}\right) dx = \frac{\delta}{2}\left[a^2x - \frac{x^5}{80p^2}\right]_{-2\sqrt{pa}}^{2\sqrt{pa}} = 2\cdot\frac{\delta}{2}\left[a^2x - \frac{x^5}{80p^2}\right]_0^{2\sqrt{pa}} = \delta\left(2a^2\sqrt{pa} - \frac{2^5p^2a^2\sqrt{pa}}{80p^2}\right)$

$= 2a^2\delta\sqrt{pa}\left(1 - \frac{16}{80}\right) = 2a^2\delta\sqrt{pa}\left(\frac{80-16}{80}\right) = 2a^2\delta\sqrt{pa}\left(\frac{64}{80}\right) = \frac{8a^2\delta\sqrt{pa}}{5}$; $M = \int dm = \delta\int_{-2\sqrt{pa}}^{2\sqrt{pa}} \left(a - \frac{x^2}{4p}\right) dx$

$= \delta\left[ax - \frac{x^3}{12p}\right]_{-2\sqrt{pa}}^{2\sqrt{pa}} = 2\cdot\delta\left[ax - \frac{x^3}{12p}\right]_0^{2\sqrt{pa}} = 2\delta\left(2a\sqrt{pa} - \frac{2^3pa\sqrt{pa}}{12p}\right) = 4a\delta\sqrt{pa}\left(1 - \frac{4}{12}\right) = 4a\delta\sqrt{pa}\left(\frac{12-4}{12}\right)$

$= \frac{8a\delta\sqrt{pa}}{3}$. So $\overline{y} = \frac{M_x}{M} = \left(\frac{8a^2\delta\sqrt{pa}}{5}\right)\left(\frac{3}{8a\delta\sqrt{pa}}\right) = \frac{3}{5}a$, as claimed.

45. Since the density is constant, its value will not affect our answers, so we can set $\delta = 1$.
A generalization of Example 6 yields $M_x = \int \widetilde{y}\, dm = \int_{\pi/2-\alpha}^{\pi/2+\alpha} a^2\sin\theta\, d\theta = a^2[-\cos\theta]_{\pi/2-\alpha}^{\pi/2+\alpha}$

$= a^2\left[-\cos\left(\frac{\pi}{2}+\alpha\right) + \cos\left(\frac{\pi}{2}-\alpha\right)\right] = a^2(\sin\alpha + \sin\alpha) = 2a^2\sin\alpha$; $M = \int dm = \int_{\pi/2-\alpha}^{\pi/2+\alpha} a\, d\theta = a[\theta]_{\pi/2-\alpha}^{\pi/2+\alpha}$

$= a\left[\left(\frac{\pi}{2}+\alpha\right) - \left(\frac{\pi}{2}-\alpha\right)\right] = 2a\alpha$. Thus, $\overline{y} = \frac{M_x}{M} = \frac{2a^2\sin\alpha}{2a\alpha} = \frac{a\sin\alpha}{\alpha}$. Now $s = a(2\alpha)$ and $a\sin\alpha = \frac{c}{2}$

$\Rightarrow c = 2a\sin\alpha$. Then $\overline{y} = \frac{a(2a\sin\alpha)}{2a\alpha} = \frac{ac}{s}$, as claimed.

46. (a) First, we note that $\overline{y} = (\text{distance from origin to } \overline{AB}) + d \Rightarrow \frac{a \sin \alpha}{\alpha} = a \cos \alpha + d \Rightarrow d = \frac{a(\sin \alpha - \alpha \cos \alpha)}{\alpha}$. Moreover, $h = a - a \cos \alpha \Rightarrow \frac{d}{h} = \frac{a(\sin \alpha - \alpha \cos \alpha)}{a(\alpha - \alpha \cos \alpha)} = \frac{\sin \alpha - \alpha \cos \alpha}{\alpha - \alpha \cos \alpha}$. The graphs below suggest that $\lim\limits_{\alpha \to 0^+} \frac{\sin \alpha - \alpha \cos \alpha}{\alpha - \alpha \cos \alpha} \approx \frac{2}{3}$.

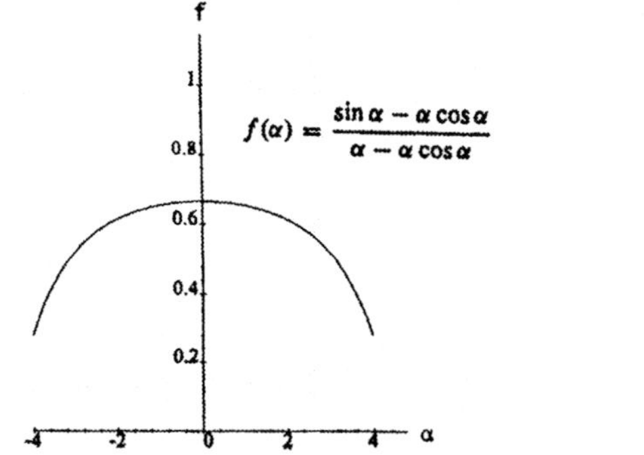

ZOOM VIEW

$f(\alpha) = \frac{\sin \alpha - \alpha \cos \alpha}{\alpha - \alpha \cos \alpha}$

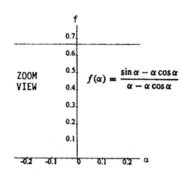

(b)

α	0.2	0.4	0.6	0.8	1.0
$f(\alpha)$	0.666222	0.664879	0.662615	0.659389	0.655145

6.5 AREAS OF SURFACES OF REVOLUTION AND THE THEOREMS OF PAPPUS

1. (a) $\frac{dy}{dx} = \sec^2 x \Rightarrow \left(\frac{dy}{dx}\right)^2 = \sec^4 x$

$\Rightarrow S = 2\pi \int_0^{\pi/4} (\tan x) \sqrt{1 + \sec^4 x}\, dx$

(c) $S \approx 3.84$

(b)

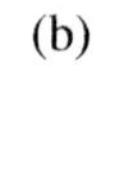

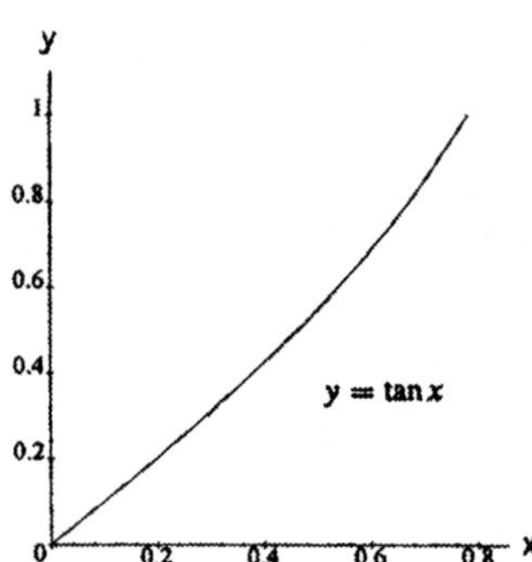

2. (a) $\frac{dy}{dx} = 2x \Rightarrow \left(\frac{dy}{dx}\right)^2 = 4x^2$

$\Rightarrow S = 2\pi \int_0^2 x^2 \sqrt{1 + 4x^2}\, dx$

(c) $S \approx 53.23$

(b)

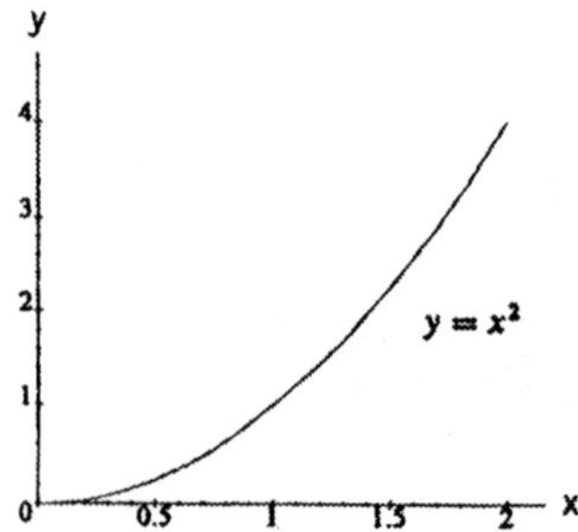

3. (a) $xy = 1 \Rightarrow x = \frac{1}{y} \Rightarrow \frac{dx}{dy} = -\frac{1}{y^2} \Rightarrow \left(\frac{dx}{dy}\right)^2 = \frac{1}{y^4}$
$\Rightarrow S = 2\pi \int_1^2 \frac{1}{y} \sqrt{1 + y^{-4}}\, dy$
(c) $S \approx 5.02$

(b)

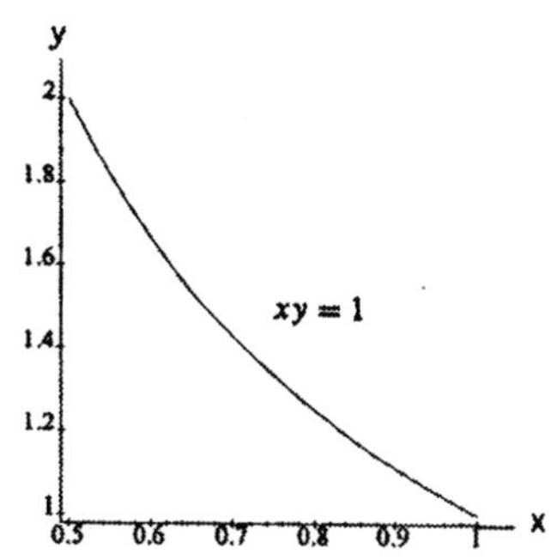

4. (a) $\frac{dx}{dy} = \cos y \Rightarrow \left(\frac{dx}{dy}\right)^2 = \cos^2 y$
$\Rightarrow S = 2\pi \int_0^{\pi} (\sin y) \sqrt{1 + \cos^2 y}\, dy$
(c) $S \approx 14.42$

(b)

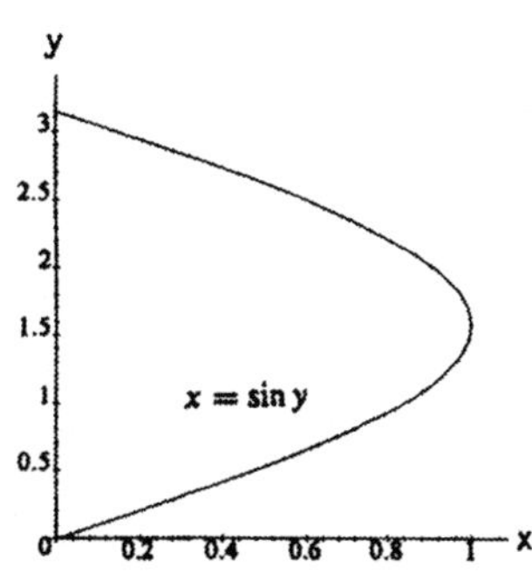

5. (a) $x^{1/2} + y^{1/2} = 3 \Rightarrow y = \left(3 - x^{1/2}\right)^2$
$\Rightarrow \frac{dy}{dx} = 2\left(3 - x^{1/2}\right)\left(-\frac{1}{2}x^{-1/2}\right)$
$\Rightarrow \left(\frac{dy}{dx}\right)^2 = \left(1 - 3x^{-1/2}\right)^2$
$\Rightarrow S = 2\pi \int_1^4 \left(3 - x^{1/2}\right)^2 \sqrt{1 + \left(1 - 3x^{-1/2}\right)^2}\, dx$
(c) $S \approx 63.37$

(b)

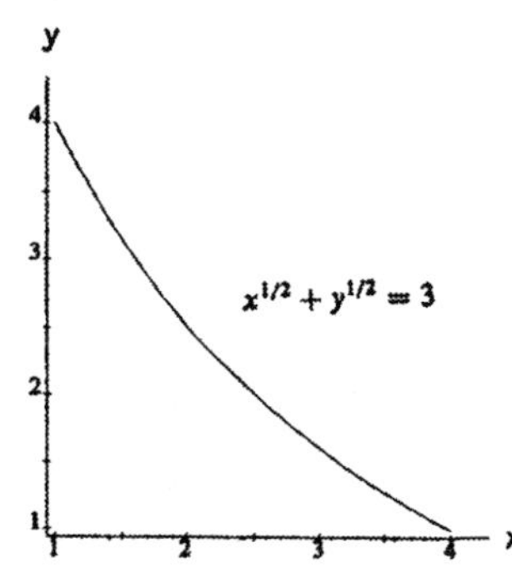

6. (a) $\frac{dx}{dy} = 1 + y^{-1/2} \Rightarrow \left(\frac{dx}{dy}\right)^2 = \left(1 + y^{-1/2}\right)^2$
$\Rightarrow S = 2\pi \int_1^2 \left(y + 2\sqrt{y}\right) \sqrt{1 + \left(1 + y^{-1/2}\right)^2}\, dx$
(c) $S \approx 51.33$

(b)

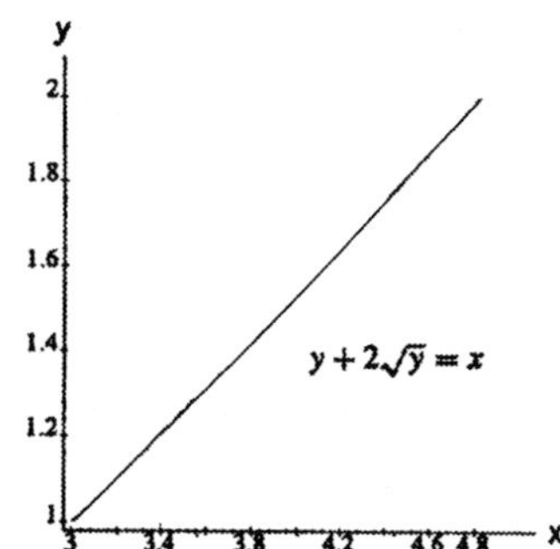

7. (a) $\frac{dx}{dy} = \tan y \Rightarrow \left(\frac{dx}{dy}\right)^2 = \tan^2 y$
$\Rightarrow S = 2\pi \int_0^{\pi/3} \left(\int_0^y \tan t \, dt\right) \sqrt{1 + \tan^2 y} \, dy$
$= 2\pi \int_0^{\pi/3} \left(\int_0^y \tan t \, dt\right) \sec y \, dy$
(c) $S \approx 2.08$

(b)

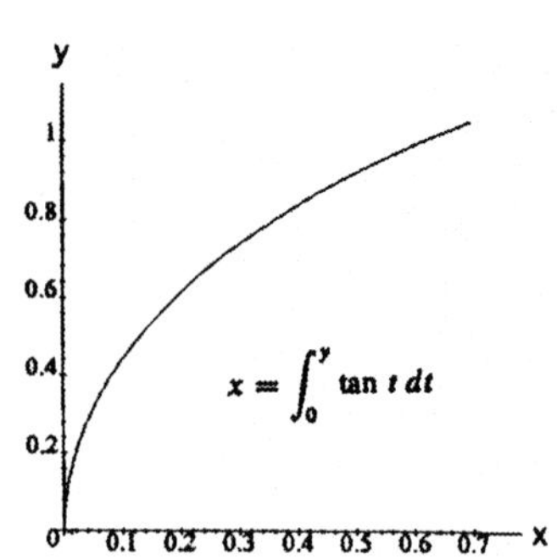

8. (a) $\frac{dy}{dx} = \sqrt{x^2 - 1} \Rightarrow \left(\frac{dy}{dx}\right)^2 = x^2 - 1$
$\Rightarrow S = 2\pi \int_1^{\sqrt{5}} \left(\int_1^x \sqrt{t^2 - 1} \, dt\right) \sqrt{1 + (x^2 - 1)} \, dx$
$= 2\pi \int_1^{\sqrt{5}} \left(\int_1^x \sqrt{t^2 - 1} \, dt\right) x \, dx$
(c) $S \approx 8.55$

(b)

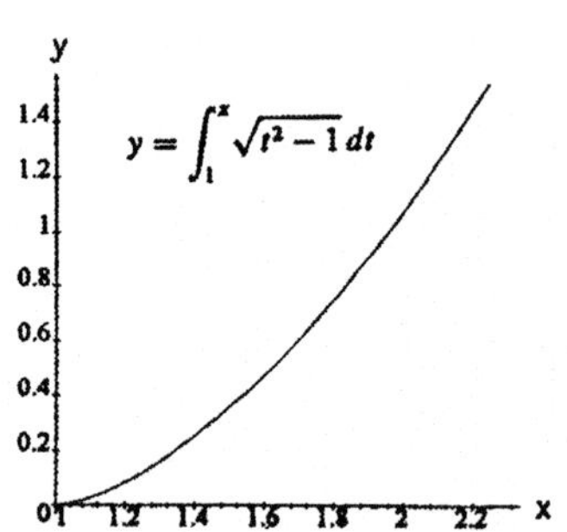

9. $y = \frac{x}{2} \Rightarrow \frac{dy}{dx} = \frac{1}{2}$; $S = \int_a^b 2\pi y \sqrt{1 + \left(\frac{dy}{dx}\right)^2} \, dx \Rightarrow S = \int_0^4 2\pi \left(\frac{x}{2}\right) \sqrt{1 + \frac{1}{4}} \, dx = \frac{\pi\sqrt{5}}{2} \int_0^4 x \, dx$
$= \frac{\pi\sqrt{5}}{2} \left[\frac{x^2}{2}\right]_0^4 = 4\pi\sqrt{5}$; Geometry formula: base circumference $= 2\pi(2)$, slant height $= \sqrt{4^2 + 2^2} = 2\sqrt{5}$
$\Rightarrow$ Lateral surface area $= \frac{1}{2}(4\pi)\left(2\sqrt{5}\right) = 4\pi\sqrt{5}$ in agreement with the integral value

10. $y = \frac{x}{2} \Rightarrow x = 2y \Rightarrow \frac{dx}{dy} = 2$; $S = \int_c^d 2\pi x \sqrt{1 + \left(\frac{dx}{dy}\right)^2} \, dy = \int_0^2 2\pi \cdot 2y\sqrt{1 + 2^2} \, dy = 4\pi\sqrt{5} \int_0^2 y \, dy = 2\pi\sqrt{5} \left[y^2\right]_0^2$
$= 2\pi\sqrt{5} \cdot 4 = 8\pi\sqrt{5}$; Geometry formula: base circumference $= 2\pi(4)$, slant height $= \sqrt{4^2 + 2^2} = 2\sqrt{5}$
$\Rightarrow$ Lateral surface area $= \frac{1}{2}(8\pi)\left(2\sqrt{5}\right) = 8\pi\sqrt{5}$ in agreement with the integral value

11. $\frac{dy}{dx} = \frac{1}{2}$; $S = \int_a^b 2\pi y \sqrt{1 + \left(\frac{dy}{dx}\right)^2} \, dx = \int_1^3 2\pi \frac{(x+1)}{2} \sqrt{1 + \left(\frac{1}{2}\right)^2} \, dx = \frac{\pi\sqrt{5}}{2} \int_1^3 (x + 1) \, dx = \frac{\pi\sqrt{5}}{2} \left[\frac{x^2}{2} + x\right]_1^3$
$= \frac{\pi\sqrt{5}}{2} \left[\left(\frac{9}{2} + 3\right) - \left(\frac{1}{2} + 1\right)\right] = \frac{\pi\sqrt{5}}{2}(4 + 2) = 3\pi\sqrt{5}$; Geometry formula: $r_1 = \frac{1}{2} + \frac{1}{2} = 1$, $r_2 = \frac{3}{2} + \frac{1}{2} = 2$,
slant height $= \sqrt{(2-1)^2 + (3-1)^2} = \sqrt{5} \Rightarrow$ Frustum surface area $= \pi(r_1 + r_2) \times$ slant height $= \pi(1 + 2)\sqrt{5}$
$= 3\pi\sqrt{5}$ in agreement with the integral value

12. $y = \frac{x}{2} + \frac{1}{2} \Rightarrow x = 2y - 1 \Rightarrow \frac{dx}{dy} = 2$; $S = \int_c^d 2\pi x \sqrt{1 + \left(\frac{dx}{dy}\right)^2} \, dy = \int_1^2 2\pi(2y - 1)\sqrt{1 + 4} \, dy = 2\pi\sqrt{5} \int_1^2 (2y - 1) \, dy$
$= 2\pi\sqrt{5} \left[y^2 - y\right]_1^2 = 2\pi\sqrt{5}[(4 - 2) - (1 - 1)] = 4\pi\sqrt{5}$; Geometry formula: $r_1 = 1$, $r_2 = 3$,
slant height $= \sqrt{(2-1)^2 + (3-1)^2} = \sqrt{5} \Rightarrow$ Frustum surface area $= \pi(1 + 3)\sqrt{5} = 4\pi\sqrt{5}$ in agreement with the integral value

13. $\frac{dy}{dx} = \frac{x^2}{3} \Rightarrow \left(\frac{dy}{dx}\right)^2 = \frac{x^4}{9} \Rightarrow S = \int_0^2 \frac{2\pi x^3}{9}\sqrt{1+\frac{x^4}{9}}\,dx;$
$\left[u = 1 + \frac{x^4}{9} \Rightarrow du = \frac{4}{9}x^3\,dx \Rightarrow \frac{1}{4}\,du = \frac{x^3}{9}\,dx;\right.$
$\left. x = 0 \Rightarrow u = 1, x = 2 \Rightarrow u = \frac{25}{9}\right]$
$\rightarrow S = 2\pi \int_1^{25/9} u^{1/2} \cdot \frac{1}{4}\,du = \frac{\pi}{2}\left[\frac{2}{3}u^{3/2}\right]_1^{25/9}$
$= \frac{\pi}{3}\left(\frac{125}{27} - 1\right) = \frac{\pi}{3}\left(\frac{125-27}{27}\right) = \frac{98\pi}{81}$

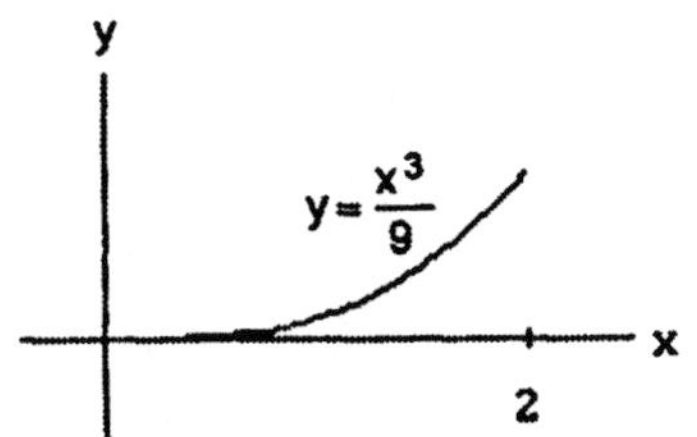

14. $\frac{dy}{dx} = \frac{1}{2}x^{-1/2} \Rightarrow \left(\frac{dy}{dx}\right)^2 = \frac{1}{4x}$
$\Rightarrow S = \int_{3/4}^{15/4} 2\pi\sqrt{x}\sqrt{1+\frac{1}{4x}}\,dx$
$= 2\pi\int_{3/4}^{15/4}\sqrt{x+\frac{1}{4}}\,dx = 2\pi\left[\frac{2}{3}\left(x+\frac{1}{4}\right)^{3/2}\right]_{3/4}^{15/4}$
$= \frac{4\pi}{3}\left[\left(\frac{15}{4}+\frac{1}{4}\right)^{3/2} - \left(\frac{3}{4}+\frac{1}{4}\right)^{3/2}\right] = \frac{4\pi}{3}\left[\left(\frac{4}{2}\right)^3 - 1\right]$
$= \frac{4\pi}{3}(8-1) = \frac{28\pi}{3}$

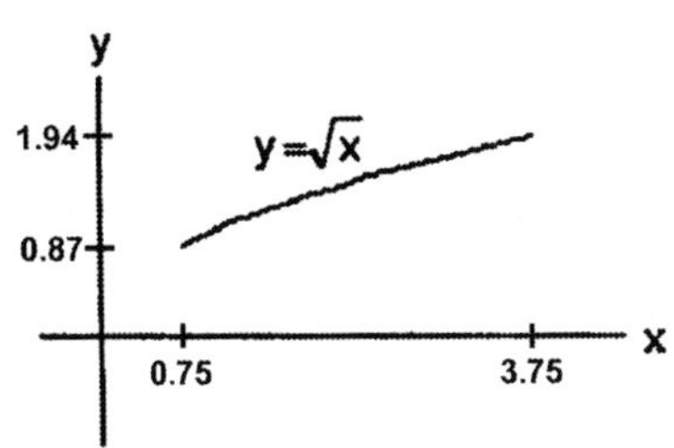

15. $\frac{dy}{dx} = \frac{1}{2}\frac{(2-2x)}{\sqrt{2x-x^2}} = \frac{1-x}{\sqrt{2x-x^2}} \Rightarrow \left(\frac{dy}{dx}\right)^2 = \frac{(1-x)^2}{2x-x^2}$
$\Rightarrow S = \int_{0.5}^{1.5} 2\pi\sqrt{2x-x^2}\sqrt{1+\frac{(1-x)^2}{2x-x^2}}\,dx$
$= 2\pi\int_{0.5}^{1.5}\sqrt{2x-x^2}\,\frac{\sqrt{2x-x^2+1-2x+x^2}}{\sqrt{2x-x^2}}\,dx$
$= 2\pi\int_{0.5}^{1.5} dx = 2\pi[x]_{0.5}^{1.5} = 2\pi$

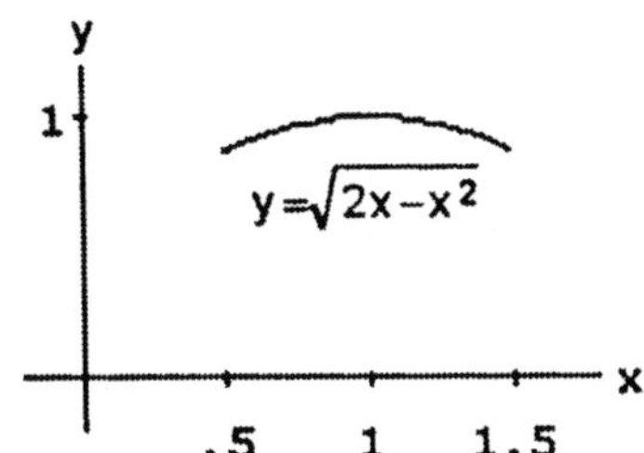

16. $\frac{dy}{dx} = \frac{1}{2\sqrt{x+1}} \Rightarrow \left(\frac{dy}{dx}\right)^2 = \frac{1}{4(x+1)}$

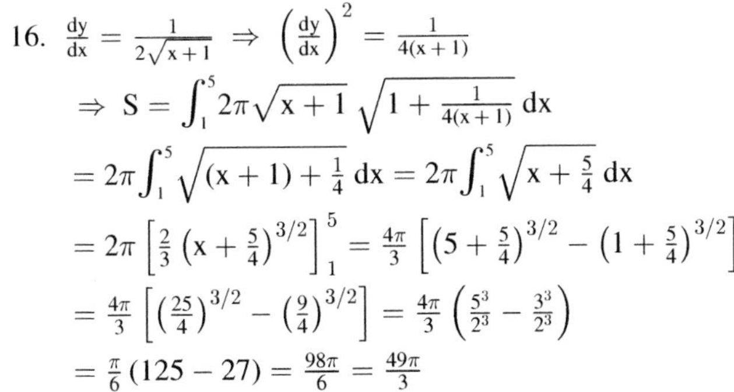

$\Rightarrow S = \int_1^5 2\pi\sqrt{x+1}\sqrt{1+\frac{1}{4(x+1)}}\,dx$
$= 2\pi\int_1^5\sqrt{(x+1)+\frac{1}{4}}\,dx = 2\pi\int_1^5\sqrt{x+\frac{5}{4}}\,dx$
$= 2\pi\left[\frac{2}{3}\left(x+\frac{5}{4}\right)^{3/2}\right]_1^5 = \frac{4\pi}{3}\left[\left(5+\frac{5}{4}\right)^{3/2} - \left(1+\frac{5}{4}\right)^{3/2}\right]$
$= \frac{4\pi}{3}\left[\left(\frac{25}{4}\right)^{3/2} - \left(\frac{9}{4}\right)^{3/2}\right] = \frac{4\pi}{3}\left(\frac{5^3}{2^3} - \frac{3^3}{2^3}\right)$
$= \frac{\pi}{6}(125-27) = \frac{98\pi}{6} = \frac{49\pi}{3}$

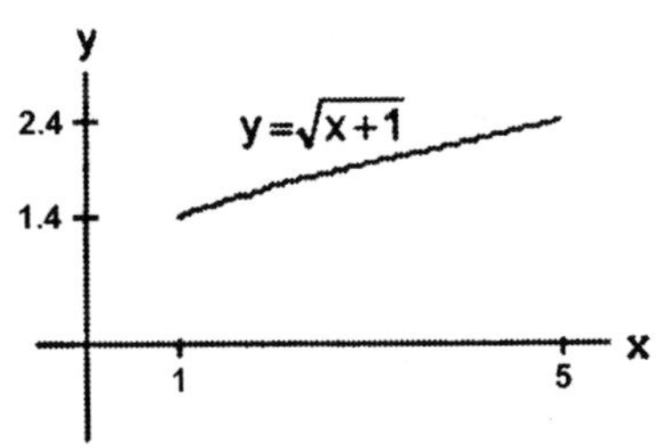

17. $\frac{dx}{dy} = y^2 \Rightarrow \left(\frac{dx}{dy}\right)^2 = y^4 \Rightarrow S = \int_0^1 \frac{2\pi y^3}{3}\sqrt{1+y^4}\,dy;$

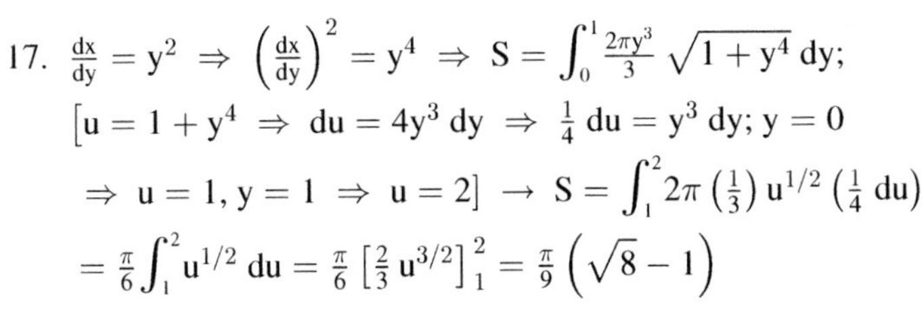

$\left[u = 1 + y^4 \Rightarrow du = 4y^3\,dy \Rightarrow \frac{1}{4}\,du = y^3\,dy;\ y = 0\right.$
$\left.\Rightarrow u = 1, y = 1 \Rightarrow u = 2\right] \rightarrow S = \int_1^2 2\pi\left(\frac{1}{3}\right)u^{1/2}\left(\frac{1}{4}\,du\right)$
$= \frac{\pi}{6}\int_1^2 u^{1/2}\,du = \frac{\pi}{6}\left[\frac{2}{3}u^{3/2}\right]_1^2 = \frac{\pi}{9}\left(\sqrt{8}-1\right)$

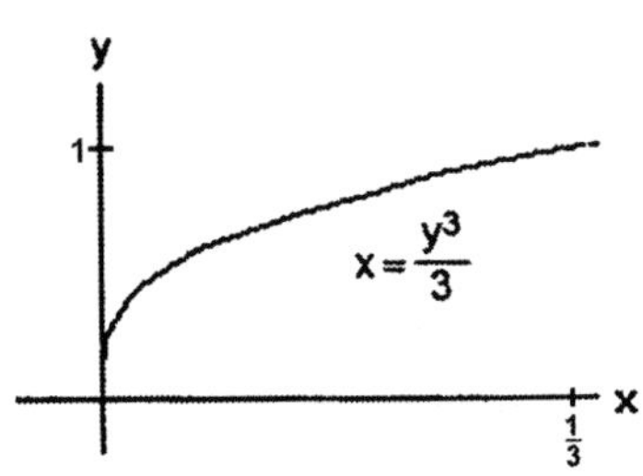

18. $x = \left(\frac{1}{3} y^{3/2} - y^{1/2}\right) \le 0$, when $1 \le y \le 3$. To get positive area, we take $x = -\left(\frac{1}{3} y^{3/2} - y^{1/2}\right)$

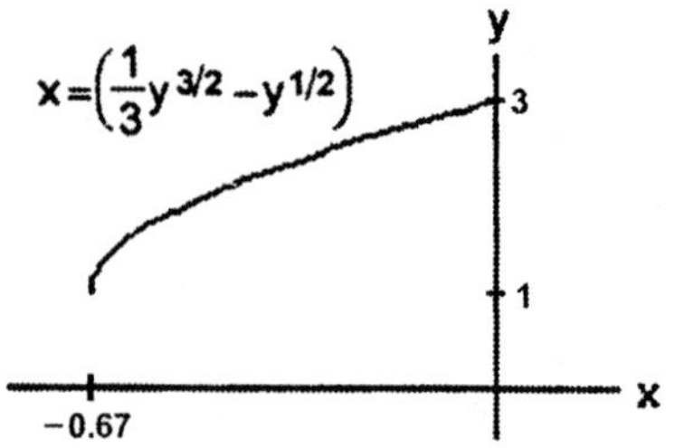

$\Rightarrow \frac{dx}{dy} = -\frac{1}{2}\left(y^{1/2} - y^{-1/2}\right) \Rightarrow \left(\frac{dx}{dy}\right)^2 = \frac{1}{4}\left(y - 2 + y^{-1}\right)$

$\Rightarrow S = -\int_1^3 2\pi\left(\frac{1}{3} y^{3/2} - y^{1/2}\right)\sqrt{1 + \frac{1}{4}\left(y - 2 + y^{-1}\right)}\, dy$

$= -2\pi \int_1^3 \left(\frac{1}{3} y^{3/2} - y^{1/2}\right)\sqrt{\frac{1}{4}\left(y + 2 + y^{-1}\right)}\, dy$

$x=\left(\frac{1}{3}y^{3/2}-y^{1/2}\right)$
y
3
1
x
−0.67

$= -2\pi \int_1^3 \left(\frac{1}{3} y^{3/2} - y^{1/2}\right)\frac{\sqrt{\left(y^{1/2}+y^{-1/2}\right)^2}}{2}\, dy = -\pi \int_1^3 y^{1/2}\left(\frac{1}{3} y - 1\right)\left(y^{1/2} + \frac{1}{y^{1/2}}\right) dy = -\pi \int_1^3 \left(\frac{1}{3} y - 1\right)(y+1)\, dy$

$= -\pi \int_1^3 \left(\frac{1}{3} y^2 - \frac{2}{3} y - 1\right) dy = -\pi\left[\frac{y^3}{9} - \frac{y^2}{3} - y\right]_1^3 = -\pi\left[\left(\frac{27}{9} - \frac{9}{3} - 3\right) - \left(\frac{1}{9} - \frac{1}{3} - 1\right)\right] = -\pi\left(-3 - \frac{1}{9} + \frac{1}{3} + 1\right)$

$= -\frac{\pi}{9}(-18 - 1 + 3) = \frac{16\pi}{9}$

19. $\frac{dx}{dy} = \frac{-1}{\sqrt{4-y}} \Rightarrow \left(\frac{dx}{dy}\right)^2 = \frac{1}{4-y} \Rightarrow S = \int_0^{15/4} 2\pi \cdot 2\sqrt{4-y}\,\sqrt{1 + \frac{1}{4-y}}\, dy = 4\pi \int_0^{15/4} \sqrt{(4-y)+1}\, dy$

$= 4\pi \int_0^{15/4} \sqrt{5-y}\, dy = -4\pi\left[\frac{2}{3}(5-y)^{3/2}\right]_0^{15/4} = -\frac{8\pi}{3}\left[\left(5 - \frac{15}{4}\right)^{3/2} - 5^{3/2}\right] = -\frac{8\pi}{3}\left[\left(\frac{5}{4}\right)^{3/2} - 5^{3/2}\right]$

$= \frac{8\pi}{3}\left(5\sqrt{5} - \frac{5\sqrt{5}}{8}\right) = \frac{8\pi}{3}\left(\frac{40\sqrt{5}-5\sqrt{5}}{8}\right) = \frac{35\pi\sqrt{5}}{3}$

20. $\frac{dx}{dy} = \frac{1}{\sqrt{2y-1}} \Rightarrow \left(\frac{dx}{dy}\right)^2 = \frac{1}{2y-1} \Rightarrow S = \int_{5/8}^1 2\pi\sqrt{2y-1}\,\sqrt{1 + \frac{1}{2y-1}}\, dy = 2\pi \int_{5/8}^1 \sqrt{(2y-1)+1}\, dy$

$= 2\pi \int_{5/8}^1 \sqrt{2}\, y^{1/2}\, dy = 2\pi\sqrt{2}\left[\frac{2}{3} y^{3/2}\right]_{5/8}^1 = \frac{4\pi\sqrt{2}}{3}\left[1^{3/2} - \left(\frac{5}{8}\right)^{3/2}\right] = \frac{4\pi\sqrt{2}}{3}\left(1 - \frac{5\sqrt{5}}{8\sqrt{8}}\right)$

$= \frac{4\pi\sqrt{2}}{3}\left(\frac{8\cdot 2\sqrt{2} - 5\sqrt{5}}{8\cdot 2\sqrt{2}}\right) = \frac{\pi}{12}\left(16\sqrt{2} - 5\sqrt{5}\right)$

21. $S = 2\pi \int_0^{\ln 2} \left(\frac{e^y + e^{-y}}{2}\right)\sqrt{1 + \left(\frac{e^y - e^{-y}}{2}\right)^2}\, dy = 2\pi \int_0^{\ln 2} \left(\frac{e^y + e^{-y}}{2}\right)\sqrt{1 + \frac{1}{4}\left(e^{2y} - 2 + e^{-2y}\right)}\, dy$

$= 2\pi \int_0^{\ln 2} \left(\frac{e^y + e^{-y}}{2}\right)\sqrt{\left(\frac{e^y + e^{-y}}{2}\right)^2}\, dy = 2\pi \int_0^{\ln 2} \left(\frac{e^y + e^{-y}}{2}\right)^2 dy = \frac{\pi}{2}\int_0^{\ln 2} \left(e^{2y} + 2 + e^{-2y}\right) dy$

$= \frac{\pi}{2}\left[\frac{1}{2} e^{2y} + 2y - \frac{1}{2} e^{-2y}\right]_0^{\ln 2} = \frac{\pi}{2}\left[\left(\frac{1}{2} e^{2\ln 2} + 2\ln 2 - \frac{1}{2} e^{-2\ln 2}\right) - \left(\frac{1}{2} + 0 - \frac{1}{2}\right)\right]$

$= \frac{\pi}{2}\left(\frac{1}{2}\cdot 4 + 2\ln 2 - \frac{1}{2}\cdot\frac{1}{4}\right) = \frac{\pi}{2}\left(2 - \frac{1}{8} + 2\ln 2\right) = \pi\left(\frac{15}{16} + \ln 2\right)$

22. $y = \frac{1}{3}\left(x^2 + 2\right)^{3/2} \Rightarrow dy = x\sqrt{x^2 + 2}\, dx \Rightarrow ds = \sqrt{1 + \left(2x^2 + x^4\right)}\, dx \Rightarrow S = 2\pi \int_0^{\sqrt{2}} x\sqrt{1 + 2x^2 + x^4}\, dx$

$= 2\pi \int_0^{\sqrt{2}} x\sqrt{\left(x^2 + 1\right)^2}\, dx = 2\pi \int_0^{\sqrt{2}} x\left(x^2 + 1\right) dx = 2\pi \int_0^{\sqrt{2}} \left(x^3 + x\right) dx = 2\pi\left[\frac{x^4}{4} + \frac{x^2}{2}\right]_0^{\sqrt{2}} = 2\pi\left(\frac{4}{4} + \frac{2}{2}\right) = 4\pi$

23. $ds = \sqrt{dx^2 + dy^2} = \sqrt{\left(y^3 - \frac{1}{4y^3}\right)^2 + 1}\, dy = \sqrt{\left(y^6 - \frac{1}{2} + \frac{1}{16y^6}\right) + 1}\, dy = \sqrt{\left(y^6 + \frac{1}{2} + \frac{1}{16y^6}\right)}\, dy$

$= \sqrt{\left(y^3 + \frac{1}{4y^3}\right)^2}\, dy = \left(y^3 + \frac{1}{4y^3}\right) dy$; $S = \int_1^2 2\pi y\, ds = 2\pi \int_1^2 y\left(y^3 + \frac{1}{4y^3}\right) dy = 2\pi \int_1^2 \left(y^4 + \frac{1}{4} y^{-2}\right) dy$

$= 2\pi\left[\frac{y^5}{5} - \frac{1}{4} y^{-1}\right]_1^2 = 2\pi\left[\left(\frac{32}{5} - \frac{1}{8}\right) - \left(\frac{1}{5} - \frac{1}{4}\right)\right] = 2\pi\left(\frac{31}{5} + \frac{1}{8}\right) = \frac{2\pi}{40}(8\cdot 31 + 5) = \frac{253\pi}{20}$

24. $y = \cos x \Rightarrow \frac{dy}{dx} = -\sin x \Rightarrow \left(\frac{dy}{dx}\right)^2 = \sin^2 x \Rightarrow S = 2\pi \int_{-\pi/2}^{\pi/2} (\cos x)\sqrt{1 + \sin^2 x}\, dx$

25. $y = \sqrt{a^2 - x^2} \Rightarrow \frac{dy}{dx} = \frac{1}{2}\left(a^2 - x^2\right)^{-1/2}(-2x) = \frac{-x}{\sqrt{a^2 - x^2}} \Rightarrow \left(\frac{dy}{dx}\right)^2 = \frac{x^2}{\left(a^2 - x^2\right)}$

$\Rightarrow S = 2\pi \int_{-a}^a \sqrt{a^2 - x^2}\,\sqrt{1 + \frac{x^2}{\left(a^2 - x^2\right)}}\, dx = 2\pi \int_{-a}^a \sqrt{\left(a^2 - x^2\right) + x^2}\, dx = 2\pi \int_{-a}^a a\, dx = 2\pi a[x]_{-a}^a$

$= 2\pi a[a - (-a)] = (2\pi a)(2a) = 4\pi a^2$

26. $y = \frac{r}{h}x \Rightarrow \frac{dy}{dx} = \frac{r}{h} \Rightarrow \left(\frac{dy}{dx}\right)^2 = \frac{r^2}{h^2} \Rightarrow S = 2\pi \int_0^h \frac{r}{h}x\sqrt{1+\frac{r^2}{h^2}}\,dx = 2\pi \int_0^h \frac{r}{h}x\sqrt{\frac{h^2+r^2}{h^2}}\,dx$
$= \frac{2\pi r}{h}\sqrt{\frac{h^2+r^2}{h^2}}\int_0^h x\,dx = \frac{2\pi r}{h^2}\sqrt{h^2+r^2}\left[\frac{x^2}{2}\right]_0^h = \frac{2\pi r}{h^2}\sqrt{h^2+r^2}\left(\frac{h^2}{2}\right) = \pi r\sqrt{h^2+r^2}$

27. The area of the surface of one wok is $S = \int_c^d 2\pi x\sqrt{1+\left(\frac{dx}{dy}\right)^2}\,dy$. Now, $x^2+y^2 = 16^2 \Rightarrow x = \sqrt{16^2-y^2}$
$\Rightarrow \frac{dx}{dy} = \frac{-y}{\sqrt{16^2-y^2}} \Rightarrow \left(\frac{dx}{dy}\right)^2 = \frac{y^2}{16^2-y^2}$; $S = \int_{-16}^{-7} 2\pi\sqrt{16^2-y^2}\sqrt{1+\frac{y^2}{16^2-y^2}}\,dy = 2\pi\int_{-16}^{-7}\sqrt{(16^2-y^2)+y^2}\,dy$
$= 2\pi\int_{-16}^{-7} 16\,dy = 32\pi \cdot 9 = 288\pi \approx 904.78\text{ cm}^2$. The enamel needed to cover one surface of one wok is $V = S \cdot 0.5\text{ mm} = S \cdot 0.05\text{ cm} = (904.78)(0.05)\text{ cm}^3 = 45.24\text{ cm}^3$. For 5000 woks, we need $5000 \cdot V = 5000 \cdot 45.24\text{ cm}^3 = (5)(45.24)\text{L} = 226.2\text{L} \Rightarrow$ 226.2 liters of each color are needed.

28. $y = \sqrt{r^2-x^2} \Rightarrow \frac{dy}{dx} = -\frac{1}{2}\frac{2x}{\sqrt{r^2-x^2}} = \frac{-x}{\sqrt{r^2-x^2}} \Rightarrow \left(\frac{dx}{dy}\right)^2 = \frac{x^2}{r^2-x^2}$; $S = 2\pi\int_a^{a+h}\sqrt{r^2-x^2}\sqrt{1+\frac{x^2}{r^2-x^2}}\,dx$
$= 2\pi\int_a^{a+h}\sqrt{(r^2-x^2)+x^2}\,dx = 2\pi r\int_a^{a+h}dx = 2\pi rh$, which is independent of a.

29. $y = \sqrt{R^2-x^2} \Rightarrow \frac{dy}{dx} = -\frac{1}{2}\frac{2x}{\sqrt{R^2-x^2}} = \frac{-x}{\sqrt{R^2-x^2}} \Rightarrow \left(\frac{dx}{dy}\right)^2 = \frac{x^2}{R^2-x^2}$; $S = 2\pi\int_a^{a+h}\sqrt{R^2-x^2}\sqrt{1+\frac{x^2}{R^2-x^2}}\,dx$
$= 2\pi\int_a^{a+h}\sqrt{(R^2-x^2)+x^2}\,dx = 2\pi R\int_a^{a+h}dx = 2\pi Rh$

30. (a) $x^2+y^2 = 45^2 \Rightarrow x = \sqrt{45^2-y^2} \Rightarrow \frac{dx}{dy} = \frac{-y}{\sqrt{45^2-y^2}} \Rightarrow \left(\frac{dx}{dy}\right)^2 = \frac{y^2}{45^2-y^2}$;
$S = \int_{-22.5}^{45} 2\pi\sqrt{45^2-y^2}\sqrt{1+\frac{y^2}{45^2-y^2}}\,dy = 2\pi\int_{-22.5}^{45}\sqrt{(45^2-y^2)+y^2}\,dy = 2\pi \cdot 45\int_{-22.5}^{45}dy$
$= (2\pi)(45)(67.5) = 6075\pi$ square feet

(b) 19,085 square feet

31. (a) $y = x \Rightarrow \left(\frac{dy}{dx}\right) = 1 \Rightarrow \left(\frac{dy}{dx}\right)^2 = 1 \Rightarrow S = 2\pi\int_{-1}^{2}|x|\sqrt{1+1}\,dx = 2\pi\int_{-1}^{0}(-x)\sqrt{2}\,dx + 2\pi\int_0^2 x\sqrt{2}\,dx$
$= -2\sqrt{2}\pi\left[\frac{x^2}{2}\right]_{-1}^{0} + 2\sqrt{2}\pi\left[\frac{x^2}{2}\right]_0^2 = -2\sqrt{2}\pi\left(0-\frac{1}{2}\right) + 2\sqrt{2}\pi(2-0) = 5\sqrt{2}\pi$

(b) $\frac{dy}{dx} = \frac{x^2}{3} \Rightarrow \left(\frac{dy}{dx}\right)^2 = \frac{x^4}{9} \Rightarrow$ by symmetry of the graph that $S = 2\int_{-\sqrt{3}}^{0} 2\pi\left(-\frac{x^3}{9}\right)\sqrt{1+\frac{x^4}{9}}\,dx$; $\left[u = 1+\frac{x^4}{9}\right.$
$\Rightarrow du = \frac{4}{9}x^3\,dx \Rightarrow -\frac{1}{4}\,du = -\frac{x^3}{9}\,dx$; $x = -\sqrt{3} \Rightarrow u = 2, x = 0 \Rightarrow u = 1\Big] \to S = 4\pi\int_2^1 u^{1/2}\left(-\frac{1}{4}\right)du$
$= -\pi\int_2^1 u^{1/2}\,du = -\pi\left[\frac{2}{3}u^{3/2}\right]_2^1 = -\pi\left(\frac{2}{3}-\frac{2}{3}\sqrt{8}\right) = \frac{2\pi}{3}\left(\sqrt{8}-1\right)$. If the absolute value bars are dropped the integral for $S = \int_{-\sqrt{3}}^{\sqrt{3}} 2\pi f(x)\,ds$ will equal zero since $\int_{-\sqrt{3}}^{\sqrt{3}} 2\pi\left(\frac{x^3}{9}\right)\sqrt{1+\frac{x^4}{9}}\,dx$ is the integral of an odd function over the symmetric interval $-\sqrt{3} \le x \le \sqrt{3}$.

32. $y = \left(1-x^{2/3}\right)^{3/2} \Rightarrow \frac{dy}{dx} = \frac{3}{2}\left(1-x^{2/3}\right)^{1/2}\left(-\frac{2}{3}x^{-1/3}\right) = -\frac{\left(1-x^{2/3}\right)^{1/2}}{x^{1/3}} \Rightarrow \left(\frac{dy}{dx}\right)^2 = \frac{1-x^{2/3}}{x^{2/3}} = \frac{1}{x^{2/3}} - 1$
$\Rightarrow S = 2\int_0^1 2\pi\left(1-x^{2/3}\right)^{3/2}\sqrt{1+\left(\frac{1}{x^{2/3}}-1\right)}\,dx = 4\pi\int_0^1\left(1-x^{2/3}\right)^{3/2}\sqrt{x^{-2/3}}\,dx$
$= 4\pi\int_0^1\left(1-x^{2/3}\right)^{3/2}x^{-1/3}\,dx$; $\left[u = 1-x^{2/3} \Rightarrow du = -\frac{2}{3}x^{-1/3}\,dx \Rightarrow -\frac{3}{2}\,du = x^{-1/3}\,dx;\right.$
$x = 0 \Rightarrow u = 1, x = 1 \Rightarrow u = 0] \to S = 4\pi\int_1^0 u^{3/2}\left(-\frac{3}{2}\,du\right) = -6\pi\left[\frac{2}{5}u^{5/2}\right]_1^0 = -6\pi\left(0-\frac{2}{5}\right) = \frac{12\pi}{5}$

33. $\frac{dx}{dt} = -\sin t$ and $\frac{dy}{dt} = \cos t \Rightarrow \sqrt{\left(\frac{dx}{dt}\right)^2+\left(\frac{dy}{dt}\right)^2} = \sqrt{(-\sin t)^2+(\cos t)^2} = 1 \Rightarrow S = \int 2\pi y\,ds$
$= \int_0^{2\pi} 2\pi(2+\sin t)(1)\,dt = 2\pi\left[2t-\cos t\right]_0^{2\pi} = 2\pi[(4\pi-1)-(0-1)] = 8\pi^2$

34. $\frac{dx}{dt} = t^{1/2}$ and $\frac{dy}{dt} = t^{-1/2} \Rightarrow \sqrt{\left(\frac{dx}{dt}\right)^2 + \left(\frac{dy}{dt}\right)^2} = \sqrt{t + t^{-1}} = \sqrt{\frac{t^2+1}{t}} \Rightarrow S = \int 2\pi x\, ds$
$= \int_0^{\sqrt{3}} 2\pi \left(\frac{2}{3} t^{3/2}\right) \sqrt{\frac{t^2+1}{t}}\, dt = \frac{4\pi}{3} \int_0^{\sqrt{3}} t\sqrt{t^2+1}\, dt$; $[u = t^2 + 1 \Rightarrow du = 2t\, dt;\ t = 0 \Rightarrow u = 1,$
$\left[t = \sqrt{3} \Rightarrow u = 4\right] \rightarrow \int_1^4 \frac{2\pi}{3} \sqrt{u}\, du = \left[\frac{4\pi}{9} u^{3/2}\right]_1^4 = \frac{28\pi}{9}$
Note: $\int_0^{\sqrt{3}} 2\pi \left(\frac{2}{3} t^{3/2}\right) \sqrt{\frac{t^2+1}{t}}\, dt$ is an improper integral but $\lim_{t \to 0^+} f(t)$ exists and is equal to 0, where
$f(t) = 2\pi \left(\frac{2}{3} t^{3/2}\right) \sqrt{\frac{t^2+1}{t}}$. Thus the discontinuity is removable: define $F(t) = f(t)$ for $t > 0$ and $F(0) = 0$
$\Rightarrow \int_0^{\sqrt{3}} F(t)\, dt = \frac{28\pi}{9}$.

35. $\frac{dx}{dt} = 1$ and $\frac{dy}{dt} = t + \sqrt{2} \Rightarrow \sqrt{\left(\frac{dx}{dt}\right)^2 + \left(\frac{dy}{dt}\right)^2} = \sqrt{1^2 + \left(t + \sqrt{2}\right)^2} = \sqrt{t^2 + 2\sqrt{2}t + 3} \Rightarrow S = \int 2\pi x\, ds$
$= \int_{-\sqrt{2}}^{\sqrt{2}} 2\pi \left(t + \sqrt{2}\right) \sqrt{t^2 + 2\sqrt{2}t + 3}\, dt$; $\left[u = t^2 + 2\sqrt{2}t + 3 \Rightarrow du = \left(2t + 2\sqrt{2}\right) dt;\ t = -\sqrt{2} \Rightarrow u = 1,\right.$
$\left. t = \sqrt{2} \Rightarrow u = 9\right] \rightarrow \int_1^9 \pi\sqrt{u}\, du = \left[\frac{2}{3}\pi u^{3/2}\right]_1^9 = \frac{2\pi}{3}(27 - 1) = \frac{52\pi}{3}$

36. $\frac{dx}{dt} = a(1 - \cos t)$ and $\frac{dy}{dt} = a \sin t \Rightarrow \sqrt{\left(\frac{dx}{dt}\right)^2 + \left(\frac{dy}{dt}\right)^2} = \sqrt{[a(1 - \cos t)]^2 + (a \sin t)^2}$
$= \sqrt{a^2 - 2a^2\cos t + a^2\cos^2 t + a^2\sin^2 t} = \sqrt{2a^2 - 2a^2\cos t} = a\sqrt{2}\sqrt{1 - \cos t} \Rightarrow S = \int 2\pi y\, ds$
$= \int_0^{2\pi} 2\pi\, a(1 - \cos t) \cdot a\sqrt{2}\sqrt{1 - \cos t}\, dt = 2\sqrt{2}\,\pi\, a^2 \int_0^{2\pi} (1 - \cos t)^{3/2}\, dt$

37. $S = 2\pi \int_0^1 y \sqrt{\left(\frac{dx}{dt}\right)^2 + \left(\frac{dy}{dt}\right)^2}\, dt = 2\pi \int_0^1 4e^{t/2} \sqrt{(e^t - 1)^2 + (2e^{t/2})^2}\, dt = 8\pi \int_0^1 e^{t/2} \sqrt{e^{2t} + 2e^t + 1}\, dt$
$= 8\pi \int_0^1 e^{t/2} \sqrt{(e^t + 1)^2}\, dt = 8\pi \int_0^1 e^{t/2}(e^t + 1)dt = 8\pi \int_0^1 \left(e^{3t/2} + e^{t/2}\right)dt = 8\pi\left(\frac{2}{3}e^{3t/2} + 2e^{t/2}\right)\Big|_0^1$
$= 8\pi\left(\left(\frac{2}{3}e^{3/2} + 2e^{1/2}\right) - \frac{8}{3}\right) = \frac{16\pi}{3}\left(e^{3/2} + 3e^{1/2} - 4\right)$

38. From Exercise 33 in Section 6.3, $\sqrt{\left(\frac{dx}{dt}\right)^2 + \left(\frac{dy}{dt}\right)^2} = \tan t \Rightarrow \text{Area} = \int 2\pi y\, ds = \int_0^{\pi/3} 2\pi \cos t \tan t\, dt = 2\pi \int_0^{\pi/3} \sin t\, dt$
$= 2\pi \left[-\cos t\right]_0^{\pi/3} = 2\pi\left[-\frac{1}{2} - (-1)\right] = \pi$

39. $\frac{dx}{dt} = 2$ and $\frac{dy}{dt} = 1 \Rightarrow \sqrt{\left(\frac{dx}{dt}\right)^2 + \left(\frac{dy}{dt}\right)^2} = \sqrt{2^2 + 1^2} = \sqrt{5} \Rightarrow S = \int 2\pi y\, ds = \int_0^1 2\pi(t + 1)\sqrt{5}\, dt$
$= 2\pi\sqrt{5}\left[\frac{t^2}{2} + t\right]_0^1 = 3\pi\sqrt{5}$. Check: slant height is $\sqrt{5} \Rightarrow$ Area is $\pi(1 + 2)\sqrt{5} = 3\pi\sqrt{5}$.

40. $\frac{dx}{dt} = h$ and $\frac{dy}{dt} = r \Rightarrow \sqrt{\left(\frac{dx}{dt}\right)^2 + \left(\frac{dy}{dt}\right)^2} = \sqrt{h^2 + r^2} \Rightarrow S = \int 2\pi y\, ds = \int_0^1 2\pi rt\sqrt{h^2 + r^2}\, dt$
$= 2\pi r\sqrt{h^2 + r^2} \int_0^1 t\, dt = 2\pi r\sqrt{h^2 + r^2} \left[\frac{t^2}{2}\right]_0^1 = \pi r\sqrt{h^2 + r^2}$. Check: slant height is $\sqrt{h^2 + r^2} \Rightarrow$ Area is $\pi r\sqrt{h^2 + r^2}$.

41. (a) An equation of the tangent line segment is (see figure) $y = f(m_k) + f'(m_k)(x - m_k)$. When $x = x_{k-1}$ we have

$r_1 = f(m_k) + f'(m_k)(x_{k-1} - m_k)$

$= f(m_k) + f'(m_k)\left(-\frac{\Delta x_k}{2}\right) = f(m_k) - f'(m_k)\,\frac{\Delta x_k}{2}$;

when $x = x_k$ we have

$r_2 = f(m_k) + f'(m_k)(x_k - m_k)$

$= f(m_k) + f'(m_k)\,\frac{\Delta x_k}{2}$;

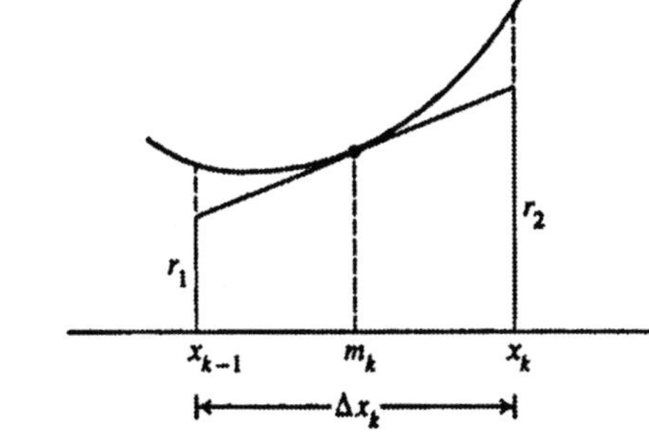

(b) $L_k^2 = (\Delta x_k)^2 + (r_2 - r_1)^2$

$= (\Delta x_k)^2 + \left[f'(m_k)\,\frac{\Delta x_k}{2} - \left(-f'(m_k)\,\frac{\Delta x_k}{2}\right)\right]^2$

$= (\Delta x_k)^2 + [f'(m_k)\Delta x_k]^2 \Rightarrow L_k = \sqrt{(\Delta x_k)^2 + [f'(m_k)\Delta x_k]^2}$, as claimed

(c) From geometry it is a fact that the lateral surface area of the frustum obtained by revolving the tangent line segment about the x-axis is given by $\Delta S_k = \pi(r_1 + r_2)L_k = \pi[2f(m_k)]\sqrt{(\Delta x_k)^2 + [f'(m_k)\Delta x_k]^2}$ using parts (a) and (b) above. Thus, $\Delta S_k = 2\pi f(m_k)\sqrt{1 + [f'(m_k)]^2}\,\Delta x_k$.

(d) $S = \lim\limits_{n\to\infty} \sum\limits_{k=1}^{n} \Delta S_k = \lim\limits_{n\to\infty} \sum\limits_{k=1}^{n} 2\pi f(m_k)\sqrt{1 + [f'(m_k)]^2}\,\Delta x_k = \int_a^b 2\pi f(x)\sqrt{1 + [f'(x)]^2}\,dx$

42. $S = \int_a^b 2\pi f(x)\,dx = \int_0^{\sqrt{3}} 2\pi \cdot \frac{x}{\sqrt{3}}\,dx = \frac{\pi}{\sqrt{3}}\left[x^2\right]_0^{\sqrt{3}} = \frac{3\pi}{\sqrt{3}} = \sqrt{3}\pi$

43. The centroid of the square is located at $(2, 2)$. The volume is $V = (2\pi)(\bar{y})(A) = (2\pi)(2)(8) = 32\pi$ and the surface area is $S = (2\pi)(\bar{y})(L) = (2\pi)(2)\left(4\sqrt{8}\right) = 32\sqrt{2}\pi$ (where $\sqrt{8}$ is the length of a side).

44. The midpoint of the hypotenuse of the triangle is $\left(\frac{3}{2}, 3\right)$ $\Rightarrow y = 2x$ is an equation of the median $\Rightarrow$ the line $y = 2x$ contains the centroid. The point $\left(\frac{3}{2}, 3\right)$ is $\frac{3\sqrt{5}}{2}$ units from the origin $\Rightarrow$ the x-coordinate of the centroid solves the equation $\sqrt{\left(x - \frac{3}{2}\right)^2 + (2x - 3)^2}$ $= \frac{\sqrt{5}}{2} \Rightarrow \left(x^2 - 3x + \frac{9}{4}\right) + (4x^2 - 12x + 9) = \frac{5}{4}$

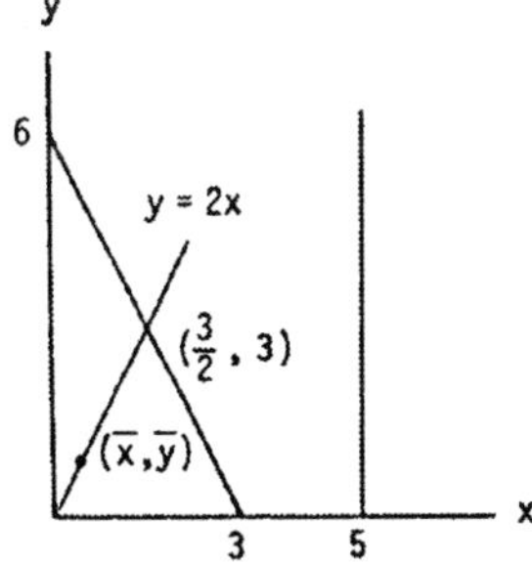

$\Rightarrow 5x^2 - 15x + 9 = -1$

$\Rightarrow x^2 - 3x + 2 = (x - 2)(x - 1) = 0 \Rightarrow \bar{x} = 1$ since the centroid must lie inside the triangle $\Rightarrow \bar{y} = 2$. By the Theorem of Pappus, the volume is $V =$ (distance traveled by the centroid)(area of the region) $= 2\pi(5 - \bar{x})\left[\frac{1}{2}(3)(6)\right]$ $= (2\pi)(4)(9) = 72\pi$

45. The centroid is located at $(2, 0) \Rightarrow V = (2\pi)(\bar{x})(A) = (2\pi)(2)(\pi) = 4\pi^2$

46. We create the cone by revolving the triangle with vertices $(0, 0)$, (h, r) and $(h, 0)$ about the x-axis (see the accompanying figure). Thus, the cone has height h and base radius r. By Theorem of Pappus, the lateral surface area swept out by the hypotenuse L is given by $S = 2\pi\bar{y}L = 2\pi\left(\frac{r}{2}\right)\sqrt{h^2 + r^2}$ $= \pi r\sqrt{r^2 + h^2}$. To calculate the volume we need the position of the centroid of the triangle. From the diagram we see that

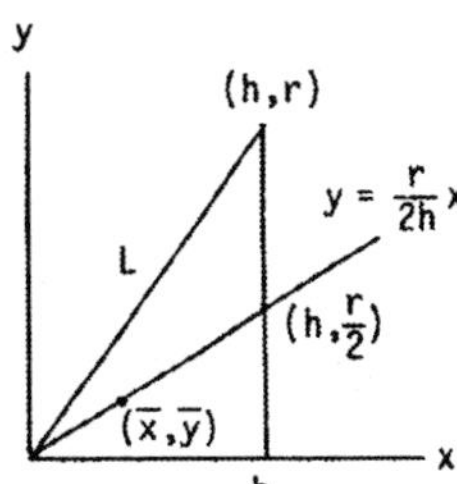

the centroid lies on the line $y = \frac{r}{2h}x$. The x-coordinate of the centroid solves the equation $\sqrt{(x - h)^2 + \left(\frac{r}{2h}x - \frac{r}{2}\right)^2}$ $= \frac{1}{3}\sqrt{h^2 + \frac{r^2}{4}} \Rightarrow \left(\frac{4h^2 + r^2}{4h^2}\right)x^2 - \left(\frac{4h^2 + r^2}{2h}\right)x + \frac{r^2}{4} + \frac{2(r^2 + 4h^2)}{9} = 0 \Rightarrow x = \frac{2h}{3}$ or $\frac{4h}{3} \Rightarrow \bar{x} = \frac{2h}{3}$, since the centroid must lie inside the triangle $\Rightarrow$ $\overline{y} = \frac{r}{2h}\,\overline{x} = \frac{r}{3}$. By the Theorem of Pappus, $V = \left[2\pi\left(\frac{r}{3}\right)\right]\left(\frac{1}{2}\,hr\right) = \frac{1}{3}\,\pi r^2 h$.

47. $S = 2\pi\,\overline{y}\,L \Rightarrow 4\pi a^2 = (2\pi\overline{y})(\pi a) \Rightarrow \overline{y} = \frac{2a}{\pi}$, and by symmetry $\overline{x} = 0$

48. $S = 2\pi\rho L \Rightarrow \left[2\pi\left(a - \frac{2a}{\pi}\right)\right](\pi a) = 2\pi a^2(\pi - 2)$

49. $V = 2\pi\,\overline{y}A \Rightarrow \frac{4}{3}\,\pi ab^2 = (2\pi\overline{y})\left(\frac{\pi ab}{2}\right) \Rightarrow \overline{y} = \frac{4b}{3\pi}$ and by symmetry $\overline{x} = 0$

50. $V = 2\pi\rho A \Rightarrow V = \left[2\pi\left(a + \frac{4a}{3\pi}\right)\right]\left(\frac{\pi a^2}{2}\right) = \frac{\pi a^3(3\pi+4)}{3}$

51. $V = 2\pi\rho\,A = (2\pi)$(area of the region) $\cdot$ (distance from the centroid to the line $y = x - a$). We must find the distance from $\left(0, \frac{4a}{3\pi}\right)$ to $y = x - a$. The line containing the centroid and perpendicular to $y = x - a$ has slope -1 and contains the point $\left(0, \frac{4a}{3\pi}\right)$. This line is $y = -x + \frac{4a}{3\pi}$. The intersection of $y = x - a$ and $y = -x + \frac{4a}{3\pi}$ is the point $\left(\frac{4a+3a\pi}{6\pi}, \frac{4a-3a\pi}{6\pi}\right)$. Thus, the distance from the centroid to the line $y = x - a$ is $\sqrt{\left(\frac{4a+3a\pi}{6\pi}\right)^2 + \left(\frac{4a}{3\pi} - \frac{4a}{6\pi} + \frac{3a\pi}{6\pi}\right)^2} = \frac{\sqrt{2}(4a+3a\pi)}{6\pi} \Rightarrow V = (2\pi)\left(\frac{\sqrt{2}(4a+3a\pi)}{6\pi}\right)\left(\frac{\pi a^2}{2}\right) = \frac{\sqrt{2}\,\pi a^3(4+3\pi)}{6}$

52. The line perpendicular to $y = x - a$ and passing through the centroid $\left(0, \frac{2a}{\pi}\right)$ has equation $y = -x + \frac{2a}{\pi}$. The intersection of the two perpendicular lines occurs when $x - a = -x + \frac{2a}{\pi} \Rightarrow x = \frac{2a+a\pi}{2\pi} \Rightarrow y = \frac{2a-a\pi}{2\pi}$. Thus the distance from the centroid to the line $y = x - a$ is $\sqrt{\left(\frac{2a+\pi a}{2} - 0\right)^2 + \left(\frac{2a-\pi a}{2} - \frac{2a}{2}\right)^2} = \frac{a(2+\pi)}{\sqrt{2\pi}}$.
Therefore, by the Theorem of Pappus the surface area is $S = 2\pi\left[\frac{a(2+\pi)}{\sqrt{2\pi}}\right](\pi a) = \sqrt{2}\pi a^2(2+\pi)$.

53. From Example 4 and Pappus's Theorem for Volumes we have the moment about the x-axis is $M_x = \overline{y}\,M$ $= \left(\frac{4a}{3\pi}\right)\left(\frac{\pi a^2}{2}\right) = \frac{2a^3}{3}$.

6.6 WORK

1. The force required to stretch the spring from its natural length of 2 m to a length of 5 m is $F(x) = kx$. The work done by F is $W = \int_0^3 F(x)\,dx = k\int_0^3 x\,dx = \frac{k}{2}\left[x^2\right]_0^3 = \frac{9k}{2}$. This work is equal to 1800 J $\Rightarrow \frac{9}{2}k = 1800$ $\Rightarrow k = 400$ N/m

2. (a) We find the force constant from Hooke's Law: $F = kx \Rightarrow k = \frac{F}{x} \Rightarrow k = \frac{800}{4} = 200$ lb/in.
   (b) The work done to stretch the spring 2 inches beyond its natural length is $W = \int_0^2 kx\,dx$ $= 200\int_0^2 x\,dx = 200\left[\frac{x^2}{2}\right]_0^2 = 200(2-0) = 400$ in $\cdot$ lb $= 33.3$ ft $\cdot$ lb
   (c) We substitute $F = 1600$ into the equation $F = 200x$ to find $1600 = 200x \Rightarrow x = 8$ in.

3. We find the force constant from Hooke's law: $F = kx$. A force of 2 N stretches the spring to 0.02 m $\Rightarrow 2 = k \cdot (0.02) \Rightarrow k = 100\,\frac{N}{m}$. The force of 4 N will stretch the rubber band y m, where $F = ky \Rightarrow y = \frac{F}{k}$ $\Rightarrow y = \frac{4N}{100\,\frac{N}{m}} \Rightarrow y = 0.04$ m $= 4$ cm. The work done to stretch the rubber band 0.04 m is $W = \int_0^{0.04} kx\,dx$ $= 100\int_0^{0.04} x\,dx = 100\left[\frac{x^2}{2}\right]_0^{0.04} = \frac{(100)(0.04)^2}{2} = 0.08$ J

4. We find the force constant from Hooke's law: $F = kx \Rightarrow k = \frac{F}{x} \Rightarrow k = \frac{90}{1} \Rightarrow k = 90\,\frac{N}{m}$. The work done to

stretch the spring 5 m beyond its natural length is $W = \int_0^5 kx\,dx = 90\int_0^5 x\,dx = 90\left[\frac{x^2}{2}\right]_0^5 = (90)\left(\frac{25}{2}\right) = 1125$ J

5. (a) We find the spring's constant from Hooke's law: $F = kx \Rightarrow k = \frac{F}{x} = \frac{21{,}714}{8-5} = \frac{21{,}714}{3} \Rightarrow k = 7238\ \frac{\text{lb}}{\text{in}}$

(b) The work done to compress the assembly the first half inch is $W = \int_0^{0.5} kx\,dx = 7238\int_0^{0.5} x\,dx$ $= 7238\left[\frac{x^2}{2}\right]_0^{0.5} = (7238)\frac{(0.5)^2}{2} = \frac{(7238)(0.25)}{2} \approx 905$ in · lb. The work done to compress the assembly the second half inch is: $W = \int_{0.5}^{1.0} kx\,dx = 7238\int_{0.5}^{1.0} x\,dx = 7238\left[\frac{x^2}{2}\right]_{0.5}^{1.0} = \frac{7238}{2}\left[1-(0.5)^2\right] = \frac{(7238)(0.75)}{2}$ ≈ 2714 in · lb

6. First, we find the force constant from Hooke's law: $F = kx \Rightarrow k = \frac{F}{x} = \frac{150}{\left(\frac{1}{16}\right)} = 16 \cdot 150 = 2{,}400\ \frac{\text{lb}}{\text{in}}$. If someone compresses the scale $x = \frac{1}{8}$ in, he/she must weigh $F = kx = 2{,}400\left(\frac{1}{8}\right) = 300$ lb. The work done to compress the scale this far is $W = \int_0^{1/8} kx\,dx = 2400\left[\frac{x^2}{2}\right]_0^{1/8} = \frac{2400}{2 \cdot 64} = 18.75$ lb · in. $= \frac{25}{16}$ ft · lb

7. The force required to haul up the rope is equal to the rope's weight, which varies steadily and is proportional to x, the length of the rope still hanging: $F(x) = 0.624x$. The work done is: $W = \int_0^{50} F(x)\,dx = \int_0^{50} 0.624x\,dx$ $= 0.624\left[\frac{x^2}{2}\right]_0^{50} = 780$ J

8. The weight of sand decreases steadily by 72 lb over the 18 ft, at 4 lb/ft. So the weight of sand when the bag is x ft off the ground is $F(x) = 144 - 4x$. The work done is: $W = \int_a^b F(x)\,dx = \int_0^{18}(144-4x)dx = \left[144x - 2x^2\right]_0^{18} = 1944$ ft · lb

9. The force required to lift the cable is equal to the weight of the cable paid out: $F(x) = (4.5)(180 - x)$ where x is the position of the car off the first floor. The work done is: $W = \int_0^{180} F(x)\,dx = 4.5\int_0^{180}(180-x)\,dx$ $= 4.5\left[180x - \frac{x^2}{2}\right]_0^{180} = 4.5\left(180^2 - \frac{180^2}{2}\right) = \frac{4.5 \cdot 180^2}{2} = 72{,}900$ ft · lb

10. Since the force is acting toward the origin, it acts opposite to the positive x-direction. Thus $F(x) = -\frac{k}{x^2}$. The work done is $W = \int_a^b -\frac{k}{x^2}\,dx = k\int_a^b -\frac{1}{x^2}\,dx = k\left[\frac{1}{x}\right]_a^b = k\left(\frac{1}{b} - \frac{1}{a}\right) = \frac{k(a-b)}{ab}$

11. The force against the piston is $F = pA$. If $V = Ax$, where x is the height of the cylinder, then $dV = A\,dx$ $\Rightarrow$ Work $= \int F\,dx = \int pA\,dx = \int_{(p_1,V_1)}^{(p_2,V_2)} p\,dV$.

12. $pV^{1.4} = c$, a constant $\Rightarrow p = cV^{-1.4}$. If $V_1 = 243$ in^3 and $p_1 = 50$ lb/in^3, then $c = (50)(243)^{1.4} = 109{,}350$ lb. Thus $W = \int_{243}^{32} 109{,}350V^{-1.4}\,dV = \left[-\frac{109{,}350}{0.4V^{0.4}}\right]_{243}^{32} = -\frac{109{,}350}{0.4}\left(\frac{1}{32^{0.4}} - \frac{1}{243^{0.4}}\right) = -\frac{109{,}350}{0.4}\left(\frac{1}{4} - \frac{1}{9}\right)$ $= -\frac{(109{,}350)(5)}{(0.4)(36)} = -37{,}968.75$ in · lb. Note that when a system is compressed, the work done by the system is negative.

13. Let r = the constant rate of leakage. Since the bucket is leaking at a constant rate and the bucket is rising at a constant rate, the amount of water in the bucket is proportional to $(20 - x)$, the distance the bucket is being raised. The leakage rate of the water is 0.8 lb/ft raised and the weight of the water in the bucket is $F = 0.8(20 - x)$. So:
$W = \int_0^{20} 0.8(20-x)\,dx = 0.8\left[20x - \frac{x^2}{2}\right]_0^{20} = 160$ ft · lb.

14. Let r = the constant rate of leakage. Since the bucket is leaking at a constant rate and the bucket is rising at a constant rate, the amount of water in the bucket is proportional to $(20 - x)$, the distance the bucket is being raised. The leakage rate of

the water is 2 lb/ft raised and the weight of the water in the bucket is $F = 2(20 - x)$. So:

$W = \int_0^{20} 2(20 - x)\, dx = 2\left[20x - \frac{x^2}{2}\right]_0^{20} = 400$ ft · lb.

Note that since the force in Exercise 14 is 2.5 times the force in Exercise 13 at each elevation, the total work is also 2.5 times as great.

15. We will use the coordinate system given.

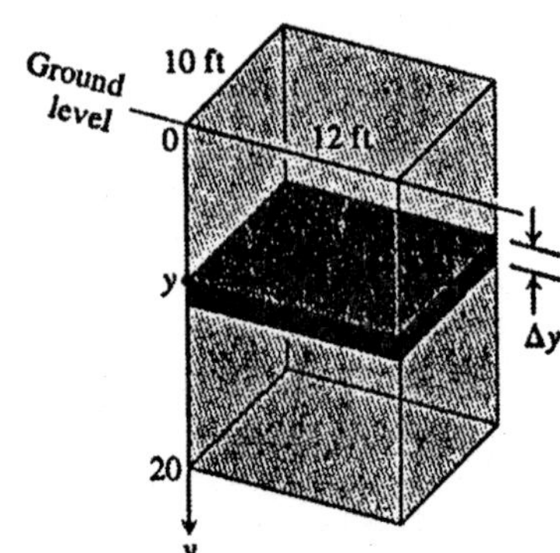

(a) The typical slab between the planes at y and $y + \Delta y$ has a volume of $\Delta V = (10)(12)\,\Delta y = 120\,\Delta y$ ft³. The force F required to lift the slab is equal to its weight: $F = 62.4\,\Delta V = 62.4 \cdot 120\,\Delta y$ lb. The distance through which F must act is about y ft, so the work done lifting the slab is about ΔW = force × distance $= 62.4 \cdot 120 \cdot y \cdot \Delta y$ ft · lb. The work it takes to lift all the water is approximately $W \approx \sum_0^{20} \Delta W = \sum_0^{20} 62.4 \cdot 120y \cdot \Delta y$ ft · lb. This is a Riemann sum for the function $62.4 \cdot 120y$ over the interval $0 \le y \le 20$. The work of pumping the tank empty is the limit of these sums:

$W = \int_0^{20} 62.4 \cdot 120y\, dy = (62.4)(120)\left[\frac{y^2}{2}\right]_0^{20} = (62.4)(120)\left(\frac{400}{2}\right) = (62.4)(120)(200) = 1{,}497{,}600$ ft · lb

(b) The time t it takes to empty the full tank with $\left(\frac{5}{11}\right)$–hp motor is $t = \frac{W}{250\,\frac{\text{ft·lb}}{\text{sec}}} = \frac{1{,}497{,}600\text{ ft·lb}}{250\,\frac{\text{ft·lb}}{\text{sec}}} = 5990.4$ sec $= 1.664$ hr $\Rightarrow t \approx 1$ hr and 40 min

(c) Following all the steps of part (a), we find that the work it takes to lower the water level 10 ft is

$W = \int_0^{10} 62.4 \cdot 120y\, dy = (62.4)(120)\left[\frac{y^2}{2}\right]_0^{10} = (62.4)(120)\left(\frac{100}{2}\right) = 374{,}400$ ft · lb and the time is $t = \frac{W}{250\,\frac{\text{ft·lb}}{\text{sec}}}$ $= 1497.6$ sec $= 0.416$ hr ≈ 25 min

(d) In a location where water weighs $62.26\ \frac{\text{lb}}{\text{ft}^3}$:

a) $W = (62.26)(24{,}000) = 1{,}494{,}240$ ft · lb.

b) $t = \frac{1{,}494{,}240}{250} = 5976.96$ sec ≈ 1.660 hr $\Rightarrow t \approx 1$ hr and 40 min

In a location where water weighs $62.59\ \frac{\text{lb}}{\text{ft}^3}$

a) $W = (62.59)(24{,}000) = 1{,}502{,}160$ ft · lb

b) $t = \frac{1{,}502{,}160}{250} = 6008.64$ sec ≈ 1.669 hr $\Rightarrow t \approx 1$ hr and 40.1 min

16. We will use the coordinate system given.

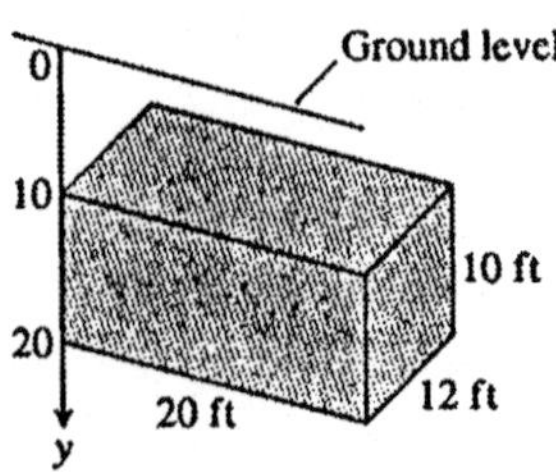

(a) The typical slab between the planes at y and $y + \Delta y$ has a volume of $\Delta V = (20)(12)\,\Delta y = 240\,\Delta y$ ft³. The force F required to lift the slab is equal to its weight: $F = 62.4\,\Delta V = 62.4 \cdot 240\,\Delta y$ lb. The distance through which F must act is about y ft, so the work done lifting the slab is about ΔW = force × distance $= 62.4 \cdot 240 \cdot y \cdot \Delta y$ ft · lb. The work it takes to lift all the water is approximately $W \approx \sum_{10}^{20} \Delta W = \sum_{10}^{20} 62.4 \cdot 240y \cdot \Delta y$ ft · lb. This is a Riemann sum for the function $62.4 \cdot 240y$ over the interval $10 \le y \le 20$. The work it takes to empty the cistern is the limit of these sums: $W = \int_{10}^{20} 62.4 \cdot 240y\, dy$ $= (62.4)(240)\left[\frac{y^2}{2}\right]_{10}^{20} = (62.4)(240)(200 - 50) = (62.4)(240)(150) = 2{,}246{,}400$ ft · lb

(b) $t = \frac{W}{275\,\frac{\text{ft·lb}}{\text{sec}}} = \frac{2{,}246{,}400\text{ ft·lb}}{275} \approx 8168.73$ sec ≈ 2.27 hours ≈ 2 hr and 16.1 min

(c) Following all the steps of part (a), we find that the work it takes to empty the tank halfway is
$W = \int_{10}^{15} 62.4 \cdot 240y\, dy = (62.4)(240)\left[\frac{y^2}{2}\right]_{10}^{15} = (62.4)(240)\left(\frac{225}{2} - \frac{100}{2}\right) = (62.4)(240)\left(\frac{125}{2}\right) = 936{,}000$ ft.
Then the time is $t = \frac{W}{275\,\frac{\text{ft}\cdot\text{lb}}{\text{sec}}} = \frac{936{,}000}{275} \approx 3403.64$ sec ≈ 56.7 min

(d) In a location where water weighs $62.26\ \frac{\text{lb}}{\text{ft}^3}$:
a) $W = (62.26)(240)(150) = 2{,}241{,}360$ ft · lb.
b) $t = \frac{2{,}241{,}360}{275} = 8150.40$ sec $= 2.264$ hours ≈ 2 hr and 15.8 min
c) $W = (62.26)(240)\left(\frac{125}{2}\right) = 933{,}900$ ft · lb; $t = \frac{933{,}900}{275} = 3396$ sec ≈ 0.94 hours ≈ 56.6 min
In a location where water weighs $62.59\ \frac{\text{lb}}{\text{ft}^3}$
a) $W = (62.59)(240)(150) = 2{,}253{,}240$ ft · lb.
b) $t = \frac{2{,}253{,}240}{275} = 8193.60$ sec $= 2.276$ hours ≈ 2 hr and 16.56 min
c) $W = (62.59)(240)\left(\frac{125}{2}\right) = 938{,}850$ ft · lb; $t = \frac{938{,}850}{275} \approx 3414$ sec ≈ 0.95 hours ≈ 56.9 min

17. The slab is a disk of area $\pi x^2 = \pi\left(\frac{y}{2}\right)^2$, thickness $\triangle y$, and height below the top of the tank $(10 - y)$. So the work to pump the oil in this slab, $\triangle W$, is $57(10 - y)\pi\left(\frac{y}{2}\right)^2$. The work to pump all the oil to the top of the tank is
$W = \int_0^{10} \frac{57\pi}{4}(10y^2 - y^3)dy = \frac{57\pi}{4}\left[\frac{10y^3}{3} - \frac{y^4}{4}\right]_0^{10} = 11{,}875\pi$ ft · lb $\approx 37{,}306$ ft · lb.

18. Each slab of oil is to be pumped to a height of 14 ft. So the work to pump a slab is $(14 - y)(\pi)\left(\frac{y}{2}\right)^2$ and since the tank is half full and the volume of the original cone is $V = \frac{1}{3}\pi r^2 h = \frac{1}{3}\pi(5^2)(10) = \frac{250\pi}{3}$ ft^3, half the volume $= \frac{250\pi}{6}$ ft^3, and with half the volume the cone is filled to a height y, $\frac{250\pi}{6} = \frac{1}{3}\pi\frac{y^2}{4}y \Rightarrow y = \sqrt[3]{500}$ ft. So $W = \int_0^{\sqrt[3]{500}} \frac{57\pi}{4}(14y^2 - y^3)\, dy$
$= \frac{57\pi}{4}\left[\frac{14y^3}{3} - \frac{y^4}{4}\right]_0^{\sqrt[3]{500}} \approx 60{,}042$ ft · lb.

19. The typical slab between the planes at y and and $y + \Delta y$ has a volume of $\Delta V = \pi(\text{radius})^2(\text{thickness}) = \pi\left(\frac{20}{2}\right)^2 \Delta y$ $= \pi \cdot 100\ \Delta y$ ft^3. The force F required to lift the slab is equal to its weight: $F = 51.2\,\Delta V = 51.2 \cdot 100\pi\,\Delta y$ lb $\Rightarrow F = 5120\pi\,\Delta y$ lb. The distance through which F must act is about $(30 - y)$ ft. The work it takes to lift all the kerosene is approximately $W \approx \sum_0^{30} \Delta W = \sum_0^{30} 5120\pi(30 - y)\,\Delta y$ ft · lb which is a Riemann sum. The work to pump the tank dry is the limit of these sums: $W = \int_0^{30} 5120\pi(30 - y)\, dy = 5120\pi\left[30y - \frac{y^2}{2}\right]_0^{30} = 5120\pi\left(\frac{900}{2}\right) = (5120)(450\pi)$
$\approx 7{,}238{,}229.48$ ft · lb

20. (Alternate Solution) Each method must pump all of the water the 15 ft to the base of the tank. Pumping to the rim requires the water to be pumped an additional 6 feet. Pumping into the bottom requires that the water be pumped an average of 3 additional feet. Thus pumping through the valve requires $\sqrt{3}$ ft$(4\pi)6$ ft^3(62.4 lb/ft^3) $\approx 14{,}115$ ft · lb less work and thus less time.

21. (a) Follow all the steps of Example 5 but make the substitution of $64.5\ \frac{\text{lb}}{\text{ft}^3}$ for $57\ \frac{\text{lb}}{\text{ft}^3}$. Then,
$W = \int_0^8 \frac{64.5\pi}{4}(10 - y)y^2\, dy = \frac{64.5\pi}{4}\left[\frac{10y^3}{3} - \frac{y^4}{4}\right]_0^8 = \frac{64.5\pi}{4}\left(\frac{10\cdot 8^3}{3} - \frac{8^4}{4}\right) = \left(\frac{64.5\pi}{4}\right)(8^3)\left(\frac{10}{3} - 2\right)$
$= \frac{64.5\pi\cdot 8^3}{3} = 21.5\pi \cdot 8^3 \approx 34{,}582.65$ ft · lb

(b) Exactly as done in Example 5 but change the distance through which F acts to distance $\approx (13 - y)$ ft. Then
$W = \int_0^8 \frac{57\pi}{4}(13 - y)y^2\, dy = \frac{57\pi}{4}\left[\frac{13y^3}{3} - \frac{y^4}{4}\right]_0^8 = \frac{57\pi}{4}\left(\frac{13\cdot 8^3}{3} - \frac{8^4}{4}\right) = \left(\frac{57\pi}{4}\right)(8^3)\left(\frac{13}{3} - 2\right) = \frac{57\pi\cdot 8^3\cdot 7}{3\cdot 4}$
$= (19\pi)(8^2)(7)(2) \approx 53.482.5$ ft · lb

22. The typical slab between the planes of y and y+Δy has a volume of about $\Delta V = \pi(\text{radius})^2(\text{thickness})$ $= \pi\left(\sqrt{y}\right)^2 \Delta y = xy\,\Delta y\ m^3$. The force F(y) is equal to the slab's weight: $F(y) = 10{,}000\ \frac{N}{m^3} \cdot \Delta V$ $= \pi 10{,}000 y\,\Delta y$ N. The height of the tank is $4^2 = 16$ m. The distance through which F(y) must act to lift the slab to the level of the top of the tank is about $(16 - y)$ m, so the work done lifting the slab is about $\Delta W = 10{,}000\pi y(16 - y)\,\Delta y\ N \cdot m$. The work done lifting all the slabs from y = 0 to y = 16 to the top is approximately $W \approx \sum_{0}^{16} 10{,}000\pi y(16 - y)\Delta y$. Taking the limit of these Riemann sums, we get

$$W = \int_0^{16} 10{,}000\pi y(16 - y)\,dy = 10{,}000\pi \int_0^{16} (16y - y^2)\,dy = 10{,}000\pi \left[\frac{16y^2}{2} - \frac{y^3}{3}\right]_0^{16} = 10{,}000\pi\left(\frac{16^3}{2} - \frac{16^3}{3}\right)$$

$$= \frac{10{,}000 \cdot \pi \cdot 16^3}{6} \approx 21{,}446{,}605.9 \text{ J}$$

23. The typical slab between the planes at y and y+Δy has a volume of about $\Delta V = \pi(\text{radius})^2(\text{thickness})$ $= \pi\left(\sqrt{25 - y^2}\right)^2 \Delta y\ m^3$. The force F(y) required to lift this slab is equal to its weight: $F(y) = 9800 \cdot \Delta V$ $= 9800\pi\left(\sqrt{25 - y^2}\right)^2 \Delta y = 9800\pi\,(25 - y^2)\,\Delta y$ N. The distance through which F(y) must act to lift the slab to the level of 4 m above the top of the reservoir is about $(4 - y)$ m, so the work done is approximately $\Delta W \approx 9800\pi\,(25 - y^2)(4 - y)\,\Delta y\ N \cdot m$. The work done lifting all the slabs from $y = -5$ m to $y = 0$ m is approximately $W \approx \sum_{-5}^{0} 9800\pi\,(25 - y^2)(4 - y)\,\Delta y\ N \cdot m$. Taking the limit of these Riemann sums, we get

$$W = \int_{-5}^{0} 9800\pi\,(25 - y^2)(4 - y)\,dy = 9800\pi \int_{-5}^{0} (100 - 25y - 4y^2 + y^3)\,dy = 9800\pi\left[100y - \frac{25}{2}y^2 - \frac{4}{3}y^3 + \frac{y^4}{4}\right]_{-5}^{0}$$

$$= -9800\pi\left(-500 - \frac{25 \cdot 25}{2} + \frac{4}{3} \cdot 125 + \frac{625}{4}\right) \approx 15{,}073{,}099.75 \text{ J}$$

24. The typical slab between the planes at y and y+Δy has a volume of about $\Delta V = \pi(\text{radius})^2(\text{thickness})$ $= \pi\left(\sqrt{100 - y^2}\right)^2 \Delta y = \pi\,(100 - y^2)\,\Delta y\ ft^3$. The force is $F(y) = \frac{56\text{ lb}}{ft^3} \cdot \Delta V = 56\pi\,(100 - y^2)\,\Delta y$ lb. The distance through which F(y) must act to lift the slab to the level of 2 ft above the top of the tank is about $(12 - y)$ ft, so the work done is $\Delta W \approx 56\pi\,(100 - y^2)(12 - y)\,\Delta y\ lb \cdot ft$. The work done lifting all the slabs from y = 0 ft to y = 10 ft is approximately $W \approx \sum_{0}^{10} 56\pi\,(100 - y^2)(12 - y)\,\Delta y\ lb \cdot ft$. Taking the limit of these Riemann sums, we get $W = \int_0^{10} 56\pi\,(100 - y^2)(12 - y)\,dy = 56\pi \int_0^{10} (100 - y^2)(12 - y)\,dy$

$$= 56\pi \int_0^{10} (1200 - 100y - 12y^2 + y^3)\,dy = 56\pi\left[1200y - \frac{100y^2}{2} - \frac{12y^3}{3} + \frac{y^4}{4}\right]_0^{10}$$

$$= 56\pi\left(12{,}000 - \frac{10{,}000}{2} - 4 \cdot 1000 + \frac{10{,}000}{4}\right) = (56\pi)\left(12 - 5 - 4 + \frac{5}{2}\right)(1000) \approx 967{,}611 \text{ ft} \cdot \text{lb}.$$

It would cost (0.5)(967,611) = 483,805¢ = \$4838.05. Yes, you can afford to hire the firm.

25. $F = m\frac{dv}{dt} = mv\frac{dv}{dx}$ by the chain rule $\Rightarrow W = \int_{x_1}^{x_2} mv\frac{dv}{dx}\,dx = m\int_{x_1}^{x_2}\left(v\frac{dv}{dx}\right)dx = m\left[\frac{1}{2}v^2(x)\right]_{x_1}^{x_2}$ $= \frac{1}{2}m\left[v^2(x_2) - v^2(x_1)\right] = \frac{1}{2}mv_2^2 - \frac{1}{2}mv_1^2$, as claimed.

26. weight $= 2$ oz $= \frac{2}{16}$ lb; mass $= \frac{\text{weight}}{32} = \frac{\frac{1}{8}}{32} = \frac{1}{256}$ slugs; $W = \left(\frac{1}{2}\right)\left(\frac{1}{256}\text{ slugs}\right)(160\text{ ft/sec})^2 \approx 50$ ft · lb

27. 90 mph $= \frac{90\text{ mi}}{1\text{ hr}} \cdot \frac{1\text{ hr}}{60\text{ min}} \cdot \frac{1\text{ min}}{60\text{ sec}} \cdot \frac{5280\text{ ft}}{1\text{ mi}} = 132$ ft/sec; $m = \frac{0.3125\text{ lb}}{32\text{ ft/sec}^2} = \frac{0.3125}{32}$ slugs; $W = \left(\frac{1}{2}\right)\left(\frac{0.3125\text{ lb}}{32\text{ ft/sec}^2}\right)(132\text{ ft/sec})^2 \approx 85.1$ ft · lb

28. weight $= 1.6$ oz $= 0.1$ lb $\Rightarrow m = \frac{0.1\text{ lb}}{32\text{ ft/sec}^2} = \frac{1}{320}$ slugs; $W = \left(\frac{1}{2}\right)\left(\frac{1}{320}\text{ slugs}\right)(280\text{ ft/sec})^2 = 122.5$ ft · lb

29. weight $= 2$ oz $= \frac{1}{8}$ lb $\Rightarrow m = \frac{\frac{1}{8}}{32}$ slugs $= \frac{1}{256}$ slugs; 124 mph $= \frac{(124)(5280)}{(60)(60)} \approx 181.87$ ft/sec; $W = \left(\frac{1}{2}\right)\left(\frac{1}{256}\text{ slugs}\right)(181.87\text{ ft/sec})^2 \approx 64.6$ ft · lb

30. weight $= 14.5$ oz $= \frac{14.5}{16}$ lb $\Rightarrow$ m $= \frac{14.5}{(16)(32)}$ slugs; W $= \left(\frac{1}{2}\right)\left(\frac{14.5}{(16)(32)}\text{ slugs}\right)(88\text{ ft/sec})^2 \approx 109.7$ ft · lb

31. weight $= 6.5$ oz $= \frac{6.5}{16}$ lb $\Rightarrow$ m $= \frac{6.5}{(16)(32)}$ slugs; W $= \left(\frac{1}{2}\right)\left(\frac{6.5}{(16)(32)}\text{ slugs}\right)(132\text{ ft/sec})^2 \approx 110.6$ ft · lb

32. $F = (18\text{ lb/ft})x \Rightarrow W = \int_0^{1/6} 18x\,dx = \left[9x^2\right]_0^{1/6} = \frac{1}{4}$ ft · lb. Now $W = \frac{1}{2}mv^2 - \frac{1}{2}mv_1^2$, where $W = \frac{1}{4}$ ft · lb, $m = \frac{\frac{1}{8}}{32} = \frac{1}{256}$ slugs and $v_1 = 0$ ft/sec. Thus, $\frac{1}{4}$ ft · lb. $= \left(\frac{1}{2}\right)\left(\frac{1}{256}\text{ slugs}\right)v^2 \Rightarrow v = 8\sqrt{2}$ ft/sec. With $v = 0$ at the top of the bearing's path and $v = 8\sqrt{2} - 32t \Rightarrow t = \frac{\sqrt{2}}{4}$ sec when the bearing is at the top of its path. The height the bearing reaches is $s = 8\sqrt{2}\,t - 16t^2 \Rightarrow$ at $t = \frac{\sqrt{2}}{4}$ the bearing reaches a height of $\left(8\sqrt{2}\right)\left(\frac{\sqrt{2}}{4}\right) - (16)\left(\frac{\sqrt{2}}{4}\right)^2 = 2$ ft

33. (a) From the diagram,

$r(y) = 60 - x = 60 - \sqrt{50^2 - (y-325)^2}$

for $325 \le y \le 375$ ft.

(b) The volume of a horizontal slice of the funnel is $\Delta V \approx \pi\left[r(y)\right]^2 \Delta y$

$= \pi\left[60 - \sqrt{50^2 - (y-325)^2}\right]^2 \Delta y$

(c) The work required to lift the single slice of water is $\Delta W \approx 62.4\Delta V(375 - y)$

$= 62.4(375 - y)\pi\left[60 - \sqrt{50^2 - (y-325)^2}\right]^2 \Delta y.$

The total work to pump our the funnel is W

$= \int_{325}^{375} 62.4(375 - y)\pi\left[60 - \sqrt{50^2 - (y-325)^2}\right]^2 dy$

$\approx 6.3358 \cdot 10^7$ ft · lb.

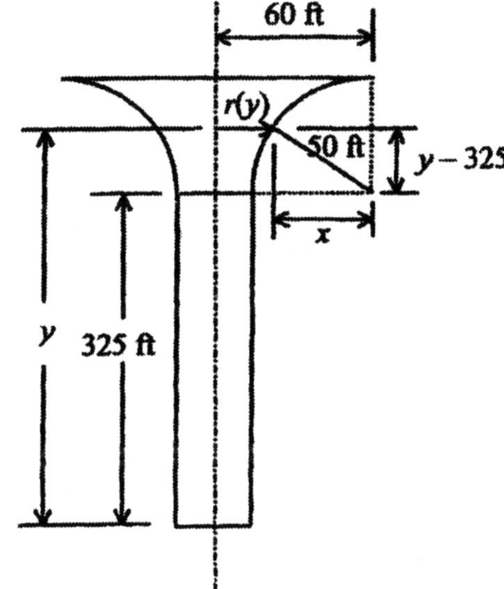

34. (a) From the result in Example 6, the work to pump out the throat is 1,353,869,354 ft · lb. Therefore, the total work required to pump out the throat and the funnel is $1{,}353{,}869{,}354 + 63{,}358{,}000 = 1{,}417227{,}354$ ft · lb.

(b) In horsepower-hours, the work required to pump out the glory hole is $\frac{1{,}417227{,}354}{1.98 \cdot 10^6} = 715.8$. Therefore, it would take $\frac{715.8\text{ hp}\cdot\text{h}}{1000\text{ hp}} = 0.7158$ hours ≈ 43 minutes.

35. We imagine the milkshake divided into thin slabs by planes perpendicular to the y-axis at the points of a partition of the interval $[0, 7]$. The typical slab between the planes at y and $y + \Delta y$ has a volume of about $\Delta V = \pi(\text{radius})^2(\text{thickness}) = \pi\left(\frac{y+17.5}{14}\right)^2 \Delta y\ \text{in}^3$. The force F(y) required to lift this slab is equal to its weight: $F(y) = \frac{4}{9}\,\Delta V = \frac{4\pi}{9}\left(\frac{y+17.5}{14}\right)^2 \Delta y$ oz. The distance through which F(y) must act to lift this slab to the level of 1 inch above the top is about $(8 - y)$ in. The work done lifting the slab is about $\Delta W = \left(\frac{4\pi}{9}\right)\frac{(y+17.5)^2}{14^2}(8 - y)\,\Delta y$ in · oz. The work done lifting all the slabs from $y = 0$ to $y = 7$ is approximately $W = \sum_0^7 \frac{4\pi}{9\cdot 14^2}(y + 17.5)^2(8 - y)\,\Delta y$ in · oz which is a Riemann sum. The work is the limit of these sums as the norm of the partition goes to zero: $W = \int_0^7 \frac{4\pi}{9\cdot 14^2}(y + 17.5)^2(8 - y)\,dy$

$= \frac{4\pi}{9\cdot 14^2}\int_0^7 (2450 - 26.25y - 27y^2 - y^3)\,dy = \frac{4\pi}{9\cdot 14^2}\left[-\frac{y^4}{4} - 9y^3 - \frac{26.25}{2}y^2 + 2450y\right]_0^7$

$= \frac{4\pi}{9\cdot 14^2}\left[-\frac{7^4}{4} - 9\cdot 7^3 - \frac{26.25}{2}\cdot 7^2 + 2450\cdot 7\right] \approx 91.32$ in · oz

36. We fill the pipe and the tank. To find the work required to fill the tank follow Example 6 with radius = 10 ft. Then $\Delta V = \pi \cdot 100\,\Delta y$ ft^3. The force required will be $F = 62.4 \cdot \Delta V = 62.4 \cdot 100\pi\,\Delta y = 6240\pi\,\Delta y$ lb. The distance through which F must act is y so the work done lifting the slab is about $\Delta W_1 = 6240\pi \cdot y \cdot \Delta y$ lb · ft. The work it takes to lift all the water into the tank is: $W_1 \approx \sum_{360}^{385} \Delta W_1 = \sum_{360}^{385} 6240\pi \cdot y \cdot \Delta y$ lb · ft. Taking the limit we end up with $W_1 = \int_{360}^{385} 6240\pi y\,dy = 6240\pi \left[\frac{y^2}{2}\right]_{360}^{385} = \frac{6240\pi}{2}\left[385^2 - 360^2\right] \approx 182{,}557{,}949$ ft · lb
To find the work required to fill the pipe, do as above, but take the radius to be $\frac{4}{2}$ in $= \frac{1}{6}$ ft.
Then $\Delta V = \pi \cdot \frac{1}{36}\,\Delta y$ ft^3 and $F = 62.4 \cdot \Delta V = \frac{62.4\pi}{36}\,\Delta y$. Also take different limits of summation and integration: $W_2 \approx \sum_{0}^{360} \Delta W_2 \Rightarrow W_2 = \int_0^{360} \frac{62.4}{36}\pi y\,dy = \frac{62.4\pi}{36}\left[\frac{y^2}{2}\right]_0^{360} = \left(\frac{62.4\pi}{36}\right)\left(\frac{360^2}{2}\right) \approx 352{,}864$ ft · lb.
The total work is $W = W_1 + W_2 \approx 182{,}557{,}949 + 352{,}864 \approx 182{,}910{,}813$ ft · lb. The time it takes to fill the tank and the pipe is Time $= \frac{W}{1650} \approx \frac{182{,}910{,}813}{1650} \approx 110{,}855$ sec ≈ 31 hr

37. Work $= \int_{6{,}370{,}000}^{35{,}780{,}000} \frac{1000\,MG}{r^2}\,dr = 1000\,MG \int_{6{,}370{,}000}^{35{,}780{,}000} \frac{dr}{r^2} = 1000\,MG\left[-\frac{1}{r}\right]_{6{,}370{,}000}^{35{,}780{,}000}$
$= (1000)(5.975 \cdot 10^{24})(6.672 \cdot 10^{-11})\left(\frac{1}{6{,}370{,}000} - \frac{1}{35{,}780{,}000}\right) \approx 5.144 \times 10^{10}$ J

38. (a) Let ρ be the x-coordinate of the second electron. Then $r^2 = (\rho - 1)^2 \Rightarrow W = \int_{-1}^{0} F(\rho)\,d\rho$
$= \int_{-1}^{0} \frac{(23\times 10^{-29})}{(\rho-1)^2}\,d\rho = -\left[\frac{23\times 10^{-29}}{\rho - 1}\right]_{-1}^{0} = (23 \times 10^{-29})\left(1 - \frac{1}{2}\right) = 11.5 \times 10^{-29}$
(b) $W = W_1 + W_2$ where W_1 is the work done against the field of the first electron and W_2 is the work done against the field of the second electron. Let ρ be the x-coordinate of the third electron. Then $r_1^2 = (\rho - 1)^2$ and $r_2^2 = (\rho + 1)^2 \Rightarrow W_1 = \int_3^5 \frac{23\times 10^{-29}}{r_1^2}\,d\rho = \int_3^5 \frac{23\times 10^{-29}}{(\rho-1)^2}\,d\rho = -23 \times 10^{-29}\left[\frac{1}{\rho-1}\right]_3^5$
$= (-23 \times 10^{-29})\left(\frac{1}{4} - \frac{1}{2}\right) = \frac{23}{4} \times 10^{-29}$, and $W_2 = \int_3^5 \frac{23\times 10^{-29}}{r_2^2}\,d\rho = \int_3^5 \frac{23\times 10^{-29}}{(\rho+1)^2}\,d\rho$
$= -23 \times 10^{-29}\left[\frac{1}{\rho+1}\right]_3^5 = (-23 \times 10^{-29})\left(\frac{1}{6} - \frac{1}{4}\right) = \frac{23\times 10^{-29}}{12}(3-2) = \frac{23}{12} \times 10^{-29}$. Therefore
$W = W_1 + W_2 = \left(\frac{23}{4} \times 10^{-29}\right) + \left(\frac{23}{12} \times 10^{-29}\right) = \frac{23}{3} \times 10^{-29} \approx 7.67 \times 10^{-29}$ J

6.7 FLUID PRESSURES AND FORCES

1. To find the width of the plate at a typical depth y, we first find an equation for the line of the plate's right-hand edge: $y = x - 5$. If we let x denote the width of the right-hand half of the triangle at depth y, then $x = 5 + y$ and the total width is $L(y) = 2x = 2(5 + y)$. The depth of the strip is $(-y)$. The force exerted by the water against one side of the plate is therefore $F = \int_{-5}^{-2} w(-y) \cdot L(y)\,dy = \int_{-5}^{-2} 62.4 \cdot (-y) \cdot 2(5 + y)\,dy$
$= 124.8 \int_{-5}^{-2} (-5y - y^2)\,dy = 124.8\left[-\frac{5}{2}y^2 - \frac{1}{3}y^3\right]_{-5}^{-2} = 124.8\left[\left(-\frac{5}{2}\cdot 4 + \frac{1}{3}\cdot 8\right) - \left(-\frac{5}{2}\cdot 25 + \frac{1}{3}\cdot 125\right)\right]$
$= (124.8)\left(\frac{105}{2} - \frac{117}{3}\right) = (124.8)\left(\frac{315 - 234}{6}\right) = 1684.8$ lb

2. An equation for the line of the plate's right-hand edge is $y = x - 3 \Rightarrow x = y + 3$. Thus the total width is $L(y) = 2x = 2(y + 3)$. The depth of the strip is $(2 - y)$. The force exerted by the water is
$F = \int_{-3}^{0} w(2 - y)L(y)\,dy = \int_{-3}^{0} 62.4 \cdot (2 - y) \cdot 2(3 + y)\,dy = 124.8\int_{-3}^{0} (6 - y - y^2)\,dy = 124.8\left[6y - \frac{y^2}{2} - \frac{y^3}{3}\right]_{-3}^{0}$
$= (-124.8)\left(-18 - \frac{9}{2} + 9\right) = (-124.8)\left(-\frac{27}{2}\right) = 1684.8$ lb

3. Using the coordinate system of Exercise 4, we find the equation for the line of the plate's right-hand edge is $y = x - 3 \Rightarrow x = y + 3$. Thus the total width is $L(y) = 2x = 2(y + 3)$. The depth of the strip changes to $(4 - y)$
$\Rightarrow F = \int_{-3}^{0} w(4 - y)L(y)\,dy = \int_{-3}^{0} 62.4 \cdot (4 - y) \cdot 2(y + 3)\,dy = 124.8\int_{-3}^{0} (12 + y - y^2)\,dy$

$= 124.8\left[12y + \frac{y^2}{2} - \frac{y^3}{3}\right]_{-3}^{0} = (-124.8)\left(-36 + \frac{9}{2} + 9\right) = (-124.8)\left(-\frac{45}{2}\right) = 2808 \text{ lb}$

4. Using the coordinate system of Exercise 4, we see that the equation for the line of the plate's right-hand edge remains the same: $y = x - 3 \Rightarrow x = 3 + y$ and $L(y) = 2x = 2(y+3)$. The depth of the strip changes to $(-y)$
$\Rightarrow F = \int_{-3}^{0} w(-y)L(y)\,dy = \int_{-3}^{0} 62.4 \cdot (-y) \cdot 2(y+3)\,dy = 124.8\int_{-3}^{0}(-y^2 - 3y)\,dy = 124.8\left[-\frac{y^3}{3} - \frac{3}{2}y^2\right]_{-3}^{0}$
$= (-124.8)\left(\frac{27}{3} - \frac{27}{2}\right) = \frac{(-124.8)(27)(2-3)}{6} = 561.6 \text{ lb}$

5. Using the coordinate system of Exercise 4, we find the equation for the line of the plate's right-hand edge to be $y = 2x - 4 \Rightarrow x = \frac{y+4}{2}$ and $L(y) = 2x = y + 4$. The depth of the strip is $(1 - y)$.
(a) $F = \int_{-4}^{0} w(1-y)L(y)\,dy = \int_{-4}^{0} 62.4 \cdot (1-y)(y+4)\,dy = 62.4\int_{-4}^{0}(4 - 3y - y^2)\,dy = 62.4\left[4y - \frac{3y^2}{2} - \frac{y^3}{3}\right]_{-4}^{0}$
$= (-62.4)\left[(-4)(4) - \frac{(3)(16)}{2} + \frac{64}{3}\right] = (-62.4)\left(-16 - 24 + \frac{64}{3}\right) = \frac{(-62.4)(-120+64)}{3} = 1164.8 \text{ lb}$
(b) $F = (-64.0)\left[(-4)(4) - \frac{(3)(16)}{2} + \frac{64}{3}\right] = \frac{(-64.0)(-120+64)}{3} \approx 1194.7 \text{ lb}$

6. Using the coordinate system given, we find an equation for the line of the plate's right-hand edge to be $y = -2x + 4$ $\Rightarrow x = \frac{4-y}{2}$ and $L(y) = 2x = 4 - y$. The depth of the strip is $(1 - y) \Rightarrow F = \int_0^1 w(1-y)(4-y)\,dy$
$= 62.4\int_0^1 (y^2 - 5y + 4)\,dy = 62.4\left[\frac{y^3}{3} - \frac{5y^2}{2} + 4y\right]_0^1$
$= (62.4)\left(\frac{1}{3} - \frac{5}{2} + 4\right) = (62.4)\left(\frac{2-15+24}{6}\right) = \frac{(62.4)(11)}{6} = 114.4 \text{ lb}$

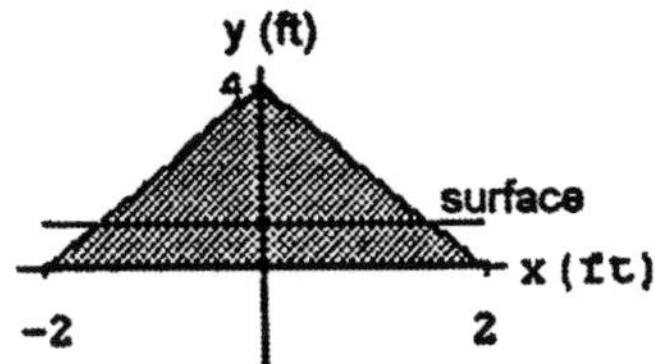

7. Using the coordinate system given in the accompanying figure, we see that the total width is $L(y) = 63$ and the depth of the strip is $(33.5 - y) \Rightarrow F = \int_0^{33} w(33.5 - y)L(y)\,dy$
$= \int_0^{33} \frac{64}{12^3} \cdot (33.5 - y) \cdot 63\,dy = \left(\frac{64}{12^3}\right)(63)\int_0^{33}(33.5 - y)\,dy$
$= \left(\frac{64}{12^3}\right)(63)\left[33.5y - \frac{y^2}{2}\right]_0^{33} = \left(\frac{64 \cdot 63}{12^3}\right)\left[(33.5)(33) - \frac{33^2}{2}\right]$
$= \frac{(64)(63)(33)(67-33)}{(2)(12^3)} = 1309 \text{ lb}$

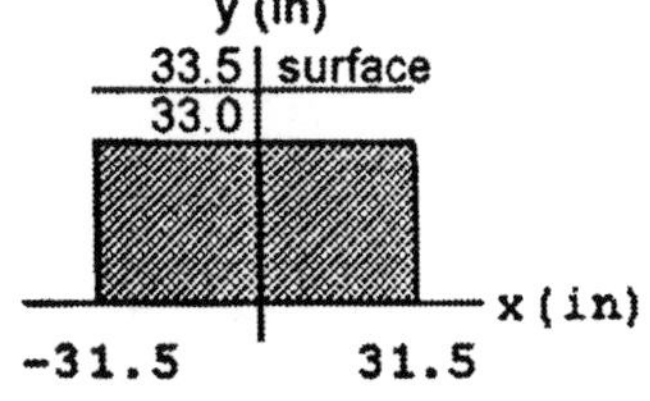

8. (a) Use the coordinate system given in the accompanying figure. The depth of the strip is $\left(\frac{11}{6} - y\right)$ ft
$\Rightarrow F = \int_0^{11/6} w\left(\frac{11}{6} - y\right)(\text{width})\,dy$
$= (62.4)(\text{width})\int_0^{11/6}\left(\frac{11}{6} - y\right)dy$
$= (62.4)(\text{width})\left[\frac{11}{6}y - \frac{y^2}{2}\right]_0^{11/6}$
$= (62.4)(\text{width})\left[\left(\frac{11}{6}\right)^2 \cdot \frac{1}{2}\right] \Rightarrow F_{\text{end}} = (62.4)(2)\left(\frac{121}{36}\right)\left(\frac{1}{2}\right) \approx 209.73 \text{ lb}$ and $F_{\text{side}} = (62.4)(4)\left(\frac{121}{36}\right)\left(\frac{1}{2}\right) \approx 419.47 \text{ lb}$

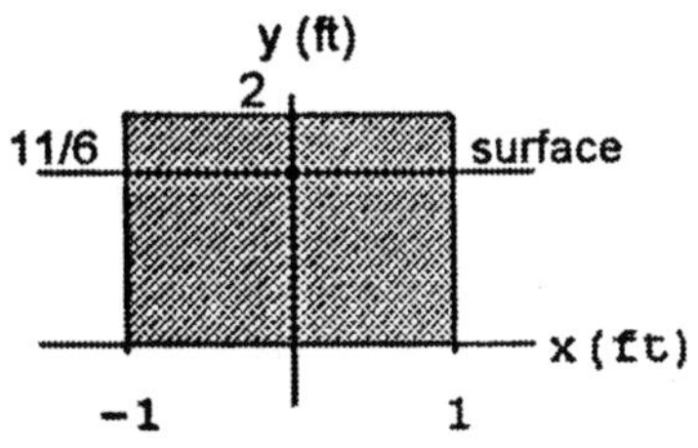

(b) Use the coordinate system given in the accompanying figure. Find Y from the condition that the entire volume of the water is conserved (no spilling): $\frac{11}{6} \cdot 2 \cdot 4 = 2 \cdot 2 \cdot Y$
$\Rightarrow Y = \frac{11}{3}$ ft. The depth of a typical strip is $\left(\frac{11}{3} - y\right)$ ft and the total width is $L(y) = 2$ ft. Thus,
$F = \int_0^{113} w\left(\frac{11}{3} - y\right)L(y)\,dy$

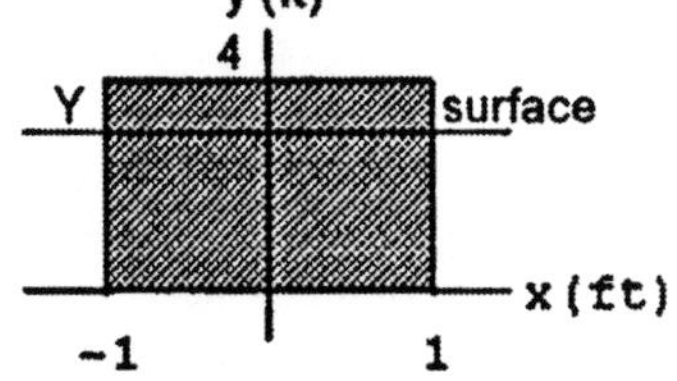

$= \int_0^{113} (62.4)\left(\frac{11}{3} - y\right) \cdot 2\,dy = (62.4)(2)\left[\frac{11}{3}y - \frac{y^2}{2}\right]_0^{11/3} = (62.4)(2)\left[\left(\frac{1}{2}\right)\left(\frac{11}{3}\right)^2\right] = \frac{(62.4)(121)}{9} \approx 838.93 \text{ lb} \Rightarrow$ the fluid force doubles.

9. Using the coordinate system given in the accompanying figure, we see that the right-hand edge is $x = \sqrt{1 - y^2}$ so the total width is $L(y) = 2x = 2\sqrt{1 - y^2}$ and the depth of the strip is $(-y)$. The force exerted by the water is therefore $F = \int_{-1}^{0} w \cdot (-y) \cdot 2\sqrt{1 - y^2}\,dy$

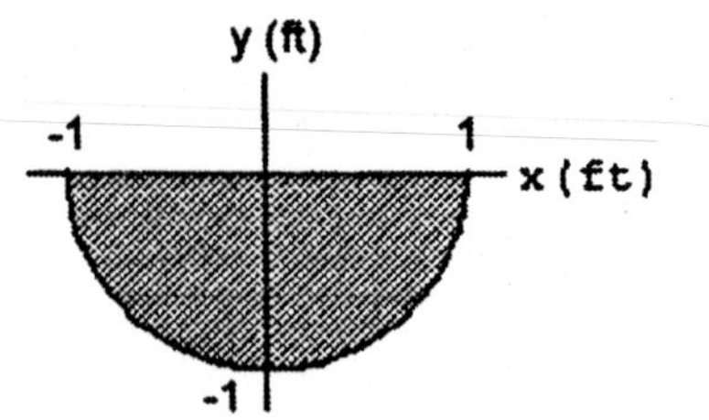

$= 62.4\int_{-1}^{0} \sqrt{1 - y^2}\,d(1 - y^2) = 62.4\left[\frac{2}{3}(1 - y^2)^{3/2}\right]_{-1}^{0} = (62.4)\left(\frac{2}{3}\right)(1 - 0) = 416 \text{ lb}$

10. Using the same coordinate system as in Exercise 15, the right-hand edge is $x = \sqrt{3^2 - y^2}$ and the total width is $L(y) = 2x = 2\sqrt{9 - y^2}$. The depth of the strip is $(-y)$. The force exerted by the milk is therefore $F = \int_{-3}^{0} w \cdot (-y) \cdot 2\sqrt{9 - y^2}\,dy = 64.5\int_{-3}^{0} \sqrt{9 - y^2}\,d(9 - y^2) = 64.5\left[\frac{2}{3}(9 - y^2)^{3/2}\right]_{-3}^{0} = (64.5)\left(\frac{2}{3}\right)(27 - 0)$ $= (64.5)(18) = 1161 \text{ lb}$

11. The coordinate system is given in the text. The right-hand edge is $x = \sqrt{y}$ and the total width is $L(y) = 2x = 2\sqrt{y}$.

(a) The depth of the strip is $(2 - y)$ so the force exerted by the liquid on the gate is $F = \int_0^1 w(2 - y)L(y)\,dy$ $= \int_0^1 50(2 - y) \cdot 2\sqrt{y}\,dy = 100\int_0^1 (2 - y)\sqrt{y}\,dy = 100\int_0^1 \left(2y^{1/2} - y^{3/2}\right) dy = 100\left[\frac{4}{3}y^{3/2} - \frac{2}{5}y^{5/2}\right]_0^1$ $= 100\left(\frac{4}{3} - \frac{2}{5}\right) = \left(\frac{100}{15}\right)(20 - 6) = 93.33 \text{ lb}$

(b) We need to solve $160 = \int_0^1 w(H - y) \cdot 2\sqrt{y}\,dy$ for h. $160 = 100\left(\frac{2H}{3} - \frac{2}{5}\right) \Rightarrow H = 3$ ft.

12. Use the coordinate system given in the accompanying figure. The total width is $L(y) = 1$.

(a) The depth of the strip is $(3 - 1) - y = (2 - y)$ ft. The force exerted by the fluid in the window is $F = \int_0^1 w(2 - y)L(y)\,dy = 62.4\int_0^1 (2 - y) \cdot 1\,dy = (62.4)\left[2y - \frac{y^2}{2}\right]_0^1 = (62.4)\left(2 - \frac{1}{2}\right) = \frac{(62.4)(3)}{2} = 93.6 \text{ lb}$

(b) Suppose that H is the maximum height to which the tank can be filled without exceeding its design limitation. This means that the depth of a typical strip is $(H - 1) - y$ and the force is $F = \int_0^1 w[(H - 1) - y]L(y)\,dy = F_{max}$, where

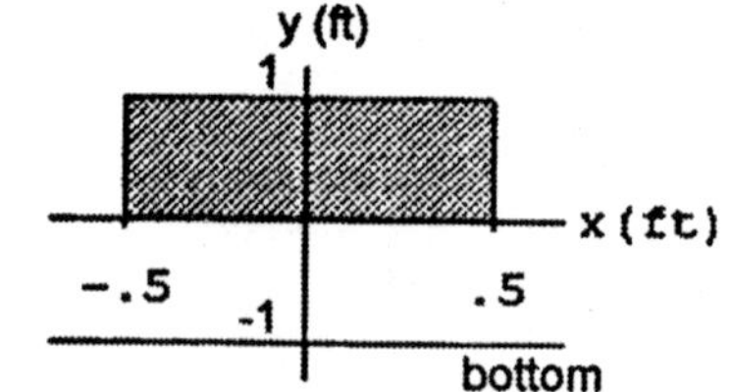

$F_{max} = 312$ lb. Thus, $F_{max} = w\int_0^1 [(H - 1) - y] \cdot 1\,dy = (62.4)\left[(H - 1)y - \frac{y^2}{2}\right]_0^1 = (62.4)\left(H - \frac{3}{2}\right)$ $= \left(\frac{62.4}{2}\right)(2H - 3) = -93.6 + 62.4H$. Then $F_{max} = -93.6 + 62.4H \Rightarrow 312 = -93.6 + 62.4H \Rightarrow H = \frac{405.6}{62.4}$ $= 6.5$ ft

13. Suppose that h is the maximum height. Using the coordinate system given in the text, we find an equation for the line of the end plate's right-hand edge is $y = \frac{5}{2}x \Rightarrow x = \frac{2}{5}y$. The total width is $L(y) = 2x = \frac{4}{5}y$ and the depth of the typical horizontal strip at level y is $(h - y)$. Then the force is $F = \int_0^h w(h - y)L(y)\,dy = F_{max}$, where $F_{max} = 6667$ lb. Hence, $F_{max} = w\int_0^h (h - y) \cdot \frac{4}{5}y\,dy = (62.4)\left(\frac{4}{5}\right)\int_0^h \left(hy - y^2\right) dy$ $= (62.4)\left(\frac{4}{5}\right)\left[\frac{hy^2}{2} - \frac{y^3}{3}\right]_0^h = (62.4)\left(\frac{4}{5}\right)\left(\frac{h^3}{2} - \frac{h^3}{3}\right) = (62.4)\left(\frac{4}{5}\right)\left(\frac{1}{6}\right)h^3 = (10.4)\left(\frac{4}{5}\right)h^3 \Rightarrow h = \sqrt[3]{\left(\frac{5}{4}\right)\left(\frac{F_{max}}{10.4}\right)}$

$= \sqrt[3]{\left(\frac{5}{4}\right)\left(\frac{6667}{10.4}\right)} \approx 9.288$ ft. The volume of water which the tank can hold is $V = \frac{1}{2}$ (Base)(Height) $\cdot$ 30, where Height $= h$ and $\frac{1}{2}$ (Base) $= \frac{2}{5}h \Rightarrow V = \left(\frac{2}{5}h^2\right)(30) = 12h^2 \approx 12(9.288)^2 \approx 1035$ ft^3.

14. (a) After 9 hours of filling there are $V = 1000 \cdot 9 = 9000$ cubic feet of water in the pool. The level of the water is $h = \frac{V}{\text{Area}}$, where Area $= 50 \cdot 30 = 1500 \Rightarrow h = \frac{9000}{1500} = 6$ ft. The depth of the typical horizontal strip at level y is then $(6 - y)$ for the coordinate system given in the text. An equation for the drain plate's right-hand edge is $y = x \Rightarrow$ total width is $L(y) = 2x = 2y$. Thus the force against the drain plate is $F = \int_0^1 w(6-y)L(y)\,dy = 62.4\int_0^1 (6-y)\cdot 2y\,dy = (62.4)(2)\int_0^1 (6y - y^2) = (62.4)(2)\left[\frac{6y^2}{2} - \frac{y^3}{3}\right]_0^1$ $= (124.8)\left(3 - \frac{1}{3}\right) = (124.8)\left(\frac{8}{3}\right) = 332.8$ lb

(b) Suppose that h is the maximum height. Then, the depth of a typical strip is $(h - y)$ and the force $F = \int_0^1 w(h-y)L(y)\,dy = F_{max}$, where $F_{max} = 520$ lb. Hence, $F_{max} = (62.4)\int_0^1 (h-y)\cdot 2y\,dy$ $= 124.8\int_0^1 (hy - y^2)\,dy = (124.8)\left[\frac{hy^2}{2} - \frac{y^3}{3}\right]_0^1 = (124.8)\left(\frac{h}{2} - \frac{1}{3}\right) = (20.8)(3h - 2) \Rightarrow \frac{520}{20.8} = 3h - 2$ $\Rightarrow h = \frac{27}{3} = 9$ ft

15. The pressure at level y is $p(y) = w \cdot y \Rightarrow$ the average pressure is $\bar{p} = \frac{1}{b}\int_0^b p(y)\,dy = \frac{1}{b}\int_0^b w \cdot y\,dy = \frac{1}{b}w\left[\frac{y^2}{2}\right]_0^b$ $= \left(\frac{w}{b}\right)\left(\frac{b^2}{2}\right) = \frac{wb}{2}$. This is the pressure at level $\frac{b}{2}$, which is the pressure at the middle of the plate.

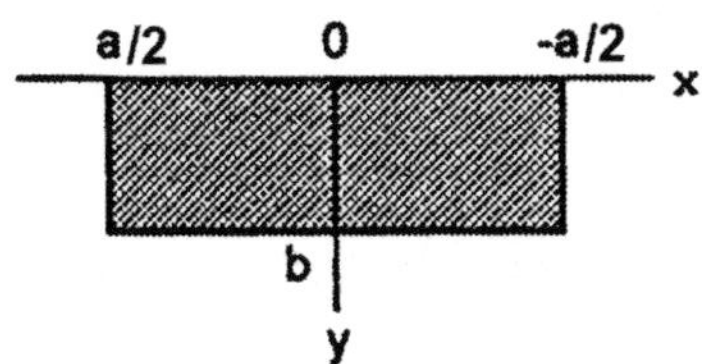

16. The force exerted by the fluid is $F = \int_0^b w(\text{depth})(\text{length})\,dy = \int_0^b w \cdot y \cdot a\,dy = (w \cdot a)\int_0^b y\,dy = (w \cdot a)\left[\frac{y^2}{2}\right]_0^b$ $= w\left(\frac{ab^2}{2}\right) = \left(\frac{wb}{2}\right)(ab) = \bar{p} \cdot$ Area, where $\bar{p}$ is the average value of the pressure (see Exercise 21).

17. When the water reaches the top of the tank the force on the movable side is $\int_{-2}^0 (62.4)\left(2\sqrt{4-y^2}\right)(-y)\,dy$ $= (62.4)\int_{-2}^0 (4-y^2)^{1/2}(-2y)\,dy = (62.4)\left[\frac{2}{3}(4-y^2)^{3/2}\right]_{-2}^0 = (62.4)\left(\frac{2}{3}\right)\left(4^{3/2}\right) = 332.8$ ft $\cdot$ lb. The force compressing the spring is $F = 100x$, so when the tank is full we have $332.8 = 100x \Rightarrow x \approx 3.33$ ft. Therefore the movable end does not reach the required 5 ft to allow drainage $\Rightarrow$ the tank will overflow.

18. (a) Using the given coordinate system we see that the total width is $L(y) = 3$ and the depth of the strip is $(3 - y)$. Thus, $F = \int_0^3 w(3-y)L(y)\,dy = \int_0^3 (62.4)(3-y)\cdot 3\,dy$ $= (62.4)(3)\int_0^3 (3-y)\,dy = (62.4)(3)\left[3y - \frac{y^2}{2}\right]_0^3$ $= (62.4)(3)\left(9 - \frac{9}{2}\right) = (62.4)(3)\left(\frac{9}{2}\right) = 842.4$ lb

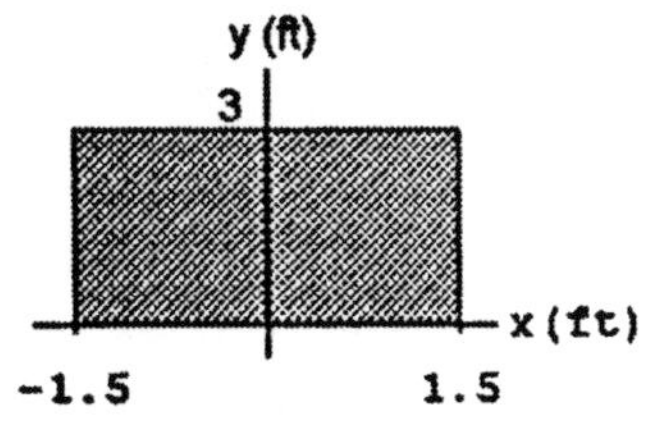

(b) Find a new water level Y such that $F_Y = (0.75)(842.4 \text{ lb}) = 631.8$ lb. The new depth of the strip is $(Y - y)$ and Y is the new upper limit of integration. Thus, $F_Y = \int_0^Y w(Y-y)L(y)\,dy$ $= 62.4\int_0^Y (Y-y)\cdot 3\,dy = (62.4)(3)\int_0^Y (Y-y)\,dy = (62.4)(3)\left[Yy - \frac{y^2}{2}\right]_0^Y = (62.4)(3)\left(Y^2 - \frac{Y^2}{2}\right)$ $= (62.4)(3)\left(\frac{Y^2}{2}\right)$. Therefore, $Y = \sqrt{\frac{2F_Y}{(62.4)(3)}} = \sqrt{\frac{1263.6}{187.2}} = \sqrt{6.75} \approx 2.598$ ft. So, $\Delta Y = 3 - Y$ $\approx 3 - 2.598 \approx 0.402$ ft ≈ 4.8 in

19. Use a coordinate system with y = 0 at the bottom of the carton and with L(y) = 3.75 and the depth of a typical strip being (7.75 − y). Then $F = \int_0^{7.75} w(7.75-y)L(y)\,dy = \left(\frac{64.5}{12^3}\right)(3.75)\int_0^{7.75}(7.75-y)\,dy = \left(\frac{64.5}{12^3}\right)(3.75)\left[7.75y - \frac{y^2}{2}\right]_0^{7.75}$
$= \left(\frac{64.5}{12^3}\right)(3.75)\frac{(7.75)^2}{2} \approx 4.2$ lb

20. The force against the base is $F_{base} = pA = whA = w \cdot h \cdot (\text{length})(\text{width}) = \left(\frac{57}{12^3}\right)(10)(5.75)(3.5) \approx 6.64$ lb.
To find the fluid force against each side, use a coordinate system with y = 0 at the bottom of the can, so that the depth of a typical strip is (10 − y): $F = \int_0^{10} w(10-y)\left(\substack{\text{width of}\\ \text{the side}}\right)dy = \left(\frac{57}{12^3}\right)\left(\substack{\text{width of}\\ \text{the side}}\right)\left[10y - \frac{y^2}{2}\right]_0^{10}$
$= \left(\frac{57}{12^3}\right)\left(\substack{\text{width of}\\ \text{the side}}\right)\left(\frac{100}{2}\right) \Rightarrow F_{end} = \left(\frac{57}{12^3}\right)(50)(3.5) \approx 5.773$ lb and $F_{side} = \left(\frac{57}{12^3}\right)(50)(5.75) \approx 9.484$ lb

21. (a) An equation of the right-hand edge is $y = \frac{3}{2}x \Rightarrow x = \frac{2}{3}y$ and $L(y) = 2x = \frac{4y}{3}$. The depth of the strip is $(3-y) \Rightarrow F = \int_0^3 w(3-y)L(y)\,dy = \int_0^3 (62.4)(3-y)\left(\frac{4}{3}y\right)dy = (62.4)\cdot\left(\frac{4}{3}\right)\int_0^3 (3y-y^2)\,dy$
$= (62.4)\left(\frac{4}{3}\right)\left[\frac{3}{2}y^2 - \frac{y^3}{3}\right]_0^3 = (62.4)\left(\frac{4}{3}\right)\left[\frac{27}{2} - \frac{27}{3}\right] = (62.4)\left(\frac{4}{3}\right)\left(\frac{27}{6}\right) = 374.4$ lb

(b) We want to find a new water level Y such that $F_Y = \frac{1}{2}(374.4) = 187.2$ lb. The new depth of the strip is (Y − y), and Y is the new upper limit of integration. Thus, $F_Y = \int_0^Y w(Y-y)L(y)\,dy$
$= 62.4\int_0^Y (Y-y)\left(\frac{4}{3}y\right)dy = (62.4)\left(\frac{4}{3}\right)\int_0^Y (Yy - y^2)\,dy = (62.4)\left(\frac{4}{3}\right)\left[Y\cdot\frac{y^2}{2} - \frac{y^3}{3}\right]_0^Y = (62.4)\left(\frac{4}{3}\right)\left(\frac{Y^3}{2} - \frac{Y^3}{3}\right)$
$= (62.4)\left(\frac{2}{9}\right)Y^3$. Therefore $Y^3 = \frac{9F_Y}{2\cdot(62.4)} = \frac{(9)(187.2)}{124.8} \Rightarrow Y = \sqrt[3]{\frac{(9)(187.2)}{124.8}} = \sqrt[3]{13.5} \approx 2.3811$ ft. So,
$\Delta Y = 3 - Y \approx 3 - 2.3811 \approx 0.6189$ ft ≈ 7.5 in. to the nearest half inch.

(c) No, it does not matter how long the trough is. The fluid pressure and the resulting force depend only on depth of the water.

22. The area of a strip of the face of height Δy and parallel to the base is $100\left(\frac{26}{24}\right)\cdot\Delta y$, where the factor of $\frac{26}{24}$ accounts for the inclination of the face of the dam. With the origin at the bottom of the dam, the force on the face is then:
$F = \int_0^{24} w(24-y)(100)\left(\frac{26}{24}\right)dy = 6760\left[24y - \frac{y^2}{2}\right]_0^{24} = 6760\left(24^2 - \frac{24^2}{2}\right) = 1{,}946{,}880$ lb.

CHAPTER 6 PRACTICE EXERCISES

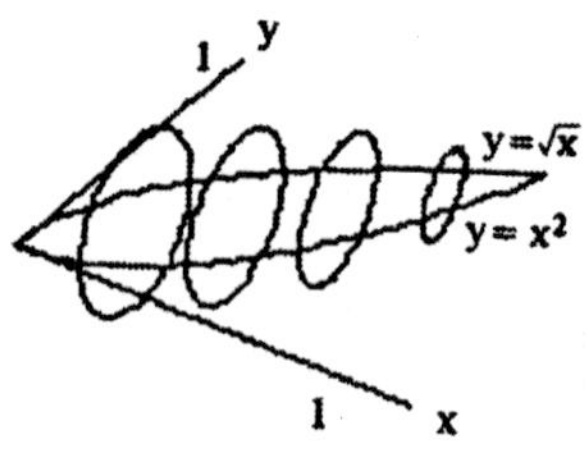

1. $A(x) = \frac{\pi}{4}(\text{diameter})^2 = \frac{\pi}{4}\left(\sqrt{x} - x^2\right)^2$
$= \frac{\pi}{4}\left(x - 2\sqrt{x}\cdot x^2 + x^4\right)$; a = 0, b = 1
$\Rightarrow V = \int_a^b A(x)\,dx = \frac{\pi}{4}\int_0^1 \left(x - 2x^{5/2} + x^4\right)dx$
$= \frac{\pi}{4}\left[\frac{x^2}{2} - \frac{4}{7}x^{7/2} + \frac{x^5}{5}\right]_0^1 = \frac{\pi}{4}\left(\frac{1}{2} - \frac{4}{7} + \frac{1}{5}\right)$
$= \frac{\pi}{4\cdot 70}(35 - 40 + 14) = \frac{9\pi}{280}$

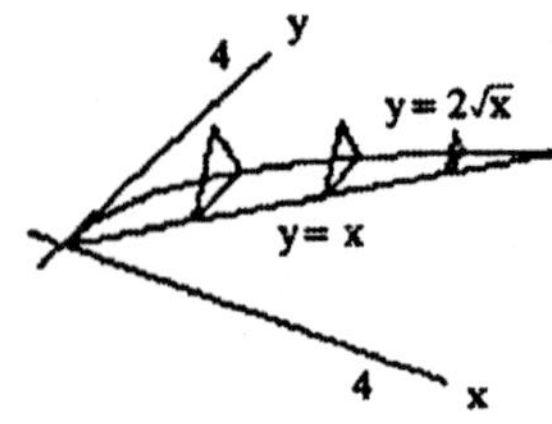

2. $A(x) = \frac{1}{2}(\text{side})^2\left(\sin\frac{\pi}{3}\right) = \frac{\sqrt{3}}{4}\left(2\sqrt{x} - x\right)^2$
$= \frac{\sqrt{3}}{4}\left(4x - 4x\sqrt{x} + x^2\right)$; a = 0, b = 4
$\Rightarrow V = \int_a^b A(x)\,dx = \frac{\sqrt{3}}{4}\int_0^4 \left(4x - 4x^{3/2} + x^2\right)dx$
$= \frac{\sqrt{3}}{4}\left[2x^2 - \frac{8}{5}x^{5/2} + \frac{x^3}{3}\right]_0^4 = \frac{\sqrt{3}}{4}\left(32 - \frac{8\cdot 32}{5} + \frac{64}{3}\right)$
$= \frac{32\sqrt{3}}{4}\left(1 - \frac{8}{5} + \frac{2}{3}\right) = \frac{8\sqrt{3}}{15}(15 - 24 + 10) = \frac{8\sqrt{3}}{15}$

3. $A(x) = \frac{\pi}{4}\,(\text{diameter})^2 = \frac{\pi}{4}\,(2\sin x - 2\cos x)^2$
$= \frac{\pi}{4} \cdot 4\,(\sin^2 x - 2\sin x\cos x + \cos^2 x)$
$= \pi(1 - \sin 2x);\ a = \frac{\pi}{4},\ b = \frac{5\pi}{4}$
$\Rightarrow\ V = \int_a^b A(x)\,dx = \pi \int_{\pi/4}^{5\pi/4} (1 - \sin 2x)\,dx$
$= \pi\left[x + \frac{\cos 2x}{2}\right]_{\pi/4}^{5\pi/4}$
$= \pi\left[\left(\frac{5\pi}{4} + \frac{\cos\frac{5\pi}{2}}{2}\right) - \left(\frac{\pi}{4} - \frac{\cos\frac{\pi}{2}}{2}\right)\right] = \pi^2$

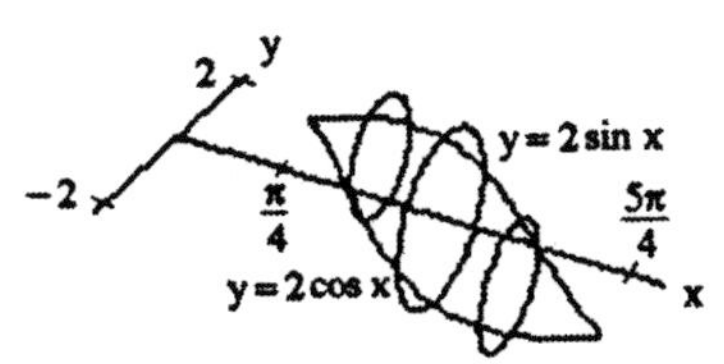

4. $A(x) = (\text{edge})^2 = \left(\left(\sqrt{6} - \sqrt{x}\right)^2 - 0\right)^2 = \left(\sqrt{6} - \sqrt{x}\right)^4 = 36 - 24\sqrt{6}\,\sqrt{x} + 36x - 4\sqrt{6}\,x^{3/2} + x^2;$
$a = 0,\ b = 6\ \Rightarrow\ V = \int_a^b A(x)\,dx = \int_0^6 \left(36 - 24\sqrt{6}\,\sqrt{x} + 36x - 4\sqrt{6}\,x^{3/2} + x^2\right) dx$
$= \left[36x - 24\sqrt{6}\cdot\frac{2}{3}\,x^{3/2} + 18x^2 - 4\sqrt{6}\cdot\frac{2}{5}\,x^{5/2} + \frac{x^3}{3}\right]_0^6 = 216 - 16\cdot\sqrt{6}\,\sqrt{6}\cdot 6 + 18\cdot 6^2 - \frac{8}{5}\,\sqrt{6}\,\sqrt{6}\cdot 6^2 + \frac{6^3}{3}$
$= 216 - 576 + 648 - \frac{1728}{5} + 72 = 360 - \frac{1728}{5} = \frac{1800 - 1728}{5} = \frac{72}{5}$

5. $A(x) = \frac{\pi}{4}\,(\text{diameter})^2 = \frac{\pi}{4}\left(2\sqrt{x} - \frac{x^2}{4}\right)^2 = \frac{\pi}{4}\left(4x - x^{5/2} + \frac{x^4}{16}\right);\ a = 0,\ b = 4\ \Rightarrow\ V = \int_a^b A(x)\,dx$
$= \frac{\pi}{4}\int_0^4 \left(4x - x^{5/2} + \frac{x^4}{16}\right) dx = \frac{\pi}{4}\left[2x^2 - \frac{2}{7}\,x^{7/2} + \frac{x^5}{5\cdot 16}\right]_0^4 = \frac{\pi}{4}\left(32 - 32\cdot\frac{8}{7} + \frac{2}{5}\cdot 32\right)$
$= \frac{32\pi}{4}\left(1 - \frac{8}{7} + \frac{2}{5}\right) = \frac{8\pi}{35}\,(35 - 40 + 14) = \frac{72\pi}{35}$

6. $A(x) = \frac{1}{2}\,(\text{edge})^2 \sin\left(\frac{\pi}{3}\right) = \frac{\sqrt{3}}{4}\left[2\sqrt{x} - \left(-2\sqrt{x}\right)\right]^2$
$= \frac{\sqrt{3}}{4}\left(4\sqrt{x}\right)^2 = 4\sqrt{3}\,x;\ a = 0,\ b = 1$
$\Rightarrow\ V = \int_a^b A(x)\,dx = \int_0^1 4\sqrt{3}\,x\,dx = \left[2\sqrt{3}\,x^2\right]_0^1$
$= 2\sqrt{3}$

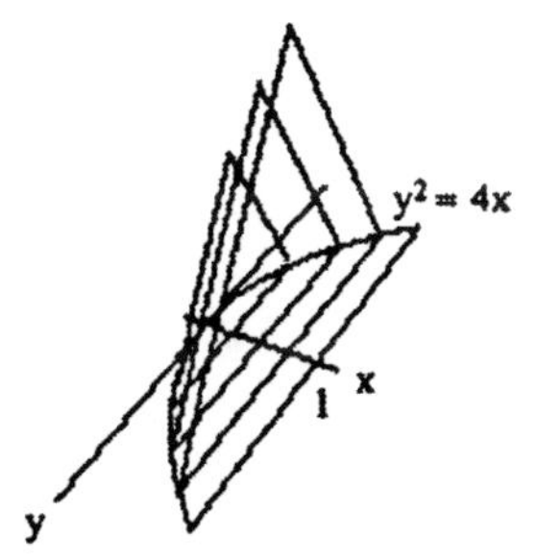

7. (a) *disk method*:

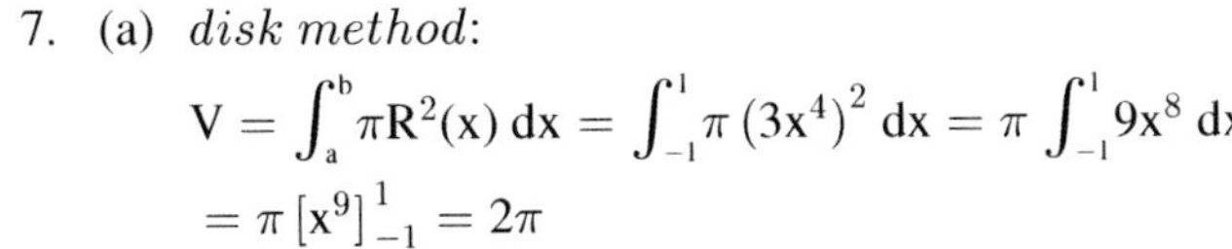

$V = \int_a^b \pi R^2(x)\,dx = \int_{-1}^1 \pi\,(3x^4)^2\,dx = \pi\int_{-1}^1 9x^8\,dx$
$= \pi\left[x^9\right]_{-1}^1 = 2\pi$

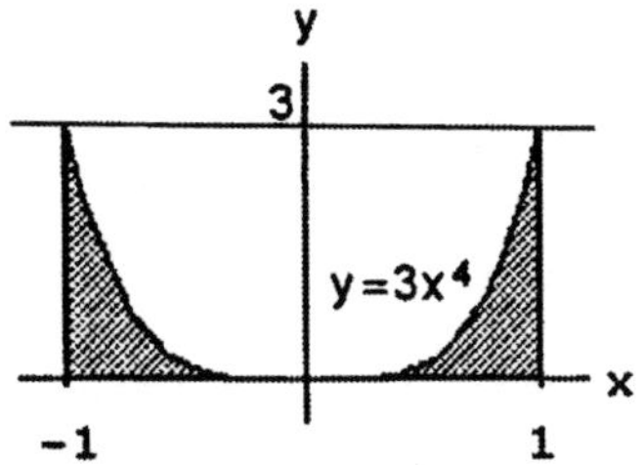

(b) *shell method*:
$V = \int_a^b 2\pi\left(\begin{smallmatrix}\text{shell}\\ \text{radius}\end{smallmatrix}\right)\left(\begin{smallmatrix}\text{shell}\\ \text{height}\end{smallmatrix}\right) dx = \int_0^1 2\pi x\,(3x^4)\,dx = 2\pi\cdot 3\int_0^1 x^5\,dx = 2\pi\cdot 3\left[\frac{x^6}{6}\right]_0^1 = \pi$
Note: The lower limit of integration is 0 rather than -1.

(c) *shell method*:
$V = \int_a^b 2\pi\left(\begin{smallmatrix}\text{shell}\\ \text{radius}\end{smallmatrix}\right)\left(\begin{smallmatrix}\text{shell}\\ \text{height}\end{smallmatrix}\right) dx = 2\pi\int_{-1}^1 (1 - x)\,(3x^4)\,dx = 2\pi\left[\frac{3x^5}{5} - \frac{x^6}{2}\right]_{-1}^1 = 2\pi\left[\left(\frac{3}{5} - \frac{1}{2}\right) - \left(-\frac{3}{5} - \frac{1}{2}\right)\right] = \frac{12\pi}{5}$

(d) *washer method*:
$R(x) = 3,\ r(x) = 3 - 3x^4 = 3\,(1 - x^4)\ \Rightarrow\ V = \int_a^b \pi\,[R^2(x) - r^2(x)]\,dx = \int_{-1}^1 \pi\left[9 - 9\,(1 - x^4)^2\right] dx$
$= 9\pi\int_{-1}^1 [1 - (1 - 2x^4 + x^8)]\,dx = 9\pi\int_{-1}^1 (2x^4 - x^8)\,dx = 9\pi\left[\frac{2x^5}{5} - \frac{x^9}{9}\right]_{-1}^1 = 18\pi\left[\frac{2}{5} - \frac{1}{9}\right] = \frac{2\pi\cdot 13}{5} = \frac{26\pi}{5}$

8. (a) *washer method:*

$R(x) = \frac{4}{x^3}$, $r(x) = \frac{1}{2} \Rightarrow V = \int_a^b \pi[R^2(x) - r^2(x)]\,dx = \int_1^2 \pi\left[\left(\frac{4}{x^3}\right)^2 - \left(\frac{1}{2}\right)^2\right]dx = \pi\left[-\frac{16}{5}x^{-5} - \frac{x}{4}\right]_1^2$

$= \pi\left[\left(\frac{-16}{5\cdot 32} - \frac{1}{2}\right) - \left(-\frac{16}{5} - \frac{1}{4}\right)\right] = \pi\left(-\frac{1}{10} - \frac{1}{2} + \frac{16}{5} + \frac{1}{4}\right) = \frac{\pi}{20}(-2 - 10 + 64 + 5) = \frac{57\pi}{20}$

(b) *shell method:*

$V = 2\pi\int_1^2 x\left(\frac{4}{x^3} - \frac{1}{2}\right)dx = 2\pi\left[-4x^{-1} - \frac{x^2}{4}\right]_1^2 = 2\pi\left[\left(-\frac{4}{2} - 1\right) - \left(-4 - \frac{1}{4}\right)\right] = 2\pi\left(\frac{5}{4}\right) = \frac{5\pi}{2}$

(c) *shell method:*

$V = 2\pi\int_a^b \left(\begin{smallmatrix}\text{shell}\\ \text{radius}\end{smallmatrix}\right)\left(\begin{smallmatrix}\text{shell}\\ \text{height}\end{smallmatrix}\right)dx = 2\pi\int_1^2 (2 - x)\left(\frac{4}{x^3} - \frac{1}{2}\right)dx = 2\pi\int_1^2 \left(\frac{8}{x^3} - \frac{4}{x^2} - 1 + \frac{x}{2}\right)dx$

$= 2\pi\left[-\frac{4}{x^2} + \frac{4}{x} - x + \frac{x^2}{4}\right]_1^2 = 2\pi\left[(-1 + 2 - 2 + 1) - \left(-4 + 4 - 1 + \frac{1}{4}\right)\right] = \frac{3\pi}{2}$

(d) *washer method:*

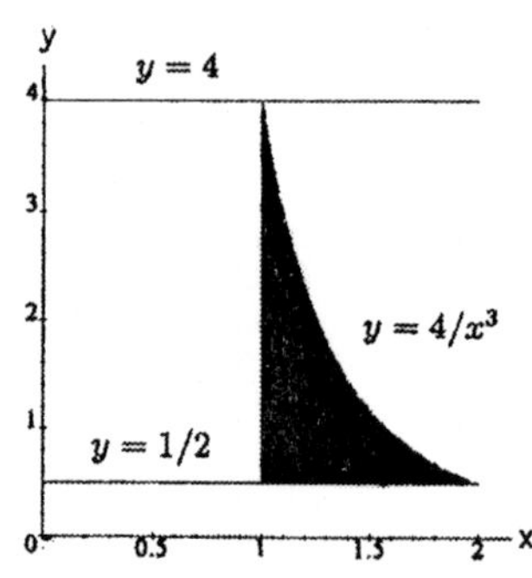

$V = \int_a^b \pi[R^2(x) - r^2(x)]\,dx$

$= \pi\int_1^2 \left[\left(\frac{7}{2}\right)^2 - \left(4 - \frac{4}{x^3}\right)^2\right]dx$

$= \frac{49\pi}{4} - 16\pi\int_1^2 (1 - 2x^{-3} + x^{-6})\,dx$

$= \frac{49\pi}{4} - 16\pi\left[x + x^{-2} - \frac{x^{-5}}{5}\right]_1^2$

$= \frac{49\pi}{4} - 16\pi\left[\left(2 + \frac{1}{4} - \frac{1}{5\cdot 32}\right) - \left(1 + 1 - \frac{1}{5}\right)\right]$

$= \frac{49\pi}{4} - 16\pi\left(\frac{1}{4} - \frac{1}{160} + \frac{1}{5}\right)$

$= \frac{49\pi}{4} - \frac{16\pi}{160}(40 - 1 + 32) = \frac{49\pi}{4} - \frac{71\pi}{10} = \frac{103\pi}{20}$

9. (a) *disk method:*

$V = \pi\int_1^5 \left(\sqrt{x - 1}\right)^2 dx = \pi\int_1^5 (x - 1)\,dx = \pi\left[\frac{x^2}{2} - x\right]_1^5$

$= \pi\left[\left(\frac{25}{2} - 5\right) - \left(\frac{1}{2} - 1\right)\right] = \pi\left(\frac{24}{2} - 4\right) = 8\pi$

(b) *washer method:*

$R(y) = 5$, $r(y) = y^2 + 1 \Rightarrow V = \int_c^d \pi\,[R^2(y) - r^2(y)]\,dy = \pi\int_{-2}^2 \left[25 - (y^2 + 1)^2\right]dy$

$= \pi\int_{-2}^2 (25 - y^4 - 2y^2 - 1)\,dy = \pi\int_{-2}^2 (24 - y^4 - 2y^2)\,dy = \pi\left[24y - \frac{y^5}{5} - \frac{2}{3}y^3\right]_{-2}^2 = 2\pi\left(24\cdot 2 - \frac{32}{5} - \frac{2}{3}\cdot 8\right)$

$= 32\pi\left(3 - \frac{2}{5} - \frac{1}{3}\right) = \frac{32\pi}{15}(45 - 6 - 5) = \frac{1088\pi}{15}$

(c) *disk method:*

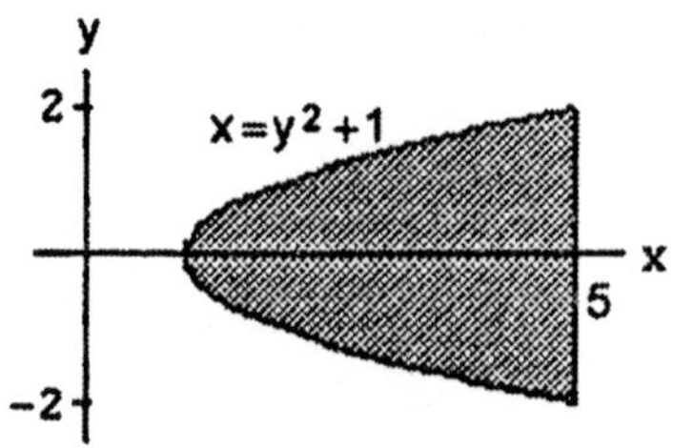

$R(y) = 5 - (y^2 + 1) = 4 - y^2$

$\Rightarrow V = \int_c^d \pi R^2(y)\,dy = \int_{-2}^2 \pi(4 - y^2)^2\,dy$

$= \pi\int_{-2}^2 (16 - 8y^2 + y^4)\,dy$

$= \pi\left[16y - \frac{8y^3}{3} + \frac{y^5}{5}\right]_{-2}^2 = 2\pi\left(32 - \frac{64}{3} + \frac{32}{5}\right)$

$= 64\pi\left(1 - \frac{2}{3} + \frac{1}{5}\right) = \frac{64\pi}{15}(15 - 10 + 3) = \frac{512\pi}{15}$

10. (a) *shell method:*

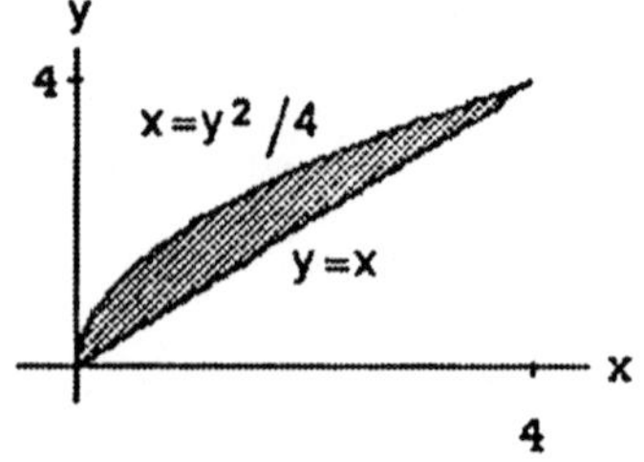

$V = \int_c^d 2\pi\left(\begin{smallmatrix}\text{shell}\\ \text{radius}\end{smallmatrix}\right)\left(\begin{smallmatrix}\text{shell}\\ \text{height}\end{smallmatrix}\right)dy = \int_0^4 2\pi y\left(y - \frac{y^2}{4}\right)dy$

$= 2\pi\int_0^4 \left(y^2 - \frac{y^3}{4}\right)dy = 2\pi\left[\frac{y^3}{3} - \frac{y^4}{16}\right]_0^4 = 2\pi\left(\frac{64}{3} - \frac{64}{4}\right)$

$= \frac{2\pi}{12}\cdot 64 = \frac{32\pi}{3}$

(b) *shell method*:

$$V=\int_a^b 2\pi\left(\begin{smallmatrix}\text{shell}\\ \text{radius}\end{smallmatrix}\right)\left(\begin{smallmatrix}\text{shell}\\ \text{height}\end{smallmatrix}\right)dx=\int_0^4 2\pi x\left(2\sqrt{x}-x\right)dx=2\pi\int_0^4\left(2x^{3/2}-x^2\right)dx=2\pi\left[\tfrac{4}{5}x^{5/2}-\tfrac{x^3}{3}\right]_0^4$$

$$=2\pi\left(\tfrac{4}{5}\cdot 32-\tfrac{64}{3}\right)=\tfrac{128\pi}{15}$$

(c) *shell method*:

$$V=\int_a^b 2\pi\left(\begin{smallmatrix}\text{shell}\\ \text{radius}\end{smallmatrix}\right)\left(\begin{smallmatrix}\text{shell}\\ \text{height}\end{smallmatrix}\right)dx=\int_0^4 2\pi(4-x)\left(2\sqrt{x}-x\right)dx=2\pi\int_0^4\left(8x^{1/2}-4x-2x^{3/2}+x^2\right)dx$$

$$=2\pi\left[\tfrac{16}{3}x^{3/2}-2x^2-\tfrac{4}{5}x^{5/2}+\tfrac{x^3}{3}\right]_0^4=2\pi\left(\tfrac{16}{3}\cdot 8-32-\tfrac{4}{5}\cdot 32+\tfrac{64}{3}\right)=64\pi\left(\tfrac{4}{3}-1-\tfrac{4}{5}+\tfrac{2}{3}\right)$$

$$=64\pi\left(1-\tfrac{4}{5}\right)=\tfrac{64\pi}{5}$$

(d) *shell method*:

$$V=\int_c^d 2\pi\left(\begin{smallmatrix}\text{shell}\\ \text{radius}\end{smallmatrix}\right)\left(\begin{smallmatrix}\text{shell}\\ \text{height}\end{smallmatrix}\right)dy=\int_0^4 2\pi(4-y)\left(y-\tfrac{y^2}{4}\right)dy=2\pi\int_0^4\left(4y-y^2-y^2+\tfrac{y^3}{4}\right)dy$$

$$=2\pi\int_0^4\left(4y-2y^2+\tfrac{y^3}{4}\right)dy=2\pi\left[2y^2-\tfrac{2}{3}y^3+\tfrac{y^4}{16}\right]_0^4=2\pi\left(32-\tfrac{2}{3}\cdot 64+16\right)=32\pi\left(2-\tfrac{8}{3}+1\right)=\tfrac{32\pi}{3}$$

11. *disk method*:

$$R(x)=\tan x,\ a=0,\ b=\tfrac{\pi}{3}\ \Rightarrow\ V=\pi\int_0^{\pi/3}\tan^2 x\,dx=\pi\int_0^{\pi/3}\left(\sec^2 x-1\right)dx=\pi[\tan x-x]_0^{\pi/3}=\frac{\pi\left(3\sqrt{3}-\pi\right)}{3}$$

12. *disk method*:

$$V=\pi\int_0^{\pi}(2-\sin x)^2\,dx=\pi\int_0^{\pi}\left(4-4\sin x+\sin^2 x\right)dx=\pi\int_0^{\pi}\left(4-4\sin x+\tfrac{1-\cos 2x}{2}\right)dx$$

$$=\pi\left[4x+4\cos x+\tfrac{x}{2}-\tfrac{\sin 2x}{4}\right]_0^{\pi}=\pi\left[\left(4\pi-4+\tfrac{\pi}{2}-0\right)-(0+4+0-0)\right]=\pi\left(\tfrac{9\pi}{2}-8\right)=\tfrac{\pi}{2}(9\pi-16)$$

13. *shell method*:

$$V=\int_0^1 2\pi\left(\begin{smallmatrix}\text{shell}\\ \text{radius}\end{smallmatrix}\right)\left(\begin{smallmatrix}\text{shell}\\ \text{height}\end{smallmatrix}\right)dx=\int_0^1 2\pi\,y\,e^{y^2}dy=\pi\,e^{y^2}\Big|_0^1=\pi(e-1)$$

14. *disk method*:

$$V=2\pi\int_0^{\pi/4}4\tan^2 x\,dx=8\pi\int_0^{\pi/4}\left(\sec^2 x-1\right)dx=8\pi[\tan x-x]_0^{\pi/4}=2\pi(4-\pi)$$

15. The material removed from the sphere consists of a cylinder and two "caps." From the diagram, the height of the cylinder is 2h, where $h^2+\left(\sqrt{3}\right)^2=2^2$, i.e. $h=1$. Thus $V_{cyl}=(2h)\pi\left(\sqrt{3}\right)^2=6\pi\ \text{ft}^3$. To get the volume of a cap, use the disk method and $x^2+y^2=2^2$: $V_{cap}=\int_1^2\pi x^2dy$
$=\int_1^2\pi(4-y^2)dy=\pi\left[4y-\tfrac{y^3}{3}\right]_1^2$
$=\pi\left[\left(8-\tfrac{8}{3}\right)-\left(4-\tfrac{1}{3}\right)\right]=\tfrac{5\pi}{3}\ \text{ft}^3$. Therefore,
$V_{removed}=V_{cyl}+2V_{cap}=6\pi+\tfrac{10\pi}{3}=\tfrac{28\pi}{3}\ \text{ft}^3$.

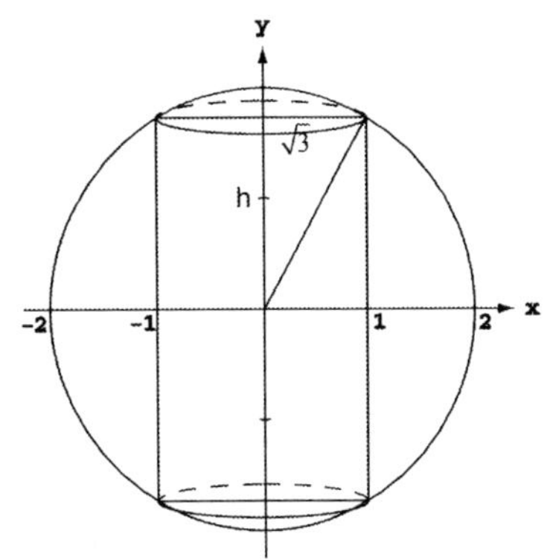

16. We rotate the region enclosed by the curve $y=\sqrt{12\left(1-\tfrac{4x^2}{121}\right)}$ and the x-axis around the x-axis. To find the volume we use the *disk* method: $V=\int_a^b\pi R^2(x)\,dx=\int_{-11/2}^{11/2}\pi\left(\sqrt{12\left(1-\tfrac{4x^2}{121}\right)}\right)^2dx=\pi\int_{-11/2}^{11/2}12\left(1-\tfrac{4x^2}{121}\right)dx$

$$=12\pi\int_{-11/2}^{11/2}\left(1-\tfrac{4x^2}{121}\right)dx=12\pi\left[x-\tfrac{4x^3}{363}\right]_{-11/2}^{11/2}=24\pi\left[\tfrac{11}{2}-\left(\tfrac{4}{363}\right)\left(\tfrac{11}{2}\right)^3\right]=132\pi\left[1-\left(\tfrac{4}{363}\right)\left(\tfrac{11^2}{4}\right)\right]$$

$$=132\pi\left(1-\tfrac{1}{3}\right)=\tfrac{264\pi}{3}=88\pi\approx 276\ \text{in}^3$$

17. $y = x^{1/2} - \frac{x^{3/2}}{3} \Rightarrow \frac{dy}{dx} = \frac{1}{2}x^{-1/2} - \frac{1}{2}x^{1/2} \Rightarrow \left(\frac{dy}{dx}\right)^2 = \frac{1}{4}\left(\frac{1}{x} - 2 + x\right) \Rightarrow L = \int_1^4 \sqrt{1 + \frac{1}{4}\left(\frac{1}{x} - 2 + x\right)}\, dx$
$\Rightarrow L = \int_1^4 \sqrt{\frac{1}{4}\left(\frac{1}{x} + 2 + x\right)}\, dx = \int_1^4 \sqrt{\frac{1}{4}\left(x^{-1/2} + x^{1/2}\right)^2}\, dx = \int_1^4 \frac{1}{2}\left(x^{-1/2} + x^{1/2}\right) dx = \frac{1}{2}\left[2x^{1/2} + \frac{2}{3}x^{3/2}\right]_1^4$
$= \frac{1}{2}\left[\left(4 + \frac{2}{3}\cdot 8\right) - \left(2 + \frac{2}{3}\right)\right] = \frac{1}{2}\left(2 + \frac{14}{3}\right) = \frac{10}{3}$

18. $x = y^{2/3} \Rightarrow \frac{dx}{dy} = \frac{2}{3}x^{-1/3} \Rightarrow \left(\frac{dx}{dy}\right)^2 = \frac{4x^{-2/3}}{9} \Rightarrow L = \int_1^8 \sqrt{1 + \left(\frac{dx}{dy}\right)^2}\, dy = \int_1^8 \sqrt{1 + \frac{4}{9x^{2/3}}}\, dy$
$= \int_1^8 \frac{\sqrt{9x^{2/3}+4}}{3x^{1/3}}\, dx = \frac{1}{3}\int_1^8 \sqrt{9x^{2/3}+4}\,\left(x^{-1/3}\right) dx$; $\left[u = 9x^{2/3} + 4 \Rightarrow du = 6y^{-1/3}\, dy;\ x = 1 \Rightarrow u = 13,\right.$
$\left. x = 8 \Rightarrow u = 40\right] \to L = \frac{1}{18}\int_{13}^{40} u^{1/2}\, du = \frac{1}{18}\left[\frac{2}{3}u^{3/2}\right]_{13}^{40} = \frac{1}{27}\left[40^{3/2} - 13^{3/2}\right] \approx 7.634$

19. $y = x^2 - \frac{\ln x}{8} \Rightarrow y' = 2x - \frac{1}{8x} \Rightarrow \sqrt{1 + (y')^2} = \sqrt{1 + \left(2x - \frac{1}{8x}\right)^2} = \sqrt{\frac{256x^4 + 32x^2 + 1}{64x^2}} = \sqrt{\frac{(16x^2+1)^2}{(8x)^2}} = \frac{16x^2+1}{8x}$
$= 2x + \frac{1}{8x} \Rightarrow \text{Length} = \int_1^2 \sqrt{1 + (y')^2}\, dx = \int_1^2 \left(2x + \frac{1}{8x}\right)dx = \left(x^2 + \frac{1}{8}\ln x\right)\Big|_1^2 = \left(4 + \frac{1}{8}\ln 2\right) - \left(1 + \frac{1}{8}\ln 1\right)$
$= 3 + \frac{1}{8}\ln 2$

20. $x = \frac{1}{12}y^3 + \frac{1}{y} \Rightarrow \frac{dx}{dy} = \frac{1}{4}y^2 - \frac{1}{y^2} \Rightarrow \left(\frac{dx}{dy}\right)^2 = \frac{1}{16}y^4 - \frac{1}{2} + \frac{1}{y^4} \Rightarrow L = \int_1^2 \sqrt{1 + \left(\frac{1}{16}y^4 - \frac{1}{2} + \frac{1}{y^4}\right)}\, dy$
$= \int_1^2 \sqrt{\frac{1}{16}y^4 + \frac{1}{2} + \frac{1}{y^4}}\, dy = \int_1^2 \sqrt{\left(\frac{1}{4}y^2 + \frac{1}{y^2}\right)^2}\, dy = \int_1^2 \left(\frac{1}{4}y^2 + \frac{1}{y^2}\right) dy = \left[\frac{1}{12}y^3 - \frac{1}{y}\right]_1^2$
$= \left(\frac{8}{12} - \frac{1}{2}\right) - \left(\frac{1}{12} - 1\right) = \frac{7}{12} + \frac{1}{2} = \frac{13}{12}$

21. $\frac{dx}{dt} = -5\sin t + 5\sin 5t$ and $\frac{dy}{dt} = 5\cos t - 5\cos 5t \Rightarrow \sqrt{\left(\frac{dx}{dt}\right)^2 + \left(\frac{dy}{dt}\right)^2}$
$= \sqrt{(-5\sin t + 5\sin 5t)^2 + (5\cos t - 5\cos 5t)^2}$
$= 5\sqrt{\sin^2 5t - 2\sin t\sin 5t + \sin^2 t + \cos^2 t - 2\cos t\cos 5t + \cos^2 5t} = 5\sqrt{2 - 2(\sin t\sin 5t + \cos t\cos 5t)}$
$= 5\sqrt{2(1 - \cos 4t)} = 5\sqrt{4\left(\frac{1}{2}\right)(1 - \cos 4t)} = 10\sqrt{\sin^2 2t} = 10|\sin 2t| = 10\sin 2t$ (since $0 \le t \le \frac{\pi}{2}$)
$\Rightarrow \text{Length} = \int_0^{\pi/2} 10\sin 2t\, dt = [-5\cos 2t]_0^{\pi/2} = (-5)(-1) - (-5)(1) = 10$

22. $\frac{dx}{dt} = 3t^2 - 12t$ and $\frac{dy}{dt} = 3t^2 + 12t \Rightarrow \sqrt{\left(\frac{dx}{dt}\right)^2 + \left(\frac{dy}{dt}\right)^2} = \sqrt{(3t^2 - 12t)^2 + (3t^2 + 12t)^2} = \sqrt{288t^2 + 18t^4}$
$= 3\sqrt{2}\,|t|\sqrt{16 + t^2} \Rightarrow \text{Length} = \int_0^1 3\sqrt{2}\,|t|\sqrt{16 + t^2}\, dt = 3\sqrt{2}\int_0^1 t\sqrt{16 + t^2}\, dt$; $\left[u = 16 + t^2 \Rightarrow du = 2t\, dt\right.$
$\left.\Rightarrow \frac{1}{2}du = t\, dt;\ t = 0 \Rightarrow u = 16;\ t = 1 \Rightarrow u = 17\right]$; $\frac{3\sqrt{2}}{2}\int_{16}^{17} \sqrt{u}\, du = \frac{3\sqrt{2}}{2}\left[\frac{2}{3}u^{3/2}\right]_{16}^{17} = \frac{3\sqrt{2}}{2}\left(\frac{2}{3}(17)^{3/2} - \frac{2}{3}(16)^{3/2}\right)$
$= \frac{3\sqrt{2}}{2}\cdot\frac{2}{3}\left((17)^{3/2} - 64\right) = \sqrt{2}\left((17)^{3/2} - 64\right) \approx 8.617.$

23. $\frac{dx}{d\theta} = -3\sin\theta$ and $\frac{dy}{d\theta} = 3\cos\theta \Rightarrow \sqrt{\left(\frac{dx}{d\theta}\right)^2 + \left(\frac{dy}{d\theta}\right)^2} = \sqrt{(-3\sin\theta)^2 + (3\cos\theta)^2} = \sqrt{3(\sin^2\theta + \cos^2\theta)} = 3$
$\Rightarrow \text{Length} = \int_0^{3\pi/2} 3\, d\theta = 3\int_0^{3\pi/2} d\theta = 3\left(\frac{3\pi}{2} - 0\right) = \frac{9\pi}{2}$

24. $x = t^2$ and $y = \frac{t^3}{3} - t,\ -\sqrt{3} \le t \le \sqrt{3} \Rightarrow \frac{dx}{dt} = 2t$ and $\frac{dy}{dt} = t^2 - 1 \Rightarrow \text{Length} = \int_{-\sqrt{3}}^{\sqrt{3}} \sqrt{(2t)^2 + (t^2 - 1)^2}\, dt$
$= \int_{-\sqrt{3}}^{\sqrt{3}} \sqrt{t^4 + 2t^2 + 1}\, dt = \int_{-\sqrt{3}}^{\sqrt{3}} \sqrt{t^4 + 2t^2 + 1}\, dt = \int_{-\sqrt{3}}^{\sqrt{3}} \sqrt{(t^2 + 1)^2}\, dt = \int_{-\sqrt{3}}^{\sqrt{3}} (t^2 + 1)\, dt = \left[\frac{t^3}{3} + t\right]_{-\sqrt{3}}^{\sqrt{3}}$
$= 4\sqrt{3}$

25. Intersection points: $3 - x^2 = 2x^2 \Rightarrow 3x^2 - 3 = 0$ $\Rightarrow 3(x-1)(x+1) = 0 \Rightarrow x = -1$ or $x = 1$. Symmetry suggests that $\overline{x} = 0$. The typical *vertical* strip has center of mass: $(\tilde{x}, \tilde{y}) = \left(x, \frac{2x^2 + (3 - x^2)}{2}\right) = \left(x, \frac{x^2+3}{2}\right)$, length: $(3 - x^2) - 2x^2 = 3(1 - x^2)$, width: dx, area: $dA = 3(1 - x^2)\,dx$, and mass: $dm = \delta \cdot dA$ $= 3\delta(1 - x^2)\,dx \Rightarrow$ the moment about the x-axis is

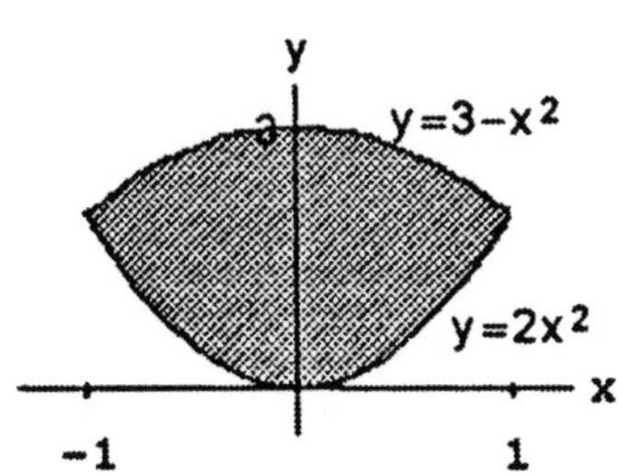

$\tilde{y}\,dm = \frac{3}{2}\delta(x^2+3)(1-x^2)\,dx = \frac{3}{2}\delta(-x^4 - 2x^2 + 3)\,dx \Rightarrow M_x = \int \tilde{y}\,dm = \frac{3}{2}\delta\int_{-1}^{1}(-x^4 - 2x^2 + 3)\,dx$

$= \frac{3}{2}\delta\left[-\frac{x^5}{5} - \frac{2x^3}{3} + 3x\right]_{-1}^{1} = 3\delta\left(-\frac{1}{5} - \frac{2}{3} + 3\right) = \frac{3\delta}{15}(-3 - 10 + 45) = \frac{32\delta}{5}$; $M = \int dm = 3\delta\int_{-1}^{1}(1 - x^2)\,dx$

$= 3\delta\left[x - \frac{x^3}{3}\right]_{-1}^{1} = 6\delta\left(1 - \frac{1}{3}\right) = 4\delta \Rightarrow \overline{y} = \frac{M_x}{M} = \frac{32\delta}{5 \cdot 4\delta} = \frac{8}{5}$. Therefore, the centroid is $(\overline{x}, \overline{y}) = \left(0, \frac{8}{5}\right)$.

26. Symmetry suggests that $\overline{x} = 0$. The typical *vertical* strip has center of mass: $(\tilde{x}, \tilde{y}) = \left(x, \frac{x^2}{2}\right)$, length: x^2, width: dx, area: $dA = x^2\,dx$, mass: $dm = \delta \cdot dA = \delta x^2\,dx$ $\Rightarrow$ the moment about the x-axis is $\tilde{y}\,dm = \frac{\delta}{2}x^2 \cdot x^2\,dx$ $= \frac{\delta}{2}x^4\,dx \Rightarrow M_x = \int \tilde{y}\,dm = \frac{\delta}{2}\int_{-2}^{2}x^4\,dx = \frac{\delta}{10}\left[x^5\right]_{-2}^{2}$

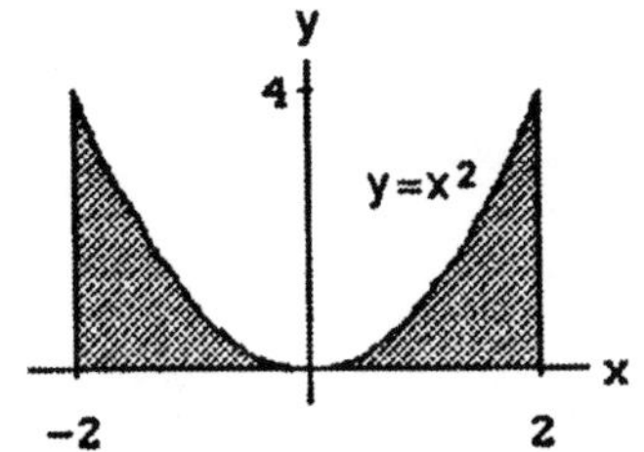

$= \frac{2\delta}{10}(2^5) = \frac{32\delta}{5}$; $M = \int dm = \delta\int_{-2}^{2}x^2\,dx = \delta\left[\frac{x^3}{3}\right]_{-2}^{2} = \frac{2\delta}{3}(2^3) = \frac{16\delta}{3} \Rightarrow \overline{y} = \frac{M_x}{M} = \frac{32 \cdot \delta \cdot 3}{5 \cdot 16 \cdot \delta} = \frac{6}{5}$. Therefore, the centroid is $(\overline{x}, \overline{y}) = \left(0, \frac{6}{5}\right)$.

27. The typical *vertical* strip has: center of mass: $(\tilde{x}, \tilde{y})$ $= \left(x, \frac{4 + \frac{x^2}{4}}{2}\right)$, length: $4 - \frac{x^2}{4}$, width: dx, area: $dA = \left(4 - \frac{x^2}{4}\right)dx$, mass: $dm = \delta \cdot dA$ $= \delta\left(4 - \frac{x^2}{4}\right)dx \Rightarrow$ the moment about the x-axis is

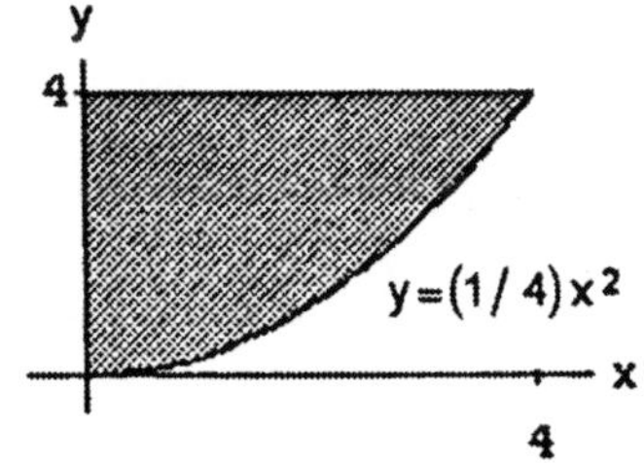

$\tilde{y}\,dm = \delta \cdot \frac{\left(4 + \frac{x^2}{4}\right)}{2}\left(4 - \frac{x^2}{4}\right)dx = \frac{\delta}{2}\left(16 - \frac{x^4}{16}\right)dx$; the moment about the y-axis is $\tilde{x}\,dm = \delta\left(4 - \frac{x^2}{4}\right) \cdot x\,dx = \delta\left(4x - \frac{x^3}{4}\right)dx$. Thus, $M_x = \int \tilde{y}\,dm = \frac{\delta}{2}\int_0^4\left(16 - \frac{x^4}{16}\right)dx$

$= \frac{\delta}{2}\left[16x - \frac{x^5}{5 \cdot 16}\right]_0^4 = \frac{\delta}{2}\left[64 - \frac{64}{5}\right] = \frac{128\delta}{5}$; $M_y = \int \tilde{x}\,dm = \delta\int_0^4\left(4x - \frac{x^3}{4}\right)dx = \delta\left[2x^2 - \frac{x^4}{16}\right]_0^4$

$= \delta(32 - 16) = 16\delta$; $M = \int dm = \delta\int_0^4\left(4 - \frac{x^2}{4}\right)dx = \delta\left[4x - \frac{x^3}{12}\right]_0^4 = \delta\left(16 - \frac{64}{12}\right) = \frac{32\delta}{3}$

$\Rightarrow \overline{x} = \frac{M_y}{M} = \frac{16 \cdot \delta \cdot 3}{32 \cdot \delta} = \frac{3}{2}$ and $\overline{y} = \frac{M_x}{M} = \frac{128 \cdot \delta \cdot 3}{5 \cdot 32 \cdot \delta} = \frac{12}{5}$. Therefore, the centroid is $(\overline{x}, \overline{y}) = \left(\frac{3}{2}, \frac{12}{5}\right)$.

28. A typical *horizontal* strip has: center of mass: $(\tilde{x}, \tilde{y}) = \left(\frac{y^2+2y}{2}, y\right)$, length: $2y - y^2$, width: dy, area: $dA = (2y - y^2)\,dy$, mass: $dm = \delta \cdot dA$ $= \delta(2y - y^2)\,dy$; the moment about the x-axis is $\tilde{y}\,dm = \delta \cdot y \cdot (2y - y^2)\,dy = \delta(2y^2 - y^3)$; the moment about the y-axis is $\tilde{x}\,dm = \delta \cdot \frac{(y^2+2y)}{2} \cdot (2y - y^2)\,dy$ $= \frac{\delta}{2}(4y^2 - y^4)\,dy \Rightarrow M_x = \int \tilde{y}\,dm = \delta\int_0^2(2y^2 - y^3)\,dy$

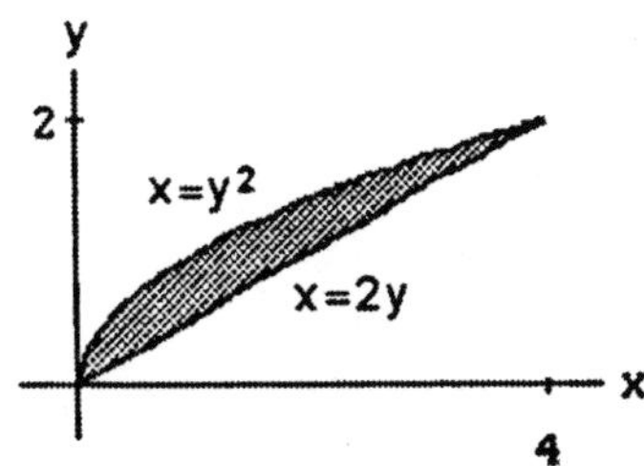

$= \delta\left[\frac{2}{3}y^3 - \frac{y^4}{4}\right]_0^2 = \delta\left(\frac{2}{3} \cdot 8 - \frac{16}{4}\right) = \delta\left(\frac{16}{3} - \frac{16}{4}\right) = \frac{\delta \cdot 16}{12} = \frac{4\delta}{3}$; $M_y = \int \tilde{x}\,dm = \frac{\delta}{2}\int_0^2(4y^2 - y^4)\,dy = \frac{\delta}{2}\left[\frac{4}{3}y^3 - \frac{y^5}{5}\right]_0^2$

$= \frac{\delta}{2}\left(\frac{4\cdot 8}{3} - \frac{32}{5}\right) = \frac{32\delta}{15}$; $M = \int dm = \delta \int_0^2 (2y - y^2)\, dy = \delta \left[y^2 - \frac{y^3}{3}\right]_0^2 = \delta\left(4 - \frac{8}{3}\right) = \frac{4\delta}{3} \Rightarrow \bar{x} = \frac{M_y}{M} = \frac{\delta \cdot 32 \cdot 3}{15 \cdot \delta \cdot 4} = \frac{8}{5}$ and $\bar{y} = \frac{M_x}{M} = \frac{4 \cdot \delta \cdot 3}{3 \cdot 4 \cdot \delta} = 1$. Therefore, the centroid is $(\bar{x}, \bar{y}) = \left(\frac{8}{5}, 1\right)$.

29. A typical horizontal strip has: center of mass: $(\tilde{x}, \tilde{y}) = \left(\frac{y^2+2y}{2}, y\right)$, length: $2y - y^2$, width: dy, area: $dA = (2y - y^2)\, dy$, mass: $dm = \delta \cdot dA = (1 + y)(2y - y^2)\, dy \Rightarrow$ the moment about the x-axis is $\tilde{y}\, dm = y(1 + y)(2y - y^2)\, dy = (2y^2 + 2y^3 - y^3 - y^4)\, dy = (2y^2 + y^3 - y^4)\, dy$; the moment about the y-axis is

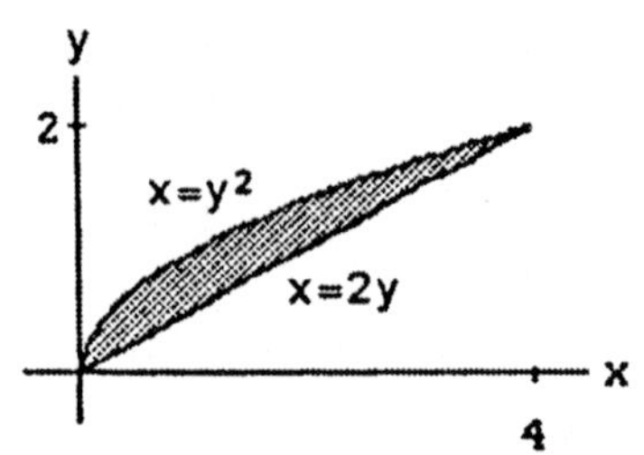

$\tilde{x}\, dm = \left(\frac{y^2+2y}{2}\right)(1 + y)(2y - y^2)\, dy = \frac{1}{2}(4y^2 - y^4)(1 + y)\, dy = \frac{1}{2}(4y^2 + 4y^3 - y^4 - y^5)\, dy$

$\Rightarrow M_x = \int \tilde{y}\, dm = \int_0^2 (2y^2 + y^3 - y^4)\, dy = \left[\frac{2}{3}y^3 + \frac{y^4}{4} - \frac{y^5}{5}\right]_0^2 = \left(\frac{16}{3} + \frac{16}{4} - \frac{32}{5}\right) = 16\left(\frac{1}{3} + \frac{1}{4} - \frac{2}{5}\right)$
$= \frac{16}{60}(20 + 15 - 24) = \frac{4}{15}(11) = \frac{44}{15}$; $M_y = \int \tilde{x}\, dm = \int_0^2 \frac{1}{2}(4y^2 + 4y^3 - y^4 - y^5)\, dy = \frac{1}{2}\left[\frac{4}{3}y^3 + y^4 - \frac{y^5}{5} - \frac{y^6}{6}\right]_0^2$
$= \frac{1}{2}\left(\frac{4\cdot 2^3}{3} + 2^4 - \frac{2^5}{5} - \frac{2^6}{6}\right) = 4\left(\frac{4}{3} + 2 - \frac{4}{5} - \frac{8}{6}\right) = 4\left(2 - \frac{4}{5}\right) = \frac{24}{5}$; $M = \int dm = \int_0^2 (1 + y)(2y - y^2)\, dy$
$= \int_0^2 (2y + y^2 - y^3)\, dy = \left[y^2 + \frac{y^3}{3} - \frac{y^4}{4}\right]_0^2 = \left(4 + \frac{8}{3} - \frac{16}{4}\right) = \frac{8}{3} \Rightarrow \bar{x} = \frac{M_y}{M} = \left(\frac{24}{5}\right)\left(\frac{3}{8}\right) = \frac{9}{5}$ and $\bar{y} = \frac{M_x}{M}$
$= \left(\frac{44}{15}\right)\left(\frac{3}{8}\right) = \frac{44}{40} = \frac{11}{10}$. Therefore, the center of mass is $(\bar{x}, \bar{y}) = \left(\frac{9}{5}, \frac{11}{10}\right)$.

30. A typical vertical strip has: center of mass: $(\tilde{x}, \tilde{y}) = \left(x, \frac{3}{2x^{3/2}}\right)$, length: $\frac{3}{x^{3/2}}$, width: dx, area: $dA = \frac{3}{x^{3/2}}\, dx$, mass: $dm = \delta \cdot dA = \delta \cdot \frac{3}{x^{3/2}}\, dx \Rightarrow$ the moment about the x-axis is $\tilde{y}\, dm = \frac{3}{2x^{3/2}} \cdot \delta \frac{3}{x^{3/2}}\, dx = \frac{9\delta}{2x^3}\, dx$; the moment about the y-axis is $\tilde{x}\, dm = x \cdot \delta \frac{3}{x^{3/2}}\, dx = \frac{3\delta}{x^{1/2}}\, dx$.

(a) $M_x = \delta \int_1^9 \frac{1}{2}\left(\frac{9}{x^3}\right) dx = \frac{9\delta}{2}\left[-\frac{x^{-2}}{2}\right]_1^9 = \frac{20\delta}{9}$; $M_y = \delta \int_1^9 x\left(\frac{3}{x^{3/2}}\right) dx = 3\delta\left[2x^{1/2}\right]_1^9 = 12\delta$;
$M = \delta \int_1^9 \frac{3}{x^{3/2}}\, dx = -6\delta\left[x^{-1/2}\right]_1^9 = 4\delta \Rightarrow \bar{x} = \frac{M_y}{M} = \frac{12\delta}{4\delta} = 3$ and $\bar{y} = \frac{M_x}{M} = \frac{\left(\frac{20\delta}{9}\right)}{4\delta} = \frac{5}{9}$

(b) $M_x = \int_1^9 \frac{x}{2}\left(\frac{9}{x^3}\right) dx = \frac{9}{2}\left[-\frac{1}{x}\right]_1^9 = 4$; $M_y = \int_1^9 x^2\left(\frac{3}{x^{3/2}}\right) dx = \left[2x^{3/2}\right]_1^9 = 52$; $M = \int_1^9 x\left(\frac{3}{x^{3/2}}\right) dx$
$= 6\left[x^{1/2}\right]_1^9 = 12 \Rightarrow \bar{x} = \frac{M_y}{M} = \frac{13}{3}$ and $\bar{y} = \frac{M_x}{M} = \frac{1}{3}$

31. $S = \int_a^b 2\pi y \sqrt{1 + \left(\frac{dy}{dx}\right)^2}\, dx$; $\frac{dy}{dx} = \frac{1}{\sqrt{2x+1}} \Rightarrow \left(\frac{dy}{dx}\right)^2 = \frac{1}{2x+1} \Rightarrow S = \int_0^3 2\pi\sqrt{2x+1}\sqrt{1 + \frac{1}{2x+1}}\, dx$
$= 2\pi \int_0^3 \sqrt{2x+1}\sqrt{\frac{2x+2}{2x+1}}\, dx = 2\sqrt{2}\pi \int_0^3 \sqrt{x+1}\, dx = 2\sqrt{2}\pi\left[\frac{2}{3}(x+1)^{3/2}\right]_0^3 = 2\sqrt{2}\pi \cdot \frac{2}{3}(8 - 1) = \frac{28\pi\sqrt{2}}{3}$

32. $S = \int_a^b 2\pi y \sqrt{1 + \left(\frac{dy}{dx}\right)^2}\, dx$; $\frac{dy}{dx} = x^2 \Rightarrow \left(\frac{dy}{dx}\right)^2 = x^4 \Rightarrow S = \int_0^1 2\pi \cdot \frac{x^3}{3}\sqrt{1 + x^4}\, dx = \frac{\pi}{6}\int_0^1 \sqrt{1 + x^4}\,(4x^3)\, dx$
$= \frac{\pi}{6}\int_0^1 \sqrt{1 + x^4}\, d(1 + x^4) = \frac{\pi}{6}\left[\frac{2}{3}(1 + x^4)^{3/2}\right]_0^1 = \frac{\pi}{9}\left[2\sqrt{2} - 1\right]$

33. $S = \int_c^d 2\pi x \sqrt{1 + \left(\frac{dx}{dy}\right)^2}\, dy$; $\frac{dx}{dy} = \frac{\left(\frac{1}{2}\right)(4 - 2y)}{\sqrt{4y - y^2}} = \frac{2 - y}{\sqrt{4y - y^2}} \Rightarrow 1 + \left(\frac{dx}{dy}\right)^2 = \frac{4y - y^2 + 4 - 4y + y^2}{4y - y^2} = \frac{4}{4y - y^2}$
$\Rightarrow S = \int_1^2 2\pi\sqrt{4y - y^2}\sqrt{\frac{4}{4y - y^2}}\, dy = 4\pi \int_1^2 dx = 4\pi$

34. $S = \int_c^d 2\pi x \sqrt{1 + \left(\frac{dx}{dy}\right)^2}\, dy$; $\frac{dx}{dy} = \frac{1}{2\sqrt{y}} \Rightarrow 1 + \left(\frac{dx}{dy}\right)^2 = 1 + \frac{1}{4y} = \frac{4y+1}{4y} \Rightarrow S = \int_2^6 2\pi\sqrt{y} \cdot \frac{\sqrt{4y+1}}{\sqrt{4y}}\, dy$
$= \pi \int_2^6 \sqrt{4y+1}\, dy = \frac{\pi}{4}\left[\frac{2}{3}(4y+1)^{3/2}\right]_2^6 = \frac{\pi}{6}(125 - 27) = \frac{\pi}{6}(98) = \frac{49\pi}{3}$

35. $x = \frac{t^2}{2}$ and $y = 2t$, $0 \le t \le \sqrt{5} \Rightarrow \frac{dx}{dt} = t$ and $\frac{dy}{dt} = 2 \Rightarrow$ Surface Area $= \int_0^{\sqrt{5}} 2\pi(2t)\sqrt{t^2+4}\,dt = \int_4^9 2\pi u^{1/2}\,du$
$= 2\pi\left[\frac{2}{3}u^{3/2}\right]_4^9 = \frac{76\pi}{3}$, where $u = t^2 + 4 \Rightarrow du = 2t\,dt$; $t = 0 \Rightarrow u = 4$, $t = \sqrt{5} \Rightarrow u = 9$

36. $x = t^2 + \frac{1}{2t}$ and $y = 4\sqrt{t}$, $\frac{1}{\sqrt{2}} \le t \le 1 \Rightarrow \frac{dx}{dt} = 2t - \frac{1}{2t^2}$ and $\frac{dy}{dt} = \frac{2}{\sqrt{t}}$

$\Rightarrow$ Surface Area $= \int_{1/\sqrt{2}}^{1} 2\pi\left(t^2 + \frac{1}{2t}\right)\sqrt{\left(2t - \frac{1}{2t^2}\right)^2 + \left(\frac{2}{\sqrt{t}}\right)^2}\,dt = 2\pi\int_{1/\sqrt{2}}^{1}\left(t^2 + \frac{1}{2t}\right)\sqrt{\left(2t + \frac{1}{2t^2}\right)^2}\,dt$
$= 2\pi\int_{1/\sqrt{2}}^{1}\left(t^2 + \frac{1}{2t}\right)\left(2t + \frac{1}{2t^2}\right)dt = 2\pi\int_{1/\sqrt{2}}^{1}\left(2t^3 + \frac{3}{2} + \frac{1}{4}t^{-3}\right)dt = 2\pi\left[\frac{1}{2}t^4 + \frac{3}{2}t - \frac{1}{8}t^{-2}\right]_{1/\sqrt{2}}^{1}$
$= 2\pi\left(2 - \frac{3\sqrt{2}}{4}\right)$

37. The equipment alone: the force required to lift the equipment is equal to its weight $\Rightarrow F_1(x) = 100$ N. The work done is $W_1 = \int_a^b F_1(x)\,dx = \int_0^{40} 100\,dx = [100x]_0^{40} = 4000$ J; the rope alone: the force required to lift the rope is equal to the weight of the rope paid out at elevation x $\Rightarrow F_2(x) = 0.8(40 - x)$. The work done is $W_2 = \int_a^b F_2(x)\,dx = \int_0^{40} 0.8(40 - x)\,dx = 0.8\left[40x - \frac{x^2}{2}\right]_0^{40} = 0.8\left(40^2 - \frac{40^2}{2}\right) = \frac{(0.8)(1600)}{2} = 640$ J; the total work is $W = W_1 + W_2 = 4000 + 640 = 4640$ J

38. The force required to lift the water is equal to the water's weight, which varies steadily from $8 \cdot 800$ lb to $8 \cdot 400$ lb over the 4750 ft elevation. When the truck is x ft off the base of Mt. Washington, the water weight is $F(x) = 8 \cdot 800 \cdot \left(\frac{2 \cdot 4750 - x}{2 \cdot 4750}\right) = (6400)\left(1 - \frac{x}{9500}\right)$ lb. The work done is $W = \int_a^b F(x)\,dx$
$= \int_0^{4750} 6400\left(1 - \frac{x}{9500}\right)dx = 6400\left[x - \frac{x^2}{2 \cdot 9500}\right]_0^{4750} = 6400\left(4750 - \frac{4750^2}{4 \cdot 4750}\right) = \left(\frac{3}{4}\right)(6400)(4750)$
$= 22{,}800{,}000$ ft · lb

39. Force constant: $F = kx \Rightarrow 20 = k \cdot 1 \Rightarrow k = 20$ lb/ft; the work to stretch the spring 1 ft is $W = \int_0^1 kx\,dx = k\int_0^1 x\,dx = \left[20\frac{x^2}{2}\right]_0^1 = 10$ ft · lb; the work to stretch the spring an additional foot is $W = \int_1^2 kx\,dx = k\int_1^2 x\,dx = 20\left[\frac{x^2}{2}\right]_1^2 = 20\left(\frac{4}{2} - \frac{1}{2}\right) = 20\left(\frac{3}{2}\right) = 30$ ft · lb

40. Force constant: $F = kx \Rightarrow 200 = k(0.8) \Rightarrow k = 250$ N/m; the 300 N force stretches the spring $x = \frac{F}{k}$ $= \frac{300}{250} = 1.2$ m; the work required to stretch the spring that far is then $W = \int_0^{1.2} F(x)\,dx = \int_0^{1.2} 250x\,dx$
$= [125x^2]_0^{1.2} = 125(1.2)^2 = 180$ J

41. We imagine the water divided into thin slabs by planes perpendicular to the y-axis at the points of a partition of the interval [0, 8]. The typical slab between the planes at y and $y + \Delta y$ has a volume of about $\Delta V = \pi(\text{radius})^2(\text{thickness})$ $= \pi\left(\frac{5}{4}y\right)^2\Delta y = \frac{25\pi}{16}y^2\,\Delta y$ ft³. The force F(y) required to lift this slab is equal to its weight: $F(y) = 62.4\,\Delta V$ $= \frac{(62.4)(25)}{16}\pi y^2\,\Delta y$ lb. The distance through which F(y) must act to lift this slab to the level 6 ft above the top is about $(6 + 8 - y)$ ft, so the work done lifting the slab is about $\Delta W = \frac{(62.4)(25)}{16}\pi y^2(14 - y)\,\Delta y$ ft · lb. The work done lifting all the slabs from $y = 0$ to $y = 8$ to the level 6 ft above the top is approximately $W \approx \sum_0^8 \frac{(62.4)(25)}{16}\pi y^2(14 - y)\,\Delta y$ ft · lb so the work to pump the water is the limit of these Riemann sums as the norm of

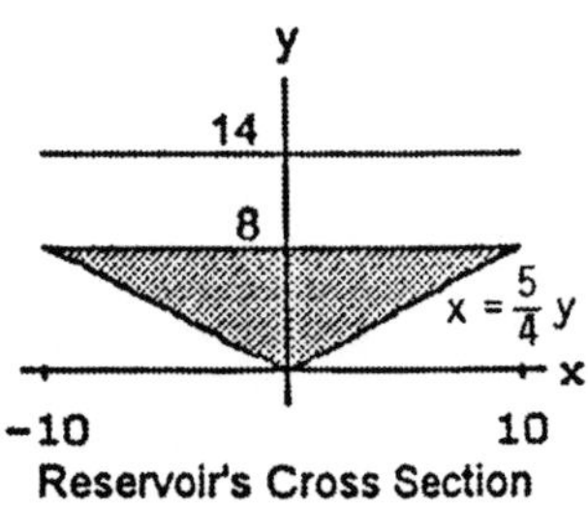

Reservoir's Cross Section

the partition goes to zero: $W = \int_0^8 \frac{(62.4)(25)}{(16)} \pi y^2(14-y)\,dy = \frac{(62.4)(25)\pi}{16} \int_0^8 (14y^2 - y^3)\,dy = (62.4)\left(\frac{25\pi}{16}\right)\left[\frac{14}{3}y^3 - \frac{y^4}{4}\right]_0^8$
$= (62.4)\left(\frac{25\pi}{16}\right)\left(\frac{14}{3}\cdot 8^3 - \frac{8^4}{4}\right) \approx 418{,}208.81$ ft · lb

42. The same as in Exercise 41, but change the distance through which F(y) must act to $(8-y)$ rather than $(6+8-y)$. Also change the upper limit of integration from 8 to 5. The integral is:
$W = \int_0^5 \frac{(62.4)(25)\pi}{16} y^2(8-y)\,dy = (62.4)\left(\frac{25\pi}{16}\right)\int_0^5 (8y^2 - y^3)\,dy = (62.4)\left(\frac{25\pi}{16}\right)\left[\frac{8}{3}y^3 - \frac{y^4}{4}\right]_0^5$
$= (62.4)\left(\frac{25\pi}{16}\right)\left(\frac{8}{3}\cdot 5^3 - \frac{5^4}{4}\right) \approx 54{,}241.56$ ft · lb

43. The tank's cross section looks like the figure in Exercise 41 with right edge given by $x = \frac{5}{10}y = \frac{y}{2}$. A typical horizontal slab has volume $\Delta V = \pi(\text{radius})^2(\text{thickness}) = \pi\left(\frac{y}{2}\right)^2 \Delta y = \frac{\pi}{4} y^2\,\Delta y$. The force required to lift this slab is its weight: $F(y) = 60 \cdot \frac{\pi}{4} y^2\,\Delta y$. The distance through which F(y) must act is $(2 + 10 - y)$ ft, so the work to pump the liquid is $W = 60\int_0^{10} \pi(12-y)\left(\frac{y^2}{4}\right) dy = 15\pi\left[\frac{12y^3}{3} - \frac{y^4}{4}\right]_0^{10} = 22{,}500\pi$ ft · lb; the time needed to empty the tank is $\frac{22{,}500 \text{ ft·lb}}{275 \text{ ft·lb/sec}} \approx 257$ sec

44. A typical horizontal slab has volume about $\Delta V = (20)(2x)\Delta y = (20)\left(2\sqrt{16-y^2}\right)\Delta y$ and the force required to lift this slab is its weight $F(y) = (57)(20)\left(2\sqrt{16-y^2}\right)\Delta y$. The distance through which F(y) must act is $(6+4-y)$ ft, so the work to pump the olive oil from the half-full tank is
$W = 57\int_{-4}^{0} (10-y)(20)\left(2\sqrt{16-y^2}\right) dy = 2880\int_{-4}^{0} 10\sqrt{16-y^2}\,dy + 1140\int_{-4}^{0} (16-y^2)^{1/2}(-2y)\,dy$
$= 22{,}800 \cdot (\text{area of a quarter circle having radius 4}) + \frac{2}{3}(1140)\left[(16-y^2)^{3/2}\right]_{-4}^{0} = (22{,}800)(4\pi) + 48{,}640$
$= 335{,}153.25$ ft · lb

45. $F = \int_a^b W \cdot \left(\begin{smallmatrix}\text{strip}\\\text{depth}\end{smallmatrix}\right) \cdot L(y)\,dy \Rightarrow F = 2\int_0^2 (62.4)(2-y)(2y)\,dy = 249.6\int_0^2 (2y - y^2)\,dy = 249.6\left[y^2 - \frac{y^3}{3}\right]_0^2$
$= (249.6)\left(4 - \frac{8}{3}\right) = (249.6)\left(\frac{4}{3}\right) = 332.8$ lb

46. $F = \int_a^b W \cdot \left(\begin{smallmatrix}\text{strip}\\\text{depth}\end{smallmatrix}\right) \cdot L(y)\,dy \Rightarrow F = \int_0^{5/6} 75\left(\frac{5}{6} - y\right)(2y+4)\,dy = 75\int_0^{5/6}\left(\frac{5}{3}y + \frac{10}{3} - 2y^2 - 4y\right) dy$
$= 75\int_0^{5/6}\left(\frac{10}{3} - \frac{7}{3}y - 2y^2\right) dy = 75\left[\frac{10}{3}y - \frac{7}{6}y^2 - \frac{2}{3}y^3\right]_0^{5/6} = (75)\left[\left(\frac{50}{18}\right) - \left(\frac{7}{6}\right)\left(\frac{25}{36}\right) - \left(\frac{2}{3}\right)\left(\frac{125}{216}\right)\right]$
$= (75)\left(\frac{25}{9} - \frac{175}{216} - \frac{250}{3\cdot 216}\right) = \left(\frac{75}{9\cdot 216}\right)(25\cdot 216 - 175\cdot 9 - 250\cdot 3) = \frac{(75)(3075)}{9\cdot 216} \approx 118.63$ lb.

47. $F = \int_a^b W \cdot \left(\begin{smallmatrix}\text{strip}\\\text{depth}\end{smallmatrix}\right) \cdot L(y)\,dy \Rightarrow F = 62.4\int_0^4 (9-y)\left(2\cdot\frac{\sqrt{y}}{2}\right) dy = 62.4\int_0^4 \left(9y^{1/2} - 3y^{3/2}\right) dy$
$= 62.4\left[6y^{3/2} - \frac{2}{5}y^{5/2}\right]_0^4 = (62.4)\left(6\cdot 8 - \frac{2}{5}\cdot 32\right) = \left(\frac{62.4}{5}\right)(48\cdot 5 - 64) = \frac{(62.4)(176)}{5} = 2196.48$ lb

48. Place the origin at the bottom of the tank. Then $F = \int_0^h W \cdot \left(\begin{smallmatrix}\text{strip}\\\text{depth}\end{smallmatrix}\right) \cdot L(y)\,dy$, h = the height of the mercury column, strip depth $= h - y$, $L(y) = 1 \Rightarrow F = \int_0^h 849(h-y)\,1\,dy = (849)\int_0^h (h-y)\,dy = 849\left[hy - \frac{y^2}{2}\right]_0^h = 849\left(h^2 - \frac{h^2}{2}\right)$
$= \frac{849}{2}h^2$. Now solve $\frac{849}{2}h^2 = 40000$ to get $h \approx 9.707$ ft. The volume of the mercury is $s^2h = 1^2 \cdot 9.707 = 9.707$ ft^3.

49. $F = w_1\int_0^6 (8-y)(2)(6-y)\,dy + w_2\int_{-6}^{0} (8-y)(2)(y+6)\,dy = 2w_1\int_0^6 (48 - 14y + y^2)\,dy + 2w_2\int_{-6}^{0} (48 + 2y - y^2)\,dy$
$= 2w_1\left[48y - 7y^2 + \frac{y^3}{3}\right]_0^6 + 2w_2\left[48y + y^2 - \frac{y^3}{3}\right]_{-6}^{0} = 216w_1 + 360w_2$

50. (a) $F = 62.4\int_0^6 (10 - y)\left[\left(8 - \frac{y}{6}\right) - \left(\frac{y}{6}\right)\right] dy$

$= \frac{62.4}{3}\int_0^6 (240 - 34y + y^2)\, dy$

$= \frac{62.4}{3}\left[240y - 17y^2 + \frac{y^3}{3}\right]_0^6 = \frac{62.4}{3}(1440 - 612 + 72)$

$= 18{,}720$ lb.

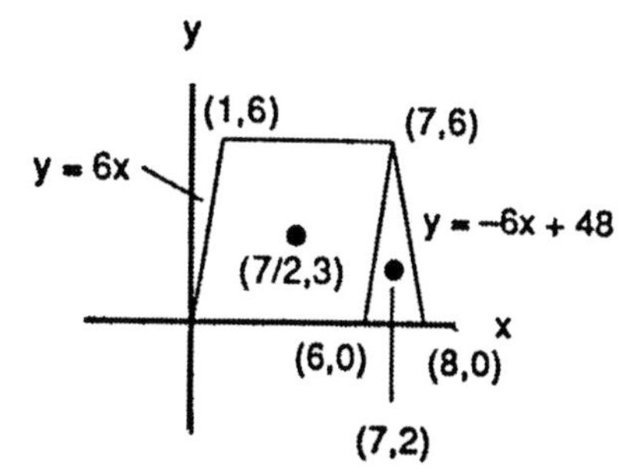

(b) The centroid $\left(\frac{7}{2}, 3\right)$ of the parallelogram is located at the intersection of $y = \frac{6}{7}x$ and $y = -\frac{6}{5}x + \frac{36}{5}$. The centroid of the triangle is located at (7, 2). Therefore, F = (62.4)(7)(36) + (62.4)(8)(6) = (300)(62.4) = 18,720 lb

CHAPTER 6 ADDITIONAL AND ADVANCED EXERCISES

1. $V = \pi\int_a^b [f(x)]^2\, dx = b^2 - ab \Rightarrow \pi\int_a^x [f(t)]^2\, dt = x^2 - ax$ for all $x > a \Rightarrow \pi[f(x)]^2 = 2x - a \Rightarrow f(x) = \sqrt{\frac{2x - a}{\pi}}$

2. $V = \pi\int_0^a [f(x)]^2\, dx = a^2 + a \Rightarrow \pi\int_0^x [f(t)]^2\, dt = x^2 + x$ for all $x > a \Rightarrow \pi[f(x)]^2 = 2x + 1 \Rightarrow f(x) = \sqrt{\frac{2x+1}{\pi}}$

3. $s(x) = Cx \Rightarrow \int_0^x \sqrt{1 + [f'(t)]^2}\, dt = Cx \Rightarrow \sqrt{1 + [f'(x)]^2} = C \Rightarrow f'(x) = \sqrt{C^2 - 1}$ for $C \ge 1$
$\Rightarrow f(x) = \int_0^x \sqrt{C^2 - 1}\, dt + k$. Then $f(0) = a \Rightarrow a = 0 + k \Rightarrow f(x) = \int_0^x \sqrt{C^2 - 1}\, dt + a \Rightarrow f(x) = x\sqrt{C^2 - 1} + a$, where $C \ge 1$.

4. (a) The graph of $f(x) = \sin x$ traces out a path from $(0, 0)$ to $(\alpha, \sin\alpha)$ whose length is $L = \int_0^\alpha \sqrt{1 + \cos^2\theta}\, d\theta$. The line segment from $(0, 0)$ to $(\alpha, \sin\alpha)$ has length $\sqrt{(\alpha - 0)^2 + (\sin\alpha - 0)^2} = \sqrt{\alpha^2 + \sin^2\alpha}$. Since the shortest distance between two points is the length of the straight line segment joining them, we have immediately that $\int_0^\alpha \sqrt{1 + \cos^2\theta}\, d\theta > \sqrt{\alpha^2 + \sin^2\alpha}$ if $0 < \alpha \le \frac{\pi}{2}$.

(b) In general, if $y = f(x)$ is continuously differentiable and $f(0) = 0$, then $\int_0^\alpha \sqrt{1 + [f'(t)]^2}\, dt > \sqrt{\alpha^2 + f^2(\alpha)}$ for $\alpha > 0$.

5. From the symmetry of $y = 1 - x^n$, n even, about the y-axis for $-1 \le x \le 1$, we have $\bar{x} = 0$. To find $\bar{y} = \frac{M_x}{M}$, we use the vertical strips technique. The typical strip has center of mass: $(\tilde{x}, \tilde{y}) = \left(x, \frac{1 - x^n}{2}\right)$, length: $1 - x^n$, width: dx, area: $dA = (1 - x^n)\, dx$, mass: $dm = 1 \cdot dA = (1 - x^n)\, dx$. The moment of the strip about the x-axis is $\tilde{y}\, dm = \frac{(1 - x^n)^2}{2}\, dx \Rightarrow M_x = \int_{-1}^1 \frac{(1 - x^n)^2}{2}\, dx = 2\int_0^1 \frac{1}{2}(1 - 2x^n + x^{2n})\, dx = \left[x - \frac{2x^{n+1}}{n+1} + \frac{x^{2n+1}}{2n+1}\right]_0^1$
$= 1 - \frac{2}{n+1} + \frac{1}{2n+1} = \frac{(n+1)(2n+1) - 2(2n+1) + (n+1)}{(n+1)(2n+1)} = \frac{2n^2 + 3n + 1 - 4n - 2 + n + 1}{(n+1)(2n+1)} = \frac{2n^2}{(n+1)(2n+1)}$.
Also, $M = \int_{-1}^1 dA = \int_{-1}^1 (1 - x^n)\, dx = 2\int_0^1 (1 - x^n)\, dx = 2\left[x - \frac{x^{n+1}}{n+1}\right]_0^1 = 2\left(1 - \frac{1}{n+1}\right) = \frac{2n}{n+1}$. Therefore, $\bar{y} = \frac{M_x}{M} = \frac{2n^2}{(n+1)(2n+1)} \cdot \frac{(n+1)}{2n} = \frac{n}{2n+1} \Rightarrow \left(0, \frac{n}{2n+1}\right)$ is the location of the centroid. As $n \to \infty$, $\bar{y} \to \frac{1}{2}$ so the limiting position of the centroid is $\left(0, \frac{1}{2}\right)$.

6. Align the telephone pole along the x-axis as shown in the accompanying figure. The slope of the top length of pole is $\frac{\left(\frac{14.5}{8\pi} - \frac{9}{8\pi}\right)}{40} = \frac{1}{8\pi} \cdot \frac{1}{40} \cdot (14.5 - 9) = \frac{5.5}{8\pi \cdot 40} = \frac{11}{8\pi \cdot 80}$. Thus, $y = \frac{9}{8\pi} + \frac{11}{8\pi \cdot 80}\, x = \frac{1}{8\pi}\left(9 + \frac{11}{80}\, x\right)$ is an equation of the line representing the top of the pole. Then, $M_y = \int_a^b x \cdot \pi y^2\, dx = \pi \int_0^{40} x \left[\frac{1}{8\pi}\left(9 + \frac{11}{80}\, x\right)\right]^2 dx$ $= \frac{1}{64\pi} \int_0^{40} x\left(9 + \frac{11}{80}\, x\right)^2 dx$; $M = \int_a^b \pi y^2\, dx$ $= \pi \int_0^{40} \left[\frac{1}{8\pi}\left(9 + \frac{11}{80}\, x\right)\right]^2 dx = \frac{1}{64\pi} \int_0^{40} \left(9 + \frac{11}{80}\, x\right)^2 dx$. Thus, $\overline{x} = \frac{M_y}{M} \approx \frac{129{,}700}{5623.3} \approx 23.06$ (using a calculator to compute the integrals). By symmetry about the x-axis, $\overline{y} = 0$ so the center of mass is about 23 ft from the top of the pole.

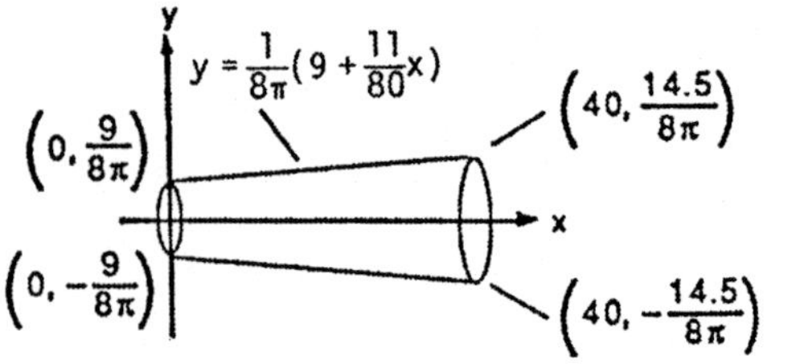

7. (a) Consider a single vertical strip with center of mass $(\tilde{x}, \tilde{y})$. If the plate lies to the right of the line, then the moment of this strip about the line $x = b$ is $(\tilde{x} - b)\, dm = (\tilde{x} - b)\, \delta\, dA \Rightarrow$ the plate's first moment about $x = b$ is the integral $\int (x - b)\delta\, dA = \int \delta x\, dA - \int \delta b\, dA = M_y - b\delta A$.

(b) If the plate lies to the left of the line, the moment of a vertical strip about the line $x = b$ is $(b - \tilde{x})\, dm = (b - \tilde{x})\, \delta\, dA \Rightarrow$ the plate's first moment about $x = b$ is $\int (b - x)\delta\, dA = \int b\delta\, dA - \int \delta x\, dA$ $= b\delta A - M_y$.

8. (a) By symmetry of the plate about the x-axis, $\overline{y} = 0$. A typical vertical strip has center of mass: $(\tilde{x}, \tilde{y}) = (x, 0)$, length: $4\sqrt{ax}$, width: dx, area: $4\sqrt{ax}\, dx$, mass: $dm = \delta\, dA = kx \cdot 4\sqrt{ax}\, dx$, for some proportionality constant k. The moment of the strip about the y-axis is $M_y = \int \tilde{x}\, dm = \int_0^a 4kx^2\sqrt{ax}\, dx$ $= 4k\sqrt{a}\int_0^a x^{5/2}\, dx = 4k\sqrt{a}\left[\frac{2}{7}\, x^{7/2}\right]_0^a = 4ka^{1/2} \cdot \frac{2}{7}\, a^{7/2} = \frac{8ka^4}{7}$. Also, $M = \int dm = \int_0^a 4kx\sqrt{ax}\, dx$ $= 4k\sqrt{a}\int_0^a x^{3/2}\, dx = 4k\sqrt{a}\left[\frac{2}{5}\, x^{5/2}\right]_0^a = 4ka^{1/2} \cdot \frac{2}{5}\, a^{5/2} = \frac{8ka^3}{5}$. Thus, $\overline{x} = \frac{M_y}{M} = \frac{8ka^4}{7} \cdot \frac{5}{8ka^3} = \frac{5}{7}\, a$ $\Rightarrow (\overline{x}, \overline{y}) = \left(\frac{5a}{7}, 0\right)$ is the center of mass.

(b) A typical horizontal strip has center of mass: $(\tilde{x}, \tilde{y}) = \left(\frac{\frac{y^2}{4a} + a}{2}, y\right) = \left(\frac{y^2 + 4a^2}{8a}, y\right)$, length: $a - \frac{y^2}{4a}$, width: dy, area: $\left(a - \frac{y^2}{4a}\right) dy$, mass: $dm = \delta\, dA = |y|\left(a - \frac{y^2}{4a}\right) dy$. Thus, $M_x = \int \tilde{y}\, dm$

$$= \int_{-2a}^{2a} y\,|y|\left(a - \frac{y^2}{4a}\right) dy = \int_{-2a}^{0} -y^2\left(a - \frac{y^2}{4a}\right) dy + \int_0^{2a} y^2\left(a - \frac{y^2}{4a}\right) dy$$

$$= \int_{-2a}^{0}\left(-ay^2 + \frac{y^4}{4a}\right) dy + \int_0^{2a}\left(ay^2 - \frac{y^4}{4a}\right) dy = \left[-\frac{a}{3}\, y^3 + \frac{y^5}{20a}\right]_{-2a}^{0} + \left[\frac{a}{3}\, y^3 - \frac{y^5}{20a}\right]_0^{2a}$$

$$= -\frac{8a^4}{3} + \frac{32a^5}{20a} + \frac{8a^4}{3} - \frac{32a^5}{20a} = 0;\ M_y = \int \tilde{x}\, dm = \int_{-2a}^{2a}\left(\frac{y^2 + 4a^2}{8a}\right)|y|\left(a - \frac{y^2}{4a}\right) dy$$

$$= \frac{1}{8a}\int_{-2a}^{2a} |y|\,(y^2 + 4a^2)\left(\frac{4a^2 - y^2}{4a}\right) dy = \frac{1}{32a^2}\int_{-2a}^{2a} |y|\,(16a^4 - y^4)\, dy$$

$$= \frac{1}{32a^2}\int_{-2a}^{0}(-16a^4y + y^5)\, dy + \frac{1}{32a^2}\int_0^{2a}(16a^4y - y^5)\, dy = \frac{1}{32a^2}\left[-8a^4y^2 + \frac{y^6}{6}\right]_{-2a}^{0} + \frac{1}{32a^2}\left[8a^4y^2 - \frac{y^6}{6}\right]_0^{2a}$$

$$= \frac{1}{32a^2}\left[8a^4 \cdot 4a^2 - \frac{64a^6}{6}\right] + \frac{1}{32a^2}\left[8a^4 \cdot 4a^2 - \frac{64a^6}{6}\right] = \frac{1}{16a^2}\left(32a^6 - \frac{32a^6}{3}\right) = \frac{1}{16a^2} \cdot \frac{2}{3}\,(32a^6) = \frac{4}{3}\, a^4;$$

$$M = \int dm = \int_{-2a}^{2a} |y|\left(\frac{4a^2 - y^2}{4a}\right) dy = \frac{1}{4a}\int_{-2a}^{2a} |y|\,(4a^2 - y^2)\, dy$$

$$= \frac{1}{4a}\int_{-2a}^{0}(-4a^2y + y^3)\, dy + \frac{1}{4a}\int_0^{2a}(4a^2y - y^3)\, dy = \frac{1}{4a}\left[-2a^2y^2 + \frac{y^4}{4}\right]_{-2a}^{0} + \frac{1}{4a}\left[2a^2y^2 - \frac{y^4}{4}\right]_0^{2a}$$

$= 2 \cdot \frac{1}{4a}\left(2a^2 \cdot 4a^2 - \frac{16a^4}{4}\right) = \frac{1}{2a}\,(8a^4 - 4a^4) = 2a^3$. Therefore, $\overline{x} = \frac{M_y}{M} = \left(\frac{4}{3}\, a^4\right)\left(\frac{1}{2a^3}\right) = \frac{2a}{3}$ and $\overline{y} = \frac{M_x}{M} = 0$ is the center of mass.

9. (a) On $[0, a]$ a typical *vertical* strip has center of mass: $(\tilde{x}, \tilde{y}) = \left(x, \frac{\sqrt{b^2 - x^2} + \sqrt{a^2 - x^2}}{2}\right)$, length: $\sqrt{b^2 - x^2} - \sqrt{a^2 - x^2}$, width: dx, area: $dA = \left(\sqrt{b^2 - x^2} - \sqrt{a^2 - x^2}\right) dx$, mass: $dm = \delta\, dA$ $= \delta\left(\sqrt{b^2 - x^2} - \sqrt{a^2 - x^2}\right) dx$. On $[a, b]$ a typical *vertical* strip has center of mass: $(\tilde{x}, \tilde{y}) = \left(x, \frac{\sqrt{b^2 - x^2}}{2}\right)$, length: $\sqrt{b^2 - x^2}$, width: dx, area: $dA = \sqrt{b^2 - x^2}\, dx$, mass: $dm = \delta\, dA = \delta\sqrt{b^2 - x^2}\, dx$. Thus, $M_x = \int \tilde{y}\, dm$

$= \int_0^a \frac{1}{2}\left(\sqrt{b^2 - x^2} + \sqrt{a^2 - x^2}\right)\delta\left(\sqrt{b^2 - x^2} - \sqrt{a^2 - x^2}\right) dx + \int_a^b \frac{1}{2}\sqrt{b^2 - x^2}\,\delta\sqrt{b^2 - x^2}\, dx$

$= \frac{\delta}{2}\int_0^a [(b^2 - x^2) - (a^2 - x^2)]\, dx + \frac{\delta}{2}\int_a^b (b^2 - x^2)\, dx = \frac{\delta}{2}\int_0^a (b^2 - a^2)\, dx + \frac{\delta}{2}\int_a^b (b^2 - x^2)\, dx$

$= \frac{\delta}{2}\left[(b^2 - a^2)\, x\right]_0^a + \frac{\delta}{2}\left[b^2 x - \frac{x^3}{3}\right]_a^b = \frac{\delta}{2}\left[(b^2 - a^2)\, a\right] + \frac{\delta}{2}\left[\left(b^3 - \frac{b^3}{3}\right) - \left(b^2 a - \frac{a^3}{3}\right)\right]$

$= \frac{\delta}{2}(ab^2 - a^3) + \frac{\delta}{2}\left(\frac{2}{3}b^3 - ab^2 + \frac{a^3}{3}\right) = \frac{\delta b^3}{3} - \frac{\delta a^3}{3} = \delta\left(\frac{b^3 - a^3}{3}\right)$; $M_y = \int \tilde{x}\, dm$

$= \int_0^a x\delta\left(\sqrt{b^2 - x^2} - \sqrt{a^2 - x^2}\right) dx + \int_a^b x\delta\sqrt{b^2 - x^2}\, dx$

$= \delta\int_0^a x\,(b^2 - x^2)^{1/2}\, dx - \delta\int_0^a x\,(a^2 - x^2)^{1/2}\, dx + \delta\int_a^b x\,(b^2 - x^2)^{1/2}\, dx$

$= \frac{-\delta}{2}\left[\frac{2(b^2 - x^2)^{3/2}}{3}\right]_0^a + \frac{\delta}{2}\left[\frac{2(a^2 - x^2)^{3/2}}{3}\right]_0^a - \frac{\delta}{2}\left[\frac{2(b^2 - x^2)^{3/2}}{3}\right]_a^b$

$= -\frac{\delta}{3}\left[(b^2 - a^2)^{3/2} - (b^2)^{3/2}\right] + \frac{\delta}{3}\left[0 - (a^2)^{3/2}\right] - \frac{\delta}{3}\left[0 - (b^2 - a^2)^{3/2}\right] = \frac{\delta b^3}{3} - \frac{\delta a^3}{3} = \frac{\delta(b^3 - a^3)}{3} = M_x$;

We calculate the mass geometrically: $M = \delta A = \delta\left(\frac{\pi b^2}{4}\right) - \delta\left(\frac{\pi a^2}{4}\right) = \frac{\delta\pi}{4}(b^2 - a^2)$. Thus, $\bar{x} = \frac{M_y}{M}$

$= \frac{\delta(b^3 - a^3)}{3} \cdot \frac{4}{\delta\pi(b^2 - a^2)} = \frac{4}{3\pi}\left(\frac{b^3 - a^3}{b^2 - a^2}\right) = \frac{4}{3\pi}\, \frac{(b - a)(a^2 + ab + b^2)}{(b - a)(b + a)} = \frac{4(a^2 + ab + b^2)}{3\pi(a + b)}$; likewise

$\bar{y} = \frac{M_x}{M} = \frac{4(a^2 + ab + b^2)}{3\pi(a + b)}$.

(b) $\lim_{b \to a} \frac{4}{3\pi}\left(\frac{a^2 + ab + b^2}{a + b}\right) = \left(\frac{4}{3\pi}\right)\left(\frac{a^2 + a^2 + a^2}{a + a}\right) = \left(\frac{4}{3\pi}\right)\left(\frac{3a^2}{2a}\right) = \frac{2a}{\pi} \Rightarrow (\bar{x}, \bar{y}) = \left(\frac{2a}{\pi}, \frac{2a}{\pi}\right)$ is the limiting position of the centroid as $b \to a$. This is the centroid of a circle of radius a (and we note the two circles coincide when $b = a$).

10. Since the area of the traingle is 36, the diagram may be labeled as shown at the right. The centroid of the triangle is $\left(\frac{a}{3}, \frac{24}{a}\right)$. The shaded portion is $144 - 36 = 108$. Write $(\underline{x}, \underline{y})$ for the centroid of the remaining region. The centroid of the whole square is obviously (6, 6). Think of the square as a sheet of uniform density, so that the centroid of the square is the average of the centroids of the two regions, weighted by area:

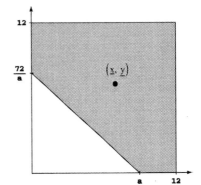

$6 = \frac{36\left(\frac{a}{3}\right) + 108(\underline{x})}{144}$ and $6 = \frac{36\left(\frac{24}{a}\right) + 108(\underline{y})}{144}$

which we solve to get $\underline{x} = 8 - \frac{a}{9}$ and $\underline{y} = \frac{8(a - 1)}{a}$. Set $\underline{x} = 7$ in. (Given). It follows that $a = 9$, whence $\underline{y} = \frac{64}{9}$ $= 7\frac{1}{9}$ in. The distances of the centroid $(\underline{x}, \underline{y})$ from the other sides are easily computed. (Note that if we set $\underline{y} = 7$ in. above, we will find $\underline{x} = 7\frac{1}{9}$.)

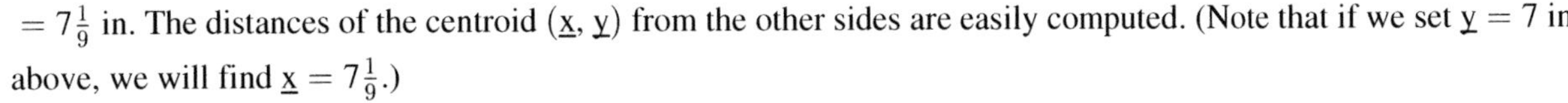

11. $y = 2\sqrt{x} \Rightarrow ds = \sqrt{\frac{1}{x} + 1}\, dx \Rightarrow A = \int_0^3 2\sqrt{x}\sqrt{\frac{1}{x} + 1}\, dx = \frac{4}{3}\left[(1 + x)^{3/2}\right]_0^3 = \frac{28}{3}$

12. This surface is a triangle having a base of $2\pi a$ and a height of $2\pi ak$. Therefore the surface area is $\frac{1}{2}(2\pi a)(2\pi ak) = 2\pi^2 a^2 k$.

13. $F = ma = t^2 \Rightarrow \frac{d^2x}{dt^2} = a = \frac{t^2}{m} \Rightarrow v = \frac{dx}{dt} = \frac{t^3}{3m} + C$; $v = 0$ when $t = 0 \Rightarrow C = 0 \Rightarrow \frac{dx}{dt} = \frac{t^3}{3m} \Rightarrow x = \frac{t^4}{12m} + C_1$; $x = 0$ when $t = 0 \Rightarrow C_1 = 0 \Rightarrow x = \frac{t^4}{12m}$. Then $x = h \Rightarrow t = (12mh)^{1/4}$. The work done is

$$W = \int F\,dx = \int_0^{(12mh)^{1/4}} F(t) \cdot \frac{dx}{dt}\,dt = \int_0^{(12mh)^{1/4}} t^2 \cdot \frac{t^3}{3m}\,dt = \frac{1}{3m}\left[\frac{t^6}{6}\right]_0^{(12mh)^{1/4}} = \left(\frac{1}{18m}\right)(12mh)^{6/4}$$

$$= \frac{(12mh)^{3/2}}{18m} = \frac{12mh \cdot \sqrt{12mh}}{18m} = \frac{2h}{3} \cdot 2\sqrt{3mh} = \frac{4h}{3}\sqrt{3mh}$$

14. Converting to pounds and feet, 2 lb/in $= \frac{2\text{ lb}}{1\text{ in}} \cdot \frac{12\text{ in}}{1\text{ ft}} = 24$ lb/ft. Thus, $F = 24x \Rightarrow W = \int_0^{1/2} 24x\,dx$ $= \left[12x^2\right]_0^{1/2} = 3$ ft · lb. Since $W = \frac{1}{2}mv_0^2 - \frac{1}{2}mv_1^2$, where $W = 3$ ft · lb, $m = \left(\frac{1}{10}\text{ lb}\right)\left(\frac{1}{32\text{ ft/sec}^2}\right)$ $= \frac{1}{320}$ slugs, and $v_1 = 0$ ft/sec, we have $3 = \left(\frac{1}{2}\right)\left(\frac{1}{320}v_0^2\right) \Rightarrow v_0^2 = 3 \cdot 640$. For the projectile height, $s = -16t^2 + v_0t$ (since $s = 0$ at $t = 0$) $\Rightarrow \frac{ds}{dt} = v = -32t + v_0$. At the top of the ball's path, $v = 0 \Rightarrow t = \frac{v_0}{32}$ and the height is $s = -16\left(\frac{v_0}{32}\right)^2 + v_0\left(\frac{v_0}{32}\right) = \frac{v_0^2}{64} = \frac{3 \cdot 640}{64} = 30$ ft.

15. The submerged triangular plate is depicted in the figure at the right. The hypotenuse of the triangle has slope -1 $\Rightarrow y - (-2) = -(x - 0) \Rightarrow x = -(y + 2)$ is an equation of the hypotenuse. Using a typical horizontal strip, the fluid pressure is $F = \int (62.4) \cdot \binom{\text{strip}}{\text{depth}} \cdot \binom{\text{strip}}{\text{length}}\,dy$

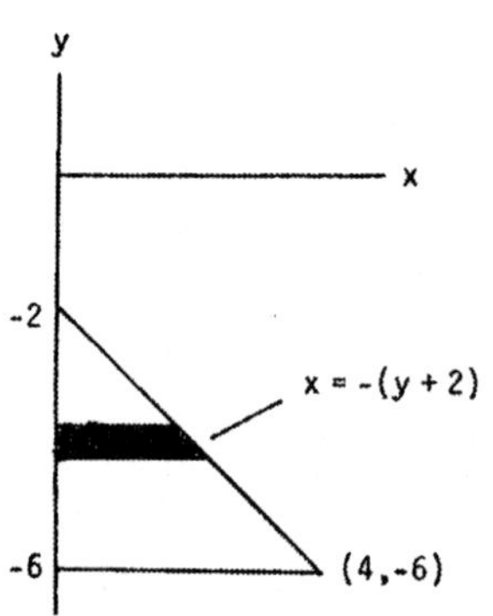

$$= \int_{-6}^{-2} (62.4)(-y)[-(y+2)]\,dy = 62.4 \int_{-6}^{-2} (y^2 + 2y)\,dy$$

$$= 62.4\left[\frac{y^3}{3} + y^2\right]_{-6}^{-2} = (62.4)\left[\left(-\frac{8}{3} + 4\right) - \left(-\frac{216}{3} + 36\right)\right]$$

$$= (62.4)\left(\frac{208}{3} - 32\right) = \frac{(62.4)(112)}{3} \approx 2329.6 \text{ lb}$$

16. Consider a rectangular plate of length ℓ and width w. The length is parallel with the surface of the fluid of weight density ω. The force on one side of the plate is $F = \omega \int_{-w}^{0} (-y)(\ell)\,dy = -\omega\ell\left[\frac{y^2}{2}\right]_{-w}^{0} = \frac{\omega\ell w^2}{2}$. The average force on one side of the plate is $F_{av} = \frac{\omega}{w}\int_{-w}^{0}(-y)dy$ $= \frac{\omega}{w}\left[-\frac{y^2}{2}\right]_{-w}^{0} = \frac{\omega w}{2}$. Therefore the force $\frac{\omega\ell w^2}{2}$ $= \left(\frac{\omega w}{2}\right)(\ell w) =$ (the average pressure up and down) · (the area of the plate).

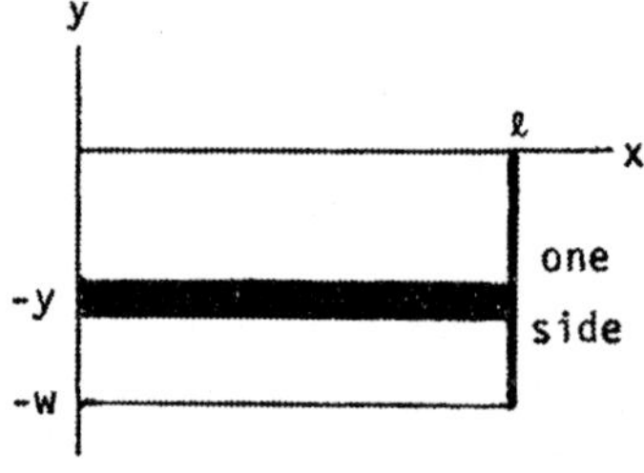

17. (a) We establish a coordinate system as shown. A typical horizontal strip has: center of pressure: $(\tilde{x}, \tilde{y})$ $= \left(\frac{b}{2}, y\right)$, length: $L(y) = b$, width: dy, area: dA $= b\,dy$, pressure: $dp = \omega\,|y|\,dA = \omega b\,|y|\,dy$

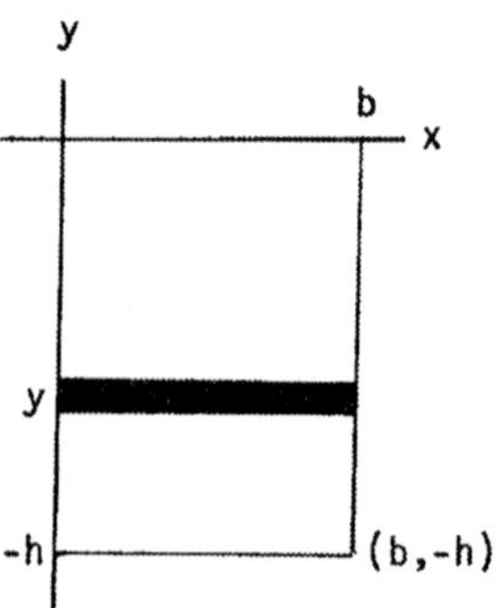

$$\Rightarrow F_x = \int \tilde{y}\,dp = \int_{-h}^{0} y \cdot \omega b\,|y|\,dy = -\omega b \int_{-h}^{0} y^2\,dy$$

$$= -\omega b\left[\frac{y^3}{3}\right]_{-h}^{0} = -\omega b\left[0 - \left(\frac{-h^3}{3}\right)\right] = \frac{-\omega b h^3}{3};$$

$$F = \int dp = \int_{-h}^{0} \omega\,|y|\,L(y)\,dy = -\omega b \int_{-h}^{0} y\,dy$$

$$= -\omega b\left[\frac{y^2}{2}\right]_{-h}^{0} = -\omega b\left[0 - \frac{h^2}{2}\right] = \frac{\omega b h^2}{2}.$$ Thus, $\bar{y} = \frac{F_x}{F} = \frac{\left(\frac{-\omega b h^3}{3}\right)}{\left(\frac{\omega b h^2}{2}\right)} = \frac{-2h}{3} \Rightarrow$ the distance below the surface is $\frac{2}{3}$h.

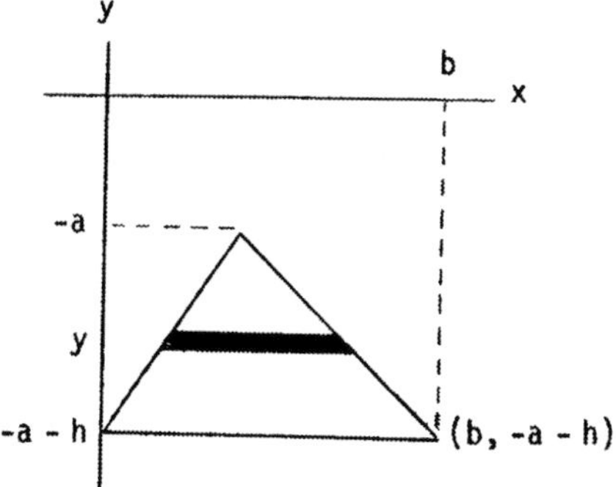

(b) A typical horizontal strip has length L(y). By similar triangles from the figure at the right, $\frac{L(y)}{b} = \frac{-y-a}{h}$ $\Rightarrow L(y) = -\frac{b}{h}(y+a)$. Thus, a typical strip has center of pressure: $(\tilde{x}, \tilde{y}) = (\tilde{x}, y)$, length: $L(y) = -\frac{b}{h}(y+a)$, width: dy, area: $dA = -\frac{b}{h}(y+a)\,dy$, pressure: $dp = \omega\,|y|\,dA = \omega(-y)\left(-\frac{b}{h}\right)(y+a)\,dy$ $= \frac{\omega b}{h}(y^2+ay)\,dy \Rightarrow F_x = \int \tilde{y}\,dp$ $= \int_{-(a+h)}^{-a} y \cdot \frac{\omega b}{h}(y^2+ay)\,dy = \frac{\omega b}{h}\int_{-(a+h)}^{-a}(y^3+ay^2)\,dy$ $= \frac{\omega b}{h}\left[\frac{y^4}{4}+\frac{ay^3}{3}\right]_{-(a+h)}^{-a}$ $= \frac{\omega b}{h}\left[\left(\frac{a^4}{4}-\frac{a^4}{3}\right)-\left(\frac{(a+h)^4}{4}-\frac{a(a+h)^3}{3}\right)\right] = \frac{\omega b}{h}\left[\frac{a^4-(a+h)^4}{4}-\frac{a^4-a(a+h)^3}{3}\right]$ $= \frac{\omega b}{12h}\left[3\left(a^4-(a^4+4a^3h+6a^2h^2+4ah^3+h^4)\right)-4\left(a^4-a\left(a^3+3a^2h+3ah^2+h^3\right)\right)\right]$ $= \frac{\omega b}{12h}\left(12a^3h+12a^2h^2+4ah^3-12a^3h-18a^2h^2-12ah^3-3h^4\right) = \frac{\omega b}{12h}\left(-6a^2h^2-8ah^3-3h^4\right)$ $= \frac{-\omega bh}{12}\left(6a^2+8ah+3h^2\right)$; $F = \int dp = \int \omega\,|y|\,L(y)\,dy = \frac{\omega b}{h}\int_{-(a+h)}^{-a}(y^2+ay)\,dy = \frac{\omega b}{h}\left[\frac{y^3}{3}+\frac{ay^2}{2}\right]_{-(a+h)}^{-a}$ $= \frac{\omega b}{h}\left[\left(\frac{-a^3}{3}+\frac{a^3}{2}\right)-\left(\frac{-(a+h)^3}{3}+\frac{a(a+h)^2}{2}\right)\right] = \frac{\omega b}{h}\left[\frac{(a+h)^3-a^3}{3}+\frac{a^3-a(a+h)^2}{2}\right]$ $= \frac{\omega b}{h}\left[\frac{a^3+3a^2h+3ah^2+h^3-a^3}{3}+\frac{a^3-(a^3+2a^2h+ah^2)}{2}\right] = \frac{\omega b}{6h}\left[2\left(3a^2h+3ah^2+h^3\right)-3\left(2a^2h+ah^2\right)\right]$ $= \frac{\omega b}{6h}\left(6a^2h+6ah^2+2h^3-6a^2h-3ah^2\right) = \frac{\omega b}{6h}\left(3ah^2+2h^3\right) = \frac{\omega bh}{6}(3a+2h)$. Thus, $\overline{y} = \frac{F_x}{F}$ $= \frac{\left(\frac{-\omega bh}{12}\right)\left(6a^2+8ah+3h^2\right)}{\left(\frac{\omega bh}{6}\right)(3a+2h)} = \left(\frac{-1}{2}\right)\left(\frac{6a^2+8ah+3h^2}{3a+2h}\right)$ $\Rightarrow$ the distance below the surface is $\frac{6a^2+8ah+3h^2}{6a+4h}$.

NOTES:

CHAPTER 7 INTEGRALS AND TRANSCENDENTAL FUNCTIONS

7.1 THE LOGARITHM DEFINED AS AN INTEGRAL

1. $\int_{-3}^{-2} \frac{1}{x}\, dx = [\ln|x|]_{-3}^{-2} = \ln 2 - \ln 3 = \ln \frac{2}{3}$

2. $\int_{-1}^{0} \frac{3}{3x-2}\, dx = [\ln|3x-2|]_{-1}^{0} = \ln 2 - \ln 5 = \ln \frac{2}{5}$

3. $\int \frac{2y}{y^2-25}\, dy = \ln|y^2 - 25| + C$

4. $\int \frac{8r}{4r^2-5}\, dr = \ln|4r^2 - 5| + C$

5. Let $u = 6 + 3\tan t \Rightarrow du = 3\sec^2 t\, dt$;
$\int \frac{3\sec^2 t}{6+3\tan t}\, dt = \int \frac{du}{u} = \ln|u| + C = \ln|6 + 3\tan t| + C$

6. Let $u = 2 + \sec y \Rightarrow du = \sec y \tan y\, dy$;
$\int \frac{\sec y \tan y}{2+\sec y}\, dy = \int \frac{du}{u} = \ln|u| + C = \ln|2 + \sec y| + C$

7. $\int \frac{dx}{2\sqrt{x}+2x} = \int \frac{dx}{2\sqrt{x}\,(1+\sqrt{x})}$; let $u = 1 + \sqrt{x} \Rightarrow du = \frac{1}{2\sqrt{x}}\, dx$; $\int \frac{dx}{2\sqrt{x}\,(1+\sqrt{x})} = \int \frac{du}{u} = \ln|u| + C$
$= \ln\left|1 + \sqrt{x}\right| + C = \ln\left(1 + \sqrt{x}\right) + C$

8. Let $u = \sec x + \tan x \Rightarrow du = (\sec x \tan x + \sec^2 x)\, dx = (\sec x)(\tan x + \sec x)\, dx \Rightarrow \sec x\, dx = \frac{du}{u}$;
$\int \frac{\sec x\, dx}{\sqrt{\ln(\sec x + \tan x)}} = \int \frac{du}{u\sqrt{\ln u}} = \int (\ln u)^{-1/2} \cdot \frac{1}{u}\, du = 2(\ln u)^{1/2} + C = 2\sqrt{\ln(\sec x + \tan x)} + C$

9. $\int_{\ln 2}^{\ln 3} e^x\, dx = [e^x]_{\ln 2}^{\ln 3} = e^{\ln 3} - e^{\ln 2} = 3 - 2 = 1$

10. $\int_{-\ln 2}^{\ln 3} e^{-x}\, dx = [-e^{-x}]_{-\ln 2}^{0} = -e^0 + e^{\ln 2} = -1 + 2 = 1$

11. $\int 8e^{(x+1)}\, dx = 8e^{(x+1)} + C$

12. $\int 2e^{(2x-1)}\, dx = e^{(2x-1)} + C$

13. $\int_{\ln 4}^{\ln 9} e^{x/2}\, dx = \left[2e^{x/2}\right]_{\ln 4}^{\ln 9} = 2\left[e^{(\ln 9)/2} - e^{(\ln 4)/2}\right] = 2\left(e^{\ln 3} - e^{\ln 2}\right) = 2(3 - 2) = 2$

14. $\int_0^{\ln 16} e^{x/4}\, dx = \left[4e^{x/4}\right]_0^{\ln 16} = 4\left(e^{(\ln 16)/4} - e^0\right) = 4\left(e^{\ln 2} - 1\right) = 4(2 - 1) = 4$

15. Let $u = r^{1/2} \Rightarrow du = \frac{1}{2} r^{-1/2}\, dr \Rightarrow 2\, du = r^{-1/2}\, dr$;
$\int \frac{e^{\sqrt{r}}}{\sqrt{r}}\, dr = \int e^{r^{1/2}} \cdot r^{-1/2}\, dr = 2\int e^u\, du = 2e^u + C = 2e^{r^{1/2}} + C = 2e^{\sqrt{r}} + C$

16. Let $u = -r^{1/2} \Rightarrow du = -\frac{1}{2} r^{-1/2}\, dr \Rightarrow -2\, du = r^{-1/2}\, dr$;
$\int \frac{e^{-\sqrt{r}}}{\sqrt{r}}\, dr = \int e^{-r^{1/2}} \cdot r^{-1/2}\, dr = -2\int e^u\, du = -2e^{-r^{1/2}} + C = -2e^{-\sqrt{r}} + C$

17. Let $u = -t^2 \Rightarrow du = -2t\, dt \Rightarrow -du = 2t\, dt$;
$\int 2te^{-t^2}\, dt = -\int e^u\, du = -e^u + C = -e^{-t^2} + C$

18. Let $u = t^4 \Rightarrow du = 4t^3\,dt \Rightarrow \frac{1}{4}\,du = t^3\,dt$;
$\int t^3 e^{t^4}\,dt = \frac{1}{4}\int e^u\,du = \frac{1}{4}e^{t^4} + C$

19. Let $u = \frac{1}{x} \Rightarrow du = -\frac{1}{x^2}\,dx \Rightarrow -du = \frac{1}{x^2}\,dx$;
$\int \frac{e^{1/x}}{x^2}\,dx = \int -e^u\,du = -e^u + C = -e^{1/x} + C$

20. Let $u = -x^{-2} \Rightarrow du = 2x^{-3}\,dx \Rightarrow \frac{1}{2}\,du = x^{-3}\,dx$;
$\int \frac{e^{-1/x^2}}{x^3}\,dx = \int e^{-x^{-2}} \cdot x^{-3}\,dx = \frac{1}{2}\int e^u\,du = \frac{1}{2}e^u + C = \frac{1}{2}e^{-x^{-2}} + C = \frac{1}{2}e^{-1/x^2} + C$

21. Let $u = \sec \pi t \Rightarrow du = \pi \sec \pi t \tan \pi t\,dt \Rightarrow \frac{du}{\pi} = \sec \pi t \tan \pi t\,dt$;
$\int e^{\sec(\pi t)}\sec(\pi t)\tan(\pi t)\,dt = \frac{1}{\pi}\int e^u\,du = \frac{e^u}{\pi} + C = \frac{e^{\sec(\pi t)}}{\pi} + C$

22. Let $u = \csc(\pi + t) \Rightarrow du = -\csc(\pi + t)\cot(\pi + t)\,dt$;
$\int e^{\csc(\pi+t)}\csc(\pi + t)\cot(\pi + t)\,dt = -\int e^u\,du = -e^u + C = -e^{\csc(\pi+t)} + C$

23. Let $u = e^v \Rightarrow du = e^v\,dv \Rightarrow 2\,du = 2e^v\,dv$; $v = \ln\frac{\pi}{6} \Rightarrow u = \frac{\pi}{6}$, $v = \ln\frac{\pi}{2} \Rightarrow u = \frac{\pi}{2}$;
$\int_{\ln(\pi/6)}^{\ln(\pi/2)} 2e^v \cos e^v\,dv = 2\int_{\pi/6}^{\pi/2} \cos u\,du = [2\sin u]_{\pi/6}^{\pi/2} = 2\left[\sin\left(\frac{\pi}{2}\right) - \sin\left(\frac{\pi}{6}\right)\right] = 2\left(1 - \frac{1}{2}\right) = 1$

24. Let $u = e^{x^2} \Rightarrow du = 2xe^{x^2}\,dx$; $x = 0 \Rightarrow u = 1$, $x = \sqrt{\ln \pi} \Rightarrow u = e^{\ln \pi} = \pi$;
$\int_0^{\sqrt{\ln \pi}} 2xe^{x^2}\cos\left(e^{x^2}\right)dx = \int_1^{\pi} \cos u\,du = [\sin u]_1^{\pi} = \sin(\pi) - \sin(1) = -\sin(1) \approx -0.84147$

25. Let $u = 1 + e^r \Rightarrow du = e^r\,dr$;
$\int \frac{e^r}{1+e^r}\,dr = \int \frac{1}{u}\,du = \ln|u| + C = \ln(1 + e^r) + C$

26. $\int \frac{1}{1+e^x}\,dx = \int \frac{e^{-x}}{e^{-x}+1}\,dx$;
let $u = e^{-x} + 1 \Rightarrow du = -e^{-x}\,dx \Rightarrow -du = e^{-x}\,dx$;
$\int \frac{e^{-x}}{e^{-x}+1}\,dx = -\int \frac{1}{u}\,du = -\ln|u| + C = -\ln(e^{-x} + 1) + C$

27. $\int_0^1 2^{-\theta}\,d\theta = \int_0^1 \left(\frac{1}{2}\right)^{\theta} d\theta = \left[\frac{\left(\frac{1}{2}\right)^{\theta}}{\ln\left(\frac{1}{2}\right)}\right]_0^1 = \frac{\frac{1}{2}}{\ln\left(\frac{1}{2}\right)} - \frac{1}{\ln\left(\frac{1}{2}\right)} = -\frac{\frac{1}{2}}{\ln\left(\frac{1}{2}\right)} = \frac{-1}{2(\ln 1 - \ln 2)} = \frac{1}{2\ln 2}$

28. $\int_{-2}^0 5^{-\theta}\,d\theta = \int_{-2}^0 \left(\frac{1}{5}\right)^{\theta} d\theta = \left[\frac{\left(\frac{1}{5}\right)^{\theta}}{\ln\left(\frac{1}{5}\right)}\right]_{-2}^0 = \frac{1}{\ln\left(\frac{1}{5}\right)} - \frac{\left(\frac{1}{5}\right)^{-2}}{\ln\left(\frac{1}{5}\right)} = \frac{1}{\ln\left(\frac{1}{5}\right)}(1 - 25) = \frac{-24}{\ln 1 - \ln 5} = \frac{24}{\ln 5}$

29. Let $u = x^2 \Rightarrow du = 2x\,dx \Rightarrow \frac{1}{2}\,du = x\,dx$; $x = 1 \Rightarrow u = 1$, $x = \sqrt{2} \Rightarrow u = 2$;
$\int_1^{\sqrt{2}} x2^{(x^2)}\,dx = \int_1^2 \left(\frac{1}{2}\right) 2^u\,du = \frac{1}{2}\left[\frac{2^u}{\ln 2}\right]_1^2 = \left(\frac{1}{2\ln 2}\right)(2^2 - 2^1) = \frac{1}{\ln 2}$

30. Let $u = x^{1/2} \Rightarrow du = \frac{1}{2}x^{-1/2}\,dx \Rightarrow 2\,du = \frac{dx}{\sqrt{x}}$; $x = 1 \Rightarrow u = 1$, $x = 4 \Rightarrow u = 2$;
$\int_1^4 \frac{2^{\sqrt{x}}}{\sqrt{x}}\,dx = \int_1^4 2^{x^{1/2}} \cdot x^{-1/2}\,dx = 2\int_1^2 2^u\,du = \left[\frac{2^{(u+1)}}{\ln 2}\right]_1^2 = \left(\frac{1}{\ln 2}\right)(2^3 - 2^2) = \frac{4}{\ln 2}$

31. Let $u = \cos t \Rightarrow du = -\sin t\,dt \Rightarrow -du = \sin t\,dt$; $t = 0 \Rightarrow u = 1$, $t = \frac{\pi}{2} \Rightarrow u = 0$;
$\int_0^{\pi/2} 7^{\cos t} \sin t\,dt = -\int_1^0 7^u\,du = \left[-\frac{7^u}{\ln 7}\right]_1^0 = \left(\frac{-1}{\ln 7}\right)(7^0 - 7) = \frac{6}{\ln 7}$

32. Let $u = \tan t \Rightarrow du = \sec^2 t\,dt$; $t = 0 \Rightarrow u = 0$, $t = \frac{\pi}{4} \Rightarrow u = 1$;
$\int_0^{\pi/4} \left(\frac{1}{3}\right)^{\tan t} \sec^2 t\,dt = \int_0^1 \left(\frac{1}{3}\right)^u du = \left[\frac{\left(\frac{1}{3}\right)^u}{\ln\left(\frac{1}{3}\right)}\right]_0^1 = \left(-\frac{1}{\ln 3}\right)\left[\left(\frac{1}{3}\right)^1 - \left(\frac{1}{3}\right)^0\right] = \frac{2}{3 \ln 3}$

33. Let $u = x^{2x} \Rightarrow \ln u = 2x \ln x \Rightarrow \frac{1}{u}\frac{du}{dx} = 2 \ln x + (2x)\left(\frac{1}{x}\right) \Rightarrow \frac{du}{dx} = 2u(\ln x + 1) \Rightarrow \frac{1}{2}\,du = x^{2x}(1 + \ln x)\,dx$;
$x = 2 \Rightarrow u = 2^4 = 16$, $x = 4 \Rightarrow u = 4^8 = 65{,}536$;
$\int_2^4 x^{2x}(1 + \ln x)\,dx = \frac{1}{2}\int_{16}^{65{,}536} du = \frac{1}{2}[u]_{16}^{65{,}536} = \frac{1}{2}(65{,}536 - 16) = \frac{65{,}520}{2} = 32{,}760$

34. Let $u = \ln x \Rightarrow du = \frac{1}{x}\,dx$; $x = 1 \Rightarrow u = 0$, $x = 2 \Rightarrow u = \ln 2$;
$\int_1^2 \frac{2^{\ln x}}{x}\,dx = \int_0^{\ln 2} 2^u\,du = \left[\frac{2^u}{\ln 2}\right]_0^{\ln 2} = \left(\frac{1}{\ln 2}\right)(2^{\ln 2} - 2^0) = \frac{2^{\ln 2} - 1}{\ln 2}$

35. $\int_0^3 \left(\sqrt{2} + 1\right) x^{\sqrt{2}}\,dx = \left[x^{(\sqrt{2}+1)}\right]_0^3 = 3^{(\sqrt{2}+1)}$

36. $\int_1^e x^{(\ln 2)-1}\,dx = \left[\frac{x^{\ln 2}}{\ln 2}\right]_1^e = \frac{e^{\ln 2} - 1^{\ln 2}}{\ln 2} = \frac{2-1}{\ln 2} = \frac{1}{\ln 2}$

37. $\int \frac{\log_{10} x}{x}\,dx = \int \left(\frac{\ln x}{\ln 10}\right)\left(\frac{1}{x}\right)dx$; $\left[u = \ln x \Rightarrow du = \frac{1}{x}\,dx\right]$
$\rightarrow \int \left(\frac{\ln x}{\ln 10}\right)\left(\frac{1}{x}\right)dx = \frac{1}{\ln 10}\int u\,du = \left(\frac{1}{\ln 10}\right)\left(\frac{1}{2}u^2\right) + C = \frac{(\ln x)^2}{2 \ln 10} + C$

38. $\int_1^4 \frac{\log_2 x}{x}\,dx = \int_1^4 \left(\frac{\ln x}{\ln 2}\right)\left(\frac{1}{x}\right)dx$; $\left[u = \ln x \Rightarrow du = \frac{1}{x}\,dx;\ x = 1 \Rightarrow u = 0,\ x = 4 \Rightarrow u = \ln 4\right]$
$\rightarrow \int_1^4 \left(\frac{\ln x}{\ln 2}\right)\left(\frac{1}{x}\right)dx = \int_0^{\ln 4}\left(\frac{1}{\ln 2}\right)u\,du = \left(\frac{1}{\ln 2}\right)\left[\frac{1}{2}u^2\right]_0^{\ln 4} = \left(\frac{1}{\ln 2}\right)\left[\frac{1}{2}(\ln 4)^2\right] = \frac{(\ln 4)^2}{2 \ln 2} = \frac{(\ln 4)^2}{\ln 4} = \ln 4$

39. $\int_1^4 \frac{\ln 2 \log_2 x}{x}\,dx = \int_1^4 \left(\frac{\ln 2}{x}\right)\left(\frac{\ln x}{\ln 2}\right)dx = \int_1^4 \frac{\ln x}{x}\,dx = \left[\frac{1}{2}(\ln x)^2\right]_1^4 = \frac{1}{2}\left[(\ln 4)^2 - (\ln 1)^2\right] = \frac{1}{2}(\ln 4)^2$
$= \frac{1}{2}(2 \ln 2)^2 = 2(\ln 2)^2$

40. $\int_1^e \frac{2 \ln 10 (\log_{10} x)}{x}\,dx = \int_1^e \frac{(\ln 10)(2 \ln x)}{(\ln 10)}\left(\frac{1}{x}\right)dx = \left[(\ln x)^2\right]_1^e = (\ln e)^2 - (\ln 1)^2 = 1$

41. $\int_0^2 \frac{\log_2 (x+2)}{x+2}\,dx = \frac{1}{\ln 2}\int_0^2 [\ln(x+2)]\left(\frac{1}{x+2}\right)dx = \left(\frac{1}{\ln 2}\right)\left[\frac{(\ln(x+2))^2}{2}\right]_0^2 = \left(\frac{1}{\ln 2}\right)\left[\frac{(\ln 4)^2}{2} - \frac{(\ln 2)^2}{2}\right]$
$= \left(\frac{1}{\ln 2}\right)\left[\frac{4(\ln 2)^2}{2} - \frac{(\ln 2)^2}{2}\right] = \frac{3}{2}\ln 2$

42. $\int_{1/10}^{10} \frac{\log_{10}(10x)}{x}\,dx = \frac{10}{\ln 10}\int_{1/10}^{10}[\ln(10x)]\left(\frac{1}{10x}\right)dx = \left(\frac{10}{\ln 10}\right)\left[\frac{(\ln(10x))^2}{20}\right]_{1/10}^{10} = \left(\frac{10}{\ln 10}\right)\left[\frac{(\ln 100)^2}{20} - \frac{(\ln 1)^2}{2}\right]$
$= \left(\frac{10}{\ln 10}\right)\left[\frac{4(\ln 10)^2}{20}\right] = 2 \ln 10$

43. $\int_0^9 \frac{2 \log_{10}(x+1)}{x+1}\,dx = \frac{2}{\ln 10}\int_0^9 \ln(x+1)\left(\frac{1}{x+1}\right)dx = \left(\frac{2}{\ln 10}\right)\left[\frac{(\ln(x+1))^2}{2}\right]_0^9 = \left(\frac{2}{\ln 10}\right)\left[\frac{(\ln 10)^2}{2} - \frac{(\ln 1)^2}{2}\right]$
$= \ln 10$

44. $\int_2^3 \frac{2 \log_2 (x-1)}{x-1}\,dx = \frac{2}{\ln 2}\int_2^3 \ln(x-1)\left(\frac{1}{x-1}\right)dx = \left(\frac{2}{\ln 2}\right)\left[\frac{(\ln(x-1))^2}{2}\right]_2^3 = \left(\frac{2}{\ln 2}\right)\left[\frac{(\ln 2)^2}{2} - \frac{(\ln 1)^2}{2}\right] = \ln 2$

45. $\int \frac{dx}{x \log_{10} x} = \int \left(\frac{\ln 10}{\ln x}\right)\left(\frac{1}{x}\right) dx = (\ln 10) \int \left(\frac{1}{\ln x}\right)\left(\frac{1}{x}\right) dx$; $\left[u = \ln x \Rightarrow du = \frac{1}{x}\, dx\right]$
$\rightarrow (\ln 10) \int \left(\frac{1}{\ln x}\right)\left(\frac{1}{x}\right) dx = (\ln 10) \int \frac{1}{u}\, du = (\ln 10) \ln |u| + C = (\ln 10) \ln |\ln x| + C$

46. $\int \frac{dx}{x (\log_8 x)^2} = \int \frac{dx}{x \left(\frac{\ln x}{\ln 8}\right)^2} = (\ln 8)^2 \int \frac{(\ln x)^{-2}}{x}\, dx = (\ln 8)^2 \frac{(\ln x)^{-1}}{-1} + C = -\frac{(\ln 8)^2}{\ln x} + C$

47. $\frac{dy}{dt} = e^t \sin (e^t - 2) \Rightarrow y = \int e^t \sin (e^t - 2)\, dt$;
let $u = e^t - 2 \Rightarrow du = e^t\, dt \Rightarrow y = \int \sin u\, du = -\cos u + C = -\cos (e^t - 2) + C$; $y(\ln 2) = 0$
$\Rightarrow -\cos \left(e^{\ln 2} - 2\right) + C = 0 \Rightarrow -\cos (2 - 2) + C = 0 \Rightarrow C = \cos 0 = 1$; thus, $y = 1 - \cos (e^t - 2)$

48. $\frac{dy}{dt} = e^{-t} \sec^2 (\pi e^{-t}) \Rightarrow y = \int e^{-t} \sec^2 (\pi e^{-t})\, dt$;
let $u = \pi e^{-t} \Rightarrow du = -\pi e^{-t}\, dt \Rightarrow -\frac{1}{\pi}\, du = e^{-t}\, dt \Rightarrow y = -\frac{1}{\pi} \int \sec^2 u\, du = -\frac{1}{\pi} \tan u + C$
$= -\frac{1}{\pi} \tan (\pi e^{-t}) + C$; $y(\ln 4) = \frac{2}{\pi} \Rightarrow -\frac{1}{\pi} \tan \left(\pi e^{-\ln 4}\right) + C = \frac{2}{\pi} \Rightarrow -\frac{1}{\pi} \tan \left(\pi \cdot \frac{1}{4}\right) + C = \frac{2}{\pi}$
$\Rightarrow -\frac{1}{\pi}(1) + C = \frac{2}{\pi} \Rightarrow C = \frac{3}{\pi}$; thus, $y = \frac{3}{\pi} - \frac{1}{\pi} \tan (\pi e^{-t})$

49. $\frac{d^2y}{dx^2} = 2e^{-x} \Rightarrow \frac{dy}{dx} = -2e^{-x} + C$; $x = 0$ and $\frac{dy}{dx} = 0 \Rightarrow 0 = -2e^0 + C \Rightarrow C = 2$; thus $\frac{dy}{dx} = -2e^{-x} + 2$
$\Rightarrow y = 2e^{-x} + 2x + C_1$; $x = 0$ and $y = 1 \Rightarrow 1 = 2e^0 + C_1 \Rightarrow C_1 = -1 \Rightarrow y = 2e^{-x} + 2x - 1 = 2(e^{-x} + x) - 1$

50. $\frac{d^2y}{dt^2} = 1 - e^{2t} \Rightarrow \frac{dy}{dt} = t - \frac{1}{2} e^{2t} + C$; $t = 1$ and $\frac{dy}{dt} = 0 \Rightarrow 0 = 1 - \frac{1}{2} e^2 + C \Rightarrow C = \frac{1}{2} e^2 - 1$; thus
$\frac{dy}{dt} = t - \frac{1}{2} e^{2t} + \frac{1}{2} e^2 - 1 \Rightarrow y = \frac{1}{2} t^2 - \frac{1}{4} e^{2t} + \left(\frac{1}{2} e^2 - 1\right) t + C_1$; $t = 1$ and $y = -1 \Rightarrow -1 = \frac{1}{2} - \frac{1}{4} e^2 + \frac{1}{2} e^2 - 1 + C_1$
$\Rightarrow C_1 = -\frac{1}{2} - \frac{1}{4} e^2 \Rightarrow y = \frac{1}{2} t^2 - \frac{1}{4} e^{2t} + \left(\frac{1}{2} e^2 - 1\right) t - \left(\frac{1}{2} + \frac{1}{4} e^2\right)$

51. $\frac{dy}{dx} = 1 + \frac{1}{x}$ at $(1, 3) \Rightarrow y = x + \ln |x| + C$; $y = 3$ at $x = 1 \Rightarrow C = 2 \Rightarrow y = x + \ln |x| + 2$

52. $\frac{d^2y}{dx^2} = \sec^2 x \Rightarrow \frac{dy}{dx} = \tan x + C$ and $1 = \tan 0 + C \Rightarrow \frac{dy}{dx} = \tan x + 1 \Rightarrow y = \int (\tan x + 1)\, dx$
$= \ln |\sec x| + x + C_1$ and $0 = \ln |\sec 0| + 0 + C_1 \Rightarrow C_1 = 0 \Rightarrow y = \ln |\sec x| + x$

53. $V = 2\pi \int_{1/2}^{2} x \left(\frac{1}{x^2}\right) dx = 2\pi \int_{1/2}^{2} \frac{1}{x}\, dx = 2\pi \left[\ln |x|\right]_{1/2}^{2} = 2\pi \left(\ln 2 - \ln \frac{1}{2}\right) = 2\pi(2 \ln 2) = \pi \ln 2^4 = \pi \ln 16$

54. $V = \pi \int_0^3 \left(\frac{9x}{\sqrt{x^3 + 9}}\right)^2 dx = 27\pi \int_0^3 dx = 27\pi \left[\ln (x^3 + 9)\right]_0^3 = 27\pi(\ln 36 - \ln 9) = 27\pi(\ln 4 + \ln 9 - \ln 9)$
$= 27\pi \ln 4 = 54\pi \ln 2$

55. $y = \frac{x^2}{8} - \ln x \Rightarrow 1 + (y')^2 = 1 + \left(\frac{x}{4} - \frac{1}{x}\right)^2 = 1 + \left(\frac{x^2 - 4}{4x}\right)^2 = \left(\frac{x^2 + 4}{4x}\right)^2 \Rightarrow L = \int_4^8 \sqrt{1 + (y')^2}\, dx$
$= \int_4^8 \frac{x^2 + 4}{4x}\, dx = \int_4^8 \left(\frac{x}{4} + \frac{1}{x}\right) dx = \left[\frac{x^2}{8} + \ln |x|\right]_4^8 = (8 + \ln 8) - (2 + \ln 4) = 6 + \ln 2$

56. $x = \left(\frac{y}{4}\right)^2 - 2 \ln \left(\frac{y}{4}\right) \Rightarrow \frac{dx}{dy} = \frac{y}{8} - \frac{2}{y} \Rightarrow 1 + \left(\frac{dx}{dy}\right)^2 = 1 + \left(\frac{y}{8} - \frac{2}{y}\right)^2 = 1 + \left(\frac{y^2 - 16}{8y}\right)^2 = \left(\frac{y^2 + 16}{8y}\right)^2$
$\Rightarrow L = \int_4^{12} \sqrt{1 + \left(\frac{dx}{dy}\right)^2}\, dy = \int_4^{12} \frac{y^2 + 16}{8y}\, dy = \int_4^{12} \left(\frac{y}{8} + \frac{2}{y}\right) dy = \left[\frac{y^2}{16} + 2 \ln y\right]_4^{12} = (9 + 2 \ln 12) - (1 + 2 \ln 4)$
$= 8 + 2 \ln 3 = 8 + \ln 9$

57. (a) $L(x) = f(0) + f'(0) \cdot x$, and $f(x) = \ln(1+x) \Rightarrow f'(x)\big|_{x=0} = \frac{1}{1+x}\Big|_{x=0} = 1 \Rightarrow L(x) = \ln 1 + 1 \cdot x \Rightarrow L(x) = x$

(b) Let $f(x) = \ln(x+1)$. Since $f''(x) = -\frac{1}{(x+1)^2} < 0$ on $[0, 0.1]$, the graph of f is concave down on this interval and the largest error in the linear approximation will occur when $x = 0.1$. This error is $0.1 - \ln(1.1) \approx 0.00469$ to five decimal places.

(c) The approximation $y = x$ for $\ln(1+x)$ is best for smaller positive values of x; in particular for $0 \le x \le 0.1$ in the graph. As x increases, so does the error $x - \ln(1+x)$. From the graph an upper bound for the error is $0.5 - \ln(1+0.5) \approx 0.095$; i.e., $|E(x)| \le 0.095$ for $0 \le x \le 0.5$. Note from the graph that $0.1 - \ln(1+0.1) \approx 0.00469$ estimates the error in replacing $\ln(1+x)$ by x over $0 \le x \le 0.1$. This is consistent with the estimate given in part (b) above.

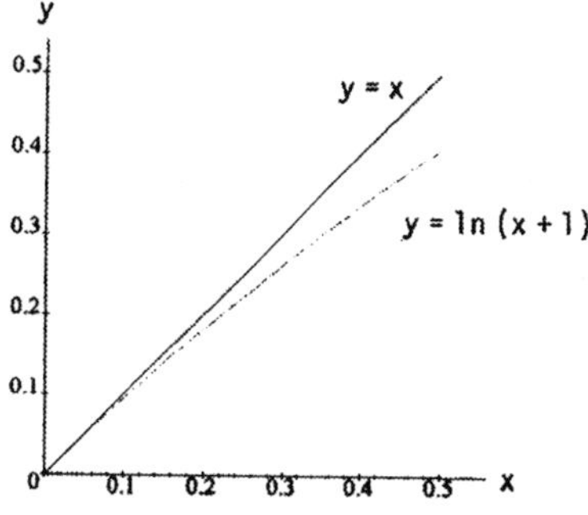

58. (a) $f(x) = e^x \Rightarrow f'(x) = e^x$; $L(x) = f(0) + f'(0)(x-0) \Rightarrow L(x) = 1 + x$

(b) $f(0) = 1$ and $L(0) = 1 \Rightarrow$ error $= 0$; $f(0.2) = e^{0.2} \approx 1.22140$ and $L(0.2) = 1.2 \Rightarrow$ error ≈ 0.02140

(c) Since $y'' = e^x > 0$, the tangent line approximation always lies below the curve $y = e^x$. Thus $L(x) = x + 1$ never overestimates e^x.

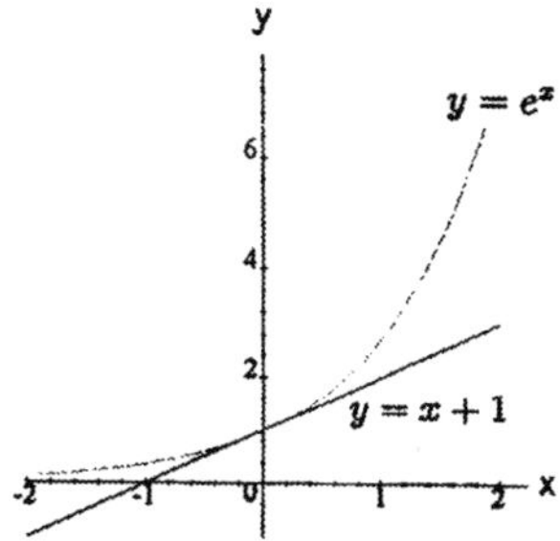

59. Note that $y = \ln x$ and $e^y = x$ are the same curve; $\int_1^a \ln x\, dx$ = area under the curve between 1 and a; $\int_0^{\ln a} e^y\, dy$ = area to the left of the curve between 0 and ln a. The sum of these areas is equal to the area of the rectangle $\Rightarrow \int_1^a \ln x\, dx + \int_0^{\ln a} e^y\, dy = a \ln a$.

60. (a) $y = e^x \Rightarrow y'' = e^x > 0$ for all $x \Rightarrow$ the graph of $y = e^x$ is always concave upward

(b) area of the trapezoid ABCD $< \int_{\ln a}^{\ln b} e^x\, dx <$ area of the trapezoid AEFD $\Rightarrow \frac{1}{2}(AB + CD)(\ln b - \ln a)$ $< \int_{\ln a}^{\ln b} e^x\, dx < \left(\frac{e^{\ln a} + e^{\ln b}}{2}\right)(\ln b - \ln a)$. Now $\frac{1}{2}(AB + CD)$ is the height of the midpoint $M = e^{(\ln a + \ln b)/2}$ since the curve containing the points B and C is linear $\Rightarrow e^{(\ln a + \ln b)/2}(\ln b - \ln a)$ $< \int_{\ln a}^{\ln b} e^x\, dx < \left(\frac{e^{\ln a} + e^{\ln b}}{2}\right)(\ln b - \ln a)$

(c) $\int_{\ln a}^{\ln b} e^x\, dx = \left[e^x\right]_{\ln a}^{\ln b} = e^{\ln b} - e^{\ln a} = b - a$, so part (b) implies that

$e^{(\ln a + \ln b)/2}(\ln b - \ln a) < b - a < \left(\frac{e^{\ln a} + e^{\ln b}}{2}\right)(\ln b - \ln a) \Rightarrow e^{(\ln a + \ln b)/2} < \frac{b-a}{\ln b - \ln a} < \frac{a+b}{2}$

$\Rightarrow e^{\ln a/2} \cdot e^{\ln b/2} < \frac{b-a}{\ln b - \ln a} < \frac{a+b}{2} \Rightarrow \sqrt{e^{\ln a}}\sqrt{e^{\ln b}} < \frac{b-a}{\ln b - \ln a} < \frac{a+b}{2} \Rightarrow \sqrt{ab} < \frac{b-a}{\ln b - \ln a} < \frac{a+b}{2}$

61. $y = \ln kx \Rightarrow y = \ln x + \ln k$; thus the graph of $y = \ln kx$ is the graph of $y = \ln x$ shifted vertically by $\ln k$, $k > 0$.

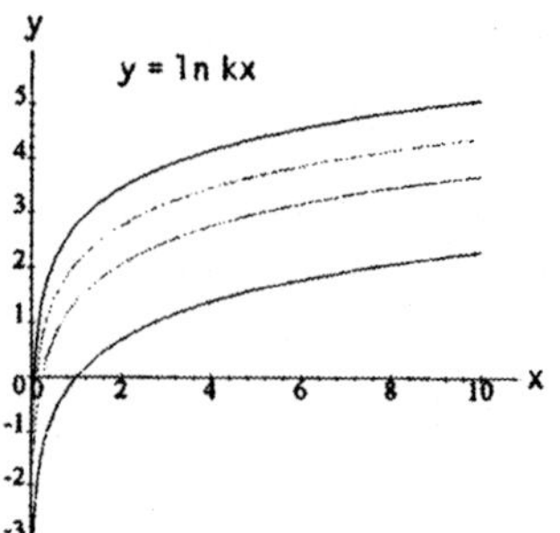

62. To turn the arches upside down we would use the formula $y = -\ln|\sin x| = \ln \frac{1}{|\sin x|}$.

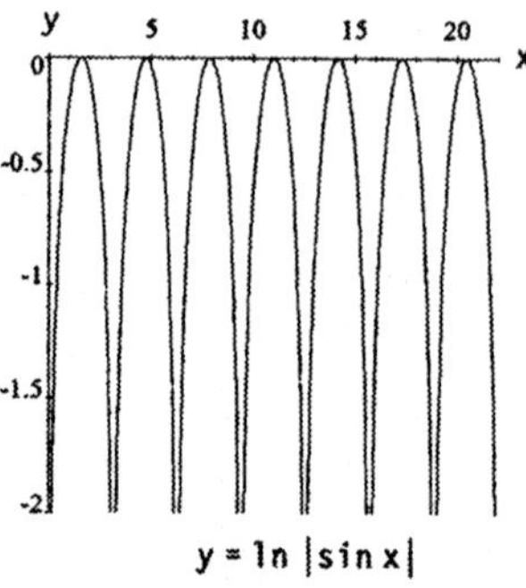

63. (a)

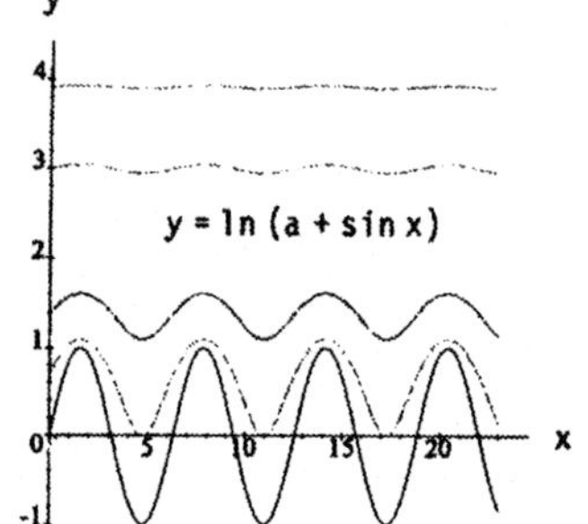

(b) $y' = \frac{\cos x}{a+\sin x}$. Since $|\sin x|$ and $|\cos x|$ are less than or equal to 1, we have for $a > 1$ $\frac{-1}{a-1} \le y' \le \frac{1}{a-1}$ for all x. Thus, $\lim_{a\to+\infty} y' = 0$ for all $x \Rightarrow$ the graph of y looks more and more horizontal as $a \to +\infty$.

64. (a) The graph of $y = \sqrt{x} - \ln x$ appears to be concave upward for all $x > 0$.

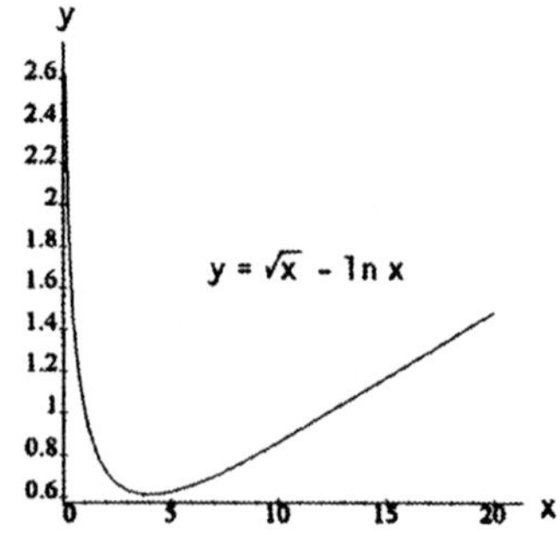

(b) $y = \sqrt{x} - \ln x \Rightarrow y' = \frac{1}{2\sqrt{x}} - \frac{1}{x} \Rightarrow y'' = -\frac{1}{4x^{3/2}} + \frac{1}{x^2} = \frac{1}{x^2}\left(-\frac{\sqrt{x}}{4} + 1\right) = 0 \Rightarrow \sqrt{x} = 4 \Rightarrow x = 16$. Thus, $y'' > 0$ if $0 < x < 16$ and $y'' < 0$ if $x > 16$ so a point of inflection exists at $x = 16$. The graph of $y = \sqrt{x} - \ln x$ closely resembles a straight line for $x \ge 10$ and it is impossible to discuss the point of inflection visually from the graph.

65. From zooming in on the graph at the right, we estimate the third root to be $x \approx -0.76666$

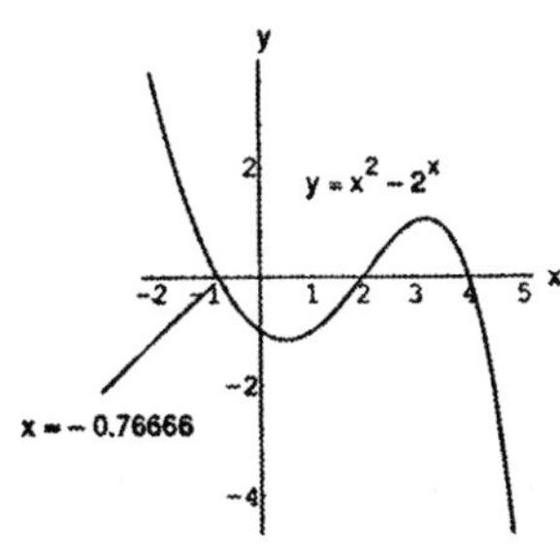

66. The functions $f(x) = x^{\ln 2}$ and $g(x) = 2^{\ln x}$ appear to have identical graphs for $x > 0$. This is no accident, because $x^{\ln 2} = e^{\ln 2 \cdot \ln x} = (e^{\ln 2})^{\ln x} = 2^{\ln x}$.

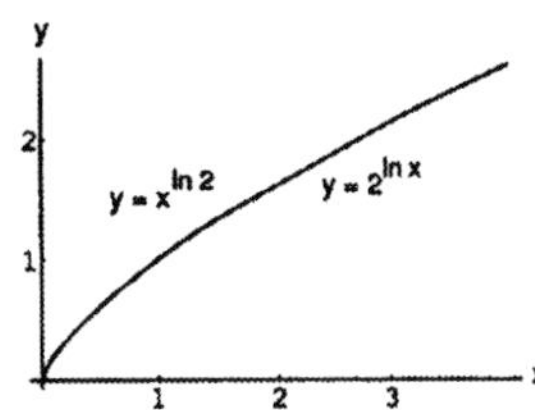

67. (a) The point of tangency is $(p, \ln p)$ and $m_{tangent} = \frac{1}{p}$ since $\frac{dy}{dx} = \frac{1}{x}$. The tangent line passes through $(0, 0) \Rightarrow$ the equation of the tangent line is $y = \frac{1}{p}x$. The tangent line also passes through$(p, \ln p) \Rightarrow \ln p = \frac{1}{p}p = 1 \Rightarrow p = e$, and the tangent line equation is $y = \frac{1}{e}x$.
(b) $\frac{d^2y}{dx^2} = -\frac{1}{x^2}$ for $x \neq 0 \Rightarrow y = \ln x$ is concave downward over its domain. Therefore, $y = \ln x$ lies below the graph of $y = \frac{1}{e}x$ for all $x > 0$, $x \neq e$, and $\ln x < \frac{x}{e}$ for $x > 0$, $x \neq e$.
(c) Multiplying by e, $e \ln x < x$ or $\ln x^e < x$.
(d) Exponentiating both sides of $\ln x^e < x$, we have $e^{\ln x^e} < e^x$, or $x^e < e^x$ for all positive $x \neq e$.
(e) Let $x = \pi$ to see that $\pi^e < e^\pi$. Therefore, e^π is bigger.

68. Using Newton's Method: $f(x) = \ln(x) - 1 \Rightarrow f'(x) = \frac{1}{x} \Rightarrow x_{n+1} = x_n - \frac{\ln(x_n) - 1}{\frac{1}{x_n}} \Rightarrow x_{n+1} = x_n\left[2 - \ln(x_n)\right]$.
Then, $x_1 = 2$, $x_2 = 2.61370564$, $x_3 = 2.71624393$, and $x_5 = 2.71828183$. Many other methods may be used. For example, graph $y = \ln x - 1$ and determine the zero of y.

69. (a) $\log_3 8 = \frac{\ln 8}{\ln 3} \approx 1.89279$
(b) $\log_7 0.5 = \frac{\ln 0.5}{\ln 7} \approx -0.35621$
(c) $\log_{20} 17 = \frac{\ln 17}{\ln 20} \approx 0.94575$
(d) $\log_{0.5} 7 = \frac{\ln 7}{\ln 0.5} \approx -2.80735$
(e) $\ln x = (\log_{10} x)(\ln 10) = 2.3 \ln 10 \approx 5.29595$
(f) $\ln x = (\log_2 x)(\ln 2) = 1.4 \ln 2 \approx 0.97041$
(g) $\ln x = (\log_2 x)(\ln 2) = -1.5 \ln 2 \approx -1.03972$
(h) $\ln x = (\log_{10} x)(\ln 10) = -0.7 \ln 10 \approx -1.61181$

70. (a) $\frac{\ln 10}{\ln 2} \cdot \log_{10} x = \frac{\ln 10}{\ln 2} \cdot \frac{\ln x}{\ln 10} = \frac{\ln x}{\ln 2} = \log_2 x$
(b) $\frac{\ln a}{\ln b} \cdot \log_a x = \frac{\ln a}{\ln b} \cdot \frac{\ln x}{\ln a} = \frac{\ln x}{\ln b} = \log_b x$

7.2 EXPONENTIAL GROWTH AND DECAY

1. (a) $y = y_0 e^{kt} \Rightarrow 0.99y_0 = y_0 e^{1000k} \Rightarrow k = \frac{\ln 0.99}{1000} \approx -0.00001$
(b) $0.9 = e^{(-0.00001)t} \Rightarrow (-0.00001)t = \ln(0.9) \Rightarrow t = \frac{\ln(0.9)}{-0.00001} \approx 10{,}536$ years
(c) $y = y_0 e^{(20{,}000)k} \approx y_0 e^{-0.2} = y_0(0.82) \Rightarrow 82\%$

2. (a) $\frac{dp}{dh} = kp \Rightarrow p = p_0 e^{kh}$ where $p_0 = 1013$; $90 = 1013e^{20k} \Rightarrow k = \frac{\ln(90) - \ln(1013)}{20} \approx -0.121$
(b) $p = 1013e^{-6.05} \approx 2.389$ millibars
(c) $900 = 1013e^{(-0.121)h} \Rightarrow -0.121h = \ln\left(\frac{900}{1013}\right) \Rightarrow h = \frac{\ln(1013) - \ln(900)}{0.121} \approx 0.977$ km

3. $\frac{dy}{dt} = -0.6y \Rightarrow y = y_0 e^{-0.6t}$; $y_0 = 100 \Rightarrow y = 100e^{-0.6t} \Rightarrow y = 100e^{-0.6} \approx 54.88$ grams when t = 1 hr

4. $A = A_0e^{kt} \Rightarrow 800 = 1000e^{10k} \Rightarrow k = \frac{\ln(0.8)}{10} \Rightarrow A = 1000e^{(\ln(0.8)/10)t}$, where A represents the amount of sugar that remains after time t. Thus after another 14 hrs, $A = 1000e^{(\ln(0.8)/10)24} \approx 585.35$ kg

5. $L(x) = L_0e^{-kx} \Rightarrow \frac{L_0}{2} = L_0e^{-18k} \Rightarrow \ln\frac{1}{2} = -18k \Rightarrow k = \frac{\ln 2}{18} \approx 0.0385 \Rightarrow L(x) = L_0e^{-0.0385x}$; when the intensity is one-tenth of the surface value, $\frac{L_0}{10} = L_0e^{-0.0385x} \Rightarrow \ln 10 = 0.0385x \Rightarrow x \approx 59.8$ ft

6. $V(t) = V_0e^{-t/40} \Rightarrow 0.1V_0 = V_0e^{-t/40}$ when the voltage is 10% of its original value $\Rightarrow t = -40\ln(0.1)$ ≈ 92.1 sec

7. $y = y_0e^{kt}$ and $y_0 = 1 \Rightarrow y = e^{kt} \Rightarrow$ at $y = 2$ and $t = 0.5$ we have $2 = e^{0.5k} \Rightarrow \ln 2 = 0.5k \Rightarrow k = \frac{\ln 2}{0.5} = \ln 4$. Therefore, $y = e^{(\ln 4)t} \Rightarrow y = e^{24\ln 4} = 4^{24} = 2.81474978 \times 10^{14}$ at the end of 24 hrs

8. $y = y_0e^{kt}$ and $y(3) = 10{,}000 \Rightarrow 10{,}000 = y_0e^{3k}$; also $y(5) = 40{,}000 = y_0e^{5k}$. Therefore $y_0e^{5k} = 4y_0e^{3k}$ $\Rightarrow e^{5k} = 4e^{3k} \Rightarrow e^{2k} = 4 \Rightarrow k = \ln 2$. Thus, $y = y_0e^{(\ln 2)t} \Rightarrow 10{,}000 = y_0e^{3\ln 2} = y_0e^{\ln 8} \Rightarrow 10{,}000 = 8y_0$ $\Rightarrow y_0 = \frac{10{,}000}{8} = 1250$

9. (a) $10{,}000e^{k(1)} = 7500 \Rightarrow e^k = 0.75 \Rightarrow k = \ln 0.75$ and $y = 10{,}000e^{(\ln 0.75)t}$. Now $1000 = 10{,}000e^{(\ln 0.75)t}$ $\Rightarrow \ln 0.1 = (\ln 0.75)t \Rightarrow t = \frac{\ln 0.1}{\ln 0.75} \approx 8.00$ years (to the nearest hundredth of a year)
(b) $1 = 10{,}000e^{(\ln 0.75)t} \Rightarrow \ln 0.0001 = (\ln 0.75)t \Rightarrow t = \frac{\ln 0.0001}{\ln 0.75} \approx 32.02$ years (to the nearest hundredth of a year)

10. (a) There are (60)(60)(24)(365) = 31,536,000 seconds in a year. Thus, assuming exponential growth, $P = 257{,}313{,}431e^{kt}$ and $257{,}313{,}432 = 257{,}313{,}431e^{(14k/31{,}536{,}000)} \Rightarrow \ln\left(\frac{257{,}313{,}432}{257{,}313{,}431}\right) = \frac{14k}{31{,}536{,}000}$ $\Rightarrow k \approx 0.0087542$
(b) $P = 257{,}313{,}431e^{(0.0087542)(15)} \approx 293{,}420{,}847$ (to the nearest integer). Answers will vary considerably with the number of decimal places retained.

11. $0.9P_0 = P_0e^k \Rightarrow k = \ln 0.9$; when the well's output falls to one-fifth of its present value $P = 0.2P_0$ $\Rightarrow 0.2P_0 = P_0e^{(\ln 0.9)t} \Rightarrow 0.2 = e^{(\ln 0.9)t} \Rightarrow \ln(0.2) = (\ln 0.9)t \Rightarrow t = \frac{\ln 0.2}{\ln 0.9} \approx 15.28$ yr

12. (a) $\frac{dp}{dx} = -\frac{1}{100}p \Rightarrow \frac{dp}{p} = -\frac{1}{100}dx \Rightarrow \ln p = -\frac{1}{100}x + C \Rightarrow p = e^{(-0.01x+C)} = e^Ce^{-0.01x} = C_1e^{-0.01x}$; $p(100) = 20.09 \Rightarrow 20.09 = C_1e^{(-0.01)(100)} \Rightarrow C_1 = 20.09e \approx 54.61 \Rightarrow p(x) = 54.61e^{-0.01x}$ (in dollars)
(b) $p(10) = 54.61e^{(-0.01)(10)} = \$49.41$, and $p(90) = 54.61e^{(-0.01)(90)} = \$22.20$
(c) $r(x) = xp(x) \Rightarrow r'(x) = p(x) + xp'(x)$; $p'(x) = -.5461e^{-0.01x} \Rightarrow r'(x)$ $= (54.61 - .5461x)e^{-0.01x}$. Thus, $r'(x) = 0$ $\Rightarrow 54.61 = .5461x \Rightarrow x = 100$. Since $r' > 0$ for any $x < 100$ and $r' < 0$ for $x > 100$, then $r(x)$ must be a maximum at $x = 100$.

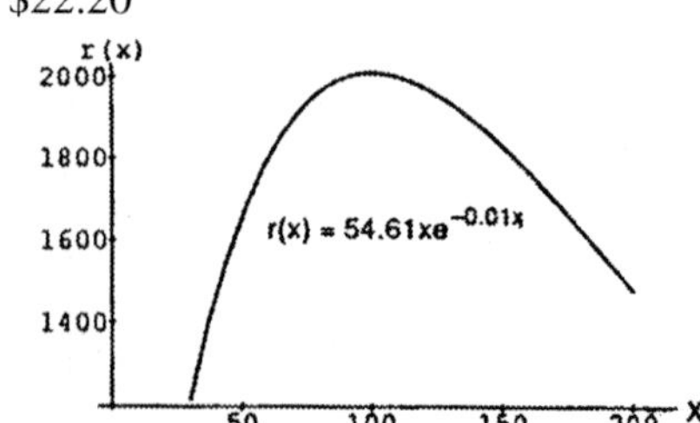

13. (a) $A_0e^{(0.04)5} = A_0e^{0.2}$
(b) $2A_0 = A_0e^{(0.04)t} \Rightarrow \ln 2 = (0.04)t \Rightarrow t = \frac{\ln 2}{0.04} \approx 17.33$ years; $3A_0 = A_0e^{(0.04)t} \Rightarrow \ln 3 = (0.04)t$ $\Rightarrow t = \frac{\ln 3}{0.04} \approx 27.47$ years

14. (a) The amount of money invested A_0 after t years is $A(t) = A_0e^t$
(b) If $A(t) = 3A_0$, then $3A_0 = A_0e^t \Rightarrow \ln 3 = t$ or $t \approx 1.099$ years

(c) At the beginning of a year the account balance is A_0e^t, while at the end of the year the balance is $A_0e^{(t+1)}$. The amount earned is $A_0e^{(t+1)} - A_0e^t = A_0e^t(e-1) \approx 1.7$ times the beginning amount.

15. $A(100) = 90{,}000 \Rightarrow 90{,}000 = 1000e^{r(100)} \Rightarrow 90 = e^{100r} \Rightarrow \ln 90 = 100r \Rightarrow r = \frac{\ln 90}{100} \approx 0.0450$ or 4.50%

16. $A(100) = 131{,}000 \Rightarrow 131{,}000 = 1000e^{100r} \Rightarrow \ln 131 = 100r \Rightarrow r = \frac{\ln 131}{100} \approx 0.04875$ or 4.875%

17. $y = y_0e^{-0.18t}$ represents the decay equation; solving $(0.9)y_0 = y_0e^{-0.18t} \Rightarrow t = \frac{\ln(0.9)}{-0.18} \approx 0.585$ days

18. $A = A_0e^{kt}$ and $\frac{1}{2}A_0 = A_0e^{139k} \Rightarrow \frac{1}{2} = e^{139k} \Rightarrow k = \frac{\ln(0.5)}{139} \approx -0.00499$; then $0.05A_0 = A_0e^{-0.00499t}$ $\Rightarrow t = \frac{\ln 0.05}{-0.00499} \approx 600$ days

19. $y = y_0e^{-kt} = y_0e^{-(k)(3/k)} = y_0e^{-3} = \frac{y_0}{e^3} < \frac{y_0}{20} = (0.05)(y_0) \Rightarrow$ after three mean lifetimes less than 5% remains

20. (a) $A = A_0e^{-kt} \Rightarrow \frac{1}{2} = e^{-2.645k} \Rightarrow k = \frac{\ln 2}{2.645} \approx 0.262$

(b) $\frac{1}{k} \approx 3.816$ years

(c) $(0.05)A = A\exp\left(-\frac{\ln 2}{2.645}t\right) \Rightarrow -\ln 20 = \left(-\frac{\ln 2}{2.645}\right)t \Rightarrow t = \frac{2.645 \ln 20}{\ln 2} \approx 11.431$ years

21. $T - T_s = (T_0 - T_s)e^{-kt}$, $T_0 = 90°C$, $T_s = 20°C$, $T = 60°C \Rightarrow 60 - 20 = 70e^{-10k} \Rightarrow \frac{4}{7} = e^{-10k}$ $\Rightarrow k = \frac{\ln\left(\frac{7}{4}\right)}{10} \approx 0.05596$

(a) $35 - 20 = 70e^{-0.05596t} \Rightarrow t \approx 27.5$ min is the total time $\Rightarrow$ it will take $27.5 - 10 = 17.5$ minutes longer to reach 35°C

(b) $T - T_s = (T_0 - T_s)e^{-kt}$, $T_0 = 90°C$, $T_s = -15°C \Rightarrow 35 + 15 = 105e^{-0.05596t} \Rightarrow t \approx 13.26$ min

22. $T - 65° = (T_0 - 65°)e^{-kt} \Rightarrow 35° - 65° = (T_0 - 65°)e^{-10k}$ and $50° - 65° = (T_0 - 65°)e^{-20k}$. Solving $-30° = (T_0 - 65°)e^{-10k}$ and $-15° = (T_0 - 65°)e^{-20k}$ simultaneously $\Rightarrow (T_0 - 65°)e^{-10k} = 2(T_0 - 65°)e^{-20k}$ $\Rightarrow e^{10k} = 2 \Rightarrow k = \frac{\ln 2}{10}$ and $-30° = \frac{T_0 - 65°}{e^{10k}} \Rightarrow -30°\left[e^{10\left(\frac{\ln 2}{10}\right)}\right] = T_0 - 65° \Rightarrow T_0 = 65° - 30°\left(e^{\ln 2}\right) = 65° - 60° = 5°$

23. $T - T_s = (T_0 - T_s)e^{-kt} \Rightarrow 39 - T_s = (46 - T_s)e^{-10k}$ and $33 - T_s = (46 - T_s)e^{-20k} \Rightarrow \frac{39-T_s}{46-T_s} = e^{-10k}$ and $\frac{33-T_s}{46-T_s} = e^{-20k} = \left(e^{-10k}\right)^2 \Rightarrow \frac{33-T_s}{46-T_s} = \left(\frac{39-T_s}{46-T_s}\right)^2 \Rightarrow (33 - T_s)(46 - T_s) = (39 - T_s)^2 \Rightarrow 1518 - 79T_s + T_s^2$ $= 1521 - 78T_s + T_s^2 \Rightarrow -T_s = 3 \Rightarrow T_s = -3°C$

24. Let x represent how far above room temperature the silver will be 15 min from now, y how far above room temperature the silver will be 120 min from now, and t_0 the time the silver will be 10°C above room temperature. We then have the following time-temperature table:

time in min.	0	20 (Now)	35	140	t_0
temperature	$T_s + 70°$	$T_s + 60°$	$T_s + x$	$T_s + y$	$T_s + 10°$

$T - T_s = (T_0 - T_s)e^{-kt} \Rightarrow (60 + T_s) - T_s = [(70 + T_s) - T_s]e^{-20k} \Rightarrow 60 = 70e^{-20k} \Rightarrow k = \left(-\frac{1}{20}\right)\ln\left(\frac{6}{7}\right)$ ≈ 0.00771

(a) $T - T_s = (T_0 - T_s)e^{-0.00771t} \Rightarrow (T_s + x) - T_s = [(70 + T_s) - T_s]e^{-(0.00771)(35)} \Rightarrow x = 70e^{-0.26985} \approx 53.44°C$

(b) $T - T_s = (T_0 - T_s)e^{-0.00771t} \Rightarrow (T_s + y) - T_s = [(70 + T_s) - T_s]e^{-(0.00771)(140)} \Rightarrow y = 70e^{-1.0794} \approx 23.79°C$

(c) $T - T_s = (T_0 - T_s)e^{-0.00771t} \Rightarrow (T_s + 10) - T_s = [(70 + T_s) - T_s]e^{-(0.00771)t_0} \Rightarrow 10 = 70e^{-0.00771t_0}$ $\Rightarrow \ln\left(\frac{1}{7}\right) = -0.00771t_0 \Rightarrow t_0 = \left(-\frac{1}{0.00771}\right)\ln\left(\frac{1}{7}\right) = 252.39 \Rightarrow 252.39 - 20 \approx 232$ minutes from now the silver will be 10°C above room temperature

25. From Example 5, the half-life of carbon-14 is 5700 yr $\Rightarrow \frac{1}{2}c_0 = c_0e^{-k(5700)} \Rightarrow k = \frac{\ln 2}{5700} \approx 0.0001216$
$\Rightarrow c = c_0e^{-0.0001216t} \Rightarrow (0.445)c_0 = c_0e^{-0.0001216t} \Rightarrow t = \frac{\ln(0.445)}{-0.0001216} \approx 6659$ years

26. From Exercise 25, $k \approx 0.0001216$ for carbon-14.
(a) $c = c_0e^{-0.0001216t} \Rightarrow (0.17)c_0 = c_0e^{-0.0001216t} \Rightarrow t \approx 14{,}571.44$ years $\Rightarrow$ 12,571 BC
(b) $(0.18)c_0 = c_0e^{-0.0001216t} \Rightarrow t \approx 14{,}101.41$ years $\Rightarrow$ 12,101 BC
(c) $(0.16)c_0 = c_0e^{-0.0001216t} \Rightarrow t \approx 15{,}069.98$ years $\Rightarrow$ 13,070 BC

27. From Exercise 25, $k \approx 0.0001216$ for carbon-14. Thus, $c = c_0e^{-0.0001216t} \Rightarrow (0.995)c_0 = c_0e^{-0.0001216t}$
$\Rightarrow t = \frac{\ln(0.995)}{-0.0001216} \approx 41$ years old

7.3 RELATIVE RATES OF GROWTH

1. (a) slower, $\lim_{x\to\infty} \frac{x+3}{e^x} = \lim_{x\to\infty} \frac{1}{e^x} = 0$
(b) slower, $\lim_{x\to\infty} \frac{x^3+\sin^2 x}{e^x} = \lim_{x\to\infty} \frac{3x^2+2\sin x\cos x}{e^x} = \lim_{x\to\infty} \frac{6x+2\cos 2x}{e^x} = \lim_{x\to\infty} \frac{6-4\sin 2x}{e^x} = 0$ by the Sandwich Theorem because $\frac{2}{e^x} \le \frac{6-4\sin 2x}{e^x} \le \frac{10}{e^x}$ for all reals and $\lim_{x\to\infty} \frac{2}{e^x} = 0 = \lim_{x\to\infty} \frac{10}{e^x}$
(c) slower, $\lim_{x\to\infty} \frac{\sqrt{x}}{e^x} = \lim_{x\to\infty} \frac{x^{1/2}}{e^x} = \lim_{x\to\infty} \frac{\left(\frac{1}{2}\right)x^{-1/2}}{e^x} = \lim_{x\to\infty} \frac{1}{2\sqrt{x}\,e^x} = 0$
(d) faster, $\lim_{x\to\infty} \frac{4^x}{e^x} = \lim_{x\to\infty} \left(\frac{4}{e}\right)^x = \infty$ since $\frac{4}{e} > 1$
(e) slower, $\lim_{x\to\infty} \frac{\left(\frac{3}{2}\right)^x}{e^x} = \lim_{x\to\infty} \left(\frac{3}{2e}\right)^x = 0$ since $\frac{3}{2e} < 1$
(f) slower, $\lim_{x\to\infty} \frac{e^{x/2}}{e^x} = \lim_{x\to\infty} \frac{1}{e^{x/2}} = 0$
(g) same, $\lim_{x\to\infty} \frac{\left(\frac{e^x}{2}\right)}{e^x} = \lim_{x\to\infty} \frac{1}{2} = \frac{1}{2}$
(h) slower, $\lim_{x\to\infty} \frac{\log_{10} x}{e^x} = \lim_{x\to\infty} \frac{\ln x}{(\ln 10)e^x} = \lim_{x\to\infty} \frac{\frac{1}{x}}{(\ln 10)e^x} = \lim_{x\to\infty} \frac{1}{(\ln 10)xe^x} = 0$

2. (a) slower, $\lim_{x\to\infty} \frac{10x^4+30x+1}{e^x} = \lim_{x\to\infty} \frac{40x^3+30}{e^x} = \lim_{x\to\infty} \frac{120x^2}{e^x} = \lim_{x\to\infty} \frac{240x}{e^x} = \lim_{x\to\infty} \frac{240}{e^x} = 0$
(b) slower, $\lim_{x\to\infty} \frac{x\ln x - x}{e^x} = \lim_{x\to\infty} \frac{x(\ln x - 1)}{e^x} = \lim_{x\to\infty} \frac{\ln x - 1 + x\left(\frac{1}{x}\right)}{e^x} = \lim_{x\to\infty} \frac{\ln x - 1 + 1}{e^x} = \lim_{x\to\infty} \frac{\ln x}{e^x}$
$= \lim_{x\to\infty} \frac{\left(\frac{1}{x}\right)}{e^x} = \lim_{x\to\infty} \frac{1}{xe^x} = 0$
(c) slower, $\lim_{x\to\infty} \frac{\sqrt{1+x^4}}{e^x} = \sqrt{\lim_{x\to\infty} \frac{1+x^4}{e^{2x}}} = \sqrt{\lim_{x\to\infty} \frac{4x^3}{2e^{2x}}} = \sqrt{\lim_{x\to\infty} \frac{12x^2}{4e^{2x}}} = \sqrt{\lim_{x\to\infty} \frac{24x}{8e^{2x}}} = \sqrt{\lim_{x\to\infty} \frac{24}{16e^{2x}}}$
$= \sqrt{0} = 0$
(d) slower, $\lim_{x\to\infty} \frac{\left(\frac{5}{2}\right)^x}{e^x} = \lim_{x\to\infty} \left(\frac{5}{2e}\right)^x = 0$ since $\frac{5}{2e} < 1$
(e) slower, $\lim_{x\to\infty} \frac{e^{-x}}{e^x} = \lim_{x\to\infty} \frac{1}{e^{2x}} = 0$
(f) faster, $\lim_{x\to\infty} \frac{xe^x}{e^x} = \lim_{x\to\infty} x = \infty$
(g) slower, since for all reals we have $-1 \le \cos x \le 1 \Rightarrow e^{-1} \le e^{\cos x} \le e^1 \Rightarrow \frac{e^{-1}}{e^x} \le \frac{e^{\cos x}}{e^x} \le \frac{e^1}{e^x}$ and also $\lim_{x\to\infty} \frac{e^{-1}}{e^x} = 0 = \lim_{x\to\infty} \frac{e^1}{e^x}$, so by the Sandwich Theorem we conclude that $\lim_{x\to\infty} \frac{e^{\cos x}}{e^x} = 0$
(h) same, $\lim_{x\to\infty} \frac{e^{x-1}}{e^x} = \lim_{x\to\infty} \frac{1}{e^{(x-x+1)}} = \lim_{x\to\infty} \frac{1}{e} = \frac{1}{e}$

3. (a) same, $\lim_{x\to\infty} \frac{x^2+4x}{x^2} = \lim_{x\to\infty} \frac{2x+4}{2x} = \lim_{x\to\infty} \frac{2}{2} = 1$
(b) faster, $\lim_{x\to\infty} \frac{x^5-x^2}{x^2} = \lim_{x\to\infty} (x^3-1) = \infty$
(c) same, $\lim_{x\to\infty} \frac{\sqrt{x^4+x^3}}{x^2} = \sqrt{\lim_{x\to\infty} \frac{x^4+x^3}{x^4}} = \sqrt{\lim_{x\to\infty} \left(1+\frac{1}{x}\right)} = \sqrt{1} = 1$

(d) same, $\lim_{x\to\infty} \frac{(x+3)^2}{x^2} = \lim_{x\to\infty} \frac{2(x+3)}{2x} = \lim_{x\to\infty} \frac{2}{2} = 1$

(e) slower, $\lim_{x\to\infty} \frac{x \ln x}{x^2} = \lim_{x\to\infty} \frac{\ln x}{x} = \lim_{x\to\infty} \frac{\left(\frac{1}{x}\right)}{1} = 0$

(f) faster, $\lim_{x\to\infty} \frac{2^x}{x^2} = \lim_{x\to\infty} \frac{(\ln 2)\, 2^x}{2x} = \lim_{x\to\infty} \frac{(\ln 2)^2\, 2^x}{2} = \infty$

(g) slower, $\lim_{x\to\infty} \frac{x^3 e^{-x}}{x^2} = \lim_{x\to\infty} \frac{x}{e^x} = \lim_{x\to\infty} \frac{1}{e^x} = 0$

(h) same, $\lim_{x\to\infty} \frac{8x^2}{x^2} = \lim_{x\to\infty} 8 = 8$

4. (a) same, $\lim_{x\to\infty} \frac{x^2+\sqrt{x}}{x^2} = \lim_{x\to\infty} \left(1 + \frac{1}{x^{3/2}}\right) = 1$

(b) same, $\lim_{x\to\infty} \frac{10x^2}{x^2} = \lim_{x\to\infty} 10 = 10$

(c) slower, $\lim_{x\to\infty} \frac{x^2 e^{-x}}{x^2} = \lim_{x\to\infty} \frac{1}{e^x} = 0$

(d) slower, $\lim_{x\to\infty} \frac{\log_{10} x^2}{x^2} = \lim_{x\to\infty} \frac{\left(\frac{\ln x^2}{\ln 10}\right)}{x^2} = \frac{1}{\ln 10} \lim_{x\to\infty} \frac{2 \ln x}{x^2} = \frac{2}{\ln 10} \lim_{x\to\infty} \frac{\left(\frac{1}{x}\right)}{2x} = \frac{1}{\ln 10} \lim_{x\to\infty} \frac{1}{x^2} = 0$

(e) faster, $\lim_{x\to\infty} \frac{x^3-x^2}{x^2} = \lim_{x\to\infty} (x-1) = \infty$

(f) slower, $\lim_{x\to\infty} \frac{\left(\frac{1}{10}\right)^x}{x^2} = \lim_{x\to\infty} \frac{1}{10^x x^2} = 0$

(g) faster, $\lim_{x\to\infty} \frac{(1.1)^x}{x^2} = \lim_{x\to\infty} \frac{(\ln 1.1)(1.1)^x}{2x} = \lim_{x\to\infty} \frac{(\ln 1.1)^2 (1.1)^x}{2} = \infty$

(h) same, $\lim_{x\to\infty} \frac{x^2+100x}{x^2} = \lim_{x\to\infty} \left(1 + \frac{100}{x}\right) = 1$

5. (a) same, $\lim_{x\to\infty} \frac{\log_3 x}{\ln x} = \lim_{x\to\infty} \frac{\left(\frac{\ln x}{\ln 3}\right)}{\ln x} = \lim_{x\to\infty} \frac{1}{\ln 3} = \frac{1}{\ln 3}$

(b) same, $\lim_{x\to\infty} \frac{\ln 2x}{\ln x} = \lim_{x\to\infty} \frac{\left(\frac{2}{2x}\right)}{\left(\frac{1}{x}\right)} = 1$

(c) same, $\lim_{x\to\infty} \frac{\ln \sqrt{x}}{\ln x} = \lim_{x\to\infty} \frac{\left(\frac{1}{2}\right) \ln x}{\ln x} = \lim_{x\to\infty} \frac{1}{2} = \frac{1}{2}$

(d) faster, $\lim_{x\to\infty} \frac{\sqrt{x}}{\ln x} = \lim_{x\to\infty} \frac{x^{1/2}}{\ln x} = \lim_{x\to\infty} \frac{\left(\frac{1}{2}\right) x^{-1/2}}{\left(\frac{1}{x}\right)} = \lim_{x\to\infty} \frac{x}{2\sqrt{x}} = \lim_{x\to\infty} \frac{\sqrt{x}}{2} = \infty$

(e) faster, $\lim_{x\to\infty} \frac{x}{\ln x} = \lim_{x\to\infty} \frac{1}{\left(\frac{1}{x}\right)} = \lim_{x\to\infty} x = \infty$

(f) same, $\lim_{x\to\infty} \frac{5 \ln x}{\ln x} = \lim_{x\to\infty} 5 = 5$

(g) slower, $\lim_{x\to\infty} \frac{\left(\frac{1}{x}\right)}{\ln x} = \lim_{x\to\infty} \frac{1}{x \ln x} = 0$

(h) faster, $\lim_{x\to\infty} \frac{e^x}{\ln x} = \lim_{x\to\infty} \frac{e^x}{\left(\frac{1}{x}\right)} = \lim_{x\to\infty} x e^x = \infty$

6. (a) same, $\lim_{x\to\infty} \frac{\log_2 x^2}{\ln x} = \lim_{x\to\infty} \frac{\left(\frac{\ln x^2}{\ln 2}\right)}{\ln x} = \frac{1}{\ln 2} \lim_{x\to\infty} \frac{\ln x^2}{\ln x} = \frac{1}{\ln 2} \lim_{x\to\infty} \frac{2 \ln x}{\ln x} = \frac{1}{\ln 2} \lim_{x\to\infty} 2 = \frac{2}{\ln 2}$

(b) same, $\lim_{x\to\infty} \frac{\log_{10} 10x}{\ln x} = \lim_{x\to\infty} \frac{\left(\frac{\ln 10x}{\ln 10}\right)}{\ln x} = \frac{1}{\ln 10} \lim_{x\to\infty} \frac{\ln 10x}{\ln x} = \frac{1}{\ln 10} \lim_{x\to\infty} \frac{\left(\frac{10}{10x}\right)}{\left(\frac{1}{x}\right)} = \frac{1}{\ln 10} \lim_{x\to\infty} 1 = \frac{1}{\ln 10}$

(c) slower, $\lim_{x\to\infty} \frac{\left(\frac{1}{\sqrt{x}}\right)}{\ln x} = \lim_{x\to\infty} \frac{1}{(\sqrt{x})(\ln x)} = 0$

(d) slower, $\lim_{x\to\infty} \frac{\left(\frac{1}{x^2}\right)}{\ln x} = \lim_{x\to\infty} \frac{1}{x^2 \ln x} = 0$

(e) faster, $\lim_{x\to\infty} \frac{x-2\ln x}{\ln x} = \lim_{x\to\infty} \left(\frac{x}{\ln x} - 2\right) = \left(\lim_{x\to\infty} \frac{x}{\ln x}\right) - 2 = \left(\lim_{x\to\infty} \frac{1}{\left(\frac{1}{x}\right)}\right) - 2 = \left(\lim_{x\to\infty} x\right) - 2 = \infty$

(f) slower, $\lim_{x\to\infty} \frac{e^{-x}}{\ln x} = \lim_{x\to\infty} \frac{1}{e^x \ln x} = 0$

(g) slower, $\lim_{x\to\infty} \frac{\ln(\ln x)}{\ln x} = \lim_{x\to\infty} \frac{\left(\frac{1/x}{\ln x}\right)}{\left(\frac{1}{x}\right)} = \lim_{x\to\infty} \frac{1}{\ln x} = 0$

(h) same, $\lim_{x\to\infty} \frac{\ln(2x+5)}{\ln x} = \lim_{x\to\infty} \frac{\left(\frac{2}{2x+5}\right)}{\left(\frac{1}{x}\right)} = \lim_{x\to\infty} \frac{2x}{2x+5} = \lim_{x\to\infty} \frac{2}{2} = \lim_{x\to\infty} 1 = 1$

7. $\lim_{x\to\infty} \frac{e^x}{e^{x/2}} = \lim_{x\to\infty} e^{x/2} = \infty \Rightarrow e^x$ grows faster than $e^{x/2}$; since for $x > e^e$ we have $\ln x > e$ and $\lim_{x\to\infty} \frac{(\ln x)^x}{e^x} = \lim_{x\to\infty} \left(\frac{\ln x}{e}\right)^x = \infty \Rightarrow (\ln x)^x$ grows faster than e^x; since $x > \ln x$ for all $x > 0$ and $\lim_{x\to\infty} \frac{x^x}{(\ln x)^x} = \lim_{x\to\infty} \left(\frac{x}{\ln x}\right)^x = \infty \Rightarrow x^x$ grows faster than $(\ln x)^x$. Therefore, slowest to fastest are: $e^{x/2}$, e^x, $(\ln x)^x$, x^x so the order is d, a, c, b

8. $\lim_{x\to\infty} \frac{(\ln 2)^x}{x^2} = \lim_{x\to\infty} \frac{(\ln(\ln 2))(\ln 2)^x}{2x} = \lim_{x\to\infty} \frac{(\ln(\ln 2))^2(\ln 2)^x}{2} = \frac{(\ln(\ln 2))^2}{2} \lim_{x\to\infty} (\ln 2)^x = 0$ $\Rightarrow (\ln 2)^x$ grows slower than x^2; $\lim_{x\to\infty} \frac{x^2}{2^x} = \lim_{x\to\infty} \frac{2x}{(\ln 2)2^x} = \lim_{x\to\infty} \frac{2}{(\ln 2)^2 2^x} = 0 \Rightarrow x^2$ grows slower than 2^x; $\lim_{x\to\infty} \frac{2^x}{e^x} = \lim_{x\to\infty} \left(\frac{2}{e}\right)^x = 0 \Rightarrow 2^x$ grows slower than e^x. Therefore, the slowest to the fastest is: $(\ln 2)^x$, x^2, 2^x and e^x so the order is c, b, a, d

9. (a) false; $\lim_{x\to\infty} \frac{x}{x} = 1$
(b) false; $\lim_{x\to\infty} \frac{x}{x+5} = \frac{1}{1} = 1$
(c) true; $x < x+5 \Rightarrow \frac{x}{x+5} < 1$ if $x > 1$ (or sufficiently large)
(d) true; $x < 2x \Rightarrow \frac{x}{2x} < 1$ if $x > 1$ (or sufficiently large)
(e) true; $\lim_{x\to\infty} \frac{e^x}{e^{2x}} = \lim_{x\to 0} \frac{1}{e^x} = 0$
(f) true; $\frac{x+\ln x}{x} = 1 + \frac{\ln x}{x} < 1 + \frac{\sqrt{x}}{x} = 1 + \frac{1}{\sqrt{x}} < 2$ if $x > 1$ (or sufficiently large)
(g) false; $\lim_{x\to\infty} \frac{\ln x}{\ln 2x} = \lim_{x\to\infty} \frac{\left(\frac{1}{x}\right)}{\left(\frac{2}{2x}\right)} = \lim_{x\to\infty} 1 = 1$
(h) true; $\frac{\sqrt{x^2+5}}{x} < \frac{\sqrt{(x+5)^2}}{x} < \frac{x+5}{x} = 1 + \frac{5}{x} < 6$ if $x > 1$ (or sufficiently large)

10. (a) true; $\frac{\left(\frac{1}{x+3}\right)}{\left(\frac{1}{x}\right)} = \frac{x}{x+3} < 1$ if $x > 1$ (or sufficiently large)
(b) true; $\frac{\left(\frac{1}{x}+\frac{1}{x^2}\right)}{\left(\frac{1}{x}\right)} = 1 + \frac{1}{x} < 2$ if $x > 1$ (or sufficiently large)
(c) false; $\lim_{x\to\infty} \frac{\left(\frac{1}{x}-\frac{1}{x^2}\right)}{\left(\frac{1}{x}\right)} = \lim_{x\to\infty} \left(1 - \frac{1}{x}\right) = 1$
(d) true; $2 + \cos x \le 3 \Rightarrow \frac{2+\cos x}{2} \le \frac{3}{2}$ if x is sufficiently large
(e) true; $\frac{e^x+x}{e^x} = 1 + \frac{x}{e^x}$ and $\frac{x}{e^x} \to 0$ as $x \to \infty \Rightarrow 1 + \frac{x}{e^x} < 2$ if x is sufficiently large
(f) true; $\lim_{x\to\infty} \frac{x\ln x}{x^2} = \lim_{x\to\infty} \frac{\ln x}{x} = \lim_{x\to\infty} \frac{\left(\frac{1}{x}\right)}{1} = 0$
(g) true; $\frac{\ln(\ln x)}{\ln x} < \frac{\ln x}{\ln x} = 1$ if x is sufficiently large
(h) false; $\lim_{x\to\infty} \frac{\ln x}{\ln(x^2+1)} = \lim_{x\to\infty} \frac{\left(\frac{1}{x}\right)}{\left(\frac{2x}{x^2+1}\right)} = \lim_{x\to\infty} \frac{x^2+1}{2x^2} = \lim_{x\to\infty} \left(\frac{1}{2} + \frac{1}{2x^2}\right) = \frac{1}{2}$

11. If f(x) and g(x) grow at the same rate, then $\lim_{x\to\infty} \frac{f(x)}{g(x)} = L \ne 0 \Rightarrow \lim_{x\to\infty} \frac{g(x)}{f(x)} = \frac{1}{L} \ne 0$. Then $\left|\frac{f(x)}{g(x)} - L\right| < 1$ if x is sufficiently large $\Rightarrow L - 1 < \frac{f(x)}{g(x)} < L + 1 \Rightarrow \frac{f(x)}{g(x)} \le |L| + 1$ if x is sufficiently large $\Rightarrow f = O(g)$. Similarly, $\frac{g(x)}{f(x)} \le \left|\frac{1}{L}\right| + 1 \Rightarrow g = O(f)$.

12. When the degree of f is less than the degree of g since in that case $\lim_{x\to\infty} \frac{f(x)}{g(x)} = 0$.

13. When the degree of f is less than or equal to the degree of g since $\lim_{x\to\infty} \frac{f(x)}{g(x)} = 0$ when the degree of f is smaller than the degree of g, and $\lim_{x\to\infty} \frac{f(x)}{g(x)} = \frac{a}{b}$ (the ratio of the leading coefficients) when the degrees are the same.

14. Polynomials of a greater degree grow at a greater rate than polynomials of a lesser degree. Polynomials of the same degree grow at the same rate.

15. $\lim_{x\to\infty} \frac{\ln(x+1)}{\ln x} = \lim_{x\to\infty} \frac{\left(\frac{1}{x+1}\right)}{\left(\frac{1}{x}\right)} = \lim_{x\to\infty} \frac{x}{x+1} = \lim_{x\to\infty} \frac{1}{1} = 1$ and $\lim_{x\to\infty} \frac{\ln(x+999)}{\ln x} = \lim_{x\to\infty} \frac{\left(\frac{1}{x+999}\right)}{\left(\frac{1}{x}\right)}$ $= \lim_{x\to\infty} \frac{x}{x+999} = 1$

16. $\lim_{x\to\infty} \frac{\ln(x+a)}{\ln x} = \lim_{x\to\infty} \frac{\left(\frac{1}{x+a}\right)}{\left(\frac{1}{x}\right)} = \lim_{x\to\infty} \frac{x}{x+a} = \lim_{x\to\infty} \frac{1}{1} = 1$. Therefore, the relative rates are the same.

17. $\lim_{x\to\infty} \frac{\sqrt{10x+1}}{\sqrt{x}} = \sqrt{\lim_{x\to\infty} \frac{10x+1}{x}} = \sqrt{10}$ and $\lim_{x\to\infty} \frac{\sqrt{x+1}}{\sqrt{x}} = \sqrt{\lim_{x\to\infty} \frac{x+1}{x}} = \sqrt{1} = 1$. Since the growth rate is transitive, we conclude that $\sqrt{10x+1}$ and $\sqrt{x+1}$ have the same growth rate (that of $\sqrt{x}$).

18. $\lim_{x\to\infty} \frac{\sqrt{x^4+x}}{x^2} = \sqrt{\lim_{x\to\infty} \frac{x^4+x}{x^4}} = 1$ and $\lim_{x\to\infty} \frac{\sqrt{x^4-x^3}}{x^2} = \sqrt{\lim_{x\to\infty} \frac{x^4-x^3}{x^4}} = 1$. Since the growth rate is transitive, we conclude that $\sqrt{x^4+x}$ and $\sqrt{x^4-x^3}$ have the same growth rate (that of x^2).

19. $\lim_{x\to\infty} \frac{x^n}{e^x} = \lim_{x\to\infty} \frac{nx^{n-1}}{e^x} = \ldots = \lim_{x\to\infty} \frac{n!}{e^x} = 0 \Rightarrow x^n = o(e^x)$ for any non-negative integer n

20. If $p(x) = a_n x^n + a_{n-1}x^{n-1} + \ldots + a_1 x + a_0$, then $\lim_{x\to\infty} \frac{p(x)}{e^x} = a_n \lim_{x\to\infty} \frac{x^n}{e^x} + a_{n-1} \lim_{x\to\infty} \frac{x^{n-1}}{e^x} + \ldots$ $+ a_1 \lim_{x\to\infty} \frac{x}{e^x} + a_0 \lim_{x\to\infty} \frac{1}{e^x}$ where each limit is zero (from Exercise 19). Therefore, $\lim_{x\to\infty} \frac{p(x)}{e^x} = 0$ $\Rightarrow e^x$ grows faster than any polynomial.

21. (a) $\lim_{x\to\infty} \frac{x^{1/n}}{\ln x} = \lim_{x\to\infty} \frac{x^{(1-n)/n}}{n\left(\frac{1}{x}\right)} = \left(\frac{1}{n}\right) \lim_{x\to\infty} x^{1/n} = \infty \Rightarrow \ln x = o\left(x^{1/n}\right)$ for any positive integer n

(b) $\ln\left(e^{17{,}000{,}000}\right) = 17{,}000{,}000 < \left(e^{17\times 10^6}\right)^{1/10^6} = e^{17} \approx 24{,}154{,}952.75$

(c) $x \approx 3.430631121 \times 10^{15}$

(d) In the interval $[3.41 \times 10^{15}, 3.45 \times 10^{15}]$ we have $\ln x = 10 \ln(\ln x)$. The graphs cross at about 3.4306311×10^{15}.

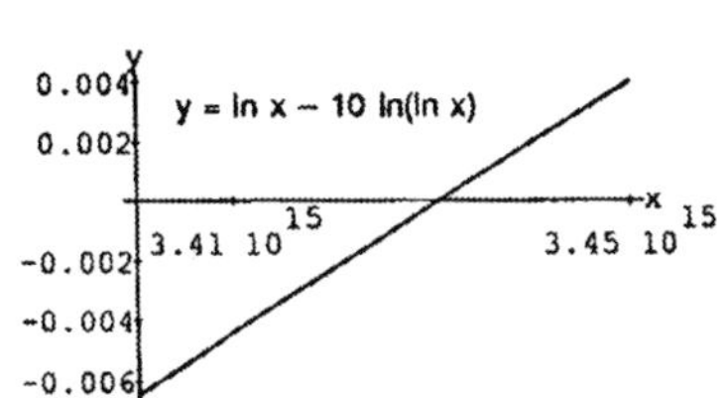

22. $\lim_{x\to\infty} \frac{\ln x}{a_n x^n + a_{n-1}x^{n-1} + \ldots + a_1 x + a_0} = \frac{\lim_{x\to\infty} \left(\frac{\ln x}{x^n}\right)}{\lim_{x\to\infty} \left(a_n + \frac{a_{n-1}}{x} + \ldots + \frac{a_1}{x^{n-1}} + \frac{a_0}{x^n}\right)} = \frac{\lim_{x\to\infty} \left[\frac{\left(\frac{1}{x}\right)}{nx^{n-1}}\right]}{a_n}$ $= \lim_{x\to\infty} \frac{1}{(a_n)(nx^n)} = 0 \Rightarrow \ln x$ grows slower than any non-constant polynomial $(n \geq 1)$

23. (a) $\lim_{n \to \infty} \frac{n \log_2 n}{n (\log_2 n)^2} = \lim_{n \to \infty} \frac{1}{\log_2 n} = 0 \Rightarrow n \log_2 n$ grows slower than $n (\log_2 n)^2$; $\lim_{n \to \infty} \frac{n \log_2 n}{n^{3/2}} = \lim_{n \to \infty} \frac{\left(\frac{\ln n}{\ln 2}\right)}{n^{1/2}}$ $= \frac{1}{\ln 2} \lim_{n \to \infty} \frac{\left(\frac{1}{n}\right)}{\left(\frac{1}{2}\right) n^{-1/2}} = \frac{2}{\ln 2} \lim_{n \to \infty} \frac{1}{n^{1/2}} = 0$ $\Rightarrow n \log_2 n$ grows slower than $n^{3/2}$. Therefore, $n \log_2 n$ grows at the slowest rate $\Rightarrow$ the algorithm that takes $O(n \log_2 n)$ steps is the most efficient in the long run.

(b)

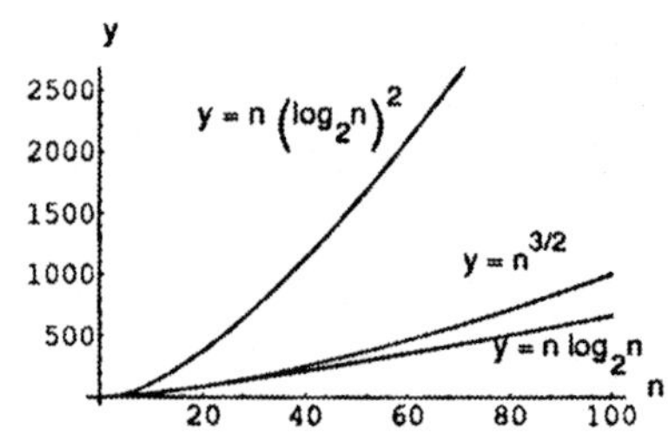

24. (a) $\lim_{n \to \infty} \frac{(\log_2 n)^2}{n} = \lim_{n \to \infty} \frac{\left(\frac{\ln n}{\ln 2}\right)^2}{n} = \lim_{n \to \infty} \frac{(\ln n)^2}{n(\ln 2)^2}$ $= \lim_{n \to \infty} \frac{2(\ln n)\left(\frac{1}{n}\right)}{(\ln 2)^2} = \frac{2}{(\ln 2)^2} \lim_{n \to \infty} \frac{\ln n}{n}$ $= \frac{2}{(\ln 2)^2} \lim_{n \to \infty} \frac{\left(\frac{1}{n}\right)}{1} = 0 \Rightarrow (\log_2 n)^2$ grows slower than n; $\lim_{n \to \infty} \frac{(\log_2 n)^2}{\sqrt{n} \log_2 n} = \lim_{n \to \infty} \frac{\log_2 n}{\sqrt{n}}$ $= \lim_{n \to \infty} \frac{\left(\frac{\ln n}{\ln 2}\right)}{n^{1/2}} = \frac{1}{\ln 2} \lim_{n \to \infty} \frac{\ln n}{n^{1/2}}$ $= \frac{1}{\ln 2} \lim_{x \to \infty} \frac{\left(\frac{1}{n}\right)}{\left(\frac{1}{2}\right) n^{-1/2}} = \frac{2}{\ln 2} \lim_{n \to \infty} \frac{1}{n^{1/2}} = 0 \Rightarrow (\log_2 n)^2$ grows slower than $\sqrt{n} \log_2 n$. Therefore $(\log_2 n)^2$ grows at the slowest rate $\Rightarrow$ the algorithm that takes $O\left((\log_2 n)^2\right)$ steps is the most efficient in the long run.

(b)

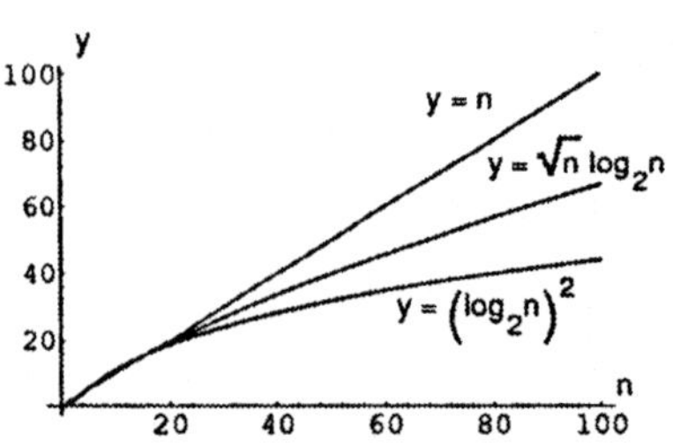

25. It could take one million steps for a sequential search, but at most 20 steps for a binary search because $2^{19} = 524{,}288 < 1{,}000{,}000 < 1{,}048{,}576 = 2^{20}$.

26. It could take 450,000 steps for a sequential search, but at most 19 steps for a binary search because $2^{18} = 262{,}144 < 450{,}000 < 524{,}288 = 2^{19}$.

7.4 HYPERBOLIC FUNCTIONS

1. $\sinh x = -\frac{3}{4} \Rightarrow \cosh x = \sqrt{1 + \sinh^2 x} = \sqrt{1 + \left(-\frac{3}{4}\right)^2} = \sqrt{1 + \frac{9}{16}} = \sqrt{\frac{25}{16}} = \frac{5}{4}$, $\tanh x = \frac{\sinh x}{\cosh x} = \frac{\left(-\frac{3}{4}\right)}{\left(\frac{5}{4}\right)} = -\frac{3}{5}$, $\coth x = \frac{1}{\tanh x} = -\frac{5}{3}$, $\operatorname{sech} x = \frac{1}{\cosh x} = \frac{4}{5}$, and $\operatorname{csch} x = \frac{1}{\sin x} = -\frac{4}{3}$

2. $\sinh x = \frac{4}{3} \Rightarrow \cosh x = \sqrt{1 + \sinh^2 x} = \sqrt{1 + \frac{16}{9}} = \sqrt{\frac{25}{9}} = \frac{5}{3}$, $\tanh x = \frac{\sinh x}{\cosh x} = \frac{\left(\frac{4}{3}\right)}{\left(\frac{5}{3}\right)} = \frac{4}{5}$, $\coth x = \frac{1}{\tanh x} = \frac{5}{4}$, $\operatorname{sech} x = \frac{1}{\cosh x} = \frac{3}{5}$, and $\operatorname{csch} x = \frac{1}{\sinh x} = \frac{3}{4}$

3. $\cosh x = \frac{17}{15}$, $x > 0 \Rightarrow \sinh x = \sqrt{\cosh^2 x - 1} = \sqrt{\left(\frac{17}{15}\right)^2 - 1} = \sqrt{\frac{289}{225} - 1} = \sqrt{\frac{64}{225}} = \frac{8}{15}$, $\tanh x = \frac{\sinh x}{\cosh x} = \frac{\left(\frac{8}{15}\right)}{\left(\frac{17}{15}\right)}$ $= \frac{8}{17}$, $\coth x = \frac{1}{\tanh x} = \frac{17}{8}$, $\operatorname{sech} x = \frac{1}{\cosh x} = \frac{15}{17}$, and $\operatorname{csch} x = \frac{1}{\sinh x} = \frac{15}{8}$

4. $\cosh x = \frac{13}{5}$, $x > 0 \Rightarrow \sinh x = \sqrt{\cosh^2 x - 1} = \sqrt{\frac{169}{25} - 1} = \sqrt{\frac{144}{25}} = \frac{12}{5}$, $\tanh x = \frac{\sinh x}{\cosh x} = \frac{\left(\frac{12}{5}\right)}{\left(\frac{13}{5}\right)} = \frac{12}{13}$, $\coth x = \frac{1}{\tanh x} = \frac{13}{12}$, $\operatorname{sech} x = \frac{1}{\cosh x} = \frac{5}{13}$, and $\operatorname{csch} x = \frac{1}{\sinh x} = \frac{5}{12}$

5. $2 \cosh (\ln x) = 2 \left(\frac{e^{\ln x} + e^{-\ln x}}{2}\right) = e^{\ln x} + \frac{1}{e^{\ln x}} = x + \frac{1}{x}$

6. $\sinh (2 \ln x) = \frac{e^{2 \ln x} - e^{-2 \ln x}}{2} = \frac{e^{\ln x^2} - e^{\ln x^{-2}}}{2} = \frac{\left(x^2 - \frac{1}{x^2}\right)}{2} = \frac{x^4 - 1}{2x^2}$

7. $\cosh 5x + \sinh 5x = \frac{e^{5x}+e^{-5x}}{2} + \frac{e^{5x}-e^{-5x}}{2} = e^{5x}$

8. $\cosh 3x - \sinh 3x = \frac{e^{3x}+e^{-3x}}{2} - \frac{e^{3x}-e^{-3x}}{2} = e^{-3x}$

9. $(\sinh x + \cosh x)^4 = \left(\frac{e^x-e^{-x}}{2} + \frac{e^x+e^{-x}}{2}\right)^4 = (e^x)^4 = e^{4x}$

10. $\ln(\cosh x + \sinh x) + \ln(\cosh x - \sinh x) = \ln(\cosh^2 x - \sinh^2 x) = \ln 1 = 0$

11. (a) $\sinh 2x = \sinh(x + x) = \sinh x \cosh x + \cosh x \sinh x = 2 \sinh x \cosh x$
(b) $\cosh 2x = \cosh(x + x) = \cosh x \cosh x + \sinh x \sin x = \cosh^2 x + \sinh^2 x$

12. $\cosh^2 x - \sinh^2 x = \left(\frac{e^x+e^{-x}}{2}\right)^2 - \left(\frac{e^x-e^{-x}}{2}\right)^2 = \frac{1}{4}\left[(e^x + e^{-x}) + (e^x - e^{-x})\right]\left[(e^x + e^{-x}) - (e^x - e^{-x})\right]$
$= \frac{1}{4}(2e^x)(2e^{-x}) = \frac{1}{4}(4e^0) = \frac{1}{4}(4) = 1$

13. $y = 6 \sinh \frac{x}{3} \Rightarrow \frac{dy}{dx} = 6\left(\cosh \frac{x}{3}\right)\left(\frac{1}{3}\right) = 2 \cosh \frac{x}{3}$

14. $y = \frac{1}{2} \sinh(2x + 1) \Rightarrow \frac{dy}{dx} = \frac{1}{2}[\cosh(2x + 1)](2) = \cosh(2x + 1)$

15. $y = 2\sqrt{t} \tanh \sqrt{t} = 2t^{1/2} \tanh t^{1/2} \Rightarrow \frac{dy}{dt} = \left[\operatorname{sech}^2\left(t^{1/2}\right)\right]\left(\frac{1}{2}t^{-1/2}\right)\left(2t^{1/2}\right) + \left(\tanh t^{1/2}\right)\left(t^{-1/2}\right)$
$= \operatorname{sech}^2 \sqrt{t} + \frac{\tanh \sqrt{t}}{\sqrt{t}}$

16. $y = t^2 \tanh \frac{1}{t} = t^2 \tanh t^{-1} \Rightarrow \frac{dy}{dt} = \left[\operatorname{sech}^2(t^{-1})\right](-t^{-2})(t^2) + (2t)(\tanh t^{-1}) = -\operatorname{sech}^2 \frac{1}{t} + 2t \tanh \frac{1}{t}$

17. $y = \ln(\sinh z) \Rightarrow \frac{dy}{dz} = \frac{\cosh z}{\sinh z} = \coth z$

18. $y = \ln(\cosh z) \Rightarrow \frac{dy}{dz} = \frac{\sinh z}{\cosh z} = \tanh z$

19. $y = (\operatorname{sech} \theta)(1 - \ln \operatorname{sech} \theta) \Rightarrow \frac{dy}{d\theta} = \left(-\frac{-\operatorname{sech} \theta \tanh \theta}{\operatorname{sech} \theta}\right)(\operatorname{sech} \theta) + (-\operatorname{sech} \theta \tanh \theta)(1 - \ln \operatorname{sech} \theta)$
$= \operatorname{sech} \theta \tanh \theta - (\operatorname{sech} \theta \tanh \theta)(1 - \ln \operatorname{sech} \theta) = (\operatorname{sech} \theta \tanh \theta)[1 - (1 - \ln \operatorname{sech} \theta)]$
$= (\operatorname{sech} \theta \tanh \theta)(\ln \operatorname{sech} \theta)$

20. $y = (\operatorname{csch} \theta)(1 - \ln \operatorname{csch} \theta) \Rightarrow \frac{dy}{d\theta} = (\operatorname{csch} \theta)\left(-\frac{-\operatorname{csch} \theta \coth \theta}{\operatorname{csch} \theta}\right) + (1 - \ln \operatorname{csch} \theta)(-\operatorname{csch} \theta \coth \theta)$
$= \operatorname{csch} \theta \coth \theta - (1 - \ln \operatorname{csch} \theta)(\operatorname{csch} \theta \coth \theta) = (\operatorname{csch} \theta \coth \theta)(1 - 1 + \ln \operatorname{csch} \theta) = (\operatorname{csch} \theta \coth \theta)(\ln \operatorname{csch} \theta)$

21. $y = \ln \cosh v - \frac{1}{2} \tanh^2 v \Rightarrow \frac{dy}{dv} = \frac{\sinh v}{\cosh v} - \left(\frac{1}{2}\right)(2 \tanh v)(\operatorname{sech}^2 v) = \tanh v - (\tanh v)(\operatorname{sech}^2 v)$
$= (\tanh v)(1 - \operatorname{sech}^2 v) = (\tanh v)(\tanh^2 v) = \tanh^3 v$

22. $y = \ln \sinh v - \frac{1}{2} \coth^2 v \Rightarrow \frac{dy}{dv} = \frac{\cosh v}{\sinh v} - \left(\frac{1}{2}\right)(2 \coth v)(-\operatorname{csch}^2 v) = \coth v + (\coth v)(\operatorname{csch}^2 v)$
$= (\coth v)(1 + \operatorname{csch}^2 v) = (\coth v)(\coth^2 v) = \coth^3 v$

23. $y = (x^2 + 1) \operatorname{sech}(\ln x) = (x^2 + 1)\left(\frac{2}{e^{\ln x}+e^{-\ln x}}\right) = (x^2 + 1)\left(\frac{2}{x+x^{-1}}\right) = (x^2 + 1)\left(\frac{2x}{x^2+1}\right) = 2x \Rightarrow \frac{dy}{dx} = 2$

24. $y = (4x^2 - 1) \operatorname{csch}(\ln 2x) = (4x^2 - 1)\left(\frac{2}{e^{\ln 2x}+e^{-\ln 2x}}\right) = (4x^2 - 1)\left(\frac{2}{2x-(2x)^{-1}}\right) = (4x^2 - 1)\left(\frac{4x}{4x^2-1}\right)$
$= 4x \Rightarrow \frac{dy}{dx} = 4$

25. $y = \sinh^{-1} \sqrt{x} = \sinh^{-1}\left(x^{1/2}\right) \Rightarrow \frac{dy}{dx} = \frac{\left(\frac{1}{2}\right)x^{-1/2}}{\sqrt{1+(x^{1/2})^2}} = \frac{1}{2\sqrt{x}\sqrt{1+x}} = \frac{1}{2\sqrt{x(1+x)}}$

26. $y = \cosh^{-1} 2\sqrt{x+1} = \cosh^{-1}\left(2(x+1)^{1/2}\right) \Rightarrow \frac{dy}{dx} = \frac{(2)\left(\frac{1}{2}\right)(x+1)^{-1/2}}{\sqrt{\left[2(x+1)^{1/2}\right]^2 - 1}} = \frac{1}{\sqrt{x+1}\sqrt{4x+3}} = \frac{1}{\sqrt{4x^2+7x+3}}$

27. $y = (1-\theta)\tanh^{-1}\theta \Rightarrow \frac{dy}{d\theta} = (1-\theta)\left(\frac{1}{1-\theta^2}\right) + (-1)\tanh^{-1}\theta = \frac{1}{1+\theta} - \tanh^{-1}\theta$

28. $y = (\theta^2 + 2\theta)\tanh^{-1}(\theta+1) \Rightarrow \frac{dy}{d\theta} = (\theta^2+2\theta)\left[\frac{1}{1-(\theta+1)^2}\right] + (2\theta+2)\tanh^{-1}(\theta+1)$
$= \frac{\theta^2+2\theta}{-\theta^2-2\theta} + (2\theta+2)\tanh^{-1}(\theta+1) = (2\theta+2)\tanh^{-1}(\theta+1) - 1$

29. $y = (1-t)\coth^{-1}\sqrt{t} = (1-t)\coth^{-1}\left(t^{1/2}\right) \Rightarrow \frac{dy}{dt} = (1-t)\left[\frac{\left(\frac{1}{2}\right)t^{-1/2}}{1-\left(t^{1/2}\right)^2}\right] + (-1)\coth^{-1}\left(t^{1/2}\right) = \frac{1}{2\sqrt{t}} - \coth^{-1}\sqrt{t}$

30. $y = (1-t^2)\coth^{-1} t \Rightarrow \frac{dy}{dt} = (1-t^2)\left(\frac{1}{1-t^2}\right) + (-2t)\coth^{-1} t = 1 - 2t\coth^{-1} t$

31. $y = \cos^{-1} x - x\,\text{sech}^{-1} x \Rightarrow \frac{dy}{dx} = \frac{-1}{\sqrt{1-x^2}} - \left[x\left(\frac{-1}{x\sqrt{1-x^2}}\right) + (1)\,\text{sech}^{-1} x\right] = \frac{-1}{\sqrt{1-x^2}} + \frac{1}{\sqrt{1-x^2}} - \text{sech}^{-1} x$
$= -\,\text{sech}^{-1} x$

32. $y = \ln x + \sqrt{1-x^2}\,\text{sech}^{-1} x = \ln x + \left(1-x^2\right)^{1/2}\text{sech}^{-1} x \Rightarrow \frac{dy}{dx}$
$= \frac{1}{x} + \left(1-x^2\right)^{1/2}\left(\frac{-1}{x\sqrt{1-x^2}}\right) + \left(\frac{1}{2}\right)\left(1-x^2\right)^{-1/2}(-2x)\,\text{sech}^{-1} x = \frac{1}{x} - \frac{1}{x} - \frac{x}{\sqrt{1-x^2}}\,\text{sech}^{-1} x = \frac{-x}{\sqrt{1-x^2}}\,\text{sech}^{-1} x$

33. $y = \text{csch}^{-1}\left(\frac{1}{2}\right)^\theta \Rightarrow \frac{dy}{d\theta} = -\frac{\left[\ln\left(\frac{1}{2}\right)\right]\left(\frac{1}{2}\right)^\theta}{\left(\frac{1}{2}\right)^\theta\sqrt{1+\left[\left(\frac{1}{2}\right)^\theta\right]^2}} = -\frac{\ln(1)-\ln(2)}{\sqrt{1+\left(\frac{1}{2}\right)^{2\theta}}} = \frac{\ln 2}{\sqrt{1+\left(\frac{1}{2}\right)^{2\theta}}}$

34. $y = \text{csch}^{-1} 2^\theta \Rightarrow \frac{dy}{d\theta} = -\frac{(\ln 2)\,2^\theta}{2^\theta\sqrt{1+\left(2^\theta\right)^2}} = \frac{-\ln 2}{\sqrt{1+2^{2\theta}}}$

35. $y = \sinh^{-1}(\tan x) \Rightarrow \frac{dy}{dx} = \frac{\sec^2 x}{\sqrt{1+(\tan x)^2}} = \frac{\sec^2 x}{\sqrt{\sec^2 x}} = \frac{\sec^2 x}{|\sec x|} = \frac{|\sec x|\,|\sec x|}{|\sec x|} = |\sec x|$

36. $y = \cosh^{-1}(\sec x) \Rightarrow \frac{dy}{dx} = \frac{(\sec x)(\tan x)}{\sqrt{\sec^2 x - 1}} = \frac{(\sec x)(\tan x)}{\sqrt{\tan^2 x}} = \frac{(\sec x)(\tan x)}{|\tan x|} = \sec x,\ 0 < x < \frac{\pi}{2}$

37. (a) If $y = \tan^{-1}(\sinh x) + C$, then $\frac{dy}{dx} = \frac{\cosh x}{1+\sinh^2 x} = \frac{\cosh x}{\cosh^2 x} = \text{sech}\, x$, which verifies the formula
(b) If $y = \sin^{-1}(\tanh x) + C$, then $\frac{dy}{dx} = \frac{\text{sech}^2 x}{\sqrt{1-\tanh^2 x}} = \frac{\text{sech}^2 x}{\text{sech}\, x} = \text{sech}\, x$, which verifies the formula

38. If $y = \frac{x^2}{2}\,\text{sech}^{-1} x - \frac{1}{2}\sqrt{1-x^2} + C$, then $\frac{dy}{dx} = x\,\text{sech}^{-1} x + \frac{x^2}{2}\left(\frac{-1}{x\sqrt{1-x^2}}\right) + \frac{2x}{4\sqrt{1-x^2}} = x\,\text{sech}^{-1} x$, which verifies the formula

39. If $y = \frac{x^2-1}{2}\coth^{-1} x + \frac{x}{2} + C$, then $\frac{dy}{dx} = x\coth^{-1} x + \left(\frac{x^2-1}{2}\right)\left(\frac{1}{1-x^2}\right) + \frac{1}{2} = x\coth^{-1} x$, which verifies the formula

40. If $y = x\tanh^{-1} x + \frac{1}{2}\ln\left(1-x^2\right) + C$, then $\frac{dy}{dx} = \tanh^{-1} x + x\left(\frac{1}{1-x^2}\right) + \frac{1}{2}\left(\frac{-2x}{1-x^2}\right) = \tanh^{-1} x$, which verifies the formula

41. $\int \sinh 2x\, dx = \frac{1}{2}\int \sinh u\, du$, where $u = 2x$ and $du = 2\,dx$
$= \frac{\cosh u}{2} + C = \frac{\cosh 2x}{2} + C$

42. $\int \sinh \frac{x}{5}\,dx = 5\int \sinh u\,du$, where $u = \frac{x}{5}$ and $du = \frac{1}{5}\,dx$
$= 5\cosh u + C = 5\cosh \frac{x}{5} + C$

43. $\int 6\cosh\left(\frac{x}{2} - \ln 3\right) dx = 12\int \cosh u\,du$, where $u = \frac{x}{2} - \ln 3$ and $du = \frac{1}{2}\,dx$
$= 12\sinh u + C = 12\sinh\left(\frac{x}{2} - \ln 3\right) + C$

44. $\int 4\cosh(3x - \ln 2)\,dx = \frac{4}{3}\int \cosh u\,du$, where $u = 3x - \ln 2$ and $du = 3\,dx$
$= \frac{4}{3}\sinh u + C = \frac{4}{3}\sinh(3x - \ln 2) + C$

45. $\int \tanh \frac{x}{7}\,dx = 7\int \frac{\sinh u}{\cosh u}\,du$, where $u = \frac{x}{7}$ and $du = \frac{1}{7}\,dx$
$= 7\ln|\cosh u| + C_1 = 7\ln\left|\cosh \frac{x}{7}\right| + C_1 = 7\ln\left|\frac{e^{x/7}+e^{-x/7}}{2}\right| + C_1 = 7\ln\left|e^{x/7} + e^{-x/7}\right| - 7\ln 2 + C_1$
$= 7\ln\left|e^{x/7} + e^{-x/7}\right| + C$

46. $\int \coth \frac{\theta}{\sqrt{3}}\,d\theta = \sqrt{3}\int \frac{\cosh u}{\sinh u}\,du$, where $u = \frac{\theta}{\sqrt{3}}$ and $du = \frac{d\theta}{\sqrt{3}}$
$= \sqrt{3}\ln|\sinh u| + C_1 = \sqrt{3}\ln\left|\sinh \frac{\theta}{\sqrt{3}}\right| + C_1 = \sqrt{3}\ln\left|\frac{e^{\theta/\sqrt{3}} - e^{-\theta/\sqrt{3}}}{2}\right| + C_1$
$= \sqrt{3}\ln\left|e^{\theta/\sqrt{3}} - e^{-\theta/\sqrt{3}}\right| - \sqrt{3}\ln 2 + C_1 = \sqrt{3}\ln\left|e^{\theta/\sqrt{3}} - e^{-\theta/\sqrt{3}}\right| + C$

47. $\int \operatorname{sech}^2\left(x - \frac{1}{2}\right) dx = \int \operatorname{sech}^2 u\,du$, where $u = \left(x - \frac{1}{2}\right)$ and $du = dx$
$= \tanh u + C = \tanh\left(x - \frac{1}{2}\right) + C$

48. $\int \operatorname{csch}^2(5 - x)\,dx = -\int \operatorname{csch}^2 u\,du$, where $u = (5 - x)$ and $du = -\,dx$
$= -(-\coth u) + C = \coth u + C = \coth(5 - x) + C$

49. $\int \frac{\operatorname{sech}\sqrt{t}\tanh\sqrt{t}}{\sqrt{t}}\,dt = 2\int \operatorname{sech} u \tanh u\,du$, where $u = \sqrt{t} = t^{1/2}$ and $du = \frac{dt}{2\sqrt{t}}$
$= 2(-\operatorname{sech} u) + C = -2\operatorname{sech}\sqrt{t} + C$

50. $\int \frac{\operatorname{csch}(\ln t)\coth(\ln t)}{t}\,dt = \int \operatorname{csch} u \coth u\,du$, where $u = \ln t$ and $du = \frac{dt}{t}$
$= -\operatorname{csch} u + C = -\operatorname{csch}(\ln t) + C$

51. $\int_{\ln 2}^{\ln 4} \coth x\,dx = \int_{\ln 2}^{\ln 4} \frac{\cosh x}{\sinh x}\,dx = \int_{3/4}^{15/8} \frac{1}{u}\,du = [\ln|u|]_{3/4}^{15/8} = \ln\left|\frac{15}{8}\right| - \ln\left|\frac{3}{4}\right| = \ln\left|\frac{15}{8}\cdot\frac{4}{3}\right| = \ln\frac{5}{2}$,
where $u = \sinh x$, $du = \cosh x\,dx$, the lower limit is $\sinh(\ln 2) = \frac{e^{\ln 2} - e^{-\ln 2}}{2} = \frac{2 - \left(\frac{1}{2}\right)}{2} = \frac{3}{4}$ and the upper limit is $\sinh(\ln 4) = \frac{e^{\ln 4} - e^{-\ln 4}}{2} = \frac{4 - \left(\frac{1}{4}\right)}{2} = \frac{15}{8}$

52. $\int_0^{\ln 2} \tanh 2x\,dx = \int_0^{\ln 2} \frac{\sinh 2x}{\cosh 2x}\,dx = \frac{1}{2}\int_1^{17/8} \frac{1}{u}\,du = \frac{1}{2}[\ln|u|]_1^{17/8} = \frac{1}{2}\left[\ln\left(\frac{17}{8}\right) - \ln 1\right] = \frac{1}{2}\ln\frac{17}{8}$, where $u = \cosh 2x$, $du = 2\sinh(2x)\,dx$, the lower limit is $\cosh 0 = 1$ and the upper limit is $\cosh(2\ln 2) = \cosh(\ln 4)$
$= \frac{e^{\ln 4} + e^{-\ln 4}}{2} = \frac{4 + \left(\frac{1}{4}\right)}{2} = \frac{17}{8}$

53. $\int_{-\ln 4}^{-\ln 2} 2e^{\theta}\cosh\theta\,d\theta = \int_{-\ln 4}^{-\ln 2} 2e^{\theta}\left(\frac{e^{\theta}+e^{-\theta}}{2}\right) d\theta = \int_{-\ln 4}^{-\ln 2} (e^{2\theta} + 1)\,d\theta = \left[\frac{e^{2\theta}}{2} + \theta\right]_{-\ln 4}^{-\ln 2}$
$= \left(\frac{e^{-2\ln 2}}{2} - \ln 2\right) - \left(\frac{e^{-2\ln 4}}{2} - \ln 4\right) = \left(\frac{1}{8} - \ln 2\right) - \left(\frac{1}{32} - \ln 4\right) = \frac{3}{32} - \ln 2 + 2\ln 2 = \frac{3}{32} + \ln 2$

54. $\int_0^{\ln 2} 4e^{-\theta} \sinh \theta \, d\theta = \int_0^{\ln 2} 4e^{-\theta} \left(\frac{e^\theta - e^{-\theta}}{2}\right) d\theta = 2 \int_0^{\ln 2} (1 - e^{-2\theta}) \, d\theta = 2 \left[\theta + \frac{e^{-2\theta}}{2}\right]_0^{\ln 2}$
$= 2\left[\left(\ln 2 + \frac{e^{-2\ln 2}}{2}\right) - \left(0 + \frac{e^0}{2}\right)\right] = 2\left(\ln 2 + \frac{1}{8} - \frac{1}{2}\right) = 2 \ln 2 + \frac{1}{4} - 1 = \ln 4 - \frac{3}{4}$

55. $\int_{-\pi/4}^{\pi/4} \cosh(\tan \theta) \sec^2 \theta \, d\theta = \int_{-1}^{1} \cosh u \, du = [\sinh u]_{-1}^{1} = \sinh(1) - \sinh(-1) = \left(\frac{e^1 - e^{-1}}{2}\right) - \left(\frac{e^{-1} - e^1}{2}\right)$
$= \frac{e - e^{-1} - e^{-1} + e}{2} = e - e^{-1}$, where $u = \tan \theta$, $du = \sec^2 \theta \, d\theta$, the lower limit is $\tan\left(-\frac{\pi}{4}\right) = -1$ and the upper limit is $\tan\left(\frac{\pi}{4}\right) = 1$

56. $\int_0^{\pi/2} 2 \sinh(\sin \theta) \cos \theta \, d\theta = 2\int_0^1 \sinh u \, du = 2 [\cosh u]_0^1 = 2(\cosh 1 - \cosh 0) = 2\left(\frac{e + e^{-1}}{2} - 1\right)$
$= e + e^{-1} - 2$, where $u = \sin \theta$, $du = \cos \theta \, d\theta$, the lower limit is $\sin 0 = 0$ and the upper limit is $\sin\left(\frac{\pi}{2}\right) = 1$

57. $\int_1^2 \frac{\cosh(\ln t)}{t} \, dt = \int_0^{\ln 2} \cosh u \, du = [\sinh u]_0^{\ln 2} = \sinh(\ln 2) - \sinh(0) = \frac{e^{\ln 2} - e^{-\ln 2}}{2} - 0 = \frac{2 - \frac{1}{2}}{2} = \frac{3}{4}$, where $u = \ln t$, $du = \frac{1}{t} \, dt$, the lower limit is $\ln 1 = 0$ and the upper limit is $\ln 2$

58. $\int_1^4 \frac{8 \cosh \sqrt{x}}{\sqrt{x}} \, dx = 16\int_1^2 \cosh u \, du = 16 [\sinh u]_1^2 = 16(\sinh 2 - \sinh 1) = 16 \left[\left(\frac{e^2 - e^{-2}}{2}\right) - \left(\frac{e - e^{-1}}{2}\right)\right]$
$= 8(e^2 - e^{-2} - e + e^{-1})$, where $u = \sqrt{x} = x^{1/2}$, $du = \frac{1}{2} x^{-1/2} dx = \frac{dx}{2\sqrt{x}}$, the lower limit is $\sqrt{1} = 1$ and the upper limit is $\sqrt{4} = 2$

59. $\int_{-\ln 2}^{0} \cosh^2\left(\frac{x}{2}\right) dx = \int_{-\ln 2}^{0} \frac{\cosh x + 1}{2} \, dx = \frac{1}{2}\int_{-\ln 2}^{0} (\cosh x + 1) \, dx = \frac{1}{2} [\sinh x + x]_{-\ln 2}^{0}$
$= \frac{1}{2} [(\sinh 0 + 0) - (\sinh(-\ln 2) - \ln 2)] = \frac{1}{2} \left[(0 + 0) - \left(\frac{e^{-\ln 2} - e^{\ln 2}}{2} - \ln 2\right)\right] = \frac{1}{2} \left[-\frac{\left(\frac{1}{2}\right) - 2}{2} + \ln 2\right]$
$= \frac{1}{2}\left(1 - \frac{1}{4} + \ln 2\right) = \frac{3}{8} + \frac{1}{2} \ln 2 = \frac{3}{8} + \ln \sqrt{2}$

60. $\int_0^{\ln 10} 4 \sinh^2\left(\frac{x}{2}\right) dx = \int_0^{\ln 10} 4\left(\frac{\cosh x - 1}{2}\right) dx = 2\int_0^{\ln 10} (\cosh x - 1) \, dx = 2 [\sinh x - x]_0^{\ln 10}$
$= 2[(\sinh(\ln 10) - \ln 10) - (\sinh 0 - 0)] = e^{\ln 10} - e^{-\ln 10} - 2 \ln 10 = 10 - \frac{1}{10} - 2 \ln 10 = 9.9 - 2 \ln 10$

61. $\sinh^{-1}\left(\frac{-5}{12}\right) = \ln\left(-\frac{5}{12} + \sqrt{\frac{25}{144} + 1}\right) = \ln\left(\frac{2}{3}\right)$

62. $\cosh^{-1}\left(\frac{5}{3}\right) = \ln\left(\frac{5}{3} + \sqrt{\frac{25}{9} - 1}\right) = \ln 3$

63. $\tanh^{-1}\left(-\frac{1}{2}\right) = \frac{1}{2} \ln\left(\frac{1 - (1/2)}{1 + (1/2)}\right) = -\frac{\ln 3}{2}$

64. $\coth^{-1}\left(\frac{5}{4}\right) = \frac{1}{2} \ln\left(\frac{(9/4)}{(1/4)}\right) = \frac{1}{2} \ln 9 = \ln 3$

65. $\operatorname{sech}^{-1}\left(\frac{3}{5}\right) = \ln\left(\frac{1 + \sqrt{1 - (9/25)}}{(3/5)}\right) = \ln 3$

66. $\operatorname{csch}^{-1}\left(-\frac{1}{\sqrt{3}}\right) = \ln\left(-\sqrt{3} + \frac{\sqrt{4/3}}{\left(1/\sqrt{3}\right)}\right) = \ln\left(-\sqrt{3} + 2\right)$

67. (a) $\int_0^{2\sqrt{3}} \frac{dx}{\sqrt{4 + x^2}} = \left[\sinh^{-1} \frac{x}{2}\right]_0^{2\sqrt{3}} = \sinh^{-1} \sqrt{3} - \sinh 0 = \sinh^{-1} \sqrt{3}$

(b) $\sinh^{-1} \sqrt{3} = \ln\left(\sqrt{3} + \sqrt{3 + 1}\right) = \ln\left(\sqrt{3} + 2\right)$

68. (a) $\int_0^{1/3} \frac{6 \, dx}{\sqrt{1 + 9x^2}} = 2\int_0^1 \frac{du}{\sqrt{a^2 + u^2}}$, where $u = 3x$, $du = 3 \, dx$, $a = 1$
$= [2 \sinh^{-1} u]_0^1 = 2(\sinh^{-1} 1 - \sinh^{-1} 0) = 2 \sinh^{-1} 1$

(b) $2 \sinh^{-1} 1 = 2 \ln\left(1 + \sqrt{1^2 + 1}\right) = 2 \ln\left(1 + \sqrt{2}\right)$

69. (a) $\int_{5/4}^{2} \frac{1}{1-x^2}\,dx = \left[\coth^{-1} x\right]_{5/4}^{2} = \coth^{-1} 2 - \coth^{-1} \frac{5}{4}$
(b) $\coth^{-1} 2 - \coth^{-1} \frac{5}{4} = \frac{1}{2}\left[\ln 3 - \ln\left(\frac{9/4}{1/4}\right)\right] = \frac{1}{2} \ln \frac{1}{3}$

70. (a) $\int_{0}^{1/2} \frac{1}{1-x^2}\,dx = \left[\tanh^{-1} x\right]_{0}^{1/2} = \tanh^{-1} \frac{1}{2} - \tanh^{-1} 0 = \tanh^{-1} \frac{1}{2}$
(b) $\tanh^{-1} \frac{1}{2} = \frac{1}{2} \ln\left(\frac{1+(1/2)}{1-(1/2)}\right) = \frac{1}{2} \ln 3$

71. (a) $\int_{1/5}^{3/13} \frac{dx}{x\sqrt{1-16x^2}} = \int_{4/5}^{12/13} \frac{du}{u\sqrt{a^2-u^2}}$, where $u = 4x$, $du = 4\,dx$, $a = 1$
$= \left[-\operatorname{sech}^{-1} u\right]_{4/5}^{12/13} = -\operatorname{sech}^{-1} \frac{12}{13} + \operatorname{sech}^{-1} \frac{4}{5}$
(b) $-\operatorname{sech}^{-1} \frac{12}{13} + \operatorname{sech}^{-1} \frac{4}{5} = -\ln\left(\frac{1+\sqrt{1-(12/13)^2}}{(12/13)}\right) + \ln\left(\frac{1+\sqrt{1-(4/5)^2}}{(4/5)}\right)$
$= -\ln\left(\frac{13+\sqrt{169-144}}{12}\right) + \ln\left(\frac{5+\sqrt{25-16}}{4}\right) = \ln\left(\frac{5+3}{4}\right) - \ln\left(\frac{13+5}{12}\right) = \ln 2 - \ln \frac{3}{2}$
$= \ln\left(2 \cdot \frac{2}{3}\right) = \ln \frac{4}{3}$

72. (a) $\int_{1}^{2} \frac{dx}{x\sqrt{4+x^2}} = \left[-\frac{1}{2} \operatorname{csch}^{-1} \left|\frac{x}{2}\right|\right]_{1}^{2} = -\frac{1}{2}\left(\operatorname{csch}^{-1} 1 - \operatorname{csch}^{-1} \frac{1}{2}\right) = \frac{1}{2}\left(\operatorname{csch}^{-1} \frac{1}{2} - \operatorname{csch}^{-1} 1\right)$
(b) $\frac{1}{2}\left(\operatorname{csch}^{-1} \frac{1}{2} - \operatorname{csch}^{-1} 1\right) = \frac{1}{2}\left[\ln\left(2 + \frac{\sqrt{5/4}}{(1/2)}\right) - \ln\left(1+\sqrt{2}\right)\right] = \frac{1}{2} \ln\left(\frac{2+\sqrt{5}}{1+\sqrt{2}}\right)$

73. (a) $\int_{0}^{\pi} \frac{\cos x}{\sqrt{1+\sin^2 x}}\,dx = \int_{0}^{0} \frac{1}{\sqrt{1+u^2}}\,du = \left[\sinh^{-1} u\right]_{0}^{0} = \sinh^{-1} 0 - \sinh^{-1} 0 = 0$, where $u = \sin x$, $du = \cos x\,dx$
(b) $\sinh^{-1} 0 - \sinh^{-1} 0 = \ln\left(0 + \sqrt{0+1}\right) - \ln\left(0 + \sqrt{0+1}\right) = 0$

74. (a) $\int_{1}^{e} \frac{dx}{x\sqrt{1+(\ln x)^2}} = \int_{0}^{1} \frac{du}{\sqrt{a^2+u^2}}$, where $u = \ln x$, $du = \frac{1}{x}\,dx$, $a = 1$
$= \left[\sinh^{-1} u\right]_{0}^{1} = \sinh^{-1} 1 - \sinh^{-1} 0 = \sinh^{-1} 1$
(b) $\sinh^{-1} 1 - \sinh^{-1} 0 = \ln\left(1 + \sqrt{1^2+1}\right) - \ln\left(0 + \sqrt{0^2+1}\right) = \ln\left(1+\sqrt{2}\right)$

75. (a) Let $E(x) = \frac{f(x)+f(-x)}{2}$ and $O(x) = \frac{f(x)-f(-x)}{2}$. Then $E(x) + O(x) = \frac{f(x)+f(-x)}{2} + \frac{f(x)-f(-x)}{2}$
$= \frac{2f(x)}{2} = f(x)$. Also, $E(-x) = \frac{f(-x)+f(-(-x))}{2} = \frac{f(x)+f(-x)}{2} = E(x) \Rightarrow E(x)$ is even, and
$O(-x) = \frac{f(-x)-f(-(-x))}{2} = -\frac{f(x)-f(-x)}{2} = -O(x) \Rightarrow O(x)$ is odd. Consequently, $f(x)$ can be written as a sum of an even and an odd function.
(b) $f(x) = \frac{f(x)+f(-x)}{2}$ because $\frac{f(x)-f(-x)}{2} = 0$ if f is even and $f(x) = \frac{f(x)-f(-x)}{2}$ because $\frac{f(x)+f(-x)}{2} = 0$ if f is odd.
Thus, if f is even $f(x) = \frac{2f(x)}{2} + 0$ and if f is odd, $f(x) = 0 + \frac{2f(x)}{2}$

76. $y = \sinh^{-1} x \Rightarrow x = \sinh y \Rightarrow x = \frac{e^y - e^{-y}}{2} \Rightarrow 2x = e^y - \frac{1}{e^y} \Rightarrow 2xe^y = e^{2y} - 1 \Rightarrow e^{2y} - 2xe^y - 1 = 0$
$\Rightarrow e^y = \frac{2x \pm \sqrt{4x^2+4}}{2} \Rightarrow e^y = x + \sqrt{x^2+1} \Rightarrow \sinh^{-1} x = y = \ln\left(x + \sqrt{x^2+1}\right)$. Since $e^y > 0$, we cannot choose $e^y = x - \sqrt{x^2+1}$ because $x - \sqrt{x^2+1} < 0$.

77. (a) $v = \sqrt{\frac{mg}{k}} \tanh\left(\sqrt{\frac{gk}{m}}\,t\right) \Rightarrow \frac{dv}{dt} = \sqrt{\frac{mg}{k}}\left[\operatorname{sech}^2\left(\sqrt{\frac{gk}{m}}\,t\right)\right]\left(\sqrt{\frac{gk}{m}}\right) = g \operatorname{sech}^2\left(\sqrt{\frac{gk}{m}}\,t\right)$.
Thus $m\frac{dv}{dt} = mg \operatorname{sech}^2\left(\sqrt{\frac{gk}{m}}\,t\right) = mg\left(1 - \tanh^2\left(\sqrt{\frac{gk}{m}}\,t\right)\right) = mg - kv^2$. Also, since $\tanh x = 0$ when $x = 0$, $v = 0$ when $t = 0$.
(b) $\lim_{t \to \infty} v = \lim_{t \to \infty} \sqrt{\frac{mg}{k}} \tanh\left(\sqrt{\frac{kg}{m}}\,t\right) = \sqrt{\frac{mg}{k}} \lim_{t \to \infty} \tanh\left(\sqrt{\frac{kg}{m}}\,t\right) = \sqrt{\frac{mg}{k}}\,(1) = \sqrt{\frac{mg}{k}}$

(c) $\sqrt{\frac{160}{0.005}} = \sqrt{\frac{160,000}{5}} = \frac{400}{\sqrt{5}} = 80\sqrt{5} \approx 178.89$ ft/sec

78. (a) $s(t) = a \cos kt + b \sin kt \Rightarrow \frac{ds}{dt} = -ak \sin kt + bk \cos kt \Rightarrow \frac{d^2s}{dt^2} = -ak^2 \cos kt - bk^2 \sin kt$
$= -k^2 (a \cos kt + b \sin kt) = -k^2 s(t) \Rightarrow$ acceleration is proportional to s. The negative constant $-k^2$ implies that the acceleration is directed toward the origin.

(b) $s(t) = a \cosh kt + b \sinh kt \Rightarrow \frac{ds}{dt} = ak \sinh kt + bk \cosh kt \Rightarrow \frac{d^2s}{dt^2} = ak^2 \cosh kt + bk^2 \sinh kt$
$= k^2 (a \cosh kt + b \sinh kt) = k^2 s(t) \Rightarrow$ acceleration is proportional to s. The positive constant k^2 implies that the acceleration is directed away from the origin.

79. $\frac{dy}{dx} = \frac{-1}{x\sqrt{1-x^2}} + \frac{x}{\sqrt{1-x^2}} \Rightarrow y = \int \frac{-1}{x\sqrt{1-x^2}}\, dx + \int \frac{x}{\sqrt{1-x^2}}\, dx \Rightarrow y = \operatorname{sech}^{-1}(x) - \sqrt{1-x^2} + C$; $x = 1$ and $y = 0 \Rightarrow C = 0 \Rightarrow y = \operatorname{sech}^{-1}(x) - \sqrt{1-x^2}$

80. To find the length of the curve: $y = \frac{1}{a} \cosh ax \Rightarrow y' = \sinh ax \Rightarrow L = \int_0^b \sqrt{1 + (\sinh ax)^2}\, dx$
$\Rightarrow L = \int_0^b \cosh ax\, dx = \left[\frac{1}{a} \sinh ax\right]_0^b = \frac{1}{a} \sinh ab$. Then the area under the curve is $A = \int_0^b \frac{1}{a} \cosh ax\, dx$
$= \left[\frac{1}{a^2} \sinh ax\right]_0^b = \frac{1}{a^2} \sinh ab = \left(\frac{1}{a}\right)\left(\frac{1}{a} \sinh ab\right)$ which is the area of the rectangle of height $\frac{1}{a}$ and length L as claimed.

81. $V = \pi \int_0^2 (\cosh^2 x - \sinh^2 x)\, dx = \pi \int_0^2 1\, dx = 2\pi$

82. $V = 2\pi \int_0^{\ln \sqrt{3}} \operatorname{sech}^2 x\, dx = 2\pi [\tanh x]_0^{\ln \sqrt{3}} = 2\pi \left[\frac{\sqrt{3} - \left(1/\sqrt{3}\right)}{\sqrt{3} + \left(1/\sqrt{3}\right)}\right] = \pi$

83. $y = \frac{1}{2} \cosh 2x \Rightarrow y' = \sinh 2x \Rightarrow L = \int_0^{\ln \sqrt{5}} \sqrt{1 + (\sinh 2x)^2}\, dx = \int_0^{\ln \sqrt{5}} \cosh 2x\, dx = \left[\frac{1}{2} \sinh 2x\right]_0^{\ln \sqrt{5}}$
$= \left[\frac{1}{2}\left(\frac{e^{2x} - e^{-2x}}{2}\right)\right]_0^{\ln \sqrt{5}} = \frac{1}{4}\left(5 - \frac{1}{5}\right) = \frac{6}{5}$

84. (a) Let the point located at $(\cosh u, 0)$ be called T. Then $A(u) =$ area of the triangle ΔOTP minus the area under the curve $y = \sqrt{x^2 - 1}$ from A to T $\Rightarrow A(u) = \frac{1}{2} \cosh u \sinh u - \int_1^{\cosh u} \sqrt{x^2 - 1}\, dx$.

(b) $A(u) = \frac{1}{2} \cosh u \sinh u - \int_1^{\cosh u} \sqrt{x^2 - 1}\, dx \Rightarrow A'(u) = \frac{1}{2}(\cosh^2 u + \sinh^2 u) - \left(\sqrt{\cosh^2 u - 1}\right)(\sinh u)$
$= \frac{1}{2} \cosh^2 u + \frac{1}{2} \sinh^2 u - \sinh^2 u = \frac{1}{2}(\cosh^2 u - \sinh^2 u) = \left(\frac{1}{2}\right)(1) = \frac{1}{2}$

(c) $A'(u) = \frac{1}{2} \Rightarrow A(u) = \frac{u}{2} + C$, and from part (a) we have $A(0) = 0 \Rightarrow C = 0 \Rightarrow A(u) = \frac{u}{2} \Rightarrow u = 2A$

85. $y = 4 \cosh \frac{x}{4} \Rightarrow 1 + \left(\frac{dy}{dx}\right)^2 = 1 + \sinh^2\left(\frac{x}{4}\right) = \cosh^2\left(\frac{x}{4}\right)$; the surface area is $S = \int_{-\ln 16}^{\ln 81} 2\pi y \sqrt{1 + \left(\frac{dy}{dx}\right)^2}\, dx$
$= 8\pi \int_{-\ln 16}^{\ln 81} \cosh^2\left(\frac{x}{4}\right) dx = 4\pi \int_{-\ln 16}^{\ln 81} \left(1 + \cosh \frac{x}{2}\right) dx = 4\pi \left[x + 2 \sinh \frac{x}{2}\right]_{-\ln 16}^{\ln 81}$
$= 4\pi \left[\left(\ln 81 + 2 \sinh\left(\frac{\ln 81}{2}\right)\right) - \left(-\ln 16 + 2 \sinh\left(\frac{-\ln 16}{2}\right)\right)\right] = 4\pi [\ln (81 \cdot 16) + 2 \sinh (\ln 9) + 2 \sinh (\ln 4)]$
$= 4\pi [\ln (9 \cdot 4)^2 + (e^{\ln 9} - e^{-\ln 9}) + (e^{\ln 4} - e^{-\ln 4})] = 4\pi \left[2 \ln 36 + \left(9 - \frac{1}{9}\right) + \left(4 - \frac{1}{4}\right)\right] = 4\pi \left(4 \ln 6 + \frac{80}{9} + \frac{15}{4}\right)$
$= 4\pi \left(4 \ln 6 + \frac{320 + 135}{36}\right) = 16\pi \ln 6 + \frac{455\pi}{9}$

86. (a) $y = \cosh x \Rightarrow ds = \sqrt{(dx)^2 + (dy)^2} = \sqrt{(dx)^2 + (\sinh^2 x)(dx)^2} = \cosh x\, dx$; $M_x = \int_{-\ln 2}^{\ln 2} y\, ds$
$= \int_{-\ln 2}^{\ln 2} \cosh x\, ds = \int_{-\ln 2}^{\ln 2} \cosh^2 x\, dx = \int_0^{\ln 2} (\cosh 2x + 1)\, dx = \left[\frac{\sinh 2x}{2} + x\right]_0^{\ln 2} = \frac{1}{4}(e^{\ln 4} - e^{-\ln 4}) + \ln 2$
$= \frac{15}{16} + \ln 2$; $M = 2 \int_0^{\ln 2} \sqrt{1 + \sinh^2 x}\, dx = 2 \int_0^{\ln 2} \cosh x\, dx = 2 [\sinh x]_0^{\ln 2} = e^{\ln 2} - e^{-\ln 2} = 2 - \frac{1}{2} = \frac{3}{2}$.

Therefore, $\overline{y} = \frac{M_x}{M} = \frac{\left(\frac{15}{16} + \ln 2\right)}{\left(\frac{3}{2}\right)} = \frac{5}{8} + \frac{\ln 4}{3}$, and by symmetry $\overline{x} = 0$.

(b) $\overline{x} = 0, \overline{y} \approx 1.09$

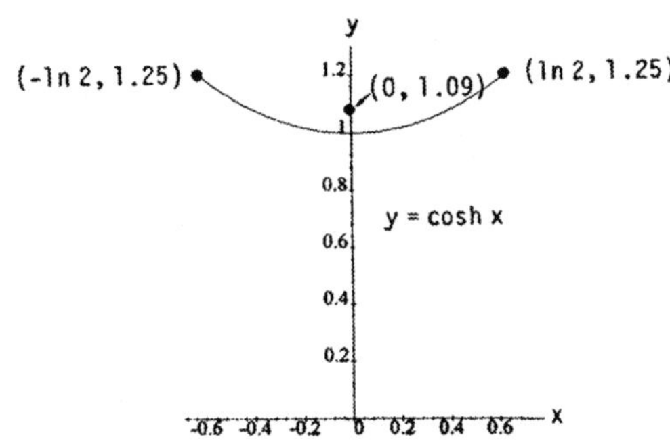

87. (a) $y = \frac{H}{w} \cosh\left(\frac{w}{H} x\right) \Rightarrow \tan\phi = \frac{dy}{dx} = \left(\frac{H}{w}\right)\left[\frac{w}{H} \sinh\left(\frac{w}{H} x\right)\right] = \sinh\left(\frac{w}{H} x\right)$

(b) The tension at P is given by $T \cos\phi = H \Rightarrow T = H \sec\phi = H\sqrt{1 + \tan^2\phi} = H\sqrt{1 + \left(\sinh \frac{w}{H} x\right)^2}$
$= H \cosh\left(\frac{w}{H} x\right) = w\left(\frac{H}{w}\right) \cosh\left(\frac{w}{H} x\right) = wy$

88. $s = \frac{1}{a} \sinh ax \Rightarrow \sinh ax = as \Rightarrow ax = \sinh^{-1} as \Rightarrow x = \frac{1}{a} \sinh^{-1} as$; $y = \frac{1}{a} \cosh ax = \frac{1}{a}\sqrt{\cosh^2 ax}$
$= \frac{1}{a}\sqrt{\sinh^2 ax + 1} = \frac{1}{a}\sqrt{a^2s^2 + 1} = \sqrt{s^2 + \frac{1}{a^2}}$

89. (a) Since the cable is 32 ft long, $s = 16$ and $x = 15$. From Exercise 88, $x = \frac{1}{a} \sinh^{-1} as \Rightarrow 15a = \sinh^{-1} 16a$
$\Rightarrow \sinh 15a = 16a$.

(b) The intersection is near $(0.042, 0.672)$.

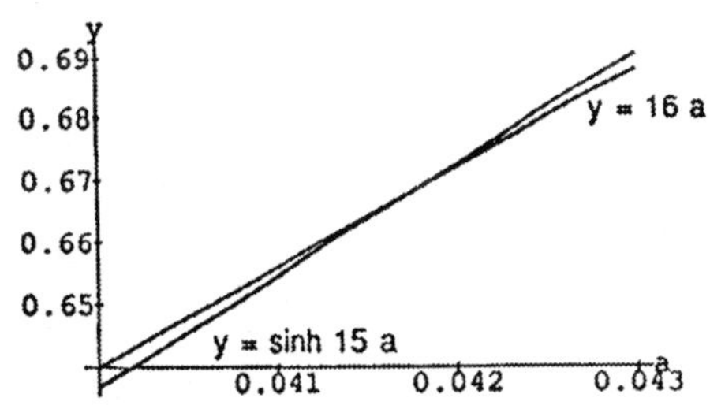

(c) Newton's method indicates that at $a \approx 0.0417525$ the curves $y = 16a$ and $y = \sinh 15a$ intersect.

(d) $T = wy \approx (2 \text{ lb})\left(\frac{1}{0.0417525}\right) \approx 47.90$ lb

(e) The sag is $\frac{1}{a}\cosh(15a) - \frac{1}{a} \approx 4.85$ ft.

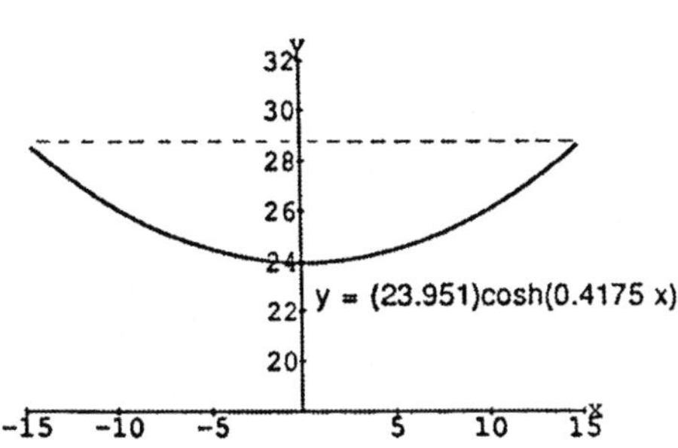

CHAPTER 7 PRACTICE EXERCISES

1. $\int e^x \sin(e^x)\, dx = \int \sin u\, du$, where $u = e^x$ and $du = e^x\, dx$
$= -\cos u + C = -\cos(e^x) + C$

2. $\int e^t \cos(3e^t - 2)\, dt = \frac{1}{3}\int \cos u\, du$, where $u = 3e^t - 2$ and $du = 3e^t\, dt$
$= \frac{1}{3} \sin u + C = \frac{1}{3} \sin(3e^t - 2) + C$

3. $\int_0^{\pi} \tan\left(\frac{x}{3}\right) dx = \int_0^{\pi} \frac{\sin\left(\frac{x}{3}\right)}{\cos\left(\frac{x}{3}\right)} dx = -3\int_1^{1/2} \frac{1}{u}\, du$, where $u = \cos\left(\frac{x}{3}\right)$, $du = -\frac{1}{3}\sin\left(\frac{x}{3}\right) dx$; $x = 0 \Rightarrow u = 1$, $x = \pi \Rightarrow u = \frac{1}{2}$

$= -3\left[\ln|u|\right]_1^{1/2} = -3\left[\ln\left|\frac{1}{2}\right| - \ln|1|\right] = -3\ln\frac{1}{2} = \ln 2^3 = \ln 8$

4. $\int_{1/6}^{1/4} 2\cot \pi x\, dx = 2\int_{1/6}^{1/4} \frac{\cos \pi x}{\sin \pi x} dx = \frac{2}{\pi}\int_{1/2}^{1/\sqrt{2}} \frac{1}{u}\, du$, where $u = \sin \pi x$, $du = \pi \cos \pi x\, dx$; $x = \frac{1}{6} \Rightarrow u = \frac{1}{2}$, $x = \frac{1}{4} \Rightarrow u = \frac{1}{\sqrt{2}}$

$= \frac{2}{\pi}\left[\ln|u|\right]_{1/2}^{1/\sqrt{2}} = \frac{2}{\pi}\left[\ln\left|\frac{1}{\sqrt{2}}\right| - \ln\left|\frac{1}{2}\right|\right] = \frac{2}{\pi}\left[\ln 1 - \frac{1}{2}\ln 2 - \ln 1 + \ln 2\right] = \frac{2}{\pi}\left[\frac{1}{2}\ln 2\right] = \frac{\ln 2}{\pi}$

5. $\int_{-\pi/2}^{\pi/6} \frac{\cos t}{1 - \sin t} dt = -\int_2^{1/2} \frac{1}{u}\, du$, where $u = 1 - \sin t$, $du = -\cos t\, dt$; $t = -\frac{\pi}{2} \Rightarrow u = 2$, $t = \frac{\pi}{6} \Rightarrow u = \frac{1}{2}$

$= -\left[\ln|u|\right]_2^{1/2} = -\left[\ln\left|\frac{1}{2}\right| - \ln|2|\right] = -\ln 1 + \ln 2 + \ln 2 = 2\ln 2 = \ln 4$

6. $\int \frac{1}{v \ln v} dv = \int \frac{1}{u}\, du$, where $u = \ln v$ and $du = \frac{1}{v}\, dv$

$= \ln|u| + C = \ln|\ln v| + C$

7. $\int \frac{\ln(x-5)}{x-5} dx = \int u\, du$, where $u = \ln(x-5)$ and $du = \frac{1}{x-5}\, dx$

$= \frac{u^2}{2} + C = \frac{[\ln(x-5)]^2}{2} + C$

8. $\int \frac{\cos(1 - \ln v)}{v} dv = -\int \cos u\, du$, where $u = 1 - \ln v$ and $du = -\frac{1}{v}\, dv$

$= -\sin u + C = -\sin(1 - \ln v) + C$

9. $\int_1^7 \frac{3}{x} dx = 3\int_1^7 \frac{1}{x} dx = 3\left[\ln|x|\right]_1^7 = 3(\ln 7 - \ln 1) = 3\ln 7$

10. $\int_1^{32} \frac{1}{5x} dx = \frac{1}{5}\int_1^{32} \frac{1}{x} dx = \frac{1}{5}\left[\ln|x|\right]_1^{32} = \frac{1}{5}(\ln 32 - \ln 1) = \frac{1}{5}\ln 32 = \ln\left(\sqrt[5]{32}\right) = \ln 2$

11. $\int_e^{e^2} \frac{1}{x\sqrt{\ln x}} dx = \int_e^{e^2} (\ln x)^{-1/2} \frac{1}{x} dx = \int_1^2 u^{-1/2}\, du$, where $u = \ln x$, $du = \frac{1}{x} dx$; $x = e \Rightarrow u = 1$, $x = e^2 \Rightarrow u = 2$

$= 2\left[u^{1/2}\right]_1^2 = 2\left(\sqrt{2} - 1\right) = 2\sqrt{2} - 2$

12. $\int_2^4 (1 + \ln t)(t \ln t)\, dt = \int_2^4 (t \ln t)(1 + \ln t)\, dt = \int_{2\ln 2}^{4\ln 4} u\, du$, where $u = t\ln t$, $du = \left((t)\left(\frac{1}{t}\right) + (\ln t)(1)\right) dt$

$= (1 + \ln t)\, dt$; $t = 2 \Rightarrow u = 2\ln 2$, $t = 4$

$\Rightarrow u = 4\ln 4$

$= \frac{1}{2}\left[u^2\right]_{2\ln 2}^{4\ln 4} = \frac{1}{2}\left[(4\ln 4)^2 - (2\ln 2)^2\right] = \frac{1}{2}\left[(8\ln 2)^2 - (2\ln 2)^2\right] = \frac{(2\ln 2)^2}{2}(16 - 1) = 30(\ln 2)^2$

13. $3^y = 2^{y+1} \Rightarrow \ln 3^y = \ln 2^{y+1} \Rightarrow y(\ln 3) = (y+1)\ln 2 \Rightarrow (\ln 3 - \ln 2)y = \ln 2 \Rightarrow \left(\ln\frac{3}{2}\right)y = \ln 2 \Rightarrow y = \frac{\ln 2}{\ln\left(\frac{3}{2}\right)}$

14. $4^{-y} = 3^{y+2} \Rightarrow \ln 4^{-y} = \ln 3^{y+2} \Rightarrow -y\ln 4 = (y+2)\ln 3 \Rightarrow -2\ln 3 = (\ln 3 + \ln 4)y \Rightarrow (\ln 12)y = -2\ln 3$

$\Rightarrow y = -\frac{\ln 9}{\ln 12}$

15. $9e^{2y} = x^2 \Rightarrow e^{2y} = \frac{x^2}{9} \Rightarrow \ln e^{2y} = \ln\left(\frac{x^2}{9}\right) \Rightarrow 2y(\ln e) = \ln\left(\frac{x^2}{9}\right) \Rightarrow y = \frac{1}{2}\ln\left(\frac{x^2}{9}\right) = \ln\sqrt{\frac{x^2}{9}} = \ln\left|\frac{x}{3}\right| = \ln|x| - \ln 3$

16. $3^y = 3\ln x \Rightarrow \ln 3^y = \ln(3\ln x) \Rightarrow y\ln 3 = \ln(3\ln x) \Rightarrow y = \frac{\ln(3\ln x)}{\ln 3} = \frac{\ln 3 + \ln(\ln x)}{\ln 3}$

17. (a) $\lim\limits_{x \to \infty} \frac{\log_2 x}{\log_3 x} = \lim\limits_{x \to \infty} \frac{\left(\frac{\ln x}{\ln 2}\right)}{\left(\frac{\ln x}{\ln 3}\right)} = \lim\limits_{x \to \infty} \frac{\ln 3}{\ln 2} = \frac{\ln 3}{\ln 2} \Rightarrow$ same rate

(b) $\lim\limits_{x \to \infty} \frac{x}{x+\left(\frac{1}{x}\right)} = \lim\limits_{x \to \infty} \frac{x^2}{x^2+1} = \lim\limits_{x \to \infty} \frac{2x}{2x} = \lim\limits_{x \to \infty} 1 = 1 \Rightarrow$ same rate

(c) $\lim\limits_{x \to \infty} \frac{\left(\frac{x}{100}\right)}{xe^{-x}} = \lim\limits_{x \to \infty} \frac{xe^x}{100x} = \lim\limits_{x \to \infty} \frac{e^x}{100} = \infty \Rightarrow$ faster

(d) $\lim\limits_{x \to \infty} \frac{x}{\tan^{-1} x} = \infty \Rightarrow$ faster

(e) $\lim\limits_{x \to \infty} \frac{\csc^{-1} x}{\left(\frac{1}{x}\right)} = \lim\limits_{x \to \infty} \frac{\sin^{-1}(x^{-1})}{x^{-1}} = \lim\limits_{x \to \infty} \frac{\frac{(-x^{-2})}{\sqrt{1-(x^{-1})^2}}}{-x^{-2}} = \lim\limits_{x \to \infty} \frac{1}{\sqrt{1-\left(\frac{1}{x^2}\right)}} = 1 \Rightarrow$ same rate

(f) $\lim\limits_{x \to \infty} \frac{\sinh x}{e^x} = \lim\limits_{x \to \infty} \frac{(e^x - e^{-x})}{2e^x} = \lim\limits_{x \to \infty} \frac{1-e^{-2x}}{2} = \frac{1}{2} \Rightarrow$ same rate

18. (a) $\lim\limits_{x \to \infty} \frac{3^{-x}}{2^{-x}} = \lim\limits_{x \to \infty} \left(\frac{2}{3}\right)^x = 0 \Rightarrow$ slower

(b) $\lim\limits_{x \to \infty} \frac{\ln 2x}{\ln x^2} = \lim\limits_{x \to \infty} \frac{\ln 2 + \ln x}{2(\ln x)} = \lim\limits_{x \to \infty} \left(\frac{\ln 2}{2\ln x} + \frac{1}{2}\right) = \frac{1}{2} \Rightarrow$ same rate

(c) $\lim\limits_{x \to \infty} \frac{10x^3+2x^2}{e^x} = \lim\limits_{x \to \infty} \frac{30x^2+4x}{e^x} = \lim\limits_{x \to \infty} \frac{60x+4}{e^x} = \lim\limits_{x \to \infty} \frac{60}{e^x} = 0 \Rightarrow$ slower

(d) $\lim\limits_{x \to \infty} \frac{\tan^{-1}\left(\frac{1}{x}\right)}{\left(\frac{1}{x}\right)} = \lim\limits_{x \to \infty} \frac{\tan^{-1}(x^{-1})}{x^{-1}} = \lim\limits_{x \to \infty} \frac{\left(\frac{-x^{-2}}{1+x^{-2}}\right)}{-x^{-2}} = \lim\limits_{x \to \infty} \frac{1}{1+\frac{1}{x^2}} = 1 \Rightarrow$ same rate

(e) $\lim\limits_{x \to \infty} \frac{\sin^{-1}\left(\frac{1}{x}\right)}{\left(\frac{1}{x^2}\right)} = \lim\limits_{x \to \infty} \frac{\sin^{-1}(x^{-1})}{x^{-2}} = \lim\limits_{x \to \infty} \frac{\left(\frac{-x^{-2}}{\sqrt{1-(x^{-1})^2}}\right)}{-2x^{-3}} = \lim\limits_{x \to \infty} \frac{x}{2\sqrt{1-\frac{1}{x^2}}} = \infty \Rightarrow$ faster

(f) $\lim\limits_{x \to \infty} \frac{\operatorname{sech} x}{e^{-x}} = \lim\limits_{x \to \infty} \frac{\left(\frac{2}{e^x+e^{-x}}\right)}{e^{-x}} = \lim\limits_{x \to \infty} \frac{2}{e^{-x}(e^x+e^{-x})} = \lim\limits_{x \to \infty} \left(\frac{2}{1+e^{-2x}}\right) = 2 \Rightarrow$ same rate

19. (a) $\frac{\left(\frac{1}{x^2}+\frac{1}{x^4}\right)}{\left(\frac{1}{x^2}\right)} = 1 + \frac{1}{x^2} \le 2$ for x sufficiently large $\Rightarrow$ true

(b) $\frac{\left(\frac{1}{x^2}+\frac{1}{x^4}\right)}{\left(\frac{1}{x^4}\right)} = x^2 + 1 > M$ for any positive integer M whenever $x > \sqrt{M} \Rightarrow$ false

(c) $\lim\limits_{x \to \infty} \frac{x}{x+\ln x} = \lim\limits_{x \to \infty} \frac{1}{1+\frac{1}{x}} = 1 \Rightarrow$ the same growth rate $\Rightarrow$ false

(d) $\lim\limits_{x \to \infty} \frac{\ln(\ln x)}{\ln x} = \lim\limits_{x \to \infty} \frac{\left[\frac{\left(\frac{1}{x}\right)}{\ln x}\right]}{\left(\frac{1}{x}\right)} = \lim\limits_{x \to \infty} \frac{1}{\ln x} = 0 \Rightarrow$ grows slower $\Rightarrow$ true

(e) $\frac{\tan^{-1} x}{1} \le \frac{\pi}{2}$ for all x $\Rightarrow$ true

(f) $\frac{\cosh x}{e^x} = \frac{1}{2}(1+e^{-2x}) \le \frac{1}{2}(1+1) = 1$ if $x > 0 \Rightarrow$ true

20. (a) $\frac{\left(\frac{1}{x^4}\right)}{\left(\frac{1}{x^2}+\frac{1}{x^4}\right)} = \frac{1}{x^2+1} \le 1$ if $x > 0 \Rightarrow$ true

(b) $\lim\limits_{x \to \infty} \frac{\left(\frac{1}{x^4}\right)}{\left(\frac{1}{x^2}+\frac{1}{x^4}\right)} = \lim\limits_{x \to \infty} \left(\frac{1}{x^2+1}\right) = 0 \Rightarrow$ true

(c) $\lim\limits_{x \to \infty} \frac{\ln x}{x+1} = \lim\limits_{x \to \infty} \frac{\left(\frac{1}{x}\right)}{1} = 0 \Rightarrow$ true

(d) $\frac{\ln 2x}{\ln x} = \frac{\ln 2}{\ln x} + 1 \le 1 + 1 = 2$ if $x \ge 2 \Rightarrow$ true

(e) $\frac{\sec^{-1} x}{1} = \frac{\cos^{-1}\left(\frac{1}{x}\right)}{1} \le \frac{\left(\frac{\pi}{2}\right)}{1} = \frac{\pi}{2}$ if $x > 1 \Rightarrow$ true

(f) $\frac{\sinh x}{e^x} = \frac{1}{2}(1-e^{-2x}) \le \frac{1}{2}$ if $x > 0 \Rightarrow$ true

21. $\frac{df}{dx} = e^x + 1 \Rightarrow \left(\frac{df^{-1}}{dx}\right)_{x=f(\ln 2)} = \frac{1}{\left(\frac{df}{dx}\right)_{x=\ln 2}} \Rightarrow \left(\frac{df^{-1}}{dx}\right)_{x=f(\ln 2)} = \frac{1}{(e^x+1)_{x=\ln 2}} = \frac{1}{2+1} = \frac{1}{3}$

22. $y = f(x) \Rightarrow y = 1 + \frac{1}{x} \Rightarrow \frac{1}{x} = y - 1 \Rightarrow x = \frac{1}{y-1} \Rightarrow f^{-1}(x) = \frac{1}{x-1}$; $f^{-1}(f(x)) = \frac{1}{\left(1+\frac{1}{x}\right)-1} = \frac{1}{\left(\frac{1}{x}\right)} = x$ and

$f(f^{-1}(x)) = 1 + \frac{1}{\left(\frac{1}{x-1}\right)} = 1 + (x-1) = x$; $\left.\frac{df^{-1}}{dx}\right|_{f(x)} = \left.\frac{-1}{(x-1)^2}\right|_{f(x)} = \frac{-1}{\left[\left(1+\frac{1}{x}\right)-1\right]^2} = -x^2$;

$f'(x) = -\frac{1}{x^2} \Rightarrow \left.\frac{df^{-1}}{dx}\right|_{f(x)} = \frac{1}{f'(x)}$

23. $y = \ln x \Rightarrow \frac{dy}{dx} = \frac{1}{x}$; $\frac{dy}{dt} = \frac{dy}{dx}\frac{dx}{dt} \Rightarrow \frac{dy}{dt} = \left(\frac{1}{x}\right)\sqrt{x} = \frac{1}{\sqrt{x}} \Rightarrow \left.\frac{dy}{dt}\right|_{e^2} = \frac{1}{e}$ m/sec

24. $y = 9e^{-x/3} \Rightarrow \frac{dy}{dx} = -3e^{-x/3}$; $\frac{dx}{dt} = \frac{(dy/dt)}{(dy/dx)} \Rightarrow \frac{dx}{dt} = \frac{\left(-\frac{1}{4}\right)\sqrt{9-y}}{-3e^{-x/3}}$; $x = 9 \Rightarrow y = 9e^{-3}$

$\Rightarrow \left.\frac{dx}{dt}\right|_{x=9} = \frac{\left(-\frac{1}{4}\right)\sqrt{9-\frac{9}{e^3}}}{\left(-\frac{3}{e^3}\right)} = \frac{1}{4}\sqrt{e^3}\sqrt{e^3-1} \approx 5$ ft/sec

25. $K = \ln(5x) - \ln(3x) = \ln 5 + \ln x - \ln 3 - \ln x = \ln 5 - \ln 3 = \ln \frac{5}{3}$

26. (a) No, there are two intersections: one at $x = 2$ and the other at $x = 4$

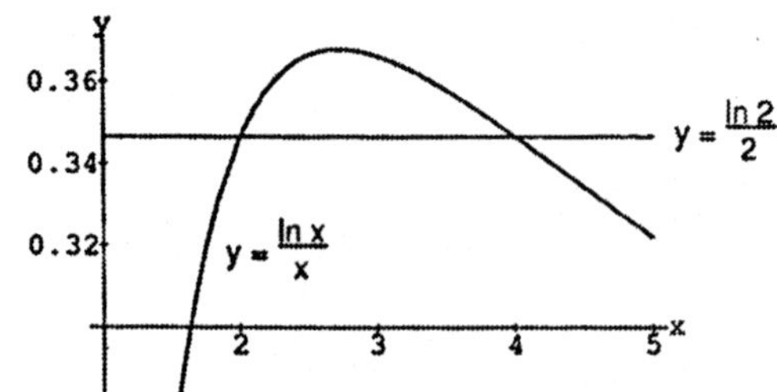

(b) Yes, because there is only one intersection

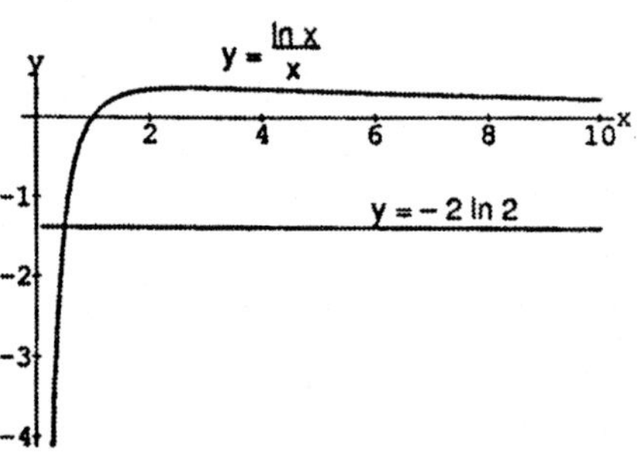

27. $\frac{\log_4 x}{\log_2 x} = \frac{\left(\frac{\ln x}{\ln 4}\right)}{\left(\frac{\ln x}{\ln 2}\right)} = \frac{\ln x}{\ln 4} \cdot \frac{\ln 2}{\ln x} = \frac{\ln 2}{\ln 4} = \frac{\ln 2}{2\ln 2} = \frac{1}{2}$

28. (a) $f(x) = \frac{\ln 2}{\ln x}$, $g(x) = \frac{\ln x}{\ln 2}$

(b) f is negative when g is negative, positive when g is positive, and undefined when g = 0; the values of f decrease as those of g increase

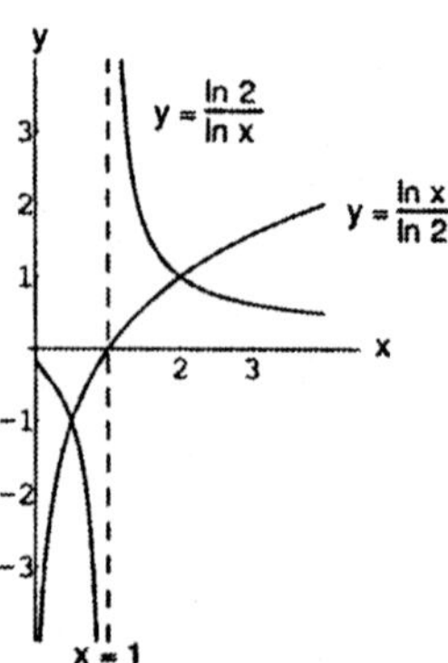

29. Since the half life is 5700 years and $A(t) = A_0e^{kt}$ we have $\frac{A_0}{2} = A_0e^{5700k} \Rightarrow \frac{1}{2} = e^{5700k} \Rightarrow \ln(0.5) = 5700k$ $\Rightarrow k = \frac{\ln(0.5)}{5700}$. With 10% of the original carbon-14 remaining we have $0.1A_0 = A_0e^{\frac{\ln(0.5)}{5700}t} \Rightarrow 0.1 = e^{\frac{\ln(0.5)}{5700}t}$ $\Rightarrow \ln(0.1) = \frac{\ln(0.5)}{5700}t \Rightarrow t = \frac{(5700)\ln(0.1)}{\ln(0.5)} \approx 18{,}935$ years (rounded to the nearest year).

30. $T - T_s = (T_o - T_s)e^{-kt} \Rightarrow 180 - 40 = (220 - 40)e^{-k/4}$, time in hours, $\Rightarrow k = -4\ln\left(\frac{7}{9}\right) = 4\ln\left(\frac{9}{7}\right) \Rightarrow 70 - 40$
$= (220 - 40)e^{-4\ln(9/7)t} \Rightarrow t = \frac{\ln 6}{4\ln\left(\frac{9}{7}\right)} \approx 1.78$ hr ≈ 107 min, the total time $\Rightarrow$ the time it took to cool from
$180°$ F to $70°$ F was $107 - 15 = 92$ min

CHAPTER 7 ADDITIONAL AND ADVANCED EXERCISES

1. $A(t) = \int_0^t e^{-x}\,dx = \left[-e^{-x}\right]_0^t = 1 - e^{-t}$, $V(t) = \pi\int_0^t e^{-2x}\,dx = \left[-\frac{\pi}{2}e^{-2x}\right]_0^t = \frac{\pi}{2}\left(1 - e^{-2t}\right)$
   (a) $\lim\limits_{t\to\infty} A(t) = \lim\limits_{t\to\infty}\left(1 - e^{-t}\right) = 1$
   (b) $\lim\limits_{t\to\infty} \frac{V(t)}{A(t)} = \lim\limits_{t\to\infty} \frac{\frac{\pi}{2}\left(1-e^{-2t}\right)}{1-e^{-t}} = \frac{\pi}{2}$
   (c) $\lim\limits_{t\to 0^+} \frac{V(t)}{A(t)} = \lim\limits_{t\to 0^+} \frac{\frac{\pi}{2}\left(1-e^{-2t}\right)}{1-e^{-t}} = \lim\limits_{t\to 0^+} \frac{\frac{\pi}{2}\left(1-e^{-t}\right)\left(1+e^{-t}\right)}{\left(1-e^{-t}\right)} = \lim\limits_{t\to 0^+} \frac{\pi}{2}\left(1+e^{-t}\right) = \pi$

2. (a) $\lim\limits_{a\to 0^+} \log_a 2 = \lim\limits_{a\to 0^+} \frac{\ln 2}{\ln a} = 0$;
   $\lim\limits_{a\to 1^-} \log_a 2 = \lim\limits_{a\to 1^-} \frac{\ln 2}{\ln a} = -\infty$;
   $\lim\limits_{a\to 1^+} \log_a 2 = \lim\limits_{a\to 1^+} \frac{\ln 2}{\ln 1} = \infty$;
   $\lim\limits_{a\to \infty} \log_a 2 = \lim\limits_{a\to \infty} \frac{\ln 2}{\ln a} = 0$

   (b)
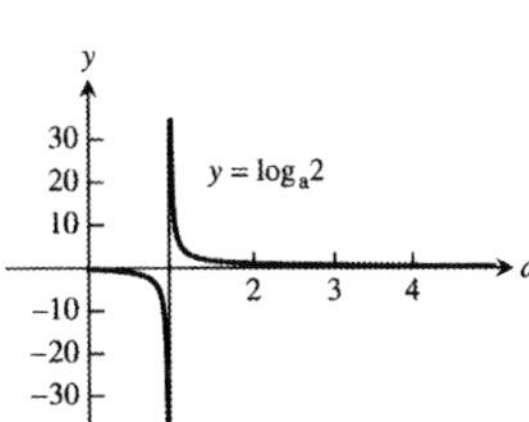

3. $f(x) = e^{g(x)} \Rightarrow f'(x) = e^{g(x)}\,g'(x)$, where $g'(x) = \frac{x}{1+x^4} \Rightarrow f'(2) = e^0\left(\frac{2}{1+16}\right) = \frac{2}{17}$

4. (a) $\frac{df}{dx} = \frac{2\ln e^x}{e^x}\cdot e^x = 2x$
   (b) $f(0) = \int_1^1 \frac{2\ln t}{t}\,dt = 0$
   (c) $\frac{df}{dx} = 2x \Rightarrow f(x) = x^2 + C$; $f(0) = 0 \Rightarrow C = 0 \Rightarrow f(x) = x^2 \Rightarrow$ the graph of f(x) is a parabola

5. (a) $g(x) + h(x) = 0 \Rightarrow g(x) = -h(x)$; also $g(x) + h(x) = 0 \Rightarrow g(-x) + h(-x) = 0 \Rightarrow g(x) - h(x) = 0$
   $\Rightarrow g(x) = h(x)$; therefore $-h(x) = h(x) \Rightarrow h(x) = 0 \Rightarrow g(x) = 0$
   (b) $\frac{f(x)+f(-x)}{2} = \frac{\left[f_E(x)+f_O(x)\right]+\left[f_E(-x)+f_O(-x)\right]}{2} = \frac{f_E(x)+f_O(x)+f_E(x)-f_O(x)}{2} = f_E(x)$;
   $\frac{f(x)-f(-x)}{2} = \frac{\left[f_E(x)+f_O(x)\right]-\left[f_E(-x)+f_O(-x)\right]}{2} = \frac{f_E(x)+f_O(x)-f_E(x)+f_O(x)}{2} = f_O(x)$
   (c) Part b $\Rightarrow$ such a decomposition is unique.

6. (a) $g(0+0) = \frac{g(0)+g(0)}{1-g(0)\,g(0)} \Rightarrow \left[1 - g^2(0)\right]g(0) = 2g(0) \Rightarrow g(0) - g^3(0) = 2g(0) \Rightarrow g^3(0) + g(0) = 0$
   $\Rightarrow g(0)\left[g^2(0) + 1\right] = 0 \Rightarrow g(0) = 0$
   (b) $g'(x) = \lim\limits_{h\to 0} \frac{g(x+h)-g(x)}{h} = \lim\limits_{h\to 0} \frac{\left[\frac{g(x)+g(h)}{1-g(x)g(h)}\right]-g(x)}{h} = \lim\limits_{h\to 0} \frac{g(x)+g(h)-g(x)+g^2(x)\,g(h)}{h\left[1-g(x)\,g(h)\right]}$
   $= \lim\limits_{h\to 0}\left[\frac{g(h)}{h}\right]\left[\frac{1+g^2(x)}{1-g(x)\,g(h)}\right] = 1\cdot\left[1 + g^2(x)\right] = 1 + g^2(x) = 1 + [g(x)]^2$
   (c) $\frac{dy}{dx} = 1 + y^2 \Rightarrow \frac{dy}{1+y^2} = dx \Rightarrow \tan^{-1} y = x + C \Rightarrow \tan^{-1}(g(x)) = x + C$; $g(0) = 0 \Rightarrow \tan^{-1} 0 = 0 + C$
   $\Rightarrow C = 0 \Rightarrow \tan^{-1}(g(x)) = x \Rightarrow g(x) = \tan x$

7. $M = \int_0^1 \frac{2}{1+x^2}\,dx = 2\left[\tan^{-1} x\right]_0^1 = \frac{\pi}{2}$ and $M_y = \int_0^1 \frac{2x}{1+x^2}\,dx = \left[\ln\left(1+x^2\right)\right]_0^1 = \ln 2 \Rightarrow \overline{x} = \frac{M_y}{M}$
   $= \frac{\ln 2}{\left(\frac{\pi}{2}\right)} = \frac{\ln 4}{\pi}$; $\overline{y} = 0$ by symmetry

8. (a) $V = \pi\int_{1/4}^4 \left(\frac{1}{2\sqrt{x}}\right)^2 dx = \frac{\pi}{4}\int_{1/4}^4 \frac{1}{x}\,dx = \frac{\pi}{4}\left[\ln|x|\right]_{1/4}^4 = \frac{\pi}{4}\left(\ln 4 - \ln\frac{1}{4}\right) = \frac{\pi}{4}\ln 16 = \frac{\pi}{4}\ln\left(2^4\right) = \pi\ln 2$

(b) $M_y = \int_{1/4}^{4} x\left(\frac{1}{2\sqrt{x}}\right) dx = \frac{1}{2}\int_{1/4}^{4} x^{1/2}\,dx = \left[\frac{1}{3}x^{3/2}\right]_{1/4}^{4} = \left(\frac{8}{3} - \frac{1}{24}\right) = \frac{64-1}{24} = \frac{63}{24}$;
$M_x = \int_{1/4}^{4} \frac{1}{2}\left(\frac{1}{2\sqrt{x}}\right)\left(\frac{1}{2\sqrt{x}}\right) dx = \frac{1}{8}\int_{1/4}^{4} \frac{1}{x}\,dx = \left[\frac{1}{8}\ln|x|\right]_{1/4}^{4} = \frac{1}{8}\ln 16 = \frac{1}{2}\ln 2$;
$M = \int_{1/4}^{4} \frac{1}{2\sqrt{x}}\,dx = \int_{1/4}^{4} \frac{1}{2}x^{-1/2}\,dx = \left[x^{1/2}\right]_{1/4}^{4} = 2 - \frac{1}{2} = \frac{3}{2}$; therefore, $\overline{x} = \frac{M_y}{M} = \left(\frac{63}{24}\right)\left(\frac{2}{3}\right) = \frac{21}{12} = \frac{7}{4}$ and
$\overline{y} = \frac{M_x}{M} = \left(\frac{1}{2}\ln 2\right)\left(\frac{2}{3}\right) = \frac{\ln 2}{3}$

9. $A(t) = A_0e^{rt}$; $A(t) = 2A_0 \Rightarrow 2A_0 = A_0e^{rt} \Rightarrow e^{rt} = 2 \Rightarrow rt = \ln 2 \Rightarrow t = \frac{\ln 2}{r} \Rightarrow t \approx \frac{.7}{r} = \frac{70}{100r} = \frac{70}{(r\%)}$

10. (a) The figure shows that $\frac{\ln e}{e} > \frac{\ln \pi}{\pi} \Rightarrow \pi \ln e > e \ln \pi \Rightarrow \ln e^{\pi} > \ln \pi^{e} \Rightarrow e^{\pi} > \pi^{e}$
(b) $y = \frac{\ln x}{x} \Rightarrow y' = \left(\frac{1}{x}\right)\left(\frac{1}{x}\right) - \frac{\ln x}{x^2} \Rightarrow \frac{1 - \ln x}{x^2}$; solving $y' = 0 \Rightarrow \ln x = 1 \Rightarrow x = e$; $y' < 0$ for $x > e$ and $y' > 0$ for $0 < x < e \Rightarrow$ an absolute maximum occurs at $x = e$

CHAPTER 8 TECHNIQUES OF INTEGRATION

8.1 BASIC INTEGRATION FORMULAS

1. $\int \frac{16x\,dx}{\sqrt{8x^2+1}}$; $\begin{bmatrix} u = 8x^2+1 \\ du = 16x\,dx \end{bmatrix} \rightarrow \int \frac{du}{\sqrt{u}} = 2\sqrt{u} + C = 2\sqrt{8x^2+1} + C$

2. $\int \frac{3\cos x\,dx}{\sqrt{1+3\sin x}}$; $\begin{bmatrix} u = 1+3\sin x \\ du = 3\cos x\,dx \end{bmatrix} \rightarrow \int \frac{du}{\sqrt{u}} = 2\sqrt{u} + C = 2\sqrt{1+3\sin x} + C$

3. $\int 3\sqrt{\sin v}\cos v\,dv$; $\begin{bmatrix} u = \sin v \\ du = \cos v\,dv \end{bmatrix} \rightarrow \int 3\sqrt{u}\,du = 3 \cdot \frac{2}{3} u^{3/2} + C = 2(\sin v)^{3/2} + C$

4. $\int \cot^3 y \csc^2 y\,dy$; $\begin{bmatrix} u = \cot y \\ du = -\csc^2 y\,dy \end{bmatrix} \rightarrow \int u^3(-du) = -\frac{u^4}{4} + C = \frac{-\cot^4 y}{4} + C$

5. $\int_0^1 \frac{16x\,dx}{8x^2+2}$; $\begin{bmatrix} u = 8x^2+2 \\ du = 16x\,dx \\ x=0 \Rightarrow u=2,\ \ x=1 \Rightarrow u=10 \end{bmatrix} \rightarrow \int_2^{10} \frac{du}{u} = [\ln|u|]_2^{10} = \ln 10 - \ln 2 = \ln 5$

6. $\int_{\pi/4}^{\pi/3} \frac{\sec^2 z\,dz}{\tan z}$; $\begin{bmatrix} u = \tan z \\ du = \sec^2 z\,dz \\ z = \frac{\pi}{4} \Rightarrow u = 1, z = \frac{\pi}{3} \Rightarrow u = \sqrt{3} \end{bmatrix} \rightarrow \int_1^{\sqrt{3}} \frac{1}{u}\,du = [\ln|u|]_1^{\sqrt{3}} = \ln\sqrt{3} - \ln 1 = \ln\sqrt{3}$

7. $\int \frac{dx}{\sqrt{x}\left(\sqrt{x}+1\right)}$; $\begin{bmatrix} u = \sqrt{x}+1 \\ du = \frac{1}{2\sqrt{x}}\,dx \\ 2\,du = \frac{dx}{\sqrt{x}} \end{bmatrix} \rightarrow \int \frac{2\,du}{u} = 2\ln|u| + C = 2\ln\left(\sqrt{x}+1\right) + C$

8. $\int \frac{dx}{x-\sqrt{x}} = \int \frac{dx}{\sqrt{x}\left(\sqrt{x}-1\right)}$; $\begin{bmatrix} u = \sqrt{x}-1 \\ du = \frac{1}{2\sqrt{x}}\,dx \\ 2\,du = \frac{dx}{\sqrt{x}} \end{bmatrix} \rightarrow \int \frac{2\,du}{u} = 2\ln|u| + C = 2\ln\left|\sqrt{x}-1\right| + C$

9. $\int \cot(3-7x)\,dx$; $\begin{bmatrix} u = 3-7x \\ du = -7\,dx \end{bmatrix} \rightarrow -\frac{1}{7}\int \cot u\,du = -\frac{1}{7}\ln|\sin u| + C = -\frac{1}{7}\ln|\sin(3-7x)| + C$

10. $\int \csc(\pi x - 1)\,dx$; $\begin{bmatrix} u = \pi x - 1 \\ du = \pi\,dx \end{bmatrix} \rightarrow \int \csc u \cdot \frac{du}{\pi} = \frac{-1}{\pi}\ln|\csc u + \cot u| + C$
$= -\frac{1}{\pi}\ln|\csc(\pi x - 1) + \cot(\pi x - 1)| + C$

11. $\int e^\theta \csc\left(e^\theta + 1\right)\,d\theta$; $\begin{bmatrix} u = e^\theta + 1 \\ du = e^\theta\,d\theta \end{bmatrix} \rightarrow \int \csc u\,du = -\ln|\csc u + \cot u| + C = -\ln\left|\csc\left(e^\theta+1\right) + \cot\left(e^\theta+1\right)\right| + C$

12. $\int \frac{\cot(3+\ln x)}{x}\,dx$; $\begin{bmatrix} u = 3+\ln x \\ du = \frac{dx}{x} \end{bmatrix} \rightarrow \int \cot u\,du = \ln|\sin u| + C = \ln|\sin(3+\ln x)| + C$

13. $\int \sec \frac{t}{3}\, dt;\ \begin{bmatrix} u = \frac{t}{3} \\ du = \frac{dt}{3} \end{bmatrix} \rightarrow \int 3 \sec u\, du = 3 \ln |\sec u + \tan u| + C = 3 \ln \left|\sec \frac{t}{3} + \tan \frac{t}{3}\right| + C$

14. $\int x \sec (x^2 - 5)\, dx;\ \begin{bmatrix} u = x^2 - 5 \\ du = 2x\, dx \end{bmatrix} \rightarrow \int \frac{1}{2} \sec u\, du = \frac{1}{2} \ln |\sec u + \tan u| + C$
$= \frac{1}{2} \ln |\sec (x^2 - 5) + \tan (x^2 - 5)| + C$

15. $\int \csc (s - \pi)\, ds;\ \begin{bmatrix} u = s - \pi \\ du = ds \end{bmatrix} \rightarrow \int \csc u\, du = -\ln |\csc u + \cot u| + C = -\ln |\csc (s - \pi) + \cot (s - \pi)| + C$

16. $\int \frac{1}{\theta^2} \csc \frac{1}{\theta}\, d\theta;\ \begin{bmatrix} u = \frac{1}{\theta} \\ du = \frac{-d\theta}{\theta^2} \end{bmatrix} \rightarrow \int -\csc u\, du = \ln |\csc u + \cot u| + C = \ln \left|\csc \frac{1}{\theta} + \cot \frac{1}{\theta}\right| + C$

17. $\int_0^{\sqrt{\ln 2}} 2xe^{x^2}\, dx;\ \begin{bmatrix} u = x^2 \\ du = 2x\, dx \\ x = 0 \Rightarrow u = 0,\ x = \sqrt{\ln 2} \Rightarrow u = \ln 2 \end{bmatrix} \rightarrow \int_0^{\ln 2} e^u\, du = [e^u]_0^{\ln 2} = e^{\ln 2} - e^0 = 2 - 1 = 1$

18. $\int_{\pi/2}^{\pi} \sin (y) e^{\cos y}\, dy;\ \begin{bmatrix} u = \cos y \\ du = -\sin y\, dy \\ y = \frac{\pi}{2} \Rightarrow u = 0,\ y = \pi \Rightarrow u = -1 \end{bmatrix} \rightarrow \int_0^{-1} -e^u\, du = \int_{-1}^{0} e^u\, du = [e^u]_{-1}^{0} = 1 - e^{-1} = \frac{e-1}{e}$

19. $\int e^{\tan v} \sec^2 v\, dv;\ \begin{bmatrix} u = \tan v \\ du = \sec^2 v\, dv \end{bmatrix} \rightarrow \int e^u\, du = e^u + C = e^{\tan v} + C$

20. $\int \frac{e^{\sqrt{t}}\, dt}{\sqrt{t}};\ \begin{bmatrix} u = \sqrt{t} \\ du = \frac{dt}{2\sqrt{t}} \end{bmatrix} \rightarrow \int 2e^u\, du = 2e^u + C = 2e^{\sqrt{t}} + C$

21. $\int 3^{x+1}\, dx;\ \begin{bmatrix} u = x + 1 \\ du = dx \end{bmatrix} \rightarrow \int 3^u\, du = \left(\frac{1}{\ln 3}\right) 3^u + C = \frac{3^{(x+1)}}{\ln 3} + C$

22. $\int \frac{2^{\ln x}}{x}\, dx;\ \begin{bmatrix} u = \ln x \\ du = \frac{dx}{x} \end{bmatrix} \rightarrow \int 2^u\, du = \frac{2^u}{\ln 2} + C = \frac{2^{\ln x}}{\ln 2} + C$

23. $\int \frac{2^{\sqrt{w}}\, dw}{2\sqrt{w}};\ \begin{bmatrix} u = \sqrt{w} \\ du = \frac{dw}{2\sqrt{w}} \end{bmatrix} \rightarrow \int 2^u\, du = \frac{2^u}{\ln 2} + C = \frac{2^{\sqrt{w}}}{\ln 2} + C$

24. $\int 10^{2\theta}\, d\theta;\ \begin{bmatrix} u = 2\theta \\ du = 2\, d\theta \end{bmatrix} \rightarrow \int \frac{1}{2} 10^u\, du = \frac{10^u}{2 \ln 10} + C = \frac{1}{2}\left(\frac{10^{2\theta}}{\ln 10}\right) + C$

25. $\int \frac{9\, du}{1 + 9u^2};\ \begin{bmatrix} x = 3u \\ dx = 3\, du \end{bmatrix} \rightarrow \int \frac{3\, dx}{1 + x^2} = 3 \tan^{-1} x + C = 3 \tan^{-1} 3u + C$

26. $\int \frac{4\, dx}{1 + (2x + 1)^2};\ \begin{bmatrix} u = 2x + 1 \\ du = 2\, dx \end{bmatrix} \rightarrow \int \frac{2\, du}{1 + u^2} = 2 \tan^{-1} u + C = 2 \tan^{-1} (2x + 1) + C$

27. $\int_0^{1/6} \frac{dx}{\sqrt{1 - 9x^2}};\ \begin{bmatrix} u = 3x \\ du = 3\, dx \\ x = 0 \Rightarrow u = 0,\ x = \frac{1}{6} \Rightarrow u = \frac{1}{2} \end{bmatrix} \rightarrow \int_0^{1/2} \frac{1}{3} \frac{du}{\sqrt{1 - u^2}} = \left[\frac{1}{3} \sin^{-1} u\right]_0^{1/2} = \frac{1}{3}\left(\frac{\pi}{6} - 0\right) = \frac{\pi}{18}$

28. $\int_0^1 \frac{dt}{\sqrt{4-t^2}} = \left[\sin^{-1} \frac{t}{2}\right]_0^1 = \sin^{-1}\left(\frac{1}{2}\right) - 0 = \frac{\pi}{6}$

29. $\int \frac{2s\,ds}{\sqrt{1-s^4}}$; $\begin{bmatrix} u = s^2 \\ du = 2s\,ds \end{bmatrix} \rightarrow \int \frac{du}{\sqrt{1-u^2}} = \sin^{-1} u + C = \sin^{-1} s^2 + C$

30. $\int \frac{2\,dx}{x\sqrt{1-4\ln^2 x}}$; $\begin{bmatrix} u = 2\ln x \\ du = \frac{2\,dx}{x} \end{bmatrix} \rightarrow \int \frac{du}{\sqrt{1-u^2}} = \sin^{-1} u + C = \sin^{-1}(2\ln x) + C$

31. $\int \frac{6\,dx}{x\sqrt{25x^2-1}} = \int \frac{6\,dx}{5x\sqrt{x^2-\frac{1}{25}}} = \frac{6}{5} \cdot 5 \sec^{-1}|5x| + C = 6\sec^{-1}|5x| + C$

32. $\int \frac{dr}{r\sqrt{r^2-9}} = \frac{1}{3}\sec^{-1}\left|\frac{r}{3}\right| + C$

33. $\int \frac{dx}{e^x+e^{-x}} = \int \frac{e^x\,dx}{e^{2x}+1}$; $\begin{bmatrix} u = e^x \\ du = e^x\,dx \end{bmatrix} \rightarrow \int \frac{du}{u^2+1} = \tan^{-1} u + C = \tan^{-1} e^x + C$

34. $\int \frac{dy}{\sqrt{e^{2y}-1}} = \int \frac{e^y\,dy}{e^y\sqrt{(e^y)^2-1}}$; $\begin{bmatrix} u = e^y \\ du = e^y\,dy \end{bmatrix} \rightarrow \int \frac{du}{u\sqrt{u^2-1}} = \sec^{-1}|u| + C = \sec^{-1} e^y + C$

35. $\int_1^{e^{\pi/3}} \frac{dx}{x\cos(\ln x)}$; $\begin{bmatrix} u = \ln x \\ du = \frac{dx}{x} \\ x = 1 \Rightarrow u = 0,\ x = e^{\pi/3} \Rightarrow u = \frac{\pi}{3} \end{bmatrix} \rightarrow \int_0^{\pi/3} \frac{du}{\cos u} = \int_0^{\pi/3} \sec u\,du = [\ln|\sec u + \tan u|]_0^{\pi/3}$
$= \ln\left|\sec\frac{\pi}{3} + \tan\frac{\pi}{3}\right| - \ln|\sec 0 + \tan 0| = \ln\left(2+\sqrt{3}\right) - \ln(1) = \ln\left(2+\sqrt{3}\right)$

36. $\int \frac{\ln x\,dx}{x+4x\ln^2 x} = \int \frac{\ln x\,dx}{x(1+4\ln^2 x)}$; $\begin{bmatrix} u = \ln^2 x \\ du = \frac{2}{x}\ln x\,dx \end{bmatrix} \rightarrow \int \frac{1}{2}\frac{du}{1+4u} = \frac{1}{8}\ln|1+4u| + C = \frac{1}{8}\ln(1+4\ln^2 x) + C$

37. $\int_1^2 \frac{8\,dx}{x^2-2x+2} = 8\int_1^2 \frac{dx}{1+(x-1)^2}$; $\begin{bmatrix} u = x-1 \\ du = dx \\ x = 1 \Rightarrow u = 0,\ x = 2 \Rightarrow u = 1 \end{bmatrix} \rightarrow 8\int_0^1 \frac{du}{1+u^2} = 8\left[\tan^{-1} u\right]_0^1$
$= 8(\tan^{-1} 1 - \tan^{-1} 0) = 8\left(\frac{\pi}{4} - 0\right) = 2\pi$

38. $\int_2^4 \frac{2\,dx}{x^2-6x+10} = 2\int_2^4 \frac{dx}{(x-3)^2+1}$; $\begin{bmatrix} u = x-3 \\ du = dx \\ x = 2 \Rightarrow u = -1,\ x = 4 \Rightarrow u = 1 \end{bmatrix} \rightarrow 2\int_{-1}^1 \frac{du}{u^2+1} = 2\left[\tan^{-1} u\right]_{-1}^1$
$= 2\left[\tan^{-1} 1 - \tan^{-1}(-1)\right] = 2\left[\frac{\pi}{4} - \left(-\frac{\pi}{4}\right)\right] = \pi$

39. $\int \frac{dt}{\sqrt{-t^2+4t-3}} = \int \frac{dt}{\sqrt{1-(t-2)^2}}$; $\begin{bmatrix} u = t-2 \\ du = dt \end{bmatrix} \rightarrow \int \frac{du}{\sqrt{1-u^2}} = \sin^{-1} u + C = \sin^{-1}(t-2) + C$

40. $\int \frac{d\theta}{\sqrt{2\theta-\theta^2}} = \int \frac{d\theta}{\sqrt{1-(\theta-1)^2}}$; $\begin{bmatrix} u = \theta-1 \\ du = d\theta \end{bmatrix} \rightarrow \int \frac{du}{\sqrt{1-u^2}} = \sin^{-1} u + C = \sin^{-1}(\theta-1) + C$

41. $\int \frac{dx}{(x+1)\sqrt{x^2+2x}} = \int \frac{dx}{(x+1)\sqrt{(x+1)^2-1}}$; $\begin{bmatrix} u = x+1 \\ du = dx \end{bmatrix} \rightarrow \int \frac{du}{u\sqrt{u^2-1}} = \sec^{-1}|u| + C = \sec^{-1}|x+1| + C,$
$|u| = |x+1| > 1$

42. $\int \frac{dx}{(x-2)\sqrt{x^2-4x+3}} = \int \frac{dx}{(x-2)\sqrt{(x-2)^2-1}}$; $\begin{bmatrix} u = x-2 \\ du = dx \end{bmatrix} \to \int \frac{du}{u\sqrt{u^2-1}} = \sec^{-1}|u| + C$
$= \sec^{-1}|x-2| + C,\ |u| = |x-2| > 1$

43. $\int (\sec x + \cot x)^2\, dx = \int (\sec^2 x + 2\sec x \cot x + \cot^2 x)\, dx = \int \sec^2 x\, dx + \int 2\csc x\, dx + \int (\csc^2 x - 1)\, dx$
$= \tan x - 2\ln|\csc x + \cot x| - \cot x - x + C$

44. $\int (\csc x - \tan x)^2\, dx = \int (\csc^2 x - 2\csc x \tan x + \tan^2 x)\, dx = \int \csc^2 x\, dx - \int 2\sec x\, dx + \int (\sec^2 x - 1)\, dx$
$= -\cot x - 2\ln|\sec x + \tan x| + \tan x - x + C$

45. $\int \csc x \sin 3x\, dx = \int (\csc x)(\sin 2x \cos x + \sin x \cos 2x)\, dx = \int (\csc x)(2\sin x \cos^2 x + \sin x \cos 2x)\, dx$
$= \int (2\cos^2 x + \cos 2x)\, dx = \int [(1 + \cos 2x) + \cos 2x]\, dx = \int (1 + 2\cos 2x)\, dx = x + \sin 2x + C$

46. $\int (\sin 3x \cos 2x - \cos 3x \sin 2x)\, dx = \int \sin(3x - 2x)\, dx = \int \sin x\, dx = -\cos x + C$

47. $\int \frac{x}{x+1}\, dx = \int \left(1 - \frac{1}{x+1}\right) dx = x - \ln|x+1| + C$

48. $\int \frac{x^2}{x^2+1}\, dx = \int \left(1 - \frac{1}{x^2+1}\right) dx = x - \tan^{-1} x + C$

49. $\int_{\sqrt{2}}^{3} \frac{2x^3}{x^2-1}\, dx = \int_{\sqrt{2}}^{3} \left(2x + \frac{2x}{x^2-1}\right) dx = \left[x^2 + \ln|x^2-1|\right]_{\sqrt{2}}^{3} = (9 + \ln 8) - (2 + \ln 1) = 7 + \ln 8$

50. $\int_{-1}^{3} \frac{4x^2-7}{2x+3}\, dx = \int_{-1}^{3} \left[(2x-3) + \frac{2}{2x+3}\right] dx = \left[x^2 - 3x + \ln|2x+3|\right]_{-1}^{3} = (9 - 9 + \ln 9) - (1 + 3 + \ln 1) = \ln 9 - 4$

51. $\int \frac{4t^3 - t^2 + 16t}{t^2+4}\, dt = \int \left[(4t-1) + \frac{4}{t^2+4}\right] dt = 2t^2 - t + 2\tan^{-1}\left(\frac{t}{2}\right) + C$

52. $\int \frac{2\theta^3 - 7\theta^2 + 7\theta}{2\theta - 5}\, d\theta = \int \left[(\theta^2 - \theta + 1) + \frac{5}{2\theta-5}\right] d\theta = \frac{\theta^3}{3} - \frac{\theta^2}{2} + \theta + \frac{5}{2}\ln|2\theta - 5| + C$

53. $\int \frac{1-x}{\sqrt{1-x^2}}\, dx = \int \frac{dx}{\sqrt{1-x^2}} - \int \frac{x\, dx}{\sqrt{1-x^2}} = \sin^{-1} x + \sqrt{1-x^2} + C$

54. $\int \frac{x + 2\sqrt{x-1}}{2x\sqrt{x-1}}\, dx = \int \frac{dx}{2\sqrt{x-1}} + \int \frac{dx}{x} = (x-1)^{1/2} + \ln|x| + C$

55. $\int_0^{\pi/4} \frac{1+\sin x}{\cos^2 x}\, dx = \int_0^{\pi/4} (\sec^2 x + \sec x \tan x)\, dx = [\tan x + \sec x]_0^{\pi/4} = \left(1 + \sqrt{2}\right) - (0 + 1) = \sqrt{2}$

56. $\int_0^{1/2} \frac{2-8x}{1+4x^2}\, dx = \int_0^{1/2} \left(\frac{2}{1+4x^2} - \frac{8x}{1+4x^2}\right) dx = \left[\tan^{-1}(2x) - \ln|1+4x^2|\right]_0^{1/2}$
$= (\tan^{-1} 1 - \ln 2) - (\tan^{-1} 0 - \ln 1) = \frac{\pi}{4} - \ln 2$

57. $\int \frac{dx}{1+\sin x} = \int \frac{(1-\sin x)}{(1-\sin^2 x)}\, dx = \int \frac{(1-\sin x)}{\cos^2 x}\, dx = \int (\sec^2 x - \sec x \tan x)\, dx = \tan x - \sec x + C$

58. $1 + \cos x = 1 + \cos\left(2 \cdot \frac{x}{2}\right) = 2\cos^2 \frac{x}{2} \Rightarrow \int \frac{dx}{1+\cos x} = \int \frac{dx}{2\cos^2\left(\frac{x}{2}\right)} = \frac{1}{2}\int \sec^2\left(\frac{x}{2}\right) dx = \tan \frac{x}{2} + C$

59. $\int \frac{1}{\sec\theta + \tan\theta}\, d\theta = \int d\theta;\ \begin{bmatrix} u = 1 + \sin\theta \\ du = \cos\theta\, d\theta \end{bmatrix} \to \int \frac{du}{u} = \ln|u| + C = \ln|1 + \sin\theta| + C$

60. $\int \frac{1}{\csc\theta+\cot\theta}\,d\theta = \int \frac{\sin\theta}{1+\cos\theta}\,d\theta;\ \begin{bmatrix} u = 1+\cos\theta \\ du = -\sin\theta\,d\theta \end{bmatrix} \to \int \frac{-du}{u} = -\ln|u| + C = -\ln|1+\cos\theta| + C$

61. $\int \frac{1}{1-\sec x}\,dx = \int \frac{\cos x}{\cos x - 1}\,dx = \int \left(1 + \frac{1}{\cos x - 1}\right) dx = \int \left(1 - \frac{1+\cos x}{\sin^2 x}\right) dx = \int \left(1 - \csc^2 x - \frac{\cos x}{\sin^2 x}\right) dx$
$= \int (1 - \csc^2 x - \csc x \cot x)\,dx = x + \cot x + \csc x + C$

62. $\int \frac{1}{1-\csc x}\,dx = \int \frac{\sin x}{\sin x - 1}\,dx = \int \left(1 + \frac{1}{\sin x - 1}\right) dx = \int \left(1 + \frac{\sin x + 1}{(\sin x - 1)(\sin x + 1)}\right) dx$
$= \int \left(1 - \frac{1+\sin x}{\cos^2 x}\right) dx = \int \left(1 - \sec^2 x - \frac{\sin x}{\cos^2 x}\right) dx = \int (1 - \sec^2 x - \sec x \tan x)\,dx = x - \tan x - \sec x + C$

63. $\int_0^{2\pi} \sqrt{\frac{1-\cos x}{2}}\,dx = \int_0^{2\pi} \left|\sin \frac{x}{2}\right| dx;\ \begin{bmatrix} \sin \frac{x}{2} \ge 0 \\ \text{for } 0 \le \frac{x}{2} \le 2\pi \end{bmatrix} \to \int_0^{2\pi} \sin\left(\frac{x}{2}\right) dx = \left[-2\cos\frac{x}{2}\right]_0^{2\pi} = -2(\cos\pi - \cos 0)$
$= (-2)(-2) = 4$

64. $\int_0^{\pi} \sqrt{1-\cos 2x}\,dx = \int_0^{\pi} \sqrt{2}\,|\sin x|\,dx;\ \begin{bmatrix} \sin x \ge 0 \\ \text{for } 0 \le x \le \pi \end{bmatrix} \to \sqrt{2}\int_0^{\pi} \sin x\,dx = \left[-\sqrt{2}\cos x\right]_0^{\pi}$
$= -\sqrt{2}(\cos\pi - \cos 0) = 2\sqrt{2}$

65. $\int_{\pi/2}^{\pi} \sqrt{1+\cos 2t}\,dt = \int_{\pi/2}^{\pi} \sqrt{2}\,|\cos t|\,dt;\ \begin{bmatrix} \cos t \le 0 \\ \text{for } \frac{\pi}{2} \le t \le \pi \end{bmatrix} \to \int_{\pi/2}^{\pi} -\sqrt{2}\cos t\,dt = \left[-\sqrt{2}\sin t\right]_{\pi/2}^{\pi}$
$= -\sqrt{2}\left(\sin\pi - \sin\frac{\pi}{2}\right) = \sqrt{2}$

66. $\int_{-\pi}^{0} \sqrt{1+\cos t}\,dt = \int_{-\pi}^{0} \sqrt{2}\left|\cos\frac{t}{2}\right| dt;\ \begin{bmatrix} \cos\frac{t}{2} \ge 0 \\ \text{for } -\pi \le t \le 0 \end{bmatrix} \to \int_{-\pi}^{0} \sqrt{2}\cos\frac{t}{2}\,dt = \left[2\sqrt{2}\sin\frac{t}{2}\right]_{-\pi}^{0}$
$= 2\sqrt{2}\left[\sin 0 - \sin\left(-\frac{\pi}{2}\right)\right] = 2\sqrt{2}$

67. $\int_{-\pi}^{0} \sqrt{1-\cos^2\theta}\,d\theta = \int_{-\pi}^{0} |\sin\theta|\,d\theta;\ \begin{bmatrix} \sin\theta \le 0 \\ \text{for } -\pi \le \theta \le 0 \end{bmatrix} \to \int_{-\pi}^{0} -\sin\theta\,d\theta = [\cos\theta]_{-\pi}^{0} = \cos 0 - \cos(-\pi)$
$= 1 - (-1) = 2$

68. $\int_{\pi/2}^{\pi} \sqrt{1-\sin^2\theta}\,d\theta = \int_{\pi/2}^{\pi} |\cos\theta|\,d\theta;\ \begin{bmatrix} \cos\theta \le 0 \\ \text{for } \frac{\pi}{2} \le \theta \le \pi \end{bmatrix} \to \int_{\pi/2}^{\pi} -\cos\theta\,d\theta = [-\sin\theta]_{\pi/2}^{\pi} = -\sin\pi + \sin\frac{\pi}{2} = 1$

69. $\int_{-\pi/4}^{\pi/4} \sqrt{\tan^2 y + 1}\,dy = \int_{-\pi/4}^{\pi/4} |\sec y|\,dy;\ \begin{bmatrix} \sec y \ge 0 \\ \text{for } -\frac{\pi}{4} \le y \le \frac{\pi}{4} \end{bmatrix} \to \int_{-\pi/4}^{\pi/4} \sec y\,dy = [\ln|\sec y + \tan y|]_{-\pi/4}^{\pi/4}$
$= \ln\left|\sqrt{2}+1\right| - \ln\left|\sqrt{2}-1\right|$

70. $\int_{-\pi/4}^{0} \sqrt{\sec^2 y - 1}\,dy = \int_{-\pi/4}^{0} |\tan y|\,dy;\ \begin{bmatrix} \tan y \le 0 \\ \text{for } -\frac{\pi}{4} \le y \le 0 \end{bmatrix} \to \int_{-\pi/4}^{0} -\tan y\,dy = [\ln|\cos y|]_{-\pi/4}^{0} = -\ln\left(\frac{1}{\sqrt{2}}\right)$
$= \ln\sqrt{2}$

71. $\int_{\pi/4}^{3\pi/4} (\csc x - \cot x)^2\,dx = \int_{\pi/4}^{3\pi/4} (\csc^2 x - 2\csc x\cot x + \cot^2 x)\,dx = \int_{\pi/4}^{3\pi/4} (2\csc^2 x - 1 - 2\csc x\cot x)\,dx$
$= [-2\cot x - x + 2\csc x]_{\pi/4}^{3\pi/4} = \left(-2\cot\frac{3\pi}{4} - \frac{3\pi}{4} + 2\csc\frac{3\pi}{4}\right) - \left(-2\cot\frac{\pi}{4} - \frac{\pi}{4} + 2\csc\frac{\pi}{4}\right)$
$= \left[-2(-1) - \frac{3\pi}{4} + 2\left(\sqrt{2}\right)\right] - \left[-2(1) - \frac{\pi}{4} + 2\left(\sqrt{2}\right)\right] = 4 - \frac{\pi}{2}$

72. $\int_0^{\pi/4} (\sec x + 4\cos x)^2\, dx = \int_0^{\pi/4} \left[\sec^2 x + 8 + 16\left(\frac{1+\cos 2x}{2}\right)\right] dx = [\tan x + 16x - 4\sin 2x]_0^{\pi/4}$
$= \left(\tan\frac{\pi}{4} + 4\pi - 4\sin\frac{\pi}{2}\right) - (\tan 0 + 0 - 4\sin 0) = 5 + 4\pi$

73. $\int \cos\theta \csc(\sin\theta)\, d\theta;\ \begin{bmatrix} u = \sin\theta \\ du = \cos\theta\, d\theta \end{bmatrix} \to \int \csc u\, du = -\ln|\csc u + \cot u| + C$
$= -\ln|\csc(\sin\theta) + \cot(\sin\theta)| + C$

74. $\int \left(1 + \frac{1}{x}\right)\cot(x + \ln x)\, dx;\ \begin{bmatrix} u = x + \ln x \\ du = \left(1 + \frac{1}{x}\right) dx \end{bmatrix} \to \int \cot u\, du = \ln|\sin u| + C = \ln|\sin(x + \ln x)| + C$

75. $\int (\csc x - \sec x)(\sin x + \cos x)\, dx = \int (1 + \cot x - \tan x - 1)\, dx = \int \cot x\, dx - \int \tan x\, dx$
$= \ln|\sin x| + \ln|\cos x| + C$

76. $\int 3\sinh\left(\frac{x}{2} + \ln 5\right) dx = \begin{bmatrix} u = \frac{x}{2} + \ln 5 \\ 2\, du = dx \end{bmatrix} = 6\int \sinh u\, du = 6\cosh u + C = 6\cosh\left(\frac{x}{2} + \ln 5\right) + C$

77. $\int \frac{6\, dy}{\sqrt{y}(1+y)};\ \begin{bmatrix} u = \sqrt{y} \\ du = \frac{1}{2\sqrt{y}}\, dy \end{bmatrix} \to \int \frac{12\, du}{1+u^2} = 12\tan^{-1} u + C = 12\tan^{-1}\sqrt{y} + C$

78. $\int \frac{dx}{x\sqrt{4x^2 - 1}} = \int \frac{2\, dx}{2x\sqrt{(2x)^2 - 1}};\ \begin{bmatrix} u = 2x \\ du = 2\, dx \end{bmatrix} \to \int \frac{du}{u\sqrt{u^2 - 1}} = \sec^{-1}|u| + C = \sec^{-1}|2x| + C$

79. $\int \frac{7\, dx}{(x-1)\sqrt{x^2 - 2x - 48}} = \int \frac{7\, dx}{(x-1)\sqrt{(x-1)^2 - 49}};\ \begin{bmatrix} u = x - 1 \\ du = dx \end{bmatrix} \to \int \frac{7\, du}{u\sqrt{u^2 - 49}} = 7 \cdot \frac{1}{7}\sec^{-1}\left|\frac{u}{7}\right| + C$
$= \sec^{-1}\left|\frac{x-1}{7}\right| + C$

80. $\int \frac{dx}{(2x+1)\sqrt{4x^2 + 4x}} = \int \frac{dx}{(2x+1)\sqrt{(2x+1)^2 - 1}};\ \begin{bmatrix} u = 2x + 1 \\ du = 2\, dx \end{bmatrix} \to \int \frac{du}{2u\sqrt{u^2 - 1}} = \frac{1}{2}\sec^{-1}|u| + C$
$= \frac{1}{2}\sec^{-1}|2x + 1| + C$

81. $\int \sec^2 t \tan(\tan t)\, dt;\ \begin{bmatrix} u = \tan t \\ du = \sec^2 t\, dt \end{bmatrix} \to \int \tan u\, du = -\ln|\cos u| + C = \ln|\sec u| + C = \ln|\sec(\tan t)| + C$

82. $\int \frac{dx}{x\sqrt{3 + x^2}} = -\frac{1}{3}\operatorname{csch}^{-1}\left|\frac{x}{\sqrt{3}}\right| + C$

83. (a) $\int \cos^3\theta\, d\theta = \int (\cos\theta)(1 - \sin^2\theta)\, d\theta;\ \begin{bmatrix} u = \sin\theta \\ du = \cos\theta\, d\theta \end{bmatrix} \to \int (1 - u^2)\, du = u - \frac{u^3}{3} + C = \sin\theta - \frac{1}{3}\sin^3\theta + C$

(b) $\int \cos^5\theta\, d\theta = \int (\cos\theta)(1 - \sin^2\theta)^2\, d\theta = \int (1 - u^2)^2\, du = \int (1 - 2u^2 + u^4)\, du = u - \frac{2}{3}u^3 + \frac{u^5}{5} + C$
$= \sin\theta - \frac{2}{3}\sin^3\theta + \frac{1}{5}\sin^5\theta + C$

(c) $\int \cos^9\theta\, d\theta = \int (\cos^8\theta)(\cos\theta)\, d\theta = \int (1 - \sin^2\theta)^4(\cos\theta)\, d\theta$

84. (a) $\int \sin^3\theta\, d\theta = \int (1 - \cos^2\theta)(\sin\theta)\, d\theta;\ \begin{bmatrix} u = \cos\theta \\ du = -\sin\theta\, d\theta \end{bmatrix} \to \int (1 - u^2)(-du) = \frac{u^3}{3} - u + C$
$= -\cos\theta + \frac{1}{3}\cos^3\theta + C$

(b) $\int \sin^5\theta\, d\theta = \int (1 - \cos^2\theta)^2(\sin\theta)\, d\theta = \int (1 - u^2)^2(-du) = \int (-1 + 2u^2 - u^4)\, du$
$= -\cos\theta + \frac{2}{3}\cos^3\theta - \frac{1}{5}\cos^5\theta + C$

(c) $\int \sin^7 \theta \, d\theta = \int (1-u^2)^3(-du) = \int (-1+3u^2-3u^4+u^6)\, du = -\cos\theta + \cos^3\theta - \frac{3}{5}\cos^5\theta + \frac{\cos^7\theta}{7} + C$

(d) $\int \sin^{13}\theta \, d\theta = \int (\sin^{12}\theta)(\sin\theta)\, d\theta = \int (1-\cos^2\theta)^6 (\sin\theta)\, d\theta$

85. (a) $\int \tan^3\theta\, d\theta = \int (\sec^2\theta - 1)(\tan\theta)\, d\theta = \int \sec^2\theta \tan\theta\, d\theta - \int \tan\theta\, d\theta = \frac{1}{2}\tan^2\theta - \int \tan\theta\, d\theta$
$= \frac{1}{2}\tan^2\theta + \ln|\cos\theta| + C$

(b) $\int \tan^5\theta\, d\theta = \int (\sec^2\theta - 1)(\tan^3\theta)\, d\theta = \int \tan^3\theta \sec^2\theta\, d\theta - \int \tan^3\theta\, d\theta = \frac{1}{4}\tan^4\theta - \int \tan^3\theta\, d\theta$

(c) $\int \tan^7\theta\, d\theta = \int (\sec^2\theta - 1)(\tan^5\theta)\, d\theta = \int \tan^5\theta \sec^2\theta\, d\theta - \int \tan^5\theta\, d\theta = \frac{1}{6}\tan^6\theta - \int \tan^5\theta\, d\theta$

(d) $\int \tan^{2k+1}\theta\, d\theta = \int (\sec^2\theta - 1)(\tan^{2k-1}\theta)\, d\theta = \int \tan^{2k-1}\theta \sec^2\theta\, d\theta - \int \tan^{2k-1}\theta\, d\theta;$

$\begin{bmatrix} u = \tan\theta \\ du = \sec^2\theta\, d\theta \end{bmatrix} \to \int u^{2k-1}\, du - \int \tan^{2k-1}\theta\, d\theta = \frac{1}{2k}u^{2k} - \int \tan^{2k-1}\theta\, d\theta = \frac{1}{2k}\tan^{2k}\theta - \int \tan^{2k-1}\theta\, d\theta$

86. (a) $\int \cot^3\theta\, d\theta = \int (\csc^2\theta - 1)(\cot\theta)\, d\theta = \int \cot\theta \csc^2\theta\, d\theta - \int \cot\theta\, d\theta = -\frac{1}{2}\cot^2\theta - \int \cot\theta\, d\theta$
$= -\frac{1}{2}\cot^2\theta - \ln|\sin\theta| + C$

(b) $\int \cot^5\theta\, d\theta = \int (\csc^2\theta - 1)(\cot^3\theta)\, d\theta = \int \cot^3\theta \csc^2\theta\, d\theta - \int \cot^3\theta\, d\theta = -\frac{1}{4}\cot^4\theta - \int \cot^3\theta\, d\theta$

(c) $\int \cot^7\theta\, d\theta = \int (\csc^2\theta - 1)(\cot^5\theta)\, d\theta = \int \cot^5\theta \csc^2\theta\, d\theta - \int \cot^5\theta\, d\theta = -\frac{1}{6}\cot^6\theta - \int \cot^5\theta\, d\theta$

(d) $\int \cot^{2k+1}\theta\, d\theta = \int (\csc^2\theta - 1)(\cot^{2k-1}\theta)\, d\theta = \int \cot^{2k-1}\theta \csc^2\theta\, d\theta - \int \cot^{2k-1}\theta\, d\theta;$

$\begin{bmatrix} u = \cot\theta \\ du = -\csc^2\theta\, d\theta \end{bmatrix} \to -\int u^{2k-1}\, du - \int \cot^{2k-1}\theta\, d\theta = -\frac{1}{2k}u^{2k} - \int \cot^{2k-1}\theta\, d\theta$
$= -\frac{1}{2k}\cot^{2k}\theta - \int \cot^{2k-1}\theta\, d\theta$

87. $A = \int_{-\pi/4}^{\pi/4} (2\cos x - \sec x)\, dx = [2\sin x - \ln|\sec x + \tan x|]_{-\pi/4}^{\pi/4}$
$= \left[\sqrt{2} - \ln\left(\sqrt{2}+1\right)\right] - \left[-\sqrt{2} - \ln\left(\sqrt{2}-1\right)\right]$
$= 2\sqrt{2} - \ln\left(\frac{\sqrt{2}+1}{\sqrt{2}-1}\right) = 2\sqrt{2} - \ln\left(\frac{\left(\sqrt{2}+1\right)^2}{2-1}\right)$
$= 2\sqrt{2} - \ln\left(3+2\sqrt{2}\right)$

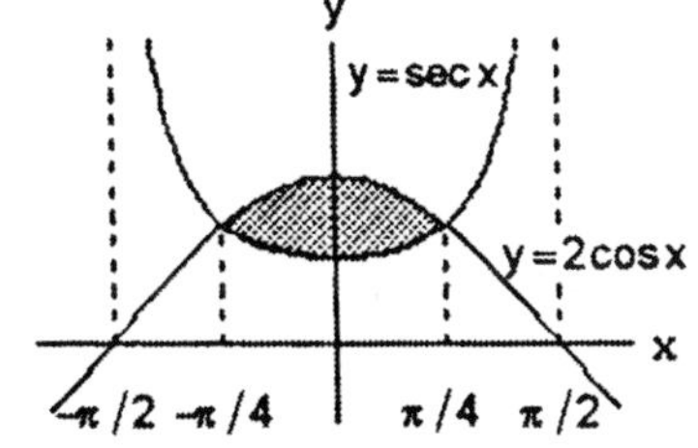

88. $A = \int_{\pi/6}^{\pi/2} (\csc x - \sin x)\, dx = [-\ln|\csc x + \cot x| + \cos x]_{\pi/6}^{\pi/2}$
$= -\ln|1+0| + \ln\left|2+\sqrt{3}\right| - \frac{\sqrt{3}}{2} = \ln\left(2+\sqrt{3}\right) - \frac{\sqrt{3}}{2}$

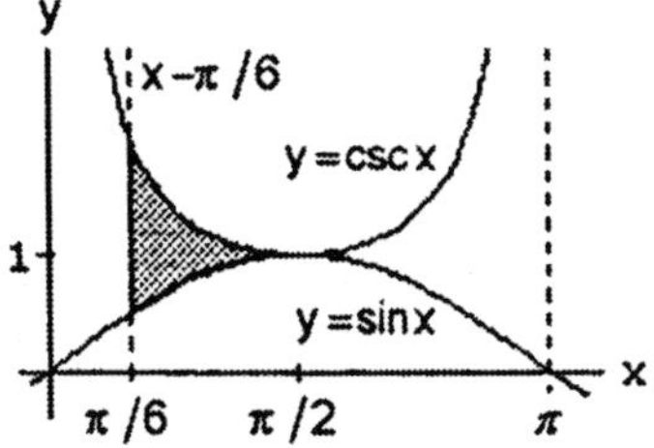

89. $V = \int_{-\pi/4}^{\pi/4} \pi(2\cos x)^2\, dx - \int_{-\pi/4}^{\pi/4} \pi\sec^2 x\, dx = 4\pi \int_{-\pi/4}^{\pi/4} \cos^2 x\, dx - \pi \int_{-\pi/4}^{\pi/4} \sec^2 x\, dx$
$= 2\pi \int_{-\pi/4}^{\pi/4} (1+\cos 2x)\, dx - \pi[\tan x]_{-\pi/4}^{\pi/4} = 2\pi\left[x + \frac{1}{2}\sin 2x\right]_{-\pi/4}^{\pi/4} - \pi[1-(-1)]$
$= 2\pi\left[\left(\frac{\pi}{4}+\frac{1}{2}\right) - \left(-\frac{\pi}{4}-\frac{1}{2}\right)\right] - 2\pi = 2\pi\left(\frac{\pi}{2}+1\right) - 2\pi = \pi^2$

90. $V = \int_{\pi/6}^{\pi/2} \pi\csc^2 x\, dx - \int_{\pi/6}^{\pi/2} \pi\sin^2 x\, dx = \pi \int_{\pi/6}^{\pi/2} \csc^2 x\, dx - \frac{\pi}{2}\int_{\pi/6}^{\pi/2} (1-\cos 2x)\, dx$
$= \pi[-\cot x]_{\pi/6}^{\pi/2} - \frac{\pi}{2}\left[x - \frac{1}{2}\sin 2x\right]_{\pi/6}^{\pi/2} = \pi\left[0 - \left(-\sqrt{3}\right)\right] - \frac{\pi}{2}\left[\left(\frac{\pi}{2}-0\right) - \left(\frac{\pi}{6} - \frac{1}{2}\cdot\frac{\sqrt{3}}{2}\right)\right]$
$= \pi\sqrt{3} - \frac{\pi}{2}\left(\frac{2\pi}{6} + \frac{\sqrt{3}}{4}\right) = \pi\left(\frac{7\sqrt{3}}{8} - \frac{\pi}{6}\right)$

91. $y = \ln(\cos x) \Rightarrow \frac{dy}{dx} = -\frac{\sin x}{\cos x} \Rightarrow \left(\frac{dy}{dx}\right)^2 = \tan^2 x = \sec^2 x - 1;\ L = \int_a^b \sqrt{1 + \left(\frac{dy}{dx}\right)^2}\, dx$
$= \int_0^{\pi/3} \sqrt{1 + (\sec^2 x - 1)}\, dx = \int_0^{\pi/3} \sec x\, dx = [\ln|\sec x + \tan x|]_0^{\pi/3} = \ln\left|2 + \sqrt{3}\right| - \ln|1 + 0| = \ln\left(2 + \sqrt{3}\right)$

92. $y = \ln(\sec x) \Rightarrow \frac{dy}{dx} = \frac{\sec x \tan x}{\sec x} \Rightarrow \left(\frac{dy}{dx}\right)^2 = \tan^2 x = \sec^2 x - 1;\ L = \int_a^b \sqrt{1 + \left(\frac{dy}{dx}\right)^2}\, dx$
$= \int_0^{\pi/4} \sec x\, dx = [\ln|\sec x + \tan x|]_0^{\pi/4} = \ln\left|\sqrt{2} + 1\right| - \ln|1 + 0| = \ln\left(\sqrt{2} + 1\right)$

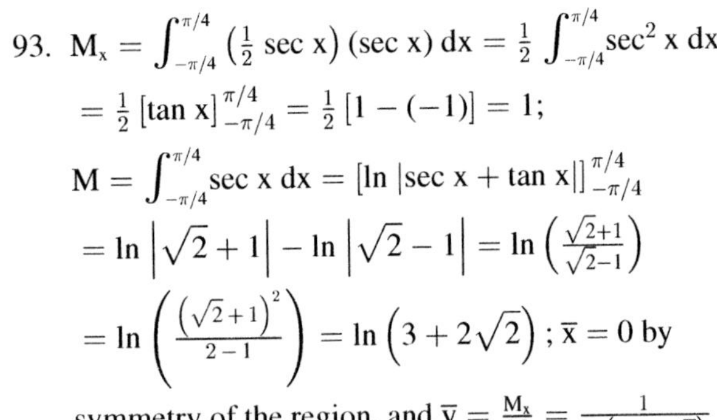

93. $M_x = \int_{-\pi/4}^{\pi/4} \left(\frac{1}{2}\sec x\right)(\sec x)\, dx = \frac{1}{2}\int_{-\pi/4}^{\pi/4} \sec^2 x\, dx$
$= \frac{1}{2}[\tan x]_{-\pi/4}^{\pi/4} = \frac{1}{2}[1 - (-1)] = 1;$
$M = \int_{-\pi/4}^{\pi/4} \sec x\, dx = [\ln|\sec x + \tan x|]_{-\pi/4}^{\pi/4}$
$= \ln\left|\sqrt{2} + 1\right| - \ln\left|\sqrt{2} - 1\right| = \ln\left(\frac{\sqrt{2}+1}{\sqrt{2}-1}\right)$
$= \ln\left(\frac{\left(\sqrt{2}+1\right)^2}{2-1}\right) = \ln\left(3 + 2\sqrt{2}\right)$; $\overline{x} = 0$ by
symmetry of the region, and $\overline{y} = \frac{M_x}{M} = \frac{1}{\ln\left(3+2\sqrt{2}\right)}$

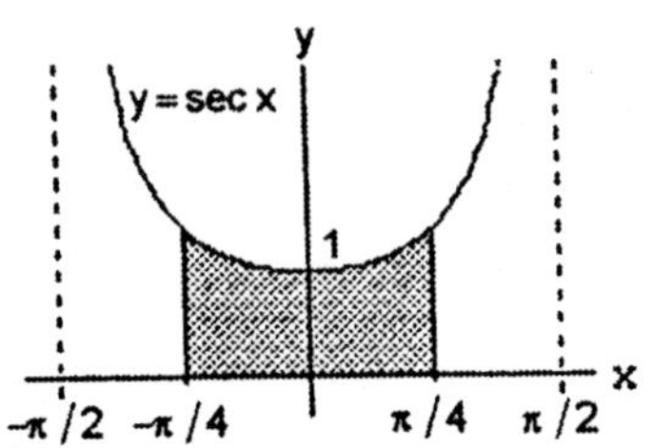

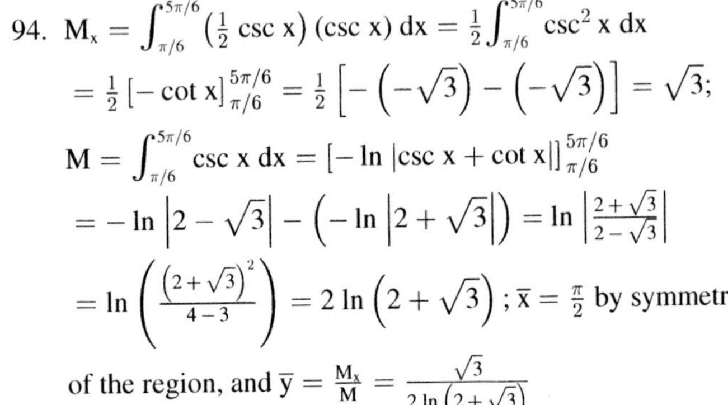

94. $M_x = \int_{\pi/6}^{5\pi/6} \left(\frac{1}{2}\csc x\right)(\csc x)\, dx = \frac{1}{2}\int_{\pi/6}^{5\pi/6} \csc^2 x\, dx$
$= \frac{1}{2}[-\cot x]_{\pi/6}^{5\pi/6} = \frac{1}{2}\left[-\left(-\sqrt{3}\right) - \left(-\sqrt{3}\right)\right] = \sqrt{3};$
$M = \int_{\pi/6}^{5\pi/6} \csc x\, dx = [-\ln|\csc x + \cot x|]_{\pi/6}^{5\pi/6}$
$= -\ln\left|2 - \sqrt{3}\right| - \left(-\ln\left|2 + \sqrt{3}\right|\right) = \ln\left|\frac{2+\sqrt{3}}{2-\sqrt{3}}\right|$
$= \ln\left(\frac{\left(2+\sqrt{3}\right)^2}{4-3}\right) = 2\ln\left(2 + \sqrt{3}\right)$; $\overline{x} = \frac{\pi}{2}$ by symmetry
of the region, and $\overline{y} = \frac{M_x}{M} = \frac{\sqrt{3}}{2\ln\left(2+\sqrt{3}\right)}$

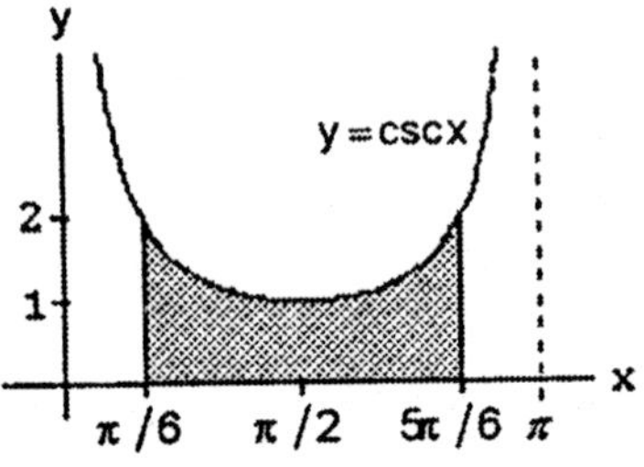

95. $\int \csc x\, dx = \int (\csc x)(1)\, dx = \int (\csc x)\left(\frac{\csc x + \cot x}{\csc x + \cot x}\right) dx = \int \frac{\csc^2 x + \csc x \cot x}{\csc x + \cot x}\, dx;$

$\left[\begin{array}{c} u = \csc x + \cot x \\ du = (-\csc x \cot x - \csc^2 x)\, dx \end{array}\right] \rightarrow \int \frac{-du}{u} = -\ln|u| + C = -\ln|\csc x + \cot x| + C$

96. $\left[(x^2 - 1)(x + 1)\right]^{-2/3} = \left[(x - 1)(x + 1)^2\right]^{-2/3} = (x - 1)^{-2/3}(x + 1)^{-4/3} = (x + 1)^{-2}\left[(x - 1)^{-2/3}(x + 1)^{2/3}\right]$
$= (x + 1)^{-2}\left(\frac{x-1}{x+1}\right)^{-2/3} = (x + 1)^{-2}\left(1 - \frac{2}{x+1}\right)^{-2/3}$

(a) $\int \left[(x^2 - 1)(x + 1)\right]^{-2/3} dx = \int (x + 1)^{-2}\left(1 - \frac{2}{x+1}\right)^{-2/3} dx;\ \left[\begin{array}{c} u = \frac{1}{x+1} \\ du = -\frac{1}{(x+1)^2}\, dx \end{array}\right]$
$\rightarrow \int -(1 - 2u)^{-2/3}\, du = \frac{3}{2}(1 - 2u)^{1/3} + C = \frac{3}{2}\left(1 - \frac{2}{x+1}\right)^{1/3} + C = \frac{3}{2}\left(\frac{x-1}{x+1}\right)^{1/3} + C$

(b) $\int \left[(x^2 - 1)(x + 1)\right]^{-2/3} dx = \int (x + 1)^{-2}\left(\frac{x-1}{x+1}\right)^{-2/3} dx;\ u = \left(\frac{x-1}{x+1}\right)^k$
$\Rightarrow du = k\left(\frac{x-1}{x+1}\right)^{k-1} \frac{[(x+1) - (x-1)]}{(x+1)^2}\, dx = 2k\frac{(x-1)^{k-1}}{(x+1)^{k+1}}\, dx;\ dx = \frac{(x+1)^2}{2k}\left(\frac{x+1}{x-1}\right)^{k-1} du$
$= \frac{(x+1)^2}{2k}\left(\frac{x-1}{x+1}\right)^{1-k} du$; then, $\int \left(\frac{x-1}{x+1}\right)^{-2/3} \frac{1}{2k}\left(\frac{x-1}{x+1}\right)^{1-k} du = \frac{1}{2k}\int \left(\frac{x-1}{x+1}\right)^{(1/3-k)} du$
$= \frac{1}{2k}\int \left(\frac{x-1}{x+1}\right)^{k(1/3k-1)} du = \frac{1}{2k}\int u^{(1/3k-1)}\, du = \frac{1}{2k}(3k)u^{1/3k} + C = \frac{3}{2}u^{1/3k} + C = \frac{3}{2}\left(\frac{x-1}{x+1}\right)^{1/3} + C$

(c) $\int[(x^2-1)(x+1)]^{-2/3}\,dx = \int (x+1)^{-2}\left(\frac{x-1}{x+1}\right)^{-2/3} dx;$

$\begin{bmatrix} u = \tan^{-1} x \\ x = \tan u \\ dx = \frac{du}{\cos^2 u} \end{bmatrix} \to \int \frac{1}{(\tan u+1)^2}\left(\frac{\tan u - 1}{\tan u + 1}\right)^{-2/3}\left(\frac{du}{\cos^2 u}\right) = \int \frac{1}{(\sin u+\cos u)^2}\left(\frac{\sin u-\cos u}{\sin u+\cos u}\right)^{-2/3} du;$

$\begin{bmatrix} \sin u + \cos u = \sin u + \sin\left(\frac{\pi}{2}-u\right) = 2\sin\frac{\pi}{4}\cos\left(u-\frac{\pi}{4}\right) \\ \sin u - \cos u = \sin u - \sin\left(\frac{\pi}{2}-u\right) = 2\cos\frac{\pi}{4}\sin\left(u-\frac{\pi}{4}\right) \end{bmatrix} \to \int \frac{1}{2\cos^2\left(u-\frac{\pi}{4}\right)}\left[\frac{\sin\left(u-\frac{\pi}{4}\right)}{\cos\left(u-\frac{\pi}{4}\right)}\right]^{-2/3} du$

$= \frac{1}{2}\int \tan^{-2/3}\left(u-\frac{\pi}{4}\right)\sec^2\left(u-\frac{\pi}{4}\right) du = \frac{3}{2}\tan^{1/3}\left(u-\frac{\pi}{4}\right) + C = \frac{3}{2}\left[\frac{\tan u - \tan\frac{\pi}{4}}{1+\tan u \tan\frac{\pi}{4}}\right]^{1/3} + C$

$= \frac{3}{2}\left(\frac{x-1}{x+1}\right)^{1/3} + C$

(d) $u = \tan^{-1}\sqrt{x} \Rightarrow \tan u = \sqrt{x} \Rightarrow \tan^2 u = x \Rightarrow dx = 2\tan u\left(\frac{1}{\cos^2 u}\right) du = \frac{2\sin u}{\cos^3 u}\,du = -\frac{2d(\cos u)}{\cos^3 u};$

$x-1 = \tan^2 u - 1 = \frac{\sin^2 u - \cos^2 u}{\cos^2 u} = \frac{1-2\cos^2 u}{\cos^2 u}; \; x+1 = \tan^2 u + 1 = \frac{\cos^2 u + \sin^2 u}{\cos^2 u} = \frac{1}{\cos^2 u};$

$\int (x-1)^{-2/3}(x+1)^{-4/3}\,dx = \int \frac{(1-2\cos^2 u)^{-2/3}}{(\cos^2 u)^{-2/3}} \cdot \frac{1}{(\cos^2 u)^{-4/3}} \cdot \frac{-2d(\cos u)}{\cos^3 u}$

$= \int (1-2\cos^2 u)^{-2/3} \cdot (-2) \cdot \cos u \cdot d(\cos u) = \frac{1}{2}\int (1-2\cos^2 u)^{-2/3} \cdot d\left(1-2\cos^2 u\right)$

$= \frac{3}{2}\left(1-2\cos^2 u\right)^{1/3} + C = \frac{3}{2}\left[\frac{\left(\frac{1-2\cos^2 u}{\cos^2 u}\right)}{\left(\frac{1}{\cos^2 u}\right)}\right]^{1/3} + C = \frac{3}{2}\left(\frac{x-1}{x+1}\right)^{1/3} + C$

(e) $u = \tan^{-1}\left(\frac{x-1}{2}\right) \Rightarrow \frac{x-1}{2} = \tan u \Rightarrow x+1 = 2(\tan u + 1) \Rightarrow dx = \frac{2\,du}{\cos^2 u} = 2d(\tan u);$

$\int (x-1)^{-2/3}(x+1)^{-4/3}\,dx = \int (\tan u)^{-2/3}(\tan u+1)^{-4/3} \cdot 2^{-2} \cdot 2 \cdot d(\tan u)$

$= \frac{1}{2}\int \left(1-\frac{1}{\tan u+1}\right)^{-2/3} d\left(1-\frac{1}{\tan u+1}\right) = \frac{3}{2}\left(1-\frac{1}{\tan u+1}\right)^{1/3} + C = \frac{3}{2}\left(1-\frac{2}{x+1}\right)^{1/3} + C$

$= \frac{3}{2}\left(\frac{x-1}{x+1}\right)^{1/3} + C$

(f) $\begin{bmatrix} u = \cos^{-1} x \\ x = \cos u \\ dx = -\sin u\,du \end{bmatrix} \to -\int \frac{\sin u\,du}{\sqrt[3]{(\cos^2 u - 1)^2(\cos u + 1)^2}} = -\int \frac{\sin u\,du}{(\sin^{4/3} u)\left(2^{2/3}\cos\frac{u}{2}\right)^{4/3}}$

$= -\int \frac{du}{(\sin u)^{1/3}\left(2^{2/3}\cos\frac{u}{2}\right)^{4/3}} = -\int \frac{du}{2\left(\sin\frac{u}{2}\right)^{1/3}\left(\cos\frac{u}{2}\right)^{5/3}} = -\frac{1}{2}\int \left(\frac{\cos\frac{u}{2}}{\sin\frac{u}{2}}\right)^{1/3} \frac{du}{\left(\cos^2\frac{u}{2}\right)}$

$= -\int \tan^{-1/3}\left(\frac{u}{2}\right) d\left(\tan\frac{u}{2}\right) = -\frac{3}{2}\tan^{2/3}\frac{u}{2} + C = \frac{3}{2}\left(-\tan^2\frac{u}{2}\right)^{1/3} + C = \frac{3}{2}\left(\frac{\cos u - 1}{\cos u + 1}\right)^{1/3} + C$

$= \frac{3}{2}\left(\frac{x-1}{x+1}\right)^{1/3} + C$

(g) $\int [(x^2-1)(x+1)]^{-2/3}\,dx; \; \begin{bmatrix} u = \cosh^{-1} x \\ x = \cosh u \\ dx = \sinh u \end{bmatrix} \to \int \frac{\sinh u\,du}{\sqrt[3]{(\cosh^2 u - 1)^2(\cosh u + 1)^2}}$

$= \int \frac{\sinh u\,du}{\sqrt[3]{(\sinh^4 u)(\cosh u+1)^2}} = \int \frac{du}{\sqrt[3]{(\sinh u)\left(4\cosh^4\frac{u}{2}\right)}} = \frac{1}{2}\int \frac{du}{\sqrt[3]{\sinh\left(\frac{u}{2}\right)\cosh^5\left(\frac{u}{2}\right)}}$

$= \int \left(\tanh\frac{u}{2}\right)^{-1/3} d\left(\tanh\frac{u}{2}\right) = \frac{3}{2}\left(\tanh\frac{u}{2}\right)^{2/3} + C = \frac{3}{2}\left(\frac{\cosh u - 1}{\cosh u + 1}\right)^{1/3} + C = \frac{3}{2}\left(\frac{x-1}{x+1}\right)^{1/3} + C$

8.2 INTEGRATION BY PARTS

1. $u = x,\ du = dx;\ dv = \sin\frac{x}{2}\,dx,\ v = -2\cos\frac{x}{2};$

$\int x\sin\frac{x}{2}\,dx = -2x\cos\frac{x}{2} - \int\left(-2\cos\frac{x}{2}\right) dx = -2x\cos\left(\frac{x}{2}\right) + 4\sin\left(\frac{x}{2}\right) + C$

2. $u = \theta,\ du = d\theta;\ dv = \cos\pi\theta\,d\theta,\ v = \frac{1}{\pi}\sin\pi\theta;$

$\int \theta\cos\pi\theta\,d\theta = \frac{\theta}{\pi}\sin\pi\theta - \int \frac{1}{\pi}\sin\pi\theta\,d\theta = \frac{\theta}{\pi}\sin\pi\theta + \frac{1}{\pi^2}\cos\pi\theta + C$

3.

		$\cos t$
t^2	$\xrightarrow{(+)}$	$\sin t$
$2t$	$\xrightarrow{(-)}$	$-\cos t$
2	$\xrightarrow{(+)}$	$-\sin t$
0		

$$\int t^2 \cos t\, dt = t^2 \sin t + 2t \cos t - 2 \sin t + C$$

4.

		$\sin x$
x^2	$\xrightarrow{(+)}$	$-\cos x$
$2x$	$\xrightarrow{(-)}$	$-\sin x$
2	$\xrightarrow{(+)}$	$\cos x$
0		

$$\int x^2 \sin x\, dx = -x^2 \cos x + 2x \sin x + 2 \cos x + C$$

5. $u = \ln x,\ du = \frac{dx}{x};\ dv = x\,dx,\ v = \frac{x^2}{2};$

$\int_1^2 x \ln x\, dx = \left[\frac{x^2}{2} \ln x\right]_1^2 - \int_1^2 \frac{x^2}{2}\frac{dx}{x} = 2 \ln 2 - \left[\frac{x^2}{4}\right]_1^2 = 2 \ln 2 - \frac{3}{4} = \ln 4 - \frac{3}{4}$

6. $u = \ln x,\ du = \frac{dx}{x};\ dv = x^3\, dx,\ v = \frac{x^4}{4};$

$\int_1^e x^3 \ln x\, dx = \left[\frac{x^4}{4} \ln x\right]_1^e - \int_1^e \frac{x^4}{4}\frac{dx}{x} = \frac{e^4}{4} - \left[\frac{x^4}{16}\right]_1^e = \frac{3e^4+1}{16}$

7. $u = \tan^{-1} y,\ du = \frac{dy}{1+y^2};\ dv = dy,\ v = y;$

$\int \tan^{-1} y\, dy = y \tan^{-1} y - \int \frac{y\,dy}{(1+y^2)} = y \tan^{-1} y - \frac{1}{2} \ln(1+y^2) + C = y \tan^{-1} y - \ln \sqrt{1+y^2} + C$

8. $u = \sin^{-1} y,\ du = \frac{dy}{\sqrt{1-y^2}};\ dv = dy,\ v = y;$

$\int \sin^{-1} y\, dy = y \sin^{-1} y - \int \frac{y\,dy}{\sqrt{1-y^2}} = y \sin^{-1} y + \sqrt{1-y^2} + C$

9. $u = x,\ du = dx;\ dv = \sec^2 x\, dx,\ v = \tan x;$

$\int x \sec^2 x\, dx = x \tan x - \int \tan x\, dx = x \tan x + \ln|\cos x| + C$

10. $\int 4x \sec^2 2x\, dx;\ [y = 2x] \rightarrow \int y \sec^2 y\, dy = y \tan y - \int \tan y\, dy = y \tan y - \ln|\sec y| + C$

$= 2x \tan 2x - \ln|\sec 2x| + C$

11.

		e^x
x^3	$\xrightarrow{(+)}$	e^x
$3x^2$	$\xrightarrow{(-)}$	e^x
$6x$	$\xrightarrow{(+)}$	e^x
6	$\xrightarrow{(-)}$	e^x
0		

$$\int x^3 e^x\, dx = x^3 e^x - 3x^2 e^x + 6x e^x - 6e^x + C = (x^3 - 3x^2 + 6x - 6)\, e^x + C$$

12.

		e^{-p}
p^4	$\xrightarrow{(+)}$	$-e^{-p}$
$4p^3$	$\xrightarrow{(-)}$	e^{-p}
$12p^2$	$\xrightarrow{(+)}$	$-e^{-p}$
$24p$	$\xrightarrow{(-)}$	e^{-p}
24	$\xrightarrow{(+)}$	$-e^{-p}$
0		

$$\int p^4 e^{-p}\,dp = -p^4e^{-p} - 4p^3e^{-p} - 12p^2e^{-p} - 24pe^{-p} - 24e^{-p} + C$$
$$= (-p^4 - 4p^3 - 12p^2 - 24p - 24)\,e^{-p} + C$$

13.

		e^x
$x^2 - 5x$	$\xrightarrow{(+)}$	e^x
$2x - 5$	$\xrightarrow{(-)}$	e^x
2	$\xrightarrow{(+)}$	e^x
0		

$$\int (x^2 - 5x)\,e^x\,dx = (x^2 - 5x)\,e^x - (2x - 5)e^x + 2e^x + C = x^2e^x - 7xe^x + 7e^x + C$$
$$= (x^2 - 7x + 7)\,e^x + C$$

14.

		e^r
$r^2 + r + 1$	$\xrightarrow{(+)}$	e^r
$2r + 1$	$\xrightarrow{(-)}$	e^r
2	$\xrightarrow{(+)}$	e^r
0		

$$\int (r^2 + r + 1)\,e^r\,dr = (r^2 + r + 1)\,e^r - (2r + 1)\,e^r + 2e^r + C$$
$$= [(r^2 + r + 1) - (2r + 1) + 2]\,e^r + C = (r^2 - r + 2)\,e^r + C$$

15.

		e^x
x^5	$\xrightarrow{(+)}$	e^x
$5x^4$	$\xrightarrow{(-)}$	e^x
$20x^3$	$\xrightarrow{(+)}$	e^x
$60x^2$	$\xrightarrow{(-)}$	e^x
$120x$	$\xrightarrow{(+)}$	e^x
120	$\xrightarrow{(-)}$	e^x
0		

$$\int x^5 e^x\,dx = x^5e^x - 5x^4e^x + 20x^3e^x - 60x^2e^x + 120xe^x - 120e^x + C$$
$$= (x^5 - 5x^4 + 20x^3 - 60x^2 + 120x - 120)\,e^x + C$$

16. e^{4t}

$t^2 \xrightarrow{(+)} \frac{1}{4}e^{4t}$

$2t \xrightarrow{(-)} \frac{1}{16}e^{4t}$

$2 \xrightarrow{(+)} \frac{1}{64}e^{4t}$

0

$$\int t^2 e^{4t}\,dt = \tfrac{t^2}{4}e^{4t} - \tfrac{2t}{16}e^{4t} + \tfrac{2}{64}e^{4t} + C = \tfrac{t^2}{4}e^{4t} - \tfrac{t}{8}e^{4t} + \tfrac{1}{32}e^{4t} + C$$
$$= \left(\tfrac{t^2}{4} - \tfrac{t}{8} + \tfrac{1}{32}\right)e^{4t} + C$$

17. $\sin 2\theta$

$\theta^2 \xrightarrow{(+)} -\frac{1}{2}\cos 2\theta$

$2\theta \xrightarrow{(-)} -\frac{1}{4}\sin 2\theta$

$2 \xrightarrow{(+)} \frac{1}{8}\cos 2\theta$

0

$$\int_0^{\pi/2} \theta^2 \sin 2\theta\,d\theta = \left[-\tfrac{\theta^2}{2}\cos 2\theta + \tfrac{\theta}{2}\sin 2\theta + \tfrac{1}{4}\cos 2\theta\right]_0^{\pi/2}$$
$$= \left[-\tfrac{\pi^2}{8}\cdot(-1) + \tfrac{\pi}{4}\cdot 0 + \tfrac{1}{4}\cdot(-1)\right] - \left[0 + 0 + \tfrac{1}{4}\cdot 1\right] = \tfrac{\pi^2}{8} - \tfrac{1}{2} = \tfrac{\pi^2-4}{8}$$

18. $\cos 2x$

$x^3 \xrightarrow{(+)} \frac{1}{2}\sin 2x$

$3x^2 \xrightarrow{(-)} -\frac{1}{4}\cos 2x$

$6x \xrightarrow{(+)} -\frac{1}{8}\sin 2x$

$6 \xrightarrow{(-)} \frac{1}{16}\cos 2x$

0

$$\int_0^{\pi/2} x^3 \cos 2x\,dx = \left[\tfrac{x^3}{2}\sin 2x + \tfrac{3x^2}{4}\cos 2x - \tfrac{3x}{4}\sin 2x - \tfrac{3}{8}\cos 2x\right]_0^{\pi/2}$$
$$= \left[\tfrac{\pi^3}{16}\cdot 0 + \tfrac{3\pi^2}{16}\cdot(-1) - \tfrac{3\pi}{8}\cdot 0 - \tfrac{3}{8}\cdot(-1)\right] - \left[0 + 0 - 0 - \tfrac{3}{8}\cdot 1\right] = -\tfrac{3\pi^2}{16} + \tfrac{3}{4} = \tfrac{3(4-\pi^2)}{16}$$

19. $u = \sec^{-1} t,\ du = \frac{dt}{t\sqrt{t^2-1}};\ dv = t\,dt,\ v = \frac{t^2}{2};$

$$\int_{2/\sqrt{3}}^{2} t\sec^{-1} t\,dt = \left[\tfrac{t^2}{2}\sec^{-1} t\right]_{2/\sqrt{3}}^{2} - \int_{2/\sqrt{3}}^{2}\left(\tfrac{t^2}{2}\right)\tfrac{dt}{t\sqrt{t^2-1}} = \left(2\cdot\tfrac{\pi}{3} - \tfrac{2}{3}\cdot\tfrac{\pi}{6}\right) - \int_{2/\sqrt{3}}^{2}\tfrac{t\,dt}{2\sqrt{t^2-1}}$$
$$= \tfrac{5\pi}{9} - \left[\tfrac{1}{2}\sqrt{t^2-1}\right]_{2/\sqrt{3}}^{2} = \tfrac{5\pi}{9} - \tfrac{1}{2}\left(\sqrt{3} - \sqrt{\tfrac{4}{3}-1}\right) = \tfrac{5\pi}{9} - \tfrac{1}{2}\left(\sqrt{3} - \tfrac{\sqrt{3}}{3}\right) = \tfrac{5\pi}{9} - \tfrac{\sqrt{3}}{3} = \tfrac{5\pi-3\sqrt{3}}{9}$$

20. $u = \sin^{-1}(x^2),\ du = \frac{2x\,dx}{\sqrt{1-x^4}};\ dv = 2x\,dx,\ v = x^2;$

$$\int_0^{1/\sqrt{2}} 2x\sin^{-1}(x^2)\,dx = \left[x^2\sin^{-1}(x^2)\right]_0^{1/\sqrt{2}} - \int_0^{1/\sqrt{2}} x^2\cdot\tfrac{2x\,dx}{\sqrt{1-x^4}} = \left(\tfrac{1}{2}\right)\left(\tfrac{\pi}{6}\right) + \int_0^{1/\sqrt{2}}\tfrac{d(1-x^4)}{2\sqrt{1-x^4}}$$
$$= \tfrac{\pi}{12} + \left[\sqrt{1-x^4}\right]_0^{1/\sqrt{2}} = \tfrac{\pi}{12} + \sqrt{\tfrac{3}{4}} - 1 = \tfrac{\pi+6\sqrt{3}-12}{12}$$

21. $I = \int e^\theta \sin\theta\,d\theta$; $[u = \sin\theta,\ du = \cos\theta\,d\theta;\ dv = e^\theta\,d\theta,\ v = e^\theta] \Rightarrow I = e^\theta\sin\theta - \int e^\theta\cos\theta\,d\theta$;
$[u = \cos\theta,\ du = -\sin\theta\,d\theta;\ dv = e^\theta\,d\theta,\ v = e^\theta] \Rightarrow I = e^\theta\sin\theta - \left(e^\theta\cos\theta + \int e^\theta\sin\theta\,d\theta\right)$
$= e^\theta\sin\theta - e^\theta\cos\theta - I + C' \Rightarrow 2I = (e^\theta\sin\theta - e^\theta\cos\theta) + C' \Rightarrow I = \frac{1}{2}(e^\theta\sin\theta - e^\theta\cos\theta) + C$, where $C = \frac{C'}{2}$ is another arbitrary constant

22. $I = \int e^{-y} \cos y\, dy$; $[u = \cos y,\ du = -\sin y\, dy;\ dv = e^{-y}\, dy,\ v = -e^{-y}]$
$\Rightarrow I = -e^{-y} \cos y - \int (-e^{-y})(-\sin y)\, dy = -e^{-y} \cos y - \int e^{-y} \sin y\, dy$; $[u = \sin y,\ du = \cos y\, dy;$
$dv = e^{-y}\, dy,\ v = -e^{-y}]$ $\Rightarrow I = -e^{-y} \cos y - \left(-e^{-y} \sin y - \int (-e^{y}) \cos y\, dy\right) = -e^{-y} \cos y + e^{-y} \sin y - I + C'$
$\Rightarrow 2I = e^{-y}(\sin y - \cos y) + C'$ $\Rightarrow I = \frac{1}{2}(e^{-y} \sin y - e^{-y} \cos y) + C$, where $C = \frac{C'}{2}$ is another arbitrary constant

23. $I = \int e^{2x} \cos 3x\, dx$; $\left[u = \cos 3x;\ du = -3 \sin 3x\, dx,\ dv = e^{2x}\, dx;\ v = \frac{1}{2} e^{2x}\right]$
$\Rightarrow I = \frac{1}{2} e^{2x} \cos 3x + \frac{3}{2} \int e^{2x} \sin 3x\, dx$; $\left[u = \sin 3x,\ du = 3 \cos 3x,\ dv = e^{2x}\, dx;\ v = \frac{1}{2} e^{2x}\right]$
$\Rightarrow I = \frac{1}{2} e^{2x} \cos 3x + \frac{3}{2}\left(\frac{1}{2} e^{2x} \sin 3x - \frac{3}{2} \int e^{2x} \cos 3x\, dx\right) = \frac{1}{2} e^{2x} \cos 3x + \frac{3}{4} e^{2x} \sin 3x - \frac{9}{4} I + C'$
$\Rightarrow \frac{13}{4} I = \frac{1}{2} e^{2x} \cos 3x + \frac{3}{4} e^{2x} \sin 3x + C'$ $\Rightarrow \frac{e^{2x}}{13}(3 \sin 3x + 2 \cos 3x) + C$, where $C = \frac{4}{13} C'$

24. $\int e^{-2x} \sin 2x\, dx$; $[y = 2x]$ $\rightarrow \frac{1}{2} \int e^{-y} \sin y\, dy = I$; $[u = \sin y,\ du = \cos y\, dy;\ dv = e^{-y}\, dy,\ v = -e^{-y}]$
$\Rightarrow I = \frac{1}{2}\left(-e^{-y} \sin y + \int e^{-y} \cos y\, dy\right)$ $[u = \cos y,\ du = -\sin y;\ dv = e^{-y}\, dy,\ v = -e^{-y}]$
$\Rightarrow I = -\frac{1}{2} e^{-y} \sin y + \frac{1}{2}\left(-e^{-y} \cos y - \int (-e^{-y})(-\sin y)\, dy\right) = -\frac{1}{2} e^{-y}(\sin y + \cos y) - I + C'$
$\Rightarrow 2I = -\frac{1}{2} e^{-y}(\sin y + \cos y) + C'$ $\Rightarrow I = -\frac{1}{4} e^{-y}(\sin y + \cos y) + C = -\frac{e^{-2x}}{4}(\sin 2x + \cos 2x) + C$, where
$C = \frac{C'}{2}$

25. $\int e^{\sqrt{3s+9}}\, ds$; $\begin{bmatrix} 3s + 9 = x^2 \\ ds = \frac{2}{3} x\, dx \end{bmatrix}$ $\rightarrow \int e^{x} \cdot \frac{2}{3} x\, dx = \frac{2}{3} \int x e^{x}\, dx$; $[u = x,\ du = dx;\ dv = e^{x}\, dx,\ v = e^{x}]$;
$\frac{2}{3} \int x e^{x}\, dx = \frac{2}{3}\left(x e^{x} - \int e^{x}\, dx\right) = \frac{2}{3}(x e^{x} - e^{x}) + C = \frac{2}{3}\left(\sqrt{3s+9}\, e^{\sqrt{3s+9}} - e^{\sqrt{3s+9}}\right) + C$

26. $u = x,\ du = dx;\ dv = \sqrt{1-x}\, dx,\ v = -\frac{2}{3}\sqrt{(1-x)^3}$;
$\int_0^1 x\sqrt{1-x}\, dx = \left[-\frac{2}{3}\sqrt{(1-x)^3}\, x\right]_0^1 + \frac{2}{3} \int_0^1 \sqrt{(1-x)^3}\, dx = \frac{2}{3}\left[-\frac{2}{5}(1-x)^{5/2}\right]_0^1 = \frac{4}{15}$

27. $u = x,\ du = dx;\ dv = \tan^2 x\, dx,\ v = \int \tan^2 x\, dx = \int \frac{\sin^2 x}{\cos^2 x}\, dx = \int \frac{1 - \cos^2 x}{\cos^2 x}\, dx = \int \frac{dx}{\cos^2 x} - \int dx$
$= \tan x - x$; $\int_0^{\pi/3} x \tan^2 x\, dx = [x(\tan x - x)]_0^{\pi/3} - \int_0^{\pi/3} (\tan x - x)\, dx = \frac{\pi}{3}\left(\sqrt{3} - \frac{\pi}{3}\right) + \left[\ln|\cos x| + \frac{x^2}{2}\right]_0^{\pi/3}$
$= \frac{\pi}{3}\left(\sqrt{3} - \frac{\pi}{3}\right) + \ln\frac{1}{2} + \frac{\pi^2}{18} = \frac{\pi\sqrt{3}}{3} - \ln 2 - \frac{\pi^2}{18}$

28. $u = \ln(x + x^2),\ du = \frac{(2x+1)\, dx}{x + x^2}$; $dv = dx,\ v = x$; $\int \ln(x + x^2)\, dx = x \ln(x + x^2) - \int \frac{2x+1}{x(x+1)} \cdot x\, dx$
$= x \ln(x + x^2) - \int \frac{(2x+1)\, dx}{x+1} = x \ln(x + x^2) - \int \frac{2(x+1) - 1}{x+1}\, dx = x \ln(x + x^2) - 2x + \ln|x+1| + C$

29. $\int \sin(\ln x)\, dx$; $\begin{bmatrix} u = \ln x \\ du = \frac{1}{x}\, dx \\ dx = e^{u}\, du \end{bmatrix}$ $\rightarrow \int (\sin u) e^{u}\, du$. From Exercise 21, $\int (\sin u) e^{u}\, du = e^{u}\left(\frac{\sin u - \cos u}{2}\right) + C$
$= \frac{1}{2}[-x \cos(\ln x) + x \sin(\ln x)] + C$

30. $\int z(\ln z)^2\,dz;\ \begin{bmatrix} u = \ln z \\ du = \frac{1}{z}\,dz \\ dz = e^u\,du \end{bmatrix} \to \int e^u \cdot u^2 \cdot e^u\,du = \int e^{2u} \cdot u^2\,du;$

$$\begin{array}{lcl} & & e^{2u} \\ u^2 & \xrightarrow{(+)} & \frac{1}{2}e^{2u} \\ 2u & \xrightarrow{(-)} & \frac{1}{4}e^{2u} \\ 2 & \xrightarrow{(+)} & \frac{1}{8}e^{2u} \\ 0 & & \end{array}$$

$$\int u^2e^{2u}\,du = \frac{u^2}{2}e^{2u} - \frac{u}{2}e^{2u} + \frac{1}{4}e^{2u} + C = \frac{e^{2u}}{4}\left[2u^2 - 2u + 1\right] + C$$
$$= \frac{z^2}{4}\left[2(\ln z)^2 - 2\ln z + 1\right] + C$$

31. (a) $u = x,\ du = dx;\ dv = \sin x\,dx,\ v = -\cos x;$

$$S_1 = \int_0^{\pi} x\sin x\,dx = [-x\cos x]_0^{\pi} + \int_0^{\pi}\cos x\,dx = \pi + [\sin x]_0^{\pi} = \pi$$

(b) $S_2 = -\int_{\pi}^{2\pi} x\sin x\,dx = -\left[[-x\cos x]_{\pi}^{2\pi} + \int_{\pi}^{2\pi}\cos x\,dx\right] = -\left[-3\pi + [\sin x]_{\pi}^{2\pi}\right] = 3\pi$

(c) $S_3 = \int_{2\pi}^{3\pi} x\sin x\,dx = [-x\cos x]_{2\pi}^{3\pi} + \int_{2\pi}^{3\pi}\cos x\,dx = 5\pi + [\sin x]_{2\pi}^{3\pi} = 5\pi$

(d) $S_{n+1} = (-1)^{n+1}\int_{n\pi}^{(n+1)\pi} x\sin x\,dx = (-1)^{n+1}\left[[-x\cos x]_{n\pi}^{(n+1)\pi} + [\sin x]_{n\pi}^{(n+1)\pi}\right]$
$= (-1)^{n+1}\left[-(n+1)\pi(-1)^n + n\pi(-1)^{n+1}\right] + 0 = (2n+1)\pi$

32. (a) $u = x,\ du = dx;\ dv = \cos x\,dx,\ v = \sin x;$

$$S_1 = -\int_{\pi/2}^{3\pi/2} x\cos x\,dx = -\left[[x\sin x]_{\pi/2}^{3\pi/2} - \int_{\pi/2}^{3\pi/2}\sin x\,dx\right] = -\left(-\frac{3\pi}{2} - \frac{\pi}{2}\right) - [\cos x]_{\pi/2}^{3\pi/2} = 2\pi$$

(b) $S_2 = \int_{3\pi/2}^{5\pi/2} x\cos x\,dx = [x\sin x]_{3\pi/2}^{5\pi/2} - \int_{3\pi/2}^{5\pi/2}\sin x\,dx = \left[\frac{5\pi}{2} - \left(-\frac{3\pi}{2}\right)\right] - [\cos x]_{3\pi/2}^{5\pi/2} = 4\pi$

(c) $S_3 = -\int_{5\pi/2}^{7\pi/2} x\cos x\,dx = -\left[[x\sin x]_{5\pi/2}^{7\pi/2} - \int_{5\pi/2}^{7\pi/2}\sin x\,dx\right] = -\left(-\frac{7\pi}{2} - \frac{5\pi}{2}\right) - [\cos x]_{5\pi/2}^{7\pi/2} = 6\pi$

(d) $S_n = (-1)^n\int_{(2n-1)\pi/2}^{(2n+1)\pi/2} x\cos x\,dx = (-1)^n\left[[x\sin x]_{(2n-1)\pi/2}^{(2n+1)\pi/2} - {}^n\int_{(2n-1)\pi/2}^{(2n+1)\pi/2}\sin x\,dx\right]$
$= (-1)^n\left[\frac{(2n+1)\pi}{2}(-1)^n - \frac{(2n-1)\pi}{2}(-1)^{n-1}\right] - [\cos x]_{(2n-1)\pi/2}^{(2n+1)\pi/2} = \frac{1}{2}(2n\pi + \pi + 2n\pi - \pi) = 2n\pi$

33. $V = \int_0^{\ln 2} 2\pi(\ln 2 - x)e^x\,dx = 2\pi\ln 2\int_0^{\ln 2} e^x\,dx - 2\pi\int_0^{\ln 2} xe^x\,dx$
$= (2\pi\ln 2)\left[e^x\right]_0^{\ln 2} - 2\pi\left(\left[xe^x\right]_0^{\ln 2} - \int_0^{\ln 2} e^x\,dx\right)$
$= 2\pi\ln 2 - 2\pi\left(2\ln 2 - \left[e^x\right]_0^{\ln 2}\right) = -2\pi\ln 2 + 2\pi = 2\pi(1 - \ln 2)$

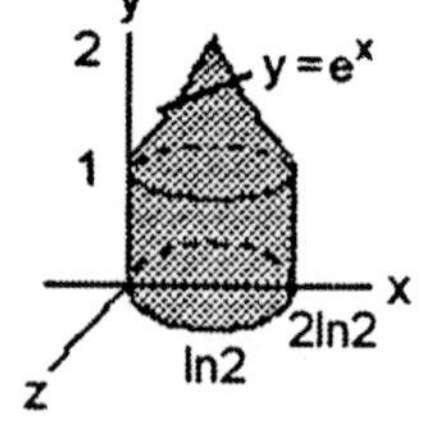

34. (a) $V = \int_0^1 2\pi xe^{-x}\,dx = 2\pi\left(\left[-xe^{-x}\right]_0^1 + \int_0^1 e^{-x}\,dx\right)$
$= 2\pi\left(-\frac{1}{e} + \left[-e^{-x}\right]_0^1\right) = 2\pi\left(-\frac{1}{e} - \frac{1}{e} + 1\right)$
$= 2\pi - \frac{4\pi}{e}$

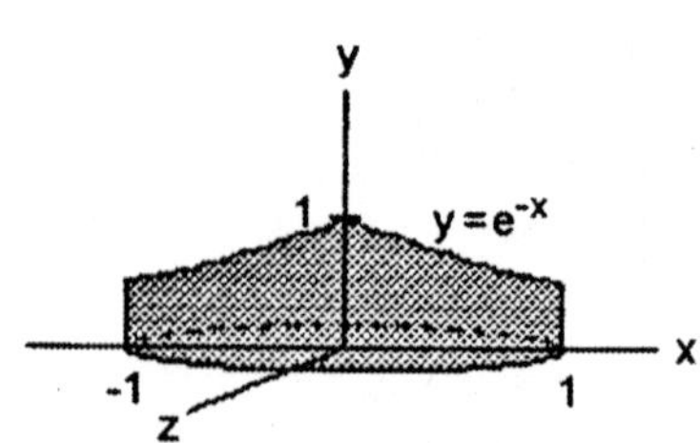

(b) $V = \int_0^1 2\pi(1-x)e^{-x}\,dx$; $u = 1 - x$, $du = -\,dx$; $dv = e^{-x}\,dx$,

$v = -e^{-x}$; $V = 2\pi\left[\left[(1-x)\left(-e^{-x}\right)\right]_0^1 - \int_0^1 e^{-x}\,dx\right]$

$= 2\pi\left[[0 - 1(-1)] + \left[e^{-x}\right]_0^1\right] = 2\pi\left(1 + \frac{1}{e} - 1\right) = \frac{2\pi}{e}$

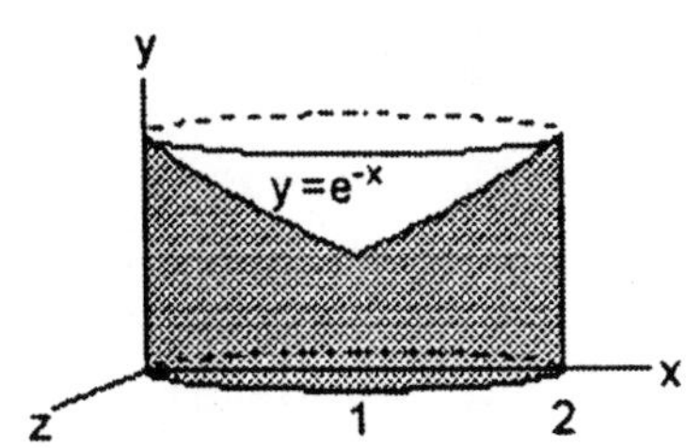

35. (a) $V = \int_0^{\pi/2} 2\pi x \cos x\,dx = 2\pi\left([x \sin x]_0^{\pi/2} - \int_0^{\pi/2} \sin x\,dx\right)$

$= 2\pi\left(\frac{\pi}{2} + [\cos x]_0^{\pi/2}\right) = 2\pi\left(\frac{\pi}{2} + 0 - 1\right) = \pi(\pi - 2)$

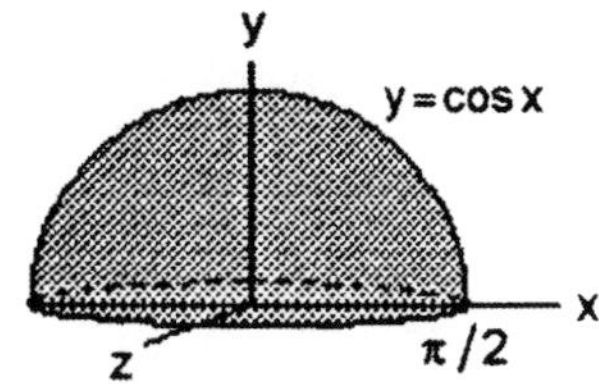

(b) $V = \int_0^{\pi/2} 2\pi\left(\frac{\pi}{2} - x\right)\cos x\,dx$; $u = \frac{\pi}{2} - x$, $du = -\,dx$; $dv = \cos x\,dx$, $v = \sin x$;

$V = 2\pi\left[\left(\frac{\pi}{2} - x\right)\sin x\right]_0^{\pi/2} + 2\pi\int_0^{\pi/2} \sin x\,dx = 0 + 2\pi[-\cos x]_0^{\pi/2} = 2\pi(0+1) = 2\pi$

36. (a) $V = \int_0^{\pi} 2\pi x(x \sin x)\,dx$;

		$\sin x$
x^2	$\xrightarrow{(+)}$	$-\cos x$
$2x$	$\xrightarrow{(-)}$	$-\sin x$
2	$\xrightarrow{(+)}$	$\cos x$
0		

$\Rightarrow V = 2\pi\int_0^{\pi} x^2 \sin x\,dx = 2\pi\left[-x^2\cos x + 2x\sin x + 2\cos x\right]_0^{\pi} = 2\pi\left(\pi^2 - 4\right)$

(b) $V = \int_0^{\pi} 2\pi(\pi - x)x\sin x\,dx = 2\pi^2\int_0^{\pi} x\sin x\,dx - 2\pi\int_0^{\pi} x^2\sin x\,dx = 2\pi^2[-x\cos x + \sin x]_0^{\pi} - \left(2\pi^3 - 8\pi\right)$

$= 8\pi$

37. $av(y) = \frac{1}{2\pi}\int_0^{2\pi} 2e^{-t}\cos t\,dt$

$= \frac{1}{\pi}\left[e^{-t}\left(\frac{\sin t - \cos t}{2}\right)\right]_0^{2\pi}$

(see Exercise 22) $\Rightarrow av(y) = \frac{1}{2\pi}\left(1 - e^{-2\pi}\right)$

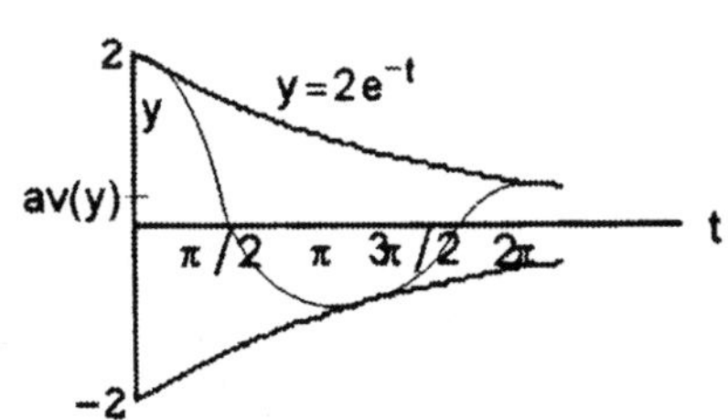

38. $av(y) = \frac{1}{2\pi}\int_0^{2\pi} 4e^{-t}(\sin t - \cos t)\,dt$

$= \frac{2}{\pi}\int_0^{2\pi} e^{-t}\sin t\,dt - \frac{2}{\pi}\int_0^{2\pi} e^{-t}\cos t\,dt$

$= \frac{2}{\pi}\left[e^{-t}\left(\frac{-\sin t - \cos t}{2}\right) - e^{-t}\left(\frac{\sin t - \cos t}{2}\right)\right]_0^{2\pi}$

$= \frac{2}{\pi}\left[-e^{-t}\sin t\right]_0^{2\pi} = 0$

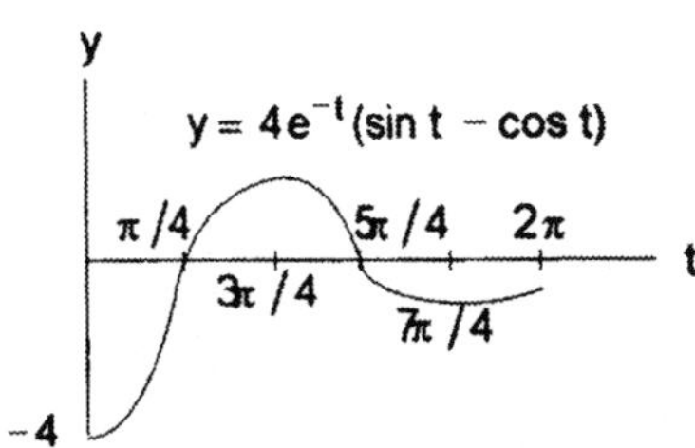

39. $I = \int x^n \cos x\,dx$; $\left[u = x^n,\ du = nx^{n-1}\,dx;\ dv = \cos x\,dx,\ v = \sin x\right]$

$\Rightarrow I = x^n \sin x - \int nx^{n-1}\sin x\,dx$

40. $I = \int x^n \sin x\,dx$; $[u = x^n,\ du = nx^{n-1}\,dx;\ dv = \sin x\,dx,\ v = -\cos x]$
$\Rightarrow I = -x^n \cos x + \int nx^{n-1}\cos x\,dx$

41. $I = \int x^n e^{ax}\,dx$; $\left[u = x^n,\ du = nx^{n-1}\,dx;\ dv = e^{ax}\,dx,\ v = \frac{1}{a}e^{ax}\right]$
$\Rightarrow I = \frac{x^n e^{ax}}{a}e^{ax} - \frac{n}{a}\int x^{n-1}e^{ax}\,dx,\ a \neq 0$

42. $I = \int (\ln x)^n\,dx$; $\left[u = (\ln x)^n,\ du = \frac{n(\ln x)^{n-1}}{x}\,dx;\ dv = 1\,dx,\ v = x\right]$
$\Rightarrow I = x(\ln x)^n - \int n(\ln x)^{n-1}\,dx$

43. $\int \sin^{-1} x\,dx = x\sin^{-1} x - \int \sin y\,dy = x\sin^{-1} x + \cos y + C = x\sin^{-1} x + \cos(\sin^{-1} x) + C$

44. $\int \tan^{-1} x\,dx = x\tan^{-1} x - \int \tan y\,dy = x\tan^{-1} x + \ln|\cos y| + C = x\tan^{-1} x + \ln|\cos(\tan^{-1} x)| + C$

45. $\int \sec^{-1} x\,dx = x\sec^{-1} x - \int \sec y\,dy = x\sec^{-1} x - \ln|\sec y + \tan y| + C$
$= x\sec^{-1} x - \ln|\sec(\sec^{-1} x) + \tan(\sec^{-1} x)| + C = x\sec^{-1} x - \ln\left|x + \sqrt{x^2 - 1}\right| + C$

46. $\int \log_2 x\,dx = x\log_2 x - \int 2^y\,dy = x\log_2 x - \frac{2^y}{\ln 2} + C = x\log_2 x - \frac{x}{\ln 2} + C$

47. Yes, $\cos^{-1} x$ is the angle whose cosine is x which implies $\sin(\cos^{-1} x) = \sqrt{1 - x^2}$.

48. Yes, $\tan^{-1} x$ is the angle whose tangent is x which implies $\sec(\tan^{-1} x) = \sqrt{1 + x^2}$.

49. (a) $\int \sinh^{-1} x\,dx = x\sinh^{-1} x - \int \sinh y\,dy = x\sinh^{-1} x - \cosh y + C = x\sinh^{-1} x - \cosh(\sinh^{-1} x) + C$;
check: $d[x\sinh^{-1} x - \cosh(\sinh^{-1} x) + C] = \left[\sinh^{-1} x + \frac{x}{\sqrt{1+x^2}} - \sinh(\sinh^{-1} x)\,\frac{1}{\sqrt{1+x^2}}\right]dx$
$= \sinh^{-1} x\,dx$

(b) $\int \sinh^{-1} x\,dx = x\sinh^{-1} x - \int x\left(\frac{1}{\sqrt{1+x^2}}\right)dx = x\sinh^{-1} x - \frac{1}{2}\int (1+x^2)^{-1/2}2x\,dx$
$= x\sinh^{-1} x - (1+x^2)^{1/2} + C$
check: $d\left[x\sinh^{-1} x - (1+x^2)^{1/2} + C\right] = \left[\sinh^{-1} x + \frac{x}{\sqrt{1+x^2}} - \frac{x}{\sqrt{1+x^2}}\right]dx = \sinh^{-1} x\,dx$

50. (a) $\int \tanh^{-1} x\,dx = x\tanh^{-1} x - \int \tanh y\,dy = x\tanh^{-1} x - \ln|\cosh y| + C$
$= x\tanh^{-1} x - \ln|\cosh(\tanh^{-1} x)| + C$;
check: $d[x\tanh^{-1} x - \ln|\cosh(\tanh^{-1} x)| + C] = \left[\tanh^{-1} x + \frac{x}{1-x^2} - \frac{\sinh(\tanh^{-1} x)}{\cosh(\tanh^{-1} x)}\,\frac{1}{1-x^2}\right]dx$
$= \left[\tanh^{-1} x + \frac{x}{1-x^2} - \frac{x}{1-x^2}\right]dx = \tanh^{-1} x\,dx$

(b) $\int \tanh^{-1} x\,dx = x\tanh^{-1} x - \int \frac{x}{1-x^2}\,dx = x\tanh^{-1} x - \frac{1}{2}\int \frac{2x}{1-x^2}\,dx = x\tanh^{-1} x + \frac{1}{2}\ln|1 - x^2| + C$
check: $d\left[x\tanh^{-1} x + \frac{1}{2}\ln|1 - x^2| + C\right] = \left[\tanh^{-1} x + \frac{x}{1-x^2} - \frac{x}{1-x^2}\right]dx = \tanh^{-1} x\,dx$

8.3 INTEGRATION OF RATIONAL FUNCTIONS BY PARTIAL FRACTIONS

1. $\frac{5x - 13}{(x-3)(x-2)} = \frac{A}{x-3} + \frac{B}{x-2} \Rightarrow 5x - 13 = A(x-2) + B(x-3) = (A+B)x - (2A+3B)$
$\Rightarrow \left.\begin{array}{c} A + B = 5 \\ 2A + 3B = 13 \end{array}\right\} \Rightarrow -B = (10 - 13) \Rightarrow B = 3 \Rightarrow A = 2$; thus, $\frac{5x-13}{(x-3)(x-2)} = \frac{2}{x-3} + \frac{3}{x-2}$

2. $\frac{5x-7}{x^2-3x+2} = \frac{5x-7}{(x-2)(x-1)} = \frac{A}{x-2} + \frac{B}{x-1} \Rightarrow 5x-7 = A(x-1) + B(x-2) = (A+B)x - (A+2B)$
$\Rightarrow \left.\begin{array}{c} A+B=5 \\ A+2B=7 \end{array}\right\} \Rightarrow B=2 \Rightarrow A=3$; thus, $\frac{5x-7}{x^2-3x+2} = \frac{3}{x-2} + \frac{2}{x-1}$

3. $\frac{x+4}{(x+1)^2} = \frac{A}{x+1} + \frac{B}{(x+1)^2} \Rightarrow x+4 = A(x+1) + B = Ax + (A+B) \Rightarrow \left.\begin{array}{c} A=1 \\ A+B=4 \end{array}\right\} \Rightarrow A=1$ and $B=3$;
thus, $\frac{x+4}{(x+1)^2} = \frac{1}{x+1} + \frac{3}{(x+1)^2}$

4. $\frac{2x+2}{x^2-2x+1} = \frac{2x+2}{(x-1)^2} = \frac{A}{x-1} + \frac{B}{(x-1)^2} \Rightarrow 2x+2 = A(x-1) + B = Ax + (-A+B) \Rightarrow \left.\begin{array}{c} A=2 \\ -A+B=2 \end{array}\right\}$
$\Rightarrow A=2$ and $B=4$; thus, $\frac{2x+2}{x^2-2x+1} = \frac{2}{x-1} + \frac{4}{(x-1)^2}$

5. $\frac{z+1}{z^2(z-1)} = \frac{A}{z} + \frac{B}{z^2} + \frac{C}{z-1} \Rightarrow z+1 = Az(z-1) + B(z-1) + Cz^2 \Rightarrow z+1 = (A+C)z^2 + (-A+B)z - B$
$\Rightarrow \left.\begin{array}{c} A+C=0 \\ -A+B=1 \\ -B=1 \end{array}\right\} \Rightarrow B=-1 \Rightarrow A=-2 \Rightarrow C=2$; thus, $\frac{z+1}{z^2(z-1)} = \frac{-2}{z} + \frac{-1}{z^2} + \frac{2}{z-1}$

6. $\frac{z}{z^3-z^2-6z} = \frac{1}{z^2-z-6} = \frac{1}{(z-3)(z+2)} = \frac{A}{z-3} + \frac{B}{z+2} \Rightarrow 1 = A(z+2) + B(z-3) = (A+B)z + (2A-3B)$
$\Rightarrow \left.\begin{array}{c} A+B=0 \\ 2A-3B=1 \end{array}\right\} \Rightarrow -5B=1 \Rightarrow B=-\frac{1}{5} \Rightarrow A=\frac{1}{5}$; thus, $\frac{z}{z^3-z^2-6z} = \frac{\frac{1}{5}}{z-3} + \frac{-\frac{1}{5}}{z+2}$

7. $\frac{t^2+8}{t^2-5t+6} = 1 + \frac{5t+2}{t^2-5t+6}$ (after long division); $\frac{5t+2}{t^2-5t+6} = \frac{5t+2}{(t-3)(t-2)} = \frac{A}{t-3} + \frac{B}{t-2}$
$\Rightarrow 5t+2 = A(t-2) + B(t-3) = (A+B)t + (-2A-3B) \Rightarrow \left.\begin{array}{c} A+B=5 \\ -2A-3B=2 \end{array}\right\} \Rightarrow -B = (10+2) = 12$
$\Rightarrow B=-12 \Rightarrow A=17$; thus, $\frac{t^2+8}{t^2-5t+6} = 1 + \frac{17}{t-3} + \frac{-12}{t-2}$

8. $\frac{t^4+9}{t^4+9t^2} = 1 + \frac{-9t^2+9}{t^4+9t^2} = 1 + \frac{-9t^2+9}{t^2(t^2+9)}$ (after long division); $\frac{-9t^2+9}{t^2(t^2+9)} = \frac{A}{t} + \frac{B}{t^2} + \frac{Ct+D}{t^2+9}$
$\Rightarrow -9t^2+9 = At(t^2+9) + B(t^2+9) + (Ct+D)t^2 = (A+C)t^3 + (B+D)t^2 + 9At + 9B$
$\Rightarrow \left.\begin{array}{c} A+C=0 \\ B+D=-9 \\ 9A=0 \\ 9B=9 \end{array}\right\} \Rightarrow A=0 \Rightarrow C=0; B=1 \Rightarrow D=-10$; thus, $\frac{t^4+9}{t^4+9t^2} = 1 + \frac{1}{t^2} + \frac{-10}{t^2+9}$

9. $\frac{1}{1-x^2} = \frac{A}{1-x} + \frac{B}{1+x} \Rightarrow 1 = A(1+x) + B(1-x)$; $x=1 \Rightarrow A=\frac{1}{2}$; $x=-1 \Rightarrow B=\frac{1}{2}$;
$\int \frac{dx}{1-x^2} = \frac{1}{2}\int \frac{dx}{1-x} + \frac{1}{2}\int \frac{dx}{1+x} = \frac{1}{2}\left[\ln|1+x| - \ln|1-x|\right] + C$

10. $\frac{1}{x^2+2x} = \frac{A}{x} + \frac{B}{x+2} \Rightarrow 1 = A(x+2) + Bx$; $x=0 \Rightarrow A=\frac{1}{2}$; $x=-2 \Rightarrow B=-\frac{1}{2}$;
$\int \frac{dx}{x^2+2x} = \frac{1}{2}\int \frac{dx}{x} - \frac{1}{2}\int \frac{dx}{x+2} = \frac{1}{2}\left[\ln|x| - \ln|x+2|\right] + C$

11. $\frac{x+4}{x^2+5x-6} = \frac{A}{x+6} + \frac{B}{x-1} \Rightarrow x+4 = A(x-1) + B(x+6)$; $x=1 \Rightarrow B=\frac{5}{7}$; $x=-6 \Rightarrow A=\frac{-2}{-7}=\frac{2}{7}$;
$\int \frac{x+4}{x^2+5x-6}\,dx = \frac{2}{7}\int \frac{dx}{x+6} + \frac{5}{7}\int \frac{dx}{x-1} = \frac{2}{7}\ln|x+6| + \frac{5}{7}\ln|x-1| + C = \frac{1}{7}\ln\left|(x+6)^2(x-1)^5\right| + C$

12. $\frac{2x+1}{x^2-7x+12} = \frac{A}{x-4} + \frac{B}{x-3} \Rightarrow 2x+1 = A(x-3) + B(x-4)$; $x=3 \Rightarrow B=\frac{7}{-1}=-7$; $x=4 \Rightarrow A=\frac{9}{1}=9$;
$\int \frac{2x+1}{x^2-7x+12}\,dx = 9\int \frac{dx}{x-4} - 7\int \frac{dx}{x-3} = 9\ln|x-4| - 7\ln|x-3| + C = \ln\left|\frac{(x-4)^9}{(x-3)^7}\right| + C$

13. $\frac{y}{y^2-2y-3} = \frac{A}{y-3} + \frac{B}{y+1} \Rightarrow y = A(y+1) + B(y-3);\ y = -1 \Rightarrow B = \frac{-1}{-4} = \frac{1}{4};\ y = 3 \Rightarrow A = \frac{3}{4};$
$\int_4^8 \frac{y\,dy}{y^2-2y-3} = \frac{3}{4}\int_4^8 \frac{dy}{y-3} + \frac{1}{4}\int_4^8 \frac{dy}{y+1} = \left[\frac{3}{4}\ln|y-3| + \frac{1}{4}\ln|y+1|\right]_4^8 = \left(\frac{3}{4}\ln 5 + \frac{1}{4}\ln 9\right) - \left(\frac{3}{4}\ln 1 + \frac{1}{4}\ln 5\right)$
$= \frac{1}{2}\ln 5 + \frac{1}{2}\ln 3 = \frac{\ln 15}{2}$

14. $\frac{y+4}{y^2+y} = \frac{A}{y} + \frac{B}{y+1} \Rightarrow y+4 = A(y+1) + By;\ y = 0 \Rightarrow A = 4;\ y = -1 \Rightarrow B = \frac{3}{-1} = -3;$
$\int_{1/2}^1 \frac{y+4}{y^2+y}\,dy = 4\int_{1/2}^1 \frac{dy}{y} - 3\int_{1/2}^1 \frac{dy}{y+1} = \left[4\ln|y| - 3\ln|y+1|\right]_{1/2}^1 = (4\ln 1 - 3\ln 2) - \left(4\ln\frac{1}{2} - 3\ln\frac{3}{2}\right)$
$= \ln\frac{1}{8} - \ln\frac{1}{16} + \ln\frac{27}{8} = \ln\left(\frac{27}{8}\cdot\frac{1}{8}\cdot 16\right) = \ln\frac{27}{4}$

15. $\frac{1}{t^3+t^2-2t} = \frac{A}{t} + \frac{B}{t+2} + \frac{C}{t-1} \Rightarrow 1 = A(t+2)(t-1) + Bt(t-1) + Ct(t+2);\ t = 0 \Rightarrow A = -\frac{1}{2};\ t = -2$
$\Rightarrow B = \frac{1}{6};\ t = 1 \Rightarrow C = \frac{1}{3};\ \int \frac{dt}{t^3+t^2-2t} = -\frac{1}{2}\int \frac{dt}{t} + \frac{1}{6}\int \frac{dt}{t+2} + \frac{1}{3}\int \frac{dt}{t-1}$
$= -\frac{1}{2}\ln|t| + \frac{1}{6}\ln|t+2| + \frac{1}{3}\ln|t-1| + C$

16. $\frac{x+3}{2x^3-8x} = \frac{A}{x} + \frac{B}{x+2} + \frac{C}{x-2} \Rightarrow \frac{1}{2}(x+3) = A(x+2)(x-2) + Bx(x-2) + Cx(x+2);\ x = 0 \Rightarrow A = \frac{3}{-8};\ x = -2$
$\Rightarrow B = \frac{1}{16};\ x = 2 \Rightarrow C = \frac{5}{16};\ \int \frac{x+3}{2x^3-8x}\,dx = -\frac{3}{8}\int \frac{dx}{x} + \frac{1}{16}\int \frac{dx}{x+2} + \frac{5}{16}\int \frac{dx}{x-2}$
$= -\frac{3}{8}\ln|x| + \frac{1}{16}\ln|x+2| + \frac{5}{16}\ln|x-2| + C = \frac{1}{16}\ln\left|\frac{(x-2)^5(x+2)}{x^6}\right| + C$

17. $\frac{x^3}{x^2+2x+1} = (x-2) + \frac{3x+2}{(x+1)^2}$ (after long division); $\frac{3x+2}{(x+1)^2} = \frac{A}{x+1} + \frac{B}{(x+1)^2} \Rightarrow 3x+2 = A(x+1) + B$
$= Ax + (A+B) \Rightarrow A = 3,\ A+B = 2 \Rightarrow A = 3,\ B = -1;\ \int_0^1 \frac{x^3\,dx}{x^2+2x+1}$
$= \int_0^1 (x-2)\,dx + 3\int_0^1 \frac{dx}{x+1} - \int_0^1 \frac{dx}{(x+1)^2} = \left[\frac{x^2}{2} - 2x + 3\ln|x+1| + \frac{1}{x+1}\right]_0^1$
$= \left(\frac{1}{2} - 2 + 3\ln 2 + \frac{1}{2}\right) - (1) = 3\ln 2 - 2$

18. $\frac{x^3}{x^2-2x+1} = (x+2) + \frac{3x-2}{(x-1)^2}$ (after long division); $\frac{3x-2}{(x-1)^2} = \frac{A}{x-1} + \frac{B}{(x-1)^2} \Rightarrow 3x-2 = A(x-1) + B$
$= Ax + (-A+B) \Rightarrow A = 3,\ -A+B = -2 \Rightarrow A = 3,\ B = 1;\ \int_{-1}^0 \frac{x^3\,dx}{x^2-2x+1}$
$= \int_{-1}^0 (x+2)\,dx + 3\int_{-1}^0 \frac{dx}{x-1} + \int_{-1}^0 \frac{dx}{(x-1)^2} = \left[\frac{x^2}{2} + 2x + 3\ln|x-1| - \frac{1}{x-1}\right]_{-1}^0$
$= \left(0 + 0 + 3\ln 1 - \frac{1}{(-1)}\right) - \left(\frac{1}{2} - 2 + 3\ln 2 - \frac{1}{(-2)}\right) = 2 - 3\ln 2$

19. $\frac{1}{(x^2-1)^2} = \frac{A}{x+1} + \frac{B}{x-1} + \frac{C}{(x+1)^2} + \frac{D}{(x-1)^2} \Rightarrow 1 = A(x+1)(x-1)^2 + B(x-1)(x+1)^2 + C(x-1)^2 + D(x+1)^2;$
$x = -1 \Rightarrow C = \frac{1}{4};\ x = 1 \Rightarrow D = \frac{1}{4}$; coefficient of $x^3 = A + B \Rightarrow A + B = 0$; constant $= A - B + C + D$
$\Rightarrow A - B + C + D = 1 \Rightarrow A - B = \frac{1}{2}$; thus, $A = \frac{1}{4} \Rightarrow B = -\frac{1}{4};\ \int \frac{dx}{(x^2-1)^2}$
$= \frac{1}{4}\int \frac{dx}{x+1} - \frac{1}{4}\int \frac{dx}{x-1} + \frac{1}{4}\int \frac{dx}{(x+1)^2} + \frac{1}{4}\int \frac{dx}{(x-1)^2} = \frac{1}{4}\ln\left|\frac{x+1}{x-1}\right| - \frac{x}{2(x^2-1)} + C$

20. $\frac{x^2}{(x-1)(x^2+2x+1)} = \frac{A}{x-1} + \frac{B}{x+1} + \frac{C}{(x+1)^2} \Rightarrow x^2 = A(x+1)^2 + B(x-1)(x+1) + C(x-1);\ x = -1$
$\Rightarrow C = -\frac{1}{2};\ x = 1 \Rightarrow A = \frac{1}{4}$; coefficient of $x^2 = A + B \Rightarrow A + B = 1 \Rightarrow B = \frac{3}{4};\ \int \frac{x^2\,dx}{(x-1)(x^2+2x+1)}$
$= \frac{1}{4}\int \frac{dx}{x-1} + \frac{3}{4}\int \frac{dx}{x+1} - \frac{1}{2}\int \frac{dx}{(x+1)^2} = \frac{1}{4}\ln|x-1| + \frac{3}{4}\ln|x+1| + \frac{1}{2(x+1)} + C$
$= \frac{\ln|(x-1)(x+1)^3|}{4} + \frac{1}{2(x+1)} + C$

21. $\frac{1}{(x+1)(x^2+1)} = \frac{A}{x+1} + \frac{Bx+C}{x^2+1} \Rightarrow 1 = A(x^2+1) + (Bx+C)(x+1);\ x = -1 \Rightarrow A = \frac{1}{2}$; coefficient of x^2
$= A + B \Rightarrow A + B = 0 \Rightarrow B = -\frac{1}{2}$; constant $= A + C \Rightarrow A + C = 1 \Rightarrow C = \frac{1}{2};\ \int_0^1 \frac{dx}{(x+1)(x^2+1)}$

$= \frac{1}{2}\int_0^1 \frac{dx}{x+1} + \frac{1}{2}\int_0^1 \frac{(-x+1)}{x^2+1}\,dx = \left[\frac{1}{2}\ln|x+1| - \frac{1}{4}\ln(x^2+1) + \frac{1}{2}\tan^{-1}x\right]_0^1$

$= \left(\frac{1}{2}\ln 2 - \frac{1}{4}\ln 2 + \frac{1}{2}\tan^{-1}1\right) - \left(\frac{1}{2}\ln 1 - \frac{1}{4}\ln 1 + \frac{1}{2}\tan^{-1}0\right) = \frac{1}{4}\ln 2 + \frac{1}{2}\left(\frac{\pi}{4}\right) = \frac{(\pi + 2\ln 2)}{8}$

22. $\frac{3t^2+t+4}{t^3+t} = \frac{A}{t} + \frac{Bt+C}{t^2+1} \Rightarrow 3t^2+t+4 = A(t^2+1) + (Bt+C)t$; $t=0 \Rightarrow A=4$; coefficient of t^2

$= A + B \Rightarrow A + B = 3 \Rightarrow B = -1$; coefficient of $t = C \Rightarrow C = 1$; $\int_1^{\sqrt{3}} \frac{3t^2+t+4}{t^3+1}\,dt$

$= 4\int_1^{\sqrt{3}} \frac{dt}{t} + \int_1^{\sqrt{3}} \frac{(-t+1)}{t^2+1}\,dt = \left[4\ln|t| - \frac{1}{2}\ln(t^2+1) + \tan^{-1}t\right]_1^{\sqrt{3}}$

$= \left(4\ln\sqrt{3} - \frac{1}{2}\ln 4 + \tan^{-1}\sqrt{3}\right) - \left(4\ln 1 - \frac{1}{2}\ln 2 + \tan^{-1}1\right) = 2\ln 3 - \ln 2 + \frac{\pi}{3} + \frac{1}{2}\ln 2 - \frac{\pi}{4}$

$= 2\ln 3 - \frac{1}{2}\ln 2 + \frac{\pi}{12} = \ln\left(\frac{9}{\sqrt{2}}\right) + \frac{\pi}{12}$

23. $\frac{y^2+2y+1}{(y^2+1)^2} = \frac{Ay+B}{y^2+1} + \frac{Cy+D}{(y^2+1)^2} \Rightarrow y^2+2y+1 = (Ay+B)(y^2+1) + Cy + D$

$= Ay^3 + By^2 + (A+C)y + (B+D) \Rightarrow A = 0, B = 1; A + C = 2 \Rightarrow C = 2; B + D = 1 \Rightarrow D = 0;$

$\int \frac{y^2+2y+1}{(y^2+1)^2}\,dy = \int \frac{1}{y^2+1}\,dy + 2\int \frac{y}{(y^2+1)^2}\,dy = \tan^{-1}y - \frac{1}{y^2+1} + C$

24. $\frac{8x^2+8x+2}{(4x^2+1)^2} = \frac{Ax+B}{4x^2+1} + \frac{Cx+D}{(4x^2+1)^2} \Rightarrow 8x^2+8x+2 = (Ax+B)(4x^2+1) + Cx + D$

$= 4Ax^3 + 4Bx^2 + (A+C)x + (B+D)$; $A = 0, B = 2; A + C = 8 \Rightarrow C = 8; B + D = 2 \Rightarrow D = 0;$

$\int \frac{8x^2+8x+2}{(4x^2+1)^2}\,dx = 2\int \frac{dx}{4x^2+1} + 8\int \frac{x\,dx}{(4x^2+1)^2} = \tan^{-1}2x - \frac{1}{4x^2+1} + C$

25. $\frac{2s+2}{(s^2+1)(s-1)^3} = \frac{As+B}{s^2+1} + \frac{C}{s-1} + \frac{D}{(s-1)^2} + \frac{E}{(s-1)^3} \Rightarrow 2s+2$

$= (As+B)(s-1)^3 + C(s^2+1)(s-1)^2 + D(s^2+1)(s-1) + E(s^2+1)$

$= [As^4 + (-3A+B)s^3 + (3A-3B)s^2 + (-A+3B)s - B] + C(s^4 - 2s^3 + 2s^2 - 2s + 1) + D(s^3 - s^2 + s - 1)$
$+ E(s^2+1)$

$= (A+C)s^4 + (-3A+B-2C+D)s^3 + (3A-3B+2C-D+E)s^2 + (-A+3B-2C+D)s + (-B+C-D+E)$

$$\Rightarrow \left.\begin{array}{l} A + C = 0 \\ -3A + B - 2C + D = 0 \\ 3A - 3B + 2C - D + E = 0 \\ -A + 3B - 2C + D = 2 \\ -B + C - D + E = 2 \end{array}\right\} \text{summing all equations} \Rightarrow 2E = 4 \Rightarrow E = 2;$$

summing eqs (2) and (3) $\Rightarrow -2B + 2 = 0 \Rightarrow B = 1$; summing eqs (3) and (4) $\Rightarrow 2A + 2 = 2 \Rightarrow A = 0; C = 0$ from eq (1); then $-1 + 0 - D + 2 = 2$ from eq (5) $\Rightarrow D = -1$;

$\int \frac{2s+2}{(s^2+1)(s-1)^3}\,ds = \int \frac{ds}{s^2+1} - \int \frac{ds}{(s-1)^2} + 2\int \frac{ds}{(s-1)^3} = -(s-1)^{-2} + (s-1)^{-1} + \tan^{-1}s + C$

26. $\frac{s^4+81}{s(s^2+9)^2} = \frac{A}{s} + \frac{Bs+C}{s^2+9} + \frac{Ds+E}{(s^2+9)^2} \Rightarrow s^4 + 81 = A(s^2+9)^2 + (Bs+C)s(s^2+9) + (Ds+E)s$

$= A(s^4 + 18s^2 + 81) + (Bs^4 + Cs^3 + 9Bs^2 + 9Cs) + Ds^2 + Es$

$= (A+B)s^4 + Cs^3 + (18A + 9B + D)s^2 + (9C + E)s + 81A \Rightarrow 81A = 81$ or $A = 1; A + B = 1 \Rightarrow B = 0;$

$C = 0; 9C + E = 0 \Rightarrow E = 0; 18A + 9B + D = 0 \Rightarrow D = -18$; $\int \frac{s^4+81}{s(s^2+9)^2}\,ds = \int \frac{ds}{s} - 18\int \frac{s\,ds}{(s^2+9)^2}$

$= \ln|s| + \frac{9}{(s^2+9)} + C$

27. $\frac{2\theta^3+5\theta^2+8\theta+4}{(\theta^2+2\theta+2)^2} = \frac{A\theta+B}{\theta^2+2\theta+2} + \frac{C\theta+D}{(\theta^2+2\theta+2)^2} \Rightarrow 2\theta^3 + 5\theta^2 + 8\theta + 4 = (A\theta+B)(\theta^2+2\theta+2) + C\theta + D$

$= A\theta^3 + (2A+B)\theta^2 + (2A+2B+C)\theta + (2B+D) \Rightarrow A = 2; 2A + B = 5 \Rightarrow B = 1; 2A + 2B + C = 8 \Rightarrow C = 2;$

$2B + D = 4 \Rightarrow D = 2$; $\int \frac{2\theta^3+5\theta^2+8\theta+4}{(\theta^2+2\theta+2)^2}\,d\theta = \int \frac{2\theta+1}{(\theta^2+2\theta+2)}\,d\theta + \int \frac{2\theta+2}{(\theta^2+2\theta+2)^2}\,d\theta$

$= \int \frac{2\theta+2}{\theta^2+2\theta+2}\,d\theta - \int \frac{d\theta}{\theta^2+2\theta+2} + \int \frac{d(\theta^2+2\theta+2)}{(\theta^2+2\theta+2)^2} = \int \frac{d(\theta^2+2\theta+2)}{\theta^2+2\theta+2} - \int \frac{d\theta}{(\theta+1)^2+1} - \frac{1}{\theta^2+2\theta+2}$

$= \frac{-1}{\theta^2+2\theta+2} + \ln(\theta^2+2\theta+2) - \tan^{-1}(\theta+1) + C$

28. $\frac{\theta^4-4\theta^3+2\theta^2-3\theta+1}{(\theta^2+1)^3} = \frac{A\theta+B}{\theta^2+1} + \frac{C\theta+D}{(\theta^2+1)^2} + \frac{E\theta+F}{(\theta^2+1)^3} \Rightarrow \theta^4-4\theta^3+2\theta^2-3\theta+1$
$= (A\theta+B)(\theta^2+1)^2 + (C\theta+D)(\theta^2+1) + E\theta + F = (A\theta+B)(\theta^4+2\theta^2+1) + (C\theta^3+D\theta^2+C\theta+D) + E\theta + F$
$= (A\theta^5+B\theta^4+2A\theta^3+2B\theta^2+A\theta+B) + (C\theta^3+D\theta^2+C\theta+D) + E\theta+F$
$= A\theta^5+B\theta^4+(2A+C)\theta^3+(2B+D)\theta^2+(A+C+E)\theta+(B+D+F) \Rightarrow A=0; B=1; 2A+C=-4$
$\Rightarrow C=-4; 2B+D=2 \Rightarrow D=0; A+C+E=-3 \Rightarrow E=1; B+D+F=1 \Rightarrow F=0;$
$\int \frac{\theta^4-4\theta^3+2\theta^2-3\theta+1}{(\theta^2+1)^3}\,d\theta = \int \frac{d\theta}{\theta^2+1} - 4\int \frac{\theta\,d\theta}{(\theta^2+1)^2} + \int \frac{\theta\,d\theta}{(\theta^2+1)^3} = \tan^{-1}\theta + 2(\theta^2+1)^{-1} - \frac{1}{4}(\theta^2+1)^{-2} + C$

29. $\frac{2x^3-2x^2+1}{x^2-x} = 2x + \frac{1}{x^2-x} = 2x + \frac{1}{x(x-1)}$; $\frac{1}{x(x-1)} = \frac{A}{x} + \frac{B}{x-1} \Rightarrow 1 = A(x-1) + Bx; x=0 \Rightarrow A=-1;$
$x=1 \Rightarrow B=1; \int \frac{2x^3-2x^2+1}{x^2-x} = \int 2x\,dx - \int \frac{dx}{x} + \int \frac{dx}{x-1} = x^2 - \ln|x| + \ln|x-1| + C = x^2 + \ln\left|\frac{x-1}{x}\right| + C$

30. $\frac{x^4}{x^2-1} = (x^2+1) + \frac{1}{x^2-1} = (x^2+1) + \frac{1}{(x+1)(x-1)}$; $\frac{1}{(x+1)(x-1)} = \frac{A}{x+1} + \frac{B}{x-1} \Rightarrow 1 = A(x-1) + B(x+1);$
$x=-1 \Rightarrow A=-\frac{1}{2}; x=1 \Rightarrow B=\frac{1}{2}; \int \frac{x^4}{x^2-1}\,dx = \int (x^2+1)\,dx - \frac{1}{2}\int \frac{dx}{x+1} + \frac{1}{2}\int \frac{dx}{x-1}$
$= \frac{1}{3}x^3 + x - \frac{1}{2}\ln|x+1| + \frac{1}{2}\ln|x-1| + C = \frac{x^3}{3} + x + \frac{1}{2}\ln\left|\frac{x-1}{x+1}\right| + C$

31. $\frac{9x^3-3x+1}{x^3-x^2} = 9 + \frac{9x^2-3x+1}{x^2(x-1)}$ (after long division); $\frac{9x^2-3x+1}{x^2(x-1)} = \frac{A}{x} + \frac{B}{x^2} + \frac{C}{x-1}$
$\Rightarrow 9x^2-3x+1 = Ax(x-1) + B(x-1) + Cx^2; x=1 \Rightarrow C=7; x=0 \Rightarrow B=-1; A+C=9 \Rightarrow A=2;$
$\int \frac{9x^3-3x+1}{x^3-x^2}\,dx = \int 9\,dx + 2\int \frac{dx}{x} - \int \frac{dx}{x^2} + 7\int \frac{dx}{x-1} = 9x + 2\ln|x| + \frac{1}{x} + 7\ln|x-1| + C$

32. $\frac{16x^3}{4x^2-4x+1} = (4x+4) + \frac{12x-4}{4x^2-4x+1}$; $\frac{12x-4}{(2x-1)^2} = \frac{A}{2x-1} + \frac{B}{(2x-1)^2} \Rightarrow 12x-4 = A(2x-1) + B$
$\Rightarrow A=6; -A+B=-4 \Rightarrow B=2; \int \frac{16x^3}{4x^2-4x+1}\,dx = 4\int (x+1)\,dx + 6\int \frac{dx}{2x-1} + 2\int \frac{dx}{(2x-1)^2}$
$= 2(x+1)^2 + 3\ln|2x-1| - \frac{1}{2x-1} + C_1 = 2x^2 + 4x + 3\ln|2x-1| - (2x-1)^{-1} + C$, where $C = 2 + C_1$

33. $\frac{y^4+y^2-1}{y^3+y} = y - \frac{1}{y(y^2+1)}$; $\frac{1}{y(y^2+1)} = \frac{A}{y} + \frac{By+C}{y^2+1} \Rightarrow 1 = A(y^2+1) + (By+C)y = (A+B)y^2 + Cy + A$
$\Rightarrow A=1; A+B=0 \Rightarrow B=-1; C=0; \int \frac{y^4+y^2-1}{y^3+y}\,dy = \int y\,dy - \int \frac{dy}{y} + \int \frac{y\,dy}{y^2+1}$
$= \frac{y^2}{2} - \ln|y| + \frac{1}{2}\ln(1+y^2) + C$

34. $\frac{2y^4}{y^3-y^2+y-1} = 2y + 2 + \frac{2}{y^3-y^2+y-1}$; $\frac{2}{y^3-y^2+y-1} = \frac{2}{(y^2+1)(y-1)} = \frac{A}{y-1} + \frac{By+C}{y^2+1}$
$\Rightarrow 2 = A(y^2+1) + (By+C)(y-1) = (Ay^2+A) + (By^2+Cy-By-C) = (A+B)y^2 + (-B+C)y + (A-C)$
$\Rightarrow A+B=0, -B+C=0$ or $C=B, A-C=A-B=2 \Rightarrow A=1, B=-1, C=-1;$
$\int \frac{2y^4}{y^3-y^2+y-1}\,dy = 2\int (y+1)\,dy + \int \frac{dy}{y-1} - \int \frac{y}{y^2+1}\,dy - \int \frac{dy}{y^2+1}$
$= (y+1)^2 + \ln|y-1| - \frac{1}{2}\ln(y^2+1) - \tan^{-1}y + C_1 = y^2 + 2y + \ln|y-1| - \frac{1}{2}\ln(y^2+1) - \tan^{-1}y + C,$
where $C = C_1 + 1$

35. $\int \frac{e^t\,dt}{e^{2t}+3e^t+2} = [e^t = y] \int \frac{dy}{y^2+3y+2} = \int \frac{dy}{y+1} - \int \frac{dy}{y+2} = \ln\left|\frac{y+1}{y+2}\right| + C = \ln\left(\frac{e^t+1}{e^t+2}\right) + C$

36. $\int \frac{e^{4t}+2e^{2t}-e^t}{e^{2t}+1}\,dt = \int \frac{e^{3t}+2e^t-1}{e^{2t}+1}e^t dt; \begin{bmatrix} y = e^t \\ dy = e^t\,dt \end{bmatrix} \to \int \frac{y^3+2y-1}{y^2+1}\,dy = \int \left(y + \frac{y-1}{y^2+1}\right) dy = \frac{y^2}{2} + \int \frac{y}{y^2+1}\,dy - \int \frac{dy}{y^2+1}$
$= \frac{y^2}{2} + \frac{1}{2}\ln(y^2+1) - \tan^{-1}y + C = \frac{1}{2}e^{2t} + \frac{1}{2}\ln(e^{2t}+1) - \tan^{-1}(e^t) + C$

37. $\int \frac{\cos y\, dy}{\sin^2 y + \sin y - 6}$; $[\sin y = t, \cos y\, dy = dt] \rightarrow \int \frac{dy}{t^2+t-6} = \frac{1}{5}\int \left(\frac{1}{t-2} - \frac{1}{t+3}\right) dt = \frac{1}{5} \ln \left|\frac{t-2}{t+3}\right| + C$
$= \frac{1}{5} \ln \left|\frac{\sin y - 2}{\sin y + 3}\right| + C$

38. $\int \frac{\sin\theta\, d\theta}{\cos^2\theta + \cos\theta - 2}$; $[\cos\theta = y] \rightarrow -\int \frac{dy}{y^2+y-2} = \frac{1}{3}\int \frac{dy}{y+2} - \frac{1}{3}\int \frac{dy}{y-1} = \frac{1}{3}\ln\left|\frac{y+2}{y-1}\right| + C = \frac{1}{3}\ln\left|\frac{\cos\theta+2}{\cos\theta-1}\right| + C$
$= \frac{1}{3}\ln\left|\frac{2+\cos\theta}{1-\cos\theta}\right| + C = -\frac{1}{3}\ln\left|\frac{\cos\theta-1}{\cos\theta+2}\right| + C$

39. $\int \frac{(x-2)^2 \tan^{-1}(2x) - 12x^3 - 3x}{(4x^2+1)(x-2)^2}\, dx = \int \frac{\tan^{-1}(2x)}{4x^2+1}\, dx - 3\int \frac{x}{(x-2)^2}\, dx$
$= \frac{1}{2}\int \tan^{-1}(2x)\, d\left(\tan^{-1}(2x)\right) - 3\int \frac{dx}{x-2} - 6\int \frac{dx}{(x-2)^2} = \frac{\left(\tan^{-1} 2x\right)^2}{4} - 3\ln|x-2| + \frac{6}{x-2} + C$

40. $\int \frac{(x+1)^2 \tan^{-1}(3x) + 9x^3 + x}{(9x^2+1)(x+1)^2}\, dx = \int \frac{\tan^{-1}(3x)}{9x^2+1}\, dx + \int \frac{x}{(x+1)^2}\, dx$
$= \frac{1}{3}\int \tan^{-1}(3x)\, d\left(\tan^{-1}(3x)\right) + \int \frac{dx}{x+1} - \int \frac{dx}{(x+1)^2} = \frac{\left(\tan^{-1} 3x\right)^2}{6} + \ln|x+1| + \frac{1}{x+1} + C$

41. $(t^2 - 3t + 2)\frac{dx}{dt} = 1$; $x = \int \frac{dt}{t^2-3t+2} = \int \frac{dt}{t-2} - \int \frac{dt}{t-1} = \ln\left|\frac{t-2}{t-1}\right| + C$; $\frac{t-2}{t-1} = Ce^x$; $t = 3$ and $x = 0$
$\Rightarrow \frac{1}{2} = C \Rightarrow \frac{t-2}{t-1} = \frac{1}{2}e^x \Rightarrow x = \ln\left|2\left(\frac{t-2}{t-1}\right)\right| = \ln|t-2| - \ln|t-1| + \ln 2$

42. $(3t^4 + 4t^2 + 1)\frac{dx}{dt} = 2\sqrt{3}$; $x = 2\sqrt{3}\int \frac{dt}{3t^4+4t^2+1} = \sqrt{3}\int \frac{dt}{t^2+\frac{1}{3}} - \sqrt{3}\int \frac{dt}{t^2+1}$
$= 3\tan^{-1}\left(\sqrt{3}t\right) - \sqrt{3}\tan^{-1} t + C$; $t = 1$ and $x = \frac{-\pi\sqrt{3}}{4} \Rightarrow -\frac{\sqrt{3}\pi}{4} = \pi - \frac{\sqrt{3}}{4}\pi + C \Rightarrow C = -\pi$
$\Rightarrow x = 3\tan^{-1}\left(\sqrt{3}t\right) - \sqrt{3}\tan^{-1} t - \pi$

43. $(t^2 + 2t)\frac{dx}{dt} = 2x + 2$; $\frac{1}{2}\int \frac{dx}{x+1} = \int \frac{dt}{t^2+2t} \Rightarrow \frac{1}{2}\ln|x+1| = \frac{1}{2}\int \frac{dt}{t} - \frac{1}{2}\int \frac{dt}{t+2} \Rightarrow \ln|x+1| = \ln\left|\frac{t}{t+2}\right| + C$;
$t = 1$ and $x = 1 \Rightarrow \ln 2 = \ln\frac{1}{3} + C \Rightarrow C = \ln 2 + \ln 3 = \ln 6 \Rightarrow \ln|x+1| = \ln 6\left|\frac{t}{t+2}\right| \Rightarrow x + 1 = \frac{6t}{t+2}$
$\Rightarrow x = \frac{6t}{t+2} - 1, t > 0$

44. $(t+1)\frac{dx}{dt} = x^2 + 1 \Rightarrow \int \frac{dx}{x^2+1} = \int \frac{dt}{t+1} \Rightarrow \tan^{-1} x = \ln|t+1| + C$; $t = 0$ and $x = \frac{\pi}{4} \Rightarrow \tan^{-1}\frac{\pi}{4} = \ln|1| + C$
$\Rightarrow C = \tan^{-1}\frac{\pi}{4} = 1 \Rightarrow \tan^{-1} x = \ln|t+1| + 1 \Rightarrow x = \tan(\ln(t+1) + 1), t > -1$

45. $V = \pi\int_{0.5}^{2.5} y^2\, dx = \pi\int_{0.5}^{2.5} \frac{9}{3x-x^2}\, dx = 3\pi\left(\int_{0.5}^{2.5}\left(-\frac{1}{x-3} + \frac{1}{x}\right)\right) dx = \left[3\pi\ln\left|\frac{x}{x-3}\right|\right]_{0.5}^{2.5} = 3\pi\ln 25$

46. $V = 2\pi\int_0^1 xy\, dx = 2\pi\int_0^1 \frac{2x}{(x+1)(2-x)}\, dx = 4\pi\int_0^1\left(-\frac{1}{3}\left(\frac{1}{x+1}\right) + \frac{2}{3}\left(\frac{1}{2-x}\right)\right) dx$
$= \left[-\frac{4\pi}{3}\left(\ln|x+1| + 2\ln|2-x|\right)\right]_0^1 = \frac{4\pi}{3}(\ln 2)$

47. $A = \int_0^{\sqrt{3}} \tan^{-1} x\, dx = \left[x\tan^{-1} x\right]_0^{\sqrt{3}} - \int_0^{\sqrt{3}} \frac{x}{1+x^2}\, dx$
$= \frac{\pi\sqrt{3}}{3} - \left[\frac{1}{2}\ln(x^2+1)\right]_0^{\sqrt{3}} = \frac{\pi\sqrt{3}}{3} - \ln 2$;
$\overline{x} = \frac{1}{A}\int_0^{\sqrt{3}} x\tan^{-1} x\, dx$
$= \frac{1}{A}\left(\left[\frac{1}{2}x^2\tan^{-1} x\right]_0^{\sqrt{3}} - \frac{1}{2}\int_0^{\sqrt{3}} \frac{x^2}{1+x^2}\, dx\right)$
$= \frac{1}{A}\left[\frac{\pi}{2} - \left[\frac{1}{2}\left(x - \tan^{-1} x\right)\right]_0^{\sqrt{3}}\right]$
$= \frac{1}{A}\left(\frac{\pi}{2} - \frac{\sqrt{3}}{2} + \frac{\pi}{6}\right) = \frac{1}{A}\left(\frac{2\pi}{3} - \frac{\sqrt{3}}{2}\right) \cong 1.10$

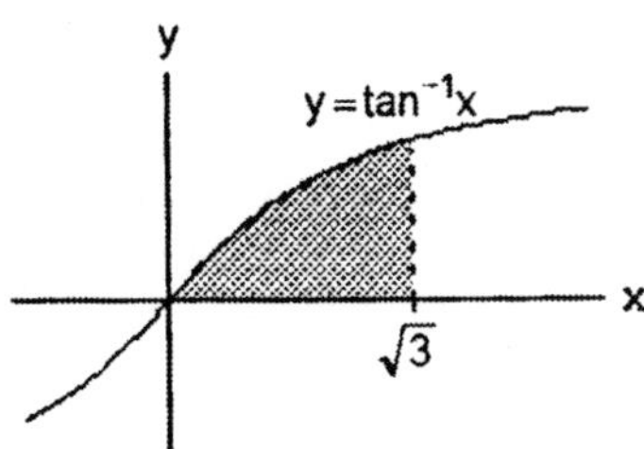

48. $A = \int_3^5 \frac{4x^2+13x-9}{x^3+2x^2-3x}\,dx = 3\int_3^5 \frac{dx}{x} - \int_3^5 \frac{dx}{x+3} + 2\int_3^5 \frac{dx}{x-1} = [3\ln|x| - \ln|x+3| + 2\ln|x-1|]_3^5 = \ln\frac{125}{9}$;
$\overline{x} = \frac{1}{A}\int_3^5 \frac{x(4x^2+13x-9)}{x^3+2x^2-3x}\,dx = \frac{1}{A}\left([4x]_3^5 + 3\int_3^5 \frac{dx}{x+3} + 2\int_3^5 \frac{dx}{x-1}\right) = \frac{1}{A}(8 + 11\ln 2 - 3\ln 6) \cong 3.90$

49. (a) $\frac{dx}{dt} = kx(N-x) \Rightarrow \int \frac{dx}{x(N-x)} = \int k\,dt \Rightarrow \frac{1}{N}\int \frac{dx}{x} + \frac{1}{N}\int \frac{dx}{N-x} = \int k\,dt \Rightarrow \frac{1}{N}\ln\left|\frac{x}{N-x}\right| = kt + C$;
$k = \frac{1}{250}$, $N = 1000$, $t = 0$ and $x = 2 \Rightarrow \frac{1}{1000}\ln\left|\frac{2}{998}\right| = C \Rightarrow \frac{1}{1000}\ln\left|\frac{x}{1000-x}\right| = \frac{t}{250} + \frac{1}{1000}\ln\left(\frac{1}{499}\right)$
$\Rightarrow \ln\left|\frac{499x}{1000-x}\right| = 4t \Rightarrow \frac{499x}{1000-x} = e^{4t} \Rightarrow 499x = e^{4t}(1000-x) \Rightarrow (499+e^{4t})\,x = 1000e^{4t} \Rightarrow x = \frac{1000e^{4t}}{499+e^{4t}}$
(b) $x = \frac{1}{2}N = 500 \Rightarrow 500 = \frac{1000e^{4t}}{499+e^{4t}} \Rightarrow 500 \cdot 499 + 500e^{4t} = 1000e^{4t} \Rightarrow e^{4t} = 499 \Rightarrow t = \frac{1}{4}\ln 499 \approx 1.55$ days

50. $\frac{dx}{dt} = k(a-x)(b-x) \Rightarrow \frac{dx}{(a-x)(b-x)} = k\,dt$
(a) $a = b$: $\int \frac{dx}{(a-x)^2} = \int k\,dt \Rightarrow \frac{1}{a-x} = kt + C$; $t = 0$ and $x = 0 \Rightarrow \frac{1}{a} = C \Rightarrow \frac{1}{a-x} = kt + \frac{1}{a}$
$\Rightarrow \frac{1}{a-x} = \frac{akt+1}{a} \Rightarrow a - x = \frac{a}{akt+1} \Rightarrow x = a - \frac{a}{akt+1} = \frac{a^2kt}{akt+1}$
(b) $a \neq b$: $\int \frac{dx}{(a-x)(b-x)} = \int k\,dt \Rightarrow \frac{1}{b-a}\int \frac{dx}{a-x} - \frac{1}{b-a}\int \frac{dx}{b-x} = \int k\,dt \Rightarrow \frac{1}{b-a}\ln\left|\frac{b-x}{a-x}\right| = kt + C$;
$t = 0$ and $x = 0 \Rightarrow \frac{1}{b-a}\ln\frac{b}{a} = C \Rightarrow \ln\left|\frac{b-x}{a-x}\right| = (b-a)kt + \ln\left(\frac{b}{a}\right) \Rightarrow \frac{b-x}{a-x} = \frac{b}{a}e^{(b-a)kt}$
$\Rightarrow x = \frac{ab\left[1 - e^{(b-a)kt}\right]}{a - be^{(b-a)kt}}$

51. (a) $\int_0^1 \frac{x^4(x-1)^4}{x^2+1}\,dx = \int_0^1 \left(x^6 - 4x^5 + 5x^4 - 4x^2 + 4 - \frac{4}{x^2+1}\right)dx = \frac{22}{7} - \pi$
(b) $\frac{\frac{22}{7}-\pi}{\pi} \cdot 100\% \cong 0.04\%$
(c) The area is less than 0.003

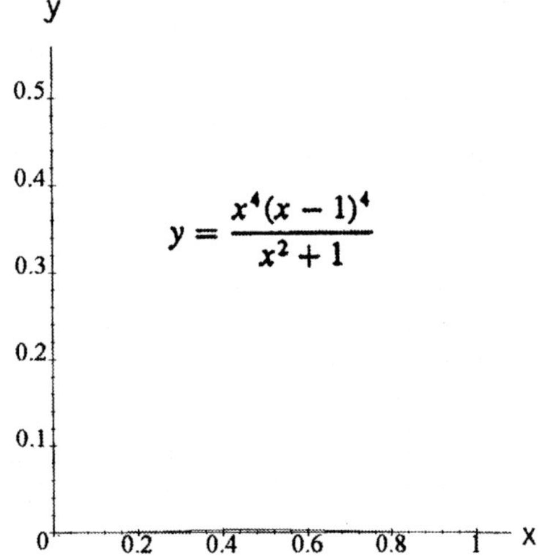

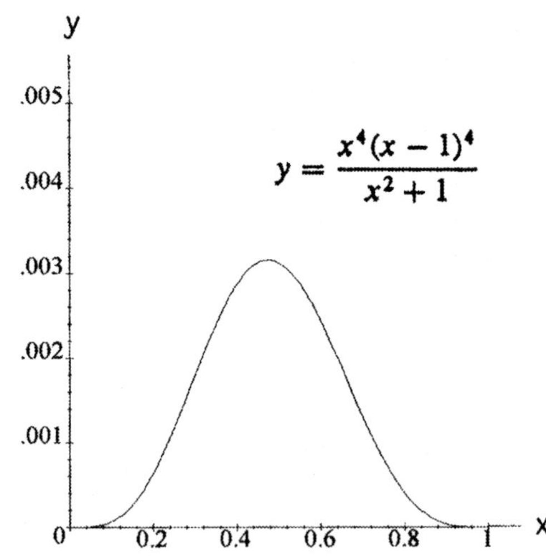

52. $P(x) = ax^2 + bx + c$, $P(0) = c = 1$ and $P'(0) = 0 \Rightarrow b = 0 \Rightarrow P(x) = ax^2 + 1$. Next,
$\frac{ax^2+1}{x^3(x-1)^2} = \frac{A}{x} + \frac{B}{x^2} + \frac{C}{x^3} + \frac{D}{x-1} + \frac{E}{(x-1)^2}$; for the integral to be a rational function, we must have $A = 0$ and $D = 0$. Thus, $ax^2 + 1 = Bx(x-1)^2 + C(x-1)^2 + Ex^3 = (B+E)x^3 + (C-2B)x^2 + (B-2C)x + C$
$\Rightarrow \left.\begin{array}{r} B + E = 0 \\ C - 2B = a \\ C = 1 \end{array}\right\} \Rightarrow E = -B$; $x = 1 \Rightarrow a + 1 = E$; therefore, $1 - 2B = a \Rightarrow 1 + 2E = a \Rightarrow 1 + 2(a+1) = a$
$\Rightarrow a = -3$

8.4 TRIGONOMETRIC INTEGRALS

1. $\int_0^{\pi/2} \sin^5 x\,dx = \int_0^{\pi/2} (\sin^2 x)^2 \sin x\,dx = \int_0^{\pi/2} (1-\cos^2 x)^2 \sin x\,dx = \int_0^{\pi/2} (1-2\cos^2 x+\cos^4 x)\sin x\,dx$
$= \int_0^{\pi/2} \sin x\,dx - \int_0^{\pi/2} 2\cos^2 x \sin x\,dx + \int_0^{\pi/2} \cos^4 x \sin x\,dx = \left[-\cos x + 2\frac{\cos^3 x}{3} - \frac{\cos^5 x}{5}\right]_0^{\pi/2}$
$= (0) - \left(-1+\frac{2}{3}-\frac{1}{5}\right) = \frac{8}{15}$

2. $\int_0^{\pi} \sin^5\left(\frac{x}{2}\right)dx$ (using Exercise 1) $= \int_0^{\pi} \sin\left(\frac{x}{2}\right)dx - \int_0^{\pi} 2\cos^2\left(\frac{x}{2}\right)\sin\left(\frac{x}{2}\right)dx + \int_0^{\pi} \cos^4\left(\frac{x}{2}\right)\sin\left(\frac{x}{2}\right)dx$
$= \left[-2\cos\left(\frac{x}{2}\right) + \frac{4}{3}\cos^3\left(\frac{x}{2}\right) - \frac{2}{5}\cos^5\left(\frac{x}{2}\right)\right]_0^{\pi} = (0) - \left(-2+\frac{4}{3}-\frac{2}{5}\right) = \frac{16}{15}$

3. $\int_{-\pi/2}^{\pi/2} \cos^3 x\,dx = \int_{-\pi/2}^{\pi/2} (\cos^2 x)\cos x\,dx = \int_{-\pi/2}^{\pi/2} (1-\sin^2 x)\cos x\,dx = \int_{-\pi/2}^{\pi/2} \cos x\,dx - \int_{-\pi/2}^{\pi/2} \sin^2 x \cos x\,dx$
$= \left[\sin x - \frac{\sin^3 x}{3}\right]_{-\pi/2}^{\pi/2} = \left(1-\frac{1}{3}\right) - \left(-1+\frac{1}{3}\right) = \frac{4}{3}$

4. $\int_0^{\pi/6} 3\cos^5 3x\,dx = \int_0^{\pi/6} (\cos^2 3x)^2 \cos 3x \cdot 3dx = \int_0^{\pi/6} (1-\sin^2 3x)^2 \cos 3x \cdot 3dx = \int_0^{\pi/6} (1-2\sin^2 3x+\sin^4 3x)\cos 3x \cdot 3dx$
$= \int_0^{\pi/6} \cos 3x \cdot 3dx - 2\int_0^{\pi/6} \sin^2 3x \cos 3x \cdot 3dx + \int_0^{\pi/6} \sin^4 3x \cos 3x \cdot 3dx = \left[\sin 3x - 2\frac{\sin^3 3x}{3} + \frac{\sin^5 3x}{5}\right]_0^{\pi/6}$
$= \left(1-\frac{2}{3}+\frac{1}{5}\right) - (0) = \frac{8}{15}$

5. $\int_0^{\pi/2} \sin^7 y\,dy = \int_0^{\pi/2} \sin^6 y \sin y\,dy = \int_0^{\pi/2} (1-\cos^2 y)^3 \sin y\,dy = \int_0^{\pi/2} \sin y\,dy - 3\int_0^{\pi/2} \cos^2 y \sin y\,dy$
$+ 3\int_0^{\pi/2} \cos^4 y \sin y\,dy - \int_0^{\pi/2} \cos^6 y \sin y\,dy = \left[-\cos y + 3\frac{\cos^3 y}{3} - 3\frac{\cos^5 y}{5} + \frac{\cos^7 y}{7}\right]_0^{\pi/2} = (0) - \left(-1+1-\frac{3}{5}+\frac{1}{7}\right) = \frac{16}{35}$

6. $\int_0^{\pi/2} 7\cos^7 t\,dt$ (using Exercise 5) $= 7\left[\int_0^{\pi/2} \cos t\,dt - 3\int_0^{\pi/2} \sin^2 t \cos t\,dt + 3\int_0^{\pi/2} \sin^4 t \cos t\,dt - \int_0^{\pi/2} \sin^6 t \cos t\,dt\right]$
$= 7\left[\sin t - 3\frac{\sin^3 t}{3} + 3\frac{\sin^5 t}{5} - \frac{\sin^7 t}{7}\right]_0^{\pi/2} = 7\left(1-1+\frac{3}{5}-\frac{1}{7}\right) - 7(0) = \frac{16}{5}$

7. $\int_0^{\pi} 8\sin^4 x\,dx = 8\int_0^{\pi} \left(\frac{1-\cos 2x}{2}\right)^2 dx = 2\int_0^{\pi} (1-2\cos 2x+\cos^2 2x)dx = 2\int_0^{\pi} dx - 2\int_0^{\pi} \cos 2x \cdot 2dx + 2\int_0^{\pi} \frac{1+\cos 4x}{2}\,dx$
$= [2x - 2\sin 2x]_0^{\pi} + \int_0^{\pi} dx + \int_0^{\pi} \cos 4x\,dx = 2\pi + \left[x + \frac{1}{4}\sin 4x\right]_0^{\pi} = 2\pi + \pi = 3\pi$

8. $\int_0^1 8\cos^4 2\pi x\,dx = 8\int_0^1 \left(\frac{1+\cos 4\pi x}{2}\right)^2 dx = 2\int_0^1 (1+2\cos 4\pi x+\cos^2 4\pi x)dx = 2\int_0^1 dx + 4\int_0^1 \cos 4\pi x\,dx + 2\int_0^1 \frac{1+\cos 8\pi x}{2}\,dx$
$= \left[2x + \frac{1}{\pi}\sin 4\pi x\right]_0^1 + \int_0^1 dx + \int_0^1 \cos 8\pi x\,dx = 2 + \left[x + \frac{1}{8\pi}\sin 8\pi x\right]_0^1 = 2 + 1 = 3$

9. $\int_{-\pi/4}^{\pi/4} 16\sin^2 x \cos^2 x\,dx = 16\int_{-\pi/4}^{\pi/4} \left(\frac{1-\cos 2x}{2}\right)\left(\frac{1+\cos 2x}{2}\right)dx = 4\int_{-\pi/4}^{\pi/4} (1-\cos^2 2x)dx = 4\int_{-\pi/4}^{\pi/4} dx - 4\int_{-\pi/4}^{\pi/4} \left(\frac{1+\cos 4x}{2}\right)dx$
$= [4x]_{-\pi/4}^{\pi/4} - 2\int_{-\pi/4}^{\pi/4} dx - 2\int_{-\pi/4}^{\pi/4} \cos 4x\,dx = \pi + \pi - \left[2x + \frac{\sin 4x}{2}\right]_{-\pi/4}^{\pi/4} = 2\pi - \left(\frac{\pi}{2} - \left(-\frac{\pi}{2}\right)\right) = \pi$

10. $\int_0^{\pi} 8\sin^4 y \cos^2 y\,dy = 8\int_0^{\pi} \left(\frac{1-\cos 2y}{2}\right)^2 \left(\frac{1+\cos 2y}{2}\right) dy = \int_0^{\pi} dy - \int_0^{\pi} \cos 2y\,dy - \int_0^{\pi} \cos^2 2y\,dy + \int_0^{\pi} \cos^3 2y\,dy$
$= \left[y - \frac{1}{2}\sin 2y\right]_0^{\pi} - \int_0^{\pi} \left(\frac{1+\cos 4y}{2}\right) dy + \int_0^{\pi} (1-\sin^2 2y)\cos 2y\,dy = \pi - \frac{1}{2}\int_0^{\pi} dy - \frac{1}{2}\int_0^{\pi} \cos 4y\,dy + \int_0^{\pi} \cos 2y\,dy$
$- \int_0^{\pi} \sin^2 2y \cos 2y\,dy = \pi + \left[-\frac{1}{2}y - \frac{1}{8}\sin 4y + \frac{1}{2}\sin 2y - \frac{1}{2}\cdot\frac{\sin^3 2y}{3}\right]_0^{\pi} = \pi - \frac{\pi}{2} = \frac{\pi}{2}$

11. $\int_0^{\pi/2} 35\sin^4 x\cos^3 x\,dx = \int_0^{\pi/2} 35\sin^4 x(1-\sin^2 x)\cos x\,dx = 35\int_0^{\pi/2}\sin^4 x\cos x\,dx - 35\int_0^{\pi/2}\sin^6 x\cos x\,dx$
$= \left[35\frac{\sin^5 x}{5} - 35\frac{\sin^7 x}{7}\right]_0^{\pi/2} = (7-5)-(0) = 2$

12. $\int_0^{\pi}\cos^2 2x\sin 2x\,dx = \left[-\frac{1}{2}\frac{\cos^3 2x}{3}\right]_0^{\pi} = -\frac{1}{6}+\frac{1}{6} = 0$

13. $\int_0^{\pi/4} 8\cos^3 2\theta\sin 2\theta\,d\theta = \left[8\left(-\frac{1}{2}\right)\frac{\cos^4 2\theta}{4}\right]_0^{\pi/4} = \left[-\cos^4 2\theta\right]_0^{\pi/4} = (0)-(-1) = 1$

14. $\int_0^{\pi/2}\sin^2 2\theta\cos^3 2\theta\,d\theta = \int_0^{\pi/2}\sin^2 2\theta(1-\sin^2 2\theta)\cos 2\theta\,d\theta = \int_0^{\pi/2}\sin^2 2\theta\cos 2\theta\,d\theta - \int_0^{\pi/2}\sin^4 2\theta\cos 2\theta\,d\theta$
$= \left[\frac{1}{2}\cdot\frac{\sin^3 2\theta}{3} - \frac{1}{2}\cdot\frac{\sin^5 2\theta}{5}\right]_0^{\pi/2} = 0$

15. $\int_0^{2\pi}\sqrt{\frac{1-\cos x}{2}}\,dx = \int_0^{2\pi}\left|\sin\frac{x}{2}\right|dx = \int_0^{2\pi}\sin\frac{x}{2}\,dx = \left[-2\cos\frac{x}{2}\right]_0^{2\pi} = 2+2 = 4$

16. $\int_0^{\pi}\sqrt{1-\cos 2x}\,dx = \int_0^{\pi}\sqrt{2}\,|\sin 2x|\,dx = \int_0^{\pi}\sqrt{2}\sin 2x\,dx = \left[-\sqrt{2}\cos 2x\right]_0^{\pi} = \sqrt{2}+\sqrt{2} = 2\sqrt{2}$

17. $\int_0^{\pi}\sqrt{1-\sin^2 t}\,dt = \int_0^{\pi}|\cos t|\,dt = \int_0^{\pi/2}\cos t\,dt - \int_{\pi/2}^{\pi}\cos t\,dt = [\sin t]_0^{\pi/2} - [\sin t]_{\pi/2}^{\pi} = 1-0-0+1 = 2$

18. $\int_0^{\pi}\sqrt{1-\cos^2\theta}\,d\theta = \int_0^{\pi}|\sin\theta|\,d\theta = \int_0^{\pi}\sin\theta\,d\theta = [-\cos\theta]_0^{\pi} = 1+1 = 2$

19. $\int_{-\pi/4}^{\pi/4}\sqrt{1+\tan^2 x}\,dx = \int_{-\pi/4}^{\pi/4}|\sec x|\,dx = \int_{-\pi/4}^{\pi/4}\sec x\,dx = [\ln|\sec x+\tan x|]_{-\pi/4}^{\pi/4} = \ln\left(\sqrt{2}+1\right) - \ln\left(\sqrt{2}-1\right)$
$= \ln\left(\frac{\sqrt{2}+1}{\sqrt{2}-1}\right) = 2\ln\left(1+\sqrt{2}\right)$

20. $\int_{-\pi/4}^{\pi/4}\sqrt{\sec^2 x-1}\,dx = \int_{-\pi/4}^{\pi/4}|\tan x|\,dx = -\int_{-\pi/4}^{0}\tan x\,dx + \int_0^{\pi/4}\tan x\,dx = [-\ln|\sec x|]_{-\pi/4}^{0} + [-\ln|\sec x|]_0^{\pi/4}$
$= -\ln(1)+\ln\sqrt{2}+\ln\sqrt{2}-\ln(1) = 2\ln\sqrt{2} = \ln 2$

21. $\int_0^{\pi/2}\theta\sqrt{1-\cos 2\theta}\,d\theta = \int_0^{\pi/2}\theta\sqrt{2}\,|\sin\theta|\,d\theta = \sqrt{2}\int_0^{\pi/2}\theta\sin\theta\,d\theta = \sqrt{2}\,[-\theta\cos\theta+\sin\theta]_0^{\pi/2} = \sqrt{2}(1) = \sqrt{2}$

22. $\int_{-\pi}^{\pi}(1-\cos^2 t)^{3/2}\,dt = \int_{-\pi}^{\pi}(\sin^2 t)^{3/2}\,dt = \int_{-\pi}^{\pi}|\sin^3 t|\,dt = -\int_{-\pi}^{0}\sin^3 t\,dt + \int_0^{\pi}\sin^3 t\,dt = -\int_{-\pi}^{0}(1-\cos^2 t)\sin t\,dt$
$+\int_0^{\pi}(1-\cos^2 t)\sin t\,dt = -\int_{-\pi}^{0}\sin t\,dt + \int_{-\pi}^{0}\cos^2 t\sin t\,dt + \int_0^{\pi}\sin t\,dt - \int_0^{\pi}\cos^2 t\sin t\,dt = \left[\cos t - \frac{\cos^3 t}{3}\right]_{-\pi}^{0}$
$+\left[-\cos t+\frac{\cos^3 t}{3}\right]_0^{\pi} = \left(1-\frac{1}{3}+1-\frac{1}{3}\right)+\left(1-\frac{1}{3}+1-\frac{1}{3}\right) = \frac{8}{3}$

23. $\int_{-\pi/3}^{0} 2\sec^3 x\,dx$; $u = \sec x$, $du = \sec x\tan x\,dx$, $dv = \sec^2 x\,dx$, $v = \tan x$;
$\int_{-\pi/3}^{0} 2\sec^3 x\,dx = [2\sec x\tan x]_{-\pi/3}^{0} - 2\int_{-\pi/3}^{0}\sec x\tan^2 x\,dx = 2\cdot 1\cdot 0 - 2\cdot 2\cdot\sqrt{3} - 2\int_{-\pi/3}^{0}\sec x\,(\sec^2 x-1)dx$
$= 4\sqrt{3} - 2\int_{-\pi/3}^{0}\sec^3 x\,dx + 2\int_{-\pi/3}^{0}\sec x\,dx;\ 2\int_{-\pi/3}^{0} 2\sec^3 x\,dx = 4\sqrt{3} + [2\ln|\sec x+\tan x|]_{-\pi/3}^{0}$
$2\int_{-\pi/3}^{0} 2\sec^3 x\,dx = 4\sqrt{3} + 2\ln|1+0| - 2\ln|2-\sqrt{3}| = 4\sqrt{3} - 2\ln\left(2-\sqrt{3}\right)$
$\int_{-\pi/3}^{0} 2\sec^3 x\,dx = 2\sqrt{3} - \ln\left(2-\sqrt{3}\right)$

24. $\int e^x \sec^3(e^x)dx;\, u = \sec(e^x),\ du = \sec(e^x)\tan(e^x)e^x dx,\ dv = \sec^2(e^x)e^x dx,\ v = \tan(e^x).$

$$\begin{aligned}\int e^x\sec^3(e^x)\,dx &= \sec(e^x)\tan(e^x) - \int \sec(e^x)\tan^2(e^x)e^x dx\\ &= \sec(e^x)\tan(e^x) - \int \sec(e^x)(\sec^2(e^x)-1)e^x dx\\ &= \sec(e^x)\tan(e^x) - \int \sec^3(e^x)e^x dx + \int \sec(e^x)e^x dx\end{aligned}$$

$$2\int e^x\sec^3(e^x)\,dx = \sec(e^x)\tan(e^x) + \ln\left|\sec(e^x)+\tan(e^x)\right| + C$$

$$\int e^x\sec^3(e^x)\,dx = \tfrac{1}{2}\left(\sec(e^x)\tan(e^x) + \ln\left|\sec(e^x)+\tan(e^x)\right|\right) + C$$

25. $\int_0^{\pi/4} \sec^4\theta\, d\theta = \int_0^{\pi/4} (1+\tan^2\theta)\sec^2\theta\, d\theta = \int_0^{\pi/4} \sec^2\theta\, d\theta + \int_0^{\pi/4} \tan^2\theta\sec^2\theta\, d\theta = \left[\tan\theta + \frac{\tan^3\theta}{3}\right]_0^{\pi/4}$
$= \left(1+\frac{1}{3}\right) - (0) = \frac{4}{3}$

26. $\int_0^{\pi/12} 3\sec^4(3x)\,dx = \int_0^{\pi/12} (1+\tan^2(3x))\sec^2(3x)3dx = \int_0^{\pi/} \sec^2(3x)3dx + \int_0^{\pi/12} \tan^2(3x)\sec^2(3x)3dx$
$= \left[\tan(3x) + \frac{\tan^3(3x)}{3}\right]_0^{\pi/12} = \left(1+\frac{1}{3}\right) - (0) = \frac{4}{3}$

27. $\int_{\pi/4}^{\pi/2} \csc^4\theta\, d\theta = \int_{\pi/4}^{\pi/2} (1+\cot^2\theta)\csc^2\theta\, d\theta = \int_{\pi/4}^{\pi/2} \csc^2\theta\, d\theta + \int_{\pi/4}^{\pi/2} \cot^2\theta\csc^2\theta\, d\theta = \left[-\cot\theta - \frac{\cot^3\theta}{3}\right]_{\pi/4}^{\pi/2}$
$= (0) - \left(-1-\frac{1}{3}\right) = \frac{4}{3}$

28. $\int_{\pi/2}^{\pi} 3\csc^4\frac{\theta}{2}\, d\theta = 3\int_{\pi/2}^{\pi} \left(1+\cot^2\frac{\theta}{2}\right)\csc^2\frac{\theta}{2}\, d\theta = 3\int_{\pi/2}^{\pi} \csc^2\frac{\theta}{2}\, d\theta + 3\int_{\pi/2}^{\pi} \cot^2\frac{\theta}{2}\csc^2\frac{\theta}{2}\, d\theta = \left[-6\cot\frac{\theta}{2} - 6\frac{\cot^3\frac{\theta}{2}}{3}\right]_{\pi/2}^{\pi}$
$= (-6\cdot 0 - 2\cdot 0) - (-6\cdot 1 - 2\cdot 1) = 8$

29. $\int_0^{\pi/4} 4\tan^3 x\, dx = 4\int_0^{\pi/4} (\sec^2 x - 1)\tan x\, dx = 4\int_0^{\pi/4} \sec^2 x\tan x\, dx - 4\int_0^{\pi/4} \tan x\, dx = \left[4\frac{\tan^2 x}{2} - 4\ln|\sec x|\right]_0^{\pi/4}$
$= 2(1) - 4\ln\sqrt{2} - 2\cdot 0 + 4\ln 1 = 2 - 2\ln 2$

30. $\int_{-\pi/4}^{\pi/4} 6\tan^4 x\, dx = 6\int_{-\pi/4}^{\pi/4} (\sec^2 x - 1)\tan^2 x\, dx = 6\int_{-\pi/4}^{\pi/4} \sec^2 x\tan^2 x\, dx - 6\int_{-\pi/4}^{\pi/4} \tan^2 x\, dx$
$= 6\int_{-\pi/4}^{\pi/4} \sec^2 x\tan^2 x\, dx - 6\int_{-\pi/4}^{\pi/4} (\sec^2 x - 1)dx = \left[6\frac{\tan^3 x}{3}\right]_{-\pi/4}^{\pi/4} - 6\int_{-\pi/4}^{\pi/4} \sec^2 x\, dx + 6\int_{-\pi/4}^{\pi/4} dx$
$= 2(1-(-1)) - [6\tan x]_{-\pi/4}^{\pi/4} + [6x]_{-\pi/4}^{\pi/4} = 4 - 6(1-(-1)) + \frac{3\pi}{2} + \frac{3\pi}{2} = 3\pi - 8$

31. $\int_{\pi/6}^{\pi/3} \cot^3 x\, dx = \int_{\pi/6}^{\pi/3} (\csc^2 x - 1)\cot x\, dx = \int_{\pi/6}^{\pi/3} \csc^2 x\cot x\, dx - \int_{\pi/6}^{\pi/3} \cot x\, dx = \left[-\frac{\cot^2 x}{2} + \ln|\csc x|\right]_{\pi/6}^{\pi/3}$
$= -\frac{1}{2}\left(\frac{1}{3} - 3\right) + \left(\ln\frac{2}{\sqrt{3}} - \ln 2\right) = \frac{4}{3} - \ln\sqrt{3}$

32. $\int_{\pi/4}^{\pi/2} 8\cot^4 t\, dt = 8\int_{\pi/4}^{\pi/2} (\csc^2 t - 1)\cot^2 t\, dt = 8\int_{\pi/4}^{\pi/2} \csc^2 t\cot^2 t\, dt - 8\int_{\pi/4}^{\pi/2} \cot^2 t\, dt$
$= -8\left[-\frac{\cot^3 t}{3}\right]_{\pi/4}^{\pi/2} - 8\int_{\pi/4}^{\pi/2} (\csc^2 t - 1)dt = -\frac{8}{3}(0-1) + [8\cot t]_{\pi/4}^{\pi/2} + [8t]_{\pi/4}^{\pi/2} = \frac{8}{3} + 8(0-1) + 4\pi - 2\pi = 2\pi - \frac{16}{3}$

33. $\int_{-\pi}^{0} \sin 3x\cos 2x\, dx = \frac{1}{2}\int_{-\pi}^{0} (\sin x + \sin 5x)\, dx = \frac{1}{2}\left[-\cos x - \frac{1}{5}\cos 5x\right]_{-\pi}^{0} = \frac{1}{2}\left(-1 - \frac{1}{5} - 1 - \frac{1}{5}\right) = -\frac{6}{5}$

34. $\int_0^{\pi/2} \sin 2x\cos 3x\, dx = \frac{1}{2}\int_0^{\pi/2} (\sin(-x) + \sin 5x)\, dx = \frac{1}{2}\left[\cos(-x) - \frac{1}{5}\cos 5x\right]_0^{\pi/2} = \frac{1}{2}(0) - \frac{1}{2}\left(1 - \frac{1}{5}\right) = -\frac{2}{5}$

35. $\int_{-\pi}^{\pi} \sin 3x \sin 3x\, dx = \frac{1}{2}\int_{-\pi}^{\pi} (\cos 0 - \cos 6x)\, dx = \frac{1}{2}\int_{-\pi}^{\pi} dx - \frac{1}{2}\int_{-\pi}^{\pi} \cos 6x\, dx = \frac{1}{2}\left[x - \frac{1}{12}\sin 6x\right]_{-\pi}^{\pi} = \frac{\pi}{2} + \frac{\pi}{2} - 0 = \pi$

36. $\int_{0}^{\pi/2} \sin x \cos x\, dx = \frac{1}{2}\int_{0}^{\pi/2} (\sin 0 + \sin 2x)\, dx = \frac{1}{2}\int_{0}^{\pi/2} \sin 2x\, dx = -\frac{1}{4}[\cos 2x]_{0}^{\pi/2} = -\frac{1}{4}(-1-1) = \frac{1}{2}$

37. $\int_{0}^{\pi} \cos 3x \cos 4x\, dx = \frac{1}{2}\int_{0}^{\pi} (\cos(-x) + \cos 7x)\, dx = \frac{1}{2}\left[-\sin(-x) + \frac{1}{7}\sin 7x\right]_{0}^{\pi} = \frac{1}{2}(0) = 0$

38. $\int_{-\pi/2}^{\pi/2} \cos 7x \cos x\, dx = \frac{1}{2}\int_{-\pi/2}^{\pi/2} (\cos 6x + \cos 8x)\, dx = \frac{1}{2}\left[\frac{1}{6}\sin 6x + \frac{1}{8}\sin 8x\right]_{-\pi/2}^{\pi/2} = 0$

39. $x = t^{2/3} \Rightarrow t^2 = x^3;\ y = \frac{t^2}{2} \Rightarrow y = \frac{x^3}{2};\ 0 \le t \le 2 \Rightarrow 0 \le x \le 2^{2/3};$

$A = \int_{0}^{2^{2/3}} 2\pi\left(\frac{x^3}{2}\right)\sqrt{1 + \frac{9}{4}x^4}\, dx;\ \left[\begin{array}{c} u = \frac{9}{4}x^4 \\ du = 9x^3 dx \end{array}\right] \to \frac{\pi}{9}\int_{0}^{9(2^{2/3})} \sqrt{1+u}\, du = \left[\frac{\pi}{9} \cdot \frac{2}{3}(1+u)^{3/2}\right]_{0}^{9(2^{2/3})}$

$= \frac{2\pi}{27}\left[\left(1 + 9(2^{2/3})\right)^{3/2} - 1\right]$

40. $y = \ln(\cos x);\ y' = \frac{-\sin x}{\cos x} = -\tan x;\ (y')^2 = \tan^2 x;\ \int_{0}^{\pi/3} \sqrt{1 + \tan^2 x}\, dx = \int_{0}^{\pi/3} |\sec x|\, dx = [\ln|\sec x + \tan x|]_{0}^{\pi/3}$

$= \ln\left(2 + \sqrt{3}\right) - \ln(1+0) = \ln\left(2 + \sqrt{3}\right)$

41. $y = \ln(\sec x);\ y' = \frac{\sec x \tan x}{\sec x} = \tan x; (y')^2 = \tan^2 x;\ \int_{0}^{\pi/4} \sqrt{1 + \tan^2 x}\, dx = \int_{0}^{\pi/4} |\sec x|\, dx = [\ln|\sec x + \tan x|]_{0}^{\pi/4}$

$= \ln\left(\sqrt{2} + 1\right) - \ln(0+1) = \ln\left(\sqrt{2}+1\right)$

42. $M = \int_{-\pi/4}^{\pi/4} \sec x\, dx = [\ln|\sec x + \tan x|]_{-\pi/4}^{\pi/4} = \ln\left(\sqrt{2}+1\right) - \ln\left|\sqrt{2} - 1\right| = \ln\frac{\sqrt{2}+1}{\sqrt{2}-1}$

$\overline{y} = \frac{1}{\ln\frac{\sqrt{2}+1}{\sqrt{2}-1}}\int_{-\pi/4}^{\pi/4} \frac{\sec^2 x}{2}\, dx = \frac{1}{2\ln\frac{\sqrt{2}+1}{\sqrt{2}-1}}[\tan x]_{-\pi/4}^{\pi/4} = \frac{1}{2\ln\frac{\sqrt{2}+1}{\sqrt{2}-1}}(1-(-1)) = \frac{1}{\ln\frac{\sqrt{2}+1}{\sqrt{2}-1}}$

$\Rightarrow (\overline{x}, \overline{y}) = \left(0, \left(\ln\frac{\sqrt{2}+1}{\sqrt{2}-1}\right)^{-1}\right)$

43. $V = \pi\int_{0}^{\pi} \sin^2 x\, dx = \pi\int_{0}^{\pi} \frac{1-\cos 2x}{2}\, dx = \frac{\pi}{2}\int_{0}^{\pi} dx - \frac{\pi}{2}\int_{0}^{\pi} \cos 2x\, dx = \frac{\pi}{2}[x]_{0}^{\pi} - \frac{\pi}{4}[\sin 2x]_{0}^{\pi} = \frac{\pi}{2}(\pi - 0) - \frac{\pi}{4}(0-0) = \frac{\pi^2}{2}$

44. $A = \int_{0}^{\pi} \sqrt{1 + \cos 4x}\, dx = \int_{0}^{\pi} \sqrt{2}\,|\cos 2x|dx = \sqrt{2}\int_{0}^{\pi/4} \cos 2x\, dx - \sqrt{2}\int_{\pi/4}^{3\pi/4} \cos 2x\, dx + \sqrt{2}\int_{3\pi/4}^{\pi} \cos 2x\, dx$

$= \frac{\sqrt{2}}{2}[\sin 2x]_{0}^{\pi/4} - \frac{\sqrt{2}}{2}[\sin 2x]_{\pi/4}^{3\pi/4} + \frac{\sqrt{2}}{2}[\sin 2x]_{3\pi/4}^{\pi} = \frac{\sqrt{2}}{2}(1-0) - \frac{\sqrt{2}}{2}(-1-1) + \frac{\sqrt{2}}{2}(0+1) = \sqrt{2} + \sqrt{2} = 2\sqrt{2}$

45. (a) $m^2 \ne n^2 \Rightarrow m + n \ne 0$ and $m - n \ne 0 \Rightarrow \int_{k}^{k+2\pi} \sin mx \sin nx\, dx = \frac{1}{2}\int_{k}^{k+2\pi} [\cos(m-n)x - \cos(m+n)x]dx$

$= \frac{1}{2}\left[\frac{1}{m-n}\sin(m-n)x - \frac{1}{m+n}\sin(m+n)x\right]_{k}^{k+2\pi}$

$= \frac{1}{2}\left(\frac{1}{m-n}\sin((m-n)(k+2\pi)) - \frac{1}{m+n}\sin((m+n)(k+2\pi))\right) - \frac{1}{2}\left(\frac{1}{m-n}\sin((m-n)k) - \frac{1}{m+n}\sin((m+n)k)\right)$

$= \frac{1}{2(m-n)}\sin((m-n)k) - \frac{1}{2(m+n)}\sin((m+n)k) - \frac{1}{2(m-n)}\sin((m-n)k) + \frac{1}{2(m+n)}\sin((m+n)k) = 0$

$\Rightarrow$ sin mx and sin nx are orthogonal.

(b) Same as part since $\frac{1}{2}\int_{k}^{k+2\pi} \cos 0\, dx = \pi$. $m^2 \ne n^2 \Rightarrow m + n \ne 0$ and $m - n \ne 0 \Rightarrow \int_{k}^{k+2\pi} \cos mx \cos nx\, dx$

$= \frac{1}{2}\int_{k}^{k+2\pi} [\cos(m-n)x + \cos(m+n)x]dx = \frac{1}{2}\left[\frac{1}{m-n}\sin(m-n)x + \frac{1}{m+n}\sin(m+n)x\right]_{k}^{k+2\pi}$

$= \frac{1}{2(m-n)}\sin((m-n)(k+2\pi)) + \frac{1}{2(m+n)}\sin((m+n)(k+2\pi)) - \frac{1}{2(m-n)}\sin((m-n)k) - \frac{1}{2(m+n)}\sin((m+n)k)$

$= \frac{1}{2(m-n)}\sin((m-n)k) + \frac{1}{2(m+n)}\sin((m+n)k) - \frac{1}{2(m-n)}\sin((m-n)k) - \frac{1}{2(m+n)}\sin((m+n)k) = 0$

$\Rightarrow$ cos mx and cos nx are orthogonal.

(c) Let m = n $\Rightarrow \sin mx \cos nx = \frac{1}{2}(\sin 0 + \sin((m+n)x))$ and $\frac{1}{2}\int_k^{k+2\pi} \sin 0\, dx = 0$ and $\frac{1}{2}\int_k^{k+2\pi} \sin((m+n)x)\, dx = 0$

$\Rightarrow$ sin mx and cos nx are orthogonal if m = n.

Let $m \neq n$.

$\int_k^{k+2\pi} \sin mx \cos nx\, dx = \frac{1}{2}\int_k^{k+2\pi} [\sin(m-n)x + \sin(m+n)x]dx = \frac{1}{2}\left[-\frac{1}{m-n}\cos(m-n)x - \frac{1}{m+n}\cos(m+n)x\right]_k^{k+2\pi}$

$= -\frac{1}{2(m-n)}\cos((m-n)(k+2\pi)) - \frac{1}{2(m+n)}\cos((m+n)(k+2\pi)) + \frac{1}{2(m-n)}\cos((m-n)k) + \frac{1}{2(m+n)}\cos((m+n)k)$

$= -\frac{1}{2(m-n)}\cos((m-n)k) - \frac{1}{2(m+n)}\cos((m+n)k) + \frac{1}{2(m-n)}\cos((m-n)k) + \frac{1}{2(m+n)}\cos((m+n)k) = 0$

$\Rightarrow$ sin mx and cos nx are orthogonal.

46. $\frac{1}{\pi}\int_{-\pi}^{\pi} f(x)\sin mx\, dx = \sum_{n=1}^{N} \frac{a_n}{\pi}\int_{-\pi}^{\pi} \sin nx \sin mx\, dx$. Since $\frac{1}{\pi}\int_{-\pi}^{\pi} \sin nx \sin mx\, dx = \begin{cases} 0 & \text{for } m \neq n \\ 1 & \text{for } m = n \end{cases}$,
the sum on the right has only one nonzero term, namely $\frac{a_m}{\pi}\int_{-\pi}^{\pi} \sin mx \sin mx\, dx = a_m$.

8.5 TRIGONOMETRIC SUBSTITUTIONS

1. $y = 3\tan\theta,\ -\frac{\pi}{2} < \theta < \frac{\pi}{2},\ dy = \frac{3\, d\theta}{\cos^2\theta},\ 9 + y^2 = 9(1+\tan^2\theta) = \frac{9}{\cos^2\theta} \Rightarrow \frac{1}{\sqrt{9+y^2}} = \frac{|\cos\theta|}{3} = \frac{\cos\theta}{3}$
(because $\cos\theta > 0$ when $-\frac{\pi}{2} < \theta < \frac{\pi}{2}$);
$\int \frac{dy}{\sqrt{9+y^2}} = 3\int \frac{\cos\theta\, d\theta}{3\cos^2\theta} = \int \frac{d\theta}{\cos\theta} = \ln|\sec\theta + \tan\theta| + C' = \ln\left|\frac{\sqrt{9+y^2}}{3} + \frac{y}{3}\right| + C' = \ln\left|\sqrt{9+y^2} + y\right| + C$

2. $\int \frac{3\, dy}{\sqrt{1+9y^2}}$; $[3y = x] \rightarrow \int \frac{dx}{\sqrt{1+x^2}}$; $x = \tan t,\ -\frac{\pi}{2} < t < \frac{\pi}{2},\ dx = \frac{dt}{\cos^2 t},\ \sqrt{1+x^2} = \frac{1}{\cos t}$;
$\int \frac{dx}{\sqrt{1+x^2}} = \int \frac{dt}{\cos^2 t\left(\frac{1}{\cos t}\right)} = \ln|\sec t + \tan t| + C = \ln\left|\sqrt{x^2+1} + x\right| + C = \ln\left|\sqrt{1+9y^2} + 3y\right| + C$

3. $\int_{-2}^{2} \frac{dx}{4+x^2} = \left[\frac{1}{2}\tan^{-1}\frac{x}{2}\right]_{-2}^{2} = \frac{1}{2}\tan^{-1}1 - \frac{1}{2}\tan^{-1}(-1) = \left(\frac{1}{2}\right)\left(\frac{\pi}{4}\right) - \left(\frac{1}{2}\right)\left(-\frac{\pi}{4}\right) = \frac{\pi}{4}$

4. $\int_0^2 \frac{dx}{8+2x^2} = \frac{1}{2}\int_0^2 \frac{dx}{4+x^2} = \frac{1}{2}\left[\frac{1}{2}\tan^{-1}\frac{x}{2}\right]_0^2 = \frac{1}{2}\left(\frac{1}{2}\tan^{-1}1 - \frac{1}{2}\tan^{-1}0\right) = \left(\frac{1}{2}\right)\left(\frac{1}{2}\right)\left(\frac{\pi}{4}\right) - 0 = \frac{\pi}{16}$

5. $\int_0^{3/2} \frac{dx}{\sqrt{9-x^2}} = \left[\sin^{-1}\frac{x}{3}\right]_0^{3/2} = \sin^{-1}\frac{1}{2} - \sin^{-1}0 = \frac{\pi}{6} - 0 = \frac{\pi}{6}$

6. $\int_0^{1/2\sqrt{2}} \frac{2\, dx}{\sqrt{1-4x^2}}$; $[t = 2x] \rightarrow \int_0^{1/2\sqrt{2}} \frac{dt}{\sqrt{1-t^2}} = \left[\sin^{-1}t\right]_0^{1/\sqrt{2}} = \sin^{-1}\frac{1}{\sqrt{2}} - \sin^{-1}0 = \frac{\pi}{4} - 0 = \frac{\pi}{4}$

7. $t = 5\sin\theta,\ -\frac{\pi}{2} < \theta < \frac{\pi}{2},\ dt = 5\cos\theta\, d\theta,\ \sqrt{25-t^2} = 5\cos\theta$;
$\int \sqrt{25-t^2}\, dt = \int (5\cos\theta)(5\cos\theta)\, d\theta = 25\int \cos^2\theta\, d\theta = 25\int \frac{1+\cos 2\theta}{2}\, d\theta = 25\left(\frac{\theta}{2} + \frac{\sin 2\theta}{4}\right) + C$
$= \frac{25}{2}(\theta + \sin\theta\cos\theta) + C = \frac{25}{2}\left[\sin^{-1}\left(\frac{t}{5}\right) + \left(\frac{t}{5}\right)\left(\frac{\sqrt{25-t^2}}{5}\right)\right] + C = \frac{25}{2}\sin^{-1}\left(\frac{t}{5}\right) + \frac{t\sqrt{25-t^2}}{2} + C$

8. $t = \frac{1}{3}\sin\theta,\ -\frac{\pi}{2} < \theta < \frac{\pi}{2},\ dt = \frac{1}{3}\cos\theta\, d\theta,\ \sqrt{1-9t^2} = \cos\theta$;
$\int \sqrt{1-9t^2}\, dt = \frac{1}{3}\int (\cos\theta)(\cos\theta)\, d\theta = \frac{1}{3}\int \cos^2\theta\, d\theta = \frac{1}{6}(\theta + \sin\theta\cos\theta) + C = \frac{1}{6}\left[\sin^{-1}(3t) + 3t\sqrt{1-9t^2}\right] + C$

9. $x = \frac{7}{2}\sec\theta,\ 0 < \theta < \frac{\pi}{2},\ dx = \frac{7}{2}\sec\theta\tan\theta\, d\theta,\ \sqrt{4x^2-49} = \sqrt{49\sec^2\theta - 49} = 7\tan\theta$;
$\int \frac{dx}{\sqrt{4x^2-49}} = \int \frac{\left(\frac{7}{2}\sec\theta\tan\theta\right) d\theta}{7\tan\theta} = \frac{1}{2}\int \sec\theta\, d\theta = \frac{1}{2}\ln|\sec\theta + \tan\theta| + C = \frac{1}{2}\ln\left|\frac{2x}{7} + \frac{\sqrt{4x^2-49}}{7}\right| + C$

10. $x = \frac{3}{5}\sec\theta,\ 0 < \theta < \frac{\pi}{2},\ dx = \frac{3}{5}\sec\theta\tan\theta\,d\theta,\ \sqrt{25x^2 - 9} = \sqrt{9\sec^2\theta - 9} = 3\tan\theta;$

$$\int \frac{5\,dx}{\sqrt{25x^2-9}} = \int \frac{5\left(\frac{3}{5}\sec\theta\tan\theta\right)d\theta}{3\tan\theta} = \int \sec\theta\,d\theta = \ln|\sec\theta + \tan\theta| + C = \ln\left|\frac{5x}{3} + \frac{\sqrt{25x^2-9}}{3}\right| + C$$

11. $y = 7\sec\theta,\ 0 < \theta < \frac{\pi}{2},\ dy = 7\sec\theta\tan\theta\,d\theta,\ \sqrt{y^2 - 49} = 7\tan\theta;$

$$\int \frac{\sqrt{y^2-49}}{y}\,dy = \int \frac{(7\tan\theta)(7\sec\theta\tan\theta)\,d\theta}{7\sec\theta} = 7\int \tan^2\theta\,d\theta = 7\int(\sec^2\theta - 1)\,d\theta = 7(\tan\theta - \theta) + C$$
$$= 7\left[\frac{\sqrt{y^2-49}}{7} - \sec^{-1}\left(\frac{y}{7}\right)\right] + C$$

12. $y = 5\sec\theta,\ 0 < \theta < \frac{\pi}{2},\ dy = 5\sec\theta\tan\theta\,d\theta,\ \sqrt{y^2 - 25} = 5\tan\theta;$

$$\int \frac{\sqrt{y^2-25}}{y^3}\,dy = \int \frac{(5\tan\theta)(5\sec\theta\tan\theta)\,d\theta}{125\sec^3\theta} = \frac{1}{5}\int \tan^2\theta\cos^2\theta\,d\theta = \frac{1}{5}\int \sin^2\theta\,d\theta = \frac{1}{10}\int(1 - \cos 2\theta)\,d\theta$$
$$= \frac{1}{10}(\theta - \sin\theta\cos\theta) + C = \frac{1}{10}\left[\sec^{-1}\left(\frac{y}{5}\right) - \left(\frac{\sqrt{y^2-25}}{y}\right)\left(\frac{5}{y}\right)\right] + C = \left[\frac{\sec^{-1}\left(\frac{y}{5}\right)}{10} - \frac{\sqrt{y^2-25}}{2y^2}\right] + C$$

13. $x = \sec\theta,\ 0 < \theta < \frac{\pi}{2},\ dx = \sec\theta\tan\theta\,d\theta,\ \sqrt{x^2 - 1} = \tan\theta;$

$$\int \frac{dx}{x^2\sqrt{x^2-1}} = \int \frac{\sec\theta\tan\theta\,d\theta}{\sec^2\theta\tan\theta} = \int \frac{d\theta}{\sec\theta} = \sin\theta + C = \frac{\sqrt{x^2-1}}{x} + C$$

14. $x = \sec\theta,\ 0 < \theta < \frac{\pi}{2},\ dx = \sec\theta\tan\theta\,d\theta,\ \sqrt{x^2 - 1} = \tan\theta;$

$$\int \frac{2\,dx}{x^3\sqrt{x^2-1}} = \int \frac{2\tan\theta\sec\theta\,d\theta}{\sec^3\theta\tan\theta} = 2\int \cos^2\theta\,d\theta = 2\int\left(\frac{1+\cos 2\theta}{2}\right)d\theta = \theta + \sin\theta\cos\theta + C$$
$$= \theta + \tan\theta\cos^2\theta + C = \sec^{-1}x + \sqrt{x^2-1}\left(\frac{1}{x}\right)^2 + C = \sec^{-1}x + \frac{\sqrt{x^2-1}}{x^2} + C$$

15. $x = 2\tan\theta,\ -\frac{\pi}{2} < \theta < \frac{\pi}{2},\ dx = \frac{2\,d\theta}{\cos^2\theta},\ \sqrt{x^2+4} = \frac{2}{\cos\theta};$

$$\int \frac{x^3\,dx}{\sqrt{x^2+4}} = \int \frac{(8\tan^3\theta)(\cos\theta)\,d\theta}{\cos^2\theta} = 8\int \frac{\sin^3\theta\,d\theta}{\cos^4\theta} = 8\int \frac{(\cos^2\theta - 1)(-\sin\theta)\,d\theta}{\cos^4\theta};$$
$$[t = \cos\theta] \to 8\int \frac{t^2-1}{t^4}\,dt = 8\int\left(\frac{1}{t^2} - \frac{1}{t^4}\right)dt = 8\left(-\frac{1}{t} + \frac{1}{3t^3}\right) + C = 8\left(-\sec\theta + \frac{\sec^3\theta}{3}\right) + C$$
$$= 8\left(-\frac{\sqrt{x^2+4}}{2} + \frac{(x^2+4)^{3/2}}{8\cdot 3}\right) + C = \frac{1}{3}(x^2+4)^{3/2} - 4\sqrt{x^2+4} + C$$

16. $x = \tan\theta,\ -\frac{\pi}{2} < \theta < \frac{\pi}{2},\ dx = \sec^2\theta\,d\theta,\ \sqrt{x^2+1} = \sec\theta;$

$$\int \frac{dx}{x^2\sqrt{x^2+1}} = \int \frac{\sec^2\theta\,d\theta}{\tan^2\theta\sec\theta} = \int \frac{\cos\theta\,d\theta}{\sin^2\theta} = -\frac{1}{\sin\theta} + C = \frac{-\sqrt{x^2+1}}{x} + C$$

17. $w = 2\sin\theta,\ -\frac{\pi}{2} < \theta < \frac{\pi}{2},\ dw = 2\cos\theta\,d\theta,\ \sqrt{4 - w^2} = 2\cos\theta;$

$$\int \frac{8\,dw}{w^2\sqrt{4-w^2}} = \int \frac{8\cdot 2\cos\theta\,d\theta}{4\sin^2\theta\cdot 2\cos\theta} = 2\int \frac{d\theta}{\sin^2\theta} = -2\cot\theta + C = \frac{-2\sqrt{4-w^2}}{w} + C$$

18. $w = 3\sin\theta,\ -\frac{\pi}{2} < \theta < \frac{\pi}{2},\ dw = 3\cos\theta\,d\theta,\ \sqrt{9 - w^2} = 3\cos\theta;$

$$\int \frac{\sqrt{9-w^2}}{w^2}\,dw = \int \frac{3\cos\theta\cdot 3\cos\theta\,d\theta}{9\sin^2\theta} = \int \cot^2\theta\,d\theta = \int\left(\frac{1-\sin^2\theta}{\sin^2\theta}\right)d\theta = \int(\csc^2\theta - 1)\,d\theta$$
$$= -\cot\theta - \theta + C = -\frac{\sqrt{9-w^2}}{w} - \sin^{-1}\left(\frac{w}{3}\right) + C$$

19. $x = \sin\theta,\ 0 \le \theta \le \frac{\pi}{3},\ dx = \cos\theta\,d\theta,\ (1 - x^2)^{3/2} = \cos^3\theta;$

$$\int_0^{\sqrt{3}/2} \frac{4x^2\,dx}{(1-x^2)^{3/2}} = \int_0^{\pi/3} \frac{4\sin^2\theta\cos\theta\,d\theta}{\cos^3\theta} = 4\int_0^{\pi/3}\left(\frac{1-\cos^2\theta}{\cos^2\theta}\right)d\theta = 4\int_0^{\pi/3}(\sec^2\theta - 1)\,d\theta$$
$$= 4[\tan\theta - \theta]_0^{\pi/3} = 4\sqrt{3} - \frac{4\pi}{3}$$

20. $x = 2\sin\theta,\ 0 \le \theta \le \frac{\pi}{6},\ dx = 2\cos\theta\,d\theta,\ (4-x^2)^{3/2} = 8\cos^3\theta;$

$$\int_0^1 \frac{dx}{(4-x^2)^{3/2}} = \int_0^{\pi/6} \frac{2\cos\theta\,d\theta}{8\cos^3\theta} = \frac{1}{4}\int_0^{\pi/6} \frac{d\theta}{\cos^2\theta} = \frac{1}{4}\left[\tan\theta\right]_0^{\pi/6} = \frac{\sqrt{3}}{12} = \frac{1}{4\sqrt{3}}$$

21. $x = \sec\theta,\ 0 < \theta < \frac{\pi}{2},\ dx = \sec\theta\tan\theta\,d\theta,\ (x^2-1)^{3/2} = \tan^3\theta;$

$$\int \frac{dx}{(x^2-1)^{3/2}} = \int \frac{\sec\theta\tan\theta\,d\theta}{\tan^3\theta} = \int \frac{\cos\theta\,d\theta}{\sin^2\theta} = -\frac{1}{\sin\theta} + C = -\frac{x}{\sqrt{x^2-1}} + C$$

22. $x = \sec\theta,\ 0 < \theta < \frac{\pi}{2},\ dx = \sec\theta\tan\theta\,d\theta,\ (x^2-1)^{5/2} = \tan^5\theta;$

$$\int \frac{x^2\,dx}{(x^2-1)^{5/2}} = \int \frac{\sec^2\theta\cdot\sec\theta\tan\theta\,d\theta}{\tan^5\theta} = \int \frac{\cos\theta}{\sin^4\theta}\,d\theta = -\frac{1}{3\sin^3\theta} + C = -\frac{x^3}{3(x^2-1)^{3/2}} + C$$

23. $x = \sin\theta,\ -\frac{\pi}{2} < \theta < \frac{\pi}{2},\ dx = \cos\theta\,d\theta,\ (1-x^2)^{3/2} = \cos^3\theta;$

$$\int \frac{(1-x^2)^{3/2}\,dx}{x^6} = \int \frac{\cos^3\theta\cdot\cos\theta\,d\theta}{\sin^6\theta} = \int \cot^4\theta\csc^2\theta\,d\theta = -\frac{\cot^5\theta}{5} + C = -\frac{1}{5}\left(\frac{\sqrt{1-x^2}}{x}\right)^5 + C$$

24. $x = \sin\theta,\ -\frac{\pi}{2} < \theta < \frac{\pi}{2},\ dx = \cos\theta\,d\theta,\ (1-x^2)^{1/2} = \cos\theta;$

$$\int \frac{(1-x^2)^{1/2}\,dx}{x^4} = \int \frac{\cos\theta\cdot\cos\theta\,d\theta}{\sin^4\theta} = \int \cot^2\theta\csc^2\theta\,d\theta = -\frac{\cot^3\theta}{3} + C = -\frac{1}{3}\left(\frac{\sqrt{1-x^2}}{x}\right)^3 + C$$

25. $x = \frac{1}{2}\tan\theta,\ -\frac{\pi}{2} < \theta < \frac{\pi}{2},\ dx = \frac{1}{2}\sec^2\theta\,d\theta,\ (4x^2+1)^2 = \sec^4\theta;$

$$\int \frac{8\,dx}{(4x^2+1)^2} = \int \frac{8\left(\frac{1}{2}\sec^2\theta\right)d\theta}{\sec^4\theta} = 4\int \cos^2\theta\,d\theta = 2(\theta + \sin\theta\cos\theta) + C = 2\tan^{-1}2x + \frac{4x}{(4x^2+1)} + C$$

26. $t = \frac{1}{3}\tan\theta,\ -\frac{\pi}{2} < \theta < \frac{\pi}{2},\ dt = \frac{1}{3}\sec^2\theta\,d\theta,\ 9t^2 + 1 = \sec^2\theta;$

$$\int \frac{6\,dt}{(9t^2+1)^2} = \int \frac{6\left(\frac{1}{3}\sec^2\theta\right)d\theta}{\sec^4\theta} = 2\int \cos^2\theta\,d\theta = \theta + \sin\theta\cos\theta + C = \tan^{-1}3t + \frac{3t}{(9t^2+1)} + C$$

27. $v = \sin\theta,\ -\frac{\pi}{2} < \theta < \frac{\pi}{2},\ dv = \cos\theta\,d\theta,\ (1-v^2)^{5/2} = \cos^5\theta;$

$$\int \frac{v^2\,dv}{(1-v^2)^{5/2}} = \int \frac{\sin^2\theta\cos\theta\,d\theta}{\cos^5\theta} = \int \tan^2\theta\sec^2\theta\,d\theta = \frac{\tan^3\theta}{3} + C = \frac{1}{3}\left(\frac{v}{\sqrt{1-v^2}}\right)^3 + C$$

28. $r = \sin\theta,\ -\frac{\pi}{2} < \theta < \frac{\pi}{2};$

$$\int \frac{(1-r^2)^{5/2}\,dr}{r^8} = \int \frac{\cos^5\theta\cdot\cos\theta\,d\theta}{\sin^8\theta} = \int \cot^6\theta\csc^2\theta\,d\theta = -\frac{\cot^7\theta}{7} + C = -\frac{1}{7}\left[\frac{\sqrt{1-r^2}}{r}\right]^7 + C$$

29. Let $e^t = 3\tan\theta,\ t = \ln(3\tan\theta),\ \tan^{-1}\left(\frac{1}{3}\right) \le \theta \le \tan^{-1}\left(\frac{4}{3}\right),\ dt = \frac{\sec^2\theta}{\tan\theta}\,d\theta,\ \sqrt{e^{2t}+9} = \sqrt{9\tan^2\theta + 9} = 3\sec\theta;$

$$\int_0^{\ln 4} \frac{e^t\,dt}{\sqrt{e^{2t}+9}} = \int_{\tan^{-1}(1/3)}^{\tan^{-1}(4/3)} \frac{3\tan\theta\cdot\sec^2\theta\,d\theta}{\tan\theta\cdot 3\sec\theta} = \int_{\tan^{-1}(1/3)}^{\tan^{-1}(4/3)} \sec\theta\,d\theta = \left[\ln|\sec\theta + \tan\theta|\right]_{\tan^{-1}(1/3)}^{\tan^{-1}(4/3)}$$

$$= \ln\left(\frac{5}{3} + \frac{4}{3}\right) - \ln\left(\frac{\sqrt{10}}{3} + \frac{1}{3}\right) = \ln 9 - \ln\left(1 + \sqrt{10}\right)$$

30. Let $e^t = \tan\theta,\ t = \ln(\tan\theta),\ \tan^{-1}\left(\frac{3}{4}\right) \le \theta \le \tan^{-1}\left(\frac{4}{3}\right),\ dt = \frac{\sec^2\theta}{\tan\theta}\,d\theta,\ 1 + e^{2t} = 1 + \tan^2\theta = \sec^2\theta;$

$$\int_{\ln(3/4)}^{\ln(4/3)} \frac{e^t\,dt}{(1+e^{2t})^{3/2}} = \int_{\tan^{-1}(3/4)}^{\tan^{-1}(4/3)} \frac{(\tan\theta)\left(\frac{\sec^2\theta}{\tan\theta}\right)d\theta}{\sec^3\theta} = \int_{\tan^{-1}(3/4)}^{\tan^{-1}(4/3)} \cos\theta\,d\theta = \left[\sin\theta\right]_{\tan^{-1}(3/4)}^{\tan^{-1}(4/3)} = \frac{4}{5} - \frac{3}{5} = \frac{1}{5}$$

31. $\int_{1/12}^{1/4} \frac{2\,dt}{\sqrt{t} + 4t\sqrt{t}};\ \left[u = 2\sqrt{t},\ du = \frac{1}{\sqrt{t}}\,dt\right] \to \int_{1/\sqrt{3}}^{1} \frac{2\,du}{1+u^2};\ u = \tan\theta,\ \frac{\pi}{6} \le \theta \le \frac{\pi}{4},\ du = \sec^2\theta\,d\theta,\ 1 + u^2 = \sec^2\theta;$

$$\int_{1/\sqrt{3}}^{1} \frac{2\,du}{1+u^2} = \int_{\pi/6}^{\pi/4} \frac{2\sec^2\theta\,d\theta}{\sec^2\theta} = \left[2\theta\right]_{\pi/6}^{\pi/4} = 2\left(\frac{\pi}{4} - \frac{\pi}{6}\right) = \frac{\pi}{6}$$

32. $y = e^{\tan\theta}, 0 \le \theta \le \frac{\pi}{4}, dy = e^{\tan\theta}\sec^2\theta\, d\theta, \sqrt{1+(\ln y)^2} = \sqrt{1+\tan^2\theta} = \sec\theta;$

$\int_1^e \frac{dy}{y\sqrt{1+(\ln y)^2}} = \int_0^{\pi/4} \frac{e^{\tan\theta}\sec^2\theta}{e^{\tan\theta}\sec\theta}\, d\theta = \int_0^{\pi/4}\sec\theta\, d\theta = [\ln|\sec\theta+\tan\theta|]_0^{\pi/4} = \ln\left(1+\sqrt{2}\right)$

33. $x = \sec\theta, 0 < \theta < \frac{\pi}{2}, dx = \sec\theta\tan\theta\, d\theta, \sqrt{x^2-1} = \sqrt{\sec^2\theta - 1} = \tan\theta;$

$\int \frac{dx}{x\sqrt{x^2-1}} = \int \frac{\sec\theta\tan\theta\, d\theta}{\sec\theta\tan\theta} = \theta + C = \sec^{-1}x + C$

34. $x = \tan\theta, dx = \sec^2\theta\, d\theta, 1 + x^2 = \sec^2\theta;$

$\int \frac{dx}{x^2+1} = \int \frac{\sec^2\theta\, d\theta}{\sec^2\theta} = \theta + C = \tan^{-1}x + C$

35. $x = \sec\theta, dx = \sec\theta\tan\theta\, d\theta, \sqrt{x^2-1} = \sqrt{\sec^2\theta-1} = \tan\theta;$

$\int \frac{x\, dx}{\sqrt{x^2-1}} = \int \frac{\sec\theta \cdot \sec\theta\tan\theta\, d\theta}{\tan\theta} = \int \sec^2\theta\, d\theta = \tan\theta + C = \sqrt{x^2-1} + C$

36. $x = \sin\theta, dx = \cos\theta\, d\theta, -\frac{\pi}{2} < \theta < \frac{\pi}{2};$

$\int \frac{dx}{\sqrt{1-x^2}} = \int \frac{\cos\theta\, d\theta}{\cos\theta} = \theta + C = \sin^{-1}x + C$

37. $x\frac{dy}{dx} = \sqrt{x^2-4}; dy = \sqrt{x^2-4}\,\frac{dx}{x}; y = \int \frac{\sqrt{x^2-4}}{x}\, dx; \begin{bmatrix} x = 2\sec\theta, 0 < \theta < \frac{\pi}{2} \\ dx = 2\sec\theta\tan\theta\, d\theta \\ \sqrt{x^2-4} = 2\tan\theta \end{bmatrix}$

$\rightarrow y = \int \frac{(2\tan\theta)(2\sec\theta\tan\theta)\, d\theta}{2\sec\theta} = 2\int \tan^2\theta\, d\theta = 2\int(\sec^2\theta - 1)\, d\theta = 2(\tan\theta - \theta) + C$

$= 2\left[\frac{\sqrt{x^2-4}}{2} - \sec^{-1}\left(\frac{x}{2}\right)\right] + C; x = 2 \text{ and } y = 0 \Rightarrow 0 = 0 + C \Rightarrow C = 0 \Rightarrow y = 2\left[\frac{\sqrt{x^2-4}}{2} - \sec^{-1}\frac{x}{2}\right]$

38. $\sqrt{x^2-9}\,\frac{dy}{dx} = 1, dy = \frac{dx}{\sqrt{x^2-9}}; y = \int \frac{dx}{\sqrt{x^2-9}}; \begin{bmatrix} x = 3\sec\theta, 0 < \theta < \frac{\pi}{2} \\ dx = 3\sec\theta\tan\theta\, d\theta \\ \sqrt{x^2-9} = 3\tan\theta \end{bmatrix} \rightarrow y = \int \frac{3\sec\theta\tan\theta\, d\theta}{3\tan\theta}$

$= \int \sec\theta\, d\theta = \ln|\sec\theta + \tan\theta| + C = \ln\left|\frac{x}{3} + \frac{\sqrt{x^2-9}}{3}\right| + C; x = 5 \text{ and } y = \ln 3 \Rightarrow \ln 3 = \ln 3 + C \Rightarrow C = 0$

$\Rightarrow y = \ln\left|\frac{x}{3} + \frac{\sqrt{x^2-9}}{3}\right|$

39. $(x^2+4)\frac{dy}{dx} = 3, dy = \frac{3\, dx}{x^2+4}; y = 3\int \frac{dx}{x^2+4} = \frac{3}{2}\tan^{-1}\frac{x}{2} + C; x = 2 \text{ and } y = 0 \Rightarrow 0 = \frac{3}{2}\tan^{-1}1 + C$

$\Rightarrow C = -\frac{3\pi}{8} \Rightarrow y = \frac{3}{2}\tan^{-1}\left(\frac{x}{2}\right) - \frac{3\pi}{8}$

40. $(x^2+1)^2\frac{dy}{dx} = \sqrt{x^2+1}, dy = \frac{dx}{(x^2+1)^{3/2}}; x = \tan\theta, dx = \sec^2\theta\, d\theta, (x^2+1)^{3/2} = \sec^3\theta;$

$y = \int \frac{\sec^2\theta\, d\theta}{\sec^3\theta} = \int \cos\theta\, d\theta = \sin\theta + C = \tan\theta\cos\theta + C = \frac{\tan\theta}{\sec\theta} + C = \frac{x}{\sqrt{x^2+1}} + C; x = 0 \text{ and } y = 1$

$\Rightarrow 1 = 0 + C \Rightarrow y = \frac{x}{\sqrt{x^2+1}} + 1$

41. $A = \int_0^3 \frac{\sqrt{9-x^2}}{3}\, dx; x = 3\sin\theta, 0 \le \theta \le \frac{\pi}{2}, dx = 3\cos\theta\, d\theta, \sqrt{9-x^2} = \sqrt{9-9\sin^2\theta} = 3\cos\theta;$

$A = \int_0^{\pi/2} \frac{3\cos\theta \cdot 3\cos\theta\, d\theta}{3} = 3\int_0^{\pi/2}\cos^2\theta\, d\theta = \frac{3}{2}[\theta + \sin\theta\cos\theta]_0^{\pi/2} = \frac{3\pi}{4}$

42. $V = \int_0^1 \pi \left(\frac{2}{1+x^2}\right)^2 dx = 4\pi \int_0^1 \frac{dx}{(x^2+1)^2}$;
$x = \tan\theta,\ dx = \sec^2\theta\, d\theta,\ x^2+1 = \sec^2\theta$;
$V = 4\pi \int_0^{\pi/4} \frac{\sec^2\theta\, d\theta}{\sec^4\theta} = 4\pi \int_0^{\pi/4} \cos^2\theta\, d\theta$
$= 2\pi \int_0^{\pi/4} (1+\cos 2\theta)\, d\theta = 2\pi \left[\theta + \frac{\sin 2\theta}{2}\right]_0^{\pi/4} = \pi\left(\frac{\pi}{2}+1\right)$

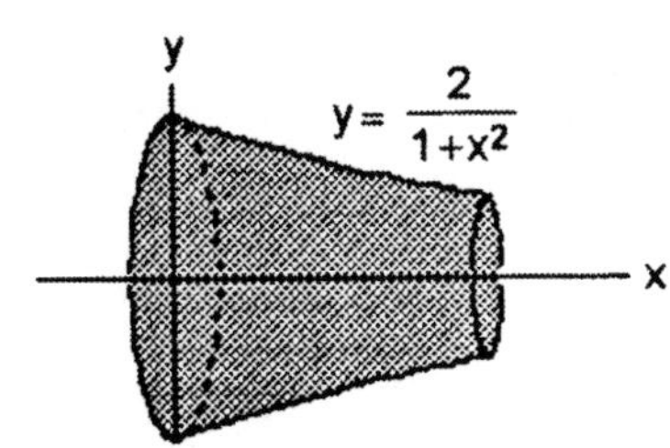

43. $\int \frac{dx}{1-\sin x} = \int \frac{\left(\frac{2\,dz}{1+z^2}\right)}{1-\left(\frac{2z}{1+z^2}\right)} = \int \frac{2\,dz}{(1-z)^2} = \frac{2}{1-z} + C = \frac{2}{1-\tan\left(\frac{x}{2}\right)} + C$

44. $\int \frac{dx}{1+\sin x+\cos x} = \int \frac{\left(\frac{2\,dz}{1+z^2}\right)}{1+\left(\frac{2z}{1+z^2}+\frac{1-z^2}{1+z^2}\right)} = \int \frac{2\,dz}{1+z^2+2z+1-z^2} = \int \frac{dz}{1+z} = \ln|1+z| + C$
$= \ln\left|\tan\left(\frac{x}{2}\right)+1\right| + C$

45. $\int_0^{\pi/2} \frac{dx}{1+\sin x} = \int_0^1 \frac{\left(\frac{2\,dz}{1+z^2}\right)}{1+\left(\frac{2z}{1+z^2}\right)} = \int_0^1 \frac{2\,dz}{(1+z)^2} = -\left[\frac{2}{1+z}\right]_0^1 = -(1-2) = 1$

46. $\int_{\pi/3}^{\pi/2} \frac{dx}{1-\cos x} = \int_{1/\sqrt{3}}^1 \frac{\left(\frac{2\,dz}{1+z^2}\right)}{1-\left(\frac{1-z^2}{1+z^2}\right)} = \int_{1/\sqrt{3}}^1 \frac{dz}{z^2} = \left[-\frac{1}{z}\right]_{1/\sqrt{3}}^1 = \sqrt{3}-1$

47. $\int_0^{\pi/2} \frac{d\theta}{2+\cos\theta} = \int_0^1 \frac{\left(\frac{2\,dz}{1+z^2}\right)}{2+\left(\frac{1-z^2}{1+z^2}\right)} = \int_0^1 \frac{2\,dz}{2+2z^2+1-z^2} = \int_0^1 \frac{2\,dz}{z^2+3} = \frac{2}{\sqrt{3}}\left[\tan^{-1}\frac{z}{\sqrt{3}}\right]_0^1 = \frac{2}{\sqrt{3}}\tan^{-1}\frac{1}{\sqrt{3}}$
$= \frac{\pi}{3\sqrt{3}} = \frac{\sqrt{3}\pi}{9}$

48. $\int_{\pi/2}^{2\pi/3} \frac{\cos\theta\, d\theta}{\sin\theta\cos\theta+\sin\theta} = \int_1^{\sqrt{3}} \frac{\left(\frac{1-z^2}{1+z^2}\right)\left(\frac{2\,dz}{1+z^2}\right)}{\left[\frac{2z(1-z^2)}{(1+z^2)^2}+\left(\frac{2z}{1+z^2}\right)\right]} = \int_1^{\sqrt{3}} \frac{2(1-z^2)\,dz}{2z-2z^3+2z+2z^3} = \int_1^{\sqrt{3}} \frac{1-z^2}{2z}\,dz$
$= \left[\frac{1}{2}\ln z - \frac{z^2}{4}\right]_1^{\sqrt{3}} = \left(\frac{1}{2}\ln\sqrt{3}-\frac{3}{4}\right) - \left(0-\frac{1}{4}\right) = \frac{\ln 3}{4} - \frac{1}{2} = \frac{1}{4}(\ln 3 - 2) = \frac{1}{2}\left(\ln\sqrt{3}-1\right)$

49. $\int \frac{dt}{\sin t-\cos t} = \int \frac{\left(\frac{2\,dz}{1+z^2}\right)}{\left(\frac{2z}{1+z^2}-\frac{1-z^2}{1+z^2}\right)} = \int \frac{2\,dz}{2z-1+z^2} = \int \frac{2\,dz}{(z+1)^2-2} = \frac{1}{\sqrt{2}}\ln\left|\frac{z+1-\sqrt{2}}{z+1+\sqrt{2}}\right| + C$
$= \frac{1}{\sqrt{2}}\ln\left|\frac{\tan\left(\frac{t}{2}\right)+1-\sqrt{2}}{\tan\left(\frac{t}{2}\right)+1+\sqrt{2}}\right| + C$

50. $\int \frac{\cos t\, dt}{1-\cos t} = \int \frac{\left(\frac{1-z^2}{1+z^2}\right)\left(\frac{2\,dz}{1+z^2}\right)}{1-\left(\frac{1-z^2}{1+z^2}\right)} = \int \frac{2(1-z^2)\,dz}{(1+z^2)^2-(1+z^2)(1-z^2)} = \int \frac{2(1-z^2)\,dz}{(1+z^2)(1+z^2-1+z^2)}$
$= \int \frac{(1-z^2)\,dz}{(1+z^2)z^2} = \int \frac{dz}{z^2(1+z^2)} - \int \frac{dz}{1+z^2} = \int \frac{dz}{z^2} - 2\int \frac{dz}{z^2+1} = -\frac{1}{z} - 2\tan^{-1} z + C = -\cot\left(\frac{t}{2}\right) - t + C$

51. $\int \sec\theta\, d\theta = \int \frac{d\theta}{\cos\theta} = \int \frac{\left(\frac{2\,dz}{1+z^2}\right)}{\left(\frac{1-z^2}{1+z^2}\right)} = \int \frac{2\,dz}{1-z^2} = \int \frac{2\,dz}{(1+z)(1-z)} = \int \frac{dz}{1+z} + \int \frac{dz}{1-z}$
$= \ln|1+z| - \ln|1-z| + C = \ln\left|\frac{1+\tan\left(\frac{\theta}{2}\right)}{1-\tan\left(\frac{\theta}{2}\right)}\right| + C$

52. $\int \csc\theta\, d\theta = \int \frac{d\theta}{\sin\theta} = \int \frac{\left(\frac{2\,dz}{1+z^2}\right)}{\left(\frac{2z}{1+z^2}\right)} = \int \frac{dz}{z} = \ln|z| + C = \ln\left|\tan\frac{\theta}{2}\right| + C$

8.6 INTEGRAL TABLES AND COMPUTER ALGEBRA SYSTEMS

1. $\int \frac{dx}{x\sqrt{x-3}} = \frac{2}{\sqrt{3}} \tan^{-1} \sqrt{\frac{x-3}{3}} + C$

(We used FORMULA 13(a) with a = 1, b = 3)

2. $\int \frac{dx}{x\sqrt{x+4}} = \frac{1}{\sqrt{4}} \ln \left| \frac{\sqrt{x+4}-\sqrt{4}}{\sqrt{x+4}+\sqrt{4}} \right| + C = \frac{1}{2} \ln \left| \frac{\sqrt{x+4}-2}{\sqrt{x+4}+2} \right| + C$

(We used FORMULA 13(b) with a = 1, b = 4)

3. $\int \frac{x\,dx}{\sqrt{x-2}} = \int \frac{(x-2)\,dx}{\sqrt{x-2}} + 2\int \frac{dx}{\sqrt{x-2}} = \int \left(\sqrt{x-2}\right)^1 dx + 2\int \left(\sqrt{x-2}\right)^{-1} dx$

$= \left(\frac{2}{1}\right) \frac{\left(\sqrt{x-2}\right)^3}{3} + 2\left(\frac{2}{1}\right) \frac{\left(\sqrt{x-2}\right)^1}{1} = \sqrt{x-2}\left[\frac{2(x-2)}{3} + 4\right] + C$

(We used FORMULA 11 with a = 1, b = −2, n = 1 and a = 1, b = −2, n = −1)

4. $\int \frac{x\,dx}{(2x+3)^{3/2}} = \frac{1}{2} \int \frac{(2x+3)\,dx}{(2x+3)^{3/2}} - \frac{3}{2} \int \frac{dx}{(2x+3)^{3/2}} = \frac{1}{2} \int \frac{dx}{\sqrt{2x+3}} - \frac{3}{2} \int \frac{dx}{\left(\sqrt{2x+3}\right)^3}$

$= \frac{1}{2} \int \left(\sqrt{2x+3}\right)^{-1} dx - \frac{3}{2} \int \left(\sqrt{2x+3}\right)^{-3} dx = \left(\frac{1}{2}\right)\left(\frac{2}{2}\right) \frac{\left(\sqrt{2x+3}\right)^1}{1} - \left(\frac{3}{2}\right)\left(\frac{2}{2}\right) \frac{\left(\sqrt{2x+3}\right)^{-1}}{(-1)} + C$

$= \frac{1}{2\sqrt{2x+3}}(2x+3+3) + C = \frac{(x+3)}{\sqrt{2x+3}} + C$

(We used FORMULA 11 with a = 2, b = 3, n = −1 and a = 2, b = 3, n = −3)

5. $\int x\sqrt{2x-3}\,dx = \frac{1}{2} \int (2x-3)\sqrt{2x-3}\,dx + \frac{3}{2} \int \sqrt{2x-3}\,dx = \frac{1}{2} \int \left(\sqrt{2x-3}\right)^3 dx + \frac{3}{2} \int \left(\sqrt{2x-3}\right)^1 dx$

$= \left(\frac{1}{2}\right)\left(\frac{2}{2}\right) \frac{\left(\sqrt{2x-3}\right)^5}{5} + \left(\frac{3}{2}\right)\left(\frac{2}{2}\right) \frac{\left(\sqrt{2x-3}\right)^3}{3} + C = \frac{(2x-3)^{3/2}}{2}\left[\frac{2x-3}{5} + 1\right] + C = \frac{(2x-3)^{3/2}(x+1)}{5} + C$

(We used FORMULA 11 with a = 2, b = −3, n = 3 and a = 2, b = −3, n = 1)

6. $\int x(7x+5)^{3/2}\,dx = \frac{1}{7} \int (7x+5)(7x+5)^{3/2}\,dx - \frac{5}{7} \int (7x+5)^{3/2}\,dx = \frac{1}{7} \int \left(\sqrt{7x+5}\right)^5 dx - \frac{5}{7} \int \left(\sqrt{7x+5}\right)^3 dx$

$= \left(\frac{1}{7}\right)\left(\frac{2}{7}\right) \frac{\left(\sqrt{7x+5}\right)^7}{7} - \left(\frac{5}{7}\right)\left(\frac{2}{7}\right) \frac{\left(\sqrt{7x+5}\right)^5}{5} + C = \left[\frac{(7x+5)^{5/2}}{49}\right]\left[\frac{2(7x+5)}{7} - 2\right] + C$

$= \left[\frac{(7x+5)^{5/2}}{49}\right]\left(\frac{14x-4}{7}\right) + C$

(We used FORMULA 11 with a = 7, b = 5, n = 5 and a = 7, b = 5, n = 3)

7. $\int \frac{\sqrt{9-4x}}{x^2}\,dx = -\frac{\sqrt{9-4x}}{x} + \frac{(-4)}{2} \int \frac{dx}{x\sqrt{9-4x}} + C$

(We used FORMULA 14 with a = −4, b = 9)

$= -\frac{\sqrt{9-4x}}{x} - 2\left(\frac{1}{\sqrt{9}}\right) \ln \left| \frac{\sqrt{9-4x}-\sqrt{9}}{\sqrt{9-4x}+\sqrt{9}} \right| + C$

(We used FORMULA 13(b) with a = −4, b = 9)

$= \frac{-\sqrt{9-4x}}{x} - \frac{2}{3} \ln \left| \frac{\sqrt{9-4x}-3}{\sqrt{9-4x}+3} \right| + C$

8. $\int \frac{dx}{x^2\sqrt{4x-9}} = -\frac{\sqrt{4x-9}}{(-9)x} + \frac{4}{18} \int \frac{dx}{x\sqrt{4x-9}} + C$

(We used FORMULA 15 with a = 4, b = −9)

$= \frac{\sqrt{4x-9}}{9x} + \left(\frac{2}{9}\right)\left(\frac{2}{\sqrt{9}}\right) \tan^{-1} \sqrt{\frac{4x-9}{9}} + C$

(We used FORMULA 13(a) with a = 4, b = 9)

$= \frac{\sqrt{4x-9}}{9x} + \frac{4}{27} \tan^{-1} \sqrt{\frac{4x-9}{9}} + C$

9. $\int x\sqrt{4x-x^2}\,dx = \int x\sqrt{2\cdot 2x - x^2}\,dx = \frac{(x+2)(2x-3\cdot 2)\sqrt{2\cdot 2\cdot x - x^2}}{6} + \frac{2^3}{2}\sin^{-1}\left(\frac{x-2}{2}\right) + C$
$= \frac{(x+2)(2x-6)\sqrt{4x-x^2}}{6} + 4\sin^{-1}\left(\frac{x-2}{2}\right) + C = \frac{(x+2)(x-3)\sqrt{4x-x^2}}{3} + 4\sin^{-1}\left(\frac{x-2}{2}\right) + C$
(We used FORMULA 51 with $a = 2$)

10. $\int \frac{\sqrt{x-x^2}}{x}\,dx = \int \frac{\sqrt{2\cdot\frac{1}{2}x - x^2}}{x}\,dx = \sqrt{2\cdot\frac{1}{2}x - x^2} + \frac{1}{2}\sin^{-1}\left(\frac{x-\frac{1}{2}}{\frac{1}{2}}\right) + C = \sqrt{x-x^2} + \frac{1}{2}\sin^{-1}(2x-1) + C$
(We used FORMULA 52 with $a = \frac{1}{2}$)

11. $\int \frac{dx}{x\sqrt{7+x^2}} = \int \frac{dx}{x\sqrt{\left(\sqrt{7}\right)^2 + x^2}} = -\frac{1}{\sqrt{7}}\ln\left|\frac{\sqrt{7}+\sqrt{\left(\sqrt{7}\right)^2+x^2}}{x}\right| + C = -\frac{1}{\sqrt{7}}\ln\left|\frac{\sqrt{7}+\sqrt{7+x^2}}{x}\right| + C$
(We used FORMULA 26 with $a = \sqrt{7}$)

12. $\int \frac{dx}{x\sqrt{7-x^2}} = \int \frac{dx}{x\sqrt{\left(\sqrt{7}\right)^2 - x^2}} = -\frac{1}{\sqrt{7}}\ln\left|\frac{\sqrt{7}+\sqrt{\left(\sqrt{7}\right)^2-x^2}}{x}\right| + C = -\frac{1}{\sqrt{7}}\ln\left|\frac{\sqrt{7}+\sqrt{7-x^2}}{x}\right| + C$
(We used FORMULA 34 with $a = \sqrt{7}$)

13. $\int \frac{\sqrt{4-x^2}}{x}\,dx = \int \frac{\sqrt{2^2-x^2}}{x}\,dx = \sqrt{2^2-x^2} - 2\ln\left|\frac{2+\sqrt{2^2-x^2}}{x}\right| + C = \sqrt{4-x^2} - 2\ln\left|\frac{2+\sqrt{4-x^2}}{x}\right| + C$
(We used FORMULA 31 with $a = 2$)

14. $\int \frac{\sqrt{x^2-4}}{x}\,dx = \int \frac{\sqrt{x^2-2^2}}{x}\,dx = \sqrt{x^2-2^2} - 2\sec^{-1}\left|\frac{x}{2}\right| + C = \sqrt{x^2-4} - 2\sec^{-1}\left|\frac{x}{2}\right| + C$
(We used FORMULA 42 with $a = 2$)

15. $\int \sqrt{25-p^2}\,dp = \int \sqrt{5^2-p^2}\,dp = \frac{p}{2}\sqrt{5^2-p^2} + \frac{5^2}{2}\sin^{-1}\frac{p}{5} + C = \frac{p}{2}\sqrt{25-p^2} + \frac{25}{2}\sin^{-1}\frac{p}{5} + C$
(We used FORMULA 29 with $a = 5$)

16. $\int q^2\sqrt{25-q^2}\,dq = \int q^2\sqrt{5^2-q^2}\,dq = \frac{5^4}{8}\sin^{-1}\left(\frac{q}{5}\right) - \frac{1}{8}q\sqrt{5^2-q^2}\,(5^2-2q^2) + C$
$= \frac{625}{8}\sin^{-1}\left(\frac{q}{5}\right) - \frac{1}{8}q\sqrt{25-q^2}\,(25-2q^2) + C$
(We used FORMULA 30 with $a = 5$)

17. $\int \frac{r^2}{\sqrt{4-r^2}}\,dr = \int \frac{r^2}{\sqrt{2^2-r^2}}\,dr = \frac{2^2}{2}\sin^{-1}\left(\frac{r}{2}\right) - \frac{1}{2}r\sqrt{2^2-r^2} + C = 2\sin^{-1}\left(\frac{r}{2}\right) - \frac{1}{2}r\sqrt{4-r^2} + C$
(We used FORMULA 33 with $a = 2$)

18. $\int \frac{ds}{\sqrt{s^2-2}} = \int \frac{ds}{\sqrt{s^2-\left(\sqrt{2}\right)^2}} = \cosh^{-1}\frac{s}{\sqrt{2}} + C = \ln\left|s + \sqrt{s^2-\left(\sqrt{2}\right)^2}\right| + C = \ln\left|s+\sqrt{s^2-2}\right| + C$
(We used FORMULA 36 with $a = \sqrt{2}$)

19. $\int \frac{d\theta}{5+4\sin 2\theta} = \frac{-2}{2\sqrt{25-16}}\tan^{-1}\left[\sqrt{\frac{5-4}{5+4}}\tan\left(\frac{\pi}{4}-\frac{2\theta}{2}\right)\right] + C = -\frac{1}{3}\tan^{-1}\left[\frac{1}{3}\tan\left(\frac{\pi}{4}-\theta\right)\right] + C$
(We used FORMULA 70 with $b = 5$, $c = 4$, $a = 2$)

20. $\int \frac{d\theta}{4+5\sin 2\theta} = \frac{-1}{2\sqrt{25-16}}\ln\left|\frac{5+4\sin 2\theta + \sqrt{25-16}\cos 2\theta}{4+5\sin 2\theta}\right| + C = -\frac{1}{6}\ln\left|\frac{5+4\sin 2\theta + 3\cos 2\theta}{4+5\sin 2\theta}\right| + C$
(We used FORMULA 71 with $a = 2$, $b = 4$, $c = 5$)

21. $\int e^{2t} \cos 3t\, dt = \frac{e^{2t}}{2^2+3^2}(2 \cos 3t + 3 \sin 3t) + C = \frac{e^{2t}}{13}(2 \cos 3t + 3 \sin 3t) + C$

(We used FORMULA 108 with $a = 2$, $b = 3$)

22. $\int e^{-3t} \sin 4t\, dt = \frac{e^{-3t}}{(-3)^2+4^2}(-3 \sin 4t - 4 \cos 4t) + C = \frac{e^{-3t}}{25}(-3 \sin 4t - 4 \cos 4t) + C$

(We used FORMULA 107 with $a = -3$, $b = 4$)

23. $\int x \cos^{-1} x\, dx = \int x^1 \cos^{-1} x\, dx = \frac{x^{1+1}}{1+1} \cos^{-1} x + \frac{1}{1+1} \int \frac{x^{1+1}\, dx}{\sqrt{1-x^2}} = \frac{x^2}{2} \cos^{-1} x + \frac{1}{2} \int \frac{x^2\, dx}{\sqrt{1-x^2}}$

(We used FORMULA 100 with $a = 1$, $n = 1$)

$= \frac{x^2}{2} \cos^{-1} x + \frac{1}{2}\left(\frac{1}{2} \sin^{-1} x\right) - \frac{1}{2}\left(\frac{1}{2} x\sqrt{1-x^2}\right) + C = \frac{x^2}{2} \cos^{-1} x + \frac{1}{4} \sin^{-1} x - \frac{1}{4} x\sqrt{1-x^2} + C$

(We used FORMULA 33 with $a = 1$)

24. $\int x \tan^{-1} x\, dx = \int x^1 \tan^{-1}(1x)\, dx = \frac{x^{1+1}}{1+1} \tan^{-1}(1x) - \frac{1}{1+1} \int \frac{x^{1+1}\, dx}{1+(1)^2x^2} = \frac{x^2}{2} \tan^{-1} x - \frac{1}{2} \int \frac{x^2\, dx}{1+x^2}$

(We used FORMULA 101 with $a = 1$, $n = 1$)

$= \frac{x^2}{2} \tan^{-1} x - \frac{1}{2} \int \left(1 - \frac{1}{1+x^2}\right) dx$ (after long division)

$= \frac{x^2}{2} \tan^{-1} x - \frac{1}{2} \int dx + \frac{1}{2} \int \frac{1}{1+x^2} dx = \frac{x^2}{2} \tan^{-1} x - \frac{1}{2} x + \frac{1}{2} \tan^{-1} x + C = \frac{1}{2}((x^2+1)\tan^{-1} x - x) + C$

25. $\int \frac{ds}{(9-s^2)^2} = \int \frac{ds}{(3^3-s^2)^2} = \frac{s}{2 \cdot 3^2 \cdot (3^2-s^2)} + \frac{1}{4 \cdot 3^3} \ln \left|\frac{s+3}{s-3}\right| + C$

(We used FORMULA 19 with $a = 3$)

$= \frac{s}{18(9-s^2)} + \frac{1}{108} \ln \left|\frac{s+3}{s-3}\right| + C$

26. $\int \frac{d\theta}{(2-\theta^2)^2} = \int \frac{d\theta}{\left[\left(\sqrt{2}\right)^2 - \theta^2\right]^2} = \frac{\theta}{2\left(\sqrt{2}\right)^2\left[\left(\sqrt{2}\right)^2 - \theta^2\right]} + \frac{1}{4\left(\sqrt{2}\right)^3} \ln \left|\frac{\theta+\sqrt{2}}{\theta-\sqrt{2}}\right| + C$

(We used FORMULA 19 with $a = \sqrt{2}$)

$= \frac{\theta}{4(2-\theta^2)} + \frac{1}{8\sqrt{2}} \ln \left|\frac{\theta+\sqrt{2}}{\theta-\sqrt{2}}\right| + C$

27. $\int \frac{\sqrt{4x+9}}{x^2}\, dx = -\frac{\sqrt{4x+9}}{x} + \frac{4}{2} \int \frac{dx}{x\sqrt{4x+9}}$

(We used FORMULA 14 with $a = 4$, $b = 9$)

$= -\frac{\sqrt{4x+9}}{x} + 2\left(\frac{1}{\sqrt{9}} \ln \left|\frac{\sqrt{4x+9}-\sqrt{9}}{\sqrt{4x+9}+\sqrt{9}}\right|\right) + C = -\frac{\sqrt{4x+9}}{x} + \frac{2}{3} \ln \left|\frac{\sqrt{4x+9}-3}{\sqrt{4x+9}+3}\right| + C$

(We used FORMULA 13(b) with $a = 4$, $b = 9$)

28. $\int \frac{\sqrt{9x-4}}{x^2}\, dx = -\frac{\sqrt{9x-4}}{x} + \frac{9}{2} \int \frac{dx}{x\sqrt{9x-4}} + C$

(We used FORMULA 14 with $a = 9$, $b = -4$)

$= -\frac{\sqrt{9x-4}}{x} + \frac{9}{2}\left(\frac{2}{\sqrt{4}} \tan^{-1} \sqrt{\frac{9x-4}{4}}\right) + C = -\frac{\sqrt{9x-4}}{x} + \frac{9}{2} \tan^{-1} \frac{\sqrt{9x-4}}{2} + C$

(We used FORMULA 13(a) with $a = 9$, $b = 4$)

29. $\int \frac{\sqrt{3t-4}}{t}\, dt = 2\sqrt{3t-4} + (-4) \int \frac{dt}{t\sqrt{3t-4}}$

(We used FORMULA 12 with $a = 3$, $b = -4$)

$= 2\sqrt{3t-4} - 4\left(\frac{2}{\sqrt{4}} \tan^{-1} \sqrt{\frac{3t-4}{4}}\right) + C = 2\sqrt{3t-4} - 4 \tan^{-1} \frac{\sqrt{3t-4}}{2} + C$

(We used FORMULA 13(a) with $a = 3$, $b = 4$)

30. $\int \frac{\sqrt{3t+9}}{t}\,dt = 2\sqrt{3t+9} + 9\int \frac{dt}{t\sqrt{3t+9}}$
(We used FORMULA 12 with a = 3, b = 9)
$= 2\sqrt{3t+9} + 9\left(\frac{1}{\sqrt{9}} \ln\left|\frac{\sqrt{3t+9}-\sqrt{9}}{\sqrt{3t+9}+\sqrt{9}}\right|\right) + C = 2\sqrt{3t+9} + 3\ln\left|\frac{\sqrt{3t+9}-3}{\sqrt{3t+9}+3}\right| + C$
(We used FORMULA 13(b) with a = 3, b = 9)

31. $\int x^2 \tan^{-1} x\,dx = \frac{x^{2+1}}{2+1}\tan^{-1} x - \frac{1}{2+1}\int \frac{x^{2+1}}{1+x^2}\,dx = \frac{x^3}{3}\tan^{-1} x - \frac{1}{3}\int \frac{x^3}{1+x^2}\,dx$
(We used FORMULA 101 with a = 1, n = 2);
$\int \frac{x^3}{1+x^2}\,dx = \int x\,dx - \int \frac{x\,dx}{1+x^2} = \frac{x^2}{2} - \frac{1}{2}\ln(1+x^2) + C \Rightarrow \int x^2\tan^{-1} x\,dx$
$= \frac{x^3}{3}\tan^{-1} x - \frac{x^2}{6} + \frac{1}{6}\ln(1+x^2) + C$

32. $\int \frac{\tan^{-1} x}{x^2}\,dx = \int x^{-2}\tan^{-1} x\,dx = \frac{x^{(-2+1)}}{(-2+1)}\tan^{-1} x - \frac{1}{(-2+1)}\int \frac{x^{(-2+1)}}{1+x^2}\,dx = \frac{x^{-1}}{(-1)}\tan^{-1} x + \int \frac{x^{-1}}{(1+x^2)}\,dx$
(We used FORMULA 101 with a = 1, n = −2);
$\int \frac{x^{-1}\,dx}{1+x^2} = \int \frac{dx}{x(1+x^2)} = \int \frac{dx}{x} - \int \frac{x\,dx}{1+x^2} = \ln|x| - \frac{1}{2}\ln(1+x^2) + C$
$\Rightarrow \int \frac{\tan^{-1} x}{x^2}\,dx = -\frac{1}{x}\tan^{-1} x + \ln|x| - \frac{1}{2}\ln(1+x^2) + C$

33. $\int \sin 3x \cos 2x\,dx = -\frac{\cos 5x}{10} - \frac{\cos x}{2} + C$
(We used FORMULA 62(a) with a = 3, b = 2)

34. $\int \sin 2x \cos 3x\,dx = -\frac{\cos 5x}{10} + \frac{\cos x}{2} + C$
(We used FORMULA 62(a) with a = 2, b = 3)

35. $\int 8\sin 4t \sin \frac{t}{2}\,dx = \frac{8}{7}\sin\left(\frac{7t}{2}\right) - \frac{8}{9}\sin\left(\frac{9t}{2}\right) + C = 8\left[\frac{\sin\left(\frac{7t}{2}\right)}{7} - \frac{\sin\left(\frac{9t}{2}\right)}{9}\right] + C$
(We used FORMULA 62(b) with a = 4, b = $\frac{1}{2}$)

36. $\int \sin \frac{t}{3} \sin \frac{t}{6}\,dt = 3\sin\left(\frac{t}{6}\right) - \sin\left(\frac{t}{2}\right) + C$
(We used FORMULA 62(b) with a = $\frac{1}{3}$, b = $\frac{1}{6}$)

37. $\int \cos \frac{\theta}{3} \cos \frac{\theta}{4}\,d\theta = 6\sin\left(\frac{\theta}{12}\right) + \frac{6}{7}\sin\left(\frac{7\theta}{12}\right) + C$
(We used FORMULA 62(c) with a = $\frac{1}{3}$, b = $\frac{1}{4}$)

38. $\int \cos \frac{\theta}{2} \cos 7\theta\,d\theta = \frac{1}{13}\sin\left(\frac{13\theta}{2}\right) + \frac{1}{15}\sin\left(\frac{15\theta}{2}\right) + C = \frac{\sin\left(\frac{13\theta}{2}\right)}{13} + \frac{\sin\left(\frac{15\theta}{2}\right)}{15} + C$
(We used FORMULA 62(c) with a = $\frac{1}{2}$, b = 7)

39. $\int \frac{x^3+x+1}{(x^2+1)^2}\,dx = \int \frac{x\,dx}{x^2+1} + \int \frac{dx}{(x^2+1)^2} = \frac{1}{2}\int \frac{d(x^2+1)}{x^2+1} + \int \frac{dx}{(x^2+1)^2}$
$= \frac{1}{2}\ln(x^2+1) + \frac{x}{2(1+x^2)} + \frac{1}{2}\tan^{-1} x + C$
(For the second integral we used FORMULA 17 with a = 1)

40. $\int \frac{x^2+6x}{(x^2+3)^2}\,dx = \int \frac{dx}{x^2+3} + \int \frac{6x\,dx}{(x^2+3)^2} - \int \frac{3\,dx}{(x^2+3)^2} = \int \frac{dx}{x^2+\left(\sqrt{3}\right)^2} + 3\int \frac{d(x^2+3)}{(x^2+3)^2} - 3\int \frac{dx}{\left[x^2+\left(\sqrt{3}\right)^2\right]^2}$

$$= \frac{1}{\sqrt{3}} \tan^{-1}\left(\frac{x}{\sqrt{3}}\right) - \frac{3}{(x^2+3)} - 3\left(\frac{x}{2\left(\sqrt{3}\right)^2\left(\left(\sqrt{3}\right)^2+x^2\right)} + \frac{1}{2\left(\sqrt{3}\right)^3}\tan^{-1}\left(\frac{x}{\sqrt{3}}\right)\right) + C$$

(For the first integral we used FORMULA 16 with $a = \sqrt{3}$; for the third integral we used FORMULA 17 with $a = \sqrt{3}$)

$$= \frac{1}{2\sqrt{3}} \tan^{-1}\left(\frac{x}{\sqrt{3}}\right) - \frac{3}{x^2+3} - \frac{x}{2(x^2+3)} + C$$

41. $\int \sin^{-1}\sqrt{x}\,dx;\ \begin{bmatrix} u = \sqrt{x} \\ x = u^2 \\ dx = 2u\,du \end{bmatrix} \to 2\int u^1 \sin^{-1} u\,du = 2\left(\frac{u^{1+1}}{1+1}\sin^{-1} u - \frac{1}{1+1}\int \frac{u^{1+1}}{\sqrt{1-u^2}}\,du\right)$

$= u^2 \sin^{-1} u - \int \frac{u^2\,du}{\sqrt{1-u^2}}$

(We used FORMULA 99 with $a = 1, n = 1$)

$= u^2 \sin^{-1} u - \left(\frac{1}{2}\sin^{-1} u - \frac{1}{2}u\sqrt{1-u^2}\right) + C = \left(u^2 - \frac{1}{2}\right)\sin^{-1} u + \frac{1}{2}u\sqrt{1-u^2} + C$

(We used FORMULA 33 with $a = 1$)

$= \left(x - \frac{1}{2}\right)\sin^{-1}\sqrt{x} + \frac{1}{2}\sqrt{x - x^2} + C$

42. $\int \frac{\cos^{-1}\sqrt{x}}{\sqrt{x}}\,dx;\ \begin{bmatrix} u = \sqrt{x} \\ x = u^2 \\ dx = 2u\,du \end{bmatrix} \to \int \frac{\cos^{-1} u}{u} \cdot 2u\,du = 2\int \cos^{-1} u\,du = 2\left(u\cos^{-1} u - \frac{1}{1}\sqrt{1-u^2}\right) + C$

(We used FORMULA 97 with $a = 1$)

$= 2\left(\sqrt{x}\cos^{-1}\sqrt{x} - \sqrt{1-x}\right) + C$

43. $\int \frac{\sqrt{x}}{\sqrt{1-x}}\,dx;\ \begin{bmatrix} u = \sqrt{x} \\ x = u^2 \\ dx = 2u\,du \end{bmatrix} \to \int \frac{u \cdot 2u}{\sqrt{1-u^2}}\,du = 2\int \frac{u^2}{\sqrt{1-u^2}}\,du = 2\left(\frac{1}{2}\sin^{-1} u - \frac{1}{2}u\sqrt{1-u^2}\right) + C$

$= \sin^{-1} u - u\sqrt{1-u^2} + C$

(We used FORMULA 33 with $a = 1$)

$= \sin^{-1}\sqrt{x} - \sqrt{x}\sqrt{1-x} + C = \sin^{-1}\sqrt{x} - \sqrt{x - x^2} + C$

44. $\int \frac{\sqrt{2-x}}{\sqrt{x}}\,dx;\ \begin{bmatrix} u = \sqrt{x} \\ x = u^2 \\ dx = 2u\,du \end{bmatrix} \to \int \frac{\sqrt{2-u^2}}{u} \cdot 2u\,du = 2\int \sqrt{\left(\sqrt{2}\right)^2 - u^2}\,du$

$= 2\left[\frac{u}{2}\sqrt{\left(\sqrt{2}\right)^2 - u^2} + \frac{\left(\sqrt{2}\right)^2}{2}\sin^{-1}\left(\frac{u}{\sqrt{2}}\right)\right] + C = u\sqrt{2-u^2} + 2\sin^{-1}\left(\frac{u}{\sqrt{2}}\right) + C$

(We used FORMULA 29 with $a = \sqrt{2}$)

$= \sqrt{2x - x^2} + 2\sin^{-1}\sqrt{\frac{x}{2}} + C$

45. $\int (\cot t)\sqrt{1 - \sin^2 t}\,dt = \int \frac{\sqrt{1-\sin^2 t}\,(\cos t)\,dt}{\sin t};\ \begin{bmatrix} u = \sin t \\ du = \cos t\,dt \end{bmatrix} \to \int \frac{\sqrt{1-u^2}\,du}{u}$

$= \sqrt{1-u^2} - \ln\left|\frac{1+\sqrt{1-u^2}}{u}\right| + C$

(We used FORMULA 31 with $a = 1$)

$= \sqrt{1-\sin^2 t} - \ln\left|\frac{1+\sqrt{1-\sin^2 t}}{\sin t}\right| + C$

46. $\int \frac{dt}{(\tan t)\sqrt{4-\sin^2 t}} = \int \frac{\cos t\, dt}{(\sin t)\sqrt{4-\sin^2 t}}$; $\begin{bmatrix} u = \sin t \\ du = \cos t\, dt \end{bmatrix} \to \int \frac{du}{u\sqrt{4-u^2}} = -\frac{1}{2} \ln \left| \frac{2+\sqrt{4-u^2}}{u} \right| + C$

(We used FORMULA 34 with a = 2)

$= -\frac{1}{2} \ln \left| \frac{2+\sqrt{4-\sin^2 t}}{\sin t} \right| + C$

47. $\int \frac{dy}{y\sqrt{3+(\ln y)^2}}$; $\begin{bmatrix} u = \ln y \\ y = e^u \\ dy = e^u\, du \end{bmatrix} \to \int \frac{e^u\, du}{e^u\sqrt{3+u^2}} = \int \frac{du}{\sqrt{3+u^2}} = \ln \left| u + \sqrt{3+u^2} \right| + C$

$= \ln \left| \ln y + \sqrt{3+(\ln y)^2} \right| + C$

(We used FORMULA 20 with $a = \sqrt{3}$)

48. $\int \frac{\cos\theta\, d\theta}{\sqrt{5+\sin^2\theta}}$; $\begin{bmatrix} u = \sin\theta \\ du = \cos\theta\, d\theta \end{bmatrix} \to \int \frac{du}{\sqrt{5+u^2}} = \ln \left| u + \sqrt{5+u^2} \right| + C = \ln \left| \sin\theta + \sqrt{5+\sin^2\theta} \right| + C$

(We used FORMULA 20 with $a = \sqrt{5}$)

49. $\int \frac{3\, dr}{\sqrt{9r^2-1}}$; $\begin{bmatrix} u = 3r \\ du = 3\, dr \end{bmatrix} \to \int \frac{du}{\sqrt{u^2-1}} = \ln \left| u + \sqrt{u^2-1} \right| + C = \ln \left| 3r + \sqrt{9r^2-1} \right| + C$

(We used FORMULA 36 with a = 1)

50. $\int \frac{3\, dy}{\sqrt{1+9y^2}}$; $\begin{bmatrix} u = 3y \\ du = 3\, dy \end{bmatrix} \to \int \frac{du}{\sqrt{1+u^2}} = \ln \left| u + \sqrt{1+u^2} \right| + C = \ln \left| 3y + \sqrt{1+9y^2} \right| + C$

(We used FORMULA 20 with a = 1)

51. $\int \cos^{-1}\sqrt{x}\, dx$; $\begin{bmatrix} t = \sqrt{x} \\ x = t^2 \\ dx = 2t\, dt \end{bmatrix} \to 2\int t \cos^{-1} t\, dt = 2\left(\frac{t^2}{2} \cos^{-1} t + \frac{1}{2} \int \frac{t^2}{\sqrt{1-t^2}}\, dt \right) = t^2 \cos^{-1} t + \int \frac{t^2}{\sqrt{1-t^2}}\, dt$

(We used FORMULA 100 with a = 1, n = 1)

$= t^2 \cos^{-1} t + \frac{1}{2} \sin^{-1} t - \frac{1}{2} t\sqrt{1-t^2} + C$

(We used FORMULA 33 with a = 1)

$= x \cos^{-1}\sqrt{x} + \frac{1}{2} \sin^{-1}\sqrt{x} - \frac{1}{2}\sqrt{x}\sqrt{1-x} + C = x \cos^{-1}\sqrt{x} + \frac{1}{2} \sin^{-1}\sqrt{x} - \frac{1}{2}\sqrt{x-x^2} + C$

52. $\int \tan^{-1}\sqrt{y}\, dy$; $\begin{bmatrix} t = \sqrt{y} \\ y = t^2 \\ dy = 2t\, dt \end{bmatrix} \to 2\int t \tan^{-1} t\, dt = 2\left[\frac{t^2}{2} \tan^{-1} t - \frac{1}{2} \int \frac{t^2}{1+t^2}\, dt \right] = t^2 \tan^{-1} t - \int \frac{t^2}{1+t^2}\, dt$

(We used FORMULA 101 with n = 1, a = 1)

$= t^2 \tan^{-1} t - \int \frac{t^2+1}{t^2+1}\, dt + \int \frac{dt}{1+t^2} = t^2 \tan^{-1} t - t + \tan^{-1} t + C = y \tan^{-1}\sqrt{y} + \tan^{-1}\sqrt{y} - \sqrt{y} + C$

53. $\int \sin^5 2x\, dx = -\frac{\sin^4 2x \cos 2x}{5 \cdot 2} + \frac{5-1}{5} \int \sin^3 2x\, dx = -\frac{\sin^4 2x \cos 2x}{10} + \frac{4}{5}\left[-\frac{\sin^2 2x \cos 2x}{3 \cdot 2} + \frac{3-1}{3} \int \sin 2x\, dx \right]$

(We used FORMULA 60 with a = 2, n = 5 and a = 2, n = 3)

$= -\frac{\sin^4 2x \cos 2x}{10} - \frac{2}{15} \sin^2 2x \cos 2x + \frac{8}{15}\left(-\frac{1}{2}\right) \cos 2x + C = -\frac{\sin^4 2x \cos 2x}{10} - \frac{2 \sin^2 2x \cos 2x}{15} - \frac{4 \cos 2x}{15} + C$

54. $\int \sin^5 \frac{\theta}{2}\, d\theta = -\frac{\sin^4 \frac{\theta}{2} \cos \frac{\theta}{2}}{5 \cdot \frac{1}{2}} + \frac{5-1}{5} \int \sin^3 \frac{\theta}{2}\, d\theta = -\frac{2}{5} \sin^4 \frac{\theta}{2} \cos \frac{\theta}{2} + \frac{4}{5}\left[-\frac{\sin^2 \frac{\theta}{2} \cos \frac{\theta}{2}}{3 \cdot \frac{1}{2}} + \frac{3-1}{3} \int \sin \frac{\theta}{2}\, d\theta \right]$

(We used FORMULA 60 with $a = \frac{1}{2}$, n = 5 and $a = \frac{1}{2}$, n = 3)

$= -\frac{2}{5} \sin^4 \frac{\theta}{2} \cos \frac{\theta}{2} - \frac{8}{15} \sin^2 \frac{\theta}{2} \cos \frac{\theta}{2} + \frac{8}{15}\left(-2 \cos \frac{\theta}{2}\right) + C = -\frac{2}{5} \sin^4 \frac{\theta}{2} \cos \frac{\theta}{2} - \frac{8}{15} \sin^2 \frac{\theta}{2} \cos \frac{\theta}{2} - \frac{16}{15} \cos \frac{\theta}{2} + C$

55. $\int 8\cos^4 2\pi t\, dt = 8\left(\frac{\cos^3 2\pi t \sin 2\pi t}{4\cdot 2\pi} + \frac{4-1}{4}\int \cos^2 2\pi t\, dt\right)$

(We used FORMULA 61 with $a = 2\pi$, $n = 4$)

$= \frac{\cos^3 2\pi t \sin 2\pi t}{\pi} + 6\left[\frac{t}{2} + \frac{\sin(2\cdot 2\pi\cdot t)}{4\cdot 2\pi}\right] + C$

(We used FORMULA 59 with $a = 2\pi$)

$= \frac{\cos^3 2\pi t \sin 2\pi t}{\pi} + 3t + \frac{3\sin 4\pi t}{4\pi} + C = \frac{\cos^3 2\pi t \sin 2\pi t}{\pi} + \frac{3\cos 2\pi t \sin 2\pi t}{2\pi} + 3t + C$

56. $\int 3\cos^5 3y\, dy = 3\left(\frac{\cos^4 3y \sin 3y}{5\cdot 3} + \frac{5-1}{5}\int \cos^3 3y\, dy\right)$

$= \frac{\cos^4 3y \sin 3y}{5} + \frac{12}{5}\left(\frac{\cos^2 3y \sin 3y}{3\cdot 3} + \frac{3-1}{3}\int \cos 3y\, dy\right)$

(We used FORMULA 61 with $a = 3$, $n = 5$ and $a = 3$, $n = 3$)

$= \frac{1}{5}\cos^4 3y \sin 3y + \frac{4}{15}\cos^2 3y \sin 3y + \frac{8}{15}\sin 3y + C$

57. $\int \sin^2 2\theta \cos^3 2\theta\, d\theta = \frac{\sin^3 2\theta \cos^2 2\theta}{2(2+3)} + \frac{3-1}{3+2}\int \sin^2 2\theta \cos 2\theta\, d\theta$

(We used FORMULA 69 with $a = 2$, $m = 3$, $n = 2$)

$= \frac{\sin^3 2\theta \cos^2 2\theta}{10} + \frac{2}{5}\int \sin^2 2\theta \cos 2\theta\, d\theta = \frac{\sin^3 2\theta \cos^2 2\theta}{10} + \frac{2}{5}\left[\frac{1}{2}\int \sin^2 2\theta\, d(\sin 2\theta)\right] = \frac{\sin^3 2\theta \cos^2 2\theta}{10} + \frac{\sin^3 2\theta}{15} + C$

58. $\int 9\sin^3\theta \cos^{3/2}\theta\, d\theta = 9\left[-\frac{\sin^2\theta \cos^{5/2}\theta}{3+\left(\frac{3}{2}\right)} + \frac{3-1}{3+\left(\frac{3}{2}\right)}\int \sin\theta \cos^{3/2}\theta\, d\theta\right]$

$= -2\sin^2\theta \cos^{5/2}\theta + 4\int \cos^{3/2}\theta \sin\theta\, d\theta$

(We used FORMULA 68 with $a = 1$, $n = 3$, $m = \frac{3}{2}$)

$= -2\sin^2\theta \cos^{5/2}\theta - 4\int \cos^{3/2}\theta\, d(\cos\theta) = -2\sin^2\theta \cos^{5/2}\theta - 4\left(\frac{2}{5}\cos^{5/2}\theta\right) + C$

$= \left(-2\cos^{5/2}\theta\right)\left(\sin^2\theta + \frac{4}{5}\right) + C$

59. $\int 2\sin^2 t \sec^4 t\, dt = \int 2\sin^2 t \cos^{-4} t\, dt = 2\left(-\frac{\sin t \cos^{-3} t}{2-4} + \frac{2-1}{2-4}\int \cos^{-4} t\, dt\right)$

(We used FORMULA 68 with $a = 1$, $n = 2$, $m = -4$)

$= \sin t \cos^{-3} t - \int \cos^{-4} t\, dt = \sin t \cos^{-3} t - \int \sec^4 t\, dt = \sin t \cos^{-3} t - \left(\frac{\sec^2 t \tan t}{4-1} + \frac{4-2}{4-1}\int \sec^2 t\, dt\right)$

(We used FORMULA 92 with $a = 1$, $n = 4$)

$= \sin t \cos^{-3} t - \left(\frac{\sec^2 t \tan t}{3}\right) - \frac{2}{3}\tan t + C = \frac{2}{3}\sec^2 t \tan t - \frac{2}{3}\tan t + C = \frac{2}{3}\tan t\left(\sec^2 t - 1\right) + C$

$= \frac{2}{3}\tan^3 t + C$

An easy way to find the integral using substitution:

$\int 2\sin^2 t \cos^{-4} t\, dt = \int 2\tan^2 t \sec^2 t\, dt = \int 2\tan^2 t\, d(\tan t) = \frac{2}{3}\tan^3 t + C$

60. $\int \csc^2 y \cos^5 y\, dy = \int \sin^{-2} y \cos^5 y\, dy = \frac{\left(\frac{1}{\sin y}\right)\cos^4 y}{5-2} + \frac{5-1}{5-2}\int \sin^{-2} y \cos^3 y\, dy$

$= \frac{\left(\frac{1}{\sin y}\right)\cos^4 y}{3} + \frac{4}{3}\left(\frac{\left(\frac{1}{\sin y}\right)\cos^2 y}{3-2} + \frac{3-1}{3-2}\int \sin^{-2} y \cos y\, dy\right)$

(We used FORMULA 69 with $n = -2$, $m = 5$, $a = 1$ and $n = -2$, $m = 3$, $a = 1$)

$= \frac{\left(\frac{1}{\sin y}\right)\cos^4 y}{3} + \frac{4}{3}\left(\frac{1}{\sin y}\right)\cos^2 y + \frac{8}{3}\int \sin^{-2} y\, d(\sin y) = \frac{\cos^4 y}{3\sin y} + \frac{4\cos^2 y}{3\sin y} - \frac{8}{3\sin y} + C$

61. $\int 4\tan^3 2x\, dx = 4\left(\frac{\tan^2 2x}{2\cdot 2} - \int \tan 2x\, dx\right) = \tan^2 2x - 4\int \tan 2x\, dx$

(We used FORMULA 86 with $n = 3$, $a = 2$)

$= \tan^2 2x - \frac{4}{2}\ln|\sec 2x| + C = \tan^2 2x - 2\ln|\sec 2x| + C$

62. $\int \tan^4\left(\frac{x}{2}\right) dx = \frac{\tan^3\left(\frac{x}{2}\right)}{\frac{1}{2}(4-1)} - \int \tan^2\left(\frac{x}{2}\right) dx = \frac{2}{3}\tan^3\left(\frac{x}{2}\right) - \int \tan^2\left(\frac{x}{2}\right) dx$

(We used FORMULA 86 with n = 4, a = $\frac{1}{2}$)

$= \frac{2}{3}\tan^3\frac{x}{2} - 2\tan\frac{x}{2} + x + C$

(We used FORMULA 84 with a = $\frac{1}{2}$)

63. $\int 8\cot^4 t\, dt = 8\left(-\frac{\cot^3 t}{3} - \int \cot^2 t\, dt\right)$

(We used FORMULA 87 with a = 1, n = 4)

$= 8\left(-\frac{1}{3}\cot^3 t + \cot t + t\right) + C$

(We used FORMULA 85 with a = 1)

64. $\int 4\cot^3 2t\, dt = 4\left[-\frac{\cot^2 2t}{2(3-1)} - \int \cot 2t\, dt\right] = -\cot^2 2t - 4\int \cot 2t\, dt$

(We used FORMULA 87 with a = 2, n = 3)

$= -\cot^2 2t - \frac{4}{2}\ln|\sin 2t| + C = -\cot^2 2t - 2\ln|\sin 2t| + C$

(We used FORMULA 83 with a = 2)

65. $\int 2\sec^3 \pi x\, dx = 2\left[\frac{\sec\pi x\tan\pi x}{\pi(3-1)} + \frac{3-2}{3-1}\int \sec\pi x\, dx\right]$

(We used FORMULA 92 with n = 3, a = π)

$= \frac{1}{\pi}\sec\pi x\tan\pi x + \frac{1}{\pi}\ln|\sec\pi x + \tan\pi x| + C$

(We used FORMULA 88 with a = π)

66. $\int \frac{1}{2}\csc^3\frac{x}{2}\, dx = \frac{1}{2}\left(-\frac{\csc\frac{x}{2}\cot\frac{x}{2}}{\frac{1}{2}(3-1)} + \frac{3-2}{3-1}\int \csc\frac{x}{2}\, dx\right)$

(We used FORMULA 93 with a = $\frac{1}{2}$, n = 3)

$= \frac{1}{2}\left[-\csc\frac{x}{2}\cot\frac{x}{2} - \ln\left|\csc\frac{x}{2} + \cot\frac{x}{2}\right|\right] + C = -\frac{1}{2}\csc\frac{x}{2}\cot\frac{x}{2} - \frac{1}{2}\ln\left|\csc\frac{x}{2} + \cot\frac{x}{2}\right| + C$

(We used FORMULA 89 with a = $\frac{1}{2}$)

67. $\int 3\sec^4 3x\, dx = 3\left[\frac{\sec^2 3x\tan 3x}{3(4-1)} + \frac{4-2}{4-1}\int \sec^2 3x\, dx\right]$

(We used FORMULA 92 with n = 4, a = 3)

$= \frac{\sec^2 3x\tan 3x}{3} + \frac{2}{3}\tan 3x + C$

(We used FORMULA 90 with a = 3)

68. $\int \csc^4\frac{\theta}{3}\, d\theta = -\frac{\csc^2\frac{\theta}{3}\cot\frac{\theta}{3}}{\frac{1}{3}(4-1)} + \frac{4-2}{4-1}\int \csc^2\frac{\theta}{3}\, d\theta$

(We used FORMULA 93 with n = 4, a = $\frac{1}{3}$)

$= -\csc^2\frac{\theta}{3}\cot\frac{\theta}{3} - \frac{2}{3}\cdot 3\cot\frac{\theta}{3} + C = -\csc^2\frac{\theta}{3}\cot\frac{\theta}{3} - 2\cot\frac{\theta}{3} + C$

(We used FORMULA 91 with a = $\frac{1}{3}$)

69. $\int \csc^5 x\, dx = -\frac{\csc^3 x\cot x}{5-1} + \frac{5-2}{5-1}\int \csc^3 x\, dx = -\frac{\csc^3 x\cot x}{4} + \frac{3}{4}\left(-\frac{\csc x\cot x}{3-1} + \frac{3-2}{3-1}\int \csc x\, dx\right)$

(We used FORMULA 93 with n = 5, a = 1 and n = 3, a = 1)

$= -\frac{1}{4}\csc^3 x\cot x - \frac{3}{8}\csc x\cot x - \frac{3}{8}\ln|\csc x + \cot x| + C$

(We used FORMULA 89 with a = 1)

70. $\int \sec^5 x\, dx = \frac{\sec^3 x\tan x}{5-1} + \frac{5-2}{5-1}\int \sec^3 x\, dx = \frac{\sec^3 x\tan x}{4} + \frac{3}{4}\left(\frac{\sec x\tan x}{3-1} + \frac{3-2}{3-1}\int \sec x\, dx\right)$

(We used FORMULA 92 with a = 1, n = 5 and a = 1, n = 3)

$= \frac{1}{4}\sec^3 x\tan x + \frac{3}{8}\sec x\tan x + \frac{3}{8}\ln|\sec x + \tan x| + C$

(We used FORMULA 88 with a = 1)

71. $\int 16x^3(\ln x)^2\,dx = 16\left[\frac{x^4(\ln x)^2}{4} - \frac{2}{4}\int x^3 \ln x\,dx\right] = 16\left[\frac{x^4(\ln x)^2}{4} - \frac{1}{2}\left[\frac{x^4(\ln x)}{4} - \frac{1}{4}\int x^3\,dx\right]\right]$
(We used FORMULA 110 with a = 1, n = 3, m = 2 and a = 1, n = 3, m = 1)
$= 16\left(\frac{x^4(\ln x)^2}{4} - \frac{x^4(\ln x)}{8} + \frac{x^4}{32}\right) + C = 4x^4(\ln x)^2 - 2x^4 \ln x + \frac{x^4}{2} + C$

72. $\int (\ln x)^3\,dx = \frac{x(\ln x)^3}{1} - \frac{3}{1}\int (\ln x)^2\,dx = x(\ln x)^3 - 3\left[\frac{x(\ln x)^2}{1} - \frac{2}{1}\int \ln x\,dx\right] = x(\ln x)^3 - 3x(\ln x)^2 + 6\left(\frac{x \ln x}{1} - \frac{1}{1}\int dx\right)$
$= x(\ln x)^3 - 3x(\ln x)^2 + 6x\ln x - 6x + C$
(We used FORMULA 110 with n = 0, a = 1 and m = 3, 2, 1)

73. $\int xe^{3x}\,dx = \frac{e^{3x}}{3^2}(3x-1) + C = \frac{e^{3x}}{9}(3x-1) + C$
(We used FORMULA 104 with a = 3)

74. $\int xe^{-2x}\,dx = \frac{e^{-2x}}{(-2)^2}(-2x-1) + C = -\frac{e^{-2x}}{4}(2x+1) + C$
(We used FORMULA 104 with a = −2)

75. $\int x^3e^{x/2}\,dx = 2x^3e^{x/2} - 3\cdot 2\int x^2e^{x/2}\,dx = 2x^3e^{x/2} - 6\left(2x^2e^{x/2} - 2\cdot 2\int xe^{x/2}\,dx\right)$
$= 2x^3e^{x/2} - 12x^2e^{x/2} + 24\cdot 4e^{x/2}\left(\frac{x}{2} - 1\right) + C = 2x^3e^{x/2} - 12x^2e^{x/2} + 96e^{x/2}\left(\frac{x}{2} - 1\right) + C$
(We used FORMULA 105 with $a = \frac{1}{2}$ twice and FORMULA 104 with $a = \frac{1}{2}$)

76. $\int x^2e^{\pi x}\,dx = \frac{1}{\pi}x^2e^{\pi x} - \frac{2}{\pi}\int xe^{\pi x}\,dx$
(We used FORMULA 105 with n = 2, a = π)
$= \frac{1}{\pi}x^2e^{\pi x} - \frac{2}{\pi\cdot\pi^2}\cdot e^{\pi x}(\pi x - 1) + C = \frac{1}{\pi}x^2e^{\pi x} - \left(\frac{2e^{\pi x}}{\pi^3}\right)(\pi x - 1) + C$
(We used FORMULA 104 with a = π)

77. $\int x^2 2^x\,dx = \frac{x^2 2^x}{\ln 2} - \frac{2}{\ln 2}\int x2^x\,dx = \frac{x^2 2^x}{\ln 2} - \frac{2}{\ln 2}\left(\frac{x2^x}{\ln 2} - \frac{1}{\ln 2}\int 2^x\,dx\right) = \frac{x^2 2^x}{\ln 2} - \frac{2}{\ln 2}\left[\frac{x2^x}{\ln 2} - \frac{2^x}{(\ln 2)^2}\right] + C$
(We used FORMULA 106 with a = 1, b = 2, n = 2, n = 1)

78. $\int x^2 2^{-x}\,dx = \frac{x^2 2^{-x}}{-\ln 2} + \frac{2}{\ln 2}\int x2^{-x}\,dx = \frac{-x^2 2^{-x}}{\ln 2} + \frac{2}{\ln 2}\left(-\frac{x2^{-x}}{\ln 2} + \frac{1}{\ln 2}\int 2^{-x}\,dx\right)$
$= -\frac{x^2 2^{-x}}{\ln 2} + \frac{2}{\ln 2}\left[\frac{x2^{-x}}{-\ln 2} - \frac{2^{-x}}{(\ln 2)^2}\right] + C$
(We used FORMULA 106 with a = −1, b = 2, n = 2, n = 1)

79. $\int x\pi^x\,dx = \frac{x\pi^x}{\ln \pi} - \frac{1}{\ln \pi}\int \pi^x\,dx = \frac{x\pi^x}{\ln \pi} - \frac{1}{\ln \pi}\left(\frac{\pi^x}{\ln \pi}\right) + C = \frac{x\pi^x}{\ln \pi} - \frac{\pi^x}{(\ln \pi)^2} + C$
(We used FORMULA 106 with n = 1, b = π, a = 1)

80. $\int x2^{\sqrt{2}x}\,dx = \frac{x2^{\sqrt{2}x}}{\sqrt{2}\ln 2} - \frac{1}{\sqrt{2}\ln 2}\int 2^{\sqrt{2}x}\,dx = \frac{x2^{x\sqrt{2}}}{\sqrt{2}\ln 2} - \frac{2^{x\sqrt{2}}}{2(\ln 2)^2} + C$
(We used FORMULA 106 with a = $\sqrt{2}$, b = 2, n = 1)

81. $\int e^t \sec^3(e^t - 1)\,dt;\ \begin{bmatrix} x = e^t - 1 \\ dx = e^t\,dt \end{bmatrix} \to \int \sec^3 x\,dx = \frac{\sec x \tan x}{3-1} + \frac{3-2}{3-1}\int \sec x\,dx$
(We used FORMULA 92 with a = 1, n = 3)
$= \frac{\sec x \tan x}{2} + \frac{1}{2}\ln|\sec x + \tan x| + C = \frac{1}{2}\left[\sec(e^t - 1)\tan(e^t - 1) + \ln|\sec(e^t - 1) + \tan(e^t - 1)|\right] + C$

82. $\int \frac{\csc^3 \sqrt{\theta}}{\sqrt{\theta}}\, d\theta;\ \left[\begin{array}{c} t=\sqrt{\theta} \\ \theta = t^2 \\ d\theta = 2t\, dt \end{array}\right] \to 2\int \csc^3 t\, dt = 2\left[-\frac{\csc t \cot t}{3-1} + \frac{3-2}{3-1}\int \csc t\, dt\right]$

(We used FORMULA 93 with a = 1, n = 3)

$= 2\left[-\frac{\csc t \cot t}{2} - \frac{1}{2}\ln|\csc t + \cot t|\right] + C = -\csc\sqrt{\theta}\cot\sqrt{\theta} - \ln\left|\csc\sqrt{\theta} + \cot\sqrt{\theta}\right| + C$

83. $\int_0^1 2\sqrt{x^2+1}\, dx;\ [x = \tan t] \to 2\int_0^{\pi/4} \sec t \cdot \sec^2 t\, dt = 2\int_0^{\pi/4} \sec^3 t\, dt = 2\left[\left[\frac{\sec t \cdot \tan t}{3-1}\right]_0^{\pi/4} + \frac{3-2}{3-1}\int_0^{\pi/4} \sec t\, dt\right]$

(We used FORMULA 92 with n = 3, a = 1)

$= [\sec t \cdot \tan t + \ln|\sec t + \tan t|]_0^{\pi/4} = \sqrt{2} + \ln\left(\sqrt{2}+1\right)$

84. $\int_0^{\sqrt{3}/2} \frac{dy}{(1-y^2)^{5/2}};\ [y = \sin x] \to \int_0^{\pi/3} \frac{\cos x\, dx}{\cos^5 x} = \int_0^{\pi/3} \sec^4 x\, dx = \left[\frac{\sec^2 x \tan x}{4-1}\right]_0^{\pi/3} + \frac{4-2}{4-1}\int_0^{\pi/3} \sec^2 x\, dx$

(We used FORMULA 92 with a = 1, n = 4)

$= \left[\frac{\sec^2 x \tan x}{3} + \frac{2}{3}\tan x\right]_0^{\pi/3} = \left(\frac{4}{3}\right)\sqrt{3} + \left(\frac{2}{3}\right)\sqrt{3} = 2\sqrt{3}$

85. $\int_1^2 \frac{(r^2-1)^{3/2}}{r}\, dr;\ [r = \sec\theta] \to \int_0^{\pi/3} \frac{\tan^3\theta}{\sec\theta}(\sec\theta\tan\theta)\, d\theta = \int_0^{\pi/3} \tan^4\theta\, d\theta = \left[\frac{\tan^3\theta}{4-1}\right]_0^{\pi/3} - \int_0^{\pi/3} \tan^2\theta\, d\theta$

$= \left[\frac{\tan^3\theta}{3} - \tan\theta + \theta\right]_0^{\pi/3} = \frac{3\sqrt{3}}{3} - \sqrt{3} + \frac{\pi}{3} = \frac{\pi}{3}$

(We used FORMULA 86 with a = 1, n = 4 and FORMULA 84 with a = 1)

86. $\int_0^{1/\sqrt{3}} \frac{dt}{(t^2+1)^{7/2}};\ [t = \tan\theta] \to \int_0^{\pi/6} \frac{\sec^2\theta\, d\theta}{\sec^7\theta} = \int_0^{\pi/6} \cos^5\theta\, d\theta = \left[\frac{\cos^4\theta \sin\theta}{5}\right]_0^{\pi/6} + \left(\frac{5-1}{5}\right)\int_0^{\pi/6} \cos^3\theta\, d\theta$

$= \left[\frac{\cos^4\theta\sin\theta}{5}\right]_0^{\pi/6} + \frac{4}{5}\left[\left[\frac{\cos^2\theta\sin\theta}{3}\right]_0^{\pi/6} + \left(\frac{3-1}{3}\right)\int_0^{\pi/6} \cos\theta\, d\theta\right]$

$= \left[\frac{\cos^4\theta\sin\theta}{5} + \frac{4}{15}\cos^2\theta\sin\theta + \frac{8}{15}\sin\theta\right]_0^{\pi/6}$

(We used FORMULA 61 with a = 1, n = 5 and a = 1, n = 3)

$= \frac{\left(\frac{\sqrt{3}}{2}\right)^4\left(\frac{1}{2}\right)}{5} + \left(\frac{4}{15}\right)\left(\frac{\sqrt{3}}{2}\right)^2\left(\frac{1}{2}\right) + \left(\frac{8}{15}\right)\left(\frac{1}{2}\right) = \frac{9}{160} + \frac{1}{10} + \frac{4}{15} = \frac{3\cdot 9 + 48 + 32\cdot 4}{480} = \frac{203}{480}$

87. $\int \frac{1}{8}\sinh^5 3x\, dx = \frac{1}{8}\left(\frac{\sinh^4 3x \cosh 3x}{5\cdot 3} - \frac{5-1}{5}\int \sinh^3 3x\, dx\right)$

$= \frac{\sinh^4 3x\cosh 3x}{120} - \frac{1}{10}\left(\frac{\sinh 3x\cosh 3x}{3\cdot 3} - \frac{3-1}{3}\int \sinh 3x\, dx\right)$

(We used FORMULA 117 with a = 3, n = 5 and a = 3, n = 3)

$= \frac{\sinh^4 3x\cosh 3x}{120} - \frac{\sinh 3x\cosh 3x}{90} + \frac{2}{30}\left(\frac{1}{3}\cosh 3x\right) + C$

$= \frac{1}{120}\sinh^4 3x\cosh 3x - \frac{1}{90}\sinh 3x\cosh 3x + \frac{1}{45}\cosh 3x + C$

88. $\int \frac{\cosh^4\sqrt{x}}{\sqrt{x}}\, dx;\ \left[\begin{array}{c} u = \sqrt{x} \\ du = \frac{dx}{2\sqrt{x}} \end{array}\right] \to 2\int \cosh^4 u\, du = 2\left(\frac{\cosh^3 u \sinh u}{4} + \frac{4-1}{4}\int \cosh^2 u\, du\right)$

$= \frac{\cosh^3 u\sinh u}{2} + \frac{3}{2}\left(\frac{\sinh 2u}{4} + \frac{u}{2}\right) + C$

(We used FORMULA 118 with a = 1, n = 4 and FORMULA 116 with a = 1)

$= \frac{1}{2}\cosh^3\sqrt{x}\sinh\sqrt{x} + \frac{3}{8}\sinh 2\sqrt{x} + \frac{3}{4}\sqrt{x} + C$

89. $\int x^2\cosh 3x\, dx = \frac{x^2}{3}\sinh 3x - \frac{2}{3}\int x\sinh 3x\, dx = \frac{x^2}{3}\sinh 3x - \frac{2}{3}\left(\frac{x}{3}\cosh 3x - \frac{1}{3}\int \cosh 3x\, dx\right)$

(We used FORMULA 122 with a = 3, n = 2 and FORMULA 121 with a = 3, n = 1)

$= \frac{x^2}{3} \sinh 3x - \frac{2x}{9} \cosh 3x + \frac{2}{27} \sinh 3x + C$

90. $\int x \sinh 5x\, dx = \frac{x}{5} \cosh 5x - \frac{1}{25} \sinh 5x + C$
(We used FORMULA 119 with $a = 5$)

91. $\int \operatorname{sech}^7 x \tanh x\, dx = -\frac{\operatorname{sech}^7 x}{7} + C$
(We used FORMULA 135 with $a = 1$, $n = 7$)

92. $\int \operatorname{csch}^3 2x \coth 2x\, dx = -\frac{\operatorname{csch}^3 2x}{3\cdot 2} + C = -\frac{\operatorname{csch}^3 2x}{6} + C$
(We used FORMULA 136 with $a = 2$, $n = 3$)

93. $u = ax + b \Rightarrow x = \frac{u-b}{a} \Rightarrow dx = \frac{du}{a}$;
$\int \frac{x\, dx}{(ax+b)^2} = \int \frac{(u-b)}{au^2} \frac{du}{a} = \frac{1}{a^2} \int \left(\frac{1}{u} - \frac{b}{u^2}\right) du = \frac{1}{a^2} \left[\ln|u| + \frac{b}{u}\right] + C = \frac{1}{a^2} \left[\ln|ax+b| + \frac{b}{ax+b}\right] + C$

94. $x = a \tan\theta \Rightarrow a^2 + x^2 = a^2 \sec^2\theta \Rightarrow 2x\, dx = 2a^2 \sec^2\theta \tan\theta \Rightarrow dx = a \sec^2\theta\, d\theta$;
$\int \frac{dx}{(a^2+x^2)^2} = \int \frac{a \sec^2\theta}{(a^2 \sec^2\theta)^2}\, d\theta = \frac{1}{a^3} \int \frac{d\theta}{\sec^2\theta} = \frac{1}{2a^3} \int (1 + \cos 2\theta)\, d\theta = \frac{1}{2a^3} \left(\theta + \frac{1}{2} \sin 2\theta\right) + C$
$= \frac{1}{2a^3} (\theta + \sin\theta\cos\theta) + C = \frac{1}{2a^3} \left[\theta + \left(\frac{\sin\theta}{\cos\theta}\right) \cos^2\theta\right] + C = \frac{1}{2a^3} \left(\theta + \frac{\tan\theta}{1+\tan^2\theta}\right) + C$
$= \frac{1}{2a^3} \left[\tan^{-1} \frac{x}{a} + \frac{x}{a\left(1 + \frac{x^2}{a^2}\right)}\right] + C = \frac{x}{2a^2(a^2+x^2)} + \frac{1}{2a^3} \tan^{-1} \frac{x}{a} + C$

95. $x = a \sin\theta \Rightarrow a^2 - x^2 = a^2 \cos^2\theta \Rightarrow -2x\, dx = -2a^2 \cos\theta \sin\theta\, d\theta \Rightarrow dx = a \cos\theta\, d\theta$;
$\int \sqrt{a^2 - x^2}\, dx = \int a \cos\theta (a \cos\theta)\, d\theta = a^2 \int \cos^2\theta\, d\theta = \frac{a^2}{2} \int (1 + \cos 2\theta)\, d\theta = \frac{a^2}{2} \left(\theta + \frac{\sin 2\theta}{2}\right) + C$
$= \frac{a^2}{2} (\theta + \cos\theta \sin\theta) + C = \frac{a^2}{2} \left(\theta + \sqrt{1 - \sin^2\theta} \cdot \sin\theta\right) + C = \frac{a^2}{2} \left(\sin^{-1} \frac{x}{a} + \frac{\sqrt{a^2 - x^2}}{a} \cdot \frac{x}{a}\right) + C$
$= \frac{a^2}{2} \sin^{-1} \frac{x}{a} + \frac{x}{2} \sqrt{a^2 - x^2} + C$

96. $x = a \sec\theta \Rightarrow x^2 - a^2 = a^2 \tan^2\theta \Rightarrow 2x\, dx = 2a^2 \tan\theta \sec^2\theta\, d\theta \Rightarrow dx = a \sec\theta \tan\theta\, d\theta$;
$\int \frac{dx}{x^2\sqrt{x^2 - a^2}} = \int \frac{a \tan\theta \sec\theta\, d\theta}{(a^2 \sec^2\theta)\, a \tan\theta} = \int \frac{d\theta}{a^2 \sec\theta} = \frac{1}{a^2} \int \cos\theta\, d\theta = \frac{1}{a^2} \sin\theta + C = \frac{1}{a^2} \sqrt{1 - \cos^2\theta} + C$
$= \left(\frac{1}{a^2}\right) \frac{\sqrt{\frac{1}{\cos^2\theta} - 1}}{\left(\frac{1}{\cos\theta}\right)} + C = \left(\frac{1}{a^2}\right) \frac{\sqrt{\sec^2\theta - 1}}{\sec\theta} + C = \left(\frac{1}{a^2}\right) \frac{\sqrt{\frac{x^2}{a^2} - 1}}{\left(\frac{x}{a}\right)} + C = \frac{\sqrt{x^2 - a^2}}{a^2 x} + C$

97. $\int x^n \sin ax\, dx = -\int x^n \left(\frac{1}{a}\right) d(\cos ax) = (\cos ax)\, x^n \left(-\frac{1}{a}\right) + \frac{1}{a} \int \cos ax \cdot n x^{n-1}\, dx$
$= -\frac{x^n}{a} \cos ax + \frac{n}{a} \int x^{n-1} \cos ax\, dx$
(We used integration by parts $\int u\, dv = uv - \int v\, du$ with $u = x^n$, $v = -\frac{1}{a} \cos ax$)

98. $\int x^n (\ln ax)^m\, dx = \int (\ln ax)^m\, d\left(\frac{x^{n+1}}{n+1}\right) = \frac{x^{n+1}(\ln ax)^m}{n+1} - \int \left(\frac{x^{n+1}}{n+1}\right) m(\ln ax)^{m-1} \left(\frac{1}{x}\right) dx$
$= \frac{x^{n+1}(\ln ax)^m}{n+1} - \frac{m}{n+1} \int x^n (\ln ax)^{m-1}\, dx$, $n \neq -1$
(We used integration by parts $\int u\, dv = uv - \int v\, du$ with $u = (\ln ax)^m$, $v = \frac{x^{n+1}}{n+1}$)

99. $\int x^n \sin^{-1} ax\, dx = \int \sin^{-1} ax\, d\left(\frac{x^{n+1}}{n+1}\right) = \frac{x^{n+1}}{n+1} \sin^{-1} ax - \int \left(\frac{x^{n+1}}{n+1}\right) \frac{a}{\sqrt{1 - (ax)^2}}\, dx$
$= \frac{x^{n+1}}{n+1} \sin^{-1} ax - \frac{a}{n+1} \int \frac{x^{n+1}\, dx}{\sqrt{1 - a^2 x^2}}$, $n \neq -1$

$\left(\text{We used integration by parts } \int u\,dv = uv - \int v\,du \text{ with } u = \sin^{-1} ax,\ v = \frac{x^{n+1}}{n+1}\right)$

100. $\int x^n \tan^{-1} ax\,dx = \int \tan^{-1} ax\,d\left(\frac{x^{n+1}}{n+1}\right) = \frac{x^{n+1}}{n+1}\tan^{-1} ax - \int \left(\frac{x^{n+1}}{n+1}\right)\frac{a}{1+(ax)^2}\,dx$

$= \frac{x^{n+1}}{n+1}\tan^{-1} ax - \frac{a}{n+1}\int \frac{x^{n+1}\,dx}{1+a^2x^2},\ n \neq -1$

$\left(\text{We used integration by parts } \int u\,dv = uv - \int v\,du \text{ with } u = \tan^{-1} ax,\ v = \frac{x^{n+1}}{n+1}\right)$

101. $S = \int_0^{\sqrt{2}} 2\pi y\sqrt{1+(y')^2}\,dx$

$= 2\pi \int_0^{\sqrt{2}} \sqrt{x^2+2}\sqrt{1+\frac{x^2}{x^2+2}}\,dx$

$= 2\sqrt{2}\pi \int_0^{\sqrt{2}} \sqrt{x^2+1}\,dx$

$= 2\sqrt{2}\pi \left[\frac{x\sqrt{x^2+1}}{2} + \frac{1}{2}\ln\left|x+\sqrt{x^2+1}\right|\right]_0^{\sqrt{2}}$

(We used FORMULA 21 with $a = 1$)

$= \sqrt{2}\pi\left[\sqrt{6} + \ln\left(\sqrt{2}+\sqrt{3}\right)\right] = 2\pi\sqrt{3} + \pi\sqrt{2}\ln\left(\sqrt{2}+\sqrt{3}\right)$

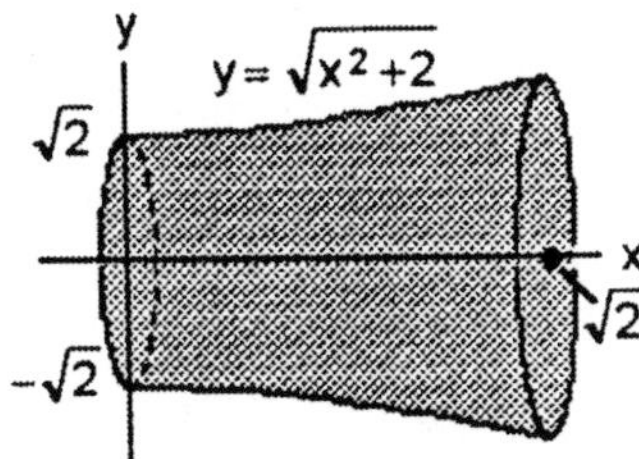

102. $L = \int_0^{\sqrt{3}/2} \sqrt{1+(2x)^2}\,dx = 2\int_0^{\sqrt{3}/2} \sqrt{\frac{1}{4}+x^2}\,dx = 2\left[\frac{x}{2}\sqrt{\frac{1}{4}+x^2} + \left(\frac{1}{4}\right)\left(\frac{1}{2}\right)\ln\left(x+\sqrt{\frac{1}{4}+x^2}\right)\right]_0^{\sqrt{3}/2}$

(We used FORMULA 2 with $a = \frac{1}{2}$)

$= \left[\frac{x}{2}\sqrt{1+4x^2} + \frac{1}{4}\ln\left(x+\frac{1}{2}\sqrt{1+4x^2}\right)\right]_0^{\sqrt{3}/2} = \frac{\sqrt{3}}{4}\sqrt{1+4\left(\frac{3}{4}\right)} + \frac{1}{4}\ln\left(\frac{\sqrt{3}}{2}+\frac{1}{2}\sqrt{1+4\left(\frac{3}{4}\right)}\right) - \frac{1}{4}\ln\frac{1}{2}$

$= \frac{\sqrt{3}}{4}(2) + \frac{1}{4}\ln\left(\frac{\sqrt{3}}{2}+1\right) + \frac{1}{4}\ln 2 = \frac{\sqrt{3}}{2} + \frac{1}{4}\ln\left(\sqrt{3}+2\right)$

103. $A = \int_0^3 \frac{dx}{\sqrt{x+1}} = \left[2\sqrt{x+1}\right]_0^3 = 2;\ \overline{x} = \frac{1}{A}\int_0^3 \frac{x\,dx}{\sqrt{x+1}}$

$= \frac{1}{A}\int_0^3 \sqrt{x+1}\,dx - \frac{1}{A}\int_0^3 \frac{dx}{\sqrt{x+1}}$

$= \frac{1}{2}\cdot\frac{2}{3}\left[(x+1)^{3/2}\right]_0^3 - 1 = \frac{4}{3};$

(We used FORMULA 11 with $a = 1$, $b = 1$, $n = 1$ and $a = 1$, $b = 1$, $n = -1$)

$\overline{y} = \frac{1}{2A}\int_0^3 \frac{dx}{x+1} = \frac{1}{4}\left[\ln(x+1)\right]_0^3 = \frac{1}{4}\ln 4 = \frac{1}{2}\ln 2 = \ln\sqrt{2}$

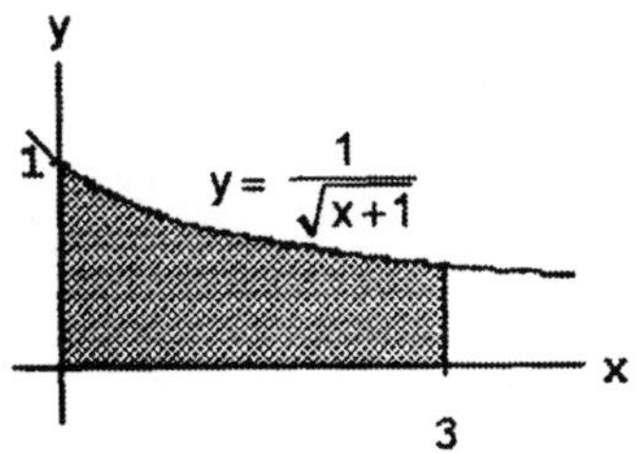

104. $M_y = \int_0^3 x\left(\frac{36}{2x+3}\right)dx = 18\int_0^3 \frac{2x+3}{2x+3}\,dx - 54\int_0^3 \frac{dx}{2x+3} = \left[18x - 27\ln|2x+3|\right]_0^3$

$= 18\cdot 3 - 27\ln 9 - (-27\ln 3) = 54 - 27\cdot 2\ln 3 + 27\ln 3 = 54 - 27\ln 3$

105. $S = 2\pi\int_{-1}^{1} x^2\sqrt{1+4x^2}\,dx;$

$\begin{bmatrix} u = 2x \\ du = 2\,dx \end{bmatrix} \rightarrow \frac{\pi}{4}\int_{-2}^{2} u^2\sqrt{1+u^2}\,du$

$= \frac{\pi}{4}\left[\frac{u}{8}(1+2u^2)\sqrt{1+u^2} - \frac{1}{8}\ln\left(u+\sqrt{1+u^2}\right)\right]_{-2}^{2}$

(We used FORMULA 22 with $a = 1$)

$= \frac{\pi}{4}\left[\frac{2}{8}(1+2\cdot 4)\sqrt{1+4} - \frac{1}{8}\ln\left(2+\sqrt{1+4}\right)\right.$

$\left. + \frac{2}{8}(1+2\cdot 4)\sqrt{1+4} + \frac{1}{8}\ln\left(-2+\sqrt{1+4}\right)\right]$

$= \frac{\pi}{4}\left[\frac{9}{2}\sqrt{5} - \frac{1}{8}\ln\left(\frac{2+\sqrt{5}}{-2+\sqrt{5}}\right)\right] \approx 7.62$

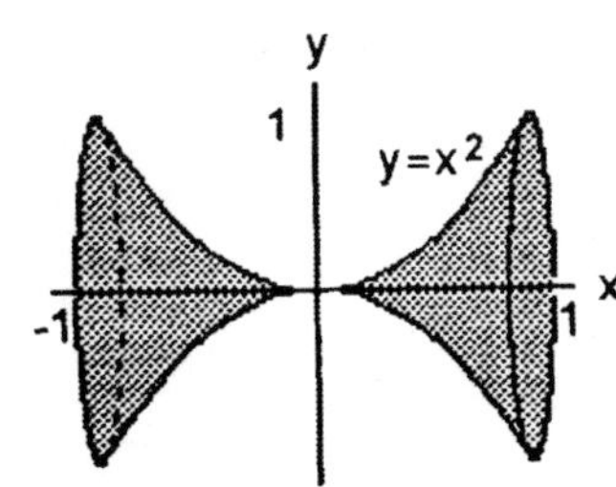

106. (a) The volume of the filled part equals the length of the tank times the area of the shaded region shown in the accompanying figure. Consider a layer of gasoline of thickness dy located at height y where $-r < y < -r + d$. The width of this layer is $2\sqrt{r^2 - y^2}$. Therefore, $A = 2\int_{-r}^{-r+d} \sqrt{r^2 - y^2}\,dy$ and $V = L \cdot A = 2L\int_{-r}^{-r+d} \sqrt{r^2 - y^2}\,dy$

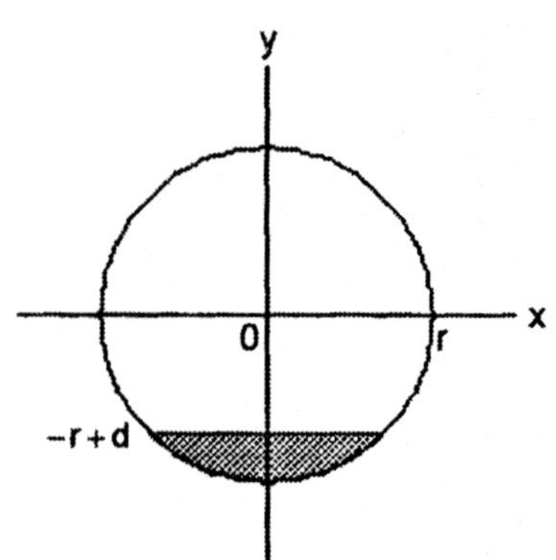

(b) $2L\int_{-r}^{-r+d} \sqrt{r^2 - y^2}\,dy = 2L\left[\frac{y\sqrt{r^2-y^2}}{2} + \frac{r^2}{2}\sin^{-1}\frac{y}{r}\right]_{-r}^{-r+d}$

(We used FORMULA 29 with $a = r$)

$= 2L\left[\frac{(d-r)}{2}\sqrt{2rd - d^2} + \frac{r^2}{2}\sin^{-1}\left(\frac{d-r}{r}\right) + \frac{r^2}{2}\left(\frac{\pi}{2}\right)\right] = 2L\left[\left(\frac{d-r}{2}\right)\sqrt{2rd - d^2} + \left(\frac{r^2}{2}\right)\left(\sin^{-1}\left(\frac{d-r}{r}\right) + \frac{\pi}{2}\right)\right]$

107. The integrand $f(x) = \sqrt{x - x^2}$ is nonnegative, so the integral is maximized by integrating over the function's entire domain, which runs from $x = 0$ to $x = 1$

$\Rightarrow \int_0^1 \sqrt{x - x^2}\,dx = \int_0^1 \sqrt{2 \cdot \frac{1}{2}x - x^2}\,dx = \left[\frac{\left(x - \frac{1}{2}\right)}{2}\sqrt{2 \cdot \frac{1}{2}x - x^2} + \frac{\left(\frac{1}{2}\right)^2}{2}\sin^{-1}\left(\frac{x - \frac{1}{2}}{\frac{1}{2}}\right)\right]_0^1$

(We used FORMULA 48 with $a = \frac{1}{2}$)

$= \left[\frac{\left(x - \frac{1}{2}\right)}{2}\sqrt{x - x^2} + \frac{1}{8}\sin^{-1}(2x - 1)\right]_0^1 = \frac{1}{8} \cdot \frac{\pi}{2} - \frac{1}{8}\left(-\frac{\pi}{2}\right) = \frac{\pi}{8}$

108. The integrand is maximized by integrating $g(x) = x\sqrt{2x - x^2}$ over the largest domain on which g is nonnegative, namely $[0, 2]$

$\Rightarrow \int_0^2 x\sqrt{2x - x^2}\,dx = \left[\frac{(x+1)(2x-3)\sqrt{2x - x^2}}{6} + \frac{1}{2}\sin^{-1}(x - 1)\right]_0^2$

(We used FORMULA 51 with $a = 1$)

$= \frac{1}{2} \cdot \frac{\pi}{2} - \frac{1}{2}\left(-\frac{\pi}{2}\right) = \frac{\pi}{2}$

CAS EXPLORATIONS

109. Example CAS commands:

Maple:

```
q1 := Int( x*ln(x), x );                    # (a)
q1 = value( q1 );
q2 := Int( x^2*ln(x), x );                  # (b)
q2 = value( q2 );
q3 := Int( x^3*ln(x), x );                  # (c)
q3 = value( q3 );
q4 := Int( x^4*ln(x), x );                  # (d)
q4 = value( q4 );
q5 := Int( x^n*ln(x), x );                  # (e)
q6 := value( q5 );
q7 := simplify(q6) assuming n::integer;
q5 = collect( factor(q7), ln(x) );
```

110. Example CAS commands:

Maple:

```
q1 := Int( ln(x)/x, x );                                # (a)
q1 = value( q1 );
q2 := Int( ln(x)/x^2, x );                              # (b)
q2 = value( q2 );
q3 := Int( ln(x)/x^3, x );                              # (c)
q3 = value( q3 );
q4 := Int( ln(x)/x^4, x );                              # (d)
q4 = value( q4 );
q5 := Int( ln(x)/x^n, x );                              # (e)
q6 := value( q5 );
q7 := simplify(q6) assuming n::integer;
q5 = collect( factor(q7), ln(x) );
```

111. Example CAS commands:

Maple:

```
q := Int( sin(x)^n/(sin(x)^n+cos(x)^n), x=0..Pi/2 );   # (a)
q = value( q );
q1 := eval( q, n=1 ):                                   # (b)
q1 = value( q1 );
for N in [1,2,3,5,7] do
  q1 := eval( q, n=N );
  print( q1 = evalf(q1) );
end do:
qq1 := PDEtools[dchange]( x=Pi/2-u, q, [u] );          # (c)
qq2 := subs( u=x, qq1 );
qq3 := q + q = q + qq2;
qq4 := combine( qq3 );
qq5 := value( qq4 );
simplify( qq5/2 );
```

109-111.Example CAS commands:

Mathematica: (functions may vary)

In Mathematica, the natural log is denoted by Log rather than Ln, Log base 10 is Log[x,10]

Mathematica does not include an arbitrary constant when computing an indefinite integral,

```
Clear[x, f, n]
f[x_]:=Log[x] / x^n
Integrate[f[x], x]
```

For exercise 111, Mathematica cannot evaluate the integral with arbitrary n. It does evaluate the integral (value is $\pi/4$ in each case) for small values of n, but for large values of n, it identifies this integral as Indeterminate

109. (e) $\int x^n \ln x\,dx = \frac{x^{n+1}\ln x}{n+1} - \frac{1}{n+1}\int x^n\,dx,\ n \neq -1$

(We used FORMULA 110 with $a = 1$, $m = 1$)

$= \frac{x^{n+1}\ln x}{n+1} - \frac{x^{n+1}}{(n+1)^2} + C = \frac{x^{n+1}}{n+1}\left(\ln x - \frac{1}{n+1}\right) + C$

110. (e) $\int x^{-n} \ln x\,dx = \frac{x^{-n+1}\ln x}{-n+1} - \frac{1}{(-n)+1}\int x^{-n}\,dx,\ n \neq 1$

(We used FORMULA 110 with $a = 1$, $m = 1$, $n = -n$)

$= \frac{x^{1-n}\ln x}{1-n} - \frac{1}{1-n}\left(\frac{x^{1-n}}{1-n}\right) + C = \frac{x^{1-n}}{1-n}\left(\ln x - \frac{1}{1-n}\right) + C$

111. (a) Neither MAPLE nor MATHEMATICA can find this integral for arbitrary n.
(b) MAPLE and MATHEMATICA get stuck at about n = 5.
(c) Let $x = \frac{\pi}{2} - u \Rightarrow dx = -du$; $x = 0 \Rightarrow u = \frac{\pi}{2}$, $x = \frac{\pi}{2} \Rightarrow u = 0$;

$I = \int_0^{\pi/2} \frac{\sin^n x\,dx}{\sin^n x + \cos^n x} = \int_{\pi/2}^0 \frac{-\sin^n\left(\frac{\pi}{2}-u\right)du}{\sin^n\left(\frac{\pi}{2}-u\right)+\cos^n\left(\frac{\pi}{2}-u\right)} = \int_0^{\pi/2} \frac{\cos^n u\,du}{\cos^n u + \sin^n u} = \int_0^{\pi/2} \frac{\cos^n x\,dx}{\cos^n x + \sin^n x}$

$\Rightarrow I + I = \int_0^{\pi/2} \left(\frac{\sin^n x + \cos^n x}{\sin^n x + \cos^n x}\right) dx = \int_0^{\pi/2} dx = \frac{\pi}{2} \Rightarrow I = \frac{\pi}{4}$

8.7 NUMERICAL INTEGRATION

1. $\int_1^2 x\,dx$

I. (a) For n = 4, $\Delta x = \frac{b-a}{n} = \frac{2-1}{4} = \frac{1}{4} \Rightarrow \frac{\Delta x}{2} = \frac{1}{8}$;
$\sum mf(x_i) = 12 \Rightarrow T = \frac{1}{8}(12) = \frac{3}{2}$;
$f(x) = x \Rightarrow f'(x) = 1 \Rightarrow f'' = 0 \Rightarrow M = 0$
$\Rightarrow |E_T| = 0$

	x_i	$f(x_i)$	m	$mf(x_i)$
x_0	1	1	1	1
x_1	5/4	5/4	2	5/2
x_2	3/2	3/2	2	3
x_3	7/4	7/4	2	7/2
x_4	2	2	1	2

(b) $\int_1^2 x\,dx = \left[\frac{x^2}{2}\right]_1^2 = 2 - \frac{1}{2} = \frac{3}{2} \Rightarrow |E_T| = \int_1^2 x\,dx - T = 0$

(c) $\frac{|E_T|}{\text{True Value}} \times 100 = 0\%$

II. (a) For n = 4, $\Delta x = \frac{b-a}{n} = \frac{2-1}{4} = \frac{1}{4} \Rightarrow \frac{\Delta x}{3} = \frac{1}{12}$;
$\sum mf(x_i) = 18 \Rightarrow S = \frac{1}{12}(18) = \frac{3}{2}$;
$f^{(4)}(x) = 0 \Rightarrow M = 0 \Rightarrow |E_S| = 0$

	x_i	$f(x_i)$	m	$mf(x_i)$
x_0	1	1	1	1
x_1	5/4	5/4	4	5
x_2	3/2	3/2	2	3
x_3	7/4	7/4	4	7
x_4	2	2	1	2

(b) $\int_1^2 x\,dx = \frac{3}{2} \Rightarrow |E_S| = \int_1^2 x\,dx - S = \frac{3}{2} - \frac{3}{2} = 0$

(c) $\frac{|E_S|}{\text{True Value}} \times 100 = 0\%$

2. $\int_1^3 (2x-1)\,dx$

I. (a) For n = 4, $\Delta x = \frac{b-a}{n} = \frac{3-1}{4} = \frac{2}{4} = \frac{1}{2} \Rightarrow \frac{\Delta x}{2} = \frac{1}{4}$;
$\sum mf(x_i) = 24 \Rightarrow T = \frac{1}{4}(24) = 6$;
$f(x) = 2x - 1 \Rightarrow f'(x) = 2 \Rightarrow f'' = 0 \Rightarrow M = 0$
$\Rightarrow |E_T| = 0$

	x_i	$f(x_i)$	m	$mf(x_i)$
x_0	1	1	1	1
x_1	3/2	2	2	4
x_2	2	3	2	6
x_3	5/2	4	2	8
x_4	3	5	1	5

(b) $\int_1^3 (2x-1)\,dx = \left[x^2 - x\right]_1^3 = (9-3) - (1-1) = 6 \Rightarrow |E_T| = \int_1^3 (2x-1)\,dx - T = 6 - 6 = 0$

(c) $\frac{|E_T|}{\text{True Value}} \times 100 = 0\%$

II. (a) For n = 4, $\Delta x = \frac{b-a}{n} = \frac{3-1}{4} = \frac{2}{4} = \frac{1}{2} \Rightarrow \frac{\Delta x}{3} = \frac{1}{6}$;
$\sum mf(x_i) = 36 \Rightarrow S = \frac{1}{6}(36) = 6$;
$f^{(4)}(x) = 0 \Rightarrow M = 0 \Rightarrow |E_S| = 0$

	x_i	$f(x_i)$	m	$mf(x_i)$
x_0	1	1	1	1
x_1	3/2	2	4	8
x_2	2	3	2	6
x_3	5/2	4	4	16
x_4	3	5	1	5

(b) $\int_1^3 (2x-1)\,dx = 6 \Rightarrow |E_S| = \int_1^3 (2x-1)\,dx - S$
$= 6 - 6 = 0$

(c) $\frac{|E_S|}{\text{True Value}} \times 100 = 0\%$

3. $\int_{-1}^1 (x^2+1)\,dx$

I. (a) For n = 4, $\Delta x = \frac{b-a}{n} = \frac{1-(-1)}{4} = \frac{2}{4} = \frac{1}{2} \Rightarrow \frac{\Delta x}{2} = \frac{1}{4}$;
$\sum mf(x_i) = 11 \Rightarrow T = \frac{1}{4}(11) = 2.75$;
$f(x) = x^2 + 1 \Rightarrow f'(x) = 2x \Rightarrow f''(x) = 2 \Rightarrow M = 2$
$\Rightarrow |E_T| \le \frac{1-(-1)}{12}\left(\frac{1}{2}\right)^2(2) = \frac{1}{12}$ or 0.08333

	x_i	$f(x_i)$	m	$mf(x_i)$
x_0	−1	2	1	2
x_1	−1/2	5/4	2	5/2
x_2	0	1	2	2
x_3	1/2	5/4	2	5/2
x_4	1	2	1	2

(b) $\int_{-1}^{1}(x^2+1)\,dx = \left[\frac{x^3}{3}+x\right]_{-1}^{1} = \left(\frac{1}{3}+1\right)-\left(-\frac{1}{3}-1\right) = \frac{8}{3} \Rightarrow E_T = \int_{-1}^{1}(x^2+1)\,dx - T = \frac{8}{3}-\frac{11}{4} = -\frac{1}{12}$

$\Rightarrow |E_T| = \left|-\frac{1}{12}\right| \approx 0.08333$

(c) $\frac{|E_T|}{\text{True Value}} \times 100 = \left(\frac{\frac{1}{12}}{\frac{8}{3}}\right) \times 100 \approx 3\%$

II. (a) For $n=4$, $\Delta x = \frac{b-a}{n} = \frac{1-(-1)}{4} = \frac{2}{4} = \frac{1}{2} \Rightarrow \frac{\Delta x}{3} = \frac{1}{6}$;

$\sum mf(x_i) = 16 \Rightarrow S = \frac{1}{6}(16) = \frac{8}{3} = 2.66667$;

$f^{(3)}(x) = 0 \Rightarrow f^{(4)}(x) = 0 \Rightarrow M = 0 \Rightarrow |E_S| = 0$

	x_i	$f(x_i)$	m	$mf(x_i)$
x_0	−1	2	1	2
x_1	−1/2	5/4	4	5
x_2	0	1	2	2
x_3	1/2	5/4	4	5
x_4	1	2	1	2

(b) $\int_{-1}^{1}(x^2+1)\,dx = \left[\frac{x^3}{3}+x\right]_{-1}^{1} = \frac{8}{3}$

$\Rightarrow |E_S| = \int_{-1}^{1}(x^2+1)\,dx - S = \frac{8}{3}-\frac{8}{3} = 0$

(c) $\frac{|E_S|}{\text{True Value}} \times 100 = 0\%$

4. $\int_{-2}^{0}(x^2-1)\,dx$

I. (a) For $n=4$, $\Delta x = \frac{b-a}{n} = \frac{0-(-2)}{4} = \frac{2}{4} = \frac{1}{2} \Rightarrow \frac{\Delta x}{2} = \frac{1}{4}$

$\sum mf(x_i) = 3 \Rightarrow T = \frac{1}{4}(3) = \frac{3}{4}$;

$f(x) = x^2-1 \Rightarrow f'(x) = 2x \Rightarrow f''(x) = 2$

$\Rightarrow M = 2 \Rightarrow |E_T| \le \frac{0-(-2)}{12}\left(\frac{1}{2}\right)^2(2) = \frac{1}{12} = 0.08333$

	x_i	$f(x_i)$	m	$mf(x_i)$
x_0	−2	3	1	3
x_1	−3/2	5/4	2	5/2
x_2	−1	0	2	0
x_3	− 1/2	−3/4	2	−3/2
x_4	0	−1	1	−1

(b) $\int_{-2}^{0}(x^2-1)\,dx = \left[\frac{x^3}{3}-x\right]_{-2}^{0} = 0-\left(-\frac{8}{3}+2\right) = \frac{2}{3} \Rightarrow E_T = \int_{-2}^{0}(x^2-1)\,dx - T = \frac{2}{3}-\frac{3}{4} = -\frac{1}{12}$

$\Rightarrow |E_T| = \frac{1}{12}$

(c) $\frac{|E_T|}{\text{True Value}} \times 100 = \left(\frac{\frac{1}{12}}{\frac{2}{3}}\right) \times 100 \approx 13\%$

II. (a) For $n=4$, $\Delta x = \frac{b-a}{n} = \frac{0-(-2)}{4} = \frac{2}{4} = \frac{1}{2}$

$\Rightarrow \frac{\Delta x}{3} = \frac{1}{6}$; $\sum mf(x_i) = 4 \Rightarrow S = \frac{1}{6}(4) = \frac{2}{3}$;

$f^{(3)}(x) = 0 \Rightarrow f^{(4)}(x) = 0 \Rightarrow M = 0 \Rightarrow |E_S| = 0$

	x_i	$f(x_i)$	m	$mf(x_i)$
x_0	−2	3	1	3
x_1	−3/2	5/4	4	5
x_2	−1	0	2	0
x_3	− 1/2	−3/4	4	−3
x_4	0	−1	1	−1

(b) $\int_{-2}^{0}(x^2-1)\,dx = \frac{2}{3} \Rightarrow |E_S| = \int_{-2}^{0}(x^2-1)\,dx - S$

$= \frac{2}{3}-\frac{2}{3} = 0$

(c) $\frac{|E_S|}{\text{True Value}} \times 100 = 0\%$

5. $\int_{0}^{2}(t^3+t)\,dt$

I. (a) For $n=4$, $\Delta x = \frac{b-a}{n} = \frac{2-0}{4} = \frac{2}{4} = \frac{1}{2}$

$\Rightarrow \frac{\Delta x}{2} = \frac{1}{4}$; $\sum mf(t_i) = 25 \Rightarrow T = \frac{1}{4}(25) = \frac{25}{4}$;

$f(t) = t^3+t \Rightarrow f'(t) = 3t^2+1 \Rightarrow f''(t) = 6t$

$\Rightarrow M = 12 = f''(2) \Rightarrow |E_T| \le \frac{2-0}{12}\left(\frac{1}{2}\right)^2(12) = \frac{1}{2}$

	t_i	$f(t_i)$	m	$mf(t_i)$
t_0	0	0	1	0
t_1	1/2	5/8	2	5/4
t_2	1	2	2	4
t_3	3/2	39/8	2	39/4
t_4	2	10	1	10

(b) $\int_{0}^{2}(t^3+t)\,dt = \left[\frac{t^4}{4}+\frac{t^2}{2}\right]_{0}^{2} = \left(\frac{2^4}{4}+\frac{2^2}{2}\right)-0 = 6 \Rightarrow |E_T| = \int_{0}^{2}(t^3+t)\,dt - T = 6-\frac{25}{4} = -\frac{1}{4} \Rightarrow |E_T| = \frac{1}{4}$

(c) $\frac{|E_T|}{\text{True Value}} \times 100 = \frac{\left|-\frac{1}{4}\right|}{6} \times 100 \approx 4\%$

II. (a) For $n=4$, $\Delta x = \frac{b-a}{n} = \frac{2-0}{4} = \frac{2}{4} = \frac{1}{2} \Rightarrow \frac{\Delta x}{3} = \frac{1}{6}$;

$\sum mf(t_i) = 36 \Rightarrow S = \frac{1}{6}(36) = 6$;

$f^{(3)}(t) = 6 \Rightarrow f^{(4)}(t) = 0 \Rightarrow M = 0 \Rightarrow |E_S| = 0$

	t_i	$f(t_i)$	m	$mf(t_i)$
t_0	0	0	1	0
t_1	1/2	5/8	4	5/2
t_2	1	2	2	4
t_3	3/2	39/8	4	39/2
t_4	2	10	1	10

(b) $\int_{0}^{2}(t^3+t)\,dt = 6 \Rightarrow |E_S| = \int_{0}^{2}(t^3+t)\,dt - S$

$= 6-6 = 0$

(c) $\frac{|E_S|}{\text{True Value}} \times 100 = 0\%$

6. $\int_{-1}^{1}(t^3+1)\,dt$

	t_i	$f(t_i)$	m	$mf(t_i)$
t_0	−1	0	1	0
t_1	−1/2	7/8	2	7/4
t_2	0	1	2	2
t_3	1/2	9/8	2	9/4
t_4	1	2	1	2

I. (a) For $n=4$, $\Delta x=\frac{b-a}{n}=\frac{1-(-1)}{4}=\frac{2}{4}=\frac{1}{2}$
$\Rightarrow \frac{\Delta x}{2}=\frac{1}{4}$; $\sum mf(t_i)=8 \Rightarrow T=\frac{1}{4}(8)=2$;
$f(t)=t^3+1 \Rightarrow f'(t)=3t^2 \Rightarrow f''(t)=6t$
$\Rightarrow M=6=f''(1) \Rightarrow |E_T|\le \frac{1-(-1)}{12}\left(\frac{1}{2}\right)^2(6)=\frac{1}{4}$

(b) $\int_{-1}^{1}(t^3+1)\,dt=\left[\frac{t^4}{4}+t\right]_{-1}^{1}=\left(\frac{1^4}{4}+1\right)-\left(\frac{(-1)^4}{4}+(-1)\right)=2 \Rightarrow |E_T|=\int_{-1}^{1}(t^3+1)\,dt-T=2-2=0$

(c) $\frac{|E_T|}{\text{True Value}}\times 100=0\%$

	t_i	$f(t_i)$	m	$mf(t_i)$
t_0	−1	0	1	0
t_1	−1/2	7/8	4	7/2
t_2	0	1	2	2
t_3	1/2	9/8	4	9/2
t_4	1	2	1	2

II. (a) For $n=4$, $\Delta x=\frac{b-a}{n}=\frac{1-(-1)}{4}=\frac{2}{4}=\frac{1}{2}$
$\Rightarrow \frac{\Delta x}{3}=\frac{1}{6}$; $\sum mf(t_i)=12 \Rightarrow S=\frac{1}{6}(12)=2$;
$f^{(3)}(t)=6 \Rightarrow f^{(4)}(t)=0 \Rightarrow M=0 \Rightarrow |E_S|=0$

(b) $\int_{-1}^{1}(t^3+1)\,dt=2 \Rightarrow |E_S|=\int_{-1}^{1}(t^3+1)\,dt-S$
$=2-2=0$

(c) $\frac{|E_S|}{\text{True Value}}\times 100=0\%$

7. $\int_{1}^{2}\frac{1}{s^2}\,ds$

	s_i	$f(s_i)$	m	$mf(s_i)$
s_0	1	1	1	1
s_1	5/4	16/25	2	32/25
s_2	3/2	4/9	2	8/9
s_3	7/4	16/49	2	32/49
s_4	2	1/4	1	1/4

I. (a) For $n=4$, $\Delta x=\frac{b-a}{n}=\frac{2-1}{4}=\frac{1}{4} \Rightarrow \frac{\Delta x}{2}=\frac{1}{8}$;
$\sum mf(s_i)=\frac{179{,}573}{44{,}100} \Rightarrow T=\frac{1}{8}\left(\frac{179{,}573}{44{,}100}\right)=\frac{179{,}573}{352{,}800}$
≈ 0.50899; $f(s)=\frac{1}{s^2} \Rightarrow f'(s)=-\frac{2}{s^3}$
$\Rightarrow f''(s)=\frac{6}{s^4} \Rightarrow M=6=f''(1)$
$\Rightarrow |E_T|\le \frac{2-1}{12}\left(\frac{1}{4}\right)^2(6)=\frac{1}{32}=0.03125$

(b) $\int_{1}^{2}\frac{1}{s^2}\,ds=\int_{1}^{2}s^{-2}\,ds=\left[-\frac{1}{s}\right]_1^2=-\frac{1}{2}-\left(-\frac{1}{1}\right)=\frac{1}{2} \Rightarrow E_T=\int_{1}^{2}\frac{1}{s^2}\,ds-T=\frac{1}{2}-0.50899=-0.00899$
$\Rightarrow |E_T|=0.00899$

(c) $\frac{|E_T|}{\text{True Value}}\times 100=\frac{0.00899}{0.5}\times 100\approx 2\%$

	s_i	$f(s_i)$	m	$mf(s_i)$
s_0	1	1	1	1
s_1	5/4	16/25	4	64/25
s_2	3/2	4/9	2	8/9
s_3	7/4	16/49	4	64/49
s_4	2	1/4	1	1/4

II. (a) For $n=4$, $\Delta x=\frac{b-a}{n}=\frac{2-1}{4}=\frac{1}{4} \Rightarrow \frac{\Delta x}{3}=\frac{1}{12}$;
$\sum mf(s_i)=\frac{264{,}821}{44{,}100} \Rightarrow S=\frac{1}{12}\left(\frac{264{,}821}{44{,}100}\right)=\frac{264{,}821}{529{,}200}$
≈ 0.50042; $f^{(3)}(s)=-\frac{24}{s^5} \Rightarrow f^{(4)}(s)=\frac{120}{s^6}$
$\Rightarrow M=120 \Rightarrow |E_S|\le \left|\frac{2-1}{180}\right|\left(\frac{1}{4}\right)^4(120)$
$=\frac{1}{384}\approx 0.00260$

(b) $\int_{1}^{2}\frac{1}{s^2}\,ds=\frac{1}{2} \Rightarrow E_S=\int_{1}^{2}\frac{1}{s^2}\,ds-S=\frac{1}{2}-0.50042=-0.00042 \Rightarrow |E_S|=0.00042$

(c) $\frac{|E_S|}{\text{True Value}}\times 100=\frac{0.0004}{0.5}\times 100\approx 0.08\%$

8. $\int_{2}^{4}\frac{1}{(s-1)^2}\,ds$

	s_i	$f(s_i)$	m	$mf(s_i)$
s_0	2	1	1	1
s_1	5/2	4/9	2	8/9
s_2	3	1/4	2	1/2
s_3	7/2	4/25	2	8/25
s_4	4	1/9	1	1/9

I. (a) For $n=4$, $\Delta x=\frac{b-a}{n}=\frac{4-2}{4}=\frac{1}{2} \Rightarrow \frac{\Delta x}{2}=\frac{1}{4}$;
$\sum mf(s_i)=\frac{1269}{450}$
$\Rightarrow T=\frac{1}{4}\left(\frac{1269}{450}\right)=\frac{1269}{1800}=0.70500$;
$f(s)=(s-1)^{-2} \Rightarrow f'(s)=-\frac{2}{(s-1)^3}$
$\Rightarrow f''(s)=\frac{6}{(s-1)^4} \Rightarrow M=6$
$\Rightarrow |E_T|\le \frac{4-2}{12}\left(\frac{1}{2}\right)^2(6)=\frac{1}{4}=0.25$

(b) $\int_{2}^{4}\frac{1}{(s-1)^2}\,ds=\left[\frac{-1}{(s-1)}\right]_2^4=\left(\frac{-1}{4-1}\right)-\left(\frac{-1}{2-1}\right)=\frac{2}{3} \Rightarrow E_T=\int_{2}^{4}\frac{1}{(s-1)^2}\,ds-T=\frac{2}{3}-0.705\approx -0.03833$
$\Rightarrow |E_T|\approx 0.03833$

(c) $\frac{|E_T|}{\text{True Value}} \times 100 = \frac{0.03833}{\left(\frac{2}{3}\right)} \times 100 \approx 6\%$

II. (a) For n = 4, $\Delta x = \frac{b-a}{n} = \frac{4-2}{4} = \frac{1}{2} \Rightarrow \frac{\Delta x}{3} = \frac{1}{6}$;

$\sum mf(s_i) = \frac{1813}{450}$

$\Rightarrow S = \frac{1}{6}\left(\frac{1813}{450}\right) = \frac{1813}{2700} \approx 0.67148$;

$f^{(3)}(s) = \frac{-24}{(s-1)^5} \Rightarrow f^{(4)}(s) = \frac{120}{(s-1)^6} \Rightarrow M = 120$

$\Rightarrow |E_S| \le \frac{4-2}{180}\left(\frac{1}{2}\right)^4(120) = \frac{1}{12} \approx 0.08333$

	s_i	$f(s_i)$	m	$mf(s_i)$
s_0	2	1	1	1
s_1	5/2	4/9	4	16/9
s_2	3	1/4	2	1/2
s_3	7/2	4/25	4	16/25
s_4	4	1/9	1	1/9

(b) $\int_2^4 \frac{1}{(s-1)^2}\,ds = \frac{2}{3} \Rightarrow E_S = \int_2^4 \frac{1}{(s-1)^2}\,ds - S \approx \frac{2}{3} - 0.67148 = -0.00481 \Rightarrow |E_S| \approx 0.00481$

(c) $\frac{|E_S|}{\text{True Value}} \times 100 = \frac{0.00481}{\left(\frac{2}{3}\right)} \times 100 \approx 1\%$

9. $\int_0^{\pi} \sin t\,dt$

I. (a) For n = 4, $\Delta x = \frac{b-a}{n} = \frac{\pi-0}{4} = \frac{\pi}{4} \Rightarrow \frac{\Delta x}{2} = \frac{\pi}{8}$;

$\sum mf(t_i) = 2 + 2\sqrt{2} \approx 4.8284$

$\Rightarrow T = \frac{\pi}{8}\left(2 + 2\sqrt{2}\right) \approx 1.89612$;

$f(t) = \sin t \Rightarrow f'(t) = \cos t \Rightarrow f''(t) = -\sin t$

$\Rightarrow M = 1 \Rightarrow |E_T| \le \frac{\pi-0}{12}\left(\frac{\pi}{4}\right)^2(1) = \frac{\pi^3}{192}$

≈ 0.16149

	t_i	$f(t_i)$	m	$mf(t_i)$
t_0	0	0	1	0
t_1	$\pi/4$	$\sqrt{2}/2$	2	$\sqrt{2}$
t_2	$\pi/2$	1	2	2
t_3	$3\pi/4$	$\sqrt{2}/2$	2	$\sqrt{2}$
t_4	π	0	1	0

(b) $\int_0^{\pi} \sin t\,dt = [-\cos t]_0^{\pi} = (-\cos \pi) - (-\cos 0) = 2 \Rightarrow |E_T| = \int_0^{\pi} \sin t\,dt - T \approx 2 - 1.89612 = 0.10388$

(c) $\frac{|E_T|}{\text{True Value}} \times 100 = \frac{0.10388}{2} \times 100 \approx 5\%$

II. (a) For n = 4, $\Delta x = \frac{b-a}{n} = \frac{\pi-0}{4} = \frac{\pi}{4} \Rightarrow \frac{\Delta x}{3} = \frac{\pi}{12}$;

$\sum mf(t_i) = 2 + 4\sqrt{2} \approx 7.6569$

$\Rightarrow S = \frac{\pi}{12}\left(2 + 4\sqrt{2}\right) \approx 2.00456$;

$f^{(3)}(t) = -\cos t \Rightarrow f^{(4)}(t) = \sin t$

$\Rightarrow M = 1 \Rightarrow |E_S| \le \frac{\pi-0}{180}\left(\frac{\pi}{4}\right)^4(1) \approx 0.00664$

	t_i	$f(t_i)$	m	$mf(t_i)$
t_0	0	0	1	0
t_1	$\pi/4$	$\sqrt{2}/2$	4	$2\sqrt{2}$
t_2	$\pi/2$	1	2	2
t_3	$3\pi/4$	$\sqrt{2}/2$	4	$2\sqrt{2}$
t_4	π	0	1	0

(b) $\int_0^{\pi} \sin t\,dt = 2 \Rightarrow E_S = \int_0^{\pi} \sin t\,dt - S \approx 2 - 2.00456 = -0.00456 \Rightarrow |E_S| \approx 0.00456$

(c) $\frac{|E_S|}{\text{True Value}} \times 100 = \frac{0.00456}{2} \times 100 \approx 0\%$

10. $\int_0^{1} \sin \pi t\,dt$

I. (a) For n = 4, $\Delta x = \frac{b-a}{n} = \frac{1-0}{4} = \frac{1}{4} \Rightarrow \frac{\Delta x}{2} = \frac{1}{8}$;

$\sum mf(t_i) = 2 + 2\sqrt{2} \approx 4.828$

$\Rightarrow T = \frac{1}{8}\left(2 + 2\sqrt{2}\right) \approx 0.60355$; $f(t) = \sin \pi t$

$\Rightarrow f'(t) = \pi \cos \pi t$

$\Rightarrow f''(t) = -\pi^2 \sin \pi t \Rightarrow M = \pi^2$

$\Rightarrow |E_T| \le \frac{1-0}{12}\left(\frac{1}{4}\right)^2(\pi^2) \approx 0.05140$

	t_i	$f(t_i)$	m	$mf(t_i)$
t_0	0	0	1	0
t_1	1/4	$\sqrt{2}/2$	2	$\sqrt{2}$
t_2	1/2	1	2	2
t_3	3/4	$\sqrt{2}/2$	2	$\sqrt{2}$
t_4	1	0	1	0

(b) $\int_0^{1} \sin \pi t\,dt = \left[-\frac{1}{\pi}\cos \pi t\right]_0^1 = \left(-\frac{1}{\pi}\cos \pi\right) - \left(-\frac{1}{\pi}\cos 0\right) = \frac{2}{\pi} \approx 0.63662 \Rightarrow |E_T| = \int_0^1 \sin \pi t\,dt - T$

$\approx \frac{2}{\pi} - 0.60355 = 0.03307$

(c) $\frac{|E_T|}{\text{True Value}} \times 100 = \frac{0.03307}{\left(\frac{2}{\pi}\right)} \times 100 \approx 5\%$

II. (a) For n = 4, $\Delta x = \frac{b-a}{n} = \frac{1-0}{4} = \frac{1}{4} \Rightarrow \frac{\Delta x}{3} = \frac{1}{12}$;

$\sum mf(t_i) = 2 + 4\sqrt{2} \approx 7.65685$

$\Rightarrow S = \frac{1}{12}\left(2 + 4\sqrt{2}\right) \approx 0.63807$;

$f^{(3)}(t) = -\pi^3 \cos \pi t \Rightarrow f^{(4)}(t) = \pi^4 \sin \pi t$

$\Rightarrow M = \pi^4 \Rightarrow |E_S| \le \frac{1-0}{180}\left(\frac{1}{4}\right)^4 (\pi^4) \approx 0.00211$

	t_i	$f(t_i)$	m	$mf(t_i)$
t_0	0	0	1	0
t_1	1/4	$\sqrt{2}/2$	4	$2\sqrt{2}$
t_2	1/2	1	2	2
t_3	3/4	$\sqrt{2}/2$	4	$2\sqrt{2}$
t_4	1	0	1	0

(b) $\int_0^1 \sin \pi t\, dt = \frac{2}{\pi} \approx 0.63662 \Rightarrow E_S = \int_0^1 \sin \pi t\, dt - S \approx \frac{2}{\pi} - 0.63807 = -0.00145 \Rightarrow |E_S| \approx 0.00145$

(c) $\frac{|E_S|}{\text{True Value}} \times 100 = \frac{0.00145}{\left(\frac{2}{\pi}\right)} \times 100 \approx 0\%$

11. (a) $n = 8 \Rightarrow \Delta x = \frac{1}{8} \Rightarrow \frac{\Delta x}{2} = \frac{1}{16}$;

$\sum mf(x_i) = 1(0.0) + 2(0.12402) + 2(0.24206) + 2(0.34763) + 2(0.43301) + 2(0.48789) + 2(0.49608)$
$+ 2(0.42361) + 1(0) = 5.1086 \Rightarrow T = \frac{1}{16}(5.1086) = 0.31929$

(b) $n = 8 \Rightarrow \Delta x = \frac{1}{8} \Rightarrow \frac{\Delta x}{3} = \frac{1}{24}$;

$\sum mf(x_i) = 1(0.0) + 4(0.12402) + 2(0.24206) + 4(0.34763) + 2(0.43301) + 4(0.48789) + 2(0.49608)$
$+ 4(0.42361) + 1(0) = 7.8749 \Rightarrow S = \frac{1}{24}(7.8749) = 0.32812$

(c) Let $u = 1 - x^2 \Rightarrow du = -2x\, dx \Rightarrow -\frac{1}{2}\, du = x\, dx$; $x = 0 \Rightarrow u = 1$, $x = 1 \Rightarrow u = 0$

$\int_0^1 x\sqrt{1-x^2}\, dx = \int_1^0 \sqrt{u}\left(-\frac{1}{2}\, du\right) = \frac{1}{2}\int_0^1 u^{1/2}\, du = \left[\frac{1}{2}\left(\frac{u^{3/2}}{\frac{3}{2}}\right)\right]_0^1 = \left[\frac{1}{3}u^{3/2}\right]_0^1 = \frac{1}{3}\left(\sqrt{1}\right)^3 - \frac{1}{3}\left(\sqrt{0}\right)^3 = \frac{1}{3}$;

$E_T = \int_0^1 x\sqrt{1-x^2}\, dx - T \approx \frac{1}{3} - 0.31929 = 0.01404$; $E_S = \int_0^1 x\sqrt{1-x^2}\, dx - S \approx \frac{1}{3} - 0.32812 = 0.00521$

12. (a) $n = 8 \Rightarrow \Delta x = \frac{3}{8} \Rightarrow \frac{\Delta x}{2} = \frac{3}{16}$;

$\sum mf(\theta_i) = 1(0) + 2(0.09334) + 2(0.18429) + 2(0.27075) + 2(0.35112) + 2(0.42443) + 2(0.49026)$
$+ 2(0.58466) + 1(0.6) = 5.3977 \Rightarrow T = \frac{3}{16}(5.3977) = 1.01207$

(b) $n = 8 \Rightarrow \Delta x = \frac{3}{8} \Rightarrow \frac{\Delta x}{3} = \frac{1}{8}$;

$\sum mf(\theta_i) = 1(0) + 4(0.09334) + 2(0.18429) + 4(0.27075) + 2(0.35112) + 4(0.42443) + 2(0.49026)$
$+ 4(0.58466) + 1(0.6) = 8.14406 \Rightarrow S = \frac{1}{8}(8.14406) = 1.01801$

(c) Let $u = 16 + \theta^2 \Rightarrow du = 2\theta\, d\theta \Rightarrow \frac{1}{2}\, du = \theta\, d\theta$; $\theta = 0 \Rightarrow u = 16$, $\theta = 3 \Rightarrow u = 16 + 3^2 = 25$

$\int_0^3 \frac{\theta}{\sqrt{16+\theta^2}}\, d\theta = \int_{16}^{25} \frac{1}{\sqrt{u}}\left(\frac{1}{2}\, du\right) = \frac{1}{2}\int_{16}^{25} u^{-1/2}\, du = \left[\frac{1}{2}\left(\frac{u^{1/2}}{\frac{1}{2}}\right)\right]_{16}^{25} = \sqrt{25} - \sqrt{16} = 1$;

$E_T = \int_0^3 \frac{\theta}{\sqrt{16+\theta^2}}\, d\theta - T \approx 1 - 1.01207 = -0.01207$; $E_S = \int_0^3 \frac{\theta}{\sqrt{16+\theta^2}}\, d\theta - S \approx 1 - 1.01801 = -0.01801$

13. (a) $n = 8 \Rightarrow \Delta x = \frac{\pi}{8} \Rightarrow \frac{\Delta x}{2} = \frac{\pi}{16}$;

$\sum mf(t_i) = 1(0.0) + 2(0.99138) + 2(1.26906) + 2(1.05961) + 2(0.75) + 2(0.48821) + 2(0.28946) + 2(0.13429)$
$+ 1(0) = 9.96402 \Rightarrow T = \frac{\pi}{16}(9.96402) \approx 1.95643$

(b) $n = 8 \Rightarrow \Delta x = \frac{\pi}{8} \Rightarrow \frac{\Delta x}{3} = \frac{\pi}{24}$;

$\sum mf(t_i) = 1(0.0) + 4(0.99138) + 2(1.26906) + 4(1.05961) + 2(0.75) + 4(0.48821) + 2(0.28946) + 4(0.13429)$
$+ 1(0) = 15.311 \Rightarrow S \approx \frac{\pi}{24}(15.311) \approx 2.00421$

(c) Let $u = 2 + \sin t \Rightarrow du = \cos t\, dt$; $t = -\frac{\pi}{2} \Rightarrow u = 2 + \sin\left(-\frac{\pi}{2}\right) = 1$, $t = \frac{\pi}{2} \Rightarrow u = 2 + \sin\frac{\pi}{2} = 3$

$\int_{-\pi/2}^{\pi/2} \frac{3\cos t}{(2+\sin t)^2}\, dt = \int_1^3 \frac{3}{u^2}\, du = 3\int_1^3 u^{-2}\, du = \left[3\left(\frac{u^{-1}}{-1}\right)\right]_1^3 = 3\left(-\frac{1}{3}\right) - 3\left(-\frac{1}{1}\right) = 2$;

$E_T = \int_{-\pi/2}^{\pi/2} \frac{3\cos t}{(2+\sin t)^2}\, dt - T \approx 2 - 1.95643 = 0.04357$; $E_S = \int_{-\pi/2}^{\pi/2} \frac{3\cos t}{(2+\sin t)^2}\, dt - S$
$\approx 2 - 2.00421 = -0.00421$

14. (a) $n = 8 \Rightarrow \Delta x = \frac{\pi}{32} \Rightarrow \frac{\Delta x}{2} = \frac{\pi}{64}$;

$\sum mf(y_i) = 1(2.0) + 2(1.51606) + 2(1.18237) + 2(0.93998) + 2(0.75402) + 2(0.60145) + 2(0.46364) + 2(0.31688) + 1(0) = 13.5488 \Rightarrow T \approx \frac{\pi}{64}(13.5488) = 0.66508$

(b) $n = 8 \Rightarrow \Delta x = \frac{\pi}{32} \Rightarrow \frac{\Delta x}{3} = \frac{\pi}{96}$;

$\sum mf(y_i) = 1(2.0) + 4(1.51606) + 2(1.18237) + 4(0.93988) + 2(0.75402) + 4(0.60145) + 2(0.46364) + 4(0.31688) + 1(0) = 20.29734 \Rightarrow S \approx \frac{\pi}{96}(20.29734) = 0.66423$

(c) Let $u = \cot y \Rightarrow du = -\csc^2 y\, dy$; $y = \frac{\pi}{4} \Rightarrow u = 1$, $y = \frac{\pi}{2} \Rightarrow u = 0$

$\int_{\pi/4}^{\pi/2} (\csc^2 y)\sqrt{\cot y}\, dy = \int_1^0 \sqrt{u}\,(-du) = \int_0^1 u^{1/2}\, du = \left[\frac{u^{3/2}}{\frac{3}{2}}\right]_0^1 = \frac{2}{3}\left(\sqrt{1}\right)^3 - \frac{2}{3}\left(\sqrt{0}\right)^3 = \frac{2}{3}$;

$E_T = \int_{\pi/4}^{\pi/2} (\csc^2 y)\sqrt{\cot y}\, dy - T \approx \frac{2}{3} - 0.66508 = 0.00159$; $E_S = \int_{\pi/4}^{\pi/2} (\csc^2 y)\sqrt{\cot y}\, dy - S \approx \frac{2}{3} - 0.66423 = 0.00244$

15. (a) $M = 0$ (see Exercise 1): Then $n = 1 \Rightarrow \Delta x = 1 \Rightarrow |E_T| = \frac{1}{12}(1)^2(0) = 0 < 10^{-4}$

(b) $M = 0$ (see Exercise 1): Then $n = 2$ (n must be even) $\Rightarrow \Delta x = \frac{1}{2} \Rightarrow |E_S| = \frac{1}{180}\left(\frac{1}{2}\right)^4(0) = 0 < 10^{-4}$

16. (a) $M = 0$ (see Exercise 2): Then $n = 1 \Rightarrow \Delta x = 2 \Rightarrow |E_T| = \frac{2}{12}(2)^2(0) = 0 < 10^{-4}$

(b) $M = 0$ (see Exercise 2): Then $n = 2$ (n must be even) $\Rightarrow \Delta x = 1 \Rightarrow |E_S| = \frac{2}{180}(1)^4(0) = 0 < 10^{-4}$

17. (a) $M = 2$ (see Exercise 3): Then $\Delta x = \frac{2}{n} \Rightarrow |E_T| \le \frac{2}{12}\left(\frac{2}{n}\right)^2(2) = \frac{4}{3n^2} < 10^{-4} \Rightarrow n^2 > \frac{4}{3}(10^4) \Rightarrow n > \sqrt{\frac{4}{3}(10^4)}$ $\Rightarrow n > 115.4$, so let $n = 116$

(b) $M = 0$ (see Exercise 3): Then $n = 2$ (n must be even) $\Rightarrow \Delta x = 1 \Rightarrow |E_S| = \frac{2}{180}(1)^4(0) = 0 < 10^{-4}$

18. (a) $M = 2$ (see Exercise 4): Then $\Delta x = \frac{2}{n} \Rightarrow |E_T| \le \frac{2}{12}\left(\frac{2}{n}\right)^2(2) = \frac{4}{3n^2} < 10^{-4} \Rightarrow n^2 > \frac{4}{3}(10^4) \Rightarrow n > \sqrt{\frac{4}{3}(10^4)}$ $\Rightarrow n > 115.4$, so let $n = 116$

(b) $M = 0$ (see Exercise 4): Then $n = 2$ (n must be even) $\Rightarrow \Delta x = 1 \Rightarrow |E_S| = \frac{2}{180}(1)^4(0) = 0 < 10^{-4}$

19. (a) $M = 12$ (see Exercise 5): Then $\Delta x = \frac{2}{n} \Rightarrow |E_T| \le \frac{2}{12}\left(\frac{2}{n}\right)^2(12) = \frac{8}{n^2} < 10^{-4} \Rightarrow n^2 > 8(10^4) \Rightarrow n > \sqrt{8(10^4)}$ $\Rightarrow n > 282.8$, so let $n = 283$

(b) $M = 0$ (see Exercise 5): Then $n = 2$ (n must be even) $\Rightarrow \Delta x = 1 \Rightarrow |E_S| = \frac{2}{180}(1)^4(0) = 0 < 10^{-4}$

20. (a) $M = 6$ (see Exercise 6): Then $\Delta x = \frac{2}{n} \Rightarrow |E_T| \le \frac{2}{12}\left(\frac{2}{n}\right)^2(6) = \frac{4}{n^2} < 10^{-4} \Rightarrow n^2 > 4(10^4) \Rightarrow n > \sqrt{4(10^4)}$ $= 200$, so let $n = 201$

(b) $M = 0$ (see Exercise 6): Then $n = 2$ (n must be even) $\Rightarrow \Delta x = 1 \Rightarrow |E_S| = \frac{2}{180}(1)^4(0) = 0 < 10^{-4}$

21. (a) $M = 6$ (see Exercise 7): Then $\Delta x = \frac{1}{n} \Rightarrow |E_T| \le \frac{1}{12}\left(\frac{1}{n}\right)^2(6) = \frac{1}{2n^2} < 10^{-4} \Rightarrow n^2 > \frac{1}{2}(10^4) \Rightarrow n > \sqrt{\frac{1}{2}(10^4)}$ $\Rightarrow n > 70.7$, so let $n = 71$

(b) $M = 120$ (see Exercise 7): Then $\Delta x = \frac{1}{n} \Rightarrow |E_S| = \frac{1}{180}\left(\frac{1}{n}\right)^4(120) = \frac{2}{3n^4} < 10^{-4} \Rightarrow n^4 > \frac{2}{3}(10^4)$ $\Rightarrow n > \sqrt[4]{\frac{2}{3}(10^4)} \Rightarrow n > 9.04$, so let $n = 10$ (n must be even)

22. (a) $M = 6$ (see Exercise 8): Then $\Delta x = \frac{2}{n} \Rightarrow |E_T| \le \frac{2}{12}\left(\frac{2}{n}\right)^2(6) = \frac{4}{n^2} < 10^{-4} \Rightarrow n^2 > 4(10^4) \Rightarrow n > \sqrt{4(10^4)}$ $\Rightarrow n > 200$, so let $n = 201$

(b) $M = 120$ (see Exercise 8): Then $\Delta x = \frac{2}{n} \Rightarrow |E_S| \le \frac{2}{180}\left(\frac{2}{n}\right)^4(120) = \frac{64}{3n^4} < 10^{-4} \Rightarrow n^4 > \frac{64}{3}(10^4)$ $\Rightarrow n > \sqrt[4]{\frac{64}{3}(10^4)} \Rightarrow n > 21.5$, so let $n = 22$ (n must be even)

23. (a) $f(x) = \sqrt{x+1} \Rightarrow f'(x) = \frac{1}{2}(x+1)^{-1/2} \Rightarrow f''(x) = -\frac{1}{4}(x+1)^{-3/2} = -\frac{1}{4\left(\sqrt{x+1}\right)^3} \Rightarrow M = \frac{1}{4\left(\sqrt{1}\right)^3} = \frac{1}{4}$.

Then $\Delta x = \frac{3}{n} \Rightarrow |E_T| \le \frac{3}{12}\left(\frac{3}{n}\right)^2\left(\frac{1}{4}\right) = \frac{9}{16n^2} < 10^{-4} \Rightarrow n^2 > \frac{9}{16}(10^4) \Rightarrow n > \sqrt{\frac{9}{16}(10^4)} \Rightarrow n > 75$, so let $n = 76$

(b) $f^{(3)}(x) = \frac{3}{8}(x+1)^{-5/2} \Rightarrow f^{(4)}(x) = -\frac{15}{16}(x+1)^{-7/2} = -\frac{15}{16\left(\sqrt{x+1}\right)^7} \Rightarrow M = \frac{15}{16\left(\sqrt{1}\right)^7} = \frac{15}{16}$. Then $\Delta x = \frac{3}{n}$

$\Rightarrow |E_S| \le \frac{3}{180}\left(\frac{3}{n}\right)^4\left(\frac{15}{16}\right) = \frac{3^5(15)}{16(180)n^4} < 10^{-4} \Rightarrow n^4 > \frac{3^5(15)(10^4)}{16(180)} \Rightarrow n > \sqrt[4]{\frac{3^5(15)(10^4)}{16(180)}} \Rightarrow n > 10.6$, so let $n = 12$ (n must be even)

24. (a) $f(x) = \frac{1}{\sqrt{x+1}} \Rightarrow f'(x) = -\frac{1}{2}(x+1)^{-3/2} \Rightarrow f''(x) = \frac{3}{4}(x+1)^{-5/2} = \frac{3}{4\left(\sqrt{x+1}\right)^5} \Rightarrow M = \frac{3}{4\left(\sqrt{1}\right)^5} = \frac{3}{4}$.

Then $\Delta x = \frac{3}{n} \Rightarrow |E_T| \le \frac{3}{12}\left(\frac{3}{n}\right)^2\left(\frac{3}{4}\right) = \frac{3^4}{48n^2} < 10^{-4} \Rightarrow n^2 > \frac{3^4(10^4)}{48} \Rightarrow n > \sqrt{\frac{3^4(10^4)}{48}} \Rightarrow n > 129.9$, so let $n = 130$

(b) $f^{(3)}(x) = -\frac{15}{8}(x+1)^{-7/2} \Rightarrow f^{(4)}(x) = \frac{105}{16}(x+1)^{-9/2} = \frac{105}{16\left(\sqrt{x+1}\right)^9} \Rightarrow M = \frac{105}{16\left(\sqrt{1}\right)^9} = \frac{105}{16}$. Then $\Delta x = \frac{3}{n}$

$\Rightarrow |E_S| \le \frac{3}{180}\left(\frac{3}{n}\right)^4\left(\frac{105}{16}\right) = \frac{3^5(105)}{16(180)n^4} < 10^{-4} \Rightarrow n^4 > \frac{3^5(105)(10^4)}{16(180)} \Rightarrow n > \sqrt[4]{\frac{3^5(105)(10^4)}{16(180)}} \Rightarrow n > 17.25$, so let $n = 18$ (n must be even)

25. (a) $f(x) = \sin(x+1) \Rightarrow f'(x) = \cos(x+1) \Rightarrow f''(x) = -\sin(x+1) \Rightarrow M = 1$. Then $\Delta x = \frac{2}{n} \Rightarrow |E_T| \le \frac{2}{12}\left(\frac{2}{n}\right)^2(1)$ $= \frac{8}{12n^2} < 10^{-4} \Rightarrow n^2 > \frac{8(10^4)}{12} \Rightarrow n > \sqrt{\frac{8(10^4)}{12}} \Rightarrow n > 81.6$, so let $n = 82$

(b) $f^{(3)}(x) = -\cos(x+1) \Rightarrow f^{(4)}(x) = \sin(x+1) \Rightarrow M = 1$. Then $\Delta x = \frac{2}{n} \Rightarrow |E_S| \le \frac{2}{180}\left(\frac{2}{n}\right)^4(1) = \frac{32}{180n^4} < 10^{-4}$ $\Rightarrow n^4 > \frac{32(10^4)}{180} \Rightarrow n > \sqrt[4]{\frac{32(10^4)}{180}} \Rightarrow n > 6.49$, so let $n = 8$ (n must be even)

26. (a) $f(x) = \cos(x+\pi) \Rightarrow f'(x) = -\sin(x+\pi) \Rightarrow f''(x) = -\cos(x+\pi) \Rightarrow M = 1$. Then $\Delta x = \frac{2}{n}$ $\Rightarrow |E_T| \le \frac{2}{12}\left(\frac{2}{n}\right)^2(1) = \frac{8}{12n^2} < 10^{-4} \Rightarrow n^2 > \frac{8(10^4)}{12} \Rightarrow n > \sqrt{\frac{8(10^4)}{12}} \Rightarrow n > 81.6$, so let $n = 82$

(b) $f^{(3)}(x) = \sin(x+\pi) \Rightarrow f^{(4)}(x) = \cos(x+\pi) \Rightarrow M = 1$. Then $\Delta x = \frac{2}{n} \Rightarrow |E_S| \le \frac{2}{180}\left(\frac{2}{n}\right)^4(1) = \frac{32}{180n^4} < 10^{-4}$ $\Rightarrow n^4 > \frac{32(10^4)}{180} \Rightarrow n > \sqrt[4]{\frac{32(10^4)}{180}} \Rightarrow n > 6.49$, so let $n = 8$ (n must be even)

27. $\frac{5}{2}(6.0 + 2(8.2) + 2(9.1)\ldots + 2(12.7) + 13.0)(30) = 15{,}990 \text{ ft}^3$.

28. (a) Using Trapezoid Rule, $\Delta x = 200 \Rightarrow \frac{\Delta x}{2} = \frac{200}{2} = 100$; $\sum mf(x_i) = 13{,}180 \Rightarrow$ Area $\approx 100(13{,}180)$ $= 1{,}318{,}000 \text{ ft}^2$. Since the average depth = 20 ft we obtain Volume $\approx 20(\text{Area}) \approx 26{,}360{,}000 \text{ ft}^3$.

(b) Now, Number of fish $= \frac{\text{Volume}}{1000} = 26{,}360$ (to the nearest fish) $\Rightarrow$ Maximum to be caught = 75% of 26,360 $= 19{,}770 \Rightarrow$ Number of licenses $= \frac{19{,}770}{20} = 988$

	x_i	$f(x_i)$	m	$mf(x_i)$
x_0	0	0	1	0
x_1	200	520	2	1040
x_2	400	800	2	1600
x_3	600	1000	2	2000
x_4	800	1140	2	2280
x_5	1000	1160	2	2320
x_6	1200	1110	2	2220
x_7	1400	860	2	1720
x_8	1600	0	1	0

29. Use the conversion 30 mph = 44 fps (ft per sec) since time is measured in seconds. The distance traveled as the car accelerates from, say, 40 mph = 58.67 fps to 50 mph = 73.33 fps in (4.5 − 3.2) = 1.3 sec is the area of the trapezoid (see figure) associated with that time interval: $\frac{1}{2}(58.67 + 73.33)(1.3) = 85.8$ ft. The total distance traveled by the Ford Mustang Cobra is the sum of all these eleven trapezoids (using $\frac{\Delta t}{2}$ and the table below):

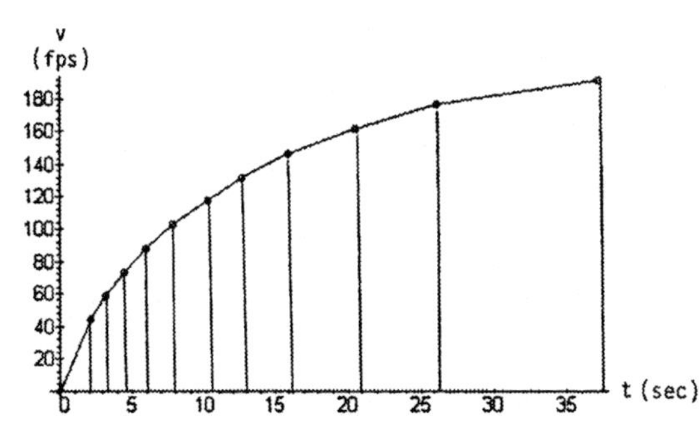

$s = (44)(1.1) + (102.67)(0.5) + (132)(0.65) + (161.33)(0.7) + (190.67)(0.95) + (220)(1.2) + (249.33)(1.25)$
$\quad + (278.67)(1.65) + (308)(2.3) + (337.33)(2.8) + (366.67)(5.45) = 5166.346 \text{ ft} \approx 0.9785 \text{ mi}$

v (mph)	0	30	40	50	60	70	80	90	100	110	120	130
v (fps)	0	44	58.67	73.33	88	102.67	117.33	132	146.67	161.33	176	190.67
t (sec)	0	2.2	3.2	4.5	5.9	7.8	10.2	12.7	16	20.6	26.2	37.1
Δt/2	0	1.1	0.5	0.65	0.7	0.95	1.2	1.25	1.65	2.3	2.8	5.45

30. Using Simpson's Rule, $\Delta x = \frac{b-a}{n} = \frac{24-0}{6} = \frac{24}{6} = 4$;
$\sum my_i = 350 \Rightarrow S = \frac{4}{3}(350) = \frac{1400}{3} \approx 466.7 \text{ in.}^2$

	x_i	y_i	m	my_i
x_0	0	0	1	0
x_1	4	18.75	4	75
x_2	8	24	2	48
x_3	12	26	4	104
x_4	16	24	2	48
x_5	20	18.75	4	75
x_6	24	0	1	0

31. Using Simpson's Rule, $\Delta x = 1 \Rightarrow \frac{\Delta x}{3} = \frac{1}{3}$;
$\sum my_i = 33.6 \Rightarrow$ Cross Section Area $\approx \frac{1}{3}(33.6)$
$= 11.2 \text{ ft}^2$. Let x be the length of the tank. Then the Volume V = (Cross Sectional Area) x = 11.2x.
Now 5000 lb of gasoline at 42 lb/ft^3
$\Rightarrow V = \frac{5000}{42} = 119.05 \text{ ft}^3$
$\Rightarrow 119.05 = 11.2x \Rightarrow x \approx 10.63$ ft

	x_i	y_i	m	my_i
x_0	0	1.5	1	1.5
x_1	1	1.6	4	6.4
x_2	2	1.8	2	3.6
x_3	3	1.9	4	7.6
x_4	4	2.0	2	4.0
x_5	5	2.1	4	8.4
x_6	6	2.1	1	2.1

32. $\frac{24}{2}[0.019 + 2(0.020) + 2(0.021) + \ldots + 2(0.031) + 0.035] = 4.2$ L

33. (a) $|E_S| \le \frac{b-a}{180}(\Delta x^4)M$; $n = 4 \Rightarrow \Delta x = \frac{\frac{\pi}{2}-0}{4} = \frac{\pi}{8}$; $|f^{(4)}| \le 1 \Rightarrow M = 1 \Rightarrow |E_S| \le \frac{(\frac{\pi}{2}-0)}{180}\left(\frac{\pi}{8}\right)^4(1) \approx 0.00021$

(b) $\Delta x = \frac{\pi}{8} \Rightarrow \frac{\Delta x}{3} = \frac{\pi}{24}$;
$\sum mf(x_i) = 10.47208705$
$\Rightarrow S = \frac{\pi}{24}(10.47208705) \approx 1.37079$

	x_i	$f(x_i)$	m	$mf(x_{1i})$
x_0	0	1	1	1
x_1	π/8	0.974495358	4	3.897981432
x_2	π/4	0.900316316	2	1.800632632
x_3	3π/8	0.784213303	4	3.136853212
x_4	π/2	0.636619772	1	0.636619772

(c) $\approx \left(\frac{0.00021}{1.37079}\right) \times 100 \approx 0.015\%$

34. (a) $\Delta x = \frac{b-a}{n} = \frac{1-0}{10} = 0.1 \Rightarrow \text{erf}(1) = \frac{2}{\sqrt{3}}\left(\frac{0.1}{3}\right)(y_0 + 4y_1 + 2y_2 + 4y_3 + \ldots + 4y_9 + y_{10})$
$\frac{2}{30\sqrt{\pi}}(e^0 + 4e^{-0.01} + 2e^{-0.04} + 4e^{-0.09} + \ldots + 4e^{-0.81} + e^{-1}) \approx 0.843$

(b) $|E_S| \le \frac{1-0}{180}(0.1)^4(12) \approx 6.7 \times 10^{-6}$

35. (a) $n = 10 \Rightarrow \Delta x = \frac{\pi - 0}{10} = \frac{\pi}{10} \Rightarrow \frac{\Delta x}{2} = \frac{\pi}{20}$;
$\sum mf(x_i) = 1(0) + 2(0.09708) + 2(0.36932) + 2(0.76248) + 2(1.19513) + 2(1.57080) + 2(1.79270)$
$+ 2(1.77912) + 2(1.47727) + 2(0.87372) + 1(0) = 19.83524 \Rightarrow T = \frac{\pi}{20}(19.83524) = 3.11571$

(b) $\pi - 3.11571 \approx 0.02588$

(c) With $M = 3.11$, we get $|E_T| \le \frac{\pi}{12}\left(\frac{\pi}{10}\right)^2(3.11) = \frac{\pi^3}{1200}(3.11) < 0.08036$

36. (a) $f''(x) = 2\cos x - x\sin x \Rightarrow f^{(3)}(x) = -3\sin x - x\cos x \Rightarrow f^{(4)}(x) = -4\cos x + x\sin x$. From the graphs shown below, $|-4\cos x + x\sin x| < 4.8$ for $0 \le x \le \pi$.

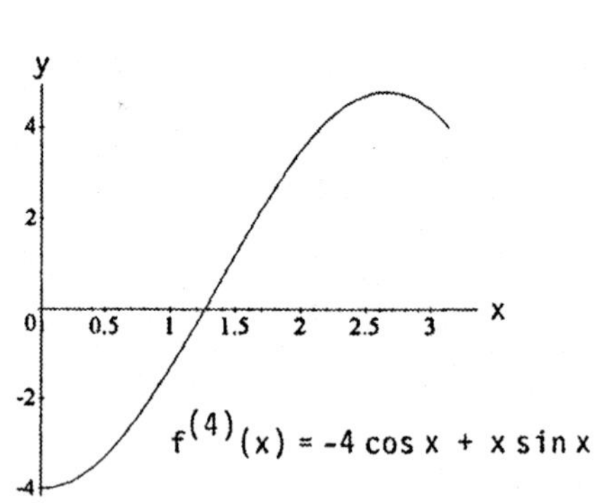

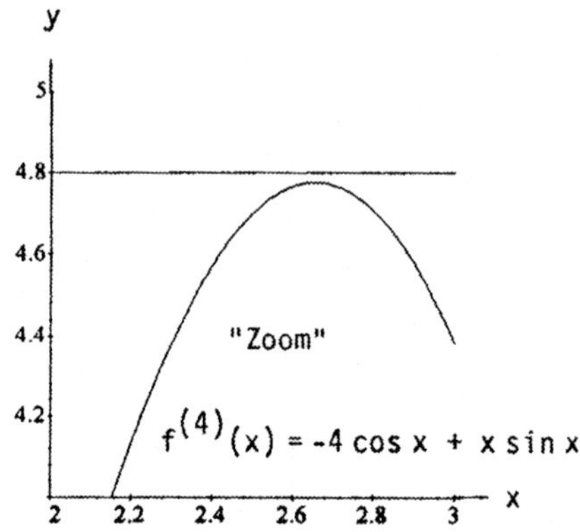

(b) $n = 10 \Rightarrow \Delta x = \frac{\pi}{10} \Rightarrow |E_S| \le \frac{\pi}{180}\left(\frac{\pi}{10}\right)^4(4.8) \approx 0.00082$

(c) $\sum mf(x_i) = 1(0) + 4(0.09708) + 2(0.36932) + 4(0.76248) + 2(1.19513) + 4(1.57080) + 2(1.79270)$
$+ 4(1.77912) + 2(1.47727) + 4(0.87372) + 1(0) = 30.0016 \Rightarrow S = \frac{\pi}{30}(30.0016) = 3.14176$

(d) $|\pi - 3.14176| \approx 0.00017$

37. $T = \frac{\Delta x}{2}(y_0 + 2y_1 + 2y_2 + 2y_3 + \ldots + 2y_{n-1} + y_n)$ where $\Delta x = \frac{b-a}{n}$ and f is continuous on [a, b]. So
$T = \frac{b-a}{n}\frac{(y_0 + y_1 + y_1 + y_2 + y_2 + \ldots + y_{n-1} + y_{n-1} + y_n)}{2} = \frac{b-a}{n}\left(\frac{f(x_0)+f(x_1)}{2} + \frac{f(x_1)+f(x_2)}{2} + \ldots + \frac{f(x_{n-1})+f(x_n)}{2}\right)$.
Since f is continuous on each interval $[x_{k-1}, x_k]$, and $\frac{f(x_{k-1})+f(x_k)}{2}$ is always between $f(x_{k-1})$ and $f(x_k)$, there is a point c_k in $[x_{k-1}, x_k]$ with $f(c_k) = \frac{f(x_{k-1})+f(x_k)}{2}$; this is a consequence of the Intermediate Value Theorem. Thus our sum is $\sum_{k=1}^{n}\left(\frac{b-a}{n}\right)f(c_k)$ which has the form $\sum_{k=1}^{n}\Delta x_k f(c_k)$ with $\Delta x_k = \frac{b-a}{n}$ for all k. This is a Riemann Sum for f on [a, b].

38. $S = \frac{\Delta x}{3}(y_0 + 4y_1 + 2y_2 + 4y_3 + \ldots + 2y_{n-2} + 4y_{n-1} + y_n)$ where n is even, $\Delta x = \frac{b-a}{n}$ and f is continuous on [a, b]. So
$S = \frac{b-a}{n}\left(\frac{y_0 + 4y_1 + y_2}{3} + \frac{y_2 + 4y_3 + y_4}{3} + \frac{y_4 + 4y_5 + y_6}{3} + \ldots + \frac{y_{n-2} + 4y_{n-1} + y_n}{3}\right)$
$= \frac{b-a}{\frac{n}{2}}\left(\frac{f(x_0)+4f(x_1)+f(x_2)}{6} + \frac{f(x_2)+4f(x_3)+f(x_4)}{6} + \frac{f(x_4)+4f(x_5)+f(x_6)}{6} + \ldots + \frac{f(x_{n-2})+4f(x_{n-1})+f(x_n)}{6}\right)$
$\frac{f(x_{2k})+4f(x_{2k+1})+f(x_{2k+2})}{6}$ is the average of the six values of the continuous function on the interval $[x_{2k}, x_{2k+2}]$, so it is between the minimum and maximum of f on this interval. By the Extreme Value Theorem for continuous functions, f takes on its maximum and minimum in this interval, so there are x_a and x_b with $x_{2k} \le x_a, x_b \le x_{2k+2}$ and $f(x_a) \le \frac{f(x_{2k})+4f(x_{2k+1})+f(x_{2k+2})}{6} \le f(x_b)$. By the Intermediate Value Theorem, there is c_k in $[x_{2k}, x_{2k+2}]$ with $f(c_k) = \frac{f(x_{2k})+4f(x_{2k+1})+f(x_{2k+2})}{6}$. So our sum has the form $\sum_{k=1}^{n/2}\Delta x_k f(c_k)$ with $\Delta x_k = \frac{b-a}{(n/2)}$, a Riemann sum for f on [a, b].

Exercises 39-42 were done using a graphing calculator with $n = 50$

39. 1.08943

40. 1.37076

41. 0.82812

42. 51.05400

43. (a) $T_{10} \approx 1.983523538$
$T_{100} \approx 1.999835504$
$T_{1000} \approx 1.999998355$

(b)

n	$\lvert E_T \rvert = 2 - T_n$
10	$0.016476462 = 1.6476462 \times 10^{-2}$
100	1.64496×10^{-4}
1000	1.646×10^{-6}

(c) $\lvert E_{T_{10n}} \rvert \approx 10^{-2} \lvert E_{T_n} \rvert$

(d) $b - a = \pi$, $(\Delta x)^2 = \frac{\pi^2}{n^2}$, $M = 1$

$\lvert E_{T_n} \rvert \le \frac{\pi}{12}\left(\frac{\pi^2}{n^2}\right) = \frac{\pi^3}{12n^2}$

$\lvert E_{T_{10n}} \rvert \le \frac{\pi^3}{12(10n)^2} \le 10^{-2} \lvert E_{T_n} \rvert$

44. (a) $S_{10} \approx 2.000109517$
$S_{100} \approx 2.000000011$
$S_{1000} \approx 2.000000000$

(b)

n	$\lvert E_S \rvert = 2 - S_n$
10	1.09517×10^{-4}
100	1.1×10^{-8}
1000	0

(c) $\lvert E_{S_{10n}} \rvert \approx 10^{-4} \lvert E_{S_n} \rvert$

(d) $b - a = \pi$, $(\Delta x)^4 = \frac{\pi^4}{n^4}$, $M = 1$

$\lvert E_{S_n} \rvert \le \frac{\pi}{180}\left(\frac{\pi^4}{n^4}\right) = \frac{\pi^5}{180n^4}$

$\lvert E_{S_{10n}} \rvert \le \frac{\pi^5}{180(10n)^4} \le 10^{-4} \lvert E_{S_n} \rvert$

45. (a) $f'(x) = 2x\cos(x^2)$, $f''(x) = 2x \cdot (-2x)\sin(x^2) + 2\cos(x^2) = -4x^2\sin(x^2) + 2\cos(x^2)$

(b)

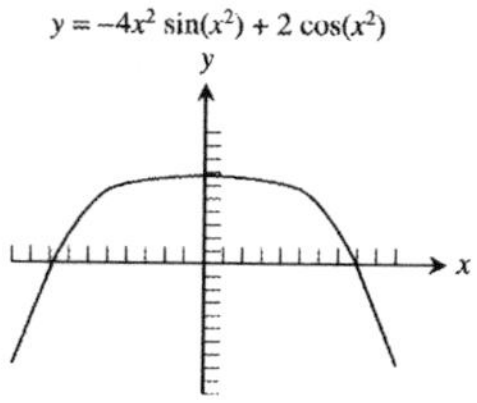

(c) The graph shows that $3 \le f''(x) \le 2$ so $\lvert f''(x) \rvert \le 3$ for $-1 \le x \le 1$.

(d) $\lvert E_T \rvert \le \frac{1-(-1)}{12}(\Delta x)^2(3) = \frac{(\Delta x)^2}{2}$

(e) For $0 < \Delta x < 0.1$, $\lvert E_T \rvert \le \frac{(\Delta x)^2}{2} \le \frac{0.1^2}{2} = 0.005 < 0.01$

(f) $n \ge \frac{1-(-1)}{\Delta x} \ge \frac{2}{0.1} = 20$

46. (a) $f'''(x) = -4x^2 \cdot 2x\cos(x^2) - 8x\sin(x^2) - 4x\sin(x^2) = -8x^3\cos(x^2) - 12x\sin(x^2)$
$f^{(4)}(x) = -8x^3 \cdot 2x\sin(x^2) - 24x^2\cos(x^2) - 12x \cdot 2x\cos(x^2) - 12\sin(x^2) = (16x^4 - 12)\sin(x^2) - 48x^2\cos(x^2)$

(b)

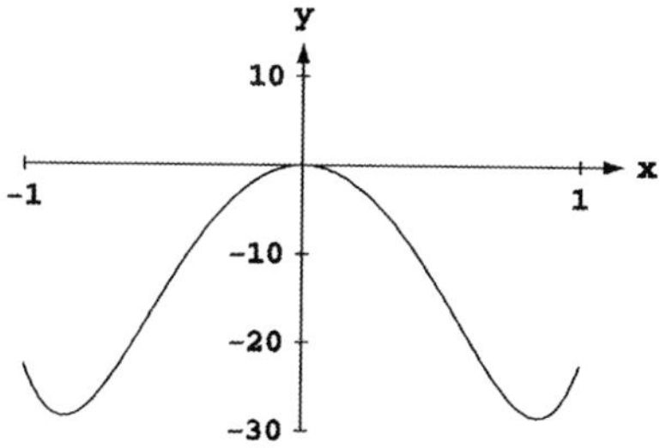

(c) The graph shows that $-30 \le f^{(4)}(x) \le 0$ so $\lvert f^{(4)}(x) \rvert \le 30$ for $-1 \le x \le 1$.

(d) $\lvert E_S \rvert \le \frac{1-(-1)}{180}(\Delta x)^4(30) = \frac{(\Delta x)^4}{3}$

(e) For $0 < \Delta x < 0.4$, $\lvert E_S \rvert \le \frac{(\Delta x)^4}{3} \le \frac{0.4^2}{3} \approx 0.00853 < 0.01$

(f) $n \ge \frac{1-(-1)}{\Delta x} \ge \frac{2}{0.4} = 5$

47. (a) Using $d = \frac{C}{\pi}$, and $A = \pi\left(\frac{d}{2}\right)^2 = \frac{C^2}{4\pi}$ yields the following areas (in square inches, rounded to the nearest tenth): 2.3, 1.6, 1.5, 2.1, 3.2, 4.8, 7.0, 9.3, 10.7, 10.7, 9.3, 6.4, 3.2

(b) If $C(y)$ is the circumference as a function of y, then the area of a cross section is

$A(y) = \pi\left(\frac{C(y)/\pi}{2}\right)^2 = \frac{C^2(y)}{4\pi}$, and the volume is $\frac{1}{4\pi}\int_0^6 C^2(y)\,dy$.

(c) $\int_0^6 A(y)\,dy = \frac{1}{4\pi}\int_0^6 C^2(y)\,dy$
$\approx \frac{1}{4\pi}\left(\frac{6-0}{24}\right)[5.4^2 + 2(4.5^2 + 4.4^2 + 5.1^2 + 6.3^2 + 7.8^2 + 9.4^2 + 10.8^2 + 11.6^2 + 11.6^2 + 10.8^2 + 9.0^2) + 6.3^2]$
$\approx 34.7\ \text{in}^3$

(d) $V = \frac{1}{4\pi}\int_0^6 C^2(y)\,dy \approx \frac{1}{4\pi}\left(\frac{6-0}{36}\right)\Big[5.4^2 + 4(4.5^2) + 2(4.4^2) + 4(5.1^2) + 2(6.3^2) + 4(7.8^2) + 2(9.4^2) + 4(10.8^2)$
$+ 2(11.6^2) + 4(11.6^2) + 2(10.8^2) + 4(9.0^2) + 6.3^2\Big] = 34.792\ \text{in}^3$

by Simpson's Rule. The Simpson's Rule estimate should be more accurate than the trapezoid estimate. The error in the Simpson's estimate is proportional to $(\Delta y)^4 = 0.0625$ whereas the error in the trapezoid estimate is proportional to $(\Delta y)^2 = 0.25$, a larger number when $\Delta y = 0.5$ in.

48. (a) Displacement Volume $V \approx \frac{\Delta x}{3}(y_0 + 4y_1 + 2y_2 + 4y_3 + \ldots + 2y_{n-2} + 4y_{n-1} + y_n)$, $x_0 = 0$, $x_n = 10 - \Delta x$,
$\Delta x = 2.54$, $n = 10 \Rightarrow \int_{x_0}^{x_n} A(x)\,dx \approx \frac{2.54}{3}\Big[0 + 4(1.07) + 2(3.84) + 4(7.82) + 2(12.20) + 4(15.18) + 2(16.14)$
$+ 4(14.00) + 2(9.21) + 4(3.24) + 0\Big] = \frac{2.54}{3}(248.02) = 209.99 \approx 210\ \text{ft}^3$.

(b) The weigth of water displaced is approximately $64 \cdot 120 = 13{,}440$ lb.

(c) The volume of a prism $= (2.54)(16.14) = 409.96 \approx 410\ \text{ft}^3$. Thus, the prismatic coefficient is $\frac{210\ \text{ft}^3}{410\ \text{ft}^3} \approx 0.51$.

49. (a) $a = 1, e = \frac{1}{2} \Rightarrow \text{Length} = 4\int_0^{\pi/2}\sqrt{1 - \frac{1}{4}\cos^2 t}\,dt$
$= 2\int_0^{\pi/2}\sqrt{4 - \cos^2 t}\,dt = \int_0^{\pi/2} f(t)\,dt$; use the Trapezoid Rule with $n = 10 \Rightarrow \Delta t = \frac{b-a}{n} = \frac{\left(\frac{\pi}{2}\right) - 0}{10}$
$= \frac{\pi}{20} \cdot \int_0^{\pi/2}\sqrt{4 - \cos^2 t}\,dt \approx \sum_{n=0}^{10} mf(x_n) = 37.3686183$
$\Rightarrow T = \frac{\Delta t}{2}(37.3686183) = \frac{\pi}{40}(37.3686183)$
$= 2.934924419 \Rightarrow \text{Length} = 2(2.934924419)$
≈ 5.870

(b) $|f''(t)| < 1 \Rightarrow M = 1$
$\Rightarrow |E_T| \le \frac{b-a}{12}(\Delta t^2 M) \le \frac{\left(\frac{\pi}{2}\right) - 0}{12}\left(\frac{\pi}{20}\right)^2 1 \le 0.0032$

	x_i	$f(x_i)$	m	$mf(x_i)$
x_0	0	1.732050808	1	1.732050808
x_1	$\pi/20$	1.739100843	2	3.478201686
x_2	$\pi/10$	1.759400893	2	3.518801786
x_3	$3\pi/20$	1.790560631	2	3.581121262
x_4	$\pi/5$	1.82906848	1	3.658136959
x_5	$\pi/4$	1.870828693	1	3.741657387
x_6	$3\pi/10$	1.911676881	2	3.823353762
x_7	$7\pi/20$	1.947791731	2	3.895583461
x_8	$2\pi/5$	1.975982919	2	3.951965839
x_9	$9\pi/20$	1.993872679	2	3.987745357
x_{10}	$\pi/2$	2	1	2

50. $\Delta x = \frac{\pi - 0}{8} = \frac{\pi}{8} \Rightarrow \frac{\Delta x}{3} = \frac{\pi}{24}$; $\sum mf(x_i) = 29.184807792$
$\Rightarrow S = \frac{\pi}{24}(29.18480779) \approx 3.82028$

	x_i	$f(x_i)$	m	$mf(x_i)$
x_0	0	1.414213562	1	1.414213562
x_1	$\pi/8$	1.361452677	4	5.445810706
x_2	$\pi/4$	1.224744871	2	2.449489743
x_3	$3\pi/8$	1.070722471	4	4.282889883
x_4	$\pi/2$	1	2	2
x_5	$5\pi/8$	1.070722471	4	4.282889883
x_6	$3\pi/4$	1.224744871	2	2.449489743
x_7	$7\pi/8$	1.361452677	4	5.445810706
x_8	π	1.414213562	1	1.414213562

51. The length of the curve $y = \sin\left(\frac{3\pi}{20}x\right)$ from 0 to 20 is: $L = \int_0^{20}\sqrt{1 + \left(\frac{dy}{dx}\right)^2}\,dx$; $\frac{dy}{dx} = \frac{3\pi}{20}\cos\left(\frac{3\pi}{20}x\right) \Rightarrow \left(\frac{dy}{dx}\right)^2$
$= \frac{9\pi^2}{400}\cos^2\left(\frac{3\pi}{20}x\right) \Rightarrow L = \int_0^{20}\sqrt{1 + \frac{9\pi^2}{400}\cos^2\left(\frac{3\pi}{20}x\right)}\,dx$. Using numerical integration we find $L \approx 21.07$ in

52. First, we'll find the length of the cosine curve: $L = \int_{-25}^{25}\sqrt{1 + \left(\frac{dy}{dx}\right)^2}\,dx$; $\frac{dy}{dx} = -\frac{25\pi}{50}\sin\left(\frac{\pi x}{50}\right)$
$\Rightarrow \left(\frac{dy}{dx}\right)^2 = \frac{\pi^2}{4}\sin^2\left(\frac{\pi x}{50}\right) \Rightarrow L = \int_{-25}^{25}\sqrt{1 + \frac{\pi^2}{4}\sin^2\left(\frac{\pi x}{50}\right)}\,dx$. Using a numerical integrator we find

$L \approx 73.1848$ ft. Surface area is: $A = \text{length} \cdot \text{width} \approx (73.1848)(300) = 21{,}955.44$ ft.
$\text{Cost} = 1.75A = (1.75)(21{,}955.44) = \$38{,}422.02$. Answers may vary slightly, depending on the numerical integration used.

53. $y = \sin x \Rightarrow \frac{dy}{dx} = \cos x \Rightarrow \left(\frac{dy}{dx}\right)^2 = \cos^2 x \Rightarrow S = \int_0^{\pi} 2\pi(\sin x)\sqrt{1+\cos^2 x}\,dx$; a numerical integration gives $S \approx 14.4$

54. $y = \frac{x^2}{4} \Rightarrow \frac{dy}{dx} = \frac{x}{2} \Rightarrow \left(\frac{dy}{dx}\right)^2 = \frac{x^2}{4} \Rightarrow S = \int_0^2 2\pi\left(\frac{x^2}{4}\right)\sqrt{1+\frac{x^2}{4}}\,dx$; a numerical integration gives $S \approx 5.28$

55. $y = x + \sin 2x \Rightarrow \frac{dy}{dx} = 1 + 2\cos 2x \Rightarrow \left(\frac{dy}{dx}\right)^2 = (1+2\cos 2x)^2$; by symmetry of the graph we have that $S = 2\int_0^{2\pi/3} 2\pi(x+\sin 2x)\sqrt{1+(1+2\cos 2x)^2}\,dx$; a numerical integration gives $S \approx 54.9$

56. $y = \frac{x}{12}\sqrt{36-x^2} \Rightarrow \frac{dy}{dx} = \frac{\sqrt{36-x^2}}{12} + \frac{x}{12}\cdot\frac{1}{2}\,\frac{(-2x)}{\sqrt{36-x^2}} = \frac{\sqrt{36-x^2}}{12} - \frac{x^2}{12\sqrt{36-x^2}} = \frac{1}{12}\,\frac{(36-x^2-x^2)}{\sqrt{36-x^2}}$
$= \frac{1}{12}\,\frac{(36-2x^2)}{\sqrt{36-x^2}} = \frac{(18-x^2)}{6\sqrt{36-x^2}} \Rightarrow \left(\frac{dy}{dx}\right)^2 = \frac{(18-x^2)^2}{36(36-x^2)} \Rightarrow S = \int_0^6 \frac{2\pi\cdot x}{12}\sqrt{36-x^2}\sqrt{1+\frac{(18-x^2)^2}{36(36-x^2)}}\,dx$
$= \int_0^6 \frac{\pi x}{6}\sqrt{(36-x^2)+\left(\frac{18-x^2}{6}\right)^2}\,dx$; using numerical integration we get $S \approx 41.8$

57. A calculator or computer numerical integrator yields $\sin^{-1} 0.6 \approx 0.643501109$.

58. A calculator or computer numerical integrator yields $\pi \approx 3.1415929$.

8.8 IMPROPER INTEGRALS

1. $\int_0^{\infty} \frac{dx}{x^2+1} = \lim_{b\to\infty} \int_0^b \frac{dx}{x^2+1} = \lim_{b\to\infty} \left[\tan^{-1} x\right]_0^b = \lim_{b\to\infty} (\tan^{-1} b - \tan^{-1} 0) = \frac{\pi}{2} - 0 = \frac{\pi}{2}$

2. $\int_1^{\infty} \frac{dx}{x^{1.001}} = \lim_{b\to\infty} \int_1^b \frac{dx}{x^{1.001}} = \lim_{b\to\infty} \left[-1000x^{-0.001}\right]_1^b = \lim_{b\to\infty} \left(\frac{-1000}{b^{0.001}} + 1000\right) = 1000$

3. $\int_0^1 \frac{dx}{\sqrt{x}} = \lim_{b\to 0^+} \int_b^1 x^{-1/2}\,dx = \lim_{b\to 0^+} \left[2x^{1/2}\right]_b^1 = \lim_{b\to 0^+} \left(2 - 2\sqrt{b}\right) = 2 - 0 = 2$

4. $\int_0^4 \frac{dx}{\sqrt{4-x}} = \lim_{b\to 4^-} \int_0^b (4-x)^{-1/2}dx = \lim_{b\to 4^-} \left[-2\sqrt{4-b} - \left(-2\sqrt{4}\right)\right] = 0 + 4 = 4$

5. $\int_{-1}^1 \frac{dx}{x^{2/3}} = \int_{-1}^0 \frac{dx}{x^{2/3}} + \int_0^1 \frac{dx}{x^{2/3}} = \lim_{b\to 0^-} \left[3x^{1/3}\right]_{-1}^b + \lim_{c\to 0^+} \left[3x^{1/3}\right]_c^1$
$= \lim_{b\to 0^-} \left[3b^{1/3} - 3(-1)^{1/3}\right] + \lim_{c\to 0^+} \left[3(1)^{1/3} - 3c^{1/3}\right] = (0+3) + (3-0) = 6$

6. $\int_{-8}^1 \frac{dx}{x^{1/3}} = \int_{-8}^0 \frac{dx}{x^{1/3}} + \int_0^1 \frac{dx}{x^{1/3}} = \lim_{b\to 0^-} \left[\frac{3}{2}x^{2/3}\right]_{-8}^b + \lim_{c\to 0^+} \left[\frac{3}{2}x^{2/3}\right]_c^1$
$= \lim_{b\to 0^-} \left[\frac{3}{2}b^{2/3} - \frac{3}{2}(-8)^{2/3}\right] + \lim_{c\to 0^+} \left[\frac{3}{2}(1)^{2/3} - \frac{3}{2}c^{2/3}\right] = \left[0 - \frac{3}{2}(4)\right] + \left(\frac{3}{2} - 0\right) = -\frac{9}{2}$

7. $\int_0^1 \frac{dx}{\sqrt{1-x^2}} = \lim_{b\to 1^-} \left[\sin^{-1} x\right]_0^b = \lim_{b\to 1^-} (\sin^{-1} b - \sin^{-1} 0) = \frac{\pi}{2} - 0 = \frac{\pi}{2}$

8. $\int_0^1 \frac{dr}{r^{0.999}} = \lim_{b\to 0^+} \left[1000r^{0.001}\right]_b^1 = \lim_{b\to 0^+} (1000 - 1000b^{0.001}) = 1000 - 0 = 1000$

9. $\int_{-\infty}^{-2} \frac{2\,dx}{x^2-1} = \int_{-\infty}^{-2} \frac{dx}{x-1} - \int_{-\infty}^{-2} \frac{dx}{x+1} = \lim_{b \to -\infty} \left[\ln|x-1|\right]_b^{-2} - \lim_{b \to -\infty} \left[\ln|x+1|\right]_b^{-2} = \lim_{b \to -\infty} \left[\ln\left|\frac{x-1}{x+1}\right|\right]_b^{-2}$
$= \lim_{b \to -\infty} \left(\ln\left|\frac{-3}{-1}\right| - \ln\left|\frac{b-1}{b+1}\right|\right) = \ln 3 - \ln\left(\lim_{b \to -\infty} \frac{b-1}{b+1}\right) = \ln 3 - \ln 1 = \ln 3$

10. $\int_{-\infty}^{2} \frac{2\,dx}{x^2+4} = \lim_{b \to -\infty} \left[\tan^{-1} \frac{x}{2}\right]_b^2 = \lim_{b \to -\infty} \left(\tan^{-1} 1 - \tan^{-1} \frac{b}{2}\right) = \frac{\pi}{4} - \left(-\frac{\pi}{2}\right) = \frac{3\pi}{4}$

11. $\int_2^{\infty} \frac{2\,dv}{v^2-v} = \lim_{b \to \infty} \left[2\ln\left|\frac{v-1}{v}\right|\right]_2^b = \lim_{b \to \infty} \left(2\ln\left|\frac{b-1}{b}\right| - 2\ln\left|\frac{2-1}{2}\right|\right) = 2\ln(1) - 2\ln\left(\frac{1}{2}\right) = 0 + 2\ln 2 = \ln 4$

12. $\int_2^{\infty} \frac{2\,dt}{t^2-1} = \lim_{b \to \infty} \left[\ln\left|\frac{t-1}{t+1}\right|\right]_2^b = \lim_{b \to \infty} \left(\ln\left|\frac{b-1}{b+1}\right| - \ln\left|\frac{2-1}{2+1}\right|\right) = \ln(1) - \ln\left(\frac{1}{3}\right) = 0 + \ln 3 = \ln 3$

13. $\int_{-\infty}^{\infty} \frac{2x\,dx}{(x^2+1)^2} = \int_{-\infty}^{0} \frac{2x\,dx}{(x^2+1)^2} + \int_0^{\infty} \frac{2x\,dx}{(x^2+1)^2}$; $\begin{bmatrix} u = x^2+1 \\ du = 2x\,dx \end{bmatrix} \to \int_{\infty}^{1} \frac{du}{u^2} + \int_1^{\infty} \frac{du}{u^2} = \lim_{b \to \infty} \left[-\frac{1}{u}\right]_b^1 + \lim_{c \to \infty} \left[-\frac{1}{u}\right]_1^c$
$= \lim_{b \to \infty} \left(-1 + \frac{1}{b}\right) + \lim_{c \to \infty} \left[-\frac{1}{c} - (-1)\right] = (-1+0) + (0+1) = 0$

14. $\int_{-\infty}^{\infty} \frac{x\,dx}{(x^2+4)^{3/2}} = \int_{-\infty}^{0} \frac{x\,dx}{(x^2+4)^{3/2}} + \int_0^{\infty} \frac{x\,dx}{(x^2+4)^{3/2}}$; $\begin{bmatrix} u = x^2+4 \\ du = 2x\,dx \end{bmatrix} \to \int_{\infty}^{4} \frac{du}{2u^{3/2}} + \int_4^{\infty} \frac{du}{2u^{3/2}}$
$= \lim_{b \to \infty} \left[-\frac{1}{\sqrt{u}}\right]_b^4 + \lim_{c \to \infty} \left[-\frac{1}{\sqrt{u}}\right]_4^c = \lim_{b \to \infty} \left(-\frac{1}{2} + \frac{1}{\sqrt{b}}\right) + \lim_{c \to \infty} \left(-\frac{1}{\sqrt{c}} + \frac{1}{2}\right) = \left(-\frac{1}{2} + 0\right) + \left(0 + \frac{1}{2}\right) = 0$

15. $\int_0^1 \frac{\theta+1}{\sqrt{\theta^2+2\theta}}\,d\theta$; $\begin{bmatrix} u = \theta^2+2\theta \\ du = 2(\theta+1)\,d\theta \end{bmatrix} \to \int_0^3 \frac{du}{2\sqrt{u}} = \lim_{b \to 0^+} \int_b^3 \frac{du}{2\sqrt{u}} = \lim_{b \to 0^+} \left[\sqrt{u}\right]_b^3 = \lim_{b \to 0^+} \left(\sqrt{3} - \sqrt{b}\right)$
$= \sqrt{3} - 0 = \sqrt{3}$

16. $\int_0^2 \frac{s+1}{\sqrt{4-s^2}}\,ds = \frac{1}{2}\int_0^2 \frac{2s\,ds}{\sqrt{4-s^2}} + \int_0^2 \frac{ds}{\sqrt{4-s^2}}$; $\begin{bmatrix} u = 4-s^2 \\ du = -2s\,ds \end{bmatrix} \to -\frac{1}{2}\int_4^0 \frac{du}{\sqrt{u}} + \lim_{c \to 2^-} \int_0^c \frac{ds}{\sqrt{4-s^2}}$
$= \lim_{b \to 0^+} \int_b^4 \frac{du}{2\sqrt{u}} + \lim_{c \to 2^-} \int_0^c \frac{ds}{\sqrt{4-s^2}} = \lim_{b \to 0^+} \left[\sqrt{u}\right]_b^4 + \lim_{c \to 2^-} \left[\sin^{-1} \frac{s}{2}\right]_0^c$
$= \lim_{b \to 0^+} \left(2 - \sqrt{b}\right) + \lim_{c \to 2^-} \left(\sin^{-1} \frac{c}{2} - \sin^{-1} 0\right) = (2-0) + \left(\frac{\pi}{2} - 0\right) = \frac{4+\pi}{2}$

17. $\int_0^{\infty} \frac{dx}{(1+x)\sqrt{x}}$; $\begin{bmatrix} u = \sqrt{x} \\ du = \frac{dx}{2\sqrt{x}} \end{bmatrix} \to \int_0^{\infty} \frac{2\,du}{u^2+1} = \lim_{b \to \infty} \int_0^b \frac{2\,du}{u^2+1} = \lim_{b \to \infty} \left[2\tan^{-1} u\right]_0^b$
$= \lim_{b \to \infty} \left(2\tan^{-1} b - 2\tan^{-1} 0\right) = 2\left(\frac{\pi}{2}\right) - 2(0) = \pi$

18. $\int_1^{\infty} \frac{dx}{x\sqrt{x^2-1}} = \int_1^2 \frac{dx}{x\sqrt{x^2-1}} + \int_2^{\infty} \frac{dx}{x\sqrt{x^2-1}} = \lim_{b \to 1^+} \int_b^2 \frac{dx}{x\sqrt{x^2-1}} + \lim_{c \to \infty} \int_2^c \frac{dx}{x\sqrt{x^2-1}}$
$= \lim_{b \to 1^+} \left[\sec^{-1}|x|\right]_b^2 + \lim_{c \to \infty} \left[\sec^{-1}|x|\right]_2^c = \lim_{b \to 1^+} \left(\sec^{-1} 2 - \sec^{-1} b\right) + \lim_{c \to \infty} \left(\sec^{-1} c - \sec^{-1} 2\right)$
$= \left(\frac{\pi}{3} - 0\right) + \left(\frac{\pi}{2} - \frac{\pi}{3}\right) = \frac{\pi}{2}$

19. $\int_0^{\infty} \frac{dv}{(1+v^2)(1+\tan^{-1} v)} = \lim_{b \to \infty} \left[\ln|1+\tan^{-1} v|\right]_0^b = \lim_{b \to \infty} \left[\ln|1+\tan^{-1} b|\right] - \ln|1+\tan^{-1} 0|$
$= \ln\left(1+\frac{\pi}{2}\right) - \ln(1+0) = \ln\left(1+\frac{\pi}{2}\right)$

20. $\int_0^{\infty} \frac{16\tan^{-1} x}{1+x^2}\,dx = \lim_{b \to \infty} \left[8\left(\tan^{-1} x\right)^2\right]_0^b = \lim_{b \to \infty} \left[8\left(\tan^{-1} b\right)^2\right] - 8\left(\tan^{-1} 0\right)^2 = 8\left(\frac{\pi}{2}\right)^2 - 8(0) = 2\pi^2$

21. $\int_{-\infty}^{0} \theta e^{\theta}\, d\theta = \lim_{b \to -\infty} \left[\theta e^{\theta} - e^{\theta}\right]_b^0 = (0 \cdot e^0 - e^0) - \lim_{b \to -\infty} \left[be^b - e^b\right] = -1 - \lim_{b \to -\infty} \left(\frac{b-1}{e^{-b}}\right)$
$= -1 - \lim_{b \to -\infty} \left(\frac{1}{-e^{-b}}\right)$ (l'Hôpital's rule for $\frac{\infty}{\infty}$ form)
$= -1 - 0 = -1$

22. $\int_0^{\infty} 2e^{-\theta} \sin\theta\, d\theta = \lim_{b \to \infty} \int_0^b 2e^{-\theta} \sin\theta\, d\theta$
$= \lim_{b \to \infty} 2\left[\frac{e^{-\theta}}{1+1}(-\sin\theta - \cos\theta)\right]_0^b$ (FORMULA 107 with a = −1, b = 1)
$= \lim_{b \to \infty} \frac{-2(\sin b + \cos b)}{2e^b} + \frac{2(\sin 0 + \cos 0)}{2e^0} = 0 + \frac{2(0+1)}{2} = 1$

23. $\int_{-\infty}^{0} e^{-|x|}\, dx = \int_{-\infty}^{0} e^x\, dx = \lim_{b \to -\infty} \left[e^x\right]_b^0 = \lim_{b \to -\infty} (1 - e^b) = (1 - 0) = 1$

24. $\int_{-\infty}^{\infty} 2xe^{-x^2}\, dx = \int_{-\infty}^{0} 2xe^{-x^2}\, dx + \int_0^{\infty} 2xe^{-x^2}\, dx = \lim_{b \to -\infty} \left[-e^{-x^2}\right]_b^0 + \lim_{c \to \infty} \left[-e^{-x^2}\right]_0^c$
$= \lim_{b \to -\infty} \left[-1 - (-e^{-b^2})\right] + \lim_{c \to \infty} \left[-e^{-c^2} - (-1)\right] = (-1 - 0) + (0 + 1) = 0$

25. $\int_0^1 x \ln x\, dx = \lim_{b \to 0^+} \left[\frac{x^2}{2} \ln x - \frac{x^2}{4}\right]_b^1 = \left(\frac{1}{2} \ln 1 - \frac{1}{4}\right) - \lim_{b \to 0^+} \left(\frac{b^2}{2} \ln b - \frac{b^2}{4}\right) = -\frac{1}{4} - \lim_{b \to 0^+} \frac{\ln b}{\left(\frac{2}{b^2}\right)} + 0$
$= -\frac{1}{4} - \lim_{b \to 0^+} \frac{\left(\frac{1}{b}\right)}{\left(-\frac{4}{b^3}\right)} = -\frac{1}{4} + \lim_{b \to 0^+} \left(\frac{b^2}{4}\right) = -\frac{1}{4} + 0 = -\frac{1}{4}$

26. $\int_0^1 (-\ln x)\, dx = \lim_{b \to 0^+} \left[x - x \ln x\right]_b^1 = \left[1 - 1 \ln 1\right] - \lim_{b \to 0^+} \left[b - b \ln b\right] = 1 - 0 + \lim_{b \to 0^+} \frac{\ln b}{\left(\frac{1}{b}\right)} = 1 + \lim_{b \to 0^+} \frac{\left(\frac{1}{b}\right)}{\left(-\frac{1}{b^2}\right)}$
$= 1 - \lim_{b \to 0^+} b = 1 - 0 = 1$

27. $\int_0^2 \frac{ds}{\sqrt{4 - s^2}} = \lim_{b \to 2^-} \left[\sin^{-1} \frac{s}{2}\right]_0^b = \lim_{b \to 2^-} \left(\sin^{-1} \frac{b}{2}\right) - \sin^{-1} 0 = \frac{\pi}{2} - 0 = \frac{\pi}{2}$

28. $\int_0^1 \frac{4r\, dr}{\sqrt{1 - r^4}} = \lim_{b \to 1^-} \left[2 \sin^{-1}(r^2)\right]_0^b = \lim_{b \to 1^-} \left[2 \sin^{-1}(b^2)\right] - 2 \sin^{-1} 0 = 2 \cdot \frac{\pi}{2} - 0 = \pi$

29. $\int_1^2 \frac{ds}{s\sqrt{s^2 - 1}} = \lim_{b \to 1^+} \left[\sec^{-1} s\right]_b^2 = \sec^{-1} 2 - \lim_{b \to 1^+} \sec^{-1} b = \frac{\pi}{3} - 0 = \frac{\pi}{3}$

30. $\int_2^4 \frac{dt}{t\sqrt{t^2 - 4}} = \lim_{b \to 2^+} \left[\frac{1}{2} \sec^{-1} \frac{t}{2}\right]_b^4 = \lim_{b \to 2^+} \left[\left(\frac{1}{2} \sec^{-1} \frac{4}{2}\right) - \frac{1}{2} \sec^{-1}\left(\frac{b}{2}\right)\right] = \frac{1}{2}\left(\frac{\pi}{3}\right) - \frac{1}{2} \cdot 0 = \frac{\pi}{6}$

31. $\int_{-1}^{4} \frac{dx}{\sqrt{|x|}} = \lim_{b \to 0^-} \int_{-1}^{b} \frac{dx}{\sqrt{-x}} + \lim_{c \to 0^+} \int_c^4 \frac{dx}{\sqrt{x}} = \lim_{b \to 0^-} \left[-2\sqrt{-x}\right]_{-1}^{b} + \lim_{c \to 0^+} \left[2\sqrt{x}\right]_c^4$
$= \lim_{b \to 0^-} \left(-2\sqrt{-b}\right) - \left(-2\sqrt{-(-1)}\right) + 2\sqrt{4} - \lim_{c \to 0^+} 2\sqrt{c} = 0 + 2 + 2 \cdot 2 - 0 = 6$

32. $\int_0^2 \frac{dx}{\sqrt{|x - 1|}} = \int_0^1 \frac{dx}{\sqrt{1 - x}} + \int_1^2 \frac{dx}{\sqrt{x - 1}} = \lim_{b \to 1^-} \left[-2\sqrt{1 - x}\right]_0^b + \lim_{c \to 1^+} \left[2\sqrt{x - 1}\right]_c^2$
$= \lim_{b \to 1^-} \left(-2\sqrt{1 - b}\right) - \left(-2\sqrt{1 - 0}\right) + 2\sqrt{2 - 1} - \lim_{c \to 1^+} \left(2\sqrt{c - 1}\right) = 0 + 2 + 2 - 0 = 4$

33. $\int_{-1}^{\infty} \frac{d\theta}{\theta^2 + 5\theta + 6} = \lim_{b \to \infty} \left[\ln\left|\frac{\theta + 2}{\theta + 3}\right|\right]_{-1}^{b} = \lim_{b \to \infty} \left[\ln\left|\frac{b + 2}{b + 3}\right|\right] - \ln\left|\frac{-1 + 2}{-1 + 3}\right| = 0 - \ln\left(\frac{1}{2}\right) = \ln 2$

34. $\int_0^{\infty} \frac{dx}{(x+1)(x^2+1)} = \lim_{b\to\infty} \left[\frac{1}{2}\ln|x+1| - \frac{1}{4}\ln(x^2+1) + \frac{1}{2}\tan^{-1}x\right]_0^b = \lim_{b\to\infty} \left[\frac{1}{2}\ln\left(\frac{x+1}{\sqrt{x^2+1}}\right) + \frac{1}{2}\tan^{-1}x\right]_0^b$
$= \lim_{b\to\infty} \left[\frac{1}{2}\ln\left(\frac{b+1}{\sqrt{b^2+1}}\right) + \frac{1}{2}\tan^{-1}b\right] - \left[\frac{1}{2}\ln\frac{1}{\sqrt{1}} + \frac{1}{2}\tan^{-1}0\right] = \frac{1}{2}\ln 1 + \frac{1}{2}\cdot\frac{\pi}{2} - \frac{1}{2}\ln 1 - \frac{1}{2}\cdot 0 = \frac{\pi}{4}$

35. $\int_0^{\pi/2} \tan\theta\, d\theta = \lim_{b\to\frac{\pi}{2}^-} [-\ln|\cos\theta|]_0^b = \lim_{b\to\frac{\pi}{2}^-} [-\ln|\cos b|] + \ln 1 = \lim_{b\to\frac{\pi}{2}^-} [-\ln|\cos b|] = +\infty$,
the integral diverges

36. $\int_0^{\pi/2} \cot\theta\, d\theta = \lim_{b\to 0^+} [\ln|\sin\theta|]_b^{\pi/2} = \ln 1 - \lim_{b\to 0^+} [\ln|\sin b|] = -\lim_{b\to 0^+} [\ln|\sin b|] = +\infty$,
the integral diverges

37. $\int_0^{\pi} \frac{\sin\theta\, d\theta}{\sqrt{\pi-\theta}}$; $[\pi-\theta = x] \to -\int_{\pi}^{0} \frac{\sin x\, dx}{\sqrt{x}} = \int_0^{\pi} \frac{\sin x\, dx}{\sqrt{x}}$. Since $0 \le \frac{\sin x}{\sqrt{x}} \le \frac{1}{\sqrt{x}}$ for all $0 \le x \le \pi$ and $\int_0^{\pi} \frac{dx}{\sqrt{x}}$ converges, then $\int_0^{\pi} \frac{\sin x}{\sqrt{x}}\, dx$ converges by the Direct Comparison Test.

38. $\int_{-\pi/2}^{\pi/2} \frac{\cos\theta\, d\theta}{(\pi-2\theta)^{1/3}}$; $\begin{bmatrix} x = \pi - 2\theta \\ \theta = \frac{\pi}{2} - \frac{x}{2} \\ d\theta = -\frac{dx}{2} \end{bmatrix} \to \int_{2\pi}^{0} \frac{-\cos\left(\frac{\pi}{2}-\frac{x}{2}\right) dx}{2x^{1/3}} = \int_0^{2\pi} \frac{\sin\left(\frac{x}{2}\right) dx}{2x^{1/3}}$. Since $0 \le \frac{\sin\frac{x}{2}}{2x^{1/3}} \le \frac{1}{2x^{1/3}}$ for all $0 \le x \le 2\pi$ and $\int_0^{2\pi} \frac{dx}{2x^{1/3}}$ converges, then $\int_0^{2\pi} \frac{\sin\frac{x}{2}\, dx}{2x^{1/3}}$ converges by the Direct Comparison Test.

39. $\int_0^{\ln 2} x^{-2}e^{-1/x}\, dx$; $\left[\frac{1}{x} = y\right] \to \int_{\infty}^{1/\ln 2} \frac{y^2e^{-y}\, dy}{-y^2} = \int_{1/\ln 2}^{\infty} e^{-y}\, dy = \lim_{b\to\infty} [-e^{-y}]_{1/\ln 2}^{b} = \lim_{b\to\infty} [-e^{-b}] - [-e^{-1/\ln 2}]$
$= 0 + e^{-1/\ln 2} = e^{-1/\ln 2}$, so the integral converges.

40. $\int_0^1 \frac{e^{-\sqrt{x}}}{\sqrt{x}}\, dx$; $[y = \sqrt{x}] \to 2\int_0^1 e^{-y}\, dy = 2 - \frac{2}{e}$, so the integral converges.

41. $\int_0^{\pi} \frac{dt}{\sqrt{t}+\sin t}$. Since for $0 \le t \le \pi$, $0 \le \frac{1}{\sqrt{t}+\sin t} \le \frac{1}{\sqrt{t}}$ and $\int_0^{\pi} \frac{dt}{\sqrt{t}}$ converges, then the original integral converges as well by the Direct Comparison Test.

42. $\int_0^1 \frac{dt}{t-\sin t}$; let $f(t) = \frac{1}{t-\sin t}$ and $g(t) = \frac{1}{t^3}$, then $\lim_{t\to 0} \frac{f(t)}{g(t)} = \lim_{t\to 0} \frac{t^3}{t-\sin t} = \lim_{t\to 0} \frac{3t^2}{1-\cos t} = \lim_{t\to 0} \frac{6t}{\sin t}$
$= \lim_{t\to 0} \frac{6}{\cos t} = 6$. Now, $\int_0^1 \frac{dt}{t^3} = \lim_{b\to 0^+} \left[-\frac{1}{2t^2}\right]_b^1 = -\frac{1}{2} - \lim_{b\to 0^+} \left[-\frac{1}{2b^2}\right] = +\infty$, which diverges $\Rightarrow \int_0^1 \frac{dt}{t-\sin t}$ diverges by the Limit Comparison Test.

43. $\int_0^2 \frac{dx}{1-x^2} = \int_0^1 \frac{dx}{1-x^2} + \int_1^2 \frac{dx}{1-x^2}$ and $\int_0^1 \frac{dx}{1-x^2} = \lim_{b\to 1^-} \left[\frac{1}{2}\ln\left|\frac{1+x}{1-x}\right|\right]_0^b = \lim_{b\to 1^-} \left[\frac{1}{2}\ln\left|\frac{1+b}{1-b}\right|\right] - 0 = \infty$, which diverges $\Rightarrow \int_0^2 \frac{dx}{1-x^2}$ diverges as well.

44. $\int_0^2 \frac{dx}{1-x} = \int_0^1 \frac{dx}{1-x} + \int_1^2 \frac{dx}{1-x}$ and $\int_0^1 \frac{dx}{1-x} = \lim_{b\to 1^-} [-\ln(1-x)]_0^b = \lim_{b\to 1^-} [-\ln(1-b)] - 0 = \infty$, which diverges $\Rightarrow \int_0^2 \frac{dx}{1-x}$ diverges as well.

45. $\int_{-1}^1 \ln|x|\, dx = \int_{-1}^0 \ln(-x)\, dx + \int_0^1 \ln x\, dx$; $\int_0^1 \ln x\, dx = \lim_{b\to 0^+} [x\ln x - x]_b^1 = [1\cdot 0 - 1] - \lim_{b\to 0^+} [b\ln b - b]$
$= -1 - 0 = -1$; $\int_{-1}^0 \ln(-x)\, dx = -1 \Rightarrow \int_{-1}^1 \ln|x|\, dx = -2$ converges.

46. $\int_{-1}^{1}(-x\ln|x|)\,dx = \int_{-1}^{0}[-x\ln(-x)]\,dx + \int_{0}^{1}(-x\ln x)\,dx = \lim_{b\to 0^+}\left[\frac{x^2}{2}\ln x - \frac{x^2}{4}\right]_b^1 - \lim_{c\to 0^+}\left[\frac{x^2}{2}\ln x - \frac{x^2}{4}\right]_c^1$
$= \left[\frac{1}{2}\ln 1 - \frac{1}{4}\right] - \lim_{b\to 0^+}\left[\frac{b^2}{2}\ln b - \frac{b^2}{4}\right] - \left[\frac{1}{2}\ln 1 - \frac{1}{4}\right] + \lim_{c\to 0^+}\left[\frac{c^2}{2}\ln c - \frac{c^2}{4}\right] = -\frac{1}{4} - 0 + \frac{1}{4} + 0 = 0 \Rightarrow$ the integral converges (see Exercise 25 for the limit calculations).

47. $\int_1^\infty \frac{dx}{1+x^3}$; $0 \le \frac{1}{x^3+1} \le \frac{1}{x^3}$ for $1 \le x < \infty$ and $\int_1^\infty \frac{dx}{x^3}$ converges $\Rightarrow \int_1^\infty \frac{dx}{1+x^3}$ converges by the Direct Comparison Test.

48. $\int_4^\infty \frac{dx}{\sqrt{x}-1}$; $\lim_{x\to\infty} \frac{\left(\frac{1}{\sqrt{x}-1}\right)}{\left(\frac{1}{\sqrt{x}}\right)} = \lim_{x\to\infty} \frac{\sqrt{x}}{\sqrt{x}-1} = \lim_{x\to\infty} \frac{1}{1-\frac{1}{\sqrt{x}}} = \frac{1}{1-0} = 1$ and $\int_4^\infty \frac{dx}{\sqrt{x}} = \lim_{b\to\infty}\left[2\sqrt{x}\right]_4^b = \infty$, which diverges $\Rightarrow \int_4^\infty \frac{dx}{\sqrt{x}-1}$ diverges by the Limit Comparison Test.

49. $\int_2^\infty \frac{dv}{\sqrt{v-1}}$; $\lim_{v\to\infty} \frac{\left(\frac{1}{\sqrt{v-1}}\right)}{\left(\frac{1}{\sqrt{v}}\right)} = \lim_{v\to\infty} \frac{\sqrt{v}}{\sqrt{v-1}} = \lim_{v\to\infty} \frac{1}{\sqrt{1-\frac{1}{v}}} = \frac{1}{\sqrt{1-0}} = 1$ and $\int_2^\infty \frac{dv}{\sqrt{v}} = \lim_{b\to\infty}\left[2\sqrt{v}\right]_2^b = \infty$, which diverges $\Rightarrow \int_2^\infty \frac{dv}{\sqrt{v-1}}$ diverges by the Limit Comparison Test.

50. $\int_0^\infty \frac{d\theta}{1+e^\theta}$; $0 \le \frac{1}{1+e^\theta} \le \frac{1}{e^\theta}$ for $0 \le \theta < \infty$ and $\int_0^\infty \frac{d\theta}{e^\theta} = \lim_{b\to\infty}\left[-e^{-\theta}\right]_0^b = \lim_{b\to\infty}\left(-e^{-b}+1\right) = 1$
$\Rightarrow \int_0^\infty \frac{d\theta}{e^\theta}$ converges $\Rightarrow \int_0^\infty \frac{d\theta}{1+e^\theta}$ converges by the Direct Comparison Test.

51. $\int_0^\infty \frac{dx}{\sqrt{x^6+1}} = \int_0^1 \frac{dx}{\sqrt{x^6+1}} + \int_1^\infty \frac{dx}{\sqrt{x^6+1}} < \int_0^1 \frac{dx}{\sqrt{x^6+1}} + \int_1^\infty \frac{dx}{x^3}$ and $\int_1^\infty \frac{dx}{x^3} = \lim_{b\to\infty}\left[-\frac{1}{2x^2}\right]_1^b$
$= \lim_{b\to\infty}\left(-\frac{1}{2b^2} + \frac{1}{2}\right) = \frac{1}{2} \Rightarrow \int_0^\infty \frac{dx}{\sqrt{x^6+1}}$ converges by the Direct Comparison Test.

52. $\int_2^\infty \frac{dx}{\sqrt{x^2-1}}$; $\lim_{x\to\infty} \frac{\left(\frac{1}{\sqrt{x^2-1}}\right)}{\left(\frac{1}{x}\right)} = \lim_{x\to\infty} \frac{x}{\sqrt{x^2-1}} = \lim_{x\to\infty} \frac{1}{\sqrt{1-\frac{1}{x^2}}} = 1$; $\int_2^\infty \frac{1}{x}\,dx = \lim_{b\to\infty}\left[\ln b\right]_2^b = \infty$, which diverges $\Rightarrow \int_2^\infty \frac{dx}{\sqrt{x^2-1}}$ diverges by the Limit Comparison Test.

53. $\int_1^\infty \frac{\sqrt{x+1}}{x^2}\,dx$; $\lim_{x\to\infty} \frac{\left(\frac{\sqrt{x}}{x^2}\right)}{\left(\frac{\sqrt{x+1}}{x^2}\right)} = \lim_{x\to\infty} \frac{\sqrt{x}}{\sqrt{x+1}} = \lim_{x\to\infty} \frac{1}{\sqrt{1+\frac{1}{x}}} = 1$; $\int_1^\infty \frac{\sqrt{x}}{x^2}\,dx = \int_1^\infty \frac{dx}{x^{3/2}}$
$= \lim_{b\to\infty}\left[-2x^{-1/2}\right]_1^b = \lim_{b\to\infty}\left(\frac{-2}{\sqrt{b}} + 2\right) = 2 \Rightarrow \int_1^\infty \frac{\sqrt{x+1}}{x^2}\,dx$ converges by the Limit Comparison Test.

54. $\int_2^\infty \frac{x\,dx}{\sqrt{x^4-1}}$; $\lim_{x\to\infty} \frac{\left(\frac{x}{\sqrt{x^4-1}}\right)}{\left(\frac{x}{\sqrt{x^4}}\right)} = \lim_{x\to\infty} \frac{\sqrt{x^4}}{\sqrt{x^4-1}} = \lim_{x\to\infty} \frac{1}{\sqrt{1-\frac{1}{x^4}}} = 1$; $\int_2^\infty \frac{x\,dx}{\sqrt{x^4}} = \int_2^\infty \frac{dx}{x} = \lim_{b\to\infty}\left[\ln x\right]_2^b = \infty$, which diverges $\Rightarrow \int_2^\infty \frac{x\,dx}{\sqrt{x^4-1}}$ diverges by the Limit Comparison Test.

55. $\int_\pi^\infty \frac{2+\cos x}{x}\,dx$; $0 < \frac{1}{x} \le \frac{2+\cos x}{x}$ for $x \ge \pi$ and $\int_\pi^\infty \frac{dx}{x} = \lim_{b\to\infty}\left[\ln x\right]_\pi^b = \infty$, which diverges
$\Rightarrow \int_\pi^\infty \frac{2+\cos x}{x}\,dx$ diverges by the Direct Comparison Test.

56. $\int_\pi^\infty \frac{1+\sin x}{x^2}\,dx$; $0 \le \frac{1+\sin x}{x^2} \le \frac{2}{x^2}$ for $x \ge \pi$ and $\int_\pi^\infty \frac{2}{x^2}\,dx = \lim_{b\to\infty}\left[-\frac{2}{x}\right]_\pi^b = \lim_{b\to\infty}\left(-\frac{2}{b} + \frac{2}{\pi}\right) = \frac{2}{\pi}$
$\Rightarrow \int_\pi^\infty \frac{2\,dx}{x^2}$ converges $\Rightarrow \int_\pi^\infty \frac{1+\sin x}{x^2}\,dx$ converges by the Direct Comparison Test.

57. $\int_4^\infty \frac{2\,dt}{t^{3/2}-1}$; $\lim_{t\to\infty} \frac{t^{3/2}}{t^{3/2}-1} = 1$ and $\int_4^\infty \frac{2\,dt}{t^{3/2}} = \lim_{b\to\infty} \left[-4t^{-1/2}\right]_4^b = \lim_{b\to\infty} \left(\frac{-4}{\sqrt{b}} + 2\right) = 2 \Rightarrow \int_4^\infty \frac{2\,dt}{t^{3/2}}$ converges $\Rightarrow \int_4^\infty \frac{2\,dt}{t^{3/2}+1}$ converges by the Limit Comparison Test.

58. $\int_2^\infty \frac{dx}{\ln x}$; $0 < \frac{1}{x} < \frac{1}{\ln x}$ for $x > 2$ and $\int_2^\infty \frac{dx}{x}$ diverges $\Rightarrow \int_2^\infty \frac{dx}{\ln x}$ diverges by the Direct Comparison Test.

59. $\int_1^\infty \frac{e^x}{x}\,dx$; $0 < \frac{1}{x} < \frac{e^x}{x}$ for $x > 1$ and $\int_1^\infty \frac{dx}{x}$ diverges $\Rightarrow \int_1^\infty \frac{e^x\,dx}{x}$ diverges by the Direct Comparison Test.

60. $\int_{e^e}^\infty \ln(\ln x)\,dx$; $[x = e^y] \to \int_e^\infty (\ln y)e^y\,dy$; $0 < \ln y < (\ln y)e^y$ for $y \ge e$ and $\int_e^\infty \ln y\,dy = \lim_{b\to\infty} [y\ln y - y]_e^b$ $= \infty$, which diverges $\Rightarrow \int_e^\infty \ln e^y\,dy$ diverges $\Rightarrow \int_{e^e}^\infty \ln(\ln x)\,dx$ diverges by the Direct Comparison Test.

61. $\int_1^\infty \frac{dx}{\sqrt{e^x - x}}$; $\lim_{x\to\infty} \frac{\left(\frac{1}{\sqrt{e^x-x}}\right)}{\left(\frac{1}{\sqrt{e^x}}\right)} = \lim_{x\to\infty} \frac{\sqrt{e^x}}{\sqrt{e^x-x}} = \lim_{x\to\infty} \frac{1}{\sqrt{1-\frac{x}{e^x}}} = \frac{1}{\sqrt{1-0}} = 1$; $\int_1^\infty \frac{dx}{\sqrt{e^x}} = \int_1^\infty e^{-x/2}\,dx$ $= \lim_{b\to\infty} \left[-2e^{-x/2}\right]_1^b = \lim_{b\to\infty} \left(-2e^{-b/2} + 2e^{-1/2}\right) = \frac{2}{\sqrt{e}} \Rightarrow \int_1^\infty e^{-x/2}\,dx$ converges $\Rightarrow \int_1^\infty \frac{dx}{\sqrt{e^x-x}}$ converges by the Limit Comparison Test.

62. $\int_1^\infty \frac{dx}{e^x - 2^x}$; $\lim_{x\to\infty} \frac{\left(\frac{1}{e^x-2^x}\right)}{\left(\frac{1}{e^x}\right)} = \lim_{x\to\infty} \frac{e^x}{e^x-2^x} = \lim_{x\to\infty} \frac{1}{1-\left(\frac{2}{e}\right)^x} = \frac{1}{1-0} = 1$ and $\int_1^\infty \frac{dx}{e^x} = \lim_{b\to\infty} \left[-e^{-x}\right]_1^b$ $= \lim_{b\to\infty} \left(-e^{-b} + e^{-1}\right) = \frac{1}{e} \Rightarrow \int_1^\infty \frac{dx}{e^x}$ converges $\Rightarrow \int_1^\infty \frac{dx}{e^x-2^x}$ converges by the Limit Comparison Test.

63. $\int_{-\infty}^\infty \frac{dx}{\sqrt{x^4+1}} = 2\int_0^\infty \frac{dx}{\sqrt{x^4+1}}$; $\int_0^\infty \frac{dx}{\sqrt{x^4+1}} = \int_0^1 \frac{dx}{\sqrt{x^4+1}} + \int_1^\infty \frac{dx}{\sqrt{x^4+1}} < \int_0^1 \frac{dx}{\sqrt{x^4+1}} + \int_1^\infty \frac{dx}{x^2}$ and $\int_1^\infty \frac{dx}{x^2} = \lim_{b\to\infty} \left[-\frac{1}{x}\right]_1^b = \lim_{b\to\infty} \left(-\frac{1}{b} + 1\right) = 1 \Rightarrow \int_{-\infty}^\infty \frac{dx}{\sqrt{x^4+1}}$ converges by the Direct Comparison Test.

64. $\int_{-\infty}^\infty \frac{dx}{e^x + e^{-x}} = 2\int_0^\infty \frac{dx}{e^x+e^{-x}}$; $0 < \frac{1}{e^x+e^{-x}} < \frac{1}{e^x}$ for $x > 0$; $\int_0^\infty \frac{dx}{e^x}$ converges $\Rightarrow 2\int_0^\infty \frac{dx}{e^x+e^{-x}}$ converges by the Direct Comparison Test.

65. (a) $\int_1^2 \frac{dx}{x(\ln x)^p}$; $[t = \ln x] \to \int_0^{\ln 2} \frac{dt}{t^p} = \lim_{b\to 0^+} \left[\frac{1}{-p+1} t^{1-p}\right]_b^{\ln 2} = \lim_{b\to 0^+} \frac{b^{1-p}}{p-1} + \frac{1}{1-p}(\ln 2)^{1-p}$
$\Rightarrow$ the integral converges for $p < 1$ and diverges for $p \ge 1$

(b) $\int_2^\infty \frac{dx}{x(\ln x)^p}$; $[t = \ln x] \to \int_{\ln 2}^\infty \frac{dt}{t^p}$ and this integral is essentially the same as in Exercise 65(a): it converges for $p > 1$ and diverges for $p \le 1$

66. $\int_0^\infty \frac{2x\,dx}{x^2+1} = \lim_{b\to\infty} \left[\ln(x^2+1)\right]_0^b = \lim_{b\to\infty} \left[\ln(b^2+1)\right] - 0 = \lim_{b\to\infty} \ln(b^2+1) = \infty \Rightarrow$ the integral $\int_{-\infty}^\infty \frac{2x}{x^2+1}\,dx$ diverges. But $\lim_{b\to\infty} \int_{-b}^b \frac{2x\,dx}{x^2+1} = \lim_{b\to\infty} \left[\ln(x^2+1)\right]_{-b}^b = \lim_{b\to\infty} \left[\ln(b^2+1) - \ln(b^2+1)\right] = \lim_{b\to\infty} \ln\left(\frac{b^2+1}{b^2+1}\right)$ $= \lim_{b\to\infty} (\ln 1) = 0$

67. $A = \int_0^\infty e^{-x}\,dx = \lim_{b\to\infty} \left[-e^{-x}\right]_0^b = \lim_{b\to\infty} \left(-e^{-b}\right) - \left(-e^{-0}\right)$ $= 0 + 1 = 1$

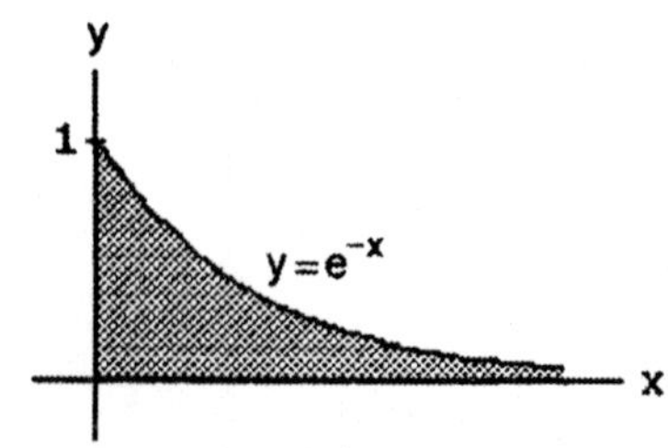

68. $\bar{x} = \frac{1}{A}\int_0^\infty xe^{-x}\,dx = \lim_{b\to\infty}\left[-xe^{-x} - e^{-x}\right]_0^b = \lim_{b\to\infty}\left(-be^{-b} - e^{-b}\right) - \left(-0\cdot e^{-0} - e^{-0}\right) = 0 + 1 = 1;$

$\bar{y} = \frac{1}{2A}\int_0^\infty \left(e^{-x}\right)^2 dx = \frac{1}{2}\int_0^\infty e^{-2x}\,dx = \lim_{b\to\infty} \frac{1}{2}\left[-\frac{1}{2}e^{-2x}\right]_0^b = \lim_{b\to\infty} \frac{1}{2}\left(-\frac{1}{2}e^{-2b}\right) - \frac{1}{2}\left(-\frac{1}{2}e^{-2\cdot 0}\right) = 0 + \frac{1}{4} = \frac{1}{4}$

69. $V = \int_0^\infty 2\pi xe^{-x}\,dx = 2\pi\int_0^\infty xe^{-x}\,dx = 2\pi \lim_{b\to\infty}\left[-xe^{-x} - e^{-x}\right]_0^b = 2\pi\left[\lim_{b\to\infty}\left(-be^{-b} - e^{-b}\right) - 1\right] = 2\pi$

70. $V = \int_0^\infty \pi\left(e^{-x}\right)^2 dx = \pi\int_0^\infty e^{-2x}\,dx = \pi \lim_{b\to\infty}\left[-\frac{1}{2}e^{-2x}\right]_0^b = \pi \lim_{b\to\infty}\left(-\frac{1}{2}e^{-2b} + \frac{1}{2}\right) = \frac{\pi}{2}$

71. $A = \int_0^{\pi/2}(\sec x - \tan x)\,dx = \lim_{b\to\frac{\pi}{2}^-}\left[\ln|\sec x + \tan x| - \ln|\sec x|\right]_0^b = \lim_{b\to\frac{\pi}{2}^-}\left(\ln\left|1 + \frac{\tan b}{\sec b}\right| - \ln|1+0|\right)$

$= \lim_{b\to\frac{\pi}{2}^-} \ln|1 + \sin b| = \ln 2$

72. (a) $V = \int_0^{\pi/2}\pi\sec^2 x\,dx - \int_0^{\pi/2}\pi\tan^2 x\,dx = \pi\int_0^{\pi/2}\left(\sec^2 x - \tan^2 x\right)dx = \int_0^{\pi/2}\pi\left[\sec^2 x - \left(\sec^2 x - 1\right)\right]dx$

$= \pi\int_0^{\pi/2} dx = \frac{\pi^2}{2}$

(b) $S_{\text{outer}} = \int_0^{\pi/2} 2\pi\sec x\sqrt{1 + \sec^2 x\tan^2 x}\,dx \ge \int_0^{\pi/2} 2\pi\sec x(\sec x\tan x)\,dx = \pi \lim_{b\to\frac{\pi}{2}^-}\left[\tan^2 x\right]_0^b$

$= \pi\left[\lim_{b\to\frac{\pi}{2}^-}\left[\tan^2 b\right] - 0\right] = \pi \lim_{b\to\frac{\pi}{2}^-}\left(\tan^2 b\right) = \infty \Rightarrow S_{\text{outer}}$ diverges; $S_{\text{inner}} = \int_0^{\pi/2} 2\pi\tan x\sqrt{1 + \sec^4 x}\,dx$

$\ge \int_0^{\pi/2} 2\pi\tan x\sec^2 x\,dx = \pi \lim_{b\to\frac{\pi}{2}^-}\left[\tan^2 x\right]_0^b = \pi\left[\lim_{b\to\frac{\pi}{2}^-}\left[\tan^2 b\right] - 0\right] = \pi \lim_{b\to\frac{\pi}{2}^-}\left(\tan^2 b\right) = \infty$

$\Rightarrow S_{\text{inner}}$ diverges

73. (a) $\int_3^\infty e^{-3x}\,dx = \lim_{b\to\infty}\left[-\frac{1}{3}e^{-3x}\right]_3^b = \lim_{b\to\infty}\left(-\frac{1}{3}e^{-3b}\right) - \left(-\frac{1}{3}e^{-3\cdot 3}\right) = 0 + \frac{1}{3}\cdot e^{-9} = \frac{1}{3}e^{-9}$

$\approx 0.0000411 < 0.000042$. Since $e^{-x^2} \le e^{-3x}$ for $x > 3$, then $\int_3^\infty e^{-x^2}\,dx < 0.000042$ and therefore $\int_0^\infty e^{-x^2}\,dx$ can be replaced by $\int_0^3 e^{-x^2}\,dx$ without introducing an error greater than 0.000042.

(b) $\int_0^3 e^{-x^2}\,dx \cong 0.88621$

74. (a) $V = \int_1^\infty \pi\left(\frac{1}{x}\right)^2 dx = \pi \lim_{b\to\infty}\left[-\frac{1}{x}\right]_1^b = \pi\left[\lim_{b\to\infty}\left(-\frac{1}{b}\right) - \left(-\frac{1}{1}\right)\right] = \pi(0+1) = \pi$

(b) When you take the limit to ∞, you are no longer modeling the real world which is finite. The comparison step in the modeling process discussed in Section 4.2 relating the mathematical world to the real world fails to hold.

75. (a)

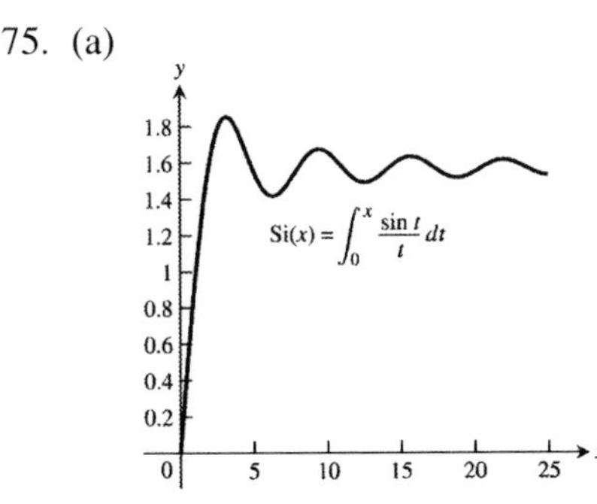

(b) > int((sin(t))/t, t=0..infinity); (answer is $\frac{\pi}{2}$)

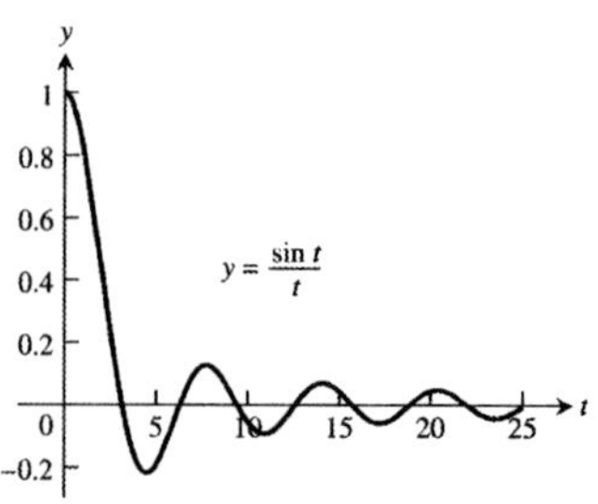

76. (a)

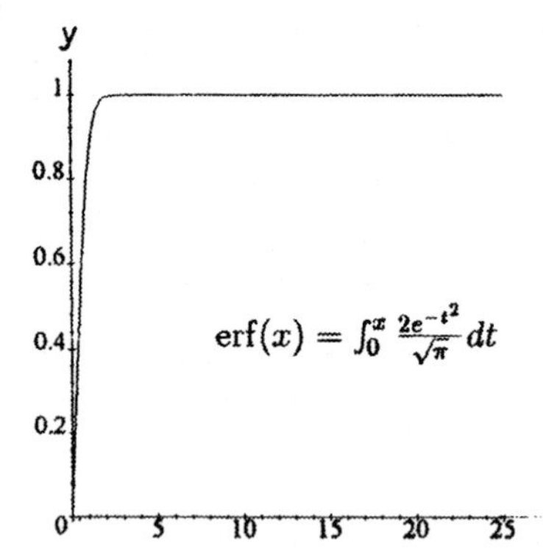

(b) > f:= 2*exp(−t^2)/sqrt(Pi);
> int(f, t=0..infinity); (answer is 1)

77. (a) $f(x) = \frac{1}{\sqrt{2\pi}} e^{-x^2/2}$

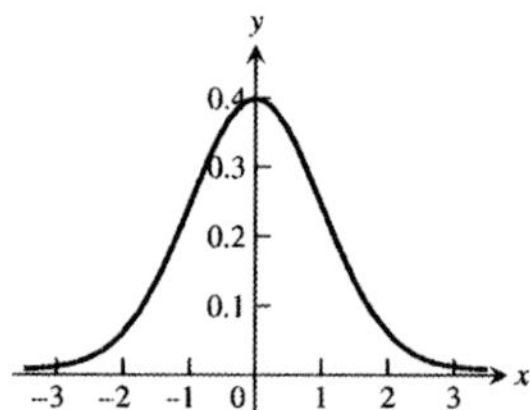

f is increasing on $(-\infty, 0]$. f is decreasing on $[0, \infty)$. f has a local maximum at $(0, f(0)) = \left(0, \frac{1}{\sqrt{2\pi}}\right)$

(b) Maple commands:
>f: = exp(−x^2/2)(sqrt(2*pi);
>int(f, x = −1..1); ≈ 0.683
>int(f, x = −2..2); ≈ 0.954
>int(f, x = −3..3); ≈ 0.997

(c) Part (b) suggests that as n increases, the integral approaches 1. We can take $\int_{-n}^{n} f(x)\,dx$ as close to 1 as we want by choosing $n > 1$ large enough. Also, we can make $\int_{n}^{\infty} f(x)\,dx$ and $\int_{-\infty}^{-n} f(x)\,dx$ as small as we want by choosing n large enough. This is because $0 < f(x) < e^{-x/2}$ for $x > 1$. (Likewise, $0 < f(x) < e^{x/2}$ for $x < -1$.)
Thus, $\int_{n}^{\infty} f(x)\,dx < \int_{n}^{\infty} e^{-x/2} dx$.
$\int_{n}^{\infty} e^{-x/2} dx = \lim_{c\to\infty} \int_{n}^{c} e^{-x/2} dx = \lim_{c\to\infty} \left[-2e^{-x/2}\right]_n^c = \lim_{c\to\infty} \left[-2e^{-c/2} + 2e^{-n/2}\right] = 2e^{-n/2}$
As $n \to \infty$, $2e^{-n/2} \to 0$, for large enough n, $\int_{n}^{\infty} f(x)\,dx$ is as small as we want. Likewise for large enough n, $\int_{-\infty}^{-n} f(x)\,dx$ is as small as we want.

78. $\int_3^{\infty} \left(\frac{1}{x-2} - \frac{1}{x}\right) dx \neq \int_3^{\infty} \frac{dx}{x-2} - \int_3^{\infty} \frac{dx}{x}$, since the left hand integral converges but both of the right hand integrals diverge.

79. (a) The statement is true since $\int_{-\infty}^{b} f(x)\,dx = \int_{-\infty}^{a} f(x)\,dx + \int_{a}^{b} f(x)\,dx$, $\int_{b}^{\infty} f(x)\,dx = \int_{a}^{\infty} f(x)\,dx - \int_{a}^{b} f(x)\,dx$ and $\int_{a}^{b} f(x)\,dx$ exists since f(x) is integrable on every interval [a, b].

(b) $\int_{-\infty}^{a} f(x)\,dx + \int_{a}^{\infty} f(x)\,dx = \int_{-\infty}^{a} f(x)\,dx + \int_{a}^{b} f(x)\,dx - \int_{a}^{b} f(x)\,dx + \int_{a}^{\infty} f(x)\,dx$
$= \int_{-\infty}^{b} f(x)\,dx + \int_{b}^{a} f(x)\,dx + \int_{a}^{\infty} f(x)\,dx = \int_{-\infty}^{b} f(x)\,dx + \int_{b}^{\infty} f(x)\,dx$

80. (a) $\int_{-\infty}^{\infty} f(x)\,dx = \int_{-\infty}^{0} f(x)\,dx + \int_{0}^{\infty} f(x)\,dx = -\int_{\infty}^{0} f(-u)\,du + \int_{0}^{\infty} f(x)\,dx$
$= \int_{0}^{\infty} f(-u)\,du + \int_{0}^{\infty} f(x)\,dx = 2\int_{0}^{\infty} f(x)\,dx$, where $u = -x$

(b) $\int_{-\infty}^{\infty} f(x)\,dx = \int_{-\infty}^{0} f(x)\,dx + \int_{0}^{\infty} f(x)\,dx = -\int_{\infty}^{0} f(-u)\,du + \int_{0}^{\infty} f(x)\,dx$
$= \int_{0}^{\infty} -f(u)\,du + \int_{0}^{\infty} f(x)\,dx = -\int_{0}^{\infty} f(x)\,dx + \int_{0}^{\infty} f(x)\,dx = 0$, where $u = -x$

81. $\int_{-\infty}^{\infty} \frac{dx}{\sqrt{x^2+1}} = \int_{-\infty}^{1} \frac{dx}{\sqrt{x^2+1}} + \int_{1}^{\infty} \frac{dx}{\sqrt{x^2+1}}$; $\int_{1}^{\infty} \frac{dx}{\sqrt{x^2+1}}$ diverges because $\lim_{x\to\infty} \frac{\left(\frac{1}{x}\right)}{\left(\frac{1}{\sqrt{x^2+1}}\right)}$
$= \lim_{x\to\infty} \frac{\sqrt{x^2+1}}{x} = \lim_{x\to\infty} \sqrt{1+\frac{1}{x^2}} = 1$ and $\int_{1}^{\infty} \frac{dx}{x}$ diverges; therefore, $\int_{-\infty}^{\infty} \frac{dx}{\sqrt{x^2+1}}$ diverges

82. $\int_{-\infty}^{\infty} \frac{1}{\sqrt{1+x^6}}\,dx$ converges, since $\int_{-\infty}^{\infty} \frac{1}{\sqrt{1+x^6}}\,dx = 2\int_{0}^{\infty} \frac{1}{\sqrt{1+x^6}}\,dx$ which was shown to converge in Exercise 51

83. $\int_{-\infty}^{\infty} \frac{dx}{e^x+e^{-x}} = \int_{-\infty}^{\infty} \frac{e^x\,dx}{e^{2x}+1}$; $\frac{e^x}{e^{2x}+1} = \frac{1}{e^x+e^{-x}} < \frac{1}{e^x}$ and $\int_{0}^{\infty} \frac{dx}{e^x} = \lim_{c\to\infty} \left[-e^{-x}\right]_0^c = \lim_{c\to\infty} \left(-e^{-c}+1\right) = 1$
$\Rightarrow \int_{-\infty}^{\infty} \frac{e^x\,dx}{e^{2x}+1} = 2\int_{0}^{\infty} \frac{dx}{e^x+e^{-x}}$ converges

84. $\int_{-\infty}^{\infty} \frac{e^{-x}\,dx}{x^2+1} = \int_{-\infty}^{-1} \frac{e^{-x}\,dx}{x^2+1} + \int_{-1}^{\infty} \frac{e^{-x}\,dx}{x^2+1}$; $\int_{-\infty}^{-1} \frac{e^{-x}\,dx}{x^2+1} = \int_{1}^{\infty} \frac{e^u\,du}{1+u^2}$, where $u = -x$, and since $\frac{e^u}{1+u^2} > \frac{1}{u}$ $(u > 1)$ and $\int_{1}^{\infty} \frac{du}{u}$ diverges, the integral $\int_{1}^{\infty} \frac{e^u\,du}{1+u^2}$ diverges $\Rightarrow \int_{-\infty}^{\infty} \frac{e^{-x}\,dx}{x^2+1}$ diverges

85. $\int_{-\infty}^{\infty} e^{-|x|}\,dx = 2\int_{0}^{\infty} e^{-x}\,dx = 2\lim_{b\to\infty} \int_{0}^{b} e^{-x}\,dx = -2\lim_{b\to\infty} \left[e^{-x}\right]_0^b = 2$, so the integral converges.

86. $\int_{-\infty}^{\infty} \frac{dx}{(x+1)^2} = \int_{-\infty}^{-2} \frac{dx}{(x+1)^2} + \int_{-2}^{-1} \frac{dx}{(x+1)^2} + \int_{-1}^{2} \frac{dx}{(x+1)^2} + \int_{2}^{\infty} \frac{dx}{(x+1)^2}$;
$\lim_{b\to -1^-} \int_{-2}^{b} \frac{dx}{(x+1)^2} = -\lim_{b\to -1^-} \left[\frac{1}{x+1}\right]_{-2}^{b} = \infty$, which diverges $\Rightarrow \int_{-\infty}^{\infty} \frac{dx}{(x+1)^2}$ diverges

87. $\int_{-\infty}^{\infty} \frac{|\sin x| + |\cos x|}{|x|+1}\,dx = 2\int_{0}^{\infty} \frac{|\sin x| + |\cos x|}{x+1}\,dx \ge 2\int_{0}^{\infty} \frac{\sin^2 x + \cos^2 x}{x+1}\,dx = 2\lim_{b\to\infty} \int_{0}^{b} \frac{dx}{x+1}\,dx$
$= 2\lim_{b\to\infty} \left[\ln|x+1|\right]_0^b = \infty$, which diverges $\Rightarrow \int_{-\infty}^{\infty} \frac{|\sin x| + |\cos x|}{|x|+1}\,dx$ diverges

88. $\int_{-\infty}^{\infty} \frac{x}{(x^2+1)(x^2+2)}\,dx = 0$ by Exercise 80(b) because the integrand is odd and the integral $\int_{0}^{\infty} \frac{x\,dx}{(x^2+1)(x^2+2)} \le \int_{0}^{\infty} \frac{dx}{x^3}$ converges

89. Example CAS commands:

<u>Maple</u>:

```
f := (x,p) -> x^p*ln(x);
domain := 0..exp(1);
fn_list := [seq( f(x,p), p=-2..2 )];
```

```
plot( fn_list, x=domain, y=-50..10, color=[red,blue,green,cyan,pink], linestyle=[1,3,4,7,9], thickness=[3,4,1,2,0],
      legend=["p= -2","p = -1","p = 0","p = 1","p = 2"], title="#89 (Section 8.8)" );
q1 := Int( f(x,p), x=domain );
q2 := value( q1 );
q3 := simplify( q2 ) assuming p>-1;
q4 := simplify( q2 ) assuming p<-1;
q5 := value( eval( q1, p=-1 ) );
i1 := q1 = piecewise( p<-1, q4, p=-1, q5, p>-1, q3 );
```

90. Example CAS commands:

Maple:

```
f := (x,p) -> x^p*ln(x);
domain := exp(1)..infinity;
fn_list := [seq( f(x,p), p=-2..2 )];
plot( fn_list, x=exp(1)..10, y=0..100, color=[red,blue,green,cyan,pink], linestyle=[1,3,4,7,9], thickness=[3,4,1,2,0],
      legend=["p = -2","p = -1","p = 0","p = 1","p = 2"], title="#90 (Section 8.8)" );
q6 := Int( f(x,p), x=domain );
q7 := value( q6 );
q8 := simplify( q7 ) assuming p>-1;
q9 := simplify( q7 ) assuming p<-1;
q10 := value( eval( q6, p=-1 ) );
i2 := q6 = piecewise( p<-1, q9, p=-1, q10, p>-1, q8 );
```

91. Example CAS commands:

Maple:

```
f := (x,p) -> x^p*ln(x);
domain := 0..infinity;
fn_list := [seq( f(x,p), p=-2..2 )];
plot( fn_list, x=0..10, y=-50..50, color=[red,blue,green,cyan,pink], linestyle=[1,3,4,7,9], thickness=[3,4,1,2,0],
      legend=["p = -2","p = -1","p = 0","p = 1","p = 2"], title="#91 (Section 8.8)" );
q11 := Int( f(x,p), x=domain ):
q11 = lhs(i1+i2);
`` = rhs(i1+i2);
`` = piecewise( p<-1, q4+q9, p=-1, q5+q10, p>-1, q3+q8 );
`` = piecewise( p<-1, -infinity, p=-1, undefined, p>-1, infinity );
```

92. Example CAS commands:

Maple:

```
f := (x,p) -> x^p*ln(abs(x));
domain := -infinity..infinity;
fn_list := [seq( f(x,p), p=-2..2 )];
plot( fn_list, x=-4..4, y=-20..10, color=[red,blue,green,cyan,pink], linestyle=[1,3,4,7,9],
      legend=["p = -2","p = -1","p = 0","p = 1","p = 2"], title="#92 (Section 8.8)" );
q12 := Int( f(x,p), x=domain );
q12p := Int( f(x,p), x=0..infinity );
q12n := Int( f(x,p), x=-infinity..0 );
q12 = q12p + q12n;
`` = simplify( q12p+q12n );
```

89-92. Example CAS commands:

Mathematica: (functions and domains may vary)

```
Clear[x, f, p]
f[x_]:= x^p Log[Abs[x]]
int = Integrate[f[x], {x, e, 100}]
int /. p → 2.5
```

In order to plot the function, a value for p must be selected.

```
p = 3;
Plot[f[x], {x, 2.72, 10}]
```

CHAPTER 8 PRACTICE EXERCISES

1. $\int x\sqrt{4x^2-9}\,dx;\ \begin{bmatrix} u = 4x^2-9 \\ du = 8x\,dx \end{bmatrix} \to \frac{1}{8}\int \sqrt{u}\,du = \frac{1}{8}\cdot\frac{2}{3}u^{3/2} + C = \frac{1}{12}\left(4x^2-9\right)^{3/2} + C$

2. $\int 6x\sqrt{3x^2+5}\,dx;\ \begin{bmatrix} u = 3x^2+5 \\ du = 6x\,dx \end{bmatrix} \to \int \sqrt{u}\,du = \frac{2}{3}u^{3/2} + C = \frac{2}{3}\left(3x^2+5\right)^{3/2} + C$

3. $\int x(2x+1)^{1/2}\,dx;\ \begin{bmatrix} u = 2x+1 \\ du = 2\,dx \end{bmatrix} \to \frac{1}{2}\int \left(\frac{u-1}{2}\right)\sqrt{u}\,du = \frac{1}{4}\left(\int u^{3/2}\,du - \int u^{1/2}\,du\right) = \frac{1}{4}\left(\frac{2}{5}u^{5/2} - \frac{2}{3}u^{3/2}\right) + C$
$= \frac{(2x+1)^{5/2}}{10} - \frac{(2x+1)^{3/2}}{6} + C$

4. $\int \frac{x}{\sqrt{1-x}}\,dx;\ \begin{bmatrix} u = 1-x \\ du = -\,dx \end{bmatrix} \to -\int \frac{(1-u)}{\sqrt{u}}\,du = \int \left(\sqrt{u} - \frac{1}{\sqrt{u}}\right)du = \frac{2}{3}u^{3/2} - 2u^{1/2} + C$
$= \frac{2}{3}(1-x)^{3/2} - 2(1-x)^{1/2} + C$

5. $\int \frac{x\,dx}{\sqrt{8x^2+1}};\ \begin{bmatrix} u = 8x^2+1 \\ du = 16x\,dx \end{bmatrix} \to \frac{1}{16}\int \frac{du}{\sqrt{u}} = \frac{1}{16}\cdot 2u^{1/2} + C = \frac{\sqrt{8x^2+1}}{8} + C$

6. $\int \frac{x\,dx}{\sqrt{9-4x^2}};\ \begin{bmatrix} u = 9-4x^2 \\ du = -8x\,dx \end{bmatrix} \to -\frac{1}{8}\int \frac{du}{\sqrt{u}} = -\frac{1}{8}\cdot 2u^{1/2} + C = -\frac{\sqrt{9-4x^2}}{4} + C$

7. $\int \frac{y\,dy}{25+y^2};\ \begin{bmatrix} u = 25+y^2 \\ du = 2y\,dy \end{bmatrix} \to \frac{1}{2}\int \frac{du}{u} = \frac{1}{2}\ln|u| + C = \frac{1}{2}\ln\left(25+y^2\right) + C$

8. $\int \frac{y^3\,dy}{4+y^4};\ \begin{bmatrix} u = 4+y^4 \\ du = 4y^3\,dy \end{bmatrix} \to \frac{1}{4}\int \frac{du}{u} = \frac{1}{4}\ln|u| + C = \frac{1}{4}\ln\left(4+y^4\right) + C$

9. $\int \frac{t^3\,dt}{\sqrt{9-4t^4}};\ \begin{bmatrix} u = 9-4t^4 \\ du = -16t^3\,dt \end{bmatrix} \to -\frac{1}{16}\int \frac{du}{\sqrt{u}} = -\frac{1}{16}\cdot 2u^{1/2} + C = -\frac{\sqrt{9-4t^4}}{8} + C$

10. $\int \frac{2t\,dt}{t^4+1};\ \begin{bmatrix} u = t^2 \\ du = 2t\,dt \end{bmatrix} \to \int \frac{du}{u^2+1} = \tan^{-1}u + C = \tan^{-1}t^2 + C$

11. $\int z^{2/3}\left(z^{5/3}+1\right)^{2/3}dz;\ \begin{bmatrix} u = z^{5/3}+1 \\ du = \frac{5}{3}z^{2/3}\,dz \end{bmatrix} \to \frac{3}{5}\int u^{2/3}\,du = \frac{3}{5}\cdot\frac{3}{5}u^{5/3} + C = \frac{9}{25}\left(z^{5/3}+1\right)^{5/3} + C$

12. $\int z^{-1/5}\left(1+z^{4/5}\right)^{-1/2} dz$; $\begin{bmatrix} u = 1 + z^{4/5} \\ du = \frac{4}{5} z^{-1/5} dz \end{bmatrix}$ $\rightarrow \frac{5}{4}\int u^{-1/2}\,du = \frac{5}{4}\cdot 2\sqrt{u} + C = \frac{5}{2}\left(1+z^{4/5}\right)^{1/2} + C$

13. $\int \frac{\sin 2\theta\, d\theta}{(1-\cos 2\theta)^2}$; $\begin{bmatrix} u = 1 - \cos 2\theta \\ du = 2\sin 2\theta\, d\theta \end{bmatrix}$ $\rightarrow \frac{1}{2}\int \frac{du}{u^2} = -\frac{1}{2u} + C = -\frac{1}{2(1-\cos 2\theta)} + C$

14. $\int \frac{\cos\theta\, d\theta}{(1+\sin\theta)^{1/2}}$; $\begin{bmatrix} u = 1 + \sin\theta \\ du = \cos\theta\, d\theta \end{bmatrix}$ $\rightarrow \int \frac{du}{u^{1/2}} = 2u^{1/2} + C = 2\sqrt{1+\sin\theta} + C$

15. $\int \frac{\sin t\, dt}{3+4\cos t}$; $\begin{bmatrix} u = 3 + 4\cos t \\ du = -4\sin t\, dt \end{bmatrix}$ $\rightarrow -\frac{1}{4}\int \frac{du}{u} = -\frac{1}{4}\ln|u| + C = -\frac{1}{4}\ln|3+4\cos t| + C$

16. $\int \frac{\cos 2t\, dt}{1+\sin 2t}$; $\begin{bmatrix} u = 1 + \sin 2t \\ du = 2\cos 2t\, dt \end{bmatrix}$ $\rightarrow \frac{1}{2}\int \frac{du}{u} = \frac{1}{2}\ln|u| + C = \frac{1}{2}\ln|1+\sin 2t| + C$

17. $\int (\sin 2x)\, e^{\cos 2x}\, dx$; $\begin{bmatrix} u = \cos 2x \\ du = -2\sin 2x\, dx \end{bmatrix}$ $\rightarrow -\frac{1}{2}\int e^u\, du = -\frac{1}{2}e^u + C = -\frac{1}{2}e^{\cos 2x} + C$

18. $\int (\sec x\tan x)\, e^{\sec x}\, dx$; $\begin{bmatrix} u = \sec x \\ du = \sec x\tan x\, dx \end{bmatrix}$ $\rightarrow \int e^u\, du = e^u + C = e^{\sec x} + C$

19. $\int e^\theta \sin(e^\theta)\cos^2(e^\theta)\, d\theta$; $\begin{bmatrix} u = \cos(e^\theta) \\ du = -\sin(e^\theta)\cdot e^\theta\, d\theta \end{bmatrix}$ $\rightarrow \int -u^2\, du = -\frac{1}{3}u^3 + C = -\frac{1}{3}\cos^3(e^\theta) + C$

20. $\int e^\theta \sec^2(e^\theta)\, d\theta$; $\begin{bmatrix} u = e^\theta \\ du = e^\theta\, d\theta \end{bmatrix}$ $\rightarrow \int \sec^2 u\, du = \tan u + C = \tan(e^\theta) + C$

21. $\int 2^{x-1}\, dx = \frac{2^{x-1}}{\ln 2} + C$

22. $\int 5^{x\sqrt{2}}\, dx = \frac{1}{\sqrt{2}}\left(\frac{5^{x\sqrt{2}}}{\ln 5}\right) + C$

23. $\int \frac{dv}{v\ln v}$; $\begin{bmatrix} u = \ln v \\ du = \frac{1}{v}\, dv \end{bmatrix}$ $\rightarrow \int \frac{du}{u} = \ln|u| + C = \ln|\ln v| + C$

24. $\int \frac{dv}{v(2+\ln v)}$; $\begin{bmatrix} u = 2 + \ln v \\ du = \frac{1}{v}\, dv \end{bmatrix}$ $\rightarrow \int \frac{du}{u} = \ln|u| + C = \ln|2+\ln v| + C$

25. $\int \frac{dx}{(x^2+1)(2+\tan^{-1}x)}$; $\begin{bmatrix} u = 2 + \tan^{-1}x \\ du = \frac{dx}{x^2+1} \end{bmatrix}$ $\rightarrow \int \frac{du}{u} = \ln|u| + C = \ln|2+\tan^{-1}x| + C$

26. $\int \frac{\sin^{-1}x\, dx}{\sqrt{1-x^2}}$; $\begin{bmatrix} u = \sin^{-1}x \\ du = \frac{dx}{\sqrt{1-x^2}} \end{bmatrix}$ $\rightarrow \int u\, du = \frac{1}{2}u^2 + C = \frac{1}{2}\left(\sin^{-1}x\right)^2 + C$

27. $\int \frac{2\, dx}{\sqrt{1-4x^2}}$; $\begin{bmatrix} u = 2x \\ du = 2\, dx \end{bmatrix}$ $\rightarrow \int \frac{du}{\sqrt{1-u^2}} = \sin^{-1}u + C = \sin^{-1}(2x) + C$

28. $\int \frac{dx}{\sqrt{49-x^2}} = \frac{1}{7}\int \frac{dx}{\sqrt{1-\left(\frac{x}{7}\right)^2}}$; $\begin{bmatrix} u = \frac{x}{7} \\ du = \frac{1}{7}\, dx \end{bmatrix}$ $\rightarrow \int \frac{du}{\sqrt{1-u^2}} = \sin^{-1}u + C = \sin^{-1}\left(\frac{x}{7}\right) + C$

29. $\int \frac{dt}{\sqrt{16-9t^2}} = \frac{1}{4}\int \frac{dt}{\sqrt{1-\left(\frac{3t}{4}\right)^2}}$; $\begin{bmatrix} u = \frac{3}{4}t \\ du = \frac{3}{4}\,dt \end{bmatrix} \rightarrow \frac{1}{3}\int \frac{du}{\sqrt{1-u^2}} = \frac{1}{3}\sin^{-1} u + C = \frac{1}{3}\sin^{-1}\left(\frac{3t}{4}\right) + C$

30. $\int \frac{dt}{\sqrt{9-4t^2}} = \frac{1}{3}\int \frac{dt}{\sqrt{1-\left(\frac{2t}{3}\right)^2}}$; $\begin{bmatrix} u = \frac{2}{3}t \\ du = \frac{2}{3}\,dt \end{bmatrix} \rightarrow \frac{1}{2}\int \frac{du}{\sqrt{1-u^2}} = \frac{1}{2}\sin^{-1} u + C = \frac{1}{2}\sin^{-1}\left(\frac{2t}{3}\right) + C$

31. $\int \frac{dt}{9+t^2} = \frac{1}{9}\int \frac{dt}{1+\left(\frac{t}{3}\right)^2}$; $\begin{bmatrix} u = \frac{1}{3}t \\ du = \frac{1}{3}\,dt \end{bmatrix} \rightarrow \frac{1}{3}\int \frac{du}{1+u^2} = \frac{1}{3}\tan^{-1} u + C = \frac{1}{3}\tan^{-1}\left(\frac{t}{3}\right) + C$

32. $\int \frac{dt}{1+25t^2}$; $\begin{bmatrix} u = 5t \\ du = 5\,dt \end{bmatrix} \rightarrow \frac{1}{5}\int \frac{du}{1+u^2} = \frac{1}{5}\tan^{-1} u + C = \frac{1}{5}\tan^{-1}(5t) + C$

33. $\int \frac{4\,dx}{5x\sqrt{25x^2-16}} = \frac{4}{25}\int \frac{dx}{x\sqrt{x^2-\frac{16}{25}}} = \frac{1}{5}\sec^{-1}\left|\frac{5x}{4}\right| + C$

34. $\int \frac{6\,dx}{x\sqrt{4x^2-9}} = 3\int \frac{dx}{x\sqrt{x^2-\frac{9}{4}}} = 2\sec^{-1}\left|\frac{2x}{3}\right| + C$

35. $\int \frac{dx}{\sqrt{4x-x^2}} = \int \frac{d(x-2)}{\sqrt{4-(x-2)^2}} = \sin^{-1}\left(\frac{x-2}{2}\right) + C$

36. $\int \frac{dx}{\sqrt{4x-x^2-3}} = \int \frac{d(x-2)}{\sqrt{1-(x-2)^2}} = \sin^{-1}(x-2) + C$

37. $\int \frac{dy}{y^2-4y+8} = \int \frac{d(y-2)}{(y-2)^2+4} = \frac{1}{2}\tan^{-1}\left(\frac{y-2}{2}\right) + C$

38. $\int \frac{dt}{t^2+4t+5} = \int \frac{d(t+2)}{(t+2)^2+1} = \tan^{-1}(t+2) + C$

39. $\int \frac{dx}{(x-1)\sqrt{x^2-2x}} = \int \frac{d(x-1)}{(x-1)\sqrt{(x-1)^2-1}} = \sec^{-1}|x-1| + C$

40. $\int \frac{dv}{(v+1)\sqrt{v^2+2v}} = \int \frac{d(v+1)}{(v+1)\sqrt{(v+1)^2-1}} = \sec^{-1}|v+1| + C$

41. $\int \sin^2 x\,dx = \int \frac{1-\cos 2x}{2}\,dx = \frac{x}{2} - \frac{\sin 2x}{4} + C$

42. $\int \cos^2 3x\,dx = \int \frac{1+\cos 6x}{2}\,dx = \frac{x}{2} + \frac{\sin 6x}{12} + C$

43. $\int \sin^3 \frac{\theta}{2}\,d\theta = \int \left(1-\cos^2 \frac{\theta}{2}\right)\left(\sin \frac{\theta}{2}\right) d\theta$; $\begin{bmatrix} u = \cos \frac{\theta}{2} \\ du = -\frac{1}{2}\sin \frac{\theta}{2}\,d\theta \end{bmatrix} \rightarrow -2\int (1-u^2)\,du = \frac{2u^3}{3} - 2u + C$
$= \frac{2}{3}\cos^3 \frac{\theta}{2} - 2\cos \frac{\theta}{2} + C$

44. $\int \sin^3 \theta \cos^2 \theta\,d\theta = \int (1-\cos^2 \theta)(\sin \theta)(\cos^2 \theta)\,d\theta$; $\begin{bmatrix} u = \cos \theta \\ du = -\sin \theta\,d\theta \end{bmatrix} \rightarrow -\int (1-u^2)u^2\,du = \int (u^4-u^2)\,du$
$= \frac{u^5}{5} - \frac{u^3}{3} + C = \frac{\cos^5 \theta}{5} - \frac{\cos^3 \theta}{3} + C$

45. $\int \tan^3 2t\, dt = \int (\tan 2t)(\sec^2 2t - 1)\, dt = \int \tan 2t \sec^2 2t\, dt - \int \tan 2t\, dt;\ \begin{bmatrix} u = 2t \\ du = 2\, dt \end{bmatrix}$
$\to \frac{1}{2}\int \tan u \sec^2 u\, du - \frac{1}{2}\int \tan u\, du = \frac{1}{4}\tan^2 u + \frac{1}{2}\ln|\cos u| + C = \frac{1}{4}\tan^2 2t + \frac{1}{2}\ln|\cos 2t| + C$
$= \frac{1}{4}\tan^2 2t - \frac{1}{2}\ln|\sec 2t| + C$

46. $\int 6 \sec^4 t\, dt = 6\int (\tan^2 t + 1)(\sec^2 t)\, dt;\ \begin{bmatrix} u = \tan t \\ du = \sec^2 t\, dt \end{bmatrix} \to 6\int (u^2 + 1)\, du = 2u^3 + 6u + C$
$= 2\tan^3 t + 6\tan t + C$

47. $\int \frac{dx}{2 \sin x \cos x} = \int \frac{dx}{\sin 2x} = \int \csc 2x\, dx = -\frac{1}{2}\ln|\csc 2x + \cot 2x| + C$

48. $\int \frac{2\, dx}{\cos^2 x - \sin^2 x} = \int \frac{2\, dx}{\cos 2x};\ \begin{bmatrix} u = 2x \\ du = 2\, dx \end{bmatrix} \to \int \frac{du}{\cos u} = \int \sec u\, du = \ln|\sec u + \tan u| + C$
$= \ln|\sec 2x + \tan 2x| + C$

49. $\int_{\pi/4}^{\pi/2} \sqrt{\csc^2 y - 1}\, dy = \int_{\pi/4}^{\pi/2} \cot y\, dy = [\ln|\sin y|]_{\pi/4}^{\pi/2} = \ln 1 - \ln \frac{1}{\sqrt{2}} = \ln \sqrt{2}$

50. $\int_{\pi/4}^{3\pi/4} \sqrt{\cot^2 t + 1}\, dt = \int_{\pi/4}^{3\pi/4} \csc t\, dt = [-\ln|\csc t + \cot t|]_{\pi/4}^{3\pi/4} = -\ln\left|\csc \frac{3\pi}{4} + \cot \frac{3\pi}{4}\right| + \ln\left|\csc \frac{\pi}{4} + \cot \frac{\pi}{4}\right|$
$= -\ln\left|\sqrt{2} - 1\right| + \ln\left|\sqrt{2} + 1\right| = \ln\left|\frac{\sqrt{2}+1}{\sqrt{2}-1}\right| = \ln\left|\frac{(\sqrt{2}+1)(\sqrt{2}+1)}{2-1}\right| = \ln\left(3 + 2\sqrt{2}\right)$

51. $\int_0^{\pi} \sqrt{1 - \cos^2 2x}\, dx = \int_0^{\pi} |\sin 2x|\, dx = \int_0^{\pi/2} \sin 2x\, dx - \int_{\pi/2}^{\pi} \sin 2x\, dx = -\left[\frac{\cos 2x}{2}\right]_0^{\pi/2} + \left[\frac{\cos 2x}{2}\right]_{\pi/2}^{\pi}$
$= -\left(-\frac{1}{2} - \frac{1}{2}\right) + \left[\frac{1}{2} - \left(-\frac{1}{2}\right)\right] = 2$

52. $\int_0^{2\pi} \sqrt{1 - \sin^2 \frac{x}{2}}\, dx = \int_0^{2\pi} \left|\cos \frac{x}{2}\right| dx = \int_0^{\pi} \cos \frac{x}{2}\, dx - \int_{\pi}^{2\pi} \cos \frac{x}{2}\, dx = \left[2 \sin \frac{x}{2}\right]_0^{\pi} - \left[2 \sin \frac{x}{2}\right]_{\pi}^{2\pi}$
$= (2 - 0) - (0 - 2) = 4$

53. $\int_{-\pi/2}^{\pi/2} \sqrt{1 - \cos 2t}\, dt = \sqrt{2}\int_{-\pi/2}^{\pi/2} |\sin t|\, dt = 2\sqrt{2}\int_0^{\pi/2} \sin t\, dt = \left[-2\sqrt{2}\cos t\right]_0^{\pi/2} = 2\sqrt{2}\,[0 - (-1)] = 2\sqrt{2}$

54. $\int_{\pi}^{2\pi} \sqrt{1 + \cos 2t}\, dt = \sqrt{2}\int_{\pi}^{2\pi} |\cos t|\, dt = -\sqrt{2}\int_{\pi}^{3\pi/2} \cos t\, dt + \sqrt{2}\int_{3\pi/2}^{2\pi} \cos t\, dt$
$= -\sqrt{2}\,[\sin t]_{\pi}^{3\pi/2} + \sqrt{2}\,[\sin t]_{3\pi/2}^{2\pi} = -\sqrt{2}(-1 - 0) + \sqrt{2}\,[0 - (-1)] = 2\sqrt{2}$

55. $\int \frac{x^2\, dx}{x^2 + 4} = x - \int \frac{4\, dx}{x^2 + 4} = x - 2\tan^{-1}\left(\frac{x}{2}\right) + C$

56. $\int \frac{x^3\, dx}{9 + x^2} = \int \left[\frac{x(x^2+9) - 9x}{x^2 + 9}\right] dx = \int \left(x - \frac{9x}{x^2+9}\right) dx = \frac{x^2}{2} - \frac{9}{2}\ln(9 + x^2) + C$

57. $\int \frac{4x^2 + 3}{2x - 1}\, dx = \int \left[(2x + 1) + \frac{4}{2x - 1}\right] dx = x + x^2 + 2\ln|2x - 1| + C$

58. $\int \frac{2x\, dx}{x - 4} = \int \left(2 + \frac{8}{x - 4}\right) dx = 2x + 8\ln|x - 4| + C$

59. $\int \frac{2y - 1}{y^2 + 4}\, dy = \int \frac{2y\, dy}{y^2 + 4} - \int \frac{dy}{y^2 + 4} = \ln(y^2 + 4) - \frac{1}{2}\tan^{-1}\left(\frac{y}{2}\right) + C$

60. $\int \frac{y+4}{y^2+1}\,dy = \int \frac{y\,dy}{y^2+1} + 4\int \frac{dy}{y^2+1} = \frac{1}{2}\ln(y^2+1) + 4\tan^{-1}y + C$

61. $\int \frac{t+2}{\sqrt{4-t^2}}\,dt = \int \frac{t\,dt}{\sqrt{4-t^2}} + 2\int \frac{dt}{\sqrt{4-t^2}} = -\sqrt{4-t^2} + 2\sin^{-1}\left(\frac{t}{2}\right) + C$

62. $\int \frac{2t^2+\sqrt{1-t^2}}{t\sqrt{1-t^2}}\,dt = \int \frac{2t\,dt}{\sqrt{1-t^2}} + \int \frac{dt}{t} = -2\sqrt{1-t^2} + \ln|t| + C$

63. $\int \frac{\tan x\,dx}{\tan x+\sec x} = \int \frac{\sin x\,dx}{\sin x+1} = \int \frac{(\sin x)(1-\sin x)}{1-\sin^2 x}\,dx = \int \frac{\sin x - 1 + \cos^2 x}{\cos^2 x}\,dx$
$= -\int \frac{d(\cos x)}{\cos^2 x} - \int \frac{dx}{\cos^2 x} + \int dx = \frac{1}{\cos x} - \tan x + x + C = x - \tan x + \sec x + C$

64. $\int \frac{\cot x\,dx}{\cot x+\csc x} = \int \frac{\cos x\,dx}{\cos x+1} = \int \frac{(\cos x)(1-\cos x)}{1-\cos^2 x}\,dx = \int \frac{\cos x - 1 + \sin^2 x}{\sin^2 x}\,dx$
$= \int \frac{d(\sin x)}{\sin^2 x} - \int \frac{dx}{\sin^2 x} + \int dx = -\frac{1}{\sin x} + \cot x + x + C = x + \cot x - \csc x + C$

65. $\int \sec(5-3x)\,dx;\ \left[\begin{array}{l} y = 5-3x \\ dy = -3\,dx \end{array}\right] \rightarrow \int \sec y \cdot \left(-\frac{dy}{3}\right) = -\frac{1}{3}\int \sec y\,dy = -\frac{1}{3}\ln|\sec y + \tan y| + C$
$= -\frac{1}{3}\ln|\sec(5-3x) + \tan(5-3x)| + C$

66. $\int x\csc(x^2+3)\,dx = \frac{1}{2}\int \csc(x^2+3)\,d(x^2+3) = -\frac{1}{2}\ln|\csc(x^2+3) + \cot(x^2+3)| + C$

67. $\int \cot\left(\frac{x}{4}\right)dx = 4\int \cot\left(\frac{x}{4}\right)d\left(\frac{x}{4}\right) = 4\ln\left|\sin\left(\frac{x}{4}\right)\right| + C$

68. $\int \tan(2x-7)\,dx = \frac{1}{2}\int \tan(2x-7)\,d(2x-7) = -\frac{1}{2}\ln|\cos(2x-7)| + C = \frac{1}{2}\ln|\sec(2x-7)| + C$

69. $\int x\sqrt{1-x}\,dx;\ \left[\begin{array}{l} u = 1-x \\ du = -\,dx \end{array}\right] \rightarrow -\int (1-u)\sqrt{u}\,du = \int \left(u^{3/2} - u^{1/2}\right)du = \frac{2}{5}u^{5/2} - \frac{2}{3}u^{3/2} + C$
$= \frac{2}{5}(1-x)^{5/2} - \frac{2}{3}(1-x)^{3/2} + C = -2\left[\frac{\left(\sqrt{1-x}\right)^3}{3} - \frac{\left(\sqrt{1-x}\right)^5}{5}\right] + C$

70. $\int 3x\sqrt{2x+1}\,dx;\ \left[\begin{array}{l} u = 2x+1 \\ du = 2\,dx \end{array}\right] \rightarrow \int 3\left(\frac{u-1}{2}\right)\sqrt{u}\cdot\frac{1}{2}\,du = \frac{3}{4}\int \left(u^{3/2} - u^{1/2}\right)du = \frac{3}{4}\cdot\frac{2}{5}u^{5/2} - \frac{3}{4}\cdot\frac{2}{3}u^{3/2} + C$
$= \frac{3}{10}(2x+1)^{5/2} - \frac{1}{2}(2x+1)^{3/2} + C = \frac{3\left(\sqrt{2x+1}\right)^5}{10} - \frac{\left(\sqrt{2x+1}\right)^3}{2} + C$

71. $\int \sqrt{z^2+1}\,dz;\ \left[\begin{array}{l} z = \tan\theta \\ dz = \sec^2\theta\,d\theta \end{array}\right] \rightarrow \int \sqrt{\tan^2\theta + 1}\cdot\sec^2\theta\,d\theta = \int \sec^3\theta\,d\theta$
$= \frac{\sec\theta\tan\theta}{3-1} + \frac{3-2}{3-1}\int \sec\theta\,d\theta$ (FORMULA 92)
$= \frac{\sin\theta}{2\cos^2\theta} + \frac{1}{2}\ln|\sec\theta + \tan\theta| + C = \frac{z\sqrt{z^2+1}}{2} + \frac{1}{2}\ln\left|z + \sqrt{1+z^2}\right| + C$

72. $\int (16+z^2)^{-3/2}\,dz;\ \left[\begin{array}{l} z = 4\tan\theta \\ dz = 4\sec^2\theta\,d\theta \end{array}\right] \rightarrow \int \frac{4\sec^2\theta\,d\theta}{64\sec^3\theta\,d\theta} = \frac{1}{16}\int \cos\theta\,d\theta = \frac{1}{16}\sin\theta + C = \frac{z}{16\sqrt{16+z^2}} + C$
$= \frac{z}{16\left(16+z^2\right)^{1/2}} + C$

73. $\int \frac{dy}{\sqrt{25+y^2}} = \frac{1}{5}\int \frac{dy}{\sqrt{1+\left(\frac{y}{5}\right)^2}} = \int \frac{du}{\sqrt{1+u^2}}, \ \left[u = \frac{y}{5}\right]; \left[\begin{array}{c} u = \tan\theta \\ du = \sec^2\theta\, d\theta \end{array}\right] \rightarrow \int \frac{\sec^2\theta\, d\theta}{\sqrt{1+\tan^2\theta}} = \int \sec\theta\, d\theta$

$= \ln|\sec\theta + \tan\theta| + C_1 = \ln\left|\sqrt{1+u^2} + u\right| + C_1 = \ln\left|\sqrt{1+\left(\frac{y}{5}\right)^2} + \frac{y}{5}\right| + C_1 = \ln\left|\frac{\sqrt{25+y^2}+y}{5}\right| + C_1$

$= \ln\left|y + \sqrt{25+y^2}\right| + C$

74. $\int \frac{dy}{\sqrt{25+9y^2}} = \frac{1}{5}\int \frac{dy}{\sqrt{1+\left(\frac{3y}{5}\right)^2}} = \frac{1}{3}\int \frac{du}{\sqrt{1+u^2}} = \frac{1}{3}\ln\left|\sqrt{1+u^2}+u\right| + C_1$ from Exercise 73

$\rightarrow \frac{1}{3}\ln\left|\sqrt{25+9y^2}+3y\right| + C$

75. $\int \frac{dx}{x^2\sqrt{1-x^2}}; \left[\begin{array}{c} x = \sin\theta \\ dx = \cos\theta\, d\theta \end{array}\right] \rightarrow \int \frac{\cos\theta\, d\theta}{\sin^2\theta\cos\theta} = \int \csc^2\theta\, d\theta = -\cot\theta + C = \frac{-\sqrt{1-x^2}}{x} + C$

76. $\int \frac{x^3\, dx}{\sqrt{1-x^2}}; \left[\begin{array}{c} x = \sin\theta \\ dx = \cos\theta\, d\theta \end{array}\right] \rightarrow \int \frac{\sin^3\theta\cos\theta\, d\theta}{\cos\theta} = \int \sin^3\theta\, d\theta = \int (1-\cos^2\theta)(\sin\theta)\, d\theta;$

$[u = \cos\theta] \rightarrow -\int (1-u^2)\, du = -u + \frac{u^3}{3} + C = -\cos\theta + \frac{1}{3}\cos^3\theta = -\sqrt{1-x^2} + \frac{1}{3}(1-x^2)^{3/2} + C$

Note: Ans $\equiv \frac{-x^2\sqrt{1-x^2}}{3} - \frac{2}{3}\sqrt{1-x^2} + C$ by another method

77. $\int \frac{x^2\, dx}{\sqrt{1-x^2}}; \left[\begin{array}{c} x = \sin\theta \\ dx = \cos\theta\, d\theta \end{array}\right] \rightarrow \int \frac{\sin^2\theta\cos\theta\, d\theta}{\cos\theta} = \int \sin^2\theta\, d\theta = \int \frac{1-\cos 2\theta}{2}\, d\theta = \frac{1}{2}\theta - \frac{1}{4}\sin 2\theta + C$

$= \frac{1}{2}\theta - \frac{1}{2}\sin\theta\cos\theta = \frac{\sin^{-1}x}{2} - \frac{x\sqrt{1-x^2}}{2} + C$

78. $\int \sqrt{4-x^2}\, dx; \left[\begin{array}{c} x = 2\sin\theta \\ dx = 2\cos\theta\, d\theta \end{array}\right] \rightarrow \int 2\cos\theta \cdot 2\cos\theta\, d\theta = 2\int (1+\cos 2\theta)\, d\theta = 2\left(\theta + \frac{1}{2}\sin 2\theta\right) + C$

$= 2\theta + 2\sin\theta\cos\theta + C = 2\sin^{-1}\left(\frac{x}{2}\right) + x\sqrt{1-\left(\frac{x}{2}\right)^2} + C = 2\sin^{-1}\left(\frac{x}{2}\right) + \frac{x\sqrt{4-x^2}}{2} + C$

79. $\int \frac{dx}{\sqrt{x^2-9}}; \left[\begin{array}{c} x = 3\sec\theta \\ dx = 3\sec\theta\tan\theta\, d\theta \end{array}\right] \rightarrow \int \frac{3\sec\theta\tan\theta\, d\theta}{\sqrt{9\sec^2\theta - 9}} = \int \frac{3\sec\theta\tan\theta\, d\theta}{3\tan\theta} = \int \sec\theta\, d\theta$

$= \ln|\sec\theta + \tan\theta| + C_1 = \ln\left|\frac{x}{3} + \sqrt{\left(\frac{x}{3}\right)^2 - 1}\right| + C_1 = \ln\left|\frac{x+\sqrt{x^2-9}}{3}\right| + C_1 = \ln\left|x + \sqrt{x^2-9}\right| + C$

80. $\int \frac{12\, dx}{(x^2-1)^{3/2}}; \left[\begin{array}{c} x = \sec\theta \\ dx = \sec\theta\tan\theta\, d\theta \end{array}\right] \rightarrow \int \frac{12\sec\theta\tan\theta\, d\theta}{\tan^3\theta} = \int \frac{12\cos\theta\, d\theta}{\sin^2\theta}; \left[\begin{array}{c} u = \sin\theta \\ du = \cos\theta\, d\theta \end{array}\right] \rightarrow \int \frac{12\, du}{u^2}$

$= -\frac{12}{u} + C = -\frac{12}{\sin\theta} + C = -\frac{12x}{\sqrt{x^2-1}} + C$

81. $\int \frac{\sqrt{w^2-1}}{w}\, dw; \left[\begin{array}{c} w = \sec\theta \\ dw = \sec\theta\tan\theta\, d\theta \end{array}\right] \rightarrow \int \left(\frac{\tan\theta}{\sec\theta}\right) \cdot \sec\theta\tan\theta\, d\theta = \int \tan^2\theta\, d\theta = \int (\sec^2\theta - 1)\, d\theta$

$= \tan\theta - \theta + C = \sqrt{w^2-1} - \sec^{-1}w + C$

82. $\int \frac{\sqrt{z^2-16}}{z}\, dz; \left[\begin{array}{c} z = 4\sec\theta \\ dz = 4\sec\theta\tan\theta\, d\theta \end{array}\right] \rightarrow \int \frac{4\tan\theta \cdot 4\sec\theta\tan\theta\, d\theta}{4\sec\theta} = 4\int \tan^2\theta\, d\theta = 4(\tan\theta - \theta) + C$

$= \sqrt{z^2-16} - 4\sec^{-1}\left(\frac{z}{4}\right) + C$

83. $u = \ln(x+1),\ du = \frac{dx}{x+1};\ dv = dx,\ v = x;$

$\int \ln(x+1)\, dx = x\ln(x+1) - \int \frac{x}{x+1}\, dx = x\ln(x+1) - \int dx + \int \frac{dx}{x+1} = x\ln(x+1) - x + \ln(x+1) + C_1$

$= (x+1)\ln(x+1) - x + C_1 = (x+1)\ln(x+1) - (x+1) + C$, where $C = C_1 + 1$

84. $u = \ln x$, $du = \frac{dx}{x}$; $dv = x^2\,dx$, $v = \frac{1}{3}x^3$;

$\int x^2 \ln x\,dx = \frac{1}{3}x^3 \ln x - \int \frac{1}{3}x^3\left(\frac{1}{x}\right)dx = \frac{x^3}{3}\ln x - \frac{x^3}{9} + C$

85. $u = \tan^{-1} 3x$, $du = \frac{3\,dx}{1+9x^2}$; $dv = dx$, $v = x$;

$\int \tan^{-1} 3x\,dx = x\tan^{-1} 3x - \int \frac{3x\,dx}{1+9x^2}$; $\begin{bmatrix} y = 1+9x^2 \\ dy = 18x\,dx \end{bmatrix} \rightarrow x\tan^{-1} 3x - \frac{1}{6}\int \frac{dy}{y}$

$= x\tan^{-1}(3x) - \frac{1}{6}\ln(1+9x^2) + C$

86. $u = \cos^{-1}\left(\frac{x}{2}\right)$, $du = \frac{-dx}{\sqrt{4-x^2}}$; $dv = dx$, $v = x$;

$\int \cos^{-1}\left(\frac{x}{2}\right)dx = x\cos^{-1}\left(\frac{x}{2}\right) + \int \frac{x\,dx}{\sqrt{4-x^2}}$; $\begin{bmatrix} y = 4-x^2 \\ dy = -2x\,dx \end{bmatrix} \rightarrow x\cos^{-1}\left(\frac{x}{2}\right) - \frac{1}{2}\int \frac{dy}{\sqrt{y}}$

$= x\cos^{-1}\left(\frac{x}{2}\right) - \sqrt{4-x^2} + C = x\cos^{-1}\left(\frac{x}{2}\right) - 2\sqrt{1-\left(\frac{x}{2}\right)^2} + C$

87.

		e^x
$(x+1)^2$	$\xrightarrow{(+)}$	e^x
$2(x+1)$	$\xrightarrow{(-)}$	e^x
2	$\xrightarrow{(+)}$	e^x
0		

$\Rightarrow \int (x+1)^2 e^x\,dx = [(x+1)^2 - 2(x+1) + 2]\,e^x + C$

88.

		$\sin(1-x)$
x^2	$\xrightarrow{(+)}$	$\cos(1-x)$
$2x$	$\xrightarrow{(-)}$	$-\sin(1-x)$
2	$\xrightarrow{(+)}$	$-\cos(1-x)$
0		

$\Rightarrow \int x^2 \sin(1-x)\,dx = x^2\cos(1-x) + 2x\sin(1-x) - 2\cos(1-x) + C$

89. $u = \cos 2x$, $du = -2\sin 2x\,dx$; $dv = e^x\,dx$, $v = e^x$;

$I = \int e^x \cos 2x\,dx = e^x\cos 2x + 2\int e^x \sin 2x\,dx$;

$u = \sin 2x$, $du = 2\cos 2x\,dx$; $dv = e^x\,dx$, $v = e^x$;

$I = e^x\cos 2x + 2\left[e^x \sin 2x - 2\int e^x \cos 2x\,dx\right] = e^x\cos 2x + 2e^x\sin 2x - 4I \Rightarrow I = \frac{e^x\cos 2x}{5} + \frac{2e^x\sin 2x}{5} + C$

90. $u = \sin 3x$, $du = 3\cos 3x\,dx$; $dv = e^{-2x}\,dx$, $v = -\frac{1}{2}e^{-2x}$;

$I = \int e^{-2x}\sin 3x\,dx = -\frac{1}{2}e^{-2x}\sin 3x + \frac{3}{2}\int e^{-2x}\cos 3x\,dx$;

$u = \cos 3x$, $du = -3\sin 3x\,dx$; $dv = e^{-2x}\,dx$, $v = -\frac{1}{2}e^{-2x}$;

$I = -\frac{1}{2}e^{-2x}\sin 3x + \frac{3}{2}\left[-\frac{1}{2}e^{-2x}\cos 3x - \frac{3}{2}\int e^{-2x}\sin 3x\,dx\right] = -\frac{1}{2}e^{-2x}\sin 3x - \frac{3}{4}e^{-2x}\cos 3x - \frac{9}{4}I$

$\Rightarrow I = \frac{4}{13}\left(-\frac{1}{2}e^{-2x}\sin 3x - \frac{3}{4}e^{-2x}\cos 3x\right) + C = -\frac{2}{13}e^{-2x}\sin 3x - \frac{3}{13}e^{-2x}\cos 3x + C$

91. $\int \frac{x\,dx}{x^2-3x+2} = \int \frac{2\,dx}{x-2} - \int \frac{dx}{x-1} = 2\ln|x-2| - \ln|x-1| + C$

92. $\int \frac{x\,dx}{x^2+4x+3} = \frac{3}{2}\int \frac{dx}{x+3} - \frac{1}{2}\int \frac{dx}{x+1} = \frac{3}{2}\ln|x+3| - \frac{1}{2}\ln|x+1| + C$

93. $\int \frac{dx}{x(x+1)^2} = \int \left(\frac{1}{x} - \frac{1}{x+1} + \frac{-1}{(x+1)^2}\right) dx = \ln|x| - \ln|x+1| + \frac{1}{x+1} + C$

94. $\int \frac{x+1}{x^2(x-1)}\, dx = \int \left(\frac{2}{x-1} - \frac{2}{x} - \frac{1}{x^2}\right) dx = 2\ln\left|\frac{x-1}{x}\right| + \frac{1}{x} + C = -2\ln|x| + \frac{1}{x} + 2\ln|x-1| + C$

95. $\int \frac{\sin\theta\, d\theta}{\cos^2\theta + \cos\theta - 2}$; $[\cos\theta = y] \rightarrow -\int \frac{dy}{y^2+y-2} = -\frac{1}{3}\int \frac{dy}{y-1} + \frac{1}{3}\int \frac{dy}{y+2} = \frac{1}{3}\ln\left|\frac{y+2}{y-1}\right| + C$
$= \frac{1}{3}\ln\left|\frac{\cos\theta+2}{\cos\theta-1}\right| + C = -\frac{1}{3}\ln\left|\frac{\cos\theta-1}{\cos\theta+2}\right| + C$

96. $\int \frac{\cos\theta\, d\theta}{\sin^2\theta + \sin\theta - 6}$; $[\sin\theta = x] \rightarrow \int \frac{dx}{x^2+x-6} = \frac{1}{5}\int \frac{dx}{x-2} - \frac{1}{5}\int \frac{dx}{x+3} = \frac{1}{5}\ln\left|\frac{\sin\theta-2}{\sin\theta+3}\right| + C$

97. $\int \frac{3x^2+4x+4}{x^3+x}\, dx = \int \frac{4}{x}\, dx - \int \frac{x-4}{x^2+1}\, dx = 4\ln|x| - \frac{1}{2}\ln(x^2+1) + 4\tan^{-1}x + C$

98. $\int \frac{4x\, dx}{x^3+4x} = \int \frac{4\, dx}{x^2+4} = 2\tan^{-1}\left(\frac{x}{2}\right) + C$

99. $\int \frac{(v+3)\, dv}{2v^3-8v} = \frac{1}{2}\int \left(-\frac{3}{4v} + \frac{5}{8(v-2)} + \frac{1}{8(v+2)}\right) dv = -\frac{3}{8}\ln|v| + \frac{5}{16}\ln|v-2| + \frac{1}{16}\ln|v+2| + C$
$= \frac{1}{16}\ln\left|\frac{(v-2)^5(v+2)}{v^6}\right| + C$

100. $\int \frac{(3v-7)\, dv}{(v-1)(v-2)(v-3)} = \int \frac{(-2)\, dv}{v-1} + \int \frac{dv}{v-2} + \int \frac{dv}{v-3} = \ln\left|\frac{(v-2)(v-3)}{(v-1)^2}\right| + C$

101. $\int \frac{dt}{t^4+4t^2+3} = \frac{1}{2}\int \frac{dt}{t^2+1} - \frac{1}{2}\int \frac{dt}{t^2+3} = \frac{1}{2}\tan^{-1}t - \frac{1}{2\sqrt{3}}\tan^{-1}\left(\frac{t}{\sqrt{3}}\right) + C = \frac{1}{2}\tan^{-1}t - \frac{\sqrt{3}}{6}\tan^{-1}\frac{t}{\sqrt{3}} + C$

102. $\int \frac{t\, dt}{t^4-t^2-2} = \frac{1}{3}\int \frac{t\, dt}{t^2-2} - \frac{1}{3}\int \frac{t\, dt}{t^2+1} = \frac{1}{6}\ln|t^2-2| - \frac{1}{6}\ln(t^2+1) + C$

103. $\int \frac{x^3+x^2}{x^2+x-2}\, dx = \int \left(x + \frac{2x}{x^2+x-2}\right) dx = \int x\, dx + \frac{2}{3}\int \frac{dx}{x-1} + \frac{4}{3}\int \frac{dx}{x+2}$
$= \frac{x^2}{2} + \frac{4}{3}\ln|x+2| + \frac{2}{3}\ln|x-1| + C$

104. $\int \frac{x^3+1}{x^3-x}\, dx = \int \left(1 + \frac{x+1}{x^3-x}\right) dx = \int \left[1 + \frac{1}{x(x-1)}\right] dx = \int dx + \int \frac{dx}{x-1} - \int \frac{dx}{x} = x + \ln|x-1| - \ln|x| + C$

105. $\int \frac{x^3+4x^2}{x^2+4x+3}\, dx = \int \left(x - \frac{3x}{x^2+4x+3}\right) dx = \int x\, dx + \frac{3}{2}\int \frac{dx}{x+1} - \frac{9}{2}\int \frac{dx}{x+3}$
$= \frac{x^2}{2} - \frac{9}{2}\ln|x+3| + \frac{3}{2}\ln|x+1| + C$

106. $\int \frac{2x^3+x^2-21x+24}{x^2+2x-8}\, dx = \int \left[(2x-3) + \frac{x}{x^2+2x-8}\right] dx = \int (2x-3)\, dx + \frac{1}{3}\int \frac{dx}{x-2} + \frac{2}{3}\int \frac{dx}{x+4}$
$= x^2 - 3x + \frac{2}{3}\ln|x+4| + \frac{1}{3}\ln|x-2| + C$

107. $\int \frac{dx}{x(3\sqrt{x+1})}$; $\begin{bmatrix} u = \sqrt{x+1} \\ du = \frac{dx}{2\sqrt{x+1}} \\ dx = 2u\, du \end{bmatrix} \rightarrow \frac{2}{3}\int \frac{u\, du}{(u^2-1)u} = \frac{1}{3}\int \frac{du}{u-1} - \frac{1}{3}\int \frac{du}{u+1} = \frac{1}{3}\ln|u-1| - \frac{1}{3}\ln|u+1| + C$
$= \frac{1}{3}\ln\left|\frac{\sqrt{x+1}-1}{\sqrt{x+1}+1}\right| + C$

108. $\int \frac{dx}{x(1+\sqrt[3]{x})}$; $\begin{bmatrix} u = \sqrt[3]{x} \\ du = \frac{dx}{3x^{2/3}} \\ dx = 3u^2\,du \end{bmatrix}$ $\rightarrow \int \frac{3u^2\,du}{u^3(1+u)} = 3\int \frac{du}{u(1+u)} = 3\ln\left|\frac{u}{u+1}\right| + C = 3\ln\left|\frac{\sqrt[3]{x}}{1+\sqrt[3]{x}}\right| + C$

109. $\int \frac{ds}{e^s - 1}$; $\begin{bmatrix} u = e^s - 1 \\ du = e^s\,ds \\ ds = \frac{du}{u+1} \end{bmatrix}$ $\rightarrow \int \frac{du}{u(u+1)} = -\int \frac{du}{u+1} + \int \frac{du}{u} = \ln\left|\frac{u}{u+1}\right| + C = \ln\left|\frac{e^s-1}{e^s}\right| + C = \ln|1 - e^{-s}| + C$

110. $\int \frac{ds}{\sqrt{e^s+1}}$; $\begin{bmatrix} u = \sqrt{e^s+1} \\ du = \frac{e^s\,ds}{2\sqrt{e^s+1}} \\ ds = \frac{2u\,du}{u^2-1} \end{bmatrix}$ $\rightarrow \int \frac{2u\,du}{u(u^2-1)} = 2\int \frac{du}{(u+1)(u-1)} = \int \frac{du}{u-1} - \int \frac{du}{u+1} = \ln\left|\frac{u-1}{u+1}\right| + C$
$= \ln\left|\frac{\sqrt{e^s+1}-1}{\sqrt{e^s+1}+1}\right| + C$

111. (a) $\int \frac{y\,dy}{\sqrt{16-y^2}} = -\frac{1}{2}\int \frac{d(16-y^2)}{\sqrt{16-y^2}} = -\sqrt{16-y^2} + C$
(b) $\int \frac{y\,dy}{\sqrt{16-y^2}}$; $[y = 4\sin x] \rightarrow 4\int \frac{\sin x \cos x\,dx}{\cos x} = -4\cos x + C = -\frac{4\sqrt{16-y^2}}{4} + C = -\sqrt{16-y^2} + C$

112. (a) $\int \frac{x\,dx}{\sqrt{4+x^2}} = \frac{1}{2}\int \frac{d(4+x^2)}{\sqrt{4+x^2}} = \sqrt{4+x^2} + C$
(b) $\int \frac{x\,dx}{\sqrt{4+x^2}}$; $[x = 2\tan y] \rightarrow \int \frac{2\tan y \cdot 2\sec^2 y\,dy}{2\sec y} = 2\int \sec y \tan y\,dy = 2\sec y + C = \sqrt{4+x^2} + C$

113. (a) $\int \frac{x\,dx}{4-x^2} = -\frac{1}{2}\int \frac{d(4-x^2)}{4-x^2} = -\frac{1}{2}\ln|4-x^2| + C$
(b) $\int \frac{x\,dx}{4-x^2}$; $[x = 2\sin\theta] \rightarrow \int \frac{2\sin\theta \cdot 2\cos\theta\,d\theta}{4\cos^2\theta} = \int \tan\theta\,d\theta = -\ln|\cos\theta| + C = -\ln\left(\frac{\sqrt{4-x^2}}{2}\right) + C$
$= -\frac{1}{2}\ln|4-x^2| + C$

114. (a) $\int \frac{t\,dt}{\sqrt{4t^2-1}} = \frac{1}{8}\int \frac{d(4t^2-1)}{\sqrt{4t^2-1}} = \frac{1}{4}\sqrt{4t^2-1} + C$
(b) $\int \frac{t\,dt}{\sqrt{4t^2-1}}$; $\left[t = \frac{1}{2}\sec\theta\right] \rightarrow \int \frac{\frac{1}{2}\sec\theta\tan\theta \cdot \frac{1}{2}\sec\theta\,d\theta}{\tan\theta} = \frac{1}{4}\int \sec^2\theta\,d\theta = \frac{\tan\theta}{4} + C = \frac{\sqrt{4t^2-1}}{4} + C$

115. $\int \frac{x\,dx}{9-x^2}$; $\begin{bmatrix} u = 9 - x^2 \\ du = -2x\,dx \end{bmatrix}$ $\rightarrow -\frac{1}{2}\int \frac{du}{u} = -\frac{1}{2}\ln|u| + C = \ln\frac{1}{\sqrt{u}} + C = \ln\frac{1}{\sqrt{9-x^2}} + C$

116. $\int \frac{dx}{x(9-x^2)} = \frac{1}{9}\int \frac{dx}{x} + \frac{1}{18}\int \frac{dx}{3-x} - \frac{1}{18}\int \frac{dx}{3+x} = \frac{1}{9}\ln|x| - \frac{1}{18}\ln|3-x| - \frac{1}{18}\ln|3+x| + C$
$= \frac{1}{9}\ln|x| - \frac{1}{18}\ln|9-x^2| + C$

117. $\int \frac{dx}{9-x^2} = \frac{1}{6}\int \frac{dx}{3-x} + \frac{1}{6}\int \frac{dx}{3+x} = -\frac{1}{6}\ln|3-x| + \frac{1}{6}\ln|3+x| + C = \frac{1}{6}\ln\left|\frac{x+3}{x-3}\right| + C$

118. $\int \frac{dx}{\sqrt{9-x^2}}$; $\begin{bmatrix} x = 3\sin\theta \\ dx = 3\cos\theta\,d\theta \end{bmatrix}$ $\rightarrow \int \frac{3\cos\theta}{3\cos\theta}\,d\theta = \int d\theta = \theta + C = \sin^{-1}\frac{x}{3} + C$

119. $\int \sin^3 x \cos^4 x\,dx = \int \cos^4 x(1-\cos^2 x)\sin x\,dx = \int \cos^4 x \sin x\,dx - \int \cos^6 x \sin x\,dx = -\frac{\cos^5 x}{5} + \frac{\cos^7 x}{7} + C$

120. $\int \cos^5 x \sin^5 x\,dx = \int \sin^5 x \cos^4 x \cos x\,dx = \int \sin^5 x\,(1-\sin^2 x)^2 \cos x\,dx$
$= \int \sin^5 x \cos x\,dx - 2\int \sin^7 x \cos x\,dx + \int \sin^9 x \cos x\,dx = \frac{\sin^6 x}{6} - \frac{2\sin^8 x}{8} + \frac{\sin^{10} x}{10} + C$

121. $\int \tan^4 x \sec^2 x\, dx = \frac{\tan^5 x}{5} + C$

122. $\int \tan^3 x \sec^3 x\, dx = \int (\sec^2 x - 1) \sec^2 x \cdot \sec x \cdot \tan x\, dx = \int \sec^4 x \cdot \sec x \cdot \tan x\, dx - \int \sec^2 x \cdot \sec x \cdot \tan x\, dx$
$= \frac{\sec^5 x}{5} - \frac{\sec^3 x}{3} + C$

123. $\int \sin 5\theta \cos 6\theta\, d\theta = \frac{1}{2}\int (\sin(-\theta) + \sin(11\theta))\, d\theta = \frac{1}{2}\int \sin(-\theta)\, d\theta + \frac{1}{2}\int \sin(11\theta)\, d\theta = \frac{1}{2}\cos(-\theta) - \frac{1}{22}\cos 11\theta + C$
$= \frac{1}{2}\cos\theta - \frac{1}{22}\cos 11\theta + C$

124. $\int \cos 3\theta \cos 3\theta\, d\theta = \frac{1}{2}\int (\cos 0 + \cos 6\theta)\, d\theta = \frac{1}{2}\int d\theta + \frac{1}{2}\int \cos 6\theta\, d\theta = \frac{1}{2}\theta + \frac{1}{12}\sin 6\theta + C$

125. $\int \sqrt{1 + \cos\left(\frac{t}{2}\right)}\, dt = \int \sqrt{2}\left|\cos \frac{t}{4}\right| dt = 4\sqrt{2}\left|\sin \frac{t}{4}\right| + C$

126. $\int e^t \sqrt{\tan^2 e^t + 1}\, dt = \int |\sec e^t|\, e^t\, dt = \ln|\sec e^t + \tan e^t| + C$

127. $|E_s| \le \frac{3-1}{180}(\Delta x)^4 M$ where $\Delta x = \frac{3-1}{n} = \frac{2}{n}$; $f(x) = \frac{1}{x} = x^{-1} \Rightarrow f'(x) = -x^{-2} \Rightarrow f''(x) = 2x^{-3} \Rightarrow f'''(x) = -6x^{-4}$
$\Rightarrow f^{(4)}(x) = 24x^{-5}$ which is decreasing on $[1, 3] \Rightarrow$ maximum of $f^{(4)}(x)$ on $[1, 3]$ is $f^{(4)}(1) = 24 \Rightarrow M = 24$. Then
$|E_s| \le 0.0001 \Rightarrow \left(\frac{3-1}{180}\right)\left(\frac{2}{n}\right)^4 (24) \le 0.0001 \Rightarrow \left(\frac{768}{180}\right)\left(\frac{1}{n^4}\right) \le 0.0001 \Rightarrow \frac{1}{n^4} \le (0.0001)\left(\frac{180}{768}\right) \Rightarrow n^4 \ge 10{,}000\left(\frac{768}{180}\right)$
$\Rightarrow n \ge 14.37 \Rightarrow n \ge 16$ (n must be even)

128. $|E_T| \le \frac{1-0}{12}(\Delta x)^2 M$ where $\Delta x = \frac{1-0}{n} = \frac{1}{n}$; $0 \le f''(x) \le 8 \Rightarrow M = 8$. Then $|E_T| \le 10^{-3} \Rightarrow \frac{1}{12}\left(\frac{1}{n}\right)^2 (8) \le 10^{-3}$
$\Rightarrow \frac{2}{3n^2} \le 10^{-3} \Rightarrow \frac{3n^2}{2} \ge 1000 \Rightarrow n^2 \ge \frac{2000}{3} \Rightarrow n \ge 25.82 \Rightarrow n \ge 26$

129. $\Delta x = \frac{b-a}{n} = \frac{\pi - 0}{6} = \frac{\pi}{6} \Rightarrow \frac{\Delta x}{2} = \frac{\pi}{12}$;
$\sum_{i=0}^{6} mf(x_i) = 12 \Rightarrow T = \left(\frac{\pi}{12}\right)(12) = \pi$;

	x_i	$f(x_i)$	m	$mf(x_i)$
x_0	0	0	1	0
x_1	$\pi/6$	1/2	2	1
x_2	$\pi/3$	3/2	2	3
x_3	$\pi/2$	2	2	4
x_4	$2\pi/3$	3/2	2	3
x_5	$5\pi/6$	1/2	2	1
x_6	π	0	1	0

$\sum_{i=0}^{6} mf(x_i) = 18$ and $\frac{\Delta x}{3} = \frac{\pi}{18} \Rightarrow$
$S = \left(\frac{\pi}{18}\right)(18) = \pi$.

	x_i	$f(x_i)$	m	$mf(x_i)$
x_0	0	0	1	0
x_1	$\pi/6$	1/2	4	2
x_2	$\pi/3$	3/2	2	3
x_3	$\pi/2$	2	4	8
x_4	$2\pi/3$	3/2	2	3
x_5	$5\pi/6$	1/2	4	2
x_6	π	0	1	0

130. $\left|f^{(4)}(x)\right| \le 3 \Rightarrow M = 3$; $\Delta x = \frac{2-1}{n} = \frac{1}{n}$. Hence $|E_s| \le 10^{-5} \Rightarrow \left(\frac{2-1}{180}\right)\left(\frac{1}{n}\right)^4 (3) \le 10^{-5} \Rightarrow \frac{1}{60n^4} \le 10^{-5} \Rightarrow n^4 \ge \frac{10^5}{60}$
$\Rightarrow n \ge 6.38 \Rightarrow n \ge 8$ (n must be even)

131. $y_{av} = \frac{1}{365-0}\int_0^{365}\left[37\sin\left(\frac{2\pi}{365}(x - 101)\right) + 25\right] dx = \frac{1}{365}\left[-37\left(\frac{365}{2\pi}\cos\left(\frac{2\pi}{365}(x - 101)\right) + 25x\right)\right]_0^{365}$
$= \frac{1}{365}\left[\left(-37\left(\frac{365}{2\pi}\right)\cos\left[\frac{2\pi}{365}(365 - 101)\right] + 25(365)\right) - \left(-37\left(\frac{365}{2\pi}\right)\cos\left[\frac{2\pi}{365}(0 - 101)\right] + 25(0)\right)\right]$
$= -\frac{37}{2\pi}\cos\left(\frac{2\pi}{365}(264)\right) + 25 + \frac{37}{2\pi}\cos\left(\frac{2\pi}{365}(-101)\right) = -\frac{37}{2\pi}\left(\cos\left(\frac{2\pi}{365}(264)\right) - \cos\left(\frac{2\pi}{365}(-101)\right)\right) + 25$

$\approx -\frac{37}{2\pi}(0.16705 - 0.16705) + 25 = 25° \text{F}$

132. $\text{av}(C_v) = \frac{1}{675-20}\int_{20}^{675}[8.27 + 10^{-5}(26T - 1.87T^2)]\,dT = \frac{1}{655}\left[8.27T + \frac{13}{10^5}T^2 - \frac{0.62333}{10^5}T^3\right]_{20}^{675}$

$\approx \frac{1}{655}[(5582.25 + 59.23125 - 1917.03194) - (165.4 + 0.052 - 0.04987)] \approx 5.434;$

$8.27 + 10^{-5}(26T - 1.87T^2) = 5.434 \Rightarrow 1.87T^2 - 26T - 283{,}600 = 0 \Rightarrow T \approx \frac{26+\sqrt{676+4(1.87)(283{,}600)}}{2(1.87)}$

$\approx 396.45° \text{C}$

133. (a) Each interval is 5 min $= \frac{1}{12}$ hour.

$\frac{1}{24}[2.5 + 2(2.4) + 2(2.3) + \ldots + 2(2.4) + 2.3] = \frac{29}{12} \approx 2.42$ gal

(b) $(60 \text{ mph})\left(\frac{12}{29} \text{ hours/gal}\right) \approx 24.83$ mi/gal

134. Using the Simpson's rule, $\triangle x = 15 \Rightarrow \frac{\triangle x}{3} = 5$;

$\sum mf(x_i) = 1211.8 \Rightarrow \text{Area} \approx (1211.8)(5) = 6059 \text{ ft}^2$;

The cost is Area $\cdot$ ($\$2.10/\text{ft}^2$) $\approx (6059 \text{ ft}^2)(\$2.10/\text{ft}^2)$

$= \$12{,}723.90 \Rightarrow$ the job cannot be done for \$11,000.

	x_i	$f(x_i)$	m	$mf(x_i)$
x_0	0	0	1	0
x_1	15	36	4	144
x_2	30	54	2	108
x_3	45	51	4	204
x_4	60	49.5	2	99
x_5	75	54	4	216
x_6	90	64.4	2	128.8
x_7	105	67.5	4	270
x_8	120	42	1	42

135. $\int_0^3 \frac{dx}{\sqrt{9-x^2}} = \lim_{b\to 3^-}\int_0^b \frac{dx}{\sqrt{9-x^2}} = \lim_{b\to 3^-}\left[\sin^{-1}\left(\frac{x}{3}\right)\right]_0^b = \lim_{b\to 3^-}\sin^{-1}\left(\frac{b}{3}\right) - \sin^{-1}\left(\frac{0}{3}\right) = \frac{\pi}{2} - 0 = \frac{\pi}{2}$

136. $\int_0^1 \ln x\,dx = \lim_{b\to 0^+}[x\ln x - x]_b^1 = (1\cdot\ln 1 - 1) - \lim_{b\to 0^+}[b\ln b - b] = -1 - \lim_{b\to 0^+}\frac{\ln b}{\left(\frac{1}{b}\right)} = -1 - \lim_{b\to 0^+}\frac{\left(\frac{1}{b}\right)}{\left(-\frac{1}{b^2}\right)}$

$= -1 + 0 = -1$

137. $\int_{-1}^1 \frac{dy}{y^{2/3}} = \int_{-1}^0 \frac{dy}{y^{2/3}} + \int_0^1 \frac{dy}{y^{2/3}} = 2\int_0^1 \frac{dy}{y^{2/3}} = 2\cdot 3\lim_{b\to 0^+}\left[y^{1/3}\right]_b^1 = 6\left(1 - \lim_{b\to 0^+} b^{1/3}\right) = 6$

138. $\int_{-2}^\infty \frac{d\theta}{(\theta+1)^{3/5}} = \int_{-2}^{-1}\frac{d\theta}{(\theta+1)^{3/5}} + \int_{-1}^2 \frac{d\theta}{(\theta+1)^{3/5}} + \int_2^\infty \frac{d\theta}{(\theta+1)^{3/5}}$ converges if each integral converges, but

$\lim_{\theta\to\infty}\frac{\theta^{3/5}}{(\theta+1)^{3/5}} = 1$ and $\int_2^\infty \frac{d\theta}{\theta^{3/5}}$ diverges $\Rightarrow \int_{-2}^\infty \frac{d\theta}{(\theta+1)^{3/5}}$ diverges

139. $\int_3^\infty \frac{2\,du}{u^2-2u} = \int_3^\infty \frac{du}{u-2} - \int_3^\infty \frac{du}{u} = \lim_{b\to\infty}\left[\ln\left|\frac{u-2}{u}\right|\right]_3^b = \lim_{b\to\infty}\left[\ln\left|\frac{b-2}{b}\right|\right] - \ln\left|\frac{3-2}{3}\right| = 0 - \ln\left(\frac{1}{3}\right) = \ln 3$

140. $\int_1^\infty \frac{3v-1}{4v^3-v^2}\,dv = \int_1^\infty\left(\frac{1}{v} + \frac{1}{v^2} - \frac{4}{4v-1}\right)dv = \lim_{b\to\infty}\left[\ln v - \frac{1}{v} - \ln(4v-1)\right]_1^b$

$= \lim_{b\to\infty}\left[\ln\left(\frac{b}{4b-1}\right) - \frac{1}{b}\right] - (\ln 1 - 1 - \ln 3) = \ln\frac{1}{4} + 1 + \ln 3 = 1 + \ln\frac{3}{4}$

141. $\int_0^\infty x^2e^{-x}\,dx = \lim_{b\to\infty}\left[-x^2e^{-x} - 2xe^{-x} - 2e^{-x}\right]_0^b = \lim_{b\to\infty}(-b^2e^{-b} - 2be^{-b} - 2e^{-b}) - (-2) = 0 + 2 = 2$

142. $\int_{-\infty}^0 xe^{3x}\,dx = \lim_{b\to-\infty}\left[\frac{x}{3}e^{3x} - \frac{1}{9}e^{3x}\right]_b^0 = -\frac{1}{9} - \lim_{b\to-\infty}\left(\frac{b}{3}e^{3b} - \frac{1}{9}e^{3b}\right) = -\frac{1}{9} - 0 = -\frac{1}{9}$

143. $\int_{-\infty}^\infty \frac{dx}{4x^2+9} = 2\int_0^\infty \frac{dx}{4x^2+9} = \frac{1}{2}\int_0^\infty \frac{dx}{x^2+\frac{9}{4}} = \frac{1}{2}\lim_{b\to\infty}\left[\frac{2}{3}\tan^{-1}\left(\frac{2x}{3}\right)\right]_0^b = \frac{1}{2}\lim_{b\to\infty}\left[\frac{2}{3}\tan^{-1}\left(\frac{2b}{3}\right)\right] - \frac{1}{3}\tan^{-1}(0)$

$= \frac{1}{2}\left(\frac{2}{3}\cdot\frac{\pi}{2}\right) - 0 = \frac{\pi}{6}$

144. $\int_{-\infty}^{\infty} \frac{4\,dx}{x^2+16} = 2\int_0^{\infty} \frac{4\,dx}{x^2+16} = 2 \lim_{b\to\infty} \left[\tan^{-1}\left(\frac{x}{4}\right)\right]_0^b = 2\left(\lim_{b\to\infty} \left[\tan^{-1}\left(\frac{b}{4}\right)\right] - \tan^{-1}(0)\right) = 2\left(\frac{\pi}{2}\right) - 0 = \pi$

145. $\lim_{\theta\to\infty} \frac{\theta}{\sqrt{\theta^2+1}} = 1$ and $\int_6^{\infty} \frac{d\theta}{\theta}$ diverges $\Rightarrow \int_6^{\infty} \frac{d\theta}{\sqrt{\theta^2+1}}$ diverges

146. $I = \int_0^{\infty} e^{-u}\cos u\,du = \lim_{b\to\infty} \left[-e^{-u}\cos u\right]_0^b - \int_0^{\infty} e^{-u}\sin u\,du = 1 + \lim_{b\to\infty} \left[e^{-u}\sin u\right]_0^b - \int_0^{\infty} (e^{-u})\cos u\,du$

$\Rightarrow I = 1 + 0 - I \Rightarrow 2I = 1 \Rightarrow I = \frac{1}{2}$ converges

147. $\int_1^{\infty} \frac{\ln z}{z}\,dz = \int_1^{e} \frac{\ln z}{z}\,dz + \int_e^{\infty} \frac{\ln z}{z}\,dz = \left[\frac{(\ln z)^2}{2}\right]_1^e + \lim_{b\to\infty} \left[\frac{(\ln z)^2}{2}\right]_e^b = \left(\frac{1^2}{2} - 0\right) + \lim_{b\to\infty} \left[\frac{(\ln b)^2}{2} - \frac{1}{2}\right]$

$= \infty \Rightarrow$ diverges

148. $0 < \frac{e^{-t}}{\sqrt{t}} \le e^{-t}$ for $t \ge 1$ and $\int_1^{\infty} e^{-t}\,dt$ converges $\Rightarrow \int_1^{\infty} \frac{e^{-t}}{\sqrt{t}}\,dt$ converges

149. $\int_{-\infty}^{\infty} \frac{2\,dx}{e^x+e^{-x}} = 2\int_0^{\infty} \frac{2\,dx}{e^x+e^{-x}} < \int_0^{\infty} \frac{4\,dx}{e^x}$ converges $\Rightarrow \int_{-\infty}^{\infty} \frac{2\,dx}{e^x+e^{-x}}$ converges

150. $\int_{-\infty}^{\infty} \frac{dx}{x^2(1+e^x)} = \int_{-\infty}^{-1} \frac{dx}{x^2(1+e^x)} + \int_{-1}^{0} \frac{dx}{x^2(1+e^x)} + \int_0^{1} \frac{dx}{x^2(1+e^x)} + \int_1^{\infty} \frac{dx}{x^2(1+e^x)}$;

$\lim_{x\to 0} \frac{\left(\frac{1}{x^2}\right)}{\left[\frac{1}{x^2(1+e^x)}\right]} = \lim_{x\to 0} \frac{x^2(1+e^x)}{x^2} = \lim_{x\to 0} (1+e^x) = 2$ and $\int_0^1 \frac{dx}{x^2}$ diverges $\Rightarrow \int_0^1 \frac{dx}{x^2(1+e^x)}$ diverges

$\Rightarrow \int_{-\infty}^{\infty} \frac{dx}{x^2(1+e^x)}$ diverges

151. $\int \frac{x\,dx}{1+\sqrt{x}}$; $\begin{bmatrix} u = \sqrt{x} \\ du = \frac{dx}{2\sqrt{x}} \end{bmatrix} \to \int \frac{u^2\cdot 2u\,du}{1+u} = \int \left(2u^2 - 2u + 2 - \frac{2}{1+u}\right) du = \frac{2}{3}u^3 - u^2 + 2u - 2\ln|1+u| + C$

$= \frac{2x^{3/2}}{3} - x + 2\sqrt{x} - 2\ln\left(1+\sqrt{x}\right) + C$

152. $\int \frac{x^3+2}{4-x^2}\,dx = -\int \left(x + \frac{4x+2}{x^2-4}\right) dx = -\int x\,dx - \frac{3}{2}\int \frac{dx}{x+2} - \frac{5}{2}\int \frac{dx}{x-2} = -\frac{x^2}{2} - \frac{3}{2}\ln|x+2| - \frac{5}{2}\ln|x-2| + C$

153. $\int \frac{dx}{x(x^2+1)^2}$; $\begin{bmatrix} x = \tan\theta \\ dx = \sec^2\theta\,d\theta \end{bmatrix} \to \int \frac{\sec^2\theta\,d\theta}{\tan\theta\sec^4\theta} = \int \frac{\cos^3\theta\,d\theta}{\sin\theta} = \int \left(\frac{1-\sin^2\theta}{\sin\theta}\right) d(\sin\theta)$

$= \ln|\sin\theta| - \frac{1}{2}\sin^2\theta + C = \ln\left|\frac{x}{\sqrt{x^2+1}}\right| - \frac{1}{2}\left(\frac{x}{\sqrt{x^2+1}}\right)^2 + C$

154. $\int \frac{\cos\sqrt{x}}{\sqrt{x}}\,dx$; $\begin{bmatrix} u = \sqrt{x} \\ du = \frac{dx}{2\sqrt{x}} \end{bmatrix} \to \int \frac{\cos u\cdot 2u\,du}{u} = 2\int \cos u\,du = 2\sin u + C = 2\sin\sqrt{x} + C$

155. $\int \frac{dx}{\sqrt{-2x-x^2}} = \int \frac{d(x+1)}{\sqrt{1-(x+1)^2}} = \sin^{-1}(x+1) + C$

156. $\int \frac{(t-1)\,dt}{\sqrt{t^2-2t}}$; $\begin{bmatrix} u = t^2 - 2t \\ du = (2t-2)\,dt = 2(t-1)\,dt \end{bmatrix} \to \frac{1}{2}\int \frac{du}{\sqrt{u}} = \sqrt{u} + C = \sqrt{t^2-2t} + C$

157. $\int \frac{du}{\sqrt{1+u^2}}$; $[u = \tan\theta] \to \int \frac{\sec^2\theta\,d\theta}{\sec\theta} = \ln|\sec\theta + \tan\theta| + C = \ln\left|\sqrt{1+u^2} + u\right| + C$

158. $\int e^t \cos e^t \, dt = \sin e^t + C$

159. $\int \frac{2 - \cos x + \sin x}{\sin^2 x} \, dx = \int 2 \csc^2 x \, dx - \int \frac{\cos x \, dx}{\sin^2 x} + \int \csc x \, dx = -2 \cot x + \frac{1}{\sin x} - \ln |\csc x + \cot x| + C$
$= -2 \cot x + \csc x - \ln |\csc x + \cot x| + C$

160. $\int \frac{\sin^2 \theta}{\cos^2 \theta} \, d\theta = \int \frac{1 - \cos^2 \theta}{\cos^2 \theta} \, d\theta = \int \sec^2 \theta \, d\theta - \int d\theta = \tan \theta - \theta + C$

161. $\int \frac{9 \, dv}{81 - v^4} = \frac{1}{2} \int \frac{dv}{v^2 + 9} + \frac{1}{12} \int \frac{dv}{3 - v} + \frac{1}{12} \int \frac{dv}{3 + v} = \frac{1}{12} \ln \left| \frac{3 + v}{3 - v} \right| + \frac{1}{6} \tan^{-1} \frac{v}{3} + C$

162. $\int \frac{\cos x \, dx}{1 + \sin^2 x} = \int \frac{d(\sin x)}{1 + \sin^2 x} = \tan^{-1} (\sin x) + C$

163.

		$\cos (2\theta + 1)$
θ	$\xrightarrow{(+)}$	$\frac{1}{2} \sin (2\theta + 1)$
1	$\xrightarrow{(-)}$	$-\frac{1}{4} \cos (2\theta + 1)$
0		

$\Rightarrow \int \theta \cos (2\theta + 1) \, d\theta = \frac{\theta}{2} \sin (2\theta + 1) + \frac{1}{4} \cos (2\theta + 1) + C$

164. $\int_2^{\infty} \frac{dx}{(x - 1)^2} = \lim_{b \to \infty} \left[\frac{1}{1 - x} \right]_2^b = \lim_{b \to \infty} \left[\frac{1}{1 - b} - (-1) \right] = 0 + 1 = 1$

165. $\int \frac{x^3 \, dx}{x^2 - 2x + 1} = \int \left(x + 2 + \frac{3x - 2}{x^2 - 2x + 1} \right) dx = \int (x + 2) \, dx + 3 \int \frac{dx}{x - 1} + \int \frac{dx}{(x - 1)^2}$
$= \frac{x^2}{2} + 2x + 3 \ln |x - 1| - \frac{1}{x - 1} + C$

166. $\int \frac{d\theta}{\sqrt{1 + \sqrt{\theta}}}; \begin{bmatrix} x = 1 + \sqrt{\theta} \\ dx = \frac{d\theta}{2\sqrt{\theta}} \\ d\theta = 2(x - 1) \, dx \end{bmatrix} \to \int \frac{2(x - 1) \, dx}{\sqrt{x}} = 2 \int \sqrt{x} \, dx - 2 \int \frac{dx}{\sqrt{x}} = \frac{4}{3} x^{3/2} - 4x^{1/2} + C$
$= \frac{4}{3} \left(1 + \sqrt{\theta} \right)^{3/2} - 4 \left(1 + \sqrt{\theta} \right)^{1/2} + C = 4 \left[\frac{\left(\sqrt{1 + \sqrt{\theta}} \right)^3}{3} - \sqrt{1 + \sqrt{\theta}} \right] + C$

167. $\int \frac{2 \sin \sqrt{x} \, dx}{\sqrt{x} \sec \sqrt{x}}; \begin{bmatrix} y = \sqrt{x} \\ dy = \frac{dx}{2\sqrt{x}} \end{bmatrix} \to \int \frac{2 \sin y \cdot 2y \, dy}{y \sec y} = \int 2 \sin 2y \, dy = -\cos (2y) + C = -\cos \left(2\sqrt{x} \right) + C$

168. $\int \frac{x^5 \, dx}{x^4 - 16} = \int \left(x + \frac{16x}{x^4 - 16} \right) dx = \frac{x^2}{2} + \int \left(\frac{2x}{x^2 - 4} - \frac{2x}{x^2 + 4} \right) dx = \frac{x^2}{2} + \ln \left| \frac{x^2 - 4}{x^2 + 4} \right| + C$

169. $\int \frac{dy}{\sin y \cos y} = \int \frac{2 \, dy}{\sin 2y} = \int 2 \csc (2y) \, dy = -\ln |\csc (2y) + \cot (2y)| + C$

170. $\int \frac{d\theta}{\theta^2 - 2\theta + 4} = \int \frac{d\theta}{(\theta - 1)^2 + 3} = \frac{\sqrt{3}}{3} \tan^{-1} \left(\frac{\theta - 1}{\sqrt{3}} \right) + C$

171. $\int \frac{\tan x}{\cos^2 x} \, dx = \int \tan x \sec^2 x \, dx = \int \tan x \cdot d(\tan x) = \frac{1}{2} \tan^2 x + C$

172. $\int \frac{dr}{(r + 1)\sqrt{r^2 + 2r}} = \int \frac{d(r + 1)}{(r + 1)\sqrt{(r + 1)^2 - 1}} = \sec^{-1} |r + 1| + C$

173. $\int \frac{(r+2)\,dr}{\sqrt{-r^2-4r}} = \int \frac{(r+2)\,dr}{\sqrt{4-(r+2)^2}}$; $\begin{bmatrix} u = 4-(r+2)^2 \\ du = -2(r+2)\,dr \end{bmatrix} \to -\int \frac{du}{2\sqrt{u}} = -\sqrt{u} + C = -\sqrt{4-(r+2)^2} + C$

174. $\int \frac{y\,dy}{4+y^4} = \frac{1}{2}\int \frac{d(y^2)}{4+(y^2)^2} = \frac{1}{4}\tan^{-1}\left(\frac{y^2}{2}\right) + C$

175. $\int \frac{\sin 2\theta\,d\theta}{(1+\cos 2\theta)^2} = -\frac{1}{2}\int \frac{d(1+\cos 2\theta)}{(1+\cos 2\theta)^2} = \frac{1}{2(1+\cos 2\theta)} + C = \frac{1}{4}\sec^2\theta + C$

176. $\int \frac{dx}{(x^2-1)^2} = \int \frac{dx}{(1-x^2)^2} = \frac{x}{2(1-x^2)} + \frac{1}{4}\ln\left|\frac{x+1}{x-1}\right| + C$ (FORMULA 19)

177. $\int_{\pi/4}^{\pi/2} \sqrt{1+\cos 4x}\,dx = -\sqrt{2}\int_{\pi/4}^{\pi/2} \cos 2x\,dx = \left[-\frac{\sqrt{2}}{2}\sin 2x\right]_{\pi/4}^{\pi/2} = \frac{\sqrt{2}}{2}$

178. $\int (15)^{2x+1}\,dx = \frac{1}{2}\int (15)^{2x+1}\,d(2x+1) = \frac{1}{2}\left(\frac{15^{2x+1}}{\ln 15}\right) + C$

179. $\int \frac{x\,dx}{\sqrt{2-x}}$; $\begin{bmatrix} y = 2-x \\ dy = -dx \end{bmatrix} \to -\int \frac{(2-y)\,dy}{\sqrt{y}} = \frac{2}{3}y^{3/2} - 4y^{1/2} + C = \frac{2}{3}(2-x)^{3/2} - 4(2-x)^{1/2} + C$
$= 2\left[\frac{\left(\sqrt{2-x}\right)^3}{3} - 2\sqrt{2-x}\right] + C$

180. $\int \frac{\sqrt{1-v^2}}{v^2}\,dv$; $[v = \sin\theta] \to \int \frac{\cos\theta \cdot \cos\theta\,d\theta}{\sin^2\theta} = \int \frac{(1-\sin^2\theta)\,d\theta}{\sin^2\theta} = \int \csc^2\theta\,d\theta - \int d\theta = \cot\theta - \theta + C$
$= -\sin^{-1} v - \frac{\sqrt{1-v^2}}{v} + C$

181. $\int \frac{dy}{y^2-2y+2} = \int \frac{d(y-1)}{(y-1)^2+1} = \tan^{-1}(y-1) + C$

182. $\int \ln\sqrt{x-1}\,dx$; $\begin{bmatrix} y = \sqrt{x-1} \\ dy = \frac{dx}{2\sqrt{x-1}} \end{bmatrix} \to \int \ln y \cdot 2y\,dy$; $u = \ln y$, $du = \frac{dy}{y}$; $dv = 2y\,dy$, $v = y^2$
$\Rightarrow \int 2y\ln y\,dy = y^2\ln y - \int y\,dy = y^2\ln y - \frac{1}{2}y^2 + C = (x-1)\ln\sqrt{x-1} - \frac{1}{2}(x-1) + C_1$
$= \frac{1}{2}[(x-1)\ln|x-1| - x] + \left(C_1 + \frac{1}{2}\right) = \frac{1}{2}[x\ln|x-1| - x - \ln|x-1|] + C$

183. $\int \theta^2\tan(\theta^3)\,d\theta = \frac{1}{3}\int \tan(\theta^3)\,d(\theta^3) = \frac{1}{3}\ln|\sec\theta^3| + C$

184. $\int \frac{x\,dx}{\sqrt{8-2x^2-x^4}} = \frac{1}{2}\int \frac{d(x^2+1)}{\sqrt{9-(x^2+1)^2}} = \frac{1}{2}\sin^{-1}\left(\frac{x^2+1}{3}\right) + C$

185. $\int \frac{z+1}{z^2(z^2+4)}\,dz = \frac{1}{4}\int \left(\frac{1}{z} + \frac{1}{z^2} - \frac{z+1}{z^2+4}\right) dz = \frac{1}{4}\ln|z| - \frac{1}{4z} - \frac{1}{8}\ln(z^2+4) - \frac{1}{8}\tan^{-1}\frac{z}{2} + C$

186. $\int x^3 e^{x^2}\,dx = \frac{1}{2}\int x^2 e^{x^2}\,d(x^2) = \frac{1}{2}\left(x^2 e^{x^2} - e^{x^2}\right) + C = \frac{(x^2-1)e^{x^2}}{2} + C$

187. $\int \frac{t\,dt}{\sqrt{9-4t^2}} = -\frac{1}{8}\int \frac{d(9-4t^2)}{\sqrt{9-4t^2}} = -\frac{1}{4}\sqrt{9-4t^2} + C$

188. $\int_0^{\pi/10} \sqrt{1+\cos 5\theta}\,d\theta = \sqrt{2}\int_0^{\pi/10} \cos\left(\frac{5\theta}{2}\right) d\theta = \frac{2\sqrt{2}}{5}\left[\sin\left(\frac{5\theta}{2}\right)\right]_0^{\pi/10} = \frac{2\sqrt{2}}{5}\left(\sin\frac{\pi}{4} - 0\right) = \frac{2}{5}$

189. $\int \frac{\cot\theta\, d\theta}{1+\sin^2\theta} = \int \frac{\cos\theta\, d\theta}{(\sin\theta)(1+\sin^2\theta)}$; $\begin{bmatrix} x = \sin\theta \\ dx = \cos\theta\, d\theta \end{bmatrix} \to \int \frac{dx}{x(1+x^2)} = \int \frac{dx}{x} - \int \frac{x\, dx}{x^2+1}$
$= \ln|\sin\theta| - \frac{1}{2}\ln(1+\sin^2\theta) + C$

190. $u = \tan^{-1} x,\ du = \frac{dx}{1+x^2}$; $dv = \frac{dx}{x^2},\ v = -\frac{1}{x}$;
$\int \frac{\tan^{-1} x\, dx}{x^2} = -\frac{1}{x}\tan^{-1} x + \int \frac{dx}{x(1+x^2)} = -\frac{1}{x}\tan^{-1} x + \int \frac{dx}{x} - \int \frac{x\, dx}{1+x^2}$
$= -\frac{1}{x}\tan^{-1} x + \ln|x| - \frac{1}{2}\ln(1+x^2) + C = -\frac{\tan^{-1} x}{x} + \ln|x| - \ln\sqrt{1+x^2} + C$

191. $\int \frac{\tan\sqrt{y}\, dy}{2\sqrt{y}}$; $[\sqrt{y} = x] \to \int \frac{\tan x \cdot 2x\, dx}{2x} = \ln|\sec x| + C = \ln|\sec\sqrt{y}| + C$

192. $\int \frac{e^t\, dt}{e^{2t}+3e^t+2}$; $[e^t = x] \to \int \frac{dx}{(x+1)(x+2)} = \int \frac{dx}{x+1} - \int \frac{dx}{x+2} = \ln|x+1| - \ln|x+2| + C$
$= \ln\left|\frac{x+1}{x+2}\right| + C = \ln\left(\frac{e^t+1}{e^t+2}\right) + C$

193. $\int \frac{\theta^2\, d\theta}{4-\theta^2} = \int \left(-1 + \frac{4}{4-\theta^2}\right) d\theta = -\int d\theta - \int \frac{d\theta}{\theta-2} + \int \frac{d\theta}{\theta+2} = -\theta - \ln|\theta-2| + \ln|\theta+2| + C$
$= -\theta + \ln\left|\frac{\theta+2}{\theta-2}\right| + C$

194. $\int \frac{1-\cos 2x}{1+\cos 2x}\, dx = \int \tan^2 x\, dx = \int (\sec^2 x - 1)\, dx = \tan x - x + C$

195. $\int \frac{\cos(\sin^{-1} x)\, dx}{\sqrt{1-x^2}}$; $\begin{bmatrix} u = \sin^{-1} x \\ du = \frac{dx}{\sqrt{1-x^2}} \end{bmatrix} \to \int \cos u\, du = \sin u + C = \sin(\sin^{-1} x) + C = x + C$

196. $\int \frac{\cos x\, dx}{\sin^3 x - \sin x} = -\int \frac{\cos x\, dx}{(\sin x)(1-\sin^2 x)} = -\int \frac{\cos x\, dx}{(\sin x)(\cos^2 x)} = -\int \frac{2\, dx}{\sin 2x} = -2\int \csc 2x\, dx$
$= \ln|\csc(2x) + \cot(2x)| + C$

197. $\int \sin\frac{x}{2}\cos\frac{x}{2}\, dx = \int \frac{1}{2}\sin\left(\frac{x}{2}+\frac{x}{2}\right) dx = \frac{1}{2}\int \sin x\, dx = -\frac{1}{2}\cos x + C$

198. $\int \frac{x^2-x+2}{(x^2+2)^2}\, dx = \int \frac{dx}{x^2+2} - \int \frac{x\, dx}{(x^2+2)^2} = \frac{1}{\sqrt{2}}\tan^{-1}\left(\frac{x}{\sqrt{2}}\right) + \frac{1}{2}(x^2+2)^{-1} + C$
$= \frac{1}{\sqrt{2}}\tan^{-1}\left(\frac{x}{\sqrt{2}}\right) + \frac{1}{2(x^2+2)} + C$

199. $\int \frac{e^t\, dt}{1+e^t} = \ln(1+e^t) + C$

200. $\int \tan^3 t\, dt = \int (\tan t)(\sec^2 t - 1)\, dt = \frac{\tan^2 t}{2} - \int \tan t\, dt = \frac{\tan^2 t}{2} - \ln|\sec t| + C$

201. $\int_1^\infty \frac{\ln y\, dy}{y^3}$; $\begin{bmatrix} x = \ln y \\ dx = \frac{dy}{y} \\ dy = e^x\, dx \end{bmatrix} \to \int_0^\infty \frac{x \cdot e^x}{e^{3x}}\, dx = \int_0^\infty x e^{-2x}\, dx = \lim_{b\to\infty} \left[-\frac{x}{2}e^{-2x} - \frac{1}{4}e^{-2x}\right]_0^b$
$= \lim_{b\to\infty} \left(\frac{-b}{2e^{2b}} - \frac{1}{4e^{2b}}\right) - \left(0 - \frac{1}{4}\right) = \frac{1}{4}$

202. $\int \frac{3+\sec^2 x+\sin x}{\tan x}\, dx = 3\int \cot x\, dx + \int \frac{\sec^2 x\, dx}{\tan x} + \int \cos x\, dx = 3\ln|\sin x| + \ln|\tan x| + \sin x + C$

203. $\int \frac{\cot v\, dv}{\ln(\sin v)} = \int \frac{\cos v\, dv}{(\sin v)\ln(\sin v)}$; $\begin{bmatrix} u = \ln(\sin v) \\ du = \frac{\cos v\, dv}{\sin v} \end{bmatrix} \to \int \frac{du}{u} = \ln|u| + C = \ln|\ln(\sin v)| + C$

204. $\int \frac{dx}{(2x-1)\sqrt{x^2-x}} = \int \frac{2\,dx}{(2x-1)\sqrt{4x^2-4x}} = \int \frac{2\,dx}{(2x-1)\sqrt{(2x-1)^2-1}}$; $\begin{bmatrix} u = 2x-1 \\ du = 2\,dx \end{bmatrix} \to \int \frac{du}{u\sqrt{u^2-1}}$
$= \sec^{-1}|u| + C = \sec^{-1}|2x-1| + C$

205. $\int e^{\ln\sqrt{x}}\,dx = \int \sqrt{x}\,dx = \frac{2}{3}x^{3/2} + C$

206. $\int e^{\theta}\sqrt{3+4e^{\theta}}\,d\theta$; $\begin{bmatrix} u = 4e^{\theta} \\ du = 4e^{\theta}\,d\theta \end{bmatrix} \to \frac{1}{4}\int\sqrt{3+u}\,du = \frac{1}{4}\cdot\frac{2}{3}(3+u)^{3/2} + C = \frac{1}{6}(3+4e^{\theta})^{3/2} + C$

207. $\int \frac{\sin 5t\,dt}{1+(\cos 5t)^2}$; $\begin{bmatrix} u = \cos 5t \\ du = -5\sin 5t\,dt \end{bmatrix} \to -\frac{1}{5}\int\frac{du}{1+u^2} = -\frac{1}{5}\tan^{-1}u + C = -\frac{1}{5}\tan^{-1}(\cos 5t) + C$

208. $\int \frac{dv}{\sqrt{e^{2v}-1}}$; $\begin{bmatrix} x = e^{v} \\ dx = e^{v}\,dv \end{bmatrix} \to \int \frac{dx}{x\sqrt{x^2-1}} = \sec^{-1}x + C = \sec^{-1}(e^{v}) + C$

209. $\int (27)^{3\theta+1}\,d\theta = \frac{1}{3}\int (27)^{3\theta+1}\,d(3\theta+1) = \frac{1}{3\ln 27}(27)^{3\theta+1} + C = \frac{1}{3}\left(\frac{27^{3\theta+1}}{\ln 27}\right) + C$

210.

		$\sin x$
x^5	$\xrightarrow{(+)}$	$-\cos x$
$5x^4$	$\xrightarrow{(-)}$	$-\sin x$
$20x^3$	$\xrightarrow{(+)}$	$\cos x$
$60x^2$	$\xrightarrow{(-)}$	$\sin x$
$120x$	$\xrightarrow{(+)}$	$-\cos x$
120	$\xrightarrow{(-)}$	$-\sin x$
0		

$\Rightarrow \int x^5\sin x\,dx = -x^5\cos x + 5x^4\sin x + 20x^3\cos x - 60x^2\sin x - 120x\cos x$
$+ 120\sin x + C$

211. $\int \frac{dr}{1+\sqrt{r}}$; $\begin{bmatrix} u = \sqrt{r} \\ du = \frac{dr}{2\sqrt{r}} \end{bmatrix} \to \int \frac{2u\,du}{1+u} = \int\left(2 - \frac{2}{1+u}\right)du = 2u - 2\ln|1+u| + C = 2\sqrt{r} - 2\ln\left(1+\sqrt{r}\right) + C$

212. $\int \frac{4x^3-20x}{x^4-10x^2+9}\,dx = \int \frac{d(x^4-10x^2+9)}{x^4-10x^2+9} = \ln|x^4-10x^2+9| + C$

213. $\int \frac{8\,dy}{y^3(y+2)} = \int\frac{dy}{y} - \int\frac{2\,dy}{y^2} + \int\frac{4\,dy}{y^3} - \int\frac{dy}{(y+2)} = \ln\left|\frac{y}{y+2}\right| + \frac{2}{y} - \frac{2}{y^2} + C$

214. $\int \frac{(t+1)\,dt}{(t^2+2t)^{2/3}}$; $\begin{bmatrix} u = t^2+2t \\ du = 2(t+1)\,dt \end{bmatrix} \to \frac{1}{2}\int\frac{du}{u^{2/3}} = \frac{1}{2}\cdot 3u^{1/3} + C = \frac{3}{2}(t^2+2t)^{1/3} + C$

215. $\int \frac{8\,dm}{m\sqrt{49m^2-4}} = \frac{8}{7}\int\frac{dm}{m\sqrt{m^2-\left(\frac{2}{7}\right)^2}} = 4\sec^{-1}\left|\frac{7m}{2}\right| + C$

216. $\int \frac{dt}{t(1+\ln t)\sqrt{(\ln t)(2+\ln t)}}$; $\begin{bmatrix} u = \ln t \\ du = \frac{dt}{t} \end{bmatrix} \to \int\frac{du}{(1+u)\sqrt{u(2+u)}} = \int\frac{du}{(u+1)\sqrt{(u+1)^2-1}}$
$= \sec^{-1}|u+1| + C = \sec^{-1}|\ln t + 1| + C$

217. If $u = \int_0^x \sqrt{1+(t-1)^4}\,dt$ and $dv = 3(x-1)^2\,dx$, then $du = \sqrt{1+(x-1)^4}\,dx$, and $v = (x-1)^3$ so integration by parts $\Rightarrow \int_0^1 3(x-1)^2 \left[\int_0^x \sqrt{1+(t-1)^4}\,dt\right] dx = \left[(x-1)^3 \int_0^x \sqrt{1+(t-1)^4}\,dt\right]_0^1$ $- \int_0^1 (x-1)^3 \sqrt{1+(x-1)^4}\,dx = \left[-\frac{1}{6}\left(1+(x-1)^4\right)^{3/2}\right]_0^1 = \frac{\sqrt{8}-1}{6}$

218. $\frac{4v^3+v-1}{v^2(v-1)(v^2+1)} = \frac{A}{v} + \frac{B}{v^2} + \frac{C}{v-1} + \frac{Dv+E}{v^2+1} \Rightarrow 4v^3 + v - 1$

$= Av(v-1)(v^2+1) + B(v-1)(v^2+1) + Cv^2(v^2+1) + (Dv+E)(v^2)(v-1)$

$v = 0$: $-1 = -B \Rightarrow B = 1$;

$v = 1$: $4 = 2C \Rightarrow C = 2$;

coefficient of v^4: $0 = A + C + D \Rightarrow A + D = -2$;

coefficient of v^3: $4 = -A + B + E - D$

coefficient of v^2: $0 = A - B + C - E \Rightarrow C - D = 4 \Rightarrow D = -2$ (summing with previous equation);

coefficient of v: $1 = -A + B \Rightarrow A = 0$;

in summary: $A = 0$, $B = 1$, $C = 2$, $D = -2$ and $E = 1$

$\Rightarrow \int_2^\infty \frac{4v^3+v-1}{v^2(v-1)(v^2+1)}\,dv = \lim_{b\to\infty} \int_2^b \left(\frac{2}{v-1} + v^{-2} + \frac{1}{1+v^2} - \frac{2v}{1+v^2}\right) dv$

$= \lim_{b\to\infty} \left[\ln(v-1)^2 - \frac{1}{v} + \tan^{-1} v - \ln(1+v^2)\right]_2^b$

$= \lim_{b\to\infty} \left[\ln\left(\frac{(b-1)^2}{1+b^2}\right) - \frac{1}{b} + \tan^{-1} b\right] - \left(\ln 1 - \frac{1}{2} + \tan^{-1} 2 - \ln 5\right) = \left(0 - 0 + \frac{\pi}{2}\right) - \left(0 - \frac{1}{2} + \tan^{-1} 2 - \ln 5\right)$

$= \frac{\pi}{2} + \ln(5) + \frac{1}{2} - \tan^{-1} 2$

219. $u = f(x)$, $du = f'(x)\,dx$; $dv = dx$, $v = x$;

$\int_{\pi/2}^{3\pi/2} f(x)\,dx = [x\,f(x)]_{\pi/2}^{3\pi/2} - \int_{\pi/2}^{3\pi/2} xf'(x)\,dx = \left[\frac{3\pi}{2} f\left(\frac{3\pi}{2}\right) - \frac{\pi}{2} f\left(\frac{\pi}{2}\right)\right] - \int_{\pi/2}^{3\pi/2} \cos x\,dx$

$= \left(\frac{3\pi b}{2} - \frac{\pi a}{2}\right) - [\sin x]_{\pi/2}^{3\pi/2} = \frac{\pi}{2}(3b-a) - [(-1) - 1] = \frac{\pi}{2}(3b-a) + 2$

220. $\int_0^a \frac{dx}{1+x^2} = [\tan^{-1} x]_0^a = \tan^{-1} a$; $\int_a^\infty \frac{dx}{1+x^2} = \lim_{b\to\infty} [\tan^{-1} x]_a^b = \lim_{b\to\infty} (\tan^{-1} b - \tan^{-1} a) = \frac{\pi}{2} - \tan^{-1} a$; therefore, $\tan^{-1} a = \frac{\pi}{2} - \tan^{-1} a \Rightarrow \tan^{-1} a = \frac{\pi}{4} \Rightarrow a = 1$ since $a > 0$.

CHAPTER 8 ADDITIONAL AND ADVANCED EXERCISES

1. $u = (\sin^{-1} x)^2$, $du = \frac{2\sin^{-1} x\,dx}{\sqrt{1-x^2}}$; $dv = dx$, $v = x$;

$\int (\sin^{-1} x)^2\,dx = x(\sin^{-1} x)^2 - \int \frac{2x\sin^{-1} x\,dx}{\sqrt{1-x^2}}$;

$u = \sin^{-1} x$, $du = \frac{dx}{\sqrt{1-x^2}}$; $dv = -\frac{2x\,dx}{\sqrt{1-x^2}}$, $v = 2\sqrt{1-x^2}$;

$-\int \frac{2x\sin^{-1} x\,dx}{\sqrt{1-x^2}} = 2(\sin^{-1} x)\sqrt{1-x^2} - \int 2\,dx = 2(\sin^{-1} x)\sqrt{1-x^2} - 2x + C$; therefore

$\int (\sin^{-1} x)^2\,dx = x(\sin^{-1} x)^2 + 2(\sin^{-1} x)\sqrt{1-x^2} - 2x + C$

2. $\frac{1}{x} = \frac{1}{x}$,

$\frac{1}{x(x+1)} = \frac{1}{x} - \frac{1}{x+1}$,

$\frac{1}{x(x+1)(x+2)} = \frac{1}{2x} - \frac{1}{x+1} + \frac{1}{2(x+2)}$,

$\frac{1}{x(x+1)(x+2)(x+3)} = \frac{1}{6x} - \frac{1}{2(x+1)} + \frac{1}{2(x+2)} - \frac{1}{6(x+3)}$,

$\frac{1}{x(x+1)(x+2)(x+3)(x+4)} = \frac{1}{24x} - \frac{1}{6(x+1)} + \frac{1}{4(x+2)} - \frac{1}{6(x+3)} + \frac{1}{24(x+4)} \Rightarrow$ the following pattern:

$\frac{1}{x(x+1)(x+2)\cdots(x+m)} = \sum_{k=0}^{m} \frac{(-1)^k}{(k!)(m-k)!(x+k)}$; therefore $\int \frac{dx}{x(x+1)(x+2)\cdots(x+m)}$

$$= \sum_{k=0}^{m} \left[\frac{(-1)^k}{(k!)(m-k)!} \ln|x+k|\right] + C$$

3. $u = \sin^{-1} x,\ du = \frac{dx}{\sqrt{1-x^2}}$; $dv = x\,dx,\ v = \frac{x^2}{2}$;

$$\int x \sin^{-1} x\,dx = \frac{x^2}{2}\sin^{-1} x - \int \frac{x^2\,dx}{2\sqrt{1-x^2}};\ \begin{bmatrix} x = \sin\theta \\ dx = \cos\theta\,d\theta \end{bmatrix} \to \int x\sin^{-1} x\,dx = \frac{x^2}{2}\sin^{-1} x - \int \frac{\sin^2\theta\cos\theta\,d\theta}{2\cos\theta}$$

$$= \frac{x^2}{2}\sin^{-1} x - \frac{1}{2}\int \sin^2\theta\,d\theta = \frac{x^2}{2}\sin^{-1} x - \frac{1}{2}\left(\frac{\theta}{2} - \frac{\sin 2\theta}{4}\right) + C = \frac{x^2}{2}\sin^{-1} x + \frac{\sin\theta\cos\theta - \theta}{4} + C$$

$$= \frac{x^2}{2}\sin^{-1} x + \frac{x\sqrt{1-x^2} - \sin^{-1} x}{4} + C$$

4. $\int \sin^{-1}\sqrt{y}\,dy;\ \begin{bmatrix} z = \sqrt{y} \\ dz = \frac{dy}{2\sqrt{y}} \end{bmatrix} \to \int 2z\sin^{-1} z\,dz$; from Exercise 3, $\int z\sin^{-1} z\,dz$

$$= \frac{z^2\sin^{-1} z}{2} + \frac{z\sqrt{1-z^2} - \sin^{-1} z}{4} + C \Rightarrow \int \sin^{-1}\sqrt{y}\,dy = y\sin^{-1}\sqrt{y} + \frac{\sqrt{y}\sqrt{1-y} - \sin^{-1}\sqrt{y}}{2} + C$$

$$= y\sin^{-1}\sqrt{y} + \frac{\sqrt{y-y^2}}{2} - \frac{\sin^{-1}\sqrt{y}}{2} + C$$

5. $\int \frac{d\theta}{1-\tan^2\theta} = \int \frac{\cos^2\theta}{\cos^2\theta - \sin^2\theta}\,d\theta = \int \frac{1+\cos 2\theta}{2\cos 2\theta}\,d\theta = \frac{1}{2}\int (\sec 2\theta + 1)\,d\theta = \frac{\ln|\sec 2\theta + \tan 2\theta| + 2\theta}{4} + C$

6. $u = \ln\left(\sqrt{x} + \sqrt{1+x}\right),\ du = \left(\frac{dx}{\sqrt{x}+\sqrt{1+x}}\right)\left(\frac{1}{2\sqrt{x}} + \frac{1}{2\sqrt{1+x}}\right) = \frac{dx}{2\sqrt{x}\sqrt{1+x}}$; $dv = dx,\ v = x$;

$$\int \ln\left(\sqrt{x}+\sqrt{1+x}\right)dx = x\ln\left(\sqrt{x}+\sqrt{1+x}\right) - \frac{1}{2}\int \frac{x\,dx}{\sqrt{x}\sqrt{1+x}};\ \frac{1}{2}\int \frac{x\,dx}{\sqrt{\left(x+\frac{1}{2}\right)^2 - \frac{1}{4}}};$$

$$\begin{bmatrix} x + \frac{1}{2} = \frac{1}{2}\sec\theta \\ dx = \frac{1}{2}\sec\theta\tan\theta\,d\theta \end{bmatrix} \to \frac{1}{4}\int \frac{(\sec\theta - 1)\cdot\sec\theta\tan\theta\,d\theta}{\left(\frac{1}{2}\tan\theta\right)} = \frac{1}{2}\int (\sec^2\theta - \sec\theta)\,d\theta$$

$$= \frac{\tan\theta - \ln|\sec\theta + \tan\theta|}{2} + C = \frac{2\sqrt{x^2+x} - \ln\left|2x+1+2\sqrt{x^2+x}\right|}{2} + C$$

$$\Rightarrow \int \ln\left(\sqrt{x}+\sqrt{1+x}\right)dx = x\ln\left(\sqrt{x}+\sqrt{1+x}\right) - \frac{2\sqrt{x^2+x} - \ln\left|2x+1+2\sqrt{x^2+x}\right|}{4} + C$$

7. $\int \frac{dt}{t-\sqrt{1-t^2}};\ \begin{bmatrix} t = \sin\theta \\ dt = \cos\theta\,d\theta \end{bmatrix} \to \int \frac{\cos\theta\,d\theta}{\sin\theta - \cos\theta} = \int \frac{d\theta}{\tan\theta - 1};\ \begin{bmatrix} u = \tan\theta \\ du = \sec^2\theta\,d\theta \\ d\theta = \frac{du}{u^2+1} \end{bmatrix} \to \int \frac{du}{(u-1)(u^2+1)}$

$$= \frac{1}{2}\int \frac{du}{u-1} - \frac{1}{2}\int \frac{du}{u^2+1} - \frac{1}{2}\int \frac{u\,du}{u^2+1} = \frac{1}{2}\ln\left|\frac{u-1}{\sqrt{u^2+1}}\right| - \frac{1}{2}\tan^{-1} u + C = \frac{1}{2}\ln\left|\frac{\tan\theta - 1}{\sec\theta}\right| - \frac{1}{2}\theta + C$$

$$= \frac{1}{2}\ln\left(t - \sqrt{1-t^2}\right) - \frac{1}{2}\sin^{-1} t + C$$

8. $\int \frac{(2e^{2x} - e^x)\,dx}{\sqrt{3e^{2x} - 6e^x - 1}};\ \begin{bmatrix} u = e^x \\ du = e^x\,dx \end{bmatrix} \to \int \frac{(2u-1)\,du}{\sqrt{3u^2 - 6u - 1}} = \frac{1}{\sqrt{3}}\int \frac{(2u-1)\,du}{\sqrt{(u-1)^2 - \frac{4}{3}}};$

$$\begin{bmatrix} u - 1 = \frac{2}{\sqrt{3}}\sec\theta \\ du = \frac{2}{\sqrt{3}}\sec\theta\tan\theta\,d\theta \end{bmatrix} \to \frac{1}{\sqrt{3}}\int \left(\frac{4}{\sqrt{3}}\sec\theta + 1\right)(\sec\theta)\,d\theta = \frac{4}{3}\int \sec^2\theta\,d\theta + \frac{1}{\sqrt{3}}\int \sec\theta\,d\theta$$

$$= \frac{4}{3}\tan\theta + \frac{1}{\sqrt{3}}\ln|\sec\theta + \tan\theta| + C_1 = \frac{4}{3}\cdot\sqrt{\frac{3}{4}(u-1)^2 - 1} + \frac{1}{\sqrt{3}}\ln\left|\frac{\sqrt{3}}{2}(u-1) + \sqrt{\frac{3}{4}(u-1)^2 - 1}\right| + C_1$$

$$= \frac{2}{3}\sqrt{3u^2 - 6u - 1} + \frac{1}{\sqrt{3}}\ln\left|u - 1 + \sqrt{(u-1)^2 - \frac{4}{3}}\right| + \left(C_1 + \frac{1}{\sqrt{3}}\ln\frac{\sqrt{3}}{2}\right)$$

$$= \frac{1}{\sqrt{3}}\left[2\sqrt{e^{2x} - 2e^x - \frac{1}{3}} + \ln\left|e^x - 1 + \sqrt{e^{2x} - 2e^x - \frac{1}{3}}\right|\right] + C$$

9. $\int \frac{1}{x^4+4}\,dx = \int \frac{1}{(x^2+2)^2 - 4x^2}\,dx = \int \frac{1}{(x^2+2x+2)(x^2-2x+2)}\,dx$

$$= \frac{1}{16}\int \left[\frac{2x+2}{x^2+2x+2} + \frac{2}{(x+1)^2+1} - \frac{2x-2}{x^2-2x+2} + \frac{2}{(x-1)^2+1}\right]dx$$

$= \frac{1}{16} \ln \left| \frac{x^2+2x+2}{x^2-2x+2} \right| + \frac{1}{8} [\tan^{-1}(x+1) + \tan^{-1}(x-1)] + C$

10. $\int \frac{1}{x^6-1}\, dx = \frac{1}{6} \int \left(\frac{1}{x-1} - \frac{1}{x+1} + \frac{x-2}{x^2-x+1} - \frac{x+2}{x^2+x+1} \right) dx$
$= \frac{1}{6} \ln \left| \frac{x-1}{x+1} \right| + \frac{1}{12} \int \left[\frac{2x-1}{x^2-x+1} - \frac{3}{\left(x-\frac{1}{2}\right)^2+\frac{3}{4}} - \frac{2x+1}{x^2+x+1} - \frac{3}{\left(x+\frac{1}{2}\right)^2+\frac{3}{4}} \right] dx$
$= \frac{1}{6} \ln \left| \frac{x-1}{x+1} \right| + \frac{1}{12} \left[\ln \left| \frac{x^2-x+1}{x^2+x+1} \right| - 2\sqrt{3} \tan^{-1}\left(\frac{2x-1}{\sqrt{3}}\right) - 2\sqrt{3} \tan^{-1}\left(\frac{2x+1}{\sqrt{3}}\right) \right] + C$

11. $\lim_{x \to \infty} \int_{-x}^{x} \sin t\, dt = \lim_{x \to \infty} [-\cos t]_{-x}^{x} = \lim_{x \to \infty} [-\cos x + \cos(-x)] = \lim_{x \to \infty} (-\cos x + \cos x) = \lim_{x \to \infty} 0 = 0$

12. $\lim_{x \to 0^+} \int_x^1 \frac{\cos t}{t^2}\, dt$; $\lim_{t \to 0^+} \frac{\left(\frac{1}{t^2}\right)}{\left(\frac{\cos t}{t^2}\right)} = \lim_{t \to 0^+} \frac{1}{\cos t} = 1 \Rightarrow \lim_{x \to 0^+} \int_x^1 \frac{\cos t}{t^2}\, dt$ diverges since $\int_0^1 \frac{dt}{t^2}$ diverges; thus

$\lim_{x \to 0^+} x \int_x^1 \frac{\cos t}{t^2}\, dt$ is an indeterminate $0 \cdot \infty$ form and we apply l'Hôpital's rule:

$\lim_{x \to 0^+} x \int_x^1 \frac{\cos t}{t^2}\, dt = \lim_{x \to 0^+} \frac{-\int_1^x \frac{\cos t}{t^2}\, dt}{\frac{1}{x}} = \lim_{x \to 0^+} \frac{-\left(\frac{\cos x}{x^2}\right)}{\left(-\frac{1}{x^2}\right)} = \lim_{x \to 0^+} \cos x = 1$

13. $\lim_{n \to \infty} \sum_{k=1}^{n} \ln \sqrt[n]{1 + \frac{k}{n}} = \lim_{n \to \infty} \sum_{k=1}^{n} \ln\left(1 + k\left(\frac{1}{n}\right)\right)\left(\frac{1}{n}\right) = \int_0^1 \ln(1+x)\, dx$; $\left[\begin{array}{c} u = 1+x,\ du = dx \\ x = 0 \Rightarrow u = 1,\ x = 1 \Rightarrow u = 2 \end{array} \right]$
$\rightarrow \int_1^2 \ln u\, du = [u \ln u - u]_1^2 = (2 \ln 2 - 2) - (\ln 1 - 1) = 2 \ln 2 - 1 = \ln 4 - 1$

14. $\lim_{n \to \infty} \sum_{k=0}^{n-1} \frac{1}{\sqrt{n^2-k^2}} = \lim_{n \to \infty} \sum_{k=0}^{n-1} \left(\frac{n}{\sqrt{n^2-k^2}}\right)\left(\frac{1}{n}\right) = \lim_{n \to \infty} \sum_{k=0}^{n-1} \left(\frac{1}{\sqrt{1-\left[k\left(\frac{1}{n}\right)\right]^2}}\right)\left(\frac{1}{n}\right)$
$= \int_0^1 \frac{1}{\sqrt{1-x^2}}\, dx = [\sin^{-1} x]_0^1 = \frac{\pi}{2}$

15. $\frac{dy}{dx} = \sqrt{\cos 2x} \Rightarrow 1 + \left(\frac{dy}{dx}\right)^2 = 1 + \cos 2x = 2\cos^2 x$; $L = \int_0^{\pi/4} \sqrt{1 + \left(\sqrt{\cos 2t}\right)^2}\, dt = \sqrt{2} \int_0^{\pi/4} \sqrt{\cos^2 t}\, dt$
$= \sqrt{2}\, [\sin t]_0^{\pi/4} = 1$

16. $\frac{dy}{dx} = \frac{-2x}{1-x^2} \Rightarrow 1 + \left(\frac{dy}{dx}\right)^2 = \frac{(1-x^2)^2+4x^2}{(1-x^2)^2} = \frac{1+2x^2+x^4}{(1-x^2)^2} = \left(\frac{1+x^2}{1-x^2}\right)^2$; $L = \int_0^{1/2} \sqrt{1 + \left(\frac{dy}{dx}\right)^2}\, dx$
$= \int_0^{1/2} \left(\frac{1+x^2}{1-x^2}\right) dx = \int_0^{1/2} \left(-1 + \frac{2}{1-x^2}\right) dx = \int_0^{1/2} \left(-1 + \frac{1}{1+x} + \frac{1}{1-x}\right) dx = \left[-x + \ln\left|\frac{1+x}{1-x}\right|\right]_0^{1/2}$
$= \left(-\frac{1}{2} + \ln 3\right) - (0 + \ln 1) = \ln 3 - \frac{1}{2}$

17. $V = \int_a^b 2\pi \left(\begin{array}{c}\text{shell}\\ \text{radius}\end{array}\right)\left(\begin{array}{c}\text{shell}\\ \text{height}\end{array}\right) dx = \int_0^1 2\pi xy\, dx$
$= 6\pi \int_0^1 x^2\sqrt{1-x}\, dx$; $\left[\begin{array}{c} u = 1-x \\ du = -dx \\ x^2 = (1-u)^2 \end{array} \right]$
$\rightarrow -6\pi \int_1^0 (1-u)^2 \sqrt{u}\, du$
$= -6\pi \int_1^0 \left(u^{1/2} - 2u^{3/2} + u^{5/2}\right) du$
$= -6\pi \left[\frac{2}{3}u^{3/2} - \frac{4}{5}u^{5/2} + \frac{2}{7}u^{7/2}\right]_1^0 = 6\pi\left(\frac{2}{3} - \frac{4}{5} + \frac{2}{7}\right)$
$= 6\pi\left(\frac{70-84+30}{105}\right) = 6\pi\left(\frac{16}{105}\right) = \frac{32\pi}{35}$

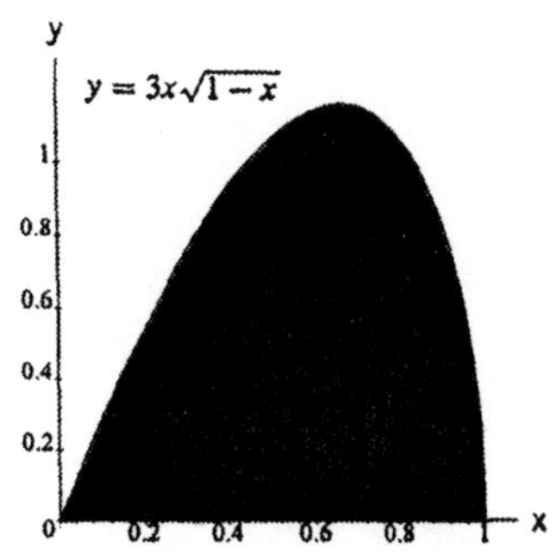

18. $V = \int_a^b \pi y^2\,dx = \pi \int_1^4 \frac{25\,dx}{x^2(5-x)}$

$= \pi \int_1^4 \left(\frac{dx}{x} + \frac{5\,dx}{x^2} + \frac{dx}{5-x}\right)$

$= \pi \left[\ln\left|\frac{x}{5-x}\right| - \frac{5}{x}\right]_1^4 = \pi\left(\ln 4 - \frac{5}{4}\right) - \pi\left(\ln\frac{1}{4} - 5\right)$

$= \frac{15\pi}{4} + 2\pi \ln 4$

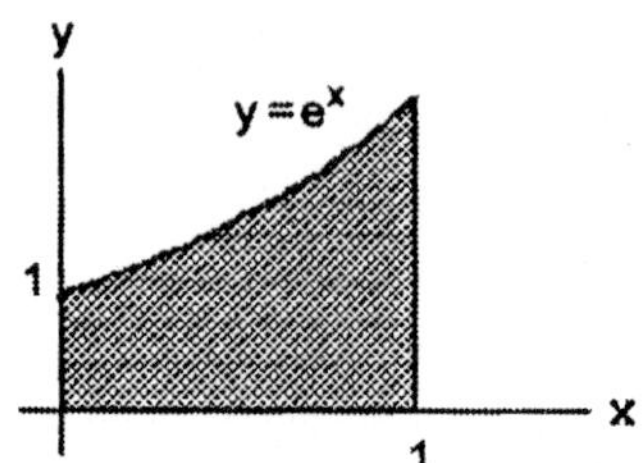

19. $V = \int_a^b 2\pi \binom{\text{shell}}{\text{radius}}\binom{\text{shell}}{\text{height}} dx = \int_0^1 2\pi x e^x\,dx$

$= 2\pi\left[xe^x - e^x\right]_0^1 = 2\pi$

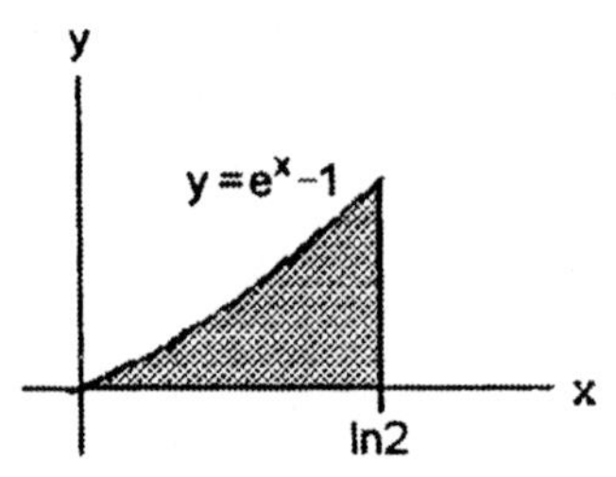

20. $V = \int_0^{\ln 2} 2\pi(\ln 2 - x)(e^x - 1)\,dx$

$= 2\pi \int_0^{\ln 2} \left[(\ln 2)e^x - \ln 2 - xe^x + x\right] dx$

$= 2\pi\left[(\ln 2)e^x - (\ln 2)x - xe^x + e^x + \frac{x^2}{2}\right]_0^{\ln 2}$

$= 2\pi\left[2\ln 2 - (\ln 2)^2 - 2\ln 2 + 2 + \frac{(\ln 2)^2}{2}\right] - 2\pi(\ln 2 + 1)$

$= 2\pi\left[-\frac{(\ln 2)^2}{2} - \ln 2 + 1\right]$

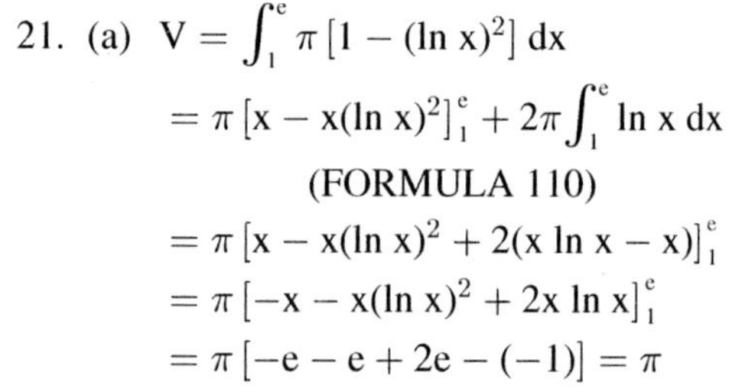

21. (a) $V = \int_1^e \pi\left[1 - (\ln x)^2\right] dx$

$= \pi\left[x - x(\ln x)^2\right]_1^e + 2\pi \int_1^e \ln x\,dx$

(FORMULA 110)

$= \pi\left[x - x(\ln x)^2 + 2(x\ln x - x)\right]_1^e$

$= \pi\left[-x - x(\ln x)^2 + 2x\ln x\right]_1^e$

$= \pi\left[-e - e + 2e - (-1)\right] = \pi$

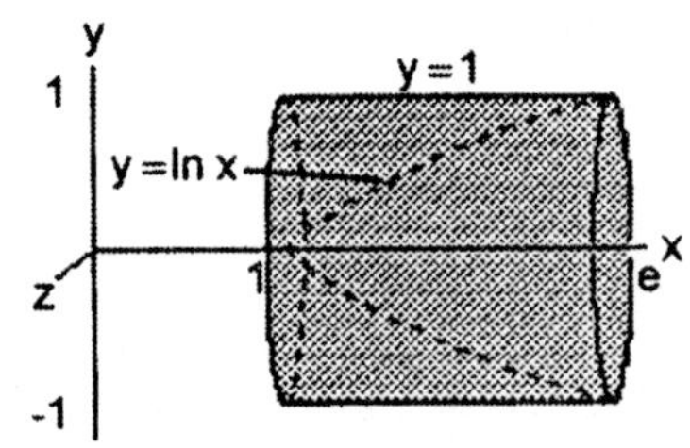

(b) $V = \int_1^e \pi(1 - \ln x)^2\,dx = \pi \int_1^e \left[1 - 2\ln x + (\ln x)^2\right] dx$

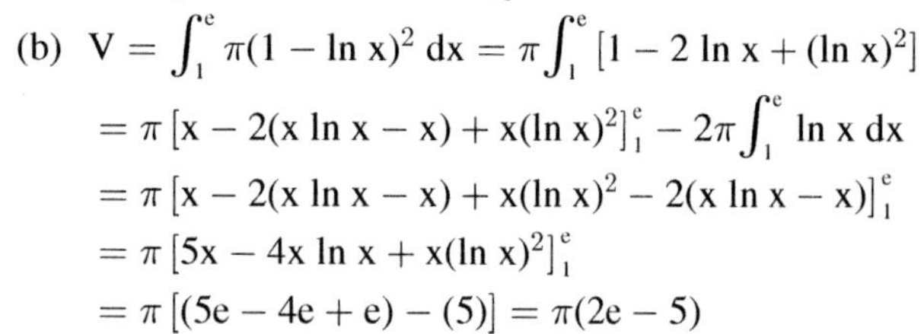

$= \pi\left[x - 2(x\ln x - x) + x(\ln x)^2\right]_1^e - 2\pi \int_1^e \ln x\,dx$

$= \pi\left[x - 2(x\ln x - x) + x(\ln x)^2 - 2(x\ln x - x)\right]_1^e$

$= \pi\left[5x - 4x\ln x + x(\ln x)^2\right]_1^e$

$= \pi\left[(5e - 4e + e) - (5)\right] = \pi(2e - 5)$

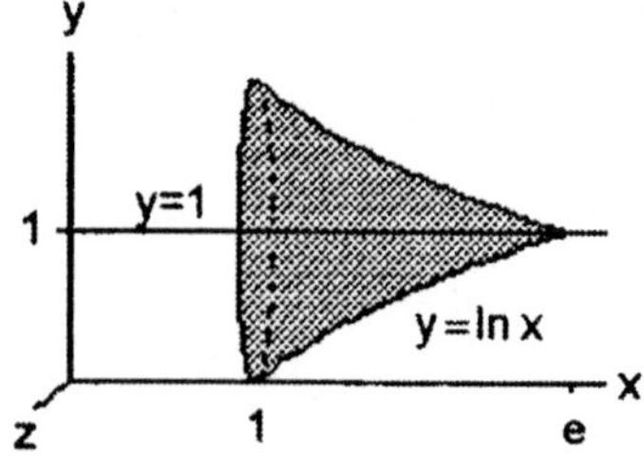

22. (a) $V = \pi \int_0^1 \left[(e^y)^2 - 1\right] dy = \pi \int_0^1 (e^{2y} - 1)\,dy = \pi\left[\frac{e^{2y}}{2} - y\right]_0^1 = \pi\left[\frac{e^2}{2} - 1 - \left(\frac{1}{2}\right)\right] = \frac{\pi(e^2 - 3)}{2}$

(b) $V = \pi \int_0^1 (e^y - 1)^2\,dy = \pi \int_0^1 (e^{2y} - 2e^y + 1)\,dy = \pi\left[\frac{e^{2y}}{2} - 2e^y + y\right]_0^1 = \pi\left[\left(\frac{e^2}{2} - 2e + 1\right) - \left(\frac{1}{2} - 2\right)\right]$

$= \pi\left(\frac{e^2}{2} - 2e + \frac{5}{2}\right) = \frac{\pi(e^2 - 4e + 5)}{2}$

23. (a) $\lim_{x \to 0^+} x\ln x = 0 \Rightarrow \lim_{x \to 0^+} f(x) = 0 = f(0) \Rightarrow$ f is continuous

(b) $V = \int_0^2 \pi x^2(\ln x)^2\,dx;\ \begin{bmatrix} u = (\ln x)^2 \\ du = (2\ln x)\frac{dx}{x} \\ dv = x^2 dx \\ v = \frac{x^3}{3} \end{bmatrix} \rightarrow \pi\left(\lim_{b\to 0^+}\left[\frac{x^3}{3}(\ln x)^2\right]_b^2 - \int_0^2\left(\frac{x^3}{3}\right)(2\ln x)\frac{dx}{x}\right)$

$= \pi\left[\left(\frac{8}{3}\right)(\ln 2)^2 - \left(\frac{2}{3}\right)\lim_{b\to 0^+}\left[\frac{x^3}{3}\ln x - \frac{x^3}{9}\right]_b^2\right] = \pi\left[\frac{8(\ln 2)^2}{3} - \frac{16(\ln 2)}{9} + \frac{16}{27}\right]$

24. $V = \int_0^1 \pi(-\ln x)^2\,dx$

$= \pi\left(\lim_{b\to 0^+}\left[x(\ln x)^2\right]_b^1 - 2\int_b^1 \ln x\,dx\right)$

$= -2\pi \lim_{b\to 0^+}\left[x\ln x - x\right]_b^1 = 2\pi$

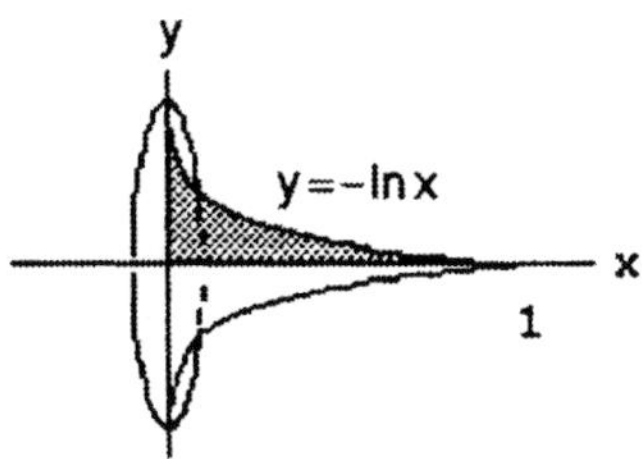

25. $M = \int_1^e \ln x\,dx = [x\ln x - x]_1^e = (e-e) - (0-1) = 1;$

$M_x = \int_1^e (\ln x)\left(\frac{\ln x}{2}\right)dx = \frac{1}{2}\int_1^e (\ln x)^2\,dx$

$= \frac{1}{2}\left(\left[x(\ln x)^2\right]_1^e - 2\int_1^e \ln x\,dx\right) = \frac{1}{2}(e-2);$

$M_y = \int_1^e x\ln x\,dx = \left[\frac{x^2\ln x}{2}\right]_1^e - \frac{1}{2}\int_1^e x\,dx$

$= \frac{1}{2}\left[x^2\ln x - \frac{x^2}{2}\right]_1^e = \frac{1}{2}\left[\left(e^2 - \frac{e^2}{2}\right) + \frac{1}{2}\right] = \frac{1}{4}(e^2+1);$

therefore, $\bar{x} = \frac{M_y}{M} = \frac{e^2+1}{4}$ and $\bar{y} = \frac{M_x}{M} = \frac{e-2}{2}$

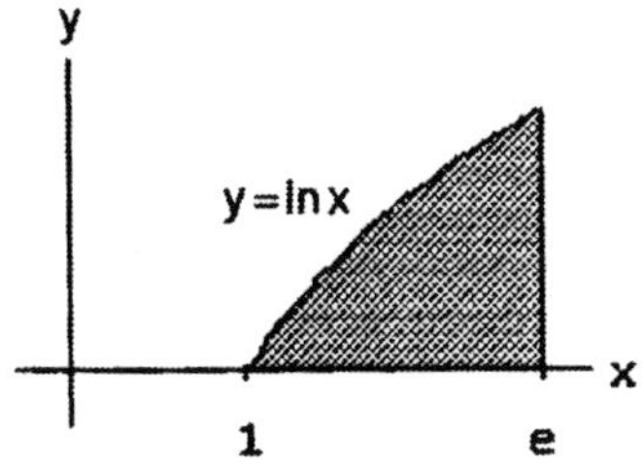

26. $M = \int_0^1 \frac{2\,dx}{\sqrt{1-x^2}} = 2\left[\sin^{-1} x\right]_0^1 = \pi;$

$M_y = \int_0^1 \frac{2x\,dx}{\sqrt{1-x^2}} = 2\left[-\sqrt{1-x^2}\right]_0^1 = 2;$

therefore, $\bar{x} = \frac{M_y}{M} = \frac{2}{\pi}$ and $\bar{y} = 0$ by symmetry

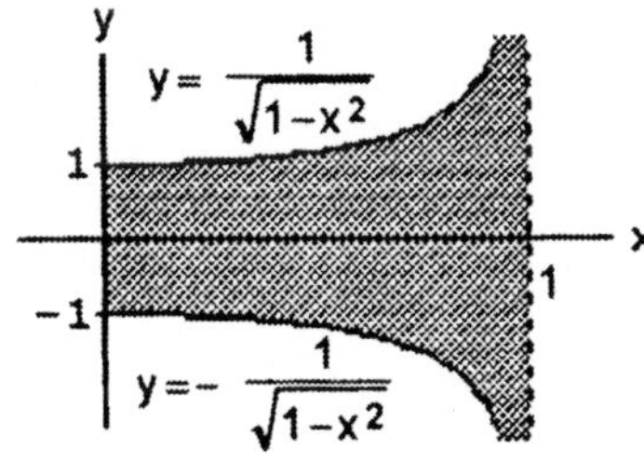

27. $L = \int_1^e \sqrt{1+\frac{1}{x^2}}\,dx = \int_1^e \frac{\sqrt{x^2+1}}{x}\,dx;\ \begin{bmatrix} x = \tan\theta \\ dx = \sec^2\theta\,d\theta \end{bmatrix} \rightarrow L = \int_{\pi/4}^{\tan^{-1} e} \frac{\sec\theta\cdot\sec^2\theta\,d\theta}{\tan\theta}$

$= \int_{\pi/4}^{\tan^{-1} e} \frac{(\sec\theta)(\tan^2\theta+1)}{\tan\theta}\,d\theta = \int_{\pi/4}^{\tan^{-1} e}(\tan\theta\sec\theta + \csc\theta)\,d\theta = \left[\sec\theta - \ln|\csc\theta + \cot\theta|\right]_{\pi/4}^{\tan^{-1} e}$

$= \left(\sqrt{1+e^2} - \ln\left|\frac{\sqrt{1+e^2}}{e} + \frac{1}{e}\right|\right) - \left[\sqrt{2} - \ln\left(1+\sqrt{2}\right)\right] = \sqrt{1+e^2} - \ln\left(\frac{\sqrt{1+e^2}}{e} + \frac{1}{e}\right) - \sqrt{2} + \ln\left(1+\sqrt{2}\right)$

28. $y = \ln x \Rightarrow 1 + \left(\frac{dx}{dy}\right)^2 = 1 + x^2 \Rightarrow S = 2\pi\int_c^d x\sqrt{1+x^2}\,dy \Rightarrow S = 2\pi\int_0^1 e^y\sqrt{1+e^{2y}}\,dy;\ \begin{bmatrix} u = e^y \\ du = e^y\,dy \end{bmatrix}$

$\rightarrow S = 2\pi\int_1^e \sqrt{1+u^2}\,du;\ \begin{bmatrix} u = \tan\theta \\ du = \sec^2\theta\,d\theta \end{bmatrix} \rightarrow 2\pi\int_{\pi/4}^{\tan^{-1} e}\sec\theta\cdot\sec^2\theta\,d\theta$

$= 2\pi\left(\frac{1}{2}\right)\left[\sec\theta\tan\theta + \ln|\sec\theta + \tan\theta|\right]_{\pi/4}^{\tan^{-1} e} = \pi\left[\left(\sqrt{1+e^2}\right)e + \ln\left|\sqrt{1+e^2}+e\right|\right] - \pi\left[\sqrt{2}\cdot 1 + \ln\left(\sqrt{2}+1\right)\right]$

$= \pi\left[e\sqrt{1+e^2} + \ln\left(\frac{\sqrt{1+e^2}+e}{\sqrt{2}+1}\right) - \sqrt{2}\right]$

29. $L = 4\int_0^1 \sqrt{1+\left(\frac{dy}{dx}\right)^2}\,dx;\ x^{2/3} + y^{2/3} = 1 \Rightarrow y = \left(1-x^{2/3}\right)^{3/2} \Rightarrow \frac{dy}{dx} = -\frac{3}{2}\left(1-x^{2/3}\right)^{1/2}\left(x^{-1/3}\right)\left(\frac{2}{3}\right)$

$\Rightarrow \left(\frac{dy}{dx}\right)^2 = \frac{1-x^{2/3}}{x^{2/3}} \Rightarrow L = 4\int_0^1 \sqrt{1+\left(\frac{1-x^{2/3}}{x^{2/3}}\right)}\,dx = 4\int_0^1 \frac{dx}{x^{1/3}} = 6\left[x^{2/3}\right]_0^1 = 6$

30. $S = 2\pi \int_{-1}^{1} f(x)\sqrt{1+[f'(x)]^2}\,dx$; $f(x) = \left(1-x^{2/3}\right)^{3/2} \Rightarrow [f'(x)]^2 + 1 = \frac{1}{x^{2/3}} \Rightarrow S = 2\pi\int_{-1}^{1}\left(1-x^{2/3}\right)^{3/2} \cdot \frac{dx}{\sqrt{x^{2/3}}}$
$= 4\pi\int_0^1 \left(1-x^{2/3}\right)^{3/2}\left(\frac{1}{x^{1/3}}\right)dx$; $\begin{bmatrix} u = x^{2/3} \\ du = \frac{2}{3}\frac{dx}{x^{1/3}} \end{bmatrix} \rightarrow 4\cdot\frac{3}{2}\pi\int_0^1 (1-u)^{3/2}\,du = -6\pi\int_0^1 (1-u)^{3/2}\,d(1-u)$
$= -6\pi\cdot\frac{2}{5}\left[(1-u)^{5/2}\right]_0^1 = \frac{12\pi}{5}$

31. $\left(\frac{dy}{dx}\right)^2 = \frac{1}{4x} \Rightarrow \frac{dy}{dx} = \frac{\pm 1}{2\sqrt{x}} \Rightarrow y = \sqrt{x}$ or $y = -\sqrt{x}$, $0 \le x \le 4$

32. The integral $\int_{-1}^{1}\sqrt{1-x^2}\,dx$ is the area enclosed by the x-axis and the semicircle $y = \sqrt{1-x^2}$. This area is half the circle's area, or $\frac{\pi}{2}$ and multiplying by 2 gives π. The length of the circular arc $y = \sqrt{1-x^2}$ from $x = -1$ to $x = 1$ is $L = \int_{-1}^{1}\sqrt{1+\left(\frac{dy}{dx}\right)^2}\,dx = \int_{-1}^{1}\sqrt{1+\left(\frac{-x}{\sqrt{1-x^2}}\right)^2}\,dx = \int_{-1}^{1}\frac{dx}{\sqrt{1-x^2}} = \frac{1}{2}(2\pi) = \pi$ since L is half the circle's circumference. In conclusion, $2\int_{-1}^{1}\sqrt{1-x^2}\,dx = \int_{-1}^{1}\frac{dx}{\sqrt{1-x^2}}$.

33. (b) $\int_{-\infty}^{\infty} e^{(x-e^x)}\,dx = \int_{-\infty}^{\infty} e^{(-e^x)}e^x\,dx$
$= \lim_{a\to-\infty}\int_a^0 e^{(-e^x)}e^x\,dx + \lim_{b\to+\infty}\int_0^b e^{(-e^x)}e^x\,dx$;
$\begin{bmatrix} u = e^x \\ du = e^x\,dx \end{bmatrix} \rightarrow$
$\lim_{a\to-\infty}\int_{e^a}^{1} e^{-u}\,du + \lim_{b\to+\infty}\int_1^{e^b} e^{-u}\,du$
$= \lim_{a\to-\infty}\left[-e^{-u}\right]_{e^a}^{1} + \lim_{b\to-\infty}\left[-e^{-u}\right]_1^{e^b}$
$= \lim_{a\to-\infty}\left[-\frac{1}{e} + e^{-(e^a)}\right] + \lim_{b\to+\infty}\left[-e^{-(e^b)} + \frac{1}{e}\right]$
$= \left(-\frac{1}{e} + e^0\right) + \left(0 + \frac{1}{e}\right) = 1$

(a)

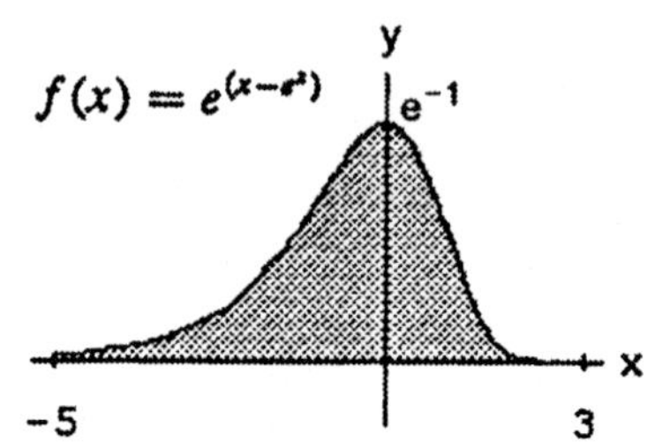

34. $u = \frac{1}{1+y}$, $du = -\frac{dy}{(1+y)^2}$; $dv = ny^{n-1}\,dy$, $v = y^n$;
$\lim_{n\to\infty}\int_0^1 \frac{ny^{n-1}}{1+y}\,dy = \lim_{n\to\infty}\left(\left[\frac{y^n}{1+y}\right]_0^1 + \int_0^1 \frac{y^n}{1+y^2}\,dy\right) = \frac{1}{2} + \lim_{n\to\infty}\int_0^1 \frac{y^n}{1+y^2}\,dy$. Now, $0 \le \frac{y^n}{1+y^2} \le y^n$
$\Rightarrow 0 \le \lim_{n\to\infty}\int_0^1 \frac{y^n}{1+y^2}\,dy \le \lim_{n\to\infty}\int_0^1 y^n\,dy = \lim_{n\to\infty}\left[\frac{y^{n+1}}{n+1}\right]_0^1 = \lim_{n\to\infty}\frac{1}{n+1} = 0 \Rightarrow \lim_{n\to\infty}\int_0^1 \frac{ny^{n-1}}{1+y}\,dy$
$= \frac{1}{2} + 0 = \frac{1}{2}$

35. $u = x^2 - a^2 \Rightarrow du = 2x\,dx$;
$\int x\left(\sqrt{x^2-a^2}\right)^n dx = \frac{1}{2}\int\left(\sqrt{u}\right)^n du = \frac{1}{2}\int u^{n/2}\,du = \frac{1}{2}\left(\frac{u^{n/2+1}}{\frac{n}{2}+1}\right) + C,\ n \ne -2$
$= \frac{u^{(n+2)/2}}{n+2} + C = \frac{\left(\sqrt{u}\right)^{n+2}}{n+2} + C = \frac{\left(\sqrt{x^2-a^2}\right)^{n+2}}{n+2} + C$

36. $\frac{\pi}{6} = \sin^{-1}\frac{1}{2} = \left[\sin^{-1}\frac{x}{2}\right]_0^1 = \int_0^1 \frac{dx}{\sqrt{4-x^2}} < \int_0^1 \frac{dx}{\sqrt{4-x^2-x^3}} < \int_0^1 \frac{dx}{\sqrt{4-2x^2}} = \frac{1}{\sqrt{2}}\int_0^{\sqrt{2}} \frac{du}{\sqrt{4-u^2}}$
$= \frac{1}{\sqrt{2}}\left[\sin^{-1}\frac{u}{2}\right]_0^{\sqrt{2}} = \frac{1}{\sqrt{2}}\sin^{-1}\frac{\sqrt{2}}{2} = \frac{1}{\sqrt{2}}\left(\frac{\pi}{4}\right) = \frac{\pi\sqrt{2}}{8}$

37. $\int_1^\infty \left(\frac{ax}{x^2+1} - \frac{1}{2x}\right) dx = \lim_{b\to\infty} \int_1^b \left(\frac{ax}{x^2+1} - \frac{1}{2x}\right) dx = \lim_{b\to\infty} \left[\frac{a}{2}\ln(x^2+1) - \frac{1}{2}\ln x\right]_1^b = \lim_{b\to\infty} \left[\frac{1}{2}\ln\frac{(x^2+1)^a}{x}\right]_1^b$
$= \lim_{b\to\infty} \frac{1}{2}\left[\ln\frac{(b^2+1)^a}{b} - \ln 2^a\right]$; $\lim_{b\to\infty} \frac{(b^2+1)^a}{b} > \lim_{b\to\infty} \frac{b^{2a}}{b} = \lim_{b\to\infty} b^{2\left(a-\frac{1}{2}\right)} = \infty$ if $a > \frac{1}{2} \Rightarrow$ the improper integral diverges if $a > \frac{1}{2}$; for $a = \frac{1}{2}$: $\lim_{b\to\infty} \frac{\sqrt{b^2+1}}{b} = \lim_{b\to\infty} \sqrt{1+\frac{1}{b^2}} = 1 \Rightarrow \lim_{b\to\infty} \frac{1}{2}\left[\ln\frac{(b^2+1)^{1/2}}{b} - \ln 2^{1/2}\right]$
$= \frac{1}{2}\left(\ln 1 - \frac{1}{2}\ln 2\right) = -\frac{\ln 2}{4}$; if $a < \frac{1}{2}$: $0 \le \lim_{b\to\infty} \frac{(b^2+1)^a}{b} < \lim_{b\to\infty} \frac{(b+1)^{2a}}{b+1} = \lim_{b\to\infty} (b+1)^{2a-1} = 0$
$\Rightarrow \lim_{b\to\infty} \ln\frac{(b^2+1)^a}{b} = -\infty \Rightarrow$ the improper integral diverges if $a < \frac{1}{2}$; in summary, the improper integral $\int_1^\infty \left(\frac{ax}{x^2+1} - \frac{1}{2x}\right) dx$ converges only when $a = \frac{1}{2}$ and has the value $-\frac{\ln 2}{4}$

38. $G(x) = \lim_{b\to\infty} \int_0^b e^{-xt}\,dt = \lim_{b\to\infty} \left[-\frac{1}{x}e^{-xt}\right]_0^b = \lim_{b\to\infty} \left(\frac{1-e^{-xb}}{x}\right) = \frac{1-0}{x} = \frac{1}{x}$ if $x > 0 \Rightarrow xG(x) = x\left(\frac{1}{x}\right)$
$= 1$ if $x > 0$

39. $A = \int_1^\infty \frac{dx}{x^p}$ converges if $p > 1$ and diverges if $p \le 1$. Thus, $p \le 1$ for infinite area. The volume of the solid of revolution about the x-axis is $V = \int_1^\infty \pi\left(\frac{1}{x^p}\right)^2 dx = \pi\int_1^\infty \frac{dx}{x^{2p}}$ which converges if $2p > 1$ and diverges if $2p \le 1$. Thus we want $p > \frac{1}{2}$ for finite volume. In conclusion, the curve $y = x^{-p}$ gives infinite area and finite volume for values of p satisfying $\frac{1}{2} < p \le 1$.

40. The area is given by the integral $A = \int_0^1 \frac{dx}{x^p}$;
$p = 1$: $A = \lim_{b\to 0^+} [\ln x]_b^1 = -\lim_{b\to 0^+} \ln b = \infty$, diverges;
$p > 1$: $A = \lim_{b\to 0^+} [x^{1-p}]_b^1 = 1 - \lim_{b\to 0^+} b^{1-p} = -\infty$, diverges;
$p < 1$: $A = \lim_{b\to 0^+} [x^{1-p}]_b^1 = 1 - \lim_{b\to 0^+} b^{1-p} = 1 - 0$, converges; thus, $p \ge 1$ for infinite area.

The volume of the solid of revolution about the x-axis is $V_x = \pi\int_0^1 \frac{dx}{x^{2p}}$ which converges if $2p < 1$ or $p < \frac{1}{2}$, and diverges if $p \ge \frac{1}{2}$. Thus, V_x is infinite whenever the area is infinite ($p \ge 1$).

The volume of the solid of revolution about the y-axis is $V_y = \pi\int_1^\infty [R(y)]^2\,dy = \pi\int_1^\infty \frac{dy}{y^{2/p}}$ which converges if $\frac{2}{p} > 1 \Leftrightarrow p < 2$ (see Exercise 39). In conclusion, the curve $y = x^{-p}$ gives infinite area and finite volume for values of p satisfying $1 \le p < 2$, as described above.

41.

e^{2x}	$(+)$	$\cos 3x$
$2e^{2x}$	$(-)$	$\frac{1}{3}\sin 3x$
$4e^{2x}$	$(+)$	$-\frac{1}{9}\cos 3x$

$I = \frac{e^{2x}}{3}\sin 3x + \frac{2e^{2x}}{9}\cos 3x - \frac{4}{9}I \Rightarrow \frac{13}{9}I = \frac{e^{2x}}{9}(3\sin 3x + 2\cos 3x) \Rightarrow I = \frac{e^{2x}}{13}(3\sin 3x + 2\cos 3x) + C$

42.

e^{3x}	$(+)$	$\sin 4x$
$3e^{3x}$	$(-)$	$-\frac{1}{4}\cos 4x$
$9e^{3x}$	$(+)$	$-\frac{1}{16}\sin 4x$

$I = -\frac{e^{3x}}{4}\cos 4x + \frac{3e^{3x}}{16}\sin 4x - \frac{9}{16}I \Rightarrow \frac{25}{16}I = \frac{e^{3x}}{16}(3\sin 4x - 4\cos 4x) \Rightarrow I = \frac{e^{3x}}{25}(3\sin 4x - 4\cos 4x) + C$

43.

$\sin 3x$	$(+)$	$\sin x$
$3\cos 3x$	$(-)$	$-\cos x$
$-9\sin 3x$	$(+)$	$-\sin x$

$I = -\sin 3x \cos x + 3\cos 3x \sin x + 9I \Rightarrow -8I = -\sin 3x \cos x + 3\cos 3x \sin x$

$\Rightarrow I = \frac{\sin 3x \cos x - 3\cos 3x \sin x}{8} + C$

44.

$\cos 5x$	$(+)$	$\sin 4x$
$-\sin 5x$	$(-)$	$-\frac{1}{4}\cos 4x$
$-25\cos 5x$	$(+)$	$-\frac{1}{16}\sin 4$

$I = -\frac{1}{4}\cos 5x \cos 4x - \frac{5}{16}\sin 5x \sin 4x + \frac{25}{16}I \Rightarrow -\frac{9}{16}I = -\frac{1}{4}\cos 5x \cos 4x - \frac{5}{16}\sin 5x \sin 4x$

$\Rightarrow I = \frac{1}{9}(4\cos 5x \cos 4x + 5\sin 5x \sin 4x) + C$

45.

e^{ax}	$(+)$	$\sin bx$
ae^{ax}	$(-)$	$-\frac{1}{b}\cos bx$
a^2e^{ax}	$(+)$	$-\frac{1}{b^2}\sin bx$

$I = -\frac{e^{ax}}{b}\cos bx + \frac{ae^{ax}}{b^2}\sin bx - \frac{a^2}{b^2}I \Rightarrow \left(\frac{a^2+b^2}{b^2}\right)I = \frac{e^{ax}}{b^2}(a\sin bx - b\cos bx)$

$\Rightarrow I = \frac{e^{ax}}{a^2+b^2}(a\sin bx - b\cos bx) + C$

46.

e^{ax}	$(+)$	$\cos bx$
ae^{ax}	$(-)$	$\frac{1}{b}\sin bx$
a^2e^{ax}	$(+)$	$-\frac{1}{b^2}\cos bx$

$I = \frac{e^{ax}}{b}\sin bx + \frac{ae^{ax}}{b^2}\cos bx - \frac{a^2}{b^2}I \Rightarrow \left(\frac{a^2+b^2}{b^2}\right)I = \frac{e^{ax}}{b^2}(a\cos bx + b\sin bx)$

$\Rightarrow I = \frac{e^{ax}}{a^2+b^2}(a\cos bx + b\sin bx) + C$

47.

$\ln(ax)$	$(+)$	1
$\frac{1}{x}$	$(+)$	x

$I = x\ln(ax) - \int\left(\frac{1}{x}\right)x\,dx = x\ln(ax) - x + C$

48.

$\ln(ax)$	$(+)$	x^2
$\frac{1}{x}$	$(+)$	$\frac{1}{3}x^3$

$I = \frac{1}{3}x^3\ln(ax) - \int\left(\frac{1}{x}\right)\left(\frac{x^3}{3}\right)dx = \frac{1}{3}x^3\ln(ax) - \frac{1}{9}x^3 + C$

49. (a) $\Gamma(1) = \int_0^\infty e^{-t}\,dt = \lim_{b\to\infty}\int_0^b e^{-t}\,dt = \lim_{b\to\infty}\left[-e^{-t}\right]_0^b = \lim_{b\to\infty}\left[-\frac{1}{e^b} - (-1)\right] = 0 + 1 = 1$

(b) $u = t^x,\ du = xt^{x-1}\,dt;\ dv = e^{-t}\,dt,\ v = -e^{-t};\ x =$ fixed positive real

$\Rightarrow \Gamma(x+1) = \int_0^\infty t^xe^{-t}\,dt = \lim_{b\to\infty}\left[-t^xe^{-t}\right]_0^b + x\int_0^\infty t^{x-1}e^{-t}\,dt = \lim_{b\to\infty}\left(-\frac{b^x}{e^b} + 0^xe^0\right) + x\Gamma(x) = x\Gamma(x)$

(c) $\Gamma(n+1) = n\Gamma(n) = n!$:

$n = 0$: $\Gamma(0+1) = \Gamma(1) = 0!$;

$n = k$: Assume $\Gamma(k+1) = k!$ for some $k > 0$;

$$\begin{aligned} n = k+1:\ \Gamma(k+1+1) &= (k+1)\,\Gamma(k+1) && \text{from part (b)} \\ &= (k+1)k! && \text{induction hypothesis} \\ &= (k+1)! && \text{definition of factorial} \end{aligned}$$

Thus, $\Gamma(n+1) = n\Gamma(n) = n!$ for every positive integer n.

50. (a) $\Gamma(x) \approx \left(\frac{x}{e}\right)^x \sqrt{\frac{2\pi}{x}}$ and $n\Gamma(n) = n!$ $\Rightarrow$ $n! \approx n\left(\frac{n}{e}\right)^n \sqrt{\frac{2\pi}{n}} = \left(\frac{n}{e}\right)^n \sqrt{2n\pi}$

(b)

n	$\left(\frac{n}{e}\right)^n \sqrt{2n\pi}$	calculator
10	3598695.619	3628800
20	2.4227868×10^{18}	2.432902×10^{18}
30	2.6451710×10^{32}	2.652528×10^{32}
40	8.1421726×10^{47}	8.1591528×10^{47}
50	3.0363446×10^{64}	3.0414093×10^{64}
60	8.3094383×10^{81}	8.3209871×10^{81}

(c)

n	$\left(\frac{n}{e}\right)^n \sqrt{2n\pi}$	$\left(\frac{n}{e}\right)^n \sqrt{2n\pi}\, e^{1/12n}$	calculator
10	3598695.619	3628810.051	3628800

NOTES:

CHAPTER 9 FURTHER APPLICATIONS OF INTEGRATION

9.1 SLOPE FIELDS AND SEPARABLE DIFFERENTIAL EQUATIONS

1. (a) $y = e^{-x} \Rightarrow y' = -e^{-x} \Rightarrow 2y' + 3y = 2(-e^{-x}) + 3e^{-x} = e^{-x}$

   (b) $y = e^{-x} + e^{-3x/2} \Rightarrow y' = -e^{-x} - \frac{3}{2}e^{-3x/2} \Rightarrow 2y' + 3y = 2\left(-e^{-x} - \frac{3}{2}e^{-3x/2}\right) + 3\left(e^{-x} + e^{-3x/2}\right) = e^{-x}$

   (c) $y = e^{-x} + Ce^{-3x/2} \Rightarrow y' = -e^{-x} - \frac{3}{2}Ce^{-3x/2} \Rightarrow 2y' + 3y = 2\left(-e^{-x} - \frac{3}{2}Ce^{-3x/2}\right) + 3\left(e^{-x} + Ce^{-3x/2}\right) = e^{-x}$

2. (a) $y = -\frac{1}{x} \Rightarrow y' = \frac{1}{x^2} = \left(-\frac{1}{x}\right)^2 = y^2$

   (b) $y = -\frac{1}{x+3} \Rightarrow y' = \frac{1}{(x+3)^2} = \left[-\frac{1}{(x+3)}\right]^2 = y^2$

   (c) $y = \frac{1}{x+C} \Rightarrow y' = \frac{1}{(x+C)^2} = \left[-\frac{1}{x+C}\right]^2 = y^2$

3. $y = \frac{1}{x}\int_1^x \frac{e^t}{t}\,dt \Rightarrow y' = -\frac{1}{x^2}\int_1^x \frac{e^t}{t}\,dt + \left(\frac{1}{x}\right)\left(\frac{e^x}{x}\right) \Rightarrow x^2y' = -\int_1^x \frac{e^t}{t}\,dt + e^x = -x\left(\frac{1}{x}\int_1^x \frac{e^t}{t}\,dt\right) + e^x = -xy + e^x$
   $\Rightarrow x^2y' + xy = e^x$

4. $y = \frac{1}{\sqrt{1+x^4}}\int_1^x \sqrt{1+t^4}\,dt \Rightarrow y' = -\frac{1}{2}\left[\frac{4x^3}{\left(\sqrt{1+x^4}\right)^3}\right]\int_1^x \sqrt{1+t^4}\,dt + \frac{1}{\sqrt{1+x^4}}\left(\sqrt{1+x^4}\right)$
   $\Rightarrow y' = \left(\frac{-2x^3}{1+x^4}\right)\left(\frac{1}{\sqrt{1+x^4}}\int_1^x \sqrt{1+t^4}\,dt\right) + 1 \Rightarrow y' = \left(\frac{-2x^3}{1+x^4}\right)y + 1 \Rightarrow y' + \frac{2x^3}{1+x^4}\cdot y = 1$

5. $y = e^{-x}\tan^{-1}(2e^x) \Rightarrow y' = -e^{-x}\tan^{-1}(2e^x) + e^{-x}\left[\frac{1}{1+(2e^x)^2}\right](2e^x) = -e^{-x}\tan^{-1}(2e^x) + \frac{2}{1+4e^{2x}}$
   $\Rightarrow y' = -y + \frac{2}{1+4e^{2x}} \Rightarrow y' + y = \frac{2}{1+4e^{2x}}$; $y(-\ln 2) = e^{-(-\ln 2)}\tan^{-1}(2e^{-\ln 2}) = 2\tan^{-1}1 = 2\left(\frac{\pi}{4}\right) = \frac{\pi}{2}$

6. $y = (x-2)e^{-x^2} \Rightarrow y' = e^{-x^2} + \left(-2xe^{-x^2}\right)(x-2) \Rightarrow y' = e^{-x^2} - 2xy$; $y(2) = (2-2)e^{-2^2} = 0$

7. $y = \frac{\cos x}{x} \Rightarrow y' = \frac{-x\sin x - \cos x}{x^2} \Rightarrow y' = -\frac{\sin x}{x} - \frac{1}{x}\left(\frac{\cos x}{x}\right) \Rightarrow y' = -\frac{\sin x}{x} - \frac{y}{x} \Rightarrow xy' = -\sin x - y$
   $\Rightarrow xy' + y = -\sin x$; $y\left(\frac{\pi}{2}\right) = \frac{\cos(\pi/2)}{(\pi/2)} = 0$

8. $y = \frac{x}{\ln x} \Rightarrow y' = \frac{\ln x - x\left(\frac{1}{x}\right)}{(\ln x)^2} \Rightarrow y' = \frac{1}{\ln x} - \frac{1}{(\ln x)^2} \Rightarrow x^2y' = \frac{x^2}{\ln x} - \frac{x^2}{(\ln x)^2} \Rightarrow x^2y' = xy - y^2$; $y(e) = \frac{e}{\ln e} = e.$

9. $2\sqrt{xy}\,\frac{dy}{dx} = 1 \Rightarrow 2x^{1/2}y^{1/2}\,dy = dx \Rightarrow 2y^{1/2}\,dy = x^{-1/2}\,dx \Rightarrow \int 2y^{1/2}\,dy = \int x^{-1/2}\,dx \Rightarrow 2\left(\frac{2}{3}y^{3/2}\right)$
   $= 2x^{1/2} + C_1 \Rightarrow \frac{2}{3}y^{3/2} - x^{1/2} = C$, where $C = \frac{1}{2}C_1$

10. $\frac{dy}{dx} = x^2\sqrt{y} \Rightarrow dy = x^2y^{1/2}\,dx \Rightarrow y^{-1/2}\,dy = x^2\,dx \Rightarrow \int y^{-1/2}\,dy = \int x^2\,dx \Rightarrow 2y^{1/2} = \frac{x^3}{3} + C$
    $\Rightarrow 2y^{1/2} - \frac{1}{3}x^3 = C$

11. $\frac{dy}{dx} = e^{x-y} \Rightarrow dy = e^xe^{-y}\,dx \Rightarrow e^y\,dy = e^x\,dx \Rightarrow \int e^y\,dy = \int e^x\,dx \Rightarrow e^y = e^x + C \Rightarrow e^y - e^x = C$

12. $\frac{dy}{dx} = 3x^2e^{-y} \Rightarrow dy = 3x^2e^{-y}dx \Rightarrow e^y\,dy = 3x^2dx \Rightarrow \int e^y\,dy = \int 3x^2dx \Rightarrow e^y = x^3 + C \Rightarrow e^y - x^3 = C$

13. $\frac{dy}{dx} = \sqrt{y}\cos^2\sqrt{y} \Rightarrow dy = \left(\sqrt{y}\cos^2\sqrt{y}\right)dx \Rightarrow \frac{\sec^2\sqrt{y}}{\sqrt{y}}dy = dx \Rightarrow \int \frac{\sec^2\sqrt{y}}{\sqrt{y}}dy = \int dx$. In the integral on the left-hand side, substitute $u = \sqrt{y} \Rightarrow du = \frac{1}{2\sqrt{y}}dy \Rightarrow 2\,du = \frac{1}{\sqrt{y}}dy$, and we have $\int \sec^2 u\,du = \int dx \Rightarrow 2\tan u = x + C$
$\Rightarrow -x + 2\tan\sqrt{y} = C$

14. $\sqrt{2xy}\,\frac{dy}{dx} = 1 \Rightarrow dy = \frac{1}{\sqrt{2xy}}\,dx \Rightarrow \sqrt{2}\sqrt{y}dy = \frac{1}{\sqrt{x}}\,dx \Rightarrow \sqrt{2}\,y^{1/2}\,dy = x^{-1/2}\,dx \Rightarrow \sqrt{2}\int y^{1/2}\,dy = \int x^{-1/2}\,dx$
$\Rightarrow \sqrt{2}\,\frac{y^{3/2}}{\frac{3}{2}}\,dy = \frac{x^{1/2}}{\frac{1}{2}} + C_1 \Rightarrow \sqrt{2}\,y^{3/2} = 3\sqrt{x} + \frac{3}{2}C_1 \Rightarrow \sqrt{2}\left(\sqrt{y}\right)^3 - 3\sqrt{x} = C$, where $C = \frac{3}{2}C_1$

15. $\sqrt{x}\,\frac{dy}{dx} = e^{y+\sqrt{x}} \Rightarrow \frac{dy}{dx} = \frac{e^y e^{\sqrt{x}}}{\sqrt{x}} \Rightarrow dy = \frac{e^y e^{\sqrt{x}}}{\sqrt{x}}dx \Rightarrow e^{-y}\,dy = \frac{e^{\sqrt{x}}}{\sqrt{x}}\,dx \Rightarrow \int e^{-y}\,dy = \int \frac{e^{\sqrt{x}}}{\sqrt{x}}\,dx$. In the integral on the right-hand side, substitute $u = \sqrt{x} \Rightarrow du = \frac{1}{2\sqrt{x}}\,dx \Rightarrow 2\,du = \frac{1}{\sqrt{x}}\,dx$, and we have $\int e^{-y}\,dy = 2\int e^u\,du$
$\Rightarrow -e^{-y} = 2e^u + C_1 \Rightarrow -e^{-y} = 2e^{\sqrt{x}} + C$, where $C = -C_1$

16. $(\sec x)\,\frac{dy}{dx} = e^{y+\sin x} \Rightarrow \frac{dy}{dx} = e^{y+\sin x}\cos x \Rightarrow dy = \left(e^y e^{\sin x}\cos x\right)dx \Rightarrow e^{-y}\,dy = e^{\sin x}\cos x\,dx$
$\Rightarrow \int e^{-y}\,dy = \int e^{\sin x}\cos x\,dx \Rightarrow -e^{-y} = e^{\sin x} + C_1 \Rightarrow e^{-y} + e^{\sin x} = C$, where $C = -C_1$

17. $\frac{dy}{dx} = 2x\sqrt{1-y^2} \Rightarrow dy = 2x\sqrt{1-y^2}dx \Rightarrow \frac{dy}{\sqrt{1-y^2}} = 2x\,dx \Rightarrow \int \frac{dy}{\sqrt{1-y^2}} = \int 2x\,dx \Rightarrow \sin^{-1} y = x^2 + C$ since $|y| < 1$
$\Rightarrow y = \sin(x^2 + C)$

18. $\frac{dy}{dx} = \frac{e^{2x-y}}{e^{x+y}} \Rightarrow dy = \frac{e^{2x-y}}{e^{x+y}}dx \Rightarrow dy = \frac{e^{2x}e^{-y}}{e^x e^y}dx = \frac{e^x}{e^{2y}}dx \Rightarrow e^{2y}\,dy = e^x\,dx \Rightarrow \int e^{2y}\,dy = \int e^x\,dx \Rightarrow \frac{e^{2y}}{2} = e^x + C_1$
$\Rightarrow e^{2y} - 2e^x = C$ where $C = 2C_1$

19. $y' = x + y \Rightarrow$ slope of 0 for the line $y = -x$.
For $x, y > 0$, $y' = x + y \Rightarrow$ slope > 0 in Quadrant I.
For $x, y < 0$, $y' = x + y \Rightarrow$ slope < 0 in Quadrant III.
For $|y| > |x|$, $y > 0$, $x < 0$, $y' = x + y \Rightarrow$ slope > 0 in Quadrant II above $y = -x$.
For $|y| < |x|$, $y > 0$, $x < 0$, $y' = x + y \Rightarrow$ slope < 0 in Quadrant II below $y = -x$.
For $|y| < |x|$, $x > 0$, $y < 0$, $y' = x + y \Rightarrow$ slope > 0 in Quadrant IV above $y = -x$.
For $|y| > |x|$, $x > 0$, $y < 0$, $y' = x + y \Rightarrow$ slope < 0 in Quadrant IV below $y = -x$.
All of the conditions are seen in slope field (d).

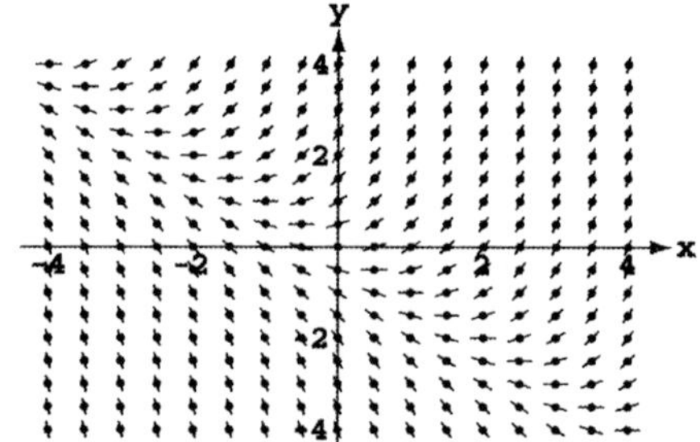

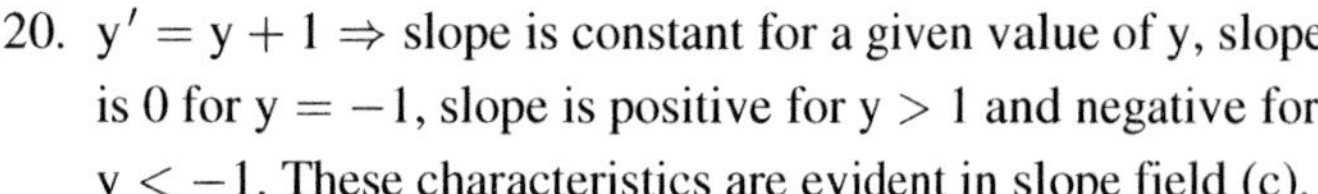
20. $y' = y + 1 \Rightarrow$ slope is constant for a given value of y, slope is 0 for $y = -1$, slope is positive for $y > 1$ and negative for $y < -1$. These characteristics are evident in slope field (c).

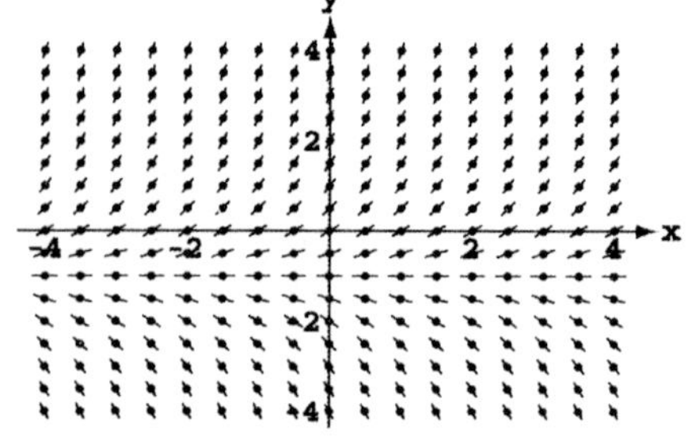

21. $y' = -\frac{x}{y} \Rightarrow$ slope $= 1$ on $y = -x$ and -1 on $y = x$.
$y' = -\frac{x}{y} \Rightarrow$ slope $= 0$ on the y-axis, excluding $(0, 0)$, and is undefined on the x-axis. Slopes are positive for $x > 0, y < 0$ and $x < 0, y > 0$ (Quadrants II and IV), otherwise negative. Field (a) is consistent with these conditions.

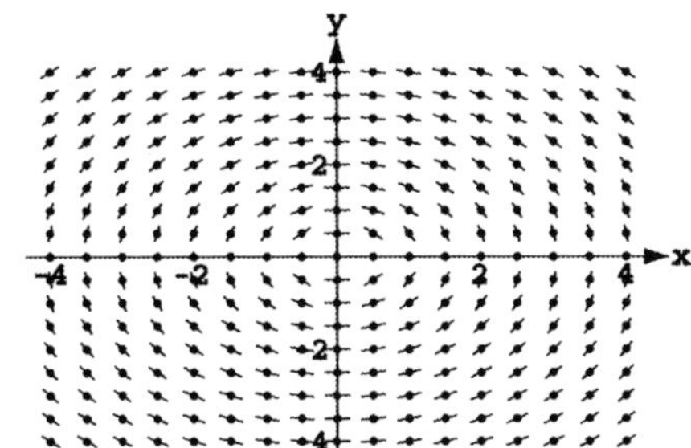

22. $y' = y^2 - x^2 \Rightarrow$ slope is 0 for $y = x$ and for $y = -x$. For $|y| > |x|$ slope is positive and for $|y| < |x|$ slope is negative. Field (b) has these characteristics.

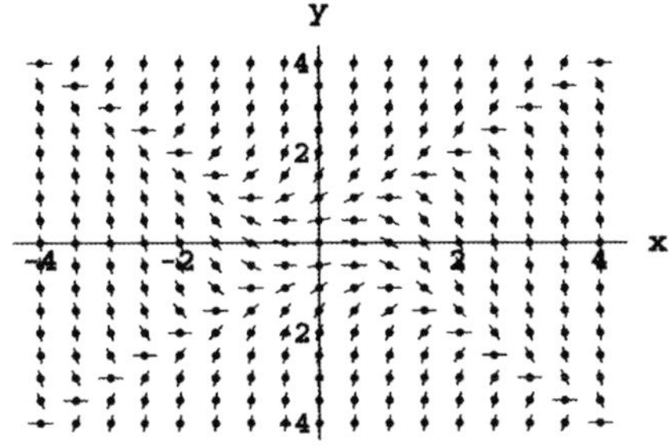

23.

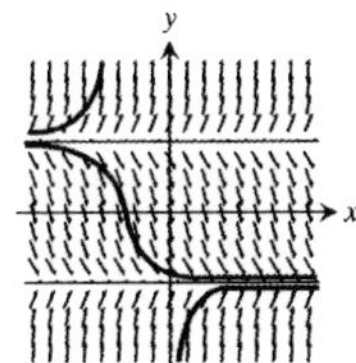

24.

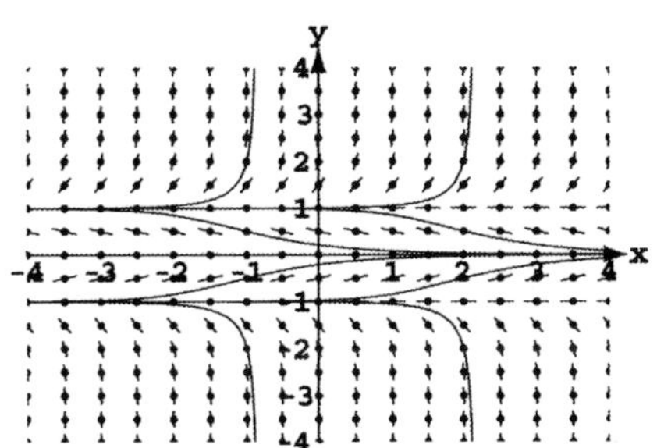

25-36. Example CAS commands:

Maple:

```
ode := diff( y(x), x ) = y(x);
icA := [0, 1];
icB := [0, 2];
icC := [0,-1];
DEplot( ode, y(x), x=0..2, [icA,icB,icC], arrows=slim, linecolor=blue, title="#25 (Section 9.1)" );
```

Mathematica:

To plot vector fields, you must begin by loading a graphics package.

```
<<Graphics`PlotField`
```

To control lengths and appearance of vectors, select the Help browser, type PlotVectorField and select Go.

```
Clear[x, y, f]
yprime = y (2 – y);
pv = PlotVectorField[{1, yprime}, {x, –5, 5}, {y, –4, 6}, Axes → True, AxesLabel → {x, y}];
```

To draw solution curves with Mathematica, you must first solve the differential equation. This will be done with the DSolve command. The y[x] and x at the end of the command specify the dependent and independent variables. The command will not work unless the y in the differential equation is referenced as y[x].

```
equation = y'[x] == y[x] (2 – y[x]) ;
initcond = y[a] == b;
sols = DSolve[{equation, initcond}, y[x], x]
vals = {{0, 1/2}, {0, 3/2}, {0, 2}, {0, 3}}
f[{a_, b_}] = sols[[1, 1, 2]];
```

```
solnset = Map[f, vals]
ps = Plot[Evaluate[solnset, {x, −5, 5}];
Show[pv, ps, PlotRange → {−4, 6}];
```

The code for problems such as 33 & 34 is similar for the direction field, but the analytical solutions involve complicated inverse functions, so the numerical solver NDSolve is used. Note that a domain interval is specified.

```
equation = y'[x] == Cos[2x − y[x]] ;
initcond = y[0] == 2;
sol = NDSolve[{equation, initcond}, y[x], {x, 0, 5}]
ps = Plot[Evaluate[y[x]/.sol, {x, 0, 5}];
N[y[x] /. sol/.x → 2]
Show[pv, ps, PlotRange → {0, 5}];
```

Solutions for 35 can be found one at a time and plots named and shown together. No direction fields here. For 36, the direction field code is similar, but the solution is found implicitly using integrations. The plot requires loading another special graphics package.

```
<<Graphics`ImplicitPlot`
Clear[x,y]
solution[c_] = Integrate[2 (y − 1), y] == Integrate[3x^2 + 4x + 2, x] + c
values = {−6, −4, −2, 0, 2, 4, 6};
solns = Map[solution, values];
ps = ImplicitPlot[solns, {x, −3, 3}, {y, −3, 3}]
Show[pv, ps]
```

25.

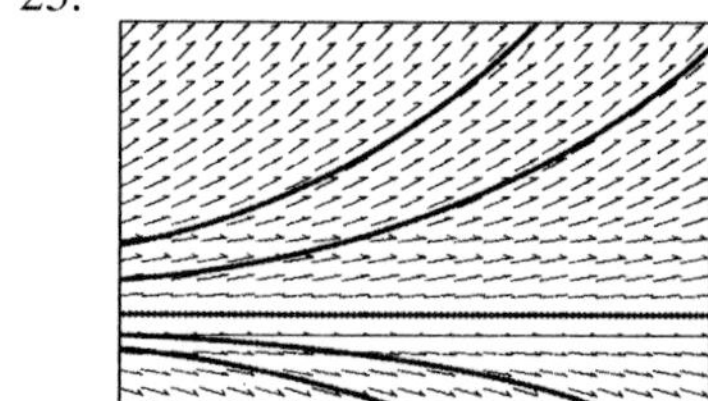

26.

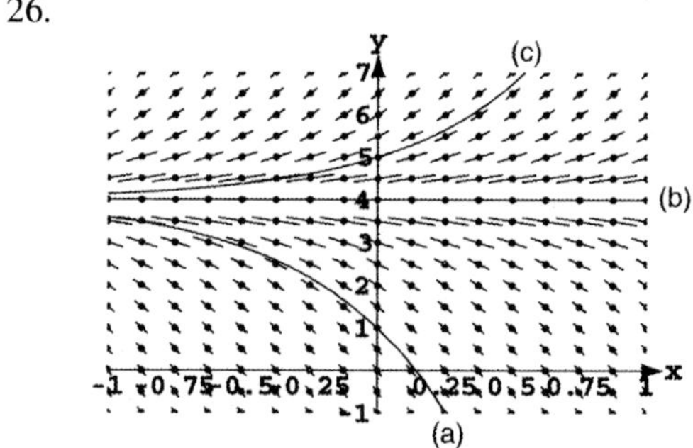

27.

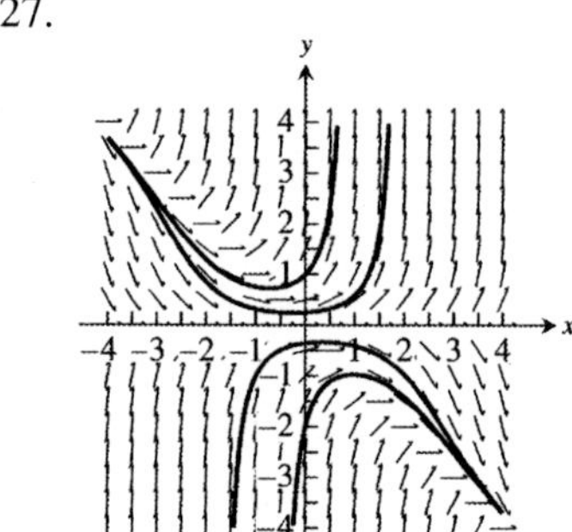

28.

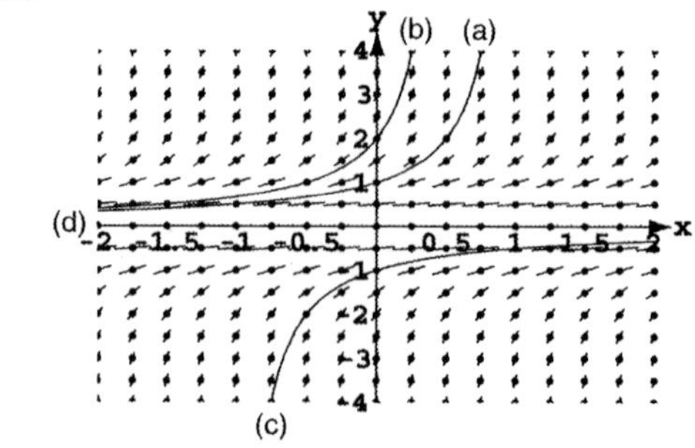

29.

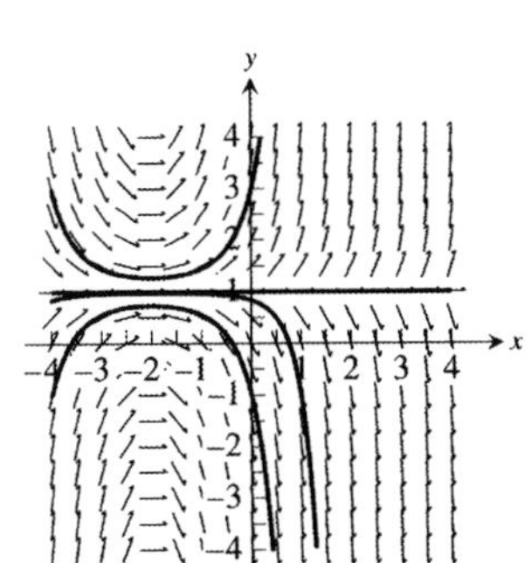

30.

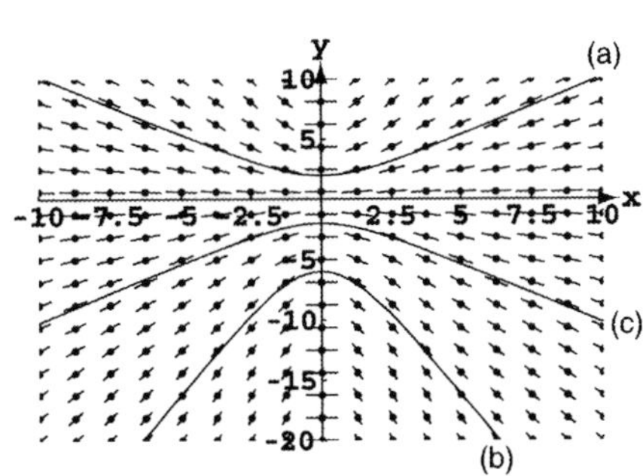

9.2 FIRST-ORDER LINEAR DIFFERENTIAL EQUATIONS

1. $x\,\frac{dy}{dx} + y = e^x \Rightarrow \frac{dy}{dx} + \left(\frac{1}{x}\right) y = \frac{e^x}{x}$, $P(x) = \frac{1}{x}$, $Q(x) = \frac{e^x}{x}$
$\int P(x)\,dx = \int \frac{1}{x}\,dx = \ln|x| = \ln x,\ x > 0 \Rightarrow v(x) = e^{\int P(x)\,dx} = e^{\ln x} = x$
$y = \frac{1}{v(x)} \int v(x)\,Q(x)\,dx = \frac{1}{x} \int x\left(\frac{e^x}{x}\right) dx = \frac{1}{x}\left(e^x + C\right) = \frac{e^x + C}{x},\ x > 0$

2. $e^x\,\frac{dy}{dx} + 2e^x y = 1 \Rightarrow \frac{dy}{dx} + 2y = e^{-x}$, $P(x) = 2$, $Q(x) = e^{-x}$
$\int P(x)\,dx = \int 2\,dx = 2x \Rightarrow v(x) = e^{\int P(x)\,dx} = e^{2x}$
$y = \frac{1}{e^{2x}} \int e^{2x} \cdot e^{-x}\,dx = \frac{1}{e^{2x}} \int e^x\,dx = \frac{1}{e^{2x}}\left(e^x + C\right) = e^{-x} + Ce^{-2x}$

3. $xy' + 3y = \frac{\sin x}{x^2},\ x > 0 \Rightarrow \frac{dy}{dx} + \left(\frac{3}{x}\right) y = \frac{\sin x}{x^3}$, $P(x) = \frac{3}{x}$, $Q(x) = \frac{\sin x}{x^3}$
$\int \frac{3}{x}\,dx = 3\ln|x| = \ln x^3,\ x > 0 \Rightarrow v(x) = e^{\ln x^3} = x^3$
$y = \frac{1}{x^3} \int x^3\left(\frac{\sin x}{x^3}\right) dx = \frac{1}{x^3} \int \sin x\,dx = \frac{1}{x^3}\left(-\cos x + C\right) = \frac{C - \cos x}{x^3},\ x > 0$

4. $y' + (\tan x)\,y = \cos^2 x,\ -\frac{\pi}{2} < x < \frac{\pi}{2} \Rightarrow \frac{dy}{dx} + (\tan x)\,y = \cos^2 x$, $P(x) = \tan x$, $Q(x) = \cos^2 x$
$\int \tan x\,dx = \int \frac{\sin x}{\cos x}\,dx = -\ln|\cos x| = \ln(\cos x)^{-1},\ -\frac{\pi}{2} < x < \frac{\pi}{2} \Rightarrow v(x) = e^{\ln(\cos x)^{-1}} = (\cos x)^{-1}$
$y = \frac{1}{(\cos x)^{-1}} \int (\cos x)^{-1} \cdot \cos^2 x\,dx = (\cos x) \int \cos x\,dx = (\cos x)(\sin x + C) = \sin x \cos x + C\cos x$

5. $x\,\frac{dy}{dx} + 2y = 1 - \frac{1}{x},\ x > 0 \Rightarrow \frac{dy}{dx} + \left(\frac{2}{x}\right) y = \frac{1}{x} - \frac{1}{x^2}$, $P(x) = \frac{2}{x}$, $Q(x) = \frac{1}{x} - \frac{1}{x^2}$
$\int \frac{2}{x}\,dx = 2\ln|x| = \ln x^2,\ x > 0 \Rightarrow v(x) = e^{\ln x^2} = x^2$
$y = \frac{1}{x^2} \int x^2\left(\frac{1}{x} - \frac{1}{x^2}\right) dx = \frac{1}{x^2} \int (x - 1)\,dx = \frac{1}{x^2}\left(\frac{x^2}{2} - x + C\right) = \frac{1}{2} - \frac{1}{x} + \frac{C}{x^2},\ x > 0$

6. $(1 + x)\,y' + y = \sqrt{x} \Rightarrow \frac{dy}{dx} + \left(\frac{1}{1+x}\right) y = \frac{\sqrt{x}}{1+x}$, $P(x) = \frac{1}{1+x}$, $Q(x) = \frac{\sqrt{x}}{1+x}$
$\int \frac{1}{1+x}\,dx = \ln(1 + x)$, since $x > 0 \Rightarrow v(x) = e^{\ln(1+x)} = 1$
$y = \frac{1}{1+x} \int (1 + x)\left(\frac{\sqrt{x}}{1+x}\right) dx = \frac{1}{1+x} \int \sqrt{x}\,dx = \left(\frac{1}{1+x}\right)\left(\frac{2}{3}x^{3/2} + C\right) = \frac{2x^{3/2}}{3(1+x)} + \frac{C}{1+x}$

7. $\frac{dy}{dx} - \frac{1}{2}y = \frac{1}{2}e^{x/2} \Rightarrow P(x) = -\frac{1}{2}$, $Q(x) = \frac{1}{2}e^{x/2} \Rightarrow \int P(x)\,dx = -\frac{1}{2}x \Rightarrow v(x) = e^{-x/2}$
$\Rightarrow y = \frac{1}{e^{-x/2}} \int e^{-x/2}\left(\frac{1}{2}e^{x/2}\right) dx = e^{x/2} \int \frac{1}{2}\,dx = e^{x/2}\left(\frac{1}{2}x + C\right) = \frac{1}{2}xe^{x/2} + Ce^{x/2}$

8. $\frac{dy}{dx} + 2y = 2xe^{-2x} \Rightarrow P(x) = 2$, $Q(x) = 2xe^{-2x} \Rightarrow \int P(x)\,dx = \int 2\,dx = 2x \Rightarrow v(x) = e^{2x}$
$\Rightarrow y = \frac{1}{e^{2x}} \int e^{2x}\left(2xe^{-2x}\right) dx = \frac{1}{e^{2x}} \int 2x\,dx = e^{-2x}\left(x^2 + C\right) = x^2e^{-2x} + Ce^{-2x}$

9. $\frac{dy}{dx} - \left(\frac{1}{x}\right) y = 2 \ln x \Rightarrow P(x) = -\frac{1}{x}, Q(x) = 2 \ln x \Rightarrow \int P(x)\, dx = -\int \frac{1}{x}\, dx = -\ln x, x > 0$
$\Rightarrow v(x) = e^{-\ln x} = \frac{1}{x} \Rightarrow y = x \int \left(\frac{1}{x}\right)(2 \ln x)\, dx = x\left[(\ln x)^2 + C\right] = x(\ln x)^2 + Cx$

10. $\frac{dy}{dx} + \left(\frac{2}{x}\right) y = \frac{\cos x}{x^2}, x > 0 \Rightarrow P(x) = \frac{2}{x}, Q(x) = \frac{\cos x}{x^2} \Rightarrow \int P(x)\, dx = \int \frac{2}{x}\, dx = 2 \ln |x| = \ln x^2, x > 0$
$\Rightarrow v(x) = e^{\ln x^2} = x^2 \Rightarrow y = \frac{1}{x^2} \int x^2 \left(\frac{\cos x}{x^2}\right) dx = \frac{1}{x^2} \int \cos x\, dx = \frac{1}{x^2}(\sin x + C) = \frac{\sin x + C}{x^2}$

11. $\frac{ds}{dt} + \left(\frac{4}{t-1}\right) s = \frac{t+1}{(t-1)^3} \Rightarrow P(t) = \frac{4}{t-1}, Q(t) = \frac{t+1}{(t-1)^3} \Rightarrow \int P(t)\, dt = \int \frac{4}{t-1}\, dt = 4 \ln |t-1| = \ln (t-1)^4$
$\Rightarrow v(t) = e^{\ln (t-1)^4} = (t-1)^4 \Rightarrow s = \frac{1}{(t-1)^4} \int (t-1)^4 \left[\frac{t+1}{(t-1)^3}\right] dt = \frac{1}{(t-1)^4} \int (t^2 - 1)\, dt$
$= \frac{1}{(t-1)^4}\left(\frac{t^3}{3} - t + C\right) = \frac{t^3}{3(t-1)^4} - \frac{t}{(t-1)^4} + \frac{C}{(t-1)^4}$

12. $(t+1)\frac{ds}{dt} + 2s = 3(t+1) + \frac{1}{(t+1)^2} \Rightarrow \frac{ds}{dt} + \left(\frac{2}{t+1}\right) s = 3 + \frac{1}{(t+1)^3} \Rightarrow P(t) = \frac{2}{t+1}, Q(t) = 3 + (t+1)^{-3}$
$\Rightarrow \int P(t)\, dt = \int \frac{2}{t+1}\, dt = 2 \ln |t+1| = \ln (t+1)^2 \Rightarrow v(t) = e^{\ln (t+1)^2} = (t+1)^2$
$\Rightarrow s = \frac{1}{(t+1)^2} \int (t+1)^2 \left[3 + (t+1)^{-3}\right] dt = \frac{1}{(t+1)^2} \int \left[3(t+1)^2 + (t+1)^{-1}\right] dt$
$= \frac{1}{(t+1)^2}\left[(t+1)^3 + \ln |t+1| + C\right] = (t+1) + (t+1)^{-2} \ln (t+1) + \frac{C}{(t+1)^2}, t > -1$

13. $\frac{dr}{d\theta} + (\cot \theta) r = \sec \theta \Rightarrow P(\theta) = \cot \theta, Q(\theta) = \sec \theta \Rightarrow \int P(\theta)\, d\theta = \int \cot \theta\, d\theta = \ln |\sin \theta| \Rightarrow v(\theta) = e^{\ln |\sin \theta|}$
$= \sin \theta$ because $0 < \theta < \frac{\pi}{2} \Rightarrow r = \frac{1}{\sin \theta} \int (\sin \theta)(\sec \theta)\, d\theta = \frac{1}{\sin \theta} \int \tan \theta\, d\theta = \frac{1}{\sin \theta}(\ln |\sec \theta| + C)$
$= (\csc \theta)(\ln |\sec \theta| + C)$

14. $\tan \theta \frac{dr}{d\theta} + r = \sin^2 \theta \Rightarrow \frac{dr}{d\theta} + \frac{r}{\tan \theta} = \frac{\sin^2 \theta}{\tan \theta} \Rightarrow \frac{dr}{d\theta} + (\cot \theta) r = \sin \theta \cos \theta \Rightarrow P(\theta) = \cot \theta, Q(\theta) = \sin \theta \cos \theta$
$\Rightarrow \int P(\theta)\, d\theta = \int \cot \theta\, d\theta = \ln |\sin \theta| = \ln (\sin \theta)$ since $0 < \theta < \frac{\pi}{2} \Rightarrow v(\theta) = e^{\ln (\sin \theta)} = \sin \theta$
$\Rightarrow r = \frac{1}{\sin \theta} \int (\sin \theta)(\sin \theta \cos \theta)\, d\theta = \frac{1}{\sin \theta} \int \sin^2 \theta \cos \theta\, d\theta = \left(\frac{1}{\sin \theta}\right)\left(\frac{\sin^3 \theta}{3} + C\right) = \frac{\sin^2 \theta}{3} + \frac{C}{\sin \theta}$

15. $\frac{dy}{dt} + 2y = 3 \Rightarrow P(t) = 2, Q(t) = 3 \Rightarrow \int P(t)\, dt = \int 2\, dt = 2t \Rightarrow v(t) = e^{2t} \Rightarrow y = \frac{1}{e^{2t}} \int 3e^{2t}\, dt$
$= \frac{1}{e^{2t}}\left(\frac{3}{2} e^{2t} + C\right); y(0) = 1 \Rightarrow \frac{3}{2} + C = 1 \Rightarrow C = -\frac{1}{2} \Rightarrow y = \frac{3}{2} - \frac{1}{2} e^{-2t}$

16. $\frac{dy}{dt} + \frac{2y}{t} = t^2 \Rightarrow P(t) = \frac{2}{t}, Q(t) = t^2 \Rightarrow \int P(t)\, dt = 2 \ln |t| \Rightarrow v(t) = e^{\ln t^2} = t^2 \Rightarrow y = \frac{1}{t^2} \int (t^2)(t^2)\, dt$
$= \frac{1}{t^2} \int t^4\, dt = \frac{1}{t^2}\left(\frac{t^5}{5} + C\right) = \frac{t^3}{5} + \frac{C}{t^2}; y(2) = 1 \Rightarrow \frac{8}{5} + \frac{C}{4} = 1 \Rightarrow C = -\frac{12}{5} \Rightarrow y = \frac{t^3}{5} - \frac{12}{5t^2}$

17. $\frac{dy}{d\theta} + \left(\frac{1}{\theta}\right) y = \frac{\sin \theta}{\theta} \Rightarrow P(\theta) = \frac{1}{\theta}, Q(\theta) = \frac{\sin \theta}{\theta} \Rightarrow \int P(\theta)\, d\theta = \ln |\theta| \Rightarrow v(\theta) = e^{\ln |\theta|} = |\theta|$
$\Rightarrow y = \frac{1}{|\theta|} \int |\theta| \left(\frac{\sin \theta}{\theta}\right) d\theta = \frac{1}{\theta} \int \theta \left(\frac{\sin \theta}{\theta}\right) d\theta$ for $\theta \neq 0 \Rightarrow y = \frac{1}{\theta} \int \sin \theta\, d\theta = \frac{1}{\theta}(-\cos \theta + C)$
$= -\frac{1}{\theta} \cos \theta + \frac{C}{\theta}; y\left(\frac{\pi}{2}\right) = 1 \Rightarrow C = \frac{\pi}{2} \Rightarrow y = -\frac{1}{\theta} \cos \theta + \frac{\pi}{2\theta}$

18. $\frac{dy}{d\theta} - \left(\frac{2}{\theta}\right) y = \theta^2 \sec \theta \tan \theta \Rightarrow P(\theta) = -\frac{2}{\theta}, Q(\theta) = \theta^2 \sec \theta \tan \theta \Rightarrow \int P(\theta)\, d\theta = -2 \ln |\theta| \Rightarrow v(\theta) = e^{-2 \ln |\theta|}$
$= \theta^{-2} \Rightarrow y = \frac{1}{\theta^{-2}} \int (\theta^{-2})(\theta^2 \sec \theta \tan \theta)\, d\theta = \theta^2 \int \sec \theta \tan \theta\, d\theta = \theta^2(\sec \theta + C) = \theta^2 \sec \theta + C\theta^2;$
$y\left(\frac{\pi}{3}\right) = 2 \Rightarrow 2 = \left(\frac{\pi^2}{9}\right)(2) + C\left(\frac{\pi^2}{9}\right) \Rightarrow C = \frac{18}{\pi^2} - 2 \Rightarrow y = \theta^2 \sec \theta + \left(\frac{18}{\pi^2} - 2\right)\theta^2$

19. $(x+1)\frac{dy}{dx} - 2(x^2+x)y = \frac{e^{x^2}}{x+1} \Rightarrow \frac{dy}{dx} - 2\left[\frac{x(x+1)}{x+1}\right]y = \frac{e^{x^2}}{(x+1)^2} \Rightarrow \frac{dy}{dx} - 2xy = \frac{e^{x^2}}{(x+1)^2} \Rightarrow P(x) = -2x,$

$Q(x) = \frac{e^{x^2}}{(x+1)^2} \Rightarrow \int P(x)\,dx = \int -2x\,dx = -x^2 \Rightarrow v(x) = e^{-x^2} \Rightarrow y = \frac{1}{e^{-x^2}}\int e^{-x^2}\left[\frac{e^{x^2}}{(x+1)^2}\right]dx$

$= e^{x^2}\int \frac{1}{(x+1)^2}\,dx = e^{x^2}\left[\frac{(x+1)^{-1}}{-1} + C\right] = -\frac{e^{x^2}}{x+1} + Ce^{x^2};\ y(0) = 5 \Rightarrow -\frac{1}{0+1} + C = 5 \Rightarrow -1 + C = 5$

$\Rightarrow C = 6 \Rightarrow y = 6e^{x^2} - \frac{e^{x^2}}{x+1}$

20. $\frac{dy}{dx} + xy = x \Rightarrow P(x) = x,\ Q(x) = x \Rightarrow \int P(x)\,dx = \int x\,dx = \frac{x^2}{2} \Rightarrow v(x) = e^{x^2/2} \Rightarrow y = \frac{1}{e^{x^2/2}}\int e^{x^2/2}\cdot x\,dx$

$= \frac{1}{e^{x^2/2}}\left(e^{x^2/2} + C\right) = 1 + \frac{C}{e^{x^2/2}};\ y(0) = -6 \Rightarrow 1 + C = -6 \Rightarrow C = -7 \Rightarrow y = 1 - \frac{7}{e^{x^2/2}}$

21. $\frac{dy}{dt} - ky = 0 \Rightarrow P(t) = -k,\ Q(t) = 0 \Rightarrow \int P(t)\,dt = \int -k\,dt = -kt \Rightarrow v(t) = e^{-kt}$

$\Rightarrow y = \frac{1}{e^{-kt}}\int \left(e^{-kt}\right)(0)\,dt = e^{kt}(0 + C) = Ce^{kt};\ y(0) = y_0 \Rightarrow C = y_0 \Rightarrow y = y_0e^{kt}$

22. (a) $\frac{dv}{dt} + \frac{k}{m}v = 0 \Rightarrow P(t) = \frac{k}{m},\ Q(t) = 0 \Rightarrow \int P(t)\,dt = \int \frac{k}{m}\,dt = \frac{k}{m}t = \frac{kt}{m} \Rightarrow v(t) = e^{kt/m}$

$\Rightarrow y = \frac{1}{e^{kt/m}}\int e^{kt/m}\cdot 0\,dt = \frac{C}{e^{kt/m}};\ v(0) = v_0 \Rightarrow \frac{C}{e^{k(0)/m}} = v_0 \Rightarrow C = v_0 \Rightarrow v = v_0\,e^{-(k/m)t}$

(b) $\frac{dv}{dt} = -\frac{k}{m}v \Rightarrow \frac{dv}{v} = -\frac{k}{m}\,dt \Rightarrow \ln v = -\frac{k}{m}t + C \Rightarrow v = e^{-(k/m)t+C} \Rightarrow v = e^{-(k/m)t}\cdot e^C$. Let $e^C = C_1$.

Then $v = \frac{1}{e^{(k/m)t}}\cdot C_1$ and $v(0) = v_0 = \frac{1}{e^{(k/m)(0)}}\cdot C_1 = C_1$. So $v = v_0\,e^{-(k/m)t}$

23. $x\int \frac{1}{x}\,dx = x(\ln|x| + C) = x\ln|x| + Cx \Rightarrow$ (b) is correct

24. $\frac{1}{\cos x}\int \cos x\,dx = \frac{1}{\cos x}(\sin x + C) = \tan x + \frac{C}{\cos x} \Rightarrow$ (b) is correct

25. Let y(t) = the amount of salt in the container and V(t) = the total volume of liquid in the tank at time t. Then, the departure rate is $\frac{y(t)}{V(t)}$ (the outflow rate).

(a) Rate entering $= \frac{2\text{ lb}}{\text{gal}}\cdot\frac{5\text{ gal}}{\text{min}} = 10$ lb/min

(b) Volume $= V(t) = 100$ gal $+ (5t$ gal $- 4t$ gal$) = (100 + t)$ gal

(c) The volume at time t is $(100 + t)$ gal. The amount of salt in the tank at time t is y lbs. So the concentration at any time t is $\frac{y}{100+t}$ lbs/gal. Then, the rate leaving $= \frac{y}{100+t}$ (lbs/gal) $\cdot$ 4 (gal/min)

$= \frac{4y}{100+t}$ lbs/min

(d) $\frac{dy}{dt} = 10 - \frac{4y}{100+t} \Rightarrow \frac{dy}{dt} + \left(\frac{4}{100+t}\right)y = 10 \Rightarrow P(t) = \frac{4}{100+t},\ Q(t) = 10 \Rightarrow \int P(t)\,dt = \int \frac{4}{100+t}\,dt$

$= 4\ln(100+t) \Rightarrow v(t) = e^{4\ln(100+t)} = (100+t)^4 \Rightarrow y = \frac{1}{(100+t)^4}\int (100+t)^4(10\,dt)$

$= \frac{10}{(100+t)^4}\left(\frac{(100+t)^5}{5} + C\right) = 2(100+t) + \frac{C}{(100+t)^4};\ y(0) = 50 \Rightarrow 2(100+0) + \frac{C}{(100+0)^4} = 50$

$\Rightarrow C = -(150)(100)^4 \Rightarrow y = 2(100+t) - \frac{(150)(100)^4}{(100+t)^4} \Rightarrow y = 2(100+t) - \frac{150}{\left(1+\frac{t}{100}\right)^4}$

(e) $y(25) = 2(100+25) - \frac{(150)(100)^4}{(100+25)^4} \approx 188.56$ lbs $\Rightarrow$ concentration $= \frac{y(25)}{\text{volume}} \approx \frac{188.6}{125} \approx 1.5$ lb/gal

26. (a) $\frac{dV}{dt} = (5-3) = 2 \Rightarrow V = 100 + 2t$

The tank is full when $V = 200 = 100 + 2t \Rightarrow t = 50$ min

(b) Let y(t) be the amount of concentrate in the tank at time t.

$\frac{dy}{dt} = \left(\frac{1}{2}\frac{\text{lb}}{\text{gal}}\right)\left(5\frac{\text{gal}}{\text{min}}\right) - \left(\frac{y}{100+2t}\frac{\text{lb}}{\text{gal}}\right)\left(3\frac{\text{gal}}{\text{min}}\right) \Rightarrow \frac{dy}{dt} = \frac{5}{2} - \frac{3}{2}\left(\frac{y}{50+t}\right) \Rightarrow \frac{dy}{dt} + \frac{3}{2(t+50)}y = \frac{5}{2}$

$Q(t) = \frac{5}{2};\ P(t) = \frac{3}{2}\left(\frac{1}{50+t}\right) \Rightarrow \int P(t)\,dt = \frac{3}{2}\int \frac{1}{t+50}\,dt = \frac{3}{2}\ln(t+50)$ since $t + 50 > 0$

$v(t) = e^{\int P(t)\,dt} = e^{\frac{3}{2}\ln(t+50)} = (t+50)^{3/2}$

$y(t) = \frac{1}{(t+50)^{3/2}} \int \frac{5}{2}(t+50)^{3/2}\,dt = (t+50)^{-3/2}\left[(t+50)^{5/2} + C\right] \Rightarrow y(t) = t + 50 + \frac{C}{(t+50)^{3/2}}$

Apply the initial condition (i.e., distilled water in the tank at $t = 0$):

$y(0) = 0 = 50 + \frac{C}{50^{3/2}} \Rightarrow C = -50^{5/2} \Rightarrow y(t) = t + 50 - \frac{50^{5/2}}{(t+50)^{3/2}}$. When the tank is full at $t = 50$,

$y(50) = 100 - \frac{50^{5/2}}{100^{3/2}} \approx 83.22$ pounds of concentrate.

27. Let y be the amount of fertilizer in the tank at time t. Then rate entering $= 1\,\frac{\text{lb}}{\text{gal}} \cdot 1\,\frac{\text{gal}}{\text{min}} = 1\,\frac{\text{lb}}{\text{min}}$ and the volume in the tank at time t is $V(t) = 100\text{ (gal)} + [1\text{ (gal/min)} - 3\text{ (gal/min)}]t\text{ min} = (100 - 2t)$ gal. Hence rate out $= \left(\frac{y}{100-2t}\right)3 = \frac{3y}{100-2t}$ lbs/min $\Rightarrow \frac{dy}{dt} = \left(1 - \frac{3y}{100-2t}\right)$ lbs/min $\Rightarrow \frac{dy}{dt} + \left(\frac{3}{100-2t}\right)y = 1$

$\Rightarrow P(t) = \frac{3}{100-2t},\ Q(t) = 1 \Rightarrow \int P(t)\,dt = \int \frac{3}{100-2t}\,dt = \frac{3\ln(100-2t)}{-2} \Rightarrow v(t) = e^{(-3\ln(100-2t))2}$

$= (100-2t)^{-3/2} \Rightarrow y = \frac{1}{(100-2t)^{-3/2}} \int (100-2t)^{-3/2}\,dt = (100-2t)^{-3/2}\left[\frac{-2(100-2t)^{-1/2}}{-2} + C\right]$

$= (100-2t) + C(100-2t)^{3/2};\ y(0) = 0 \Rightarrow [100 - 2(0)] + C[100 - 2(0)]^{3/2} \Rightarrow C(100)^{3/2} = -100$

$\Rightarrow C = -(100)^{-1/2} = -\frac{1}{10} \Rightarrow y = (100-2t) - \frac{(100-2t)^{3/2}}{10}$. Let $\frac{dy}{dt} = 0 \Rightarrow \frac{dy}{dt} = -2 - \frac{\left(\frac{3}{2}\right)(100-2t)^{1/2}(-2)}{10}$

$= -2 + \frac{3\sqrt{100-2t}}{10} = 0 \Rightarrow 20 = 3\sqrt{100-2t} \Rightarrow 400 = 9(100-2t) \Rightarrow 400 = 900 - 18t \Rightarrow -500 = -18t$

$\Rightarrow t \approx 27.8$ min, the time to reach the maximum. The maximum amount is then

$y(27.8) = [100 - 2(27.8)] - \frac{[100-2(27.8)]^{3/2}}{10} \approx 14.8$ lb

28. Let $y = y(t)$ be the amount of carbon monoxide (CO) in the room at time t. The amount of CO entering the room is $\left(\frac{4}{100} \times \frac{3}{10}\right) = \frac{12}{1000}$ ft^3/min, and the amount of CO leaving the room is $\left(\frac{y}{4500}\right)\left(\frac{3}{10}\right) = \frac{y}{15{,}000}$ ft^3/min. Thus, $\frac{dy}{dt} = \frac{12}{1000} - \frac{y}{15{,}000} \Rightarrow \frac{dy}{dt} + \frac{1}{15{,}000}y = \frac{12}{1000} \Rightarrow P(t) = \frac{1}{15{,}000},\ Q(t) = \frac{12}{1000} \Rightarrow v(t) = e^{t/15{,}000}$

$\Rightarrow y = \frac{1}{e^{t/15{,}000}} \int \frac{12}{1000} e^{t/15{,}000}\,dt \Rightarrow y = e^{-t/15{,}000}\left(\frac{12 \cdot 15{,}000}{1000} e^{t/15{,}000} + C\right) = e^{-t/15{,}000}\left(180e^{t/15{,}000} + C\right);$

$y(0) = 0 \Rightarrow 0 = 1(180 + C) \Rightarrow C = -180 \Rightarrow y = 180 - 180e^{-t/15{,}000}$. When the concentration of CO is 0.01% in the room, the amount of CO satisfies $\frac{y}{4500} = \frac{.01}{100} \Rightarrow y = 0.45$ ft^3. When the room contains this amount we have $0.45 = 180 - 180e^{-t/15{,}000} \Rightarrow \frac{179.55}{180} = e^{-t/15{,}000} \Rightarrow t = -15{,}000\ln\left(\frac{179.55}{180}\right) \approx 37.55$ min.

29. Steady State $= \frac{V}{R}$ and we want $i = \frac{1}{2}\left(\frac{V}{R}\right) \Rightarrow \frac{1}{2}\left(\frac{V}{R}\right) = \frac{V}{R}\left(1 - e^{-Rt/L}\right) \Rightarrow \frac{1}{2} = 1 - e^{-Rt/L} \Rightarrow -\frac{1}{2} = -e^{-Rt/L}$

$\Rightarrow \ln\frac{1}{2} = -\frac{Rt}{L} \Rightarrow -\frac{L}{R}\ln\frac{1}{2} = t \Rightarrow t = \frac{L}{R}\ln 2$ sec

30. (a) $\frac{di}{dt} + \frac{R}{L}i = 0 \Rightarrow \frac{1}{i}\,di = -\frac{R}{L}\,dt \Rightarrow \ln i = -\frac{Rt}{L} + C_1 \Rightarrow i = e^{C_1}e^{-Rt/L} = Ce^{-Rt/L};\ i(0) = I \Rightarrow I = C$

$\Rightarrow i = Ie^{-Rt/L}$ amp

(b) $\frac{1}{2}I = Ie^{-Rt/L} \Rightarrow e^{-Rt/L} = \frac{1}{2} \Rightarrow -\frac{Rt}{L} = \ln\frac{1}{2} = -\ln 2 \Rightarrow t = \frac{L}{R}\ln 2$ sec

(c) $t = \frac{L}{R} \Rightarrow i = Ie^{(-Rt/L)(L/R)} = Ie^{-1}$ amp

31. (a) $t = \frac{3L}{R} \Rightarrow i = \frac{V}{R}\left(1 - e^{(-R/L)(3L/R)}\right) = \frac{V}{R}\left(1 - e^{-3}\right) \approx 0.9502\,\frac{V}{R}$ amp, or about 95% of the steady state value

(b) $t = \frac{2L}{R} \Rightarrow i = \frac{V}{R}\left(1 - e^{(-R/L)(2L/R)}\right) = \frac{V}{R}\left(1 - e^{-2}\right) \approx 0.8647\,\frac{V}{R}$ amp, or about 86% of the steady state value

32. (a) $\frac{di}{dt} + \frac{R}{L}i = \frac{V}{L} \Rightarrow P(t) = \frac{R}{L},\ Q(t) = \frac{V}{L} \Rightarrow \int P(t)\,dt = \int \frac{R}{L}\,dt = \frac{Rt}{L} \Rightarrow v(t) = e^{Rt/L}$

$\Rightarrow i = \frac{1}{e^{Rt/L}} \int e^{Rt/L}\left(\frac{V}{L}\right)dt = \frac{1}{e^{Rt/L}}\left[\frac{L}{R}e^{Rt/L}\left(\frac{V}{L}\right) + C\right] = \frac{V}{R} + Ce^{-(R/L)t}$

(b) $i(0) = 0 \Rightarrow \frac{V}{R} + C = 0 \Rightarrow C = -\frac{V}{R} \Rightarrow i = \frac{V}{R} - \frac{V}{R}e^{-Rt/L}$

(c) $i = \frac{V}{R} \Rightarrow \frac{di}{dt} = 0 \Rightarrow \frac{di}{dt} + \frac{R}{L}i = 0 + \left(\frac{R}{L}\right)\left(\frac{V}{R}\right) = \frac{V}{L} \Rightarrow i = \frac{V}{R}$ is a solution of Eq. (11); $i = Ce^{-(R/L)t}$

33. $y' - y = -y^2$; we have $n = 2$, so let $u = y^{1-2} = y^{-1}$. Then $y = u^{-1}$ and $\frac{du}{dx} = -1y^{-2}\frac{dy}{dx} \Rightarrow \frac{dy}{dx} = -y^2\frac{du}{dx}$

$\Rightarrow -u^{-2}\frac{du}{dx} - u^{-1} = -u^{-2} \Rightarrow \frac{du}{dx} + u = 1$. With $e^{\int dx} = e^x$ as the integrating factor, we have

$e^x\left(\frac{du}{dx}+u\right)=\frac{d}{dx}(e^xu)=e^x$. Integrating, we get $e^xu=e^x+C \Rightarrow u=1+\frac{C}{e^x}=\frac{1}{y} \Rightarrow y=\frac{1}{1+\frac{C}{e^x}}=\frac{e^x}{e^x+C}$

34. $y'-y=xy^2$; we have $n=2$, so let $u=y^{-1}$. Then $y=u^{-1}$ and $\frac{du}{dx}=-y^{-2}\frac{dy}{dx} \Rightarrow \frac{dy}{dx}=-y^2\frac{du}{dx}=-u^{-2}\frac{du}{dx}$.
Substituting: $-u^{-2}\frac{du}{dx}-u^{-1}=xu^{-2} \Rightarrow \frac{du}{dx}+u=-x$. Using $e^{\int dx}=e^x$ as an integrating factor:
$e^x\left(\frac{du}{dx}+u\right)=\frac{d}{dx}(e^xu)=-x\,e^x \Rightarrow e^xu=e^x(1-x)+C \Rightarrow u=\frac{e^x(1-x)+C}{e^x} \Rightarrow y=u^{-1}=\frac{e^x}{e^x-xe^x+C}$

35. $xy'+y=y^{-2} \Rightarrow y'+\left(\frac{1}{x}\right)y=\left(\frac{1}{x}\right)y^{-2}$. Let $u=y^{1-(-2)}=y^3 \Rightarrow y=u^{1/3}$ and $y^{-2}=u^{-2/3}$.
$\frac{du}{dx}=3y^2\frac{dy}{dx} \Rightarrow y'=\frac{dy}{dx}=\left(\frac{1}{3}\right)\left(\frac{du}{dx}\right)(y^{-2})=\left(\frac{1}{3}\right)\left(\frac{du}{dx}\right)\left(u^{-2/3}\right)$. Thus we have
$\left(\frac{1}{3}\right)\left(\frac{du}{dx}\right)\left(u^{-2/3}\right)+\left(\frac{1}{x}\right)u^{1/3}=\left(\frac{1}{x}\right)u^{-2/3} \Rightarrow \frac{du}{dx}+\left(\frac{3}{x}\right)u=\left(\frac{3}{x}\right)1$. The integrating factor, $v(x)$, is
$e^{\int\frac{3}{x}dx}=e^{3\ln x}=e^{\ln x^3}=x^3$. Thus $\frac{d}{dx}(x^3u)=\left(\frac{3}{x}\right)x^3=3x^2 \Rightarrow x^3u=x^3+C \Rightarrow u=1+\frac{C}{x^3}=y^3$
$\Rightarrow y=\left(1+\frac{C}{x^3}\right)^{1/3}$

36. $x^2y'+2xy=y^3 \Rightarrow y'+\left(\frac{2}{x}\right)y=\left(\frac{1}{x^2}\right)y^3$. $P(x)=\left(\frac{2}{x}\right)$, $Q(x)=\left(\frac{1}{x^2}\right)$, $n=3$. Let $u=y^{1-3}=y^{-2}$.
Substituting gives $\frac{du}{dx}+(-2)\left(\frac{2}{x}\right)u=-2\left(\frac{1}{x^2}\right) \Rightarrow \frac{du}{dx}+\left(\frac{-4}{x}\right)u=\frac{-2}{x^2}$. Let the integrating factor, $v(x)$, be
$e^{\int\left(\frac{-4}{x}\right)dx}=e^{\ln x^{-4}}=x^{-4}$. Thus $\frac{d}{dx}(x^{-4}u)=-2x^{-6} \Rightarrow x^{-4}u=\frac{2}{5}x^{-5}+C \Rightarrow u=\frac{2}{5x}+Cx^4=y^{-2}$
$\Rightarrow y=\left(\frac{2}{5x}+Cx^4\right)^{-1/2}$

9.3 EULER'S METHOD

1. $y_1=y_0+\left(1-\frac{y_0}{x_0}\right)dx=-1+\left(1-\frac{-1}{2}\right)(.5)=-0.25$,
$y_2=y_1+\left(1-\frac{y_1}{x_1}\right)dx=-0.25+\left(1-\frac{-0.25}{2.5}\right)(.5)=0.3$,
$y_3=y_2+\left(1-\frac{y_2}{x_2}\right)dx=0.3+\left(1-\frac{0.3}{3}\right)(.5)=0.75$;
$\frac{dy}{dx}+\left(\frac{1}{x}\right)y=1 \Rightarrow P(x)=\frac{1}{x},\ Q(x)=1 \Rightarrow \int P(x)\,dx=\int\frac{1}{x}\,dx=\ln|x|=\ln x,\ x>0 \Rightarrow v(x)=e^{\ln x}=x$
$\Rightarrow y=\frac{1}{x}\int x\cdot 1\,dx=\frac{1}{x}\left(\frac{x^2}{2}+C\right);\ x=2,\ y=-1 \Rightarrow -1=1+\frac{C}{2} \Rightarrow C=-4 \Rightarrow y=\frac{x}{2}-\frac{4}{x}$
$\Rightarrow y(3.5)=\frac{3.5}{2}-\frac{4}{3.5}=\frac{4.25}{7}\approx 0.6071$

2. $y_1=y_0+x_0(1-y_0)\,dx=0+1(1-0)(.2)=.2$,
$y_2=y_1+x_1(1-y_1)\,dx=.2+1.2(1-.2)(.2)=.392$,
$y_3=y_2+x_2(1-y_2)\,dx=.392+1.4(1-.392)(.2)=.5622$;
$\frac{dy}{1-y}=x\,dx \Rightarrow -\ln|1-y|=\frac{x^2}{2}+C;\ x=1,\ y=0 \Rightarrow -\ln 1=\frac{1}{2}+C \Rightarrow C=-\frac{1}{2} \Rightarrow \ln|1-y|=-\frac{x^2}{2}+\frac{1}{2}$
$\Rightarrow y=1-e^{(1-x^2)/2} \Rightarrow y(1.6)\approx .5416$

3. $y_1=y_0+(2x_0y_0+2y_0)\,dx=3+[2(0)(3)+2(3)](.2)=4.2$,
$y_2=y_1+(2x_1y_1+2y_1)\,dx=4.2+[2(.2)(4.2)+2(4.2)](.2)=6.216$,
$y_3=y_2+(2x_2y_2+2y_2)\,dx=6.216+[2(.4)(6.216)+2(6.216)](.2)=9.6969$;
$\frac{dy}{dx}=2y(x+1) \Rightarrow \frac{dy}{y}=2(x+1)\,dx \Rightarrow \ln|y|=(x+1)^2+C;\ x=0,\ y=3 \Rightarrow \ln 3=1+C \Rightarrow C=\ln 3-1$
$\Rightarrow \ln y=(x+1)^2+\ln 3-1 \Rightarrow y=e^{(x+1)^2+\ln 3-1}=e^{\ln 3}e^{x^2+2x}=3e^{x(x+2)} \Rightarrow y(.6)\approx 14.2765$

4. $y_1=y_0+y_0^2(1+2x_0)\,dx=1+1^2[1+2(-1)](.5)=.5$,
$y_2=y_1+y_1^2(1+2x_1)\,dx=.5+(.5)^2[1+2(-.5)](.5)=.5$,
$y_3=y_2+y_2^2(1+2x_2)\,dx=.5+(.5)^2[1+2(0)](.5)=.625$;
$\frac{dy}{y^2}=(1+2x)\,dx \Rightarrow -\frac{1}{y}=x+x^2+C;\ x=-1,\ y=1 \Rightarrow -1=-1+(-1)^2+C \Rightarrow C=-1 \Rightarrow \frac{1}{y}=1-x-x^2$
$\Rightarrow y=\frac{1}{1-x-x^2} \Rightarrow y(.5)=\frac{1}{1-.5-(.5)^2}=4$

5. $y_1 = y_0 + 2x_0 e^{x_0^2}\,dx = 2 + 2(0)(.1) = 2,$
$y_2 = y_1 + 2x_1 e^{x_1^2}\,dx = 2 + 2(.1)\,e^{.1^2}(.1) = 2.0202,$
$y_3 = y_2 + 2x_2 e^{x_2^2}\,dx = 2.0202 + 2(.2)\,e^{.2^2}(.1) = 2.0618,$
$dy = 2xe^{x^2}\,dx \Rightarrow y = e^{x^2} + C;\ y(0) = 2 \Rightarrow 2 = 1 + C \Rightarrow C = 1 \Rightarrow y = e^{x^2} + 1 \Rightarrow y(.3) = e^{.3^2} + 1 \approx 2.0942$

6. $y_1 = y_0 + (y_0 + e^{x_0} - 2)\,dx = 2 + (2 + e^0 - 2)\,(.5) = 2.5,$
$y_2 = y_1 + (y_1 + e^{x_1} - 2)\,dx = 2.5 + (2.5 + e^{.5} - 2)\,(.5) = 3.5744,$
$y_3 = y_2 + (y_2 + e^{x_2} - 2)\,dx = 3.5744 + (3.5744 + e^1 - 2)\,(.5) = 5.7207;$
$\frac{dy}{dx} - y = e^x - 2 \Rightarrow P(x) = -1, Q(x) = e^x - 2 \Rightarrow \int P(x)\,dx = -x \Rightarrow v(x) = e^{-x} \Rightarrow y = \frac{1}{e^{-x}} \int e^{-x}(e^x - 2)\,dx$
$= e^x(x + 2e^{-x} + C);\ y(0) = 2 \Rightarrow 2 = 2 + C \Rightarrow C = 0 \Rightarrow y = xe^x + 2 \Rightarrow y(1.5) = 1.5e^{1.5} + 2 \approx 8.7225$

7. $y_1 = 1 + 1(.2) = 1.2,$
$y_2 = 1.2 + (1.2)(.2) = 1.44,$
$y_3 = 1.44 + (1.44)(.2) = 1.728,$
$y_4 = 1.728 + (1.728)(.2) = 2.0736,$
$y_5 = 2.0736 + (2.0736)(.2) = 2.48832;$
$\frac{dy}{y} = dx \Rightarrow \ln y = x + C_1 \Rightarrow y = Ce^x;\ y(0) = 1 \Rightarrow 1 = Ce^0 \Rightarrow C = 1 \Rightarrow y = e^x \Rightarrow y(1) = e \approx 2.7183$

8. $y_1 = 2 + \left(\frac{2}{1}\right)(.2) = 2.4,$
$y_2 = 2.4 + \left(\frac{2.4}{1.2}\right)(.2) = 2.8,$
$y_3 = 2.8 + \left(\frac{2.8}{1.4}\right)(.2) = 3.2,$
$y_4 = 3.2 + \left(\frac{3.2}{1.6}\right)(.2) = 3.6,$
$y_5 = 3.6 + \left(\frac{3.6}{1.8}\right)(.2) = 4;$
$\frac{dy}{y} = \frac{dx}{x} \Rightarrow \ln y = \ln x + C \Rightarrow y = kx;\ y(1) = 2 \Rightarrow 2 = k \Rightarrow y = 2x \Rightarrow y(2) = 4$

9. $y_1 = -1 + \left[\frac{(-1)^2}{\sqrt{1}}\right](.5) = -.5,$
$y_2 = -.5 + \left[\frac{(-.5)^2}{\sqrt{1.5}}\right](.5) = -.39794,$
$y_3 = -.39794 + \left[\frac{(-.39794)^2}{\sqrt{2}}\right](.5) = -.34195,$
$y_4 = -.34195 + \left[\frac{(-.34195)^2}{\sqrt{2.5}}\right](.5) = -.30497,$
$y_5 = -.27812,\ y_6 = -.25745,\ y_7 = -.24088,\ y_8 = -.2272;$
$\frac{dy}{y^2} = \frac{dx}{\sqrt{x}} \Rightarrow -\frac{1}{y} = 2\sqrt{x} + C;\ y(1) = -1 \Rightarrow 1 = 2 + C \Rightarrow C = -1 \Rightarrow y = \frac{1}{1 - 2\sqrt{x}} \Rightarrow y(5) = \frac{1}{1 - 2\sqrt{5}} \approx -.2880$

10. $y_1 = 1 + (1 - e^0)\left(\frac{1}{3}\right) = 1,$
$y_2 = 1 + \left(1 - e^{2/3}\right)\left(\frac{1}{3}\right) = 0.68408,$
$y_3 = 0.68408 + \left(0.68408 - e^{4/3}\right)\left(\frac{1}{3}\right) = -0.35245,$
$y_4 = -0.35245 + \left(-0.35245 - e^{6/3}\right)\left(\frac{1}{3}\right) = -2.93295,$
$y_5 = -2.93295 + \left(-2.93295 - e^{8/3}\right)\left(\frac{1}{3}\right) = -8.70790,$
$y_6 = -8.7079 + \left(-8.7079 - e^{10/3}\right)\left(\frac{1}{3}\right) = -20.95441;$
$y' - y = -e^{2x} \Rightarrow P(x) = -1, Q(x) = -e^{2x} \Rightarrow \int P(x)\,dx = -x \Rightarrow v(x) = e^{-x} \Rightarrow y = \frac{1}{e^{-x}} \int e^{-x}(-e^{2x})\,dx$
$= e^x(-e^x + C);\ y(0) = 1 \Rightarrow 1 = -1 + C \Rightarrow C = 2 \Rightarrow y = -e^{2x} + 2e^x \Rightarrow y(2) = -e^4 + 2e^2 \approx -39.8200$

11. Let $z_n = y_{n-1} + 2y_{n-1}(x_{n-1} + 1)dx$ and $y_n = y_{n-1} + (y_{n-1}(x_{n-1} + 1) + z_n(x_n + 1))dx$ with $x_0 = 0$, $y_0 = 3$, and $dx = 0.2$. The exact solution is $y = 3e^{x(x+2)}$. Using a programmable calculator or a spreadsheet (I used a spreadsheet) gives the values in the following table.

x	z	y-approx	y-exact	Error
0	---	3	3	0
0.2	4.2	4.608	4.658122	0.050122
0.4	6.81984	7.623475	7.835089	0.211614
0.6	11.89262	13.56369	14.27646	0.712777

12. Let $z_n = y_{n-1} + x_{n-1}(1 - y_{n-1})dx$ and $y_n = y_{n-1} + \left(\frac{x_{n-1}(1 - y_{n-1}) + x_n(1 - z_n)}{2}\right)dx$ with $x_0 = 1$, $y_0 = 0$, and $dx = 0.2$.
The exact solution is $y = 1 - e^{(1-x^2)/2}$. Using a programmable calculator or a spreadsheet (I used a spreadsheet) gives the values in the following table.

x	z	y-approx	y-exact	Error
1	---	0	0	0
1.2	0.2	1.196	0.197481	0.001481
1.4	0.38896	0.378026	0.381217	0.003191
1.6	0.552178	0.536753	0.541594	0.004841

13. $\frac{dy}{dx} = 2xe^{x^2}, y(0) = 2 \Rightarrow y_{n+1} = y_n + 2x_ne^{x_n^2}dx = y_n + 2x_ne^{x_n^2}(0.1) = y_n + 0.2x_ne^{x_n^2}$
On a TI-92 Plus calculator home screen, type the following commands:
2 STO > y: 0 STO > x: y (enter)
y + 0.2*x*e^(x^2) STO > y: x + 0.1 STO > x: y (enter, 10 times)
The last value displayed gives $y_{\text{Euler}}(1) \approx 3.45835$
The exact solution: $dy = 2xe^{x^2}dx \Rightarrow y = e^{x^2} + C; y(0) = 2 = e^0 + C \Rightarrow C = 1 \Rightarrow y = 1 + e^{x^2}$
$\Rightarrow y_{\text{exact}}(1) = 1 + e \approx 3.71828$

14. $\frac{dy}{dx} = y + e^x - 2, y(0) = 2 \Rightarrow y_{n+1} = y_n + (y_n + e^{x_n} - 2)dx = y_n + 0.5(y_n + e^{x_n} - 2)$
On a TI-92 Plus calculator home screen, type the following commands:
2 STO > y: 0 STO > x: y (enter)
y + 0.5*(y + e^x − 2) STO > y: x + 0.5 STO > x: y (enter, 4 times)
The last value displayed gives $y_{\text{Euler}}(2) \approx 9.82187$
The exact solution: $\frac{dy}{dx} - y = e^x - 2 \Rightarrow P(x) = 1, Q(x) = e^x - 2 \Rightarrow \int P(x)\,dx = -x \Rightarrow v(x) = e^{-x}$
$\Rightarrow y = \frac{1}{e^{-x}}\int e^{-x}(e^x - 2)dx = e^x(x + 2e^{-x} + C); y(0) = 2 \Rightarrow 2 = 2 + C \Rightarrow C = 0$
$\Rightarrow y = xe^x + 2 \Rightarrow y_{\text{exact}}(2) = 2e^2 + 2 \approx 16.7781$

15. $\frac{dy}{dx} = \frac{\sqrt{x}}{y}, y > 0, y(0) = 1 \Rightarrow y_{n+1} = y_n + \frac{\sqrt{x_n}}{y_n}dx = y_n + \frac{\sqrt{x_n}}{y_n}(0.1) = y_n + 0.1\frac{\sqrt{x_n}}{y_n}$
On a TI-92 Plus calculator home screen, type the following commands:
1 STO > y: 0 STO > x: y (enter)
y + 0.1*($\sqrt{x}$/y) STO > y: x + 0.1 STO > x: y (enter, 10 times)
The last value displayed gives $y_{\text{Euler}}(1) \approx 1.5000$
The exact solution: $dy = \frac{\sqrt{x}}{y}dx \Rightarrow y\,dy = \sqrt{x}\,dx \Rightarrow \frac{y^2}{2} = \frac{2}{3}x^{3/2} + C; \frac{(y(0))^2}{2} = \frac{1^2}{2} = \frac{1}{2} = \frac{2}{3}(0)^{3/2} + C \Rightarrow C = \frac{1}{2}$
$\Rightarrow \frac{y^2}{2} = \frac{2}{3}x^{3/2} + \frac{1}{2} \Rightarrow y = \sqrt{\frac{4}{3}x^{3/2} + 1} \Rightarrow y_{\text{exact}}(1) = \sqrt{\frac{4}{3}(1)^{3/2} + 1} \approx 1.5275$

16. $\frac{dy}{dx} = 1 + y^2, y(0) = 0 \Rightarrow y_{n+1} = y_n + (1 + y_n^2)dx = y_n + (1 + y_n^2)(0.1) = y_n + 0.1(1 + y_n^2)$
On a TI-92 Plus calculator home screen, type the following commands:
0 STO > y: 0 STO > x: y (enter)
y + 0.1*(1 + y^2) STO > y: x + 0.1 STO > x: y (enter, 10 times)
The last value displayed gives $y_{\text{Euler}}(1) \approx 1.3964$
The exact solution: $dy = (1 + y^2)dx \Rightarrow \frac{dy}{1+y^2} = dx \Rightarrow \tan^{-1}y = x + C; \tan^{-1}y(0) = \tan^{-1}0 = 0 = 0 + C \Rightarrow C = 0$

$\Rightarrow \tan^{-1} y = x \Rightarrow y = \tan x \Rightarrow y_{exact}(1) = \tan 1 \approx 1.5574$

17. (a) $\frac{dy}{dx} = 2y^2(x-1) \Rightarrow \frac{dy}{y^2} = 2(x-1)dx \Rightarrow \int y^{-2}dy = \int (2x-2)dx \Rightarrow -y^{-1} = x^2 - 2x + C$
Initial value: $y(2) = -\frac{1}{2} \Rightarrow 2 = 2^2 - 2(2) + C \Rightarrow C = 2$
Solution: $-y^{-1} = x^2 - 2x + 2$ or $y = -\frac{1}{x^2-2x+2}$
$y(3) = -\frac{1}{3^2-2(3)+2} = -\frac{1}{5} = -0.2$
(b) To find the approximation, set $y_1 = 2y^2(x-1)$ and use EULERT with initial values $x = 2$ and $y = -\frac{1}{2}$ and step size 0.2 for 5 Points. This gives $y(3) \approx -0.1851$; error ≈ 0.0149.
(c) Use step size 0.1 for 10 points. This gives $y(3) \approx -0.1929$; error ≈ 0.0071.
(d) Use step size 0.05 for 20 points. This gives $y(3) \approx -0.1965$; error ≈ 0.0035.

18. (a) $\frac{dy}{dx} = y - 1 \Rightarrow \int \frac{dy}{y-1} = \int dx \Rightarrow \ln|y-1| = x + C \Rightarrow |y-1| = e^{x+C} \Rightarrow y - 1 = \pm e^C e^x \Rightarrow y = Ae^x + 1$
Initial value: $y(0) = 3 \Rightarrow 3 = Ae^0 + 1 \Rightarrow A = 2$
Solution: $y = 2e^x + 1$
$y(1) = 2e + 1 \approx 6.4366$
(b) To find the approximation, set $y_1 = y - 1$ and use a graphing calculator or CAS with initial values $x = 0$ and $y = 3$ and step size 0.2 for 5 Points. This gives $y(1) \approx 5.9766$; error ≈ 0.4599
(c) Use step size 0.1 for 10 points. This gives $y(1) \approx 6.1875$; error ≈ 0.2491.
(d) Use step size 0.05 for 20 points. This gives $y(1) \approx 6.3066$; error ≈ 0.1300.

19. The exact solution is $y = \frac{-1}{x^2-2x+2}$, so $y(3) = -0.2$. To find the approximation, let $z_n = y_{n-1} + 2y_{n-1}^2(x_{n-1}-1)dx$ and $y_n = y_{n-1} + (y_{n-1}^2(x_{n-1}-1) + z_n^2(x_n^2-1))dx$ with initial values $x_0 = 2$ and $y_0 = -\frac{1}{2}$. Use a spreadsheet, graphing calculator, or CAS as indicated in parts (a) through (d).
(a) Use $dx = 0.2$ with 5 steps to obtain $y(3) \approx -0.2024 \Rightarrow$ error ≈ 0.0024.
(b) Use $dx = 0.1$ with 10 steps to obtain $y(3) \approx -0.2005 \Rightarrow$ error ≈ 0.0005.
(c) Use $dx = 0.05$ with 20 steps to obtain $y(3) \approx -0.2001 \Rightarrow$ error ≈ 0.0001.
(d) Each time the step size is cut in half, the error is reduced to approximately one-fourth of what it was for the larger step size.

20. The exact solution is $y = 2e^x + 1$, so $y(1) = 2e + 1 \approx 6.4366$. To find the approximation, let $z_n = y_{n-1} + (y_{n-1} - 1)dx$ and $y_n = y_{n-1} + \left(\frac{y_{n-1}+z_n-2}{2}\right)dx$ with initial value $y_n = 3$. Use a spreadsheet, graphing calculator, or CAS as indicated in parts (a) through (d).
(a) Use $dx = 0.2$ with 5 steps to obtain $y(1) \approx 6.4054 \Rightarrow$ error ≈ 0.0311.
(b) Use $dx = 0.1$ with 10 steps to obtain $y(1) \approx 6.4282 \Rightarrow$ error ≈ 0.0084
(c) Use $dx = 0.05$ with 20 steps to obtain $y(1) \approx 6.4344 \Rightarrow$ error ≈ 0.0022
(d) Each time the step size is cut in half, the error is reduced to approximately one-fourth of what it was for the larger step size.

13-16. Example CAS commands:

Maple:

```
ode := diff( y(x), x ) = 2*x*exp(x^2);ic := y(0)=2;
xstar := 1;
dx := 0.1;
approx := dsolve( {ode,ic}, y(x), numeric, method=classical[foreuler], stepsize=dx ):
approx(xstar);
exact := dsolve( {ode,ic}, y(x) );
eval( exact, x=xstar );
```

```
evalf( % );
```

17. Example CAS commands:

Maple:

```
ode := diff( y(x), x ) = 2*y(x)*(x-1);ic := y(2)=-1/2;
xstar := 3;
exact := dsolve( {ode,ic}, y(x) );                  # (a)
eval( exact, x=xstar );
evalf( % );
approx1 := dsolve( {ode,ic}, y(x),                  # (b)
            numeric, method=classical[foreuler], stepsize=0.2 ):
approx1(xstar);
approx2 := dsolve( {ode,ic}, y(x),                  # (c)
            numeric, method=classical[foreuler], stepsize=0.1 ):
approx2(xstar);
approx3 := dsolve( {ode,ic}, y(x),                  # (d)
            numeric, method=classical[foreuler], stepsize=0.05 ):
approx3(xstar);
```

19. Example CAS commands:

Maple:

```
ode := diff( y(x), x ) = 2*y(x)*(x-1);ic := y(2)=-1/2;
xstar := 3;
approx1 := dsolve( {ode,ic}, y(x),              # (a)
            numeric, method=classical[heunform], stepsize=0.2 ):
approx1(xstar);
approx2 := dsolve( {ode,ic}, y(x),              # (b)
            numeric, method=classical[heunform], stepsize=0.1 ):
approx2(xstar);
approx3 := dsolve( {ode,ic}, y(x),              # (c)
            numeric, method=classical[heunform], stepsize=0.05 ):
approx3(xstar);
```

21. Example CAS commands:

Maple:

```
ode := diff( y(x), x ) = x + y(x);ic := y(0)=-7/10;
x0 := -4;x1 := 4;y0 := -4; y1 := 4;
b := 1;
P1 := DEplot( ode, y(x), x=x0..x1, y=y0..y1, arrows=thin, title="#21(a) (Section 9.3)" ):
P1;
Ygen := unapply( rhs(dsolve( ode, y(x) )), x,_C1 );                                # (b)
P2 := seq( plot( Ygen(x,c), x=x0..x1, y=y0..y1, color=blue ), c=-2..2 ):           # (c)
display( [P1,P2], title="#21(c) (Section 9.3)" );
CC := solve( Ygen(0,C)=rhs(ic), C );                                                # (d)
Ypart := Ygen(x,CC);
P3 := plot( Ypart, x=0..b, title="#21(d) (Section 9.3)" ):
P3;
euler4 := dsolve( {ode,ic}, numeric, method=classical[foreuler], stepsize=(x1-x0)/4 ):   # (e)
P4 := odeplot( euler4, [x,y(x)], x=0..b, numpoints=4, color=blue ):
```

```
display( [P3,P4], title="#21(e) (Section 9.3)" );
euler8 := dsolve( {ode,ic}, numeric, method=classical[foreuler], stepsize=(x1-x0)/8 ):   # (f)
P5 := odeplot( euler8, [x,y(x)], x=0..b, numpoints=8, color=green ):
euler16 := dsolve( {ode,ic}, numeric, method=classical[foreuler], stepsize=(x1-x0)/16 ):
P6 := odeplot( euler16, [x,y(x)], x=0..b, numpoints=16, color=pink ):
euler32 := dsolve( {ode,ic}, numeric, method=classical[foreuler], stepsize=(x1-x0)/32 ):
P7 := odeplot( euler32, [x,y(x)], x=0..b, numpoints=32, color=cyan ):
display( [P3,P4,P5,P6,P7], title="#21(f) (Section 9.3)" );
<< N | h        | `percent error` >,                                                   # (g)
 < 4 | (x1-x0)/ 4 | evalf[5]( abs(1-eval(y(x),euler4(b))/eval(Ypart,x=b))*100 ) >,
 < 8 | (x1-x0)/ 8 | evalf[5]( abs(1-eval(y(x),euler8(b))/eval(Ypart,x=b))*100 ) >,
 < 16 | (x1-x0)/16 | evalf[5]( abs(1-eval(y(x),euler16(b))/eval(Ypart,x=b))*100 ) >,
 < 32 | (x1-x0)/32 | evalf[5]( abs(1-eval(y(x),euler32(b))/eval(Ypart,x=b))*100 ) > >;
```

13-24. Example CAS commands:

Mathematica: (assigned functions, step sizes, and values for initial conditions may vary)

For exercises 13 - 20, find the exact solution as follows. Set up two error lists.

```
Clear[x, y, f]
f[x_,y_]:= 2 y^2 (x – 1)
a = 2; b = –1/2;
xstar = 3;
desol=DSolve[{y'[x] == f[x, y[x]], y[a] == b}, y[x], x] //Simplify
actual[x_] = desol[[1, 1, 2]];
{xstar, actual[xstar]}
errorlisteuler = { };
errorlisteulerimp = { };
pa = Plot[actual[x], {x, a, xstar}]
```

Euler's method with error at x*. The **Do** command is used with a sequence of commands that are repeated n times.

```
a = 2; b = -1/2;
dx = 0.2;
xstar = 3; n = (xstar – a) /dx;
solnslist = {{a,b}};
Do[ {new = b +  f[a,b] dx, a = a + dx, b = new, AppendTo[solnslist, {a,b}]},{n}]
solnslist
error= actual[xstar] – solnslist[[n, 2]]
relativeerror= error / actual[xstar]
AppendTo[errorlisteuler, error]
pe = ListPlot[solnslist, PlotStyle → {Hue[.4], PointSize[0.02]}]
Show[pa, pe]
```

Rerun with different values for dx, starting from largest to smallest. After doing this, observe what happens to the error as the step size decreases by entering the input command: errorlisteuler

Improved Euler's method. with error at x*

```
a = 2; b = –1/2;
dx = 0.2;
xstar = 3; n = (xstar – a) /dx;
solnslist = {{a,b}};
Do[{new1 = b + f[a,b] dx, new2 = b + (f[a, b] + f[a+dx, new1])/2 dx, a = a + dx, b = new2,
        AppendTo[solnslist, {a,b}]},{n}]
solnslist
error= actual[xstar] – solnslist[[n, 2]
```

```
relativeerror= error / actual[xstar]
AppendTo[errorlisteulerimp, error]
peimp = ListPlot[solnslist, PlotStyle → {Hue[.8], PointSize[0.02]}]
Show[pa, peimp]
```

Rerun with different values for dx, starting from largest to smallest. After doing this, observe what happens to the error as the step size decreases by entering the input command: errorlisteulerimp

You can also type Show[pa, pe, peimp]. This would be appropriate for a fixed value of dx with each method.

You can also make a list of relative errors.

Problems 21 - 24 involve use of code from section 9.1 together with the above code for Euler's method.

9.4 GRAPHICAL SOUTIONS OF AUTONOMOUS DIFFERENTIAL EQUATIONS

1. $y' = (y+2)(y-3)$
   (a) $y = -2$ is a stable equilibrium value and $y = 3$ is an unstable equilibrium.
   (b) $y'' = (2y-1)y' = 2(y+2)\left(y-\frac{1}{2}\right)(y-3)$

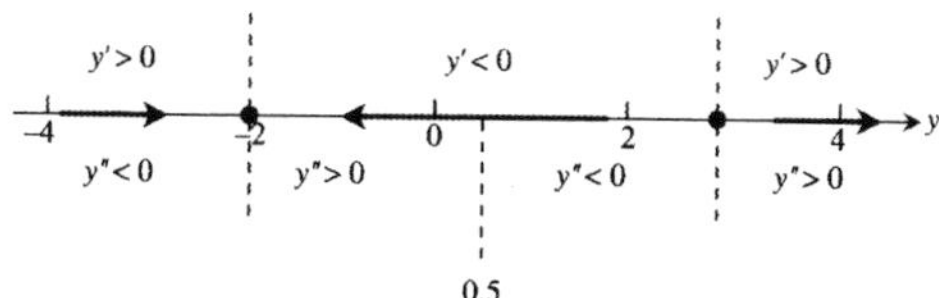

   (c)

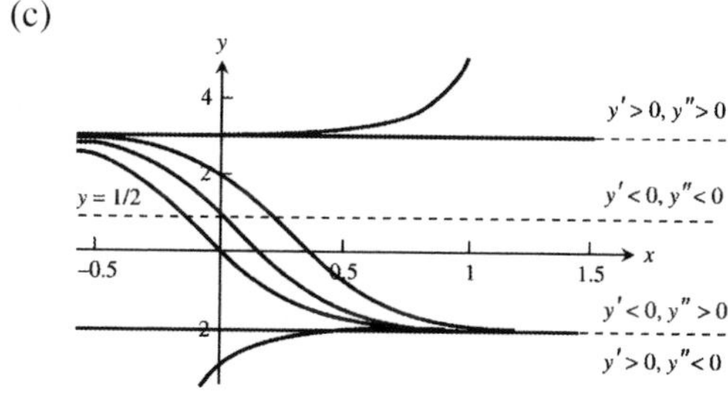

2. $y' = (y+2)(y-2)$
   (a) $y = -2$ is a stable equilibrium value and $y = 2$ is an unstable equilibrium.
   (b) $y'' = 2yy' = 2(y+2)y(y-2)$

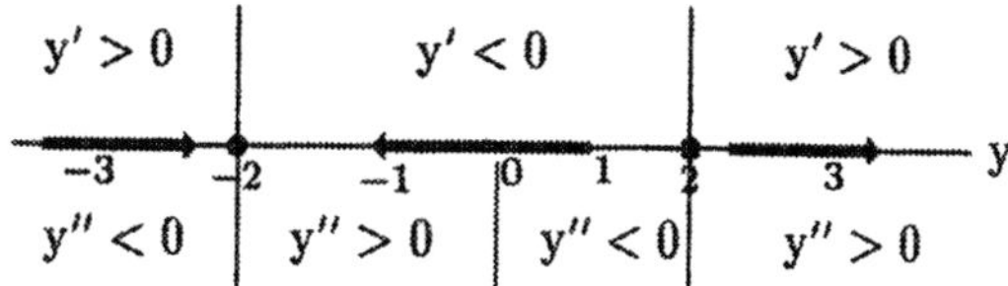

   (c)

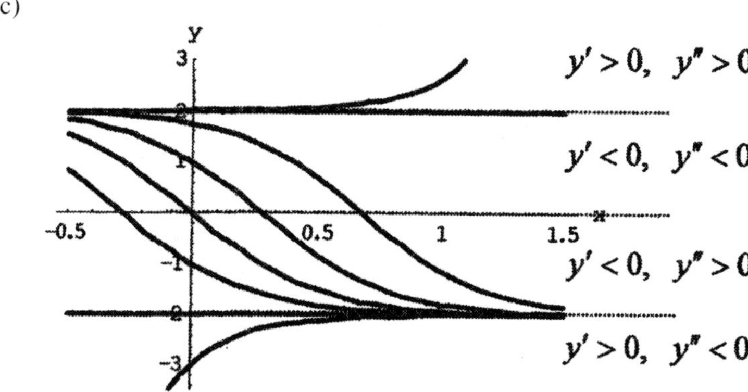

3. $y' = y^3 - y = (y+1)y(y-1)$
   (a) $y = -1$ and $y = 1$ is an unstable equilibrium and $y = 0$ is a stable equilibrium value.
   (b) $y'' = (3y^2 - 1)y' = 3(y+1)\left(y + \frac{1}{\sqrt{3}}\right)y\left(y - \frac{1}{\sqrt{3}}\right)(y-1)$

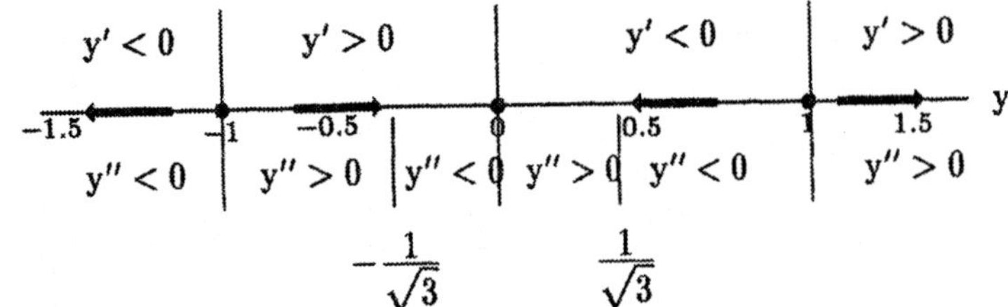

   (c)

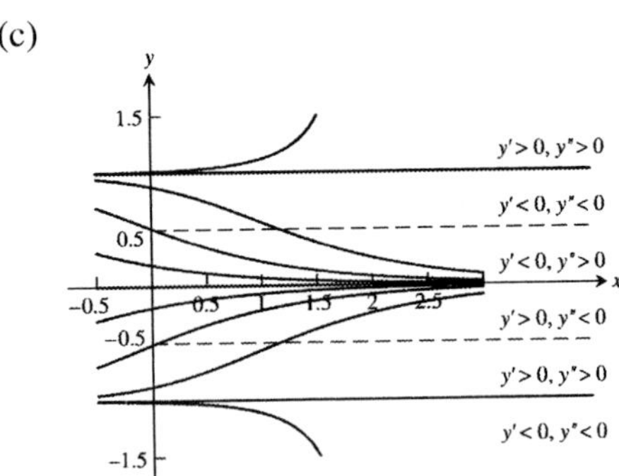

4. $y' = y(y-2)$
   (a) $y = 0$ is a stable equilibrium value and $y = 2$ is an unstable equilibrium.
   (b) $y'' = (2y - 2)y' = 2y(y-1)(y-2)$

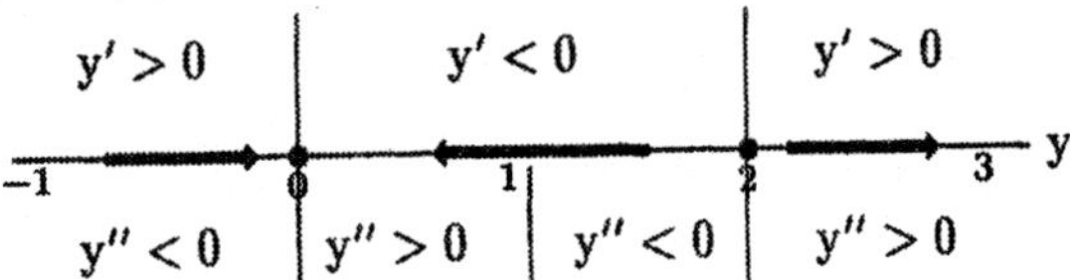

   (c)

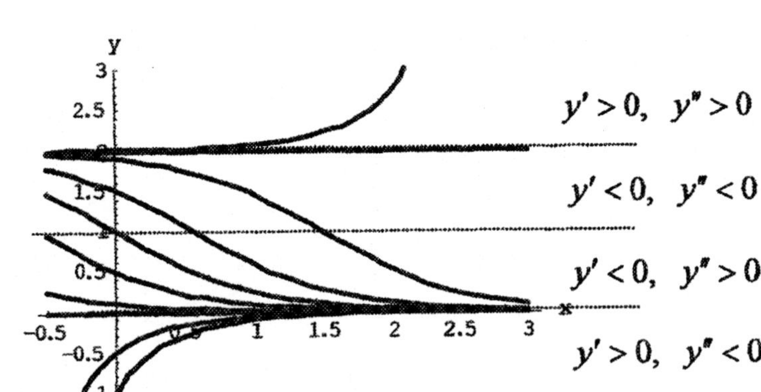

5. $y' = \sqrt{y}, y > 0$
   (a) There are no equilibrium values.
   (b) $y'' = \frac{1}{2\sqrt{y}}\, y' = \frac{1}{2\sqrt{y}}\sqrt{y} = \frac{1}{2}$

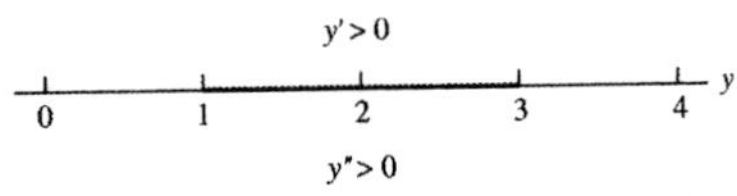

(c)

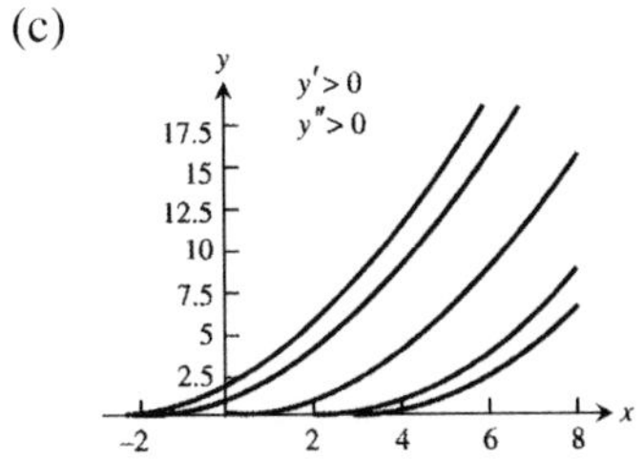

6. $y' = y - \sqrt{y},\ y > 0$
   (a) $y = 1$ is an unstable equilibrium.
   (b) $y'' = \left(1 - \frac{1}{2\sqrt{y}}\right) y' = \left(1 - \frac{1}{2\sqrt{y}}\right)\left(y - \sqrt{y}\right) = \left(\sqrt{y} - \frac{1}{2}\right)\left(\sqrt{y} - 1\right)$

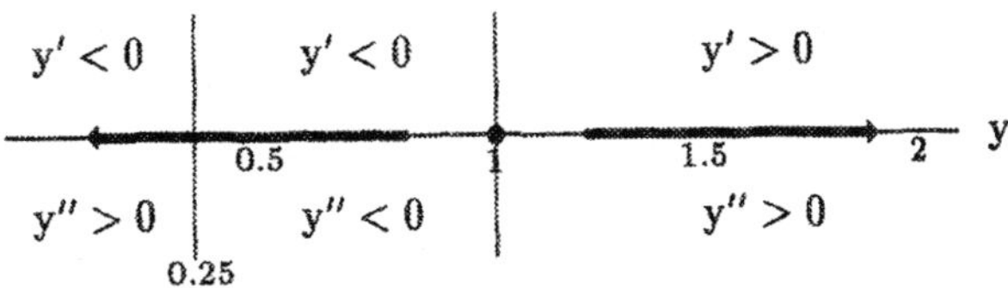

   (c)

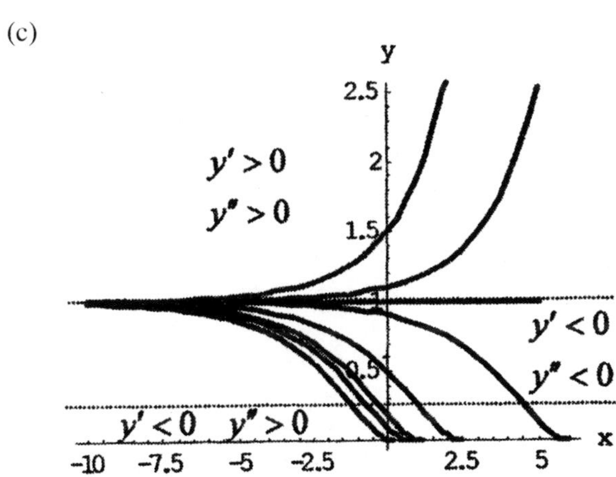

7. $y' = (y - 1)(y - 2)(y - 3)$
   (a) $y = 1$ and $y = 3$ is an unstable equilibrium and $y = 2$ is a stable equilibrium value.
   (b) $y'' = (3y^2 - 12y + 11)(y - 1)(y - 2)(y - 3) = 3(y - 1)\left(y - \frac{6 - \sqrt{3}}{3}\right)(y - 2)\left(y - \frac{6 + \sqrt{3}}{3}\right)(y - 3)$

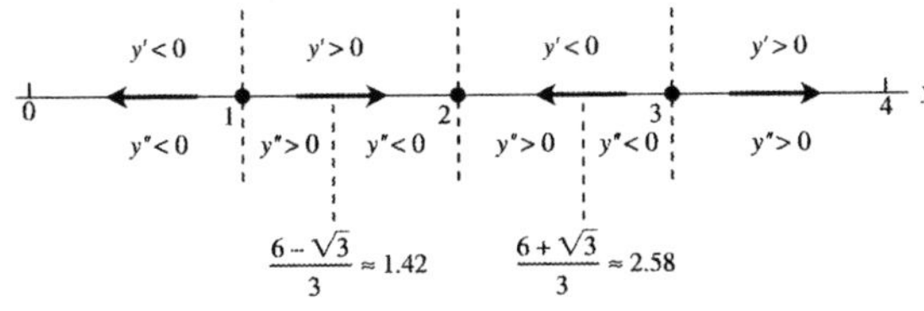

   (c)

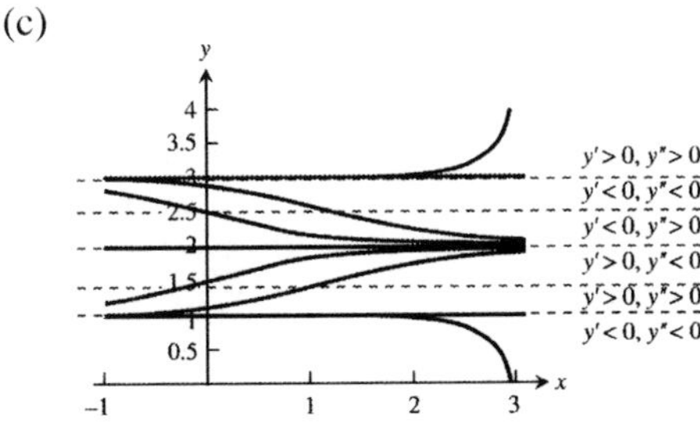

8. $y' = y^3 - y^2 = y^2(y-1)$
   (a) $y = 0$ and $y = 1$ is an unstable equilibrium.
   (b) $y'' = (3y^2 - 2y)(y^3 - y^2) = y^3(3y-2)(y-1)$

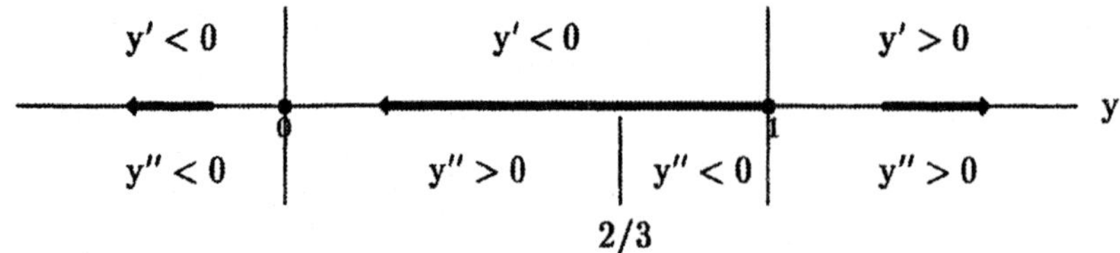

   (c)

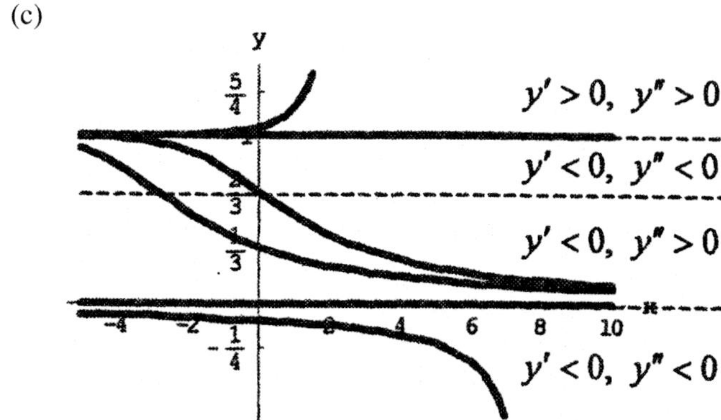

9. $\frac{dP}{dt} = 1 - 2P$ has a stable equilibrium at $P = \frac{1}{2}$. $\frac{d^2P}{dt^2} = -2\frac{dP}{dt} = -2(1-2P)$

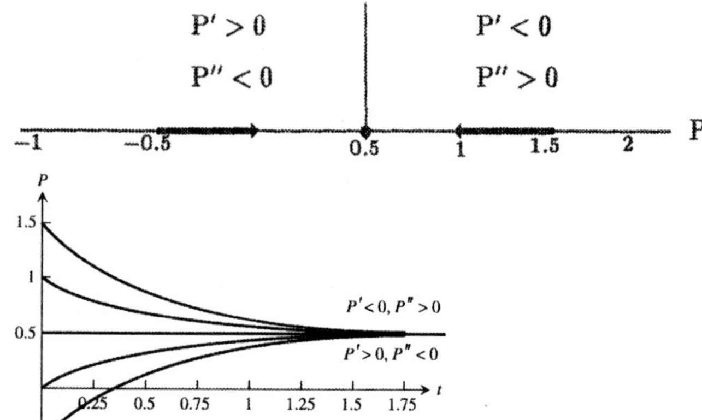

10. $\frac{dP}{dt} = P(1-2P)$ has an unstable equilibrium at $P = 0$ and a stable equilibrium at $P = \frac{1}{2}$.
$\frac{d^2P}{dt^2} = (1-4P)\frac{dP}{dt} = P(1-4P)(1-2P)$

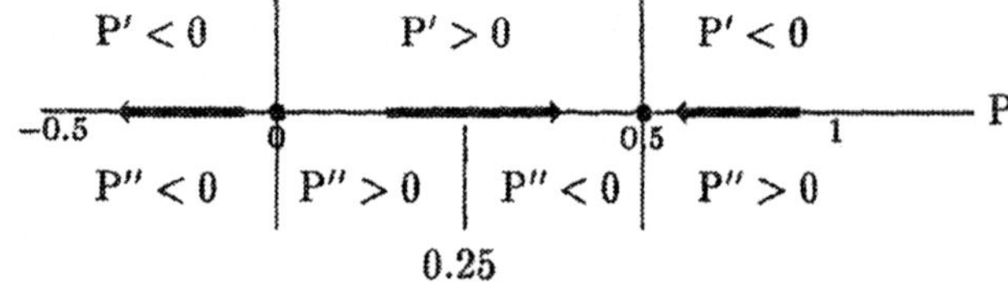

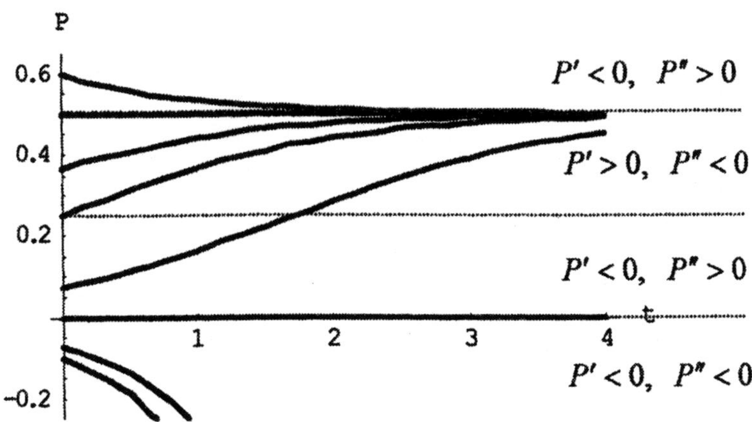

11. $\frac{dP}{dt} = 2P(P-3)$ has a stable equilibrium at $P = 0$ and an unstable equilibrium at $P = 3$.
$\frac{d^2P}{dt^2} = 2(2P-3)\frac{dP}{dt} = 4P(2P-3)(P-3)$

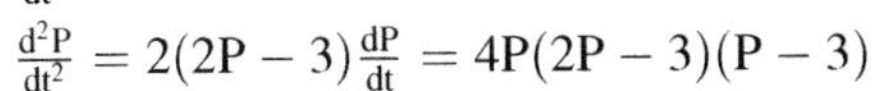

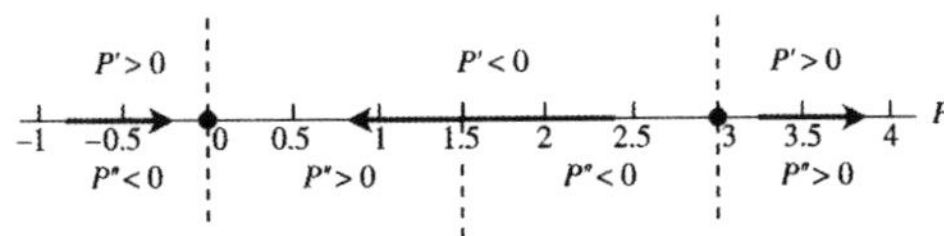

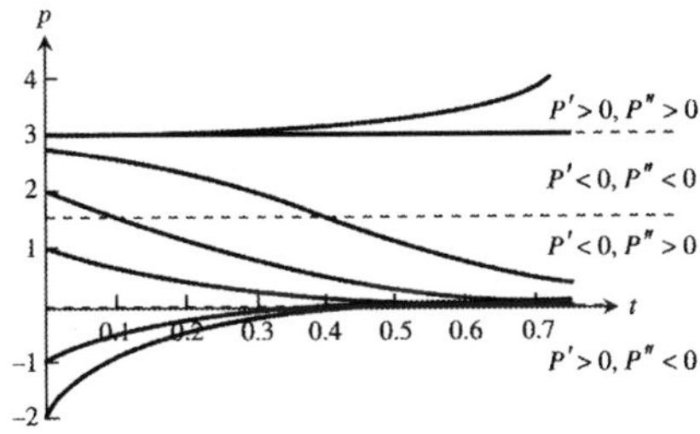

12. $\frac{dP}{dt} = 3P(1-P)\left(P-\frac{1}{2}\right)$ has a stable equilibria at $P = 0$ and $P = 1$ an unstable equilibrium at $P = \frac{1}{2}$.
$\frac{d^2P}{dt^2} = -\frac{3}{2}(6P^2 - 6P+1)\frac{dP}{dt} = \frac{3}{2}P\left(P - \frac{3-\sqrt{3}}{6}\right)\left(P-\frac{1}{2}\right)\left(P - \frac{3+\sqrt{3}}{6}\right)(P-1)$

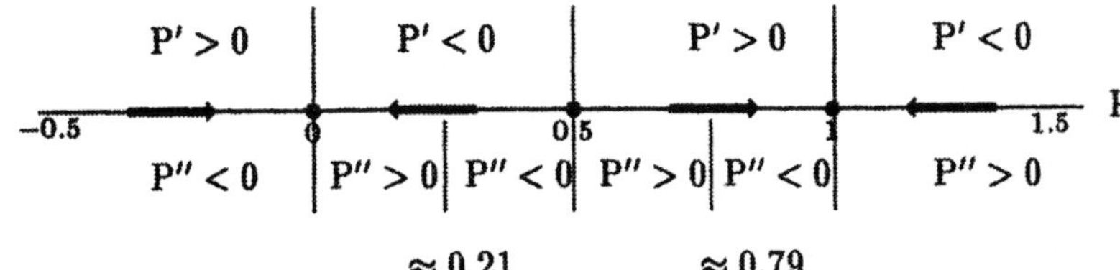

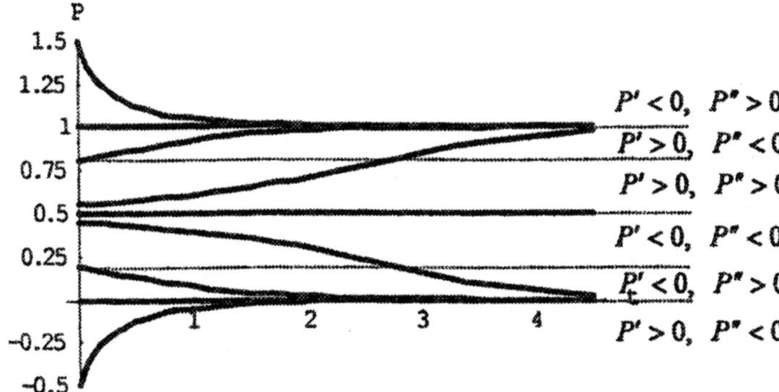

13.

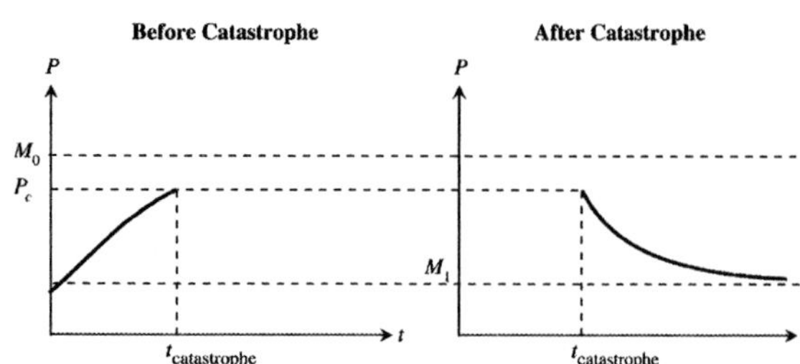

Before the catastrophe, the population exhibits logistic growth and $P(t) \to M_0$, the stable equilibrium. After the catastrophe, the population declines logistically and $P(t) \to M_1$, the new stable equilibrium.

14. $\frac{dP}{dt} = rP(M - P)(P - m)$, r, M, m > 0

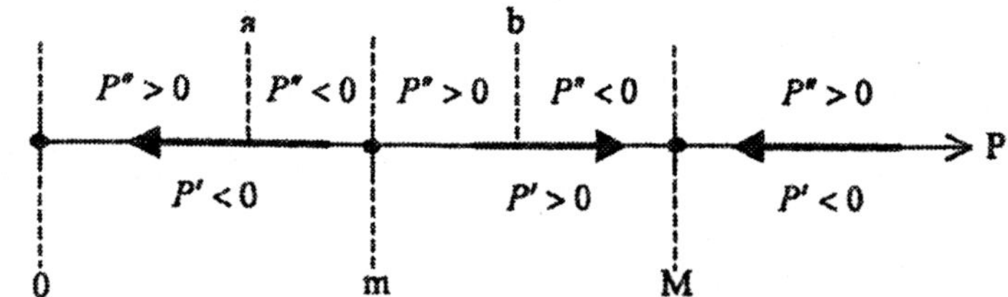

The model has 3 equilibrium points. The rest point $P = 0$, $P = M$ are asymptotically stable while $P = m$ is unstable. For initial populations greater than m, the model predicts P approaches M for large t. For initial populations less than m, the model predicts extinction. Points of inflection occur at $P = a$ and $P = b$ where $a = \frac{1}{3}\left[M + m - \sqrt{M^2 - mM + m^2}\right]$ and $b = \frac{1}{3}\left[M + m + \sqrt{M^2 - mM + m^2}\right]$.

(a) The model is reasonable in the sense that if $P < m$, then $P \to 0$ as $t \to \infty$; if $m < P < M$, then $P \to M$ as $t \to \infty$; if $P > M$, then $P \to M$ as $t \to \infty$.

(b) It is different if the population falls below m, for then $P \to 0$ as $t \to \infty$ (extinction). If is probably a more realistic model for that reason because we know some populations have become extinct after the population level became too low.

(c) For $P > M$ we see that $\frac{dP}{dt} = rP(M - P)(P - m)$ is negative. Thus the curve is everywhere decreasing. Moreover, $P \equiv M$ is a solution to the differential equation. Since the equation satisfies the existence and uniqueness conditions, solution trajectories cannot cross. Thus, $P \to M$ as $t \to \infty$.

(d) See the initial discussion above.

(e) See the initial discussion above.

15. $\frac{dv}{dt} = g - \frac{k}{m}v^2$, g, k, m > 0 and $v(t) \geq 0$

Equilibrium: $\frac{dv}{dt} = g - \frac{k}{m}v^2 = 0 \Rightarrow v = \sqrt{\frac{mg}{k}}$

Concavity: $\frac{d^2v}{dt^2} = -2\left(\frac{k}{m}v\right)\frac{dv}{dt} = -2\left(\frac{k}{m}v\right)\left(g - \frac{k}{m}v^2\right)$

(a)

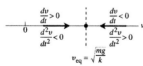

$\frac{dv}{dt} > 0$ $\quad \frac{dv}{dt} < 0$

0 $\quad$ v

$\frac{d^2v}{dt^2} < 0$ $\quad \frac{d^2v}{dt^2} > 0$

$v_{eq} = \sqrt{\frac{mg}{k}}$

(b)

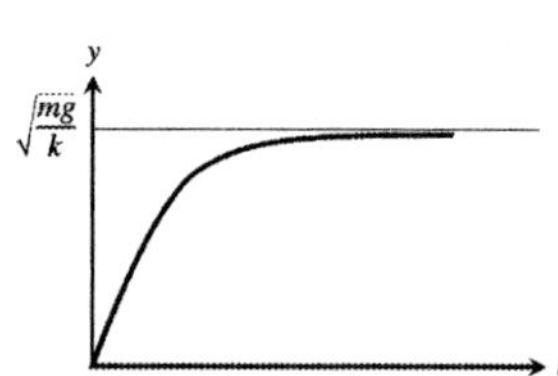

(c) $v_{terminal} = \sqrt{\frac{160}{0.005}} = 178.9\ \frac{ft}{s} = 122$ mph

16. $F = F_p - F_r$
$ma = mg - k\sqrt{v}$
$\frac{dv}{dt} = g - \frac{k}{m}\sqrt{v},\ v(0) = v_0$
Thus, $\frac{dv}{dt} = 0$ implies $v = \left(\frac{mg}{k}\right)^2$, the terminal velocity. If $v_0 < \left(\frac{mg}{k}\right)^2$, the object will fall faster and faster, approaching the terminal velocity; if $v_0 > \left(\frac{mg}{k}\right)^2$, the object will slow down to the terminal velocity.

17. $F = F_p - F_r$
$ma = 50 - 5|v|$
$\frac{dv}{dt} = \frac{1}{m}(50 - 5|v|)$
The maximum velocity occurs when $\frac{dv}{dt} = 0$ or $v = 10\ \frac{ft}{sec}$.

18. (a) The model seems reasonable because the rate of spread of a piece of information, an innovation, or a cultural fad is proportional to the product of the number of individuals who have it (X) and those who do not (N − X). When X is small, there are only a few individuals to spread the item so the rate of spread is slow. On the other hand, when (N − X) is small the rate of spread will be slow because there are only a few indiciduals who can receive it during the interval of time. The rate of spread will be fastest when both X and (N − X) are large because then there are a lot of individuals to spread the item and a lot of individuals to receive it.

(b) There is a stable equilibrium at $X = N$ and an unstable equilibrium at $X = 0$.
$\frac{d^2X}{dt^2} = k\frac{dX}{dt}(N - X) - kX\frac{dX}{dt} = k^2X(N - X)(N - 2X) \Rightarrow$ inflection points at $X = 0$, $X = \frac{N}{2}$, and $X = N$.

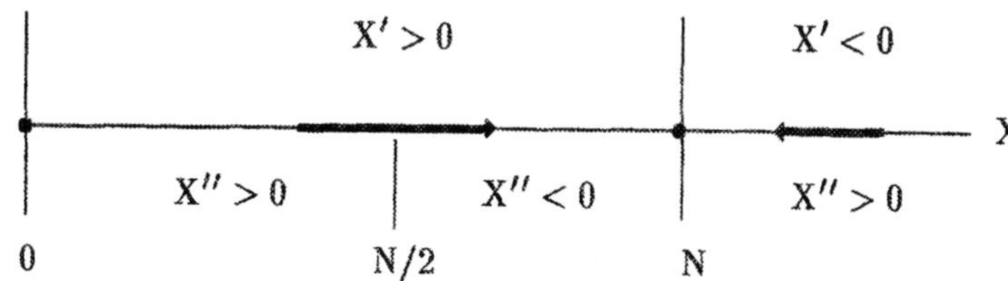

(c)

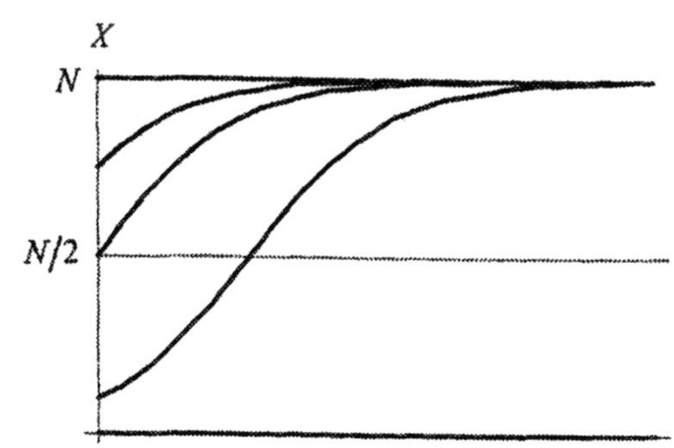

(d) The spread rate is most rapid when $x = \frac{N}{2}$. Eventually all of the people will receive the item.

19. $L\frac{di}{dt} + Ri = V \Rightarrow \frac{di}{dt} = \frac{V}{L} - \frac{R}{L}i = \frac{R}{L}\left(\frac{V}{R} - i\right)$, $V, L, R > 0$
Equilibrium: $\frac{di}{dt} = \frac{R}{L}\left(\frac{V}{R} - i\right) = 0 \Rightarrow i = \frac{V}{R}$
Concavity: $\frac{d^2i}{dt^2} = -\left(\frac{R}{L}\right)\frac{di}{dt} = -\left(\frac{R}{L}\right)^2\left(\frac{V}{R} - i\right)$
Phase Line:

$\frac{di}{dt} > 0$ $\quad$ $\frac{di}{dt} < 0$ $\quad i$
0 $\quad$ $\frac{d^2i}{dt^2} < 0$ $\quad$ $\frac{d^2i}{dt^2} > 0$
$i_{eq} = \frac{V}{R}$

If the switch is closed at $t = 0$, then $i(0) = 0$, and the graph of the solution looks like this:

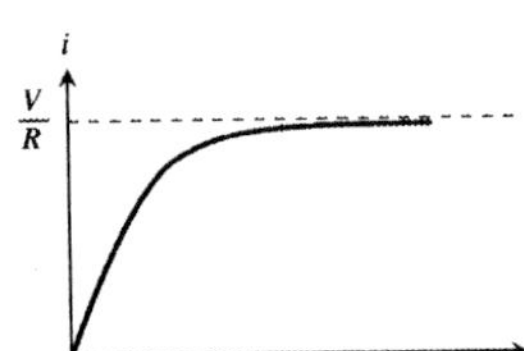

As $t \to \infty$, it $\to i_{\text{steady state}} = \frac{V}{R}$. (In the steady state condition, the self-inductance acts like a simple wire connector and, as a result, the current throught the resistor can be calculated using the familiar version of Ohm's Law.)

20. (a) Free body diagram of he pearl:

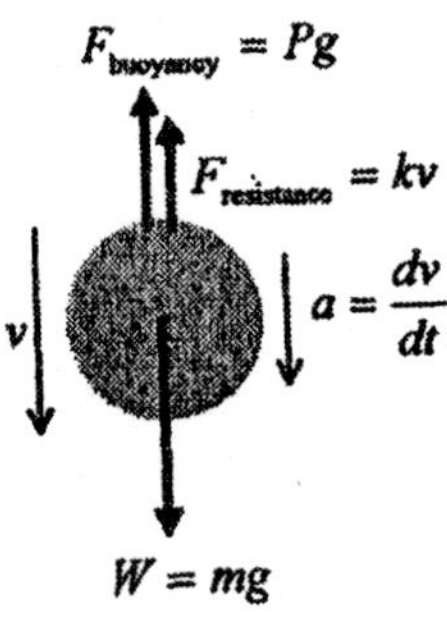

(b) Use Newton's Second Law, summing forces in the direction of the acceleration:
$mg - Pg - kv = ma \Rightarrow \frac{dv}{dt} = \left(\frac{m-P}{m}\right)g - \frac{k}{m}v.$

(c) Equilibrium: $\frac{dv}{dt} = \frac{k}{m}\left(\frac{(m-P)g}{k} - v\right) = 0$

$\Rightarrow v_{\text{terminal}} = \frac{(m-P)g}{k}$

Concavity: $\frac{d^2v}{dt^2} = -\frac{k}{m}\frac{dv}{dt} = -\left(\frac{k}{m}\right)^2\left(\frac{(m-P)g}{k} - v\right)$

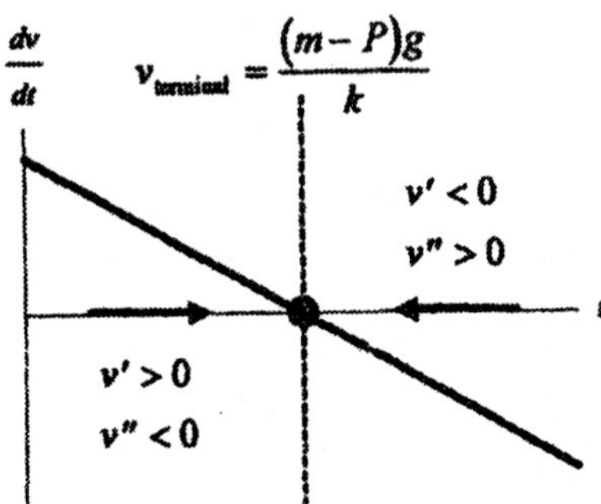

(d)

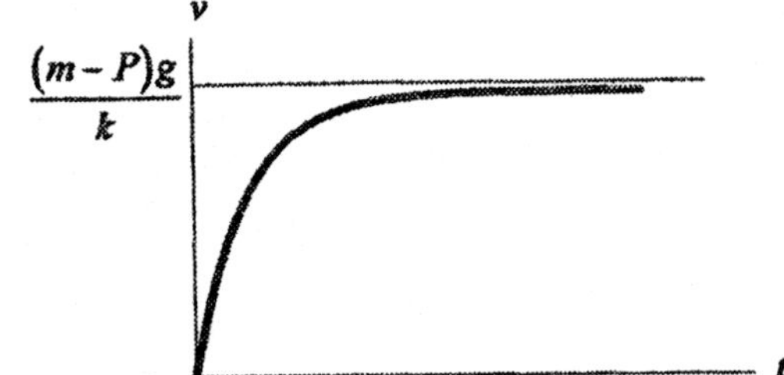

(e) The terminal velocity of the pearl is $\frac{(m-P)g}{k}$.

9.5 APPLICATIONS OF FIRST ORDER DIFFERENTIAL EQUATIONS

1. Note that the total mass is $66 + 7 = 73$ kg, therefore, $v = v_0e^{-(k/m)t} \Rightarrow v = 9e^{-3.9t/73}$

(a) $s(t) = \int 9e^{-3.9t/73}dt = -\frac{2190}{13}e^{-3.9t/73} + C$

Since $s(0) = 0$ we have $C = \frac{2190}{13}$ and $\lim_{t\to\infty} s(t) = \lim_{t\to\infty} \frac{2190}{13}\left(1 - e^{-3.9t/73}\right) = \frac{2190}{13} \approx 168.5$

The cyclist will coast about 168.5 meters.

(b) $1 = 9e^{-3.9t/73} \Rightarrow \frac{3.9t}{73} = \ln 9 \Rightarrow t = \frac{73 \ln 9}{3.9} \approx 41.13$ sec

It will take about 41.13 seconds.

2. $v = v_0 e^{-(k/m)t} \Rightarrow v = 9e^{-(59{,}000/51{,}000{,}000)t} \Rightarrow v = 9e^{-59t/51{,}000}$

(a) $s(t) = \int 9e^{-59t/51{,}000}dt = -\frac{459{,}0000}{59}e^{-59t/51{,}000} + C$

Since $s(0) = 0$ we have $C = \frac{459{,}0000}{59}$ and $\lim_{t\to\infty} s(t) = \lim_{t\to\infty} \frac{459{,}0000}{59}\left(1 - e^{-59t/51{,}000}\right) = \frac{459{,}0000}{59} \approx 7780$ m

The ship will coast about 7780 m, or 7.78 km.

(b) $1 = 9e^{-59t/51{,}000} \Rightarrow \frac{59t}{51{,}000} = \ln 9 \Rightarrow t = \frac{51{,}000 \ln 9}{59} \approx 1899.3$ sec

It will take about 31.65 minutes.

3. The total distance traveled $= \frac{v_0 m}{k} \Rightarrow \frac{(2.75)(39.92)}{k} = 4.91 \Rightarrow k = 22.36$. Therefore, the distance traveled is given by the function $s(t) = 4.91\left(1 - e^{-(22.36/39.92)t}\right)$. The graph shows s(t) and the data points.

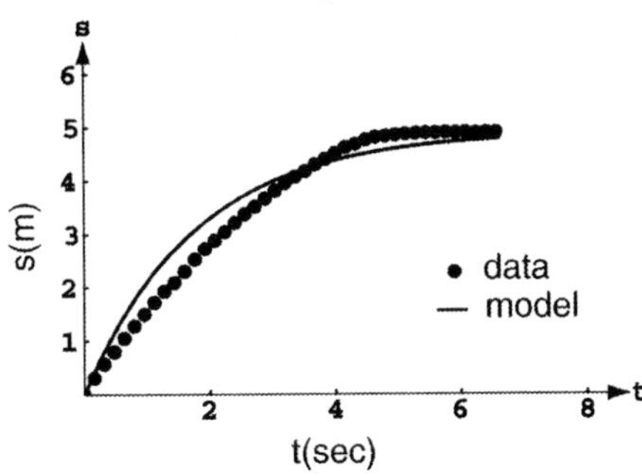

4. $\frac{v_0 m}{k} =$ coasting distance $\Rightarrow \frac{(0.80)(49.90)}{k} = 1.32 \Rightarrow k = \frac{998}{33}$

We know that $\frac{v_0 m}{k} = 1.32$ and $\frac{k}{m} = \frac{998}{33(49.9)} = \frac{20}{33}$.

Using Equation 3, we have: $s(t) = \frac{v_0 m}{k}\left(1 - e^{-(k/m)t}\right) = 1.32\left(1 - e^{-20t/33}\right) \approx 1.32\left(1 - e^{-0.606t}\right)$

5. (a) $\frac{dP}{dt} = 0.0015P(150 - P) = \frac{0.255}{150}P(150 - P) = \frac{k}{M}P(M - P)$

Thus, $k = 0.255$ and $M = 150$, and $P = \frac{M}{1 + Ae^{-kt}} = \frac{150}{1 + Ae^{-0.255t}}$

Initial condition: $P(0) = 6 \Rightarrow 6 = \frac{150}{1 + Ae^0} \Rightarrow 1 + A = 25 \Rightarrow A = 24$

Formula: $P = \frac{150}{1 + 24e^{-0.255t}}$

(b) $100 = \frac{150}{1 + 24e^{-0.255t}} \Rightarrow 1 + 24e^{-0.255t} = \frac{3}{2} \Rightarrow 24e^{-0.255t} = \frac{1}{2} \Rightarrow e^{-0.255t} = \frac{1}{48} \Rightarrow -0.255t = -\ln 48$

$\Rightarrow t = \frac{\ln 48}{0.255} \approx 17.21$ weeks

$125 = \frac{150}{1 + 24e^{-0.255t}} \Rightarrow 1 + 24e^{-0.255t} = \frac{6}{5} \Rightarrow 24e^{-0.255t} = \frac{1}{5} \Rightarrow e^{-0.255t} = \frac{1}{120} \Rightarrow -0.255t = -\ln 120$

$\Rightarrow t = \frac{\ln 120}{0.255} \approx 21.28$

It will take about 17.21 weeks to reach 100 guppies, and about 21.28 weeks to reach 125 guppies.

6. (a) $\frac{dP}{dt} = 0.0004P(250 - P) = \frac{0.1}{250}P(150 - P) = \frac{k}{M}P(M - P)$

Thus, $k = 0.1$ and $M = 250$, and $P = \frac{M}{1 + Ae^{-kt}} = \frac{250}{1 + Ae^{-0.1t}}$

Initial condition: $P(0) = 28$, where $t = 0$ represents the year 1970

$28 = \frac{250}{1 + Ae^0} \Rightarrow 28(1 + A) = 250 \Rightarrow A = \frac{250}{28} - 1 = \frac{111}{14} \approx 7.9286$

Formula: $P = \frac{250}{1 + \frac{111}{14}e^{-0.1t}}$ or approximately $P = \frac{250}{1 + 7.9286e^{-0.1t}}$

(b) The population P(t) will round to 250 when $P(t) \geq 249.5 \Rightarrow 249.5 = \frac{250}{1 + \frac{111}{14}e^{-0.1t}} \Rightarrow 249.5\left(1 + \frac{111}{14}e^{-0.1t}\right) = 250$

$\Rightarrow \frac{(249.5)(111e^{-0.1t})}{14} = 0.5 \Rightarrow e^{-0.1t} = \frac{14}{55{,}389} \Rightarrow -0.1t = \ln \frac{14}{55{,}389} \Rightarrow t = 10\,(\ln 55{,}389 - \ln 14) \approx 82.8$.

It will take about 83 years.

7. (a) Using the general solution form Example 2, part (c),

$\frac{dy}{dt} = (0.08875 \times 10^{-7})(8 \times 10^7 - y)y \Rightarrow y(t) = \frac{M}{1 + Ae^{-rMt}} = \frac{8 \times 10^7}{1 + Ae^{-(0.08875)(8)t}} = \frac{8 \times 10^7}{1 + Ae^{-0.71t}}$

Apply the initial condition:
$y(0) = 1.6 \times 10^7 = \frac{8\times10^7}{1+A} \Rightarrow \frac{8}{1.6} - 1 = 4 \Rightarrow y(1) = \frac{8\times10^7}{1+4e^{-0.71}} \approx 2.69671 \times 10^7$ kg.

(b) $y(t) = 4 \times 10^7 = \frac{8\times10^7}{1+4e^{-0.71t}} \Rightarrow 4e^{-0.71t} = 1 \Rightarrow t = -\frac{\ln\left(\frac{1}{4}\right)}{0.71} \approx 1.95253$ years.

8. (a) If a part of the population leaves or is removed from the environment (e.g., a preserve or a region) each year, then c would represent the rate of reduction of the population due to this removal and/or migration. When grizzly bears become a nuisance (e.g., feeding on livestock) or threaten human safety, they are often relocated to other areas or even eliminated, but only after relocation efforts fail. In addition, bears are killed, sometimes accidently and sometimes maliciously. For an environment that has a capacity of about 100 bears, a realistic value for c would probably be between 0 and 4.

(b)

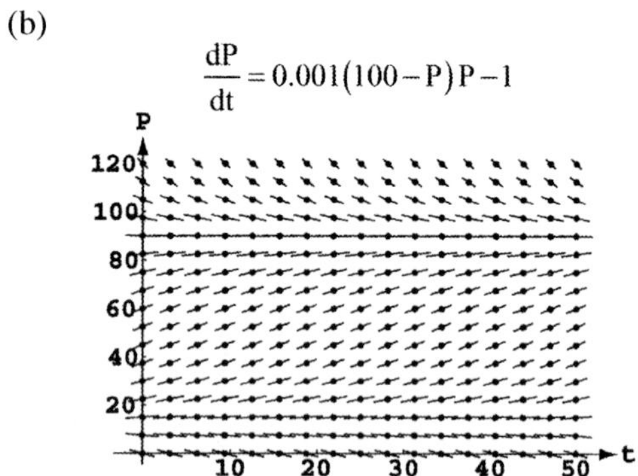

Equilibrium solutions: $\frac{dP}{dt} = 0 = 0.001(100 - P)P - 1 \Rightarrow P^2 - 100P + 1000 = 0 \Rightarrow P_{eq} \approx 11.27$ (unstable) and $P_{eq} \approx 88.73$ (stable)

(c)

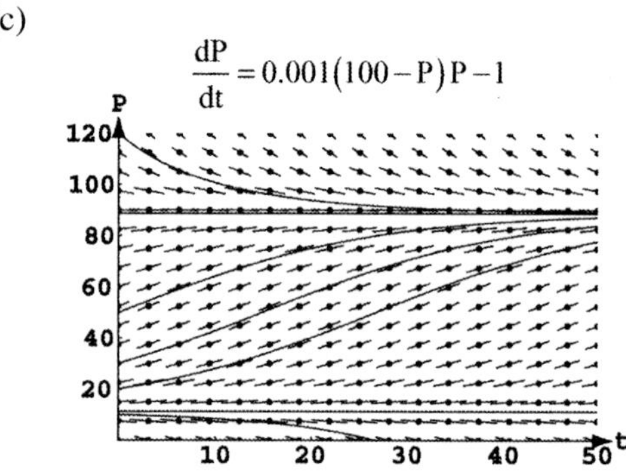

For $0 < P(0) \le 11$, the bear population will eventually disappear, for $12 \le P(0) \le 88$, the population will grow to about 89, the population will remain at about 89, and for $P(0) > 89$, the population will decrease to about 89 bears.

9. (a) $\frac{dy}{dt} = 1 + y \Rightarrow dy = (1 + y)dt \Rightarrow \frac{dy}{1+y} = dt \Rightarrow \ln|1 + y| = t + C_1 \Rightarrow e^{\ln|1+y|} = e^{t+C_1} \Rightarrow |1 + y| = e^t e^{C_1}$
$1 + y = \pm C_2 e^t \Rightarrow y = Ce^t - 1$, where $C_2 = e^{C_1}$ and $C = \pm C_2$. Apply the initial condition: $y(0) = 1 = Ce^0 - 1$
$\Rightarrow C = 2 \Rightarrow y = 2e^t - 1$.

(b) $\frac{dy}{dt} = 0.5(400 - y)y \Rightarrow dy = 0.5(400 - y)y\,dt \Rightarrow \frac{dy}{(400-y)y} = 0.5\,dt$. Using the partial fraction decomposition in Example 2, part (c), we obtain $\frac{1}{400}\left(\frac{1}{y} + \frac{1}{400-y}\right)dy = 0.5\,dt \Rightarrow \left(\frac{1}{y} + \frac{1}{400-y}\right)dy = 200\,dt$
$\Rightarrow \int\left(\frac{1}{y} + \frac{1}{400-y}\right)dy = \int 200\,dt \Rightarrow \ln|y| - \ln|y - 400| = 200t + C_1 \Rightarrow \ln\left|\frac{y}{y-400}\right| = 200t + C_1$
$\Rightarrow e^{\ln\left|\frac{y}{y-400}\right|} = e^{200t+C_1} = e^{200t}e^{C_1} \Rightarrow \left|\frac{y}{y-400}\right| = C_2e^{200t}$ (where $C_2 = e^{C_1}$) $\Rightarrow \frac{y}{y-400} = \pm C_2e^{200t}$
$\Rightarrow \frac{y}{y-400} = Ce^{200t}$ (where $C = \pm C_2$) $\Rightarrow y = Ce^{200t}y - 400\,Ce^{200t} \Rightarrow (1 - Ce^{200t})y = -400\,Ce^{200t}$
$\Rightarrow y = \frac{400\,Ce^{200t}}{Ce^{200t} - 1} \Rightarrow y = \frac{400}{1 - \frac{1}{C}e^{-200t}} = \frac{400}{1 + Ae^{-200t}}$, where $A = -\frac{1}{C}$. Apply the initial condition:
$y(0) = 2 = \frac{400}{1 + Ae^0} \Rightarrow A = 199 \Rightarrow y(t) = \frac{400}{1 + 199e^{-200t}}$

10. $\frac{dP}{dt} = r(M - P)P \Rightarrow dP = r(M - P)P\,dt \Rightarrow \frac{dP}{(M-P)P} = r\,dt$. Using the partial fraction decomposition in Example 6, part (c), we obtain $\frac{1}{M}\left(\frac{1}{P} + \frac{1}{M-P}\right)dP = r\,dt \Rightarrow \left(\frac{1}{P} + \frac{1}{M-P}\right)dP = rM\,dt \Rightarrow \int\left(\frac{1}{P} - \frac{1}{P-M}\right)dP = \int rM\,dt$
$\Rightarrow \ln|P| - \ln|P - M| = (rM)\,t + C_1 \Rightarrow \ln\left|\frac{P}{P-M}\right| = (rM)\,t + C_1 \Rightarrow e^{\ln\left|\frac{P}{P-M}\right|} = e^{(rM)t+C_1} = e^{(rM)t}e^{C_1}$
$\Rightarrow \left|\frac{P}{P-M}\right| = C_2e^{(rM)t}$ (where $C_2 = e^{C_1}$) $\Rightarrow \frac{P}{P-M} = \pm C_2e^{(rM)t} \Rightarrow \frac{P}{P-M} = Ce^{(rM)t}$ (where $C = \pm C_2$)
$\Rightarrow P = Ce^{(rM)t}P - MCe^{(rM)t} \Rightarrow \left(1 - Ce^{(rM)t}\right)P = -MCe^{(rM)t} \Rightarrow P = \frac{MCe^{(rM)t}}{Ce^{(rM)t} - 1} \Rightarrow P = \frac{M}{1 - \frac{1}{C}e^{-(rM)t}}$
$\Rightarrow P = \frac{M}{1 - Ae^{-(rM)t}}$, where $A = -\frac{1}{C}$.

11. (a) $\frac{dP}{dt} = kP^2 \Rightarrow \int P^{-2}dP = \int k\,dt \Rightarrow -P^{-1} = kt + C \Rightarrow P = \frac{-1}{kt + C}$
Initial condition: $P(0) = P_0 \Rightarrow P_0 = -\frac{1}{C} \Rightarrow C = \frac{-1}{P_0}$
Solution: $P = -\frac{1}{kt - (1/P_0)} = \frac{P_0}{1 - kP_0t}$
(b) There is a vertical asymptote at $t = \frac{1}{kP_0}$

12. (a) $\frac{dP}{dt} = r(M - P)(P - m) \Rightarrow \frac{dP}{dt} = r(1200 - P)(P - 100) \Rightarrow \frac{1}{(1200-P)(P-100)}\frac{dP}{dt} = r \Rightarrow \frac{1100}{(1200-P)(P-100)}\frac{dP}{dt} = 1100\,r$
$\Rightarrow \frac{(P-100)+(1200-P)}{(1200-P)(P-100)}\frac{dP}{dt} = 1100\,r \Rightarrow \left(\frac{1}{1200-P} + \frac{1}{P-100}\right)\frac{dP}{dt} = 1100\,r \Rightarrow \left(\frac{1}{1200-P} + \frac{1}{P-100}\right)dP = 1100\,r\,dt$
$\Rightarrow \int\left(\frac{1}{1200-P} + \frac{1}{P-100}\right)dP = \int 1100\,r\,dt \Rightarrow -\ln(1200 - P) + \ln(P - 100) = 1100\,rt + C_1$
$\Rightarrow \ln\left|\frac{P-100}{1200-P}\right| = 1100\,rt + C_1 \Rightarrow \ln\left|\frac{P-100}{1200-P}\right| = 1100\,rt + C_1 \Rightarrow \frac{P-100}{1200-P} = \pm e^{C_1}e^{1100\,rt} \Rightarrow \frac{P-100}{1200-P} = Ce^{1100\,rt}$
where $C = \pm e^{C_1} \Rightarrow P - 100 = 1200Ce^{1100\,rt} - CPe^{1100\,rt} \Rightarrow P(1 + Ce^{1100\,rt}) = 1200Ce^{1100\,rt} + 100$
$\Rightarrow P = \frac{1200Ce^{1100\,rt} + 100}{Ce^{1100\,rt} + 1} = \frac{1200 + \frac{100}{C}e^{-1100\,rt}}{1 + \frac{1}{C}e^{-1100\,rt}} \Rightarrow P = \frac{1200 + 100Ae^{-1100\,rt}}{1 + Ae^{-1100\,rt}}$ where $A = \frac{1}{C}$.
(b) Apply the initial condition: $300 = \frac{1200 + 100A}{1 + A} \Rightarrow 300 + 300A = 1200 + 100A \Rightarrow A = \frac{9}{2} \Rightarrow P = \frac{2400 + 900Ae^{-1100\,rt}}{2 + 9e^{-1100\,rt}}$.
(Note that $P \to 1200$ as $t \to \infty$.)
(c) $\frac{dP}{dt} = r(M - P)(P - m) \Rightarrow \frac{1}{(M-P)(P-m)}\frac{dP}{dt} = r \Rightarrow \frac{M-m}{(M-P)(P-m)}\frac{dP}{dt} = r(M - m) \Rightarrow \frac{(P-m)+(M-P)}{(M-P)(P-m)}\frac{dP}{dt} = r(M - m)$
$\Rightarrow \left(\frac{1}{M-P} + \frac{1}{P-m}\right)\frac{dP}{dt} = r(M - m) \Rightarrow \int\left(\frac{1}{M-P} + \frac{1}{P-m}\right)dP = \int r(M - m)dt$
$\Rightarrow -\ln(M - P) + \ln(P - m) = (M - m)\,rt + C_1 \Rightarrow \ln\left|\frac{P-m}{M-P}\right| = (M - m)\,rt + C_1 \Rightarrow \frac{P-m}{M-P} = \pm e^{C_1}e^{(M-m)\,rt}$
$\Rightarrow \frac{P-m}{M-P} = Ce^{(M-m)\,rt}$ where $C = \pm e^{C_1} \Rightarrow P - m = MCe^{(M-m)\,rt} - CPe^{(M-m)\,rt}$
$\Rightarrow P\left(1 + Ce^{(M-m)\,rt}\right) = MCe^{(M-m)\,rt} + m \Rightarrow P = \frac{MCe^{(M-m)\,rt} + m}{Ce^{(M-m)\,rt} + 1} \Rightarrow P = \frac{M + \frac{m}{C}e^{-(M-m)\,rt}}{1 + \frac{1}{C}e^{-(M-m)\,rt}} \Rightarrow P = \frac{M + mAe^{-(M-m)\,rt}}{1 + Ae^{-(M-m)\,rt}}$
$A = \frac{1}{C}$.
Apply the initial condition $P(0) = P_0$
$P_0 = \frac{M+mA}{1+A} \Rightarrow P_0 + P_0A = M + mA \Rightarrow A = \frac{M - P_0}{P_0 - m} \Rightarrow P = \frac{M(P_0 - m) + m(M - P_0)e^{-(M-m)\,rt}}{(P_0 - m) + (M - P_0)e^{-(M-m)\,rt}}$
(Note that $P \to M$ as $t \to \infty$ provided $P_0 > m$.)

13. $y = mx \Rightarrow \frac{y}{x} = m \Rightarrow \frac{xy' - y}{x^2} = 0 \Rightarrow y' = \frac{y}{x}$. So for orthogonals: $\frac{dy}{dx} = -\frac{x}{y} \Rightarrow y\,dy = -x\,dx \Rightarrow \frac{y^2}{2} + \frac{x^2}{2} = C$
$\Rightarrow x^2 + y^2 = C_1$

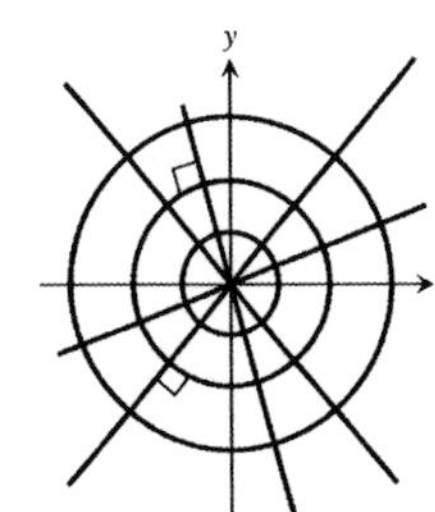

14. $y = cx^2 \Rightarrow \frac{y}{x^2} = c \Rightarrow \frac{x^2y' - 2xy}{x^4} = 0 \Rightarrow x^2y' = 2xy$
$\Rightarrow y' = \frac{2y}{x}$. So for the orthogonals: $\frac{dy}{dx} = -\frac{x}{2y}$
$\Rightarrow 2ydy = -xdx \Rightarrow y^2 = -\frac{x^2}{2} + C \Rightarrow y = \pm\sqrt{\frac{x^2}{2} + C}$,
$C > 0$

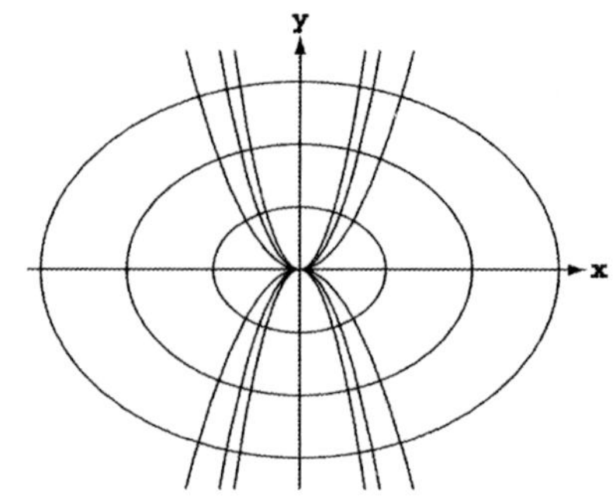

15. $kx^2 + y^2 = 1 \Rightarrow 1 - y^2 = kx^2 \Rightarrow \frac{1-y^2}{x^2} = k$
$\Rightarrow \frac{x^2(2y)y' - (1-y^2)2x}{x^4} = 0 \Rightarrow -2yx^2y' = (1-y^2)(2x)$
$\Rightarrow y' = \frac{(1-y^2)(2x)}{-2xy^2} = \frac{(1-y^2)}{-xy}$. So for the orthogonals:
$\frac{dy}{dx} = \frac{xy}{1-y^2} \Rightarrow \frac{(1-y^2)}{y}dy = x\,dx \Rightarrow \ln y - \frac{y^2}{2} = \frac{x^2}{2} + C$

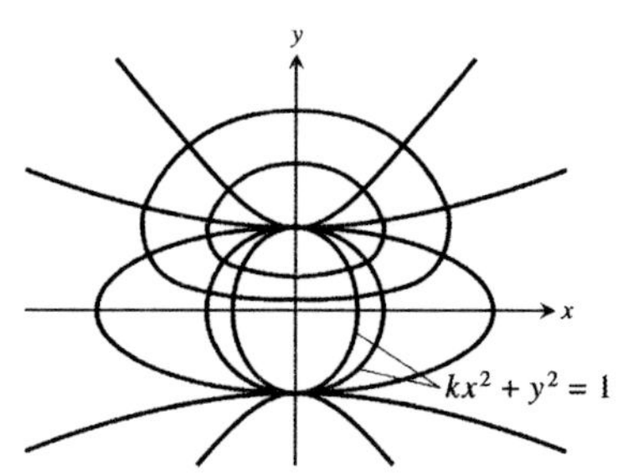

16. $2x^2 + y^2 = c^2 \Rightarrow 4x + 2yy' = 0 \Rightarrow y' = -\frac{4x}{2y} = -\frac{2x}{y}$. For orthogonals: $\frac{dy}{dx} = \frac{y}{2x} \Rightarrow \frac{dy}{y} = \frac{dx}{2x} \Rightarrow \ln y = \frac{1}{2}\ln x + C$
$\Rightarrow \ln y = \ln x^{1/2} + \ln C_1 \Rightarrow y = C_1\,|x|^{1/2}$

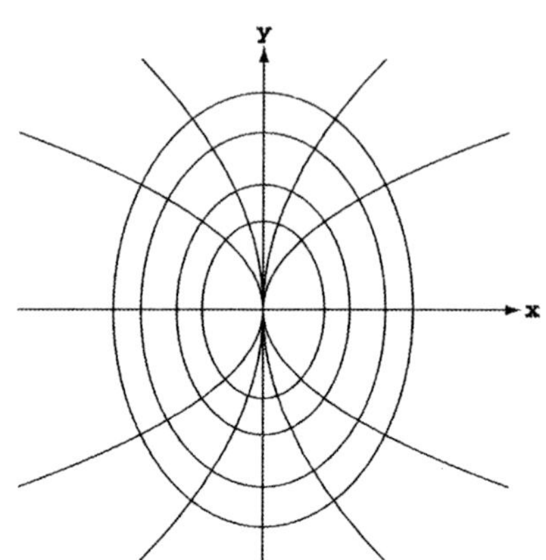

17. $y = ce^{-x} \Rightarrow \frac{y}{e^{-x}} = c \Rightarrow \frac{e^{-x}y' - y(e^{-x})(-1)}{(e^{-x})^2} = 0$
$\Rightarrow e^{-x}y' = -ye^{-x} \Rightarrow y' = -y$. So for the orthogonals:
$\frac{dy}{dx} = \frac{1}{y} \Rightarrow y\,dy = dx \Rightarrow \frac{y^2}{2} = x + C$
$\Rightarrow y^2 = 2x + C_1 \Rightarrow y = \pm\sqrt{2x + C_1}$

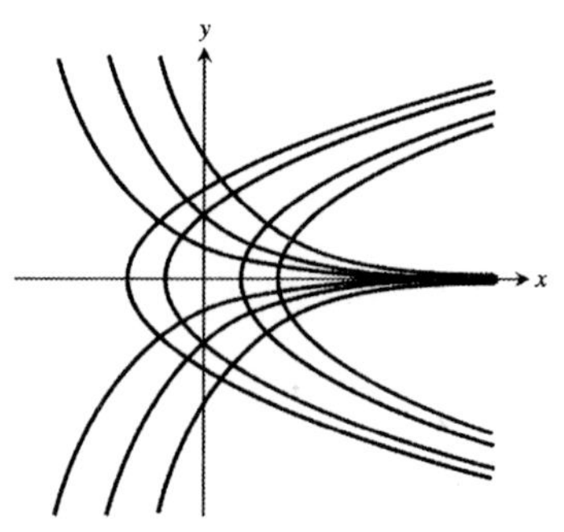

18. $y = e^{kx} \Rightarrow \ln y = kx \Rightarrow \frac{\ln y}{x} = k \Rightarrow \frac{x\left(\frac{1}{y}\right)y' - \ln y}{x^2} = 0$
$\Rightarrow \left(\frac{x}{y}\right)y' - \ln y = 0 \Rightarrow y' = \frac{y\ln y}{x}$. So for the orthogonals:
$\frac{dy}{dx} = \frac{-x}{y\ln y} \Rightarrow y\ln y\,dy = -x\,dx$
$\Rightarrow \frac{1}{2}y^2\ln y - \frac{1}{4}(y^2) = \left(-\frac{1}{2}x^2\right) + C$
$\Rightarrow y^2\ln y - \frac{y^2}{2} = -x^2 + C_1$

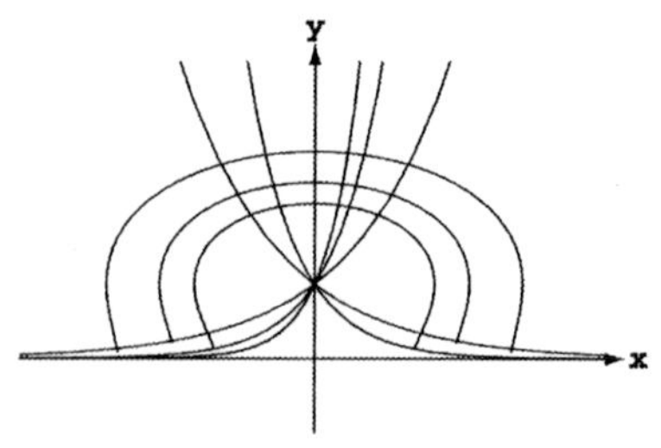

19. $2x^2 + 3y^2 = 5$ and $y^2 = x^3$ intersect at $(1, 1)$. Also, $2x^2 + 3y^2 = 5 \Rightarrow 4x + 6y\,y' = 0 \Rightarrow y' = -\frac{4x}{6y} \Rightarrow y'(1, 1) = -\frac{2}{3}$
$y_1^2 = x^3 \Rightarrow 2y_1y_1' = 3x^2 \Rightarrow y_1' = \frac{3x^2}{2y_1} \Rightarrow y_1'(1, 1) = \frac{3}{2}$. Since $y' \cdot y_1' = \left(-\frac{2}{3}\right)\left(\frac{3}{2}\right) = -1$, the curves are orthogonal.

20. (a) $x\,dx + y\,dy = 0 \Rightarrow \frac{x^2}{2} + \frac{y^2}{2} = C$ is the general equation of the family with slope $y' = -\frac{x}{y}$. For the orthogonals: $y' = \frac{y}{x} \Rightarrow \frac{dy}{y} = \frac{dx}{x} \Rightarrow \ln y = \ln x + C$ or $y = C_1 x$ (where $C_1 = e^C$) is the general equation of the orthogonals.

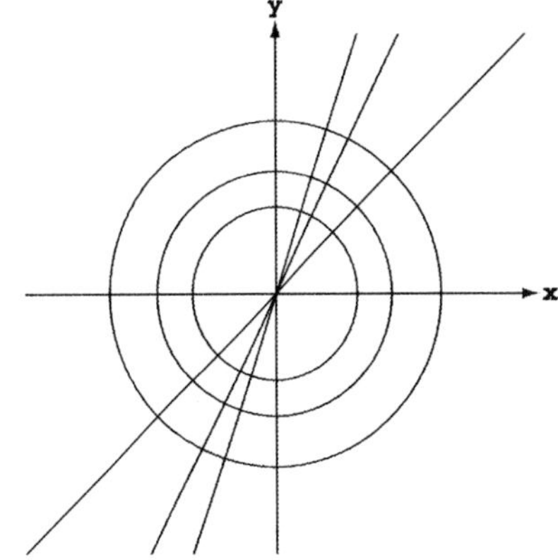

(b) $x\,dy - 2y\,dx = 0 \Rightarrow 2y\,dx = x\,dy \Rightarrow \frac{dy}{2y} = \frac{dx}{x}$
$\Rightarrow \frac{1}{2}\left(\frac{dy}{y}\right) = \frac{dx}{x} \Rightarrow \frac{1}{2}\ln y = \ln x + C \Rightarrow y = C_1x^2$ is the equation for the solution family.
$\frac{1}{2}\ln y - \ln x = C \Rightarrow \frac{1}{2}\frac{y'}{y} - \frac{1}{x} = 0 \Rightarrow y' = \frac{2y}{x}$
$\Rightarrow$ slope of orthogonals is $\frac{dy}{dx} = -\frac{x}{2y}$
$\Rightarrow 2y\,dy = -x\,dx \Rightarrow y^2 = -\frac{x^2}{2} + C$ is the general equation of the orthogonals.

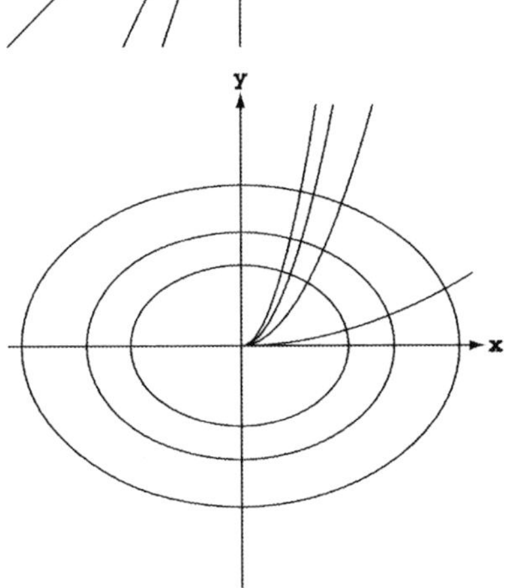

21. $y^2 = 4a^2 - 4ax$ and $y^2 = 4b^2 + 4bx \Rightarrow$ (at intersection) $4a^2 - 4ax = 4b^2 + 4bx \Rightarrow a^2 - b^2 = x(a + b)$
$\Rightarrow (a + b)(a - b) = (a + b)x \Rightarrow x = a - b$. Now, $y^2 = 4a^2 - 4a(a - b) = 4a^2 - 4a^2 + 4ab = 4ab \Rightarrow y = \pm 2\sqrt{ab}$.
Thus the intersections are at $\left(a - b,\ \pm 2\sqrt{ab}\right)$. So, $y^2 = 4a^2 - 4ax \Rightarrow y_1' = -\frac{4a}{2y}$ which are equal to $-\frac{4a}{2\left(2\sqrt{ab}\right)}$ and $-\frac{4a}{2\left(-2\sqrt{ab}\right)} = -\sqrt{\frac{a}{b}}$ and $\sqrt{\frac{a}{b}}$ at the intersections. Also, $y^2 = 4b^2 + 4bx \Rightarrow y_2' = \frac{4b}{2y}$ which are equal to $\frac{4b}{2\left(2\sqrt{ab}\right)}$ and $\frac{4b}{2\left(-2\sqrt{ab}\right)} = -\sqrt{\frac{b}{a}}$ and $\sqrt{\frac{b}{a}}$ at the intersections. $(y_1') \cdot (y_2') = -1$. Thus the curves are orthogonal.

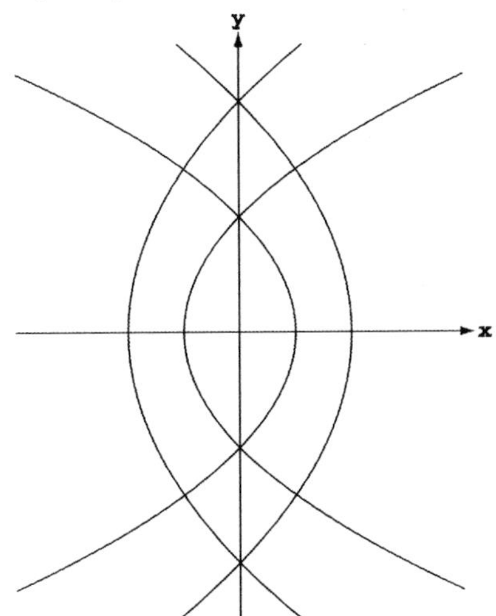

CHAPTER 9 PRACTICE EXERCISES

1. $\frac{dy}{dx} = \sqrt{y}\cos^2\sqrt{y} \Rightarrow \frac{dy}{\sqrt{y}\cos^2\sqrt{y}} = dx \Rightarrow 2\tan\sqrt{y} = x + C \Rightarrow y = \left(\tan^{-1}\left(\frac{x+C}{2}\right)\right)^2$

2. $y' = \frac{3y(x+1)^2}{y-1} \Rightarrow \frac{(y-1)}{y}dy = 3(x+1)^2dx \Rightarrow y - \ln y = (x+1)^3 + C$

3. $yy' = \sec(y^2)\sec^2x \Rightarrow \frac{y\,dy}{\sec(y^2)} = \sec^2x\,dx \Rightarrow \frac{\sin(y^2)}{2} = \tan x + C \Rightarrow \sin(y^2) = 2\tan x + C_1$

4. $y\cos^2(x)\,dy + \sin x\,dx = 0 \Rightarrow y\,dy = -\frac{\sin x}{\cos^2(x)}dx \Rightarrow \frac{y^2}{2} = -\frac{1}{\cos(x)} + C \Rightarrow y = \pm\sqrt{\frac{-2}{\cos(x)} + C_1}$

5. $y' = xe^y\sqrt{x-2} \Rightarrow e^{-y}dy = x\sqrt{x-2}\,dx \Rightarrow -e^{-y} = \frac{2(x-2)^{3/2}(3x+4)}{15} + C \Rightarrow e^{-y} = \frac{-2(x-2)^{3/2}(3x+4)}{15} - C$
$\Rightarrow -y = \ln\left[\frac{-2(x-2)^{3/2}(3x+4)}{15} - C\right] \Rightarrow y = -\ln\left[\frac{-2(x-2)^{3/2}(3x+4)}{15} - C\right]$

6. $y' = xye^{x^2} \Rightarrow \frac{dy}{y} = e^{x^2}x\,dx \Rightarrow \ln y = \frac{1}{2}e^{x^2} + C$

7. $\sec x\,dy + x\cos^2y\,dx = 0 \Rightarrow \frac{dy}{\cos^2y} = -\frac{x\,dx}{\sec x} \Rightarrow \tan y = -\cos x - x\sin x + C$

8. $2x^2dx - 3\sqrt{y}\csc x\,dy = 0 \Rightarrow 3\sqrt{y}\,dy = \frac{2x^2}{\csc x}dx \Rightarrow 2y^{3/2} = 2(2-x^2)\cos x + 4x\sin x + C$
$\Rightarrow y^{3/2} = (2-x^2)\cos x + 2x\sin x + C_1$

9. $y' = \frac{e^y}{xy} \Rightarrow ye^{-y}dy = \frac{dx}{x} \Rightarrow (y+1)e^{-y} = -\ln|x| + C$

10. $y' = xe^{x-y}\csc y \Rightarrow y' = \frac{xe^x}{e^y}\csc y \Rightarrow \frac{e^y}{\csc y}dy = xe^xdx \Rightarrow \frac{e^y}{2}(\sin y - \cos y) = (x-1)e^x + C$

11. $x(x-1)dy - y\,dx = 0 \Rightarrow x(x-1)dy = y\,dx \Rightarrow \frac{dy}{y} = \frac{dx}{x(x-1)} \Rightarrow \ln y = \ln(x-1) - \ln(x) + C$
$\Rightarrow \ln y = \ln(x-1) - \ln(x) + \ln C_1 \Rightarrow \ln y = \ln\left(\frac{C_1(x-1)}{x}\right) \Rightarrow y = \frac{C_1(x-1)}{x}$

12. $y' = (y^2-1)(x^{-1}) \Rightarrow \frac{dy}{y^2-1} = \frac{dx}{x} \Rightarrow \frac{\ln\left(\frac{y-1}{y+1}\right)}{2} = \ln x + C \Rightarrow \ln\left(\frac{y-1}{y+1}\right) = 2\ln x + \ln C_1 \Rightarrow \frac{y-1}{y+1} = C_1x^2$

13. $2y' - y = xe^{x/2} \Rightarrow y' - \frac{1}{2}y = \frac{x}{2}e^{x/2}.$
$p(x) = -\frac{1}{2},\ v(x) = e^{\int\left(-\frac{1}{2}\right)dx} = e^{-x/2}.$
$e^{-x/2}y' - \frac{1}{2}e^{-x/2}y = \left(e^{-x/2}\right)\left(\frac{x}{2}\right)\left(e^{x/2}\right) = \frac{x}{2} \Rightarrow \frac{d}{dx}\left(e^{-x/2}y\right) = \frac{x}{2} \Rightarrow e^{-x/2}y = \frac{x^2}{4} + C \Rightarrow y = e^{x/2}\left(\frac{x^2}{4} + C\right)$

14. $\frac{y'}{2} + y = e^{-x}\sin x \Rightarrow y' + 2y = 2e^{-x}\sin x.$
$p(x) = 2,\ v(x) = e^{\int 2dx} = e^{2x}.$
$e^{2x}y' + 2e^{2x}y = 2e^{2x}e^{-x}\sin x = 2e^x\sin x \Rightarrow \frac{d}{dx}(e^{2x}y) = 2e^x\sin x \Rightarrow e^{2x}y = e^x(\sin x - \cos x) + C$
$\Rightarrow y = e^{-x}(\sin x - \cos x) + Ce^{-2x}$

15. $xy' + 2y = 1 - x^{-1} \Rightarrow y' + \left(\frac{2}{x}\right)y = \frac{1}{x} - \frac{1}{x^2}.$
$v(x) = e^{2\int\frac{dx}{x}} = e^{2\ln x} = e^{\ln x^2} = x^2.$
$x^2y' + 2xy = x - 1 \Rightarrow \frac{d}{dx}(x^2y) = x - 1 \Rightarrow x^2y = \frac{x^2}{2} - x + C \Rightarrow y = \frac{1}{2} - \frac{1}{x} + \frac{C}{x^2}$

16. $xy' - y = 2x\ln x \Rightarrow y' - \left(\frac{1}{x}\right)y = 2\ln x.$
$v(x) = e^{-\int \frac{dx}{x}} = e^{-\ln x} = \frac{1}{x}.\ \left(\frac{1}{x}\right)y' - \left(\frac{1}{x}\right)^2 y = \frac{2}{x}\ln x \Rightarrow$
$\frac{d}{dx}\left(\frac{1}{x}\cdot y\right) = \frac{2}{x}\ln x \Rightarrow \frac{1}{x}\cdot y = [\ln x]^2 + C \Rightarrow y = x[\ln x]^2 + Cx$

17. $(1+e^x)dy + (ye^x + e^{-x})dx = 0 \Rightarrow (1+e^x)y' + e^x y = -e^{-x} \Rightarrow y' = \frac{e^x}{1+e^x}y = \frac{-e^{-x}}{(1+e^x)}.$
$v(x) = e^{\int \frac{e^x dx}{(1+e^x)}} = e^{\ln(e^x+1)} = e^x + 1.$
$(e^x+1)y' + (e^x+1)\left(\frac{e^x}{1+e^x}\right)y = \frac{-e^{-x}}{(1+e^x)}(e^x+1) \Rightarrow \frac{d}{dx}[(e^x+1)y] = -e^{-x} \Rightarrow (e^x+1)y = e^{-x} + C$
$\Rightarrow y = \frac{e^{-x}+C}{e^x+1} = \frac{e^{-x}+C}{1+e^x}$

18. $\frac{dx}{dy} + x - 4ye^y = 0 \Rightarrow x' + x = 4ye^y.$ Let $v(y) = e^{\int dy} = e^y.$ Then $e^y x' + xe^y = 4ye^{2y} \Rightarrow \frac{d}{dy}(xe^y) = 4ye^{2y}$
$\Rightarrow xe^y = (2y-1)e^{2y} + C \Rightarrow x = (2y-1)e^y + Ce^{-y}$

19. $(x+3y^2)\,dy + y\,dx = 0 \Rightarrow x\,dy + y\,dx = -3y^2dy \Rightarrow \frac{d}{dx}(xy) = -3y^2dy \Rightarrow xy = -y^3 + C$

20. $y\,dx + (3x - y^{-2}\cos y)\,dx = 0 \Rightarrow x' + \left(\frac{3}{y}\right)x = y^{-3}\cos y.$ Let $v(y) = e^{\int \frac{3dy}{y}} = e^{3\ln y} = e^{\ln y^3} = y^3.$
Then $y^3x' + 3y^2x = \cos y$ and $y^3x = \int \cos y\,dy = \sin y + C.$ So $x = y^{-3}(\sin y + C)$

21. $\frac{dy}{dx} = e^{-x-y-2} \Rightarrow e^y dy = e^{-(x+2)}dx \Rightarrow e^y = -e^{-(x+2)} + C.$ We have $y(0) = -2,$ so $e^{-2} = -e^{-2} + C \Rightarrow C = 2e^{-2}$ and
$e^y = -e^{-(x+2)} + 2e^{-2} \Rightarrow y = \ln\left(-e^{-(x+2)} + 2e^{-2}\right)$

22. $\frac{dy}{dx} = \frac{y\ln y}{1+x^2} \Rightarrow \frac{dy}{y\ln y} = \frac{dx}{1+x^2} \Rightarrow \ln(\ln y) = \tan^{-1}(x) + C \Rightarrow y = e^{e^{\tan^{-1}(x)+C}}.$ We have $y(0) = e^2 \Rightarrow e^2 = e^{e^{\tan^{-1}(0)+C}}$
$\Rightarrow e^{\tan^{-1}(0)+C} = 2 \Rightarrow \tan^{-1}(0) + C = \ln 2 \Rightarrow 0 + C = \ln 2 \Rightarrow C = \ln 2 \Rightarrow y = e^{e^{\tan^{-1}(x)+\ln 2}}$

23. $(x+1)\frac{dy}{dx} + 2y = x \Rightarrow y' + \left(\frac{2}{x+1}\right)y = \frac{x}{x+1}.$ Let $v(x) = e^{\int \frac{2}{x+1}dx} = e^{2\ln(x+1)} = e^{\ln(x+1)^2} = (x+1)^2.$
So $y'(x+1)^2 + \frac{2}{(x+1)}(x+1)^2y = \frac{x}{(x+1)}(x+1)^2 \Rightarrow \frac{d}{dx}\left[y(x+1)^2\right] = x(x+1) \Rightarrow y(x+1)^2 = \int x(x+1)dx$
$\Rightarrow y(x+1)^2 = \frac{x^3}{3} + \frac{x^2}{2} + C \Rightarrow y = (x+1)^{-2}\left(\frac{x^3}{3} + \frac{x^2}{2} + C\right).$ We have $y(0) = 1 \Rightarrow 1 = C.$ So
$y = (x+1)^{-2}\left(\frac{x^3}{3} + \frac{x^2}{2} + 1\right)$

24. $x\frac{dy}{dx} + 2y = x^2 + 1 \Rightarrow y' + \left(\frac{2}{x}\right)y = x + \frac{1}{x}.$ Let $v(x) = e^{\int\left(\frac{2}{x}\right)dx} = e^{\ln x^2} = x^2.$ So $x^2y' + 2xy = x^3 + x$
$\Rightarrow \frac{d}{dx}(x^2y) = x^3 + x \Rightarrow x^2y = \frac{x^4}{4} + \frac{x^2}{2} + C \Rightarrow y = \frac{x^2}{4} + \frac{C}{x^2} + \frac{1}{2}.$ We have $y(1) = 1 \Rightarrow 1 = \frac{1}{4} + C + \frac{1}{2} \Rightarrow C = \frac{1}{4}.$
So $y = \frac{x^2}{4} + \frac{1}{4x^2} + \frac{1}{2} = \frac{x^4+2x^2+1}{4x^2}$

25. $\frac{dy}{dx} + 3x^2y = x^2.$ Let $v(x) = e^{\int 3x^2dx} = e^{x^3}.$ So $e^{x^3}y' + 3x^2e^{x^3}y = x^2e^{x^3} \Rightarrow \frac{d}{dx}\left(e^{x^3}y\right) = x^2e^{x^3} \Rightarrow e^{x^3}y = \frac{1}{3}e^{x^3} + C.$
We have $y(0) = -1 \Rightarrow e^{0^3}(-1) = \frac{1}{3}e^{0^3} + C \Rightarrow -1 = \frac{1}{3} + C \Rightarrow C = -\frac{4}{3}$ and $e^{x^3}y = \frac{1}{3}e^{x^3} - \frac{4}{3} \Rightarrow y = \frac{1}{3} - \frac{4}{3}e^{-x^3}$

26. $xdy + (y - \cos x)dx = 0 \Rightarrow xy' + y - \cos x = 0 \Rightarrow y' + \left(\frac{1}{x}\right)y = \frac{\cos x}{x}.$ Let $v(x) = e^{\int \frac{1}{x}dx} = e^{\ln x} = x.$
So $xy' + x\left(\frac{1}{x}\right)y = \cos x \Rightarrow \frac{d}{dx}(xy) = \cos x \Rightarrow xy = \int \cos x\,dx \Rightarrow xy = \sin x + C.$ We have $y\left(\frac{\pi}{2}\right) = 0 \Rightarrow \left(\frac{\pi}{2}\right)0 = 1 + C$
$\Rightarrow C = -1.$ So $xy = -1 + \sin x \Rightarrow y = \frac{-1+\sin x}{x}$

27. $x\,dy - \left(y + \sqrt{y}\right)dx = 0 \Rightarrow \frac{dy}{(y+\sqrt{y})} = \frac{dx}{x} \Rightarrow 2\ln\left(\sqrt{y}+1\right) = \ln x + C.$ We have $y(1) = 1 \Rightarrow 2\ln\left(\sqrt{1}+1\right) = \ln 1 + C$
$\Rightarrow 2\ln 2 = C = \ln 2^2 = \ln 4.$ So $2\ln\left(\sqrt{y}+1\right) = \ln x + \ln 4 = \ln(4x) \Rightarrow \ln\left(\sqrt{y}+1\right) = \frac{1}{2}\ln(4x) = \ln(4x)^{1/2}$

$\Rightarrow e^{\ln(\sqrt{y}+1)} = e^{\ln(4x)^{1/2}} \Rightarrow \sqrt{y} + 1 = 2\sqrt{x} \Rightarrow y = \left(2\sqrt{x} - 1\right)^2$

28. $y^{-2}\frac{dx}{dy} = \frac{e^x}{e^{2x}+1} \Rightarrow \frac{e^{2x}+1}{e^x}dx = \frac{dy}{y^{-2}} \Rightarrow \frac{y^3}{3} = e^x - e^{-x} + C$. We have $y(0) = 1 \Rightarrow \frac{(1)^3}{3} = e^0 - e^0 + C \Rightarrow C = \frac{1}{3}$. So $\frac{y^3}{3} = e^x - e^{-x} + \frac{1}{3} \Rightarrow y^3 = 3(e^x - e^{-x}) + 1 \Rightarrow y = [3(e^x - e^{-x}) + 1\,]^{1/3}$

29. $xy' + (x-2)y = 3x^3e^{-x} \Rightarrow y' + \left(\frac{x-2}{x}\right)y = 3x^2e^{-x}$. Let $v(x) = e^{\int\left(\frac{x-2}{x}\right)dx} = e^{x - 2\ln x} = \frac{e^x}{x^2}$. So $\frac{e^x}{x^2}y' + \frac{e^x}{x^2}\left(\frac{x-2}{x}\right)y = 3 \Rightarrow \frac{d}{dx}\left(y \cdot \frac{e^x}{x^2}\right) = 3 \Rightarrow y \cdot \frac{e^x}{x^2} = 3x + C$. We have $y(1) = 0 \Rightarrow 0 = 3(1) + C \Rightarrow C = -3$ $\Rightarrow y \cdot \frac{e^x}{x^2} = 3x - 3 \Rightarrow y = x^2e^{-x}(3x - 3)$

30. $y\,dx + (3x - xy + 2)dy = 0 \Rightarrow \frac{dx}{dy} + \frac{3x - xy + 2}{y} = 0 \Rightarrow \frac{dx}{dy} + \frac{3x}{y} - x = -\frac{2}{y} \Rightarrow \frac{dx}{dy} + \left(\frac{3}{y} - 1\right)x = -\frac{2}{y}$.
$P(y) = \frac{3}{y} - 1 \Rightarrow \int P(y)dy = 3\ln y - y \Rightarrow v(y) = e^{3\ln y - y} = y^3e^{-y}$
$y^3e^{-y}x' + y^3e^{-y}\left(\frac{3}{y} - 1\right)x = -2y^2e^{-y} \Rightarrow y^3e^{-y}x = \int -2y^2e^{-y}dy = 2e^{-y}(y^2 + 2y + 2) + C$
$\Rightarrow y^3 = \frac{2(y^2+2y+2)+Ce^y}{x}$. We have $y(2) = -1 \Rightarrow -1 = \frac{2(1-2+2)+Ce^{-1}}{2} \Rightarrow C = -4e$ and
$\Rightarrow y^3 = \frac{2(y^2+2y+2)-4e^{y+1}}{x}$

31. To find the approximate values let $y_n = y_{n-1} + (y_{n-1} + \cos x_{n-1})(0.1)$ with $x_0 = 0$, $y_0 = 0$, and 20 steps. Use a spreadsheet, graphing calculator, or CAS to obtain the values in the following table.

x	y
0	0
0.1	0.1000
0.2	0.2095
0.3	0.3285
0.4	0.4568
0.5	0.5946
0.6	0.7418
0.7	0.8986
0.8	1.0649
0.9	1.2411
1.0	1.4273

x	y
1.1	1.6241
1.2	1.8319
1.3	2.0513
1.4	2.2832
1.5	2.5285
1.6	2.7884
1.7	3.0643
1.8	3.3579
1.9	3.6709
2.0	4.0057

32. To find the approximate values let $z_n = y_{n-1} + ((2 - y_{n-1})(2\,x_{n-1} + 3))(0.1)$ and $y_n = y_{n-1} + \left(\frac{(2-y_{n-1})(2\,x_{n-1}+3) + (2-z_n)(2\,x_n+3)}{2}\right)(0.1)$ with initial values $x_0 = -3$, $y_0 = 1$, and 20 steps. Use a spreadsheet, graphing calculator, or CAS to obtain the values in the following table.

x	y
−3	1
−2.9	0.6680
−2.8	0.2599
−2.7	−0.2294
−2.6	−0.8011
−2.5	−1.4509
−2.4	−2.1687
−2.3	−2.9374
−2.2	−3.7333
−2.1	−4.5268
−2.0	−5.2840

x	y
−1.9	−5.9686
−1.8	−6.5456
−1.7	−6.9831
−1.6	−7.2562
−1.5	−7.3488
−1.4	−7.2553
−1.3	−6.9813
−1.2	−6.5430
−1.1	−5.9655
−1.0	−5.2805

33. To estimate y(3), let $z_n = y_{n-1} + \left(\frac{x_{n-1} - 2y_{n-1}}{x_{n-1} + 1}\right)(0.05)$ and $y_n = y_{n-1} + \frac{1}{2}\left(\frac{x_{n-1} - 2y_{n-1}}{x_{n-1} + 1} + \frac{x_n - 2z_n}{x_n + 1}\right)(0.05)$ with initial values $x_0 = 0$, $y_0 = 1$, and 60 steps. Use a spreadsheet, graphing calculator, or CAS to obtain $y(3) \approx 0.9063$.

34. To estimate y(4), let $z_n = y_{n-1} + \left(\frac{x_{n-1}^2 - 2y_{n-1} + 1}{x_{n-1}}\right)(0.05)$ with initial values $x_0 = 1$, $y_0 = 1$, and 60 steps. Use a spreadsheet, graphing calculator, or CAS to obtain $y(4) \approx 4.4974$.

35. Let $y_n = y_{n-1} + \left(\frac{1}{e^{x_{n-1} + y_{n-1} + 2}}\right)(dx)$ with starting values $x_0 = 0$ and $y_0 = 2$, and steps of 0.1 and -0.1. Use a spreadsheet, programmable calculator, or CAS to generate the following graphs.

(a)

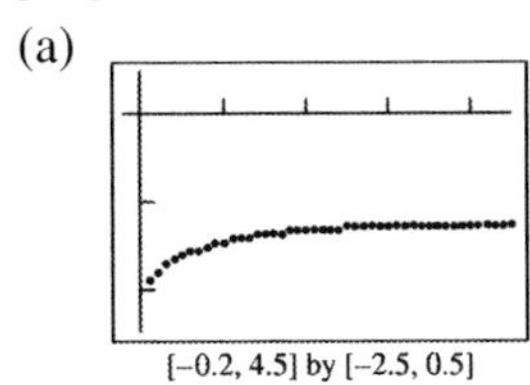

[−0.2, 4.5] by [−2.5, 0.5]

(b) Note that we choose a small interval of x-values because the y-values decrease very rapidly and our calculator cannot handle the calculations for $x \le -1$. (This occurs because the analytic solution is $y = -2 + \ln(2 - e^{-x})$, which has an asymptote at $x = -\ln 2 \approx 0.69$. Obviously, the Euler approximations are misleading for $x \le -0.7$.)

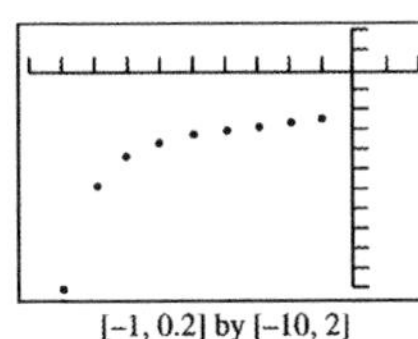

[−1, 0.2] by [−10, 2]

36. Let $z_n = y_{n-1} - \left(\frac{x_{n-1}^2 + y_{n-1}}{e^{y_{n-1}} + x_{n-1}}\right)(dx)$ and $y_n = y_{n-1} + \frac{1}{2}\left(\frac{x_{n-1}^2 + y_{n-1}}{e^{y_{n-1}} + x_{n-1}} + \frac{x_n^2 + z_n}{e^{z_n} + x_n}\right)(dx)$ with starting values $x_0 = 0$ and $y_0 = 0$, and steps of 0.1 and -0.1. Use a spreadsheet, programmable calculator, or CAS to generate the following graphs.

(a)

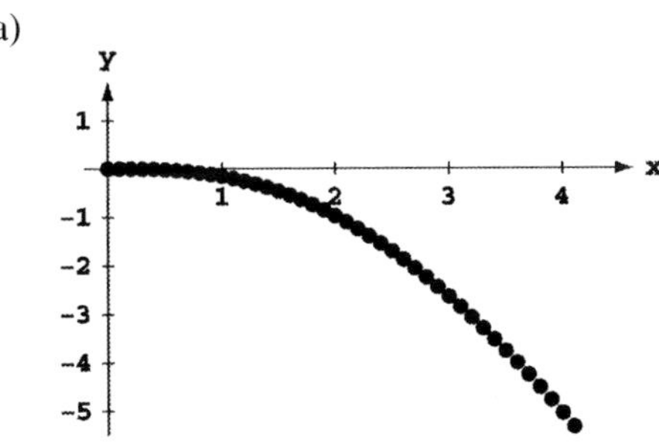

(b)

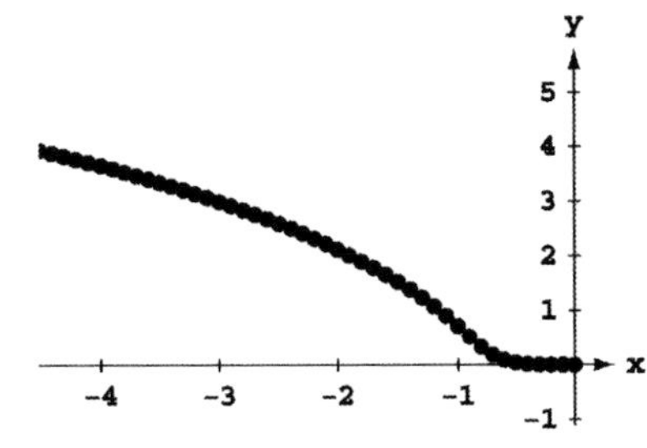

37.

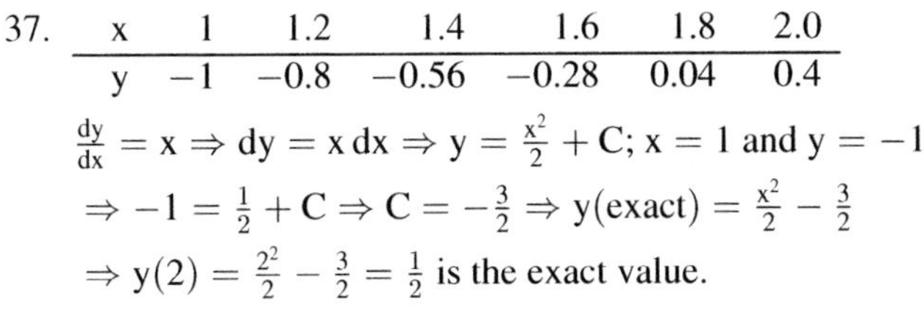

x	1	1.2	1.4	1.6	1.8	2.0
y	−1	−0.8	−0.56	−0.28	0.04	0.4

$\frac{dy}{dx} = x \Rightarrow dy = x\,dx \Rightarrow y = \frac{x^2}{2} + C$; $x = 1$ and $y = -1$

$\Rightarrow -1 = \frac{1}{2} + C \Rightarrow C = -\frac{3}{2} \Rightarrow y(\text{exact}) = \frac{x^2}{2} - \frac{3}{2}$

$\Rightarrow y(2) = \frac{2^2}{2} - \frac{3}{2} = \frac{1}{2}$ is the exact value.

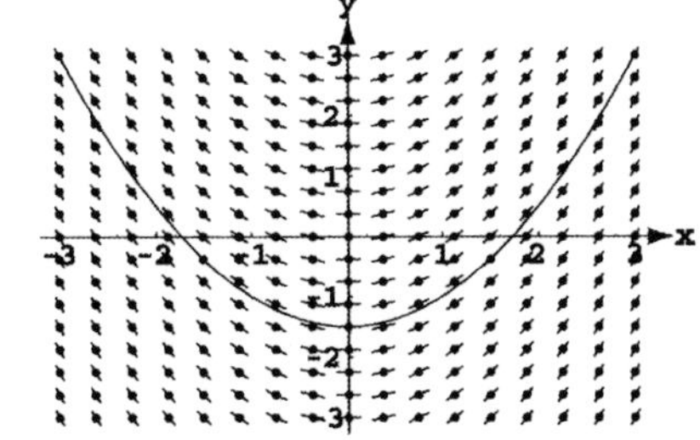

38.

x	1	1.2	1.4	1.6	1.8	2.0
y	−1	−0.8	−0.6333	−0.4904	−0.3654	−0.2544

$\frac{dy}{dx} = \frac{1}{x} \Rightarrow dy = \frac{1}{x}dx \Rightarrow y = \ln|x| + C$; $x = 1$ and $y = -1$
$\Rightarrow -1 = \ln 1 + C \Rightarrow C = -1 \Rightarrow y(\text{exact}) = \ln|x| - 1$
$\Rightarrow y(2) = \ln 2 - 1 \approx -0.3069$ is the exact value.

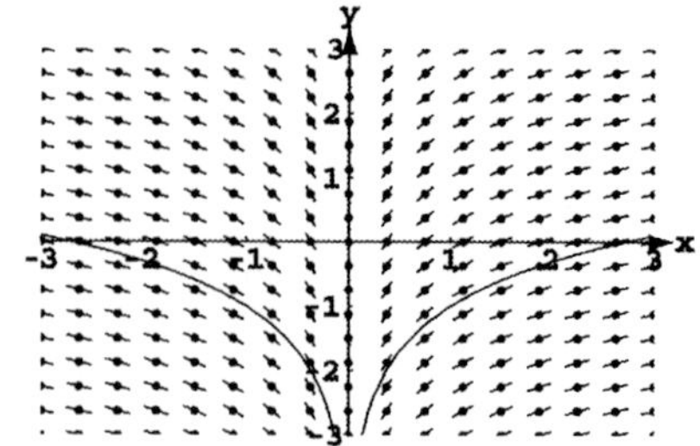

39.

x	1	1.2	1.4	1.6	1.8	2.0
y	−1	−1.2	−0.488	−1.9046	−2.5141	−3.4192

$\frac{dy}{dx} = xy \Rightarrow \frac{dy}{y} = x\,dx \Rightarrow \ln|y| = \frac{x^2}{2} + C$
$\Rightarrow y = e^{\frac{x^2}{2}+C} = e^{\frac{x^2}{2}} \cdot e^C = C_1 e^{\frac{x^2}{2}}$; $x = 1$ and $y = -1$
$\Rightarrow -1 = C_1 e^{1/2} \Rightarrow C_1 = -e^{1/2} y(\text{exact}) = -e^{1/2} \cdot e^{\frac{x^2}{2}}$
$= -e^{(x^2-1)/2} \Rightarrow y(2) = -e^{3/2} \approx -4.4817$ is the exact value.

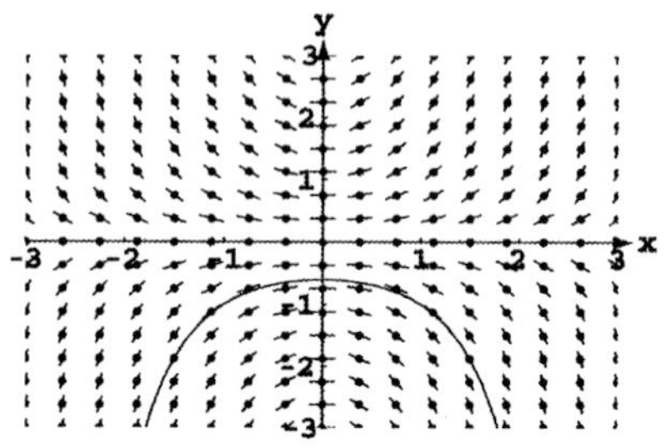

40.

x	1	1.2	1.4	1.6	1.8	2.0
y	−1	−1.2	−1.3667	−1.5130	−1.6452	−1.7688

$\frac{dy}{dx} = \frac{1}{y} \Rightarrow y\,dy = dx \Rightarrow \frac{y^2}{2} = x + C$; $x = 1$ and $y = -1$
$\frac{1}{2} = 1 + C \Rightarrow C = -\frac{1}{2} \Rightarrow y^2 = 2x - 1$
$\Rightarrow y(\text{exact}) = \sqrt{2x-1} \Rightarrow y(2) = -\sqrt{3} \approx -1.7321$ is the exact value.

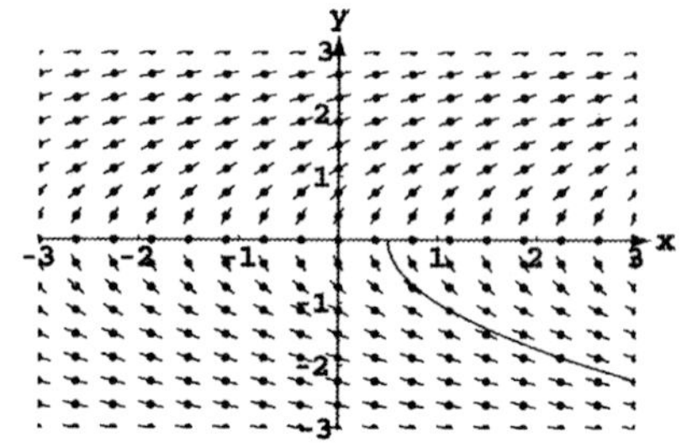

41. $\frac{dy}{dx} = y^2 - 1 \Rightarrow y' = (y+1)(y-1)$. We have $y' = 0 \Rightarrow (y+1) = 0, (y-1) = 0 \Rightarrow y = -1, 1$.

(a) Equilibrium points are −1 (stable) and 1 (unstable)

(b) $y' = y^2 - 1 \Rightarrow y'' = 2yy' \Rightarrow y'' = 2y(y^2-1) = 2y(y+1)(y-1)$. So $y'' = 0 \Rightarrow y = 0, y = -1, y = 1$.

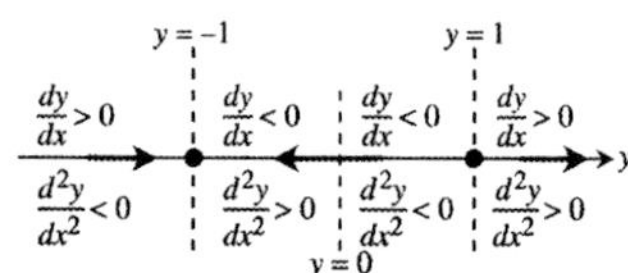

(c)

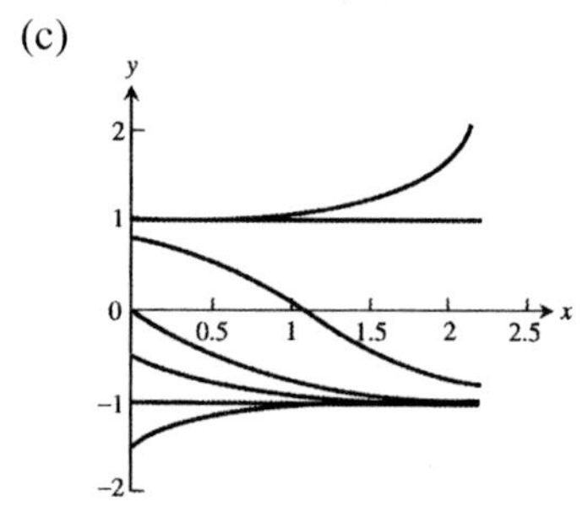

42. $\frac{dy}{dx} = y - y^2 \Rightarrow y' = y(1-y)$. We have $y' = 0 \Rightarrow y(1-y) = 0 \Rightarrow y = 0, 1 - y = 0 \Rightarrow y = 0, 1$.

(a) The equilibrium points are 0 and 1. So, 0 is unstable and 1 is stable.

(b) Let ⟶ = increasing, ⟵ = decreasing.

$y' < 0$ ⟵ (0) ⟶ $y' > 0$ (1) ⟵ $y' < 0$ ⟶ y

$y' = y - y^2 \Rightarrow y'' = y' - 2yy' \Rightarrow y'' = (y - y^2) - 2y(y - y^2) = y - y^2 - 2y^2 + 2y^3 \Rightarrow y'' = 2y^3 - 3y^2 + y$
$= y(2y^2 - 3y + 1) \Rightarrow y'' = y(2y-1)(y-1)$. So, $y'' = 0 \Rightarrow y = 0, 2y - 1 = 0, y - 1 = 0 \Rightarrow y = 0, y = \frac{1}{2}$,

$y = 1$.

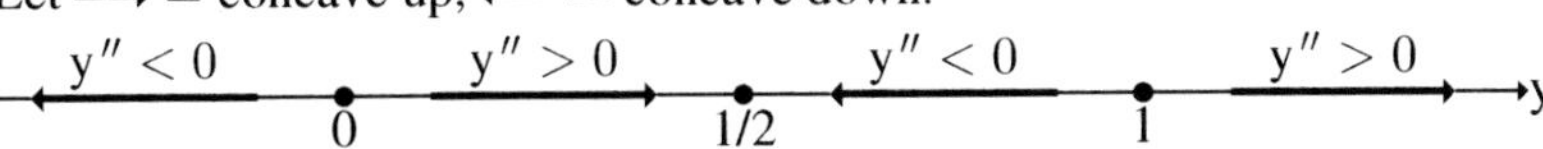

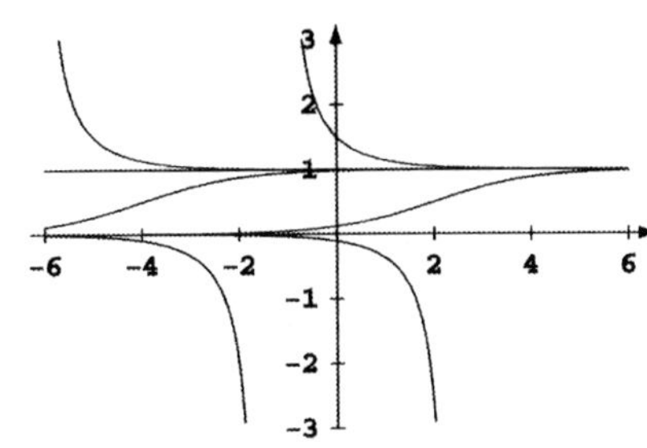

43. (a) Force = Mass times Acceleration (Newton's Second Law) or $F = ma$. Let $a = \frac{dv}{dt} = \frac{dv}{ds} \cdot \frac{ds}{dt} = v\frac{dv}{ds}$. Then $ma = -mgR^2s^{-2} \Rightarrow a = -gR^2s^{-2} \Rightarrow v\frac{dv}{ds} = -gR^2s^{-2} \Rightarrow v\,dv = -gR^2s^{-2}ds \Rightarrow \int v\,dv = \int -gR^2s^{-2}ds$ $\Rightarrow \frac{v^2}{2} = \frac{gR^2}{s} + C_1 \Rightarrow v^2 = \frac{2gR^2}{s} + 2C_1 = \frac{2gR^2}{s} + C$. When $t = 0$, $v = v_0$ and $s = R \Rightarrow v_0^2 = \frac{2gR^2}{R} + C$ $\Rightarrow C = v_0^2 - 2gR \Rightarrow v^2 = \frac{2gR^2}{s} + v_0^2 - 2gR$

(b) If $v_0 = \sqrt{2gR}$, then $v^2 = \frac{2gR^2}{s} \Rightarrow v = \sqrt{\frac{2gR^2}{s}}$, since $v \geq 0$ if $v_0 \geq \sqrt{2gR}$. Then $\frac{ds}{dt} = \frac{\sqrt{2gR^2}}{\sqrt{s}} \Rightarrow \sqrt{s}\,ds = \sqrt{2gR^2}\,dt$ $\Rightarrow \int s^{1/2}ds = \int \sqrt{2gR^2}\,dt \Rightarrow \frac{2}{3}s^{3/2} = \sqrt{2gR^2}t + C_1 \Rightarrow s^{3/2} = \left(\frac{3}{2}\sqrt{2gR^2}\right)t + C$; $t = 0$ and $s = R$ $\Rightarrow R^{3/2} = \left(\frac{3}{2}\sqrt{2gR^2}\right)(0) + C \Rightarrow C = R^{3/2} \Rightarrow s^{3/2} = \left(\frac{3}{2}\sqrt{2gR^2}\right)t + R^{3/2} = \left(\frac{3}{2}R\sqrt{2g}\right)t + R^{3/2}$ $= R^{3/2}\left[\left(\frac{3}{2}R^{-1/2}\sqrt{2g}\right)t + 1\right] = R^{3/2}\left[\left(\frac{3\sqrt{2gR}}{2R}\right)t + 1\right] = R^{3/2}\left[\left(\frac{3v_0}{2R}\right)t + 1\right] \Rightarrow s = R\left[1 + \left(\frac{3v_0}{2R}\right)t\right]^{2/3}$

44. $\frac{v_0 m}{k}$ = coasting distance $\Rightarrow \frac{(0.86)(30.84)}{k} = 0.97 \Rightarrow k \approx 27.343$. $s(t) = \frac{v_0 m}{k}\left(1 - e^{-(k/m)t}\right) \Rightarrow s(t) = 0.97\left(1 - e^{-(27.343/30.84)t}\right)$ $\Rightarrow s(t) = 0.97(1 - e^{-0.8866t})$. A graph of the model is shown superimposed on a graph of the data.

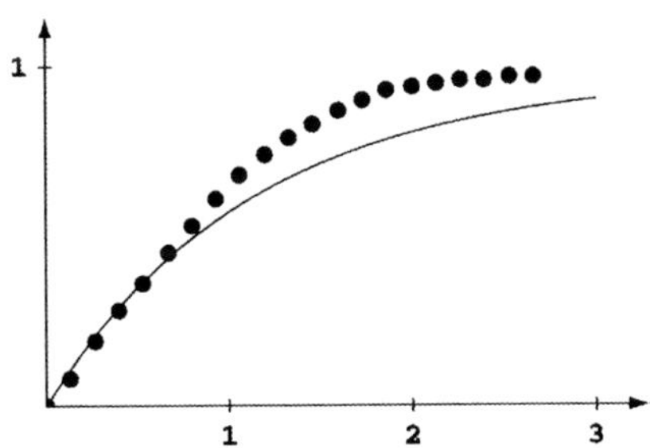

CHAPTER 9 ADDITIONAL AND ADVANCED EXERCISES

1. (a) $\frac{dy}{dt} = k\frac{A}{V}(c - y) \Rightarrow dy = -k\frac{A}{V}(y - c)dt \Rightarrow \frac{dy}{y - c} = -k\frac{A}{V}dt \Rightarrow \int \frac{dy}{y - c} = -\int k\frac{A}{V}dt \Rightarrow \ln|y - c| = -k\frac{A}{V}t + C_1$ $\Rightarrow y - c = \pm e^{C_1}e^{-k\frac{A}{V}t}$. Apply the initial condition, $y(0) = y_0 \Rightarrow y_0 = c + C \Rightarrow C = y_0 - c$ $\Rightarrow y = c + (y_0 - c)e^{-k\frac{A}{V}t}$.

(b) Steady state solution: $y_\infty = \lim\limits_{t\to\infty} y(t) = \lim\limits_{t\to\infty}\left[c + (y_0 - c)e^{-k\frac{A}{V}t}\right] = c + (y_0 - c)(0) = c$

2. Measure the amounts of oxygen involved in mL. Then the inflow of oxygen is 1000 mL/min (Assumed: it will take 5 minutes to deliver the 5L = 5000mL); the amount of oxygen at $t = 0$ is 210 mL; letting $A =$ the amount of oxygen in the flask, the concentration at time t is A mL/L; the outflow rate of oxygen is A mL/L (lb/sec). The rate of change in A, $\frac{dA}{dt}$, equals the rate of gain (1000 mL/min) minus rate of loss (A mL/min). Thus: $\frac{dA}{dt} = 1000 - A \Rightarrow \frac{dA}{1000 - A} = dt \Rightarrow \ln(A - 1000) = -t + C \Rightarrow A - 1000 = Ce^{-t}$. At $t = 0$, $A = 210$, so $C = -790$ and $A = 1000 - 790e^{-t}$. Thus, $A(5) = 1000 - 790e^{-5} \approx 994.7$ mL. The concentration is $\frac{994.7\text{ mL}}{1000\text{ mL}} = 99.47\%$.

3. The amount of CO_2 in the room at time t is $A(t)$. The rate of change in the amount of CO_2, $\frac{dA}{dt}$ is the rate of internal production (R_1) plus the inflow rate (R_2) minus the outflow rate (R_3).

$R_1 = \left(20\ \frac{\text{breaths/min}}{\text{student}}\right)(30\text{ students})\left(\frac{100}{1728}\text{ ft}^3\right)\left(0.04\ \frac{\text{ft}^3\,CO_2}{\text{ft}^3}\right) \approx 1.39\ \frac{\text{ft}^3\,CO_2}{\text{min}}$

$R_2 = \left(1000\ \frac{\text{ft}^3}{\text{min}}\right)\left(0.0004\ \frac{\text{ft}^3\,CO_2}{\text{min}}\right) = 0.4\ \frac{\text{ft}^3\,CO_2}{\text{min}}$

$R_3 = \left(\frac{A}{10{,}000}\right)1000 = 0.1A\ \frac{\text{ft}^3\,CO_2}{\text{min}}$

$\frac{dA}{dt} = 1.39 + 0.4 - 0.1A = 1.79 - 0.1A \Rightarrow A' + 0.1A = 1.79$. Let $v(t) = e^{\int 0.1\,dt}$. We have $\frac{d}{dt}\left(Ae^{\int 0.1\,dt}\right) = 1.79e^{\int 0.1\,dt}$ $\Rightarrow Ae^{0.1t} = \int 1.79e^{0.1t}dt = 17.9e^{0.1t} + C$. At $t = 0$, $A = (10{,}000)(0.0004) = 4\text{ ft}^3\ CO_2 \Rightarrow C = -13.9$ $\Rightarrow A = 17.9 - 13.9e^{-0.1t}$. So $A(60) = 17.9 - 13.9e^{-0.1(60)} \approx 17.87\text{ ft}^3$ of CO_2 in the $10{,}000\text{ ft}^3$ room. The percent of CO_2 is $\frac{17.87}{10{,}000} \times 100 = 0.18\%$

4. $\frac{d(mv)}{dt} = F + (v+u)\frac{dm}{dt} \Rightarrow F = \frac{d(mv)}{dt} - (v+u)\frac{dm}{dt} \Rightarrow F = m\frac{dv}{dt} + v\frac{dm}{dt} - v\frac{dm}{dt} - u\frac{dm}{dt} \Rightarrow F = m\frac{dv}{dt} - u\frac{dm}{dt}$.
$\frac{dm}{dt} = -b \Rightarrow m = -|b|t + C$. At $t = 0$, $m = m_0$, so $C = m_0$ and $m = m_0 - |b|t$.
Thus, $F = (m_0 - |b|t)\frac{dv}{dt} - u|b| = -(m_0 - |b|t)|g| \Rightarrow \frac{dv}{dt} = -g + \frac{u|b|}{m_0 - |b|t} \Rightarrow v = -gt - u\ln\left(\frac{m_0 - |b|t}{m_0}\right) + C_1$
$v = 0$ at $t = 0 \Rightarrow C_1 = 0$. So $v = -gt - u\ln\left(\frac{m_0 - |b|t}{m_0}\right) = \frac{dy}{dt} \Rightarrow y = \int\left[-gt - u\ln\left(\frac{m_0 - |b|t}{m_0}\right)\right]dt$ and $u = c$, $y = 0$ at $t = 0 \Rightarrow y = -\frac{1}{2}gt^2 + c\left[t + \left(\frac{m_0 - |b|t}{|b|}\right)\ln\left(\frac{m_0 - |b|t}{m_0}\right)\right]$

5. (a) Let y be any function such that $v(x)y = \int v(x)Q(x)\,dx + C$, $v(x) = e^{\int P(x)\,dx}$. Then $\frac{d}{dx}(v(x)\cdot y) = v(x)\cdot y' + y\cdot v'(x) = v(x)Q(x)$. We have $v(x) = e^{\int P(x)\,dx} \Rightarrow v'(x) = \ = e^{\int P(x)\,dx}P(x) = v(x)P(x)$. Thus $v(x)\cdot y' + y\cdot v(x)\,P(x) = v(x)Q(x) \Rightarrow y' + y\,P(x) = Q(x) \Rightarrow$ the given y is a solution.

(b) If v and Q are continuous on $[a, b]$ and $x \in (a, b)$, then $\frac{d}{dx}\left[\int_{x_0}^{x} v(t)Q(t)\,dt\right] = v(x)Q(x)$ $\Rightarrow \int_{x_0}^{x} v(t)Q(t)\,dt = \int v(x)Q(x)\,dx$. So $C = y_0v(x_0) - \int v(x)Q(x)\,dx$. From part (a), $v(x)y = \int v(x)Q(x)\,dx + C$. Substituting for C: $v(x)y = \int v(x)Q(x)\,dx + y_0v(x_0) - \int v(x)Q(x)\,dx \Rightarrow v(x)y = y_0v(x_0)$ when $x = x_0$.

6. (a) $y' + P(x)y = 0$, $y(x_0) = 0$. Use $v(x) = e^{\int P(x)\,dx}$ as an integrating factor. Then $\frac{d}{dx}(v(x)y) = 0 \Rightarrow v(x)y = C$ $\Rightarrow y = Ce^{-\int P(x)\,dx}$ and $y_1 = C_1e^{-\int P(x)\,dx}$, $y_2 = C_2e^{-\int P(x)\,dx}$, $y_1(x_0) = y_2(x_0) = 0$, $y_1 - y_2 = (C_1 - C_2)e^{-\int P(x)\,dx}$ $= C_3e^{-\int P(x)\,dx}$ and $y_1 - y_2 = 0 - 0 = 0$. So $y_1 - y_2$ is a solution to $y' + P(x)y = 0$ with $y(x_0) = 0$.

(b) $\frac{d}{dx}(v(x)[y_1(x) - y_2(x)]) = \frac{d}{dx}\left(e^{\int P(x)\,dx}\left[e^{-\int P(x)\,dx}(C_1 - C_2)\right]\right) = \frac{d}{dx}(C_1 - C_2) = \frac{d}{dx}(C_3) = 0$.
$\int \frac{d}{dx}(v(x)[y_1(x) - y_2(x)])dx = (v(x)[y_1(x) - y_2(x)]) = \int 0\,dx = C$

(c) $y_1 = C_1e^{-\int P(x)\,dx}$, $y_2 = C_2e^{-\int P(x)\,dx}$, $y = y_1 - y_2$. So $y(x_0) = 0 \Rightarrow C_1e^{-\int P(x)\,dx} - C_2e^{-\int P(x)\,dx} = 0$ $\Rightarrow C_1 - C_2 = 0 \Rightarrow C_1 = C_2 \Rightarrow y_1(x) = y_2(x)$ for $a < x < b$.

CHAPTER 10 CONIC SECTIONS AND POLAR COORDINATES

10.1 CONIC SECTIONS AND QUADRATIC EQUATIONS

1. $x = \frac{y^2}{8} \Rightarrow 4p = 8 \Rightarrow p = 2$; focus is $(2, 0)$, directrix is $x = -2$

2. $x = -\frac{y^2}{4} \Rightarrow 4p = 4 \Rightarrow p = 1$; focus is $(-1, 0)$, directrix is $x = 1$

3. $y = -\frac{x^2}{6} \Rightarrow 4p = 6 \Rightarrow p = \frac{3}{2}$; focus is $\left(0, -\frac{3}{2}\right)$, directrix is $y = \frac{3}{2}$

4. $y = \frac{x^2}{2} \Rightarrow 4p = 2 \Rightarrow p = \frac{1}{2}$; focus is $\left(0, \frac{1}{2}\right)$, directrix is $y = -\frac{1}{2}$

5. $\frac{x^2}{4} - \frac{y^2}{9} = 1 \Rightarrow c = \sqrt{4+9} = \sqrt{13} \Rightarrow$ foci are $\left(\pm\sqrt{13}, 0\right)$; vertices are $(\pm 2, 0)$; asymptotes are $y = \pm\frac{3}{2}x$

6. $\frac{x^2}{4} + \frac{y^2}{9} = 1 \Rightarrow c = \sqrt{9-4} = \sqrt{5} \Rightarrow$ foci are $\left(0, \pm\sqrt{5}\right)$; vertices are $(0, \pm 3)$

7. $\frac{x^2}{2} + y^2 = 1 \Rightarrow c = \sqrt{2-1} = 1 \Rightarrow$ foci are $(\pm 1, 0)$; vertices are $\left(\pm\sqrt{2}, 0\right)$

8. $\frac{y^2}{4} - x^2 = 1 \Rightarrow c = \sqrt{4+1} = \sqrt{5} \Rightarrow$ foci are $\left(0, \pm\sqrt{5}\right)$; vertices are $(0, \pm 2)$; asymptotes are $y = \pm 2x$

9. $y^2 = 12x \Rightarrow x = \frac{y^2}{12} \Rightarrow 4p = 12 \Rightarrow p = 3$;
focus is $(3, 0)$, directrix is $x = -3$

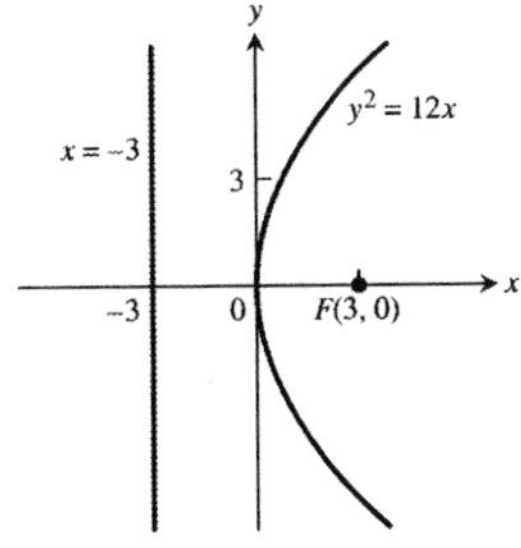

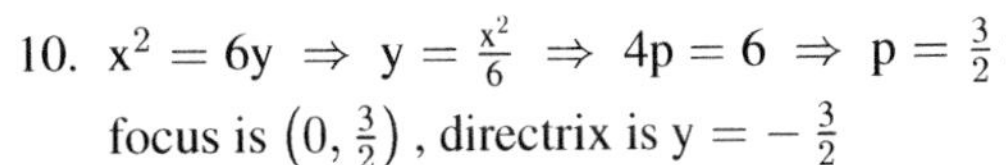

10. $x^2 = 6y \Rightarrow y = \frac{x^2}{6} \Rightarrow 4p = 6 \Rightarrow p = \frac{3}{2}$;
focus is $\left(0, \frac{3}{2}\right)$, directrix is $y = -\frac{3}{2}$

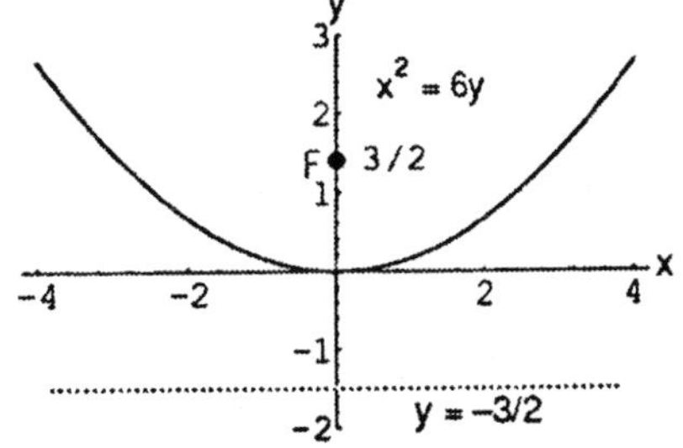

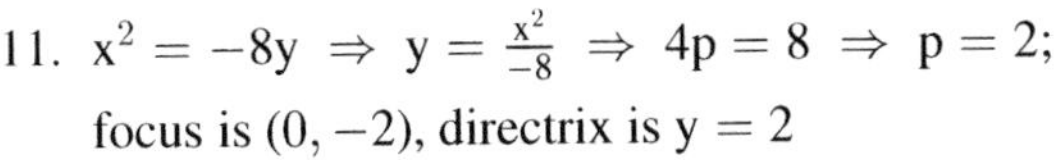

11. $x^2 = -8y \Rightarrow y = \frac{x^2}{-8} \Rightarrow 4p = 8 \Rightarrow p = 2$;
focus is $(0, -2)$, directrix is $y = 2$

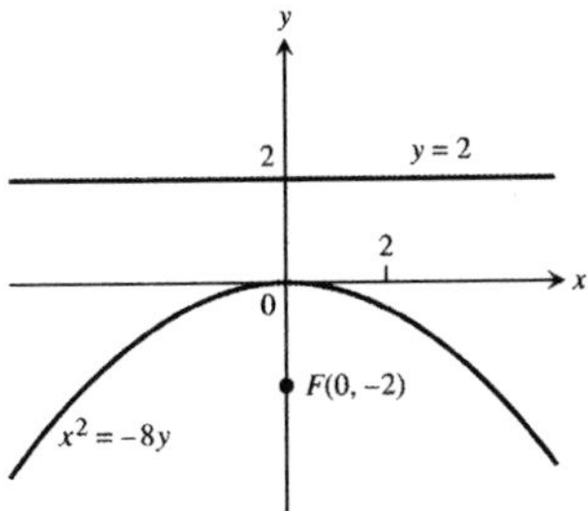

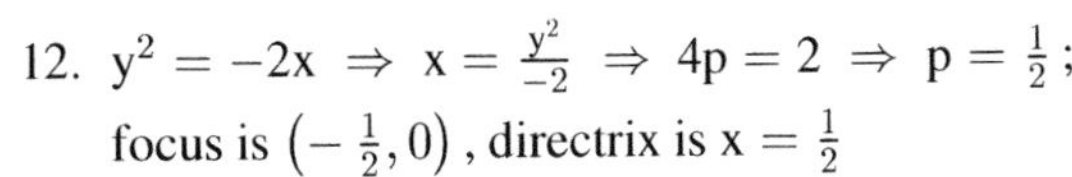

12. $y^2 = -2x \Rightarrow x = \frac{y^2}{-2} \Rightarrow 4p = 2 \Rightarrow p = \frac{1}{2}$;
focus is $\left(-\frac{1}{2}, 0\right)$, directrix is $x = \frac{1}{2}$

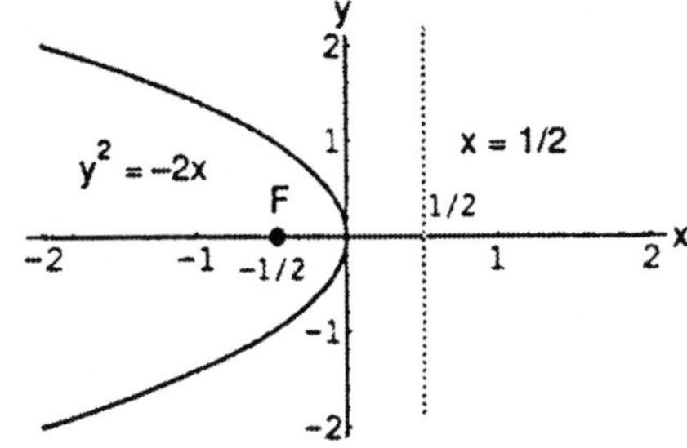

13. $y = 4x^2 \Rightarrow y = \frac{x^2}{\left(\frac{1}{4}\right)} \Rightarrow 4p = \frac{1}{4} \Rightarrow p = \frac{1}{16}$; focus is $\left(0, \frac{1}{16}\right)$, directrix is $y = -\frac{1}{16}$

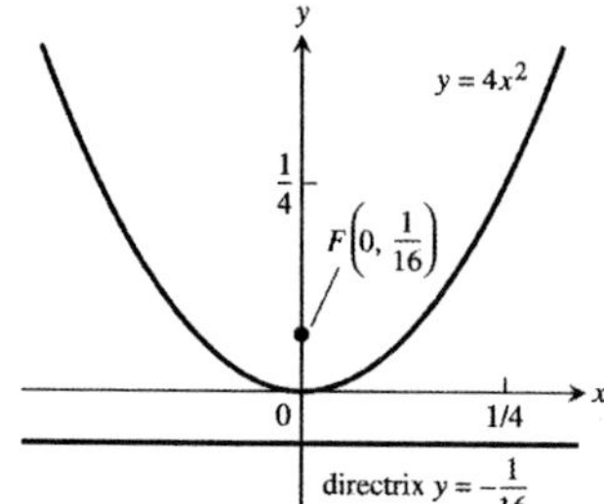

14. $y = -8x^2 \Rightarrow y = -\frac{x^2}{\left(\frac{1}{8}\right)} \Rightarrow 4p = \frac{1}{8} \Rightarrow p = \frac{1}{32}$; focus is $\left(0, -\frac{1}{32}\right)$, directrix is $y = \frac{1}{32}$

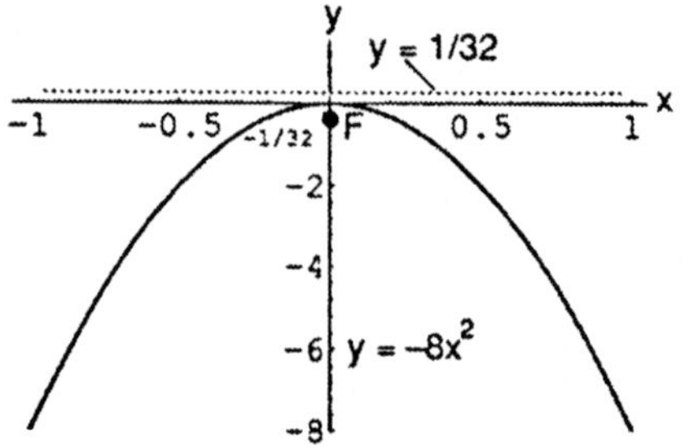

15. $x = -3y^2 \Rightarrow x = -\frac{y^2}{\left(\frac{1}{3}\right)} \Rightarrow 4p = \frac{1}{3} \Rightarrow p = \frac{1}{12}$; focus is $\left(-\frac{1}{12}, 0\right)$, directrix is $x = \frac{1}{12}$

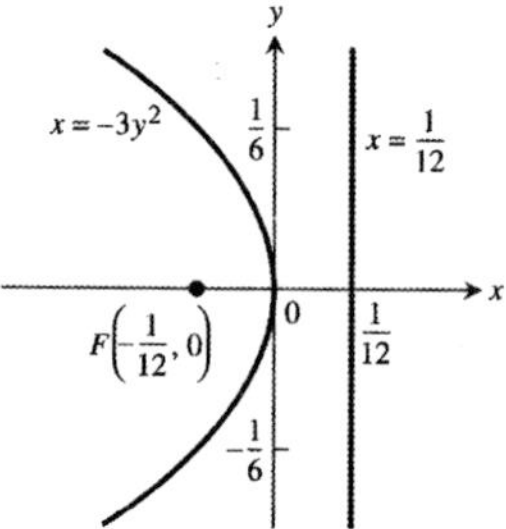

16. $x = 2y^2 \Rightarrow x = \frac{y^2}{\left(\frac{1}{2}\right)} \Rightarrow 4p = \frac{1}{2} \Rightarrow p = \frac{1}{8}$; focus is $\left(\frac{1}{8}, 0\right)$, directrix is $x = -\frac{1}{8}$

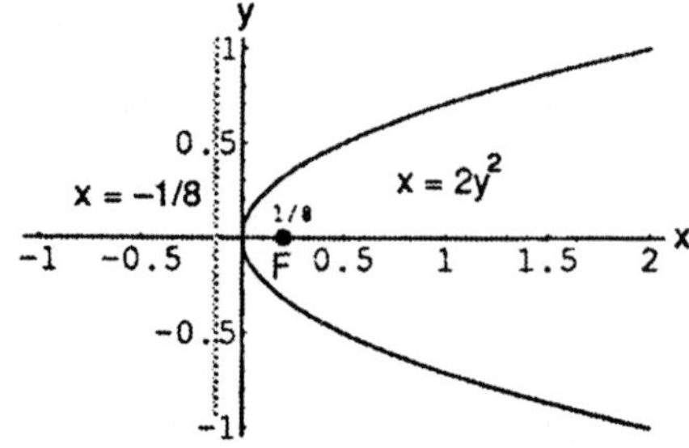

17. $16x^2 + 25y^2 = 400 \Rightarrow \frac{x^2}{25} + \frac{y^2}{16} = 1$
$\Rightarrow c = \sqrt{a^2 - b^2} = \sqrt{25 - 16} = 3$

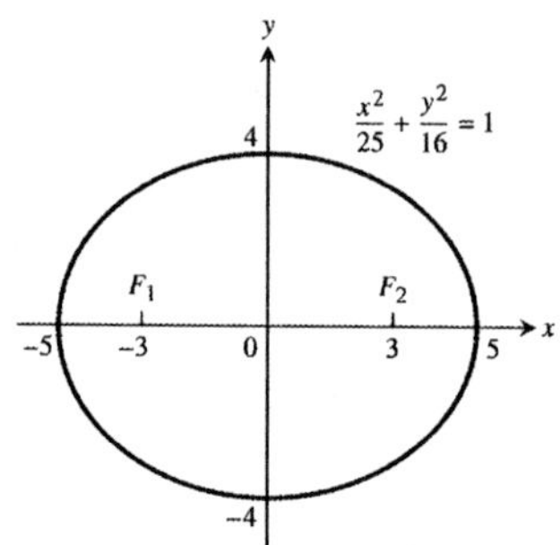

18. $7x^2 + 16y^2 = 112 \Rightarrow \frac{x^2}{16} + \frac{y^2}{7} = 1$
$\Rightarrow c = \sqrt{a^2 - b^2} = \sqrt{16 - 7} = 3$

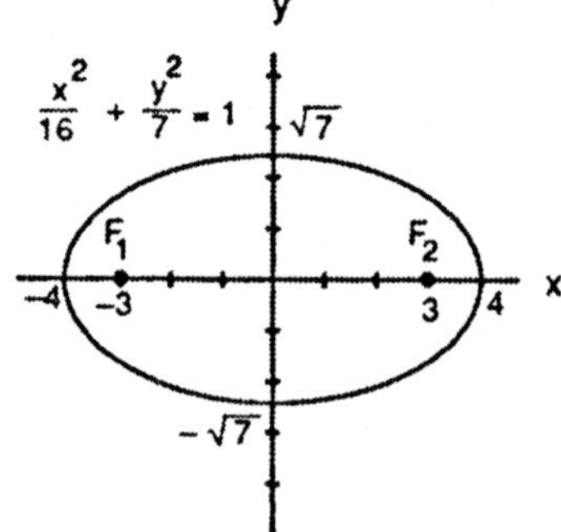

19. $2x^2 + y^2 = 2 \Rightarrow x^2 + \frac{y^2}{2} = 1$
$\Rightarrow c = \sqrt{a^2 - b^2} = \sqrt{2 - 1} = 1$

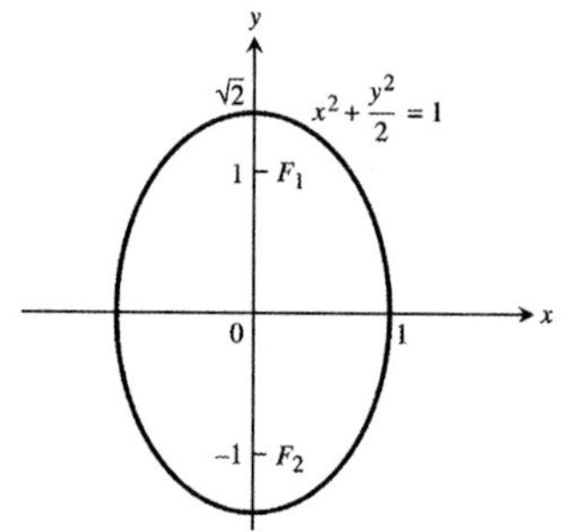

20. $2x^2 + y^2 = 4 \Rightarrow \frac{x^2}{2} + \frac{y^2}{4} = 1$
$\Rightarrow c = \sqrt{a^2 - b^2} = \sqrt{4 - 2} = \sqrt{2}$

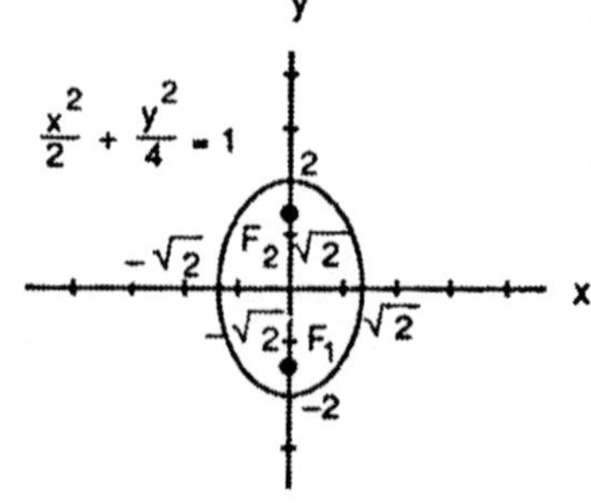

21. $3x^2 + 2y^2 = 6 \Rightarrow \frac{x^2}{2} + \frac{y^2}{3} = 1$
$\Rightarrow c = \sqrt{a^2 - b^2} = \sqrt{3 - 2} = 1$

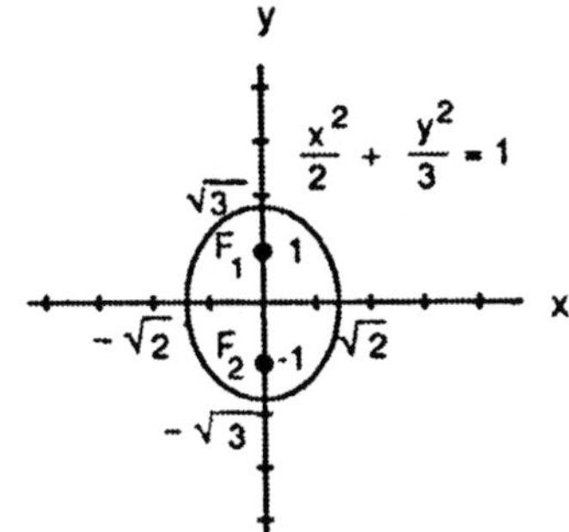

22. $9x^2 + 10y^2 = 90 \Rightarrow \frac{x^2}{10} + \frac{y^2}{9} = 1$
$\Rightarrow c = \sqrt{a^2 - b^2} = \sqrt{10 - 9} = 1$

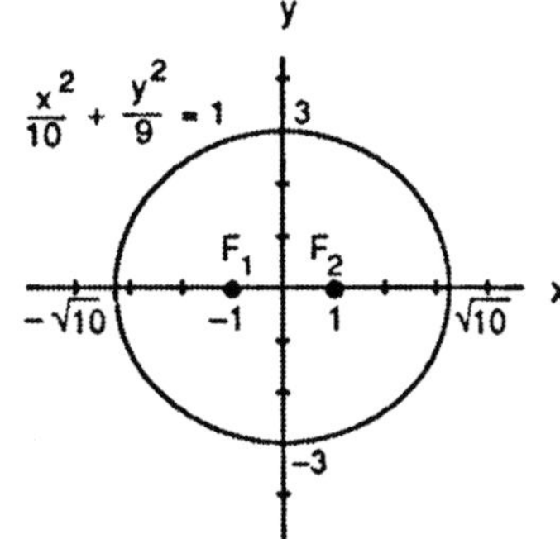

23. $6x^2 + 9y^2 = 54 \Rightarrow \frac{x^2}{9} + \frac{y^2}{6} = 1$
$\Rightarrow c = \sqrt{a^2 - b^2} = \sqrt{9 - 6} = \sqrt{3}$

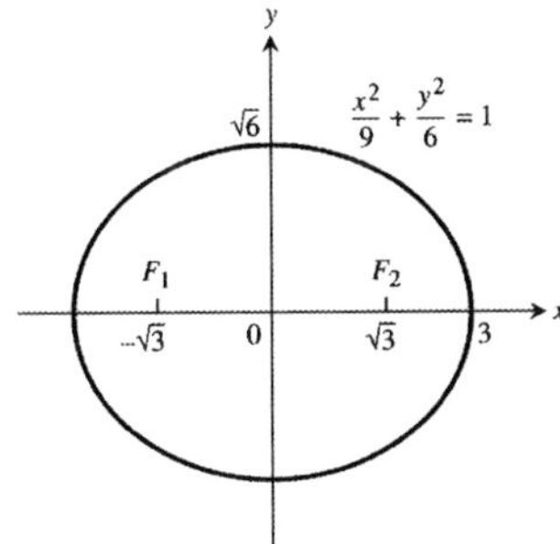

24. $169x^2 + 25y^2 = 4225 \Rightarrow \frac{x^2}{25} + \frac{y^2}{169} = 1$
$\Rightarrow c = \sqrt{a^2 - b^2} = \sqrt{169 - 25} = 12$

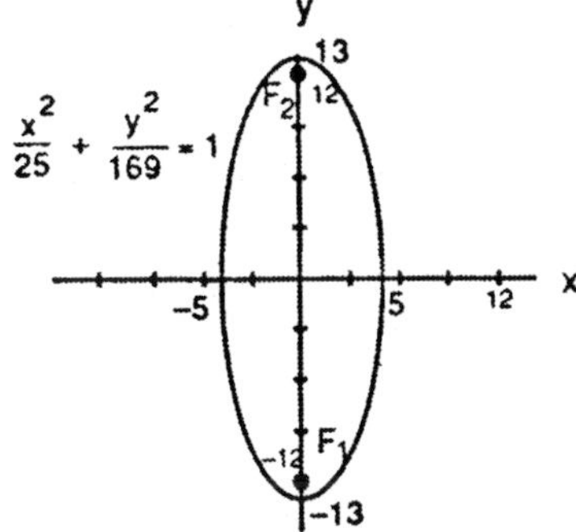

25. Foci: $\left(\pm\sqrt{2}, 0\right)$, Vertices: $(\pm 2, 0) \Rightarrow a = 2, c = \sqrt{2} \Rightarrow b^2 = a^2 - c^2 = 4 - \left(\sqrt{2}\right)^2 = 2 \Rightarrow \frac{x^2}{4} + \frac{y^2}{2} = 1$

26. Foci: $(0, \pm 4)$, Vertices: $(0, \pm 5) \Rightarrow a = 5, c = 4 \Rightarrow b^2 = 25 - 16 = 9 \Rightarrow \frac{x^2}{9} + \frac{y^2}{25} = 1$

27. $x^2 - y^2 = 1 \Rightarrow c = \sqrt{a^2 + b^2} = \sqrt{1 + 1} = \sqrt{2}$;
asymptotes are $y = \pm x$

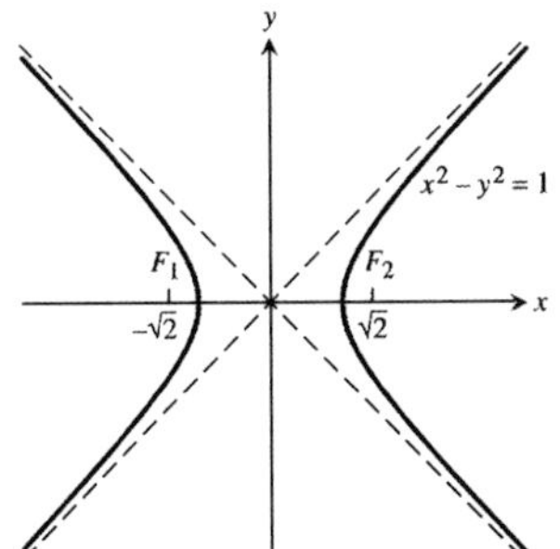

28. $9x^2 - 16y^2 = 144 \Rightarrow \frac{x^2}{16} - \frac{y^2}{9} = 1$
$\Rightarrow c = \sqrt{a^2 + b^2} = \sqrt{16 + 9} = 5$;
asymptotes are $y = \pm \frac{3}{4}x$

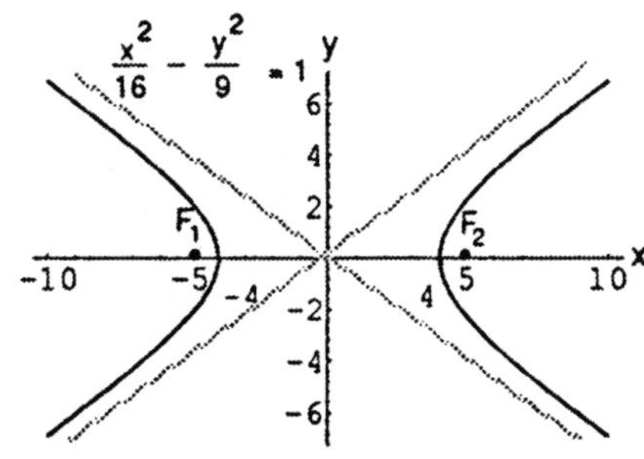

29. $y^2 - x^2 = 8 \Rightarrow \frac{y^2}{8} - \frac{x^2}{8} = 1 \Rightarrow c = \sqrt{a^2 + b^2}$
$= \sqrt{8 + 8} = 4$; asymptotes are $y = \pm x$

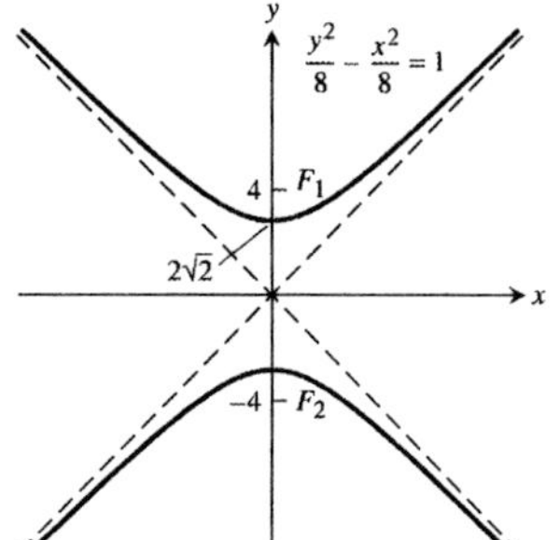

30. $y^2 - x^2 = 4 \Rightarrow \frac{y^2}{4} - \frac{x^2}{4} = 1 \Rightarrow c = \sqrt{a^2 + b^2}$
$= \sqrt{4 + 4} = 2\sqrt{2}$; asymptotes are $y = \pm x$

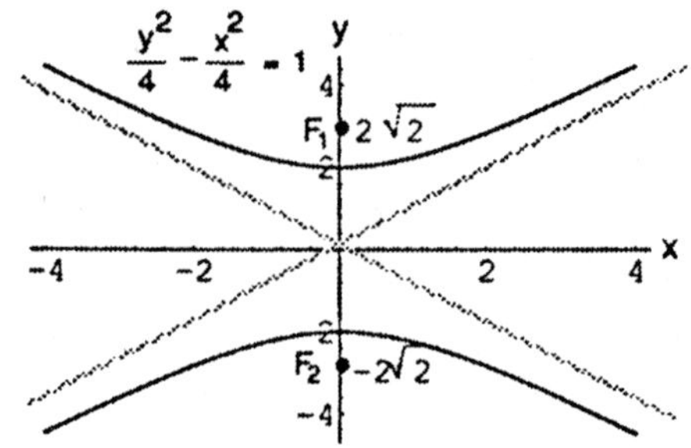

31. $8x^2 - 2y^2 = 16 \Rightarrow \frac{x^2}{2} - \frac{y^2}{8} = 1 \Rightarrow c = \sqrt{a^2 + b^2}$
$= \sqrt{2 + 8} = \sqrt{10}$; asymptotes are $y = \pm 2x$

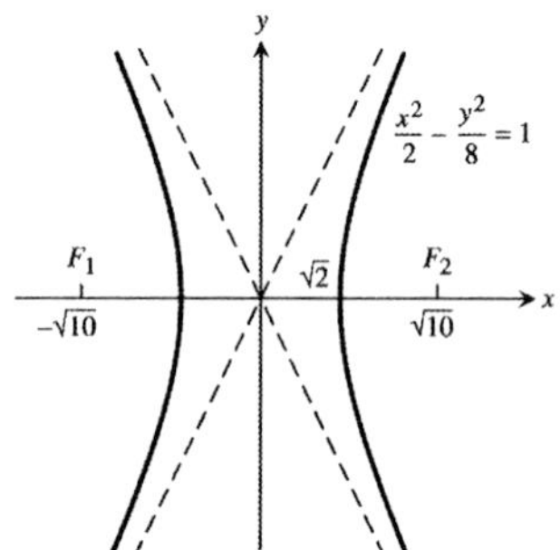

32. $y^2 - 3x^2 = 3 \Rightarrow \frac{y^2}{3} - x^2 = 1 \Rightarrow c = \sqrt{a^2 + b^2}$
$= \sqrt{3 + 1} = 2$; asymptotes are $y = \pm \sqrt{3}x$

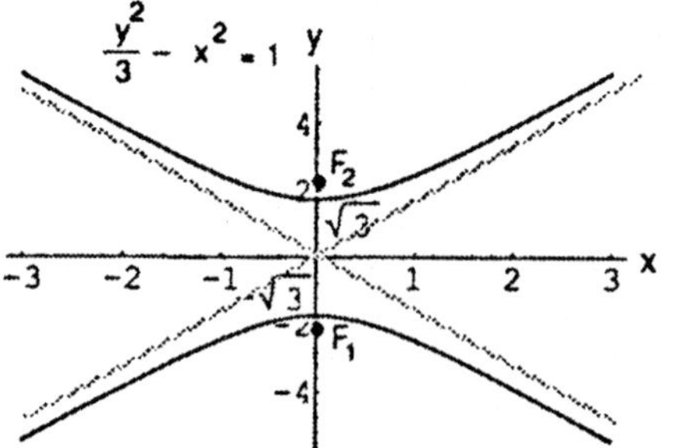

33. $8y^2 - 2x^2 = 16 \Rightarrow \frac{y^2}{2} - \frac{x^2}{8} = 1 \Rightarrow c = \sqrt{a^2 + b^2}$
$= \sqrt{2 + 8} = \sqrt{10}$; asymptotes are $y = \pm \frac{x}{2}$

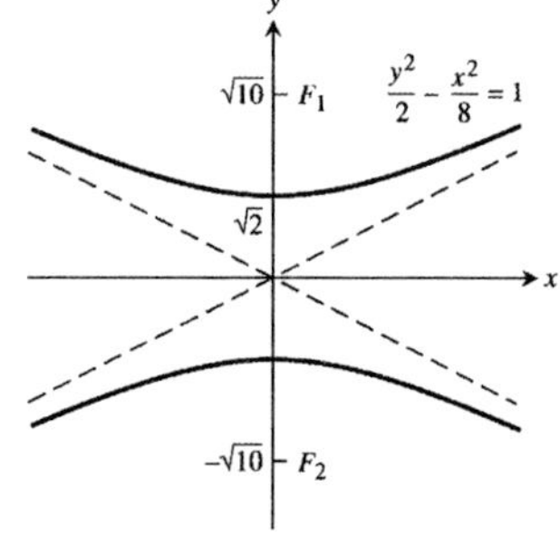

34. $64x^2 - 36y^2 = 2304 \Rightarrow \frac{x^2}{36} - \frac{y^2}{64} = 1 \Rightarrow c = \sqrt{a^2 + b^2}$
$= \sqrt{36 + 64} = 10$; asymptotes are $y = \pm \frac{4}{3}$

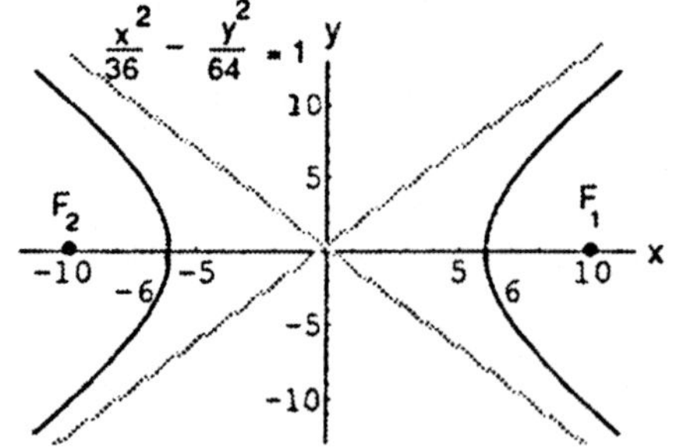

35. Foci: $\left(0, \pm \sqrt{2}\right)$, Asymptotes: $y = \pm x \Rightarrow c = \sqrt{2}$ and $\frac{a}{b} = 1 \Rightarrow a = b \Rightarrow c^2 = a^2 + b^2 = 2a^2 \Rightarrow 2 = 2a^2$
$\Rightarrow a = 1 \Rightarrow b = 1 \Rightarrow y^2 - x^2 = 1$

36. Foci: $(\pm 2, 0)$, Asymptotes: $y = \pm \frac{1}{\sqrt{3}} x \Rightarrow c = 2$ and $\frac{b}{a} = \frac{1}{\sqrt{3}} \Rightarrow b = \frac{a}{\sqrt{3}} \Rightarrow c^2 = a^2 + b^2 = a^2 + \frac{a^2}{3} = \frac{4a^2}{3}$
$\Rightarrow 4 = \frac{4a^2}{3} \Rightarrow a^2 = 3 \Rightarrow a = \sqrt{3} \Rightarrow b = 1 \Rightarrow \frac{x^2}{3} - y^2 = 1$

37. Vertices: $(\pm 3, 0)$, Asymptotes: $y = \pm \frac{4}{3} x \Rightarrow a = 3$ and $\frac{b}{a} = \frac{4}{3} \Rightarrow b = \frac{4}{3}(3) = 4 \Rightarrow \frac{x^2}{9} - \frac{y^2}{16} = 1$

38. Vertices: $(0, \pm 2)$, Asymptotes: $y = \pm \frac{1}{2} x \Rightarrow a = 2$ and $\frac{a}{b} = \frac{1}{2} \Rightarrow b = 2(2) = 4 \Rightarrow \frac{y^2}{4} - \frac{x^2}{16} = 1$

39. (a) $y^2 = 8x \Rightarrow 4p = 8 \Rightarrow p = 2 \Rightarrow$ directrix is $x = -2$, focus is $(2, 0)$, and vertex is $(0, 0)$; therefore the new directrix is $x = -1$, the new focus is $(3, -2)$, and the new vertex is $(1, -2)$

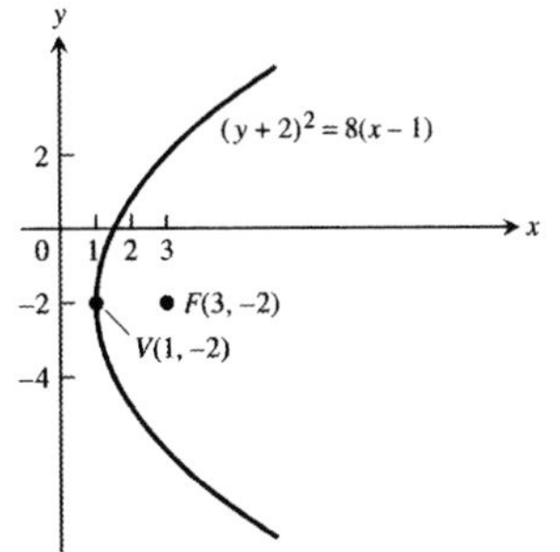

40. (a) $x^2 = -4y \Rightarrow 4p = 4 \Rightarrow p = 1 \Rightarrow$ directrix is $y = 1$, focus is $(0, -1)$, and vertex is $(0, 0)$; therefore the new directrix is $y = 4$, the new focus is $(-1, 2)$, and the new vertex is $(-1, 3)$

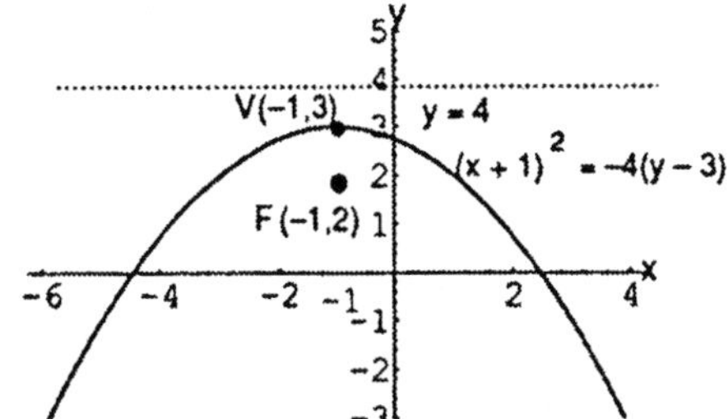

41. (a) $\frac{x^2}{16} + \frac{y^2}{9} = 1 \Rightarrow$ center is $(0, 0)$, vertices are $(-4, 0)$ and $(4, 0)$; $c = \sqrt{a^2 - b^2} = \sqrt{7} \Rightarrow$ foci are $\left(\sqrt{7}, 0\right)$ and $\left(-\sqrt{7}, 0\right)$; therefore the new center is $(4, 3)$, the new vertices are $(0, 3)$ and $(8, 3)$, and the new foci are $\left(4 \pm \sqrt{7}, 3\right)$

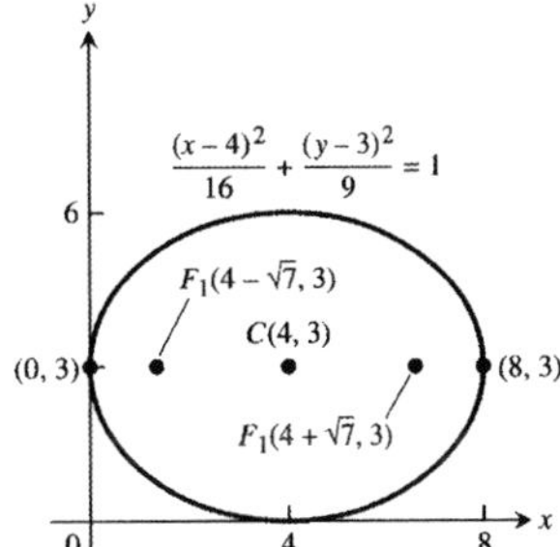

42. (a) $\frac{x^2}{9} + \frac{y^2}{25} = 1 \Rightarrow$ center is $(0, 0)$, vertices are $(0, 5)$ and $(0, -5)$; $c = \sqrt{a^2 - b^2} = \sqrt{16} = 4 \Rightarrow$ foci are $(0, 4)$ and $(0, -4)$; therefore the new center is $(-3, -2)$, the new vertices are $(-3, 3)$ and $(-3, -7)$, and the new foci are $(-3, 2)$ and $(-3, -6)$

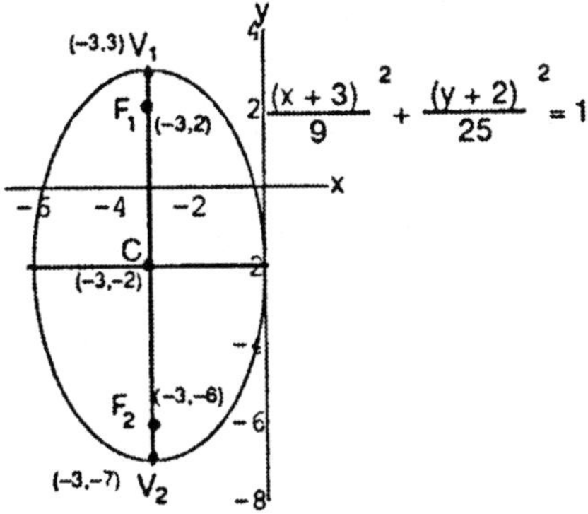

43. (a) $\frac{x^2}{16} - \frac{y^2}{9} = 1 \Rightarrow$ center is $(0, 0)$, vertices are $(-4, 0)$ and $(4, 0)$, and the asymptotes are $\frac{x}{4} = \pm \frac{y}{3}$ or $y = \pm \frac{3x}{4}$; $c = \sqrt{a^2 + b^2} = \sqrt{25} = 5 \Rightarrow$ foci are $(-5, 0)$ and $(5, 0)$; therefore the new center is $(2, 0)$, the new vertices are $(-2, 0)$ and $(6, 0)$, the new foci are $(-3, 0)$ and $(7, 0)$, and the new asymptotes are $y = \pm \frac{3(x-2)}{4}$

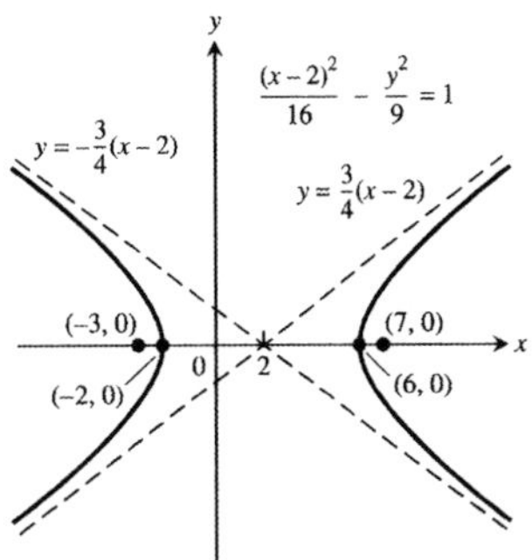

44. (a) $\frac{y^2}{4} - \frac{x^2}{5} = 1 \Rightarrow$ center is $(0, 0)$, vertices are $(0, -2)$ and $(0, 2)$, and the asymptotes are $\frac{y}{2} = \pm \frac{x}{\sqrt{5}}$ or $y = \pm \frac{2x}{\sqrt{5}}$; $c = \sqrt{a^2 + b^2} = \sqrt{9} = 3 \Rightarrow$ foci are $(0, 3)$ and $(0, -3)$; therefore the new center is $(0, -2)$, the new vertices are $(0, -4)$ and $(0, 0)$, the new foci are $(0, 1)$ and $(0, -5)$, and the new asymptotes are $y + 2 = \pm \frac{2x}{\sqrt{5}}$

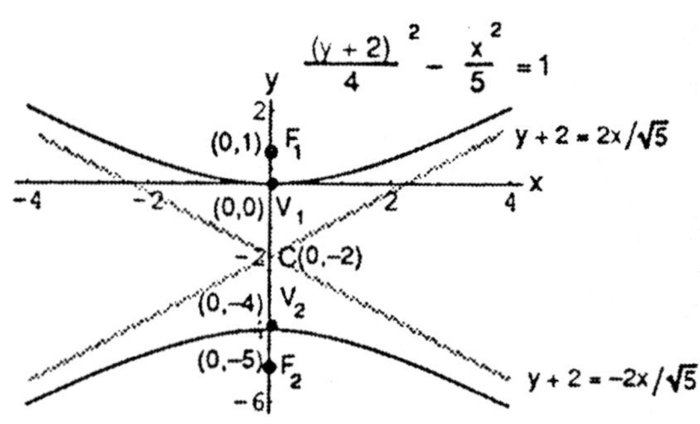

45. $y^2 = 4x \Rightarrow 4p = 4 \Rightarrow p = 1 \Rightarrow$ focus is $(1, 0)$, directrix is $x = -1$, and vertex is $(0, 0)$; therefore the new vertex is $(-2, -3)$, the new focus is $(-1, -3)$, and the new directrix is $x = -3$; the new equation is $(y + 3)^2 = 4(x + 2)$

46. $y^2 = -12x \Rightarrow 4p = 12 \Rightarrow p = 3 \Rightarrow$ focus is $(-3, 0)$, directrix is $x = 3$, and vertex is $(0, 0)$; therefore the new vertex is $(4, 3)$, the new focus is $(1, 3)$, and the new directrix is $x = 7$; the new equation is $(y - 3)^2 = -12(x - 4)$

47. $x^2 = 8y \Rightarrow 4p = 8 \Rightarrow p = 2 \Rightarrow$ focus is $(0, 2)$, directrix is $y = -2$, and vertex is $(0, 0)$; therefore the new vertex is $(1, -7)$, the new focus is $(1, -5)$, and the new directrix is $y = -9$; the new equation is $(x - 1)^2 = 8(y + 7)$

48. $x^2 = 6y \Rightarrow 4p = 6 \Rightarrow p = \frac{3}{2} \Rightarrow$ focus is $\left(0, \frac{3}{2}\right)$, directrix is $y = -\frac{3}{2}$, and vertex is $(0, 0)$; therefore the new vertex is $(-3, -2)$, the new focus is $\left(-3, -\frac{1}{2}\right)$, and the new directrix is $y = -\frac{7}{2}$; the new equation is $(x + 3)^2 = 6(y + 2)$

49. $\frac{x^2}{6} + \frac{y^2}{9} = 1 \Rightarrow$ center is $(0, 0)$, vertices are $(0, 3)$ and $(0, -3)$; $c = \sqrt{a^2 - b^2} = \sqrt{9 - 6} = \sqrt{3} \Rightarrow$ foci are $\left(0, \sqrt{3}\right)$ and $\left(0, -\sqrt{3}\right)$; therefore the new center is $(-2, -1)$, the new vertices are $(-2, 2)$ and $(-2, -4)$, and the new foci are $\left(-2, -1 \pm \sqrt{3}\right)$; the new equation is $\frac{(x+2)^2}{6} + \frac{(y+1)^2}{9} = 1$

50. $\frac{x^2}{2} + y^2 = 1 \Rightarrow$ center is $(0, 0)$, vertices are $\left(\sqrt{2}, 0\right)$ and $\left(-\sqrt{2}, 0\right)$; $c = \sqrt{a^2 - b^2} = \sqrt{2 - 1} = 1 \Rightarrow$ foci are $(-1, 0)$ and $(1, 0)$; therefore the new center is $(3, 4)$, the new vertices are $\left(3 \pm \sqrt{2}, 4\right)$, and the new foci are $(2, 4)$ and $(4, 4)$; the new equation is $\frac{(x-3)^2}{2} + (y - 4)^2 = 1$

51. $\frac{x^2}{3} + \frac{y^2}{2} = 1 \Rightarrow$ center is $(0, 0)$, vertices are $\left(\sqrt{3}, 0\right)$ and $\left(-\sqrt{3}, 0\right)$; $c = \sqrt{a^2 - b^2} = \sqrt{3 - 2} = 1 \Rightarrow$ foci are $(-1, 0)$ and $(1, 0)$; therefore the new center is $(2, 3)$, the new vertices are $\left(2 \pm \sqrt{3}, 3\right)$, and the new foci are $(1, 3)$ and $(3, 3)$; the new equation is $\frac{(x-2)^2}{3} + \frac{(y-3)^2}{2} = 1$

52. $\frac{x^2}{16} + \frac{y^2}{25} = 1 \Rightarrow$ center is $(0, 0)$, vertices are $(0, 5)$ and $(0, -5)$; $c = \sqrt{a^2 - b^2} = \sqrt{25 - 16} = 3 \Rightarrow$ foci are $(0, 3)$ and $(0, -3)$; therefore the new center is $(-4, -5)$, the new vertices are $(-4, 0)$ and $(-4, -10)$, and the new foci are $(-4, -2)$ and $(-4, -8)$; the new equation is $\frac{(x+4)^2}{16} + \frac{(y+5)^2}{25} = 1$

53. $\frac{x^2}{4} - \frac{y^2}{5} = 1 \Rightarrow$ center is $(0, 0)$, vertices are $(2, 0)$ and $(-2, 0)$; $c = \sqrt{a^2 + b^2} = \sqrt{4 + 5} = 3 \Rightarrow$ foci are $(3, 0)$ and $(-3, 0)$; the asymptotes are $\pm \frac{x}{2} = \frac{y}{\sqrt{5}} \Rightarrow y = \pm \frac{\sqrt{5}x}{2}$; therefore the new center is $(2, 2)$, the new vertices are $(4, 2)$ and $(0, 2)$, and the new foci are $(5, 2)$ and $(-1, 2)$; the new asymptotes are $y - 2 = \pm \frac{\sqrt{5}(x-2)}{2}$; the new

equation is $\frac{(x-2)^2}{4} - \frac{(y-2)^2}{5} = 1$

54. $\frac{x^2}{16} - \frac{y^2}{9} = 1 \Rightarrow$ center is $(0,0)$, vertices are $(4,0)$ and $(-4,0)$; $c = \sqrt{a^2+b^2} = \sqrt{16+9} = 5 \Rightarrow$ foci are $(-5,0)$ and $(5,0)$; the asymptotes are $\pm\frac{x}{4} = \frac{y}{3} \Rightarrow y = \pm\frac{3x}{4}$; therefore the new center is $(-5,-1)$, the new vertices are $(-1,-1)$ and $(-9,-1)$, and the new foci are $(-10,-1)$ and $(0,-1)$; the new asymptotes are $y+1 = \pm\frac{3(x+5)}{4}$; the new equation is $\frac{(x+5)^2}{16} - \frac{(y+1)^2}{9} = 1$

55. $y^2 - x^2 = 1 \Rightarrow$ center is $(0,0)$, vertices are $(0,1)$ and $(0,-1)$; $c = \sqrt{a^2+b^2} = \sqrt{1+1} = \sqrt{2} \Rightarrow$ foci are $\left(0, \pm\sqrt{2}\right)$; the asymptotes are $y = \pm x$; therefore the new center is $(-1,-1)$, the new vertices are $(-1,0)$ and $(-1,-2)$, and the new foci are $\left(-1, -1\pm\sqrt{2}\right)$; the new asymptotes are $y+1 = \pm(x+1)$; the new equation is $(y+1)^2 - (x+1)^2 = 1$

56. $\frac{y^2}{3} - x^2 = 1 \Rightarrow$ center is $(0,0)$, vertices are $\left(0,\sqrt{3}\right)$ and $\left(0,-\sqrt{3}\right)$; $c = \sqrt{a^2+b^2} = \sqrt{3+1} = 2 \Rightarrow$ foci are $(0,2)$ and $(0,-2)$; the asymptotes are $\pm x = \frac{y}{\sqrt{3}} \Rightarrow y = \pm\sqrt{3}x$; therefore the new center is $(1,3)$, the new vertices are $\left(1, 3\pm\sqrt{3}\right)$, and the new foci are $(1,5)$ and $(1,1)$; the new asymptotes are $y-3 = \pm\sqrt{3}(x-1)$; the new equation is $\frac{(y-3)^2}{3} - (x-1)^2 = 1$

57. $x^2 + 4x + y^2 = 12 \Rightarrow x^2 + 4x + 4 + y^2 = 12 + 4 \Rightarrow (x+2)^2 + y^2 = 16$; this is a circle: center at $C(-2,0)$, $a = 4$

58. $2x^2 + 2y^2 - 28x + 12y + 114 = 0 \Rightarrow x^2 - 14x + 49 + y^2 + 6y + 9 = -57 + 49 + 9 \Rightarrow (x-7)^2 + (y+3)^2 = 1$; this is a circle: center at $C(7,-3)$, $a = 1$

59. $x^2 + 2x + 4y - 3 = 0 \Rightarrow x^2 + 2x + 1 = -4y + 3 + 1 \Rightarrow (x+1)^2 = -4(y-1)$; this is a parabola: $V(-1,1)$, $F(-1,0)$

60. $y^2 - 4y - 8x - 12 = 0 \Rightarrow y^2 - 4y + 4 = 8x + 12 + 4 \Rightarrow (y-2)^2 = 8(x+2)$; this is a parabola: $V(-2,2)$, $F(0,2)$

61. $x^2 + 5y^2 + 4x = 1 \Rightarrow x^2 + 4x + 4 + 5y^2 = 5 \Rightarrow (x+2)^2 + 5y^2 = 5 \Rightarrow \frac{(x+2)^2}{5} + y^2 = 1$; this is an ellipse: the center is $(-2,0)$, the vertices are $\left(-2\pm\sqrt{5}, 0\right)$; $c = \sqrt{a^2-b^2} = \sqrt{5-1} = 2 \Rightarrow$ the foci are $(-4,0)$ and $(0,0)$

62. $9x^2 + 6y^2 + 36y = 0 \Rightarrow 9x^2 + 6(y^2 + 6y + 9) = 54 \Rightarrow 9x^2 + 6(y+3)^2 = 54 \Rightarrow \frac{x^2}{6} + \frac{(y+3)^2}{9} = 1$; this is an ellipse: the center is $(0,-3)$, the vertices are $(0,0)$ and $(0,-6)$; $c = \sqrt{a^2-b^2} = \sqrt{9-6} = \sqrt{3} \Rightarrow$ the foci are $\left(0, -3\pm\sqrt{3}\right)$

63. $x^2 + 2y^2 - 2x - 4y = -1 \Rightarrow x^2 - 2x + 1 + 2(y^2 - 2y + 1) = 2 \Rightarrow (x-1)^2 + 2(y-1)^2 = 2$ $\Rightarrow \frac{(x-1)^2}{2} + (y-1)^2 = 1$; this is an ellipse: the center is $(1,1)$, the vertices are $\left(1\pm\sqrt{2}, 1\right)$; $c = \sqrt{a^2-b^2} = \sqrt{2-1} = 1 \Rightarrow$ the foci are $(2,1)$ and $(0,1)$

64. $4x^2 + y^2 + 8x - 2y = -1 \Rightarrow 4(x^2 + 2x + 1) + y^2 - 2y + 1 = 4 \Rightarrow 4(x+1)^2 + (y-1)^2 = 4$ $\Rightarrow (x+1)^2 + \frac{(y-1)^2}{4} = 1$; this is an ellipse: the center is $(-1,1)$, the vertices are $(-1,3)$ and

$(-1,-1)$; $c = \sqrt{a^2 - b^2} = \sqrt{4-1} = \sqrt{3} \Rightarrow$ the foci are $\left(-1, 1 \pm \sqrt{3}\right)$

65. $x^2 - y^2 - 2x + 4y = 4 \Rightarrow x^2 - 2x + 1 - (y^2 - 4y + 4) = 1 \Rightarrow (x-1)^2 - (y-2)^2 = 1$; this is a hyperbola: the center is $(1,2)$, the vertices are $(2,2)$ and $(0,2)$; $c = \sqrt{a^2 + b^2} = \sqrt{1+1} = \sqrt{2} \Rightarrow$ the foci are $\left(1 \pm \sqrt{2}, 2\right)$; the asymptotes are $y - 2 = \pm(x-1)$

66. $x^2 - y^2 + 4x - 6y = 6 \Rightarrow x^2 + 4x + 4 - (y^2 + 6y + 9) = 1 \Rightarrow (x+2)^2 - (y+3)^2 = 1$; this is a hyperbola: the center is $(-2,-3)$, the vertices are $(-1,-3)$ and $(-3,-3)$; $c = \sqrt{a^2 + b^2} = \sqrt{1+1} = \sqrt{2} \Rightarrow$ the foci are $\left(-2 \pm \sqrt{2}, -3\right)$; the asymptotes are $y + 3 = \pm(x+2)$

67. $2x^2 - y^2 + 6y = 3 \Rightarrow 2x^2 - (y^2 - 6y + 9) = -6 \Rightarrow \frac{(y-3)^2}{6} - \frac{x^2}{3} = 1$; this is a hyperbola: the center is $(0,3)$, the vertices are $\left(0, 3 \pm \sqrt{6}\right)$; $c = \sqrt{a^2 + b^2} = \sqrt{6+3} = 3 \Rightarrow$ the foci are $(0,6)$ and $(0,0)$; the asymptotes are $\frac{y-3}{\sqrt{6}} = \pm\frac{x}{\sqrt{3}} \Rightarrow y = \pm\sqrt{2}x + 3$

68. $y^2 - 4x^2 + 16x = 24 \Rightarrow y^2 - 4(x^2 - 4x + 4) = 8 \Rightarrow \frac{y^2}{8} - \frac{(x-2)^2}{2} = 1$; this is a hyperbola: the center is $(2,0)$, the vertices are $\left(2, \pm\sqrt{8}\right)$; $c = \sqrt{a^2 + b^2} = \sqrt{8+2} = \sqrt{10} \Rightarrow$ the foci are $\left(2, \pm\sqrt{10}\right)$; the asymptotes are $\frac{y}{\sqrt{8}} = \pm\frac{x-2}{\sqrt{2}} \Rightarrow y = \pm 2(x-2)$

69.

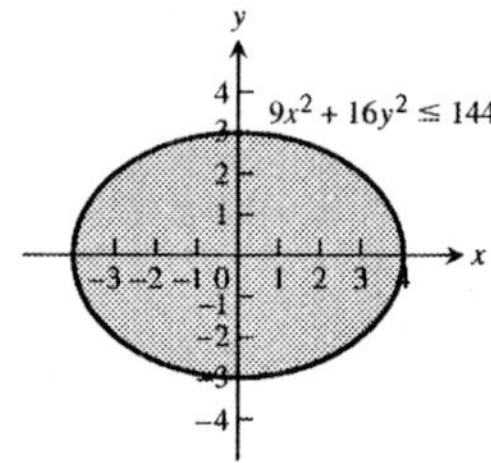

70.

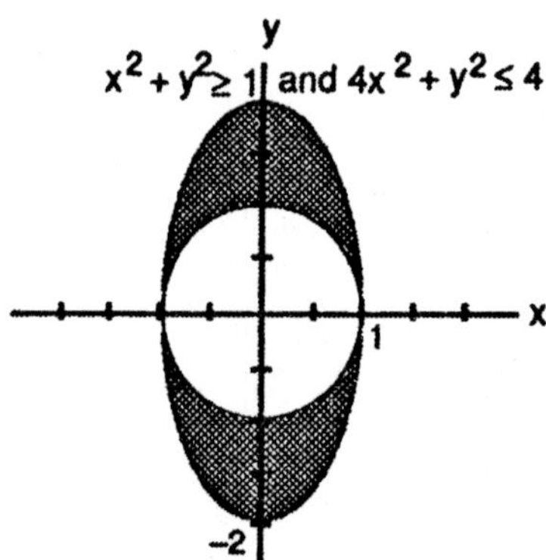

71.

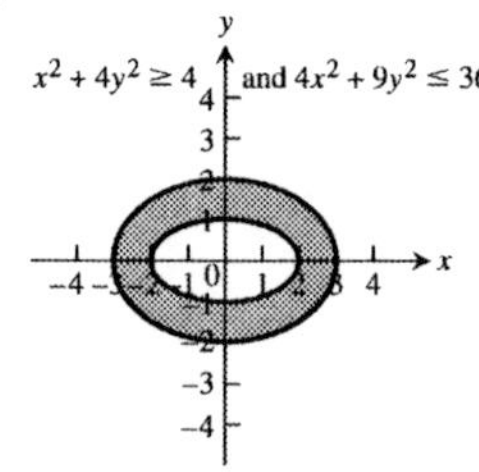

72.

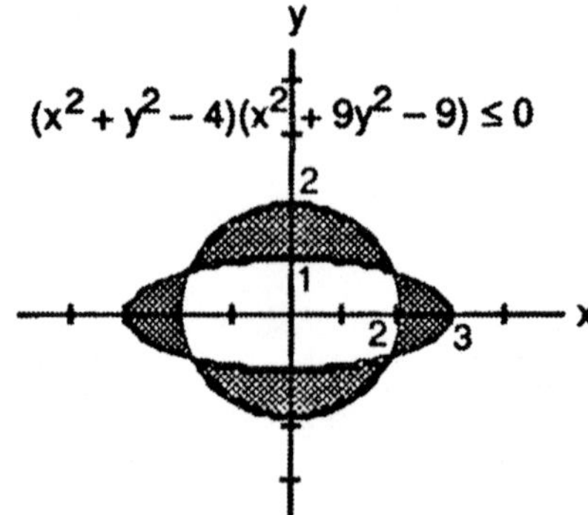

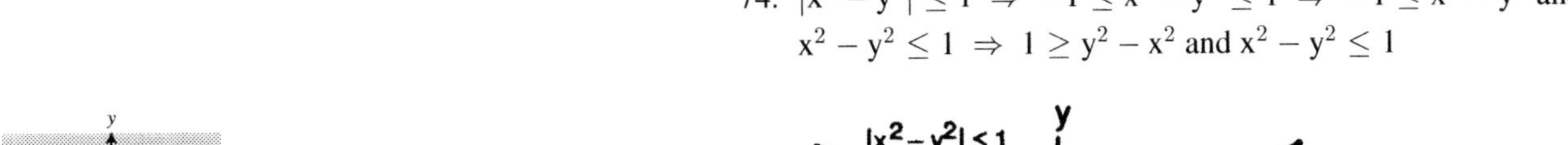

73.

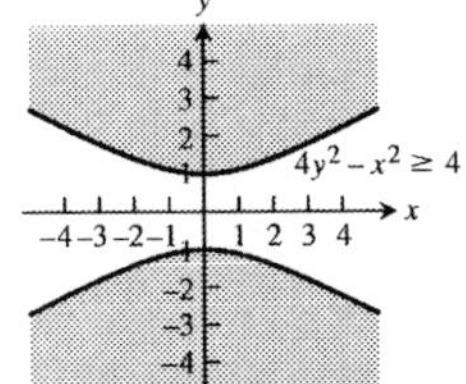

74. $|x^2 - y^2| \leq 1 \Rightarrow -1 \leq x^2 - y^2 \leq 1 \Rightarrow -1 \leq x^2 - y^2$ and $x^2 - y^2 \leq 1 \Rightarrow 1 \geq y^2 - x^2$ and $x^2 - y^2 \leq 1$

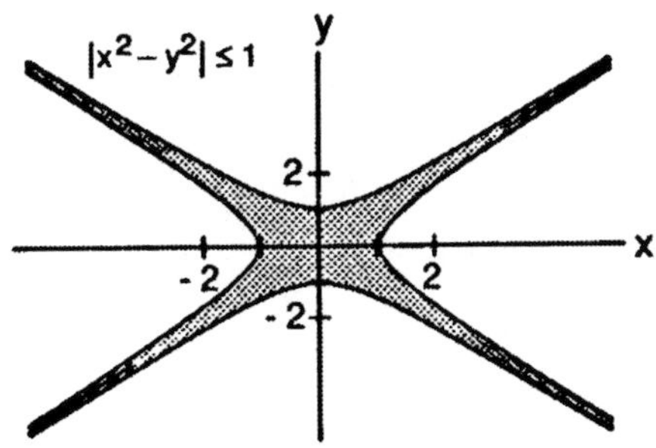

75. Volume of the Parabolic Solid: $V_1 = \int_0^{b/2} 2\pi x \left(h - \frac{4h}{b^2} x^2\right) dx = 2\pi h \int_0^{b/2} \left(x - \frac{4x^3}{b^2}\right) dx = 2\pi h \left[\frac{x^2}{2} - \frac{x^4}{b^2}\right]_0^{b/2}$
$= \frac{\pi h b^2}{8}$; Volume of the Cone: $V_2 = \frac{1}{3}\pi\left(\frac{b}{2}\right)^2 h = \frac{1}{3}\pi\left(\frac{b^2}{4}\right) h = \frac{\pi h b^2}{12}$; therefore $V_1 = \frac{3}{2} V_2$

76. $y = \int \frac{w}{H} x\, dx = \frac{w}{H}\left(\frac{x^2}{2}\right) + C = \frac{wx^2}{2H} + C$; $y = 0$ when $x = 0 \Rightarrow 0 = \frac{w(0)^2}{2H} + C \Rightarrow C = 0$; therefore $y = \frac{wx^2}{2H}$ is the equation of the cable's curve

77. A general equation of the circle is $x^2 + y^2 + ax + by + c = 0$, so we will substitute the three given points into this equation and solve the resulting system: $\left.\begin{array}{r} a \quad\;\; + c = -1 \\ b + c = -1 \\ 2a + 2b + c = -8 \end{array}\right\} \Rightarrow c = \frac{4}{3}$ and $a = b = -\frac{7}{3}$; therefore $3x^2 + 3y^2 - 7x - 7y + 4 = 0$ represents the circle

78. A general equation of the circle is $x^2 + y^2 + ax + by + c = 0$, so we will substitute each of the three given points into this equation and solve the resulting system: $\left.\begin{array}{r} 2a + 3b + c = -13 \\ 3a + 2b + c = -13 \\ -4a + 3b + c = -25 \end{array}\right\} \Rightarrow a = 2, b = 2$, and $c = -23$; therefore $x^2 + y^2 + 2x + 2y - 23 = 0$ represents the circle

79. $r^2 = (-2-1)^2 + (1-3)^2 = 13 \Rightarrow (x+2)^2 + (y-1)^2 = 13$ is an equation of the circle; the distance from the center to $(1.1, 2.8)$ is $\sqrt{(-2-1.1)^2 + (1-2.8)^2} = \sqrt{12.85} < \sqrt{13}$, the radius $\Rightarrow$ the point is inside the circle

80. $(x-2)^2 + (y-1)^2 = 5 \Rightarrow 2(x-2) + 2(y-1)\frac{dy}{dx} = 0 \Rightarrow \frac{dy}{dx} = -\frac{x-2}{y-1}$; $y = 0 \Rightarrow (x-2)^2 + (0-1)^2 = 5$
$\Rightarrow (x-2)^2 = 4 \Rightarrow x = 4$ or $x = 0 \Rightarrow$ the circle crosses the x-axis at $(4, 0)$ and $(0, 0)$; $x = 0$
$\Rightarrow (0-2)^2 + (y-1)^2 = 5 \Rightarrow (y-1)^2 = 1 \Rightarrow y = 2$ or $y = 0 \Rightarrow$ the circle crosses the y-axis at $(0, 2)$ and $(0, 0)$.
At $(4, 0)$: $\frac{dy}{dx} = -\frac{4-2}{0-1} = 2 \Rightarrow$ the tangent line is $y = 2(x-4)$ or $y = 2x - 8$
At $(0, 0)$: $\frac{dy}{dx} = -\frac{0-2}{0-1} = -2 \Rightarrow$ the tangent line is $y = -2x$
At $(0, 2)$: $\frac{dy}{dx} = -\frac{0-2}{2-1} = 2 \Rightarrow$ the tangent line is $y - 2 = 2x$ or $y = 2x + 2$

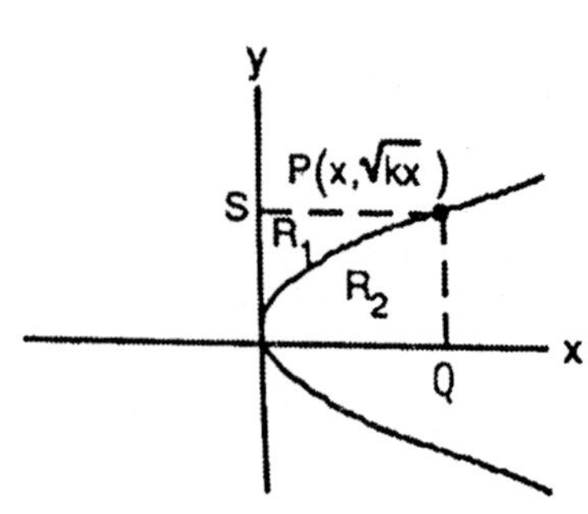

81. (a) $y^2 = kx \Rightarrow x = \frac{y^2}{k}$; the volume of the solid formed by revolving R_1 about the y-axis is $V_1 = \int_0^{\sqrt{kx}} \pi \left(\frac{y^2}{k}\right)^2 dy = \frac{\pi}{k^2} \int_0^{\sqrt{kx}} y^4\, dy = \frac{\pi x^2 \sqrt{kx}}{5}$; the volume of the right circular cylinder formed by revolving PQ about the y-axis is $V_2 = \pi x^2 \sqrt{kx} \Rightarrow$ the volume of the solid formed by revolving R_2 about the y-axis is $V_3 = V_2 - V_1 = \frac{4\pi x^2 \sqrt{kx}}{5}$. Therefore we can see the ratio of V_3 to V_1 is 4:1.

(b) The volume of the solid formed by revolving R_2 about the x-axis is $V_1 = \int_0^x \pi \left(\sqrt{kt}\right)^2 dt = \pi k \int_0^x t\, dt = \frac{\pi k x^2}{2}$. The volume of the right circular cylinder formed by revolving PS about the x-axis is $V_2 = \pi \left(\sqrt{kx}\right)^2 x = \pi k x^2 \Rightarrow$ the volume of the solid formed by revolving R_1 about the x-axis is $V_3 = V_2 - V_1 = \pi k x^2 - \frac{\pi k x^2}{2} = \frac{\pi k x^2}{2}$. Therefore the ratio of V_3 to V_1 is 1:1.

82. Let $P_1(-p, y_1)$ be any point on $x = -p$, and let $P(x, y)$ be a point where a tangent intersects $y^2 = 4px$. Now $y^2 = 4px \Rightarrow 2y \frac{dy}{dx} = 4p \Rightarrow \frac{dy}{dx} = \frac{2p}{y}$; then the slope of a tangent line from P_1 is $\frac{y - y_1}{x - (-p)} = \frac{dy}{dx} = \frac{2p}{y}$ $\Rightarrow y^2 - yy_1 = 2px + 2p^2$. Since $x = \frac{y^2}{4p}$, we have $y^2 - yy_1 = 2p\left(\frac{y^2}{4p}\right) + 2p^2 \Rightarrow y^2 - yy_1 = \frac{1}{2}y^2 + 2p^2$ $\Rightarrow \frac{1}{2}y^2 - yy_1 - 2p^2 = 0 \Rightarrow y = \frac{2y_1 \pm \sqrt{4y_1^2 + 16p^2}}{2} = y_1 \pm \sqrt{y_1^2 + 4p^2}$. Therefore the slopes of the two tangents from P_1 are $m_1 = \frac{2p}{y_1 + \sqrt{y_1^2 + 4p^2}}$ and $m_2 = \frac{2p}{y_1 - \sqrt{y_1^2 + 4p^2}} \Rightarrow m_1 m_2 = \frac{4p^2}{y_1^2 - (y_1^2 + 4p^2)} = -1$ $\Rightarrow$ the lines are perpendicular

83. Let $y = \sqrt{1 - \frac{x^2}{4}}$ on the interval $0 \le x \le 2$. The area of the inscribed rectangle is given by $A(x) = 2x\left(2\sqrt{1 - \frac{x^2}{4}}\right) = 4x\sqrt{1 - \frac{x^2}{4}}$ (since the length is 2x and the height is 2y) $\Rightarrow A'(x) = 4\sqrt{1 - \frac{x^2}{4}} - \frac{x^2}{\sqrt{1 - \frac{x^2}{4}}}$. Thus $A'(x) = 0 \Rightarrow 4\sqrt{1 - \frac{x^2}{4}} - \frac{x^2}{\sqrt{1 - \frac{x^2}{4}}} = 0 \Rightarrow 4\left(1 - \frac{x^2}{4}\right) - x^2 = 0 \Rightarrow x^2 = 2$ $\Rightarrow x = \sqrt{2}$ (only the positive square root lies in the interval). Since $A(0) = A(2) = 0$ we have that $A\left(\sqrt{2}\right) = 4$ is the maximum area when the length is $2\sqrt{2}$ and the height is $\sqrt{2}$.

84. (a) Around the x-axis: $9x^2 + 4y^2 = 36 \Rightarrow y^2 = 9 - \frac{9}{4}x^2 \Rightarrow y = \pm\sqrt{9 - \frac{9}{4}x^2}$ and we use the positive root $\Rightarrow V = 2\int_0^2 \pi \left(\sqrt{9 - \frac{9}{4}x^2}\right)^2 dx = 2\int_0^2 \pi \left(9 - \frac{9}{4}x^2\right) dx = 2\pi \left[9x - \frac{3}{4}x^3\right]_0^2 = 24\pi$

(b) Around the y-axis: $9x^2 + 4y^2 = 36 \Rightarrow x^2 = 4 - \frac{4}{9}y^2 \Rightarrow x = \pm\sqrt{4 - \frac{4}{9}y^2}$ and we use the positive root $\Rightarrow V = 2\int_0^3 \pi \left(\sqrt{4 - \frac{4}{9}y^2}\right)^2 dy = 2\int_0^3 \pi \left(4 - \frac{4}{9}y^2\right) dy = 2\pi \left[4y - \frac{4}{27}y^3\right]_0^3 = 16\pi$

85. $9x^2 - 4y^2 = 36 \Rightarrow y^2 = \frac{9x^2 - 36}{4} \Rightarrow y = \pm\frac{3}{2}\sqrt{x^2 - 4}$ on the interval $2 \le x \le 4 \Rightarrow V = \int_2^4 \pi \left(\frac{3}{2}\sqrt{x^2 - 4}\right)^2 dx$ $= \frac{9\pi}{4}\int_2^4 (x^2 - 4)\, dx = \frac{9\pi}{4}\left[\frac{x^3}{3} - 4x\right]_2^4 = \frac{9\pi}{4}\left[\left(\frac{64}{3} - 16\right) - \left(\frac{8}{3} - 8\right)\right] = \frac{9\pi}{4}\left(\frac{56}{3} - 8\right) = \frac{3\pi}{4}(56 - 24) = 24\pi$

86. $x^2 - y^2 = 1 \Rightarrow x = \pm\sqrt{1 + y^2}$ on the interval $-3 \le y \le 3 \Rightarrow V = \int_{-3}^3 \pi \left(\sqrt{1 + y^2}\right)^2 dy = 2\int_0^3 \pi \left(\sqrt{1 + y^2}\right)^2 dy$ $= 2\pi \int_0^3 (1 + y^2)\, dy = 2\pi \left[y + \frac{y^3}{3}\right]_0^3 = 24\pi$

87. Let $y = \sqrt{16 - \frac{16}{9}x^2}$ on the interval $-3 \le x \le 3$. Since the plate is symmetric about the y-axis, $\bar{x} = 0$. For a vertical strip: $(\tilde{x}, \tilde{y}) = \left(x, \frac{\sqrt{16 - \frac{16}{9}x^2}}{2}\right)$, length $= \sqrt{16 - \frac{16}{9}x^2}$, width $= dx \Rightarrow$ area $= dA = \sqrt{16 - \frac{16}{9}x^2}\,dx$ $\Rightarrow$ mass $= dm = \delta\,dA = \delta\sqrt{16 - \frac{16}{9}x^2}\,dx$. Moment of the strip about the x-axis:
$\tilde{y}\,dm = \frac{\sqrt{16 - \frac{16}{9}x^2}}{2}\left(\delta\sqrt{16 - \frac{16}{9}x^2}\right)dx = \delta\left(8 - \frac{8}{9}x^2\right)dx$ so the moment of the plate about the x-axis is
$M_x = \int \tilde{y}\,dm = \int_{-3}^{3} \delta\left(8 - \frac{8}{9}x^2\right)dx = \delta\left[8x - \frac{8}{27}x^3\right]_{-3}^{3} = 32\delta$; also the mass of the plate is
$M = \int_{-3}^{3} \delta\sqrt{16 - \frac{16}{9}x^2}\,dx = \int_{-3}^{3} 4\delta\sqrt{1 - \left(\frac{1}{3}x\right)^2}\,dx = 4\delta\int_{-1}^{1} 3\sqrt{1 - u^2}\,du$ where $u = \frac{x}{3} \Rightarrow 3\,du = dx$; $x = -3$ $\Rightarrow u = -1$ and $x = 3 \Rightarrow u = 1$. Hence, $4\delta\int_{-1}^{1} 3\sqrt{1 - u^2}\,du = 12\delta\int_{-1}^{1}\sqrt{1 - u^2}\,du$
$= 12\delta\left[\frac{1}{2}\left(u\sqrt{1 - u^2} + \sin^{-1}u\right)\right]_{-1}^{1} = 6\pi\delta \Rightarrow \bar{y} = \frac{M_x}{M} = \frac{32\delta}{6\pi\delta} = \frac{16}{3\pi}$. Therefore the center of mass is $\left(0, \frac{16}{3\pi}\right)$.

88. $y = \sqrt{x^2 + 1} \Rightarrow \frac{dy}{dx} = \frac{1}{2}(x^2 + 1)^{-1/2}(2x) = \frac{x}{\sqrt{x^2 + 1}} \Rightarrow \left(\frac{dy}{dx}\right)^2 = \frac{x^2}{x^2 + 1} \Rightarrow \sqrt{1 + \left(\frac{dy}{dx}\right)^2} = \sqrt{1 + \frac{x^2}{x^2 + 1}}$
$= \sqrt{\frac{2x^2 + 1}{x^2 + 1}} \Rightarrow S = \int_0^{\sqrt{2}} 2\pi y\sqrt{1 + \left(\frac{dy}{dx}\right)^2}\,dx = \int_0^{\sqrt{2}} 2\pi\sqrt{x^2 + 1}\sqrt{\frac{2x^2 + 1}{x^2 + 1}}\,dx = \int_0^{\sqrt{2}} 2\pi\sqrt{2x^2 + 1}\,dx$;
$\begin{bmatrix} u = \sqrt{2}x \\ du = \sqrt{2}\,dx \end{bmatrix} \rightarrow \frac{2\pi}{\sqrt{2}}\int_0^2 \sqrt{u^2 + 1}\,du = \frac{2\pi}{\sqrt{2}}\left[\frac{1}{2}\left(u\sqrt{u^2 + 1} + \ln\left(u + \sqrt{u^2 + 1}\right)\right)\right]_0^2 = \frac{\pi}{\sqrt{2}}\left[2\sqrt{5} + \ln\left(2 + \sqrt{5}\right)\right]$

89. $\frac{dr_A}{dt} = \frac{dr_B}{dt} \Rightarrow \frac{d}{dt}(r_A - r_B) = 0 \Rightarrow r_A - r_B = C$, a constant $\Rightarrow$ the points P(t) lie on a hyperbola with foci at A and B

90. (a) $\tan\beta = m_L \Rightarrow \tan\beta = f'(x_0)$ where $f(x) = \sqrt{4px}$;
$f'(x) = \frac{1}{2}(4px)^{-1/2}(4p) = \frac{2p}{\sqrt{4px}} \Rightarrow f'(x_0) = \frac{2p}{\sqrt{4px_0}}$
$= \frac{2p}{y_0} \Rightarrow \tan\beta = \frac{2p}{y_0}$.

(b) $\tan\phi = m_{FP} = \frac{y_0 - 0}{x_0 - p} = \frac{y_0}{x_0 - p}$

(c) $\tan\alpha = \frac{\tan\phi - \tan\beta}{1 + \tan\phi\tan\beta} = \frac{\left(\frac{y_0}{x_0 - p} - \frac{2p}{y_0}\right)}{1 + \left(\frac{y_0}{x_0 - p}\right)\left(\frac{2p}{y_0}\right)}$
$= \frac{y_0^2 - 2p(x_0 - p)}{y_0(x_0 - p + 2p)} = \frac{4px_0 - 2px_0 + 2p^2}{y_0(x_0 + p)} = \frac{2p(x_0 + p)}{y_0(x_0 + p)} = \frac{2p}{y_0}$

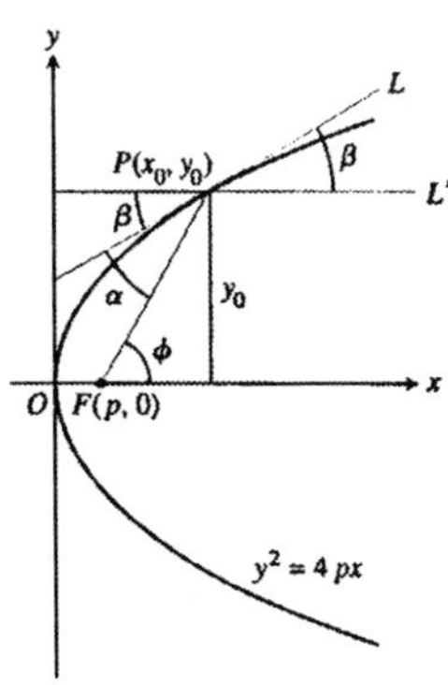

91. PF will always equal PB because the string has constant length AB = FP + PA = AP + PB.

92. (a) In the labeling of the accompanying figure we have $\frac{y}{1} = \tan t$ so the coordinates of A are $(1, \tan t)$. The coordinates of P are therefore $(1 + r, \tan t)$. Since $1^2 + y^2 = (OA)^2$, we have $1^2 + \tan^2 t = (1 + r)^2$
$\Rightarrow 1 + r = \sqrt{1 + \tan^2 t} = \sec t \Rightarrow r = \sec t - 1$.
The coordinates of P are therefore $(x, y) = (\sec t, \tan t)$
$\Rightarrow x^2 - y^2 = \sec^2 t - \tan^2 t = 1$

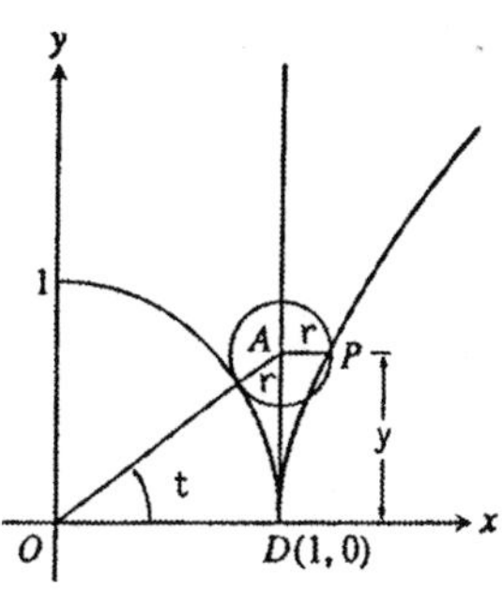

(b) In the labeling of the accompany figure the coordinates of A are (cos t, sin t), the coordinates of C are (1, tan t), and the coordinates of P are (1 + d, tan t). By similar triangles, $\frac{d}{AB} = \frac{OC}{OA} \Rightarrow \frac{d}{1-\cos t} = \frac{\sqrt{1+\tan^2 t}}{1}$ $\Rightarrow d = (1 - \cos t)(\sec t) = \sec t - 1$. The coordinates of P are therefore (sec t, tan t) and P moves on the hyperbola $x^2 - y^2 = 1$ as in part (a).

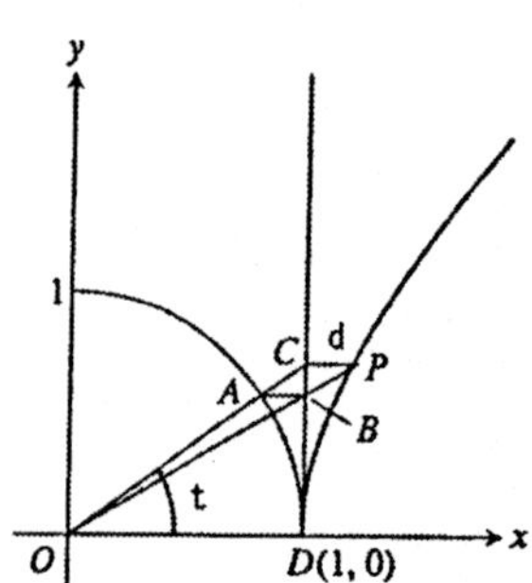

93. $x^2 = 4py$ and $y = p \Rightarrow x^2 = 4p^2 \Rightarrow x = \pm 2p$. Therefore the line $y = p$ cuts the parabola at points $(-2p, p)$ and $(2p, p)$, and these points are $\sqrt{[2p-(-2p)]^2 + (p-p)^2} = 4p$ units apart.

94. $\lim\limits_{x\to\infty}\left(\frac{b}{a}x - \frac{b}{a}\sqrt{x^2-a^2}\right) = \frac{b}{a}\lim\limits_{x\to\infty}\left(x - \sqrt{x^2-a^2}\right) = \frac{b}{a}\lim\limits_{x\to\infty}\left[\frac{\left(x-\sqrt{x^2-a^2}\right)\left(x+\sqrt{x^2-a^2}\right)}{x+\sqrt{x^2-a^2}}\right]$

$= \frac{b}{a}\lim\limits_{x\to\infty}\left[\frac{x^2-(x^2-a^2)}{x+\sqrt{x^2-a^2}}\right] = \frac{b}{a}\lim\limits_{x\to\infty}\left[\frac{a^2}{x+\sqrt{x^2-a^2}}\right] = 0$

10.2 CLASSIFYING CONIC SECTIONS BY ECCENTRICITY

1. $16x^2 + 25y^2 = 400 \Rightarrow \frac{x^2}{25} + \frac{y^2}{16} = 1 \Rightarrow c = \sqrt{a^2-b^2}$
$= \sqrt{25-16} = 3 \Rightarrow e = \frac{c}{a} = \frac{3}{5}$; $F(\pm 3, 0)$;
directrices are $x = 0 \pm \frac{a}{e} = \pm \frac{5}{\left(\frac{3}{5}\right)} = \pm \frac{25}{3}$

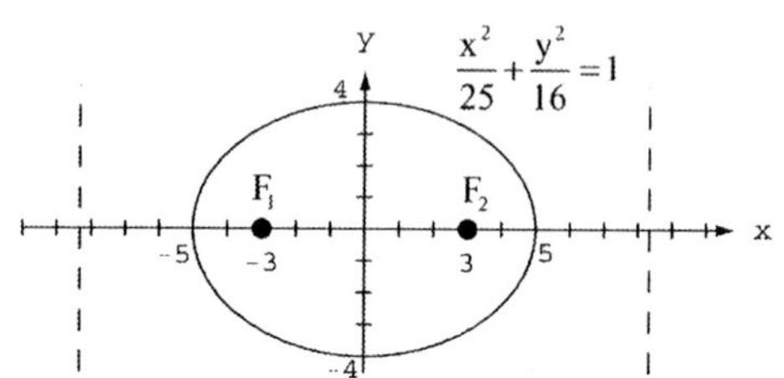

2. $7x^2 + 16y^2 = 112 \Rightarrow \frac{x^2}{16} + \frac{y^2}{7} = 1 \Rightarrow c = \sqrt{a^2-b^2}$
$= \sqrt{16-7} = 3 \Rightarrow e = \frac{c}{a} = \frac{3}{4}$; $F(\pm 3, 0)$;
directrices are $x = 0 \pm \frac{a}{e} = \pm \frac{4}{\left(\frac{3}{4}\right)} = \pm \frac{16}{3}$

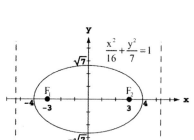

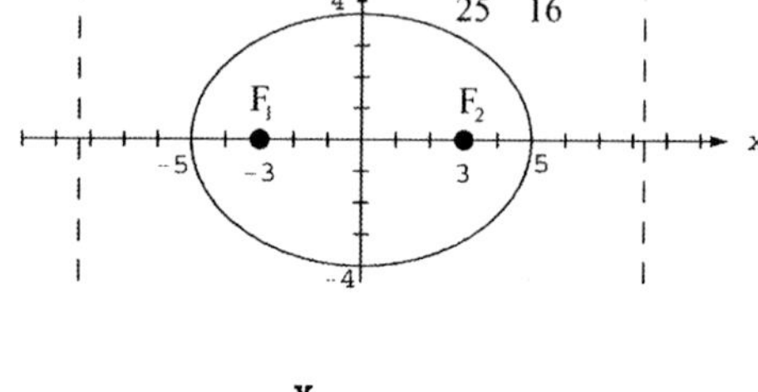

3. $2x^2 + y^2 = 2 \Rightarrow x^2 + \frac{y^2}{2} = 1 \Rightarrow c = \sqrt{a^2-b^2}$
$= \sqrt{2-1} = 1 \Rightarrow e = \frac{c}{a} = \frac{1}{\sqrt{2}}$; $F(0, \pm 1)$;
directrices are $y = 0 \pm \frac{a}{e} = \pm \frac{\sqrt{2}}{\left(\frac{1}{\sqrt{2}}\right)} = \pm 2$

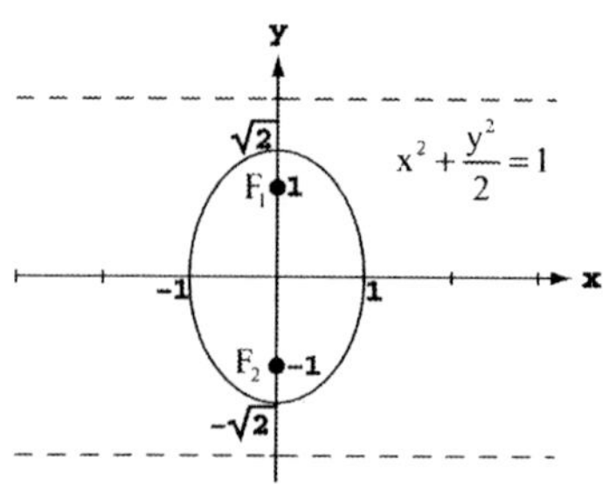

4. $2x^2 + y^2 = 4 \Rightarrow \frac{x^2}{2} + \frac{y^2}{4} = 1 \Rightarrow c = \sqrt{a^2 - b^2}$
$= \sqrt{4-2} = \sqrt{2} \Rightarrow e = \frac{c}{a} = \frac{\sqrt{2}}{2}$; $F\left(0, \pm\sqrt{2}\right)$;
directrices are $y = 0 \pm \frac{a}{e} = \pm \frac{2}{\left(\frac{\sqrt{2}}{2}\right)} = \pm 2\sqrt{2}$

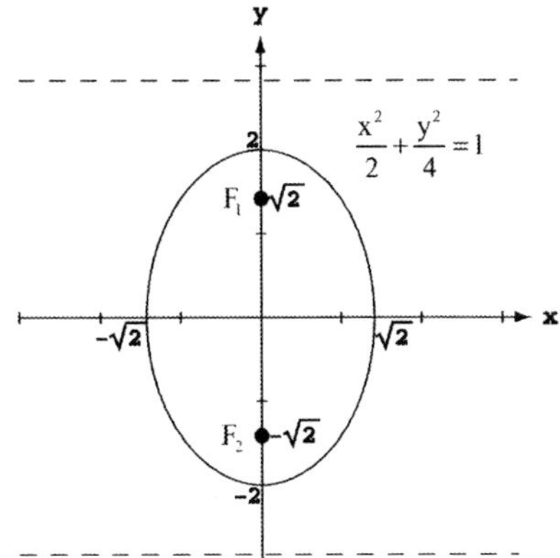

5. $3x^2 + 2y^2 = 6 \Rightarrow \frac{x^2}{2} + \frac{y^2}{3} = 1 \Rightarrow c = \sqrt{a^2 - b^2}$
$= \sqrt{3-2} = 1 \Rightarrow e = \frac{c}{a} = \frac{1}{\sqrt{3}}$; $F(0, \pm 1)$;
directrices are $y = 0 \pm \frac{a}{e} = \pm \frac{\sqrt{3}}{\left(\frac{1}{\sqrt{3}}\right)} = \pm 3$

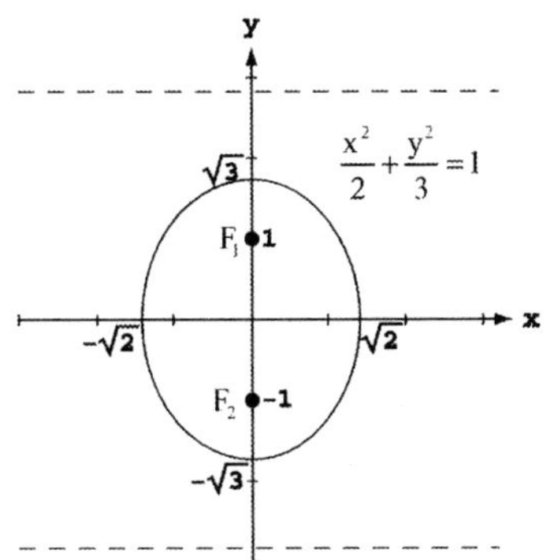

6. $9x^2 + 10y^2 = 90 \Rightarrow \frac{x^2}{10} + \frac{y^2}{9} = 1 \Rightarrow c = \sqrt{a^2 - b^2}$
$= \sqrt{10-9} = 1 \Rightarrow e = \frac{c}{a} = \frac{1}{\sqrt{10}}$; $F(\pm 1, 0)$;
directrices are $x = 0 \pm \frac{a}{e} = \pm \frac{\sqrt{10}}{\left(\frac{1}{\sqrt{10}}\right)} = \pm 10$

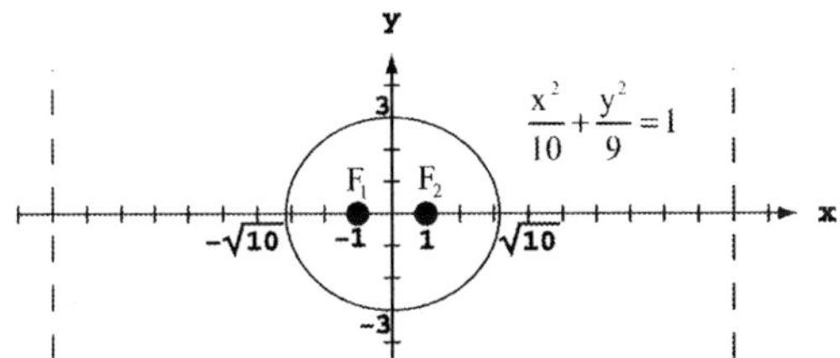

7. $6x^2 + 9y^2 = 54 \Rightarrow \frac{x^2}{9} + \frac{y^2}{6} = 1 \Rightarrow c = \sqrt{a^2 - b^2}$
$= \sqrt{9-6} = \sqrt{3} \Rightarrow e = \frac{c}{a} = \frac{\sqrt{3}}{3}$; $F\left(\pm\sqrt{3}, 0\right)$;
directrices are $x = 0 \pm \frac{a}{e} = \pm \frac{3}{\left(\frac{\sqrt{3}}{3}\right)} = \pm 3\sqrt{3}$

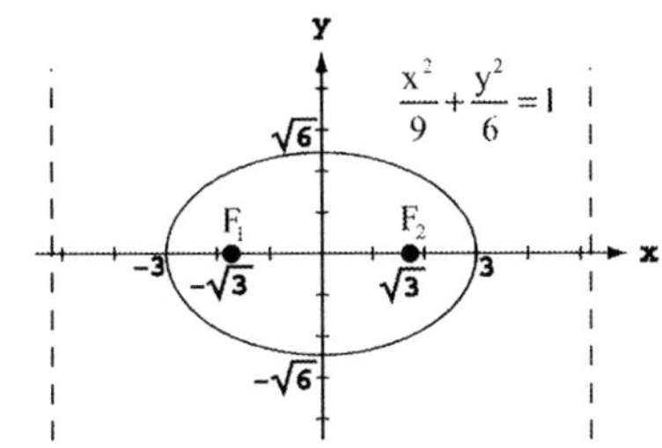

8. $169x^2 + 25y^2 = 4225 \Rightarrow \frac{x^2}{25} + \frac{y^2}{169} = 1 \Rightarrow c = \sqrt{a^2 - b^2}$
$= \sqrt{169 - 25} = 12 \Rightarrow e = \frac{c}{a} = \frac{12}{13}$; $F(0, \pm 12)$;
directrices are $y = 0 \pm \frac{a}{e} = \pm \frac{13}{\left(\frac{12}{13}\right)} = \pm \frac{169}{12}$

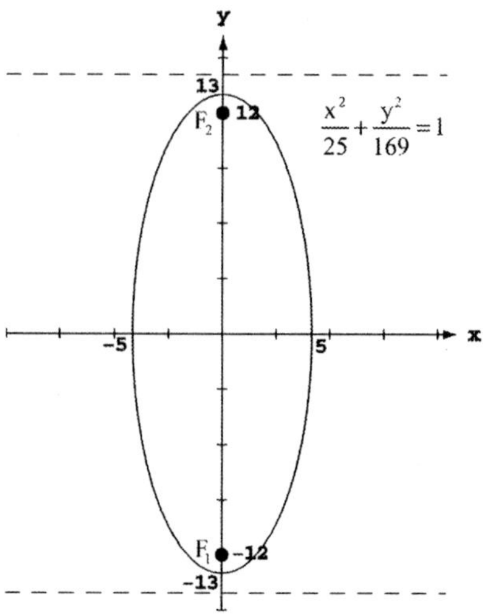

9. Foci: $(0, \pm 3)$, $e = 0.5 \Rightarrow c = 3$ and $a = \frac{c}{e} = \frac{3}{0.5} = 6 \Rightarrow b^2 = 36 - 9 = 27 \Rightarrow \frac{x^2}{27} + \frac{y^2}{36} = 1$

10. Foci: $(\pm 8, 0)$, $e = 0.2 \Rightarrow c = 8$ and $a = \frac{c}{e} = \frac{8}{0.2} = 40 \Rightarrow b^2 = 1600 - 64 = 1536 \Rightarrow \frac{x^2}{1600} + \frac{y^2}{1536} = 1$

11. Vertices: $(0, \pm 70)$, $e = 0.1 \Rightarrow a = 70$ and $c = ae = 70(0.1) = 7 \Rightarrow b^2 = 4900 - 49 = 4851 \Rightarrow \frac{x^2}{4851} + \frac{y^2}{4900} = 1$

12. Vertices: $(\pm 10, 0)$, $e = 0.24 \Rightarrow a = 10$ and $c = ae = 10(0.24) = 2.4 \Rightarrow b^2 = 100 - 5.76 = 94.24$
$\Rightarrow \frac{x^2}{100} + \frac{y^2}{94.24} = 1$

13. Focus: $\left(\sqrt{5}, 0\right)$, Directrix: $x = \frac{9}{\sqrt{5}} \Rightarrow c = ae = \sqrt{5}$ and $\frac{a}{e} = \frac{9}{\sqrt{5}} \Rightarrow \frac{ae}{e^2} = \frac{9}{\sqrt{5}} \Rightarrow \frac{\sqrt{5}}{e^2} = \frac{9}{\sqrt{5}} \Rightarrow e^2 = \frac{5}{9}$
$\Rightarrow e = \frac{\sqrt{5}}{3}$. Then $PF = \frac{\sqrt{5}}{3} PD \Rightarrow \sqrt{\left(x - \sqrt{5}\right)^2 + (y - 0)^2} = \frac{\sqrt{5}}{3}\left|x - \frac{9}{\sqrt{5}}\right| \Rightarrow \left(x - \sqrt{5}\right)^2 + y^2 = \frac{5}{9}\left(x - \frac{9}{\sqrt{5}}\right)^2$
$\Rightarrow x^2 - 2\sqrt{5}x + 5 + y^2 = \frac{5}{9}\left(x^2 - \frac{18}{\sqrt{5}}x + \frac{81}{5}\right) \Rightarrow \frac{4}{9}x^2 + y^2 = 4 \Rightarrow \frac{x^2}{9} + \frac{y^2}{4} = 1$

14. Focus: $(4, 0)$, Directrix: $x = \frac{16}{3} \Rightarrow c = ae = 4$ and $\frac{a}{e} = \frac{16}{3} \Rightarrow \frac{ae}{e^2} = \frac{16}{3} \Rightarrow \frac{4}{e^2} = \frac{16}{3} \Rightarrow e^2 = \frac{3}{4} \Rightarrow e = \frac{\sqrt{3}}{2}$. Then
$PF = \frac{\sqrt{3}}{2} PD \Rightarrow \sqrt{(x - 4)^2 + (y - 0)^2} = \frac{\sqrt{3}}{2}\left|x - \frac{16}{3}\right| \Rightarrow (x - 4)^2 + y^2 = \frac{3}{4}\left(x - \frac{16}{3}\right)^2 \Rightarrow x^2 - 8x + 16 + y^2$
$= \frac{3}{4}\left(x^2 - \frac{32}{3}x + \frac{256}{9}\right) \Rightarrow \frac{1}{4}x^2 + y^2 = \frac{16}{3} \Rightarrow \frac{x^2}{\left(\frac{64}{3}\right)} + \frac{y^2}{\left(\frac{16}{3}\right)} = 1$

15. Focus: $(-4, 0)$, Directrix: $x = -16 \Rightarrow c = ae = 4$ and $\frac{a}{e} = 16 \Rightarrow \frac{ae}{e^2} = 16 \Rightarrow \frac{4}{e^2} = 16 \Rightarrow e^2 = \frac{1}{4} \Rightarrow e = \frac{1}{2}$. Then
$PF = \frac{1}{2} PD \Rightarrow \sqrt{(x + 4)^2 + (y - 0)^2} = \frac{1}{2}|x + 16| \Rightarrow (x + 4)^2 + y^2 = \frac{1}{4}(x + 16)^2 \Rightarrow x^2 + 8x + 16 + y^2$
$= \frac{1}{4}(x^2 + 32x + 256) \Rightarrow \frac{3}{4}x^2 + y^2 = 48 \Rightarrow \frac{x^2}{64} + \frac{y^2}{48} = 1$

16. Focus: $\left(-\sqrt{2}, 0\right)$, Directrix: $x = -2\sqrt{2} \Rightarrow c = ae = \sqrt{2}$ and $\frac{a}{e} = 2\sqrt{2} \Rightarrow \frac{ae}{e^2} = 2\sqrt{2} \Rightarrow \frac{\sqrt{2}}{e^2} = 2\sqrt{2} \Rightarrow e^2 = \frac{1}{2}$
$\Rightarrow e = \frac{1}{\sqrt{2}}$. Then $PF = \frac{1}{\sqrt{2}} PD \Rightarrow \sqrt{\left(x + \sqrt{2}\right)^2 + (y - 0)^2} = \frac{1}{\sqrt{2}}\left|x + 2\sqrt{2}\right| \Rightarrow \left(x + \sqrt{2}\right)^2 + y^2$
$= \frac{1}{2}\left(x + 2\sqrt{2}\right)^2 \Rightarrow x^2 + 2\sqrt{2}x + 2 + y^2 = \frac{1}{2}\left(x^2 + 4\sqrt{2}x + 8\right) \Rightarrow \frac{1}{2}x^2 + y^2 = 2 \Rightarrow \frac{x^2}{4} + \frac{y^2}{2} = 1$

17. $e = \frac{4}{5} \Rightarrow$ take $c = 4$ and $a = 5$; $c^2 = a^2 - b^2$ $\Rightarrow 16 = 25 - b^2 \Rightarrow b^2 = 9 \Rightarrow b = 3$; therefore $\frac{x^2}{25} + \frac{y^2}{9} = 1$

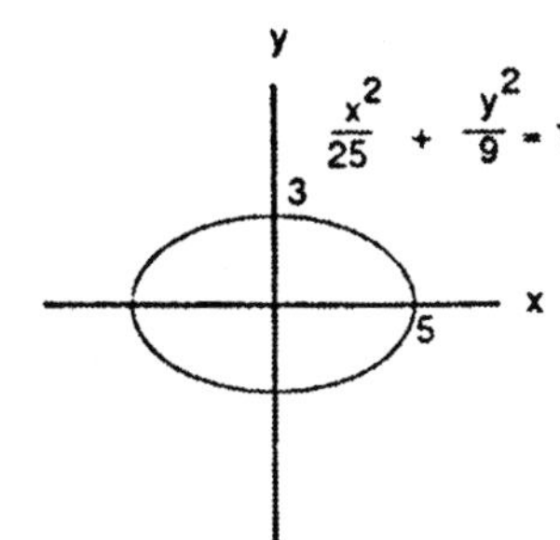

18. The eccentricity e for Pluto is 0.25 $\Rightarrow e = \frac{c}{a} = 0.25 = \frac{1}{4}$ $\Rightarrow$ take $c = 1$ and $a = 4$; $c^2 = a^2 - b^2 \Rightarrow 1 = 16 - b^2$ $\Rightarrow b^2 = 15 \Rightarrow b = \sqrt{15}$; therefore, $\frac{x^2}{16} + \frac{y^2}{15} = 1$ is a model of Pluto's orbit.

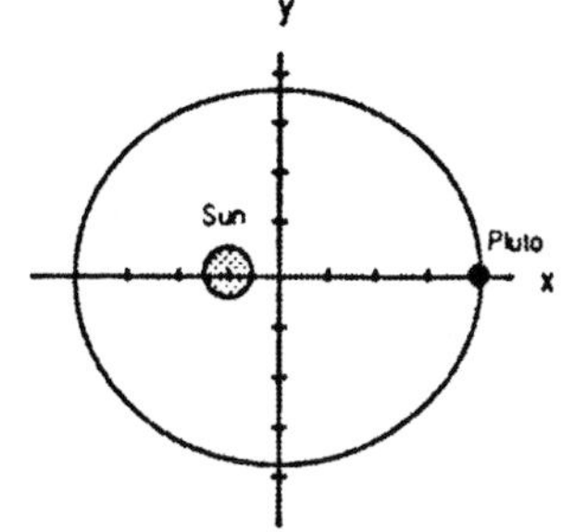

19. One axis is from $A(1, 1)$ to $B(1, 7)$ and is 6 units long; the other axis is from $C(3, 4)$ to $D(-1, 4)$ and is 4 units long. Therefore $a = 3$, $b = 2$ and the major axis is vertical. The center is the point $C(1, 4)$ and the ellipse is given by $\frac{(x-1)^2}{4} + \frac{(y-4)^2}{9} = 1$; $c^2 = a^2 - b^2 = 3^2 - 2^2 = 5$ $\Rightarrow c = \sqrt{5}$; therefore the foci are $F\left(1, 4 \pm \sqrt{5}\right)$, the eccentricity is $e = \frac{c}{a} = \frac{\sqrt{5}}{3}$, and the directrices are $y = 4 \pm \frac{a}{e} = 4 \pm \frac{3}{\left(\frac{\sqrt{5}}{3}\right)} = 4 \pm \frac{9\sqrt{5}}{5}$.

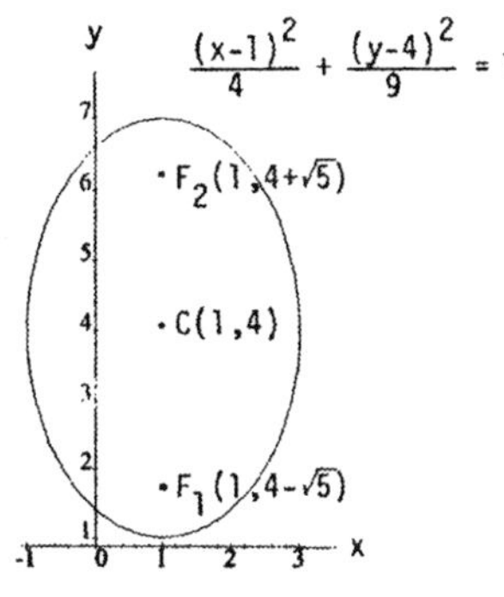

20. Using $PF = e \cdot PD$, we have $\sqrt{(x-4)^2 + y^2} = \frac{2}{3}|x - 9| \Rightarrow (x-4)^2 + y^2 = \frac{4}{9}(x-9)^2 \Rightarrow x^2 - 8x + 16 + y^2$ $= \frac{4}{9}(x^2 - 18x + 81) \Rightarrow \frac{5}{9}x^2 + y^2 = 20 \Rightarrow 5x^2 + 9y^2 = 180$ or $\frac{x^2}{36} + \frac{y^2}{20} = 1$.

21. The ellipse must pass through $(0, 0) \Rightarrow c = 0$; the point $(-1, 2)$ lies on the ellipse $\Rightarrow -a + 2b = -8$. The ellipse is tangent to the x-axis $\Rightarrow$ its center is on the y-axis, so $a = 0$ and $b = -4 \Rightarrow$ the equation is $4x^2 + y^2 - 4y = 0$. Next, $4x^2 + y^2 - 4y + 4 = 4 \Rightarrow 4x^2 + (y - 24)^2 = 4 \Rightarrow x^2 + \frac{(y-2)^2}{4} = 1 \Rightarrow a = 2$ and $b = 1$ (now using the standard symbols) $\Rightarrow c^2 = a^2 - b^2 = 4 - 1 = 3 \Rightarrow c = \sqrt{3} \Rightarrow e = \frac{c}{a} = \frac{\sqrt{3}}{2}$.

22. We first prove a result which we will use: let m_1, and m_2 be two nonparallel, nonperpendicular lines. Let α be the acute angle between the lines. Then $\tan\alpha = \frac{m_1 - m_2}{1 + m_1 m_2}$. To see this result, let θ_1 be the angle of inclination of the line with slope m_1, and θ_2 be the angle of inclination of the line with slope m_2. Assume $m_1 > m_2$. Then $\theta_1 > \theta_2$ and we have $\alpha = \theta_1 - \theta_2$. Then $\tan\alpha = \tan(\theta_1 - \theta_2)$ $= \frac{\tan\theta_1 - \tan\theta_2}{1 + \tan\theta_1 \tan\theta_2} = \frac{m_1 - m_2}{1 + m_1 m_2}$, since $m_1 = \tan\theta_1$ and and $m_2 = \tan\theta_2$.

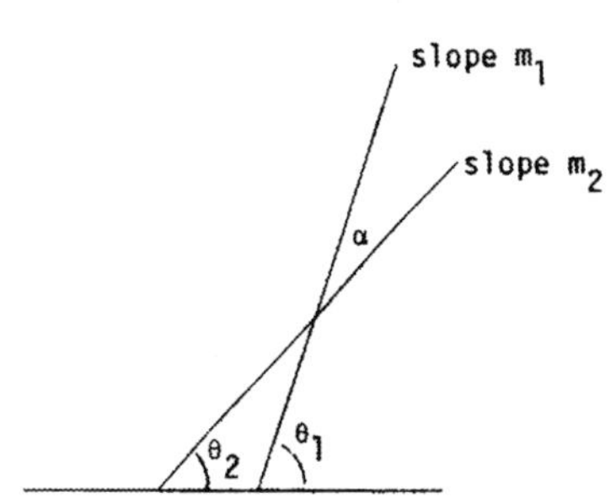

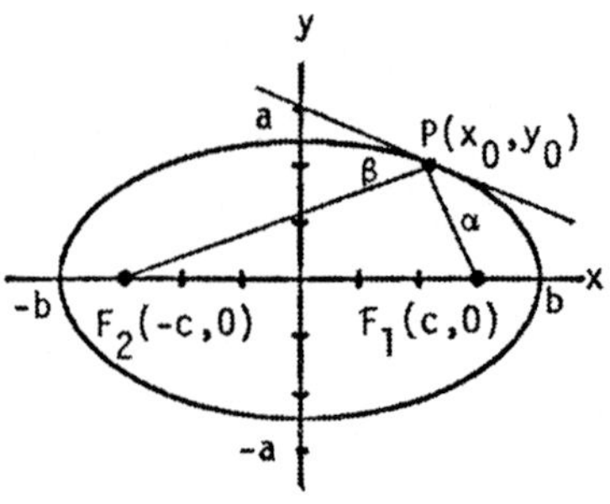

Now we prove the reflective property of ellipses (see the accompanying figure): If $\frac{x^2}{a^2}+\frac{y^2}{b^2}=1$, then $b^2x^2+a^2y^2=a^2b^2$ and $y=\frac{b}{a}\sqrt{a^2-x^2} \Rightarrow y'=\frac{-bx}{a\sqrt{a^2-x^2}}$.
Let $P(x_0,y_0)$ be any point on the ellipse $\Rightarrow y'(x_0)=\frac{-bx_0}{a\sqrt{a^2-x_0^2}}=\frac{-b^2x_0}{a^2y_0}$. Let $F_1(c,0)$ and $F_2(-c,0)$ be the foci. Then $m_{PF_1}=\frac{y_0}{x_0-c}$ and $m_{PF_2}=\frac{y_0}{x_0+c}$. Let α and β be the angles between the tangent line and PF_1 and PF_2, respectively. Then

$$\tan\alpha=\frac{\left(-\frac{b^2x_0}{a^2y_0}-\frac{y_0}{x_0-c}\right)}{\left(1-\frac{b^2x_0y_0}{a^2y_0(x_0-c)}\right)}=\frac{-b^2x_0^2+b^2x_0c-a^2y_0^2}{a^2y_0x_0-a^2y_0c-b^2x_0y_0}=\frac{b^2x_0c-(b^2x_0^2+a^2y_0^2)}{-a^2y_0c+(a^2-b^2)\,x_0y_0}=\frac{b^2x_0c-a^2b^2}{-a^2y_0c+c^2x_0y_0}=\frac{b^2}{cy_0}.$$

Similarly, $\tan\beta=\frac{b^2}{cy_0}$. Since $\tan\alpha=\tan\beta$, and α and β are both less than 90°, we have $\alpha=\beta$.

23. $x^2-y^2=1 \Rightarrow c=\sqrt{a^2+b^2}=\sqrt{1+1}=\sqrt{2} \Rightarrow e=\frac{c}{a}=\frac{\sqrt{2}}{1}=\sqrt{2}$; asymptotes are $y=\pm x$; $F\left(\pm\sqrt{2},0\right)$; directrices are $x=0\pm\frac{a}{e}=\pm\frac{1}{\sqrt{2}}$

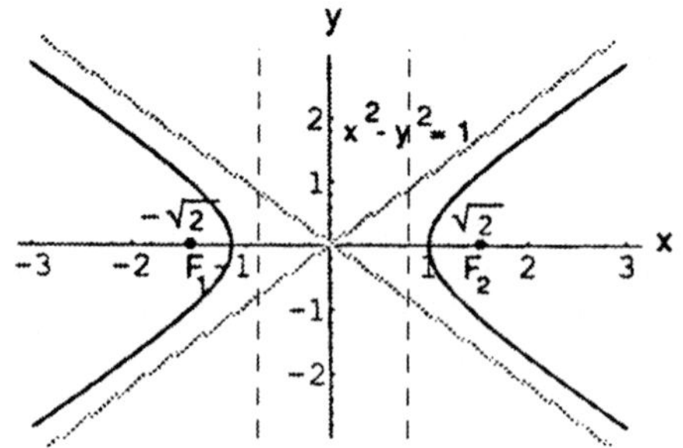

24. $9x^2-16y^2=144 \Rightarrow \frac{x^2}{16}-\frac{y^2}{9}=1 \Rightarrow c=\sqrt{a^2+b^2}=\sqrt{16+9}=5 \Rightarrow e=\frac{c}{a}=\frac{5}{4}$; asymptotes are $y=\pm\frac{3}{4}x$; $F(\pm 5,0)$; directrices are $x=0\pm\frac{a}{e}=\pm\frac{16}{5}$

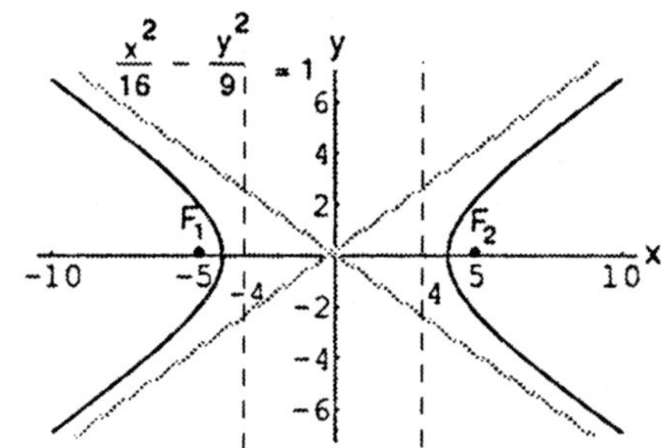

25. $y^2-x^2=8 \Rightarrow \frac{y^2}{8}-\frac{x^2}{8}=1 \Rightarrow c=\sqrt{a^2+b^2}=\sqrt{8+8}=4 \Rightarrow e=\frac{c}{a}=\frac{4}{\sqrt{8}}=\sqrt{2}$; asymptotes are $y=\pm x$; $F(0,\pm 4)$; directrices are $y=0\pm\frac{a}{e}=\pm\frac{\sqrt{8}}{\sqrt{2}}=\pm 2$

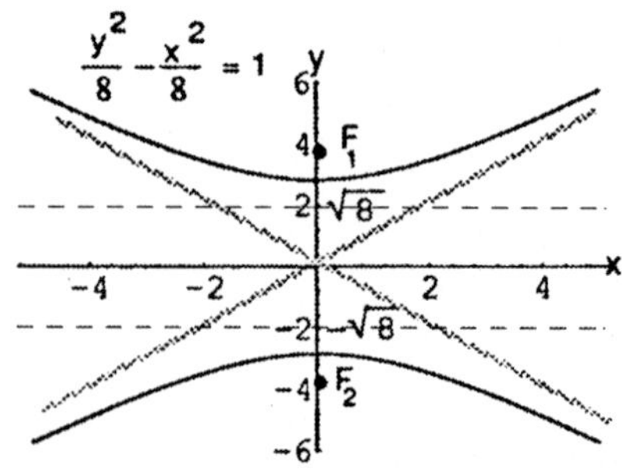

26. $y^2 - x^2 = 4 \Rightarrow \frac{y^2}{4} - \frac{x^2}{4} = 1 \Rightarrow c = \sqrt{a^2 + b^2}$ $= \sqrt{4+4} = 2\sqrt{2} \Rightarrow e = \frac{c}{a} = \frac{2\sqrt{2}}{2} = \sqrt{2}$; asymptotes are $y = \pm x$; $F\left(0, \pm 2\sqrt{2}\right)$; directrices are $y = 0 \pm \frac{a}{e}$ $= \pm \frac{2}{\sqrt{2}} = \pm\sqrt{2}$

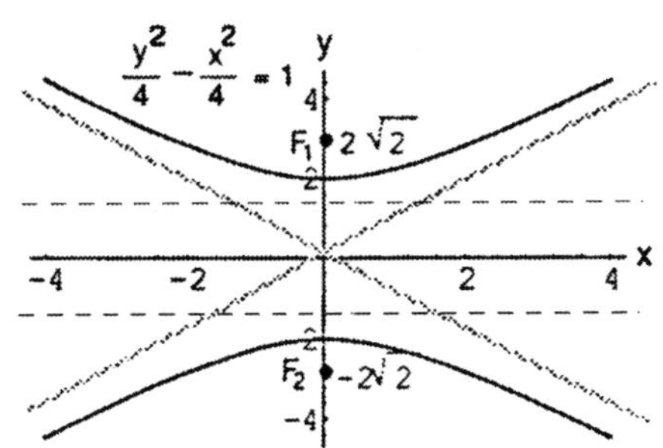

27. $8x^2 - 2y^2 = 16 \Rightarrow \frac{x^2}{2} - \frac{y^2}{8} = 1 \Rightarrow c = \sqrt{a^2 + b^2}$ $= \sqrt{2+8} = \sqrt{10} \Rightarrow e = \frac{c}{a} = \frac{\sqrt{10}}{\sqrt{2}} = \sqrt{5}$; asymptotes are $y = \pm 2x$; $F\left(\pm\sqrt{10}, 0\right)$; directrices are $x = 0 \pm \frac{a}{e}$ $= \pm \frac{\sqrt{2}}{\sqrt{5}} = \pm \frac{2}{\sqrt{10}}$

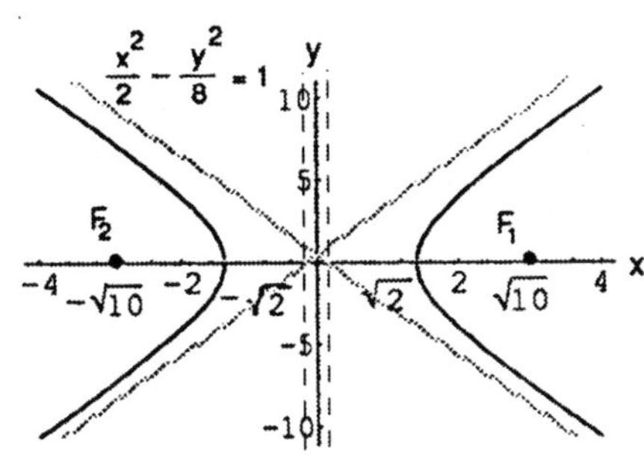

28. $y^2 - 3x^2 = 3 \Rightarrow \frac{y^2}{3} - x^2 = 1 \Rightarrow c = \sqrt{a^2 + b^2}$ $= \sqrt{3+1} = 2 \Rightarrow e = \frac{c}{a} = \frac{2}{\sqrt{3}}$; asymptotes are $y = \pm\sqrt{3}\,x$; $F(0, \pm 2)$; directrices are $y = 0 \pm \frac{a}{e}$ $= \pm \frac{\sqrt{3}}{\left(\frac{2}{\sqrt{3}}\right)} = \pm \frac{3}{2}$

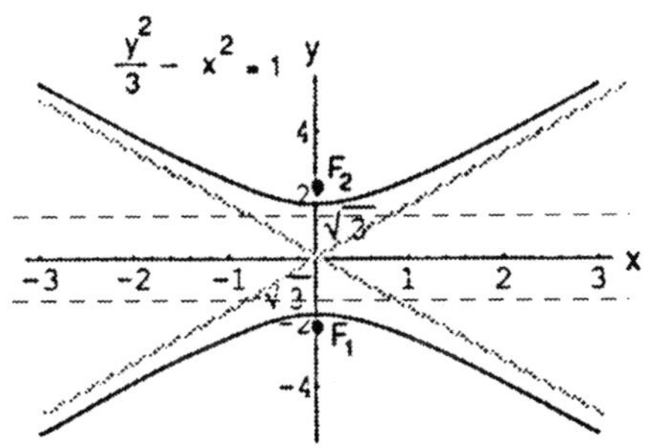

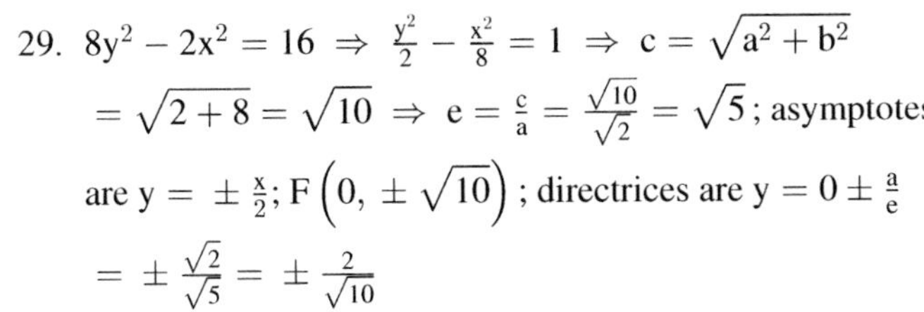
29. $8y^2 - 2x^2 = 16 \Rightarrow \frac{y^2}{2} - \frac{x^2}{8} = 1 \Rightarrow c = \sqrt{a^2 + b^2}$ $= \sqrt{2+8} = \sqrt{10} \Rightarrow e = \frac{c}{a} = \frac{\sqrt{10}}{\sqrt{2}} = \sqrt{5}$; asymptotes are $y = \pm \frac{x}{2}$; $F\left(0, \pm\sqrt{10}\right)$; directrices are $y = 0 \pm \frac{a}{e}$ $= \pm \frac{\sqrt{2}}{\sqrt{5}} = \pm \frac{2}{\sqrt{10}}$

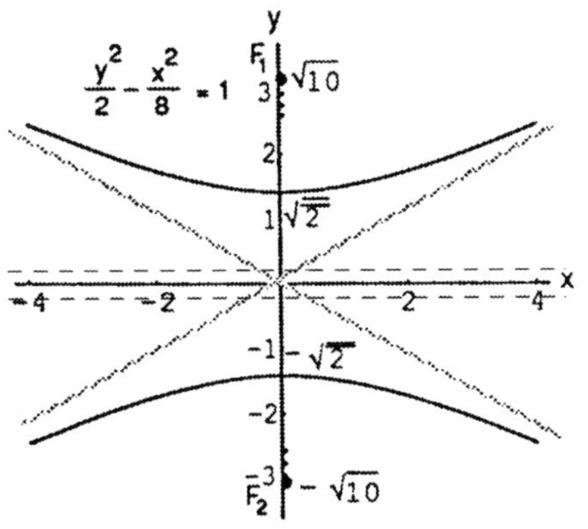

30. $64x^2 - 36y^2 = 2304 \Rightarrow \frac{x^2}{36} - \frac{y^2}{64} = 1 \Rightarrow c = \sqrt{a^2 + b^2}$ $= \sqrt{36+64} = 10 \Rightarrow e = \frac{c}{a} = \frac{10}{6} = \frac{5}{3}$; asymptotes are $y = \pm\frac{4}{3}x$; $F(\pm 10, 0)$; directrices are $x = 0 \pm \frac{a}{e}$ $= \pm \frac{6}{\left(\frac{5}{3}\right)} = \pm \frac{18}{5}$

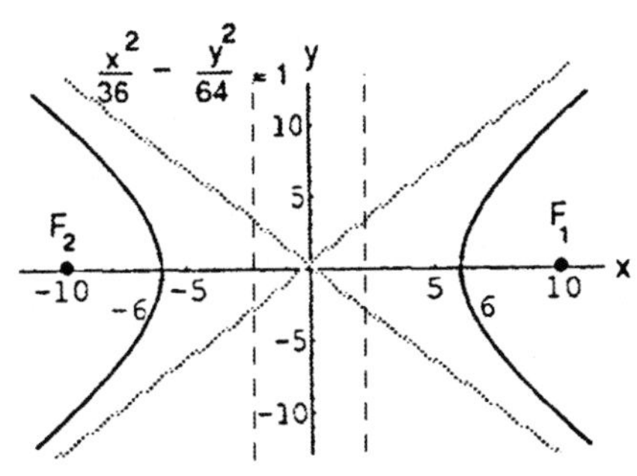

31. Vertices $(0, \pm 1)$ and $e = 3 \Rightarrow a = 1$ and $e = \frac{c}{a} = 3 \Rightarrow c = 3a = 3 \Rightarrow b^2 = c^2 - a^2 = 9 - 1 = 8 \Rightarrow y^2 - \frac{x^2}{8} = 1$

32. Vertices $(\pm 2, 0)$ and $e = 2 \Rightarrow a = 2$ and $e = \frac{c}{a} = 2 \Rightarrow c = 2a = 4 \Rightarrow b^2 = c^2 - a^2 = 16 - 4 = 12 \Rightarrow \frac{x^2}{4} - \frac{y^2}{12} = 1$

33. Foci $(\pm 3, 0)$ and $e = 3 \Rightarrow c = 3$ and $e = \frac{c}{a} = 3 \Rightarrow c = 3a \Rightarrow a = 1 \Rightarrow b^2 = c^2 - a^2 = 9 - 1 = 8 \Rightarrow x^2 - \frac{y^2}{8} = 1$

34. Foci $(0, \pm 5)$ and $e = 1.25 \Rightarrow c = 5$ and $e = \frac{c}{a} = 1.25 = \frac{5}{4} \Rightarrow c = \frac{5}{4}a \Rightarrow 5 = \frac{5}{4}a \Rightarrow a = 4 \Rightarrow b^2 = c^2 - a^2$ $= 25 - 16 = 9 \Rightarrow \frac{y^2}{16} - \frac{x^2}{9} = 1$

35. Focus $(4, 0)$ and Directrix $x = 2 \Rightarrow c = ae = 4$ and $\frac{a}{e} = 2 \Rightarrow \frac{ae}{e^2} = 2 \Rightarrow \frac{4}{e^2} = 2 \Rightarrow e^2 = 2 \Rightarrow e = \sqrt{2}$. Then $PF = \sqrt{2}\,PD \Rightarrow \sqrt{(x-4)^2 + (y-0)^2} = \sqrt{2}\,|x-2| \Rightarrow (x-4)^2 + y^2 = 2(x-2)^2 \Rightarrow x^2 - 8x + 16 + y^2$ $= 2(x^2 - 4x + 4) \Rightarrow -x^2 + y^2 = -8 \Rightarrow \frac{x^2}{8} - \frac{y^2}{8} = 1$

36. Focus $\left(\sqrt{10}, 0\right)$ and Directrix $x = \sqrt{2} \Rightarrow c = ae = \sqrt{10}$ and $\frac{a}{e} = \sqrt{2} \Rightarrow \frac{ae}{e^2} = \sqrt{2} \Rightarrow \frac{\sqrt{10}}{e^2} = \sqrt{2} \Rightarrow e^2 = \sqrt{5}$ $\Rightarrow e = \sqrt[4]{5}$. Then $PF = \sqrt[4]{5}\,PD \Rightarrow \sqrt{\left(x - \sqrt{10}\right)^2 + (y-0)^2} = \sqrt[4]{5}\left|x - \sqrt{2}\right| \Rightarrow \left(x - \sqrt{10}\right)^2 + y^2$ $= \sqrt{5}\left(x - \sqrt{2}\right)^2 \Rightarrow x^2 - 2\sqrt{10}\,x + 10 + y^2 = \sqrt{5}\left(x^2 - 2\sqrt{2}\,x + 2\right) \Rightarrow \left(1 - \sqrt{5}\right)x^2 + y^2 = 2\sqrt{5} - 10$ $\Rightarrow \frac{\left(1-\sqrt{5}\right)x^2}{2\sqrt{5} - 10} + \frac{y^2}{2\sqrt{5} - 10} = 1 \Rightarrow \frac{x^2}{2\sqrt{5}} - \frac{y^2}{10 - 2\sqrt{5}} = 1$

37. Focus $(-2, 0)$ and Directrix $x = -\frac{1}{2} \Rightarrow c = ae = 2$ and $\frac{a}{e} = \frac{1}{2} \Rightarrow \frac{ae}{e^2} = \frac{1}{2} \Rightarrow \frac{2}{e^2} = \frac{1}{2} \Rightarrow e^2 = 4 \Rightarrow e = 2$. Then $PF = 2PD \Rightarrow \sqrt{(x+2)^2 + (y-0)^2} = 2\left|x + \frac{1}{2}\right| \Rightarrow (x+2)^2 + y^2 = 4\left(x + \frac{1}{2}\right)^2 \Rightarrow x^2 + 4x + 4 + y^2$ $= 4\left(x^2 + x + \frac{1}{4}\right) \Rightarrow -3x^2 + y^2 = -3 \Rightarrow x^2 - \frac{y^2}{3} = 1$

38. Focus $(-6, 0)$ and Directrix $x = -2 \Rightarrow c = ae = 6$ and $\frac{a}{e} = 2 \Rightarrow \frac{ae}{e^2} = 2 \Rightarrow \frac{6}{e^2} = 2 \Rightarrow e^2 = 3 \Rightarrow e = \sqrt{3}$. Then $PF = \sqrt{3}\,PD \Rightarrow \sqrt{(x+6)^2 + (y-0)^2} = \sqrt{3}\,|x+2| \Rightarrow (x+6)^2 + y^2 = 3(x+2)^2 \Rightarrow x^2 + 12x + 36 + y^2$ $= 3(x^2 + 4x + 4) \Rightarrow -2x^2 + y^2 = -24 \Rightarrow \frac{x^2}{12} - \frac{y^2}{24} = 1$

39. $\sqrt{(x-1)^2 + (y+3)^2} = \frac{3}{2}|y-2| \Rightarrow x^2 - 2x + 1 + y^2 + 6y + 9 = \frac{9}{4}(y^2 - 4y + 4) \Rightarrow 4x^2 - 5y^2 - 8x + 60y + 4 = 0$ $\Rightarrow 4(x^2 - 2x + 1) - 5(y^2 - 12y + 36) = -4 + 4 - 180 \Rightarrow \frac{(y-6)^2}{36} - \frac{(x-1)^2}{45} = 1$

40. $c^2 = a^2 + b^2 \Rightarrow b^2 = c^2 - a^2$; $e = \frac{c}{a} \Rightarrow c = ea \Rightarrow c^2 = e^2a^2 \Rightarrow b^2 = e^2a^2 - a^2 = a^2(e^2 - 1)$; thus, $\frac{x^2}{a^2} - \frac{y^2}{b^2} = 1 \Rightarrow \frac{x^2}{a^2} - \frac{y^2}{a^2(e^2-1)} = 1$; the asymptotes of this hyperbola are $y = \pm(e^2 - 1)x \Rightarrow$ as e increases, the absolute values of the slopes of the asymptotes increase and the hyperbola approaches a straight line.

41. To prove the reflective property for hyperbolas: $\frac{x^2}{a^2} - \frac{y^2}{b^2} = 1 \Rightarrow a^2y^2 = b^2x^2 - a^2b^2$ and $\frac{dy}{dx} = \frac{xb^2}{ya^2}$. Let $P(x_0, y_0)$ be a point of tangency (see the accompanying figure). The slope from P to $F(-c, 0)$ is $\frac{y_0}{x_0 + c}$ and from P to $F_2(c, 0)$ it is $\frac{y_0}{x_0 - c}$. Let the tangent through P meet the x-axis in point A, and define the angles $\angle F_1PA = \alpha$ and $\angle F_2PA = \beta$. We will show that $\tan\alpha = \tan\beta$. From the preliminary result in Exercise 22,

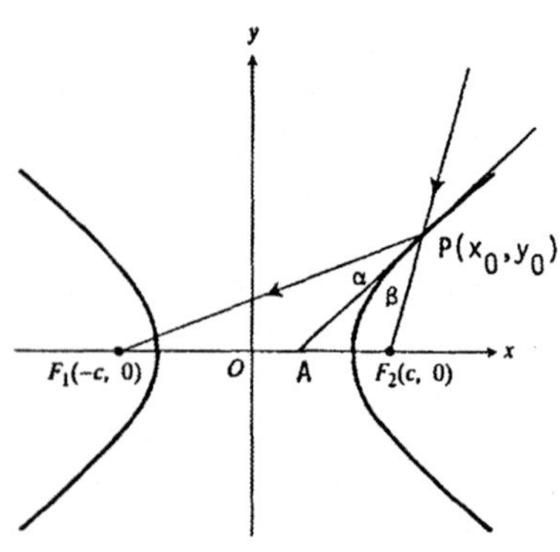

$\tan\alpha = \frac{\left(\frac{x_0b^2}{y_0a^2} - \frac{y_0}{x_0+c}\right)}{1+\left(\frac{x_0b^2}{y_0a^2}\right)\left(\frac{y_0}{x_0+c}\right)} = \frac{x_0^2b^2 + x_0b^2c - y_0^2a^2}{x_0y_0a^2 + y_0a^2c + x_0y_0b^2} = \frac{a^2b^2 + x_0b^2c}{x_0y_0c^2 + y_0a^2c} = \frac{b^2}{y_0c}$. In a similar manner,

$\tan\beta = \frac{\left(\frac{y_0}{x_0-c} - \frac{x_0b^2}{y_0a^2}\right)}{1+\left(\frac{y_0}{x_0-c}\right)\left(\frac{x_0b^2}{y_0a^2}\right)} = \frac{b^2}{y_0c}$. Since $\tan\alpha = \tan\beta$, and α and β are acute angles, we have $\alpha = \beta$.

42. From the accompanying figure, a ray of light emanating from the focus A that met the parabola at P would be reflected from the hyperbola as if it came directly from B (Exercise 41). The same light ray would be reflected off the ellipse to pass through B. Thus BPC is a straight line. Let β be the angle of incidence of the light ray on the hyperbola. Let α be the angle of incidence of the light ray on the ellipse. Note that $\alpha + \beta$ is the angle between the tangent lines to the ellipse and hyperbola at P. Since BPC is a straight line, $2\alpha + 2\beta = 180°$. Thus $\alpha + \beta = 90°$.

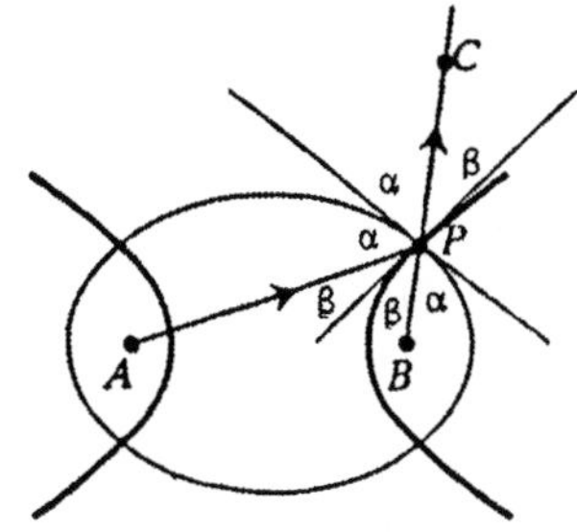

10.3 QUADRATIC EQUATIONS AND ROTATIONS

1. $x^2 - 3xy + y^2 - x = 0 \Rightarrow B^2 - 4AC = (-3)^2 - 4(1)(1) = 5 > 0 \Rightarrow$ Hyperbola

2. $3x^2 - 18xy + 27y^2 - 5x + 7y = -4 \Rightarrow B^2 - 4AC = (-18)^2 - 4(3)(27) = 0 \Rightarrow$ Parabola

3. $3x^2 - 7xy + \sqrt{17}y^2 = 1 \Rightarrow B^2 - 4AC = (-7)^2 - 4(3)\sqrt{17} \approx -0.477 < 0 \Rightarrow$ Ellipse

4. $2x^2 - \sqrt{15}\,xy + 2y^2 + x + y = 0 \Rightarrow B^2 - 4AC = \left(-\sqrt{15}\right)^2 - 4(2)(2) = -1 < 0 \Rightarrow$ Ellipse

5. $x^2 + 2xy + y^2 + 2x - y + 2 = 0 \Rightarrow B^2 - 4AC = 2^2 - 4(1)(1) = 0 \Rightarrow$ Parabola

6. $2x^2 - y^2 + 4xy - 2x + 3y = 6 \Rightarrow B^2 - 4AC = 4^2 - 4(2)(-1) = 24 > 0 \Rightarrow$ Hyperbola

7. $x^2 + 4xy + 4y^2 - 3x = 6 \Rightarrow B^2 - 4AC = 4^2 - 4(1)(4) = 0 \Rightarrow$ Parabola

8. $x^2 + y^2 + 3x - 2y = 10 \Rightarrow B^2 - 4AC = 0^2 - 4(1)(1) = -4 < 0 \Rightarrow$ Ellipse (circle)

9. $xy + y^2 - 3x = 5 \Rightarrow B^2 - 4AC = 1^2 - 4(0)(1) = 1 > 0 \Rightarrow$ Hyperbola

10. $3x^2 + 6xy + 3y^2 - 4x + 5y = 12 \Rightarrow B^2 - 4AC = 6^2 - 4(3)(3) = 0 \Rightarrow$ Parabola

11. $3x^2 - 5xy + 2y^2 - 7x - 14y = -1 \Rightarrow B^2 - 4AC = (-5)^2 - 4(3)(2) = 1 > 0 \Rightarrow$ Hyperbola

12. $2x^2 - 4.9xy + 3y^2 - 4x = 7 \Rightarrow B^2 - 4AC = (-4.9)^2 - 4(2)(3) = 0.01 > 0 \Rightarrow$ Hyperbola

13. $x^2 - 3xy + 3y^2 + 6y = 7 \Rightarrow B^2 - 4AC = (-3)^2 - 4(1)(3) = -3 < 0 \Rightarrow$ Ellipse

14. $25x^2 + 21xy + 4y^2 - 350x = 0 \Rightarrow B^2 - 4AC = 21^2 - 4(25)(4) = 41 > 0 \Rightarrow$ Hyperbola

15. $6x^2 + 3xy + 2y^2 + 17y + 2 = 0 \Rightarrow B^2 - 4AC = 3^2 - 4(6)(2) = -39 < 0 \Rightarrow$ Ellipse

16. $3x^2 + 12xy + 12y^2 + 435x - 9y + 72 = 0 \Rightarrow B^2 - 4AC = 12^2 - 4(3)(12) = 0 \Rightarrow$ Parabola

17. $\cot 2\alpha = \frac{A-C}{B} = \frac{0}{1} = 0 \Rightarrow 2\alpha = \frac{\pi}{2} \Rightarrow \alpha = \frac{\pi}{4}$; therefore $x = x' \cos\alpha - y' \sin\alpha$,
$y = x' \sin\alpha + y' \cos\alpha \Rightarrow x = x'\frac{\sqrt{2}}{2} - y'\frac{\sqrt{2}}{2},\ y = x'\frac{\sqrt{2}}{2} + y'\frac{\sqrt{2}}{2}$
$\Rightarrow \left(\frac{\sqrt{2}}{2}x' - \frac{\sqrt{2}}{2}y'\right)\left(\frac{\sqrt{2}}{2}x' + \frac{\sqrt{2}}{2}y'\right) = 2 \Rightarrow \frac{1}{2}x'^2 - \frac{1}{2}y'^2 = 2 \Rightarrow x'^2 - y'^2 = 4 \Rightarrow$ Hyperbola

18. $\cot 2\alpha = \frac{A-C}{B} = \frac{1-1}{1} = 0 \Rightarrow 2\alpha = \frac{\pi}{2} \Rightarrow \alpha = \frac{\pi}{4}$; therefore $x = x' \cos\alpha - y' \sin\alpha$,
$y = x' \sin\alpha + y' \cos\alpha \Rightarrow x = x'\frac{\sqrt{2}}{2} - y'\frac{\sqrt{2}}{2},\ y = x'\frac{\sqrt{2}}{2} + y'\frac{\sqrt{2}}{2}$
$\Rightarrow \left(\frac{\sqrt{2}}{2}x' - \frac{\sqrt{2}}{2}y'\right)^2 + \left(\frac{\sqrt{2}}{2}x' + \frac{\sqrt{2}}{2}y'\right)\left(\frac{\sqrt{2}}{2}x' - \frac{\sqrt{2}}{2}y'\right) + \left(\frac{\sqrt{2}}{2}x' + \frac{\sqrt{2}}{2}y'\right)^2 = 1$
$\Rightarrow \frac{1}{2}x'^2 - x'y' + \frac{1}{2}y'^2 + \frac{1}{2}x'^2 - \frac{1}{2}y'^2 + \frac{1}{2}x'^2 + x'y' + \frac{1}{2}y'^2 = 1 \Rightarrow \frac{3}{2}x'^2 + \frac{1}{2}y'^2 = 1 \Rightarrow 3x'^2 + y'^2 = 2 \Rightarrow$ Ellipse

19. $\cot 2\alpha = \frac{A-C}{B} = \frac{3-1}{2\sqrt{3}} = \frac{1}{\sqrt{3}} \Rightarrow 2\alpha = \frac{\pi}{3} \Rightarrow \alpha = \frac{\pi}{6}$; therefore $x = x' \cos\alpha - y' \sin\alpha$,
$y = x' \sin\alpha + y' \cos\alpha \Rightarrow x = \frac{\sqrt{3}}{2}x' - \frac{1}{2}y',\ y = \frac{1}{2}x' + \frac{\sqrt{3}}{2}y'$
$\Rightarrow 3\left(\frac{\sqrt{3}}{2}x' - \frac{1}{2}y'\right)^2 + 2\sqrt{3}\left(\frac{\sqrt{3}}{2}x' + \frac{1}{2}y'\right)\left(\frac{1}{2}x' + \frac{\sqrt{3}}{2}y'\right) + \left(\frac{1}{2}x' + \frac{\sqrt{3}}{2}y'\right)^2 - 8\left(\frac{\sqrt{3}}{2}x' - \frac{1}{2}y'\right)$
$+ 8\sqrt{3}\left(\frac{1}{2}x' + \frac{\sqrt{3}}{2}y'\right) = 0 \Rightarrow 4x'^2 + 16y' = 0 \Rightarrow$ Parabola

20. $\cot 2\alpha = \frac{A-C}{B} = \frac{1-2}{-\sqrt{3}} = \frac{1}{\sqrt{3}} \Rightarrow 2\alpha = \frac{\pi}{3} \Rightarrow \alpha = \frac{\pi}{6}$; therefore $x = x' \cos\alpha - y' \sin\alpha$,
$y = x' \sin\alpha + y' \cos\alpha \Rightarrow x = \frac{\sqrt{3}}{2}x' - \frac{1}{2}y',\ y = \frac{1}{2}x' + \frac{\sqrt{3}}{2}y'$
$\Rightarrow \left(\frac{\sqrt{3}}{2}x' - \frac{1}{2}y'\right)^2 - \sqrt{3}\left(\frac{\sqrt{3}}{2}x' - \frac{1}{2}y'\right)\left(\frac{1}{2}x' + \frac{\sqrt{3}}{2}y'\right) + 2\left(\frac{1}{2}x' + \frac{\sqrt{3}}{2}y'\right)^2 = 1 \Rightarrow \frac{1}{2}x'^2 + \frac{5}{2}y'^2 = 1$
$\Rightarrow x'^2 + 5y'^2 = 2 \Rightarrow$ Ellipse

21. $\cot 2\alpha = \frac{A-C}{B} = \frac{1-1}{-2} = 0 \Rightarrow 2\alpha = \frac{\pi}{2} \Rightarrow \alpha = \frac{\pi}{4}$; therefore $x = x' \cos\alpha - y' \sin\alpha$,
$y = x' \sin\alpha + y' \cos\alpha \Rightarrow x = \frac{\sqrt{2}}{2}x' - \frac{\sqrt{2}}{2}y',\ y = \frac{\sqrt{2}}{2}x' + \frac{\sqrt{2}}{2}y'$
$\Rightarrow \left(\frac{\sqrt{2}}{2}x' - \frac{\sqrt{2}}{2}y'\right)^2 - 2\left(\frac{\sqrt{2}}{2}x' - \frac{\sqrt{2}}{2}y'\right)\left(\frac{\sqrt{2}}{2}x' + \frac{\sqrt{2}}{2}y'\right) + \left(\frac{\sqrt{2}}{2}x' + \frac{\sqrt{2}}{2}y'\right)^2 = 2 \Rightarrow y'^2 = 1$
$\Rightarrow$ Parallel horizontal lines

22. $\cot 2\alpha = \frac{A-C}{B} = \frac{3-1}{-2\sqrt{3}} = -\frac{1}{\sqrt{3}} \Rightarrow 2\alpha = \frac{2\pi}{3} \Rightarrow \alpha = \frac{\pi}{3}$; therefore $x = x' \cos\alpha - y' \sin\alpha$,
$y = x' \sin\alpha + y' \cos\alpha \Rightarrow x = \frac{1}{2}x' - \frac{\sqrt{3}}{2}y',\ y = \frac{\sqrt{3}}{2}x' + \frac{1}{2}y'$
$\Rightarrow 3\left(\frac{1}{2}x' - \frac{\sqrt{3}}{2}y'\right)^2 - 2\sqrt{3}\left(\frac{1}{2}x' - \frac{\sqrt{3}}{2}y'\right)\left(\frac{\sqrt{3}}{2}x' + \frac{1}{2}y'\right) + \left(\frac{\sqrt{3}}{2}x' + \frac{1}{2}y'\right)^2 = 1 \Rightarrow 4y'^2 = 1$
$\Rightarrow$ Parallel horizontal lines

23. $\cot 2\alpha = \frac{A-C}{B} = \frac{\sqrt{2}-\sqrt{2}}{2\sqrt{2}} = 0 \Rightarrow 2\alpha = \frac{\pi}{2} \Rightarrow \alpha = \frac{\pi}{4}$; therefore $x = x' \cos\alpha - y' \sin\alpha$,
$y = x' \sin\alpha + y' \cos\alpha \Rightarrow x = \frac{\sqrt{2}}{2}x' - \frac{\sqrt{2}}{2}y',\ y = \frac{\sqrt{2}}{2}x' + \frac{\sqrt{2}}{2}y'$
$\Rightarrow \sqrt{2}\left(\frac{\sqrt{2}}{2}x' - \frac{\sqrt{2}}{2}y'\right)^2 + 2\sqrt{2}\left(\frac{\sqrt{2}}{2}x' - \frac{\sqrt{2}}{2}y'\right)\left(\frac{\sqrt{2}}{2}x' + \frac{\sqrt{2}}{2}y'\right) + \sqrt{2}\left(\frac{\sqrt{2}}{2}x' + \frac{\sqrt{2}}{2}y'\right)^2$
$- 8\left(\frac{\sqrt{2}}{2}x' - \frac{\sqrt{2}}{2}y'\right) + 8\left(\frac{\sqrt{2}}{2}x' + \frac{\sqrt{2}}{2}y'\right) = 0 \Rightarrow 2\sqrt{2}x'^2 + 8\sqrt{2}y' = 0 \Rightarrow$ Parabola

24. $\cot 2\alpha = \frac{A-C}{B} = \frac{0-0}{1} = 0 \Rightarrow 2\alpha = \frac{\pi}{2} \Rightarrow \alpha = \frac{\pi}{4}$; therefore $x = x' \cos\alpha - y' \sin\alpha$,
$y = x' \sin\alpha + y' \cos\alpha \Rightarrow x = \frac{\sqrt{2}}{2}x' - \frac{\sqrt{2}}{2}y',\ y = \frac{\sqrt{2}}{2}x' + \frac{\sqrt{2}}{2}y'$

$\Rightarrow \left(\frac{\sqrt{2}}{2}x' - \frac{\sqrt{2}}{2}y'\right)\left(\frac{\sqrt{2}}{2}x' + \frac{\sqrt{2}}{2}y'\right) - \left(\frac{\sqrt{2}}{2}x' + \frac{\sqrt{2}}{2}y'\right) - \left(\frac{\sqrt{2}}{2}x' - \frac{\sqrt{2}}{2}y'\right) + 1 = 0 \Rightarrow x'^2 - y'^2 - 2\sqrt{2}\,x' + 2$
$= 0 \Rightarrow$ Hyperbola

25. $\cot 2\alpha = \frac{A-C}{B} = \frac{3-3}{2} = 0 \Rightarrow 2\alpha = \frac{\pi}{2} \Rightarrow \alpha = \frac{\pi}{4}$; therefore $x = x'\cos\alpha - y'\sin\alpha$,
$y = x'\sin\alpha + y'\cos\alpha \Rightarrow x = \frac{\sqrt{2}}{2}x' - \frac{\sqrt{2}}{2}y'$, $y = \frac{\sqrt{2}}{2}x' + \frac{\sqrt{2}}{2}y'$
$\Rightarrow 3\left(\frac{\sqrt{2}}{2}x' - \frac{\sqrt{2}}{2}y'\right)^2 + 2\left(\frac{\sqrt{2}}{2}x' - \frac{\sqrt{2}}{2}y'\right)\left(\frac{\sqrt{2}}{2}x' + \frac{\sqrt{2}}{2}y'\right) + 3\left(\frac{\sqrt{2}}{2}x' + \frac{\sqrt{2}}{2}y'\right)^2 = 19 \Rightarrow 4x'^2 + 2y'^2 = 19$
$\Rightarrow$ Ellipse

26. $\cot 2\alpha = \frac{A-C}{B} = \frac{3-(-1)}{4\sqrt{3}} = \frac{1}{\sqrt{3}} \Rightarrow 2\alpha = \frac{\pi}{3} \Rightarrow \alpha = \frac{\pi}{6}$; therefore $x = x'\cos\alpha - y'\sin\alpha$,
$y = x'\sin\alpha + y'\cos\alpha \Rightarrow x = \frac{\sqrt{3}}{2}x' - \frac{1}{2}y'$, $y = \frac{1}{2}x' + \frac{\sqrt{3}}{2}y'$
$\Rightarrow 3\left(\frac{\sqrt{3}}{2}x' - \frac{1}{2}y'\right)^2 + 4\sqrt{3}\left(\frac{\sqrt{3}}{2}x' - \frac{1}{2}y'\right)\left(\frac{1}{2}x' + \frac{\sqrt{3}}{2}y'\right) - \left(\frac{1}{2}x' + \frac{\sqrt{3}}{2}y'\right)^2 = 7 \Rightarrow 5x'^2 - 3y'^2 = 7$
$\Rightarrow$ Hyperbola

27. $\cot 2\alpha = \frac{14-2}{16} = \frac{3}{4} \Rightarrow \cos 2\alpha = \frac{3}{5}$ (if we choose 2α in Quadrant I); thus $\sin\alpha = \sqrt{\frac{1-\cos 2\alpha}{2}} = \sqrt{\frac{1-\left(\frac{3}{5}\right)}{2}} = \frac{1}{\sqrt{5}}$
and $\cos\alpha = \sqrt{\frac{1+\cos 2\alpha}{2}} = \sqrt{\frac{1+\left(\frac{3}{5}\right)}{2}} = \frac{2}{\sqrt{5}}$ (or $\sin\alpha = \frac{2}{\sqrt{5}}$ and $\cos\alpha = \frac{-1}{\sqrt{5}}$)

28. $\cot 2\alpha = \frac{A-C}{B} = \frac{4-1}{-4} = -\frac{3}{4} \Rightarrow \cos 2\alpha = -\frac{3}{5}$ (if we choose 2α in Quadrant II); thus $\sin\alpha = \sqrt{\frac{1-\cos 2\alpha}{2}}$
$= \sqrt{\frac{1-\left(-\frac{3}{5}\right)}{2}} = \frac{2}{\sqrt{5}}$ and $\cos\alpha = \sqrt{\frac{1+\cos 2\alpha}{2}} = \sqrt{\frac{1+\left(-\frac{3}{5}\right)}{2}} = \frac{1}{\sqrt{5}}$ (or $\sin\alpha = \frac{1}{\sqrt{5}}$ and $\cos\alpha = \frac{-2}{\sqrt{5}}$)

29. $\tan 2\alpha = \frac{-1}{1-3} = \frac{1}{2} \Rightarrow 2\alpha \approx 26.57° \Rightarrow \alpha \approx 13.28° \Rightarrow \sin\alpha \approx 0.23$, $\cos\alpha \approx 0.97$; then $A' \approx 0.9$, $B' \approx 0.0$,
$C' \approx 3.1$, $D' \approx 0.7$, $E' \approx -1.2$, and $F' = -3 \Rightarrow 0.9\,x'^2 + 3.1\,y'^2 + 0.7x' - 1.2y' - 3 = 0$, an ellipse

30. $\tan 2\alpha = \frac{1}{2-(-3)} = \frac{1}{5} \Rightarrow 2\alpha \approx 11.31° \Rightarrow \alpha \approx 5.65° \Rightarrow \sin\alpha \approx 0.10$, $\cos\alpha \approx 1.00$; then $A' \approx 2.1$, $B' \approx 0.0$,
$C' \approx -3.1$, $D' \approx 3.0$, $E' \approx -0.3$, and $F' = -7 \Rightarrow 2.1\,x'^2 - 3.1\,y'^2 + 3.0x' - 0.3y' - 7 = 0$, a hyperbola

31. $\tan 2\alpha = \frac{-4}{1-4} = \frac{4}{3} \Rightarrow 2\alpha \approx 53.13° \Rightarrow \alpha \approx 26.57° \Rightarrow \sin\alpha \approx 0.45$, $\cos\alpha \approx 0.89$; then $A' \approx 0.0$, $B' \approx 0.0$,
$C' \approx 5.0$, $D' \approx 0$, $E' \approx 0$, and $F' = -5 \Rightarrow 5.0\,y'^2 - 5 = 0$ or $y' = \pm 1.0$, parallel lines

32. $\tan 2\alpha = \frac{-12}{2-18} = \frac{3}{4} \Rightarrow 2\alpha \approx 36.87° \Rightarrow \alpha \approx 18.43° \Rightarrow \sin\alpha \approx 0.32$, $\cos\alpha \approx 0.95$; then $A' \approx 0.0$, $B' \approx 0.0$,
$C' \approx 20.1$, $D' \approx 0$, $E' \approx 0$, and $F' = -49 \Rightarrow 20.1\,y'^2 - 49 = 0$, parallel lines

33. $\tan 2\alpha = \frac{5}{3-2} = 5 \Rightarrow 2\alpha \approx 78.69° \Rightarrow \alpha \approx 39.35° \Rightarrow \sin\alpha \approx 0.63$, $\cos\alpha \approx 0.77$; then $A' \approx 5.0$, $B' \approx 0.0$,
$C' \approx -0.05$, $D' \approx -5.0$, $E' \approx -6.2$, and $F' = -1 \Rightarrow 5.0\,x'^2 - 0.05\,y'^2 - 5.0x' - 6.2y' - 1 = 0$, a hyperbola

34. $\tan 2\alpha = \frac{7}{2-9} = -1 \Rightarrow 2\alpha \approx -45.00° \Rightarrow \alpha \approx -22.5° \Rightarrow \sin\alpha \approx -0.38$, $\cos\alpha \approx 0.92$; then $A' \approx 0.5$, $B' \approx 0.0$,
$C' \approx 10.4$, $D' \approx 18.4$, $E' \approx 7.6$, and $F' = -86 \Rightarrow 0.5\,x'^2 + 10.4(y')^2 + 18.4x' + 7.6y' - 86 = 0$, an ellipse

35. $\alpha = 90° \Rightarrow x = x'\cos 90° - y'\sin 90° = -y'$ and $y = x'\sin 90° + y'\cos 90° = x'$

(a) $\frac{x'^2}{b^2} + \frac{y'^2}{a^2} = 1$ (b) $\frac{y'^2}{a^2} - \frac{x'^2}{b^2} = 1$ (c) $x'^2 + y'^2 = a^2$

(d) $y = mx \Rightarrow y - mx = 0 \Rightarrow D = -m$ and $E = 1$; $\alpha = 90° \Rightarrow D' = 1$ and $E' = m \Rightarrow my' + x' = 0 \Rightarrow y' = -\frac{1}{m}x'$

(e) $y = mx + b \Rightarrow y - mx - b = 0 \Rightarrow D = -m$ and $E = 1$; $\alpha = 90° \Rightarrow D' = 1$, $E' = m$ and $F' = -b$
$\Rightarrow my' + x' - b = 0 \Rightarrow y' = -\frac{1}{m}x' + \frac{b}{m}$

36. $\alpha = 180° \Rightarrow x = x' \cos 180° - y' \sin 180° = -x'$ and $y = x' \sin 180° + y' \cos 180° = -y'$

(a) $\frac{x'^2}{a^2} + \frac{y'^2}{b^2} = 1$ (b) $\frac{x'^2}{a^2} - \frac{y'^2}{b^2} = 1$ (c) $x'^2 + y'^2 = a^2$

(d) $y = mx \Rightarrow y - mx = 0 \Rightarrow D = -m$ and $E = 1$; $\alpha = 180° \Rightarrow D' = m$ and $E' = -1 \Rightarrow -y' + mx' = 0 \Rightarrow y' = mx'$

(e) $y = mx + b \Rightarrow y - mx - b = 0 \Rightarrow D = -m$ and $E = 1$; $\alpha = 180° \Rightarrow D' = m$, $E' = -1$ and $F' = -b$ $\Rightarrow -y' + mx' - b = 0 \Rightarrow y' = mx' - b$

37. (a) $A' = \cos 45° \sin 45° = \left(\frac{\sqrt{2}}{2}\right)\left(\frac{\sqrt{2}}{2}\right) = \frac{1}{2}$, $B' = 0$, $C' = -\cos 45° \sin 45° = -\frac{1}{2}$, $F' = -1$ $\Rightarrow \frac{1}{2}x'^2 - \frac{1}{2}y'^2 = 1 \Rightarrow x'^2 - y'^2 = 2$

(b) $A' = \frac{1}{2}$, $C' = -\frac{1}{2}$ (see part (a) above), $D' = E' = B' = 0$, $F' = -a \Rightarrow \frac{1}{2}x'^2 - \frac{1}{2}y'^2 = a \Rightarrow x'^2 - y'^2 = 2a$

38. $xy = 2 \Rightarrow x'^2 - y'^2 = 4 \Rightarrow \frac{x'^2}{4} - \frac{y'^2}{4} = 1$ (see Exercise 37(b)) $\Rightarrow a = 2$ and $b = 2 \Rightarrow c = \sqrt{4+4} = 2\sqrt{2}$ $\Rightarrow e = \frac{c}{a} = \frac{2\sqrt{2}}{2} = \sqrt{2}$

39. Yes, the graph is a hyperbola: with $AC < 0$ we have $-4AC > 0$ and $B^2 - 4AC > 0$.

40. The one curve that meets all three of the stated criteria is the ellipse $x^2 + 4xy + 5y^2 - 1 = 0$. The reasoning: The symmetry about the origin means that $(-x, -y)$ lies on the graph whenever (x, y) does. Adding $Ax^2 + Bxy + Cy^2 + Dx + Ey + F = 0$ and $A(-x)^2 + B(-x)(-y) + C(-y)^2 + D(-x) + E(-y) + F = 0$ and dividing the result by 2 produces the equivalent equation $Ax^2 + Bxy + Cy^2 + F = 0$. Substituting $x = 1$, $y = 0$ (because the point $(1, 0)$ lies on the curve) shows further that $A = -F$. Then $-Fx^2 + Bxy + Cy^2 + F = 0$. By implicit differentiation, $-2Fx + By + Bxy' + 2Cyy' = 0$, so substituting $x = -2$, $y = 1$, and $y' = 0$ (from Property 3) gives $4F + B = 0 \Rightarrow B = -4F \Rightarrow$ the conic is $-Fx^2 - 4Fxy + Cy^2 + F = 0$. Now substituting $x = -2$ and $y = 1$ again gives $-4F + 8F + C + F = 0 \Rightarrow C = -5F \Rightarrow$ the equation is now $-Fx^2 - 4Fxy - 5Fy^2 + F = 0$. Finally, dividing through by $-F$ gives the equation $x^2 + 4xy + 5y^2 - 1 = 0$.

41. Let α be any angle. Then $A' = \cos^2\alpha + \sin^2\alpha = 1$, $B' = 0$, $C' = \sin^2\alpha + \cos^2\alpha = 1$, $D' = E' = 0$ and $F' = -a^2$ $\Rightarrow x'^2 + y'^2 = a^2$.

42. If $A = C$, then $B' = B\cos 2\alpha + (C - A)\sin 2\alpha = B\cos 2\alpha$. Then $\alpha = \frac{\pi}{4} \Rightarrow 2\alpha = \frac{\pi}{2} \Rightarrow B' = B\cos\frac{\pi}{2} = 0$ so the xy-term is eliminated.

43. (a) $B^2 - 4AC = 4^2 - 4(1)(4) = 0$, so the discriminant indicates this conic is a parabola

(b) The left-hand side of $x^2 + 4xy + 4y^2 + 6x + 12y + 9 = 0$ factors as a perfect square: $(x + 2y + 3)^2 = 0$ $\Rightarrow x + 2y + 3 = 0 \Rightarrow 2y = -x - 3$; thus the curve is a degenerate parabola (i.e., a straight line).

44. (a) $B^2 - 4AC = 6^2 - 4(9)(1) = 0$, so the discriminant indicates this conic is a parabola

(b) The left-hand side of $9x^2 + 6xy + y^2 - 12x - 4y + 4 = 0$ factors as a perfect square: $(3x + y - 2)^2 = 0$ $\Rightarrow 3x + y - 2 = 0 \Rightarrow y = -3x + 2$; thus the curve is a degenerate parabola (i.e., a straight line).

45. (a) $B^2 - 4AC = 1 - 4(0)(0) = 1 \Rightarrow$ hyperbola

(b) $xy + 2x - y = 0 \Rightarrow y(x-1) = -2x \Rightarrow y = \frac{-2x}{x-1}$

(c) $y = \frac{-2x}{x-1} \Rightarrow \frac{dy}{dx} = \frac{2}{(x-1)^2}$ and we want $\frac{-1}{\left(\frac{dy}{dx}\right)} = -2$, the slope of $y = -2x \Rightarrow -2 = -\frac{(x-1)^2}{2}$ $\Rightarrow (x-1)^2 = 4 \Rightarrow x = 3$ or $x = -1$; $x = 3$ $\Rightarrow y = -3 \Rightarrow (3, -3)$ is a point on the hyperbola where the line with slope $m = -2$ is normal $\Rightarrow$ the line is $y + 3 = -2(x-3)$ or $y = -2x + 3$; $x = -1 \Rightarrow y = -1 \Rightarrow (-1, -1)$ is a point on the hyperbola where the line with slope $m = -2$ is normal $\Rightarrow$ the line is $y + 1 = -2(x+1)$ or $y = -2x - 3$

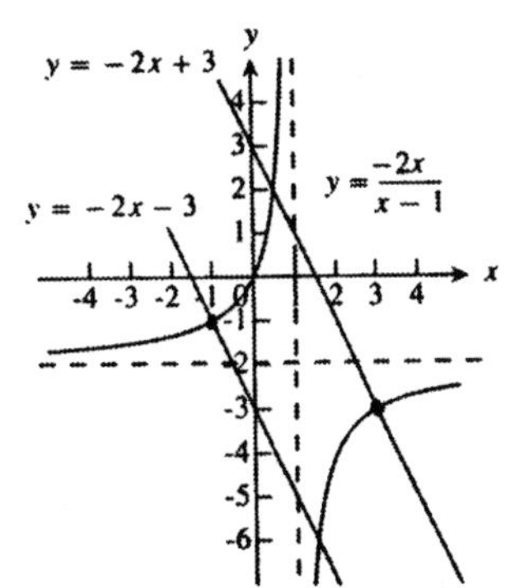

46. (a) False: let $A = C = 1$, $B = 2 \Rightarrow B^2 - 4AC = 0 \Rightarrow$ parabola

(b) False: see part (a) above

(c) True: $AC < 0 \Rightarrow -4AC > 0 \Rightarrow B^2 - 4AC > 0 \Rightarrow$ hyperbola

47. Assume the ellipse has been rotated to eliminate the xy-term $\Rightarrow$ the new equation is $A'x'^2 + C'y'^2 = 1 \Rightarrow$ the semi-axes are $\sqrt{\frac{1}{A'}}$ and $\sqrt{\frac{1}{C'}} \Rightarrow$ the area is $\pi\left(\sqrt{\frac{1}{A'}}\right)\left(\sqrt{\frac{1}{C'}}\right) = \frac{\pi}{\sqrt{A'C'}} = \frac{2\pi}{\sqrt{4A'C'}}$. Since $B^2 - 4AC$ $= B'^2 - 4A'C' = -4A'C'$ (because $B' = 0$) we find that the area is $\frac{2\pi}{\sqrt{4AC - B^2}}$ as claimed.

48. (a) $A' + C' = (A\cos^2\alpha + B\cos\alpha\sin\alpha + C\sin^2\alpha) + (A\sin^2\alpha - B\cos\alpha\sin\alpha + C\sin^2\alpha)$
$= A(\cos^2\alpha + \sin^2\alpha) + C(\sin^2\alpha + \cos^2\alpha) = A + C$

(b) $D'^2 + E'^2 = (D\cos\alpha + E\sin\alpha)^2 + (-D\sin\alpha + E\cos\alpha)^2 = D^2\cos^2\alpha + 2DE\cos\alpha\sin\alpha + E^2\sin^2\alpha$
$+ D^2\sin^2\alpha - 2DE\sin\alpha\cos\alpha + E^2\cos^2\alpha = D^2(\cos^2\alpha + \sin^2\alpha) + E^2(\sin^2\alpha + \cos^2\alpha) = D^2 + E^2$

49. $B'^2 - 4A'C'$

$= (B\cos 2\alpha + (C-A)\sin 2\alpha)^2 - 4(A\cos^2\alpha + B\cos\alpha\sin\alpha + C\sin^2\alpha)(A\sin^2\alpha - B\cos\alpha\sin\alpha + C\cos^2\alpha)$

$= B^2\cos^2 2\alpha + 2B(C-A)\sin 2\alpha\cos 2\alpha + (C-A)^2\sin^2 2\alpha - 4A^2\cos^2\alpha\sin^2\alpha + 4AB\cos^3\alpha\sin\alpha$
$- 4AC\cos^4\alpha - 4AB\cos\alpha\sin^3\alpha + 4B^2\cos^2\alpha\sin^2\alpha - 4BC\cos^3\alpha\sin\alpha - 4AC\sin^4\alpha + 4BC\cos\alpha\sin^3\alpha$
$- 4C^2\cos^2\alpha\sin^2\alpha$

$= B^2\cos^2 2\alpha + 2BC\sin 2\alpha\cos 2\alpha - 2AB\sin 2\alpha\cos 2\alpha + C^2\sin^2 2\alpha - 2AC\sin^2 2\alpha + A^2\sin^2 2\alpha$
$- 4A^2\cos^2\alpha\sin^2\alpha + 4AB\cos^3\alpha\sin\alpha - 4AC\cos^4\alpha - 4AB\cos\alpha\sin^3\alpha + B^2\sin^2 2\alpha - 4BC\cos^3\alpha\sin\alpha$
$- 4AC\sin^4\alpha + 4BC\cos\alpha\sin^3\alpha - 4C^2\cos^2\alpha\sin^2\alpha$

$= B^2 + 2BC(2\sin\alpha\cos\alpha)(\cos^2\alpha - \sin^2\alpha) - 2AB(2\sin\alpha\cos\alpha)(\cos^2\alpha - \sin^2\alpha) + C^2(4\sin^2\alpha\cos^2\alpha)$
$- 2AC(4\sin^2\alpha\cos^2\alpha) + A^2(4\sin^2\alpha\cos^2\alpha) - 4A^2\cos^2\alpha\sin^2\alpha + 4AB\cos^3\alpha\sin\alpha - 4AC\cos^4\alpha$
$- 4AB\cos\alpha\sin^3\alpha - 4BC\cos^3\alpha\sin\alpha - 4AC\sin^4\alpha + 4BC\cos\alpha\sin^3\alpha - 4C^2\cos^2\alpha\sin^2\alpha$

$= B^2 - 8AC\sin^2\alpha\cos^2\alpha - 4AC\cos^4\alpha - 4AC\sin^4\alpha$

$= B^2 - 4AC(\cos^4\alpha + 2\sin^2\alpha\cos^2\alpha + \sin^4\alpha)$

$= B^2 - 4AC(\cos^2\alpha + \sin^2\alpha)^2$

$= B^2 - 4AC$

10.4 CONICS AND PARAMETRIC EQUATIONS; THE CYCLOID

1. $x = \cos t,\ y = \sin t,\ 0 \le t \le \pi$
$\Rightarrow \cos^2 t + \sin^2 t = 1 \Rightarrow x^2 + y^2 = 1$

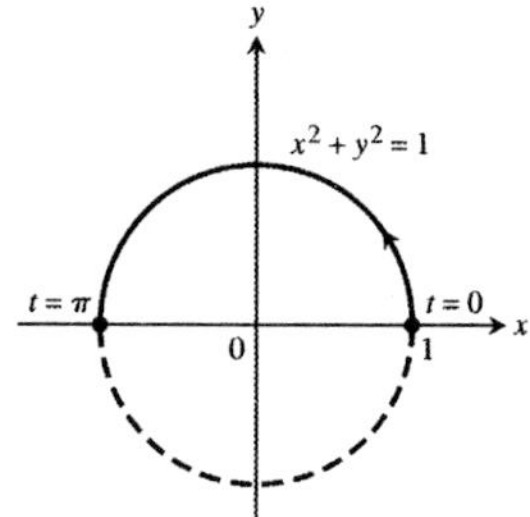

2. $x = \sin(2\pi(1 - t)),\ y = \cos(2\pi(1 - t)),\ 0 \le t \le 1$
$\Rightarrow \sin^2(2\pi(1 - t)) + \cos^2(2\pi(1 - t)) = 1$
$\Rightarrow x^2 + y^2 = 1$

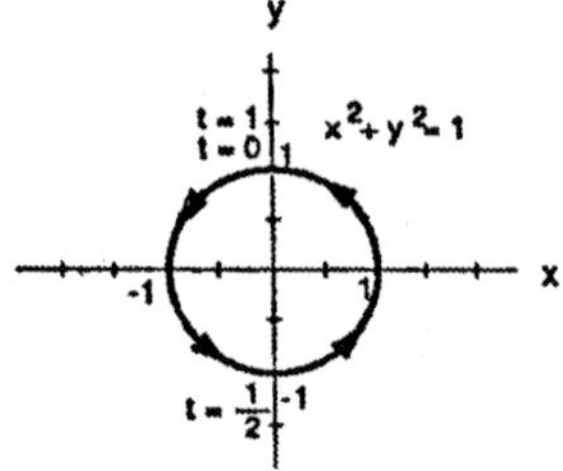

3. $x = 4\cos t,\ y = 5\sin t,\ 0 \le t \le \pi$
$\Rightarrow \frac{16\cos^2 t}{16} + \frac{25\sin^2 t}{25} = 1 \Rightarrow \frac{x^2}{16} + \frac{y^2}{25} = 1$

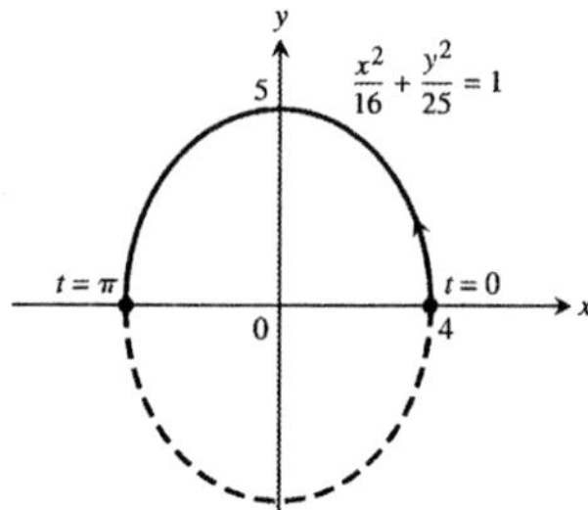

4. $x = 4\sin t,\ y = 5\cos t,\ 0 \le t \le 2\pi$
$\Rightarrow \frac{16\sin^2 t}{16} + \frac{25\cos^2 t}{25} = 1 \Rightarrow \frac{x^2}{16} + \frac{y^2}{25} = 1$

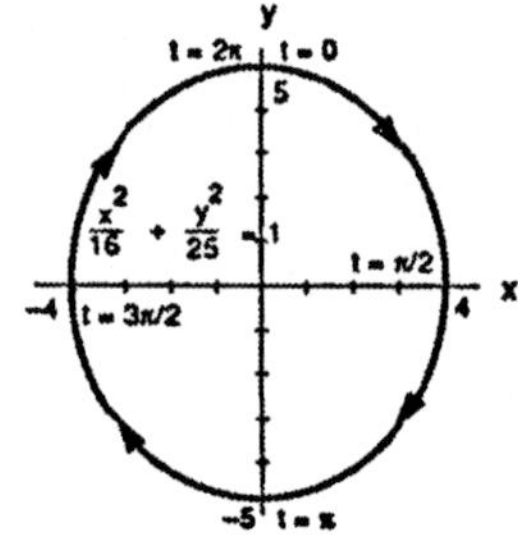

5. $x = t,\ y = \sqrt{t},\ t \ge 0 \Rightarrow y = \sqrt{x}$

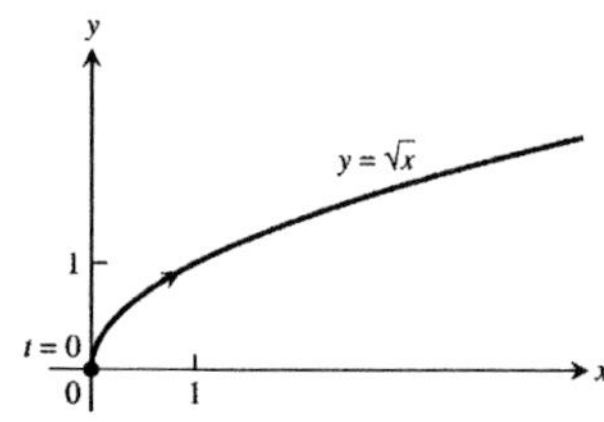

6. $x = \sec^2 t - 1,\ y = \tan t,\ -\frac{\pi}{2} < t < \frac{\pi}{2}$
$\Rightarrow \sec^2 t - 1 = \tan^2 t \Rightarrow x = y^2$

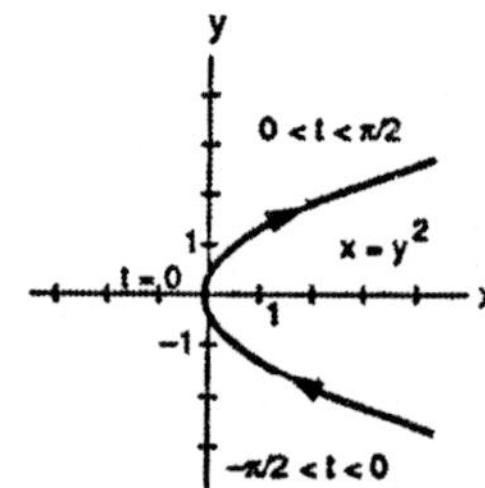

7. $x = -\sec t,\ y = \tan t,\ -\frac{\pi}{2} < t < \frac{\pi}{2}$
$\Rightarrow \sec^2 t - \tan^2 t = 1 \Rightarrow x^2 - y^2 = 1$

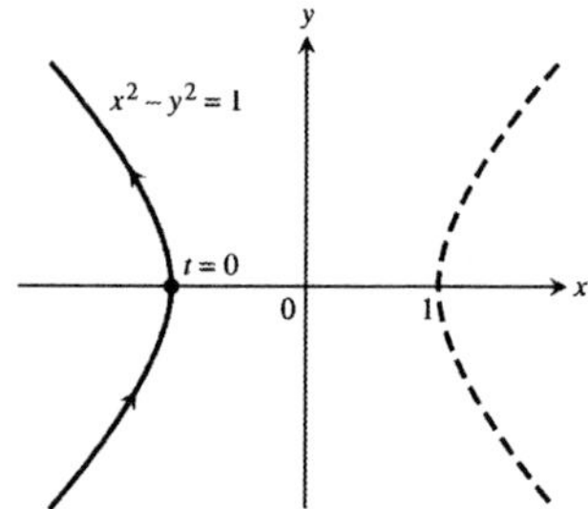

8. $x = \csc t,\ y = \cot t,\ 0 < t < \pi$
$\Rightarrow 1 + \cot^2 t = \csc^2 t \Rightarrow 1 + y^2 = x^2 \Rightarrow x^2 - y^2 = 1$

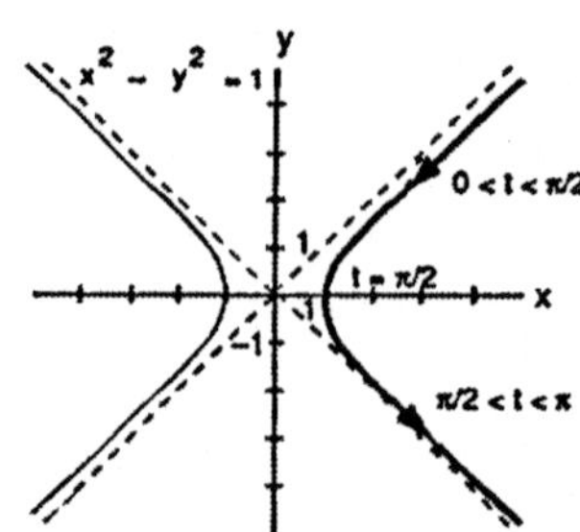

9. $x = t,\ y = \sqrt{4 - t^2},\ 0 \le t \le 2$
$\Rightarrow\ y = \sqrt{4 - x^2}$

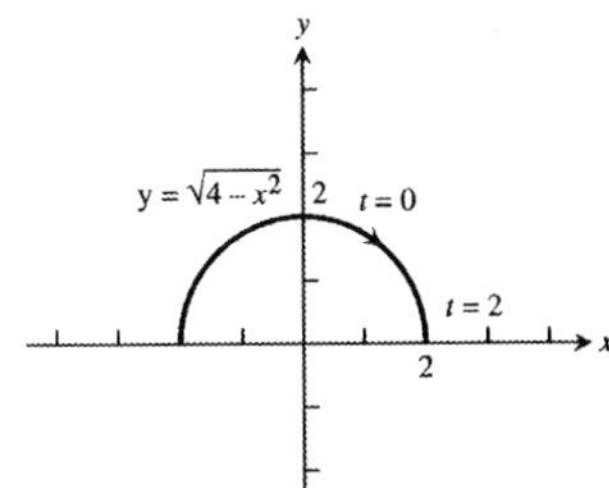

10. $x = t^2,\ y = \sqrt{t^4 + 1},\ t \ge 0$
$\Rightarrow\ y = \sqrt{x^2 + 1},\ x \ge 0$

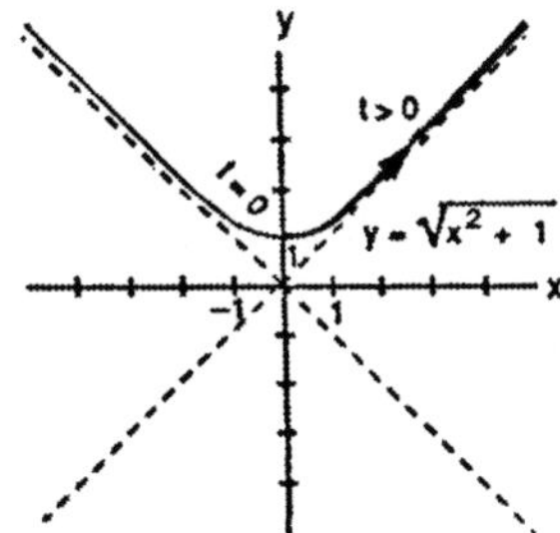

11. $x = -\cosh t,\ y = \sinh t,\ -\infty < 1 < \infty$
$\Rightarrow\ \cosh^2 t - \sinh^2 t = 1\ \Rightarrow\ x^2 - y^2 = 1$

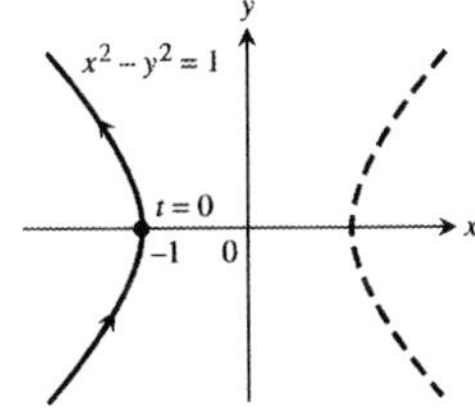

12. $x = 2\sinh t,\ y = 2\cosh t,\ -\infty < t < \infty$
$\Rightarrow\ 4\cosh^2 t - 4\sinh^2 t = 4\ \Rightarrow\ y^2 - x^2 = 4$

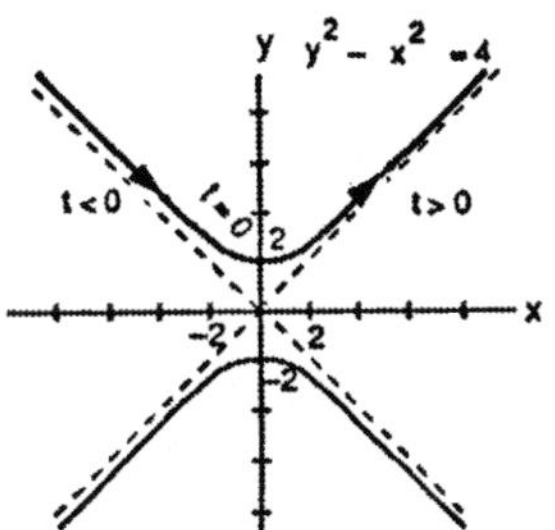

13. Arc PF = Arc AF since each is the distance rolled and
$\frac{\text{Arc PF}}{b} = \angle FCP \Rightarrow \text{Arc PF} = b(\angle FCP);\ \frac{\text{Arc AF}}{a} = \theta$
$\Rightarrow \text{Arc AF} = a\theta \Rightarrow a\theta = b(\angle FCP) \Rightarrow \angle FCP = \frac{a}{b}\,\theta;$
$\angle OCG = \frac{\pi}{2} - \theta;\ \angle OCG = \angle OCP + \angle PCE$
$= \angle OCP + \left(\frac{\pi}{2} - \alpha\right)$. Now $\angle OCP = \pi - \angle FCP$
$= \pi - \frac{a}{b}\,\theta$. Thus $\angle OCG = \pi - \frac{a}{b}\,\theta + \frac{\pi}{2} - \alpha \Rightarrow \frac{\pi}{2} - \theta$
$= \pi - \frac{a}{b}\,\theta + \frac{\pi}{2} - \alpha \Rightarrow \alpha = \pi - \frac{a}{b}\,\theta + \theta = \pi - \left(\frac{a-b}{b}\,\theta\right)$.

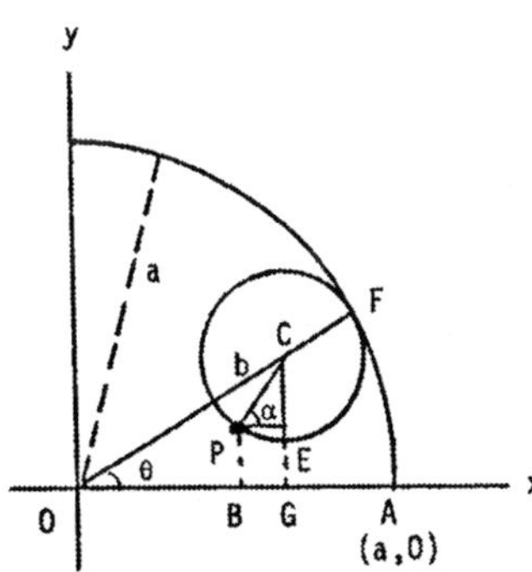

Then $x = OG - BG = OG - PE = (a - b)\cos\theta - b\cos\alpha = (a - b)\cos\theta - b\cos\left(\pi - \frac{a-b}{b}\,\theta\right)$
$= (a - b)\cos\theta + b\cos\left(\frac{a-b}{b}\,\theta\right)$. Also $y = EG = CG - CE = (a - b)\sin\theta - b\sin\alpha$
$= (a - b)\sin\theta - b\sin\left(\pi - \frac{a-b}{b}\,\theta\right) = (a - b)\sin\theta - b\sin\left(\frac{a-b}{b}\,\theta\right)$. Therefore
$x = (a - b)\cos\theta + b\cos\left(\frac{a-b}{b}\,\theta\right)$ and $y = (a - b)\sin\theta - b\sin\left(\frac{a-b}{b}\,\theta\right)$.
If $b = \frac{a}{4}$, then $x = \left(a - \frac{a}{4}\right)\cos\theta + \frac{a}{4}\cos\left(\frac{a - \left(\frac{a}{4}\right)}{\left(\frac{a}{4}\right)}\,\theta\right)$
$= \frac{3a}{4}\cos\theta + \frac{a}{4}\cos 3\theta = \frac{3a}{4}\cos\theta + \frac{a}{4}(\cos\theta\cos 2\theta - \sin\theta\sin 2\theta)$
$= \frac{3a}{4}\cos\theta + \frac{a}{4}((\cos\theta)(\cos^2\theta - \sin^2\theta) - (\sin\theta)(2\sin\theta\cos\theta))$
$= \frac{3a}{4}\cos\theta + \frac{a}{4}\cos^3\theta - \frac{a}{4}\cos\theta\sin^2\theta - \frac{2a}{4}\sin^2\theta\cos\theta$
$= \frac{3a}{4}\cos\theta + \frac{a}{4}\cos^3\theta - \frac{3a}{4}(\cos\theta)(1 - \cos^2\theta) = a\cos^3\theta;$
$y = \left(a - \frac{a}{4}\right)\sin\theta - \frac{a}{4}\sin\left(\frac{a - \left(\frac{a}{4}\right)}{\left(\frac{a}{4}\right)}\,\theta\right) = \frac{3a}{4}\sin\theta - \frac{a}{4}\sin 3\theta = \frac{3a}{4}\sin\theta - \frac{a}{4}(\sin\theta\cos 2\theta + \cos\theta\sin 2\theta)$
$= \frac{3a}{4}\sin\theta - \frac{a}{4}((\sin\theta)(\cos^2\theta - \sin^2\theta) + (\cos\theta)(2\sin\theta\cos\theta))$
$= \frac{3a}{4}\sin\theta - \frac{a}{4}\sin\theta\cos^2\theta + \frac{a}{4}\sin^3\theta - \frac{2a}{4}\cos^2\theta\sin\theta$
$= \frac{3a}{4}\sin\theta - \frac{3a}{4}\sin\theta\cos^2\theta + \frac{a}{4}\sin^3\theta$
$= \frac{3a}{4}\sin\theta - \frac{3a}{4}(\sin\theta)(1 - \sin^2\theta) + \frac{a}{4}\sin^3\theta = a\sin^3\theta.$

14. P traces a hypocycloid where the larger radius is 2a and the smaller is a $\Rightarrow$ $x = (2a - a)\cos\theta + a\cos\left(\frac{2a-a}{a}\theta\right) = 2a\cos\theta$, $0 \le \theta \le 2\pi$, and $y = (2a - a)\sin\theta - a\sin\left(\frac{2a-a}{a}\theta\right) = a\sin\theta - a\sin\theta = 0$. Therefore P traces the diameter of the circle back and forth as θ goes from 0 to 2π.

15. Draw line AM in the figure and note that $\angle AMO$ is a right angle since it is an inscribed angle which spans the diameter of a circle. Then $AN^2 = MN^2 + AM^2$. Now, $OA = a$, $\frac{AN}{a} = \tan t$, and $\frac{AM}{a} = \sin t$. Next $MN = OP$ $\Rightarrow OP^2 = AN^2 - AM^2 = a^2\tan^2 t - a^2\sin^2 t$ $\Rightarrow OP = \sqrt{a^2\tan^2 t - a^2\sin^2 t}$ $= (a\sin t)\sqrt{\sec^2 t - 1} = \frac{a\sin^2 t}{\cos t}$. In triangle BPO, $x = OP\sin t = \frac{a\sin^3 t}{\cos t} = a\sin^2 t\tan t$ and $y = OP\cos t = a\sin^2 t$ $\Rightarrow$ $x = a\sin^2 t\tan t$ and $y = a\sin^2 t$.

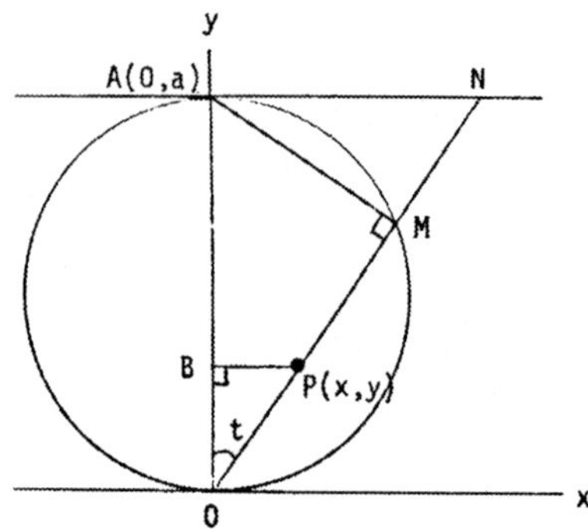

16. Let the x-axis be the line the wheel rolls along with the y-axis through a low point of the trochoid (see the accompanying figure).

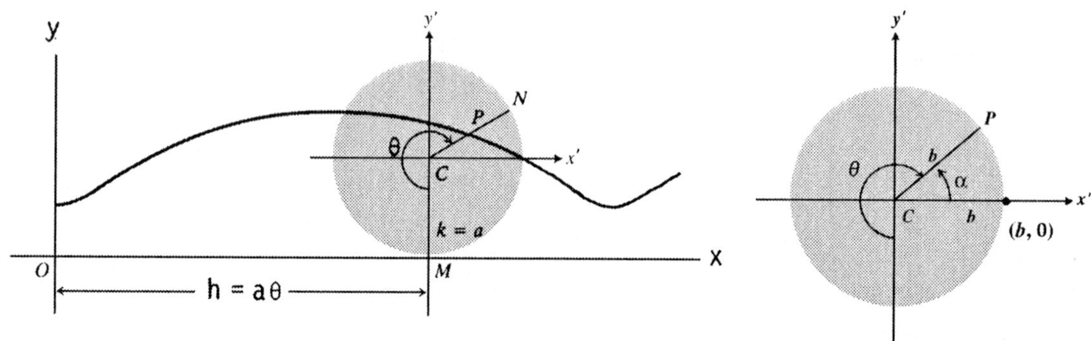

Let θ denote the angle through which the wheel turns. Then $h = a\theta$ and $k = a$. Next introduce $x'y'$-axes parallel to the xy-axes and having their origin at the center C of the wheel. Then $x' = b\cos\alpha$ and $y' = b\sin\alpha$, where $\alpha = \frac{3\pi}{2} - \theta$. It follows that $x' = b\cos\left(\frac{3\pi}{2} - \theta\right) = -b\sin\theta$ and $y' = b\sin\left(\frac{3\pi}{2} - \theta\right) = -b\cos\theta$ $\Rightarrow$ $x = h + x' = a\theta - b\sin\theta$ and $y = k + y' = a - b\cos\theta$ are parametric equations of the trochoid.

17. $D = \sqrt{(x-2)^2 + \left(y - \frac{1}{2}\right)^2}$ $\Rightarrow$ $D^2 = (x-2)^2 + \left(y - \frac{1}{2}\right)^2 = (t-2)^2 + \left(t^2 - \frac{1}{2}\right)^2$ $\Rightarrow$ $D^2 = t^4 - 4t + \frac{17}{4}$ $\Rightarrow$ $\frac{d(D^2)}{dt} = 4t^3 - 4 = 0$ $\Rightarrow$ $t = 1$. The second derivative is always positive for $t \ne 0$ $\Rightarrow$ $t = 1$ gives a local minimum for D^2 (and hence D) which is an absolute minimum since it is the only extremum $\Rightarrow$ the closest point on the parabola is (1, 1).

18. $D = \sqrt{\left(2\cos t - \frac{3}{4}\right)^2 + (\sin t - 0)^2}$ $\Rightarrow$ $D^2 = \left(2\cos t - \frac{3}{4}\right)^2 + \sin^2 t$ $\Rightarrow$ $\frac{d(D^2)}{dt} = 2\left(2\cos t - \frac{3}{4}\right)(-2\sin t) + 2\sin t\cos t = (-2\sin t)\left(3\cos t - \frac{3}{2}\right) = 0$ $\Rightarrow$ $-2\sin t = 0$ or $3\cos t - \frac{3}{2} = 0$ $\Rightarrow$ $t = 0, \pi$ or $t = \frac{\pi}{3}, \frac{5\pi}{3}$. Now $\frac{d^2(D^2)}{dt^2} = -6\cos^2 t + 3\cos t + 6\sin^2 t$ so that $\frac{d^2(D^2)}{dt^2}(0) = -3$ $\Rightarrow$ relative maximum, $\frac{d^2(D^2)}{dt^2}(\pi) = -9$ $\Rightarrow$ relative maximum, $\frac{d^2(D^2)}{dt^2}\left(\frac{\pi}{3}\right) = \frac{9}{2}$ $\Rightarrow$ relative minimum, and $\frac{d^2(D^2)}{dt^2}\left(\frac{5\pi}{3}\right) = \frac{9}{2}$ $\Rightarrow$ relative minimum. Therefore both $t = \frac{\pi}{3}$ and $t = \frac{5\pi}{3}$ give points on the ellipse closest to the point $\left(\frac{3}{4}, 0\right)$ $\Rightarrow$ $\left(1, \frac{\sqrt{3}}{2}\right)$ and $\left(1, -\frac{\sqrt{3}}{2}\right)$ are the desired points.

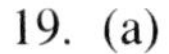

19. (a)

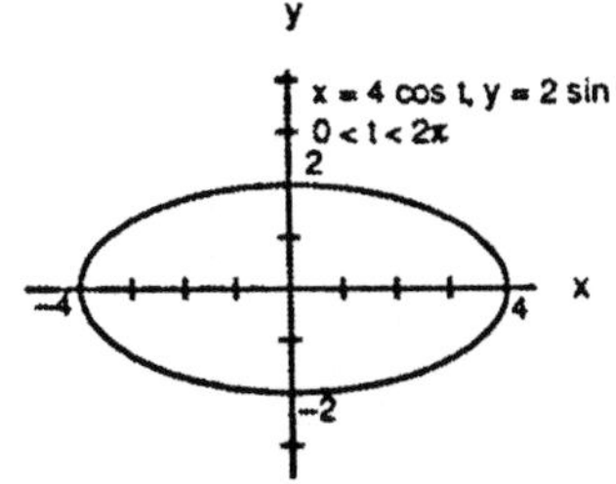

(b)

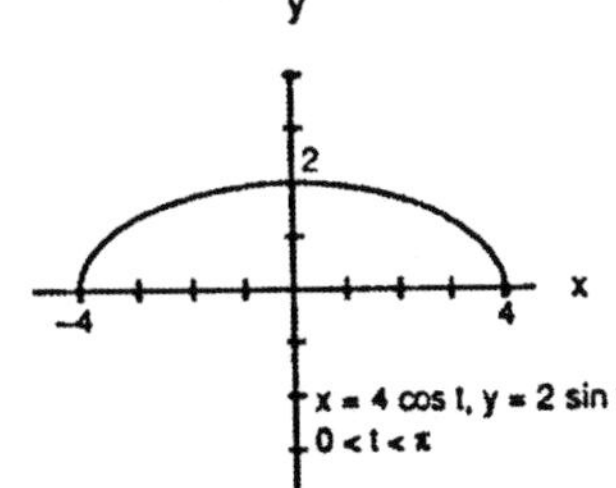

(c)

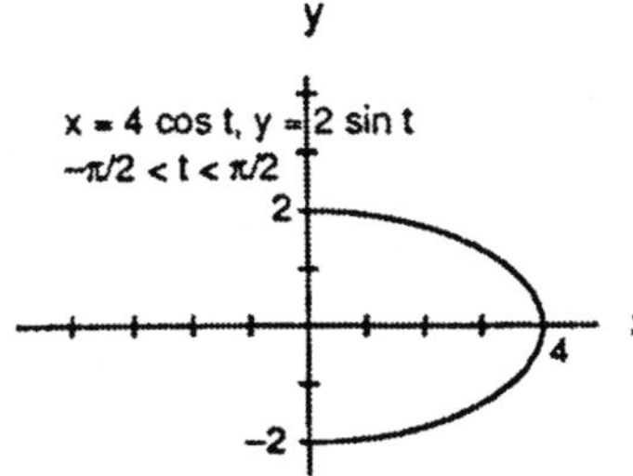

20. (a)

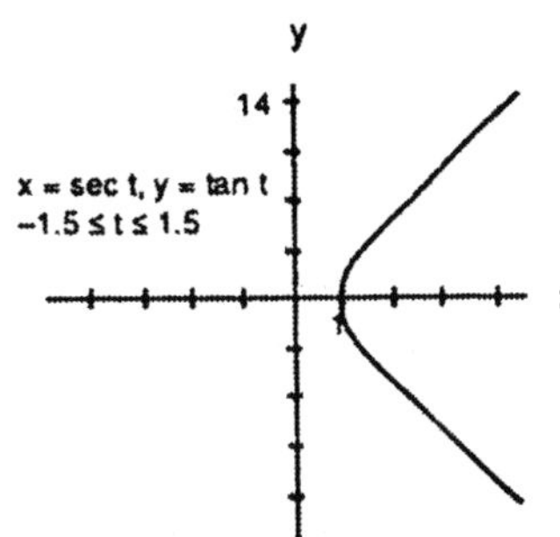

(b)

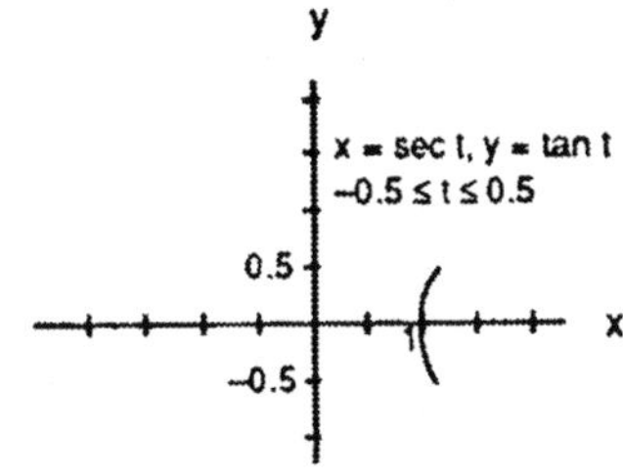

(c)

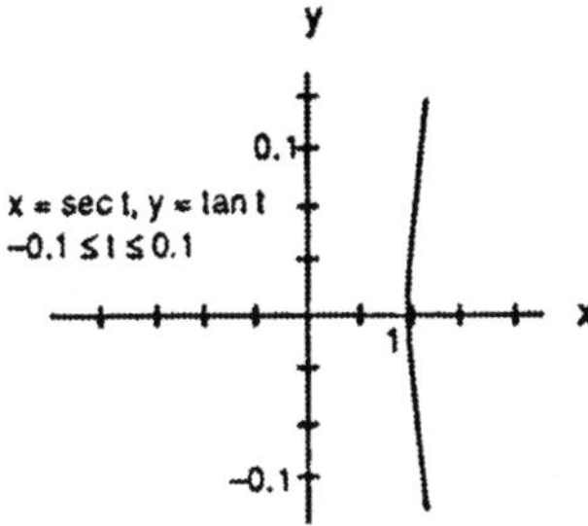

21.

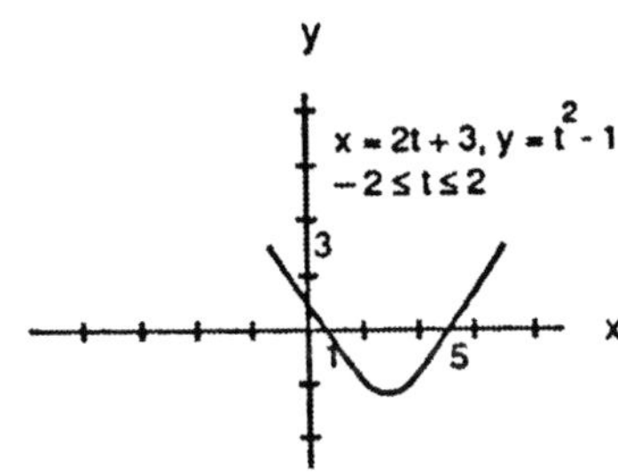

22. (a)

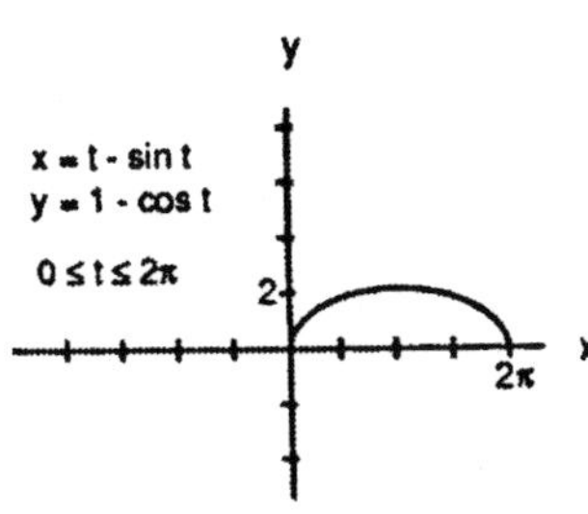

(b)

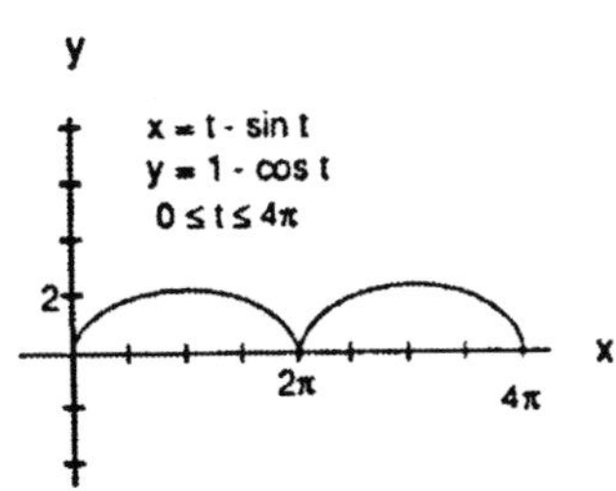

(c)

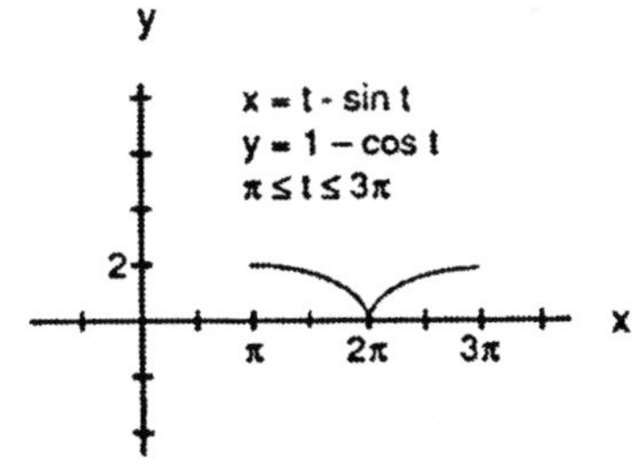

23. (a)

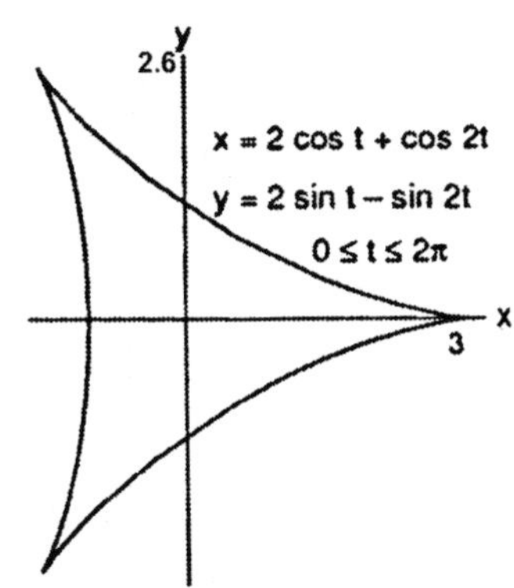

(b)

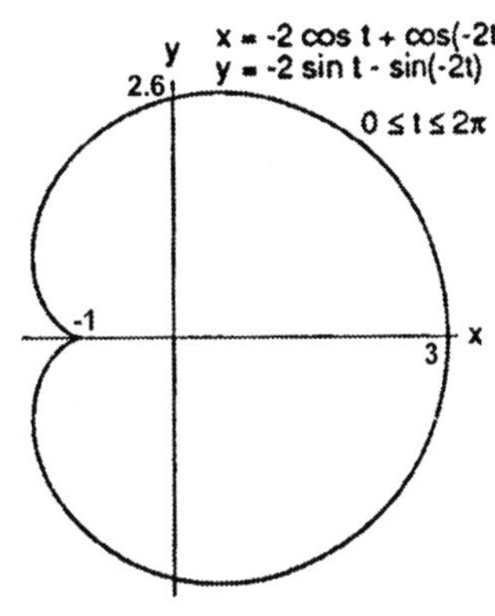

24. (a)

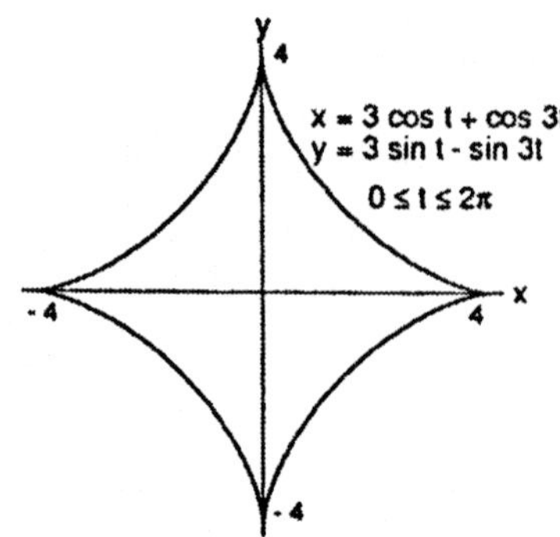

(b)

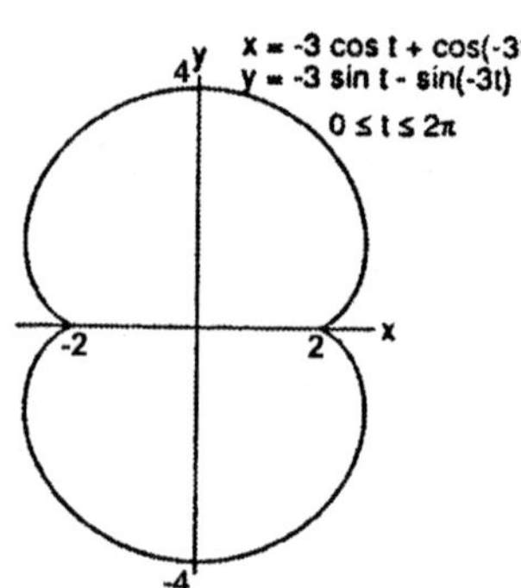

25. (a)

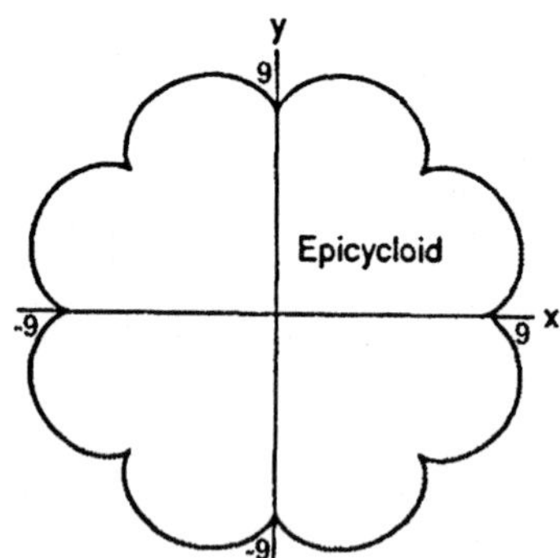

(b)

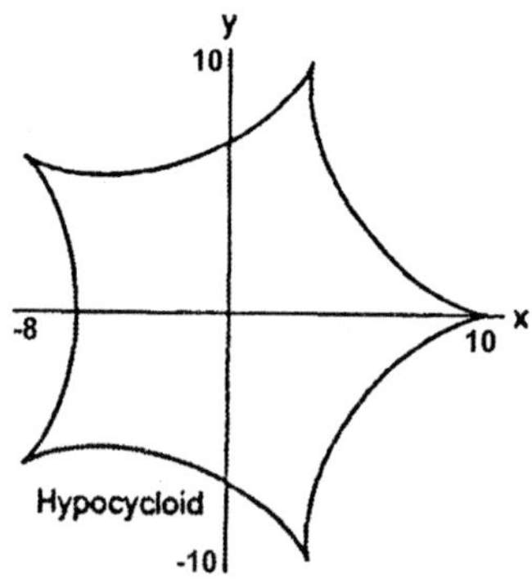

(c)

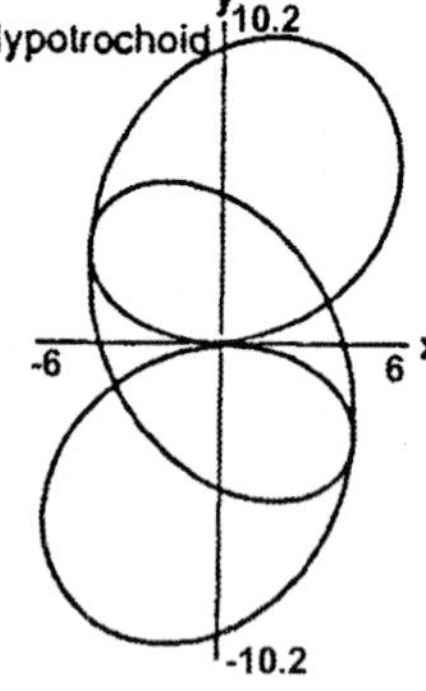

26. (a)

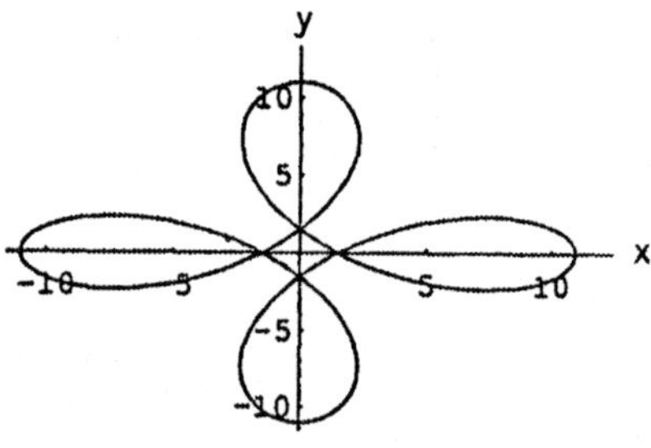

$x = 6 \cos t + 5 \cos 3t$, $y = 6 \sin t - 5 \sin 3t$, $0 \le t \le 2\pi$

(b)

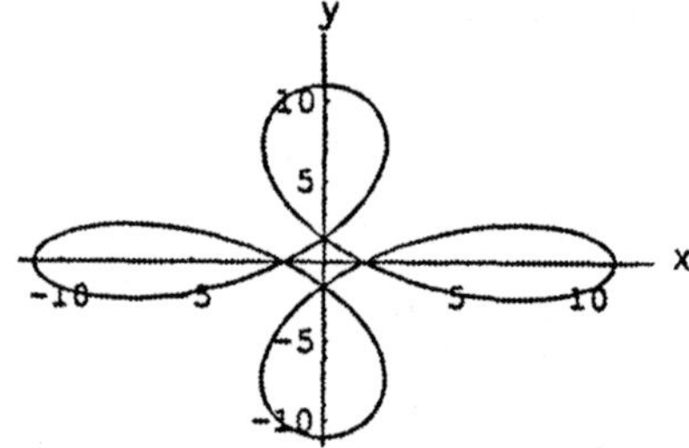

$x = 6 \cos 2t + 5 \cos 6t$, $y = 6 \sin 2t - 5 \sin 6t$, $0 \le t \le \pi$

(c)

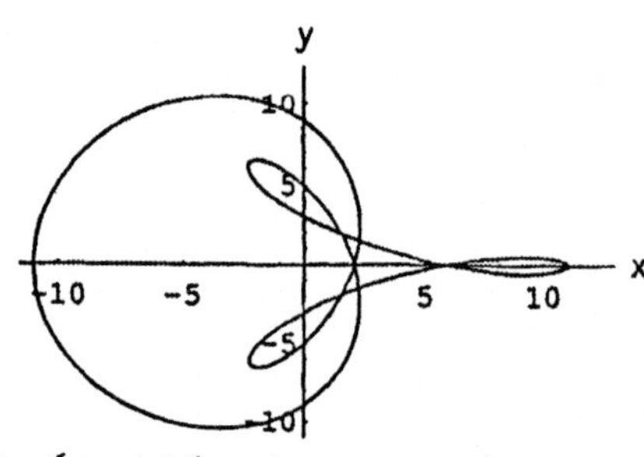

$x = 6 \cos t + 5 \cos 3t$, $y = 6 \sin 2t - 5 \sin 3t$, $0 \le t \le 2\pi$

(d)

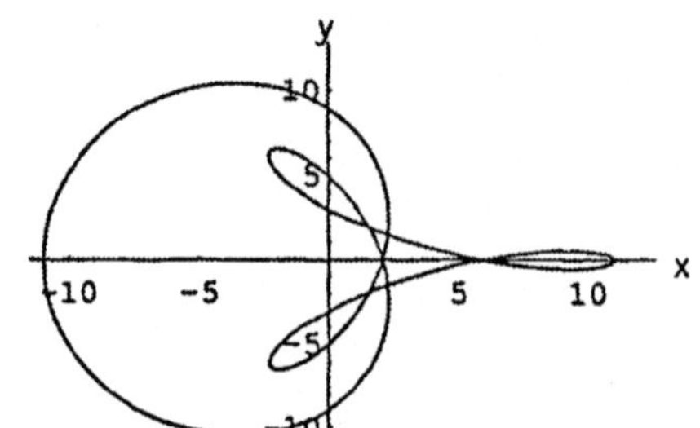

$x = 6 \cos 2t + 5 \cos 6t$, $y = 6 \sin 4t - 5 \sin 6t$, $0 \le t \le \pi$

10.5 POLAR COORDINATES

1. a, e; b, g; c, h; d, f

2. a, f; b, h; c, g; d, e

3. (a) $\left(2, \frac{\pi}{2} + 2n\pi\right)$ and $\left(-2, \frac{\pi}{2} + (2n+1)\pi\right)$, n an integer
   (b) $(2, 2n\pi)$ and $(-2, (2n+1)\pi)$, n an integer
   (c) $\left(2, \frac{3\pi}{2} + 2n\pi\right)$ and $\left(-2, \frac{3\pi}{2} + (2n+1)\pi\right)$, n an integer
   (d) $(2, (2n+1)\pi)$ and $(-2, 2n\pi)$, n an integer

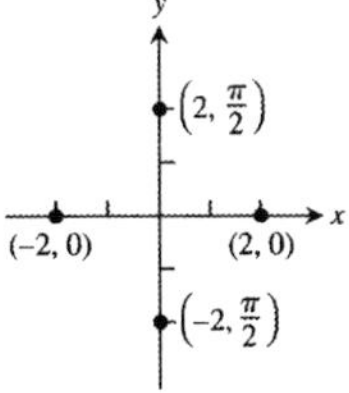

4. (a) $\left(3, \frac{\pi}{4} + 2n\pi\right)$ and $\left(-3, \frac{5\pi}{4} + 2n\pi\right)$, n an integer
   (b) $\left(-3, \frac{\pi}{4} + 2n\pi\right)$ and $\left(3, \frac{5\pi}{4} + 2n\pi\right)$, n an integer
   (c) $\left(3, -\frac{\pi}{4} + 2n\pi\right)$ and $\left(-3, \frac{3\pi}{4} + 2n\pi\right)$, n an integer
   (d) $\left(-3, -\frac{\pi}{4} + 2n\pi\right)$ and $\left(3, \frac{3\pi}{4} + 2n\pi\right)$, n an integer

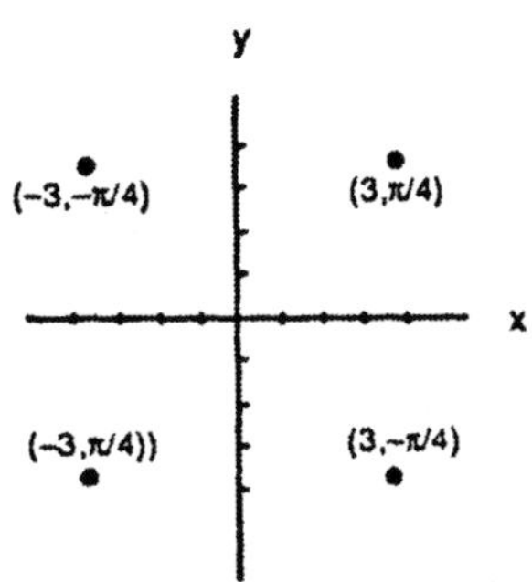

5. (a) $x = r\cos\theta = 3\cos 0 = 3$, $y = r\sin\theta = 3\sin 0 = 0 \Rightarrow$ Cartesian coordinates are $(3, 0)$
   (b) $x = r\cos\theta = -3\cos 0 = -3$, $y = r\sin\theta = -3\sin 0 = 0 \Rightarrow$ Cartesian coordinates are $(-3, 0)$
   (c) $x = r\cos\theta = 2\cos\frac{2\pi}{3} = -1$, $y = r\sin\theta = 2\sin\frac{2\pi}{3} = \sqrt{3} \Rightarrow$ Cartesian coordinates are $\left(-1, \sqrt{3}\right)$
   (d) $x = r\cos\theta = 2\cos\frac{7\pi}{3} = 1$, $y = r\sin\theta = 2\sin\frac{7\pi}{3} = \sqrt{3} \Rightarrow$ Cartesian coordinates are $\left(1, \sqrt{3}\right)$
   (e) $x = r\cos\theta = -3\cos\pi = 3$, $y = r\sin\theta = -3\sin\pi = 0 \Rightarrow$ Cartesian coordinates are $(3, 0)$
   (f) $x = r\cos\theta = 2\cos\frac{\pi}{3} = 1$, $y = r\sin\theta = 2\sin\frac{\pi}{3} = \sqrt{3} \Rightarrow$ Cartesian coordinates are $\left(1, \sqrt{3}\right)$
   (g) $x = r\cos\theta = -3\cos 2\pi = -3$, $y = r\sin\theta = -3\sin 2\pi = 0 \Rightarrow$ Cartesian coordinates are $(-3, 0)$
   (h) $x = r\cos\theta = -2\cos\left(-\frac{\pi}{3}\right) = -1$, $y = r\sin\theta = -2\sin\left(-\frac{\pi}{3}\right) = \sqrt{3} \Rightarrow$ Cartesian coordinates are $\left(-1, \sqrt{3}\right)$

6. (a) $x = \sqrt{2}\cos\frac{\pi}{4} = 1$, $y = \sqrt{2}\sin\frac{\pi}{4} = 1 \Rightarrow$ Cartesian coordinates are $(1, 1)$
   (b) $x = 1\cos 0 = 1$, $y = 1\sin 0 = 0 \Rightarrow$ Cartesian coordinates are $(1, 0)$
   (c) $x = 0\cos\frac{\pi}{2} = 0$, $y = 0\sin\frac{\pi}{2} = 0 \Rightarrow$ Cartesian coordinates are $(0, 0)$
   (d) $x = -\sqrt{2}\cos\left(\frac{\pi}{4}\right) = -1$, $y = -\sqrt{2}\sin\left(\frac{\pi}{4}\right) = -1 \Rightarrow$ Cartesian coordinates are $(-1, -1)$
   (e) $x = -3\cos\frac{5\pi}{6} = \frac{3\sqrt{3}}{2}$, $y = -3\sin\frac{5\pi}{6} = -\frac{3}{2} \Rightarrow$ Cartesian coordinates are $\left(\frac{3\sqrt{3}}{2}, -\frac{3}{2}\right)$
   (f) $x = 5\cos\left(\tan^{-1}\frac{4}{3}\right) = 3$, $y = 5\sin\left(\tan^{-1}\frac{4}{3}\right) = 4 \Rightarrow$ Cartesian coordinates are $(3, 4)$
   (g) $x = -1\cos 7\pi = 1$, $y = -1\sin 7\pi = 0 \Rightarrow$ Cartesian coordinates are $(1, 0)$
   (h) $x = 2\sqrt{3}\cos\frac{2\pi}{3} = -\sqrt{3}$, $y = 2\sqrt{3}\sin\frac{2\pi}{3} = 3 \Rightarrow$ Cartesian coordinates are $\left(-\sqrt{3}, 3\right)$

7.

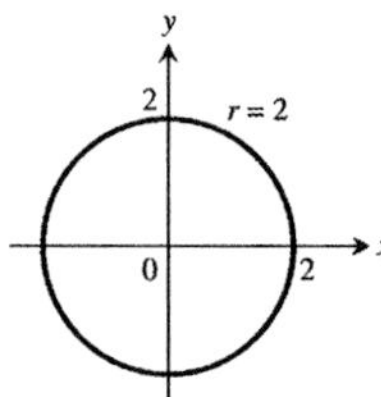

8.

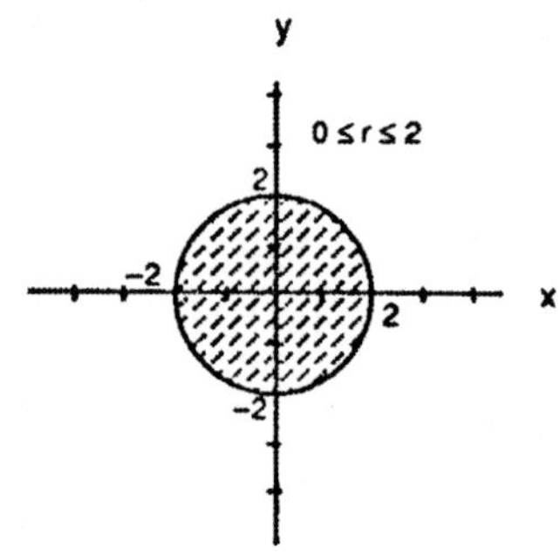

9.

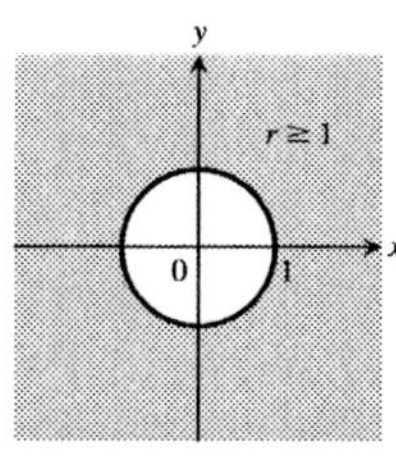

10.

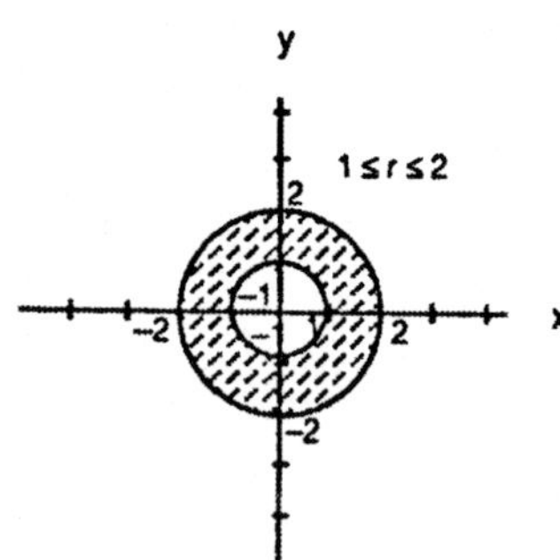

11.

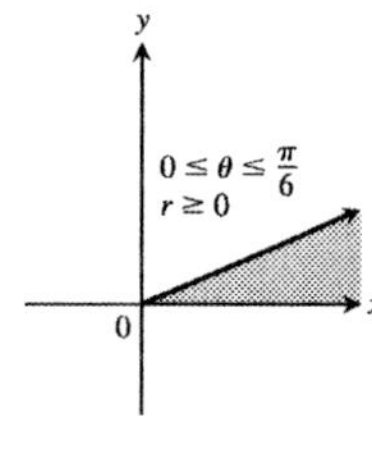

12.

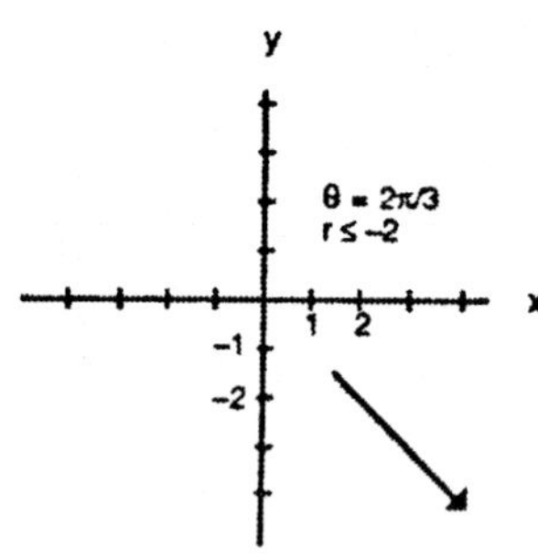

13.

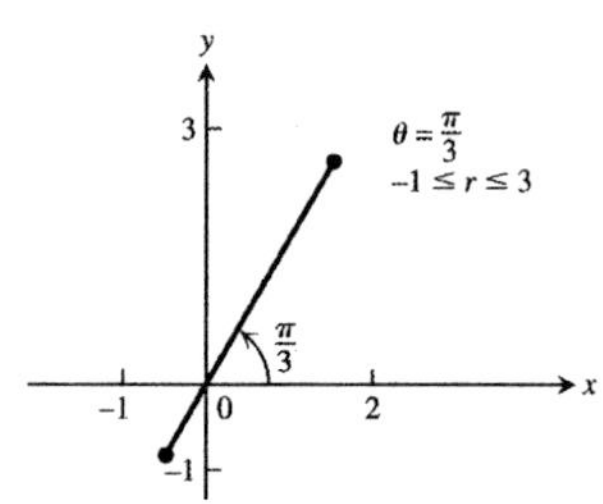

14.

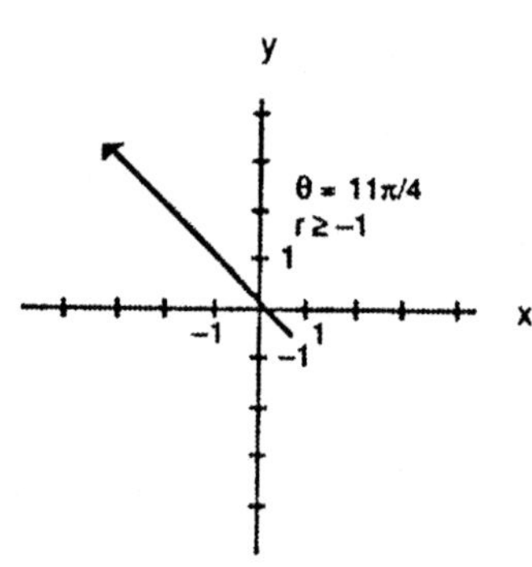

15.

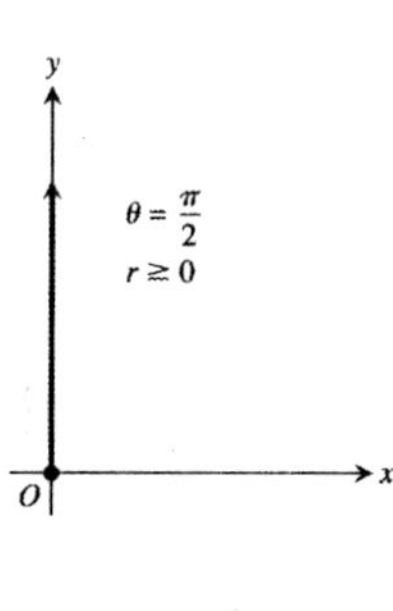

16.

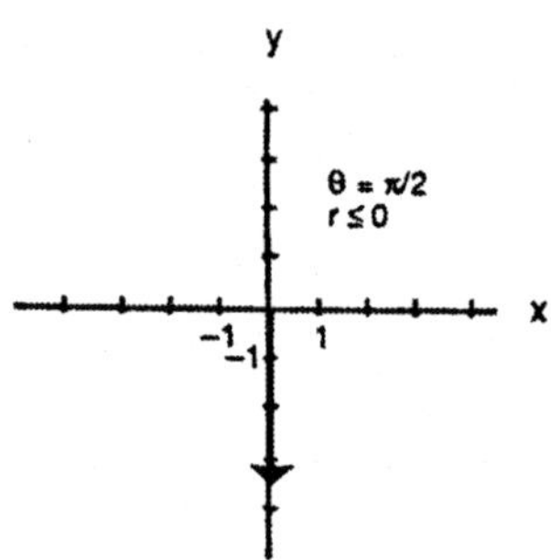

17.

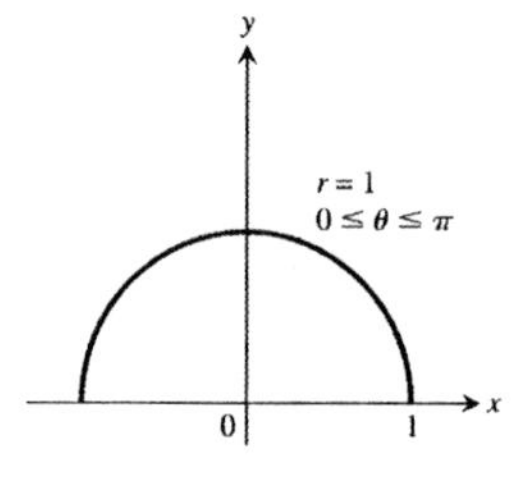

18.

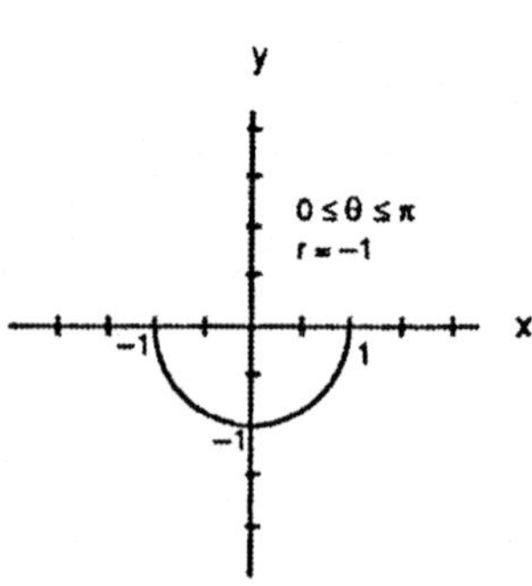

19.

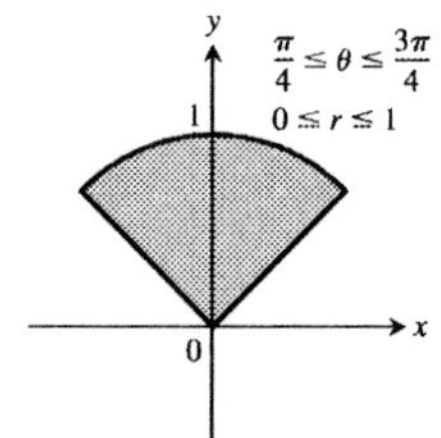

20.

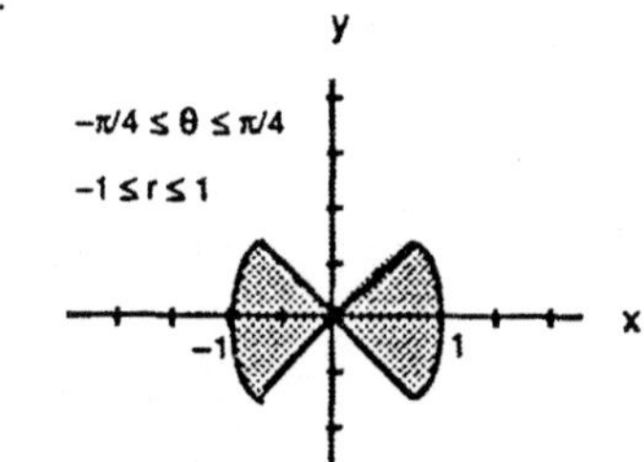

21.

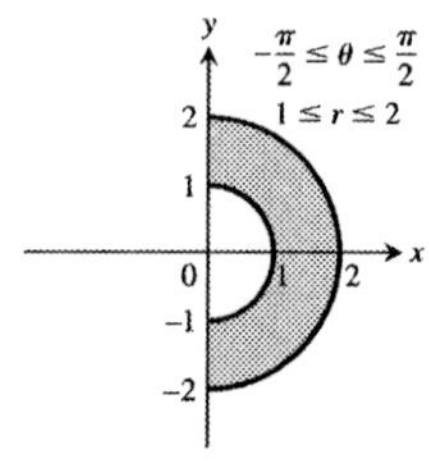

22.

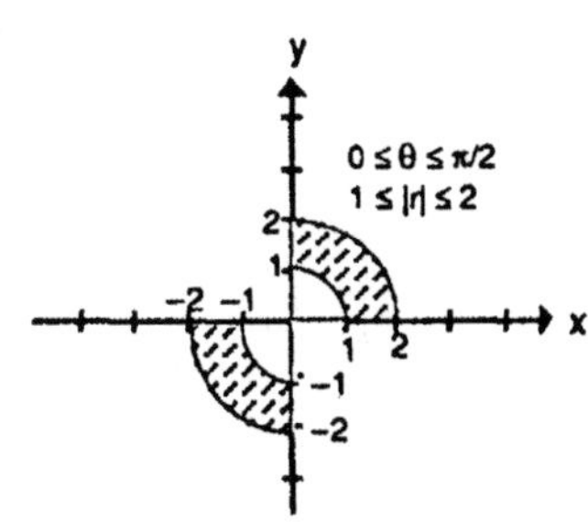

23. $r \cos \theta = 2 \Rightarrow x = 2$, vertical line through $(2, 0)$

24. $r \sin \theta = -1 \Rightarrow y = -1$, horizontal line through $(0, -1)$

25. $r \sin \theta = 0 \Rightarrow y = 0$, the x-axis

26. $r \cos \theta = 0 \Rightarrow x = 0$, the y-axis

27. $r = 4 \csc \theta \Rightarrow r = \frac{4}{\sin \theta} \Rightarrow r \sin \theta = 4 \Rightarrow y = 4$, a horizontal line through $(0, 4)$

28. $r = -3 \sec \theta \Rightarrow r = \frac{-3}{\cos \theta} \Rightarrow r \cos \theta = -3 \Rightarrow x = -3$, a vertical line through $(-3, 0)$

29. $r \cos \theta + r \sin \theta = 1 \Rightarrow x + y = 1$, line with slope $m = -1$ and intercept $b = 1$

30. $r \sin \theta = r \cos \theta \Rightarrow y = x$, line with slope $m = 1$ and intercept $b = 0$

31. $r^2 = 1 \Rightarrow x^2 + y^2 = 1$, circle with center $C = (0, 0)$ and radius 1

32. $r^2 = 4r \sin \theta \Rightarrow x^2 + y^2 = 4y \Rightarrow x^2 + y^2 - 4y + 4 = 4 \Rightarrow x^2 + (y - 2)^2 = 4$, circle with center $C = (0, 2)$ and radius 2

33. $r = \frac{5}{\sin \theta - 2 \cos \theta} \Rightarrow r \sin \theta - 2r \cos \theta = 5 \Rightarrow y - 2x = 5$, line with slope $m = 2$ and intercept $b = 5$

34. $r^2 \sin 2\theta = 2 \Rightarrow 2r^2 \sin \theta \cos \theta = 2 \Rightarrow (r \sin \theta)(r \cos \theta) = 1 \Rightarrow xy = 1$, hyperbola with focal axis $y = x$

35. $r = \cot \theta \csc \theta = \left(\frac{\cos \theta}{\sin \theta}\right)\left(\frac{1}{\sin \theta}\right) \Rightarrow r \sin^2 \theta = \cos \theta \Rightarrow r^2 \sin^2 \theta = r \cos \theta \Rightarrow y^2 = x$, parabola with vertex $(0, 0)$ which opens to the right

36. $r = 4 \tan \theta \sec \theta \Rightarrow r = 4\left(\frac{\sin \theta}{\cos^2 \theta}\right) \Rightarrow r \cos^2 \theta = 4 \sin \theta \Rightarrow r^2 \cos^2 \theta = 4r \sin \theta \Rightarrow x^2 = 4y$, parabola with vertex $= (0, 0)$ which opens upward

37. $r = (\csc \theta)\, e^{r \cos \theta} \Rightarrow r \sin \theta = e^{r \cos \theta} \Rightarrow y = e^x$, graph of the natural exponential function

38. $r \sin \theta = \ln r + \ln \cos \theta = \ln (r \cos \theta) \Rightarrow y = \ln x$, graph of the natural logarithm function

39. $r^2 + 2r^2 \cos \theta \sin \theta = 1 \Rightarrow x^2 + y^2 + 2xy = 1 \Rightarrow x^2 + 2xy + y^2 = 1 \Rightarrow (x + y)^2 = 1 \Rightarrow x + y = \pm 1$, two parallel straight lines of slope -1 and y-intercepts $b = \pm 1$

40. $\cos^2 \theta = \sin^2 \theta \Rightarrow r^2 \cos^2 \theta = r^2 \sin^2 \theta \Rightarrow x^2 = y^2 \Rightarrow |x| = |y| \Rightarrow \pm x = y$, two perpendicular lines through the origin with slopes 1 and -1, respectively.

41. $r^2 = -4r \cos \theta \Rightarrow x^2 + y^2 = -4x \Rightarrow x^2 + 4x + y^2 = 0 \Rightarrow x^2 + 4x + 4 + y^2 = 4 \Rightarrow (x + 2)^2 + y^2 = 4$, a circle with center $C(-2, 0)$ and radius 2

42. $r^2 = -6r\sin\theta \Rightarrow x^2 + y^2 = -6y \Rightarrow x^2 + y^2 + 6y = 0 \Rightarrow x^2 + y^2 + 6y + 9 = 9 \Rightarrow x^2 + (y+3)^2 = 9$, a circle with center $C(0, -3)$ and radius 3

43. $r = 8\sin\theta \Rightarrow r^2 = 8r\sin\theta \Rightarrow x^2 + y^2 = 8y \Rightarrow x^2 + y^2 - 8y = 0 \Rightarrow x^2 + y^2 - 8y + 16 = 16$ $\Rightarrow x^2 + (y-4)^2 = 16$, a circle with center $C(0, 4)$ and radius 4

44. $r = 3\cos\theta \Rightarrow r^2 = 3r\cos\theta \Rightarrow x^2 + y^2 = 3x \Rightarrow x^2 + y^2 - 3x = 0 \Rightarrow x^2 - 3x + \frac{9}{4} + y^2 = \frac{9}{4}$ $\Rightarrow \left(x - \frac{3}{2}\right)^2 + y^2 = \frac{9}{4}$, a circle with center $C\left(\frac{3}{2}, 0\right)$ and radius $\frac{3}{2}$

45. $r = 2\cos\theta + 2\sin\theta \Rightarrow r^2 = 2r\cos\theta + 2r\sin\theta \Rightarrow x^2 + y^2 = 2x + 2y \Rightarrow x^2 - 2x + y^2 - 2y = 0$ $\Rightarrow (x-1)^2 + (y-1)^2 = 2$, a circle with center $C(1, 1)$ and radius $\sqrt{2}$

46. $r = 2\cos\theta - \sin\theta \Rightarrow r^2 = 2r\cos\theta - r\sin\theta \Rightarrow x^2 + y^2 = 2x - y \Rightarrow x^2 - 2x + y^2 + y = 0$ $\Rightarrow (x-1)^2 + \left(y + \frac{1}{2}\right)^2 = \frac{5}{4}$, a circle with center $C\left(1, -\frac{1}{2}\right)$ and radius $\frac{\sqrt{5}}{2}$

47. $r\sin\left(\theta + \frac{\pi}{6}\right) = 2 \Rightarrow r\left(\sin\theta\cos\frac{\pi}{6} + \cos\theta\sin\frac{\pi}{6}\right) = 2 \Rightarrow \frac{\sqrt{3}}{2}r\sin\theta + \frac{1}{2}r\cos\theta = 2 \Rightarrow \frac{\sqrt{3}}{2}y + \frac{1}{2}x = 2$ $\Rightarrow \sqrt{3}y + x = 4$, line with slope $m = -\frac{1}{\sqrt{3}}$ and intercept $b = \frac{4}{\sqrt{3}}$

48. $r\sin\left(\frac{2\pi}{3} - \theta\right) = 5 \Rightarrow r\left(\sin\frac{2\pi}{3}\cos\theta - \cos\frac{2\pi}{3}\sin\theta\right) = 5 \Rightarrow \frac{\sqrt{3}}{2}r\cos\theta + \frac{1}{2}r\sin\theta = 5 \Rightarrow \frac{\sqrt{3}}{2}x + \frac{1}{2}y = 5$ $\Rightarrow \sqrt{3}x + y = 10$, line with slope $m = -\sqrt{3}$ and intercept $b = 10$

49. $x = 7 \Rightarrow r\cos\theta = 7$

50. $y = 1 \Rightarrow r\sin\theta = 1$

51. $x = y \Rightarrow r\cos\theta = r\sin\theta \Rightarrow \theta = \frac{\pi}{4}$

52. $x - y = 3 \Rightarrow r\cos\theta - r\sin\theta = 3$

53. $x^2 + y^2 = 4 \Rightarrow r^2 = 4 \Rightarrow r = 2$ or $r = -2$

54. $x^2 - y^2 = 1 \Rightarrow r^2\cos^2\theta - r^2\sin^2\theta = 1 \Rightarrow r^2(\cos^2\theta - \sin^2\theta) = 1 \Rightarrow r^2\cos 2\theta = 1$

55. $\frac{x^2}{9} + \frac{y^2}{4} = 1 \Rightarrow 4x^2 + 9y^2 = 36 \Rightarrow 4r^2\cos^2\theta + 9r^2\sin^2\theta = 36$

56. $xy = 2 \Rightarrow (r\cos\theta)(r\sin\theta) = 2 \Rightarrow r^2\cos\theta\sin\theta = 2 \Rightarrow 2r^2\cos\theta\sin\theta = 4 \Rightarrow r^2\sin 2\theta = 4$

57. $y^2 = 4x \Rightarrow r^2\sin^2\theta = 4r\cos\theta \Rightarrow r\sin^2\theta = 4\cos\theta$

58. $x^2 + xy + y^2 = 1 \Rightarrow x^2 + y^2 + xy = 1 \Rightarrow r^2 + r^2\sin\theta\cos\theta = 1 \Rightarrow r^2(1 + \sin\theta\cos\theta) = 1$

59. $x^2 + (y-2)^2 = 4 \Rightarrow x^2 + y^2 - 4y + 4 = 4 \Rightarrow x^2 + y^2 = 4y \Rightarrow r^2 = 4r\sin\theta \Rightarrow r = 4\sin\theta$

60. $(x-5)^2 + y^2 = 25 \Rightarrow x^2 - 10x + 25 + y^2 = 25 \Rightarrow x^2 + y^2 = 10x \Rightarrow r^2 = 10r\cos\theta \Rightarrow r = 10\cos\theta$

61. $(x-3)^2 + (y+1)^2 = 4 \Rightarrow x^2 - 6x + 9 + y^2 + 2y + 1 = 4 \Rightarrow x^2 + y^2 = 6x - 2y - 6 \Rightarrow r^2 = 6r\cos\theta - 2r\sin\theta - 6$

62. $(x+2)^2 + (y-5)^2 = 16 \Rightarrow x^2 + 4x + 4 + y^2 - 10y + 25 = 16 \Rightarrow x^2 + y^2 = -4x + 10y - 13 \Rightarrow r^2$ $= -4r\cos\theta + 10r\sin\theta - 13$

63. $(0, \theta)$ where θ is any angle

64. (a) $x = a \Rightarrow r \cos\theta = a \Rightarrow r = \frac{a}{\cos\theta} \Rightarrow r = a \sec\theta$

(b) $y = b \Rightarrow r \sin\theta = b \Rightarrow r = \frac{b}{\sin\theta} \Rightarrow r = b \csc\theta$

10.6 GRAPHING IN POLAR COORDINATES

1. $1 + \cos(-\theta) = 1 + \cos\theta = r \Rightarrow$ symmetric about the x-axis; $1 + \cos(-\theta) \neq -r$ and $1 + \cos(\pi - \theta) = 1 - \cos\theta \neq r \Rightarrow$ not symmetric about the y-axis; therefore not symmetric about the origin

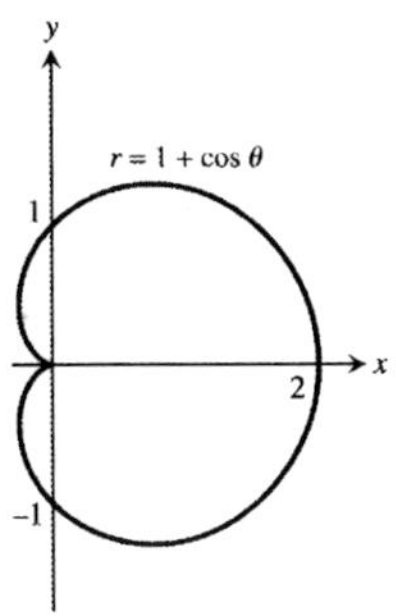

2. $2 - 2\cos(-\theta) = 2 - 2\cos\theta = r \Rightarrow$ symmetric about the x-axis; $2 - 2\cos(-\theta) \neq -r$ and $2 - 2\cos(\pi - \theta) = 2 + 2\cos\theta \neq r \Rightarrow$ not symmetric about the y-axis; therefore not symmetric about the origin

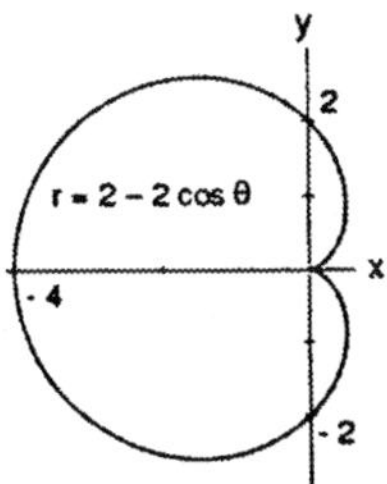

3. $1 - \sin(-\theta) = 1 + \sin\theta \neq r$ and $1 - \sin(\pi - \theta) = 1 - \sin\theta \neq -r \Rightarrow$ not symmetric about the x-axis; $1 - \sin(\pi - \theta) = 1 - \sin\theta = r \Rightarrow$ symmetric about the y-axis; therefore not symmetric about the origin

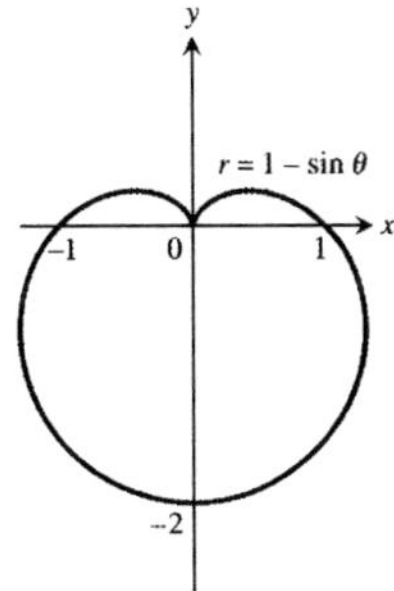

4. $1 + \sin(-\theta) = 1 - \sin\theta \neq r$ and $1 + \sin(\pi - \theta) = 1 + \sin\theta \neq -r \Rightarrow$ not symmetric about the x-axis; $1 + \sin(\pi - \theta) = 1 + \sin\theta = r \Rightarrow$ symmetric about the y-axis; therefore not symmetric about the origin

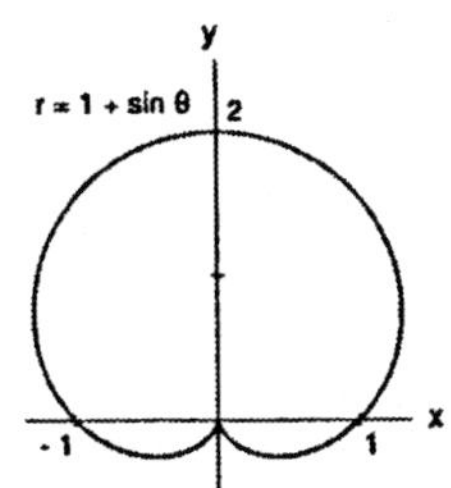

5. $2 + \sin(-\theta) = 2 - \sin\theta \neq r$ and $2 + \sin(\pi - \theta) = 2 + \sin\theta \neq -r \Rightarrow$ not symmetric about the x-axis; $2 + \sin(\pi - \theta) = 2 + \sin\theta = r \Rightarrow$ symmetric about the y-axis; therefore not symmetric about the origin

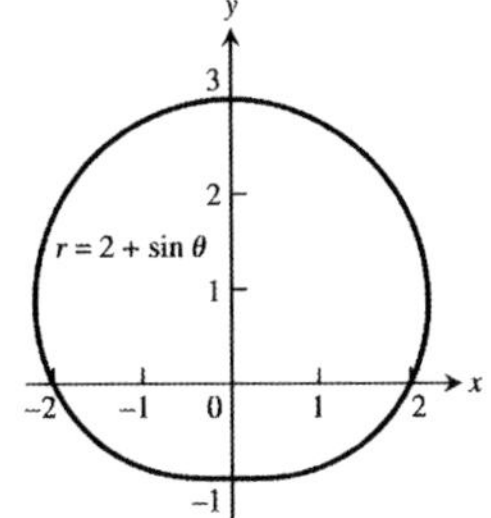

6. $1 + 2\sin(-\theta) = 1 - 2\sin\theta \neq r$ and $1 + 2\sin(\pi - \theta)$ $= 1 + 2\sin\theta \neq -r \Rightarrow$ not symmetric about the x-axis; $1 + 2\sin(\pi - \theta) = 1 + 2\sin\theta = r \Rightarrow$ symmetric about the y-axis; therefore not symmetric about the origin

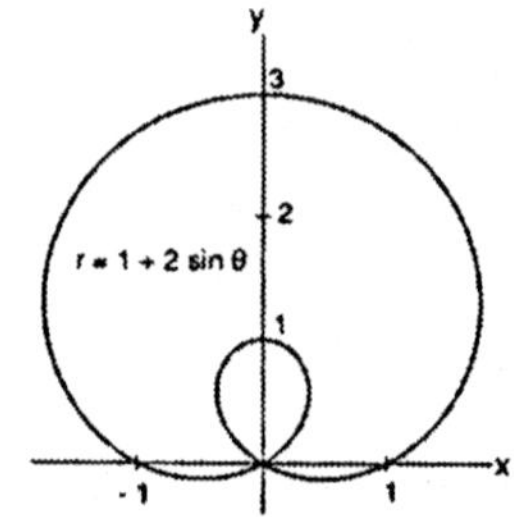

7. $\sin\left(-\frac{\theta}{2}\right) = -\sin\left(\frac{\theta}{2}\right) = -r \Rightarrow$ symmetric about the y-axis; $\sin\left(\frac{2\pi-\theta}{2}\right) = \sin\left(\frac{\theta}{2}\right)$, so the graph is symmetric about the x-axis, and hence the origin.

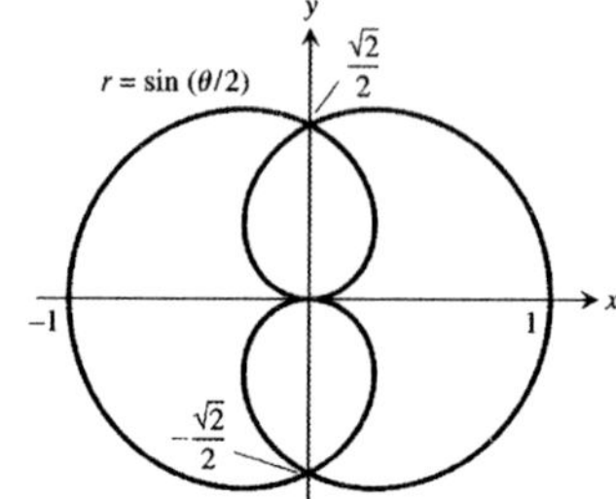

8. $\cos\left(-\frac{\theta}{2}\right) = \cos\left(\frac{\theta}{2}\right) = r \Rightarrow$ symmetric about the x-axis; $\cos\left(\frac{2\pi-\theta}{2}\right) = \cos\left(\frac{\theta}{2}\right)$, so the graph is symmetric about the y-axis, and hence the origin.

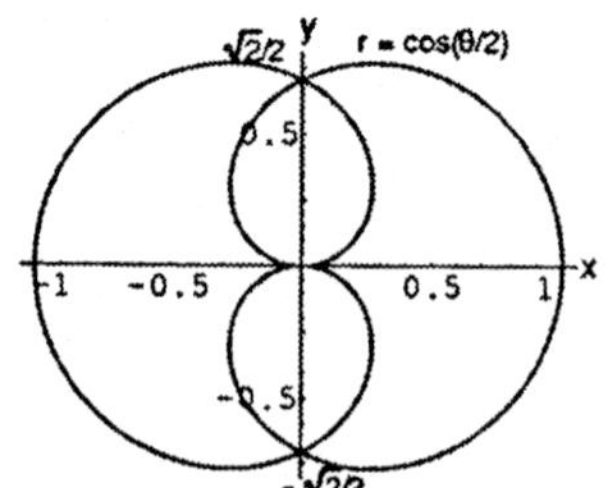

9. $\cos(-\theta) = \cos\theta = r^2 \Rightarrow (r, -\theta)$ and $(-r, -\theta)$ are on the graph when (r, θ) is on the graph $\Rightarrow$ symmetric about the x-axis and the y-axis; therefore symmetric about the origin

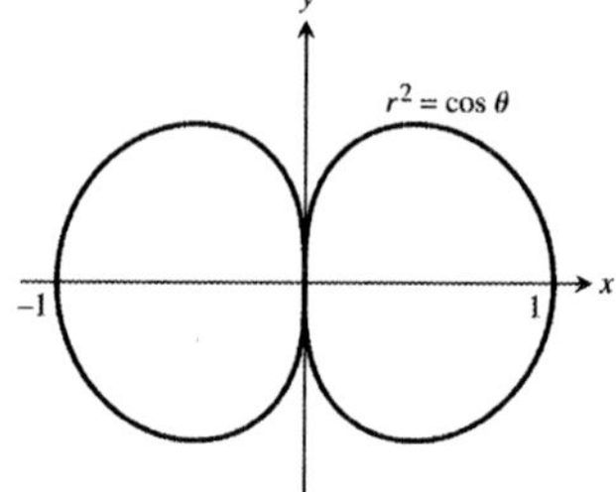

10. $\sin(\pi - \theta) = \sin\theta = r^2 \Rightarrow (r, \pi - \theta)$ and $(-r, \pi - \theta)$ are on the graph when (r, θ) is on the graph $\Rightarrow$ symmetric about the y-axis and the x-axis; therefore symmetric about the origin

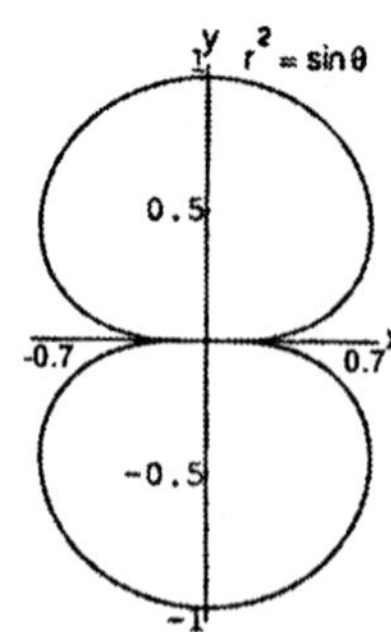

11. $-\sin(\pi-\theta) = -\sin\theta = r^2 \Rightarrow (r, \pi-\theta)$ and $(-r, \pi-\theta)$ are on the graph when (r, θ) is on the graph $\Rightarrow$ symmetric about the y-axis and the x-axis; therefore symmetric about the origin

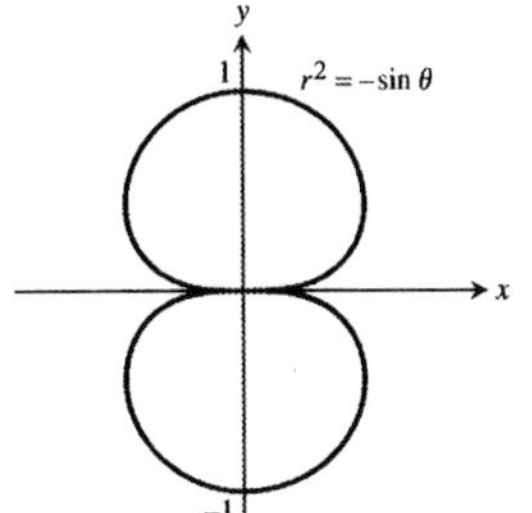

12. $-\cos(-\theta) = -\cos\theta = r^2 \Rightarrow (r, -\theta)$ and $(-r, -\theta)$ are on the graph when (r, θ) is on the graph $\Rightarrow$ symmetric about the x-axis and the y-axis; therefore symmetric about the origin

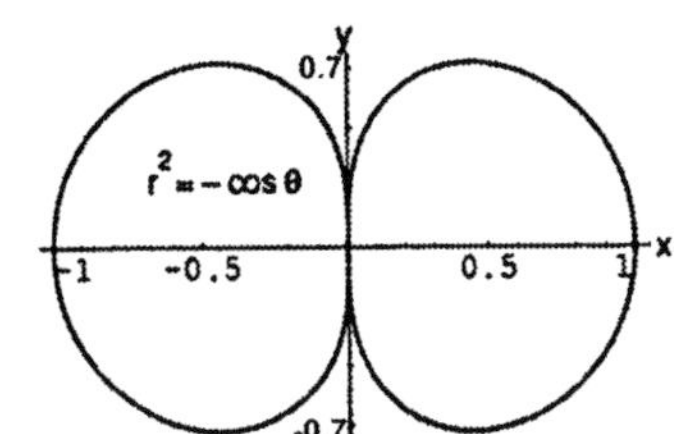

13. Since $(\pm r, -\theta)$ are on the graph when (r, θ) is on the graph $\left((\pm r)^2 = 4\cos 2(-\theta) \Rightarrow r^2 = 4\cos 2\theta\right)$, the graph is symmetric about the x-axis and the y-axis $\Rightarrow$ the graph is symmetric about the origin

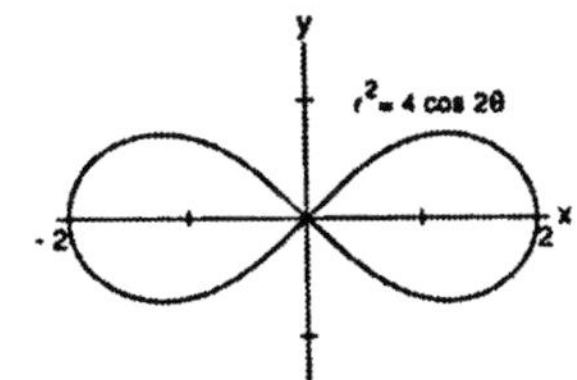

14. Since (r, θ) on the graph $\Rightarrow (-r, \theta)$ is on the graph $\left((\pm r)^2 = 4\sin 2\theta \Rightarrow r^2 = 4\sin 2\theta\right)$, the graph is symmetric about the origin. But $4\sin 2(-\theta) = -4\sin 2\theta \neq r^2$ and $4\sin 2(\pi-\theta) = 4\sin(2\pi - 2\theta) = 4\sin(-2\theta) = -4\sin 2\theta \neq r^2 \Rightarrow$ the graph is not symmetric about the x-axis; therefore the graph is not symmetric about the y-axis

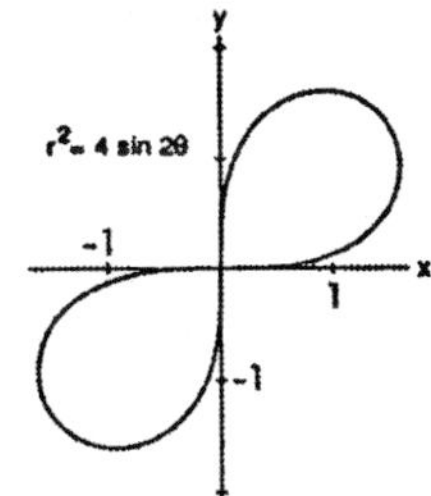

15. Since (r, θ) on the graph $\Rightarrow (-r, \theta)$ is on the graph $\left((\pm r)^2 = -\sin 2\theta \Rightarrow r^2 = -\sin 2\theta\right)$, the graph is symmetric about the origin. But $-\sin 2(-\theta) = -(-\sin 2\theta)$ $\sin 2\theta \neq r^2$ and $-\sin 2(\pi-\theta) = -\sin(2\pi - 2\theta)$ $= -\sin(-2\theta) = -(-\sin 2\theta) = \sin 2\theta \neq r^2 \Rightarrow$ the graph is not symmetric about the x-axis; therefore the graph is not symmetric about the y-axis

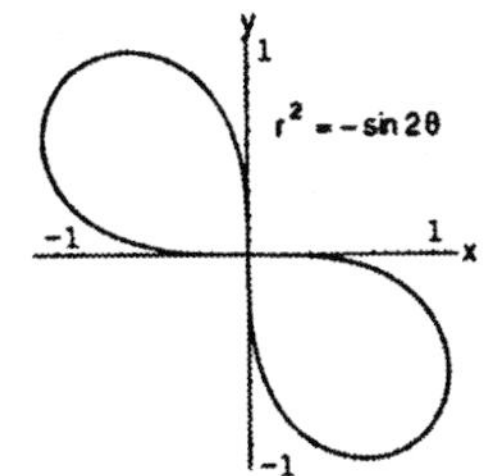

16. Since$(\pm r, -\theta)$ are on the graph when (r, θ) is on the graph $\left((\pm r)^2 = -\cos 2(-\theta) \Rightarrow r^2 = -\cos 2\theta\right)$, the graph is symmetric about the x-axis and the y-axis $\Rightarrow$ the graph is symmetric about the origin.

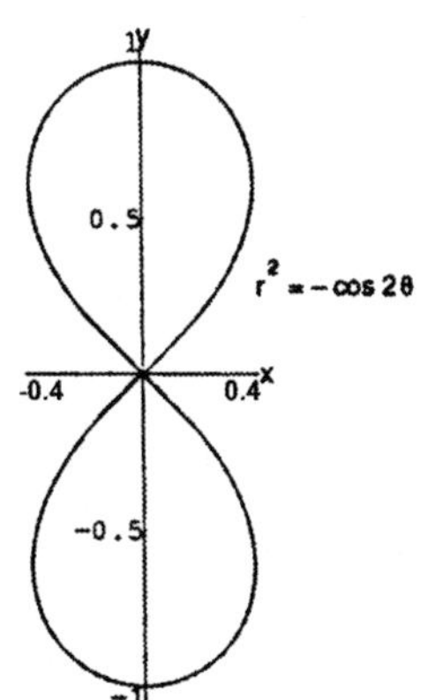

17. $\theta = \frac{\pi}{2} \Rightarrow r = -1 \Rightarrow \left(-1, \frac{\pi}{2}\right)$, and $\theta = -\frac{\pi}{2} \Rightarrow r = -1$
$\Rightarrow \left(-1, -\frac{\pi}{2}\right)$; $r' = \frac{dr}{d\theta} = -\sin\theta$; Slope $= \frac{r'\sin\theta + r\cos\theta}{r'\cos\theta - r\sin\theta}$
$= \frac{-\sin^2\theta + r\cos\theta}{-\sin\theta\cos\theta - r\sin\theta} \Rightarrow$ Slope at $\left(-1, \frac{\pi}{2}\right)$ is
$\frac{-\sin^2\left(\frac{\pi}{2}\right) + (-1)\cos\frac{\pi}{2}}{-\sin\frac{\pi}{2}\cos\frac{\pi}{2} - (-1)\sin\frac{\pi}{2}} = -1$; Slope at $\left(-1, -\frac{\pi}{2}\right)$ is
$\frac{-\sin^2\left(-\frac{\pi}{2}\right) + (-1)\cos\left(-\frac{\pi}{2}\right)}{-\sin\left(-\frac{\pi}{2}\right)\cos\left(-\frac{\pi}{2}\right) - (-1)\sin\left(-\frac{\pi}{2}\right)} = 1$

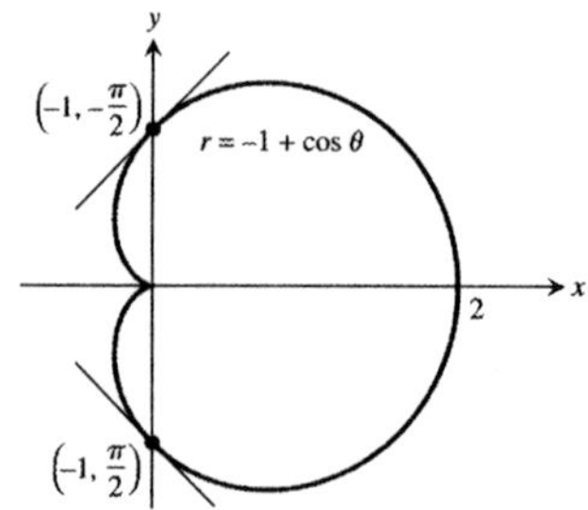

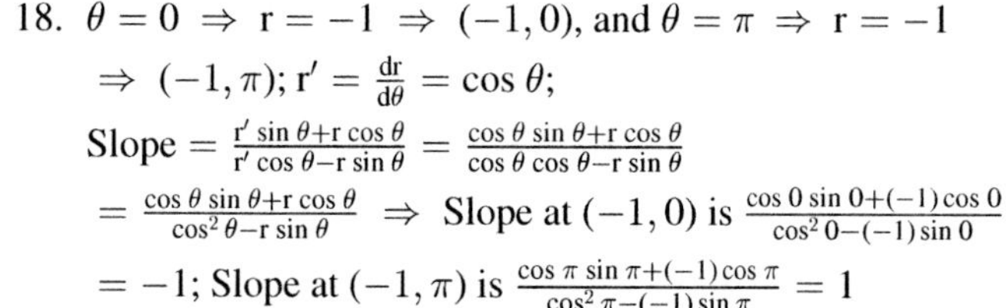

18. $\theta = 0 \Rightarrow r = -1 \Rightarrow (-1, 0)$, and $\theta = \pi \Rightarrow r = -1$
$\Rightarrow (-1, \pi)$; $r' = \frac{dr}{d\theta} = \cos\theta$;
Slope $= \frac{r'\sin\theta + r\cos\theta}{r'\cos\theta - r\sin\theta} = \frac{\cos\theta\sin\theta + r\cos\theta}{\cos\theta\cos\theta - r\sin\theta}$
$= \frac{\cos\theta\sin\theta + r\cos\theta}{\cos^2\theta - r\sin\theta} \Rightarrow$ Slope at $(-1, 0)$ is $\frac{\cos 0\sin 0 + (-1)\cos 0}{\cos^2 0 - (-1)\sin 0}$
$= -1$; Slope at $(-1, \pi)$ is $\frac{\cos\pi\sin\pi + (-1)\cos\pi}{\cos^2\pi - (-1)\sin\pi} = 1$

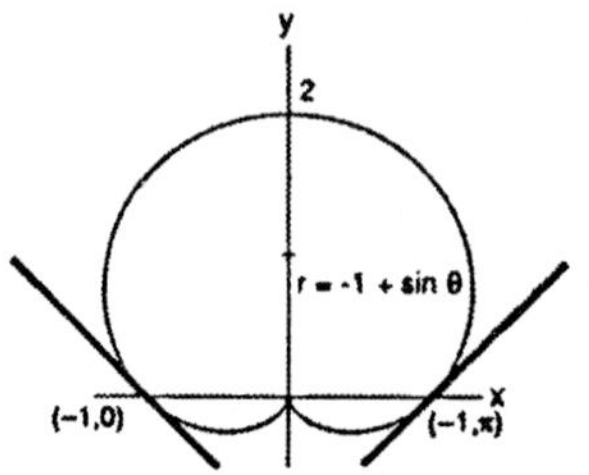

19. $\theta = \frac{\pi}{4} \Rightarrow r = 1 \Rightarrow \left(1, \frac{\pi}{4}\right)$; $\theta = -\frac{\pi}{4} \Rightarrow r = -1$
$\Rightarrow \left(-1, -\frac{\pi}{4}\right)$; $\theta = \frac{3\pi}{4} \Rightarrow r = -1 \Rightarrow \left(-1, \frac{3\pi}{4}\right)$;
$\theta = -\frac{3\pi}{4} \Rightarrow r = 1 \Rightarrow \left(1, -\frac{3\pi}{4}\right)$;
$r' = \frac{dr}{d\theta} = 2\cos 2\theta$;
Slope $= \frac{r'\sin\theta + r\cos\theta}{r'\cos\theta - r\sin\theta} = \frac{2\cos 2\theta\sin\theta + r\cos\theta}{2\cos 2\theta\cos\theta - r\sin\theta}$
$\Rightarrow$ Slope at $\left(1, \frac{\pi}{4}\right)$ is $\frac{2\cos\left(\frac{\pi}{2}\right)\sin\left(\frac{\pi}{4}\right) + (1)\cos\left(\frac{\pi}{4}\right)}{2\cos\left(\frac{\pi}{2}\right)\cos\left(\frac{\pi}{4}\right) - (1)\sin\left(\frac{\pi}{4}\right)} = -1$;

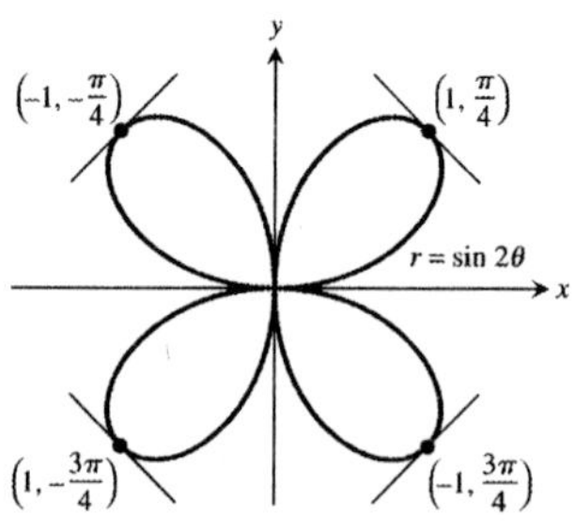

Slope at $\left(-1, -\frac{\pi}{4}\right)$ is $\frac{2\cos\left(-\frac{\pi}{2}\right)\sin\left(-\frac{\pi}{4}\right) + (-1)\cos\left(-\frac{\pi}{4}\right)}{2\cos\left(-\frac{\pi}{2}\right)\cos\left(-\frac{\pi}{4}\right) - (-1)\sin\left(-\frac{\pi}{4}\right)} = 1$;

Slope at $\left(-1, \frac{3\pi}{4}\right)$ is $\frac{2\cos\left(\frac{3\pi}{2}\right)\sin\left(\frac{3\pi}{4}\right) + (-1)\cos\left(\frac{3\pi}{4}\right)}{2\cos\left(\frac{3\pi}{2}\right)\cos\left(\frac{3\pi}{4}\right) - (-1)\sin\left(\frac{3\pi}{4}\right)} = 1$;

Slope at $\left(1, -\frac{3\pi}{4}\right)$ is $\frac{2\cos\left(-\frac{3\pi}{2}\right)\sin\left(-\frac{3\pi}{4}\right) + (1)\cos\left(-\frac{3\pi}{4}\right)}{2\cos\left(-\frac{3\pi}{2}\right)\cos\left(-\frac{3\pi}{4}\right) - (1)\sin\left(-\frac{3\pi}{4}\right)} = -1$

20. $\theta = 0 \Rightarrow r = 1 \Rightarrow (1, 0)$; $\theta = \frac{\pi}{2} \Rightarrow r = -1 \Rightarrow \left(-1, \frac{\pi}{2}\right)$;
$\theta = -\frac{\pi}{2} \Rightarrow r = -1 \Rightarrow \left(-1, -\frac{\pi}{2}\right)$; $\theta = \pi \Rightarrow r = 1$
$\Rightarrow (1, \pi)$; $r' = \frac{dr}{d\theta} = -2\sin 2\theta$;

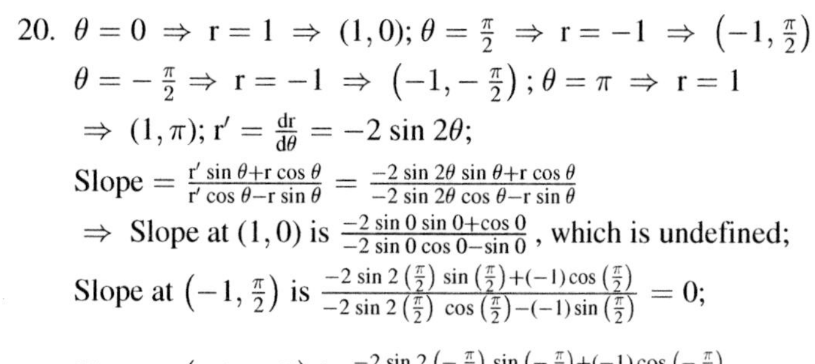

Slope $= \frac{r'\sin\theta + r\cos\theta}{r'\cos\theta - r\sin\theta} = \frac{-2\sin 2\theta\sin\theta + r\cos\theta}{-2\sin 2\theta\cos\theta - r\sin\theta}$
$\Rightarrow$ Slope at $(1, 0)$ is $\frac{-2\sin 0\sin 0 + \cos 0}{-2\sin 0\cos 0 - \sin 0}$, which is undefined;
Slope at $\left(-1, \frac{\pi}{2}\right)$ is $\frac{-2\sin 2\left(\frac{\pi}{2}\right)\sin\left(\frac{\pi}{2}\right) + (-1)\cos\left(\frac{\pi}{2}\right)}{-2\sin 2\left(\frac{\pi}{2}\right)\cos\left(\frac{\pi}{2}\right) - (-1)\sin\left(\frac{\pi}{2}\right)} = 0$;

Slope at $\left(-1, -\frac{\pi}{2}\right)$ is $\frac{-2\sin 2\left(-\frac{\pi}{2}\right)\sin\left(-\frac{\pi}{2}\right) + (-1)\cos\left(-\frac{\pi}{2}\right)}{-2\sin 2\left(-\frac{\pi}{2}\right)\cos\left(-\frac{\pi}{2}\right) - (-1)\sin\left(-\frac{\pi}{2}\right)} = 0$;
Slope at $(1, \pi)$ is $\frac{-2\sin 2\pi\sin\pi + \cos\pi}{-2\sin 2\pi\cos\pi - \sin\pi}$, which is undefined

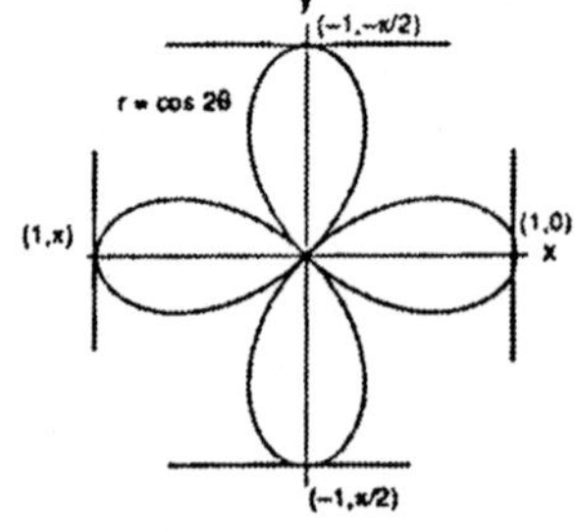

21. (a)

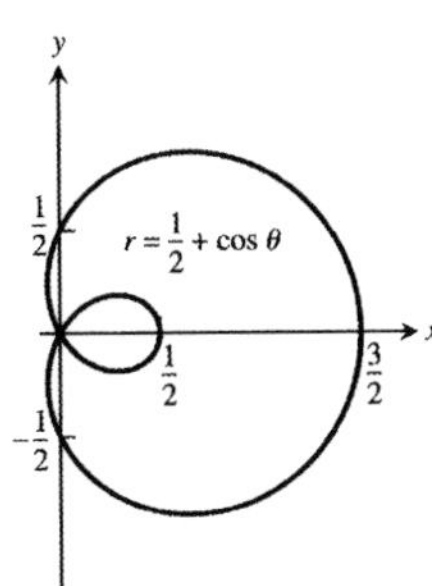

(b)

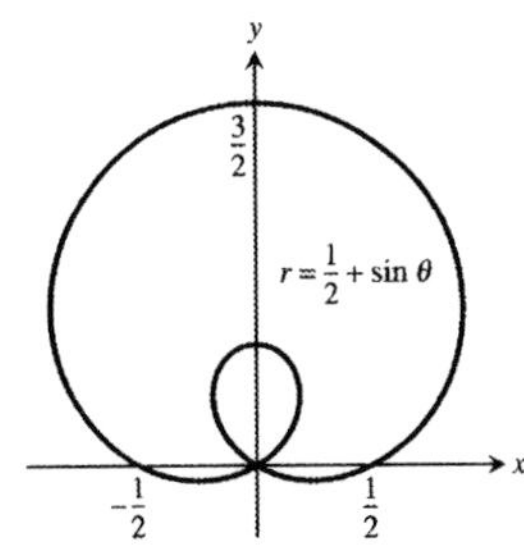

22. (a)

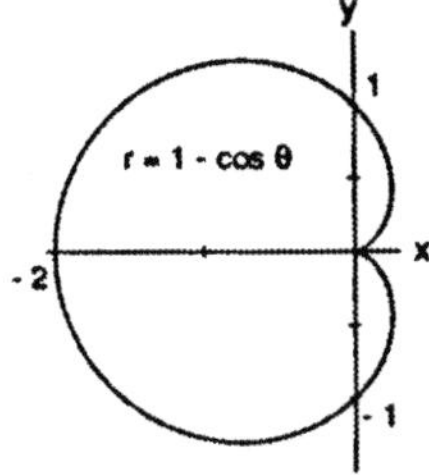

(b)

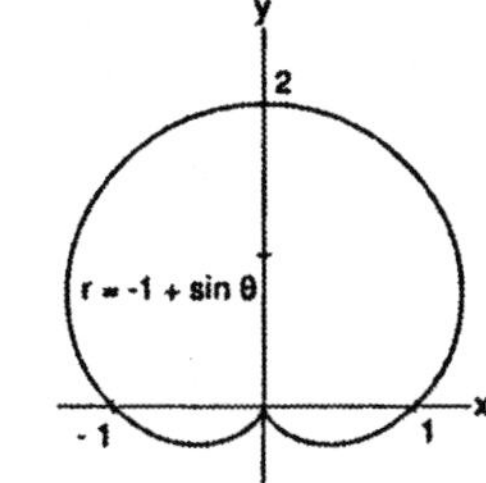

23. (a)

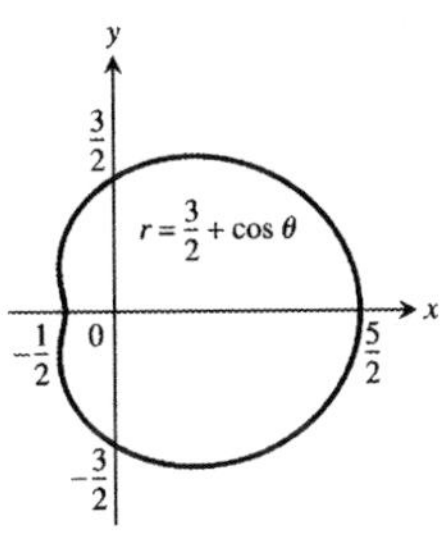

(b)

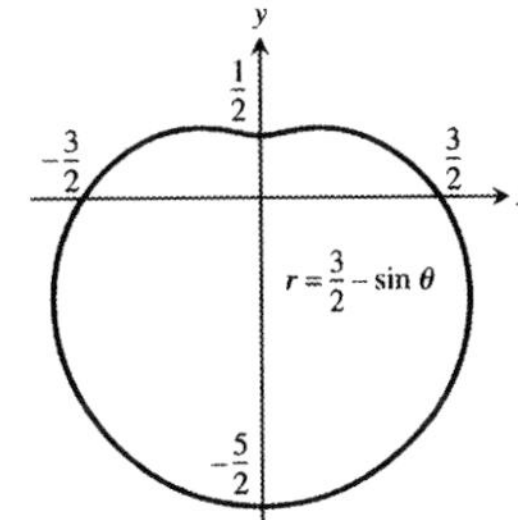

24. (a)

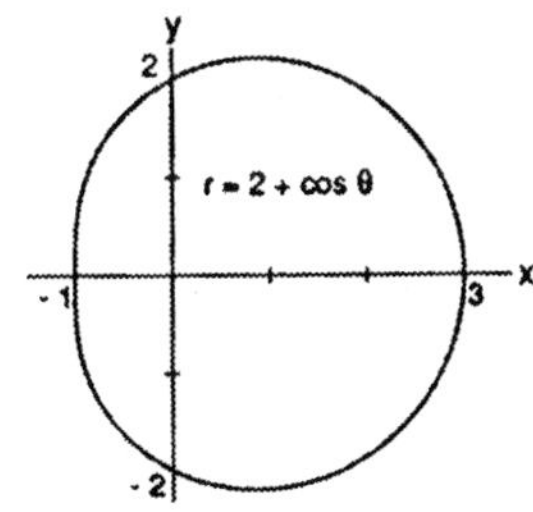

(b)

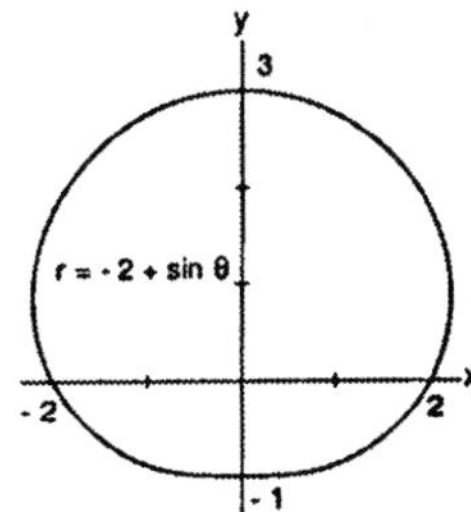

25.

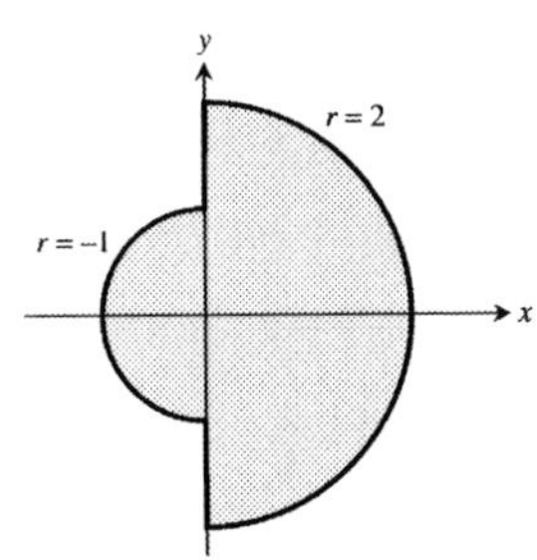

26. $r = 2 \sec \theta \Rightarrow r = \frac{2}{\cos \theta} \Rightarrow r \cos \theta = 2 \Rightarrow x = 2$

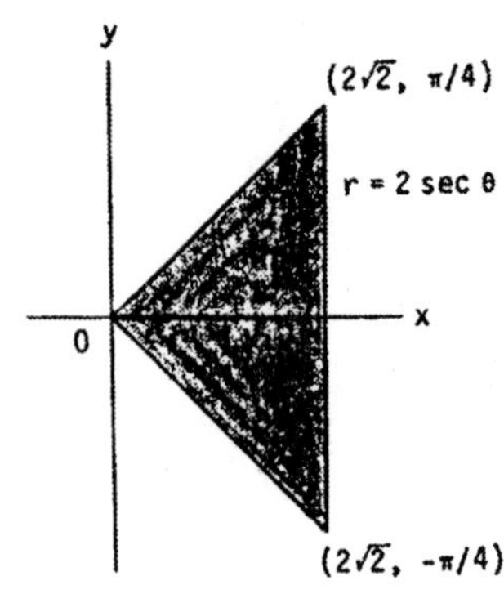

27.

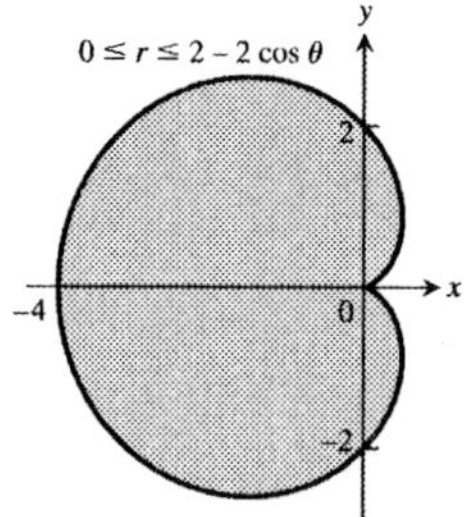

28.

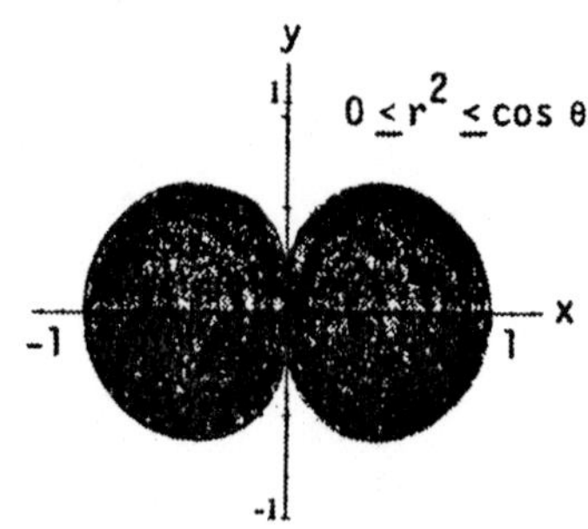

29. $\left(2, \frac{3\pi}{4}\right)$ is the same point as $\left(-2, -\frac{\pi}{4}\right)$; $r = 2 \sin 2\left(-\frac{\pi}{4}\right) = 2 \sin\left(-\frac{\pi}{2}\right) = -2 \Rightarrow \left(-2, -\frac{\pi}{4}\right)$ is on the graph $\Rightarrow \left(2, \frac{3\pi}{4}\right)$ is on the graph

30. $\left(\frac{1}{2}, \frac{3\pi}{2}\right)$ is the same point as $\left(-\frac{1}{2}, \frac{\pi}{2}\right)$; $r = -\sin\left(\frac{\left(\frac{\pi}{2}\right)}{3}\right) = -\sin \frac{\pi}{6} = -\frac{1}{2} \Rightarrow \left(-\frac{1}{2}, \frac{\pi}{2}\right)$ is on the graph $\Rightarrow \left(\frac{1}{2}, \frac{3\pi}{2}\right)$ is on the graph

31. $1 + \cos \theta = 1 - \cos \theta \Rightarrow \cos \theta = 0 \Rightarrow \theta = \frac{\pi}{2}, \frac{3\pi}{2}$ $\Rightarrow r = 1$; points of intersection are $\left(1, \frac{\pi}{2}\right)$ and $\left(1, \frac{3\pi}{2}\right)$. The point of intersection $(0, 0)$ is found by graphing.

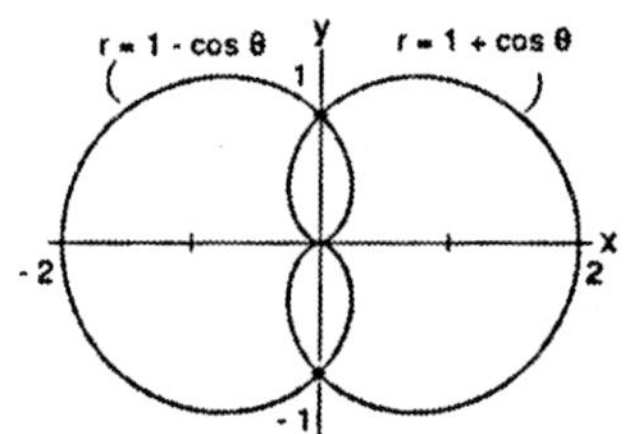

32. $1 + \sin \theta = 1 - \sin \theta \Rightarrow \sin \theta = 0 \Rightarrow \theta = 0, \pi \Rightarrow r = 1$; points of intersection are $(1, 0)$ and $(1, \pi)$. The point of intersection $(0, 0)$ is found by graphing.

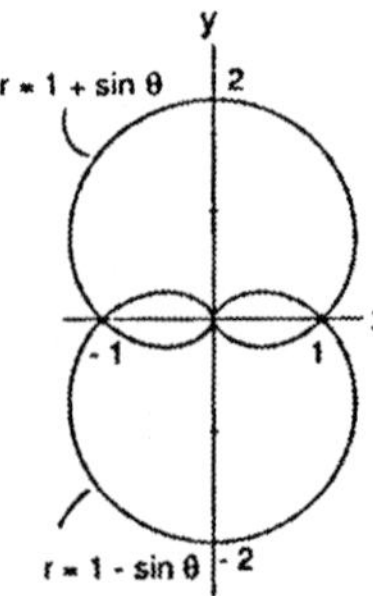

33. $2 \sin \theta = 2 \sin 2\theta \Rightarrow \sin \theta = \sin 2\theta \Rightarrow \sin \theta$
$= 2 \sin \theta \cos \theta \Rightarrow \sin \theta - 2 \sin \theta \cos \theta = 0$
$\Rightarrow (\sin \theta)(1 - 2 \cos \theta) = 0 \Rightarrow \sin \theta = 0$ or $\cos \theta = \frac{1}{2}$
$\Rightarrow \theta = 0, \pi, \frac{\pi}{3}$, or $-\frac{\pi}{3}$; $\theta = 0$ or $\pi \Rightarrow r = 0$,
$\theta = \frac{\pi}{3} \Rightarrow r = \sqrt{3}$, and $\theta = -\frac{\pi}{3} \Rightarrow r = -\sqrt{3}$; points of intersection are $(0, 0)$, $\left(\sqrt{3}, \frac{\pi}{3}\right)$, and $\left(-\sqrt{3}, -\frac{\pi}{3}\right)$

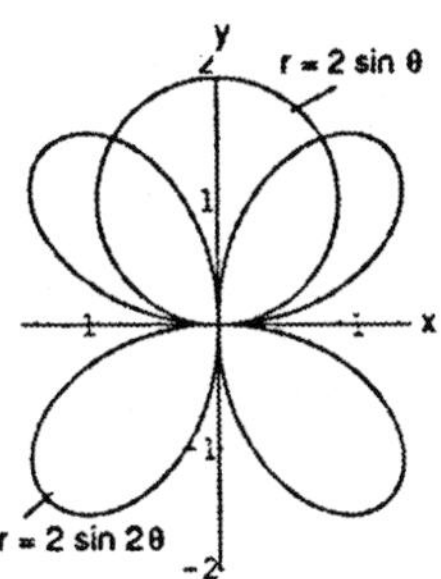

34. $\cos \theta = 1 - \cos \theta \Rightarrow 2 \cos \theta = 1 \Rightarrow \cos \theta = \frac{1}{2}$
$\Rightarrow \theta = \frac{\pi}{3}, -\frac{\pi}{3} \Rightarrow r = \frac{1}{2}$; points of intersection are $\left(\frac{1}{2}, \frac{\pi}{3}\right)$ and $\left(\frac{1}{2}, -\frac{\pi}{3}\right)$. The point $(0, 0)$ is found by graphing.

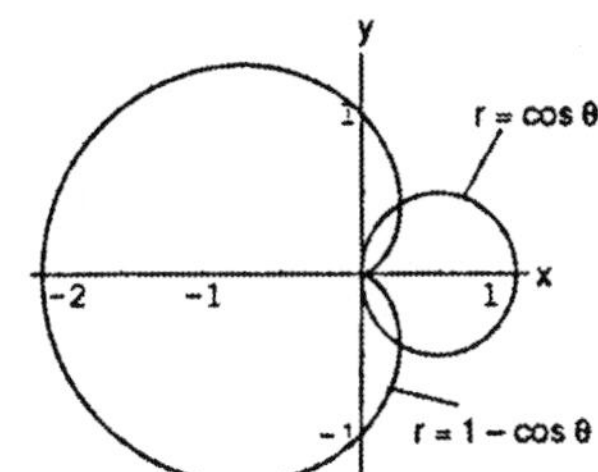

35. $\left(\sqrt{2}\right)^2 = 4 \sin \theta \Rightarrow \frac{1}{2} = \sin \theta \Rightarrow \theta = \frac{\pi}{6}, \frac{5\pi}{6}$; points of intersection are $\left(\sqrt{2}, \frac{\pi}{6}\right)$ and $\left(\sqrt{2}, \frac{5\pi}{6}\right)$. The points $\left(\sqrt{2}, -\frac{\pi}{6}\right)$ and $\left(\sqrt{2}, -\frac{5\pi}{6}\right)$ are found by graphing.

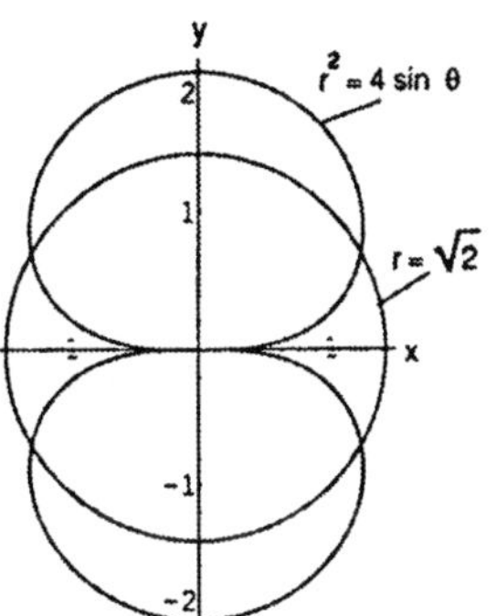

36. $\sqrt{2} \sin \theta = \sqrt{2} \cos \theta \Rightarrow \sin \theta = \cos \theta \Rightarrow \theta = \frac{\pi}{4}, \frac{5\pi}{4}$;
$\theta = \frac{\pi}{4} \Rightarrow r^2 = 1 \Rightarrow r = \pm 1$ and $\theta = \frac{5\pi}{4} \Rightarrow r^2 = -1$
$\Rightarrow$ no solution for r; points of intersection are $\left(\pm 1, \frac{\pi}{4}\right)$.
The points $(0, 0)$ and $\left(\pm 1, \frac{3\pi}{4}\right)$ are found by graphing.

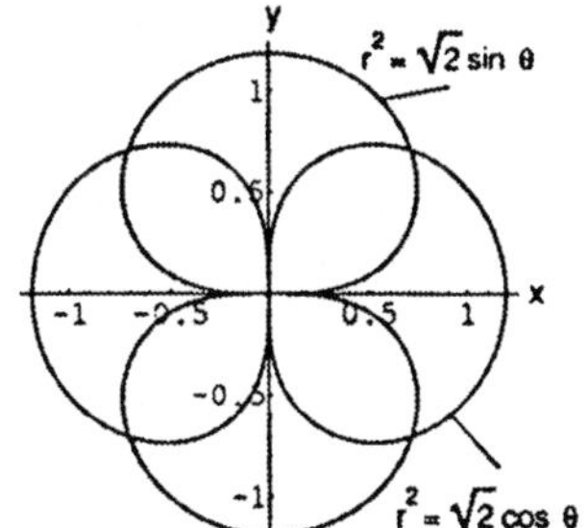

37. $1 = 2 \sin 2\theta \Rightarrow \sin 2\theta = \frac{1}{2} \Rightarrow 2\theta = \frac{\pi}{6}, \frac{5\pi}{6}, \frac{13\pi}{6}, \frac{17\pi}{6}$
$\Rightarrow \theta = \frac{\pi}{12}, \frac{5\pi}{12}, \frac{13\pi}{12}, \frac{17\pi}{12}$; points of intersection are $\left(1, \frac{\pi}{12}\right), \left(1, \frac{5\pi}{12}\right), \left(1, \frac{13\pi}{12}\right)$, and $\left(1, \frac{17\pi}{12}\right)$. No other points are found by graphing.

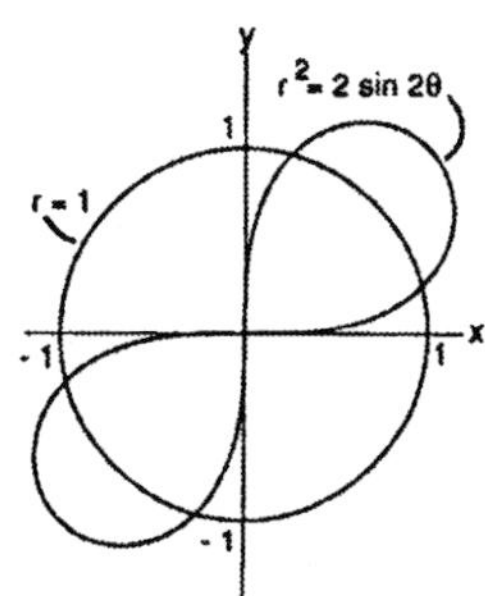

38. $\sqrt{2}\cos 2\theta = \sqrt{2}\sin 2\theta \Rightarrow \cos 2\theta = \sin 2\theta$
$\Rightarrow 2\theta = \frac{\pi}{4}, \frac{5\pi}{4}, \frac{9\pi}{4}, \frac{13\pi}{4} \Rightarrow \theta = \frac{\pi}{8}, \frac{5\pi}{8}, \frac{9\pi}{8}, \frac{13\pi}{8}$;
$\theta = \frac{\pi}{8}, \frac{9\pi}{8} \Rightarrow r^2 = 1 \Rightarrow r = \pm 1; \theta = \frac{5\pi}{8}, \frac{13\pi}{8}$
$\Rightarrow r^2 = -1 \Rightarrow$ no solution for r; points of intersection are $\left(1, \frac{\pi}{8}\right)$ and $\left(1, \frac{9\pi}{8}\right)$. The point of intersection $(0, 0)$ is found by graphing.

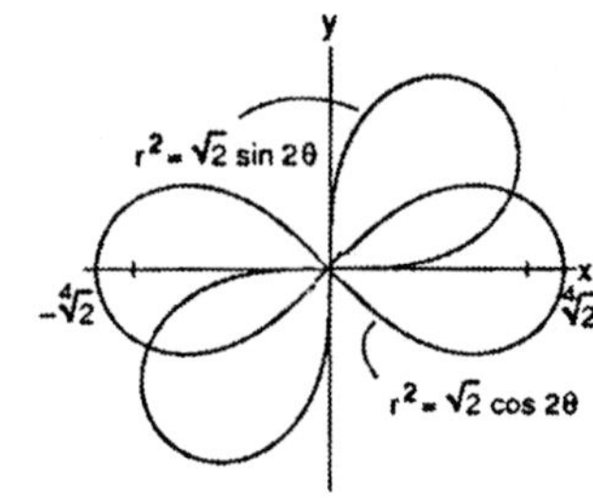

39. $r^2 = \sin 2\theta$ and $r^2 = \cos 2\theta$ are generated completely for $0 \le \theta \le \frac{\pi}{2}$. Then $\sin 2\theta = \cos 2\theta \Rightarrow 2\theta = \frac{\pi}{4}$ is the only solution on that interval $\Rightarrow \theta = \frac{\pi}{8} \Rightarrow r^2 = \sin 2\left(\frac{\pi}{8}\right) = \frac{1}{\sqrt{2}}$
$\Rightarrow r = \pm \frac{1}{\sqrt[4]{2}}$; points of intersection are $\left(\pm \frac{1}{\sqrt[4]{2}}, \frac{\pi}{8}\right)$.
The point of intersection $(0, 0)$ is found by graphing.

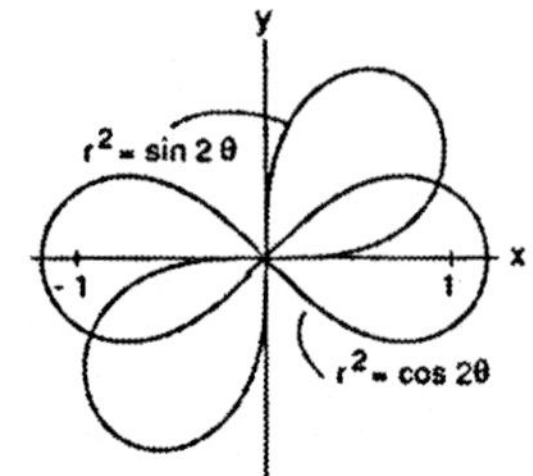

40. $1 - \sin\frac{\theta}{2} = 1 + \cos\frac{\theta}{2} \Rightarrow -\sin\frac{\theta}{2} = \cos\frac{\theta}{2} \Rightarrow \frac{\theta}{2} = \frac{3\pi}{4}, \frac{7\pi}{4}$
$\Rightarrow \theta = \frac{3\pi}{2}, \frac{7\pi}{2}; \theta = \frac{3\pi}{2} \Rightarrow r = 1 + \cos\frac{3\pi}{4} = 1 - \frac{\sqrt{2}}{2}$;
$\theta = \frac{7\pi}{2} \Rightarrow r = 1 + \cos\frac{7\pi}{4} = 1 + \frac{\sqrt{2}}{2}$; points of intersection are $\left(1 - \frac{\sqrt{2}}{2}, \frac{3\pi}{2}\right)$ and $\left(1 + \frac{\sqrt{2}}{2}, \frac{7\pi}{2}\right)$. The three points of intersection $(0, 0)$ and $\left(1 \pm \frac{\sqrt{2}}{2}, \frac{\pi}{2}\right)$ are found by graphing and symmetry.

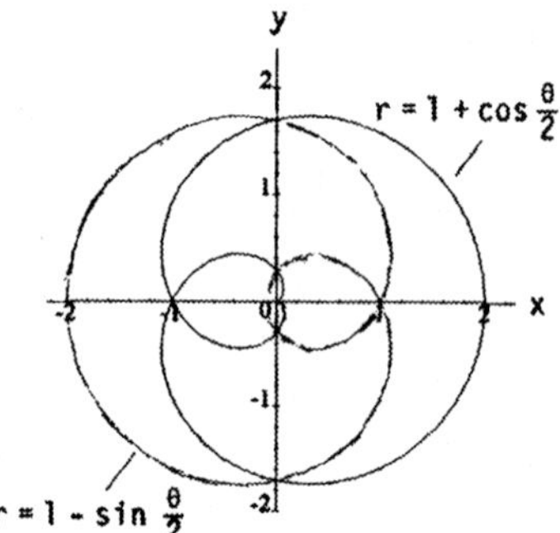

41. $1 = 2\sin 2\theta \Rightarrow \sin 2\theta = \frac{1}{2} \Rightarrow 2\theta = \frac{\pi}{6}, \frac{5\pi}{6}, \frac{13\pi}{6}, \frac{17\pi}{6}$
$\Rightarrow \theta = \frac{\pi}{12}, \frac{5\pi}{12}, \frac{13\pi}{12}, \frac{17\pi}{12}$; points of intersection are $\left(1, \frac{\pi}{12}\right), \left(1, \frac{5\pi}{12}\right), \left(1, \frac{13\pi}{12}\right)$, and $\left(1, \frac{17\pi}{12}\right)$. The points of intersection $\left(1, \frac{7\pi}{12}\right), \left(1, \frac{11\pi}{12}\right), \left(1, \frac{19\pi}{12}\right)$ and $\left(1, \frac{23\pi}{12}\right)$ are found by graphing and symmetry.

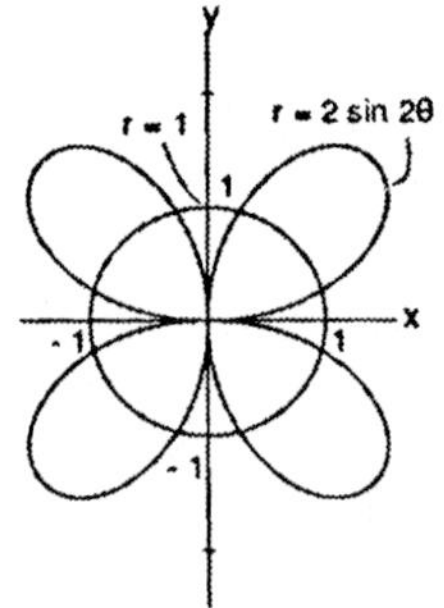

42. $r^2 = 2\sin 2\theta$ is completely generated on $0 \le \theta \le \frac{\pi}{2}$ so that $1 = 2\sin 2\theta \Rightarrow \sin 2\theta = \frac{1}{2} \Rightarrow 2\theta = \frac{\pi}{6}, \frac{5\pi}{6} \Rightarrow \theta = \frac{\pi}{12}, \frac{5\pi}{12}$; points of intersection are $\left(1, \frac{\pi}{12}\right)$ and $\left(1, \frac{5\pi}{12}\right)$. The points of intersection $\left(-1, \frac{\pi}{12}\right)$ and $\left(-1, \frac{5\pi}{12}\right)$ are found by graphing.

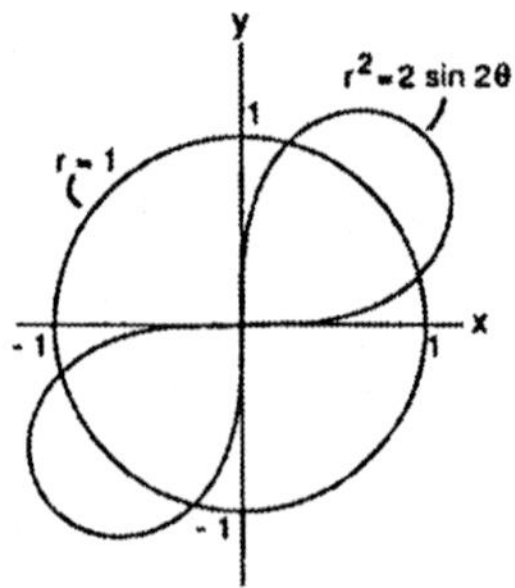

43. Note that (r, θ) and $(-r, \theta + \pi)$ describe the same point in the plane. Then $r = 1 - \cos\theta \Leftrightarrow -1 - \cos(\theta + \pi)$ $= -1 - (\cos\theta\cos\pi - \sin\theta\sin\pi) = -1 + \cos\theta = -(1 - \cos\theta) = -r$; therefore (r, θ) is on the graph of $r = 1 - \cos\theta \Leftrightarrow (-r, \theta + \pi)$ is on the graph of $r = -1 - \cos\theta \Rightarrow$ the answer is (a).

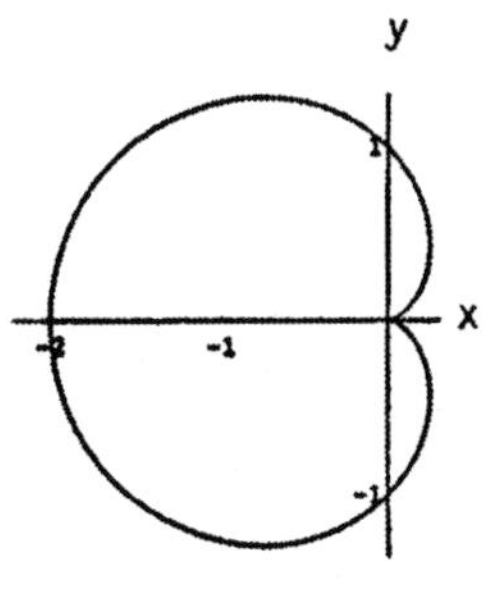

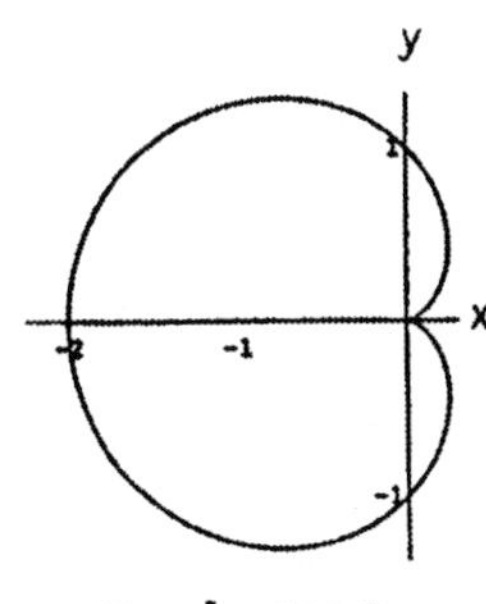

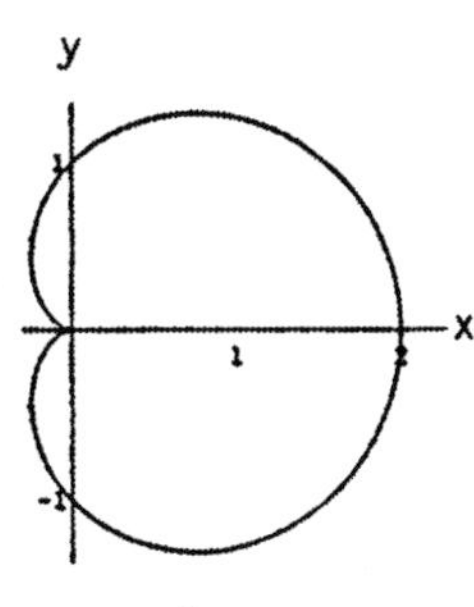

44. Note that (r, θ) and $(-r, \theta + \pi)$ describe the same point in the plane. Then $r = \cos 2\theta \Leftrightarrow -\sin\left(2(\theta + \pi)\right) + \frac{\pi}{2}\right)$ $= -\sin\left(2\theta + \frac{5\pi}{2}\right) = -\sin(2\theta)\cos\left(\frac{5\pi}{2}\right) - \cos(2\theta)\sin\left(\frac{5\pi}{2}\right) = -\cos 2\theta = -r$; therefore (r, θ) is on the graph of $r = -\sin\left(2\theta + \frac{\pi}{2}\right) \Rightarrow$ the answer is (a).

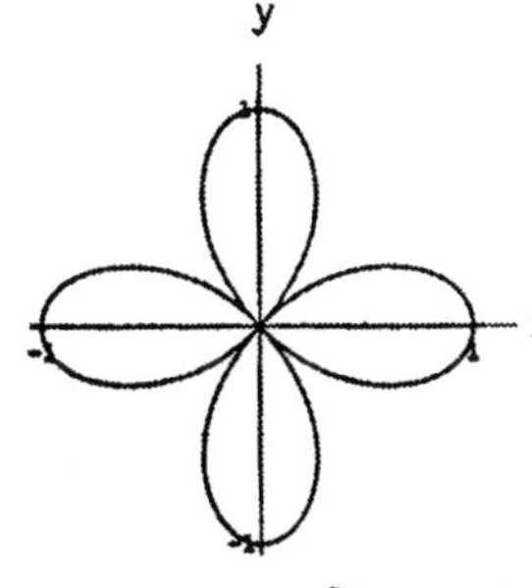

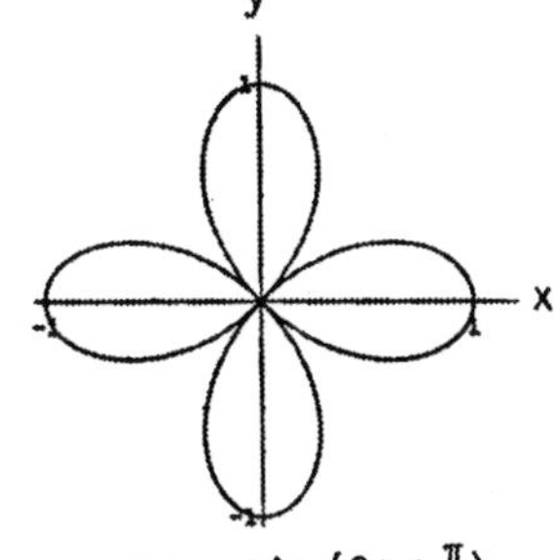

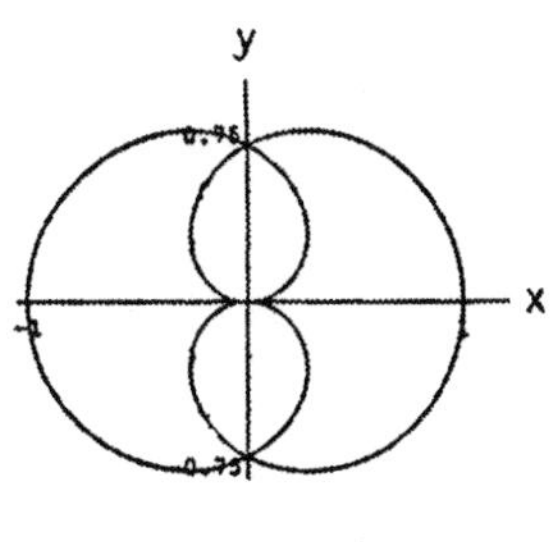

45.

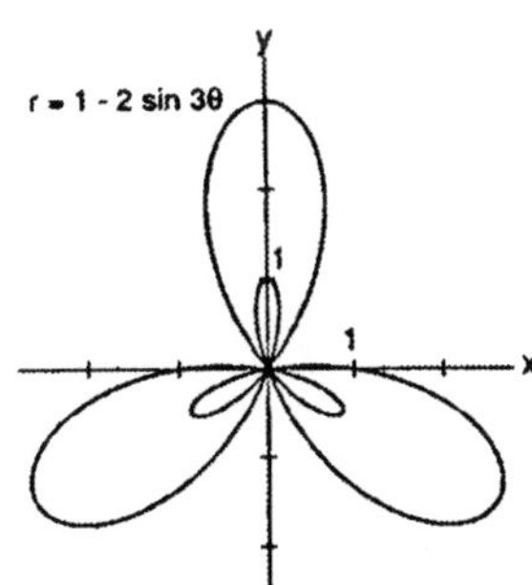

46.

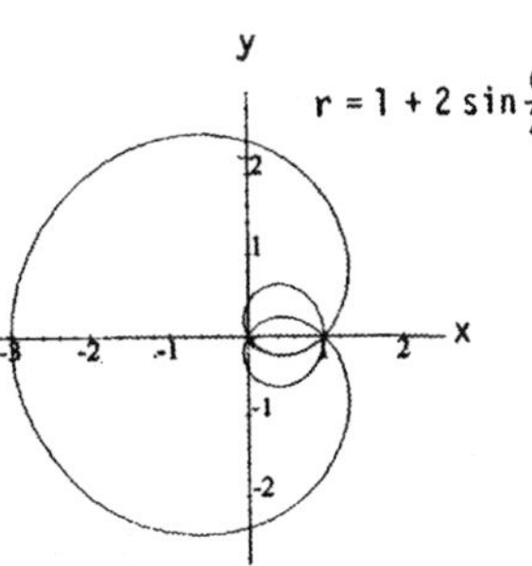

47. (a)

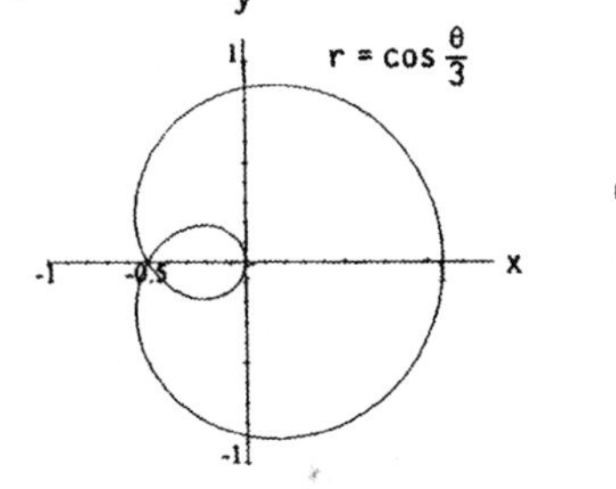

(b)

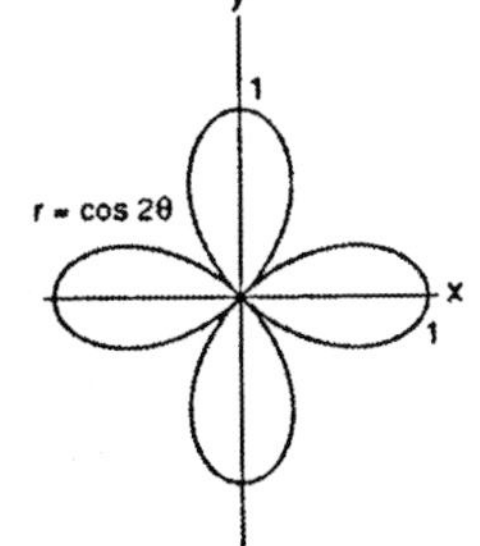

(c)

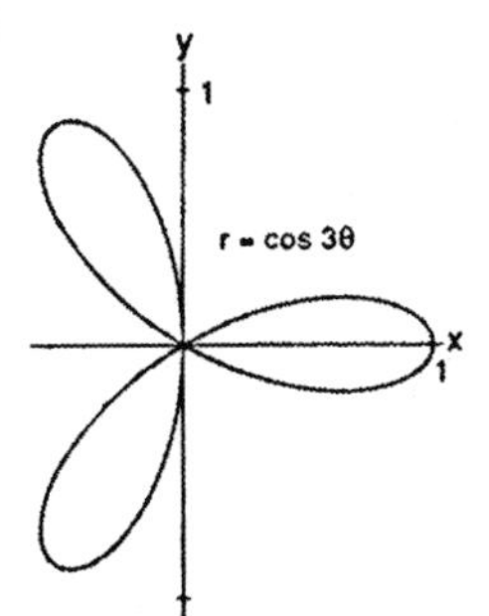

(d)

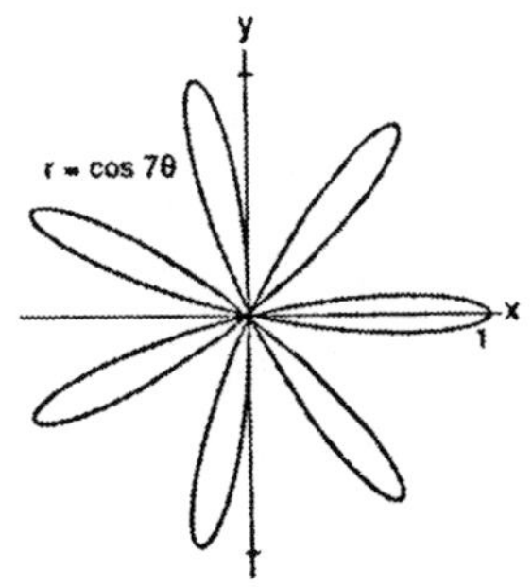

48. (a)

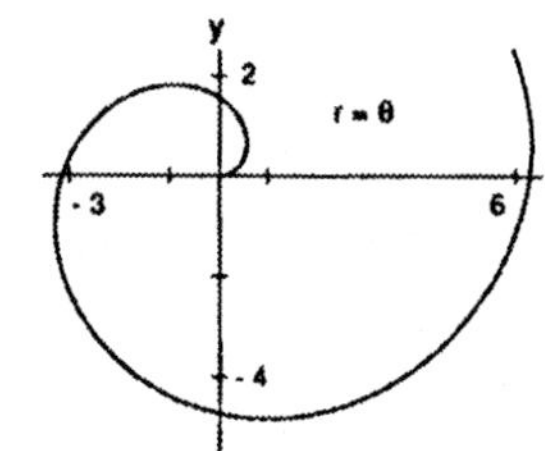

(b)

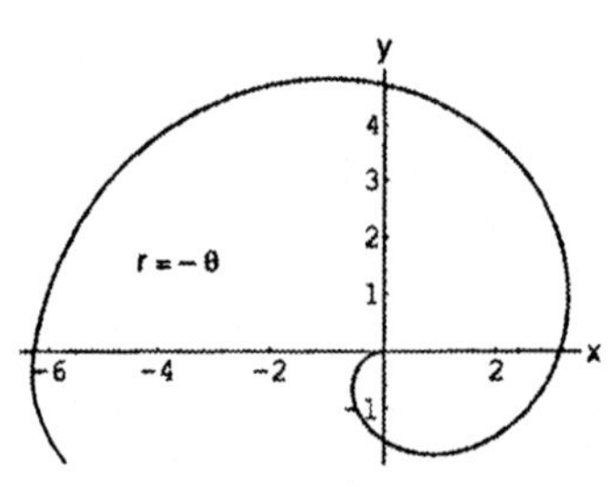

(c)

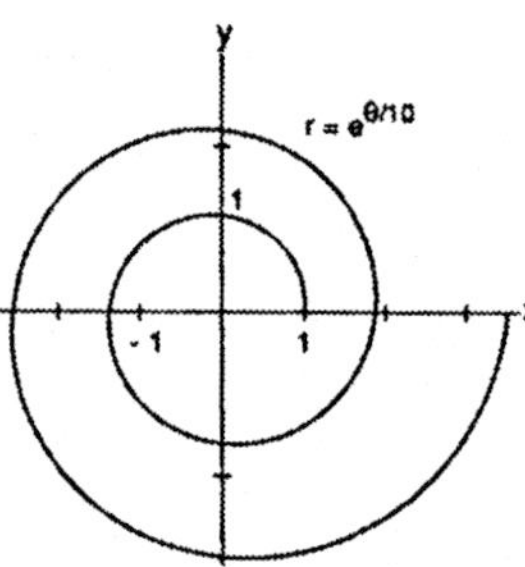

(d)

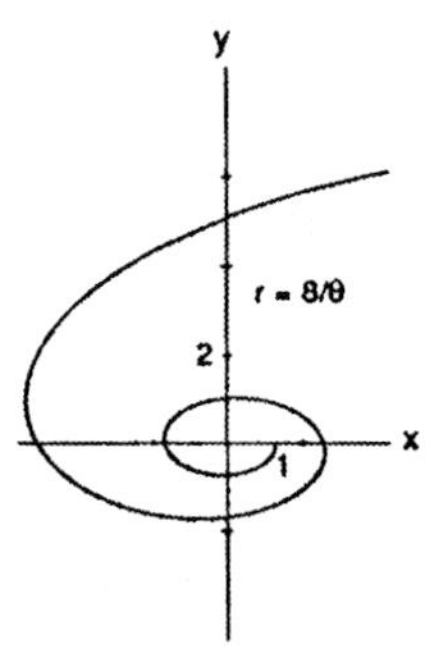

(e)

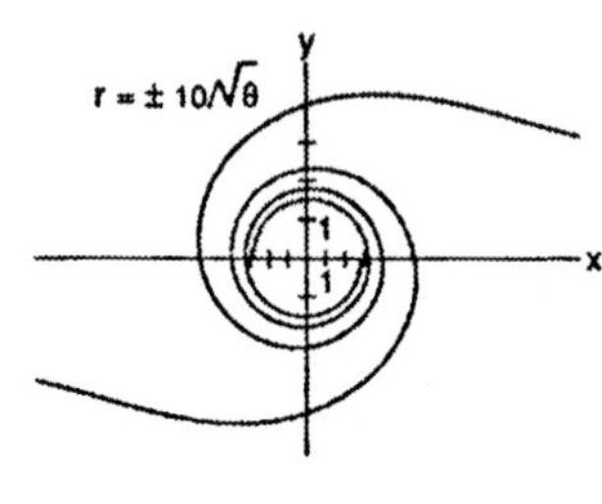

49. (a) $r^2 = -4\cos\theta \Rightarrow \cos\theta = -\frac{r^2}{4}$; $r = 1 - \cos\theta \Rightarrow r = 1 - \left(-\frac{r^2}{4}\right) \Rightarrow 0 = r^2 - 4r + 4 \Rightarrow (r-2)^2 = 0$ $\Rightarrow r = 2$; therefore $\cos\theta = -\frac{2^2}{4} = -1 \Rightarrow \theta = \pi \Rightarrow (2, \pi)$ is a point of intersection

(b) $r = 0 \Rightarrow 0^2 = 4\cos\theta \Rightarrow \cos\theta = 0 \Rightarrow \theta = \frac{\pi}{2}, \frac{3\pi}{2} \Rightarrow \left(0, \frac{\pi}{2}\right)$ or $\left(0, \frac{3\pi}{2}\right)$ is on the graph; $r = 0 \Rightarrow 0 = 1 - \cos\theta$ $\Rightarrow \cos\theta = 1 \Rightarrow \theta = 0 \Rightarrow (0,0)$ is on the graph. Since $(0,0) = \left(0, \frac{\pi}{2}\right)$ for polar coordinates, the graphs intersect at the origin.

50. (a) Let $r = f(\theta)$ be symmetric about the x-axis and the y-axis. Then (r, θ) on the graph $\Rightarrow (r, -\theta)$ is on the graph because of symmetry about the x-axis. Then $(-r, -(-\theta)) = (-r, \theta)$ is on the graph because of symmetry about the y-axis. Therefore $r = f(\theta)$ is symmetric about the origin.

(b) Let $r = f(\theta)$ be symmetric about the x-axis and the origin. Then (r, θ) on the graph $\Rightarrow (r, -\theta)$ is on the graph because of symmetry about the x-axis. Then $(-r, -\theta)$ is on the graph because of symmetry about the origin. Therefore $r = f(\theta)$ is symmetric about the y-axis.

(c) Let $r = f(\theta)$ be symmetric about the y-axis and the origin. Then (r, θ) on the graph $\Rightarrow (-r, -\theta)$ is on the graph because of symmetry about the y-axis. Then $(-(-r), -\theta) = (r, -\theta)$ is on the graph because of symmetry about the origin. Therefore $r = f(\theta)$ is symmetric about the x-axis.

51. The maximum width of the petal of the rose which lies along the x-axis is twice the largest y value of the curve on the interval $0 \le \theta \le \frac{\pi}{4}$. So we wish to maximize $2y = 2r\sin\theta = 2\cos 2\theta \sin\theta$ on $0 \le \theta \le \frac{\pi}{4}$. Let $f(\theta) = 2\cos 2\theta \sin\theta = 2(1 - 2\sin^2\theta)(\sin\theta) = 2\sin\theta - 4\sin^3\theta \Rightarrow f'(\theta) = 2\cos\theta - 12\sin^2\theta\cos\theta$. Then $f'(\theta) = 0 \Rightarrow 2\cos\theta - 12\sin^2\theta\cos\theta = 0 \Rightarrow (\cos\theta)(1 - 6\sin^2\theta) = 0 \Rightarrow \cos\theta = 0$ or $1 - 6\sin^2\theta = 0 \Rightarrow \theta = \frac{\pi}{2}$ or $\sin\theta = \frac{\pm 1}{\sqrt{6}}$. Since we want $0 \le \theta \le \frac{\pi}{4}$, we choose $\theta = \sin^{-1}\left(\frac{1}{\sqrt{6}}\right) \Rightarrow f(\theta) = 2\sin\theta - 4\sin^3\theta$ $= 2\left(\frac{1}{\sqrt{6}}\right) - 4 \cdot \frac{1}{6\sqrt{6}} = \frac{2\sqrt{6}}{9}$. We can see from the graph of $r = \cos 2\theta$ that a maximum does occur in the interval $0 \le \theta \le \frac{\pi}{4}$. Therefore the maximum width occurs at $\theta = \sin^{-1}\left(\frac{1}{\sqrt{6}}\right)$, and the maximum width is $\frac{2\sqrt{6}}{9}$.

52. We wish to maximize $y = r\sin\theta = 2(1 + \cos\theta)(\sin\theta) = 2\sin\theta + 2\sin\theta\cos\theta$. Then $\frac{dy}{d\theta} = 2\cos\theta + 2(\sin\theta)(-\sin\theta) + 2\cos\theta\cos\theta = 2\cos\theta - 2\sin^2\theta + 2\cos^2\theta = 2\cos\theta + 4\cos^2\theta - 2$; thus $\frac{dy}{d\theta} = 0 \Rightarrow 4\cos^2\theta + 2\cos\theta - 2 = 0 \Rightarrow 2\cos^2\theta + \cos\theta - 1 = 0 \Rightarrow (2\cos\theta - 1)(\cos\theta + 1) = 0 \Rightarrow \cos\theta = \frac{1}{2}$ or $\cos\theta = -1 \Rightarrow \theta = \frac{\pi}{3}, \frac{5\pi}{3}, \pi$. From the graph, we can see that the maximum occurs in the first quadrant so we choose $\theta = \frac{\pi}{3}$. Then $y = 2\sin\frac{\pi}{3} + 2\sin\frac{\pi}{3}\cos\frac{\pi}{3} = \frac{3\sqrt{3}}{2}$. The x-coordinate of this point is $x = r\cos\frac{\pi}{3}$ $= 2\left(1 + \cos\frac{\pi}{3}\right)\left(\cos\frac{\pi}{3}\right) = \frac{3}{2}$. Thus the maximum height is $h = \frac{3\sqrt{3}}{2}$ occurring at $x = \frac{3}{2}$.

10.7 AREA AND LENGTHS IN POLAR COORDINATES

1. $A = \int_0^{2\pi} \frac{1}{2}(4 + 2\cos\theta)^2\, d\theta = \int_0^{2\pi} \frac{1}{2}(16 + 16\cos\theta + 4\cos^2\theta)\, d\theta = \int_0^{2\pi} \left[8 + 8\cos\theta + 2\left(\frac{1+\cos 2\theta}{2}\right)\right] d\theta$
$= \int_0^{2\pi} (9 + 8\cos\theta + \cos 2\theta)\, d\theta = \left[9\theta + 8\sin\theta + \frac{1}{2}\sin 2\theta\right]_0^{2\pi} = 18\pi$

2. $A = \int_0^{2\pi} \frac{1}{2}[a(1 + \cos\theta)]^2\, d\theta = \int_0^{2\pi} \frac{1}{2}a^2(1 + 2\cos\theta + \cos^2\theta)\, d\theta = \frac{1}{2}a^2 \int_0^{2\pi} \left(1 + 2\cos\theta + \frac{1+\cos 2\theta}{2}\right) d\theta$
$= \frac{1}{2}a^2 \int_0^{2\pi} \left(\frac{3}{2} + 2\cos\theta + \frac{1}{2}\cos 2\theta\right) d\theta = \frac{1}{2}a^2\left[\frac{3}{2}\theta + 2\sin\theta + \frac{1}{4}\sin 2\theta\right]_0^{2\pi} = \frac{3}{2}\pi a^2$

3. $A = 2\int_0^{\pi/4} \frac{1}{2}\cos^2 2\theta\, d\theta = \int_0^{\pi/4} \frac{1+\cos 4\theta}{2}\, d\theta = \frac{1}{2}\left[\theta + \frac{\sin 4\theta}{4}\right]_0^{\pi/4} = \frac{\pi}{8}$

4. $A = 2\int_{-\pi/4}^{\pi/4} \frac{1}{2}(2a^2\cos 2\theta)\, d\theta = 2a^2 \int_{-\pi/4}^{\pi/4} \cos 2\theta\, d\theta = 2a^2\left[\frac{\sin 2\theta}{2}\right]_{-\pi/4}^{\pi/4} = 2a^2$

5. $A = \int_0^{\pi/2} \frac{1}{2}(4\sin 2\theta)\, d\theta = \int_0^{\pi/2} 2\sin 2\theta\, d\theta = [-\cos 2\theta]_0^{\pi/2} = 2$

6. $A = (6)(2)\int_0^{\pi/6} \frac{1}{2}(2\sin 3\theta)\, d\theta = 12\int_0^{\pi/6} \sin 3\theta\, d\theta = 12\left[-\frac{\cos 3\theta}{3}\right]_0^{\pi/6} = 4$

7. $r = 2\cos\theta$ and $r = 2\sin\theta \Rightarrow 2\cos\theta = 2\sin\theta$
$\Rightarrow \cos\theta = \sin\theta \Rightarrow \theta = \frac{\pi}{4}$; therefore
$A = 2\int_0^{\pi/4} \frac{1}{2}(2\sin\theta)^2\, d\theta = \int_0^{\pi/4} 4\sin^2\theta\, d\theta$
$= \int_0^{\pi/4} 4\left(\frac{1-\cos 2\theta}{2}\right) d\theta = \int_0^{\pi/4} (2 - 2\cos 2\theta)\, d\theta$
$= [2\theta - \sin 2\theta]_0^{\pi/4} = \frac{\pi}{2} - 1$

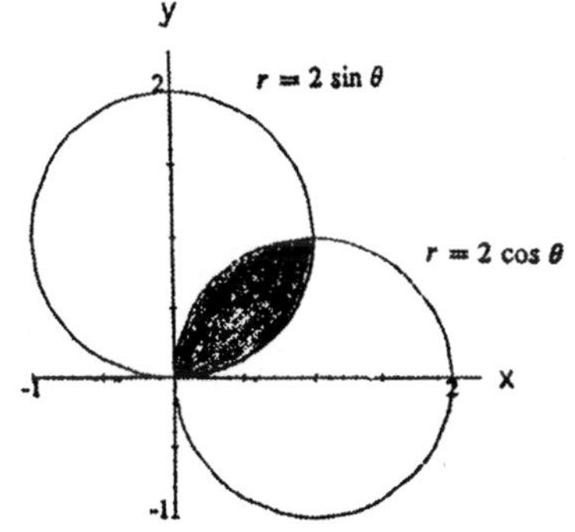

8. $r = 1$ and $r = 2\sin\theta \Rightarrow 2\sin\theta = 1 \Rightarrow \sin\theta = \frac{1}{2}$
$\Rightarrow \theta = \frac{\pi}{6}$ or $\frac{5\pi}{6}$; therefore
$A = \pi(1)^2 - \int_{\pi/6}^{5\pi/6} \frac{1}{2}\left[(2\sin\theta)^2 - 1^2\right] d\theta$
$= \pi - \int_{\pi/6}^{5\pi/6} \left(2\sin^2\theta - \frac{1}{2}\right) d\theta$
$= \pi - \int_{\pi/6}^{5\pi/6} \left(1 - \cos 2\theta - \frac{1}{2}\right) d\theta$
$= \pi - \int_{\pi/6}^{5\pi/6} \left(\frac{1}{2} - \cos 2\theta\right) d\theta = \pi - \left[\frac{1}{2}\theta - \frac{\sin 2\theta}{2}\right]_{\pi/6}^{5\pi/6}$
$= \pi - \left(\frac{5\pi}{12} - \frac{1}{2}\sin\frac{5\pi}{3}\right) + \left(\frac{\pi}{12} - \frac{1}{2}\sin\frac{\pi}{3}\right) = \frac{4\pi - 3\sqrt{3}}{6}$

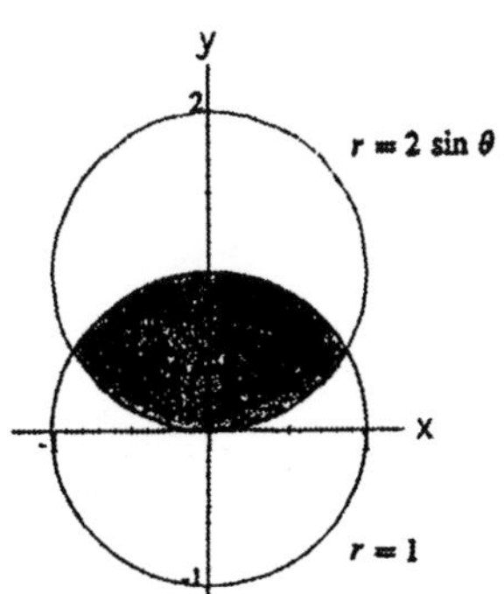

9. $r = 2$ and $r = 2(1 - \cos\theta) \Rightarrow 2 = 2(1 - \cos\theta)$
$\Rightarrow \cos\theta = 0 \Rightarrow \theta = \pm\frac{\pi}{2}$; therefore

$$A = 2\int_0^{\pi/2} \tfrac{1}{2}[2(1-\cos\theta)]^2\,d\theta + \tfrac{1}{2}\text{area of the circle}$$
$$= \int_0^{\pi/2} 4(1 - 2\cos\theta + \cos^2\theta)\,d\theta + \left(\tfrac{1}{2}\pi\right)(2)^2$$
$$= \int_0^{\pi/2} 4\left(1 - 2\cos\theta + \tfrac{1+\cos 2\theta}{2}\right)d\theta + 2\pi$$
$$= \int_0^{\pi/2} (4 - 8\cos\theta + 2 + 2\cos 2\theta)\,d\theta + 2\pi$$
$$= [6\theta - 8\sin\theta + \sin 2\theta]_0^{\pi/2} + 2\pi = 5\pi - 8$$

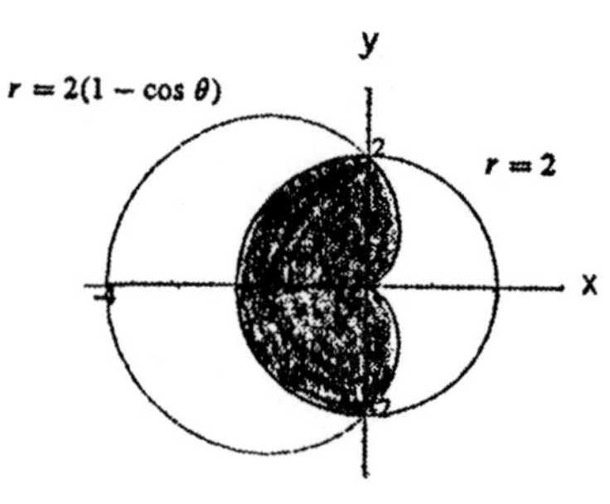

10. $r = 2(1 - \cos\theta)$ and $r = 2(1 + \cos\theta) \Rightarrow 1 - \cos\theta$
$= 1 + \cos\theta \Rightarrow \cos\theta = 0 \Rightarrow \theta = \frac{\pi}{2}$ or $\frac{3\pi}{2}$; the graph also
gives the point of intersection $(0, 0)$; therefore

$$A = 2\int_0^{\pi/2} \tfrac{1}{2}[2(1-\cos\theta)]^2\,d\theta + 2\int_{\pi/2}^{\pi} \tfrac{1}{2}[2(1+\cos\theta)]^2\,d\theta$$
$$= \int_0^{\pi/2} 4(1 - 2\cos\theta + \cos^2\theta)\,d\theta$$
$$\quad + \int_{\pi/2}^{\pi} 4(1 + 2\cos\theta + \cos^2\theta)\,d\theta$$
$$= \int_0^{\pi/2} 4\left(1 - 2\cos\theta + \tfrac{1+\cos 2\theta}{2}\right)d\theta + \int_{\pi/2}^{\pi} 4\left(1 + 2\cos\theta + \tfrac{1+\cos 2\theta}{2}\right)d\theta$$
$$= \int_0^{\pi/2} (6 - 8\cos\theta + 2\cos 2\theta)\,d\theta + \int_{\pi/2}^{\pi} (6 + 8\cos\theta + 2\cos 2\theta)\,d\theta$$
$$= [6\theta - 8\sin\theta + \sin 2\theta]_0^{\pi/2} + [6\theta + 8\sin\theta + \sin 2\theta]_{\pi/2}^{\pi} = 6\pi - 16$$

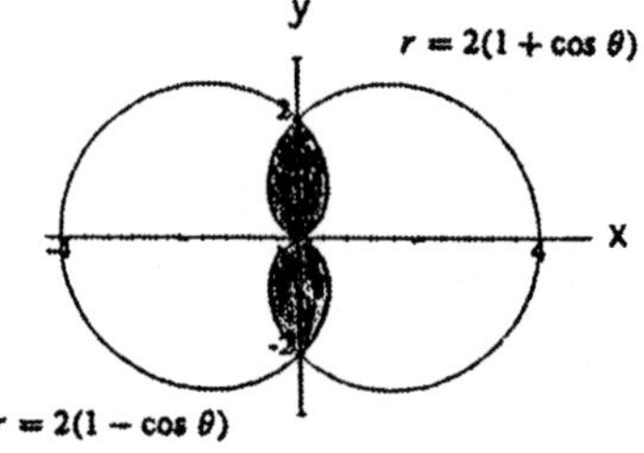

11. $r = \sqrt{3}$ and $r^2 = 6\cos 2\theta \Rightarrow 3 = 6\cos 2\theta \Rightarrow \cos 2\theta = \frac{1}{2}$
$\Rightarrow \theta = \frac{\pi}{6}$ (in the 1st quadrant); we use symmetry of the
graph to find the area, so

$$A = 4\int_0^{\pi/6} \left[\tfrac{1}{2}(6\cos 2\theta) - \tfrac{1}{2}\left(\sqrt{3}\right)^2\right] d\theta$$
$$= 2\int_0^{\pi/6} (6\cos 2\theta - 3)\,d\theta = 2\,[3\sin 2\theta - 3\theta]_0^{\pi/6}$$
$$= 3\sqrt{3} - \pi$$

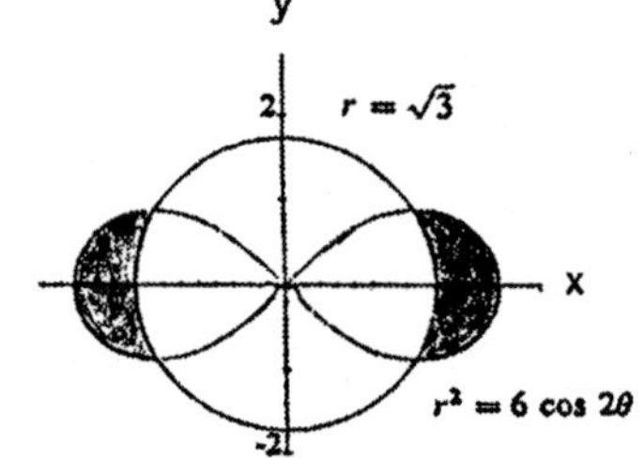

12. $r = 3a\cos\theta$ and $r = a(1 + \cos\theta) \Rightarrow 3a\cos\theta = a(1 + \cos\theta)$
$\Rightarrow 3\cos\theta = 1 + \cos\theta \Rightarrow \cos\theta = \frac{1}{2} \Rightarrow \theta = \frac{\pi}{3}$ or $-\frac{\pi}{3}$;
the graph also gives the point of intersection $(0, 0)$; therefore

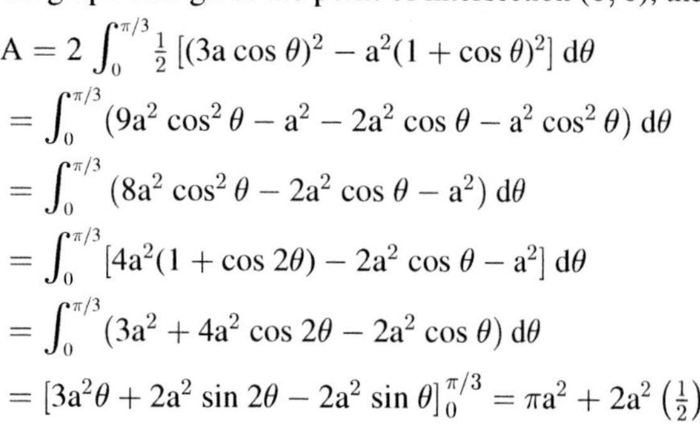

$$A = 2\int_0^{\pi/3} \tfrac{1}{2}\left[(3a\cos\theta)^2 - a^2(1+\cos\theta)^2\right] d\theta$$
$$= \int_0^{\pi/3} (9a^2\cos^2\theta - a^2 - 2a^2\cos\theta - a^2\cos^2\theta)\,d\theta$$
$$= \int_0^{\pi/3} (8a^2\cos^2\theta - 2a^2\cos\theta - a^2)\,d\theta$$
$$= \int_0^{\pi/3} [4a^2(1 + \cos 2\theta) - 2a^2\cos\theta - a^2]\,d\theta$$
$$= \int_0^{\pi/3} (3a^2 + 4a^2\cos 2\theta - 2a^2\cos\theta)\,d\theta$$

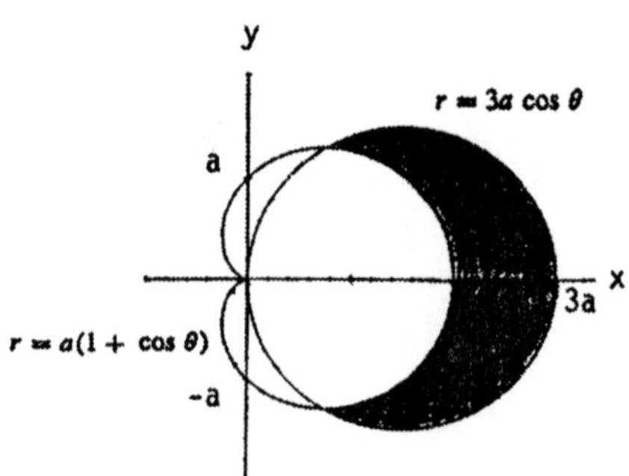

$$= [3a^2\theta + 2a^2\sin 2\theta - 2a^2\sin\theta]_0^{\pi/3} = \pi a^2 + 2a^2\left(\tfrac{1}{2}\right) - 2a^2\left(\tfrac{\sqrt{3}}{2}\right) = a^2\left(\pi + 1 - \sqrt{3}\right)$$

13. $r = 1$ and $r = -2\cos\theta \Rightarrow 1 = -2\cos\theta \Rightarrow \cos\theta = -\frac{1}{2}$
$\Rightarrow \theta = \frac{2\pi}{3}$ in quadrant II; therefore
$A = 2\int_{2\pi/3}^{\pi} \frac{1}{2}\left[(-2\cos\theta)^2 - 1^2\right] d\theta = \int_{2\pi/3}^{\pi} (4\cos^2\theta - 1)\, d\theta$
$= \int_{2\pi/3}^{\pi} [2(1+\cos 2\theta) - 1]\, d\theta = \int_{2\pi/3}^{\pi} (1 + 2\cos 2\theta)\, d\theta$
$= [\theta + \sin 2\theta]_{2\pi/3}^{\pi} = \frac{\pi}{3} + \frac{\sqrt{3}}{2}$

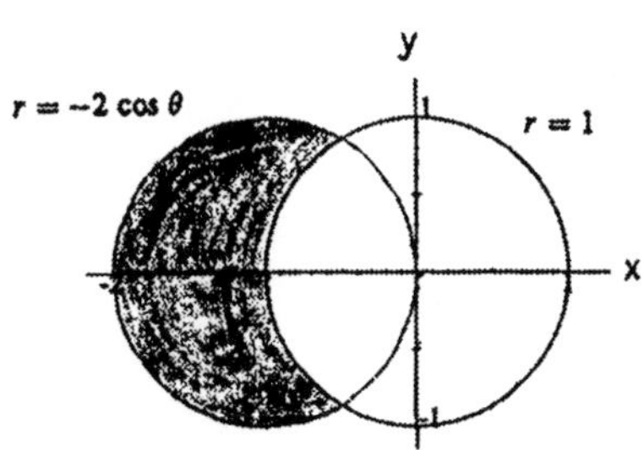

14. (a) $A = 2\int_0^{2\pi/3} \frac{1}{2}(2\cos\theta + 1)^2\, d\theta = \int_0^{2\pi/3} (4\cos^2\theta + 4\cos\theta + 1)\, d\theta = \int_0^{2\pi/3} [2(1+\cos 2\theta) + 4\cos\theta + 1]\, d\theta$
$= \int_0^{2\pi/3} (3 + 2\cos 2\theta + 4\cos\theta)\, d\theta = [3\theta + \sin 2\theta + 4\sin\theta]_0^{2\pi/3} = 2\pi - \frac{\sqrt{3}}{2} + \frac{4\sqrt{3}}{2} = 2\pi + \frac{3\sqrt{3}}{2}$

(b) $A = \left(2\pi + \frac{3\sqrt{3}}{2}\right) - \left(\pi - \frac{3\sqrt{3}}{2}\right) = \pi + 3\sqrt{3}$ (from 14(a) above and Example 2 in the text)

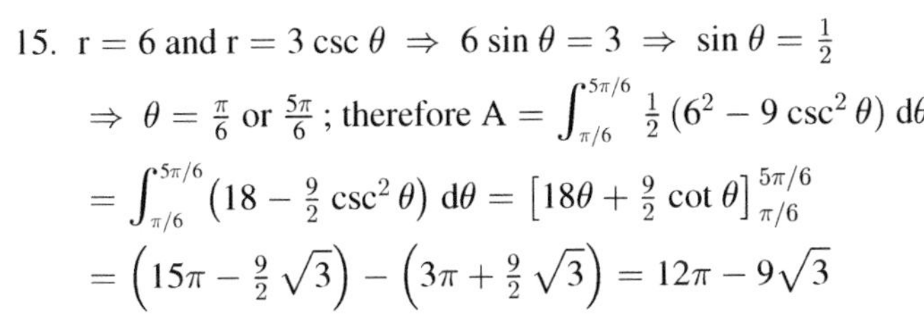

15. $r = 6$ and $r = 3\csc\theta \Rightarrow 6\sin\theta = 3 \Rightarrow \sin\theta = \frac{1}{2}$
$\Rightarrow \theta = \frac{\pi}{6}$ or $\frac{5\pi}{6}$; therefore $A = \int_{\pi/6}^{5\pi/6} \frac{1}{2}(6^2 - 9\csc^2\theta)\, d\theta$
$= \int_{\pi/6}^{5\pi/6} \left(18 - \frac{9}{2}\csc^2\theta\right) d\theta = \left[18\theta + \frac{9}{2}\cot\theta\right]_{\pi/6}^{5\pi/6}$
$= \left(15\pi - \frac{9}{2}\sqrt{3}\right) - \left(3\pi + \frac{9}{2}\sqrt{3}\right) = 12\pi - 9\sqrt{3}$

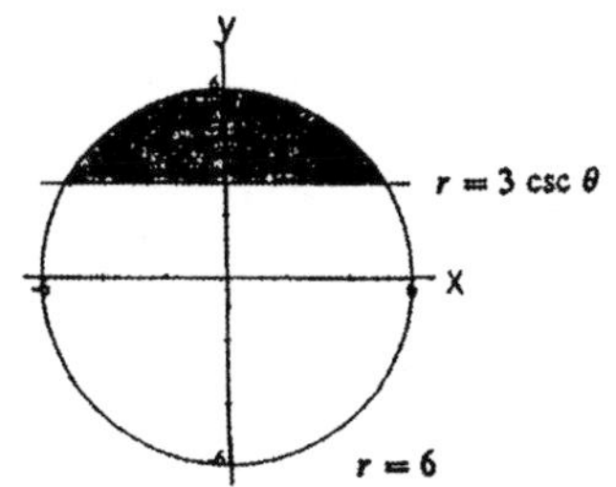

16. $r^2 = 6\cos 2\theta$ and $r = \frac{3}{2}\sec\theta \Rightarrow \frac{9}{4}\sec^2\theta = 6\cos 2\theta \Rightarrow \frac{9}{24} = \cos^2\theta\cos 2\theta \Rightarrow \frac{3}{8} = (\cos^2\theta)(2\cos^2\theta - 1)$
$\Rightarrow \frac{3}{8} = 2\cos^4\theta - \cos^2\theta \Rightarrow 2\cos^4\theta - \cos^2\theta - \frac{3}{8} = 0 \Rightarrow 16\cos^4\theta - 8\cos^2\theta - 3 = 0$
$\Rightarrow (4\cos^2\theta + 1)(4\cos^2\theta - 3) = 0 \Rightarrow \cos^2\theta = \frac{3}{4}$ or $\cos^2\theta = -\frac{1}{4} \Rightarrow \cos\theta = \pm\frac{\sqrt{3}}{2}$ (the second equation has no real roots) $\Rightarrow \theta = \frac{\pi}{6}$ (in the first quadrant); thus $A = 2\int_0^{\pi/6} \frac{1}{2}\left(6\cos 2\theta - \frac{9}{4}\sec^2\theta\right) d\theta = \int_0^{\pi/6} \left(6\cos 2\theta - \frac{9}{4}\sec^2\theta\right) d\theta$
$= \left[3\sin 2\theta - \frac{9}{4}\tan\theta\right]_0^{\pi/6} = 3\left(\frac{\sqrt{3}}{2}\right) - \frac{9}{4\sqrt{3}} = \frac{3\sqrt{3}}{2} - \frac{3\sqrt{3}}{4} = \frac{3\sqrt{3}}{4}$

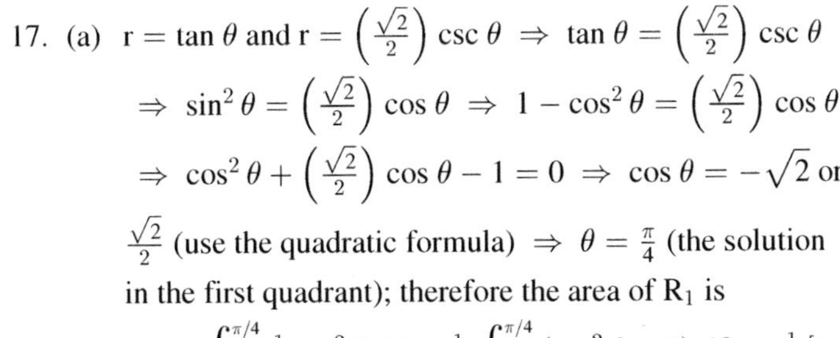

17. (a) $r = \tan\theta$ and $r = \left(\frac{\sqrt{2}}{2}\right)\csc\theta \Rightarrow \tan\theta = \left(\frac{\sqrt{2}}{2}\right)\csc\theta$
$\Rightarrow \sin^2\theta = \left(\frac{\sqrt{2}}{2}\right)\cos\theta \Rightarrow 1 - \cos^2\theta = \left(\frac{\sqrt{2}}{2}\right)\cos\theta$
$\Rightarrow \cos^2\theta + \left(\frac{\sqrt{2}}{2}\right)\cos\theta - 1 = 0 \Rightarrow \cos\theta = -\sqrt{2}$ or $\frac{\sqrt{2}}{2}$ (use the quadratic formula) $\Rightarrow \theta = \frac{\pi}{4}$ (the solution in the first quadrant); therefore the area of R_1 is
$A_1 = \int_0^{\pi/4} \frac{1}{2}\tan^2\theta\, d\theta = \frac{1}{2}\int_0^{\pi/4} (\sec^2\theta - 1)\, d\theta = \frac{1}{2}[\tan\theta - \theta]_0^{\pi/4} = \frac{1}{2}\left(\tan\frac{\pi}{4} - \frac{\pi}{4}\right) = \frac{1}{2} - \frac{\pi}{8}$;
$AO = \left(\frac{\sqrt{2}}{2}\right)\csc\frac{\pi}{2} = \frac{\sqrt{2}}{2}$ and $OB = \left(\frac{\sqrt{2}}{2}\right)\csc\frac{\pi}{4} = 1 \Rightarrow AB = \sqrt{1^2 - \left(\frac{\sqrt{2}}{2}\right)^2} = \frac{\sqrt{2}}{2}$
$\Rightarrow$ the area of R_2 is $A_2 = \frac{1}{2}\left(\frac{\sqrt{2}}{2}\right)\left(\frac{\sqrt{2}}{2}\right) = \frac{1}{4}$; therefore the area of the region shaded in the text is $2\left(\frac{1}{2} - \frac{\pi}{8} + \frac{1}{4}\right) = \frac{3}{2} - \frac{\pi}{4}$. Note: The area must be found this way since no common interval generates the region. For example, the interval $0 \le \theta \le \frac{\pi}{4}$ generates the arc OB of $r = \tan\theta$ but does not generate the segment AB of the line $r = \frac{\sqrt{2}}{2}\csc\theta$. Instead the interval generates the half-line from B to $+\infty$ on the line $r = \frac{\sqrt{2}}{2}\csc\theta$.

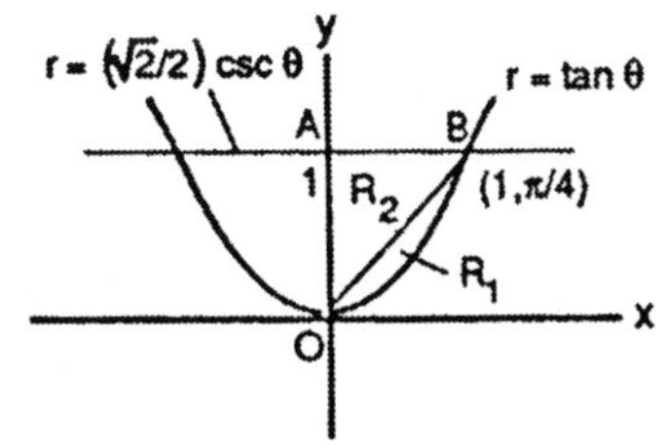

(b) $\lim_{\theta \to \pi/2^-} \tan\theta = \infty$ and the line $x = 1$ is $r = \sec\theta$ in polar coordinates; then $\lim_{\theta \to \pi/2^-} (\tan\theta - \sec\theta)$
$= \lim_{\theta \to \pi/2^-} \left(\frac{\sin\theta}{\cos\theta} - \frac{1}{\cos\theta}\right) = \lim_{\theta \to \pi/2^-} \left(\frac{\sin\theta - 1}{\cos\theta}\right) = \lim_{\theta \to \pi/2^-} \left(\frac{\cos\theta}{-\sin\theta}\right) = 0 \Rightarrow r = \tan\theta$ approaches $r = \sec\theta$ as $\theta \to \frac{\pi^-}{2} \Rightarrow r = \sec\theta$ (or $x = 1$) is a vertical asymptote of $r = \tan\theta$. Similarly, $r = -\sec\theta$

(or x = -1) is a vertical asymptote of r = $\tan\theta$.

18. It is not because the circle is generated twice from $\theta = 0$ to 2π. The area of the cardioid is
$A = 2\int_0^{\pi} \frac{1}{2}(\cos\theta + 1)^2\, d\theta = \int_0^{\pi} (\cos^2\theta + 2\cos\theta + 1)\, d\theta = \int_0^{\pi} \left(\frac{1+\cos 2\theta}{2} + 2\cos\theta + 1\right) d\theta$
$= \left[\frac{3\theta}{2} + \frac{\sin 2\theta}{4} + 2\sin\theta\right]_0^{\pi} = \frac{3\pi}{2}$. The area of the circle is $A = \pi\left(\frac{1}{2}\right)^2 = \frac{\pi}{4} \Rightarrow$ the area requested is actually
$\frac{3\pi}{2} - \frac{\pi}{4} = \frac{5\pi}{4}$

19. $r = \theta^2, 0 \le \theta \le \sqrt{5} \Rightarrow \frac{dr}{d\theta} = 2\theta$; therefore Length $= \int_0^{\sqrt{5}} \sqrt{(\theta^2)^2 + (2\theta)^2}\, d\theta = \int_0^{\sqrt{5}} \sqrt{\theta^4 + 4\theta^2}\, d\theta$
$= \int_0^{\sqrt{5}} |\theta| \sqrt{\theta^2 + 4}\, d\theta =$ (since $\theta \ge 0$) $\int_0^{\sqrt{5}} \theta\sqrt{\theta^2+4}\, d\theta$; $\big[u = \theta^2 + 4 \Rightarrow \frac{1}{2}\, du = \theta\, d\theta;\ \theta = 0 \Rightarrow u = 4,$
$\theta = \sqrt{5} \Rightarrow u = 9\big] \rightarrow \int_4^9 \frac{1}{2}\sqrt{u}\, du = \frac{1}{2}\left[\frac{2}{3}u^{3/2}\right]_4^9 = \frac{19}{3}$

20. $r = \frac{e^{\theta}}{\sqrt{2}}, 0 \le \theta \le \pi \Rightarrow \frac{dr}{d\theta} = \frac{e^{\theta}}{\sqrt{2}}$; therefore Length $= \int_0^{\pi} \sqrt{\left(\frac{e^{\theta}}{\sqrt{2}}\right)^2 + \left(\frac{e^{\theta}}{\sqrt{2}}\right)^2}\, d\theta = \int_0^{\pi} \sqrt{2\left(\frac{e^{2\theta}}{2}\right)}\, d\theta$
$= \int_0^{\pi} e^{\theta}\, d\theta = \left[e^{\theta}\right]_0^{\pi} = e^{\pi} - 1$

21. $r = 1 + \cos\theta \Rightarrow \frac{dr}{d\theta} = -\sin\theta$; therefore Length $= \int_0^{2\pi} \sqrt{(1+\cos\theta)^2 + (-\sin\theta)^2}\, d\theta$
$= 2\int_0^{\pi} \sqrt{2 + 2\cos\theta}\, d\theta = 2\int_0^{\pi} \sqrt{\frac{4(1+\cos\theta)}{2}}\, d\theta = 4\int_0^{\pi} \sqrt{\frac{1+\cos\theta}{2}}\, d\theta = 4\int_0^{\pi} \cos\left(\frac{\theta}{2}\right) d\theta = 4\left[2\sin\frac{\theta}{2}\right]_0^{\pi} = 8$

22. $r = a\sin^2\frac{\theta}{2}, 0 \le \theta \le \pi, a > 0 \Rightarrow \frac{dr}{d\theta} = a\sin\frac{\theta}{2}\cos\frac{\theta}{2}$; therefore Length $= \int_0^{\pi} \sqrt{\left(a\sin^2\frac{\theta}{2}\right)^2 + \left(a\sin\frac{\theta}{2}\cos\frac{\theta}{2}\right)^2}\, d\theta$
$= \int_0^{\pi} \sqrt{a^2\sin^4\frac{\theta}{2} + a^2\sin^2\frac{\theta}{2}\cos^2\frac{\theta}{2}}\, d\theta = \int_0^{\pi} a\left|\sin\frac{\theta}{2}\right| \sqrt{\sin^2\frac{\theta}{2} + \cos^2\frac{\theta}{2}}\, d\theta =$ (since $0 \le \theta \le \pi$) $a\int_0^{\pi} \sin\left(\frac{\theta}{2}\right) d\theta$
$= \left[-2a\cos\frac{\theta}{2}\right]_0^{\pi} = 2a$

23. $r = \frac{6}{1+\cos\theta}, 0 \le \theta \le \frac{\pi}{2} \Rightarrow \frac{dr}{d\theta} = \frac{6\sin\theta}{(1+\cos\theta)^2}$; therefore Length $= \int_0^{\pi/2} \sqrt{\left(\frac{6}{1+\cos\theta}\right)^2 + \left(\frac{6\sin\theta}{(1+\cos\theta)^2}\right)^2}\, d\theta$
$= \int_0^{\pi/2} \sqrt{\frac{36}{(1+\cos\theta)^2} + \frac{36\sin^2\theta}{(1+\cos\theta)^4}}\, d\theta = 6\int_0^{\pi/2} \left|\frac{1}{1+\cos\theta}\right| \sqrt{1 + \frac{\sin^2\theta}{(1+\cos\theta)^2}}\, d\theta$
$=$ (since $\frac{1}{1+\cos\theta} > 0$ on $0 \le \theta \le \frac{\pi}{2}$) $6\int_0^{\pi/2} \left(\frac{1}{1+\cos\theta}\right) \sqrt{\frac{1 + 2\cos\theta + \cos^2\theta + \sin^2\theta}{(1+\cos\theta)^2}}\, d\theta$
$= 6\int_0^{\pi/2} \left(\frac{1}{1+\cos\theta}\right) \sqrt{\frac{2+2\cos\theta}{(1+\cos\theta)^2}}\, d\theta = 6\sqrt{2}\int_0^{\pi/2} \frac{d\theta}{(1+\cos\theta)^{3/2}} = 6\sqrt{2}\int_0^{\pi/2} \frac{d\theta}{\left(2\cos^2\frac{\theta}{2}\right)^{3/2}} = 3\int_0^{\pi/2} \left|\sec^3\frac{\theta}{2}\right| d\theta$
$= 3\int_0^{\pi/2} \sec^3\frac{\theta}{2}\, d\theta = 6\int_0^{\pi/4} \sec^3 u\, du =$ (use tables) $6\left(\left[\frac{\sec u \tan u}{2}\right]_0^{\pi/4} + \frac{1}{2}\int_0^{\pi/4} \sec u\, du\right)$
$= 6\left(\frac{1}{\sqrt{2}} + \left[\frac{1}{2}\ln|\sec u + \tan u|\right]_0^{\pi/4}\right) = 3\left[\sqrt{2} + \ln\left(1 + \sqrt{2}\right)\right]$

24. $r = \frac{2}{1-\cos\theta}, \frac{\pi}{2} \le \theta \le \pi \Rightarrow \frac{dr}{d\theta} = \frac{-2\sin\theta}{(1-\cos\theta)^2}$; therefore Length $= \int_{\pi/2}^{\pi} \sqrt{\left(\frac{2}{1-\cos\theta}\right)^2 + \left(\frac{-2\sin\theta}{(1-\cos\theta)^2}\right)^2}\, d\theta$
$= \int_{\pi/2}^{\pi} \sqrt{\frac{4}{(1-\cos\theta)^2}\left(1 + \frac{\sin^2\theta}{(1-\cos\theta)^2}\right)}\, d\theta = \int_{\pi/2}^{\pi} \left|\frac{2}{1-\cos\theta}\right| \sqrt{\frac{(1-\cos\theta)^2 + \sin^2\theta}{(1-\cos\theta)^2}}\, d\theta$
$=$ (since $1 - \cos\theta \ge 0$ on $\frac{\pi}{2} \le \theta \le \pi$) $2\int_{\pi/2}^{\pi} \left(\frac{1}{1-\cos\theta}\right) \sqrt{\frac{1 - 2\cos\theta + \cos^2\theta + \sin^2\theta}{(1-\cos\theta)^2}}\, d\theta$
$= 2\int_{\pi/2}^{\pi} \left(\frac{1}{1-\cos\theta}\right) \sqrt{\frac{2-2\cos\theta}{(1-\cos\theta)^2}}\, d\theta = 2\sqrt{2}\int_{\pi/2}^{\pi} \frac{d\theta}{(1-\cos\theta)^{3/2}} = 2\sqrt{2}\int_{\pi/2}^{\pi} \frac{d\theta}{\left(2\sin^2\frac{\theta}{2}\right)^{3/2}} = \int_{\pi/2}^{\pi} \left|\csc^3\frac{\theta}{2}\right| d\theta$
$= \int_{\pi/2}^{\pi} \csc^3\left(\frac{\theta}{2}\right) d\theta =$ (since $\csc\frac{\theta}{2} \ge 0$ on $\frac{\pi}{2} \le \theta \le \pi$) $2\int_{\pi/4}^{\pi/2} \csc^3 u\, du =$ (use tables)

$2\left(\left[-\frac{\csc u \cot u}{2}\right]_{\pi/4}^{\pi/2} + \frac{1}{2}\int_{\pi/4}^{\pi/2} \csc u \, du\right) = 2\left(\frac{1}{\sqrt{2}} - \left[\frac{1}{2}\ln|\csc u + \cot u|\right]_{\pi/4}^{\pi/2}\right) = 2\left[\frac{1}{\sqrt{2}} + \frac{1}{2}\ln\left(\sqrt{2}+1\right)\right]$

$= \sqrt{2} + \ln\left(1+\sqrt{2}\right)$

25. $r = \cos^3\frac{\theta}{3} \Rightarrow \frac{dr}{d\theta} = -\sin\frac{\theta}{3}\cos^2\frac{\theta}{3}$; therefore Length $= \int_0^{\pi/4}\sqrt{\left(\cos^3\frac{\theta}{3}\right)^2 + \left(-\sin\frac{\theta}{3}\cos^2\frac{\theta}{3}\right)^2}\, d\theta$

$= \int_0^{\pi/4}\sqrt{\cos^6\left(\frac{\theta}{3}\right) + \sin^2\left(\frac{\theta}{3}\right)\cos^4\left(\frac{\theta}{3}\right)}\, d\theta = \int_0^{\pi/4}\left(\cos^2\frac{\theta}{3}\right)\sqrt{\cos^2\left(\frac{\theta}{3}\right) + \sin^2\left(\frac{\theta}{3}\right)}\, d\theta = \int_0^{\pi/4}\cos^2\left(\frac{\theta}{3}\right) d\theta$

$= \int_0^{\pi/4}\frac{1+\cos\left(\frac{2\theta}{3}\right)}{2}\, d\theta = \frac{1}{2}\left[\theta + \frac{3}{2}\sin\frac{2\theta}{3}\right]_0^{\pi/4} = \frac{\pi}{8} + \frac{3}{8}$

26. $r = \sqrt{1+\sin 2\theta},\ 0 \le \theta \le \pi\sqrt{2} \Rightarrow \frac{dr}{d\theta} = \frac{1}{2}(1+\sin 2\theta)^{-1/2}(2\cos 2\theta) = (\cos 2\theta)(1+\sin 2\theta)^{-1/2}$; therefore

Length $= \int_0^{\pi\sqrt{2}}\sqrt{(1+\sin 2\theta) + \frac{\cos^2 2\theta}{(1+\sin 2\theta)}}\, d\theta = \int_0^{\pi\sqrt{2}}\sqrt{\frac{1+2\sin 2\theta + \sin^2 2\theta + \cos^2 2\theta}{1+\sin 2\theta}}\, d\theta$

$= \int_0^{\pi\sqrt{2}}\sqrt{\frac{2+2\sin 2\theta}{1+\sin 2\theta}}\, d\theta = \int_0^{\pi\sqrt{2}}\sqrt{2}\, d\theta = \left[\sqrt{2}\,\theta\right]_0^{\pi\sqrt{2}} = 2\pi$

27. $r = \sqrt{1+\cos 2\theta} \Rightarrow \frac{dr}{d\theta} = \frac{1}{2}(1+\cos 2\theta)^{-1/2}(-2\sin 2\theta)$; therefore Length $= \int_0^{\pi\sqrt{2}}\sqrt{(1+\cos 2\theta) + \frac{\sin^2 2\theta}{(1+\cos 2\theta)}}\, d\theta$

$= \int_0^{\pi\sqrt{2}}\sqrt{\frac{1+2\cos 2\theta + \cos^2 2\theta + \sin^2 2\theta}{1+\cos 2\theta}}\, d\theta = \int_0^{\pi\sqrt{2}}\sqrt{\frac{2+2\cos 2\theta}{1+\cos 2\theta}}\, d\theta = \int_0^{\pi\sqrt{2}}\sqrt{2}\, d\theta = \left[\sqrt{2}\,\theta\right]_0^{\pi\sqrt{2}} = 2\pi$

28. (a) $r = a \Rightarrow \frac{dr}{d\theta} = 0$; Length $= \int_0^{2\pi}\sqrt{a^2+0^2}\, d\theta = \int_0^{2\pi}|a|\, d\theta = [a\theta]_0^{2\pi} = 2\pi a$

(b) $r = a\cos\theta \Rightarrow \frac{dr}{d\theta} = -a\sin\theta$; Length $= \int_0^{\pi}\sqrt{(a\cos\theta)^2 + (-a\sin\theta)^2}\, d\theta = \int_0^{\pi}\sqrt{a^2\left(\cos^2\theta + \sin^2\theta\right)}\, d\theta$

$= \int_0^{\pi}|a|\, d\theta = [a\theta]_0^{\pi} = \pi a$

(c) $r = a\sin\theta \Rightarrow \frac{dr}{d\theta} = a\cos\theta$; Length $= \int_0^{\pi}\sqrt{(a\cos\theta)^2 + (a\sin\theta)^2}\, d\theta = \int_0^{\pi}\sqrt{a^2\left(\cos^2\theta + \sin^2\theta\right)}\, d\theta$

$= \int_0^{\pi}|a|\, d\theta = [a\theta]_0^{\pi} = \pi a$

29. $r = \sqrt{\cos 2\theta},\ 0 \le \theta \le \frac{\pi}{4} \Rightarrow \frac{dr}{d\theta} = \frac{1}{2}(\cos 2\theta)^{-1/2}(-\sin 2\theta)(2) = \frac{-\sin 2\theta}{\sqrt{\cos 2\theta}}$; therefore Surface Area

$= \int_0^{\pi/4}(2\pi r\cos\theta)\sqrt{\left(\sqrt{\cos 2\theta}\right)^2 + \left(\frac{-\sin 2\theta}{\sqrt{\cos 2\theta}}\right)^2}\, d\theta = \int_0^{\pi/4}\left(2\pi\sqrt{\cos 2\theta}\right)(\cos\theta)\sqrt{\cos 2\theta + \frac{\sin^2 2\theta}{\cos 2\theta}}\, d\theta$

$= \int_0^{\pi/4}\left(2\pi\sqrt{\cos 2\theta}\right)(\cos\theta)\sqrt{\frac{1}{\cos 2\theta}}\, d\theta = \int_0^{\pi/4} 2\pi\cos\theta\, d\theta = [2\pi\sin\theta]_0^{\pi/4} = \pi\sqrt{2}$

30. $r = \sqrt{2}e^{\theta/2},\ 0 \le \theta \le \frac{\pi}{2} \Rightarrow \frac{dr}{d\theta} = \sqrt{2}\left(\frac{1}{2}\right)e^{\theta/2} = \frac{\sqrt{2}}{2}e^{\theta/2}$; therefore Surface Area

$= \int_0^{\pi/2}\left(2\pi\sqrt{2}e^{\theta/2}\right)(\sin\theta)\sqrt{\left(\sqrt{2}e^{\theta/2}\right)^2 + \left(\frac{\sqrt{2}}{2}e^{\theta/2}\right)^2}\, d\theta = \int_0^{\pi/2}\left(2\pi\sqrt{2}e^{\theta/2}\right)(\sin\theta)\sqrt{2e^{\theta} + \frac{1}{2}e^{\theta}}\, d\theta$

$= \int_0^{\pi/2}\left(2\pi\sqrt{2}e^{\theta/2}\right)(\sin\theta)\sqrt{\frac{5}{2}e^{\theta}}\, d\theta = \int_0^{\pi/2}\left(2\pi\sqrt{2}e^{\theta/2}\right)(\sin\theta)\left(\frac{\sqrt{5}}{\sqrt{2}}e^{\theta/2}\right) d\theta = 2\pi\sqrt{5}\int_0^{\pi/2} e^{\theta}\sin\theta\, d\theta$

$= 2\pi\sqrt{5}\left[\frac{e^{\theta}}{2}(\sin\theta - \cos\theta)\right]_0^{\pi/2} = \pi\sqrt{5}\left(e^{\pi/2}+1\right)$ where we integrated by parts

31. $r^2 = \cos 2\theta \Rightarrow r = \pm\sqrt{\cos 2\theta}$; use $r = \sqrt{\cos 2\theta}$ on $\left[0, \frac{\pi}{4}\right] \Rightarrow \frac{dr}{d\theta} = \frac{1}{2}(\cos 2\theta)^{-1/2}(-\sin 2\theta)(2) = \frac{-\sin 2\theta}{\sqrt{\cos 2\theta}}$;

therefore Surface Area $= 2\int_0^{\pi/4}\left(2\pi\sqrt{\cos 2\theta}\right)(\sin\theta)\sqrt{\cos 2\theta + \frac{\sin^2 2\theta}{\cos 2\theta}}\, d\theta = 4\pi\int_0^{\pi/4}\sqrt{\cos 2\theta}\,(\sin\theta)\sqrt{\frac{1}{\cos 2\theta}}\, d\theta$

$= 4\pi\int_0^{\pi/4}\sin\theta\, d\theta = 4\pi[-\cos\theta]_0^{\pi/4} = 4\pi\left[-\frac{\sqrt{2}}{2} - (-1)\right] = 2\pi\left(2-\sqrt{2}\right)$

32. $r = 2a\cos\theta \Rightarrow \frac{dr}{d\theta} = -2a\sin\theta$; therefore Surface Area $= \int_0^\pi 2\pi(2a\cos\theta)(\cos\theta)\sqrt{(2a\cos\theta)^2 + (-2a\sin\theta)^2}\,d\theta$
$= 4a\pi\int_0^\pi (\cos^2\theta)\sqrt{4a^2(\cos^2\theta + \sin^2\theta)}\,d\theta = 8a\pi\int_0^\pi (\cos^2\theta)\,|a|\,d\theta = 8a^2\pi\int_0^\pi \cos^2\theta\,d\theta$
$= 8a^2\pi\int_0^\pi \left(\frac{1+\cos 2\theta}{2}\right)d\theta = 4a^2\pi\int_0^\pi (1+\cos 2\theta)\,d\theta = 4a^2\pi\left[\theta + \frac{1}{2}\sin 2\theta\right]_0^\pi = 4a^2\pi^2$

33. Let $r = f(\theta)$. Then $x = f(\theta)\cos\theta \Rightarrow \frac{dx}{d\theta} = f'(\theta)\cos\theta - f(\theta)\sin\theta \Rightarrow \left(\frac{dx}{d\theta}\right)^2 = [f'(\theta)\cos\theta - f(\theta)\sin\theta]^2$
$= [f'(\theta)]^2\cos^2\theta - 2f'(\theta)f(\theta)\sin\theta\cos\theta + [f(\theta)]^2\sin^2\theta$; $y = f(\theta)\sin\theta \Rightarrow \frac{dy}{d\theta} = f'(\theta)\sin\theta + f(\theta)\cos\theta$
$\Rightarrow \left(\frac{dy}{d\theta}\right)^2 = [f'(\theta)\sin\theta + f(\theta)\cos\theta]^2 = [f'(\theta)]^2\sin^2\theta + 2f'(\theta)f(\theta)\sin\theta\cos\theta + [f(\theta)]^2\cos^2\theta$. Therefore
$\left(\frac{dx}{d\theta}\right)^2 + \left(\frac{dy}{d\theta}\right)^2 = [f'(\theta)]^2(\cos^2\theta + \sin^2\theta) + [f(\theta)]^2(\cos^2\theta + \sin^2\theta) = [f'(\theta)]^2 + [f(\theta)]^2 = r^2 + \left(\frac{dr}{d\theta}\right)^2$.
Thus, $L = \int_\alpha^\beta \sqrt{\left(\frac{dx}{d\theta}\right)^2 + \left(\frac{dy}{d\theta}\right)^2}\,d\theta = \int_\alpha^\beta \sqrt{r^2 + \left(\frac{dr}{d\theta}\right)^2}\,d\theta$.

34. (a) $r_{av} = \frac{1}{2\pi - 0}\int_0^{2\pi} a(1-\cos\theta)\,d\theta = \frac{a}{2\pi}[\theta - \sin\theta]_0^{2\pi} = a$
(b) $r_{av} = \frac{1}{2\pi - 0}\int_0^{2\pi} a\,d\theta = \frac{1}{2\pi}[a\theta]_0^{2\pi} = a$
(c) $r_{av} = \frac{1}{\left(\frac{\pi}{2}\right) - \left(-\frac{\pi}{2}\right)}\int_{-\pi/2}^{\pi/2} a\cos\theta\,d\theta = \frac{1}{\pi}[a\sin\theta]_{-\pi/2}^{\pi/2} = \frac{2a}{\pi}$

35. $r = 2f(\theta)$, $\alpha \le \theta \le \beta \Rightarrow \frac{dr}{d\theta} = 2f'(\theta) \Rightarrow r^2 + \left(\frac{dr}{d\theta}\right)^2 = [2f(\theta)]^2 + [2f'(\theta)]^2 \Rightarrow$ Length $= \int_\alpha^\beta \sqrt{4[f(\theta)]^2 + 4[f'(\theta)]^2}\,d\theta$
$= 2\int_\alpha^\beta \sqrt{[f(\theta)]^2 + [f'(\theta)]^2}\,d\theta$ which is twice the length of the curve $r = f(\theta)$ for $\alpha \le \theta \le \beta$.

36. Again $r = 2f(\theta) \Rightarrow r^2 + \left(\frac{dr}{d\theta}\right)^2 = [2f(\theta)]^2 + [2f'(\theta)]^2 \Rightarrow$ Surface Area $= \int_\alpha^\beta 2\pi[2f(\theta)\sin\theta]\sqrt{4[f(\theta)]^2 + 4[f'(\theta)]^2}\,d\theta$
$= 4\int_\alpha^\beta 2\pi[f(\theta)\sin\theta]\sqrt{[f(\theta)]^2 + [f'(\theta)]^2}\,d\theta$ which is four times the area of the surface generated by revolving $r = f(\theta)$ about the x-axis for $\alpha \le \theta \le \beta$.

37. $\overline{x} = \frac{\frac{2}{3}\int_0^{2\pi} r^3\cos\theta\,d\theta}{\int_0^{2\pi} r^2\,d\theta} = \frac{\frac{2}{3}\int_0^{2\pi}[a(1+\cos\theta)]^3(\cos\theta)\,d\theta}{\int_0^{2\pi}[a(1+\cos\theta)]^2\,d\theta} = \frac{\frac{2}{3}a^3\int_0^{2\pi}(1+3\cos\theta+3\cos^2\theta+\cos^3\theta)(\cos\theta)\,d\theta}{a^2\int_0^{2\pi}(1+2\cos\theta+\cos^2\theta)\,d\theta}$
$= \frac{\frac{2}{3}a\int_0^{2\pi}\left[\cos\theta + 3\left(\frac{1+\cos 2\theta}{2}\right) + 3(1-\sin^2\theta)(\cos\theta) + \left(\frac{1+\cos 2\theta}{2}\right)^2\right]d\theta}{\int_0^{2\pi}\left[1+2\cos\theta+\left(\frac{1+\cos 2\theta}{2}\right)\right]d\theta}$ = (After considerable algebra using
the identity $\cos^2 A = \frac{1+\cos 2A}{2}$) $\frac{a\int_0^{2\pi}\left(\frac{15}{12}+\frac{8}{3}\cos\theta+\frac{4}{3}\cos 2\theta - 2\cos\theta\sin^2\theta + \frac{1}{12}\cos 4\theta\right)d\theta}{\int_0^{2\pi}\left(\frac{3}{2}+2\cos\theta+\frac{1}{2}\cos 2\theta\right)d\theta}$
$= \frac{a\left[\frac{15}{12}\theta + \frac{8}{3}\sin\theta + \frac{2}{3}\sin 2\theta - \frac{2}{3}\sin^3\theta + \frac{1}{48}\sin 4\theta\right]_0^{2\pi}}{\left[\frac{3}{2}\theta + 2\sin\theta + \frac{1}{4}\sin 2\theta\right]_0^{2\pi}} = \frac{a\left(\frac{15}{6}\pi\right)}{3\pi} = \frac{5}{6}a$;
$\overline{y} = \frac{\frac{2}{3}\int_0^{2\pi} r^3\sin\theta\,d\theta}{\int_0^{2\pi} r^2\,d\theta} = \frac{\frac{2}{3}\int_0^{2\pi}[a(1+\cos\theta)]^3(\sin\theta)\,d\theta}{3\pi}$; $\left[u = a(1+\cos\theta) \Rightarrow -\frac{1}{a}du = \sin\theta\,d\theta;\ \theta = 0 \Rightarrow u = 2a;\right.$
$\left.\theta = 2\pi \Rightarrow u = 2a\right] \rightarrow \frac{\frac{2}{3}\int_{2a}^{2a} -\frac{1}{a}u^3\,du}{3\pi} = \frac{0}{3\pi} = 0$. Therefore the centroid is $(\overline{x}, \overline{y}) = \left(\frac{5}{6}a, 0\right)$

38. $\int_0^\pi r^2\,d\theta = \int_0^\pi a^2\,d\theta = [a^2\theta]_0^\pi = a^2\pi$; $\overline{x} = \frac{\frac{2}{3}\int_0^\pi r^3\cos\theta\,d\theta}{\int_0^\pi r^2\,d\theta} = \frac{\frac{2}{3}\int_0^\pi a^3\cos\theta\,d\theta}{a^2\pi} = \frac{\frac{2}{3}a^3[\sin\theta]_0^\pi}{a^2\pi} = \frac{0}{a^2\pi} = 0$;
$\overline{y} = \frac{\frac{2}{3}\int_0^\pi r^3\sin\theta\,d\theta}{\int_0^\pi r^2\,d\theta} = \frac{\frac{2}{3}\int_0^\pi a^3\sin\theta\,d\theta}{a^2\pi} = \frac{\frac{2}{3}a^3[-\cos\theta]_0^\pi}{a^2\pi} = \frac{\left(\frac{4}{3}\right)a^3}{a^2\pi} = \frac{4a}{3\pi}$. Therefore the centroid is $(\overline{x}, \overline{y}) = \left(0, \frac{4a}{3\pi}\right)$.

10.8 CONIC SECTIONS IN POLAR COORDINATES

1. $r\cos\left(\theta - \frac{\pi}{6}\right) = 5 \Rightarrow r\left(\cos\theta\cos\frac{\pi}{6} + \sin\theta\sin\frac{\pi}{6}\right) = 5 \Rightarrow \frac{\sqrt{3}}{2}r\cos\theta + \frac{1}{2}r\sin\theta = 5 \Rightarrow \frac{\sqrt{3}}{2}x + \frac{1}{2}y = 5 \Rightarrow \sqrt{3}x + y = 10 \Rightarrow y = -\sqrt{3}x + 10$

2. $r\cos\left(\theta - \frac{3\pi}{4}\right) = 2 \Rightarrow r\left(\cos\theta\cos\frac{3\pi}{4} + \sin\theta\sin\frac{3\pi}{4}\right) = 2 \Rightarrow -\frac{\sqrt{2}}{2}r\cos\theta + \frac{\sqrt{2}}{2}r\sin\theta = 2 \Rightarrow -\frac{\sqrt{2}}{2}x + \frac{\sqrt{2}}{2}y = 2 \Rightarrow -\sqrt{2}x + \sqrt{2}y = 4 \Rightarrow y = x + 2\sqrt{2}$

3. $r\cos\left(\theta - \frac{4\pi}{3}\right) = 3 \Rightarrow r\left(\cos\theta\cos\frac{4\pi}{3} + \sin\theta\sin\frac{4\pi}{3}\right) = 3 \Rightarrow -\frac{1}{2}r\cos\theta - \frac{\sqrt{3}}{2}r\sin\theta = 3 \Rightarrow -\frac{1}{2}x - \frac{\sqrt{3}}{2}y = 3 \Rightarrow x + \sqrt{3}y = -6 \Rightarrow y = -\frac{\sqrt{3}}{3}x - 2\sqrt{3}$

4. $r\cos\left(\theta - \left(-\frac{\pi}{4}\right)\right) = 4 \Rightarrow r\cos\left(\theta + \frac{\pi}{4}\right) = 4 \Rightarrow r\left(\cos\theta\cos\frac{\pi}{4} - \sin\theta\sin\frac{\pi}{4}\right) = 4 \Rightarrow \frac{\sqrt{2}}{2}r\cos\theta - \frac{\sqrt{2}}{2}r\sin\theta = 4 \Rightarrow \frac{\sqrt{2}}{2}x - \frac{\sqrt{2}}{2}y = 4 \Rightarrow \sqrt{2}x - \sqrt{2}y = 8 \Rightarrow y = x - 4\sqrt{2}$

5. $r\cos\left(\theta - \frac{\pi}{4}\right) = \sqrt{2} \Rightarrow r\left(\cos\theta\cos\frac{\pi}{4} + \sin\theta\sin\frac{\pi}{4}\right) = \sqrt{2} \Rightarrow \frac{1}{\sqrt{2}}r\cos\theta + \frac{1}{\sqrt{2}}r\sin\theta = \sqrt{2} \Rightarrow \frac{1}{\sqrt{2}}x + \frac{1}{\sqrt{2}}y = \sqrt{2} \Rightarrow x + y = 2 \Rightarrow y = 2 - x$

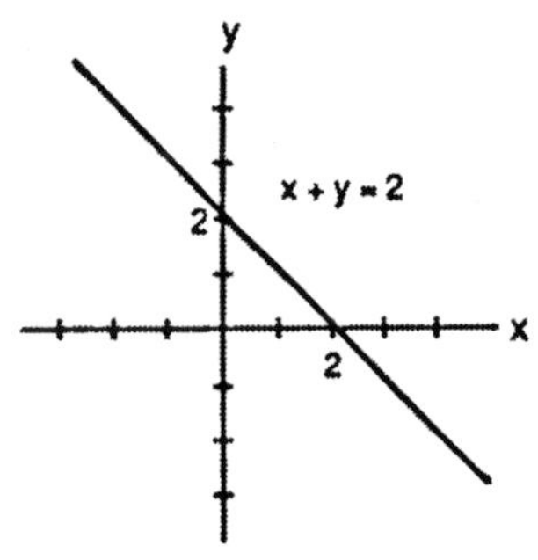

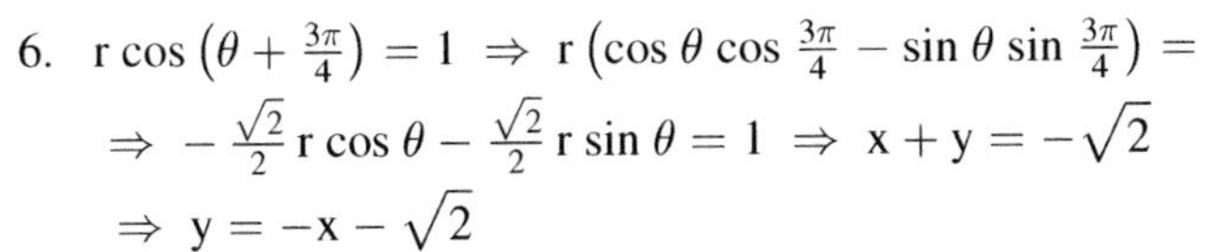

6. $r\cos\left(\theta + \frac{3\pi}{4}\right) = 1 \Rightarrow r\left(\cos\theta\cos\frac{3\pi}{4} - \sin\theta\sin\frac{3\pi}{4}\right) = 1 \Rightarrow -\frac{\sqrt{2}}{2}r\cos\theta - \frac{\sqrt{2}}{2}r\sin\theta = 1 \Rightarrow x + y = -\sqrt{2} \Rightarrow y = -x - \sqrt{2}$

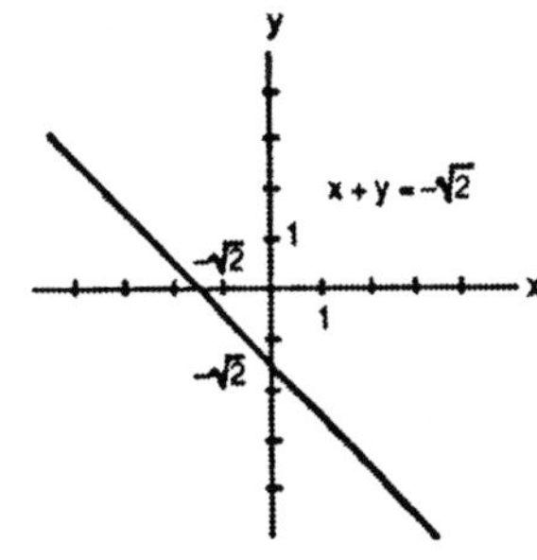

7. $r\cos\left(\theta - \frac{2\pi}{3}\right) = 3 \Rightarrow r\left(\cos\theta\cos\frac{2\pi}{3} + \sin\theta\sin\frac{2\pi}{3}\right) = 3 \Rightarrow -\frac{1}{2}r\cos\theta + \frac{\sqrt{3}}{2}r\sin\theta = 3 \Rightarrow -\frac{1}{2}x + \frac{\sqrt{3}}{2}y = 3 \Rightarrow -x + \sqrt{3}y = 6 \Rightarrow y = \frac{\sqrt{3}}{3}x + 2\sqrt{3}$

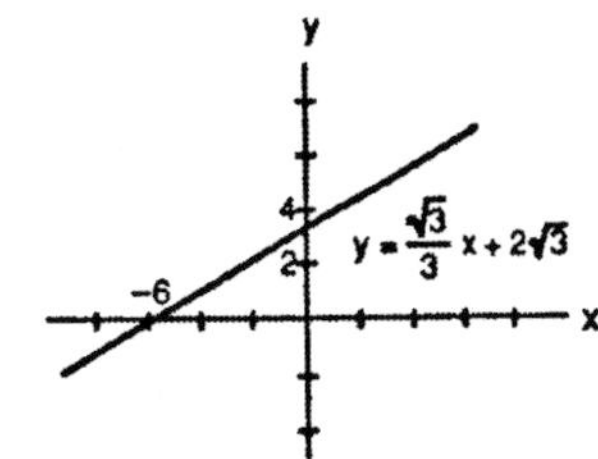

8. $r\cos\left(\theta+\frac{\pi}{3}\right)=2 \Rightarrow r\left(\cos\theta\cos\frac{\pi}{3}-\sin\theta\sin\frac{\pi}{3}\right)=2$
$\Rightarrow \frac{1}{2}r\cos\theta-\frac{\sqrt{3}}{2}r\sin\theta=2 \Rightarrow \frac{1}{2}x-\frac{\sqrt{3}}{2}y=2$
$\Rightarrow x-\sqrt{3}y=4 \Rightarrow y=\frac{\sqrt{3}}{3}x-\frac{4\sqrt{3}}{3}$

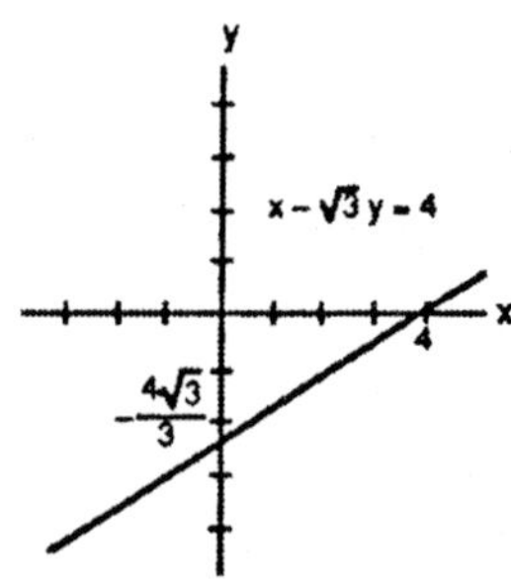

9. $\sqrt{2}x+\sqrt{2}y=6 \Rightarrow \sqrt{2}r\cos\theta+\sqrt{2}r\sin\theta=6 \Rightarrow r\left(\frac{\sqrt{2}}{2}\cos\theta+\frac{\sqrt{2}}{2}\sin\theta\right)=3 \Rightarrow r\left(\cos\frac{\pi}{4}\cos\theta+\sin\frac{\pi}{4}\sin\theta\right)$
$=3 \Rightarrow r\cos\left(\theta-\frac{\pi}{4}\right)=3$

10. $\sqrt{3}x-y=1 \Rightarrow \sqrt{3}r\cos\theta-r\sin\theta=1 \Rightarrow r\left(\frac{\sqrt{3}}{2}\cos\theta-\frac{1}{2}\sin\theta\right)=\frac{1}{2} \Rightarrow r\left(\cos\frac{\pi}{6}\cos\theta-\sin\frac{\pi}{6}\sin\theta\right)$
$=\frac{1}{2} \Rightarrow r\cos\left(\theta+\frac{\pi}{6}\right)=\frac{1}{2}$

11. $y=-5 \Rightarrow r\sin\theta=-5 \Rightarrow -r\sin\theta=5 \Rightarrow r\sin(-\theta)=5 \Rightarrow r\cos\left(\frac{\pi}{2}-(-\theta)\right)=5 \Rightarrow r\cos\left(\theta+\frac{\pi}{2}\right)=5$

12. $x=-4 \Rightarrow r\cos\theta=-4 \Rightarrow -r\cos\theta=4 \Rightarrow r\cos(\theta-\pi)=4$

13. $r=2(4)\cos\theta=8\cos\theta$

14. $r=-2(1)\sin\theta=-2\sin\theta$

15. $r=2\sqrt{2}\sin\theta$

16. $r=-2\left(\frac{1}{2}\right)\cos\theta=-\cos\theta$

17.

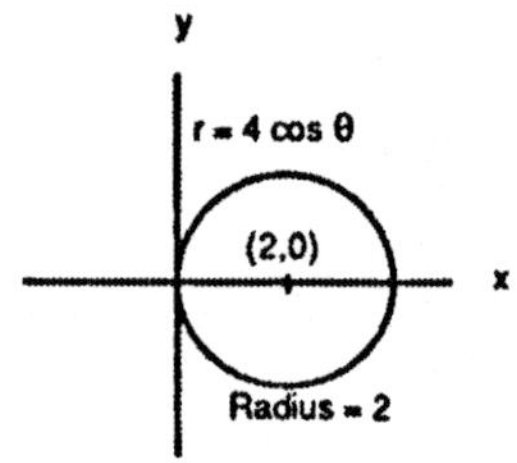

18.

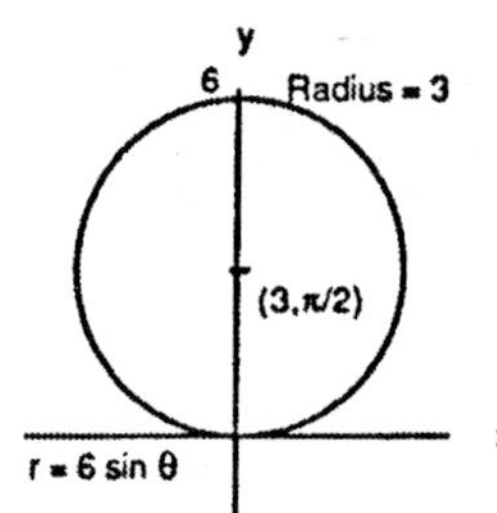

19.

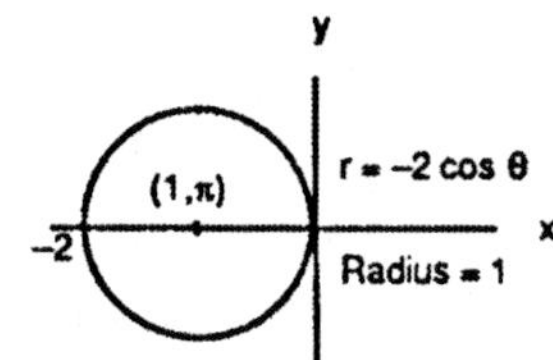

20.

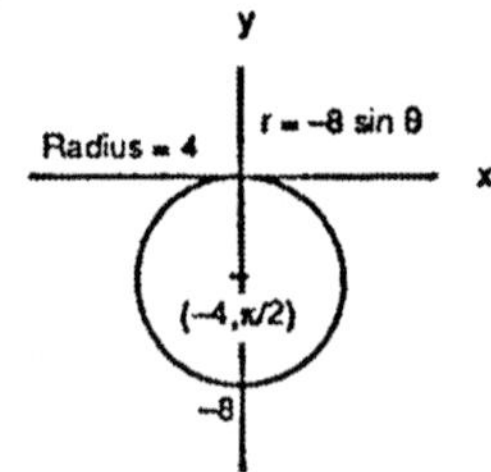

21. $(x-6)^2+y^2=36 \Rightarrow C=(6,0),\ a=6$
$\Rightarrow r=12\cos\theta$ is the polar equation

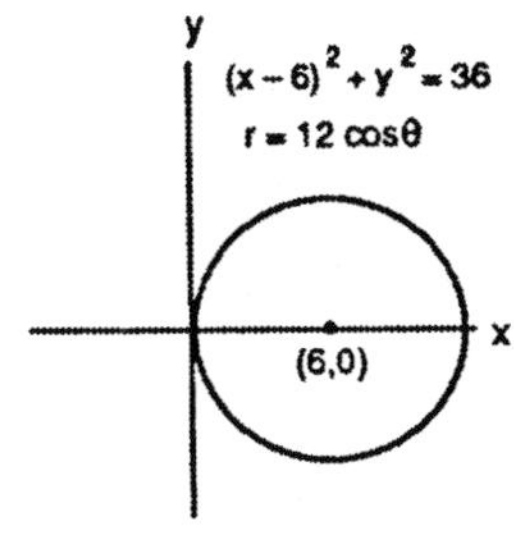

22. $(x+2)^2+y^2=4 \Rightarrow C=(-2,0),\ a=2$
$\Rightarrow r=-4\cos\theta$ is the polar equation

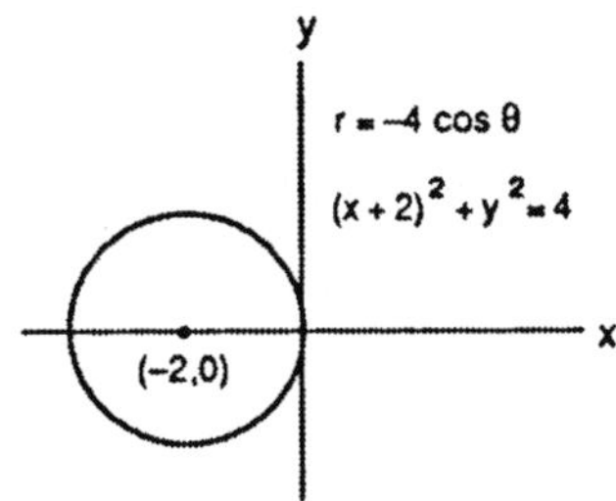

23. $x^2+(y-5)^2=25 \Rightarrow C=(0,5),\ a=5$
$\Rightarrow r=10\sin\theta$ is the polar equation

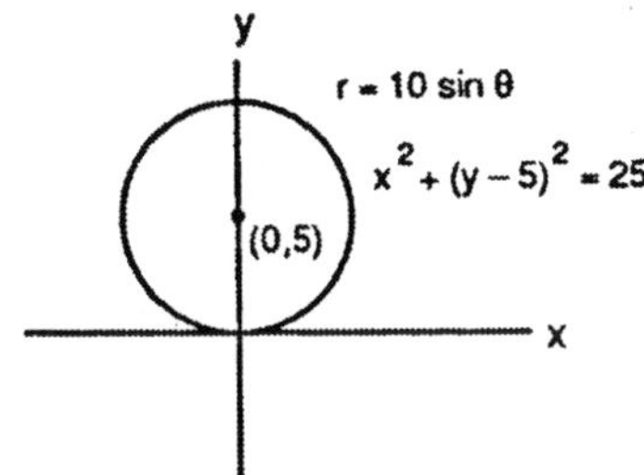

24. $x^2+(y+7)^2=49 \Rightarrow C=(0,-7),\ a=7$
$\Rightarrow r=-14\sin\theta$ is the polar equation

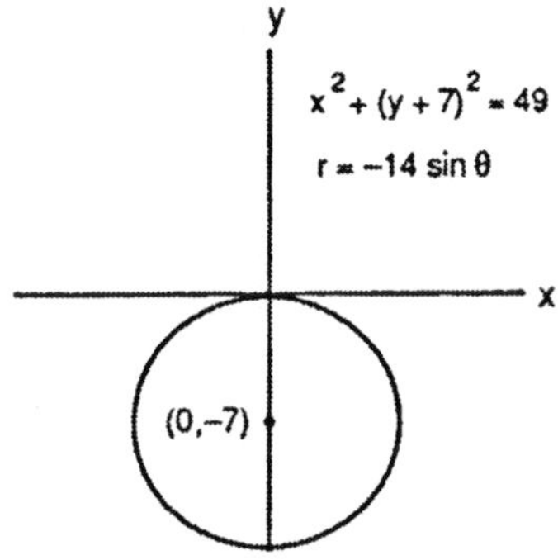

25. $x^2+2x+y^2=0 \Rightarrow (x+1)^2+y^2=1$
$\Rightarrow C=(-1,0),\ a=1 \Rightarrow r=-2\cos\theta$ is the polar equation

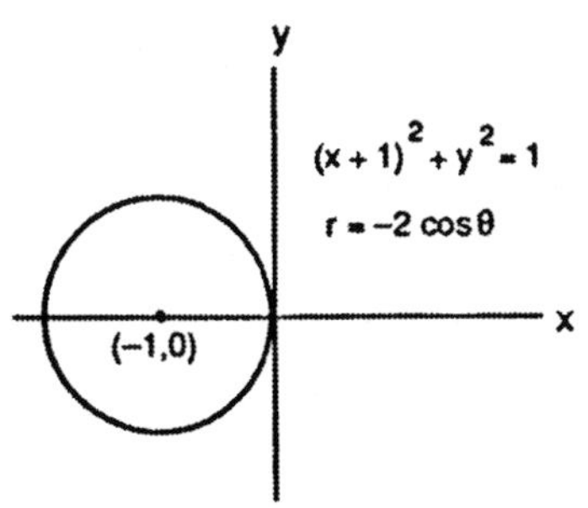

26. $x^2-16x+y^2=0 \Rightarrow (x-8)^2+y^2=64$
$\Rightarrow C=(8,0),\ a=8 \Rightarrow r=16\cos\theta$ is the polar equation

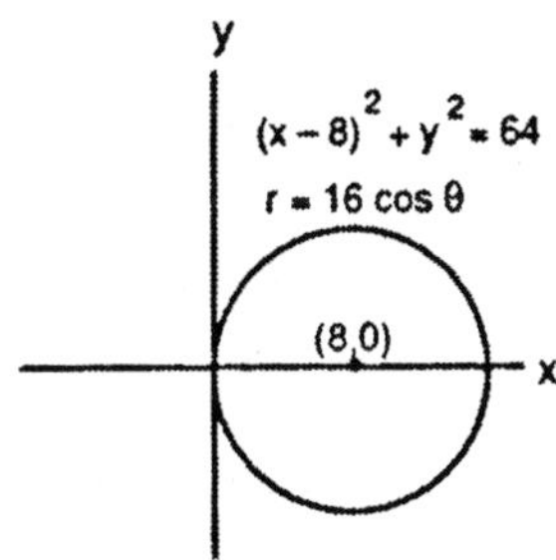

27. $x^2+y^2+y=0 \Rightarrow x^2+\left(y+\frac{1}{2}\right)^2=\frac{1}{4}$
$\Rightarrow C=\left(0,-\frac{1}{2}\right),\ a=\frac{1}{2} \Rightarrow r=-\sin\theta$ is the polar equation

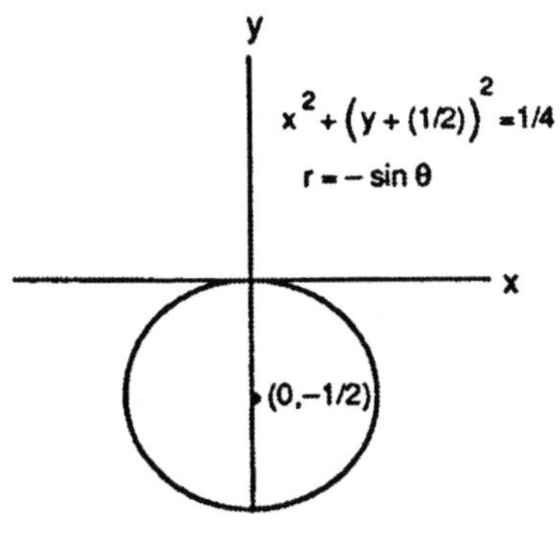

28. $x^2+y^2-\frac{4}{3}y=0 \Rightarrow x^2+\left(y-\frac{2}{3}\right)^2=\frac{4}{9}$
$\Rightarrow C=\left(0,\frac{2}{3}\right),\ a=\frac{2}{3} \Rightarrow r=\frac{4}{3}\sin\theta$ is the polar equation

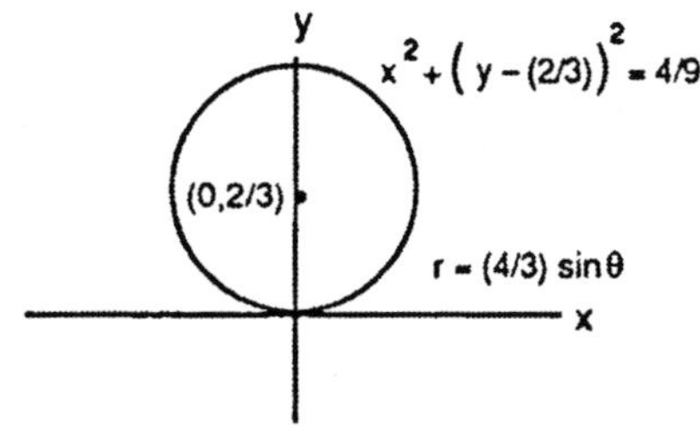

29. $e = 1, x = 2 \Rightarrow k = 2 \Rightarrow r = \frac{2(1)}{1+(1)\cos\theta} = \frac{2}{1+\cos\theta}$

30. $e = 1, y = 2 \Rightarrow k = 2 \Rightarrow r = \frac{2(1)}{1+(1)\sin\theta} = \frac{2}{1+\sin\theta}$

31. $e = 5, y = -6 \Rightarrow k = 6 \Rightarrow r = \frac{6(5)}{1-5\sin\theta} = \frac{30}{1-5\sin\theta}$

32. $e = 2, x = 4 \Rightarrow k = 4 \Rightarrow r = \frac{4(2)}{1+2\cos\theta} = \frac{8}{1+2\cos\theta}$

33. $e = \frac{1}{2}, x = 1 \Rightarrow k = 1 \Rightarrow r = \frac{\left(\frac{1}{2}\right)(1)}{1+\left(\frac{1}{2}\right)\cos\theta} = \frac{1}{2+\cos\theta}$

34. $e = \frac{1}{4}, x = -2 \Rightarrow k = 2 \Rightarrow r = \frac{\left(\frac{1}{4}\right)(2)}{1-\left(\frac{1}{4}\right)\cos\theta} = \frac{2}{4-\cos\theta}$

35. $e = \frac{1}{5}, x = -10 \Rightarrow k = 10 \Rightarrow r = \frac{\left(\frac{1}{5}\right)(10)}{1-\left(\frac{1}{5}\right)\sin\theta} = \frac{10}{5-\sin\theta}$

36. $e = \frac{1}{3}, y = 6 \Rightarrow k = 6 \Rightarrow r = \frac{\left(\frac{1}{3}\right)(6)}{1+\left(\frac{1}{3}\right)\sin\theta} = \frac{6}{3+\sin\theta}$

37. $r = \frac{1}{1+\cos\theta} \Rightarrow e = 1, k = 1 \Rightarrow x = 1$

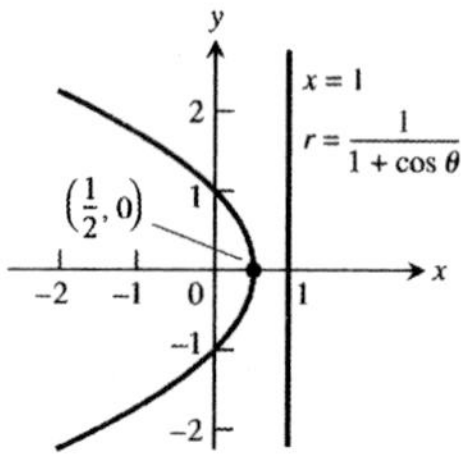

38. $r = \frac{6}{2+\cos\theta} = \frac{3}{1+\left(\frac{1}{2}\right)\cos\theta} \Rightarrow e = \frac{1}{2}, k = 6 \Rightarrow x = 6;$
$a\left(1-e^2\right) = ke \Rightarrow a\left[1-\left(\frac{1}{2}\right)^2\right] = 3 \Rightarrow \frac{3}{4}a = 3$
$\Rightarrow a = 4 \Rightarrow ea = 2$

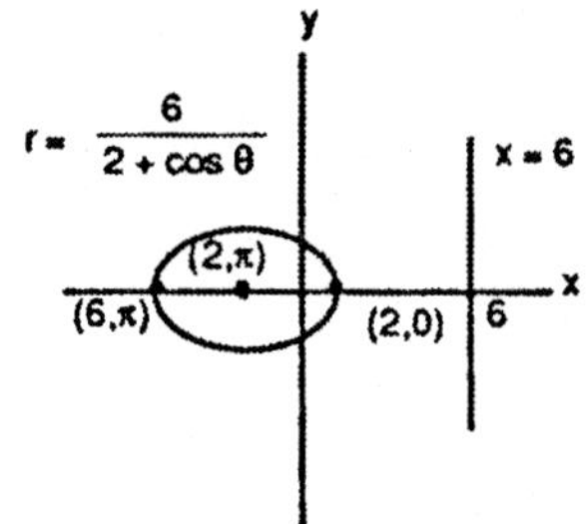

39. $r = \frac{25}{10-5\cos\theta} \Rightarrow r = \frac{\left(\frac{25}{10}\right)}{1-\left(\frac{5}{10}\right)\cos\theta} = \frac{\left(\frac{5}{2}\right)}{1-\left(\frac{1}{2}\right)\cos\theta}$
$\Rightarrow e = \frac{1}{2}, k = 5 \Rightarrow x = -5; a\left(1-e^2\right) = ke$
$\Rightarrow a\left[1-\left(\frac{1}{2}\right)^2\right] = \frac{5}{2} \Rightarrow \frac{3}{4}a = \frac{5}{2} \Rightarrow a = \frac{10}{3} \Rightarrow ea = \frac{5}{3}$

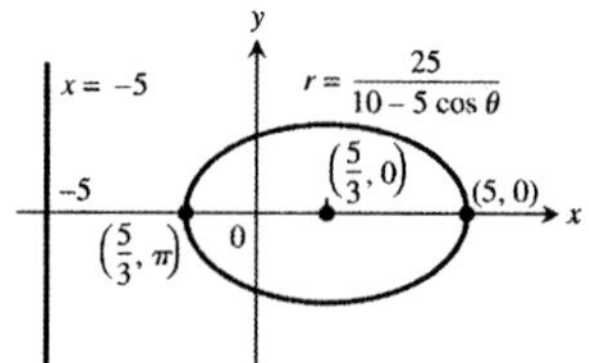

40. $r = \frac{4}{2-2\cos\theta} \Rightarrow r = \frac{2}{1-\cos\theta} \Rightarrow e = 1, k = 2 \Rightarrow x = -2$

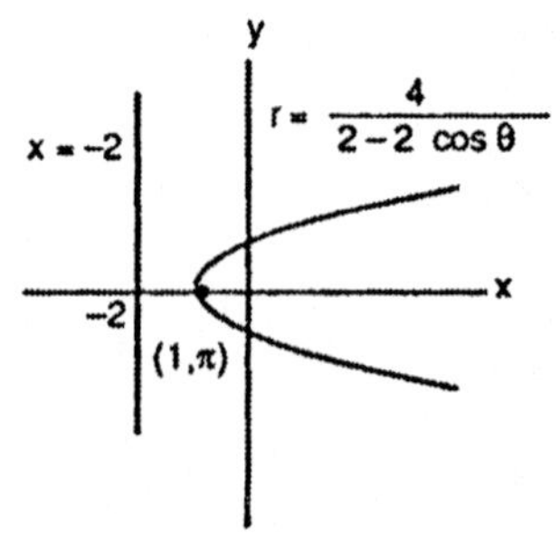

41. $r = \frac{400}{16+8\sin\theta} \Rightarrow r = \frac{\left(\frac{400}{16}\right)}{1+\left(\frac{8}{16}\right)\sin\theta} \Rightarrow r = \frac{25}{1+\left(\frac{1}{2}\right)\sin\theta}$
$e = \frac{1}{2}$, $k = 50 \Rightarrow y = 50$; $a\left(1-e^2\right) = ke$
$\Rightarrow a\left[1-\left(\frac{1}{2}\right)^2\right] = 25 \Rightarrow \frac{3}{4}a = 25 \Rightarrow a = \frac{100}{3}$
$\Rightarrow ea = \frac{50}{3}$

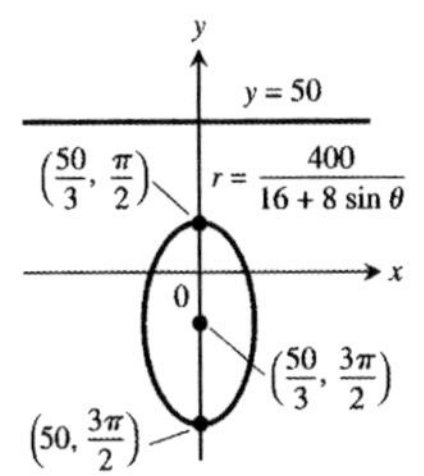

42. $r = \frac{12}{3+3\sin\theta} \Rightarrow r = \frac{4}{1+\sin\theta} \Rightarrow e = 1$,
$k = 4 \Rightarrow y = 4$

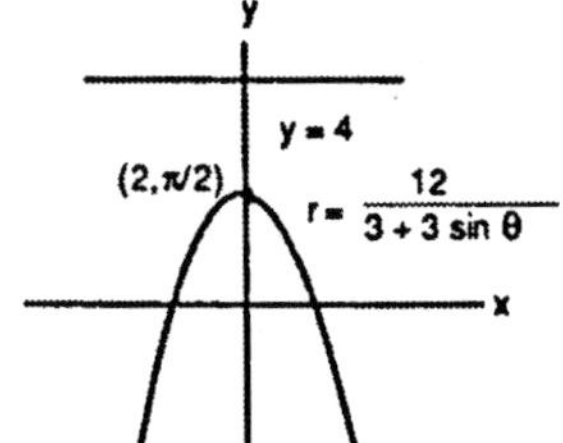

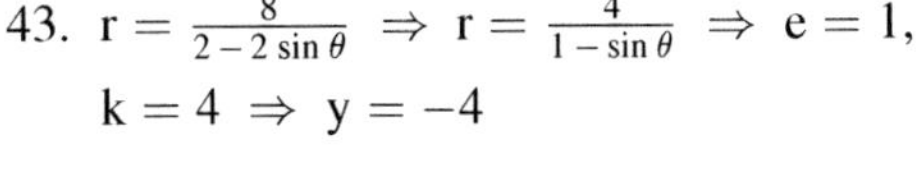
43. $r = \frac{8}{2-2\sin\theta} \Rightarrow r = \frac{4}{1-\sin\theta} \Rightarrow e = 1$,
$k = 4 \Rightarrow y = -4$

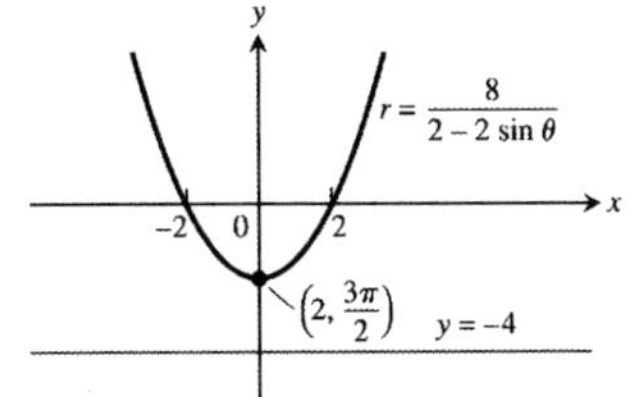

44. $r = \frac{4}{2-\sin\theta} \Rightarrow r = \frac{2}{1-\left(\frac{1}{2}\right)\sin\theta} \Rightarrow e = \frac{1}{2}, k = 4$
$\Rightarrow y = -4$; $a\left(1-e^2\right) = ke \Rightarrow a\left[1-\left(\frac{1}{2}\right)^2\right] = 2$
$\Rightarrow \frac{3}{4}a = 2 \Rightarrow a = \frac{8}{3} \Rightarrow ea = \frac{4}{3}$

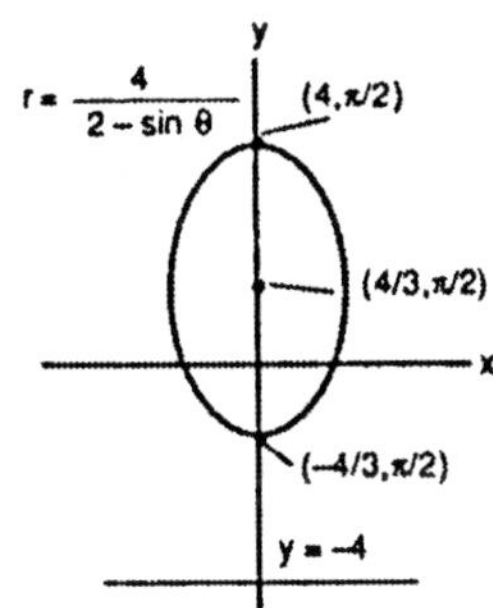

45.

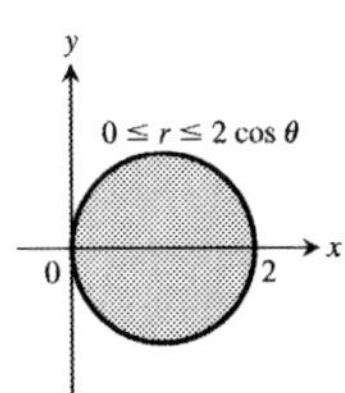

46.

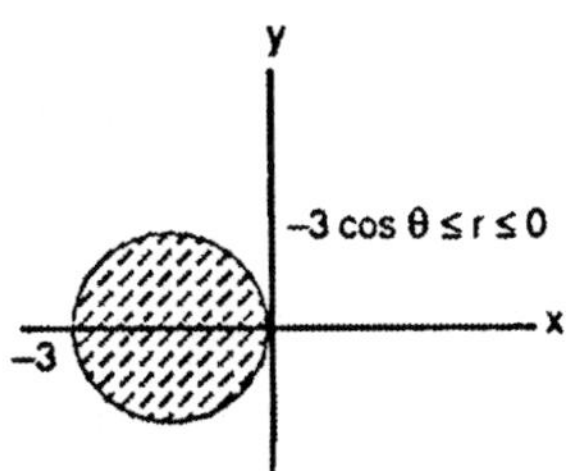

47.

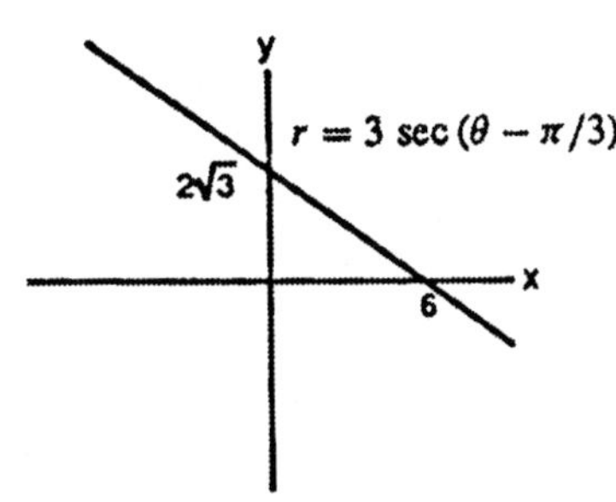

48.

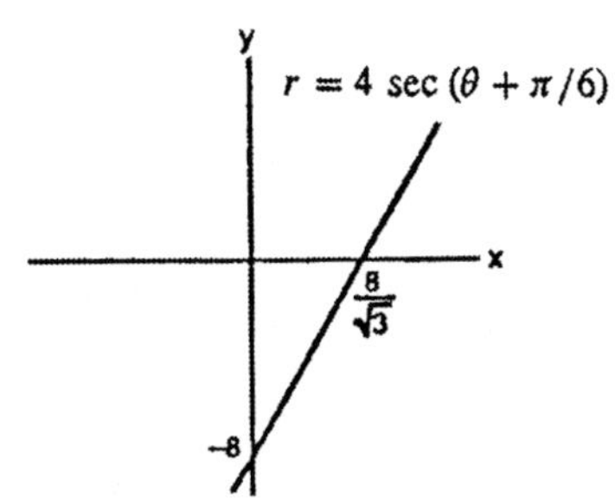

49.

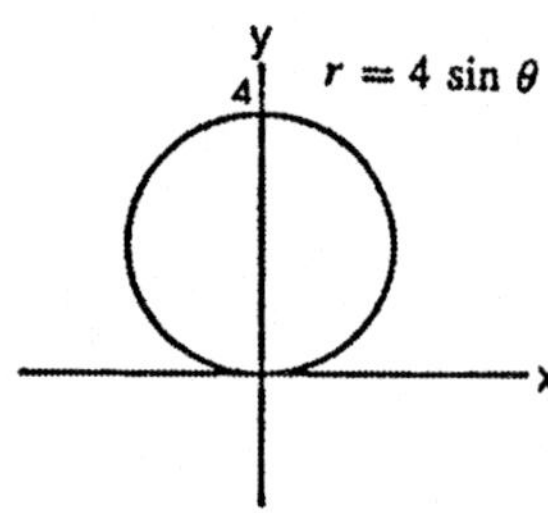

50.

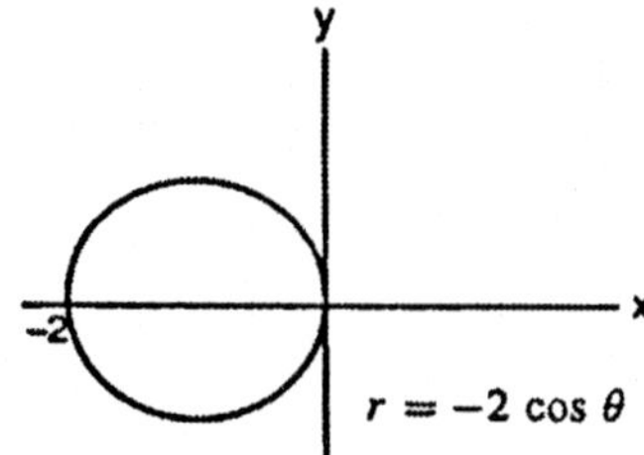

51.

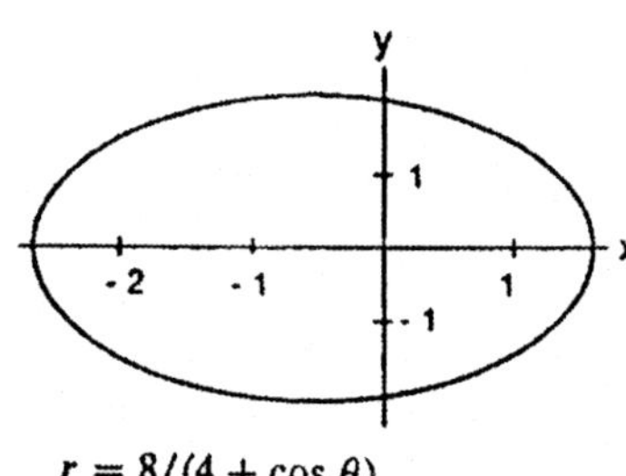

52.

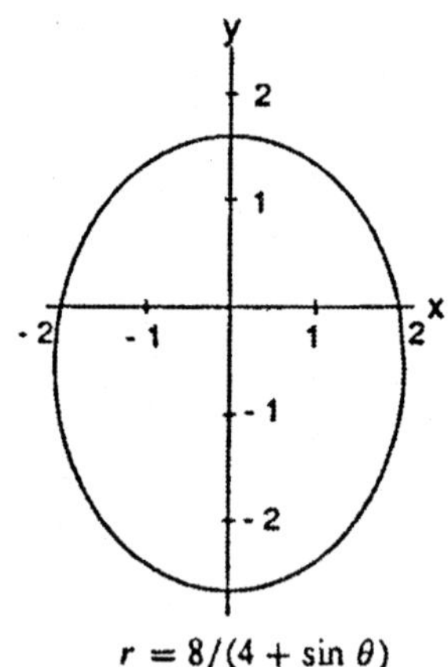

53.

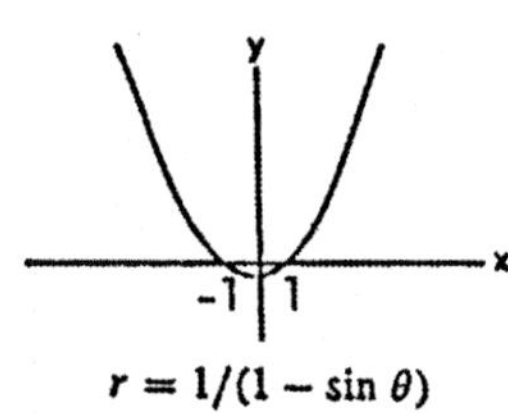

54.

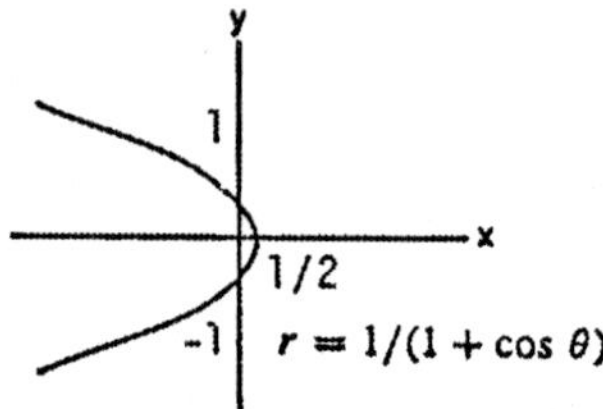

55.

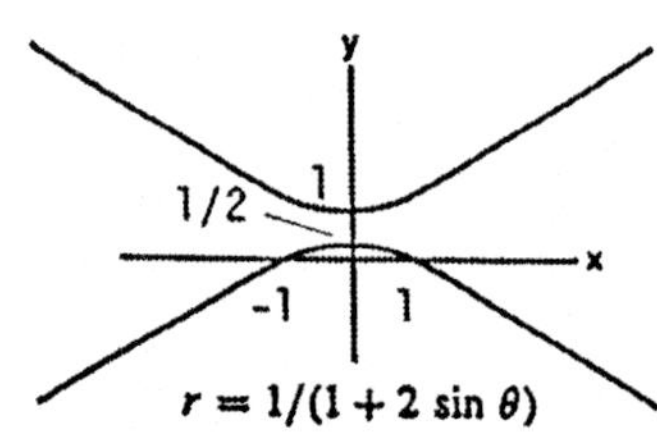

56.

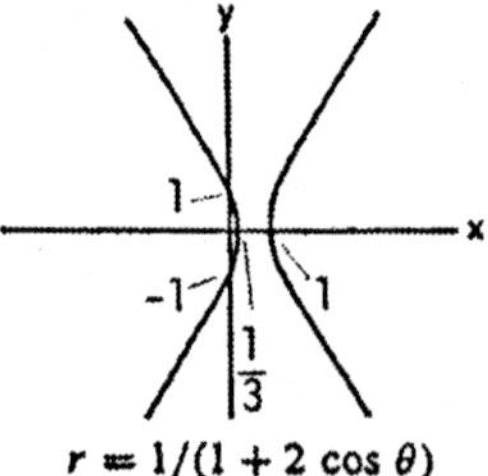

57. (a) Perihelion = a − ae = a(1 − e), Aphelion = ea + a = a(1 + e)

(b)

Planet	Perihelion	Aphelion
Mercury	0.3075 AU	0.4667 AU
Venus	0.7184 AU	0.7282 AU
Earth	0.9833 AU	1.0167 AU
Mars	1.3817 AU	1.6663 AU
Jupiter	4.9512 AU	5.4548 AU
Saturn	9.0210 AU	10.0570 AU
Uranus	18.2977 AU	20.0623 AU
Neptune	29.8135 AU	30.3065 AU
Pluto	29.6549 AU	49.2251 AU

58. Mercury: $r = \frac{(0.3871)(1-0.2056^2)}{1+0.2056 \cos\theta} = \frac{0.3707}{1+0.2056 \cos\theta}$

Venus: $r = \frac{(0.7233)(1-0.0068^2)}{1+0.0068 \cos\theta} = \frac{0.7233}{1+0.0068 \cos\theta}$

Earth: $r = \frac{1(1-0.0167^2)}{1+0.0167 \cos\theta} = \frac{0.9997}{1+0.0617 \cos\theta}$

Mars: $r = \frac{(1.524)(1-0.0934^2)}{1+0.0934 \cos\theta} = \frac{1.511}{1+0.0934 \cos\theta}$

Jupiter: $r = \frac{(5.203)(1-0.0484^2)}{1+0.0484 \cos\theta} = \frac{5.191}{1+0.0484 \cos\theta}$

Saturn: $r = \frac{(9.539)(1-0.0543^2)}{1+0.0543 \cos\theta} = \frac{9.511}{1+0.0543 \cos\theta}$

Uranus: $r = \frac{(19.18)(1-0.0460^2)}{1+0.0460 \cos\theta} = \frac{19.14}{1+0.0460 \cos\theta}$

Neptune: $r = \frac{(30.06)(1-0.0082^2)}{1+0.0082 \cos\theta} = \frac{30.06}{1+0.0082 \cos\theta}$

59. (a) $r = 4\sin\theta \Rightarrow r^2 = 4r\sin\theta \Rightarrow x^2 + y^2 = 4y$; $r = \sqrt{3}\sec\theta \Rightarrow r = \frac{\sqrt{3}}{\cos\theta} \Rightarrow r\cos\theta = \sqrt{3}$ $\Rightarrow x = \sqrt{3}$; $x = \sqrt{3} \Rightarrow \left(\sqrt{3}\right)^2 + y^2 = 4y$ $\Rightarrow y^2 - 4y + 3 = 0 \Rightarrow (y-3)(y-1) = 0 \Rightarrow y = 3$ or $y = 1$. Therefore in Cartesian coordinates, the points of intersection are $\left(\sqrt{3}, 3\right)$ and $\left(\sqrt{3}, 1\right)$. In polar coordinates, $4\sin\theta = \sqrt{3}\sec\theta \Rightarrow 4\sin\theta\cos\theta = \sqrt{3}$ $\Rightarrow 2\sin\theta\cos\theta = \frac{\sqrt{3}}{2} \Rightarrow \sin 2\theta = \frac{\sqrt{3}}{2} \Rightarrow 2\theta = \frac{\pi}{3}$ or $\frac{2\pi}{3} \Rightarrow \theta = \frac{\pi}{6}$ or $\frac{\pi}{3}$; $\theta = \frac{\pi}{6} \Rightarrow r = 2$, and $\theta = \frac{\pi}{3}$ $\Rightarrow r = 2\sqrt{3} \Rightarrow \left(2, \frac{\pi}{6}\right)$ and $\left(2\sqrt{3}, \frac{\pi}{3}\right)$ are the points of intersection in polar coordinates.

(b)

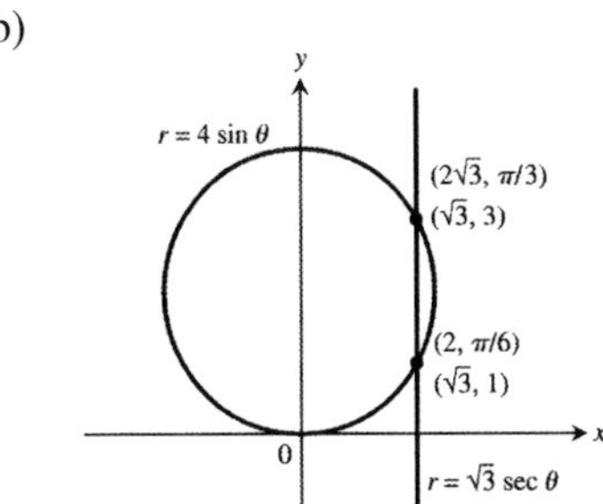

60. (a) $r = 8\cos\theta \Rightarrow r^2 = 8r\cos\theta \Rightarrow x^2 + y^2 = 8x$ $\Rightarrow x^2 - 8x + y^2 = 0 \Rightarrow (x-4)^2 + y^2 = 16$; $r = 2\sec\theta \Rightarrow r = \frac{2}{\cos\theta} \Rightarrow r\cos\theta = 2$ $\Rightarrow x = 2$; $x = 2 \Rightarrow 2^2 - 8(2) + y^2 = 0$ $\Rightarrow y^2 = 12 \Rightarrow y = \pm 2\sqrt{3}$. Therefore $\left(2, \pm 2\sqrt{3}\right)$ are the points of intersection in Cartesian coordinates. In polar coordinates, $8\cos\theta = 2\sec\theta \Rightarrow 8\cos^2\theta = 2$ $\Rightarrow \cos^2\theta = \frac{1}{4} \Rightarrow \cos\theta = \pm\frac{1}{2} \Rightarrow \theta = \frac{\pi}{3}, \frac{2\pi}{3}, \frac{4\pi}{3}$, or $\frac{5\pi}{3}$; $\theta = \frac{\pi}{3}$ and $\frac{5\pi}{3} \Rightarrow r = 4$, and $\theta = \frac{2\pi}{3}$ and $\frac{4\pi}{3}$ $\Rightarrow r = -4 \Rightarrow \left(4, \frac{\pi}{3}\right)$ and $\left(4, \frac{5\pi}{3}\right)$ are the points of intersection in polar coordinates. The points $\left(-4, \frac{2\pi}{3}\right)$ and $\left(-4, \frac{4\pi}{3}\right)$ are the same points.

(b)

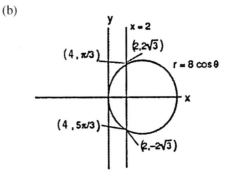

61. $r\cos\theta = 4 \Rightarrow x = 4 \Rightarrow k = 4$: parabola $\Rightarrow e = 1 \Rightarrow r = \frac{4}{1+\cos\theta}$

62. $r\cos\left(\theta - \frac{\pi}{2}\right) = 2 \Rightarrow r\left(\cos\theta\cos\frac{\pi}{2} + \sin\theta\sin\frac{\pi}{2}\right) = 2 \Rightarrow r\sin\theta = 2 \Rightarrow y = 2 \Rightarrow k = 2$: parabola $\Rightarrow e = 1$ $\Rightarrow r = \frac{2}{1+\sin\theta}$

63. (a) Let the ellipse be the orbit, with the Sun at one focus. Then $r_{max} = a + c$ and $r_{min} = a - c \Rightarrow \frac{r_{max} - r_{min}}{r_{max} + r_{min}}$ $= \frac{(a+c)-(a-c)}{(a+c)+(a-c)} = \frac{2c}{2a} = \frac{c}{a} = e$

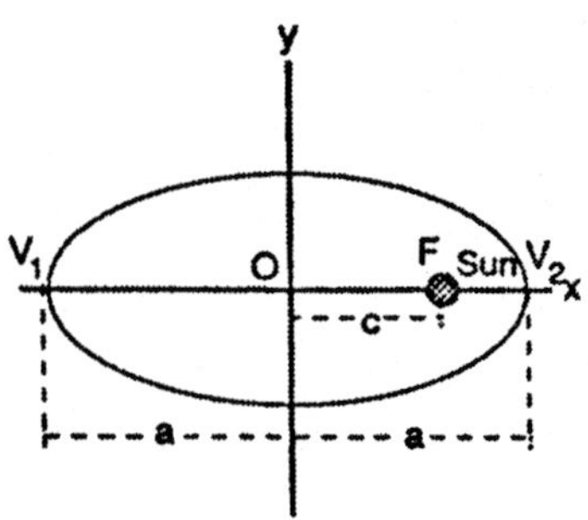

(b) Let F_1, F_2 be the foci. Then $PF_1 + PF_2 = 10$ where P is any point on the ellipse. If P is a vertex, then $PF_1 = a + c$ and $PF_2 = a - c$ $\Rightarrow (a + c) + (a - c) = 10$ $\Rightarrow 2a = 10 \Rightarrow a = 5$. Since $e = \frac{c}{a}$ we have $0.2 = \frac{c}{5}$ $\Rightarrow c = 1.0 \Rightarrow$ the pins should be 2 inches apart.

64. $e = 0.97$, Major axis $= 36.18$ AU $\Rightarrow a = 18.09$, Minor axis $= 9.12$ AU $\Rightarrow b = 4.56$ (1 AU $\approx 1.49 \times 10^8$ km)

(a) $r = \frac{ke}{1+e\cos\theta} = \frac{a(1-e^2)}{1+e\cos\theta} = \frac{(18.09)[1-(0.97)^2]}{1+0.97\cos\theta} = \frac{1.07}{1+0.97\cos\theta}$ AU

(b) $\theta = 0 \Rightarrow r = \frac{1.07}{1+0.97} \approx 0.5431$ AU $\approx 8.09 \times 10^7$ km

(c) $\theta = \pi \Rightarrow r = \frac{1.07}{1-0.97} \approx 35.7$ AU $\approx 5.32 \times 10^9$ km

65. $x^2 + y^2 - 2ay = 0 \Rightarrow (r\cos\theta)^2 + (r\sin\theta)^2 - 2ar\sin\theta = 0$ $\Rightarrow r^2\cos^2\theta + r^2\sin^2\theta - 2ar\sin\theta = 0 \Rightarrow r^2 = 2ar\sin\theta$ $\Rightarrow r = 2a\sin\theta$

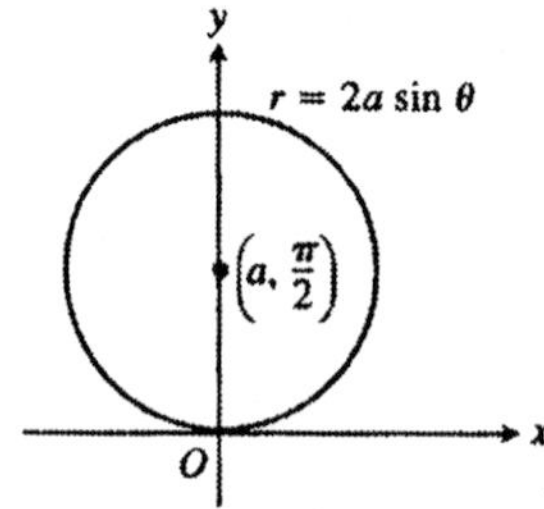

66. $y^2 = 4ax + 4a^2 \Rightarrow (r\sin\theta)^2 = 4ar\cos\theta + 4a^2 \Rightarrow r^2\sin^2\theta$ $= 4ar\cos\theta + 4a^2 \Rightarrow r^2(1 - \cos^2\theta) = 4ar\cos\theta + 4a^2$ $\Rightarrow r^2 - r^2\cos^2\theta = 4ar\cos\theta + 4a^2 \Rightarrow r^2$ $= r^2\cos^2\theta + 4ar\cos\theta + 4a^2 \Rightarrow r^2 = (r\cos\theta + 2a)^2$ $\Rightarrow r = \pm(r\cos\theta + 2a) \Rightarrow r - r\cos\theta = 2a$ or $r + r\cos\theta = -2a \Rightarrow r = \frac{2a}{1-\cos\theta}$ or $r = \frac{-2a}{1+\cos\theta}$; the equations have the same graph, which is a parabola opening to the right

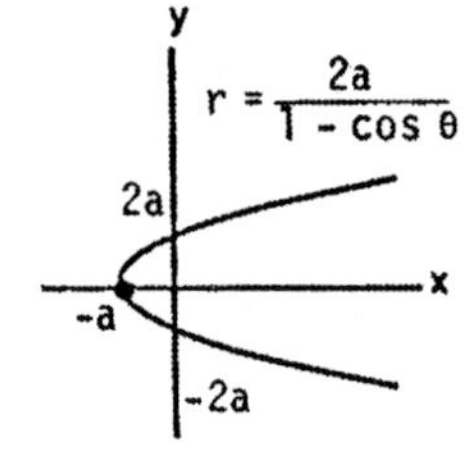

67. $x\cos\alpha + y\sin\alpha = p \Rightarrow r\cos\theta\cos\alpha + r\sin\theta\sin\alpha = p$ $\Rightarrow r(\cos\theta\cos\alpha + \sin\theta\sin\alpha) = p \Rightarrow r\cos(\theta - \alpha) = p$

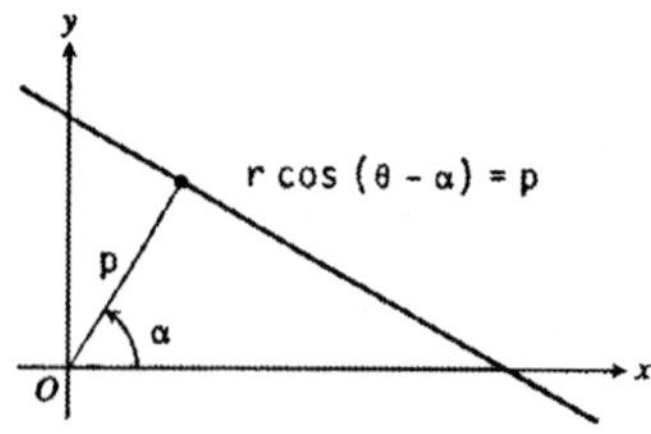

68. $(x^2 + y^2)^2 + 2ax(x^2 + y^2) - a^2y^2 = 0$ $\Rightarrow (r^2)^2 + 2a(r\cos\theta)(r^2) - a^2(r\sin\theta)^2 = 0$ $\Rightarrow r^4 + 2ar^3\cos\theta - a^2r^2\sin^2\theta = 0$ $\Rightarrow r^2[r^2 + 2ar\cos\theta - a^2(1 - \cos^2\theta)] = 0$ (assume $r \neq 0$) $\Rightarrow r^2 + 2ar\cos\theta - a^2 + a^2\cos^2\theta = 0$ $\Rightarrow (r^2 + 2ar\cos\theta + a^2\cos^2\theta) - a^2 = 0$ $\Rightarrow (r + a\cos\theta)^2 = a^2 \Rightarrow r + a\cos\theta = \pm a$ $\Rightarrow r = a(1 - \cos\theta)$ or $r = -a(1 + \cos\theta)$;

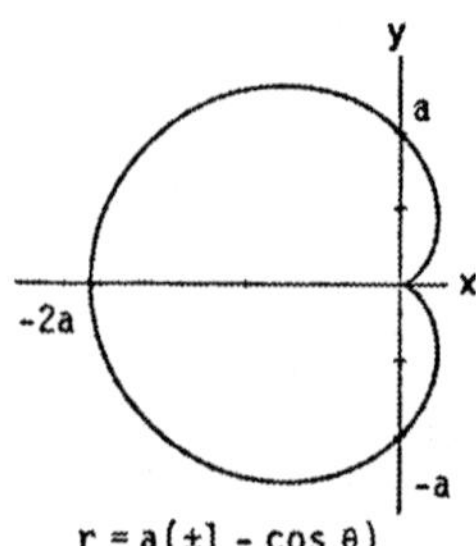

the equations have the same graph, which is a cardioid

69 - 70. Example CAS commands:

Maple:

```
with( plots );#69
f := (r,k,e) -> k*e/(1+e*cos(theta));
elist := [3/4,1,5/4];                          # (a)
P1 := seq( plot( f(r,-2,e), theta=-Pi..Pi, coords=polar ), e=elist ):
display( [P1], insequence=true, view=[-20..20,-20..20], title="#69(a) (Section 10.8)\nk=-2" );
P2 := seq( plot( f(r,2,e), theta=-Pi..Pi, coords=polar ), e=elist ):
display( [P2], insequence=true, view=[-20..20,-20..20], title="#69(a) (Section 10.8)\nk=2" );
elist2 := [7/6,5/4,4/3,3/2,2,3,5,10,20];       # (b)
P3 := seq( plot( f(r,-1,e), theta=-Pi..Pi, coords=polar ), e=elist2 ):
display( [P3], insequence=true, view=[-20..20,-20..20], title="#69(b) (Section 10.8)\nk=-1, e>1" );
elist3 := [1/2,1/3,1/4,1/10,1/20];
P4 := seq( plot( f(r,-1,e), theta=-Pi..Pi, coords=polar ), e=elist3 ):
display( [P4], insequence=true, title="#69(b) (Section 10.8)\nk=-1, e<1" );
klist := -5..-1;                               # (c)
P5 := seq( plot( f(r,k,1/2), theta=-Pi..Pi, coords=polar ), k=klist ):
display( [P5], insequence=true, title="#69(c) (Section 10.8)\ne=1/2, k<0" );
P6 := seq( plot( f(r,k,1), theta=-Pi..Pi, coords=polar ), k=klist ):
display( [P6], insequence=true, view=[-4..50,-50..50], title="#69(c) (Section 10.8)\ne=1, k<0" );
P7 := seq( plot( f(r,k,2), theta=-Pi..Pi, coords=polar ), k=klist ):
display( [P5], insequence=true, title="#69(c) (Section 10.8)\ne=2, k<0" );
```

Mathematica: (assigned function and values for parameters and bounds may vary):

To do polar plots in Mathematica, it is necessary to first load a graphics package

In the **PolarPlot** command, it is assumed that the variable r is given as a function of the variable θ.

```
<<Graphics`Graphics`
f[θ_, k_,ec_]:= (k ec) / (1 + ec Cos[θ])
PolarPlot[{ f[θ, −2, 3/4], f[θ, −2, 1], f[θ, −2, 5/4]}, {θ, 0, 2π}, PlotRange → {−20, 20},
      PlotStyle → {RGBColor[1, 0, 0], RGBColor[0, 1, 0], RGBColor[0, 0, 1]}];
PolarPlot[{ f[θ, −1, 1], f[θ, −2, 1], f[θ, −3, 1], f[θ, −4, 1], f[θ, −5, 1]}, {θ, 0, 2π}, PlotRange → {−20, 20},
      PlotStyle →
      {RGBColor[1, 0, 0], RGBColor[0, 1, 0], RGBColor[0, 0, 1,, RGBColor[.5, .5, 0], RGBColor[0, .5, .5]}];
```

The limitation on the range is primarily needed when plotting hyperbolas.

Problem 70 can be done in a similar fashion.

CHAPTER 10 PRACTICE EXERCISES

1. $x^2 = -4y \Rightarrow y = -\frac{x^2}{4} \Rightarrow 4p = 4 \Rightarrow p = 1$; therefore Focus is $(0, -1)$, Directrix is $y = 1$

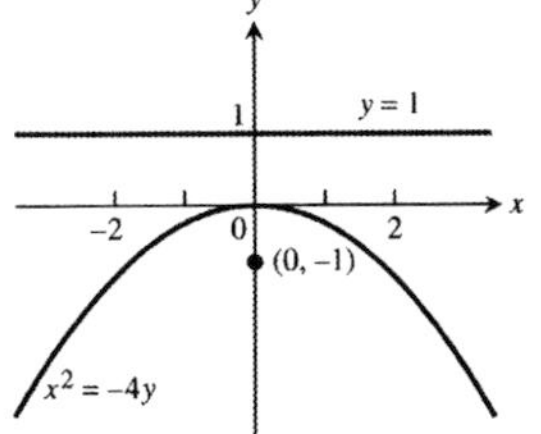

2. $x^2 = 2y \Rightarrow \frac{x^2}{2} = y \Rightarrow 4p = 2 \Rightarrow p = \frac{1}{2}$; therefore Focus is $\left(0, \frac{1}{2}\right)$; Directrix is $y = -\frac{1}{2}$

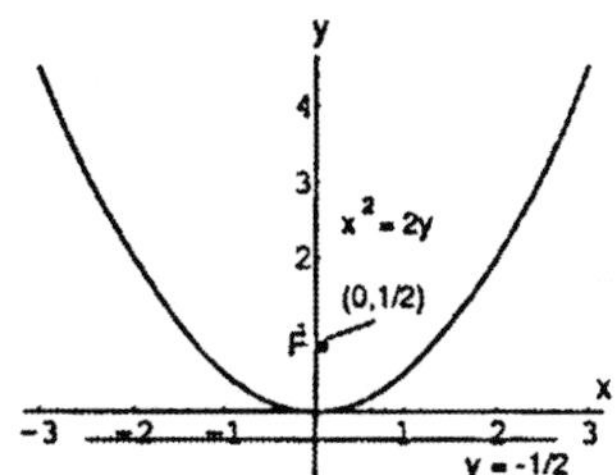

3. $y^2 = 3x \Rightarrow x = \frac{y^2}{3} \Rightarrow 4p = 3 \Rightarrow p = \frac{3}{4}$; therefore Focus is $\left(\frac{3}{4}, 0\right)$, Directrix is $x = -\frac{3}{4}$

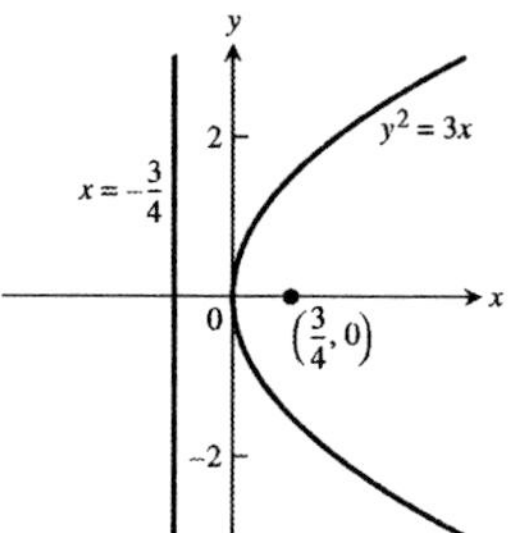

4. $y^2 = -\frac{8}{3}x \Rightarrow x = -\frac{y^2}{\left(\frac{8}{3}\right)} \Rightarrow 4p = \frac{8}{3} \Rightarrow p = \frac{2}{3}$; therefore Focus is $\left(-\frac{2}{3}, 0\right)$, Directrix is $x = \frac{2}{3}$

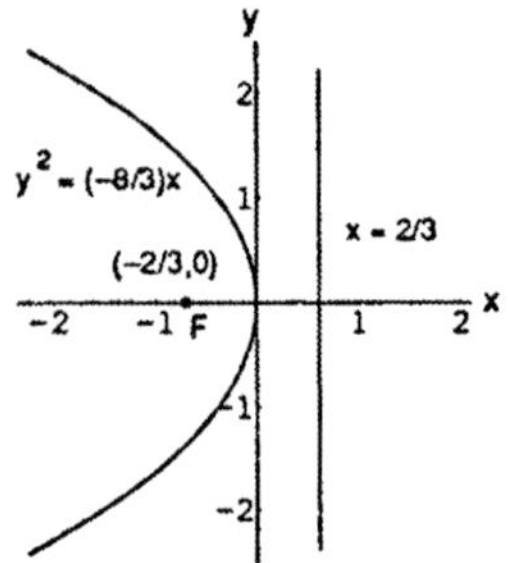

5. $16x^2 + 7y^2 = 112 \Rightarrow \frac{x^2}{7} + \frac{y^2}{16} = 1$
$\Rightarrow c^2 = 16 - 7 = 9 \Rightarrow c = 3; e = \frac{c}{a} = \frac{3}{4}$

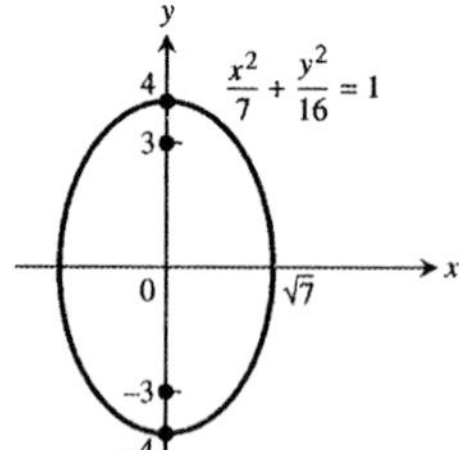

6. $x^2 + 2y^2 = 4 \Rightarrow \frac{x^2}{4} + \frac{y^2}{2} = 1 \Rightarrow c^2 = 4 - 2 = 2$
$\Rightarrow c = \sqrt{2}; e = \frac{c}{a} = \frac{\sqrt{2}}{2}$

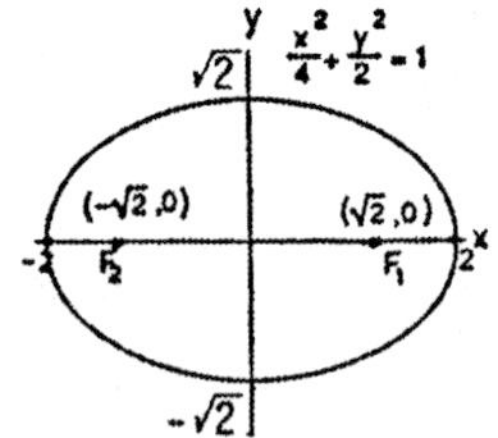

7. $3x^2 - y^2 = 3 \Rightarrow x^2 - \frac{y^2}{3} = 1 \Rightarrow c^2 = 1 + 3 = 4$
$\Rightarrow c = 2; e = \frac{c}{a} = \frac{2}{1} = 2$; the asymptotes are
$y = \pm\sqrt{3}x$

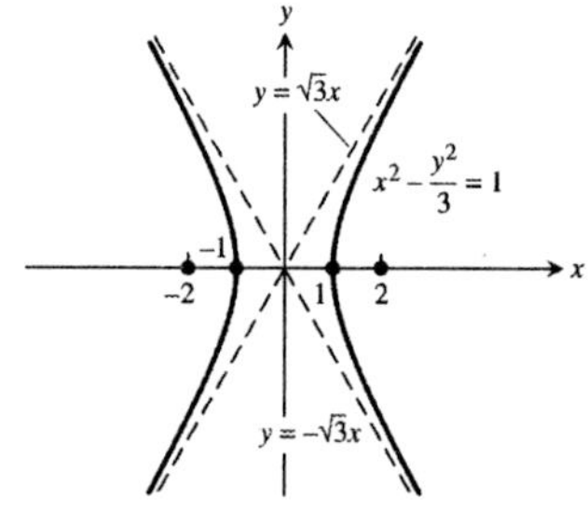

8. $5y^2 - 4x^2 = 20 \Rightarrow \frac{y^2}{4} - \frac{x^2}{5} = 1 \Rightarrow c^2 = 4 + 5 = 9$
$\Rightarrow c = 3, e = \frac{c}{a} = \frac{3}{2}$; the asymptotes are $y = \pm\frac{2}{\sqrt{5}}x$

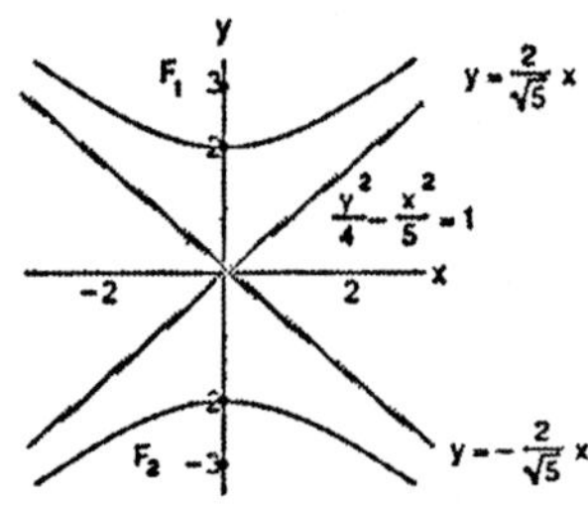

9. $x^2 = -12y \Rightarrow -\frac{x^2}{12} = y \Rightarrow 4p = 12 \Rightarrow p = 3 \Rightarrow$ focus is $(0, -3)$, directrix is $y = 3$, vertex is $(0, 0)$; therefore new vertex is $(2, 3)$, new focus is $(2, 0)$, new directrix is $y = 6$, and the new equation is $(x - 2)^2 = -12(y - 3)$

10. $y^2 = 10x \Rightarrow \frac{y^2}{10} = x \Rightarrow 4p = 10 \Rightarrow p = \frac{5}{2} \Rightarrow$ focus is $\left(\frac{5}{2}, 0\right)$, directrix is $x = -\frac{5}{2}$, vertex is $(0, 0)$; therefore new vertex is $\left(-\frac{1}{2}, -1\right)$, new focus is $(2, -1)$, new directrix is $x = -3$, and the new equation is $(y + 1)^2 = 10\left(x + \frac{1}{2}\right)$

11. $\frac{x^2}{9} + \frac{y^2}{25} = 1 \Rightarrow a = 5$ and $b = 3 \Rightarrow c = \sqrt{25 - 9} = 4 \Rightarrow$ foci are $(0, \pm 4)$, vertices are $(0, \pm 5)$, center is $(0, 0)$; therefore the new center is $(-3, -5)$, new foci are $(-3, -1)$ and $(-3, -9)$, new vertices are $(-3, -10)$ and $(-3, 0)$, and the new equation is $\frac{(x+3)^2}{9} + \frac{(y+5)^2}{25} = 1$

12. $\frac{x^2}{169} + \frac{y^2}{144} = 1 \Rightarrow a = 13$ and $b = 12 \Rightarrow c = \sqrt{169 - 144} = 5 \Rightarrow$ foci are $(\pm 5, 0)$, vertices are $(\pm 13, 0)$, center is $(0, 0)$; therefore the new center is $(5, 12)$, new foci are $(10, 12)$ and $(0, 12)$, new vertices are $(18, 12)$ and

$(-8, 12)$, and the new equation is $\frac{(x-5)^2}{169} + \frac{(y-12)^2}{144} = 1$

13. $\frac{y^2}{8} - \frac{x^2}{2} = 1 \Rightarrow a = 2\sqrt{2}$ and $b = \sqrt{2} \Rightarrow c = \sqrt{8+2} = \sqrt{10} \Rightarrow$ foci are $\left(0, \pm\sqrt{10}\right)$, vertices are $\left(0, \pm 2\sqrt{2}\right)$, center is $(0, 0)$, and the asymptotes are $y = \pm 2x$; therefore the new center is $\left(2, 2\sqrt{2}\right)$, new foci are $\left(2, 2\sqrt{2} \pm \sqrt{10}\right)$, new vertices are $\left(2, 4\sqrt{2}\right)$ and $(2, 0)$, the new asymptotes are $y = 2x - 4 + 2\sqrt{2}$ and $y = -2x + 4 + 2\sqrt{2}$; the new equation is $\frac{\left(y - 2\sqrt{2}\right)^2}{8} - \frac{(x-2)^2}{2} = 1$

14. $\frac{x^2}{36} - \frac{y^2}{64} = 1 \Rightarrow a = 6$ and $b = 8 \Rightarrow c = \sqrt{36+64} = 10 \Rightarrow$ foci are $(\pm 10, 0)$, vertices are $(\pm 6, 0)$, the center is $(0, 0)$ and the asymptotes are $\frac{y}{8} = \pm \frac{x}{6}$ or $y = \pm \frac{4}{3}x$; therefore the new center is $(-10, -3)$, the new foci are $(-20, -3)$ and $(0, -3)$, the new vertices are $(-16, -3)$ and $(-4, -3)$, the new asymptotes are $y = \frac{4}{3}x + \frac{31}{3}$ and $y = -\frac{4}{3}x - \frac{49}{3}$; the new equation is $\frac{(x+10)^2}{36} - \frac{(y+3)^2}{64} = 1$

15. $x^2 - 4x - 4y^2 = 0 \Rightarrow x^2 - 4x + 4 - 4y^2 = 4 \Rightarrow (x-2)^2 - 4y^2 = 4 \Rightarrow \frac{(x-2)^2}{4} - y^2 = 1$, a hyperbola; $a = 2$ and $b = 1 \Rightarrow c = \sqrt{1+4} = \sqrt{5}$; the center is $(2, 0)$, the vertices are $(0, 0)$ and $(4, 0)$; the foci are $\left(2 \pm \sqrt{5}, 0\right)$ and the asymptotes are $y = \pm \frac{x-2}{2}$

16. $4x^2 - y^2 + 4y = 8 \Rightarrow 4x^2 - y^2 + 4y - 4 = 4 \Rightarrow 4x^2 - (y-2)^2 = 4 \Rightarrow x^2 - \frac{(y-2)^2}{4} = 1$, a hyperbola; $a = 1$ and $b = 2 \Rightarrow c = \sqrt{1+4} = \sqrt{5}$; the center is $(0, 2)$, the vertices are $(1, 2)$ and $(-1, 2)$, the foci are $\left(\pm\sqrt{5}, 2\right)$ and the asymptotes are $y = \pm 2x + 2$

17. $y^2 - 2y + 16x = -49 \Rightarrow y^2 - 2y + 1 = -16x - 48 \Rightarrow (y-1)^2 = -16(x+3)$, a parabola; the vertex is $(-3, 1)$; $4p = 16 \Rightarrow p = 4 \Rightarrow$ the focus is $(-7, 1)$ and the directrix is $x = 1$

18. $x^2 - 2x + 8y = -17 \Rightarrow x^2 - 2x + 1 = -8y - 16 \Rightarrow (x-1)^2 = -8(y+2)$, a parabola; the vertex is $(1, -2)$; $4p = 8 \Rightarrow p = 2 \Rightarrow$ the focus is $(1, -4)$ and the directrix is $y = 0$

19. $9x^2 + 16y^2 + 54x - 64y = -1 \Rightarrow 9(x^2 + 6x) + 16(y^2 - 4y) = -1 \Rightarrow 9(x^2 + 6x + 9) + 16(y^2 - 4y + 4) = 144$ $\Rightarrow 9(x+3)^2 + 16(y-2)^2 = 144 \Rightarrow \frac{(x+3)^2}{16} + \frac{(y-2)^2}{9} = 1$, an ellipse; the center is $(-3, 2)$; $a = 4$ and $b = 3$ $\Rightarrow c = \sqrt{16-9} = \sqrt{7}$; the foci are $\left(-3 \pm \sqrt{7}, 2\right)$; the vertices are $(1, 2)$ and $(-7, 2)$

20. $25x^2 + 9y^2 - 100x + 54y = 44 \Rightarrow 25(x^2 - 4x) + 9(y^2 + 6y) = 44 \Rightarrow 25(x^2 - 4x + 4) + 9(y^2 + 6y + 9) = 225$ $\Rightarrow \frac{(x-2)^2}{9} + \frac{(y+3)^2}{25} = 1$, an ellipse; the center is $(2, -3)$; $a = 5$ and $b = 3 \Rightarrow c = \sqrt{25-9} = 4$; the foci are $(2, 1)$ and $(2, -7)$; the vertices are $(2, 2)$ and $(2, -8)$

21. $x^2 + y^2 - 2x - 2y = 0 \Rightarrow x^2 - 2x + 1 + y^2 - 2y + 1 = 2 \Rightarrow (x-1)^2 + (y-1)^2 = 2$, a circle with center $(1, 1)$ and radius $= \sqrt{2}$

22. $x^2 + y^2 + 4x + 2y = 1 \Rightarrow x^2 + 4x + 4 + y^2 + 2y + 1 = 6 \Rightarrow (x+2)^2 + (y+1)^2 = 6$, a circle with center $(-2, -1)$ and radius $= \sqrt{6}$

23. $B^2 - 4AC = 1 - 4(1)(1) = -3 < 0 \Rightarrow$ ellipse

24. $B^2 - 4AC = 4^2 - 4(1)(4) = 0 \Rightarrow$ parabola

25. $B^2 - 4AC = 3^2 - 4(1)(2) = 1 > 0 \Rightarrow$ hyperbola

26. $B^2 - 4AC = 2^2 - 4(1)(-2) = 12 > 0 \Rightarrow$ hyperbola

27. $x^2 - 2xy + y^2 = 0 \Rightarrow (x - y)^2 = 0 \Rightarrow x - y = 0$ or $y = x$, a straight line

28. $B^2 - 4AC = (-3)^2 - 4(1)(4) = -7 < 0 \Rightarrow$ ellipse

29. $B^2 - 4AC = 1^2 - 4(2)(2) = -15 < 0 \Rightarrow$ ellipse; $\cot 2\alpha = \frac{A-C}{B} = 0 \Rightarrow 2\alpha = \frac{\pi}{2} \Rightarrow \alpha = \frac{\pi}{4}$; $x = \frac{\sqrt{2}}{2}x' - \frac{\sqrt{2}}{2}y'$ and $y = \frac{\sqrt{2}}{2}x' + \frac{\sqrt{2}}{2}y' \Rightarrow 2\left(\frac{\sqrt{2}}{2}x' - \frac{\sqrt{2}}{2}y'\right)^2 + \left(\frac{\sqrt{2}}{2}x' - \frac{\sqrt{2}}{2}y'\right)\left(\frac{\sqrt{2}}{2}x' + \frac{\sqrt{2}}{2}y'\right) + 2\left(\frac{\sqrt{2}}{2}x' + \frac{\sqrt{2}}{2}y'\right)^2 - 15 = 0$
$\Rightarrow 5x'^2 + 3y'^2 = 30$

30. $B^2 - 4AC = 2^2 - 4(3)(3) = -32 < 0 \Rightarrow$ ellipse; $\cot 2\alpha = \frac{A-C}{B} = 0 \Rightarrow 2\alpha = \frac{\pi}{2} \Rightarrow \alpha = \frac{\pi}{4}$; $x = \frac{\sqrt{2}}{2}x' - \frac{\sqrt{2}}{2}y'$ and $y = \frac{\sqrt{2}}{2}x' + \frac{\sqrt{2}}{2}y' \Rightarrow 3\left(\frac{\sqrt{2}}{2}x' - \frac{\sqrt{2}}{2}y'\right)^2 + 2\left(\frac{\sqrt{2}}{2}x' - \frac{\sqrt{2}}{2}y'\right)\left(\frac{\sqrt{2}}{2}x' + \frac{\sqrt{2}}{2}y'\right) + 3\left(\frac{\sqrt{2}}{2}x' + \frac{\sqrt{2}}{2}y'\right)^2 = 19$
$\Rightarrow 4x'^2 + 2y'^2 = 19$

31. $B^2 - 4AC = \left(2\sqrt{3}\right)^2 - 4(1)(-1) = 16 \Rightarrow$ hyperbola; $\cot 2\alpha = \frac{A-C}{B} = \frac{1}{\sqrt{3}} \Rightarrow 2\alpha = \frac{\pi}{3} \Rightarrow \alpha = \frac{\pi}{6}$; $x = \frac{\sqrt{3}}{2}x' - \frac{1}{2}y'$ and $y = \frac{1}{2}x' + \frac{\sqrt{3}}{2}y' \Rightarrow \left(\frac{\sqrt{3}}{2}x' - \frac{1}{2}y'\right)^2 + 2\sqrt{3}\left(\frac{\sqrt{3}}{2}x' - \frac{1}{2}y'\right)\left(\frac{1}{2}x' + \frac{\sqrt{3}}{2}y'\right) - \left(\frac{1}{2}x' + \frac{\sqrt{3}}{2}y'\right)^2 = 4$
$\Rightarrow 2x'^2 - 2y'^2 = -4 \Rightarrow y'^2 - x'^2 = 2$

32. $B^2 - 4AC = (-3)^2 - 4(1)(1) = 5 > 0 \Rightarrow$ hyperbola; $\cot 2\alpha = \frac{A-C}{B} = 0 \Rightarrow 2\alpha = \frac{\pi}{2} \Rightarrow \alpha = \frac{\pi}{4}$; $x = \frac{\sqrt{2}}{2}x' - \frac{\sqrt{2}}{2}y'$ and $y = \frac{\sqrt{2}}{2}x' + \frac{\sqrt{2}}{2}y' \Rightarrow \left(\frac{\sqrt{2}}{2}x' - \frac{\sqrt{2}}{2}y'\right)^2 - 3\left(\frac{\sqrt{2}}{2}x' - \frac{\sqrt{2}}{2}y'\right)\left(\frac{\sqrt{2}}{2}x' + \frac{\sqrt{2}}{2}y'\right) + \left(\frac{\sqrt{2}}{2}x' + \frac{\sqrt{2}}{2}y'\right)^2 = 5$
$\Rightarrow \frac{5}{2}y'^2 - \frac{1}{2}x'^2 = 5$ or $5y'^2 - x'^2 = 10$

33. $x = \frac{1}{2}\tan t$ and $y = \frac{1}{2}\sec t \Rightarrow x^2 = \frac{1}{4}\tan^2 t$ and $y^2 = \frac{1}{4}\sec^2 t \Rightarrow 4x^2 = \tan^2 t$ and $4y^2 = \sec^2 t \Rightarrow 4x^2 + 1 = 4y^2 \Rightarrow 4y^2 - 4x^2 = 1$

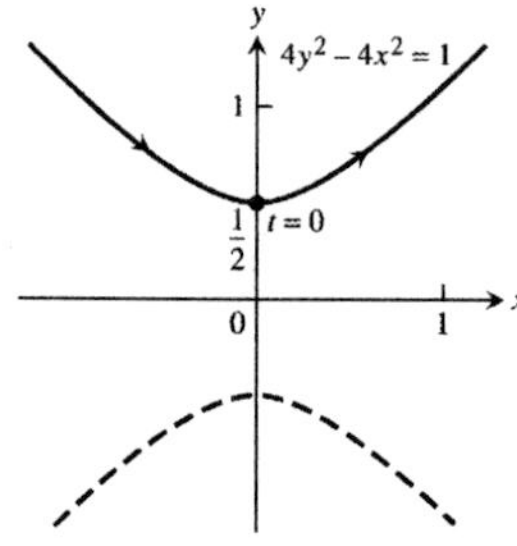

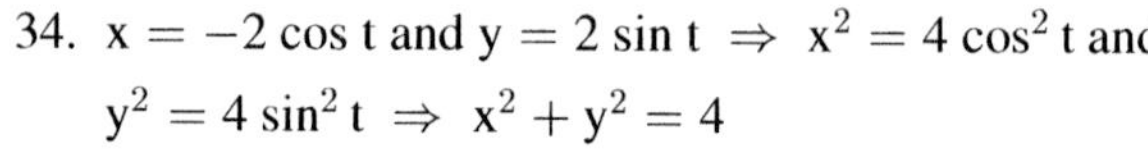
34. $x = -2\cos t$ and $y = 2\sin t \Rightarrow x^2 = 4\cos^2 t$ and $y^2 = 4\sin^2 t \Rightarrow x^2 + y^2 = 4$

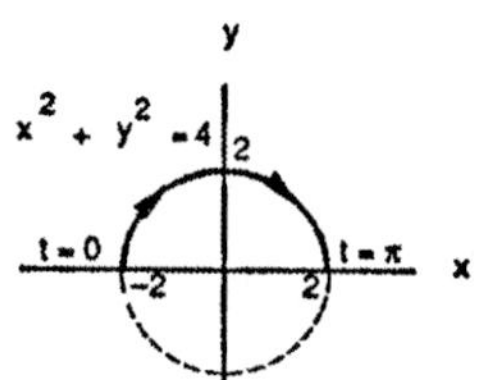

35. $x = -\cos t$ and $y = \cos^2 t \Rightarrow y = (-x)^2 = x^2$

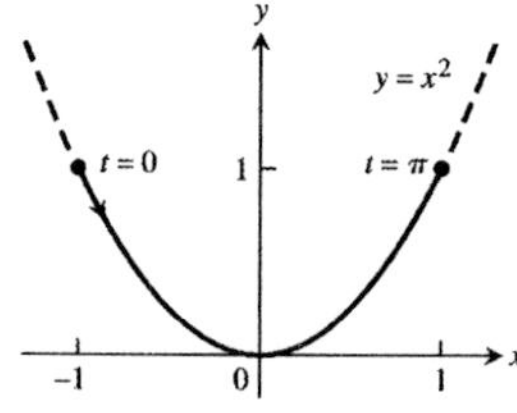

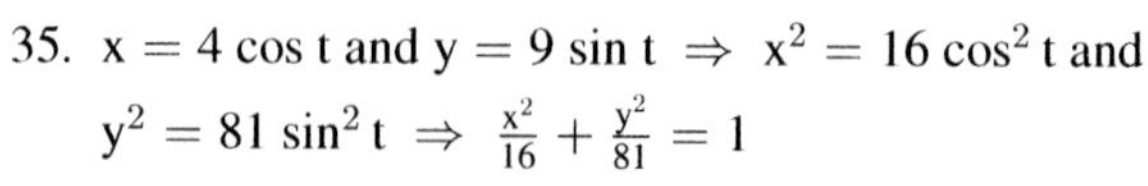
35. $x = 4\cos t$ and $y = 9\sin t \Rightarrow x^2 = 16\cos^2 t$ and $y^2 = 81\sin^2 t \Rightarrow \frac{x^2}{16} + \frac{y^2}{81} = 1$

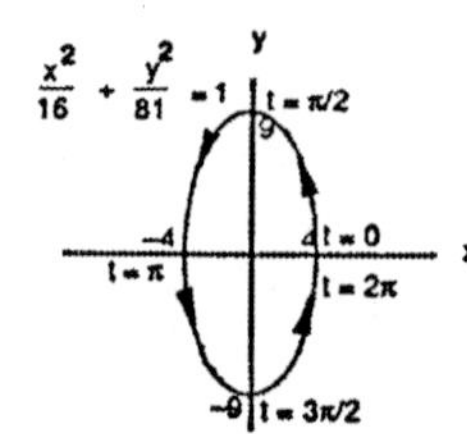

37.

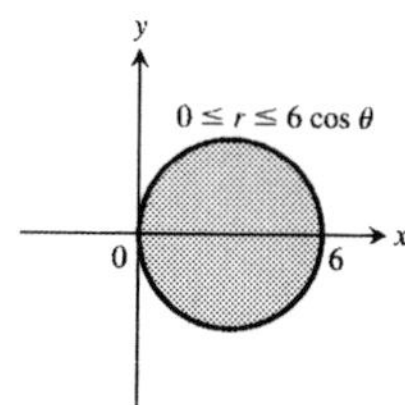

38.

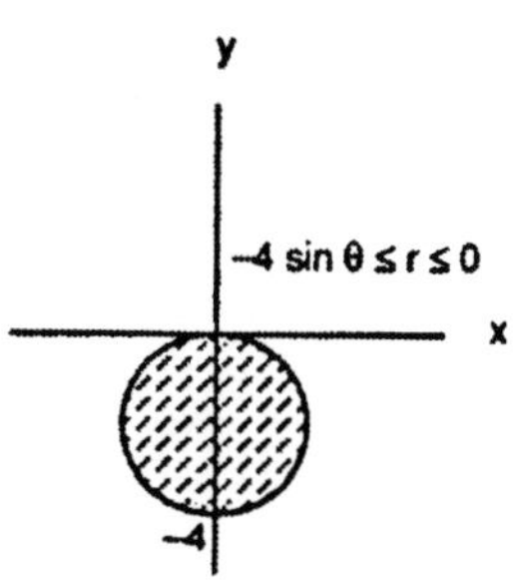

39. d

40. e

41. l

42. f

43. k

44. h

45. i

46. j

47. $r = \sin\theta$ and $r = 1 + \sin\theta \Rightarrow \sin\theta = 1 + \sin\theta \Rightarrow 0 = 1$ so no solutions exist. There are no points of intersection found by solving the system. The point of intersection $(0, 0)$ is found by graphing.

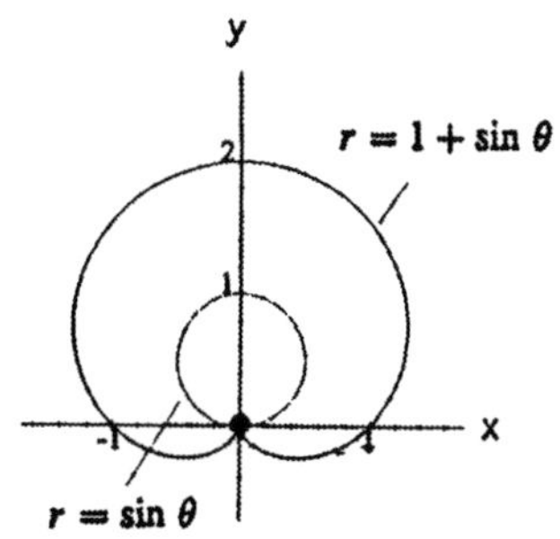

48. $r = \cos\theta$ and $r = 1 - \cos\theta \Rightarrow \cos\theta = 1 - \cos\theta$ $\Rightarrow \cos\theta = \frac{1}{2} \Rightarrow \theta = \frac{\pi}{3}, -\frac{\pi}{3}; \theta = \frac{\pi}{3} \Rightarrow r = \frac{1}{2}; \theta = -\frac{\pi}{3}$ $\Rightarrow r = \frac{1}{2}$. The points of intersection are $\left(\frac{1}{2}, \frac{\pi}{3}\right)$ and $\left(\frac{1}{2}, -\frac{\pi}{3}\right)$. The point of intersection $(0, 0)$ is found by graphing.

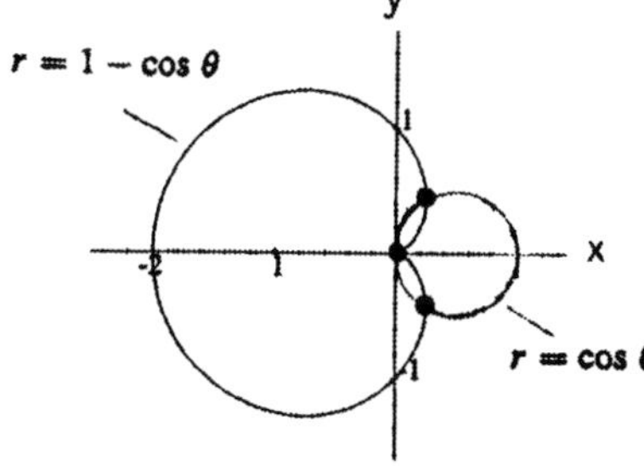

49. $r = 1 + \cos\theta$ and $r = 1 - \cos\theta \Rightarrow 1 + \cos\theta = 1 - \cos\theta$ $\Rightarrow 2\cos\theta = 0 \Rightarrow \cos\theta = 0 \Rightarrow \theta = \frac{\pi}{2}, \frac{3\pi}{2}; \theta = \frac{\pi}{2}$ or $\frac{3\pi}{2}$ $\Rightarrow r = 1$. The points of intersection are $\left(1, \frac{\pi}{2}\right)$ and $\left(1, \frac{3\pi}{2}\right)$. The point of intersection $(0, 0)$ is found by graphing.

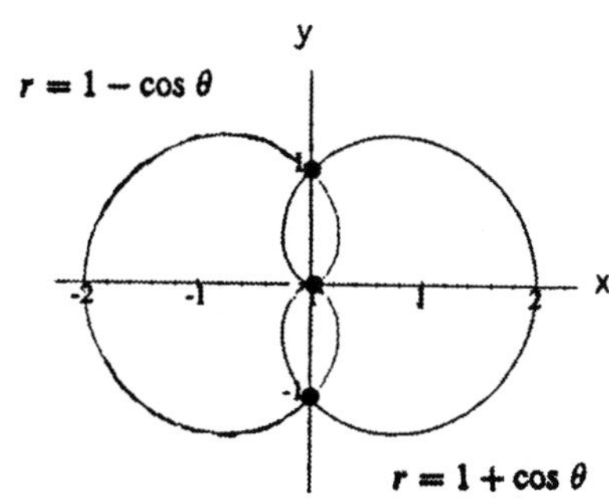

50. $r = 1 + \sin\theta$ and $r = 1 - \sin\theta \Rightarrow 1 + \sin\theta = 1 - \sin\theta$
$\Rightarrow 2\sin\theta = 0 \Rightarrow \sin\theta = 0 \Rightarrow \theta = 0, \pi;\ \theta = 0$ or π
$\Rightarrow r = 1$. The points of intersection are $(1, 0)$ and $(1, \pi)$.
The point of intersection $(0, 0)$ is found by graphing.

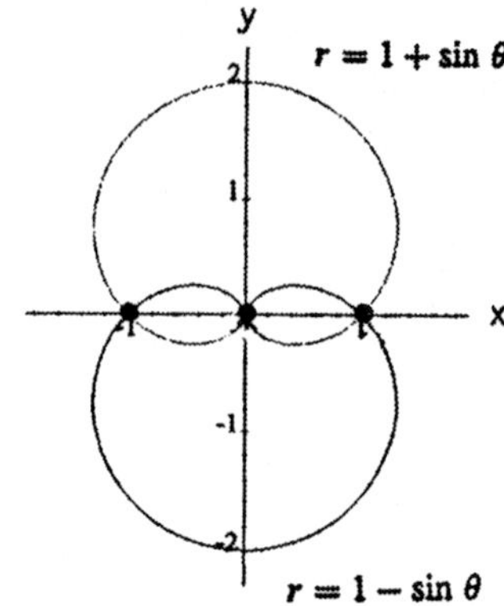

51. $r = 1 + \sin\theta$ and $r = -1 + \sin\theta$ intersect at all points of $r = 1 + \sin\theta$ because the graphs coincide. This can be seen by graphing them.

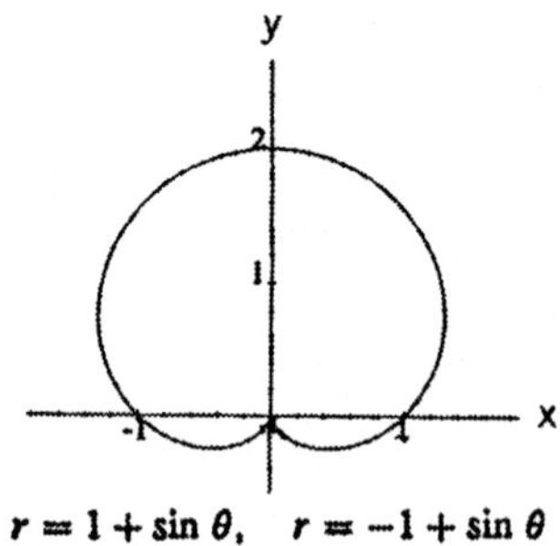

52. $r = 1 + \cos\theta$ and $r = -1 + \cos\theta$ intersect at all points of $r = 1 + \cos\theta$ because the graphs coincide. This can be seen by graphing them.

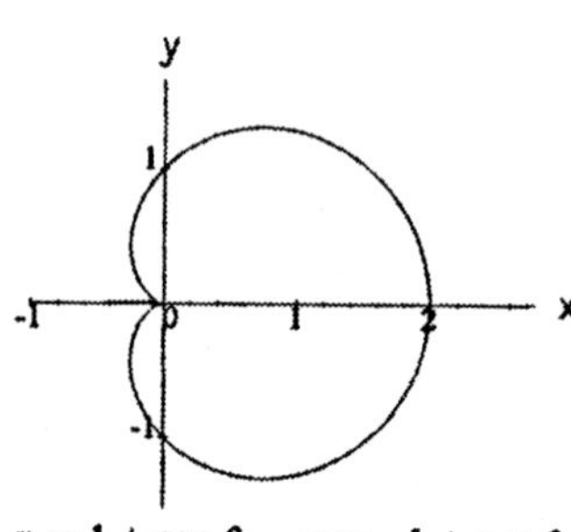

53. $r = \sec\theta$ and $r = 2\sin\theta \Rightarrow \sec\theta = 2\sin\theta$
$\Rightarrow 1 = 2\sin\theta\cos\theta \Rightarrow 1 = \sin 2\theta \Rightarrow 2\theta = \frac{\pi}{2} \Rightarrow \theta = \frac{\pi}{4}$
$\Rightarrow r = 2\sin\frac{\pi}{4} = \sqrt{2} \Rightarrow$ the point of intersection is
$\left(\sqrt{2}, \frac{\pi}{4}\right)$. No other points of intersection exist.

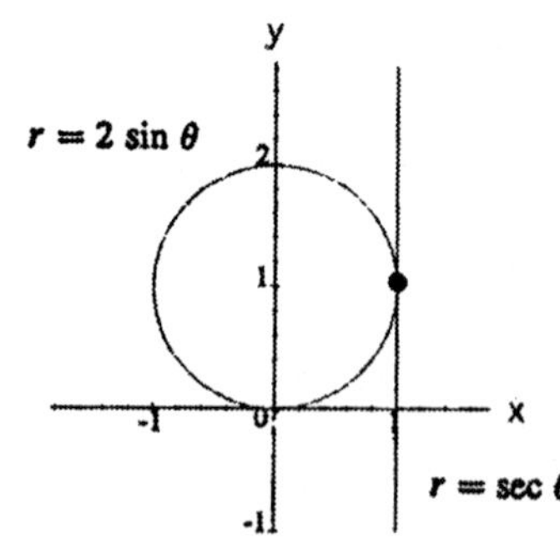

54. $r = -2\csc\theta$ and $r = -4\cos\theta \Rightarrow -2\csc\theta = -4\cos\theta$
$\Rightarrow 1 = 2\sin\theta\cos\theta \Rightarrow 1 = \sin 2\theta \Rightarrow 2\theta = \frac{\pi}{2}, \frac{5\pi}{2}$
$\Rightarrow \theta = \frac{\pi}{4}, \frac{5\pi}{4};\ \theta = \frac{\pi}{4} \Rightarrow r = -4\cos\frac{\pi}{4} = -2\sqrt{2};$
$\theta = \frac{5\pi}{4} \Rightarrow r = -4\cos\frac{5\pi}{4} = 2\sqrt{2}$. The point of intersection is $\left(2\sqrt{2}, \frac{5\pi}{4}\right)$ and the point $\left(-2\sqrt{2}, \frac{\pi}{4}\right)$ is the same point.

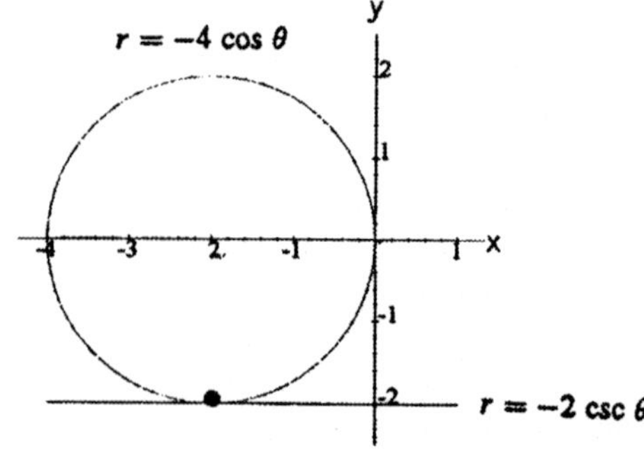

55. $r\cos\left(\theta+\frac{\pi}{3}\right)=2\sqrt{3} \Rightarrow r\left(\cos\theta\cos\frac{\pi}{3}-\sin\theta\sin\frac{\pi}{3}\right)$
$=2\sqrt{3} \Rightarrow \frac{1}{2}r\cos\theta-\frac{\sqrt{3}}{2}r\sin\theta=2\sqrt{3}$
$\Rightarrow r\cos\theta-\sqrt{3}r\sin\theta=4\sqrt{3} \Rightarrow x-\sqrt{3}y=4\sqrt{3}$
$\Rightarrow y=\frac{\sqrt{3}}{3}x-4$

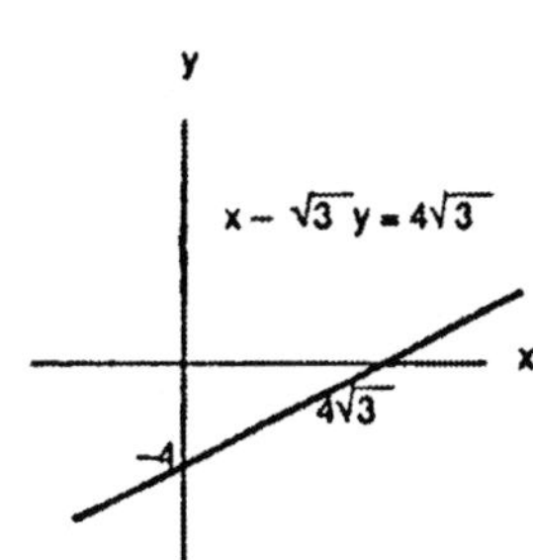

56. $r\cos\left(\theta-\frac{3\pi}{4}\right)=\frac{\sqrt{2}}{2} \Rightarrow r\left(\cos\theta\cos\frac{3\pi}{4}+\sin\theta\sin\frac{3\pi}{4}\right)$
$=\frac{\sqrt{2}}{2} \Rightarrow -\frac{\sqrt{2}}{2}r\cos\theta+\frac{\sqrt{2}}{2}r\sin\theta=\frac{\sqrt{2}}{2} \Rightarrow -x+y=1$
$\Rightarrow y=x+1$

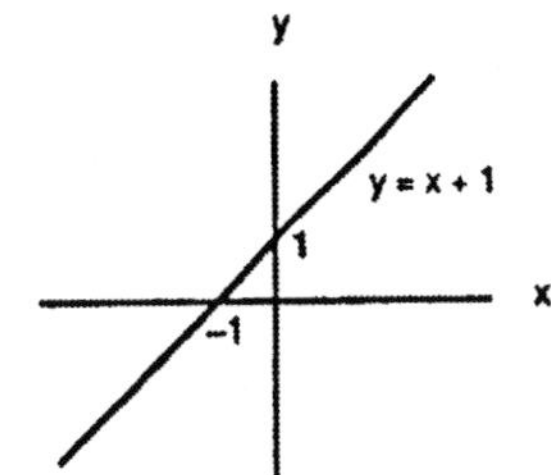

57. $r=2\sec\theta \Rightarrow r=\frac{2}{\cos\theta} \Rightarrow r\cos\theta=2 \Rightarrow x=2$

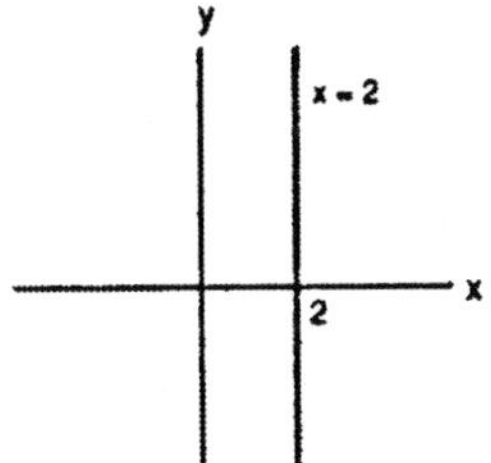

58. $r=-\sqrt{2}\sec\theta \Rightarrow r\cos\theta=-\sqrt{2} \Rightarrow x=-\sqrt{2}$

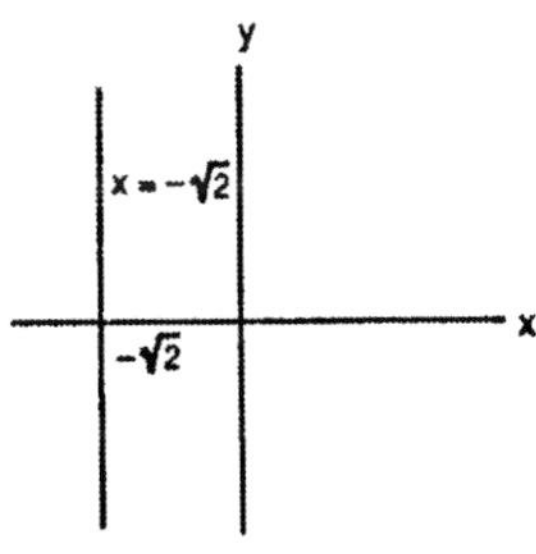

59. $r=-\frac{3}{2}\csc\theta \Rightarrow r\sin\theta=-\frac{3}{2} \Rightarrow y=-\frac{3}{2}$

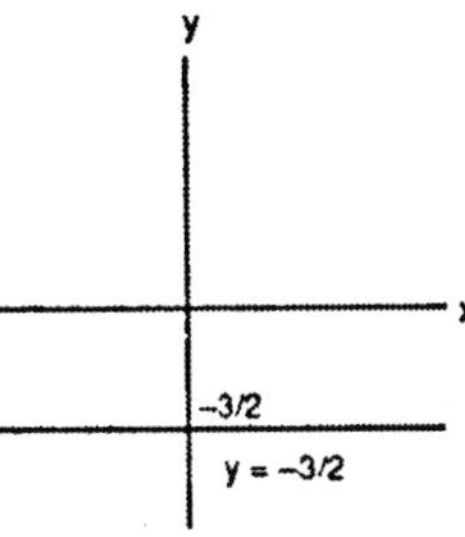

60. $r=3\sqrt{3}\csc\theta \Rightarrow r\sin\theta=3\sqrt{3} \Rightarrow y=3\sqrt{3}$

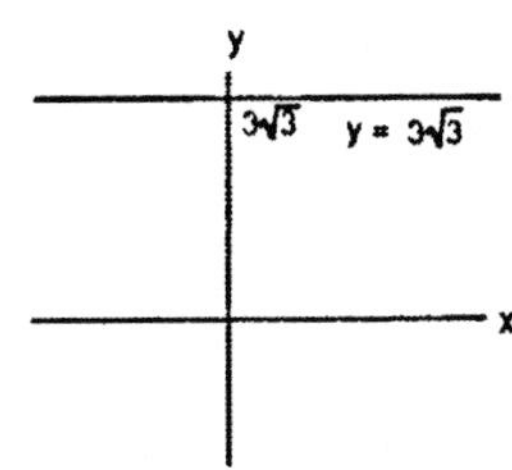

61. $r = -4 \sin\theta \Rightarrow r^2 = -4r \sin\theta \Rightarrow x^2 + y^2 + 4y = 0$ $\Rightarrow x^2 + (y+2)^2 = 4$; circle with center $(0, -2)$ and radius 2.

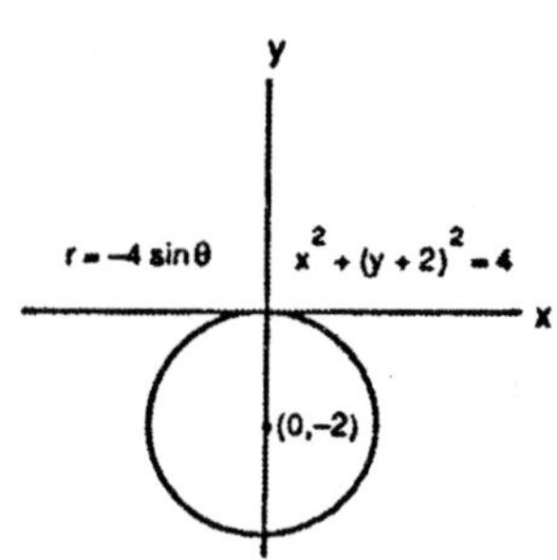

62. $r = 3\sqrt{3} \sin\theta \Rightarrow r^2 = 3\sqrt{3}\, r \sin\theta$ $\Rightarrow x^2 + y^2 - 3\sqrt{3}\, y = 0 \Rightarrow x^2 + \left(y - \frac{3\sqrt{3}}{2}\right)^2 = \frac{27}{4}$; circle with center $\left(0, \frac{3\sqrt{3}}{2}\right)$ and radius $\frac{3\sqrt{3}}{2}$

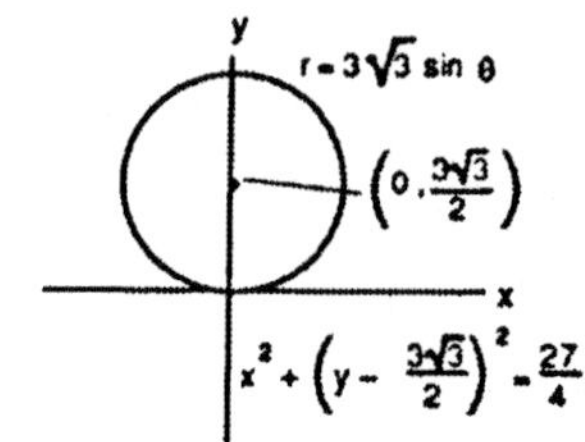

63. $r = 2\sqrt{2} \cos\theta \Rightarrow r^2 = 2\sqrt{2}\, r \cos\theta$ $\Rightarrow x^2 + y^2 - 2\sqrt{2}\, x = 0 \Rightarrow \left(x - \sqrt{2}\right)^2 + y^2 = 2$; circle with center $\left(\sqrt{2}, 0\right)$ and radius $\sqrt{2}$

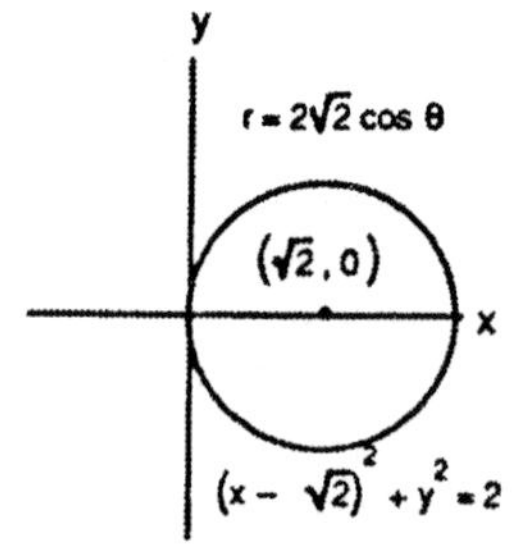

64. $r = -6 \cos\theta \Rightarrow r^2 = -6r \cos\theta \Rightarrow x^2 + y^2 + 6x = 0$ $\Rightarrow (x+3)^2 + y^2 = 9$; circle with center $(-3, 0)$ and radius 3

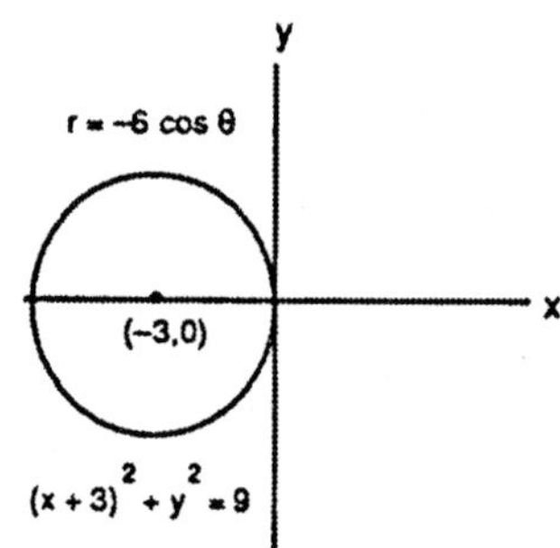

65. $x^2 + y^2 + 5y = 0 \Rightarrow x^2 + \left(y + \frac{5}{2}\right)^2 = \frac{25}{4} \Rightarrow C = \left(0, -\frac{5}{2}\right)$ and $a = \frac{5}{2}$; $r^2 + 5r \sin\theta = 0 \Rightarrow r = -5 \sin\theta$

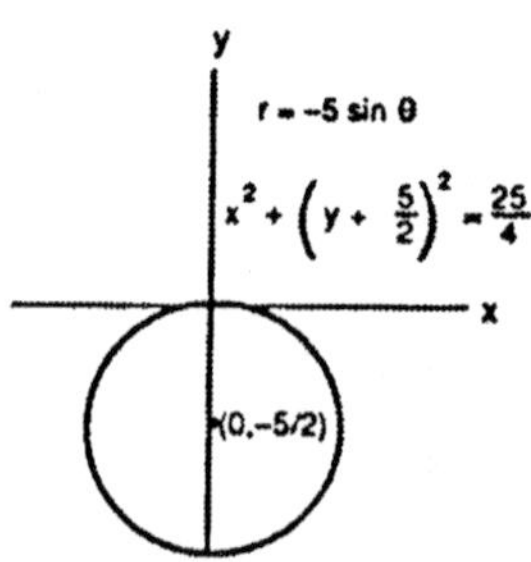

66. $x^2 + y^2 - 2y = 0 \Rightarrow x^2 + (y-1)^2 = 1 \Rightarrow C = (0, 1)$ and $a = 1$; $r^2 - 2r \sin\theta = 0 \Rightarrow r = 2 \sin\theta$

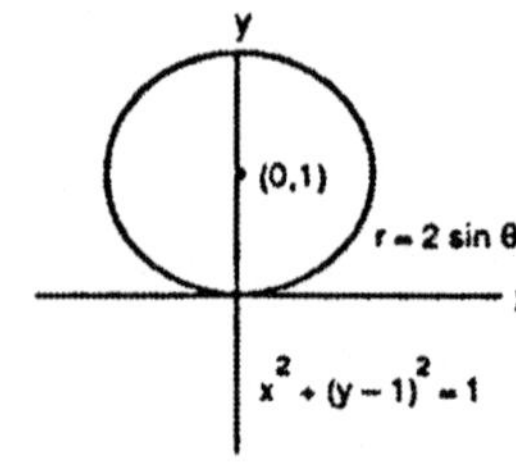

67. $x^2 + y^2 - 3x = 0 \Rightarrow \left(x - \frac{3}{2}\right)^2 + y^2 = \frac{9}{4} \Rightarrow C = \left(\frac{3}{2}, 0\right)$ and $a = \frac{3}{2}$; $r^2 - 3r\cos\theta = 0 \Rightarrow r = 3\cos\theta$

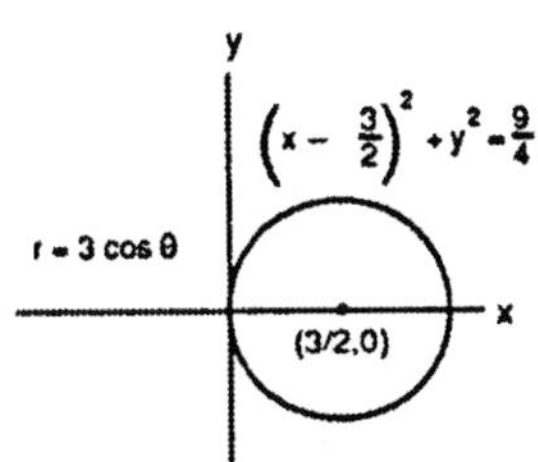

68. $x^2 + y^2 + 4x = 0 \Rightarrow (x + 2)^2 + y^2 = 4 \Rightarrow C = (-2, 0)$ and $a = 2$; $r^2 + 4r\cos\theta = 0 \Rightarrow r = -4\cos\theta$

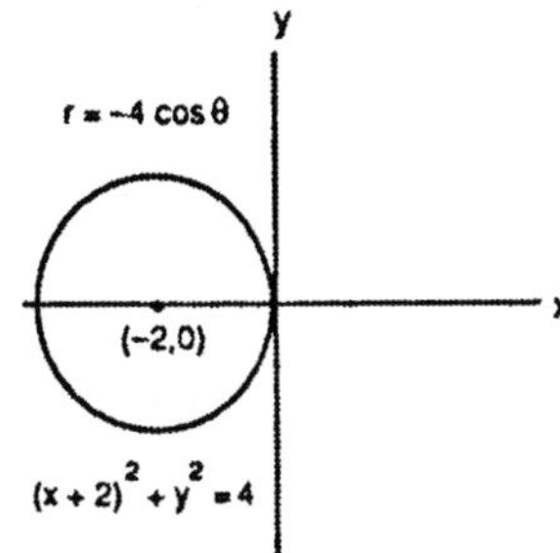

69. $r = \frac{2}{1 + \cos\theta} \Rightarrow e = 1 \Rightarrow$ parabola with vertex at $(1, 0)$

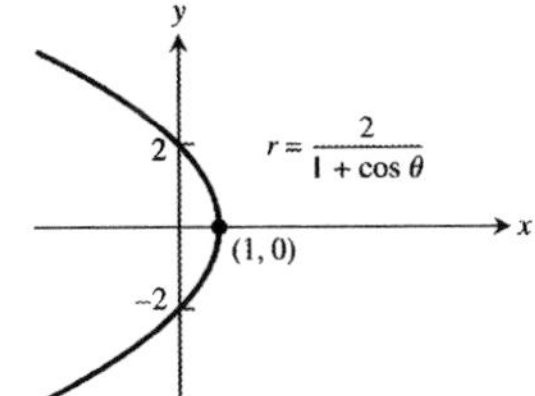

70. $r = \frac{8}{2 + \cos\theta} \Rightarrow r = \frac{4}{1 + \left(\frac{1}{2}\right)\cos\theta} \Rightarrow e = \frac{1}{2} \Rightarrow$ ellipse; $ke = 4 \Rightarrow \frac{1}{2}k = 4 \Rightarrow k = 8$; $k = \frac{a}{e} - ea \Rightarrow 8 = \frac{a}{\left(\frac{1}{2}\right)} - \frac{1}{2}a$ $\Rightarrow a = \frac{16}{3} \Rightarrow ea = \left(\frac{1}{2}\right)\left(\frac{16}{3}\right) = \frac{8}{3}$; therefore the center is $\left(\frac{8}{3}, \pi\right)$; vertices are $(8, \pi)$ and $\left(\frac{8}{3}, 0\right)$

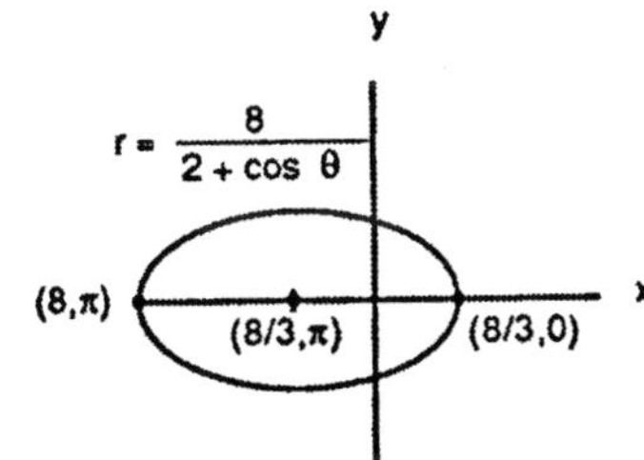

71. $r = \frac{6}{1 - 2\cos\theta} \Rightarrow e = 2 \Rightarrow$ hyperbola; $ke = 6 \Rightarrow 2k = 6$ $\Rightarrow k = 3 \Rightarrow$ vertices are $(2, \pi)$ and $(6, \pi)$

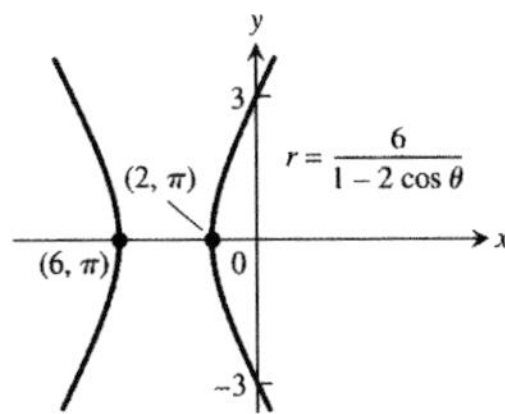

72. $r = \frac{12}{3 + \sin\theta} \Rightarrow r = \frac{4}{1 + \left(\frac{1}{3}\right)\sin\theta} \Rightarrow e = \frac{1}{3}$; $ke = 4$ $\Rightarrow \frac{1}{3}k = 4 \Rightarrow k = 12$; $a(1 - e^2) = 4 \Rightarrow a\left[1 - \left(\frac{1}{3}\right)^2\right]$ $= 4 \Rightarrow a = \frac{9}{2} \Rightarrow ea = \left(\frac{1}{3}\right)\left(\frac{9}{2}\right) = \frac{3}{2}$; therefore the center is $\left(\frac{3}{2}, \frac{3\pi}{2}\right)$; vertices are $\left(3, \frac{\pi}{2}\right)$ and $\left(6, \frac{3\pi}{2}\right)$

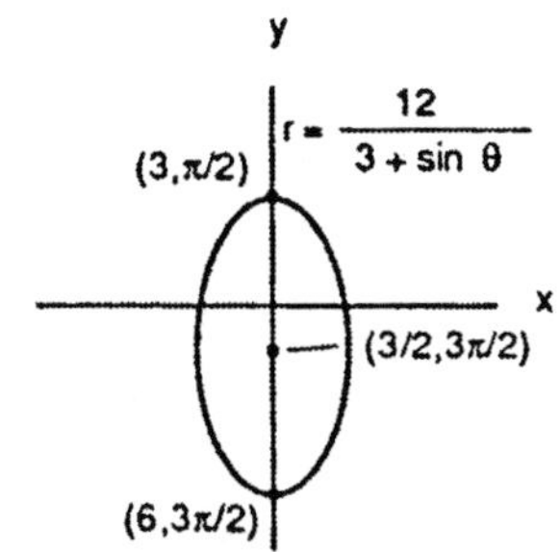

73. $e = 2$ and $r\cos\theta = 2 \Rightarrow x = 2$ is directrix $\Rightarrow k = 2$; the conic is a hyperbola; $r = \frac{ke}{1 + e\cos\theta} \Rightarrow r = \frac{(2)(2)}{1 + 2\cos\theta}$ $\Rightarrow r = \frac{4}{1 + 2\cos\theta}$

74. $e = 1$ and $r \cos\theta = -4 \Rightarrow x = -4$ is directrix $\Rightarrow k = 4$; the conic is a parabola; $r = \frac{ke}{1 - e\cos\theta} \Rightarrow r = \frac{(4)(1)}{1 - \cos\theta}$
$\Rightarrow r = \frac{4}{1 - \cos\theta}$

75. $e = \frac{1}{2}$ and $r \sin\theta = 2 \Rightarrow y = 2$ is directrix $\Rightarrow k = 2$; the conic is an ellipse; $r = \frac{ke}{1 + e\sin\theta} \Rightarrow r = \frac{(2)\left(\frac{1}{2}\right)}{1 + \left(\frac{1}{2}\right)\sin\theta}$
$\Rightarrow r = \frac{2}{2 + \sin\theta}$

76. $e = \frac{1}{3}$ and $r \sin\theta = -6 \Rightarrow y = -6$ is directrix $\Rightarrow k = 6$; the conic is an ellipse; $r = \frac{ke}{1 - e\sin\theta} \Rightarrow r = \frac{(6)\left(\frac{1}{3}\right)}{1 - \left(\frac{1}{3}\right)\sin\theta}$
$\Rightarrow r = \frac{6}{3 - \sin\theta}$

77. $A = 2\int_0^{\pi} \frac{1}{2} r^2\, d\theta = \int_0^{\pi} (2 - \cos\theta)^2\, d\theta = \int_0^{\pi} (4 - 4\cos\theta + \cos^2\theta)\, d\theta = \int_0^{\pi} \left(4 - 4\cos\theta + \frac{1 + \cos 2\theta}{2}\right) d\theta$
$= \int_0^{\pi} \left(\frac{9}{2} - 4\cos\theta + \frac{\cos 2\theta}{2}\right) d\theta = \left[\frac{9}{2}\theta - 4\sin\theta + \frac{\sin 2\theta}{4}\right]_0^{\pi} = \frac{9}{2}\pi$

78. $A = \int_0^{\pi/3} \frac{1}{2}(\sin^2 3\theta)\, d\theta = \int_0^{\pi/3} \left(\frac{1 - \cos 6\theta}{2}\right) d\theta = \frac{1}{4}\left[\theta - \frac{1}{6}\sin 6\theta\right]_0^{\pi/3} = \frac{\pi}{12}$

79. $r = 1 + \cos 2\theta$ and $r = 1 \Rightarrow 1 = 1 + \cos 2\theta \Rightarrow 0 = \cos 2\theta \Rightarrow 2\theta = \frac{\pi}{2} \Rightarrow \theta = \frac{\pi}{4}$; therefore
$A = 4\int_0^{\pi/4} \frac{1}{2}\left[(1 + \cos 2\theta)^2 - 1^2\right] d\theta = 2\int_0^{\pi/4} (1 + 2\cos 2\theta + \cos^2 2\theta - 1)\, d\theta$
$= 2\int_0^{\pi/4} \left(2\cos 2\theta + \frac{1}{2} + \frac{\cos 4\theta}{2}\right) d\theta = 2\left[\sin 2\theta + \frac{1}{2}\theta + \frac{\sin 4\theta}{8}\right]_0^{\pi/4} = 2\left(1 + \frac{\pi}{8} + 0\right) = 2 + \frac{\pi}{4}$

80. The circle lies interior to the cardioid (see the graphs in Exercises 61 and 63). Thus,
$A = 2\int_{-\pi/2}^{\pi/2} \frac{1}{2}\left[2(1 + \sin\theta)\right]^2 d\theta - \pi$ (the integral is the area of the cardioid minus the area of the circle)
$= \int_{-\pi/2}^{\pi/2} 4(1 + 2\sin\theta + \sin^2\theta)\, d\theta - \pi = \int_{-\pi/2}^{\pi/2} (6 + 8\sin\theta - 2\cos 2\theta)\, d\theta - \pi = \left[6\theta - 8\cos\theta - \sin 2\theta\right]_{-\pi/2}^{\pi/2} - \pi$
$= [3\pi - (-3\pi)] - \pi = 5\pi$

81. $r = -1 + \cos\theta \Rightarrow \frac{dr}{d\theta} = -\sin\theta$; Length $= \int_0^{2\pi} \sqrt{(-1 + \cos\theta)^2 + (-\sin\theta)^2}\, d\theta = \int_0^{2\pi} \sqrt{2 - 2\cos\theta}\, d\theta$
$= \int_0^{2\pi} \sqrt{\frac{4(1 - \cos\theta)}{2}}\, d\theta = \int_0^{2\pi} 2\sin\frac{\theta}{2}\, d\theta = \left[-4\cos\frac{\theta}{2}\right]_0^{2\pi} = (-4)(-1) - (-4)(1) = 8$

82. $r = 2\sin\theta + 2\cos\theta,\ 0 \le \theta \le \frac{\pi}{2} \Rightarrow \frac{dr}{d\theta} = 2\cos\theta - 2\sin\theta$; $r^2 + \left(\frac{dr}{d\theta}\right)^2 = (2\sin\theta + 2\cos\theta)^2 + (2\cos\theta - 2\sin\theta)^2$
$= 8(\sin^2\theta + \cos^2\theta) = 8 \Rightarrow L = \int_0^{\pi/2} \sqrt{8}\, d\theta = \left[2\sqrt{2}\,\theta\right]_0^{\pi/2} = 2\sqrt{2}\left(\frac{\pi}{2}\right) = \pi\sqrt{2}$

83. $r = 8\sin^3\left(\frac{\theta}{3}\right),\ 0 \le \theta \le \frac{\pi}{4} \Rightarrow \frac{dr}{d\theta} = 8\sin^2\left(\frac{\theta}{3}\right)\cos\left(\frac{\theta}{3}\right)$; $r^2 + \left(\frac{dr}{d\theta}\right)^2 = \left[8\sin^3\left(\frac{\theta}{3}\right)\right]^2 + \left[8\sin^2\left(\frac{\theta}{3}\right)\cos\left(\frac{\theta}{3}\right)\right]^2$
$= 64\sin^4\left(\frac{\theta}{3}\right) \Rightarrow L = \int_0^{\pi/4} \sqrt{64\sin^4\left(\frac{\theta}{3}\right)}\, d\theta = \int_0^{\pi/4} 8\sin^2\left(\frac{\theta}{3}\right) d\theta = \int_0^{\pi/4} 8\left[\frac{1 - \cos\left(\frac{2\theta}{3}\right)}{2}\right] d\theta$
$= \int_0^{\pi/4} \left[4 - 4\cos\left(\frac{2\theta}{3}\right)\right] d\theta = \left[4\theta - 6\sin\left(\frac{2\theta}{3}\right)\right]_0^{\pi/4} = 4\left(\frac{\pi}{4}\right) - 6\sin\left(\frac{\pi}{6}\right) - 0 = \pi - 3$

84. $r = \sqrt{1 + \cos 2\theta} \Rightarrow \frac{dr}{d\theta} = \frac{1}{2}(1 + \cos 2\theta)^{-1/2}(-2\sin 2\theta) = \frac{-\sin 2\theta}{\sqrt{1 + \cos 2\theta}} \Rightarrow \left(\frac{dr}{d\theta}\right)^2 = \frac{\sin^2 2\theta}{1 + \cos 2\theta}$
$\Rightarrow r^2 + \left(\frac{dr}{d\theta}\right)^2 = 1 + \cos 2\theta + \frac{\sin^2 2\theta}{1 + \cos 2\theta} = \frac{(1 + \cos 2\theta)^2 + \sin^2 2\theta}{1 + \cos 2\theta} = \frac{1 + 2\cos 2\theta + \cos^2 2\theta + \sin^2 2\theta}{1 + \cos 2\theta}$
$= \frac{2 + 2\cos 2\theta}{1 + \cos 2\theta} = 2 \Rightarrow L = \int_{-\pi/2}^{\pi/2} \sqrt{2}\, d\theta = \sqrt{2}\left[\frac{\pi}{2} - \left(-\frac{\pi}{2}\right)\right] = \sqrt{2}\,\pi$

85. $r = \sqrt{\cos 2\theta} \Rightarrow \frac{dr}{d\theta} = \frac{-\sin 2\theta}{\sqrt{\cos 2\theta}}$; Surface Area $= \int_0^{\pi/4} 2\pi(r \sin\theta)\sqrt{r^2 + \left(\frac{dr}{d\theta}\right)^2}\, d\theta$
$= \int_0^{\pi/4} 2\pi\sqrt{\cos 2\theta}\,(\sin\theta)\sqrt{\cos 2\theta + \frac{\sin^2 2\theta}{\cos 2\theta}}\, d\theta = \int_0^{\pi/4} 2\pi\sqrt{\cos 2\theta}\,(\sin\theta)\sqrt{\frac{1}{\cos 2\theta}}\, d\theta = \int_0^{\pi/4} 2\pi \sin\theta\, d\theta$
$= [2\pi(-\cos\theta)]_0^{\pi/4} = 2\pi\left(1 - \frac{\sqrt{2}}{2}\right) = \left(2 - \sqrt{2}\right)\pi$

86. $r^2 = \sin 2\theta \Rightarrow 2r\frac{dr}{d\theta} = 2\cos 2\theta \Rightarrow r\frac{dr}{d\theta} = \cos 2\theta$; Surface Area $= 2\int_0^{\pi/2} 2\pi(r\cos\theta)\sqrt{r^2 + \left(\frac{dr}{d\theta}\right)^2}\, d\theta$
$= 2\int_0^{\pi/2} 2\pi(\cos\theta)\sqrt{r^4 + \left(r\frac{dr}{dt}\right)^2}\, d\theta = 2\int_0^{\pi/2} 2\pi(\cos\theta)\sqrt{(\sin 2\theta)^2 + (\cos 2\theta)^2}\, d\theta = 2\int_0^{\pi/2} 2\pi\cos\theta\, d\theta$
$= 2\,[2\pi\sin\theta]_0^{\pi/2} = 4\pi$

87. (a) Around the x-axis: $9x^2 + 4y^2 = 36 \Rightarrow y^2 = 9 - \frac{9}{4}x^2 \Rightarrow y = \pm\sqrt{9 - \frac{9}{4}x^2}$ and we use the positive root:
$V = 2\int_0^2 \pi\left(\sqrt{9 - \frac{9}{4}x^2}\right)^2 dx = 2\int_0^2 \pi\left(9 - \frac{9}{4}x^2\right) dx = 2\pi\left[9x - \frac{3}{4}x^3\right]_0^2 = 24\pi$
(b) Around the y-axis: $9x^2 + 4y^2 = 36 \Rightarrow x^2 = 4 - \frac{4}{9}y^2 \Rightarrow x = \pm\sqrt{4 - \frac{4}{9}y^2}$ and we use the positive root:
$V = 2\int_0^3 \pi\left(\sqrt{4 - \frac{4}{9}y^2}\right)^2 dy = 2\int_0^3 \pi\left(4 - \frac{4}{9}y^2\right) dy = 2\pi\left[4y - \frac{4}{27}y^3\right]_0^3 = 16\pi$

88. $9x^2 - 4y^2 = 36,\ x = 4 \Rightarrow y^2 = \frac{9x^2 - 36}{4} \Rightarrow y = \frac{3}{2}\sqrt{x^2 - 4};\ V = \int_2^4 \pi\left(\frac{3}{2}\sqrt{x^2 - 4}\right)^2 dx = \frac{9\pi}{4}\int_2^4 (x^2 - 4)\, dx$
$= \frac{9\pi}{4}\left[\frac{x^3}{3} - 4x\right]_2^4 = \frac{9\pi}{4}\left[\left(\frac{64}{3} - 16\right) - \left(\frac{8}{3} - 8\right)\right] = \frac{9\pi}{4}\left(\frac{56}{3} - \frac{24}{3}\right) = \frac{3\pi}{4}(32) = 24\pi$

89. Each portion of the wave front reflects to the other focus, and since the wave front travels at a constant speed as it expands, the different portions of the wave arrive at the second focus simultaneously, from all directions, causing a spurt at the second focus.

90. The velocity of the signals is $v = 980$ ft/ms. Let t_1 be the time it takes for the signal to go from A to S. Then
$d_1 = 980t_1$ and $d_2 = 980(t_1 + 1400)$
$\Rightarrow d_2 - d_1 = 980(1400) = 1.372 \times 10^6$ ft or 259.8 miles.
The ship is 259.8 miles closer to A than to B.
The difference of the distances is always constant (259.8 miles) so the ship is traveling along a branch of a hyperbola with foci at the two towers. The branch is the one having tower A as its focus.

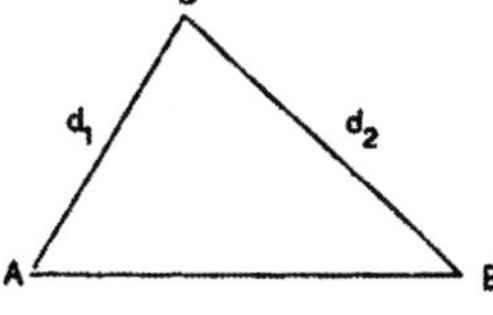

91. The time for the bullet to hit the target remains constant, say $t = t_0$. Let the time it takes for sound to travel from the target to the listener be t_2. Since the listener hears the sounds simultaneously, $t_1 = t_0 + t_2$ where t_1 is the time for the sound to travel from the rifle to the listener. If v is the velocity of sound, then $vt_1 = vt_0 + vt_2$ or $vt_1 - vt_2 = vt_0$. Now vt_1 is the distance from the rifle to the listener and vt_2 is the distance from the target to the listener. Therefore the difference of the distances is constant since vt_0 is constant so the listener is on a branch of a hyperbola with foci at the rifle and the target. The branch is the one with the target as focus.

92. Let (r_1, θ_1) be a point on the graph where $r_1 = a\theta_1$. Let (r_2, θ_2) be on the graph where $r_2 = a\theta_2$ and $\theta_2 = \theta_1 + 2\pi$. Then r_1 and r_2 lie on the same ray on consecutive turns of the spiral and the distance between the two points is $r_2 - r_1 = a\theta_2 - a\theta_1 = a(\theta_2 - \theta_1) = 2\pi a$, which is constant.

93. (a) $r = \frac{k}{1+e\cos\theta} \Rightarrow r + er\cos\theta = k \Rightarrow \sqrt{x^2+y^2} + ex = k \Rightarrow \sqrt{x^2+y^2} = k - ex \Rightarrow x^2 + y^2$
$= k^2 - 2kex + e^2x^2 \Rightarrow x^2 - e^2x^2 + y^2 + 2kex - k^2 = 0 \Rightarrow (1-e^2)\,x^2 + y^2 + 2kex - k^2 = 0$

(b) $e = 0 \Rightarrow x^2 + y^2 - k^2 = 0 \Rightarrow x^2 + y^2 = k^2 \Rightarrow$ circle;
$0 < e < 1 \Rightarrow e^2 < 1 \Rightarrow e^2 - 1 < 0 \Rightarrow B^2 - 4AC = 0^2 - 4(1-e^2)(1) = 4(e^2-1) < 0 \Rightarrow$ ellipse;
$e = 1 \Rightarrow B^2 - 4AC = 0^2 - 4(0)(1) = 0 \Rightarrow$ parabola;
$e > 1 \Rightarrow e^2 > 1 \Rightarrow B^2 - 4AC = 0^2 - 4(1-e^2)(1) = 4e^2 - 4 > 0 \Rightarrow$ hyperbola

94. (a) The length of the major axis is 300 miles + 8000 miles + 1000 miles = 2a $\Rightarrow$ a = 4650 miles. If the center of the earth is one focus and the distance from the center of the earth to the satellite's low point is 4300 miles (half the diameter plus the distance above the North Pole), then the distance from the center of the ellipse to the focus (center of the earth) is 4650 miles − 4300 miles = 350 miles = c. Therefore $e = \frac{c}{a} = \frac{350 \text{ miles}}{4650 \text{ miles}} = \frac{7}{93}$.

(b) $r = \frac{a(1-e^2)}{1+e\cos\theta} \Rightarrow r = \frac{4650\left[1-\left(\frac{7}{93}\right)^2\right]}{\left(1+\frac{7}{93}\cos\theta\right)} = \frac{430{,}000}{93+7\cos\theta}$ mile

CHAPTER 10 ADDITIONAL AND ADVANCED EXERCISES

1. Directrix x = 3 and focus (4, 0) $\Rightarrow$ vertex is $\left(\frac{7}{2}, 0\right)$
$\Rightarrow p = \frac{1}{2} \Rightarrow$ the equation is $x - \frac{7}{2} = \frac{y^2}{2}$

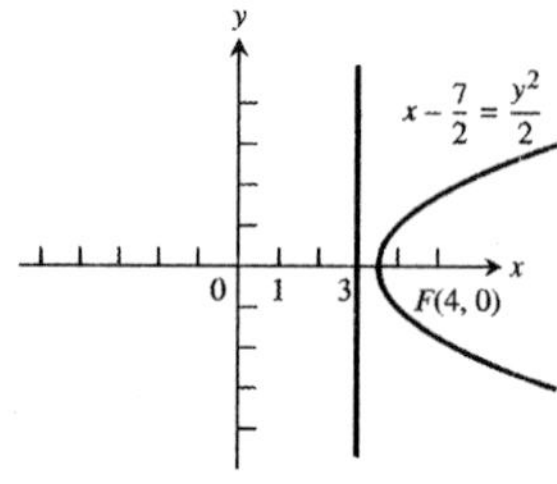

2. $x^2 - 6x - 12y + 9 = 0 \Rightarrow x^2 - 6x + 9 = 12y \Rightarrow \frac{(x-3)^2}{12} = y \Rightarrow$ vertex is (3, 0) and p = 3 $\Rightarrow$ focus is (3, 3) and the directrix is y = −3

3. $x^2 = 4y \Rightarrow$ vertex is (0, 0) and p = 1 $\Rightarrow$ focus is (0, 1); thus the distance from P(x, y) to the vertex is $\sqrt{x^2+y^2}$ and the distance from P to the focus is $\sqrt{x^2+(y-1)^2} \Rightarrow \sqrt{x^2+y^2} = 2\sqrt{x^2+(y-1)^2}$
$\Rightarrow x^2 + y^2 = 4\left[x^2 + (y-1)^2\right] \Rightarrow x^2 + y^2 = 4x^2 + 4y^2 - 8y + 4 \Rightarrow 3x^2 + 3y^2 - 8y + 4 = 0$, which is a circle

4. Let the segment a + b intersect the y-axis in point A and intersect the x-axis in point B so that PB = b and PA = a (see figure). Draw the horizontal line through P and let it intersect the y-axis in point C. Let $\angle PBO = \theta$
$\Rightarrow \angle APC = \theta$. Then $\sin\theta = \frac{y}{b}$ and $\cos\theta = \frac{x}{a}$
$\Rightarrow \frac{x^2}{a^2} + \frac{y^2}{b^2} = \cos^2\theta + \sin^2\theta = 1$.

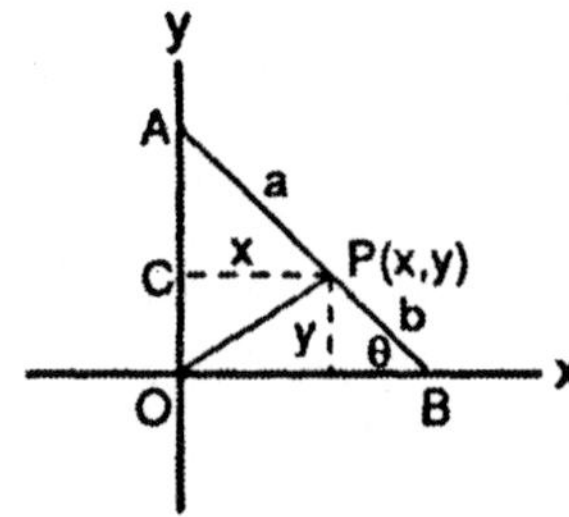

5. Vertices are (0, ± 2) $\Rightarrow$ a = 2; $e = \frac{c}{a} \Rightarrow 0.5 = \frac{c}{2} \Rightarrow c = 1 \Rightarrow$ foci are (0, ± 1)

6. Let the center of the ellipse be (x, 0); directrix x = 2, focus (4, 0), and $e = \frac{2}{3} \Rightarrow \frac{a}{e} - c = 2 \Rightarrow \frac{a}{e} = 2 + c$
$\Rightarrow a = \frac{2}{3}(2+c)$. Also $c = ae = \frac{2}{3}a \Rightarrow a = \frac{2}{3}\left(2 + \frac{2}{3}a\right) \Rightarrow a = \frac{4}{3} + \frac{4}{9}a \Rightarrow \frac{5}{9}a = \frac{4}{3} \Rightarrow a = \frac{12}{5}$; $x - 2 = \frac{a}{e}$
$\Rightarrow x - 2 = \left(\frac{12}{5}\right)\left(\frac{3}{2}\right) = \frac{18}{5} \Rightarrow x = \frac{28}{5} \Rightarrow$ the center is $\left(\frac{28}{5}, 0\right)$; $x - 4 = c \Rightarrow c = \frac{28}{5} - 4 = \frac{8}{5}$ so that $c^2 = a^2 - b^2$
$= \left(\frac{12}{5}\right)^2 - \left(\frac{8}{5}\right)^2 = \frac{80}{25}$; therefore the equation is $\frac{\left(x-\frac{28}{5}\right)^2}{\left(\frac{144}{25}\right)} + \frac{y^2}{\left(\frac{80}{25}\right)} = 1$ or $\frac{25\left(x-\frac{28}{5}\right)^2}{144} + \frac{5y^2}{16} = 1$

7. Let the center of the hyperbola be $(0, y)$.

(a) Directrix $y = -1$, focus $(0, -7)$ and $e = 2 \Rightarrow c - \frac{a}{e} = 6 \Rightarrow \frac{a}{e} = c - 6 \Rightarrow a = 2c - 12$. Also $c = ae = 2a$ $\Rightarrow a = 2(2a) - 12 \Rightarrow a = 4 \Rightarrow c = 8$; $y - (-1) = \frac{a}{e} = \frac{4}{2} = 2 \Rightarrow y = 1 \Rightarrow$ the center is $(0, 1)$; $c^2 = a^2 + b^2$ $\Rightarrow b^2 = c^2 - a^2 = 64 - 16 = 48$; therefore the equation is $\frac{(y-1)^2}{16} - \frac{x^2}{48} = 1$

(b) $e = 5 \Rightarrow c - \frac{a}{e} = 6 \Rightarrow \frac{a}{e} = c - 6 \Rightarrow a = 5c - 30$. Also, $c = ae = 5a \Rightarrow a = 5(5a) - 30 \Rightarrow 24a = 30 \Rightarrow a = \frac{5}{4}$ $\Rightarrow c = \frac{25}{4}$; $y - (-1) = \frac{a}{e} = \frac{\left(\frac{5}{4}\right)}{5} = \frac{1}{4} \Rightarrow y = -\frac{3}{4} \Rightarrow$ the center is $\left(0, -\frac{3}{4}\right)$; $c^2 = a^2 + b^2 \Rightarrow b^2 = c^2 - a^2$ $= \frac{625}{16} - \frac{25}{16} = \frac{75}{2}$; therefore the equation is $\frac{\left(y+\frac{3}{4}\right)^2}{\left(\frac{25}{16}\right)} - \frac{x^2}{\left(\frac{75}{2}\right)} = 1$ or $\frac{16\left(y+\frac{3}{4}\right)^2}{25} - \frac{2x^2}{75} = 1$

8. The center is $(0, 0)$ and $c = 2 \Rightarrow 4 = a^2 + b^2 \Rightarrow b^2 = 4 - a^2$. The equation is $\frac{y^2}{a^2} - \frac{x^2}{b^2} = 1 \Rightarrow \frac{49}{a^2} - \frac{144}{b^2} = 1$ $\Rightarrow \frac{49}{a^2} - \frac{144}{(4-a^2)} = 1 \Rightarrow 49\left(4 - a^2\right) - 144a^2 = a^2\left(4 - a^2\right) \Rightarrow 196 - 49a^2 - 144a^2 = 4a^2 - a^4 \Rightarrow a^4 - 197a^2 + 196$ $= 0 \Rightarrow \left(a^2 - 196\right)\left(a^2 - 1\right) = 0 \Rightarrow a = 14$ or $a = 1$; $a = 14 \Rightarrow b^2 = 4 - (14)^2 < 0$ which is impossible; $a = 1$ $\Rightarrow b^2 = 4 - 1 = 3$; therefore the equation is $y^2 - \frac{x^2}{3} = 1$

9. (a) $b^2x^2 + a^2y^2 = a^2b^2 \Rightarrow \frac{dy}{dx} = -\frac{b^2x}{a^2y}$; at (x_1, y_1) the tangent line is $y - y_1 = \left(-\frac{b^2x_1}{a^2y_1}\right)(x - x_1)$ $\Rightarrow a^2yy_1 + b^2xx_1 = b^2x_1^2 + a^2y_1^2 = a^2b^2 \Rightarrow b^2xx_1 + a^2yy_1 - a^2b^2 = 0$

(b) $b^2x^2 - a^2y^2 = a^2b^2 \Rightarrow \frac{dy}{dx} = \frac{b^2x}{a^2y}$; at (x_1, y_1) the tangent line is $y - y_1 = \left(\frac{b^2x_1}{a^2y_1}\right)(x - x_1)$ $\Rightarrow b^2xx_1 - a^2yy_1 = b^2x_1^2 - a^2y_1^2 = a^2b^2 \Rightarrow b^2xx_1 - a^2yy_1 - a^2b^2 = 0$

10. $Ax^2 + Bxy + Cy^2 + Dx + Ey + F = 0$ has the derivative $\frac{dy}{dx} = \frac{-2Ax - By - D}{Bx + 2Cy + E}$; at (x_1, y_1) the tangent line is $y - y_1 = \left(\frac{-2Ax_1 - By_1 - D}{Bx_1 + 2Cy_1 + E}\right)(x - x_1) \Rightarrow Byx_1 + 2Cyy_1 + Ey - By_1x_1 - 2Cy_1^2 - Ey_1$ $= -2Axx_1 - Bxy_1 - Dx + 2Ax_1^2 + Bx_1y_1 + Dx_1 \Rightarrow 2Axx_1 + B(yx_1 + xy_1) + 2Cyy_1 + Dx - Dx_1 + Ey - Ey_1$ $= 2Ax_1^2 + 2Bx_1y_1 + 2Cy_1^2$. Now add $2Dx_1 + 2Ey_1$ to both sides of this last equation, divide the result by 2, and represent the constant value on the right by $-F$ to get:

$Axx_1 + B\left(\frac{yx_1 + xy_1}{2}\right) + Cyy_1 + D\left(\frac{x + x_1}{2}\right) + E\left(\frac{y + y_1}{2}\right) = -F$

11.

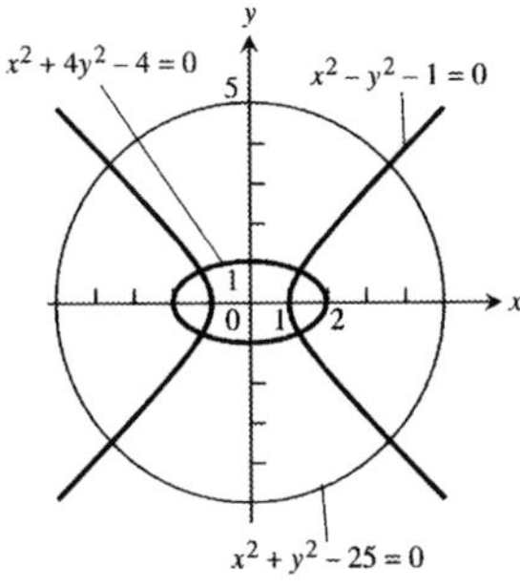

12.

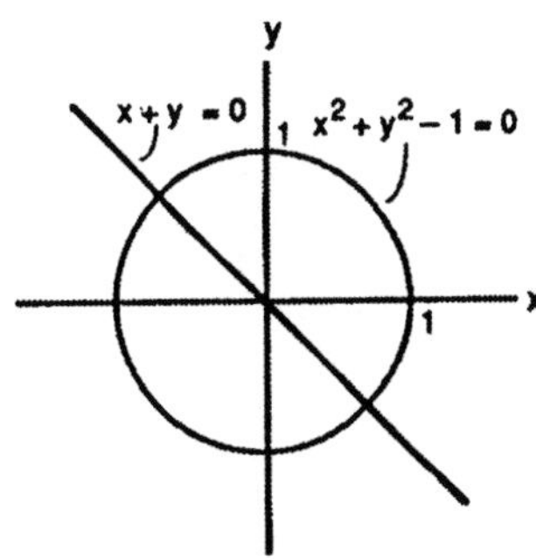

13.

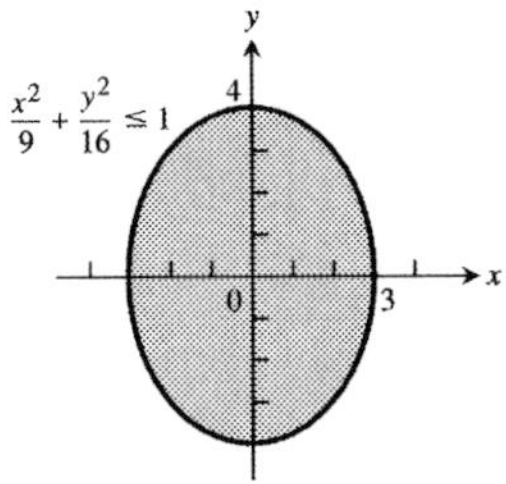

14.

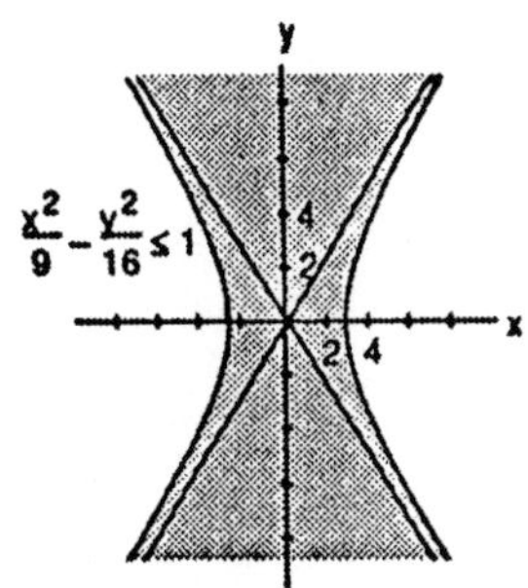

15. $(9x^2 + 4y^2 - 36)(4x^2 + 9y^2 - 16) \le 0$
$\Rightarrow 9x^2 + 4y^2 - 36 \le 0$ and $4x^2 + 9y^2 - 16 \ge 0$
or $9x^2 + 4y^2 - 36 \ge 0$ and $4x^2 + 9y^2 - 16 \le 0$

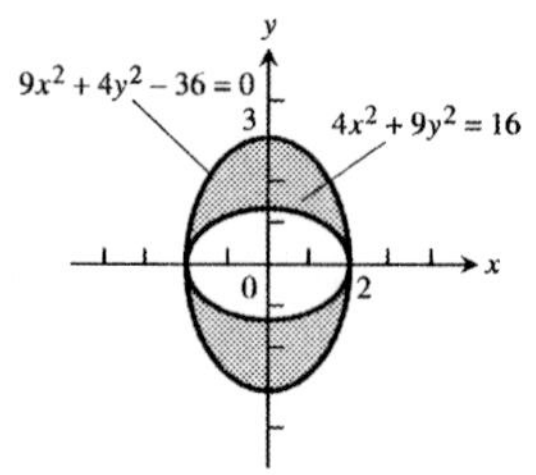

16. $(9x^2 + 4y^2 - 36)(4x^2 + 9y^2 - 16) > 0$, which is the complement of the set in Exercise 15

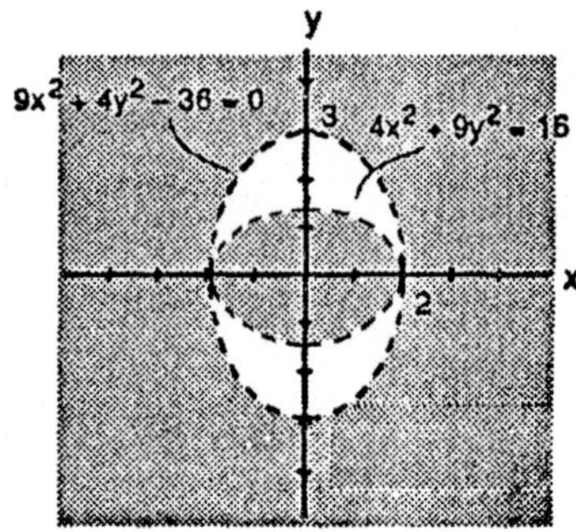

17. $x^4 - (y^2 - 9)^2 = 0 \Rightarrow x^2 - (y^2 - 9) = 0$ or
$x^2 + (y^2 - 9) = 0 \Rightarrow y^2 - x^2 = 9$ or $x^2 + y^2 = 9$

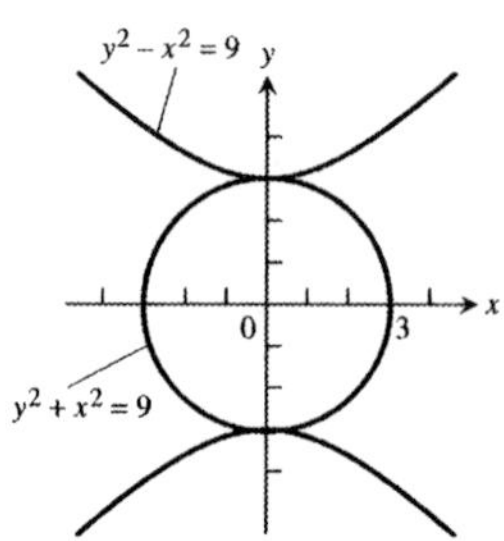

18. $x^2 + xy + y^2 < 3 \Rightarrow \tan 2\alpha = \frac{1}{1-1}$ which is undefined
$\Rightarrow 2\alpha = 90° \Rightarrow \alpha = 45° \Rightarrow A'$
$= \cos^2 45° + \cos 45° \sin 45° + \sin^2 45° = \frac{3}{2}$, $B' = 0$,
$C' = \sin^2 45° - \sin 45° \cos 45° + \cos^2 45° = \frac{1}{2}$
$\Rightarrow \frac{3}{2} x'^2 + \frac{1}{2} y'^2 < 3$ which is the interior of a rotated ellipse

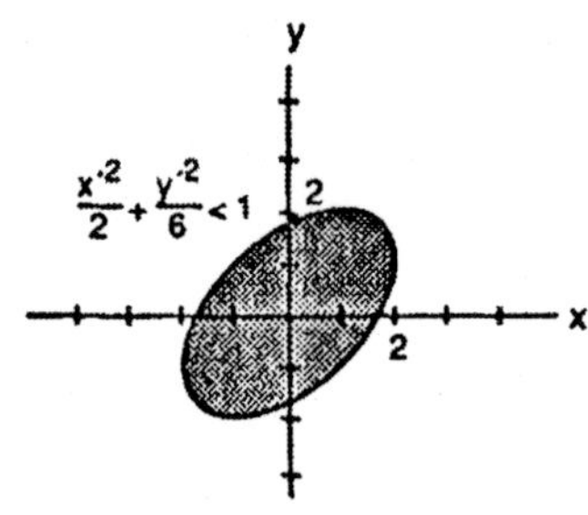

19. Arc PF = Arc AF since each is the distance rolled;
$\angle PCF = \frac{\text{Arc PF}}{b} \Rightarrow \text{Arc PF} = b(\angle PCF)$; $\theta = \frac{\text{Arc AF}}{a}$
$\Rightarrow \text{Arc AF} = a\theta \Rightarrow a\theta = b(\angle PCF) \Rightarrow \angle PCF = \left(\frac{a}{b}\right)\theta$;
$\angle OCB = \frac{\pi}{2} - \theta$ and $\angle OCB = \angle PCF - \angle PCE$
$= \angle PCF - \left(\frac{\pi}{2} - \alpha\right) = \left(\frac{a}{b}\right)\theta - \left(\frac{\pi}{2} - \alpha\right) \Rightarrow \frac{\pi}{2} - \theta$
$= \left(\frac{a}{b}\right)\theta - \left(\frac{\pi}{2} - \alpha\right) \Rightarrow \frac{\pi}{2} - \theta = \left(\frac{a}{b}\right)\theta - \frac{\pi}{2} + \alpha$
$\Rightarrow \alpha = \pi - \theta - \left(\frac{a}{b}\right)\theta \Rightarrow \alpha = \pi - \left(\frac{a+b}{b}\right)\theta$.

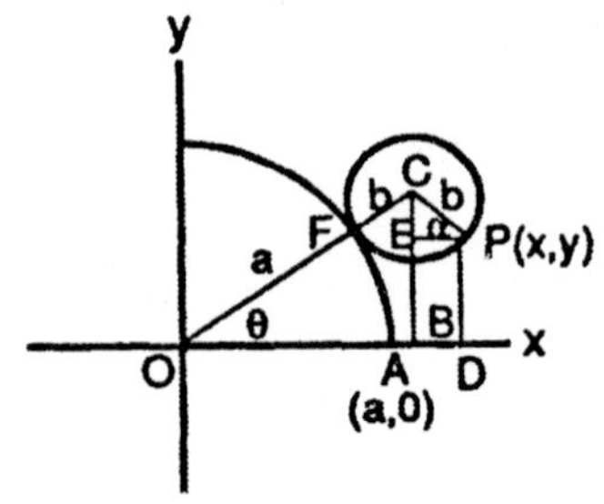

Now $x = OB + BD = OB + EP = (a+b)\cos\theta + b\cos\alpha = (a+b)\cos\theta + b\cos\left(\pi - \left(\frac{a+b}{b}\right)\theta\right)$
$= (a+b)\cos\theta + b\cos\pi\cos\left(\left(\frac{a+b}{b}\right)\theta\right) + b\sin\pi\sin\left(\left(\frac{a+b}{b}\right)\theta\right) = (a+b)\cos\theta - b\cos\left(\left(\frac{a+b}{b}\right)\theta\right)$ and
$y = PD = CB - CE = (a+b)\sin\theta - b\sin\alpha = (a+b)\sin\theta - b\sin\left(\left(\frac{a+b}{b}\right)\theta\right)$
$= (a+b)\sin\theta - b\sin\pi\cos\left(\left(\frac{a+b}{b}\right)\theta\right) + b\cos\pi\sin\left(\left(\frac{a+b}{b}\right)\theta\right) = (a+b)\sin\theta - b\sin\left(\left(\frac{a+b}{b}\right)\theta\right)$;
therefore $x = (a+b)\cos\theta - b\cos\left(\left(\frac{a+b}{b}\right)\theta\right)$ and $y = (a+b)\sin\theta - b\sin\left(\left(\frac{a+b}{b}\right)\theta\right)$

20. (a) $x = a(t - \sin t) \Rightarrow \frac{dx}{dt} = a(1 - \cos t)$ and let $\delta = 1 \Rightarrow dm = dA = y\,dx = y\left(\frac{dx}{dt}\right) dt$
$= a(1 - \cos t)\,a(1 - \cos t)\,dt = a^2(1 - \cos t)^2\,dt$; then $A = \int_0^{2\pi} a^2(1 - \cos t)^2\,dt$
$= a^2 \int_0^{2\pi} (1 - 2\cos t + \cos^2 t)\,dt = a^2 \int_0^{2\pi} \left(1 - 2\cos t + \frac{1}{2} + \frac{1}{2}\cos 2t\right) dt = a^2\left[\frac{3}{2}t - 2\sin t + \frac{\sin 2t}{4}\right]_0^{2\pi}$
$= 3\pi a^2$; $\widetilde{x} = x = a(t - \sin t)$ and $\widetilde{y} = \frac{1}{2}y = \frac{1}{2}a(1 - \cos t) \Rightarrow M_x = \int \widetilde{y}\,dm = \int \widetilde{y}\,\delta\,dA$
$= \int_0^{2\pi} \frac{1}{2}a(1 - \cos t)\,a^2(1 - \cos t)^2\,dt = \frac{1}{2}a^3 \int_0^{2\pi} (1 - \cos t)^3\,dt = \frac{a^3}{2}\int_0^{2\pi} (1 - 3\cos t + 3\cos^2 t - \cos^3 t)\,dt$
$= \frac{a^3}{2}\int_0^{2\pi} \left[1 - 3\cos t + \frac{3}{2} + \frac{3\cos 2t}{2} - (1 - \sin^2 t)(\cos t)\right] dt = \frac{a^3}{2}\left[\frac{5}{2}t - 3\sin t + \frac{3\sin 2t}{4} - \sin t + \frac{\sin^3 t}{3}\right]_0^{2\pi}$
$= \frac{5\pi a^3}{2}$. Therefore $\overline{y} = \frac{M_x}{M} = \frac{\left(\frac{5\pi a^3}{2}\right)}{3\pi a^2} = \frac{5}{6}a$. Also, $M_y = \int \widetilde{x}\,dm = \int \widetilde{x}\,\delta\,dA$
$= \int_0^{2\pi} a(t - \sin t)\,a^2(1 - \cos t)^2\,dt = a^3 \int_0^{2\pi} (t - 2t\cos t + t\cos^2 t - \sin t + 2\sin t\cos t - \sin t\cos^2 t)\,dt$
$= a^3\left[\frac{t^2}{2} - 2\cos t - 2t\sin t + \frac{1}{4}t^2 + \frac{1}{8}\cos 2t + \frac{t}{4}\sin 2t + \cos t + \sin^2 t + \frac{\cos^3 t}{3}\right]_0^{2\pi} = 3\pi^2 a^3$. Thus
$\overline{x} = \frac{M_y}{M} = \frac{3\pi^2 a^3}{3\pi a^2} = \pi a \Rightarrow \left(\pi a, \frac{5}{6}a\right)$ is the center of mass.

(b) $x = \frac{2}{3}t^{3/2} \Rightarrow \frac{dx}{dt} = t^{1/2}$ and $y = 2t^{1/2} \Rightarrow \frac{dy}{dt} = t^{-1/2}$; let $\delta = 1 \Rightarrow dm = dA = y\,dx = y\left(\frac{dx}{dt}\right) dt$
$\Rightarrow \left(2t^{1/2}\right)\left(t^{1/2}\right) dt = 2t\,dt$; $\widetilde{x} = x = \frac{2}{3}t^{3/2}$ and $\widetilde{y} = \frac{y}{2} = t^{1/2} \Rightarrow M_x = \int \widetilde{y}\,dm = \int_0^{\sqrt{3}} t^{1/2}\,(2t\,dt)$
$= \int_0^{\sqrt{3}} 2t^{3/2}\,dt = \left[\frac{4}{5}t^{5/2}\right]_0^{\sqrt{3}} = \frac{12}{5}\sqrt[4]{3}$. Also, $M_y = \int \widetilde{x}\,dm = \int \widetilde{x}\,dA = \int_0^{\sqrt{3}} \frac{2}{3}t^{3/2}(2t\,dt)$
$= \int_0^{\sqrt{3}} \frac{4}{3}t^{5/2}\,dt = \left[\frac{8}{21}t^{7/2}\right]_0^{\sqrt{3}} = \frac{8}{7}\sqrt[4]{27}$.

21. (a) $x = e^{2t}\cos t$ and $y = e^{2t}\sin t \Rightarrow x^2 + y^2 = e^{4t}\cos^2 t + e^{4t}\sin^2 t = e^{4t}$. Also $\frac{y}{x} = \frac{e^{2t}\sin t}{e^{2t}\cos t} = \tan t$
$\Rightarrow t = \tan^{-1}\left(\frac{y}{x}\right) \Rightarrow x^2 + y^2 = e^{4\tan^{-1}(y/x)}$ is the Cartesian equation. Since $r^2 = x^2 + y^2$ and
$\theta = \tan^{-1}\left(\frac{y}{x}\right)$, the polar equation is $r^2 = e^{4\theta}$ or $r = e^{2\theta}$ for $r > 0$

(b) $ds^2 = r^2\,d\theta^2 + dr^2$; $r = e^{2\theta} \Rightarrow dr = 2e^{2\theta}\,d\theta$
$\Rightarrow ds^2 = r^2\,d\theta^2 + \left(2e^{2\theta}\,d\theta\right)^2 = \left(e^{2\theta}\right)^2 d\theta^2 + 4e^{4\theta}\,d\theta^2$
$= 5e^{4\theta}\,d\theta^2 \Rightarrow ds = \sqrt{5}\,e^{2\theta}\,d\theta \Rightarrow L = \int_0^{2\pi} \sqrt{5}\,e^{2\theta}\,d\theta$
$= \left[\frac{\sqrt{5}\,e^{2\theta}}{2}\right]_0^{2\pi} = \frac{\sqrt{5}}{2}\left(e^{4\pi} - 1\right)$

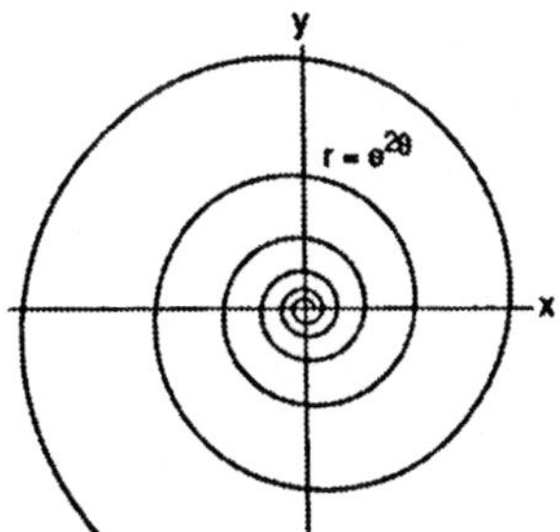

22. $r = 2\sin^3\left(\frac{\theta}{3}\right) \Rightarrow dr = 2\sin^2\left(\frac{\theta}{3}\right)\cos\left(\frac{\theta}{3}\right) d\theta \Rightarrow ds^2 = r^2\,d\theta^2 + dr^2 = \left[2\sin^3\left(\frac{\theta}{3}\right)\right]^2 d\theta^2 + \left[2\sin^2\left(\frac{\theta}{3}\right)\cos\left(\frac{\theta}{3}\right) d\theta\right]^2$
$= 4\sin^6\left(\frac{\theta}{3}\right) d\theta^2 + 4\sin^4\left(\frac{\theta}{3}\right)\cos^2\left(\frac{\theta}{3}\right) d\theta^2 = \left[4\sin^4\left(\frac{\theta}{3}\right)\right]\left[\sin^2\left(\frac{\theta}{3}\right) + \cos^2\left(\frac{\theta}{3}\right)\right] d\theta^2 = 4\sin^4\left(\frac{\theta}{3}\right) d\theta^2$
$\Rightarrow ds = 2\sin^2\left(\frac{\theta}{3}\right) d\theta$. Then $L = \int_0^{3\pi} 2\sin^2\left(\frac{\theta}{3}\right) d\theta = \int_0^{3\pi} \left[1 - \cos\left(\frac{2\theta}{3}\right)\right] d\theta = \left[\theta - \frac{3}{2}\sin\left(\frac{2\theta}{3}\right)\right]_0^{3\pi} = 3\pi$

23. $r = 1 + \cos\theta$ and $S = \int 2\pi\rho\,ds$, where $\rho = y = r\sin\theta$; $ds = \sqrt{r^2\,d\theta^2 + dr^2}$
$= \sqrt{(1 + \cos\theta)^2\,d\theta^2 + \sin^2\theta\,d\theta^2}\ \sqrt{1 + 2\cos\theta + \cos^2\theta + \sin^2\theta}\,d\theta = \sqrt{2 + 2\cos\theta}\,d\theta = \sqrt{4\cos^2\left(\frac{\theta}{2}\right)}\,d\theta$
$= 2\cos\left(\frac{\theta}{2}\right) d\theta$ since $0 \le \theta \le \frac{\pi}{2}$. Then $S = \int_0^{\pi/2} 2\pi(r\sin\theta)\cdot 2\cos\left(\frac{\theta}{2}\right) d\theta = \int_0^{\pi/2} 4\pi(1 + \cos\theta)\cdot\sin\theta\cos\left(\frac{\theta}{2}\right) d\theta$
$= \int_0^{\pi/2} 4\pi\left[2\cos^2\left(\frac{\theta}{2}\right)\right]\left[2\sin\left(\frac{\theta}{2}\right)\cos\left(\frac{\theta}{2}\right)\cos\left(\frac{\theta}{2}\right)\right] d\theta = \int_0^{\pi/2} 16\pi\cos^4\left(\frac{\theta}{2}\right)\sin\left(\frac{\theta}{2}\right) d\theta = \left[\frac{-32\pi\cos^5\left(\frac{\theta}{2}\right)}{5}\right]_0^{\pi/2}$
$= \frac{(-32\pi)\left(\frac{\sqrt{2}}{2}\right)^5}{5} - \left(-\frac{32\pi}{5}\right) = \frac{32\pi - 4\pi\sqrt{2}}{5}$

24. The region in question is the figure eight in the middle. The arc of $r = 2a \sin^2\left(\frac{\theta}{2}\right)$ in the first quadrant gives $\frac{1}{4}$ of that region. Therefore the area is $A = 4\int_0^{\pi/2} \frac{1}{2} r^2\, d\theta$

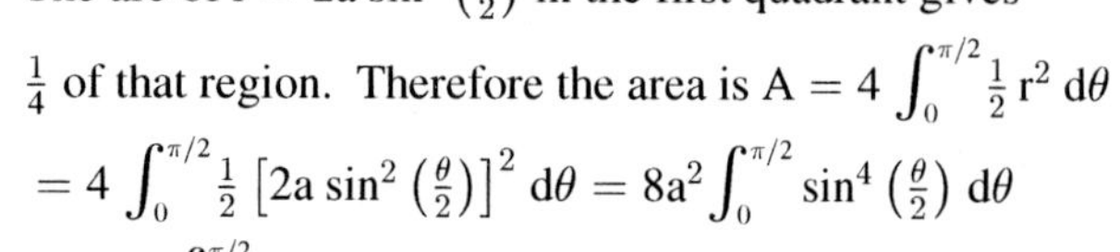

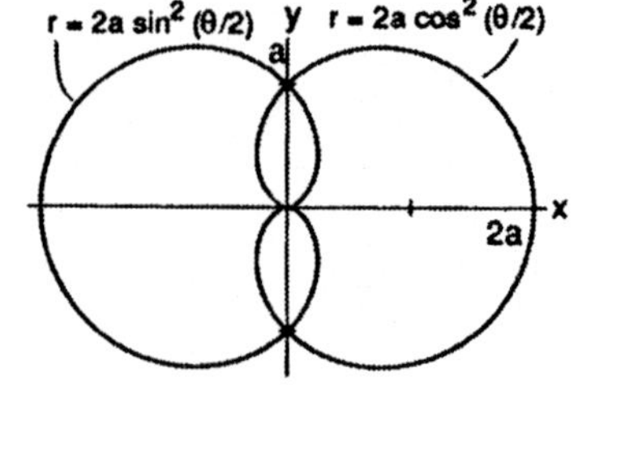

$= 4\int_0^{\pi/2} \frac{1}{2}\left[2a \sin^2\left(\frac{\theta}{2}\right)\right]^2 d\theta = 8a^2\int_0^{\pi/2} \sin^4\left(\frac{\theta}{2}\right) d\theta$

$= 8a^2\int_0^{\pi/2} \sin^2\left(\frac{\theta}{2}\right)\left[1 - \cos^2\left(\frac{\theta}{2}\right)\right] d\theta$

$= 8a^2\int_0^{\pi/2} \left[\sin^2\left(\frac{\theta}{2}\right) - \sin^2\left(\frac{\theta}{2}\right)\cos^2\left(\frac{\theta}{2}\right)\right] d\theta$

$= 8a^2\int_0^{\pi/2} \left(\frac{1-\cos\theta}{2} - \frac{\sin^2\theta}{4}\right) d\theta$

$= 2a^2\int_0^{\pi/2} \left(2 - 2\cos\theta - \frac{1-\cos 2\theta}{2}\right) d\theta = a^2\int_0^{\pi/2} (3 - 4\cos\theta + \cos 2\theta)\, d\theta = a^2\left[3\theta - 4\sin\theta + \frac{1}{2}\sin 2\theta\right]_0^{\pi/2}$

$= a^2\left(\frac{3\pi}{2} - 4\right)$

25. $e = 2$ and $r\cos\theta = 2 \Rightarrow x = 2$ is the directrix $\Rightarrow k = 2$; the conic is a hyperbola with $r = \frac{ke}{1+e\cos\theta}$

$\Rightarrow r = \frac{(2)(2)}{1+2\cos\theta} = \frac{4}{1+2\cos\theta}$

26. $e = 1$ and $r\cos\theta = -4 \Rightarrow x = -4$ is the directrix $\Rightarrow k = 4$; the conic is a parabola with $r = \frac{ke}{1-e\cos\theta}$

$\Rightarrow r = \frac{(4)(1)}{1-\cos\theta} = \frac{4}{1-\cos\theta}$

27. $e = \frac{1}{2}$ and $r\sin\theta = 2 \Rightarrow y = 2$ is the directrix $\Rightarrow k = 2$; the conic is an ellipse with $r = \frac{ke}{1+e\sin\theta}$

$\Rightarrow r = \frac{2\left(\frac{1}{2}\right)}{1+\left(\frac{1}{2}\right)\sin\theta} = \frac{2}{2+\sin\theta}$

28. $e = \frac{1}{3}$ and $r\sin\theta = -6 \Rightarrow y = -6$ is the directrix $\Rightarrow k = 6$; the conic is an ellipse with $r = \frac{ke}{1-e\sin\theta}$

$\Rightarrow r = \frac{6\left(\frac{1}{3}\right)}{1-\left(\frac{1}{3}\right)\sin\theta} = \frac{6}{3-\sin\theta}$

29. The length of the rope is $L = 2x + 2c + y \geq 8c$.

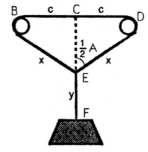

(a) The angle A ($\angle BED$) occurs when the distance $CF = \ell$ is maximized. Now $\ell = \sqrt{x^2 - c^2} + y$

$\Rightarrow \ell = \sqrt{x^2 - c^2} + L - 2x - 2c$

$\Rightarrow \frac{d\ell}{dx} = \frac{1}{2}\left(x^2 - c^2\right)^{-1/2}(2x) - 2 = \frac{x}{\sqrt{x^2 - c^2}} - 2.$

Thus $\frac{d\ell}{dx} = 0 \Rightarrow \frac{x}{\sqrt{x^2 - c^2}} - 2 = 0 \Rightarrow x = 2\sqrt{x^2 - c^2}$

$\Rightarrow x^2 = 4x^2 - 4c^2 \Rightarrow 3x^2 = 4c^2 \Rightarrow \frac{c^2}{x^2} = \frac{3}{4}$

$\Rightarrow \frac{c}{x} = \frac{\sqrt{3}}{2}$. Since $\frac{c}{x} = \sin\frac{A}{2}$ we have $\sin\frac{A}{2} = \frac{\sqrt{3}}{2}$

$\Rightarrow \frac{A}{2} = 60^\circ \Rightarrow A = 120^\circ$

(b) If the ring is fixed at E (i.e., y is held constant) and E is moved to the right, for example, the rope will slip around the pegs so that BE lengthens and DE becomes shorter $\Rightarrow$ BE + ED is always $2x = L - y - 2c$, which is constant $\Rightarrow$ the point E lies on an ellipse with the pegs as foci.

(c) Minimal potential energy occurs when the weight is at its lowest point $\Rightarrow$ E is at the intersection of the ellipse and its minor axis.

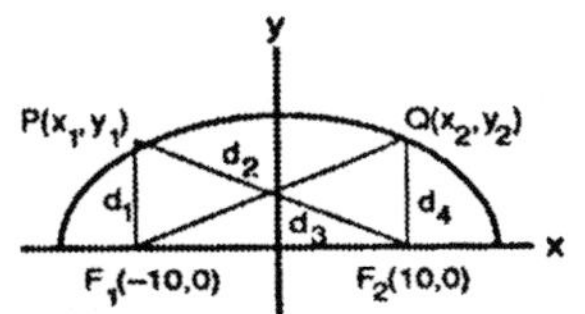

30. $\frac{d_1}{c} + \frac{d_2}{c} = \frac{30}{c} \Rightarrow d_1 + d_2 = 30$; $\frac{d_3}{c} + \frac{d_4}{c} = \frac{30}{c}$ $\Rightarrow d_3 + d_4 = 30$. Therefore P and Q lie on an ellipse with F_1 and F_2 as foci. Now $2a = d_1 + d_2 = 30 \Rightarrow a = 15$ and the focal distance is $10 \Rightarrow b^2 = 15^2 - 10^2 = 125$ $\Rightarrow$ an equation of the ellipse is $\frac{x^2}{225} + \frac{y^2}{125} = 1$. Next $x_2 = x_1 + v_0 t = x_1 + v_0\left(\frac{10}{v_0}\right) = x_1 + 10$. If the plane is flying level, then P and Q must be symmetric to the y-axis $\Rightarrow x_1 = -x_2 \Rightarrow x_2 = -x_2 + 10$ $\Rightarrow x_2 = 5 \Rightarrow \frac{5^2}{225} + \frac{y_2^2}{125} = 1 \Rightarrow y_2^2 = \frac{1000}{9} \Rightarrow y_2 = \frac{10\sqrt{10}}{3}$ since y_2 must be positive. Therefore the position of the plane is $\left(5, \frac{10\sqrt{10}}{3}\right)$ where the origin $(0, 0)$ is located midway between the two stations.

31. If the vertex is $(0, 0)$, then the focus is $(p, 0)$. Let $P(x, y)$ be the present position of the comet. Then $\sqrt{(x-p)^2 + y^2} = 4 \times 10^7$. Since $y^2 = 4px$ we have $\sqrt{(x-p)^2 + 4px} = 4 \times 10^7 \Rightarrow (x-p)^2 + 4px = 16 \times 10^{14}$. Also, $x - p = 4 \times 10^7 \cos 60° = 2 \times 10^7 \Rightarrow x = p + 2 \times 10^7$. Therefore $(2 \times 10^7)^2 + 4p(p + 2 \times 10^7) = 16 \times 10^{14}$ $\Rightarrow 4 \times 10^{14} + 4p^2 + 8p \times 10^7 = 16 \times 10^{14} \Rightarrow 4p^2 + 8p \times 10^7 - 12 \times 10^{14} = 0 \Rightarrow p^2 + 2p \times 10^7 - 3 \times 10^{14} = 0$ $\Rightarrow (p + 3 \times 10^7)(p - 10^7) = 0 \Rightarrow p = -3 \times 10^7$ or $p = 10^7$. Since p is positive we obtain $p = 10^7$ miles.

32. $x = 2t$ and $y = t^2 \Rightarrow y = \frac{x^2}{4}$; let $D = \sqrt{(x-0)^2 + \left(\frac{x^2}{4} - 3\right)^2} = \sqrt{x^2 + \frac{x^4}{16} - \frac{3}{2}x^2 + 9} = \sqrt{\frac{x^4}{16} - \frac{1}{2}x^2 + 9}$ $= \frac{1}{4}\sqrt{x^4 - 8x^2 + 144}$ be the distance from any point on the parabola to $(0, 3)$. We want to minimize D. Then $\frac{dD}{dx} = \frac{1}{8}(x^4 - 8x^2 + 144)^{-1/2}(4x^3 - 16x) = \frac{\left(\frac{1}{2}\right)x^3 - 2x}{\sqrt{x^4 - 8x^2 + 144}} = 0 \Rightarrow \frac{1}{2}x^3 - 2x = 0 \Rightarrow x^3 - 4x = 0 \Rightarrow x = 0$ or $x = \pm 2$. Now $x = 0 \Rightarrow y = 0$ and $x = \pm 2 \Rightarrow y = 1$. The distance from $(0, 0)$ to $(0, 3)$ is $D = 3$. The distance from $(\pm 2, 1)$ to $(0, 3)$ is $D = \sqrt{(\pm 2)^2 + (1-3)^2} = 2\sqrt{2}$ which is less than 3. Therefore the points closest to $(0, 3)$ are $(\pm 2, 1)$.

33. $\cot 2\alpha = \frac{A-C}{B} = 0 \Rightarrow \alpha = 45°$ is the angle of rotation $\Rightarrow A' = \cos^2 45° + \cos 45° \sin 45° + \sin^2 45° = \frac{3}{2}$, $B' = 0$, and $C' = \sin^2 45° - \sin 45° \cos 45° + \cos^2 45° = \frac{1}{2} \Rightarrow \frac{3}{2}x'^2 + \frac{1}{2}y'^2 = 1 \Rightarrow b = \sqrt{\frac{2}{3}}$ and $a = \sqrt{2} \Rightarrow c^2 = a^2 - b^2$ $= 2 - \frac{2}{3} = \frac{4}{3} \Rightarrow c = \frac{2}{\sqrt{3}}$. Therefore the eccentricity is $e = \frac{c}{a} = \frac{\left(\frac{2}{\sqrt{3}}\right)}{\sqrt{2}} = \sqrt{\frac{2}{3}} \approx 0.82$.

34. The angle of rotation is $\alpha = \frac{\pi}{4} \Rightarrow A' = \sin\frac{\pi}{4}\cos\frac{\pi}{4} = \frac{1}{2}$, $B' = 0$, and $C' = -\sin\frac{\pi}{4}\cos\frac{\pi}{4} = -\frac{1}{2} \Rightarrow \frac{x'^2}{2} - \frac{y'^2}{2} = 1$ $\Rightarrow a = \sqrt{2}$ and $b = \sqrt{2} \Rightarrow c^2 = a^2 + b^2 = 4 \Rightarrow c = 2$. Therefore the eccentricity is $e = \frac{c}{a} = \frac{2}{\sqrt{2}} = \sqrt{2}$.

35. $\sqrt{x} + \sqrt{y} = 1 \Rightarrow x + 2\sqrt{xy} + y = 1 \Rightarrow 2\sqrt{xy} = 1 - (x + y) \Rightarrow 4xy = 1 - 2(x + y) + (x + y)^2$ $\Rightarrow 4xy = x^2 + 2xy + y^2 - 2x - 2y + 1 \Rightarrow x^2 - 2xy + y^2 - 2x - 2y + 1 = 0 \Rightarrow B^2 - 4AC = (-2)^2 - 4(1)(1) = 0$ $\Rightarrow$ the curve is part of a parabola

36. $\alpha = \frac{\pi}{4} \Rightarrow A' = 2\sin\frac{\pi}{4}\cos\frac{\pi}{4} = 1$, $B' = 0$, $C' = -2\sin\frac{\pi}{4}\cos\frac{\pi}{4} = -1$, $D' = -\sqrt{2}\sin\frac{\pi}{4} = -1$, $E' = -\sqrt{2}\cos\frac{\pi}{4}$ $= -1$, $F' = 2 \Rightarrow x'^2 - y'^2 - x' - y' + 2 = 0 \Rightarrow \left(x'^2 - x'\right) - \left(y'^2 + y'\right) = -2 \Rightarrow \left(x'^2 - x' + \frac{1}{4}\right) - \left(y'^2 + y' + \frac{1}{4}\right)$ $= -2 \Rightarrow \frac{\left(y' + \frac{1}{2}\right)^2}{2} - \frac{\left(x' - \frac{1}{2}\right)^2}{2} = 1$. The center is $(x', y') = \left(\frac{1}{2}, -\frac{1}{2}\right) \Rightarrow x = \frac{1}{2}\cos\frac{\pi}{4} - \left(-\frac{1}{2}\right)\sin\frac{\pi}{4} = \frac{\sqrt{2}}{2}$ and $y = \frac{1}{2}\sin\frac{\pi}{4} - \frac{1}{2}\cos\frac{\pi}{4} = 0$ or the center is $(x, y) = \left(\frac{\sqrt{2}}{2}, 0\right)$. Next $a = \sqrt{2} \Rightarrow$ the vertices are $(x', y') = \left(\frac{1}{2}, \sqrt{2} - \frac{1}{2}\right)$ and $\left(\frac{1}{2}, -\sqrt{2} - \frac{1}{2}\right) \Rightarrow x = \frac{1}{2}\cos\frac{\pi}{4} - \left(\sqrt{2} - \frac{1}{2}\right)\sin\frac{\pi}{4} = \frac{\sqrt{2}}{2} - 1$ and $y = \frac{1}{2}\sin\frac{\pi}{4} + \left(\sqrt{2} - \frac{1}{2}\right)\cos\frac{\pi}{4} = 1$ or $(x, y) = \left(\frac{\sqrt{2}}{2} - 1, 1\right)$ is one vertex, and $x = \frac{1}{2}\cos\frac{\pi}{4} - \left(-\sqrt{2} - \frac{1}{2}\right)\sin\frac{\pi}{4}$

$= \frac{\sqrt{2}}{2} + 1$ and $y = \frac{1}{2}\sin\frac{\pi}{4} + \left(-\sqrt{2} - \frac{1}{2}\right)\sin\frac{\pi}{4} = -1$ or $(x, y) = \left(\frac{\sqrt{2}}{2} + 1, -1\right)$ is the other vertex. Also $c^2 = 2 + 2 = 4 \Rightarrow c = 2 \Rightarrow$ the foci are $(x', y') = \left(\frac{1}{2}, \frac{3}{2}\right)$ and $\left(\frac{1}{2}, -\frac{5}{2}\right) \Rightarrow x = \frac{1}{2}\cos\frac{\pi}{4} - \frac{3}{2}\sin\frac{\pi}{4} = -\frac{\sqrt{2}}{2}$ and $y = \frac{1}{2}\sin\frac{\pi}{4} + \frac{3}{2}\cos\frac{\pi}{4} = \sqrt{2}$ or $(x, y) = \left(-\frac{\sqrt{2}}{2}, \sqrt{2}\right)$ is one focus, and $x = \frac{1}{2}\cos\frac{\pi}{4} + \frac{5}{2}\sin\frac{\pi}{4} = \frac{3\sqrt{2}}{2}$ and $y = \frac{1}{2}\sin\frac{\pi}{4} - \frac{5}{2}\cos\frac{\pi}{4} = -\sqrt{2}$ or $(x, y) = \left(\frac{3\sqrt{2}}{2}, -\sqrt{2}\right)$ is the other focus. The asymptotes are $y' + \frac{1}{2} = \pm\left(x' - \frac{1}{2}\right)$ in the rotated system. Since $x = \frac{1}{\sqrt{2}}x' - \frac{1}{\sqrt{2}}y'$ and $y = \frac{1}{\sqrt{2}}x' + \frac{1}{\sqrt{2}}y' \Rightarrow x + y = \frac{2}{\sqrt{2}}x'$ $\Rightarrow \frac{\sqrt{2}}{2}x + \frac{\sqrt{2}}{2}y = x'$ and $x - y = -\frac{2}{\sqrt{2}}y' \Rightarrow -\frac{\sqrt{2}}{2}x + \frac{\sqrt{2}}{2}y = y'$; the asymptotes are $-\frac{\sqrt{2}}{2}x + \frac{\sqrt{2}}{2}y + \frac{1}{2} = \pm\left(\frac{\sqrt{2}}{2}x + \frac{\sqrt{2}}{2}y - \frac{1}{2}\right) \Rightarrow$ the asymptotes are $-\sqrt{2}x + 1 = 0$ or $x = \frac{1}{\sqrt{2}}$ and $\sqrt{2}y = 0$ or $y = 0$. Finally, the x'-axis is the line through $\left(\frac{\sqrt{2}}{2}, 0\right)$ with a slope of 1 (recall that $\alpha = \frac{\pi}{4}$) $\Rightarrow y = x - \frac{\sqrt{2}}{2}$. The y'-axis is the line through $\left(\frac{\sqrt{2}}{2}, 0\right)$ with a slope of $-1 \Rightarrow y = -x + \frac{\sqrt{2}}{2}$.

37. (a) The equation of a parabola with focus $(0, 0)$ and vertex $(a, 0)$ is $r = \frac{2a}{1 + \cos\theta}$ and rotating this parabola through $\alpha = 45°$ gives $r = \frac{2a}{1 + \cos\left(\theta - \frac{\pi}{4}\right)}$.

(b) Foci at $(0, 0)$ and $(2, 0) \Rightarrow$ the center is $(1, 0) \Rightarrow a = 3$ and $c = 1$ since one vertex is at $(4, 0)$. Then $e = \frac{c}{a} = \frac{1}{3}$. For ellipses with one focus at the origin and major axis along the x-axis we have $r = \frac{a(1 - e^2)}{1 - e\cos\theta} = \frac{3\left(1 - \frac{1}{9}\right)}{1 - \left(\frac{1}{3}\right)\cos\theta} = \frac{8}{3 - \cos\theta}$.

(c) Center at $\left(2, \frac{\pi}{2}\right)$ and focus at $(0, 0) \Rightarrow c = 2$; center at $\left(2, \frac{\pi}{2}\right)$ and vertex at $\left(1, \frac{\pi}{2}\right) \Rightarrow a = 1$. Then $e = \frac{c}{a} = \frac{2}{1} = 2$. Also $k = ae - \frac{a}{e} = (1)(2) - \frac{1}{2} = \frac{3}{2}$. Therefore $r = \frac{ke}{1 + e\sin\theta} = \frac{\left(\frac{3}{2}\right)(2)}{1 + 2\sin\theta} = \frac{3}{1 + 2\sin\theta}$.

38. Let (d_1, θ_1) and (d_2, θ_2) be the polar coordinates of P_1 and P_2, respectively. Then $\theta_2 = \theta_1 + \pi$, and we have $d_1 = \frac{3}{2 + \cos\theta_1}$ and $d_2 = \frac{3}{2 + \cos(\theta_1 + \pi)}$. Therefore $\frac{1}{d_1} + \frac{1}{d_2} = \frac{2 + \cos\theta_1}{3} + \frac{2 + \cos(\theta_1 + \pi)}{3} = \frac{4 + \cos\theta_1 + \cos\theta_1\cos\pi - \sin\theta_1\sin\pi}{3} = \frac{4}{3}$.

39. Arc PT = Arc TO since each is the same distance rolled. Now Arc PT = a(∠TAP) and Arc TO = a(∠TBO) ⇒ ∠TAP = ∠TBO. Since AP = a = BO we have that ΔADP is congruent to ΔBCO ⇒ CO = DP ⇒ OP is parallel to AB ⇒ ∠TBO = ∠TAP = θ. Then OPDC is a square ⇒ r = CD = AB − AD − CB = AB − 2CB ⇒ $r = 2a - 2a\cos\theta = 2a(1 - \cos\theta)$, which is the polar equation of a cardioid.

40. Note first that the point P traces out a circular arc as the door closes until the second door panel PQ is tangent to the circle. This happens when P is located at $\left(\frac{1}{\sqrt{2}}, \frac{1}{\sqrt{2}}\right)$, since ∠OPQ is 90° at that time. Thus the curve is the circle $x^2 + y^2 = 1$ for $0 \le x \le \frac{1}{\sqrt{2}}$. When $x \ge \frac{1}{\sqrt{2}}$, the second door panel is tangent to the curve at P. Now let t represent ∠POQ so that as t runs from $\frac{\pi}{2}$ to 0, the door closes. The coordinates of P are given by $(\cos t, \sin t)$, and the coordinates of Q by $(2\cos t, 0)$ (since triangle POQ is isosceles). Therefore at a fixed instant of time t, the slope of the line formed by the second panel PQ is $m = \frac{\Delta y}{\Delta x} = \frac{\sin t - 0}{\cos t - 2\cos t} = -\tan t \Rightarrow$ the tangent line PQ is $y - 0 = (-\tan t)(x - 2\cos t) \Rightarrow y = (-\tan t)x + 2\sin t$. Now, to find an equation of the curve for $\frac{1}{\sqrt{2}} \le x \le 1$, we want to find, for <u>fixed</u> x, the largest value of y as t ranges over the interval $0 \le t \le \frac{\pi}{4}$. We solve $\frac{dy}{dt} = 0 \Rightarrow (-\sec^2 t)x + 2\cos t = 0 \Rightarrow (-\sec^2 t)x = -2\cos t \Rightarrow x = 2\cos^3 t$. (Note that $\frac{d^2y}{dt^2} = (-2\sec^2 t\tan t)x - 2\sin t < 0$ on $0 \le t \le \frac{\pi}{2}$, so a maximum occurs for y.) Now $x = 2\cos^3 t \Rightarrow$ the

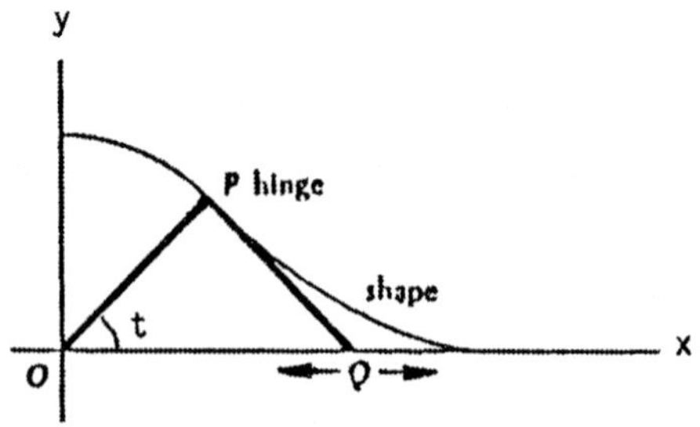

corresponding y value is $y = (-\tan t)(2\cos^3 t) + 2\sin t = -2\sin t\cos^2 t + 2\sin t = (2\sin t)(-\cos^2 t + 1)$ $= 2\sin^3 t$. Therefore parametric equations for the path of the curve are given by $x = 2\cos^3 t$ and $y = 2\sin^3 t$ for $0 \le t \le \frac{\pi}{4}$. In Cartesian coordinates, we have the curve $x^{2/3} + y^{2/3} = (2\cos^3 t)^{2/3} + (2\sin^3 t)^{2/3}$ $= 2^{2/3}(\cos^2 t + \sin^2 t) = 2^{2/3} \Rightarrow$ the curve traced out by the door is given by

$$\left.\begin{array}{ll} x^2 + y^2 = 1 & \text{for } 0 \le x \le \frac{1}{\sqrt{2}} \\ x^{2/3} + y^{2/3} = 2^{2/3} & \text{for } \frac{1}{\sqrt{2}} \le x \le 1 \end{array}\right\}$$

41. $\beta = \psi_2 - \psi_1 \Rightarrow \tan\beta = \tan(\psi_2 - \psi_1) = \frac{\tan\psi_2 - \tan\psi_1}{1 + \tan\psi_2 \tan\psi_1}$; the curves will be orthogonal when $\tan\beta$ is undefined, or when $\tan\psi_2 = \frac{-1}{\tan\psi_1} \Rightarrow \frac{r}{g'(\theta)} = \frac{-1}{\left[\frac{r}{f'(\theta)}\right]}$ $\Rightarrow r^2 = -f'(\theta)\,g'(\theta)$

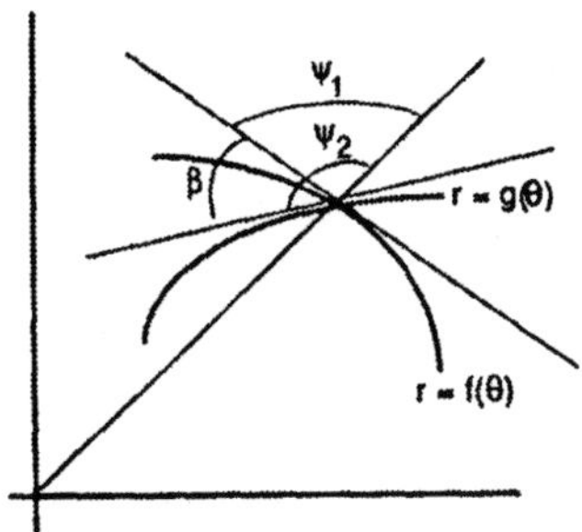

42. $r = \sin^4\left(\frac{\theta}{4}\right) \Rightarrow \frac{dr}{d\theta} = \sin^3\left(\frac{\theta}{4}\right)\cos\left(\frac{\theta}{4}\right) \Rightarrow \tan\psi = \frac{\sin^4\left(\frac{\theta}{4}\right)}{\sin^3\left(\frac{\theta}{4}\right)\cos\left(\frac{\theta}{4}\right)} = \tan\left(\frac{\theta}{4}\right)$

43. $r = 2a\sin 3\theta \Rightarrow \frac{dr}{d\theta} = 6a\cos 3\theta \Rightarrow \tan\psi = \frac{r}{\left(\frac{dr}{d\theta}\right)} = \frac{2a\sin 3\theta}{6a\cos 3\theta} = \frac{1}{3}\tan 3\theta$; when $\theta = \frac{\pi}{6}$, $\tan\psi = \frac{1}{3}\tan\frac{\pi}{2}$ $\Rightarrow \psi = \frac{\pi}{2}$

44. (a)

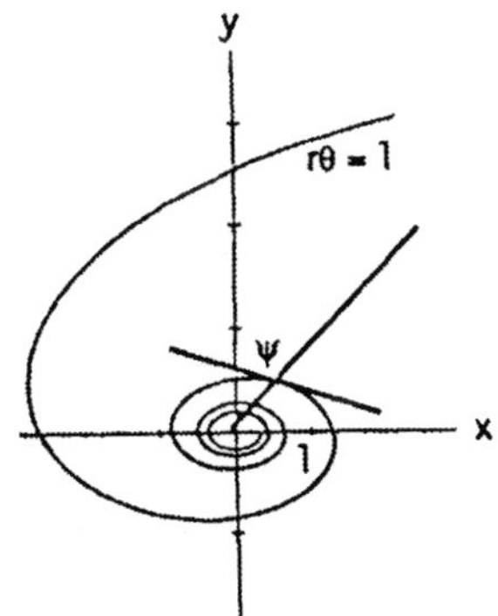

(b) $r\theta = 1 \Rightarrow r = \theta^{-1} \Rightarrow \frac{dr}{d\theta} = -\theta^{-2} \Rightarrow \tan\psi\big|_{\theta=1} = \frac{\theta^{-1}}{-\theta^{-2}} = -\theta \Rightarrow \lim_{\theta\to\infty} \tan\psi = -\infty$ $\Rightarrow \psi \to \frac{\pi}{2}$ from the right as the spiral winds in around the origin.

45. $\tan\psi_1 = \frac{\sqrt{3}\cos\theta}{-\sqrt{3}\sin\theta} = -\cot\theta$ is $-\frac{1}{\sqrt{3}}$ at $\theta = \frac{\pi}{3}$; $\tan\psi_2 = \frac{\sin\theta}{\cos\theta} = \tan\theta$ is $\sqrt{3}$ at $\theta = \frac{\pi}{3}$; since the product of these slopes is -1, the tangents are perpendicular

46. $a(1 + \cos\theta) = 3a\cos\theta \Rightarrow 1 = 2\cos\theta \Rightarrow \cos\theta = \frac{1}{2}$ or $\theta = \frac{\pi}{3}$; $\tan\psi_1 = \frac{a(1+\cos\theta)}{-a\sin\theta}$ is $-\sqrt{3}$ at $\theta = \frac{\pi}{3}$; $\tan\psi_2 = \frac{3a\cos\theta}{-3a\sin\theta}$ is $-\frac{1}{\sqrt{3}}$ at $\theta = \frac{\pi}{3}$. Then

$$\tan\beta = \frac{-\frac{1}{\sqrt{3}} - \left(-\sqrt{3}\right)}{1 + \left(-\frac{1}{\sqrt{3}}\right)\left(-\sqrt{3}\right)} = \frac{-\frac{1}{\sqrt{3}} + \sqrt{3}}{2} = \frac{1}{\sqrt{3}} \Rightarrow \beta = \frac{\pi}{6}$$

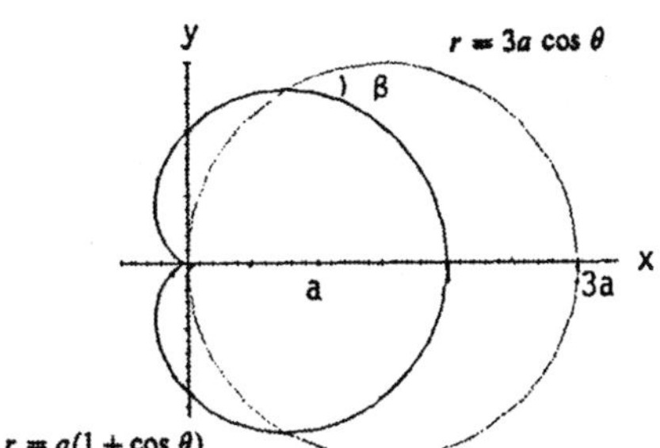

47. $r_1 = \frac{1}{1-\cos\theta} \Rightarrow \frac{dr_1}{d\theta} = -\frac{\sin\theta}{(1-\cos\theta)^2}$; $r_2 = \frac{3}{1+\cos\theta} \Rightarrow \frac{dr_2}{d\theta} = \frac{3\sin\theta}{(1+\cos\theta)^2}$; $\frac{1}{1-\cos\theta} = \frac{3}{1+\cos\theta}$ $\Rightarrow 1+\cos\theta = 3-3\cos\theta \Rightarrow 4\cos\theta = 2 \Rightarrow \cos\theta = \frac{1}{2} \Rightarrow \theta = \pm\frac{\pi}{3} \Rightarrow r_1 = r_2 = 2 \Rightarrow$ the curves intersect at the points $\left(2, \pm\frac{\pi}{3}\right)$; $\tan\psi_1 = \frac{\left(\frac{1}{1-\cos\theta}\right)}{\left[\frac{-\sin\theta}{(1-\cos\theta)^2}\right]} = -\frac{1-\cos\theta}{\sin\theta}$ is $-\frac{1}{\sqrt{3}}$ at $\theta = \frac{\pi}{3}$; $\tan\psi_2 = \frac{\left(\frac{3}{1+\cos\theta}\right)}{\left[\frac{3\sin\theta}{(1+\cos\theta)^2}\right]} = \frac{1+\cos\theta}{\sin\theta}$ is $\sqrt{3}$ at $\theta = \frac{\pi}{3}$; therefore $\tan\beta$ is undefined at $\theta = \frac{\pi}{3}$ since $1+\tan\psi_1\tan\psi_2 = 1+\left(-\frac{1}{\sqrt{3}}\right)\left(\sqrt{3}\right) = 0 \Rightarrow \beta = \frac{\pi}{2}$; $\tan\psi_1\big|_{\theta=-\pi/3} = -\frac{1-\cos\left(-\frac{\pi}{3}\right)}{\sin\left(-\frac{\pi}{3}\right)} = \frac{1}{\sqrt{3}}$ and $\tan\psi_2\big|_{\theta=-\pi/3} = \frac{1+\cos\left(-\frac{\pi}{3}\right)}{\sin\left(-\frac{\pi}{3}\right)} = -\sqrt{3} \Rightarrow \tan\beta$ is also undefined at $\theta = -\frac{\pi}{3} \Rightarrow \beta = \frac{\pi}{2}$

48. (a) We need $\psi+\theta = \pi$, so that $\tan\psi = \tan(\pi-\theta)$ $= -\tan\theta$. Now $\tan\psi = \frac{r}{\left(\frac{dr}{d\theta}\right)} = \frac{a(1+\cos\theta)}{-a\sin\theta}$ $= -\tan\theta = -\frac{\sin\theta}{\cos\theta} \Rightarrow \cos\theta+\cos^2\theta = \sin^2\theta$ $\Rightarrow \cos\theta+\cos^2\theta = 1-\cos^2\theta$ $\Rightarrow 2\cos^2\theta+\cos\theta-1 = 0$ $\Rightarrow \cos\theta = \frac{1}{2}$ or $\cos\theta = -1$; $\cos\theta = \frac{1}{2} \Rightarrow \theta = \pm\frac{\pi}{3}$ $\Rightarrow r = \frac{3a}{2}$; $\cos\theta = -1 \Rightarrow \theta = \pi \Rightarrow r = 0$. Therefore the points where the tangent line is horizontal are $\left(\frac{3a}{2}, \pm\frac{\pi}{3}\right)$ and $(0, \pi)$.

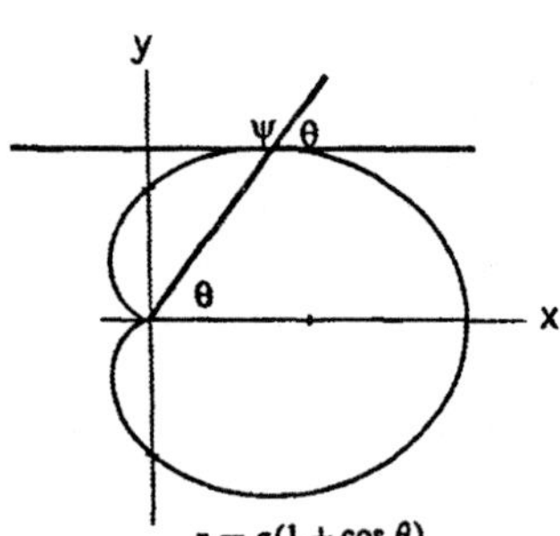

(b) We need $\psi+\theta = \frac{\pi}{2}$ so that $\tan\psi = \tan\left(\frac{\pi}{2}-\theta\right) = \cot\theta$. Thus $\tan\psi = \frac{r}{\left(\frac{dr}{d\theta}\right)} = \frac{a(1+\cos\theta)}{-a\sin\theta} = \cot\theta$ $= \frac{\cos\theta}{\sin\theta} \Rightarrow \sin\theta+\sin\theta\cos\theta = -\sin\theta\cos\theta \Rightarrow \cos\theta = -\frac{1}{2}$ or $\sin\theta = 0$; $\cos\theta = -\frac{1}{2} \Rightarrow \theta = \pm\frac{2\pi}{3}$ $\Rightarrow r = \frac{a}{2}$; $\sin\theta = 0 \Rightarrow \theta = 0$ (not π, see part (a)) $\Rightarrow r = 2a$. Therefore the points where the tangent line is vertical are $\left(\frac{a}{2}, \pm\frac{2\pi}{3}\right)$ and $(2a, 0)$.

49. $r_1 = \frac{a}{1+\cos\theta} \Rightarrow \frac{dr_1}{d\theta} = \frac{a\sin\theta}{(1+\cos\theta)^2}$ and $r_2 = \frac{b}{1-\cos\theta} \Rightarrow \frac{dr_2}{d\theta} = -\frac{b\sin\theta}{(1-\cos\theta)^2}$; then $\tan\psi_1 = \frac{\left(\frac{a}{1+\cos\theta}\right)}{\left[\frac{a\sin\theta}{(1+\cos\theta)^2}\right]} = \frac{1+\cos\theta}{\sin\theta}$ and $\tan\psi_2 = \frac{\left(\frac{b}{1-\cos\theta}\right)}{\left[\frac{-b\sin\theta}{(1-\cos\theta)^2}\right]} = \frac{1-\cos\theta}{-\sin\theta} \Rightarrow 1+\tan\psi_1\tan\psi_2$ $= 1+\left(\frac{1+\cos\theta}{\sin\theta}\right)\left(\frac{1-\cos\theta}{-\sin\theta}\right) = 1-\frac{1-\cos^2\theta}{\sin^2\theta} = 0 \Rightarrow \beta$ is undefined $\Rightarrow$ the parabolas are orthogonal at each point of intersection

50. $\tan\psi = \frac{r}{\left(\frac{dr}{d\theta}\right)} = \frac{a(1-\cos\theta)}{a\sin\theta}$ is 1 at $\theta = \frac{\pi}{2} \Rightarrow \psi = \frac{\pi}{4}$

51. $r = 3\sec\theta \Rightarrow r = \frac{3}{\cos\theta}$; $\frac{3}{\cos\theta} = 4+4\cos\theta \Rightarrow 3 = 4\cos\theta+4\cos^2\theta \Rightarrow (2\cos\theta+3)(2\cos\theta-1) = 0$ $\Rightarrow \cos\theta = \frac{1}{2}$ or $\cos\theta = -\frac{3}{2} \Rightarrow \theta = \frac{\pi}{3}$ or $\frac{5\pi}{3}$ (the second equation has no solutions); $\tan\psi_2 = \frac{4(1+\cos\theta)}{-4\sin\theta}$ $= -\frac{1+\cos\theta}{\sin\theta}$ is $-\sqrt{3}$ at $\frac{\pi}{3}$ and $\tan\psi_1 = \frac{3\sec\theta}{3\sec\theta\tan\theta} = \cot\theta$ is $\frac{1}{\sqrt{3}}$ at $\frac{\pi}{3}$. Then $\tan\beta$ is undefined since $1+\tan\psi_1\tan\psi_2 = 1+\left(\frac{1}{\sqrt{3}}\right)\left(-\sqrt{3}\right) = 0 \Rightarrow \beta = \frac{\pi}{2}$. Also, $\tan\psi_2\big|_{5\pi/3} = \sqrt{3}$ and $\tan\psi_1\big|_{5\pi/3} = -\frac{1}{\sqrt{3}}$ $\Rightarrow 1+\tan\psi_1\tan\psi_2 = 1+\left(-\frac{1}{\sqrt{3}}\right)\left(\sqrt{3}\right) = 0 \Rightarrow \tan\beta$ is also undefined $\Rightarrow \beta = \frac{\pi}{2}$.

52. $\tan\psi = \frac{a\tan\left(\frac{\theta}{2}\right)}{\frac{a}{2}\sec^2\left(\frac{\theta}{2}\right)} = 1$ at $\theta = \frac{\pi}{2} \Rightarrow \psi = \frac{\pi}{4}$; $m_{tan} = \tan(\theta+\psi) = \tan\frac{3\pi}{4} = -1$

53. $\frac{1}{1-\cos\theta} = \frac{1}{1-\sin\theta} \Rightarrow 1-\cos\theta = 1-\sin\theta \Rightarrow \cos\theta = \sin\theta \Rightarrow \theta = \frac{\pi}{4}$; $\tan\psi_1 = \frac{\left(\frac{1}{1-\cos\theta}\right)}{\left[\frac{-\sin\theta}{(1-\cos\theta)^2}\right]} = \frac{1-\cos\theta}{-\sin\theta}$; $\tan\psi_2 = \frac{\left(\frac{1}{1-\sin\theta}\right)}{\left[\frac{\cos\theta}{(1-\sin\theta)^2}\right]} = \frac{1-\sin\theta}{\cos\theta}$. Thus at $\theta = \frac{\pi}{4}$, $\tan\psi_1 = \frac{1-\cos\left(\frac{\pi}{4}\right)}{-\sin\left(\frac{\pi}{4}\right)} = 1-\sqrt{2}$ and

$\tan \psi_2 = \frac{1 - \sin\left(\frac{\pi}{4}\right)}{\cos\left(\frac{\pi}{4}\right)} = \sqrt{2} - 1$. Then $\tan \beta = \frac{\left(\sqrt{2}-1\right) - \left(1 - \sqrt{2}\right)}{1 + \left(\sqrt{2}-1\right)\left(1-\sqrt{2}\right)} = \frac{2\sqrt{2}-2}{2\sqrt{2}-2} = 1 \Rightarrow \beta = \frac{\pi}{4}$

54. (a)

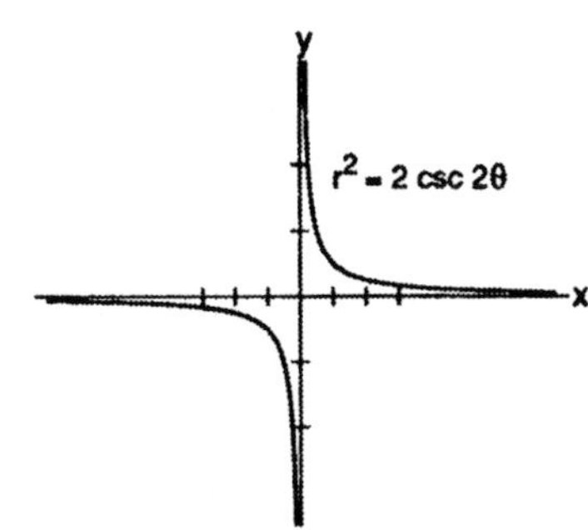

(b) $r^2 = 2 \csc 2\theta = \frac{2}{\sin 2\theta} = \frac{2}{2 \sin\theta \cos\theta}$
$\Rightarrow r^2 \sin\theta \cos\theta = 1 \Rightarrow xy = 1$, a hyperbola

(c) At $\theta = \frac{\pi}{4}$, $x = y = 1 \Rightarrow \frac{dy}{dx} = -\frac{1}{x^2} = -1$
$= m_{\tan} \Rightarrow \phi = \frac{3\pi}{4} \Rightarrow \psi = \phi - \theta = \frac{3\pi}{4} - \frac{\pi}{4} = \frac{\pi}{2}$

55. (a) $\tan \alpha = \frac{r}{\left(\frac{dr}{d\theta}\right)} \Rightarrow \frac{dr}{r} = \frac{d\theta}{\tan\alpha} \Rightarrow \ln r = \frac{\theta}{\tan\alpha} + C$ (by integration) $\Rightarrow r = Be^{\theta/(\tan\alpha)}$ for some constant B;

$A = \frac{1}{2}\int_{\theta_1}^{\theta_2} B^2 e^{2\theta/(\tan\alpha)}\, d\theta = \left[\frac{B^2 (\tan\alpha) e^{2\theta/(\tan\alpha)}}{4}\right]_{\theta_1}^{\theta_2} = \frac{\tan\alpha}{4}\left[B^2 e^{2\theta_2/(\tan\alpha)} - B^2 e^{2\theta_1/(\tan\alpha)}\right]$

$= \frac{\tan\alpha}{4}\left(r_2^2 - r_1^2\right)$ since $r_2^2 = B^2 e^{2\theta_2/(\tan\alpha)}$ and $r_1^2 = B^2 e^{2\theta_1/(\tan\alpha)}$; constant of proportionality $K = \frac{\tan\alpha}{4}$

(b) $\tan\alpha = \frac{r}{\left(\frac{dr}{d\theta}\right)} \Rightarrow \frac{dr}{d\theta} = \frac{r}{\tan\alpha} \Rightarrow \left(\frac{dr}{d\theta}\right)^2 = \frac{r^2}{\tan^2\alpha} \Rightarrow r^2 + \left(\frac{dr}{d\theta}\right)^2 = r^2 + \frac{r^2}{\tan^2\alpha} = r^2\left(\frac{\tan^2\alpha + 1}{\tan^2\alpha}\right)$

$= r^2\left(\frac{\sec^2\alpha}{\tan^2\alpha}\right) \Rightarrow \text{Length} = \int_{\theta_1}^{\theta_2} r\left(\frac{\sec\alpha}{\tan\alpha}\right) d\theta = \int_{\theta_1}^{\theta_2} Be^{\theta/(\tan\alpha)} \cdot \frac{\sec\alpha}{\tan\alpha}\, d\theta = \left[B(\sec\alpha)e^{\theta/(\tan\alpha)}\right]_{\theta_1}^{\theta_2}$

$= (\sec\alpha)\left[Be^{\theta_2/(\tan\alpha)} - Be^{\theta_1(\tan\alpha)}\right] = K(r_2 - r_1)$ where $K = \sec\alpha$ is the constant of proportionality

56. $r^2 \sin 2\theta = 2a^2 \Rightarrow r^2 \sin\theta\cos\theta = a^2 \Rightarrow xy = a^2$ and $\frac{dy}{dx} = -\frac{a^2}{x^2}$. If $P(x_1, y_1)$ is a point on the curve, the tangent line is $y - y_1 = -\frac{a^2}{x_1^2}(x - x_1)$, so the tangent line crosses the x-axis when $y = 0 \Rightarrow -y_1 = -\frac{a^2}{x_1^2}(x - x_1)$
$\Rightarrow \frac{x_1^2 y_1}{a^2} = x - x_1 \Rightarrow x = \frac{x_1^2 y_1}{a^2} + x_1 = x_1 + x_1 = 2x_1$
since $\frac{x_1 y_1}{a^2} = 1$. Let Q be $(2x_1, 0)$. Then

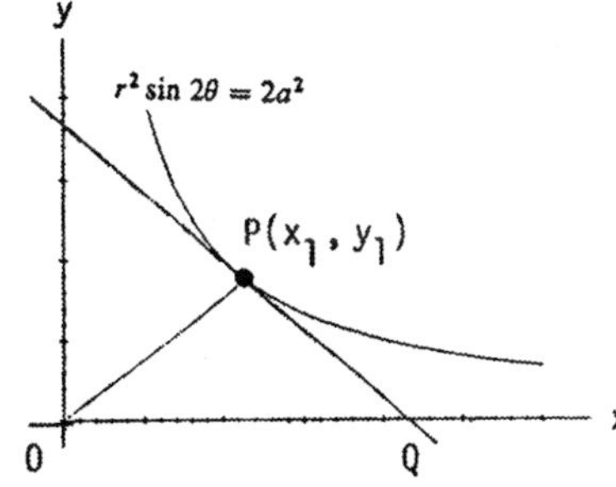

$PQ = \sqrt{(2x_1 - x_1)^2 + (0 - y_1)^2} = \sqrt{x_1^2 + y_1^2}$ and
$OP = r = \sqrt{(x_1 - 0)^2 + (y_1 - 0)^2} = \sqrt{x_1^2 + y_1^2} \Rightarrow OP = PQ$ and the triangle is isosceles.

NOTES:

CHAPTER 11 INFINITE SEQUENCES AND SERIES

11.1 SEQUENCES

1. $a_1 = \frac{1-1}{1^2} = 0$, $a_2 = \frac{1-2}{2^2} = -\frac{1}{4}$, $a_3 = \frac{1-3}{3^2} = -\frac{2}{9}$, $a_4 = \frac{1-4}{4^2} = -\frac{3}{16}$

2. $a_1 = \frac{1}{1!} = 1$, $a_2 = \frac{1}{2!} = \frac{1}{2}$, $a_3 = \frac{1}{3!} = \frac{1}{6}$, $a_4 = \frac{1}{4!} = \frac{1}{24}$

3. $a_1 = \frac{(-1)^2}{2-1} = 1$, $a_2 = \frac{(-1)^3}{4-1} = -\frac{1}{3}$, $a_3 = \frac{(-1)^4}{6-1} = \frac{1}{5}$, $a_4 = \frac{(-1)^5}{8-1} = -\frac{1}{7}$

4. $a_1 = 2 + (-1)^1 = 1$, $a_2 = 2 + (-1)^2 = 3$, $a_3 = 2 + (-1)^3 = 1$, $a_4 = 2 + (-1)^4 = 3$

5. $a_1 = \frac{2}{2^2} = \frac{1}{2}$, $a_2 = \frac{2^2}{2^3} = \frac{1}{2}$, $a_3 = \frac{2^3}{2^4} = \frac{1}{2}$, $a_4 = \frac{2^4}{2^5} = \frac{1}{2}$

6. $a_1 = \frac{2-1}{2} = \frac{1}{2}$, $a_2 = \frac{2^2-1}{2^2} = \frac{3}{4}$, $a_3 = \frac{2^3-1}{2^3} = \frac{7}{8}$, $a_4 = \frac{2^4-1}{2^4} = \frac{15}{16}$

7. $a_1 = 1$, $a_2 = 1 + \frac{1}{2} = \frac{3}{2}$, $a_3 = \frac{3}{2} + \frac{1}{2^2} = \frac{7}{4}$, $a_4 = \frac{7}{4} + \frac{1}{2^3} = \frac{15}{8}$, $a_5 = \frac{15}{8} + \frac{1}{2^4} = \frac{31}{16}$, $a_6 = \frac{63}{32}$,
$a_7 = \frac{127}{64}$, $a_8 = \frac{255}{128}$, $a_9 = \frac{511}{256}$, $a_{10} = \frac{1023}{512}$

8. $a_1 = 1$, $a_2 = \frac{1}{2}$, $a_3 = \frac{\left(\frac{1}{2}\right)}{3} = \frac{1}{6}$, $a_4 = \frac{\left(\frac{1}{6}\right)}{4} = \frac{1}{24}$, $a_5 = \frac{\left(\frac{1}{24}\right)}{5} = \frac{1}{120}$, $a_6 = \frac{1}{720}$, $a_7 = \frac{1}{5040}$, $a_8 = \frac{1}{40{,}320}$,
$a_9 = \frac{1}{362{,}880}$, $a_{10} = \frac{1}{3{,}628{,}800}$

9. $a_1 = 2$, $a_2 = \frac{(-1)^2(2)}{2} = 1$, $a_3 = \frac{(-1)^3(1)}{2} = -\frac{1}{2}$, $a_4 = \frac{(-1)^4\left(-\frac{1}{2}\right)}{2} = -\frac{1}{4}$, $a_5 = \frac{(-1)^5\left(-\frac{1}{4}\right)}{2} = \frac{1}{8}$,
$a_6 = \frac{1}{16}$, $a_7 = -\frac{1}{32}$, $a_8 = -\frac{1}{64}$, $a_9 = \frac{1}{128}$, $a_{10} = \frac{1}{256}$

10. $a_1 = -2$, $a_2 = \frac{1\cdot(-2)}{2} = -1$, $a_3 = \frac{2\cdot(-1)}{3} = -\frac{2}{3}$, $a_4 = \frac{3\cdot\left(-\frac{2}{3}\right)}{4} = -\frac{1}{2}$, $a_5 = \frac{4\cdot\left(-\frac{1}{2}\right)}{5} = -\frac{2}{5}$, $a_6 = -\frac{1}{3}$,
$a_7 = -\frac{2}{7}$, $a_8 = -\frac{1}{4}$, $a_9 = -\frac{2}{9}$, $a_{10} = -\frac{1}{5}$

11. $a_1 = 1$, $a_2 = 1$, $a_3 = 1 + 1 = 2$, $a_4 = 2 + 1 = 3$, $a_5 = 3 + 2 = 5$, $a_6 = 8$, $a_7 = 13$, $a_8 = 21$, $a_9 = 34$, $a_{10} = 55$

12. $a_1 = 2$, $a_2 = -1$, $a_3 = -\frac{1}{2}$, $a_4 = \frac{\left(-\frac{1}{2}\right)}{-1} = \frac{1}{2}$, $a_5 = \frac{\left(\frac{1}{2}\right)}{\left(-\frac{1}{2}\right)} = -1$, $a_6 = -2$, $a_7 = 2$, $a_8 = -1$, $a_9 = -\frac{1}{2}$, $a_{10} = \frac{1}{2}$

13. $a_n = (-1)^{n+1}$, $n = 1, 2, \ldots$

14. $a_n = (-1)^n$, $n = 1, 2, \ldots$

15. $a_n = (-1)^{n+1}n^2$, $n = 1, 2, \ldots$

16. $a_n = \frac{(-1)^{n+1}}{n^2}$, $n = 1, 2, \ldots$

17. $a_n = n^2 - 1$, $n = 1, 2, \ldots$

18. $a_n = n - 4$, $n = 1, 2, \ldots$

19. $a_n = 4n - 3$, $n = 1, 2, \ldots$

20. $a_n = 4n - 2$, $n = 1, 2, \ldots$

21. $a_n = \frac{1 + (-1)^{n+1}}{2}$, $n = 1, 2, \ldots$

22. $a_n = \frac{n - \frac{1}{2} + (-1)^n\left(\frac{1}{2}\right)}{2} = \left\lfloor \frac{n}{2} \right\rfloor$, $n = 1, 2, \ldots$

23. $\lim\limits_{n\to\infty} 2 + (0.1)^n = 2 \Rightarrow$ converges (Theorem 5, #4)

24. $\lim_{n\to\infty} \frac{n+(-1)^n}{n} = \lim_{n\to\infty} 1 + \frac{(-1)^n}{n} = 1 \Rightarrow$ converges

25. $\lim_{n\to\infty} \frac{1-2n}{1+2n} = \lim_{n\to\infty} \frac{\left(\frac{1}{n}\right)-2}{\left(\frac{1}{n}\right)+2} = \lim_{n\to\infty} \frac{-2}{2} = -1 \Rightarrow$ converges

26. $\lim_{n\to\infty} \frac{2n+1}{1-3\sqrt{n}} = \lim_{n\to\infty} \frac{2\sqrt{n}+\left(\frac{1}{\sqrt{n}}\right)}{\left(\frac{1}{\sqrt{n}}-3\right)} = -\infty \Rightarrow$ diverges

27. $\lim_{n\to\infty} \frac{1-5n^4}{n^4+8n^3} = \lim_{n\to\infty} \frac{\left(\frac{1}{n^4}\right)-5}{1+\left(\frac{8}{n}\right)} = -5 \Rightarrow$ converges

28. $\lim_{n\to\infty} \frac{n+3}{n^2+5n+6} = \lim_{n\to\infty} \frac{n+3}{(n+3)(n+2)} = \lim_{n\to\infty} \frac{1}{n+2} = 0 \Rightarrow$ converges

29. $\lim_{n\to\infty} \frac{n^2-2n+1}{n-1} = \lim_{n\to\infty} \frac{(n-1)(n-1)}{n-1} = \lim_{n\to\infty} (n-1) = \infty \Rightarrow$ diverges

30 $\lim_{n\to\infty} \frac{1-n^3}{70-4n^2} = \lim_{n\to\infty} \frac{\left(\frac{1}{n^2}\right)-n}{\left(\frac{70}{n^2}\right)-4} = \infty \Rightarrow$ diverges

31. $\lim_{n\to\infty} (1+(-1)^n)$ does not exist $\Rightarrow$ diverges

32. $\lim_{n\to\infty} (-1)^n\left(1-\frac{1}{n}\right)$ does not exist $\Rightarrow$ diverges

33. $\lim_{n\to\infty} \left(\frac{n+1}{2n}\right)\left(1-\frac{1}{n}\right) = \lim_{n\to\infty} \left(\frac{1}{2}+\frac{1}{2n}\right)\left(1-\frac{1}{n}\right) = \frac{1}{2} \Rightarrow$ converges

34. $\lim_{n\to\infty} \left(2-\frac{1}{2^n}\right)\left(3+\frac{1}{2^n}\right) = 6 \Rightarrow$ converges

35. $\lim_{n\to\infty} \frac{(-1)^{n+1}}{2n-1} = 0 \Rightarrow$ converges

36. $\lim_{n\to\infty} \left(-\frac{1}{2}\right)^n = \lim_{n\to\infty} \frac{(-1)^n}{2^n} = 0 \Rightarrow$ converges

37. $\lim_{n\to\infty} \sqrt{\frac{2n}{n+1}} = \sqrt{\lim_{n\to\infty} \frac{2n}{n+1}} = \sqrt{\lim_{n\to\infty} \left(\frac{2}{1+\frac{1}{n}}\right)} = \sqrt{2} \Rightarrow$ converges

38. $\lim_{n\to\infty} \frac{1}{(0.9)^n} = \lim_{n\to\infty} \left(\frac{10}{9}\right)^n = \infty \Rightarrow$ diverges

39. $\lim_{n\to\infty} \sin\left(\frac{\pi}{2}+\frac{1}{n}\right) = \sin\left(\lim_{n\to\infty}\left(\frac{\pi}{2}+\frac{1}{n}\right)\right) = \sin\frac{\pi}{2} = 1 \Rightarrow$ converges

40. $\lim_{n\to\infty} n\pi\cos(n\pi) = \lim_{n\to\infty} (n\pi)(-1)^n$ does not exist $\Rightarrow$ diverges

41. $\lim_{n\to\infty} \frac{\sin n}{n} = 0$ because $-\frac{1}{n} \le \frac{\sin n}{n} \le \frac{1}{n} \Rightarrow$ converges by the Sandwich Theorem for sequences

42. $\lim_{n\to\infty} \frac{\sin^2 n}{2^n} = 0$ because $0 \le \frac{\sin^2 n}{2^n} \le \frac{1}{2^n} \Rightarrow$ converges by the Sandwich Theorem for sequences

43. $\lim_{n\to\infty} \frac{n}{2^n} = \lim_{n\to\infty} \frac{1}{2^n \ln 2} = 0 \Rightarrow$ converges (using l'Hôpital's rule)

44. $\lim_{n\to\infty} \frac{3^n}{n^3} = \lim_{n\to\infty} \frac{3^n \ln 3}{3n^2} = \lim_{n\to\infty} \frac{3^n(\ln 3)^2}{6n} = \lim_{n\to\infty} \frac{3^n(\ln 3)^3}{6} = \infty \Rightarrow$ diverges (using l'Hôpital's rule)

45. $\lim_{n\to\infty} \frac{\ln(n+1)}{\sqrt{n}} = \lim_{n\to\infty} \frac{\left(\frac{1}{n+1}\right)}{\left(\frac{1}{2\sqrt{n}}\right)} = \lim_{n\to\infty} \frac{2\sqrt{n}}{n+1} = \lim_{n\to\infty} \frac{\left(\frac{2}{\sqrt{n}}\right)}{1+\left(\frac{1}{n}\right)} = 0 \Rightarrow$ converges

46. $\lim\limits_{n \to \infty} \frac{\ln n}{\ln 2n} = \lim\limits_{n \to \infty} \frac{\left(\frac{1}{n}\right)}{\left(\frac{2}{2n}\right)} = 1 \Rightarrow$ converges

47. $\lim\limits_{n \to \infty} 8^{1/n} = 1 \Rightarrow$ converges (Theorem 5, #3)

48. $\lim\limits_{n \to \infty} (0.03)^{1/n} = 1 \Rightarrow$ converges (Theorem 5, #3)

49. $\lim\limits_{n \to \infty} \left(1 + \frac{7}{n}\right)^n = e^7 \Rightarrow$ converges (Theorem 5, #5)

50. $\lim\limits_{n \to \infty} \left(1 - \frac{1}{n}\right)^n = \lim\limits_{n \to \infty} \left[1 + \frac{(-1)}{n}\right]^n = e^{-1} \Rightarrow$ converges (Theorem 5, #5)

51. $\lim\limits_{n \to \infty} \sqrt[n]{10n} = \lim\limits_{n \to \infty} 10^{1/n} \cdot n^{1/n} = 1 \cdot 1 = 1 \Rightarrow$ converges (Theorem 5, #3 and #2)

52. $\lim\limits_{n \to \infty} \sqrt[n]{n^2} = \lim\limits_{n \to \infty} \left(\sqrt[n]{n}\right)^2 = 1^2 = 1 \Rightarrow$ converges (Theorem 5, #2)

53. $\lim\limits_{n \to \infty} \left(\frac{3}{n}\right)^{1/n} = \frac{\lim\limits_{n \to \infty} 3^{1/n}}{\lim\limits_{n \to \infty} n^{1/n}} = \frac{1}{1} = 1 \Rightarrow$ converges (Theorem 5, #3 and #2)

54. $\lim\limits_{n \to \infty} (n+4)^{1/(n+4)} = \lim\limits_{x \to \infty} x^{1/x} = 1 \Rightarrow$ converges; (let $x = n + 4$, then use Theorem 5, #2)

55. $\lim\limits_{n \to \infty} \frac{\ln n}{n^{1/n}} = \frac{\lim\limits_{n \to \infty} \ln n}{\lim\limits_{n \to \infty} n^{1/n}} = \frac{\infty}{1} = \infty \Rightarrow$ diverges (Theorem 5, #2)

56. $\lim\limits_{n \to \infty} [\ln n - \ln(n+1)] = \lim\limits_{n \to \infty} \ln\left(\frac{n}{n+1}\right) = \ln\left(\lim\limits_{n \to \infty} \frac{n}{n+1}\right) = \ln 1 = 0 \Rightarrow$ converges

57. $\lim\limits_{n \to \infty} \sqrt[n]{4^n n} = \lim\limits_{n \to \infty} 4\sqrt[n]{n} = 4 \cdot 1 = 4 \Rightarrow$ converges (Theorem 5, #2)

58. $\lim\limits_{n \to \infty} \sqrt[n]{3^{2n+1}} = \lim\limits_{n \to \infty} 3^{2+(1/n)} = \lim\limits_{n \to \infty} 3^2 \cdot 3^{1/n} = 9 \cdot 1 = 9 \Rightarrow$ converges (Theorem 5, #3)

59. $\lim\limits_{n \to \infty} \frac{n!}{n^n} = \lim\limits_{n \to \infty} \frac{1 \cdot 2 \cdot 3 \cdots (n-1)(n)}{n \cdot n \cdot n \cdots n \cdot n} \le \lim\limits_{n \to \infty} \left(\frac{1}{n}\right) = 0$ and $\frac{n!}{n^n} \ge 0 \Rightarrow \lim\limits_{n \to \infty} \frac{n!}{n^n} = 0 \Rightarrow$ converges

60. $\lim\limits_{n \to \infty} \frac{(-4)^n}{n!} = 0 \Rightarrow$ converges (Theorem 5, #6)

61. $\lim\limits_{n \to \infty} \frac{n!}{10^{6n}} = \lim\limits_{n \to \infty} \frac{1}{\left(\frac{(10^6)^n}{n!}\right)} = \infty \Rightarrow$ diverges (Theorem 5, #6)

62. $\lim\limits_{n \to \infty} \frac{n!}{2^n 3^n} = \lim\limits_{n \to \infty} \frac{1}{\left(\frac{6^n}{n!}\right)} = \infty \Rightarrow$ diverges (Theorem 5, #6)

63. $\lim\limits_{n \to \infty} \left(\frac{1}{n}\right)^{1/(\ln n)} = \lim\limits_{n \to \infty} \exp\left(\frac{1}{\ln n} \ln\left(\frac{1}{n}\right)\right) = \lim\limits_{n \to \infty} \exp\left(\frac{\ln 1 - \ln n}{\ln n}\right) = e^{-1} \Rightarrow$ converges

64. $\lim\limits_{n \to \infty} \ln\left(1 + \frac{1}{n}\right)^n = \ln\left(\lim\limits_{n \to \infty} \left(1 + \frac{1}{n}\right)^n\right) = \ln e = 1 \Rightarrow$ converges (Theorem 5, #5)

65. $\lim\limits_{n \to \infty} \left(\frac{3n+1}{3n-1}\right)^n = \lim\limits_{n \to \infty} \exp\left(n \ln\left(\frac{3n+1}{3n-1}\right)\right) = \lim\limits_{n \to \infty} \exp\left(\frac{\ln(3n+1) - \ln(3n-1)}{\frac{1}{n}}\right)$

$= \lim_{n\to\infty} \exp\left(\frac{\frac{3}{3n+1} - \frac{3}{3n-1}}{\left(-\frac{1}{n^2}\right)}\right) = \lim_{n\to\infty} \exp\left(\frac{6n^2}{(3n+1)(3n-1)}\right) = \exp\left(\frac{6}{9}\right) = e^{2/3} \Rightarrow$ converges

66. $\lim_{n\to\infty} \left(\frac{n}{n+1}\right)^n = \lim_{n\to\infty} \exp\left(n \ln\left(\frac{n}{n+1}\right)\right) = \lim_{n\to\infty} \exp\left(\frac{\ln n - \ln(n+1)}{\left(\frac{1}{n}\right)}\right) = \lim_{n\to\infty} \exp\left(\frac{\frac{1}{n} - \frac{1}{n+1}}{\left(-\frac{1}{n^2}\right)}\right)$
$= \lim_{n\to\infty} \exp\left(-\frac{n^2}{n(n+1)}\right) = e^{-1} \Rightarrow$ converges

67. $\lim_{n\to\infty} \left(\frac{x^n}{2n+1}\right)^{1/n} = \lim_{n\to\infty} x\left(\frac{1}{2n+1}\right)^{1/n} = x \lim_{n\to\infty} \exp\left(\frac{1}{n} \ln\left(\frac{1}{2n+1}\right)\right) = x \lim_{n\to\infty} \exp\left(\frac{-\ln(2n+1)}{n}\right)$
$= x \lim_{n\to\infty} \exp\left(\frac{-2}{2n+1}\right) = xe^0 = x,\ x > 0 \Rightarrow$ converges

68. $\lim_{n\to\infty} \left(1 - \frac{1}{n^2}\right)^n = \lim_{n\to\infty} \exp\left(n \ln\left(1 - \frac{1}{n^2}\right)\right) = \lim_{n\to\infty} \exp\left(\frac{\ln\left(1 - \frac{1}{n^2}\right)}{\left(\frac{1}{n}\right)}\right) = \lim_{n\to\infty} \exp\left[\frac{\left(\frac{2}{n^3}\right)/\left(1 - \frac{1}{n^2}\right)}{\left(-\frac{1}{n^2}\right)}\right]$
$= \lim_{n\to\infty} \exp\left(\frac{-2n}{n^2-1}\right) = e^0 = 1 \Rightarrow$ converges

69. $\lim_{n\to\infty} \frac{3^n \cdot 6^n}{2^{-n} \cdot n!} = \lim_{n\to\infty} \frac{36^n}{n!} = 0 \Rightarrow$ converges (Theorem 5, #6)

70. $\lim_{n\to\infty} \frac{\left(\frac{10}{11}\right)^n}{\left(\frac{9}{10}\right)^n + \left(\frac{11}{12}\right)^n} = \lim_{n\to\infty} \frac{\left(\frac{12}{11}\right)^n \left(\frac{10}{11}\right)^n}{\left(\frac{12}{11}\right)^n \left(\frac{9}{10}\right)^n + \left(\frac{12}{11}\right)^n \left(\frac{11}{12}\right)^n} = \lim_{n\to\infty} \frac{\left(\frac{120}{121}\right)^n}{\left(\frac{108}{110}\right)^n + 1} = 0 \Rightarrow$ converges
(Theorem 5, #4)

71. $\lim_{n\to\infty} \tanh n = \lim_{n\to\infty} \frac{e^n - e^{-n}}{e^n + e^{-n}} = \lim_{n\to\infty} \frac{e^{2n} - 1}{e^{2n} + 1} = \lim_{n\to\infty} \frac{2e^{2n}}{2e^{2n}} = \lim_{n\to\infty} 1 = 1 \Rightarrow$ converges

72. $\lim_{n\to\infty} \sinh(\ln n) = \lim_{n\to\infty} \frac{e^{\ln n} - e^{-\ln n}}{2} = \lim_{n\to\infty} \frac{n - \left(\frac{1}{n}\right)}{2} = \infty \Rightarrow$ diverges

73. $\lim_{n\to\infty} \frac{n^2 \sin\left(\frac{1}{n}\right)}{2n-1} = \lim_{n\to\infty} \frac{\sin\left(\frac{1}{n}\right)}{\left(\frac{2}{n} - \frac{1}{n^2}\right)} = \lim_{n\to\infty} \frac{-\left(\cos\left(\frac{1}{n}\right)\right)\left(\frac{1}{n^2}\right)}{\left(-\frac{2}{n^2} + \frac{2}{n^3}\right)} = \lim_{n\to\infty} \frac{-\cos\left(\frac{1}{n}\right)}{-2 + \left(\frac{2}{n}\right)} = \frac{1}{2} \Rightarrow$ converges

74. $\lim_{n\to\infty} n\left(1 - \cos\frac{1}{n}\right) = \lim_{n\to\infty} \frac{\left(1 - \cos\frac{1}{n}\right)}{\left(\frac{1}{n}\right)} = \lim_{n\to\infty} \frac{\left[\sin\left(\frac{1}{n}\right)\right]\left(\frac{1}{n^2}\right)}{\left(\frac{1}{n^2}\right)} = \lim_{n\to\infty} \sin\left(\frac{1}{n}\right) = 0 \Rightarrow$ converges

75. $\lim_{n\to\infty} \tan^{-1} n = \frac{\pi}{2} \Rightarrow$ converges

76. $\lim_{n\to\infty} \frac{1}{\sqrt{n}} \tan^{-1} n = 0 \cdot \frac{\pi}{2} = 0 \Rightarrow$ converges

77. $\lim_{n\to\infty} \left(\frac{1}{3}\right)^n + \frac{1}{\sqrt{2^n}} = \lim_{n\to\infty} \left(\left(\frac{1}{3}\right)^n + \left(\frac{1}{\sqrt{2}}\right)^n\right) = 0 \Rightarrow$ converges (Theorem 5, #4)

78. $\lim_{n\to\infty} \sqrt[n]{n^2 + n} = \lim_{n\to\infty} \exp\left[\frac{\ln(n^2+n)}{n}\right] = \lim_{n\to\infty} \exp\left(\frac{2n+1}{n^2+n}\right) = e^0 = 1 \Rightarrow$ converges

79. $\lim_{n\to\infty} \frac{(\ln n)^{200}}{n} = \lim_{n\to\infty} \frac{200(\ln n)^{199}}{n} = \lim_{n\to\infty} \frac{200 \cdot 199 (\ln n)^{198}}{n} = \ldots = \lim_{n\to\infty} \frac{200!}{n} = 0 \Rightarrow$ converges

80. $\lim_{n\to\infty} \frac{(\ln n)^5}{\sqrt{n}} = \lim_{n\to\infty} \left[\frac{\left(\frac{5(\ln n)^4}{n}\right)}{\left(\frac{1}{2\sqrt{n}}\right)}\right] = \lim_{n\to\infty} \frac{10(\ln n)^4}{\sqrt{n}} = \lim_{n\to\infty} \frac{80(\ln n)^3}{\sqrt{n}} = \ldots = \lim_{n\to\infty} \frac{3840}{\sqrt{n}} = 0 \Rightarrow$ converges

81. $\lim_{n \to \infty} \left(n - \sqrt{n^2 - n}\right) = \lim_{n \to \infty} \left(n - \sqrt{n^2 - n}\right)\left(\frac{n + \sqrt{n^2 - n}}{n + \sqrt{n^2 - n}}\right) = \lim_{n \to \infty} \frac{n}{n + \sqrt{n^2 - n}} = \lim_{n \to \infty} \frac{1}{1 + \sqrt{1 - \frac{1}{n}}}$
$= \frac{1}{2} \Rightarrow$ converges

82. $\lim_{n \to \infty} \frac{1}{\sqrt{n^2 - 1} - \sqrt{n^2 + n}} = \lim_{n \to \infty} \left(\frac{1}{\sqrt{n^2 - 1} - \sqrt{n^2 + n}}\right)\left(\frac{\sqrt{n^2 - 1} + \sqrt{n^2 + n}}{\sqrt{n^2 - 1} + \sqrt{n^2 + n}}\right) = \lim_{n \to \infty} \frac{\sqrt{n^2 - 1} + \sqrt{n^2 + n}}{-1 - n}$
$= \lim_{n \to \infty} \frac{\sqrt{1 - \frac{1}{n^2}} + \sqrt{1 + \frac{1}{n}}}{\left(-\frac{1}{n} - 1\right)} = -2 \Rightarrow$ converges

83. $\lim_{n \to \infty} \frac{1}{n} \int_1^n \frac{1}{x}\, dx = \lim_{n \to \infty} \frac{\ln n}{n} = \lim_{n \to \infty} \frac{1}{n} = 0 \Rightarrow$ converges (Theorem 5, #1)

84. $\lim_{n \to \infty} \int_1^n \frac{1}{x^p}\, dx = \lim_{n \to \infty} \left[\frac{1}{1-p} \frac{1}{x^{p-1}}\right]_1^n = \lim_{n \to \infty} \frac{1}{1-p}\left(\frac{1}{n^{p-1}} - 1\right) = \frac{1}{p-1}$ if $p > 1 \Rightarrow$ converges

85. $1, 1, 2, 4, 8, 16, 32, \ldots = 1, 2^0, 2^1, 2^2, 2^3, 2^4, 2^5, \ldots \Rightarrow x_1 = 1$ and $x_n = 2^{n-2}$ for $n \geq 2$

86. (a) $1^2 - 2(1)^2 = -1$, $3^2 - 2(2)^2 = 1$; let $f(a, b) = (a + 2b)^2 - 2(a + b)^2 = a^2 + 4ab + 4b^2 - 2a^2 - 4ab - 2b^2$
$= 2b^2 - a^2$; $a^2 - 2b^2 = -1 \Rightarrow f(a, b) = 2b^2 - a^2 = 1$; $a^2 - 2b^2 = 1 \Rightarrow f(a, b) = 2b^2 - a^2 = -1$

(b) $r_n^2 - 2 = \left(\frac{a + 2b}{a + b}\right)^2 - 2 = \frac{a^2 + 4ab + 4b^2 - 2a^2 - 4ab - 2b^2}{(a + b)^2} = \frac{-(a^2 - 2b^2)}{(a + b)^2} = \frac{\pm 1}{y_n^2} \Rightarrow r_n = \sqrt{2 \pm \left(\frac{1}{y_n}\right)^2}$

In the first and second fractions, $y_n \geq n$. Let $\frac{a}{b}$ represent the $(n - 1)$th fraction where $\frac{a}{b} \geq 1$ and $b \geq n - 1$ for n a positive integer ≥ 3. Now the nth fraction is $\frac{a + 2b}{a + b}$ and $a + b \geq 2b \geq 2n - 2 \geq n \Rightarrow y_n \geq n$. Thus, $\lim_{n \to \infty} r_n = \sqrt{2}$.

87. (a) $f(x) = x^2 - 2$; the sequence converges to $1.414213562 \approx \sqrt{2}$
(b) $f(x) = \tan(x) - 1$; the sequence converges to $0.7853981635 \approx \frac{\pi}{4}$
(c) $f(x) = e^x$; the sequence $1, 0, -1, -2, -3, -4, -5, \ldots$ diverges

88. (a) $\lim_{n \to \infty} n f\left(\frac{1}{n}\right) = \lim_{\Delta x \to 0^+} \frac{f(\Delta x)}{\Delta x} = \lim_{\Delta x \to 0^+} \frac{f(0 + \Delta x) - f(0)}{\Delta x} = f'(0)$, where $\Delta x = \frac{1}{n}$
(b) $\lim_{n \to \infty} n \tan^{-1}\left(\frac{1}{n}\right) = f'(0) = \frac{1}{1 + 0^2} = 1$, $f(x) = \tan^{-1} x$
(c) $\lim_{n \to \infty} n\left(e^{1/n} - 1\right) = f'(0) = e^0 = 1$, $f(x) = e^x - 1$
(d) $\lim_{n \to \infty} n \ln\left(1 + \frac{2}{n}\right) = f'(0) = \frac{2}{1 + 2(0)} = 2$, $f(x) = \ln(1 + 2x)$

89. (a) If $a = 2n + 1$, then $b = \left\lfloor \frac{a^2}{2} \right\rfloor = \left\lfloor \frac{4n^2 + 4n + 1}{2} \right\rfloor = \left\lfloor 2n^2 + 2n + \frac{1}{2} \right\rfloor = 2n^2 + 2n$, $c = \left\lceil \frac{a^2}{2} \right\rceil = \left\lceil 2n^2 + 2n + \frac{1}{2} \right\rceil$
$= 2n^2 + 2n + 1$ and $a^2 + b^2 = (2n + 1)^2 + \left(2n^2 + 2n\right)^2 = 4n^2 + 4n + 1 + 4n^4 + 8n^3 + 4n^2$
$= 4n^4 + 8n^3 + 8n^2 + 4n + 1 = \left(2n^2 + 2n + 1\right)^2 = c^2$.

(b) $\lim_{a \to \infty} \frac{\left\lfloor \frac{a^2}{2} \right\rfloor}{\left\lceil \frac{a^2}{2} \right\rceil} = \lim_{a \to \infty} \frac{2n^2 + 2n}{2n^2 + 2n + 1} = 1$ or $\lim_{a \to \infty} \frac{\left\lfloor \frac{a^2}{2} \right\rfloor}{\left\lceil \frac{a^2}{2} \right\rceil} = \lim_{a \to \infty} \sin\theta = \lim_{\theta \to \pi/2} \sin\theta = 1$

90. (a) $\lim_{n \to \infty} (2n\pi)^{1/(2n)} = \lim_{n \to \infty} \exp\left(\frac{\ln 2n\pi}{2n}\right) = \lim_{n \to \infty} \exp\left(\frac{\left(\frac{2\pi}{2n\pi}\right)}{2}\right) = \lim_{n \to \infty} \exp\left(\frac{1}{2n}\right) = e^0 = 1$;

$n! \approx \left(\frac{n}{e}\right)^n \sqrt{2n\pi}$, Stirlings approximation $\Rightarrow \sqrt[n]{n!} \approx \left(\frac{n}{e}\right)(2n\pi)^{1/(2n)} \approx \frac{n}{e}$ for large values of n

(b)

n	$\sqrt[n]{n!}$	$\frac{n}{e}$
40	15.76852702	14.71517765
50	19.48325423	18.39397206
60	23.19189561	22.07276647

91. (a) $\lim_{n \to \infty} \frac{\ln n}{n^c} = \lim_{n \to \infty} \frac{\left(\frac{1}{n}\right)}{cn^{c-1}} = \lim_{n \to \infty} \frac{1}{cn^c} = 0$

(b) For all $\epsilon > 0$, there exists an $N = e^{-(\ln \epsilon)/c}$ such that $n > e^{-(\ln \epsilon)/c} \Rightarrow \ln n > -\frac{\ln \epsilon}{c} \Rightarrow \ln n^c > \ln\left(\frac{1}{\epsilon}\right)$
$\Rightarrow n^c > \frac{1}{\epsilon} \Rightarrow \frac{1}{n^c} < \epsilon \Rightarrow \left|\frac{1}{n^c} - 0\right| < \epsilon \Rightarrow \lim_{n \to \infty} \frac{1}{n^c} = 0$

92. Let $\{a_n\}$ and $\{b_n\}$ be sequences both converging to L. Define $\{c_n\}$ by $c_{2n} = b_n$ and $c_{2n-1} = a_n$, where $n = 1, 2, 3, \ldots$. For all $\epsilon > 0$ there exists N_1 such that when $n > N_1$ then $|a_n - L| < \epsilon$ and there exists N_2 such that when $n > N_2$ then $|b_n - L| < \epsilon$. If $n > 1 + 2\max\{N_1, N_2\}$, then $|c_n - L| < \epsilon$, so $\{c_n\}$ converges to L.

93. $\lim_{n \to \infty} n^{1/n} = \lim_{n \to \infty} \exp\left(\frac{1}{n} \ln n\right) = \lim_{n \to \infty} \exp\left(\frac{1}{n}\right) = e^0 = 1$

94. $\lim_{n \to \infty} x^{1/n} = \lim_{n \to \infty} \exp\left(\frac{1}{n} \ln x\right) = e^0 = 1$, because x remains fixed while n gets large

95. Assume the hypotheses of the theorem and let ϵ be a positive number. For all ϵ there exists a N_1 such that when $n > N_1$ then $|a_n - L| < \epsilon \Rightarrow -\epsilon < a_n - L < \epsilon \Rightarrow L - \epsilon < a_n$, and there exists a N_2 such that when $n > N_2$ then $|c_n - L| < \epsilon \Rightarrow -\epsilon < c_n - L < \epsilon \Rightarrow c_n < L + \epsilon$. If $n > \max\{N_1, N_2\}$, then $L - \epsilon < a_n \le b_n \le c_n < L + \epsilon \Rightarrow |b_n - L| < \epsilon \Rightarrow \lim_{n \to \infty} b_n = L$.

96. Let $\epsilon > 0$. We have f continuous at L $\Rightarrow$ there exists δ so that $|x - L| < \delta \Rightarrow |f(x) - f(L)| < \epsilon$. Also, $a_n \to L \Rightarrow$ there exists N so that for $n > N$ $|a_n - L| < \delta$. Thus for $n > N$, $|f(a_n) - f(L)| < \epsilon \Rightarrow f(a_n) \to f(L)$.

97. $a_{n+1} \ge a_n \Rightarrow \frac{3(n+1)+1}{(n+1)+1} > \frac{3n+1}{n+1} \Rightarrow \frac{3n+4}{n+2} > \frac{3n+1}{n+1} \Rightarrow 3n^2 + 3n + 4n + 4 > 3n^2 + 6n + n + 2$
$\Rightarrow 4 > 2$; the steps are reversible so the sequence is nondecreasing; $\frac{3n+1}{n+1} < 3 \Rightarrow 3n + 1 < 3n + 3$
$\Rightarrow 1 < 3$; the steps are reversible so the sequence is bounded above by 3

98. $a_{n+1} \ge a_n \Rightarrow \frac{(2(n+1)+3)!}{((n+1)+1)!} > \frac{(2n+3)!}{(n+1)!} \Rightarrow \frac{(2n+5)!}{(n+2)!} > \frac{(2n+3)!}{(n+1)!} \Rightarrow \frac{(2n+5)!}{(2n+3)!} > \frac{(n+2)!}{(n+1)!}$
$\Rightarrow (2n+5)(2n+4) > n + 2$; the steps are reversible so the sequence is nondecreasing; the sequence is not bounded since $\frac{(2n+3)!}{(n+1)!} = (2n+3)(2n+2)\cdots(n+2)$ can become as large as we please

99. $a_{n+1} \le a_n \Rightarrow \frac{2^{n+1}3^{n+1}}{(n+1)!} \le \frac{2^n 3^n}{n!} \Rightarrow \frac{2^{n+1}3^{n+1}}{2^n 3^n} \le \frac{(n+1)!}{n!} \Rightarrow 2 \cdot 3 \le n + 1$ which is true for $n \ge 5$; the steps are reversible so the sequence is decreasing after a_5, but it is not nondecreasing for all its terms; $a_1 = 6$, $a_2 = 18$, $a_3 = 36$, $a_4 = 54$, $a_5 = \frac{324}{5} = 64.8 \Rightarrow$ the sequence is bounded from above by 64.8

100. $a_{n+1} \ge a_n \Rightarrow 2 - \frac{2}{n+1} - \frac{1}{2^{n+1}} \ge 2 - \frac{2}{n} - \frac{1}{2^n} \Rightarrow \frac{2}{n} - \frac{2}{n+1} \ge \frac{1}{2^{n+1}} - \frac{1}{2^n} \Rightarrow \frac{2}{n(n+1)} \ge -\frac{1}{2^{n+1}}$; the steps are reversible so the sequence is nondecreasing; $2 - \frac{2}{n} - \frac{1}{2^n} \le 2 \Rightarrow$ the sequence is bounded from above

101. $a_n = 1 - \frac{1}{n}$ converges because $\frac{1}{n} \to 0$ by Example 1; also it is a nondecreasing sequence bounded above by 1

102. $a_n = n - \frac{1}{n}$ diverges because $n \to \infty$ and $\frac{1}{n} \to 0$ by Example 1, so the sequence is unbounded

103. $a_n = \frac{2^n - 1}{2^n} = 1 - \frac{1}{2^n}$ and $0 < \frac{1}{2^n} < \frac{1}{n}$; since $\frac{1}{n} \to 0$ (by Example 1) $\Rightarrow \frac{1}{2^n} \to 0$, the sequence converges; also it is a nondecreasing sequence bounded above by 1

104. $a_n = \frac{2^n - 1}{3^n} = \left(\frac{2}{3}\right)^n - \frac{1}{3^n}$; the sequence converges to 0 by Theorem 5, #4

105. $a_n = ((-1)^n + 1)\left(\frac{n+1}{n}\right)$ diverges because $a_n = 0$ for n odd, while for n even $a_n = 2\left(1 + \frac{1}{n}\right)$ converges to 2; it diverges by definition of divergence

106. $x_n = \max\{\cos 1, \cos 2, \cos 3, \dots, \cos n\}$ and $x_{n+1} = \max\{\cos 1, \cos 2, \cos 3, \dots, \cos(n+1)\} \ge x_n$ with $x_n \le 1$ so the sequence is nondecreasing and bounded above by 1 $\Rightarrow$ the sequence converges.

107. If $\{a_n\}$ is nonincreasing with lower bound M, then $\{-a_n\}$ is a nondecreasing sequence with upper bound $-M$. By Theorem 1, $\{-a_n\}$ converges and hence $\{a_n\}$ converges. If $\{a_n\}$ has no lower bound, then $\{-a_n\}$ has no upper bound and therefore diverges. Hence, $\{a_n\}$ also diverges.

108. $a_n \ge a_{n+1} \Leftrightarrow \frac{n+1}{n} \ge \frac{(n+1)+1}{n+1} \Leftrightarrow n^2 + 2n + 1 \ge n^2 + 2n \Leftrightarrow 1 \ge 0$ and $\frac{n+1}{n} \ge 1$; thus the sequence is nonincreasing and bounded below by 1 $\Rightarrow$ it converges

109. $a_n \ge a_{n+1} \Leftrightarrow \frac{1+\sqrt{2n}}{\sqrt{n}} \ge \frac{1+\sqrt{2(n+1)}}{\sqrt{n+1}} \Leftrightarrow \sqrt{n+1} + \sqrt{2n^2+2n} \ge \sqrt{n} + \sqrt{2n^2+2n} \Leftrightarrow \sqrt{n+1} \ge \sqrt{n}$ and $\frac{1+\sqrt{2n}}{\sqrt{n}} \ge \sqrt{2}$; thus the sequence is nonincreasing and bounded below by $\sqrt{2} \Rightarrow$ it converges

110. $a_n \ge a_{n+1} \Leftrightarrow \frac{1-4^n}{2^n} \ge \frac{1-4^{n+1}}{2^{n+1}} \Leftrightarrow 2^{n+1} - 2^{n+1}4^n \ge 2^n - 2^n 4^{n+1} \Leftrightarrow 2^{n+1} - 2^n \ge 2^{n+1}4^n - 2^n 4^{n+1}$ $\Leftrightarrow 2 - 1 \ge 2 \cdot 4^n - 4^{n+1} \Leftrightarrow 1 \ge 4^n(2-4) \Leftrightarrow 1 \ge (-2)\cdot 4^n$; thus the sequence is nonincreasing. However, $a_n = \frac{1}{2^n} - \frac{4^n}{2^n} = \frac{1}{2^n} - 2^n$ which is not bounded below so the sequence diverges

111. $\frac{4^{n+1}+3^n}{4^n} = 4 + \left(\frac{3}{4}\right)^n$ so $a_n \ge a_{n+1} \Leftrightarrow 4 + \left(\frac{3}{4}\right)^n \ge 4 + \left(\frac{3}{4}\right)^{n+1} \Leftrightarrow \left(\frac{3}{4}\right)^n \ge \left(\frac{3}{4}\right)^{n+1} \Leftrightarrow 1 \ge \frac{3}{4}$ and $4 + \left(\frac{3}{4}\right)^n \ge 4$; thus the sequence is nonincreasing and bounded below by 4 $\Rightarrow$ it converges

112. $a_1 = 1$, $a_2 = 2 - 3$, $a_3 = 2(2-3) - 3 = 2^2 - (2^2 - 1)\cdot 3$, $a_4 = 2\left(2^2 - (2^2-1)\cdot 3\right) - 3 = 2^3 - (2^3 - 1)\,3$, $a_5 = 2\left[2^3 - (2^3-1)\,3\right] - 3 = 2^4 - (2^4 - 1)\,3, \dots, a_n = 2^{n-1} - (2^{n-1} - 1)\,3 = 2^{n-1} - 3\cdot 2^{n-1} + 3$ $= 2^{n-1}(1-3) + 3 = -2^n + 3$; $a_n \ge a_{n+1} \Leftrightarrow -2^n + 3 \ge -2^{n+1} + 3 \Leftrightarrow -2^n \ge -2^{n+1} \Leftrightarrow 1 \le 2$ so the sequence is nonincreasing but not bounded below and therefore diverges

113. Let $0 < M < 1$ and let N be an integer greater than $\frac{M}{1-M}$. Then $n > N \Rightarrow n > \frac{M}{1-M} \Rightarrow n - nM > M$ $\Rightarrow n > M + nM \Rightarrow n > M(n+1) \Rightarrow \frac{n}{n+1} > M$.

114. Since M_1 is a least upper bound and M_2 is an upper bound, $M_1 \le M_2$. Since M_2 is a least upper bound and M_1 is an upper bound, $M_2 \le M_1$. We conclude that $M_1 = M_2$ so the least upper bound is unique.

115. The sequence $a_n = 1 + \frac{(-1)^n}{2}$ is the sequence $\frac{1}{2}, \frac{3}{2}, \frac{1}{2}, \frac{3}{2}, \dots$. This sequence is bounded above by $\frac{3}{2}$, but it clearly does not converge, by definition of convergence.

116. Let L be the limit of the convergent sequence $\{a_n\}$. Then by definition of convergence, for $\frac{\epsilon}{2}$ there corresponds an N such that for all m and n, $m > N \Rightarrow |a_m - L| < \frac{\epsilon}{2}$ and $n > N \Rightarrow |a_n - L| < \frac{\epsilon}{2}$. Now $|a_m - a_n| = |a_m - L + L - a_n| \le |a_m - L| + |L - a_n| < \frac{\epsilon}{2} + \frac{\epsilon}{2} = \epsilon$ whenever $m > N$ and $n > N$.

117. Given an $\epsilon > 0$, by definition of convergence there corresponds an N such that for all $n > N$, $|L_1 - a_n| < \epsilon$ and $|L_2 - a_n| < \epsilon$. Now $|L_2 - L_1| = |L_2 - a_n + a_n - L_1| \le |L_2 - a_n| + |a_n - L_1| < \epsilon + \epsilon = 2\epsilon$. $|L_2 - L_1| < 2\epsilon$ says that the difference between two fixed values is smaller than any positive number 2ϵ. The only nonnegative number smaller than every positive number is 0, so $|L_1 - L_2| = 0$ or $L_1 = L_2$.

118. Let k(n) and i(n) be two order-preserving functions whose domains are the set of positive integers and whose ranges are a subset of the positive integers. Consider the two subsequences $a_{k(n)}$ and $a_{i(n)}$, where $a_{k(n)} \to L_1$, $a_{i(n)} \to L_2$ and $L_1 \neq L_2$. Thus $\left|a_{k(n)} - a_{i(n)}\right| \to |L_1 - L_2| > 0$. So there does not exist N such that for all m, n > N $\Rightarrow |a_m - a_n| < \epsilon$. So by Exercise 116, the sequence $\{a_n\}$ is not convergent and hence diverges.

119. $a_{2k} \to L \Leftrightarrow$ given an $\epsilon > 0$ there corresponds an N_1 such that $[2k > N_1 \Rightarrow |a_{2k} - L| < \epsilon]$. Similarly, $a_{2k+1} \to L \Leftrightarrow [2k+1 > N_2 \Rightarrow |a_{2k+1} - L| < \epsilon]$. Let $N = \max\{N_1, N_2\}$. Then $n > N \Rightarrow |a_n - L| < \epsilon$ whether n is even or odd, and hence $a_n \to L$.

120. Assume $a_n \to 0$. This implies that given an $\epsilon > 0$ there corresponds an N such that $n > N \Rightarrow |a_n - 0| < \epsilon$ $\Rightarrow |a_n| < \epsilon \Rightarrow ||a_n|| < \epsilon \Rightarrow ||a_n| - 0| < \epsilon \Rightarrow |a_n| \to 0$. On the other hand, assume $|a_n| \to 0$. This implies that given an $\epsilon > 0$ there corresponds an N such that for n > N, $||a_n| - 0| < \epsilon \Rightarrow ||a_n|| < \epsilon \Rightarrow |a_n| < \epsilon$ $\Rightarrow |a_n - 0| < \epsilon \Rightarrow a_n \to 0$.

121. $\left|\sqrt[n]{0.5} - 1\right| < 10^{-3} \Rightarrow -\frac{1}{1000} < \left(\frac{1}{2}\right)^{1/n} - 1 < \frac{1}{1000} \Rightarrow \left(\frac{999}{1000}\right)^n < \frac{1}{2} < \left(\frac{1001}{1000}\right)^n \Rightarrow n > \frac{\ln\left(\frac{1}{2}\right)}{\ln\left(\frac{999}{1000}\right)} \Rightarrow n > 692.8$ $\Rightarrow N = 692;\ a_n = \left(\frac{1}{2}\right)^{1/n}$ and $\lim_{n \to \infty} a_n = 1$

122. $\left|\sqrt[n]{n} - 1\right| < 10^{-3} \Rightarrow -\frac{1}{1000} < n^{1/n} - 1 < \frac{1}{1000} \Rightarrow \left(\frac{999}{1000}\right)^n < n < \left(\frac{1001}{1000}\right)^n \Rightarrow n > 9123 \Rightarrow N = 9123;$ $a_n = \sqrt[n]{n} = n^{1/n}$ and $\lim_{n \to \infty} a_n = 1$

123. $(0.9)^n < 10^{-3} \Rightarrow n \ln(0.9) < -3 \ln 10 \Rightarrow n > \frac{-3 \ln 10}{\ln(0.9)} \approx 65.54 \Rightarrow N = 65;\ a_n = \left(\frac{9}{10}\right)^n$ and $\lim_{n \to \infty} a_n = 0$

124. $\frac{2^n}{n!} < 10^{-7} \Rightarrow n! > 2^n 10^7$ and by calculator experimentation, $n > 14 \Rightarrow N = 14;\ a_n = \frac{2^n}{n!}$ and $\lim_{n \to \infty} a_n = 0$

125. (a) $f(x) = x^2 - a \Rightarrow f'(x) = 2x \Rightarrow x_{n+1} = x_n - \frac{x_n^2 - a}{2x_n} \Rightarrow x_{n+1} = \frac{2x_n^2 - (x_n^2 - a)}{2x_n} = \frac{x_n^2 + a}{2x_n} = \frac{\left(x_n + \frac{a}{x_n}\right)}{2}$

(b) $x_1 = 2$, $x_2 = 1.75$, $x_3 = 1.732142857$, $x_4 = 1.73205081$, $x_5 = 1.732050808$; we are finding the positive number where $x^2 - 3 = 0$; that is, where $x^2 = 3$, $x > 0$, or where $x = \sqrt{3}$.

126. $x_1 = 1.5$, $x_2 = 1.416666667$, $x_3 = 1.414215686$, $x_4 = 1.414213562$, $x_5 = 1.414213562$; we are finding the positive number $x^2 - 2 = 0$; that is, where $x^2 = 2$, $x > 0$, or where $x = \sqrt{2}$.

127. $x_1 = 1$, $x_2 = 1 + \cos(1) = 1.540302306$, $x_3 = 1.540302306 + \cos(1 + \cos(1)) = 1.570791601$, $x_4 = 1.570791601 + \cos(1.570791601) = 1.570796327 = \frac{\pi}{2}$ to 9 decimal places. After a few steps, the arc (x_{n-1}) and line segment $\cos(x_{n-1})$ are nearly the same as the quarter circle.

128. (a) $S_1 = 6.815$, $S_2 = 6.4061$, $S_3 = 6.021734$, $S_4 = 5.66042996$, $S_5 = 5.320804162$, $S_6 = 5.001555913$, $S_7 = 4.701462558$, $S_8 = 4.419374804$, $S_9 = 4.154212316$, $S_{10} = 3.904959577$, $S_{11} = 3.670662003$, $S_{12} = 3.450422282$ so it will take Ford about 12 years to catch up

(b) $x \approx 11.8$

129-140. Example CAS Commands:

Maple:

```
with( Student[Calculus1] );
f := x -> sin(x);
a := 0;
b := Pi;
```

```
plot( f(x), x=a..b, title="#23(a) (Section 5.1)" );
N := [ 100, 200, 1000 ];                              # (b)
for n in N do
  Xlist := [ a+1.*(b-a)/n*i $ i=0..n ];
  Ylist := map( f, Xlist );
end do:
for n in N do                                         # (c)
  Avg[n] := evalf(add(y,y=Ylist)/nops(Ylist));
end do;
avg := FunctionAverage( f(x), x=a..b, output=value );
evalf( avg );
FunctionAverage(f(x),x=a..b,output=plot);     # (d)
fsolve( f(x)=avg, x=0.5 );
fsolve( f(x)=avg, x=2.5 );
fsolve( f(x)=Avg[1000], x=0.5 );
fsolve( f(x)=Avg[1000], x=2.5 );
```

Mathematica: (sequence functions may vary):

```
Clear[a, n]
a[n_]; = n^(1/n)
first25= Table[N[a[n]],{n, 1, 25}]
Limit[a[n], n → 8]
```

The last command (Limit) will not always work in Mathematica. You could also explore the limit by enlarging your table to more than the first 25 values.

If you know the limit (1 in the above example), to determine how far to go to have all further terms within 0.01 of the limit, do the following.

```
Clear[minN, lim]
lim= 1
Do[{diff=Abs[a[n] – lim], If[diff < .01, {minN= n, Abort[]}]}, {n, 2, 1000}]
minN
```

For sequences that are given recursively, the following code is suggested. The portion of the command a[n_]:=a[n] stores the elements of the sequence and helps to streamline computation.

```
Clear[a, n]
a[1]= 1;
a[n_]; = a[n]= a[n – 1] + (1/5)^(n–1)
first25= Table[N[a[n]], {n, 1, 25}]
```

The limit command does not work in this case, but the limit can be observed as 1.25.

```
Clear[minN, lim]
lim= 1.25
Do[{diff=Abs[a[n] – lim], If[diff < .01, {minN= n, Abort[]}]}, {n, 2, 1000}]
minN
```

141. Example CAS Commands:

Maple:

```
with( Student[Calculus1] );
A := n->(1+r/m)*A(n-1) + b;
A(0) := A0;
A(0) := 1000; r := 0.02015; m := 12; b := 50;              # (a)
pts1 := [seq( [n,A(n)], n=0..99 )]:
plot( pts1, style=point, title="#141(a) (Section 11.1)");
```

```
A(60);
The sequence { A[n] } is not unbounded;
limit( A[n], n=infinity ) = infinity.
A(0) := 5000; r := 0.0589; m := 12; b := -50;                 # (b)
pts1 := [seq( [n,A(n)], n=0..99 )]:
plot( pts1, style=point, title="#141(b) (Section 11.1)");
A(60);
pts1 := [seq( [n,A(n)], n=0..199 )]:
plot( pts1, style=point, title="#141(b) (Section 11.1)");
# This sequence is not bounded, and diverges to -infinity:
limit( A[n], n=infinity ) = -infinity.
A(0) := 5000; r := 0.045; m := 4; b := 0;                     # (c)
for n from 1 while A(n)<20000 do end do; n;
```

It takes 31 years (124 quarters) for the investment to grow to $20,000 when the interest rate is 4.5%, compounded quarterly.

```
r := 0.0625;
for n from 1 while A(n)<20000 do end do; n;
```

When the interest rate increases to 6.25% (compounded quarterly), it takes only 22.5 years for the balance to reach $20,000.

```
B := k -> (1+r/m)^k * (A(0)+m*b/r) - m*b/r;                  # (d)
A(0) := 1000.; r := 0.02015; m := 12; b := 50;
for k from 0 to 49 do
  printf( "%5d  %9.2f  %9.2f  %9.2f\n", k, A(k), B(k), B(k)-A(k) );
end do;
A(0) := 'A(0)'; r := 'r'; m := 'm'; b := 'b'; n := 'n';
eval( AA(n+1) - ((1+r/m)*AA(n) + b), AA=B );
simplify( % );
```

142. Example CAS Commands:

Maple:

```
r := 3/4.;                         # (a)
for k in $1..9 do
  A := k/10.;
  L := [0,A];
  for n from 1 to 99 do
    A := r*A*(1-A);
    L := L, [n,A];
  end do;
  pt[r,k/10] := [L];
end do:
plot( [seq( pt[r,a], a=[($1..9)/10] )], style=point, title="#142(a) (Section 11.1)" );
R1 := [1.1, 1.2, 1.5, 2.5, 2.8, 2.9];                         # (b)
for r in R1 do
  for k in $1..9 do
    A := k/10.;
    L := [0,A];
    for n from 1 to 99 do
      A := r*A*(1-A);
      L := L, [n,A];
```

```
    end do;
    pt[r,k/10] := [L];
  end do:
  t := sprintf("#142(b) (Section 11.1)\nr = %f", r);
  P[r] := plot( [seq( pt[r,a], a=[($1..9)/10] )], style=point, title=t );
end do:
display( [seq(P[r], r=R1)], insequence=true );
R2 := [3.05, 3.1, 3.2, 3.3, 3.35, 3.4];                    # (c)
for r in R2 do
  for k in $1..9 do
    A := k/10.;
    L := [0,A];
    for n from 1 to 99 do
      A := r*A*(1-A);
      L := L, [n,A];
    end do;
    pt[r,k/10] := [L];
  end do:
  t := sprintf("#142(c) (Section 11.1)\nr = %f", r);
  P[r] := plot( [seq( pt[r,a], a=[($1..9)/10] )], style=point, title=t );
end do:
display( [seq(P[r], r=R2)], insequence=true );
R3 := [3.46, 3.47, 3.48, 3.49, 3.5, 3.51, 3.52, 3.53, 3.542, 3.544, 3.546, 3.548];          # (d)
for r in R3 do
  for k in $1..9 do
    A := k/10.;
    L := [0,A];
    for n from 1 to 199 do
      A := r*A*(1-A);
      L := L, [n,A];
    end do;
    pt[r,k/10] := [L];
  end do:
  t := sprintf("#142(d) (Section 11.1)\nr = %f", r);
  P[r] := plot( [seq( pt[r,a], a=[($1..9)/10] )], style=point, title=t );
end do:
display( [seq(P[r], r=R3)], insequence=true );
R4 := [3.5695];                              # (e)
for r in R4 do
  for k in $1..9 do
    A := k/10.;
    L := [0,A];
    for n from 1 to 299 do
      A := r*A*(1-A);
      L := L, [n,A];
    end do;
    pt[r,k/10] := [L];
  end do:
  t := sprintf("#142(e) (Section 11.1)\nr = %f", r);
```

```
  P[r] := plot( [seq( pt[r,a], a=[($1..9)/10] )], style=point, title=t );
end do:
display( [seq(P[r], r=R4)], insequence=true );
R5 := [3.65];                                              # (f)
for r in R5 do
  for k in $1..9 do
    A := k/10.;
    L := [0,A];
    for n from 1 to 299 do
      A := r*A*(1-A);
      L := L, [n,A];
    end do;
    pt[r,k/10] := [L];
  end do:
  t := sprintf("#142(f) (Section 11.1)\nr = %f", r);
  P[r] := plot( [seq( pt[r,a], a=[($1..9)/10] )], style=point, title=t );
end do:
display( [seq(P[r], r=R5)], insequence=true );
R6 := [3.65, 3.75];                                        # (g)
for r in R6 do
  for a in [0.300, 0.301, 0.600, 0.601 ] do
    A := a;
    L := [0,a];
    for n from 1 to 299 do
      A := r*A*(1-A);
      L := L, [n,A];
    end do;
    pt[r,a] := [L];
  end do:
  t := sprintf("#142(g) (Section 11.1)\nr = %f", r);
  P[r] := plot( [seq( pt[r,a], a=[0.300, 0.301, 0.600, 0.601] )], style=point, title=t );
end do:
display( [seq(P[r], r=R6)], insequence=true );
```

11.2 INFINITE SERIES

1. $s_n = \frac{a(1-r^n)}{(1-r)} = \frac{2\left(1-\left(\frac{1}{3}\right)^n\right)}{1-\left(\frac{1}{3}\right)} \Rightarrow \lim_{n\to\infty} s_n = \frac{2}{1-\left(\frac{1}{3}\right)} = 3$

2. $s_n = \frac{a(1-r^n)}{(1-r)} = \frac{\left(\frac{9}{100}\right)\left(1-\left(\frac{1}{100}\right)^n\right)}{1-\left(\frac{1}{100}\right)} \Rightarrow \lim_{n\to\infty} s_n = \frac{\left(\frac{9}{100}\right)}{1-\left(\frac{1}{100}\right)} = \frac{1}{11}$

3. $s_n = \frac{a(1-r^n)}{(1-r)} = \frac{1-\left(-\frac{1}{2}\right)^n}{1-\left(-\frac{1}{2}\right)} \Rightarrow \lim_{n\to\infty} s_n = \frac{1}{\left(\frac{3}{2}\right)} = \frac{2}{3}$

4. $s_n = \frac{1-(-2)^n}{1-(-2)}$, a geometric series where $|r| > 1 \Rightarrow$ divergence

5. $\frac{1}{(n+1)(n+2)} = \frac{1}{n+1} - \frac{1}{n+2} \Rightarrow s_n = \left(\frac{1}{2}-\frac{1}{3}\right) + \left(\frac{1}{3}-\frac{1}{4}\right) + \ldots + \left(\frac{1}{n+1}-\frac{1}{n+2}\right) = \frac{1}{2} - \frac{1}{n+2} \Rightarrow \lim_{n\to\infty} s_n = \frac{1}{2}$

6. $\frac{5}{n(n+1)} = \frac{5}{n} - \frac{5}{n+1} \Rightarrow s_n = \left(5 - \frac{5}{2}\right) + \left(\frac{5}{2} - \frac{5}{3}\right) + \left(\frac{5}{3} - \frac{5}{4}\right) + \ldots + \left(\frac{5}{n-1} - \frac{5}{n}\right) + \left(\frac{5}{n} - \frac{5}{n+1}\right) = 5 - \frac{5}{n+1}$
$\Rightarrow \lim_{n \to \infty} s_n = 5$

7. $1 - \frac{1}{4} + \frac{1}{16} - \frac{1}{64} + \ldots$, the sum of this geometric series is $\frac{1}{1 - \left(-\frac{1}{4}\right)} = \frac{1}{1 + \left(\frac{1}{4}\right)} = \frac{4}{5}$

8. $\frac{1}{16} + \frac{1}{64} + \frac{1}{256} + \ldots$, the sum of this geometric series is $\frac{\left(\frac{1}{16}\right)}{1 - \left(\frac{1}{4}\right)} = \frac{1}{12}$

9. $\frac{7}{4} + \frac{7}{16} + \frac{7}{64} + \ldots$, the sum of this geometric series is $\frac{\left(\frac{7}{4}\right)}{1 - \left(\frac{1}{4}\right)} = \frac{7}{3}$

10. $5 - \frac{5}{4} + \frac{5}{16} - \frac{5}{64} + \ldots$, the sum of this geometric series is $\frac{5}{1 - \left(-\frac{1}{4}\right)} = 4$

11. $(5 + 1) + \left(\frac{5}{2} + \frac{1}{3}\right) + \left(\frac{5}{4} + \frac{1}{9}\right) + \left(\frac{5}{8} + \frac{1}{27}\right) + \ldots$, is the sum of two geometric series; the sum is
$\frac{5}{1 - \left(\frac{1}{2}\right)} + \frac{1}{1 - \left(\frac{1}{3}\right)} = 10 + \frac{3}{2} = \frac{23}{2}$

12. $(5 - 1) + \left(\frac{5}{2} - \frac{1}{3}\right) + \left(\frac{5}{4} - \frac{1}{9}\right) + \left(\frac{5}{8} - \frac{1}{27}\right) + \ldots$, is the difference of two geometric series; the sum is
$\frac{5}{1 - \left(\frac{1}{2}\right)} - \frac{1}{1 - \left(\frac{1}{3}\right)} = 10 - \frac{3}{2} = \frac{17}{2}$

13. $(1 + 1) + \left(\frac{1}{2} - \frac{1}{5}\right) + \left(\frac{1}{4} + \frac{1}{25}\right) + \left(\frac{1}{8} - \frac{1}{125}\right) + \ldots$, is the sum of two geometric series; the sum is
$\frac{1}{1 - \left(\frac{1}{2}\right)} + \frac{1}{1 + \left(\frac{1}{5}\right)} = 2 + \frac{5}{6} = \frac{17}{6}$

14. $2 + \frac{4}{5} + \frac{8}{25} + \frac{16}{125} + \ldots = 2\left(1 + \frac{2}{5} + \frac{4}{25} + \frac{8}{125} + \ldots\right)$; the sum of this geometric series is $2\left(\frac{1}{1 - \left(\frac{2}{5}\right)}\right) = \frac{10}{3}$

15. $\frac{4}{(4n-3)(4n+1)} = \frac{1}{4n-3} - \frac{1}{4n+1} \Rightarrow s_n = \left(1 - \frac{1}{5}\right) + \left(\frac{1}{5} - \frac{1}{9}\right) + \left(\frac{1}{9} - \frac{1}{13}\right) + \ldots + \left(\frac{1}{4n-7} - \frac{1}{4n-3}\right)$
$+ \left(\frac{1}{4n-3} - \frac{1}{4n+1}\right) = 1 - \frac{1}{4n+1} \Rightarrow \lim_{n \to \infty} s_n = \lim_{n \to \infty} \left(1 - \frac{1}{4n+1}\right) = 1$

16. $\frac{6}{(2n-1)(2n+1)} = \frac{A}{2n-1} + \frac{B}{2n+1} = \frac{A(2n+1) + B(2n-1)}{(2n-1)(2n+1)} \Rightarrow A(2n+1) + B(2n-1) = 6$
$\Rightarrow (2A + 2B)n + (A - B) = 6 \Rightarrow \begin{cases} 2A + 2B = 0 \\ A - B = 6 \end{cases} \Rightarrow \begin{cases} A + B = 0 \\ A - B = 6 \end{cases} \Rightarrow 2A = 6 \Rightarrow A = 3$ and $B = -3$. Hence,
$\sum_{n=1}^{k} \frac{6}{(2n-1)(2n+1)} = 3 \sum_{n=1}^{k} \left(\frac{1}{2n-1} - \frac{1}{2n+1}\right) = 3\left(\frac{1}{1} - \frac{1}{3} + \frac{1}{3} - \frac{1}{5} + \frac{1}{5} - \frac{1}{7} + \ldots - \frac{1}{2(k-1)+1} + \frac{1}{2k-1} - \frac{1}{2k+1}\right)$
$= 3\left(1 - \frac{1}{2k+1}\right) \Rightarrow$ the sum is $\lim_{k \to \infty} 3\left(1 - \frac{1}{2k+1}\right) = 3$

17. $\frac{40n}{(2n-1)^2(2n+1)^2} = \frac{A}{(2n-1)} + \frac{B}{(2n-1)^2} + \frac{C}{(2n+1)} + \frac{D}{(2n+1)^2}$
$= \frac{A(2n-1)(2n+1)^2 + B(2n+1)^2 + C(2n+1)(2n-1)^2 + D(2n-1)^2}{(2n-1)^2(2n+1)^2}$
$\Rightarrow A(2n-1)(2n+1)^2 + B(2n+1)^2 + C(2n+1)(2n-1)^2 + D(2n-1)^2 = 40n$
$\Rightarrow A\left(8n^3 + 4n^2 - 2n - 1\right) + B\left(4n^2 + 4n + 1\right) + C\left(8n^3 - 4n^2 - 2n + 1\right) = D\left(4n^2 - 4n + 1\right) = 40n$
$\Rightarrow (8A + 8C)n^3 + (4A + 4B - 4C + 4D)n^2 + (-2A + 4B - 2C - 4D)n + (-A + B + C + D) = 40n$
$\Rightarrow \begin{cases} 8A + 8C = 0 \\ 4A + 4B - 4C + 4D = 0 \\ -2A + 4B - 2C - 4D = 40 \\ -A + B + C + D = 0 \end{cases} \Rightarrow \begin{cases} 8A + 8C = 0 \\ A + B - C + D = 0 \\ -A + 2B - C - 2D = 20 \\ -A + B + C + D = 0 \end{cases} \Rightarrow \begin{cases} B + D = 0 \\ 2B - 2D = 20 \end{cases} \Rightarrow 4B = 20 \Rightarrow B = 5$
and $D = -5 \Rightarrow \begin{cases} A + C = 0 \\ -A + 5 + C - 5 = 0 \end{cases} \Rightarrow C = 0$ and $A = 0$. Hence, $\sum_{n=1}^{k} \left[\frac{40n}{(2n-1)^2(2n+1)^2}\right]$

$= 5\sum_{n=1}^{k}\left[\frac{1}{(2n-1)^2} - \frac{1}{(2n+1)^2}\right] = 5\left(\frac{1}{1} - \frac{1}{9} + \frac{1}{9} - \frac{1}{25} + \frac{1}{25} - \ldots - \frac{1}{(2(k-1)+1)^2} + \frac{1}{(2k-1)^2} - \frac{1}{(2k+1)^2}\right)$

$= 5\left(1 - \frac{1}{(2k+1)^2}\right) \Rightarrow$ the sum is $\lim_{n\to\infty} 5\left(1 - \frac{1}{(2k+1)^2}\right) = 5$

18. $\frac{2n+1}{n^2(n+1)^2} = \frac{1}{n^2} - \frac{1}{(n+1)^2} \Rightarrow s_n = \left(1 - \frac{1}{4}\right) + \left(\frac{1}{4} - \frac{1}{9}\right) + \left(\frac{1}{9} - \frac{1}{16}\right) + \ldots + \left[\frac{1}{(n-1)^2} - \frac{1}{n^2}\right] + \left[\frac{1}{n^2} - \frac{1}{(n+1)^2}\right]$

$\Rightarrow \lim_{n\to\infty} s_n = \lim_{n\to\infty} \left[1 - \frac{1}{(n+1)^2}\right] = 1$

19. $s_n = \left(1 - \frac{1}{\sqrt{2}}\right) + \left(\frac{1}{\sqrt{2}} - \frac{1}{\sqrt{3}}\right) + \left(\frac{1}{\sqrt{3}} - \frac{1}{\sqrt{4}}\right) + \ldots + \left(\frac{1}{\sqrt{n-1}} + \frac{1}{\sqrt{n}}\right) + \left(\frac{1}{\sqrt{n}} - \frac{1}{\sqrt{n+1}}\right) = 1 - \frac{1}{\sqrt{n+1}}$

$\Rightarrow \lim_{n\to\infty} s_n = \lim_{n\to\infty} \left(1 - \frac{1}{\sqrt{n+1}}\right) = 1$

20. $s_n = \left(\frac{1}{2} - \frac{1}{2^{1/2}}\right) + \left(\frac{1}{2^{1/2}} - \frac{1}{2^{1/3}}\right) + \left(\frac{1}{2^{1/3}} - \frac{1}{2^{1/4}}\right) + \ldots + \left(\frac{1}{2^{1/(n-1)}} - \frac{1}{2^{1/n}}\right) + \left(\frac{1}{2^{1/n}} - \frac{1}{2^{1/(n+1)}}\right) = \frac{1}{2} - \frac{1}{2^{1/(n+1)}}$

$\Rightarrow \lim_{n\to\infty} s_n = \frac{1}{2} - \frac{1}{1} = -\frac{1}{2}$

21. $s_n = \left(\frac{1}{\ln 3} - \frac{1}{\ln 2}\right) + \left(\frac{1}{\ln 4} - \frac{1}{\ln 3}\right) + \left(\frac{1}{\ln 5} - \frac{1}{\ln 4}\right) + \ldots + \left(\frac{1}{\ln(n+1)} - \frac{1}{\ln n}\right) + \left(\frac{1}{\ln(n+2)} - \frac{1}{\ln(n+1)}\right)$

$= -\frac{1}{\ln 2} + \frac{1}{\ln(n+2)} \Rightarrow \lim_{n\to\infty} s_n = -\frac{1}{\ln 2}$

22. $s_n = [\tan^{-1}(1) - \tan^{-1}(2)] + [\tan^{-1}(2) - \tan^{-1}(3)] + \ldots + [\tan^{-1}(n-1) - \tan^{-1}(n)]$

$+ [\tan^{-1}(n) - \tan^{-1}(n+1)] = \tan^{-1}(1) - \tan^{-1}(n+1) \Rightarrow \lim_{n\to\infty} s_n = \tan^{-1}(1) - \frac{\pi}{2} = \frac{\pi}{4} - \frac{\pi}{2} = -\frac{\pi}{4}$

23. convergent geometric series with sum $\frac{1}{1-\left(\frac{1}{\sqrt{2}}\right)} = \frac{\sqrt{2}}{\sqrt{2}-1} = 2 + \sqrt{2}$

24. divergent geometric series with $|r| = \sqrt{2} > 1$

25. convergent geometric series with sum $\frac{\left(\frac{3}{2}\right)}{1-\left(-\frac{1}{2}\right)} = 1$

26. $\lim_{n\to\infty} (-1)^{n+1} n \neq 0 \Rightarrow$ diverges

27. $\lim_{n\to\infty} \cos(n\pi) = \lim_{n\to\infty} (-1)^n \neq 0 \Rightarrow$ diverges

28. $\cos(n\pi) = (-1)^n \Rightarrow$ convergent geometric series with sum $\frac{1}{1-\left(-\frac{1}{5}\right)} = \frac{5}{6}$

29. convergent geometric series with sum $\frac{1}{1-\left(\frac{1}{e^2}\right)} = \frac{e^2}{e^2-1}$

30. $\lim_{n\to\infty} \ln\frac{1}{n} = -\infty \neq 0 \Rightarrow$ diverges

31. convergent geometric series with sum $\frac{2}{1-\left(\frac{1}{10}\right)} - 2 = \frac{20}{9} - \frac{18}{9} = \frac{2}{9}$

32. convergent geometric series with sum $\frac{1}{1-\left(\frac{1}{x}\right)} = \frac{x}{x-1}$

33. difference of two geometric series with sum $\frac{1}{1-\left(\frac{2}{3}\right)} - \frac{1}{1-\left(\frac{1}{3}\right)} = 3 - \frac{3}{2} = \frac{3}{2}$

34. $\lim_{n\to\infty} \left(1 - \frac{1}{n}\right)^n = \lim_{n\to\infty} \left(1 + \frac{-1}{n}\right)^n = e^{-1} \neq 0 \Rightarrow$ diverges

35. $\lim_{n \to \infty} \frac{n!}{1000^n} = \infty \neq 0 \Rightarrow$ diverges

36. $\lim_{n \to \infty} \frac{n^n}{n!} = \lim_{n \to \infty} \frac{n \cdot n \cdots n}{1 \cdot 2 \cdots n} > \lim_{n \to \infty} n = \infty \Rightarrow$ diverges

37. $\sum_{n=1}^{\infty} \ln\left(\frac{n}{n+1}\right) = \sum_{n=1}^{\infty} [\ln(n) - \ln(n+1)] \Rightarrow s_n = [\ln(1) - \ln(2)] + [\ln(2) - \ln(3)] + [\ln(3) - \ln(4)] + \ldots$
$+ [\ln(n-1) - \ln(n)] + [\ln(n) - \ln(n+1)] = \ln(1) - \ln(n+1) = -\ln(n+1) \Rightarrow \lim_{n \to \infty} s_n = -\infty, \Rightarrow$ diverges

38. $\lim_{n \to \infty} a_n = \lim_{n \to \infty} \ln\left(\frac{n}{2n+1}\right) = \ln\left(\frac{1}{2}\right) \neq 0 \Rightarrow$ diverges

39. convergent geometric series with sum $\frac{1}{1 - \left(\frac{e}{\pi}\right)} = \frac{\pi}{\pi - e}$

40. divergent geometric series with $|r| = \frac{e^\pi}{\pi^e} \approx \frac{23.141}{22.459} > 1$

41. $\sum_{n=0}^{\infty} (-1)^n x^n = \sum_{n=0}^{\infty} (-x)^n$; $a = 1, r = -x$; converges to $\frac{1}{1-(-x)} = \frac{1}{1+x}$ for $|x| < 1$

42. $\sum_{n=0}^{\infty} (-1)^n x^{2n} = \sum_{n=0}^{\infty} (-x^2)^n$; $a = 1, r = -x^2$; converges to $\frac{1}{1+x^2}$ for $|x| < 1$

43. $a = 3, r = \frac{x-1}{2}$; converges to $\frac{3}{1 - \left(\frac{x-1}{2}\right)} = \frac{6}{3-x}$ for $-1 < \frac{x-1}{2} < 1$ or $-1 < x < 3$

44. $\sum_{n=0}^{\infty} \frac{(-1)^n}{2} \left(\frac{1}{3 + \sin x}\right)^n = \sum_{n=0}^{\infty} \frac{1}{2} \left(\frac{-1}{3 + \sin x}\right)^n$; $a = \frac{1}{2}, r = \frac{-1}{3 + \sin x}$; converges to $\frac{\left(\frac{1}{2}\right)}{1 - \left(\frac{-1}{3 + \sin x}\right)}$
$= \frac{3 + \sin x}{2(4 + \sin x)} = \frac{3 + \sin x}{8 + 2 \sin x}$ for all x $\left(\text{since } \frac{1}{4} \leq \frac{1}{3 + \sin x} \leq \frac{1}{2} \text{ for all x}\right)$

45. $a = 1, r = 2x$; converges to $\frac{1}{1-2x}$ for $|2x| < 1$ or $|x| < \frac{1}{2}$

46. $a = 1, r = -\frac{1}{x^2}$; converges to $\frac{1}{1 - \left(\frac{-1}{x^2}\right)} = \frac{x^2}{x^2+1}$ for $\left|\frac{1}{x^2}\right| < 1$ or $|x| > 1$.

47. $a = 1, r = -(x+1)^n$; converges to $\frac{1}{1 + (x+1)} = \frac{1}{2+x}$ for $|x+1| < 1$ or $-2 < x < 0$

48. $a = 1, r = \frac{3-x}{2}$; converges to $\frac{1}{1 - \left(\frac{3-x}{2}\right)} = \frac{2}{x-1}$ for $\left|\frac{3-x}{2}\right| < 1$ or $1 < x < 5$

49. $a = 1, r = \sin x$; converges to $\frac{1}{1 - \sin x}$ for $x \neq (2k+1)\frac{\pi}{2}$, k an integer

50. $a = 1, r = \ln x$; converges to $\frac{1}{1 - \ln x}$ for $|\ln x| < 1$ or $e^{-1} < x < e$

51. $0.\overline{23} = \sum_{n=0}^{\infty} \frac{23}{100} \left(\frac{1}{10^2}\right)^n = \frac{\left(\frac{23}{100}\right)}{1 - \left(\frac{1}{100}\right)} = \frac{23}{99}$

52. $0.\overline{234} = \sum_{n=0}^{\infty} \frac{234}{1000} \left(\frac{1}{10^3}\right)^n = \frac{\left(\frac{234}{1000}\right)}{1 - \left(\frac{1}{1000}\right)} = \frac{234}{999}$

53. $0.\overline{7} = \sum_{n=0}^{\infty} \frac{7}{10} \left(\frac{1}{10}\right)^n = \frac{\left(\frac{7}{10}\right)}{1 - \left(\frac{1}{10}\right)} = \frac{7}{9}$

54. $0.\overline{d} = \sum_{n=0}^{\infty} \frac{d}{10} \left(\frac{1}{10}\right)^n = \frac{\left(\frac{d}{10}\right)}{1 - \left(\frac{1}{10}\right)} = \frac{d}{9}$

55. $0.0\overline{6} = \sum_{n=0}^{\infty} \left(\frac{1}{10}\right)\left(\frac{6}{10}\right)\left(\frac{1}{10}\right)^n = \frac{\left(\frac{6}{100}\right)}{1 - \left(\frac{1}{10}\right)} = \frac{6}{90} = \frac{1}{15}$

56. $1.\overline{414} = 1 + \sum_{n=0}^{\infty} \frac{414}{1000}\left(\frac{1}{10^3}\right)^n = 1 + \frac{\left(\frac{414}{1000}\right)}{1-\left(\frac{1}{1000}\right)} = 1 + \frac{414}{999} = \frac{1413}{999}$

57. $1.24\overline{123} = \frac{124}{100} + \sum_{n=0}^{\infty} \frac{123}{10^5}\left(\frac{1}{10^3}\right)^n = \frac{124}{100} + \frac{\left(\frac{123}{10^5}\right)}{1-\left(\frac{1}{10^3}\right)} = \frac{124}{100} + \frac{123}{10^5 - 10^2} = \frac{124}{100} + \frac{123}{99{,}900} = \frac{123{,}999}{99{,}900} = \frac{41{,}333}{33{,}300}$

58. $3.\overline{142857} = 3 + \sum_{n=0}^{\infty} \frac{142{,}857}{10^6}\left(\frac{1}{10^6}\right)^n = 3 + \frac{\left(\frac{142{,}857}{10^6}\right)}{1-\left(\frac{1}{10^6}\right)} = 3 + \frac{142{,}857}{10^6 - 1} = \frac{3{,}142{,}854}{999{,}999} = \frac{116{,}402}{37{,}037}$

59. (a) $\sum_{n=-2}^{\infty} \frac{1}{(n+4)(n+5)}$ (b) $\sum_{n=0}^{\infty} \frac{1}{(n+2)(n+3)}$ (c) $\sum_{n=5}^{\infty} \frac{1}{(n-3)(n-2)}$

60. (a) $\sum_{n=-1}^{\infty} \frac{5}{(n+2)(n+3)}$ (b) $\sum_{n=3}^{\infty} \frac{5}{(n-2)(n-1)}$ (c) $\sum_{n=20}^{\infty} \frac{5}{(n-19)(n-18)}$

61. (a) one example is $\frac{1}{2} + \frac{1}{4} + \frac{1}{8} + \frac{1}{16} + \ldots = \frac{\left(\frac{1}{2}\right)}{1-\left(\frac{1}{2}\right)} = 1$

(b) one example is $-\frac{3}{2} - \frac{3}{4} - \frac{3}{8} - \frac{3}{16} - \ldots = \frac{\left(-\frac{3}{2}\right)}{1-\left(\frac{1}{2}\right)} = -3$

(c) one example is $1 - \frac{1}{2} - \frac{1}{4} - \frac{1}{8} - \frac{1}{16} - \ldots$; the series $\frac{k}{2} + \frac{k}{4} + \frac{k}{8} + \ldots = \frac{\left(\frac{k}{2}\right)}{1-\left(\frac{1}{2}\right)} = k$ where k is any positive or negative number.

62. The series $\sum_{n=0}^{\infty} k\left(\frac{1}{2}\right)^{n+1}$ is a geometric series whose sum is $\frac{\left(\frac{k}{2}\right)}{1-\left(\frac{1}{2}\right)} = k$ where k can be any positive or negative number.

63. Let $a_n = b_n = \left(\frac{1}{2}\right)^n$. Then $\sum_{n=1}^{\infty} a_n = \sum_{n=1}^{\infty} b_n = \sum_{n=1}^{\infty} \left(\frac{1}{2}\right)^n = 1$, while $\sum_{n=1}^{\infty} \left(\frac{a_n}{b_n}\right) = \sum_{n=1}^{\infty} (1)$ diverges.

64. Let $a_n = b_n = \left(\frac{1}{2}\right)^n$. Then $\sum_{n=1}^{\infty} a_n = \sum_{n=1}^{\infty} b_n = \sum_{n=1}^{\infty} \left(\frac{1}{2}\right)^n = 1$, while $\sum_{n=1}^{\infty} (a_n b_n) = \sum_{n=1}^{\infty} \left(\frac{1}{4}\right)^n = \frac{1}{3} \neq AB$.

65. Let $a_n = \left(\frac{1}{4}\right)^n$ and $b_n = \left(\frac{1}{2}\right)^n$. Then $A = \sum_{n=1}^{\infty} a_n = \frac{1}{3}$, $B = \sum_{n=1}^{\infty} b_n = 1$ and $\sum_{n=1}^{\infty} \left(\frac{a_n}{b_n}\right) = \sum_{n=1}^{\infty} \left(\frac{1}{2}\right)^n = 1 \neq \frac{A}{B}$.

66. Yes: $\sum \left(\frac{1}{a_n}\right)$ diverges. The reasoning: $\sum a_n$ converges $\Rightarrow a_n \to 0 \Rightarrow \frac{1}{a_n} \to \infty \Rightarrow \sum \left(\frac{1}{a_n}\right)$ diverges by the nth-Term Test.

67. Since the sum of a finite number of terms is finite, adding or subtracting a finite number of terms from a series that diverges does not change the divergence of the series.

68. Let $A_n = a_1 + a_2 + \ldots + a_n$ and $\lim_{n \to \infty} A_n = A$. Assume $\sum (a_n + b_n)$ converges to S. Let $S_n = (a_1 + b_1) + (a_2 + b_2) + \ldots + (a_n + b_n) \Rightarrow S_n = (a_1 + a_2 + \ldots + a_n) + (b_1 + b_2 + \ldots + b_n)$ $\Rightarrow b_1 + b_2 + \ldots + b_n = S_n - A_n \Rightarrow \lim_{n \to \infty} (b_1 + b_2 + \ldots + b_n) = S - A \Rightarrow \sum b_n$ converges. This contradicts the assumption that $\sum b_n$ diverges; therefore, $\sum (a_n + b_n)$ diverges.

69. (a) $\frac{2}{1-r} = 5 \Rightarrow \frac{2}{5} = 1 - r \Rightarrow r = \frac{3}{5}; 2 + 2\left(\frac{3}{5}\right) + 2\left(\frac{3}{5}\right)^2 + \ldots$

(b) $\frac{\left(\frac{13}{2}\right)}{1-r} = 5 \Rightarrow \frac{13}{10} = 1 - r \Rightarrow r = -\frac{3}{10}; \frac{13}{2} - \frac{13}{2}\left(\frac{3}{10}\right) + \frac{13}{2}\left(\frac{3}{10}\right)^2 - \frac{13}{2}\left(\frac{3}{10}\right)^3 + \ldots$

70. $1 + e^b + e^{2b} + \ldots = \frac{1}{1-e^b} = 9 \Rightarrow \frac{1}{9} = 1 - e^b \Rightarrow e^b = \frac{8}{9} \Rightarrow b = \ln\left(\frac{8}{9}\right)$

71. $s_n = 1 + 2r + r^2 + 2r^3 + r^4 + 2r^5 + \ldots + r^{2n} + 2r^{2n+1}$, $n = 0, 1, \ldots$
$\Rightarrow s_n = (1 + r^2 + r^4 + \ldots + r^{2n}) + (2r + 2r^3 + 2r^5 + \ldots + 2r^{2n+1}) \Rightarrow \lim_{n \to \infty} s_n = \frac{1}{1-r^2} + \frac{2r}{1-r^2}$
$= \frac{1+2r}{1-r^2}$, if $|r^2| < 1$ or $|r| < 1$

72. $L - s_n = \frac{a}{1-r} - \frac{a(1-r^n)}{1-r} = \frac{ar^n}{1-r}$

73. distance $= 4 + 2\left[(4)\left(\frac{3}{4}\right) + (4)\left(\frac{3}{4}\right)^2 + \ldots\right] = 4 + 2\left(\frac{3}{1-\left(\frac{3}{4}\right)}\right) = 28$ m

74. time $= \sqrt{\frac{4}{4.9}} + 2\sqrt{\left(\frac{4}{4.9}\right)\left(\frac{3}{4}\right)} + 2\sqrt{\left(\frac{4}{4.9}\right)\left(\frac{3}{4}\right)^2} + 2\sqrt{\left(\frac{4}{4.9}\right)\left(\frac{3}{4}\right)^3} + \ldots = \sqrt{\frac{4}{4.9}} + 2\sqrt{\frac{4}{4.9}}\left[\sqrt{\frac{3}{4}} + \sqrt{\left(\frac{3}{4}\right)^2} + \ldots\right]$
$= \frac{2}{\sqrt{4.9}} + \left(\frac{4}{\sqrt{4.9}}\right)\left[\frac{\sqrt{\frac{3}{4}}}{1-\sqrt{\frac{3}{4}}}\right] = \frac{2}{\sqrt{4.9}} + \left(\frac{4}{\sqrt{4.9}}\right)\left(\frac{\sqrt{3}}{2-\sqrt{3}}\right) = \frac{\left(4-2\sqrt{3}\right)+4\sqrt{3}}{\sqrt{4.9}\left(2-\sqrt{3}\right)} = \frac{4+2\sqrt{3}}{\sqrt{4.9}\left(2-\sqrt{3}\right)} \approx 12.58$ sec

75. area $= 2^2 + \left(\sqrt{2}\right)^2 + (1)^2 + \left(\frac{1}{\sqrt{2}}\right)^2 + \ldots = 4 + 2 + 1 + \frac{1}{2} + \ldots = \frac{4}{1-\frac{1}{2}} = 8$ m^2

76. area $= 2\left[\frac{\pi\left(\frac{1}{2}\right)^2}{2}\right] + 4\left[\frac{\pi\left(\frac{1}{4}\right)^2}{2}\right] + 8\left[\frac{\pi\left(\frac{1}{8}\right)^2}{2}\right] + \ldots = \pi\left(\frac{1}{4} + \frac{1}{8} + \frac{1}{16} + \ldots\right) = \pi\left(\frac{\left(\frac{1}{4}\right)}{1-\left(\frac{1}{2}\right)}\right) = \frac{\pi}{2}$

77. (a) $L_1 = 3, L_2 = 3\left(\frac{4}{3}\right), L_3 = 3\left(\frac{4}{3}\right)^2, \ldots, L_n = 3\left(\frac{4}{3}\right)^{n-1} \Rightarrow \lim_{n \to \infty} L_n = \lim_{n \to \infty} 3\left(\frac{4}{3}\right)^{n-1} = \infty$

(b) Using the fact that the area of an equilateral triangle of side length s is $\frac{\sqrt{3}}{4}s^2$, we see that $A_1 = \frac{\sqrt{3}}{4}$,
$A_2 = A_1 + 3\left(\frac{\sqrt{3}}{4}\right)\left(\frac{1}{3}\right)^2 = \frac{\sqrt{3}}{4} + \frac{\sqrt{3}}{12}$, $A_3 = A_2 + 3(4)\left(\frac{\sqrt{3}}{4}\right)\left(\frac{1}{3^2}\right)^2 = \frac{\sqrt{3}}{4} + \frac{\sqrt{3}}{12} + \frac{\sqrt{3}}{27}$,
$A_4 = A_3 + 3(4)^2\left(\frac{\sqrt{3}}{4}\right)\left(\frac{1}{3^3}\right)^2$, $A_5 = A_4 + 3(4)^3\left(\frac{\sqrt{3}}{4}\right)\left(\frac{1}{3^4}\right)^2, \ldots,$
$A_n = \frac{\sqrt{3}}{4} + \sum_{k=2}^{n} 3(4)^{k-2}\left(\frac{\sqrt{3}}{4}\right)\left(\frac{1}{3^2}\right)^{k-1} = \frac{\sqrt{3}}{4} + \sum_{k=2}^{n} 3\sqrt{3}(4)^{k-3}\left(\frac{1}{9}\right)^{k-1} = \frac{\sqrt{3}}{4} + 3\sqrt{3}\left(\sum_{k=2}^{n} \frac{4^{k-3}}{9^{k-1}}\right)$.
$\lim_{n \to \infty} A_n = \lim_{n \to \infty} \left(\frac{\sqrt{3}}{4} + 3\sqrt{3}\left(\sum_{k=2}^{n} \frac{4^{k-3}}{9^{k-1}}\right)\right) = \frac{\sqrt{3}}{4} + 3\sqrt{3}\left(\frac{\frac{1}{36}}{1-\frac{4}{9}}\right) = \frac{\sqrt{3}}{4} + 3\sqrt{3}\left(\frac{1}{20}\right) = \frac{2\sqrt{3}}{5}$

78. Each term of the series $\sum_{n=1}^{\infty} \frac{1}{n^2}$ represents the area of one of the squares shown in the figure, and all of the squares lie inside the rectangle of width 1 and length $\sum_{n=0}^{\infty} \left(\frac{1}{2}\right)^n = \frac{1}{1-\frac{1}{2}} = 2$. Since the squares do not fill the rectangle completely, and the area of the rectangle is 2, we have $\sum_{n=1}^{\infty} \frac{1}{n^2} < 2$.

11.3 THE INTEGRAL TEST

1. converges; a geometric series with $r = \frac{1}{10} < 1$

2. converges; a geometric series with $r = \frac{1}{e} < 1$

3. diverges; by the nth-Term Test for Divergence, $\lim_{n \to \infty} \frac{n}{n+1} = 1 \neq 0$

4. diverges by the Integral Test; $\int_1^n \frac{5}{x+1}\,dx = 5\ln(n+1) - 5\ln 2 \Rightarrow \int_1^\infty \frac{5}{x+1}\,dx \to \infty$

5. diverges; $\sum_{n=1}^{\infty} \frac{3}{\sqrt{n}} = 3\sum_{n=1}^{\infty} \frac{1}{\sqrt{n}}$, which is a divergent p-series $(p = \frac{1}{2})$

6. converges; $\sum_{n=1}^{\infty} \frac{-2}{n\sqrt{n}} = -2\sum_{n=1}^{\infty} \frac{1}{n^{3/2}}$, which is a convergent p-series $(p = \frac{3}{2})$

7. converges; a geometric series with $r = \frac{1}{8} < 1$

8. diverges; $\sum_{n=1}^{\infty} \frac{-8}{n} = -8\sum_{n=1}^{\infty} \frac{1}{n}$ and since $\sum_{n=1}^{\infty} \frac{1}{n}$ diverges, $-8\sum_{n=1}^{\infty} \frac{1}{n}$ diverges

9. diverges by the Integral Test: $\int_2^n \frac{\ln x}{x}\,dx = \frac{1}{2}(\ln^2 n - \ln 2) \Rightarrow \int_2^\infty \frac{\ln x}{x}\,dx \to \infty$

10. diverges by the Integral Test: $\int_2^\infty \frac{\ln x}{\sqrt{x}}\,dx$; $\begin{bmatrix} t = \ln x \\ dt = \frac{dx}{x} \\ dx = e^t\,dt \end{bmatrix} \to \int_{\ln 2}^\infty te^{t/2}\,dt = \lim_{b \to \infty} \left[2te^{t/2} - 4e^{t/2}\right]_{\ln 2}^{b}$
$= \lim_{b \to \infty} \left[2e^{b/2}(b-2) - 2e^{(\ln 2)/2}(\ln 2 - 2)\right] = \infty$

11. converges; a geometric series with $r = \frac{2}{3} < 1$

12. diverges; $\lim_{n \to \infty} \frac{5^n}{4^n + 3} = \lim_{n \to \infty} \frac{5^n \ln 5}{4^n \ln 4} = \lim_{n \to \infty} \left(\frac{\ln 5}{\ln 4}\right)\left(\frac{5}{4}\right)^n = \infty \neq 0$

13. diverges; $\sum_{n=0}^{\infty} \frac{-2}{n+1} = -2\sum_{n=0}^{\infty} \frac{1}{n+1}$, which diverges by the Integral Test

14. diverges by the Integral Test: $\int_1^n \frac{dx}{2x-1} = \frac{1}{2}\ln(2n-1) \to \infty$ as $n \to \infty$

15. diverges; $\lim_{n \to \infty} a_n = \lim_{n \to \infty} \frac{2^n}{n+1} = \lim_{n \to \infty} \frac{2^n \ln 2}{1} = \infty \neq 0$

16. diverges by the Integral Test: $\int_1^n \frac{dx}{\sqrt{x}\left(\sqrt{x}+1\right)}$; $\begin{bmatrix} u = \sqrt{x} + 1 \\ du = \frac{dx}{\sqrt{x}} \end{bmatrix} \to \int_2^{\sqrt{n}+1} \frac{du}{u} = \ln\left(\sqrt{n}+1\right) - \ln 2$
$\to \infty$ as $n \to \infty$

17. diverges; $\lim_{n \to \infty} \frac{\sqrt{n}}{\ln n} = \lim_{n \to \infty} \frac{\left(\frac{1}{2\sqrt{n}}\right)}{\left(\frac{1}{n}\right)} = \lim_{n \to \infty} \frac{\sqrt{n}}{2} = \infty \neq 0$

18. diverges; $\lim_{n \to \infty} a_n = \lim_{n \to \infty} \left(1 + \frac{1}{n}\right)^n = e \neq 0$

19. diverges; a geometric series with $r = \frac{1}{\ln 2} \approx 1.44 > 1$

20. converges; a geometric series with $r = \frac{1}{\ln 3} \approx 0.91 < 1$

21. converges by the Integral Test: $\int_3^{\infty} \frac{\left(\frac{1}{x}\right)}{(\ln x)\sqrt{(\ln x)^2 - 1}}\, dx$; $\left[\begin{array}{c} u = \ln x \\ du = \frac{1}{x}\, dx \end{array}\right] \rightarrow \int_{\ln 3}^{\infty} \frac{1}{u\sqrt{u^2 - 1}}\, du$
$= \lim_{b \to \infty} \left[\sec^{-1} |u|\right]_{\ln 3}^{b} = \lim_{b \to \infty} \left[\sec^{-1} b - \sec^{-1} (\ln 3)\right] = \lim_{b \to \infty} \left[\cos^{-1}\left(\frac{1}{b}\right) - \sec^{-1} (\ln 3)\right]$
$= \cos^{-1}(0) - \sec^{-1}(\ln 3) = \frac{\pi}{2} - \sec^{-1}(\ln 3) \approx 1.1439$

22. converges by the Integral Test: $\int_1^{\infty} \frac{1}{x(1 + \ln^2 x)}\, dx = \int_1^{\infty} \frac{\left(\frac{1}{x}\right)}{1 + (\ln x)^2}\, dx$; $\left[\begin{array}{c} u = \ln x \\ du = \frac{1}{x}\, dx \end{array}\right] \rightarrow \int_0^{\infty} \frac{1}{1+u^2}\, du$
$= \lim_{b \to \infty} \left[\tan^{-1} u\right]_0^b = \lim_{b \to \infty} \left(\tan^{-1} b - \tan^{-1} 0\right) = \frac{\pi}{2} - 0 = \frac{\pi}{2}$

23. diverges by the nth-Term Test for divergence; $\lim_{n \to \infty} n \sin\left(\frac{1}{n}\right) = \lim_{n \to \infty} \frac{\sin\left(\frac{1}{n}\right)}{\left(\frac{1}{n}\right)} = \lim_{x \to 0} \frac{\sin x}{x} = 1 \neq 0$

24. diverges by the nth-Term Test for divergence; $\lim_{n \to \infty} n \tan\left(\frac{1}{n}\right) = \lim_{n \to \infty} \frac{\tan\left(\frac{1}{n}\right)}{\left(\frac{1}{n}\right)} = \lim_{n \to \infty} \frac{\left(-\frac{1}{n^2}\right)\sec^2\left(\frac{1}{n}\right)}{\left(-\frac{1}{n^2}\right)}$
$= \lim_{n \to \infty} \sec^2\left(\frac{1}{n}\right) = \sec^2 0 = 1 \neq 0$

25. converges by the Integral Test: $\int_1^{\infty} \frac{e^x}{1+e^{2x}}\, dx$; $\left[\begin{array}{c} u = e^x \\ du = e^x\, dx \end{array}\right] \rightarrow \int_e^{\infty} \frac{1}{1+u^2}\, du = \lim_{n \to \infty} \left[\tan^{-1} u\right]_e^b$
$= \lim_{b \to \infty} \left(\tan^{-1} b - \tan^{-1} e\right) = \frac{\pi}{2} - \tan^{-1} e \approx 0.35$

26. converges by the Integral Test: $\int_1^{\infty} \frac{2}{1+e^x}\, dx$; $\left[\begin{array}{c} u = e^x \\ du = e^x\, dx \\ dx = \frac{1}{u}\, du \end{array}\right] \rightarrow \int_e^{\infty} \frac{2}{u(1+u)}\, du = \int_e^{\infty} \left(\frac{2}{u} - \frac{2}{u+1}\right) du$
$= \lim_{b \to \infty} \left[2 \ln \frac{u}{u+1}\right]_e^b = \lim_{b \to \infty} 2 \ln\left(\frac{b}{b+1}\right) - 2 \ln\left(\frac{e}{e+1}\right) = 2 \ln 1 - 2 \ln\left(\frac{e}{e+1}\right) = -2 \ln\left(\frac{e}{e+1}\right) \approx 0.63$

27. converges by the Integral Test: $\int_1^{\infty} \frac{8 \tan^{-1} x}{1+x^2}\, dx$; $\left[\begin{array}{c} u = \tan^{-1} x \\ du = \frac{dx}{1+x^2} \end{array}\right] \rightarrow \int_{\pi/4}^{\pi/2} 8u\, du = \left[4u^2\right]_{\pi/4}^{\pi/2} = 4\left(\frac{\pi^2}{4} - \frac{\pi^2}{16}\right) = \frac{3\pi^2}{4}$

28. diverges by the Integral Test: $\int_1^{\infty} \frac{x}{x^2+1}\, dx$; $\left[\begin{array}{c} u = x^2 + 1 \\ du = 2x\, dx \end{array}\right] \rightarrow \frac{1}{2}\int_2^{\infty} \frac{du}{4} = \lim_{b \to \infty} \left[\frac{1}{2} \ln u\right]_2^b$
$= \lim_{b \to \infty} \frac{1}{2}(\ln b - \ln 2) = \infty$

29. converges by the Integral Test: $\int_1^{\infty} \operatorname{sech} x\, dx = 2 \lim_{b \to \infty} \int_1^b \frac{e^x}{1+(e^x)^2}\, dx = 2 \lim_{b \to \infty} \left[\tan^{-1} e^x\right]_1^b$
$= 2 \lim_{b \to \infty} \left(\tan^{-1} e^b - \tan^{-1} e\right) = \pi - 2 \tan^{-1} e \approx 0.71$

30. converges by the Integral Test: $\int_1^{\infty} \operatorname{sech}^2 x\, dx = \lim_{b \to \infty} \int_1^b \operatorname{sech}^2 x\, dx = \lim_{b \to \infty} \left[\tanh x\right]_1^b = \lim_{b \to \infty} (\tanh b - \tanh 1)$
$= 1 - \tanh 1 \approx 0.76$

31. $\int_1^{\infty} \left(\frac{a}{x+2} - \frac{1}{x+4}\right) dx = \lim_{b \to \infty} \left[a \ln |x+2| - \ln |x+4|\right]_1^b = \lim_{b \to \infty} \ln \frac{(b+2)^a}{b+4} - \ln\left(\frac{3^a}{5}\right)$;
$\lim_{b \to \infty} \frac{(b+2)^a}{b+4} = a \lim_{b \to \infty} (b+2)^{a-1} = \begin{cases} \infty, & a > 1 \\ 1, & a = 1 \end{cases}$ $\Rightarrow$ the series converges to $\ln\left(\frac{5}{3}\right)$ if $a = 1$ and diverges to ∞ if $a > 1$. If $a < 1$, the terms of the series eventually become negative and the Integral Test does not apply. From that point on, however, the series behaves like a negative multiple of the harmonic series, and so it diverges.

32. $\int_3^{\infty}\left(\frac{1}{x-1}-\frac{2a}{x+1}\right)dx = \lim_{b\to\infty}\left[\ln\left|\frac{x-1}{(x+1)^{2a}}\right|\right]_3^b = \lim_{b\to\infty}\ln\frac{b-1}{(b+1)^{2a}} - \ln\left(\frac{2}{4^{2a}}\right)$; $\lim_{b\to\infty}\frac{b-1}{(b+1)^{2a}}$

$= \lim_{b\to\infty}\frac{1}{2a(b+1)^{2a-1}} = \begin{cases}1, & a=\frac{1}{2}\\ \infty, & a<\frac{1}{2}\end{cases} \Rightarrow$ the series converges to $\ln\left(\frac{4}{2}\right) = \ln 2$ if $a=\frac{1}{2}$ and diverges to ∞ if

if $a<\frac{1}{2}$. If $a>\frac{1}{2}$, the terms of the series eventually become negative and the Integral Test does not apply. From that point on, however, the series behaves like a negative multiple of the harmonic series, and so it diverges.

33. (a)

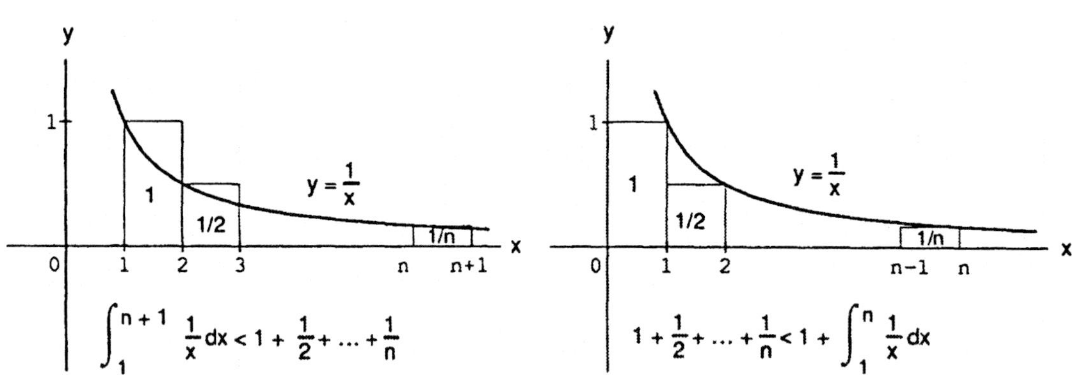

(b) There are $(13)(365)(24)(60)(60)\left(10^9\right)$ seconds in 13 billion years; by part (a) $s_n \le 1+\ln n$ where $n = (13)(365)(24)(60)(60)\left(10^9\right) \Rightarrow s_n \le 1+\ln\left((13)(365)(24)(60)(60)\left(10^9\right)\right)$
$= 1+\ln(13)+\ln(365)+\ln(24)+2\ln(60)+9\ln(10) \approx 41.55$

34. No, because $\sum_{n=1}^{\infty}\frac{1}{nx} = \frac{1}{x}\sum_{n=1}^{\infty}\frac{1}{n}$ and $\sum_{n=1}^{\infty}\frac{1}{n}$ diverges

35. Yes. If $\sum_{n=1}^{\infty} a_n$ is a divergent series of positive numbers, then $\left(\frac{1}{2}\right)\sum_{n=1}^{\infty} a_n = \sum_{n=1}^{\infty}\left(\frac{a_n}{2}\right)$ also diverges and $\frac{a_n}{2} < a_n$.

There is no "smallest" divergent series of positive numbers: for any divergent series $\sum_{n=1}^{\infty} a_n$ of positive numbers $\sum_{n=1}^{\infty}\left(\frac{a_n}{2}\right)$ has smaller terms and still diverges.

36. No, if $\sum_{n=1}^{\infty} a_n$ is a convergent series of positive numbers, then $2\sum_{n=1}^{\infty} a_n = \sum_{n=1}^{\infty} 2a_n$ also converges, and $2a_n \ge a_n$.

There is no "largest" convergent series of positive numbers.

37. Let $A_n = \sum_{k=1}^{n} a_k$ and $B_n = \sum_{k=1}^{n} 2^k a_{\left(2^k\right)}$, where $\{a_k\}$ is a nonincreasing sequence of positive terms converging to 0. Note that $\{A_n\}$ and $\{B_n\}$ are nondecreasing sequences of positive terms. Now,

$B_n = 2a_2 + 4a_4 + 8a_8 + \ldots + 2^n a_{\left(2^n\right)} = 2a_2 + (2a_4+2a_4) + (2a_8+2a_8+2a_8+2a_8) + \ldots$
$+ \underbrace{\left(2a_{\left(2^n\right)} + 2a_{\left(2^n\right)} + \ldots + 2a_{\left(2^n\right)}\right)}_{2^{n-1}\text{ terms}} \le 2a_1 + 2a_2 + (2a_3+2a_4) + (2a_5+2a_6+2a_7+2a_8) + \ldots$

$+ \left(2a_{\left(2^{n-1}\right)} + 2a_{\left(2^{n-1}+1\right)} + \ldots + 2a_{\left(2^n\right)}\right) = 2A_{\left(2^n\right)} \le 2\sum_{k=1}^{\infty} a_k$. Therefore if $\sum a_k$ converges,

then $\{B_n\}$ is bounded above $\Rightarrow \sum 2^k a_{\left(2^k\right)}$ converges. Conversely,

$A_n = a_1 + (a_2+a_3) + (a_4+a_5+a_6+a_7) + \ldots + a_n < a_1 + 2a_2 + 4a_4 + \ldots + 2^n a_{\left(2^n\right)} = a_1 + B_n < a_1 + \sum_{k=1}^{\infty} 2^k a_{\left(2^k\right)}$.

Therefore, if $\sum_{k=1}^{\infty} 2^k a_{(2^k)}$ converges, then $\{A_n\}$ is bounded above and hence converges.

38. (a) $a_{(2^n)} = \frac{1}{2^n \ln(2^n)} = \frac{1}{2^n \cdot n(\ln 2)} \Rightarrow \sum_{n=2}^{\infty} 2^n a_{(2^n)} = \sum_{n=2}^{\infty} 2^n \frac{1}{2^n \cdot n(\ln 2)} = \frac{1}{\ln 2} \sum_{n=2}^{\infty} \frac{1}{n}$, which diverges $\Rightarrow \sum_{n=2}^{\infty} \frac{1}{n \ln n}$ diverges.

(b) $a_{(2^n)} = \frac{1}{2^{np}} \Rightarrow \sum_{n=1}^{\infty} 2^n a_{(2^n)} = \sum_{n=1}^{\infty} 2^n \cdot \frac{1}{2^{np}} = \sum_{n=1}^{\infty} \frac{1}{(2^n)^{p-1}} = \sum_{n=1}^{\infty} \left(\frac{1}{2^{p-1}}\right)^n$, a geometric series that converges if $\frac{1}{2^{p-1}} < 1$ or $p > 1$, but diverges if $p \le 1$.

39. (a) $\int_2^{\infty} \frac{dx}{x(\ln x)^p}$; $\left[\begin{array}{l} u = \ln x \\ du = \frac{dx}{x} \end{array}\right] \to \int_{\ln 2}^{\infty} u^{-p}\, du = \lim_{b \to \infty} \left[\frac{u^{-p+1}}{-p+1}\right]_{\ln 2}^{b} = \lim_{b \to \infty} \left(\frac{1}{1-p}\right)\left[b^{-p+1} - (\ln 2)^{-p+1}\right]$

$= \begin{cases} \frac{1}{p-1}(\ln 2)^{-p+1}, & p > 1 \\ \infty, & p < 1 \end{cases} \Rightarrow$ the improper integral converges if $p > 1$ and diverges if $p < 1$. For $p = 1$: $\int_2^{\infty} \frac{dx}{x \ln x} = \lim_{b \to \infty} \left[\ln(\ln x)\right]_2^b = \lim_{b \to \infty} \left[\ln(\ln b) - \ln(\ln 2)\right] = \infty$, so the improper integral diverges if $p = 1$.

(b) Since the series and the integral converge or diverge together, $\sum_{n=2}^{\infty} \frac{1}{n(\ln n)^p}$ converges if and only if $p > 1$.

40. (a) $p = 1 \Rightarrow$ the series diverges

(b) $p = 1.01 \Rightarrow$ the series converges

(c) $\sum_{n=2}^{\infty} \frac{1}{n(\ln n^3)} = \frac{1}{3} \sum_{n=2}^{\infty} \frac{1}{n(\ln n)}$; $p = 1 \Rightarrow$ the series diverges

(d) $p = 3 \Rightarrow$ the series converges

41. (a) From Fig. 11.8 in the text with $f(x) = \frac{1}{x}$ and $a_k = \frac{1}{k}$, we have $\int_1^{n+1} \frac{1}{x}\, dx \le 1 + \frac{1}{2} + \frac{1}{3} + \ldots + \frac{1}{n}$ $\le 1 + \int_1^n f(x)\, dx \Rightarrow \ln(n+1) \le 1 + \frac{1}{2} + \frac{1}{3} + \ldots + \frac{1}{n} \le 1 + \ln n \Rightarrow 0 \le \ln(n+1) - \ln n$ $\le \left(1 + \frac{1}{2} + \frac{1}{3} + \ldots + \frac{1}{n}\right) - \ln n \le 1$. Therefore the sequence $\left\{\left(1 + \frac{1}{2} + \frac{1}{3} + \ldots + \frac{1}{n}\right) - \ln n\right\}$ is bounded above by 1 and below by 0.

(b) From the graph in Fig. 11.8(a) with $f(x) = \frac{1}{x}$, $\frac{1}{n+1} < \int_n^{n+1} \frac{1}{x}\, dx = \ln(n+1) - \ln n$ $\Rightarrow 0 > \frac{1}{n+1} - [\ln(n+1) - \ln n] = \left(1 + \frac{1}{2} + \frac{1}{3} + \ldots + \frac{1}{n+1} - \ln(n+1)\right) - \left(1 + \frac{1}{2} + \frac{1}{3} + \ldots + \frac{1}{n} - \ln n\right)$. If we define $a_n = 1 + \frac{1}{2} = \frac{1}{3} + \frac{1}{n} - \ln n$, then $0 > a_{n+1} - a_n \Rightarrow a_{n+1} < a_n \Rightarrow \{a_n\}$ is a decreasing sequence of nonnegative terms.

42. $e^{-x^2} \le e^{-x}$ for $x \ge 1$, and $\int_1^{\infty} e^{-x}\, dx = \lim_{b \to \infty} \left[-e^{-x}\right]_1^b = \lim_{b \to \infty} \left(-e^{-b} + e^{-1}\right) = e^{-1} \Rightarrow \int_1^{\infty} e^{-x^2}\, dx$ converges by the Comparison Test for improper integrals $\Rightarrow \sum_{n=0}^{\infty} e^{-n^2} = 1 + \sum_{n=1}^{\infty} e^{-n^2}$ converges by the Integral Test.

11.4 COMPARISON TESTS

1. diverges by the Limit Comparison Test (part 1) when compared with $\sum_{n=1}^{\infty} \frac{1}{\sqrt{n}}$, a divergent p-series:

$$\lim_{n \to \infty} \frac{\left(\frac{1}{2\sqrt{n} + \sqrt[3]{n}}\right)}{\left(\frac{1}{\sqrt{n}}\right)} = \lim_{n \to \infty} \frac{\sqrt{n}}{2\sqrt{n} + \sqrt[3]{n}} = \lim_{n \to \infty} \left(\frac{1}{2 + n^{-1/6}}\right) = \frac{1}{2}$$

2. diverges by the Direct Comparison Test since $n + n + n > n + \sqrt{n} + 0 \Rightarrow \frac{3}{n+\sqrt{n}} > \frac{1}{n}$, which is the nth term of the divergent series $\sum_{n=1}^{\infty} \frac{1}{n}$ or use Limit Comparison Test with $b_n = \frac{1}{n}$

3. converges by the Direct Comparison Test; $\frac{\sin^2 n}{2^n} \le \frac{1}{2^n}$, which is the nth term of a convergent geometric series

4. converges by the Direct Comparison Test; $\frac{1+\cos n}{n^2} \le \frac{2}{n^2}$ and the p-series $\sum \frac{1}{n^2}$ converges

5. diverges since $\lim_{n\to\infty} \frac{2n}{3n-1} = \frac{2}{3} \neq 0$

6. converges by the Limit Comparison Test (part 1) with $\frac{1}{n^{3/2}}$, the nth term of a convergent p-series:
$$\lim_{n\to\infty} \frac{\left(\frac{n+1}{n^2\sqrt{n}}\right)}{\left(\frac{1}{n^{3/2}}\right)} = \lim_{n\to\infty} \left(\frac{n+1}{n}\right) = 1$$

7. converges by the Direct Comparison Test; $\left(\frac{n}{3n+1}\right)^n < \left(\frac{n}{3n}\right)^n = \left(\frac{1}{3}\right)^n$, the nth term of a convergent geometric series

8. converges by the Limit Comparison Test (part 1) with $\frac{1}{n^{3/2}}$, the nth term of a convergent p-series:
$$\lim_{n\to\infty} \frac{\left(\frac{1}{n^{3/2}}\right)}{\left(\frac{1}{\sqrt{n^3+2}}\right)} = \lim_{n\to\infty} \sqrt{\frac{n^3+2}{n^3}} = \lim_{n\to\infty} \sqrt{1+\frac{2}{n^3}} = 1$$

9. diverges by the Direct Comparison Test; $n > \ln n \Rightarrow \ln n > \ln \ln n \Rightarrow \frac{1}{n} < \frac{1}{\ln n} < \frac{1}{\ln (\ln n)}$ and $\sum_{n=3}^{\infty} \frac{1}{n}$ diverges

10. diverges by the Limit Comparison Test (part 3) when compared with $\sum_{n=2}^{\infty} \frac{1}{n}$, a divergent p-series:
$$\lim_{n\to\infty} \frac{\left(\frac{1}{(\ln n)^2}\right)}{\left(\frac{1}{n}\right)} = \lim_{n\to\infty} \frac{n}{(\ln n)^2} = \lim_{n\to\infty} \frac{1}{2(\ln n)\left(\frac{1}{n}\right)} = \frac{1}{2}\lim_{n\to\infty} \frac{n}{\ln n} = \frac{1}{2}\lim_{n\to\infty} \frac{1}{\left(\frac{1}{n}\right)} = \frac{1}{2}\lim_{n\to\infty} n = \infty$$

11. converges by the Limit Comparison Test (part 2) when compared with $\sum_{n=1}^{\infty} \frac{1}{n^2}$, a convergent p-series:
$$\lim_{n\to\infty} \frac{\left[\frac{(\ln n)^2}{n^3}\right]}{\left(\frac{1}{n^2}\right)} = \lim_{n\to\infty} \frac{(\ln n)^2}{n} = \lim_{n\to\infty} \frac{2(\ln n)\left(\frac{1}{n}\right)}{1} = 2\lim_{n\to\infty} \frac{\ln n}{n} = 0$$

12. converges by the Limit Comparison Test (part 2) when compared with $\sum_{n=1}^{\infty} \frac{1}{n^2}$, a convergent p-series:
$$\lim_{n\to\infty} \frac{\left[\frac{(\ln n)^3}{n^3}\right]}{\left(\frac{1}{n^2}\right)} = \lim_{n\to\infty} \frac{(\ln n)^3}{n} = \lim_{n\to\infty} \frac{3(\ln n)^2\left(\frac{1}{n}\right)}{1} = 3\lim_{n\to\infty} \frac{(\ln n)^2}{n} = 3\lim_{n\to\infty} \frac{2(\ln n)\left(\frac{1}{n}\right)}{1} = 6\lim_{n\to\infty} \frac{\ln n}{n}$$
$$= 6 \cdot 0 = 0$$

13. diverges by the Limit Comparison Test (part 3) with $\frac{1}{n}$, the nth term of the divergent harmonic series:
$$\lim_{n\to\infty} \frac{\left[\frac{1}{\sqrt{n}\ln n}\right]}{\left(\frac{1}{n}\right)} = \lim_{n\to\infty} \frac{\sqrt{n}}{\ln n} = \lim_{n\to\infty} \frac{\left(\frac{1}{2\sqrt{n}}\right)}{\left(\frac{1}{n}\right)} = \lim_{n\to\infty} \frac{\sqrt{n}}{2} = \infty$$

14. converges by the Limit Comparison Test (part 2) with $\frac{1}{n^{5/4}}$, the nth term of a convergent p-series:

$$\lim_{n\to\infty} \frac{\left[\frac{(\ln n)^2}{n^{3/2}}\right]}{\left(\frac{1}{n^{5/4}}\right)} = \lim_{n\to\infty} \frac{(\ln n)^2}{n^{1/4}} = \lim_{n\to\infty} \frac{\left(\frac{2\ln n}{n}\right)}{\left(\frac{1}{4n^{3/4}}\right)} = 8 \lim_{n\to\infty} \frac{\ln n}{n^{1/4}} = 8 \lim_{n\to\infty} \frac{\left(\frac{1}{n}\right)}{\left(\frac{1}{4n^{3/4}}\right)} = 32 \lim_{n\to\infty} \frac{1}{n^{1/4}} = 32 \cdot 0 = 0$$

15. diverges by the Limit Comparison Test (part 3) with $\frac{1}{n}$, the nth term of the divergent harmonic series:

$$\lim_{n\to\infty} \frac{\left(\frac{1}{1+\ln n}\right)}{\left(\frac{1}{n}\right)} = \lim_{n\to\infty} \frac{n}{1+\ln n} = \lim_{n\to\infty} \frac{1}{\left(\frac{1}{n}\right)} = \lim_{n\to\infty} n = \infty$$

16. diverges by the Limit Comparison Test (part 3) with $\frac{1}{n}$, the nth term of the divergent harmonic series:

$$\lim_{n\to\infty} \frac{\left(\frac{1}{(1+\ln n)^2}\right)}{\left(\frac{1}{n}\right)} = \lim_{n\to\infty} \frac{n}{(1+\ln n)^2} = \lim_{n\to\infty} \frac{1}{\left[\frac{2(1+\ln n)}{n}\right]} = \lim_{n\to\infty} \frac{n}{2(1+\ln n)} = \lim_{n\to\infty} \frac{1}{\left(\frac{2}{n}\right)} = \lim_{n\to\infty} \frac{n}{2} = \infty$$

17. diverges by the Integral Test: $\int_2^\infty \frac{\ln(x+1)}{x+1}\,dx = \int_{\ln 3}^\infty u\,du = \lim_{b\to\infty} \left[\frac{1}{2}u^2\right]_{\ln 3}^{b} = \lim_{b\to\infty} \frac{1}{2}\left(b^2 - \ln^2 3\right) = \infty$

18. diverges by the Limit Comparison Test (part 3) with $\frac{1}{n}$, the nth term of the divergent harmonic series:

$$\lim_{n\to\infty} \frac{\left(\frac{1}{1+\ln^2 n}\right)}{\left(\frac{1}{n}\right)} = \lim_{n\to\infty} \frac{n}{1+\ln^2 n} = \lim_{n\to\infty} \frac{1}{\left(\frac{2\ln n}{n}\right)} = \lim_{n\to\infty} \frac{n}{2\ln n} = \lim_{n\to\infty} \frac{1}{\left(\frac{2}{n}\right)} = \lim_{n\to\infty} \frac{n}{2} = \infty$$

19. converges by the Direct Comparison Test with $\frac{1}{n^{3/2}}$, the nth term of a convergent p-series: $n^2 - 1 > n$ for $n \ge 2 \Rightarrow n^2(n^2-1) > n^3 \Rightarrow n\sqrt{n^2-1} > n^{3/2} \Rightarrow \frac{1}{n^{3/2}} > \frac{1}{n\sqrt{n^2-1}}$ or use Limit Comparison Test with $\frac{1}{n^2}$.

20. converges by the Direct Comparison Test with $\frac{1}{n^{3/2}}$, the nth term of a convergent p-series: $n^2 + 1 > n^2$ $\Rightarrow n^2 + 1 > \sqrt{n}n^{3/2} \Rightarrow \frac{n^2+1}{\sqrt{n}} > n^{3/2} \Rightarrow \frac{\sqrt{n}}{n^2+1} < \frac{1}{n^{3/2}}$ or use Limit Comparison Test with $\frac{1}{n^{3/2}}$.

21. converges because $\sum_{n=1}^{\infty} \frac{1-n}{n2^n} = \sum_{n=1}^{\infty} \frac{1}{n2^n} + \sum_{n=1}^{\infty} \frac{-1}{2^n}$ which is the sum of two convergent series: $\sum_{n=1}^{\infty} \frac{1}{n2^n}$ converges by the Direct Comparison Test since $\frac{1}{n2^n} < \frac{1}{2^n}$, and $\sum_{n=1}^{\infty} \frac{-1}{2^n}$ is a convergent geometric series

22. converges by the Direct Comparison Test: $\sum_{n=1}^{\infty} \frac{n+2^n}{n^2 2^n} = \sum_{n=1}^{\infty} \left(\frac{1}{n2^n} + \frac{1}{n^2}\right)$ and $\frac{1}{n2^n} + \frac{1}{n^2} \le \frac{1}{2^n} + \frac{1}{n^2}$, the sum of the nth terms of a convergent geometric series and a convergent p-series

23. converges by the Direct Comparison Test: $\frac{1}{3^{n-1}+1} < \frac{1}{3^{n-1}}$, which is the nth term of a convergent geometric series

24. diverges; $\lim_{n\to\infty} \left(\frac{3^{n-1}+1}{3^n}\right) = \lim_{n\to\infty} \left(\frac{1}{3} + \frac{1}{3^n}\right) = \frac{1}{3} \ne 0$

25. diverges by the Limit Comparison Test (part 1) with $\frac{1}{n}$, the nth term of the divergent harmonic series:

$$\lim_{n\to\infty} \frac{\left(\sin\frac{1}{n}\right)}{\left(\frac{1}{n}\right)} = \lim_{x\to 0} \frac{\sin x}{x} = 1$$

26. diverges by the Limit Comparison Test (part 1) with $\frac{1}{n}$, the nth term of the divergent harmonic series:

$$\lim_{n \to \infty} \frac{\left(\tan \frac{1}{n}\right)}{\left(\frac{1}{n}\right)} = \lim_{n \to \infty} \left(\frac{1}{\cos \frac{1}{n}}\right) \frac{\left(\sin \frac{1}{n}\right)}{\left(\frac{1}{n}\right)} = \lim_{x \to 0} \left(\frac{1}{\cos x}\right)\left(\frac{\sin x}{x}\right) = 1 \cdot 1 = 1$$

27. converges by the Limit Comparison Test (part 1) with $\frac{1}{n^2}$, the nth term of a convergent p-series:

$$\lim_{n \to \infty} \frac{\left(\frac{10n+1}{n(n+1)(n+2)}\right)}{\left(\frac{1}{n^2}\right)} = \lim_{n \to \infty} \frac{10n^2+n}{n^2+3n+2} = \lim_{n \to \infty} \frac{20n+1}{2n+3} = \lim_{n \to \infty} \frac{20}{2} = 10$$

28. converges by the Limit Comparison Test (part 1) with $\frac{1}{n^2}$, the nth term of a convergent p-series:

$$\lim_{n \to \infty} \frac{\left(\frac{5n^3-3n}{n^2(n-2)\left(n^2+5\right)}\right)}{\left(\frac{1}{n^2}\right)} = \lim_{n \to \infty} \frac{5n^3-3n}{n^3-2n^2+5n-10} = \lim_{n \to \infty} \frac{15n^2-3}{3n^2-4n+5} = \lim_{n \to \infty} \frac{30n}{6n-4} = 5$$

29. converges by the Direct Comparison Test: $\frac{\tan^{-1} n}{n^{1.1}} < \frac{\frac{\pi}{2}}{n^{1.1}}$ and $\sum_{n=1}^{\infty} \frac{\frac{\pi}{2}}{n^{1.1}} = \frac{\pi}{2} \sum_{n=1}^{\infty} \frac{1}{n^{1.1}}$ is the product of a convergent p-series and a nonzero constant

30. converges by the Direct Comparison Test: $\sec^{-1} n < \frac{\pi}{2} \Rightarrow \frac{\sec^{-1} n}{n^{1.3}} < \frac{\left(\frac{\pi}{2}\right)}{n^{1.3}}$ and $\sum_{n=1}^{\infty} \frac{\left(\frac{\pi}{2}\right)}{n^{1.3}} = \frac{\pi}{2} \sum_{n=1}^{\infty} \frac{1}{n^{1.3}}$ is the product of a convergent p-series and a nonzero constant

31. converges by the Limit Comparison Test (part 1) with $\frac{1}{n^2}$: $\lim_{n \to \infty} \frac{\left(\frac{\coth n}{n^2}\right)}{\left(\frac{1}{n^2}\right)} = \lim_{n \to \infty} \coth n = \lim_{n \to \infty} \frac{e^n+e^{-n}}{e^n-e^{-n}}$ $= \lim_{n \to \infty} \frac{1+e^{-2n}}{1-e^{-2n}} = 1$

32. converges by the Limit Comparison Test (part 1) with $\frac{1}{n^2}$: $\lim_{n \to \infty} \frac{\left(\frac{\tanh n}{n^2}\right)}{\left(\frac{1}{n^2}\right)} = \lim_{n \to \infty} \tanh n = \lim_{n \to \infty} \frac{e^n-e^{-n}}{e^n+e^{-n}}$ $= \lim_{n \to \infty} \frac{1-e^{-2n}}{1+e^{-2n}} = 1$

33. diverges by the Limit Comparison Test (part 1) with $\frac{1}{n}$: $\lim_{n \to \infty} \frac{\left(\frac{1}{n\sqrt[n]{n}}\right)}{\left(\frac{1}{n}\right)} = \lim_{n \to \infty} \frac{1}{\sqrt[n]{n}} = 1.$

34. converges by the Limit Comparison Test (part 1) with $\frac{1}{n^2}$: $\lim_{n \to \infty} \frac{\left(\frac{\sqrt[n]{n}}{n^2}\right)}{\left(\frac{1}{n^2}\right)} = \lim_{n \to \infty} \sqrt[n]{n} = 1$

35. $\frac{1}{1+2+3+\ldots+n} = \frac{1}{\left(\frac{n(n+1)}{2}\right)} = \frac{2}{n(n+1)}$. The series converges by the Limit Comparison Test (part 1) with $\frac{1}{n^2}$:

$$\lim_{n \to \infty} \frac{\left(\frac{2}{n(n+1)}\right)}{\left(\frac{1}{n^2}\right)} = \lim_{n \to \infty} \frac{2n^2}{n^2+n} = \lim_{n \to \infty} \frac{4n}{2n+1} = \lim_{n \to \infty} \frac{4}{2} = 2.$$

36. $\frac{1}{1+2^2+3^2+\ldots+n^2} = \frac{1}{\frac{n(n+1)(2n+1)}{6}} = \frac{6}{n(n+1)(2n+1)} \le \frac{6}{n^3} \Rightarrow$ the series converges by the Direct Comparison Test

37. (a) If $\lim_{n \to \infty} \frac{a_n}{b_n} = 0$, then there exists an integer N such that for all $n > N$, $\left|\frac{a_n}{b_n} - 0\right| < 1 \Rightarrow -1 < \frac{a_n}{b_n} < 1$ $\Rightarrow a_n < b_n$. Thus, if $\sum b_n$ converges, then $\sum a_n$ converges by the Direct Comparison Test.

(b) If $\lim_{n \to \infty} \frac{a_n}{b_n} = \infty$, then there exists an integer N such that for all $n > N$, $\frac{a_n}{b_n} > 1 \Rightarrow a_n > b_n$. Thus, if $\sum b_n$ diverges, then $\sum a_n$ diverges by the Direct Comparison Test.

38. Yes, $\sum_{n=1}^{\infty} \frac{a_n}{n}$ converges by the Direct Comparison Test because $\frac{a_n}{n} < a_n$

39. $\lim_{n \to \infty} \frac{a_n}{b_n} = \infty \Rightarrow$ there exists an integer N such that for all $n > N$, $\frac{a_n}{b_n} > 1 \Rightarrow a_n > b_n$. If $\sum a_n$ converges, then $\sum b_n$ converges by the Direct Comparison Test

40. $\sum a_n$ converges $\Rightarrow \lim_{n \to \infty} a_n = 0 \Rightarrow$ there exists an integer N such that for all $n > N$, $0 \le a_n < 1 \Rightarrow a_n^2 < a_n$ $\Rightarrow \sum a_n^2$ converges by the Direct Comparison Test

41. Example CAS commands:

<u>Maple</u>:

```
a := n -> 1./n^3/sin(n)^2;
s := k -> sum( a(n), n=1..k );                                   # (a)]
limit( s(k), k=infinity );
pts := [seq( [k,s(k)], k=1..100 )]:                              # (b)
plot( pts, style=point, title="#41(b) (Section 11.4)" );
pts := [seq( [k,s(k)], k=1..200 )]:                              # (c)
plot( pts, style=point, title="#41(c) (Section 11.4)" );
pts := [seq( [k,s(k)], k=1..400 )]:                              # (d)
plot( pts, style=point, title="#41(d) (Section 11.4)" );
evalf( 355/113 );
```

<u>Mathematica</u>:

```
Clear[a, n, s, k, p]
a[n_]:= 1 / ( n^3 Sin[n]^2 )
s[k_]= Sum[ a[n], {n, 1, k}]
points[p_]:= Table[{k, N[s[k]]}, {k, 1, p}]
points[100]
ListPlot[points[100]]
points[200]
ListPlot[points[200]
points[400]
ListPlot[points[400], PlotRange → All]
```

To investigate what is happening around k = 355, you could do the following.

```
N[355/113]
N[π − 355/113]
Sin[355]//N
a[355]//N
N[s[354]]
N[s[355]]
N[s[356]]
```

11.5 THE RATIO AND ROOT TESTS

1. converges by the Ratio Test: $\lim\limits_{n\to\infty} \frac{a_{n+1}}{a_n} = \lim\limits_{n\to\infty} \frac{\left[\frac{(n+1)^{\sqrt{2}}}{2^{n+1}}\right]}{\left[\frac{n^{\sqrt{2}}}{2^n}\right]} = \lim\limits_{n\to\infty} \frac{(n+1)^{\sqrt{2}}}{2^{n+1}} \cdot \frac{2^n}{n^{\sqrt{2}}}$

$= \lim\limits_{n\to\infty} \left(1+\frac{1}{n}\right)^{\sqrt{2}} \left(\frac{1}{2}\right) = \frac{1}{2} < 1$

2. converges by the Ratio Test: $\lim\limits_{n\to\infty} \frac{a_{n+1}}{a_n} = \lim\limits_{n\to\infty} \frac{\left(\frac{(n+1)^2}{e^{n+1}}\right)}{\left(\frac{n^2}{e^n}\right)} = \lim\limits_{n\to\infty} \frac{(n+1)^2}{e^{n+1}} \cdot \frac{e^n}{n^2} = \lim\limits_{n\to\infty} \left(1+\frac{1}{n}\right)^2 \left(\frac{1}{e}\right) = \frac{1}{e} < 1$

3. diverges by the Ratio Test: $\lim\limits_{n\to\infty} \frac{a_{n+1}}{a_n} = \lim\limits_{n\to\infty} \frac{\left(\frac{(n+1)!}{e^{n+1}}\right)}{\left(\frac{n!}{e^n}\right)} = \lim\limits_{n\to\infty} \frac{(n+1)!}{e^{n+1}} \cdot \frac{e^n}{n!} = \lim\limits_{n\to\infty} \frac{n+1}{e} = \infty$

4. diverges by the Ratio Test: $\lim\limits_{n\to\infty} \frac{a_{n+1}}{a_n} = \lim\limits_{n\to\infty} \frac{\left(\frac{(n+1)!}{10^{n+1}}\right)}{\left(\frac{n!}{10^n}\right)} = \lim\limits_{n\to\infty} \frac{(n+1)!}{10^{n+1}} \cdot \frac{10^n}{n!} = \lim\limits_{n\to\infty} \frac{n}{10} = \infty$

5. converges by the Ratio Test: $\lim\limits_{n\to\infty} \frac{a_{n+1}}{a_n} = \lim\limits_{n\to\infty} \frac{\left(\frac{(n+1)^{10}}{10^{n+1}}\right)}{\left(\frac{n^{10}}{10^n}\right)} = \lim\limits_{n\to\infty} \frac{(n+1)^{10}}{10^{n+1}} \cdot \frac{10^n}{n^{10}} = \lim\limits_{n\to\infty} \left(1+\frac{1}{n}\right)^{10} \left(\frac{1}{10}\right)$

$= \frac{1}{10} < 1$

6. diverges; $\lim\limits_{n\to\infty} a_n = \lim\limits_{n\to\infty} \left(\frac{n-2}{n}\right)^n = \lim\limits_{n\to\infty} \left(1+\frac{-2}{n}\right)^n = e^{-2} \neq 0$

7. converges by the Direct Comparison Test: $\frac{2+(-1)^n}{(1.25)^n} = \left(\frac{4}{5}\right)^n [2+(-1)^n] \le \left(\frac{4}{5}\right)^n(3)$ which is the nth term of a convergent geometric series

8. converges; a geometric series with $|r| = \left|-\frac{2}{3}\right| < 1$

9. diverges; $\lim\limits_{n\to\infty} a_n = \lim\limits_{n\to\infty} \left(1-\frac{3}{n}\right)^n = \lim\limits_{n\to\infty} \left(1+\frac{-3}{n}\right)^n = e^{-3} \approx 0.05 \neq 0$

10. diverges; $\lim\limits_{n\to\infty} a_n = \lim\limits_{n\to\infty} \left(1-\frac{1}{3n}\right)^n = \lim\limits_{n\to\infty} \left(1+\frac{\left(-\frac{1}{3}\right)}{n}\right)^n = e^{-1/3} \approx 0.72 \neq 0$

11. converges by the Direct Comparison Test: $\frac{\ln n}{n^3} < \frac{n}{n^3} = \frac{1}{n^2}$ for $n \ge 2$, the nth term of a convergent p-series.

12. converges by the nth-Root Test: $\lim\limits_{n\to\infty} \sqrt[n]{a_n} = \lim\limits_{n\to\infty} \sqrt[n]{\frac{(\ln n)^n}{n^n}} = \lim\limits_{n\to\infty} \frac{((\ln n)^n)^{1/n}}{(n^n)^{1/n}} = \lim\limits_{n\to\infty} \frac{\ln n}{n}$

$= \lim\limits_{n\to\infty} \frac{\left(\frac{1}{n}\right)}{1} = 0 < 1$

13. diverges by the Direct Comparison Test: $\frac{1}{n} - \frac{1}{n^2} = \frac{n-1}{n^2} > \frac{1}{2}\left(\frac{1}{n}\right)$ for $n > 2$ or by the Limit Comparison Test (part 1) with $\frac{1}{n}$.

14. converges by the nth-Root Test: $\lim\limits_{n\to\infty} \sqrt[n]{a_n} = \lim\limits_{n\to\infty} \sqrt[n]{\left(\frac{1}{n}-\frac{1}{n^2}\right)^n} = \lim\limits_{n\to\infty} \left(\left(\frac{1}{n}-\frac{1}{n^2}\right)^n\right)^{1/n}$

$= \lim\limits_{n\to\infty} \left(\frac{1}{n}-\frac{1}{n^2}\right) = 0 < 1$

15. diverges by the Direct Comparison Test: $\frac{\ln n}{n} > \frac{1}{n}$ for $n \geq 3$

16. converges by the Ratio Test: $\lim_{n \to \infty} \frac{a_{n+1}}{a_n} = \lim_{n \to \infty} \frac{(n+1)\ln(n+1)}{2^{n+1}} \cdot \frac{2^n}{n \ln(n)} = \frac{1}{2} < 1$

17. converges by the Ratio Test: $\lim_{n \to \infty} \frac{a_{n+1}}{a_n} = \lim_{n \to \infty} \frac{(n+2)(n+3)}{(n+1)!} \cdot \frac{n!}{(n+1)(n+2)} = 0 < 1$

18. converges by the Ratio Test: $\lim_{n \to \infty} \frac{a_{n+1}}{a_n} = \lim_{n \to \infty} \frac{(n+1)^3}{e^{n+1}} \cdot \frac{e^n}{n^3} = \frac{1}{e} < 1$

19. converges by the Ratio Test: $\lim_{n \to \infty} \frac{a_{n+1}}{a_n} = \lim_{n \to \infty} \frac{(n+4)!}{3!\,(n+1)!\,3^{n+1}} \cdot \frac{3!\,n!\,3^n}{(n+3)!} = \lim_{n \to \infty} \frac{n+4}{3(n+1)} = \frac{1}{3} < 1$

20. converges by the Ratio Test: $\lim_{n \to \infty} \frac{a_{n+1}}{a_n} = \lim_{n \to \infty} \frac{(n+1)2^{n+1}(n+2)!}{3^{n+1}(n+1)!} \cdot \frac{3^n n!}{n2^n(n+1)!}$
$= \lim_{n \to \infty} \left(\frac{n+1}{n}\right)\left(\frac{2}{3}\right)\left(\frac{n+2}{n+1}\right) = \frac{2}{3} < 1$

21. converges by the Ratio Test: $\lim_{n \to \infty} \frac{a_{n+1}}{a_n} = \lim_{n \to \infty} \frac{(n+1)!}{(2n+3)!} \cdot \frac{(2n+1)!}{n!} = \lim_{n \to \infty} \frac{n+1}{(2n+3)(2n+2)} = 0 < 1$

22. converges by the Ratio Test: $\lim_{n \to \infty} \frac{a_{n+1}}{a_n} = \lim_{n \to \infty} \frac{(n+1)!}{(n+1)^{n+1}} \cdot \frac{n^n}{n!} = \lim_{n \to \infty} \left(\frac{n}{n+1}\right)^n = \lim_{n \to \infty} \frac{1}{\left(\frac{n+1}{n}\right)^n}$
$= \lim_{n \to \infty} \frac{1}{\left(1+\frac{1}{n}\right)^n} = \frac{1}{e} < 1$

23. converges by the Root Test: $\lim_{n \to \infty} \sqrt[n]{a_n} = \lim_{n \to \infty} \sqrt[n]{\frac{n}{(\ln n)^n}} = \lim_{n \to \infty} \frac{\sqrt[n]{n}}{\ln n} = \lim_{n \to \infty} \frac{1}{\ln n} = 0 < 1$

24. converges by the Root Test: $\lim_{n \to \infty} \sqrt[n]{a_n} = \lim_{n \to \infty} \sqrt[n]{\frac{n}{(\ln n)^{n/2}}} = \lim_{n \to \infty} \frac{\sqrt[n]{n}}{\sqrt{\ln n}} = \frac{\lim_{n\to\infty} \sqrt[n]{n}}{\lim_{n\to\infty} \sqrt{\ln n}} = 0 < 1$
$\left(\lim_{n \to \infty} \sqrt[n]{n} = 1\right)$

25. converges by the Direct Comparison Test: $\frac{n!\,\ln n}{n(n+2)!} = \frac{\ln n}{n(n+1)(n+2)} < \frac{n}{n(n+1)(n+2)} = \frac{1}{(n+1)(n+2)} < \frac{1}{n^2}$
which is the nth-term of a convergent p-series

26. diverges by the Ratio Test: $\lim_{n \to \infty} \frac{a_{n+1}}{a_n} = \lim_{n \to \infty} \frac{3^{n+1}}{(n+1)^3\,2^{n+1}} \cdot \frac{n^3 2^n}{3^n} = \lim_{n \to \infty} \frac{n^3}{(n+1)^3}\left(\frac{3}{2}\right) = \frac{3}{2} > 1$

27. converges by the Ratio Test: $\lim_{n \to \infty} \frac{a_{n+1}}{a_n} = \lim_{n \to \infty} \frac{\left(\frac{1+\sin n}{n}\right) a_n}{a_n} = 0 < 1$

28. converges by the Ratio Test: $\lim_{n \to \infty} \frac{a_{n+1}}{a_n} = \lim_{n \to \infty} \frac{\left(\frac{1+\tan^{-1} n}{n}\right) a_n}{a_n} = \lim_{n \to \infty} \frac{1+\tan^{-1} n}{n} = 0$ since the numerator approaches $1 + \frac{\pi}{2}$ while the denominator tends to ∞

29. diverges by the Ratio Test: $\lim_{n \to \infty} \frac{a_{n+1}}{a_n} = \lim_{n \to \infty} \frac{\left(\frac{3n-1}{2n+1}\right) a_n}{a_n} = \lim_{n \to \infty} \frac{3n-1}{2n+1} = \frac{3}{2} > 1$

30. diverges; $a_{n+1} = \frac{n}{n+1} a_n \Rightarrow a_{n+1} = \left(\frac{n}{n+1}\right)\left(\frac{n-1}{n} a_{n-1}\right) \Rightarrow a_{n+1} = \left(\frac{n}{n+1}\right)\left(\frac{n-1}{n}\right)\left(\frac{n-2}{n-1} a_{n-2}\right)$
$\Rightarrow a_{n+1} = \left(\frac{n}{n+1}\right)\left(\frac{n-1}{n}\right)\left(\frac{n-2}{n-1}\right)\cdots\left(\frac{1}{2}\right) a_1 \Rightarrow a_{n+1} = \frac{a_1}{n+1} \Rightarrow a_{n+1} = \frac{3}{n+1}$, which is a constant times the general term of the diverging harmonic series

31. converges by the Ratio Test: $\lim_{n \to \infty} \frac{a_{n+1}}{a_n} = \lim_{n \to \infty} \frac{\left(\frac{2}{n}\right) a_n}{a_n} = \lim_{n \to \infty} \frac{2}{n} = 0 < 1$

32. converges by the Ratio Test: $\lim_{n\to\infty} \frac{a_{n+1}}{a_n} = \lim_{n\to\infty} \frac{\left(\frac{\sqrt[n]{n}}{2}\right) a_n}{a_n} = \lim_{n\to\infty} \frac{\sqrt[n]{n}}{n} = \frac{1}{2} < 1$

33. converges by the Ratio Test: $\lim_{n\to\infty} \frac{a_{n+1}}{a_n} = \lim_{n\to\infty} \frac{\left(\frac{1+\ln n}{n}\right) a_n}{a_n} = \lim_{n\to\infty} \frac{1+\ln n}{n} = \lim_{n\to\infty} \frac{1}{n} = 0 < 1$

34. $\frac{n+\ln n}{n+10} > 0$ and $a_1 = \frac{1}{2} \Rightarrow a_n > 0$; $\ln n > 10$ for $n > e^{10} \Rightarrow n + \ln n > n + 10 \Rightarrow \frac{n+\ln n}{n+10} > 1$
$\Rightarrow a_{n+1} = \frac{n+\ln n}{n+10} a_n > a_n$; thus $a_{n+1} > a_n \ge \frac{1}{2} \Rightarrow \lim_{n\to\infty} a_n \ne 0$, so the series diverges by the nth-Term Test

35. diverges by the nth-Term Test: $a_1 = \frac{1}{3}$, $a_2 = \sqrt[2]{\frac{1}{3}}$, $a_3 = \sqrt[3]{\sqrt[2]{\frac{1}{3}}} = \sqrt[6]{\frac{1}{3}}$, $a_4 = \sqrt[4]{\sqrt[3]{\sqrt[2]{\frac{1}{3}}}} = \sqrt[4!]{\frac{1}{3}}, \ldots,$
$a_n = \sqrt[n!]{\frac{1}{3}} \Rightarrow \lim_{n\to\infty} a_n = 1$ because $\left\{\sqrt[n!]{\frac{1}{3}}\right\}$ is a subsequence of $\left\{\sqrt[n]{\frac{1}{3}}\right\}$ whose limit is 1 by Table 8.1

36. converges by the Direct Comparison Test: $a_1 = \frac{1}{2}$, $a_2 = \left(\frac{1}{2}\right)^2$, $a_3 = \left(\left(\frac{1}{2}\right)^2\right)^3 = \left(\frac{1}{2}\right)^6$, $a_4 = \left(\left(\frac{1}{2}\right)^6\right)^4 = \left(\frac{1}{2}\right)^{24}, \ldots$
$\Rightarrow a_n = \left(\frac{1}{2}\right)^{n!} < \left(\frac{1}{2}\right)^n$ which is the nth-term of a convergent geometric series

37. converges by the Ratio Test: $\lim_{n\to\infty} \frac{a_{n+1}}{a_n} = \lim_{n\to\infty} \frac{2^{n+1}(n+1)!\,(n+1)!}{(2n+2)!} \cdot \frac{(2n)!}{2^n n!\, n!} = \lim_{n\to\infty} \frac{2(n+1)(n+1)}{(2n+2)(2n+1)}$
$= \lim_{n\to\infty} \frac{n+1}{2n+1} = \frac{1}{2} < 1$

38. diverges by the Ratio Test: $\lim_{n\to\infty} \frac{a_{n+1}}{a_n} = \lim_{n\to\infty} \frac{(3n+3)!}{(n+1)!\,(n+2)!\,(n+3)!} \cdot \frac{n!\,(n+1)!\,(n+2)!}{(3n)!}$
$= \lim_{n\to\infty} \frac{(3n+3)(3+2)(3n+1)}{(n+1)(n+2)(n+3)} = \lim_{n\to\infty} 3\left(\frac{3n+2}{n+2}\right)\left(\frac{3n+1}{n+3}\right) = 3 \cdot 3 \cdot 3 = 27 > 1$

39. diverges by the Root Test: $\lim_{n\to\infty} \sqrt[n]{a_n} \equiv \lim_{n\to\infty} \sqrt[n]{\frac{(n!)^n}{(n^n)^2}} = \lim_{n\to\infty} \frac{n!}{n^2} = \infty > 1$

40. converges by the Root Test: $\lim_{n\to\infty} \sqrt[n]{\frac{(n!)^n}{n^{n^2}}} = \lim_{n\to\infty} \sqrt[n]{\frac{(n!)^n}{(n^n)^n}} = \lim_{n\to\infty} \frac{n!}{n^n} = \lim_{n\to\infty} \left(\frac{1}{n}\right)\left(\frac{2}{n}\right)\left(\frac{3}{n}\right)\cdots\left(\frac{n-1}{n}\right)\left(\frac{n}{n}\right)$
$\le \lim_{n\to\infty} \frac{1}{n} = 0 < 1$

41. converges by the Root Test: $\lim_{n\to\infty} \sqrt[n]{a_n} = \lim_{n\to\infty} \sqrt[n]{\frac{n^n}{2^{n^2}}} = \lim_{n\to\infty} \frac{n}{2^n} = \lim_{n\to\infty} \frac{1}{2^n \ln 2} = 0 < 1$

42. diverges by the Root Test: $\lim_{n\to\infty} \sqrt[n]{a_n} = \lim_{n\to\infty} \sqrt[n]{\frac{n^n}{(2^n)^2}} = \lim_{n\to\infty} \frac{n}{4} = \infty > 1$

43. converges by the Ratio Test: $\lim_{n\to\infty} \frac{a_{n+1}}{a_n} = \lim_{n\to\infty} \frac{1\cdot 3\cdot\cdots\cdot(2n-1)(2n+1)}{4^{n+1}2^{n+1}(n+1)!} \cdot \frac{4^n\, 2^n\, n!}{1\cdot 3\cdot\cdots\cdot(2n-1)}$
$= \lim_{n\to\infty} \frac{2n+1}{(4\cdot 2)(n+1)} = \frac{1}{4} < 1$

44. converges by the Ratio Test: $a_n = \frac{1\cdot 3\cdots(2n-1)}{(2\cdot 4\cdots 2n)(3^n+1)} = \frac{1\cdot 2\cdot 3\cdot 4\cdots(2n-1)(2n)}{(2\cdot 4\cdots 2n)^2(3^n+1)} = \frac{(2n)!}{(2^n n!)^2(3^n+1)}$
$\Rightarrow \lim_{n\to\infty} \frac{(2n+2)!}{[2^{n+1}(n+1)!]^2(3^{n+1}+1)} \cdot \frac{(2^n n!)^2(3^n+1)}{(2n)!} = \lim_{n\to\infty} \frac{(2n+1)(2n+2)(3^n+1)}{2^2(n+1)^2(3^{n+1}+1)}$
$= \lim_{n\to\infty} \left(\frac{4n^2+6n+2}{4n^2+8n+4}\right)\frac{(1+3^{-n})}{(3+3^{-n})} = 1 \cdot \frac{1}{3} = \frac{1}{3} < 1$

45. Ratio: $\lim_{n\to\infty} \frac{a_{n+1}}{a_n} = \lim_{n\to\infty} \frac{1}{(n+1)^p} \cdot \frac{n^p}{1} = \lim_{n\to\infty} \left(\frac{n}{n+1}\right)^p = 1^p = 1 \Rightarrow$ no conclusion

Root: $\lim_{n\to\infty} \sqrt[n]{a_n} = \lim_{n\to\infty} \sqrt[n]{\frac{1}{n^p}} = \lim_{n\to\infty} \frac{1}{(\sqrt[n]{n})^p} = \frac{1}{(1)^p} = 1 \Rightarrow$ no conclusion

46. Ratio: $\lim_{n \to \infty} \frac{a_{n+1}}{a_n} = \lim_{n \to \infty} \frac{1}{(\ln(n+1))^p} \cdot \frac{(\ln n)^p}{1} = \left[\lim_{n \to \infty} \frac{\ln n}{\ln(n+1)}\right]^p = \left[\lim_{n \to \infty} \frac{\left(\frac{1}{n}\right)}{\left(\frac{1}{n+1}\right)}\right]^p = \left(\lim_{n \to \infty} \frac{n+1}{n}\right)^p$

$= (1)^p = 1 \Rightarrow$ no conclusion

Root: $\lim_{n \to \infty} \sqrt[n]{a_n} = \lim_{n \to \infty} \sqrt[n]{\frac{1}{(\ln n)^p}} = \frac{1}{\left(\lim_{n\to\infty} (\ln n)^{1/n}\right)^p}$; let $f(n) = (\ln n)^{1/n}$, then $\ln f(n) = \frac{\ln(\ln n)}{n}$

$\Rightarrow \lim_{n \to \infty} \ln f(n) = \lim_{n \to \infty} \frac{\ln(\ln n)}{n} = \lim_{n \to \infty} \frac{\left(\frac{1}{n \ln n}\right)}{1} = \lim_{n \to \infty} \frac{1}{n \ln n} = 0 \Rightarrow \lim_{n \to \infty} (\ln n)^{1/n}$

$= \lim_{n \to \infty} e^{\ln f(n)} = e^0 = 1$; therefore $\lim_{n \to \infty} \sqrt[n]{a_n} = \frac{1}{\left(\lim_{n\to\infty} (\ln n)^{1/n}\right)^p} = \frac{1}{(1)^p} = 1 \Rightarrow$ no conclusion

47. $a_n \le \frac{n}{2^n}$ for every n and the series $\sum_{n=1}^{\infty} \frac{n}{2^n}$ converges by the Ratio Test since $\lim_{n \to \infty} \frac{(n+1)}{2^{n+1}} \cdot \frac{2^n}{n} = \frac{1}{2} < 1$

$\Rightarrow \sum_{n=1}^{\infty} a_n$ converges by the Direct Comparison Test

11.6 ALTERNATING SERIES, ABSOLUTE AND CONDITIONAL CONVERGENCE

1. converges absolutely $\Rightarrow$ converges by the Absolute Convergence Test since $\sum_{n=1}^{\infty} |a_n| = \sum_{n=1}^{\infty} \frac{1}{n^2}$ which is a convergent p-series

2. converges absolutely $\Rightarrow$ converges by the Absolute Convergence Test since $\sum_{n=1}^{\infty} |a_n| = \sum_{n=1}^{\infty} \frac{1}{n^{3/2}}$ which is a convergent p-series

3. diverges by the nth-Term Test since for $n > 10 \Rightarrow \frac{n}{10} > 1 \Rightarrow \lim_{n \to \infty} \left(\frac{n}{10}\right)^n \ne 0 \Rightarrow \sum_{n=1}^{\infty} (-1)^{n+1} \left(\frac{n}{10}\right)^n$ diverges

4. diverges by the nth-Term Test since $\lim_{n \to \infty} \frac{10^n}{n^{10}} = \lim_{n \to \infty} \frac{10^n (\ln 10)^{10}}{10!} = \infty$ (after 10 applications of L'Hôpital's rule)

5. converges by the Alternating Series Test because $f(x) = \ln x$ is an increasing function of $x \Rightarrow \frac{1}{\ln x}$ is decreasing $\Rightarrow u_n \ge u_{n+1}$ for $n \ge 1$; also $u_n \ge 0$ for $n \ge 1$ and $\lim_{n \to \infty} \frac{1}{\ln n} = 0$

6. converges by the Alternating Series Test since $f(x) = \frac{\ln x}{x} \Rightarrow f'(x) = \frac{1 - \ln x}{x^2} < 0$ when $x > e \Rightarrow f(x)$ is decreasing $\Rightarrow u_n \ge u_{n+1}$; also $u_n \ge 0$ for $n \ge 1$ and $\lim_{n \to \infty} u_n = \lim_{n \to \infty} \frac{\ln n}{n} = \lim_{n \to \infty} \frac{\left(\frac{1}{n}\right)}{1} = 0$

7. diverges by the nth-Term Test since $\lim_{n \to \infty} \frac{\ln n}{\ln n^2} = \lim_{n \to \infty} \frac{\ln n}{2 \ln n} = \lim_{n \to \infty} \frac{1}{2} = \frac{1}{2} \ne 0$

8. converges by the Alternating Series Test since $f(x) = \ln(1 + x^{-1}) \Rightarrow f'(x) = \frac{-1}{x(x+1)} < 0$ for $x > 0 \Rightarrow f(x)$ is decreasing $\Rightarrow u_n \ge u_{n+1}$; also $u_n \ge 0$ for $n \ge 1$ and $\lim_{n \to \infty} u_n = \lim_{n \to \infty} \ln\left(1 + \frac{1}{n}\right) = \ln\left(\lim_{n \to \infty} \left(1 + \frac{1}{n}\right)\right) = \ln 1 = 0$

9. converges by the Alternating Series Test since $f(x) = \frac{\sqrt{x}+1}{x+1} \Rightarrow f'(x) = \frac{1 - x - 2\sqrt{x}}{2\sqrt{x}(x+1)^2} < 0 \Rightarrow f(x)$ is decreasing $\Rightarrow u_n \ge u_{n+1}$; also $u_n \ge 0$ for $n \ge 1$ and $\lim_{n \to \infty} u_n = \lim_{n \to \infty} \frac{\sqrt{n}+1}{n+1} = 0$

10. diverges by the nth-Term Test since $\lim_{n\to\infty} \frac{3\sqrt{n+1}}{\sqrt{n}+1} = \lim_{n\to\infty} \frac{3\sqrt{1+\frac{1}{n}}}{1+\left(\frac{1}{\sqrt{n}}\right)} = 3 \neq 0$

11. converges absolutely since $\sum_{n=1}^{\infty} |a_n| = \sum_{n=1}^{\infty} \left(\frac{1}{10}\right)^n$ a convergent geometric series

12. converges absolutely by the Direct Comparison Test since $\left|\frac{(-1)^{n+1}(0.1)^n}{n}\right| = \frac{1}{(10)^n n} < \left(\frac{1}{10}\right)^n$ which is the nth term of a convergent geometric series

13. converges conditionally since $\frac{1}{\sqrt{n}} > \frac{1}{\sqrt{n+1}} > 0$ and $\lim_{n\to\infty} \frac{1}{\sqrt{n}} = 0 \Rightarrow$ convergence; but $\sum_{n=1}^{\infty} |a_n| = \sum_{n=1}^{\infty} \frac{1}{n^{1/2}}$ is a divergent p-series

14. converges conditionally since $\frac{1}{1+\sqrt{n}} > \frac{1}{1+\sqrt{n+1}} > 0$ and $\lim_{n\to\infty} \frac{1}{1+\sqrt{n}} = 0 \Rightarrow$ convergence; but $\sum_{n=1}^{\infty} |a_n| = \sum_{n=1}^{\infty} \frac{1}{1+\sqrt{n}}$ is a divergent series since $\frac{1}{1+\sqrt{n}} \geq \frac{1}{2\sqrt{n}}$ and $\sum_{n=1}^{\infty} \frac{1}{n^{1/2}}$ is a divergent p-series

15. converges absolutely since $\sum_{n=1}^{\infty} |a_n| = \sum_{n=1}^{\infty} \frac{n}{n^3+1}$ and $\frac{n}{n^3+1} < \frac{1}{n^2}$ which is the nth-term of a converging p-series

16. diverges by the nth-Term Test since $\lim_{n\to\infty} \frac{n!}{2^n} = \infty$

17. converges conditionally since $\frac{1}{n+3} > \frac{1}{(n+1)+3} > 0$ and $\lim_{n\to\infty} \frac{1}{n+3} = 0 \Rightarrow$ convergence; but $\sum_{n=1}^{\infty} |a_n|$ $= \sum_{n=1}^{\infty} \frac{1}{n+3}$ diverges because $\frac{1}{n+3} \geq \frac{1}{4n}$ and $\sum_{n=1}^{\infty} \frac{1}{n}$ is a divergent series

18. converges absolutely because the series $\sum_{n=1}^{\infty} \left|\frac{\sin n}{n^2}\right|$ converges by the Direct Comparison Test since $\left|\frac{\sin n}{n^2}\right| \leq \frac{1}{n^2}$

19. diverges by the nth-Term Test since $\lim_{n\to\infty} \frac{3+n}{5+n} = 1 \neq 0$

20. converges conditionally since $f(x) = \ln x$ is an increasing function of $x \Rightarrow \frac{1}{3\ln x} = \frac{1}{\ln(x^3)}$ is decreasing $\Rightarrow \frac{1}{3\ln n} > \frac{1}{3\ln(n+1)} > 0$ for $n \geq 2$ and $\lim_{n\to\infty} \frac{1}{3\ln n} = 0 \Rightarrow$ convergence; but $\sum_{n=2}^{\infty} |a_n| = \sum_{n=2}^{\infty} \frac{1}{\ln(n^3)}$ $= \sum_{n=2}^{\infty} \frac{1}{3\ln n}$ diverges because $\frac{1}{3\ln n} > \frac{1}{3n}$ and $\sum_{n=2}^{\infty} \frac{1}{n}$ diverges

21. converges conditionally since $f(x) = \frac{1}{x^2} + \frac{1}{x} \Rightarrow f'(x) = -\left(\frac{2}{x^3} + \frac{1}{x^2}\right) < 0 \Rightarrow f(x)$ is decreasing and hence $u_n > u_{n+1} > 0$ for $n \geq 1$ and $\lim_{n\to\infty} \left(\frac{1}{n^2} + \frac{1}{n}\right) = 0 \Rightarrow$ convergence; but $\sum_{n=1}^{\infty} |a_n| = \sum_{n=1}^{\infty} \frac{1+n}{n^2}$ $= \sum_{n=1}^{\infty} \frac{1}{n^2} + \sum_{n=1}^{\infty} \frac{1}{n}$ is the sum of a convergent and divergent series, and hence diverges

22. converges absolutely by the Direct Comparison Test since $\left|\frac{(-2)^{n+1}}{n+5^n}\right| = \frac{2^{n+1}}{n+5^n} < 2\left(\frac{2}{5}\right)^n$ which is the nth term of a convergent geometric series

23. converges absolutely by the Ratio Test: $\lim_{n \to \infty} \left(\frac{u_{n+1}}{u_n}\right) = \lim_{n \to \infty} \left[\frac{(n+1)^2 \left(\frac{2}{3}\right)^{n+1}}{n^2 \left(\frac{2}{3}\right)^n}\right] = \frac{2}{3} < 1$

24. diverges by the nth-Term Test since $\lim_{n \to \infty} a_n = \lim_{n \to \infty} 10^{1/n} = 1 \neq 0$

25. converges absolutely by the Integral Test since $\int_1^{\infty} (\tan^{-1} x)\left(\frac{1}{1+x^2}\right) dx = \lim_{b \to \infty} \left[\frac{(\tan^{-1} x)^2}{2}\right]_1^b$
$= \lim_{b \to \infty} \left[(\tan^{-1} b)^2 - (\tan^{-1} 1)^2\right] = \frac{1}{2}\left[\left(\frac{\pi}{2}\right)^2 - \left(\frac{\pi}{4}\right)^2\right] = \frac{3\pi^2}{32}$

26. converges conditionally since $f(x) = \frac{1}{x \ln x} \Rightarrow f'(x) = -\frac{[\ln(x)+1]}{(x \ln x)^2} < 0 \Rightarrow f(x)$ is decreasing
$\Rightarrow u_n > u_{n+1} > 0$ for $n \geq 2$ and $\lim_{n \to \infty} \frac{1}{n \ln n} = 0 \Rightarrow$ convergence; but by the Integral Test,
$\int_2^{\infty} \frac{dx}{x \ln x} = \lim_{b \to \infty} \int_2^b \left(\frac{\left(\frac{1}{x}\right)}{\ln x}\right) dx = \lim_{b \to \infty} \left[\ln(\ln x)\right]_2^b = \lim_{b \to \infty} \left[\ln(\ln b) - \ln(\ln 2)\right] = \infty$
$\Rightarrow \sum_{n=1}^{\infty} |a_n| = \sum_{n=1}^{\infty} \frac{1}{n \ln n}$ diverges

27. diverges by the nth-Term Test since $\lim_{n \to \infty} \frac{n}{n+1} = 1 \neq 0$

28. converges conditionally since $f(x) = \frac{\ln x}{x - \ln x} \Rightarrow f'(x) = \frac{\left(\frac{1}{x}\right)(x - \ln x) - (\ln x)\left(1 - \frac{1}{x}\right)}{(x - \ln x)^2}$
$= \frac{1 - \left(\frac{\ln x}{x}\right) - \ln x + \left(\frac{\ln x}{x}\right)}{(x - \ln x)^2} = \frac{1 - \ln x}{(x - \ln x)^2} < 0 \Rightarrow u_n \geq u_{n+1} > 0$ when $n > e$ and $\lim_{n \to \infty} \frac{\ln n}{n - \ln n}$
$= \lim_{n \to \infty} \frac{\left(\frac{1}{n}\right)}{1 - \left(\frac{1}{n}\right)} = 0 \Rightarrow$ convergence; but $n - \ln n < n \Rightarrow \frac{1}{n - \ln n} > \frac{1}{n} \Rightarrow \frac{\ln n}{n - \ln n} > \frac{1}{n}$ so that
$\sum_{n=1}^{\infty} |a_n| = \sum_{n=1}^{\infty} \frac{\ln n}{n - \ln n}$ diverges by the Direct Comparison Test

29. converges absolutely by the Ratio Test: $\lim_{n \to \infty} \left(\frac{u_{n+1}}{u_n}\right) = \lim_{n \to \infty} \frac{(100)^{n+1}}{(n+1)!} \cdot \frac{n!}{(100)^n} = \lim_{n \to \infty} \frac{100}{n+1} = 0 < 1$

30. converges absolutely since $\sum_{n=1}^{\infty} |a_n| = \sum_{n=1}^{\infty} \left(\frac{1}{5}\right)^n$ is a convergent geometric series

31. converges absolutely by the Direct Comparison Test since $\sum_{n=1}^{\infty} |a_n| = \sum_{n=1}^{\infty} \frac{1}{n^2 + 2n + 1}$ and
$\frac{1}{n^2 + 2n + 1} < \frac{1}{n^2}$ which is the nth-term of a convergent p-series

32. converges absolutely since $\sum_{n=1}^{\infty} |a_n| = \sum_{n=1}^{\infty} \left(\frac{\ln n}{\ln n^2}\right)^n = \sum_{n=1}^{\infty} \left(\frac{\ln n}{2 \ln n}\right)^n = \sum_{n=1}^{\infty} \left(\frac{1}{2}\right)^n$ is a convergent
geometric series

33. converges absolutely since $\sum_{n=1}^{\infty} |a_n| = \sum_{n=1}^{\infty} \left|\frac{(-1)^n}{n\sqrt{n}}\right| = \sum_{n=1}^{\infty} \frac{1}{n^{3/2}}$ is a convergent p-series

34. converges conditionally since $\sum_{n=1}^{\infty} \frac{\cos n\pi}{n} = \sum_{n=1}^{\infty} \frac{(-1)^n}{n}$ is the convergent alternating harmonic series, but
$\sum_{n=1}^{\infty} |a_n| = \sum_{n=1}^{\infty} \frac{1}{n}$ diverges

35. converges absolutely by the Root Test: $\lim\limits_{n\to\infty} \sqrt[n]{|a_n|} = \lim\limits_{n\to\infty} \left(\frac{(n+1)^n}{(2n)^n}\right)^{1/n} = \lim\limits_{n\to\infty} \frac{n+1}{2n} = \frac{1}{2} < 1$

36. converges absolutely by the Ratio Test: $\lim\limits_{n\to\infty} \left|\frac{a_{n+1}}{a_n}\right| = \lim\limits_{n\to\infty} \frac{((n+1)!)^2}{((2n+2)!)} \cdot \frac{(2n)!}{(n!)^2} = \lim\limits_{n\to\infty} \frac{(n+1)^2}{(2n+2)(2n+1)} = \frac{1}{4} < 1$

37. diverges by the nth-Term Test since $\lim\limits_{n\to\infty} |a_n| = \lim\limits_{n\to\infty} \frac{(2n)!}{2^n n!\, n} = \lim\limits_{n\to\infty} \frac{(n+1)(n+2)\cdots(2n)}{2^n n}$
$= \lim\limits_{n\to\infty} \frac{(n+1)(n+2)\cdots(n+(n-1))}{2^{n-1}} > \lim\limits_{n\to\infty} \left(\frac{n+1}{2}\right)^{n-1} = \infty \neq 0$

38. converges absolutely by the Ratio Test: $\lim\limits_{n\to\infty} \left|\frac{a_{n+1}}{a_n}\right| = \lim\limits_{n\to\infty} \frac{(n+1)!\,(n+1)!\,3^{n+1}}{(2n+3)!} \cdot \frac{(2n+1)!}{n!\,n!\,3^n}$
$= \lim\limits_{n\to\infty} \frac{(n+1)^2\, 3}{(2n+2)(2n+3)} = \frac{3}{4} < 1$

39. converges conditionally since $\frac{\sqrt{n+1}-\sqrt{n}}{1} \cdot \frac{\sqrt{n+1}+\sqrt{n}}{\sqrt{n+1}+\sqrt{n}} = \frac{1}{\sqrt{n+1}+\sqrt{n}}$ and $\left\{\frac{1}{\sqrt{n+1}+\sqrt{n}}\right\}$ is a decreasing sequence of positive terms which converges to 0 $\Rightarrow \sum\limits_{n=1}^{\infty} \frac{(-1)^n}{\sqrt{n+1}+\sqrt{n}}$ converges; but $\sum\limits_{n=1}^{\infty} |a_n| = \sum\limits_{n=1}^{\infty} \frac{1}{\sqrt{n+1}+\sqrt{n}}$ diverges by the Limit Comparison Test (part 1) with $\frac{1}{\sqrt{n}}$; a divergent p-series:
$\lim\limits_{n\to\infty} \left(\frac{\frac{1}{\sqrt{n+1}+\sqrt{n}}}{\frac{1}{\sqrt{n}}}\right) = \lim\limits_{n\to\infty} \frac{\sqrt{n}}{\sqrt{n+1}+\sqrt{n}} = \lim\limits_{n\to\infty} \frac{1}{\sqrt{1+\frac{1}{n}}+1} = \frac{1}{2}$

40. diverges by the nth-Term Test since $\lim\limits_{n\to\infty} \left(\sqrt{n^2+n}-n\right) = \lim\limits_{n\to\infty} \left(\sqrt{n^2+n}-n\right) \cdot \left(\frac{\sqrt{n^2+n}+n}{\sqrt{n^2+n}+n}\right)$
$= \lim\limits_{n\to\infty} \frac{n}{\sqrt{n^2+n}+n} = \lim\limits_{n\to\infty} \frac{1}{\sqrt{1+\frac{1}{n}}+1} = \frac{1}{2} \neq 0$

41. diverges by the nth-Term Test since $\lim\limits_{n\to\infty} \left(\sqrt{n+\sqrt{n}}-\sqrt{n}\right) = \lim\limits_{n\to\infty} \left[\left(\sqrt{n+\sqrt{n}}-\sqrt{n}\right)\left(\frac{\sqrt{n+\sqrt{n}}+\sqrt{n}}{\sqrt{n+\sqrt{n}}+\sqrt{n}}\right)\right]$
$= \lim\limits_{n\to\infty} \frac{\sqrt{n}}{\sqrt{n+\sqrt{n}}+\sqrt{n}} = \lim\limits_{n\to\infty} \frac{1}{\sqrt{1+\frac{1}{\sqrt{n}}}+1} = \frac{1}{2} \neq 0$

42. converges conditionally since $\left\{\frac{1}{\sqrt{n}+\sqrt{n+1}}\right\}$ is a decreasing sequence of positive terms converging to 0
$\Rightarrow \sum\limits_{n=1}^{\infty} \frac{(-1)^n}{\sqrt{n}+\sqrt{n+1}}$ converges; but $\lim\limits_{n\to\infty} \frac{\left(\frac{1}{\sqrt{n}+\sqrt{n+1}}\right)}{\left(\frac{1}{\sqrt{n}}\right)} = \lim\limits_{n\to\infty} \frac{\sqrt{n}}{\sqrt{n}+\sqrt{n+1}} = \lim\limits_{n\to\infty} \frac{1}{1+\sqrt{1+\frac{1}{n}}} = \frac{1}{2}$
so that $\sum\limits_{n=1}^{\infty} \frac{1}{\sqrt{n}+\sqrt{n+1}}$ diverges by the Limit Comparison Test with $\sum\limits_{n=1}^{\infty} \frac{1}{\sqrt{n}}$ which is a divergent p-series

43. converges absolutely by the Direct Comparison Test since $\operatorname{sech}(n) = \frac{2}{e^n+e^{-n}} = \frac{2e^n}{e^{2n}+1} < \frac{2e^n}{e^{2n}} = \frac{2}{e^n}$ which is the nth term of a convergent geometric series

44. converges absolutely by the Limit Comparison Test (part 1): $\sum\limits_{n=1}^{\infty} |a_n| = \sum\limits_{n=1}^{\infty} \frac{2}{e^n-e^{-n}}$
Apply the Limit Comparison Test with $\frac{1}{e^n}$, the n-th term of a convergent geometric series:
$\lim\limits_{n\to\infty} \left(\frac{\frac{2}{e^n-e^{-n}}}{\frac{1}{e^n}}\right) = \lim\limits_{n\to\infty} \frac{2e^n}{e^n-e^{-n}} = \lim\limits_{n\to\infty} \frac{2}{1-e^{-2n}} = 2$

45. $|\text{error}| < \left|(-1)^6\left(\frac{1}{5}\right)\right| = 0.2$

46. $|\text{error}| < \left|(-1)^6\left(\frac{1}{10^5}\right)\right| = 0.00001$

47. $|\text{error}| < \left|(-1)^6 \frac{(0.01)^5}{5}\right| = 2 \times 10^{-11}$

48. $|\text{error}| < |(-1)^4 t^4| = t^4 < 1$

49. $\frac{1}{(2n)!} < \frac{5}{10^6} \Rightarrow (2n)! > \frac{10^6}{5} = 200{,}000 \Rightarrow n \geq 5 \Rightarrow 1 - \frac{1}{2!} + \frac{1}{4!} - \frac{1}{6!} + \frac{1}{8!} \approx 0.54030$

50. $\frac{1}{n!} < \frac{5}{10^6} \Rightarrow \frac{10^6}{5} < n! \Rightarrow n \geq 9 \Rightarrow 1 - 1 + \frac{1}{2!} - \frac{1}{3!} + \frac{1}{4!} - \frac{1}{5!} + \frac{1}{6!} - \frac{1}{7!} + \frac{1}{8!} \approx 0.367881944$

51. (a) $a_n \geq a_{n+1}$ fails since $\frac{1}{3} < \frac{1}{2}$

(b) Since $\sum_{n=1}^{\infty} |a_n| = \sum_{n=1}^{\infty} \left[\left(\frac{1}{3}\right)^n + \left(\frac{1}{2}\right)^n\right] = \sum_{n=1}^{\infty} \left(\frac{1}{3}\right)^n + \sum_{n=1}^{\infty} \left(\frac{1}{2}\right)^n$ is the sum of two absolutely convergent series, we can rearrange the terms of the original series to find its sum:
$\left(\frac{1}{3} + \frac{1}{9} + \frac{1}{27} + \ldots\right) - \left(\frac{1}{2} + \frac{1}{4} + \frac{1}{8} + \ldots\right) = \frac{\left(\frac{1}{3}\right)}{1-\left(\frac{1}{3}\right)} - \frac{\left(\frac{1}{2}\right)}{1-\left(\frac{1}{2}\right)} = \frac{1}{2} - 1 = -\frac{1}{2}$

52. $s_{20} = 1 - \frac{1}{2} + \frac{1}{3} - \frac{1}{4} + \ldots + \frac{1}{19} - \frac{1}{20} \approx 0.6687714032 \Rightarrow s_{20} + \frac{1}{2} \cdot \frac{1}{21} \approx 0.692580927$

53. The unused terms are $\sum_{j=n+1}^{\infty} (-1)^{j+1} a_j = (-1)^{n+1}(a_{n+1} - a_{n+2}) + (-1)^{n+3}(a_{n+3} - a_{n+4}) + \ldots$
$= (-1)^{n+1}\left[(a_{n+1} - a_{n+2}) + (a_{n+3} - a_{n+4}) + \ldots\right]$. Each grouped term is positive, so the remainder has the same sign as $(-1)^{n+1}$, which is the sign of the first unused term.

54. $s_n = \frac{1}{1\cdot 2} + \frac{1}{2\cdot 3} + \frac{1}{3\cdot 4} + \ldots + \frac{1}{n(n+1)} = \sum_{k=1}^{n} \frac{1}{k(k+1)} = \sum_{k=1}^{n} \left(\frac{1}{k} - \frac{1}{k+1}\right)$
$= \left(1 - \frac{1}{2}\right) + \left(\frac{1}{2} - \frac{1}{3}\right) + \left(\frac{1}{3} - \frac{1}{4}\right) + \left(\frac{1}{4} - \frac{1}{5}\right) + \ldots + \left(\frac{1}{n} - \frac{1}{n+1}\right)$ which are the first 2n terms of the first series, hence the two series are the same. Yes, for
$s_n = \sum_{k=1}^{n} \left(\frac{1}{k} - \frac{1}{k+1}\right) = \left(1 - \frac{1}{2}\right) + \left(\frac{1}{2} - \frac{1}{3}\right) + \left(\frac{1}{3} - \frac{1}{4}\right) + \left(\frac{1}{4} - \frac{1}{5}\right) + \ldots + \left(\frac{1}{n-1} - \frac{1}{n}\right) + \left(\frac{1}{n} - \frac{1}{n+1}\right) = 1 - \frac{1}{n+1}$
$\Rightarrow \lim_{n \to \infty} s_n = \lim_{n \to \infty} \left(1 - \frac{1}{n+1}\right) = 1 \Rightarrow$ both series converge to 1. The sum of the first $2n + 1$ terms of the first series is $\left(1 - \frac{1}{n+1}\right) + \frac{1}{n+1} = 1$. Their sum is $\lim_{n \to \infty} s_n = \lim_{n \to \infty} \left(1 - \frac{1}{n+1}\right) = 1$.

55. Theorem 16 states that $\sum_{n=1}^{\infty} |a_n|$ converges $\Rightarrow \sum_{n=1}^{\infty} a_n$ converges. But this is equivalent to $\sum_{n=1}^{\infty} a_n$ diverges $\Rightarrow \sum_{n=1}^{\infty} |a_n|$ diverges.

56. $|a_1 + a_2 + \ldots + a_n| \leq |a_1| + |a_2| + \ldots + |a_n|$ for all n; then $\sum_{n=1}^{\infty} |a_n|$ converges $\Rightarrow \sum_{n=1}^{\infty} a_n$ converges and these imply that $\left|\sum_{n=1}^{\infty} a_n\right| \leq \sum_{n=1}^{\infty} |a_n|$

57. (a) $\sum_{n=1}^{\infty} |a_n + b_n|$ converges by the Direct Comparison Test since $|a_n + b_n| \leq |a_n| + |b_n|$ and hence $\sum_{n=1}^{\infty} (a_n + b_n)$ converges absolutely

(b) $\sum_{n=1}^{\infty} |b_n|$ converges $\Rightarrow \sum_{n=1}^{\infty} -b_n$ converges absolutely; since $\sum_{n=1}^{\infty} a_n$ converges absolutely and $\sum_{n=1}^{\infty} -b_n$ converges absolutely, we have $\sum_{n=1}^{\infty} [a_n + (-b_n)] = \sum_{n=1}^{\infty} (a_n - b_n)$ converges absolutely by part (a)

(c) $\sum_{n=1}^{\infty} |a_n|$ converges $\Rightarrow |k| \sum_{n=1}^{\infty} |a_n| = \sum_{n=1}^{\infty} |k a_n|$ converges $\Rightarrow \sum_{n=1}^{\infty} k a_n$ converges absolutely

58. If $a_n = b_n = (-1)^n \frac{1}{\sqrt{n}}$, then $\sum_{n=1}^{\infty} (-1)^n \frac{1}{\sqrt{n}}$ converges, but $\sum_{n=1}^{\infty} a_n b_n = \sum_{n=1}^{\infty} \frac{1}{n}$ diverges

59. $s_1 = -\frac{1}{2}$, $s_2 = -\frac{1}{2} + 1 = \frac{1}{2}$,

$s_3 = -\frac{1}{2} + 1 - \frac{1}{4} - \frac{1}{6} - \frac{1}{8} - \frac{1}{10} - \frac{1}{12} - \frac{1}{14} - \frac{1}{16} - \frac{1}{18} - \frac{1}{20} - \frac{1}{22} \approx -0.5099$,

$s_4 = s_3 + \frac{1}{3} \approx -0.1766$,

$s_5 = s_4 - \frac{1}{24} - \frac{1}{26} - \frac{1}{28} - \frac{1}{30} - \frac{1}{32} - \frac{1}{34} - \frac{1}{36} - \frac{1}{38} - \frac{1}{40} - \frac{1}{42} - \frac{1}{44} \approx -0.512$,

$s_6 = s_5 + \frac{1}{5} \approx -0.312$,

$s_7 = s_6 - \frac{1}{46} - \frac{1}{48} - \frac{1}{50} - \frac{1}{52} - \frac{1}{54} - \frac{1}{56} - \frac{1}{58} - \frac{1}{60} - \frac{1}{62} - \frac{1}{64} - \frac{1}{66} \approx -0.51106$

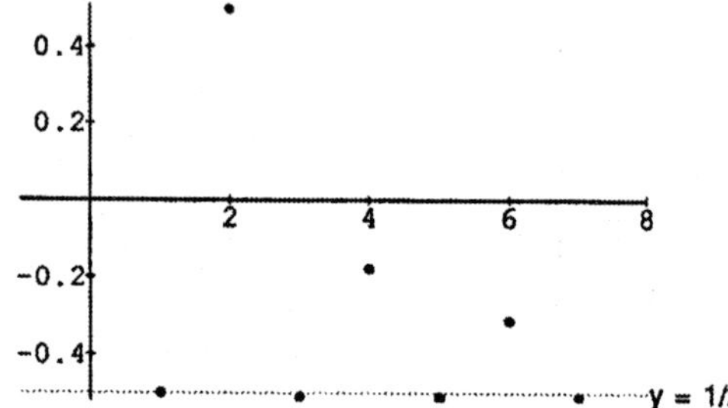

60. (a) Since $\sum |a_n|$ converges, say to M, for $\epsilon > 0$ there is an integer N_1 such that $\left|\sum_{n=1}^{N_1-1} |a_n| - M\right| < \frac{\epsilon}{2}$

$\Leftrightarrow \left|\sum_{n=1}^{N_1-1} |a_n| - \left(\sum_{n=1}^{N_1-1} |a_n| + \sum_{n=N_1}^{\infty} |a_n|\right)\right| < \frac{\epsilon}{2} \Leftrightarrow \left|-\sum_{n=N_1}^{\infty} |a_n|\right| < \frac{\epsilon}{2} \Leftrightarrow \sum_{n=N_1}^{\infty} |a_n| < \frac{\epsilon}{2}$. Also, $\sum a_n$ converges to L $\Leftrightarrow$ for $\epsilon > 0$ there is an integer N_2 (which we can choose greater than or equal to N_1) such that $|s_{N_2} - L| < \frac{\epsilon}{2}$. Therefore, $\sum_{n=N_1}^{\infty} |a_n| < \frac{\epsilon}{2}$ and $|s_{N_2} - L| < \frac{\epsilon}{2}$.

(b) The series $\sum_{n=1}^{\infty} |a_n|$ converges absolutely, say to M. Thus, there exists N_1 such that $\left|\sum_{n=1}^{k} |a_n| - M\right| < \epsilon$ whenever $k > N_1$. Now all of the terms in the sequence $\{|b_n|\}$ appear in $\{|a_n|\}$. Sum together all of the terms in $\{|b_n|\}$, in order, until you include all of the terms $\{|a_n|\}_{n=1}^{N_1}$, and let N_2 be the largest index in the sum $\sum_{n=1}^{N_2} |b_n|$ so obtained. Then $\left|\sum_{n=1}^{N_2} |b_n| - M\right| < \epsilon$ as well $\Rightarrow \sum_{n=1}^{\infty} |b_n|$ converges to M.

61. (a) If $\sum_{n=1}^{\infty} |a_n|$ converges, then $\sum_{n=1}^{\infty} a_n$ converges and $\frac{1}{2}\sum_{n=1}^{\infty} a_n + \frac{1}{2}\sum_{n=1}^{\infty} |a_n| = \sum_{n=1}^{\infty} \frac{a_n + |a_n|}{2}$ converges where $b_n = \frac{a_n + |a_n|}{2} = \begin{cases} a_n, & \text{if } a_n \ge 0 \\ 0, & \text{if } a_n < 0 \end{cases}$.

(b) If $\sum_{n=1}^{\infty} |a_n|$ converges, then $\sum_{n=1}^{\infty} a_n$ converges and $\frac{1}{2}\sum_{n=1}^{\infty} a_n - \frac{1}{2}\sum_{n=1}^{\infty} |a_n| = \sum_{n=1}^{\infty} \frac{a_n - |a_n|}{2}$ converges where $c_n = \frac{a_n - |a_n|}{2} = \begin{cases} 0, & \text{if } a_n \ge 0 \\ a_n, & \text{if } a_n < 0 \end{cases}$.

62. The terms in this conditionally convergent series were not added in the order given.

63. Here is an example figure when N = 5. Notice that $u_3 > u_2 > u_1$ and $u_3 > u_5 > u_4$, but $u_n \geq u_{n+1}$ for $n \geq 5$.

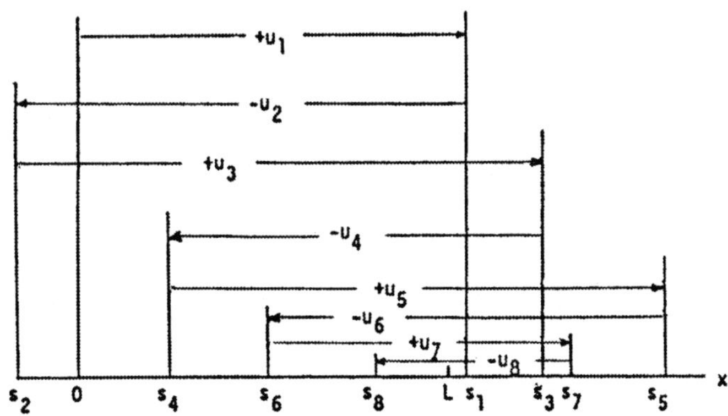

11.7 POWER SERIES

1. $\lim_{n\to\infty} \left|\frac{u_{n+1}}{u_n}\right| < 1 \Rightarrow \lim_{n\to\infty} \left|\frac{x^{n+1}}{x^n}\right| < 1 \Rightarrow |x| < 1 \Rightarrow -1 < x < 1$; when $x = -1$ we have $\sum_{n=1}^{\infty} (-1)^n$, a divergent series; when $x = 1$ we have $\sum_{n=1}^{\infty} 1$, a divergent series

   (a) the radius is 1; the interval of convergence is $-1 < x < 1$
   (b) the interval of absolute convergence is $-1 < x < 1$
   (c) there are no values for which the series converges conditionally

2. $\lim_{n\to\infty} \left|\frac{u_{n+1}}{u_n}\right| < 1 \Rightarrow \lim_{n\to\infty} \left|\frac{(x+5)^{n+1}}{(x+5)^n}\right| < 1 \Rightarrow |x+5| < 1 \Rightarrow -6 < x < -4$; when $x = -6$ we have $\sum_{n=1}^{\infty} (-1)^n$, a divergent series; when $x = -4$ we have $\sum_{n=1}^{\infty} 1$, a divergent series

   (a) the radius is 1; the interval of convergence is $-6 < x < -4$
   (b) the interval of absolute convergence is $-6 < x < -4$
   (c) there are no values for which the series converges conditionally

3. $\lim_{n\to\infty} \left|\frac{u_{n+1}}{u_n}\right| < 1 \Rightarrow \lim_{n\to\infty} \left|\frac{(4x+1)^{n+1}}{(4x+1)^n}\right| < 1 \Rightarrow |4x+1| < 1 \Rightarrow -1 < 4x+1 < 1 \Rightarrow -\frac{1}{2} < x < 0$; when $x = -\frac{1}{2}$ we have $\sum_{n=1}^{\infty} (-1)^n(-1)^n = \sum_{n=1}^{\infty} (-1)^{2n} = \sum_{n=1}^{\infty} 1^n$, a divergent series; when $x = 0$ we have $\sum_{n=1}^{\infty} (-1)^n(1)^n$ $= \sum_{n=1}^{\infty} (-1)^n$, a divergent series

   (a) the radius is $\frac{1}{4}$; the interval of convergence is $-\frac{1}{2} < x < 0$
   (b) the interval of absolute convergence is $-\frac{1}{2} < x < 0$
   (c) there are no values for which the series converges conditionally

4. $\lim_{n\to\infty} \left|\frac{u_{n+1}}{u_n}\right| < 1 \Rightarrow \lim_{n\to\infty} \left|\frac{(3x-2)^{n+1}}{n+1} \cdot \frac{n}{(3x-2)^n}\right| < 1 \Rightarrow |3x-2| \lim_{n\to\infty} \left(\frac{n}{n+1}\right) < 1 \Rightarrow |3x-2| < 1$ $\Rightarrow -1 < 3x-2 < 1 \Rightarrow \frac{1}{3} < x < 1$; when $x = \frac{1}{3}$ we have $\sum_{n=1}^{\infty} \frac{(-1)^n}{n}$ which is the alternating harmonic series and is conditionally convergent; when $x = 1$ we have $\sum_{n=1}^{\infty} \frac{1}{n}$, the divergent harmonic series

   (a) the radius is $\frac{1}{3}$; the interval of convergence is $\frac{1}{3} \leq x < 1$
   (b) the interval of absolute convergence is $\frac{1}{3} < x < 1$

(c) the series converges conditionally at $x = \frac{1}{3}$

5. $\lim_{n\to\infty} \left|\frac{u_{n+1}}{u_n}\right| < 1 \Rightarrow \lim_{n\to\infty} \left|\frac{(x-2)^{n+1}}{10^{n+1}} \cdot \frac{10^n}{(x-2)^n}\right| < 1 \Rightarrow \frac{|x-2|}{10} < 1 \Rightarrow |x-2| < 10 \Rightarrow -10 < x-2 < 10$

$\Rightarrow -8 < x < 12$; when $x = -8$ we have $\sum_{n=1}^{\infty} (-1)^n$, a divergent series; when $x = 12$ we have $\sum_{n=1}^{\infty} 1$, a divergent series

(a) the radius is 10; the interval of convergence is $-8 < x < 12$
(b) the interval of absolute convergence is $-8 < x < 12$
(c) there are no values for which the series converges conditionally

6. $\lim_{n\to\infty} \left|\frac{u_{n+1}}{u_n}\right| < 1 \Rightarrow \lim_{n\to\infty} \left|\frac{(2x)^{n+1}}{(2x)^n}\right| < 1 \Rightarrow \lim_{n\to\infty} |2x| < 1 \Rightarrow |2x| < 1 \Rightarrow -\frac{1}{2} < x < \frac{1}{2}$; when $x = -\frac{1}{2}$ we have $\sum_{n=1}^{\infty} (-1)^n$, a divergent series; when $x = \frac{1}{2}$ we have $\sum_{n=1}^{\infty} 1$, a divergent series

(a) the radius is $\frac{1}{2}$; the interval of convergence is $-\frac{1}{2} < x < \frac{1}{2}$
(b) the interval of absolute convergence is $-\frac{1}{2} < x < \frac{1}{2}$
(c) there are no values for which the series converges conditionally

7. $\lim_{n\to\infty} \left|\frac{u_{n+1}}{u_n}\right| < 1 \Rightarrow \lim_{n\to\infty} \left|\frac{(n+1)x^{n+1}}{(n+3)} \cdot \frac{(n+2)}{nx^n}\right| < 1 \Rightarrow |x| \lim_{n\to\infty} \frac{(n+1)(n+2)}{(n+3)(n)} < 1 \Rightarrow |x| < 1$

$\Rightarrow -1 < x < 1$; when $x = -1$ we have $\sum_{n=1}^{\infty} (-1)^n \frac{n}{n+2}$, a divergent series by the nth-term Test; when $x = 1$ we have $\sum_{n=1}^{\infty} \frac{n}{n+2}$, a divergent series

(a) the radius is 1; the interval of convergence is $-1 < x < 1$
(b) the interval of absolute convergence is $-1 < x < 1$
(c) there are no values for which the series converges conditionally

8. $\lim_{n\to\infty} \left|\frac{u_{n+1}}{u_n}\right| < 1 \Rightarrow \lim_{n\to\infty} \left|\frac{(x+2)^{n+1}}{n+1} \cdot \frac{n}{(x+2)^n}\right| < 1 \Rightarrow |x+2| \lim_{n\to\infty} \left(\frac{n}{n+1}\right) < 1 \Rightarrow |x+2| < 1$

$\Rightarrow -1 < x+2 < 1 \Rightarrow -3 < x < -1$; when $x = -3$ we have $\sum_{n=1}^{\infty} \frac{1}{n}$, a divergent series; when $x = -1$ we have $\sum_{n=1}^{\infty} \frac{(-1)^n}{n}$, a convergent series

(a) the radius is 1; the interval of convergence is $-3 < x \le -1$
(b) the interval of absolute convergence is $-3 < x < -1$
(c) the series converges conditionally at $x = -1$

9. $\lim_{n\to\infty} \left|\frac{u_{n+1}}{u_n}\right| < 1 \Rightarrow \lim_{n\to\infty} \left|\frac{x^{n+1}}{(n+1)\sqrt{n+1}\,3^{n+1}} \cdot \frac{n\sqrt{n}\,3^n}{x^n}\right| < 1 \Rightarrow \frac{|x|}{3}\left(\lim_{n\to\infty} \frac{n}{n+1}\right)\left(\sqrt{\lim_{n\to\infty} \frac{n}{n+1}}\right) < 1$

$\Rightarrow \frac{|x|}{3}(1)(1) < 1 \Rightarrow |x| < 3 \Rightarrow -3 < x < 3$; when $x = -3$ we have $\sum_{n=1}^{\infty} \frac{(-1)^n}{n^{3/2}}$, an absolutely convergent series; when $x = 3$ we have $\sum_{n=1}^{\infty} \frac{1}{n^{3/2}}$, a convergent p-series

(a) the radius is 3; the interval of convergence is $-3 \le x \le 3$
(b) the interval of absolute convergence is $-3 \le x \le 3$
(c) there are no values for which the series converges conditionally

10. $\lim_{n\to\infty} \left|\frac{u_{n+1}}{u_n}\right| < 1 \Rightarrow \lim_{n\to\infty} \left|\frac{(x-1)^{n+1}}{\sqrt{n+1}} \cdot \frac{\sqrt{n}}{(x-1)^n}\right| < 1 \Rightarrow |x-1| \sqrt{\lim_{n\to\infty} \frac{n}{n+1}} < 1 \Rightarrow |x-1| < 1$

$\Rightarrow -1 < x - 1 < 1 \Rightarrow 0 < x < 2$; when $x = 0$ we have $\sum_{n=1}^{\infty} \frac{(-1)^n}{n^{1/2}}$, a conditionally convergent series; when $x = 2$ we have $\sum_{n=1}^{\infty} \frac{1}{n^{1/2}}$, a divergent series

(a) the radius is 1; the interval of convergence is $0 \le x < 2$
(b) the interval of absolute convergence is $0 < x < 2$
(c) the series converges conditionally at $x = 0$

11. $\lim_{n \to \infty} \left|\frac{u_{n+1}}{u_n}\right| < 1 \Rightarrow \lim_{n \to \infty} \left|\frac{x^{n+1}}{(n+1)!} \cdot \frac{n!}{x^n}\right| < 1 \Rightarrow |x| \lim_{n \to \infty} \left(\frac{1}{n+1}\right) < 1$ for all x
(a) the radius is ∞; the series converges for all x
(b) the series converges absolutely for all x
(c) there are no values for which the series converges conditionally

12. $\lim_{n \to \infty} \left|\frac{u_{n+1}}{u_n}\right| < 1 \Rightarrow \lim_{n \to \infty} \left|\frac{3^{n+1} x^{n+1}}{(n+1)!} \cdot \frac{n!}{3^n x^n}\right| < 1 \Rightarrow 3|x| \lim_{n \to \infty} \left(\frac{1}{n+1}\right) < 1$ for all x
(a) the radius is ∞; the series converges for all x
(b) the series converges absolutely for all x
(c) there are no values for which the series converges conditionally

13. $\lim_{n \to \infty} \left|\frac{u_{n+1}}{u_n}\right| < 1 \Rightarrow \lim_{n \to \infty} \left|\frac{x^{2n+3}}{(n+1)!} \cdot \frac{n!}{x^{2n+1}}\right| < 1 \Rightarrow x^2 \lim_{n \to \infty} \left(\frac{1}{n+1}\right) < 1$ for all x
(a) the radius is ∞; the series converges for all x
(b) the series converges absolutely for all x
(c) there are no values for which the series converges conditionally

14. $\lim_{n \to \infty} \left|\frac{u_{n+1}}{u_n}\right| < 1 \Rightarrow \lim_{n \to \infty} \left|\frac{(2x+3)^{2n+3}}{(n+1)!} \cdot \frac{n!}{(2x+3)^{2n+1}}\right| < 1 \Rightarrow (2x+3)^2 \lim_{n \to \infty} \left(\frac{1}{n+1}\right) < 1$ for all x
(a) the radius is ∞; the series converges for all x
(b) the series converges absolutely for all x
(c) there are no values for which the series converges conditionally

15. $\lim_{n \to \infty} \left|\frac{u_{n+1}}{u_n}\right| < 1 \Rightarrow \lim_{n \to \infty} \left|\frac{x^{n+1}}{\sqrt{(n+1)^2+3}} \cdot \frac{\sqrt{n^2+3}}{x^n}\right| < 1 \Rightarrow |x| \sqrt{\lim_{n \to \infty} \frac{n^2+3}{n^2+2n+4}} < 1 \Rightarrow |x| < 1$
$\Rightarrow -1 < x < 1$; when $x = -1$ we have $\sum_{n=1}^{\infty} \frac{(-1)^n}{\sqrt{n^2+3}}$, a conditionally convergent series; when $x = 1$ we have $\sum_{n=1}^{\infty} \frac{1}{\sqrt{n^2+3}}$, a divergent series
(a) the radius is 1; the interval of convergence is $-1 \le x < 1$
(b) the interval of absolute convergence is $-1 < x < 1$
(c) the series converges conditionally at $x = -1$

16. $\lim_{n \to \infty} \left|\frac{u_{n+1}}{u_n}\right| < 1 \Rightarrow \lim_{n \to \infty} \left|\frac{x^{n+1}}{\sqrt{(n+1)^2+3}} \cdot \frac{\sqrt{n^2+3}}{x^n}\right| < 1 \Rightarrow |x| \sqrt{\lim_{n \to \infty} \frac{n^2+3}{n^2+2n+4}} < 1 \Rightarrow |x| < 1$
$\Rightarrow -1 < x < 1$; when $x = -1$ we have $\sum_{n=1}^{\infty} \frac{1}{\sqrt{n^2+3}}$, a divergent series; when $x = 1$ we have $\sum_{n=1}^{\infty} \frac{(-1)^n}{\sqrt{n^2+3}}$, a conditionally convergent series
(a) the radius is 1; the interval of convergence is $-1 < x \le 1$
(b) the interval of absolute convergence is $-1 < x < 1$
(c) the series converges conditionally at $x = 1$

17. $\lim_{n\to\infty}\left|\frac{u_{n+1}}{u_n}\right| < 1 \Rightarrow \lim_{n\to\infty}\left|\frac{(n+1)(x+3)^{n+1}}{5^{n+1}}\cdot\frac{5^n}{n(x+3)^n}\right| < 1 \Rightarrow \frac{|x+3|}{5}\lim_{n\to\infty}\left(\frac{n+1}{n}\right) < 1 \Rightarrow \frac{|x+3|}{5} < 1$

$\Rightarrow |x+3| < 5 \Rightarrow -5 < x+3 < 5 \Rightarrow -8 < x < 2$; when $x = -8$ we have $\sum_{n=1}^{\infty}\frac{n(-5)^n}{5^n} = \sum_{n=1}^{\infty}(-1)^n n$, a divergent series; when $x = 2$ we have $\sum_{n=1}^{\infty}\frac{n5^n}{5^n} = \sum_{n=1}^{\infty} n$, a divergent series

(a) the radius is 5; the interval of convergence is $-8 < x < 2$
(b) the interval of absolute convergence is $-8 < x < 2$
(c) there are no values for which the series converges conditionally

18. $\lim_{n\to\infty}\left|\frac{u_{n+1}}{u_n}\right| < 1 \Rightarrow \lim_{n\to\infty}\left|\frac{(n+1)x^{n+1}}{4^{n+1}(n^2+2n+2)}\cdot\frac{4^n(n^2+1)}{nx^n}\right| < 1 \Rightarrow \frac{|x|}{4}\lim_{n\to\infty}\left|\frac{(n+1)(n^2+1)}{n(n^2+2n+2)}\right| < 1 \Rightarrow |x| < 4$

$\Rightarrow -4 < x < 4$; when $x = -4$ we have $\sum_{n=1}^{\infty}\frac{n(-1)^n}{n^2+1}$, a conditionally convergent series; when $x = 4$ we have $\sum_{n=1}^{\infty}\frac{n}{n^2+1}$, a divergent series

(a) the radius is 4; the interval of convergence is $-4 \le x < 4$
(b) the interval of absolute convergence is $-4 < x < 4$
(c) the series converges conditionally at $x = -4$

19. $\lim_{n\to\infty}\left|\frac{u_{n+1}}{u_n}\right| < 1 \Rightarrow \lim_{n\to\infty}\left|\frac{\sqrt{n+1}\,x^{n+1}}{3^{n+1}}\cdot\frac{3^n}{\sqrt{n}\,x^n}\right| < 1 \Rightarrow \frac{|x|}{3}\sqrt{\lim_{n\to\infty}\left(\frac{n+1}{n}\right)} < 1 \Rightarrow \frac{|x|}{3} < 1 \Rightarrow |x| < 3$

$\Rightarrow -3 < x < 3$; when $x = -3$ we have $\sum_{n=1}^{\infty}(-1)^n\sqrt{n}$, a divergent series; when $x = 3$ we have $\sum_{n=1}^{\infty}\sqrt{n}$, a divergent series

(a) the radius is 3; the interval of convergence is $-3 < x < 3$
(b) the interval of absolute convergence is $-3 < x < 3$
(c) there are no values for which the series converges conditionally

20. $\lim_{n\to\infty}\left|\frac{u_{n+1}}{u_n}\right| < 1 \Rightarrow \lim_{n\to\infty}\left|\frac{\sqrt[n+1]{n+1}\,(2x+5)^{n+1}}{\sqrt[n]{n}\,(2x+5)^n}\right| < 1 \Rightarrow |2x+5|\lim_{n\to\infty}\left(\frac{\sqrt[n+1]{n+1}}{\sqrt[n]{n}}\right) < 1$

$\Rightarrow |2x+5|\left(\frac{\lim_{t\to\infty}\sqrt[t]{t}}{\lim_{n\to\infty}\sqrt[n]{n}}\right) < 1 \Rightarrow |2x+5| < 1 \Rightarrow -1 < 2x+5 < 1 \Rightarrow -3 < x < -2$; when $x = -3$ we have $\sum_{n=1}^{\infty}(-1)^n\sqrt[n]{n}$, a divergent series since $\lim_{n\to\infty}\sqrt[n]{n} = 1$; when $x = -2$ we have $\sum_{n=1}^{\infty}\sqrt[n]{n}$, a divergent series

(a) the radius is $\frac{1}{2}$; the interval of convergence is $-3 < x < -2$
(b) the interval of absolute convergence is $-3 < x < -2$
(c) there are no values for which the series converges conditionally

21. $\lim_{n\to\infty}\left|\frac{u_{n+1}}{u_n}\right| < 1 \Rightarrow \lim_{n\to\infty}\left|\frac{\left(1+\frac{1}{n+1}\right)^{n+1}x^{n+1}}{\left(1+\frac{1}{n}\right)^n x^n}\right| < 1 \Rightarrow |x|\left(\frac{\lim_{t\to\infty}\left(1+\frac{1}{t}\right)^t}{\lim_{n\to\infty}\left(1+\frac{1}{n}\right)^n}\right) < 1 \Rightarrow |x|\left(\frac{e}{e}\right) < 1 \Rightarrow |x| < 1$

$\Rightarrow -1 < x < 1$; when $x = -1$ we have $\sum_{n=1}^{\infty}(-1)^n\left(1+\frac{1}{n}\right)^n$, a divergent series by the nth-Term Test since $\lim_{n\to\infty}\left(1+\frac{1}{n}\right)^n = e \ne 0$; when $x = 1$ we have $\sum_{n=1}^{\infty}\left(1+\frac{1}{n}\right)^n$, a divergent series

(a) the radius is 1; the interval of convergence is $-1 < x < 1$
(b) the interval of absolute convergence is $-1 < x < 1$
(c) there are no values for which the series converges conditionally

22. $\lim_{n\to\infty} \left|\frac{u_{n+1}}{u_n}\right| < 1 \Rightarrow \lim_{n\to\infty} \left|\frac{\ln(n+1)x^{n+1}}{x^n \ln n}\right| < 1 \Rightarrow |x| \lim_{n\to\infty} \left|\frac{\left(\frac{1}{n+1}\right)}{\left(\frac{1}{n}\right)}\right| < 1 \Rightarrow |x| \lim_{n\to\infty} \left(\frac{n}{n+1}\right) < 1 \Rightarrow |x| < 1$

$\Rightarrow -1 < x < 1$; when $x = -1$ we have $\sum_{n=1}^{\infty} (-1)^n \ln n$, a divergent series by the nth-Term Test since

$\lim_{n\to\infty} \ln n \neq 0$; when $x = 1$ we have $\sum_{n=1}^{\infty} \ln n$, a divergent series

(a) the radius is 1; the interval of convergence is $-1 < x < 1$
(b) the interval of absolute convergence is $-1 < x < 1$
(c) there are no values for which the series converges conditionally

23. $\lim_{n\to\infty} \left|\frac{u_{n+1}}{u_n}\right| < 1 \Rightarrow \lim_{n\to\infty} \left|\frac{(n+1)^{n+1}x^{n+1}}{n^n x^n}\right| < 1 \Rightarrow |x| \left(\lim_{n\to\infty} \left(1 + \frac{1}{n}\right)^n\right)\left(\lim_{n\to\infty} (n+1)\right) < 1$

$\Rightarrow e|x| \lim_{n\to\infty} (n+1) < 1 \Rightarrow$ only $x = 0$ satisfies this inequality

(a) the radius is 0; the series converges only for $x = 0$
(b) the series converges absolutely only for $x = 0$
(c) there are no values for which the series converges conditionally

24. $\lim_{n\to\infty} \left|\frac{u_{n+1}}{u_n}\right| < 1 \Rightarrow \lim_{n\to\infty} \left|\frac{(n+1)!(x-4)^{n+1}}{n!(x-4)^n}\right| < 1 \Rightarrow |x-4| \lim_{n\to\infty} (n+1) < 1 \Rightarrow$ only $x = 4$ satisfies this inequality

(a) the radius is 0; the series converges only for $x = 4$
(b) the series converges absolutely only for $x = 4$
(c) there are no values for which the series converges conditionally

25. $\lim_{n\to\infty} \left|\frac{u_{n+1}}{u_n}\right| < 1 \Rightarrow \lim_{n\to\infty} \left|\frac{(x+2)^{n+1}}{(n+1)2^{n+1}} \cdot \frac{n2^n}{(x+2)^n}\right| < 1 \Rightarrow \frac{|x+2|}{2} \lim_{n\to\infty} \left(\frac{n}{n+1}\right) < 1 \Rightarrow \frac{|x+2|}{2} < 1 \Rightarrow |x+2| < 2$

$\Rightarrow -2 < x+2 < 2 \Rightarrow -4 < x < 0$; when $x = -4$ we have $\sum_{n=1}^{\infty} \frac{-1}{n}$, a divergent series; when $x = 0$ we have

$\sum_{n=1}^{\infty} \frac{(-1)^{n+1}}{n}$, the alternating harmonic series which converges conditionally

(a) the radius is 2; the interval of convergence is $-4 < x \le 0$
(b) the interval of absolute convergence is $-4 < x < 0$
(c) the series converges conditionally at $x = 0$

26. $\lim_{n\to\infty} \left|\frac{u_{n+1}}{u_n}\right| < 1 \Rightarrow \lim_{n\to\infty} \left|\frac{(-2)^{n+1}(n+2)(x-1)^{n+1}}{(-2)^n(n+1)(x-1)^n}\right| < 1 \Rightarrow 2|x-1| \lim_{n\to\infty} \left(\frac{n+2}{n+1}\right) < 1 \Rightarrow 2|x-1| < 1$

$\Rightarrow |x-1| < \frac{1}{2} \Rightarrow -\frac{1}{2} < x - 1 < \frac{1}{2} \Rightarrow \frac{1}{2} < x < \frac{3}{2}$; when $x = \frac{1}{2}$ we have $\sum_{n=1}^{\infty}(n+1)$, a divergent series; when $x = \frac{3}{2}$

we have $\sum_{n=1}^{\infty}(-1)^n(n+1)$, a divergent series

(a) the radius is $\frac{1}{2}$; the interval of convergence is $\frac{1}{2} < x < \frac{3}{2}$
(b) the interval of absolute convergence is $\frac{1}{2} < x < \frac{3}{2}$
(c) there are no values for which the series converges conditionally

27. $\lim_{n\to\infty} \left|\frac{u_{n+1}}{u_n}\right| < 1 \Rightarrow \lim_{n\to\infty} \left|\frac{x^{n+1}}{(n+1)(\ln(n+1))^2} \cdot \frac{n(\ln n)^2}{x^n}\right| < 1 \Rightarrow |x| \left(\lim_{n\to\infty} \frac{n}{n+1}\right)\left(\lim_{n\to\infty} \frac{\ln n}{\ln(n+1)}\right)^2 < 1$

$\Rightarrow |x|(1)\left(\lim_{n\to\infty} \frac{\left(\frac{1}{n}\right)}{\left(\frac{1}{n+1}\right)}\right)^2 < 1 \Rightarrow |x| \left(\lim_{n\to\infty} \frac{n+1}{n}\right)^2 < 1 \Rightarrow |x| < 1 \Rightarrow -1 < x < 1$; when $x = -1$ we have

$\sum_{n=1}^{\infty} \frac{(-1)^n}{n(\ln n)^2}$ which converges absolutely; when $x = 1$ we have $\sum_{n=1}^{\infty} \frac{1}{n(\ln n)^2}$ which converges

(a) the radius is 1; the interval of convergence is $-1 \le x \le 1$

(b) the interval of absolute convergence is $-1 \le x \le 1$

(c) there are no values for which the series converges conditionally

28. $\lim_{n\to\infty} \left|\frac{u_{n+1}}{u_n}\right| < 1 \Rightarrow \lim_{n\to\infty} \left|\frac{x^{n+1}}{(n+1)\ln(n+1)} \cdot \frac{n\ln(n)}{x^n}\right| < 1 \Rightarrow |x| \left(\lim_{n\to\infty} \frac{n}{n+1}\right)\left(\lim_{n\to\infty} \frac{\ln(n)}{\ln(n+1)}\right) < 1$
$\Rightarrow |x|(1)(1) < 1 \Rightarrow |x| < 1 \Rightarrow -1 < x < 1$; when $x = -1$ we have $\sum_{n=2}^{\infty} \frac{(-1)^n}{n\ln n}$, a convergent alternating series; when $x = 1$ we have $\sum_{n=2}^{\infty} \frac{1}{n\ln n}$ which diverges by Exercise 38, Section 11.3

(a) the radius is 1; the interval of convergence is $-1 \le x < 1$

(b) the interval of absolute convergence is $-1 < x < 1$

(c) the series converges conditionally at $x = -1$

29. $\lim_{n\to\infty} \left|\frac{u_{n+1}}{u_n}\right| < 1 \Rightarrow \lim_{n\to\infty} \left|\frac{(4x-5)^{2n+3}}{(n+1)^{3/2}} \cdot \frac{n^{3/2}}{(4x-5)^{2n+1}}\right| < 1 \Rightarrow (4x-5)^2 \left(\lim_{n\to\infty} \frac{n}{n+1}\right)^{3/2} < 1 \Rightarrow (4x-5)^2 < 1$
$\Rightarrow |4x-5| < 1 \Rightarrow -1 < 4x-5 < 1 \Rightarrow 1 < x < \frac{3}{2}$; when $x = 1$ we have $\sum_{n=1}^{\infty} \frac{(-1)^{2n+1}}{n^{3/2}} = \sum_{n=1}^{\infty} \frac{-1}{n^{3/2}}$ which is absolutely convergent; when $x = \frac{3}{2}$ we have $\sum_{n=1}^{\infty} \frac{(1)^{2n+1}}{n^{3/2}}$, a convergent p-series

(a) the radius is $\frac{1}{4}$; the interval of convergence is $1 \le x \le \frac{3}{2}$

(b) the interval of absolute convergence is $1 \le x \le \frac{3}{2}$

(c) there are no values for which the series converges conditionally

30. $\lim_{n\to\infty} \left|\frac{u_{n+1}}{u_n}\right| < 1 \Rightarrow \lim_{n\to\infty} \left|\frac{(3x+1)^{n+2}}{2n+4} \cdot \frac{2n+2}{(3x+1)^{n+1}}\right| < 1 \Rightarrow |3x+1| \lim_{n\to\infty} \left(\frac{2n+2}{2n+4}\right) < 1 \Rightarrow |3x+1| < 1$
$\Rightarrow -1 < 3x+1 < 1 \Rightarrow -\frac{2}{3} < x < 0$; when $x = -\frac{2}{3}$ we have $\sum_{n=1}^{\infty} \frac{(-1)^{n+1}}{2n+1}$, a conditionally convergent series; when $x = 0$ we have $\sum_{n=1}^{\infty} \frac{(1)^{n+1}}{2n+1} = \sum_{n=1}^{\infty} \frac{1}{2n+1}$, a divergent series

(a) the radius is $\frac{1}{3}$; the interval of convergence is $-\frac{2}{3} \le x < 0$

(b) the interval of absolute convergence is $-\frac{2}{3} < x < 0$

(c) the series converges conditionally at $x = -\frac{2}{3}$

31. $\lim_{n\to\infty} \left|\frac{u_{n+1}}{u_n}\right| < 1 \Rightarrow \lim_{n\to\infty} \left|\frac{(x+\pi)^{n+1}}{\sqrt{n+1}} \cdot \frac{\sqrt{n}}{(x+\pi)^n}\right| < 1 \Rightarrow |x+\pi| \lim_{n\to\infty} \left|\sqrt{\frac{n}{n+1}}\right| < 1$
$\Rightarrow |x+\pi| \sqrt{\lim_{n\to\infty} \left(\frac{n}{n+1}\right)} < 1 \Rightarrow |x+\pi| < 1 \Rightarrow -1 < x+\pi < 1 \Rightarrow -1-\pi < x < 1-\pi$;
when $x = -1-\pi$ we have $\sum_{n=1}^{\infty} \frac{(-1)^n}{\sqrt{n}} = \sum_{n=1}^{\infty} \frac{(-1)^n}{n^{1/2}}$, a conditionally convergent series; when $x = 1-\pi$ we have $\sum_{n=1}^{\infty} \frac{1^n}{\sqrt{n}} = \sum_{n=1}^{\infty} \frac{1}{n^{1/2}}$, a divergent p-series

(a) the radius is 1; the interval of convergence is $(-1-\pi) \le x < (1-\pi)$

(b) the interval of absolute convergence is $-1-\pi < x < 1-\pi$

(c) the series converges conditionally at $x = -1-\pi$

32. $\lim_{n\to\infty} \left|\frac{u_{n+1}}{u_n}\right| < 1 \Rightarrow \lim_{n\to\infty} \left|\frac{\left(x-\sqrt{2}\right)^{2n+3}}{2^{n+1}} \cdot \frac{2^n}{\left(x-\sqrt{2}\right)^{2n+1}}\right| < 1 \Rightarrow \frac{\left(x-\sqrt{2}\right)^2}{2} \lim_{n\to\infty} |1| < 1$
$\Rightarrow \frac{\left(x-\sqrt{2}\right)^2}{2} < 1 \Rightarrow \left(x-\sqrt{2}\right)^2 < 2 \Rightarrow \left|x-\sqrt{2}\right| < \sqrt{2} \Rightarrow -\sqrt{2} < x-\sqrt{2} < \sqrt{2} \Rightarrow 0 < x < 2\sqrt{2}$; when $x = 0$ we have $\sum_{n=1}^{\infty} \frac{\left(-\sqrt{2}\right)^{2n+1}}{2^n} = -\sum_{n=1}^{\infty} \frac{2^{n+1/2}}{2^n} = -\sum_{n=1}^{\infty} \sqrt{2}$ which diverges since $\lim_{n\to\infty} a_n \ne 0$; when $x = 2\sqrt{2}$

we have $\sum_{n=1}^{\infty} \frac{(\sqrt{2})^{2n+1}}{2^n} = \sum_{n=1}^{\infty} \frac{2^{n+1/2}}{2^n} = \sum_{n=1}^{\infty} \sqrt{2}$, a divergent series

(a) the radius is $\sqrt{2}$; the interval of convergence is $0 < x < 2\sqrt{2}$

(b) the interval of absolute convergence is $0 < x < 2\sqrt{2}$

(c) there are no values for which the series converges conditionally

33. $\lim_{n \to \infty} \left|\frac{u_{n+1}}{u_n}\right| < 1 \Rightarrow \lim_{n \to \infty} \left|\frac{(x-1)^{2n+2}}{4^{n+1}} \cdot \frac{4^n}{(x-1)^{2n}}\right| < 1 \Rightarrow \frac{(x-1)^2}{4} \lim_{n \to \infty} |1| < 1 \Rightarrow (x-1)^2 < 4 \Rightarrow |x-1| < 2$ $\Rightarrow -2 < x-1 < 2 \Rightarrow -1 < x < 3$; at $x = -1$ we have $\sum_{n=0}^{\infty} \frac{(-2)^{2n}}{4^n} = \sum_{n=0}^{\infty} \frac{4^n}{4^n} = \sum_{n=0}^{\infty} 1$, which diverges; at $x = 3$ we have $\sum_{n=0}^{\infty} \frac{2^{2n}}{4^n} = \sum_{n=0}^{\infty} \frac{4^n}{4^n} = \sum_{n=0}^{\infty} 1$, a divergent series; the interval of convergence is $-1 < x < 3$; the series $\sum_{n=0}^{\infty} \frac{(x-1)^{2n}}{4^n} = \sum_{n=0}^{\infty} \left(\left(\frac{x-1}{2}\right)^2\right)^n$ is a convergent geometric series when $-1 < x < 3$ and the sum is $\frac{1}{1-\left(\frac{x-1}{2}\right)^2} = \frac{1}{\left[\frac{4-(x-1)^2}{4}\right]} = \frac{4}{4-x^2+2x-1} = \frac{4}{3+2x-x^2}$

34. $\lim_{n \to \infty} \left|\frac{u_{n+1}}{u_n}\right| < 1 \Rightarrow \lim_{n \to \infty} \left|\frac{(x+1)^{2n+2}}{9^{n+1}} \cdot \frac{9^n}{(x+1)^{2n}}\right| < 1 \Rightarrow \frac{(x+1)^2}{9} \lim_{n \to \infty} |1| < 1 \Rightarrow (x+1)^2 < 9 \Rightarrow |x+1| < 3$ $\Rightarrow -3 < x+1 < 3 \Rightarrow -4 < x < 2$; when $x = -4$ we have $\sum_{n=0}^{\infty} \frac{(-3)^{2n}}{9^n} = \sum_{n=0}^{\infty} 1$ which diverges; at $x = 2$ we have $\sum_{n=0}^{\infty} \frac{3^{2n}}{9^n} = \sum_{n=0}^{\infty} 1$ which also diverges; the interval of convergence is $-4 < x < 2$; the series $\sum_{n=0}^{\infty} \frac{(x+1)^{2n}}{9^n} = \sum_{n=0}^{\infty} \left(\left(\frac{x+1}{3}\right)^2\right)^n$ is a convergent geometric series when $-4 < x < 2$ and the sum is $\frac{1}{1-\left(\frac{x+1}{3}\right)^2} = \frac{1}{\left[\frac{9-(x+1)^2}{9}\right]} = \frac{9}{9-x^2-2x-1} = \frac{9}{8-2x-x^2}$

35. $\lim_{n \to \infty} \left|\frac{u_{n+1}}{u_n}\right| < 1 \Rightarrow \lim_{n \to \infty} \left|\frac{(\sqrt{x}-2)^{n+1}}{2^{n+1}} \cdot \frac{2^n}{(\sqrt{x}-2)^n}\right| < 1 \Rightarrow |\sqrt{x}-2| < 2 \Rightarrow -2 < \sqrt{x}-2 < 2 \Rightarrow 0 < \sqrt{x} < 4$ $\Rightarrow 0 < x < 16$; when $x = 0$ we have $\sum_{n=0}^{\infty} (-1)^n$, a divergent series; when $x = 16$ we have $\sum_{n=0}^{\infty} (1)^n$, a divergent series; the interval of convergence is $0 < x < 16$; the series $\sum_{n=0}^{\infty} \left(\frac{\sqrt{x}-2}{2}\right)^n$ is a convergent geometric series when $0 < x < 16$ and its sum is $\frac{1}{1-\left(\frac{\sqrt{x}-2}{2}\right)} = \frac{1}{\left(\frac{2-\sqrt{x}+2}{2}\right)} = \frac{2}{4-\sqrt{x}}$

36. $\lim_{n \to \infty} \left|\frac{u_{n+1}}{u_n}\right| < 1 \Rightarrow \lim_{n \to \infty} \left|\frac{(\ln x)^{n+1}}{(\ln x)^n}\right| < 1 \Rightarrow |\ln x| < 1 \Rightarrow -1 < \ln x < 1 \Rightarrow e^{-1} < x < e$; when $x = e^{-1}$ or e we obtain the series $\sum_{n=0}^{\infty} 1^n$ and $\sum_{n=0}^{\infty} (-1)^n$ which both diverge; the interval of convergence is $e^{-1} < x < e$; $\sum_{n=0}^{\infty} (\ln x)^n = \frac{1}{1-\ln x}$ when $e^{-1} < x < e$

37. $\lim_{n \to \infty} \left|\frac{u_{n+1}}{u_n}\right| < 1 \Rightarrow \lim_{n \to \infty} \left|\left(\frac{x^2+1}{3}\right)^{n+1} \cdot \left(\frac{3}{x^2+1}\right)^n\right| < 1 \Rightarrow \frac{(x^2+1)}{3} \lim_{n \to \infty} |1| < 1 \Rightarrow \frac{x^2+1}{3} < 1 \Rightarrow x^2 < 2$ $\Rightarrow |x| < \sqrt{2} \Rightarrow -\sqrt{2} < x < \sqrt{2}$; at $x = \pm\sqrt{2}$ we have $\sum_{n=0}^{\infty} (1)^n$ which diverges; the interval of convergence is $-\sqrt{2} < x < \sqrt{2}$; the series $\sum_{n=0}^{\infty} \left(\frac{x^2+1}{3}\right)^n$ is a convergent geometric series when $-\sqrt{2} < x < \sqrt{2}$ and its sum is $\frac{1}{1-\left(\frac{x^2+1}{3}\right)} = \frac{1}{\left(\frac{3-x^2-1}{3}\right)} = \frac{3}{2-x^2}$

38. $\lim_{n\to\infty}\left|\frac{u_{n+1}}{u_n}\right| < 1 \Rightarrow \lim_{n\to\infty}\left|\frac{(x^2-1)^{n+1}}{2^{n+1}}\cdot\frac{2^n}{(x^2+1)^n}\right| < 1 \Rightarrow |x^2-1| < 2 \Rightarrow -\sqrt{3} < x < \sqrt{3}$; when $x = \pm\sqrt{3}$ we have $\sum_{n=0}^{\infty} 1^n$, a divergent series; the interval of convergence is $-\sqrt{3} < x < \sqrt{3}$; the series $\sum_{n=0}^{\infty}\left(\frac{x^2-1}{2}\right)^n$ is a convergent geometric series when $-\sqrt{3} < x < \sqrt{3}$ and its sum is $\frac{1}{1-\left(\frac{x^2-1}{2}\right)} = \frac{1}{\left(\frac{2-(x^2-1)}{2}\right)} = \frac{2}{3-x^2}$

39. $\lim_{n\to\infty}\left|\frac{(x-3)^{n+1}}{2^{n+1}}\cdot\frac{2^n}{(x-3)^n}\right| < 1 \Rightarrow |x-3| < 2 \Rightarrow 1 < x < 5$; when $x = 1$ we have $\sum_{n=1}^{\infty}(1)^n$ which diverges; when $x = 5$ we have $\sum_{n=1}^{\infty}(-1)^n$ which also diverges; the interval of convergence is $1 < x < 5$; the sum of this convergent geometric series is $\frac{1}{1+\left(\frac{x-3}{2}\right)} = \frac{2}{x-1}$. If $f(x) = 1 - \frac{1}{2}(x-3) + \frac{1}{4}(x-3)^2 + \ldots + \left(-\frac{1}{2}\right)^n(x-3)^n + \ldots$ $= \frac{2}{x-1}$ then $f'(x) = -\frac{1}{2} + \frac{1}{2}(x-3) + \ldots + \left(-\frac{1}{2}\right)^n n(x-3)^{n-1} + \ldots$ is convergent when $1 < x < 5$, and diverges when $x = 1$ or 5. The sum for $f'(x)$ is $\frac{-2}{(x-1)^2}$, the derivative of $\frac{2}{x-1}$.

40. If $f(x) = 1 - \frac{1}{2}(x-3) + \frac{1}{4}(x-3)^2 + \ldots + \left(-\frac{1}{2}\right)^n(x-3)^n + \ldots = \frac{2}{x-1}$ then $\int f(x)\,dx$ $= x - \frac{(x-3)^2}{4} + \frac{(x-3)^3}{12} + \ldots + \left(-\frac{1}{2}\right)^n\frac{(x-3)^{n+1}}{n+1} + \ldots$. At $x = 1$ the series $\sum_{n=1}^{\infty}\frac{-2}{n+1}$ diverges; at $x = 5$ the series $\sum_{n=1}^{\infty}\frac{(-1)^n 2}{n+1}$ converges. Therefore the interval of convergence is $1 < x \le 5$ and the sum is $2\ln|x-1| + (3 - \ln 4)$, since $\int \frac{2}{x-1}\,dx = 2\ln|x-1| + C$, where $C = 3 - \ln 4$ when $x = 3$.

41. (a) Differentiate the series for sin x to get $\cos x = 1 - \frac{3x^2}{3!} + \frac{5x^4}{5!} - \frac{7x^6}{7!} + \frac{9x^8}{9!} - \frac{11x^{10}}{11!} + \ldots$ $= 1 - \frac{x^2}{2!} + \frac{x^4}{4!} - \frac{x^6}{6!} + \frac{x^8}{8!} - \frac{x^{10}}{10!} + \ldots$. The series converges for all values of x since $\lim_{n\to\infty}\left|\frac{x^{2n+2}}{(2n+2)!}\cdot\frac{(2n)!}{x^{2n}}\right| = x^2\lim_{n\to\infty}\left(\frac{1}{(2n+1)(2n+2)}\right) = 0 < 1$ for all x.

(b) $\sin 2x = 2x - \frac{2^3x^3}{3!} + \frac{2^5x^5}{5!} - \frac{2^7x^7}{7!} + \frac{2^9x^9}{9!} - \frac{2^{11}x^{11}}{11!} + \ldots = 2x - \frac{8x^3}{3!} + \frac{32x^5}{5!} - \frac{128x^7}{7!} + \frac{512x^9}{9!} - \frac{2048x^{11}}{11!} + \ldots$

(c) $2\sin x\cos x = 2\left[(0\cdot 1) + (0\cdot 0 + 1\cdot 1)x + \left(0\cdot\frac{-1}{2} + 1\cdot 0 + 0\cdot 1\right)x^2 + \left(0\cdot 0 - 1\cdot\frac{1}{2} + 0\cdot 0 - 1\cdot\frac{1}{3!}\right)x^3\right.$ $+ \left(0\cdot\frac{1}{4!} + 1\cdot 0 - 0\cdot\frac{1}{2} - 0\cdot\frac{1}{3!} + 0\cdot 1\right)x^4 + \left(0\cdot 0 + 1\cdot\frac{1}{4!} + 0\cdot 0 + \frac{1}{2}\cdot\frac{1}{3!} + 0\cdot 0 + 1\cdot\frac{1}{5!}\right)x^5$ $\left.+ \left(0\cdot\frac{1}{6!} + 1\cdot 0 + 0\cdot\frac{1}{4!} + 0\cdot\frac{1}{3!} + 0\cdot\frac{1}{2} + 0\cdot\frac{1}{5!} + 0\cdot 1\right)x^6 + \ldots\right] = 2\left[x - \frac{4x^3}{3!} + \frac{16x^5}{5!} - \ldots\right]$ $= 2x - \frac{2^3x^3}{3!} + \frac{2^5x^5}{5!} - \frac{2^7x^7}{7!} + \frac{2^9x^9}{9!} - \frac{2^{11}x^{11}}{11!} + \ldots$

42. (a) $\frac{d}{x}(e^x) = 1 + \frac{2x}{2!} + \frac{3x^2}{3!} + \frac{4x^3}{4!} + \frac{5x^4}{5!} + \ldots = 1 + x + \frac{x^2}{2!} + \frac{x^3}{3!} + \frac{x^4}{4!} + \ldots = e^x$; thus the derivative of e^x is e^x itself

(b) $\int e^x\,dx = e^x + C = x + \frac{x^2}{2} + \frac{x^3}{3!} + \frac{x^4}{4!} + \frac{x^5}{5!} + \ldots + C$, which is the general antiderivative of e^x

(c) $e^{-x} = 1 - x + \frac{x^2}{2!} - \frac{x^3}{3!} + \frac{x^4}{4!} - \frac{x^5}{5!} + \ldots$; $e^{-x}\cdot e^x = 1\cdot 1 + (1\cdot 1 - 1\cdot 1)x + \left(1\cdot\frac{1}{2!} - 1\cdot 1 + \frac{1}{2!}\cdot 1\right)x^2$ $+ \left(1\cdot\frac{1}{3!} - 1\cdot\frac{1}{2!} + \frac{1}{2!}\cdot 1 - \frac{1}{3!}\cdot 1\right)x^3 + \left(1\cdot\frac{1}{4!} - 1\cdot\frac{1}{3!} + \frac{1}{2!}\cdot\frac{1}{2!} - \frac{1}{3!}\cdot 1 + \frac{1}{4!}\cdot 1\right)x^4$ $+ \left(1\cdot\frac{1}{5!} - 1\cdot\frac{1}{4!} + \frac{1}{2!}\cdot\frac{1}{3!} - \frac{1}{3!}\cdot\frac{1}{2!} + \frac{1}{4!}\cdot 1 - \frac{1}{5!}\cdot 1\right)x^5 + \ldots = 1 + 0 + 0 + 0 + 0 + 0 + \ldots$

43. (a) $\ln|\sec x| + C = \int \tan x\,dx = \int\left(x + \frac{x^3}{3} + \frac{2x^5}{15} + \frac{17x^7}{315} + \frac{62x^9}{2835} + \ldots\right)dx$ $= \frac{x^2}{2} + \frac{x^4}{12} + \frac{x^6}{45} + \frac{17x^8}{2520} + \frac{31x^{10}}{14,175} + \ldots + C$; $x = 0 \Rightarrow C = 0 \Rightarrow \ln|\sec x| = \frac{x^2}{2} + \frac{x^4}{12} + \frac{x^6}{45} + \frac{17x^8}{2520} + \frac{31x^{10}}{14,175} + \ldots$, converges when $-\frac{\pi}{2} < x < \frac{\pi}{2}$

(b) $\sec^2 x = \frac{d(\tan x)}{dx} = \frac{d}{dx}\left(x + \frac{x^3}{3} + \frac{2x^5}{15} + \frac{17x^7}{315} + \frac{62x^9}{2835} + \ldots\right) = 1 + x^2 + \frac{2x^4}{3} + \frac{17x^6}{45} + \frac{62x^8}{315} + \ldots$, converges when $-\frac{\pi}{2} < x < \frac{\pi}{2}$

(c) $\sec^2 x = (\sec x)(\sec x) = \left(1 + \frac{x^2}{2} + \frac{5x^4}{24} + \frac{61x^6}{720} + \ldots\right)\left(1 + \frac{x^2}{2} + \frac{5x^4}{24} + \frac{61x^6}{720} + \ldots\right)$
$= 1 + \left(\frac{1}{2} + \frac{1}{2}\right)x^2 + \left(\frac{5}{24} + \frac{1}{4} + \frac{5}{24}\right)x^4 + \left(\frac{61}{720} + \frac{5}{48} + \frac{5}{48} + \frac{61}{720}\right)x^6 + \ldots$
$= 1 + x^2 + \frac{2x^4}{3} + \frac{17x^6}{45} + \frac{62x^8}{315} + \ldots, -\frac{\pi}{2} < x < \frac{\pi}{2}$

44. (a) $\ln|\sec x + \tan x| + C = \int \sec x\, dx = \int \left(1 + \frac{x^2}{2} + \frac{5x^4}{24} + \frac{61x^6}{720} + \ldots\right) dx$
$= x + \frac{x^3}{6} + \frac{x^5}{24} + \frac{61x^7}{5040} + \frac{277x^9}{72{,}576} + \ldots + C;\ x = 0 \Rightarrow C = 0 \Rightarrow \ln|\sec x + \tan x|$
$= x + \frac{x^3}{6} + \frac{x^5}{24} + \frac{61x^7}{5040} + \frac{277x^9}{72{,}576} + \ldots$, converges when $-\frac{\pi}{2} < x < \frac{\pi}{2}$

(b) $\sec x \tan x = \frac{d(\sec x)}{dx} = \frac{d}{dx}\left(1 + \frac{x^2}{2} + \frac{5x^4}{24} + \frac{61x^6}{720} + \ldots\right) = x + \frac{5x^3}{6} + \frac{61x^5}{120} + \frac{277x^7}{1008} + \ldots$, converges when $-\frac{\pi}{2} < x < \frac{\pi}{2}$

(c) $(\sec x)(\tan x) = \left(1 + \frac{x^2}{2} + \frac{5x^4}{24} + \frac{61x^6}{720} + \ldots\right)\left(x + \frac{x^3}{3} + \frac{2x^5}{15} + \frac{17x^7}{315} + \ldots\right)$
$= x + \left(\frac{1}{3} + \frac{1}{2}\right)x^3 + \left(\frac{2}{15} + \frac{1}{6} + \frac{5}{24}\right)x^5 + \left(\frac{17}{315} + \frac{1}{15} + \frac{5}{72} + \frac{61}{720}\right)x^7 + \ldots = x + \frac{5x^3}{6} + \frac{61x^5}{120} + \frac{277x^7}{1008} + \ldots,$
$-\frac{\pi}{2} < x < \frac{\pi}{2}$

45. (a) If $f(x) = \sum_{n=0}^{\infty} a_n x^n$, then $f^{(k)}(x) = \sum_{n=k}^{\infty} n(n-1)(n-2)\cdots(n-(k-1))\, a_n x^{n-k}$ and $f^{(k)}(0) = k!a_k$
$\Rightarrow a_k = \frac{f^{(k)}(0)}{k!}$; likewise if $f(x) = \sum_{n=0}^{\infty} b_n x^n$, then $b_k = \frac{f^{(k)}(0)}{k!} \Rightarrow a_k = b_k$ for every nonnegative integer k

(b) If $f(x) = \sum_{n=0}^{\infty} a_n x^n = 0$ for all x, then $f^{(k)}(x) = 0$ for all x $\Rightarrow$ from part (a) that $a_k = 0$ for every nonnegative integer k

46. $\frac{1}{1-x} = 1 + x + x^2 + x^3 + x^4 + \ldots \Rightarrow x\left[\frac{1}{(1-x)^2}\right] = x\left(1 + 2x + 3x^2 + 4x^3 + \ldots\right) \Rightarrow \frac{x}{(1-x)^2}$
$= x + 2x^2 + 3x^3 + 4x^4 + \ldots \Rightarrow x\left[\frac{1+x}{(1-x)^3}\right] = x\left(1 + 4x + 9x^2 + 16x^3 + \ldots\right) \Rightarrow \frac{x+x^2}{(1-x)^3}$
$= x + 4x^2 + 9x^3 + 16x^4 + \ldots \Rightarrow \frac{\left(\frac{1}{2} + \frac{1}{4}\right)}{\left(\frac{1}{8}\right)} = \frac{1}{2} + \frac{4}{4} + \frac{9}{8} + \frac{16}{16} + \ldots \Rightarrow \sum_{n=1}^{\infty} \frac{n^2}{2^n} = 6$

47. The series $\sum_{n=1}^{\infty} \frac{x^n}{n}$ converges conditionally at the left-hand endpoint of its interval of convergence $[-1, 1)$; the series $\sum_{n=1}^{\infty} \frac{x^n}{(n^2)}$ converges absolutely at the left-hand endpoint of its interval of convergence $[-1, 1]$

48. Answers will vary. For instance:

(a) $\sum_{n=1}^{\infty} \left(\frac{x}{3}\right)^n$ (b) $\sum_{n=1}^{\infty} (x+1)^n$ (c) $\sum_{n=1}^{\infty} \left(\frac{x-3}{2}\right)^n$

11.8 TAYLOR AND MACLAURIN SERIES

1. $f(x) = \ln x$, $f'(x) = \frac{1}{x}$, $f''(x) = -\frac{1}{x^2}$, $f'''(x) = \frac{2}{x^3}$; $f(1) = \ln 1 = 0$, $f'(1) = 1$, $f''(1) = -1$, $f'''(1) = 2 \Rightarrow P_0(x) = 0$, $P_1(x) = (x-1)$, $P_2(x) = (x-1) - \frac{1}{2}(x-1)^2$, $P_3(x) = (x-1) - \frac{1}{2}(x-1)^2 + \frac{1}{3}(x-1)^3$

2. $f(x) = \ln(1+x)$, $f'(x) = \frac{1}{1+x} = (1+x)^{-1}$, $f''(x) = -(1+x)^{-2}$, $f'''(x) = 2(1+x)^{-3}$; $f(0) = \ln 1 = 0$, $f'(0) = \frac{1}{1} = 1$, $f''(0) = -(1)^{-2} = -1$, $f'''(0) = 2(1)^{-3} = 2 \Rightarrow P_0(x) = 0$, $P_1(x) = x$, $P_2(x) = x - \frac{x^2}{2}$, $P_3(x) = x - \frac{x^2}{2} + \frac{x^3}{3}$

3. $f(x)=\frac{1}{x}=x^{-1}$, $f'(x)=-x^{-2}$, $f''(x)=2x^{-3}$, $f'''(x)=-6x^{-4}$; $f(2)=\frac{1}{2}$, $f'(2)=-\frac{1}{4}$, $f''(2)=\frac{1}{4}$, $f'''(x)=-\frac{3}{8}$
$\Rightarrow P_0(x)=\frac{1}{2}$, $P_1(x)=\frac{1}{2}-\frac{1}{4}(x-2)$, $P_2(x)=\frac{1}{2}-\frac{1}{4}(x-2)+\frac{1}{8}(x-2)^2$,
$P_3(x)=\frac{1}{2}-\frac{1}{4}(x-2)+\frac{1}{8}(x-2)^2-\frac{1}{16}(x-2)^3$

4. $f(x)=(x+2)^{-1}$, $f'(x)=-(x+2)^{-2}$, $f''(x)=2(x+2)^{-3}$, $f'''(x)=-6(x+2)^{-4}$; $f(0)=(2)^{-1}=\frac{1}{2}$, $f'(0)=-(2)^{-2}$
$=-\frac{1}{4}$, $f''(0)=2(2)^{-3}=\frac{1}{4}$, $f'''(0)=-6(2)^{-4}=-\frac{3}{8}$ $\Rightarrow P_0(x)=\frac{1}{2}$, $P_1(x)=\frac{1}{2}-\frac{x}{4}$, $P_2(x)=\frac{1}{2}-\frac{x}{4}+\frac{x^2}{8}$,
$P_3(x)=\frac{1}{2}-\frac{x}{4}+\frac{x^2}{8}-\frac{x^3}{16}$

5. $f(x)=\sin x$, $f'(x)=\cos x$, $f''(x)=-\sin x$, $f'''(x)=-\cos x$; $f\left(\frac{\pi}{4}\right)=\sin\frac{\pi}{4}=\frac{\sqrt{2}}{2}$, $f'\left(\frac{\pi}{4}\right)=\cos\frac{\pi}{4}=\frac{\sqrt{2}}{2}$,
$f''\left(\frac{\pi}{4}\right)=-\sin\frac{\pi}{4}=-\frac{\sqrt{2}}{2}$, $f'''\left(\frac{\pi}{4}\right)=-\cos\frac{\pi}{4}=-\frac{\sqrt{2}}{2}$ $\Rightarrow P_0=\frac{\sqrt{2}}{2}$, $P_1(x)=\frac{\sqrt{2}}{2}+\frac{\sqrt{2}}{2}\left(x-\frac{\pi}{4}\right)$,
$P_2(x)=\frac{\sqrt{2}}{2}+\frac{\sqrt{2}}{2}\left(x-\frac{\pi}{4}\right)-\frac{\sqrt{2}}{4}\left(x-\frac{\pi}{4}\right)^2$, $P_3(x)=\frac{\sqrt{2}}{2}+\frac{\sqrt{2}}{2}\left(x-\frac{\pi}{4}\right)-\frac{\sqrt{2}}{4}\left(x-\frac{\pi}{4}\right)^2-\frac{\sqrt{2}}{12}\left(x-\frac{\pi}{4}\right)^3$

6. $f(x)=\cos x$, $f'(x)=-\sin x$, $f''(x)=-\cos x$, $f'''(x)=\sin x$; $f\left(\frac{\pi}{4}\right)=\cos\frac{\pi}{4}=\frac{1}{\sqrt{2}}$,
$f'\left(\frac{\pi}{4}\right)=-\sin\frac{\pi}{4}=-\frac{1}{\sqrt{2}}$, $f''\left(\frac{\pi}{4}\right)=-\cos\frac{\pi}{4}=-\frac{1}{\sqrt{2}}$, $f'''\left(\frac{\pi}{4}\right)=\sin\frac{\pi}{4}=\frac{1}{\sqrt{2}}$ $\Rightarrow P_0(x)=\frac{1}{\sqrt{2}}$,
$P_1(x)=\frac{1}{\sqrt{2}}-\frac{1}{\sqrt{2}}\left(x-\frac{\pi}{4}\right)$, $P_2(x)=\frac{1}{\sqrt{2}}-\frac{1}{\sqrt{2}}\left(x-\frac{\pi}{4}\right)-\frac{1}{2\sqrt{2}}\left(x-\frac{\pi}{4}\right)^2$,
$P_3(x)=\frac{1}{\sqrt{2}}-\frac{1}{\sqrt{2}}\left(x-\frac{\pi}{4}\right)-\frac{1}{2\sqrt{2}}\left(x-\frac{\pi}{4}\right)^2+\frac{1}{6\sqrt{2}}\left(x-\frac{\pi}{4}\right)^3$

7. $f(x)=\sqrt{x}=x^{1/2}$, $f'(x)=\left(\frac{1}{2}\right)x^{-1/2}$, $f''(x)=\left(-\frac{1}{4}\right)x^{-3/2}$, $f'''(x)=\left(\frac{3}{8}\right)x^{-5/2}$; $f(4)=\sqrt{4}=2$,
$f'(4)=\left(\frac{1}{2}\right)4^{-1/2}=\frac{1}{4}$, $f''(4)=\left(-\frac{1}{4}\right)4^{-3/2}=-\frac{1}{32}$, $f'''(4)=\left(\frac{3}{8}\right)4^{-5/2}=\frac{3}{256}$ $\Rightarrow P_0(x)=2$, $P_1(x)=2+\frac{1}{4}(x-4)$,
$P_2(x)=2+\frac{1}{4}(x-4)-\frac{1}{64}(x-4)^2$, $P_3(x)=2+\frac{1}{4}(x-4)-\frac{1}{64}(x-4)^2+\frac{1}{512}(x-4)^3$

8. $f(x)=(x+4)^{1/2}$, $f'(x)=\left(\frac{1}{2}\right)(x+4)^{-1/2}$, $f''(x)=\left(-\frac{1}{4}\right)(x+4)^{-3/2}$, $f'''(x)=\left(\frac{3}{8}\right)(x+4)^{-5/2}$; $f(0)=(4)^{1/2}=2$,
$f'(0)=\left(\frac{1}{2}\right)(4)^{-1/2}=\frac{1}{4}$, $f''(0)=\left(-\frac{1}{4}\right)(4)^{-3/2}=-\frac{1}{32}$, $f'''(0)=\left(\frac{3}{8}\right)(4)^{-5/2}=\frac{3}{256}$ $\Rightarrow P_0(x)=2$,
$P_1(x)=2+\frac{1}{4}x$, $P_2(x)=2+\frac{1}{4}x-\frac{1}{64}x^2$, $P_3(x)=2+\frac{1}{4}x-\frac{1}{64}x^2+\frac{1}{512}x^3$

9. $e^x=\sum_{n=0}^{\infty}\frac{x^n}{n!}\Rightarrow e^{-x}=\sum_{n=0}^{\infty}\frac{(-x)^n}{n!}=1-x+\frac{x^2}{2!}-\frac{x^3}{3!}+\frac{x^4}{4!}-\ldots$

10. $e^x=\sum_{n=0}^{\infty}\frac{x^n}{n!}\Rightarrow e^{x/2}=\sum_{n=0}^{\infty}\frac{\left(\frac{x}{2}\right)^n}{n!}=1+\frac{x}{2}+\frac{x^2}{4\cdot 2!}+\frac{x^3}{2^3\cdot 3!}+\frac{x^4}{2^4\cdot 4!}+\ldots$

11. $f(x)=(1+x)^{-1}\Rightarrow f'(x)=-(1+x)^{-2}$, $f''(x)=2(1+x)^{-3}$, $f'''(x)=-3!(1+x)^{-4}\Rightarrow\ldots f^{(k)}(x)$
$=(-1)^k k!(1+x)^{-k-1}$; $f(0)=1$, $f'(0)=-1$, $f''(0)=2$, $f'''(0)=-3!$, $\ldots$, $f^{(k)}(0)=(-1)^k k!$
$\Rightarrow\frac{1}{1+x}=1-x+x^2-x^3+\ldots=\sum_{n=0}^{\infty}(-x)^n=\sum_{n=0}^{\infty}(-1)^n x^n$

12. $f(x)=(1-x)^{-1}\Rightarrow f'(x)=(1-x)^{-2}$, $f''(x)=2(1-x)^{-3}$, $f'''(x)=3!(1-x)^{-4}\Rightarrow\ldots f^{(k)}(x)$
$=k!(1-x)^{-k-1}$; $f(0)=1$, $f'(0)=1$, $f''(0)=2$, $f'''(0)=3!$, $\ldots$, $f^{(k)}(0)=k!$
$\Rightarrow\frac{1}{1-x}=1+x+x^2+x^3+\ldots=\sum_{n=0}^{\infty}x^n$

13. $\sin x=\sum_{n=0}^{\infty}\frac{(-1)^n x^{2n+1}}{(2n+1)!}\Rightarrow\sin 3x=\sum_{n=0}^{\infty}\frac{(-1)^n(3x)^{2n+1}}{(2n+1)!}=\sum_{n=0}^{\infty}\frac{(-1)^n 3^{2n+1}x^{2n+1}}{(2n+1)!}=3x-\frac{3^3x^3}{3!}+\frac{3^5x^5}{5!}-\ldots$

14. $\sin x = \sum_{n=0}^{\infty} \frac{(-1)^n x^{2n+1}}{(2n+1)!} \Rightarrow \sin \frac{x}{2} = \sum_{n=0}^{\infty} \frac{(-1)^n \left(\frac{x}{2}\right)^{2n+1}}{(2n+1)!} = \sum_{n=0}^{\infty} \frac{(-1)^n x^{2n+1}}{2^{2n+1}(2n+1)!} = \frac{x}{2} - \frac{x^3}{2^3 \cdot 3!} + \frac{x^5}{2^5 \cdot 5!} + \ldots$

15. $7 \cos(-x) = 7 \cos x = 7 \sum_{n=0}^{\infty} \frac{(-1)^n x^{2n}}{(2n)!} = 7 - \frac{7x^2}{2!} + \frac{7x^4}{4!} - \frac{7x^6}{6!} + \ldots$, since the cosine is an even function

16. $\cos x = \sum_{n=0}^{\infty} \frac{(-1)^n x^{2n}}{(2n)!} \Rightarrow 5 \cos \pi x = 5 \sum_{n=0}^{\infty} \frac{(-1)^n (\pi x)^{2n}}{(2n)!} = 5 - \frac{5\pi^2 x^2}{2!} + \frac{5\pi^4 x^4}{4!} - \frac{5\pi^6 x^6}{6!} + \ldots$

17. $\cosh x = \frac{e^x + e^{-x}}{2} = \frac{1}{2}\left[\left(1 + x^2 + \frac{x^2}{2!} + \frac{x^3}{3!} + \frac{x^4}{4!} + \ldots\right) + \left(1 - x + \frac{x^2}{2!} - \frac{x^3}{3!} + \frac{x^4}{4!} - \ldots\right)\right] = 1 + \frac{x^2}{2!} + \frac{x^4}{4!} + \frac{x^6}{6!} + \ldots$
$= \sum_{n=0}^{\infty} \frac{x^{2n}}{(2n)!}$

18. $\sinh x = \frac{e^x - e^{-x}}{2} = \frac{1}{2}\left[\left(1 + x + \frac{x^2}{2!} + \frac{x^3}{3!} + \frac{x^4}{4!} + \ldots\right) - \left(1 - x + \frac{x^2}{2!} - \frac{x^3}{3!} + \frac{x^4}{4!} - \ldots\right)\right] = x + \frac{x^3}{3!} + \frac{x^5}{5!} + \frac{x^6}{6!} + \ldots$
$= \sum_{n=0}^{\infty} \frac{x^{2n+1}}{(2n+1)!}$

19. $f(x) = x^4 - 2x^3 - 5x + 4 \Rightarrow f'(x) = 4x^3 - 6x^2 - 5, f''(x) = 12x^2 - 12x, f'''(x) = 24x - 12, f^{(4)}(x) = 24$
$\Rightarrow f^{(n)}(x) = 0$ if $n \geq 5$; $f(0) = 4, f'(0) = -5, f''(0) = 0, f'''(0) = -12, f^{(4)}(0) = 24, f^{(n)}(0) = 0$ if $n \geq 5$
$\Rightarrow x^4 - 2x^3 - 5x + 4 = 4 - 5x - \frac{12}{3!}x^3 + \frac{24}{4!}x^4 = x^4 - 2x^3 - 5x + 4$ itself

20. $f(x) = (x+1)^2 \Rightarrow f'(x) = 2(x+1); f''(x) = 2 \Rightarrow f^{(n)}(x) = 0$ if $n \geq 3$; $f(0) = 1, f'(0) = 2, f''(0) = 2, f^{(n)}(0) = 0$ if $n \geq 3 \Rightarrow (x+1)^2 = 1 + 2x + \frac{2}{2!}x^2 = 1 + 2x + x^2$

21. $f(x) = x^3 - 2x + 4 \Rightarrow f'(x) = 3x^2 - 2, f''(x) = 6x, f'''(x) = 6 \Rightarrow f^{(n)}(x) = 0$ if $n \geq 4$; $f(2) = 8, f'(2) = 10,$
$f''(2) = 12, f'''(2) = 6, f^{(n)}(2) = 0$ if $n \geq 4 \Rightarrow x^3 - 2x + 4 = 8 + 10(x-2) + \frac{12}{2!}(x-2)^2 + \frac{6}{3!}(x-2)^3$
$= 8 + 10(x-2) + 6(x-2)^2 + (x-2)^3$

22. $f(x) = 2x^3 + x^2 + 3x - 8 \Rightarrow f'(x) = 6x^2 + 2x + 3, f''(x) = 12x + 2, f'''(x) = 12 \Rightarrow f^{(n)}(x) = 0$ if $n \geq 4$; $f(1) = -2,$
$f'(1) = 11, f''(1) = 14, f'''(1) = 12, f^{(n)}(1) = 0$ if $n \geq 4 \Rightarrow 2x^3 + x^2 + 3x - 8$
$= -2 + 11(x-1) + \frac{14}{2!}(x-1)^2 + \frac{12}{3!}(x-1)^3 = -2 + 11(x-1) + 7(x-1)^2 + 2(x-1)^3$

23. $f(x) = x^4 + x^2 + 1 \Rightarrow f'(x) = 4x^3 + 2x, f''(x) = 12x^2 + 2, f'''(x) = 24x, f^{(4)}(x) = 24, f^{(n)}(x) = 0$ if $n \geq 5$;
$f(-2) = 21, f'(-2) = -36, f''(-2) = 50, f'''(-2) = -48, f^{(4)}(-2) = 24, f^{(n)}(-2) = 0$ if $n \geq 5 \Rightarrow x^4 + x^2 + 1$
$= 21 - 36(x+2) + \frac{50}{2!}(x+2)^2 - \frac{48}{3!}(x+2)^3 + \frac{24}{4!}(x+2)^4 = 21 - 36(x+2) + 25(x+2)^2 - 8(x+2)^3 + (x+2)^4$

24. $f(x) = 3x^5 - x^4 + 2x^3 + x^2 - 2 \Rightarrow f'(x) = 15x^4 - 4x^3 + 6x^2 + 2x, f''(x) = 60x^3 - 12x^2 + 12x + 2,$
$f'''(x) = 180x^2 - 24x + 12, f^{(4)}(x) = 360x - 24, f^{(5)}(x) = 360, f^{(n)}(x) = 0$ if $n \geq 6$; $f(-1) = -7,$
$f'(-1) = 23, f''(-1) = -82, f'''(-1) = 216, f^{(4)}(-1) = -384, f^{(5)}(-1) = 360, f^{(n)}(-1) = 0$ if $n \geq 6$
$\Rightarrow 3x^5 - x^4 + 2x^3 + x^2 - 2 = -7 + 23(x+1) - \frac{82}{2!}(x+1)^2 + \frac{216}{3!}(x+1)^3 - \frac{384}{4!}(x+1)^4 + \frac{360}{5!}(x+1)^5$
$= -7 + 23(x+1) - 41(x+1)^2 + 36(x+1)^3 - 16(x+1)^4 + 3(x+1)^5$

25. $f(x) = x^{-2} \Rightarrow f'(x) = -2x^{-3}, f''(x) = 3!\,x^{-4}, f'''(x) = -4!\,x^{-5} \Rightarrow f^{(n)}(x) = (-1)^n(n+1)!\,x^{-n-2};$
$f(1) = 1, f'(1) = -2, f''(1) = 3!, f'''(1) = -4!, f^{(n)}(1) = (-1)^n(n+1)! \Rightarrow \frac{1}{x^2}$
$= 1 - 2(x-1) + 3(x-1)^2 - 4(x-1)^3 + \ldots = \sum_{n=0}^{\infty} (-1)^n(n+1)(x-1)^n$

26. $f(x) = \frac{x}{1-x} \Rightarrow f'(x) = (1-x)^{-2}, f''(x) = 2(1-x)^{-3}, f'''(x) = 3!\,(1-x)^{-4} \Rightarrow f^{(n)}(x) = n!\,(1-x)^{-n-1};$
$f(0) = 0, f'(0) = 1, f''(0) = 2, f'''(0) = 3! \Rightarrow \frac{x}{1-x} = x + x^2 + x^3 + \ldots = \sum_{n=0}^{\infty} x^{n+1}$

27. $f(x) = e^x \Rightarrow f'(x) = e^x, f''(x) = e^x \Rightarrow f^{(n)}(x) = e^x; f(2) = e^2, f'(2) = e^2, \ldots f^{(n)}(2) = e^2$
$\Rightarrow e^x = e^2 + e^2(x-2) + \frac{e^2}{2}(x-2)^2 + \frac{e^3}{3!}(x-2)^3 + \ldots = \sum_{n=0}^{\infty} \frac{e^2}{n!}(x-2)^n$

28. $f(x) = 2^x \Rightarrow f'(x) = 2^x \ln 2, f''(x) = 2^x(\ln 2)^2, f'''(x) = 2^x(\ln 2)^3 \Rightarrow f^{(n)}(x) = 2^x(\ln 2)^n; f(1) = 2, f'(1) = 2\ln 2,$
$f''(1) = 2(\ln 2)^2, f'''(1) = 2(\ln 2)^3, \ldots, f^{(n)}(1) = 2(\ln 2)^n$
$\Rightarrow 2^x = 2 + (2\ln 2)(x-1) + \frac{2(\ln 2)^2}{2}(x-1)^2 + \frac{2(\ln 2)^3}{3!}(x-1)^3 + \ldots = \sum_{n=0}^{\infty} \frac{2(\ln 2)^n(x-1)^n}{n!}$

29. If $e^x = \sum_{n=0}^{\infty} \frac{f^{(n)}(a)}{n!}(x-a)^n$ and $f(x) = e^x$, we have $f^{(n)}(a) = e^a$ f or all $n = 0, 1, 2, 3, \ldots$
$\Rightarrow e^x = e^a\left[\frac{(x-a)^0}{0!} + \frac{(x-a)^1}{1!} + \frac{(x-a)^2}{2!} + \ldots\right] = e^a\left[1 + (x-a) + \frac{(x-a)^2}{2!} + \ldots\right]$ at $x = a$

30. $f(x) = e^x \Rightarrow f^{(n)}(x) = e^x$ for all $n \Rightarrow f^{(n)}(1) = e$ for all $n = 0, 1, 2, \ldots$
$\Rightarrow e^x = e + e(x-1) + \frac{e}{2!}(x-1)^2 + \frac{e}{3!}(x-1)^3 + \ldots = e\left[1 + (x-1) + \frac{(x-1)^2}{2!} + \frac{(x-1)^3}{3!} + \ldots\right]$

31. $f(x) = f(a) + f'(a)(x-a) + \frac{f''(a)}{2}(x-a)^2 + \frac{f'''(a)}{3!}(x-a)^3 + \ldots \Rightarrow f'(x)$
$= f'(a) + f''(a)(x-a) + \frac{f'''(a)}{3!}3(x-a)^2 + \ldots \Rightarrow f''(x) = f''(a) + f'''(a)(x-a) + \frac{f^{(4)}(a)}{4!}4\cdot 3(x-a)^2 + \ldots$
$\Rightarrow f^{(n)}(x) = f^{(n)}(a) + f^{(n+1)}(a)(x-a) + \frac{f^{(n+2)}(a)}{2}(x-a)^2 + \ldots$
$\Rightarrow f(a) = f(a) + 0, f'(a) = f'(a) + 0, \ldots, f^{(n)}(a) = f^{(n)}(a) + 0$

32. $E(x) = f(x) - b_0 - b_1(x-a) - b_2(x-a)^2 - b_3(x-a)^3 - \ldots - b_n(x-a)^n$
$\Rightarrow 0 = E(a) = f(a) - b_0 \Rightarrow b_0 = f(a)$; from condition (b),
$\lim_{x \to a} \frac{f(x) - f(a) - b_1(x-a) - b_2(x-a)^2 - b_3(x-a)^3 - \ldots - b_n(x-a)^n}{(x-a)^n} = 0$
$\Rightarrow \lim_{x \to a} \frac{f'(x) - b_1 - 2b_2(x-a) - 3b_3(x-a)^2 - \ldots - nb_n(x-a)^{n-1}}{n(x-a)^{n-1}} = 0$
$\Rightarrow b_1 = f'(a) \Rightarrow \lim_{x \to a} \frac{f''(x) - 2b_2 - 3!\,b_3(x-a) - \ldots - n(n-1)b_n(x-a)^{n-2}}{n(n-1)(x-a)^{n-2}} = 0$
$\Rightarrow b_2 = \frac{1}{2}f''(a) \Rightarrow \lim_{x \to a} \frac{f'''(x) - 3!\,b_3 - \ldots - n(n-1)(n-2)b_n(x-a)^{n-3}}{n(n-1)(n-2)(x-a)^{n-3}} = 0$
$= b_3 = \frac{1}{3!}f'''(a) \Rightarrow \lim_{x \to a} \frac{f^{(n)}(x) - n!\,b_n}{n!} = 0 \Rightarrow b_n = \frac{1}{n!}f^{(n)}(a)$; therefore,
$g(x) = f(a) + f'(a)(x-a) + \frac{f''(a)}{2!}(x-a)^2 + \ldots + \frac{f^{(n)}(a)}{n!}(x-a)^n = P_n(x)$

33. $f(x) = \ln(\cos x) \Rightarrow f'(x) = -\tan x$ and $f''(x) = -\sec^2 x; f(0) = 0, f'(0) = 0, f''(0) = -1$
$\Rightarrow L(x) = 0$ and $Q(x) = -\frac{x^2}{2}$

34. $f(x) = e^{\sin x} \Rightarrow f'(x) = (\cos x)e^{\sin x}$ and $f''(x) = (-\sin x)e^{\sin x} + (\cos x)^2 e^{\sin x}; f(0) = 1, f'(0) = 1,$
$f''(0) = 1 \Rightarrow L(x) = 1 + x$ and $Q(x) = 1 + x + \frac{x^2}{2}$

35. $f(x) = (1-x^2)^{-1/2} \Rightarrow f'(x) = x(1-x^2)^{-3/2}$ and $f''(x) = (1-x^2)^{-3/2} + 3x^2(1-x^2)^{-5/2}; f(0) = 1,$
$f'(0) = 0, f''(0) = 1 \Rightarrow L(x) = 1$ and $Q(x) = 1 + \frac{x^2}{2}$

36. $f(x) = \cosh x \Rightarrow f'(x) = \sinh x$ and $f''(x) = \cosh x; f(0) = 1, f'(0) = 0, f''(0) = 1 \Rightarrow L(x) = 1$ and $Q(x) = 1 + \frac{x^2}{2}$

37. $f(x) = \sin x \Rightarrow f'(x) = \cos x$ and $f''(x) = -\sin x$; $f(0) = 0$, $f'(0) = 1$, $f''(0) = 0 \Rightarrow L(x) = x$ and $Q(x) = x$

38. $f(x) = \tan x \Rightarrow f'(x) = \sec^2 x$ and $f''(x) = 2\sec^2 x \tan x$; $f(0) = 0$, $f'(0) = 1$, $f'' = 0 \Rightarrow L(x) = x$ and $Q(x) = x$

11.9 CONVERGENCE OF TAYLOR SERIES; ERROR ESTIMATES

1. $e^x = 1 + x + \frac{x^2}{2!} + \ldots = \sum_{n=0}^{\infty} \frac{x^n}{n!} \Rightarrow e^{-5x} = 1 + (-5x) + \frac{(-5x)^2}{2!} + \ldots = 1 - 5x + \frac{5^2x^2}{2!} - \frac{5^3x^3}{3!} + \ldots = \sum_{n=0}^{\infty} \frac{(-1)^n 5^n x^n}{n!}$

2. $e^x = 1 + x + \frac{x^2}{2!} + \ldots = \sum_{n=0}^{\infty} \frac{x^n}{n!} \Rightarrow e^{-x/2} = 1 + \left(\frac{-x}{2}\right) + \frac{\left(-\frac{x}{2}\right)^2}{2!} + \ldots = 1 - \frac{x}{2} + \frac{x^2}{2^2 2!} - \frac{x^3}{2^3 3!} + \ldots$
$= \sum_{n=0}^{\infty} \frac{(-1)^n x^n}{2^n n!}$

3. $\sin x = x - \frac{x^3}{3!} + \frac{x^5}{5!} - \ldots = \sum_{n=0}^{\infty} \frac{(-1)^n x^{2n+1}}{(2n+1)!} \Rightarrow 5 \sin(-x) = 5\left[(-x) - \frac{(-x)^3}{3!} + \frac{(-x)^5}{5!} - \ldots\right]$
$= \sum_{n=0}^{\infty} \frac{5(-1)^{n+1} x^{2n+1}}{(2n+1)!}$

4. $\sin x = x - \frac{x^3}{3!} + \frac{x^5}{5!} - \ldots = \sum_{n=0}^{\infty} \frac{(-1)^n x^{2n+1}}{(2n+1)!} \Rightarrow \sin \frac{\pi x}{2} = \frac{\pi x}{2} - \frac{\left(\frac{\pi x}{2}\right)^3}{3!} + \frac{\left(\frac{\pi x}{2}\right)^5}{5!} - \frac{\left(\frac{\pi x}{2}\right)^7}{7!} + \ldots$
$= \sum_{n=0}^{\infty} \frac{(-1)^n \pi^{2n+1} x^{2n+1}}{2^{2n+1}(2n+1)!}$

5. $\cos x = \sum_{n=0}^{\infty} \frac{(-1)^n x^{2n}}{(2n)!} \Rightarrow \cos\sqrt{x+1} = \sum_{n=0}^{\infty} \frac{(-1)^n \left[(x+1)^{1/2}\right]^{2n}}{(2n)!} = \sum_{n=0}^{\infty} \frac{(-1)^n (x+1)^n}{(2n)!} = 1 - \frac{x+1}{2!} + \frac{(x+1)^2}{4!} - \frac{(x+1)^3}{6!} + \ldots$

6. $\cos x = \sum_{n=0}^{\infty} \frac{(-1)^n x^{2n}}{(2n)!} \Rightarrow \cos\left(\frac{x^{3/2}}{\sqrt{2}}\right) = \cos\left(\left(\frac{x^3}{2}\right)^{1/2}\right) = \sum_{n=0}^{\infty} \frac{(-1)^n \left(\left(\frac{x^3}{2}\right)^{1/2}\right)^{2n}}{(2n)!} = \sum_{n=0}^{\infty} \frac{(-1)^n x^{3n}}{2^n (2n)!}$
$= 1 - \frac{x^3}{2 \cdot 2!} + \frac{x^6}{2^2 \cdot 4!} - \frac{x^9}{2^3 \cdot 6!} + \ldots$

7. $e^x = \sum_{n=0}^{\infty} \frac{x^n}{n!} \Rightarrow xe^x = x\left(\sum_{n=0}^{\infty} \frac{x^n}{n!}\right) = \sum_{n=0}^{\infty} \frac{x^{n+1}}{n!} = x + x^2 + \frac{x^3}{2!} + \frac{x^4}{3!} + \frac{x^5}{4!} + \ldots$

8. $\sin x = \sum_{n=0}^{\infty} \frac{(-1)^n x^{2n+1}}{(2n+1)!} \Rightarrow x^2 \sin x = x^2\left(\sum_{n=0}^{\infty} \frac{(-1)^n x^{2n+1}}{(2n+1)!}\right) = \sum_{n=0}^{\infty} \frac{(-1)^n x^{2n+3}}{(2n+1)!} = x^3 - \frac{x^5}{3!} + \frac{x^7}{5!} - \frac{x^9}{7!} + \ldots$

9. $\cos x = \sum_{n=0}^{\infty} \frac{(-1)^n x^{2n}}{(2n)!} \Rightarrow \frac{x^2}{2} - 1 + \cos x = \frac{x^2}{2} - 1 + \sum_{n=0}^{\infty} \frac{(-1)^n x^{2n}}{(2n)!} = \frac{x^2}{2} - 1 + 1 - \frac{x^2}{2} + \frac{x^4}{4!} - \frac{x^6}{6!} + \frac{x^8}{8!} - \frac{x^{10}}{10!} + \ldots$
$= \frac{x^4}{4!} - \frac{x^6}{6!} + \frac{x^8}{8!} - \frac{x^{10}}{10!} + \ldots = \sum_{n=2}^{\infty} \frac{(-1)^n x^{2n}}{(2n)!}$

10. $\sin x = \sum_{n=0}^{\infty} \frac{(-1)^n x^{2n+1}}{(2n+1)!} \Rightarrow \sin x - x + \frac{x^3}{3!} = \left(\sum_{n=0}^{\infty} \frac{(-1)^n x^{2n+1}}{(2n+1)!}\right) - x + \frac{x^3}{3!}$
$= \left(x - \frac{x^3}{3!} + \frac{x^5}{5!} - \frac{x^7}{7!} + \frac{x^9}{9!} - \frac{x^{11}}{11!} + \ldots\right) - x + \frac{x^3}{3!} = \frac{x^5}{5!} - \frac{x^7}{7!} + \frac{x^9}{9!} - \frac{x^{11}}{11!} + \ldots = \sum_{n=2}^{\infty} \frac{(-1)^n x^{2n+1}}{(2n+1)!}$

11. $\cos x = \sum_{n=0}^{\infty} \frac{(-1)^n x^{2n}}{(2n)!} \Rightarrow x \cos \pi x = x \sum_{n=0}^{\infty} \frac{(-1)^n (\pi x)^{2n}}{(2n)!} = \sum_{n=0}^{\infty} \frac{(-1)^n \pi^{2n} x^{2n+1}}{(2n)!} = x - \frac{\pi^2 x^3}{2!} + \frac{\pi^4 x^5}{4!} - \frac{\pi^6 x^7}{6!} + \ldots$

12. $\cos x = \sum_{n=0}^{\infty} \frac{(-1)^n x^{2n}}{(2n)!} \Rightarrow x^2 \cos(x^2) = x^2 \sum_{n=0}^{\infty} \frac{(-1)^n (x^2)^{2n}}{(2n)!} = \sum_{n=0}^{\infty} \frac{(-1)^n x^{4n+2}}{(2n)!} = x^2 - \frac{x^6}{2!} + \frac{x^{10}}{4!} - \frac{x^{14}}{6!} + \ldots$

13. $\cos^2 x = \frac{1}{2} + \frac{\cos 2x}{2} = \frac{1}{2} + \frac{1}{2} \sum_{n=0}^{\infty} \frac{(-1)^n (2x)^{2n}}{(2n)!} = \frac{1}{2} + \frac{1}{2}\left[1 - \frac{(2x)^2}{2!} + \frac{(2x)^4}{4!} - \frac{(2x)^6}{6!} + \frac{(2x)^8}{8!} - \ldots\right]$

$= 1 - \frac{(2x)^2}{2\cdot 2!} + \frac{(2x)^4}{2\cdot 4!} - \frac{(2x)^6}{2\cdot 6!} + \frac{(2x)^8}{2\cdot 8!} - \ldots = 1 + \sum_{n=1}^{\infty} \frac{(-1)^n (2x)^{2n}}{2\cdot(2n)!} = 1 + \sum_{n=1}^{\infty} \frac{(-1)^n\, 2^{2n-1}\, x^{2n}}{(2n)!}$

14. $\sin^2 x = \left(\frac{1-\cos 2x}{2}\right) = \frac{1}{2} - \frac{1}{2}\cos 2x = \frac{1}{2} - \frac{1}{2}\left(1 - \frac{(2x)^2}{2!} + \frac{(2x)^4}{4!} - \frac{(2x)^6}{6!} + \ldots\right) = \frac{(2x)^2}{2\cdot 2!} - \frac{(2x)^4}{2\cdot 4!} + \frac{(2x)^6}{2\cdot 6!} - \ldots$

$= \sum_{n=1}^{\infty} \frac{(-1)^{n+1} (2x)^{2n}}{2\cdot(2n)!} = \sum_{n=1}^{\infty} \frac{(-1)^n\, 2^{2n-1}\, x^{2n}}{(2n)!}$

15. $\frac{x^2}{1-2x} = x^2\left(\frac{1}{1-2x}\right) = x^2 \sum_{n=0}^{\infty} (2x)^n = \sum_{n=0}^{\infty} 2^n x^{n+2} = x^2 + 2x^3 + 2^2x^4 + 2^3x^5 + \ldots$

16. $x \ln(1+2x) = x \sum_{n=1}^{\infty} \frac{(-1)^{n-1}(2x)^n}{n} = \sum_{n=1}^{\infty} \frac{(-1)^{n-1} 2^n x^{n+1}}{n} = 2x^2 - \frac{2^2x^3}{2} + \frac{2^3x^4}{4} - \frac{2^4x^5}{5} + \ldots$

17. $\frac{1}{1-x} = \sum_{n=0}^{\infty} x^n = 1 + x + x^2 + x^3 + \ldots \Rightarrow \frac{d}{dx}\left(\frac{1}{1-x}\right) = \frac{1}{(1-x)^2} = 1 + 2x + 3x^2 + \ldots = \sum_{n=1}^{\infty} nx^{n-1}$

$= \sum_{n=0}^{\infty} (n+1)x^n$

18. $\frac{2}{(1-x)^3} = \frac{d^2}{dx^2}\left(\frac{1}{1-x}\right) = \frac{d}{dx}\left(\frac{1}{(1-x)^2}\right) = \frac{d}{dx}(1 + 2x + 3x^2 + \ldots) = 2 + 6x + 12x^2 + \ldots = \sum_{n=2}^{\infty} n(n-1)x^{n-2}$

$= \sum_{n=0}^{\infty} (n+2)(n+1)x^n$

19. By the Alternating Series Estimation Theorem, the error is less than $\frac{|x|^5}{5!} \Rightarrow |x|^5 < (5!)(5 \times 10^{-4})$
$\Rightarrow |x|^5 < 600 \times 10^{-4} \Rightarrow |x| < \sqrt[5]{6 \times 10^{-2}} \approx 0.56968$

20. If $\cos x = 1 - \frac{x^2}{2}$ and $|x| < 0.5$, then the error is less than $\left|\frac{(.5)^4}{24}\right| = 0.0026$, by Alternating Series Estimation Theorem; since the next term in the series is positive, the approximation $1 - \frac{x^2}{2}$ is too small, by the Alternating Series Estimation Theorem

21. If $\sin x = x$ and $|x| < 10^{-3}$, then the error is less than $\frac{(10^{-3})^3}{3!} \approx 1.67 \times 10^{-10}$, by Alternating Series Estimation Theorem; The Alternating Series Estimation Theorem says $R_2(x)$ has the same sign as $-\frac{x^3}{3!}$. Moreover, $x < \sin x$
$\Rightarrow 0 < \sin x - x = R_2(x) \Rightarrow x < 0 \Rightarrow -10^{-3} < x < 0.$

22. $\sqrt{1+x} = 1 + \frac{x}{2} - \frac{x^2}{8} + \frac{x^3}{16} - \ldots$. By the Alternating Series Estimation Theorem the $|\text{error}| < \left|\frac{-x^2}{8}\right| < \frac{(0.01)^2}{8}$
$= 1.25 \times 10^{-5}$

23. $|R_2(x)| = \left|\frac{e^c x^3}{3!}\right| < \frac{3^{(0.1)}(0.1)^3}{3!} < 1.87 \times 10^{-4}$, where c is between 0 and x

24. $|R_2(x)| = \left|\frac{e^c x^3}{3!}\right| < \frac{(0.1)^3}{3!} = 1.67 \times 10^{-4}$, where c is between 0 and x

25. $|R_4(x)| < \left|\frac{\cosh c}{5!} x^5\right| = \left|\frac{e^c+e^{-c}}{2} \frac{x^5}{5!}\right| < \frac{1.65+\frac{1}{1.65}}{2} \cdot \frac{(0.5)^5}{5!} = (1.13)\,\frac{(0.5)^5}{5!} \approx 0.000294$

26. If we approximate e^h with $1 + h$ and $0 \le h \le 0.01$, then $|\text{error}| < \left|\frac{e^c h^2}{2}\right| \le \frac{e^{0.01} h \cdot h}{2} \le \left(\frac{e^{0.01}(0.01)}{2}\right) h$
$= 0.00505h < 0.006h = (0.6\%)h$, where c is between 0 and h.

27. $|R_1| = \left|\frac{1}{(1+c)^2} \frac{x^2}{2!}\right| < \frac{x^2}{2} = \left|\frac{x}{2}\right| |x| < .01\,|x| = (1\%)\,|x| \Rightarrow \left|\frac{x}{2}\right| < .01 \Rightarrow 0 < |x| < .02$

28. $\tan^{-1} x = x - \frac{x^3}{3} + \frac{x^5}{5} - \frac{x^7}{7} + \ldots \Rightarrow \frac{\pi}{4} = \tan^{-1} 1 = 1 - \frac{1}{3} + \frac{1}{5} - \frac{1}{7} + \ldots$; $|\text{error}| < \frac{1}{2n+1} < .01$
$\Rightarrow 2n + 1 > 100 \Rightarrow n > 49$

29. (a) $\sin x = x - \frac{x^3}{3!} + \frac{x^5}{5!} - \frac{x^7}{7!} + \ldots \Rightarrow \frac{\sin x}{x} = 1 - \frac{x^2}{3!} + \frac{x^4}{5!} - \frac{x^6}{7!} + \ldots$, $s_1 = 1$ and $s_2 = 1 - \frac{x^2}{6}$; if L is the sum of the series representing $\frac{\sin x}{x}$, then by the Alternating Series Estimation Theorem, $L - s_1 = \frac{\sin x}{x} - 1 < 0$ and $L - s_2 = \frac{\sin x}{x} - \left(1 - \frac{x^2}{6}\right) > 0$. Therefore $1 - \frac{x^2}{6} < \frac{\sin x}{x} < 1$

(b) The graph of $y = \frac{\sin x}{x}$, $x \ne 0$, is bounded below by the graph of $y = 1 - \frac{x^2}{6}$ and above by the graph of $y = 1$ as derived in part (a).

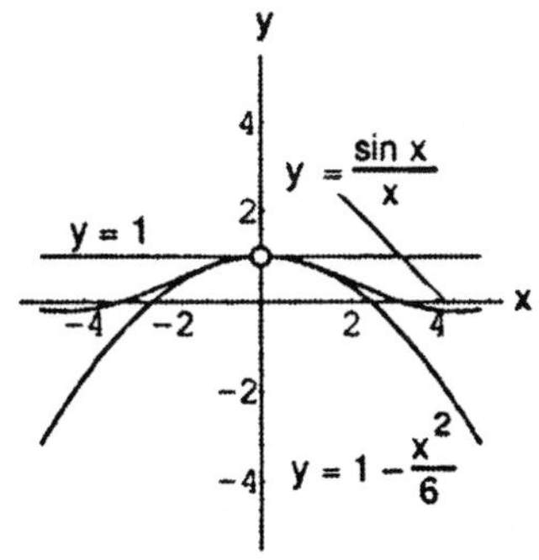

30. (a) $\cos x = 1 - \frac{x^2}{2!} + \frac{x^4}{4!} - \frac{x^6}{6!} + \ldots \Rightarrow 1 - \cos x = \frac{x^2}{2!} - \frac{x^4}{4!} + \frac{x^6}{6!} - \frac{x^8}{8!} + \ldots \Rightarrow \frac{1 - \cos x}{x^2} = \frac{1}{2} - \frac{x^2}{4!} + \frac{x^4}{6!} - \frac{x^6}{8!} + \ldots$;
if L is the sum of the series representing $\frac{1-\cos x}{x^2}$, then by the Alternating Series Estimation Theorem $L - s_1 = \frac{1-\cos x}{x^2} - \frac{1}{2} < 0$ and $\frac{1-\cos x}{x^2} - \left(\frac{1}{2} - \frac{x^2}{4!}\right) > 0$. Therefore $\frac{1}{2} - \frac{x^2}{24} < \frac{1-\cos x}{x^2} < \frac{1}{2}$.

(b) The graph of $y = \frac{1-\cos x}{x^2}$ is bounded below by the graph of $y = \frac{1}{2} - \frac{x^2}{24}$ and above by the graph of $y = \frac{1}{2}$ as indicated in part (a).

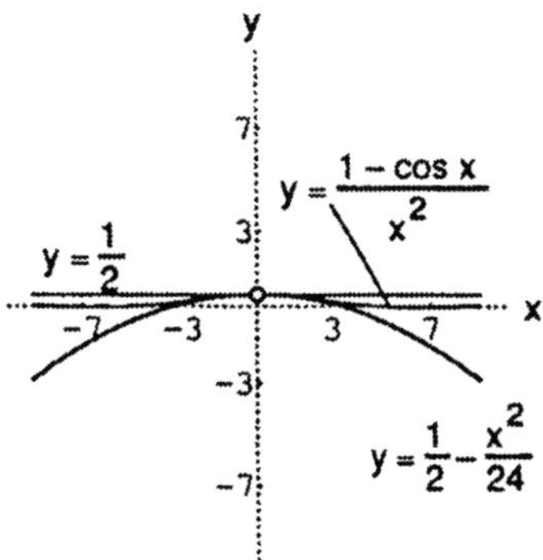

31. $\sin x$ when $x = 0.1$; the sum is $\sin(0.1) \approx 0.099833417$

32. $\cos x$ when $x = \frac{\pi}{4}$; the sum is $\cos\left(\frac{\pi}{4}\right) = \frac{1}{\sqrt{2}} \approx 0.707106781$

33. $\tan^{-1} x$ when $x = \frac{\pi}{3}$; the sum is $\tan^{-1}\left(\frac{\pi}{3}\right) \approx 0.808448$

34. $\ln(1 + x)$ when $x = \pi$; the sum is $\ln(1 + \pi) \approx 1.421080$

35. $e^x \sin x = 0 + x + x^2 + x^3\left(-\frac{1}{3!}+\frac{1}{2!}\right) + x^4\left(-\frac{1}{3!}+\frac{1}{3!}\right) + x^5\left(\frac{1}{5!}-\frac{1}{2!}\frac{1}{3!}+\frac{1}{4!}\right) + x^6\left(\frac{1}{5!}-\frac{1}{3!}\frac{1}{3!}+\frac{1}{5!}\right) + \ldots$
$= x + x^2 + \frac{1}{3}x^3 - \frac{1}{30}x^5 - \frac{1}{90}x^6 + \ldots$

36. $e^x \cos x = 1 + x + x^2\left(-\frac{1}{2!}+\frac{1}{2!}\right) + x^3\left(-\frac{1}{2!}+\frac{1}{3!}\right) + x^4\left(\frac{1}{4!}-\frac{1}{2!}\frac{1}{2!}+\frac{1}{4!}\right) + x^5\left(\frac{1}{4!}-\frac{1}{2!}\frac{1}{3!}+\frac{1}{5!}\right) + \ldots$
$= 1 + x - \frac{1}{3}x^3 - \frac{1}{6}x^4 - \frac{1}{30}x^5 + \ldots$

37. $\sin^2 x = \left(\frac{1-\cos 2x}{2}\right) = \frac{1}{2} - \frac{1}{2}\cos 2x = \frac{1}{2} - \frac{1}{2}\left(1 - \frac{(2x)^2}{2!} + \frac{(2x)^4}{4!} - \frac{(2x)^6}{6!} + \ldots\right) = \frac{2x^2}{2!} - \frac{2^3x^4}{4!} + \frac{2^5x^6}{6!} - \ldots$
$\Rightarrow \frac{d}{dx}(\sin^2 x) = \frac{d}{dx}\left(\frac{2x^2}{2!} - \frac{2^3x^4}{4!} + \frac{2^5x^6}{6!} - \ldots\right) = 2x - \frac{(2x)^3}{3!} + \frac{(2x)^5}{5!} - \frac{(2x)^7}{7!} + \ldots \Rightarrow 2\sin x\cos x$
$= 2x - \frac{(2x)^3}{3!} + \frac{(2x)^5}{5!} - \frac{(2x)^7}{7!} + \ldots = \sin 2x$, which checks

38. $\cos^2 x = \cos 2x + \sin^2 x = \left(1 - \frac{(2x)^2}{2!} + \frac{(2x)^4}{4!} - \frac{(2x)^6}{6!} + \frac{(2x)^8}{8!} + \ldots\right) + \left(\frac{2x^2}{2!} - \frac{2^3x^4}{4!} + \frac{2^5x^6}{6!} - \frac{2^7x^8}{8!} + \ldots\right)$
$= 1 - \frac{2x^2}{2!} + \frac{2^3x^4}{4!} - \frac{2^5x^6}{6!} + \ldots = 1 - x^2 + \frac{1}{3}x^4 - \frac{2}{45}x^6 + \frac{1}{315}x^8 - \ldots$

39. A special case of Taylor's Theorem is $f(b) = f(a) + f'(c)(b-a)$, where c is between a and b $\Rightarrow$ $f(b) - f(a) = f'(c)(b-a)$, the Mean Value Theorem.

40. If f(x) is twice differentiable and at $x = a$ there is a point of inflection, then $f''(a) = 0$. Therefore, $L(x) = Q(x) = f(a) + f'(a)(x-a)$.

41. (a) $f'' \le 0$, $f'(a) = 0$ and $x = a$ interior to the interval I $\Rightarrow f(x) - f(a) = \frac{f''(c_2)}{2}(x-a)^2 \le 0$ throughout I $\Rightarrow f(x) \le f(a)$ throughout I $\Rightarrow$ f has a local maximum at $x = a$

(b) similar reasoning gives $f(x) - f(a) = \frac{f''(c_2)}{2}(x-a)^2 \ge 0$ throughout I $\Rightarrow f(x) \ge f(a)$ throughout I $\Rightarrow$ f has a local minimum at $x = a$

42. $f(x) = (1-x)^{-1} \Rightarrow f'(x) = (1-x)^{-2} \Rightarrow f''(x) = 2(1-x)^{-3} \Rightarrow f^{(3)}(x) = 6(1-x)^{-4}$
$\Rightarrow f^{(4)}(x) = 24(1-x)^{-5}$; therefore $\frac{1}{1-x} \approx 1 + x + x^2 + x^3$. $|x| < 0.1 \Rightarrow \frac{10}{11} < \frac{1}{1-x} < \frac{10}{9} \Rightarrow \left|\frac{1}{(1-x)^5}\right| < \left(\frac{10}{9}\right)^5$
$\Rightarrow \left|\frac{x^4}{(1-x)^5}\right| < x^4\left(\frac{10}{9}\right)^5 \Rightarrow$ the error $e_3 \le \left|\frac{\max f^{(4)}(x)\, x^4}{4!}\right| < (0.1)^4\left(\frac{10}{9}\right)^5 = 0.00016935 < 0.00017$, since $\left|\frac{f^{(4)}(x)}{4!}\right| = \left|\frac{1}{(1-x)^5}\right|$.

43. (a) $f(x) = (1+x)^k \Rightarrow f'(x) = k(1+x)^{k-1} \Rightarrow f''(x) = k(k-1)(1+x)^{k-2}$; $f(0) = 1$, $f'(0) = k$, and $f''(0) = k(k-1)$ $\Rightarrow Q(x) = 1 + kx + \frac{k(k-1)}{2}x^2$

(b) $|R_2(x)| = \left|\frac{3\cdot2\cdot1}{3!}x^3\right| < \frac{1}{100} \Rightarrow |x^3| < \frac{1}{100} \Rightarrow 0 < x < \frac{1}{100^{1/3}}$ or $0 < x < .21544$

44. (a) Let $P = x + \pi \Rightarrow |x| = |P - \pi| < .5 \times 10^{-n}$ since P approximates π accurate to n decimals. Then, $P + \sin P = (\pi + x) + \sin(\pi + x) = (\pi + x) - \sin x = \pi + (x - \sin x) \Rightarrow |(P + \sin P) - \pi|$ $= |\sin x - x| \le \frac{|x|^3}{3!} < \frac{0.125}{3!} \times 10^{-3n} < .5 \times 10^{-3n} \Rightarrow P + \sin P$ gives an approximation to π correct to 3n decimals.

45. If $f(x) = \sum_{n=0}^{\infty} a_n x^n$, then $f^{(k)}(x) = \sum_{n=k}^{\infty} n(n-1)(n-2)\cdots(n-k+1)a_n x^{n-k}$ and $f^{(k)}(0) = k!\,a_k$ $\Rightarrow a_k = \frac{f^{(k)}(0)}{k!}$ for k a nonnegative integer. Therefore, the coefficients of f(x) are identical with the corresponding coefficients in the Maclaurin series of f(x) and the statement follows.

46. Note: f even $\Rightarrow f(-x) = f(x) \Rightarrow -f'(-x) = f'(x) \Rightarrow f'(-x) = -f'(x) \Rightarrow f'$ odd;
f odd $\Rightarrow f(-x) = -f(x) \Rightarrow -f'(-x) = -f'(x) \Rightarrow f'(-x) = f'(x) \Rightarrow f'$ even;
also, f odd $\Rightarrow f(-0) = f(0) \Rightarrow 2f(0) = 0 \Rightarrow f(0) = 0$

(a) If f(x) is even, then any odd-order derivative is odd and equal to 0 at $x = 0$. Therefore, $a_1 = a_3 = a_5 = \ldots = 0$; that is, the Maclaurin series for f contains only even powers.

(b) If f(x) is odd, then any even-order derivative is odd and equal to 0 at $x = 0$. Therefore, $a_0 = a_2 = a_4 = \ldots = 0$; that is, the Maclaurin series for f contains only odd powers.

47. (a) Suppose f(x) is a continuous periodic function with period p. Let x_0 be an arbitrary real number. Then f assumes a minimum m_1 and a maximum m_2 in the interval $[x_0, x_0 + p]$; i.e., $m_1 \le f(x) \le m_2$ for all x in $[x_0, x_0 + p]$. Since f is periodic it has exactly the same values on all other intervals $[x_0 + p, x_0 + 2p]$, $[x_0 + 2p, x_0 + 3p], \ldots$, and $[x_0 - p, x_0]$, $[x_0 - 2p, x_0 - p], \ldots$, and so forth. That is, for all real numbers $-\infty < x < \infty$ we have $m_1 \le f(x) \le m_2$. Now choose $M = \max\{|m_1|, |m_2|\}$. Then $-M \le -|m_1| \le m_1 \le f(x) \le m_2 \le |m_2| \le M \Rightarrow |f(x)| \le M$ for all x.

(b) The dominate term in the nth order Taylor polynomial generated by cos x about x = a is $\frac{\sin(a)}{n!}(x-a)^n$ or $\frac{\cos(a)}{n!}(x-a)^n$. In both cases, as |x| increases the absolute value of these dominate terms tends to ∞, causing the graph of $P_n(x)$ to move away from cos x.

48. (b) $\tan^{-1} x = x - \frac{x^3}{3} + \frac{x^5}{5} - \ldots \Rightarrow \frac{x - \tan^{-1} x}{x^3} = \frac{1}{3} - \frac{x^2}{5} + \ldots$; from the Alternating Series Estimation Theorem, $\frac{x - \tan^{-1} x}{x^3} - \frac{1}{3} < 0$ $\Rightarrow \frac{x - \tan^{-1} x}{x^3} - \left(\frac{1}{3} - \frac{x^2}{5}\right) > 0 \Rightarrow \frac{1}{3} < \frac{x - \tan^{-1} x}{x^3} < \frac{1}{3} - \frac{x^2}{5}$; therefore, the $\lim\limits_{x \to 0} \frac{x - \tan^{-1} x}{x^3} = \frac{1}{3}$

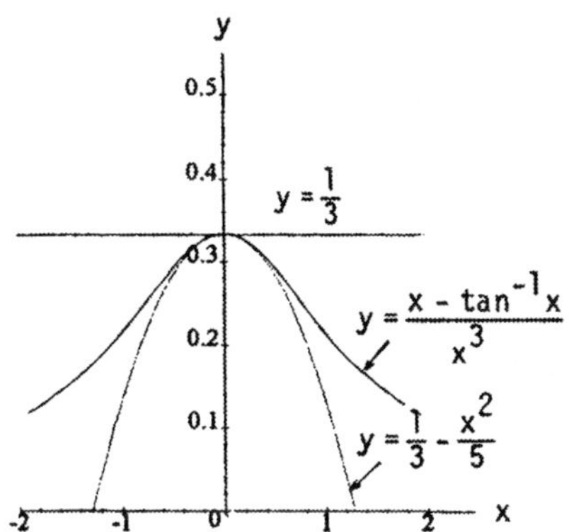

49. (a) $e^{-i\pi} = \cos(-\pi) + i\sin(-\pi) = -1 + i(0) = -1$

(b) $e^{i\pi/4} = \cos\left(\frac{\pi}{4}\right) + i\sin\left(\frac{\pi}{4}\right) = \frac{1}{\sqrt{2}} + \frac{i}{\sqrt{2}} = \left(\frac{1}{\sqrt{2}}\right)(1+i)$

(c) $e^{-i\pi/2} = \cos\left(-\frac{\pi}{2}\right) + i\sin\left(-\frac{\pi}{2}\right) = 0 + i(-1) = -i$

50. $e^{i\theta} = \cos\theta + i\sin\theta \Rightarrow e^{-i\theta} = e^{i(-\theta)} = \cos(-\theta) + i\sin(-\theta) = \cos\theta - i\sin\theta$;
$e^{i\theta} + e^{-i\theta} = \cos\theta + i\sin\theta + \cos\theta - i\sin\theta = 2\cos\theta \Rightarrow \cos\theta = \frac{e^{i\theta} + e^{-i\theta}}{2}$;
$e^{i\theta} - e^{-i\theta} = \cos\theta + i\sin\theta - (\cos\theta - i\sin\theta) = 2i\sin\theta \Rightarrow \sin\theta = \frac{e^{i\theta} - e^{-i\theta}}{2i}$

51. $e^x = 1 + x + \frac{x^2}{2!} + \frac{x^3}{3!} + \frac{x^4}{4!} + \ldots \Rightarrow e^{i\theta} = 1 + i\theta + \frac{(i\theta)^2}{2!} + \frac{(i\theta)^3}{3!} + \frac{(i\theta)^4}{4!} + \ldots$ and
$e^{-i\theta} = 1 - i\theta + \frac{(-i\theta)^2}{2!} + \frac{(-i\theta)^3}{3!} + \frac{(-i\theta)^4}{4!} + \ldots = 1 - i\theta + \frac{(i\theta)^2}{2!} - \frac{(i\theta)^3}{3!} + \frac{(i\theta)^4}{4!} - \ldots$

$$\Rightarrow \frac{e^{i\theta} + e^{-i\theta}}{2} = \frac{\left(1 + i\theta + \frac{(i\theta)^2}{2!} + \frac{(i\theta)^3}{3!} + \frac{(i\theta)^4}{4!} + \ldots\right) + \left(1 - i\theta + \frac{(i\theta)^2}{2!} - \frac{(i\theta)^3}{3!} + \frac{(i\theta)^4}{4!} - \ldots\right)}{2}$$
$= 1 - \frac{\theta^2}{2!} + \frac{\theta^4}{4!} - \frac{\theta^6}{6!} + \ldots = \cos\theta$;

$$\frac{e^{i\theta} - e^{-i\theta}}{2i} = \frac{\left(1 + i\theta + \frac{(i\theta)^2}{2!} + \frac{(i\theta)^3}{3!} + \frac{(i\theta)^4}{4!} + \ldots\right) - \left(1 - i\theta + \frac{(i\theta)^2}{2!} - \frac{(i\theta)^3}{3!} + \frac{(i\theta)^4}{4!} - \ldots\right)}{2i}$$
$= \theta - \frac{\theta^3}{3!} + \frac{\theta^5}{5!} - \frac{\theta^7}{7!} + \ldots = \sin\theta$

52. $e^{i\theta} = \cos\theta + i\sin\theta \Rightarrow e^{-i\theta} = e^{i(-\theta)} = \cos(-\theta) + i\sin(-\theta) = \cos\theta - i\sin\theta$

(a) $e^{i\theta} + e^{-i\theta} = (\cos\theta + i\sin\theta) + (\cos\theta - i\sin\theta) = 2\cos\theta \Rightarrow \cos\theta = \frac{e^{i\theta} + e^{-i\theta}}{2} = \cosh i\theta$

(b) $e^{i\theta} - e^{-i\theta} = (\cos\theta + i\sin\theta) - (\cos\theta - i\sin\theta) = 2i\sin\theta \Rightarrow i\sin\theta = \frac{e^{i\theta} - e^{-i\theta}}{2} = \sinh i\theta$

53. $e^x \sin x = \left(1 + x + \frac{x^2}{2!} + \frac{x^3}{3!} + \frac{x^4}{4!} + \ldots\right)\left(x - \frac{x^3}{3!} + \frac{x^5}{5!} - \frac{x^7}{7!} + \ldots\right)$

$= (1)x + (1)x^2 + \left(-\frac{1}{6} + \frac{1}{2}\right)x^3 + \left(-\frac{1}{6} + \frac{1}{6}\right)x^4 + \left(\frac{1}{120} - \frac{1}{12} + \frac{1}{24}\right)x^5 + \ldots = x + x^2 + \frac{1}{3}x^3 - \frac{1}{30}x^5 + \ldots;$

$e^x \cdot e^{ix} = e^{(1+i)x} = e^x(\cos x + i \sin x) = e^x \cos x + i(e^x \sin x) \Rightarrow e^x \sin x$ is the series of the imaginary part

of $e^{(1+i)x}$ which we calculate next; $e^{(1+i)x} = \sum_{n=0}^{\infty} \frac{(x+ix)^n}{n!} = 1 + (x + ix) + \frac{(x+ix)^2}{2!} + \frac{(x+ix)^3}{3!} + \frac{(x+ix)^4}{4!} + \ldots$

$= 1 + x + ix + \frac{1}{2!}(2ix^2) + \frac{1}{3!}(2ix^3 - 2x^3) + \frac{1}{4!}(-4x^4) + \frac{1}{5!}(-4x^5 - 4ix^5) + \frac{1}{6!}(-8ix^6) + \ldots \Rightarrow$ the imaginary part

of $e^{(1+i)x}$ is $x + \frac{2}{2!}x^2 + \frac{2}{3!}x^3 - \frac{4}{5!}x^5 - \frac{8}{6!}x^6 + \ldots = x + x^2 + \frac{1}{3}x^3 - \frac{1}{30}x^5 - \frac{1}{90}x^6 + \ldots$ in agreement with our

product calculation. The series for $e^x \sin x$ converges for all values of x.

54. $\frac{d}{dx}\left(e^{(a+ib)}\right) = \frac{d}{dx}\left[e^{ax}(\cos bx + i \sin bx)\right] = ae^{ax}(\cos bx + i \sin bx) + e^{ax}(-b \sin bx + bi \cos bx)$

$= ae^{ax}(\cos bx + i \sin bx) + bie^{ax}(\cos bx + i \sin bx) = ae^{(a+ib)x} + ibe^{(a+ib)x} = (a + ib)e^{(a+ib)x}$

55. (a) $e^{i\theta_1}e^{i\theta_2} = (\cos\theta_1 + i \sin\theta_1)(\cos\theta_2 + i \sin\theta_2) = (\cos\theta_1\cos\theta_2 - \sin\theta_1\sin\theta_2) + i(\sin\theta_1\cos\theta_2 + \sin\theta_2\cos\theta_1)$

$= \cos(\theta_1 + \theta_2) + i \sin(\theta_1 + \theta_2) = e^{i(\theta_1+\theta_2)}$

(b) $e^{-i\theta} = \cos(-\theta) + i \sin(-\theta) = \cos\theta - i \sin\theta = (\cos\theta - i \sin\theta)\left(\frac{\cos\theta + i \sin\theta}{\cos\theta + i \sin\theta}\right) = \frac{1}{\cos\theta + i \sin\theta} = \frac{1}{e^{i\theta}}$

56. $\frac{a-bi}{a^2+b^2}e^{(a+bi)x} + C_1 + iC_2 = \left(\frac{a-bi}{a^2+b^2}\right)e^{ax}(\cos bx + i \sin bx) + C_1 + iC_2$

$= \frac{e^{ax}}{a^2+b^2}(a \cos bx + ia \sin bx - ib \cos bx + b \sin bx) + C_1 + iC_2$

$= \frac{e^{ax}}{a^2+b^2}[(a \cos bx + b \sin bx) + (a \sin bx - b \cos bx)i] + C_1 + iC_2$

$= \frac{e^{ax}(a \cos bx + b \sin bx)}{a^2+b^2} + C_1 + \frac{ie^{ax}(a \sin bx - b \cos bx)}{a^2+b^2} + iC_2;$

$e^{(a+bi)x} = e^{ax}e^{ibx} = e^{ax}(\cos bx + i \sin bx) = e^{ax}\cos bx + ie^{ax}\sin bx$, so that given

$\int e^{(a+bi)x}\,dx = \frac{a-bi}{a^2+b^2}e^{(a+bi)x} + C_1 + iC_2$ we conclude that $\int e^{ax}\cos bx\,dx = \frac{e^{ax}(a \cos bx + b \sin bx)}{a^2+b^2} + C_1$

and $\int e^{ax}\sin bx\,dx = \frac{e^{ax}(a \sin bx - b \cos bx)}{a^2+b^2} + C_2$

57-62. Example CAS commands:

Maple:

```
f := x -> 1/sqrt(1+x);
x0 := -3/4;
x1 := 3/4;
# Step 1:
plot( f(x), x=x0..x1, title="Step 1: #57 (Section 11.9)" );
# Step 2:
P1 := unapply( TaylorApproximation(f(x), x = 0, order=1), x );
P2 := unapply( TaylorApproximation(f(x), x = 0, order=2), x );
P3 := unapply( TaylorApproximation(f(x), x = 0, order=3), x );
# Step 3:
D2f := D(D(f));
D3f := D(D(D(f)));
D4f := D(D(D(D(f))));
plot( [D2f(x),D3f(x),D4f(x)], x=x0..x1, thickness=[0,2,4], color=[red,blue,green], title="Step 3: #57 (Section 11.9)" );
c1 := x0;
M1 := abs( D2f(c1) );
c2 := x0;
M2 := abs( D3f(c2) );
c3 := x0;
M3 := abs( D4f(c3) );
# Step 4:
```

```
R1 := unapply( abs(M1/2!*(x-0)^2), x );
R2 := unapply( abs(M2/3!*(x-0)^3), x );
R3 := unapply( abs(M3/4!*(x-0)^4), x );
plot( [R1(x),R2(x),R3(x)], x=x0..x1, thickness=[0,2,4], color=[red,blue,green], title="Step 4: #57 (Section 11.9)" );
# Step 5:
E1 := unapply( abs(f(x)-P1(x)), x );
E2 := unapply( abs(f(x)-P2(x)), x );
E3 := unapply( abs(f(x)-P3(x)), x );
plot( [E1(x),E2(x),E3(x),R1(x),R2(x),R3(x)], x=x0..x1, thickness=[0,2,4], color=[red,blue,green],
     linestyle=[1,1,1,3,3,3], title="Step 5: #57 (Section 11.9)" );
# Step 6:
TaylorApproximation( f(x), view=[x0..x1,DEFAULT], x=0, output=animation, order=1..3 );
L1 := fsolve( abs(f(x)-P1(x))=0.01, x=x0/2 );              # (a)
R1 := fsolve( abs(f(x)-P1(x))=0.01, x=x1/2 );
L2 := fsolve( abs(f(x)-P2(x))=0.01, x=x0/2 );
R2 := fsolve( abs(f(x)-P2(x))=0.01, x=x1/2 );
L3 := fsolve( abs(f(x)-P3(x))=0.01, x=x0/2 );
R3 := fsolve( abs(f(x)-P3(x))=0.01, x=x1/2 );
plot( [E1(x),E2(x),E3(x),0.01], x=min(L1,L2,L3)..max(R1,R2,R3), thickness=[0,2,4,0], linestyle=[0,0,0,2],
     color=[red,blue,green,black], view=[DEFAULT,0..0.01], title="#57(a) (Section 11.9)" );
abs(`f(x)`-`P`[1](x) ) <= evalf( E1(x0) );                 # (b)
abs(`f(x)`-`P`[2](x) ) <= evalf( E2(x0) );
abs(`f(x)`-`P`[3](x) ) <= evalf( E3(x0) );
```

Mathematica: (assigned function and values for a, b, c, and n may vary)

```
Clear[x, f, c]
f[x_]= (1 + x)^(3/2)
{a, b}= {−1/2, 2};
pf=Plot[ f[x], {x, a, b}];
poly1[x_]=Series[f[x], {x,0,1}]//Normal
poly2[x_]=Series[f[x], {x,0,2}]//Normal
poly3[x_]=Series[f[x], {x,0,3}]//Normal
Plot[{f[x], poly1[x], poly2[x], poly3[x]}, {x, a, b},
        PlotStyle → {RGBColor[1,0,0], RGBColor[0,1,0], RGBColor[0,0,1], RGBColor[0,.5,.5]}];
```

The above defines the approximations. The following analyzes the derivatives to determine their maximum values.

```
f''[c]
Plot[f''[x], {x, a, b}];
f'''[c]
Plot[f'''[x], {x, a, b}];
f''''[c]
Plot[f''''[x], {x, a, b}];
```

Noting the upper bound for each of the above derivatives occurs at x = a, the upper bounds m1, m2, and m3 can be defined and bounds for remainders viewed as functions of x.

```
m1=f''[a]
m2=-f'''[a]
m3=f''''[a]
r1[x_]=m1 x^2 /2!
Plot[r1[x], {x, a, b}];
r2[x_]=m2 x^3 /3!
Plot[r2[x], {x, a, b}];
```

r3[x_]=m3 x^4 /4!
Plot[r3[x], {x, a, b}];

A three dimensional look at the error functions, allowing both c and x to vary can also be viewed. Recall that c must be a value between 0 and x, so some points on the surfaces where c is not in that interval are meaningless.

Plot3D[f''[c] x^2 /2!, {x, a, b}, {c, a, b}, PlotRange → All]
Plot3D[f'''[c] x^3 /3!, {x, a, b}, {c, a, b}, PlotRange → All]
Plot3D[f''''[c] x^4 /4!, {x, a, b}, {c, a, b}, PlotRange → All]

11.10 APPLICATIONS OF POWER SERIES

1. $(1+x)^{1/2} = 1 + \frac{1}{2}x + \frac{\left(\frac{1}{2}\right)\left(-\frac{1}{2}\right)x^2}{2!} + \frac{\left(\frac{1}{2}\right)\left(-\frac{1}{2}\right)\left(-\frac{3}{2}\right)x^3}{3!} + \ldots = 1 + \frac{1}{2}x - \frac{1}{8}x^2 + \frac{1}{16}x^3 - \ldots$

2. $(1+x)^{1/3} = 1 + \frac{1}{3}x + \frac{\left(\frac{1}{3}\right)\left(-\frac{2}{3}\right)x^2}{2!} + \frac{\left(\frac{1}{3}\right)\left(-\frac{2}{3}\right)\left(-\frac{5}{3}\right)x^3}{3!} + \ldots = 1 + \frac{1}{3}x - \frac{1}{9}x^2 + \frac{5}{81}x^3 - \ldots$

3. $(1-x)^{-1/2} = 1 - \frac{1}{2}(-x) + \frac{\left(-\frac{1}{2}\right)\left(-\frac{3}{2}\right)(-x)^2}{2!} + \frac{\left(-\frac{1}{2}\right)\left(-\frac{3}{2}\right)\left(-\frac{5}{2}\right)(-x)^3}{3!} + \ldots = 1 + \frac{1}{2}x + \frac{3}{8}x^2 + \frac{5}{16}x^3 + \ldots$

4. $(1-2x)^{1/2} = 1 + \frac{1}{2}(-2x) + \frac{\left(\frac{1}{2}\right)\left(-\frac{1}{2}\right)(-2x)^2}{2!} + \frac{\left(\frac{1}{2}\right)\left(-\frac{1}{2}\right)\left(-\frac{3}{2}\right)(-2x)^3}{3!} + \ldots = 1 - x - \frac{1}{2}x^2 - \frac{1}{2}x^3 - \ldots$

5. $\left(1+\frac{x}{2}\right)^{-2} = 1 - 2\left(\frac{x}{2}\right) + \frac{(-2)(-3)\left(\frac{x}{2}\right)^2}{2!} + \frac{(-2)(-3)(-4)\left(\frac{x}{2}\right)^3}{3!} + \ldots = 1 - x + \frac{3}{4}x^2 - \frac{1}{2}x^3$

6. $\left(1-\frac{x}{2}\right)^{-2} = 1 - 2\left(-\frac{x}{2}\right) + \frac{(-2)(-3)\left(-\frac{x}{2}\right)^2}{2!} + \frac{(-2)(-3)(-4)\left(-\frac{x}{2}\right)^3}{3!} + \ldots = 1 + x + \frac{3}{4}x^2 + \frac{1}{2}x^3 + \ldots$

7. $\left(1+x^3\right)^{-1/2} = 1 - \frac{1}{2}x^3 + \frac{\left(-\frac{1}{2}\right)\left(-\frac{3}{2}\right)\left(x^3\right)^2}{2!} + \frac{\left(-\frac{1}{2}\right)\left(-\frac{3}{2}\right)\left(-\frac{5}{2}\right)\left(x^3\right)^3}{3!} + \ldots = 1 - \frac{1}{2}x^3 + \frac{3}{8}x^6 - \frac{5}{16}x^9 + \ldots$

8. $\left(1+x^2\right)^{-1/3} = 1 - \frac{1}{3}x^2 + \frac{\left(-\frac{1}{3}\right)\left(-\frac{4}{3}\right)\left(x^2\right)^2}{2!} + \frac{\left(-\frac{1}{3}\right)\left(-\frac{4}{3}\right)\left(-\frac{7}{3}\right)\left(x^2\right)^3}{3!} + \ldots = 1 - \frac{1}{3}x^2 + \frac{2}{9}x^4 - \frac{14}{81}x^6 + \ldots$

9. $\left(1+\frac{1}{x}\right)^{1/2} = 1 + \frac{1}{2}\left(\frac{1}{x}\right) + \frac{\left(\frac{1}{2}\right)\left(-\frac{1}{2}\right)\left(\frac{1}{x}\right)^2}{2!} + \frac{\left(\frac{1}{2}\right)\left(-\frac{1}{2}\right)\left(-\frac{3}{2}\right)\left(\frac{1}{x}\right)^3}{3!} + \ldots = 1 + \frac{1}{2x} - \frac{1}{8x^2} + \frac{1}{16x^3} + \ldots$

10. $\left(1-\frac{2}{x}\right)^{1/3} = 1 + \frac{1}{3}\left(-\frac{2}{x}\right) + \frac{\left(\frac{1}{3}\right)\left(-\frac{2}{3}\right)\left(-\frac{2}{x}\right)^2}{2!} + \frac{\left(\frac{1}{3}\right)\left(-\frac{2}{3}\right)\left(-\frac{5}{3}\right)\left(-\frac{2}{x}\right)^3}{3!} + \ldots = 1 - \frac{2}{3x} - \frac{4}{9x^2} - \frac{40}{81x^3} - \ldots$

11. $(1+x)^4 = 1 + 4x + \frac{(4)(3)x^2}{2!} + \frac{(4)(3)(2)x^3}{3!} + \frac{(4)(3)(2)x^4}{4!} = 1 + 4x + 6x^2 + 4x^3 + x^4$

12. $\left(1+x^2\right)^3 = 1 + 3x^2 + \frac{(3)(2)\left(x^2\right)^2}{2!} + \frac{(3)(2)(1)\left(x^2\right)^3}{3!} = 1 + 3x^2 + 3x^4 + x^6$

13. $(1-2x)^3 = 1 + 3(-2x) + \frac{(3)(2)(-2x)^2}{2!} + \frac{(3)(2)(1)(-2x)^3}{3!} = 1 - 6x + 12x^2 - 8x^3$

14. $\left(1-\frac{x}{2}\right)^4 = 1 + 4\left(-\frac{x}{2}\right) + \frac{(4)(3)\left(-\frac{x}{2}\right)^2}{2!} + \frac{(4)(3)(2)\left(-\frac{x}{2}\right)^3}{3!} + \frac{(4)(3)(2)(1)\left(-\frac{x}{2}\right)^4}{4!} = 1 - 2x + \frac{3}{2}x^2 - \frac{1}{2}x^3 + \frac{1}{16}x^4$

15. Assume the solution has the form $y = a_0 + a_1x + a_2x^2 + \ldots + a_{n-1}x^{n-1} + a_nx^n + \ldots$
$\Rightarrow \frac{dy}{dx} = a_1 + 2a_2x + \ldots + na_nx^{n-1} + \ldots$
$\Rightarrow \frac{dy}{dx} + y = (a_1 + a_0) + (2a_2 + a_1)x + (3a_3 + a_2)x^2 + \ldots + (na_n + a_{n-1})x^{n-1} + \ldots = 0$
$\Rightarrow a_1 + a_0 = 0$, $2a_2 + a_1 = 0$, $3a_3 + a_2 = 0$ and in general $na_n + a_{n-1} = 0$. Since $y = 1$ when $x = 0$ we have $a_0 = 1$. Therefore $a_1 = -1$, $a_2 = \frac{-a_1}{2\cdot 1} = \frac{1}{2}$, $a_3 = \frac{-a_2}{3} = -\frac{1}{3\cdot 2}$, …, $a_n = \frac{-a_{n-1}}{n} = \frac{(-1)^n}{n!}$

$\Rightarrow y = 1 - x + \frac{1}{2}x^2 - \frac{1}{3!}x^3 + \ldots + \frac{(-1)^n}{n!}x^n + \ldots = \sum_{n=0}^{\infty} \frac{(-1)^n x^n}{n!} = e^{-x}$

16. Assume the solution has the form $y = a_0 + a_1x + a_2x^2 + \ldots + a_{n-1}x^{n-1} + a_nx^n + \ldots$
$\Rightarrow \frac{dy}{dx} = a_1 + 2a_2x + \ldots + na_nx^{n-1} + \ldots$
$\Rightarrow \frac{dy}{dx} - 2y = (a_1 - 2a_0) + (2a_2 - 2a_1)x + (3a_3 - 2a_2)x^2 + \ldots + (na_n - 2a_{n-1})x^{n-1} + \ldots = 0$
$\Rightarrow a_1 - 2a_0 = 0,\ 2a_2 - 2a_1 = 0,\ 3a_3 - 2a_2 = 0$ and in general $na_n - 2a_{n-1} = 0$. Since $y = 1$ when $x = 0$ we have $a_0 = 1$. Therefore $a_1 = 2a_0 = 2(1) = 2,\ a_2 = \frac{2}{2}a_1 = \frac{2}{2}(2) = \frac{2^2}{2},\ a_3 = \frac{2}{3}a_2 = \frac{2}{3}\left(\frac{2^2}{2}\right) = \frac{2^3}{3\cdot 2}, \ldots,$
$a_n = \left(\frac{2}{n}\right)a_{n-1} = \left(\frac{2}{n}\right)\left(\frac{2^{n-1}}{n-1}\right)a_{n-2} = \frac{2^n}{n!} \Rightarrow y = 1 + 2x + \frac{2^2}{2}x^2 + \frac{2^3}{3!}x^3 + \ldots + \frac{2^n}{n!}x^n + \ldots$
$= 1 + (2x) + \frac{(2x)^2}{2!} + \frac{(2x)^3}{3!} + \ldots + \frac{(2x)^n}{n!} + \ldots = \sum_{n=0}^{\infty} \frac{(2x)^n}{n!} = e^{2x}$

17. Assume the solution has the form $y = a_0 + a_1x + a_2x^2 + \ldots + a_{n-1}x^{n-1} + a_nx^n + \ldots$
$\Rightarrow \frac{dy}{dx} = a_1 + 2a_2x + \ldots + na_nx^{n-1} + \ldots$
$\Rightarrow \frac{dy}{dx} - y = (a_1 - a_0) + (2a_2 - a_1)x + (3a_3 - a_2)x^2 + \ldots + (na_n - a_{n-1})x^{n-1} + \ldots = 1$
$\Rightarrow a_1 - a_0 = 1,\ 2a_2 - a_1 = 0,\ 3a_3 - a_2 = 0$ and in general $na_n - a_{n-1} = 0$. Since $y = 0$ when $x = 0$ we have $a_0 = 0$. Therefore $a_1 = 1,\ a_2 = \frac{a_1}{2} = \frac{1}{2},\ a_3 = \frac{a_2}{3} = \frac{1}{3\cdot 2},\ a_4 = \frac{a_3}{4} = \frac{1}{4\cdot 3\cdot 2}, \ldots, a_n = \frac{a_{n-1}}{n} = \frac{1}{n!}$
$\Rightarrow y = 0 + 1x + \frac{1}{2}x^2 + \frac{1}{3\cdot 2}x^3 + \frac{1}{4\cdot 3\cdot 2}x^4 + \ldots + \frac{1}{n!}x^n + \ldots$
$= \left(1 + 1x + \frac{1}{2}x^2 + \frac{1}{3\cdot 2}x^3 + \frac{1}{4\cdot 3\cdot 2}x^4 + \ldots + \frac{1}{n!}x^n + \ldots\right) - 1 = \sum_{n=0}^{\infty} \frac{x^n}{n!} - 1 = e^x - 1$

18. Assume the solution has the form $y = a_0 + a_1x + a_2x^2 + \ldots + a_{n-1}x^{n-1} + a_nx^n + \ldots$
$\Rightarrow \frac{dy}{dx} = a_1 + 2a_2x + \ldots + na_nx^{n-1} + \ldots$
$\Rightarrow \frac{dy}{dx} + y = (a_1 + a_0) + (2a_2 + a_1)x + (3a_3 + a_2)x^2 + \ldots + (na_n + a_{n-1})x^{n-1} + \ldots = 1$
$\Rightarrow a_1 + a_0 = 1,\ 2a_2 + a_1 = 0,\ 3a_3 + a_2 = 0$ and in general $na_n + a_{n-1} = 0$. Since $y = 2$ when $x = 0$ we have $a_0 = 2$. Therefore $a_1 = 1 - a_0 = -1,\ a_2 = \frac{-a_1}{2\cdot 1} = \frac{1}{2},\ a_3 = \frac{-a_2}{3} = -\frac{1}{3\cdot 2}, \ldots, a_n = \frac{-a_{n-1}}{n} = \frac{(-1)^n}{n!}$
$\Rightarrow y = 2 - x + \frac{1}{2}x^2 - \frac{1}{3\cdot 2}x^3 + \ldots + \frac{(-1)^n}{n!}x^n + \ldots = 1 + \left(1 - x + \frac{1}{2}x^2 - \frac{1}{3\cdot 2}x^3 + \ldots + \frac{(-1)^n}{n!}x^n + \ldots\right)$
$= 1 + \sum_{n=0}^{\infty} \frac{(-1)^n x^n}{n!} = 1 + e^{-x}$

19. Assume the solution has the form $y = a_0 + a_1x + a_2x^2 + \ldots + a_{n-1}x^{n-1} + a_nx^n + \ldots$
$\Rightarrow \frac{dy}{dx} = a_1 + 2a_2x + \ldots + na_nx^{n-1} + \ldots$
$\Rightarrow \frac{dy}{dx} - y = (a_1 - a_0) + (2a_2 - a_1)x + (3a_3 - a_2)x^2 + \ldots + (na_n - a_{n-1})x^{n-1} + \ldots = x$
$\Rightarrow a_1 - a_0 = 0,\ 2a_2 - a_1 = 1,\ 3a_3 - a_2 = 0$ and in general $na_n - a_{n-1} = 0$. Since $y = 0$ when $x = 0$ we have $a_0 = 0$. Therefore $a_1 = 0,\ a_2 = \frac{1 + a_1}{2} = \frac{1}{2},\ a_3 = \frac{a_2}{3} = \frac{1}{3\cdot 2},\ a_4 = \frac{a_3}{4} = \frac{1}{4\cdot 3\cdot 2}, \ldots, a_n = \frac{a_{n-1}}{n} = \frac{1}{n!}$
$\Rightarrow y = 0 + 0x + \frac{1}{2}x^2 + \frac{1}{3\cdot 2}x^3 + \frac{1}{4\cdot 3\cdot 2}x^4 + + \ldots + \frac{1}{n!}x^n + \ldots$
$= \left(1 + 1x + \frac{1}{2}x^2 + \frac{1}{3\cdot 2}x^3 + \frac{1}{4\cdot 3\cdot 2}x^4 + \ldots + \frac{1}{n!}x^n + \ldots\right) - 1 - x = \sum_{n=0}^{\infty} \frac{x^n}{n!} - 1 - x = e^x - x - 1$

20. Assume the solution has the form $y = a_0 + a_1x + a_2x^2 + \ldots + a_{n-1}x^{n-1} + a_nx^n + \ldots$
$\Rightarrow \frac{dy}{dx} = a_1 + 2a_2x + \ldots + na_nx^{n-1} + \ldots$
$\Rightarrow \frac{dy}{dx} + y = (a_1 + a_0) + (2a_2 + a_1)x + (3a_3 + a_2)x^2 + \ldots + (na_n + a_{n-1})x^{n-1} + \ldots = 2x$
$\Rightarrow a_1 + a_0 = 0,\ 2a_2 + a_1 = 2,\ 3a_3 + a_2 = 0$ and in general $na_n + a_{n-1} = 0$. Since $y = -1$ when $x = 0$ we have $a_0 = -1$. Therefore $a_1 = 1,\ a_2 = \frac{2 - a_1}{2} = \frac{1}{2},\ a_3 = \frac{-a_2}{3} = -\frac{1}{3\cdot 2}, \ldots, a_n = \frac{-a_{n-1}}{n} = \frac{(-1)^n}{n!}$
$\Rightarrow y = -1 + 1x + \frac{1}{2}x^2 - \frac{1}{3\cdot 2}x^3 + \ldots + \frac{(-1)^n}{n!}x^n + \ldots$

$= \left(1 - 1x + \frac{1}{2}x^2 - \frac{1}{3\cdot2}x^3 + \ldots + \frac{(-1)^n}{n!}x^n + \ldots\right) - 2 + 2x = \sum_{n=0}^{\infty} \frac{(-1)^n x^n}{n!} - 2 + 2x = e^{-x} + 2x - 2$

21. $y' - xy = a_1 + (2a_2 - a_0)x + (3a_3 - a_1)x + \ldots + (na_n - a_{n-2})x^{n-1} + \ldots = 0 \Rightarrow a_1 = 0, 2a_2 - a_0 = 0, 3a_3 - a_1 = 0,$ $4a_4 - a_2 = 0$ and in general $na_n - a_{n-2} = 0$. Since $y = 1$ when $x = 0$, we have $a_0 = 1$. Therefore $a_2 = \frac{a_0}{2} = \frac{1}{2}$, $a_3 = \frac{a_1}{3} = 0$, $a_4 = \frac{a_2}{4} = \frac{1}{2\cdot4}$, $a_5 = \frac{a_3}{5} = 0, \ldots, a_{2n} = \frac{1}{2\cdot4\cdot6\cdots2n}$ and $a_{2n+1} = 0$

$\Rightarrow y = 1 + \frac{1}{2}x^2 + \frac{1}{2\cdot4}x^4 + \frac{1}{2\cdot4\cdot6}x^6 + \ldots + \frac{1}{2\cdot4\cdot6\cdots2n}x^{2n} + \ldots = \sum_{n=0}^{\infty} \frac{x^{2n}}{2^n n!} = \sum_{n=0}^{\infty} \frac{\left(\frac{x^2}{2}\right)^n}{n!} = e^{x^2/2}$

22. $y' - x^2y = a_1 + 2a_2x + (3a_3 - a_0)x^2 + (4a_4 - a_1)x^3 + \ldots + (na_n - a_{n-3})x^{n-1} + \ldots = 0 \Rightarrow a_1 = 0, a_2 = 0,$ $3a_3 - a_0 = 0, 4a_4 - a_1 = 0$ and in general $na_n - a_{n-3} = 0$. Since $y = 1$ when $x = 0$, we have $a_0 = 1$. Therefore $a_3 = \frac{a_0}{3} = \frac{1}{3}$, $a_4 = \frac{a_1}{4} = 0$, $a_5 = \frac{a_2}{5} = 0$, $a_6 = \frac{a_3}{6} = \frac{1}{3\cdot6}, \ldots, a_{3n} = \frac{1}{3\cdot6\cdot9\cdots3n}$, $a_{3n+1} = 0$ and $a_{3n+2} = 0$

$\Rightarrow y = 1 + \frac{1}{3}x^3 + \frac{1}{3\cdot6}x^6 + \frac{1}{3\cdot6\cdot9}x^9 + \ldots + \frac{1}{3\cdot6\cdot9\cdots3n}x^{3n} + \ldots = \sum_{n=0}^{\infty} \frac{x^{3n}}{3^n n!} = \sum_{n=0}^{\infty} \frac{\left(\frac{x^3}{3}\right)^n}{n!} = e^{x^3/3}$

23. $(1 - x)y' - y = (a_1 - a_0) + (2a_2 - a_1 - a_1)x + (3a_3 - 2a_2 - a_2)x^2 + (4a_4 - 3a_3 - a_3)x^3 + \ldots$ $+ (na_n - (n-1)a_{n-1} - a_{n-1})x^{n-1} + \ldots = 0 \Rightarrow a_1 - a_0 = 0, 2a_2 - 2a_1 = 0, 3a_3 - 3a_2 = 0$ and in general $(na_n - na_{n-1}) = 0$. Since $y = 2$ when $x = 0$, we have $a_0 = 2$. Therefore $a_1 = 2, a_2 = 2, \ldots, a_n = 2 \Rightarrow y = 2 + 2x + 2x^2 + \ldots = \sum_{n=0}^{\infty} 2x^n = \frac{2}{1-x}$

24. $(1 + x^2)\,y' + 2xy = a_1 + (2a_2 + 2a_0)x + (3a_3 + 2a_1 + a_1)x^2 + (4a_4 + 2a_2 + 2a_2)x^3 + \ldots + (na_n + na_{n-2})x^{n-1} + \ldots$ $= 0 \Rightarrow a_1 = 0, 2a_2 + 2a_0 = 0, 3a_3 + 3a_1 = 0, 4a_4 + 4a_2 = 0$ and in general $na_n + na_{n-2} = 0$. Since $y = 3$ when $x = 0$, we have $a_0 = 3$. Therefore $a_2 = -3, a_3 = 0, a_4 = 3, \ldots, a_{2n+1} = 0, a_{2n} = (-1)^n 3$

$\Rightarrow y = 3 - 3x^2 + 3x^4 - \ldots = \sum_{n=0}^{\infty} 3(-1)^n x^{2n} = \sum_{n=0}^{\infty} 3\left(-x^2\right)^n = \frac{3}{1+x^2}$

25. $y = a_0 + a_1x + a_2x^2 + \ldots + a_nx^n + \ldots \Rightarrow y'' = 2a_2 + 3\cdot2a_3x + \ldots + n(n-1)a_nx^{n-2} + \ldots \Rightarrow y'' - y$ $= (2a_2 - a_0) + (3\cdot2a_3 - a_1)x + (4\cdot3a_4 - a_2)x^2 + \ldots + (n(n-1)a_n - a_{n-2})x^{n-2} + \ldots = 0 \Rightarrow 2a_2 - a_0 = 0,$ $3\cdot2a_3 - a_1 = 0, 4\cdot3a_4 - a_2 = 0$ and in general $n(n-1)a_n - a_{n-2} = 0$. Since $y' = 1$ and $y = 0$ when $x = 0$, we have $a_0 = 0$ and $a_1 = 1$. Therefore $a_2 = 0$, $a_3 = \frac{1}{3\cdot2}$, $a_4 = 0$, $a_5 = \frac{1}{5\cdot4\cdot3\cdot2}, \ldots, a_{2n+1} = \frac{1}{(2n+1)!}$ and $a_{2n} = 0 \Rightarrow y = x + \frac{1}{3!}x^3 + \frac{1}{5!}x^5 + \ldots = \sum_{n=0}^{\infty} \frac{x^{2n+1}}{(2n+1)!} = \sinh x$

26. $y = a_0 + a_1x + a_2x^2 + \ldots + a_nx^n + \ldots \Rightarrow y'' = 2a_2 + 3\cdot2a_3x + \ldots + n(n-1)a_nx^{n-2} + \ldots \Rightarrow y'' + y$ $= (2a_2 + a_0) + (3\cdot2a_3 + a_1)x + (4\cdot3a_4 + a_2)x^2 + \ldots + (n(n-1)a_n + a_{n-2})x^{n-2} + \ldots = 0 \Rightarrow 2a_2 + a_0 = 0,$ $3\cdot2a_3 + a_1 = 0, 4\cdot3a_4 + a_2 = 0$ and in general $n(n-1)a_n + a_{n-2} = 0$. Since $y' = 0$ and $y = 1$ when $x = 0$, we have $a_0 = 1$ and $a_1 = 0$. Therefore $a_2 = -\frac{1}{2}$, $a_3 = 0$, $a_4 = \frac{1}{4\cdot3\cdot2}$, $a_5 = 0, \ldots, a_{2n+1} = 0$ and $a_{2n} = \frac{(-1)^n}{(2n)!}$

$\Rightarrow y = 1 - \frac{1}{2}x^2 + \frac{1}{4!}x^4 - \ldots = \sum_{n=0}^{\infty} \frac{(-1)^n x^{2n}}{(2n)!} = \cos x$

27. $y = a_0 + a_1x + a_2x^2 + \ldots + a_nx^n + \ldots \Rightarrow y'' = 2a_2 + 3\cdot2a_3x + \ldots + n(n-1)a_nx^{n-2} + \ldots \Rightarrow y'' + y$ $= (2a_2 + a_0) + (3\cdot2a_3 + a_1)x + (4\cdot3a_4 + a_2)x^2 + \ldots + (n(n-1)a_n + a_{n-2})x^{n-2} + \ldots = x \Rightarrow 2a_2 + a_0 = 0,$ $3\cdot2a_3 + a_1 = 1, 4\cdot3a_4 + a_2 = 0$ and in general $n(n-1)a_n + a_{n-2} = 0$. Since $y' = 1$ and $y = 2$ when $x = 0$, we have $a_0 = 2$ and $a_1 = 1$. Therefore $a_2 = -1$, $a_3 = 0$, $a_4 = \frac{1}{4\cdot3}$, $a_5 = 0, \ldots, a_{2n} = -2\cdot\frac{(-1)^{n+1}}{(2n)!}$ and $a_{2n+1} = 0 \Rightarrow y = 2 + x - x^2 + 2\cdot\frac{x^4}{4!} + \ldots = 2 + x - 2\sum_{n=1}^{\infty} \frac{(-1)^{n+1}x^{2n}}{(2n)!} = x + \cos 2x$

28. $y = a_0 + a_1x + a_2x^2 + \ldots + a_nx^n + \ldots \Rightarrow y'' = 2a_2 + 3 \cdot 2a_3x + \ldots + n(n-1)a_nx^{n-2} + \ldots \Rightarrow y'' - y$ $= (2a_2 - a_0) + (3 \cdot 2a_3 - a_1)x + (4 \cdot 3a_4 - a_2)x^2 + \ldots + (n(n-1)a_n - a_{n-2})x^{n-2} + \ldots = x \Rightarrow 2a_2 - a_0 = 0,$ $3 \cdot 2a_3 - a_1 = 1, 4 \cdot 3a_4 - a_2 = 0$ and in general $n(n-1)a_n - a_{n-2} = 0$. Since $y' = 2$ and $y = -1$ when $x = 0$, we have $a_0 = -1$ and $a_1 = 2$. Therefore $a_2 = \frac{-1}{2}, a_3 = \frac{1}{2}, a_4 = \frac{-1}{2 \cdot 3 \cdot 4}, a_5 = \frac{1}{5 \cdot 4 \cdot 2} = \frac{3}{5!}, \ldots, a_{2n} = \frac{-1}{(2n)!}$ and $a_{2n+1} = \frac{3}{(2n+1)!} \Rightarrow y = -1 + 2x - \frac{1}{2}x^2 + \frac{3}{3!}x^3 - \ldots = -1 + 2x - \sum_{n=1}^{\infty} \frac{x^{2n}}{(2n)!} + \sum_{n=1}^{\infty} \frac{3x^{2n+1}}{(2n+1)!}$

29. $y = a_0 + a_1(x-2) + a_2(x-2)^2 + \ldots + a_n(x-2)^n + \ldots$ $\Rightarrow y'' = 2a_2 + 3 \cdot 2a_3(x-2) + \ldots + n(n-1)a_n(x-2)^{n-2} + \ldots \Rightarrow y'' - y$ $= (2a_2 - a_0) + (3 \cdot 2a_3 - a_1)(x-2) + (4 \cdot 3a_4 - a_2)(x-2)^2 + \ldots + (n(n-1)a_n - a_{n-2})(x-2)^{n-2} + \ldots = -x$ $= -(x-2) - 2 \Rightarrow 2a_2 - a_0 = -2, 3 \cdot 2a_3 - a_1 = -1$, and $n(n-1)a_n - a_{n-2} = 0$ for $n > 3$. Since $y = 0$ when $x = 2$, we have $a_0 = 0$, and since $y' = -2$ when $x = 2$, we have $a_1 = -2$. Therefore $a_2 = -1, a_3 = -\frac{1}{2}, a_4 = \frac{1}{4 \cdot 3}(-1) = \frac{-2}{4 \cdot 3 \cdot 2 \cdot 1},$ $a_5 = \frac{1}{5 \cdot 4}\left(-\frac{1}{2}\right) = \frac{-3}{5 \cdot 4 \cdot 3 \cdot 2 \cdot 1}, \ldots, a_{2n} = \frac{-2}{(2n)!}$, and $a_{2n+1} = \frac{-3}{(2n+1)!}$. Since $a_1 = -2$, we have $a_1(x-2) = (-2)(x-2)$ and $(-2)(x-2) = (-3+1)(x-2) = (-3)(x-2) + (1)(x-2) = x - 2 - 3(x-2)$.
$\Rightarrow y = x - 2 - 3(x-2) - \frac{2}{2!}(x-2)^2 - \frac{3}{3!}(x-2)^3 - \frac{2}{4!}(x-2)^4 - \frac{3}{5!}(x-2)^5 - \ldots$
$\Rightarrow y = x - 2 - \frac{2}{2!}(x-2)^2 - \frac{2}{4!}(x-2)^4 - \ldots - 3(x-2)x - \frac{3}{3!}(x-2)^3 - \frac{3}{5!}(x-2)^5 - \ldots$
$\Rightarrow y = x - 2\sum_{n=0}^{\infty} \frac{(x-2)^{2n}}{(2n)!} - 3\sum_{n=0}^{\infty} \frac{(x-2)^{2n+1}}{(2n+1)!}$

30. $y'' - x^2y = 2a_2 + 6a_3x + (4 \cdot 3a_4 - a_0)x^2 + \ldots + (n(n-1)a_n - a_{n-4})x^{n-2} + \ldots = 0 \Rightarrow 2a_2 = 0, 6a_3 = 0,$ $4 \cdot 3a_4 - a_0 = 0, 5 \cdot 4a_5 - a_1 = 0$, and in general $n(n-1)a_n - a_{n-4} = 0$. Since $y' = b$ and $y = a$ when $x = 0$, we have $a_0 = a, a_1 = b, a_2 = 0, a_3 = 0, a_4 = \frac{a}{3 \cdot 4}, a_5 = \frac{b}{4 \cdot 5}, a_6 = 0, a_7 = 0, a_8 = \frac{a}{3 \cdot 4 \cdot 7 \cdot 8}, a_9 = \frac{b}{4 \cdot 5 \cdot 8 \cdot 9}$
$\Rightarrow y = a + bx + \frac{a}{3 \cdot 4}x^4 + \frac{b}{4 \cdot 5}x^5 + \frac{a}{3 \cdot 4 \cdot 7 \cdot 8}x^8 + \frac{b}{4 \cdot 5 \cdot 8 \cdot 9}x^9 + \ldots$

31. $y'' + x^2y = 2a_2 + 6a_3x + (4 \cdot 3a_4 + a_0)x^2 + \ldots + (n(n-1)a_n + a_{n-4})x^{n-2} + \ldots = x \Rightarrow 2a_2 = 0, 6a_3 = 1,$ $4 \cdot 3a_4 + a_0 = 0, 5 \cdot 4a_5 + a_1 = 0$, and in general $n(n-1)a_n + a_{n-4} = 0$. Since $y' = b$ and $y = a$ when $x = 0$, we have $a_0 = a$ and $a_1 = b$. Therefore $a_2 = 0, a_3 = \frac{1}{2 \cdot 3}, a_4 = -\frac{a}{3 \cdot 4}, a_5 = -\frac{b}{4 \cdot 5}, a_6 = 0, a_7 = \frac{-1}{2 \cdot 3 \cdot 6 \cdot 7}$
$\Rightarrow y = a + bx + \frac{1}{2 \cdot 3}x^3 - \frac{a}{3 \cdot 4}x^4 - \frac{b}{4 \cdot 5}x^5 - \frac{1}{2 \cdot 3 \cdot 6 \cdot 7}x^7 + \frac{ax^8}{3 \cdot 4 \cdot 7 \cdot 8} + \frac{bx^9}{4 \cdot 5 \cdot 8 \cdot 9} + \ldots$

32. $y'' - 2y' + y = (2a_2 - 2a_1 + a_0) + (2 \cdot 3a_3 - 4a_2 + a_1)x + (3 \cdot 4a_4 - 2 \cdot 3a_3 + a_2)x^2 + \ldots$ $+ ((n-1)na_n - 2(n-1)a_{n-1} + a_{n-2})x^{n-2} + \ldots = 0 \Rightarrow 2a_2 - 2a_1 + a_0 = 0, 2 \cdot 3a_3 - 4a_2 + a_1 = 0,$ $3 \cdot 4a_4 - 2 \cdot 3a_3 + a_2 = 0$ and in general $(n-1)na_n - 2(n-1)a_{n-1} + a_{n-2} = 0$. Since $y' = 1$ and $y = 0$ when when $x = 0$, we have $a_0 = 0$ and $a_1 = 1$. Therefore $a_2 = 1, a_3 = \frac{1}{2}, a_4 = \frac{1}{6}, a_5 = \frac{1}{24}$ and $a_n = \frac{1}{(n-1)!}$
$\Rightarrow y = x + x^2 + \frac{1}{2}x^3 + \frac{1}{6}x^4 + \frac{1}{24}x^5 + \ldots = \sum_{n=1}^{\infty} \frac{x^n}{(n-1)!} = \sum_{n=0}^{\infty} \frac{x^{n+1}}{n!} = x\sum_{n=0}^{\infty} \frac{x^n}{n!} = xe^x$

33. $\int_0^{0.2} \sin x^2\, dx = \int_0^{0.2} \left(x^2 - \frac{x^6}{3!} + \frac{x^{10}}{5!} - \ldots\right) dx = \left[\frac{x^3}{3} - \frac{x^7}{7 \cdot 3!} + \ldots\right]_0^{0.2} \approx \left[\frac{x^3}{3}\right]_0^{0.2} \approx 0.00267$ with error $|E| \le \frac{(.2)^7}{7 \cdot 3!} \approx 0.0000003$

34. $\int_0^{0.2} \frac{e^{-x} - 1}{x}\, dx = \int_0^{0.2} \frac{1}{x}\left(1 - x + \frac{x^2}{2!} - \frac{x^3}{3!} + \frac{x^4}{4!} - \ldots - 1\right) dx = \int_0^{0.2} \left(-1 + \frac{x}{2} - \frac{x^2}{6} + \frac{x^3}{24} - \ldots\right) dx$ $= \left[-x + \frac{x^2}{4} - \frac{x^3}{18} + \ldots\right]_0^{0.2} \approx -0.19044$ with error $|E| \le \frac{(0.2)^4}{96} \approx 0.00002$

35. $\int_0^{0.1} \frac{1}{\sqrt{1 + x^4}}\, dx = \int_0^{0.1} \left(1 - \frac{x^4}{2} + \frac{3x^8}{8} - \ldots\right) dx = \left[x - \frac{x^5}{10} + \ldots\right]_0^{0.1} \approx [x]_0^{0.1} \approx 0.1$ with error $|E| \le \frac{(0.1)^5}{10} = 0.000001$

36. $\int_0^{0.25} \sqrt[3]{1+x^2}\,dx = \int_0^{0.25} \left(1 + \frac{x^2}{3} - \frac{x^4}{9} + \ldots\right) dx = \left[x + \frac{x^3}{9} - \frac{x^5}{45} + \ldots\right]_0^{0.25} \approx \left[x + \frac{x^3}{9}\right]_0^{0.25} \approx 0.25174$ with error $|E| \le \frac{(0.25)^5}{45} \approx 0.0000217$

37. $\int_0^{0.1} \frac{\sin x}{x}\,dx = \int_0^{0.1} \left(1 - \frac{x^2}{3!} + \frac{x^4}{5!} - \frac{x^6}{7!} + \ldots\right) dx = \left[x - \frac{x^3}{3\cdot 3!} + \frac{x^5}{5\cdot 5!} - \frac{x^7}{7\cdot 7!} + \ldots\right]_0^{0.1} \approx \left[x - \frac{x^3}{3\cdot 3!} + \frac{x^5}{5\cdot 5!}\right]_0^{0.1}$ ≈ 0.0999444611, $|E| \le \frac{(0.1)^7}{7\cdot 7!} \approx 2.8 \times 10^{-12}$

38. $\int_0^{0.1} \exp(-x^2)\,dx = \int_0^{0.1} \left(1 - x^2 + \frac{x^4}{2!} - \frac{x^6}{3!} + \frac{x^8}{4!} - \ldots\right) dx = \left[x - \frac{x^3}{3} + \frac{x^5}{10} + \frac{x^7}{42} + \ldots\right]_0^{0.1} \approx \left[x - \frac{x^3}{3} + \frac{x^5}{10} - \frac{x^7}{42}\right]_0^{0.1}$ ≈ 0.0996676643, $|E| \le \frac{(0.1)^9}{216} \approx 4.6 \times 10^{-12}$

39. $(1+x^4)^{1/2} = (1)^{1/2} + \frac{\left(\frac{1}{2}\right)}{1}(1)^{-1/2}(x^4) + \frac{\left(\frac{1}{2}\right)\left(-\frac{1}{2}\right)}{2!}(1)^{-3/2}(x^4)^2 + \frac{\left(\frac{1}{2}\right)\left(-\frac{1}{2}\right)\left(-\frac{3}{2}\right)}{3!}(1)^{-5/2}(x^4)^3$ $+ \frac{\left(\frac{1}{2}\right)\left(-\frac{1}{2}\right)\left(-\frac{3}{2}\right)\left(-\frac{5}{2}\right)}{4!}(1)^{-7/2}(x^4)^4 + \ldots = 1 + \frac{x^4}{2} - \frac{x^8}{8} + \frac{x^{12}}{16} - \frac{5x^{16}}{128} + \ldots$ $\Rightarrow \int_0^{0.1} \left(1 + \frac{x^4}{2} - \frac{x^8}{8} + \frac{x^{12}}{16} - \frac{5x^{16}}{128} + \ldots\right) dx \approx \left[x + \frac{x^5}{10}\right]_0^{0.1} \approx 0.100001$, $|E| \le \frac{(0.1)^9}{72} \approx 1.39 \times 10^{-11}$

40. $\int_0^1 \left(\frac{1-\cos x}{x^2}\right) dx = \int_0^1 \left(\frac{1}{2} - \frac{x^2}{4!} + \frac{x^4}{6!} - \frac{x^6}{8!} + \frac{x^8}{10!} - \ldots\right) dx \approx \left[\frac{x}{2} - \frac{x^3}{3\cdot 4!} + \frac{x^5}{5\cdot 6!} - \frac{x^7}{7\cdot 8!} + \frac{x^9}{9\cdot 10!}\right]_0^1$ ≈ 0.4863853764, $|E| \le \frac{1}{11\cdot 12!} \approx 1.9 \times 10^{-10}$

41. $\int_0^1 \cos t^2\,dt = \int_0^1 \left(1 - \frac{t^4}{2} + \frac{t^8}{4!} - \frac{t^{12}}{6!} + \ldots\right) dt = \left[t - \frac{t^5}{10} + \frac{t^9}{9\cdot 4!} - \frac{t^{13}}{13\cdot 6!} + \ldots\right]_0^1 \Rightarrow |\text{error}| < \frac{1}{13\cdot 6!} \approx .00011$

42. $\int_0^1 \cos\sqrt{t}\,dt = \int_0^1 \left(1 - \frac{t}{2} + \frac{t^2}{4!} - \frac{t^3}{6!} + \frac{t^4}{8!} - \ldots\right) dt = \left[t - \frac{t^2}{4} + \frac{t^3}{3\cdot 4!} - \frac{t^4}{4\cdot 6!} + \frac{t^5}{5\cdot 8!} - \ldots\right]_0^1$ $\Rightarrow |\text{error}| < \frac{1}{5\cdot 8!} \approx 0.000004960$

43. $F(x) = \int_0^x \left(t^2 - \frac{t^6}{3!} + \frac{t^{10}}{5!} - \frac{t^{14}}{7!} + \ldots\right) dt = \left[\frac{t^3}{3} - \frac{t^7}{7\cdot 3!} + \frac{t^{11}}{11\cdot 5!} - \frac{t^{15}}{15\cdot 7!} + \ldots\right]_0^x \approx \frac{x^3}{3} - \frac{x^7}{7\cdot 3!} + \frac{x^{11}}{11\cdot 5!}$ $\Rightarrow |\text{error}| < \frac{1}{15\cdot 7!} \approx 0.000013$

44. $F(x) = \int_0^x \left(t^2 - t^4 + \frac{t^6}{2!} - \frac{t^8}{3!} + \frac{t^{10}}{4!} - \frac{t^{12}}{5!} + \ldots\right) dt = \left[\frac{t^3}{3} - \frac{t^5}{5} + \frac{t^7}{7\cdot 2!} - \frac{t^9}{9\cdot 3!} + \frac{t^{11}}{11\cdot 4!} - \frac{t^{13}}{13\cdot 5!} + \ldots\right]_0^x$ $\approx \frac{x^3}{3} - \frac{x^5}{5} + \frac{x^7}{7\cdot 2!} - \frac{x^9}{9\cdot 3!} + \frac{x^{11}}{11\cdot 4!} \Rightarrow |\text{error}| < \frac{1}{13\cdot 5!} \approx 0.00064$

45. (a) $F(x) = \int_0^x \left(t - \frac{t^3}{3} + \frac{t^5}{5} - \frac{t^7}{7} + \ldots\right) dt = \left[\frac{t^2}{2} - \frac{t^4}{12} + \frac{t^6}{30} - \ldots\right]_0^x \approx \frac{x^2}{2} - \frac{x^4}{12} \Rightarrow |\text{error}| < \frac{(0.5)^6}{30} \approx .00052$

(b) $|\text{error}| < \frac{1}{33\cdot 34} \approx .00089$ when $F(x) \approx \frac{x^2}{2} - \frac{x^4}{3\cdot 4} + \frac{x^6}{5\cdot 6} - \frac{x^8}{7\cdot 8} + \ldots + (-1)^{15}\frac{x^{32}}{31\cdot 32}$

46. (a) $F(x) = \int_0^x \left(1 - \frac{t}{2} + \frac{t^2}{3} - \frac{t^3}{4} + \ldots\right) dt = \left[t - \frac{t^2}{2\cdot 2} + \frac{t^3}{3\cdot 3} - \frac{t^4}{4\cdot 4} + \frac{t^5}{5\cdot 5} - \ldots\right]_0^x \approx x - \frac{x^2}{2^2} + \frac{x^3}{3^2} - \frac{x^4}{4^2} + \frac{x^5}{5^2}$ $\Rightarrow |\text{error}| < \frac{(0.5)^6}{6^2} \approx .00043$

(b) $|\text{error}| < \frac{1}{32^2} \approx .00097$ when $F(x) \approx x - \frac{x^2}{2^2} + \frac{x^3}{3^2} - \frac{x^4}{4^2} + \ldots + (-1)^{31}\frac{x^{31}}{31^2}$

47. $\frac{1}{x^2}\left(e^x - (1+x)\right) = \frac{1}{x^2}\left(\left(1 + x + \frac{x^2}{2} + \frac{x^3}{3!} + \ldots\right) - 1 - x\right) = \frac{1}{2} + \frac{x}{3!} + \frac{x^2}{4!} + \ldots \Rightarrow \lim_{x\to 0} \frac{e^x - (1+x)}{x^2}$ $= \lim_{x\to 0}\left(\frac{1}{2} + \frac{x}{3!} + \frac{x^2}{4!} + \ldots\right) = \frac{1}{2}$

48. $\frac{1}{x}\left(e^x - e^{-x}\right) = \frac{1}{x}\left[\left(1 + x + \frac{x^2}{2!} + \frac{x^3}{3!} + \frac{x^4}{4!} + \ldots\right) - \left(1 - x + \frac{x^2}{2!} - \frac{x^3}{3!} + \frac{x^4}{4!} - \ldots\right)\right] = \frac{1}{x}\left(2x + \frac{2x^3}{3!} + \frac{2x^5}{5!} + \frac{2x^7}{7!} + \ldots\right)$
$= 2 + \frac{2x^2}{3!} + \frac{2x^4}{5!} + \frac{2x^6}{7!} + \ldots \Rightarrow \lim\limits_{x \to 0} \frac{e^x - e^{-x}}{x} = \lim\limits_{x \to \infty}\left(2 + \frac{2x^2}{3!} + \frac{2x^4}{5!} + \frac{2x^6}{7!} + \ldots\right) = 2$

49. $\frac{1}{t^4}\left(1 - \cos t - \frac{t^2}{2}\right) = \frac{1}{t^4}\left[1 - \frac{t^2}{2} - \left(1 - \frac{t^2}{2} + \frac{t^4}{4!} - \frac{t^6}{6!} + \ldots\right)\right] = -\frac{1}{4!} + \frac{t^2}{6!} - \frac{t^4}{8!} + \ldots \Rightarrow \lim\limits_{t \to 0} \frac{1 - \cos t - \left(\frac{t^2}{2}\right)}{t^4}$
$= \lim\limits_{t \to 0}\left(-\frac{1}{4!} + \frac{t^2}{6!} - \frac{t^4}{8!} + \ldots\right) = -\frac{1}{24}$

50. $\frac{1}{\theta^5}\left(-\theta + \frac{\theta^3}{6} + \sin\theta\right) = \frac{1}{\theta^5}\left(-\theta + \frac{\theta^3}{6} + \theta - \frac{\theta^3}{3!} + \frac{\theta^5}{5!} - \ldots\right) = \frac{1}{5!} - \frac{\theta^2}{7!} + \frac{\theta^4}{9!} - \ldots \Rightarrow \lim\limits_{\theta \to 0} \frac{\sin\theta - \theta + \left(\frac{\theta^3}{6}\right)}{\theta^5}$
$= \lim\limits_{\theta \to 0}\left(\frac{1}{5!} - \frac{\theta^2}{7!} + \frac{\theta^4}{9!} - \ldots\right) = \frac{1}{120}$

51. $\frac{1}{y^3}\left(y - \tan^{-1} y\right) = \frac{1}{y^3}\left[y - \left(y - \frac{y^3}{3} + \frac{y^5}{5} - \ldots\right)\right] = \frac{1}{3} - \frac{y^2}{5} + \frac{y^4}{7} - \ldots \Rightarrow \lim\limits_{y \to 0} \frac{y - \tan^{-1} y}{y^3} = \lim\limits_{y \to 0}\left(\frac{1}{3} - \frac{y^2}{5} + \frac{y^4}{7} - \ldots\right)$
$= \frac{1}{3}$

52. $\frac{\tan^{-1} y - \sin y}{y^3 \cos y} = \frac{\left(y - \frac{y^3}{3} + \frac{y^5}{5} - \ldots\right) - \left(y - \frac{y^3}{3!} + \frac{y^5}{5!} - \ldots\right)}{y^3 \cos y} = \frac{\left(-\frac{y^3}{6} + \frac{23y^5}{5!} - \ldots\right)}{y^3 \cos y} = \frac{\left(-\frac{1}{6} + \frac{23y^2}{5!} - \ldots\right)}{\cos y}$
$\Rightarrow \lim\limits_{y \to 0} \frac{\tan^{-1} y - \sin y}{y^3 \cos y} = \lim\limits_{y \to 0} \frac{\left(-\frac{1}{6} + \frac{23y^2}{5!} - \ldots\right)}{\cos y} = -\frac{1}{6}$

53. $x^2\left(-1 + e^{-1/x^2}\right) = x^2\left(-1 + 1 - \frac{1}{x^2} + \frac{1}{2x^4} - \frac{1}{6x^6} + \ldots\right) = -1 + \frac{1}{2x^2} - \frac{1}{6x^4} + \ldots \Rightarrow \lim\limits_{x \to \infty} x^2\left(e^{-1/x^2} - 1\right)$
$= \lim\limits_{x \to \infty}\left(-1 + \frac{1}{2x^2} - \frac{1}{6x^4} + \ldots\right) = -1$

54. $(x + 1)\sin\left(\frac{1}{x+1}\right) = (x + 1)\left(\frac{1}{x+1} - \frac{1}{3!(x+1)^3} + \frac{1}{5!(x+1)^5} - \ldots\right) = 1 - \frac{1}{3!(x+1)^2} + \frac{1}{5!(x+1)^4} - \ldots$
$\Rightarrow \lim\limits_{x \to \infty} (x + 1)\sin\left(\frac{1}{x+1}\right) = \lim\limits_{x \to \infty}\left(1 - \frac{1}{3!(x+1)^2} + \frac{1}{5!(x+1)^4} - \ldots\right) = 1$

55. $\frac{\ln(1 + x^2)}{1 - \cos x} = \frac{\left(x^2 - \frac{x^4}{2} + \frac{x^6}{3} - \ldots\right)}{1 - \left(1 - \frac{x^2}{2!} + \frac{x^4}{4!} - \ldots\right)} = \frac{\left(1 - \frac{x^2}{2} + \frac{x^4}{3} - \ldots\right)}{\left(\frac{1}{2!} - \frac{x^2}{4!} + \ldots\right)} \Rightarrow \lim\limits_{x \to 0} \frac{\ln(1 + x^2)}{1 - \cos x} = \lim\limits_{x \to 0} \frac{\left(1 - \frac{x^2}{2} + \frac{x^4}{3} - \ldots\right)}{\left(\frac{1}{2!} - \frac{x^2}{4!} + \ldots\right)} = 2! = 2$

56. $\frac{x^2 - 4}{\ln(x - 1)} = \frac{(x - 2)(x + 2)}{\left[(x - 2) - \frac{(x-2)^2}{2} + \frac{(x-2)^3}{3} - \ldots\right]} = \frac{x + 2}{\left[1 - \frac{x-2}{2} + \frac{(x-2)^2}{3} - \ldots\right]} \Rightarrow \lim\limits_{x \to 2} \frac{x^2 - 4}{\ln(x - 1)}$
$= \lim\limits_{x \to 2} \frac{x+2}{\left[1 - \frac{x-2}{2} + \frac{(x-2)^2}{3} - \ldots\right]} = 4$

57. $\ln\left(\frac{1+x}{1-x}\right) = \ln(1 + x) - \ln(1 - x) = \left(x - \frac{x^2}{2} + \frac{x^3}{3} - \frac{x^4}{4} + \ldots\right) - \left(-x - \frac{x^2}{2} - \frac{x^3}{3} - \frac{x^4}{4} - \ldots\right) = 2\left(x + \frac{x^3}{3} + \frac{x^5}{5} + \ldots\right)$

58. $\ln(1 + x) = x - \frac{x^2}{2} + \frac{x^3}{3} - \frac{x^4}{4} + \ldots + \frac{(-1)^{n-1}x^n}{n} + \ldots \Rightarrow |\text{error}| = \left|\frac{(-1)^{n-1}x^n}{n}\right| = \frac{1}{n10^n}$ when $x = 0.1$;
$\frac{1}{n10^n} < \frac{1}{10^8} \Rightarrow n10^n > 10^8$ when $n \ge 8 \Rightarrow$ 7 terms

59. $\tan^{-1} x = x - \frac{x^3}{3} + \frac{x^5}{5} - \frac{x^7}{7} + \frac{x^9}{9} - \ldots + \frac{(-1)^{n-1}x^{2n-1}}{2n-1} + \ldots \Rightarrow |\text{error}| = \left|\frac{(-1)^{n-1}x^{2n-1}}{2n-1}\right| = \frac{1}{2n-1}$ when $x = 1$;
$\frac{1}{2n-1} < \frac{1}{10^3} \Rightarrow n > \frac{1001}{2} = 500.5 \Rightarrow$ the first term not used is the 501st $\Rightarrow$ we must use 500 terms

60. $\tan^{-1} x = x - \frac{x^3}{3} + \frac{x^5}{5} - \frac{x^7}{7} + \frac{x^9}{9} - \ldots + \frac{(-1)^{n-1}x^{2n-1}}{2n-1} + \ldots$ and $\lim_{n \to \infty} \left| \frac{x^{2n+1}}{2n+1} \cdot \frac{2n-1}{x^{2n-1}} \right| = x^2 \lim_{n \to \infty} \left| \frac{2n-1}{2n+1} \right| = x^2$

$\Rightarrow \tan^{-1} x$ converges for $|x| < 1$; when $x = -1$ we have $\sum_{n=1}^{\infty} \frac{(-1)^n}{2n-1}$ which is a convergent series; when $x = 1$ we have $\sum_{n=1}^{\infty} \frac{(-1)^{n-1}}{2n-1}$ which is a convergent series $\Rightarrow$ the series representing $\tan^{-1} x$ diverges for $|x| > 1$

61. $\tan^{-1} x = x - \frac{x^3}{3} + \frac{x^5}{5} - \frac{x^7}{7} + \frac{x^9}{9} - \ldots + \frac{(-1)^{n-1}x^{2n-1}}{2n-1} + \ldots$ and when the series representing $48 \tan^{-1}\left(\frac{1}{18}\right)$ has an error less than $\frac{1}{3} \cdot 10^{-6}$, then the series representing the sum $48 \tan^{-1}\left(\frac{1}{18}\right) + 32 \tan^{-1}\left(\frac{1}{57}\right) - 20 \tan^{-1}\left(\frac{1}{239}\right)$ also has an error of magnitude less than 10^{-6}; thus

$|\text{error}| = 48 \frac{\left(\frac{1}{18}\right)^{2n-1}}{2n-1} < \frac{1}{3 \cdot 10^6} \Rightarrow n \geq 4$ using a calculator $\Rightarrow$ 4 terms

62. $\ln(\sec x) = \int_0^x \tan t\, dt = \int_0^x \left(t + \frac{t^3}{3} + \frac{2t^5}{15} + \ldots\right) dt \approx \frac{x^2}{2} + \frac{x^4}{12} + \frac{x^6}{45} + \ldots$

63. (a) $(1 - x^2)^{-1/2} \approx 1 + \frac{x^2}{2} + \frac{3x^4}{8} + \frac{5x^6}{16} \Rightarrow \sin^{-1} x \approx x + \frac{x^3}{6} + \frac{3x^5}{40} + \frac{5x^7}{112}$; Using the Ratio Test:

$\lim_{n \to \infty} \left| \frac{1 \cdot 3 \cdot 5 \cdots (2n-1)(2n+1)x^{2n+3}}{2 \cdot 4 \cdot 6 \cdots (2n)(2n+2)(2n+3)} \cdot \frac{2 \cdot 4 \cdot 6 \cdots (2n)(2n+1)}{1 \cdot 3 \cdot 5 \cdots (2n-1)x^{2n+1}} \right| < 1 \Rightarrow x^2 \lim_{n \to \infty} \left| \frac{(2n+1)(2n+1)}{(2n+2)(2n+3)} \right| < 1$

$\Rightarrow |x| < 1 \Rightarrow$ the radius of convergence is 1. See Exercise 69.

(b) $\frac{d}{dx}(\cos^{-1} x) = -(1 - x^2)^{-1/2} \Rightarrow \cos^{-1} x = \frac{\pi}{2} - \sin^{-1} x \approx \frac{\pi}{2} - \left(x + \frac{x^3}{6} + \frac{3x^5}{40} + \frac{5x^7}{112}\right) \approx \frac{\pi}{2} - x - \frac{x^3}{6} - \frac{3x^5}{40} - \frac{5x^7}{112}$

64. (a) $(1 + t^2)^{-1/2} \approx (1)^{-1/2} + \left(-\frac{1}{2}\right)(1)^{-3/2}(t^2) + \frac{\left(-\frac{1}{2}\right)\left(-\frac{3}{2}\right)(1)^{-5/2}(t^2)^2}{2!} + \frac{\left(-\frac{1}{2}\right)\left(-\frac{3}{2}\right)\left(-\frac{5}{2}\right)(1)^{-7/2}(t^2)^3}{3!}$

$= 1 - \frac{t^2}{2} + \frac{3t^4}{2^2 \cdot 2!} - \frac{3 \cdot 5t^6}{2^3 \cdot 3!} \Rightarrow \sinh^{-1} x \approx \int_0^x \left(1 - \frac{t^2}{2} + \frac{3t^4}{8} - \frac{5t^6}{16}\right) dt = x - \frac{x^3}{6} + \frac{3x^5}{40} - \frac{5x^7}{112}$

(b) $\sinh^{-1}\left(\frac{1}{4}\right) \approx \frac{1}{4} - \frac{1}{384} + \frac{3}{40{,}960} = 0.24746908$; the error is less than the absolute value of the first unused term, $\frac{5x^7}{112}$, evaluated at $t = \frac{1}{4}$ since the series is alternating $\Rightarrow |\text{error}| < \frac{5\left(\frac{1}{4}\right)^7}{112} \approx 2.725 \times 10^{-6}$

65. $\frac{-1}{1+x} = -\frac{1}{1-(-x)} = -1 + x - x^2 + x^3 - \ldots \Rightarrow \frac{d}{dx}\left(\frac{-1}{1+x}\right) = \frac{1}{1+x^2} = \frac{d}{dx}(-1 + x - x^2 + x^3 - \ldots)$
$= 1 - 2x + 3x^2 - 4x^3 + \ldots$

66. $\frac{1}{1-x^2} = 1 + x^2 + x^4 + x^6 + \ldots \Rightarrow \frac{d}{dx}\left(\frac{1}{1-x^2}\right) = \frac{2x}{(1-x^2)^2} = \frac{d}{dx}(1 + x^2 + x^4 + x^6 + \ldots) = 2x + 4x^3 + 6x^5 + \ldots$

67. Wallis' formula gives the approximation $\pi \approx 4\left[\frac{2 \cdot 4 \cdot 4 \cdot 6 \cdot 6 \cdot 8 \cdots (2n-2) \cdot (2n)}{3 \cdot 3 \cdot 5 \cdot 5 \cdot 7 \cdot 7 \cdots (2n-1) \cdot (2n-1)}\right]$ to produce the table

n	$\sim \pi$
10	3.221088998
20	3.181104886
30	3.167880758
80	3.151425420
90	3.150331383
93	3.150049112
94	3.149959030
95	3.149870848
100	3.149456425

At $n = 1929$ we obtain the first approximation accurate to 3 decimals: 3.141999845. At $n = 30{,}000$ we still do not obtain accuracy to 4 decimals: 3.141617732, so the convergence to π is very slow. Here is a Maple CAS procedure to produce these approximations:

```
pie :=
  proc(n)
  local i,j;
    a(2) := evalf(8/9);
    for i from 3 to n do a(i) := evalf(2*(2*i-2)*i/(2*i-1)^2*a(i-1)) od;
    [[j,4*a(j)] $ (j = n-5 .. n)]
  end
```

68. $\ln 1 = 0$; $\ln 2 = \ln \frac{1+\left(\frac{1}{3}\right)}{1-\left(\frac{1}{3}\right)} \approx 2\left(\frac{1}{3} + \frac{\left(\frac{1}{3}\right)^3}{3} + \frac{\left(\frac{1}{3}\right)^5}{5} + \frac{\left(\frac{1}{3}\right)^7}{7}\right) \approx 0.69314$; $\ln 3 = \ln 2 + \ln\left(\frac{3}{2}\right) = \ln 2 + \ln \frac{1+\left(\frac{1}{5}\right)}{1-\left(\frac{1}{5}\right)}$
$\approx \ln 2 + 2\left(\frac{1}{5} + \frac{\left(\frac{1}{5}\right)^3}{3} + \frac{\left(\frac{1}{5}\right)^5}{5} + \frac{\left(\frac{1}{5}\right)^7}{7}\right) \approx 1.09861$; $\ln 4 = 2\ln 2 \approx 1.38628$; $\ln 5 = \ln 4 + \ln\left(\frac{5}{4}\right) = \ln 4 + \ln \frac{1+\left(\frac{1}{9}\right)}{1-\left(\frac{1}{9}\right)}$
≈ 1.60943; $\ln 6 = \ln 2 + \ln 3 \approx 1.79175$; $\ln 7 = \ln 6 + \ln\left(\frac{7}{6}\right) = \ln 6 + \ln \frac{1+\left(\frac{1}{13}\right)}{1-\left(\frac{1}{13}\right)} \approx 1.94591$; $\ln 8 = 3\ln 2$
≈ 2.07944; $\ln 9 = 2\ln 3 \approx 2.19722$; $\ln 10 = \ln 2 + \ln 5 \approx 2.30258$

69. $(1-x^2)^{-1/2} = (1+(-x^2))^{-1/2} = (1)^{-1/2} + \left(-\frac{1}{2}\right)(1)^{-3/2}(-x^2) + \frac{\left(-\frac{1}{2}\right)\left(-\frac{3}{2}\right)(1)^{-5/2}(-x^2)^2}{2!}$
$+ \frac{\left(-\frac{1}{2}\right)\left(-\frac{3}{2}\right)\left(-\frac{5}{2}\right)(1)^{-7/2}(-x^2)^3}{3!} + \ldots = 1 + \frac{x^2}{2} + \frac{1\cdot 3x^4}{2^2\cdot 2!} + \frac{1\cdot 3\cdot 5x^6}{2^3\cdot 3!} + \ldots = 1 + \sum_{n=1}^{\infty} \frac{1\cdot 3\cdot 5\cdots(2n-1)x^{2n}}{2^n\cdot n!}$
$\Rightarrow \sin^{-1} x = \int_0^x (1-t^2)^{-1/2}\,dt = \int_0^x \left(1 + \sum_{n=1}^{\infty} \frac{1\cdot 3\cdot 5\cdots(2n-1)x^{2n}}{2^n\cdot n!}\right) dt = x + \sum_{n=1}^{\infty} \frac{1\cdot 3\cdot 5\cdots(2n-1)x^{2n+1}}{2\cdot 4\cdots(2n)(2n+1)}$,
where $|x| < 1$

70. $\left[\tan^{-1} t\right]_x^{\infty} = \frac{\pi}{2} - \tan^{-1} x = \int_x^{\infty} \frac{dt}{1+t^2} = \int_x^{\infty} \left[\frac{\left(\frac{1}{t^2}\right)}{1+\left(\frac{1}{t^2}\right)}\right] dt = \int_x^{\infty} \frac{1}{t^2}\left(1 - \frac{1}{t^2} + \frac{1}{t^4} - \frac{1}{t^6} + \ldots\right) dt$
$= \int_x^{\infty} \left(\frac{1}{t^2} - \frac{1}{t^4} + \frac{1}{t^6} - \frac{1}{t^8} + \ldots\right) dt = \lim_{b\to\infty} \left[-\frac{1}{t} + \frac{1}{3t^3} - \frac{1}{5t^5} + \frac{1}{7t^7} - \ldots\right]_x^b = \frac{1}{x} - \frac{1}{3x^3} + \frac{1}{5x^5} - \frac{1}{7x^7} + \ldots$
$\Rightarrow \tan^{-1} x = \frac{\pi}{2} - \frac{1}{x} + \frac{1}{3x^3} - \frac{1}{5x^5} + \ldots$, $x > 1$; $\left[\tan^{-1} t\right]_{-\infty}^{x} = \tan^{-1} x + \frac{\pi}{2} = \int_{-\infty}^{x} \frac{dt}{1+t^2}$
$= \lim_{b\to -\infty} \left[-\frac{1}{t} + \frac{1}{3t^3} - \frac{1}{5t^5} + \frac{1}{7t^7} - \ldots\right]_b^x = -\frac{1}{x} + \frac{1}{3x^3} - \frac{1}{5x^5} + \frac{1}{7x^7} - \ldots \Rightarrow \tan^{-1} x = -\frac{\pi}{2} - \frac{1}{x} + \frac{1}{3x^3} - \frac{1}{5x^5} + \ldots$,
$x < -1$

71. (a) $\tan\left(\tan^{-1}(n+1) - \tan^{-1}(n-1)\right) = \frac{\tan(\tan^{-1}(n+1)) - \tan(\tan^{-1}(n-1))}{1+\tan(\tan^{-1}(n+1))\tan(\tan^{-1}(n-1))} = \frac{(n+1)-(n-1)}{1+(n+1)(n-1)} = \frac{2}{n^2}$

(b) $\sum_{n=1}^{N} \tan^{-1}\left(\frac{2}{n^2}\right) = \sum_{n=1}^{N} \left[\tan^{-1}(n+1) - \tan^{-1}(n-1)\right] = \left(\tan^{-1} 2 - \tan^{-1} 0\right) + \left(\tan^{-1} 3 - \tan^{-1} 1\right)$
$+ \left(\tan^{-1} 4 - \tan^{-1} 2\right) + \ldots + \left(\tan^{-1}(N+1) - \tan^{-1}(N-1)\right) = \tan^{-1}(N+1) + \tan^{-1} N - \frac{\pi}{4}$

(c) $\sum_{n=1}^{\infty} \tan^{-1}\left(\frac{2}{n^2}\right) = \lim_{n\to\infty} \left[\tan^{-1}(N+1) + \tan^{-1} N - \frac{\pi}{4}\right] = \frac{\pi}{2} + \frac{\pi}{2} - \frac{\pi}{4} = \frac{3\pi}{4}$

11.11 FOURIER SERIES

1. $a_0 = \frac{1}{2\pi}\int_0^{2\pi} 1\,dx = 1$, $a_k = \frac{1}{\pi}\int_0^{2\pi} \cos kx\,dx = \frac{1}{\pi}\left[\frac{\sin kx}{k}\right]_0^{2\pi} = 0$, $b_k = \frac{1}{\pi}\int_0^{2\pi} \sin kx\,dx = \frac{1}{\pi}\left[-\frac{\cos kx}{k}\right]_0^{2\pi} = 0$.
Thus, the Fourier series for $f(x)$ is 1.

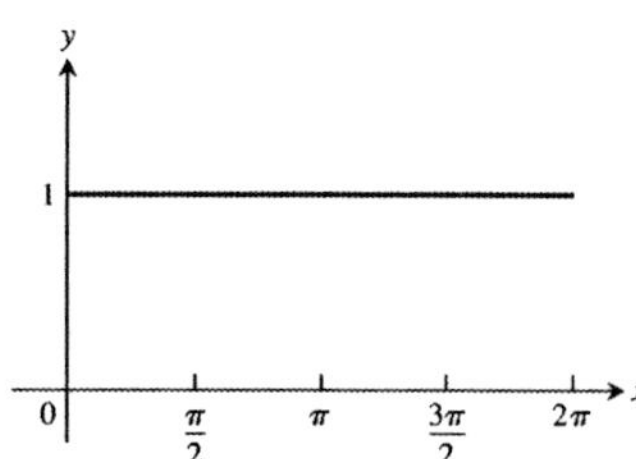

2. $a_0 = \frac{1}{2\pi}\left[\int_0^{\pi} 1\,dx + \int_{\pi}^{2\pi} -1\,dx\right] = 0$, $a_k = \frac{1}{\pi}\left[\int_0^{\pi} \cos kx\,dx - \int_{\pi}^{2\pi} \cos kx\,dx\right] = \frac{1}{\pi}\left[\frac{\sin kx}{k}\Big|_0^{\pi} - \frac{\sin kx}{k}\Big|_{\pi}^{2\pi}\right] = 0$,

$b_k = \frac{1}{\pi}\left[\int_0^{\pi} \sin kx\,dx - \int_{\pi}^{2\pi} \sin kx\,dx\right] = \frac{1}{\pi}\left[-\frac{\cos kx}{k}\Big|_0^{\pi} + \frac{\cos kx}{k}\Big|_{\pi}^{2\pi}\right] = \frac{1}{k\pi}\left[(-\cos k\pi + 1) + (\cos 2\pi k - \cos \pi k)\right]$

$= \frac{1}{k\pi}(2 - 2\cos k\pi) = \begin{cases} \frac{4}{k\pi}, & \text{k odd} \\ 0, & \text{k even} \end{cases}$.

Thus, the Fourier series for f(x) is $\frac{4}{\pi}\left[\sin x + \frac{\sin 3x}{3} + \frac{\sin 5x}{5} + \ldots\right]$.

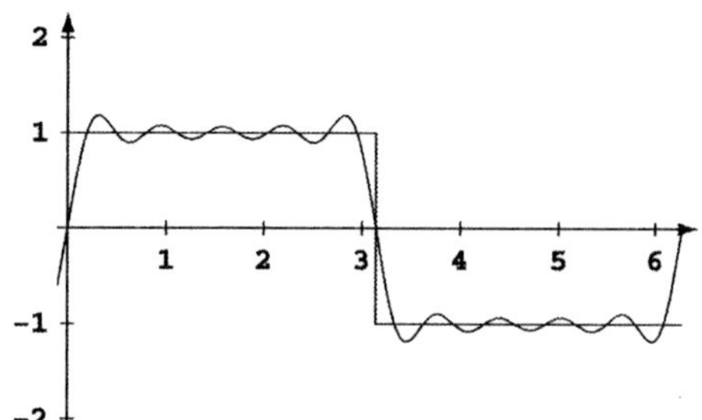

3. $a_0 = \frac{1}{2\pi}\left[\int_0^{\pi} x\,dx + \int_{\pi}^{2\pi} (x - 2\pi)\,dx\right] = \frac{1}{2\pi}\left[\frac{1}{2}\pi^2 + \frac{1}{2}(4\pi^2 - \pi^2) - 2\pi^2\right] = 0$. Note,

$\int_{\pi}^{2\pi} (x - 2\pi)\cos kx\,dx = -\int_0^{\pi} u \cos ku\,du$ (Let $u = 2\pi - x$). So $a_k = \frac{1}{\pi}\left[\int_0^{\pi} x \cos kx\,dx + \int_{\pi}^{2\pi} (x - 2\pi)\cos kx\,dx\right] = 0$.

Note, $\int_{\pi}^{2\pi} (x - 2\pi)\sin kx\,dx = \int_0^{\pi} u \sin ku\,du$ (Let $u = 2\pi - x$). So $b_k = \frac{1}{\pi}\left[\int_0^{\pi} x \sin kx\,dx + \int_{\pi}^{2\pi} (x - 2\pi)\sin kx\,dx\right]$

$= \frac{2}{\pi}\int_0^{\pi} x \sin kx\,dx = \frac{2}{\pi}\left[-\frac{x}{k}\cos kx + \frac{1}{k^2}\sin kx\right]_0^{\pi} = -\frac{2}{k}\cos k\pi = \frac{2}{k}(-1)^{k+1}$.

Thus, the Fourier series for f(x) is $\sum_{k=1}^{\infty}(-1)^{k+1}\frac{2\sin kx}{k}$.

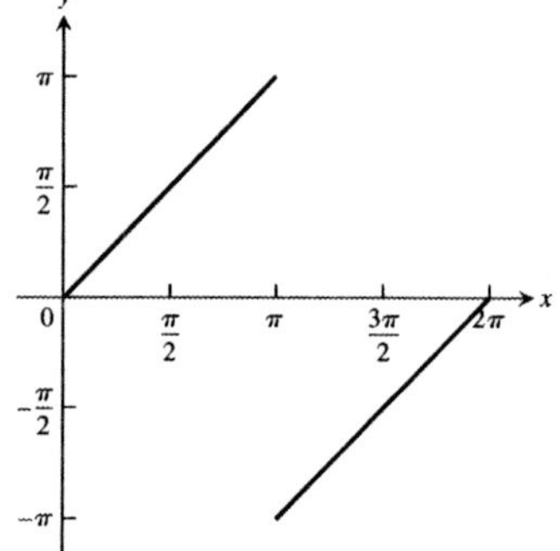

4. $a_0 = \frac{1}{2\pi}\int_0^{2\pi} f(x)\,dx = \frac{1}{2\pi}\int_0^{\pi} x^2\,dx = \frac{1}{6}\pi^2$, $a_k = \frac{1}{\pi}\int_0^{2\pi} f(x)\cos kx\,dx = \frac{1}{\pi}\int_0^{\pi} x^2 \cos kx\,dx$

$= \frac{1}{\pi}\left[\left(\frac{x^2}{k} - \frac{2}{k^3}\right)\sin kx + \frac{2}{k^2}x\cos kx\right]_0^{\pi} = \frac{2}{k^2}\cos k\pi = (-1)^k\left(\frac{2}{k^2}\right)$, $b_k = \frac{1}{\pi}\int_0^{2\pi} f(x)\sin kx\,dx = \frac{1}{\pi}\int_0^{\pi} x^2 \sin kx\,dx =$

$= \frac{1}{\pi}\left[\left(\frac{2}{k^3} - \frac{x^2}{k}\right)\cos kx + \frac{2}{k^2}x\sin kx\right]_0^{\pi} = \frac{1}{\pi}\left[\left(\frac{2}{k^3} - \frac{\pi^2}{k}\right)(-1)^k - \frac{2}{k^3}\right] = \frac{1}{\pi}\left[\left((-1)^k - 1\right)\frac{2}{k^3}\right] - \frac{\pi}{k}(-1)^k$

$= \begin{cases} -\frac{4}{\pi k^3} + \frac{\pi}{k}, & \text{k odd} \\ -\frac{\pi}{k}, & \text{k even} \end{cases}.$

Thus, the Fourier series for f(x) is $\frac{1}{6}\pi^2 - 2\cos x + \left(\frac{\pi^2-4}{\pi}\right)\sin x + \frac{1}{2}\cos 2x - \frac{\pi}{2}\sin 2x - \frac{2}{9}\cos 3x + \left(\frac{9\pi^2-4}{27\pi}\right)\sin 3x + \ldots$

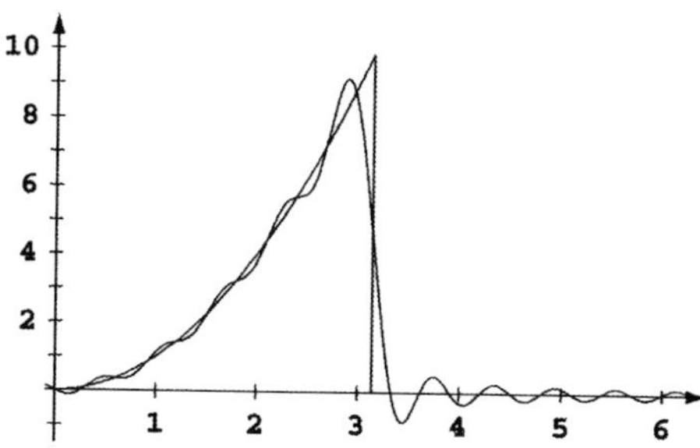

5. $a_0 = \frac{1}{2\pi}\int_0^{2\pi} e^x\,dx = \frac{1}{2\pi}(e^{2\pi}-1)$, $a_k = \frac{1}{\pi}\int_0^{2\pi} e^x \cos kx\,dx = \frac{1}{\pi}\left[\frac{e^x}{1+k^2}(\cos kx + k\sin kx)\right]_0^{2\pi} = \frac{e^{2\pi}-1}{\pi(1+k^2)}$,
$b_k = \frac{1}{\pi}\int_0^{2\pi} e^x \sin kx\,dx = \frac{1}{\pi}\left[\frac{e^x}{1+k^2}(\sin kx - k\cos kx)\right]_0^{2\pi} = \frac{k(1-e^{2\pi})}{\pi(1+k^2)}$.
Thus, the Fourier series for f(x) is $\frac{1}{2\pi}(e^{2\pi}-1) + \frac{e^{2\pi}-1}{\pi}\sum_{k=1}^{\infty}\left(\frac{\cos kx}{1+k^2} - \frac{k\sin kx}{1+k^2}\right)$.

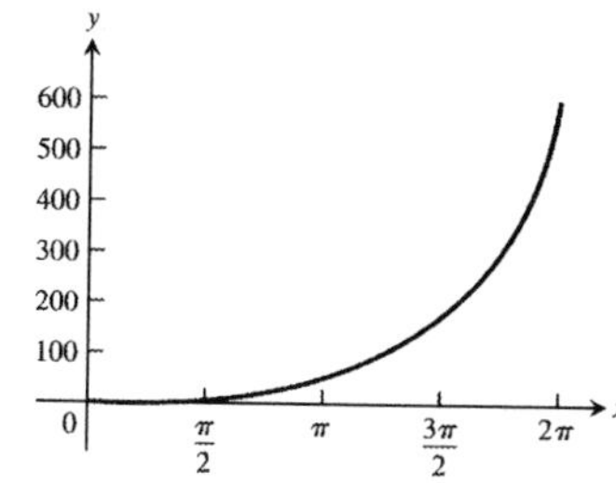

6. $a_0 = \frac{1}{2\pi}\int_0^{2\pi} f(x)\,dx = \frac{1}{2\pi}\int_0^{\pi} e^x\,dx = \frac{e^\pi-1}{2\pi}$, $a_k = \frac{1}{\pi}\int_0^{2\pi} f(x)\cos kx\,dx = \frac{1}{\pi}\int_0^{\pi} e^x\cos kx\,dx = \frac{1}{\pi}\left[\frac{e^x}{1+k^2}(\cos kx + k\sin kx)\right]_0^{\pi}$
$= \frac{1}{\pi(1+k^2)}\left[e^\pi(-1)^k - 1\right] = \begin{cases} \frac{-(1+e^\pi)}{\pi(1+k^2)}, & \text{k odd} \\ \frac{e^\pi-1}{\pi(1+k^2)}, & \text{k even} \end{cases}$. $b_k = \frac{1}{\pi}\int_0^{2\pi} f(x)\sin kx\,dx = \frac{1}{\pi}\int_0^{\pi} e^x \sin kx\,dx$
$= \frac{1}{\pi}\left[\frac{e^x}{1+k^2}(\sin kx - k\cos kx)\right]_0^{\pi} = \frac{-k}{\pi(1+k^2)}\left[e^\pi(-1)^k - 1\right] = \begin{cases} \frac{k(1+e^\pi)}{\pi(1+k^2)}, & \text{k odd} \\ \frac{1-e^\pi}{\pi(1+k^2)}, & \text{k even} \end{cases}$.
Thus, the Fourier series for f(x) is
$\frac{e^\pi-1}{2\pi} - \frac{(1+e^\pi)}{2\pi}\cos x + \frac{(1+e^\pi)}{2\pi}\sin x + \frac{e^\pi-1}{5\pi}\cos 2x + \frac{2(1-e^\pi)}{5\pi}\sin 2x - \frac{(1+e^\pi)}{10\pi}\cos 3x + \frac{3(1+e^\pi)}{10\pi}\sin 3x + \ldots$

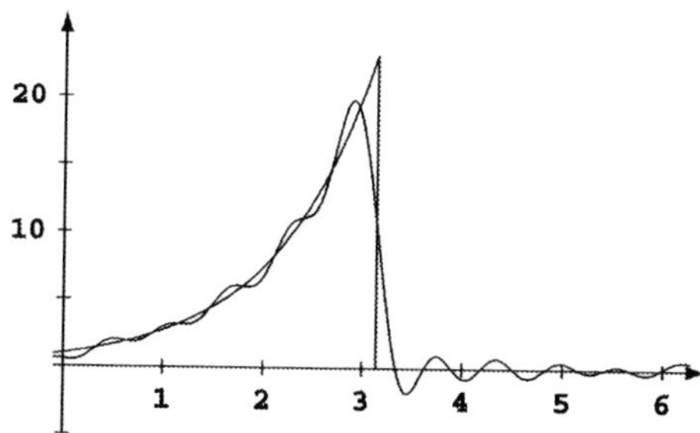

7. $a_0 = \frac{1}{2\pi}\int_0^{2\pi} f(x)\,dx = \frac{1}{2\pi}\int_0^{2\pi}\cos x\,dx = 0$, $a_k = \frac{1}{\pi}\int_0^{2\pi}\cos x\cos kx\,dx = \begin{cases} \frac{1}{\pi}\left[\frac{\sin(k-1)x}{2(k-1)} + \frac{\sin(k+1)x}{2(k+1)}\right]_0^{\pi}, & k \neq 1 \\ \frac{1}{\pi}\left[\frac{1}{2}x + \frac{1}{4}\sin 2x\right]_0^{\pi}, & k = 1 \end{cases}$

$= \begin{cases} 0, & k \neq 1 \\ \frac{1}{2}, & k = 1 \end{cases}.$

$b_k = \frac{1}{\pi}\int_0^{2\pi} \cos x \sin kx\, dx = \begin{cases} -\frac{1}{\pi}\left[\frac{\cos(k-1)x}{2(k-1)} + \frac{\cos(k+1)x}{2(k+1)}\right]_0^{\pi}, & k \neq 1 \\ -\frac{1}{4\pi}\cos 2x\Big|_0^{\pi}, & k = 1 \end{cases} = \begin{cases} 0, & k \text{ odd} \\ \frac{2k}{\pi(k^2-1)}, & k \text{ even} \end{cases}.$

Thus, the Fourier series for f(x) is $\frac{1}{2}\cos x + \sum_{k \text{ even}} \frac{2k}{\pi(k^2-1)}\sin kx$.

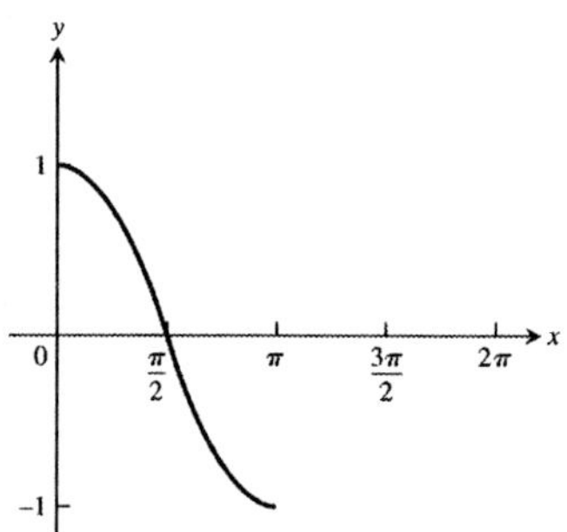

8. $a_0 = \frac{1}{2\pi}\int_0^{2\pi} f(x)\, dx = \frac{1}{2\pi}\left[\int_0^{\pi} 2\, dx + \int_{\pi}^{2\pi} -x\, dx\right] = 1 - \frac{3}{4}\pi,\ a_k = \frac{1}{\pi}\int_0^{2\pi} f(x)\cos kx\, dx$

$= \frac{1}{\pi}\left[\int_0^{\pi} 2\cos kx\, dx + \int_{\pi}^{2\pi} -x\cos kx\, dx\right] = -\frac{1}{\pi}\left[\frac{\cos kx}{k^2} + \frac{x\sin kx}{k}\right]_{\pi}^{2\pi} = \frac{-1+(-1)^k}{\pi k^2} = \begin{cases} -\frac{2}{\pi k^2}, & k \text{ odd} \\ 0, & k \text{ even} \end{cases}.$

$b_k = \frac{1}{\pi}\int_0^{2\pi} f(x)\sin kx\, dx = \frac{1}{\pi}\left[\int_0^{\pi} 2\sin kx\, dx + \int_{\pi}^{2\pi} -x\sin kx\, dx\right] = \frac{1}{\pi}\left[-\frac{2}{k}\cos kx\Big|_0^{\pi} + \left(\frac{x\cos kx}{k} - \frac{\sin kx}{k^2}\right)\Big|_{\pi}^{2\pi}\right]$

$= \begin{cases} \frac{1}{k}\left(\frac{4}{\pi} + 3\right), & k \text{ odd} \\ \frac{1}{k}, & k \text{ even} \end{cases}.$

Thus, the Fourier series for f(x) is $1 - \frac{3}{4}\pi - \frac{2}{\pi}\cos x + \left(\frac{4}{\pi} + 3\right)\sin x + \frac{1}{2}\sin 2x - \frac{2}{9\pi}\cos 3x + \frac{1}{3}\left(\frac{4}{\pi} + 3\right)\sin 3x + \ldots$.

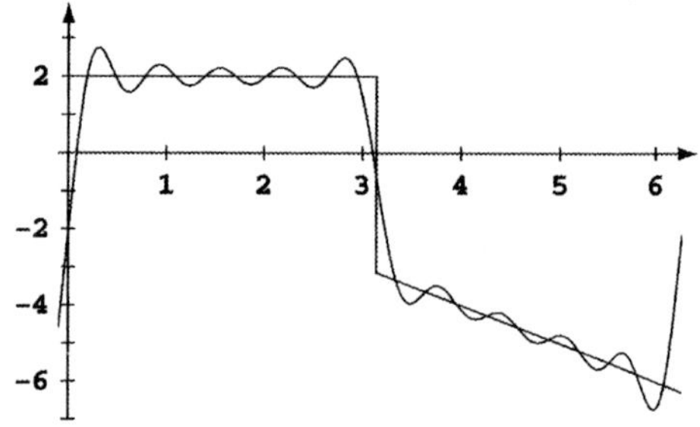

9. $\int_0^{2\pi} \cos px\, dx = \frac{1}{p}\sin px\Big|_0^{2\pi} = 0$ if $p \neq 0$.

10. $\int_0^{2\pi} \sin px\, dx = -\frac{1}{p}\cos px\Big|_0^{2\pi} = -\frac{1}{p}[1-1] = 0$ if $p \neq 0$.

11. $\int_0^{2\pi} \cos px \cos qx\, dx = \int_0^{2\pi} \frac{1}{2}[\cos(p+q)x + \cos(p-q)x]dx = \frac{1}{2}\left[\frac{1}{p+q}\sin(p+q)x + \frac{1}{p-q}\sin(p-q)x\right]_0^{2\pi} = 0$ if $p \neq q$.
If $p = q$ then $\int_0^{2\pi} \cos px \cos qx\, dx = \int_0^{2\pi} \cos^2 px\, dx = \int_0^{2\pi} \frac{1}{2}(1 + \cos 2px)\, dx = \frac{1}{2}\left(x + \frac{1}{2p}\sin 2px\right)\Big|_0^{2\pi} = \pi$.

12. $\int_0^{2\pi} \sin px \sin qx\, dx = \int_0^{2\pi} \frac{1}{2}[\cos(p-q)x - \cos(p+q)x]dx = \frac{1}{2}\left[\frac{1}{p-q}\sin(p-q)x - \frac{1}{p+q}\sin(p+q)x\right]_0^{2\pi} = 0$ if $p \neq q$.
If $p = q$ then $\int_0^{2\pi} \sin px \sin qx\, dx = \int_0^{2\pi} \sin^2 px\, dx = \int_0^{2\pi} \frac{1}{2}(1 - \cos 2px)\, dx = \frac{1}{2}\left(x - \frac{1}{2p}\sin 2px\right)\Big|_0^{2\pi} = \pi$.

13. $\int_0^{2\pi} \sin px \cos qx\, dx = \int_0^{2\pi} \frac{1}{2}[\sin(p+q)x + \sin(p-q)x\,]dx = -\frac{1}{2}\left[\frac{1}{p+q}\cos(p+q)x + \frac{1}{p-q}\cos(p-q)x\,\right]_0^{2\pi}$
$= -\frac{1}{2}\left[(1-1)\frac{1}{p+q} + (1-1)\frac{1}{p-q}\right] = 0.$ If $p = q$ then $\int_0^{2\pi} \sin px \cos qx\, dx = \int_0^{2\pi} \sin px \cos px\, dx = \int_0^{2\pi} \frac{1}{2}\sin 2px\, dx$
$= -\frac{1}{4\pi}\cos 2px\Big|_0^{2\pi} = -\frac{1}{4\pi}(1-1) = 0.$

14. Yes. Note that if f is continuous at c, then the expression $\frac{f(c^+)+f(c^-)}{2} = f(c)$ since $f(c^+) = \lim_{x\to c^+} f(x) = f(c)$ and $f(c^-) = \lim_{x\to c^-} f(x) = f(c)$. Now since the sum of two piecewise continuous functions on $[0, 2\pi]$ is also continuous on $[0, 2\pi]$, the function $f+g$ satisfies the hypothesis of Theorem 24, and so its Fourier series converges to $\frac{(f+g)(c^+)+(f+g)(c^-)}{2}$ for $0 < c < 2\pi$. Let $s_f(x)$ denote the Fourier series for $f(x)$. Then for any c in the interval $(0, 2\pi)$
$s_{f+g}(c) = \frac{(f+g)(c^+)+(f+g)(c^-)}{2} = \frac{1}{2}\left[\lim_{x\to c^+}(f+g)(x) + \lim_{x\to c^-}(f+g)(x)\right] = \frac{1}{2}\left[\lim_{x\to c^+} f(x) + \lim_{x\to c^+} g(x) + \lim_{x\to c^-} f(x) + \lim_{x\to c^-} g(x)\right]$
$= \frac{1}{2}[(f(c^+)+g(c^+)) + (f(c^-)+g(c^-))] = s_f(c) + s_g(c)$, since f and g satisfy the hypothesis of Theorem 24.

15. (a) $f(x)$ is piecewise continuous on $[0, 2\pi]$ and $f'(x) = 1$ for all $x \neq \pi \Rightarrow f'(x)$ is piecewise continuous on $[0, 2\pi]$. Then by Theorem 24, the Fourier series for $f(x)$ converges to $f(x)$ for all $x \neq \pi$ and converges to $\frac{1}{2}(f(\pi^+) + f(\pi^-))$ $= \frac{1}{2}(-\pi + \pi) = 0$ at $x = \pi$.

(b) The Fourier series for $f(x)$ is $\sum_{k=1}^{\infty}(-1)^{k+1}\frac{2\sin kx}{k}$. If we differentiate this series term by term we get the series $\sum_{k=1}^{\infty}(-1)^{k+1} 2\cos kx$, which diverges by the nth term test for divergence for any x since $\lim_{k\to\infty}(-1)^{k+1} 2\cos kx \neq 0$.

16. Since the Fourier series in discontinuous at $x = \pi$, by Theorem 24, the Fourier series will converge to $\frac{f(c^+)+f(c^-)}{2}$. Thus, at $x = \pi$ we have $\frac{f(\pi^+)+f(\pi^-)}{2} = \frac{1}{6}\pi^2 - 2\cos x + \left(\frac{\pi^2-4}{\pi}\right)\sin x + \frac{1}{2}\cos 2x - \frac{\pi}{2}\sin 2x - \frac{2}{9}\cos 3x + \left(\frac{9\pi^2-4}{27\pi}\right)\sin 3x + \ldots.$
$\Rightarrow \frac{0+\pi^2}{2} = \frac{1}{6}\pi^2 - 2\cos\pi + \left(\frac{\pi^2-4}{\pi}\right)\sin\pi + \frac{1}{2}\cos 2\pi - \frac{\pi}{2}\sin 2\pi - \frac{2}{9}\cos 3\pi + \left(\frac{9\pi^2-4}{27\pi}\right)\sin 3\pi + \ldots$
$\Rightarrow \frac{0+\pi^2}{2} = \frac{1}{6}\pi^2 + 2 + \frac{1}{2} + \frac{2}{9} + \ldots = \frac{1}{6}\pi^2 + 2\left(1 + \frac{1}{4} + \frac{1}{9} + \ldots\right) = \frac{1}{6}\pi^2 + 2\sum_{n=1}^{\infty}\frac{1}{n^2} \Rightarrow \frac{\pi^2}{2} = \frac{\pi^2}{6} + 2\sum_{n=1}^{\infty}\frac{1}{n^2}$
$\frac{\pi^2}{2} - \frac{\pi^2}{6} = 2\sum_{n=1}^{\infty}\frac{1}{n^2} \Rightarrow \frac{\pi^2}{3} = 2\sum_{n=1}^{\infty}\frac{1}{n^2} \Rightarrow \frac{\pi^2}{6} = \sum_{n=1}^{\infty}\frac{1}{n^2}.$

CHAPTER 11 PRACTICE EXERCISES

1. converges to 1, since $\lim_{n\to\infty} a_n = \lim_{n\to\infty}\left(1 + \frac{(-1)^n}{n}\right) = 1$

2. converges to 0, since $0 \le a_n \le \frac{2}{\sqrt{n}}$, $\lim_{n\to\infty} 0 = 0$, $\lim_{n\to\infty}\frac{2}{\sqrt{n}} = 0$ using the Sandwich Theorem for Sequences

3. converges to -1, since $\lim_{n\to\infty} a_n = \lim_{n\to\infty}\left(\frac{1-2^n}{2^n}\right) = \lim_{n\to\infty}\left(\frac{1}{2^n} - 1\right) = -1$

4. converges to 1, since $\lim_{n\to\infty} a_n = \lim_{n\to\infty}[1 + (0.9)^n] = 1 + 0 = 1$

5. diverges, since $\left\{\sin\frac{n\pi}{2}\right\} = \{0, 1, 0, -1, 0, 1, \ldots\}$

6. converges to 0, since $\{\sin n\pi\} = \{0, 0, 0, \ldots\}$

7. converges to 0, since $\lim_{n\to\infty} a_n = \lim_{n\to\infty}\frac{\ln n^2}{n} = 2\lim_{n\to\infty}\frac{\left(\frac{1}{n}\right)}{1} = 0$

8. converges to 0, since $\lim_{n\to\infty} a_n = \lim_{n\to\infty} \frac{\ln(2n+1)}{n} = \lim_{n\to\infty} \frac{\left(\frac{2}{2n+1}\right)}{1} = 0$

9. converges to 1, since $\lim_{n\to\infty} a_n = \lim_{n\to\infty} \left(\frac{n+\ln n}{n}\right) = \lim_{n\to\infty} \frac{1+\left(\frac{1}{n}\right)}{1} = 1$

10. converges to 0, since $\lim_{n\to\infty} a_n = \lim_{n\to\infty} \frac{\ln(2n^3+1)}{n} = \lim_{n\to\infty} \frac{\left(\frac{6n^2}{2n^3+1}\right)}{1} = \lim_{n\to\infty} \frac{12n}{6n^2} = \lim_{n\to\infty} \frac{2}{n} = 0$

11. converges to e^{-5}, since $\lim_{n\to\infty} a_n = \lim_{n\to\infty} \left(\frac{n-5}{n}\right)^n = \lim_{n\to\infty} \left(1 + \frac{(-5)}{n}\right)^n = e^{-5}$ by Theorem 5

12. converges to $\frac{1}{e}$, since $\lim_{n\to\infty} a_n = \lim_{n\to\infty} \left(1 + \frac{1}{n}\right)^{-n} = \lim_{n\to\infty} \frac{1}{\left(1+\frac{1}{n}\right)^n} = \frac{1}{e}$ by Theorem 5

13. converges to 3, since $\lim_{n\to\infty} a_n = \lim_{n\to\infty} \left(\frac{3^n}{n}\right)^{1/n} = \lim_{n\to\infty} \frac{3}{n^{1/n}} = \frac{3}{1} = 3$ by Theorem 5

14. converges to 1, since $\lim_{n\to\infty} a_n = \lim_{n\to\infty} \left(\frac{3}{n}\right)^{1/n} = \lim_{n\to\infty} \frac{3^{1/n}}{n^{1/n}} = \frac{1}{1} = 1$ by Theorem 5

15. converges to ln 2, since $\lim_{n\to\infty} a_n = \lim_{n\to\infty} n\left(2^{1/n} - 1\right) = \lim_{n\to\infty} \frac{2^{1/n} - 1}{\left(\frac{1}{n}\right)} = \lim_{n\to\infty} \frac{\left[\frac{\left(-2^{1/n} \ln 2\right)}{n^2}\right]}{\left(\frac{-1}{n^2}\right)} = \lim_{n\to\infty} 2^{1/n} \ln 2$
$= 2^0 \cdot \ln 2 = \ln 2$

16. converges to 1, since $\lim_{n\to\infty} a_n = \lim_{n\to\infty} \sqrt[n]{2n+1} = \lim_{n\to\infty} \exp\left(\frac{\ln(2n+1)}{n}\right) = \lim_{n\to\infty} \exp\left(\frac{\frac{2}{2n+1}}{1}\right) = e^0 = 1$

17. diverges, since $\lim_{n\to\infty} a_n = \lim_{n\to\infty} \frac{(n+1)!}{n!} = \lim_{n\to\infty} (n+1) = \infty$

18. converges to 0, since $\lim_{n\to\infty} a_n = \lim_{n\to\infty} \frac{(-4)^n}{n!} = 0$ by Theorem 5

19. $\frac{1}{(2n-3)(2n-1)} = \frac{\left(\frac{1}{2}\right)}{2n-3} - \frac{\left(\frac{1}{2}\right)}{2n-1} \Rightarrow s_n = \left[\frac{\left(\frac{1}{2}\right)}{3} - \frac{\left(\frac{1}{2}\right)}{5}\right] + \left[\frac{\left(\frac{1}{2}\right)}{5} - \frac{\left(\frac{1}{2}\right)}{7}\right] + \ldots + \left[\frac{\left(\frac{1}{2}\right)}{2n-3} - \frac{\left(\frac{1}{2}\right)}{2n-1}\right] = \frac{\left(\frac{1}{2}\right)}{3} - \frac{\left(\frac{1}{2}\right)}{2n-1}$
$\Rightarrow \lim_{n\to\infty} s_n = \lim_{n\to\infty} \left[\frac{1}{6} - \frac{\left(\frac{1}{2}\right)}{2n-1}\right] = \frac{1}{6}$

20. $\frac{-2}{n(n+1)} = \frac{-2}{n} + \frac{2}{n+1} \Rightarrow s_n = \left(\frac{-2}{2} + \frac{2}{3}\right) + \left(\frac{-2}{3} + \frac{2}{4}\right) + \ldots + \left(\frac{-2}{n} + \frac{2}{n+1}\right) = -\frac{2}{2} + \frac{2}{n+1} \Rightarrow \lim_{n\to\infty} s_n$
$= \lim_{n\to\infty} \left(-1 + \frac{2}{n+1}\right) = -1$

21. $\frac{9}{(3n-1)(3n+2)} = \frac{3}{3n-1} - \frac{3}{3n+2} \Rightarrow s_n = \left(\frac{3}{2} - \frac{3}{5}\right) + \left(\frac{3}{5} - \frac{3}{8}\right) + \left(\frac{3}{8} - \frac{3}{11}\right) + \ldots + \left(\frac{3}{3n-1} - \frac{3}{3n+2}\right)$
$= \frac{3}{2} - \frac{3}{3n+2} \Rightarrow \lim_{n\to\infty} s_n = \lim_{n\to\infty} \left(\frac{3}{2} - \frac{3}{3n+2}\right) = \frac{3}{2}$

22. $\frac{-8}{(4n-3)(4n+1)} = \frac{-2}{4n-3} + \frac{2}{4n+1} \Rightarrow s_n = \left(\frac{-2}{9} + \frac{2}{13}\right) + \left(\frac{-2}{13} + \frac{2}{17}\right) + \left(\frac{-2}{17} + \frac{2}{21}\right) + \ldots + \left(\frac{-2}{4n-3} + \frac{2}{4n+1}\right)$
$= -\frac{2}{9} + \frac{2}{4n+1} \Rightarrow \lim_{n\to\infty} s_n = \lim_{n\to\infty} \left(-\frac{2}{9} + \frac{2}{4n+1}\right) = -\frac{2}{9}$

23. $\sum_{n=0}^{\infty} e^{-n} = \sum_{n=0}^{\infty} \frac{1}{e^n}$, a convergent geometric series with $r = \frac{1}{e}$ and $a = 1 \Rightarrow$ the sum is $\frac{1}{1-\left(\frac{1}{e}\right)} = \frac{e}{e-1}$

24. $\sum_{n=1}^{\infty} (-1)^n \frac{3}{4^n} = \sum_{n=0}^{\infty} \left(-\frac{3}{4}\right)\left(\frac{-1}{4}\right)^n$ a convergent geometric series with $r = -\frac{1}{4}$ and $a = \frac{-3}{4} \Rightarrow$ the sum is $\frac{\left(-\frac{3}{4}\right)}{1-\left(\frac{-1}{4}\right)} = -\frac{3}{5}$

25. diverges, a p-series with $p = \frac{1}{2}$

26. $\sum_{n=1}^{\infty} \frac{-5}{n} = -5 \sum_{n=1}^{\infty} \frac{1}{n}$, diverges since it is a nonzero multiple of the divergent harmonic series

27. Since $f(x) = \frac{1}{x^{1/2}} \Rightarrow f'(x) = -\frac{1}{2x^{3/2}} < 0 \Rightarrow f(x)$ is decreasing $\Rightarrow a_{n+1} < a_n$, and $\lim_{n \to \infty} a_n = \lim_{n \to \infty} \frac{1}{\sqrt{n}} = 0$, the series $\sum_{n=1}^{\infty} \frac{(-1)^n}{\sqrt{n}}$ converges by the Alternating Series Test. Since $\sum_{n=1}^{\infty} \frac{1}{\sqrt{n}}$ diverges, the given series converges conditionally.

28. converges absolutely by the Direct Comparison Test since $\frac{1}{2n^3} < \frac{1}{n^3}$ for $n \geq 1$, which is the nth term of a convergent p-series

29. The given series does not converge absolutely by the Direct Comparison Test since $\frac{1}{\ln(n+1)} > \frac{1}{n+1}$, which is the nth term of a divergent series. Since $f(x) = \frac{1}{\ln(x+1)} \Rightarrow f'(x) = -\frac{1}{(\ln(x+1))^2(x+1)} < 0 \Rightarrow f(x)$ is decreasing $\Rightarrow a_{n+1} < a_n$, and $\lim_{n \to \infty} a_n = \lim_{n \to \infty} \frac{1}{\ln(n+1)} = 0$, the given series converges conditionally by the Alternating Series Test.

30. $\int_2^{\infty} \frac{1}{x(\ln x)^2}\,dx = \lim_{b \to \infty} \int_2^b \frac{1}{x(\ln x)^2}\,dx = \lim_{b \to \infty} \left[-(\ln x)^{-1}\right]_2^b = -\lim_{b \to \infty} \left(\frac{1}{\ln b} - \frac{1}{\ln 2}\right) = \frac{1}{\ln 2} \Rightarrow$ the series converges absolutely by the Integral Test

31. converges absolutely by the Direct Comparison Test since $\frac{\ln n}{n^3} < \frac{n}{n^3} = \frac{1}{n^2}$, the nth term of a convergent p-series

32. diverges by the Direct Comparison Test for $e^{n^n} > n \Rightarrow \ln\left(e^{n^n}\right) > \ln n \Rightarrow n^n > \ln n \Rightarrow \ln n^n > \ln(\ln n)$ $\Rightarrow n \ln n > \ln(\ln n) \Rightarrow \frac{\ln n}{\ln(\ln n)} > \frac{1}{n}$, the nth term of the divergent harmonic series

33. $\lim_{n \to \infty} \frac{\left(\frac{1}{n\sqrt{n^2+1}}\right)}{\left(\frac{1}{n^2}\right)} = \sqrt{\lim_{n \to \infty} \frac{n^2}{n^2+1}} = \sqrt{1} = 1 \Rightarrow$ converges absolutely by the Limit Comparison Test

34. Since $f(x) = \frac{3x^2}{x^3+1} \Rightarrow f'(x) = \frac{3x(2-x^3)}{(x^3+1)^2} < 0$ when $x \geq 2 \Rightarrow a_{n+1} < a_n$ for $n \geq 2$ and $\lim_{n \to \infty} \frac{3n^2}{n^3+1} = 0$, the series converges by the Alternating Series Test. The series does not converge absolutely: By the Limit Comparison Test, $\lim_{n \to \infty} \frac{\left(\frac{3n^2}{n^3+1}\right)}{\left(\frac{1}{n}\right)} = \lim_{n \to \infty} \frac{3n^3}{n^3+1} = 3$. Therefore the convergence is conditional.

35. converges absolutely by the Ratio Test since $\lim_{n \to \infty} \left[\frac{n+2}{(n+1)!} \cdot \frac{n!}{n+1}\right] = \lim_{n \to \infty} \frac{n+2}{(n+1)^2} = 0 < 1$

36. diverges since $\lim_{n \to \infty} a_n = \lim_{n \to \infty} \frac{(-1)^n(n^2+1)}{2n^2+n-1}$ does not exist

37. converges absolutely by the Ratio Test since $\lim_{n \to \infty} \left[\frac{3^{n+1}}{(n+1)!} \cdot \frac{n!}{3^n}\right] = \lim_{n \to \infty} \frac{3}{n+1} = 0 < 1$

38. converges absolutely by the Root Test since $\lim_{n\to\infty} \sqrt[n]{a_n} = \lim_{n\to\infty} \sqrt[n]{\frac{2^n 3^n}{n^n}} = \lim_{n\to\infty} \frac{6}{n} = 0 < 1$

39. converges absolutely by the Limit Comparison Test since $\lim_{n\to\infty} \frac{\left(\frac{1}{n^{3/2}}\right)}{\left(\frac{1}{\sqrt{n(n+1)(n+2)}}\right)} = \sqrt{\lim_{n\to\infty} \frac{n(n+1)(n+2)}{n^3}} = 1$

40. converges absolutely by the Limit Comparison Test since $\lim_{n\to\infty} \frac{\left(\frac{1}{n^2}\right)}{\left(\frac{1}{n\sqrt{n^2-1}}\right)} = \sqrt{\lim_{n\to\infty} \frac{n^2(n^2-1)}{n^4}} = 1$

41. $\lim_{n\to\infty} \left|\frac{u_{n+1}}{u_n}\right| < 1 \Rightarrow \lim_{n\to\infty} \left|\frac{(x+4)^{n+1}}{(n+1)3^{n+1}} \cdot \frac{n3^n}{(x+4)^n}\right| < 1 \Rightarrow \frac{|x+4|}{3} \lim_{n\to\infty} \left(\frac{n}{n+1}\right) < 1 \Rightarrow \frac{|x+4|}{3} < 1$
$\Rightarrow |x+4| < 3 \Rightarrow -3 < x+4 < 3 \Rightarrow -7 < x < -1$; at $x = -7$ we have $\sum_{n=1}^{\infty} \frac{(-1)^n 3^n}{n3^n} = \sum_{n=1}^{\infty} \frac{(-1)^n}{n}$, the alternating harmonic series, which converges conditionally; at $x = -1$ we have $\sum_{n=1}^{\infty} \frac{3^n}{n3^n} = \sum_{n=1}^{\infty} \frac{1}{n}$, the divergent harmonic series

(a) the radius is 3; the interval of convergence is $-7 \le x < -1$
(b) the interval of absolute convergence is $-7 < x < -1$
(c) the series converges conditionally at $x = -7$

42. $\lim_{n\to\infty} \left|\frac{u_{n+1}}{u_n}\right| < 1 \Rightarrow \lim_{n\to\infty} \left|\frac{(x-1)^{2n}}{(2n+1)!} \cdot \frac{(2n-1)!}{(x-1)^{2n-2}}\right| < 1 \Rightarrow (x-1)^2 \lim_{n\to\infty} \frac{1}{(2n)(2n+1)} = 0 < 1$, which holds for all x

(a) the radius is ∞; the series converges for all x
(b) the series converges absolutely for all x
(c) there are no values for which the series converges conditionally

43. $\lim_{n\to\infty} \left|\frac{u_{n+1}}{u_n}\right| < 1 \Rightarrow \lim_{n\to\infty} \left|\frac{(3x-1)^{n+1}}{(n+1)^2} \cdot \frac{n^2}{(3x-1)^n}\right| < 1 \Rightarrow |3x-1| \lim_{n\to\infty} \frac{n^2}{(n+1)^2} < 1 \Rightarrow |3x-1| < 1$
$\Rightarrow -1 < 3x-1 < 1 \Rightarrow 0 < 3x < 2 \Rightarrow 0 < x < \frac{2}{3}$; at $x = 0$ we have $\sum_{n=1}^{\infty} \frac{(-1)^{n-1}(-1)^n}{n^2} = \sum_{n=1}^{\infty} \frac{(-1)^{2n-1}}{n^2}$
$= -\sum_{n=1}^{\infty} \frac{1}{n^2}$, a nonzero constant multiple of a convergent p-series, which is absolutely convergent; at $x = \frac{2}{3}$ we have $\sum_{n=1}^{\infty} \frac{(-1)^{n-1}(1)^n}{n^2} = \sum_{n=1}^{\infty} \frac{(-1)^{n-1}}{n^2}$, which converges absolutely

(a) the radius is $\frac{1}{3}$; the interval of convergence is $0 \le x \le \frac{2}{3}$
(b) the interval of absolute convergence is $0 \le x \le \frac{2}{3}$
(c) there are no values for which the series converges conditionally

44. $\lim_{n\to\infty} \left|\frac{u_{n+1}}{u_n}\right| < 1 \Rightarrow \lim_{n\to\infty} \left|\frac{n+2}{2n+3} \cdot \frac{(2x+1)^{n+1}}{2^{n+1}} \cdot \frac{2n+1}{n+1} \cdot \frac{2^n}{(2x+1)^n}\right| < 1 \Rightarrow \frac{|2x+1|}{2} \lim_{n\to\infty} \left|\frac{n+2}{2n+3} \cdot \frac{2n+1}{n+1}\right| < 1$
$\Rightarrow \frac{|2x+1|}{2}(1) < 1 \Rightarrow |2x+1| < 2 \Rightarrow -2 < 2x+1 < 2 \Rightarrow -3 < 2x < 1 \Rightarrow -\frac{3}{2} < x < \frac{1}{2}$; at $x = -\frac{3}{2}$ we have $\sum_{n=1}^{\infty} \frac{n+1}{2n+1} \cdot \frac{(-2)^n}{2^n} = \sum_{n=1}^{\infty} \frac{(-1)^n(n+1)}{2n+1}$ which diverges by the nth-Term Test for Divergence since $\lim_{n\to\infty} \left(\frac{n+1}{2n+1}\right) = \frac{1}{2} \ne 0$; at $x = \frac{1}{2}$ we have $\sum_{n=1}^{\infty} \frac{n+1}{2n+1} \cdot \frac{2^n}{2^n} = \sum_{n=1}^{\infty} \frac{n+1}{2n+1}$, which diverges by the nth-Term Test

(a) the radius is 1; the interval of convergence is $-\frac{3}{2} < x < \frac{1}{2}$
(b) the interval of absolute convergence is $-\frac{3}{2} < x < \frac{1}{2}$
(c) there are no values for which the series converges conditionally

45. $\lim_{n\to\infty}\left|\frac{u_{n+1}}{u_n}\right| < 1 \Rightarrow \lim_{n\to\infty}\left|\frac{x^{n+1}}{(n+1)^{n+1}}\cdot\frac{n^n}{x^n}\right| < 1 \Rightarrow |x|\lim_{n\to\infty}\left|\left(\frac{n}{n+1}\right)^n\left(\frac{1}{n+1}\right)\right| < 1 \Rightarrow \frac{|x|}{e}\lim_{n\to\infty}\left(\frac{1}{n+1}\right) < 1$

$\Rightarrow \frac{|x|}{e}\cdot 0 < 1$, which holds for all x

(a) the radius is ∞; the series converges for all x
(b) the series converges absolutely for all x
(c) there are no values for which the series converges conditionally

46. $\lim_{n\to\infty}\left|\frac{u_{n+1}}{u_n}\right| < 1 \Rightarrow \lim_{n\to\infty}\left|\frac{x^{n+1}}{\sqrt{n+1}}\cdot\frac{\sqrt{n}}{x^n}\right| < 1 \Rightarrow |x|\lim_{n\to\infty}\sqrt{\frac{n}{n+1}} < 1 \Rightarrow |x| < 1$; when $x = -1$ we have $\sum_{n=1}^{\infty}\frac{(-1)^n}{\sqrt{n}}$, which converges by the Alternating Series Test; when $x = 1$ we have $\sum_{n=1}^{\infty}\frac{1}{\sqrt{n}}$, a divergent p-series

(a) the radius is 1; the interval of convergence is $-1 \le x < 1$
(b) the interval of absolute convergence is $-1 < x < 1$
(c) the series converges conditionally at $x = -1$

47. $\lim_{n\to\infty}\left|\frac{u_{n+1}}{u_n}\right| < 1 \Rightarrow \lim_{n\to\infty}\left|\frac{(n+2)x^{2n+1}}{3^{n+1}}\cdot\frac{3^n}{(n+1)x^{2n-1}}\right| < 1 \Rightarrow \frac{x^2}{3}\lim_{n\to\infty}\left(\frac{n+2}{n+1}\right) < 1 \Rightarrow -\sqrt{3} < x < \sqrt{3}$;
the series $\sum_{n=1}^{\infty} -\frac{n+1}{\sqrt{3}}$ and $\sum_{n=1}^{\infty}\frac{n+1}{\sqrt{3}}$, obtained with $x = \pm\sqrt{3}$, both diverge

(a) the radius is $\sqrt{3}$; the interval of convergence is $-\sqrt{3} < x < \sqrt{3}$
(b) the interval of absolute convergence is $-\sqrt{3} < x < \sqrt{3}$
(c) there are no values for which the series converges conditionally

48. $\lim_{n\to\infty}\left|\frac{u_{n+1}}{u_n}\right| < 1 \Rightarrow \lim_{n\to\infty}\left|\frac{(x-1)^{2n+3}}{2n+3}\cdot\frac{2n+1}{(x-1)^{2n+1}}\right| < 1 \Rightarrow (x-1)^2\lim_{n\to\infty}\left(\frac{2n+1}{2n+3}\right) < 1 \Rightarrow (x-1)^2(1) < 1$
$\Rightarrow (x-1)^2 < 1 \Rightarrow |x-1| < 1 \Rightarrow -1 < x-1 < 1 \Rightarrow 0 < x < 2$; at $x = 0$ we have $\sum_{n=1}^{\infty}\frac{(-1)^n(-1)^{2n+1}}{2n+1}$
$= \sum_{n=1}^{\infty}\frac{(-1)^{3n+1}}{2n+1} = \sum_{n=1}^{\infty}\frac{(-1)^{n-1}}{2n+1}$ which converges conditionally by the Alternating Series Test and the fact that $\sum_{n=1}^{\infty}\frac{1}{2n+1}$ diverges; at $x = 2$ we have $\sum_{n=1}^{\infty}\frac{(-1)^n(1)^{2n+1}}{2n+1} = \sum_{n=1}^{\infty}\frac{(-1)^n}{2n+1}$, which also converges conditionally

(a) the radius is 1; the interval of convergence is $0 \le x \le 2$
(b) the interval of absolute convergence is $0 < x < 2$
(c) the series converges conditionally at $x = 0$ and $x = 2$

49. $\lim_{n\to\infty}\left|\frac{u_{n+1}}{u_n}\right| < 1 \Rightarrow \lim_{n\to\infty}\left|\frac{\operatorname{csch}(n+1)x^{n+1}}{\operatorname{csch}(n)x^n}\right| < 1 \Rightarrow |x|\lim_{n\to\infty}\left|\frac{\left(\frac{2}{e^{n+1}-e^{-n-1}}\right)}{\left(\frac{2}{e^n-e^{-n}}\right)}\right| < 1$

$\Rightarrow |x|\lim_{n\to\infty}\left|\frac{e^{-1}-e^{-2n-1}}{1-e^{-2n-2}}\right| < 1 \Rightarrow \frac{|x|}{e} < 1 \Rightarrow -e < x < e$; the series $\sum_{n=1}^{\infty}(\pm e)^n \operatorname{csch} n$, obtained with $x = \pm e$, both diverge since $\lim_{n\to\infty}(\pm e)^n \operatorname{csch} n \ne 0$

(a) the radius is e; the interval of convergence is $-e < x < e$
(b) the interval of absolute convergence is $-e < x < e$
(c) there are no values for which the series converges conditionally

50. $\lim_{n\to\infty}\left|\frac{u_{n+1}}{u_n}\right| < 1 \Rightarrow \lim_{n\to\infty}\left|\frac{x^{n+1}\coth(n+1)}{x^n\coth(n)}\right| < 1 \Rightarrow |x|\lim_{n\to\infty}\left|\frac{1+e^{-2n-2}}{1-e^{-2n-2}}\cdot\frac{1-e^{-2n}}{1+e^{-2n}}\right| < 1 \Rightarrow |x| < 1$
$\Rightarrow -1 < x < 1$; the series $\sum_{n=1}^{\infty}(\pm 1)^n \coth n$, obtained with $x = \pm 1$, both diverge since $\lim_{n\to\infty}(\pm 1)^n \coth n \ne 0$

(a) the radius is 1; the interval of convergence is $-1 < x < 1$
(b) the interval of absolute convergence is $-1 < x < 1$

(c) there are no values for which the series converges conditionally

51. The given series has the form $1 - x + x^2 - x^3 + \ldots + (-x)^n + \ldots = \frac{1}{1+x}$, where $x = \frac{1}{4}$; the sum is $\frac{1}{1+\left(\frac{1}{4}\right)} = \frac{4}{5}$

52. The given series has the form $x - \frac{x^2}{2} + \frac{x^3}{3} - \ldots + (-1)^{n-1}\frac{x^n}{n} + \ldots = \ln(1+x)$, where $x = \frac{2}{3}$; the sum is $\ln\left(\frac{5}{3}\right) \approx 0.510825624$

53. The given series has the form $x - \frac{x^3}{3!} + \frac{x^5}{5!} - \ldots + (-1)^n \frac{x^{2n+1}}{(2n+1)!} + \ldots = \sin x$, where $x = \pi$; the sum is $\sin \pi = 0$

54. The given series has the form $1 - \frac{x^2}{2!} + \frac{x^4}{4!} - \ldots + (-1)^n \frac{x^{2n}}{(2n)!} + \ldots = \cos x$, where $x = \frac{\pi}{3}$; the sum is $\cos\frac{\pi}{3} = \frac{1}{2}$

55. The given series has the form $1 + x + \frac{x^2}{2!} + \frac{x^2}{3!} + \ldots + \frac{x^n}{n!} + \ldots = e^x$, where $x = \ln 2$; the sum is $e^{\ln(2)} = 2$

56. The given series has the form $x - \frac{x^3}{3} + \frac{x^5}{5} - \ldots + (-1)^n \frac{x^{2n-1}}{(2n-1)} + \ldots = \tan^{-1} x$, where $x = \frac{1}{\sqrt{3}}$; the sum is $\tan^{-1}\left(\frac{1}{\sqrt{3}}\right) = \frac{\pi}{6}$

57. Consider $\frac{1}{1-2x}$ as the sum of a convergent geometric series with $a = 1$ and $r = 2x \Rightarrow \frac{1}{1-2x} = 1 + (2x) + (2x)^2 + (2x)^3 + \ldots = \sum_{n=0}^{\infty} (2x)^n = \sum_{n=0}^{\infty} 2^n x^n$ where $|2x| < 1 \Rightarrow |x| < \frac{1}{2}$

58. Consider $\frac{1}{1+x^3}$ as the sum of a convergent geometric series with $a = 1$ and $r = -x^3 \Rightarrow \frac{1}{1+x^3} = \frac{1}{1-(-x^3)} = 1 + (-x^3) + (-x^3)^2 + (-x^3)^3 + \ldots = \sum_{n=0}^{\infty} (-1)^n x^{3n}$ where $|-x^3| < 1 \Rightarrow |x^3| < 1 \Rightarrow |x| < 1$

59. $\sin x = \sum_{n=0}^{\infty} \frac{(-1)^n x^{2n+1}}{(2n+1)!} \Rightarrow \sin \pi x = \sum_{n=0}^{\infty} \frac{(-1)^n (\pi x)^{2n+1}}{(2n+1)!} = \sum_{n=0}^{\infty} \frac{(-1)^n \pi^{2n+1} x^{2n+1}}{(2n+1)!}$

60. $\sin x = \sum_{n=0}^{\infty} \frac{(-1)^n x^{2n+1}}{(2n+1)!} \Rightarrow \sin \frac{2x}{3} = \sum_{n=0}^{\infty} \frac{(-1)^n \left(\frac{2x}{3}\right)^{2n+1}}{(2n+1)!} = \sum_{n=0}^{\infty} \frac{(-1)^n 2^{2n+1} x^{2n+1}}{3^{2n+1}(2n+1)!}$

61. $\cos x = \sum_{n=0}^{\infty} \frac{(-1)^n x^{2n}}{(2n)!} \Rightarrow \cos\left(x^{5/2}\right) = \sum_{n=0}^{\infty} \frac{(-1)^n \left(x^{5/2}\right)^{2n}}{(2n)!} = \sum_{n=0}^{\infty} \frac{(-1)^n x^{5n}}{(2n)!}$

62. $\cos x = \sum_{n=0}^{\infty} \frac{(-1)^n x^{2n}}{(2n)!} \Rightarrow \cos\sqrt{5x} = \cos\left((5x)^{1/2}\right) = \sum_{n=0}^{\infty} \frac{(-1)^n \left((5x)^{1/2}\right)^{2n}}{(2n)!} = \sum_{n=0}^{\infty} \frac{(-1)^n 5^n x^n}{(2n)!}$

63. $e^x = \sum_{n=0}^{\infty} \frac{x^n}{n!} \Rightarrow e^{(\pi x/2)} = \sum_{n=0}^{\infty} \frac{\left(\frac{\pi x}{2}\right)^n}{n!} = \sum_{n=0}^{\infty} \frac{\pi^n x^n}{2^n n!}$

64. $e^x = \sum_{n=0}^{\infty} \frac{x^n}{n!} \Rightarrow e^{-x^2} = \sum_{n=0}^{\infty} \frac{\left(-x^2\right)^n}{n!} = \sum_{n=0}^{\infty} \frac{(-1)^n x^{2n}}{n!}$

65. $f(x) = \sqrt{3+x^2} = (3+x^2)^{1/2} \Rightarrow f'(x) = x\left(3+x^2\right)^{-1/2} \Rightarrow f''(x) = -x^2\left(3+x^2\right)^{-3/2} + \left(3+x^2\right)^{-1/2}$ $\Rightarrow f'''(x) = 3x^3\left(3+x^2\right)^{-5/2} - 3x\left(3+x^2\right)^{-3/2}$; $f(-1) = 2$, $f'(-1) = -\frac{1}{2}$, $f''(-1) = -\frac{1}{8} + \frac{1}{2} = \frac{3}{8}$, $f'''(-1) = -\frac{3}{32} + \frac{3}{8} = \frac{9}{32} \Rightarrow \sqrt{3+x^2} = 2 - \frac{(x+1)}{2 \cdot 1!} + \frac{3(x+1)^2}{2^3 \cdot 2!} + \frac{9(x+1)^3}{2^5 \cdot 3!} + \ldots$

66. $f(x) = \frac{1}{1-x} = (1-x)^{-1} \Rightarrow f'(x) = (1-x)^{-2} \Rightarrow f''(x) = 2(1-x)^{-3} \Rightarrow f'''(x) = 6(1-x)^{-4};\ f(2) = -1,\ f'(2) = 1,$
$f''(2) = -2,\ f'''(2) = 6 \Rightarrow \frac{1}{1-x} = -1 + (x-2) - (x-2)^2 + (x-2)^3 - \ldots$

67. $f(x) = \frac{1}{x+1} = (x+1)^{-1} \Rightarrow f'(x) = -(x+1)^{-2} \Rightarrow f''(x) = 2(x+1)^{-3} \Rightarrow f'''(x) = -6(x+1)^{-4};\ f(3) = \frac{1}{4},$
$f'(3) = -\frac{1}{4^2},\ f''(3) = \frac{2}{4^3},\ f'''(2) = \frac{-6}{4^4} \Rightarrow \frac{1}{x+1} = \frac{1}{4} - \frac{1}{4^2}(x-3) + \frac{1}{4^3}(x-3)^2 - \frac{1}{4^4}(x-3)^3 + \ldots$

68. $f(x) = \frac{1}{x} = x^{-1} \Rightarrow f'(x) = -x^{-2} \Rightarrow f''(x) = 2x^{-3} \Rightarrow f'''(x) = -6x^{-4};\ f(a) = \frac{1}{a},\ f'(a) = -\frac{1}{a^2},\ f''(a) = \frac{2}{a^3},$
$f'''(a) = \frac{-6}{a^4} \Rightarrow \frac{1}{x} = \frac{1}{a} - \frac{1}{a^2}(x-a) + \frac{1}{a^3}(x-a)^2 - \frac{1}{a^4}(x-a)^3 + \ldots$

69. Assume the solution has the form $y = a_0 + a_1x + a_2x^2 + \ldots + a_{n-1}x^{n-1} + a_nx^n + \ldots$
$\Rightarrow \frac{dy}{dx} = a_1 + 2a_2x + \ldots + na_nx^{n-1} + \ldots \Rightarrow \frac{dy}{dx} + y$
$= (a_1 + a_0) + (2a_2 + a_1)x + (3a_3 + a_2)x^2 + \ldots + (na_n + a_{n-1})x^{n-1} + \ldots = 0 \Rightarrow a_1 + a_0 = 0,\ 2a_2 + a_1 = 0,$
$3a_3 + a_2 = 0$ and in general $na_n + a_{n-1} = 0$. Since $y = -1$ when $x = 0$ we have $a_0 = -1$. Therefore $a_1 = 1,$
$a_2 = \frac{-a_1}{2\cdot 1} = -\frac{1}{2},\ a_3 = \frac{-a_2}{3} = \frac{1}{3\cdot 2},\ a_4 = \frac{-a_3}{4} = -\frac{1}{4\cdot 3\cdot 2},\ \ldots,\ a_n = \frac{-a_{n-1}}{n} = \frac{-1}{n}\frac{(-1)^n}{(n-1)!} = \frac{(-1)^{n+1}}{n!}$
$\Rightarrow y = -1 + x - \frac{1}{2}x^2 + \frac{1}{3\cdot 2}x^3 - \ldots + \frac{(-1)^{n+1}}{n!}x^n + \ldots = -\sum_{n=0}^{\infty} \frac{(-1)^n x^n}{n!} = -e^{-x}$

70. Assume the solution has the form $y = a_0 + a_1x + a_2x^2 + \ldots + a_{n-1}x^{n-1} + a_nx^n + \ldots$
$\Rightarrow \frac{dy}{dx} = a_1 + 2a_2x + \ldots + na_nx^{n-1} + \ldots \Rightarrow \frac{dy}{dx} - y$
$= (a_1 - a_0) + (2a_2 - a_1)x + (3a_3 - a_2)x^2 + \ldots + (na_n - a_{n-1})x^{n-1} + \ldots = 0 \Rightarrow a_1 - a_0 = 0,\ 2a_2 - a_1 = 0,$
$3a_3 - a_2 = 0$ and in general $na_n - a_{n-1} = 0$. Since $y = -3$ when $x = 0$ we have $a_0 = -3$. Therefore $a_1 = -3,$
$a_2 = \frac{a_1}{2} = \frac{-3}{2},\ a_3 = \frac{a_2}{3} = \frac{-3}{3\cdot 2},\ a_n = \frac{a_{n-1}}{n} = \frac{-3}{n!} \Rightarrow y = -3 - 3x - \frac{3}{2\cdot 1}x^2 - \frac{3}{3\cdot 2}x^3 - \ldots - \frac{-3}{n!}x^n + \ldots$
$= -3\left(1 + x + \frac{x^2}{2!} + \frac{x^3}{3!} + \ldots + \frac{x^n}{n!} + \ldots\right) = -3\sum_{n=0}^{\infty} \frac{x^n}{n!} = -3e^x$

71. Assume the solution has the form $y = a_0 + a_1x + a_2x^2 + \ldots + a_{n-1}x^{n-1} + a_nx^n + \ldots$
$\Rightarrow \frac{dy}{dx} = a_1 + 2a_2x + \ldots + na_nx^{n-1} + \ldots \Rightarrow \frac{dy}{dx} + 2y$
$= (a_1 + 2a_0) + (2a_2 + 2a_1)x + (3a_3 + 2a_2)x^2 + \ldots + (na_n + 2a_{n-1})x^{n-1} + \ldots = 0$. Since $y = 3$ when $x = 0$ we have $a_0 = 3$. Therefore $a_1 = -2a_0 = -2(3) = -3(2),\ a_2 = -\frac{2}{2}a_1 = -\frac{2}{2}(-2\cdot 3) = 3\left(\frac{2^2}{2}\right),\ a_3 = -\frac{2}{3}a_2$
$= -\frac{2}{3}\left[3\left(\frac{2^2}{2}\right)\right] = -3\left(\frac{2^3}{3\cdot 2}\right),\ \ldots,\ a_n = \left(-\frac{2}{n}\right)a_{n-1} = \left(-\frac{2}{n}\right)\left(3\left(\frac{(-1)^{n-1}2^{n-1}}{(n-1)!}\right)\right) = 3\left(\frac{(-1)^n 2^n}{n!}\right)$
$\Rightarrow y = 3 - 3(2x) + 3\frac{(2)^2}{2}x^2 - 3\frac{(2)^3}{3\cdot 2}x^3 + \ldots + 3\frac{(-1)^n 2^n}{n!}x^n + \ldots$
$= 3\left[1 - (2x) + \frac{(2x)^2}{2!} - \frac{(2x)^3}{3!} + \ldots + \frac{(-1)^n(2x)^n}{n!} + \ldots\right] = 3\sum_{n=0}^{\infty} \frac{(-1)^n(2x)^n}{n!} = 3e^{-2x}$

72. Assume the solution has the form $y = a_0 + a_1x + a_2x^2 + \ldots + a_{n-1}x^{n-1} + a_nx^n + \ldots$
$\Rightarrow \frac{dy}{dx} = a_1 + 2a_2x + \ldots + na_nx^{n-1} + \ldots \Rightarrow \frac{dy}{dx} + y$
$= (a_1 + a_0) + (2a_2 + a_1)x + (3a_3 + a_2)x^2 + \ldots + (na_n + a_{n-1})x^{n-1} + \ldots = 1 \Rightarrow a_1 + a_0 = 1,\ 2a_2 + a_1 = 0,$
$3a_3 + a_2 = 0$ and in general $na_n + a_{n-1} = 0$ for $n > 1$. Since $y = 0$ when $x = 0$ we have $a_0 = 0$. Therefore
$a_1 = 1 - a_0 = 1,\ a_2 = \frac{-a_1}{2\cdot 1} = -\frac{1}{2},\ a_3 = \frac{-a_2}{3} = \frac{1}{3\cdot 2},\ a_4 = \frac{-a_3}{4} = -\frac{1}{4\cdot 3\cdot 2},\ \ldots,\ a_n$
$= \frac{-a_{n-1}}{n} = \left(\frac{-1}{n}\right)\frac{(-1)^n}{(n-1)!} = \frac{(-1)^{n+1}}{n!} \Rightarrow y = 0 + x - \frac{1}{2}x^2 + \frac{1}{3\cdot 2}x^3 - \ldots + \frac{(-1)^{n+1}}{n!}x^n + \ldots$
$= -1\left[1 - x + \frac{1}{2}x^2 - \frac{1}{3\cdot 2}x^3 - \ldots + \frac{(-1)^n}{n!}x^n + \ldots\right] + 1 = -\sum_{n=0}^{\infty} \frac{(-1)^n x^n}{n!} + 1 = 1 - e^{-x}$

73. Assume the solution has the form $y = a_0 + a_1x + a_2x^2 + \ldots + a_{n-1}x^{n-1} + a_nx^n + \ldots$
$\Rightarrow \frac{dy}{dx} = a_1 + 2a_2x + \ldots + na_nx^{n-1} + \ldots \Rightarrow \frac{dy}{dx} - y$
$= (a_1 - a_0) + (2a_2 - a_1)x + (3a_3 - a_2)x^2 + \ldots + (na_n - a_{n-1})x^{n-1} + \ldots = 3x \Rightarrow a_1 - a_0 = 0,\ 2a_2 - a_1 = 3,$

$3a_3 - a_2 = 0$ and in general $na_n - a_{n-1} = 0$ for $n > 2$. Since $y = -1$ when $x = 0$ we have $a_0 = -1$. Therefore $a_1 = -1$, $a_2 = \frac{3+a_1}{2} = \frac{2}{2}$, $a_3 = \frac{a_2}{3} = \frac{2}{3\cdot 2}$, $a_4 = \frac{a_3}{4} = \frac{2}{4\cdot 3\cdot 2}$, ..., $a_n = \frac{a_{n-1}}{n} = \frac{2}{n!}$
$\Rightarrow y = -1 - x + \left(\frac{2}{2}\right)x^2 + \frac{3}{3\cdot 2}x^3 + \frac{2}{4\cdot 3\cdot 2}x^4 + \ldots + \frac{2}{n!}x^n + \ldots$
$= 2\left(1 + x + \frac{1}{2}x^2 + \frac{1}{3\cdot 2}x^3 + \frac{1}{4\cdot 3\cdot 2}x^4 + \ldots + \frac{1}{n!}x^n + \ldots\right) - 3 - 3x = 2\sum_{n=0}^{\infty}\frac{x^n}{n!} - 3 - 3x = 2e^x - 3x - 3$

74. Assume the solution has the form $y = a_0 + a_1x + a_2x^2 + \ldots + a_{n-1}x^{n-1} + a_nx^n + \ldots$
$\Rightarrow \frac{dy}{dx} = a_1 + 2a_2x + \ldots + na_nx^{n-1} + \ldots \Rightarrow \frac{dy}{dx} + y$
$= (a_1 + a_0) + (2a_2 + a_1)x + (3a_3 + a_2)x^2 + \ldots + (na_n + a_{n-1})x^{n-1} + \ldots = x \Rightarrow a_1 + a_0 = 0,\ 2a_2 + a_1 = 1,$
$3a_3 + a_2 = 0$ and in general $na_n + a_{n-1} = 0$ for $n > 2$. Since $y = 0$ when $x = 0$ we have $a_0 = 0$. Therefore $a_1 = 0$, $a_2 = \frac{1-a_1}{2} = \frac{1}{2}$, $a_3 = \frac{-a_2}{3} = -\frac{1}{3\cdot 2}$, ..., $a_n = \frac{-a_{n-1}}{n} = \frac{(-1)^n}{n!}$
$\Rightarrow y = 0 - 0x + \frac{1}{2}x^2 - \frac{1}{3\cdot 2}x^3 + \ldots + \frac{(-1)^n}{n!}x^n + \ldots = \left(1 - x + \frac{1}{2}x^2 - \frac{1}{3\cdot 2}x^3 + \ldots + \frac{(-1)^n}{n!}x^n + \ldots\right) - 1 + x$
$= \sum_{n=0}^{\infty}\frac{(-1)^nx^n}{n!} - 1 + x = e^{-x} + x - 1$

75. Assume the solution has the form $y = a_0 + a_1x + a_2x^2 + \ldots + a_{n-1}x^{n-1} + a_nx^n + \ldots$
$\Rightarrow \frac{dy}{dx} = a_1 + 2a_2x + \ldots + na_nx^{n-1} + \ldots \Rightarrow \frac{dy}{dx} - y$
$= (a_1 - a_0) + (2a_2 - a_1)x + (3a_3 - a_2)x^2 + \ldots + (na_n - a_{n-1})x^{n-1} + \ldots = x \Rightarrow a_1 - a_0 = 0,\ 2a_2 - a_1 = 1,$
$3a_3 - a_2 = 0$ and in general $na_n - a_{n-1} = 0$ for $n > 2$. Since $y = 1$ when $x = 0$ we have $a_0 = 1$. Therefore $a_1 = 1$, $a_2 = \frac{1+a_1}{2} = \frac{2}{2}$, $a_3 = \frac{a_2}{3} = \frac{2}{3\cdot 2}$, $a_4 = \frac{a_3}{4} = \frac{2}{4\cdot 3\cdot 2}$, ..., $a_n = \frac{a_{n-1}}{n} = \frac{2}{n!}$
$\Rightarrow y = 1 + x + \left(\frac{2}{2}\right)x^2 + \frac{2}{3\cdot 2}x^3 + \frac{2}{4\cdot 2\cdot 2}x^4 + \ldots + \frac{2}{n!}x^n + \ldots$
$= 2\left(1 + x + \frac{1}{2}x^2 + \frac{1}{3\cdot 2}x^3 + \frac{1}{4\cdot 3\cdot 2}x^4 + \ldots + \frac{1}{n!}x^n + \ldots\right) - 1 - x = 2\sum_{n=0}^{\infty}\frac{x^n}{n!} - 1 - x = 2e^x - x - 1$

76. Assume the solution has the form $y = a_0 + a_1x + a_2x^2 + \ldots + a_{n-1}x^{n-1} + a_nx^n + \ldots$
$\Rightarrow \frac{dy}{dx} = a_1 + 2a_2x + \ldots + na_nx^{n-1} + \ldots \Rightarrow \frac{dy}{dx} - y$
$= (a_1 - a_0) + (2a_2 - a_1)x + (3a_3 - a_2)x^2 + \ldots + (na_n - a_{n-1})x^{n-1} + \ldots = -x \Rightarrow a_1 - a_0 = 0,\ 2a_2 - a_1 = -1,$
$3a_3 - a_2 = 0$ and in general $na_n - a_{n-1} = 0$ for $n > 2$. Since $y = 2$ when $x = 0$ we have $a_0 = 2$. Therefore $a_1 = 2$, $a_2 = \frac{-1+a_1}{2} = \frac{1}{2}$, $a_3 = \frac{a_2}{3} = \frac{1}{3\cdot 2}$, $a_4 = \frac{a_3}{4} = \frac{1}{4\cdot 3\cdot 2}$, ..., $a_n = \frac{a_{n-1}}{n} = \frac{1}{n!}$
$\Rightarrow y = 2 + 2x + \frac{1}{2}x^2 + \frac{1}{3\cdot 2}x^3 + \frac{1}{4\cdot 3\cdot 2}x^4 + \ldots + \frac{1}{n!}x^n + \ldots$
$= \left(1 + x + \frac{1}{2}x^2 + \frac{1}{3\cdot 2}x^3 + \frac{1}{4\cdot 3\cdot 2}x^4 + \ldots + \frac{1}{n!}x^n + \ldots\right) + 1 + x = \sum_{n=0}^{\infty}\frac{x^n}{n!} + 1 + x = e^x + x + 1$

77. $\int_0^{1/2}\exp(-x^3)\,dx = \int_0^{1/2}\left(1 - x^3 + \frac{x^6}{2!} - \frac{x^9}{3!} + \frac{x^{12}}{4!} + \ldots\right)dx = \left[x - \frac{x^4}{4} + \frac{x^7}{7\cdot 2!} - \frac{x^{10}}{10\cdot 3!} + \frac{x^{13}}{13\cdot 4!} - \ldots\right]_0^{1/2}$
$\approx \frac{1}{2} - \frac{1}{2^4\cdot 4} + \frac{1}{2^7\cdot 7\cdot 2!} - \frac{1}{2^{10}\cdot 10\cdot 3!} + \frac{1}{2^{13}\cdot 13\cdot 4!} - \frac{1}{2^{16}\cdot 16\cdot 5!} \approx 0.484917143$

78. $\int_0^1 x\sin(x^3)\,dx = \int_0^1 x\left(x^3 - \frac{x^9}{3!} + \frac{x^{15}}{5!} - \frac{x^{21}}{7!} + \frac{x^{27}}{9!} + \ldots\right)dx = \int_0^1\left(x^4 - \frac{x^{10}}{3!} + \frac{x^{16}}{5!} - \frac{x^{22}}{7!} + \frac{x^{28}}{9!} - \ldots\right)dx$
$= \left[\frac{x^5}{5} - \frac{x^{11}}{11\cdot 3!} + \frac{x^{17}}{17\cdot 5!} - \frac{x^{23}}{23\cdot 7!} + \frac{x^{29}}{29\cdot 9!} - \ldots\right]_0^1 \approx 0.185330149$

79. $\int_1^{1/2}\frac{\tan^{-1}x}{x}\,dx = \int_1^{1/2}\left(1 - \frac{x^2}{3} + \frac{x^4}{5} - \frac{x^6}{7} + \frac{x^8}{9} - \frac{x^{10}}{11} + \ldots\right)dx = \left[x - \frac{x^3}{9} + \frac{x^5}{25} - \frac{x^7}{49} + \frac{x^9}{81} - \frac{x^{11}}{121} + \ldots\right]_0^{1/2}$
$\approx \frac{1}{2} - \frac{1}{9\cdot 2^3} + \frac{1}{5^2\cdot 2^5} - \frac{1}{7^2\cdot 2^7} + \frac{1}{9^2\cdot 2^9} - \frac{1}{11^2\cdot 2^{11}} + \frac{1}{13^2\cdot 2^{13}} - \frac{1}{15^2\cdot 2^{15}} + \frac{1}{17^2\cdot 2^{17}} - \frac{1}{19^2\cdot 2^{19}} + \frac{1}{21^2\cdot 2^{21}}$
≈ 0.4872223583

80. $\int_0^{1/64} \frac{\tan^{-1} x}{\sqrt{x}}\,dx = \int_0^{1/64} \frac{1}{\sqrt{x}}\left(x - \frac{x^3}{3} + \frac{x^5}{5} - \frac{x^7}{7} + \ldots\right) dx = \int_0^{1/64} \left(x^{1/2} - \frac{1}{3}x^{5/2} + \frac{1}{5}x^{9/2} - \frac{1}{7}x^{13/2} + \ldots\right) dx$
$= \left[\frac{2}{3}x^{3/2} - \frac{2}{21}x^{7/2} + \frac{2}{55}x^{11/2} - \frac{2}{105}x^{15/2} + \ldots\right]_0^{1/64} = \left(\frac{2}{3\cdot 8^3} - \frac{2}{21\cdot 8^7} + \frac{2}{55\cdot 8^{11}} - \frac{2}{105\cdot 8^{15}} + \ldots\right) \approx 0.0013020379$

81. $\lim\limits_{x \to 0} \frac{7 \sin x}{e^{2x} - 1} = \lim\limits_{x \to 0} \frac{7\left(x - \frac{x^3}{3!} + \frac{x^5}{5!} - \ldots\right)}{\left(2x + \frac{2^2x^2}{2!} + \frac{2^3x^3}{3!} + \ldots\right)} = \lim\limits_{x \to 0} \frac{7\left(1 - \frac{x^2}{3!} + \frac{x^4}{5!} - \ldots\right)}{\left(2 + \frac{2^2x}{2!} + \frac{2^3x^2}{3!} + \ldots\right)} = \frac{7}{2}$

82. $\lim\limits_{\theta \to 0} \frac{e^{\theta} - e^{-\theta} - 2\theta}{\theta - \sin\theta} = \lim\limits_{\theta \to 0} \frac{\left(1 + \theta + \frac{\theta^2}{2!} + \frac{\theta^3}{3!} + \ldots\right) - \left(1 - \theta + \frac{\theta^2}{2!} - \frac{\theta^3}{3!} + \ldots\right) - 2\theta}{\theta - \left(\theta - \frac{\theta^3}{3!} + \frac{\theta^5}{5!} - \ldots\right)} = \lim\limits_{\theta \to 0} \frac{2\left(\frac{\theta^3}{3!} + \frac{\theta^5}{5!} + \ldots\right)}{\left(\frac{\theta^3}{3!} - \frac{\theta^5}{5!} + \ldots\right)}$
$= \lim\limits_{\theta \to 0} \frac{2\left(\frac{1}{3!} + \frac{\theta^2}{5!} + \ldots\right)}{\left(\frac{1}{3!} - \frac{\theta^2}{5!} + \ldots\right)} = 2$

83. $\lim\limits_{t \to 0} \left(\frac{1}{2 - 2\cos t} - \frac{1}{t^2}\right) = \lim\limits_{t \to 0} \frac{t^2 - 2 + 2\cos t}{2t^2(1 - \cos t)} = \lim\limits_{t \to 0} \frac{t^2 - 2 + 2\left(1 - \frac{t^2}{2} + \frac{t^4}{4!} - \ldots\right)}{2t^2\left(1 - 1 + \frac{t^2}{2} - \frac{t^4}{4!} + \ldots\right)} = \lim\limits_{t \to 0} \frac{2\left(\frac{t^4}{4!} - \frac{t^6}{6!} + \ldots\right)}{\left(t^4 - \frac{2t^6}{4!} + \ldots\right)}$
$= \lim\limits_{t \to 0} \frac{2\left(\frac{1}{4!} - \frac{t^2}{6!} + \ldots\right)}{\left(1 - \frac{2t^2}{4!} + \ldots\right)} = \frac{1}{12}$

84. $\lim\limits_{h \to 0} \frac{\left(\frac{\sin h}{h}\right) - \cos h}{h^2} = \lim\limits_{h \to 0} \frac{\left(1 - \frac{h^2}{3!} + \frac{h^4}{5!} - \ldots\right) - \left(1 - \frac{h^2}{2!} + \frac{h^4}{4!} - \ldots\right)}{h^2}$
$= \lim\limits_{h \to 0} \frac{\left(\frac{h^2}{2!} - \frac{h^2}{3!} + \frac{h^4}{5!} - \frac{h^4}{4!} + \frac{h^6}{6!} - \frac{h^6}{7!} + \ldots\right)}{h^2} = \lim\limits_{h \to 0} \left(\frac{1}{2!} - \frac{1}{3!} + \frac{h^2}{5!} - \frac{h^2}{4!} + \frac{h^4}{6!} - \frac{h^4}{7!} + \ldots\right) = \frac{1}{3}$

85. $\lim\limits_{z \to 0} \frac{1 - \cos^2 z}{\ln(1 - z) + \sin z} = \lim\limits_{z \to 0} \frac{1 - \left(1 - z^2 + \frac{z^4}{3} - \ldots\right)}{\left(-z - \frac{z^2}{2} - \frac{z^3}{3} - \ldots\right) + \left(z - \frac{z^3}{3!} + \frac{z^5}{5!} - \ldots\right)} = \lim\limits_{z \to 0} \frac{\left(z^2 - \frac{z^4}{3} + \ldots\right)}{\left(-\frac{z^2}{2} - \frac{2z^3}{3} - \frac{z^4}{4} - \ldots\right)}$
$= \lim\limits_{z \to 0} \frac{\left(1 - \frac{z^2}{3} + \ldots\right)}{\left(-\frac{1}{2} - \frac{2z}{3} - \frac{z^2}{4} - \ldots\right)} = -2$

86. $\lim\limits_{y \to 0} \frac{y^2}{\cos y - \cosh y} = \lim\limits_{y \to 0} \frac{y^2}{\left(1 - \frac{y^2}{2} + \frac{y^4}{4!} - \frac{y^6}{6!} + \ldots\right) - \left(1 + \frac{y^2}{2!} + \frac{y^4}{4!} + \frac{y^6}{6!} + \ldots\right)} = \lim\limits_{y \to 0} \frac{y^2}{\left(-\frac{2y^2}{2} - \frac{2y^6}{6!} - \ldots\right)}$
$= \lim\limits_{y \to 0} \frac{1}{\left(-1 - \frac{2y^4}{6!} - \ldots\right)} = -1$

87. $\lim\limits_{x \to 0} \left(\frac{\sin 3x}{x^3} + \frac{r}{x^2} + s\right) = \lim\limits_{x \to 0} \left[\frac{\left(3x - \frac{(3x)^3}{6} + \frac{(3x)^5}{120} - \ldots\right)}{x^3} + \frac{r}{x^2} + s\right] = \lim\limits_{x \to 0} \left(\frac{3}{x^2} - \frac{9}{2} + \frac{81x^2}{40} + \ldots + \frac{r}{x^2} + s\right) = 0$
$\Rightarrow \frac{r}{x^2} + \frac{3}{x^2} = 0$ and $s - \frac{9}{2} = 0 \Rightarrow r = -3$ and $s = \frac{9}{2}$

88. (a) $\csc x \approx \frac{1}{x} + \frac{x}{6} \Rightarrow \csc x \approx \frac{6 + x^2}{6x} \Rightarrow \sin x \approx \frac{6x}{6 + x^2}$

(b) The approximation $\sin x \approx \frac{6x}{6 + x^2}$ is better than $\sin x \approx x$.

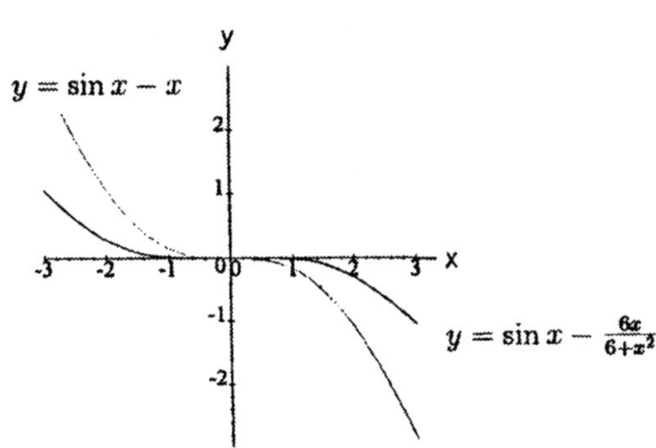

89. (a) $\sum_{n=1}^{\infty}\left(\sin\frac{1}{2n}-\sin\frac{1}{2n+1}\right)=\left(\sin\frac{1}{2}-\sin\frac{1}{3}\right)+\left(\sin\frac{1}{4}-\sin\frac{1}{5}\right)+\left(\sin\frac{1}{6}-\sin\frac{1}{7}\right)+\ldots+\left(\sin\frac{1}{2n}-\sin\frac{1}{2n+1}\right)$ $+\ldots=\sum_{n=2}^{\infty}(-1)^n\sin\frac{1}{n}$; $f(x)=\sin\frac{1}{x}\Rightarrow f'(x)=\frac{-\cos\left(\frac{1}{x}\right)}{x^2}<0$ if $x\geq 2\Rightarrow \sin\frac{1}{n+1}<\sin\frac{1}{n}$, and $\lim_{n\to\infty}\sin\frac{1}{n}=0\Rightarrow\sum_{n=2}^{\infty}(-1)^n\sin\frac{1}{n}$ converges by the Alternating Series Test

(b) $|\text{error}|<\left|\sin\frac{1}{42}\right|\approx 0.02381$ and the sum is an underestimate because the remainder is positive

90. (a) $\sum_{n=1}^{\infty}\left(\tan\frac{1}{2n}-\tan\frac{1}{2n+1}\right)=\sum_{n=2}^{\infty}(-1)^n\tan\frac{1}{n}$ (see Exercise 89); $f(x)=\tan\frac{1}{x}\Rightarrow f'(x)=\frac{-\sec^2\left(\frac{1}{x}\right)}{x^2}<0$ $\Rightarrow\tan\frac{1}{n+1}<\tan\frac{1}{n}$, and $\lim_{n\to\infty}\tan\frac{1}{n}=0\Rightarrow\sum_{n=2}^{\infty}(-1)^n\tan\frac{1}{n}$ converges by the Alternating Series Test

(b) $|\text{error}|<\left|\tan\frac{1}{42}\right|\approx 0.02382$ and the sum is an underestimate because the remainder is positive

91. $\lim_{n\to\infty}\left|\frac{2\cdot5\cdot8\cdots(3n-1)(3n+2)x^{n+1}}{2\cdot4\cdot6\cdots(2n)(2n+2)}\cdot\frac{2\cdot4\cdot6\cdots(2n)}{2\cdot5\cdot8\cdots(3n-1)x^n}\right|<1\Rightarrow|x|\lim_{n\to\infty}\left|\frac{3n+2}{2n+2}\right|<1\Rightarrow|x|<\frac{2}{3}$ $\Rightarrow$ the radius of convergence is $\frac{2}{3}$

92. $\lim_{n\to\infty}\left|\frac{3\cdot5\cdot7\cdots(2n+1)(2n+3)(x-1)^{n+1}}{4\cdot9\cdot14\cdots(5n-1)(5n+4)}\cdot\frac{4\cdot9\cdot14\cdots(5n-1)}{3\cdot5\cdot7\cdots(2n+1)x^n}\right|<1\Rightarrow|x|\lim_{n\to\infty}\left|\frac{2n+3}{5n+4}\right|<1\Rightarrow|x|<\frac{5}{2}$ $\Rightarrow$ the radius of convergence is $\frac{5}{2}$

93. $\sum_{k=2}^{n}\ln\left(1-\frac{1}{k^2}\right)=\sum_{k=2}^{n}\left[\ln\left(1+\frac{1}{k}\right)+\ln\left(1-\frac{1}{k}\right)\right]=\sum_{k=2}^{n}[\ln(k+1)-\ln k+\ln(k-1)-\ln k]$ $=[\ln 3-\ln 2+\ln 1-\ln 2]+[\ln 4-\ln 3+\ln 2-\ln 3]+[\ln 5-\ln 4+\ln 3-\ln 4]+[\ln 6-\ln 5+\ln 4-\ln 5]$ $+\ldots+[\ln(n+1)-\ln n+\ln(n-1)-\ln n]=[\ln 1-\ln 2]+[\ln(n+1)-\ln n]$ after cancellation $\Rightarrow\sum_{k=2}^{n}\ln\left(1-\frac{1}{k^2}\right)=\ln\left(\frac{n+1}{2n}\right)\Rightarrow\sum_{k=2}^{\infty}\ln\left(1-\frac{1}{k^2}\right)=\lim_{n\to\infty}\ln\left(\frac{n+1}{2n}\right)=\ln\frac{1}{2}$ is the sum

94. $\sum_{k=2}^{n}\frac{1}{k^2-1}=\frac{1}{2}\sum_{k=2}^{n}\left(\frac{1}{k-1}-\frac{1}{k+1}\right)=\frac{1}{2}\left[\left(\frac{1}{1}-\frac{1}{3}\right)+\left(\frac{1}{2}-\frac{1}{4}\right)+\left(\frac{1}{3}-\frac{1}{5}\right)+\left(\frac{1}{4}-\frac{1}{6}\right)+\ldots+\left(\frac{1}{n-2}-\frac{1}{n}\right)\right.$ $\left.+\left(\frac{1}{n-1}-\frac{1}{n+1}\right)\right]=\frac{1}{2}\left(\frac{1}{1}+\frac{1}{2}-\frac{1}{n}-\frac{1}{n+1}\right)=\frac{1}{2}\left(\frac{3}{2}-\frac{1}{n}-\frac{1}{n+1}\right)=\frac{1}{2}\left[\frac{3n(n+1)-2(n+1)-2n}{2n(n+1)}\right]=\frac{3n^2-n-2}{4n(n+1)}$ $\Rightarrow\sum_{k=2}^{\infty}\frac{1}{k^2-1}=\lim_{n\to\infty}\frac{1}{2}\left(\frac{3}{2}-\frac{1}{n}-\frac{1}{n+1}\right)=\frac{3}{4}$

95. (a) $\lim_{n\to\infty}\left|\frac{1\cdot4\cdot7\cdots(3n-2)(3n+1)x^{3n+3}}{(3n+3)!}\cdot\frac{(3n)!}{1\cdot4\cdot7\cdots(3n-2)x^{3n}}\right|<1\Rightarrow|x^3|\lim_{n\to\infty}\frac{(3n+1)}{(3n+1)(3n+2)(3n+3)}$ $=|x^3|\cdot 0<1\Rightarrow$ the radius of convergence is ∞

(b) $y=1+\sum_{n=1}^{\infty}\frac{1\cdot4\cdot7\cdots(3n-2)}{(3n)!}x^{3n}\Rightarrow\frac{dy}{dx}=\sum_{n=1}^{\infty}\frac{1\cdot4\cdot7\cdots(3n-2)}{(3n-1)!}x^{3n-1}$

$\Rightarrow\frac{d^2y}{dx^2}=\sum_{n=1}^{\infty}\frac{1\cdot4\cdot7\cdots(3n-2)}{(3n-2)!}x^{3n-2}=x+\sum_{n=2}^{\infty}\frac{1\cdot4\cdot7\cdots(3n-5)}{(3n-3)!}x^{3n-2}$

$=x\left(1+\sum_{n=1}^{\infty}\frac{1\cdot4\cdot7\cdots(3n-2)}{(3n)!}x^{3n}\right)=xy+0\Rightarrow a=1$ and $b=0$

96. (a) $\frac{x^2}{1+x}=\frac{x^2}{1-(-x)}=x^2+x^2(-x)+x^2(-x)^2+x^2(-x)^3+\ldots=x^2-x^3+x^4-x^5+\ldots=\sum_{n=2}^{\infty}(-1)^n x^n$ which converges absolutely for $|x|<1$

(b) $x=1\Rightarrow\sum_{n=2}^{\infty}(-1)^n x^n=\sum_{n=2}^{\infty}(-1)^n$ which diverges

97. Yes, the series $\sum_{n=1}^{\infty} a_n b_n$ converges as we now show. Since $\sum_{n=1}^{\infty} a_n$ converges it follows that $a_n \to 0 \Rightarrow a_n < 1$ for $n >$ some index $N \Rightarrow a_n b_n < b_n$ for $n > N \Rightarrow \sum_{n=1}^{\infty} a_n b_n$ converges by the Direct Comparison Test with $\sum_{n=1}^{\infty} b_n$

98. No, the series $\sum_{n=1}^{\infty} a_n b_n$ might diverge (as it would if a_n and b_n both equaled n) or it might converge (as it would if a_n and b_n both equaled $\frac{1}{n}$).

99. $\sum_{n=1}^{\infty} (x_{n+1} - x_n) = \lim_{n \to \infty} \sum_{k=1}^{\infty} (x_{k+1} - x_k) = \lim_{n \to \infty} (x_{n+1} - x_1) = \lim_{n \to \infty} (x_{n+1}) - x_1 \Rightarrow$ both the series and sequence must either converge or diverge.

100. It converges by the Limit Comparison Test since $\lim_{n \to \infty} \frac{\left(\frac{a_n}{1+a_n}\right)}{a_n} = \lim_{n \to \infty} \frac{1}{1+a_n} = 1$ because $\sum_{n=1}^{\infty} a_n$ converges and so $a_n \to 0$.

101. Newton's method gives $x_{n+1} = x_n - \frac{(x_n-1)^{40}}{40\,(x_n-1)^{39}} = \frac{39}{40}\, x_n + \frac{1}{40}$, and if the sequence $\{x_n\}$ has the limit L, then $L = \frac{39}{40} L + \frac{1}{40} \Rightarrow L = 1$ and $\{x_n\}$ converges since $\left|\frac{f(x) f''(x)}{[f'(x)]^2}\right| = \frac{39}{40} < 1$

102. $\sum_{n=1}^{\infty} \frac{a_n}{n} = a_1 + \frac{a_2}{2} + \frac{a_3}{3} + \frac{a_4}{4} + \ldots \geq a_1 + \left(\frac{1}{2}\right) a_2 + \left(\frac{1}{3} + \frac{1}{4}\right) a_4 + \left(\frac{1}{5} + \frac{1}{6} + \frac{1}{7} + \frac{1}{8}\right) a_8$
$+ \left(\frac{1}{9} + \frac{1}{10} + \frac{1}{11} + \ldots + \frac{1}{16}\right) a_{16} + \ldots \geq \frac{1}{2} (a_2 + a_4 + a_8 + a_{16} + \ldots)$ which is a divergent series

103. $a_n = \frac{1}{\ln n}$ for $n \geq 2 \Rightarrow a_2 \geq a_3 \geq a_4 \geq \ldots$, and $\frac{1}{\ln 2} + \frac{1}{\ln 4} + \frac{1}{\ln 8} + \ldots = \frac{1}{\ln 2} + \frac{1}{2 \ln 2} + \frac{1}{3 \ln 2} + \ldots$
$= \frac{1}{\ln 2} \left(1 + \frac{1}{2} + \frac{1}{3} + \ldots\right)$ which diverges so that $1 + \sum_{n=2}^{\infty} \frac{1}{n \ln n}$ diverges by the Integral Test.

104. (a) $T = \frac{\left(\frac{1}{2}\right)}{2} \left(0 + 2\left(\frac{1}{2}\right)^2 e^{1/2} + e\right) = \frac{1}{8} e^{1/2} + \frac{1}{4} e \approx 0.885660616$

(b) $x^2 e^x = x^2 \left(1 + x + \frac{x^2}{2} + \ldots\right) = x^2 + x^3 + \frac{x^4}{2} + \ldots \Rightarrow \int_0^1 \left(x^2 + x^3 + \frac{x^4}{2}\right) dx = \left[\frac{x^3}{3} + \frac{x^4}{4} + \frac{x^5}{10}\right]_0^1 = \frac{41}{60} = 0.68\overline{333}$

(c) If the second derivative is positive, the curve is concave upward and the polygonal line segments used in the trapezoidal rule lie above the curve. The trapezoidal approximation is therefore greater than the actual area under the graph.

(d) All terms in the Maclaurin series are positive. If we truncate the series, we are omitting positive terms and hence the estimate is too small.

(e) $\int_0^1 x^2 e^x\, dx = \left[x^2 e^x - 2x e^x + 2e^x\right]_0^1 = e - 2e + 2e - 2 = e - 2 \approx 0.7182818285$

105. $a_0 = \frac{1}{2\pi} \int_0^{2\pi} f(x)\, dx = \frac{1}{2\pi} \int_\pi^{2\pi} 1\, dx = \frac{1}{2}$, $a_k = \frac{1}{\pi} \int_0^{2\pi} f(x) \cos kx\, dx = \frac{1}{\pi} \int_\pi^{2\pi} \cos kx\, dx = 0$.
$b_k = \frac{1}{\pi} \int_0^{2\pi} f(x) \sin kx\, dx = \frac{1}{\pi} \int_\pi^{2\pi} \sin kx\, dx = -\frac{\cos kx}{\pi k}\Big|_\pi^{2\pi} = -\frac{1}{\pi k}\left(1 - (-1)^k\right) = \begin{cases} -\frac{2}{\pi k}, & k \text{ odd} \\ 0, & k \text{ even} \end{cases}$.

Thus, the Fourier series of f(x) is $\frac{1}{2} - \sum_{k \text{ odd}} \frac{2}{\pi k} \sin kx$

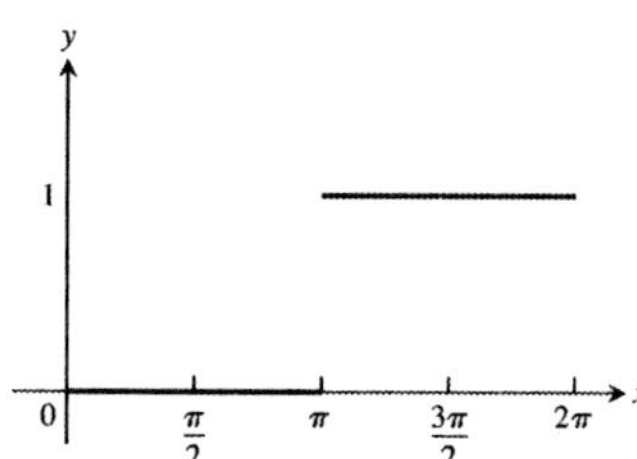

106. $a_0 = \frac{1}{2\pi}\left[\int_0^{\pi} x\, dx + \int_{\pi}^{2\pi} 1\, dx\right] = \frac{1}{2} + \frac{1}{4}\pi$, $a_k = \frac{1}{\pi}\left[\int_0^{\pi} x \cos kx\, dx + \int_{\pi}^{2\pi} \cos kx\, dx\right] = \frac{1}{\pi}\left[\frac{\cos kx}{k^2} + \frac{x \sin kx}{k}\right]_0^{\pi}$

$= \frac{1}{\pi k^2}\left((-1)^k - 1\right) = \begin{cases} -\frac{2}{\pi k^2}, & \text{k odd} \\ 0, & \text{k even} \end{cases}$.

$b_k = \frac{1}{\pi}\left[\int_0^{\pi} x \sin kx\, dx + \int_{\pi}^{2\pi} \sin kx\, dx\right] = \frac{1}{\pi}\left[\frac{\sin kx}{k^2} - \frac{x \cos kx}{k}\right]_0^{\pi} - \frac{\cos kx}{\pi k}\Big|_{\pi}^{2\pi} = \frac{(-1)^{k+1}}{k} - \frac{1}{\pi k}\left(1 - (-1)^k\right)$

$= \begin{cases} \frac{1}{k}\left(1 - \frac{2}{\pi}\right), & \text{k odd} \\ -\frac{1}{k}, & \text{k even} \end{cases}$.

Thus, the Fourier series of f(x) is $\frac{1}{2} + \frac{1}{4}\pi - \frac{2}{\pi}\cos x + \left(1 - \frac{2}{\pi}\right)\sin x - \frac{1}{2}\sin 2x - \frac{2}{9\pi}\cos 3x + \frac{1}{3}\left(1 - \frac{2}{\pi}\right)\sin 3x + \ldots$

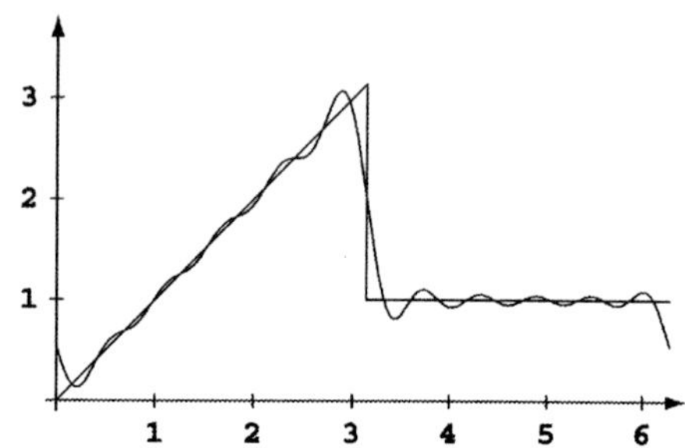

107. $a_0 = \frac{1}{2\pi}\left[\int_0^{\pi} (\pi - x)\, dx + \int_{\pi}^{2\pi} (x - 2\pi)\, dx\right] = \frac{1}{2\pi}\left[\int_0^{\pi} (\pi - x)\, dx - \int_0^{\pi} (\pi - u)\, du\right] = 0$ where we used the substitution $u = x - \pi$ in the second integral. We have $a_k = \frac{1}{\pi}\left[\int_0^{\pi} (\pi - x) \cos kx\, dx + \int_{\pi}^{2\pi} (x - 2\pi) \cos kx\, dx\right]$. Using the substitution $u = x - \pi$ in the second integral gives $\int_{\pi}^{2\pi} (x - 2\pi) \cos kx\, dx = \int_0^{\pi} -(\pi - u) \cos(ku + k\pi)\, du$

$= \begin{cases} \int_0^{\pi} (\pi - u) \cos ku\, du, & \text{k odd} \\ \int_0^{\pi} -(\pi - u) \cos ku\, du, & \text{k even} \end{cases}$.

Thus, $a_k = \begin{cases} \frac{2}{\pi}\int_0^{\pi} (\pi - x) \cos kx\, dx, & \text{k odd} \\ 0, & \text{k even} \end{cases}$.

Now, since k is odd, letting $v = \pi - x \Rightarrow \frac{2}{\pi}\int_0^{\pi} (\pi - x) \cos kx\, dx = -\frac{2}{\pi}\int_0^{\pi} v \cos kv\, dv = -\frac{2}{\pi}\left(-\frac{2}{k^2}\right) = \frac{4}{\pi k^2}$, k odd. (See Exercise 106). So, $a_k = \begin{cases} \frac{4}{\pi k^2}, & \text{k odd} \\ 0, & \text{k even} \end{cases}$.

Using similar techniques we see that $b_k = \begin{cases} \frac{2}{\pi}\int_0^{\pi} (\pi - u) \sin ku\, du, & \text{k odd} \\ 0, & \text{k even} \end{cases} = \begin{cases} \frac{2}{k}, & \text{k odd} \\ 0, & \text{k even} \end{cases}$.

Thus, the Fourier series of f(x) is $\sum_{\text{k odd}} \left(\frac{4}{\pi k^2}\cos kx + \frac{2}{k}\sin kx\right)$.

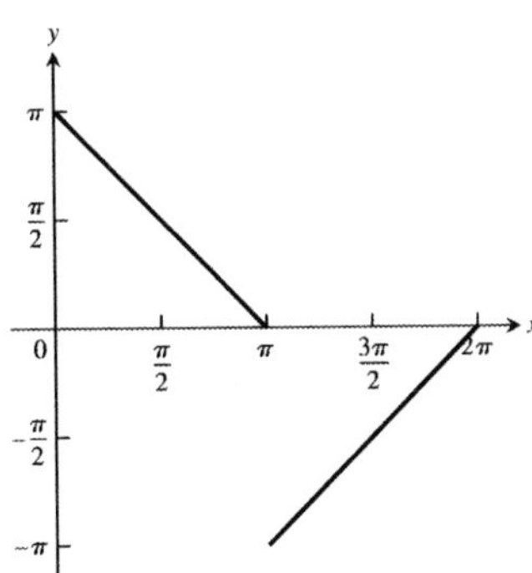

108. $a_0 = \frac{1}{2\pi}\int_0^{2\pi} |\sin x|\, dx = \frac{1}{\pi}\int_0^{\pi} \sin x\, dx = \frac{2}{\pi}$. We have $a_k = \frac{1}{\pi}\int_0^{2\pi} |\sin x| \cos kx\, dx$

$= \frac{1}{\pi}\left[\int_0^{\pi} \sin x \cos kx\, dx - \int_{\pi}^{2\pi} \sin x \cos kx\, dx\right]$. Using techniques similar to those used in Exercise 107, we find

$a_k = \begin{cases} 0, & k \text{ odd} \\ \frac{2}{\pi}\int_0^{\pi} \sin x \cos kx\, dx, & k \text{ even} \end{cases} = \begin{cases} 0, & k \text{ odd} \\ \frac{-4}{(k^2-1)\pi}, & k \text{ even} \end{cases}$.

$b_k = \frac{1}{\pi}\int_0^{2\pi} |\sin x| \sin kx\, dx = \frac{1}{\pi}\left[\int_0^{\pi} \sin x \sin kx\, dx - \int_{\pi}^{2\pi} \sin x \sin kx\, dx\right] = \begin{cases} 0, & k \text{ odd} \\ \frac{2}{\pi}\int_0^{\pi} \sin x \sin kx\, dx, & k \text{ even} \end{cases} = 0$

for all k.

Thus, the Fourier series of f(x) is $\frac{2}{\pi} + \sum_{k \text{ even}} \left(\frac{-4}{(k^2-1)\pi}\cos kx\right)$.

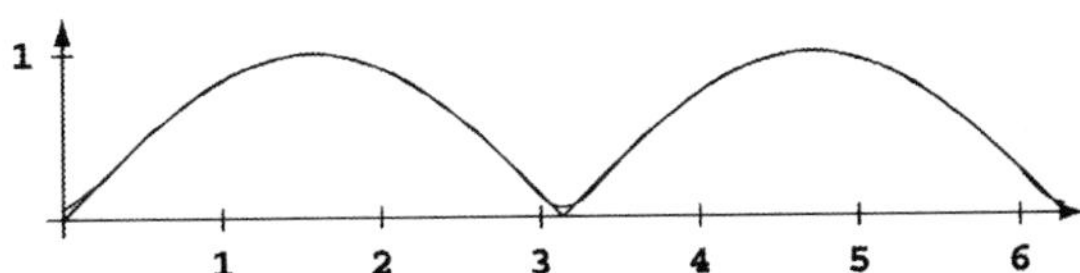

CHAPTER 11 ADDITIONAL AND ADVANCED EXERCISES

1. converges since $\frac{1}{(3n-2)^{(2n+1)/2}} < \frac{1}{(3n-2)^{3/2}}$ and $\sum_{n=1}^{\infty} \frac{1}{(3n-2)^{3/2}}$ converges by the Limit Comparison Test:

$\lim_{n \to \infty} \frac{\left(\frac{1}{n^{3/2}}\right)}{\left(\frac{1}{(3n-2)^{3/2}}\right)} = \lim_{n \to \infty} \left(\frac{3n-2}{n}\right)^{3/2} = 3^{3/2}$

2. converges by the Integral Test: $\int_1^{\infty} (\tan^{-1} x)^2 \frac{dx}{x^2+1} = \lim_{b \to \infty} \left[\frac{(\tan^{-1} x)^3}{3}\right]_1^b = \lim_{b \to \infty} \left[\frac{(\tan^{-1} b)^3}{3} - \frac{\pi^3}{192}\right]$

$= \left(\frac{\pi^3}{24} - \frac{\pi^3}{192}\right) = \frac{7\pi^3}{192}$

3. diverges by the nth-Term Test since $\lim_{n \to \infty} a_n = \lim_{n \to \infty} (-1)^n \tanh n = \lim_{b \to \infty} (-1)^n \left(\frac{1-e^{-2n}}{1+e^{-2n}}\right) = \lim_{n \to \infty} (-1)^n$

does not exist

4. converges by the Direct Comparison Test: $n! < n^n \Rightarrow \ln(n!) < n \ln(n) \Rightarrow \frac{\ln(n!)}{\ln(n)} < n$

$\Rightarrow \log_n(n!) < n \Rightarrow \frac{\log_n(n!)}{n^3} < \frac{1}{n^2}$, which is the nth-term of a convergent p-series

5. converges by the Direct Comparison Test: $a_1 = 1 = \frac{12}{(1)(3)(2)^2}$, $a_2 = \frac{1 \cdot 2}{3 \cdot 4} = \frac{12}{(2)(4)(3)^2}$, $a_3 = \left(\frac{2 \cdot 3}{4 \cdot 5}\right)\left(\frac{1 \cdot 2}{3 \cdot 4}\right)$

$= \frac{12}{(3)(5)(4)^2}$, $a_4 = \left(\frac{3 \cdot 4}{5 \cdot 6}\right)\left(\frac{2 \cdot 3}{4 \cdot 5}\right)\left(\frac{1 \cdot 2}{3 \cdot 4}\right) = \frac{12}{(4)(6)(5)^2}, \ldots \Rightarrow 1 + \sum_{n=1}^{\infty} \frac{12}{(n+1)(n+3)(n+2)^2}$ represents the given series and $\frac{12}{(n+1)(n+3)(n+2)^2} < \frac{12}{n^4}$, which is the nth-term of a convergent p-series

6. converges by the Ratio Test: $\lim_{n\to\infty} \frac{a_{n+1}}{a_n} = \lim_{n\to\infty} \frac{n}{(n-1)(n+1)} = 0 < 1$

7. diverges by the nth-Term Test since if $a_n \to L$ as $n \to \infty$, then $L = \frac{1}{1+L} \Rightarrow L^2 + L - 1 = 0 \Rightarrow L = \frac{-1\pm\sqrt{5}}{2}$ $\neq 0$

8. Split the given series into $\sum_{n=1}^{\infty} \frac{1}{3^{2n+1}}$ and $\sum_{n=1}^{\infty} \frac{2n}{3^{2n}}$; the first subseries is a convergent geometric series and the second converges by the Root Test: $\lim_{n\to\infty} \sqrt[n]{\frac{2n}{3^{2n}}} = \lim_{n\to\infty} \frac{\sqrt[n]{2}\,\sqrt[n]{n}}{9} = \frac{1\cdot 1}{9} = \frac{1}{9} < 1$

9. $f(x) = \cos x$ with $a = \frac{\pi}{3} \Rightarrow f\left(\frac{\pi}{3}\right) = 0.5, f'\left(\frac{\pi}{3}\right) = -\frac{\sqrt{3}}{2}, f''\left(\frac{\pi}{3}\right) = -0.5, f'''\left(\frac{\pi}{3}\right) = \frac{\sqrt{3}}{2}, f^{(4)}\left(\frac{\pi}{3}\right) = 0.5;$
$\cos x = \frac{1}{2} - \frac{\sqrt{3}}{2}\left(x - \frac{\pi}{3}\right) - \frac{1}{4}\left(x - \frac{\pi}{3}\right)^2 + \frac{\sqrt{3}}{12}\left(x - \frac{\pi}{3}\right)^3 + \ldots$

10. $f(x) = \sin x$ with $a = 2\pi \Rightarrow f(2\pi) = 0, f'(2\pi) = 1, f''(2\pi) = 0, f'''(2\pi) = -1, f^{(4)}(2\pi) = 0, f^{(5)}(2\pi) = 1,$
$f^{(6)}(2\pi) = 0, f^{(7)}(2\pi) = -1; \sin x = (x - 2\pi) - \frac{(x-2\pi)^3}{3!} + \frac{(x-2\pi)^5}{5!} - \frac{(x-2\pi)^7}{7!} + \ldots$

11. $e^x = 1 + x + \frac{x^2}{2!} + \frac{x^3}{3!} + \ldots$ with $a = 0$

12. $f(x) = \ln x$ with $a = 1 \Rightarrow f(1) = 0, f'(1) = 1, f''(1) = -1, f'''(1) = 2, f^{(4)}(1) = -6;$
$\ln x = (x-1) - \frac{(x-1)^2}{2} + \frac{(x-1)^3}{3} - \frac{(x-1)^4}{4} + \ldots$

13. $f(x) = \cos x$ with $a = 22\pi \Rightarrow f(22\pi) = 1, f'(22\pi) = 0, f''(22\pi) = -1, f'''(22\pi) = 0, f^{(4)}(22\pi) = 1,$
$f^{(5)}(22\pi) = 0, f^{(6)}(22\pi) = -1; \cos x = 1 - \frac{1}{2}(x - 22\pi)^2 + \frac{1}{4!}(x - 22\pi)^4 - \frac{1}{6!}(x - 22\pi)^6 + \ldots$

14. $f(x) = \tan^{-1} x$ with $a = 1 \Rightarrow f(1) = \frac{\pi}{4}, f'(1) = \frac{1}{2}, f''(1) = -\frac{1}{2}, f'''(1) = \frac{1}{2};$
$\tan^{-1} x = \frac{\pi}{4} + \frac{(x-1)}{2} - \frac{(x-1)^2}{4} + \frac{(x-1)^3}{12} + \ldots$

15. Yes, the sequence converges: $c_n = (a^n + b^n)^{1/n} \Rightarrow c_n = b\left(\left(\frac{a}{b}\right)^n + 1\right)^{1/n} \Rightarrow \lim_{n\to\infty} c_n = \ln b + \lim_{n\to\infty} \frac{\ln\left(\left(\frac{a}{b}\right)^n + 1\right)}{n}$
$= \ln b + \lim_{n\to\infty} \frac{\left(\frac{a}{b}\right)^n \ln\left(\frac{a}{b}\right)}{\left(\frac{a}{b}\right)^n + 1} = \ln b + \frac{0\cdot\ln\left(\frac{a}{b}\right)}{0+1} = \ln b$ since $0 < a < b$. Thus, $\lim_{n\to\infty} c_n = e^{\ln b} = b$.

16. $1 + \frac{2}{10} + \frac{3}{10^2} + \frac{7}{10^3} + \frac{2}{10^4} + \frac{3}{10^5} + \frac{7}{10^6} + \ldots = 1 + \sum_{n=1}^{\infty} \frac{2}{10^{3n-2}} + \sum_{n=1}^{\infty} \frac{3}{10^{3n-1}} + \sum_{n=1}^{\infty} \frac{7}{10^{3n}}$
$= 1 + \sum_{n=0}^{\infty} \frac{2}{10^{3n+1}} + \sum_{n=0}^{\infty} \frac{3}{10^{3n+2}} + \sum_{n=0}^{\infty} \frac{7}{10^{3n+3}} = 1 + \frac{\left(\frac{2}{10}\right)}{1-\left(\frac{1}{10}\right)^3} + \frac{\left(\frac{3}{10^2}\right)}{1-\left(\frac{1}{10}\right)^3} + \frac{\left(\frac{7}{10^3}\right)}{1-\left(\frac{1}{10}\right)^3}$
$= 1 + \frac{200}{999} + \frac{30}{999} + \frac{7}{999} = \frac{999+237}{999} = \frac{412}{333}$

17. $s_n = \sum_{k=0}^{n-1} \int_k^{k+1} \frac{dx}{1+x^2} \Rightarrow s_n = \int_0^1 \frac{dx}{1+x^2} + \int_1^2 \frac{dx}{1+x^2} + \ldots + \int_{n-1}^n \frac{dx}{1+x^2} \Rightarrow s_n = \int_0^n \frac{dx}{1+x^2}$
$\Rightarrow \lim_{n\to\infty} s_n = \lim_{n\to\infty} (\tan^{-1} n - \tan^{-1} 0) = \frac{\pi}{2}$

18. $\lim_{n\to\infty} \left|\frac{u_{n+1}}{u_n}\right| = \lim_{n\to\infty} \left|\frac{(n+1)x^{n+1}}{(n+2)(2x+1)^{n+1}} \cdot \frac{(n+1)(2x+1)^n}{nx^n}\right| = \lim_{n\to\infty} \left|\frac{x}{2x+1} \cdot \frac{(n+1)^2}{n(n+2)}\right| = \left|\frac{x}{2x+1}\right| < 1$
$\Rightarrow |x| < |2x+1|$; if $x > 0$, $|x| < |2x+1| \Rightarrow x < 2x+1 \Rightarrow x > -1$; if $-\frac{1}{2} < x < 0$, $|x| < |2x+1|$
$\Rightarrow -x < 2x+1 \Rightarrow 3x > -1 \Rightarrow x > -\frac{1}{3}$; if $x < -\frac{1}{2}$, $|x| < |2x+1| \Rightarrow -x < -2x-1 \Rightarrow x < -1$. Therefore,

the series converges absolutely for $x < -1$ and $x > -\frac{1}{3}$.

19. (a) Each A_{n+1} fits into the corresponding upper triangular region, whose vertices are: $(n, f(n) - f(n+1))$, $(n+1, f(n+1))$ and $(n, f(n))$ along the line whose slope is $f(n+1) - f(n)$. All the A_n's fit into the first upper triangular region whose area is $\frac{f(1)-f(2)}{2} \Rightarrow \sum_{n=1}^{\infty} A_n < \frac{f(1)-f(2)}{2}$

(b) If $A_k = \frac{f(k+1)+f(k)}{2} - \int_k^{k+1} f(x)\,dx$, then

$$\sum_{k=1}^{n-1} A_k = \frac{f(1)+f(2)+f(2)+f(3)+f(3)+\ldots+f(n-1)+f(n)}{2} - \int_1^2 f(x)\,dx - \int_2^3 f(x)\,dx - \ldots - \int_{n-1}^{n} f(x)\,dx$$

$$= \frac{f(1)+f(n)}{2} + \sum_{k=2}^{n-1} f(k) - \int_1^n f(x)\,dx \Rightarrow \sum_{k=1}^{n-1} A_k = \sum_{k=1}^{n} f(k) - \frac{f(1)+f(n)}{2} - \int_1^n f(x)\,dx < \frac{f(1)-f(2)}{2}$$, from part (a). The sequence $\left\{\sum_{k=1}^{n-1} A_k\right\}$ is bounded above and increasing, so it converges and the limit in question must exist.

(c) Let $L = \lim_{n\to\infty} \left[\sum_{k=1}^{n} f(k) - \int_1^n f(x)\,dx - \frac{1}{2}(f(1)+f(n))\right]$, which exists by part (b). Since f is positive and decreasing $\lim_{n\to\infty} f(n) = M \ge 0$ exists. Thus $\lim_{n\to\infty} \left[\sum_{k=1}^{n} f(k) - \int_1^n f(x)\,dx\right] = L + \frac{1}{2}(f(1)+M)$.

20. The number of triangles removed at stage n is 3^{n-1}; the side length at stage n is $\frac{b}{2^{n-1}}$; the area of a triangle at stage n is $\frac{\sqrt{3}}{4}\left(\frac{b}{2^{n-1}}\right)^2$.

(a) $\frac{\sqrt{3}}{4} b^2 + 3\frac{\sqrt{3}}{4}\left(\frac{b^2}{2^2}\right) + 3^2\frac{\sqrt{3}}{4}\left(\frac{b^2}{2^4}\right) + 3^3\frac{\sqrt{3}}{4}\left(\frac{b^2}{2^6}\right) + \ldots = \frac{\sqrt{3}}{4} b^2 \sum_{n=0}^{\infty} \frac{3^n}{2^{2n}} = \frac{\sqrt{3}}{4} b^2 \sum_{n=0}^{\infty} \left(\frac{3}{4}\right)^n$

(b) a geometric series with sum $\frac{\left(\frac{\sqrt{3}}{4} b^2\right)}{1-\left(\frac{3}{4}\right)} = \sqrt{3}b^2$

(c) No; for instance,the three vertices of the original triangle are not removed. However the total area removed is $\sqrt{3}b^2$ which equals the area of the original triangle. Thus the set of points not removed has area 0.

21. (a) No, the limit does not appear to depend on the value of the constant a

(b) Yes, the limit depends on the value of b

(c) $s = \left(1 - \frac{\cos\left(\frac{a}{n}\right)}{n}\right)^n \Rightarrow \ln s = \frac{\ln\left(1 - \frac{\cos\left(\frac{a}{n}\right)}{n}\right)}{\left(\frac{1}{n}\right)} \Rightarrow \lim_{n\to\infty} \ln s = \frac{\left(\frac{1}{1-\frac{\cos\left(\frac{a}{n}\right)}{n}}\right)\left(\frac{-\frac{a}{n}\sin\left(\frac{a}{n}\right)+\cos\left(\frac{a}{n}\right)}{n^2}\right)}{\left(-\frac{1}{n^2}\right)}$

$= \lim_{n\to\infty} \frac{\frac{a}{n}\sin\left(\frac{a}{n}\right) - \cos\left(\frac{a}{n}\right)}{1 - \frac{\cos\left(\frac{a}{n}\right)}{n}} = \frac{0-1}{1-0} = -1 \Rightarrow \lim_{n\to\infty} s = e^{-1} \approx 0.3678794412$; similarly,

$\lim_{n\to\infty} \left(1 - \frac{\cos\left(\frac{a}{n}\right)}{bn}\right)^n = e^{-1/b}$

22. $\sum_{n=1}^{\infty} a_n$ converges $\Rightarrow \lim_{n\to\infty} a_n = 0$; $\lim_{n\to\infty} \left[\left(\frac{1+\sin a_n}{2}\right)^n\right]^{1/n} = \lim_{n\to\infty} \left(\frac{1+\sin a_n}{2}\right) = \frac{1+\sin\left(\lim_{n\to\infty} a_n\right)}{2} = \frac{1+\sin 0}{2}$ $= \frac{1}{2} \Rightarrow$ the series converges by the nth-Root Test

23. $\lim_{n\to\infty} \left|\frac{u_{n+1}}{u_n}\right| < 1 \Rightarrow \lim_{n\to\infty} \left|\frac{b^{n+1}x^{n+1}}{\ln(n+1)} \cdot \frac{\ln n}{b^n x^n}\right| < 1 \Rightarrow |bx| < 1 \Rightarrow -\frac{1}{b} < x < \frac{1}{b} = 5 \Rightarrow b = \pm\frac{1}{5}$

24. A polynomial has only a finite number of nonzero terms in its Taylor series, but the functions sin x, ln x and e^x have infinitely many nonzero terms in their Taylor expansions.

25. $\lim\limits_{x \to 0} \frac{\sin(ax) - \sin x - x}{x^3} = \lim\limits_{x \to 0} \frac{\left(ax - \frac{a^3x^3}{3!} + \ldots\right) - \left(x - \frac{x^3}{3!} + \ldots\right) - x}{x^3}$
$= \lim\limits_{x \to 0} \left[\frac{a-2}{x^2} - \frac{a^3}{3!} + \frac{1}{3!} - \left(\frac{a^5}{5!} - \frac{1}{5!}\right)x^2 + \ldots\right]$ is finite if $a - 2 = 0 \Rightarrow a = 2$;
$\lim\limits_{x \to 0} \frac{\sin 2x - \sin x - x}{x^3} = -\frac{2^3}{3!} + \frac{1}{3!} = -\frac{7}{6}$

26. $\lim\limits_{x \to 0} \frac{\cos ax - b}{2x^2} = -1 \Rightarrow \lim\limits_{x \to 0} \frac{\left(1 - \frac{a^2x^2}{2} + \frac{a^4x^4}{4!} - \ldots\right) - b}{2x^2} = -1 \Rightarrow \lim\limits_{x \to 0} \left(\frac{1-b}{2x^2} - \frac{a^2}{4} + \frac{a^2x^2}{48} - \ldots\right) = -1$
$\Rightarrow b = 1$ and $a = \pm 2$

27. (a) $\frac{u_n}{u_{n+1}} = \frac{(n+1)^2}{n^2} = 1 + \frac{2}{n} + \frac{1}{n^2} \Rightarrow C = 2 > 1$ and $\sum\limits_{n=1}^{\infty} \frac{1}{n^2}$ converges

(b) $\frac{u_n}{u_{n+1}} = \frac{n+1}{n} = 1 + \frac{1}{n} + \frac{0}{n^2} \Rightarrow C = 1 \le 1$ and $\sum\limits_{n=1}^{\infty} \frac{1}{n}$ diverges

28. $\frac{u_n}{u_{n+1}} = \frac{2n(2n+1)}{(2n-1)^2} = \frac{4n^2+2n}{4n^2-4n+1} = 1 + \frac{\left(\frac{6}{4}\right)}{n} + \frac{5}{4n^2-4n+1} = 1 + \frac{\left(\frac{3}{2}\right)}{n} + \frac{\left[\frac{5n^2}{(4n^2-4n+1)}\right]}{n^2}$ after long division
$\Rightarrow C = \frac{3}{2} > 1$ and $|f(n)| = \frac{5n^2}{4n^2-4n+1} = \frac{5}{\left(4 - \frac{4}{n} + \frac{1}{n^2}\right)} \le 5 \Rightarrow \sum\limits_{n=1}^{\infty} u_n$ converges by Raabe's Test

29. (a) $\sum\limits_{n=1}^{\infty} a_n = L \Rightarrow a_n^2 \le a_n \sum\limits_{n=1}^{\infty} a_n = a_n L \Rightarrow \sum\limits_{n=1}^{\infty} a_n^2$ converges by the Direct Comparison Test

(b) converges by the Limit Comparison Test: $\lim\limits_{n \to \infty} \frac{\left(\frac{a_n}{1-a_n}\right)}{a_n} = \lim\limits_{n \to \infty} \frac{1}{1-a_n} = 1$ since $\sum\limits_{n=1}^{\infty} a_n$ converges and therefore $\lim\limits_{x \to \infty} a_n = 0$

30. If $0 < a_n < 1$ then $|\ln(1 - a_n)| = -\ln(1 - a_n) = a_n + \frac{a_n^2}{2} + \frac{a_n^3}{3} + \ldots < a_n + a_n^2 + a_n^3 + \ldots = \frac{a_n}{1-a_n}$, a positive term of a convergent series, by the Limit Comparison Test and Exercise 29b

31. $(1 - x)^{-1} = 1 + \sum\limits_{n=1}^{\infty} x^n$ where $|x| < 1 \Rightarrow \frac{1}{(1-x)^2} = \frac{d}{dx}(1 - x)^{-1} = \sum\limits_{n=1}^{\infty} nx^{n-1}$ and when $x = \frac{1}{2}$ we have
$4 = 1 + 2\left(\frac{1}{2}\right) + 3\left(\frac{1}{2}\right)^2 + 4\left(\frac{1}{2}\right)^3 + \ldots + n\left(\frac{1}{2}\right)^{n-1} + \ldots$

32. (a) $\sum\limits_{n=1}^{\infty} x^{n+1} = \frac{x^2}{1-x} \Rightarrow \sum\limits_{n=1}^{\infty} (n+1)x^n = \frac{2x-x^2}{(1-x)^2} \Rightarrow \sum\limits_{n=1}^{\infty} n(n+1)x^{n-1} = \frac{2}{(1-x)^3} \Rightarrow \sum\limits_{n=1}^{\infty} n(n+1)x^n = \frac{2x}{(1-x)^3}$
$\Rightarrow \sum\limits_{n=1}^{\infty} \frac{n(n+1)}{x^n} = \frac{\frac{2}{x}}{\left(1 - \frac{1}{x}\right)^3} = \frac{2x^2}{(x-1)^3}$, $|x| > 1$

(b) $x = \sum\limits_{n=1}^{\infty} \frac{n(n+1)}{x^n} \Rightarrow x = \frac{2x^2}{(x-1)^3} \Rightarrow x^3 - 3x^2 + x - 1 = 0 \Rightarrow x = 1 + \left(1 + \frac{\sqrt{57}}{9}\right)^{1/3} + \left(1 - \frac{\sqrt{57}}{9}\right)^{1/3}$
≈ 2.769292, using a CAS or calculator

33. The sequence $\{x_n\}$ converges to $\frac{\pi}{2}$ from below so $\epsilon_n = \frac{\pi}{2} - x_n > 0$ for each n. By the Alternating Series Estimation Theorem $\epsilon_{n+1} \approx \frac{1}{3!}(\epsilon_n)^3$ with $|\text{error}| < \frac{1}{5!}(\epsilon_n)^5$, and since the remainder is negative this is an overestimate $\Rightarrow 0 < \epsilon_{n+1} < \frac{1}{6}(\epsilon_n)^3$.

34. Yes, the series $\sum\limits_{n=1}^{\infty} \ln(1 + a_n)$ converges by the Direct Comparison Test: $1 + a_n < 1 + a_n + \frac{a_n^2}{2!} + \frac{a_n^3}{3!} + \ldots$
$\Rightarrow 1 + a_n < e^{a_n} \Rightarrow \ln(1 + a_n) < a_n$

35. (a) $\frac{1}{(1-x)^2} = \frac{d}{dx}\left(\frac{1}{1-x}\right) = \frac{d}{dx}(1 + x + x^2 + x^3 + \ldots) = 1 + 2x + 3x^2 + 4x^3 + \ldots = \sum_{n=1}^{\infty} nx^{n-1}$

(b) from part (a) we have $\sum_{n=1}^{\infty} n\left(\frac{5}{6}\right)^{n-1}\left(\frac{1}{6}\right) = \left(\frac{1}{6}\right)\left[\frac{1}{1-\left(\frac{5}{6}\right)}\right]^2 = 6$

(c) from part (a) we have $\sum_{n=1}^{\infty} np^{n-1}q = \frac{q}{(1-p)^2} = \frac{q}{q^2} = \frac{1}{q}$

36. (a) $\sum_{k=1}^{\infty} p_k = \sum_{k=1}^{\infty} 2^{-k} = \frac{\left(\frac{1}{2}\right)}{1-\left(\frac{1}{2}\right)} = 1$ and $E(x) = \sum_{k=1}^{\infty} kp_k = \sum_{k=1}^{\infty} k2^{-k} = \frac{1}{2}\sum_{k=1}^{\infty} k2^{1-k} = \left(\frac{1}{2}\right)\frac{1}{\left[1-\left(\frac{1}{2}\right)\right]^2} = 2$ by Exercise 35(a)

(b) $\sum_{k=1}^{\infty} p_k = \sum_{k=1}^{\infty} \frac{5^{k-1}}{6^k} = \frac{1}{5}\sum_{k=1}^{\infty}\left(\frac{5}{6}\right)^k = \left(\frac{1}{5}\right)\left[\frac{\left(\frac{5}{6}\right)}{1-\left(\frac{5}{6}\right)}\right] = 1$ and $E(x) = \sum_{k=1}^{\infty} kp_k = \sum_{k=1}^{\infty} k\,\frac{5^{k-1}}{6^k} = \frac{1}{6}\sum_{k=1}^{\infty} k\left(\frac{5}{6}\right)^{k-1}$
$= \left(\frac{1}{6}\right)\frac{1}{\left[1-\left(\frac{5}{6}\right)\right]^2} = 6$

(c) $\sum_{k=1}^{\infty} p_k = \sum_{k=1}^{\infty} \frac{1}{k(k+1)} = \sum_{k=1}^{\infty}\left(\frac{1}{k} - \frac{1}{k+1}\right) = \lim_{k\to\infty}\left(1 - \frac{1}{k+1}\right) = 1$ and $E(x) = \sum_{k=1}^{\infty} kp_k = \sum_{k=1}^{\infty} k\left(\frac{1}{k(k+1)}\right)$
$= \sum_{k=1}^{\infty}\frac{1}{k+1}$, a divergent series so that E(x) does not exist

37. (a) $R_n = C_0e^{-kt_0} + C_0e^{-2kt_0} + \ldots + C_0e^{-nkt_0} = \frac{C_0e^{-kt_0}\left(1-e^{-nkt_0}\right)}{1-e^{-kt_0}} \Rightarrow R = \lim_{n\to\infty} R_n = \frac{C_0e^{-kt_0}}{1-e^{-kt_0}} = \frac{C_0}{e^{kt_0}-1}$

(b) $R_n = \frac{e^{-1}(1-e^{-n})}{1-e^{-1}} \Rightarrow R_1 = e^{-1} \approx 0.36787944$ and $R_{10} = \frac{e^{-1}(1-e^{-10})}{1-e^{-1}} \approx 0.58195028$;
$R = \frac{1}{e-1} \approx 0.58197671$; $R - R_{10} \approx 0.00002643 \Rightarrow \frac{R-R_{10}}{R} < 0.0001$

(c) $R_n = \frac{e^{-.1}\left(1-e^{-.1n}\right)}{1-e^{-.1}}$, $\frac{R}{2} = \frac{1}{2}\left(\frac{1}{e^{.1}-1}\right) \approx 4.7541659$; $R_n > \frac{R}{2} \Rightarrow \frac{1-e^{-.1n}}{e^{.1}-1} > \left(\frac{1}{2}\right)\left(\frac{1}{e^{.1}-1}\right)$
$\Rightarrow 1 - e^{-n/10} > \frac{1}{2} \Rightarrow e^{-n/10} < \frac{1}{2} \Rightarrow -\frac{n}{10} < \ln\left(\frac{1}{2}\right) \Rightarrow \frac{n}{10} > -\ln\left(\frac{1}{2}\right) \Rightarrow n > 6.93 \Rightarrow n = 7$

38. (a) $R = \frac{C_0}{e^{kt_0}-1} \Rightarrow Re^{kt_0} = R + C_0 = C_H \Rightarrow e^{kt_0} = \frac{C_H}{C_L} \Rightarrow t_0 = \frac{1}{k}\ln\left(\frac{C_H}{C_L}\right)$

(b) $t_0 = \frac{1}{0.05}\ln e = 20$ hrs

(c) Give an initial dose that produces a concentration of 2 mg/ml followed every $t_0 = \frac{1}{0.02}\ln\left(\frac{2}{0.5}\right) \approx 69.31$ hrs by a dose that raises the concentration by 1.5 mg/ml

(d) $t_0 = \frac{1}{0.2}\ln\left(\frac{0.1}{0.03}\right) = 5\ln\left(\frac{10}{3}\right) \approx 6$ hrs

39. The convergence of $\sum_{n=1}^{\infty}|a_n|$ implies that $\lim_{n\to\infty}|a_n| = 0$. Let $N > 0$ be such that $|a_n| < \frac{1}{2} \Rightarrow 1 - |a_n| > \frac{1}{2}$
$\Rightarrow \frac{|a_n|}{1-|a_n|} < 2|a_n|$ for all $n > N$. Now $|\ln(1+a_n)| = \left|a_n - \frac{a_n^2}{2} + \frac{a_n^3}{3} - \frac{a_n^4}{4} + \ldots\right| \le |a_n| + \left|\frac{a_n^2}{2}\right| + \left|\frac{a_n^3}{3}\right| + \left|\frac{a_n^4}{4}\right| + \ldots$
$< |a_n| + |a_n|^2 + |a_n|^3 + |a_n|^4 + \ldots = \frac{|a_n|}{1-|a_n|} < 2|a_n|$. Therefore $\sum_{n=1}^{\infty}\ln(1+a_n)$ converges by the Direct Comparison Test since $\sum_{n=1}^{\infty}|a_n|$ converges.

40. $\sum_{n=3}^{\infty}\frac{1}{n\ln n(\ln(\ln n))^p}$ converges if $p > 1$ and diverges otherwise by the Integral Test: when $p = 1$ we have
$\lim_{b\to\infty}\int_3^b \frac{dx}{x\ln x(\ln(\ln x))} = \lim_{b\to\infty}\left[\ln(\ln(\ln x))\right]_3^b = \infty$; when $p \ne 1$ we have $\lim_{b\to\infty}\int_3^b \frac{dx}{x\ln x(\ln(\ln x))^p}$
$= \lim_{b\to\infty}\left[\frac{(\ln(\ln x))^{-p+1}}{1-p}\right]_3^b = \begin{cases} \frac{(\ln(\ln 3))^{-p+1}}{1-p}, & \text{if } p > 1 \\ \infty, & \text{if } p < 1 \end{cases}$

41. (a) $s_{2n+1} = \frac{c_1}{1} + \frac{c_2}{2} + \frac{c_3}{3} + \ldots + \frac{c_{2n+1}}{2n+1} = \frac{t_1}{1} + \frac{t_2 - t_1}{2} + \frac{t_3 - t_2}{3} + \ldots + \frac{t_{2n+1} - t_{2n}}{2n+1}$

$= t_1\left(1 - \frac{1}{2}\right) + t_2\left(\frac{1}{2} - \frac{1}{3}\right) + \ldots + t_{2n}\left(\frac{1}{2n} - \frac{1}{2n+1}\right) + \frac{t_{2n+1}}{2n+1} = \sum_{k=1}^{2n} \frac{t_k}{k(k+1)} + \frac{t_{2n+1}}{2n+1}.$

(b) $\{c_n\} = \{(-1)^n\} \Rightarrow \sum_{n=1}^{\infty} \frac{(-1)^n}{n}$ converges

(c) $\{c_n\} = \{1, -1, -1, 1, 1, -1, -1, 1, 1, \ldots\} \Rightarrow$ the series $1 - \frac{1}{2} - \frac{1}{3} + \frac{1}{4} + \frac{1}{5} - \frac{1}{6} - \frac{1}{7} + \ldots$ converges

42. (a) $(1 - t + t^2 - t^3 + \ldots + (-1)^n t^n)(1 + t) = 1 - t + t^2 - t^3 + \ldots + (-1)^n t^n + t - t^2 + t^3 - t^4 + \ldots + (-1)^n t^{n+1}$

$= 1 + (-1)^n t^{n+1} \Rightarrow 1 - t + t^2 - t^3 + \ldots + (-1)^n t^n - \frac{(-1)^n t^{n+1}}{1+t} = \frac{1}{1+t}$

(b) $\int_0^x \frac{1}{1+t}\,dt = \int_0^x \left[1 - t + t^2 + \ldots + (-1)^n t^n + \frac{(-1)^{n+1} t^{n+1}}{1+t}\right] dt \Rightarrow [\ln|1+t|]_0^x$

$= \left[t - \frac{t^2}{2} + \frac{t^3}{3} + \ldots + \frac{(-1)^n t^{n+1}}{n+1}\right]_0^x + \int_0^x \frac{(-1)^{n+1} t^{n+1}}{n+1}\,dt \Rightarrow \ln|1+x|$

$= x - \frac{x^2}{2} + \frac{x^3}{3} - \ldots + \frac{(-1)^n x^{n+1}}{n+1} + R_{n+1}$, where $R_{n+1} = \int_0^x \frac{(-1)^{n+1} t^{n+1}}{n+1}\,dt$

(c) $x > 0$ and $R_{n+1} = (-1)^{n+1} \int_0^x \frac{t^{n+1}}{1+t}\,dt \Rightarrow |R_{n+1}| = \int_0^x \frac{t^{n+1}}{1+t}\,dt \le \int_0^x t^{n+1}\,dt = \frac{x^{n+2}}{n+2}$

(d) $-1 < x < 0$ and $R_{n+1} = (-1)^{n+1} \int_0^x \frac{t^{n+1}}{1+t}\,dt \Rightarrow |R_{n+1}| = \left|\int_0^x \frac{t^{n+1}}{1+t}\,dt\right| \le \int_0^x \left|\frac{t^{n+1}}{1+t}\right| dt$

$\le \int_0^x \frac{|t|^{n+1}}{1-|x|}\,dx = \frac{|x|^{n+2}}{(1-|x|)(n+2)}$ since $|1+t| \ge 1 - |x|$

(e) From part (d) we have $|R_{n+1}| \le \frac{|x|^{n+2}}{(1-|x|)(n+2)} \Rightarrow$ the given series converges since

$\lim_{n \to \infty} \frac{|x|^{n+2}}{(1-|x|)(n+2)} = 0 \Rightarrow |R_{n+1}| \to 0$ when $|x| < 1$. If $x = 1$, by part (c) $|R_{n+1}| \le \frac{|x|^{n+2}}{n+2} = \frac{1}{n+2} \to 0.$

Thus the given series converges to $\ln(1 + x)$ for $-1 < x \le 1$.